INSTRUCTOR'S EDITION
SECOND EDITION

Experiencing Introductory and Intermediate Algebra

D1310753

JoAnne Thomasson
Pellissippi State Technical Community College

Bob Pesut
Pellissippi State Technical Community College

Prentice Hall

Pearson Education, Inc.
Upper Saddle River, New Jersey 07458

Senior Acquisitions Editor: *Paul Murphy*
Editor in Chief: *Christine Hoag*
Project Manager/Development Editor: *Elaine Page*
Production Editor: *Lynn Savino Wendel*
Vice President/Director of Production and Manufacturing: *David W. Riccardi*
Senior Managing Editor: *Linda Mihatov Behrens*
Assistant Managing Editor: *Bayani Mendoza De Leon*
Executive Managing Editor: *Kathleen Schiaparelli*
Assistant Managing Editor, Math Media Production: *John Matthews*
Manufacturing Buyer: *Alan Fischer*
Manufacturing Manager: *Trudy Pisciotti*
Executive Marketing Manager: *Eilish Collins Main*
Marketing Assistant: *Annett Uebel*
Media Project Manager, Developmental Math: *Audra J. Walsh*
Editorial Assistant: *Heather Balderson*
Art Director: *Jonathan Boylan*
Cover and Interior Designer: *Maureen Eide*
Art Editor: *Thomas Benfatti*
Creative Director: *Carole Anson*
Director of Creative Services: *Paul Belfanti*
Director, Image Resource Center: *Melinda Reo*
Manager, Rights and Permissions: *Zina Arabia*
Image Permission Coordinator: *Debbie Hewitson*
Photo Researcher: *Melinda Alexander*
Image Specialist: *Beth Boyd Brenzel*
Cover Photo Credits: CORBIS
Art Studio: *Precision Graphics*

© 2003, 1999 by Pearson Education, Inc.
Pearson Education, Inc.
Upper Saddle River, New Jersey 07458

Printed in the United States of America

10 9 8 7 6 5 4 3 2 1

ISBN 0-13-009294-0

Student edition ISBN 0-13-035682-4

Pearson Education Ltd., London
Pearson Education Australia Pty. Limited, Sydney
Pearson Education Singapore, Pte. Ltd
Pearson Education North Asia Ltd, Hong Kong
Pearson Education Canada, Ltd, Toronto
Pearson Educación de Mexico, S.A. de C.V.
Pearson Education—Japan, Tokyo
Pearson Education Malaysia, Pte. Ltd

Interest Formulas

Simple Interest

$I = Prt$

I = simple interest

P = principal

r = rate of interest

t = number of time periods

$A = P + I$

A = simple interest amount

$I = A - P$

I = interest

Compounded Interest

$A = P(1 + r)^t$

A = compounded amount

P = principal

r = interest rate per time period

t = number of time periods

Continuously Compounded Interest

$A = Pe^{rt}$

A = amount compounded continuously

P = principal

r = annual interest rate compounded continuously

t = number of years

$e \approx 2.718$ (irrational number)

Temperature Formulas

Fahrenheit Temperature (F)

$F = \dfrac{9}{5}C + 32$ C = Celsius temperature

Celsius Temperature (C)

$C = \dfrac{5}{9}(F - 32)$ F = Fahrenheit temperature

Other Formulas

Distance Traveled

$d = rt$

d = distance traveled

r = rate (speed)

t = time traveled

Vertical Position

$s = -16t^2 + v_0 t + s_0$

s = position in feet above ground level

t = time in seconds

v_0 = initial velocity in feet per second

s_0 = initial position in feet above ground level

Pendulum

$T = 2\pi\sqrt{\dfrac{L}{32}}$

T = period in seconds

L = length in feet of suspension

Quadratic Formula

$x = \dfrac{-b \pm \sqrt{b^2 - 4ac}}{2a}$

$a, b,$ and c are real numbers

$a \neq 0$ and $ax^2 + bx + c = 0$

Distance between Two Points

$d = \sqrt{(x_2 - x_1)^2 + (y_2 - y_1)^2}$

(x_1, y_1) and (x_2, y_2) are coordinates of two points

Products of Polynomials

Two Binomial Factors (FOIL Method)

$(a + b)(c + d) = ac + ad + bc + bd$

Squaring a Binomial

$(a + b)^2 = a^2 + 2ab + b^2$

$(a - b)^2 = a^2 - 2ab + b^2$

Sum and Difference of the Same Two Terms

$(a + b)(a - b) = a^2 - b^2$

Other Special Products

$(a + b)(a^2 - ab + b^2) = a^3 + b^3$

$(a - b)(a^2 + ab + b^2) = a^3 - b^3$

Thanks to our spouses and families
for their continued encouragement and support
while writing this text.

Jack	*Gretchen*
Tracy	*Lauren*
Tommy	*Katherine*
Shawn	*Tracy*
Cameron	*Jim*
Caitlin	*Emma*
Chloe	*B.P.*
J.T.	

Contents

8 Polynomial Functions 531

9 Exponents and Polynomials 593

10 Factoring 649

14 | Exponential and Logarithmic Functions and Equations   973

Preface

The second edition of *Experiencing Introductory and Intermediate Algebra* continues to embrace the goal of promoting a new approach to teaching and learning developmental mathematics. This approach combines a traditional model with the reform movements presented in the National Council of Teachers of Mathematics (NCTM) standards and the American Mathematical Association of Two-Year Colleges (AMATYC) standards. The NCTM goals state that in our present technological society, students should learn to value mathematics, reason and communicate mathematically, become confident of their mathematical abilities, and become mathematical problem solvers. The AMATYC standards for intellectual development state that students will model real-world situations, connect mathematics with other disciplines, and use appropriate technology.

In this second edition, we have incorporated recommendations and suggestions from instructors and reviewers of the text. Instructors who currently use the text valued the real-world application feature and encouraged us to expand it. At the same time, they recommended that the text be streamlined to reduce its volume. The contents of this edition are still organized by families of functions, according to the AMATYC standards. Consequently, the first seven chapters of the text focus on linear expressions, equations, and functions. The next four chapters, Chapters 8 through 11, focus on polynomial expressions, equations, and functions. Chapter 12 presents rational expressions, equations, and functions, Chapter 13 examines radical expressions, equations, and functions, and Chapter 14 features exponential and logarithmic expressions, equations, and functions.

We have condensed the discussion of the real-number system into a review of pre-algebra numeric topics that may be taught as a whole or by sections of choice. We have combined Chapters 1 and 2 of the previous edition into one chapter, postponing the discussion of rational exponents to a later chapter. Also, the coverage of radicals in the combined chapter is limited to square roots and cube roots.

In the previous edition, exponents and polynomials were presented together with factoring, all within one chapter. We have now separated this material into two chapters. The first of the two, Chapter 9, focuses on exponents and polynomial operations. We have added a separate section on polynomial division that outlines polynomial long division in greater detail.

Feedback from reviewers indicated a need for expanding and strengthening the discussion of factoring, so we have created a separate chapter on the subject, Chapter 10. In this new chapter, we offer more examples, more exercises, and a different ordering of topics. In the previous edition, there was a separate chapter on complex numbers. In this edition, complex numbers and equations with imaginary solutions have been placed at the end of Chapter 13, following the discussion of radical expressions, equations, and functions. We believe that that is the appropriate place for complex numbers for two major reasons: students will have just learned the algebra of radicals, and the new location provides a perfect opportunity to revisit the quadratic formula, thereby reinforcing their understanding of the concept. However, the section on complex numbers and equations with imaginary solutions has been written as a stand-alone section, and if an instructor so chooses, it could be presented after the discussion of the quadratic formula in Chapter 11.

The first half of the text presents a balanced discussion of algebraic, numerical, and graphical methods for solving linear equations, so that students have a solid understanding of what the concepts represent. In the second half, the discussion of polynomial equations continues to utilize the same algebraic, numerical, and graphical techniques. This approach provides an opportunity for students who enter the sequence at that late point to gain an understanding of those methods. However, as we progress further into the second half of the text, the emphasis increasingly is on algebraic methods. Numerical and graphical methods are used only for checking solutions, rather than obtaining them. This way of teaching the topics will strengthen the students' algebraic skills for their subsequent math courses.

The new feature in the second edition is the inclusion of a project at the end of each chapter. The project enriches the study of the material presented in the chapter and provides connections to other areas of mathematics and other disciplines. Students may be asked to research the history of mathematical topics, collect and interpret data for use in their mathematical modeling activities, and build on the applications they have studied. The companion Web site provides support for those instructors who need to access data.

Approach

We have carefully written *Experiencing Introductory and Intermediate Algebra* in a positive manner to help students build confidence in their ability to do algebra. After completing the course, students should be able to do all of the following:

- Model real-world situations.
- Reason mathematically and develop convincing mathematical arguments.
- Use an appropriate method—numeric, graphic, or algebraic—to solve problems.
- Connect algebra to other disciplines.
- Communicate mathematically.
- Use appropriate technology.
- Work collaboratively in groups.

To teach these skills, we introduce a problem-solving procedure in Chapter 4 and use this approach throughout the text. Numeric, graphic, and algebraic approaches to solving problems are described, and students are encouraged to choose that method which is appropriate to solve their problems. Every section of the text addresses real-world situations, so students can see reasons for learning algebra and can connect up what they learn with other disciplines, both inside and outside of mathematics. Students are asked to discover mathematical ideas on their own, to strengthen their mathematical reasoning skills, and to communicate their results. We then explain these results mathematically to reinforce the concepts the students have found.

Content

The text is written for a two-semester course in beginning and intermediate algebra. However, it is also flexible enough for use in a one-semester course. In both courses, topics are covered with a minimal amount of repetition. In Chapter 1, we introduce the set of real numbers, develop the properties of the real-number system, and present the rules for operations on real numbers. We complete these numeric topics with discussions of integer exponents, scientific notation, and radicals.

After completing the numeric foundation, we introduce variables, algebraic expressions, and equations in Chapter 2. There, we discuss geometric formulas and other formulas used in the first seven chapters of the text. This early introduction to these formulas allows us to integrate geometric and other applications throughout the book. In Chapter 3, we examine additional topics needed for the study of algebra: ordered pairs, relations, functions, and graphs. This early discussion of functions supports the structure of the remainder of the text, which focuses on the study of various families of functions.

Chapters 4, 5, 6, and 7 cover topics related to linear functions. In Chapter 4, we begin the explicit study of algebra by solving linear equations in one variable and absolute-value equations. Here and throughout the text, we teach how to solve equations numerically, graphically, and algebraically. Chapter 5 focuses on linear equations in two variables and on functions. Chapter 6 presents methods for solving systems of linear equations in two variables, emphasizing solutions by graphing, by substitution, and by elimination. Inequalities and solutions of linear inequalities are discussed in Chapter 7. We recommend the first seven chapters as a beginning algebra text.

We designed Chapter 8 as the first chapter in the second semester of study, allowing for a review of graphical methods and the concept of a function. The remainder of the text follows a standard pattern consisting of the introduction of a family of functions, rules for operating with the expressions that define the functions, and methods for solving related equations numerically, graphically, and algebraically. We follow this pattern in discussing polynomial functions in Chapter 8, polynomial expressions in Chapter 9, factoring polynomials in Chapter 10, and polynomial equations and inequalities in Chapter 11.

In Chapter 11, we solve quadratic equations numerically, graphically, and algebraically by factoring, by completing the square, and by using the quadratic formula. This arrangement allows for closure on quadratic equations and enables the student to choose an appropriate method for solving equations by examining the equation given.

In Chapter 12, we describe rational functions, operations with rational expressions, and the solution of rational equations. We complete the coverage of radical functions, expressions, and equations in Chapter 13, along with functions, expressions, and equations having rational exponents. The chapter also includes a section on the complex-number system, describing equations in one variable with complex solutions. This section is designed to stand on its own or to be incorporated into earlier chapters if desired. Finally, Chapter 14 focuses on inverse functions, exponential functions, and logarithmic functions. It also presents a discussion of the properties of exponents and logarithms and the methods for solving exponential equations and logarithmic equations in one variable.

Pedagogy

Use of Technology

Graphing calculators allow students the freedom to experiment with and explore mathematical ideas. Using graphing calculators helps boost confidence and increase motivation. Skills such as estimating, computing, graphing, and analyzing data can be developed and reinforced with the use of a calculator. When students are relieved of tedious computations, they can focus on processes instead. They can also go beyond the limitations of traditional paper-and-pencil work and deal directly with real-world numbers.

Students should learn not only *how* to use technology, but also *when* to do so. The text assumes that all students have a TI-83 Plus graphing calculator available for use at all times. This requirement will minimize the amount of time necessary to demonstrate particular calculator functions. Technology boxes in the text are designed as stand-alone discussions of topics of interest and present the keystrokes required to produce selected TI-83 Plus calculator screens. Additional calculator activities and instructions are included in the calculator exercises at the end of each section. Other calculator activities are available on the companion Web site for the text.

Multiple Approaches

Throughout the text, concepts are developed using *numeric*, *graphic*, *algebraic*, and *verbal* approaches. The *numeric* presentation emphasizes tables of values, constructed either manually or by using a calculator. *Graphical* techniques follow naturally from the numeric methods. The *algebraic* approach is introduced and supported by the numeric and graphic methods. Students are encouraged to decide which approach is the most appropriate for solving particular problems. They are also challenged to express their solutions both *orally* and in *writing*.

Interactive and Collaborative Learning

We believe students should learn to read, write, and speak mathematically. We have written this text at a level that developmental readers can understand. Several features, including the discovery boxes, the checkup exercises at the end of each objective, the writing exercises at the end of every section, the reflections in each chapter review, the chapter tests, the cumulative review, and the chapter projects, ask students to write mathematically.

Experiencing Mathematics

The graphing calculator enables students to explore and experiment with mathematical ideas as they discover algebraic concepts in the discovery boxes. After a student discovers a concept, the text explains why the concept applies. Students develop a sense of ownership of the algebraic principles through this discovery process. As a result, the students acquire a better understanding of the reasoning behind the mathematics.

To help students keep a positive attitude toward mathematics while experiencing algebra, we provide frequent helping hands, or study tips, to reinforce skills. Students review their skills at the end of each chapter in the section-by-section review and in the chapter review, as well as in the various cumulative reviews. In addition, each chapter test begins with a test-taking tip designed to further bolster students' confidence and improve their ability to perform well on tests.

Connection with Other Experiences

For a meaningful experience in this course, students must make a connection between algebra and the world around them. To help students make this connection, we begin each section of the text with a real-world application and solve this application before the section ends. Each section also presents a list of objectives, one of which is to model real-world situations by using concepts discussed in that section. Calculator exercises, chapter projects, and writing exercises within each set of section exercises often involve connections to disciplines and fields outside of mathematics, as well as to areas of applied mathematics, such as geometry, probability, and statistics.

Supplements for the Instructor

Instructor's Edition (0-13-009294-0)
Includes answers to all section exercises, calculator exercises, and end-of-chapter exercises.

Instructor's Solutions Manual (0-13-009445-5)
Includes solutions to all section exercises, calculator exercises, and end-of-chapter exercises as well as answers to the chapter projects.

Instructor's Resource Manual with Tests (0-13-009446-3)
This supplement contains 6 sample tests for each chapter, for a total of 84 tests. Also included are four sample final exams that cover the cumulative content from the entire textbook and answers to all tests.

MathPro Explorer 4.0 Network Version (IBM/Mac) (0-13-047651-X)

MathPro5 Tutorial Software (0-13-047654-4)

TestGen-EQ (0-13-009447-1)

WebCT (0-13-009449-8)

Course Compass (0-13-009452-8)

Blackboard (0-13-009453-6)

Supplements for the Student

Student Solutions Manual (0-13-009444-7)
This supplement provides full solutions to odd-numbered section exercises, odd-numbered calculator exercises, and all exercises within the section-by-section review, chapter review, chapter test, and cumulative review at the end of each chapter.

Lecture Videos on VHS (0-13-047640-4) and CD (0-13-047656-0)
Full lectures appear at the chapter level on videotape and also are digitized on CD.

MathPro Explorer 4.0 Student Version (0-13-047652-8)

MathPro5 Tutorial Software (0-13-009448-X)

Companion Web Site www.prenhall.com/thomasson_intro_intermediate

Prentice Hall Math Tutor Center (0-13-009451-X)

Acknowledgments

This text was completed with the help of many individuals who offered encouragement, suggestions, and criticisms. We would like to thank the following individuals who reviewed the text:

Linda Bastian	*Portland Community College*
Gale Brewer	*Amarillo College*
Tina Cannon	*Tuscalum College*
Edie Carter	*Amarillo College*
Delaine Cochran	*Indiana University, Southeast*
Julie Davis	*Amarillo College*
Susan Fleming	*Virginia Highlands Community College*
Laurette Foster	*Prairie View A & M University*
Randall Gallaher	*Lewis and Clark Community College*
Tracey Hoy	*College of Lake County*
Kandace Kling	*Portland Community College*
Marva Lucas	*Middle Tennessee State University*
Lou Ann Mahaney	*Tarrant County College, Northeast*
Joyce Martin	*East Central College*
Scott McDaniel	*Middle Tennesee State University*
David Platt	*Front Range Community College*
Ilga Ross	*Portland Community College*
Michael Scroggins	*Lewis and Clark Community College*
Pansy Waycaster	*Southwest Virginia Community College*
Gerald Williams	*San Juan College*
Karl Zilm	*Lewis and Clark Community College*

To the students who participated in the various research projects, class tests, and tests of new materials, your involvement has been invaluable. To the mathematics faculty at Pellissippi State Technical Community College who supported us in this project, we offer our thanks. Many faculty members throughout the college encouraged our undertaking of this project, and we want to acknowledge their support. We wish to thank the college administration and faculty for their recognition of our efforts. In particular, we want to thank Nancy Williams, who was the first college employee to encourage us to author this text, and Dougal Moore, who was very helpful in reviewing the second edition.

We also want to thank the editorial, production, and marketing staff at Prentice Hall for their patience, assistance, and contributions to the project.

Karin Wagner has always supported the project. We thank Paul Murphy, Elaine Page and Lynn Savino Wendel, who guided us through the production process to bring the second edition of the text to fruition. We thank Patrick Kelley, who was the first individual to persuade us to undertake the effort to author this text.

Last, but certainly not least, we thank our spouses, children, family, and friends for the encouragement, support, and sacrifices they have made in order to allow us to complete the project.

JoAnne Thomasson
Bob Pesut
Pellissippi State Technical Community College

Chapter 1

Real Numbers

In this chapter, we introduce the set of rational numbers and write numerical expressions that use them. We evaluate these expressions by means of the basic operations of addition, subtraction, multiplication, and division. We also evaluate expressions containing exponents and radicals, which often include irrational numbers. Together, these two categories of numbers form the set of real numbers, the basis for most of this text.

Positive and negative numbers occur in many situations. One such situation involves temperature. Positive temperatures are temperatures above zero degrees. Negative temperatures are temperatures below zero degrees. However, temperatures may be measured on different scales, such as Fahrenheit and Celsius. Measuring temperatures on different scales will result in different values for a given temperature. We conclude this chapter with a project to illustrate this feature.

1.1 APPLICATION

Human body temperature is about 98°F. According to the *Guinness Book of World Records*, the coldest inhabited place in the Northern Hemisphere is the Siberian village of Oymyakon (called the "Cold Pole"). The unofficial recorded temperature there is the same as the opposite of the human body temperature. Write and evaluate an expression for the temperature of Oymyakon.

We will discuss this application further. See page 11.

1.1 Rational Numbers and the Number Line

OBJECTIVES

1 Identify a number as a member of the set of natural numbers, whole numbers, integers, or rational numbers.

2 Graph rational numbers on a number line.

3 Determine the order of rational numbers, and use the symbols $=$, $<$, $>$, $\leq$, and $\geq$ to complete an equation or inequality.

4 Evaluate the absolute value of a rational number.

5 Evaluate the opposite of a rational number.

6 Model real-world situations with rational numbers.

1.1.1 Identifying Rational Numbers and Their Subsets

A **set** is a group or collection of objects. The objects, which we call **members** or **elements** of the set, could be anything, such as state capitals of the United States, baseball players who have hit over 500 home runs in one season, or stars in the sky. In mathematics, we often discuss sets of numbers.

A set with no members is called an **empty set** or **null set**. For example, the set of baseball players who have hit over 500 home runs in one season is an empty set.

If it is possible to list all members of a set, it is called a **finite set**, such as the state capitals of the United States. If it is not possible to list all members of a set, it is called an **infinite set**, such as the set of points on a line.

We indicate a set by enclosing its members in a pair of braces: { }. If the set is infinite, we include enough members to show the pattern of the set. Then we add three dots, called an **ellipsis**, to indicate that the pattern continues. The following are examples of sets of numbers:

$A = \{\,\}$ Empty set A (no members)

$B = \{1, 3, 5, 7, 9\}$ Finite set B (odd numbers between 0 and 10)

$C = \{1, 3, 5, \ldots\}$ Infinite set C (odd numbers greater than or equal to 1)

A set that is contained in another set is called a **subset** of the latter set. In the three previous sets, set B is a subset of set C. In this section, we discuss the set of rational numbers and several of its subsets.

The first set of numbers we consider is the set N of **natural numbers**, also called **counting numbers**.

$$N = \{1, 2, 3, \ldots\}$$

If we extend the set of natural numbers by including 0 as an element, we have the set W of **whole numbers**:

$$W = \{0, 1, 2, 3, \ldots\}$$

To extend the set of whole numbers, we include the opposites of all the natural numbers. The **opposite** of a natural number is a negative number and is written with a negative sign, $-$, in front of the number.

The opposite of 1 is negative 1, or -1.

The opposite of 2 is negative 2, or -2.

The opposites of all natural numbers form a set of negative numbers, $\{-1, -2, -3, \ldots\}$, or, since the order does not matter, $\{\ldots, -3, -2, -1\}$. If we combine this set of negative numbers with the set of whole numbers, we have the set Z of **integers**:

$$Z = \{\ldots -3, -2, -1, 0, 1, 2, 3, \ldots\}$$

To visualize these sets, we use a **number line**. We construct a number line by drawing a line and labeling equally spaced intervals with numbers in consecutive order. We can represent the integers on a number line like this:

If we examine the number line of integers, we see a space between each pair of consecutive integers. The numbers located between each pair include fractions and decimals. Fractions are simply ratios of integers, while some decimals are equivalent to fractions. We can name many fractions and decimals between 0 and 1, such as $\frac{1}{8}, \frac{1}{4}, \frac{1}{3}, 0.5, \frac{2}{3}, 0.75$, and $\frac{7}{8}$. On the number line, they look like this:

In fact, if we could name and locate all fractions and decimals between 0 and 1, the entire space between 0 and 1 would appear to be filled with numbers.

We can extend the set of integers by including all fractions and their decimal equivalents. This gives us the set R of **rational numbers**.

> Rational numbers are numbers that can be written as a fraction (a ratio of integers), excluding the possibility of a zero denominator.

All integers are rational numbers, because we may write them with a denominator of 1. Thus,

$$6 = \frac{6}{1} \qquad -5 = \frac{-5}{1}$$

All terminating decimals are rational numbers, such as

$$0.25 = \frac{25}{100} = \frac{1}{4} \qquad 0.9 = \frac{9}{10} \qquad 0.85 = \frac{85}{100} = \frac{17}{20}$$

All repeating decimals are rational numbers. For example,

$$0.3333\ldots = \frac{1}{3} \qquad 0.5454\ldots = \frac{6}{11} \qquad 0.8333\ldots = \frac{5}{6}$$

Other forms of decimals will be discussed later.

All of these examples of rational numbers are written in fractional notation in simplest form. **Fractional notation** (in simplest form) is a form of a rational number written as a ratio of an integer numerator to a nonzero integer denominator, with the numerator and denominator having no common factors other than 1.

$$\frac{3 \quad \text{numerator}}{4 \quad \text{denominator}}$$ is in fractional notation in simplest form because 3 and 4 are integers that have no common factors other than 1.

In summary, since the set of rational numbers includes all possible ratios of integers (excluding a denominator of 0), it includes all integers, which includes all whole numbers, which includes all counting numbers, as shown in the following chart:

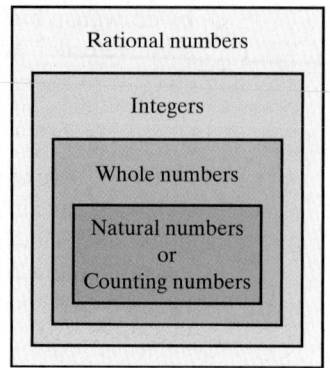

EXAMPLE I Identify the possible sets of numbers (natural numbers, whole numbers, integers, or rational numbers) to which the given number belongs.

a. 123   **b.** 0   **c.** −386   **d.** 36.25   **e.** $-\dfrac{2}{3}$ **f.** $2\dfrac{3}{4}$ **g.** $-3\dfrac{1}{4}$

Solution

a. natural number, whole number, integer, rational number
b. whole number, integer, rational number
c. integer, rational number
d. rational number   **e.** rational number
f. rational number   **g.** rational number

You will need to know how to enter rational numbers into your calculator. First, you must be able to distinguish between the negative symbol key $\boxed{(-)}$ and the subtraction sign key that is an operation sign $\boxed{-}$. Next, you will need to know how to enter fractions and have the calculator return an answer in the form of a fraction.

T E C H N O L O G Y

RATIONAL NUMBERS AND OPPOSITES

Enter $-\dfrac{2}{3}, 2\dfrac{3}{4}, -3\dfrac{1}{4}$.

```
-2/3▶Frac
              -2/3
2+3/4▶Frac
              11/4
-(3+1/4)▶Frac
              -13/4
```

Figure 1.1

For Figure 1.1,

Enter the fraction, $-\frac{2}{3}$, as a quotient, and choose to have your answer displayed as a fraction by selecting ▶ FRAC, option 1, under the MATH menu. The negative symbol is $\boxed{(-)}$.

Enter the mixed number, $2\frac{3}{4}$, as a sum of the whole number, 2, and the fraction, $\frac{3}{4}$. Choose to have the answer displayed as a fraction.

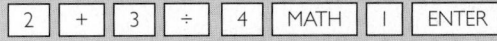

Enter the negative mixed number as the opposite of a sum of the whole number, 3, and the fraction, $\frac{1}{4}$. Use $\boxed{(-)}$ as an opposite sign. Choose to have the answer displayed as a fraction.

$\boxed{(-)}\ \boxed{(}\ \boxed{3}\ \boxed{+}\ \boxed{1}\ \boxed{\div}\ \boxed{4}\ \boxed{)}\ \boxed{MATH}\ \boxed{1}\ \boxed{ENTER}$

Note that the mixed-number answers are displayed as improper fractions. Directions for changing an improper fraction to a mixed number are given in Section 1.2, "Calculator Exercises."

 **HELPING HAND** In keying a negative number into your calculator, the negative sign must always be outside the parentheses containing the value.

 1.1.1 Checkup

In exercises 1–4, consider the following numbers, which may belong to one or more of the subsets of the rational numbers:

$$15, -3, 0, \frac{1}{3}, \frac{6}{3}, \frac{8}{3}, -\frac{5}{3}, 1 \text{ billion}, -15.4, 25.75, -180$$

1. Which of these numbers are natural numbers?
2. Which are rational numbers?
3. Which are integers?
4. Which are whole numbers?
5. Define *set, empty set, finite set,* and *infinite set.*
6. Describe the following sets of numbers: the integers; the whole numbers; the natural numbers; the rational numbers.

1.1.2 Graphing Rational Numbers

We used a number line to visualize sets of numbers. We can also use a number line to understand relationships between numbers. To **graph**, or **plot**, a number on a number line, we place a dot on the line at the location of the number.

EXAMPLE 2 Graph $2, \frac{3}{4}, -1.5, \frac{7}{4}$, and $-2\frac{1}{4}$ on a number line.

Solution
Construct a number line. Label equally spaced intervals with integers in increasing order. Place a dot on the number line at the location where the number appears. Thus,

2 is located 2 units to the right of 0.
$\frac{3}{4}$ is located $\frac{3}{4}$ units to the right of 0.
-1.5 is located 1.5 units to the left of 0, or 0.5 $\left(\frac{1}{2}\right)$ unit to the left of -1.
$\frac{7}{4}$ is located $1\frac{3}{4}$ units to the right of 0, or $\frac{1}{4}$ unit to the left of 2.
$-2\frac{1}{4}$ is located $2\frac{1}{4}$ units to the left of 0, or $\frac{1}{4}$ unit to the left of -2.

1.1.2 Checkup

Graph the following numbers on a number line:

$$-2, 4.75, 7, -5\frac{1}{4}, 0, -8.6, 3\frac{1}{3}, \frac{3}{5}, -\frac{1}{3}$$

1.1.3 Ordering Rational Numbers

Two numbers have the same value, or are **equal** to each other, if they are located in the same position on a number line. For example, 3.5 and $3\frac{1}{2}$ have the same location on the number line.

$$3.5 = 3\frac{1}{2} \qquad \text{3.5 is equal to } 3\frac{1}{2}.$$

Numbers that are not of equal value can be compared by using their locations on a number line. The farther a number is located to the right, the larger the number is. The farther a number is located to the left, the smaller the number is. We use **inequality (order) symbols** of "is greater than," $>$, and "is less than," $<$, to write these statements. For example, use $>$ or $<$ to compare

$$7 \underline{\qquad} 2$$
$$2 \underline{\qquad} 7$$

In comparing 7 and 2 on a number line, we notice that 7 is to the right of 2; therefore, 7 is greater than 2.

$$7 > 2$$

We can also see that 2 is to the left of 7. Therefore, 2 is less than 7.

$$2 < 7$$

HELPING HAND We can think of the inequality symbol as an arrow pointing to the smaller number.

The two inequalities, $7 > 2$ and $2 < 7$, are **equivalent**, or mean the same thing. To write equivalent inequalities, the numbers are exchanged and the inequality symbol is reversed. (The arrow is still pointing to the smaller number.)

EXAMPLE 3 Use $>$, $<$, or $=$ to compare the following numbers:

a. $2 \rule{1cm}{0.4pt} -5$ **b.** $0 \rule{1cm}{0.4pt} -\dfrac{1}{2}$

c. $-6 \rule{1cm}{0.4pt} -2$ **d.** $-1\dfrac{1}{4} \rule{1cm}{0.4pt} -1.25$

Solution

a. $2 \underline{\ >\ } -5$ 2 is to the right of -5. Therefore, $2 > -5$.

b. $0 \underline{\ >\ } -\dfrac{1}{2}$ 0 is to the right of $-\frac{1}{2}$. Therefore, $0 > -\frac{1}{2}$.

c. $-6 \underline{\ <\ } -2$ -6 is to the left of -2. Therefore, $-6 < -2$.

d. $-1\dfrac{1}{4} \underline{\ =\ } -1.25$ $-1\frac{1}{4}$ and -1.25 are located in the same position on the number line. Therefore, the numbers are equal.

If we combine each of the inequality symbols with the equals symbol, we obtain the inequality symbols $\geq$ (is greater than or equal to) and $\leq$ (is less than or equal to). The following inequalities are true:

$$-1 \leq 0 \qquad \text{-1 is less than or equal to 0.}$$

$$-1 \leq \frac{-5}{5} \qquad \text{-1 is less than or equal to $\frac{-5}{5}$.}$$

We may also write a compound inequality containing more than one inequality symbol:

$$-1 \leq 0 < 5 \qquad \text{-1 is less than or equal to 0, and 0 is less than 5.}$$

On a number line, we will observe that 0 is between the values of -1 and 5.

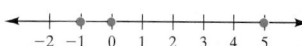

The conventional way to write this form of a compound inequality is to use the "is less than" or the "is less than or is equal to" symbol. This is easy to remember, because we are reading the number line from left to right. Therefore, at times we will need to write an equivalent compound inequality.

EXAMPLE 4 Write equivalent order relations for the following statements:

a. $0.6 < 0.85$ **b.** $-0.5 = -\dfrac{1}{2}$ **c.** $-2 \geq -2.5 \geq -3$

Solution

a. $0.85 > 0.6$ Exchange the numbers and replace $<$ with $>$.

b. $-\dfrac{1}{2} = -0.5$ Exchange the numbers. The $=$ does not change.

c. $-3 \le -2.5 \le -2$ Exchange the numbers and replace $\ge$ with $\le$.

1.1.3 Checkup

Use one of the symbols $>, <,$ or $=$ to compare the two numbers.

1. $0 \underline{\hspace{1cm}} -2$ **2.** $1\dfrac{1}{2} \underline{\hspace{1cm}} 1.5$ **3.** $-2.3 \underline{\hspace{1cm}} -3.5$

Write an equivalent order relation for each statement.

4. $5 < 15$ **5.** $6 > -2$ **6.** $2.75 = \dfrac{11}{4}$ **7.** $4 > 0 \ge -3$

1.1.4 Evaluating Absolute Values

We are now ready to consider mathematical expressions. An **expression** is a combination of numbers and mathematical operations. (By a mathematical operation, we mean addition, subtraction, multiplication, or division.) The quantity $2 + 3$ is a mathematical expression, as is the quantity 2×3. We **evaluate** expressions by finding their numerical value. The expression $2 + 3$ has a value of 5, and the expression 2×3 has a value of 6.

The distance (number of units) a number is from 0 on the number line is called the **absolute value** of the number. The absolute value of a number is always nonnegative (positive or zero), because distance is always measured in nonnegative units.

To write an absolute-value expression, enclose the number in a set of vertical bars, | |. Do not confuse these bars with parentheses, ( ), brackets, [ ], or braces, { }.

$$|6| = 6 \qquad \text{The absolute value of 6 is 6.}$$
$$|-6| = 6 \qquad \text{The absolute value of } -6 \text{ is 6.}$$

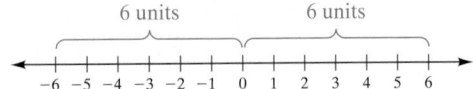

EXAMPLE 5   Evaluate.

a. $|3.5|$ **b.** $\left|-\dfrac{1}{2}\right|$ **c.** $|0|$

Solution

a. $|3.5| = 3.5$ 3.5 is 3.5 units from 0.

b. $\left|-\dfrac{1}{2}\right| = \dfrac{1}{2}$ $-\frac{1}{2}$ is $\frac{1}{2}$ unit from 0.

c. $|0| = 0$ 0 is 0 units from 0.

The checks for Example 5a and 5b are shown in Figure 1.2.

TECHNOLOGY

ABSOLUTE VALUE

Evaluate $|3.5|$, $\left| -\dfrac{1}{2} \right|$.

```
abs(3.5)
              3.5
abs(-1/2)▶Frac
              1/2
```

Figure 1.2

For Figure 1.2,

To enter an absolute-value expression on your calculator, use the abs function found under the MATH NUM menu option 1. Close the set of parentheses that are opened for you to enclose each of the numbers, 3.5 and $-\frac{1}{2}$, in a set of parentheses.

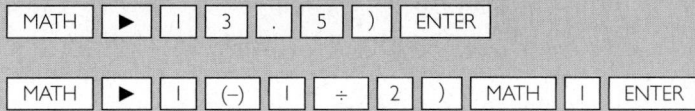

 1.1.4 Checkup

In exercises 1–3, evaluate the absolute-value expressions.

1. $|-15|$ **2.** $|-3.3|$ **3.** $\left| \dfrac{2}{7} \right|$ **4.** What does the absolute value of a number represent?

1.1.5 Evaluating Opposites

Previously, we defined the opposite of a natural number to be a negative number. More generally, two numbers with different signs, but with the same absolute value, meaning the same distance from 0, are called opposites. The opposite of a positive number is a negative number. The opposite of a negative number is a positive number. To write an opposite expression, enclose the expression in a set of parentheses, ( ), with a negative sign in front of it. In the preceding discussion, 6 and -6 are both 6 units from 0; therefore, 6 and -6 are opposites.

$$-(6) = -6 \qquad \text{The opposite of 6 is } -6.$$
$$-(-6) = 6 \qquad \text{The opposite of } -6 \text{ is } 6.$$

EXAMPLE 6 Evaluate:

a. $-\left(1\dfrac{1}{2}\right)$ **b.** $-\left| \dfrac{3}{4} \right|$ **c.** $-(-(75))$

Solution

Check

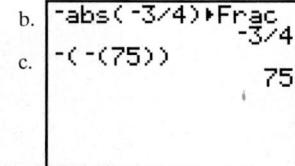

a. $-\left(1\dfrac{1}{2}\right) = -1\dfrac{1}{2}$ $-1\frac{1}{2}$ is the same distance from 0 as $1\frac{1}{2}$.

b. $-\left|-\dfrac{3}{4}\right| = -\dfrac{3}{4}$ Evaluate the absolute value of $-\frac{3}{4}$ and obtain $\frac{3}{4}$. Then take the opposite of $\frac{3}{4}$ and obtain $-\frac{3}{4}$.

c. $-(-(75)) = 75$ Evaluate the opposite of 75 and obtain -75. Next, the opposite of -75 is 75. In other words, the opposite of the opposite of a number is the number itself.

Figure 1.3

 HELPING HAND In evaluating an expression with more than one set of parentheses, you should always work from the inside out. This procedure will be explained in more detail later.

The checks for Example 6b and 6c are shown in Figure 1.3. The opposite symbol is $\boxed{(-)}$.

✓ 1.1.5 Checkup

Evaluate.

1. $-\left(3\dfrac{1}{3}\right)$ **2.** $-\left(-\dfrac{1}{2}\right)$ **3.** $-(-(15))$ **4.** $-|35|$ **5.** $-\left|-\dfrac{4}{7}\right|$

1.1.6 Modeling the Real World

In our daily lives, many things can be expressed as rational numbers. For example, positive and negative numbers are used to define geographic elevations above or below sea level and temperatures above or below zero degrees. Many other situations also lend themselves to being expressed as rational numbers.

EXAMPLE 7   Mount McKinley is 20,320 feet above sea level. Rounded to the nearest 10 feet, the depth of Agulhas Basin in the Indian Ocean is the same distance below sea level. Write an expression, using an opposite symbol, to represent the depth of Agulhas Basin. Evaluate the expression.

Write an expression, using an absolute-value symbol and a rational number, to represent the distance Agulhas Basin is from sea level. Evaluate the expression.

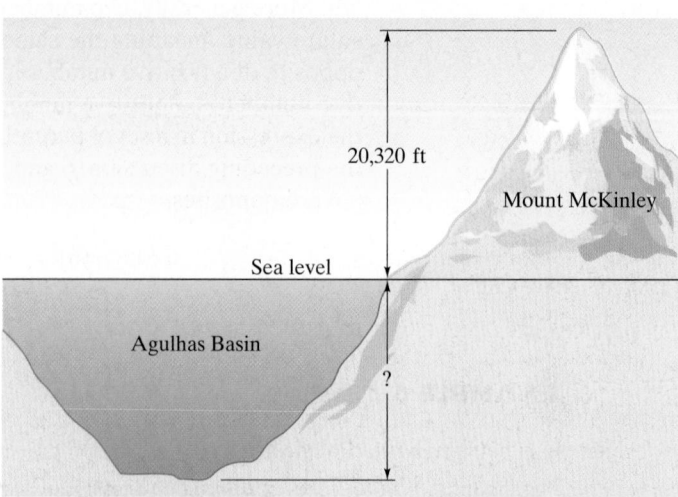

Solution

The depth of Agulhas Basin is the opposite of the height of Mount McKinley. Thus,

$$-(20{,}320) = -20{,}320$$

Agulhas Basin is approximately $-20{,}320$ feet. The absolute-value symbol means the distance from 0, or, in this case, sea level. Therefore,

$$|-20{,}320| = 20{,}320$$

Agulhas Basin is approximately 20,320 feet below sea level.

APPLICATION

Human body temperature is about 98°F. According to the *Guinness Book of World Records,* the coldest inhabited place in the Northern Hemisphere is the Siberian village of Oymyakon (called the "Cold Pole"). The unofficial recorded temperature there is the same as the opposite of the human body temperature. Write and evaluate an expression for the temperature of Oymyakon.

Discussion

Oymyakon has a temperature that is the opposite of the human body temperature, 98°F. Therefore,

$$-(98) = -98$$

The unofficial temperature in Oymyakon is $-98°F$.

✓ 1.1.6 Checkup

1. The coldest temperature (Fahrenheit) on record in the United States was 80 degrees below zero ($-80°F$) at Prospect Creek Camp in the Endicott Mountains of northern Alaska on January 23, 1971. Write an expression, using an absolute-value symbol and a rational number, to represent the number of degrees Fahrenheit that this temperature is away from zero.

2. Between 1995 and 1998, the population of the state of Ohio increased by 0.6%. During the same period, the population of the state of West Virginia experienced the opposite change. Write an expression, using an opposite symbol, for the percentage change in West Virginia's population.

3. In Super Bowl XXXV, Baltimore's quarterback was sacked three times for a total of 20 yards. New York's quarterback was sacked four times for a total of 26 yards. Representing gains by positive numbers and losses by negative numbers, write rational-number representations of each of these situations.

1.1 Exercises

Identify the possible sets of numbers (natural, whole, integer, or rational) to which each number belongs.

1. a. -15 **b.** 29 **c.** 1 million   **d.** $\dfrac{3}{7}$ **e.** $-4\dfrac{1}{3}$

2. a. 278 **b.** $-15\dfrac{2}{3}$ **c.** 5 billion   **d.** -14.2 **e.** $5\dfrac{1}{2}$

3. a. 0 **b.** 12.75 **c.** $2\dfrac{4}{9}$ **d.** -8.35 **e.** $\dfrac{4}{5}$

4. a. $\dfrac{9}{3}$ **b.** -199 **c.** $-\dfrac{11}{15}$ **d.** 0.009 **e.** $\dfrac{3}{2}$

5. On the number line, between what two integers would you graph the rational number $\frac{17}{3}$?

6. On the number line, between what two integers would you graph the rational number $-\frac{3}{2}$?

7. Graph the numbers on a number line.

$$-2, 1.5, \frac{1}{2}, -3.5, \frac{9}{4}, -1\frac{1}{4}, -\frac{4}{5}, 4$$

8. Graph the numbers on a number line.

$$\frac{7}{2}, -2.25, -1.5, 0.5, 2, -2\frac{1}{4}, -1, -\frac{1}{5}$$

Use one of the symbols $>$, $<$, or $=$ to compare the two numbers.

9. -9 _____ -5

10. -7 _____ -4

11. 0 _____ -6

12. 0 _____ 11

13. 5 _____ 3

14. 12 _____ 9

15. $-\frac{1}{2}$ _____ $-\frac{2}{5}$

16. $-\frac{1}{3}$ _____ $-\frac{2}{7}$

17. $\frac{3}{7}$ _____ $\frac{7}{10}$

18. $\frac{3}{8}$ _____ $\frac{2}{7}$

19. $2\frac{3}{5}$ _____ $-2\frac{3}{5}$

20. $-4\frac{3}{7}$ _____ $4\frac{3}{7}$

21. $1\frac{4}{5}$ _____ 1.8

22. $2\frac{2}{5}$ _____ 2.4

23. -3.7 _____ -5.8

24. -6.2 _____ -4.6

25. 1.7 _____ 3.2

26. 4.1 _____ 3.9

27. $|-7|$ _____ 7

28. 6 _____ $|-6|$

29. $|-3.5|$ _____ -3.5

30. -14.3 _____ $|14.3|$

31. $\left|\frac{1}{2}\right|$ _____ $-\left(-\frac{1}{2}\right)$

32. $-\left(-\frac{3}{4}\right)$ _____ $\left|-\frac{3}{4}\right|$

33. $-|2|$ _____ -2

34. -5 _____ $-|5|$

Write an equivalent order relation for each statement.

35. $0.7 > 0.295$

36. $-2.56 < 0.06$

37. $0 \geq -1 > -5$

38. $25 > 24 \geq 20$

39. $\frac{2}{5} = 0.4$

40. $-0.86 = -\frac{43}{50}$

In exercises 41–56, evaluate.

41. $|15.34|$

42. $|25|$

43. $|-15.34|$

44. $|-25|$

45. $\left|-\left(-3\frac{1}{3}\right)\right|$

46. $|-(-25)|$

47. $-|23|$

48. $-|25|$

49. $-|-23|$

50. $-|-25|$

51. $-(15)$

52. $-\left(\frac{1}{2}\right)$

53. $-(-25)$

54. $-(-35)$

55. $-\left(-\left(-\frac{3}{2}\right)\right)$

56. $-\left(-\left(-\frac{5}{4}\right)\right)$

57. According to *Money* magazine, one of the top funds in its list of the Money 100 for 2001 was the Clipper Fund, with a 26.97% return on investment for one year. At the same time The SSgA Growth & Income Fund was listed as having approximately the opposite one-year return.

   a. Write an expression, using an opposite symbol, to represent the one-year return for the SSgA Growth & Income Fund.

   b. Write an expression, using an absolute-value symbol and a rational number, to represent the percentage that the SSgA Growth & Income Fund lost on investment for one year.

58. The Money 100 list of top funds for 2001 reported that the Vanguard Windsor-II fund had a one-year return on investment of 5.82%. The Torray Fund has approximately the opposite one-year return.

   a. Write an expression, using an opposite symbol, to represent the one-year return for the Torray Fund.

   b. Write an expression, using an absolute-value symbol and a rational number, to represent the percentage that the Torray Fund lost on investment for one year.

59. In a Monopoly game, a player gets $45 from the sale of a stock on one play. On another play, a player is assessed $115 for street repairs for a hotel. On a third play, a player collects $200 due to a bank error in her favor. Write rational numbers for each of these situations, using positive numbers for money received and negative numbers for money paid out.

60. On Monday, S & S stock rose 1.25 points; on Tuesday, it dropped 0.75 point; on Wednesday, it dropped 2 points; on Thursday, it rose 1.75 points; and on Friday, it rose 1.50 points. Write rational numbers for each of these situations, using positive numbers for rises in stock prices and negative numbers for drops in stock prices.

61. In 1988, Belayneh Dinsamo of Ethiopia set the world record for a man running a marathon. If a runner beats this record, the time is recorded as the number of seconds below the record, written as a negative number. In the year 2000, Antonio Pinto of Portugal beat Dinsamo's record by 14 seconds. Write an expression, using an opposite symbol, to represent Pinto's time.

62. In 1994, Uta Pippig of Germany set the world record for a woman running a marathon. In 1998, Tegla Loroupe of Kenya beat this record by 58 seconds. Write an expression, using an opposite symbol, to represent Loroupe's time.

63. On a true–false test, an instructor awards five points for every correct answer, deducts five points for each incorrect answer, and deducts two points for each unanswered question. Write rational numbers for the situations in which (a) a student answers a question correctly, (b) a student answers a question incorrectly, and (c) a student leaves a question unanswered.

64. In a game of Jeopardy, write rational numbers for the situations in which a player (a) answers a $100 question correctly and (b) answers a $200 question incorrectly.

65. The lowest temperature recorded in the conterminous 48 United States, 70 degrees below zero Fahrenheit ($-70°F$), was observed at Rogers Pass in Montana in 1954. Write an expression, using absolute-value symbols and a rational number, to represent how many degrees this temperature was below zero.

66. In February, 2000, Tiger Woods won the AT&T Pebble Beach National Pro-Am with a score that was 15 under par, posted as a score of -15. Write an expression, using absolute-value symbols and a rational number, to represent how many strokes his score was below par.

1.1 Calculator Exercises

The number line is used in statistical analysis to display data. In statistics, such a graph is called a **dot plot**. In plotting data on a dot plot, if a number occurs more than once, dots are placed one above another, at the same location on the number line to indicate that a value occurs a multiple number of times. As an example, suppose the following sets of numbers were collected for a statistical study on Fahrenheit temperatures:

$32°, 34°, 35°, 34°, 33°, 37°, 31°, 38°, 42°, 39°, 32°, 33°,$
$30°, 31°, 35°, 36°, 33°, 38°, 39°, 35°$

The dot plot for these data is shown in the following diagram:

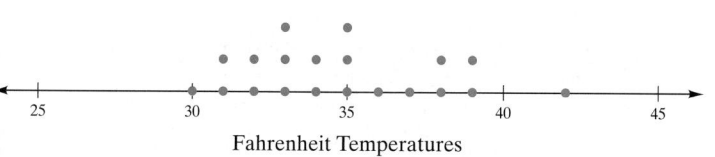

Fahrenheit Temperatures

The plot can be graphed on the calculator using a program. To key the program into your calculator, press [PRGM], scroll over to NEW, and press [ENTER]. The calculator will ask for the name of the program. Since the calculator will be set to ALPHA mode, just type the letters of the name. Each time you press [ENTER], the calculator will move to the next line of the program. Most of the instructions can be found under the [PRGM] key in the edit mode. Just search through the menus until you find the instruction you are looking for, and select it by pressing [ENTER]. The instruction will appear in the program listing. Instructions not found under the [PRGM] menu are identified to the left of the following listing of the program:

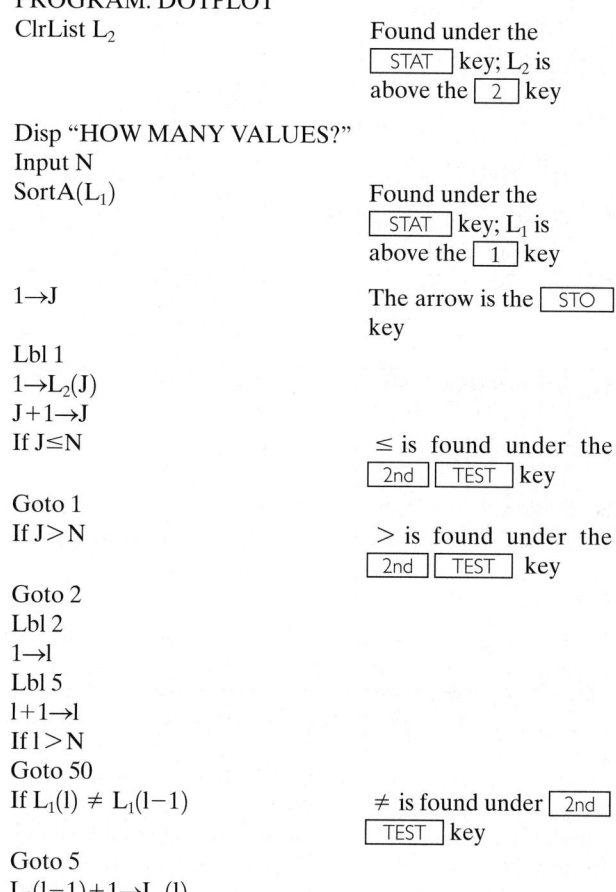

PROGRAM: DOTPLOT
ClrList L_2 — Found under the [STAT] key; L_2 is above the [2] key

Disp "HOW MANY VALUES?"
Input N
SortA(L_1) — Found under the [STAT] key; L_1 is above the [1] key

$1 \rightarrow J$ — The arrow is the [STO] key

Lbl 1
$1 \rightarrow L_2(J)$
$J+1 \rightarrow J$
If $J \leq N$ — $\leq$ is found under the [2nd] [TEST] key

Goto 1
If $J > N$ — $>$ is found under the [2nd] [TEST] key

Goto 2
Lbl 2
$1 \rightarrow I$
Lbl 5
$I+1 \rightarrow I$
If $I > N$
Goto 50
If $L_1(I) \neq L_1(I-1)$ — $\neq$ is found under [2nd] [TEST] key

Goto 5
$L_2(I-1)+1 \rightarrow L_2(I)$
If $I \leq N-1$

Goto 5
Lbl 50
PlotsOff

        Found under `2nd`
        `STATPLOT`

Plot1(Scatter, L_1, L_2, $\square$)

        Found under `2nd`
        `STATPLOT`;
        Scatter is found in `2nd`
        `STATPLOT`, TYPE;
        The symbol is found
        there under MARK

$L_1(1)-1\rightarrow$Xmin

        Xmin is found under
        `VARS`, Window

$L_1(N)+1\rightarrow$Xmax

        All Window settings are
        found under
        `VARS`, Window

$5\rightarrow$Xscl
$-5\rightarrow$Ymin
$N/2\rightarrow$Ymax
$2\rightarrow$Yscl
DispGraph

The screen displays for this program are as follows:

```
PROGRAM:DOTPLOT
:ClrList L2
:Disp "HOW MANY
VALUES?"
:Input N
:SortA(L1)
:1→J
:Lbl 1
```

```
PROGRAM:DOTPLOT
:1→L2(J)
:J+1→J
:If J≤N
:Goto 1
:If J>N
:Goto 2
:Lbl 2
```

```
PROGRAM:DOTPLOT
:1→I
:Lbl 5
:I+1→I
:If I>N
:Goto 50
:If L1(I)≠L1(I-1
)
```

```
PROGRAM:DOTPLOT
:Goto 5
:L2(I-1)+1→L2(I)

:If I≤N-1
:Goto 5
:Lbl 50
:PlotsOff
```

```
PROGRAM:DOTPLOT
:Plot1(Scatter,L
1,L2,□)
:L1(1)-1→Xmin
:L1(N)+1→Xmax
:5→Xscl
: -5→Ymin
:N/2→Ymax
```

```
PROGRAM:DOTPLOT
:2→Yscl
:DispGraph
:
:
:
:
:
```

To use the program, first store the data in L_1 of the calculator:

- Press `STAT`
- Choose Edit
- Move the cursor to list L_1
- Clear the list by placing the cursor on the title, press `CLEAR`, and move the cursor down.
- Enter the values in L_1, pressing `ENTER` after each value.

Keep track of how many values were keyed into L_1.

Press `Y =`, and clear out any equations stored there.

Next press `PRGM`, choose DOTPLOT, and press `ENTER` to begin running the program. The program will ask you how many values you entered into L1. Type the number in, and press `ENTER`. The program will display the dot plot. To identify points press the `TRACE` key and use the arrows to move from one point to another.

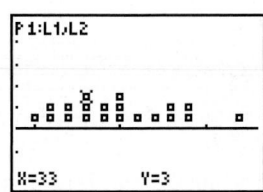

Try drawing dot plots for the following sets of data, and check them with the calculator program:

1. The St. Louis Rams won Super Bowl XXXIV by defeating the Tennessee Titans in an exciting game. The number of points scored in each of the Rams' regular-season games is as follows:

   27 35 38 42 41 34 21 27 35 23 43 34 30 31 34 31

2. Tiger Woods won the 2000 U.S. Open with a score of 272. The following are the scores of the top finishers in the tournament, including Tiger's.

   272 287 287 288 289 289 290 291 291 291 291 292 292
   292 292 293 293 293 293 293 293

1.1 Writing Exercises

How do you feel about mathematics? Has it always been an easy subject for you, or have you always struggled with it? This is your chance to tell your instructor about your feelings toward the subject. Send an e-mail message or write a short letter to your instructor about your feelings toward math. Be sure to give reasons for your feelings—either good or bad past experiences with a math class, a math instructor, or whatever. Suggest to your instructor what would be helpful to you and what would cause you problems in the class. Describe any personal situations that might affect your study of mathematics, such as work, family responsibilities, or course load. Explain how you plan to handle these situations.

1.2 Addition and Subtraction

OBJECTIVES

1 Evaluate the sum of two rational numbers.

2 Evaluate the difference of two rational numbers.

3 Evaluate sums and differences involving more than two rational numbers.

4 Model real-world situations by using the sum or difference of rational numbers.

APPLICATION

In 1999, Mount Everest, the highest point in Tibet, was estimated to be 8850 meters above sea level. The lowest point in China, Turpan Pendi, is 154 meters below sea level. Determine the range of the terrain (the difference of the highest and lowest points).

After completing this section, we will discuss this application further. See page 25.

1.2.1 Evaluating Sums of Rational Numbers

When we add two rational numbers, we obtain a **sum**. The numbers that we add are called **addends**. We will find these terms helpful as we discuss the rules for adding rational numbers.

$$\underset{\text{addend + addend = sum}}{8 \quad + \quad 4 \quad = 12} \qquad \underset{\text{addend + addend = sum}}{4 \quad + \quad 8 \quad = 12}$$

Note that the order in which we add two rational numbers does not change the sum.

COMMUTATIVE LAW FOR ADDITION

Changing the order of two rational-number addends does not change the sum.

To evaluate an expression with grouping symbols, operate on the grouping symbols first.

$$(6 + 2) + 4 = 8 + 4 = 12 \qquad 6 + (2 + 4) = 6 + 6 = 12$$

Note that changing the grouping of three rational-number addends does not change the sum.

ASSOCIATIVE LAW FOR ADDITION

Changing the grouping of three rational-number addends does not change the sum.

In this text, we use the calculator to do simple problems quickly, so that we can use the results to discover rules of mathematics. The rules of mathematics do not depend on the calculator, and the rules we discover are more basic and fundamental in mathematics than are calculator operations. In fact, the people who designed and built your calculator used those rules to do so.

When we add two nonzero rational numbers, there are two possible combinations: The two numbers may have the same signs (**like signs**) or different signs (**unlike signs**). In order to discover rules for adding these combinations, complete the following sets of exercises with your calculator. (Remember to use the $\boxed{(-)}$ for a negative number.)

Discovery 1

Adding Rational Numbers with Like Signs

Evaluate each expression, and compare the results obtained in the left column with the corresponding results in the right column.

1. a. $6 + 2 = $ _____ **b.** $|6| + |2| = $ _____

2. a. $3 + 9 = $ _____ **b.** $|3| + |9| = $ _____

3. a. $-6 + (-2) = $ _____ **b.** $|-6| + |-2| = $ _____

4. a. $-3 + (-9) = $ _____ **b.** $|-3| + |-9| = $ _____

State a rule for writing the sum of two rational numbers with like signs.

In the column on the left of the first two exercises, we add two positive rational numbers. Each sum results in a positive number. In the column on the right, we add the absolute values of the addends. The results are the same as on the left.

In the column on the left of the last two exercises, we add two negative rational numbers. Each sum results in a negative number. In the column on the right, we add the absolute values of the addends. The results are the opposite of those on the left.

Discovery 2

Adding Rational Numbers with Unlike Signs

Evaluate each expression, and compare the results obtained in the left column with the corresponding results in the right column.

1. a. $6 + (-2) = $ _____ **b.** Since $|6| > |-2|$, then $|6| - |-2| = $ _____

2. a. $3 + (-9) = $ _____ **b.** Since $|-9| > |3|$, then $|-9| - |3| = $ _____

3. a. $-6 + 2 = $ _____ **b.** Since $|-6| > |2|$, then $|-6| - |2| = $ _____

4. a. $-3 + 9 = $ _____ **b.** Since $|9| > |-3|$, then $|9| - |-3| = $ _____

State a rule for writing the sum of a positive and a negative rational number.

In the column on the left, we add a positive and a negative rational number. Each sum results in a number whose sign is the same as the sign of the addend with the larger absolute value. In the column on the right, we subtract the smaller absolute-value addend from the larger absolute-value addend. The results are the same as the absolute value of the sum in the left column.

ADDITION OF RATIONAL NUMBERS

To add two rational numbers with like signs (both positive or both negative),

- Add the absolute values of the addends.
- The sign of the sum is the sign of the addends.

To add two rational numbers with unlike signs (one positive and one negative),

- Subtract the smaller absolute-value addend from the larger absolute-value addend.
- The sign of the sum is the sign of the addend with the larger absolute value.

In addition to the commutative and associative properties, two other properties of addition should be noted. The first property we will discuss is the identity property of 0. Adding 0 to any rational number results in the number itself. In other words, when we add 0 to any rational number, the number's "identity" does not change. Therefore, 0 is called the **additive identity**.

$$0 + 4 = 4$$

The second property we use in addition is the additive-inverse property. When we add a rational number and its opposite, the sum is 0. Therefore, the opposite is sometimes called the **additive inverse** of a rational number.

$$6 + (-6) = 0$$

To help you understand the addition rules we have just discovered, let's add rational numbers by using a number line. This will help you visualize how the signs of the addends determine the sign of the sum. We will do three of the exercises in the previous discovery sets on a number line.

To perform addition on a number line, we start at the **origin**, 0, we move to the right if an addend is positive, and we move to the left if an addend is negative.

Adding Two Positive Rational Numbers

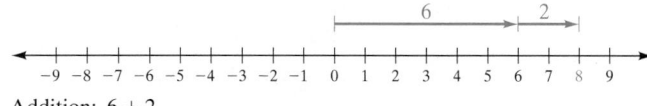

Addition: $6 + 2$

The sum is 8. The result is positive, because both moves were in the positive direction. The absolute value of this sum is the sum of the absolute values of the addends, because both moves were in the same direction from 0.

Adding Two Negative Rational Numbers

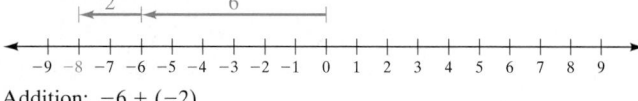

Addition: $-6 + (-2)$

The sum is -8. The result is negative, because both moves were in the negative direction. The absolute value of this sum is the sum of the absolute values of the addends, because both moves were in the same direction from 0.

Adding a Positive and a Negative Rational Number

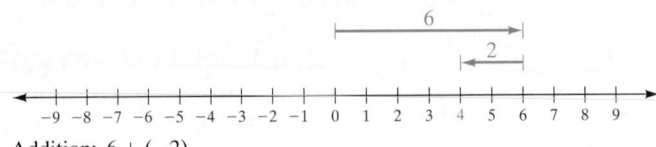

Addition: $6 + (-2)$

The sum is 4. The result is positive, because, although the moves were in opposite directions, the larger move was in the positive direction. The absolute value of this sum is the difference in the absolute values of the addends, because the moves were in different directions and overlapped each other.

We are now ready to use the addition rules we have discovered and illustrated. Let's begin with examples of integers to prepare us for the more complicated exercises that follow.

EXAMPLE 1 Add and check.

a. $8 + (-2)$ **b.** $-6 + (-5)$ **c.** $-7 + 3$

Solution   **Check**

a. $8 + (-2) = 6$
Unlike signs: The sum is positive because $|8| > |-2|$ and 8 is positive. The absolute value of the sum is $|8| - |-2| = 6$. Therefore, the sum is positive 6.

a. $\boxed{8+ {}^-2}$ 6
b. $-6+ {}^-5$ -11
c. $-7+3$ -4

b. $-6 + (-5) = -11$
Like signs: The sum is the same sign as the addends: negative. The absolute value of the sum is $|-6| + |-5| = 11$. Therefore, the sum is negative 11.

c. $-7 + 3 = -4$
Unlike signs: The sum is negative because $|-7| > |3|$ and -7 is negative. The absolute value of the sum is $|-7| - |3| = 4$. Therefore, the sum is negative 4.

EXAMPLE 2 Add and check.

a. $-4.89 + 6.4$ **b.** $-12.5 + (-2)$ **c.** $\dfrac{1}{2} + \left(-\dfrac{2}{3}\right)$ **d.** $-1\dfrac{2}{3} + 2\dfrac{3}{5}$

Solution

a. $-4.89 + 6.4 = 1.51$
Unlike signs: The sum is positive because $|6.4| > |-4.89|$ and 6.4 is positive. The absolute value of the sum is $|6.4| - |-4.89| = 1.51$. Therefore, the sum is positive 1.51.

b. $-12.5 + (-2) = -14.5$
Like signs: The sum is the same sign as the addends: negative. The absolute value of the sum is $|-12.5| + |-2| = 14.5$. Therefore, the sum is negative 14.5.

c. $\dfrac{1}{2} + \left(-\dfrac{2}{3}\right) = \dfrac{3}{6} + \left(-\dfrac{4}{6}\right)$
$= -\dfrac{1}{6}$
Unlike signs: First, convert the fractions to their lowest common denominator: $\frac{1}{2} = \frac{3}{6}$ and $-\frac{2}{3} = -\frac{4}{6}$. The sum is negative because $|-\frac{4}{6}| > |\frac{3}{6}|$ and $-\frac{4}{6}$ is negative. The absolute value of the sum is $|-\frac{4}{6}| - |\frac{3}{6}| = \frac{1}{6}$. Therefore, the sum is negative $\frac{1}{6}$.

d. $-1\dfrac{2}{3} + 2\dfrac{3}{5} = -1\dfrac{10}{15} + 2\dfrac{9}{15}$

Unlike signs: Convert the fractions to their lowest common denominator: 15.

$= -\dfrac{25}{15} + \dfrac{39}{15}$

$= \dfrac{14}{15}$

Convert the mixed numbers to improper fractions. The sum is positive because $|\frac{39}{15}| > |\frac{-25}{15}|$ and $\frac{39}{15}$ is positive. The absolute value of the sum is $|\frac{39}{15}| - |\frac{-25}{15}| = \frac{14}{15}$. Therefore, the sum is positive $\frac{14}{15}$.

The checks for Examples 2c and 2d are shown in Figure 1.4.

TECHNOLOGY

ADDING RATIONAL NUMBERS

Add. c. $\dfrac{1}{2} + \left(-\dfrac{2}{3}\right)$ d. $-1\dfrac{2}{3} + 2\dfrac{3}{5}$

```
c. (1/2)+(-2/3)▶Fra
   c
              -1/6
d. -(1+2/3)+(2+3/5)
   ▶Frac
             14/15
```

Figure 1.4

For Figure 1.4,

Enter the fractions, $\frac{1}{2}$ and $-\frac{2}{3}$, in parentheses for proper grouping and ease in reading. Choose to have the answer displayed as a fraction. This option, ▶ FRAC, is located under the MATH menu option 1.

| (| 1 | ÷ | 2 |) | + | (| (−) | 2 | ÷ | 3 |) | MATH | 1 | ENTER |

Mixed numbers are written as the sum of the whole number and the fraction. Negative mixed numbers have the negative sign outside the parentheses. Choose to have the answer displayed as a fraction by using option 1 under the MATH menu.

| (−) | (| 1 | + | 2 | ÷ | 3 |) | + | (| 2 | + | 3 | ÷ | 5 |) | MATH |
| 1 | ENTER |

✓ 1.2.1 Checkup

Add and check.

1. $-43 + 77$ **2.** $-8.75 + 3$ **3.** $0 + 4.5$ **4.** $-1\dfrac{1}{3} + \left(-2\dfrac{2}{3}\right)$ **5.** $-\dfrac{5}{7} + \dfrac{3}{4}$

6. When you use your calculator to add mixed numbers, it is necessary to enclose the number in parentheses, with any negative signs outside the parentheses. Explain why this is so.

1.2.2 Evaluating Differences of Rational Numbers

When two rational numbers are subtracted, we obtain a **difference**. The numbers that we subtract are called the **minuend** and **subtrahend**. We do not use these terms often, but they are useful for describing the process of subtraction.

$$10 \;-\; 4 \;=\; 6 \qquad\qquad 4 \;-\; 10 \;=\; -6$$
minuend − subtrahend = difference   minuend − subtrahend = difference

Note that the order in which we subtract does change the difference. Subtraction is *not* commutative.

Once again, we will use the calculator to explore the rules underlying mathematical operations. This time, we investigate the subtraction of rational numbers.

When we subtract two nonzero rational numbers, we have any of three combinations. The two numbers may be both positive, both negative, or a positive number and a negative number. Complete the following sets of exercises with your calculator. (Remember to use $\boxed{(-)}$ for a negative number and $\boxed{-}$ for subtraction.)

Discovery 3

Subtracting Two Rational Numbers

Evaluate each expression, and compare the results obtained in the left column with the corresponding results in the right column.

1. a. $6 - 2 =$ _____   **b.** $6 + (-2) =$ _____

2. a. $3 - 9 =$ _____   **b.** $3 + (-9) =$ _____

3. a. $-6 - (-2) =$ _____   **b.** $-6 + 2 =$ _____

4. a. $-3 - (-9) =$ _____   **b.** $-3 + 9 =$ _____

5. a. $6 - (-2) =$ _____   **b.** $6 + 2 =$ _____

6. a. $-3 - 9 =$ _____   **b.** $-3 + (-9) =$ _____

For exercises 1 and 2, state a rule for writing the difference of two positive rational numbers.

For exercises 3 and 4, state a rule for writing the difference of two negative rational numbers.

For exercises 5 and 6, state a rule for writing the difference of a positive and a negative rational number.

Note that the same rule was used to describe each of the three sets in the discovery.

In the column on the left, we subtract two rational numbers. In the column on the right, we add the minuend and the opposite of the subtrahend. The results are the same in both columns.

SUBTRACTION OF RATIONAL NUMBERS

To subtract two rational numbers, add the minuend to the opposite of the subtrahend.

To check the rules of subtraction that we discovered, we need to remember that subtraction is related to addition. Therefore, we can write a subtraction expression as a related addition expression:

$$8 \quad - \quad 3 \quad = 5 \text{ is related to } 3 \quad + \quad 5 \quad = \quad 8.$$

minuend − subtrahend = difference   subtrahend + difference = minuend

We will check examples from the previous discovery sets using this relationship.

$$6 - 2 = \underline{\quad} \text{ is related to } 2 + \underline{\quad} = 6$$

The missing difference is 4.

$$-6 - (-2) = \underline{\quad} \text{ is related to } -2 + \underline{\quad} = -6$$

The missing difference is −4.

$$6 - (-2) = \underline{\quad} \text{ is related to } -2 + \underline{\quad} = 6$$

The missing difference is 8.

We are now ready to use the subtraction rules we have discovered and illustrated. We first perform subtraction on integers in order to prepare ourselves for the more complicated problems that follow.

EXAMPLE 3   Subtract and check.

a. $8 - 12$ **b.** $-9 - (-6)$ **c.** $-7 - 3$ **d.** $4 - (-12)$

Solution   **Check**

a. $8 - 12 = 8 + (-12)$ Change the subtraction to   a. $\boxed{\begin{array}{l} \text{8-12} \\ \text{-4} \end{array}}$
         $= -4$ addition and take the oppo-   b. $\boxed{\begin{array}{l} \text{-9- -6} \\ \text{-3} \end{array}}$
                                site of 12. Then add.

b. $-9 - (-6) = -9 + 6$ Change the subtraction to
              $= -3$ addition and take the oppo-
                               site of −6. Then add.

c. $-7 - 3 = -7 + (-3)$ Change the subtraction to   c. $\boxed{\begin{array}{l} \text{-7-3} \\ \text{-10} \end{array}}$
            $= -10$ addition and take the oppo-   d. $\boxed{\begin{array}{l} \text{4- -12} \\ \text{16} \end{array}}$
                               site of 3. Then add.

d. $4 - (-12) = 4 + 12$ Change the subtraction to
             $= 16$ addition and take the oppo-
                               site of −12. Then add.

EXAMPLE 4   Subtract and check.

a. $9.6 - (-3.8)$ **b.** $0 - (-8.9)$ **c.** $-\dfrac{2}{5} - \dfrac{1}{3}$ **d.** $\dfrac{1}{4} - \left(-1\dfrac{1}{2}\right)$

Solution

a. $9.6 - (-3.8) = 9.6 + 3.8$ Add the opposite of −3.8.
              $= 13.4$

b. $0 - (-8.9) = 0 + 8.9$ Add the opposite of −8.9.
            $= 8.9$

c. $-\dfrac{2}{5} - \dfrac{1}{3} = -\dfrac{2}{5} + \left(-\dfrac{1}{3}\right)$ Add the opposite of $\frac{1}{3}$.

$$= -\dfrac{6}{15} + \left(-\dfrac{5}{15}\right)$$

Change to the lowest common denominator, LCD, of 5 and 3.

$$= -\dfrac{11}{15}$$

d. $\dfrac{1}{4} - \left(-1\dfrac{1}{2}\right) = \dfrac{1}{4} + 1\dfrac{1}{2}$ *Add the opposite of $-1\frac{1}{2}$.*

$\qquad\qquad\qquad = \dfrac{1}{4} + 1\dfrac{2}{4}$ *Change to the LCD of 4 and 2.*

$\qquad\qquad\qquad = 1\dfrac{3}{4}$

The checks for Examples 5a and 5d are shown in Figure 1.5.

TECHNOLOGY

SUBTRACTING RATIONAL NUMBERS

Subtract. a. $9.6 - (-3.8)$ d. $\dfrac{1}{4} - \left(-1\dfrac{1}{2}\right)$

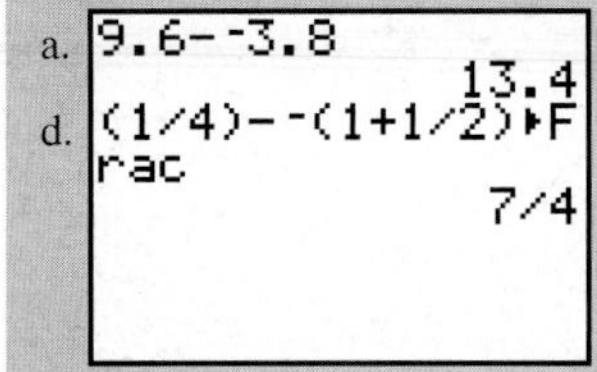

```
a. 9.6--3.8
                 13.4
d. (1/4)--(1+1/2)▶F
   rac
                  7/4
```

Figure I.5

For Figure 1.5,

Negative integers and decimals may be entered without the parentheses. Note the difference in the negative, $\boxed{(-)}$, and the minus, $\boxed{-}$.

$\boxed{9}\;\boxed{.}\;\boxed{6}\;\boxed{-}\;\boxed{(-)}\;\boxed{3}\;\boxed{.}\;\boxed{8}\;\boxed{\text{ENTER}}$

Enter the fraction, $\frac{1}{4}$, in parentheses for proper grouping and ease in reading.

Negative mixed numbers have the negative sign outside the parentheses. Choose to have the answer displayed as a fraction. This option, ▶FRAC, is located under the MATH menu option 1.

$\boxed{(}\;\boxed{1}\;\boxed{\div}\;\boxed{4}\;\boxed{)}\;\boxed{-}\;\boxed{(-)}\;\boxed{(}\;\boxed{1}\;\boxed{+}\;\boxed{1}\;\boxed{\div}\;\boxed{2}\;\boxed{)}\;\boxed{\text{MATH}}$
$\boxed{1}\;\boxed{\text{ENTER}}$

 1.2.2 Checkup

Subtract and check.

1. $12 - 36$ **2.** $-43 - (-77)$ **3.** $-8.75 - 3$ **4.** $0 - 4.5$ **5.** $-\dfrac{5}{7} - \dfrac{3}{4}$ **6.** $1\dfrac{1}{3} - \left(-2\dfrac{2}{3}\right)$

1.2.3 Evaluating Sums and Differences

To perform addition and subtraction operations in the same exercise, first change all subtractions to equivalent additions. Then add all the numbers from left to right.

EXAMPLE 5 Evaluate.

a. $13 + 15 + (-17) + 14 + (-16)$

b. $7.3 - 4.23 + 5 - (-7.9) + (-8.75)$

c. $-\dfrac{2}{3} + \left(-\dfrac{1}{6}\right) + \dfrac{3}{8} - \dfrac{5}{12}$

Solution

a.
$$
\begin{aligned}
13 + 15 + (-17) + 14 + (-16) &= 28 + (-17) + 14 + (-16) \\
&= 11 + 14 + (-16) \\
&= 25 + (-16) \\
&= 9
\end{aligned}
$$

Add the numbers from left to right.

Another way to do Example 5a is to determine the sum of the positive numbers and the sum of the negative numbers:

$$13 + 15 + 14 = 42$$ Sum the positive numbers.

$$-17 + -16 = -33$$ Sum the negative numbers.

Then add these results:

$$42 + (-33) = 9$$

b. $7.3 - 4.23 + 5 - (-7.9) + (-8.75) =$ Rewrite subtraction as addition.

$7.3 + (-4.23) + 5 + 7.9 + (-8.75)$ Add from left to right.

$$
\begin{aligned}
&= 3.07 + 5 + 7.9 + (-8.75) \\
&= 8.07 + 7.9 + (-8.75) \\
&= 15.97 + (-8.75) \\
&= 7.22
\end{aligned}
$$

c.
$$-\dfrac{2}{3} + \left(-\dfrac{1}{6}\right) + \dfrac{3}{8} - \dfrac{5}{12} = -\dfrac{2}{3} + \left(-\dfrac{1}{6}\right) + \dfrac{3}{8} + \left(-\dfrac{5}{12}\right)$$ Rewrite subtraction as addition.

$$= -\dfrac{16}{24} + \left(-\dfrac{4}{24}\right) + \dfrac{9}{24} + \left(-\dfrac{10}{24}\right)$$ Change to the LCD fo 24.

$$= -\dfrac{20}{24} + \dfrac{9}{24} + \left(-\dfrac{10}{24}\right)$$ Add from left to right.

$$= -\dfrac{11}{24} + \left(-\dfrac{10}{24}\right)$$

$$= -\dfrac{21}{24} = -\dfrac{7}{8}$$ Reduce to lowest terms.

1.2.3 Checkup

Evaluate.

1. $-13 + 52 - (-2) - 13 + (-21)$

2. $\dfrac{1}{4} + \left(-\dfrac{3}{5}\right) - \left(-\dfrac{1}{2}\right) - \dfrac{3}{10} + \left(-\dfrac{3}{20}\right)$

3. $12.3 + (-11.5) - 3.7 - (-23.1) + 2$

1.2.4 Modeling the Real World

The addition and subtraction rules for rational numbers are the same rules that govern situations involving addition and subtraction in our daily lives. Operations that involve totals or balances, such as totaling an amount spent in a store or balancing a bank account, can be written as rational expressions. Once we've decided how to represent real data as rational expressions, we add or subtract the expressions.

When we model real-world situations, it is best to always follow a logical set of steps.

MODELING REAL-WORLD SITUATIONS
To model a real-world situation,

- Read and understand the situation.
- Write an expression with the information given.
- Evaluate the expression.
- Check the value obtained.
- Write an answer to the question, using complete sentences.

EXAMPLE 6   Lorenzo is the owner of a handcrafted-furniture business. He is balancing his accounting records of transactions for the day. He took in three sales: $123.00, $798.00, and $563.00. He paid out to suppliers $699.38 for wood and $76.93 for varnish. He also received a credit from the phone company for $12.75 for an overpayment of his last bill. What is his account balance at the end of the day?

Solution
First, understand the situation. We begin with the sum of the positive numbers that represent amounts taken in from sales. We subtract from these positive numbers the amount paid out to each supplier. A credit is a return of an amount paid out. We subtract (a return) a negative amount (amount paid out).

Second, write an expression:

$$\overset{\text{sales}}{123.00} + \overset{\text{sales}}{798.00} + \overset{\text{sales}}{563.00} - \overset{\text{supplies}}{699.38} - \overset{\text{supplies}}{76.93} - \overset{\text{credit}}{(-12.75)}$$

Third, evaluate the expression. Change all subtractions to additions and the number following each subtraction symbol to its opposite. Note that subtracting a credit is the same as adding a positive amount. The result is

$$123.00 + 798.00 + 563.00 + (-699.38) + (-76.93) + 12.75$$

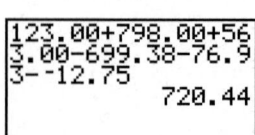

Next, add all the positive numbers and all the negative numbers, and find the sum of the two sums:

$$123.00 + 798.00 + 563.00 + 12.75 = 1496.75$$
$$-699.38 + (-76.93) = -776.31$$
$$1496.75 + (-776.31) = 720.44$$

Check your results.

Finally, answer the question.

Lorenzo had a balance of $720.44 at the end of the day.

APPLICATION

Mount Everest, the highest point in Tibet, was estimated in 1999 to be 8850 meters above sea level. The lowest point in China, Turpan Pendi, is 154 meters below sea level. Determine the range of the terrain (the difference of the highest and lowest points).

Discussion

The elevation of Mount Everest is 8850 meters, and that of Turpan Pendi is −154 meters. The range is determined by the difference of the highest point and the lowest points.

$$8850 - (-154) = 8850 + 154 = 9004$$

The range of the terrain in China is 9004 meters.

✓ 1.2.4 Checkup

1. At the start of the month, Beverly's checking account showed a balance of $897.63. During the month, she added to her account by making two deposits: one for $355.00 and another for $572.00. During the same month, she decreased her account by making two withdrawals: one for $120.00 and another for $300.00. She wrote three checks for $185.23, $104.50, and $231.97. However, she stopped payment on the last check, thereby subtracting the deduction from her account. The bank charged $10.00 to stop payment on the check. Write an addition-and-subtraction problem to represent these transactions. What is the current balance in Beverly's account?

2. The December 2000 precipitation for Lake Michigan was $1\frac{7}{10}$ inches. The average December precipitation from 1900 to 1995 was $1\frac{91}{100}$ inches. Determine how much lower the December 2000 precipitation was from the average.

1.2 Exercises

Add and check.

1. $-7 + 9$
2. $-5 + 6$
3. $-9 + (-2)$
4. $(-2) + (-3)$
5. $5 + (-7)$
6. $7 + (-9)$
7. $32 + (-579)$
8. $703 + (-21)$
9. $2.7 + 3.96$
10. $0.06 + 3.1$
11. $1.2 + (-2.5)$
12. $-5.5 + 2.7$
13. $-2.73 + 4.1$
14. $3.9 + (-1.81)$
15. $-1.1 + (-2.27)$
16. $-3.5 + (-7.9)$
17. $-\frac{3}{5} + \left(-\frac{1}{2}\right)$
18. $-\frac{1}{4} + \left(-\frac{3}{7}\right)$
19. $-\frac{7}{9} + \frac{1}{6}$
20. $-\frac{3}{8} + \frac{1}{12}$
21. $\frac{2}{3} + \left(-\frac{2}{9}\right)$
22. $\frac{5}{6} + \left(-\frac{1}{3}\right)$
23. $-2\frac{3}{4} + 3\frac{2}{3}$
24. $-1\frac{4}{5} + 4\frac{2}{3}$

Subtract and check.

25. $-7 - 9$
26. $-10 - 2$
27. $-9 - (-2)$
28. $-12 - (-6)$
29. $5 - (-13)$
30. $12 - (-4)$
31. $32 - 579$
32. $703 - 21$
33. $1.2 - (-2.5)$
34. $3.9 - (-1.81)$
35. $-1.1 - (-2.27)$
36. $-3.5 - (-7.9)$
37. $2.7 - 3.96$
38. $0.06 - 3.1$
39. $-\frac{3}{5} - \left(-\frac{1}{2}\right)$
40. $-\frac{1}{4} - \left(-\frac{3}{7}\right)$
41. $-\frac{7}{9} - \frac{1}{6}$
42. $-\frac{3}{8} - \frac{1}{12}$
43. $\frac{2}{3} - \left(-\frac{7}{9}\right)$
44. $\frac{5}{6} - \left(-\frac{1}{3}\right)$
45. $\frac{3}{7} - 3$
46. $\frac{3}{5} - 2$
47. $-5 - \left(-1\frac{4}{5}\right)$
48. $-7 - \left(-3\frac{2}{3}\right)$
49. $3\frac{2}{3} - 5$
50. $4\frac{1}{4} - 6$

Evaluate and check.

51. $17 + (-23) + 0 + 13$

52. $32 + 0 + (-14) + 72$

53. $-\dfrac{1}{2} + \dfrac{1}{3} + \left(-\dfrac{1}{4}\right)$

54. $\dfrac{2}{5} + \left(-\dfrac{1}{2}\right) + \left(-\dfrac{3}{4}\right)$

55. $17 - (-23) - (-16) - 13$

56. $32 - (-55) - (-14) - 72$

57. $1.2 - (-2.31) - (-5.7)$

58. $4.3 - (-2.3) - (-9.72)$

59. $-\dfrac{1}{2} - \dfrac{1}{3} - \left(-\dfrac{1}{4}\right)$

60. $\dfrac{2}{3} - \left(-\dfrac{2}{3}\right) - \left(-\dfrac{3}{4}\right)$

61. $1124 - (-924) - 2305 - (-1156) - (-109)$

62. $3562 - (-901) - (-805) - (-3231) - 1020$

63. $23 + 56 - 34 + (-12) - 68 - (-31)$

64. $132 - (-239) + (-141) - 53 + 75 - 18$

65. $1\dfrac{1}{5} - 2\dfrac{3}{10} + \dfrac{4}{5} - \left(-\dfrac{7}{10}\right) + \left(-\dfrac{3}{5}\right)$

66. $1\dfrac{1}{3} + 2\dfrac{3}{5} - 5\dfrac{2}{3} - \left(-1\dfrac{2}{5}\right) + \dfrac{4}{15}$

67. $3.75 - 1.2 + (-1.09) - (-0.76) + 13.13$

68. $-1.08 + 5.7 - (-0.05) - 4.37 + (-1.11) - 2$

Write and evaluate an expression to represent each of the situations that follow. Use positive and negative numbers to represent gains and losses accordingly. Interpret the result.

69. In 1998, projections of the resident populations for states showed that the state of California grew by 2,881,000 people over its 1990 census of 29,786,000. What was the 1998 projected population for California?

70. In 1998, projections of the resident populations for states showed that the state of Rhode Island decreased by 15,000 people from its 1990 census of 1,003,000. What was the 1998 projected population for Rhode Island?

71. According to the *Guinness Book of World Records,* one of the greatest temperature ranges in the world occurs in Siberia. Temperatures at Verkhoyansk have ranged from $-90°F$ to $98°F$. What is the range of these temperatures? (The range is calculated as the high value minus the low value.)

72. One of the greatest temperature variations recorded in one day occurred at Browning, Montana, from $44°F$ to $-56°F$, on January 23–24, 1916. What was the range of temperatures that day?

73. The mean surface temperature of the Moon has been reported to be $130°C$ during the lunar day and $-180°C$ at night. What is the change in mean surface temperature from lunar day to night?

74. The mean surface temperature of the planet Mercury has been reported to be $350°C$ during the day and $-170°C$ during the night. What is the change in mean surface temperature during these times?

75. The highest point in the United States is Mount McKinley, Alaska, which has an elevation of 20,320 feet above sea level. The lowest point in the United States is Death Valley, California, which has an elevation of 282 feet below sea level (-282 feet). Find the range between the highest point and the lowest point.

76. The highest point in the state of Louisiana is Driskill Mountain, which has an elevation of 163 meters above sea level. The lowest point in Louisiana is New Orleans, which has an elevation of 2 meters below sea level. Find the range between these highest and lowest points.

77. A cookie recipe requires $\frac{1}{2}$ cup of granulated sugar, $\frac{1}{4}$ cup of brown sugar, 1 cup of sifted all-purpose flour, 1 cup of semisweet chocolate pieces, and $\frac{1}{2}$ cup of chopped nuts. How many cups of dry ingredients does the recipe use?

78. A quilting pattern requires $1\frac{1}{4}$ yards of solid blue material, $\frac{3}{4}$ yard of yellow floral material, $\frac{1}{2}$ yard of green floral material, and $2\frac{1}{4}$ yards of navy plaid material. How many yards of material does the pattern use?

79. Karin is a saleswoman with a weekly quota of $5000 in sales. She keeps track of her sales by noting how much she is above her quota (a positive number) or how much she is below her quota (a negative number). Last month, her weekly sales in relation to her quota were $255 below, $375 above, $575 below, and $1525 above. What was

Karin's overall standing for the month in terms of her quota? Is she above or below quota? How can you tell?

80. For 6 weeks, Karin's sales in relation to her quota were $385 above, $285 above, $555 below, $405 below, $265 above, and $575 above. She keeps track of her sales by noting sales above quota as a positive number and sales below quota as a negative number. What was her overall performance in sales? Is she above or below quota for the six-week period?

81. Lindsay's parents opened a college savings account at the credit union. The initial deposit was $1500. For each of the next three months, they deposited $150. However, they withdrew $75 for a savings bond, $500 for a stock investment opportunity, and $200 for a municipal bond. The credit union paid the account $12 in interest for the three-month period. What is the balance of Lindsay's account?

82. At Mallory's birth, her grandparents opened a savings account for her by making an initial deposit of $500. Mallory's parents added deposits of $75, $50, and $100, received as gifts from relatives. Her parents withdrew $125 to make a premium payment on a new insurance policy. The bank added $35 interest at the end of the first year. What is the balance in the account?

83. Rosie makes and sells T-shirts. On one morning, she sold three shirts at $19.95 each, paid $25.00 to purchase dyes, paid the electric bill of $59.27, and refunded the price of one shirt that was the wrong size. What was her balance at the end of that morning?

84. Richard runs a bookstore. In one hour, he sold a book for $39.95, another for $27.95, and a third for $19.99. He paid out $175.00 for the monthly rent and paid $29.95 to the newspaper to run an advertisement for him. A customer returned a book for a refund of $14.95. What was his balance at the end of that hour?

85. An interesting application of the skills of adding and subtracting numbers occurs in the reading of parts diagrams for a machine shop. Many times, the diagrams list only the essential measurements and assume that the user of the diagram can obtain the remaining measurements through addition and subtraction. The drawing in Figure 1.6 is an example of such a diagram. It is a drawing of a part called a *taper*, which a machinist produces. All measurements are in inches. Write and evaluate an addition or subtraction expression to determine the length of A in the drawing.

86. For the taper in exercise 85, write and evaluate an addition or subtraction expression to determine the lengths of B and C in the drawing.

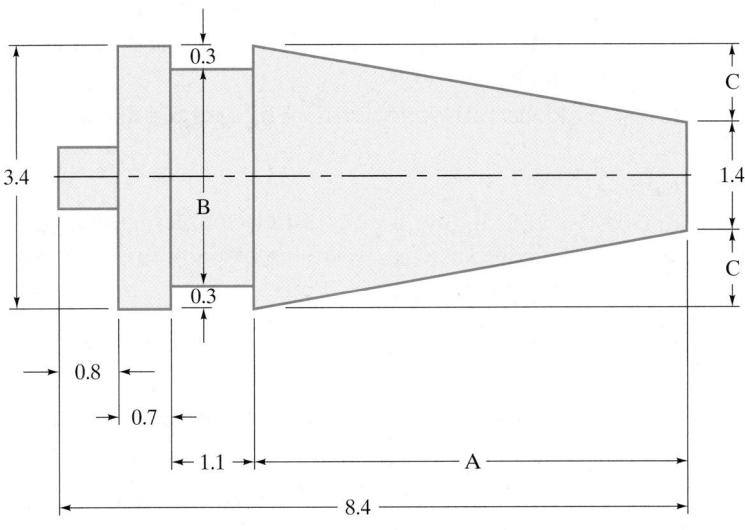

Figure 1.6

1.2 Calculator Exercises

Converting an Improper Fraction to a Mixed Number

In performing operations with mixed numbers, some calculators may present the result as an improper fraction. If the calculator does not provide an option to convert the improper fraction to a mixed number, the following procedure may be used:

1. Convert the improper fraction to a decimal by dividing on the calculator.

2. Subtract the integer part of the decimal (the part to the left of the decimal point) from the decimal on the calculator.

3. Convert the remaining part of the decimal (the part to the right of the decimal point) to a fraction on the calculator.

4. Add the integer part back to the fraction part to obtain the mixed number.

For example, convert the improper fraction $\frac{43}{15}$ to a mixed number.

1. $\frac{43}{15}$ is 2.86666666 ....

2. 2.8666666666 ... $-\ 2 = 0.8666666666 ...$

3. 0.8666666666 ... [MATH] [I] [ENTER] yields $\frac{13}{15}$.

4. $\frac{13}{15}$ added to 2 yields $2\frac{13}{15}$.

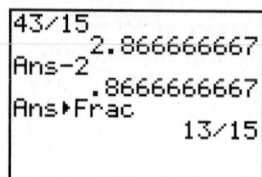

Convert the following improper fractions to mixed numbers, using the preceding approach:

1. $\dfrac{55}{7}$ 2. $-\dfrac{295}{113}$ 3. $\dfrac{1227}{487}$ 4. $-\dfrac{108}{19}$

1.2 Writing Exercises

Write your own definitions for the following mathematical terms used in this section: sum, addends, difference, minuend, subtrahend.

1.3   Multiplication and Division

OBJECTIVES

1 Evaluate the product of two rational numbers.

2 Evaluate the quotient of two rational numbers.

3 Evaluate products and quotients involving more than two rational numbers.

4 Evaluate sums, differences, products, and quotients involving more than two rational numbers.

5 Model real-world situations by using the sum or difference of rational numbers.

APPLICATION

Fifty-two percent of the U.S. national debt consists of marketable loans such as bills, notes, and bonds. In 1999, the national debt was $5.6 trillion. Determine the amount of the debt that is made up of marketable loans.

After completing this section, we will discuss this application further. See page 40.

1.3.1 Evaluating Products of Rational Numbers

When two rational numbers are multiplied, we obtain a **product**. The numbers that we multiply are called **factors**. We will use these terms as we discuss multiplying rational numbers.

$$6 \times 3 = 18 \qquad 6 \cdot 3 = 18 \qquad (6)(3) = 18$$

factor × factor = product   factor · factor = product   (factor)(factor) = product

The order in which we multiply two rational numbers does not change the product.

$$6 \cdot 2 = 12 \qquad 2 \cdot 6 = 12$$

COMMUTATIVE LAW FOR MULTIPLICATION

Changing the order of two rational-number factors does not change the product.

Also, changing the grouping when multiplying three rational numbers does not change the product.

$$(6 \cdot 2) \cdot 4 = 12 \cdot 4 = 48 \qquad 6 \cdot (2 \cdot 4) = 6 \cdot 8 = 48$$

ASSOCIATIVE LAW FOR MULTIPLICATION
Changing the grouping of three rational-number factors does not change the product.

As before, we use the calculator to perform mathematical operations so that we can discover the rules that govern these operations. In this case, we want to discover the rules for multiplication of rational numbers.

When we multiply two rational numbers, there are two possible combinations: The two numbers may have like signs; the two numbers may have unlike signs. Complete the following set of exercises with your calculator.

 Discovery 4

Multiplying Two Rational Numbers

Evaluate each expression, and compare the results obtained in the left column with the corresponding results in the right column.

1. a. $3 \cdot 2 =$ _____   **b.** $|3| \cdot |2| =$ _____

2. a. $-3 \cdot (-2) =$ _____   **b.** $|-3| \cdot |-2| =$ _____

3. a. $3 \cdot (-2) =$ _____   **b.** $|3| \cdot |-2| =$ _____

4. a. $-3 \cdot 2 =$ _____   **b.** $|-3| \cdot |2| =$ _____

For exercises 1 and 2, state a rule for writing the product of two rational numbers with like signs.

For exercises 3 and 4, state a rule for writing the product of two rational numbers with unlike signs.

In the column on the left, when we multiply two rational numbers with like signs, each product results in a positive number. In the column on the right, we multiply the absolute values of the factors. The results are the same as on the left.

In the column on the left, when we multiply two rational numbers with unlike signs, each product results in a negative number. In the column on the right, we multiply the absolute values of the factors. The results are the opposite of those on the left.

MULTIPLICATION OF RATIONAL NUMBERS
To multiply two rational numbers with like signs (both positive or both negative),

- Multiply the absolute values of the factors.
- The sign of the product is positive.

To multiply rational numbers with unlike signs (one positive and one negative),

- Multiply the absolute values of the factors.
- The sign of the product is negative.

Four additional properties of multiplication should be noted. The first property we discuss is the **identity property of 1**. Multiplying any rational number by 1 results in the number itself. Therefore, 1 is called the **multiplicative identity**.

$$(6)(1) = 6$$

An extension of this property is multiplying a rational number by -1. The product is the opposite of the rational number. Therefore, the **multiplication property of -1** states that the product of -1 and a rational number is the opposite of the rational number itself.

$$6(-1) = -6$$

Another property we use in multiplication is the **multiplicative inverse property**. When we multiply a nonzero rational number by its **reciprocal**, the product is 1. Therefore, the reciprocal is sometimes called the **multiplicative inverse** of a rational number.

$$-\frac{2}{3} \cdot \left(-\frac{3}{2}\right) = 1$$

Thus, $-\frac{2}{3}$ and $-\frac{3}{2}$ are reciprocals or multiplicative inverses.

The result of multiplying any rational number by 0 is 0. This property is referred to as the **multiplication property of 0**.

$$0 \cdot 5 = 0$$

To understand the rules of multiplication that we have just discovered, we need to remember that multiplication is a shortcut for repeated addition. If we work two examples in the preceding discovery sets, the result is

$$3 \cdot 2 = 2 + 2 + 2 = 6$$
three two's

$$3 \cdot (-2) = (-2) + (-2) + (-2) = -6$$
three negative two's

A problem occurs when we try to multiply two negative numbers.

$$-3 \cdot (-2) = $$ *How do we write negative two a negative three times?*

Let's look at the following pattern and see if we can infer that this product is really 6:

first factor decreases by one

$$2(-2) = -4$$
$$1(-2) = -2$$ *product increases by two*
$$0(-2) = 0$$
$$-1(-2) = 2$$
$$-2(-2) = 4$$
$$-3(-2) = \underline{\quad}$$ *(Using the preceding pattern, we see that this answer must be 6.)*

Therefore, $-3(-2) = 6$.

We are now ready to use the multiplication rules we have discovered and illustrated. Once again, we begin with examples of integers.

EXAMPLE 1 Multiply.

a. $8(-2)$ **b.** $-6(-5)$ **c.** $-7 \cdot 3$

Solution   **Check**

a. $8(-2) = -16$ Unlike signs: The product is   a. $8*-2$
                       negative. The product of the ab- -16
                       solute values of the factors is b. $-6*-5$
                       $|8| \cdot |-2| = 16$. Therefore, the 30
                       product is negative 16.   c. $-7*3$
                                                                       -21

b. $-6(-5) = 30$ Like signs: The product is posi-
                       tive. The product of the ab-
                       solute values of the factors is
                       $|-6| \cdot |-5| = 30$. Therefore,
                       the product is positive 30.

c. $-7 \cdot 3 = -21$ Unlike signs: The product is negative. The product of the absolute
                       values of the factors is $|-7| \cdot |3| = 21$. Therefore, the product is
                       negative 21.

EXAMPLE 2   Multiply.

a. $(-4.89)(-6.4)$ **b.** $(-8.9)(0)$ **c.** $\dfrac{1}{2} \cdot -\dfrac{1}{3}$ **d.** $-1\dfrac{2}{3} \cdot 2\dfrac{3}{5}$

Solution

a. $(-4.89)(-6.4) = 31.296$ Like signs: The product is positive. The product of the ab-
                                  solute values of the factors is $|-4.89| \cdot |-6.4| = 31.296$.
                                  Therefore, the product is positive 31.296.

b. $(-8.9)(0) = 0$ The product of a number and 0 is 0.

c. $\dfrac{1}{2} \cdot -\dfrac{1}{3} = -\dfrac{1}{6}$ Unlike signs: The product is negative. The product of the
                                  absolute values of the factors is $|\frac{1}{2}| \cdot |\frac{-1}{3}| = \frac{1}{6}$. Therefore,
                                  the product is negative $\frac{1}{6}$.

d. $-1\dfrac{2}{3} \cdot 2\dfrac{3}{5} = -\dfrac{5}{3} \cdot \dfrac{13}{5}$ Unlike signs: The product is negative. Change the mixed
                                  numbers to improper fractions and multiply. The product
$\qquad = -\dfrac{65}{15}$ of the absolute values of the factors is $|-\frac{5}{3}| \cdot |\frac{13}{5}| = \frac{65}{15}$.
                                  Therefore, the product is negative $\frac{65}{15}$.

$\qquad = -\dfrac{13}{3}$ Reduce the fraction and change to a mixed number.

$\qquad = -4\dfrac{1}{3}$

The checks for Examples 2a and 2d are shown in Figure 1.7.

T E C H N O L O G Y

MULTIPLYING RATIONAL NUMBERS

Multiply. a. $(-4.89)(-6.4)$ d. $-1\dfrac{2}{3} \cdot 2\dfrac{3}{5}$

a. $-4.89*-6.4$
           31.296
d. $-(1+2/3)*(2+3/5)$
   ▸Frac
           $-13/3$

Figure 1.7

(Continued on page 32)

For Figure 1.7,

There is no need to place the decimal numbers in parentheses. You may use the multiplication sign instead.

Place the negative mixed number as a sum in parentheses with the negative outside the grouping. Choose to have your answer displayed as a fraction. This option, ▶ FRAC, is located under the MATH menu option 1.

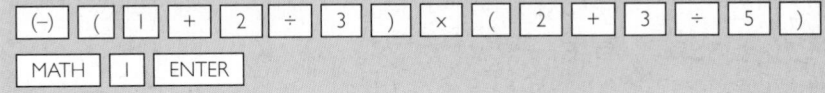

 1.3.1 Checkup

Multiply.

1. $36 \cdot 2$ **2.** $-\frac{1}{3} \cdot \left(-\frac{3}{4}\right)$ **3.** $(-7.2)(0.02)$ **4.** $2\frac{1}{4} \cdot \left(-1\frac{1}{3}\right)$

1.3.2 Discovering Division Rules

When two rational numbers are divided, we obtain a **quotient**. The numbers we divide are called the **dividend** and the **divisor**. We will use these terms when we discuss the division of rational numbers.

$$15 \div 3 = 5 \qquad 3\overline{)15}\,^{5} \qquad \frac{15}{3} = 5$$

dividend ÷ divisor = quotient $\qquad$ quotient / divisor)dividend $\qquad$ dividend/divisor = quotient

When we divide two rational numbers, there are the same two combinations—like and unlike signs—as in the case of the multiplication rules. We will add a discovery for the rules of division involving 0. Complete the following sets of exercises with your calculator.

 Discovery 5

Dividing Two Rational Numbers

Evaluate each expression, and compare the results obtained in the left column with the corresponding results in the right column.

1. a. $8 \div 2 =$ _____   **b.** $|8| \div |2| =$ _____
2. a. $-8 \div (-2) =$ _____   **b.** $|-8| \div |-2| =$ _____
3. a. $8 \div (-2) =$ _____   **b.** $|8| \div |-2| =$ _____
4. a. $-8 \div 2 =$ _____   **b.** $|-8| \div |2| =$ _____

In exercises 1 and 2, state a rule for writing the quotient of two rational numbers with like signs.

In exercises 3 and 4, state a rule for writing the quotient of two rational numbers with unlike signs.

In the column on the left, when we divide two rational numbers with like signs, each quotient is a positive number. In the column on the right, we divide the absolute values of the dividend and divisor. The results are the same as on the left.

In the column on the left, when we divide two rational numbers with unlike signs, each quotient is a negative number. In the column on the right, we divide the absolute values of the dividend and divisor. The results are the opposite of those on the left.

Discovery 6

Division Involving Zero

Evaluate each expression.

1. $8 \div 0 =$ _____   **2.** $-8 \div 0 =$ _____
3. $0 \div 2 =$ _____   **4.** $0 \div (-5) =$ _____
5. $0 \div 0 =$ _____

In exercises 1 and 2, state a rule for dividing a rational number by 0.
In exercises 3 and 4, state a rule for dividing 0 by a rational number other than 0.
In exercise 5, state a rule for dividing 0 by 0.

Dividing a nonzero rational number by 0 results in an error on the calculator. That is, the quotient cannot be found when the divisor is 0. We call this quotient **undefined**.

Dividing 0 by any rational number other than 0 results in 0. Dividing 0 by 0 results in an error. (The quotient is **indeterminate**.)

HELPING HAND "Undefined" and "indeterminate" do not mean the same thing. This is explained on the next page.

DIVISION OF RATIONAL NUMBERS

To divide two rational numbers with like signs (both positive or both negative),

- Divide the absolute values of the numbers.
- The sign of the quotient is positive.

To divide two rational numbers with unlike signs (one positive and one negative),

- Divide the absolute values of the numbers.
- The sign of the quotient is negative.

To divide 0 by a nonzero rational number, the result is 0.
To divide a rational number by 0 is undefined.
To divide 0 by 0 is indeterminate.

To check the rules of division that we have just discovered, we need to remember that division is related to multiplication. Therefore, we can write any division expression as a related multiplication expression. For example,

$$12 \div 4 = 3 \qquad \text{is related to} \qquad 4 \cdot 3 = 12$$

dividend ÷ divisor = quotient   divisor quotient = dividend

If we use this fact, we can find a quotient by completing a multiplication expression. We will work examples from the previous discovery sets to understand this concept.

$$8 \div 2 = \underline{\hspace{1cm}} \text{ is related to } 2 \cdot \underline{\hspace{1cm}} = 8.$$

The missing quotient must be 4.

$$-8 \div -2 = \underline{\hspace{1cm}} \text{ is related to } -2 \cdot \underline{\hspace{1cm}} = -8.$$

The missing quotient must be 4.

$$8 \div -2 = \underline{\hspace{1cm}} \text{ is related to } -2 \cdot \underline{\hspace{1cm}} = 8.$$

The missing quotient must be −4.

$$-8 \div 2 = \underline{\hspace{1cm}} \text{ is related to } 2 \cdot \underline{\hspace{1cm}} = -8.$$

The missing quotient must be −4.

The divisions involving zero are especially important to understand. Study the following very carefully:

$$8 \div 0 = \underline{\hspace{1cm}} \text{ is related to } 0 \cdot \underline{\hspace{1cm}} = 8.$$

The missing quotient cannot be found, because 0 times any number results in 0, not 8. Therefore, we say the quotient is undefined.

$$0 \div 2 = \underline{\hspace{1cm}} \text{ is related to } 2 \cdot \underline{\hspace{1cm}} = 0.$$

The missing quotient must be 0.

$$0 \div 0 = \underline{\hspace{1cm}} \text{ is related to } 0 \cdot \underline{\hspace{1cm}} = 0.$$

The missing quotient could be any number. Since it is impossible to determine only one number that correctly completes this statement, we say the quotient is indeterminate.

We are now ready to use the division rules we have discovered and illustrated. Let's begin with examples of integers.

EXAMPLE 3   Divide.

a. $8 \div (-4)$ **b.** $\dfrac{-6}{-2}$ **c.** $\dfrac{2}{-6}$

Solution

a. $8 \div (-4) = -2$

Unlike signs: The quotient is negative. The quotient of the absolute values of the numbers is $|8| \div |-4| = 2$. Therefore, the quotient is negative 2.

b. $\dfrac{-6}{-2} = 3$

Like signs: The quotient is positive. The quotient of the absolute values of the numbers is $|-6| \div |-2| = 3$. Therefore, the quotient is positive 3.

c. $\dfrac{2}{-6} = -\dfrac{1}{3}$

Unlike signs: The quotient is negative. The quotient of the absolute values of the numbers is $|2| \div |-6| = 0.\overline{33}$, or $\frac{1}{3}$. Therefore, the quotient is negative $0.\overline{33}$, or negative $\frac{1}{3}$.

Check

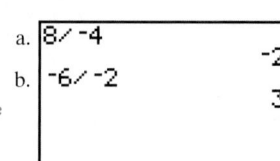

a.
```
8/-4
                -2
```
b.
```
-6/-2
                 3
```

c.
```
2/-6
       -.3333333333
Ans►Frac
              -1/3
```

Even though it is correct to do so, there is no need to place the numbers in parentheses; enter the numbers with the division sign between them for faster entry.

EXAMPLE 4   Divide.

a. $1.2 \div (-3)$ **b.** $\dfrac{-12}{-0.8}$ **c.** $\dfrac{0}{-0.34}$ **d.** $-8.9 \div 0$

Solution

a. $1.2 \div (-3) = -0.4$ Unlike signs: The quotient is negative. The quotient of the absolute values of the numbers is $|1.2| \div |-3| = 0.4$. Therefore, the quotient is negative 0.4.

b. $\dfrac{-12}{-0.8} = 15$ Like signs: The quotient is positive. The quotient of the absolute values of the numbers is $|-12| \div |-0.8| = 15$. Therefore, the quotient is positive 15.

c. $\dfrac{0}{-0.34} = 0$ The quotient of 0 divided by a rational number other than 0 is 0.

d. $-8.9 \div 0 = $ undefined   The quotient for a rational number divided by 0 is undefined.

The checks for Examples 4c and 4d are shown in Figures 1.8a and 1.8b.

T E C H N O L O G Y

DIVIDING RATIONAL NUMBERS

Divide.   c. $\dfrac{0}{-0.34}$ d. $-8.9 \div 0$

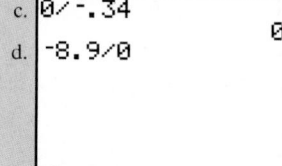

Figure 1.8a

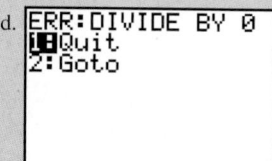

Figure 1.8b

For Figures 1.8a and 1.8b,

There is no need to enter the whole number 0 in the decimal -0.34.

| 0 | ÷ | (-) | . | 3 | 4 | ENTER |

Since it is not possible to divide by 0, an error is displayed on the screen when you enter the expression.

| (-) | 8 | . | 9 | ÷ | 0 | ENTER |

Press | ENTER | to quit the error screen and return to the default home screen.

We must remember that, to divide proper fractions, we change the division symbol to a multiplication symbol and change the divisor to its reciprocal. We then multiply the results.

EXAMPLE 5   Divide.

  a. $-\dfrac{3}{4} \div \left(-\dfrac{1}{2}\right)$ **b.** $-1\dfrac{2}{3} \div 2\dfrac{3}{5}$ **c.** $\dfrac{2}{3} \div (-3)$

Solution   **Check**

a. $-\dfrac{3}{4} \div \left(-\dfrac{1}{2}\right) = -\dfrac{3}{4} \cdot \left(-\dfrac{2}{1}\right)$ Change division to multiplication and change the divisor to its reciprocal. Use the rules of multiplication. The quotient is positive $\frac{3}{2}$.

$\qquad\qquad = \dfrac{6}{4}$

$\qquad\qquad = \dfrac{3}{2}$

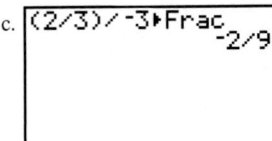

a. (-3/4)/(-1/2)▸Frac
                  3/2
b. -(1+2/3)/(2+3/5)
   ▸Frac
                -25/39

b. $-1\dfrac{2}{3} \div 2\dfrac{3}{5} = \dfrac{-5}{3} \div \dfrac{13}{5}$ Change the mixed numbers to equivalent improper fractions. Change the division to multiplication and change the divisor to its reciprocal. Use the rules of multiplication. The quotient is negative $\frac{25}{39}$.

$\qquad\qquad = \dfrac{-5}{3} \cdot \dfrac{5}{13}$

$\qquad\qquad = -\dfrac{25}{39}$

c. $\dfrac{2}{3} \div (-3) = \dfrac{2}{3} \div \left(\dfrac{-3}{1}\right)$ Change the whole number to a fraction: $-3 = \frac{-3}{1}$. Change the division to multiplication and change the divisor to its reciprocal, $\frac{-1}{3}$. Use the rules of multiplication. The quotient is negative $\frac{2}{9}$.

$\qquad\qquad = \dfrac{2}{3} \cdot \left(\dfrac{-1}{3}\right)$

$\qquad\qquad = -\dfrac{2}{9}$

c. (2/3)/-3▸Frac
                -2/9

Enter fractions in sets of parentheses for proper grouping and ease in reading.

✓ 1.3.2 Checkup

Divide and check.

1. a. $\dfrac{-12}{-4}$ **b.** $\dfrac{63}{-21}$ **c.** $\dfrac{-3}{-5}$

2. a. $36.9 \div (-9)$ **b.** $\dfrac{-18}{-0.5}$ **c.** $0 \div (-13.257)$

3. a. $-\dfrac{1}{3} \div \left(-\dfrac{3}{4}\right)$ **b.** $2\dfrac{1}{4} \div \left(-1\dfrac{1}{3}\right)$ **c.** $-\dfrac{1}{5} \div 0$

1.3.3 Evaluating Products and Quotients

We again use the calculator to perform mathematical operations so that we can discover underlying rules. In this case, we want to discover the rules for multiplication of multiple factors. Complete the following exercises with your calculator.

Discovery 7

Multiplying Three or More Nonzero Rational Numbers

Evaluate each expression, and compare the results obtained in the left column with the results in the right column.

1. one negative factor
$-6 \cdot 3 \cdot 4 \cdot 1 = $ _____

2. two negative factors
$-6 \cdot (-3) \cdot 4 \cdot 1 = $ _____

3. three negative factors
$-6 \cdot (-3) \cdot (-4) \cdot 1 = $ _____

4. four negative factors
$-6 \cdot (-3) \cdot (-4) \cdot (-1) = $ _____

State a rule for determining the sign of the product of three or more rational numbers.

In the column on the left, an odd number of negative factors results in a negative product. In the column on the right, an even number of negative factors results in a positive product. This discovery is an extension of the rules of multiplication of nonzero factors.

To perform multiplication and division operations in the same problem, perform the operations in order from left to right.

EXAMPLE 6 Evaluate and check.

a. $\left(\dfrac{1}{3}\right)\left(-\dfrac{5}{8}\right)\left(-\dfrac{5}{7}\right)$ **b.** $(-380)(257)(0)(25)$

c. $(-3.6) \div (9)(-4.76) \div (-0.2)$

d. $\left(-\dfrac{2}{3}\right) \div \left(\dfrac{1}{3}\right) \div \left(-\dfrac{1}{4}\right)\left(\dfrac{3}{8}\right)$

Solution

a. $\left(\dfrac{1}{3}\right)\left(-\dfrac{5}{8}\right)\left(-\dfrac{5}{7}\right) = \dfrac{25}{168}$ This is an even number of negative factors (two). The product is positive. Multiply the absolute values of the factors from left to right.

b. $(-380)(257)(0)(25) = 0$ A factor of 0 results in a product of 0.

c. $(-3.6) \div (9)(-4.76) \div (-0.2)$ Perform operations from left to right.
$= -0.4(-4.76) \div (-0.2)$
$= 1.904 \div (-0.2)$
$= -9.52$

d. $\left(-\dfrac{2}{3}\right) \div \left(\dfrac{1}{3}\right) \div \left(-\dfrac{1}{4}\right)\left(\dfrac{3}{8}\right)$

$= \left(\dfrac{-2}{3}\right) \cdot \left(\dfrac{3}{1}\right) \cdot \left(\dfrac{-4}{1}\right)\left(\dfrac{3}{8}\right)$ Change division to multiplication, and write the fractions following the division signs (the divisors) as their reciprocals. Perform multiplication rules from left to right.

$= \dfrac{72}{24} = 3$

Check

a. ```
(1/3)(-5/8)(-5/7
)▶Frac
              25/168
```
b. ```
-380*257*0*25
                   0
```
c. ```
-3.6/9*-4.76/-.2
             -9.52
```
d. ```
(-2/3)/(1/3)/(-1
/4)(3/8)▶Frac
                 3
```

HELPING HAND Some calculators use **implied multiplication**. This means that if two numbers are written together without a sign between them, the multiplication is completed first, not in order from left to right. (See "Calculator Exercises" at the end of this section.)

 1.3.3 Checkup

Evaluate. Check, using your calculator.

1. $(-2)(3)(-1)(-2)(4)$

2. $\left(1\frac{5}{7}\right)\left(-3\frac{5}{9}\right)\left(-\frac{3}{4}\right)\left(-\frac{2}{3}\right)$

3. $(2.5)(-3.5)(0)(5.6)(0)(3.9)$

4. $(-12)(-3) \div (-6)(15) \div 9(25) \div (-2)$

5. $(-18.9) \div (-9)(2.24) \div (-0.4)$

1.3.4 Evaluating Addition, Subtraction, Multiplication, and Division

We are now ready to evaluate expressions that involve the four operations of addition, subtraction, multiplication, and division. To do this, we must establish a few rules, so that we all obtain the same results. Complete the following exercises to discover the order of operations for addition, subtraction, multiplication, and division.

 Discovery 8

Order of Operations

Consider the expression $6 \div 2 + 1 \cdot 3 - 5$. Complete the following possible methods of evaluation:

1. Evaluate in order from left to right.

2. First, evaluate all additions and subtractions in order from left to right. Then, evaluate all multiplications and divisions in order from left to right.

3. First, evaluate all multiplications and divisions in order from left to right. Then, evaluate all additions and subtractions in order from left to right.

4. Enter the expression into your calculator.

Write a rule for performing the order of operations by comparing the calculator value with the values obtained in Exercises 1, 2, and 3.

The calculator result is 1. The third set of directions also resulted in 1. Therefore, the calculator is programmed to evaluate the expression in the same order as described in the third set of directions; it performed all multiplications and divisions before additions and subtractions. The calculator evaluated the expression according to the **order of operations**.

ORDER OF OPERATIONS
To evaluate an expression involving the four basic operations,

• Evaluate multiplication and division from left to right.
• Evaluate addition and subtraction from left to right.

 **HELPING HAND** The rule expressing the order of operations does not state that multiplication is completed before division (or addition before subtraction). It does state that multiplication and division are completed in the same pass through the expression, in order from left to right. Likewise, addition and subtraction are completed in the same pass in the order they appear from left to right.

EXAMPLE 7 Evaluate, using the rule for the order of operations.

$$\text{a. } 15 + 7 - 6 \div 3 - 2 \cdot 8 \qquad \text{b. } -12(-3) - 6(4) + 8$$

Solution

Check

a.
$$
\begin{aligned}
15 + 7 - 6 \div 3 - 2 \cdot 8 &= 15 + 7 - 2 - 2 \cdot 8 &&\text{Divide.} \\
&= 15 + 7 - 2 - 16 &&\text{Multiply.} \\
&= 22 - 2 - 16 &&\text{Add.} \\
&= 20 - 16 &&\text{Subtract.} \\
&= 4
\end{aligned}
$$

a.
```
15+7-6/3-2*8
              4
```
b.
```
-12(-3)-6(4)+8
             20
```

b.
$$
\begin{aligned}
-12(-3) - 6(4) + 8 &= 36 - 6(4) + 8 &&\text{Multiply.} \\
&= 36 - 24 + 8 &&\text{Multiply.} \\
&= 12 + 8 &&\text{Subtract.} \\
&= 20 &&\text{Add.}
\end{aligned}
$$

 ### 1.3.4 Checkup

Evaluate, using the rule for the order of operations.
1. $11 + 9 - 4 \div 2 - 3 \cdot 4$ 2. $-6(-5) - 2(3) + 7$
3. When following the order of operations, you must first do all multiplications and divisions before you do any addition or subtraction. Does this mean that you do all of the multiplications first and then go back and do all of the divisions? Explain.

1.3.5 Modeling the Real World

In our daily lives, many situations can be expressed as products of rational numbers. Whenever there is a repeated process, such as writing a check for the same amount several weeks in a row or paying rent every month for a year, we may represent the amount as a multiplication expression. Also, when we take a "percentage of" or "fractional part of" a value, we write a multiplication expression. In these cases, the rules for multiplication of rational numbers may be applied to obtain a product.

Many situations in our daily lives also involve the division of rational numbers. When we compare two quantities, we use a ratio. **A ratio** is the quotient of the two quantities. When the ratio compares two different kinds of measure, we call the ratio a **rate**. The rate (or speed) at which you travel in an automobile is given in miles per hour or kilometers per hour. The word *per* means division.

EXAMPLE 8 Greenland has the lowest (population) density (people per square mile) in the world. In 2000, Greenland's population was estimated to be 59,827, and its land area is 848,484 square miles. Determine its density.

Solution
The density, in people per square mile, is found by dividing the number of people by the area in square miles.

$$\frac{59827 \text{ people}}{848484 \text{ miles}} \approx 0.07 \frac{\text{person}}{\text{mile}}$$

The density of Greenland is about 0.07 person per square mile.

EXAMPLE 9 The human heart pumps 4 liters of blood in 90 seconds. Determine the human heart rate in liters per second. How many liters are pumped in 1 minute?

Solution
The heart rate is given in liters per second, so we will divide the number of liters by the number of seconds.

$$\frac{4 \text{ liters}}{90 \text{ seconds}} = \frac{2 \text{ liters}}{45 \text{ seconds}}$$

The human heart rate is $\frac{2}{45}$ liters per second. In one minute, or 60 seconds, the heart will pump 60 times the rate per second.

$$60 \text{ seconds} \cdot \frac{2 \text{ liters}}{45 \text{ seconds}} = \frac{120}{45} \text{ liters} = \frac{8}{3} \text{ liters} = 2\frac{2}{3} \text{ liters}$$

In one minute, the heart will pump $2\frac{2}{3}$ liters of blood.

APPLICATION

Fifty-two percent of the U.S. national debt consists of marketable loans, such as bills, notes, and bonds, that can be traded. In 1999, the national debt was $5.6 trillion. Determine the amount of the debt that is made up of marketable loans.

Discussion

A debt corresponds to a negative number. Therefore, the national debt of $5.6 trillion may be represented as −5.6. If 52% of this debt consists of marketable loans, determine the amount of the debt ceiling that is made up of marketable loans by multiplying the debt ceiling by the percentage, written as a decimal.

$$52\% \text{ of } -5.6 = 0.52 \times -5.6 = -2.912$$
$$52\% = 0.52$$

This translates into a national debt of $2.912 trillion in marketable loans, or $2,912,000,000,000.

1.3.5 Checkup

1. Napoleon sent a message from Rome to Paris using a semaphore system (signaling with flags from mountaintop to mountaintop). It took approximately 4 hours to send the message a distance of 700 miles. Determine the rate of transmission of the message in miles per hour. At the same rate, how far could you send a message in $2\frac{1}{2}$ hours, using this system?

2. For the year 2000, the *Statistical Abstract of the United States* projected the percentage distribution of the U.S. population by race to be that shown on the following bar chart (persons of Hispanic origin may be of any race and thus are not projected separately):

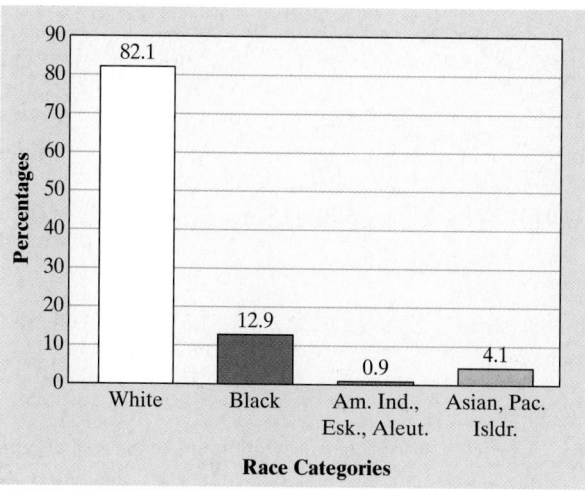

Race Categories

The *Abstract* also projects the total U.S. population in the year 2000 to be 274,634,000. Use this total to determine the size of the population projected to be black.

3. According to the 2001 *World Almanac*, the population of Australia was 19,164,000, while its land area measures 2,968,000 square miles. Determine the density (people per square mile) of Australia.

4. The caseload in a hospital ward is 15 patients for two nurses. Determine the rate of assignment of cases per nurse. How many cases can a ward handle if it employs 12 nurses?

1.3 Exercises

Multiply and check.

1. $-3 \cdot (-8)$
2. $-4 \cdot (-7)$
3. $5 \cdot (-3)$
4. $9 \cdot (-4)$
5. $-2 \cdot 10$
6. $-8 \cdot 6$
7. $45 \cdot (-3)$
8. $56 \cdot (-4)$
9. $-32 \cdot (-4)$
10. $-55 \cdot (-11)$
11. $0 \cdot (-15)$
12. $-2 \cdot 0$
13. $(-1.7)(-0.2)$
14. $(-34.2)(-2)$
15. $(24.3)(0.3)$
16. $(0.5)(10)$
17. $(-0.25)(50)$
18. $(-5.7)(0.19)$
19. $\left(\frac{2}{5}\right)\left(\frac{25}{48}\right)$
20. $\left(\frac{11}{17}\right)\left(\frac{2}{3}\right)$
21. $\left(1\frac{2}{3}\right)\left(-\frac{3}{4}\right)$
22. $\left(-1\frac{1}{5}\right)\left(-\frac{5}{6}\right)$
23. $\left(-\frac{4}{7}\right)\left(\frac{3}{16}\right)$
24. $\left(-\frac{15}{17}\right)\left(\frac{5}{9}\right)$

Divide and check.

25. $15 \div (-3)$
26. $32 \div (-4)$
27. $-32 \div (-4)$
28. $-55 \div (-11)$
29. $27 \div 3$
30. $16 \div 2$
31. $0 \div (-15)$
32. $0 \div 12$
33. $26 \div (-0.13)$
34. $-54 \div 0.6$
35. $-1.7 \div (-0.2)$
36. $0.5 \div 10$
37. $-2.7 \div (-2.7)$
38. $3.4 \div (-3.4)$
39. $\frac{5}{-25}$
40. $\frac{-3}{6}$
41. $\frac{0.88}{-1.1}$
42. $\frac{-0.25}{50}$
43. $\frac{2}{5} \div \frac{25}{48}$
44. $\frac{11}{17} \div \frac{2}{3}$
45. $1\frac{2}{3} \div \left(-\frac{3}{4}\right)$
46. $2\frac{1}{5} \div \left(-\frac{1}{3}\right)$
47. $-\frac{1}{3} \div \left(-\frac{3}{7}\right)$
48. $-\frac{2}{5} \div \left(-\frac{5}{9}\right)$
49. $-\frac{4}{7} \div \frac{3}{16}$
50. $-\frac{15}{17} \div \frac{5}{9}$
51. $\frac{3}{5} \div 0$
52. $\frac{1}{2} \div 0$
53. $-\frac{2}{3} \div \left(-\frac{2}{3}\right)$
54. $-\frac{2}{3} \div \frac{3}{2}$
55. $\frac{8}{9} \div 4$
56. $-\frac{7}{8} \div 2$
57. $14 \div \left(-\frac{1}{3}\right)$
58. $-21 \div \frac{3}{2}$

Evaluate and check.

59. $(-2)(-3)(-4)(-10)(20)$

60. $(-5)(-4)(6)(-100)(-5)$

61. $\left(-\dfrac{1}{5}\right)\left(-\dfrac{2}{3}\right)\left(-\dfrac{4}{5}\right)\left(\dfrac{1}{2}\right)$

62. $\left(-\dfrac{5}{7}\right)\left(\dfrac{14}{25}\right)\left(-\dfrac{2}{5}\right)\left(\dfrac{1}{3}\right)$

63. $(5.2)(-0.1)(-2.2)$

64. $(-2.9)(-1.1)(0.2)$

65. $(14)(0)(-35)(0)(-312)$

66. $(-1.4)(0)(3.76)(0)(-45.2)$

67. $(-15)(4) \div (-3)(12) \div 3(-10)$

68. $(-40) \div (-8)(9)(-12)$

69. $(-3.3)(2.7) \div (-11)(0.6)$

70. $(15.5) \div (-0.5)(-3.3) \div 11$

71. $\left(\dfrac{2}{3}\right)\left(-\dfrac{5}{8}\right) \div \left(-\dfrac{5}{16}\right)$

72. $\left(\dfrac{7}{22}\right) \div \left(-\dfrac{7}{11}\right)\left(-\dfrac{1}{2}\right)(4)$

73. $15 + 9 \div 3 - 7 \cdot 6 \div 2$

74. $21 - 12 \div 3 + 6 \cdot 4 \div 8$

75. $2.7 + 5.6 - 16 \div 4 - 3 \cdot 2$

76. $2.2 + (-1.5) - 27 \div 9 + 5 \cdot 3$

77. $4.3(3) - 5(1.6) + 42.9 \div 3$

78. $2(3.4) - 4.1(2) + 16.8 \div 2.1$

Write and evaluate a multiplication or division expression for each situation.

79. The deductions from Sara's paycheck typically amount to 22% of her gross pay. If she grosses $645 every 2 weeks, how much can she expect to have deducted? Deductions are represented by negative numbers. Over a 12-week period, what can she expect her total deduction to be?

80. Ron bets a total of $5 per race on nine races each day at the horse track. Bets are represented as negative numbers, since this is money Ron pays out. If he attends a week's worth of races (six days, since the horses don't race on Mondays), how much money will he bet in the week?

81. Sammy is paid 5 cents for each flyer he distributes. He can distribute 40 flyers per hour. He is permitted to distribute flyers for $1\frac{1}{2}$ hours each weekday. He does so on three weekdays. How much will he earn for the work?

82. Melanie earns 25 cents for her scout troop for each box of cookies she sells. She averages 20 boxes of cookies sold each time she works at the cookie booth in the local mall. She is permitted to work the cookie booth six different times. How much money did she make for her troop?

83. George averages 50 miles per hour on his business trip. He drives approximately $8\frac{1}{2}$ hours each day, and his drive takes him four days to complete. Approximately how many miles did George drive to get to his destination?

84. Michele hikes for $9\frac{1}{2}$ hours per day on the Appalachian Trail. She hikes at an average pace of $1\frac{3}{4}$ miles per hour. How far can she hike in five days?

85. A grocery store receives a shipment of canned vegetables. The vegetables arrive in cases with 24 cans in each case. Each can of vegetables will retail for 59 cents. If the grocery store receives 14 cases of vegetables, what will be the total retail value of the shipment?

86. A distributor packages cleaner concentrate in 12-ounce bottles. The concentrate is packaged 18 bottles to a case. An order for 12 cases of concentrate is received. It costs the distributor 35 cents per ounce to obtain the concentrate. What costs will the distributor incur to obtain the concentrate for this order?

87. Al's car averages 19.4 miles to a gallon of gas. If Al plans to drive from Nashville, Tennessee, to Washington, DC, a distance of 659 miles, approximately how many gallons of gas will he use?

88. Al completes his car trip of 659 miles and finds that he has used 38 gallons of gas. What was his gas mileage for the trip?

89. How many 40-fluid-ounce bottles of fabric softener can be filled from a production run of 47,500 fluid ounces, assuming that nothing is lost to spillage?

90. How many 98-ounce boxes of laundry detergent can be filled from a production run of 2800 pounds of detergent, assuming that nothing is lost to spillage? (*Note*: Each pound of detergent is equivalent to 16 ounces of detergent.)

91. To promote a town fund-raiser, Smallville sets up a clock that ticks off the seconds until the event begins. If the clock starts at 16,000,000 seconds, how many days will it be until the event kicks off?

92. Pioneer Village uses a clock to tick off the seconds until its bicentennial celebration will begin. If the clock starts at 20,000,000 seconds, how many weeks will it be until the event kicks off?

93. Billie wants to buy cabinets to hold her CD collection. She owns 335 CDs. If each cabinet has three drawers and each drawer holds 20 CDs, how many cabinets will she need to buy in order to store her entire collection?

94. Bruce has a collection of 650 LPs (long-playing records). He will buy record cabinets to store his collection. Each cabinet has three shelves, and each shelf holds 125 LPs. How many cabinets will Bruce need to buy for his collection?

95. In 1999, the company with the largest advertising expenditures was General Motors Corporation, which spent $4,040,374,000 for the 12-month period. If this amount was evenly distributed throughout the year, what would you estimate was the expenditure for a 3-month quarter?

96. Aerialist Philippe Petit walked a 1350-foot-long tightrope between two buildings in a time of 50 minutes. At the same rate of speed, how many minutes would you expect him to take to walk a tightrope of 1000 feet?

97. In a recommended portfolio allocation plan based on an investor risk profile, the following pie chart is recommended for a conservative investor:

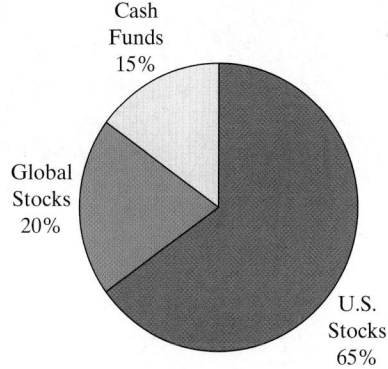

Suppose such an investor planned to invest $15,000. How much should he or she invest in U.S. stocks?

98. Using the chart in Exercise 97, state how much a conservative investor should invest in global stocks if he or she plans to invest a total of $45,000.

99. A computer-aided design (CAD) operator needs 20 minutes to plot 3 drawings. Determine the rate in minutes per drawing. How long will it take to plot 17 drawings of the same size and detail?

100. A metal joint 8 feet long requires 45 rivets. Determine the rate, in rivets per foot. How many rivets are required in a joint 5 feet long?

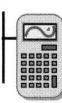

1.3 Calculator Exercises

Use your calculator to perform the following operations (in the first exercise, type parentheses to indicate the multiplication; in the second exercise, use the multiplication key on the calculator to indicate the multiplication):

1. $50 \div 5(2)$ **2.** $50 \div 5 \cdot 2$

The correct answer is 20, but some calculators interpret the parentheses to mean that the multiplication operation is to be done before the division operation. In exercise 1, if your calculator returned an answer of 5 instead of 20, the calculator is programmed for implied multiplication. This means any multiplication indicated by using parentheses is performed first. This violates our rule to perform the operations in order from

left to right when we have an expression involving both multiplication and division. Calculators that use implied multiplication have reasons for doing so. If your calculator gave you an incorrect answer, you must be careful always to use the multiplication key to indicate multiplication, unless you wish the multiplication to take precedence over operating from left to right.

Use your calculator to perform the following operations, and then determine the solution without using the calculator to see if your calculator is following implied multiplication rules.

3. $81 \div 9(-3)$ **4.** $81 \div 9 \cdot (-3)$ **5.** $100 \div 25(4)$

6. $100 \div 25 \cdot 4$ **7.** $2 \div 2(5)$ **8.** $2 \div 2 \cdot 5$

1.3 Writing Exercises

Being able to recognize words, phrases, or situations that imply mathematical operations will increase your skills in solving word problems.

1. List as many words or phrases as you can that imply multiplication.

2. List as many words or phrases as you can that imply division.

3. Pick three of the words that imply multiplication, and write a sentence for each that illustrates the use of the word.

4. Pick three of the words that imply division, and write a sentence for each that illustrates the use of the word.

1.4   Exponents and Roots

OBJECTIVES

1 Evaluate exponential expressions with nonnegative integer exponents.

2 Evaluate square roots and cube roots.

3 Graph real numbers on a number line.

4 Model real-world situations by using exponents and roots.

APPLICATION

In the town of Marostica in northern Italy, a game of chess using live people as pieces is played every two years. The event celebrates a game played in 1454 between two suitors for a lady's hand, in which the winner could claim the lady as his bride. The board on which the living chess game is played measures eight squares by eight squares, and each square measures 10 feet by 10 feet. What is the area of the board?

After completing this section, we will discuss this application further. See page 56.

1.4.1 Evaluating Exponential Expressions with a Nonnegative Integer Exponent

When a number is repeated as a factor, the product may be written in exponential form instead of as a multiplicative expression. The repeated factor is the **base** of the expression. The number of times the base is repeated as a factor is written as an **exponent**.

$$2 \cdot 2 \cdot 2 \cdot 2 \cdot 2 \cdot 2 = \quad 2^6$$

2 repeated 6 times   base^{exponent}

The base is 2 and the exponent is 6.

In the preceding example, $2 \cdot 2 \cdot 2 \cdot 2 \cdot 2 \cdot 2$ is called an **expanded form**, and 2^6 is called an **exponential form**, or an **exponential expression**.

For the moment, we will examine expressions with integer exponents greater than 1 and with bases that are positive numbers or 0.

To evaluate any such exponential expression, write it in expanded form and then multiply the factors.

$2^2 = 2 \cdot 2 = 4$ 2 (raised) to the second power is 4, or 2 squared is 4.

$2^3 = 2 \cdot 2 \cdot 2 = 8$ 2 (raised) to the third power is 8, or 2 cubed is 8.

$2^6 = 2 \cdot 2 \cdot 2 \cdot 2 \cdot 2 \cdot 2 = 64$ 2 (raised) to the sixth power is 64. (All powers except 2 and 3 are read in this manner.)

EXAMPLE 1   Write in expanded form and evaluate.

a. $\left(\dfrac{2}{3}\right)^2$ **b.** 1.5^3 **c.** 0^4 **d.** $\left(9\dfrac{1}{3}\right)^5$

Solution

a. $\left(\dfrac{2}{3}\right)^2 = \dfrac{2}{3} \cdot \dfrac{2}{3} = \dfrac{4}{9}$ $\frac{2}{3}$ is repeated as a factor two times.

b. $1.5^3 = (1.5)(1.5)(1.5) = 3.375$ 1.5 is repeated as a factor three times.

c. $0^4 = 0 \cdot 0 \cdot 0 \cdot 0 = 0$ 0 is repeated as a factor four times.

d. $\left(9\dfrac{1}{3}\right)^5 = \left(\dfrac{28}{3}\right)\left(\dfrac{28}{3}\right)\left(\dfrac{28}{3}\right)\left(\dfrac{28}{3}\right)\left(\dfrac{28}{3}\right)$ $9\frac{1}{3} = \frac{28}{3}$ and is repeated as a factor five times.

$$= \dfrac{17,210,368}{243}$$

To evaluate an exponential expression on your calculator, it is not necessary to enter the expression in expanded form. The calculator has special keys for exponents. The checks for Example 1a, 1b, and 1d are shown in Figures 1.9a, 1.9b, and 1.9c, respectively.

TECHNOLOGY

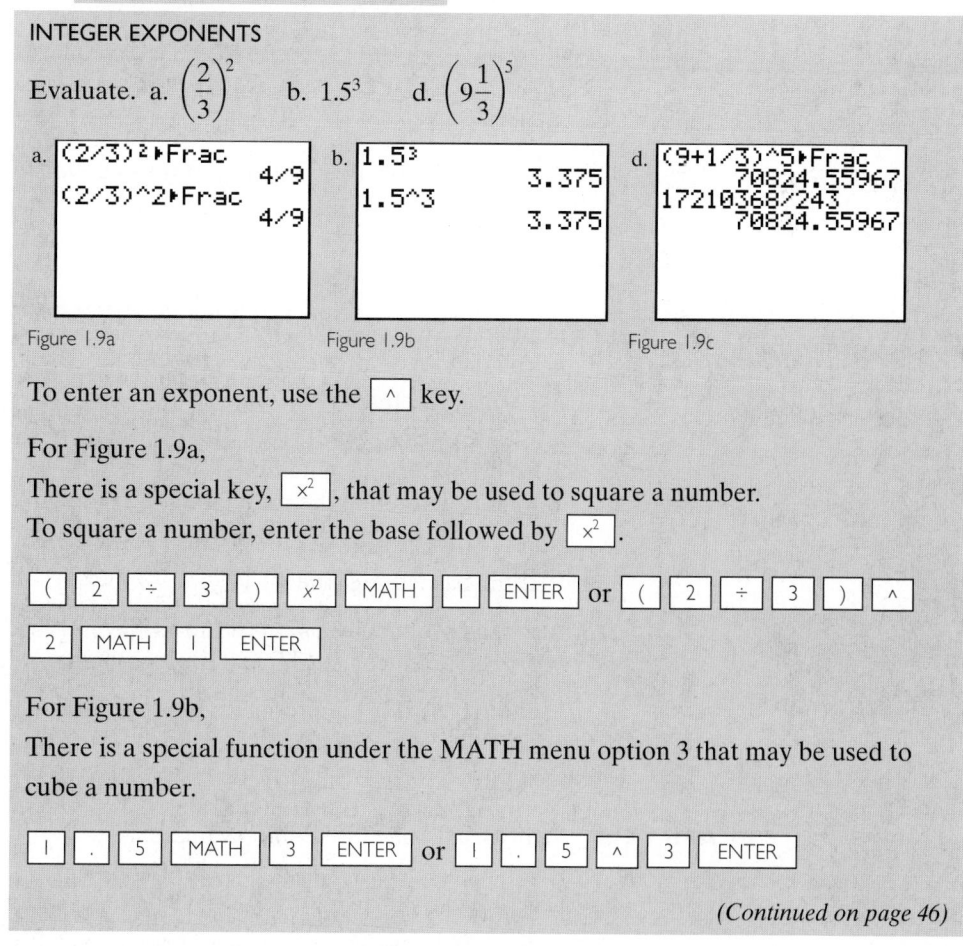

INTEGER EXPONENTS

Evaluate. a. $\left(\dfrac{2}{3}\right)^2$ b. 1.5^3 d. $\left(9\dfrac{1}{3}\right)^5$

a.
```
(2/3)²▶Frac
            4/9
(2/3)^2▶Frac
            4/9
```

b.
```
1.5³
            3.375
1.5^3
            3.375
```

d.
```
(9+1/3)^5▶Frac
       70824.55967
17210368/243
       70824.55967
```

Figure 1.9a   Figure 1.9b   Figure 1.9c

To enter an exponent, use the [^] key.

For Figure 1.9a,

There is a special key, [x^2], that may be used to square a number.

To square a number, enter the base followed by [x^2].

[( ] [2] [÷] [3] [ )] [x^2] [MATH] [1] [ENTER] **or** [( ] [2] [÷] [3] [ )] [^]

[2] [MATH] [1] [ENTER]

For Figure 1.9b,

There is a special function under the MATH menu option 3 that may be used to cube a number.

[1] [.] [5] [MATH] [3] [ENTER] **or** [1] [.] [5] [^] [3] [ENTER]

(Continued on page 46)

For Figure 1.9c,

Enter the mixed number, $9\frac{1}{3}$, as a sum of 9 and $\frac{1}{3}$ enclosed in parentheses.

Use the $\boxed{\wedge}$ for inserting the exponent.

Since the result is not a fraction, enter the fraction answer that we are checking, $\frac{17210368}{243}$, and compare the decimal values.

$\boxed{1}\,\boxed{7}\,\boxed{2}\,\boxed{1}\,\boxed{0}\,\boxed{3}\,\boxed{6}\,\boxed{8}\,\boxed{\div}\,\boxed{2}\,\boxed{4}\,\boxed{3}\,\boxed{\text{ENTER}}$

If the repeated factor in an exponential expression is a negative number, write the factor in parentheses as the base. For example,

$$-2 \cdot -2 \cdot -2 \cdot -2 \cdot -2 \cdot -2 = (-2)^6$$

−2 repeated six times   base^{exponent}

The parentheses are very important here. If no parentheses are used, the base is not considered to be a negative number. For example,

$$-2^6 = -(2 \cdot 2 \cdot 2 \cdot 2 \cdot 2 \cdot 2) = -64$$

-2^6 means the opposite of "2 to the sixth power."

$$(-2)^6 = -2 \cdot -2 \cdot -2 \cdot -2 \cdot -2 \cdot -2 = 64$$

$(-2)^6$ means "−2 to the sixth power."

Therefore, $(-2)^6 \neq -2^6$ because $(-2)^6 = 64$ and $-2^6 = -64$.

HELPING HAND It is very important to understand the difference between -2^6 and $(-2)^6$. Always determine the base of the exponential expression before expanding.

-2^6 2 is the base.

$(-2)^6$ −2 is the base.

EVALUATING EXPRESSIONS WITH INTEGER EXPONENTS > 1

To evaluate an exponential expression with an integer exponent greater than 1, determine the product of its expanded form.

If the base is positive, the product is positive.

If the base is negative:

• The product is positive if the exponent is even.
• The product is negative if the exponent is odd.

If the base is 0, the product is 0.

EXAMPLE 2 Write in expanded form and evaluate.

a. $(-1.8)^5$ **b.** $\left(-\frac{3}{4}\right)^4$ **c.** $(-1)^6$ **d.** -1^6

Solution   **Check**

a. $(-1.8)^5 = -1.8 \cdot -1.8 \cdot -1.8 \cdot -1.8 \cdot -1.8$

a. $(-1.8)^5$
                                        -18.89568

The base is -1.8.

b. $(-3/4)^4 \blacktriangleright$Frac
                                        $81/256$

$= -18.89568$

b. $\left(-\dfrac{3}{4}\right)^4 = \left(-\dfrac{3}{4}\right)\left(-\dfrac{3}{4}\right)\left(-\dfrac{3}{4}\right)\left(-\dfrac{3}{4}\right) = \dfrac{81}{256}$

The base is $-\frac{3}{4}$.

c. $(-1)^6 = (-1)(-1)(-1)(-1)(-1)(-1) = 1$

c. $(-1)^6$
                                        1

The base is -1.

d. -1^6
                                        -1

d. $-1^6 = -(1 \cdot 1 \cdot 1 \cdot 1 \cdot 1 \cdot 1) = -1$

The base is 1.

The expanded form of an expression with an exponent of 1 or 0 is not obvious. We need to find a second method for evaluating such expressions.

Complete the following exercises with your calculator.

Discovery 9

Expressions with an Exponent of 1 or 0

Evaluate each expression.

1. 10^1 **2.** $(-10)^1$ **3.** 0^1 **4.** -10^0 **5.** $(-10)^0$ **6.** 0^0

State a rule for evaluating an exponential expression with an exponent of 1.

State a rule for evaluating an exponential expression with an exponent of 0.

An exponential expression with an exponent of 1 evaluates to the base number.
    An exponential expression with an exponent of 0 and a nonzero base is 1. If the base is 0 and the exponent is 0, the expression is indeterminate.

EVALUATING WITH EXPONENTS OF 1 OR 0

The value of an exponential expression with an exponent of 1 is equal to the base number.

The value of an exponential expression with an exponent of 0 is

• 1 if the base is not 0.
• indeterminate if the base is 0.

To understand these rules, examine the following pattern:

$$10^4 = 10 \cdot 10 \cdot 10 \cdot 10$$
$$10^3 = 10 \cdot 10 \cdot 10 \qquad \text{Divide by 10.}$$

Exponent decreases by 1. $\quad 10^2 = 10 \cdot 10$

$$10^1 = \underline{\hspace{1cm}}$$
$$10^0 = \underline{\hspace{1cm}}$$

Using the preceding pattern, we obtain $10^1 = \underline{10}$ and $10^0 = \underline{1}$. (If we divide by 10, we get 1.) We will explain why 0^0 is indeterminate later in the text.

EXAMPLE 3 Evaluate.

a. 18^0 b. $(-18)^0$ c. -18^0 d. $(-15)^1$ e. -15^1

Solution

a. $18^0 = 1$ The base is 18.

b. $(-18)^0 = 1$ The base is −18.

c. $-18^0 = -1$ The base is 18.

d. $(-15)^1 = -15$ The base is −15.

e. $-15^1 = -15$ The base is 15.

1.4.1 Checkup

In exercises 1–5, evaluate.

1. 1.3^2 2. 0^6 3. $\left(\dfrac{-2}{5}\right)^5$

4. a. $(-6)^2$ b. -6^2 c. $(-6)^3$ d. -6^3

5. a. 7^1 b. $(-7)^1$ c. -7^1 d. 7^0 e. $(-7)^0$ f. -7^0

6. In exercise 4, explain the difference between the expressions in a and b, and discuss the impact of this difference on the final answers. Also, explain the difference between c and d, and discuss the impact (or lack thereof) of this difference on the final answers.

7. When the base of an exponential expression is negative, how can you tell whether the value of the exponential expression will be positive or negative?

1.4.2 Evaluating Roots

Previously, we discussed squaring a number, such as $3^2 = 9$. The result 9 is called a perfect square. A **perfect square** is defined to be the result of squaring a rational number.

If we need to reverse this operation—that is, go from the square of a number back to the number itself—we take the **square root** of the number. Thus, 3 is the square root of 9, because $3^2 = 9$. We call 3 the **positive square root** or **principal square root**.

Because $(-3)^2 = 9$ also, we refer to −3 as being the **negative square root** of 9.

$$\sqrt{9} = 3 \qquad \text{The positive square root of 9 is 3.}$$
$$-\sqrt{9} = -3 \qquad \text{The negative square root of 9 is negative 3.}$$

Square roots of perfect squares simplify to rational numbers.

Now, let's find the square root of a negative number, such as −9. To find a value for $\sqrt{-9}$, we must determine what number to square in order to obtain −9. Since squaring a positive number results in a positive number ($3 \cdot 3 = 9$) and squaring a negative number results in a positive number ($-3 \cdot -3 = 9$), we know we cannot find a real number whose square is −9 (a negative number). Therefore, $\sqrt{-9}$ cannot be evaluated in the real-number system. We will introduce the *complex-number system*, in which we can evaluate $\sqrt{-9}$, in a later chapter.

EVALUATING SQUARE ROOTS

To evaluate the positive or principal root of a positive number that is a perfect square,

- Determine the positive number whose square results in the perfect square.

To evaluate the negative root of a positive number that is a perfect square,

- Determine the negative number whose square results in the perfect square.

The square root of 0 is 0.

The positive and negative square roots of a negative number are not defined among the real numbers.

EXAMPLE 4 Evaluate.

a. $\sqrt{64}$ b. $\sqrt{1.44}$ c. $\sqrt{\dfrac{4}{9}}$ d. $-\sqrt{25}$ e. $\sqrt{-25}$ f. $\sqrt{0}$

Solution

a. $\sqrt{64} = 8$

The positive square root of 64 is 8, because $8^2 = 64$.

b. $\sqrt{1.44} = 1.2$

The positive square root of 1.44 is 1.2, because $1.2^2 = 1.44$.

c. $\sqrt{\dfrac{4}{9}} = \dfrac{2}{3}$

The positive square root of $\frac{4}{9}$ is $\frac{2}{3}$, because $\left(\frac{2}{3}\right)^2 = \frac{4}{9}$.

d. $-\sqrt{25} = -5$

The negative square root of 25 is -5, because $(-5)^2 = 25$.

e. $\sqrt{-25}$ is not a real number.

There is no real number whose square is -25 (a negative number).

f. $\sqrt{0} = 0$

The square root of 0 is 0, because $0^2 = 0$.

The checks for Example 1c, 1d, and 1e are shown in Figure 1.10a and Figure 1.10b.

TECHNOLOGY

SQUARE ROOTS

Evaluate. c. $\sqrt{\dfrac{4}{9}}$ d. $-\sqrt{25}$ e. $\sqrt{-25}$

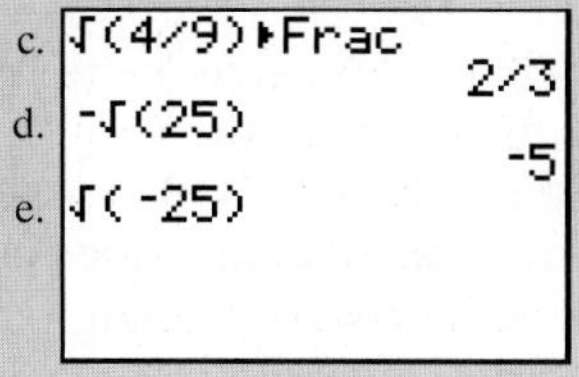

```
c. √(4/9)►Frac
                2/3
d. -√(25)
                 -5
e. √(-25)
```

```
e. ERR:NONREAL ANS
   1:Quit
   2:Goto
```

Figure 1.10a

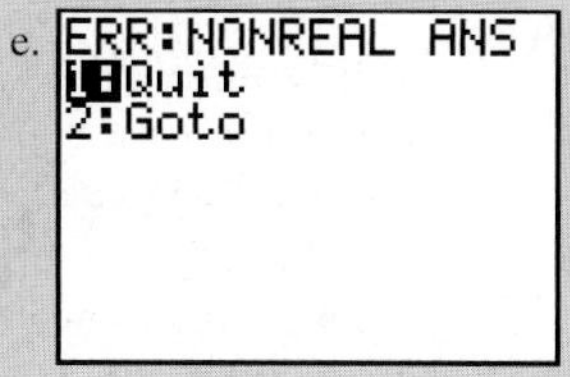

Figure 1.10b

(Continued on page 50)

For Figure 1.10a and Figure 1.10b,

To enter a square root, enter [2nd] [√]. Close the parentheses that are opened for you in order to enclose the number in a set of parentheses.

[2nd] [√] [4] [÷] [9] [)] [MATH] [1] [ENTER]
[(-)] [2nd] [√] [2] [5] [)] [ENTER]

Since the square root of a negative number is not defined as a real number, if your calculator is in Real mode (the default mode), an error will be displayed when you enter the expression.

[2nd] [√] [(-)] [2] [5] [)] [ENTER]

Press [ENTER] to quit the error screen and return to the default home screen.

Let's find a value for the square root of a positive rational number that is not a perfect square. Remember, perfect squares are 1, 4, 9, and so on. A number such as 2 is between the perfect squares 1 and 4. Therefore, the value of $\sqrt{2}$ should be between the values of $\sqrt{1}$ and $\sqrt{4}$, or between 1 and 2. Rather than experiment to find $\sqrt{2}$, it is easier to use a calculator.

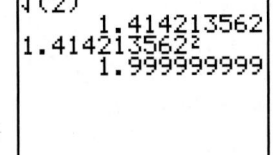

On a calculator, we find that $\sqrt{2}$ is given by 1.414213562. If we check, we get $(1.414213562)^2 = 1.999999999$. This is a close approximation, but not exactly 2. In fact, we cannot find a rational number whose square is exactly equal to 2.

$$\sqrt{2} \approx 1.414 \qquad \textit{Rounded to the nearest thousandth}$$

The fact that we cannot find a rational number whose square is equal to 2 means that the value of $\sqrt{2}$ is not a rational number. Thus, we cannot write it as a ratio of integers. Therefore, we call $\sqrt{2}$ an irrational number. An **irrational number** is a number that cannot be written as a ratio of integers. **Real numbers** are the set of all rational and irrational numbers.

EXAMPLE 5   Determine between what two integers the values of the following square roots lie, and then estimate the square roots to the nearest thousandth on your calculator.

a. $\sqrt{3}$ **b.** $-\sqrt{5}$ **c.** $\sqrt{99}$

Calculator Solution

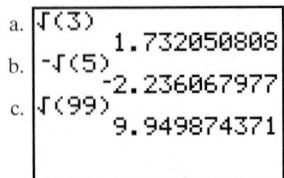

a. $\sqrt{3}$ is between $\sqrt{1} = 1$ and $\sqrt{4} = 2$, $\sqrt{3} \approx 1.732$.

b. $-\sqrt{5}$ is between $-\sqrt{4} = -2$ and $-\sqrt{9} = -3$, $-\sqrt{5} \approx -2.236$.

c. $\sqrt{99}$ is between $\sqrt{81} = 9$ and $\sqrt{100} = 10$, $\sqrt{99} \approx 9.950$.

The square root of a number, defined in the first part of this objective, is one example of a radical expression. In the expression $\sqrt{9}$, the symbol $\sqrt{}$ is called a **radical sign**, and 9 is the **radicand**. $\sqrt{9}$ is a **radical expression**.

We also can define the roots of expressions with exponents larger than 2. For example, we may want to reverse cubing a number and call the result a **cube root**. In order to use the same notation as with square roots, we will need to add an index to the radical sign. An **index** is the power we are reversing. To write a cube root, in which we are reversing a power of 3, we would write the following:

$$\sqrt[3]{64}$$ What number multiplied as a factor 3 times is 64?

$$\underset{\text{index}}{}\sqrt[]{\underset{\text{radicand}}{}}$$

To evaluate $\sqrt[3]{64}$, we determine that $4^3 = 64$. To evaluate $\sqrt[3]{-64}$, we determine $(-4)^3 = -64$.

$$\sqrt[3]{64} = 4$$ The cube root of 64 is 4, because $4^3 = 64$.
$$\sqrt[3]{-64} = -4$$ The cube root of -64 is -4, because $(-4)^3 = -64$.

HELPING HAND We can evaluate cube roots of negative numbers in the real-number system.

EVALUATING CUBE ROOTS
To evaluate a cube root of a number that is a perfect cube,

* Determine the number whose cube results in the perfect cube. If the radicand is positive, the number is positive. If the radicand is negative, the number is negative.

To evaluate a cube root of a number that is not a perfect cube,

* Determine a number whose cube is approximately the value of the radicand. A calculator may be needed to find this number.

The value of a radical expression with a radicand of 0 is 0.

EXAMPLE 6 Evaluate, rounding to the nearest thousandth.

a. $\sqrt[3]{-27}$ b. $\sqrt[3]{30}$ c. $\sqrt[3]{\dfrac{1}{27}}$

Solution

a. $\sqrt[3]{-27} = -3$ b. $\sqrt[3]{30} \approx 3.107232506 \approx 3.107$ c. $\sqrt[3]{\dfrac{1}{27}} = \dfrac{1}{3}$

The checks for Example 6 are shown in Figure 1.11.

T E C H N O L O G Y

CUBE ROOTS

Evaluate. a. $\sqrt[3]{-27}$ b. $\sqrt[3]{30}$ c. $\sqrt[3]{\dfrac{1}{27}}$

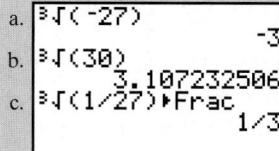

```
a. ³√(-27)
                  -3
b. ³√(30)
         3.107232506
c. ³√(1/27)▶Frac
                1/3
```

Figure 1.11

(Continued on page 52)

For Figure 1.11,
The cube root is located under the MATH menu option 4. Remember to close the set of parentheses that is opened for you.

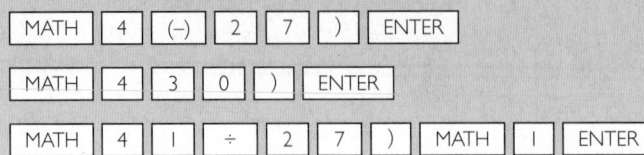

 1.4.2 Checkup

Evaluate.

1. **a.** $\sqrt{49}$ **b.** $\sqrt{0.81}$ **c.** $\sqrt{\dfrac{25}{36}}$

   d. $-\sqrt{16}$ **e.** $\sqrt{-16}$

Determine between what two integers the value of each square root lies. Then estimate, using your calculator.

2. **a.** $\sqrt{17}$ **b.** $-\sqrt{15}$

Evaluate, rounding to the nearest thousandth.

3. **a.** $\sqrt[3]{125}$ **b.** $\sqrt[3]{-64}$ **c.** $\sqrt[3]{\dfrac{8}{125}}$ **d.** $\sqrt[3]{9}$

4. Explain the difference between the principal root of a number and the negative root of a number.

5. What is meant by an irrational number?

6. What is the index of a radical expression? What is the radicand of a radical expression?

1.4.3 Graphing Real Numbers

In the first section of this text, we discussed the set of rational numbers. A rational number is any number that may be written as a ratio of integers, excluding a zero denominator. We located rational numbers as points on a number line. However, there are points on a number line that we did not identify. These are the irrational numbers. Irrational numbers are numbers that cannot be written as a ratio of integers.

The set of real numbers is the set of all rational and irrational numbers. The set of real numbers makes up a number line. Examples of irrational numbers are as follows:

$0.13133133313333\ldots$	(Only terminating and repeating decimals can be written as rational numbers.)
π	$\pi \approx 3.141592654$. The value of π is defined to be the ratio of the circumference of a circle to its diameter. (This decimal representation continues without terminating or repeating.)
$\sqrt{2}$	$\sqrt{2} \approx 1.414213562$. (This decimal representation continues without terminating or repeating.)
$-\sqrt{2}$	$-\sqrt{2} \approx -1.414213562$. (This decimal representation continues without terminating or repeating.)
$\sqrt[3]{5}$	$\sqrt[3]{5} \approx 1.709975947$. (This decimal representation continues without terminating or repeating.)

Just as in the first section, we can graph irrational numbers on the real-number line. To graph an irrational number, we need to evaluate the number to determine its approximate value, in order to place it correctly on the line.

EXAMPLE 7 Graph the real numbers 3, $\sqrt{3}$, $-\sqrt{3}$, $\sqrt{36}$, $\sqrt[3]{3}$, and π on a number line. Label the points.

Solution

3 is located 3 units to the right of 0.

$\sqrt{3} \approx 1.732050808$, which is approximately 1.7 units to the right of 0.

$-\sqrt{3} \approx -1.732050808$, which is approximately 1.7 units to the left of 0.

$\sqrt{36} = 6$, which is 6 units to the right of 0.

$\sqrt[3]{3} \approx 1.44224957$, which is approximately 1.4 units to the right of 0.

$\pi \approx 3.141592654$, which is approximately 3.1 units to the right of 0.

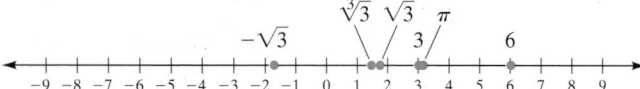

At times, it may be necessary to graph an infinite set of real numbers on the number line. For example, we can graph all of the real numbers between -1 and 3, including 3. We say that -1 is the **lower bound** of the set of numbers and 3 is the **upper bound** of the set. We graph -1 with an open dot because it is not included in the given set of numbers. We graph 3 with a closed dot because it is included in the set of numbers. We graph a solid line between the two bounds.

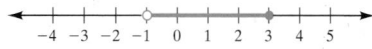

All real numbers between −1 and 3, including 3

Suppose we want to graph all of the real numbers greater than 5. We say that 5 is the lower bound of the set of numbers. We graph 5 with an open dot because it is not a member of the set of numbers being graphed, and we graph a solid line to the right of 5 to include all real numbers greater than 5. We say that the set of numbers increases without bound because it has no upper bound.

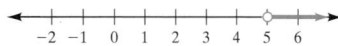

All real numbers greater than 5

When we graph all of the real numbers less than or equal to 5, the upper bound of the set of numbers is 5. We graph 5 with a closed dot because it is a member of the set of numbers being graphed, and we graph a solid line to the left of 5 to include all real numbers less than 5. We say that the set of numbers decreases without bound because it has no lower bound.

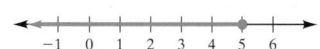

All real numbers less than or equal to 5

EXAMPLE 8 Complete the following table:

Description	Number line
All real numbers between 2 and 5, inclusive	
All real numbers less than 0	
All real numbers greater than or equal to −2	

Solution
Remember, an open dot means that the indicated number is not included in the given set, and a closed dot means that the indicated number is included in the given set.

Description	Number line
All real numbers between 2 and 5, inclusive	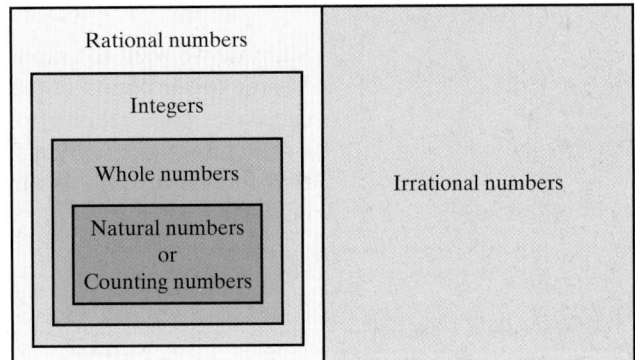
All real numbers less than 0	
All real numbers greater than or equal to −2	

In summary, we have now discussed the entire set of real numbers with its subsets. We have identified natural numbers, whole numbers, and integers as subsets of rational numbers. However, rational numbers are not a subset of irrational numbers. They are mutually exclusive sets: Any number that belongs to one of them does not belong to the other. To visualize this relationship, see Figure 1.12.

Real numbers

Rational numbers

Integers

Whole numbers

Natural numbers
or
Counting numbers

Irrational numbers

Figure 1.12

 1.4.3 Checkup

1. Graph the following real numbers on a number line, and label the points:

$$\sqrt{16}, -\sqrt{4}, \sqrt[3]{70}, -\sqrt{60}, 5, -\pi, \frac{175}{99}, \sqrt{0}, \sqrt[3]{64}$$

2. Complete the following table:

Description	Number line
All real numbers between −1 and 2, including −1	
All real numbers between −1 and 2, inclusive	
All real numbers between −1 and 2	
All real numbers greater than −1	
All real numbers less than or equal to 2	

3. Explain the difference between irrational numbers and rational numbers.

1.4.4 Modeling the Real World

In geometry, we evaluate the area and volume of geometric figures using positive integer exponents. To determine the area of a square, we square the length of its side. This is why raising a number to the second power is referred to as *the number squared*. The length of a side consists of a unit of measure, as well as a number. Therefore, if the number is squared, the unit of measure is also squared. For example, if we have a square of side 2 ft, the area is 4 ft^2. To determine the volume of a cube, we cube the length of each edge—hence *a number cubed*. As before, we cube the number as well as the unit of measurement. For a cube of edge 2 ft, the volume is 8 ft^3.

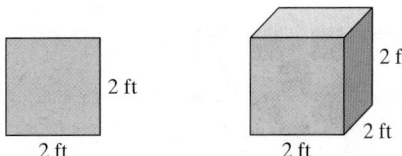

Since we determined the area of a square by squaring the length of its side, we determine the length of the side of a square by taking the square root of its area, including the number and the unit of measure. We also determined the volume of a cube by cubing the length of its edge. Likewise, to determine the length of an edge of a cube, we take the cube root of the volume, including the number and the unit of measure.

EXAMPLE 9   Ernö Rubik, professor at Budapest School of Commercial Art in Hungary, developed a cube with different colored sides divided into three rows and three columns. Each row and column slides up 360 degrees. The object of the game involving Rubik's cube is to rotate the cube until each face is multicolored and then return the cube to its original state. In the United States, Ideal Toys began manufacturing the cube in the late 1970s.

The volume of the original cube is 166.375 cubic centimeters. Determine the length of an edge of Rubik's cube. Determine the length of an edge of one of the small cubes that make up the large cube. Determine the volume of the small cube.

Solution
The length of the edge of the cube is the cube root of its volume.

$$\sqrt[3]{166.375 \text{ cm}^3} = 5.5 \text{ cm}$$

Rubik's cube is 5.5 cm on an edge.

Since there are three rows and three columns on a side, we divide the length of an edge by 3 to determine the length of the edge of the small cube.

$$\frac{5.5}{3} = 1.8\overline{3} \text{ cm}$$

Each small cube is $1.8\overline{3}$ centimeters on an edge.

The volume of the small cube is the length of an edge cubed.

$$(1.8\overline{3} \text{ cm})^3 \approx 6.16 \text{ cm}^3 \qquad \text{Do not round your answer until your final calculation.}$$

Each small cube's volume is about 6.16 cubic centimeters.

APPLICATION

In the town of Marostica in northern Italy, a game of chess using live people as pieces is played every two years. The event celebrates a game played in 1454 between two suitors for a lady's hand, in which the winner could claim the lady as his bride. The board on which the living chess game is played measures eight squares by eight squares, and each square measures 10 feet by 10 feet. What is the area of the board?

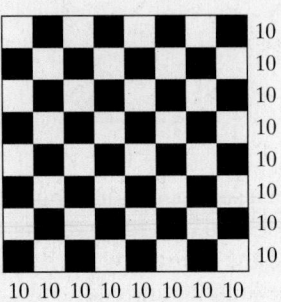

Discussion

To find the area of the board, first find the area of each individual square on the board by squaring the length of its side. This gives us $10 \cdot 10$, or 10^2. Then, multiply this area, by the number of squares on the board.

Since there are eight squares in a row and eight rows of squares, multiply 8 times 8 to find the total number of squares, or $8 \cdot 8 = 8^2$.

Therefore, the total area is

$$(10^2)(8^2) = 6400$$

The area of the board is 6400 ft^2.

✓ 1.4.4 Checkup

1. After the popularity of the Rubik's cube waned, another cube, called Rubik's Revenge, was produced. This cube has an additional row and column on each face. The volume of the cube is approximately 395 cubic centimeters. Determine the length of an edge of Rubik's Revenge. The center square is designed so that the color combinations of its four small squares can be varied. Determine the length of an edge of this center square consisting of four small squares. What is the volume of the center cube?

2. The Shedd Aquarium in Chicago has an exhibit of beluga whales that live in the Oceanarium's Secluded Bay, a tank that holds 400,000 gallons of filtered salt water. This many gallons equate to a tank that would hold approximately 53,000 cubic feet of water. If the tank were in the shape of a cube, what would be the measure of each edge?

3. The board game known in Japan as Go is played throughout the Orient. It is a game with simple rules, but extremely complicated strategies. The board for the game measures 18 squares by 18 squares. Boards may vary in size, but one board for the game has each square measuring about 0.95 inch by 0.95 inch. What is the area of the Go board?

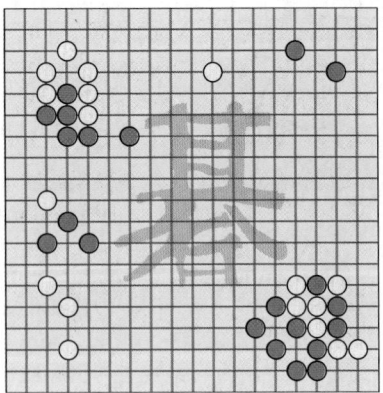

1.4 Exercises

Write in expanded form and evaluate the following exponential expressions:

1. 3^4 **2.** 2^6 **3.** $(-3)^4$ **4.** $(-2)^6$ **5.** -3^4

6. -2^6 **7.** $(-4)^3$ **8.** $(-3)^5$ **9.** -4^3 **10.** -3^5

11. $(-2.5)^2$ **12.** $(-0.4)^2$ **13.** $-\left(-\dfrac{3}{7}\right)^2$ **14.** $-\left(-\dfrac{2}{3}\right)^4$ **15.** $\left(1\dfrac{1}{3}\right)^3$

16. $\left(1\dfrac{1}{2}\right)^3$ **17.** $\left(3\dfrac{2}{11}\right)^4$ **18.** $\left(7\dfrac{21}{29}\right)^7$ **19.** 0^8 **20.** 0^5

21. 1^{32} **22.** 1^{10} **23.** $(-1)^{29}$ **24.** $(-1)^9$ **25.** 1256^1

26. $\left(-\dfrac{4}{17}\right)^1$ **27.** 1256^0 **28.** $(-34.601)^0$ **29.** -4721^0 **30.** -8.23^0

31. $(-325)^0$ **32.** $(-10.8)^0$

Evaluate the following square roots:

33. $\sqrt{36}$ **34.** $\sqrt{81}$ **35.** $\sqrt{256}$ **36.** $\sqrt{324}$ **37.** $-\sqrt{25}$ **38.** $-\sqrt{49}$

39. $\sqrt{0.64}$ **40.** $\sqrt{0.81}$ **41.** $-\sqrt{\dfrac{16}{9}}$ **42.** $-\sqrt{\dfrac{36}{49}}$ **43.** $\sqrt{1}$ **44.** $-\sqrt{1}$

45. $-\sqrt{0}$ **46.** $\sqrt{0}$ **47.** $\sqrt{-16}$ **48.** $\sqrt{-100}$

Determine between what two integers the values of the square roots lie. Then estimate the square roots using your calculator, rounding to three decimal places.

49. $\sqrt{10}$ **50.** $\sqrt{22}$ **51.** $-\sqrt{3}$ **52.** $-\sqrt{14}$

Evaluate the cube roots. Express your answers in fractional notation or round decimals to the nearest thousandth, as appropriate.

53. $\sqrt[3]{64}$ **54.** $\sqrt[3]{8}$ **55.** $\sqrt[3]{1728}$ **56.** $\sqrt[3]{9261}$ **57.** $\sqrt[3]{1234}$

58. $\sqrt[3]{4321}$ **59.** $\sqrt[3]{-125}$ **60.** $\sqrt[3]{-64}$ **61.** $\sqrt[3]{\dfrac{1}{8}}$ **62.** $\sqrt[3]{-\dfrac{729}{1331}}$

63. Graph the following numbers on a number line, and label the points:

$$\sqrt[3]{25},\ -\sqrt{45},\ \sqrt{\dfrac{9}{16}},\ -\sqrt{16},\ -\sqrt{\dfrac{8}{3}},\ \sqrt[3]{8},\ \sqrt{75}$$

64. Graph the following numbers on a number line, and label the points:

$$-\sqrt{9},\ \sqrt{-63},\ -\sqrt{\dfrac{7}{3}},\ \sqrt{17},\ \sqrt{\dfrac{25}{4}},\ \sqrt[3]{64},\ \sqrt[3]{-64}$$

65. Complete the following table:

Description	Number line
All real numbers between −1.5 and 3, including −1.5	
All real numbers between −1.5 and 3, including 3	
All real numbers between −1.5 and 3, inclusive	
All real numbers between −1.5 and 3	
All real numbers greater than −1.5	
All real numbers less than or equal to 3	

66. Complete the following table:

Description	Number line
All real numbers between −1 and 2.5, including −1	
All real numbers between −1 and 2.5, including 2.5	
All real numbers between −1 and 2.5, inclusive	
All real numbers between −1 and 2.5	
All real numbers greater than or equal to 2.5	
All real numbers less than or equal to −1	

67. The square base of the Great Pyramid, the tomb of Pharaoh Cheops, measures 755 feet on each of its four sides. What is the area of ground covered by the pyramid?

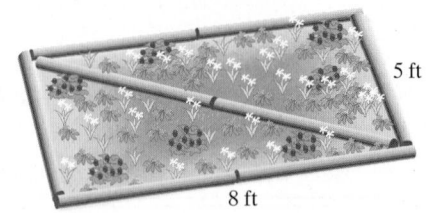

5 ft

8 ft

68. Three squares are nested, one within another. The innermost square is to be painted light blue. It measures 6 feet on a side. The middle square is to be painted white and measures 10 feet on a side. The outer square is to be painted dark blue and measures 15 feet on a side. How many square feet of surface will be painted with each color?

69. Jennie is putting in a flower garden in her backyard. The garden will be rectangular, with a length of 8 feet and a width of 5 feet. She wants to divide it in half diagonally with landscaping logs. How many feet of logs does she need for the diagonal? (In order to figure this, Jennie must square the length and width, add the two squares, and then take the square root of the sum.)

70. Jennie decides to enlarge her flower garden from exercise 69. She plans to have a length of 15 feet and a width of 8 feet. Now how many feet of logs does she need for the diagonal?

71. Instructions for planting a shrub require that a hole be dug in the shape of a cube with a volume of 3 cubic feet. What are the dimensions of the hole?

72. A pit for a luau is shaped like a cube with a volume of 15 cubic feet. What are the dimensions of the pit?

73. A card table has a surface in the shape of a square with an area of 12 square feet. What are the dimensions of the surface?

74. A square goldfish pond has a surface area of 50 square feet and a depth of 2 feet. What are the dimensions of the pond?

75. The Taj Mahal is one of the most beautiful buildings in the world. It is designed with a high central dome surrounded by small chambers arranged about two intersecting axes so that all four sides of the center structure are the same length. The center structure covers an area of about 377,000 square feet. What is the length of a side of this structure?

76. For the Taj Mahal, described in exercise 75, what is the length of the diagonal of the center structure, measured from one corner to the opposite corner? (Recall that the length of the diagonal of a square is the square root of the sum of the squares of its two adjacent sides.)

77. St. Peter's Basilica in Vatican City has a dome above the papal altar that spans an area of about 19,400 square feet. The radius of the domed area is the distance from the center of the circle to its outer edge. To find the radius of this circular area, you must divide the area by the mathematical constant π and then take the square root of the quotient. What is the radius of this domed area?

78. One of the largest domes ever constructed is in the Louisiana Superdome. The area of floor covered by the dome measures about 363,000 square feet. Use the calculation method in exercise 77 to determine the radius of the domed area.

1.4 Calculator Exercises

Part 1. Relating perfect squares and perfect cubes with roots.

In this section, we discussed perfect squares and perfect cubes. If you construct a table of perfect squares and perfect cubes, you can use the table to identify square roots and cube roots of numbers that are perfect squares or perfect cubes. The table that follows is such a table. Complete all columns and rows.

Number, n	Square, n^2	Cube, n^3
1	1	1
2	4	8
3	9	27
4	16	64
5	25	
6	36	
7		
8		
9		
10		
11		
12		
13		
14		
15		
16		
17		
18		
19		
20		

Now use the table to determine the following values:

1. $\sqrt{196}$ **2.** $\sqrt{529}$ **3.** $\sqrt[3]{4913}$ **4.** $\sqrt[3]{9261}$

Explain the relationship between raising a number to a power and taking a root of a number.

Part 2. Using the calculator to determine other roots.

You have seen how to use the square-root key on your calculator to evaluate square roots. You have also seen how to use the MATH key on your calculator to evaluate cube roots. There is another option under the MATH key that can be used to evaluate square roots and cube roots: option 5. Note that this option has the symbol $\sqrt[x]{}$ beside it. This means that if you decide to use the option, you must first specify what the index is, then choose the option, and then specify the radicand.

As an example, suppose you wish to know the square root of 161.29. You can determine the square root using the square-root key or the MATH key as shown in the following screen:

```
√(161.29)
            12.7
2×√161.29
            12.7
```

Next, suppose you wish to know the cube root of 5000. You can determine the cube root using option 4 or option 5 of the MATH key as shown in this screen:

```
³√(5000)
        17.09975947
3×√5000
        17.09975947
```

Now, you probably would not choose to use option 5 to the MATH key for square roots or cube roots, since it may seem simpler to choose the other method you have learned. However, in later chapters we will be working with expressions in which the index of the radicand is larger than 3. In other words, you may want to know what number is raised to the fourth power, the fifth power, or some other power to yield a certain value. Then option 5 of the MATH key will be handy. As a final example, suppose you want to know what number has to be raised to the sixth power to yield 2,985,984. To determine this, you can use option 5 as follows:

```
6 ×√2985984
            12
```

Use option 5 of the MATH key to evaluate the following radicals:

1. $\sqrt[5]{57,392}$ 2. $-\sqrt[4]{37,652}$ 3. $\sqrt[4]{\pi}$

4. $\sqrt[6]{\dfrac{64}{729}}$ 5. $\sqrt[7]{2.5}$ 6. $\sqrt[5]{-391.35393}$

1.4 Writing Exercises

Part 1.
Determine the resulting sign of the exponential expressions. Do not evaluate the expressions.

1. $(-55)^8$ 2. $(-85)^{12}$ 3. -55^8 4. -85^{12}
5. $(-55)^5$ 6. $(-85)^7$ 7. -55^5 8. -85^7

Discuss the rules you used to determine the resulting signs for the preceding exercises.

Part 2.
In some radical expressions, the radicand may be negative or positive. In other radical expressions, the radicand must always be positive. Describe the conditions that determine when each of the two situations applies.

1.5 Scientific Notation

OBJECTIVES

1 Evaluate exponential expressions with negative integer exponents.

2 Write equivalent scientific notation, standard notation, and calculator notation.

3 Write and evaluate numerical representations for real-world data, and express the results in the desired notation.

APPLICATION

Common table salt consists of atoms in the shape of a face-centered cubic structure. The volume of this cube is about 1.68×10^{-28} m^3. Find the length of the edge of the cube.

After completing this section, we will discuss this application further. See page 67.

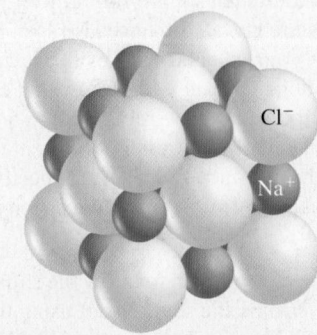

Cl$^-$

Na$^+$

1.5.1 Evaluating Exponential Expressions with a Negative Integer Exponent

If an exponential expression has a negative exponent, it is impossible to determine the number of times to repeat the base when one is writing in expanded form. Therefore, we must discover an alternative method to evaluate these exponential expressions. Complete the following sets of exercises with your calculator.

Discovery 10

Nonzero Integer Bases with Negative Integer Exponents

Evaluate each expression, and compare the results obtained in the left column with the corresponding results in the right column.

1. a. $10^{-1} = $ _____   **b.** $\left(\dfrac{1}{10}\right)^{1} = $ _____

2. a. $10^{-2} = $ _____   **b.** $\left(\dfrac{1}{10}\right)^{2} = $ _____

3. a. $10^{-3} = $ _____   **b.** $\left(\dfrac{1}{10}\right)^{3} = $ _____

4. a. $\left(\dfrac{1}{10}\right)^{-1} = $ _____   **b.** $10^{1} = $ _____

5. a. $\left(\dfrac{1}{10}\right)^{-2} = $ _____   **b.** $10^{2} = $ _____

6. a. $\left(\dfrac{1}{10}\right)^{-3} = $ _____   **b.** $10^{3} = $ _____

State a rule for evaluating an exponential expression with a nonzero integer base and a negative integer exponent.

In the column on the left, we evaluate an exponential expression with a nonzero integer base and a negative integer exponent. In the column on the right, we evaluate an exponential expression consisting of a base that is the reciprocal of the base on the left and an exponent that is the opposite of the exponent on the left. The results are the same in both columns.

EVALUATING EXPRESSIONS WITH NEGATIVE EXPONENTS

To evaluate an exponential expression with a nonzero base and a negative integer exponent,

- Rewrite the expression as the reciprocal of the base with the opposite of the exponent.
- Evaluate the new expression.

An exponential expression with a zero base and a negative integer exponent is undefined.

To understand these rules, examine the following pattern:

$$10^4 = 10 \cdot 10 \cdot 10 \cdot 10$$

Exponent decreases by 1. $\qquad 10^3 = 10 \cdot 10 \cdot 10 \qquad$ *Divide by 10.*

$$10^2 = 10 \cdot 10$$

$$10^1 = 10$$

$$10^0 = 1$$

$$10^{-1} = \underline{\qquad}$$

$$10^{-2} = \underline{\qquad}$$

Using the preceding pattern, $10^{-1} = \frac{1}{10}$ and $10^{-2} = \frac{1}{100}$.

It is impossible to evaluate an exponential expression with a negative exponent and a zero base, because if we take the reciprocal of 0, we obtain $\frac{1}{0}$, which is undefined.

$$0^{-3} = \left(\frac{1}{0}\right)^3, \text{ which is undefined.}$$

EXAMPLE 1 Write an equivalent exponential expression having a positive exponent. Evaluate.

a. $(-10)^{-1}$ **b.** $(-10)^{-2}$ **c.** $\left(-\dfrac{1}{10}\right)^{-1}$ **d.** $\left(-\dfrac{1}{10}\right)^{-2}$

Solution **Check**

a. $(-10)^{-1} = \left(-\dfrac{1}{10}\right)^1$ *Reciprocal of −10 is $-\frac{1}{10}$.*

$\qquad = -\dfrac{1}{10}$

a.
```
(-10)^-1▶Frac
              -1/10
```
b.
```
(-10)^-2▶Frac
              1/100
```

b. $(-10)^{-2} = \left(-\dfrac{1}{10}\right)^2$ *Reciprocal of −10 is $-\frac{1}{10}$.*

$\qquad = \dfrac{1}{100}$

c. $\left(-\dfrac{1}{10}\right)^{-1} = (-10)^1$ *Reciprocal of $-\frac{1}{10}$ is −10.*

$\qquad = -10$

c.
```
(-1/10)^-1
               -10
```
d.
```
(-1/10)^-2
               100
```

d. $\left(-\dfrac{1}{10}\right)^{-2} = (-10)^2$ *Reciprocal of $-\frac{1}{10}$ is −10.*

$\qquad = 100$

1.5.1 Checkup

In exercises 1–4, write an equivalent exponential expression having a positive exponent. Evaluate.

1. 3^{-1} **2.** $\left(\dfrac{1}{3}\right)^{-2}$ **3.** $(-3)^{-1}$ **4.** $\left(-\dfrac{1}{3}\right)^{-2}$

5. What is the effect of a negative exponent on an exponential expression?

6. If an exponential expression has a negative exponent, what is the restriction on the value of the base?

1.5.2 Writing Equivalent Scientific Notation, Standard Notation, and Calculator Notation

Numbers are usually written in **standard notation**.

5,000,000	Five million
−3,458,000,000	negative three billion, four hundred fifty-eight million
0.000034	thirty-four millionths

However, very large and very small numbers are often written in scientific notation. **Scientific notation** is an expression written as the product of a number whose absolute value is between 1 and 10, including 1, and an integer power of 10. We call the number whose absolute value is between 1 and 10, including 1, the **numerical factor**. Since calculators display a limited number of digits, very large and very small numbers are written in an abbreviated scientific notation. We will call this form calculator notation. **Calculator notation** is a display consisting of the numerical factor, followed by "E," followed by the integer exponent of 10.

Standard Notation	*Scientific Notation*	*Calculator Notation*
	numerical factor $\times$ 10$^{\text{integer exponent}}$	numerical factor E integer exponent
5,000,000 =	5×10^6 =	5 E6
−3,458,000,000 =	-3.458×10^9 =	−3.458 E9
0.000034 =	3.4×10^{-5} =	3.4 E−5

HELPING HAND Remember that calculator notation is not an appropriate notation for your written answer.

CONVERTING FROM STANDARD TO SCIENTIFIC NOTATION

To convert a number from standard notation to scientific notation, write a product of the numerical factor and an exponential expression.

Determine the numerical factor.

• Move the decimal point in the number so that the resulting number, the numerical factor, has an absolute value between 1 and 10, including 1.

Determine the exponential expression.

• The exponent of the base 10 is positive if the decimal was moved to the left, negative if the decimal was moved to the right, and 0 if the decimal was not moved.

• The absolute value of the exponent is the value of the number of places the decimal was moved.

EXAMPLE 2   Write in scientific notation.

a. 65,780,000,000,000   **b.** −65,780,000,000,000

c. 0.00000000002895   **d.** −0.00000000002895

Solution   **Check**

a. $65,780,000,000,000 = 6.578 \times 10^{13}$ *The decimal was moved 13 places to the left.*

a.
```
65780000000000
        6.578E13
```

b. $-65,780,000,000,000 = -6.578 \times 10^{13}$ *The decimal was moved 13 places to the left.*

b.
```
-65780000000000
        -6.578E13
```

c. $0.00000000002895 = 2.895 \times 10^{-11}$ *The decimal was moved 11 places to the right.*

c.
```
.00000000002895
       2.895E-11
```

d. $-0.00000000002895 = -2.895 \times 10^{-11}$ *The decimal was moved 11 places to the right.*

d.
```
-.00000000002895
      -2.895E-11
```

The calculator will automatically display very large or very small numbers in calculator notation. Other numbers can be displayed by changing the default setting on your calculator. See Section 1.5 Calculator Exercises at the end of this section.

HELPING HAND In scientific notation, numbers with absolute values greater than or equal to 10 have positive exponents. Numbers with absolute values less than 1 have negative exponents.

CONVERTING FROM SCIENTIFIC TO STANDARD NOTATION

To write a number in standard form if the exponent is positive,

- Move the decimal point in the numerical factor to the right the number of places the absolute value of the exponent denotes.

To write a number in standard form if the exponent is negative,

- Move the decimal point in the numerical factor to the left the number of places the absolute value of the exponent denotes.

If the exponent is 0, a number in standard form is the numerical factor.

EXAMPLE 3   Write in standard notation.

a. 3.86×10^7 **b.** -3.86×10^7 **c.** 7.4×10^{-3} **d.** -7.4×10^{-3}

Solution

a. $3.86 \times 10^7 = 38,600,000$ **b.** $-3.86 \times 10^7 = -38,600,000$

Move the decimal seven places to the right.   *Move the decimal seven places to the right.*

c. $7.4 \times 10^{-3} = 0.0074$ **d.** $-7.4 \times 10^{-3} = -0.0074$

Move the decimal three places to the left.   *Move the decimal three places to the left.*

CONVERTING FROM CALCULATOR TO SCIENTIFIC NOTATION

To convert a number from calculator notation to scientific notation, write a product of a numerical factor and an exponential expression.

- The numerical factor is the number displayed before the "E."
- The exponential expression consists of a base 10 and an exponent whose value is the number after the "E."

EXAMPLE 4   Write in scientific and standard notation.

a. 3.86 E7   **b.** 7.4 E−3

Solution

a. $3.86 \text{ E7} = 3.86 \times 10^7$ $= 38,600,000$

  3.86 is the numerical factor.   Move the decimal point
  7 is the exponent.   seven places to the right.

b. $7.4 \text{ E}-3 = 7.4 \times 10^{-3}$ $= 0.0074$

  7.4 is the numerical factor.   Move the decimal point
  −3 is the exponent.   three places to the left.

EXAMPLE 5   Complete the table.

Standard Notation	Scientific Notation	Calculator Notation
−2,340,000,000		
0.000000234		
		5.76E8
	-5.76×10^8	
		−1.4E−6
	1.4×10^{-6}	

Solution

Standard Notation	Scientific Notation	Calculator Notation
−2,340,000,000	-2.34×10^9	−2.34E9
0.000000234	2.34×10^{-7}	2.34E−7
576,000,000	5.76×10^8	5.76E8
−576,000,000	-5.76×10^8	−5.76E8
−0.0000014	-1.4×10^{-6}	−1.4E−6
0.0000014	1.4×10^{-6}	1.4E−6

1.5.2 Checkup

Complete the following table:

	Standard Notation	Scientific Notation	Calculator Notation
1.			6.5E7
2.			−3.12E−4
3.		-8.33×10^7	
4.		9.3×10^{-3}	
5.	0.00365		
6.	−2,340		

1.5.3 Modeling the Real World

Scientific notation is used most of the time in dealing with very large and very small numbers. Scientific notation is common in areas such as astronomy, which deals with large distances and huge masses; environmental science, which deals with global populations and resources; biology and chemistry, which deal with very small cells and molecules; and economics, which deals with large amounts of money and financial data.

EXAMPLE 6 According to the U.S. Bureau of the Census, the nation's population in 1998 was 270,561,000. Write a numerical representation of this number in scientific notation.

Solution

$$270,561,000 = 2.70561 \times 10^8$$

Move the decimal eight places to the left to obtain a number between 1 and 10. Therefore, the exponent of 10 is 8.

The U.S. population was about 2.7×10^8 in 1998.

EXAMPLE 7 Common table salt consists of atoms of the elements sodium (Na) and chlorine (Cl), arranged with alternate atoms at the corners of a cube. The distance between neighboring atoms is 2.76×10^{-10} meter. Approximate the length of the edge of the cube.

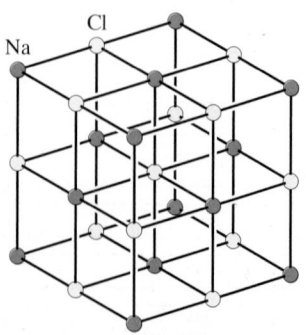

Solution
The length of the cube is about twice the distance between the neighboring atoms.

$$2(2.76 \times 10^{-10}) = 2(2.76 \times 10^{-10}) = (2 \cdot 2.76) \times 10^{-10} \qquad \text{Associative property}$$
$$= 5.52 \times 10^{-10}$$

The length of the edge of the cube is about 5.52×10^{-10} meter. In chemistry, we define an angstrom, Å, as 1×10^{-10} meter. Therefore, the distance along the edge of the cube is about 5.52 Å.

APPLICATION

Common table salt consists of atoms in the shape of a face-centered cubic structure. The volume of this cube is about 1.68×10^{-28} m^3. Find the length of the edge of the cube.

Discussion

To determine the length of the edge of a cube, we take the cube root of the volume.

$$\sqrt[3]{1.68 \times 10^{-28} \text{m}^3} \approx 5.52 \times 10^{-10} \text{ m}$$

The length of the edge is approximately 5.52×10^{-10} m.

Note: This is the same value obtained in Example 7.

✓ 1.5.3 Checkup

1. The speed of light in a vacuum is accepted to be 2.99792458×10^8 meters per second. Write a numerical representation of this number in standard notation. What do you think would be a useful approximation to the speed of light?

2. The amount of heat needed to change 125 grams of water at 100°C to steam at 100°C is about 67,500 calories. Write a numerical representation, in scientific notation, of the number of calories needed to change twice as much water to steam.

3. The biggest satellite of any planet in the Solar System is Ganymede, which orbits Jupiter. It has a mass of 1.46×10^{23} tons, approximately twice the mass of the Earth's Moon. Approximate the mass of the Earth's Moon.

1.5 Exercises

Write an equivalent exponential expression having a positive exponent. Evaluate.

1. 2^{-1} **2.** 5^{-1} **3.** $\left(\dfrac{1}{2}\right)^{-2}$ **4.** $\left(\dfrac{1}{5}\right)^{-2}$ **5.** $(-2)^{-1}$ **6.** $(-5)^{-1}$ **7.** $\left(-\dfrac{1}{2}\right)^{-2}$ **8.** $\left(-\dfrac{1}{5}\right)^{-2}$

Complete the following table:

	Standard Notation	Scientific Notation	Calculator Notation
9.	23,450,000,000		
10.	18,300,000,000,000,000,000		
11.	−0.000006591		
12.	−0.00000000072193		
13.	3.6943		
14.	9.98031		
15.			−7.1103E5
16.			−1.005E11
17.			1.966E−2
18.			5.555E−6
19.			−9.95E0
20.			−8.103E0

(Continued on page 68)

Number	Standard Notation	Scientific Notation	Calculator Notation
Ten thousand			
One hundred thousand			
One million			
Ten million			
One hundred million			
One billion			
One trillion			
One quadrillion			
One quintillion			
One sextillion			
One septillion			
One octillion			
One nonillion			
One decillion			

Explain the relationship between the number of zeroes in the number you obtain and the value of the exponent in scientific notation for the numbers listed.

1.6 Properties of Real Numbers and Order of Operations

OBJECTIVES

1 Discover, identify, and illustrate the distributive laws of multiplication over addition and subtraction.

2 Evaluate expressions using the order of operations.

3 Model real-world situations by using expressions and evaluate the expressions by means of the order of operations.

APPLICATION

Federal guidelines on the identification of obesity in adults state that the recommended body weight in pounds of a woman who is 5 feet, 6 inches (or 66 inches), tall is found from the expression

$$2.2[45.5 + 2.3(66 - 60)]$$

Determine this weight.

After completing this section, we will discuss this application further. See page 77.

In previous sections, we have written equivalent expressions for rational numbers. We discussed properties of rational numbers that help us write additional equivalent expressions and understand the ones we have already written. These properties will be useful later on, when we study algebra.

All of the properties of rational numbers that we have discussed are also properties of real numbers. The following is a summary of those properties:

PROPERTIES OF REAL NUMBERS

Identity Property of 0
The sum of a real number and 0 is the number itself.

Identity Property of 1
The product of a real number and 1 is the real number itself.

Multiplication Property of -1
The product of a real number and -1 is the opposite of the number itself.

Multiplication Property of 0
The product of a real number and 0 is 0.

Additive Inverse Property
The sum of a real number and its opposite, or additive inverse, is 0.

Multiplicative Inverse Property
The product of a nonzero real number and its reciprocal, or multiplicative inverse, is 1.

Commutative Law for Addition
Changing the order of two real-number addends does not change the sum.

Commutative Law for Multiplication
Changing the order of two real-number factors does not change the product.

Associative Law for Addition
Changing the grouping of three real-number addends does not change the sum.

Associative Law for Multiplication
Changing the grouping of three real-number factors does not change the product.

We will find that we can discuss properties of a real number even if we don't know its value. In fact, we will sometimes be able to use these properties to figure out the value of an unknown number from information about it.

1.6.1 Identifying the Distributive Law

All of the properties previously described in this section have dealt with only one operation at a time, such as all additions or all multiplications in an expression. The next property involves both multiplication and addition. Complete the following set of exercises with your calculator.

Discovery 11

Combining Multiplication and Addition

Evaluate each expression, and compare the results obtained in the left column with the corresponding results in the right column.

1. a. $2(6 + 4) =$ _____   **b.** $2(6) + 2(4) =$ _____
2. a. $-2(-6 + 4) =$ _____   **b.** $-2(-6) + (-2)(4) =$ _____

In your own words, state a rule for combining multiplication and addition.

In the column on the left, we multiply a sum of real-number addends by a real-number factor. In the column on the right, we multiply each addend by the real-number factor and then add the products. The results are the same in both columns.

This rule is one of the four distributive laws:

DISTRIBUTIVE LAWS

Multiplication over Addition
The product of a real-number factor and a sum of real-number addends is the same as the sum of the products of the factor and each addend.

Multiplication over Subtraction
The product of a real-number factor and a difference of real numbers is the same as the difference of the product of the factor and the minuend less the product of the factor and the subtrahend.

Division over Addition
The quotient of a real-number divisor and a sum of real-number addends is the same as the sum of the quotients of the divisor and each addend.

Division over Subtraction
The quotient of a real-number divisor and a difference of real numbers is the same as the difference of the quotient of the divisor and the minuend less the quotient of the divisor and the subtrahend.

Combining the multiplication property of -1 and the distributive law results in a very important property. The following example illustrates this property:

$$-(3 + 6) = -1(3 + 6) \qquad \text{Multiplication property of } -1 \text{ (in reverse)}$$
$$= -1(3) + (-1)(6) \qquad \text{Distributive law}$$
$$= -(3) + [-(6)] \qquad \text{Multiplication property of } -1$$

In this example, we see that the opposite of a sum of real numbers is the same as the sum of the opposite of each addend.

$$-(3 + 6) = -(9) = -9 \qquad \text{and} \qquad -(3) + [-(6)] = -3 + (-6) = -9$$

Note: This is also true for taking the opposite of a difference.

$$-(3 - 6) = -(3) - [-(6)] = -3 + 6$$

OPPOSITE OF A SUM

The opposite of a sum of real numbers is the same as the sum of the opposites of each addend.

EXAMPLE 1 Write an equivalent expression for the following expressions, using the distributive law or the opposite-of-a-sum property:

a. $9(3 - 4)$ **b.** $(4 + 2)3$ **c.** $\dfrac{15 + 5}{5}$ **d.** $-(3 + 6)$

Solution

a. $9(3 - 4) = 9(3) - 9(4)$ Distributive law
Multiply each addend by the factor 9.

b. $(4 + 2)3 = 4(3) + 2(3)$ Distributive law
Multiply each addend by the factor 3.

c. $\dfrac{15 + 5}{5} = \dfrac{15}{5} + \dfrac{5}{5}$ Distributive law
Divide each addend by the divisor 5.

d. $-(3 + 6) = -(3) + [-(6)]$ or $-3 - 6$ Opposite-of-a-sum property
Take the opposite of each addend.

The reverse of the distributive property is called **factoring**. The following are some examples of factoring:

$4(3) + 4(2) = 4(3 + 2)$ The real-number factor, 4, is called the common factor of the addends.

$2(3) - 2(4) = 2(3 - 4)$ The common factor is 2.

EXAMPLE 2   Factor the expressions by writing an equivalent expression, using the distributive law in reverse. Then evaluate the equivalent expression.

a. $3(5.1) + 3(4.9)$ **b.** $5\left(\dfrac{1}{2}\right) - 3\left(\dfrac{1}{2}\right)$

Solution

a. $3(5.1) + 3(4.9) = 3(5.1 + 4.9)$ The common factor is 3.
$\qquad\qquad\qquad\quad = 3(10)$
$\qquad\qquad\qquad\quad = 30$

b. $5\left(\dfrac{1}{2}\right) - 3\left(\dfrac{1}{2}\right) = \dfrac{1}{2}(5 - 3)$ The common factor is $\frac{1}{2}$.
$\qquad\qquad\qquad\quad = \dfrac{1}{2}(2)$
$\qquad\qquad\qquad\quad = 1$

1.6.1 Checkup

For each of the following, write an equivalent expression, using the distributive law or the opposite-of-a-sum property.

1. $5(3 + 7)$ **2.** $(17 - 25)5$ **3.** $\dfrac{-46 + 62}{2}$ **4.** $-(-5 + 9)$

Factor by writing an equivalent expression, using the distributive law in reverse. Then evaluate the equivalent expression.

5. $5(1.2) + 5(1.8)$ **6.** $17\left(\dfrac{1}{3}\right) - 23\left(\dfrac{1}{3}\right)$

7. Explain the terms *factoring* and *common factor*.

1.6.2 Evaluating Expressions by Using the Order of Operations

With the basic arithmetic operations, we use parentheses, brackets, braces, absolute-value symbols, fraction bars, and radicals as grouping symbols. These grouping symbols change the order of operations of the arithmetic operations (addition, subtraction, multiplication, and division). To evaluate an expression involving grouping symbols, perform the operations within the grouping symbols first. If more than one pair of grouping symbols is present, perform the innermost operation first.

ORDER OF OPERATIONS

To evaluate an expression, perform operations in the following order:

- Perform all operations within grouping symbols. If more than one grouping symbol is present, perform the innermost operation first and work outward.
- Evaluate exponents and roots.
- Evaluate multiplication and division from left to right.
- Evaluate addition and subtraction from left to right.

A phrase we can use to help us remember the correct order of operations is "Please Excuse My Dear Aunt Sally." The first letter of each word corresponds to an operation.

Please	**P**arentheses or groupings
Excuse	**E**xponents and roots
My	**M**ultiply ⎫
Dear	**D**ivide ⎭ and
Aunt	**A**dd ⎫
Sally	**S**ubtract ⎭ and

HELPING HAND Remember that we perform multiplication and division at the same time, from left to right. Then we perform addition and subtraction at the same time, from left to right.

EXAMPLE 3 Evaluate using the order of operations.

a. $4(-3 + 6) - 7(8 - 2)$ **b.** $-(3^2 - 7) + 8(-2 - 4)$

c. $\dfrac{16 - 2^2 + 3}{3 + 2}$ **d.** $2\{[2(3 - 4) + 7] - 8\} + 3(-5)$

e. $-\sqrt{25 + 11} + 6$ **f.** $-|3.5 - 4.26 - 5|$

Solution

a.
$$4(-3 + 6) - 7(8 - 2) = 4(3) - 7(8 - 2) \qquad \text{Add within parentheses.}$$
$$= 4(3) - 7(6) \qquad \text{Subtract within parentheses.}$$
$$= 12 - 7(6) \qquad \text{Multiply.}$$
$$= 12 - 42 \qquad \text{Multiply and subtract.}$$
$$= -30$$

b.
$$-(3^2 - 7) + 8(-2 - 4) = -(9 - 7) + 8(-2 - 4) \qquad \text{Evaluate the exponent.}$$
$$= -(2) + 8(-2 - 4) \qquad \text{Subtract within parentheses.}$$
$$= -(2) + 8(-6) \qquad \text{Subtract within parentheses.}$$
$$= -2 + 8(-6) \qquad \text{Take the opposite. (Multiply by } -1.)$$
$$= -2 + -48 \qquad \text{Multiply and add.}$$
$$= -50$$

c. $\dfrac{16 - 2^2 + 3}{3 + 2} = \dfrac{16 - 4 + 3}{3 + 2}$ Evaluate the exponent.

$= \dfrac{12 + 3}{3 + 2}$ Subtract.

$= \dfrac{15}{5} = 3$ Add both groups separated by the fraction bar and divide.

d. $2\{[2(3 - 4) + 7] - 8\} + 3(-5) = 2\{[2(-1) + 7] - 8\} + 3(-5)$ Subtract within parentheses.

$= 2\{[-2 + 7] - 8\} + 3(-5)$ Multiply.

$= 2\{5 - 8\} + 3(-5)$ Add within brackets.

$= 2\{-3\} + 3(-5)$ Subtract.

$= -6 - 15$ Multiply and subtract.

$= -21$

e. $-\sqrt{25 + 11} + 6 = -\sqrt{36} + 6$ Add within the radical sign.

$= -6 + 6$

$= 0$

f. $-|3.5 - 4.26 - 5| = -|-5.76|$ Subtract within absolute values.

$= -(5.76)$ Evaluate the absolute value, and then take the opposite.

$= -5.76$

In entering these examples into your calculator, it may be necessary to add additional parentheses for grouping. Remember to use parentheses for all grouping symbols. This is explained in Section 1.6, Calculator Exercises. The checks for Examples 3c, 3d, and 3e are shown in Figure 1.13. ●

TECHNOLOGY

ORDER OF OPERATIONS

Evaluate. c. $\dfrac{16 - 2^2 + 3}{3 + 2}$ d. $2\{[2(3 - 4) + 7] - 8\} + 3(-5)$

e. $-\sqrt{25 + 11} + 6$

```
c. (16-2²+3)/(3+2)
                    3
d. 2((2(3-4)+7)-8)+
   3(-5)
e.                -21
   -√(25+11)+6
                   0
```

Figure 1.13

For Figure 1.13,

It is necessary to enter both the numerator and the denominator in a set of parentheses to ensure proper grouping.

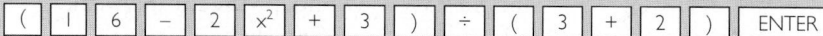

$(\;|\; 6 \;-\; 2 \;x^2 \;+\; 3 \;) \;\div\; (\;3 \;+\; 2 \;) \;|\; \text{ENTER}$

(Continued on page 76)

Use sets of parentheses instead of brackets and braces.

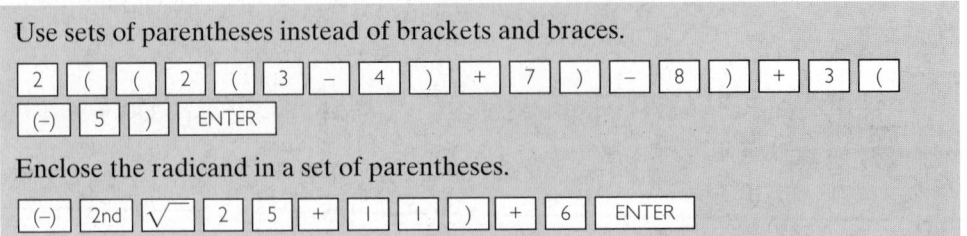

Enclose the radicand in a set of parentheses.

 1.6.2 Checkup

In Exercises 1–5, evaluate, using the order of operations.

1. $5(-4 + 7) - 3(9 - 5)$ **2.** $-(4^3 - 15) + 6(-5 - 3)$

3. $\dfrac{27 - 4^2 + 5}{7 + 1}$ **4.** $-\sqrt{9 + 16} + 1$

5. $-2\{4[3 - (5 - 7)] + 6 \div 3\} + 7(-3)$

6. What does the memory device "Please Excuse My Dear Aunt Sally" help you to remember? Explain.

1.6.3 Modeling the Real World

In using real numbers to evaluate real-world data, the same rules for the order of operations apply. Be careful in determining what grouping symbols, if any, are needed to describe the situation correctly, since grouping symbols can change the order of operations. It is often a good idea to check your answer and see whether it seems reasonable to you. For example, if you calculate an average test score and your answer is higher than any of the individual scores, you've made an error somewhere.

EXAMPLE 4   The daily low temperatures (in degrees Celsius) for the week of October 15–21, 1999, for Greenwood, Canada, are given in the following chart:

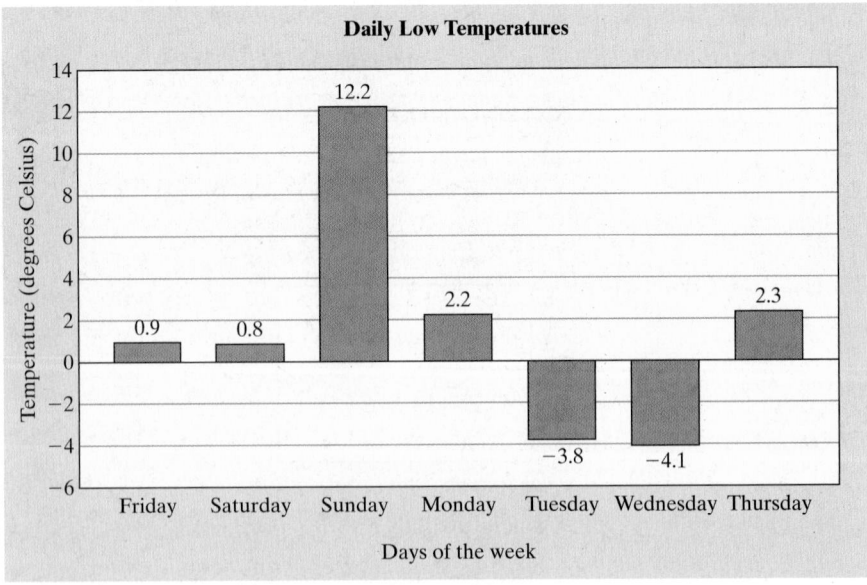

Find the average temperature for the week.

Solution

In statistics, the average is calculated by adding up all of the values and dividing by the count of the values. Therefore, the average weekly low temperature is found by adding the seven low temperatures and then dividing the sum by 7, the number of temperatures that were added.

$$\frac{0.9 + 0.8 + 12.2 + 2.2 + (-3.8) + (-4.1) + 2.3}{7} = \frac{10.5}{7} = 1.5$$

The average low temperature for the week was 1.5 degrees Celsius.

APPLICATION

Federal guidelines on the identification of obesity in adults state that the recommended body weight in pounds of a woman who is 5 feet, 6 inches (or 66 inches) tall is found from the expression

$$2.2[45.5 + 2.3(66 - 60)]$$

Determine this weight.

Discussion

$$2.2[45.5 + 2.3(66 - 60)] = 2.2[45.5 + 2.3(6)]$$
$$= 2.2[45.5 + 13.8]$$
$$= 2.2[59.3]$$
$$= 130.46$$

The recommended weight for a woman 5 feet, 6 inches is about 130 pounds.

✓ 1.6.3 Checkup

1. Federal guidelines on the identification of obesity in adult men state that the recommended body weight in pounds of a man who is 6 feet (or 72 inches) tall is found from the expression

$$2.2[50 + 2.3(72 - 60)]$$

Determine this weight.

2. U.S. coal production for the years 1994 to 1999 is shown in the following chart:

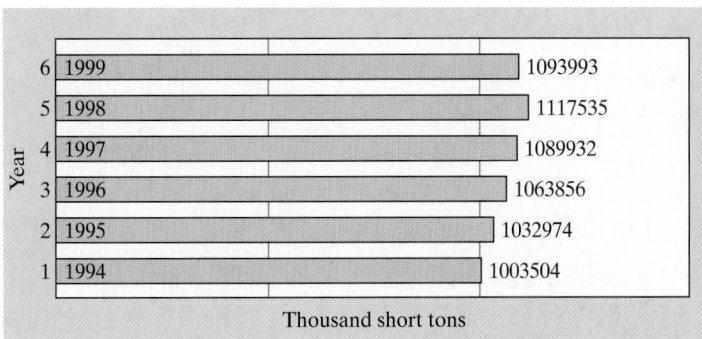

Year	Thousand short tons
6 1999	1093993
5 1998	1117535
4 1997	1089932
3 1996	1063856
2 1995	1032974
1 1994	1003504

All measurements are in thousand short tons. (A short ton is 2,000 pounds.) Find the average coal production per year for the 1994–99 period.

1.6 Exercises

Write an equivalent expression using the distributive law of real numbers. Check on your calculator.

1. $\left(\frac{3}{8}\right)\left(\frac{5}{7} - \frac{1}{9}\right)$ 2. $-23(31 + 25)$ 3. $2.7(-1.5 + 3.2)$ 4. $\left(1\frac{1}{4}\right)\left(5\frac{1}{2} - 3\frac{1}{8}\right)$

5. $\frac{217 - 175}{7}$ 6. $\frac{-78 + 108}{6}$

Write an equivalent expression using the opposite-of-a-sum property of real numbers. Check on your calculator.

7. $-(15 + 19.3)$ 8. $-(5.7 + 0.06)$ 9. $-\left(-\frac{6}{7} - \frac{5}{9}\right)$ 10. $-(-51 - 19)$

11. $-\left(1\frac{1}{7} - 2\frac{1}{5}\right)$ 12. $-(89 - 17)$ 13. $-(19.37 + 15.043)$ 14. $-\left(2\frac{1}{4} + 3\frac{5}{8}\right)$

Factor, using the distributive law of real numbers in reverse. Check on your calculator.

15. $15(17) + 15(23)$

16. $81(22) + 81(8)$

17. $-3(14) - 3(21)$

18. $-13(21) - 13(16)$

19. $6(1.2) - 6(1.3)$

20. $5(4.5) - 5(7)$

21. $3\left(\dfrac{2}{3}\right) - 9\left(\dfrac{2}{3}\right)$

22. $2\left(\dfrac{3}{4}\right) - 10\left(\dfrac{3}{4}\right)$

Evaluate, using the order of operations. Check on your calculator.

23. $-(6^2 - 12) + 5(-3 - 8)$

24. $-(8^2 - 15) + 4(-9 - 11)$

25. $[4(7 - 5) + 2] - 9$

26. $[6(8 - 2) + 1] - 7$

27. $-\sqrt{36 + 64} + 5$

28. $-\sqrt{81 + 144} - 2$

29. $\dfrac{18 - 4^2 + 7}{2 + 1}$

30. $\dfrac{25 - 3^3 + 14}{1 + 5}$

31. $-5(39 - 4^2) - 2(23 - 11)$

32. $-6(100 - 9^2) - (15 - 23)$

33. $4(15 - 8) + 31(14 - 11)$

34. $7(22 - 19) + 16(12 - 14)$

35. $-|5.2 - 31.3 + 3.95|$

36. $-\left|3\dfrac{1}{3} - 7\dfrac{5}{6} + 1\dfrac{1}{2}\right|$

37. $6^2 + 12 \div (-2) - 12 \cdot (-4)$

38. $4^3 + 57 \div (-3) - 7 \cdot (-3)$

39. $\left(\dfrac{2}{3}\right)^2 \div \left(\dfrac{1}{3} + \dfrac{1}{2}\right) \cdot \left(\dfrac{8}{9}\right)$

40. $\left(\dfrac{3}{4}\right)^2 \div \left(\dfrac{1}{2} + \dfrac{1}{6}\right) \cdot \left(\dfrac{2}{5}\right)$

41. $\left(\dfrac{1}{5}\right) \cdot \left(\dfrac{15}{22}\right) \div \left(\dfrac{1}{11} + \dfrac{1}{33}\right) - \left(\dfrac{1}{3}\right)^2$

42. $\left(\dfrac{1}{3}\right) \cdot \left(\dfrac{6}{7}\right) \div \left(\dfrac{1}{7} + \dfrac{1}{2}\right) - \left(\dfrac{1}{3}\right)^2$

43. $100 - (24 + 7^2 - 5) \cdot 3 + 102 \div 2$

44. $214 - (5 + 11^2 - 16) \cdot 4 + 55 \div 11$

45. $2\{3[5 - 2(3 + 4)] + 9 \div 3\} - 9(-8)$

46. $15(7) - 5\{3[8 + 2(5 - 9)] - 14 \div 7\}$

47. $\dfrac{29 + 3 - 2^3}{2^2 + 2^3}$

48. $\dfrac{15 + 3 \cdot 9 - 6}{5 + 2^2}$

49. $15 - \sqrt{2^2 + 3 \cdot 7}$

50. $-\sqrt{9^2 + 19} + 3 \cdot 6$

51. $\dfrac{4 + 3 \cdot 9 - 5^2 - 2 \cdot 3}{6^2 - 5}$

52. $\dfrac{15 - 2^4 + 1^2}{3 \cdot 9 - 5^2}$

53. $\dfrac{8^2 + 3 \cdot 12}{5 - 4 \cdot 6 + 2 \cdot 3^2 + 1^2}$

54. $\dfrac{45 - 7 \cdot 5}{3^3 - 1 + 2 \cdot 19 - 8^2}$

Use the order of operations to answer the following questions:

55. Federal guidelines on the identification of obesity in adult women state that the recommended body weight in pounds of a woman who is 5 feet, 3 inches (or 63 inches) tall is found from the expression

$$2.2[45.5 + 2.3(63 - 60)]$$

Determine this weight in pounds.

56. The same federal guidelines in exercise 55 state that the recommended body weight of a woman who is 5 feet, 9 inches (69 inches) tall is found from the expression

$$2.2[45.5 + 2.3(69 - 60)]$$

Determine this weight in pounds.

57. The federal guidelines also state that the recommended body weight of a man who is 6 feet, 3 inches (75 inches) tall is found from the expression

$$2.2[50 + 2.3(75 - 60)]$$

Determine the man's weight in pounds.

58. The same federal guidelines state that the recommended body weight of a man who is 5 feet, 10 inches (70 inches), tall is found from the expression

$$2.2[50 + 2.3(70 - 60)]$$

Determine this man's weight in pounds.

59. An electrical engineer must calculate the current (in amperes) of a three-phase household electrical system as equal to the power delivered (in kilowatts) divided by the voltage (in volts) times $\sqrt{3}$. If the power is 120 kilowatts and the line carries 400 volts, calculate the current. Round your answer to thousandths.

$$\text{Current} = \dfrac{120}{400\sqrt{3}}$$

60. In exercise 59, refigure the current if the power is 100 kilowatts and the line carries 200 volts. Round your answer to thousandths.

61. A car left a skid mark of 50 feet before crashing. On the basis of research data, police estimated the speed of the car (in miles per hour) as $2\sqrt{5 \cdot 50}$. What is the estimated speed of the car? Round your answer to the nearest mile per hour.

62. According to the same calculation method as in exercise 61, the estimated speed of a car that left a skid mark of 120 feet is $2\sqrt{5 \cdot 120}$. What is this speed to the nearest mile per hour?

63. When the pavement is wet, police adjust the calculations in exercises 61 and 62. They estimate the speed of a car that left a 300-foot skid mark to be $(3.25)\sqrt{300}$. What is the speed to the nearest mile per hour?

64. If a car left a 450-foot skid mark on wet pavement, police would estimate its speed as $(3.25)\sqrt{450}$. Was the car going faster than the speed limit of 65 mph?

65. If a nozzle has a diameter of 1.5 inches and the static pressure of the water is 40 pounds per square inch, then the flow rate of water through the nozzle is calculated as $(2.97)(1.5^2)\sqrt{40}$ gallons per minute. What is this flow rate to the nearest gallon per minute?

66. According to the calculation method from exercise 65, the flow rate of water through a 2-inch nozzle at 50 pounds per square inch is calculated as $(2.97)(2^2)\sqrt{50}$ gallons per minute. What would this flow rate equal to the nearest gallon per minute?

67. Index returns for the year 2000 are shown in the following chart:

68. The daily low temperatures (in degrees Celsius) for the week of October 15–21, 1999, for Truro, Canada, are given in the following bar chart:

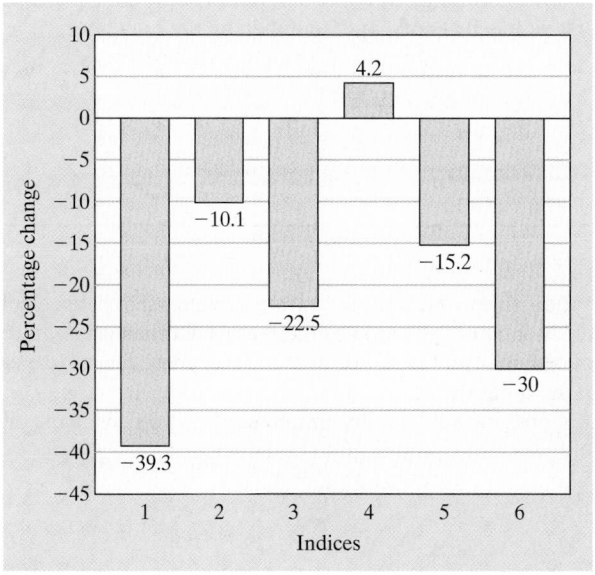

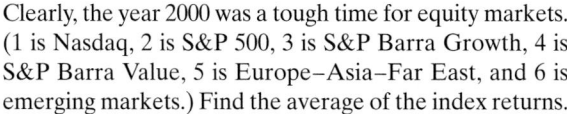

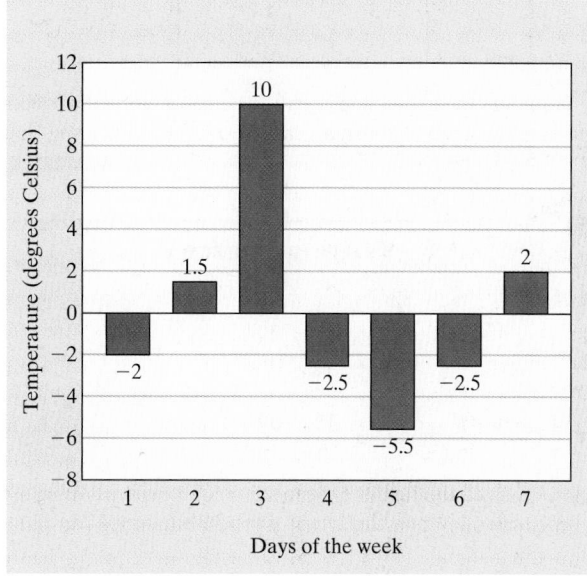

Clearly, the year 2000 was a tough time for equity markets. (1 is Nasdaq, 2 is S&P 500, 3 is S&P Barra Growth, 4 is S&P Barra Value, 5 is Europe–Asia–Far East, and 6 is emerging markets.) Find the average of the index returns.

Find the average daily low temperature for the week.

I.6 Calculator Exercises

In entering information from an exercise involving an order of operations into your calculator, you must be careful not to use the calculator keys for braces and brackets. These keys serve a special function in the calculator and are not meant to be used as grouping symbols. Rather, you should use only the parentheses keys as grouping symbols. If you have one grouping nested within another, the calculator will process your operations properly if you use parentheses for both groupings—the outer one and the nested one.

As an example, suppose you wish to evaluate the expression

$$2\{3(2 + 4) - 5[3(14 - 10) + 7] + 11\} - 4(5 - 6) + 8$$

First, use pencil and paper to prove to yourself that the correct answer is −120.

Then key the expression into the calculator, using parentheses for every grouping symbol. The result is as follows:

Next, key the expression into the calculator, using the braces and bracket keys. The result is a syntax error, indicating that

you are not using the keys for their intended functions. The following two screens show your error:

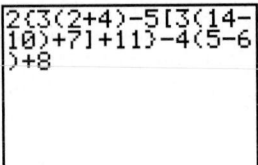

Now that you know you must use only the parentheses keys for grouping symbols, work out the following exercises, first with pencil and paper and then on the calculator:

1. $7 - \{5[2 + 9(4 - 6) - 12] + 8(17 - 12 \div 6)\}$

2. $\dfrac{4 - [2(8 + 3) - 5]}{5 + 2\{3[4 - (8 - 2)] + 11\}}$

I.6 Writing Exercise

Evaluate.

1. $\dfrac{35 + 77}{7}$ **2.** $\dfrac{1}{7}(35 + 77)$

3. $\dfrac{1}{7}(35) + \dfrac{1}{7}(77)$ **4.** $\dfrac{35}{7} + \dfrac{77}{7}$

Notice that all the results are the same. The expressions must be equivalent. While this is not a proof, but simply an illus-

tration, the problems demonstrate a relationship between division and multiplication by reciprocals. Explain what this relationship is, and explain how the first exercise relates to each of the other three.

Construct a similar illustration in which you divide the difference of two numbers by a real number.

CHAPTER I SUMMARY

After completing this chapter, you should be able to define the following key terms in your own words:

1.1
set
members
elements
empty set
null set
finite set
infinite set

ellipsis
subset
natural numbers
counting numbers
whole numbers
opposites
integers
number line

rational numbers
fractional notation
graph
plot
equal
inequality symbol
order symbol
equivalent

expression
evaluate
absolute value

1.2
sum
addends
additive identity
additive inverse
origin
difference
minuend
subtrahend

1.3
product
factor
identity property of 1
multiplicative identity
multiplication property of -1
multiplicative inverse property
multiplicative inverse

multiplication property of 0
quotient
dividend
divisor
undefined
indeterminate
reciprocals
implied multiplication
order of operators
ratio
rate

1.4
base
exponent
expanded form
exponential form
exponential expression
perfect square
square root
positive square root
principal square root

negative square root
irrational number
real number
radical sign
radicand
radical expression
cube root
index
lower bound
upper bound

1.5
standard notation
scientific notation
numerical factor
calculator notation

1.6
factoring
common factor
order of operations

Reflections

1. What is a rational number? What types of numbers are included in the set of rational numbers?

2. Explain what is meant by the absolute value of a number.

3. What is meant by the opposite of a number?

4. How do you use absolute values to add two numbers with like signs? with unlike signs? How do you use absolute values to subtract two numbers that may have like or unlike signs?

5. How do you use absolute values to multiply two numbers that have like signs? unlike signs? How do you use absolute values to divide two such numbers?

6. In an exponential expression, what is the effect of a negative integer exponent? What is the effect of a zero exponent? What is the effect of an exponent equal to 1?

7. What is the principal root of a number?

8. When will a radical expression have a positive value? When will it have a negative value? When will it not have a value in the set of real numbers?

9. Why is it important to follow the order of operations in evaluating a mathematical expression?

CHAPTER 1 SECTION-BY-SECTION REVIEW

1.1

Recall	Examples
Counting numbers, N Whole numbers, W Integers, Z Rational numbers, R, are numbers that can be written as a ratio of two integers, excluding the possibility of a zero denominator.	1, 2, 3 0, 1, 2, 3 $-2, -1, 0, 1, 2$ $-\dfrac{2}{3}, 0.75, -5$
To graph or plot a number on a number line, place a dot at the location of the number.	Graph -2 and 1.

Recall	Examples
Inequality symbols are $<$, $>$, $\leq$, and $\geq$.	$-2 < 0$ $-5 > -10$ $-3 \leq 0 \leq 5$
Absolute value is the distance a number is from 0.	$\lvert -3 \rvert = 3$ $\lvert 3 \rvert = 3$
Opposites are two numbers that have different signs but the same absolute values.	3 and -3 are opposites. $-(3) = -3$ $-(-3) = 3$

Express your answers in fractional notation or round decimals to the nearest thousandth as appropriate.

1. Identify the possible sets of numbers (natural numbers, whole numbers, integers, rational numbers) to which each of the following numbers belong:

$$-15, 1000, 0, -2000, 13, -\frac{15}{17}, 12.97, 3\frac{5}{8}, \frac{12}{4}$$

2. a. Graph the following numbers on a number line:

$$-1.1, \frac{3}{4}, -2\frac{1}{2}, 0, 2.5, 1\frac{1}{2}, 4, -3$$

b. Graph all real numbers between -2 and 3, including 3, on the number line.

Use one of the symbols $>$, $<$, or $=$ to compare the two numbers.

3. -7 ____ -9

4. $\dfrac{3}{11}$ ____ $\dfrac{1}{3}$

5. $\dfrac{27}{75}$ ____ 0.36

6. 2035 ____ 491

7. 12.304 ____ 12.344

8. -28 ____ 13

9. 331 ____ -331

10. -1.34 ____ -1.04

11. $-\dfrac{3}{4}$ ____ $-\dfrac{5}{8}$

12. $\lvert -10 \rvert$ ____ 10

13. $-\lvert 10 \rvert$ ____ 10

14. $-(-2)$ ____ $\lvert -2 \rvert$

Write an equivalent order relation for each statement.

15. $-23 > -49 \geq -100$

16. $\dfrac{252}{529} < \dfrac{13}{17}$

17. $-34.58 = -\dfrac{1729}{50}$

Evaluate.

18. $-\left\lvert -\dfrac{17}{33} \right\rvert$

19. $\lvert -(-67) \rvert$

20. $-(-32.698)$

21. $-(-(-(257)))$

22. Vijay Singh won the Masters Golf Tournament in 2000 at the Augusta National Golf Course. He scored par the first round, five under par the second round, two under par the third round, and three under par the fourth round. Write a rational-number representation for each of his scores, using positive numbers for scores above par and negative numbers for scores below par.

23. In 1999, NASA lost a $125,000,000 spacecraft, the *Mars Climate Orbiter*. According to NASA, the loss was likely due to the fact that the company which built the spacecraft failed to convert English units of measurement to metric ones. Write a rational-number representation for NASA's loss.

1.2

Recall	Examples
Addition • To add two rational numbers with like signs, add the absolute values of the addends. The sum has the sign of the addends. • To add two rational numbers with unlike signs, subtract the absolute values of the addends. The sum has the sign of the addend with the larger absolute value.	$3 + 4 = 7$ $-3 + (-4) = -7$ $3 + (-4) = -1$ $-3 + 4 = 1$

Recall	Examples
Laws for Addition • Commutative law for addition: Changing the order of two numbers when adding the numbers does not change the sum. • Associative law for addition: Changing the grouping of three numbers when adding the numbers does not change the sum.	$6 + (-2) = -2 + 6$ $-6 + (2 + 4) = (-6 + 2) + 4$
Subtraction To subtract two rational numbers, add the minuend to the opposite of the subtrahend.	$3 - 4 = 3 + (-4) = -1$ $3 - (-4) = 3 + 4 = 7$ $-3 - 4 = -3 + (-4) = -7$ $-3 - (-4) = -3 + 4 = 1$

Add, and then check, using your calculator.

24. $-27 + 13$ **25.** $-1123 + (-3406)$ **26.** $-5\frac{2}{9} + 4\frac{5}{9}$ **27.** $\frac{102}{43} + \left(-\frac{29}{75}\right)$ **28.** $235,407 + (-571,004)$

29. $12.097 + 1.92$ **30.** $\frac{32}{77} + \frac{50}{99}$ **31.** $-3.57 + (-41.098)$ **32.** $0 + (-123)$ **33.** $15.28 + (-15.28)$

Subtract, and then check, using your calculator.

34. $-35 - (-61)$ **35.** $-\frac{3}{8} - \frac{4}{7}$ **36.** $15.6 - 18$ **37.** $0 - 3.97$ **38.** $-2.3 - (-2.3)$

39. $3\frac{5}{9} - 5\frac{2}{9}$ **40.** $4,079,321 - 7,056,902$ **41.** $-473 - 2091$ **42.** $-\frac{34}{75} - \left(-\frac{203}{275}\right)$ **43.** $101.02 - (-23.9)$

44. $553 - (-392)$ **45.** $3.5 - 4.7 - (-8.2) - (-10.1) - 4.9 - (-16.7)$

46. $1 - 2 - (-3) - (-4) - 5 - 6 - (-7) - 8 - (-9)$ **47.** $\frac{3}{2} - \frac{1}{4} - \left(-\frac{8}{3}\right) - \frac{5}{6} - \left(-\frac{7}{2}\right) - 3$

Evaluate.

48. $31 - 16 + (-23) - 45 - (-37) + 52 + (-83)$ **49.** $3.7 - 6.83 + 5.5 + (-9.02) - 0.8 - (-15.2)$

50. $\frac{4}{5} - \frac{7}{3} + \frac{8}{9} - \left(-\frac{4}{15}\right) + \left(-\frac{2}{3}\right) + 3$

Write and evaluate an expression for each situation. Interpret the result.

51. Cleta had a beginning balance of $735.66 in her checking account. She wrote checks for $276.12, $187.05, and $68.57. She made deposits of $75.00, $185.00, and $50.00. The bank deducted a monthly service charge of $4.65 and also charged her $12.00 for her order of blank checks. What was the closing balance on her account?

52. The highest point in California is the top of Mount Whitney, with an elevation of 14,494 feet. The lowest point is in Death Valley, with an elevation of −282 feet. Determine the range of the elevation from the highest to the lowest point.

1.3

Recall	Examples
Multiplication • To multiply two rational numbers with like signs, multiply the absolute values of the factors. The product is positive. • To multiply two rational numbers with unlike signs, multiply the absolute values of the factors. The product is negative. • The product of any rational number and 0 is 0.	$3 \cdot 4 = 12$ $-3 \cdot (-4) = 12$ $3 \cdot (-4) = -12$ $-3 \cdot 4 = -12$ $3 \cdot 0 = 0$

(Continued on page 84)

Recall	Examples
Laws for Multiplication • Commutative law for multiplication: Changing the order of two numbers when multiplying the numbers does not change the product. • Associative law for multiplication: Changing the grouping of three numbers when multiplying the numbers does not change the product.	$6 \cdot (-2) = -2 \cdot 6$ $-6(2 \cdot 4) = (-6 \cdot 2)4$
Division • To divide two rational numbers with like signs, divide the absolute values of the numbers. The quotient is positive. • To divide two rational numbers with unlike signs, divide the absolute values of the numbers. The quotient is negative. • The quotient of 0 divided by any nonzero rational number is 0. • The quotient of any nonzero rational number divided by 0 is undefined. • The quotient of 0 divided by 0 is indeterminate.	$12 \div 3 = 4$ $-12 \div (-3) = 4$ $12 \div (-4) = -3$ $-12 \div 4 = -3$ $0 \div 4 = 0.$ $4 \div 0$ is undefined. $0 \div 0$ is indeterminate.
Order of Operations Perform operations from left to right in the following order: • Multiplication and division • Addition and subtraction	$12 - 6 \div 2 + 3 \cdot (-5)$ $= 12 - 3 + 3 \cdot (-5)$ $= 12 - 3 + (-15)$ $= -6$

Multiply, and then check, using your calculator.

53. $(-13)(-6)$ **54.** $(21)(-5)$ **55.** $\left(-7\frac{2}{5}\right)\left(-\frac{5}{7}\right)$ **56.** $(23.05)(-0.04)$

57. $0 \cdot (-11)$ **58.** $(-1) \cdot 25$ **59.** $\left(\frac{23}{171}\right)\left(\frac{33}{230}\right)$ **60.** $(-2.04)(-4.12)$

61. $2905 \cdot 1197$ **62.** $\frac{-13}{29} \cdot \frac{58}{169}$ **63.** $765 \cdot (-835)$ **64.** $(43.996)(0)$

65. $(-32)(20)(-1)(5)(-2)(10)$ **66.** $(-1.1)(0.2)(-4)(-10)(-0.8)$

67. $\left(-\frac{1}{3}\right)\left(\frac{6}{7}\right)\left(-\frac{14}{15}\right)\left(-\frac{25}{8}\right)$ **68.** $(-54)(21)(0)(32)(0)(-25)$

Divide, and then check, using your calculator.

69. $220 \div (-4)$ **70.** $-78 \div (-3)$ **71.** $\frac{5}{9} \div \left(\frac{-2}{3}\right)$ **72.** $-10.557 \div 2.3$

73. $0 \div 25$ **74.** $-2 \div 0$ **75.** $-13.7 \div (-1)$ **76.** $-1,363,443 \div 2539$

77. $\frac{143,883}{657}$ **78.** $\frac{-87}{121} \div \frac{-58}{99}$ **79.** $\frac{65}{323} \div \frac{91}{247}$ **80.** $7.32864 \div 1.056$

Evaluate.

81. $(-25) \div (-5)(12)(-3) \div (-9)(-5)$ **82.** $(4.2)(-3.2) \div (1.6)(0.2) \div (-0.4)(-2.2)$

83. $\left(\frac{5}{12}\right)\left(-\frac{6}{25}\right) \div \left(-\frac{3}{10}\right)\left(\frac{15}{17}\right) \div \left(-\frac{5}{13}\right)$ **84.** $17 + 31 - 20 \div 2 - 15 \cdot (-3)$

85. $-2(-4.8) - 5(1.7) + 9.2$

Write and evaluate a multiplication and/or division expression for each situation. Interpret the result.

86. A national charity distributes the following pie chart to as-

sure its contributors that most of the donations go directly into helping those whom the charity serves:

Fund Raising 8%

Management & Gen. Expenses 9%

Program Services 83%

If you donated $125 to the charity, how much will go directly to program services?

87. A National Geographic Society ad stated, "One in seven adult Americans can't find the U.S. on a world map." The Bureau of the Census reported that the resident population 20 years and older in 1998 was 192,631,000. How many of these residents could not find the United States on a world map?

88. Clarence records his gasoline usage on a trip from Louisville, Kentucky, to Cleveland, Ohio, to see the Rock and Roll Hall of Fame. He uses 18.7 gallons of gas to drive 345 miles. What was his average mileage for the trip?

1.4

Recall	Examples
Integer Exponents Greater than 1 • To evaluate, determine the product of the expanded form.	$5^3 = 5 \cdot 5 \cdot 5 = 125$ $(-5)^3 = -5 \cdot -5 \cdot -5 = -125$
Exponent of 1 • The value is the base number.	$5^1 = 5 \qquad (-5)^1 = -5$
Exponent of 0 • If the base is nonzero, the value is 1. • If the base is 0, the expression is indeterminate.	$5^0 = 1 \qquad (-5)^0 = 1$ 0^0 is indeterminate.
Square Roots • The principal square root is positive and must have a radicand that is positive. • The negative square root is negative and must have a radicand that is positive. • A square root is not defined in the real numbers if the radicand is negative.	$\sqrt{25} = 5 \qquad \sqrt{24} \approx 4.899$ $-\sqrt{36} = -6$ $\sqrt{-12}$ is not a real number.
Cube Roots The value of a cube root is • Positive if the radicand is positive. • Negative if the radicand is negative.	$\sqrt[3]{8} = 2$ $\sqrt[3]{-83} \approx -4.362$
Number-line graph of infinite sets of real numbers • Determine the upper and lower bounds if possible. • Graph an open dot if the bound is not included in the set of real numbers. Graph a closed dot if the bound is included in the set of real numbers.	Graph on a number line. All real numbers between -1 and 4 All real numbers less than 4

(Continued on page 86)

Recall	Examples
• Draw a solid line to the left of the number to include values less than the upper bound. Draw a line to the right of the lower bound to include numbers greater than the lower bound.	All real numbers greater than or equal to 4 $\begin{array}{c}\longleftarrow\ \ |\ \ \bullet\!\!-\!\!|\!\!-\!\!|\!\!-\!\!|\!\!-\!\!|\ \ \longrightarrow\\ \quad 3\ \ 4\ \ 5\ \ 6\ \ 7\end{array}$

Evaluate.

89. 2^8 **90.** $(-3)^5$ **91.** $(-3)^4$ **92.** 1.2^2 **93.** $\left(1\frac{1}{3}\right)^3$ **94.** 0^{10}

95. 1^{15} **96.** $(-1)^{18}$ **97.** $(-1)^{21}$ **98.** $\left(-\frac{3}{4}\right)^6$ **99.** $\left(2\frac{1}{3}\right)^5$ **100.** $(-15)^0$

101. -15^0 **102.** -15^1 **103.** 3.079^1 **104.** $\left(\frac{13}{23}\right)^0$ **105.** 1^0 **106.** 0^0

107. 1^1 **108.** 0^1 **109.** $\sqrt{81}$ **110.** $\sqrt{0.64}$ **111.** $\sqrt{\frac{9}{25}}$ **112.** $-\sqrt{49}$

113. $\sqrt{-16}$ **114.** $\sqrt{1.2769}$ **115.** $-\sqrt{470.89}$ **116.** $-\sqrt{\frac{576}{1369}}$ **117.** $\sqrt{15}$ **118.** $\sqrt[3]{27}$

119. $\sqrt[3]{0.125}$ **120.** $\sqrt[3]{\frac{64}{729}}$ **121.** $\sqrt[3]{-27000}$ **122.** $-\sqrt[3]{10}$

123. Graph the following numbers on a number line, and label the points:

$$9,\ \sqrt{9},\ -\sqrt{16},\ \sqrt{18},\ -\sqrt{68},\ \pi,\ -\sqrt{7.29},\ \sqrt[3]{140}$$

124. Complete the following table:

Description	Number line
All real numbers between -0.5 and 1, including -0.5	
All real numbers between -0.5 and 1, including 1	
All real numbers between -0.5 and 1, inclusive	
All real numbers between -0.5 and 1	
All real numbers greater than or equal to 1	
All real numbers less than or equal to -0.5	

125. The diamond in a baseball field is a square-shaped plot that has an area of 729 square meters. What are the dimensions of the diamond?

126. A box for facial tissues is a cube that will be marketed as a bathroom boutique design. The box has a volume of $91\frac{1}{8}$ cubic inches. What are the dimensions of the box?

1.5

Recall	Examples
Integer Exponents Less than 0 • If the base is nonzero, rewrite the expression as the reciprocal of the base with the opposite of the exponent. Then evaluate the new expression. • If the base is 0, the expression is undefined.	$(5)^{-2}=\left(\frac{1}{5}\right)^2=\left(\frac{1}{5}\right)\left(\frac{1}{5}\right)=\frac{1}{25}$ $\left(-\frac{1}{5}\right)^{-2}=(-5)^2=(-5)(-5)=25$ $0^{-3}=\left(\frac{1}{0}\right)^3$ is undefined.

Recall	Examples
Converting Standard Notation to Scientific Notation Write a product of the numerical factor and an exponential expression. • Move the decimal point in the number so that the resulting numerical factor has an absolute value between 1 and 10, including 1. • Determine the exponential expression. The exponent of the base 10 is positive if the decimal was moved to the left, negative if the decimal was moved to the right, and 0 if the decimal was not moved. • The absolute value of the exponent is the value of the number of places the decimal was moved.	$3,540,000 = 3.54 \times 10^6$ $0.00034 = 3.4 \times 10^{-4}$
Converting Scientific Notation to Standard Notation Move the decimal point in the numerical factor the number of places the exponent denotes (its absolute value). • The decimal moves to the right if the exponent is positive. • The decimal moves to the left if the exponent is negative.	$5.2 \times 10^4 = 52,000$ $-3.81 \times 10^{-4} = -0.000381$
Converting Calculator Notation to Scientific Notation Write a product of the numerical factor and an exponential expression. • The numerical factor is the number displayed before the "E." • The exponential factor has a base of 10 and an exponent whose value is the number after the "E."	$-5.6E6 = -5.6 \times 10^6$ $4.38E-6 = 4.38 \times 10^{-6}$

Write an equivalent exponential expression having a positive exponent. Evaluate.

127. 4^{-1} **128.** $\left(\dfrac{1}{4}\right)^{-1}$ **129.** $\left(\dfrac{1}{4}\right)^{-2}$ **130.** $(-4)^{-2}$

Complete the following table:

	Standard Notation	Scientific Notation	Calculator Notation
131.	0.000000189		
132.	$-27,085,000,000$		
133.		5.89×10^{11}	
134.			$-7.093 \text{ E}-5$

135. In 1998, the total U.S. export of goods was reported as \$682,100,000,000. Write this number in scientific notation.

136. In 1998, the total U.S. import of the goods was reported as 9.119×10^{11} U.S. dollars. Write this number in standard notation. Using the information in exercise 135, by how many dollars do U.S. imports exceed U.S. exports?

137. Red blood cells of mammals measure about 8×10^{-6} meters in diameter. Write this number in standard notation. The radius of the cell is the diameter divided by 2. What is the radius of the blood cell?

1.6

Recall	Examples
Distributive Laws • The product of a real-number factor and a sum (or difference) of real number addends is the same as the sum (or difference) of the products of the factor and each addend.	$2(3 + 4) = 2(3) + 2(4) = 14$ $-3(-2 - 5) = -3(-2) - (-3)(5) = 21$

(Continued on page 88)

Recall	Examples
• The quotient of a real-number divisor and a sum (or difference) of real number addends is the same as the sum (or difference) of the quotients of the divisor and each addend (minuend/subtrahend).	$\dfrac{3+5}{2} = \dfrac{3}{2} + \dfrac{5}{2} = \dfrac{8}{2} = 4$ $\dfrac{3-5}{2} = \dfrac{3}{2} - \dfrac{5}{2} = -\dfrac{2}{2} = -1$
Opposite-of-a-Sum Property The opposite of the sum of real numbers is the same as the sum of the opposites.	$-(3+4) = -3 + (-4) = -7$ $-(5-7) = -5 - (-7) = -5 + 7 = 2$
Factoring Factoring is doing the reverse of the distributive laws.	$7(2) + 7(8) = 7(2+8) = 7(10) = 70$
Order of Operations Perform operations from left to right in the following order: grouping symbols, powers and radicals, multiplication and division, addition and subtraction.	$2\{3[2-(3-2^2)]\} =$ $2\{3[2-(3-4)]\} =$ $2\{3[2-(-1)]\} =$ $2\{3[3]\} =$ $2\{9\} = 18$

Write an equivalent expression using the distributive law or the opposite-of-a-sum property, and then check on your calculator.

138. $-2.6(-1.9 + 3.2)$ **139.** $\dfrac{5}{6}\left(-\dfrac{3}{5} - \dfrac{4}{15}\right)$ **140.** $\dfrac{1687 - 1372}{7}$ **141.** $-(2.7 + 3.09)$ **142.** $-[32 + (-51)]$

Factor. Check for equivalence.

143. $21(18) + 21(42)$ **144.** $5(97) - 5(17)$ **145.** $5.6(18) + 5.6(-8)$ **146.** $121\left(\dfrac{3}{4}\right) - 161\left(\dfrac{3}{4}\right)$

Evaluate exercises 147–153, using the order of operations.

147. $41(-73 + 65) - (52 - 46)$ **148.** $-(27 - 4^2) - 16(-5 - 3)$ **149.** $\dfrac{191 + 104 - 11^2}{5^2 + 3^3 - 46}$

150. $[15 + 21(14 - 18)] - 7^2$ **151.** $22 - \sqrt{9^2 - 45} + 5$ **152.** $\dfrac{8 \cdot 9 - 2 \cdot 6^2}{125 - 7^2}$

153. $\dfrac{78 - 3 \cdot 17}{5^2 - 4 \cdot 6 - 1^3}$

154. Federal guidelines on the identification of obesity in adult men state that the lean body weight (in pounds) of a man who is 5 feet, 10 inches tall and weighs 184 pounds is found from the expression

$$2.2[0.5(184) - 4.1(184^2 \div 70^2)]$$

Determine this weight.

155. The median weekly earnings of males 25 years and older are given in the following table for the years (1) 1985, (2) 1990, (3) 1995, and (4) 1998.

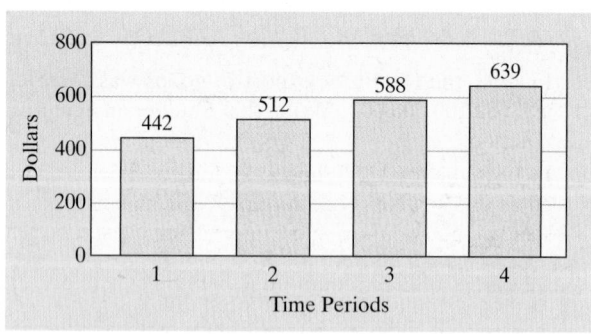

The data were published in the *Statistical Abstract of the United States*. Find the average of the earnings for the four periods shown.

156. Mary is making a tablecloth for a rectangular table. The tablecloth measures 7 feet by 5 feet. Mary will buy lace to sew around the edge of the tablecloth and also to criss-cross it diagonally. Write an expression and evaluate it to find how much lace Mary will need to trim the edge of the tablecloth and to sew the two diagonals. (Remember that the diagonal is equal to the square root of the sum of the squares of the two sides of the tablecloth.)

CHAPTER I CHAPTER REVIEW

Complete the following table:

	Standard Notation	Scientific Notation	Calculator Notation
1.	355,400,000,000,000,000		
2.	−0.000092		
3.		7.94×10^{-12}	
4.			−6.876 E4

Factor. Check for equivalence on your calculator.

5. $42(73) - 42(23)$ **6.** $-31(18) - 31(42)$

Use one of the symbols $>$, $<$, *or* $=$ *to compare the two numbers.*

7. $75 \_\_\_\_ 59$ **8.** $3.54 \_\_\_\_ 3.65$ **9.** $\dfrac{31}{51} \_\_\_\_ \dfrac{10}{17}$ **10.** $-142 \_\_\_\_ -105$ **11.** $-\dfrac{21}{59} \_\_\_\_ -\dfrac{29}{59}$

12. $\dfrac{3}{8} \_\_\_\_ 0.375$ **13.** $-3.7 \_\_\_\_ 0.53$ **14.** $0 \_\_\_\_ -197$ **15.** $31 \_\_\_\_ 0$

Evaluate.

16. $-\left|\dfrac{9}{25}\right|$ **17.** $|-(-102)|$ **18.** $-(0.085)$ **19.** $-549 + (-908)$

20. $3.07 + (-2.9)$ **21.** $-\dfrac{1}{8} + \left(-1\dfrac{1}{4}\right)$ **22.** $-67,853 + 80,000$ **23.** $0.005 + 0.05$

24. $\dfrac{17}{65} + \dfrac{8}{91}$ **25.** $-576 - (-394)$ **26.** $0.52 - 3$ **27.** $0 - \dfrac{3}{11}$

28. $-4.07 - (-5.1)$ **29.** $\dfrac{13}{25} - \dfrac{2}{5}$ **30.** $4500 - (-45)$ **31.** $\left(-\dfrac{12}{55}\right)\left(-\dfrac{5}{6}\right)$

32. $\left(-4\dfrac{3}{7}\right)\left(\dfrac{14}{31}\right)$ **33.** $0 \cdot (-8.05)$ **34.** $(-1)(-88)$ **35.** $\left(\dfrac{17}{72}\right)\left(\dfrac{9}{34}\right)$

36. $(-31.6)(-1.001)$ **37.** $4000 \cdot 1200$ **38.** $\dfrac{4}{13} \div \left(-\dfrac{12}{13}\right)$ **39.** $0 \div 0$

40. $-21 \div 0$ **41.** $56.8 \div (-1)$ **42.** $-9.02 \div (-1.1)$ **43.** $\dfrac{-8}{55} \div \left(\dfrac{-4}{17}\right)$

44. $5.5 \div 2.2$ **45.** $\dfrac{184,008}{902}$ **46.** $\dfrac{2}{5} \div \dfrac{56}{75}$

47. $25 + 63 - 48 \div 3 - 22 \cdot (-8)$ **48.** $2 - \{4[3 + 2(7 - 5)] - 18 \div 3\}$

49. $-(34 - 7^2) - 21(-8 - 4)$ **50.** $96 - \sqrt{11^2 - 21} - 36$

51. $\dfrac{412 + 204 - 4^2}{5^2 + 35}$ **52.** $\dfrac{10 \cdot 25 - 2 \cdot 5^3}{275 - 3^5}$ **53.** $\dfrac{35 - 6 \cdot 22}{6^3 - 4 \cdot 50 - 4^2}$ **54.** 13^2

55. $(3.4)^7$ **56.** $(-5)^3$ **57.** $(-5)^4$ **58.** $\left(-1\dfrac{2}{5}\right)^3$

59. $(-1)^{24}$ **60.** $(-1)^{35}$ **61.** 3^4 **62.** $-\left(\dfrac{1}{3}\right)^5$

63. $(-2.1)^4$ **64.** 22^0 **65.** $(-22)^0$ **66.** -22^0

67. 22^1

68. -22^1

69. $(-22)^1$

70. 0^0

71. 0^1

72. $(2,333,145)^0$

73. $(65.02)^1$

74. $(-0.84)^0$

75. $\left(\dfrac{79}{95}\right)^0$

76. $(-0.0004)^0$

77. 10^{-2}

78. $(-10)^{-2}$

79. $\left(\dfrac{1}{10}\right)^{-1}$

80. $\left(-\dfrac{1}{10}\right)^{-1}$

81. $\sqrt{2500}$

82. $\sqrt{-400}$

83. $-\sqrt{\dfrac{441}{1444}}$

84. $\sqrt{1.44}$

85. $-\sqrt{\dfrac{5}{400}}$

86. $-\sqrt{2235.6}$

87. $\sqrt[3]{-64}$

88. $\sqrt{\dfrac{81}{16}}$

89. $-\sqrt[3]{274.625}$

90. $-\sqrt[3]{25}$

91. $\sqrt{2.56}$

92. $\sqrt[3]{-2\dfrac{5}{7}}$

93. $\sqrt[3]{\dfrac{27}{64}}$

94. $-\sqrt{6.5536}$

95. $7.3 - 38.6 + 5.5 + (-2.09) - 8 - (-2.51)$

96. $\dfrac{3}{7} - \dfrac{17}{21} + \dfrac{11}{14} - \left(-\dfrac{5}{6}\right) + \left(-\dfrac{35}{42}\right) + 7$

97. $(-55)(12)(-2)(9)(-3)(1)$

98. $\left(-\dfrac{14}{22}\right)\left(\dfrac{8}{21}\right)\left(-\dfrac{11}{4}\right)\left(-\dfrac{3}{5}\right)$

99. $(-11.2)(3.1)(0)(-9.4)(-1)(-7.5)$

100. $(13.1)(-4.2) \div (2.62)(0.5) \div (-0.7)(-1.1)$

101. $\left(\dfrac{9}{14}\right)\left(-\dfrac{7}{18}\right) \div \left(-\dfrac{3}{4}\right)\left(\dfrac{6}{7}\right) \div \left(-\dfrac{1}{21}\right)$

In exercises 102–104, write an equivalent order relation for each statement.

102. $76 > 75 \geq 74$

103. $-\dfrac{2}{5} < \dfrac{1}{7}$

104. $0.52 = \dfrac{13}{25}$

105. a. Graph the following numbers on a number line, and label the points:

$$-7,\ \sqrt{7},\ -\sqrt{7},\ \sqrt{25},\ \sqrt[3]{0},\ \sqrt{81},\ -\sqrt[3]{64},\ \sqrt{\dfrac{9}{25}},\ \dfrac{-2}{3},\ 2\dfrac{1}{2},\ -2.7$$

b. Graph all real numbers between -2 and 1.5, including -2, on the number line.

106. The film *Who Framed Roger Rabbit?* has the longest list of credits of any film, with 763 names. It takes $6\frac{1}{2}$ minutes to run the credits. On the average, how many credits run each minute?

107. A 10-gallon hat actually holds only $\frac{3}{4}$ gallon of water. How many gallons of water will fill ten 10-gallon hats?

108. Captain Kirk of the starship *Enterprise* had 430 people in his crew. Captain Picard, his successor on the *Enterprise,* had 1012 crew members and civilians under his command. Captain Janeway of the *Voyager* had 127 crew members at one point. How many more people did Captain Picard command than Captain Kirk? How many fewer people did Captain Janeway command than Captain Kirk?

109. In the National Football League, the home team is required to provide 24 new footballs for each game. If the league has 31 teams, and if each team plays eight home games a season, how many footballs are needed?

110. In early March of 2001, the stock market was experiencing a severe downturn. The following chart presents the changes in value of the indexes at the close of trading for the week:

Index	Dow	NASDAQ	S&P 500	Russell 2000
Change	-213.63	-115.95	-31.32	-7.84

Determine the average of the four indexes.

111. Barium crystallizes in a body-centered cube with a volume of 1.303×10^{-28} cubic meter. Determine the length of each side of the cube.

CHAPTER 1 TEST

 TEST-TAKING TIPS

Many students feel that they should pull an "all-nighter" the night before a test, cramming as much study into that time as they can. An all-nighter, however, will not help you during a test. What it will do is tire you out and raise your anxiety level. It would be better to prepare for a test gradually by studying for an hour or two on each of several nights leading up to the day of the test. Then, the night before the test, just do a cursory review of the material you have been studying. Next, get a good night's sleep. If you can't sleep, try engaging in some physical activity to relax you. You will do better. Remember, your brain keeps working on the material, even when you are not consciously studying.

Have a healthy breakfast and take a walk before going to class. In class, changing the rate and pattern of your breathing to a deep, slow rhythm will calm your body and ease your mind. Before you start a test, take at least three deep, slow breaths, and blow out as much air as you can. This will help relax you and clear your mind for the test.

Use one of the symbols $>$, $<$, *or* $=$ *to compare the numbers.*

1. $\dfrac{4}{9}$ ____ $\dfrac{5}{6}$

2. -18 ____ -23

3. $\dfrac{17}{25}$ ____ 0.68

Write an equivalent order relation for each statement.

4. $\dfrac{5}{16} > \dfrac{1}{4} \geq \dfrac{1}{5}$

5. $-3.2 < -2.3$

6. $\dfrac{3}{50} = 0.06$

In exercises 7–26, evaluate. Express your answers in the same form as the original exercise wherever possible, and round decimals to the nearest thousandth.

7. $|-13.37|$

8. $-|-20|$

9. $59 + (-95)$

10. $-\dfrac{17}{95} + \dfrac{4}{19}$

11. $4.378 - 7.98$

12. $\left(-\dfrac{3}{4}\right)\left(-\dfrac{8}{15}\right)$

13. $-6985 - (-2576)$

14. $45.78 \cdot (-1)$

15. $(-37{,}562)(456)$

16. $-413.9 + (-597.65)$

17. $-819 \div (-9)$

18. $-\dfrac{44}{57} \div \dfrac{11}{19}$

19. $15.9 \div 0$

20. $0 \div (-53)$

21. $2\dfrac{3}{7} \div 1\dfrac{3}{14}$

22. $-12.05 - 2.4$

23. $(-23)(-4)(0)(-17)(0)(-45)$

24. $\left(-\dfrac{5}{6}\right)\left(-\dfrac{3}{7}\right)\left(\dfrac{1}{2}\right)\left(-\dfrac{2}{15}\right)$

25. $4.3 + (-0.1) - (-2) + (-1.1)$

26. $(-42)(22) \div 77(-4) \div (-16)$

27. a. Graph the following numbers on a number line, and label the points: $-3.5, 3, \dfrac{1}{2}, 2\dfrac{3}{4}, -1\dfrac{1}{2}, \sqrt[3]{100}, \sqrt{25}$

  b. Graph all real numbers between -3 and 5, including -3, on a number line.

Write in scientific notation.

28. $5{,}239{,}000{,}000{,}000{,}000$ **29.** -0.00000203

Evaluate exercises 30–44.

30. $(1.5)^2$

31. $\left(\dfrac{4}{3}\right)^4$

32. 1^9

33. 0^{12}

34. $(-4)^3$

35. $(-4.008)^1$

36. 0^0

37. $(-10)^{-1}$

38. $(-3)^0$

39. $\sqrt{\dfrac{36}{121}}$

40. $-\sqrt{3.6}$

41. $-[51.3 - (-20.9)]$

42. $\dfrac{-609 + 928}{29}$

43. $\sqrt[3]{17^2 - 6 \cdot 12 - 1} + 126 \div 9$

44. $\dfrac{2(5^2 + 3^2) - 8^2 - 2^2}{3.65}$

45. A flower garden in the shape of a square measures 9.5 feet on a side. A catalog offers an assortment of flowers to fill a garden that has 68 square feet of space. Write an expression and find the amount of excess space the garden will have.

46. A carton of milk contains 1.89 liters. The milk is labeled as 2% reduced-fat milk. How many liters of the milk are fat? If there are eight servings of milk in the container, how many liters of fat are contained in each serving?

47. The gross domestic product (GDP) of the United States was estimated to be 8.511×10^{12} dollars in 1998. At the same time, the GDP of Russia was estimated to be 5.934×10^{11} U.S. dollars. By how much did the GDP of the United States exceed that of Russia? Write your answer in scientific notation and in standard notation.

48. Explain how you would determine the sign of the product of two rational numbers.

49. How would you determine the sign of the product of three or more rational numbers?

Chapter 1

Project

Part I

Your company is closing the office where you currently work. You have been given a choice to transfer to one of the following locations:

Washington, D.C.   Seattle, WA   Minneapolis, MN
Boston, MA   New York City, NY   Detroit, MI

Determine your top two choices out of these possibilities. Your major concern is cold weather. To help you make a decision, complete the table that follows for each of your choices.

1. You will need to determine the average low temperature, in degrees Fahrenheit, for each month of the year and for the two locations that you prefer. You can find this information from various sources, such as the internet or the library. (Make a copy of your source.) To convert the low temperatures from degrees Fahrenheit to degrees Celsius, you must first subtract 32 and then multiply the result by $\frac{5}{9}$. Round your answers to the nearest whole degree. Construct two tables using the table headings shown.

Average Monthly Low Temperatures for _____

Month	Temperature (in Degrees Fahrenheit)	Subtract 32	Multiply by $\frac{5}{9}$ Temperature (in Degrees Celsius)
January			
⋮			
December			

2. Use the 12 measures listed in each table to calculate the annual average low temperature in degrees Fahrenheit and in degrees Celsius for both locations. Round your answers to the nearest whole degree.

Part II

A **dot plot** consists of a graph in which each data value is plotted as a point (or dot) along the number line. Dots that represent the same value occurring repeatedly are stacked above one another. This form of graph is used in statistical studies to understand the distribution of a set of data values. Plot the 12 measures listed in the table in Part I to depict the distribution of temperatures in degrees Fahrenheit and then in degrees Celsius. Refer to Section 1.1, Calculator Exercises, for more information and a calculator program to check your dot plot.

Part III

Now prepare a one-paragraph summary of your recommended choice of location. Use the information you produced in Parts I and II to justify your decision.

Chapter 2

Variables, Expressions, Equations, and Formulas

In the previous chapter, we described the set of real numbers and the rules of arithmetic that apply to real numbers. In this chapter, we generalize arithmetic by introducing algebra. We define variables, algebraic expressions, and equations, and we discuss geometric formulas and other formulas. As before, we use real-world data in each discussion.

The word *algebra* comes from the Arabic word *al-jabr*, which appears in the title of one of the earliest algebra books known. The book was written about 825 A.D., which is about the same time as the setting for the stories of the Arabian Nights. The book's author was an astronomer and mathematician named Muhammed ibn Musa al-Khwarizmi, and the full title may be translated as "the science of restoring and reduction." We will see in this chapter how the Arabian mathematicians solved equations by moving terms around ("restoring") and combining like terms ("reduction").

Many of the formulas that we discuss in this chapter were written in ancient times and are accompanied by an interesting history. As you proceed through the chapter, you may want to research the history behind the different formulas introduced. We will conclude the chapter with a project that features one of these historical events, the search for a value of π.

2.1 APPLICATION

The area of a rectangle is found by multiplying its length by its width. The perimeter of a rectangle is twice the sum of its length and width. Write an algebraic expression for the area and perimeter of a rectangle.

According to the *American Football Rules,* the dimensions of a rectangular playing field for football are 120 yards long and $53\frac{1}{3}$ yards wide. Determine the area and perimeter of a football field.

We will discuss this application further. See page 100.

2.1 Variables and Algebraic Expressions

OBJECTIVES

1 Translate word expressions into algebraic expressions.
2 Evaluate algebraic expressions.
3 Model real-world situations using algebraic expressions.

2.1.1 Translating Word Expressions

In the first chapter, we worked with **numeric expressions**, which are combinations of numbers and mathematical operations, such as

$$14 + 5, \quad -28 - 3, \quad 23 \cdot 5, \quad 45 \div (-5)$$

In algebra, we work with algebraic expressions. Examples of algebraic expressions are

$$14 + x, \quad y - 3, \quad 23a, \quad \frac{s}{t}$$

In an algebraic expression, we use letters as symbols. A symbol representing only one number is called a **constant**. A symbol that can represent more than one number is called a **variable**. In the expression $14 + x$, 14 is a constant and x is a variable. In the expression πd, π is a constant representing one number (approximately 3.14), and d is a variable representing the diameter of a circle. We define an **algebraic expression** as an expression containing variables.

In this section, we will translate word expressions into algebraic expressions. In order to do this, we need to know that certain words translate into operation symbols. Table 2.1 lists examples of these words and their translations.

TABLE 2.1

Addition (+)	Subtraction (−)	Multiplication (·)	Division (÷)
add	subtract	multiply	divide
sum	difference	product	quotient
plus	minus	times	divided by
increased by	decreased by	multiplied by	divided into
more than	less than	of	per
total	less	twice	ratio
addends	taken from	double	dividend
	net	triple	divisor
	minuend	factors	
	subtrahend		

To write an algebraic expression,

- Define the variable or variables.
- Translate the words into numbers and symbols.

EXAMPLE 1   Translate each word expression into an algebraic expression.

a. 12 dollars less than the original price

b. 25% of the retail price

c. half of the sum of the length of the larger base and the length of the smaller base

d. the product of the length, width, and height measurements

Solution

a. Let x = the original price
$$x - 12$$
"Less than" translates into subtraction. Be careful with the order.

b. Let p = the retail price — Define the variable.
$$25\% = 0.25$$
$$0.25 \cdot p \quad \text{or} \quad 0.25p$$
"Of" translates into multiplication.

c. Let B = length of the large base — Define two variables.
$$b = \text{length of the small base}$$
$$\frac{1}{2} \cdot (B + b) \quad \text{or} \quad \frac{1}{2}(B + b)$$
"Of" translates into multiplication. "Sum" translates into addition.

d. Let L = length — Define three variables.
$$W = \text{width}$$
$$H = \text{height}$$
$$L \cdot W \cdot H, \quad \text{or} \quad LWH$$
"Product" translates into multiplication.

 HELPING HAND You may omit the times sign in multiplication expressions involving variables or parentheses. Capital letters are different variables than their corresponding lower case letters.

✓ 2.1.1 Checkup

In exercises 1–3, translate each word expression into an algebraic expression.

1. One-half the product of the base and the height of a triangle

2. One hundred dollars more than the previous price of the coat

3. Six less than the product of a number and 8

4. Explain the difference between basic arithmetic and algebra

2.1.2 Evaluating Algebraic Expressions

An algebraic expression can have different values, depending on what value the variable has. To determine the value of the expression, we **substitute** the value of the variable into the expression. For instance, if the variable x has the value 5, then the expression $2x$ has the value 10. If x has the value -3, then the expression $2x$ has the value -6. This process is called **evaluating** the algebraic expression.

> To evaluate an algebraic expression,
>
> - Substitute the given values for the variables, always enclosing the substituted values in parentheses.
> - Evaluate the resulting numeric expression, following the order of operations.

EXAMPLE 2   Evaluate $b^2 - 4ac$ for

a. $a = 1, b = 3,$ and $c = 2$ **b.** $a = 1, b = -5,$ and $c = 2$

Solution

a. $b^2 - 4ac = (3)^2 - 4(1)(2)$ Substitute values.

$\qquad\qquad = 9 - 8$

$\qquad\qquad = 1$

b. $b^2 - 4ac = (-5)^2 - 4(1)(2)$ Substitute values.

$\qquad\qquad = 25 - 8$

$\qquad\qquad = 17$

The checks for Example 2 are shown in Figures 2.1a and 2.1b.

TECHNOLOGY

EVALUATING EXPRESSIONS

Evaluate $b^2 - 4ac$ for

a. $a = 1, b = 3,$ and $c = 2$ b. $a = 1, b = -5,$ and $c = 2$

```
(3)²-4(1)(2)
                1
(-5)²-4(1)(2)
               17
```

Figure 2.1a

```
1→A:3→B:2→C:B²-4
AC
                1
1→A:-5→B:2→C:B²-
4AC
               17
```

Figure 2.1b

For Figure 2.1a,

To evaluate an algebraic expression, we may substitute the value(s) in place of the variable(s) and evaluate the numeric expression. Be sure to enclose the values substituted in parentheses.

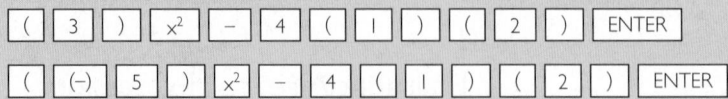

For Figure 2.1b,

To evaluate an algebraic expression, we may store the value for the variables and enter the algebraic expression. For ease in editing, we will enter all of these commands as one entry. To do so, we separate each entry with a colon, ALPHA : .

Store the three values for *a*, *b*, and *c* separated by colons.

| | STO▶ | ALPHA | A | ALPHA | : | 3 | STO▶ | ALPHA | B |

| ALPHA | : | 2 | STO▶ | ALPHA | C | ALPHA | : |

Enter the expression.

| ALPHA | B | x² | − | 4 | ALPHA | A | ALPHA | C | ENTER |

In order to enter the second expression without retyping, recall the previous entry and edit it. Press 2nd ENTRY to recall the previous entry. Then edit the previous entry, using the arrow keys in combination with delete, DEL , and insert 2nd INS .

2nd ENTRY Move the cursor to the left, using the arrow keys. Place the cursor on top of the 3 and delete 3, DEL . Insert the new value for *b*, −5, 2nd INS (−) 5 ENTER .

EXAMPLE 3 For $x = 3$, evaluate the following expressions and check on your calculator.

      a. $-x$ **b.** $-(-x)$ **c.** x^2 **d.** $-x^2$

Solution   **Check**

Note: The calculator has a special key for *x*, X,T,θ,n .

a. $-x$ for $x = 3$
      $-x = -(3) = -3$ Substitute 3 for x.
                                The opposite of 3 is −3.

b. $-(-x)$ for $x = 3$
      $-(-x) = -(-(3)) = 3$ Substitute 3 for x.
                                  The opposite of the
                                  opposite of 3 is 3.

c. x^2 for $x = 3$
      $x^2 = (3)^2 = 9$ Substitute 3 for x.

d. $-x^2$ for $x = 3$
      $-x^2 = -(3)^2 = -(9) = -9$ Substitute 3 for x.
                                  The opposite of 9 is −9.

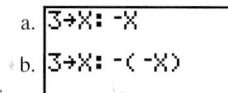

a. `3→X: -X`   -3
b. `3→X: -( -X)`   3

c. `3→X: X²`   9
d. `3→X: -X²`   -9

 2.1.2 Checkup

1. Evaluate $\sqrt{a^2 + b^2}$ for
   a. $a = 3$ and $b = 4$ **b.** $a = 2.4$ and $b = 3.2$ **c.** $a = \dfrac{2}{3}$ and $b = \dfrac{8}{9}$
2. For $a = -12$, evaluate
   a. $2a$ **b.** $-2a$ **c.** $-(-2a)$ **d.** $2a^2$ **e.** $-2a^2$
3. Explain what it means to evaluate an algebraic expression.

2.1.3 Modeling the Real World

Algebraic expressions are helpful for evaluating properties of geometric figures, such as areas and perimeters. For example, once you know the expression for the perimeter or area of a rectangle in terms of the lengths of its sides, you can evaluate the perimeter or area of any rectangle, no matter how big or small, given the lengths of its sides. Always remember that whatever operations you perform on the numbers when you evaluate the expression must also be performed on the units of measurement involved. For example,

$$3 \text{ ft} + 2 \text{ ft} = 5 \text{ ft}$$

$$(3 \text{ ft})(2 \text{ ft}) = 6 \text{ ft}^2$$

EXAMPLE 4   The daily charge for renting a 24-foot U-Haul truck is $39.95 plus $0.49 per mile driven, with a $5.00 nonrefundable charge. Write an algebraic expression for the cost of renting a truck for x days and driving it y miles. Determine the cost of renting a truck for three days and driving it 543 miles.

Solution

Let x = the number of days rented   *Define the variables.*

$\quad\quad y$ = the number of miles driven

$39.95x + 0.49y + 5.00$ *"Plus" translates to addition.*
                                     "Per mile" means to multiply.

An expression for the cost of rental is $39.95x + 0.49y + 5.00$ dollars.

$39.95x + 0.49y + 5.00 = 39.95(3) + 0.49(543) + 5.00$ *Substitute given*
$\quad\quad\quad\quad\quad\quad\quad = 390.92$ *values.*

The cost of renting a 24-foot truck for three days and driving it 543 miles is $390.92.

APPLICATION

The area of a rectangle is found by multiplying its length by its width. The perimeter of a rectangle is twice the sum of its length and width. Write an algebraic expression for the area and perimeter of a rectangle.

According to the *American Football Rules*, the dimensions of a rectangular playing field for football are 120 yards long and $53\frac{1}{3}$ yards wide. Determine the area and perimeter of a football field.

Discussion

Let L = length of a rectangle and W = width of a rectangle.

LW is an expression for the area of the rectangle. $2(L + W)$ is an expression for the perimeter of the rectangle.

Area

$LW = (120 \text{ yd})\left(53\frac{1}{3}\text{ yd}\right)$ *Substitute values.*

$\quad\quad = 6400 \text{ yd}^2$

Perimeter

$2(L + W) = 2\left(120 \text{ yd} + 53\frac{1}{3}\text{ yd}\right)$ *Substitute values.*

$\quad\quad\quad\quad = 346\frac{2}{3}\text{ yd}$

The area of a football field is 6400 square yards, and the perimeter is $346\frac{2}{3}$ yards.

2.1.3 Checkup

1. The daily charge for renting a 15-kW diesel generator is $110. Write an algebraic expression for the cost of renting the generator for *d* days. Determine the cost of renting the generator for 3.5 days.

2. The length of a diagonal of a rectangle is equal to the square root of the sum of the squares of the length and width of the rectangle. Write an algebraic expression for the length of the diagonal of a rectangle. Determine the length of the diagonal of a playing field for football that is 100 yards long and $53\frac{1}{3}$ yards wide.

2.1 Exercises

Translate each word expression into an algebraic expression.

1. Three-fourths of the total price of a gallon of milk

2. Six percent of the amount invested

3. A total amount divided by 15

4. The quotient of 55 and a count

5. The product of 2.5 and a number, decreased by the quotient of 19.59 and the number

6. The square of a number, increased by the product of 3 times the number

7. Twenty-five subtracted from the product of 12 and a number

8. Three subtracted from the product of a number and 8

9. Eighty dollars more than triple the cost of a chair

10. The total of three quiz grades divided by 3

11. One-third of the difference of the length of a rectangle and 5

12. Five less than twice the length of a side of a square

13. Four more than the product of a number and 5

14. Six more than one-third of the number of adults in a room

15. The sum of two numbers, increased by twice the product of the numbers

16. The product of a number and 3, decreased by the quotient of 10 and the number

17. Double a number divided by the sum of the number and 5

18. Eight less than the product of two numbers

19. Joe has $10,000 to invest in two separate mutual funds. If he invests *x* dollars in the first fund, write an algebraic expression for the amount he has left to invest in the second fund.

20. The number of adults and children that attended a school play was 456. If we let *a* represent the number of adults, write an algebraic expression for the number of children attending the play.

21. For $x = -3$, evaluate
    a. $-x$ **b.** $-(-x)$ **c.** x^2 **d.** $-x^2$

22. For $x = -1.5$, evaluate
    a. $-x$ **b.** $-(-x)$ **c.** x^2 **d.** $-x^2$

Evaluate $3x + 5$ for

23. $x = -5$ **24.** $x = -13$ **25.** $x = \frac{2}{3}$ **26.** $x = 2\frac{2}{3}$ **27.** $x = -2.7$ **28.** $x = -1.06$

Evaluate $18 - 3z$ for

29. $z = -12.07$ **30.** $z = -0.9$ **31.** $z = 23$ **32.** $z = 33$ **33.** $z = -\frac{5}{6}$ **34.** $z = -5\frac{4}{9}$

Evaluate $\frac{1}{2}bh$ for

35. $b = 56$ and $h = 14$ **36.** $b = 12.8$ and $h = 10.5$ **37.** $b = \frac{8}{5}$ and $h = \frac{16}{3}$

38. $b = 12$ and $h = 11$ **39.** $b = 6.8$ and $h = 4.2$ **40.** $b = \frac{4}{3}$ and $h = \frac{5}{2}$

Evaluate $x^2 + 2x + 9$ for

41. $x = -7$ **42.** $x = -8$ **43.** $x = 2.5$ **44.** $x = 1.6$

Evaluate $\sqrt{4x^2 - 20x + 25} + 8$ for

45. $x = 3$ **46.** $x = 5$ **47.** $x = -4$ **48.** $x = -1$

Evaluate $|4.5 - 3.1b|$ for

49. $b = 2$ **50.** $b = 3$ **51.** $b = -2$ **52.** $b = -3$

Evaluate $\dfrac{-b + \sqrt{b^2 - 4ac}}{2a}$ for

53. $a = 1, b = -2, c = -15$ **54.** $a = 1, b = 5, c = -14$ **55.** $a = 2, b = 5, c = -12$ **56.** $a = 3, b = 11, c = -4$

57. At a charity fund-raiser, adult tickets were sold for $6 each and children's tickets were sold for $2 each. Write an algebraic expression for the total amount of money raised from the sale of tickets. How much money was raised if the fund-raiser sold 235 adult tickets and 380 children's tickets?

58. Katie works part time at two different jobs. The first job pays her $9.50 per hour, while the second job pays her $12.25 per hour. Write an algebraic expression for the total amount of money she will earn on both jobs. How much will she earn if she works 15 hours on the first job and 12 hours on the second job?

59. Pablo receives a commission of $200 for each personal computer system he sells and a commission of $50 for each piece of ancillary equipment he sells. He has a fixed monthly expense of $750. Write an algebraic expression for his monthly profit. How much is his profit if he sells 19 personal computer systems and 12 pieces of ancillary equipment in a given month?

60. Shondra is paid a base rate of $1200 per month as a tutor. In addition, she receives $8 per hour spent tutoring and $6 per hour spent administering tests. Write an algebraic expression for her total monthly earnings. How much does she earn if she spent 125 hours tutoring and 30 hours administering tests in a given month?

61. The distance around a circle (circumference) is calculated as the product of π and the diameter of the circle. Write the algebraic expression used to find the circumference of a circle. What is the circumference of a circular wading pool with a diameter of 12 feet? Round your answer to the nearest tenth.

62. The circumference of a circle is calculated as the product of π and twice the radius of the circle. Write the algebraic expression used to find the circumference of a circle. What is the circumference of a circular dining table with a radius of 3 feet? Round your answer to the nearest tenth.

63. The daily charge for renting a minivan is $64.99 per day plus $0.29 per mile driven. Write an algebraic expression for renting the minivan for d days and driving it m miles. Determine the cost of renting the minivan for eight days and driving it 1250 miles.

64. The daily charge for renting a luxury sedan is $76.99 per day plus $0.29 per mile driven. Write an algebraic expression for renting the luxury sedan for d days and driving it m miles. Determine the cost of renting the luxury sedan for three days and driving it 175 miles.

65. The front of a box fan has the shape of a square. The area of the face is determined as the square of the length of one of its sides. The perimeter of the face is determined as the length of one of its sides multiplied by four. Write algebraic expressions for the area and perimeter of the face of the box fan. Determine the area and perimeter of the face of the fan when its sides measure 21 inches long.

66. The front of a clock has the shape of a triangle. The area of the face of the clock is determined as the product of the length of its base and its height, divided by 2. The perimeter of the face of the clock is determined as the sum of the lengths of its three sides. Write algebraic expressions for the area and perimeter of the face of the clock. Determine the area and perimeter of the face of the clock when its base measures 10 inches, its height measures 13 inches, and the other two sides measure 14 inches each.

67. The sales tax rate is 8.5%. Write an algebraic expression for the tax on a coat whose retail price is a given number of dollars. Write an algebraic expression for the total cost of the coat. What is the total cost of a coat that retails for $149.00?

68. The recommended tip for a restaurant server is 15%. Write an algebraic expression for the tip on a bill for a given number of dollars. Write an expression for the total of the bill and the tip. If a bill totals $87.58 before the tip, what is the total bill, including the tip?

2.1 Calculator Exercises

Evaluate $a^2 + b^2$ for

1. $a = 7$ and $b = 9$ **2.** $a = 0.8$ and $b = 0.6$

3. $a = \dfrac{9}{5}$ and $b = \dfrac{12}{5}$ **4.** $a = 6$ and $b = 5$

5. $a = 0.4$ and $b = 0.3$ **6.** $a = \dfrac{15}{7}$ and $b = \dfrac{8}{7}$

Now use your calculator to evaluate $(a + b)^2$ for the values given in Exercises 1–6.

The results are different. Explain the differences between the two expressions, $a^2 + b^2$ and $(a + b)^2$. Can you suggest why they yield different results when they are evaluated for the same values of the variables?

2.1 Writing Exercises

Sometimes we need to translate an algebraic expression into an appropriate word expression. Most of the time, more than one translation will be correct. For example, $3x + 7$ translates into "the product of 3 and a number, increased by 7," "7 more than 3 times a number," or "the sum of triple a number and 7." Write an appropriate word expression for the following four algebraic expressions.

1. $\dfrac{x}{y} + 3$ **2.** $2\pi r$ **3.** $a^2 + b^2$ **4.** $H - L$

5. An algebraic expression for the average of three grades is given by $\dfrac{a + b + c}{3}$, where the letters represent the three grades. Write a word expression for the average of the grades.

6. The gas mileage of a car is determined by $\dfrac{e - b}{g}$, where e is the ending odometer reading, b is the beginning odometer reading, and g is the number of gallons of gas used. Write a word expression for calculating the gas mileage of the car.

2.2 Simplifying Algebraic Expressions

OBJECTIVES

1 Identify the terms, coefficients, and like terms of an algebraic expression.
2 Simplify algebraic expressions by combining like terms.
3 Simplify algebraic expressions by removing parentheses and combining like terms.
4 Model real-world situations by using algebraic expressions.

APPLICATION

The area of a trapezoid is sometimes written as $\frac{1}{2}h(b + B)$, where h is the height of the trapezoid, B is the length of the base, and b is the length of the top. Write an equivalent form for this expression.

The Great Wall of China has a trapezoidal cross section, with average dimensions of 25 feet for the base, 20 feet for the top, and 30 feet for the height. Use the equivalent form for the area of a trapezoid to determine the area of a cross section of the wall.

After completing this section, we will discuss this application further. See page 109.

Great Wall of China

2.2.1 Understanding the Terminology of Algebraic Expressions

In the previous section, we evaluated algebraic expressions. At times, we do not know the values for the variables in an algebraic expression. However, we would still like to write the expression in its simplest equivalent form. We call this **simplifying** the algebraic expression.

Before we begin simplifying, we need to define the words used with algebraic expressions. A **term** is a number, a variable, or a product of a number and one or more variables. A **variable term** is a term that contains one or more variables. A **constant term** is a term that does not contain a variable. The **numerical coefficient** (often called, simply, **coefficient**) is the numerical factor—both sign and number—of the term.

Algebraic expression	Number of terms	Variable terms	Constant terms	Coefficients
$6x^2 + 7x + 5$	3	$6x^2, 7x$	5	$6, 7, 5$
$6x^2 + 7x - 5$ or $6x^2 + 7x + (-5)$	3	$6x^2, 7x$	-5	$6, 7, -5$

HELPING HAND The constant term is also a coefficient.

Grouping symbols affect the number of terms of an algebraic expression.

$$6x^2 + (7x - 5) \qquad \text{Terms are } 6x^2 \text{ and } (7x - 5).$$

HELPING HAND The terms of an expression can be identified as the addends of the expression. Therefore, the terms are separated by plus signs that are not inside grouping symbols.

EXAMPLE 1 Complete the following table:

	Algebraic expression	Number of terms	Variable terms	Constant terms	Coefficients
a.	$-x^2y - 9xy + x$				
b.	$\dfrac{x}{3} + x - 4$				
c.	$5x(x + 1) - 2(x + 1)$				

Solution

	Algebraic expression	Number of terms	Variable terms	Constant terms	Coefficients
a.	$-x^2y - 9xy + x$	3	$-x^2y, \ -9xy, x$	none	$-1, -9, 1$
b.	$\dfrac{x}{3} + x - 4$	3	$\dfrac{x}{3}, x$	-4	$\dfrac{1}{3}, 1, -4$
c.	$5x(x + 1) - 2(x + 1)$	2	$5x(x + 1), -2(x + 1)$	none	$5, -2$

Note:

a. $-x^2y = -1x^2y$ and $x = 1x$

b. $\dfrac{x}{3} = \dfrac{1}{3} \cdot x = \dfrac{1}{3}x$

c. $5x(x+1) - 2(x+1) = 5x(x+1) + (-2)(x+1)$

The products (terms) are separated by a plus sign.

Two terms are called **like terms** if both are constant terms or both contain the same variables with the same exponents. The expression $6x^2 + 7x - 5$ has no like terms. Notice that the variable terms both have the same variable, x, but the variables have different exponents. However, in the expression $6x^2 + 7x^2 - 5$ there are two like terms, $6x^2$ and $7x^2$, because both variable terms have the same variables with the same exponents.

EXAMPLE 2 Identify the like terms.

a. $x^2 - 4 - y^2 + 6$ **b.** $6x^2 + 7x^2y - 8x^2y + 4x^2y - y^2 - 5xy$

Solution

a. -4 and 6 are like terms. (They are both constants.)

b. $7x^2y$, $-8x^2y$, and $4x^2y$ are like terms. (They are variable terms with the same variables with the same exponents.)

 2.2.1 Checkup

Complete the table.

Algebraic expression	Number of terms	Variable terms	Constant terms	Coefficients	Like terms
1. $3y^2 + 9y + 8$					
2. $3p(p+8) - 5(p+8)$					
3. $3x^2 + 3xy + 6y$ $- 5xy + 7y - 2$					

4. How can you identify the terms of an algebraic expression?

5. When are two terms of an algebraic expression like terms?

6. All constant terms in an algebraic expression are coefficients, but not all coefficients are constant terms. Explain.

2.2.2 Simplifying by Combining Like Terms

To simplify an algebraic expression, we **combine like terms**. In order to do this, we will use the distributive law and factor out the common variables in the like terms.

$$3x + 4x = (3+4)x \qquad \text{Factor out the common factor of x.}$$
$$= 7x \qquad \text{Add.}$$

In simplifying a more complicated algebraic expression, first use the associative and commutative properties for addition to rearrange the terms before combining like terms.

$$6x^2 + 8x - 7x^2 + x - 2 = 6x^2 + (-7x^2) + 8x + x \quad - 2 \qquad \text{Rearrange terms.}$$
$$= [6 + (-7)]x^2 + (8 + 1)x - 2 \qquad \text{Factor.}$$
$$= \qquad -1x^2 \quad + \quad 9x \quad - 2 \qquad \text{Simplify.}$$
$$= \qquad -x^2 \quad + \quad 9x \quad - 2 \qquad \text{It is not necessary to write the } -1 \text{ coefficient.}$$

HELPING HAND We must be careful with the signs of the coefficients when we rearrange terms.

> To simplify an algebraic expression by combining like terms,
>
> - Rearrange terms if necessary.
> - The sum of the coefficients of the like terms is the coefficient of the simplified term.
> - The common variable factor for the variable terms does not change in the simplified expression.

It is not necessary to write a coefficient of 1 or −1 with a variable term. Also, it is not necessary to write a variable term with a coefficient of 0 or a constant term of 0.

EXAMPLE 3   Simplify.

a. $2a^3 + 3a^2 - a^3 - 3a^2 - 3$ **b.** $2.3xy - 4.4x + 2xy + 6.56x$

c. $\dfrac{x}{2} + \dfrac{3x}{4} - \dfrac{2x^2}{3} - x^2$

Solution

a. $2a^3 + 3a^2 - a^3 - 3a^2 - 3 = 2a^3 - a^3 + 3a^2 - 3a^2 - 3$ Rearrange terms.
$$= 1a^3 \qquad + \quad 0a^2 \quad - 3 \qquad \text{Combine like terms.}$$
$$= a^3 \qquad\qquad\qquad - 3 \qquad \text{It is not necessary to write the coefficient 1 or the } 0a^2.$$

b. $2.3xy - 4.4x + 2xy + 6.56x = 2.3xy + 2xy - 4.4x + 6.56x$ Rearrange terms.
$$= 4.3xy \qquad + \qquad 2.16x \qquad \text{Combine like terms.}$$

c. $\dfrac{x}{2} + \dfrac{3x}{4} - \dfrac{2x^2}{3} - x^2 = \dfrac{1}{2}x + \dfrac{3}{4}x - \dfrac{2}{3}x^2 - x^2$
$$= \dfrac{2}{4}x + \dfrac{3}{4}x - \dfrac{2}{3}x^2 - \dfrac{3}{3}x^2 \qquad \text{Find the least common denominator for like terms.}$$
$$= \dfrac{5}{4}x - \dfrac{5}{3}x^2 \qquad \text{Combine like terms.}$$

HELPING HAND We do not write improper fraction coefficients as mixed numbers.

2.2.2 Checkup

Simplify exercises 1–3.

1. $4x^3 + 3x^2 - 5x + 3 - 2x^3 - x^2 + x - 3$

2. $1.7xy + 2.5xz - 3.9yz + 4.3xy - 2.5xz + 1.9yz$

3. $\dfrac{7x}{3} + \dfrac{y}{3} - \dfrac{x}{4} + \dfrac{3y}{4}$

4. What does it mean to collect like terms?

2.2.3 Simplifying Algebraic Expressions with Parentheses

In order to simplify algebraic expressions with multiplication, we need to remember the associative law for multiplication.

$$6(3x) = (6 \cdot 3)x \qquad \text{Associative law for multiplication}$$
$$= 18x$$

In products involving variables, we multiply numeric coefficients.

In dividing expressions with variable terms by a number, we divide the numeric coefficient of the term by the number.

$$\frac{24x}{8} = \frac{24}{8} \cdot x$$
$$= 3x$$

Algebraic expressions may have parentheses or other grouping symbols. Therefore, in order to simplify these expressions, we must apply the distributive law to remove the grouping symbols.

$$-(4x + 2) = -4x - 2 \qquad \text{Distributive law}$$
$$3(4x - 1) = 3(4x) - 3(1) \qquad \text{Distributive law}$$
$$= 12x - 3 \qquad \text{Multiply.}$$
$$\frac{6x - 12}{3} = \frac{6x}{3} - \frac{12}{3} \qquad \text{Distributive law}$$
$$= 2x - 4 \qquad \text{Divide.}$$

To simplify algebraic expressions involving mixed operations, use the properties of real numbers and the order of operations.

EXAMPLE 4 Simplify.

a. $2(3x + 4y) + 5(2x - y)$

b. $4(2a - b) - (6a + 10b)$

c. $[10(x + 2) - 5] + [8 + 4(2x - 6)]$

d. $3\{[2(3x - 4) + 7] - [2(5x + 1) - 6]\}$

e. $\dfrac{5(2x + 4) - 7(x + 2)}{2x - (3x + 3) + x}$

Solution

a. $2(3x + 4y) + 5(2x - y) = 6x + 8y + 10x - 5y \qquad \text{Distributive law}$
$$= 16x + 3y \qquad \text{Combine like terms.}$$

b. $4(2a - b) - (6a + 10b) = 8a - 4b - 6a - 10b$ Distributive law

$$= 2a - 14b$$ Combine like terms.

c. $[10(x + 2) - 5] + [8 + 4(2x - 6)] = [10x + 20 - 5] + [8 + 8x - 24]$

Distributive law

$$= [10x + 15] \quad + [8x - 16]$$

Combine like terms.

$$= 10x + 15 \quad + 8x - 16$$

Remove brackets (addition).

$$= 18x - 1$$ Combine like terms.

d. $3\{[2(3x - 4) + 7] - [2(5x + 1) - 6]\} = 3\{[6x - 8 + 7] - [10x + 2 - 6]\}$

Distributive law

$$= 3\{[6x - 1] \quad - [10x - 4]\}$$

Combine like terms

$$= 3\{6x - 1 \quad - 10x + 4\}$$

Opposite-of-a-sum property

$$= 3\{-4x + 3\}$$ Combine like terms.

$$= -12x + 9$$ Distributive law

e. $\dfrac{5(2x + 4) - 7(x + 2)}{2x - (3x + 3) + x} = \dfrac{10x + 20 - 7x - 14}{2x - 3x - 3 + x}$ Apply the distributive law in the numerator and the opposite-of-a-sum property in the denominator.

$$= \dfrac{3x + 6}{-3}$$ Combine like terms.

$$= \dfrac{3x}{-3} + \dfrac{6}{-3}$$ Distributive law

$$= -x - 2$$ Divide.

2.2.3 Checkup

Simplify exercises 1–4.

1. $3(6p + 7q) + 4(8p - 5q)$

2. $-4(5a + 2b + c) - 2(-a + 3b)$

3. $4\{[2(7p - 3) + 6] - [9(p - 4) - 15]\}$

4. $\dfrac{9(3x - 7) - 8(6x + 12)}{x - 2(3x + 2) + 1 + 5x}$

5. An algebraic expression may have groupings nested within one another. How should you proceed to remove the grouping symbols?

2.2.4 Modeling the Real World

Algebraic expressions can be written for many important business concepts. **Revenue** is the amount of money that is received from sales. A business has **costs**, sometimes made up of **fixed costs**, such as rent, utilities, and setup costs, and **variable costs**, which are the costs that arise for each unit produced. An expression for the **profit** of a business is determined by subtracting the total cost from the revenue. The **break-even point** in a business venture is determined when the revenue is equal to the total cost. At this point, the business does not realize a profit or a loss. The business will realize a profit if the amount of revenue exceeds the amount of the total cost. The business will realize a loss if the total cost exceeds the amount of revenue.

Many algebraic expressions used to describe real-world situations can be simplified by applying the associative or distributive laws. In fact, you will find that calculations are often easier to do if you first simplify the algebraic expression before substituting numbers, instead of substituting numbers first and then trying to simplify.

EXAMPLE 5 A manufacturer of mountain bicycles determines that the cost of the materials is $24.74 per bicycle plus an additional overhead cost of $125.98 per day. Each bicycle sells for $195.99. What is the daily profit for x bicycles? What is the daily profit for 10 bicycles?

Solution

The profit is determined by subtracting the total cost from the revenue. The total revenue is $195.99 (the selling price per bicycle) times x (the number of bicycles), or $195.99x$. The total daily cost is $125.98 (overhead or fixed cost) plus $24.74 (the cost per bicycle) times x (the number of bicycles), or $125.98 + 24.74x$. Therefore, the daily profit (the revenue minus the total cost) is $195.99x - (125.98 + 24.74x)$.

$$195.99x - (125.98 + 24.74x) = 195.99x - 125.98 - 24.74x \qquad \text{Distribute.}$$
$$= 171.25x - 125.98 \qquad \text{Simplify.}$$

The daily profit for x bicycles is $171.25x - 125.98$ dollars.

The daily profit for 10 bicycles is found by substituting 10 for x in the expression.

$$171.25x - 125.98 = 171.25(10) - 125.98$$
$$= 1586.52$$

The daily profit for 10 bicycles is $1586.52.

APPLICATION

The area of a trapezoid is sometimes written as $\frac{1}{2}h(b + B)$, where h is the height of the trapezoid, B is the length of the base, and b is the length of the top. Write an equivalent form for this expression.

The Great Wall of China has a trapezoidal cross section, with average dimensions of 25 feet for the base, 20 feet for the top, and 30 feet for the height. Use the equivalent form for the area of a trapezoid to determine the area of a cross section of the wall.

(Continued on page 110)

Discussion

$$\frac{1}{2}h(b + B) = \frac{1}{2}hb + \frac{1}{2}hB \qquad \text{Distribute } \tfrac{1}{2}h.$$

An equivalent form of the area of a trapezoid is $\frac{1}{2}hb + \frac{1}{2}hB$.

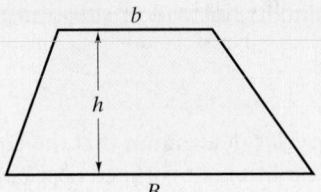

To determine the area of the cross section of the Great Wall of China, substitute 25 for B, 20 for b, and 30 for h.

$$\frac{1}{2}hb + \frac{1}{2}hB = \frac{1}{2}(30)(20) + \frac{1}{2}(30)(25)$$
$$= 300 + 375$$
$$= 675$$

The area of the cross section of the Great Wall of China is about 675 square feet.

 2.2.4 Checkup

1. An artist spends an average of $85 on paints and canvas for each painting she produces. She pays $225 per month to rent a studio for her work. In a given month, she produces and sells x paintings for an average price of $500 each. What is her profit for the month? What is her profit if she produces and sells six paintings?

2. The perimeter of a rectangular figure is sometimes expressed as twice the sum of its length and width. Write an algebraic expression for the perimeter. Write an equivalent expression without using parentheses. Evaluate each expression to find the amount of fringe needed to finish the edges on a rectangular tablecloth if the tablecloth measures 52 inches by 90 inches.

2.2 Exercises

Without simplifying, (a) determine the number of terms, (b) list the constant terms, (c) list the variable terms, (d) list the coefficient of each term, and (e) identify like terms, for each expression.

1. $2x^2 - 6x + x + 12$

2. $3y^5 + y^3 - y + 1 + 4y^3$

3. $3.4a - 11.2b - 0.3a$

4. $\frac{2}{5}m + \frac{1}{10}n - \frac{3}{10}m + \frac{1}{5}n$

5. $3m(n - 5) + 6(n - 5)$

6. $3p(p + q) - q(p + q)$

7. $x^2 + 3xy - y^2 + 7$

8. $a^2 - b^2 + 6ab - 3$

Simplify.

9. $5x + 9 - 13x + 17 - 12 + 9x$

10. $22a + 31 - 12a + 51 + 27 + 42a$

11. $2x^3 + 7x^2 - 2x + 8 - x^3 - 7x^2 + 3x - 2$

12. $5h^3 + 2h^2 + h - 10 - 4h^2 + 2h^3 - h + 1$

13. $3.05a + 6.29b - 1.18a + 0.49b$

14. $19.92z - 47.08x + 3.076x - 2.572z$

15. $\frac{1}{6}x + \frac{2}{9} - \frac{2}{3}x + \frac{5}{18}$

16. $\frac{3}{4}z - \frac{5}{16} + \frac{7}{8}z + \frac{17}{32}$

17. $6x^3 + 3x^2y - 5xy^2 + 3y^3 - 5x^2y + xy^2 + x^3 + 6y^3$

18. $9d^5 - 7d^3e^2 + 8d^2e^3 - 15e^5 + 6d^2e^3 - 3d^3e^2 + d^5 - e^5$

19. $\frac{5x}{7} + \frac{y}{6} + \frac{5x}{6} - \left(\frac{-2y}{7}\right)$

20. $\frac{-7a}{11} + \frac{10b}{11} - \frac{5a}{9} - \left(\frac{-4b}{9}\right)$

21. $5.3x + 1.4 + (3.4 - 1.7x)$

22. $1.9y + 3.7 + (4.8y - 7.9)$

23. $(45x - 112) + (21x + 33)$

24. $(48 - 22t) + (12t - 23)$

25. $(235 - 12y) - (307 + 31y)$

26. $(483z - 79) - (12 - 235z)$

27. $(x + y + 4z) - (2x - 5y + z)$

28. $(41a + 21b + c) - (a - 12b - 13c)$

29. $-(1.8y - 3.5z) + (4.1y - 2.7z)$

30. $-(2.07t + 5.2r) + (0.15t - 7.6r)$

31. $(a + b) - (a + b) + (a + b) - (a + b)$

32. $(x - y) + (x - y) + (x - y) - (x - y)$

33. $(-x + y) + (x - y)$

34. $-(a - b) - (-a + b)$

35. $(a + 5b) - (-a + 5b)$

36. $-(5m - 10n) + (-5m - 10n)$

37. $-15(2x + 3)$

38. $-12(5z + 1)$

39. $-4(-5z - 14)$

40. $-2(-a - 205)$

41. $2.2(3.5x - 7.3)$

42. $3.6(1.1a - 0.8)$

43. $72\left(-\dfrac{7}{6}m + \dfrac{49}{72}\right)$

44. $55\left(-\dfrac{12}{55}z + 6\right)$

45. $-48\left(-\dfrac{5}{12}b + \dfrac{3}{16}\right)$

46. $-42\left(\dfrac{5}{14}y - \dfrac{20}{21}\right)$

47. $\dfrac{36x + 60}{12}$

48. $\dfrac{-90m - 45n}{15}$

49. $\dfrac{20.4b - 3.4c}{-6.8}$

50. $\dfrac{-2.88m + 9.6n}{-9.6}$

51. $\dfrac{96a + 24b - 115c}{8}$

52. $\dfrac{-75x - 17y + 125z}{5}$

53. $11(-3a + 2b - 4c) - 8(5a - 7b + 2c)$

54. $-31(x - 7y + 9z) + 5(-4x + 7y + 6z)$

55. $-4.6(2x - 5y) + 9.9(5x - 3y)$

56. $16.2(5p - 3q) - 3.8(4p - 8q)$

57. $\dfrac{3}{8}\left(-\dfrac{4}{9}p - \dfrac{2}{9}q\right) + \dfrac{2}{3}\left(\dfrac{7}{8}p - \dfrac{6}{7}q\right)$

58. $-\dfrac{5}{16}\left(\dfrac{2}{5}a + \dfrac{8}{15}b\right) - \dfrac{2}{3}\left(\dfrac{1}{2}a - \dfrac{1}{6}b\right)$

59. $[15 - 2(3x + 6y - 10) + 4x] + [6x + 2(8y - 12)]$

60. $-[23x + 5(2x - 6y) + 9] + [4y - 3(6x + y) - 10]$

61. $2[-5a + 3(2b - 4c) + 15] - [7(2a + 6b - c) + 12]$

62. $-4[5(6x + 4y) + 7z - 22] - [55 - 3(x + y - z)]$

63. $6\{2[x + 2(3y - 4z)] - [x - y + 3(y + 2z)]\}$

64. $2\{3[a + 2(b - 4c) + 3b] - [a + 2(b + c) + 6c]\}$

65. $\dfrac{8(5a + 7c) - 6(2a + 4c)}{4}$

66. $\dfrac{3(x + 2y) + 5(3x - 9y)}{3}$

67. $\dfrac{2.6m + 3(1.2m - 2.6n) - (4.8m + 7.4n) - 1.6n}{3m + 2(-2m + 1) + m}$

68. $\dfrac{2(1.1a + 4.6b) - 3(5.72a - 6.2b)}{(a + 1) - (a - 1)}$

69. Shirley and Joan filed an expense report for a business trip. Shirley rented a car for $21.99 per day plus $0.29 per mile driven. Each was given a daily budget of $120.00 per day for food and lodging and a miscellaneous total trip allowance of $50.00. Their trip lasted x days and they drove y miles. Their total expense for the trip was $21.99x + 0.29y + 2(120x + 50)$ dollars. Simplify the expense to an equivalent expression without parentheses. What was the total expense for a trip that lasted four days and covered 625 miles?

70. On a business trip, Jim and Dave rented a car at $37.99 per day plus $0.29 per mile. Each of the men was allowed $95.00 per day for food and lodging. They were each given a miscellaneous spending allowance of $75.00 for the trip, which lasted d days and covered m miles. Their total expense was $37.99d + 0.29m + 2(95d + 75)$ dollars. Simplify the expense to an equivalent expression without parentheses. What was the total expense if the trip lasted three days and covered 750 miles?

71. Franklin's Electrical Services sends two electricians and three apprentices out to wire a new house that is under construction. Each electrician is paid $38.00 per hour, and then $22.00 is added for travel expenses. The three apprentices working with the crew are each paid $16.00 per hour, but are not given the travel payment. The cost of a job lasting h hours is $2(38h + 22) + 3(16h)$ dollars. Simplify the cost to an equivalent expression without parentheses. What is the cost of a job lasting seven hours?

72. Green's Lawn Service quoted a price of $33.50 per application to fertilize lawns in a neighborhood. Several homes in the neighborhood agreed to the service and contracted for three treatments for the season. In addition, the homeowners all agreed to one application of a lime treatment for $49.95 each. The cost for caring for the lawns of h homes in the neighborhood is $3(33.50h) + 49.95h$ dollars. Simplify the cost to an equivalent expression without parentheses. How much is the contract for if 24 homes in the neighborhood subscribe to the service?

73. Carl's Carpet Cleaners charges a fee of $20.00 per visit plus $1.55 per square yard of carpeting to clean carpets. The materials used for the cleaning average $0.65 per square yard of carpeting cleaned, and the cost of each trip averages $6.50. What is the profit for a job where c square yards of carpet are cleaned? What is the profit for a job cleaning 250 square yards of carpet?

74. Pete's Plumbing Company charges $35.00 for each visit to a home. In addition, Pete charges $55.00 an hour for the work he and his helper do on the job. He pays his helper $25.00 an hour. The trip to the house costs him an average of $12.00. What is the profit he makes for a job requiring h hours of work? What is the profit for a job that takes 3.5 hours to complete?

75. Speedy Mail Delivery charges $5.00 plus $1.50 per ounce to deliver a letter overnight. It costs the company about $2.25 for each letter delivered. What is the profit for delivering a letter weighing w ounces? What is the profit for delivering a letter weighing 4 ounces?

76. Tillie's Typing Service charges $10.00 plus $2.00 per page for each manuscript Tillie types. She spends about $3.00 to set up each manuscript and $0.50 per page on supplies. What is her profit for typing a manuscript having p pages? What is the profit for a manuscript having 112 pages?

77. Laurie purchases some compact discs at $9.99 each. The sales clerk adds 8% sales tax to the price of the discs. What is the total cost if Laurie purchases x compact discs? If Laurie pays with a $50.00 bill, what is her change? If Laurie buys three compact discs, what is her cost and what is her change?

78. Mary Lynn paints and sells T-shirts. The shirts cost her $4.50 each. After she paints them, she sells them at craft shows for $15.00 each. She spends $18.50 for the paint kit that she uses for all the T-shirts that she paints and sells. What is her profit for painting and selling x shirts. What is her profit if she sells 22 shirts at the craft show?

2.2 Calculator Exercises

Can the TI-83 Plus calculator simplify $2x + 3x$? To see if it can, type $2x + 3x$ into your calculator and press $\boxed{\text{ENTER}}$. The calculator does not display $5x$. Next, type $5x$ into your calculator. You should see the calculator gives you a numerical value that is the same for both expressions. If you compare the result on your calculator with that of other students, you will most likely find that they also have numerical values displayed, and theirs are different from yours. The reason for this is that when you type an expression into the calculator, it evaluates the expression instead of simplifying it. When it

evaluates the expression, it uses whatever value happened to be stored in the x location from the previous use. The calculator cannot perform algebraic simplifications; it can only perform evaluations. Therefore, you cannot use the calculator to perform algebraic operations. However, you can use the calculator to check your arithmetic when you add coefficients to collect like terms or when you multiply or divide coefficients while applying the distributive law to remove parentheses. Use the calculator this way to check your work on simplifying the expressions that follow.

Simplify. Use your calculator to check your arithmetic.

1. $12.078x + 2.093 - 17.42x - 13.9035$

2. $(2579x - 4302) - (1087x - 306)$

3. $\dfrac{10}{13}x - \dfrac{5}{52}y - \dfrac{17}{20}x - \dfrac{7}{13}y$

4. $\left(2\dfrac{11}{25}\right)x + 5\dfrac{17}{30} - \left(3\dfrac{13}{15}\right)x - 3\dfrac{23}{75}$

5. $(1.0009x + 0.0004) - (0.0909x - 1.0031)$

6. $-(935.3376x + 701.315) - (83.027x - 581.9534)$

7. $3.995x + 12.083 - 2.995x - 9.083$

2.2 Writing Exercises

1. Consider the two terms $3a^4b^3$ and $4a^4b^3$. Are the terms like terms? Explain.

2. Now consider the two terms $3a^4b^3$ and $4a^3b^4$. Are the terms like terms? Explain.

3. What should you look for in order to find like terms?

2.3 Equations

OBJECTIVES

1 Identify expressions and equations.

2 Determine whether a number is a solution of an algebraic equation.

3 Write equations for word statements.

4 Model real-world situations by using equations.

In 1905, Albert Einstein suggested his famous equation that relates energy and mass. The theory of relativity states that "energy equals mass multiplied by the velocity of light squared." Write an equation for this famous relationship, using E for energy, m for mass, and c for the velocity of light.

After completing this section, we will discuss this application further. See page 118.

2.3.1 Identifying Expressions and Equations

In the first part of this text, we discussed expressions. We evaluate an expression by finding a numeric value for the expression. If the expression contains variables, we first substitute a value for each of the variables and then find a numeric value for the expression.

In this section, we will form a mathematical statement by combining two expressions with an equals sign. This kind of statement is called an equation. Therefore, an **equation** is a mathematical statement asserting that two expressions have the same (or equal) value.

We may equate two numeric expressions:

$$2 + 3 = 9 - 4$$

$$14 - (-3) + 8(4) \div 2 = (-5)(-6) + 7$$

Or we may equate two algebraic expressions:

$$4x - 3 = 3x + 2$$

$$3x + 2y - z = 2x - y + 2z$$

It is very important to understand the difference between an expression and an equation.

EXAMPLE 1 Determine whether each of the following is an expression or an equation.

a. $6 + x - y + 3z$ **b.** $6 + x = y + 3z$ **c.** $P = a + b + c$

Solution

a. $6 + x - y + 3z$ is an expression.   (There is no equals sign.)

b. $6 + x = y + 3z$ is an equation.   (There is an equals sign.)

c. $P = a + b + c$ is an equation.   (There is an equals sign.)

2.3.1 Checkup

In exercises 1–4, determine whether each of the following is an expression or an equation:

1. $3x + 6 = 17$ **2.** $5x - 35$

3. $2L + 2W$ **4.** $\frac{1}{3}x + 3 = 2$

5. Explain the difference between an algebraic expression and an algebraic equation.

2.3.2 Determining Solutions of Algebraic Equations

Equations may be true or false. If the expression on the left side of the equation and the expression on the right side of the equation have the same value, then the equation is true. If the two expressions have different values, the equation is false.

Here is an example, taken from the preceding part of this section:

$$14 - (-3) + 8(4) \div 2 = (-5)(-6) + 7$$

$14 + 3 \quad + 8(4) \div 2$	$30 \quad + 7$
$14 + 3 \quad + 32 \div 2$	37
$14 + 3 \quad + 16$	
33	

The two expressions have different values ($33 \neq 37$). Therefore, the equation is false. (Did you notice this when you first saw this equation? It's not always easy to tell.)

If the expressions in an equation contain a variable, we need to substitute a value for the variable in order to be able to evaluate the expressions. However, not all values of the variable will make the equation true. If there is a value for the variable that makes a true equation, that value is called a **solution** of the equation.

SOLUTION OF AN ALGEBRAIC EQUATION

A solution of an algebraic equation is a value for the variable that will result in a true equation.

To determine whether a number is a solution of an algebraic equation,

* Substitute the possible solution for the variable.
* Evaluate the expression on the left side of the equation and the expression on the right side of the equation.
* Determine whether the expressions are equal.

If the expressions are equal, then the value substituted for the variable is a solution.

For example, determine whether $x = 5$ is a solution of $4x - 3 = 3x + 2$.

$4x \quad - 3 =$	$3x + 2$	
$4(5) - 3$	$3(5) + 2$	Substitute 5 for x.
$20 - 3$	$15 + 2$	
17	17	

Since $17 = 17$, the equation is true for $x = 5$. Therefore, 5 is a solution of the equation.

On a calculator, as shown in Figure 2.2a, each expression results in 17. Therefore, 5 is a solution of the equation.

The result is 1 in Figure 2.2b, which indicates that the equation is true when x is 5; therefore, 5 is a solution of the equation.

TECHNOLOGY

TESTING ALGEBRAIC EQUATIONS

Determine whether $x = 5$ is a solution of the equation $4x - 3 = 3x + 2$.

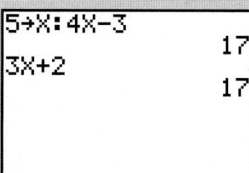

Figure 2.2a

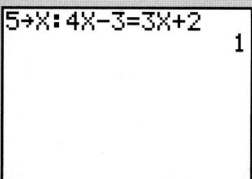

Figure 2.2b

For Figure 2.2a,

Check the value given by evaluating the expression on the left and the expression on the right.

Store the given value for the variable, $x = 5$.

| 5 | STO▶ | X,T,θ,n |

Separate the entries.

| ALPHA | : |

Enter the left side and the right side separately.

| 4 | X,T,θ,n | – | 3 | ENTER |
| 3 | X,T,θ,n | + | 2 | ENTER |

Since $17 = 17$, $x = 5$ is a solution.

For Figure 2.2b,

Check the value by using the TEST function of the calculator.

Store the given value for the variable, $x = 5$.

| 5 | STO▶ | X,T,θ,n |

Separate the two entries.

| ALPHA | : |

Enter the equation. The equals sign is under TEST menu option 1.

| 4 | X,T,θ,n | – | 3 | 2nd | TEST | I | 3 | X,T,θ,n | + | 2 | ENTER |

The calculator returns a 1 to indicate that the equation is true and returns a 0 to indicate that the equation is false. Since the calculator returned a 1, $x = 5$ is a solution of the equation.

Note: On some calculators, a solution may wrongly produce a 0 (for "false") if the calculator is working with nonterminating decimals (fractions). In this case, evaluate both expressions separately and visually check for equivalence.

EXAMPLE 2 Determine whether the given value is a solution of the equation.

a. $4(x - 2) + 7x = 6x + 2x + 5x$ for $x = -4$

b. $8 - 3a^2 - (a + 6) = (a + 2) - 2(2a + 4)$ for $a = 1$

Solution

Check

a. $4(x - 2) + 7x = 6x + 2x + 5x$ for $x = -4$

$$
\begin{array}{c|c}
4(x - 2) \quad + 7x & = \quad 6x \quad + \quad 2x \quad + \quad 5x \\
\hline
4[(-4) - 2] + 7(-4) & 6(-4) + 2(-4) + 5(-4) \\
4\,(-6) \qquad + 7(-4) & -24 \;+\; (-8) + (-20) \\
-24 \qquad\quad + (-28) & -52 \\
\qquad\qquad -52 &
\end{array}
$$

The solution is -4, because $-52 = -52$.

b. $8 - 3a^2 - (a + 6) = (a + 2) - 2(2a + 4)$ for $a = 1$

$$
\begin{array}{c|c}
8 - \quad 3a^2 \; - (a + 6) & = \quad (a + 2) - 2(2a + 4) \\
\hline
8 - 3(1)^2 - (1 + 6) & (1 + 2) - 2[2(1) + 4] \\
8 - \quad 3 \; - \quad 7 & 3 \quad - 2(\; 2 \; + 4) \\
\qquad\qquad -2 & 3 \quad - 2(\; 6 \;) \\
& 3 \quad - 12 \\
& -9
\end{array}
$$

The solution is not 1, because $-2 \neq -9$.

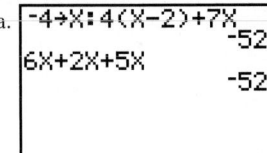

a.

```
-4→X:4(X-2)+7X
               -52
6X+2X+5X
               -52
```

b.

```
1→A:8-3A²-(A+6)=
(A+2)-2(2A+4)
               0
```

The calculator display of 0 means false.

 2.3.2 Checkup

Determine whether the given value is a solution of the equation.

1. $3x - 5 = 8x - 2$ for $x = -\dfrac{3}{5}$ **2.** $3c^2 + 5c - 2 = 0$ for $c = 2$

2.3.3 Converting Word Statements into Equations

In solving real-life mathematics problems, we often need to convert word statements into equations. To do this, we translate each expression in the word statement into numbers and symbols, as we did in previous sections. The only difference is that we now equate the two expressions by using an equals sign.

The following are some words and expressions that translate into an equals sign:

equals, is equal to, is the same as, results in, is, was, will be, becomes, gives, is equivalent to

To write an equation from a word statement,

- Define the variable or variables.
- Write two algebraic expressions for the word expressions.
- Join the expressions with an equals sign.

For example,

The sum of twice a number and 3 is the same as the difference of 3 times the number and 5.

Let n = a number   First define a variable.

$$2n + 3 \qquad\qquad = \qquad\qquad 3n - 5$$

Sum of twice a number and 3   is the same as   Difference of three times the number and 5

EXAMPLE 3   Write an equation for the word statement.
The sum of 0.32 times the number of miles and 25 is the same as 0.40 times the number of miles.

Solution
The sum of 0.32 times the number of miles and 25 is the same as 0.40 times the number of miles.

Let x = the number of miles

$$0.32x + 25 = 0.40x \qquad \text{"Sum" means addition.}$$

2.3.3 Checkup

Write an equation for the word statement.
The sum of 1500 and 15 times the number of cartridges is equal to 25 times the number of cartridges.

2.3.4 Modeling the Real World

One of the most important steps in solving real-world mathematical problems of any kind is correctly translating the given word statements into mathematical equations. This is usually the first step in solving problems in physics and chemistry, medicine and biology, economics and business, ecology and computer science—in fact, any area of human activity that involves mathematical relationships. You need to learn various concepts in these fields so that you know what variables and relationships are available to you. But once you have the right equation, you can usually figure out what you know and what you need to find in order to solve the problem.

EXAMPLE 4   Jesse plans to produce and sell get-well baskets. He estimates that each basket will cost $2.25, and it will cost $4.50 per basket to fill them. The cost of advertising is $15.00. If a basket sells for $9.99, write an equation needed to find the break-even point.

Solution
Let x = the number of baskets produced and sold
The total revenue is $9.99 per basket, or $9.99x$.
The total cost is $2.25 per basket, plus $4.50 per basket, plus $15.00, or

$$2.25x + 4.50x + 15.00 = 6.75x + 15.00$$

A company is said to break even when the total revenue equals the total cost.

$$9.99x = 6.75x + 15.00$$

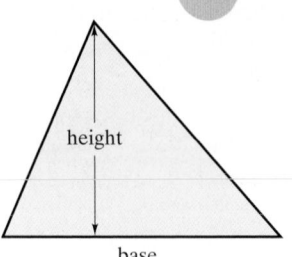

EXAMPLE 5   The area of a triangle is the product of one-half its base and its height. Write an equation for this statement.

Solution
Let A = area
b = base
h = height

$$A = \frac{1}{2}bh$$ "Product" means multiply.

APPLICATION

In 1905, Albert Einstein suggested his famous equation that relates energy and mass. The theory of relativity states that "energy equals mass multiplied by the velocity of light squared." Write an equation for this famous relationship, using E for energy, m for mass, and c for the velocity of light.

Discussion

Let E = energy   Define the variables.
m = mass
c = velocity of light

$$E \quad = \quad mc^2$$

Energy   equals   Mass multiplied by the velocity of light squared

---

✓ 2.3.4 Checkup

1. An enterprising student rents a 26-foot truck from U-Rent-It Truck Rentals. The cost of renting the truck is $141 per day. The student will use the truck to help other students move from a dormitory to off-campus housing. Each student will be charged $49 to help pack his or her belongings onto the truck and to deliver them to their new housing. Write an equation to determine the break-even point for each day's business.

2. In 1826, Georg Simon Ohm studied the effects of resistance in limiting the flow of electricity. He discovered that, for a given resistance, the current is directly proportional to the applied voltage. That is, the resistance, R, of a given conductor can be calculated as the voltage, V, divided by the current, I. Write an equation for this famous relationship.

2.3 Exercises

Determine whether each item is an expression or an equation.

1. $3x - 15 = 6$

2. $5x + 9 - 2x$

3. $15x^3 + 7x^2 - 2x + 4$

4. $3x^2 - 6x + 1 = 0$

5. $0.5x^2 - 2.8 = 0.7$

6. $1.8x^2 - 7.6x - 0.3$

7. $\frac{2}{3}x + \frac{6}{7} = 0$

8. $\frac{4}{9}x^3 - \frac{5}{7}$

9. $\frac{4}{9}x^2 - \frac{3}{5}x - \frac{1}{4}$

10. $\frac{1}{2} = \frac{x}{15}$

Determine whether the given value is a solution of the equation.

11. $2x + 4 = 10$ for $x = 3$

12. $-3a - 2 = -16$ for $a = 6$

13. $5y - 7 = 9$ for $y = 3$

14. $-12z + 15 = 50$ for $z = -3$

15. $6a + 5 = 3a + 17$ for $a = 3$

16. $5p + 16 = 6p - 12$ for $p = 28$

17. $9z - 23 = 6z - 29$ for $z = -2$

18. $8a + 26 = -5a + 39$ for $a = 1$

19. $3(x - 5) + 9 = 4(6x - 5) - 7$ for $x = 1$

20. $5y + 2(y - 7) = 6 + 3(y - 1)$ for $y = 4$

21. $2[3(x - 4) - 6] + x = 3x$ for $x = 8$

22. $3x - [5(2x + 1) + 3] = x$ for $x = -1$

23. $x^2 + 5 = 33 - 3x$ for $x = 4$

24. $2x^2 - 20 = 3x + 34$ for $x = 7$

Write an equation for each word statement.

25. The sum of a number and 6 is 15.

26. The product of 2 and a number is 12.

27. Five less than twice the number of children equals the sum of the number of adults and 2.

28. The quotient of the total annual cost divided by 12 is 2200 dollars.

29. The square of a number, less 21, is equal to 100.

30. Double a number, less 100, is equivalent to the sum of the number and 50.

31. Twice the sum of a number and the square of 5 is equal to the sum of the number and 100.

32. Half of a number increased by 60 is 200.

33. Twice the sum of a number and 2 is equal to the sum of 4 and the product of 2 times the number.

34. Three times the difference of a number and 6 is equal to the difference of triple the number and 18.

35. The sum of 17 and the quotient of a number divided by 2 is equal to the sum of 4 and the product of 3 and the number.

36. The sum of the square of a number and the number is equivalent to twice the number.

37. The perimeter of a semicircular sector is equal to the diameter added to the product of π and the radius. (This product is one-half the circumference of a circle.)

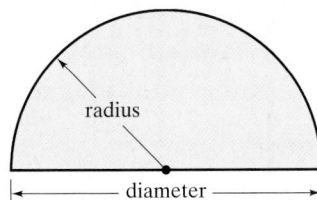

38. The average rate of speed is equal to the distance traveled divided by the time elapsed.

39. The interest, I, paid on a continuously compounded loan is equal to the difference of the compounded amount, A, less the principal, P.

40. The total amount of a loan, A, is the sum of the principal, P, plus the interest, I.

41. The sum of the number of nickels and the number of dimes is equal to twice the number of nickels.

42. The number of quarters is equal to five more than twice the number of dimes.

43. The sum of 5% of the dollars invested in one account and 7% of the dollars invested in another account is equal to the total interest earned on the investments, $176.

44. The total interest earned on investments, $256, is the sum of 8% of the dollars invested in one account and 10% of the dollars invested in a second account.

45. One angle measures 10 degrees more than triple the measure of another angle.

46. The length of one side of a triangle is 6 inches more than twice the length of a second side.

47. A solenoid is constructed by winding wire on an iron core. An electric current passing through the wire produces a magnetic induction in the core very similar to the magnetic field seen in a bar magnet. The magnetic intensity, H, is equal to the product of the number of turns, N, and the current, I, divided by the length of the solenoid, L. Write an equation to represent this relation.

48. In physics, the number of moles, n, contained in a gas is equal to its mass, m, divided by its molecular mass, M. Write an equation expressing this relation.

2.3 Calculator Exercises

Check the following two equations, using your calculator's TEST key:

1. $10 + 42 \div 7x = 4x$ when $x = 3$

2. $10 + 42 \div (7x) = 4x$ when $x = 3$

If the calculator indicates that the first equation is false and that the second is true, you should pay attention to the importance of the parentheses. When you use the calculator to check a solution or to evaluate an expression, it is important to know when you wish to divide by a product of a number and a variable and when you wish to divide by a number and then multiply by a variable. Students often want to divide by a product of a number and a variable, but fail to recognize the need for the parentheses.

3. Rewrite Exercise 1, using fraction notation.

4. Rewrite Exercise 2, using fraction notation.

5. Discuss the differences in the answers to Exercises 3 and 4.

2.3 Writing Exercise

In this section, you were instructed regarding the difference between expressions and equations. Write a short description of what is meant by an algebraic expression and an algebraic equation. Describe the defining difference between an expression and an equation. Give your own examples of an algebraic expression and an algebraic equation.

2.4 Formulas

OBJECTIVES

1 Evaluate two-dimensional geometric formulas.

2 Evaluate three-dimensional geometric formulas.

3 Evaluate angle formulas.

4 Evaluate other formulas.

5 Model real-world situations by using formulas.

APPLICATION

Diane plans to remove the carpet in her extra bedroom (br 2 in the house plan) and have it replaced with hardwood flooring. If the hardwood floor that she chose costs $3.31 per board foot installed, determine the cost of replacing the carpet with hardwood flooring, excluding the closet floor {1 board foot = 1.2 (or $1\frac{1}{5}$) times the square footage}.

After completing this section, we will discuss this application further. See page 128.

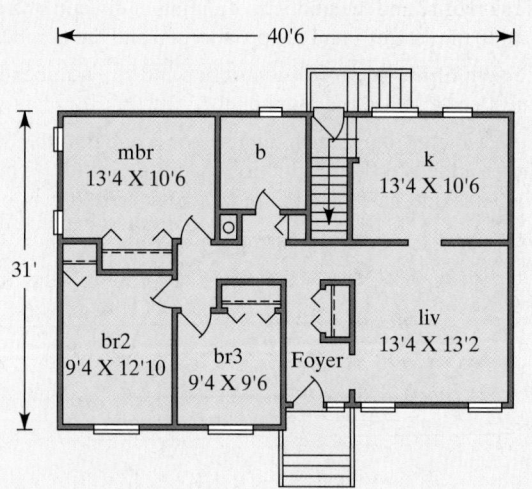

In Example 5 of section 2.3, we wrote a special type of equation called a formula. A **formula** is an equation used to find a numeric value for an unknown variable. We know the values of all the variables in the formula, except one. For example, the formula in Example 5 may be used to find the area, A, of a triangle when the base, b, and height, h, are known.

$$A = \frac{1}{2} bh \quad \text{for } b = \text{base and } h = \text{height}$$

To find a numeric value for the area, A, we substitute values for b and h. We say we *evaluate* the formula.

To evaluate a formula,

- Write the formula.
- Substitute the values for the variables.
- Evaluate the expression containing the substituted values.

For example, evaluate $A = \frac{1}{2}bh$ for $b = 12$ inches and $h = 5$ inches.

$$A = \frac{1}{2}bh \qquad \text{Write the formula.}$$

$$A = \frac{1}{2}(12)(5) \qquad \text{Substitute.}$$

$$A = 30 \text{ square inches} \qquad \text{Evaluate.}$$

The area of the triangle is 30 square inches.

In this section, we will introduce you to the formulas that you need in order to work exercises in the first seven chapters of this text. Many of these formulas have been discussed in previous sections. However, the formulas have not been presented using variables. In this section, you will be asked to evaluate each of these formulas with the information given. In later sections, you will be asked to use the formulas to write equations expressing given information.

2.4.1 Evaluating Two-Dimensional Geometric Formulas

Two-dimensional figures have properties of perimeter (circumference) and area. **Perimeter**, P, is the distance around a closed two-dimensional figure. **Circumference**, C, is the distance around a circle. **Area**, A, is the amount of surface covered by a two-dimensional figure.

The following are the two-dimensional figures we will use in this text, together with formulas for their perimeters and areas, as well as other useful formulas:

Triangle

$$P = a + b + c$$

$$A = \frac{1}{2}bh$$

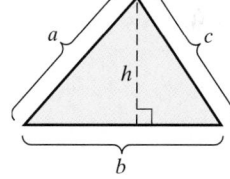

Rectangle

$$P = 2L + 2W$$

$$A = LW$$

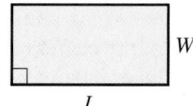

Square

$$P = 4s$$

$$A = s^2$$

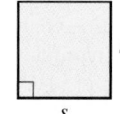

Parallelogram

$$A = bh$$

$$P = 2a + 2b$$

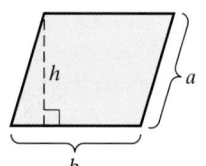

Circle

$$d = 2r$$

$$r = \frac{1}{2}d$$

$$C = \pi d \quad \text{or} \quad C = 2\pi r$$

$$A = \pi r^2$$

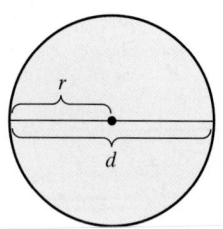

π
3.141592654

HELPING HAND The calculator value for π is more precise than the noncalculator value (3.14) we used earlier. In this text, you need to use the more precise value, unless told otherwise. (See Example 1.) To be as accurate as the calculator, we would need to use a value for π with more decimal places. To determine this value, we can enter $\boxed{\text{2nd}}$ $\boxed{\pi}$ on the calculator, which will then display a value for π rounded to nine decimal places.

EXAMPLE I A soccer field has a center circle that is used for kickoffs. The team taking the kick is allowed as many players inside the circle as it wishes. The length of the radius of the circle depends on the age of the players. For players six years and under, the radius is 5 yards. Find the exact area of the circle, and approximate this area to the nearest tenth of a square yard.

Solution

$A = \pi r^2$	*Write the formula.*	5→R: πR²
$A = \pi(5)^2$	*Substitute.*	78.53981634
$A = 25\pi$	*Evaluate.*	
$A \approx 78.5$		

The exact area is 25π square yards. The approximate area is therefore 78.5 square yards.

 2.4.1 Checkup

1. For soccer players 9 years and under and 10 years and under, the radius of the center circle of the soccer field is 8 yards. What is the exact area of this circle? What is the area approximated to the nearest tenth of a square yard?

2. For the center circle in Exercise 1, determine the exact circumference. What is the circumference approximated to the nearest tenth of a yard?

3. What is a formula?

4. What is the difference between the perimeter of a two-dimensional figure and the area of a two-dimensional figure?

2.4.2 Evaluating Three-Dimensional Geometric Formulas

Three-dimensional figures have properties of volume and surface area. **Volume**, V, is the amount of space enclosed in a three-dimensional figure. **Surface area**, S, is the total area of all exposed surfaces of a three-dimensional figure.

The following are the three-dimensional figures we will use in this text, together with the formulas for their volume and surface areas:

Rectangular solid

$V = LWH$

$S = 2LW + 2WH + 2LH$

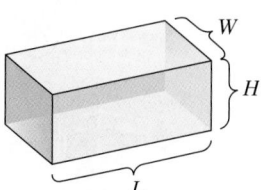

Cube

$V = s^3$

$S = 6s^2$

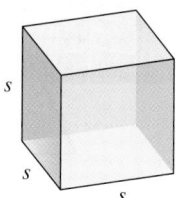

Right circular cylinder

$V = \pi r^2 h$

$S = 2\pi r^2 + 2\pi rh$

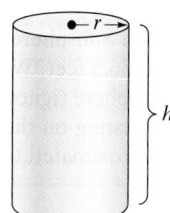

Sphere

$V = \dfrac{4}{3}\pi r^3$

$S = 4\pi r^2$

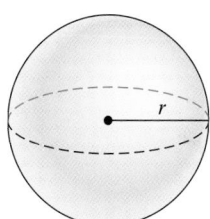

Right circular cone

$V = \dfrac{1}{3}\pi r^2 h$

$S = \pi r\sqrt{r^2 + h^2} + \pi r^2$

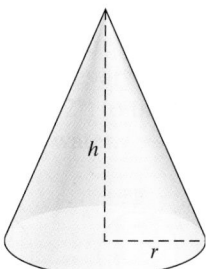

EXAMPLE 2 The 1982 World's Fair centerpiece, the Sunsphere, is a 266-foot-tall steel tower topped by reflective bronze-coated glass windows in the shape of a sphere. The radius of the sphere is 36.5 feet. Approximate the volume to the nearest tenth of a cubic foot.

Sometimes it is useful to convert temperatures measured on one scale to the other scale. The following formulas enable us to do this:

Celsius to Fahrenheit

$$F = \frac{9}{5}C + 32$$

Fahrenheit to Celsius

$$C = \frac{5}{9}(F - 32)$$

EXAMPLE 4 The average human body temperature is 98.6°F. Convert this temperature to Celsius.

Solution

$$C = \frac{5}{9}(F - 32) \qquad \text{Write the formula.}$$

$$C = \frac{5}{9}(98.6 - 32) \qquad \text{Substitute.}$$

$$C = 37 \qquad \text{Evaluate.}$$

The average human body temperature is 37°C.

Interest, I, is the amount of money a lender charges for borrowing and for investing money. In order to determine the amount of interest, we need to know the **principal**, P, or amount borrowed or invested, the **interest rate**, r, and the period of **time**, t, for which the loan or investment is made.

Interest may be calculated by several different methods. **Simple interest** is interest based on the principal alone. **Compound interest** is based on taking the interest accumulated in one period and adding it to the principal in order to determine the interest applicable in the next period. **Continuously compounded interest** is regular compound interest with very short periods, so that the addition of interest occurs on a continuous basis. In this objective, we will work only with simple interest. Later in the text we will discuss other, more complicated, interest calculations.

Simple interest (I)

$$I = Prt \qquad P = \text{principal}, r = \text{rate of interest}, t = \text{number of periods of time}$$
$$A = P + I \qquad A = \text{amount of simple interest}, P = \text{principal}, I = \text{interest}$$

EXAMPLE 5 According to the American Association of Community Colleges, the total of the average annual tuition and fees is $1518. Emin applied for a student loan with an annual compounded interest rate of $8\frac{1}{4}\%$. Four years later, the payoff amount would be $2084.41. However, his grandfather offered to lend him the same amount for 9% per year simple interest. Determine the better choice for Emin if he intends to pay off the loan after four years.

Solution
First, determine the amount of simple interest.

$$I = Prt$$
$$I = (1518)(0.09)(4)$$
$$I = 546.48$$

Next, determine the total payoff: the amount borrowed plus the amount of simple interest.

$$A = P + I$$
$$A = 1518 + 546.48$$
$$A = 2064.48$$

Emin will pay about $2084.41 for the student loan and $2064.48 to his grand-father. His better choice is to borrow from his grandfather.

Another formula we will use often allows us to find the distance traveled, d, if we are given the rate (speed), r, and time traveled, t.

Distance traveled (d)

$$d = rt \qquad r = \text{rate (speed)}, t = \text{time traveled}$$

Two related formulas are $r = d/t$ and $t = d/r$.

EXAMPLE 6 The X-34, a single-engine rocket plane, can reach altitudes of up to 250,000 feet and travel up to eight times the speed of sound. (Sound travels at 1070 feet per second.) Estimate the length of time it would take for the plane to reach its maximum altitude if it is launched from an L-1011 airliner at an altitude of 40,000 feet and travels at its maximum speed.

Solution
The distance traveled is the difference in the maximum altitude and the alti-tude at which the plane was launched, or $250,000 - 40,000 = 210,000$ feet. The speed is eight times the speed of sound, or $8(1070) = 8560$ feet per second.

$$t = \frac{d}{r} \qquad \text{Write the formula.}$$

$$t = \frac{210,000}{8560} \qquad \text{Substitute.}$$

$$t \approx 24.5$$

The X-34 will reach its maximum altitude in about 24.5 seconds from a launch at 40,000 feet.

2.4.4 Checkup

1. Scientists believe that the temperature at the center of the Earth's core could be 7200°F. Convert this tempera-ture to Celsius.

2. When the astronauts landed on the Moon, they found a world that was airless, waterless, and devoid of life. Temperatures on the Moon range from up to 134°C on the bright side to −170°C on the dark side. Convert these temperatures to Fahrenheit.

3. Find the amount of simple interest on a loan of $10,000 at a rate of interest of 8.5% per year for two years.

4. A bluefin tuna can swim about 73 feet per second for short periods. At this rate, how far can the tuna swim in two minutes? If one mile is equal to 5280 feet, convert the distance to miles.

2.4.5 Modeling the Real World

Many objects in the real world are not perfect geometric shapes. For instance, the Earth is a little wider at the equator than it is through the North and South Poles, so it is not exactly a sphere. Similarly, most rooms in a house tend to be a little out of perfect alignment and so are not exactly rectangular solids. But we can usually approximate real objects by geometrical shapes and use the formulas for volumes, areas, and perimeters introduced earlier in this section. The errors involved in these approximations are often small enough for us to ignore.

EXAMPLE 7

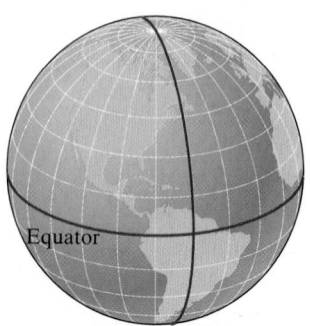

North Pole

Equator

South Pole

a. The equatorial diameter of the Earth is approximately 12,756.34 km. Determine the Earth's equatorial circumference (assuming that our planet is a sphere).

b. The polar diameter of the Earth is approximately 12,713.54 km. Determine the Earth's polar circumference.

c. What is the difference between the equatorial circumference and the polar circumference?

Calculator Solution

a. The equatorial diameter:

$$C = \pi d$$
$$C \approx \pi(12{,}756.34)$$
$$C \approx 40{,}075.224$$

The equatorial circumference is approximately 40,075.224 km.

b. The polar diameter:

$$C = \pi d$$
$$C \approx \pi(12{,}713.54)$$
$$C \approx 39{,}940.764$$

The polar circumference is approximately 39,940.764 km.

```
a. π(12756.34)
          40075.22403
b. π(12713.54)
          39940.76387
```

c. The difference is $40{,}075.224 - 39{,}940.764 = 134.460$ km.

Although the difference seems large to us, in comparison to the total circumference of the Earth, it is not. Therefore, approximating the Earth as a sphere is a reasonable approximation for most purposes. A more accurate model would be necessary for calculating the orbit of a satellite.

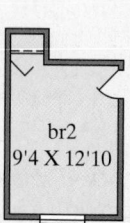

APPLICATION

Diane plans to remove the carpet in her extra bedroom (br 2 in the house plan) and have it replaced with hardwood flooring. If the hardwood floor that she chose costs $3.31 per board foot installed, determine the cost of replacing the carpet with hardwood flooring, excluding the closet floor {1 board foot = 1.2 (or $1\frac{1}{5}$) times the square footage}.

Discussion

The dimensions of the extra bedroom are 9 feet, 4 inches (length), by 12 feet, 10 inches (width). Since we are measuring board feet, we will need to convert these figures to feet. (Remember, 12 inches equal 1 foot!)

br2
9'4 X 12'10

$$L = 9 \text{ feet, } 4 \text{ inches} = 9\frac{4}{12} = 9\frac{1}{3} \text{ feet.}$$

$$W = 12 \text{ feet, } 10 \text{ inches} = 12\frac{10}{12} = 12\frac{5}{6} \text{ feet.}$$

$$A = LW \qquad \text{Area of a rectangle}$$

$$A = \left(9\frac{1}{3}\right)\left(12\frac{5}{6}\right) \qquad \text{Substitute.}$$

$$A = \frac{1078}{9}$$

$$A = 119\frac{7}{9} \text{ square feet}$$

Since 1 board foot = 1.2 (or $1\frac{1}{5}$) times the square footage, the number of board feet is

$$119\frac{7}{9} \cdot 1\frac{1}{5} = \frac{6468}{45} = 143\frac{11}{15}$$

It costs $3.31 per board foot to install the flooring.

The total cost is therefore

$$143\frac{11}{15} \cdot 3.31 \approx 475.76$$

It will cost Diane about $475.76 to install hardwood flooring in her extra bedroom.

2.4.5 Checkup

1. The equatorial diameter of the planet Mercury is approximately 4880 km. Determine the equatorial circumference of Mercury, assuming that the planet is a sphere.

2. According to the *Guinness Book of World Records*, the largest corrugated cardboard box was constructed by Kappa Van Dam of the Netherlands in 1999. The box measured 22.9 feet by 8.5 feet by 7.9 feet. Find the volume and surface area of the box.

2.4 Exercises

In exercises 1–10, find the area and perimeter of each two-dimensional geometric figure.

1. A triangle has a base of 39 inches, a height of 24 inches, and sides of 40 inches and 25 inches.

2. A triangle has a height of 16 millimeters, a base of 75 millimeters, and sides of 20 and 65 millimeters.

3. A rectangle has a length of 6 centimeters and a width of 4 centimeters.

4. A rectangle's length is 8 feet and its width is 5 feet.

5. A square's side measures $2\frac{1}{2}$ feet.

6. A square has sides measuring 5.6 meters each.

7. A parallelogram has a base of 68 meters, a height of 45 meters, and a side of 53 meters.

8. The base of a parallelogram measures 95 inches, its height measures 39 inches, and its side measures 89 inches.

9. A circle has a radius of $5\frac{1}{4}$ inches. Round your answers to the nearest tenth.

10. The radius of a circle measures $2\frac{3}{4}$ feet. Round your answers to the nearest tenth.

11. John wants to order outdoor carpeting to cover his patio, which has the shape of a rectangle. The length measures 5 yards and the width measures 4 yards. How many square yards of space will John need to cover? Carpeting is on sale at $6.99 per square yard. How much will it cost for the carpeting?

12. A corner of Ginger's yard is fenced in as a dog pen. The corner is shaped like a triangle with a base of 12 feet and a height of 9 feet. How many square feet of space will the triangle contain?

13. Gretchen made a square tablecloth measuring 52 inches on a side. What is the area of coverage? How much fringe material will she need for the perimeter.

14. Marcos built a square raised deck in his yard. The deck was 15 feet on a side. How many square feet of space did the deck contain? Marcos wants to run latticework around the deck, leaving 4 feet along one side for stairs. How many feet of latticework will he need for this?

15. It is estimated that grizzly bears searching for food range over an area equivalent to a circle with a radius of 16.5 miles. Over how large an area in square miles do they roam? What is the circumference of this circular area?

16. Polar bears are the world's largest land carnivores, and it is estimated that, in their search for food, they roam over an area equivalent to a circle with a radius of 80 miles. Over how large an area in square miles do they roam? What is the perimeter (circumference) of this circular area?

17. When Levittown, New York, was developed as the first mass-housing suburb in the country following World War II, the size of the standard lot was 60 feet by 100 feet. What is the area of this standard lot? What is the perimeter of the lot?

18. Linda has designed a circular area with a fountain in the center for her yard. The circular area has a radius of 5 feet. What is the square footage of this area? How many feet will the border of the area be?

In exercises 19–28, find the volume and surface area of each three-dimensional geometric solid.

19. A box has a length of 5 feet, a width of 3 feet, and a height of 1 foot.

20. A chest has a width of 4 feet, a length of 6 feet, and a height of 3 feet.

21. A cube has a side measuring 7.5 inches.

22. Each side of a cube measures 8.2 centimeters.

23. A right circular can has a height of 5 inches and a diameter of 3 inches. Round your answers to the nearest tenth.

24. A can of vegetable shortening has a height of 6 inches and a diameter of 5 inches. Round your answers to the nearest tenth.

25. A sphere has a diameter of 20 centimeters. Round your answers to the nearest whole number.

26. A ball has a radius of 5 inches. Round your answers to the nearest whole number.

27. A right cone has a height of 0.75 foot and a radius of 0.25 foot. Round your answers to the nearest thousandth.

28. A right cone's height measures 8 centimeters and its radius measures 2 centimeters. Round your answer to the nearest hundredth.

29. Jim built a toy box for his grandson. He made the box 4 feet long, 2 feet wide, and 2 feet high. How many cubic feet of toys will the box hold? Jim painted all six sides of the exterior with a high-gloss enamel. How many square feet of surface area did he paint? (He did not paint the interior, since he lined it with canvas.) If one pint of paint covers 20 square feet, how many pints of paint did Jim need for one coat?

30. A cube of ice is to be carved into a swan. The cube measures 2.5 feet on a side. How many cubic feet of ice does it contain? What is the total surface area of the cube of ice?

31. Danny bought a spherical tank to store propane gas for his outdoor grill. The tank was listed as having a radius of 10 inches. How many cubic inches of propane gas will the tank hold? How many square inches of surface area does the tank have? Round your answers to the nearest tenth.

32. A packing crate is rectangular, with a width of 2 feet, a length of 4 feet, and a height of 2 feet. What is the volume of the crate? How many square feet of wood does the outer surface contain?

33. Tracy constructed a doll display case out of clear acrylic panes. She built the case in the shape of a cube. If the case is 18 inches on a side, how many cubic inches of space does it contain? How many square inches of acrylic does the surface of the cubic measure?

34. The crystal base of a table lamp has the shape of a right circular cylinder, with a height of 20 inches and a radius of 4 inches. How many cubic inches of colored pellets will it take to fill the cylinder? What is the surface area of the base before any holes are made in it? Round your answers to the nearest tenth.

35. A paperweight is shaped as a right cone with a height of 7 inches and a radius measuring 3.5 inches. What is the volume of the paperweight? Round your answer to the nearest hundredth.

36. A cake-decorating kit has a squeeze bag in the shape of a right cone. The bag has a height of 9 inches and a radius of 3 inches. What is the volume of the bag?

37. Three holes will be bored into a steel plate. They form a triangle, with one angle measuring 30 degrees and another measuring 65 degrees. What is the measure of the third angle?

38. Three stakes in the ground mark off a triangular plot. One angle of the plot measures 55 degrees and another measures 65 degrees. What is the measure of the third angle?

39. A disco ball at a club in Los Angeles has a diameter of 95.25 inches. What is the radius of the ball? Determine the volume of the ball. Find its surface area. The ball is covered with 6,900 mirrors in the shape of squares measuring 2 inches on a side. How many square inches of the surface is covered with mirrors?

40. In early 2001, space shuttle *Atlantis*'s astronauts installed the $1.4 billion Destiny laboratory on the international space station. Destiny is an aluminum cylinder 28 feet long and 14 feet in diameter. Find the volume of this laboratory. The master bedroom in the house plan shown earlier in this chapter has dimensions of approximately 13 feet by 10 feet by 8 feet high. Find the volume of this room. Divide the volume of the Destiny laboratory by the volume of the room to see how many rooms would be equivalent to the laboratory. What do you now think of the size of the lab?

41. Two angles are complementary. One angle measures 65 degrees. Find the measure of the other angle.

42. An angle measures 71 degrees. What is its complement?

43. Two angles are supplementary. One angle measures 65 degrees. Find the measure of the other angle.

44. What is the supplement of an angle that measures 124 degrees?

45. Two angles of a triangle measure 33 degrees and 68 degrees. Find the measure of the other angle.

46. One angle of a triangle measures 47 degrees and another angle measures 88 degrees. What is the measure of the third angle?

47. The *pitch* of a roof measures the angle that the roof makes with a horizontal line. The angle that the roof makes with a vertical line is the complement of the pitch. What is the pitch of the roof if its complement measures 50 degrees?

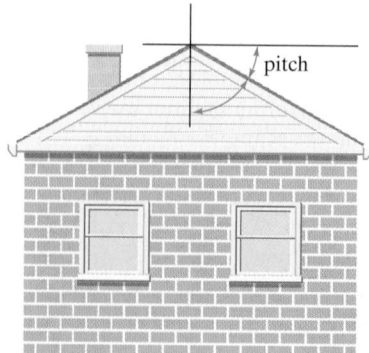

48. A ramp connects a loading dock to a driveway. The ramp forms a 70-degree angle with the vertical edge of the dock. The angle the ramp forms with the horizontal driveway is the complement of this angle. What is the measure of the angle formed with the horizontal driveway?

49. A guy wire forms two angles with the ground. The angles are supplementary. If one of the angles measures 45 degrees, what is the measure of the other angle?

50. An anchoring rope on a tent forms two supplementary angles with the ground. If one of the angles measures 60 degrees, what is the measure of the other angle?

51. Weather station KLOW in Seattle reported a daily low temperature of 25°C. What is the corresponding Fahrenheit temperature?

52. Todd the Weather Guy reported a daily high temperature of 31°C in Atlanta. Convert to the corresponding Fahrenheit temperature.

53. Marti told her friend that the high temperature for the day in Miami was 92°F. Convert this to Celsius measure.

54. The weather channel reported that the low temperature for the day in Boston was 66°F. What was the corresponding Celsius temperature?

55. At sea level, the boiling point of water is 100°C. What is the corresponding Fahrenheit temperature?

56. The melting point of gold is 1063°C. Convert to Fahrenheit temperature.

57. In Denver, the boiling point of water is 95°C. Convert to Fahrenheit temperature.

58. The boiling point of gold is 2966°C. What is the corresponding Fahrenheit temperature?

59. The melting point of mercury is −37.97°F. What is the corresponding Celsius temperature?

60. The boiling point of mercury is 673.84°F. Convert to Celsius temperature.

61. The average high temperature in Rio de Janeiro in January is 84°F. What is the Celsius temperature?

62. The average low temperature in Rio de Janeiro in January is 73°F. Convert to Celsius temperature.

63. JoAnne borrowed $2500 for 1 year at 6.5% simple interest to buy a laptop computer. How much interest did she pay?

64. Jack borrowed $5500 for 1 year at 7% simple interest to buy a used pickup truck. Find the interest on the loan.

65. How much simple interest is earned on $500 at a 7% annual interest rate for 1 month ($\frac{1}{12}$ of a year)?

66. What will be the simple interest and total amount of $100 invested at an 8% annual interest rate for 3 months ($\frac{1}{4}$ of a year)?

67. How much simple interest would be paid on a loan of $1200 at an 8% annual interest rate for 6 months?

68. What is the simple interest on a loan of $2400 at a 12% annual interest rate for 6 months?

69. The drive from home to college took $9\frac{1}{2}$ hours at an average speed of 55 miles per hour. What was the distance covered?

70. The drive to the beach took 8.25 hours at an average speed of 62 miles per hour. How many miles was the trip?

71. In 1911, Ray Harroun won the first Indianapolis 500 with an average speed of 74.602 miles per hour in 6 hours, 42 minutes, and 8 seconds. At this speed, how far did he drive in the first 3 hours of the race?

72. In 2000, Juan Montoya won the Indianapolis 500 with an average speed of 167.607 miles per hour in 2 hours, 58 minutes, and 59 seconds. At this speed, how far did he drive in the first 2 hours of the race?

2.4 Calculator Exercises

It is possible to write a calculator program to perform calculations for the geometric formulas we have studied. As an example, if you enter the following steps into your calculator, you will have created a program to calculate the area, circumference, and diameter of a circle when a number for the radius is entered into the calculator:

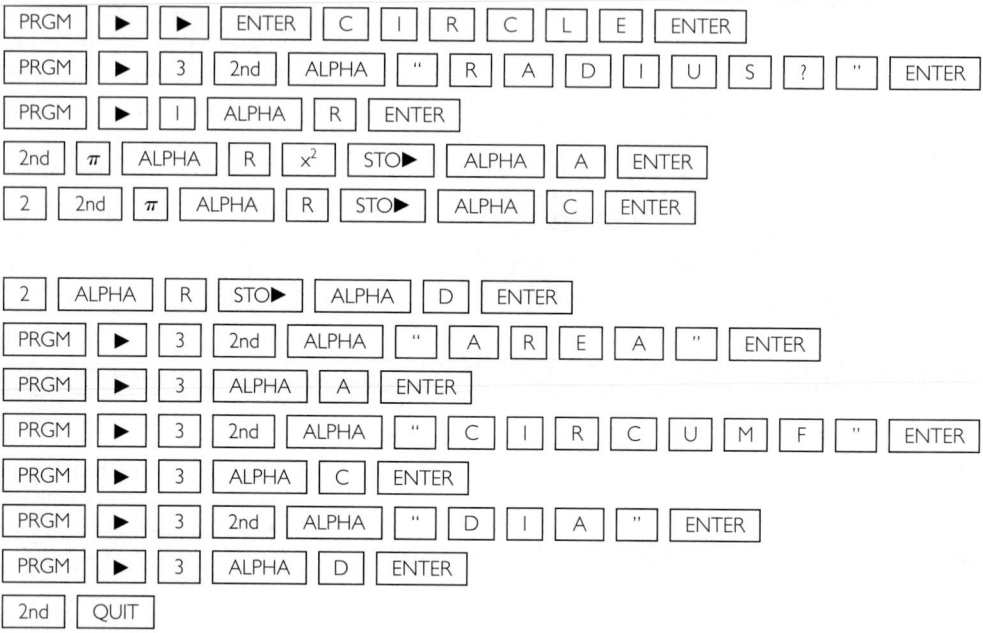

If you have entered the program correctly, you should see the displays shown in Figures 2.3a and 2.3b.

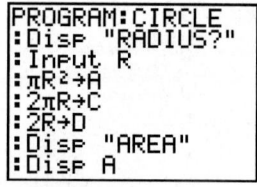

Figure 2.3a

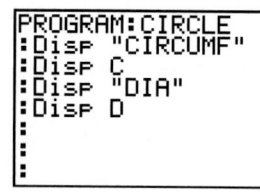

Figure 2.3b

To run the program, perform the following steps:

Enter [PRGM][1], where "1" is the number of the program in your list.

Press [ENTER], and the calculator will ask for a value for the radius.

Enter a value, and then press [ENTER].

The calculator will display the three measures.

To run the program for another value of the radius, press [ENTER].

Then, repeat the preceding steps.

To quit, enter [2nd][QUIT].

As a check, if you stored a value of 3 for the radius, the calculator should have displayed an area of 28.27433388, a circumference of 18.84955592, and a diameter of 6. You can use this program as a model to develop other programs for the various formulas presented in this section.

2.4 Writing Exercise

In this section, you learned formulas that are used to measure various properties of two-dimensional and three-dimensional geometric shapes. Write a short definition of what is meant by two-dimensional and three-dimensional shapes. Discuss what can be measured for each, and illustrate your discussion with examples. Explain what distinguishes one from the other.

CHAPTER 2 SUMMARY

After completing this chapter, you should be able to define in your own words the following key terms:

2.1
numeric expression
constant
variable
algebraic expression
substitute
evaluate

2.2
simplify
term
variable term
constant term
numerical coefficient

coefficient
like terms
combining like terms

2.3
equation
solution

2.4
formula
perimeter
circumference
area
volume

surface area
angle
right angle
complementary angles
supplementary angles
interest
principal
interest rate
time
simple interest
compounded interest
continuously compounded interest

Reflections

1. Explain the difference between simplifying an algebraic expression and evaluating an algebraic expression.
2. Define the terms of an algebraic expression. Define like terms in an algebraic expression.
3. Explain the difference between an algebraic expression and an algebraic equation.
4. How can you tell that the given values for the variables in an algebraic equation represent a solution of the equation?
5. What are formulas and why do we use them?
6. What properties of two-dimensional figures are usually represented by formulas?
7. What properties of three-dimensional figures are usually represented by formulas?

CHAPTER 2 SECTION-BY-SECTION REVIEW

2.1

Recall	Examples
Translate into an algebraic expression. • Define the variables. • Translate the words into numbers and symbols.	One third of the sum of three test grades Let a = the first test grade $\quad\quad b$ = the second test grade $\quad\quad c$ = the third test grade $\dfrac{1}{3}(a + b + c)$
Evaluate an algebraic expression. • Substitute the values for the variables, enclosing them in parentheses. • Evaluate the resulting numeric expression.	Evaluate $-x^2 + 2x - 3$ for $x = -2$. $-x^2 + 2x - 3 = -(-2)^2 + 2(-2) - 3$ $\quad\quad\quad\quad\quad\quad\quad = -11$

In exercises 1–3, translate each word expression into an algebraic expression.

1. **a.** Four more than the product of a number and 55
   b. The product of 4 more than a number and 55

2. **a.** Three-fourths of the sum of a number and 35
   b. The sum of three-fourths of a number and 35

3. **a.** Twenty less than twice a number
   b. Twenty less twice a number

4. The total cost of a car is the sum of the down payment of $2500 and the product of monthly payments of $275 for *n* months. Write an algebraic expression for the total cost of the car.

5. Pat's hourly rate of pay for a job is the quotient of his total earnings of $650 divided by the hours he worked, *h*. Write an algebraic expression for Pat's hourly rate of pay.

6. A telephone salesperson is paid $200 per week plus $5.50 for every customer who purchases the product. What is the person's weekly pay if *k* customers agree to purchase?

7. A solution of water and antifreeze is 40% antifreeze. If there are *x* liters of the solution, write an algebraic expression for the number of liters of antifreeze in the solution. Then write an algebraic expression for the number of liters of water in the solution.

8. An auto trip lasted two days and covered 758 miles of travel. If, on the first day of the trip, *x* miles were traveled, write an algebraic expression for the number of miles covered on the second day of the trip.

9. A dress was placed on sale at a discount of 25%. If the dress originally sold for *x* dollars, write an algebraic expression for the amount of discount of the dress. Write an expression for the sale price of the dress.

For $x = -\frac{2}{3}$, evaluate the following expressions:

10. $-x$ 11. $-(-x)$ 12. x^2

13. $-x^2$ 14. $|-x|$

Evaluate $5x - 25$ for

15. $x = 8$ 16. $x = -4.6$

Evaluate $\sqrt{a^2 + b^2}$ for

17. $a = 9, b = 12$ 18. $a = 4.5, b = 6$

Evaluate $\dfrac{12y - 84}{6y + 36}$ for

19. $y = 9$ 20. $y = -6$

Evaluate $|2x^2 - 45x - 75| - 2$ for

21. $x = 0.1$ 22. $x = 10$

23. The volume of a box is equal to the product of the measures of the length, width, and height of the box.
   a. Write an algebraic expression for the volume of a box.
   b. The Longaberger® Company's home office building is really a basket! It is located in Newark, Ohio. According to information provided by the company's public relations office, the base of the building is 192 feet long and 126 feet wide. Not including the handles, the height of the building is about 103 feet high. While the building is not exactly a rectangular box shape, assume that it is and determine its volume.

2.2

Recall	Examples
Combine like terms. • The sum of the coefficients of like terms is the coefficient of the simplified term. • The common variable factor is the variable factor of the simplified term.	Combine like terms. $2x^2 + 3x - 1 - x^2 - 3x + 4 = 2x^2 - x^2 + 3x - 3x - 1 + 4$ $= x^2 + 3$

Recall	Examples
Simplify expressions. • Remove parentheses by using the distributive law. • Combine like terms of the resulting expression.	Simplify. $(2a + b - 5) - (a + b + 3) = 2a + b - 5 - a - b - 3$ $= a - 8$ $2[-3(x + 1) + 4(2x - 1)] = 2[-3x - 3 + 8x - 4]$ $= 2[5x - 7]$ $= 10x - 14$

Without simplifying the algebraic expression, complete the following table:

Algebraic expression	Number of terms	Constant terms	Variable terms	Coefficients of terms	Like terms
24. $3x - 2y + 4x + 9y$					
25. $2a^2 - a + 3a^2 - 5a^3$					
26. $2.4x + 5.1 + 6.2x$					
27. $4a(a + b) - b(a + b)$					

Simplify.

28. $2.4z + 1.7z - 3.9z$ **29.** $17x + 51 + 26x - 86 - 19x - 7$ **30.** $\dfrac{3}{4}x + \dfrac{5}{8}y - \dfrac{2}{3}x + \dfrac{1}{4}y$

31. $15x^2 - 14xy + 12y^2 - 23 + 42xy - 7y^2 + 21 - 6x^2$ **32.** $(3a + 4b) - (-2a - 6b) + (a - b) - (-a + b)$

33. $-2(3.8a - 4.7b)$ **34.** $-105\left(\dfrac{6}{7}x - \dfrac{4}{15}y\right)$ **35.** $2x(3x - 17y)$

36. $\dfrac{27a - 36b + 15c}{9}$ **37.** $12(7x + 9y) + 15(3x - 7y)$ **38.** $3[-2(x + 3y) - 5] - [3(2x + y) + 16]$

39. $\dfrac{25(2a - 6b) + 5(4b - 3c) + 75}{4a - 2(2a - 3) - 1}$

40. In gym class, Tom can do x push-ups without stopping. His friend Charles can do 5 less than twice the number Tom can do. Jim can do 7 more than Tom can do. An algebraic expression for the total number of push-ups all three students can do is $x + (2x - 5) + (x + 7)$. Simplify this expression. What is the total number if Tom can do 45 push-ups?

41. Drucilla and Esmeralda attended a mathematics conference. Drucilla rented a car at $45.00 per day and $0.20 per mile. Each woman received a daily allowance of $125.00 for lodging and meals and a miscellaneous spending allowance of $20.00 per day. If the trip lasted d days and covered a distance of m miles, the total cost of the trip was $45d + 0.20m + 2(125d + 20d)$ dollars. Simplify this expression. What is the cost of a trip that took four days if the distance covered was 345 miles?

Write an expression for each situation. Then simplify and evaluate.

42. Lincoln School students are selling cans of salted peanuts and cans of a bridge mix for a fund-raiser. The peanuts costs $1.25 per can and the bridge mix costs $1.50 per can. The school will sell both for $2.75 per can. What is the profit for selling x cans of peanuts and y cans of the bridge mix? Evaluate the profit if the school sells 220 cans of peanuts and 480 cans of the bridge mix.

43. At the Summer Olympics gift shop, Katie bought some commemorative pins for $5 each. The clerk added 6% sales tax to the cost. Katie paid with a $20 bill. If Katie bought x pins, what was her change? Determine Katie's change if she bought three pins.

44. Margaret buys unfinished frames, which she decorates and sells. She pays $3.25 for each frame. After she decorates them, she sells them at craft fairs for $12.00 each. She spends $52.65 for paint and lace to decorate the frames, which is enough for all the frames she can make and sell. Write an algebraic expression for Margaret's net profit, given that she can sell x frames. What will her profit be if she makes and sells 32 frames?

2.3

Recall	Examples
To determine whether a number is a solution of an algebraic equation • Substitute the possible solution for the variable. • Evaluate both expressions. • If the expressions have equal values, then the number is a solution.	Is 3 a solution of $2x - 5 = 3x - 8$? $$\begin{array}{c c} 2x - 5 & = \quad 3x - 8 \\ \hline 2(3) - 5 & 3(3) - 8 \\ 6 - 5 & 9 - 8 \\ 1 & 1 \end{array}$$ The value 3 is a solution because $1 = 1$.
Translate into algebraic equations. • Define the variables. • Translate the words into numbers and symbols.	The test average of three grades is one third of the sum of the grades. Let A = the test average $\quad a$ = the first test grade $\quad b$ = the second test grade $\quad c$ = the third test grade $$A = \frac{1}{3}(a + b + c)$$

Determine whether each item is an expression or an equation.

45. $2x + 7y - x$

46. $5x = 3y - x$

Determine whether the given value is a solution of the equation.

47. $15y - 35 = 12$ for $y = 3$

48. $8a + 2(3a - 7) = 11a - 32$ for $a = -6$

49. $x^3 - 25x = 2x^2 - 3x - 35$ for $x = 5$

50. $2(x - 3.4) = x - 5$ for $x = 1.8$

51. $x - \dfrac{2}{3} = -\dfrac{1}{9}$ for $x = \dfrac{4}{9}$

In exercises 52–56, define all variables.

52. The sum of 5 and the product of 4 and a number is equivalent to the sum of 65 and the quotient of the number divided by 4. Write an equation that represents this relationship.

53. A writer is advanced $1500 to purchase a personal computer system to be used to author a pamphlet that is to be published by a publishing company. If the writer earns royalties of $1.35 for each pamphlet sold, write an equation to determine the break-even point for the writer.

54. The total interest earned, $1500, is the sum of 6% of the dollars invested in one account and 8% of the dollars invested in another account. Write an equation to represent this situation.

55. Write an equation which says that the dollars invested in one account added to the dollars invested in a second account total $30,000.

56. An equation is used to determine the optimum gauge of wire for a hammer dulcimer. The equation states that the fundamental frequency, f, of the wire is calculated as the square root of the quotient of tension, T, in the wire divided by the mass per unit length of wire, m. The square root is then divided by two times the length of the wire, L. Write the equation in question.

2.4

Recall	Examples
To evaluate a formula • Write the formula. • Substitute the value for the variables. • Evaluate the resulting numeric expression.	Evaluate $A = \pi r^2$ for $r = 8$ feet. $A = \pi r^2$ $A = \pi(8 \text{ ft})^2$ $A = 64\pi$ square feet (exact answer) $A \approx 201.06$ square feet (rounded answer)

Find the area and perimeter of each two-dimensional figure.

57. A triangle has a base of 26 m, a height of 16 m, and sides of 22 m and 20 m.

58. A rectangle has a width of 20 inches and a length of 44 inches.

59. A square measures 15 cm on a side.

60. A parallelogram has a base of 10.0 m, a height of 6.5 m, and a side of 7.0 m.

61. A circle has a radius of 12.2 feet.

In exercises 62–66, find the volume and surface area of each three-dimensional figure.

62. A box has a length of 35 inches, a width of 9 inches, and a height of 21 inches.

63. A cubical carton measures 14.6 cm on each side.

64. A right circular cylinder has a height of 54 inches and a radius of 18 inches.

65. A ball has a radius of 32.6 cm.

66. A right circular cone has a height of 7 cm and a radius of 4 cm.

67. Two angles are supplementary. One angle measures 58 degrees. What is the measure of the other angle?

68. Two angles are complementary. One angle measures 58 degrees. What is the measure of the other angle?

69. One angle of a triangle measures 67 degrees. Another angle of the triangle measures 88 degrees. What is the measure of the third angle of the triangle?

70. Dan wants to purchase sod for his backyard. The yard is rectangular, with a width of 60 feet and a length of 85 feet. How much sod should he order? If he fences in the yard, how many linear feet of fencing should he order if he plans a 3-foot gate and attaches the fence to the back of the house, which measures 35 feet across?

71. Amelia marks off a circular plot in the yard for a garden. The plot has a radius of 8 feet. In the center she places a circular tile on which a birdbath will sit. The tile has a diameter of 2 feet. How many square feet of the garden will there be for flowers?

72. Big Ed's Pizza Shop offers 10-inch (diameter) pizzas for $6.75 or 14-inch pizzas for $13.50. Since two of the 10-inch pizzas cost as much as one 14-inch pizza, which is a better deal, two 10-inch pizzas or one 14-inch pizza?

73. A garden shop delivers a truckload of mulch. The bed of the truck measures 8 feet by 5 feet by 2 feet high. How many cubic feet of mulch will the truck hold?

74. What is the surface area of a spherical storage tank with a diameter of 6 feet?

75. Find the simple interest on a loan of $850 at 12.5% per year for one year. What is the total amount of the loan, including the interest?

76. Convert 50°C to Fahrenheit.

77. Convert 80°F to Celsius.

78. LuAnn's automobile trip took $5\frac{3}{4}$ hours of driving time, with an average speed of travel of 62 miles per hour. What was the distance traveled?

CHAPTER 2 CHAPTER REVIEW

Without simplifying the algebraic expressions, complete the table.

Algebriac expression	Number of terms	Constant terms	Variable terms	Coefficients of terms	Like terms
1. $12x + y - z + 23$					
2. $3(a - 2) + 5(b - 4) + 75$					
3. $12 - 7x + 14x - 18 + x$					
4. $b^2 + 2b - 3b^2 + 6b + b^3$					

Evaluate for $x = -18$.

5. x^2 **6.** $-x^2$ **7.** $(-x)^2$ **8.** $-(-x)^2$

Evaluate $\sqrt{12y + 20} - 16$ for

9. $y = 8$ **10.** $y = -\frac{1}{3}$ **11.** $y = -5$

Evaluate $\dfrac{7x + 84}{2x - 3}$ *for*

12. $x = -12$ **13.** $x = 3$ **14.** $x = 1.5$

Evaluate $-\dfrac{b}{2a}$ *for*

15. $a = 1, b = -4$ **16.** $a = 2, b = 5$ **17.** $a = \dfrac{1}{2}, b = -3$

Determine whether the given value is a solution of the equation.

18. $3x - 7 = x + 1$ for $x = 4$

19. $5x + 17 = 10x + 5$ for $x = 3$

20. $3x + 17 = 2(x - 5)$ for $x = -27$

21. $2.1x - 1.9 = 0.6x - 4.6$ for $x = -1.8$

22. $\dfrac{3}{4}\left(x - \dfrac{8}{9}\right) = \dfrac{1}{2}\left(x + \dfrac{2}{3}\right)$ for $x = 5\dfrac{1}{3}$

23. $3x^2 - 6x - 10 = x^2 + x - 5$ for $x = -5$

Simplify.

24. $12h + 9h - 4h$ **25.** $6m + 22 - m - 12 + 3m$

26. $3x - 35 + 4y - 5x - 6y + 7x + 27 + 17y + 22x$

27. $3x^4 + 5x - 7x^2 + 12x^4 - 17x - 34x + x^3 - 1$

28. $(6.2a + 5.3b) + (4.7a - 1.9b)$ **29.** $-(27y - 15)$

30. $5g + 8 - (g + 4)$ **31.** $(-2x + 4y - 7z) - (-x + 6y + 8z)$ **32.** $50\left(\dfrac{11}{25}a + \dfrac{33}{50}b\right)$

33. $\dfrac{104x - 156y + 30z}{13}$ **34.** $\dfrac{-18x + 24y - 36z}{-6}$ **35.** $4(3.9x - 11.1y) + 7(2.9x - 0.7y)$

36. $12[-3(2a - 5b) + 9] + 8[-9(a + 13) - 6(b - 12)]$ **37.** $\dfrac{14.4(2x + 5) - 21.6(5x - 2) + 7.2(x - 1)}{3x - 4(x + 8) + x + 34.4}$

38. Javan paid for a purchase of x DVDs with a \$100 bill. If the DVDs sold for \$19.95 each, write an algebraic expression for the change he got back. What was his change if he purchased three DVDs?

39. Apples sell for \$0.89 per pound and bananas sell for \$0.49 per pound. If Heather buys x pounds of apples and y pounds of bananas, write an algebraic expression for the total cost of her purchase. What was her total cost if she purchased 4 pounds of apples and 2 pounds of bananas?

40. Write an equation for the following word statement: The sum of the square of a number and twice the number is equal to the number increased by 306.

41. Lakeetha charges a flat fee of \$225 plus an hourly fee of \$45 for consulting services while writing computer programs for a hospital auditing system. Write an algebraic expression for the amount of money she will make for one of her consulting contracts. How much will she earn for this assignment if she spends 120 hours developing the program?

42. Carmen sets up a savings account with an initial deposit of \$500. She has \$145 directly deposited to the account each week. She has the bank automatically deduct \$15 per week from her savings account for a Christmas club account. Write an algebraic expression for the amount of money she has in her savings account after n weeks. How much money is in the checking account after 15 weeks? If Carmen withdrew \$625 during that period, what is the balance of her account?

43. In a sales competition, each salesperson receives a flat fee of \$100, plus \$25 for each appliance sold. Beatrice has sold x appliances. Marie sold five less than twice what Beatrice sold. Ann sold one-half of what Beatrice sold. Magdalene sold the same as Beatrice. The total money earned by the women is $(100 + 25x) + [100 + 25(2x - 5)] + [100 + 25(\frac{x}{2})] + (100 + 25x)$ dollars. Simplify this expression. What was the total money earned if Beatrice sold 20 appliances?

44. Chum bought a circular above ground swimming pool for his backyard. The pool has a radius of 10 feet. If he fills the pool to a depth of 4.5 feet, how many cubic feet of water will the pool hold?

45. If the temperature reads 96°F, what is the Celsius reading?

46. How much money will you have if you invest \$18,500 at an interest rate of 5.5% compounded annually over a period of 10 years?

47. What is the distance Randy traveled if he bicycled for 1.25 hours at an average speed of 13 miles per hour?

48. Find the volume and surface area of a circular can that has a height of 1.5 inches and a radius of 1.625 inches.

49. A used book store buys paperbacks at a cost of \$0.50 per book. The store then resells the used paperbacks for \$2.00 each. The store has a weekly overhead cost for rent and utilities of \$175.00. If the store buys and resells x books, write an expression for the profit from the sales. What is the profit if the store buys and sells 250 books in a week?

50. In biology, the speed of an enzymatic reaction is often described by a formula according to which the velocity of the reaction, v, is equal to the product of a and x, divided by the sum of x and k. (a is the maximum reaction velocity, x is the concentration of the substrate, and k is the concentration of the substrate when the velocity is half of the maximum velocity.) Write an equation that represents the speed of the reaction.

51. The diameter of the wire in a standard paper clip is 0.04 inch. If the wire is 4 inches long, use the formula for a right circular cylinder to determine the volume of the wire.

In exercises 52–55, determine whether the item is an equation or an expression.

52. $6x + 3 = 2(x - 5)$ **53.** $6x + 3 - 2(x - 5)$

54. $\dfrac{5}{2x - 1} + \dfrac{7}{x + 2}$ **55.** $2.4x - 1.8 = 4.9 - 1.7x$

56. What is the measure of the complementary angle of an angle that measures 85 degrees?

57. An angle measures 43 degrees. What is the measure of its supplementary angle?

58. Find the measure of the third angle of a triangle if the other two angles measure 31 degrees and 58 degrees.

CHAPTER 2 TEST

TEST-TAKING TIPS

When you take a test, start out by first reading the instructions. Many students don't do this and, after working on a test for a while, find that they have not done what was asked for on the test. It is also good advice not to do more than has been asked. When you go beyond the instructions, you waste valuable time. You also open yourself up to making more mistakes, and the grader may deduct points for erroneous results, even though they were not required.

1. Jessica invested $3500 into two interest-earning accounts. If she invested x dollars into the first account, write an algebraic expression for the amount she invested into the second account.

2. Write an algebraic expression for the sales tax on a purchase when 8% sales tax is added to the selling price of x dollars. Write an algebraic expression for the total cost of the purchase. What will the total cost be if the selling price is $15.00?

3. The length of a rectangle is 5 inches less than three times the width of the rectangle. Write an equation for this statement.

4. In statistics, a standard score, z, is calculated as the ratio of the difference between the raw score, x, and the mean, m, divided by the standard deviation, s. Write an equation for this calculation.

Evaluate each algebraic expression for the given value.

5. $\sqrt{4x^2 - 20x + 25} + 8$ for $x = 2$

6. $\dfrac{-18x + 54}{x - 8}$ for $x = -1$

Evaluate for $x = -6$.

7. x^2 **8.** $-x^2$ **9.** $(-x)^2$

Without simplifying, consider the algebraic expression

$$y^3 - 5y^2 + 15y - 3 + 7y^2 - 12 + 4y + 6y^3$$

10. How many terms are in the expression?   **11.** List the variable terms.   **12.** List the constant terms.

13. List the coefficients of the terms.   **14.** List the like terms.

Simplify exercises 15–18.

15. $\dfrac{2x}{3} + \dfrac{5y}{6} - \dfrac{8}{9} + \dfrac{x}{6} + \dfrac{7y}{9} + \dfrac{1}{3}$ **16.** $-(5p + 2q) - (-9p + q) + (p + q) - (-p - q)$

17. $\dfrac{25x - 45}{-5}$ **18.** $5[2(x + 3) - 4(2x + 1)]$

In exercises 19 and 20, determine whether the given value is a solution of the equation.

19. $-7x - 4 = 6x + 9$ for $x = -2$

20. $8x^2 + 40x + 45 = 2x^2 - 2x - 27$ for $x = -4$

21. Orhan builds a toolbox that is 4 feet long, 2 feet wide, and 1.5 feet high. What is the volume of the box? What is the outside surface area?

22. If Tracy makes a single deposit of $2000 into a savings account that earns simple interest at 5.5% annually, how much will the account contain in 40 years?

23. What does the supplementary angle of a 78-degree angle measure?

24. Find the area and perimeter of a rectangle with a length of 6 meters and a width of 2.5 meters.

25. What is the Fahrenheit temperature if the Celsius temperature is 25°C?

26. Explain the difference between the area and the perimeter of a two-dimensional figure.

Project

The formulas we use in geometry are the results of centuries of study by mathematicians and scientists. The development of a method for determining the value of π has a history as part of the "three famous problems" that the Greeks pondered. One of those problems was finding a square whose area is the same as that of a given circle. Early attempts were empirical—that is, experimental—in nature. In the Ahmes Papyrus (circa 1550 B.C.), the area of a circle with diameter d was approximated by a square with sides of length $(d - \frac{1}{9}d)$, so the area of the circle was calculated as $(d - \frac{1}{9}d)^2$. If this calculation is compared with our formula for the area of a circle, $A = \pi r^2$, it can be shown that the ancient formula is equivalent to a value of $\pi = 3.1605$. Other methods were also used.

Part I

We will conduct an empirical study similar to those used in ancient times to evaluate π. We know that the circumference and diameter of a circle are related by the formula $C = \pi d$, or $\pi = \frac{C}{d}$. In order to use this formula to estimate π, you will need a long piece of string, a meter stick, and three different circular objects.

1. Wrap a string around the top of the circular object and mark the circumference. Lay the string on a meter stick and measure the marked length (the circumference).

2. Place the string across the top of the circular object through its center, marking the distance, and measure the string (the diameter).

3. Determine the ratio of the circumference to the diameter or the quotient of the circumference divided by the diameter, rounded to six decimal places.

   Complete the following table with your data:

Object	Circumference	Diameter	Circumference ÷ diameter

Now determine the average of the three ratios.
The average ratio of the circumference to the diameter is _____.
The answers in the last column should be close to the same value, as well as to the average of these ratios. In fact, if you measured very accurately, the number should be approximately the value of π. Therefore, the real number π is defined as the ratio of the circumference of a circle to its diameter.

Part II

You can research the history of the development of methods for estimating the irrational number π in the library and on the Internet. Find a reference that discusses the history of π, and write a one-page paper summarizing your findings. Be sure to list your reference source.

Part III

At the beginning of this chapter, there is a photo of the Dome of the Rock. This beautiful structure has a circular dome in its center. Search the Internet to find other photos of the interior and exterior of the structure. Print one of the photos, and include it with your report on this structure. Find information on the measurements of the dome. If you are able to find the measurements, use them in the formulas for a circle, and apply the formulas to describe the size of the dome. Write a one-page summary of your findings on the beautiful Dome of the Rock.

Part IV

The real number π has many other applications in mathematics. Do research in the library or on the Internet to determine a different application or formula that involves π. Write a one-page summary describing the application that uses this mathematical constant. Cite your reference source in your paper.

In Chapter 2, we saw that giving a value to one or more variables in an algebraic expression or equation usually determines the value of other variables. For example, if we know the temperature in degrees Fahrenheit, we are able to determine the temperature in degrees Celsius. An equation or formula relates these two variables; therefore, changing the value of the variable for Fahrenheit temperature changes the value of the variable representing the Celsius temperature.

In this chapter, we will visualize this relationship in a different kind of way. We will use tables to help us organize our data. We will introduce the concept of a graph of an equation and show how it translates into a picture we can see. Finally, we will use the graphs we produce to understand the relationship between two variables in more depth.

Many important relationships between two variables are found in business and science. Tables and graphs help us visualize and analyze these relationships. We will end the chapter with a project in which you will collect the hourly temperature for one day and discuss the relationship between time of day and temperature.

Relations, Functions, and Graphs

3.1 APPLICATION

According to end-of-the-year reports for the year 2000, General Motors (GM) had the largest total revenue in the United States, about $185 billion. The following table gives a 10-year summary of revenue from GM sales:

Year	2000	1999	1998	1997	1996	1995	1994	1993	1992	1991
Sales (in billions of dollars)	185	177	161	166	158	164	151	134	129	120

Write a set of ordered pairs that represent these data.

We will discuss this application further. See page 151.

3.1 Tables of Values, Ordered Pairs, and Relations

1 Create tables of values for given data.

2 Write ordered pairs for given data.

3 Identify the domain and range of a relation.

4 Create tables of values for real-world data.

3.1.1 Creating Tables of Values

Sometimes we need to evaluate a formula or an equation more than once. It is often convenient to organize our information in a table of values. A **table of values** is a table with at least two columns. One column lists the values substituted for the variable. The second column lists the values obtained for the unknown variable when the formula is evaluated.

For example, suppose a biologist needs to convert Celsius temperatures into Fahrenheit temperatures. Instead of using the conversion formula each time, she decides to make a reference table of all temperatures she is using. A portion of this table will look like the following:

Celsius Temperature (°C)	Fahrenheit Temperature (°F)
20	68
21	69.8
22	71.6
23	73.4

To complete the next three entries in the table, we will extend it to include a middle column (to show our work). The first column is labeled with the independent variable. The **independent variable** is the variable for which we are substituting values. The second column is labeled with the formula or equation we are evaluating. The third column is labeled with the dependent variable. The **dependent variable** is the variable that is determined by the substitutions we make. We refer to the information in the table as **data**.

To construct a table of values, set up a three-column table.

- The first column is labeled with the independent variable.
- The second column is labeled with the formula or equation needed to find the unknown variable.
- The third column is labeled with the dependent variable.

Complete the table.

- Enter a number in the first column.
- Substitute the number from the first column into the formula or equation in the second column, and evaluate the results.
- Enter the results in the third column.

For example, complete the next three entries in the Celsius-to-Fahrenheit temperature table.

°C	$F = \dfrac{9}{5}C + 32$	°F
24	$F = \dfrac{9}{5}(24) + 32$	75.2
25	$F = \dfrac{9}{5}(25) + 32$	77
26	$F = \dfrac{9}{5}(26) + 32$	78.8

You can construct a table of values with the use of your calculator. The calculator has two modes, auto and ask, for constructing a table of values.

TECHNOLOGY

TABLE OF VALUES USING AUTO MODE

Set the calculator to automatically generate a table of values for $y = \frac{9}{5}x + 32$, given that $x = \{24, 25, 26\}$.

Figure 3.1a   Figure 3.1b   Figure 3.1c

For Figure 3.1a,
Set up the table.

| 2nd | | TBLSET |

Set a minimum value for the independent variable, x.

| 2 | | 4 | | ENTER | (minimum value 24)

Set the size of increments to be added to the independent variable.

| 1 | | ENTER | (increments of 1)

Set the calculator to perform the evaluations automatically.

| ENTER | | ▼ | | ENTER |

For Figure 3.1b,
Enter the formula in terms of x for the first y.

| Y= | | (| | 9 | | ÷ | | 5 | |) | | X,θ,n | | + | | 3 | | 2 |

For Figure 3.1c,
View the table.

| 2nd | | TABLE |

You may view additional entries in the table by using the up or down arrow keys.

TECHNOLOGY

TABLE OF VALUES USING ASK MODE

Set the calculator to generate a table of values, asking for x-values from the user and automatically performing the calculations, for $y = \frac{9}{5}x + 32$, given that $x = \{24, 25, 26\}$.

Figure 3.2a

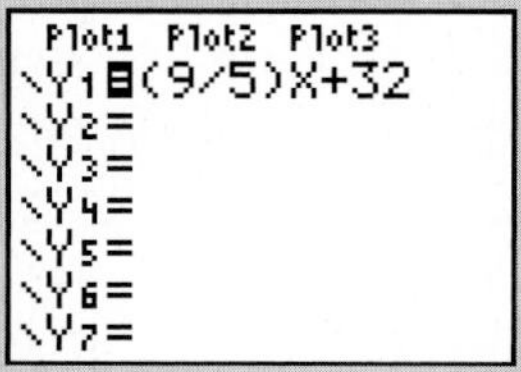

Figure 3.2b

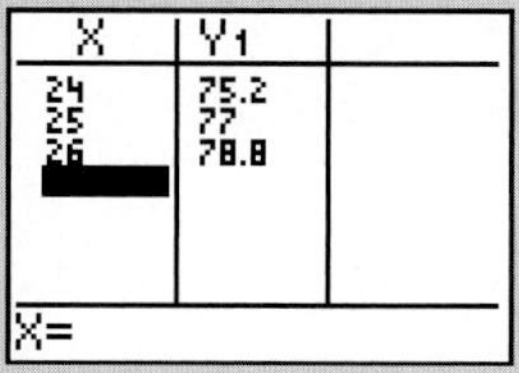

Figure 3.2c

For Figure 3.2a,
Set up the table.

| 2nd | | TBLSET |

Set the table to ask mode for the independent variable, x. (Ignore the first two entries.)

| ▼ | | ▼ | | ▶ | | ENTER | | ▼ | | ENTER |

For Figure 3.2b,
Enter the formula in terms of x for the first y.

| Y= | | (| | 9 | | ÷ | | 5 | |) | | X,T,θ,n | | + | | 3 | | 2 |

For Figure 3.2c,
View the table.

| 2nd | | TABLE |

Enter the values for x.

| 2 | | 4 | | ENTER | | 2 | | 5 | | ENTER | | 2 | | 6 | | ENTER |

EXAMPLE 1
a. Construct a table of values for the circumference of a circle, given the set of values {5, 10, 15, 20, 25} for the diameter, d.

b. Construct a table of values for $y = 3x$, given the set of values {0, 5, 7} for x.

Solution

Check

a.

d	$C = \pi d$	C
5	$C = \pi(5)$	15.708
10	$C = \pi(10)$	31.416
15	$C = \pi(15)$	47.124
20	$C = \pi(20)$	62.832
25	$C = \pi(25)$	78.540

X	Y₁
5	15.708
10	31.416
15	47.124
20	62.832
25	78.54

X=

Figure 3.3a

The calculator table in Figure 3.3a was generated in ask mode.

b.

x	$y = 3x$	y
0	$y = 3(0)$	0
5	$y = 3(5)$	15
7	$y = 3(7)$	21

X	Y₁
0	0
5	15
7	21

X=

Figure 3.3b

The calculator table in Figure 3.3b was generated in ask mode.

3.1.1 Checkup

In exercises 1–2, construct a table of values for each equation and check your results on your calculator.

1. Convert Fahrenheit temperatures to Celsius temperatures with the formula $C = \frac{5}{9}(F - 32)$, given the set of values {86, 77, 68, 59, 50, 41, 32} for F.

2. Construct a table of values of $a = -6b + 7$, given the set of values $\{-3, -\frac{1}{6}, 0, \frac{2}{3}, 2, 3, 5\}$ for b.

3. What is a table of values, and how can it help you when you work with an equation or a formula?

3.1.2 Writing Ordered Pairs

Another way to organize the data contained in an equation is to write ordered pairs. An **ordered pair** consists of two numbers in parentheses, separated by a comma. The first number in the ordered pair is the value of the independent variable. The second number in the ordered pair is the value of the dependent variable. The order of the numbers is very important.

For example, the previous temperature conversion table yields the following ordered pairs:

°C	°F	Independent Variable (C,	Dependent Variable F)
20	68	(20,	68)
21	69.8	(21,	69.8)
22	71.6	(22,	71.6)
23	73.4	(23,	73.4)

Ordered pairs are a common and useful way of organizing data for the purpose of graphing an equation, as we will see in section 3.2.

EXAMPLE 2 Write ordered pairs for the data found in Example 1.

Solution

a. $(5, 15.708), (10, 31.416), (15, 47.124), (20, 62.832), (25, 78.540)$
b. $(0, 0), (5, 15), (7, 21)$

3.1.2 Checkup

1. Write ordered pairs for the data found in exercises 1 and 2 of 3.1.1 Checkup.
2. In writing an ordered pair for a set of data, why is the order of the numbers important?

3.1.3 Identifying the Domain and Range of a Relation

A set of ordered pairs is a **relation**. For example,
The set of ordered pairs

$$T = \{(20, 68), (21, 69.8), (22, 71.6), (23, 73.4)\}$$

that the biologist used for temperature conversion is a relation.

The list of ordered pairs in a relation can be finite or infinite. A **finite** list has a definite number of ordered pairs. An **infinite** list does not have a definite number of ordered pairs.

$$A = \{(1, 1), (1, 2), (1, 3)\} \qquad \text{Finite}$$
$$B = \{\dots, (-2, -1), (-1, 0), (0, 1), (1, 2), (2, 3), \dots\} \qquad \text{Infinite}$$

A relation can also be written as an equation with two variables. We can use the equation to determine the ordered pairs by substituting a value for the independent variable and determining a value for the dependent variable.

$y = 2x$ for $x = \{1, 2, 3\}$ is a relation of a finite set of ordered pairs, $\{(1, 2), (2, 4), (3, 6)\}$.

HELPING HAND A table of values will help us determine the ordered pairs when an equation is given.

The set of all possible values for the independent variable is called the **domain** of the relation. The set of all possible values for the dependent variable is called the **range** of the relation. For example, using the previous relations, we have

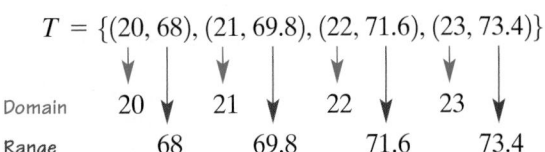

$$T = \{(20, 68), (21, 69.8), (22, 71.6), (23, 73.4)\}$$

Domain   20   21   22   23

Range   68   69.8   71.6   73.4

The domain is $\{20, 21, 22, 23\}$.
The range is $\{68, 69.8, 71.6, 73.4\}$.

$A = \{(1, 1), (1, 2), (1, 3)\}$
The domain is $\{1\}$.
The range is $\{1, 2, 3\}$.

$B = \{\ldots, (-2, -1), (-1, 0), (0, 1), (1, 2), (2, 3), \ldots\}$
The domain is $\{\ldots, -2, -1, 0, 1, 2, \ldots\}$, or all integers.
The range is $\{\ldots, -1, 0, 1, 2, 3, \ldots\}$, or all integers.

$y = 2x$ for $x = \{1, 2, 3\}$
The domain is $\{1, 2, 3\}$.
The range is $\{2, 4, 6\}$.

EXAMPLE 3   Determine the domain and range for the following relations, assuming that x is the independent variable and y is the dependent variable:
a. $\{(1, 3), (2, 5), (6, 5)\}$ **b.** $\{(1, 2), (2, 2), (3, 2), \ldots\}$
c. $y = 3x + 2$ for $x = \{-1, 0, 1\}$

Solution

a. domain $\{1, 2, 6\}$
      range $\{3, 5\}$ There is no need to write more than one 5.

b. domain $\{1, 2, 3, \ldots\}$, or all counting numbers
      range $\{2\}$

c. A table of values for $y = 3x + 2$ will help us determine the ordered pairs.

x	$y = 3x + 2$	y	
-1	$y = 3(-1) + 2$	-1	$(-1, -1)$
0	$y = 3(0) + 2$	2	$(0, 2)$
1	$y = 3(1) + 2$	5	$(1, 5)$

    domain $\{-1, 0, 1\}$
    range $\{-1, 2, 5\}$

 3.1.3 Checkup

In exercises 1–3, determine the domain and range for each relation. Assume that x is the independent variable.
  1. $\{(5, 15), (10, 30), (15, 45)\}$ **2.** $\{(0, 4), (1, 6), (2, 8), (3, 10), \ldots\}$ **3.** $y = x + 15$ for $x = \{0.5, 1.5, 2.5\}$

4. Explain the terms *domain* and *range* of a relation.
5. What is the difference between a finite list of ordered pairs and an infinite list of ordered pairs?

3.1.4 Modeling the Real World

Real-life situations can often be modeled as relations. Some relations are more convenient to describe by tables of values, other relations can best be described by sets of ordered pairs, and still others are most appropriately described by equations. You need to be familiar with all these forms of relations in order to describe a situation in the most useful manner.

EXAMPLE 4 The daily charge for renting a 24-foot U-Haul truck is $39.95, plus $0.49 per mile driven, with a $5.00 nonrefundable deposit.

a. Write an equation for the cost of renting a 24-foot U-Haul truck for one day.

b. Complete a table of values for the cost of renting a 24-foot U-Haul truck for one day for 100 miles, 200 miles, and 300 miles.

Solution

a. Let m = the number of miles driven and c = the cost of a one-day rental.

$$c = 39.95 + 0.49m + 5.00$$
$$c = 44.95 + 0.49m$$

b. Add a middle column to the table to show your work.

Number of Miles m	$c = 44.95 + 0.49m$	Cost of a One-Day Rental c
100	$c = 44.95 + 0.49(100)$	93.95
200	$c = 44.95 + 0.49(200)$	142.95
300	$c = 44.95 + 0.49(300)$	191.95

EXAMPLE 5 Latoya plans to borrow $2500 from her grandmother for her tuition payment. Her grandmother will charge her simple interest of 7% per year. Determine the amount of interest Latoya will need to repay if she repays the loan after one, two, three, or four years.

Solution

Use the simple-interest formula, $I = PRT$, where I = simple interest, P = principal or amount borrowed, R = rate of loan per year, and T = time in years. Substitute the given values for the principal, $2500, and rate, 7% = 0.07, to obtain an equation.

$$I = PRT$$
$$I = 2500(0.07)T$$

Since Latoya needs to know several amounts, a table will help us organize our information. The independent variable is the time in years, and the dependent variable is the amount of interest.

T	I = 2500(0.07) T	I
1	I = 2500(0.07)(1)	175
2	I = 2500(0.07)(2)	350
3	I = 2500(0.07)(3)	525
4	I = 2500(0.07)(4)	700

Latoya will need to repay $175 in interest for one year, $350 for two years, $525 for three years, and $700 for four years.

APPLICATION

According to end-of-the-year reports for the year 2000, General Motors (GM) had the largest total revenue in the United States, about $185 billion. The following table gives a 10-year summary of revenue from GM sales:

Year	2000	1999	1998	1997	1996	1995	1994	1993	1992	1991
Sales (in billions of dollars)	185	177	161	166	158	164	151	134	129	120

Write a set of ordered pairs that represent these data.

Discussion

Let x = year and y = revenue from sales in billions of dollars. Then we have

(1991, 120), (1992, 129), (1993, 134), (1994, 151), (1995, 164),
(1996, 158), (1997, 166), (1998, 161), (1999, 177), (2000, 185)

A second method that will enable us to use smaller numbers is to let x = number of years after 1991 and y = revenue from sales in billions of dollars. Then $x = 0$ corresponds to 1991, $x = 1$ corresponds to 1992, $x = 2$ corresponds to 1993, and so on, and we have

(0, 120), (1, 129), (2, 134), (3, 151), (4, 164), (5, 158), (6, 166), (7, 161), (8, 177), (9, 185)

3.1.4 Checkup

1. Sven attended a teachers' conference. His school reimbursed him $410 for plane fare, $95 per day for lodging, and $36 per day for meals.
   a. Write an equation for the total reimbursement, r, in terms of the number of days, n, for the trip.
   b. Construct a table showing the possible reimbursements for a trip that lasts from two through five days.
   c. List the ordered pairs found in the table for part b.

2. The volume of a videotape cabinet is calculated with the formula $V = LWH$. The cabinet has a width $W = 12$ inches and a length $L = 23$ inches. Determine the volume of the cabinet if the height, H, is 29.5, 38.5, or 47.5 inches.

3. The U.S. Patent and Trademark Office reported the following summary of number of patents issued (in thousands—for example, 99.2 represents 99,200) for each year between 1990 and 1997:

Year	1990	1991	1992	1993	1994	1995	1996	1997
Patents Issued	99.2	106.8	107.4	109.7	113.6	113.8	121.7	124.1

Write a set of ordered pairs to represent these data.

3.1 Exercises

Express your answers in fractional form or round decimals to the nearest thousandth, as appropriate. Construct a table of values, given each set of values for x.

1. $y = 5x + 4$, given $\{-2, -1, 0, 1, 2, 3\}$ for x.

2. $y = -8x - 6$, given $\{-3, -2, -1, 0, 1\}$ for x.

3. $y = \frac{3}{5}x - 2$, given $\{-15, -10, -5, 0, 5, 10, 15\}$ for x.

4. $y = \frac{7}{9}x + 3$, given $\{-18, -9, 0, 9, 18, 27, 36\}$ for x.

5. $y = 2.3x + 1.6$, given $\{-2, -1, 0, 1, 2\}$ for x.

6. $y = -4.8x - 9.2$, given $\{-3, -1, 0, 1, 3\}$ for x.

7. $y = \frac{1}{3}(x + 7)$, given $\{-1, -4, -7, -10, -13\}$ for x.

8. $y = \frac{1}{6}(x - 2)$, given $\{-10, -4, 2, 8, 14\}$ for x.

Write a table of values for each equation. Select five values for the independent variable in each table.

9. $y = 6x - 8$

10. $y = -11x + 15$

11. $y = \frac{2}{7}x - 2$

12. $y = -\frac{3}{8}x + 5$

13. $y = -4.6x + 2.1$

14. $y = 10.6x - 0.8$

15. $y = \frac{1}{4}(3x - 2)$

16. $y = \frac{3}{8}(x - 5)$

In exercises 17–28, write a table of values for each equation with the given domain.

17. $y = 12x - 13$, where x is an even integer between -5 and 5.

18. $y = -9x + 12$, where x is an odd integer between -4 and 4.

19. $z = \frac{1}{3}y + 5$, where y is an integer multiple of 3 between -7 and 7.

20. $p = \frac{7}{8}q - 4$, where q is an integer multiple of 8 between -17 and 17.

21. $a = 14.2b + 5.7$, for a domain of integer values between -3 and 3.

22. $m = 1.9n - 3.7$, for a domain of integer values between -4 and 2.

23. $y = 2x^2 + 3x + 1$, with a domain of $\{-3, -2, -1, 0, 1, 2, 3\}$.

24. $y = -3x^2 + 11x - 10$, with a domain of $\{-2, -1, 0, 1, 2, 3, 4\}$.

25. $y = (2x - 3)(3x + 4)$, for integer values of x between -3 and 3.

26. $y = (4x + 1)(-x + 2)$, for integer values of x between 4 and 10.

27. $y = \frac{3x + 7}{x - 1}$, for odd integer values of x between -4 and 4.

28. $y = \frac{-5x + 2}{-x + 3}$, for odd integer values of x between -2 and 6.

29. Chameeka took a vacation by automobile. She drove at an average speed of 55 miles per hour each day.
   a. Write an equation for the total number of miles she traveled.
   b. Complete a table of values for the total number of miles Chameeka traveled if she drove for 4 hours, 8 hours, 12 hours, 16 hours, and 20 hours.

30. During an auction, the successful bid for place settings of china was $15.50 for each four-piece setting.
   a. Write an equation for the total cost of purchasing a given number of four-piece place settings at the bid price.

b. Complete a table of values for the total cost of purchasing place settings if the number of four-piece settings purchased was 2, 4, 6, 8, 10, or 12.

Write tables of values for these formulas with their domains.

31. The volume of a box whose width is 2 feet, whose length is 4 feet, and whose height takes on the values {1, 3, 5, 7, 9} feet.

32. The volume of a cube whose edge measures {1, 2, 3, 4, 5, 6} feet.

33. The amount of interest earned on an investment of $5000 invested at 4.5% simple interest for a time *t*, where *t* starts at 1 year and is repeatedly increased by 1 year until it reaches 12 years.

34. The amount of interest on a loan of $3000 at a rate of simple interest of 7% for a time *t*, where *t* starts at 1 year and is incremented by 2 years until it reaches 15 years.

35. Ohm's law in electricity is $I = V \div R$, where *I* is current (measured in amperes), *V* is voltage (measured in volts), and *R* is resistance (measured in ohms). Find the current when the voltage is 9 volts and the resistance assumes values from 1 ohm to 9 ohms in increments of 1 ohm.

36. Ohm's law in electricity may also be stated as $V = I \cdot R$, where *V* is voltage (measured in volts), *I* is current (measured in amperes), and *R* is resistance (measured in ohms). Find the voltage when the current is 5 amps and the resistance varies as a natural number between 1 and 10 ohms, inclusive.

Write ordered pairs.

37. $y = 1.2x + 4$ when $x = -2, -1, 0, 1, 2$.

38. $y = 6 - 0.3x$ when $x = -4, -2, 0, 2, 4$.

39. $q = 1 - p$ when $p = \frac{1}{6}, \frac{1}{5}, \frac{1}{4}, \frac{1}{3}, \frac{1}{2}$.

40. $S = \frac{n(n+1)}{2}$ when $n = 1, 2, 3, 4, 5$.

41. The perimeter of a square when the side measures 2, 4, 6, 8, 10 inches.

42. The change received when paying with a $10.00 bill for a purchase of $4.00, $6.00, $8.00, $9.50.

43. The profits after taxes for all U.S. manufacturing corporations (in billions of dollars), according to the U.S. Bureau of the Census, is presented in the following table for the years 1990 through 1998:

Year	1990	1991	1992	1993	1994	1995	1996	1997	1998
Profits after tax	110	66	22	83	175	198	225	244	238

Write ordered pairs for the data.

44. The average monthly bill (in dollars) for cellular phones, according to the Cellular Telecommunications Industry Association, is presented in the following table:

Year	1990	1992	1993	1994	1995	1996	1997	1998
Average bill	80.90	68.68	61.48	56.21	51.00	47.70	42.78	39.43

Write ordered pairs for the data.

In exercises 45–60, determine the domain and range for each relation. Assume that x is the independent variable.

45. $R = \{(3, 15.8), (5, 17.8), (7, 19.8), (9, 21.8)\}$

46. $U = \{(11, 8), (15, 12), (19, 16), (23, 20)\}$

47. $S = \{(4, -3), (4, -1), (4, 1), (4, 3), \ldots\}$

48. $H = \{(3, -1), (2, -1), (1, -1), (0, -1), (-1, -1), \ldots\}$

49. $T = \{\ldots, (2, -2), (2, -1), (2, 0), (2, 1), (2, 2), \ldots\}$

50. $I = \{\ldots, (-2, 0), (-1, 0), (0, 0), (1, 0), (2, 0), \ldots\}$

51. $y = 4x - 5$ for $x = \{2, 4, 6\}$

52. $y = 3x + 5$ for $x = \{10, 20, 30\}$

53. $y = 6 - x$, where *x* assumes integer values between −1 and 7.

54. $y = -x - 8$, where *x* assumes integer values between −5 and 5.

55. $y = x^2 + 1$, where *x* assumes the values {0, 0.5, 1, 1.5, 2, 2.5, 3, 3.5, 4}

56. $y = x^2 - 2$, where *x* assumes the values {0, 0.5, 1, 1.5, 2, 2.5, 3, 3.5, 4}

57. $y = \sqrt{x - 2}$, for $x = \{2, 3, 6, 11, 18, 27\}$

58. $y = \sqrt{x + 2}$, for $x = \{-2, -1, 2, 7, 14, 23\}$

59. $y = \frac{6}{x - 2}$, for $x = \{-3, -1, 1, 3, 5, 7\}$

60. $y = \frac{10}{4 - x}$, for $x = \{-3, -1, 1, 3, 5, 7\}$

61. A diver jumped off a 50-foot cliff. His position, s feet above water, t seconds after the jump is given by the equation $s = -16t^2 + 50$. Construct a table of values of the diver's position above the water, where t assumes the values 0, 0.5, 1, 1.5 seconds.

62. A toy rocket is shot upward from the ground with an initial velocity of 225 feet per second. The position of the rocket, s feet above the ground, t seconds after blastoff is given by the equation $s = -16t^2 + 225t$. Construct a table of values of the position of the rocket, where t assumes integer values between 0 and 14, inclusive.

63. Rebecca is traveling on the interstate highway at a constant speed of 65 miles per hour. Write ordered pairs for her distance traveled at the end of each hour for a four-hour trip.

64. David is driving his truck between two turnpike entrances that are 195 miles apart. Write ordered pairs for his speed to the nearest mile per hour if he completes the trip in 2.5, 2.75, 3, 3.25, or 3.5 hours. (The Highway Patrol could use this information to determine whether David exceeded the speed limit on his trip.)

65. An executive lease for an efficiency apartment requires a nonrefundable deposit of $225 and a weekly charge of $175.
  a. Write an equation for the cost of renting the apartment.
  b. Complete a table of values for the cost of renting the apartment for 1, 2, 3, or 4 weeks.

66. A contract for renting a reception hall for a wedding requires a nonrefundable deposit of $300 plus an hourly charge of $75.
  a. Write an equation for the cost of renting the reception hall.
  b. Complete a table of values for the cost of renting the hall for 2, 2.5, 3, 3.5, or 4 hours.

3.1 Calculator Exercises

When you set up a table on a calculator, you have the option of setting the independent variable, x, to Auto or to Ask. In this section, you have seen that if you set x to Auto and specify the beginning value and the increment value, the calculator will automatically generate a table. You have also seen that if you set x to Ask, you can input whatever value you want for x, after which the calculator will display the corresponding value for y. These two settings are shown in the following diagrams:

What happens when the dependent variable, y, is set to Ask? The two settings are shown in these diagrams:

```
TABLE SETUP
 TblStart=0
 ⌂Tbl=1
Indpnt: Auto Ask
Depend: Auto Ask
```

```
TABLE SETUP
 TblStart=0
 ⌂Tbl=1
Indpnt: Auto Ask
Depend: Auto Ask
```

In this case, the calculator table will not display a value of y until you select the cell of the table for which you want the value calculated. To do so, you must move the cursor to the column for y, select the cell that corresponds to the value of x displayed, and press ENTER. The calculator will display a corresponding value of y for this value of x. If you have both options set to Ask, then you must first enter the values you want to use for x, then move to the column for y, and, finally, select the cell that you want evaluated.

In this chapter, we see no use for this feature, but it is discussed here since many students ask about it out of curiosity. Just remember that we will use the Ask option only for the independent variable, x, but we will always set the dependent variable, y, to the Auto option.

3.1 Writing Exercise

In this section, you learned several concepts dealing with relations between two variables. Write a brief explanation of each of the following concepts:

1. Describe the various ways in which to define the relation between two variables.

2. Explain dependent and independent variables.

3. Explain the domain and the range of a relation.

3.2 Rectangular Coordinate System and Graphing

OBJECTIVES

1 Construct a coordinate plane.

2 Graph relations.

3 Identify the domain and range of a relation from its graph.

4 Interpret graphs of real-world data.

APPLICATION

The following graph illustrates the revenue from sales, cost of sales, and gross operating profit for General Motors Corporation in the years 1996 through 1999:

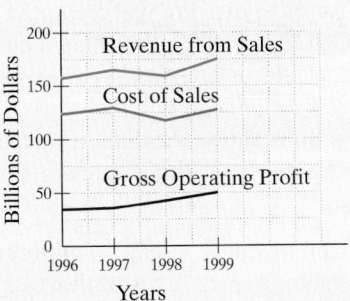

Approximate the domain and range of the revenue, cost, and profit.

After completing this section, we will discuss this application further. See page 169.

3.2.1 Constructing a Rectangular Coordinate System

We have seen how to organize data in a table of values and in sets of ordered pairs. A third way to organize data is to use a two-dimensional graph. In Chapter 1, we graphed numbers on a real-number line. Now we want to graph ordered pairs, consisting of two numbers. We use a **rectangular coordinate system**, or **Cartesian coordinate system**. The word *Cartesian* comes from the name of the great French philosopher and mathematician René Descartes (1596–1650). Sometimes, we simply refer to the system as a **coordinate plane**. A coordinate system combines two real-number lines, perpendicular to each other and intersecting at 0 on each line. Remember, perpendicular lines are lines that intersect at right angles (90°).

The horizontal number line, or horizontal axis, is often called the **x-axis**. The vertical number line, or vertical axis, is often called the **y-axis**. These two lines intersect at a point called the **origin**. We place an arrow at the end of each of the axes to indicate its positive direction.

We can draw lines perpendicular to the axes through the locations of all the integers on the two number lines. This network of lines forms a rectangular grid, which may be divided by the x-axis and y-axis into four regions called **quadrants**. The quadrants are labeled with Roman numerals in a counterclockwise direction, beginning with the upper right-hand region.

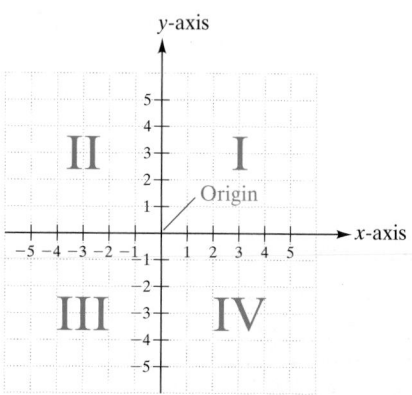

We can set up a coordinate plane on the calculator. The calculator has several choices of screens built in for us to use. We will call these screens *default* screens. We can also set the screen to a setting other than a default screen. At times, we will need to change calculator screens in order to view a picture more easily.

The Technology section that follows shows the default calculator coordinate planes. Below each screen in parentheses is the window setting. It is written in the form

(*x* minimum value, *x* maximum value, *x* scale, *y* minimum value,
y maximum value, *y* scale, *x* resolution)

A shorter version may not include the *x* scale, *y* scale, or *x* resolution:

(*x* minimum value, *x* maximum value, *y* minimum value, *y* maximum value)

TECHNOLOGY

DEFAULT GRAPH SCREENS

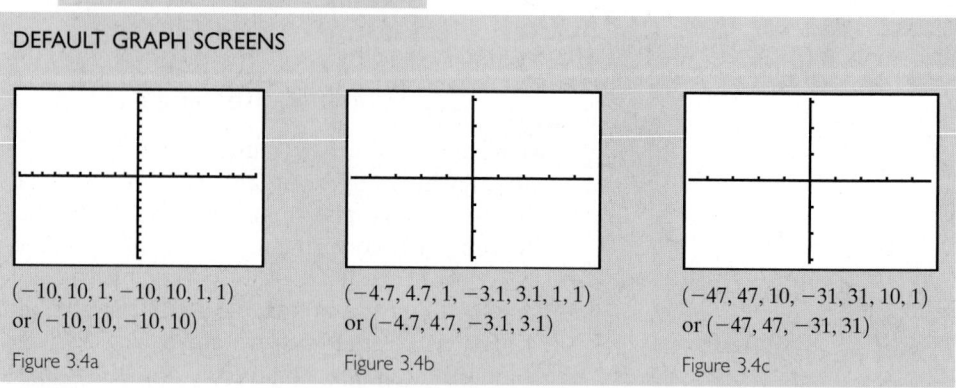

(−10, 10, 1, −10, 10, 1, 1)
or (−10, 10, −10, 10)

Figure 3.4a

(−4.7, 4.7, 1, −3.1, 3.1, 1, 1)
or (−4.7, 4.7, −3.1, 3.1)

Figure 3.4b

(−47, 47, 10, −31, 31, 10, 1)
or (−47, 47, −31, 31)

Figure 3.4c

For Figure 3.4a,

For a standard screen, choose the Zstandard screen. Enter | ZOOM | | 6 | on your calculator.

For Figure 3.4b,

For a decimal screen, choose the Zdecimal screen. Enter | ZOOM | | 4 | on your calculator.

For Figure 3.4c,

For an integer screen (centered), choose the Zstandard screen and, then choose Zinteger screen. Enter | ZOOM | | 6 | | ZOOM | | 8 | | ENTER | on your calculator. (Choosing the Zstandard screen first centers the origin on the screen.)

Note: To view the screen settings, enter | WINDOW |.

The default screens may not give us a good picture. We can change the default calculator settings to values of our choice.

For example, set the screen to the following setting, as shown in Figure 3.5b:

$$(-20, 20, 10, -100, 100, 10, 1)$$

TECHNOLOGY

SETTING GRAPH SCREENS

Set the calculator graph screen to $(-20, 20, 10, -100, 100, 10, 1)$.

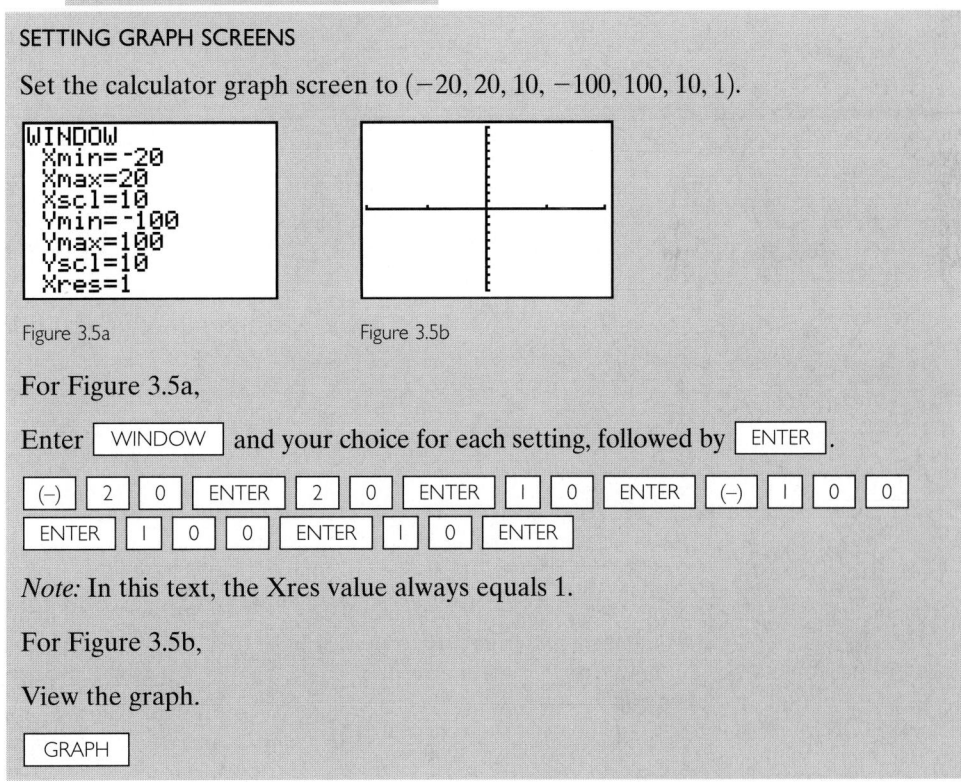

Figure 3.5a

Figure 3.5b

For Figure 3.5a,

Enter | WINDOW | and your choice for each setting, followed by | ENTER |.

| (−) | | 2 | | 0 | | ENTER | | 2 | | 0 | | ENTER | | 1 | | 0 | | ENTER | | (−) | | 1 | | 0 | | 0 |
| ENTER | | 1 | | 0 | | 0 | | ENTER | | 1 | | 0 | | ENTER |

Note: In this text, the Xres value always equals 1.

For Figure 3.5b,

View the graph.

| GRAPH |

EXAMPLE 1 **a.** Construct a coordinate plane on graph paper, and label the *x*-axis from −4 to 4 and the *y*-axis from −3 to 3, both in increments of 1. This plane is similar to the decimal screen on your calculator. On your calculator, set up a default decimal screen.

b. Construct a coordinate plane on graph paper. Label the *x*-axis from −140 to 140 in increments of 30. Label the *y*-axis from −120 to 120 in increments of 40. On your calculator, set up the same window.

Solution

a.

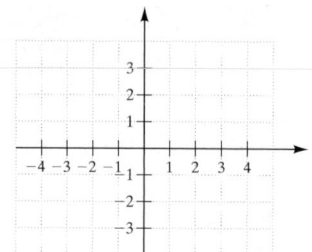

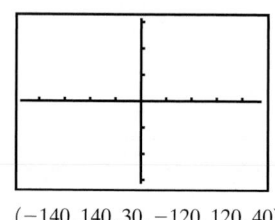

$(-4.7, 4.7, -3.1, 3.1)$

b.

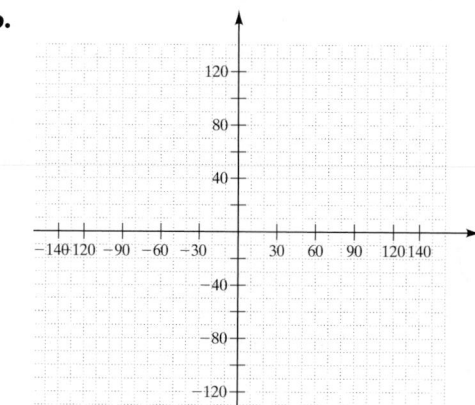

$(-140, 140, 30, -120, 120, 40)$

 HELPING HAND Be very careful to number lines, not the spaces between the lines.

3.2.1 Checkup

1. Construct a coordinate system on graph paper, and label the *x*-axis from −10 to 10 and the *y*-axis from −10 to 10, both in increments of 1 unit. To what setting on the ZOOM button of your calculator does this scale compare? On your calculator, set up the screen and check the window format to see that you have selected your scales correctly.

2. Construct a coordinate system on graph paper, and label the *x*-axis from −50 to 50 and the *y*-axis from −30 to 30, both in increments of 10 units. To what setting on the ZOOM button of your calculator does this scale compare? On your calculator, set up the screen and check the

window format to see that you have selected your scales correctly.

3. Define each of the following terms:
   a. coordinate plane
   b. axes
   c. origin
   d. quadrants

4. List the default choices on your calculator for graphing within the Cartesian coordinate system and explain what each setting does for graphing.

3.2.2 Graphing Relations

Locations in the coordinate plane are written as ordered pairs. The numbers in an ordered pair are called the **coordinates** of the point at that location. Each coordinate corresponds to the distance of the point from the *x*-axis or *y*-axis. The first number in the ordered pair is often called the **x-coordinate**. This corresponds to the distance of the point from the *y*-axis, which we measure along the *x*-axis. Similarly, the second number in the ordered pair is often called the **y-coordinate**, which corresponds to the distance of the point from the *x*-axis, as measured along the *y*-axis.

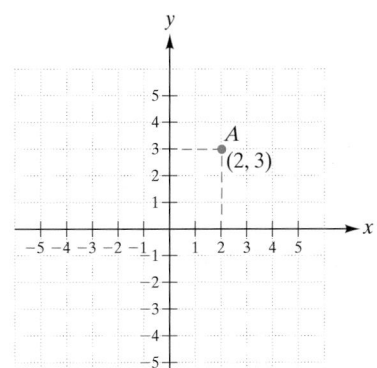

> To graph an ordered pair (or plot a point), we place a dot at its location on the coordinate plane.
>
> - First, locate the *x*-coordinate on the *x*-axis.
> - Second, locate the *y*-coordinate on the *y*-axis.
> - Place a dot at the intersection of the two lines that are perpendicular to the axes and that go through these locations on the axes.
> - Label the point with its name or coordinates.

For example, to graph the ordered pair $A(2, 3)$, locate the *x*-coordinate, 2, on the *x*-axis and the *y*-coordinate, 3, on the *y*-axis. The intersection of the two lines through these locations is the location of the dot for the ordered pair $(2, 3)$. Label the point by its name, *A*.

EXAMPLE 2 Graph and label each set of ordered pairs on a coordinate plane. Note the difference in the locations of the pairs of points.

a. $A(1, 3)$ and $B(3, 1)$
b. $C(-2, 3)$ and $D(3, -2)$
c. $E(2, 0)$ and $F(0, 2)$

Solution

a.

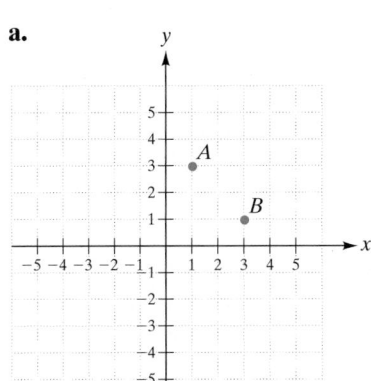

b.

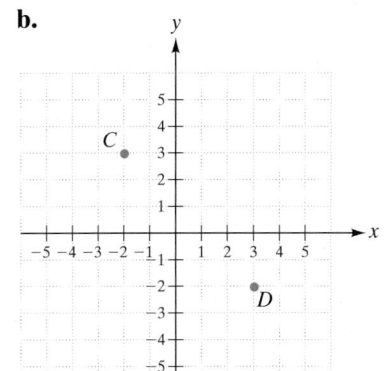

c.

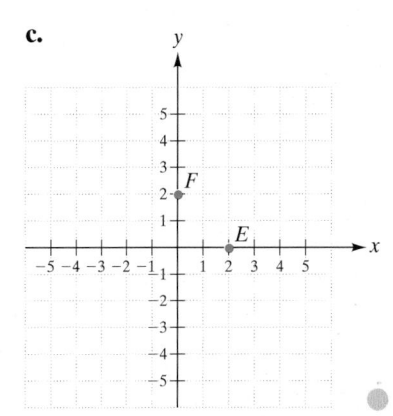

EXAMPLE 3 Using the given graph, state the coordinates of the points shown.

Solution

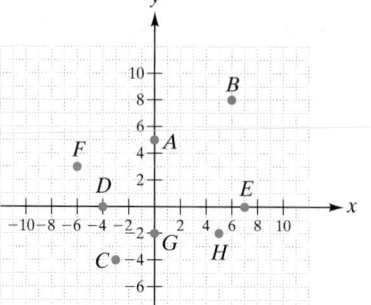

$A(0,5)$ $B(6,8)$ $C(-3,-4)$ $D(-4,0)$
$E(7,0)$ $F(-6,3)$ $G(0,-2)$ $H(5,-2)$

At times it may be necessary to locate a point in the coordinate plane approximately, without actually plotting it. In such cases, it is useful to know how the coordinates of points vary from one quadrant to another and what coordinates correspond to points located on the axes. Complete the following sets of exercises on your calculator.

 Discovery 1

Signs of the Coordinates in Each Quadrant

Set your calculator to the integer screen setting. Use your arrow keys to locate five points in each of the four quadrants. Note that the coordinates of the points are displayed on the screen. Write the ordered pairs.

Quadrant I	Quadrant II	Quadrant III	Quadrant IV
(___,___)	(___,___)	(___,___)	(___,___)
(___,___)	(___,___)	(___,___)	(___,___)
(___,___)	(___,___)	(___,___)	(___,___)
(___,___)	(___,___)	(___,___)	(___,___)
(___,___)	(___,___)	(___,___)	(___,___)

Write a rule for a condition on the signs of the coordinates in each quadrant.

In quadrant I, both coordinates are positive; in quadrant II, the x-coordinate is negative and the y-coordinate is positive; in quadrant III, both coordinates are negative; and in quadrant IV, the x-coordinate is positive and the y-coordinate is negative.

Discovery 2

Location of the Zero Coordinate on the Axes

Set your calculator to the integer screen setting. Use your arrow keys to locate five points on each of the axes. Write the ordered pairs.

x-Axis	y-Axis
(___,___)	(___,___)
(___,___)	(___,___)
(___,___)	(___,___)

(___,___)   (___,___)
(___,___)   (___,___)

Write a rule for a condition on the numbers in an ordered pair for any point on the x-axis or the y-axis.

On the x-axis, the y-coordinate is always 0, and on the y-axis, the x-coordinate is always 0.

SIGNS OF COORDINATES IN THE PLANE
In quadrant I, the x-coordinate is always positive and the y-coordinate is always positive. $(+, +)$
In quadrant II, the x-coordinate is always negative and the y-coordinate is always positive. $(-, +)$
In quadrant III, the x-coordinate is always negative and the y-coordinate is always negative. $(-, -)$
In quadrant IV, the x-coordinate is always positive and the y-coordinate is always negative. $(+, -)$
On the x-axis, the y-coordinate is always 0. $(x, 0)$
On the y-axis, the x-coordinate is always 0. $(0, y)$

EXAMPLE 4 Determine the location of the following points by quadrant or by axis:

a. $(24, 39)$ **b.** $(-24, 39)$ **c.** $(24, -39)$
d. $(-24, -39)$ **e.** $(0, 39)$ **f.** $(-24, 0)$

Solution

a. quadrant I   Coordinates are $(+, +)$.
b. quadrant II   Coordinates are $(-, +)$.
c. quadrant IV   Coordinates are $(+, -)$.
d. quadrant III   Coordinates are $(-, -)$.
e. y-axis   The x-coordinate is 0.
f. x-axis   The y-coordinate is 0.

Now we know how to locate a point, represented by an ordered pair, on the coordinate plane. Our next step is to graph a set of ordered pairs—that is, a relation—on the coordinate plane. In fact, since we have shown how to represent a relation by a set of ordered pairs, a table of values, or an equation, we can graph any of these forms in the same fashion.

To graph a relation in list form,

• Plot each ordered pair in the relation.

To graph a relation in table form,

• Write a set of ordered pairs and plot each ordered pair.

EXAMPLE 5 Graph each relation.

a. $A = \{(1, 1), (1, 2), (1, 3)\}$

b.

x	y
-10	-20
-5	-10
0	0
5	10
10	20

Solution

a. Plot each ordered pair.

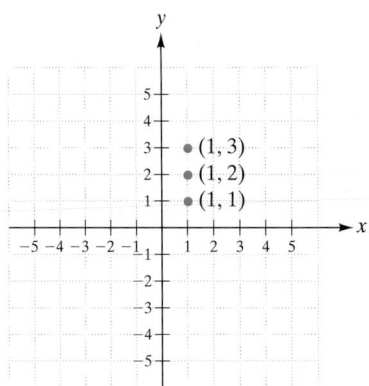

b. Plot the set of ordered pairs from the table.

$$\{(-10, -20), (-5, -10), (0, 0), (5, 10), (10, 20)\}$$

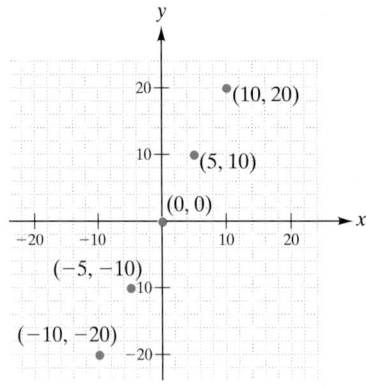

To graph a relation in equation form if the domain is given,

- Set up a table of values for the given domain.
- Plot the set of ordered pairs that the table determines.

To graph a relation in equation form if the domain is not given,

- Choose values for the independent variable that will result in a real-number value for the dependent variable. This will determine a sample of the possible ordered pairs.
- Plot the set of ordered pairs that the equation determines.

EXAMPLE 6 Graph each relation.

　a. $y = 2x$ for $x = \{1, 2, 3\}$　　**b.** $y = 2x + 3$

Solution

a. In order to graph the relation $y = 2x$ for $x = \{1, 2, 3\}$, we set up a table of values and graph the ordered pairs.

x	$y = 2x$	y
1	$y = 2(1)$	2
2	$y = 2(2)$	4
3	$y = 2(3)$	6

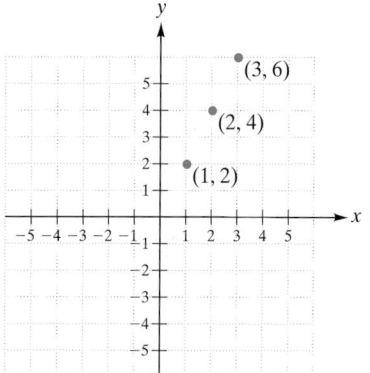

b. If a relation is represented by an equation such as $y = 2x + 3$ and the values for x are not given, we find a set of sample ordered pairs. We choose values for x in the domain of the relation. For example, we can choose $x = 1$, 1.25, $1\frac{1}{2}$, and 2 and set up a table of values.

x	$y = 2x + 3$	y
1	$y = 2(1) + 3$	5
1.25	$y = 2(1.25) + 3$	5.5
$1\frac{1}{2}$	$y = 2\left(1\frac{1}{2}\right) + 3$	6
2	$y = 2(2) + 3$	7

When we graph the sample ordered pairs, we see a straight line being formed. In fact, if we could graph all the ordered pairs that satisfy this relation, a solid line would be formed. Therefore, to complete the graph of the relation, draw a line through the sample points. An arrow indicates that the pattern continues.

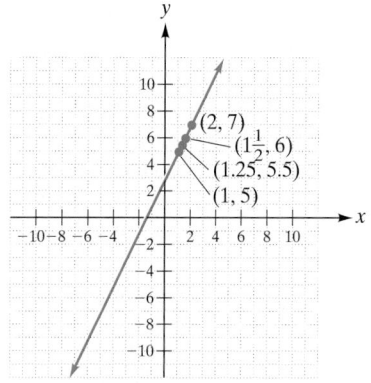

Our calculator is designed to graph relations such as $y = 2x + 3$. ●

T E C H N O L O G Y

GRAPH A RELATION

Graph the relation $y = 2x + 3$.

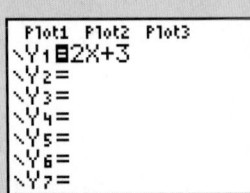

Figure 3.6a

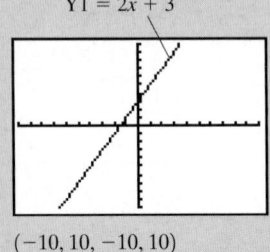

$(-10, 10, -10, 10)$

Figure 3.6b

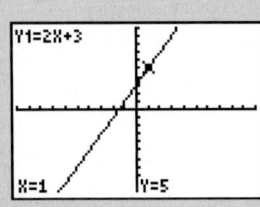

$(-10, 10, -10, 10)$

Figure 3.6c

For Figure 3.6a,

Enter the equation into the calculator in the $Y =$ menu.

| Y = | 2 | X,T,θ,n | + | 3 |

For Figure 3.6b,

Set the calculator to the desired screen setting and graph.
We will use the default standard screen.

| ZOOM | 6 |

For Figure 3.6c,

To view the points on the graph, we trace the graph and use the left and right arrow keys to move along the graph. To see the coordinates of a point that is not traced, enter the value of the independent variable, and then press ENTER . For example, we graphed the point $(1, 5)$ in Example 6b.

| TRACE | 1 | ENTER |

Not all relations will graph as lines. Some will graph as smooth curves. Viewing a graph on the calculator will help us determine the number of ordered pairs that are needed to draw curves for these relations. There is no magic number, so we will have to judge each relation individually. In this chapter, you will be given a window setting to use in your calculator. However, you should become familiar with the shape of the curve that certain equations graph, because in later chapters you will need to choose a window setting for yourself.

EXAMPLE 7 Graph $y = x^2 + 2x - 8$.

Solution

If we view the curve graphed by the relation on the calculator, we see that a minimum of five points would suggest the curve. We will need to trace the graph or look at the table of values in order to graph this relation.

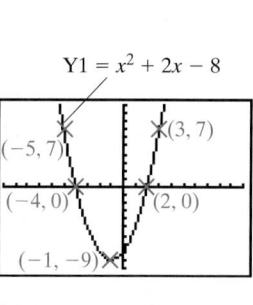

$Y1 = x^2 + 2x - 8$

$(-5, 7)$

$(3, 7)$

$(-4, 0)$

$(2, 0)$

$(-1, -9)$

$(-10, 10, -10, 10)$

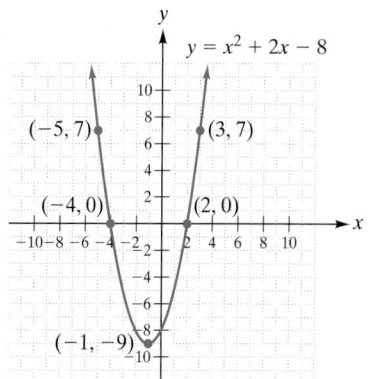

$y = x^2 + 2x - 8$

$(-5, 7)$ $(3, 7)$

$(-4, 0)$ $(2, 0)$

$(-1, -9)$

3.2.2 Checkup

In exercises 1–2, graph and label each set of ordered pairs on a coordinate plane.

1. $E(-5, -6)$ and $F(-6, -5)$
2. $G(3, -3)$ and $H(-3, 3)$
3. In exercises 1–2, what is the effect of swapping the two values of the coordinate pair? What is the effect of having a negative sign on the x-coordinate? What is the effect of having a negative sign on the y-coordinate?
4. State the coordinates of the points graphed in Figure 3.7.

5. Determine the location of each point by quadrant or by axis.

   a. $(-16, 95)$ b. $(123, 135)$ c. $(0.001, -1.009)$

   d. $\left(-3\frac{1}{3}, -2\frac{1}{5}\right)$ e. $(0, 0)$ f. $(-3.6, 0)$

   g. $\left(0, \frac{2}{9}\right)$

Graph each relation.

6. $R = \{(-2, 4), (0, 4), (3, 4), (7, 4)\}$
7. $y = -2x$ for $x = \{-2, 0, 2\}$
8. $y = -3x + 5$
9. $y = 10 - 3x - x^2$

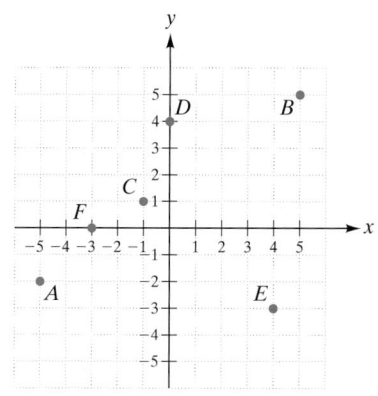

Figure 3.7

3.2.3 Identifying the Domain and Range

We can determine the domain and range of a relation directly from its graph.

To determine the domain of a relation from its graph, examine the graph to see what values of the independent variable (the x-coordinates) are used to draw the graph.

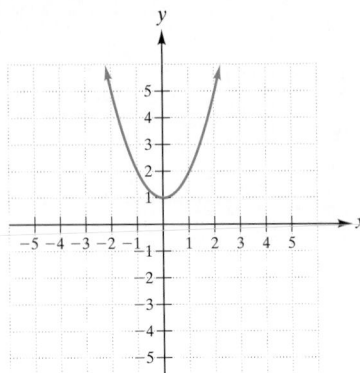

Figure 3.8

To determine the range of a relation from its graph, write a set of values of the dependent variable (the *y*-coordinates) used in the graph.

For example, determine the domain and range of the relation whose graph is shown in Figure 3.8.

We see that the graph includes points for every possible value of *x*, as indicated in the following diagram:

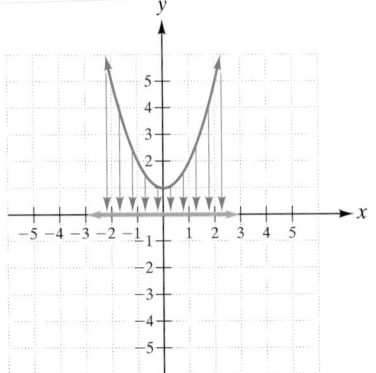

Thus, the domain is the set of all real numbers.

We see that the possible *y*-values are all greater than or equal to 1, as indicated in the following diagram:

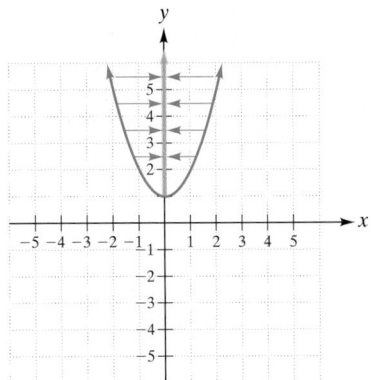

Thus, the range is the set of all real numbers *y* greater than or equal to 1 ($y \geq 1$).

EXAMPLE 8 Determine the domain and range of each relation.

a.

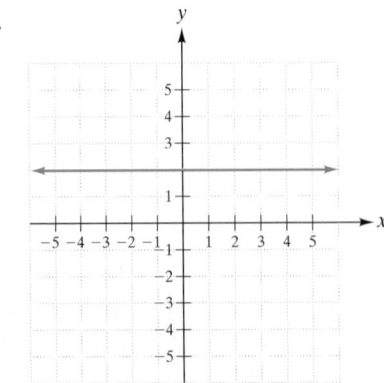

b.

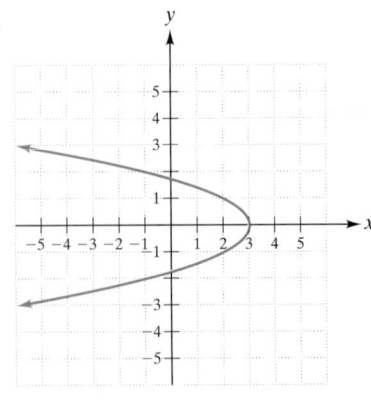

c.

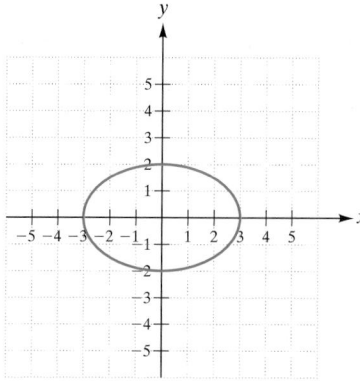

Solution

a.

Domain	Range

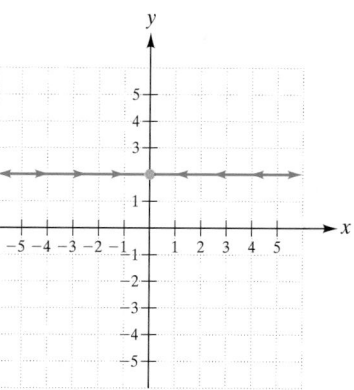

All real numbers $y = 2$

b.

Domain	Range

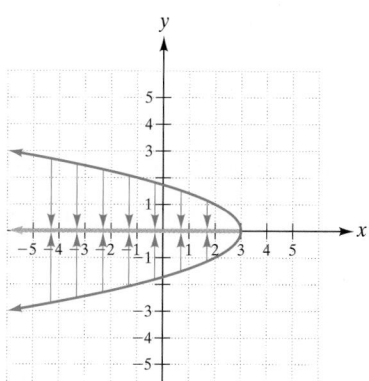   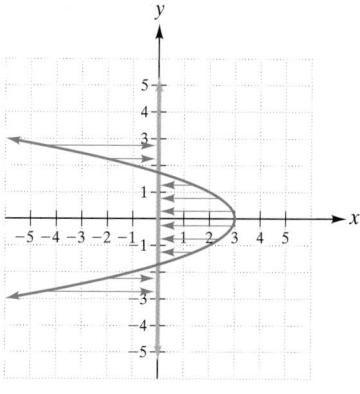

$x \leq 3$ all real numbers

c.

Domain	Range

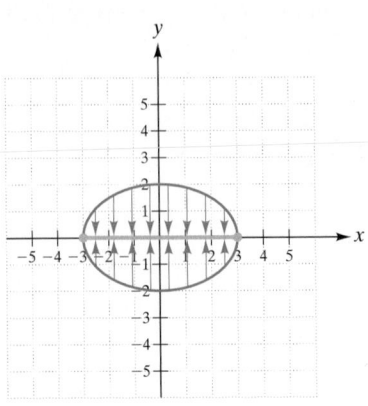

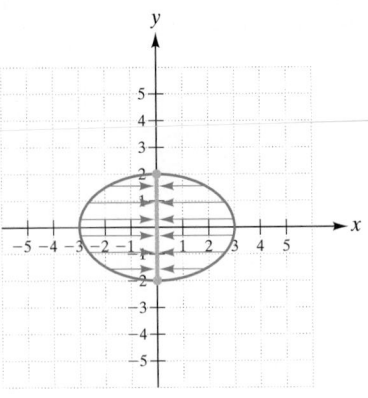

all real numbers
between and
including −3 and 3

$-3 \le x \le 3$

all real numbers
between and
including −2 and 2

$-2 \le x \le 2$

 3.2.3 Checkup

Determine the domain and range of each relation.

1.

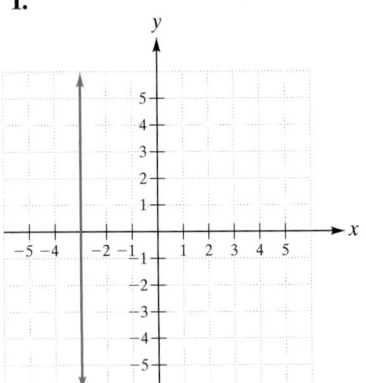

2.

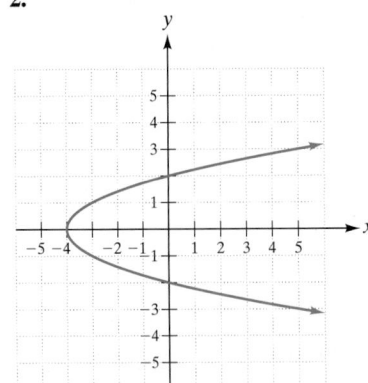

3.

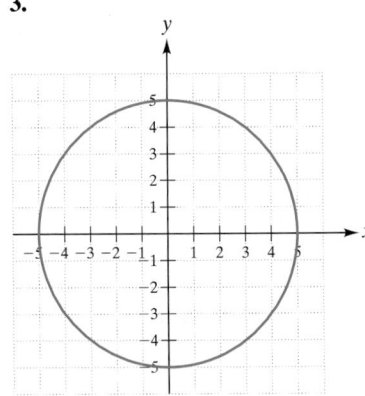

3.2.4 Modeling the Real World

Many important relations are found in business, science, and social studies—in fact, in almost any area of activity. Graphs are an important tool to help us visualize these relations. However, you should be aware that real situations sometimes impose restrictions on the domain and range of relations. For example, if the independent variable represents time or the number of units sold, then the domain must be limited to nonnegative numbers. If the dependent variable represents cost or height above the ground, then the range must be limited to nonnegative numbers.

EXAMPLE 9 Buddy dropped a pair of pliers from the top of a 30-foot power pole.

a. A relation for the height in feet above ground level of the pliers, s, in terms of time, t, in seconds, is given by $s = -16t^2 + 30$. Graph this relation as $Y1 = -16x^2 + 30$, using $(-2, 2, 1, -30, 30, 10, 1)$ as the graphing window on your calculator.

b. Approximate the domain and range of the relation.

Solution

a.

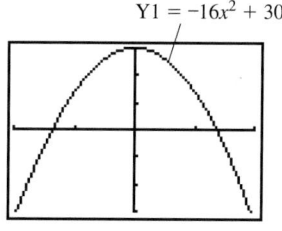

$Y1 = -16x^2 + 30$

$(-2, 2, -30, 30)$

b. The independent variable is time, t, so it must be nonnegative. The dependent variable, s, is height above the ground, so it must also be nonnegative. If we trace the function and view the t-values, we see that the domain is all real numbers greater than or equal to 0 and less than or equal to approximately 1.4, or $0 \le t \le 1.4$. If we trace the function and view the s-values, we see that the range is all real numbers greater than or equal to 0 and less than or equal to 30, or $0 \le s \le 30$.

APPLICATION

The following graph illustrates the revenue from sales, cost of sales, and gross operating profit for General Motors in the years 1996 through 1999:

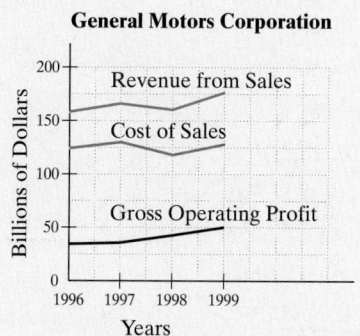

General Motors Corporation

Revenue from Sales

Cost of Sales

Gross Operating Profit

Billions of Dollars

Years

Discussion

The domain of the revenue, cost, and profit is {1996, 1997, 1998, 1999}.

The range of the revenue from sales is about $158 million to $177 million. The range of the cost of sales is about $118 million to $130 million. The range of the gross operating profit is about $34 million to $50 million.

Approximate the domain and range of the revenue, cost, and profit.

3.2.4 Checkup

1. A toy rocket is shot upward from ground level with an initial velocity of 250 feet per second.
   a. A relation for the height in feet above ground level of the rocket, s, in terms of time, t, in seconds, is given by $s = -16t^2 + 250t$. Graph the relation using $(-20, 20, 5, -1000, 1000, 100, 1)$ as the graphing window on your calculator.
   b. Approximate the domain and range of the relation.
2. A quality-control engineer recorded the numbers of non-defective parts coming off an assembly line that produced 75 parts per day. The data for each day are shown in the following figure:

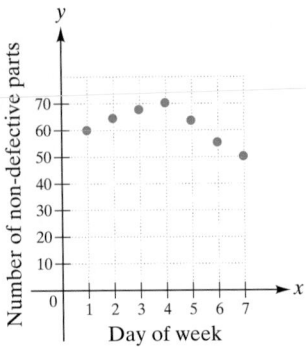

Determine the domain and range of the relation.

3.2 Exercises

Graph and label each ordered pair on a coordinate plane.

1. $A(-7, -5)$ and $B(-5, -7)$
2. $C(-2, -4)$ and $D(-4, -2)$
3. $E(4, 9)$ and $F(9, 4)$
4. $G(2, 6)$ and $H(6, 2)$
5. $I(-5, 5)$ and $J(5, -5)$
6. $K(-4, 4)$ and $L(4, -4)$
7. $M(2, -1)$ and $N(-1, 2)$
8. $O(3, -4)$ and $P(-4, 3)$
9. $Q(3, -1)$ and $R(-3, 1)$
10. $S(2, -7)$ and $T(-2, 7)$
11. $U(-8, -2)$ and $V(8, 2)$
12. $W(3, 5)$ and $X(-3, -5)$
13. $A(1.2, 2.4)$
14. $B(3.4, 1.2)$
15. $C(-4.5, -2.6)$
16. $D(-3.6, -2.5)$
17. $E(-2.4, 2.1)$
18. $F(4.1, -2.4)$
19. $G(1.8, -2.7)$
20. $H(-2.7, 1.8)$

Determine the location of each point by quadrant or by axis. Do not graph the point.

21. $(-21, 35)$
22. $(-33, -35)$
23. $(4, 96)$
24. $(28, -92)$
25. $(-3, -19)$
26. $(-75, 22)$
27. $(0, -31)$
28. $(94, 0)$
29. $(90, -100)$
30. $(2, 29)$
31. $(24, 0)$
32. $(0, -36)$
33. $(-19, 0)$
34. $(0, 0)$
35. $(0.05, 1.003)$
36. $(-2.09, 8.6)$
37. $(0, 3.7)$
38. $(-9.3, 0)$
39. $\left(\dfrac{13}{27}, -\dfrac{11}{19}\right)$
40. $\left(-\dfrac{57}{105}, -\dfrac{17}{21}\right)$
41. $\left(-\dfrac{53}{100}, -\dfrac{39}{100}\right)$
42. $\left(\dfrac{20}{33}, 0\right)$
43. $\left(0, \dfrac{28}{51}\right)$
44. $\left(3\dfrac{1}{7}, -1\dfrac{4}{5}\right)$

45. State the coordinates of the points graphed in Figure 3.9.
46. State the coordinates of the points graphed in Figure 3.10.

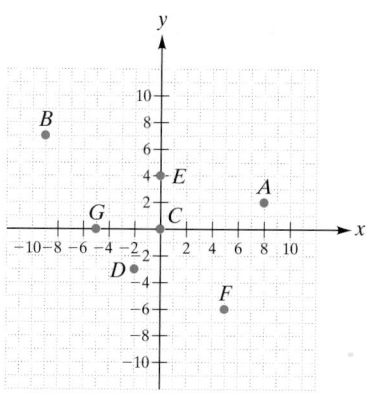

Figure 3.9

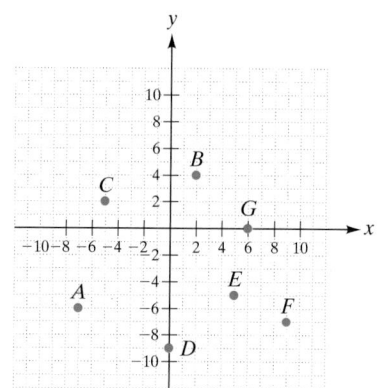

Figure 3.10

Graph each relation.

47. $A = \{(-3, -3), (-2, -2), (-1, -1), (0, 0), (1, 1), (2, 2), (3, 3)\}$

48. $B = \{(-2.5, -2), (-1.5, -1), (-0.5, 0.5), (0.5, 1), (1.5, 2)\}$

49. $C = \{(-4, 4), (-2, 2), (0, 0), (2, -2), (4, -4)\}$

50. $D = \{(0, 2), (1, 3), (2, 4), (3, 5), (4, 6)\}$

51. $E = \{(5, -3), (5, -1), (5, 1), (5, 3)\}$

52. $F = \{(-4, -2), (-2, -2), (0, -2), (2, -2), (4, -2)\}$

53.

x	y
-2	-1
-1	0
0	1
1	2
2	3

54.

x	y
-2	3
-1	2
0	1
1	0
2	-1

55.

x	y
-2	-6
-1	-5
0	-4
1	-3
2	-2
3	-1

56.

x	y
-4	-2
-2	-1
0	0
2	1
4	2

57. $y = 12x - 15$ for $x = \{-1, 0, 1\}$

58. $y = 14x$ for $x = \{-2, -1, 1, 2\}$

59. $y = \dfrac{1}{2}x + 3$ for $x = \{-4, -2, 0, 2, 4\}$ **60.** $y = -\dfrac{3}{4}x + 5$ for $x = \{-8, -4, 0, 4, 8\}$ **61.** $y = -10x + 9$

62. $y = -x - 1$ **63.** $y = -\dfrac{2}{3}x - 2$ **64.** $y = \dfrac{2}{5}x + \dfrac{1}{5}$ **65.** $y = 2x^2 - 5$

66. $y = -3x^2 - 1$ **67.** $y = |2x|$ **68.** $y = |-4x|$

Determine the domain and the range of each relation.

69.

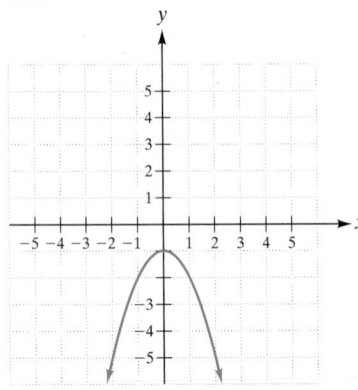

70.

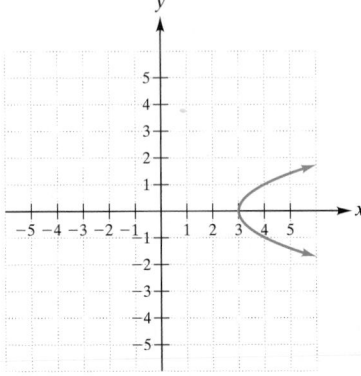

71.

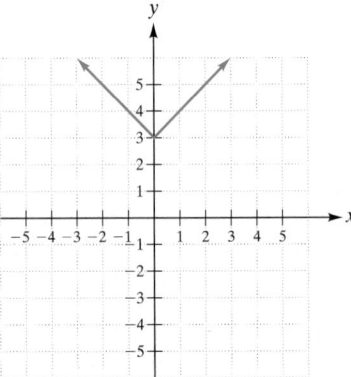

72.

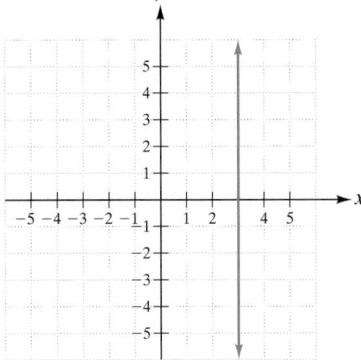

73. The following chart shows the percentage of votes cast for the Democratic and Republican candidates for president for the years 1960 through 2000, where x is the number of years since 1960:

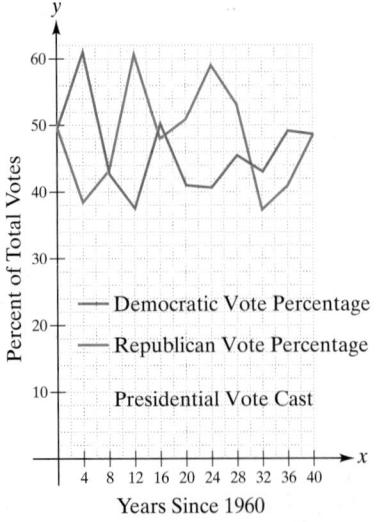

Democratic Vote Percentage

Republican Vote Percentage

Presidential Vote Cast

Years Since 1960

Determine the domain and the range of the relation if the Democratic voting percentage is the relation of interest.

74. The following chart shows violent crime rates in the United States for the years 1987 through 1997, where x is the number of years since 1987:

Violent and Property Crimes (1987–1997)

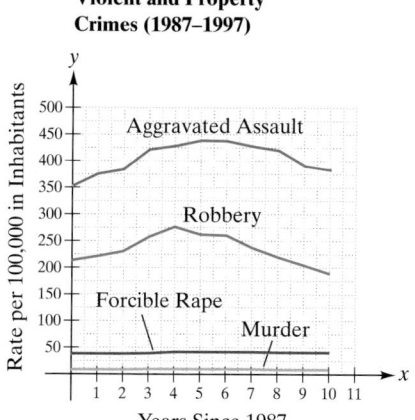

Years Since 1987

Determine the domain and range if the relation of interest is aggravated assaults.

75. An egg is dropped from a height of 100 feet.

  a. A relation for the height in feet above ground level of the egg, s, in terms of time, t, in seconds, is given by $s = -16t^2 + 100$. Graph the relation, using $(-4.7, 4.7, 1, -124, 124, 40, 1)$ as the graphing window on your calculator.

  b. State the domain and range of the relation.

76. A pendulum has a length of x feet.

  a. A relation for the period, T, in seconds required for one complete back-and-forth swing of the pendulum is given by $T = 2\pi\sqrt{\dfrac{x}{32}}$. Graph the relation, using $(-4.7, 4.7, 1, -3.1, 3.1, 1, 1)$ as the graphing window on your calculator.

  b. State the domain and range of the relation.

3.2 Calculator Exercises

A. Comparing View Screens on the Calculator

To give you an idea of what the various screen settings on a calculator mean, first draw a coordinate plane on a large sheet of graph paper. Place the origin of the coordinate system in the middle of the graph paper, and mark the x-axis in units of 5 from -50 to 50. Also, mark the y-axis in units of 5 from -35 to 35. Make the graph as large as the paper will allow. You will use this graph to draw boxes for the various calculator settings.

Clear the ⎡ Y = ⎤ screen. Then set your calculator to the decimal setting, ZDecimal, option 4.

Press ⎡ WINDOW ⎤ to view the setting's limits. Plot the limits on the graph and draw a box. This box represents the portion of a graphed relation you can view with the current setting.

Now set the calculator to the standard setting, ZStandard, option 6. Press ⎡ WINDOW ⎤ to view this setting's limits. Plot the limits on the same graph and draw another box. This box represents the portion of a graphed relation you can view with the current setting. You should see that you can view much more of a graphed relation with this setting.

Next, set the calculator to the integer setting, ZInteger, option 8. Be sure to press ⎡ ENTER ⎤ after choosing option 8,

in order to move to the new setting. Press ⎡ WINDOW ⎤ to view this setting's limits. Plot these limits on the same graph and draw a third box. This box represents the portion of a graphed relation you can view with the current setting. Once again, you should see that you can view much more of a graphed relation with this setting.

So if you use a larger setting, you see more of a graph. However, if you use a smaller setting, you may see more detail in a smaller portion of the graph. It's up to you to become familiar enough with these settings in order to make a decision regarding which setting you should use. Be aware that there are other settings you can use. You can set the screen to any setting you choose, and you can also zoom in or zoom out on a setting to see more detail or more of the graph. Try the various settings on the following equations, and decide which is best for viewing them:

1. $y = 2x - 3$

2. $y = 2x^2 - 3x + 1$

3. $y = 2x^3 - 3x^2 + x - 4$

B. Transferring Graphs from the Calculator to Paper

Set your calculator screen to the decimal setting. Then graph the following relations on this setting:

1. $y = 0.6x - 1.2$ **2.** $y = -0.5x + 2.2$

3. $y = |x| - 2$ **4.** $y = |x - 2|$

5. $y = x - 2$

Find the coordinate pairs for each relation using the domain $\{-2, -1, 0, 1, 2\}$. Now plot the points on graph paper, using the coordinate plane. Then sketch the graph connecting these points.

6. Consider the keystrokes needed to enter exercises 3–5 into your calculator. What is the important difference that distinguishes these three exercises? Pay attention to this important difference as you use this calculator method.

3.2 Writing Exercise

In this section, you learned that sometimes a graph is a finite set of points and at other times the graph is an infinite set of points represented by a curve. Explain what must be true about the domain and range of the relation for each type of graph. In your explanation, give an example of a relation that yields each type of graph.

3.3 Functions and Function Notation

OBJECTIVES

1 Determine whether relations written in list form are functions.
2 Determine whether relations represented as graphs are functions.
3 Evaluate functions written in function notation.
4 Use real-world data to determine functions.

APPLICATION

Using data from the 1996–1999 end-of-the-year reports, financial analysts for General Motors determined that the company's profit, in millions of dollars, could be determined by the function $P(x) = 1021x^2 + 2327x + 33,837$ for x years after 1996. Estimate the amount of profit for the year 2000.

After completing this section, we will discuss this application further. See page 181.

3.3.1 Identifying Functions from Ordered Pairs

A function is a special type of relation that we frequently use in mathematics. A **function** is a relation in which every element in the domain corresponds to one and only one element in the range.

A relation

$$M = \{(1, 2), (2, 4), (3, 5), (4, 6)\}$$

written in list form, is a function, because we match every element in the domain, $\{1, 2, 3, 4\}$, to only one element in the range, $\{2, 4, 5, 6\}$.

Domain	Range
1 ⟶	2
2 ⟶	4
3 ⟶	5
4 ⟶	6

A relation

$$N = \{(1, 2), (1, 3), (2, 4)\}$$

written in list form, is not a function, because we match one element, 1, in the domain $\{1, 2\}$ to more than one element, 2 and 3, in the range $\{2, 3, 4\}$.

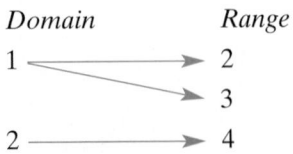

Domain   *Range*

EXAMPLE 1 Determine whether each relation is a function.

a. $\{(-5, 10), (-3, -6), (3, 6), (5, 10)\}$
b. $\{(3, 6), (-3, 6), (5, 10), (-5, 10)\}$
c. $\{(3, 6), (3, -6), (5, 10), (5, -10)\}$

Solution

a.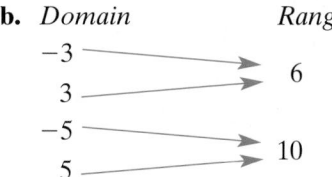

Domain   Range

$\{(3, 6), (-3, -6), (5, 10), (-5, -10)\}$ is a function, because every element in the domain corresponds to only one element in the range. The first co-ordinate is not repeated in the set of ordered pairs.

b.

Domain   Range

$\{(3, 6), (-3, 6), (5, 10), (-5, 10)\}$ is a function, because every element in the domain corresponds to only one element in the range. The first coor-dinate is not repeated in the set of ordered pairs. The repeated elements in the range, 6 and 10, do not matter.

c.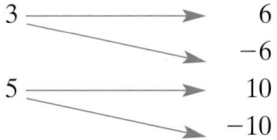

Domain   Range

$\{(3, 6), (3, -6), (5, 10), (5, -10)\}$ is not a function, because an element in the domain, 3, corresponds to two elements in the range, 6 and −6. The first coordinate, 3, is repeated in the set of ordered pairs. Likewise, the element 5 in the domain corresponds to two elements in the range.

3.3.1 Checkup

In exercises 1–4, determine whether each relation is a function.

1. $A = \{(-2, 3), (-1, 5), (0, 7), (1, 5), (2, 3)\}$ 2. $B = \{(-3, 5), (-1, 5), (1, 5), (3, 5)\}$
3. $C = \{(-3, -2), (-3, 2), (-1, -3), (-1, 3)\}$ 4. $D = \{(2, -1), (2, 0), (2, 1), (2, 2)\}$
5. What should you look for in a relation to determine whether the relation is a function?

3.3.2 Identifying Functions from Graphs

Relations are sometimes represented as graphs. Therefore, a function may be represented as a graph. We need to determine a method for deciding whether a relation represented as a graph is a function. Complete the following set of exercises.

Discovery 3

Graphs of Functions

The following are graphs of functions:

1.

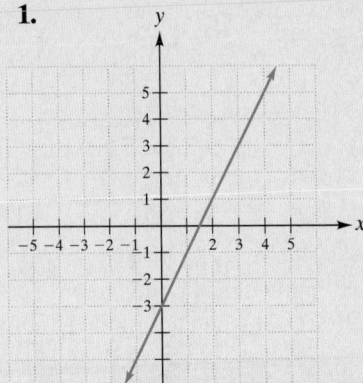

2.

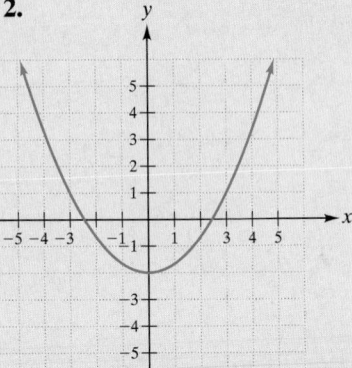

3.

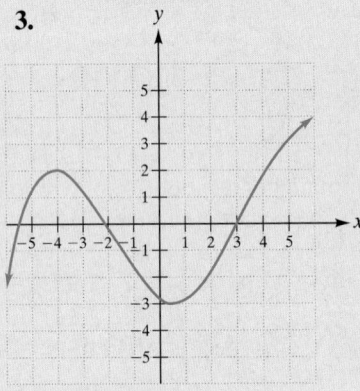

If possible, draw a vertical line through more than one point on the graph. Write a rule for determining a function from the graph of a relation by drawing a line through points on the graph.

Check your rule on the following two graphs of relations that are not functions:

4.

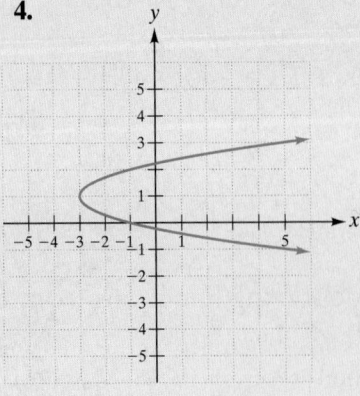

5.

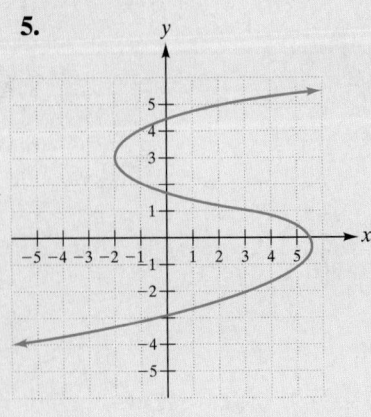

In the set of functions, a vertical line does not cross the graph of a function more than once. In the check, we see that a vertical line crossed the graphs of the relations that were not functions more than once.

VERTICAL-LINE TEST

If a vertical line can be drawn such that it intersects the graph of a relation more than once, the graph does not represent a function. If the vertical line intersects the graph only once, then the graph represents a function.

We know that this rule is a valid test because if we graph two distinct ordered pairs with the same x-coordinate, they will lie on the same vertical line, and a function does not have two ordered pairs with the same x-coordinate. For example, graph the relation $\{(4, 3), (4, -5)\}$, which is not a function.

Examples of applying the vertical-line test are given in the following figures:

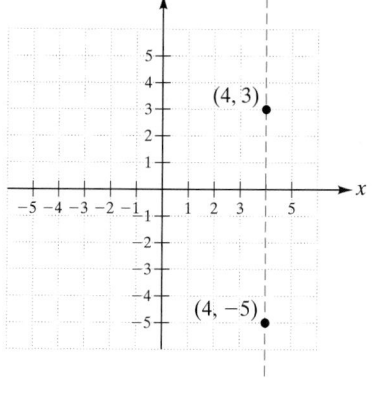

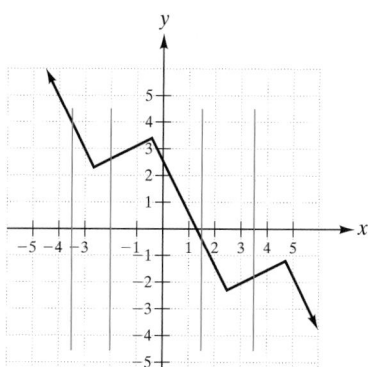

This graph represents a function. All possible vertical lines cross the graph only once.

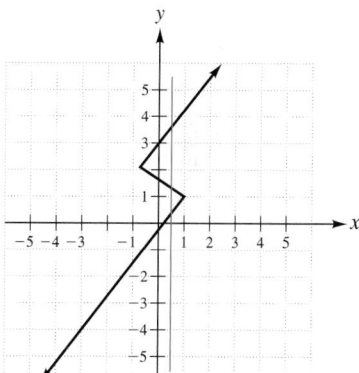

This graph does not represent a function. The vertical line drawn is one of many such lines that cross the graph more than once.

EXAMPLE 2   Determine whether each graph represents a function.

a.

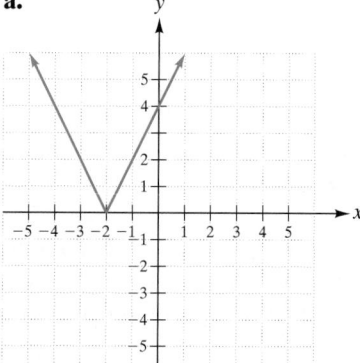

b.

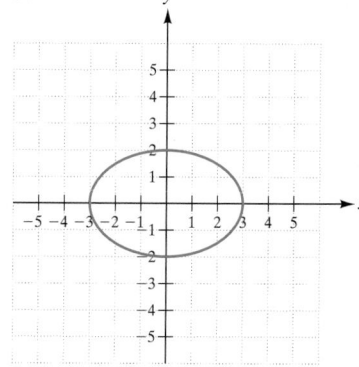

Solution

a. The graph represents a function, because all possible vertical lines cross the graph only once.

b. The graph does not represent a function, because a vertical line can be drawn to cross the graph more than once.

✓ **3.3.2** Checkup

Determine whether each graph represents a function.

1.

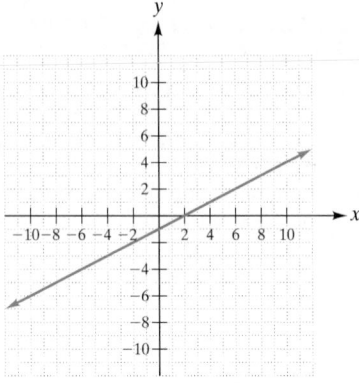

2.

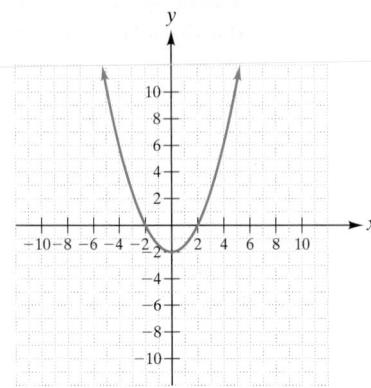

3.

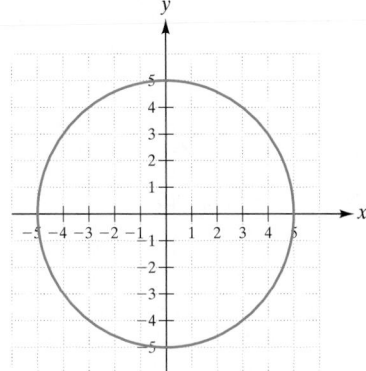

4.

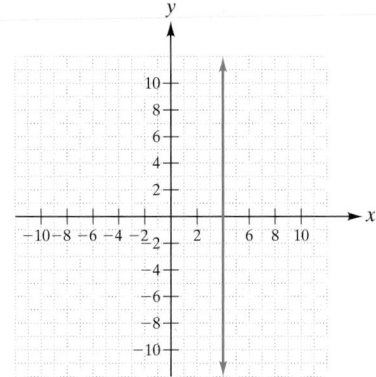

5. Explain how to check the graph of a relation to determine whether it is a function.

3.3.3 Function Notation

A function is a relation, so it may be represented in all of the same ways as a relation. That is, a function may be written as a list of ordered pairs, in equation form, or as a graph. However, there is a notation specific to functions called function notation. **Function notation** is written as an equation with the name of the function, the name of the independent variable in parentheses, and an expression to be evaluated to determine the dependent variable. For example,

$$y = x + 2 \quad \text{Relation} \qquad\qquad u = 3t - 5 \quad \text{Relation}$$
$$f(x) = x + 2 \quad \text{Function notation} \qquad g(t) = 3t - 5 \quad \text{Function notation}$$

The latter is read "f of x equals x plus 2."

The latter is read "g of t equals $3t$ minus 5."

Caution: The notation $f(x)$ does not mean multiplication of f by x.

HELPING HAND Function notation is very similar to the equation form of a relation. The difference is that in function notation the y is replaced with the name of the function and the name of the independent variable in parentheses.

We can evaluate a function written in function notation just as we evaluate a function written as an equation to determine ordered pairs.

> To evaluate a function for a given number,
>
> • Substitute the number for the independent variable in the equation.
> • Perform the indicated operations.

For example, evaluate the function $y = x + 2$ for $x = 3$.
This means

$$\text{Given } f(x) = x + 2, \text{ find } f(3). \qquad \text{Function notation}$$
$$f(3) = 3 + 2 \qquad \text{Substitute 3 for the variable, } x.$$
$$f(3) = 5 \qquad \text{Evaluate.}$$

The ordered pair is written $(x, f(x))$, or $(3, 5)$.

> To evaluate a function for an algebraic expression,
>
> • Substitute the expression for the independent variable in the equation.
> • Simplify the expression.

For example, given $g(t) = 3t - 5$, find $g(a)$.

$$g(t) = 3t - 5 \qquad \text{Function}$$
$$g(a) = 3(a) - 5 \qquad \text{Substitute the expression for the variable, } t.$$
$$g(a) = 3a - 5$$

Given $g(t) = 3t - 5$, find $g(a + 1)$.

$$g(t) = 3t - 5 \qquad \text{Function}$$
$$g(a + 1) = 3(a + 1) - 5 \qquad \text{Substitute the expression } a + 1 \text{ for the variable, } t.$$
$$g(a + 1) = 3a + 3 - 5 \qquad \text{Distribute and simplify.}$$
$$g(a + 1) = 3a - 2$$

HELPING HAND We cannot evaluate $g(a)$ and $g(a + 1)$ on a calculator, because we are evaluating a function for an algebraic expression.

EXAMPLE 3 Evaluate each function for the given number or expression.

a. Given $f(x) = x^2 + x + 5$, find $f(2)$ and $f(-2)$.
b. Given $g(v) = \sqrt{v + 7} + 1$, find $g(8)$ and $g(-8)$.
c. Given $f(x) = 2x - 4$, find $f(a)$ and $f(a + h)$.

Solution

a. Given $f(x) = x^2 + x + 5$, find $f(2)$ and $f(-2)$.

$f(2) = (2)^2 + 2 + 5$	$f(-2) = (-2)^2 + (-2) + 5$	Substitute.
$f(2) = 4 + 2 + 5$	$f(-2) = 4 + (-2) + 5$	Evaluate.
$f(2) = 11$	$f(-2) = 7$	

b. Given $g(v) = \sqrt{v + 7} + 1$, find $g(8)$ and $g(-8)$.

$g(8) = \sqrt{8 + 7} + 1$	$g(-8) = \sqrt{-8 + 7} + 1$	Substitute.
$g(8) = \sqrt{15} + 1$	$g(-8) = \sqrt{-1} + 1$	Evaluate.
$g(8) \approx 4.873$	$g(-8)$ is not a real number.	

c. Given $f(x) = 2x - 4$, find $f(a)$ and $f(a + h)$.

$f(a) = 2a - 4$	$f(a + h) = 2(a + h) - 4$	Substitute.
	$f(a + h) = 2a + 2h - 4$	Distribute.

The solutions for 3a and 3b can be checked using your calculator. However, 3c cannot be evaluated on a calculator.

3.3.3 Checkup

In exercises 1–3, evaluate each function for the given number or expression.

1. Given $f(x) = \sqrt{21 - 15x} - 17$, find $f(4)$ and $f(-4)$.

2. Given $h(z) = 3z^3 - 2z^2 + 4z - 8$, find $h(1), h(0),$ and $h(-1)$.

3. Given $g(x) = 4x - 8$, find $g(b), g(1),$ and $g(b + 1)$.

4. Explain what the notation $f(x)$ means.

3.3.4 Modeling the Real World

In real-world applications, a function is defined as a process or rule that results in exactly one **output value** for each **input value**. Many business applications involve pairs of values wherein the value of the independent variable (or input value) results in a single value for the dependent variable (or output value). Therefore, these business relations are functions. Given a business function, you can determine the output value by substituting the input value into the function. Some of the most common business relations that we will write as functions are revenue, $R(x)$, cost, $C(x)$, and profit, $P(x)$.

EXAMPLE 4 Dougal plans to decorate sweatshirts. The paint costs $14.75. The sweatshirts cost $7.14 each.

a. Write a cost function, $C(x)$, for the cost of x sweatshirts.

b. Determine the cost of two sweatshirts, using the cost function in part a.

Solution

a. Let x = the number of sweatshirts decorated

The cost is 14.75 plus 7.14 times the number of sweatshirts, x.

$$C(x) = 14.75 + 7.14x$$

b. To find the cost of two sweatshirts, find $C(2)$.

$$C(2) = 14.75 + 7.14(2)$$
$$C(2) = 29.03$$

Two sweatshirts cost Dougal $29.03 to produce.

APPLICATION

Using data from the 1996–1999 end-of-the-year reports, financial analysts for General Motors determined that the company's profit, in millions of dollars, could be determined by the function $P(x) = 1021x^2 + 2327x + 33{,}837$ for x years after 1996. Estimate the amount of profit for the year 2000.

$$P(x) = 1021x^2 + 2327x + 33{,}837$$
$$P(4) = 1021(4)^2 + 2327(4) + 33{,}837$$
$$P(4) = 59{,}481$$

In the year 2000, General Motors' profit was approximately $59,481,000,000.

Discussion

The year 2000 corresponds to $x = 4$ (four years after 1996).

3.3.4 Checkup

1. The charge for renting a chain saw is $10 per day or fraction of a day, with a $5 flat fee for resharpening the saw.
   a. Write a function that represents the charge for renting the saw for d days.
   b. Determine the cost of renting the saw for four days.

2. Using data from the same end-of-year reports as mentioned in the previous application, financial analysts de-

termined that the income taxes, in millions of dollars, paid by General Motors could be estimated by the function $T(x) = 577x^2 - 1274x + 1734$ for x years after 1996. Estimate the amount of income tax for the year 2000.

3.3 Exercises

Determine whether each relation is a function.

1. $A = \{(-2, 1), (-2, 3), (2, -3), (2,-1)\}$

2. $B = \{(1.2, 2.4), (1.4, 2.8), (1.6, 3.2), (1.8, 3.6)\}$

3. $C = \{(1.1, 1.1), (2.2, 2.2), (3.3, 3.3), (4.4, 4.4), (5.5, 5.5)\}$

4. $D = \{(1, 9), (2, 8), (3, 7), (4, 6), (5, 5)\}$

5. $E = \{(6, -1), (6, -3), (6, -5), (6, -7)\}$

6. $F = \left\{ \left(\frac{1}{2}, 2\right), \left(\frac{1}{2}, 3\right), \left(\frac{1}{2}, 4\right), \left(\frac{1}{2}, 5\right) \right\}$

7. $G = \left\{ \left(-1, \frac{2}{3}\right), \left(-2, \frac{2}{3}\right), \left(-3, \frac{2}{3}\right), \left(-4, \frac{2}{3}\right) \right\}$

8. $H = \{(8, -2), (7, -2), (6, -2), (5, -2), (4, -2)\}$

9.

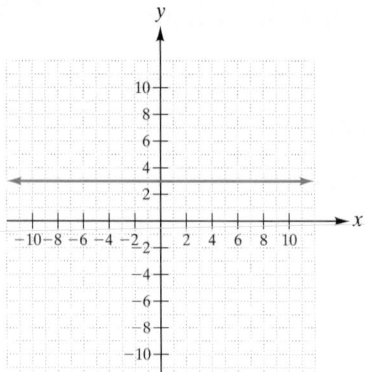

10.

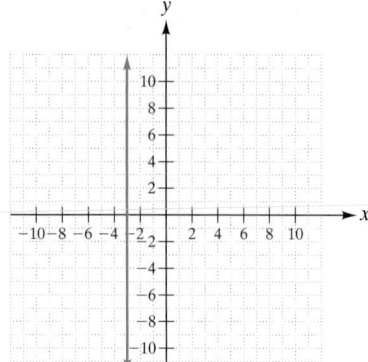

11.

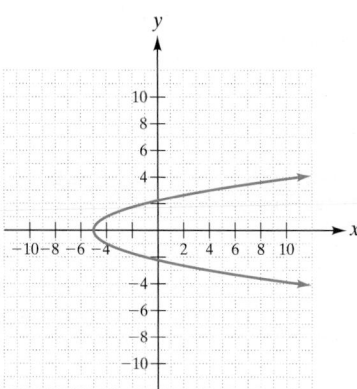

12.

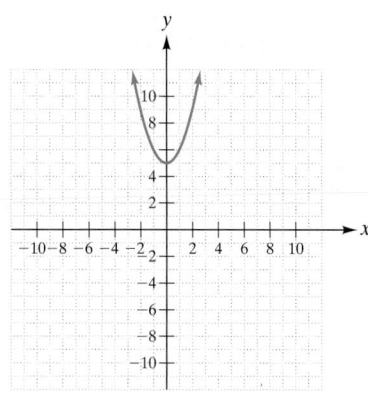

13.

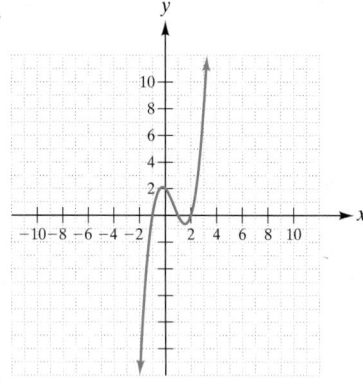

14.

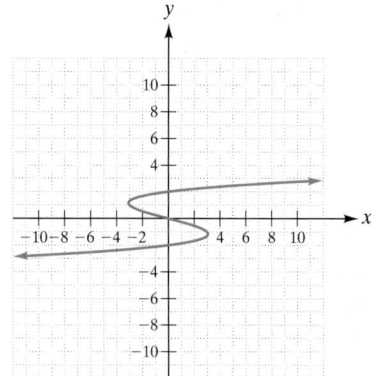

Given $f(x) = 20x + 12$, find

15. $f(5)$

16. $f(50)$

17. $f(-7)$

18. $f(-17)$

19. $f(2.4)$

20. $f(-2.4)$

21. $f\left(-\dfrac{1}{4}\right)$

22. $f\left(\dfrac{1}{10}\right)$

23. $f(a)$

24. $f(-a)$

25. $f(h + 2)$

26. $f(h - 2)$

27. $f(a - 4)$

28. $f(4 - a)$

29. $f(a + h)$

30. $f(x + h)$

Given $h(x) = 2x^2 - 4x + 5$, find

31. $h(7)$

32. $h(1)$

33. $h(-4)$

34. $h(-1)$

35. $h(-1.1)$

36. $h(0.1)$

37. $h\left(-\dfrac{2}{5}\right)$

38. $h\left(\dfrac{2}{5}\right)$

39. $h\left(2\dfrac{3}{5}\right)$

40. $h\left(-7\dfrac{1}{2}\right)$

41. $h(b)$

42. $h(-b)$

Given $g(x) = |-3x + 9|$, *find*

43. $g(5)$ **44.** $g(0)$ **45.** $g(-5)$ **46.** $g(-1)$

47. $g(4.5)$ **48.** $g(0.1)$ **49.** $g(-4.5)$ **50.** $g(-0.01)$

51. $g\left(\dfrac{2}{3}\right)$ **52.** $g\left(-\dfrac{2}{3}\right)$ **53.** $g\left(-4\dfrac{2}{3}\right)$ **54.** $g\left(4\dfrac{2}{3}\right)$

Given $F(x) = \sqrt{x + 15} + 21$, *find*

55. $F(85)$ **56.** $F(34)$ **57.** $F(-6)$ **58.** $F(-11)$

59. $F(-25)$ **60.** $F(-100)$ **61.** $F(5.25)$ **62.** $F(6.16)$

63. $F\left(-2\dfrac{3}{4}\right)$ **64.** $F\left(2\dfrac{16}{25}\right)$

65. Goodbuy Television's production process for manufacturing television sets has a fixed cost of $1500 for setting up a production run. Materials and labor to produce the sets cost $35 per television. Write a cost function for a production run. What is the cost of a production run that produces 400 televisions?

66. Creaky Car Company is considering setting up a new line in its production plant. The setup cost for each run on the line is estimated to be $25,000. The cost of materials and labor to produce parts is $550 per unit. Write a cost function for a production run. If the company can produce 3000 units on one run of the line, what will be the cost of production?

67. Fixed costs associated with selling CD players amount to $470 for advertising and counter space. The players sell for $125 each. Write a profit function (sales minus costs) for selling the players. What is the profit when 400 players are sold?

68. Truck rental costs to deliver parts to a customer average $185 per delivery. The parts are sold for $1200 per lot. Write a profit function (sales minus costs) for one shipment of a given number of lots. What will be the profit for one delivery of 50 lots of parts to a customer?

69. Trucks-4-U offers to rent a truck for a drop-off fee of $39, plus a daily rental fee of $25. Write a cost function for renting a truck for a given number of days. What will be the charge for renting the truck for three days?

70. Susie Seller is paid $475 per week, plus a commission of $165, for each major furniture sale she completes. Write a pay function for her week's pay, given that she completes a certain number of major sales. What will her week's pay be if she makes four major sales this week?

71. Handi Parking charges a fee of $2.50, plus $1.00 for each half hour of parking. Write a cost function for parking for a given number of half-hour increments. What is the charge for parking a car for $3\frac{1}{2}$ hours?

72. Party Palace rents a party room for $140 per evening and charges $18.50 per person for food and refreshments. Write a cost function to rent the room for a party for a given number of guests. What will be the charge for a party of 75 guests?

73. A Musical Mastery bus tour has 25 seats to sell. Each seat on the tour costs $175. However, in order to entice more customers to sign up for the tour, the company advertises that it will reduce the price $3 for each seat filled on the bus. The revenue function, given that x customers sign up for the tour is $R(x) = 175x - 3x^2$. How much revenue will the company collect if 22 customers make reservations for the tour?

74. The landlord of Midrose Place has 40 apartments to rent. In the past, a rent of $375 per month has been low enough that all the apartments will be rented. For each $25 increase in rent, one additional apartment will become vacant. The function for the total monthly rental receipts given the rent increases by x increments of $25 is $R(x) = 15,000 + 625x - 25x^2$. What will be the total monthly rental receipts if the landlord raises the rent $75?

75. Recycled CDs, Incorporated, offers a choice of five used CDs for $25, with each additional CD costing $4. Write a cost function for purchasing five or more CDs. What will be the cost of buying 12 used CDs?

76. Comix Collectors Club offers a sale of 10 comic books for $35, with each additional comic book selling for $1.50. Write a cost function for purchasing 10 or more comic books. What will be the cost of purchasing 18 comic books?

77. Using data from the president's budget for fiscal year 2001, analysts estimate that the percentage of households owning their own homes is given by the function $p(x) = -0.05x^2 + 1.21x + 60.08$, where x is the number of years after 1990. Use this function to estimate the percentage of households owning their own homes in the year 2001.

78. Using data from the president's budget for fiscal year 2001, analysts estimate that the discretionary anticrime budget (in billions of dollars) is given by the function $B(x) = 0.08x^2 + 0.07x + 13.62$, where x is the number of years after 1990. Use this function to estimate the anticrime budget for the year 2001. Given that the president budgeted 29.0 billion dollars for the year 2001, do you think he followed the estimation function? Explain.

3.3 Calculator Exercises

The table function of your calculator can be helpful when a function is to be evaluated repeatedly.

As an example, if

$$f(x) = x^3 + x^2 + x + 1$$

this expression can be stored in the calculator, and the "ask" feature can be used to generate evaluations as needed. The keystrokes to do so are as follows:

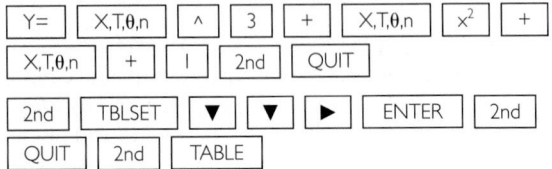

To evaluate the function for a given value, enter the value into the calculator and press [ENTER].

Thus, $f(-5) = -104$. Use this setup to find the following values of the function:

1. $f(65)$ **2.** $f(-83)$ **3.** $f(\pi)$

4. $f(\sqrt{2})$ **5.** $f(5634)$ **6.** $f(-\pi)$

Given $F(x) = \sqrt{x^2 + 8x + 16}$, find

7. $F(-8)$ **8.** $F(0)$ **9.** $F(-4)$

10. $F(0.8)$ **11.** $F(-6.3)$ **12.** $F\left(\dfrac{3}{4}\right)$

Given $h(x) = \dfrac{5}{x - 5}$, find

13. $h(10)$ **14.** $h(-5)$ **15.** $h(20)$

16. $h(5.5)$ **17.** $h(5)$ **18.** $h\left(\dfrac{1}{5}\right)$

3.3 Writing Exercise

"Every function is a relation, but not every relation is a function." Discuss this statement, explaining what it means and why it is true. Include examples to illustrate your explanation.

3.4 Analyzing Graphs

OBJECTIVES

1 Identify the intercepts of graphs.

2 Identify the maxima and minima of graphs.

3 Identify the intersection of two graphs.

4 Analyze graphs containing real-world data.

APPLICATION

The total net income of a company is the sum of the company's income from ongoing and discontinued operations, plus all other positive or negative adjustments. The total net income of y billion dollars for the x years after 1991 for General Motors is represented in the following graph:

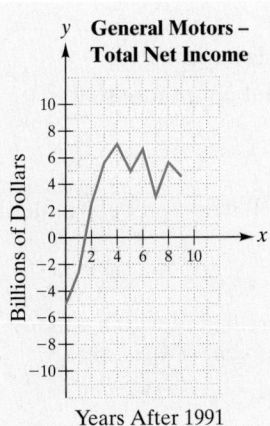

General Motors – Total Net Income

Billions of Dollars

Years After 1991

Approximate the x-intercept and discuss its relevance.

After completing this section, we will discuss this application further. See page 195.

Now we want to analyze a graph of a relation or a function by visually determining its characteristics. We first need to define several terms that will help us describe the graph.

3.4.1 Identifying Intercepts

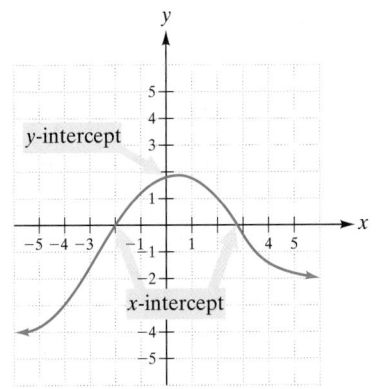

A graph may or may not cross the x-axis. If it does, it may cross the x-axis once or several times. An **x-intercept** is the location where a graph touches or crosses the x-axis. There may be more than one x-intercept for a function. Similarly, a **y-intercept** is the location where a graph touches or crosses the y-axis. There may be more than one y-intercept for a relation, but not for a function.

In section 3.2, we discovered that the y-coordinate is 0 for all points located on the x-axis. Therefore, the x-intercept will always have a y-coordinate of 0. Similarly, the x-coordinate of a point located on the y-axis is 0. Hence, the y-intercept must have an x-coordinate of 0.

HELPING HAND The x-intercept and the y-intercept are always written as ordered pairs.

EXAMPLE 1 Determine the x-intercepts and the y-intercepts for the following graphs:

a.

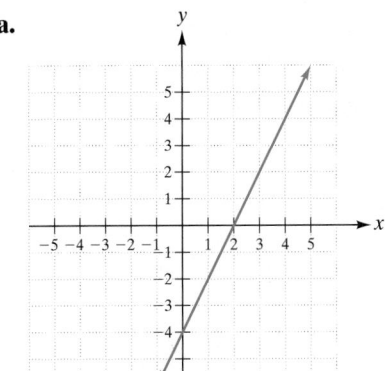

b.

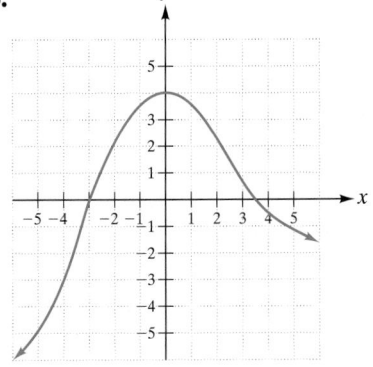

Solution

a. The *x*-intercept is $(2, 0)$. The *y*-intercept is $(0, -4)$.

b. The *x*-intercepts are $(-3, 0)$ and approximately $(\frac{7}{2}, 0)$. The *y*-intercept is $(0, 4)$.

The calculator has features that will help you determine the *y*-intercept of a graph.

EXAMPLE 2   Graph $y = x^2 - 1$ on a decimal screen and with a window setting of $(-5, 10, -5, 10)$. Then determine the *y*-intercept of the curve.

Calculator Solution
The *y*-intercept is $(0, -1)$.
See Figure 3.11a and Figure 3.11b.

T E C H N O L O G Y

Y-INTERCEPT

Determine the *y*-intercept of the graph of $y = x^2 - 1$.

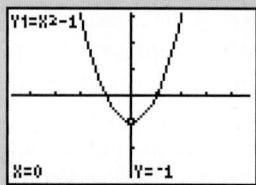
$(-4.7, 4.7, -3.1, 3.1)$
Figure 3.11a

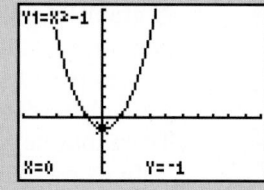
$(-5, 10, -5, 10)$
Figure 3.11b

For Figure 3.11a,
Enter the relation $y = x^2 - 1$ in Y1.

| Y= | X,T,θ,n | x² | − | 1 |

Set the window to the default decimal screen, and graph the relation.

| ZOOM | 4 |

Trace the graph to determine the *y*-intercept. Since the decimal screen is centered in the window, the first point traced is the *y*-intercept.

| TRACE |

For Figure 3.11b,
Enter the relation in Y1.

| Y= | X,T,θ,n | x² | − | 1 |

Set the window to the desired window setting, and graph the relation.

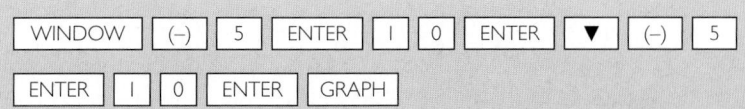

Trace the graph to determine the *y*-intercept. Since the graph is not centered in the window, the first point traced is not the *y*-intercept. Ask for the *y*-coordinate when *x* = 0.

The *y*-intercept is $(0, -1)$.

The calculator also has a feature to help us find the *x*-intercepts of a graph.

EXAMPLE 3 Graph $y = x^3 + 4.05x^2 + 3.15x$ on a decimal screen, and determine the *x*-intercepts of the curve.

Calculator Solution
The *x*-intercepts are $(0, 0)$, $(-3, 0)$, and $(-1.05, 0)$. See Figure 3.12a, Figure 3.12b, and Figure 3.12c.

T E C H N O L O G Y

X-INTERCEPTS

Determine the *x*-intercepts of the graph of $y = x^3 + 4.05x^2 + 3.15x$.

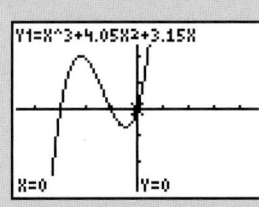

$(-4.7, 4.7, -3.1, 3.1)$

Figure 3.12a

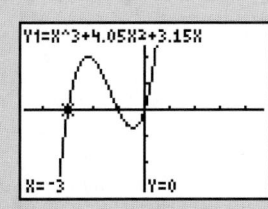

$(-4.7, 4.7, -3.1, 3.1)$

Figure 3.12b

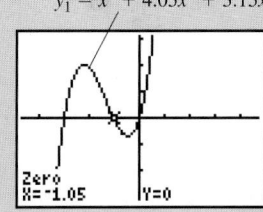

$(-4.7, 4.7, -3.1, 3.1)$

Figure 3.12c

For Figure 3.12a and Figure 3.12b,
Enter the relation $y = x^3 + 4.05x^2 + 3.15x$ in Y1.

Set the window to the default decimal screen, and graph the relation.

| ZOOM | 4 |

(Continued on page 188)

Trace the graph to determine the *x*-intercept—that is, the points on the graph where $y = 0$.

TRACE

For Figure 3.12c,
One of the *x*-intercepts cannot be found by tracing the graph. To find the *x*-intercept, choose ZERO, option 2, under the CALC menu. Press
2nd CALC 2 . Trace the graph to the left side of the intercept, called the *left bound*, and press ENTER . Move the cursor to the right of the intercept, called the *right bound*. Press ENTER . Move the cursor as close to the intercept as possible, and press ENTER . The calculator will display the coordinates of the missing *x*-intercept.
The *x*-intercepts are $(0, 0)$, $(-3, 0)$, and $(-1.05, 0)$.

✓ 3.4.1 Checkup

Determine the x-intercepts and the y-intercepts of each graph.

1.

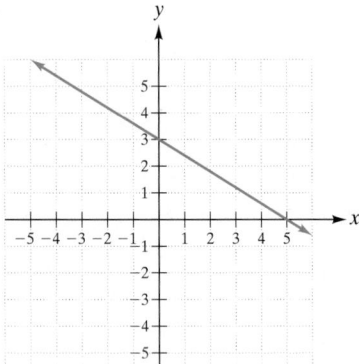

2.

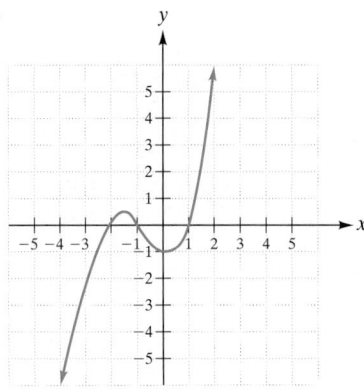

In exercises 3 and 4, graph each function on your calculator, and determine the x-intercepts and the y-intercepts.

3. $y = x^2 - 4$ **4.** $y = \dfrac{2}{3}x^3 - \dfrac{5}{6}x^2 - \dfrac{17}{6}x + 3$

5. If the graph of a relation has *x*-intercepts and *y*-intercepts, where do they occur?

6. Which coordinate of an *x*-intercept has a value of 0? Which coordinate of a *y*-intercept has a value of 0?

3.4.2 Determining Maxima and Minima

A function is said to be **increasing** if the values of the function increase as the values of the independent variable increase. A function is said to be **decreasing** if the values of the function decrease as the values of the independent variable increase. A function is **constant** if its values do not change as the values of the independent variable increase.

A function value is called a **relative maximum** (plural, *maxima*) if it is larger than the function values of its neighboring points. A function value is called a **relative minimum** (plural, *minima*) if it is smaller than the function values of its neighboring points.

We can visualize these definitions by tracing the graph of a function. Complete the following set of exercises on your calculator.

Discovery 4

Increasing and Decreasing Functions, Relative Maximum, and Relative Minimum

Graph $f(x) = x^2 - 4x$ on an integer window.

1. Trace the function from left to right (the *x*-values are increasing), and determine the following function values:

   $f(-2) =$ _____ $f(-1) =$ _____ $f(0) =$ _____
   $f(1) =$ _____ $f(2) =$ _____ $f(3) =$ _____
   $f(4) =$ _____ $f(5) =$ _____ $f(6) =$ _____

2. The function values first _____ and then _____ as the *x*-values increase. (Insert *decrease* or *increase*.)

3. Tracing the graph from left to right, we find that the graph first _____ and then _____. (insert *falls* or *rises*.)

Write a rule to determine, from a graph, whether the function is increasing or decreasing.

4. The function has a relative _____. (Insert *maximum* or *minimum*.)

5. The graph has one _____ point. (Insert *low* or *high*.)

Write a rule to determine, from a graph, whether the function has a relative maximum or minimum.

Tracing the graph from left to right, we see that the graph first falls and then rises. As we trace the graph from left to right, we see that the *x*-values are increasing and the function values are first decreasing and then increasing.

The graph has one low point, $(2, -4)$. At this low point on the graph, the function has a relative minimum value of $f(2) = -4$. This is also where the graph changes from falling to rising and where the function changes from increasing to decreasing.

Similarly, a function would have a relative maximum value at a high point on its graph. The graph would change from rising to falling, and the function would change from increasing to decreasing.

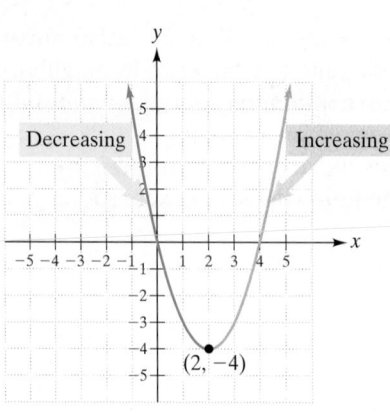

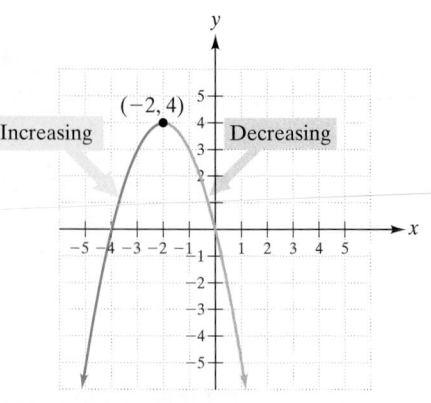

Relative minimum of $y = -4$ at $x = 2$.   Relative maximum of $y = 4$ at $x = -2$.

EXAMPLE 4 Use the following graph to answer the questions that follow:

    a. What is the relative maximum value of the function that is graphed?

    b. What is the relative minimum value of the function that is graphed?

    c. For what x-values is the function increasing?

    d. For what x-values is the function decreasing?

Solution

    a. The graph does not have a high point.

         The function does not have a relative maximum.

    b. The graph has a low point at $(1, 2)$.

         The function has relative minimum of $y = 2$ at $x = 1$.

    c. The graph is rising to the right of $x = 1$.

         The function is increasing for $x > 1$.

    d. The graph is falling to the left of $x = 1$.

         The function is decreasing for $x < 1$.

HELPING HAND In analyzing a function, it is easier to first determine any relative maxima and minima and then determine where the function is increasing or decreasing.

We can also determine an approximate relative maximum or relative minimum on our calculator.

EXAMPLE 5 Graph $g(x) = x^3 + 2x^2 - 4x + 4$ on a window $(-4.7, 4.7, -31, 31)$.

    a. What are the relative maxima of the function?

    b. What are the relative minima of the function?

    c. For what x-values is the function increasing?

    d. For what x-values is the function decreasing?

Calculator Solution
See Figure 3. 13a and Figure 3. 13b.

a. The function has a relative maximum of $y = 12$ at $x = -2$ (a high point).

b. The function has a relative minimum of $y = 2.5$ at $x = 0.7$ (a low point).

c. The graph is rising to the left of $x = -2$ and to the right of $x \approx 0.7$.
The function is increasing for $x < -2$ and $x > 0.7$.

d. The graph is falling between $x = -2$ and $x \approx 0.7$.
The function is decreasing for $-2 < x < 0.7$.

HELPING HAND Note that the x-values incorporate all real numbers, the domain of the function.

TECHNOLOGY

RELATIVE MAXIMUM AND RELATIVE MINIMUM

Determine the relative maximum and relative minimum of the function
$g(x) = x^3 + 2x^2 - 4x + 4$.

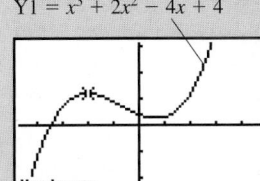

$(-4.7, 4.7, -31, 31)$

Figure 3.13a

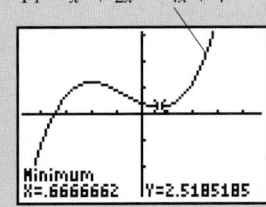

$(-4.7, 4.7, -31, 31)$

Figure 3.13b

For Figure 3.13a,
First graph the function as Y1.

A high point on the graph, $(-2, 12)$, can be found by tracing the graph.
The calculator will estimate a maximum function value between two given values called the left bound and the right bound. Choose MAXIMUM under the CALC function, option 4, by pressing | 2nd | | CALC | | 4 |. Move the cursor to the left of the high point and press | ENTER |. Move the cursor to the right of the high point and press | ENTER |. Move as close as possible to the high point and press | ENTER |. Note that the approximation is not exact.

For Figure 3.13b,
First graph the function as Y1.

A low point on the graph cannot be found by tracing the graph.

(Continued on page 192)

The calculator will estimate a minimum function value between two given values called the left bound and the right bound. Choose MINIMUM under the CALC function, option 3, by pressing [2nd] [CALC] [3]. Move the cursor to the left of the low point and press [ENTER]. Move the cursor to the right of the low point and press [ENTER]. Move as close as possible to the low point and press [ENTER].

The function has a relative maximum of $y = 12$ at $x = -2$.
The function has a relative minimum of $y \approx 2.5$ at $x \approx 0.7$.

3.4.2 Checkup

1. Use the following graph to answer the questions that follow:

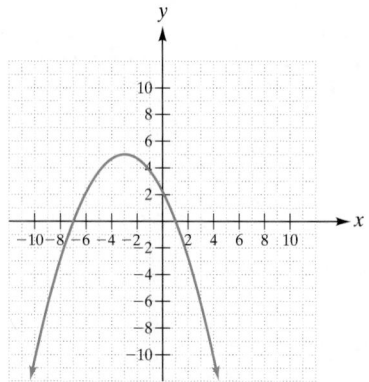

a. What are the relative maxima of the function that is graphed?
b. What are the relative minima of the function that is graphed?
c. For what x-values is the function increasing?
d. For what x-values is the function decreasing?

In exercises 2–4, use your calculator to determine (a) the relative maxima, (b) the relative minima, (c) between what x-values each function is increasing, and (d) between what x-values each function is decreasing. Use the indicated calculator window.

2. $y = -|x + 1|$; decimal screen.
3. $y = 1 - 4x + 2x^2$; decimal screen.
4. $y = 3x^4 - 14x^3 + 54x - 3$; $(-4.7, 4.7, 1, -62, 62, 20, 1)$.
5. What does it mean to say that a function is increasing or decreasing over a certain part of its domain?
6. Define *relative maximum* and *relative minimum*.

3.4.3 Determining Intersections

The **intersection** of two graphs is the location where the two graphs cross. There may be more than one point of intersection. The intersection is significant because at that point the x-values of the two graphs are equal and the *y*-values of the two graphs are equal. This will turn out to be important as we continue our discussion.

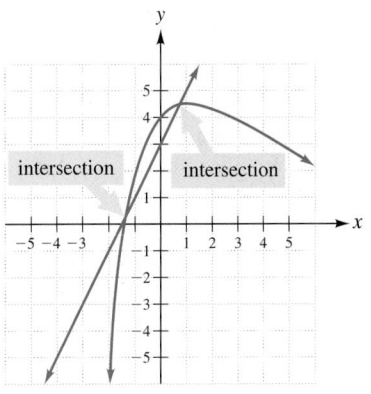

EXAMPLE 6   Determine the point of intersection of the graphs shown in the accompanying diagram.

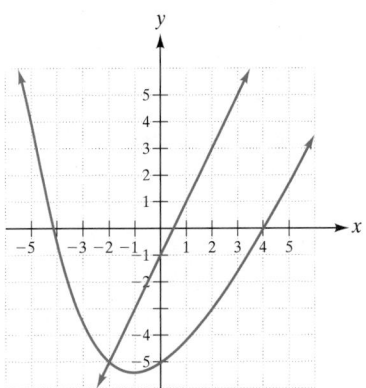

Solution
The point of intersection of the graphs is $(-2, -5)$.

EXAMPLE 7   The calculator has a feature to determine the point of intersection of two graphs.

Graph the functions $y = 2x$ and $y = x + 1$ on a decimal screen. Determine the point of intersection.

Calculator Solution
The point of intersection of the graphs is $(1, 2)$. See Figure 3.14b.

TECHNOLOGY

INTERSECTION OF TWO GRAPHS

Determine the point of intersection of the graphs of $y = 2x$ and $y = x + 1$.

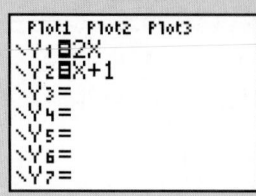

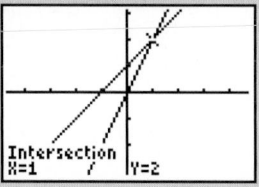

$(-4.7, 4.7, -3.1, 3.1)$

Figure 3.14a   Figure 3.14b

For Figure 3.14a,

Enter $y = 2x$ as Y1.   Move to Y2.   Enter $y = x + 1$ as Y2.

| Y= | 2 | X,T,θ,n | ▼ or | ENTER | | X,T,θ,n | + | 1 |

For Figure 3.14b,

Graph the curve on a decimal screen.

| ZOOM | 4 |

Trace the graphs to find their intersection by using the left and right arrow keys. To move between the graphs, use the up and down arrow keys. The point of intersection is $(1, 2)$.

To check the point of intersection, use INTERSECT under the CALC menu, option 5, by pressing | 2nd | CALC | 5 |. Move the cursor to the closest location to the intersection on the first graph, and press | ENTER |. Move the cursor to the closest location to the intersection on the second graph, and press | ENTER |. Move as close as possible to the intersection point, and press | ENTER |.

The point of intersection is $(1, 2)$.

✔ 3.4.3 Checkup

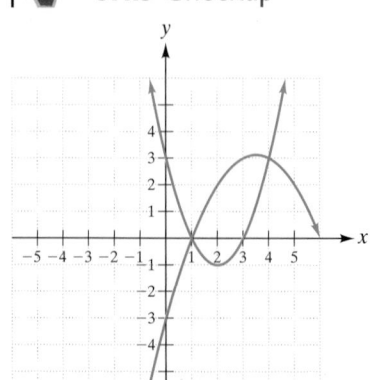

1. Determine the point or points of intersection of the graphs.

In exercises 2 and 3, find the points or point of intersection of the graphs of the two functions.

2. $y = 2x - 15$ and $y = -2x + 5$

3. $y = 1.5(x - 1)$ and $y = x^2 - 4$

4. What is the term for the location at which two graphs cross? What is important about the coordinates of the point of crossing?

3.4.4 Modeling the Real World

We have seen that graphs are a very important way of representing functions that model real-world situations. One reason for this is that they make it easier to visualize things such as maximum and minimum values of functions or intersections of functions. However, remember that dealing with real-world situations, the answers you get must make practical sense as well as mathematical sense. For example, suppose your function represents the number of T-shirts you need to sell to make a profit, and you find that your profit is a maximum if you sell 7.89 T-shirts. This means that you need to sell 8 T-shirts, since fractions of a T-shirt are not part of your sales model.

EXAMPLE 8 Dougal plans to decorate sweatshirts and sell them for a profit. She spends $14.75 for paint. The sweatshirts cost $7.14 each. The appliques cost $3.96 per shirt. She plans to sell the shirts for $20.00 each.

a. Write a cost function, $C(x)$, for the cost of the x sweatshirts.

b. Write a revenue function, $R(x)$, for the amount collected from the sale of x sweatshirts.

c. Graph the cost and revenue functions on a calculator. Use the following settings for your window: (0, 10, 1, 0, 62, 10, 1). Determine the intersection of the two graphs. Interpret the meaning of the coordinates of the intersection.

Solution

a. Let x = the number of sweatshirts decorated and sold

The sweatshirts cost $7.14 per sweatshirt, or $7.14x$.

The appliques cost $3.96 per sweatshirt, or $3.96x$.

The paint costs $14.75.

$$C(x) = 7.14x + 3.96x + 14.75, \quad \text{or} \quad C(x) = 11.10x + 14.75$$

b. The sweatshirts sell for $20.00 each, or $20.00x$.

$$R(x) = 20.00x \quad \text{or} \quad R(x) = 20x$$

c. The intersection of the graphs is (1.657, 33.146). Therefore, Dougal must make and sell approximately 1.657 sweatshirts for approximately $33.15 in order to break even (equal cost and revenue). Actually, Dougal must make and sell 2 sweatshirts, since it is impossible to sell 1.657 sweatshirts.

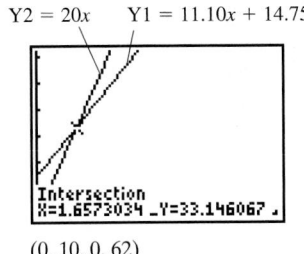

Y2 = 20x   Y1 = 11.10x + 14.75

Intersection
X=1.6573034 Y=33.146067

(0, 10, 0, 62)

APPLICATION

The total net income of a company is the sum of the company's income from ongoing and discontinued operations, plus all other positive or negative adjustments. The total net income of y billion dollars for the x years after 1991 for General Motors is represented in the following graph:

(Continued on page 196)

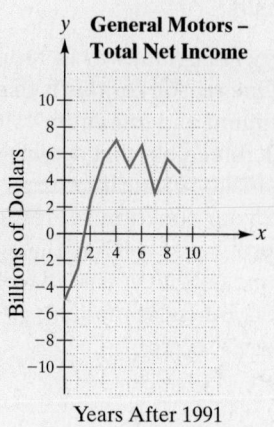

Years After 1991

Approximate the *x*-intercept and discuss its relevance.

Discussion

According to the graph, the total net income function has an *x*-intercept between $(1, 0)$ and $(2, 0)$, or about $(1.5, 0)$. General Motors Corporation had a negative net income (a loss) when $x = 1$ (the year 1992) and a positive net income (a profit) when $x = 2$ (the year 1993).

 3.4.4 Checkup

1. Aquarius packages the new U.S. Mint's Washington Quarters States Collection quarters in plastic containers and resells them. He pays $0.35 for each container and obtains the quarters at a local bank at their face value of $0.25 each. He spends $35.00 to place an advertisement in the local trading newspaper and sells the packaged quarters for $1.00 each.
   a. Write a cost function, $C(x)$, for the cost of packaging and advertising the sale of *x* quarters.
   b. Write a revenue function, $R(x)$, for the amount of money collected from selling *x* quarters.
   c. Graph the cost and revenue functions on a calculator, using $(0, 150, 10, 0, 150, 10, 1)$ as the window. Determine the intersection point and interpret its meaning.

2. The president's budget for fiscal year 2001 reports the unified surplus, *y*, in billions of dollars, for *x* years since 1992 as shown in the following graph (the unified surplus includes payments into and out of the Social Security fund to determine all government receipts and spending):

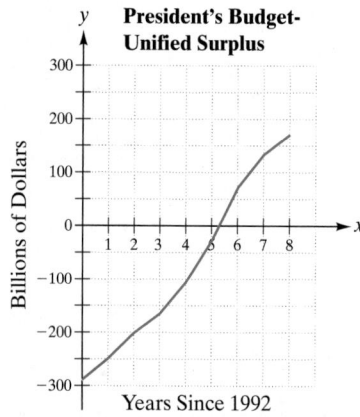

Years Since 1992

Approximate the *x*-intercept and explain its meaning.

3.4 Exercises

Use each graph to determine the following:

   a. *x-intercepts*   **b.** *y-intercepts*   **c.** *relative maxima*   **d.** *relative minima*

   e. *x-values for which the function is increasing*   **f.** *x-values for which the function is decreasing*

1.

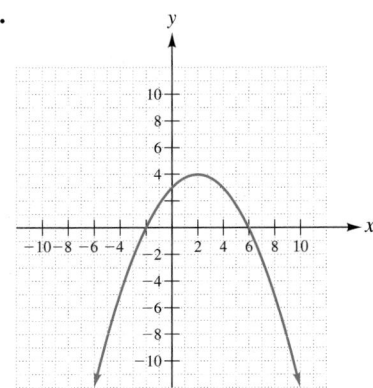

2.

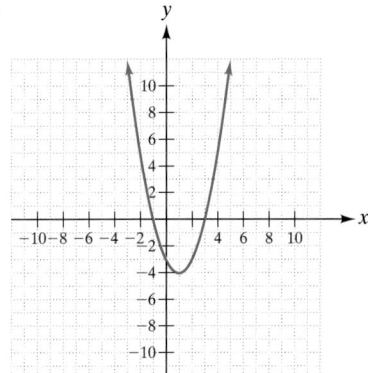

3.

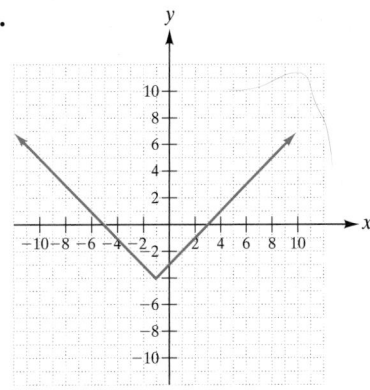

4.

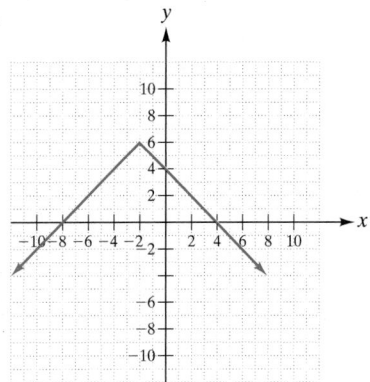

Graph each function and determine the x-intercepts and the y-intercepts.

5. $y = 3x - 6$

6. $y = 4x + 8$

7. $y = \frac{1}{2}x + 1$

8. $y = \frac{2}{3}x + 4$

9. $y = 1.2x - 6$

10. $y = 0.2x + 1$

11. $f(x) = -12x + 24$

12. $F(x) = 15x - 45$

13. $f(x) = 9x + 15$

14. $G(x) = -8x + 36$

15. $y = x^2 - 9$

16. $y = x^2 - 16$

17. $y = x^2 + 6x + 9$

18. $y = x^2 - 4x + 4$

19. $y = 4x^2 + 4x + 1$

20. $y = 4x^2 - 12x + 9$

21. $g(x) = x^2 + 10x - 3$

22. $f(x) = 0.4x^2 - 0.4x - 6.5$

23. $H(x) = x^2 - 5x - 24$

24. $g(x) = 2x^2 + 13x - 70$

25. $y = x^3 + x^2 - 2x$

26. $y = x^3 + 4x^2 + 3x$

27. $f(x) = x^3 + 2x^2 - x - 2$

28. $f(x) = x^3 + x^2 - 4x - 4$

29. $h(x) = |x| - 6$

30. $f(x) = |2x| - 6$

31. $y = |2x - 3| - 1$

32. $y = 5 - |3x + 1|$

33. $y = |x^2 - 2| - 1$

34. $y = |x^2 - 3| - 1$

Graph each function. Determine any relative minima and relative maxima. Then determine the x-values for which the function is increasing and the x-values for which the function is decreasing.

35. $y = 2x + 8$

36. $y = 4 - x$

37. $f(x) = 3 - 2x$

38. $g(x) = x - 2$

39. $y = 1 - x^2$

40. $y = x^2 + 1$

41. $g(x) = x^2 + 4x + 3$

42. $h(x) = 4x - 5 - x^2$

43. $y = |x + 3|$

44. $y = |x| + 3$

45. $f(x) = -|x + 3|$

46. $p(x) = -|x| + 3$

Given each pair of functions, find the point or points of intersection of their graphs.

47. $y = 3x - 5$ and $y = -2x + 15$

48. $y = x - 7$ and $y = -x + 9$

49. $f(x) = 2x + 7$ and $g(x) = -x + 1$

50. $f(x) = -x - 3$ and $g(x) = 2x$

51. $y = -5x + 2$ and $y = 3x + 8$

52. $y = -4x$ and $y = -x + 8$

53. $r(x) = 5x - 7$ and $c(x) = 12$

54. $p(x) = 5$ and $q(x) = -3x + 14$

55. $y = 3$ and $y = -x^2 + 4$

56. $y = x^2 - 16$ and $y = 9$

57. $f(x) = 2x^2 - 4x + 5$ and $g(x) = 4x - 1$

58. $m(x) = -2x^2 - 8x - 2$ and $n(x) = 2x + 6$

59. $y = \frac{1}{4}x^2 - 2$ and $y = \frac{1}{2}x$

60. $y = \frac{1}{3}x^2 - 6$ and $y = x$

61. $y = |x| - 5$ and $y = 2$

62. $y = |2x + 8| - 12$ and $y = 4$

63. A promotion to sell laboratory equipment to a school offers each item for a regular price of $400. However, to stimulate sales, the price for each piece will be reduced by $10 times the number of pieces sold, up to a maximum of 25 pieces. A function for the total cost of purchasing x pieces of equipment is given by $c(x) = x(400 - 10x)$, or $c(x) = 400x - 10x^2$. Graph the function, using a window of $(0, 94, 10, -3100, 6200, 1000, 1)$. Determine any relative minima or relative maxima. Then determine for which x-values the function is increasing and for which x-values the function is decreasing.

64. Roadrunners Bus Company charges $100 per person for a weekend bus excursion. To promote reservations, the company offers to reduce the price per person by $1 times the number of people who take the trip. The bus can hold up to 65 people. A function for the revenue if x people take the trip is given by $r(x) = x(100 - 1x)$, or $r(x) = 100x - x^2$. Graph the function, using a window of $(0, 188, 10, -3100, 6200, 1000, 1)$. Determine any relative minima or relative maxima. Then determine for which x-values the function is increasing and for which x-values the function is decreasing.

65. To operate her craft booth, Jillie has a fixed daily expense of $50. She collects a fee of $5 on each craft sold. (That is, items are priced to sell at $5 above wholesale cost.) Write a function for Jillie's daily profit after expenses are subtracted from revenue when x items are sold. Graph the function, using a window of $(0, 94, 10, -310, 620, 100, 1)$. Determine any relative minima or relative maxima. Then determine for which x-values the function is increasing and for which x-values the function is decreasing.

66. Sierra starts college with a scholarship of $50,000. She withdraws $5,000 per semester for tuition and expenses. Write a function representing the amount of money remaining in the scholarship fund after x semesters of withdrawals. Graph the function, using a window of $(0, 18.8, 2, -31000, 62000, 10000, 1)$. Determine any relative minima or relative maxima. Then determine for which x-values the function is increasing and for which x-values the function is decreasing.

Write a function for each situation and use it to answer the questions.

67. Charlie's Container Company can build containers at a cost of $4 per container, with a setup cost of $50 per run. The company sells the containers for $10 each.
  a. Write a cost function for the cost of making x containers in one run.
  b. Write a revenue function for the revenue received for selling x containers.
  c. Graph the two functions on the same coordinate plane and determine their point of intersection. Use a window of $(0, 9.4, 1, 0, 124, 20, 1)$.
  d. Interpret what this point of intersection tells you.

68. Jim's Carvings, Incorporated, pays a carver $12 per carving plus a bonus of a $100 for signing up to supply carvings to the company. The company retails the carvings for $24 each.
  a. Write a cost function for the cost of obtaining x carvings from a carver.
  b. Write a revenue function for the money received from selling x carvings.
  c. Graph the two functions and locate the point of intersection of the graphs. Use a window of $(0, 18.8, 2, 0, 620, 100, 1)$.
  d. What does the point of intersection indicate?

69. Tatyana's employers make a deal with her to encourage her in her college work. They offer her a bonus of $200 plus $50 for each credit hour with a passing grade. Alternatively, if she prefers, instead of the $200, they will pay her $75 for each credit hour with a passing grade.
  a. Write functions to represent how much Tatyana will receive under each offer if she receives x credit hours with passing grades.
  b. Graph the two functions on the same coordinate plane and determine their point of intersection. Use a window of $(0, 18.8, 2, 0, 930, 150, 1)$.
  c. Explain what the point of intersection means to Tatyana.

70. Dandylawn Mowing Service offers Khalid a job. The service gives him a choice of two payment methods. He can earn a base pay of $25 per week plus $10 for each lawn he mows, or he can earn $15 per lawn mowed with no base pay.

   a. Write functions to represent how much Khalid will receive under each payment plan if he mows x lawns a week.

   b. Graph the two functions on the same coordinate plane and determine their point of intersection. Use a window of (0, 9.4, 1, 0, 124, 20, 1).

   c. What does the point of intersection represent to Khalid?

71. The total net income, y million dollars, for x years after 1995, for Yahoo! Inc., is shown in the following graph:

Total Net Income–Yahoo! Inc.

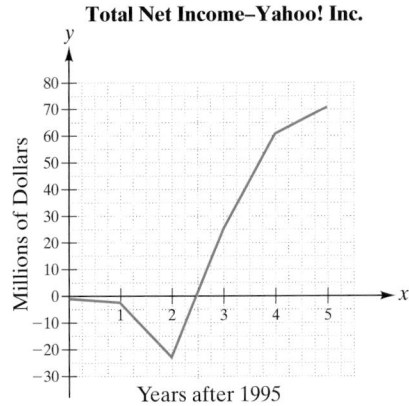

Years after 1995

Estimate the x-intercept and interpret its meaning.

72. The total net income, y million dollars, for x years after 1993, for Northwest Airlines Corporation is shown in the following graph:

Total Net Income–Northwest Airlines Corporation

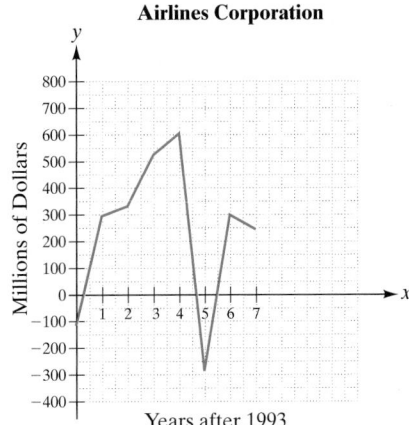

Years after 1993

Estimate the x-intercepts and interpret their meanings.

3.4 Calculator Exercises

Graph the following two functions, using the integer setting for the coordinate screen.

$$y = \frac{2}{3}x^2 - 6 \quad \text{and} \quad y = -2x - 6$$

Can you tell by looking at the screen how many times the line crosses the curve? It is difficult to see, but it does cross twice. Try changing the setting by using the ZOOM button. First try ⎡ZOOM⎤ ⎡4⎤. Is this better? Can you find the points of intersection now? Next, try ⎡ZOOM⎤ ⎡6⎤. Is this better? If not, try ⎡ZOOM⎤ ⎡8⎤ ⎡ENTER⎤. You will learn more about the ⎡ZOOM⎤ button later, but you should see that it can help

you see the graph better if you learn how to vary the setting. Remember, you can't get lost when you experiment with the ⎡ZOOM⎤ key. You can always get back to the integer setting simply by pressing ⎡ZOOM⎤ ⎡6⎤, waiting until the graph is drawn, and then pressing ⎡ZOOM⎤ ⎡8⎤ ⎡ENTER⎤.

Experiment with the ZOOM setting to find the points of intersection of each pair of functions.

1. $g(x) = \frac{1}{3}x + 3$ and $f(x) = \frac{1}{4}x^2 - 4$

2. $y = |x| - 6$ and $y = -|x| + 4$

3. $y = x^2 - 18$ and $y = -x^2 + 54$

3.4 Writing Exercise

In analyzing graphs of functions, we stated that there may be more than one x-intercept for a function, but that there must not be more than one y-intercept for a function. Explain what we mean when we refer to x-intercepts and y-intercepts, and then explain why a function can have several x-intercepts but only one y-intercept. (*Hint:* Think about the vertical-line test for checking whether a graph is the graph of a function.) Draw an example of a function that has two x-intercepts; one that has three x-intercepts; one with four x-intercepts.

CHAPTER 3 SUMMARY

After completing this chapter, you should be able to define in your own words the following key terms:

3.1
table of values
independent variable
dependent variable
data
ordered pair
relation
finite
infinite
domain
range

3.2
rectangular coordinate system
Cartesian coordinate system

coordinate plane
x-axis
y-axis
origin
quadrants
coordinates
x-coordinate
y-coordinate

3.3
function
vertical-line test
function notation
input value
output value

3.4
x-intercept
y-intercept
increasing
decreasing
constant
relative maximum
relative minimum
intersection

Reflections

1. What is the difference between a dependent variable and an independent variable?
2. Explain the importance of order in an ordered pair of values.
3. In mathematics, what is a relation? What is the domain of a relation? What is the range of a relation?
4. Describe the Cartesian coordinate system and discuss its use.
5. What is a function? Are all relations functions? Explain.
6. What is an intercept of a graph?
7. Describe what a relative minimum or a relative maximum of a graph represents.
8. When two relations are graphed on the same coordinate system, what is the significance of any points of intersection?

CHAPTER 3 SECTION-BY-SECTION REVIEW

Express your answers in fractional form, or round to nearest thousandth, as appropriate.

3.1

Recall	Examples
Table of values • First column: values of independent variable • Second column: equation • Third column: values of dependent variable	Construct a table of values for $y = x + 4$ when $x = \{1, 2, 3\}$.

x	$y = x + 4$	y
1	$y = 1 + 4$	5
2	$y = 2 + 4$	6
3	$y = 3 + 4$	7

Recall	Examples
Ordered pairs • Two numbers enclosed in parentheses and separated by a comma. • First number is the value of the independent variable. • Second number is the value of the dependent variable.	Ordered pairs for the previous example are $(1, 5)$, $(2, 6)$, and $(3, 7)$.
Domain • Set of all possible values of the independent variable.	The domain of the previous example is $\{1, 2, 3\}$.
Range • Set of all possible values of the dependent variable.	The range of the previous example is $\{5, 6, 7\}$.

Construct a table of values for each equation, given the set of values for the independent variable.

1. $b = -2a + 7$, given $\{-3, -2, -1, 0, 1, 2, 3\}$ for a.

2. $y = \dfrac{2}{3}x + 4$, given $\{9, 6, 3, 0, -3, -6, -9\}$ for x.

3. $y = 0.4x - 1.2$, given $\{-3, -2, -1, 0, 1, 2, 3\}$ for x.

4. $y = 5x^3 - 3x^2 + 2x - 22$, given $\{-18, -7, 0, 6, 21, 22.5\}$ for x.

5. $y = |x^2 - 6x + 5|$, given $\{-2, -1, 0, 1, 2, 3\}$ for x.

6. $y = 3.6x^2 + 1.5x - 14.2$, given $\{-2.7, -1.9, -0.6, 0, 0.8, 1.5, 2.4\}$ for x.

Construct a table of values for each equation. Select five values for the independent variable, x, in each table.

7. $y = -1.6x + 4.5$ **8.** $y = \dfrac{3}{2}x - 6$ **9.** $y = |2x - 9|$

In exercises 10–12, construct a table of values for each equation with the given domain.

10. $y = (5x + 2)(x - 4)$, for odd integer values of x between -6 and 6.

11. $y = \dfrac{3}{5}x + 8$, where x is an integer multiple of 5 between -20 and 20.

12. $y = 17.1x - 12.9$, where x is an integer value between -4 and 4.

13. In the spring of 2001, the price of unleaded gasoline was posted as $1.649 per gallon.
   a. Write an equation for the total cost of purchasing a given number of gallons of gasoline.
   b. Complete a table of values for the total cost of purchasing 5, 10, 15, and 20 gallons of gasoline.

Construct a table of values for each application with the given domain.

14. The area of a circle for even integer values of the radius between 2 and 12 inches.

15. The measure of an angle with a complementary angle measuring $\{10, 20, 30, 40, 45\}$ degrees.

16. The amount of simple interest on an investment of $2000 at a rate of interest of 6% for t years, where t assumes integer values between 1 and 5.

17. The Celsius temperature when the temperature assumes values of $\{-23, -14, 0, 41, 50, 59, 100\}$ degrees Fahrenheit.

Write ordered pairs.

18. $y = 7 - 3x$ when $x = -10, -5, 0, 5, 10$.

19. $y = \sqrt{x + 8}$ when $x = -8, -7, -4, 1, 8$.

20. $d = \dfrac{2}{3}c - 1$ when $c = -6, -3, 0, 3, 6$.

21. The radius of a circle whose diameter measures 2, 4, 6, 8, and 10 inches.

22. The U.S. Patent and Trademark Office reported the following summary of the number of trademarks issued (in thousands, so that 86.9 represents 86,900) for each year between 1993 and 1997:

Year	1993	1994	1995	1996	1997
Trademarks	86.9	70.1	92.5	98.6	119.9

Write a set of ordered pairs that represents this information.

In exercises 23–28, determine the domain and the range of each relation. Assume that x is the independent variable.

23. $P = \{(1, 2), (3, 6), (5, 10), (7, 14), (9, 18)\}$

24. $C = \{\ldots, (-6, 6), (-4, 4), (-2, 2), (0, 0), (2, -2), (4, -4), (6, -6), \ldots\}$

25. $y = 4(x + 5) - 1$ for $x = \{-5, -4, -3, -2, -1\}$

26. $y = x^2 + 2.5$, for $x = \{0, 0.5, 1, 1.5, 2\}$

27. $y = \sqrt{5 + x}$, for $x = \{-5, -4, -3, -2, -1, 0\}$

28. $y = \dfrac{12}{1 - x}$, for $x = \{-4, -2, 0, 2, 4\}$

29. $T = 2\pi\sqrt{\dfrac{L}{32}}$, where T is the period of a pendulum that has a suspension of L feet. (L is the independent variable.) Construct a table of values of the period, where L assumes the values 1, 8, 16, 24, and 32 feet.

3.2

Recall	Examples
Graph a relation. • Plot all points given • If no points are given, set up a table of values and plot the ordered pairs determined from the table.	Graph $y = x^2 - 3$. Determine a sample set of ordered pairs. Let $x = \{-2, -1, 0, 1, 2\}$.

x	$y = x^2 - 3$	y
-2	$y = (-2)^2 - 3$	1
-1	$y = (-1)^2 - 3$	-2
0	$y = (0)^2 - 3$	-3
1	$y = (1)^2 - 3$	-2
2	$y = (2)^2 - 3$	1

Domain	
Domain • Determine all values that can be graphed for the independent variable.	The domain of the previous example is all real numbers.
Range • Determine all values that can be graphed for the dependent variable.	The range of the previous example is all real numbers greater than or equal to -3, $y \geq -3$.

Graph and label each ordered pair on a coordinate plane.

30. $A(3, 2)$ **31.** $B(4, -3)$ **32.** $C(-3, 2)$ **33.** $D(-4, -3)$ **34.** $E(0, 5)$ **35.** $F(-5, 0)$

36. State the coordinates of the points graphed in Figure 3.15.

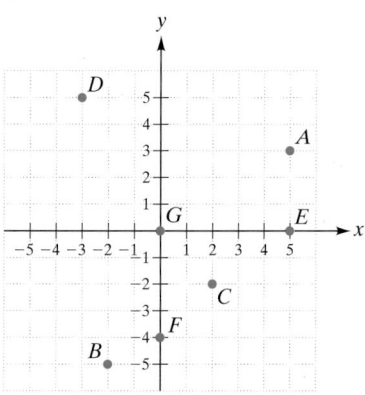

Figure 3.15

Using Figure 3.15, state the quadrant in which each plotted point lies or the axis on which the point lies.

37. A	**38.** B	**39.** C	**40.** D
41. E	**42.** F	**43.** G	

Graph each relation.

44. $S = \{(0, -4), (1, -3), (2, -2), (3, -1), (4, 0), (5, 1)\}$

45.

x	y
-1	-5
0	-3
1	-1
2	1
3	3
4	5

46. $y = 3 - 2x$ **47.** $T = \{(-5, 3), (-3, 3), (-1, 3), (1, 3), (3, 3), (5, 3)\}$

Determine the domain and range of each relation.

48.

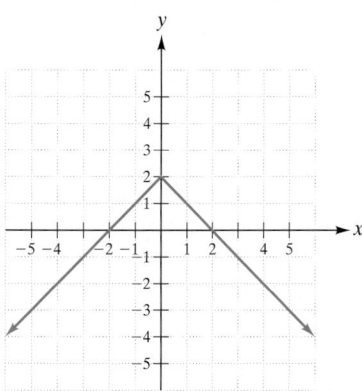

49.

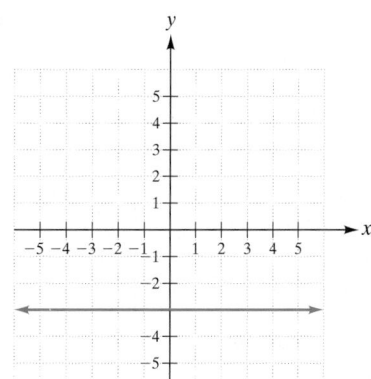

50. The U.S. energy supply and disposition are shown in the following figure:

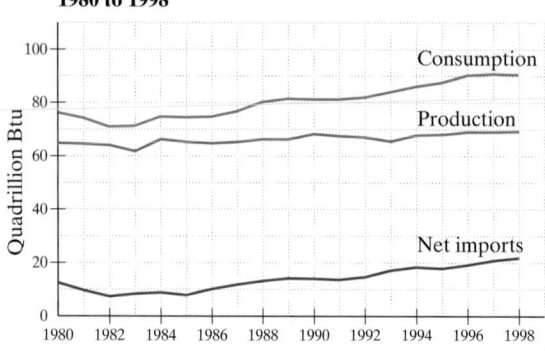

Energy Production, Trade, and Consumption: 1980 to 1998

Source: U.S. Census Bureau, Statistical Abstract of the United States: 1999

Consider the line for consumption as the relation of interest. Determine the domain and range of this relation.

51. The weekly charge for a family to send its children to a child-care center is $40 plus $10 per child.
   a. Write a relation for the weekly charge, y, in terms of the number of children, x.
   b. Graph the relation, using $(-4.7, 4.7, 1, -31, 93, 10, 1)$ as the graphing window on your calculator.
   c. State the domain and range of the relation.

52. A square has sides measuring x feet.
   a. Write a relation for the area, y square feet, in terms of the length of a side, x feet.
   b. Graph the relation, using $(-4.7, 4.7, 1, -6.2, 31, 10, 1)$ as the graphing window on your calculator.
   c. State the domain and range of the relation.

3.3

Recall	Examples
Vertical-line test • A graph represents a function if all possible vertical lines intersect or touch the graph only once.	$y = f(x)$
Function notation • An equation in function notation consists of the name of the function, the name of the independent variable, and an expression to be evaluated to determine the dependent variable. • To evaluate a function for a given value, substitute the value for the independent variable.	Given $f(x) = -x^2 + 2x - 1$, evaluate $f(-3)$ and $f(a)$. $f(-3) = -(-3)^2 + 2(-3) - 1$ $\quad$ $f(a) = -(a)^2 + 2(a) - 1$ $f(-3) = -16$ $\qquad\qquad$ $f(a) = -a^2 + 2a - 1$

Determine whether each relation is a function.

53. $S = \{(-2, 3), (-1, 5), (0, 7), (-1, 9), (-2, 11)\}$ **54.** $T = \{(-2, 3), (-1, 5), (0, 7), (1, 5), (2, 3)\}$

55.

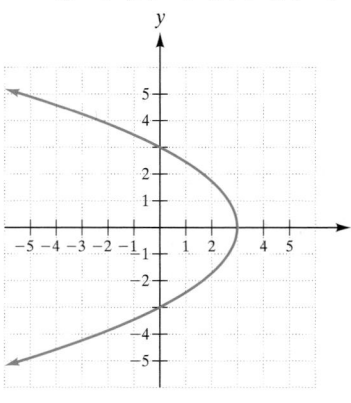

56.

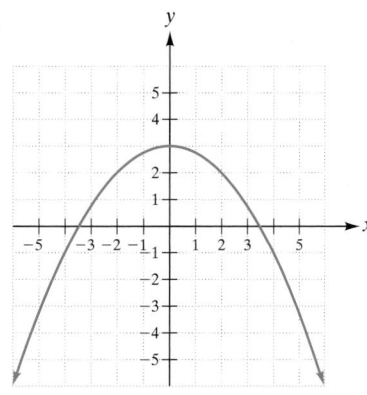

Given $f(x) = -4x + 13$, find

57. $f(13)$ **58.** $f(-21)$ **59.** $f(2.5)$ **60.** $f(-3.7)$ **61.** $f(3 + h)$ **62.** $f(-b)$

Given $g(x) = 5x^2 + x - 4$, find

63. $g(3)$ **64.** $g(-2)$ **65.** $g(0.5)$ **66.** $g(a)$ **67.** $g(-a)$ **68.** $g\left(-\dfrac{1}{4}\right)$

Given $S(x) = \sqrt{2x + 3} - 5$, find

69. $S(3)$ **70.** $S(11)$ **71.** $S(59)$

72. A process requires $4500 to set up for a production run. The cost of labor and materials to produce a single widget is $17. Write a cost function for a production run. What is the cost of producing 1200 widgets on one production run?

73. A learning institute will conduct a training session at your company for a fee of $1500 plus a charge of $125 per person attending. Write a function for the total charge of a training session. What will be the charge for a training session for 20 employees?

74. Dmitri charges $1.50 to paint faces at a church carnival. He purchased his supplies for $15.00. Write a profit function for painting faces. How much profit will Dmitri make for the church if he paints 135 faces?

75. Using data from a financial services organization, analysts can estimate the net annual sales (in billions of dollars) for Wal-Mart Stores by the function $S(x) = 2.1x^2 + 5.7x + 83.9$ for x years after 1995. Use this function to estimate the net sales for the year 2000.

3.4

Recall	Examples
y-intercept • Location where a graph touches or intersects the *y*-axis.	*y*-intercept $(0, 4)$

(Continued on page 206)

Recall	Examples
x-intercept • Location where a graph touches or intersects the *x*-axis.	 *x*-intercept $(-2, 0)$ *x*-intercept $(0, 0)$
Relative minimum • A function value that is smaller than its neighboring function values.	 $(1, -3)$ Relative minimum of $y = -3$ at $x = 1$.
Relative maximum • A function value that is larger than its neighboring function values.	 $(-1, 2)$ Relative maximum of $y = 2$ at $x = -1$.

Recall	Examples
Decreasing function • A function is decreasing if the function values decrease as the values of the independent variable increase.	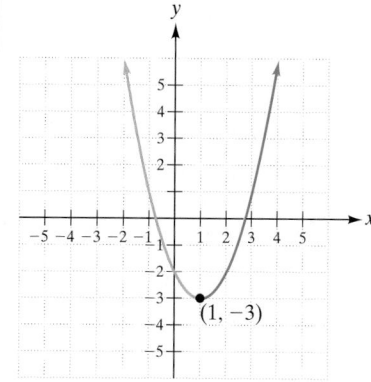 The function is decreasing for $x < 1$.
Increasing function • A function is increasing if the function values increase as the values of the independent variable increase.	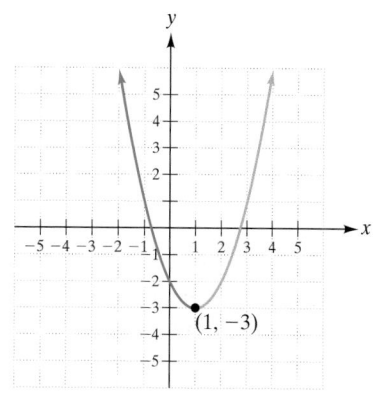 The function is increasing for $x > 1$.
Intersection of two graphs • The point where two graphs cross.	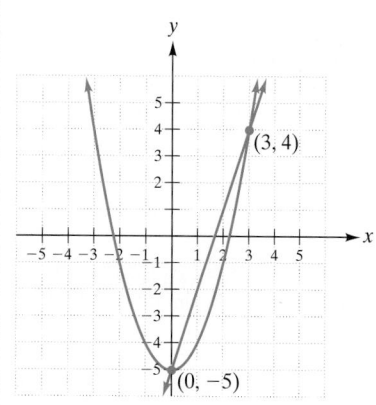 Intersections are $(0, -5)$ and $(3, 4)$.

76. Use the graph to determine the following:

   a. x-intercepts

   b. y-intercepts

   c. relative maxima

   d. relative minima

   e. x-values for which the function is increasing

   f. x-values for which the function is decreasing

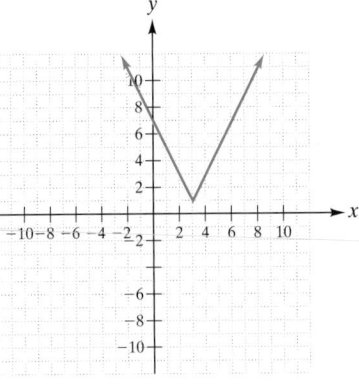

Graph each function and determine the x-intercepts and the y-intercepts.

77. $y = 3x + 9$ **78.** $y = \dfrac{3}{4}x - 9$

79. $y = x^2 - 0.36$ **80.** $y = |x| - 4$

Graph each function. Determine any relative minima and relative maxima. Then determine the values of x for which the function is increasing and the values of x for which it is decreasing.

81. $h(x) = 6 - 2x$ **82.** $y = 3 - x^2$ **83.** $y = |x| + 2$ **84.** $f(x) = |x^2 - 1|$

What are the points of intersection of the graphs of each pair of functions?

85. $y = 2x - 2$ and $y = -\dfrac{1}{3}x + 5$ **86.** $y = x^2 - 6$ and $y = x$ **87.** $f(x) = |x + 5|$ and $g(x) = 2$

88. An office contracts to purchase no more than eight desks that ordinarily sell for $325 each. To encourage multiple sales, the desks are offered at a price of $\$(325 - 15x)$ each, where x is the number of desks purchased. The total cost of purchasing x desks is given by $c(x) = x(325 - 15x)$, or $c(x) = 325x - 15x^2$. Graph this function, using a window of $(0, 9.4, 1, -3100, 3100, 500, 1)$. Determine any relative minima or relative maxima. Then determine for which x-values the function is increasing and for which x-values the function is decreasing.

89. The selling price of a horn switch wiring harness for a car is listed at $10.45. Write a revenue function for x harnesses sold. Graph the function. Determine any relative minima or relative maxima. Then determine for which x-values the function is increasing and for which x-values the function is decreasing.

90. A bank account has a beginning balance of $216.00. There is no activity in the account, except for a monthly service charge of $4.50. Write a function that represents the balance of the account after x months of inactivity. Graph the function. Determine any relative minima or relative maxima. Then determine for which x-values the function is increasing and for which x-values the function is decreasing.

91. Hans can choose to receive a tuition reimbursement stipend of $400 plus $65 per credit hour for each credit hour passed, or he can instead choose to receive $100 per credit hour passed without a stipend.

   a. Write functions that represent how much Hans will receive under each option.

   b. Graph the two functions and determine their point of intersection. Use a window of $(0, 18.8, 1, 0, 3100, 500, 1)$.

   c. What does the point of intersection represent to Hans?

92. The cost of production of certain items includes a setup cost of $500 plus a cost of $12 per item for labor and materials. The finished items sell for $25 each.

   a. Write functions that represent the total cost and total revenue associated with producing and selling x items.

   b. Graph the two functions and determine their point of intersection. Use a window of $(0, 94, 10, 0, 1550, 500, 1)$.

   c. Interpret the point of intersection in business terms.

CHAPTER 3 CHAPTER REVIEW

Given $f(x) = -x + 9,$ *find*

  1. $f(9)$ **2.** $f(-9)$ **3.** $f(1.8)$ **4.** $f(-2.7)$ **5.** $f(-b)$ **6.** $f(1 + h)$

Given $g(x) = x^2 - 3x - 4,$ *find*

  7. $g(4)$ **8.** $g(-1)$ **9.** $g(1.5)$ **10.** $g(v)$ **11.** $g(-v)$ **12.** $g\left(-\dfrac{2}{3}\right)$

Given $S(x) = \sqrt{6x - 8}$, *find*

13. $S(4)$ **14.** $S(12)$ **15.** $S(44)$

Determine whether each relation is a function.

16. $P = \{(3, 8), (2, 6), (1, 4), (0, 2), (-1, 0)\}$ **17.** $Q = \{(2, -3), (2, -2), (2, -1), (2, 0), (2, 1)\}$

Graph each function and determine the x-intercepts and the y-intercepts. Determine any relative minima and relative maxima. Then determine the values of x for which the function is increasing and the values of x for which it is decreasing.

18. $y = 4.8x - 1.2$ **19.** $y = \dfrac{2}{5}x + 4$ **20.** $y = x^2 - 1.21$ **21.** $y = 2 - |x|$

Determine the points of intersection of the graphs of each pair of functions.

22. $y = 2x + 2$ and $y = -2x - 10$ **23.** $y = x^2$ and $y = 3x$ **24.** $f(x) = |2x|$ and $g(x) = x + 3$

Determine the domain and the range of each relation. Assume that x is the independent variable.

25. $A = \{(2, 1), (4, 2), (6, 3), (8, 4), (10, 5)\}$ **26.** $B = \{\dots, (-6, 3), (-4, 3), (-2, 3), (0, 3), (2, 3), (4, 3), (6, 3), \dots\}$
27. $y = x^2$ for $x = \{-5, -4, -3, -2, -1\}$ **28.** $y = x^2 - 1.5$ for $x = \{0, 0.5, 1, 1.5, 2\}$

Write ordered pairs.

29. $y = 12 - 8x$ when $x = -6, -3, 0, 3, 6$. **30.** $y = \sqrt{10 - 3x}$ when $x = 3, 2, 1, 0, -1, -2$.

31. $t = \dfrac{4}{7}s + 5$ when $s = -7, 0, 7, 14, 21$.

Construct a table of values for each relation.

32. $y = (3x - 5)(2x + 1)$, for even integer values of x between -5 and 5.

33. $y = \dfrac{3}{4}x - 5$, where x is an integer multiple of 4 between -13 and 13.

34. $y = 15.8 - 4.7x$, where x is an integer value between -3 and 3.

35. $y = 4x^2 - 17x - 15$, with a domain of $\left\{-2, -\dfrac{3}{4}, 0, \dfrac{3}{4}, 5\right\}$.

36. $y = |1 - 2x - 3x^2|$, given $\{-6, -3, 0, 3, 6, 9\}$ for x.
37. $y = 4.6x^2 + 2.8x + 10.4$, given $\{-3.7, -2.2, -0.7, 0, 0.8, 2.3, 3.8\}$ for x.

Construct a table of values for each relation. Select three values for the independent variable, x, in each table.

38. $y = 4.5x - 1.6$ **39.** $y = \dfrac{1}{4}x + 3$ **40.** $y = |3x - 10|$

41. Write ordered pairs for the circumferences of circles whose radii measure $\frac{1}{4}, \frac{1}{2}, 1, \frac{3}{2},$ and 2 inches.

Construct a table of values for each relation.

42. The area of a square for odd integer values of the length of a side between 1 and 7 feet.

43. The measure of the second of two supplementary angles, where one angle measures $\{30, 60, 90, 120, 150)$ degrees.

44. The simple interest on an investment of $2000 at a rate of interest of 6% for t years, where t assumes integer values between 1 and 5.

45. The Fahrenheit temperature when the temperature assumes values of $\{-10, -5, 0, 5, 10, 15, 20, 25\}$ degrees Celsius.

46. Computer zip disks are on sale for $7.95 each.
   a. Write an equation for the total cost of purchasing a given number of disks.
   b. Complete a table of values for the total cost of purchasing x disks, where x assumes integer values between 0 and 6.

47. The setup costs for a production run for a certain item are $2500. The labor and materials needed to produce a single production item cost $12. Write a function for the total cost of a production run. What will be the total cost of a production run of 1650 items?

48. A rental firm charges $15 to rent a grinder, plus $2 for each hour of rental. Write a function for the total rental cost. What is the cost for renting the grinder for 10 hours?

49. A caterer will arrange an awards luncheon for your employees. He charges $275.00 to rent his party room and $9.50 per person for the luncheon. Write a function for the total charge. What will be the charge for a luncheon for 135 employees?

50. A charity basketball game charges $4 admission. Expenses for the game total $185. Write a function for the profit. What is the profit on 310 admissions?

Write a function to represent each application, graph the function, and determine the values of x for which the function is increasing and decreasing.

51. A storage tank holds 250 gallons of fluid when full. The tank dispenses fluid at a rate of 3.5 gallons per minute. Given a tank that was full, write a function for the amount of fluid remaining in the tank after x minutes of dispensing fluid. Graph the function. Determine any relative minima or relative maxima. Determine x-values for which the function is increasing or decreasing.

52. Jillian's grandparents deposited $1000 into a savings account at her birth. From then on, they deposited $50 per month into the account. Write a function for the total amount deposited into the account by her grandparents after x months. Graph the function. Determine any relative minima or relative maxima. Determine x-values for which the function is increasing or decreasing.

53. A contest winner has the option of receiving $25,000 cash initially plus an annual payment of $5000 per year, or no initial cash payment but an annual payment of $6000 per year.

  a. Write functions to represent how much money the winner will receive under each option.

  b. Graph the two functions and determine their point of intersection. Use a window of (0, 37.6, 4, 0, 248,000, 4000, 1).

  c. What does the point of intersection represent to the winner?

54. A retailer purchases appliances at a cost of $22 per appliance plus a total shipping charge of $600 for each lot ordered. She then sells the appliances for $75 apiece.

  a. Write functions that represent her total acquisition cost for a lot of x appliances and her total revenue for selling x appliances.

  b. Graph the two functions and determine their point of intersection. Use a window of (0, 47, 5, 0, 3100, 500, 1).

  c. Interpret the point of intersection in terms of business decisions.

CHAPTER 3 TEST

A⁺   TEST-TAKING TIPS

Before you start a test, you should "dump your brain." By this, we mean that you should use scratch paper to list all the important information that you will need for the test — formulas, terms, tips, and so on. This gives you something to refer to while taking the test. Then you should scan the test questions and pick out those problems that you think will be the easiest to do. Do those problems first. This will accomplish two things. First, it will allow you to expend your energies on the parts that you know best while you are still fresh. Second, it will build your confidence while you are taking the test and will help you relax knowing that you can do the work. Save the difficult problems for last. Even then, don't spend a lot of time on any one problem. If you can't seem to get a handle on it, go on to another problem. Your subconscious will continue to work on the first problem, and when you return to it, you may find that the solution will come easily.

1. Construct a table of values for $y = 2x^2 + 17x - 9$, given $\{-9, 0, 3\}$ as the set of values for x.

2. Consider the relation $y = |2x - 3| - 1$, where x is the independent variable.

  a. Graph the relation.

  b. What are the relative maxima of the relation?

  c. What are the relative minima of the relation?

  d. For what values of x is the relation increasing?

  e. For what values of x is the relation decreasing?

  f. What are the x-intercepts of the relation?

  g. What are the y-intercepts of the relation?

3. Consider the following relation:

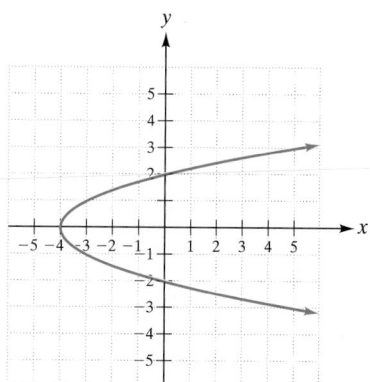

    a. Is this relation a function? Justify your answer.

    b. What is the domain of the relation?

    c. What is the range of the relation?

4. Identify the coordinates of the points in Figure 3.16.

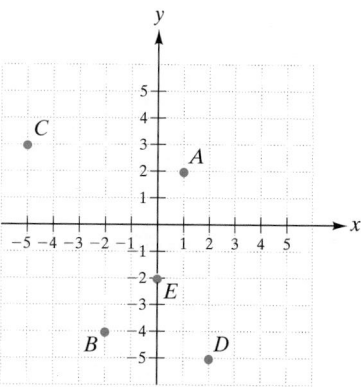

Figure 3.16

5. In the figure, in which quadrant does point D lie?

In exercises 6–8, graph the relations. Label all points used.

6. $y = 2x^2 - 8x$ **7.** $y = \dfrac{3}{4}x - 2$ **8.** $y = \sqrt{7 - x} + 1$

In exercises 9–12, for the function $f(x) = \frac{1}{2}x + 6$, find

9. $f(4)$ **10.** $f(-6)$ **11.** $f(a)$ **12.** $f(2 + a)$

13. A production process has a setup cost of \$450.00. The cost of labor and materials to produce a single item is \$21.50. Write a function that represents the cost of producing x items in one production run.

14. For the production process described in exercise 13, what is the cost of producing 250 items in one production run?

15. Find the intersection of the graphs of $y = 3x - 10$ and $y = -x - 2$.

16. What is meant by the term *ordered pair*, and why is the word *ordered* important?

Chapter 3

Project

The temperature outside your home or school usually fluctuates throughout the day and night. In this project, you will record samples from the daily temperature and discuss various aspects of your data.

You will need a Calculator-Based Laboratory (CBL) with temperature probe and your TI-83 Plus calculator. First, load the DAYTEMP program from the Prentice Hall companion Web site. Next, connect the TI-83 Plus calculator to the CBL that has a temperature probe in Chan 1. The program is designed to record the temperature every hour for 24 hours. It will store in the TI-83 Plus the number of hours, x, after you start the program for L_3 and the temperature, y, in degrees Celsius for L_4. The program will also graph the results on your calculator. Record the time that you start the program. Allow the program to run for a 24-hour period outside.

If you do not have time or equipment to collect the hourly temperatures, you may choose to use projected hourly temperatures. A sample set of data points may also be found on the companion Web site.

Part I

Use the data collected to complete the exercises that follow. Let x be the number of hours after data collection began. Let y be the temperature in degrees Celsius.

1. Write ordered pairs for the set of data collected.

2. Use the data collected to sketch a graph of the daily temperatures. Label all points and the axes.

3. Determine the sets of numbers for the domain and range of this relation. Explain the meaning of these sets of numbers.

4. Does your data represent a function?

5. Does the graph have an x-intercept? If so, explain its meaning. If not, explain why not.

6. Does the graph have a y-intercept? If so, explain its meaning. If not, explain why not.

7. **a.** Determine any relative minima. Where do they occur? Convert each relative minimum to time of day and explain its meaning.

   b. Determine the absolute minimum—the lowest temperature of the day. Is this value also a relative minimum? For what x-value does it occur? Convert the absolute minimum to time of day.

   c. Determine any relative maxima. Where do they occur? Convert each relative maximum to time of day and explain its meaning.

   d. Determine the absolute maximum—the highest temperature of the day. Is this value also a relative maximum? For what x-value does it occur? Convert the absolute maximum to time of day.

8. Between what x-values is the temperature increasing? Between what x-values is it decreasing? Convert the x-values to time of day and explain their meaning.

Part II

Now convert the temperature to Fahrenheit (to two decimal places), and repeat the previous set of exercises with your new data. Let x be the number of hours after data collection began. Let y be the temperature in degrees Fahrenheit.

1. Write ordered pairs for the set of data collected.

2. Use the data collected to sketch a graph of the daily temperatures. Label all points and the axes.

3. Determine the sets of numbers for the domain and range of this relation. Explain the meaning of these sets of numbers.

4. Does your data represent a function?

5. Does the graph have an x-intercept? If so, explain its meaning. If not, explain why not.

6. Does the graph have a y-intercept? If so, explain its meaning. If not, explain why not.

7. a. Determine any relative minima. Where do they occur? Convert each relative minimum to time of day and explain its meaning.

b. Determine the absolute minimum—the lowest temperature of the day. Is this value also a relative minimum? For what x-value does it occur? Convert the absolute minimum to time of day.

c. Determine any relative maxima. Where do they occur? Convert each relative maximum to time of day and explain its meaning.

d. Determine the absolute maximum—the highest temperature of the day. Is this value also a relative maximum? For what x-value does it occur? Convert the absolute maximum to time of day.

8. Between what x-values is the temperature increasing? Between what x-values is it decreasing? Convert the x-values to time of day and explain their meaning.

CHAPTERS 1–3 CUMULATIVE REVIEW

Consider the following real numbers:

$$-\frac{2}{3}, \quad 0, \quad 12, \quad 1\frac{4}{5}, \quad -0.33, \quad \sqrt{7}$$

1. Which numbers are whole numbers?

2. Which numbers are integers?

3. Which numbers are rational numbers?

4. Which numbers are irrational numbers?

Use one of the symbols $>$, $<$, or $=$ to compare the numbers.

5. $\dfrac{3}{8}$ _____ $\dfrac{1}{3}$ **6.** $\dfrac{2}{3}$ _____ 0.66 **7.** -2.8 _____ -1.6

8. Graph the following numbers on a number line, and label the points:

$$-3.1 \qquad -\frac{1}{2} \qquad 2\frac{3}{4} \qquad \sqrt{5} \qquad -\sqrt{25}$$

Evaluate. Express your answers in fractional notation or round decimals to the nearest thousandth, as appropriate.

9. $-28 + 13$

10. $4.8 - 7.36$

11. $-87 \div (-29)$

12. $-\dfrac{5}{8} - \dfrac{2}{3}$

13. $-2\dfrac{3}{4} \div 1\dfrac{3}{7}$

14. $\left(-\dfrac{2}{3}\right)\left(\dfrac{3}{8}\right)\left(-\dfrac{7}{16}\right)\left(\dfrac{9}{10}\right)$

15. $(12.96)(-4.8)$

16. $(14)(0)(5)(-6)$

17. $(-12)(16) \div 4(-2)$

18. $14 + (-7) + 22 - 16 - (-18)$

19. $-[3.8 - (-2.4)]$

20. $\dfrac{2(3^2 + 7) - 2^5}{3.18}$

21. $-|12 - 20|$

22. $\sqrt{\dfrac{16}{25}}$

23. $-\sqrt{1.2}$

24. $\sqrt{-16}$

25. $\sqrt[3]{1\dfrac{13}{81}}$

26. 14^0

27. 1^{12}

28. 0^0

29. -8^4

30. $(-8)^4$

31. $\left(\dfrac{1}{10}\right)^{-2}$

Write in scientific notation.

32. 0.00000305 **33.** $-4,235,600$

Write in standard notation.

34. 3.56×10^{-2} **35.** 6.78×10^8

36. Evaluate the expression $\sqrt{-x^2 + 5x - 2} + 5$ for $x = 3$.

37. Consider the algebraic expression $a^3 - 2a^2 + a - 2a^3 + 7a - 5$.

    a. How many terms are in the expression?

    b. List the variable terms.

    c. List the constant terms.

    d. Simplify the expression.

Simplify exercises 38–40.

38. $-(3y + 2z) + (4y - 2z) - (-3y - 5z)$

39. $\dfrac{3x}{4} + \dfrac{5y}{8} - \dfrac{1}{16} - \dfrac{3x}{4} + \dfrac{y}{8} - \dfrac{5}{6}$

40. $2[8 + 3(x - 4) - 2(3x + 1)]$

41. Determine whether -2 is a solution of the equation $-x^2 + 3x + 8 = -3x$.

In exercises 42–46, consider the relation $y = 2x^2 + 3$, where x is the independent variable.

42. Graph the relation.

43. What is the domain of the relation? What is the range of the relation?

44. Is this relation a function? Justify your answer.

45. Determine the relative minima if possible.

46. Determine the *x*-values for which the relation is increasing.

47. For the function $f(x) = \frac{1}{3}x - 5$, find

    a. $f(9)$ **b.** $f(3 + h)$

48. Determine the volume of a rectangular solid with a length of 3.5 feet, a width of 2.25 feet, and a height of 1.75 feet.

49. Kelsie invests $500 for four years with simple interest applied annually at 5.5%. Find the total amount of interest she will receive. What is her total investment amount?

50. The Christmas House produces Christmas decorations. The setup cost for a certain ornament is $35.00. The cost of labor and materials per ornament is $2.80. Write a function to represent the cost of producing *x* ornaments in one production run. What is the cost of producing 150 ornaments in one production run?

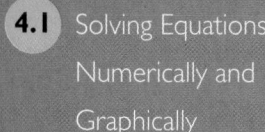

Chapter 4

Linear Equations in One Variable

In Chapter 3 we saw how to describe relations using tables, equations, and graphs. In this chapter we present methods for solving equations—that is, using the description of a relation to find values of a variable that make the equation true. We work here with linear equations in one variable, which have fewer complications than more general kinds of equations. However, the three different methods we present—numeric, graphic, and algebraic—will apply to other kinds of equations in later chapters, as well as to solving real-world problems.

The real-world situations that we will examine involve linear relationships and may be solved using a linear equation in one variable. These linear relationships are found in many aspects of our lives including analyzing the weather, renting an automobile, setting up a business, or making an investment. Many geometric relationships can also be linear. We conclude the chapter with a project illustrating a famous geometric relationship, the golden ratio, that has a very interesting history, beginning with the Egyptians, who thought the ratio was sacred. Indeed, some of their hieroglyphics have proportions based on the golden ratio.

4.1 APPLICATION

The average high temperature for Washington, DC, in the month of May is 76°F. If the high temperatures for the first six days of the third week of May are 73°, 75°, 77°, 75°, 76°, and 75°, what temperature is needed on the last day of the week to obtain a weekly average that is the same as the monthly average high temperature for May?

We will discuss this application further. See page 231.

4.1 Solving Equations Numerically and Graphically

OBJECTIVES
1. Identify linear equations in one variable.
2. Solve linear equations numerically.
3. Solve linear equations graphically.
4. Identify linear equations that are contradictions and linear equations that are identities numerically and graphically.
5. Model real-world situations by using linear equations, and solve them numerically or graphically.

4.1.1 Linear Equations in One Variable

In this text, we will be solving various kinds of equations. In this chapter, we will solve **linear equations in one variable**. These equations can be written in a particular (standard) form.

> **STANDARD FORM FOR A LINEAR EQUATION IN ONE VARIABLE**
> A linear equation in one variable (linear equation) is an equation that can be written in the form
>
> $$ax + b = 0, \text{ where } a \text{ and } b \text{ are real numbers and } a \neq 0.$$

For example,

1.	$2x + 5$	$= 0$	Standard form: $ax + b = 0$
2.	$2x - 5$	$= 6$	not standard form
3.	$x + 2 + 3(x - 4)$	$= 3x - 8$	not standard form

The first equation is in the exact form $ax + b = 0$, where $a = 2$ and $b = 5$. The last two equations can be written in this form with algebraic manipulations that we will learn later in the chapter.

 **HELPING HAND** Note that the variable x is raised to the first power. This *must* be the case for a linear equation in one variable.

Examples of nonlinear equations that contain variables raised to powers other than 1, variables in denominators of fractions, or roots of variables are listed here. We will solve these equations later in the text.

1. $2x^2 = 4$ x raised to the second power

2. $\dfrac{2}{x} + 1 = 0$ x in the denominator of a fraction

3. $2\sqrt{x} + 5 = x - 3$ x in the radicand of a radical expression

Until we learn algebraic manipulations, we will identify a linear equation in one variable, x, as an equation consisting of two expressions. Each of these expressions can be simplified to the form $ax + b$, where $a \neq 0$ in at least one of the expressions and the coefficient a is not the same in each expression. Using the previous linear examples, we have

Form: $ax + b$ $ax + b$

1.	$2x + 5 = 0x + 0$	$a = 2, b = 5; a = 0, b = 0$
2.	$2x + (-5) = 0x + 6$	$a = 2, b = -5; a = 0, b = 6$
3.	$4x + (-10) = 3x + (-8)$	$a = 4, b = -10; a = 3, b = -8$

These equations are called linear equations because the graphs of the functions defined by the two expressions in the equation turn out to be straight lines. Let's test this statement on the calculator. Complete the following set of exercises with your calculator.

 Discovery 1

Graphs of Functions Defined by Expressions in a Linear Equation

On a standard screen, graph the following functions, determined from the given linear equation:

1. $3x - 8 = 6$ **2.** $-2x + 5 = -3x - 4$

    Y1 $= 3x - 8$ Y2 $= 6$ Y1 $= -2x + 5$ Y2 $= -3x - 4$

Describe the characteristic of the graph of the function defined by the expression in Y1.
Describe the characteristic of the graph of the function defined by the expression in Y2.

On an integer screen, graph the following functions, determined from the given nonlinear equation:

3. $\dfrac{2}{x} + x^3 = x$

    Y1 $= \dfrac{2}{x} + x^3$ Y2 $= x$

Describe the characteristic of the graph of the function defined by the expression in Y1.
Describe the characteristic of the graph of the function defined by the expression in Y2.

For linear equations, we see that the graphs of the functions defined by the expressions are straight lines.

EXAMPLE 1   Identify each equation as linear or nonlinear. Check your results graphically on your calculator.

   a. $5x + (x - 2) = 3(x - 2)$ **b.** $\sqrt[4]{x + 2} = 3x - 7$

   c. $\dfrac{x}{5} + 3 = 6x$ **d.** $2x^2 = 4$

Solution

   a. Linear, because the equation simplifies to $6x - 2 = 3x - 6$, with both expressions in the form $ax + b$.

   b. Nonlinear, because the radical expression has a variable in its radicand.

   c. Linear, because the equation simplifies to $\frac{1}{5}x + 3 = 6x$, with both expressions in the form $ax + b$.

   d. Nonlinear, because x has an exponent of 2.

Check

Check your results graphically on your calculator. You may graph the expression on the left side of the equation and the expression on the right side of the equation in the same window.

a. $Y1 = 5x + (x - 2)$ $Y2 = 3(x - 2)$

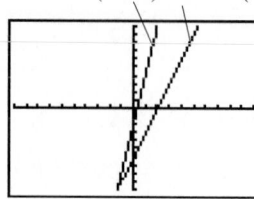

$(-10, 10, -10, 10)$
Linear, because the functions graphed are both lines.

b. $Y2 = 3x - 7$ $Y1 = \sqrt[4]{x + 2}$

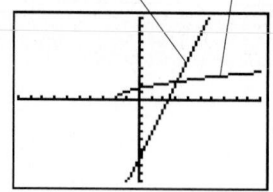

$(-10, 10, -10, 10)$
Nonlinear, because the graph of the function Y1, defined by the radical expression containing a variable in the radicand, is not a line.

c. $Y2 = 6x$ $Y1 = \frac{x}{5} + 3$

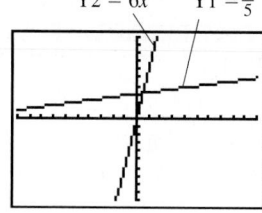

$(-10, 10, -10, 10)$
Linear, because the functions graphed are both lines.

d. $Y1 = 2x^2$ $Y2 = 4$

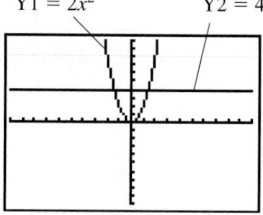

$(-10, 10, -10, 10)$
Nonlinear, because the graph of Y1, defined by the expression containing x^2, is not a line.

4.1.1 Checkup

In exercises 1–5, identify each equation as linear or nonlinear.

1. $5x - (x - 7) = x - 5$

2. $4.1x - 2.3 = 5.3x^2 + 0.6$

3. $\frac{1}{7}x - \frac{3}{7} = \frac{5}{14}x + \frac{1}{2}$

4. $2x + 8 = \sqrt{x} + 1$

5. $\frac{1}{x^2} + 12 = 5$

6. What is meant by the standard form of a linear equation in one variable?

4.1.2 Solving Equations Numerically

Previously, we determined whether a number was a solution of an equation by substituting the number for the variable and evaluating the two resulting expressions in the equation. If the expressions were equivalent, then the number substituted was called a solution. The set of all possible solutions is called a **solution set**. For example, determine whether 0 is a solution of $3x - 9 = -2x + 6$.

$3x - 9 =$	$-2x + 6$	
$3(0) - 9$	$-2(0) + 6$	Substitute 0 for x.
$0 - 9$	$0 + 6$	Simplify.
-9	6	

Therefore, 0 is not a solution, because the resulting values for the expressions, -9 and 6, are not equal.

To find a solution of the equation $3x - 9 = -2x + 6$, we could continue to substitute values for x until we find a number that results in equivalent values for each expression in the equation. To do this, it is convenient to use a table. You can see what we mean by trying the following Discovery exercise.

 ## Discovery 2

Numerical Solutions

To solve the equation $3x - 9 = -2x + 6$, complete the extended table of values shown, compare the values obtained, and determine the difference of the values.

x	$3x - 9$	$=$	$-2x + 6$		
0	-9		6	$-9 < 6$	$-9 - 6 = -15$
1					
2					
3					
4					

Write a rule to determine the solution of an equation from a table of values.

The solution is the number in the first column that, when substituted for the variable, results in two equal expressions. In comparing the values obtained for the expressions, we see that the first expression is first less than, then equal to, and then greater than, the second expression. The difference of the value of the first expression and the value of the second expression changes from negative to zero to positive.

SOLVING A LINEAR EQUATION NUMERICALLY

To solve a linear equation numerically,
Set up an extended table of values.

- The first column is labeled with the independent variable.
- The second column is labeled with the expression on the left side of the equation.
- The third column is labeled with the expression on the right side of the equation.

Complete the table.

- Substitute values for the independent variable.
- Evaluate the second and third columns.
- Continue evaluating until the values of the two expressions (the numbers in the second and third columns) are equal.

The value of the independent variable (the number in the first column) substituted to find the equivalent expressions is the solution.

For example, solve $2x + 3 = x + 5$ numerically for integer solutions.
The following is a sample table to determine an integer solution:

x	$2x + 3$	$=$	$x + 5$		
0	$2(0) + 3$		$0 + 5$	$3 < 5$	$3 - 5 = -2$
	3		5		
1	$2(1) + 3$		$1 + 5$	$5 < 6$	$5 - 6 = -1$
	5		6		
2	$2(2) + 3$		$2 + 5$	$7 = 7$	$7 - 7 = 0$
	7		7		
3	$2(3) + 3$		$3 + 5$	$9 > 8$	$9 - 8 = 1$
	9		8		
4	$2(4) + 3$		$4 + 5$	$11 > 9$	$11 - 9 = 2$
	11		9		

When 2 is substituted for the variable x, the two expressions are equivalent ($7 = 7$). Therefore, 2 is the solution of the linear equation $2x + 3 = x + 5$.

Note that in comparing the values obtained for the two expressions, the first expression is less than the second expression when $x = 0$ and $x = 1$ (values less than the solution), but greater than the second expression when $x = 3$ and $x = 4$ (values greater than the solution).

Note also that the difference of the value obtained for the first expression and the value obtained for the second expression is negative for $x = 0$ and $x = 1$, but is positive for $x = 3$ and $x = 4$.

To solve $2x + 3 = x + 5$ numerically for integer solutions on a calculator, complete the steps in the technology box that follows. As shown in Figure 4.1c, the solution is 2, because when $x = 2$, Y1 and Y2 have equal values.

TECHNOLOGY

SOLVING EQUATIONS NUMERICALLY

Solve $2x + 3 = x + 5$ numerically for integer solutions.

```
TABLE SETUP
 TblStart=0
 △Tbl=1
Indpnt: Auto Ask
Depend: Auto Ask
```

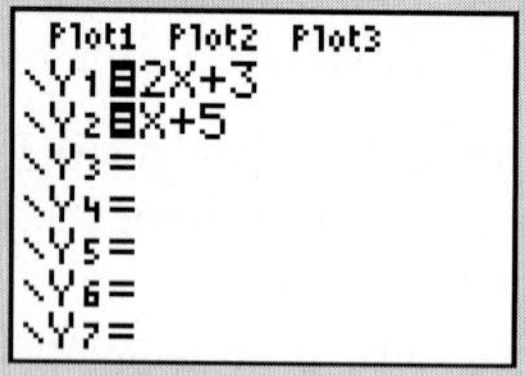

```
Plot1 Plot2 Plot3
\Y1■2X+3
\Y2■X+5
\Y3=
\Y4=
\Y5=
\Y6=
\Y7=
```

X	Y1	Y2		
0	3	5	$3 < 5$	$3 - 5 = -2$
1	5	6	$5 < 6$	$5 - 6 = -1$
2	7	7	$7 = 7$	$7 - 7 = 0$
3	9	8	$9 > 8$	$9 - 8 = 1$
4	11	9	$11 > 9$	$11 - 9 = 2$
5	13	10	$13 > 10$	$13 - 10 = 3$
6	15	11	$15 > 11$	$15 - 11 = 4$

X=0

Figure 4.1a   Figure 4.1b 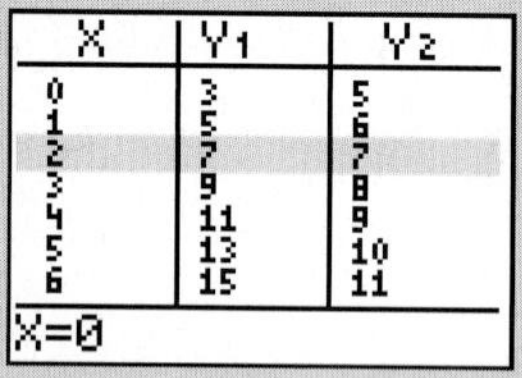   Figure 4.1c

Rename the independent variable x if necessary.

For Figure 4.1a,

Set up the table.

| 2nd | | TBLSET |

Set up the first column for the independent variable, x, by setting a minimum integer value, 0, and increments of 1 for integers.

| 0 | ENTER | (Minimum number in the table is 0.)

| 1 | ENTER | (Independent variable values are increasing by 1.)

Set the calculator to perform the operations automatically.

| ENTER | ▼ | ENTER |

For Figure 4.1b,

Set up the second column to be the expression on the left side by entering the left expression of the equation, $2x + 3$, in Y1.

| Y= | 2 | X,T,θ,n | + | 3 | ENTER |

Set up the third column to be the expression on the right side by entering the right expression of the equation, $x + 5$, in Y2.

| X,T,θ,n | + | 5 |

For Figure 4.1c,

View the table.

| 2nd | TABLE |

Move beyond the screen to view additional rows by using the up and down arrows. The solution is the x-value that results in equal Y1 and Y2 values. The solution of $2x + 3 = x + 5$ is 2 because $7 = 7$.

Linear equations may have a noninteger solution. Try solving this next equation with your calculator.

Discovery 3

Linear Equations with Noninteger Solutions

$(5x + 4) - 2(3x + 1) = 2(x - 7)$ does not have an integer solution. Complete the table of values, compare the values obtained, and determine their difference.

x	$(5x + 4) - 2(3x + 1)$	$= 2(x - 7)$		
3	-1	-8	$-1 > -8$	$-1 - (-8) = 7$
4				
5				
6				
7				

Write a rule for determining when the solution of an equation is between two integers given in a table of values.

The expression on the left is greater than the expression on the right for x-values of 3, 4, and 5. The expression on the left is less than the expression on the right for x-values of 6 and 7. Therefore, the expression on the left is equal to the expression on the right at some x-value between 5 and 6. The solution is thus noninteger. (We will need a different method to find this solution.)

Note that the differences are positive for the x-values of 3, 4, and 5. The differences are negative for the x-values of 6 and 7.

EXAMPLE 2 Solve numerically if possible.

a. $3a + 5 = 2a$ **b.** $6x - (4x + 3) = 7 - 3x$

c. $4 - 5x - (3x + 2) = 7 - x$

Solution

a. Rewrite the equation in terms of x—that is, $3x + 5 = 2x$.

Let $Y1 = 3x + 5$ and $Y2 = 2x$.

A sample table is shown in Figure 4.2a.

X	Y₁	Y₂		
-7	-16	-14	$-16 < -14$	$-16 - (-14) = -2$
-6	-13	-12	$-13 < -12$	$-13 - (-12) = -1$
-5	-10	-10	$-10 = -10$	$-10 - (-10) = 0$
-4	-7	-8	$-7 > -8$	$-7 - (-8) = 1$
-3	-4	-6	$-4 > -6$	$-4 - (-6) = 2$
-2	-1	-4	$-1 > -4$	$-1 - (-4) = 3$
-1	2	-2	$2 > -2$	$2 - (-2) = 4$

X= -7

Figure 4.2a

The solution is -5 because, when -5 is substituted for the variable a in both expressions, the results are equivalent: $-10 = -10$.

b. A sample table is shown in Figure 4.2b.

X	Y₁	Y₂		
0	-3	7	$-3 < 7$	$-3 - 7 = -10$
1	-1	4	$-1 < 4$	$-1 - 4 = -5$
2	1	1	$1 = 1$	$1 - 1 = 0$
3	3	-2	$3 > -2$	$3 - (-2) = 5$
4	5	-5	$5 > -5$	$5 - (-5) = 10$
5	7	-8	$7 > -8$	$7 - (-8) = 15$
6	9	-11	$9 > -11$	$9 - (-11) = 20$

X=0

$Y1 = 6x - (4x + 3)$ $Y2 = 7 - 3x$
Figure 4.2b

The solution is 2 because, when 2 is substituted for the variable x in both expressions, the results are equivalent: $1 = 1$.

c. A sample table is shown in Figure 4.2c.

X	Y₁	Y₂		
-2	18	9	$18 > 9$	$18 - 9 = 9$
-1	10	8	$10 > 8$	$10 - 8 = 2$
0	2	7	$2 < 7$	$2 - 7 = -5$
1	-6	6	$-6 < 6$	$-6 - 6 = -12$
2	-14	5	$-14 < 5$	$-14 - 5 = -19$
3	-22	4	$-22 < 4$	$-22 - 4 = -26$
4	-30	3	$-30 < 3$	$-30 - 3 = -33$

X= -2

$Y1 = 4 - 5x - (3x + 2)$ $Y2 = 7 - x$
Figure 4.2c

The expression on the left is greater than the expression on the right for $x = -1$ and less than the expression on the right for $x = 0$. The solution is noninteger and between -1 and 0. We will need another method to find the solution of this equation.

4.1.2 Checkup

Solve exercises 1–4 numerically if possible. If the solution is a noninteger one, indicate between what two integers it is located.

1. $3x + 6 = 0$ **2.** $2b - 3 = 10 - 3b$

3. $\dfrac{1}{2}(3 + 2x) = 2x - \dfrac{1}{2}$

4. $8.40 + 0.50x = 10.90 + 0.30x$

5. When solving a linear equation numerically, how can you tell between what two integers a noninteger solution will be located?

4.1.3 Solving Equations Graphically

A second method of solving an equation is to graph two functions. The functions to be graphed are written using each expression in the equation as a rule for one of the functions. For example, for the equation $3x - 9 = -2x + 6$, we write the two functions $Y1 = 3x - 9$ and $Y2 = -2x + 6$. Try it yourself in the next Discovery exercise. Complete the exercise on your calculator.

 Discovery 4

Graphical Solutions

To solve the equation $3x - 9 = -2x + 6$, graph the functions $Y1 = 3x - 9$ and $Y2 = -2x + 6$. Label the point of intersection of the graphs.

Write a rule for determining the solution of an equation from the graph of the two functions.

Write a rule for determining the numeric value of each expression when the equation is evaluated at its solution.

The solution of the equation is the x-coordinate of the point of intersection. The y-coordinate of the point of intersection is the value of each expression in the equation when x is replaced by the solution.

SOLVING A LINEAR EQUATION GRAPHICALLY
To solve a linear equation graphically,

- Write two functions, using each expression in the equation as a rule.
- Graph both functions on the same coordinate plane by plotting points found in the table of values and connecting the points with a line, to include all values in the domains of the functions.
- Determine the point of intersection of the lines.

The solution of the equation is the x-coordinate of the point of intersection of the two lines.

The y-coordinate of the point of intersection of the two lines is the value obtained for both expressions when the equation is evaluated with the solution.

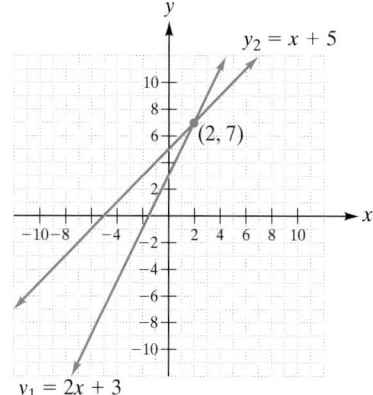

Figure 4.3

For example, solve $2x + 3 = x + 5$ graphically.
Let $y_1 = 2x + 3$ and $y_2 = x + 5$, and graph the equations.

The solution is 2, because the x-coordinate of the intersection is 2, as shown in Figure 4.3.

To solve $2x + 3 = x + 5$ graphically on your calculator, complete the steps in the next technology box.

As shown in Figure 4.4b, the solution is 2, because the x-coordinate of the intersection is 2.

T E C H N O L O G Y

SOLVING EQUATIONS GRAPHICALLY

Solve $2x + 3 = x + 5$ graphically.

$$y_1 = 2x + 3 \qquad y_2 = x + 5$$

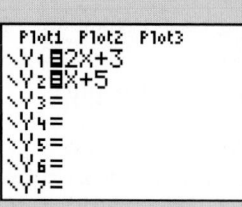

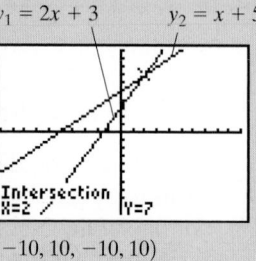

$(-10, 10, -10, 10)$

Figure 4.4a　　　　　　　Figure 4.4b　　　　　　　Figure 4.4c

Rename the independent variable x if necessary.

For Figure 4.4a,

Enter the expression on the left side of the equation, $2x + 3$, as Y1.

| Y= | 2 | X,T,θ,n | + | 3 | ENTER |

Enter the expression on the right side of the equation, $x + 5$, as Y2.

| X,T,θ,n | + | 5 |

For Figure 4.4b,

Graph the equations. (In this case, we will use the standard window.)

| ZOOM | 6 |

Find the intersection of the graphs. First trace the graph.

| TRACE |

Use the arrow keys to find the intersection.

If the intersection cannot be found by tracing, use Intersect, option 5, under the CALC menu.

| 2nd | CALC | 5 | ENTER | ENTER | ENTER |

For Figure 4.4c,

The solution is the x-value of the intersection point and is stored in x. The y-coordinate of the point of intersection is the value obtained for both the left side (Y1) and the right side (Y2) and is also stored. We can use this feature to check whether Y1 equals Y2.

Quit the graph screen and enter x.

| 2nd | QUIT | X,T,θ,n | ENTER |

Enter Y1 and Y2.

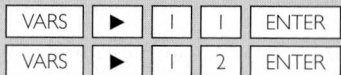

Since $x = 2$ when $7 = 7$ (or Y1 = Y2), the solution of $2x + 3 = x + 5$ is 2.

EXAMPLE 3   Solve graphically.

a. $3a + 5 = 2a$ b. $6x - (4x + 3) = 7 - 3x$

c. $(5x + 4) - 2(3x + 1) = 2(x - 7)$

Calculator Graphic Solution

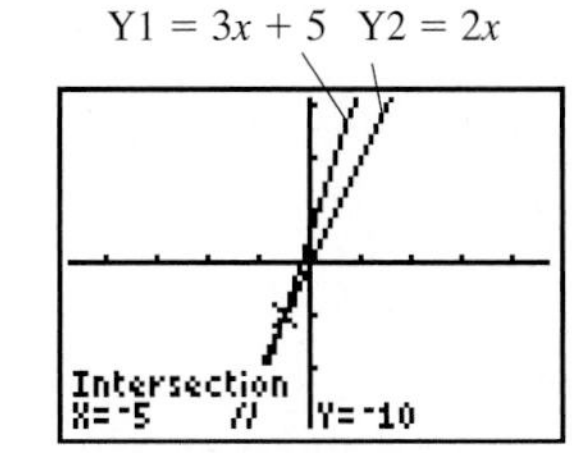

Y1 = 3x + 5   Y2 = 2x

Intersection
X=-5   //   Y=-10

$(-47, 47, -31, 31)$

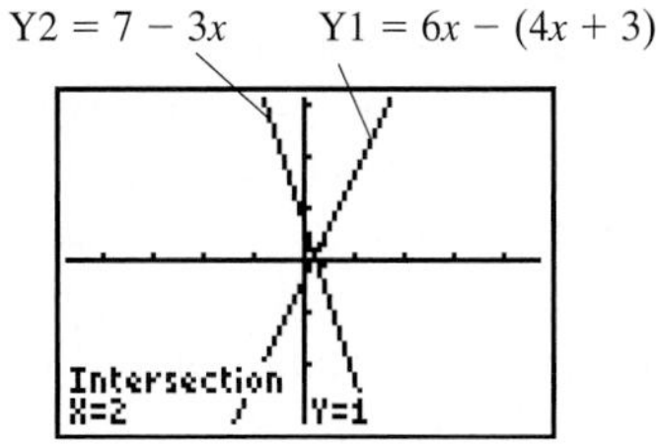

Y2 = 7 − 3x   Y1 = 6x − (4x + 3)

Intersection
X=2   /   Y=1

$(-47, 47, -31, 31)$

a.   To graph the equation with your calculator, change the variable a to x.

$$3x + 5 = 2x$$
$$\text{Let } Y1 = 3x + 5$$
$$Y2 = 2x$$

The solution is -5, because the x-coordinate of the intersection is -5.

b.   When using your calculator to graph you do not have to simplify the expressions on the left side or right side of the equation.

$$\text{Let } Y1 = 6x - (4x + 3)$$
$$Y2 = 7 - 3x$$

The solution is 2, because the x-coordinate of the intersection is 2.

c. $(5x + 4) - 2(3x + 1) = 2(x - 7)$

$$\text{Let } Y1 = (5x + 4) - 2(3x + 1)$$
$$Y2 = 2(x - 7)$$

There is no integer solution. Use the Intersect function on your calculator to find an approximate solution.

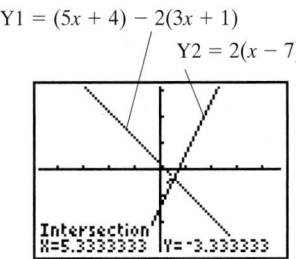

Y1 = (5x + 4) − 2(3x + 1)

Y2 = 2(x − 7)

Intersection
X=5.3333333   Y=-3.333333

Figure 4.5a

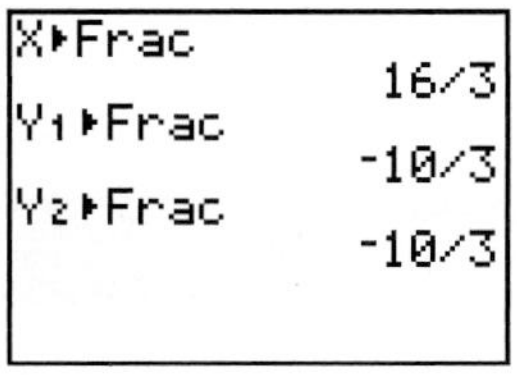

X▸Frac
                16/3
Y1▸Frac
                -10/3
Y2▸Frac
                -10/3

Figure 4.5b

As shown in Figures 4.5a and 4.5b, the solution is $5.\overline{3}$ or $\frac{16}{3}$.

 4.1.3 Checkup

Solve exercises 1–4 graphically.

1. $3x + 6 = 0$ **2.** $2b - 3 = 10 - 3b$

3. $\frac{1}{2}(3 + 2x) = 2x - \frac{1}{2}$

4. $8.40 + 0.50x = 10.90 + 0.30x$

5. Explain why the x-coordinate of the point of intersection of the two functions defined from the original equation represents the solution of the original equation.

4.1.4 Identifying Contradictions and Identities

The linear equations in Examples 2 and 3 have one solution. This is not always true. Try solving the following equations with your calculator.

Discovery 5

Linear Equations with No Solution

1. Solve $2x + 5 = 2x + 10$ numerically by completing a table of values.

   Write a rule explaining how to solve the equation by viewing its table of values.

2. Solve $2x + 5 = 2x + 10$ graphically. Sketch the graph.

   Write a rule explaining how to solve the equation by viewing its graph.

Viewing the table of values, we find that the expression on the left is always five less than the expression on the right. The two expressions will never be equal. The equation does not appear to have a solution. Such an equation is called a contradiction. A **contradiction** is an equation with no solution.

The two graphs do not appear to intersect. Therefore, there is no ordered pair common to both functions, which means that there is no solution of the equation. The equation is a contradiction. (The lines are parallel.)

HELPING HAND The equation $2x + 5 = 2.0000001x + 10$ looks as if it has no solution when we examine a table of values. The table has a constant difference of 5 between the expressions on the left and right. However, other methods will give us a noninteger solution. This fact emphasizes the need to know other methods to check our findings.

For example, if we graphically solve $2x + 5 = 2.000001x + 10$, the lines appear parallel when they are not. However, if we move between the two graphs with the up and down arrows, we can see that the difference of the y-coordinates is not constant for all x-coordinates.

Discovery 6

Linear Equations with Many Solutions

1. Solve $2x + 5 = (x + 3) + (x + 2)$ numerically by completing a table of values.

   Write a rule explaining how to solve the equation by viewing its table of values.

2. Solve $2x + 5 = (x + 3) + (x + 2)$ graphically. Sketch the graph.

   Write a rule explaining how to solve the equation by viewing its graph.

Viewing the table of values, we find that the two expressions are equal for every value of the independent variable evaluated. (This is also true for any other real-number value chosen for x.) The solution set of the equation is the set of all real numbers. Such an equation is called an identity. An **identity** is an equation for which all permissible replacements of the variable result in a true equation.

Although it seems as if there is only one graph on the screen, actually there are two, but they are the same line. Therefore, all ordered pairs on the graph are common to both functions, and all their x-coordinates are solutions of the equation. The solution set is the set of all real numbers (the domain of the functions graphed). The equation is an identity. (The lines coincide.)

EXAMPLE 4 Solve graphically and numerically.

a. $4(x - 2) = 4x - 8$ **b.** $3x + 4 = 2x + (x + 10)$

Calculator Solution

a. Let Y1 = $4(x - 2) = 4x - 8$
        Y2 = $4x - 8$

Y1 = $4(x - 2)$

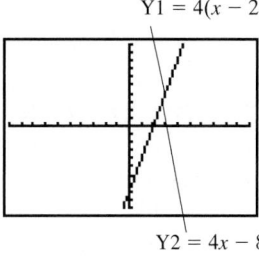

Y2 = $4x - 8$

$(-10, 10, -10, 10)$

The two graphs are the same. The solution set is the set of all real numbers

X	Y₁	Y₂		
0	-8	-8	$-8 = -8$	
1	-4	-4	$-4 = -4$	
2	0	0	$0 = 0$	
3	4	4	$4 = 4$	
4	8	8	$8 = 8$	
5	12	12	$12 = 12$	
6	16	16	$16 = 16$	

X=0

The expressions are always equal. The solution set is the set of all real numbers.

HELPING HAND The solution set of a linear equation is the set of all real numbers when the simplified expression on the left side and the simplified expression on the right side are the same.

b. Let Y1 = $3x + 4$
        Y2 = $2x + (x + 10)$

Y1 = $3x + 4$

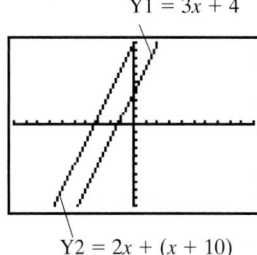

Y2 = $2x + (x + 10)$

$(-10, 10, -10, 10)$

The two graphs do not intersect. The table of values confirms this fact. The expressions will never be equal. There is no solution.

X	Y₁	Y₂		
0	4	10	$4 < 10$	$4 - 10 = -6$
1	7	13	$7 < 13$	$7 - 13 = -6$
2	10	16	$10 < 16$	$10 - 16 = -6$
3	13	19	$13 < 19$	$13 - 19 = -6$
4	16	22	$16 < 22$	$16 - 22 = -6$
5	19	25	$19 < 25$	$19 - 25 = -6$
6	22	28	$22 < 28$	$22 - 28 = -6$

X=0

The difference of the values of the two expressions is always the same. The expressions will never be equal. There is no solution.

We can summarize the preceding discoveries and exercises.

NUMERICAL SOLUTIONS OF A LINEAR EQUATION

To solve a linear equation numerically for integer solutions, set up an extended table of values. One of four possibilities will occur:

An integer solution exists. The solution is the integer in the first column that corresponds to equal values in the second and third columns.

(Continued on page 230)

A noninteger solution exists. The values in the second column change in order from less than to greater than, or from greater than to less than, the values in the third column. The noninteger solution is between the two integers in the first column that correspond to this change.

No solution exists. In the second and third columns, all of the differences of the values are the same.

An infinite number of solutions exist. In the second and third columns, all of the corresponding values are equal.

In conclusion, a linear equation may be solved numerically by using a table of values. However, if the solution is noninteger, it will be difficult to find by that method. The graphic method will solve noninteger equations.

GRAPHICAL SOLUTIONS OF A LINEAR EQUATION

To solve a linear equation graphically, graph the two functions defined by the expressions on the left and right sides of the equation. One of three possibilities will occur:

One solution exists. The graphs intersect. The solution is the x-coordinate of the point of intersection.

No solution exists. The graphs are parallel.

An infinite number of solutions exist. The graphs coincide.

4.1.4 Checkup

Solve exercises 1–2 numerically and graphically if possible.

1. $(a - 1) + (a - 3) = 2(a - 2)$

2. $(x + 1) + (3x + 4) = 3(x - 1) + x$

3. In solving a linear equation numerically and graphically, what results indicate that the equation may have no solution? What do we call a linear equation that has no solution?

4. In solving a linear equation numerically and graphically, what results indicate that the equation has many solutions? What do we call a linear equation that has many solutions?

4.1.5 Modeling the Real World

Linear equations in one variable are very common descriptions of real-world situations. In practical terms, you don't always need to solve these equations—for example, you may know that a $20 bill is going to cover the cost of filling your car with gasoline at your local service station, regardless of whether it will take 12 gallons, 13 gallons, or whatever your gas tank will hold. But if you need to know how much money you'll need for gas when you drive across the country, or if you're comparing the costs of two different car rental plans, the equation-solving methods discussed in this section can help you make the right decision.

EXAMPLE 5 Jacques rented a carpet shampooer for $28.50 per day, including all supplies. Jill rented a floor buffer for $18.25 per day, plus $20.50 for the wax. They kept the shampooer and buffer for the same number of days and spent the same amount. Determine how many days the shampooer and buffer were rented and the amount each person spent.

Solution

Let x = the number of days rented

The shampooer rental was $28.50 per day for x days, or $28.50x$. The buffer rental was $18.25 per day for x days plus $20.50, or $18.25x + 20.50$.

$$28.50x = 18.25x + 20.50$$

Numeric

Set up a table of values as shown in the accompanying figure.

The solution is 2, because $x = 2$ corresponds to the value of 57, where Y1 and Y2 are equal. The shampooer and buffer were rented for two days. Jacques and Jill spent $57.00 each.

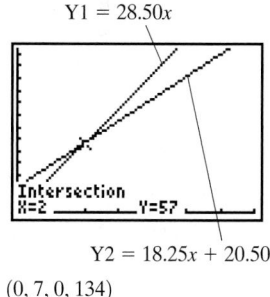

X	Y1	Y2
1	28.5	38.75
2	57	57
3	85.5	75.25
4	114	93.5
5	142.5	111.75
6	171	130
7	199.5	148.25

X=1

Y1 = 28.50x
Y2 = 18.25x + 20.50

Graphic

Graph the two functions Y1 = 28.50x and Y2 = 18.25x + 20.50 as shown in the accompanying figure.

The solution is 2, because 2 is the x-coordinate of the point of intersection. The shampooer and buffer were rented for two days. Jacques and Jill spent $57.00 each.

Y1 = 28.50x

Intersection
X=2   Y=57

Y2 = 18.25x + 20.50

$(0, 7, 0, 134)$

APPLICATION

The average high temperature for Washington, DC, in the month of May is 76°F. If the high temperatures for the first six days of the third week of May are 73°, 75°, 77°, 75°, 76°, and 75°, what temperature is needed on the last day of the week to obtain a weekly average that is the same as the monthly average high temperature for May?

Discussion

Let x = the high temperature for the last day of the week. The average temperature is found by dividing the sum of the temperatures by the number of temperatures, namely, 7.

$$\frac{73 + 75 + 77 + 75 + 76 + 75 + x}{7} = 76$$

Graphic Solution

Y2 = 76

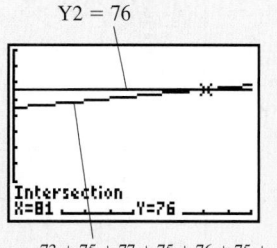

Intersection
X=81   Y=76

$Y1 = \dfrac{73 + 75 + 77 + 75 + 76 + 75 + x}{7}$

$(0, 100, 0, 100)$
The solution is 81.

Numeric Solution

X	Y1	Y2
79	75.714	76
80	75.857	76
81	76	76
82	76.143	76
83	76.286	76
84	76.429	76
85	76.571	76

X=79

$Y1 = \dfrac{73 + 75 + 77 + 75 + 76 + 75 + x}{7}$

Y2 = 76

The solution is 81.

The last day of the week, the temperature must reach 81°F to obtain a weekly average high temperature of 76°F.

4.1.5 Checkup

1. The cost of producing decorated baskets consists of a setup cost of $150 for materials and a cost of $5 per basket for labor. The baskets sell for $20 each. Determine the number of baskets for which the cost of production equals the revenue received.

2. Phillipe scored 83, 88, 91, 90, and 92 on his first five algebra exams. He is trying to earn an average of 90 in the course. What is the score he must get on the next test to achieve an average of 90?

4.1 Exercises

Identify each equation as linear or nonlinear.

1. $6x - 55 = x + 72$

2. $5(x - 3) = -(2x - 1)$

3. $4x^2 + 5 = 2x - 6$

4. $70x - 48 = 150x + 102$

5. $\dfrac{7}{9}z - \dfrac{2}{3} = 0$

6. $4.7x^2 - 5.3 = 4.4x + 0.7$

7. $\sqrt[3]{4x + 16} = 27$

8. $x^{1/3} = 64$

9. $3(2x - 5) = x + 3(x - 9)$

10. $235x - 476 = 0$

Use the calculator screens first to write the equation being solved and then to determine the solution or solutions (if any) of the equation.

11.

```
Plot1 Plot2 Plot3
\Y1■2X-7
\Y2■X+2
\Y3=
\Y4=
\Y5=
\Y6=
\Y7=
```

X	Y1	Y2
6	5	8
7	7	9
8	9	10
9	11	11
10	13	12
11	15	13
12	17	14

X=12

12.

```
Plot1 Plot2 Plot3
\Y1■(1/2)X-3
\Y2■X-1
\Y3=
\Y4=
\Y5=
\Y6=
\Y7=
```

X	Y1	Y2
-6	-6	-7
-5	-5.5	-6
-4	-5	-5
-3	-4.5	-4
-2	-4	-3
-1	-3.5	-2
0	-3	-1

X=-6

13.

```
Plot1 Plot2 Plot3
\Y1■.5X+1.25
\Y2■.5(X+2.5)
\Y3=
\Y4=
\Y5=
\Y6=
\Y7=
```

X	Y1	Y2
0	1.25	1.25
1	1.75	1.75
2	2.25	2.25
3	2.75	2.75
4	3.25	3.25
5	3.75	3.75
6	4.25	4.25

X=0

14.

```
Plot1 Plot2 Plot3
\Y1▪(2/3)X+1
\Y2▪X+2-(1/3)(X+
3)
\Y3=
\Y4=
\Y5=
\Y6=
```

X	Y1	Y2
-1	.33333	.33333
0	1	1
1	1.6667	1.6667
2	2.3333	2.3333
3	3	3
4	3.6667	3.6667
5	4.3333	4.3333

X= -1

15.

```
Plot1 Plot2 Plot3
\Y1▪X-(4.5-.5X)
\Y2▪1.5(X+2)
\Y3=
\Y4=
\Y5=
\Y6=
\Y7=
```

X	Y1	Y2
0	-4.5	3
1	-3	4.5
2	-1.5	6
3	0	7.5
4	1.5	9
5	3	10.5
6	4.5	12

X=0

16.

```
Plot1 Plot2 Plot3
\Y1▪X+3(X-5)
\Y2▪2(2X-3)
\Y3=
\Y4=
\Y5=
\Y6=
\Y7=
```

X	Y1	Y2
0	-15	-6
1	-11	-2
2	-7	2
3	-3	6
4	1	10
5	5	14
6	9	18

X=0

17.

```
Plot1 Plot2 Plot3
\Y1▪(1/3)X+1
\Y2▪(3/2)X-1
\Y3=
\Y4=
\Y5=
\Y6=
\Y7=
```

X	Y1	Y2
0	1	-1
1	1.3333	.5
2	1.6667	2
3	2	3.5
4	2.3333	5
5	2.6667	6.5
6	3	8

X=0

18.

```
Plot1 Plot2 Plot3
\Y1▪X-1
\Y2▪3X+4
\Y3=
\Y4=
\Y5=
\Y6=
\Y7=
```

X	Y1	Y2
-6	-7	-14
-5	-6	-11
-4	-5	-8
-3	-4	-5
-2	-3	-2
-1	-2	1
0	-1	4

X= -6

Solve numerically, if possible. Otherwise, solve graphically.

19. $2x - 7 = 35 - x$

20. $4z + 7 = 3z + 12$

21. $3(2x + 11) = 3(5 + x)$

22. $3(x + 10) = -2(x + 5)$

23. $6.8a + 4.3 = 2.6a + 33.7$

24. $\frac{1}{2}(x + 6) = \frac{1}{4}(x + 16) - 1$

25. $7(x + 10) + 15 = 6(x + 15) + (x - 5)$

26. $3(x - 4) = 2(x + 6) - 3(x + 7)$

27. $(a - 4) - (a + 4) = (a + 3) - (a - 2)$

28. $4(x + 1) - 2(x + 3) = 3(x + 1) - (x + 3)$

29. $3.5(z - 1) = 7(0.5z + 0.6) + 2$

30. $2(2.9x - 2.3) = 4.6(x - 1) + 1.2x$

31. $\frac{4}{5}(x - 1) = 6\left(\frac{1}{15}x - \frac{1}{10}\right)$

32. $\frac{1}{2}(x + 1) + \frac{1}{4}(x + 1) = \frac{3}{4}(x + 1)$

Use the calculator screens first to write the equation being solved and then to determine the solution or solutions (if any) of the equation.

33.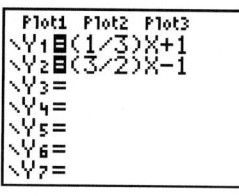

```
Plot1 Plot2 Plot3
\Y1▪3X+2
\Y2▪4-X
\Y3=
\Y4=
\Y5=
\Y6=
\Y7=
```

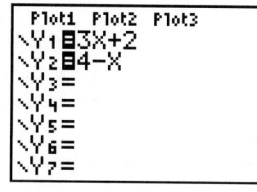

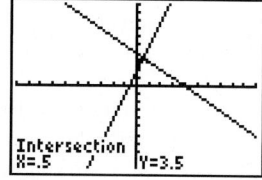

Intersection
X=.5   Y=3.5

34.

```
Plot1 Plot2 Plot3
\Y1▪.4X-1.5
\Y2▪1-.6X
\Y3=
\Y4=
\Y5=
\Y6=
\Y7=
```

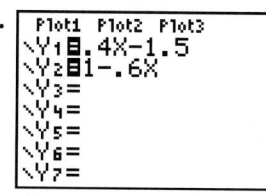

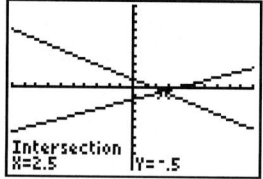

Intersection
X=2.5   Y= -.5

35.

```
Plot1 Plot2 Plot3
\Y1▪(1/2)X+5
\Y2▪4-.5(6-X)
\Y3=
\Y4=
\Y5=
\Y6=
\Y7=
```

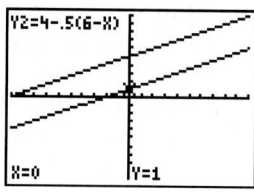

Y2=4-.5(6-X)

X=0   Y=1

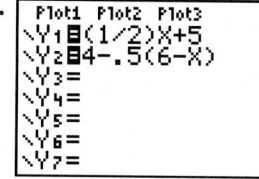

Y1=(1/2)X+5

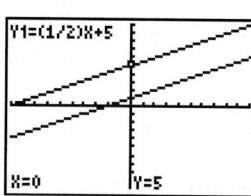

X=0   Y=5

36.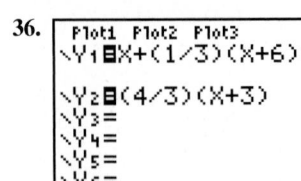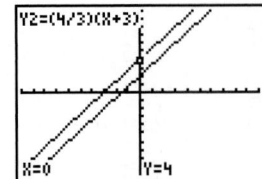

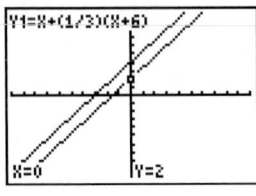

Solve graphically.

37. $x + 6 = 9 + 2x$

38. $(4x - 1) + (x - 6) = 3(2x + 1) - (x + 10)$

39. $(x + 4) + (x + 2) = (x - 1) + (x - 3)$

40. $2x - 8 = x - 6$

41. $2(x + 3) = 3(x - 1) - (x - 9)$

42. $2 - 3x = 4 - x$

43. $1.7x - 22.2 = 13.8 - 0.7x$

44. $5(0.5x + 0.3) = 0.1(25x + 15)$

45. $2.2(x - 1) + 1.7x = 3.5(x + 1) + 0.4x$

46. $7.3x + 23.7 = 2.6x - 13.9$

47. $\frac{4}{5}x + \frac{1}{5} = \frac{1}{5}x + 2$

48. $\frac{17}{24}x + \frac{1}{3} = \frac{1}{4}x + \frac{2}{3}$

49. $\frac{2}{3}(x + 1) - \frac{1}{3} = \frac{1}{3}(x + 1) + \frac{1}{3}x$

50. $\frac{4}{13}x + \frac{17}{4} = \frac{4}{13}x - \frac{7}{2}$

Solve each real world application numerically or graphically.

51. The Rent-a-Ride car rental company will lease a compact car for $49.95 per day, with unlimited mileage. The Rent-R-Wheels car rental company offers the same car for $29.95 per day plus $0.25 per mile. Determine the number of miles driven that will make the cost of a one-day rental the same for the two offers.

52. Ercille has the choice of paying a flat fee of $25.00 or of paying a fee of $10.00 plus $0.75 per page to have a paper typed. How many pages must the paper be in order to pay the same price for both offers?

53. A shoe factory has a daily setup cost of $280. The cost of materials and labor for each pair of shoes produced is $8. The factory sells the shoes at wholesale for $22 a pair. How many pairs of shoes should be produced each day to break even (to have the cost of production equal the revenue from sales)?

54. Handyman Pete offers to paint your garage for $149. Handywoman Gladys offers to do the same job for $45 plus $13 per hour. How many hours will it take Gladys to complete the job in order to do it for the same cost as Pete?

55. On a business trip, Ingrid spent $28, $19, $22, and $27 for meals during the first four days of the trip. How much can she spend on the fifth day if she must keep her average daily expense for meals at $25?

56. For the first four weeks of her diet, Caitlin lost 3 pounds, 2 pounds, 3 pounds, and 1 pound. How much must she lose during the fifth week in order to average a loss of 2 pounds per week?

57. An area rug is 3 feet longer than its width. What are its dimensions if the perimeter of the rug is 26 feet?

58. Mr. Castorini is fitting a piece of pipe into a water line. He must bend the pipe so that one end is four times as long as the other end. How long is each end if the total length of the pipe is 14 inches?

59. Charlene will paper a room for a setup charge of $25, plus $9 per roll of wallpaper and $5 per roll to pay her assistant. Greta does not have an assistant and will do the same work for a setup charge of $25 plus $14 per roll, claiming that she can work faster and more efficiently alone. How many rolls of paper will the job require if Charlene's charge and Greta's charge for the job are the same?

60. Handi-Man Rentals will rent an auger for a flat fee of $20 plus $12 per hour. Tool-Time Rentals will rent the same auger for a flat fee of $25 plus $12 per hour. For what number of hours will the total rental be the same for the two firms?

61. The average high temperature for Destin, Florida, in the month of May is 84°F. The high temperatures for the previous six days were 84°, 85°, 85°, 85°, 85°, and 86°. What must the temperature be on the seventh day to achieve a weekly average that is the same as the monthly average high temperature?

62. The average high temperature for Juneau, Alaska, in the month of May is 55°F. The high temperatures for the previous four days were 55°, 55°, 56°, and 58°. What should the temperature be on the fifth day to realize a weekly average that is the same as the monthly average high temperature?

4.1 Calculator Exercises

A. Using the Zoom Feature to Find Intersection Points

Use your calculator to solve exercises 1 and 2 graphically with the integer screen setting. Even though the lines do not cross on the integer screen, they look as if they will cross if extended. You can still use the intersection method to find the solution. Experiment with other screen settings to find out whether you can see the intersection point. Press ZOOM 3 ENTER to see more of the graph. Afterward, be sure to reset to the integer screen before attempting the second exercise.

1. $10x - 156 = 108 - 2x$ **2.** $9.2x + 55.8 = 1.4x - 37.8$

In exercises 3 and 4, use the standard screen setting, ZOOM 6 , to graph each equation as two functions. The lines do not appear on the screen, since they are outside the domain and range shown. To see the lines after graphing with the standard screen, repeatedly use the zoom-out feature, ZOOM 3 ENTER , until the lines appear. Now trace toward the point of intersection, and then use the CALC key to find the intersection.

3. $7x + 450 = 2x + 1700$ **4.** $12x + 800 = 8x - 1200$

Before starting another exercise, always reset your screen selection after you have used the ZOOM feature.

B. Using the Test Key to Solve Linear Equations Numerically

When storing the left and right side of the equation in Y1 and Y2, you can also store the condition Y1 = Y2 in Y3. Do this by moving the cursor down to Y3 after entering the functions for Y1 and Y2. Then enter VARS ► 1 1 2nd TEST 1 VARS ► 1 2 . Now when you scan the table to find a solution, you can scan the column for Y3, and when you find a row containing a 1, this means that the equation Y1 = Y2 is true, and the value of x for this row is the solution of the equation. Try this additional step on some of the exercises in this section.

4.1 Writing Exercises

1. When using the numerical method to solve a linear equation in one variable, you construct a table of integer values. When you then search the table for a solution, one of four situations can arise:
   a. There are no solutions of the equation.
   b. There is one integer solution of the equation.
   c. There is one noninteger solution of the equation.
   d. The solution set for the equation is the set of all real numbers in the domain of the relation.

For each of these outcomes, describe what you would see when you examine the table of values in search of solutions. For example, if there are no solutions of the equa-

tion, what would you see when you examine the columns for the left and right sides of the equation?

2. When you graphically solve a linear equation in one variable, one of three situations can arise:
   a. There are no solutions of the equation.
   b. There is one solution of the equation.
   c. The solution set for the equation is the set of all real numbers in the domain of the relation.

For each of these outcomes, describe what you would see when you examine the graph in search of solutions.

4.2 Solving Equations by Using Addition and Multiplication

OBJECTIVES
1 Solve linear equations algebraically by using the addition property of equations.
2 Solve linear equations algebraically by using the multiplication property of equations.
3 Model real-world situations by using linear equations, and solve the equations algebraically.

APPLICATION

Only 2% of U.S. tornadoes reach "violent" intensity: F4 or F5 on the Fujita Scale of Tornado Intensity. From 1950 to 1999, an average of about 16 "violent" tornadoes were reported annually in the United States. Approxi-

mate the annual average number of U.S. tornadoes from 1950 to 1999.

After completing this section, we will discuss this application further. See page 243.

Earlier in the text, we defined equivalent expressions to be two expressions with the same value. Similarly, **equivalent equations** are two equations that have exactly the same solutions. In order to write equivalent equations, we will need to know the properties of equations.

4.2.1 Solving Linear Equations by Using Addition

The addition property of equations is used to write equivalent equations. Let's see if we can discover how it works. Complete the following set of exercises.

 Discovery 7

Addition Property of Equations

Given the equation $7 = 7$, add 2 to both expressions.

$$7 = 7$$

$7 + 2$	$7 + 2$
9	9

1. Given the equation $7 = 7$, add -2 to both expressions.
2. Given the equation $6 + 1 = 4 + 3$, add 2 to both expressions.
3. Given the equation $6 + 1 = 4 + 3$, add -2 to both expressions.

Write a rule for the addition property of equations.

In each of the preceding exercises, we began with an equation and added the same number to both expressions. The resulting expressions remained equal in value.

This property holds true for subtraction as well, because subtraction is defined to be adding the opposite of a number.

ADDITION PROPERTY OF EQUATIONS
Given expressions a, b, and c,

$$\text{if } a = b, \text{ then } a + c = b + c \text{ and } a - c = b - c.$$

To illustrate why the addition property is true, we can think of an equation as a balanced scale. Each expression weighs the same amount. If an equal weight is added to or subtracted from each side of a balanced scale, the scale will remain balanced. The same is true of a balanced equation. That is, if an expression is added to or subtracted from the equal expressions, the expressions remain equal.

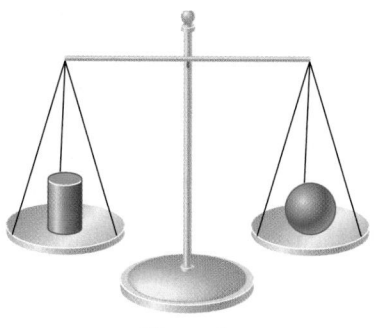

If $a = b$

then $a + c = b + c$

We are now ready to use the addition property of equations to solve algebraically a linear equation consisting of algebraic expressions. Our goal is to find a value for the variable that will make the equation true.

Let's begin with a simple equation: $x - 3 = 5$. We know that the solution of the equation is 8, because if we replace x with 8, the result is $5 = 5$, a true equation. Therefore, we begin with $x - 3 = 5$ and should end with the equation $x = 8$ (the solution). The step in the middle involves the addition property.

Solve $x - 3 = 5$.

We want to isolate the variable x on one side of the equation to determine the solution. To do this, we want to eliminate the term -3. Therefore, we must add the opposite of the term, 3, to -3 and obtain 0. However, we must add the same value to both sides of the equation in order to produce an equivalent equation.

$$x - 3 = 5$$
$$x - 3 + 3 = 5 + 3 \quad \text{Add 3 to both sides.}$$
$$x = 8 \quad \text{Combine like terms.}$$

The solution is 8.

Similarly, solve $x + 3 = 5$.

We want to isolate the variable x on one side of the equation to determine the solution. To do this, we want to eliminate the term 3. We must add the opposite of the term, -3 (or subtract 3), to 3 and obtain 0. We must add (or subtract) the same value to both sides of the equation.

$$x + 3 = 5$$
$$x + 3 - 3 = 5 - 3 \quad \text{Subtract 3 from both sides.}$$
$$x = 2 \quad \text{Combine like terms.}$$

The solution is 2.

SOLVING A LINEAR EQUATION BY USING THE ADDITION PROPERTY OF EQUATIONS

To solve a linear equation by using the addition property, first simplify both expressions in the equation. Then isolate the variable on one side of the equation.

- Add a term if a term is subtracted from the variable, and then combine like terms.

or

- Subtract a term if a term is added to the variable, and then combine like terms.

Check the solution by substituting, solving numerically, or solving graphically.

EXAMPLE 1   Solve each equation algebraically. Check your solution.

a. $5 + (x - 7) = 9$

b. $14 = (3x - 5) - (2x + 4)$

Solution

a. Algebraic

$$5 + (x - 7) = 9$$

$5 + x - 7 = 9$	*Remove parentheses.*
$x - 2 = 9$	*Combine like terms.*
$x - 2 + 2 = 9 + 2$	*Add 2 to both sides.*
$x = 11$	*Combine like terms.*

Substitution Check

$$\frac{5 + (x - 7) = 9}{5 + (11 - 7) \;\Big|\; 9}$$
$$9$$

The solution is 11.

Numeric

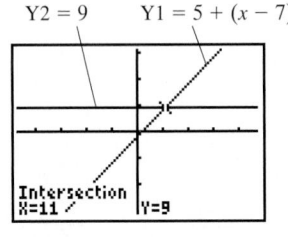

Y1 = 5 + (x − 7)
Y2 = 9

Graphic

Y2 = 9   Y1 = 5 + (x − 7)

Intersection
X=11   Y=9

(−47, 47, −31, 31)

b. Algebraic

$$14 = (3x - 5) - (2x + 4)$$

$14 = 3x - 5 - 2x - 4$	*Remove parentheses.*
$14 = x - 9$	*Combine like terms.*
$14 + 9 = x - 9 + 9$	*Add 9 to both sides.*
$23 = x \quad \text{or} \quad x = 23$	

Substitution Check

$$\frac{14 = (3x - 5) - (2x + 4)}{14 \;\Big|\; (3(23) - 5) - (2(23) + 4)}$$
$$14$$

The solution is 23.

Numeric

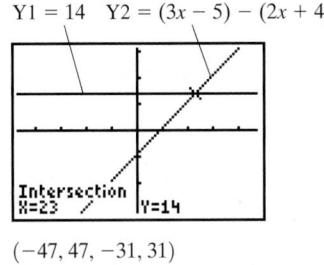

Y1 = 14
Y2 = (3x − 5) − (2x + 4)

Graphic

Y1 = 14   Y2 = (3x − 5) − (2x + 4)

Intersection
X=23   Y=14

(−47, 47, −31, 31)

 4.2.1 Checkup

In exercises 1–3, use the addition property of equations to solve each equation algebraically. Check your solutions.

1. $\left(\dfrac{1}{3}x + \dfrac{3}{5}\right) + \left(\dfrac{2}{3}x - \dfrac{1}{5}\right) = \dfrac{4}{7}$

2. $6b - (5b + 12) = 3$

3. $8 - (5 - x) = 25$

4. What does it mean to solve a linear equation by isolating the variable? How does the addition property of equations enable you to do this?

4.2.2 Solving Linear Equations by Using Multiplication

Another property of equations involves multiplication. It is very similar to the addition property of equations, which is not too surprising, since multiplication is based on repeated additions of numbers. Let's see if we can discover how the multiplication property works. Complete the following set of exercises.

Discovery 8

Multiplication Property of Equations

Given the equation $7 = 7$, multiply both expressions by 2.

$$7 = 7$$

$7 \cdot 2$	$7 \cdot 2$
14	14

1. Given the equation $7 = 7$, multiply both expressions by -2.

2. Given the equation $6 + 1 = 4 + 3$, multiply both expressions by 2.

3. Given the equation $6 + 1 = 4 + 3$, multiply both expressions by -2.

Write a rule for the multiplication property of equations.

In each of the preceding exercises, we began with an equation and multiplied both expressions by the same number. The resulting expressions remained equal in value.

This property holds true for division as well, because division is defined to be multiplication by the reciprocal of a number.

MULTIPLICATION PROPERTY OF EQUATIONS

Given expressions a, b, and c,

$$\text{if } a = b, \text{ then } a \cdot c = b \cdot c \text{ and } a \div c = b \div c \text{ (when } c \neq 0\text{)}.$$

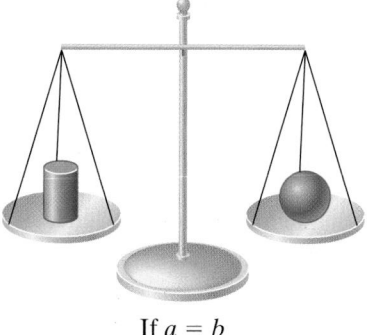

If $a = b$

To illustrate why the multiplication property is true, we can think of an equation as a balanced scale. Each expression weighs the same amount. If a weight is multiplied by a value on each side of a balanced scale, the scale remains balanced. The same is true of a balanced equation. That is, if each expression is multiplied by an equal expression, the resulting expressions remain equal.

We are now ready to use the multiplication property of equations to solve a linear equation, such as $3x = 15$. We know that the solution of the equation is 5, because if we replace x with 5, the result is $15 = 15$, a true equation. Therefore, we begin with $3x = 15$ and should end with the equation $x = 5$ (the solution). The step in the middle involves the multiplication property.

Solve $3x = 15$.

We want to isolate the variable x on one side of the equation to determine the solution. To do this, we want to eliminate the factor 3. Therefore, we must divide by 3 and obtain a factor 1. However, we must divide both sides of the equation by the same value in order to produce an equivalent equation.

$$3x = 15$$

$$\frac{3x}{3} = \frac{15}{3} \qquad \text{Divide both sides by 3.}$$

$$x = 5 \qquad \text{Simplify.}$$

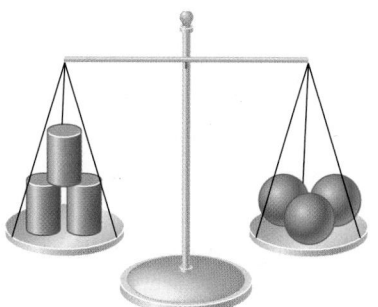

then $a \cdot c = b \cdot c$

The solution is 5.

Similarly, solve $\frac{x}{3} = 5$.

$$3\left(\frac{x}{3}\right) = 3(5) \qquad \text{Multiply both sides by 3.}$$

$$x = 15 \qquad \text{Simplify.}$$

The solution is 15.

SOLVING A LINEAR EQUATION BY USING THE MULTIPLICATION PROPERTY OF EQUATIONS

To solve a linear equation by using the multiplication property, first simplify both expressions in the equation. Then isolate the variable on one side of the equation.

- Multiply by an algebraic expression if the variable is divided by an algebraic expression.

or

- Divide by an algebraic expression if the variable is multiplied by an algebraic expression.

Check the solution by substituting, solving numerically, or solving graphically.

EXAMPLE 2 Solve each equation algebraically. Check your solution.

a. $-x = 9$ **b.** $46 = (4x + 3) + 3(2x - 1)$

c. $\frac{3}{4}x = 9$

Solution

a. Algebraic

$$-x = 9 \qquad \text{The coefficient of } -x \text{ is } -1.$$

$$\frac{-x}{-1} = \frac{9}{-1} \qquad \text{Divide both sides by } -1.$$

$$x = -9 \qquad \text{Simplify.}$$

The solution is -9.

Substitution Check

$$-x = 9$$

$$-(-9) \,\big|\, 9$$

$$9$$

The solution, -9, checks.

b. Algebraic

$$46 = (4x + 3) + 3(2x - 1)$$

$$46 = 4x + 3 + 6x - 3 \qquad \text{Remove parentheses and distribute.}$$

$$46 = 10x \qquad \text{Combine like terms.}$$

$$\frac{46}{10} = \frac{10x}{10} \qquad \text{Divide both sides by 10.}$$

$$4.6 = x \quad \text{or} \quad x = 4.6$$

The solution is 4.6.

Graphic

$\text{Y1} = 46 \quad \text{Y2} = (4x + 3) + 3(2x - 1)$

Intersection
X=4.6   Y=46

$(-47, 47, -62, 62)$

The integer setting $(-47, 47, -31, 31)$ will not result in two graphs, because the range is not large enough for the equation $\text{Y1} = 46$ to be seen. Therefore, we need to double the range of the integer screen. The solution is 4.6.

c. Algebraic

$$\frac{3}{4}x = 9$$

$$\frac{3}{4}x \div \frac{3}{4} = 9 \div \frac{3}{4} \qquad \text{Divide both sides by } \tfrac{3}{4}.$$

$$x = \frac{9}{1} \cdot \frac{4}{3} \qquad \text{Simplify.}$$

$$x = \frac{36}{3}$$

$$x = 12$$

Numeric

X	Y₁	Y₂
9	6.75	9
10	7.5	9
11	8.25	9
12	9	9
13	9.75	9
14	10.5	9
15	11.25	9

X=9

$Y1 = \dfrac{3}{4}x \qquad Y2 = 9$

The solution is 12.

The solution is 12.

Note: This example could have been solved in a different way. Dividing by a fraction is equivalent to multiplying by the fraction's reciprocal. For example,

$$\frac{3}{4}x = 9$$

$$\frac{3}{4}x \cdot \frac{4}{3} = 9 \cdot \frac{4}{3} \qquad \text{Multiply both sides by the reciprocal of } \tfrac{3}{4}, \text{ or } \tfrac{4}{3}.$$

$$x = 12 \qquad \text{Simplify.}$$

The solution remains 12.

4.2.2 Checkup

Use the multiplication property of equations to solve each equation algebraically. Check your solutions.

1. $5x - 6x = 325$ **2.** $\dfrac{5}{11}x = -15$ **3.** $-4a = -\dfrac{8}{15}$ **4.** $8(5x + 12) - 3(32 - 4x) = 260$

4.2.3 Modeling the Real World

Solving linear equations by using the addition and multiplication properties of equations is the first big step in using the power of algebra to solve real-world problems. We will see this throughout the rest of this chapter. Remember, as in all real-world problem solving, whatever operation you do on a number must also be done on the number's unit of measurement. You know, for example, that 24 ÷ 4 = 6, but remember also that 24 sq ft ÷ 4 ft = 6 ft.

EXAMPLE 3 Heat bursts are an odd atmospheric event that occurs in thunderstorms. In Glasgow, MT, on September 9, 1994, the temperature at 5:02 A.M. was 67°F. A heat burst from a thunderstorm shot the temperature up to 93°F at 5:17 A.M. Write an equation to determine the number of degrees Fahrenheit the temperature rose.

Solution

Algebraic

Let x = the number of degrees Fahrenheit the temperature rose

The heat burst temperature is the sum of the temperature before the increase and the number of degrees the temperature rose, or $67 + x = 93$.

$$67 + x = 93$$ Write an equation.

$$67 + x - 67 = 93 - 67$$ Subtract 67 from both sides.

$$x = 26$$

$$\frac{67 + \quad x = 93}{67 + 26 \quad | \quad 93}$$
$$93 \quad |$$

The solution is 26.

The temperature rose 26°F due to the heat burst.

EXAMPLE 4   David's small dog needs a pen with an area of 156 square feet for proper exercise. David wants to use 12 feet of the back of his house as the width of the pen. If he is constructing a rectangular pen, what is the length he needs?

Solution

Algebraic

Let x = length of the pen

Substitution Check

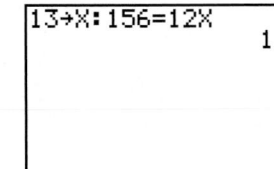

The solution is 13.

The area formula is $A = LW$.

$$156 = x \cdot 12$$ Substitute known values.

$$\frac{156}{12} = \frac{x \cdot 12}{12}$$ Divide both sides by 12.

$$13 = x$$

The length will be 13 feet.

EXAMPLE 5   How many liters of a 45% alcohol solution does a nurse need if she wants the solution to contain 100 liters of pure alcohol?

Solution

Algebraic

Let L = number of liters of 45% alcohol solution
    Since the amount of alcohol in the solution is 45% of L liters, an expression for this would be 0.45 L. This must equal 100 liters.

$$0.45L = 100$$

$$\frac{0.45L}{0.45} = \frac{100}{0.45}$$ Divide both sides by 0.45.

$$L \approx 222.2$$ Simplify.

Graphic

Y2 = 100   Y1 = 0.45x

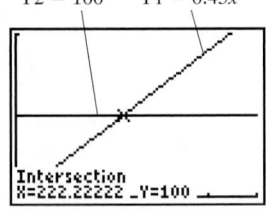

(0, 500, 0, 200)
The solution is about 222.2.

The nurse will need approximately 222.2 liters of the 45% solution.

APPLICATION

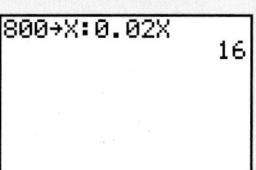

Only 2% of U.S. tornadoes reach "violent" intensity: F4 or F5 on the Fujita Scale of Tornado Intensity. From 1950 to 1999, an average of about 16 "violent" tornadoes were reported annually in the United States. Approximate the annual average number of U.S. tornadoes from 1950 to 1999.

Discussion

Let x = the annual average number of tornadoes reported from 1955 to 1999
Two percent of the U.S. tornadoes, $0.02x$, are equal to 16, or $0.02x = 16$.

Algebraic Solution

$$0.02x = 16 \qquad \text{Write an equation.}$$
$$\frac{0.02x}{0.02} = \frac{16}{0.02} \qquad \text{Divide both sides by 0.02.}$$
$$x = 800$$

Substitution Check

```
800→X:0.02X
              16
```

Since 16 = 16, the solution is 800.

The annual average number of U.S. tornadoes from 1955 to 1999 was approximately 800.

 4.2.3 Checkup

Write an equation and solve algebraically.

1. The gravitational pull on the surface of the Moon is about one-sixth as strong as it is on Earth. An astronaut and his space suit would weigh about 60 pounds on the surface of the Moon. Determine the weight of the astronaut and his suit on Earth.

2. The average height of the adult male is approximately 69 inches. Big Bird of TV's "Sesame Street" is 29 inches taller than the average adult male. Determine Big Bird's height.

3. Cheryl plans to put a square flower garden in her backyard. She has 38 linear feet of landscaping bricks to use to border the garden. What will be the length of each side?

4. A triangle has an area of $8\frac{1}{4}$ square feet. If the base of the triangle measures three feet, what is its height?

4.2 Exercises

Use the addition property of equations to solve algebraically.

1. $x + 33 = 51$
2. $y + 73 = -31$
3. $75 = a - 41$
4. $x - 123 = -47$
5. $-4.91 = y + 3.07$
6. $a - 0.153 = -4.759$
7. $y - \dfrac{1}{6} = -\dfrac{1}{6}$
8. $a - \dfrac{13}{18} = \dfrac{5}{6}$
9. $27 + (x - 13) = 11$
10. $8 = (12 + a) - 54$
11. $(13.9 + x) + 0.88 = -2.07$
12. $(x - 14.75) - 10.5 = -2.65$
13. $\left(x - \dfrac{3}{10}\right) - \dfrac{2}{5} = -3\dfrac{1}{2}$
14. $4\dfrac{1}{2} = \left(\dfrac{5}{14} + b\right) - \dfrac{3}{7}$
15. $(5x - 2) - (4x + 7) = 27$
16. $42 = (36x - 21) - (35x + 60)$
17. $\left(\dfrac{1}{3}x + \dfrac{1}{8}\right) + \left(\dfrac{3}{4} + \dfrac{2}{3}x\right) = -\dfrac{3}{16}$
18. $\left(\dfrac{5}{7}z + \dfrac{3}{5}\right) - \left(\dfrac{3}{10} - \dfrac{2}{7}z\right) = \dfrac{13}{15}$
19. $(3x + 76) - (2x - 45) = 31$
20. $(16x + 15) - (15x + 16) = -1$

Use the multiplication property of equations to solve algebraically. Round decimal answers to the nearest thousandth.

21. $-324 = -4y$
22. $7x = -434$
23. $-5.1x = 0.102$
24. $-3.7y = 13.69$
25. $-3\dfrac{1}{3}x = -1\dfrac{1}{3}$
26. $\dfrac{1}{5}a = -3\dfrac{1}{5}$

27. $\dfrac{x}{4} = 1.22$

28. $\dfrac{x}{17.3} = -4$

29. $-x = 57$

30. $-x = -16\dfrac{1}{5}$

31. $57 = 2x + 17x$

32. $4a - 5a = -17$

33. $18.22x - 12.9x = -12.76$

34. $-121 = 2.2x + 9.9x$

35. $\dfrac{5}{14} = \dfrac{9}{14}a + \dfrac{3}{7}a$

36. $\dfrac{1}{5}x - \dfrac{1}{7}x = -\dfrac{2}{49}$

37. $2(3x + 6) + 3(x - 4) = 126$

38. $(a + 17) - (2a + 17) = 41$

39. $4.8(a + 3) + 2.4(a - 6) = -7.2$

40. $2.2(x + 3.7) + 7.4(x - 1.1) = 60.48$

41. $3\left(\dfrac{1}{2}x - \dfrac{3}{4}\right) - 18\left(x - \dfrac{1}{8}\right) = 0$

42. $\dfrac{1}{4}(x + 2) - \dfrac{1}{6}(x + 3) = -\dfrac{1}{12}$

Write an equation and solve algebraically.

43. A box of Flakies cereal is marked to contain 10 servings of 1 cup each. Four servings have been used so far. How many servings remain in the box?

44. Jerome's diet limits his daily intake to 1200 calories. If Jerome has taken diet drinks for breakfast and lunch, each containing 250 calories, how many calories can he have for the remainder of the day?

45. Chuck's net paycheck was $1784.26 and the deductions amounted to $567.32. What was his gross pay? (*Hint:* Gross pay less deductions equals net pay.)

46. A jacket that originally sold for $129.95 was marked down to a sale price of $88.49. What was the amount of the markdown?

47. A recipe calls for $5\frac{1}{2}$ cups of flour. Glenda has $3\frac{3}{4}$ cups of flour in her canister. How much flour must she borrow from a neighbor to make the recipe?

48. In order to control overtime costs, Acme Industries limits its employees to 12 hours of overtime per week. If Ali has worked $3\frac{1}{3}$ hours of overtime on Monday and $4\frac{1}{4}$ hours of overtime on Wednesday, how many overtime hours remaining in the week can he work to reach the limit?

49. Tameka was charged $54.32 for a dress that was priced at $49.95. She had forgotten that sales tax would be added to the selling price. How much money in sales tax did Tameka pay on the purchase?

50. Employment for the Tennessee Valley Authority was officially recorded as 28,392 employees in 1990. By 1998, the employment for this federal agency had dropped to 13,818 employees. What was the amount of reduction in the workforce over the eight-year period?

51. A rectangular room measures 18 feet by 22 feet. Karla has 35 feet of wallpaper border to put around the room at the ceiling. How much more border must she buy in order to complete the job? If she buys two rolls of border that contain 20 feet each, will she have enough?

52. The selling price of a house is $125,000. A down payment of $32,000 is required to purchase the home. How much of a loan will be needed to purchase the home?

53. In an amateur talent contest, 264 of the paid admissions were adults. This figure represented 55% of the total number of paid admissions. How many paid admissions were there?

54. What was the selling price of a suit if the 8.75% sales tax amounted to $21.88?

55. If Erika's class receives $2.50 for each packet of gourmet coffee it sells, how many packets must the class sell if it wishes to make $1450 to purchase a computer?

56. How many months will it take to pay off a loan of $3150 (with interest already included) if monthly payments are $175?

57. Jane and her two children received $45,240 as their portion of a probated estate. If they received $\frac{3}{5}$ of the estate, how much was the estate worth?

58. Angelo was charged $4.49 for $\frac{3}{4}$ pound of baked ham. What was the selling price per pound of the ham?

59. The floor in Colonel Mustard's library is in the shape of a parallelogram. The base of the parallelogram from one end of the library to the other is 20 feet. If the covering on the floor is 350 square feet, what is the perpendicular distance (height) across the library?

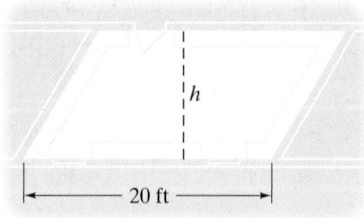

60. In designing a rectangular storage bin, the base is limited to measurements 6 feet long by 4 feet wide. How high must the bin be if it must hold 120 cubic feet?

61. A cylindrical tank is needed to hold 300 cubic feet of water. If the radius of the tank is 4 feet, what must its height be, rounded to the nearest foot?

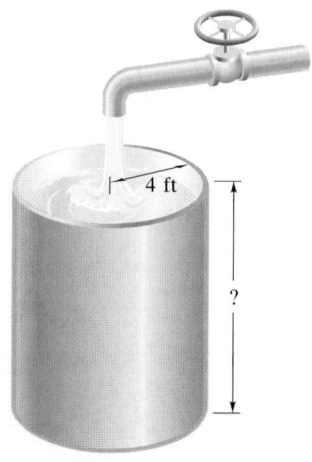

4 ft

?

62. What is the radius (rounded to the nearest inch) of a circle that has a circumference of 100 inches?

63. How much must you place into savings if you wish to earn $864 in simple interest at 4.5% per year over three years?

64. What was the average speed for a trip of 855 miles that took 13 hours to complete?

65. If each of the five partners in a firm earned $12,730 last quarter in profits, what was the firm's quarterly profits?

66. If seven people share a lottery prize, and each receives $6,570,000, what was the jackpot, rounded to the nearest million dollars?

67. Connie earned $40.50 commission on her sales for Monday. If her commission is 3% of sales, what were her sales for that day?

68. Kitty receives 4% commission on sales of her records. If she received $4800 in commissions last year, what was the value of her record sales?

69. At the start of an Iditarod dog-sled race from Anchorage to Nome, 12 dogs must be on the towline. If this number represents 75% of the maximum number of dogs a musher can have at the start of the race, what is the maximum number of dogs allowed?

70. The greatest land mountain range in the world is the Himalayas, which contain 88% of the world's peaks that are over 24,000 feet high. If the Himalayas contain 96 of these peaks, how many are there worldwide?

71. As of March, 1999, the population of the United States was estimated to exceed that of the 1990 census by 23,300,000 people. If the 1990 census reported 248,700,000 people, what is the 1999 estimated population?

72. The highest continuously active volcano in the world is Cotopaxi volcano in Ecuador. The tallest mountain in South America is Aconcagua, which, at a height of 22,831 feet, is 3,431 feet taller than the volcano. What is the height of the volcano?

73. In late 1999, Lucent Technologies announced a record-breaking development of a transistor with a minute length of 50 nanometers. (A nanometer is one billionth of a meter.) This is approximately 2,000 times thinner than the width of a single human hair. Approximate the width of a single human hair.

74. The highest recorded temperature created by man is 950,000,000°F, claimed to be 30 times hotter than the center of the Sun. It was created in 1994 at the Princeton Plasma Physics Laboratory. Approximate the temperature of the center of the Sun. (You wonder how they can measure these things!)

75. How many gallons of a 70% antifreeze solution does a mechanic need if he wants the solution to contain 4 gallons of pure antifreeze?

76. How many pints of a 20% insecticide solution does a gardener need if she wants the solution to contain 0.5 pint of pure insecticide?

4.2 Calculator Exercises

The calculator has a special feature that can be used to solve equations. In order to use this feature, you must enter the following instruction into the calculator:

solve(expression, variable, guess)

In this instruction, "expression" represents the equation's left side after the equation has been rewritten to equal zero on the right side, "variable" instructs the calculator as to which variable is being solved for, and "guess" is a reasonable guess that the calculator uses as a starting point for solving the equation. As an example, suppose you wish to solve $2(3x - 6) + 3(x + 4) = 126$. Begin by rewriting the equation

as $2(3x - 6) + 3(x + 4) - 126 = 0$. Then type the following instruction into the calculator:

solve $(2(3x - 6) + 3(x + 4) - 126, x, 10)$

The calculator will return a value of 14, the solution of the original equation.

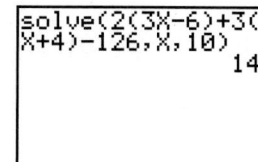

The keystrokes needed to enter the instruction are [2nd] [CATALOG] [▼] (repeatedly until you reach the solve instruction), then [ENTER], then the expression followed by [,] [X,T,θ,n] [,] the guess, and, finally, [)]. Press [ENTER] to execute the instruction.

Following are some applications that you can solve by this method. Write a linear equation to represent each situation, and solve the equation by using the "solve" instruction. Refer to Chapter 2 for the formulas needed in these exercises.

1. A circular fence is to be placed around a swimming pool. The radius of the enclosed area will be 75 feet. Two gates will be placed at opposite ends of the pool. Each gate measures 5 feet wide. How many linear feet of fencing will be required to surround the area? Round your answer to the nearest tenth of a foot.

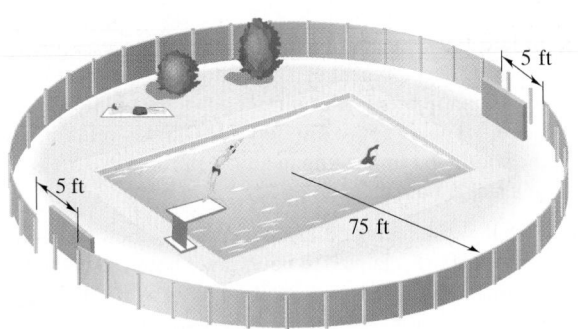

5 ft

5 ft

75 ft

2. A circle has a diameter of 25.8 inches. By how many inches is the circumference larger than the diameter?

3. If a stock account grew by $531\frac{1}{4}$ points for 250 shares of stock, what was the increase per share of stock?

4. How many pieces of tubing, each measuring $5\frac{3}{8}$ inches long, can be cut from a piece measuring $34\frac{1}{2}$ inches long?

5. If a car averages 18.5 miles per gallon for highway driving, approximately how many gallons of fuel will be needed for a trip of 220 miles?

6. In laying brick, a rule of thumb is that 6.5 bricks are needed for each square foot of wall. How many square feet of wall can be constructed from a pallet containing 800 bricks?

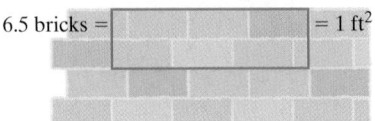

6.5 bricks = [ ] = 1 ft²

7. The label on a package of Choc-o-Block cookies states that one serving of the cookies contains 2.5 grams of fat, representing 4% of the daily recommended amount of fat. Using this information, calculate the daily recommended amount of fat.

8. A bakery states that one serving of its Sweetie-Goo cookies contains 3 grams of fat, representing 5% of the daily recommended amount of fat. Calculate the daily recommended amount from this information. Does your answer agree with that of problem 7? If not, how do you explain the difference?

9. To measure the velocity of water flow in a stream, a ball is thrown into the stream, and the time it takes to travel 250 feet is measured. Using the formula $d = rt$, find the speed (in feet per second) if the ball took 15 seconds to travel the 250 feet.

4.2 Writing Exercise

You have learned two properties of equations that can be used to help you solve linear equations algebraically: the addition property and the multiplication property. Explain how you would decide when to use each of the properties. In your explanation, state what characteristic you would look for in an equation that would signal the property to use. There are four different characteristics of an equation to consider. In your explanation, give examples of equations with each of the characteristics.

4.3 Solving Equations by Using a Combination of Properties

OBJECTIVES

1 Solve linear equations algebraically by using a combination of properties of equations.

2 Model real-world situations by using linear equations, and solve the equations algebraically.

APPLICATION

Crickets are called the poor man's thermometer, because temperature directly affects their rate of activity. According to weather folklore, the temperature can be related to the number of cricket chirps in a given period. Several variations of this relationship may be found. One relationship uses field crickets to determine the Fahrenheit temperature. To do so, one calculates the average number of chirps in 15 seconds (the number of chirps per minute, divided by 4) and adds 37. If the temperature is 77°F, how many chirps will a field cricket produce in one minute?

After completing this section, we will discuss this application further. See page 254.

4.3.1 Using Combinations of Properties of Equations

The addition property and the multiplication property are the keys to solving all linear equations. Now that we have mastered the basics, we are ready to solve more complicated linear equations. In order to do this, we will need to apply combinations of the properties of equations. Since there are several different ways to solve linear equations, we will set up a few rules so that at least in the beginning we are performing the same steps. When we become more sure of ourselves, we may follow these steps in different orders and obtain the same results.

SOLVING A LINEAR EQUATION BY USING A COMBINATION OF PROPERTIES OF EQUATIONS

To solve a linear equation by using a combination of properties of equations,

- Simplify both expressions in the equation (preferably leaving them without fractions).
- Isolate the variable to one side of the equation (preferably the left side) by using the addition property of equations.
- Isolate the constants to the other side (preferably the right side) of the equation by using the addition property of equations.
- Reduce the coefficient of the variable to 1 by using the multiplication property of equations.

Check the solution by substituting, solving numerically, or solving graphically.

The process of simplifying equations by performing operations on both sides that will isolate the variable on one side is the basis of solving all equations, not just linear ones. Therefore, this process is important to remember.

EXAMPLE 1   Use a combination of the properties of equations to solve the following equations algebraically, and then check your solutions:

a. $2x - 3 = 7$ **b.** $6x + 5 = 2x + 25$ **c.** $5x + 4 = 2(3x - 8)$

Algebraic Solution

a. $2x - 3 = 7$

$2x - 3 + 3 = 7 + 3$ Add 3 to both sides.

$2x = 10$ Simplify.

$$\frac{2x}{2} = \frac{10}{2} \qquad \text{Divide both sides by 2.}$$

$$x = 5 \qquad \text{Simplify.}$$

The solution is 5.
The check is left to you.

b.
$$6x + 5 = 2x + 25$$

$$6x + 5 - 2x = 2x + 25 - 2x \qquad \text{Subtract 2x from both sides.}$$

$$4x + 5 = 25 \qquad \text{Simplify.}$$

$$4x + 5 - 5 = 25 - 5 \qquad \text{Subtract 5 from both sides.}$$

$$4x = 20 \qquad \text{Simplify.}$$

$$\frac{4x}{4} = \frac{20}{4} \qquad \text{Divide both sides by 4.}$$

$$x = 5 \qquad \text{Simplify.}$$

The solution is 5.
The check is left to you.

c.
$$5x + 4 = 2(3x - 8)$$

$$5x + 4 = 6x - 16 \qquad \text{Distribute.}$$

$$5x + 4 - 6x = 6x - 16 - 6x \qquad \text{Subtract 6x from both sides.}$$

$$-x + 4 = -16 \qquad \text{Simplify.}$$

$$-x + 4 - 4 = -16 - 4 \qquad \text{Subtract 4 from both sides.}$$

$$-x = -20 \qquad \text{Simplify.}$$

$$\frac{-x}{-1} = \frac{-20}{-1} \qquad \text{Divide both sides by } -1 \text{ (the coefficient of x).}$$

$$x = 20 \qquad \text{Simplify.}$$

The solution is 20.
The check is left to you.

The linear equations in the previous section and in Example 1 of this section all have one solution. However, from the first section in this chapter, we know that that is not always the case: Linear equations may be contradictions or identities. Let's see what happens when we apply our rules in those cases. Complete the following set of exercises.

Discovery 9

Linear Equations with No Solution

Solve algebraically the previous example of a linear equation with no solution: $2x + 5 = 2x + 10$.

Write a rule that explains why the equation has no solution.

When we attempt to isolate the variable term to one side of the equation, that term is eliminated from both sides of the equation. The result is a false equation. Therefore, there is no solution. The equation is a contradiction.

Discovery 10

Linear Equations with Many Solutions

Solve algebraically the previous example of a linear equation with many solutions: $2x + 5 = (x + 3) + (x + 2)$.

Write a rule that explains why the equation has many solutions.

When we attempt to isolate the variable term to one side, that term is eliminated from the equation. The result is a true equation. Therefore, the solution set is the set of all real numbers. The equation is an identity.

EXAMPLE 2 Solve algebraically.

a. $4(x - 2) = 4x - 8$ **b.** $3x + 4 = 2x + (x + 10)$

Algebraic Solution

a.
$$4(x - 2) = 4x - 8$$
$$4x - 8 = 4x - 8 \qquad \text{Distribute 4.}$$
$$4x - 8 - 4x = 4x - 8 - 4x \qquad \text{Subtract 4x from both sides.}$$
$$-8 = -8 \qquad \text{Simplify.}$$

The original equation is an identity. We can confirm this by a numeric or graphic check.

Since this is a true equation, the solution is all possible values for x, or the set of all real numbers.

b.
$$3x + 4 = 2x + (x + 10)$$
$$3x + 4 = 2x + x + 10 \qquad \text{Remove parentheses.}$$
$$3x + 4 = 3x + 10 \qquad \text{Simplify.}$$
$$3x + 4 - 3x = 3x + 10 - 3x \qquad \text{Subtract 3x from both sides.}$$
$$4 = 10 \qquad \text{Simplify.}$$

The original equation is a contradiction. We can confirm this by a numeric or graphic check.

Since this is a false equation, there is no solution.

If an equation has fractional coefficients in the terms, we must be very careful in applying the rules. For example, consider the following algebraic solution:

$$\frac{3}{4}x + \frac{5}{6} = \frac{5}{3}$$
$$\frac{3}{4}x + \frac{5}{6} - \frac{5}{6} = \frac{5}{3} - \frac{5}{6} \qquad \text{Subtract } \tfrac{5}{6} \text{ from both sides.}$$
$$\frac{3}{4}x = \frac{10}{6} - \frac{5}{6} \qquad \text{Simplify and change to LCD.}$$
$$\frac{3}{4}x = \frac{5}{6} \qquad \text{Simplify.}$$

$$\frac{\overset{1}{\cancel{4}}}{\underset{1}{\cancel{3}}}\left(\frac{\overset{1}{\cancel{3}}}{\underset{1}{\cancel{4}}}x\right) = \frac{\overset{2}{\cancel{4}}}{3}\left(\frac{5}{\cancel{6}}\right)$$

Multiply both sides by $\frac{4}{3}$.

$$x = \frac{10}{9}$$

The solution is $\frac{10}{9}$.

If an equation has several fractional coefficients in the terms, the multiplication property of equations allows us to solve an equation in an easier way than dealing with the fractions. To clear fractional (decimal) coefficients from terms in an equation, multiply by the least common denominator for all the fractional coefficients. For example,

$$\frac{3}{4}x + \frac{5}{6} = \frac{5}{3}$$

$$12\left(\frac{3}{4}x + \frac{5}{6}\right) = 12\left(\frac{5}{3}\right)$$

Multiply both sides by 12 (LCD of all the fractional coefficients).

$$\overset{3}{\cancel{12}}\left(\frac{3}{\cancel{4}}x\right) + \overset{2}{\cancel{12}}\left(\frac{5}{\cancel{6}}\right) = \overset{4}{\cancel{12}}\left(\frac{5}{\cancel{3}}\right)$$

Distribute 12.

$$9x + 10 = 20$$

Simplify.

$$9x + 10 - 10 = 20 - 10$$

Subtract 10 from both sides.

$$9x = 10$$

Simplify.

$$\frac{9x}{9} = \frac{10}{9}$$

Divide both sides by 9.

$$x = \frac{10}{9}$$

The solution remains $\frac{10}{9}$.

EXAMPLE 3   Solve algebraically. Check your solutions.

a. $-\frac{2}{3}x + \frac{7}{5} = -\frac{5}{6}x$ **b.** $\frac{3}{8}\left(x + \frac{1}{4}\right) = \frac{5}{6}x + 2$

c. $0.25x - 2.75 = 0.1x + 2$

Algebraic Solution

a.
$$-\frac{2}{3}x + \frac{7}{5} = -\frac{5}{6}x$$

$$30\left(-\frac{2}{3}x + \frac{7}{5}\right) = 30\left(-\frac{5}{6}x\right)$$

Multiply both sides by 30 (LCD).

$$\overset{10}{\cancel{30}}\left(-\frac{2}{\cancel{3}}x\right) + \overset{6}{\cancel{30}}\left(\frac{7}{\cancel{5}}\right) = \overset{5}{\cancel{30}}\left(-\frac{5}{\cancel{6}}x\right)$$

Distribute 30.

$$-20x + 42 = -25x$$

Simplify.

$$-20x + 42 + 20x = -25x + 20x$$

Add 20x to both sides because the right expression did not have a constant term and this will save steps.

$$42 = -5x \qquad \text{Simplify.}$$

$$\frac{42}{-5} = \frac{-5x}{-5} \qquad \text{Divide both sides by } -5.$$

$$-\frac{42}{5} = x \qquad \text{Simplify.}$$

The solution is $-\frac{42}{5}$.

The check is left to you.

b. $\dfrac{3}{8}\left(x + \dfrac{1}{4}\right) = \dfrac{5}{6}x + 2$

HELPING HAND The first fraction in the expression on the left is a factor, not a term. It will be simpler if we distribute before we eliminate the fractions.

$$\frac{3}{8}x + \frac{3}{32} = \frac{5}{6}x + 2 \qquad \text{Distribute.}$$

$$96\left(\frac{3}{8}x + \frac{3}{32}\right) = 96\left(\frac{5}{6}x + 2\right) \qquad \text{Multiply both sides by 96 (LCD).}$$

$$\overset{12}{96}\left(\frac{3}{8}x\right) + \overset{3}{96}\left(\frac{3}{32}\right) = \overset{16}{96}\left(\frac{5}{6}x\right) + 96(2) \qquad \text{Distribute.}$$

$$36x + 9 = 80x + 192 \qquad \text{Simplify.}$$

$$36x + 9 - 80x = 80x + 192 - 80x \qquad \text{Subtract 80x from both sides.}$$

$$-44x + 9 = 192 \qquad \text{Simplify.}$$

$$-44x + 9 - 9 = 192 - 9 \qquad \text{Subtract 9 from both sides.}$$

$$-44x = 183 \qquad \text{Simplify.}$$

$$\frac{-44x}{-44} = \frac{183}{-44} \qquad \text{Divide both sides by } -44.$$

$$x = -\frac{183}{44} \qquad \text{Simplify.}$$

The solution is $-\frac{183}{44}$.

The check is left to you.

c. $0.25x - 2.75 = 0.1x + 2$

HELPING HAND Decimals are equivalent to fractions. The LCD is determined by the place value of each decimal.

$$100(0.25x - 2.75) = 100(0.1x + 2) \qquad \text{Multiply both sides by 100 (LCD).}$$

$$100(0.25x) - 100(2.75) = 100(0.1x) + 100(2) \qquad \text{Distribute.}$$

$$25x - 275 = 10x + 200 \qquad \text{Simplify.}$$

$$25x - 275 - 10x = 10x + 200 - 10x \qquad \text{Subtract 10x from both sides.}$$

$$15x - 275 = 200 \qquad \text{Simplify.}$$

$$15x - 275 + 275 = 200 + 275 \qquad \text{Add 275 to both sides.}$$

$$15x = 475 \qquad \text{Simplify.}$$

$$\frac{15x}{15} = \frac{475}{15}$$ Divide both sides by 15.

$$x \approx 31.667$$ Simplify.

The solution is approximately 31.667.

The check is left to you.

ALGEBRAIC SOLUTIONS OF LINEAR EQUATIONS

To solve a linear equation algebraically, use the properties of equations to isolate the variable. One of three possibilities will occur:

One solution exists.
No solution exists. Solving results in a contradiction.
An infinite number of solutions exist. Solving results in an identity.

 4.3.1 Checkup

In exercises 1–5, use a combination of the properties of equations to solve algebraically. Check your solutions.

1. $2x - 5 = 4(x + 2)$ **2.** $6(x - 3) + 23 = 6x + 5$

3. $x + 5(2x - 7) = 10(x + 4) + (x + 5)$

4. $\frac{1}{2}\left(x + \frac{3}{4}\right) = \frac{1}{3}x - \frac{1}{8}$ **5.** $1.2x - 4 = 0.8x + 1.2$

6. If a linear equation has parentheses in it, what should you do first when attempting to solve it?

7. If a linear equation has coefficients that are fractions or decimals, what can you do to make the equation easier to solve?

4.3.2 Modeling the Real World

Real-world situations tend to be complicated. But many situations can be expressed or approximated by linear equations, and now we know all we need to know in order to solve them. Remember that once you've solved an equation, you will need to look at your solution and see if it makes sense in terms of the original situation. Is it possible to have a negative solution? Is a fractional result realistic? Don't forget to ask yourself these kinds of questions.

EXAMPLE 4   Cameron has invested $6000 at 8% annual simple interest for one year. He needs to have earned a total of $635 in interest at the end of one year. How much additional money should he invest at 5% annual simple interest?

Solution

Let x = the amount to be invested at 5%

At 8% annual rate	At 5% annual rate
$I = PRT$	$I = PRT$
$I = 6000(0.08)(1)$	$I = x(0.05)(1)$

The total interest will be the sum of the two interests.

$$6000(0.08)(1) + x(0.05)(1) = 635$$
$$480 + 0.05x = 635$$

Algebraic

$$480 + 0.05x = 635$$
$$480 + 0.05x - 480 = 635 - 480 \quad \text{Subtract 480.}$$
$$0.05x = 155$$
$$\frac{0.05x}{0.05} = \frac{155}{0.05} \quad \text{Divide by 0.05.}$$
$$x = 3100$$

Graphic

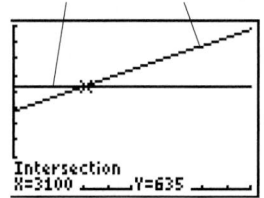

Y2 = 635 Y1 = 480 + 0.05x

Intersection
X=3100   Y=635

(0, 10,000, 0, 1000)
The solution is 3100.

Cameron needs to invest $3100 at 5% simple interest in order to have earned $635 in interest at the end of the year.

EXAMPLE 5 Charles plans to sell flower boxes for $5.00 each. He estimates that the cost of the wood is $3.25 per box. The other materials needed cost $1.75 per box. An advertisement costs $3.50. Determine the break-even point. (That is, determine when the revenue and the cost are equal.)

Solution
Let x = number of flower boxes
The revenue is $5.00 per flower box, or $5.00x$.
The cost is $3.25 per flower box, plus $1.75 per flower box, plus $3.50, or $3.25x + 1.75x + 3.50$.

Algebraic

$$5.00x = 3.25x + 1.75x + 3.50$$
$$5.00x = 5.00x + 3.50 \quad \text{Simplify.}$$
$$5.00x - 5.00x = 5.00x + 3.50 - 5.00x \quad \text{Subtract}$$
$$0 = 3.50 \quad \text{5.00x from both sides.}$$

Numeric

X	Y₁	Y₂
0	0	3.5
1	5	8.5
2	10	13.5
3	15	18.5
4	20	23.5
5	25	28.5
6	30	33.5

X=0

Y1 = $5.00x$
Y2 = $3.25x + 1.75x + 3.50$
Note that the left side is always 3.5 less than the right side.

This is a contradiction. There is no solution. Charles will not break even.
   If Charles examines his estimates, he will see that the cost of wood and materials is $5.00 per flower box, his selling price. Therefore, when he adds the cost of the advertisement to the cost of the wood and materials, his cost exceeds his selling price. He must raise the price of his flower boxes in order to break even.

EXAMPLE 6 A pharmacist needs a 40% alcohol solution. If she plans to mix 30 cubic centimeters (cc) of a 20% alcohol solution with a 70% alcohol solution, how many cc of the 70% solution does she use? How many cc are in the 40% solution?

Solution

Let x = number of cc of 70% alcohol solution

$x + 30$ = number of cc of 40% alcohol solution

The 40% solution is a mixture of the 20% alcohol solution and the 70% alcohol solution. Therefore, we add the amount of alcohol in the 20% solution, which is 20% of 30, and the amount of alcohol in the 70% solution, which is 70% of x. The result is the amount of alcohol in the 40% solution, 40% of $(x + 30)$.

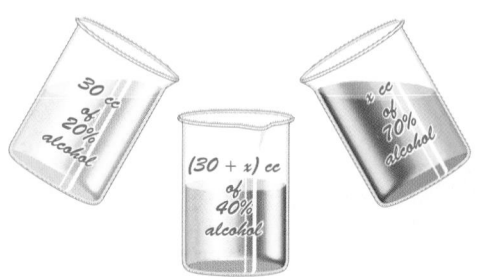

Algebraic

$$0.2(30) + 0.7x = 0.4(x + 30)$$

$$6 + 0.7x = 0.4x + 12 \qquad \text{Simplify.}$$

$$6 + 0.7x - 0.4x = 0.4x + 12 - 0.4x \qquad \text{Subtract 0.4x from both sides.}$$

$$6 + 0.3x = 12 \qquad \text{Simplify.}$$

$$6 + 0.3x - 6 = 12 - 6 \qquad \text{Subtract 6 from both sides.}$$

$$0.3x = 6 \qquad \text{Simplify.}$$

$$\frac{0.3x}{0.3} = \frac{6}{0.3} \qquad \text{Divide both sides by 0.3.}$$

$$x = 20 \qquad \text{Simplify.}$$

Numeric

X	Y1	Y2
17	17.9	18.8
18	18.6	19.2
19	19.3	19.6
20	20	20
21	20.7	20.4
22	21.4	20.8
23	22.1	21.2

X=17

Y1 = 0.2 (30) + 0.72
Y2 = 0.4 (x + 30)

The solution is 20.

When 20 cc of the 70% alcohol solution is added to 30 cc of the 20% alcohol solution, the result will be 50 cc of the needed 40% alcohol solution.

APPLICATION

Crickets are called the poor man's thermometer, because temperature directly affects their rate of activity. According to weather folklore, the temperature can be related to the number of cricket chirps in a given period. Several variations of this relationship may be found. One relationship uses field crickets to determine the Fahrenheit temperature. To do so, one calculates the average number of chirps in 15 seconds (the number of chirps per minute, divided by 4) and adds 37. If the temperature is 77°F, how many chirps will a field cricket produce in one minute?

Discussion

Let x = the number of chirps per minute

$\dfrac{x}{4}$ = the average number of chirps in 15 seconds

$\dfrac{x}{4} + 37$ = the sum of 37 and the average number of chirps in 15 seconds

The sum is equal to 77, or $\dfrac{x}{4} + 37 = 77$.

Algebraic

$$\frac{x}{4} + 37 = 77$$ *Write the equation.*

$$\frac{x}{4} + 37 - 37 = 77 - 37$$ *Subtract 37 from both sides.*

$$\frac{x}{4} = 40$$ *Simplify.*

$$\frac{x}{4} \cdot 4 = 40 \cdot 4$$ *Multiply both sides by 4.*

$$x = 160$$ *Simplify.*

Numeric

X	Y1	Y2
158	76.5	77
159	76.75	77
160	77	77
161	77.25	77
162	77.5	77
163	77.75	77
164	78	77

X=158

Y1 = $\frac{x}{4}$ + 37   Y2 = 77

The solution is 160.

When the temperature is 77°F, the field cricket chirps at a rate of 160 chirps per minute.

4.3.2 Checkup

Solve algebraically.

1. A furniture store advertises that you may purchase a bedroom suite with a down payment of $50, no interest charges, and 36 months to pay the balance. How much will the monthly payments be if the bedroom suite sells for $2750?

2. Gladys is offered a job to help prepare handcrafted baskets for a bazaar. She will be paid $50.00 per day plus $5.00 per item sold. She must purchase all of her supplies herself. For each day's order, she buys $35.00 worth of dried flowers and $15.00 worth of ribbon. In addition, each basket costs her $3.75, and each basket lining costs her $1.25. Determine Gladys's break-even point.

3. If *x* cubic centimeters (cc) of a 40% alcohol solution is mixed with 20 cc of a 70% alcohol solution, a 60% alcohol solution is obtained. How many cc of the 40% alcohol solution were used?

4.3 Exercises

Solve algebraically and check. If necessary, round decimal answers to one more decimal place than that shown in the original exercise.

1. $4x + 8 = 0$
2. $5x - 15 = 0$
3. $-x - 41 = 3$
4. $15 - x = 3$
5. $-3x + 7 = 7$
6. $36 = 6x + 9$
7. $15.17 = 5.9x - 4.3$
8. $-4.22x - 0.4 = -21.5$
9. $-9.2x - 4.3 = -70.54$
10. $-6.3p + 1.5 = -4.8$
11. $-\frac{5}{9}b + \frac{11}{12} = \frac{23}{36}$
12. $\frac{5}{12}z + \frac{1}{6} = \frac{4}{9}$
13. $-2\frac{2}{3}z - 3\frac{1}{2} = -8\frac{5}{6}$
14. $-\frac{1}{9}y - \frac{5}{18} = -\frac{1}{6}$
15. $5x + 6 = x + 126$
16. $-5x - 18 = 2x - 4$
17. $27x - 49 = -12x - 10$
18. $2x + 17 = 17 - 4x$
19. $156z - 210 = 47z + 662$
20. $728a + 958 = 116a - 878$
21. $4x - (3x + 5) = x - 5$
22. $7x + 4 = 3x + 2(2x + 2)$
23. $6x - (x + 1) = 5x + 7$
24. $6x + (2x + 7) = 2(4x + 9)$
25. $5(0.3x + 8.7) = 1.5x + 43.5$
26. $3.4x + 8.8 = 0.2(17x + 5)$
27. $5.5x = 1.2x + 3.3(x - 2)$
28. $0.2x - 1.4 = 0.2(x - 7)$
29. $\frac{3}{4}x + 6 = \frac{1}{2}x + \frac{1}{4}x$
30. $\frac{1}{3}x + \frac{3}{5} = \frac{1}{2}x - \frac{2}{5}$
31. $3x - \frac{1}{4} = \left(x + \frac{1}{2}\right) + \left(2x + \frac{1}{3}\right)$
32. $\frac{3}{8}x + 2 = \frac{1}{8}x + 4$
33. $11x - 12 = 7(3x - 6) - 2(x + 9)$
34. $2(2x - 9) + 3(3x - 4) = 4(x - 3) + 9(x - 2)$
35. $7x - 5(3x + 9) = -2(4x + 35)$
36. $5(x + 3) - 2x = 2(x + 11) - (2x - 23)$

Use the multiplication property of equations to clear the fractions from each equation. Then solve algebraically.

37. $\frac{1}{4}x + \frac{5}{9} = \frac{5}{6}$

38. $\frac{3}{7}x + 2 = \frac{3}{4}x - 1$

39. $3x + \frac{1}{4} = 2x + \frac{7}{36}$

40. $\frac{9}{11}y - \frac{17}{22} = \frac{4}{11}y + \frac{1}{22}$

41. $\frac{2}{5}b - 12 = \frac{2}{3}b + 20$

42. $\frac{7}{9}x - 15 = \frac{4}{9}x - 37$

43. $\frac{3}{4}\left(x + \frac{4}{5}\right) = -\frac{7}{8}x - \frac{2}{5}$

44. $\frac{3}{4}\left(20p + \frac{1}{2}\right) = 13p - \frac{3}{8}$

Use the multiplication property of equations to clear the decimals from each equation. Then solve algebraically.

45. $0.05x + 10.5 = 0.15x - 0.25$

46. $0.01x + 0.11 = 0.47 - 0.09x$

47. $21.1x + 0.46 = 10.9x + 0.46$

48. $-1.05x - 15.41 = 2.55x - 47.09$

49. $15.2y - 175.43 = -2.4y - 176.31$

50. $81 = 0.5(120 - x) + 0.8x$

Solve algebraically.

51. Happy Harpo's car rental firm offers a weekly special for vacationers: You can rent a car for seven days, paying $49.95 plus a charge of $0.12 per mile. If you have budgeted $200 for travel, how many miles is the maximum you would be able to drive in one week under this plan?

52. Elegant Eydie's car rental firm will rent you a luxury car for $119.95 per week and $0.22 per mile. If you budgeted $500 for a week's travel, how many miles is the maximum you could drive?

53. A furniture store will carry your loan without interest for 24 months on any purchase if you pay $200 down. What would the monthly payments be for a living-room set selling for $2252?

54. During a special promotion, Frugal Frieda's furniture company offers to carry your loan without interest for 30 months, with no down payment required. Not only that, but they will give you an instant rebate of $150 on the purchase to apply to the account. What would the monthly charges be on a purchase of $4350?

55. If 4 liters of a 10% vinegar solution is mixed with x liters of a 30% vinegar solution, the result is a 25% vinegar solution. How many liters of the 30% vinegar solution were mixed?

56. Five gallons of a 10% glucose solution is mixed with x gallons of a 40% glucose solution to obtain a 20% glucose solution. How many gallons of the 40% glucose solution were used?

57. Gina learned that her average weekly earnings are $25.00 more than twice her brother's average weekly earnings. If her weekly earnings average $730.10, what are her brother's average weekly earnings?

58. Dean's stock investments yielded a profit for the year that was $3565 less than 1.5 times his profit last year. If his profit this year was $9917, what was his profit last year?

59. Acme Sales pays each member of its sales staff $150 per week plus a commission of 10% of the person's sales. The company deducts 2% of the person's sales from its payment to the sales staff to cover overhead expenses. On the other hand, Mega Sales pays each of its salespeople $200 per week plus a commission of 8% of the person's sales. Mega deducts $50 per week from the payment to its salespeople to cover overhead expenses. For what level of sales will the two companies pay the same weekly amount to their salespeople?

60. During its festival promotion, the Discount Mall will rent a booth to retailers for $285 plus a daily charge of $25. The company also adds to this a daily charge of $20 for janitorial and other support. A license to sell at the Discount Mall costs $15. At the same time, the Christmas Mall next door will rent a booth to retailers for $255 plus a daily charge of $45. There is no charge for support labor, and a license to sell at the Christmas Mall costs $45. Find the number of days for which the two rental options have the same total cost.

61. The sales force at a car dealership has a choice of two pay plans. The first plan pays the salesperson $300 per week plus 4% of the total sales for the week. The second plan pays the salesperson $700 per week plus 4% of the total sales in excess of $5000. Given that sales must exceed $5000, for what value of sales will the two plans be equal?

62. The Flexi-Rental firm will rent a car with a choice of plans. The first plan charges $39.50 plus $0.20 per mile. The second plan charges $49.50 plus $0.20 per mile for every mile over 100 miles. For what number of miles will the two plans cost the same?

63. Nathan budgets $2000 per school year for fees and books at his college. He has a part-time job that will earn him $600 toward his school expenses. A trust fund his parents set up for him contains $20,000 and pays 4.5% simple interest per year. How much additional money must be invested in another trust fund that pays 8% simple interest per year to meet Nathan's budget at the end of the year?

64. Max budgets $5800 per school year for out-of-state fees and books at his college. He has $3,000 saved for these expenses, as well as a mutual fund that has $35,000 invested. The mutual fund pays Max 6.5% simple interest per year. How much additional money does Max need to invest in a savings account that pays 5% simple interest per year in order to meet his budget at the end of the year?

65. To calculate Fahrenheit temperature using the chirps of tree crickets, one calculates the average number of chirps in 7 seconds (the number of chirps in 1 minute, divided by 8.6) and adds 46. How many chirps per minute will you count if the temperature is 81°F?

66. To calculate Fahrenheit temperature using the chirps of snowy tree crickets, one calculates the average number of chirps in 14 seconds (the number of chirps in 1 minute, divided by 4.3) and adds 42. How many chirps per minute would you count if the temperature is 95°F?

4.3 Calculator Exercises

In Section 4.2, we saw how to use the "solve" command as an alternative method for solving a linear equation. The method first required that you rewrite the equation so that its left-hand side was set equal to zero. This can be time consuming and lead to errors. You can obtain the same result if you take the original equation you wish to solve and store the left-hand side in Y1 and the right-hand side in Y2. Then enter the "solve" command as follows:

Solve (Y1-Y2, x, guess)

Remember, the Y1 and Y2 symbols can be keyed by pressing

| VARS | ▶ | | | | for Y1 and | VARS | ▶ | | | 2 | for Y2.

After finishing one problem, you need only type in the new expressions for Y1 and Y2, and you will be ready to use the same solve command for the next problem.

Write a linear equation which represents each of the applications that follow. Then use the solve command on your calculator to solve the equation.

1. A door-to-door cosmetics salesperson is charged a 3.75% fee on total sales plus $25.00 to participate in the program. If the fee for one month was $49.48, what were the total sales for that month?

2. A nurse must administer 620 grains of a medication to a patient. She has one 200-grain tablet that she must use, and the remainder is in 120-grain tablets. How many of the 120-grain tablets must she administer?

3. A carpenter wants to create a $4\frac{1}{4}$-inch-thick tabletop by layering sheets of different woods on a 2-inch base. The sheets are each $\frac{3}{4}$ inches thick. How many sheets of wood does he need?

4.3 Writing Exercise

At the beginning of this section, we said that you should follow a few rules in solving linear equations, using a combination of properties of equations. The rules first suggested that you use the addition property to isolate the variable to one side of the equation and the constant term to the other side and then use the multiplication property to complete the solution. Explain the advantages of solving in this order, as opposed to using the multiplication principle first. Describe difficulties you might encounter in reversing the order. Since every rule has its exceptions, describe situations in which you might want to apply the multiplication principle first. (*Hint:* In some examples in this section, the multiplication principle was used first.) Finish your discussion by explaining the steps you will follow in applying these properties of equations to solve a linear equation, and illustrate with an example.

4.4 Solving Equations or Formulas for a Specified Variable

OBJECTIVES

1 Solve equations or formulas for a specified variable.

2 Model real-world situations by using linear equations, and then solve the equations.

APPLICATION

Temperature can be recorded in Celsius and Fahrenheit scales. If you know the conversion formula to calculate Celsius temperatures from Fahrenheit temperatures is $C = \frac{5}{9}(F - 32)$, solve the formula for F to obtain a new formula to calculate Fahrenheit temperatures from Celsius temperatures.

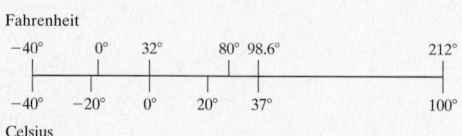

After completing this section, we will discuss this application further. See page 262.

4.4.1 Solving Equations for a Variable

We defined the term *formula* in an earlier chapter. A formula is a statement equating an unknown variable with an expression that can be evaluated to find a value for the unknown variable. For example, $A = LW$ is a formula used to find a value for the variable A(area) when values for L(length) and W(width) are known for a rectangle.

If $L = 12$ feet and $W = 6$ feet, then

$$A = LW$$
$$A = 12 \cdot 6 \qquad \text{Substitute.}$$
$$A = 72 \text{ square feet} \qquad \text{Evaluate.}$$

However, we may know values for A and W, but not for L.
If $A = 48$ square feet and $W = 8$ feet, then

$$A = LW$$
$$48 = L \cdot 8 \qquad \text{Substitute.}$$
$$\frac{48}{8} = \frac{L \cdot 8}{8} \qquad \text{Divide both sides by 8.}$$
$$6 = L \qquad \text{Simplify.}$$
$$\text{or} \quad L = 6 \text{ feet}$$

When this type of evaluation must be repeated several times, we need a new formula for L in terms of A and W. That is, we need to solve the formula $A = LW$ for the variable L. We do this by using the same properties of equations that we used in the last two sections.

For example, given $A = LW$, solve for L.

$$A = LW$$
$$\frac{A}{W} = \frac{LW}{W} \qquad \text{Divide both sides by } W.$$
$$\frac{A}{W} = L \quad \text{or} \quad L = \frac{A}{W} \qquad \text{Simplify.}$$

We now have a formula for L in terms of A and W.

HELPING HAND The conventional form for writing an equation or formula is to write the variable we are solving for on the left side. Recall that in Chapter 1 we wrote equivalent order relations. For equations or formulas, we simply exchange the expressions on each side.

SOLVING AN EQUATION OR FORMULA FOR A VARIABLE
To solve an equation or formula for a variable by using the properties of equations,

- Clear the equation or formula of fractions.
- Isolate the term(s) involving the desired variable to one side, using the addition property of equations.
- Use the addition property of equations to collect all the other terms to the other side.
- Collect like terms in both expressions. If the terms containing the desired variable are unlike, factor out the desired variable.
- Use the multiplication property of equations to reduce the coefficient of the desired variable to 1.

HELPING HAND Since we are using a process we have previously learned, with the only difference being that we are using variables instead of numbers, we can substitute numbers for the variables and solve the equation first. This will help us "see" the steps to use for the variables.

EXAMPLE 1 Solve each formula for the indicated variable.

a. Perimeter of a rectangle, $P = 2L + 2W$, for L.

b. Area of a triangle, $A = \dfrac{1}{2}bh$, for b.

c. Average of three grades, $A = \dfrac{a + b + c}{3}$, for c.

d. Area of a trapezoid, $A = \dfrac{1}{2}hb + \dfrac{1}{2}hB$, for h.

Solution

a. Perimeter of a rectangle, $P = 2L + 2W$, for L.

$$P = 2L + 2W$$
$$P - 2W = 2L + 2W - 2W \qquad \text{Subtract } 2W \text{ from both sides.}$$
$$P - 2W = 2L \qquad \text{Simplify.}$$
$$\frac{P - 2W}{2} = \frac{2L}{2} \qquad \text{Divide both sides by 2.}$$
$$\frac{P - 2W}{2} = L \quad \text{or} \quad L = \frac{P - 2W}{2} \qquad \text{Simplify.}$$

We could continue to simplify by dividing each term of the expression by 2, yielding $L = \frac{P}{2} - W$.

b. Area of a triangle, $A = \dfrac{1}{2}bh$, for b.

$$A = \frac{1}{2}bh$$
$$2A = 2\left(\frac{1}{2}bh\right) \qquad \text{Multiply both sides by 2 to clear the fraction.}$$

$$2A = bh \qquad \text{Simplify.}$$

$$\frac{2A}{h} = \frac{bh}{h} \qquad \text{Divide both sides by } h.$$

$$\frac{2A}{h} = b \quad \text{or} \quad b = \frac{2A}{h} \qquad \text{Simplify.}$$

c. Average of three grades, $A = \dfrac{a + b + c}{3}$, for c.

$$A = \frac{a + b + c}{3}$$

$$3A = 3\left(\frac{a + b + c}{3}\right) \qquad \text{Multiply both sides by 3 to clear the fraction.}$$

$$3A = a + b + c \qquad \text{Simplify.}$$

$$3A - a - b = a + b + c - a - b \qquad \text{Subtract } a \text{ and } b \text{ from both sides.}$$

$$3A - a - b = c \qquad \text{Simplify.}$$

$$\text{or} \quad c = 3A - a - b$$

d. Area of a trapezoid, $A = \dfrac{1}{2}hb + \dfrac{1}{2}hB$, for h.

$$A = \frac{1}{2}hb + \frac{1}{2}hB$$

$$2A = 2\left(\frac{1}{2}hb + \frac{1}{2}hB\right) \qquad \text{Multiply both sides by 2 to clear the fractions.}$$

$$2A = 2\left(\frac{1}{2}hb\right) + 2\left(\frac{1}{2}hB\right) \qquad \text{Distribute.}$$

$$2A = hb + hB \qquad \text{Simplify.}$$

$$2A = h(b + B) \qquad \text{Since we are solving for } h, \text{ which is a factor in two unlike terms, we factor out the common factor } h.$$

$$\frac{2A}{b + B} = \frac{h(b + B)}{b + B} \qquad \text{Divide both sides by } (b + B).$$

$$\frac{2A}{b + B} = h \qquad \text{Simplify.}$$

$$\text{or} \quad h = \frac{2A}{b + B}$$

EXAMPLE 2 Solve for y in terms of x.

a. $3x - y = 6$ **b.** $4x + 2y = 6$ **c.** $y - 3 = \dfrac{1}{2}(x + 1)$

Solution

a.
$$3x - y = 6$$

$$3x - y - 3x = 6 - 3x \qquad \text{Subtract 3x from both sides.}$$

$$-y = 6 - 3x \qquad \text{Simplify.}$$

$$\frac{-y}{-1} = \frac{6 - 3x}{-1}$$ Divide both sides by -1.

$$y = -6 + 3x$$ Distribute and simplify.

or $$y = 3x - 6$$ Rearrange the expression on the right with the variable term first.

HELPING HAND The conventional form for writing an equation when y is equated to an expression containing a variable term and a constant term is to write the right side of the equation with the variable term followed by the constant term.

b. $$4x + 2y = 6$$

$$4x + 2y - 4x = 6 - 4x$$ Subtract $4x$ from both sides.

$$2y = 6 - 4x$$ Simplify.

$$\frac{2y}{2} = \frac{6 - 4x}{2}$$ Divide both sides by 2.

$$y = 3 - 2x$$ Distribute and simplify.

or $$y = -2x + 3$$ Rearrange the expression on the right with the variable term first.

c. $$y - 3 = \frac{1}{2}(x + 1)$$

$$y - 3 = \frac{1}{2}x + \frac{1}{2}$$ Distribute.

$$y - 3 + 3 = \frac{1}{2}x + \frac{1}{2} + 3$$ Add 3 to both sides.

$$y = \frac{1}{2}x + \frac{7}{2}$$ Simplify.

4.4.1 Checkup

Solve each formula for the indicated variable.

1. Area of a triangle, $A = \frac{1}{2}bh$, for h.

2. Average of four grades, $A = \dfrac{a + b + c + d}{4}$, for a.

Solve for y in terms of x.

3. $44x + 22y = 55$ **4.** $y + 5 = \frac{7}{3}(x + 6)$

4.4.2 Modeling the Real World

Many important formulas are used to describe real-world relationships in geometry, science, and business—indeed, just about any subject you can think of. But once you understand how to solve a formula for any variable, you don't need to memorize all the different forms of the same equation. For example, if you know that speed is defined as distance traveled divided by time taken traveling, $r = \frac{d}{t}$, you can solve for d and come up with the distance formula, $d = rt$. Rearranging formulas in this way is a very useful skill to have.

EXAMPLE 3 Linda's class wants to decorate and sell t-shirts. Linda can buy the t-shirts for $4 each and the paint for $12.

a. Write an equation for the cost, c to decorate n t-shirts.

b. Use your equation to find a new equation for the number of t-shirts, n, in terms of the cost, c.

c. Use the equation obtained in part b to determine the number of t-shirts the class can make for $150, $250, and $350.

Solution

a. Let c = total cost

   n = number of t-shirts decorated

   $c = 4n + 12$

b. Solve $c = 4n + 12$ for n.

$$c = 4n + 12$$

$$c - 12 = 4n + 12 - 12 \qquad \text{Subtract 12 from both sides.}$$

$$c - 12 = 4n \qquad \text{Simplify.}$$

$$\frac{c - 12}{4} = \frac{4n}{4} \qquad \text{Divide both sides by 4.}$$

$$\frac{1}{4}c - 3 = n \qquad \text{Simplify.}$$

$$n = \frac{1}{4}c - 3 \qquad \text{Write in conventional form.}$$

c. A table of values will give us the number of t-shirts for different values.

c	$n = \dfrac{1}{4}c - 3$	n
150	$n = \dfrac{1}{4}(150) - 3$	$34\dfrac{1}{2}$
250	$n = \dfrac{1}{4}(250) - 3$	$59\dfrac{1}{2}$
350	$n = \dfrac{1}{4}(350) - 3$	$84\dfrac{1}{2}$

Note: It is not possible to make the 0.5 shirt. To stay within the budget, Linda cannot round up to 35 shirts. Therefore, Linda's class can make 34 shirts for $150, 59 shirts for $250, and 84 shirts for $350.

APPLICATION

Temperature can be recorded in Celsius and Fahrenheit scales. If you know the conversion formula to calculate Celsius temperatures from Fahrenheit temperatures is $C = \frac{5}{9}(F - 32)$, solve the formula for F to obtain a new formula to calculate Fahrenheit temperatures from Celsius temperatures.

Discussion

$$C = \frac{5}{9}(F - 32)$$

$$C = \frac{5}{9}F - \frac{5}{9} \cdot 32 \qquad \text{Distribute.}$$

$$C = \frac{5}{9}F - \frac{160}{9}$$

$$9C = 9\left(\frac{5}{9}F - \frac{160}{9}\right)$$ Multiply both sides by 9.

$$9C = 5F - 160$$ Simplify.

$$9C + 160 = 5F - 160 + 160$$ Add 160 to both sides.

$$9C + 160 = 5F$$ Simplify.

$$\frac{9C + 160}{5} = \frac{5F}{5}$$ Divide both sides by 5.

$$\frac{9}{5}C + 32 = F$$ Distribute.

$$\text{or } F = \frac{9}{5}C + 32$$

Two formulas used to convert between Celsius and Fahrenheit temperatures are $C = \frac{5}{9}(F - 32)$ and $F = \frac{9}{5}C + 32$.

 4.4.2 Checkup

1. Happy Harpo's car rental firm offers a weekly special for vacationers: You can rent a car for seven days, paying $49.95 plus a charge of $0.12 per mile.

   a. Write an equation for the cost of renting the car and driving x miles during the week.

   b. Use your equation to find a new equation for the number of miles driven, x, in terms of the cost, c.

   c. Use the equation in part b to find how many miles is the maximum you can drive if your vacation budget is

$200. (Compare your answer with that which you obtained for exercise 51 in section 4.3.)

   d. Use the equation in part b to find how many miles is the maximum you can drive if your budget is $150; if your budget is $250; if your budget is $500.

2. Solve the Fahrenheit temperature formula, $F = \frac{9}{5}C + 32$, for the Celsius temperature, C.

4.4 Exercises

Solve each formula for the indicated variable.

1. The perimeter of a square, $P = 4s$, for s.
2. The area of a parallelogram, $A = bh$, for b.
3. The circumference of a circle, $C = \pi d$, for d.
4. The circumference of a circle, $C = 2\pi r$, for r.
5. The volume of a rectangular solid, $V = LWH$, for L.
6. The volume of a rectangular solid, $V = LWH$, for W.
7. The surface area of a rectangular solid, $S = 2LW + 2LH + 2WH$, for L.
8. The surface area of a rectangular solid, $S = 2LW + 2LH + 2WH$, for H.
9. The volume of a cylinder, $V = \pi r^2 h$, for h.
10. The surface area of a cylinder, $S = 2\pi r^2 + 2\pi rh$, for h.
11. Simple interest, $I = Prt$, for P.

12. Simple interest, $I = Prt$, for r.
13. Velocity of an object falling from rest, $v = gt$, where g is the acceleration and t is the time, for g.
14. Velocity of an object falling from rest, $v = gt$, for t.
15. Current in a circuit, $I = \dfrac{V}{R}$, where V is the voltage and R is the resistance, for R.
16. Einstein's law of mass–energy equivalence, $E = mc^2$, where c is a constant (the speed of light), for m, the mass.
17. From statistics, the standardized variable, $z = \dfrac{x - m}{s}$, for m, the mean.
18. The standardized variable, $z = \dfrac{x - m}{s}$, for s, the standard deviation.

Solve for y in terms of x.

19. $4x + 3y = 0$ 20. $5x + 15y = 0$ 21. $-5x + 10y = 0$ 22. $6x - 18y = 0$
23. $-x - y = 0$ 24. $-x + y = 0$ 25. $5x + 4y = 20$ 26. $-4x - 4y = 12$

27. $-x - y = 7$
28. $13x - y = 13$
29. $7x - 14y = -28$
30. $8x + 7y = -56$
31. $-x + y = -1$
32. $-21x + 7y = -14$
33. $y - 5 = 4(x - 6)$
34. $y - 1 = -4(x - 3)$
35. $y + 6 = -2(x - 7)$
36. $y - 8 = 3(x + 9)$
37. $y + 2 = -1(x + 4)$
38. $y + 12 = -5(x + 10)$
39. $y - 4 = \frac{2}{3}(x + 9)$
40. $y + 6 = -\frac{3}{4}(x - 8)$
41. $y + \frac{5}{9} = -\frac{2}{3}\left(x - \frac{1}{3}\right)$
42. $y - \frac{7}{18} = \frac{4}{9}\left(x - \frac{1}{2}\right)$

43. With a down payment of $200, Jenny purchased a living-room suite, agreeing to make monthly payments of $85, with no interest charges, until the balance is paid up. Write an equation for Jenny's total payments, P, to pay off the bill in m months. Find a new equation for the number of months needed to pay off the bill. How long will it take to pay off a purchase of $2240? How long will it take to pay off a purchase of $1200?

44. Richard charges a flat fee of $75.00 per job plus $35.00 per hour to do interior painting. Write an equation for the total cost, c, of a job that takes h hours to complete. Find a new equation for the number of hours worked on a particular job that costs c dollars. How many hours did Richard work if the job cost $495.00? How many hours did he work if the job cost $652.50?

45. Mildred is planning a Halloween party for her kindergarten class. She has already spent $12.50 on decorations for the room. Write an equation for her total cost of the party, T, if she spends c dollars for favors and treats on each of the 22 students in the class. Find a new equation for the amount of money she can spend on each child so that she spends T dollars on the party. How much can she spend on each child if she has $35.00 for the party? How much can she spend on each child if she has $50.00 for the party?

46. The soccer team is planning a fund-raiser to buy equipment and uniforms for its team. The equipment will cost $175. Write an equation for the total amount needed, T, if the team spends c dollars for each uniform, given that there are 14 members on the team. Find a new equation for the amount spent on each uniform if the team is limited to spending T dollars total. How much will each member receive for uniforms if the team raises $675? How much will each member receive if the team raises $1000?

47. Ted's boss's weekly earnings average $75 more than three times Ted's average weekly earnings. Write an equation for the boss's average weekly earnings, B, in terms of Ted's weekly earnings, T. Find a new equation for Ted's weekly earnings if we know what his boss' earnings are. What are Ted's weekly earnings if his boss averages $725 per week? What are Ted's weekly earnings if his boss averages $1275 per week?

48. In a gender discrimination suit, it was alleged that the average hourly wage of male employees was $0.75 less than 1.5 times the average hourly wage of female employees. Write an equation for the average hourly wage of male employees, M, given that the average hourly wage of female employees is F dollars. Find a new equation for the average hourly wage of female employees if the average hourly wage of male employees is known. What would be the average hourly wage of female employees if male employees average $12.45 per hour? What would be the average hourly wage for females if the male employees average $21.00 per hour?

49. You can rent a tree stump grinder for a flat fee of $85 plus $185 per day. Write an equation to represent the total cost, C, of renting the equipment for d days. Find a new equation for the number of days you can rent the equipment for a fixed cost, C. For how many days can the equipment be rented if you want to limit the total cost to $270? For how many days' rental will the cost be limited to $825?

50. Carol is taking a motor trip around parts of Alaska for her vacation. She has already driven 50 miles from Seward. If she averages 40 miles per hour, write an equation for her total distance traveled, D, after driving h more hours. Find a new equation for the number of driving hours remaining if the total distance of the trip is known. How many driving hours remain if the total distance is about 510 miles to Fairbanks? How many driving hours remain if the total distance is about 130 miles to Anchorage?

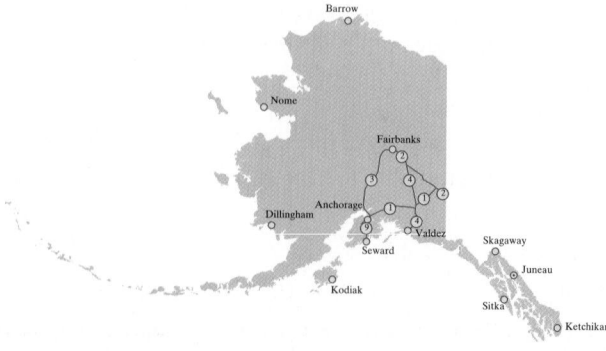

51. A storage bin must be built to fit into a corner of a shed. The bin is limited to being 3 feet wide and 5 feet long. Write an equation for the volume of the bin, V, in terms of its height, h. Find a new equation for the height of the bin in terms of its volume. What should the height be if the volume of the bin must be 60 cubic feet? What should the height be if the volume of the bin must be 100 cubic feet?

52. For the storage bin in exercise 51, write an equation for the total surface area, S, of the bin (including a bottom, top, and four sides) if the width is 3 feet, the length is 5 feet, and the height is h feet. Find a new equation for the height if the total surface area is known. What is the height if the total surface area is 86 square feet? What is the height if the total surface area is 120 square feet?

4.4 Calculator Exercises

In Chapter 2, various methods for calculating interest on an investment were discussed. One such method, continuously compounded interest, will be discussed in this calculator exercise. An investment of P dollars continuously compounded for t years at an annual interest rate of r percent will accrue to an amount

$$A = Pe^{rt}$$

where e is the irrational number that is approximated by 2.718281828.... Find the key for this number on your calculator.

We will use the preceding formula to determine what amount will accrue to an investment of $10,000 at the end of a given number of years when the interest rate is 4.5% per year. Substituting these values into the formula gives us

$$10000 = Pe^{0.045t}$$

as the relation for this particular investment situation. Solve this equation for P. Store the new equation in your calculator, using the $\boxed{Y=}$ key. To do this, you must replace P by y and t by x in your new equation. Then use the table feature of your calculator to complete the following table:

A	r	t	P
$10,000	4.5%	5	
$10,000	4.5%	7	
$10,000	4.5%	10	
$10,000	4.5%	12	

Next, change the amount, A, to $25,000 and the interest rate to 7%. Determine the new equation for P, store it in your calculator, and complete the following table:

A	r	t	P
$25,000	7%	5	
$25,000	7%	7	
$25,000	7%	10	
$25,000	7%	12	

What do these tables tell you about the relationship between the time over which the investment accrues interest and how much money you must initially invest in order to accumulate a certain amount of money? This type of analysis should impress you with the power of mathematics to help people make important decisions pertaining to managing their finances.

4.4 Writing Exercise

In this section, you learned to take a formula and, by applying the properties of equations, write a new formula by solving for one of the variables. Discuss some of the reasons that you would want to use this technique. Illustrate your reasons with an example in which you start with one of the formulas you have worked with, and use it to solve for another variable. Explain why you are solving for the variable; that is, give reasons as to why you need to solve for another variable. Complete your example by specifying some values for the variables in the formula and calculating the value for the variable for which you solved initially.

4.5 More Real-World Models

OBJECTIVES

1 Model real-world situations involving consecutive integers.
2 Model real-world situations involving geometric figures.
3 Model real-world situations involving interest.
4 Model real-world situations involving a percentage of increase or decrease.

APPLICATION

The Vietnam Veterans Memorial was designed by a 21-year-old student, Maya Ying Lin. The memorial consists of two walls, each $246\frac{2}{3}$ feet in length. The walls meet at an angle of 125.2 degrees, with one wall pointing to the Washington Monument and one pointing to the Lincoln Memorial. How many degrees should one of the walls be rotated for the walls to form a straight line? The area enclosed by the walls is in the shape of a triangle with two equal angles. Determine the measure of these angles.

After completing this section, we will discuss this application further. See page 273.

We now know three methods for solving a linear equation: numeric, graphic, and algebraic. We have also solved real-world models with each of these methods. However, numerous other types of models can be solved with such methods.

In this section, we will discuss other types of real-world models involving linear equations. We will also examine some classic algebra exercises that are common in traditional algebra texts. These exercises are not as applicable to the real world as previous exercises in this text, but the ideas needed to solve them will help strengthen your problem-solving skills.

Remember, there is no set way to work all exercises. Also, once a linear equation is written for an exercise, it may be solved with any of the methods we have learned.

MODELING REAL-WORLD SITUATIONS
To model a real-world situation,

• Read and understand the situation. A drawing is often helpful.

- Define a variable for the unknown quantity. Define other quantities in terms of the variable if possible.
- Write an equation with the information given.
- Solve the equation, either numerically, graphically, or algebraically.
- Check the solution with the original problem.
- Write an answer for the question asked, using complete sentences.

The following examples illustrate types of situations that require a linear equation for their solution.

4.5.1 Consecutive-Integer Models

We discussed the set of integers in Chapter 1. Remember,

$$Z = \{\ldots, -3, -2, -1, 0, 1, 2, 3, \ldots\}.$$

Now we need to define new terminology that is used in this section. **Consecutive integers** are integers in increasing order with no integers between them. **Consecutive even integers** are even integers in increasing order with no even integers between them. **Consecutive odd integers** are odd integers in increasing order with no odd integers between them. For example,

$$1, 2, 3, 4 \text{ are consecutive integers.}$$
$$-2, 0, 2, 4 \text{ are consecutive even integers.}$$
$$-7, -5, -3, -1 \text{ are consecutive odd integers.}$$

Consecutive integers may be represented with variables. For example, write consecutive integers in terms of the first integer.

Let x = the first integer

The second consecutive integer is 1 more than the first, or $x + 1$.

The third consecutive integer is 1 more than the second, or $(x + 1) + 1$, or $x + 2$.

The fourth consecutive integer is 1 more than the third, or $(x + 2) + 1$, or $x + 3$.

Write consecutive even integers in terms of the first even integer.

Let x = the first even integer

The second consecutive even integer is 2 more than the first, or $x + 2$.

The third consecutive even integer is 2 more than the second, or $(x + 2) + 2$, or $x + 4$.

The fourth consecutive even integer is 2 more than the third, or $(x + 4) + 2$, or $x + 6$.

Write consecutive odd integers in terms of the first odd integer.

Let x = the first odd integer

The second consecutive odd integer is 2 more than the first, or $x + 2$.

The third consecutive odd integer is 2 more than the second, or $(x + 2) + 2$, or $x + 4$.

The fourth consecutive odd integer is 2 more than the third, or $(x + 4) + 2$, or $x + 6$.

HELPING HAND Note that the only difference in the consecutive even- and odd-integer representation is the beginning definition for the variable, x, because both even and odd integers are every other integer.

EXAMPLE 1 Del is wiring an electrical appliance and has to cut a wire into three pieces. The first piece must be an integer length, and each consecutive piece must be 1 inch longer than the preceding one. He measures the wire to be 27 inches. What are the lengths of the three pieces of wire?

Solution
The wire is to be cut into three consecutive integer lengths.

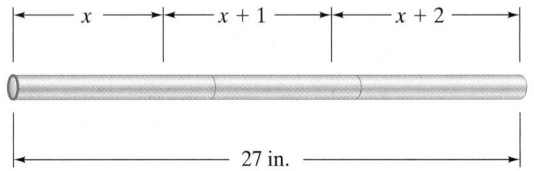

Let $x =$ integer length of the first piece
$x + 1 =$ integer length of the second piece
$x + 2 =$ integer length of the third piece

The sum of the integers will be 27 inches, which is the total length.

Algebraic

$$x + (x + 1) + (x + 2) = 27$$
$$3x + 3 = 27 \qquad \text{Simplify.}$$
$$3x + 3 - 3 = 27 - 3 \qquad \text{Subtract 3 from both sides.}$$
$$3x = 24 \qquad \text{Simplify.}$$
$$\frac{3x}{3} = \frac{24}{3} \qquad \text{Divide both sides by 3.}$$
$$x = 8 \qquad \text{Simplify.}$$

Numeric

X	Y₁	Y₂
5	18	27
6	21	27
7	24	27
8	27	27
9	30	27
10	33	27
11	36	27

X=8

Y1 = $x + (x + 1) + (x + 2)$
Y2 = 27

The solution is 8.

Find the other integers (lengths of pieces) by substituting the solution into the other expressions.

Second piece: $x + 1 = 8 + 1 = 9$
Third piece: $x + 2 = 8 + 2 = 10$

To check the solution, we first determine that the values are indeed consecutive integers. Then we total the lengths to see whether the sum is 27 inches: $8 + 9 + 10 = 27$. It is, so the solution is correct.

The three pieces should be cut 8, 9, and 10 inches in length.

✓ 4.5.1 Checkup

1. The sum of three consecutive even integers is 42. Find the integers.

2. Jim is building a scale model for his architecture class. He has a 25-inch piece of balsa wood that he wants to cut into five pieces so that each piece is 1 inch longer than the preceding piece. What are the five lengths he needs to cut?

3. What are the differences in the terms *consecutive integers*, *consecutive even integers*, and *consecutive odd integers*?

4.5.2 Geometric-Formula Models

Geometric formulas are helpful in all kinds of everyday situations. You often need to use these formulas whenever you have to find a measurement of a length, an angle, or an area. The measurement you want to find is your unknown variable, and the geometric formula is the linear equation you want to solve. You will find that a diagram of the situation can help you set up the equation correctly.

EXAMPLE 2   Shawn wants to add baseboards to his study. He measures the dimensions of the study and finds that the perimeter is 47 feet. When Shawn arrives at the home-repair store, he discovers a "manager's special" on carpets. He decides to carpet the floor of his study before adding the baseboards. However, he does not have the room's dimensions. All he can remember is that the rectangular study has a perimeter of 47 feet and the length is $4\frac{1}{2}$ feet more than the width. Determine the area of the floor in the study.

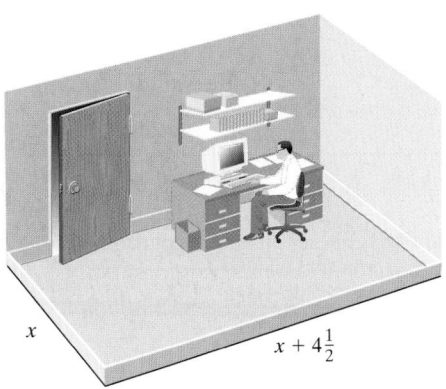

Solution

A figure will help us see the problem.

$$\text{Let } x = \text{width of study}$$
$$x + 4\tfrac{1}{2} = \text{length of study}$$

To find the area of the floor, Shawn must know the dimensions of the study. He can find them by using the perimeter formula for a rectangle.

Algebraic

$$P = 2L + 2W$$

$$47 = 2\left(x + 4\frac{1}{2}\right) + 2(x) \qquad \text{Substitute.}$$

$$47 = 2x + 9 + 2x \qquad \text{Distribute.}$$

$$47 = 4x + 9 \qquad \text{Simplify.}$$

$$47 - 9 = 4x + 9 - 9 \qquad \text{Subtract 9 from both sides.}$$

$$38 = 4x \qquad \text{Simplify.}$$

$$\frac{38}{4} = \frac{4x}{4} \qquad \text{Divide both sides by 4.}$$

$$9\frac{1}{2} = x \qquad \text{Simplify.}$$

Graphic

Y1 = 47   Y2 = $2(x + 4\frac{1}{2}) + 2x$

Intersection
X=9.5 _____ Y=47

$(0, 50, 0, 100)$
The width is 9.5 feet.

The width is $9\frac{1}{2}$ feet. To find the length, substitute the solution into the expression: $x + 4\frac{1}{2} = 9\frac{1}{2} + 4\frac{1}{2} = 14$. The length is 14 feet. To check this solution, we check the perimeter formula.

$$P = 2L \ + \ 2W$$

$$47 \ \bigg| \ 2(14) + 2\left(9\frac{1}{2}\right)$$

$$28 \ + \ 19$$

$$47$$

The solution checks. The dimensions of the study are $9\frac{1}{2}$ feet by 14 feet.

The area of the rectangular floor is found by substituting the dimensions for the variables into the area formula.

$$A = LW$$

$$A = (14)\left(9\frac{1}{2}\right)$$

$$A = 133 \text{ square feet}$$

The area of the floor is 133 square feet.

4.5.2 Checkup

Carlos is building a sandbox for his son, making it triangular to fit into a corner of the patio. He wants two sides to be of equal length and the third side to be 1.5 feet longer than the others. If he has enough material for a perimeter of 12 feet, how long will each side be?

4.5.3 Interest Models

We saw in section 2.4 that there are several ways to calculate interest on a loan. Compound interest involves exponential functions, but simple interest is a linear relationship. Situations involving calculations of interest or principal at various rates of interest are common and important in business and finance.

EXAMPLE 3   Dennis is building a trailer. He plans to sell it for $2500 when he is finished. He is borrowing the money needed to build the trailer from his father. His father charges 18% simple interest for one year. What is the most Dennis can borrow and know that he will have enough to pay his father at the end of one year?

Solution
Let x = the amount borrowed
Using the simple-interest formula, we obtain

$$I = PRT$$

$$I = x(0.18)(1) \qquad \text{x = amount borrowed, R = 0.18, T = 1}$$

$$I = 0.18x$$

Dennis will need to repay the amount borrowed, x, plus simple interest, $0.18x$. The maximum amount should be $2500. Therefore, $x + 0.18x = 2500$.

Algebraic

$$x + 0.18x = 2500$$

$$1.18x = 2500 \qquad \text{Combine like terms.}$$

$$\frac{1.18x}{1.18} = \frac{2500}{1.18} \qquad \text{Divide both sides by 1.18.}$$

$$x \approx 2118.64$$

Graphic

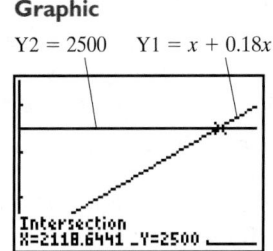

(0, 2500, 0, 3500)
The solution is about 2118.64.

To check this solution, add the amount borrowed plus interest to see whether we get 2500.

$$2118.64 + 0.18(2118.64) \approx 2500 \text{ (rounded to the nearest cent).}$$

Therefore, the solution is correct.

Dennis can borrow $2118.64 from his father and repay it at the end of one year.

EXAMPLE 4 Nguyen received an inheritance of $10,000 from his great-uncle's estate. He invested part of the money in a simple-interest account paying 9% annually. He invested the remainder of the money in a savings account paying 4.75% simple interest annually. If he earned $815 at the end of one year, how much did he invest in each account?

Solution

$$\text{Let} \qquad a = \text{amount invested at 9\% interest}$$

$$10{,}000 - a = \text{amount invested at 4.75\% interest}$$

Use the simple-interest formula to determine the amount of interest for each account.

9% account	4.75% account	
$I = PRT$	$I = PRT$	
$I = a(0.09)(1)$	$I = (10{,}000 - a)(0.0475)(1)$	Substitute.
$I = 0.09a$	$I = (10{,}000 - a)(0.0475)$	Simplify.
	$I = 475 - 0.0475a$	Distribute.

The sum of the interest is $815, or $0.09a + (475 - 0.0475a) = 815$.

Algebraic

$$0.09a + (475 - 0.0475a) = 815$$

$$0.0425a + 475 = 815 \qquad \text{Simplify.}$$

$$0.0425a + 475 - 475 = 815 - 475 \qquad \begin{array}{l}\text{Subtract 475}\\\text{from both sides.}\end{array}$$

$$0.0425a = 340 \qquad \text{Simplify.}$$

$$\frac{0.0425a}{0.0425} = \frac{340}{0.0425} \qquad \begin{array}{l}\text{Divide both sides}\\\text{by 0.0425.}\end{array}$$

$$a = 8000 \qquad \text{Simplify.}$$

Graphic

Y2 = 815   Y1 = 0.09x + (475 − 0.0475x)

Intersection
X=8000   Y=815

(0, 10000, 0, 1000)
The solution is 8000.

The amount deposited in the 9% account is $8000.

The amount deposited in the 4.75% account is

$$10{,}000 - a = 10{,}000 - 8000 = 2000.$$

To check, determine the amount of interest from each account. Their sum is 815.

$$(0.09)(8000) + (0.0475)(2000) = 720 + 95 = 815$$

Nguyen deposited $8000 in the 9% account and $2000 in the 4.75% account.

4.5.3 Checkup

1. When applying for a short-term loan at many firms, you must ask for the amount plus simple interest at the time you apply. If you apply for a loan of $3000 that includes the interest at 9% simple interest for one year, how much money will you actually receive? How much interest was paid on the loan?

2. Zeke would like to borrow $5000 at simple interest for one year. He is not able to borrow the total amount from one loan agency, so he must borrow part of it at 7% simple interest and the remainder at 8.5% simple interest. If his total interest payment is to be $365, how much will he borrow at each interest rate?

4.5.4 Percentage-of-Increase-or-Decrease Models

One of the most common linear relations in business and consumer finances involves calculating the percentage increase or decrease of some value. The value may be a cost, a sales price, a profit, or some other quantity. In any case, the method of calculating the increased or decreased value (or the calculation of the value before the increase or decrease) is the same.

EXAMPLE 5 Caitlin paid $39.99 for a battery-and-charger set. The Internet advertisement claimed that this was a savings of $9.96, or 20%. Determine the original list price. The advertisement also claimed that the original list price was $49.95. Is the advertisement correct?

List Price: $~~$49.95~~
Our Price: $39.99
You Save: $9.96 (20%)
Availability: Usually
ships within 24 hours.

Solution

Let x = the original list price
$0.20x$ = the amount of savings (20% of the original price)

The sale price, $39.99, is the original price minus the savings, or $x - 0.20x = 39.99$.

Algebraic

$$x - 0.20x = 39.99$$
$$0.80x = 39.99 \quad \text{Combine like terms.}$$
$$\frac{0.80x}{0.80} = \frac{39.99}{0.80} \quad \text{Divide both sides by 0.80.}$$
$$x = 49.9875$$

Graphic

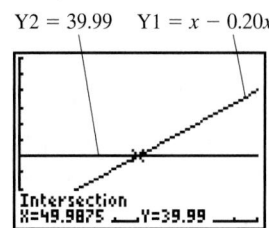

$(0, 100, 0, 100)$
The solution is about 49.99.

If Caitlin saved 20%, then the original list price was $49.99, so she saved $49.99 − $39.99, or $10.00. The advertisement stated the original list price incorrectly. There is an error of $0.04.

APPLICATION

The Vietnam Veterans Memorial was designed by a 21-year-old student, Maya Ying Lin. The memorial is made of two walls, each $246\frac{2}{3}$ feet in length. The walls meet at an angle of 125.2 degrees, with one wall pointing to the Washington Monument and one pointing to the Lincoln Memorial. How many degrees should one of the walls be rotated for the walls to form a straight line? The area enclosed by the walls is in the shape of a triangle with two equal angles. Determine the measure of these angles.

Algebraic Solution

$$125.2 + x = 180$$
$$125.2 + x - 125.2 = 180 - 125.2 \quad \text{Subtract 125.2.}$$
$$x = 54.8$$

In order to form a straight line, one wall must be rotated 54.8°.

The area enclosed by the walls is in the shape of an isosceles triangle (a triangle with two equal sides). The sum of the angles of a triangle is 180°.

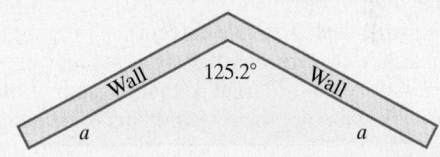

Let a = the measure of each equal angle

$$a + a + 125.2 = 180$$

$$a + a + 125.2 = 180$$
$$2a + 125.2 = 180 \quad \text{Combine like terms.}$$
$$2a + 125.2 - 125.2 = 180 - 125.2 \quad \text{Subtract 125.2.}$$
$$2a = 54.8$$
$$\frac{2a}{2} = \frac{54.8}{2} \quad \text{Divide both sides by 2.}$$
$$a = 27.4$$

The two equal angles are each 27.4°.

Discussion

If the two walls form a straight line (or 180°), the existing angle of 125.2° is supplementary to the angle through which one of the walls must be rotated.

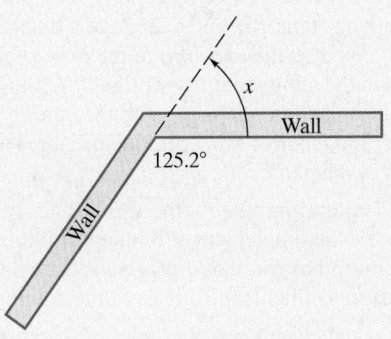

Let x = the measure of the angle
The sum of the angles is 180°.
$125.2 + x = 180$

4.5.4 Checkup

1. Tillie's Travel Agency adds a 10% surcharge for handling motel reservations for a firm. If the charge for a room is $68.75, what was the charge before the surcharge was added?

2. Two roads meet at a Y-intersection. The angle between the roads is 56°. A traffic engineer wants to design intersections that are perpendicular. Find the complement of the given angle to determine how much rotation will be needed on one of the roads.

4.5 Exercises

1. A patient's medication is to be increased in consecutive even-integer dosages in three administrations, for a total of 24 grains of medication. How many grains of medication will be administered each time?

2. A patient's medication is to be decreased in consecutive integer amounts over five administrations, for a total of 70 cc of medication. How many cc will be administered each time?

3. A lottery has four stages of winning. At each stage, the number of winners is increased in consecutive odd numbers. If there are 24 prizes in all, how many will be awarded at each stage?

4. Lottery winnings will be distributed over a five-day period, with the number of winners each day increasing by consecutive even integers. If the lottery will distribute 30 prizes, how many will be awarded each day?

5. An instructor ranks her eight students according to their level of classroom participation and then assigns them consecutive integer numbers as grades. If the total number of points for the eight students is 676, what were the lowest and highest grades given?

6. A movie rating scale has five categories, ranging from "strongly dislike" to "strongly like." Consecutive even integers are assigned left to right so that the sum of the integers is zero. What are the five numbers to be assigned to the scale?

7. The sides of an equilateral triangle are of equal length. What is the length of a side if the perimeter of the triangle is $29\frac{1}{4}$ inches?

8. The second side of a triangle is twice as long as the first side. The third side is 1 centimeter shorter than the second side. If the perimeter is $11\frac{1}{2}$ centimeters, how long is each side?

9. By definition, an isosceles triangle has at least two sides of equal length. What are the lengths of the sides of the triangle if the third side is two-thirds as long as each of the equal sides and the perimeter is 16 feet?

10. If the third side of an isosceles triangle is half as long as each of the equal sides and the perimeter is 38 meters, how long is each side?

11. The perimeter of a rectangle is 400 yards. What are the dimensions of the rectangle if the length is 30 yards more than the width?

12. If the perimeter of a square is 1 inch, what are the dimensions of the square?

13. The width of a rectangle is 55% of the length. What are the dimensions of the rectangle if the perimeter is 294.5 centimeters?

14. The length of a rectangle is 2 inches more than twice its width. If its perimeter is 52 inches, what are the dimensions of the rectangle?

15. Karl wants to build a dog run for his German shepherd. He wants the length of the run to be five times as long as the width. If he has 96 feet of fencing, what should the dimensions of the dog run be? How many square feet of yard will the run cover?

16. See exercise 15. If Karl builds a square dog run with the fencing, how long would each side be? Will this give him more square feet of yard for the run than in exercise 15?

17. A privacy fence meets the side of a house at an angle of 138°. What is the measure of the angle behind the fence that is the supplement of this angle? If the owner plants a garden behind the fence and the garden is in the shape of a right triangle, what are the measures of the three angles of the garden?

18. A loading ramp meets the ground at an angle of 165°. What is the supplement of this angle? If the triangle below the ramp has the shape of a right triangle, what are the measures of the three angles of the triangle?

19. E-Z Loan Agency writes a loan agreement stipulating that the agency will be reimbursed for the amount to be borrowed plus simple interest. What is the amount of money borrowed if $4500 is to be paid back on a one-year simple-interest loan at 12.5% interest? How much interest was paid on the loan?

20. E-Z-R Loan Agency writes a loan agreement stipulating that $1308 is to be repaid for a simple-interest loan for one year at 9%. How much is being borrowed? How much interest is being paid on the loan?

21. A mutual fund will pay 9% simple interest on an investment for one year. If you want to receive a total amount of $5000 at the end of that year, how much must you invest?

22. An investment fund offers 12% simple interest on an investment for one year. How much should be invested to receive a total amount of $12,500 at the end of the year?

23. A wise and successful businessman wants to establish an endowment of $500,000 with a community college. (What a great guy!) He invests money into a simple-interest account for a year. If the account pays 10% simple interest, how much should he invest, rounded to the nearest thousand dollars, in order to be able to establish the endowment?

24. A municipal bond earns 7.8% simple interest annually. How much should be invested in order to have a total amount of $8000 at the end of a year?

25. Megan received $15,000 from a probated estate. She invested part of the money in a simple-interest account paying 8% and the rest in another simple-interest account paying 6.5%. If she earned $1117.50 in interest at the end of the year, how much did she invest in each account?

26. Zach received $15,000 from his great-great-grandfather's estate. He invested part of his money in a corporate bond that paid 9.2% simple interest annually and the rest in a savings account paying 3.5% simple interest annually. If he earned $952.20 interest in one year, how much did he invest in each instrument?

27. The Bedrock Savings and Loan Company will pay 7% simple interest on an investment of P dollars for one year. Write an equation for the total amount, A, in a savings account after a year. Find a new equation for the amount to be invested if you know the amount desired after one year. How much should be invested if you wish to have $1350 in an account at the end of the year? How much should be invested if you desire to have $2500 at the end of the year?

28. Friendly Dan's Loan Agency charges 11% simple interest on a loan. Write an equation for the total payback, P, if you take out a loan of L dollars for one year. Find a new equation for the amount of money borrowed if the total amount of the payback is known. How much will you receive as a loan if the payback is $1332? How much will you receive as a loan if the payback is $800?

29. The sale price of a dress that was reduced by 20% is $68. What was the original selling price of the dress?

30. The sale price of a CD player is $110.49 after a reduction of 15%. What was the original selling price? What was the amount of the reduction?

31. A boutique sells all of its items at a 60% markup over the cost of the items from their suppliers. If an item sells for $19.95, what was the cost to the boutique?

32. An antiques dealer purchases a restored radio and resells it for $350, which represents a 75% markup of the cost. How much did the dealer pay for the radio?

33. An artist notices that the item she sold to a boutique for $9.50 is selling there for $17.10. What is the markup percentage at this boutique?

34. If an antiques dealer sells a vintage radio for $385 and paid $275 for it, what was the markup percentage?

35. A TV set is on sale for $195. The store claims that this is a 25% savings over the regular price. What is the regular price of the TV?

36. A pro shop has its golf clubs on sale, with 20% off the regular price. If a golf club is on sale for $59, what is the regular price of the club?

37. Slippery Sam claims that he can save you 35% of the suggested retail price (SRP) of a car. If he is willing to sell you the car for $12,500, what should the SRP be?

38. What would be the suggested retail price of a car that is on sale for $8995 if this represents a 15%-off sale?

39. A clothing store reduces the original price of a coat by 10% for a sale. Write an equation for the sale price of the coat, y, if the original price is x dollars. Find a new equation for the original price if the sale price is known. What is the original price of a coat that is on sale for $53.96? What is the original price of a coat that is on sale for $98.95?

40. In the country of Erehwon, the sales tax on a purchase is 8.75%. Write an expression for the total cost of a sale, y, if the subtotal before tax is x dollars. Find a new equation for the subtotal if the total cost, including tax, is known. What is the subtotal when the total cost is $27.19? What is the subtotal when the total cost is $143.55?

41. Denzel received a cost-of-living increase. His new hourly wage is $13.75 per hour. If the cost-of-living index was 2.2%, what was his hourly wage before the increase?

42. Joan received a 6% merit increase, which raised her annual salary to $32,500. What was her salary before the increase?

43. For large parties, many restaurants automatically add a 15% gratuity to the bill, rather than allowing the customers to add their own gratuity. If the total bill, including gratuity, is $143.24, what was the bill before the gratuity was added?

44. What is the cost before gratuity of a dinner party if the bill plus 15% gratuity is $262.95? How much was the gratuity?

45. Dylan designed a bay window for her kitchen. The bay window has five equally wide windows. Here is Dylan's sketch:

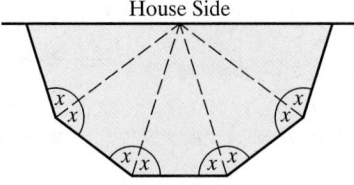

House Side

The sketch shows that the bay window has the shape of five triangles of the same shape and size. In order to cut the trim for the windows, she will need to set her saw to bevel the edges of the trim. To do so, Dylan must know the measure of the angles labeled x on the sketch. Determine the measure of these angles.

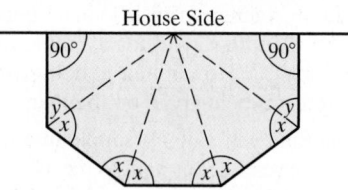

46. Dylan altered the design for her kitchen bay window from that shown in exercise 45. Her new design is shown in the following sketch:

The change she made is that the sides which touch the house now make a 90° angle with the house. She still must determine the angles labeled x and y on the sketch. Determine the measures of these angles.

4.5 Calculator Exercises

Many calculators have a list feature that can be used when you wish to evaluate a formula for more than one value. As an example, a mutual fund will pay 9% simple interest for an investment for one year. How much should you invest if you wish to receive the following amounts after a year: {\$900, \$1800, \$2700}? To solve this problem, let x = the investment P, use the formula $I = Prt$, and substitute the list for the interest I, 9% for the interest rate r, and 1 (year) for the time t.

$$\{900, 1800, 2700\} = (x)(0.09)(1)$$

Solving for x yields

$$x = \frac{\{900, 1800, 2700\}}{0.09}$$

You can use the braces on the calculator to enter this equation, and the calculator will return a set of values in braces, representing the principal invested for each of the interest values.

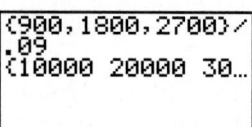

The calculator display is interpreted to mean that \$10,000 must be invested to earn \$900 interest, \$20,000 to earn \$1,800 interest, and so on. Use the calculator arrows to scroll left and right.

Use this technique in the following exercises:

1. Find the Fahrenheit temperature, given {50, 55, 60, 65} as the Celsius temperatures.

2. Find the Celsius temperature, given {0, 25, 50, 75, 100} as the Fahrenheit temperatures.

3. Find the complement of an angle whose measure is {15, 30, 45, 60} degrees.

4. Find the distance traveled at a rate of 60 miles per hour, given a time of {1, 1.5, 2, 2.5} hours driven.

4.5 Writing Exercise

Many business mathematics textbooks use the techniques presented here to solve application problems. Find a textbook that presents an illustration of a business application similar to any of the problems we have studied. It may be a problem dealing with cost-of-living increases, markups or markdowns in selling prices, investments, or something else. Make a photocopy of the page in the text discussing the application. Write a one-paragraph critique of the illustration. In your critique, state whether you think the author did a good job in presenting the application. Was it clear or was it confusing? If the presentation was not good, suggest what you might have done to present the example differently. Turn in your critique with the photocopy and the library call number of the textbook.

4.6 Solving Linear Absolute-Value Equations

OBJECTIVES

1 Identify linear absolute-value equations in one variable.

2 Solve linear absolute-value equations numerically and graphically.

3 Solve linear absolute-value equations algebraically.

4 Model real-world situations by using linear absolute-value equations, and then solve the equations.

APPLICATION

The difference of the tallest mountain on the North American continent, Mount McKinley, and the tallest mountain on the South American continent, Aconcagua, is 765 meters. Mount McKinley is 6194 meters. Determine the height of Aconcagua.

After completing this section, we will discuss this application further. See page 284.

In this section, we will solve a linear equation in one variable that contains an absolute-value expression. Before we begin, we need to review the definition of absolute value. An absolute-value expression is defined to be the distance of the expression from 0 (section 1.1). For example, $|3| = 3$ and $|-3| = 3$. Remember, the result is always nonnegative.

4.6.1 Linear Absolute-Value Equations in One Variable

STANDARD FORM FOR A LINEAR ABSOLUTE-VALUE EQUATION IN ONE VARIABLE

A linear absolute-value equation in one variable is an equation written in the form $|ax + b| = c$, where a, b, and c are real numbers and $a \neq 0$.

For example, the following are linear absolute-value equations in one variable:

1. $|2x + 3| = 5$ Standard form: $|ax + b| = c$
2. $|2x + 3| - 6 = 5$ Not standard form
3. $4|2x + 3| = 0$ Not standard form

The first equation is in the exact form $|ax + b| = c$, with $a = 2$, $b = 3$, and $c = 5$. The last two equations can be solved for the absolute-value expression by using the properties of equations. For example,

$|2x + 3| - 6 = 5$,   or $|2x + 3| = 11$ Add 6 to both sides of the equation.

$4|2x + 3| = 0$,   or $|2x + 3| = 0$ Divide both sides of the equation by 4.

These equations are called **linear absolute-value equations in one variable** because of the characteristics of the graphs of the functions defined by the two expressions in each equation. To see what we mean, let's graph some absolute-value expressions on the calculator. Complete the following set of exercises on your calculator.

Discovery 11

Graphs of the Functions Defined by Expressions in a Linear Absolute-Value Equation

On an integer screen, graph the following functions, determined from the given linear absolute-value equation:

1. $|2x + 3| = 5$ 2. $|2x - 3| = -5$

    Y1 $= |2x + 3|$ Y2 $= 5$ Y1 $= |2x - 3|$ Y2 $= -5$

Explain the characteristic of the graphs of the functions defined by the absolute-value expression, Y1.   *(Continued on page 278)*

Explain the characteristic of the graphs of the functions defined by the constant, Y2.

On an integer screen, graph the following functions, determined from the given linear absolute-value equation:

3. $|2x + 3| - 6 = 5$ **4.** $4|2x + 3| = 0$
     Y1 $= |2x + 3| - 6$ Y2 $= 5$ Y1 $= 4|2x + 3|$ Y2 $= 0$

Explain the characteristic of the graphs of the functions defined by the absolute-value expression, Y1.

Explain the characteristic of the graphs of the functions defined by the constant, Y2.

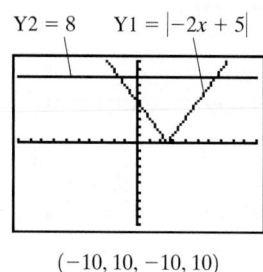

Y2 $= 8$ Y1 $= |-2x + 5|$

$(-10, 10, -10, 10)$

For the linear absolute-value equations in the form $|ax + b| = c$, the graph of the function defined by the absolute-value expression is a V-shaped graph formed by two line segments, and the graph of the constant function is a horizontal line. For the linear absolute-value equations not in $|ax + b| = c$ form, the graph of the function defined by the expression containing the absolute-value expression is formed by lines.

For example, the equation $|-2x + 5| = 8$ is a linear absolute-value equation. The graph of the function Y1 $= |-2x + 5|$ is a V-shaped graph made up of two line segments. The graph of the function Y2 $= 8$ is a line.

EXAMPLE 1   Identify each equation as a linear absolute-value equation or a nonlinear absolute value equation. Check your results graphically.

a. $-|3x - 5| + 8 = 0$ **b.** $|2x^3 + 3| = 5$

Algebraic Solution

a. $-|3x - 5| + 8 = 0$ Linear absolute value, because it simplifies to $|3x - 5| = 8$.

b. $|2x^3 + 3| = 5$ Nonlinear absolute value, because the x term is raised to the third power.

Graphic Check

a. Y2 $= 0$ Y1 $= -|3x - 5| + 8$

$(-47, 47, -31, 31)$

Note: The graph of Y2 $= 0$ is the x-axis.

The equation is a linear absolute-value equation, because the functions graphed both consist of line segments.

b. Y2 = 5   Y1 = $|2x^3 + 3|$

$(-10, 10, -10, 10)$

The equation is a nonlinear absolute-value equation, because the functions graphed are not both made up of line segments. The graph of the function defined by the absolute-value expression is not a line, but a curve.

We will solve linear absolute-value equations by using the three methods we used with linear equations: numerical, graphical, and algebraic.

4.6.1 Checkup

In exercises 1–4, identify each equation as a linear absolute-value equation or a nonlinear absolute-value equation.

1. $9 - 2|2x + 1| = 8$ **2.** $-2 = |3 - x|$

3. $|5 - 2x^2| = 10$ **4.** $|x| = 1$

5. What does the absolute value of a number represent?

6. What distinguishes a linear absolute-value equation from an absolute-value equation that is nonlinear?

4.6.2 Solving Absolute-Value Equations Graphically

Let's solve some linear absolute-value equations graphically and numerically. We will use a calculator to find the solutions of three linear absolute-value equations. Complete the following set of exercises on your calculator.

Discovery 12

Solving a Linear Absolute-Value Equation

Solve each equation graphically and check your solution numerically.

1. a. $|x + 4| = 3$ **2. a.** $|x + 4| = -3$ **3. a.** $|x + 4| = 0$

   b. $|3 - x| = 2$ **b.** $|3 - x| = -2$ **b.** $|3 - x| = 0$

Write a rule for the number of solutions of a linear absolute-value equation when the absolute-value expression equals a positive number, a negative number, and 0.

In the first exercise, in which the absolute-value expression equals a positive number, the equation has two solutions. In the second exercise, in which the absolute-value expression equals a negative number, the equation has no solution. In the third exercise, in which the absolute-value expression equals 0, the equation has one solution.

NUMBER OF SOLUTIONS OF A LINEAR ABSOLUTE-VALUE EQUATION

To determine the number of solutions of a linear absolute-value equation, write the equation in the form $|ax + b| = c$.

- If the constant, c, is positive, there are two solutions.
- If the constant, c, is negative, there are no solutions.
- If the constant, c, is 0, there is one solution.

We know this is true because the absolute values of a positive number and its opposite (a negative number) equal the same positive number. Hence, two solutions are possible if c is a positive number. However, an absolute value cannot equal a negative number. Hence, no solution is possible if c is negative. Also, the absolute value of 0 equals 0. Hence, there is only one solution possible if c is 0.

HELPING HAND Numerical solutions are sometimes hard to find. It is much easier to solve absolute-value equations graphically and then check the solutions numerically.

EXAMPLE 2 Solve graphically and check numerically.

$$\textbf{a.} \ |2x| + 8 = 8 \qquad \textbf{b.} \ \left|\frac{x-2}{3}\right| = 5 \qquad \textbf{c.} \ 3|-2x+1| - 6 = 9$$

Solution

Graphic

a. Y2 = 8   Y1 = $|2x|$ + 8

Numeric

a.

The solution is 0.

Y1 = $|2x|$ + 8
Y2 = 8
The solution is 0.

b. Y2 = 5   Y1 = $\left|\frac{x-2}{3}\right|$

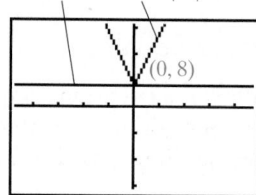

The solutions are −13 and 17.

Y1 = $\left|\dfrac{x-2}{3}\right|$ Y2 = 5

The solutions are −13 and 17.

c. Y2 = 9   Y1 = $3|-2x+1| - 6$

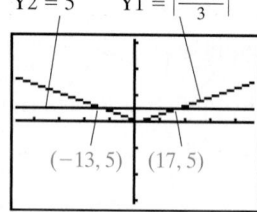

The solutions are −2 and 3.

Y1 = $3|-2x+1| - 6$
Y2 = 9
The solutions are −2 and 3.

4.6.2 Checkup

Solve exercises 1–5 graphically. Check numerically.

1. $|3x + 1| = 4$ **2.** $|3x| + 1 = 4$

3. $\left|\dfrac{x + 1}{2}\right| = 3$ **4.** $|x + 8| + 6 = 1$

5. $2|-3x + 2| - 3 = 11$

6. How do you determine the solutions of a linear absolute-value equation when you solve the equation graphically?

4.6.3 Solving Equations Algebraically

We are now ready to solve linear absolute-value equations algebraically. We begin with the rule we found earlier in this section.

> **SOLVING A LINEAR ABSOLUTE-VALUE EQUATION ALGEBRAICALLY**
> First, determine the number of solutions of an absolute-value equation by writing it in the form $|ax + b| = c$.
> - If the constant, c, is positive, there are two solutions.
> - If the constant, c, is negative, there are no solutions.
> - If the constant, c, is 0, there is one solution.
>
> Second, write and evaluate an equation or equations needed to find the solution(s).
> - If there are two solutions, then
> $$ax + b = c \quad \text{or} \quad ax + b = -c$$
> - If there is one solution, then
> $$ax + b = 0 \quad \text{since} \quad c = 0$$

For example, $|x| = 3$ has two solutions, because the absolute-value expression is equal to a positive number. The two solutions are found when the expression within the absolute-value symbols, x, equals -3 or 3. Therefore, $x = -3$ or $x = 3$.

$|x| = -3$ has no solution, because the absolute-value expression equals a negative number.

$|x| = 0$ has one solution, because the absolute-value expression equals 0. The one solution is found when the expression within the absolute-value symbols, x, equals 0. Therefore, $x = 0$.

EXAMPLE 3 Solve algebraically.

a. $|x + 4| = 3$ **b.** $|x + 4| = -3$ **c.** $|x + 4| = 0$

Solution
Algebraic

a. $|x + 4| = 3$
There are two solutions, because the absolute-value expression equals a positive number.

$$x + 4 = 3 \qquad \text{or} \qquad x + 4 = -3$$
$$x + 4 - 4 = 3 - 4 \qquad x + 4 - 4 = -3 - 4$$
$$x = -1 \qquad\qquad x = -7$$

The two solutions are -1 and -7.
The check is left for you.

b. $|x + 4| = -3$

There is no solution, because the absolute-value expression equals a negative number.

c. $|x + 4| = 0$

There is one solution, because the absolute-value expression equals 0.

$$x + 4 = 0$$
$$x + 4 - 4 = 0 - 4$$
$$x = -4$$

The solution is -4.

The check is left for you.

Now let's solve the equations of Example 2 algebraically.

EXAMPLE 4 Solve algebraically. (Remember to isolate the absolute-value expression first.)

a. $|2x| + 8 = 8$ **b.** $\left|\dfrac{x - 2}{3}\right| = 5$ **c.** $3|-2x + 1| - 6 = 9$

Solution

Algebraic

a. $|2x| + 8 = 8$

$|2x| + 8 - 8 = 8 - 8$ Subtract 8 from both sides to isolate the absolute-value expression.

$|2x| = 0$ Simplify.

$2x = 0$ Rewrite for one solution.

$\dfrac{2x}{2} = \dfrac{0}{2}$ Divide both sides by 2.

$x = 0$ Simplify.

The solution is 0.

The check is left for you.

b. $\left|\dfrac{x - 2}{3}\right| = 5$

$\dfrac{x - 2}{3} = 5$ or $\dfrac{x - 2}{3} = -5$ Rewrite for two solutions.

$3\left(\dfrac{x - 2}{3}\right) = 3(5)$ $3\left(\dfrac{x - 2}{3}\right) = 3(-5)$ Multiply both sides by 3.

$x - 2 = 15$ $x - 2 = -15$ Simplify.

$x - 2 + 2 = 15 + 2$ $x - 2 + 2 = -15 + 2$ Add 2 to both sides.

$x = 17$ $x = -13$ Simplify.

The solutions are 17 and -13.

The check is left for you.

c. $3|-2x + 1| - 6 = 9$

$3|-2x + 1| - 6 + 6 = 9 + 6$ Add 6 to both sides to isolate the absolute-value expression.

$3|-2x + 1| = 15$ Simplify.

$\dfrac{3|-2x + 1|}{3} = \dfrac{15}{3}$ Divide both sides by 3.

$|-2x + 1| = 5$ Simplify.

$$-2x + 1 = 5 \quad \text{or} \quad -2x + 1 = -5 \quad \text{Rewrite for two solutions.}$$

$$-2x + 1 - 1 = 5 - 1 \quad -2x + 1 - 1 = -5 - 1 \quad \text{Subtract 1 from both sides.}$$

$$-2x = 4 \quad\quad\quad -2x = -6 \quad \text{Simplify.}$$

$$\frac{-2x}{-2} = \frac{4}{-2} \quad\quad\quad \frac{-2x}{-2} = \frac{-6}{-2} \quad \text{Divide both sides by } -2.$$

$$x = -2 \quad\quad\quad x = 3 \quad \text{Simplify.}$$

The solutions are −2 and 3.
The check is left for you.

4.6.3 Checkup

Solve exercises 1–6 algebraically.

1. $|x - 9| = 0$ **2.** $|3x + 1| = 4$

3. $|3x| + 1 = 4$ **4.** $\left|\dfrac{x + 1}{2}\right| = 3$

5. $|x + 8| + 6 = 1$ **6.** $2|-3x + 2| - 3 = 11$

7. Describe how you would solve a linear absolute-value equation algebraically.

4.6.4 Modeling the Real World

Linear absolute-value equations can describe several kinds of real-world situations. In particular, absolute values are used to describe the size, or magnitude, of many quantities that can be positive or negative in direction. Among these quantities are distance, speed, force, and other so-called vectors. Absolute values can also describe the magnitude of financial transactions, without regard to whether the money is a credit (positive) or a debit (negative). However, remember that when you solve linear absolute-value equations, the answers you get must make sense in terms of the real situation. Sometimes you have to throw away an answer because it doesn't describe the physical situation correctly.

EXAMPLE 5 A plastic disposable hospital bowl is advertised to have an outer diameter of 5.146 inches with a tolerance (allowance for error) of plus or minus 0.015 inch. What is the least and greatest possible length of the diameter?

TOP VIEW

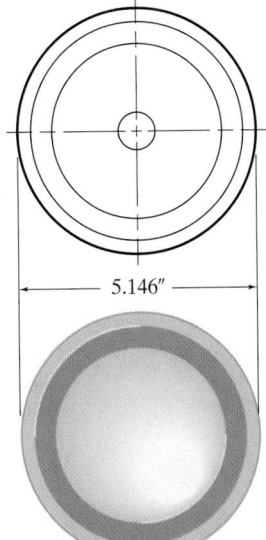

5.146″

Solution

Algebraic

Let x = the actual length of the diameter

Since we do not know which length is greater—the actual length or the length given on the package—we will take the absolute value of the difference of the lengths and equate it to the absolute value of the tolerance.

Substitution Check

```
5.161→X:abs(X-5.
146)=0.015
               1
5.131→X:abs(X-5.
146)=0.015
               1
```

$$|x - 5.146| = |\pm 0.015|$$
$$|x - 5.146| = 0.015 \quad \text{Take the absolute value of the right side.}$$

Write equations for the two possible solutions.

$$x - 5.146 = 0.015$$
$$x - 5.146 + 5.146 = 0.015 + 5.146$$
$$x = 5.161$$

or

$$x - 5.146 = -0.015$$
$$x - 5.146 + 5.146 = -0.015 + 5.146$$
$$x = 5.131$$

The greatest diameter measurement is 5.161 inches. The least diameter measurement is 5.131 inches.

APPLICATION

The difference of the tallest mountain on the North American continent, Mount McKinley, and the tallest mountain on the South American continent, Aconcagua, is 765 meters. Mount McKinley is 6194 meters. Determine the height of Aconcagua.

Discussion

Let x = the height of Aconcagua in meters Since we are not told which mountain is taller, we will need to take the absolute value of the difference, $|6194 - x|$, and set it equal to 765.

$$|6194 - x| = 765$$

Algebraic Solution

$$6194 - x = 765 \qquad \text{or} \qquad 6194 - x = -765$$
$$6194 - x - 6194 = 765 - 6194 \qquad 6194 - x - 6194 = -765 - 6194$$
$$-x = -5429 \qquad\qquad -x = -6959$$
$$\frac{-x}{-1} = \frac{-5429}{-1} \qquad\qquad \frac{x}{-1} = \frac{-6959}{-1}$$
$$x = 5429 \qquad\qquad x = 6959$$

Aconcagua is either 5429 meters high (if it is lower than Mt. McKinley) or 6959 meters high (if it is higher than Mt. McKinley). More information is needed to determine a final answer.

Since we are taking the absolute value of the difference, we could subtract in either order and obtain the same results. Instead of $|6194 - x| = 765$,

we could use $|x - 6194|$. Check for yourself. We can see why we obtain the same solution if we notice that $6194 - x$ is the opposite of $x - 6194$. (The sign of each term is the opposite of that of the corresponding term.) The absolute value of an expression and the absolute value of its opposite are equal.

✓ 4.6.4 Checkup

An auto manufacturer designs a sports car to comfortably seat a driver who is 6 feet tall (72 inches), give or take 3 inches. Another way of stating this is that the absolute value of the difference between 72 inches and a person's height, h, in inches, equals 3. Write a linear absolute-value equation, and solve it to find the greatest and smallest heights for drivers who will comfortably sit in the car.

4.6 Exercises

Solve graphically. Check numerically.

1. $|x| = 9$ **2.** $|-x| = 9$ **3.** $|x + 1| = 3$ **4.** $|x - 3| = 1$

5. $|6x + 6| = 12$ **6.** $|3x - 9| = 6$ **7.** $|-x - 1| = 1$ **8.** $|-4x - 8| = 4$

9. $|x + 21| = 26$ **10.** $|x - 19| = 29$ **11.** $|-x - 19| = 22$ **12.** $|-x - 1| = 18$

13. $|x - 2.17| = 6.09$ **14.** $|-x + 22.83| = 0.899$ **15.** $\left|x + \dfrac{2}{5}\right| = \dfrac{11}{15}$ **16.** $\left|x + \dfrac{7}{8}\right| = 6\dfrac{9}{16}$

17. $2|x - 15| + 25 = 15$ **18.** $3|2x + 12| + 14 = 14$ **19.** $7|x - 32| + 21 = 21$ **20.** $4|3x - 6| + 19 = 13$

21. $3|x + 5| - 7 = 4$ **22.** $3|2x - 5| - 16 = 2$

Solve algebraically.

23. $|x| = 138$

24. $|z| = 2400$

25. $|a| = -512$

26. $|m| = 0.009$

27. $|b| = 41.67$

28. $|s| = -22\frac{17}{21}$

29. $|c| = 14\frac{5}{9}$

30. $|x| = \frac{11}{12}$

31. $|x + 578| = 286$

32. $|z + 303| = 471$

33. $|x - 721| = 1942$

34. $|x - 651| = 124$

35. $5|x - 29| - 46 = 74$

36. $4|2x - 18| - 16 = 84$

37. $2|x + 41| + 96 = 48$

38. $3|x + 62| + 76 = 32$

39. $\left|\dfrac{x - 14}{7}\right| = 8$

40. $\left|\dfrac{2x + 3}{5}\right| = 14$

41. $|-3x - 6| - 18 = -6$

42. $|-2x - 16| - 34 = -17$

43. $-4|x + 12| = -16$

44. $-3|-2x + 8| = -15$

Write a linear absolute-value equation for each situation and solve.

45. The maximum depth of Lake Superior differs from that of Lake Michigan by 410 feet. If Lake Michigan has a maximum depth of 923 feet, what are the possible values for the maximum depth of Lake Superior? Given that Lake Superior is deeper, what is its maximum depth?

46. The largest lake in the world is the Caspian Sea, which was incorrectly called a sea by the Romans because of its salty water. One of the most beautiful lakes in the world is Lake Baikal in Russia. The difference in maximum depths of these two lakes is 1952 feet. Given that Lake Baikal has a maximum depth of 5315 feet, what are the possible values for the maximum depth of the Caspian Sea? Given further that Lake Baikal is the deepest lake in the world, what is the maximum depth of the Caspian Sea?

47. Ralph Raffish manufactures sports clothes to fit the average male height, give or take 6 inches. If the average male is 5 feet 9 inches tall, what are the maximum and minimum heights the clothes will fit?

48. Rexus Luxury Cars designs its cars to comfortably seat a driver whose height differs by 4 inches from the average height of buyers of the cars. If the average height of buyers of the cars is established by a customer survey to be 6

feet 1 inch, what are the limits on heights of drivers the car would comfortably seat?

49. A political survey reports that 42% of the voters support an issue, with a margin of error of 3%. What are the minimum and maximum percentages that may occur?

50. A poll reports that 48% of the voters of a city and county support unification of the city and county charters, with a margin of error of 4.5%. If the poll is accurate, what are the minimum and maximum percentages that could occur in the election?

51. If a bathroom scale has a margin of error of 2 pounds, what is the range for Dee's actual weight if the scale displays a weight of 132 pounds?

52. A scale in a meat market has an allowable margin of error of 0.05 pound. What is the range of weights for a package of hamburger if the scale displays a weight of one pound?

53. The tolerance in cutting a piece of pipe is $\frac{1}{4}$ inch. What are the possible lengths of a piece of pipe that is to be cut to a length of $5\frac{1}{2}$ inches?

54. A metal rod is to be cut to a length of 25.5 inches, with a tolerance of 0.125 inch. What are the possible lengths of the rod?

55. The most extreme score in a set of test scores differs from the average by 22 points. If the average of the test scores is 71 points, what are the possible values for the most extreme score?

56. The range of scores for a test was 28 points. If one of the extreme scores was 66, what are the possible limits for the scores? (Note that 66 could be either the highest score or the lowest score on the test.)

57. The difference in the ages of Barbra and her movie costar is 12 years. If Barbra's age is 47, what are the possible pairs of values for their ages?

58. The difference in sales of two brands of soda pop averages 2,500,000 bottles per month. If one brand sells 12,100,000 bottles of soda pop per month, what are the possible values for the monthly sales of the other brand?

4.6 Calculator Exercises

Some linear absolute-value equations are extremely difficult to solve algebraically, particularly if there is more than one absolute-value expression in the equation. It is then that the graph-

ical method of solving is handy. Following are some such exercises. Solve them graphically. Check your answers numerically.

1. $|x + 2| = |x - 3|$

2. $3|x + 2| - 5 = 3|x + 2| + 5$

3. $|x + 6| - 9 = -|x + 3| + 8$

4. $|x + 2| = -3x + 8$

5. $|x + 3| + |x - 3| = 15$

Could you have solved any of these exercises algebraically? If so, which ones? For those you could not solve algebraically, why couldn't you?

4.6 Writing Exercise

You have seen both ways to solve linear absolute-value equations: graphically and algebraically. Which method did you prefer? Write a brief summary of which method you prefer and why you prefer it. Try to think of situations in which you would be forced to use one method over the other.

CHAPTER 4 SUMMARY

After completing this chapter, you should be able to define in your own words the following key terms.

4.1
linear equation in one variable
solution set
contradiction
identity

4.5
consecutive integers
consecutive even integers
consecutive odd integers

4.6
linear absolute-value equation in one
  variable

4.2
equivalent equations

Reflections

1. Explain how the addition property of equations is used to solve a linear equation in one variable.

2. Explain how the multiplication property of equations is used to solve a linear equation in one variable.

3. How can you graphically solve a linear equation in one variable?

4. How can you numerically solve a linear equation in one variable?

5. Explain how you can tell that a linear equation in one variable has no solution.

6. Explain how you can tell that a linear equation in one variable has many solutions.

7. How does a linear absolute-value equation in one variable differ from a linear equation in one variable?

8. What must you do differently when algebraically solving a linear absolute-value equation in one variable, compared with solving a linear equation in one variable?

CHAPTER 4 SECTION-BY-SECTION REVIEW

4.1

Recall	Examples
Linear equation in one variable • An equation that can be written in the form $ax + b = 0$, where a and b are real numbers and $a \neq 0$.	Linear equations in one variable $\frac{3}{4}x + 5 = 0$; $2x - 5 = -4x - 7$; $x = 9$
Solving numerically • Complete a three column table of values—one column for the independent variable, one for the left side of the equation, and one for the right side of the equation. • The solution is the value of the independent variable that results in equal values for the left and right sides of the equation.	Solve $2x + 1 = 3x - 5$ numerically. <table><tr><td>x</td><td>$2x + 1$</td><td>$=$</td><td>$3x - 5$</td></tr><tr><td>5</td><td>11</td><td></td><td>10</td></tr><tr><td>6</td><td>13</td><td></td><td>13</td></tr><tr><td>7</td><td>15</td><td></td><td>16</td></tr><tr><td>8</td><td>17</td><td></td><td>19</td></tr></table>The solution is 6, because $13 = 13$.

Recall	Examples
Solving graphically • Graph the two functions defined by the left side of the equation and the right side of the equation. • The solution is the x-coordinate of the point of intersection of the two graphs.	Solve $3x + 2 = 5x - 2$ graphically. 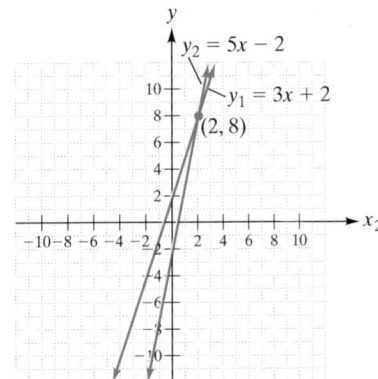 The solution is 2, because the intersection is $(2, 8)$.

Identify each equation as linear or nonlinear.

1. $4x^2 - 2x + 1 = 0$ **2.** $5x + 3 = x - 4$ **3.** $5.7x - 8.2(x + 4.6) = 0$

4. $\sqrt{x} + 2 = 3x - 6$ **5.** $\dfrac{1}{8}x + \dfrac{3}{4} = \dfrac{11}{16}$ **6.** $\dfrac{3}{x} + 5 = 12x$

Use the calculator screens to write the equation being solved and to determine the solution of the equation.

7.
```
Plot1 Plot2 Plot3
\Y1◻(3/4)(X+7)-5

\Y2◻(1/3)(X+12)
\Y3=
\Y4=
\Y5=
\Y6=
```

X	Y1	Y2
6	4.75	6
7	5.5	6.3333
8	6.25	6.6667
9	7	7
10	7.75	7.3333
11	8.5	7.6667
12	9.25	8

X=12

11. $14x + 12 = 11(x - 5) + 60$

13. $2(2x + 1) + x = 3(2x - 1) - (x + 1)$

Use the calculator screens to write the equation being solved and to determine the solution of the equation.

14.
```
Plot1 Plot2 Plot3
\Y1◻2.3(X-5.6)+4

\Y2◻3X-11.33
\Y3=
\Y4=
\Y5=
\Y6=
```

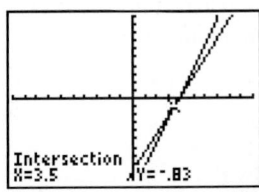

Intersection
X=3.5   Y=-.83

Solve numerically for an integer solution if possible.

8. $4x + 7 = 2x - 5$

9. $2.4x - 9.6 = 4.8$

10. $\dfrac{3}{5}x - \dfrac{7}{10} = \dfrac{1}{5}x + \dfrac{1}{2}$

12. $3(x - 2) - 1 = 4(x - 1) - (x + 3)$

Solve exercises 15–19 graphically.

15. $2x - 2 = -x + 4$

16. $\dfrac{1}{2}x - 2 = -\dfrac{1}{3}x - \dfrac{11}{3}$

17. $(x + 3) + (x + 1) = 3(x + 1) - (x - 1)$

18. $(x + 6) - 3(x + 1) = (2x + 5) - 2(2x + 3)$

19. $1.2x + 0.72 = -2.1x + 8.64$

20. Theresa offers to rake and clean the wooded area of your backyard for a flat fee of $90, or for a fee of $25 plus $6.50 for each hour worked. Determine how many hours of work will make the two offers equivalent.

21. A church receives weekly donations of $2200, $1750, and $1885 for the first three weeks of the month. How much does the church need to receive in the fourth week to meet its average weekly goal of $2000 in donations?

22. Two sides of a triangular worktable surface are of equal length. The third side measures 3 feet longer than each of the other two sides. If the perimeter of the surface is 26.25 feet, what are the lengths of each side?

4.2

Recall	Examples
Solving algebraically by using the addition property of equations • Simplify both expressions in the equation. • To both sides of the equation, add if a term is subtracted from the variable or subtract if a term is added to the variable. • Check the solution.	Solve $3x - 4 - 2x = 3$ algebraically. $3x - 4 - 2x = 3$ $x - 4 = 3$ $x - 4 + 4 = 3 + 4$ $x = 7$ Check: $3x - 4 - 2x = 3$ $3(7) - 4 - 2(7) \mid 3$ $ 3 \mid$ The solution checks.
Solving algebraically by using the multiplication property of equations • Simplify both expressions in the equation. • To both sides of the equation, multiply if the variable is divided by a number or divide if the variable is multiplied by a number. • Check the solution.	Solve $2x + 3x = 1$ algebraically. $2x + 3x = 1$ $5x = 1$ $\dfrac{5x}{5} = \dfrac{1}{5}$ $x = \dfrac{1}{5}$ Check: $2x + 3x = 1$ $2\left(\dfrac{1}{5}\right) + 3\left(\dfrac{1}{5}\right) \mid 1$ $ 1 \mid$ The solution checks.

Solve algebraically.

23. $41 + x = 67$

24. $y - \dfrac{7}{13} = \dfrac{11}{39}$

25. $5 - (2 - x) = 1$

26. $0.59(z - 1) + 0.41(z + 2) = 3.163$

27. $-4x = 272$

28. $45.86z = -1765.61$

29. $\dfrac{a}{7} = 15$

30. $\dfrac{4}{5}x = \dfrac{64}{125}$

31. $15x - 16x = -12$

32. $-y = 2.98$

33. $4(2x - 6) + 6(x + 4) = 49$

Write an equation for each situation and solve algebraically.

34. Total attendance at a benefit luncheon was 247 people. If 189 people purchased tickets, how many tickets were given as complimentary passes?

35. If a graduating class had 154 males, representing 55% of the class, how many students graduated in total?

36. Students at an elementary school are selling coupon books for $15 each as a fund-raiser. If the goal of the school is to raise $80,000 for a computerized classroom, how many books must the students sell?

37. Miyoshi was to receive $\frac{3}{7}$ of the proceeds from the sale of an estate. If she received $18,270, what were the proceeds of the sale?

38. The largest electric current achieved by scientists occurred at the Oak Ridge National Laboratory in 1996. The current of 1,000 amperes per square centimeter ordinarily carried by household wires would equal only 0.05 percent of the current achieved by the scientists. Write a linear equation and solve to determine the current achieved by the scientists.

39. How many quarts of a 35% liquid fertilizer solution does a farmer need if he wants the solution to contain 4 quarts of pure liquid fertilizer?

4.3

Recall	Example
Solving algebraically • Simplify both expressions in the equation (preferably without fractions).	Solve $\frac{2}{3}x + \frac{1}{2} = x - \frac{1}{4}$ algebraically. $$\frac{2}{3}x + \frac{1}{2} = x - \frac{1}{4}$$ $$12\left(\frac{2}{3}x + \frac{1}{2}\right) = 12\left(x - \frac{1}{4}\right)$$ $$\overset{4}{\cancel{12}}\left(\frac{2}{\cancel{3}}x\right) + \overset{6}{\cancel{12}}\left(\frac{1}{\cancel{2}}\right) = 12(x) - \overset{3}{\cancel{12}}\left(\frac{1}{\cancel{4}}\right)$$
• Isolate the variable to one side of the equation and the constants to the other side of the equation by using the addition property of equations. • Reduce the coefficient of the variable to 1 by using the multiplication property of equations.	$$8x + 6 = 12x - 3$$ $$8x + 6 - 12x = 12x - 3 - 12x$$ $$-4x + 6 = -3$$ $$-4x + 6 - 6 = -3 - 6$$ $$-4x = -9$$ $$\frac{-4x}{-4} = \frac{-9}{-4}$$ $$x = \frac{9}{4}$$
• Check the solution.	Check: $$\frac{2}{3}x + \frac{1}{2} = x - \frac{1}{4}$$ $$\begin{array}{c|c} \frac{2}{\cancel{3}}\left(\frac{\cancel{9}^{\,3}}{\cancel{4}^{\,2}}\right) + \frac{1}{2} & \frac{9}{4} - \frac{1}{4} \\ \hline 2 & 2 \end{array}$$ The solution checks.

Solve algebraically.

40. $3x + 7 = 4x + 21$ **41.** $14 - 2x = 5x$ **42.** $8.7x + 4.33 = -2.4x - 33.41$

43. $\frac{5}{7}a + \frac{11}{14} = \frac{2}{7}a$ **44.** $2(x + 5) - (x + 6) = 2(x + 2) - x$ **45.** $3(x - 4) + 2(x + 1) = 5x + 10$

Write an equation for each situation and solve algebraically.

46. Your business trip allows $250.00 for reimbursement of car rental. If you rent a car for a flat fee of $49.95 per week plus $0.22 per mile, what is the maximum number of miles you can drive without exceeding your allowance?

47. Acme Industries invests $150,000 in new equipment, which the company will depreciate evenly over a seven-year period. At the end of that time, Acme expects to sell the equipment for scrap for $25,000. What will the annual depreciation for the equipment be? (*Hint:* Depreciation plus scrap value equals the investment.)

48. One contractor offers to paint your home for a fee of $175 plus $35 per hour. Another contractor offers to do the same job for $100 plus $40 per hour. For what number of hours will the two offers be the same?

49. Erin has invested $8000 into a mutual fund that pays her 9% simple interest at the end of one year. How much should she invest in a second mutual fund that pays 8% simple interest per year if she wishes to have a total of $1600 in simple interest at the end of the year?

4.4

Recall	Example
Solving an equation for a variable • Clear the equation of fractions. • Isolate the terms involving the desired variable to one side of the equation and all other terms to the other side of the equation by using the addition property of equations. • If the variable being solved for is in unlike terms, factor out the variable. • Reduce the coefficient of the variable to 1 by using the multiplication property of equations.	Solve $S = 2LW + 2WH + 2LH$ for L. $$S = 2LW + 2WH + 2LH$$ $$S - 2WH = 2LW + 2WH + 2LH - 2WH$$ $$S - 2WH = 2LW + 2LH$$ $$S - 2WH = L(2W + 2H)$$ $$\frac{S - 2WH}{2W + 2H} = \frac{L(2W + 2H)}{2W + 2H}$$ $$\frac{S - 2WH}{2W + 2H} = L$$ or $\qquad L = \dfrac{S - 2WH}{2W + 2H}$

Solve exercises 50–52 for the indicated variable.

50. $A = \dfrac{1}{2}h(b + B)$ for h **51.** $S = 2LW + 2WH + 2LH$ for W **52.** $\dfrac{3}{4}x - \dfrac{5}{8}y = \dfrac{11}{12}$ for y

53. An annuity will pay you $6000 immediately and an annual payment of $8000. Write an equation for the total amount, A, that you will receive from the annuity over n years. Use the equation to solve for n, the number of years the annuity will last for a particular amount. Use the new equation to find how long the annuity will last if the amount is $78,000. How long will the annuity last if the amount is $126,000?

4.5

Recall	Example
Modeling real-world situations • Define a variable. • Write an equation. • Solve the equation. • Check the solution. • Write an answer to the question asked.	The numbers of people per square mile in Mongolia, Namibia, Australia, and Iceland are consecutive integers. The sum of these integers is 22. Determine the density (people per square mile) of these countries. Let x = the density of Mongolia $x + 1$ = the density of Namibia $x + 2$ = the density of Australia $x + 3$ = the density of Iceland $x + (x + 1) + (x + 2) + (x + 3) = 22$ $\qquad\qquad 4x + 6 = 22$ $\qquad\qquad 4x + 6 - 6 = 22 - 6$ $\qquad\qquad\qquad 4x = 16$ $\qquad\qquad\qquad \dfrac{4x}{4} = \dfrac{16}{4}$ $\qquad\qquad\qquad x = 4$ The consecutive integers are 4, 5, 6, and 7. Check: The sum of $4 + 5 + 6 + 7 = 22$. Mongolia has 4 people per square mile. Namibia has 5 people per square mile. Australia has 6 people per square mile. Iceland has 7 people per square mile.

54. Rob was cutting three lengths of wire from a 3-foot piece. He can't find the instructions that tell how many inches each piece should measure, but he remembers that they were consecutive even integers. What are the lengths he should cut? What would the lengths be if the lengths were consecutive integers instead of consecutive even integers?

55. Two lines meet at an angle of 48°. What is the measure of the complement of this angle? If the complement is one angle of a triangle that has two other angles with equal measure, what do each of the other two angles measure?

56. Noah wants to build a rectangular holding pen for some animals. He wants the length of the pen to be three times the width. He will install two gates on the pen at opposite ends. One gate is 4 feet wide and the other is 6 feet wide. If he has 230 feet of fencing, what should the dimensions of the pen be?

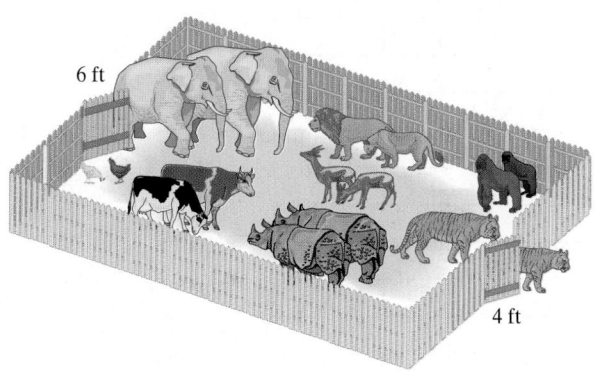

6 ft

4 ft

57. For how many years should you invest $5000 at 7.5% simple interest per year if you want to earn $2250 in interest?

58. How much should you invest at 7.5% simple interest if you want to earn $2250 in 3 years?

59. The sale price of a suit was $210. The sale was advertised as a 20%-off sale. Determine the original price of the suit.

60. If the markup on a signed art print is 30% and the print sells for $32.50, how much was the artist paid for it?

4.6

Recall	Examples
Linear absolute-value equation in one variable • An equation that can be written in the form $\|ax + b\| = c$, where a, b, and c are real numbers and $a \neq 0$.	Linear absolute-value equation in one variable $\|2x - 5\| = 4$; $\dfrac{\|x - 2\|}{3} = 1$
Solving graphically a linear absolute-value equation in one variable • Graph the two functions defined by the left side of the equation and the right side of the equation. • The solution(s) are the x-coordinates of the points of intersection of the graphs.	Solve $\|2x - 1\| = 5$. Graphic Solution $y_1 = \|2x - 1\|$ $y_2 = 5$ $(-2, 5)$ $(3, 5)$ The solutions are -2 and 3.
Solving algebraically a linear absolute value equation in one variable • Isolate the absolute value expression and determine the number of solutions to the equation. • Write and evaluate an equation(s) needed to determine the solution(s).	Solve $\|2x - 1\| = 5$. Algebraic Solution $\|2x - 1\| = 5$ $2x - 1 = 5 \qquad \text{or} \qquad 2x - 1 = -5$ $2x - 1 + 1 = 5 + 1 \qquad 2x - 1 + 1 = -5 + 1$ $2x = 6 \qquad\qquad 2x = -4$ $\dfrac{2x}{2} = \dfrac{6}{2} \qquad\qquad \dfrac{2x}{2} = \dfrac{-4}{2}$ $x = 3 \qquad\qquad x = -2$

(Continued on page 292)

Recall	Examples
Checking numerically a linear absolute-value equation in one variable	Check numerically that the solutions of $\|2x - 1\| = 5$ are -2 and 3.
• Set up a table of values.	Numeric Check
• The solution(s) are the x-values that correspond to equal values for the two expressions.	

x	$\|2x - 1\|$	$= 5$
-3	7	5
-2	5	5
-1	3	5
0	1	5
1	1	5
2	3	5
3	5	5
4	7	5

The solutions are -2 and 3.

Solve algebraically and check graphically.

61. $|a - 7| = 0$ **62.** $|-b + 12| = 4$ **63.** $|c - 2| = -4$ **64.** $|2x - 7| = 8$

65. $2|x - 7| - 4 = 8$ **66.** $5|2x - 7| + 10 = 8$ **67.** $2\left|\dfrac{x - 1}{2}\right| - 5 = -2$

Write a linear absolute-value equation that represents each situation and solve.

68. The tolerance on a machining job is 0.04 mm. If the basic dimension of the part is 62.79 mm, what are the permissible limits on the part?

69. A voter survey shows that 49% of the voters are in favor of a consolidation proposal, with a margin of error of 4%. What are the limits on the percentage of voters who favor the proposal?

CHAPTER 4 CHAPTER REVIEW

Solve graphically.

1. $3(x + 2) - 2(x - 1) = (2x + 5) - (x - 3)$ **2.** $(2x + 1) - (3x - 7) = (x + 5) - 2(x - 2)$

3. $x + 1 = 2x + 5$ **4.** $\dfrac{1}{3}x + 3 = 6 - \dfrac{2}{3}x$

5. $1.2x - 6.12 = -2.2x + 4.42$

Solve each equation numerically for an integer solution.

6. $4x - 5 = 7 - 2x$ **7.** $14(x + 6) - 17 = 12(x + 5)$

8. $1.5x + 5.5 = -2.4x - 6.2$ **9.** $2(2x - 3) + 3(x + 1) = 6(x - 1) + (x + 3)$

10. $\dfrac{1}{3}x + \dfrac{14}{3} = \dfrac{5}{2} - \dfrac{3}{4}x$ **11.** $4(x - 2) - (x - 1) = 3x + 1$

Identify each equation as linear or nonlinear.

12. $5 - \dfrac{3}{7}x = \dfrac{2}{3}x + 2$ **13.** $2x - 4 = 1 + \dfrac{9}{x}$ **14.** $-2x^3 + 3x = x - 5$

15. $15 = 2(x - 7) - (x - 2)$ **16.** $3.9(x - 1.2) - 6.7x = 0$ **17.** $\sqrt[3]{x + 1} - 5 = 4x$

Solve algebraically.

18. $\dfrac{z}{29} = -12$ **19.** $\dfrac{5}{22} = \dfrac{25}{33}y$ **20.** $59a = 1888$

21. $-174.243 = 2.41x$

22. $2.3x - 3.3x = 14$

23. $-b = 14.59$

24. $3(x - 8) + 8(x + 3) = 77$

25. $z + 193 = -251$

26. $\frac{13}{17} + a = \frac{5}{51}$

27. $0.92(x - 2) + 0.08(x + 1) = 5.73$

28. $6.2x + 5.67 = 4.9x + 16.98$

29. $24.96 - 3.9a = 0$

30. $\frac{2}{3}x - \frac{3}{4} = -\frac{5}{6}$

31. $\frac{4}{9}y + \frac{11}{18} = \frac{5}{6}y$

32. $7x - 4 = 3x + 20$

33. $7x = 15 - 3x$

34. $2(x + 2) + (x + 1) = 5(x + 1) - 2x$

35. $4(2x - 1) = 3(2x + 1) + 2(x + 1)$

Solve.

36. $|2x + 6| = 0$

37. $|4 - z| = 4$

38. $|2x - 1| = -4$

39. $|5x + 2| = 12$

40. $5|x - 7| = 15$

41. $3\left|\dfrac{x + 1}{4}\right| - 9 = -6$

Solve each equation for the indicated variable.

42. $y - 6 = \frac{2}{3}(x + 9)$ for y

43. $\frac{1}{9}x + \frac{2}{3}y = \frac{1}{6}$ for y

44. $S = 2\pi r^2 + 2\pi rh$ for h

45. $I = PRT$ for P

Write a linear absolute-value equation for each situation and solve.

46. The difference in the heights of the Peachtree Tower and the Bank of America Tower, both in Atlanta, is 253 feet. If the Peachtree Tower is 770 feet high, what are the two possible heights for the Bank of America Tower? Given that the Peachtree Tower is not as tall, which height is correct for the Bank of America Tower?

47. An instructor decides to give grades of *C* to student scores that differ from the class average by 5 or fewer points. If the class average is 72 points, what are the limits for a *C* grade?

Write a linear equation for each situation and solve.

48. If the sales tax on a purchase was $5.41, which represents 8.25% tax, what was the subtotal of the purchases before taxes? What was the total bill, including sales tax?

49. Chuck was given an advance of $5,650 for developing software for a publishing firm. The amount of the advance represented $\frac{2}{5}$ of the total amount he would receive for his work. How much money does Chuck expect to earn in total for this project?

50. A part-time employee earns $13.25 per hour working in a science laboratory. If her gross earnings last year were $16,562.50, how many hours did she work?

51. A newspaper delivery service pays its employees $15.00 per day plus $0.25 for each house along the delivery route that purchases a newspaper. To how many houses does an employee need to deliver papers in order to earn $45.00 per day?

52. Equipment purchased for $250,000 is to be sold for a scrap value of $15,000 after 20 years. What will be the annual depreciation rate for the equipment? (*Hint:* Depreciation plus scrap value equals investment.)

53. Laurie wants to buy a personal computer. She has saved $985. If the computer sells for $1399, how much more does she need to save before she can buy the computer?

54. Instructions for a model-plane kit require a piece of balsa wood to be cut into four pieces, each 1 inch longer than the previous piece. If the wood is 26 inches long, how long will each piece be? What lengths would the strips be if each piece were 2 inches longer than the preceding piece, allowing for 2 inches of scrap?

55. A triangle has one angle measuring 62°. The other two angles are equal in measure. What are the measures of the other two angles? What is the measure of the supplement of each of the two equal angles?

56. If an ink-jet printer was marked down 15% to a sale price of $118.95, what was the price before the markdown?

57. At what simple-interest rate should you invest $8000 for three years in order to earn $1560 in interest?

58. How much should you invest at 4.5% simple interest for a year to earn $562.50 in interest?

59. The length of a room is one and one-half times its width. If the perimeter of the room is 84 feet, what are its dimensions?

60. If the price of a television set with 8.75% sales tax was $325.16, what was the price of the set before tax? How much was the sales tax?

61. A plumber charges you $40.00 per visit plus $55.50 for each hour worked on a job. Write an equation for the total amount, A, that you will pay for a job that requires h hours of labor. Use the equation to solve for h, the hours worked, in terms of the total amount for the job. Use the new equation to find how many hours a job costing $178.75 lasted. If the plumber is paid $95.50, for how long did he work?

62. A designer rug is rectangular, with diagonals running from one corner to the opposite corner. If the diagonal makes a 35-degree angle with the length of the rug, how large an angle does it make with the width of the rug?

63. How many gallons of a 60% antifreeze solution does a mechanic need if she wants the solution to contain 2 gallons of pure antifreeze?

CHAPTER 4 TEST

 **TEST-TAKING TIPS**

When you review for an exam, list all the major concepts the exam will cover. Locate exercises that illustrate each of the concepts. Think of how you would recognize them if they were randomly ordered on the exam. Spend more of your study time on those exercises in which you know you are weak. For extra practice, seek out additional exercises in the text that you have not worked out. Try to find someone to study with so that you can help each other with your weaknesses. If you can't complete some exercises, see your instructor or go to a tutoring center well before test time to seek extra help. If you do this, the exam should not be a surprise to you, it won't throw you, and your confidence level will be high.

Identify each equation as linear or nonlinear.

1. $3.14x + 9.07 = 5.72x$

2. $5x = 12 + \dfrac{19}{x}$

3. $4x + 21 = 5x^2$

4. $4(x - 6) = 3(5 - x) + 12$

Solve.

5. $2(2x - 5) - 2(2 - x) = 6(x + 1) + 1$

6. $2(x + 5) = -3(x + 1) - 2$

7. $1.41(x + 5.08) + 1.17x + 0.00102 = -3.46x - 5.39334$

8. $(x + 1) - 4(x - 1) = 3(2 - x) - 1$

9. $\dfrac{4}{5}x + \dfrac{31}{10} = \dfrac{-4}{3}x + \dfrac{41}{6}$

10. $|x + 3| = 7$

11. $3|2x - 9| + 14 = 5$

Write a linear equation for each situation and solve.

12. A piece of wire measures 45 inches. For splicing purposes, it will be cut into three pieces, each of which is 1 inch longer than the preceding piece. In what lengths should the three pieces be cut?

13. Ricardo's dad is loaning him the money to buy a used car, which sells for $2470. Ricardo will use the $850 he has saved as a down payment and plans to pay his dad back in 12 months. His dad will not charge him any interest on the loan. What will be Ricardo's monthly payment to his dad?

14. If a stereo is on sale at 25% off the retail price and its sale price is $179.95, what was the price before it went on sale?

15. Solve $P = 2L + 2W$ for W, where P is the perimeter of a rectangle with length L and width W. Use the formula to find the width of a rectangle whose length is 14.8 inches and whose perimeter measures 44.8 inches.

16. One angle of a triangle measures 42°. The other two angles are equal in measure. What is the measure of the other angles? What is the measure of the supplement of the angles that have equal measure?

17. Fruit drinks are a mixture of fruit juice and water. How many liters of a drink containing 60% apple juice must be mixed with 500 liters of a drink containing 20% cranberry juice to make a drink containing 50% cranberry–apple juice?

18. Annie works at the Book City bookstore. She was offered a choice of payment plans. She could earn $1500 per month, or she could earn $1200 per month plus 4% of all her sales for the month. For what value of sales would the two pay plans be equal?

19. Serene Landscaping will clear the wooded area behind Candise's house for a labor charge of $25 per hour and a flat fee of $50 to haul away the brush. Evergreen Yard Service offers to do the same job for a labor charge of $30 per hour and no extra charge for hauling away the brush. For what number of hours will the two plans cost the same?

20. The margin of error in a voter poll is 3%. If the survey suggests that 52% of the voters support a new tax levy, what are the limits on the true percentage of voters who will support the levy?

21. Describe how to solve a linear equation in one variable graphically. Explain in detail how to locate the solution.

Project

Graphing-calculator technology makes it easy to study the golden ratio, one of the most interesting graphical and numerical concepts in algebra. The golden ratio can be found by dividing a line segment into two parts such that the length of the smaller part divided by the length of the larger part is the same as the length of the larger part divided by the length of the line segment.

Let s represent the length of the shorter part, and let the unit length 1 represent the length of the longer part. The following is a sketch of the line segment:

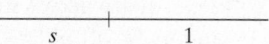

Algebraically, the golden ratio is

$$\frac{\text{shorter length}}{\text{longer length}} = \frac{\text{longer length}}{\text{total length}}$$

$$\frac{s}{1} = \frac{1}{s+1}$$

This equation will be solved later in the text, but for now just accept it that one of its solutions is $s = \frac{\sqrt{5}-1}{2}$.

Since the ratio of the shorter part to the longer part is s to 1, or $\frac{\sqrt{5}-1}{2}$ to 1, it follows that the shorter part is $\frac{\sqrt{5}-1}{2}$ times the length of the longer part.

Part I

Many philosophers, artists, mathematicians, architects, musicians, and others have been intrigued by the golden ratio and have used it in their undertakings. As an example of how you might use it, suppose you want to construct a rectangle whose length and width are pleasing to the eye. Furthermore, suppose the perimeter of the rectangle is fixed at 50 centimeters. Complete the following steps to determine the length and the width of the rectangle needed:

1. Define the variable x to be the length of the rectangle—the longer side.

2. Write an expression for the width of the rectangle—the shorter side— using s as defined above and x.

3. Draw a rectangle and label its length and width in terms of x.

4. Write an expression for the perimeter of your rectangle (using the expressions for length and width developed in steps 1 and 2), and set the expression equal to the value given in order to obtain an equation.

5. Solve the equation you found in step 4 for the length, x.

6. Approximate the value of the length to one decimal place (by substituting the value for s).

7. Find the width, using the expression from step 2. Substitute the values for s and x, and then round your answer to one decimal place.

8. Check your answer to see that the perimeter is in fact 50 centimeters. If it is not, explain.

Next, construct a rectangle with the dimensions you have determined. Is the rectangle's shape pleasing to the eye? Does the perimeter check?

Part II

Here is another interesting fact: If you mark off a square in your rectangle, with a side measuring the same as the width of the rectangle, the resulting inner rectangle also has dimensions in the golden ratio. You can continue marking off squares in each inner rectangle to obtain another golden rectangle. The resulting picture should be an aesthetically pleasing modern work of art. To do this, hold your rectangle with the width, or shorter side, up. Mark off a square across the top of the rectangle. Turn the rectangle clockwise, and mark off the next square across the top of the remaining rectangle. Continue this procedure and mark off at least five squares. Do you like the pattern? The golden ratio can be tried with other geometric shapes as well.

Part III

Now let's try to generate a table of dimensions of rectangles whose lengths and widths are in the golden ratio. Complete the following table, rounding your answers to one decimal place:

Length	Width	Perimeter
5 cm		
10 cm		
15 cm		
20 cm		
25 cm		
30 cm		
35 cm		
40 cm		

(*Hint:* You can use your calculator to fill in the table. Begin by storing the expression for the golden ratio, s, under the letter S in the calculator. Then set Y1, the width, equal to SX and Y2 equal to the formula for the perimeter, in terms of S and x. Finally, use the table feature of the calculator to obtain the values needed for the table.)

Part IV

As stated earlier, the golden ratio has a long history of use because of its aesthetic properties. The Egyptians thought that the golden ratio was sacred, and it can be found in the design of their temples, pyramids, and artwork. Even some Egyptian hieroglyphics have proportions based upon the golden ratio. Leonardo da Vinci's drawings often have overlays of rectangles with the golden ratio. The golden ratio also may be seen in many of the rectangles used by Piet Mondrian in his form of art called neoplasticism. The golden ratio can be found in many of the dimensions of the Parthenon, the famous Greek temple.

The design of the United Nations building in New York City is said to have windows in the shape of the golden ratio. The music of Beethoven and Mozart are said to have pieces that divide into parts exactly according to the golden ratio. Renaissance writers called it the "divine proportion."

However, there is also controversy about the golden ratio. Is it really as pleasing as is claimed? Does some of the architecture, such as the Greek and Egyptian, conform with the golden ratio as a result of erosion?

As a final task in this project, find a reference on the golden ratio. You may go to the library to search, or use the Internet. Write a short summary of your findings. Be sure to document your reference sources.

Chapter 5

Linear Equations and Functions

In Chapter 4 we studied linear equations in one variable, learning to solve them by substituting a suitable value for the variable. In this chapter we examine linear equations in two variables and discuss methods for solving them. Some of the same ideas we learned before will apply here, too. We will also see several new ideas, such as the slope of a line, which is a very important and powerful tool in analyzing equations. With these tools, we will be able to study the relationships between two lines and between two equations. As a result, we will learn how to write linear equations to predict new information. This is an important use of basic mathematics in the real world.

Situations that can be described by linear equations (or functions) often occur in the areas of business and science. One important application of using linear equations to predict new information is illustrated in the chapter project. In the project, you are asked to write linear models from data found in the U.S. Census *Statistical Abstract*. Using the linear model of your choosing, you are requested to predict future events. Another application included in the project explores the relationship between distance and time of an object that may be moving.

5.1 APPLICATION

In 1974, an SR-71 Blackbird jet plane was flown from New York City to London in 1 hour and 55 minutes, the fastest transatlantic flight ever. If we let d represent the plane's distance from London, then an equation representing that distance is $d = 3462 - 1807t$, where t is the flight time in hours. Graph the equation and interpret the intercepts of the graph.

We will discuss this application further. See page 314.

5.1 Graphing by Using Ordered Pairs

OBJECTIVES

1 Identify linear equations in two variables.
2 Graph linear equations in two variables by using a set of ordered pairs.
3 Algebraically determine the *y*-intercept and *x*-intercept of a graph.
4 Graph linear equations whose graphs have only one intercept.
5 Graph linear equations in two variables by using the intercept method.
6 Model real-world situations by using linear graphs.

5.1.1 Identifying Linear Equations in Two Variables

In the last chapter, we solved linear equations in one variable. We used this type of equation to solve a problem with one unknown quantity or a problem that can be written in terms of one unknown quantity. However, if we need to solve a problem with two unknown quantities, then we may use a **linear equation in two variables**.

> **STANDARD FORM FOR A LINEAR EQUATION IN TWO VARIABLES**
> The standard form for a linear equation in two variables is $ax + by = c$, where a, b, and c are real numbers and a and b are not both equal to 0.

For example, the following are equations in standard form:

$$ax + by = c$$

$2x + 5y = 2$	$a = 2$	$b = 5$	$c = 2$
$x - y = 0$	$a = 1$	$b = -1$	$c = 0$
$2x = 7$	$a = 2$	$b = 0$	$c = 7$
$-3y = 1$	$a = 0$	$b = -3$	$c = 1$

Note: The last two equations, when either $a = 0$ or $b = 0$, are linear equations in one variable, which we discussed in Chapter 4. In the current chapter, they are treated as special cases of linear equations in two variables.

The relation $y = -2x + 5$ is a linear equation in two variables because it can be rearranged into standard form by the properties of equations. For example,

$$y = -2x + 5$$
$$y + 2x = -2x + 5 + 2x \qquad \text{Add 2x to both sides.}$$
$$y + 2x = 5 \qquad \text{Simplify the right side.}$$
$$2x + y = 5 \qquad \text{Rearrange the left side.}$$

These equations are called linear equations because their graphs are straight lines. For example, the graph of $y = -2x + 5$ is linear, as shown in the figure at the left.

All linear equations in two variables are relations. The relation $y = -2x + 5$ is also a function. (The graph passes the vertical line test.) Therefore, it may be written in function notation as $f(x) = -2x + 5$ and is called a **linear function**. (Remember, not all relations are functions.)

Y1 = −2x + 5

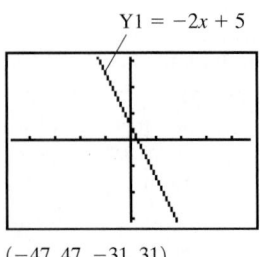

$(-47, 47, -31, 31)$

EXAMPLE 1 Identify each equation as linear or nonlinear. Express each linear equation in standard form.

a. $2x^2 + 3y = 4$ **b.** $y = -3x$
c. $x = 0$ **d.** $y = 2x + \sqrt{5}$

Solution

a. $2x^2 + 3y = 4$ is a nonlinear equation, because the x term is squared.

b. $y = -3x$ is a linear equation in two variables. Writing the equation in standard form proceeds as follows:

$$y = -3x$$
$$y + 3x = -3x + 3x$$
$$y + 3x = 0$$
$$3x + y = 0$$

c. $x = 0$ is a linear equation in two variables. It is written in standard form as $1x + 0y = 0$.

d. $y = 2x + \sqrt{5}$ is a linear equation in two variables. To put the equation in standard form, we write

$$y = 2x + \sqrt{5}$$
$$-2x + y = \sqrt{5}$$

 **HELPING HAND** Be careful to examine an equation with a radical expression. If the radical has a radicand containing a variable term, the equation is nonlinear. However, if the radicand contains only a constant term, the equation is linear.

Check your results for parts a, b, and d on your calculator. First solve for y and then graph the equation. Note that you cannot graph part c on your calculator.

a.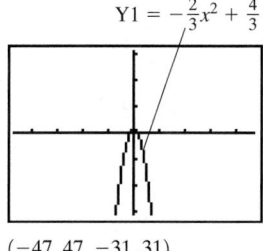

$(-47, 47, -31, 31)$
nonlinear equation

b.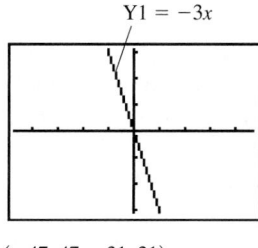

$(-47, 47, -31, 31)$
linear equation

d.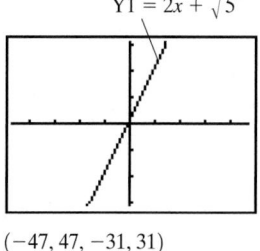

$(-47, 47, -31, 31)$
linear equation

✓ 5.1.1 Checkup

In exercises 1–5, identify each equation as linear or nonlinear. Express each linear equation in standard form.

1. $x = 3y^2 + 12$ **2.** $y = -\dfrac{3}{8}$

3. $y = 8.2x - 3.6$ **4.** $\sqrt{3}x = \sqrt{2}y$

5. $3\sqrt{x} + 2y = 0$

6. Why is the equation $ax + by = c$ called a linear equation?

5.1.2 Graphing by Using Ordered Pairs

We determine a solution of a linear equation in one variable by substituting a value for the variable. If the result is a true statement, then the value is a solution. Similarly, we determine a solution of a linear equation in two variables by substituting values for each of the two variables (an ordered pair). If the result is a true statement, then the ordered pair is a solution of the equation. We say the solution satisfies the equation.

> **SOLUTION OF A LINEAR EQUATION IN TWO VARIABLES**
> An ordered pair (x, y) is a solution of a linear equation in two variables if the values of the coordinates, when substituted for their corresponding variables, result in a true equation.

Some linear equations in one variable have more than one solution. A linear equation in two variables always has more than one solution. To determine the solutions of a linear equation in two variables, we will use a table of values, as we did with relations in Chapter 3.

> To determine solutions of a linear equation in two variables, x and y,
>
> • Solve the equation for y.
> • Set up an extended table of values.
> • Complete the table with at least three values for x.

EXAMPLE 2   Given $x - y = 3$, determine three ordered-pair solutions.

Solution

Solve for y.

$$x - y = 3$$
$$x - y - x = 3 - x$$
$$-y = 3 - x$$
$$\frac{-y}{-1} = \frac{3 - x}{-1}$$
$$y = -3 + x \quad \text{or} \quad y = x - 3$$

Complete the table of values. We will use 0, 1, and 2 for x. (Note that we may choose any value for x that is in the domain of the relation.)

x	$y = x - 3$	y	(x, y)
0	$y = 0 - 3$ $y = -3$	-3	$(0, -3)$
1	$y = 1 - 3$ $y = -2$	-2	$(1, -2)$
2	$y = 2 - 3$ $y = -1$	-1	$(2, -1)$

Therefore, $(0, -3)$, $(1, -2)$, and $(2, -1)$ are three possible solutions of the equation.

On your calculator, enter the equation in Y1, set up the table for integers, and view the table. The result is a table of possible ordered-pair solutions. Note that more than three ordered-pair solutions are shown.

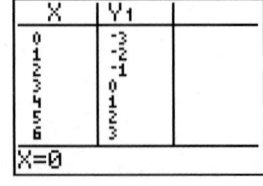

Y1 = $x - 3$

The preceding linear equation in two variables has an infinite number of possible ordered-pair solutions. To illustrate these solutions, we will use a graph. Remember that

1. Every solution of an equation can be represented by a point on its graph.
2. Every point on a graph represents a solution of its equation.

> To graph a linear equation in two variables using ordered pairs,
>
> • Graph at least two ordered-pair solutions found in a table of values.
> • Connect the points with a straight line.
>
> A third ordered pair should be used as a checkpoint. Label the coordinates of the points graphed.

EXAMPLE 3   Graph $x - y = 3$.

Solution

Solve for y.

$$y = x - 3$$

Graphing the ordered pairs found in the previous table gives three points: $(0, -3)$, $(1, -2)$, and $(2, -1)$. Connecting the points with a straight line will locate other possible solutions.

On your calculator, enter the expression in Y1 and graph it on a standard screen.

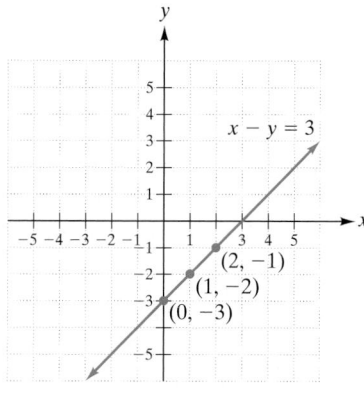

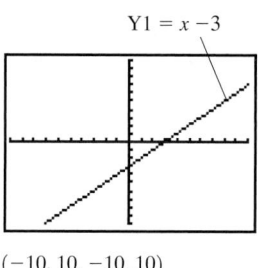

$(-10, 10, -10, 10)$

EXAMPLE 4   Graph $4x - 2y = 6$.

Solution

Solve for y.

$$4x - 2y = 6$$
$$4x - 2y - 4x = 6 - 4x$$
$$-2y = -4x + 6$$
$$\frac{-2y}{-2} = \frac{-4x + 6}{-2}$$
$$y = 2x - 3$$

Set up a table of values. We will use $-3, 0,$ and 2 for x.

x	$y = 2x - 3$	y	(x, y)
-3	$y = 2(-3) - 3$	-9	$(-3, -9)$
0	$y = 2(0) - 3$	-3	$(0, -3)$
2	$y = 2(2) - 3$	1	$(2, 1)$

Graph the ordered pairs and connect them with a straight line. Check the graph on your calculator.

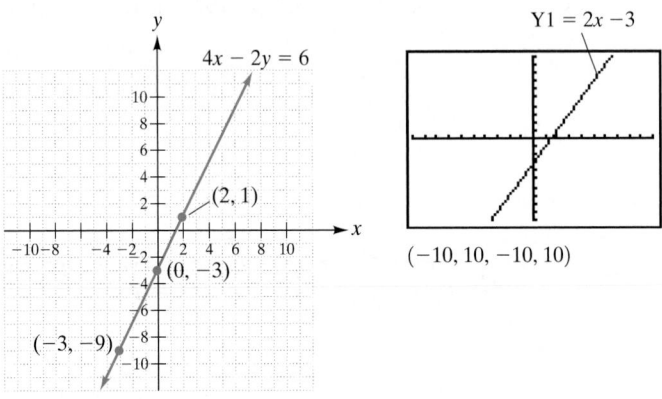

$(-10, 10, -10, 10)$

EXAMPLE 5   Graph $p(x) = \frac{2}{3}x + 1$.

Solution

$$p(x) = \frac{2}{3}x + 1$$

Set up a table. Choose any number for x. We'll use multiples of 3—that is, $-3, 0,$ and 3—because we are multiplying these numbers by a fraction with a denominator of 3. The result will be an integer.

x	$p(x) = \frac{2}{3}x + 1$	$p(x)$	$(x, p(x))$
-3	$p(x) = \frac{2}{3}(-3) + 1$ $p(x) = -2 + 1$ $p(x) = -1$	-1	$(-3, -1)$
0	$p(x) = \frac{2}{3}(0) + 1$ $p(x) = 1$	1	$(0, 1)$
3	$p(x) = \frac{2}{3}(3) + 1$ $p(x) = 2 + 1$ $p(x) = 3$	3	$(3, 3)$

Therefore, $(-3, -1), (0, 1),$ and $(3, 3)$ are three possible solutions of the linear equation (function).

Graph the ordered pairs and connect them with a straight line. Check your graph on your calculator.

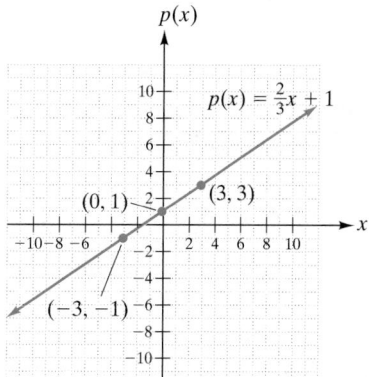

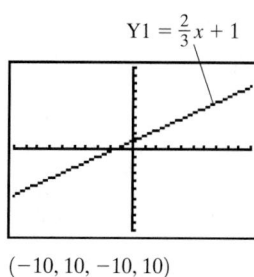

$(-10, 10, -10, 10)$

 5.1.2 Checkup

In exercises 1–6, determine three ordered-pair solutions and graph each equation.

1. $x = 6$

2. $4x - y = 1$

3. $g(x) = \dfrac{3}{5}x + 2$

4. $y = -1.5x - 4$

5. $x + y = 4$

6. $h(x) = -\dfrac{2}{3}x + 5$

7. How many solutions are there for a linear equation in two variables?

5.1.3 Algebraically Determining the *y*-Intercept and *x*-Intercept of a Graph

Special solutions of a linear equation in two variables are the *y*-intercept and *x*-intercept of the graph of the equation. Remember that we discussed these points in Chapter 3. The *y*-intercept is the point where a graph touches or crosses the *y*-axis. The *x*-coordinate of this point is 0. Similarly, the *x*-intercept is the point where a graph touches or crosses the *x*-axis. The *y*-coordinate of this point is 0. For example, for the linear equation $x - y = 3$, the *x*-intercept is $(3, 0)$ and the *y*-intercept is $(0, -3)$. On your calculator, the intercepts are the same.

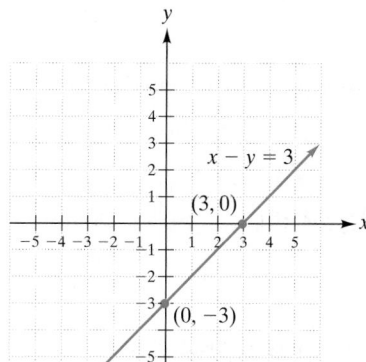

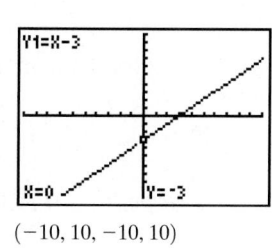

$(-10, 10, -10, 10)$

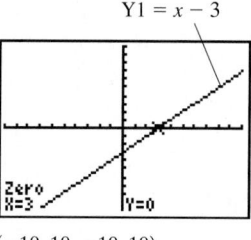

$(-10, 10, -10, 10)$

HELPING HAND It is important to remember that the y-coordinate of the x-intercept is always 0 (because it is on the x-axis). Also, the x-coordinate of the y-intercept is always 0 (because it is on the y-axis).

> To determine algebraically the y-intercept of a graph from its linear equation,
>
> - Substitute 0 for x.
> - Solve for y.
> - Write an ordered pair.
>
> To determine algebraically the x-intercept of a graph from its linear equation,
>
> - Substitute 0 for y.
> - Solve for x.
> - Write an ordered pair.

For example, given the linear equation $x - y = 3$, find the y-intercept and the x-intercept algebraically.

To determine the y-intercept, we substitute 0 for x and solve for y.

$$x - y = 3$$
$$0 - y = 3$$
$$-y = 3$$
$$\frac{-y}{-1} = \frac{3}{-1}$$
$$y = -3$$

Therefore, the y-coordinate is -3. The y-intercept is $(0, -3)$.

To determine the x-intercept, we substitute 0 for y and solve for x.

$$x - y = 3$$
$$x - 0 = 3$$
$$x = 3$$

Therefore, the x-coordinate is 3. The x-intercept is $(3, 0)$.

There is another way to determine the graph's y-intercept from its linear equation. To see what this method is, complete the following set of exercises on your calculator.

Discovery 1

y-intercepts

Graph each linear equation and label the y-intercept.

1. $y = x - 5$ **2.** $y = x + 5$

3. $y = x - 10$ **4.** $y = x + 10$

Write a rule to determine the y-coordinate of the y-intercept of a graph from its linear equation.

The y-coordinate of the y-intercept of the graph is the same as the constant term in the equation when the equation is solved for y.

To determine algebraically the y-intercept of a graph from its linear equation,

- Solve the equation for y.
- The constant term is the y-coordinate of the y-intercept.
- Write an ordered pair. Remember, the x-coordinate is 0.

For example, determine the y-intercept of the graph of the linear equation $x - y = 3$.

Solve for y and obtain $y = x - 3$. The constant term is -3, so the y-coordinate of the y-intercept is -3. The y-intercept is $(0, -3)$.

EXAMPLE 6 Determine algebraically the x-intercept and y-intercept of the graph of the linear equation $2x + y = 5$.

Solution

To determine the x-coordinate of the x-intercept, substitute 0 for y in the equation and solve for x.

$$2x + y = 5$$
$$2x + 0 = 5$$
$$2x = 5$$
$$\frac{2x}{2} = \frac{5}{2}$$
$$x = \frac{5}{2}$$

The x-intercept is $\left(\frac{5}{2}, 0\right)$.

To determine the y-coordinate of the y-intercept, substitute 0 for x in the equation and solve for y.

$$2x + y = 5$$
$$2(0) + y = 5$$
$$y = 5$$

The y-intercept is $(0, 5)$.

The alternative way to determine the y-intercept is to solve the equation for y.

$$2x + y = 5$$
$$2x + y - 2x = 5 - 2x$$
$$y = 5 - 2x$$
$$y = -2x + 5$$

The y-coordinate of the y-intercept is 5. The y-intercept is $(0, 5)$.

✓ 5.1.3 Checkup

Determine algebraically the x-intercept and y-intercept of the graph of each equation.

1. $6x + y = 12$ **2.** $y = 5x + 11$

In exercises 3–4, solve for y and state the y-intercept of each equation.

3. $7x - y = 15$ **4.** $4x + 3y = 24$

5. Describe how you would determine the x-intercept and y-intercept of a graph of a linear equation in two variables from the graph. Then describe how you would determine these two points algebraically.

5.1.4 Linear Graphs with One Intercept

Previously, we defined the standard form of a linear equation in two variables, $ax + by = c$, and stated that both a and b are not equal to 0. Now let us discuss the cases when a, b, or c equals 0. Let's see what some of these graphs look like. First, we will let $c = 0$. Complete the following set of exercises on your calculator.

Discovery 2

Linear Equations in Two Variables, $ax + by = c$, where $c = 0$

Graph the following linear equations in two variables and label the x-intercept and y-intercept:

1. $x + y = 0$ **2.** $-2x + 3y = 0$

Write a rule for determining when the graph of an equation has one point that is both the x-intercept and y-intercept.

Note that the equations are in the standard form $ax + by = c$, with $c = 0$. We see that the graphs of the equations have the same point for the x-intercept and the y-intercept.

The graph of $ax + by = 0$ has one intercept at the origin, $(0, 0)$.

If only $a = 0$, we have a special case of a linear equation in two variables. The equation is of the form $0x + by = c$, or $by = c$. An example of such an equation is $2y = 8$. Set up a table of values for this equation.

First, solve the equation for y.

$$2y = 8 \quad \text{or} \quad 0x + 2y = 8$$
$$2y = 0x + 8$$
$$y = 0x + 4$$

Since the coefficient of the x-term is 0, we will always obtain $y = 4$ for any x-value.

x	$y = 0x + 4$	y
0	$y = 0(0) + 4$	4
1	$y = 0(1) + 4$	4
2	$y = 0(2) + 4$	4

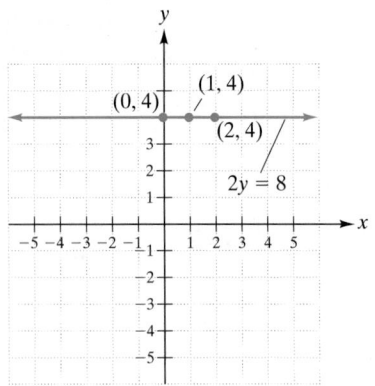

A graph corresponding to the table is shown. There is only one intercept, the y-intercept, $(0, 4)$. The line is horizontal.

The graph of $y = k$ is a horizontal line with a y-intercept of $(0, k)$.

If only $b = 0$, we have a second special case: an equation of the form $ax + 0y = c$, or $ax = c$.
Set up a table of values for $2x = -12$.

First, we cannot solve the equation for y, since the coefficient of y is 0. Therefore, we solve the equation for x.

$$2x = -12 \quad \text{or} \quad 2x + 0y = -12$$
$$2x = 0y - 12$$
$$x = 0y - 6$$

Now, if we substitute values for x other than -6, the result will be a false statement. Therefore, to obtain ordered pairs, we substitute values for y. Since the coefficient of y is 0, we will always obtain $x = -6$ for any y-value.

x	$x = 0y - 6$	y
-6	$x = 0(0) - 6$	0
-6	$x = 0(1) - 6$	1
-6	$x = 0(2) - 6$	2

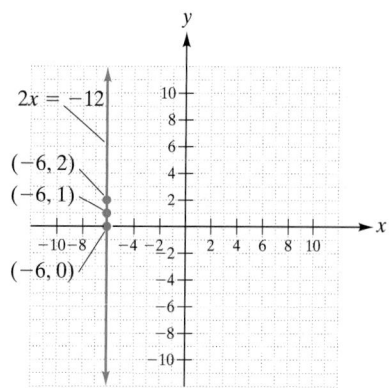

A graph corresponding to the table is shown. There is only one intercept, the x-intercept, $(-6, 0)$. The line is vertical.

This relation is not a function. In fact, a vertical line is the only case in which a linear equation in two variables is not a function. These equations cannot be graphed on a calculator by using the $\boxed{\text{Y=}}$ menu.

The graph of $x = h$ is a vertical line with an x-intercept of $(h, 0)$.

EXAMPLE 7   Determine the intercept of the graph of each linear equation in two variables. Then graph the equation.

a. $y - 3 = 0$ **b.** $x = y$ **c.** $2x - 6 = 0$

Solution

a. The equation does not have an x-variable. The coefficient of the x-term is 0. Solve for y.

$$y - 3 = 0$$
$$y = 3$$

The graph is a horizontal line with a y-intercept of $(0, 3)$.

b. In standard form, the equation is $x - y = 0$. Since $c = 0$, the x-intercept and the y-intercept are $(0, 0)$. To complete the graph, determine two additional solutions.

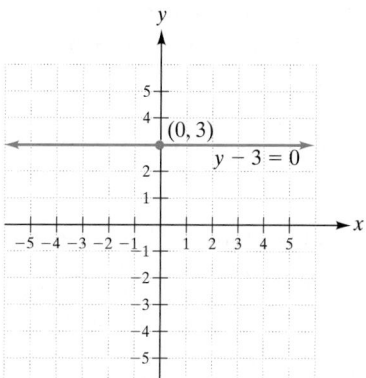

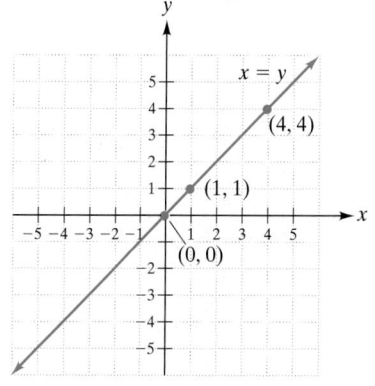

c. The equation does not have a y-variable. The coefficient of the y-term is 0. Solve for x.

$$2x - 6 = 0$$
$$x = 3$$

The graph is a vertical line with x-intercept of $(3, 0)$.

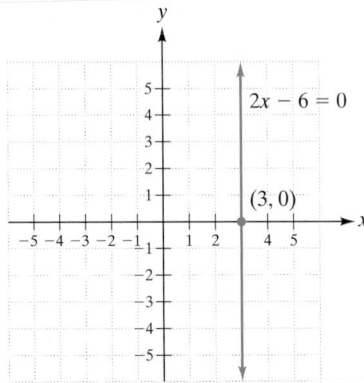

5.1.4 Checkup

In exercises 1–3, determine the intercepts of the graph of each linear equation. Then graph the equation as a check.

1. $\dfrac{1}{2}x = 8$ **2.** $2x + 4y = 0$ **3.** $5y = 3$

4. When will the graph of a linear equation have only an x-intercept?

5. When will the graph of a linear equation have only a y-intercept?

6. When will the x-intercept and the y-intercept of the graph of a linear equation be at the same point? What will the coordinates of the point be?

5.1.5 Graphing Linear Equations in Two Variables by Using the Intercept Method

Now we are ready to graph a linear equation in two variables by using the intercept method. We can use this method only if the graph has two distinct points as the x-intercept and y-intercept. In other words, the equation written in standard form must not have a, b, or c equal to 0.

The intercept method of graphing involves determining the two intercepts and connecting them with a straight line.

To graph a linear equation in two variables by using the intercept method,

- Determine the x-intercept.
- Determine the y-intercept.
- Plot the intercepts and label their coordinates.
- Connect the intercepts with a straight line.
- Check the graph by locating a third point on it, and determine that the coordinates of that point are a solution of the equation.

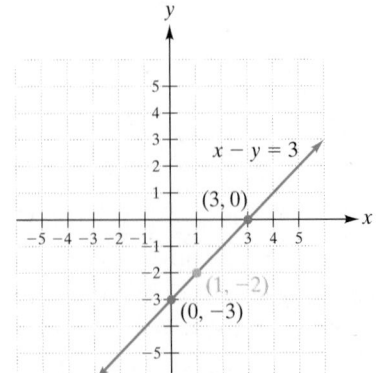

For example, graph the linear equation $x - y = 3$, using the intercept method.

Graph the two intercepts previously found for the equation, $(3, 0)$ and $(0, -3)$, and connect them with a straight line.

To check the graph, determine a point on it, and check whether that point is a solution of the equation. Let's check the point $(1, -2)$.

$$\begin{array}{c|c} x - y = 3 \\ \hline 1 - (-2) & 3 \\ 1 + 2 \\ 3 \end{array}$$

Since $3 = 3$, $(1, -2)$ is a solution. The graph checks.

EXAMPLE 8   Graph the linear equation $3x + 4y = 12$, using the intercept method.

Solution

Determine the y-intercept.
Solve for y.

$$3x + 4y = 12$$
$$3x + 4y - 3x = 12 - 3x$$
$$4y = 12 - 3x$$
$$\frac{4y}{4} = \frac{12 - 3x}{4}$$
$$y = \frac{12}{4} - \frac{3x}{4}$$
$$y = -\frac{3}{4}x + 3$$

An alternative method is to substitute
0 for x and solve for y.

$$3(0) + 4y = 12$$
$$4y = 12$$
$$\frac{4y}{4} = \frac{12}{4}$$
$$y = 3$$

The y-coordinate is the constant 3. The x-coordinate is 0. The y-intercept is $(0, 3)$.

Determine the x-intercept.
Substitute 0 for y and solve for x.

$$3x + 4(0) = 12$$
$$3x = 12$$
$$\frac{3x}{3} = \frac{12}{3}$$
$$x = 4$$

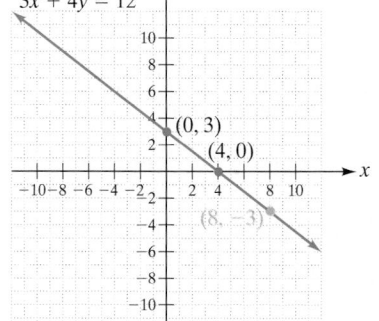

The x-intercept is $(4, 0)$.

Graph the two intercepts and connect the points with a straight line.

Check the graph by checking a point on the line. One of the integer ordered pairs located on the graph is $(8, -3)$.

$$\begin{array}{c|c} 3x + 4y = 12 \\ \hline 3(8) + 4(-3) & 12 \\ 24 + (-12) \\ 12 \end{array}$$

Since $12 = 12$, $(8, -3)$ is a solution. The graph checks.

HELPING HAND When the intercepts are not integer pairs, it is difficult to graph them accurately. It is better to use some other method.

5.1.5 Checkup

In exercises 1–2, graph each equation, using the intercept method. Check your graphs by using a third point. Label the intercepts and third point on each graph.

1. $x + y = 4$ **2.** $2x - 5y = 10$

3. The intercept method of graphing a linear equation works only when none of the coefficients a, b, and c are equal to zero. Explain why this method does not work if a, b, or c equals zero.

4. Given $ax + by = c$, if $a = 0$ and $c = 0$, what is unusual about the graph of the equation?

5. Given $ax + by = c$, if $b = 0$ and $c = 0$, what is unusual about the graph of the equation?

5.1.6 Modeling the Real World

Graphs of linear equations in two variables are highly useful for representing real-world data, because you need only two ordered pairs of data to determine the straight-line graph of such an equation. Then you can use the graph to determine additional data that satisfy the original equation. So you go easily from knowing two solutions of the equation to knowing as many solutions as you want.

An important point to keep in mind is that real-world data often have practical limitations. A linear equation may be accurate for a period of time, but then the situation may change. This is why the domain and range of a function are so important—they tell you when the relationship is valid and when it is not.

EXAMPLE 9   Mike began his new job as supervisor on a lamp production line. The week before he began, the crew produced 25 lamps. During his first week, 30 lamps were produced. During the second week of Mike's supervision, 35 lamps were produced.

a. Let x = the number of weeks Mike supervised

   y = the number of lamps produced during the week

   Write three ordered pairs from the information given.

b. Assuming that the same scenario continues, graph a linear representation of lamp production.

c. Using the graph, predict what production will be during week 5.

d. Using the graph, predict what production will be during the 15th week. Is this reasonable to expect? Explain.

Solution

a. $(0, 25), (1, 30), (2, 35)$

b.

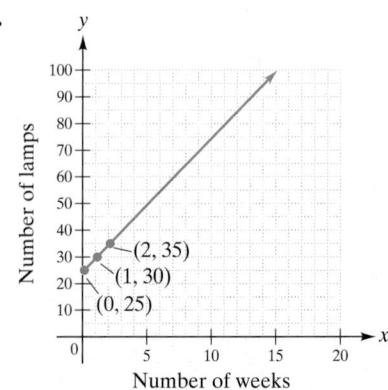

c. The point (5, 50) is on the graph. Therefore, 50 lamps will be produced during week 5.

d. The point (15, 100) is on the graph. Therefore, 100 lamps will be produced during the 15th week. This is unlikely, though, unless additional people are hired or new equipment is added. The prediction capabilities are limited in this scenario.

Using the intercept method to graph an equation is often simpler than using other methods, because you can find the intercepts by just substituting 0 for the variables in the equation. When the equation is based on a real-world relationship, you may get an extra bonus because the intercepts may be significant points themselves. For example, if your graph shows a relationship between distance d, and time t, then the d-intercept occurs where time (the t-coordinate) is 0—that is, the starting distance. The t-intercept (if there is one) occurs where distance (the d-coordinate) is 0. It is usually worthwhile to find the intercepts of a real-world graph and see what information they can give you.

EXAMPLE 10   Myletta plans to sell her clay flowerpots at a craft fair. A booth costs $50.00 to rent, and she estimates that each pot costs $3.00 to make. If Myletta sells the pots for $10.00, write a linear function for the profit she will obtain. Graph the function by the intercept method. Explain the meaning of the intercepts.

Solution

Let x = the number of flowerpots made and sold
$P(x)$ = the profit from the sale of the flowerpots

The profit is equal to the revenue, $10.00x$, minus the cost, $3.00x + 50.00$.

$$P(x) = 10.00x - (3.00x + 50.00)$$
$$P(x) = 10.00x - 3.00x - 50.00$$
$$P(x) = 7.00x - 50.00$$

The $P(x)$-intercept is $(0, -50.00)$.

To determine the x-intercept, let $P(x) = 0$ and solve for x.

$$P(x) = 7.00x - 50.00$$
$$0 = 7.00x - 50.00$$
$$0 - 7.00x = 7.00x - 7.00x - 50.00$$
$$-7.00x = -50.00$$
$$\frac{-7.00x}{-7.00} = \frac{-50.00}{-7.00}$$
$$x = \frac{50}{7}$$
$$x \approx 7.14$$

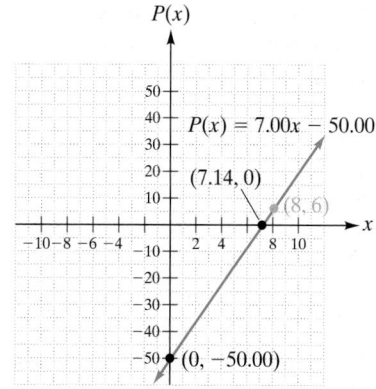

$P(x)$

$P(x) = 7.00x - 50.00$

$(7.14, 0)$

$(8, 6)$

$(0, -50.00)$

The x-intercept is about $(7.14, 0)$.

Check the graph by checking a point on the line. One integer ordered pair located on the graph is the point $(8, 6)$.

$$P(x) = 7.00x - 50.00$$

$$
\begin{array}{c|l}
6 & 7.00(8) - 50.00 \\
 & 56.00 - 50.00 \\
 & \quad\quad 6.00
\end{array}
$$

The coordinate pair $(8, 6)$ is a solution because $6 = 6.00$. The graph checks.

The $P(x)$-intercept is $(0, -50.00)$. That is, when 0 flowerpots are made and sold, the profit is $-\$50.00$; that is, there is a loss of $50.00. The x-intercept is about $(7.14, 0)$. When 7.14 pots are made and sold, the profit is $0.00. (Myletta will break even.) However, it is not possible to make and sell 7.14 pots, so she must make and sell 8 flowerpots in order to break even (or, in this case, realize a small profit).

APPLICATION

In 1974, an SR-71 Blackbird jet plane was flown from New York City to London in 1 hour and 55 minutes, the fastest transatlantic flight ever. If we let d represent the plane's distance from London, then an equation representing that distance is $d = 3462 - 1807t$, where t is the flight time in hours. Graph the equation and interpret the intercepts of the graph.

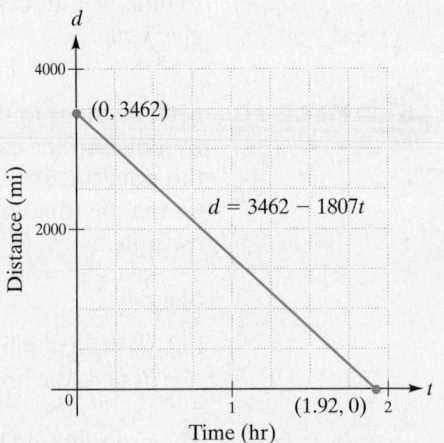

Discussion

Let t = time (in hours) in flight
d = plane's distance (in miles) from London
The equation $d = 3462 - 1807t$ is solved for d. The constant term 3462 is the d-coordinate of the d-intercept. The d-intercept is $(0, 3462)$.

We determine the t-intercept by substituting 0 for d and solving for t.

$$0 = 3462 - 1807t$$

$$1807t = 3462$$

$$\frac{1807t}{1807} = \frac{3462}{1807}$$

$$t \approx 1.92$$

The t-intercept is approximately $(1.92, 0)$.

Graph the two intercepts and connect the points with a straight line.

The coordinates of the d-intercept, $(0, 3462)$, mean that at a time of 0 hours (just before the plane begins its flight), the plane is 3462 miles from London. This is the distance from New York City to London.

The coordinates of the t-intercept, approximately $(1.92, 0)$, mean that at a time of approximately 1.92 hours, the plane will be 0 miles from London. In other words, the plane will arrive in London in approximately 1.92 hours.

✔ 5.1.6 Checkup

1. Dan's Delivery Service charges $3 to deliver a package weighing 1 pound. The charge for a package weighing 3 pounds is $7, while the charge for delivering a package weighing 5 pounds is $11. Let x represent the weight of the package and y represent the delivery charge.

   a. Write three ordered pairs from the information given.

   b. Assuming that a linear relationship exists, graph the relation given by the information.

   c. Using the graph, predict what the cost would be for a package weighing 8 pounds.

   d. What would you say are limitations on the domain of the relation? Do you think Dan would put a limit on the weight of packages he delivers, or would you expect the relationship you graphed to continue for all possible weights of packages? Would the range of the relation have any limits? Explain.

2. In 1974, an SR-71 Blackbird jet plane was flown from London to Los Angeles in 3 hours and 48 minutes. If we let *d* represent the plane's distance from Los Angeles, an equation for that distance is $d = 5645 - 1486t$, where *t* is the flight time in hours. Graph the equation and interpret the intercepts of the graph. From the graph, estimate how long it will take the aircraft to fly 3500 miles (the approximate distance from London to New York City).

3. Don conducts an abstract algebra preparatory class for students who are returning to graduate school. The setup cost for the class is $700 to rent space. Don spends $35 per student for materials for the class. He charges the students $175 to take the class. Write a linear function for the profit that Don will realize. Graph the function by the intercept method. Explain the meaning of the intercepts.

5.1 Exercises

Identify each equation as linear or nonlinear. Express each linear equation in standard form.

1. $5x + 7y = 35$ **2.** $\sqrt{3}x + 2y = 6$ **3.** $-4\sqrt{x} + y = 8$ **4.** $6x^2 + 2y = 12$
5. $x^2 + y^2 = 1$ **6.** $5x - \sqrt{y} = -10$ **7.** $2x - 5 = 0$ **8.** $7y = 2y + 14$

Determine three ordered-pair solutions of each linear equation, and graph the equation.

9. $x - 6y = 12$ **10.** $x + 7y = 21$ **11.** $p(x) = \dfrac{4x + 1}{3}$ **12.** $q(x) = \dfrac{2x - 3}{5}$

13. $y = 8$ **14.** $4y - 16 = 0$ **15.** $r(x) = -\dfrac{3}{4}x + 4$ **16.** $s(x) = \dfrac{5}{6}x - 3$

17. $y = 2.8x - 1.6$ **18.** $y = 4.7 - 1.9x$ **19.** $y = \dfrac{3x - 5}{2}$ **20.** $y = \dfrac{3x - 2}{5}$

21. $3x + y - 4 = x + 2y - 3$ **22.** $7x - 2y + 8 = 5x - 3y + 1$ **23.** $5y = -20$ **24.** $3y + 7 = 25$

Determine the intercepts of each graph.

25.

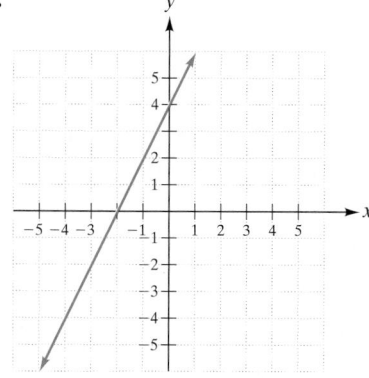

26.

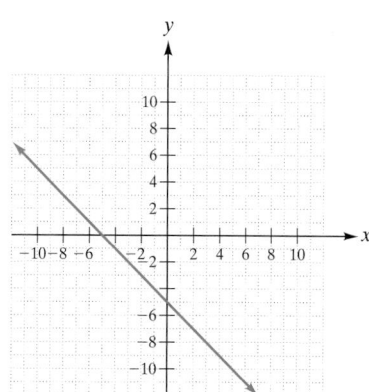

27.

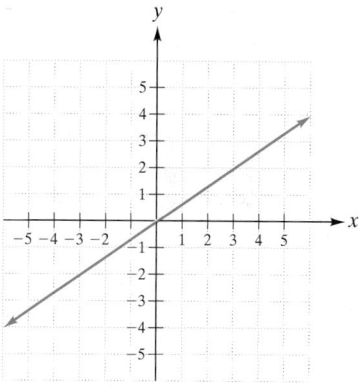

28.

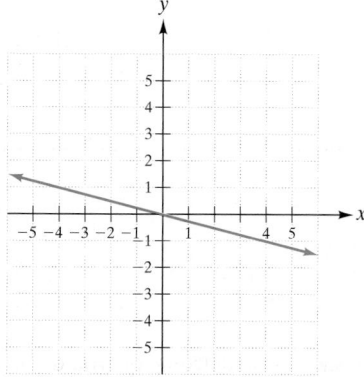

29.

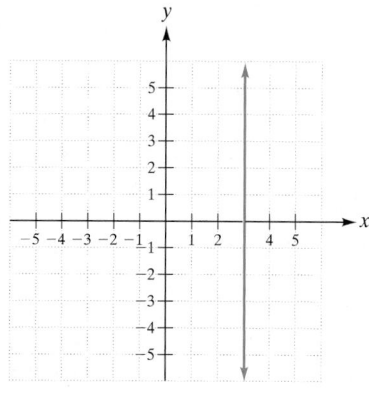

30.

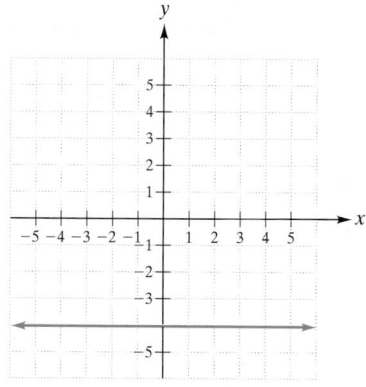

Determine algebraically the intercepts of the graph of each linear equation.

31. $3x + 5y = 12$ **32.** $7x + 9y = 63$ **33.** $4x - 7y = 14$ **34.** $x - y = 31$ **35.** $-2x - 9y = 27$

36. $-12x - 8y = 36$ **37.** $6x + 9y - 36 = 0$ **38.** $5x + 3y + 45 = 0$ **39.** $3x + 7y = 0$ **40.** $16x + 3y = 0$

41. $6x - 8 = 2x + 32$ **42.** $3x - 12 = x - 2$ **43.** $y = 3y - 22$ **44.** $6y = y - 25$

Solve each linear equation for y to determine the y-intercept.

45. $12x - y = 24$ **46.** $x + 9y = 18$ **47.** $y = 5(x - 3)$ **48.** $y = 3(x - 7) + 5$ **49.** $5x - 15y = 0$

50. $14x = 42y$ **51.** $3y = 12y + 18$ **52.** $18 - 5y = y$ **53.** $y + 5 = 5$ **54.** $3(y + 7) - 4 = 17$

55. $-17.6x + 2.2y = 19.8$ **56.** $12.6x - 6.3y = 50.4$ **57.** $x = 12y$ **58.** $100y = -5x$

In exercises 59–68, graph each equation, using the intercept method. Check your graphs by using a third point. Label the intercepts and third point on each graph.

59. $3x + 5y = 30$ **60.** $x + 3y = 33$ **61.** $4x - 3y = 24$ **62.** $5x - 7y = 70$ **63.** $x - y = 9$

64. $-x + y = 9$ **65.** $-x - y = 9$ **66.** $x + y = 9$ **67.** $2x - 7y = -14$ **68.** $8x - 3y = 24$

69. An assembly line is used to pack boxes of candy. When only one person is available, the assembly line cannot operate. When two people are working, they can pack 5 boxes per minute. When four people are working, they can pack 15 boxes per minute.
  a. Let x be the number of persons working the assembly line, and let y be the number of boxes of candy packed per minute. Write three ordered pairs from the information given.
  b. Assuming that the relation continues for other numbers of workers, graph the information.
  c. Use the graph to predict how many boxes per minute would be packed by a crew of five people.
  d. Are there any limitations on the domain and range of this relation? Is it reasonable to assume that as the number of workers increases, the number of boxes packed per minute could still be estimated by the graph? Explain your answer.

70. On an examination, Alex missed none of the questions and received a score of 100. Beth missed 6 questions and scored an 85. Chiyo missed 10 questions and scored 75.
  a. Let x be the number of questions missed, and let y be the score on the test. Write three ordered pairs from the information given.
  b. Assuming a linear relation, graph the information.
  c. Use the graph to predict what score students will receive if they miss 12 questions on the test.
  d. For what domain would it make sense to use this relation?

71. Itsu is measuring the borders around equilateral triangles for a science project. He finds that a triangle with a side of 2 inches has a border of 6 inches, one with a side of 3.5 inches has a border of 10.5 inches, and one with a side of 10.5 inches has a border of 31.5 inches.

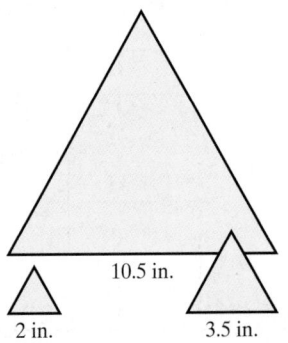

10.5 in.

2 in.   3.5 in.

  a. Let x be the length of a side of the equilateral triangle and y be the length of the border around the triangle. Write three ordered pairs from the information given.
  b. Graph a linear representation of the information.
  c. Predict what the border would be for an equilateral triangle with a side of 4 inches.
  d. Is it reasonable to assume that the relation represented by the graph would work for any equilateral triangle? Explain your answer.

72. Carla measured the perimeters of several rectangles, all of which had the same width, but differing lengths. The perimeter of a rectangle with a length of 15 cm was 50 cm. Another rectangle with a length of 25 cm had a perimeter of 70 cm. A third with a length of 10 cm had a perimeter of 40 cm.

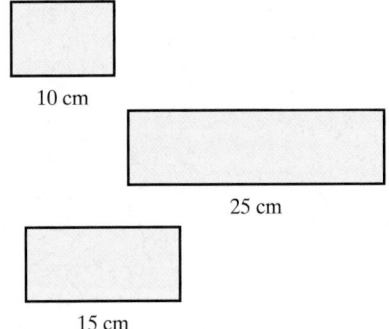

10 cm

25 cm

15 cm

  a. Let x be the length of the rectangle and y be the perimeter. Write three ordered pairs from the information given.
  b. Graph a linear representation of the information.
  c. Using the graph, what would the perimeter be for a rectangle with the same width and with a length of 20 cm?
  d. Would this relation hold for all rectangles that had the same width and varying lengths? Explain your answer.

73. Toasty Toasters determines that the fixed cost of manufacturing toasters is $250.00 per day and the variable cost of manufacturing is $3.75 per toaster. If $D(x)$ is the daily cost of manufacturing x toasters, then $D(x) = 3.75x + 250$.
  a. What would be the cost of manufacturing 20 toasters per day?
  b. What would be the cost of manufacturing 30 toasters per day?
  c. Is it true that the cost of manufacturing 25 toasters per day would be less than $350.00?

d. Construct a table of values showing the cost of manufacturing x toasters per day, where x begins at 0 and increases up to 50 in increments of 5 toasters per day.

74. Appalachian Crafts agrees to pay Anne $300 per month to provide pottery exclusively to the company's shop for sale and further agrees to pay her $25 for each item sold. If $P(x)$ represents the amount paid to Anne for a month and x represents the number of pieces of art sold, then $P(x) = 25x + 300$.

    a. How much will Anne earn if the shop sells 12 items in a month?

    b. What will Anne earn in a month when 5 items sell?

    c. If the shop sells 8 items, will Anne earn $500?

    d. Construct a table of values to show Anne's monthly earnings for selling x items in the shop, where x begins at 0 and increases up to 20 in increments of 2 items.

75. The fixed daily cost of operating a water park is $1000, and the cost to provide services is $4.50 per customer. The water park charges $18.75 per person for admission. Write a linear function for the profit that the water park will realize. Graph the function by the intercept method. Explain the meaning of the intercepts.

76. The fixed daily cost of operating a bowling alley is $200, and the cost of operating the lanes is $0.65 per game bowled. The bowling alley charges $2.00 per game bowled. Write a linear function for the profit that the bowling alley will obtain. Graph the function by the intercept method. Explain the meaning of the intercepts.

77. A Boeing 767 jet aircraft flies from Atlanta to Los Angeles in 4 hours and 34 minutes. If d represents the plane's distance from Los Angeles, an equation representing that distance is $d = 1944 - 425.7t$, where t is the flight time in hours. Graph the equation and interpret the intercepts of the graph. From the graph, approximate how long it will take to fly 1000 miles from Atlanta.

78. A Boeing 747 jet aircraft flies from New York City to London, England, in 8 hours and 55 minutes. If d represents the plane's distance from London, an equation representing that distance is $d = 3471 - 389.3t$, where t is the flight time in hours. Graph the equation and interpret the intercepts of the graph. From the graph, estimate how long it will take to fly 2000 miles from New York City.

 ## 5.1 Calculator Exercises

If you purchase applications for the TI-83 Plus from Texas Instruments, you can obtain an application that enables you to graph vertical lines easily on the calculator. Once you obtain the applications, check the procedure that follows.

To use the applications, press the APPS key on the calculator. Choose option 5 from the menu. A logo will appear. Press any key to continue. The calculator will display the menu for the Y= key. To change to the X = format, move the cursor to the top of the display, and highlight $x =$ in the first location. When you press ENTER , the calculator will change the displays to X = displays. The following screens are displayed:

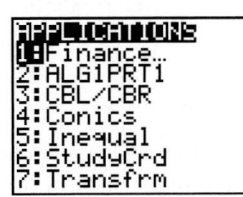

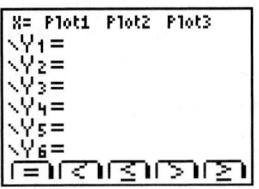

To graph the line for the equation $2x - 3 = x + 5$, first solve the equation for x. You will get $x = 8$. Then store this equation in the first location. Next, proceed to the graph as you normally would by using the $\boxed{\text{ZOOM}}$ key or setting the window and using the $\boxed{\text{GRAPH}}$ key. $\boxed{\text{ZOOM}}$ $\boxed{6}$ yields the following graph:

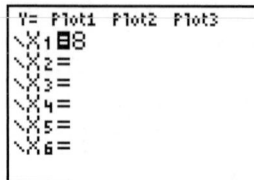 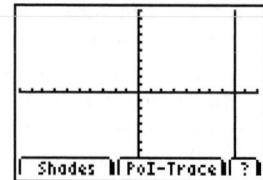

Ignore the choices that are displayed at the bottom of the screen. They cannot be used for our current work. However, they will be useful to us later on, when we want to study the behavior of several lines at the same time.

To return to the previous settings for the calculator, press $\boxed{\text{APPS}}$, choose option 5 again, and then choose option 2 to quit the application. You should get into the habit of doing this when you are through using the application.

Try using this graphing procedure on the following equations:

1. $5(x - 1) = 2x - 4$ 2. $\dfrac{3}{4}x + 1 = 7$

3. $4.5x + 1.2 = 2.3x - 5.4$

5.1 Writing Exercises

1. The standard form for a linear equation in two variables is

$$ax + by = c$$

where a, b, and c are real numbers and a and b are not both equal to 0. Give examples in which a or b are equal to 0. Then explain why the equation cannot have a and b both equal to 0.

2. It is important that you fully understand the relationship between the solutions of a linear equation in two variables and the graph of the linear equation. Explain what a graph represents with respect to a linear equation in two variables. Explain what a graph can tell you about a linear equation in two variables. Exactly what is the purpose of graphing a linear equation in two variables?

5.2 Graphing by Using the Slope–Intercept Method

OBJECTIVES

1 Determine the slope of a line from its graph.

2 Determine the slope of a line, given two points on the line.

3 Determine the slope of a line from its linear equation in two variables.

4 Graph linear equations in two variables by using the slope–intercept method.

5 Determine the average rate of change, given real-world data. Model real-world situations by using graphs.

APPLICATION

Linear depreciation diminishes the value of an asset by a fixed amount each period until the net value is zero. This is the simplest depreciation calculation to make for tax purposes.

If you purchase a computer for $2000 to use in your business and estimate its useful life to be five years, determine the amount of depreciation per year (the slope of a line).

Given x years of use and the value of the computer in y dollars, determine the y-intercept and graph the equation.

After completing this section, we will discuss this application further. See page 335.

5.2.1 Determining the Slope of a Line from Its Graph

The graphs of linear equations may appear to rise or fall from left to right. Or they may be horizontal or vertical.

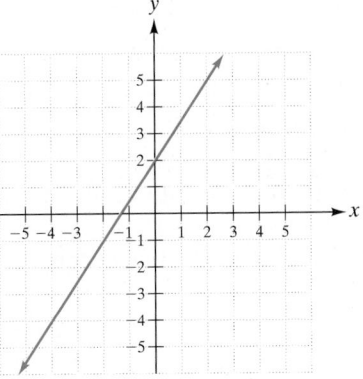

rises (Linear function is increasing.)

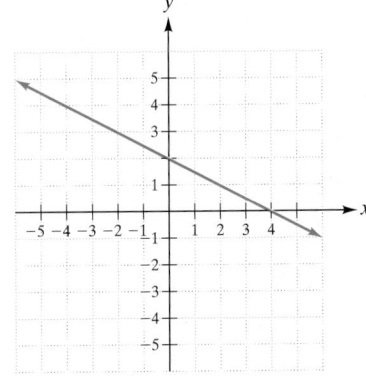

falls (Linear function is decreasing.)

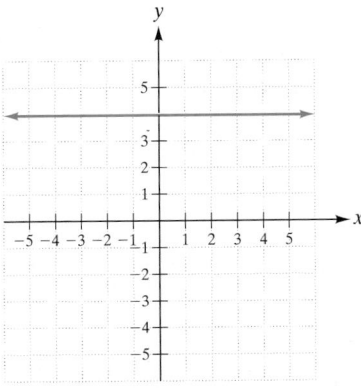

horizontal (Linear function is constant.)

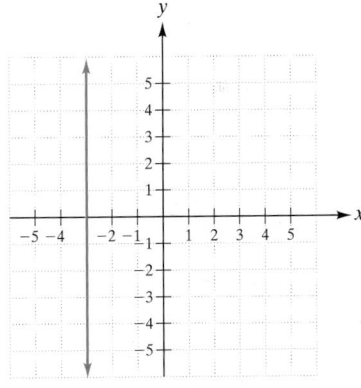

falls (This graph does not represent a function.)

Of two graphs that rise or fall, one may be more steep or less steep than the other.

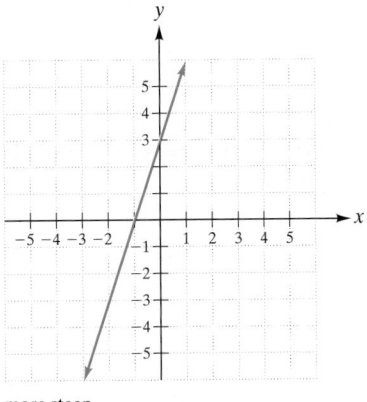

more steep

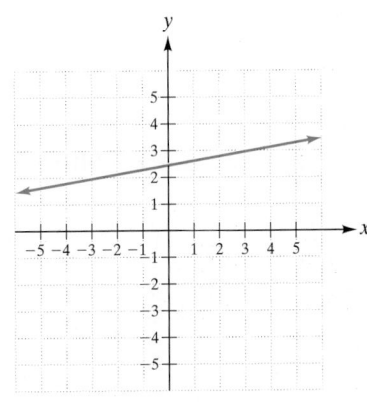

less steep

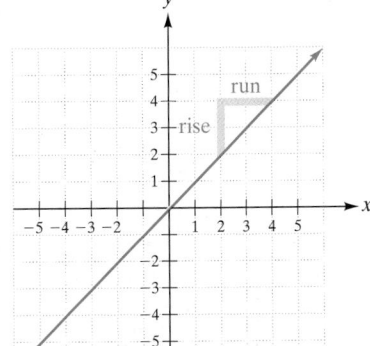

We describe the graph of a line by its steepness, which we call the slope of the line. We calculate this slope by determining the change in vertical distance of the line (called the **rise**) that corresponds to a change in horizontal distance of the line (called the **run**). The **slope** of a line is defined as the ratio of the amount of rise to the amount of run. (Students quickly learn this definition as "rise over run.")

SLOPE OF A LINE
The slope of a line is the ratio of the amount of rise to the amount of run.

$$\frac{\text{amount of rise}}{\text{amount of run}}$$

To determine the slope of a line graphically,

- Locate two points on the line whose coordinates are integers.
- Draw two legs needed to complete a right triangle with the given ordered pairs as vertices.
- Determine the length of the two legs drawn. (One leg is the rise and the other is the run.)
- Write a ratio of the rise to run, $\frac{\text{rise}}{\text{run}}$.

For example, determine graphically the slope of the line shown below.

Two possible integer ordered pairs are located, the triangle legs drawn, and the leg lengths determined. The slope is $\frac{\text{rise}}{\text{run}} = \frac{2}{5}$.

Note: The right triangle could have been drawn differently, but would result in the same slope. Also, the slope of the line remains the same for any set of ordered pairs located on the line, including noninteger ones. This is an important characteristic of a straight line.

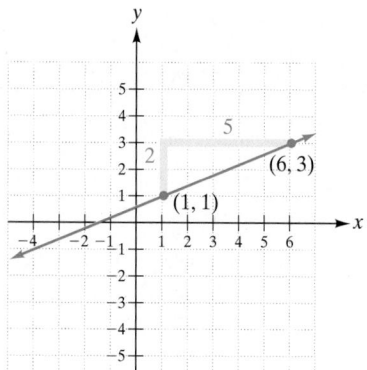
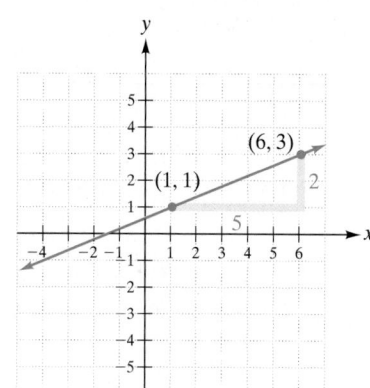
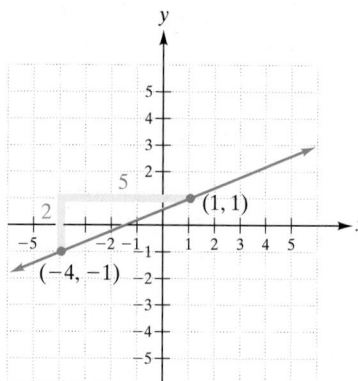

EXAMPLE 1 Determine graphically the slope of each line.

a.

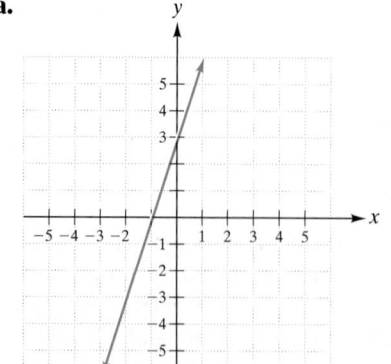

b.

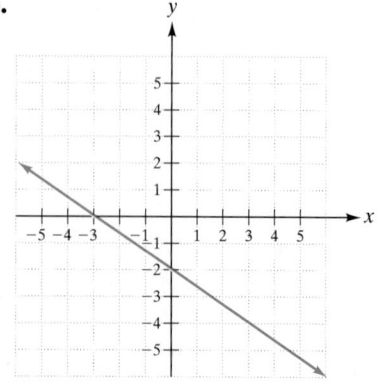

c.

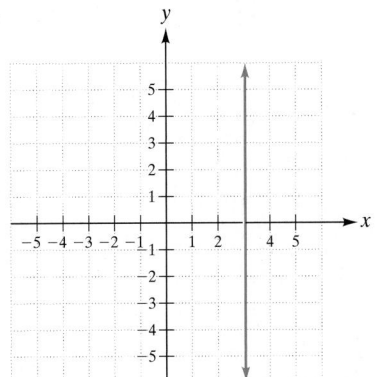

d.

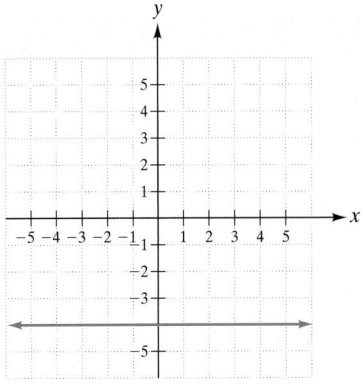

Solution

a.

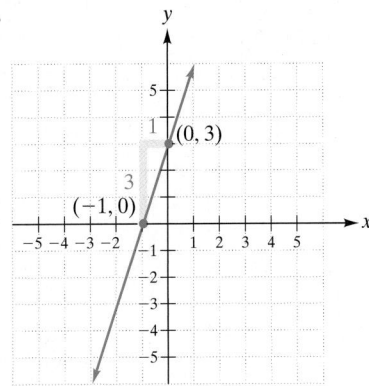

b.

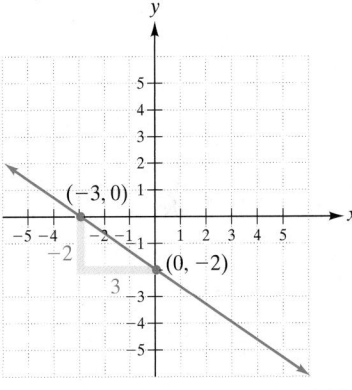

The slope is $\frac{3}{1} = 3$.

The slope is $\frac{-2}{3}$.

No right triangles can be drawn for Examples 1c and 1d.

c. The slope is $\frac{\text{any number}}{0}$; that is, the slope is undefined, because division by 0 is undefined.

d. The slope is $\frac{0}{\text{any number}}$, or 0.

Example 1 shows us that different lines can have slopes that are positive, negative, 0, or undefined. Let's take a closer look at these different slopes. Complete the following set of exercises to discover different types of slopes.

Discovery 3

Types of Slopes

Determine the slopes of the following graphs:

1.

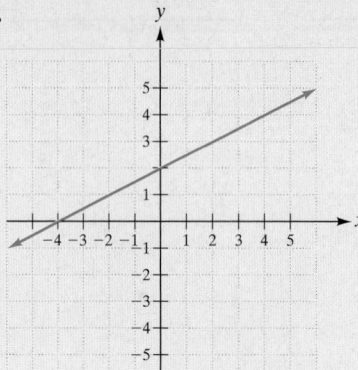

2.

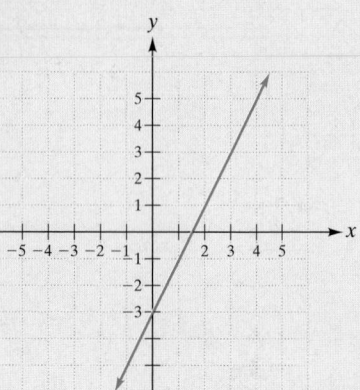

3.

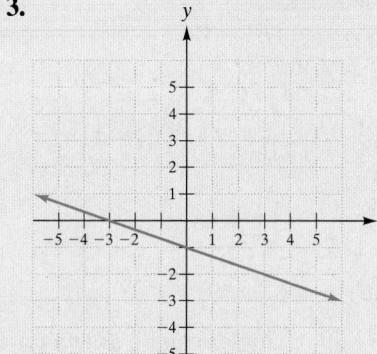

4.

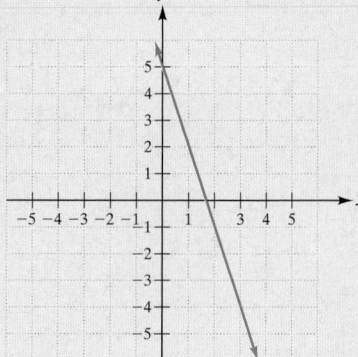

5.

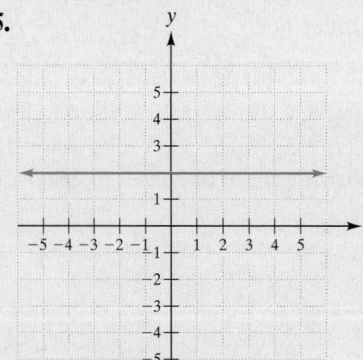

6.

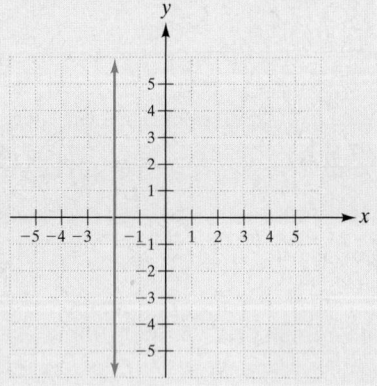

Choose the correct answer.

7. In exercises 1 and 2, the linear function is increasing. The slopes have a *positive/negative* value. Viewing the graphs from left to right, the graphs both *rise/fall*.

8. In exercises 3 and 4, the linear function is decreasing. The slopes have a *positive/negative* value. Viewing the graphs from left to right, the graphs both *rise/fall*.

9. In exercise 5, the linear function is constant. The slope is *0/undefined*. The graph is a *vertical/horizontal* line.

10. In exercise 6, the graph does not represent a function. The slope is *0/undefined*. The graph is a *vertical/horizontal* line.

11. In observing the absolute value of the slope, we see that the larger the absolute value, the *more/less* steep is the graph.

A graph with a positive slope rises from left to right. The function it represents is increasing. A graph with a negative slope falls from left to right. The function it represents is decreasing. A horizontal line has a slope of 0. The function is constant. A vertical line has an undefined slope. It does not represent a function. The larger the absolute value of the slope, the steeper is the graph of the line.

EXAMPLE 2   Determine the slope of each of the graphs that follow. In each case, determine whether the graph represents a function. If the graph represents a function, is the function increasing, decreasing, or constant?

a.

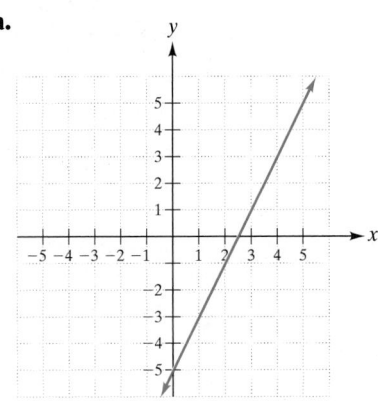

b.

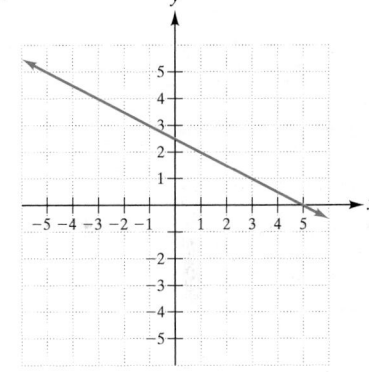

c.

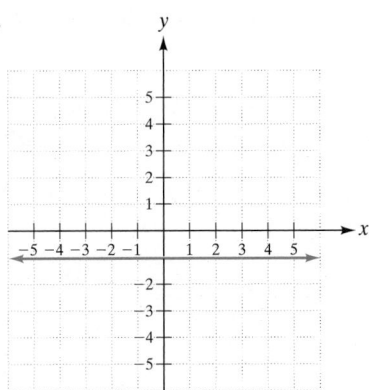

d.

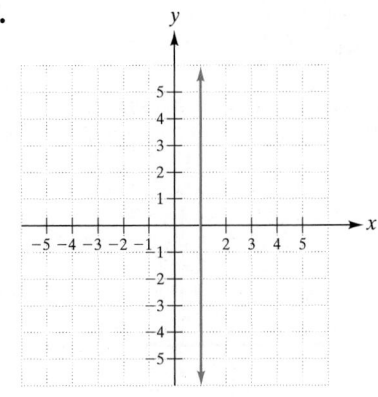

Solution

a. According to the labeled graph, the slope is $\frac{2}{1}$, or 2. The graph represents a function. (It passes the vertical-line test.) The function is increasing (the graph is rising from left to right).

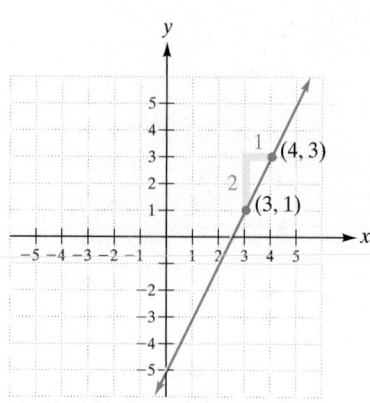

b. According to the labeled graph, the slope is $-\frac{1}{2}$. The graph represents a function. The function is decreasing (the graph is falling from left to right).

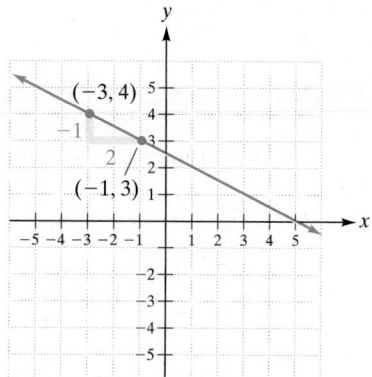

c. The slope of a horizontal line is 0. The graph represents a function. The function is constant.

d. The slope of a vertical line is undefined. The graph does not represent a function.

 5.2.1 Checkup

In exercises 1–4, determine the slope of the graph. Determine whether the graph represents a function. If the graph represents a function, is the function increasing, decreasing, or constant?

1.

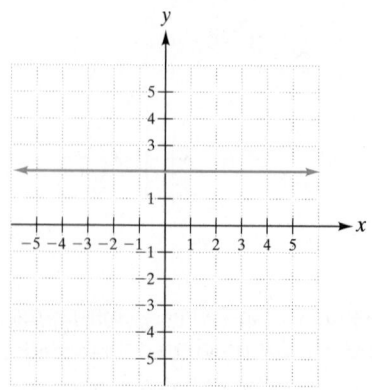

2.

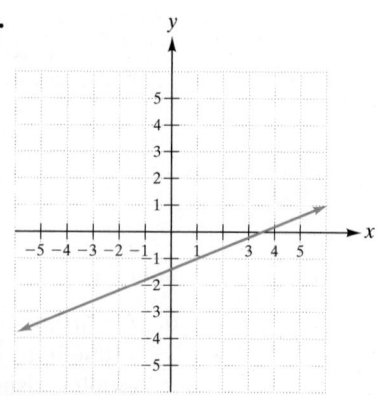

3.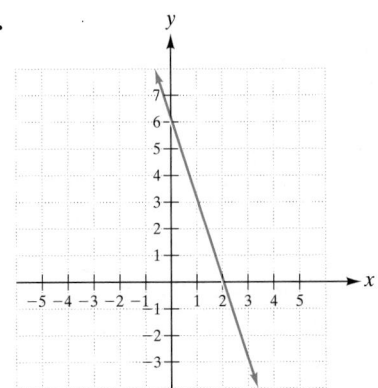

4.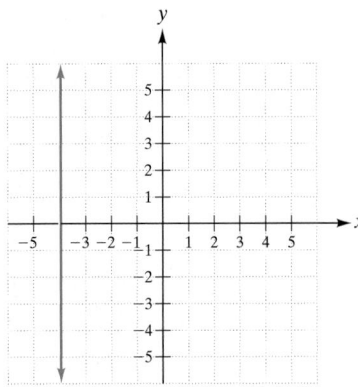

5. If you know the slope of a linear equation, what can you say about its graph?

5.2.2 Determining the Slope of a Line Given Two Points on the Line

The slope of a line is a very important characteristic of the linear equation represented by that line. In fact, there are times when we might want to determine the slope of a line even when the graph is not drawn for us. For example, suppose we know only two points on the line. We could plot the points, draw the graph, and then determine the slope. But sometimes it's easier to find the slope without drawing the graph. To see how to do this, complete the following set of exercises.

 Discovery 4

Slope Formula

1. Locate and label the points $(1, 3)$ and $(5, 4)$ on a graph.
   Draw a line connecting the points.
   Draw the legs of a right triangle needed to determine the slope of the line, and label each length.

2. The rise of the graph is _____.

3. The run of the graph is _____.

4. The difference of the y-coordinates of the ordered pairs is
   $4 - 3 =$ _____.

5. The difference of the x-coordinates of the ordered pairs is
   $5 - 1 =$ _____.

6. The slope of the graph is _____.

Write a rule to determine the slope of a graph from the coordinates of two ordered pairs.

The rise of the graph and the difference of the y-coordinates of the ordered pairs are the same. The run of the graph and the difference of the x-coordinates of the ordered pairs are the same.

Before we write a formula for the slope of a graph, we need to define some notation. The traditional symbol for the slope is m. To label the coordinates of the two ordered pairs, we use subscripts to distinguish the coordinates of the different points. For example, (x_1, y_1) and (x_2, y_2) distinguish two different ordered pairs. The numbers 1 and 2 are called **subscripts**. A subscript is written to the right and below a variable. The coordinates of the first point are written as (x_1, y_1). The coordinates of the second point are written as (x_2, y_2).

SLOPE OF A LINE THROUGH TWO GIVEN POINTS
Given two ordered pairs, (x_1, y_1) and (x_2, y_2),

$$m = \frac{\text{rise}}{\text{run}} = \frac{\text{difference in } y\text{-coordinates}}{\text{difference in } x\text{-coordinates}} = \frac{y_2 - y_1}{x_2 - x_1}$$

Since $m = \dfrac{\text{change in } y}{\text{change in } x}$, this is often written as $m = \dfrac{\Delta y}{\Delta x}$.

 **HELPING HAND** Do not mix up the order of coordinates during the subtraction. It is very important to subtract both coordinates of the first point from both coordinates of the second point.

EXAMPLE 3   Determine the slope of a line containing the given points.

a. $(1, 3)$ and $(5, 4)$ **b.** $(9, 2)$ and $(9, -1)$ **c.** $(3, -4)$ and $(2, -4)$

Solution

a. $x_1 = 1, y_1 = 3$
$x_2 = 5, y_2 = 4$

$m = \dfrac{y_2 - y_1}{x_2 - x_1}$

$m = \dfrac{4 - 3}{5 - 1}$

$m = \dfrac{1}{4}$

b. $x_1 = 9, y_1 = 2$
$x_2 = 9, y_2 = -1$

$m = \dfrac{y_2 - y_1}{x_2 - x_1}$

$m = \dfrac{-1 - 2}{9 - 9}$

$m = \dfrac{-3}{0}$

m is undefined.

c. $x_1 = 3, y_1 = -4$
$x_2 = 2, y_2 = -4$

$m = \dfrac{y_2 - y_1}{x_2 - x_1}$

$m = \dfrac{-4 - (-4)}{2 - 3}$

$m = \dfrac{0}{-1}$

$m = 0$

 **HELPING HAND** The order of labeling the points does not matter. In Example 3a, the results would be the same if $(x_1, y_1) = (5, 4)$ and $(x_2, y_2) = (1, 3)$. Try this for yourself.

 ### 5.2.2 Checkup

Determine the slope of the line containing the given points.

1. $(-3, -2)$ and $(1, 4)$ **2.** $(4, 2)$ and $(-1, 2)$ **3.** $(2, 1)$ and $(2, -4)$

5.2.3 Determining the Slope of a Line from Its Linear Equation

We may also need to determine the slope of a line when a graph is not given and only the linear equation is known. Even though we could graph the line from the equation and then determine the slope, it is easier to determine the slope algebraically without drawing the graph. Complete the following discovery to see how to find the slope from the known equation.

Discovery 5

Determining Slope from a Linear Equation

1. Graph the given linear equations in two variables.

   Label two integer coordinate points.

   a. $y = 2x + 4$ **b.** $y = -2x$ **c.** $y = \dfrac{1}{2}x - 5$

2. Determine the slope of each of the preceding lines.

3. Determine the coefficient of the x-term in each of the equations in part 1.

Write a rule to determine the slope of the graph from a linear equation in two variables.

The slope of each graph is the same as the coefficient of the x-term in the linear equation when it is solved for y.

SLOPE OF A LINE, GIVEN A LINEAR EQUATION

Given a linear equation in two variables solved for the dependent variable y, the slope of the line represented by the equation is the coefficient of the independent variable x.

To determine the slope of a line from its linear equation,

- Solve the equation for y.
- The coefficient of the x-term is the slope of the graph.

For example, determine the slope of the line for the linear equation $-2x + 3y = 5$.

Solve for y.

$$-2x + 3y = 5$$
$$-2x + 3y + 2x = 5 + 2x$$
$$3y = 5 + 2x$$
$$3y = 2x + 5$$
$$\frac{3y}{3} = \frac{2x + 5}{3}$$
$$y = \frac{2}{3}x + \frac{5}{3}$$

The slope of the line is the coefficient of the x-term, or $\frac{2}{3}$. Recall that in the previous section we learned that the constant term indicates that the y-intercept is $(0, \frac{5}{3})$.

TWO SPECIAL CASES OF THE SLOPE OF A LINE GIVEN A LINEAR EQUATION

Given a linear equation in two variables of the form $x = h$, the graph is a vertical line with an undefined slope.

Given a linear equation in two variables of the form $y = k$, the graph is a horizontal line with a zero slope.

HELPING HAND A slope of 0 and an undefined slope are not the same.

EXAMPLE 4 Determine by inspection the slope and y-intercept of the line for the given linear equation.

a. $f(x) = -3x + 7$ **b.** $6y = 19$ **c.** $x - 4 = 10$

Solution

a. The coefficient of the x-term is -3. Therefore, $m = -3$. The constant term is 7. The y-intercept is $(0, 7)$.

b. Solve for y.

$$\frac{6y}{6} = \frac{19}{6}$$

$$y = \frac{19}{6} \quad \text{or} \quad y = 0x + \frac{19}{6}$$

The coefficient of the x term is 0. Therefore, $m = 0$. The line has a y-intercept of $(0, \frac{19}{6})$.

c. Solve for x because there are no y's.

$$x - 4 = 10$$
$$x - 4 + 4 = 10 + 4$$
$$x = 14$$

Therefore, m is undefined, because $x = 14$ is a vertical line. The line has an x-intercept of $(14, 0)$.

5.2.3 Checkup

In exercises 1–4, determine by inspection the slope and y-intercept of the line for each linear equation.

1. $t(x) = -\frac{7}{11}x + 13$ **2.** $8x + 3y = 12$

3. $2y - 5 = 3y + 1$ **4.** $4x - 3 = x + 6$

5. How can you determine the slope of the line for a linear equation by inspecting the equation? How can you determine the y-intercept of the line for a linear equation by inspecting the equation?

5.2.4 Graphing Linear Equations by Using the Slope–Intercept Method

Earlier, we said that the slope of a line is an important characteristic of a linear equation. One reason for its importance is that if you know the slope of a line, you can graph the line even if you know only one point on it, instead of two.

To graph a line when you know both a point on the line and the slope of the line,

- Plot the known point and label its coordinates.
- Count the rise and run from the point you have located. For a positive rise, count upward; for a negative rise, count downward. For a positive run, count to the right; for a negative run, count to the left.
- Place a second point where the next ordered pair is located. Label the coordinates of the point.
- Draw a straight line connecting the two points.

For example, graph a line that contains the point $(2, -1)$ and has a slope of $\frac{3}{5}$.

Plot the point $(2, -1)$. Locate a second point by counting the slope (rise of 3 and run of 5). Draw a straight line connecting the two points.

HELPING HAND The slope can be counted using different triangles, as shown in the following diagrams:

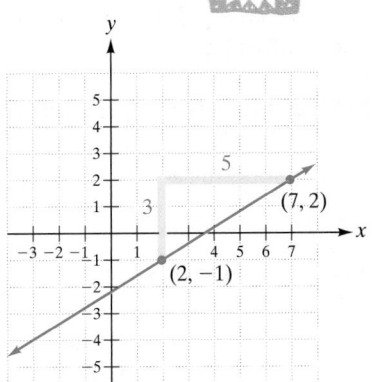 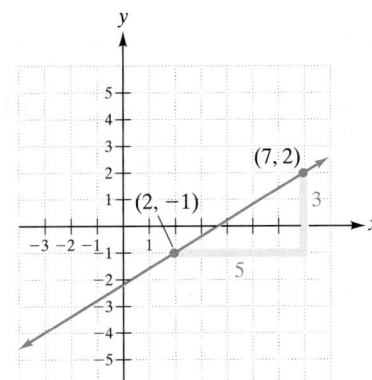 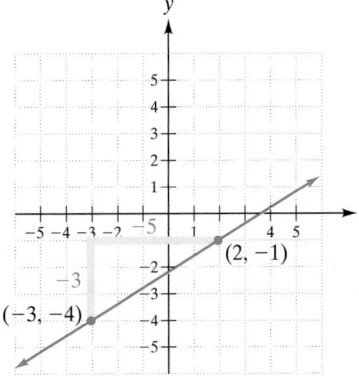

Remember that $\frac{-3}{-5} = \frac{3}{5}$.

EXAMPLE 5 Graph a line that contains the given point and has the given slope. Label two points.

a. $(3, 1)$; $m = -2$ **b.** $(3, 4)$; $m = 0$

c. $(-5, -2)$; m is undefined.

Solution

Locate the given point and determine the rise and run.

a. 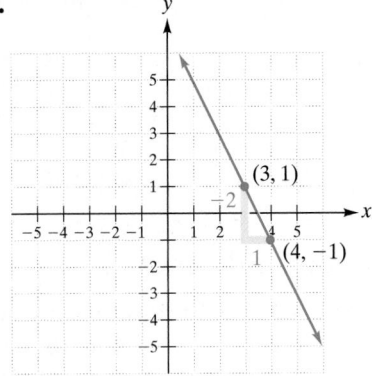   **b.**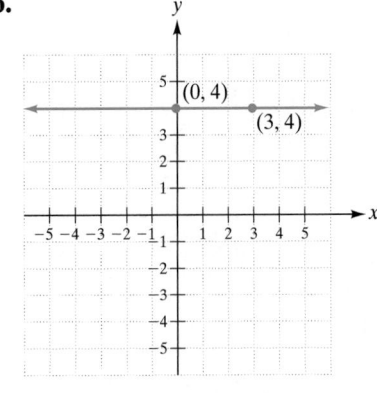

Note: The line is horizontal because $m = 0$.

c.

(graph showing vertical line through points $(-5, 0)$ and $(-5, -2)$)

Note: The line is vertical because m is undefined.

We have found that if a linear equation in two variables is solved for y, the slope of its graph is the coefficient of the x-term. The constant term is the y-coordinate of the y-intercept. Therefore, we have a linear equation in two variables in what we call the **slope–intercept form**.

SLOPE–INTERCEPT FORM FOR A LINEAR EQUATION IN TWO VARIABLES

The slope–intercept form for a linear equation in two variables is

$$y = mx + b$$

where m is the slope of the graphed line and b is the y-coordinate of the y-intercept of the graph.

We have enough information from the slope–intercept equation to graph a line.

> To graph a linear equation in two variables, using the slope–intercept method,
>
> - Solve the equation for y.
> - Determine the slope and y-coordinate of the y-intercept from the equation.
> - Plot the y-intercept and label its coordinates.
> - Locate the next point on the line by counting the slope (rise over run). Label the coordinates of the point.
> - Draw a straight line through the two points.

For example, graph the linear equation $3x + y = -1$.
First, solve for y.

$$3x + y = -1$$
$$3x + y - 3x = -1 - 3x$$
$$y = -1 - 3x$$
$$y = -3x - 1$$

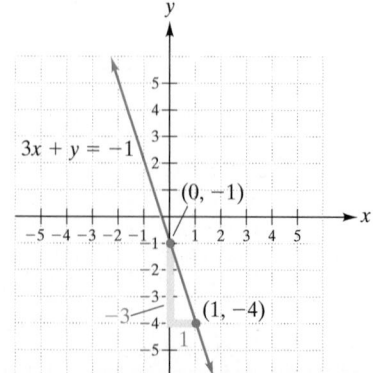

Therefore, $m = -3$ and $b = -1$.
The y-intercept is $(0, -1)$ and the slope is -3, or $\frac{-3}{1}$.
Locate the y-intercept and count the slope. Label the coordinates of both points. Draw a straight line through the two points.

EXAMPLE 6 Graph the given linear equation, using the slope–intercept method. Check by graphing on your calculator.

a. $b(x) = -3x + 7$ **b.** $3x - 4y = -4$ **c.** $6y = 19$

Solution

a. $b(x) = -3x + 7$

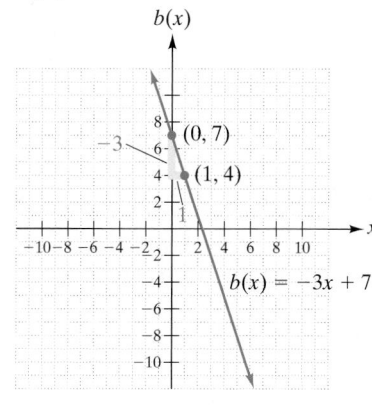

Check

The slope is −3, or $\frac{-3}{1}$. Starting at the y-intercept, $(0, 7)$, move three units down and one unit to the right.

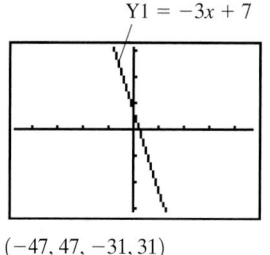

$Y1 = -3x + 7$

$(-47, 47, -31, 31)$

b. Solve for y. $3x - 4y = -4$

$$-4y = -3x - 4$$

$$y = \tfrac{3}{4}x + 1$$

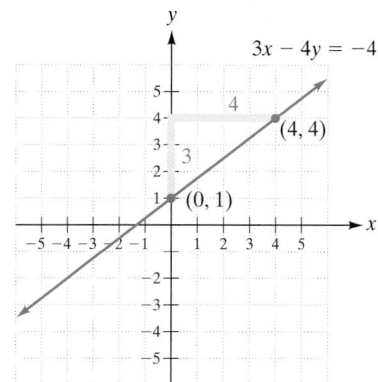

The slope is $\frac{3}{4}$. Starting at the y-intercept, $(0, 1)$, move three units up and four units to the right.

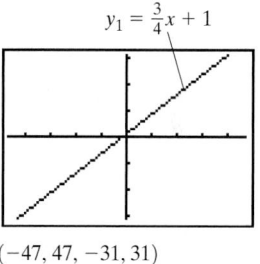

$y_1 = \tfrac{3}{4}x + 1$

$(-47, 47, -31, 31)$

c. Solve for y. $6y = 19$

$$y = \frac{19}{6}$$

$$y = \frac{19}{6} \quad \text{or} \quad y = 0x + \frac{19}{6}$$

The line is horizontal because the slope is zero. The y-intercept is $(0, \frac{19}{6})$.

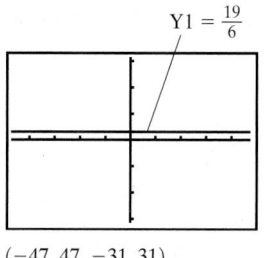

$Y1 = \dfrac{19}{6}$

$(-47, 47, -31, 31)$

5.2.4 Checkup

Graph a line that contains the given point and has the given slope.

1. $(-4, -3); m = \dfrac{2}{3}$

2. $(-2, 4); m = -1$

3. $(6, 0); m$ is undefined.

4. $(3, 6); m = 0$

Graph exercises 5–7, using the slope–intercept method. Check by graphing on your calculator.

5. $v(x) = \dfrac{4}{3}x - 2$

6. $5x + 2y = 4$

7. $6y = 2y + 7$

8. If it takes two points to draw a line, explain why we can still draw a line if we are given only one point and the slope of the line.

5.2.5 Modeling the Real World

The slope has many applications. One real-world application is the **grade** of a road, a measure of how steep the road is. We usually represent a grade as a percent. For example, a 4% grade means that for every vertical distance of 100 feet, the road drops or rises 4 feet. Note that we do not use a positive or a negative percent, because we do not know the orientation of the person viewing the road.

100 ft

4 ft

EXAMPLE 7 The U.S. Army Humvee can drive on road grades rising 30 feet vertically over a horizontal distance of 50 feet. Find the grade of the road.

Solution

The grade of the terrain is the slope, written as a percent.

$$m = \frac{\text{rise}}{\text{run}} = \frac{30}{50} = 0.6 = 60\%$$

The grade of the road for the Humvee is 60%.

When we use a linear equation to represent a real-world situation, the slope can be a very interesting, as well as important, quantity. The slope is the change in the dependent variable produced by a unit change in the independent variable. A common situation occurs when the independent variable represents time. In this case, the slope tells us how the dependent variable changes over a period of time. We call this the **average rate of change** of the dependent variable. From the average rate of change, we can sometimes predict what value a quantity will have in the future or what value it had in the past. For a non-linear relation, we have to change our idea of the slope.

EXAMPLE 8 According to the National Association of Realtors, in 1990 the median price of an existing home was $95,500. In 2000, the median price of an existing home was $139,000. Find the average rate of change per year in the median cost of existing homes from 1990 to 2000.

Solution

Let x = the year
y = the median cost of an existing home

We are given the data points (1990, 95,500) and (2000, 139,000). To determine the average rate of change from 1990 to 2000, we must use $x_1 = 1990$, $y_1 = 95,500$, $x_2 = 2000$, and $y_2 = 139,000$. The order is important in this case.

$$m = \frac{y_2 - y_1}{x_2 - x_1}$$

$$m = \frac{139,000 - 95,500}{2000 - 1990} \quad \frac{\text{cost in dollars}}{\text{years}}$$

$$m = 4350$$

The average rate of change in the mean cost of an existing home from 1990 to 2000 was $4350 per year.

A common average-rate-of-change situation that we encounter daily is the rate at which an object's distance changes with respect to time. That is, the speed at which the object moves. The following example discusses one of these situations. Further investigations are presented in Part III of the Chapter 5 Project.

EXAMPLE 9 Malcolm walked to school from his home, a distance of 2 miles. After attending his 50-minute class, he ran back home at a rate of twice his walking rate. Let t = the time in minutes of the trip and $d(t)$ = the distance from Malcolm's home in miles. Choose the graph that best illustrates Malcolm's trip.

a.
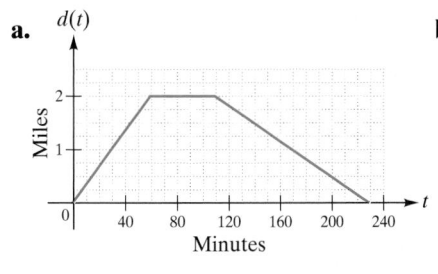

b.
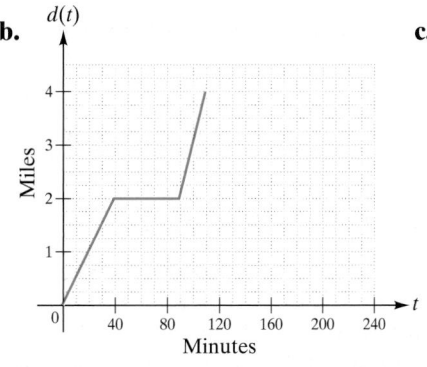

c.
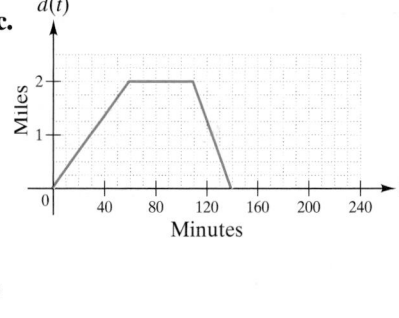

Solution

To interpret the graph, we know that the average rate of change of distance with respect to time, Malcolm's speed, is shown by the slope of the line. The faster the speed, the steeper is the line. The distance function is increasing when Malcolm is moving away from his home (the slope is positive), decreasing when Malcolm is moving toward his home (the slope is negative), and constant when Malcolm is moving neither toward nor away from his home (the slope is zero). Therefore, the graph should have three linear segments. The first segment should have a positive slope. The second segment should have a zero slope for 50 minutes. The third segment should have a negative slope and be twice as steep as the first segment.

a. In the first graph, the first segment has a positive slope. In this segment, Malcolm moved at an average speed of 1 mile per 30 minutes. The second segment has a zero slope; here, Malcolm moved at an average speed of 0 miles per minute. The third segment has a negative slope, indicating that Malcolm moved at an average speed of 1 mile per 60 minutes; that is, the slope is half as steep as that of the first segment. In other words, the graph shows that Malcolm's return trip is slower than his trip to school.

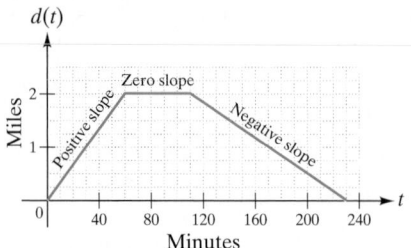

b. In the second graph, the first segment has a positive slope. In this segment, Malcolm moved at an average speed of 1 mile per 20 minutes. The second segment has a zero slope; Malcolm moved at an average speed of 0 miles per minute during this segment. The third segment has a positive slope, indicating that Malcolm progressed at an average speed of 1 mile per 10 minutes. The graph shows that Malcolm's distance from home at the end of his trip is 4 miles. In other words, Malcolm did not return home.

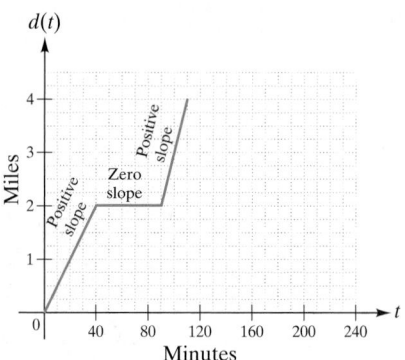

c. In the third graph, the first segment has a positive slope. In this segment, Malcolm moved at an average speed of 1 mile per 30 minutes. The second segment has a zero slope; here, Malcolm moved at an average speed of 0 miles per minute. The third segment has a negative slope, indicating that Malcolm moved at an average speed of 1 mile per 15 minutes; that is, the slope is twice as steep here as in the first segment. This graph illustrates Malcolm's trip.

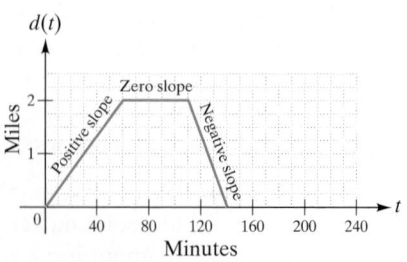

APPLICATION

Linear depreciation diminishes the value of an asset by a fixed amount each period until the net value is zero. This is the simplest depreciation calculation to make for tax purposes.

If you purchase a computer for $2000 to use in your business and estimate its useful life to be five years, determine the amount of depreciation per year (the slope of a line). Given x years of use and the value of the computer in y dollars, determine the y-intercept and graph the equation.

Discussion

The fixed amount of depreciation is $2000 divided by 5 years, or $400 per year. The slope of the line is -400, because the computer is depreciating, or decreasing in value.

To graph this situation, we need to determine the y-intercept. Let $x = 0$ and determine the corresponding y-value. At 0 years after purchase, the value of the computer is $2000. The y-intercept is $(0, 2000)$.

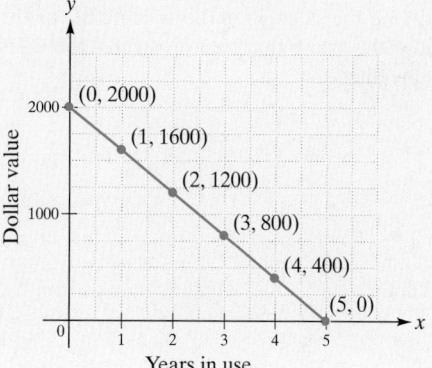

 5.2.5 Checkup

1. A Humvee can drive off the road, rising 30 feet vertically over a horizontal distance of 75 feet. Find the grade of this terrain.

2. According to the *New York Times* almanac, in 1990 there were 9444 commercial radio stations in the United States. In 1999, there were 10,540 commercial radio stations. Find the average rate of change per year in the number of radio stations from 1990 to 1999.

3. Margaret was walking home from the grocery store, which was 1 mile from her home. After walking a quarter of a mile, she stopped to talk with a neighbor who was working in her yard. After talking for 4 minutes, Margaret doubled her walking speed to finish her trip home. Which graph best illustrates Margaret's walk home? Explain your choice.

a.

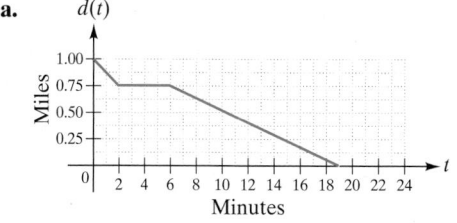

b.

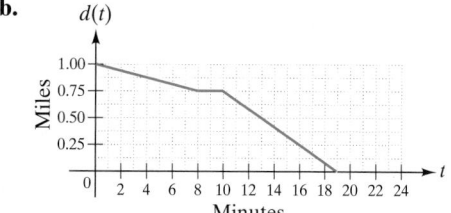

c.

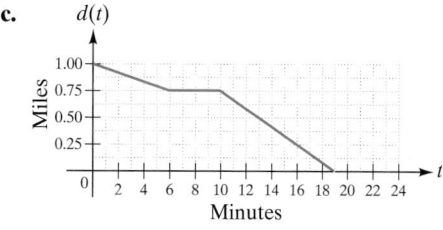

4. A company purchased a machine for $135,000. The company used simple linear depreciation for the machine for tax purposes. If the company estimated the useful life of the machine to be 15 years, determine the amount of depreciation per year.

5. Another common rate-of-change problem relates to the sale of a product. When a product is first introduced on

the market, sales steadily increase for a period of time, then they may level off, and finally they may decrease as the product loses popularity or is replaced with another product. Suppose a product experiences a steady increase in sales for the first four years that it is marketed, then sales remain level for three more years, and finally sales decrease steadily at a rate that is twice that of the initial increase. Let x be the years during which the product is marketed and y be the sales, in millions of items, of the product. Choose the graph that best illustrates the sales history of the product.

b.

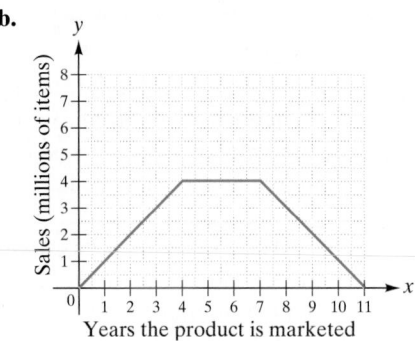

a.

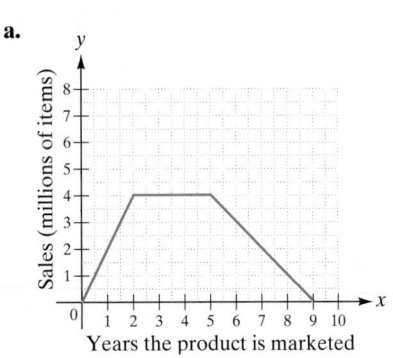

c.

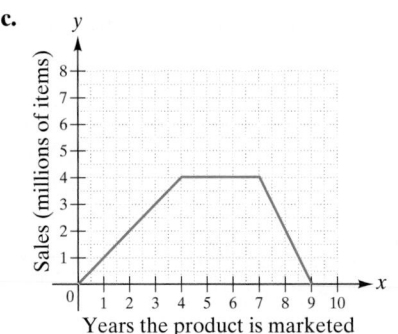

5.2 Exercises

Determine the slope of the graph. Determine whether the graph represents a function. If it does, is the function increasing, decreasing, or constant?

1.

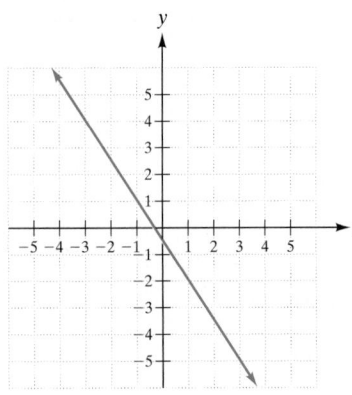

2.

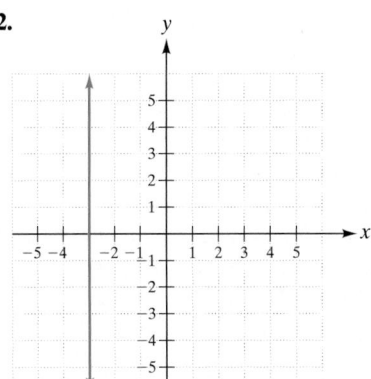

3.

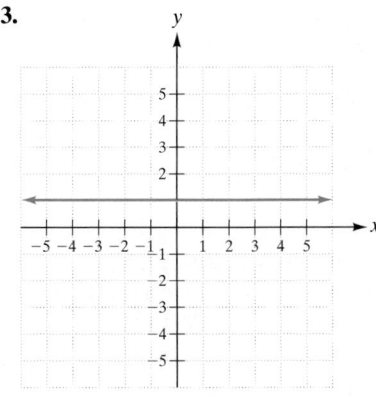

4.

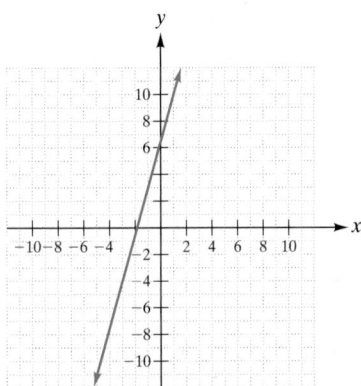

5.

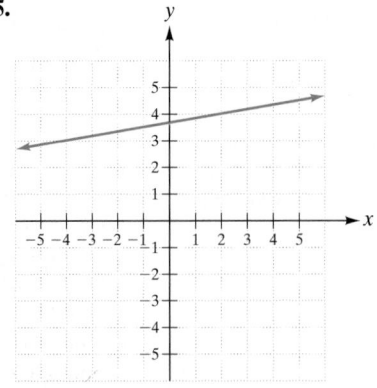

6.

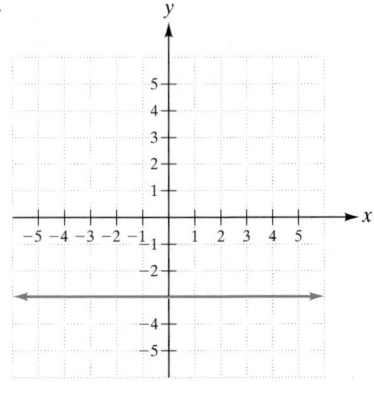

7.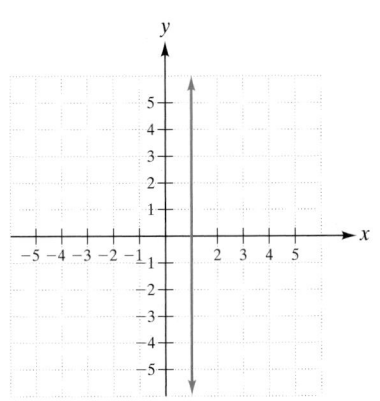

8.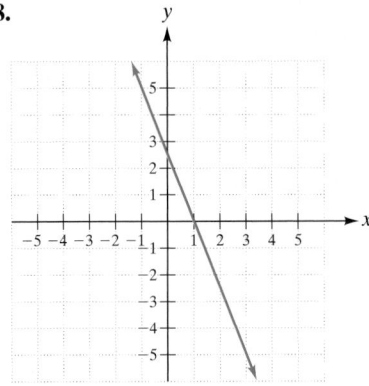

Determine the slope of the line containing the given points.

9. $(-7, -2)$ and $(5, 6)$
10. $(3, -1)$ and $(-2, 8)$
11. $(-12, -9)$ and $(4, -9)$
12. $(-8, 9)$ and $(-8, -5)$

13. $(0, 3)$ and $(0, 8)$
14. $(1, -6)$ and $(8, 9)$
15. $(6, -4)$ and $(7, -6)$
16. $(-5, 0)$ and $(7, 0)$

17. $(0, 4)$ and $(5, 0)$
18. $(-3, 0)$ and $(0, 5)$
19. $(11.5, -9.2)$ and $(6.9, 18.4)$
20. $(-3.8, 1.9)$ and $(5.7, 5.7)$

21. $\left(\dfrac{1}{2}, \dfrac{3}{4}\right)$ and $\left(-\dfrac{1}{2}, -\dfrac{5}{6}\right)$
22. $\left(-\dfrac{3}{7}, \dfrac{1}{4}\right)$ and $\left(\dfrac{5}{14}, -\dfrac{7}{8}\right)$

Determine by inspection the slope and the y-intercept of the line for each equation.

23. $y = 21x + 15$
24. $y = -19x + 28$
25. $y = 5.95x - 2.01$
26. $y = 14.8 - 3.6x$

27. $y = 85,600 - 1255x$
28. $y = 45x + 1250$
29. $16x - 4y = 64$
30. $24x + 3y = 39$

31. $7y + 18 = 2(y + 6) - 4$
32. $12(y + 2) = 5(2y - 4)$
33. $\dfrac{3}{2}x - \dfrac{3}{5}y = \dfrac{21}{10}$
34. $-\dfrac{5}{3}x + \dfrac{5}{4}y = \dfrac{25}{12}$

35. $x = -4\dfrac{7}{8}$
36. $\dfrac{5}{9}x = -15$

Graph a line that contains the given point and has the given slope.

37. $(8, 3)$; $m = \dfrac{4}{7}$
38. $(-7, -6)$; $m = \dfrac{5}{6}$
39. $(-10, 4)$; $m = -\dfrac{5}{9}$
40. $(9, -10)$; $m = -\dfrac{4}{5}$

41. $(5, 7)$; $m = 4$
42. $(-8, 7)$; $m = -6$
43. $(0, 9)$; $m = 0$
44. $(0, 9)$; m is undefined.

45. $(9, 0)$; m is undefined.
46. $(9, 0)$; $m = 0$

Graph exercises 47–58, using the slope–intercept method. Check by graphing on your calculator.

47. $y = \dfrac{5}{3}x - 4$
48. $y = -\dfrac{3}{4}x + 6$
49. $16x - 8y = 40$
50. $-21x + 7y = -35$

51. $7x + 2y = -16$
52. $9x - 4y = 24$
53. $14y + 21 = 6y + 5$
54. $31y - 19 = 26y + 6$

55. $5y = 150x + 350$
56. $6y = 300x - 360$
57. $f(x) = 0.3x - 1.2$
58. $m(x) = -0.8x + 2.7$

59. A car dealer advertises that its off-road vehicles can drive on terrain rising 56 feet vertically over a horizontal distance of 160 feet. Find the grade of the advertised terrain.

60. A motor bike drives on a terrain rising 31 feet vertically over a horizontal distance of 124 feet. What is the grade of the terrain?

The pitch of a roof is really the measure of the steepness of the roof. Often the pitch is reported as a fraction whose denominator is 12. A roof that rises 3.6 inches over a horizontal distance of 12 inches is said to have a pitch of 0.30, or 30%.

61. What is the pitch of a roof that rises 3.3 inches over a horizontal distance of 12 inches?

62. Find the pitch of a roof that rises 5 inches over a horizontal distance of 12 inches.

63. A troop of scouts hiked for three hours to reach a work site that was a distance of 6 miles from their camp. They worked at the site for one hour before starting back to the camp. On the way back, they hiked at the same rate of speed for one hour, then rested for an hour, and then completed the return hike two hours later. Which graph best illustrates their journey? Explain your choice.

a. $d(t)$

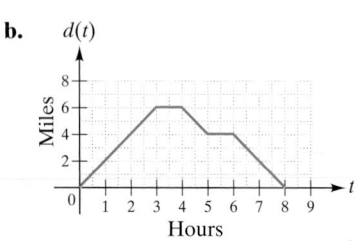

b. $d(t)$

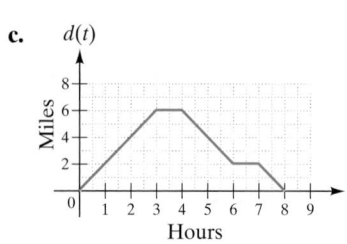

c. $d(t)$

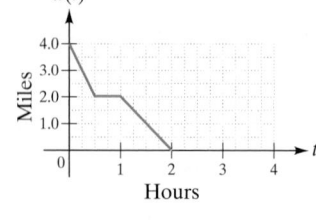

64. Stephanie lived four miles away from the sport center at which she trained. She left the center, jogging toward home. After a half hour, she stopped and spent another half hour cooling down and visiting with her jogging friend. She then walked back to the sport center at a pace that was half as fast as she jogged. Which graph best illustrates her distances from home? Explain your choice.

a. $d(t)$

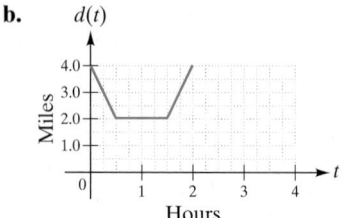

b. $d(t)$

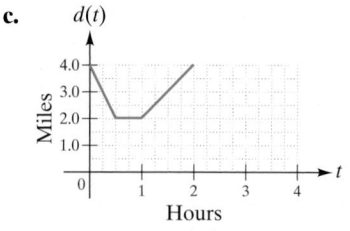

c. $d(t)$

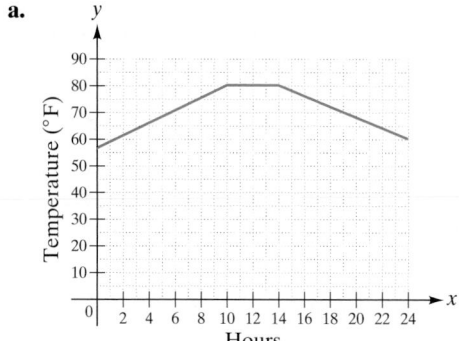

65. Yesterday, the temperature rose steadily from a low of 56°F to a high of 80°F over a 10-hour period. The temperature held constant for 4 hours and then dropped steadily to a low of 60°F over the next 10 hours. Choose the graph that best illustrates this temperature variation.

a. y

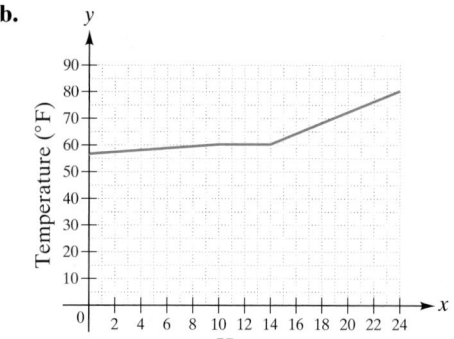

b. y

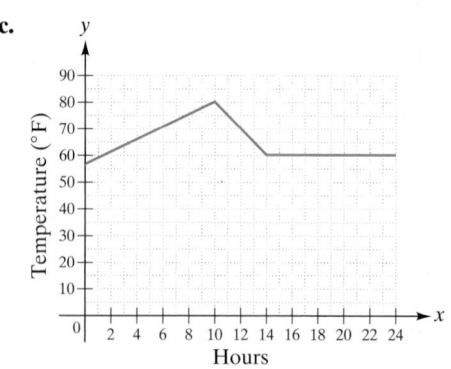

c. y

66. The enrollment at a community college grew steadily for the first six years from an initial enrollment of 1500. The enrollment remained constant at 4000 for the next three years. The enrollment then increased steadily for the next six years at a rate that was half as fast as the beginning rate. Choose the graph that best illustrates this enrollment change.

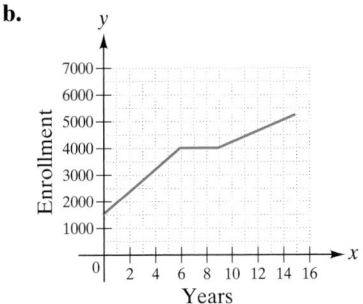

a.

b.

c.

67. A minivan is purchased new for $19,800 by a child-care center that will use straight-line depreciation to depreciate the van over the next five years. Determine the depreciation per year.

68. A lawn service company purchased a pickup truck for $26,800 and used straight-line depreciation to depreciate the truck over the next eight years. Determine the depreciation per year.

69. In 1997, the net assets of the Fidelity Magellan Fund were $51.243 billion, and in 2001 they were $80.190 billion. Find the average rate of change per year of the net assets of this fund.

70. According to the *World Almanac and Book of Facts*, the life expectancy of U.S. males increased from 47.3 years in 1900 to 73.9 years in 1998. Find the average rate of change of life expectancy per year for U.S. males.

In business applications, supply and demand functions are often used to study the economics associated with marketing products. The studies associate the price at which you sell a product with the available supply or the quantity demanded of the product. (For example, when VCRs first were placed on the market, the supply was low, causing the price to be high, and the demand was low. Later, the supply increased, the price decreased, and demand rose.) Exercises 71 and 72 deal with supply and demand functions.

71. Suppose the daily demand for a product is given by

$$D(p) = 80 - \frac{4}{5}p$$

where $D(p)$ is the quantity demanded each day by consumers when the price is p dollars per item.
   a. Graph this equation, using the slope–intercept method.
   b. Use the graph to determine what the demand for the product will be when the price is $10 per item; $20 per item; $40 per item; $64 per item.
   c. Explain in general what is happening with the demand for the product as the price increases.
   d. Are there any practical limits on the price per item, p, for this equation, that would result in an unreasonable price? Explain.

72. Suppose the daily supply of a product is given by

$$S(p) = \frac{5}{8}p$$

where $S(p)$ is the quantity supplied each day when the price is p dollars per item.
   a. Graph this equation, using the slope–intercept method.
   b. What will the daily supply be if the price is $10 per item; $20 per item; $40 per item; $64 per item?
   c. Explain in general what is happening with the supply for the product as the price increases.
   d. Are there any practical limits on the price per item for this equation? Explain.

5.2 Calculator Exercises

Complete the table that follows without graphing. (Remember to first solve the equation for y to obtain the slope–intercept form.) Then check your conclusions by graphing each equation on your calculator.

Equation	$y = mx + b$		Conclusions	
	Slope m	Constant b	Graph's Inclination ↗ or ↘	Graph's y-Intercept
$y = 3x + 6$				
$y = -2x + 7$				
$y = -x - 3$				
$y = 4x - 1$				
$5x - 3y = 9$				
$4x + 5y = 10$				
$y = \dfrac{7}{8}x - \dfrac{3}{4}$				
$y = -1.7x + 3.2$				

5.2 Writing Exercises

You have learned three methods for graphing a linear equation in two variables: using a table of values of coordinate pairs, using the intercept method, and using the slope–intercept method. State which method you would use for each of the following exercises and why you would use that method.

1. $x + 2y = 4$ **2.** $y = -x + 3$ **3.** $3x + 5y = -15$ **4.** $y = \dfrac{2}{3}x - 1$

5. $7x + y = 18$ **6.** $y = 2.5x + 4.5$ **7.** $3y = 2y - 3$ **8.** $5x - 7 = 3x + 1$

5.3 Coinciding, Parallel, and Perpendicular Lines

OBJECTIVES

1 Determine whether two linear graphs are coinciding, parallel, intersecting only, or intersecting and perpendicular by inspecting their corresponding equations.

2 Determine from real-world data whether two linear graphs are coinciding, parallel, intersecting only, or intersecting and perpendicular.

APPLICATION

In 1992, Brian J. Whipp and Susan A. Ward of the School of Medicine at UCLA wrote a paper entitled "Will Women Soon Outrun Men?"[1] In a prestigious science journal, the authors graphed the world record track times for both men and women. They noted that the slope of the graph for women was larger than the slope of the graph for men. Using this fact, they predicted that women would eventually overtake men in track competition in the year 1998. In critiquing this article that very year, Randall Woods used current data to write a linear equation for men and one for women, given year x and mean speed y (in meters per minute). The data on

running resulted in the equations $y = 0.534x - 430$ for men and $y = 4.58x - 8800$ for women. According to these equations, will the women overtake the men and outrun them? If so, when? What is wrong with these conclusions?

After completing this section, we will discuss this application further. See page 348.

[1]"Will Women Soon Outrun Men?"; Nature, January 2, 1992, Vol. 355[25].

5.3.1 Determining the Relationship of Two Lines

From previous graphing, we know that the graphs of two lines may have one of three relationships. Two **coinciding** lines have all points in common and are said to be **coincident**. Two **parallel** lines have no points in common. Two **intersecting** lines have one common point. In this section, we will see how the concept of the slope enables us to tell which of these three cases we have, just from analyzing the equations of the lines.

Let's first look at the characteristics of coinciding lines. Complete the following set of exercises on your calculator to determine whether the pairs of linear equations graph as coinciding lines.

Discovery 6

Coinciding Lines

1. Graph the given pairs of linear equations.
   a. $-2x + y = 2$
   $y = 2x + 2$
   b. $2x - 4y = 4$
   $y = \frac{1}{2}x - 1$
   c. $3x + y = -3$
   $y = -3x - 3$

2. Determine the slope and y-coordinate of the y-intercept for each graph.
   a. $-2x + y = 2$
   $y = 2x + 2$
   b. $2x - 4y = 4$
   $y = \frac{1}{2}x - 1$
   c. $3x + y = -3$
   $y = -3x - 3$
   Choose the correct answers.

3. The lines graphed are *coinciding/parallel/intersecting/intersecting and perpendicular*.

4. The slopes, m, in each pair of linear equations are *equal/not equal*.

5. The y-coordinates of the y-intercepts, b, in each pair of linear equations are *equal/not equal*.

Write a rule for determining that the graphs of two linear equations are coinciding.

If two coinciding lines have all points in common, then the slopes between each pair of points are equal and the y-coordinates of the y-intercepts in each pair are equal.

COINCIDING LINES

The graphs of two linear equations are coinciding if the graphs have equal slopes, m, and equal y-coordinates of the y-intercepts, b.

To determine whether two nonvertical lines are coinciding by inspecting their equations,

- Solve both equations for y.
- Determine the slope, m, and y-coordinate of the y-intercept, b, for each equation.

Nonvertical coinciding lines have equal slopes (m) and equal y-coordinates of the y-intercepts (b).

To determine whether two vertical lines are coinciding by inspecting their equations,

- Solve the equations for x. Vertical coinciding lines have the same constant term.

Now let's take a look at the characteristics of parallel lines. Complete the following set of exercises on your calculator to determine whether the two linear equations graph as parallel lines.

 Discovery 7

Parallel Lines

1. Graph the given pairs of linear equations.
   a. $-2x + y = 4$ **b.** $x - 2y = 4$ **c.** $y = -3x - 5$
        $y = 2x - 5$ $2x - 4y = -12$ $3x + y = 6$

2. Determine the slope and y-coordinate of the y-intercept for each graph.
   a. $-2x + y = 4$ **b.** $x - 2y = 4$ **c.** $y = -3x - 5$
        $y = 2x - 5$ $2x - 4y = -12$ $3x + y = 6$
   Choose the correct answers.

3. The lines graphed are *coinciding/parallel/intersecting/intersecting and perpendicular.*

4. The slopes, m, in each pair of linear equations in two variables are *equal/not equal.*

5. The y-coordinates of the y-intercepts, b, in each pair of linear equations in two variables are *equal/not equal.*

Write a rule for determining that the graphs of two linear equations in two variables are parallel.

In each pair of lines graphed, the two lines are parallel. Also, the slopes in each pair are equal. The y-coordinates of the y-intercepts in each pair are not equal.

PARALLEL LINES

The graphs of two linear equations are parallel if the graphs have equal slopes, m, and unequal y-coordinates of the y-intercepts, b.

To determine whether two nonvertical lines are parallel by inspecting their equations,

- Solve the equations for y.
- Determine the slope, m, and y-coordinate of the y-intercept, b, for each equation.

Nonvertical parallel lines have equal slopes (m) and unequal y-coordinates of the y-intercepts (b).

To determine whether two vertical lines are parallel by inspecting their equations,

- Solve the equations for x.

Vertical parallel lines have different constant terms.

HELPING HAND Parallel lines never intersect, so they must have a different y-intercept (or x-intercept for vertical lines) and the same slope.

The next Discovery exercise explores the characteristics of intersecting lines. Complete the following set of exercises on your calculator to determine whether the two linear equations graph as intersecting lines.

Discovery 8

Intersecting Lines

1. Graph the given pairs of linear equations.

   a. $-2x + y = 2$ **b.** $x - 2y = 4$ **c.** $-3x - y = 6$
        $y = 3x - 1$ $4x - 4y = -12$ $-3x + y = -6$

2. Determine the slope and y-coordinate of the y-intercept for each graph.

   a. $-2x + y = 2$ **b.** $x - 2y = 4$ **c.** $-3x - y = 6$
        $y = 3x - 1$ $4x - 4y = -12$ $-3x + y = -6$

Choose the correct answers.

3. The lines graphed are *coinciding/parallel/intersecting/intersecting and perpendicular*.

4. The slopes, m, in each pair of linear equations in two variables are *equal/not equal*.

Write a rule for determining that the graphs of two linear equations in two variables are intersecting.

In each pair of lines graphed, the two lines are intersecting. Also, the slopes in each pair are unequal.

INTERSECTING LINES
The graphs of two linear equations are intersecting if the graphs have unequal slopes, m.

Perpendicular lines are a special case of intersecting lines. Two intersecting lines that form four right angles are called **perpendicular** lines. Complete the following set of exercises on your calculator to investigate this special case of intersecting lines.

Discovery 9

Perpendicular Lines

1. Graph the given pairs of linear equations.
   a. $-2x + 3y = 6$ **b.** $x - 2y = 4$ **c.** $-3x - y = 3$

   $y = -\dfrac{3}{2}x - 2$ $-8x - 4y = -16$ $-x + 3y = -15$

2. Determine the slope and y-coordinate of the y-intercept for each graph.
   a. $-2x + 3y = 6$ **b.** $x - 2y = 4$ **c.** $-3x - y = 3$

   $y = -\dfrac{3}{2}x - 2$ $-8x - 4y = -16$ $-x + 3y = -15$

   Choose the correct answers.

3. The lines graphed are *coinciding/parallel/intersecting/intersecting and perpendicular*.

4. The slopes, m, in each pair of linear equations in two variables are *equal/not equal*.

5. The two slopes, m, in each pair of linear equations in two variables are reciprocals and have *the same/opposite* sign.

Write a rule for determining that the graphs of two linear equations in two variables are intersecting and perpendicular.

   In each pair of lines graphed, the lines are intersecting and perpendicular. The slopes are opposite reciprocals of each other.

INTERSECTING AND PERPENDICULAR LINES
The graphs of two linear equations are intersecting and perpendicular if the graphs have slopes, m, that are opposite reciprocals of each other.

To determine whether two nonvertical lines are intersecting or are intersecting and perpendicular by inspecting their equations,

- Solve the equations for y.
- Determine the slope, m, and y-coordinate of the y-intercept, b, of each equation.

Intersecting lines have unequal slopes (m). Nonvertical intersecting and perpendicular lines have slopes (m) that are opposite reciprocals of each other.

HELPING HAND The product of the slopes of nonvertical perpendicular lines is -1. For example, $m = \frac{2}{3}$ and $m = -\frac{3}{2}$ are slopes of perpendicular lines, and $\left(\frac{2}{3}\right)\left(-\frac{3}{2}\right) = -1$. This relationship can be used as a test for perpendicular lines.

   A special case of perpendicular lines is when one line is vertical ($x = h$) and the other line is horizontal ($y = k$).
   All of the relationships we have just discovered are summarized in the next box.

DETERMINING THE RELATIONSHIP BETWEEN TWO LINES FROM THEIR LINEAR EQUATIONS IN TWO VARIABLES

Two linear equations written in slope–intercept form, $y = mx + b$, have graphs that are

- **Coinciding (or Coincident) lines** if the equations have equal slopes (m) and equal y-coordinates of the y-intercepts (b).
- **Parallel lines** if the equations have equal slopes (m) and nonequal y-coordinates of the y-intercepts (b).
- **Intersecting lines** if the equations have nonequal slopes (m).
- **Intersecting and perpendicular lines** if the equations have slopes (m) that are opposite reciprocals of each other.

Two linear equations written in the form $x = h$ have graphs that are

- **Coinciding (or Coincident) lines** if the equations have the same constant term.
- **Parallel lines** if the equations have different constant terms.

Two linear equations, one in the form $x = h$, and the other in the form $y = k$, have graphs that are a special case of

- **Intersecting and perpendicular lines**.

EXAMPLE 1   Determine by inspection whether the graphs of the given pairs of linear equations are coinciding, parallel, intersecting only, or intersecting and perpendicular.

a. $y = 2x + 4$
   $-y = -2x - 4$

b. $2x - y = 6$
   $2x + y = 6$

c. $2x + 3y = 1$
   $3x - 2y = 1$

d. $f(x) = 2x - 5$
   $g(x) = 2x + 5$

e. $x = 5$
   $3x + 4 = 7$

f. $y = 5$
   $x = 5$

Solution

a. Solve the equations for y.

$$y = 2x + 4 \qquad m = 2 \qquad b = 4$$
$$y = 2x + 4 \qquad m = 2 \qquad b = 4$$

The slopes (m) are equal. The y-coordinates (b) are equal. The lines are coinciding.

b. Solve the equations for y.

$$y = 2x - 6 \qquad m = 2 \qquad b = -6$$
$$y = -2x + 6 \qquad m = -2 \qquad b = 6$$

The lines are intersecting at one point, because the slopes (m) are not equal: $2 \neq -2$. However, even though the slopes have opposite signs, the slopes are not reciprocals of each other; therefore, the lines are not perpendicular.

c. Solve the equations for y.

$$y = -\frac{2}{3}x + \frac{1}{3} \qquad m = -\frac{2}{3} \qquad b = \frac{1}{3}$$

$$y = \frac{3}{2}x - \frac{1}{2} \qquad m = \frac{3}{2} \qquad b = -\frac{1}{2}$$

The lines are intersecting and perpendicular, because the slopes (m) are opposite reciprocals of each other: $-\frac{2}{3}$ and $\frac{3}{2}$. The product of the slopes is -1.

d.

$$f(x) = 2x - 5 \qquad m = 2 \qquad b = -5$$
$$g(x) = 2x + 5 \qquad m = 2 \qquad b = 5$$

The slopes (m) are equal, but the y-coordinates (b) are not equal. The lines are parallel.

e. Both linear equations, $x = 5$ and $x = 1(3x + 4 = 7)$, graph vertical lines. The lines are parallel.

f. The lines are intersecting and perpendicular, because one line is vertical and the other is horizontal.

Graph the two lines on your calculator to check their relationship. *Note:* An integer or decimal screen (square screen) is necessary to determine perpendicular lines by inspection. Other screens may skew the picture. For instance, the check for Example 1c is shown here on an integer screen and on the standard screen.

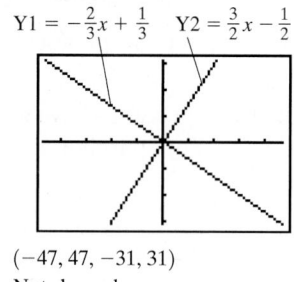

$Y1 = -\frac{2}{3}x + \frac{1}{3}$ $Y2 = \frac{3}{2}x - \frac{1}{2}$

$(-47, 47, -31, 31)$
Not skewed

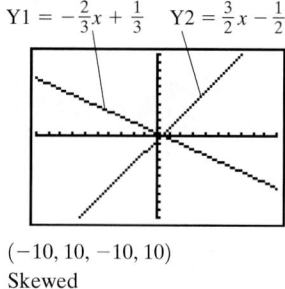

$Y1 = -\frac{2}{3}x + \frac{1}{3}$ $Y2 = \frac{3}{2}x - \frac{1}{2}$

$(-10, 10, -10, 10)$
Skewed

5.3.1 Checkup

In exercises 1–6, determine by inspection whether the graphs of the given pairs of linear equations are coinciding, parallel, intersecting only, or intersecting and perpendicular.

1. $4x - 5y = 10$
  $5x + 4y = 4$

2. $2x - 3y = 9$
  $2x - 3y = -9$

3. $5x - y = -1$
  $3x + 2y = -4$

4. $2y = 14$
  $y + 3 = 0$

5. $2x = 6$
  $y + 3 = 5$

6. $y = \frac{2}{3}x - 1$
  $2x - 3y = 3$

7. All perpendicular lines are intersecting, but not all intersecting lines are perpendicular. Explain.

8. Assume that you have two equations, $y = m_1x + b_1$ and $y = m_2x + b_2$. (Note that the numbers 1 and 2 that appear in the equations are called *subscripts*. They are used only to indicate that the values of m may differ from one another and so may the values of b. Subscripts have no mathematical meaning in the sense that exponents do.) When you graph the two equations, you will obtain two lines. The lines may be intersecting only, intersecting and perpendicular, parallel, or coinciding. Complete the following table to identify the relationship between m_1 and m_2 and between b_1 and b_2 for each situation:

Situation	Relationship between slopes m_1 and m_2	Relationship between y-intercept values b_1 and b_2
Graphs coincide.	$m_1 = m_2$	$b_1 = b_2$
Graphs are parallel.		
Graphs intersect only.		
Graphs intersect and are perpendicular.		

5.3.2 Modeling the Real World

How useful is it to know that two graphed lines are coinciding, parallel, or intersecting? If the lines represent equations based on real-world situations, this knowledge can be very useful. For example, economists often graph sales revenue and production costs on the same coordinate system. If the two lines intersect, the point of intersection is called the **break-even point**. The values of the coordinates of this point correspond to how much money you have to make in sales in order to equal the money you spend in production costs. If the two lines don't intersect, you may never be able to earn back in sales the money you spent as costs. This is certainly an important thing to know. In other situations, the point of intersection of two lines may correspond to when one moving object is going to catch up to another. Even without knowing just where the point of intersection is, it's often useful to know from the equations whether there is a point of intersection at all.

EXAMPLE 2 Amy sells wooden Christmas ornaments at a fair booth. The paint costs $0.89 per ornament, the ribbon costs $0.11 per ornament, and the unpainted ornaments cost $0.50 each. The fair charges $25.00 to set up a booth.

a. Write an equation for the total cost, y, required to produce and sell x ornaments.

b. Last year Amy sold the ornaments for $1.50 each. If she plans to sell the ornaments at the same price this year, write an equation for the total revenue, y, for selling x ornaments.

c. Will there be a break-even point (an intersection of the two graphs in parts a and b)? Explain.

d. Write an equation for the total revenue, y, received for selling x ornaments if Amy decides to sell the ornaments for $2.00. Will there be a break-even point? Explain.

e. Graph the equations on your calculator, using the window (0, 100, 10, 0, 250, 10, 1). Determine the break-even point. Interpret your answer.

Solution

a. The cost per ornament will be $1.50.

$$0.89 + 0.11 + 0.50 = 1.50$$

The total cost will be $1.50 per ornament plus the booth cost of $25.00.
Let y = total cost
x = number of ornaments

$$y = 1.50x + 25.00$$

b. The total revenue will be $1.50 per ornament.
Let y = total revenue
x = number of ornaments

$$y = 1.50x$$

c. Using the equations $y = 1.50x + 25.00$ and $y = 1.50x$, we see that the linear equations have equal slopes: $1.50 = 1.50$. They also have unequal y-coordinates of the y-intercepts: $25.00 \neq 0$. Therefore, the lines are parallel and do not intersect, so there is no break-even point.

d. The total revenue will be $2.00 per ornament.

Let y = total revenue

x = number of ornaments

$$y = 2.00x$$

Using the equations $y = 1.50x + 25.00$ and $y = 2.00x$, we see that the linear equations have unequal slopes: $1.50 \neq 2.00$. The lines will intersect at one point, and that will be the break-even point.

e. $Y1 = 1.50x + 25.00$ $Y2 = 2.00x$

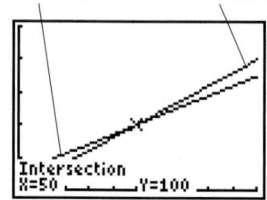

$(0, 100, 10, 0, 250, 10, 1)$

The intersection is $(50, 100)$. Amy will break even and begin to make a profit if she sells 50 ornaments for a total revenue of $100.

APPLICATION

In 1992, Brian J. Whipp and Susan A. Ward of the School of Medicine at UCLA wrote a paper entitled "Will Women Soon Outrun Men?" In a prestigious science journal, the authors graphed the world record track times for both men and women. They noted that the slope of the graph for women was larger than the slope of the graph for men. Using this fact, they predicted that women would eventually overtake men in track competition in the year 1998. In critiquing this article that very year, Randall Woods used current data to write a linear equation for men and one for women, given year x and mean speed y (in meters per minute). The data on running resulted in the equations $y = 0.534x - 430$ for the men and $y = 4.58x - 8800$ for women. According to these equations, will the women overtake the men and outrun them? If so, when? What is wrong with these conclusions?

Discussion

Given that the two equations have different slopes, we know that the equations will graph intersecting lines. The intersection point of these two graphs will determine the answer to the question, "Will Women Soon Outrun Men?"

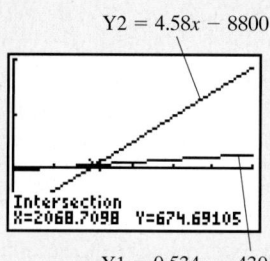

$Y2 = 4.58x - 8800$

$Y1 = 0.534x - 430$

$(0, 6000, 1000, -10,000, 20,000, 10,000)$

According to the graph, women will outrun men in the year 2068. This conclusion is based on the fact that there are no physical limitations on running and that the rate of increase will continue beyond the time of the most recent data used. Note that both equations would graph lines with a negative y-intercept, meaning that when time was 0 years, the mean speed of the runners was negative. This has no physical meaning.

✓ 5.3.2 Checkup

1. Tim has a lawn care business. He spends an average of $8.50 per lawn for fertilizer and an average of $6.50 per lawn for labor to spread the fertilizer. He paid $85 for the spreader.

a. Write an equation for the total cost, y, of providing Tim's service for x lawns.

b. Tim charges $20 per lawn for the service. Write an equation for the total revenue, y, from providing the service for x lawns.

c. Will there be a break-even point? Explain.

d. Write an equation for the total revenue, y, from providing the service for x lawns if Tim charges $15 per lawn. Will there be a break-even point? Explain.

2. Using data from *Statistical Abstracts of the United States*, we can represent the percentage of males who listed their marital status as married by the function $m(x) = -0.411x + 71.19$, where x is the number of years since 1960. Likewise, we can represent the percentage of females who listed their marital status as married by the function $f(x) = -0.423x + 67.02$. Will the graphs of these functions ever intersect? Explain.

5.3 Exercises

In exercises 1–26, determine by inspection whether the graphs of the given pairs of linear equations are (a) coinciding, (b) parallel, (c) intersecting only, or (d) both intersecting and perpendicular. Check by graphing.

1. $3x - 2y = 5(y + 7)$
$7x = 3(1 - y)$

2. $4x + 5 = 0$
$2(y - 1) = 6 - y$

3. $x = 4(y - 3)$
$x = 4(y + 5)$

4. $x - 6 = 0$
$x + 6 = 0$

5. $4x - y = 6$
$2x - y + 3 = 0$

6. $x + y = 0$
$x + 4y = 8$

7. $x = 2(y - 7)$
$y = \frac{1}{2}x + 7$

8. $2x - 5 = x + 5$
$x - 6 = 4$

9. $5x + y = -6$
$3x + y = 0$

10. $y = 3x - 4$
$2x = 3y + 6$

11. $4x + y = 8$
$4x + y + 2 = 0$

12. $y = 4x - 5$
$y + 1 = 4(x - 1)$

13. $y - 5 = 0$
$2x + 6 = x + 9$

14. $5x + y = 3$
$x = 5y$

15. $2y - 3 = 13$
$y + 1 = 4$

16. $x = -6$
$x + 3y = 3$

17. $x + 3 = 0$
$x - 5 = 0$

18. $14x - 2y = -1$
$7x - y = 6$

19. $2x - 9 = 0$
$x - 4 = 5 - x$

20. $y = 1$
$3(y + 3) = 2(y + 4)$

21. $3(y - 3) = 1$
$5y = 10 + 2y$

22. $y + 4 = 0$
$2x - 5y = 15$

23. $x = 2$
$y = 2x - 1$

24. $x + y = 9$
$y = -(x + 3)$

25. $y - 3 = 0$
$2x + 3y = 0$

26. $y + 3 = 0$
$y + 6 = 0$

27. Brook can buy candy bars wholesale for 15 cents each. He wants to resell them in his dormitory to earn some spending money. He hires his roommate Eric to sell the candy for him, paying Eric $2.50 per day, plus 10 cents per candy bar sold. The dorm's resident advisor charges Brook $1.00 a day for permission to sell the candy at the dorm.

a. Write an equation for the total cost, y, in terms of x candy bars sold by Eric in one day.

b. Write an equation for the total revenue, y, in terms of x candy bars sold when Eric sells the candy bars for 25 cents each.

c. Find the break-even point, if there is one, when Brook will start making a profit for each day Eric sells candy for him.

d. Write an equation for the total revenue, y, in terms of x candy bars sold when Eric sells the candy bars three for a dollar.

e. Find the break-even point, if there is one, when Brook will start making a profit for each day Eric sells candy for him at this price.

f. What will be the break-even point if Brook sells his own candy bars for 25 cents each?

g. What will be the break-even point if Brook sells his own candy bars three for a dollar?

h. What do you think Brook should do?

28. Joe wants to start a business selling restored antique radios. He can rent a counter in the local antiques mall for $200 per month. Miscellaneous costs average $85 per month. From his past bookkeeping records, he figures that he has purchased antique radios for an average of $70 each. He also figures that the cost of his labor and of parts required to restore the radios average $130 for each radio. (For the calculator graphs, use a window of $(0, 20, 1, 0, 2000, 100, 1)$.)

a. Write a function for the total cost, y, of setting up shop to restore and sell x radios in a given month.

b. Write an equation for the total revenue, y, received if Joe sells x radios at $200 each.

c. Find the break-even point, if there is one, when Joe will start making a profit each month.

d. Write an equation for the total revenue if Joe sells his radios at $300 each.

e. Find the break-even point, if there is one, when Joe will start making a profit each month at this price.

f. What will be the break-even point if Joe sells his radios for $250, reduces his counter rental to $125 per month by moving, and decreases his miscellaneous costs to $50 per month?

g. What do you think Joe should do?

29. Laurie can rent a car for $35.00 per day plus $0.25 per mile from Krazy Kar Rental, or she can rent a car for $60.00 per day with unlimited miles from Rational Car Rental. Write equations that represent her cost for one day of rental from each company if she drives the car x miles. Inspect the equations to determine whether their graphs will intersect. Explain. If the graphs do intersect, graph and find the intersection point, and interpret what it represents.

30. Jim can rent a stump grinder for a day for a flat fee of $65 from Rent All, Inc., or he can rent a grinder for a fee of $15 plus $10 per hour from Best Rental. Write equations that represent his cost for one day of rental from each company if he rents the grinder for x hours. Will the graphs of the equations intersect? Explain. If they do, graph and find the intersection point. Explain what the point represents.

31. Football superstar Archie receives a pass at the 50-yard line. He races in for a touchdown at a speed of 10 feet per second. Write an equation for the distance he will travel in x seconds. Defensive superstar Speedie is behind the 50-yard line. He races to catch Archie at a speed of 15 feet per second and crosses the 50-yard line 4 seconds after Archie. Write an equation for the distance Speedie will travel beyond the 50-yard line. Use both equations to determine whether Speedie will catch Archie before he reaches the end zone, which is 150 feet away.

32. In the Great Automobile Race of 1895, Villainous Victor left the starting line in Paris at 12 noon and averaged 18 miles per hour with his automobile. Trueheart Tom left the starting line 45 minutes later because of startup problems, but he was able to average 21.5 miles per hour with his gasoline-powered carriage. Write equations to determine the distance each man traveled, where x is the number of hours since 12 noon. Use the equations to determine at what distance Tom will overtake Victor, assuming that neither has car trouble in the race.

33. From data compiled by the U.S. Internal Revenue Service, the number of returns filed by corporations engaged in wholesale and retail trade can be estimated by the equation $y = 21.33x + 1017$, where x is the number of years since 1990 and y is the number of returns in thousands. At the same time, the number of returns filed by corporations engaged in services is estimated by the equation $y = 99.66x + 963$. Will the graphs of the two equations intersect? If so, find the intersection point and explain what it represents. If not, explain why.

34. From data collected by the U.S. Census Bureau, the number of employees in the wholesale trade industry is given by the equation $y = 0.27x + 19.8$, where x is the number of years since 1990 and y is the number of employees in millions. The number of employees in the services industry is given by the equation $y = 1.17x + 28.8$. Will the graphs of the two equations intersect? If so, find the intersection point and explain what it represents. If not, explain why.

35. Kate has a choice to make. As a Christmas present, her grandfather offers to give her each year an amount of money equal to $50 plus twice her age, or she can receive a fixed amount of $75 each year. Write two equations that represent the amounts of money she will receive when she is x years old. Given that she is eight years old, write a third equation that represents her age. Will the graph of this equation intersect the graphs of the other two equations? Explain. Will it be perpendicular to either of the first two equations' graphs? What do the intersection points of the three graphs represent?

36. A newspaper accepts classified ads under two payment plans. The first plan charges $0.26 per word for the ad. The second plan charges a fee of $8.00 plus $0.02 per word. Write two equations that represent the cost of placing an ad of x words under each plan. Given that your ad contains 30 words, write a third equation that represents the length of the ad. Will the graph of this equation intersect the graphs of the first two equations? Explain. Will it be perpendicular to either of the first two equations' graphs? Interpret the intersection points.

5.3 Calculator Exercises

Let's explore settings of the window for graphing a linear equation in two variables.

1. Clear out any equations in the ⟨ Y= ⟩ key. Use the ⟨ ZOOM ⟩ key to set the coordinate system to the standard setting. What are the window settings for this choice?

2. Set the coordinate system to the integer setting. What are the window settings for this choice?

3. Reset the integer setting by adding 47 to Xmin and Xmax and 31 to Ymin and Ymax. Then press $\boxed{\text{GRAPH}}$. Use the arrow keys to move the cursor. Do you see that you have been able to reset the coordinate system and still maintain integer pairs for x and y?

4. Repeat steps 1 and 2, and then reset the window by multiplying each setting by 10. Leave Xres set to 1. Press $\boxed{\text{GRAPH}}$. Move the cursor. Are the x- and y-values still integer pairs?

Are you beginning to see how you can adjust the window to be any domain and range that is appropriate for your needs while still maintaining integer pairs? Experiment with other settings. Remember that you can always go back to $\boxed{\text{ZOOM}}$ $\boxed{6}$ $\boxed{\text{ZOOM}}$ $\boxed{8}$ $\boxed{\text{ENTER}}$ to get back to your familiar settings.

Use the preceding techniques to experiment with setting the windows for a calculator graph to analyze the following break-even exercises. Determine which pairs of equations will result in intersecting graphs, which will be parallel, which will be perpendicular, and which will be coinciding. For graphs that intersect, find the point of intersection.

5. Cost function: $y = 300x + 450$
Revenue function: $y = 375x$

6. Cost function: $y = 1250 + 725x$
Revenue function: $y = 725x$

7. Cost function: $y = 1800$
Revenue function: $y = 125x$

5.3 Writing Exercises

Suppose that you intend to graph two linear equations, given by $y = m_1 + b_1 x$ and $y = m_2 + b_2 x$.

Before graphing the equations, you study them to compare the values of their slopes and the y-coordinates of their y-intercepts. Explain what these values can indicate to you. Next, complete the following matching exercise:

_____ **1.** Nonvertical coinciding

_____ **2.** Vertical coinciding

_____ **3.** Nonvertical parallel

_____ **4.** Vertical parallel

_____ **5.** Intersecting

_____ **6.** Nonvertical perpendicular

_____ **7.** Special perpendicular

a. $m_1 = m_2, b_1 \neq b_2$

b. $m_1 \cdot m_2 = -1$ or $m_2 = -\dfrac{1}{m_1}$

c. $m_1 = m_2, b_1 = b_2$

d. $m_1 \neq m_2$

e. $x = c, y = d$

f. $x = c, x = d, c = d$

g. $x = c, x = d, c \neq d$

5.4 Writing Linear Equations from Given Data

OBJECTIVES

1 Write a linear equation in two variables, given the slope and y-intercept of the graph of the equation.

2 Write a linear equation in two variables, given the slope of the graph of the equation and the coordinates of a point through which the graph passes.

3 Write a linear equation in two variables, given the coordinates of two points through which the graph of the equation passes.

4 Write a linear equation in two variables, given the coordinates of a point through which the graph of the equation passes and a description of the line in terms of a second equation.

5 Model real-world situations by using linear equations in two variables.

APPLICATION

In the science fiction movie *Star Games 23: The Comet Strikes Jasper*, astronomers discover the comet Hale–Farewell traveling straight toward the planet Jasper. They measure the speed of the comet as 20,000 km/hr and its position as 3.0×10^7 km from Jasper. Assuming that the comet continues on its course in a straight line at the same speed, write an equation for its distance from Jasper and determine when it will hit the planet.

After completing this section, we will discuss this application further. See page 362.

So far, we have determined graphs for linear equations in two variables. However, suppose we are given information about a graph and need to determine its corresponding equation. Now we will see how to do that.

5.4.1 Writing a Linear Equation Given a Slope and *y*-Intercept

We already know the slope–intercept form of a linear equation, $y = mx + b$. Therefore, if we are given the slope, m, and the y-coordinate of the y-intercept, b, of a line, we can use that form to write an equation.

For example, write a linear equation in two variables for a line that has a slope of $\frac{1}{2}$ and a y-intercept of $(0, 3)$.

$$m = \tfrac{1}{2} \text{ and } b = 3 \ (y\text{-coordinate of the } y\text{-intercept})$$

Using the slope–intercept form of a linear equation, we have $y = mx + b$,

$$y = \frac{1}{2}x + 3 \qquad \text{Substitute values for } m \text{ and } b.$$

The slope and y-intercept may also be determined from a linear graph. Therefore, we can write an equation if we are given such a graph.

To write a linear equation in two variables from a given nonvertical graph,

- Determine the y-coordinate of the y-intercept (b).
- Determine the slope of the line (m) by counting the rise and the run between two integer coordinate points.
- Write a linear equation in two variables by substituting the values for m and b into the slope–intercept form of the equation $y = mx + b$.

To write a linear equation in two variables from a given vertical graph, write an equation in the form $x = h$, where h is the x-coordinate of the x-intercept of the graph.

EXAMPLE 1 Write a linear equation in two variables for the given graph or from the given information.

a.

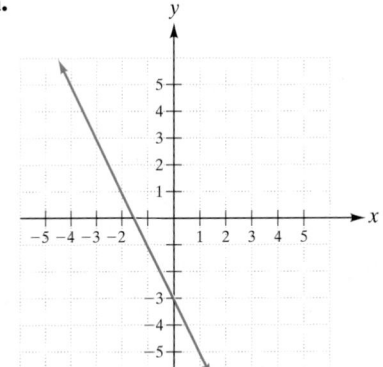

b.

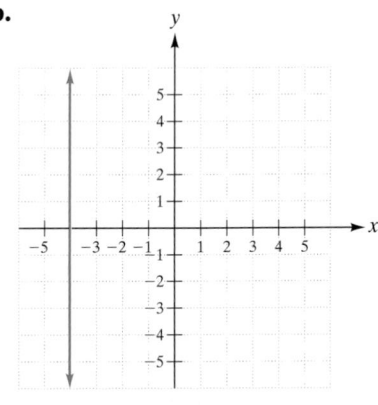

c.

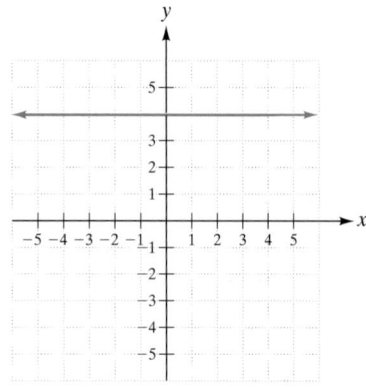

d. $m = 1; b = 0$ **e.** $m = 0; b = 3$

Solution

The graphs are labeled with the information needed to write an equation.

a.

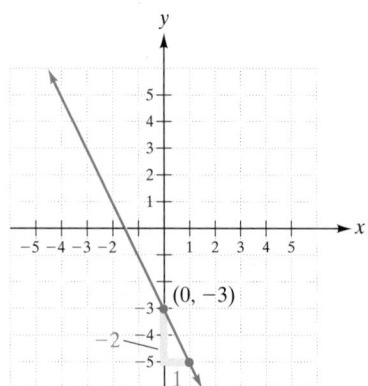

$m = \dfrac{-2}{1} = -2 \quad b = -3$

$y = mx + b$

$y = -2x - 3$

b.

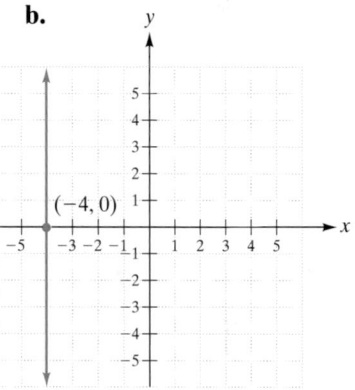

Vertical line with $h = -4$

$x = -4$

c.

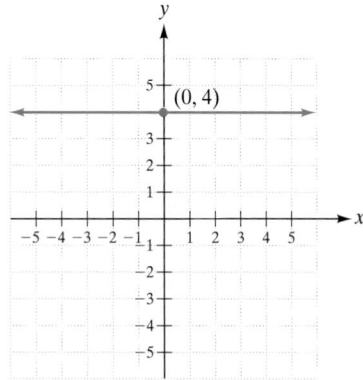

$m = 0 \quad b = 4$

$y = mx + b$

$y = 0x + 4$

$y = 4$

d. $m = 1; b = 0; y = mx + b$ **e.** $m = 0; b = 3; y = mx + b$

$$y = 1x + 0$$ $$y = 0x + 3$$

$$y = x$$ $$y = 3$$

 5.4.1 Checkup

In exercises 1–6, write a linear equation in two variables for the given graph or from the given information.

1.

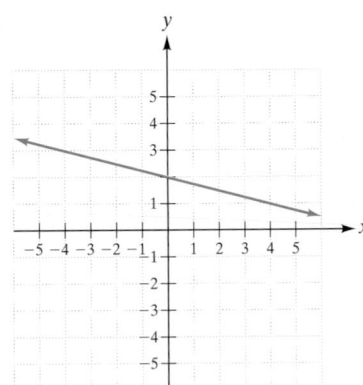

2.

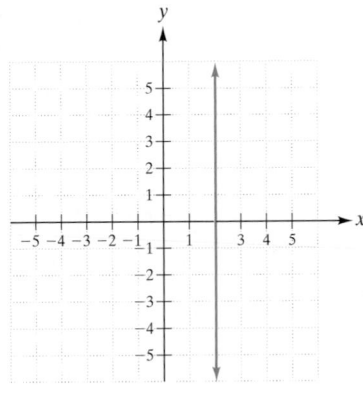

3.

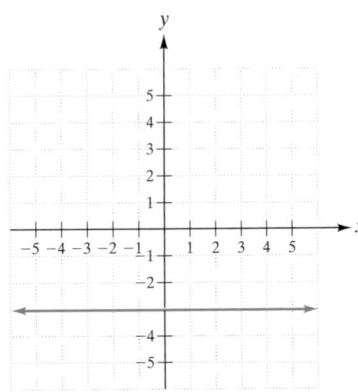

4.

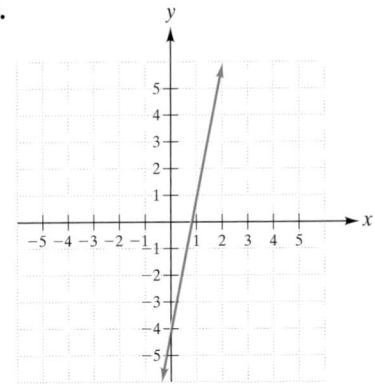

5. $m = 4; b = -1$

6. $m = -0.4; b = 3.2$

7. Why is $y = mx + b$ called the slope–intercept form of a linear equation?

5.4.2 Writing a Linear Equation from a Given Point and Slope

Sometimes we know the slope of a line and the coordinates of a point through which the line passes (x_1, y_1), and we want to write a linear equation in two variables for the line. To do so, we need a form for a linear equation in two variables that involves the slope and the coordinates of a point.

To write such a form for a linear equation in two variables, we will use a revised form of the slope formula, $m = \dfrac{y_2 - y_1}{x_2 - x_1}$. Instead of using two points on the line, we will write the variables x and y and use only the coordinates of one point, (x_1, y_1).

$$m = \frac{y_2 - y_1}{x_2 - x_1}$$

$$m = \frac{y - y_1}{x - x_1} \qquad \text{Use x and y instead of } x_2 \text{ and } y_2.$$

$$m(x - x_1) = \frac{y - y_1}{x - x_1}(x - x_1) \qquad \text{Multiply both sides by } (x - x_1).$$

$$m(x - x_1) = y - y_1 \qquad \text{Simplify.}$$

$$y - y_1 = m(x - x_1) \qquad \text{Write an equivalent equation.}$$

This form of a linear equation in two variables is called the **point–slope form**.

> **POINT-SLOPE FORM FOR A LINEAR EQUATION IN TWO VARIABLES**
>
> The point–slope form for a linear equation in two variables is $y - y_1 = m(x - x_1)$, where m is the slope of the line and (x_1, y_1) are the coordinates of a point located on the line.

We can use this point–slope form to write a linear equation in two variables, given the slope of a line and a point through which the line passes.

EXAMPLE 2   Write a linear equation in two variables for a line with a slope of $\frac{1}{2}$ and containing the point $(-3, 2)$.

Solution
Use the point–slope form for a linear equation. Let $m = \frac{1}{2}$, $x_1 = -3$, and $y_1 = 2$.

$$y - y_1 = m(x - x_1)$$

$$y - 2 = \frac{1}{2}[x - (-3)]$$

$$y - 2 = \frac{1}{2}(x + 3)$$

$$y - 2 = \frac{1}{2}x + \frac{3}{2}$$

$$y = \frac{1}{2}x + \frac{3}{2} + 2$$

$$y = \frac{1}{2}x + \frac{7}{2}$$

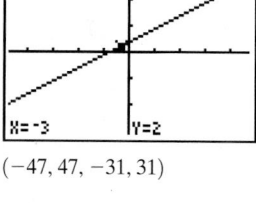

$(-47, 47, -31, 31)$

First, examine the equation to check the slope. Second, graph the equation on your calculator to determine that the given point is on the line.

$$y = mx + b$$

$$y = \frac{1}{2}x + \frac{7}{2}$$

By examination of the equation, the slope is $\frac{1}{2}$. On the calculator, we see that the graph of the equation passes through the point $(-3, 2)$.

HELPING HAND In the writing exercises at the end of this section, another method for finding the equation of a line, given its slope and a point, is presented.

5.4.2 Checkup

In exercises 1–2, write a linear equation in two variables for a line with the given slope and containing the given ordered pair. Graph the equation to check your solution.

1. $m = 2; (-1, 1)$ **2.** $m = \dfrac{3}{4}; (2, -3)$

3. Why is $y - y_1 = m(x - x_1)$ called the point–slope form of a linear equation?

5.4.3 Writing a Linear Equation for Two Given Points

If we know two ordered pairs (x_1, y_1) and (x_2, y_2) on a line and want to write a linear equation in two variables for the line, we must first determine the slope of the line and then use the point–slope form for a linear equation.

> To write a linear equation in two variables, given two data points,
>
> - Determine the slope of the line (m), using the slope formula.
> - Write a linear equation in two variables by substituting the values for m and (x_1, y_1) into the point–slope form of the equation, $y - y_1 = m(x - x_1)$.

EXAMPLE 3   Write a linear equation in two variables for a line containing the given ordered pairs. Graph the equation on your calculator to check your solution.

a. $(1, 3)$ and $(2, 7)$ **b.** $(3, 0)$ and $(4, 5)$
c. $(-1, 4)$ and $(2, 4)$ **d.** $(-1, 5)$ and $(-1, -2)$

Solution

a. Let $x_1 = 1$, $y_1 = 3$, $x_2 = 2$, and $y_2 = 7$

$$m = \frac{y_2 - y_1}{x_2 - x_1}$$

$$m = \frac{7 - 3}{2 - 1}$$

$$m = \frac{4}{1}$$

$$m = 4$$

Now substitute one of the given ordered pairs, $(1, 3)$, and the slope, 4, into the point–slope form.

$$y - y_1 = m(x - x_1)$$
$$y - 3 = 4(x - 1) \qquad (x_1, y_1) = (1, 3) \quad m = 4$$
$$y - 3 = 4x - 4$$
$$y = 4x - 1$$

It does not matter which ordered-pair coordinates are used. The results are the same.

$$y - y_1 = m(x - x_1)$$
$$y - 7 = 4(x - 2) \qquad (x_1, y_1) = (2, 7) \quad m = 4$$
$$y - 7 = 4x - 8$$
$$y = 4x - 1$$

Graph this equation on your calculator to check that the line contains both points. Enter the equation in Y1, graph the equation, and trace the graph to locate the given points.

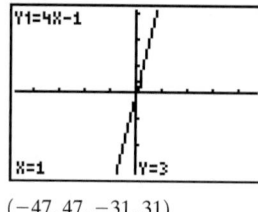
$(-47, 47, -31, 31)$

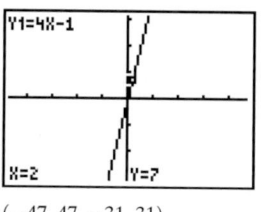
$(-47, 47, -31, 31)$

b. Determine the slope from the two given points.
Let $x_1 = 3$, $y_1 = 0$, $x_2 = 4$, and $y_2 = 5$

$$m = \frac{y_2 - y_1}{x_2 - x_1}$$

$$m = \frac{5 - 0}{4 - 3}$$

$$m = \frac{5}{1}$$

$$m = 5$$

Use the point–slope form. We will use $(3, 0)$ as the point. Therefore, $x_1 = 3$ and $y_1 = 0$.

$$y - y_1 = m(x - x_1)$$
$$y - 0 = 5(x - 3)$$
$$y - 0 = 5x - 15$$
$$y = 5x - 15$$

Check

$Y1 = 5x - 15$

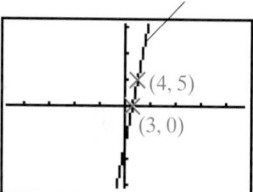
$(-47, 47, -31, 31)$

c. Determine the slope from the two given points.
Let $x_1 = -1$, $y_1 = 4$, $x_2 = 2$, and $y_2 = 4$

$$m = \frac{y_2 - y_1}{x_2 - x_1}$$

$$m = \frac{4 - 4}{2 - (-1)}$$

$$m = \frac{0}{3}$$

$$m = 0$$

The line is a horizontal line. The constant term is 4, because both ordered pairs have a y-coordinate of 4. The equation is $y = 4$.

If we had not noticed that the line was horizontal, we could have written the equation using the point–slope form. Use $(2, 4)$ as the point.

$$y - y_1 = m(x - x_1)$$
$$y - 4 = 0(x - 2)$$
$$y - 4 = 0x - 0$$
$$y - 4 = 0$$
$$y = 4$$

Check

$Y1 = 4$

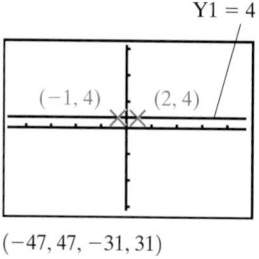
$(-47, 47, -31, 31)$

d. Determine the slope from the two given points.
Let $x_1 = -1$, $y_1 = 5$, $x_2 = -1$, and $y_2 = -2$

$$m = \frac{y_2 - y_1}{x_2 - x_1}$$

$$m = \frac{-2 - 5}{-1 - (-1)}$$

$$m = \frac{-7}{0}$$

m is undefined.

The line is a vertical line. The constant term of the equation is -1, because both ordered pairs have an x-coordinate of -1. The equation is $x = -1$. A calculator check is inappropriate.

 5.4.3 Checkup

In exercises 1–3, write a linear equation in two variables for a line containing the given ordered pairs. Graph the equation to check that it is correct.

1. $(2, 3)$ and $(-1, 1)$

2. $\left(-2, \dfrac{3}{2}\right)$ and $\left(4, \dfrac{3}{2}\right)$

3. $\left(-\dfrac{5}{2}, \dfrac{9}{2}\right)$ and $\left(-\dfrac{5}{2}, 2\right)$

4. We know that if you are given two points, you can draw a line between them. How can you find the equation of the line that connects them?

5.4.4 Writing a Linear Equation for a Graph Described by a Second Equation

In some situations, we need to write a linear equation in two variables when its graph is described in terms of a second equation. The details may vary from case to case, but the procedure always involves finding the slope and a point on the line.

> To write a linear equation in two variables in slope–intercept form when its graph is described in terms of a second equation,
> Determine the slope of the line.
>
> - If the two lines have the same slope or are parallel, use the slope of the given equation.
> - If the two lines are perpendicular, use the opposite reciprocal of the slope of the given equation.
>
> Write the equation.
>
> - Substitute the values for the slope and the given point into the point–slope form of the equation.
> - Solve for y.
>
> Check the equation.
>
> - Verify that the given conditions have been satisfied.
> - A graph may be needed to check the conditions.

EXAMPLE 4   Write a linear equation in two variables for a line determined by the given information. Graph the equation on your calculator to check that the equation is correct.

a. A line passes through the point $(-1, -3)$ and is parallel to the graph of $2x + 3y = -2$.

b. A line passes through the point $(3, 5)$ and is perpendicular to the graph of $x - 2y = 1$.

Solution

a. Parallel lines have equal slopes. The slope of the given line is found by solving the equation for y to obtain the slope–intercept form, $y = mx + b$.

$$2x + 3y = -2$$
$$2x + 3y - 2x = -2 - 2x$$
$$3y = -2 - 2x$$
$$3y = -2x - 2$$
$$\frac{3y}{3} = \frac{-2x - 2}{3}$$
$$y = -\frac{2}{3}x - \frac{2}{3}$$

The slope of the given line is $-\frac{2}{3}$. A parallel line will have the same slope. Use the point–slope form with $-\frac{2}{3}$ as the slope and $(-1, -3)$ as the point.

$$y - y_1 = m(x - x_1)$$

Check

$$y - (-3) = -\frac{2}{3}[x - (-1)]$$

$\text{Y2} = -\frac{2}{3}x - \frac{2}{3}$

$(-1, -3)$

$\text{Y1} = -\frac{2}{3}x - \frac{11}{3}$

$$y + 3 = -\frac{2}{3}(x + 1)$$

$$y + 3 = -\frac{2}{3}x - \frac{2}{3}$$

$$y + 3 - 3 = -\frac{2}{3}x - \frac{2}{3} - 3$$

$$y = -\frac{2}{3}x - \frac{2}{3} - \frac{9}{3}$$

$$y = -\frac{2}{3}x - \frac{11}{3}$$

b. Perpendicular lines have opposite reciprocal slopes. The slope of the given line is found by solving the equation for y to obtain the slope–intercept form, $y = mx + b$.

$$x - 2y = 1$$
$$x - 2y - x = 1 - x$$
$$-2y = 1 - x$$
$$\frac{-2y}{-2} = \frac{1 - x}{-2}$$
$$y = \frac{-1}{2} + \frac{1}{2}x$$
$$y = \frac{1}{2}x - \frac{1}{2}$$

The slope of the given line is $\frac{1}{2}$. The slope of a line perpendicular to the given line is -2 (the opposite reciprocal of $\frac{1}{2}$).

Use the point–slope form. Use -2 as the slope and $(3, 5)$ as the point.

$$y - y_1 = m(x - x_1)$$
$$y - 5 = -2(x - 3)$$
$$y - 5 = -2x + 6$$
$$y - 5 + 5 = -2x + 6 + 5$$
$$y = -2x + 11$$

Check

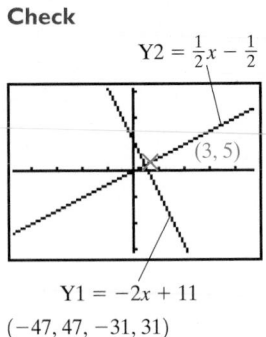

$Y2 = \frac{1}{2}x - \frac{1}{2}$

$(3, 5)$

$Y1 = -2x + 11$

$(-47, 47, -31, 31)$

 5.4.4 Checkup

In exercises 1 and 2, write a linear equation in two variables for a line determined by the given information. Graph the equation on your calculator to check that the equation is correct.

1. A line passes through the point $(3, 2)$ and is parallel to the graph of $3x + 2y = 3$.
2. A line passes through the point $(1, -1)$ and is perpendicular to the graph of $2x - 5y = 15$.

3. If you must find the equation of a line that is parallel to the graph of another equation, how can you determine which slope to use for your equation?

4. If you must find the equation of a line that is perpendicular to the graph of another equation, how can you determine which slope to use for your equation?

5.4.5 Modeling the Real World

Many real-world situations involve complicated relationships described by equations that are difficult to solve. But often, you can use a linear equation in two variables to approximate a solution. All you need to graph the linear equation is a point on the line and the slope of the line; then you can use the graph to find an approximate solution to your original problem. This is usually a lot easier to do than solving a more complicated equation, and frequently the approximate solution is reasonably close to the actual solution.

EXAMPLE 5   According to the National Safety Council, unintentional-injury deaths (accidental deaths) have decreased at a rate of 486 deaths per year over a 10-year period, from 97,100 in 1988.

a. Write a linear equation to determine the number of unintentional-injury deaths, y, in terms of x years after 1988.

b. Use the equation to determine the number of unintentional-injury deaths in 1998. According to the same report, 92,200 unintentional-injury deaths were reported in 1998. Compare this figure with the answer you obtained from your equation.

c. Use your equation to predict the number of unintentional-injury deaths expected in 2008.

Solution

a. The slope of the graph is decreasing at a rate of 486 per year, or $m = -486$. Since 1988 is 0 years after 1988, the y-intercept is $(0, 97,100)$, or $b = 97,100$.

$$y = mx + b$$
$$y = -486x + 97,100$$

b. The year 1998 is 10 years after 1988; therefore, $x = 10$.

$$y = -486x + 97,100$$
$$y = -486(10) + 97,100$$
$$y = 92,240$$

The equation yields 40 deaths more than the actual reported figure of 92,200.

c. The year 2008 is 20 years after 1988; therefore, $x = 20$.

$$y = -486x + 97,100$$
$$y = -486(20) + 97,100$$
$$y = 87,380$$

The equation estimates that in 2008 the number of unintentional-injury deaths will be 87,380, assuming that the rate of decrease continues to hold.

EXAMPLE 6 Many formulas are found by writing equations to fit given data points. An example of this is the temperature conversion formulas. Two known reference points on the Celsius and Fahrenheit temperature scales are the freezing and boiling points of water. Water freezes at 0 degrees Celsius and 32 degrees Fahrenheit. Water boils at 100 degrees Celsius and 212 degrees Fahrenheit.

a. Write a formula (equation) for the Fahrenheit temperature (F) in terms of the Celsius temperature (C).

b. Using the formula you derived in part a, determine the temperature in degrees Fahrenheit if the temperature is 20 degrees Celsius.

Solution

a. The two data points (C, F) are $(0, 32)$ and $(100, 212)$. First, determine the slope.

$$m = \frac{F_2 - F_1}{C_2 - C_1}$$

$$m = \frac{212 - 32}{100 - 0}$$

$$m = \frac{180}{100}$$

$$m = \frac{9}{5}$$

Use the point–slope form to write an equation. Write the equation in terms of C and F instead of x and y.

$$F - F_1 = m(C - C_1)$$

$$F - 32 = \frac{9}{5}(C - 0)$$

$$F - 32 = \frac{9}{5}C - 0$$

$$F = \frac{9}{5}C + 32$$

b. Substitute 20 for C in the equation just determined.

$$F = \frac{9}{5}C + 32$$

$$F = \frac{9}{5}(20) + 32$$

$$F = 36 + 32$$

$$F = 68$$

The temperature is 68 degrees Fahrenheit when the Celsius reading is 20 degrees.

APPLICATION

In the science fiction movie *Star Games 23: The Comet Strikes Jasper,* astronomers discover the comet Hale–Farewell traveling straight toward the planet Jasper. They measure the speed of the comet as 20,000 km/hr and its position as 3.0×10^7 km from Jasper. Assuming that the comet continues on its course in a straight line at the same speed, write an equation for its distance from Jasper and determine when it will hit the planet.

Discussion

Let x = the travel time (in hours) of the comet
$\quad\ y$ = the distance (in kilometers) the comet is from Jasper

The average rate of change of the distance is decreasing at a rate of 20,000 km/hr (the speed). Therefore, $m = -20{,}000$ or -2.0×10^4.

At the time of 0 hours, the comet's distance from Jasper is 3.0×10^7 km. This gives us a data point of $(0, 3.0 \times 10^7)$. Therefore, $b = 3.0 \times 10^7$.

$y = mx + b$ Slope–intercept form

$y = (-2.0 \times 10^4)x + (3.0 \times 10^7)$ Substitute values for m and b.

The comet will hit Jasper when $y = 0$, because its distance from Jasper, y, will then be 0 km. Therefore, substitute 0 for y and solve for x. (This point is the x-intercept.)

$$y = (-2.0 \times 10^4)x + (3.0 \times 10^7)$$
$$0 = (-2.0 \times 10^4)x + (3.0 \times 10^7)$$
$$0 + (2.0 \times 10^4)x = (-2.0 \times 10^4)x + (3.0 \times 10^7) + (2.0 \times 10^4)x$$
$$(2.0 \times 10^4)x = 3.0 \times 10^7$$
$$\frac{(2.0 \times 10^4)x}{2.0 \times 10^4} = \frac{3.0 \times 10^7}{2.0 \times 10^4}$$
$$x = 1.5 \times 10^3$$
$$x = 1500$$

The comet will hit Jasper in 1500 hours.

The preceding example was science fiction, but astronomers actually did a calculation like that for Comet Shoemaker–Levy, which slammed into Jupiter on July 16, 1994. Of course, the actual calculation was more complicated because the comet wasn't moving at a constant speed or in a straight line, but even so, astronomers were able to pinpoint the time of collision to within minutes for each part of the "string-of-pearls" comet. For example, on July 12, a table of anticipated collision times was published in the *New York Times*. The predicted time for the largest piece, Fragment G, to hit Jupiter was 3:24 A.M. Eastern time. On July 19, the *New York Times* reported that the time of impact for Fragment G on July 16 was 3:28 A.M. Eastern Daylight time, only 4 minutes off the predicted time.

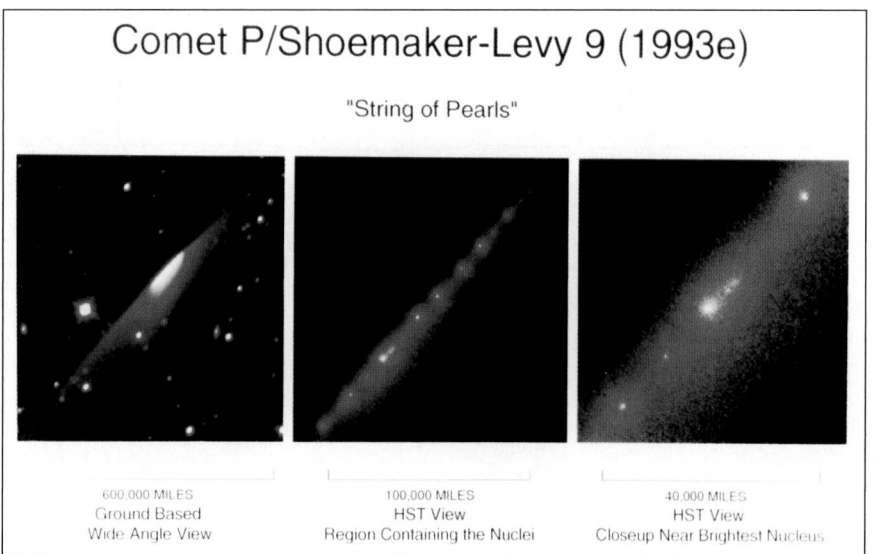

Comet P/Shoemaker-Levy 9 (1993e)

"String of Pearls"

600,000 MILES	100,000 MILES	40,000 MILES
Ground Based	HST View	HST View
Wide Angle View	Region Containing the Nuclei	Closeup Near Brightest Nucleus

 5.4.5 Checkup

1. In 1990, the accidental death rate was 37.0 deaths per 100,000 population. In 1997, the rate was 34.4 per 100,000 population. Let x represent the number of years since 1990, and let y represent the accidental death rate per 100,000 population.
   a. Find the slope of the line connecting the two points $(0, 37.0)$ and $(7, 34.4)$.
   b. Write an equation to determine the number of accidental deaths per 100,000 population after 1990.
   c. Use the equation to predict the number of accidental deaths per 100,000 population in 1995. The actual number recorded was 35.5 deaths per 100,000 population. Was your prediction close?

   d. What would your equation predict as the rate of accidental deaths for 2000?
2. If we reverse the variables in Example 6, the two data points (F, C) will become $(32, 0)$ and $(212, 100)$.
   a. Determine the new slope of the relation between F and C.
   b. Use the point–slope form to write a new equation for C in terms of F.
   c. Using the formula, find the Celsius temperature if the temperature is 75 degrees Fahrenheit.

5.4 Exercises

Write a linear equation for each graph.

1.

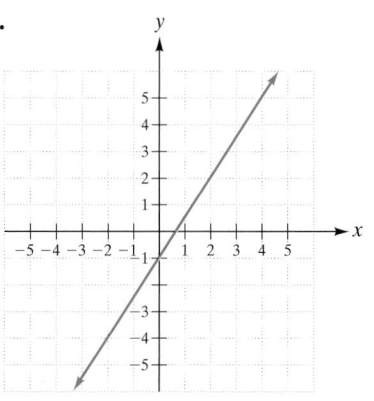

2.

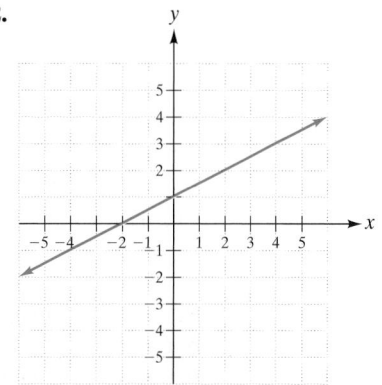

3.

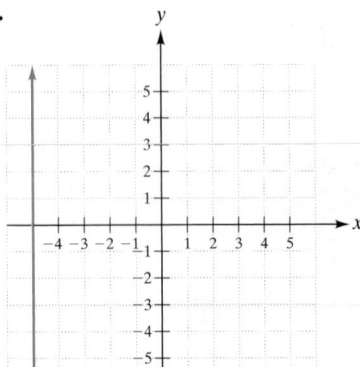

4.

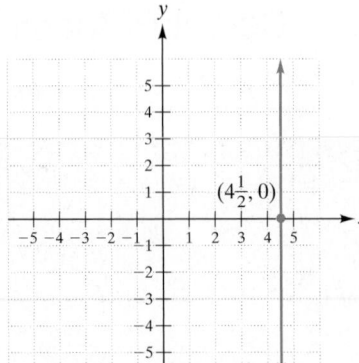

$(4\frac{1}{2}, 0)$

5.

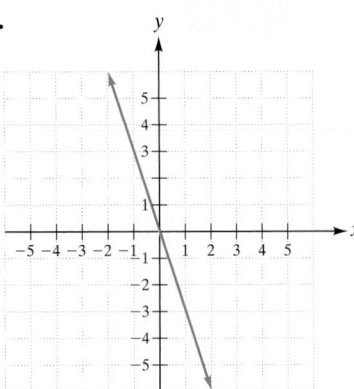

6.

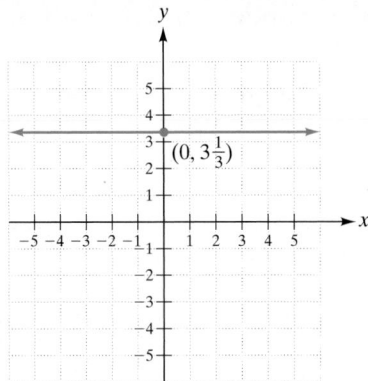

$(0, 3\frac{1}{3})$

7.

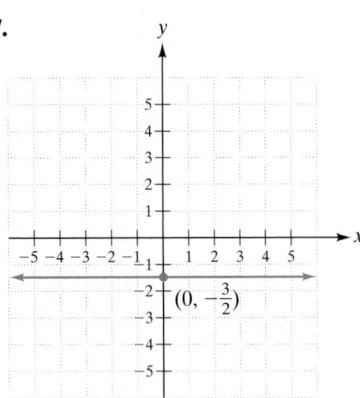

$(0, -\frac{3}{2})$

8.

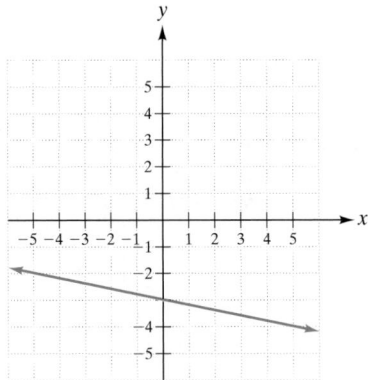

Write a linear equation for the given information.

9. $m = -\dfrac{2}{5}, b = 4$ **10.** $m = -\dfrac{1}{7}, b = -9$ **11.** $m = \dfrac{5}{9}, b = 0$ **12.** $m = \dfrac{1}{7}, b = -1$

13. $m = 4, b = -\dfrac{3}{4}$ **14.** $m = 11, b = \dfrac{1}{2}$ **15.** $m = -4.1, b = 0.5$ **16.** $m = -6.2, b = -2.2$

17. $m = 0, b = -33$ **18.** $m = -4, b = 0$

Write a linear equation in two variables for a line with the given slope and containing the given ordered pair. Graph the equation to check that it is correct.

19. $m = \dfrac{2}{3}, (3, -3)$ **20.** $m = -2, (4, 0)$ **21.** $m = -3, (0, 4)$ **22.** $m = \dfrac{4}{3}, (-6, -3)$

23. $m = -1.7, (3, -1.5)$ **24.** $m = 1.4, (1.5, -1.2)$

Write a linear equation for a line containing the given ordered pairs. Graph the equation to check that it is correct.

25. $(-1, 1)$ and $(1, -2)$ **26.** $(8, 6)$ and $(1, 6)$ **27.** $(-1, -2)$ and $(-1, 5)$

28. $(1, 6)$ and $(2, 1)$ **29.** $(-1, 1)$ and $(-2, -1)$ **30.** $(-3, 2)$ and $(4, 4)$

31. $(-2, 2)$ and $(4, 2)$

32. $(2, 9)$ and $(2, 1)$

33. $\left(4\frac{1}{2}, 5\frac{1}{4}\right)$ and $(1, 4)$

34. $\left(-2, -1\frac{1}{3}\right)$ and $\left(2\frac{1}{2}, 1\right)$

35. $\left(-1\frac{1}{3}, 2\right)$ and $(0, 0)$

36. $\left(2\frac{1}{2}, 5\frac{1}{2}\right)$ and $\left(3\frac{1}{2}, 1\right)$

37. $(0.5, 0)$ and $(-0.8, 4.2)$

38. $(-4, -1)$ and $(-5.1, -4.5)$

39. $(2.4, 2.8)$ and $(-2.6, -2.2)$

40. $(1, 6)$ and $(1.5, -3.5)$

In exercises 41–60, write a linear equation for a line determined by the given information. Check by graphing the equations.

41. A line passes through the point $(8, 7)$ and has the same slope as the line whose equation is $3x - 8y = 32$.

42. A line passes through the point $(6, 2)$ and has the same slope as the line whose equation is $5x + 3y = 15$.

43. A line passes through the point $(4, 0)$ and has the same slope as the line whose equation is $x + 2y = 7$.

44. A line passes through the point $(0, -3)$ and has the same slope as the line whose equation is $2x - y = 4$.

45. A line whose y-intercept has a y-coordinate of -5 has the same slope as the line whose equation is $y = 3x + 4$.

46. A line whose y-intercept has a y-coordinate of 2 has the same slope as the line whose equation is $y = -3x + 5$.

47. A line passes through the point $\left(\frac{4}{9}, -\frac{5}{6}\right)$ and is parallel to the graph of $y = 3x - 10$.

48. A line passes through the point $\left(\frac{2}{3}, -\frac{1}{6}\right)$ and is parallel to the graph of $y = 2x + 5$.

49. A line parallel to the graph of $y = -1.2x + 3.5$ passes through the point $(4, -2)$.

50. A line parallel to the graph of $y = -0.8x + 2.4$ passes through the point $(9, -3)$.

51. A line parallel to the graph of $2x + 4y = 5$ passes through the origin.

52. A line parallel to the graph of $x - 3y = 6$ passes through the origin.

53. A line passes through the point $(3.6, 5.8)$ and is perpendicular to the line whose equation is $y = 3x + 12$.

54. A line passes through the point $(-3.6, 1.8)$ and is perpendicular to the graph of $y = -6x$.

55. A line passes through the point $(15, -30)$ and is perpendicular to the graph of $y = 5x - 1$.

56. A line passes through the point $(-10, 8)$ and is perpendicular to the graph of $y = 2x + 5$.

57. A line passes through the origin and is perpendicular to the line whose equation is $y = -\frac{2}{3}x - 1$.

58. A line passes through the origin and is perpendicular to the line whose equation is $y = \frac{5}{4}x + 2$.

59. A line whose x-intercept has an x-coordinate of 3 is perpendicular to the line whose equation is $3x + 2y = 4$.

60. A line whose x-intercept has an x-coordinate of -2 is perpendicular to the line whose equation is $5x - 4y = 4$.

61. According to data collected by the U.S. Centers for Disease Control and Prevention, the total number of AIDS cases reported has been decreasing since 1994, when 77,103 cases were reported. In 1995, the number of cases reported was 70,864. Let x represent the number of years after 1994, and let y represent the number of AIDS cases reported.

  a. Find the slope of the line connecting the points $(0, 77,103)$ and $(1, 70,864)$.

  b. Write an equation to estimate the number of AIDS cases reported x years after 1994.

  c. What would you estimate the number of cases to be for 1997? The reported number of cases was 58,443. Is your estimate close?

  d. What does the equation predict for the number of cases in the year 2005?

  e. In what year would the number of cases become close to 0? Is this a realistic prediction? Explain.

62. The U.S. Centers for Disease Control and Prevention reported 13,802 AIDS cases in females in 1994. In 1995, the number of cases in females declined slightly, to 13,413. Let x represent the number of years after 1994, and let y represent the number of AIDS cases reported in females.

  a. Find the slope of the line connecting $(0, 13,802)$ and $(1, 13,413)$.

  b. Write an equation to estimate the number of female AIDS cases reported x years after 1994.

  c. Estimate the number of female AIDS cases reported for 1997. Compare your estimate with the reported number of 12,747. Did the equation come close?

  d. What does the equation predict will be the number of female AIDS cases reported in the year 2005?

  e. In what year would the number of cases become close to 0? Is this a realistic prediction? Explain.

63. In scientific work, the Kelvin temperature scale measures temperatures from absolute zero, which is the lowest possible temperature. Temperatures on this scale have a linear relationship with the Celsius scale. Given the points (C, K) of $(0, 273)$ and $(-10, 263)$, use the slope–intercept form to write the relationship between the two scales. What Kelvin temperature corresponds to 100°C?

64. The Rankine temperature scale is used in engineering practice. Temperatures on this scale have a linear relationship with the Fahrenheit scale. Given the points (F, R) of $(32, 492)$ and $(180, 640)$, use the point–slope form to determine the relationship between the two scales. What Rankine temperature corresponds to 75°F?

The tables in exercises 65 and 66 were reported to be generated from linear equations. Use the data in the tables to find the equations, and then verify the data points.

65.

x	1.2	1.6	2.0	2.4	2.8
y	15.92	16.96	18.00	19.04	20.08

66.

x	1.0	1.5	2.0	2.5	3.0
y	17.8	15.9	14.0	12.1	10.2

67. A consumer article reported that a car weighing 2500 pounds averages 40 miles per gallon of gasoline, while a car weighing 3500 pounds averages 35 miles per gallon. Write a linear equation relating the gas mileage, y, to the weight of the car, x.

68. A consumer article reported that a car weighing 5000 pounds averages 15 miles per gallon of gasoline, while a car weighing 4500 pounds averages 19 miles per gallon. Write a linear equation relating the gas mileage to the weight of the car. How does this equation compare with that of exercise 67? Can you explain any differences?

69. In one city, a company ran radio advertisements 6 times per week and had sales of 5000 units. In another city, the company ran the radio advertisements 15 times per week and had sales of 14,000 units. Write a linear equation that relates sales, y, to the number of times the advertisements were run each week, x. What would you predict the sales would be if the advertisements were run 10 times per week?

70. A study of aging found that people who were 65 years old tended to live an additional 16.5 years on the average. Persons who were 79 years old tended to live an average of 8.4 years longer. Write a linear equation that relates the average additional years of life, y, to a person's age, x. What would you predict is the remaining life span for a 70-year-old person?

71. At New Futures Community College, enrollment has increased at a rate of 120 students per year. In 1990, the enrollment of the college was 5470 students. Write a linear equation to determine the enrollment of the college x years after 1990. Use the equation to determine the enrollment of the college in 1996. If the actual enrollment in 1996 was 6225, did the equation provide a good estimate of this number? Use the equation to predict what the enrollment will be in the year 2010 if the increase continues.

72. At New Endeavors Computer Systems, business has been declining at a rate of $250,000 per year. The company's business total for 1993 was $4,375,000. Write a linear equation to determine the amount of business the company has x years after 1990. Use the equation to determine the amount of business in 1990. How does this compare with the actual amount of $5,065,000? Use the equation to predict what the amount of business will be in the year 2010 if the decline continues.

73. Between 1980 and 1985, the economic loss from motor vehicle accidents increased by about $3.78 billion per year, as reported by the National Safety Council. The loss for 1980 was $57.1 billion.
  a. Write a linear equation to determine the economic loss, y, in terms of x years after 1980.
  b. Use the equation to estimate the economic loss for 1997. According to data from the National Safety Council, the economic loss in 1997 was 123.7 billions of dollars. Compare this figure with your answer.
  c. Use the equation to predict what the economic loss will be in 2005.

74. The U.S. National Center for Education Statistics reported that in 1990 school expenditures for public elementary and secondary schools totaled $271 trillion and was increasing at a rate of about $6 trillion per year.
  a. Write a linear equation to determine these expenditures, y, in terms of x years after 1990.
  b. Use the equation to estimate the expenditure for 1998. According to the data reported, the approximate expenditure for 1998 was $324 trillion. Compare this figure with your answer.
  c. Use the equation to predict what the expenditures will be in 2005.

5.4 Calculator Exercises

Graphing calculators have statistical procedures that can be used to find the equation of a line passing through two points. While this is not the intent of the procedures, they can be used to write such an equation. As an example, suppose you wish to find the equation of the line that passes through the points (3, 55) and (1, 10).

a. Clear any data from the lists L1 and L2, which is where the coordinate pairs will be stored.

| STAT | 4 | 2nd | LI | , | 2nd | L2 | ENTER |

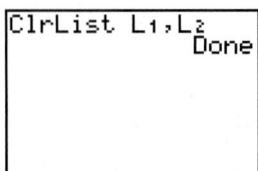

b. Store the coordinate pairs in your calculator, using L1 for the *x*-coordinate and L2 for the *y*-coordinate.

| STAT | | |

Use the arrow keys to move to the L1 column.

| 3 | ENTER | | | ENTER | ▶ | 55 | ENTER | 10 | ENTER | 2nd | QUIT |

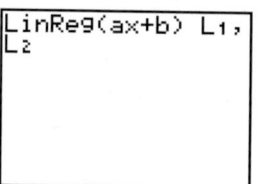

c. Calculate the "linear regression," which will provide the equation.

| STAT | ▶ | 4 | 2nd | L1 | , | 2nd | L2 | ENTER |

This instruction directs the calculator to find the *x*-coordinates in list L1 and the *y*-coordinates in list L2 and to come up with the equation. When the ENTER command is keyed, the calculator displays the result.

d. The equation for the line passing through the two points is given by

$$y = 22.5x - 12.5$$

Use this method to find the equations of the lines passing through the given pairs of points. Check the solutions by graphing or by substituting the coordinate pairs into the equations.

1. $(12, 925)$ and $(72, 4225)$ **2.** $(16, -351)$ and $(40, 417)$ **3.** $(0, 6.4)$ and $(36.4, 103.59)$

4. $\left(1, \dfrac{17}{4}\right)$ and $\left(3, \dfrac{57}{4}\right)$ **5.** $(-5, 21)$ and $(-9, 7)$ **6.** $(16, -2)$ and $(7, 4)$

5.4 Writing Exercises

In this section, you learned that if you were given the slope of a line (m) and a point (x_1, y_1) that the line passes through, you can find the equation of the line by substituting the values for m, x_1, and y_1 into the point–slope form,

$$y - y_1 = m(x - x_1)$$

and simplifying the resulting equation. However, if you substitute the values for m, x_1 (for *x*), and y_1 (for *y*) into the slope–intercept form,

$$y = mx + b$$

you can solve for b. Then, when you substitute the values for m and b back into the slope–intercept form, you will obtain the same equation as you did using the point–slope form.

Try both methods to find the equation of a line with a slope of $\frac{3}{4}$ and passing through $(4, 1)$. Explain why you think both methods yield the same equation. Then state which method you prefer, and explain why you prefer that method.

CHAPTER 5 SUMMARY

After completing this chapter, you should be able to define in your own words the following key terms.

5.1
linear equation in two variables
linear function

5.2
rise
run
slope

subscripts
slope–intercept form
grade
average rate of change

5.3
coinciding
coincident

parallel
intersecting
perpendicular
break-even point

5.4
point–slope form

Reflections

1. How can you determine algebraically the x-intercept and the y-intercept of the graph of a linear equation?
2. If the graph of a linear equation is a horizontal line, what can you say about its intercepts? If the graph of a linear equation is a vertical line, what can you say about its intercepts?
3. Explain how to graph a linear equation in two variables by using the intercept method.
4. What does the slope of a line measure? How can you determine the slope of a line?
5. Describe what the graph of a linear equation in two variables will look like if
   a. the slope of the line is positive.　　**b.** the slope of the line is negative.
   c. the slope of the line is 0.　　**d.** the slope of the line is undefined.
6. How would you graph a linear equation in two variables using the slope–intercept method?
7. What can you say about the slopes of two lines if
   a. the lines are parallel?　　　　　　　　**b.** the lines are intersecting only?
   c. the lines are intersecting and perpendicular?　　**d.** the lines are coinciding?
8. How can you write an equation for a line if it is described by
   a. its slope and y-intercept?　　**b.** its slope and a point on the line?
   c. two points on the line?　　**d.** a point on the line and the equation of another line parallel to the given line?
   e. a point on the line and the equation of another line perpendicular to the given line?

CHAPTER 5 SECTION-BY-SECTION REVIEW

5.1

Recall	Examples
Linear equation in two variables • The standard form is $ax + by = c$, where a, b, and c are real numbers and a and b are not both 0.	Linear equations in two variables $2x + 4y = 5$ $x = 7$ $y = \dfrac{2}{3}$
Graph by using ordered pairs • Determine three ordered-pair solutions of the equation, using a table of values. • Graph and label the three ordered-pair solutions. • Connect the graphed points with a straight line, and add arrows to the ends of the line.	Graph $-2x + 4y = 1$. Solve for y. $y = \dfrac{1}{2}x + \dfrac{1}{4}$ Set up a table. $$\begin{array}{c\|c\|c} x & y = \frac{1}{2}x + \frac{1}{4} & y \\ \hline -2 & y = \frac{1}{2}(-2) + \frac{1}{4} & -\frac{3}{4} \\ 0 & y = \frac{1}{2}(0) + \frac{1}{4} & \frac{1}{4} \\ 2 & y = \frac{1}{2}(2) + \frac{1}{4} & \frac{5}{4} \end{array}$$
y-intercept • To determine the y-coordinate of the y-intercept, b, algebraically, substitute 0 for x and solve for y. • Or solve the equation for y. The constant term is the y-coordinate of the y-intercept, b, where $y = mx + b$. • Write an ordered pair $(0, b)$.	Determine the y-intercept of the graph, $2x + y = 6$. Let $x = 0$.　　Solve for y. $2x + y = 6$　　$2x + y = 6$ $2(0) + y = 6$　　$y = -2x + 6$ $y = 6$　　or $b = 6$ The y-intercept is $(0, 6)$.

Recall	Examples
x-intercept • To determine the *x*-intercept algebraically, substitute 0 for *y* and solve for *x*. • Write an ordered pair with the second coordinate 0 and the value obtained for *x* as the first coordinate.	Determine the *x*-intercept of the graph for $2x + y = 6$. Let $y = 0$. $2x + y = 6$ $2x + (0) = 6$ $2x = 6$ $x = 3$ The *x*-intercept is $(3, 0)$.
Graph by using the intercept method • Determine the *x*-intercept and *y*-intercept. • Graph the intercepts. • Draw a line connecting the two points with arrows on both ends. • Check a third ordered-pair solution from the graph.	Graph $2x + y = 6$ by using the intercept method. Use the *x*-intercept and *y*-intercept found in the previous two examples, and graph. Use the point $(1, 4)$ to check for correctness. $2x + y = 6$ $2(1) + 4 \mid 6$ 6 The solution checks.	

Identify each equation as linear or nonlinear. Express each linear equation in standard form.

1. $y = 0.6x + 2.3$ **2.** $y = 4x^2 - 2$

3. $5x - 3y + 7 = x - y + 9$ **4.** $5y - 12 = 7 - y$

Determine three solutions for each equation, and graph the equations.

5. $12x + 6y = 48$ **6.** $y = \dfrac{8}{13}x - 7$ **7.** $y = -9$ **8.** $9x - y = 12$

9. $y = -\dfrac{4}{3}x + 2$ **10.** $5y - 2 = y - 10$ **11.** $2x + 16 = x + 18$

Determine the intercepts of each graph of each linear equation. If possible check your answers with your calculator.

12. $8x + 12y = -24$ **13.** $y = \dfrac{2}{5}x + 4$ **14.** $2y - 4 = y - 7$ **15.** $2(x - 1) + 5 = 9$

Solve each equation for y to determine the y-intercept.

16. $9x - 3y = -12$ **17.** $6x + 2y = x$ **18.** $3y + 2 = 2(y - 4)$

Graph each equation by using the intercept method. Label the points.

19. $7x + 11y = 77$ **20.** $-x - y = 2$ **21.** $7.40x + 14.80y = 29.60$

22. Noriko translates Japanese texts into English as a consultant to a firm that does business in Japan. She is paid $85 to translate 10 pages of text on one job and $160 to translate 20 pages of text on another job.
   a. Let *x* represent the number of pages translated on a particular job, and let *y* represent Noriko's pay for the job. Graph a representation of the information.
   b. Assume that the information is from a linear relation, and connect the points. From the graph, how much will Noriko receive for a job that is 30 pages long?

23. The rental cost for a copy machine is $25 per week. It costs $0.02 to make a copy of a page of a document. The copy center charges its customers $0.04 per copy made. Write a linear function for the weekly profit that the copy center will realize. Graph the function by the intercept method. Explain the meaning of the intercepts.

24. A McDonnell Douglas MD80J aircraft can fly from Chicago to Los Angeles in 4 hours and 25 minutes. If we let *y* represent the plane's distance from Chicago, then an equation representing that distance is $y = 1749 - 396x$, where *x* is the flight time in hours. Graph the equation and interpret the intercepts of the graph. From the graph, estimate how long it will take to fly 1000 miles from Chicago.

25. Kari was awarded a scholarship for a year of college. The amount awarded was $5200, to be paid out as $100 weekly. Let x represent the number of weeks payment has been made, and let y represent the balance of the scholarship account.

    a. Write a linear equation in two variables for the balance of the account after x weeks of payments.

    b. Without graphing, determine the intercepts for a graph of the equation. Interpret what the intercepts represent.

    c. Graph the equation by using the intercept method.

    d. From the graph, determine how much money remains in the account after 32 weeks of payments.

5.2

Recall	Examples
Slope of a line • Given a graph, locate two integer coordinate points on the graph, draw a right triangle, count the rise and the run, and write the ratio, $\frac{\text{rise}}{\text{run}}$. • Given two ordered pairs, (x_1, y_1) and (x_2, y_2), use the slope formula, $m = \dfrac{y_2 - y_1}{x_2 - x_1}$ • Given an equation of the line, solve for y. The coordinate of the x-term is the slope.	Determine the slope of the given graph. 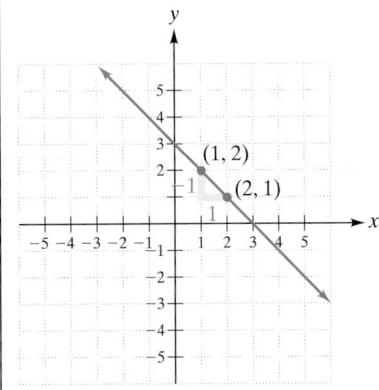 The slope is $\frac{-1}{1} = -1$. Determine the slope of a line passing through the points $(1, 2)$ and $(5, -2)$. $m = \dfrac{y_2 - y_1}{x_2 - x_1}$ $m = \dfrac{-2 - 2}{5 - 1}$ $m = \dfrac{-4}{4} = -1$ Determine the slope of the line, given $y = -x + 3$. The slope is the coefficient of the x-term, or -1.
Graph by using the slope–intercept method • Graph and label the y-intercept. • Count the slope $\left(\frac{\text{rise}}{\text{run}}\right)$. Graph and label the point determined. • For greater accuracy, count the slope from the second point, graph and label a third point. • Connect the points to draw the line.	Graph $y = -2x + 6$ by using the slope–intercept method. The slope intercept form is $y = mx + b$. The slope, m, is -2, or $\frac{-2}{1}$, and the y-intercept is $(0, 6)$. 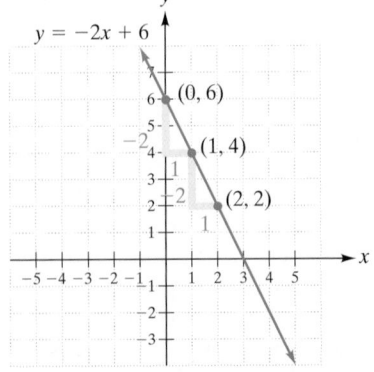

In exercises 26–29, determine the slope of the graph. Determine whether the graph represents a function. If the graph represents a function, is the function increasing, decreasing, or constant?

26.

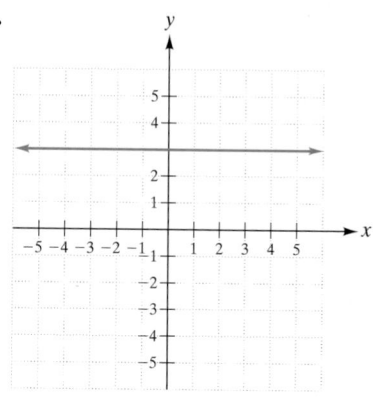

27.

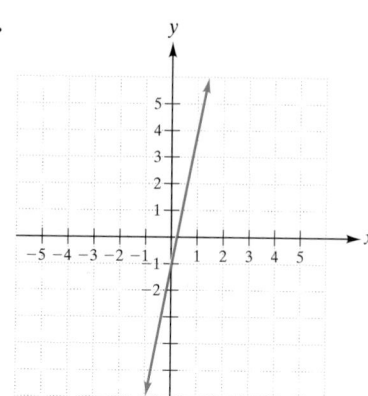

28.

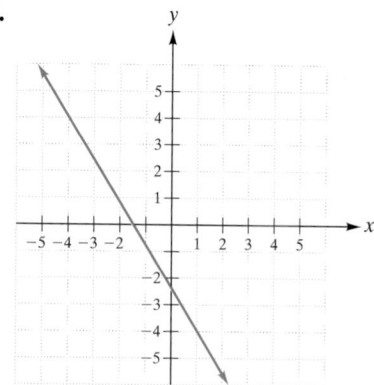

29.

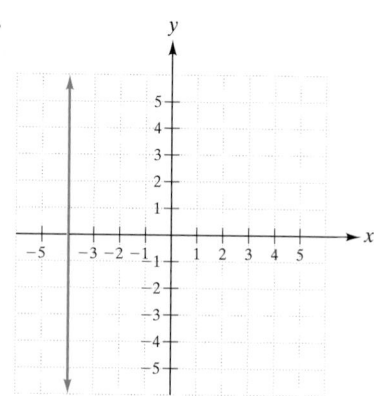

Determine the slope of the line containing the given points.

30. $(-5, -3)$ and $(1, 0)$ **31.** $(-7, 7)$ and $(-4, 2)$ **32.** $(4, -3)$ and $(10, -3)$ **33.** $(4, -3)$ and $(4, 3)$

Determine by inspection the slope and the y-intercept of the line for the given equation.

34. $y = 23x - 51$ **35.** $6x + 5y = 12$ **36.** $4(y - 2) = 3(2y - 1) + 4$

Graph a line that contains the given point and has the given slope.

37. $(-2, 3)$; $m = \dfrac{5}{9}$ **38.** $(3, -2)$; $m = -3$ **39.** $(-2, -2)$; $m = 0$ **40.** $(-2, -2)$; m is undefined.

Graph each equation by using the slope–intercept method. Label the points.

41. $y = -\dfrac{5}{3}x + 2$ **42.** $3x + 2y = 12$ **43.** $m(x) = \dfrac{5}{8}x$ **44.** $y = -4$

45. What is the grade of a road that rises 4 feet over a horizontal distance of 160 feet?

46. A professional photographer purchases a digital camera to use in her business. If she uses simple linear deprecia-tion for a camera that cost $485 and estimates its useful life to be 4 years, determine the amount of depreciation per year.

47. According to data from the U.S. Bureau of the Census, the number of farms in the United States has decreased steadily since 1940, but has held relatively constant since 1980. Which of the following charts best illustrates these changes?

a.

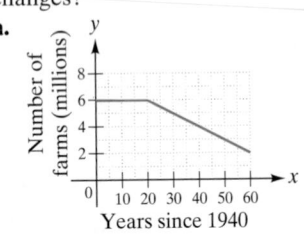

b.

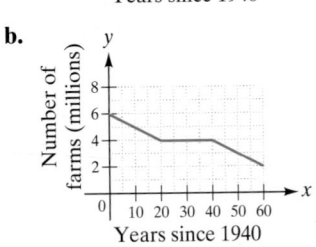

c.

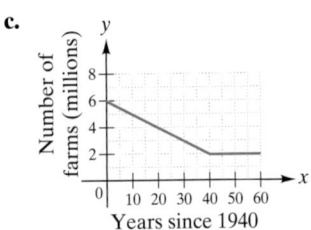

48. Janet left home to walk to the shopping mall that was 2 miles from her home. After walking a mile, she stopped to visit with her sister. She visited for an hour and then continued her walk to the mall. She spent a half hour shopping and then realized that she needed to hurry home, so she quickened her returning pace of walking to twice that of her initial pace. Which graph best illustrates Janet's journey? Explain your choice.

a.

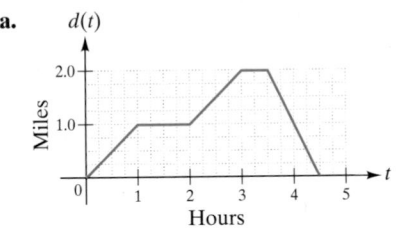

b.

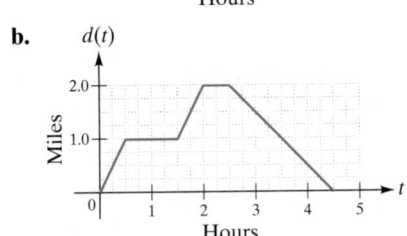

c.

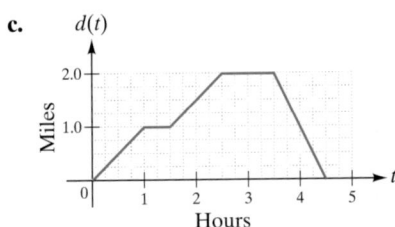

5.3

Recall	Examples
Coinciding lines • Lines that have equal slopes and the same y-intercept.	When solved for y, equations for two coinciding lines are $y = 2x - 3$ and $y = 2x - 3$.
Parallel lines • Lines that have equal slopes, but different y-intercepts.	When solved for y, equations for two parallel lines are $y = 2x - 5$ and $y = 2x - 3$.
Intersecting lines • Lines that have different slopes.	When solved for y, equations for two intersecting lines are $y = 2x - 3$ and $y = -2x + 3$.
Intersecting and perpendicular lines • Lines that have opposite reciprocal slopes.	When solved for y, equations for two intersecting and perpendicular lines are $y = 2x - 3$ and $y = -\frac{1}{2}x - 4$.

Determine by inspection whether the graphs of each pair of equations are coinciding, parallel, intersecting only, or both intersecting and perpendicular. Check by graphing.

49. $y = 2x + 6$
$3y - x = 15$

50. $2y - 2x = y + 3x + 2$
$5x - y = -2$

51. $4x - 20 = 0$
$2x + 3 = x + 4$

52. $2x + y = 4$
$y = -2(x + 2)$

53. $3(y + 2) = 2(y + 4)$ **54.** $5y - 4(y + 3) = -10$ **55.** $2x - 3y = -9$ **56.** $y = x + 7$
$2(x - 2) = 0$ $\qquad$ $y - 1 = -2(x - 1)$ $\qquad$ $3x + 2y = 6$ $\qquad$ $y = -x + 7$

57. J. R. decides to go into business buying and selling used
graphing calculators to students. He spends $25 having
posters printed to hang on student bulletin boards and
pays $35 to place an ad in the student newspaper. He will
buy used calculators at $30 each. Since he is too busy to
sell the calculators, he pays Brandon a flat fee of $25 to
help him, with the promise of paying him $5 for each cal-
culator he sells.

a. Let x represent the number of calculators bought and
resold, and let y represent the cost of selling calculators. Write a linear equation for the total cost of selling x calculators.

b. If Brandon sells the calculators at $35 each, write a
linear equation for the total revenue of selling x calculators.

c. Graph the two equations and find the break-even
point, if there is one, at which the revenue equals the
cost of selling x calculators.

d. If Brandon sells the calculators at $60 each, write a new
linear equation for the total revenue of selling x calculators.

e. Graph the original cost function and the new revenue
function, and determine the new break-even point if
there is one.

f. What would you recommend that J. R. do?

58. From data collected by the U.S. Bureau of the Census,
the college population of males in millions can be estimated by the equation $y = 0.9x + 91$, where x is the
number of years since 1987 and y is the number of male
college students in millions. Likewise, the college population of females in millions can be estimated by the equation $y = 0.9x + 99$, where x is the number of years since
1987 and y is the number of female college students in
millions. What can you say about the graphs of these two
equations? Justify your answer.

5.4

Recall	Examples
Write an equation of a line • Given a slope and the y-intercept, use the slope–intercept form of the equation, $y = mx + b$.	Write an equation of a line with a slope of -3 and a y-intercept of $(0, 4)$. $y = mx + b$ $y = -3x + 4$
• Given a point and the slope of a line, use the point–slope form of the equation, $y - y_1 = m(x - x_1)$.	Write an equation of a line with a slope of -3 and passing through $(-1, 4)$. $y - y_1 = m(x - x_1)$ $y - 4 = -3(x - (-1))$ $y - 4 = -3(x + 1)$ $y - 4 = -3x - 3$ $y - 4 + 4 = -3x - 3 + 4$ $y = -3x + 1$
• Given two points on the line, first determine the slope by using the formula $m = \dfrac{y_2 - y_1}{x_2 - x_1}$, and then use the point–slope form of the equation, $y - y_1 = m(x - x_1)$.	Write an equation of a line passing through $(-1, 4)$ and $(2, 3)$. $m = \dfrac{y_2 - y_1}{x_2 - x_1}$ $\qquad$ $y - y_1 = m(x - x_1)$ $m = \dfrac{3 - 4}{2 - (-1)}$ $\qquad$ $y - 3 = -\dfrac{1}{3}(x - 2)$ $m = -\dfrac{1}{3}$ $\qquad$ $y - 3 = -\dfrac{1}{3}x + \dfrac{2}{3}$ $y - 3 + 3 = -\dfrac{1}{3}x + \dfrac{2}{3} + 3$ $y = -\dfrac{1}{3}x + \dfrac{11}{3}$

59. Write a linear equation for the graph.

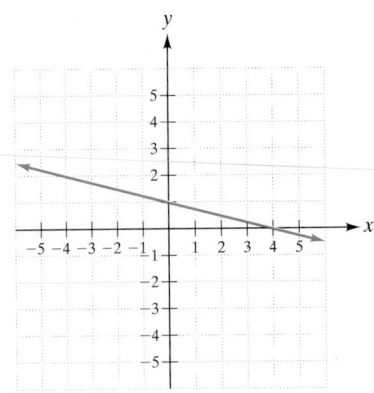

In exercises 60–69, write an equation for a line that satisfies each of the given conditions.

60. The line has a y-intercept of $(0, 3)$ and a slope of -2.

61. The line passes through $(0, -2)$ with a slope of $\frac{3}{5}$.

62. The line is horizontal with a y-intercept of $(0, -3.5)$.

63. The line is vertical with an x-intercept of $(2.6, 0)$.

64. The line has a slope of -5 and passes through $(2, -3)$.

65. The line passes through the points $(9, 5)$ and $(-2, 2)$.

66. The line passes through $(4, 6)$ and has a slope of $\frac{1}{3}$.

67. The line passes through $(1, 1)$ and is parallel to the graph of $y = 4x + 5$.

68. The line passes through $(2, 4)$ and is perpendicular to the graph of $y = 2x - 1$.

69. Holly Carton researched the sales of her latest record for the 10 weeks after its release. She found the sales (measured in thousands of records sold) for each week to be given by the following table:

Week	1	2	3	4	5	6	7	8	9	10
Records Sold	12	15	19	26	28	29	30	30	29	28

a. Use the information for week 2 and week 8 to find a linear equation relating sales to the number of weeks since the release of the record. The two coordinate pairs would be $(2, 15)$ and $(8, 30)$.

b. What does the equation predict sales to be for week 5? Is this close to the actual figure of 28?

70. The U.S. Bureau of Labor Statistics reported that there were 90.4 million employees in nonfarm industries in 1980 and that there were 109.4 million such employees in 1990. Let x represent the number of years after 1980, and let y represent the number of nonfarm employees.

a. Find the slope of the line connecting $(0, 90.4)$ and $(10, 109.4)$.

b. Write an equation to estimate the number of nonfarm employees x years after 1980.

c. Estimate the number of nonfarm employees for 1998. Compare this with the actual reported number of 125.8 million. How well did your equation do in estimating the number?

d. What does the equation predict that the number of nonfarm employees will be in 2005? Does this number seem reasonable to you? Explain.

CHAPTER 5 REVIEW

Graph each linear equation by using a table of values. Label the points.

1. $3(y - 5) = y + 7$ **2.** $14x + 7y = 7$ **3.** $y = -\dfrac{7}{5}x - 10$

Graph a line containing the given point with the given slope.

4. $(2, 6)$; $m = 0$ **5.** $(2, 6)$; m is undefined.   **6.** $(1, 4)$; $m = -2$ **7.** $(0, 3)$; $m = \dfrac{2}{7}$

Graph each equation by using the slope–intercept method. Label the points.

8. $y = \dfrac{-3}{8}x$ **9.** $y = 7$ **10.** $5x - 4y = 12$

Graph each equation by using the intercept method. Label the points.

11. $15.50x + 21.70y = 108.50$ **12.** $12x - 24y = -48$ **13.** $-x + y = -6$

Determine by inspection whether the graphs of each pair of linear equations are coinciding, parallel, intersecting only, or both intersecting and perpendicular.

14. $2x - 3 = 1$
$4x - 3y = 6$

15. $5x - 4y = 8$
$4x - 5y = -15$

16. $y = 5(x - 1)$
$2(x - 1) = 7x - y + 3$

17. $5(x + 1) = 15$
$3y + 1 = 10$

18. $y = -2x + 3$
$2(x + y) = y + 3$

19. $y = x + 3$
$y = -x - 4$

20. $5y = 20$
$2(y + 3) = -2$

Determine three solutions for each equation, and graph the equations.

21. $y = 8$

22. $8x - 9y = -72$

23. $t(x) = \dfrac{9}{11}x - 8$

Identify each equation as linear or nonlinear. Express each linear equation in standard form.

24. $y = x$

25. $y = 2x^2 - 8$

26. $y = 1.3x - 0.5$

27. $y = 15x$

28. $8y - 5x = 21$

29. $2x + 3y - 1 = y^2 + x$

Determine by inspection the slope and the y-intercept of the line for each linear equation.

30. $x + 4(x - 2) = 2$

31. $x - 3(y + 2) = 4(x + 1) - y$

32. $12x - 4y = 8$

33. $2y - 6 = 4(y - 2) + 2$

34. $y = 13x - 15$

35. $y = -5.03x + 7.92$

Determine the slope of the line containing the given points.

36. $(5, 5)$ and $(8, 5)$

37. $(-3, -3)$ and $(-3, 3)$

38. $(-4, 4)$ and $(-1, -5)$

39. $\left(-\dfrac{2}{5}, \dfrac{4}{7}\right)$ and $\left(-\dfrac{4}{7}, \dfrac{1}{5}\right)$

In exercises 40–48, write an equation for a line that satisfies the given conditions.

40. The line contains the points $(4, -2)$ and $(5, 2)$.

41. The line passes through $(-1, -1)$ and has a slope of $-\frac{1}{4}$.

42. The line passes through $(0, 5)$ with a slope of $-\frac{2}{3}$.

43. The line is vertical with an x-intercept of $(4.1, 0)$.

44. The line passes through $(-2, -2)$ and is parallel to the line for $y = -3x + 6$.

45. The line is horizontal with a y-intercept of $(0, 8)$.

46. The line has a slope of 4 and passes through the point $(1.2, 8)$.

47. The line has a y-intercept of $(0, -2)$ and a slope of 3.

48. The line passes through $(0, 2)$ and is perpendicular to the line for $y = \frac{-1}{2}x - 1$.

49. Frank earned money as a freelance umpire for intramural school baseball games. He was paid $24 for a game that lasted 7 innings. For another game that went 9 innings, he was paid $28. Denote his pay as y and the number of innings he worked as x.
   a. Graph the two coordinate pairs.
   b. Assume that the information is from a linear relation, and connect the points. How much would Frank receive for a game that lasted 8 innings? How much for a game that lasted 10 innings?

50. A storage tank contains 180 gallons of liquid at the start of a production run. Liquid is being added to the tank at a rate of 10 gallons per hour. At the same time, liquid is being drained from the tank and bottled at the rate of 15 gallons per hour. Let y represent the amount of liquid in the tank, and let x represent the number of hours of production.
   a. Write a linear equation in two variables for the amount of liquid in the tank after x hours of production.
   b. Without graphing, determine the intercepts of a graph of the equation. Interpret what the intercepts represent.
   c. Graph the equation by using the intercept method.
   d. From the graph, determine how much liquid remains in the tank after 16 hours of production.

51. A sports trainer charges his clients $20 per visit plus $50 per hour to come to their homes and conduct fitness training sessions.
   a. Let x represent the number of hours for the training session, and let y represent the total charge for the session. Write an equation to represent the trainer's total charge for a session.
   b. How much would the trainer earn if he trained for 2 hours?
   c. How much would he earn for a session that lasted $1\frac{1}{2}$ hours?

52. Shannon set up a business to sell hair bows at a kiosk in the mall during the holiday season. The cost of renting the kiosk is $75.00 per week. She had to purchase supplies that cost $150.00. In addition, the materials used to make each hair bow cost her $0.75 each.
   a. Write a linear equation to determine Shannon's total cost, $c(x)$, in terms of the number of hair bows, x, she makes and sells during the week.
   b. If she sells the hair bows for $2.00 each, write a linear equation to represent her revenue, $r(x)$, for selling x bows during the week.

c. Graph the two equations, and find the break-even point, if there is one, at which revenue equals cost.

d. Will Shannon make a profit if she sells 150 hair bows during the week? What if she sells 200 bows? What is the minimum number of bows she must sell to avoid a loss?

53. National Motors, Incorporated, advertises that its all-terrain vehicle, the Conqueror, can climb a hill that rises 60 feet vertically over a horizontal distance of 125 feet. Find the grade of the hill.

54. Find the pitch of a roof that rises 2 inches over a horizontal distance of 12 inches.

55. A first edition of a novel was purchased in 1980 for $15.00. The novel was sold to a book dealer in 2000 for $275.00. Find the average rate of change per year in the value of the novel between 1980 and 2000.

56. Jonathan started a new teaching position. His first teaching assignment lasted 6 months, and then he was on summer hiatus for 2 months with no pay. He returned after the hiatus and assumed additional responsibilities that increased his rate of pay. Choose which of the following graphs best represents Jonathan's cumulative pay for the 12-month period:

a.

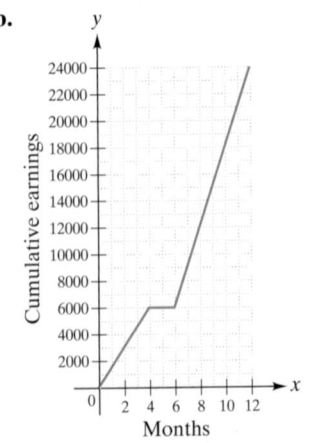

b.

c.

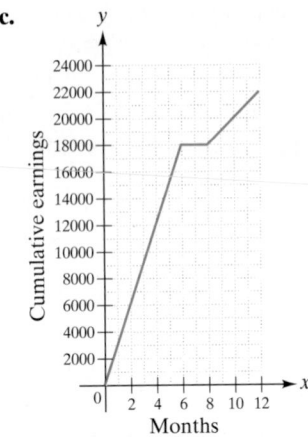

57. The U.S. Energy Information Administration reported that in 1992 carbon dioxide emissions totaled 1361 million metric tons. In 1993, they totaled 1394 million metric tons. Let x represent the number of years after 1992, and let y represent the emissions level.

a. Find the slope of the line connecting $(0, 1361)$ and $(1, 1394)$.

b. Write an equation to estimate the emissions level x years after 1992.

c. Estimate the emissions level for the years 1994 through 1997. Compare your estimates with the actual levels of 1414, 1428, 1479, and 1501, respectively, for those years. How well did your equation do in estimating these levels?

d. What does the equation predict the emissions level will be in 2005?

58. Beckie operates a picture-framing operation out of her home. She spends $35 per week to advertise in the local paper. She spends an average of $6.50 per picture for framing and matting materials. She charges her customers $40 for each picture she frames. Write a linear function for the weekly profit that Beckie can realize. Graph the function by the intercept method. Explain the meaning of the intercepts.

59. Dennis jogged for two miles, rested for a half hour, and then walked briskly home at a pace that was half as fast as his jogging pace. Which of the following graphs illustrates his exercise routine? Explain your choice.

a.

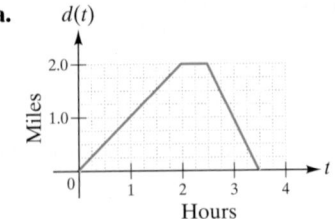

b.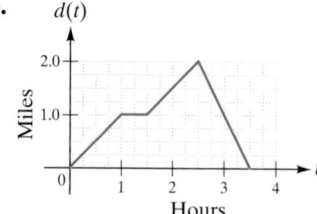

c.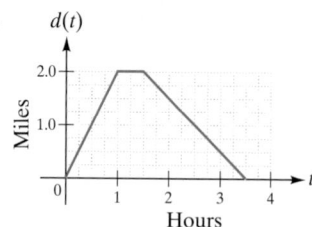

CHAPTER 5 TEST

TEST-TAKING TIPS

Rehearse for a test just as you would rehearse for a speech in speech class. By this, we mean "practice, practice, practice." If you have practice tests available, you should study them. Aim for 100% understanding. If you don't understand an exercise, get help immediately. Start practicing at least a week before the test, and do so each day for a reasonable length of time. If you find that you are weak on certain topics, practice with exercises on those topics. Make index card notes summarizing examples that you can review each day. These are like the flash cards you may have used when you were a youngster. Rehearsing moves information from your short-term memory to your long-term memory. It is the only way you can achieve the kind of recall you will need to reach your goal of 100% mastery of the material. Remember, if you don't set this goal, you will have no incentive to reach it!

Identify each equation as linear or nonlinear.

1. $5x - 7y = 2(x + 1) - 3$ **2.** $y = \dfrac{11}{12}x - 5$

3. $y^2 - 16 = 0$ **4.** $7x + 2y = 3x^2 + 3$

Graph the equation by using a table of values. Label the points.

5. $2x - 8y = 0$ **6.** $3x + 2y = 7$

Graph by using the intercept method. Label the points.

7. $12x + 15y = 60$ **8.** $2(x - 3) = x + 1$ **9.** $4y - 3 = 2y + 9$

Graph by using the slope–intercept method. Label the points.

10. $g(x) = 3x - 5$ **11.** $3x - 4y = 4$

Determine the slope of the line containing the two points.

12. $(0, -1)$ and $(-1, -1)$ **13.** $(4.8, 0.6)$ and $(-1.5, 2.4)$

14. $(-3, -4)$ and $(1, 0)$ **15.** $(-2, 2)$ and $(-2, 8)$

Determine whether the graphs of each pair of equations are coinciding, parallel, intersecting only, or both intersecting and perpendicular.

16. $y = -8x + 1$ **17.** $9(y - 2) + 2x = 7x$ **18.** $3x - y = 2$ **19.** $5y - 2 = 3y + 2$
     $y = x - 8$ $4x = 9(y + 4) - x$ $y - x = 2(x - 1)$ $3x = 2(x + 2)$

Determine the slope and the y-intercept of the graph of each linear equation.

20. $4y = 3x + 12$ **21.** $7x + 7y = 21$

In exercises 22–25, write an equation of a line that satisfies each of the following conditions:

22. A line passes through $(2, 1)$ and is perpendicular to the graph of $y = \frac{3}{5}x$.

23. A line passes through $(0, 7)$ with a slope of 9.

24. A line passes through $(-1, 2)$ with a slope of 6.

25. A line passes through $(2, 3)$ and $(-4, 1)$.

26. Determine three integer solutions for the following graph. Determine the slope of the graph. Determine whether the graph represents a function. If the graph represents a function, is the function increasing, decreasing, or constant?

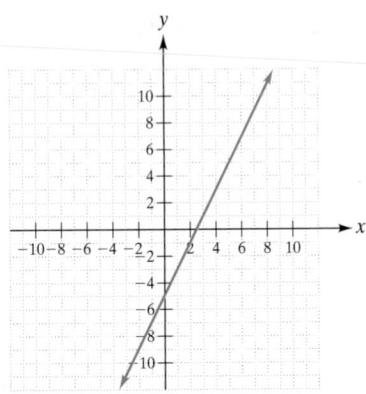

27. A novelty shop purchases a candy-vending machine for $4000. For tax purposes, the shop uses simple linear de-preciation and depreciates the machine over a useful life of six years. Determine the amount of depreciation per year.

28. The *World Almanac and Book of Facts* states that in 1940 the typical farm size was 174 acres, in 1980 it was 426 acres, and since then it has remained around 430 acres. Find the average rate of change per year of farm size be-tween 1940 and 1980.

29. Marty is paying off a $1450 loan at the rate of $150 per month. Write a linear equation for his balance, y, in terms of the number of payments, x, he has made. Graph the equation and interpret the intercepts. Find the first value of x for which his loan drops below $1000. Explain what this value represents.

30. What is the difference between graphing a linear equation by the intercept method and graphing it by the slope–intercept method?

Chapter 5

Project

Suppose that you are working for an automobile manufacturing company. The company president (your instructor) wants to know the following:

1. The approximate number of new passenger cars that will be sold in the United States in the year 2010.

2. The year in which the sale of new passenger cars will decline to 8 million.

3. The average decrease in the number of new passenger cars sold in the United States per year.

You will be assigned a group to complete this report. Each group will present it's results to the company president and the board directors (the class) in a 10-minute or less talk.

The U.S. Bureau of the Census releases information about the United States in an annual publication called the *Statistical Abstract of the United States*. This publication is available both on-line and in print form and contains information gathered from a variety of sources, including the American Automobile Manufacturers Association and Ward's Communications. Using information from one of these sources, find the latest data available on new-car sales.

Part I

Complete the following exercises to help you justify your answers to the preceding questions:

1. Use data from the latest four years. Let $x =$ the number of years after the earliest of the four years and $N(x) =$ the number of new passenger cars. (For example, if the earliest year is 1995, then $x = 0$ for 1995, $x = 1$ for 1996, and $x = 2$ for 1997.) Complete a table of values for the independent variable x and the dependent variable $N(x)$.

2. Graph the four coordinate pairs found in the table in exercise 1. Do the points appear to lie on a straight line?

3. Using the four data points two at a time, we can determine six different pairs of points. List the six pairs.

4. Write an equation of a line through each pair of points. Name the six equations as $N_1(x)$, $N_2(x)$, $N_3(x)$, $N_4(x)$, $N_5(x)$, and $N_6(x)$.

5. Graph $N_1(x)$. On the same graph, plot and label the four coordinate pairs in exercise 1.

6. Complete the following table. Note that column three is determined by substituting values of x in column one into the function $N_1(x)$.

Number of years after _____ x	Number of new automobiles in thousands $N(x)$	Estimated number of new automobiles $N_1(x)$	Difference between estimate and given value $|N(x) - N_1(x)|$
		Total difference	

7. Repeat exercises 5 and 6 five times, replacing $N_1(x)$ with $N_2(x)$, $N_3(x)$, $N_4(x)$, $N_5(x)$, and $N_6(x)$.

8. Compare the total differences from the six tables. The equation that results in the least total difference may be considered the line of best fit. Statisticians use a similar, but more complicated, process to determine a line of best fit. State the equation that you determined to be the line of best fit.

9. Use the equation from exercise 8 to predict the answers to the president's questions.

Part II

The TI-83 Plus has built-in statistical features to find the best-fitting line, called a linear regression line. In 5.4 Calculator Exercises, we used this feature to write an equation for two points. In order to use this feature for more than two points, you will need to enter the values of x in L1 and the corresponding values of $N(x)$ in L2. Be sure that the pairs of numbers match in the two lists. Then calculate LinReg $(ax + b)$.

1. Write an equation for $N_7(x)$ by using this statistical feature.

2. Repeat exercise 6 in Part I replacing $N_1(x)$ with $N_7(x)$.

3. Would you consider $N_7(x)$ to be a better fit to the data than the equation found in Part I? If so, why?

4. Use the information that you have gathered to predict the answers to the president's questions.

Part III

In this project we will use the TI-83 Plus with a Ranger program and a Calculator Based Ranger (CBR) to collect data involving the relationship of distance walked with respect to time walked.

In order to run the Ranger program,

a. Connect the CBR to the TI-83 Plus.

b. Under programs select RANGER.

c. Under the MAIN MENU select 2: SETUP/SAMPE.

d. With the up or down arrow key select a line and press ENTER to change the settings. When the setting are correct arrow to the top of the page and press ENTER to start. The screen settings should read as shown:

> REALTIME: YES
> TIME(S): 15
> DISPLAY: DIST
> BEGIN ON: [ENTER]
> SMOOTHING: NONE
> UNITS: FEET

e. Press ENTER to start the program.

f. Place the CBR on a table. On the floor mark a distance in front of the CBR in feet beginning with 3 feet for a distance of 20 feet.

g. When you are ready to collect your data, press ENTER again to start the data collection. The CBR will begin to make clicking noises. The calculator will collect data points for 15 seconds.

h. The calculator will display a graph having time in seconds, T, on the horizontal axis and distance in feet, D, on the vertical axis.

i. When you are ready to collect additional data, press ENTER and choose 5 for REPEAT SAMPLE.

You will need one person designated as a walker and one person to control the calculator and CBR.

For each of the given situations described below, complete parts a–e.

a. On your paper, set up a table of values to describe the situation. Use integer coordinates $\{0, 1, 2, \dots, 15\}$ for the independent variable T and determine the dependent variable D.

b. Graph the situation.

c. Perform the situation described using the CBR, the Ranger program, and your TI-83.

d. Compare your graph to the graph on your calculator.

e. Identify the slope of the line as either positive or negative and approximate its value. Determine the y-intercept of the graph. Write an equation for the line.

1. The walker should stand on the 3-foot line. When told to begin, walk away from the CBR at a constant rate of one foot per second for 15 seconds.

2. The walker should stand on the 20-foot line. When told to begin, walk toward from the CBR at a constant rate of one foot per second for 15 seconds.

3. The walker should stand on the 5-foot line. When told to begin, the walker should remain on the 10-foot line for 15 seconds.

Let's see how well you can predict the graph. Sketch a graph for the following situations without using the CBR. Then perform the situation described and check your sketch with the calculator graph.

4. The walker should stand on the 10-foot line. When told to begin, walk at a constant rate of one foot per second toward the CBR for 5 seconds, remain at this position for 5 seconds, and the walk at a constant rate of one foot per second away from the CBR.

5. The walker should stand on the 3-foot line. When told to begin, walk at a constant rate of two foot per second away the CBR for 5 seconds, remain at this position for 5 seconds, and then continue to walk at a constant rate of one foot per second away from the CBR.

CHAPTERS 1–5 CUMULATIVE REVIEW

1. Use one of the symbols $<$, $>$, or $=$ to compare the numbers.

a. $\dfrac{5}{9}$ ____ $\dfrac{2}{3}$ **b.** $\dfrac{8}{5}$ ____ 1.6 **c.** -5.4 ____ -6.5

2. Graph the following numbers on a number line, and label the points:

$$-\sqrt{9},\ \sqrt{7},\ 3\frac{2}{3},\ -\frac{3}{4},\ 6.4,\ -5,\ \pi$$

Evaluate. Express your answers in fractional notation, or round decimals to the nearest thousandths, as appropriate.

3. $5.7 - 4.68$

4. $-|17 - 29|$

5. $\sqrt{\dfrac{49}{81}}$

6. $-\sqrt{2.5}$

7. $\sqrt{-4}$

8. $-51 \div (-17)$

9. $-\dfrac{3}{7} - \dfrac{4}{5}$

10. $(-8)(21) \div 3(-7)$

11. $\dfrac{4(5^2 - 9) - 14}{7 + 3}$

12. $15 + (-8) + 13 - 12 - (-7)$

13. $-[14.3 - (-2.68)]$

14. $\left(-\dfrac{3}{5}\right)\left(\dfrac{5}{9}\right)\left(-\dfrac{10}{21}\right)\left(\dfrac{7}{8}\right)$

15. $25°$

16. -3^4

17. $(-3)^4$

18. $\left(\dfrac{2}{3}\right)^{-2}$

Simplify exercises 19–20.

19. $-(4a - 2b) + (7a - 4b) - (-5a - b)$

20. $2[3(2x + 3y) - (4x + y)] + 7x$

21. Consider the algebraic expression

$$x^3 - 2x^2 + 7x - 5x^3 + 2x - 4$$

  a. How many terms are in the expression?
  b. List the variable terms.
  c. List the constant terms.
  d. List the coefficients.
  e. List the like terms.

22. Is $x = -3$ a solution of $x^2 + 5x - 3 = 5x + 6$?

Consider the relation $y = 3x^2 - 1$, where x is the independent variable.

23. Graph the relation.

24. What is the domain of the relation? What is the range of the relation?

25. Is this relation a function? Justify your answer.

26. For which values of x is the relation increasing?

27. What is the maximum or minimum of the relation?

28. For the function $f(x) = 3x + 7$, find
  a. $f(-4)$ **b.** $f(-4 + h)$

Solve.

29. $5x - 3 = x - 7$

30. $2(x - 2) + 1 = 3(x + 6) + 2(x - 7)$

31. $(7x + 4) - (x + 6) = 2(3x - 1)$

32. $4(1.2x + 2) - 0.2(24x + 5) = 6$

33. $|5x - 1| = 6$

Graph using the Cartesian coordinate system. Label enough points to determine the graph.

34. $f(x) = -2x + 6$

35. $8x + 2y = -4$

36. $3x - 4 = 5$

37. $2(y + 4) - 6 = 4$

Determine whether the graphs of the pair of equations are co-inciding, parallel, intersecting only, or both intersecting and perpendicular.

38. $4x + 2y = 8$ **39.** $3x + 2y = -2$
$\ \ y = -2x + 6$ $\ -2x + 3y = 3$

Determine the slope of a line that satisfies the following conditions:

40. The line passes through $(-2, 3)$ and $(-3, -1)$.

41. $y = 3x - 6$

42. $x = 4$

In exercises 43 and 44, write an equation for a line that satisfies the following conditions:

43. The line passes through $(1, -2)$ and $(-3, 4)$.

44. The line passes through $(2, 3)$ and is perpendicular to the graph of $4x - y = 1$.

45. Solve for z: $A = \dfrac{x + y + z}{3}$

46. A dress is on sale for $87.49. If the sale price represents a discount of 30% off the original price, write an equation and solve it to find the original price of the dress.

47. Caroline borrowed money for college tuition at a rate of 7.5% simple interest for one year. If she paid $1612.50 to pay off the loan, write an equation and solve it to find the amount of money she borrowed.

48. An independent producer of music CD's pays $300 to setup a production process to manufacture the disks. It costs $2.50 for materials and labor to produce each CD. The CD's are sold for $15 each. Write an equation to find the break-even point. Will the producer make a profit if 20 CD's are made and sold? Will he make a profit if 30 CD's are made and sold?

49. In 1994, the enrollment of students at the local community college was 7,500. In 1999, the enrollment had grown to 9,500 students. Let x represent the number of years after 1994, and let y represent the enrollment in a given year. Write two coordinate pairs for these data.
   a. Find the slope of the line connecting the two coordinate pairs.
   b. Write an equation to estimate the number of students enrolled x years after 1994.
   c. Estimate the number of students enrolled for the year 2000.
   d. What does the equation predict the enrollment will be in the year 2010? Does this number seem reasonable to you? Explain.

50. A car-leasing firm purchases a luxury automobile for $38,500. The firm depreciates the vehicle over a four-year period and then sells the vehicle for $12,500. If the firm uses simple linear depreciation, determine the amount of depreciation per year.

Chapter 6

Systems of Linear Equations

We have now seen several methods for solving linear equations in two variables. But many situations in mathematics require more than one equation to describe the relation between two (or three) variables. In this chapter, we examine methods for solving systems of linear equations in two variables and systems of linear equations in three variables. Just like methods for solving single linear equations, these methods provide graphical solutions and algebraic solutions. Under the category of algebraic solutions, we will study two methods: substitution and elimination. Then we will apply those methods to solving several types of real-world problems.

In this chapter, we encounter many types of real-world situations that involve systems of equations. Examples of these systems are found in business, economics, science, and social science.

One such application involves travel both in the air and on the water when the rate of travel is determined by more than one force. In the project at the end of the chapter, you will be asked to determine the average rate of travel (speed) of an aircraft in flight between the east and west coasts. This speed not only determines how fast you arrive at your destination, but also determines other factors. One such other factor is the aircraft's point of no return: the point in the flight when the time it takes the plane to return to its point of departure is equal to the time it takes the plane to continue to its destination.

6.1 APPLICATION

In earlier chapters, we discussed two different temperature scales: Fahrenheit and Celsius. We know that the freezing point of water is 32 degrees on the Fahrenheit scale and 0 degrees on the Celsius scale. The boiling point of water is 212 degrees Fahrenheit and 100 degrees Celsius. Is there a temperature for which both scales indicate the same temperature? If so, determine this temperature.

We will discuss this application further. See page 394.

6.1 Solving Systems of Two Equations Graphically

OBJECTIVES
1 Determine whether a given ordered pair is a solution of a system of linear equations.
2 Solve systems of linear equations graphically.
3 Model real-world situations by using systems of linear equations, and solve the equations graphically.

6.1.1 Determining a Solution of a System of Linear Equations

In Chapter 5, we discussed how to solve a linear equation in two variables. But it often happens that the relationships in a situation are described by more than one equation at a time. What do we do then? How do we solve several equations at one time? The first step is to define the situation mathematically.

SYSTEM OF LINEAR EQUATIONS IN TWO VARIABLES
A system of linear equations is a set of two or more linear equations.

For example, a system of linear equations in two variables may be

$$-2x - 3y = 1$$
$$3x - y = -7$$

To determine a solution of a system of linear equations in two variables, we need to find an ordered pair that satisfies *all* of the equations in the system. That is, when the ordered-pair coordinates are substituted into all of the equations, the resulting equations are true.

SOLUTION OF A SYSTEM OF LINEAR EQUATIONS
An ordered pair is a solution of a system of linear equations if it is a solution of every equation in the system.

EXAMPLE 1 Determine whether each ordered pair is a solution of the given system of linear equations.

a. $(-2, 1)$; $-2x - 3y = 1$ **b.** $(-3, -2)$; $x = -3$
              $3x - y = -7$ $y = 2$

Solution

a. Substitute the values for the coordinates.

$-2x - 3y = 1$		$3x - y = -7$	
$-2(-2) - 3(1)$	1	$3(-2) - 1$	-7
$4 - 3$		$-6 - 1$	
1		-7	

Both equations are true. Therefore, $(-2, 1)$ is a solution of the system.

Check

On your calculator, the solution is determined by storing -2 for x and 1 for y and evaluating the expressions in the equations.

```
-2→X:1→Y: -2X-3Y
                1
3X-Y
               -7
```

The solution is $(-2, 1)$ because when $x = -2$ and $y = 1$, the equations are true.

b. Substitute the values for the coordinates.

$$x = -3 \qquad y = 2$$

$$\begin{array}{c|c} \hline -3 & -3 \end{array} \qquad \begin{array}{c|c} \hline -2 & 2 \end{array}$$

The second equation is false, because $-2 \neq 2$. Therefore, $(-3, -2)$ is not a solution of the system.

6.1.1 Checkup

In exercises 1 and 2, determine whether each ordered pair is a solution of the given system of linear equations.

1. $(3, 1)$; $2x - 3y = 3$
$ \quad x + 2y = 6$

2. $(5, 0)$; $y = 3x - 15$
$ \quad x - 5 = 0$

3. How does the solution of a system of linear equations differ from the solution of a single linear equation?

4. What must be true of a coordinate pair in order for it to be a solution of a system of linear equations?

6.1.2 Solving Systems of Linear Equations Graphically

We know that we can represent the solutions of a linear equation in two variables by graphing. Therefore, if we graph both of the linear equations in a system, we should be able to determine the solution of the system. Let's see how this works. Complete the following set of exercises on your calculator.

 Discovery 1

System of Linear Equations with One Solution

Solve the given system of linear equations by graphing both equations on the same integer screen.

$$y = 2x + 1$$
$$y = 4x - 3$$

1. The point of intersection is _____.
2. Substitute the coordinates of the intersection point into both equations of the system. The solution of the system is _____.

Write a rule for solving a system of linear equations graphically.

The coordinates of the point of intersection of the graphs form an ordered-pair solution of the system.

SOLVING A SYSTEM OF LINEAR EQUATIONS GRAPHICALLY

To solve a system of linear equations graphically, graph the equations on the same coordinate plane. The coordinates of the point of intersection give the ordered-pair solution.

EXAMPLE 2 Solve each system of linear equations graphically.

a. $-2x - 3y = 1$
$ \quad 3x - y = -7$

b. $x = -3$
$ \quad y = 2$

Graphic Solution

Check

a. Graph both equations of the system on the same coordinate plane.

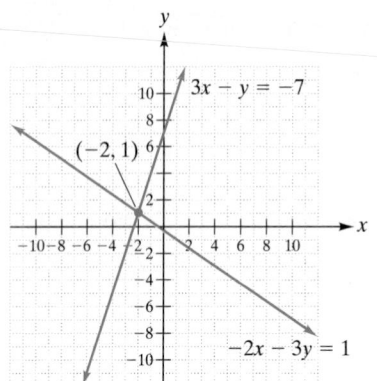

The solution is $(-2, 1)$.

To solve the system on your calculator, solve each equation for y and graph the equations.

$$Y1 = -\tfrac{2}{3}x - \tfrac{1}{3} \qquad Y2 = 3x + 7$$

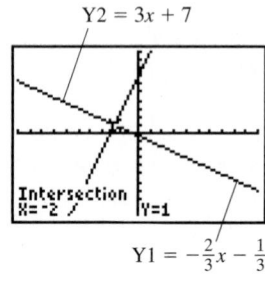

$$Y1 = -\tfrac{2}{3}x - \tfrac{1}{3}$$

$(-10, 10, -10, 10)$

The solution is $(-2, 1)$.

b. Graph both equations of the system on the same coordinate plane.

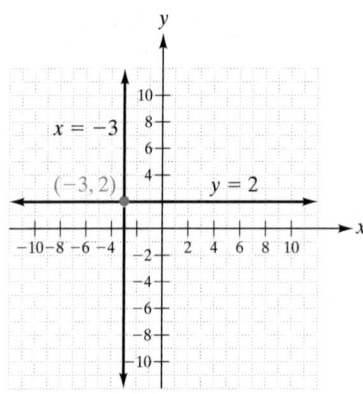

The solution is $(-3, 2)$.

The system contains the equation $x = -3$, which is not a function. Therefore, our calculator won't graph this equation. However, if you wish to check on the calculator, see the directions in 6.1 Calculator Exercises.

Both of the systems in Example 2 had one solution. We call a system with at least one solution a **consistent** system. A system with no solution is called an **inconsistent** system.

In Chapter 5, we determined that two linear graphs can be parallel or coinciding, as well as intersect in one point. We call equations whose graphs coincide **dependent**. We call equations whose graphs do not coincide **independent**.

Both of the systems we solved in Example 2 are consistent, and the equations are independent. Now let's look at inconsistent systems and systems with equations that are dependent. Complete the following sets of exercises on your calculator.

Discovery 2

Inconsistent System of Linear Equations

Solve the given system of linear equations by graphing both equations on the same integer screen.

$$-2x + y = 1$$
$$-4x + 2y = 10$$

Write a rule for determining, by graphing, that a system of linear equations is inconsistent—in other words, has no solution.

The two graphs are parallel and do not intersect. Since the graphs do not intersect, there is no ordered-pair solution of the system. The system is inconsistent.

We saw in Chapter 5 that two equations graph as parallel lines if their slopes (m) are the same and the y-coordinates of their y-intercepts (b) are different. If the equations in the system are both solved for y (that is, they are put into slope–intercept form), the system has no solution if the equations have the same slope (m) and different y-coordinates of their y-intercepts (b). Therefore, it is not necessary to actually graph the equations to determine that they have no solution.

Discovery 3

System of Dependent Linear Equations

Solve the given system of linear equations by graphing both equations on the same integer screen.

$$-3x - y = -2$$
$$y = -3x + 2$$

Write a rule for determining, by graphing, that a system of linear equations consists of dependent equations—in other words, that it has an infinite number of solutions.

The two graphs are coinciding. Since the graphs coincide at all of their points, all ordered pairs that satisfy one equation also satisfy the second equation in the system. The number of solutions is infinite. In order to represent all these solutions, we need to determine all ordered pairs (x, y) that satisfy one of the linear equations. Since both equations in the system are the same, the solutions are the ordered pairs that satisfy either equation. The equations in this system are dependent.

We also saw in Chapter 5 that two equations graph as coinciding lines if they have the same slope (m) and the same y-coordinates of their y-intercepts (b). If the equations in the system are both solved for y (that is, they are put into slope–intercept form), the system has an infinite number of solutions if the equations have the same slope (m) and the same y-coordinates of their y-intercepts (b). Therefore, it is not necessary to actually graph the equations to determine that they have an infinite number of solutions.

EXAMPLE 3   Determine whether the given systems of linear equations have no solution or infinitely many solutions. If the system has an infinite number of solutions, describe the solution set.

a. $y = -x - 3$ **b.** $y = \dfrac{2}{3}x - \dfrac{1}{3}$ **c.** $x = -7$ **d.** $y = 5$

$y = -x + 2$ $3y = 2x - 1$ $2x = -2$ $2y - 10 = 0$

Solution

a. The slopes are equal $(-1 = -1)$, and the y-coordinates of the y-intercepts are different $(-3 \neq 2)$. The graphs will be parallel. There is no solution of the system.

b. Solve the second equation for y: $y = \frac{2}{3}x - \frac{1}{3}$. The slopes are equal $(\frac{2}{3} = \frac{2}{3})$, and the y-coordinates of the y-intercepts are equal $(-\frac{1}{3} = -\frac{1}{3})$. The graphs will thus be coinciding, and there are an infinite number of solutions of the system. The solution set is the set of all ordered pairs (x, y) that satisfy $y = \frac{2}{3}x - \frac{1}{3}$.

c. The equations will graph as vertical lines (equal undefined slopes) with different x-coordinates of the x-intercepts $(-7 \neq -1)$. The graphs will be parallel. There is no solution of the system.

d. Solve the second equation for y: $y = 5$. The equations will graph as horizontal lines (equal slopes of $0 = 0$) and equal y-coordinates of y-intercepts $(5 = 5)$. The graphs will thus be coinciding, and there are an infinite number of solutions of the system. The solution set is the set of all ordered pairs (x, y) that satisfy $y = 5$.

Check

Use your calculator to check your results. Solve the equations for y, if necessary, and graph both equations of the system on the same window. In part c, the system contains two vertical lines. The calculator is not appropriate for this system.

a. $Y2 = -x - 3$

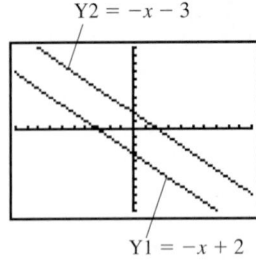

$Y1 = -x + 2$

$(-10, 10, -10, 10)$
The graphs are parallel.
There is no solution set.

b. $Y1 = \frac{2}{3}x - \frac{1}{3}$

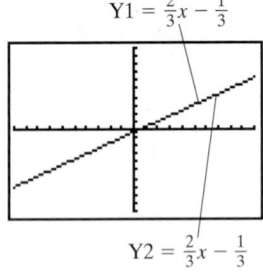

$Y2 = \frac{2}{3}x - \frac{1}{3}$

$(-10, 10, -10, 10)$
The graphs are coinciding.
Thus, there are an infinite
number of solutions of the
system. The solution set is
the set of all ordered pairs
(x, y) that satisfy $y = \frac{2}{3}x - \frac{1}{3}$.

d. $Y1 = 5$

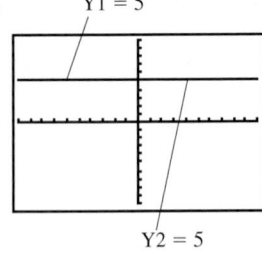

$Y2 = 5$

$(-10, 10, -10, 10)$
The graphs are coinciding.
Thus, there are an infinite
number of solutions of the
system. The solution set is
the set of all ordered pairs
(x, y) that satisfy $y = 5$.

GRAPHICAL SOLUTIONS OF A SYSTEM OF LINEAR EQUATIONS

In solving a system of linear equations in two variables graphically, one of three possibilities will occur.

1. The equations have different slopes (m). The graphs intersect.

One solution exists: the ordered-pair intersection.

The system is consistent.

The equations are independent.

2. The equations have the same slope (m) and different y-coordinates of their y-intercepts (b). The graphs are parallel.

No solution exists.

The system is inconsistent.

The equations are independent.

3. The equations have the same slope (m) and the same y-coordinates of their y-intercepts (b). The graphs coincide.

An infinite number of solutions exist. The solution set is the set of all ordered pairs (x, y) that satisfy either equation of the system.

The system is consistent.

The equations are dependent.

6.1.2 Checkup

Solve each system of equations graphically. Check your solution.

1. $y = -3x + 4$
  $x + y = 2$

2. $y = -2x + 3$
  $x = 4$

In exercises 3 and 4, determine whether each system of equations has no solution or infinitely many solutions. If the system has an infinite number of solutions, describe the solution set.

3. $y = 2x + 3$
  $y - x + 1 = x + 4$

4. $y = -3$
  $y + 4 = 7$

5. What is meant by a dependent system of linear equations? an independent system? a consistent system? an inconsistent system?

6. You have learned definitions that describe a system of equations as being consistent or inconsistent. You have also learned definitions that describe equations as being dependent or independent. Use your understanding of these definitions to complete the table that follows. When you fill in each cell of the table, draw a diagram of what the graphs of the equations in the system will look like. (One cell of the table is completed for you as an illustration.)

	Consistent system	Inconsistent system
Dependent equations	The graphs of the two equations will be *coinciding* lines.	Can a system of equations be inconsistent and at the same time consist of dependent equations?_____
Independent equations	The graphs of the two equations will be _____ lines.	The graphs of the two equations will be _____ lines.

6.1.3 Modeling the Real World

There are two common situations in which graphing two linear equations in two variables at the same time is very helpful. One situation is to compare two different relationships. For example, you might be offered two different car rental plans. Which one should you choose? To find out, you could graph the two equations describing the plans and locate their intersection. Beyond that point, one graph is higher than the other, meaning that the plan it represents will cost more. If you need to rent the car for longer than the time corresponding to the intersection point, you want to choose the plan whose graph is lower *beyond* that time. Likewise, if you need to rent the car for less than the time corresponding to the intersection point, you want to choose the plan whose graph is lower *before* that time.

EXAMPLE 4 Tamara is flying into an airport for a one-day business trip. Avis offered her a rental of a midsize automobile for $33.95 per day with unlimited mileage. Thrifty offered her a rental of a midsize automobile for $29.98 for the first 300 miles and $0.10 per mile for all miles over 300. Write a system of equations and solve them. Interpret your results.

Solution

We have two systems of equations.
The first system of equations is for 300 miles or less.
Let x = the number of mile driven
   $C(x)$ = the cost of a one-day rental

$$C(x) = 33.95 \qquad \text{Avis rental}$$
$$C(x) = 29.98 \qquad \text{Thrifty rental}$$

For this system of equations, the cost of rental will not be equal. The cost of the rental from Thrifty will always be less.
   The second system of equations is for miles over 300.
   Let x = the number of miles driven over 300
      $C(x)$ = the cost of a one-day rental

$$C(x) = 33.95 \qquad\qquad \text{Avis rental}$$
$$C(x) = 29.98 + 0.10x \qquad \text{Thrifty rental}$$

Graph the system.

Y2 = 29.98 + 0.10x

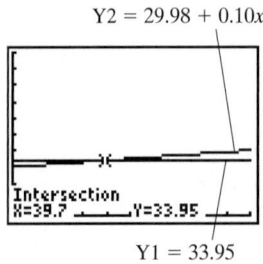

Y1 = 33.95

$(0, 100, 0, 100)$

The intersection point is $(39.7, 33.95)$.

The two plans will be equal if Tamara drives 300 plus 39.7 miles, or 339.7 miles. If Tamara plans to drive less than 339.7 miles, Thrifty rental will cost less. If she plans to drive more than 339.7 miles, then Avis is her best buy.

A second situation occurs when the two variables in an equation are related in more than one way. For example, you can design many different rectangular rooms that will have the same perimeter. If the perimeter of a room is 50 feet, the sides of the room could be 10 feet by 15 feet, 12 feet by 13 feet, or many other dimensions. But if you have to make the length of the room equal to twice the width of the room, then only one pair of dimensions will work. (Can you guess what the dimensions are? Take a look at Example 5.)

EXAMPLE 5 Leopoldo has 48 feet of baseboard. He plans to use all of the baseboard in his new office. He would like to build the rectangular room such that the length of the room is twice the width. Find the dimensions that Leopoldo should use for his new office if the opening for the door is 2 feet.

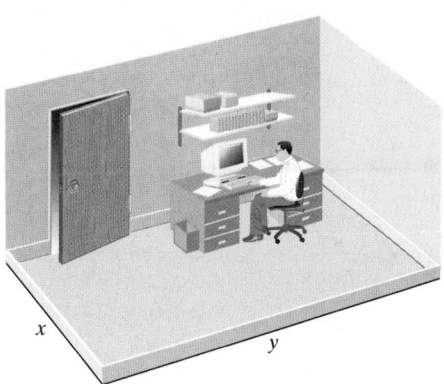

Solution

Let x = width of the room
y = length of the room

Since Leopoldo has 48 feet of baseboard and the opening for the door is 2 feet, we know the perimeter of the room: $P = 48 + 2 = 50$ feet. We also know a formula for the perimeter of a rectangle: $P = 2L + 2W$. If we substitute 50 for P in the formula, we obtain one equation.

$$P = 2L + 2W$$
$$50 = 2y + 2x$$

Since the room's length is to be twice the room's width, a second equation is

$$y = 2x$$

To solve graphically the system of equations

$$50 = 2y + 2x$$
$$y = 2x$$

we graph both equations on the same coordinate system.
The solution is $(8\frac{1}{3}, 16\frac{2}{3})$.

Leopoldo should build the office $8\frac{1}{3}$ feet by $16\frac{2}{3}$ feet.

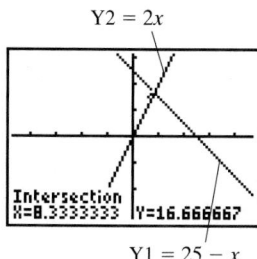

Y2 = 2x

Intersection
X=8.3333333 Y=16.66666?

Y1 = 25 − x

$(-47, 47, -31, 31)$

APPLICATION

In earlier chapters, we discussed two different temperature scales: Fahrenheit and Celsius. We know that the freezing point of water is 32 degrees on the Fahrenheit scale and 0 degrees on the Celsius scale. The boiling point of water is 212 degrees Fahrenheit and 100 degrees Celsius. Is there a temperature for which both scales indicate the same temperature? If so, determine this temperature.

Discussion

Let x = the temperature for which both scales are equal

$f(x)$ = the temperature in degrees Fahrenheit

$c(x)$ = the temperature in degrees Celsius

Using the temperature formulas, we obtain the following system of equations:

$$f(x) = \frac{9}{5}x + 32$$

$$c(x) = \frac{5}{9}(x - 32)$$

On your calculator, graph each equation in the system.

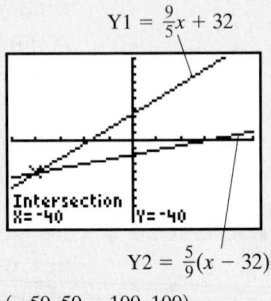

$Y1 = \frac{9}{5}x + 32$

Intersection
X=-40   Y=-40

$Y2 = \frac{5}{9}(x - 32)$

$(-50, 50, -100, 100)$

The intersection of the two graphs is $(-40, -40)$. The temperature will be equal on both scales at 40 degrees below zero.

✔ 6.1.3 Checkup

1. Tyler is renting a truck for a move and has two choices. A Penske Truck Rental shop will rent him a truck for one day for $49.95 with unlimited miles. A Discount Truck Rental will rent him a truck for one day for $36.23 and charge him $0.15 per mile for all miles over 200 miles. Write two systems of equations and solve each of them to determine the number of miles that make both offers equal. Interpret your results.

2. Nola's rectangular vegetable garden is surrounded by 64 feet of fencing. Next year, she will enlarge the garden by doubling the width and increasing the length by 15 feet. She will need 118 feet of fencing to surround the larger garden. Write a system of equations and solve them to find the dimensions of Nola's current garden and her future garden.

3. The percent of both males and females completing four or more years of college has increased since 1990. However, the percent of females is increasing at a much faster rate than the percent of males. The percent of males, $m(x)$, completing four or more years of college x years after 1990 can be estimated by $m(x) = 0.31x + 24.04$. The percent of females, $f(x)$, completing four or more years of college x years after 1990 can be estimated by $f(x) = 0.52x + 17.95$. Assuming that these rates continue into the future, determine the year in which the percent of males and females completing four or more years of college will be equal.

6.1 Exercises

Determine whether each ordered pair is a solution of the given system of linear equations.

1. $(3, 4)$; $y = x + 1$
   $2y = x + 5$

2. $(-1, -6)$; $y = 3x - 3$
   $y + 6 = 0$

3. $\left(\frac{4}{5}, \frac{6}{5}\right)$; $y = \frac{1}{4}x + 1$
   $y = \frac{1}{2}x + \frac{4}{5}$

4. $\left(7, \frac{2}{5}\right)$; $y = \frac{1}{7}x - \frac{3}{5}$
   $7y = x + \frac{28}{5}$

5. $(0.3, 2.1)$; $3y = x + 6$
   $10y = 25$

6. $(0.8, 0.5)$; $y = 2x - 1$
   $10y = 10x - 2$

7. $\left(\frac{2}{3}, \frac{5}{7}\right)$; $7y + 2 = 7$ **8.** $\left(\frac{1}{8}, \frac{1}{8}\right)$; $y = x$ **9.** $(5, -2)$; $2y = -x$

       $5y + 3 = 0$ $8y = 1$ $x = 5$

10. $(-4, -8)$; $y = -2x$

        $x = -4$

Solve each system of linear equations graphically. If the system has infinitely many solutions, describe the solution set.

11. $2x + 3y = -6$ **12.** $-2x - 6y = -6$ **13.** $a + b = 4$ **14.** $5x - 3y = 15$

      $x - 4y = 8$ $x + 3y = 3$ $a - 2b = 7$ $4x + y = 12$

15. $x + 2y = 6$ **16.** $2p + 4q = 8$ **17.** $x - y = -1$ **18.** $2x - 3y = -6$

      $x + 2y = 2$ $3p + q = -8$ $3x - 3y = -3$ $-2x + 3y = -6$

19. $y = \frac{1}{2}x + 3$ **20.** $y = \frac{2}{3}x + \frac{1}{3}$ **21.** $y = 3x - 1$ **22.** $y = -\frac{2}{3}x - 2$

      $y = \frac{-3}{2}x + 9$ $y = -\frac{2}{3}x + 1$ $y = -2x + 4$ $6y = -4x - 12$

23. $y = 2x + 2$ **24.** $y = 3x - 6$ **25.** $y = \frac{1}{2}x + 2$ **26.** $y = -3x + 2$

      $y = 2x - 3$ $y = -5x + 2$ $6y = 3x + 12$ $y = -3x$

27. $3x + 2y = 12$ **28.** $x + 4 = 0$ **29.** $3y - 2 = 2y + 2$ **30.** $-x + 2y = 4$

      $y - 2 = 1$ $3x + 2 = -4$ $x - 1 = 0$ $3x - 2 = 4$

31. $y - 1 = 0$ **32.** $5x + 10 = -5$ **33.** $y = 3x - 1$ **34.** $y = -2x + 3$

      $y + 3 = 0$ $2y + 7 = 5$ $6x = -5 - 12y$ $12x = 8y - 17$

35. $y = x - 7$ **36.** $y = \frac{1}{2}x + 3$ **37.** $y = 1 + 4x$ **38.** $y = -2x - 4$

      $x = y + 7$ $x + 2y = 6$ $y = 4x - 2$ $y = 6 - 2x$

39. Ahmet needs to rent a moving van. Turtle Rental will rent him a van for $39.95 plus $0.25 per mile for every mile over 200 miles. Snail Rental will rent him a similar van for $79.95 with unlimited miles.

    a. Write a system of linear equations in two variables, relating the cost of rental in terms of the miles driven.

    b. Using a window of $(0, 940, 100, 0, 620, 100, 1)$, determine a solution of the system on your calculator.

    c. Interpret the solution in terms of which rental option would be better for Ahmet.

    d. If Ahmet plans a short move of less than 300 miles, which firm should he rent from?

    e. If he plans a longer move of 600 miles or more, which firm should he rent from?

40. In exercise 39, Ahmet had to choose between two rental plans. Rework the exercise, assuming that Turtle Rental will rent a truck for $49.95 with unlimited miles and that Snail Rental will rent a truck for $19.95 plus $0.40 per mile.

41. An advertising firm needs to decide whether to use pre-sorted or regular first-class mailing in one of its promotions. To use pre-sorted mailing, the company must buy a permit that costs $125 and then pay 23 cents per piece of mail. If the firm uses regular first-class mailing, it will pay 34 cents per piece of mail.

    a. Write a system of linear equations in two variables to represent the cost, y, of mailing x pieces of mail under each option.

    b. Graph the system on your calculator, using a window of $(0, 1500, 100, 0, 1000, 100, 1)$. Determine the solution of the system of equations.

    c. When will it be more cost effective to use the pre-sorted mailing? Explain your answer.

    d. When will it be more cost effective to use the first-class mailing? Explain your answer.

42. A dinner coupon book sells for $50.00. With the book, you save $4.00 on each dinner purchased. The average cost of the dinners is $17.50.

    a. Write a system of linear equations in two variables to represent the total cost, y, of purchasing x dinners with and without the coupon book.

    b. Solve the system of equations.

    c. When will it be beneficial to purchase the coupon book?

    d. When will it be more cost effective not to purchase the coupon book? Explain.

43. Kelly spends a total of $500 per month for rent and car payments. Her rent payment is three times as large as her car payment. Write a system of equations to represent these facts, with x representing the car payment and y representing the rent payment. Solve the system. Explain.

44. Jim is designing an apartment complex for a client. The client wants the complex to generate exactly $7800 per month. The complex will have one-bedroom apartments renting for $330 and two-bedroom apartments renting for $450. The complex will have 20 units. Write a system of equations to represent this situation. Solve the system to determine how many apartments of each size should be built.

45. Jenny works 40 hours per week at two part-time jobs. One job pays $7.00 per hour and the other pays $8.20 per hour. She wishes to earn exactly $298 to meet her budgeted expenses. Use a system of equations to find how many hours Jenny should work on each job to meet her budget.

46. Hugh earned $316 per week, working part time at two jobs. One job paid $3.00 more per hour than the other job. Hugh worked 18 hours per week at the lower paying job and 22 hours per week at the higher paying job. Use a system of equations to find the hourly wage Hugh earned for each job.

47. The Shriners added an assembly room to their temple for gatherings. The room was rectangular, with a length that was 25 feet less than twice the width. The perimeter of the room was 550 feet. Write and solve a system of equations to find the dimensions of the room.

48. The Knights of Columbus expanded their council home by adding an assembly hall. The width of the hall was 5 feet more than half the length. The perimeter of the hall was 310 feet. Write and solve a system of equations to find the dimensions of the room.

49. Reba had one more than twice as many Top Ten song hits as Shania. Together, they had 22 Top Ten hits. Write and solve a system of equations to find how many Top Ten hits each had.

50. An heirloom tea set had 21 pieces consisting of saucers and cups. There were five more saucers than cups. Write and solve a system of equations to determine how many cups and saucers were in the set.

51. Female participation in high school athletic programs is increasing at a rate that is higher than that for males, according to data published by the National Federation of State High School Associations. The number of males participating, in millions, is estimated by $m(x) = 0.056x + 3.354$, where x is the number of years since 1990. Likewise, the number of females participating, in millions, is estimated by $f(x) = 0.103x + 1.842$. Assuming that these rates continue into the future, determine the year in which the participation by males and females will be equal.

52. The number of foreign visitors to the United States, in millions, can be estimated by $f(x) = 0.77x + 42.00$, where x is the number of years since 1990. Likewise, the number of U.S. travelers to foreign countries, in millions, can be estimated by $u(x) = 1.58x + 41.56$. These equations were developed with the use of data from the U.S. Department of Commerce. Determine the year in which the number of U.S. travelers to foreign countries will equal the number of foreign travelers to the United States.

 6.1 Calculator Exercises

The calculator does not easily graph equations of the form $x = c$, since these are not functions. However, the calculator can still be used to help you solve a system of linear equations when one of the equations in the system is of that form. There are two ways in which to do this. One uses the TRACE feature of the calculator, and the other uses the APPS feature of the calculator. Each of these is described next.

A. Using the TRACE feature

• Solve the equation that is a function of y. Store this function in Y1.

• Graph the function stored.

• Solve the equation that is not a function, $x = c$, and enter the value of c from this equation into the calculator.

• Trace the function.

• The coordinate pair displayed on the screen represents the solution of the system of equations.

As an example, solve the following system of linear equations:

$$2x + 3y = 8$$
$$4x - 1 = 6$$

Solution:

• The first equation yields $y = \dfrac{-2}{3}x + \dfrac{8}{3}$.

• Store this equation in Y1.

• The second equation yields $x = \dfrac{7}{4}$, or $x = 1.75$.

• Trace the first function and enter the value $\dfrac{7}{4}$ for x.

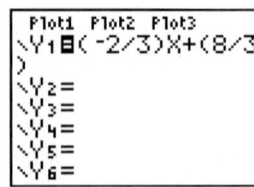 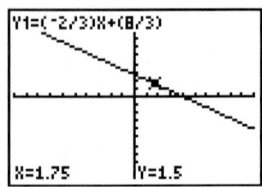

The calculator traces to the solution of the system, which is $(1.75, 1.5)$, or $\left(\dfrac{7}{4}, \dfrac{3}{2}\right)$.

B. Using the APPS feature

In the previous chapter, we explained that if you purchase applications for the TI-83 Plus from Texas Instruments, you can obtain an application that enables you to graph vertical lines on the calculator quite easily. If you have a system of equations in which one of the equations graphs to a vertical line, you can use the APPS key to graph the system. For example, solve the following system of linear equations:

$$2x + 3y = 8$$
$$4x - 1 = 6$$

- Set up the calculator to graph the equation that is a function.

- To use the applications, press the | APPS | key on the calculator.

- Choose the inequal option from the menu. A logo will appear.

- Press any key to continue. The calculator will display the menu for the | Y= | key.

- To change to the X = format, move the cursor to the top of the screen and highlight x = in the first location.

- When you press | ENTER |, the calculator will change the displays to X = displays.

- Enter the value obtained when you solve the second equation for *x* in the first location.

- Proceed to the graph as you normally would, by using the | ZOOM | key or setting the window and using the | GRAPH | key. | ZOOM | | 6 | yields the following graph:

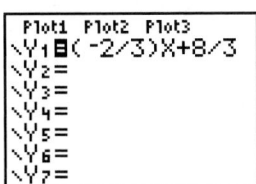

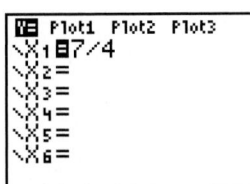
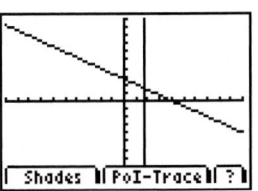

- To find the intersection point of the system, press | ALPHA | | TRACE | or | ALPHA | | ZOOM |. The calculator will display the point of intersection of the two graphs.

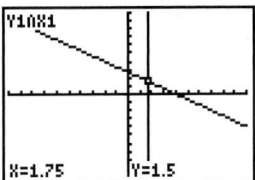

To return to the previous settings for the calculator, press | APPS |, choose option 5 again, and then choose option 2 to quit the application. You should get into the habit of doing this whenever you are through with an application.

Try either of these methods on the following systems of equations:

1. $5x + 2y = 5$
$x + 7 = 10$

2. $5x - y = -19$
$5x + 20 = 13$

3. $\frac{1}{2}x + \frac{1}{3}y = 7$
$2x - 7 = x + 1$

4. $x + 5 = 5x - 4$
$3y + 5 = y - 2$

6.1 Writing Exercises

1. In this section, the following statement was made: "We can represent the solutions of a linear equation in two variables by graphing." Explain what this statement means. In your explanation, discuss the relationship between ordered-pair solutions of a linear equation in two variables and points on the graph of the equation. Explain why the graph is the best way to indicate solutions of such an equation.

2. In the exercises in this section, sometimes it would be best to use the intercept method to graph the system of equations, other times it would be best to use the slope–intercept method, and still other times it might be best to use the table-of-values method or the method of horizontal and vertical lines. First, recall what each of these methods requires. Then look at the equations given in the exercises, and see if you can tell why a particular method would be best for a given form. Write a short paragraph explaining how the form of the equation might lead you to use that method.

6.2 Solving Systems of Two Equations by Using Substitution

OBJECTIVES

1 Solve systems of linear equations by using substitution.

2 Model real-world situations by using systems of linear equations, and solve the equations by substitution.

APPLICATION

According to the 2000 census, the receipts of the federal government, in trillions of dollars, x years after 1990 may be approximated by $R(x) = 0.087x + 0.993$. Federal government outlays, or expenditures, in trillions of dollars, x years after 1990 may be approximated by $E(x) = 0.05x + 1.257$. Determine the year when the receipts equal the expenditures.

After completing this section, we will discuss this application further. See page 405.

6.2.1 Solving Systems of Linear Equations by Using Substitution

Sometimes an ordered-pair solution of a system of linear equations may not have integer coordinates. Solving these systems graphically may be difficult. Therefore, algebraic methods are needed. One such method is the **substitution method**.

To understand the substitution method, let's review the process of substitution. We have evaluated equations when we know values of the variables by substituting each value for a variable and simplifying the equation. Let's apply this idea to a system of equations.

Solve the system. $\quad 4x + y = 2$
$$x = 3$$

Using substitution, substitute 3 for x in the first equation.

$4x + y = 2$
$4(3) + y = 2 \qquad$ Substitute 3 for x.
$12 + y = 2 \qquad$ Simplify.

Solve for the variable y.

$12 + y - 12 = 2 - 12 \qquad$ Subtract 12 from both sides.
$y = -10 \qquad$ Simplify.

The solution must have $x = 3$ and $y = -10$ as coordinate values. The ordered-pair solution is $(3, -10)$.

A more complicated example occurs when x equals an expression instead of a number. For example, solve the system. $\quad 4x + y = 2$
$$x = y + 3$$

Since $x = y + 3$, substitute $(y + 3)$ for x in the first equation.

HELPING HAND Be careful to include parentheses around the expression being substituted.

$$4x + y = 2$$
$$4(y + 3) + y = 2 \qquad \text{Substitute } y + 3 \text{ for } x.$$
$$4y + 12 + y = 2 \qquad \text{Distribute 4.}$$
$$5y + 12 = 2 \qquad \text{Simplify.}$$

Solve for the variable y.

$$5y + 12 - 12 = 2 - 12 \qquad \text{Subtract 12 from both sides.}$$
$$5y = -10 \qquad \text{Simplify.}$$
$$\frac{5y}{5} = \frac{-10}{5} \qquad \text{Divide both sides by 5.}$$
$$y = -2 \qquad \text{Simplify.}$$

To determine the x-coordinate, substitute -2 for y in the second equation of the system, and solve for x.

$$x = y + 3$$
$$x = -2 + 3 \qquad \text{Substitute } -2 \text{ for } y.$$
$$x = 1 \qquad \text{Simplify.}$$

The solution is $(1, -2)$.

SOLVING A SYSTEM OF LINEAR EQUATIONS IN TWO VARIABLES BY USING SUBSTITUTION

To solve a system of linear equations in two variables by using substitution,

- Solve one of the equations for a variable. (Preferably, choose a variable with a numerical coefficient of 1.)
- Substitute the expression for the variable found in the first step for the variable in the other equation of the system.
- Solve the resulting equation for the second variable.
- Substitute the value obtained for the second variable into one of the original equations.
- Solve for the remaining variable.

The ordered-pair solution is determined by the values obtained in solving for each individual variable.

HELPING HAND Remember to check your solution by substituting it into the original equations or by solving graphically.

EXAMPLE I Solve each system of linear equations in two variables by using substitution. Check the solution.

a. $2y = -6$ **b.** $4x + y = 2$
    $-x - 2y = 1$ $x - 2y = 3$

Algebraic Solution

a. Solve the first equation, $2y = -6$, for y: $y = -3$.
    Substitute -3 for y in the second equation.

$$-x - 2y = 1$$

$$-x - 2(-3) = 1 \qquad \text{Substitute } -3 \text{ for } y.$$

$$-x + 6 = 1 \qquad \text{Simplify.}$$

Solve for x.

$$-x + 6 - 6 = 1 - 6 \qquad \text{Subtract 6 from both sides.}$$

$$-x = -5 \qquad \text{Simplify.}$$

$$\frac{-x}{-1} = \frac{-5}{-1} \qquad \text{Divide both sides by } -1.$$

$$x = 5 \qquad \text{Simplify.}$$

The solution is $(5, -3)$.

Check by substituting the solution back into the original equations.

$2y = -6$	$-x - 2y = 1$	**Check**
$2(-3) \mid -6$	$-(5) - 2(-3) \mid 1$	
$-6 \mid$	$-5 + 6$	
	$1 \mid$	

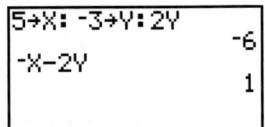

The solution checks in both equations.

b. First, we must solve one of the equations for a variable. If we solve the second equation, $x - 2y = 3$, for x, we obtain $x = 2y + 3$. Using substitution, we substitute $(2y + 3)$ for x in the first equation, $4x + y = 2$.

$$4x + y = 2$$

$$4(2y + 3) + y = 2 \qquad \text{Substitute } 2y + 3 \text{ for } x.$$

$$8y + 12 + y = 2 \qquad \text{Distribute 4.}$$

$$9y + 12 = 2 \qquad \text{Simplify.}$$

Solve for y.

$$9y + 12 - 12 = 2 - 12 \qquad \text{Subtract 12 from both sides.}$$

$$9y = -10 \qquad \text{Simplify.}$$

$$\frac{9y}{9} = -\frac{10}{9} \qquad \text{Divide both sides by 9.}$$

$$y = -\frac{10}{9} \qquad \text{Simplify.}$$

Next, substitute $-\frac{10}{9}$ for y in the second equation of the system, and solve for x.

$$x - 2y = 3$$

$$x - 2\left(-\frac{10}{9}\right) = 3 \qquad \text{Substitute } -\frac{10}{9} \text{ for } y.$$

$$x + \frac{20}{9} = 3 \qquad \text{Simplify.}$$

$$x + \frac{20}{9} - \frac{20}{9} = 3 - \frac{20}{9} \qquad \text{Subtract } \frac{20}{9} \text{ from both sides.}$$

$$x = \frac{27}{9} - \frac{20}{9} \qquad \text{Simplify.}$$

Check

Y1 = −4x + 2

Intersection
X=.7777778 Y=-1.111111

Y2 = $\frac{1}{2}x - \frac{3}{2}$

$(-10, 10, -10, 10)$

$$x = \frac{7}{9}$$ Simplify.

The solution is $(\frac{7}{9}, -\frac{10}{9})$.

To check your solution graphically, solve both equations for y. The intersection point is the solution.

Each of the systems in Example 1 had one solution. However, we know from the previous section that a system may have no solution or an infinite number of solutions. Let's try substitution on these kinds of systems. Complete the following sets of exercises.

Discovery 4

Inconsistent System of Linear Equations

Solve the given system of linear equations by using substitution.

$$-2x + y = 1$$
$$-4x + 2y = 3$$

Write a rule for determining, by using the substitution method, that a system of linear equations is inconsistent—in other words, that it has no solution.

A system has no solution if, after substitution, the resulting equation is a contradiction.

HELPING HAND It is important to state that an inconsistent system has no solution.

Discovery 5

System of Dependent Linear Equations

Solve the given system of linear equations by using substitution.

$$-3x - y = -2$$
$$y = -3x + 2$$

Write a rule for determining, by using the substitution method, that a system of linear equations has dependent equations—in other words, an infinite number of solutions.

A system has an infinite number of solutions if, after substitution, the resulting equation is an identity.

HELPING HAND It is important to describe the solution set for a system of dependent equations as all ordered pairs (x, y) that satisfy the equations.

EXAMPLE 2 Using substitution, determine whether each system of equations has no solution or infinitely many solutions. If the system has an infinite number of solutions, describe the solutions. Check by solving graphically.

a. $y = -x - 3$ **b.** $y = \frac{2}{3}x - \frac{1}{3}$

$y = -x + 2$ $3y = 2x - 1$

Algebraic Solution

Check

a. Since the first equation is solved for y, substitute the expression $(-x - 3)$ for y in the second equation.

$$y = -x + 2$$
$$-x - 3 = -x + 2 \qquad \text{Substitute } (-x - 3) \text{ for } y.$$

Solve for x.

$$-x - 3 + x = -x + 2 + x \qquad \text{Add } x \text{ to both sides.}$$
$$-3 = 2 \qquad \text{Simplify.}$$

$$Y2 = -x + 2$$
$$Y1 = -x - 3$$
$$(-10, 10, -10, 10)$$

The result is a contradiction. There is no solution of the system.
Checking by solving graphically results in two parallel lines.

b. Since the first equation is solved for y, substitute the expression $(\frac{2}{3}x - \frac{1}{3})$ for y in the second equation.

$$3y = 2x - 1$$
$$3\left(\frac{2}{3}x - \frac{1}{3}\right) = 2x - 1 \qquad \text{Substitute } (\frac{2}{3}x - \frac{1}{3}) \text{ for } y.$$
$$2x - 1 = 2x - 1 \qquad \text{Distribute 3.}$$

Solve for x.

$$2x - 1 - 2x = 2x - 1 - 2x \qquad \text{Subtract } 2x \text{ from both sides.}$$
$$-1 = -1 \qquad \text{Simplify.}$$

Check

$$Y1 = \frac{2}{3}x - \frac{1}{3}$$
$$Y2 = \frac{2}{3}x - \frac{1}{3}$$
$$(-47, 47, -31, 31)$$

The result is an identity. There are an infinite number of solutions. The solutions are all ordered pairs (x, y) that satisfy $y = \frac{2}{3}x - \frac{1}{3}$.
Checking by solving graphically results in two coinciding lines.

The following summary describes the process of algebraically solving a system of linear equations by using substitution:

SOLVING A SYSTEM OF LINEAR EQUATIONS BY USING SUBSTITUTION

In solving a system of linear equations in two variables by using substitution, one of three possibilities will occur.

1. A value for both variables will be determined.

One solution exists: the ordered-pair coordinates.

The system is consistent.

The equations are independent.

2. The substitution will result in a contradiction.

No solution exists.

The system is inconsistent.

The equations are independent.

3. The substitution will result in an identity.

An infinite number of solutions exist. The solution set is the set of all ordered pairs (x, y) that satisfy either equation of the system.

The system is consistent.

The equations are dependent.

6.2.1 Checkup

Solve each system of linear equations in two variables by using substitution. Check your solution.

1. $2x - y = 5$
$\quad 4x = 1$

2. $-x + 5y = -12$
$\quad 2x + y = -2$

3. $10x - 2y = 6$
$\quad y = 5x - 3$

4. $y = \dfrac{1}{2}x - 7$
$\quad 2y = x + 6$

In exercises 3 and 4, determine by substitution whether each system of equations has no solution or infinitely many solutions. If the system has an infinite number of solutions, describe the solution set.

5. In solving a system of equations by using the substitution method, one of three outcomes can occur. These are listed in the table that follows. Complete the table to describe the system of equations for each outcome. One row of the table is completed to help you.

Outcome	Number of solutions	Is the system consistent or inconsistent?	Are the equations dependent or independent?
Able to solve the system for a coordinate pair.	One solution exists.	The system is consistent.	The equations are independent.
The system results in an identity.			
The system results in a contradiction.			

 Modeling the Real World

The substitution method for solving a system of linear equations works best when one of the equations is relatively easy to solve for one variable. This often happens in geometry problems, where you can frequently solve a geometric formula for a side of a rectangle or the radius of a circle in terms of other variables and numbers.

EXAMPLE 3   If two angles are complementary and one angle is 6 more than twice the other, find the measures of the angles.

Algebraic Solution

Let x = the measure of the first angle
$\quad y$ = the measure of the complementary angle

Since the angles are complementary, the sum of the angle measurements is 90 degrees. We write the following equation:

$$x + y = 90$$

One angle is 6 more than twice the other.

$$x = 2y + 6$$

The system to be solved is

(1) $x + y = 90$
(2) $x = 2y + 6$

We substitute $(2y + 6)$ for x in the first equation and solve for y.

(1) $\qquad x + y = 90$

$\qquad (2y + 6) + y = 90$ $\qquad$ Substitute $2y + 6$ for x.

$\qquad 3y + 6 = 90$ $\qquad$ Simplify.

$\qquad 3y + 6 - 6 = 90 - 6$ $\qquad$ Subtract 6.

$\qquad 3y = 84$

$\qquad \dfrac{3y}{3} = \dfrac{84}{3}$ $\qquad$ Divide both sides by 3.

$\qquad y = 28$ $\qquad$ Simplify.

Substitute 28 for y in the second equation.

(2) $x = 2y + 6$

$\quad x = 2(28) + 6$ $\qquad$ Substitute 28 for y.

$\quad x = 62$ $\qquad$ Simplify.

The check is left to you.

The complementary angles have measures of 28 degrees and 62 degrees. ●

When we write a system of equations intended to describe real-world situations, we may use function notation. This often is the case for the business applications of total revenue and total cost. To determine algebraically the break-even point, when the total revenue equals the total cost (that is, there is no profit or loss), the substitution method works best.

EXAMPLE 4 The owner of Bateman Camera and Video needs to determine the number of videotapes he must produce to break even before he decides to sell this new line of products. He determined that the fixed cost of producing the videotapes would be $9000 and that there would be a variable cost of $15 per video produced. He plans to sell the videos for $45 each.

a. Write a function for the total cost of producing the videotapes.

b. Write a function for the total revenue received from selling the videotapes.

c. Write a system of equations and determine the number of videotapes that must be produced and sold to break even.

Solution

a. Let x = the number of videotapes produced and sold
$\qquad C(x)$ = the total cost of producing x videotapes

The total cost equals the variable cost plus the fixed cost.

$$C(x) = 15x + 9000$$

b. Let $R(x)$ = the total revenue received from selling x videotapes

The total revenue equals the selling price times the number of videotapes sold.

$$R(x) = 45x$$

c. Since we want $C(x)$ and $R(x)$ to be equal, we can change the function notation to relation form, using y for both the total cost and the total revenue.

(1) $y = 15x + 9000$

(2) $y = 45x$

(1) $45x = 15x + 9000$ Substitute 45x for y in the first equation.

$45x - 15x = 15x + 9000 - 15x$ Subtract 15x from both sides.

$30x = 9000$ Simplify.

$$\frac{30x}{30} = \frac{9000}{30}$$ Divide both sides by 30.

$x = 300$ Simplify.

Substitute 300 for x in the second equation.

$$y = 45x$$
$$y = 45(300)$$
$$y = 13,500$$

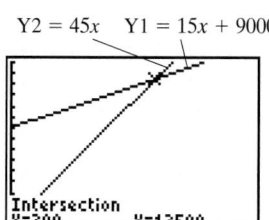

Y2 = 45x   Y1 = 15x + 9000

Intersection
X=300 ___ . ___ Y=13500

We can check the solution (300, 13,500) by solving graphically.
The owner will break even if he produces and sells 300 videotapes.

APPLICATION

According to the 2000 census, the receipts of the federal government, in trillions of dollars, x years after 1990 may be approximated by $R(x) = 0.087x + 0.993$. Federal government outlays, or expenditures, in trillions of dollars, x years after 1990 may be approximated by $E(x) = 0.05x + 1.257$. Determine the year when the receipts equal the expenditures.

Discussion

To determine when the receipts equal the expenditures, we will write and solve a system of equations.

(1) $R(x) = 0.087x + 0.993$ or $y = 0.087x + 0.993$

(2) $E(x) = 0.05x + 1.257$ $y = 0.05x + 1.257$

$0.05x + 1.257 = 0.087x + 0.993$ Substitute 0.05x + 1.257 for y in the first equation.

$0.05x + 1.257 - 0.087x = 0.087x + 0.993 - 0.087x$ Subtract 0.087x.

$-0.037x + 1.257 = 0.993$ Simplify.

$-0.037x + 1.257 - 1.257 = 0.993 - 1.257$ Subtract 1.257.

$-0.037x = -0.264$ Simplify.

$$\frac{-0.037x}{-0.037} = \frac{-0.264}{-0.037}$$ Divide by −0.037.

$x \approx 7.1$ Simplify.

Note that there is no need to determine $R(x)$ and $E(x)$ to answer the question asked.

The receipts will equal the expenditures about 7.1 years after 1990 (or $1990 + 7.1 = 1997.1$).

We will need to round up to the next year, 1998.

6.2.2 Checkup

1. Two angles are supplementary. The larger angle measures 15 degrees more than twice the smaller angle. Find the measures of the two angles.

2. Deanna set up a bakery in her home to bake German chocolate cakes to sell to local restaurants. She spent $245.00 on supplies such as pans, trays, storage, etc. She estimated that each cake she sold cost her $2.50 to produce. She sold the cakes to restaurants for $20.00 each.

   a. Write a function for the total cost of producing the cakes.

   b. Write a function for the total revenue received from selling the cakes.

   c. Write a system of equations and determine the number of cakes Deanna produced and sold to break even.

3. According to data from the U.S. Social Security Administration, the number of male workers (in millions) with Social Security insured status can be estimated by the function $m(x) = 0.775x + 86.3$, where x is the number of years after 1990. Likewise, the number of female workers (in millions) with Social Security insured status can be estimated by the function $f(x) = 1.121x + 77.3$. Determine the year in which the number of insured female workers will equal the number of insured male workers, assuming that the functions still describe the real world.

6.2 Exercises

Solve each system of linear equations by using substitution. Check the solution. If a system has an infinite number of solutions, describe the solution set.

1. $x - 2y = 26$
  $5x + 10y = -10$

2. $2x - 3y = 5$
  $2y + 6 = 8$

3. $y = 3x - 15$
  $x - 5 = 0$

4. $3x + 5y = 15$
  $4x - 2y = 7$

5. $x - 5y = 20$
  $2y + 3 = -7$

6. $y = -4x + 3$
  $5(x - 1) = 0$

7. $3y = x + 6$
  $x - 3y = 9$

8. $y = 2x + 3$
  $y - x + 1 = x + 4$

9. $y = x - 3$
  $2x - y = 19$

10. $y = -3x + 4$
  $5x - 5y = 20$

11. $y = x + 1$
  $5x - 10y = 3$

12. $y = 3x - 3$
  $5(y - 5) = 20$

13. $4y = x + 4$
  $5y = 2x + 11$

14. $y = \frac{1}{7}x - \frac{3}{5}$
  $7y = x + \frac{28}{5}$

15. $3y = x + 6$
  $10y = 25$

16. $y = 2x - 1$
  $y = x - 4$

17. $y = -3x + 2$
  $2y + 2x = 2 + y - x$

18. $y = -x - 3$
  $2x + y = 1$

19. $2x + y = -3$
  $y = -0.5x + 3$

20. $2x + 3y = -6$
  $x - 4y = 8$

21. $5x - 3y = -13$
  $4x + y = 27$

22. $x + y = 4$
  $x - 2y = 7$

23. $x + 2y = -28$
  $3x + y = -9$

24. $x + 2y = 6$
  $x + 2y = 2$

25. $x - y = -1$
  $3x - 3y = -3$

26. $-2x - 6y = -6$
  $x + 3y = 3$

27. $y = 3x - 1$
  $2x + y = 4$

28. $y = 3x - 6$
  $5x + y = 2$

29. $y = \frac{1}{2}x + 3$
  $3x + 2y = -2$

30. $y = \frac{3}{4}x + 1$
  $2x + 3y = 37$

31. $y = \frac{1}{2}x + 2$
  $6y = 3x + 12$

32. $x = -\frac{3}{2}y - 3$
  $4x + 6y = -12$

33. $y = 2x + 2$
  $2x - y = 3$

34. $3x + y = 2$
  $y = -3x$

35. $3x + 2y = 12$
  $y - 2 = 1$

36. $x - 2y = -4$
  $3x - 2 = 4$

37. $3x - y = 5$
  $x + 2y = 4$

38. $3x - 2y = 8$
  $y = -2x + 3$

39. $y = x - 7$
  $x = y + 7$

40. $2y = x + 3$
  $x + 2y = 6$

41. $y = 1 + 4x$
  $4x - y = 2$

42. $y = -2x - 4$
  $2x + y = 6$

43. $5x + 7y = 35$
  $2x - 5y = 53$

44. $2x - 5y = 8$
  $3x + 4y = 35$

Write a system of equations for each situation, and solve by using substitution.

45. Two angles are complementary. The larger angle measures 10 degrees less than four times the smaller angle. Find the measures of the angles.

46. Two angles are supplementary. The difference of their measures is 40 degrees. Find the measures of the angles.

47. Two angles of a triangle measure the same. The third angle measures 20 degrees more than the sum of the other two angles. How much does each angle measure?

48. Two angles of a triangle are the same size. The third angle is 15 degrees larger than each of the other two. What is the size of each angle?

49. The radius of a large circle is 5 inches more than twice the radius of a smaller circle. The larger circle has a circumference of 283 inches. Find the radius of the smaller circle and the radius of the larger circle, to the nearest inch.

50. A large circle has a circumference of 163 cm. Its radius is 10 cm less than three times the radius of a smaller circle. What is the radius of each circle, to the nearest cm?

51. Mr. McDonald had 15 acres of farmland, some of which he planted in corn and the rest in soybeans. The number of acres planted in corn was equivalent to 25 acres of farmland, less the number of acres planted in soybeans. How many acres were planted in each crop?

52. A field has 12 acres of farmland, part of which is planted in wheat and the rest in alfalfa. The number of acres of wheat equals one-half the difference of a 24-acre field, less twice the acreage of the alfalfa. How many acres are planted in each crop?

53. The Homestore sells toaster ovens for $49 each, retail price. The wholesale cost to stock the ovens is $22 each. The fixed cost associated with acquiring the ovens, storing them in inventory, using shelf space, and advertising the ovens for sale is $2500.
   a. Write a function for the total cost of stocking the ovens for sale.
   b. Write a function for the total revenue received from selling the ovens.
   c. Write a system of equations, and determine the number of ovens that must be sold to break even.

54. Handyman Depot sells ceiling fans for a retail price of $89 each. The wholesale cost to stock the fans is $35 each. The cost of ordering the fans from the manufacturer, storing them in inventory, using shelf space, and advertising the fans for sale is $3600.
   a. Write a function for the total revenue received from selling the fans.
   b. Write a function for the total cost of stocking the fans for sale.
   c. Write a system of equations, and determine the number of fans that must be sold to break even.

55. The net stock value (in billions of current dollars) of the metal-mining industry is estimated by $M(x) = 1.09x + 28.9$, where x is the number of years after 1990. Likewise, the net stock value (in billions of current dollars) of the coal-mining industry is estimated by $C(x) = 0.78x + 31.8$. These relations are based upon data reported by the U.S. Bureau of Economic Analysis. Determine the year when the two net stock values will be equal if the two trends are assumed to continue. How does your result compare with the following table of actual values?

Year	1990	1995	1996	1997
Metal-mining net stock value	29	34	35	37
Coal-mining net stock value	32	35	36	38

56. The net stock value (in billions of current dollars) of the industrial-machinery-and-equipment industry is estimated by $I(x) = 3.17x + 109.7$, where x is the number of years after 1990. At the same time, the net stock value (in billions of current dollars) of the electronic-and-other-electric-equipment industry is estimated by $E(x) = 6.57x + 89.7$, based upon data reported by the U.S. Bureau of Economic Analysis. In what year will the two net stock values be equal, assuming that the trends continue? How does your result compare with the following table of values reported?

Year	1990	1995	1996	1997
Net stock value of industrial machinery and equipment	110	125	128	133
Net stock value of electronic and other electric equipment	91	118	129	139

57. The total receipts of the United States for 2000 were estimated to be 1.9563×10^{12} dollars. Some of these receipts were from individual income taxes, and the rest were from other sources, such as corporate income taxes, social security taxes, excise taxes, and miscellaneous taxes. The receipts from other sources were 5.31×10^{10} dollars more than the receipts from individual income taxes. Determine how much was received from individual income taxes and how much was received from other sources.

58. Helium atoms have a much smaller diameter than cesium atoms. The arithmetic average of the two diameters is 2.75×10^{-10} m. The difference in size of the two diameters is 4.5×10^{-10} m. Determine the two diameters.

6.2 Calculator Exercises

The calculator can help you with the arithmetic when you are solving a system of equations by the substitution method, but it cannot do much more. However if you wish to check your solution by solving the system graphically, the calculator can be quite useful. Next is a shortcut that can save you effort when you use the calculator to solve a system graphically.

Whenever you use the calculator to solve a system of linear equations graphically, you must first solve functions for y in order to store them in the calculator. In solving functions for y, it is not necessary to simplify the expressions, since the calculator can handle them in nonsimplified form. This can save you time and keep you from making mistakes in the process. If the function is given to you in standard form, $ax + by = c$, just solve for y without simplifying and store $y = (c - ax) \div b$ in Y1. Do likewise for Y2. As an example, solve the following system of linear equations:

$$\frac{3}{8}x + \frac{7}{16}y = \frac{2303}{5280}$$

$$x + \frac{9}{17}y = \frac{137}{170}$$

We will store the following equations in Y1 and Y2:

$$Y1 = \left(\frac{2303}{5280} - \frac{3}{8}x\right) \div \left(\frac{7}{16}\right)$$

$$Y2 = \left(\frac{137}{170} - x\right) \div \left(\frac{9}{17}\right)$$

Note that you can almost do this step of solving for y in your head as you enter it into the calculator. Note also that we did not simplify these two equations. When we graph the two equations and find the intersection, the calculator displays the values of x and y as follows:

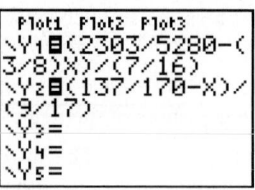
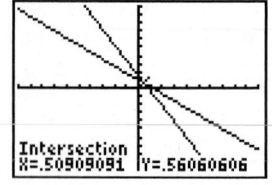

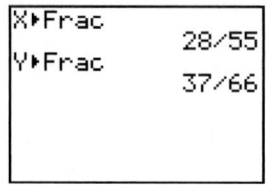

To convert the answers to fractions, simply enter the letters X and Y, followed by the $\boxed{\text{MATH}}$ $\boxed{\text{I}}$ command, and the calculator will display the fractions.

Try solving the following systems by using the substitution method. Then use the calculator to check your solution graphically by following the preceding example.

1. $15.80x + y = 2655.10$
    $18.40x + 73.20y = 19361.22$
2. $4.055x - 8.752y = -42.949$
    $x + 0.405y = 2.225$
3. $\frac{7}{13}x + \frac{5}{17}y = \frac{2}{3}$
    $x - \frac{5}{34}y = \frac{19}{42}$
4. $\frac{28}{65}x + \frac{51}{56}y = \frac{43}{35}$
    $\frac{35}{39}x - y = \frac{61}{51}$

6.2 Writing Exercises

You have learned two methods for solving systems of equations: the graphical method and the substitution method. Decide which method is easier to use to solve each system. List the reasons for your decision. Then solve the equations, using the method you chose.

1. $y = 0.0803x + 1.0507$
    $y = -0.8532x + 1.9842$
2. $y = 85x + 300$
    $450x - y = 286{,}225$

6.3 Solving Systems of Two Equations by Using Elimination

OBJECTIVES   1 Solve systems of linear equations by using elimination.

2 Model real-world situations by using systems of equations, and solve the equations by elimination.

APPLICATION

According to the 2000 *Forbes* magazine Celebrity 100 list, the combined annual earnings of the highest paid TV writer, Larry David, and TV producer, David E. Kelley, were $360 million. The writer's earnings less twice the producer's earnings were $6 million. Determine the annual earnings of Larry David and David E. Kelley.

After completing this section, we will discuss this application further. See page 417.

6.3.1 Solving Systems of Linear Equations by Using Elimination

Sometimes neither equation in a system can be solved for one variable in any simple way. Therefore, a second algebraic method is useful. We call this second method for solving a system of linear equations the **elimination method**.

In order to understand the elimination method, we will review the addition property of equations. This property states that we may add equivalent expressions to both sides of an equation, and the result is an equivalent equation; that is,

If $a = b$ and $c = d$, then $a + c = b + d$ for any expressions $a, b, c,$ and d.

Remember that in Chapter 4 we illustrated why this is true with a balanced scale representing the equation. Now we begin with two equations (balanced scales).

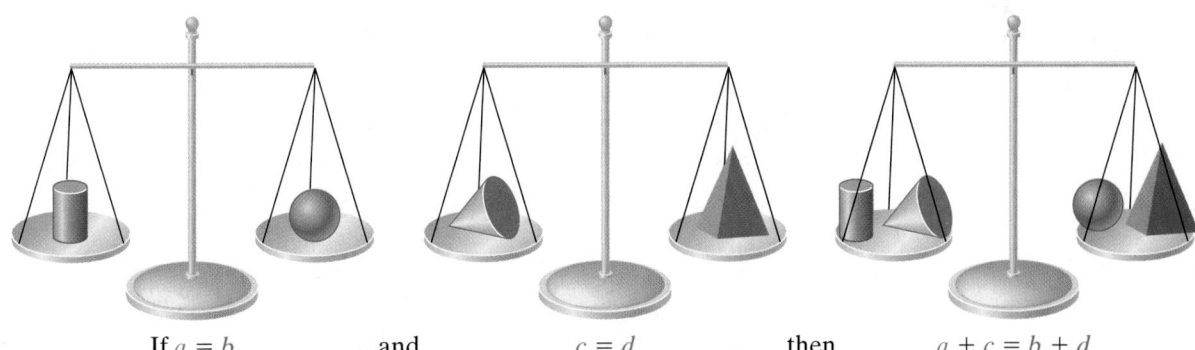

If $a = b$ and $c = d$ then $a + c = b + d$

We will apply the addition property to a system of equations to obtain a new equation with one variable eliminated.

Solve the system. $2x - y = 8$
$y = 4$

Using the addition property of equations, we obtain

If $2x - y = 8$ and $y = 4$, then $2x - y + y = 8 + 4$.

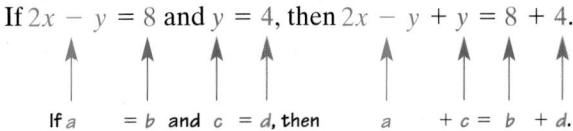

We add to each member of the first equation the corresponding member of the second equation. In other words, we add y to the left side of the first equation, $2x - y$, and 4 to the right side of the first equation, 8.

$$2x - y = 8$$
$$y = 4 \qquad \text{Add corresponding members of the equations.}$$
$$\overline{2x \quad\;\; = 12} \qquad \text{Simplify.}$$

$$\frac{2x}{2} = \frac{12}{2} \qquad \text{Divide both sides by 2.}$$

$$x = 6 \qquad \text{Simplify.}$$

Since $y = 4$, the solution of the system is $(6, 4)$.

This process works very nicely if one of the variables is eliminated when we apply the addition property. However, in many cases, we will need to write equivalent equations for the system before we apply the addition property. For example, solve the system. $\quad 2x - y = -1$
$$4x - y = 3$$

If we add $4x - y$ to the expression on the left side of the first equation, $2x - y$, we obtain $6x - 2y$. No variable is eliminated. However, we should see that if the y-term were positive in one expression and negative in the other expression, the y-variable would have been eliminated. In order to get a positive y, we need to multiply one of the variable expressions by -1 to change the sign of y. To do this, we apply the multiplication property of equations and multiply *both* sides of the equation by the same value, -1.

(1) $\quad 2x - y = -1$
(2) $\quad 4x - y = 3$
(1) $\quad 2x - y = -1$
(2) $\quad -1(4x - y) = -1(3) \qquad \text{Multiply both sides by } -1.$

 **HELPING HAND** Even though the second equation is the only equation changed in this process, rewrite the entire system each time an equation changes. Doing this helps you keep track of what the system presently looks like.

(1) $\quad 2x - y = -1$
(2) $\quad \underline{-4x + y = -3}$
$$-2x \quad\;\; = -4 \qquad \text{Add corresponding members of the equation.}$$

$$\frac{-2x}{-2} = \frac{-4}{-2} \qquad \text{Divide both sides by } -2.$$

$$x = 2 \qquad \text{Simplify.}$$

Substitute 2 for x in the first equation to find the y-value of the solution.

(1) $\quad 2x - y = -1$
$$2(2) - y = -1 \qquad \text{Substitute 2 for x.}$$
$$4 - y = -1 \qquad \text{Simplify.}$$
$$4 - y - 4 = -1 - 4 \qquad \text{Subtract 4 from both sides.}$$
$$-y = -5 \qquad \text{Simplify.}$$

$$\frac{-y}{-1} = \frac{-5}{-1} \qquad \text{Divide both sides by } -1.$$

$$y = 5 \qquad \text{Simplify.}$$

The solution of the system is $(2, 5)$.

In some cases, we will need to rewrite both equations in the system before applying the addition property to eliminate a variable. For example, solve the system.

$$2x + 3y = 5$$
$$3x + 2y = -5$$

We see that no variable is eliminated if we apply the addition property of equations. Also, the x-variable will not be eliminated if we multiply either expression by an integer. The same is true for the y-variable. However, the least common multiple of the coefficients of the variables could be found and then used to determine a number needed to multiply each equation in the system. For example, if we want to eliminate the x-variable in the system, we determine the least common multiple, 6, of the x-coefficients, 2 and 3. To eliminate the x-variable, we want one expression to have 6 and the other expression to have -6 as the coefficient of x. One way to accomplish this is by multiplying the members of the first equation by 3 and the members of the second equation by -2.

$(1)\ 2x + 3y = 5$

$(2)\ 3x + 2y = -5$

$(1)\ 3(2x + 3y) = 3(5)$ Multiply both sides by 3.

$(2)\ -2(3x + 2y) = -2(-5)$ Multiply both sides by -2.

$(1)\quad 6x + 9y = 15$

$(2)\ \underline{-6x - 4y = 10}$

$\qquad\qquad 5y = 25$ Add corresponding members of the equation.

$$\frac{5y}{5} = \frac{25}{5} \qquad \text{Divide both sides by 5.}$$

$$y = 5 \qquad \text{Simplify.}$$

Substitute 5 for y in the first equation to find the x-value of the solution.

$(1)\qquad 2x + 3y = 5$

$\qquad\quad 2x + 3(5) = 5$ Substitute 5 for y.

$\qquad\quad 2x + 15 = 5$ Simplify.

$\quad 2x + 15 - 15 = 5 - 15$ Subtract 15 from both sides.

$\qquad\qquad 2x = -10$ Simplify.

$$\frac{2x}{2} = \frac{-10}{2} \qquad \text{Divide both sides by 2.}$$

$$x = -5 \qquad \text{Simplify.}$$

The solution of the system is $(-5, 5)$.

SOLVING A SYSTEM OF LINEAR EQUATIONS IN TWO VARIABLES BY USING ELIMINATION

To solve a system of linear equations in two variables by using elimination,

- Write both equations in standard form.
- Multiply the members of the equation(s) by a number, if necessary, to obtain coefficients that are opposites.
- Add the expressions in the second equation to the corresponding expressions in the first equation, eliminating one of the variables.
- Solve the resulting equation for the remaining variable.
- Substitute the value obtained in one of the original equations, and solve for the other variable.

The ordered-pair solution is determined by the values obtained in solving for each individual variable. Check the solution.

EXAMPLE 1   Solve each system of linear equations in two variables by using elimination. Check the solution by substituting it into the original equations.

a. $2x = -3y + 1$ **b.** $3x + 2y = 1$ **c.** $\dfrac{1}{2}x + \dfrac{2}{3}y = 1$

$\quad\ \ x + 2y = -1$ $\quad 2x - 5y = -2$ $\quad\dfrac{1}{4}x - \dfrac{1}{5}y = -\dfrac{1}{10}$

Algebraic Solution

a. (1) $2x = -3y + 1$

$\quad$ (2) $x + 2y = -1$

Write the first equation in standard form.

$$2x + 3y = -3y + 1 + 3y \qquad \text{Add 3y to both sides of the equation.}$$
$$2x + 3y = 1 \qquad\qquad\qquad \text{Simplify.}$$

The system with equations in standard form is as follows:

$\quad$ (1) $2x + 3y = 1$

$\quad$ (2) $x + 2y = -1$

$\quad$ (1) $2x + 3y = 1$

$\quad$ (2) $-2(x + 2y) = -2(-1)$ $\qquad$ Multiply both sides by -2.

$\quad$ (1) $\quad\ 2x + 3y = 1$

$\quad$ (2) $\underline{-2x - 4y = 2}$

$\qquad\qquad\quad -y = 3 \qquad\qquad\qquad$ Add corresponding members of the equations.

$$\frac{-y}{-1} = \frac{3}{-1} \qquad\qquad \text{Divide both sides by } -1.$$

$\qquad\qquad\quad\ y = -3 \qquad\qquad\qquad$ Simplify.

Substitute -3 for y in the second equation to find the x-value of the solution.

$\quad$ (2) $\qquad x + 2y = -1$

$\qquad\qquad x + 2(-3) = -1 \qquad$ Substitute -3 for y.

$\qquad\qquad\quad\ x - 6 = -1 \qquad\qquad$ Simplify.

$\qquad\quad x - 6 + 6 = -1 + 6 \qquad$ Add 6 to both sides.

$\qquad\qquad\qquad\quad x = 5 \qquad\qquad\qquad$ Simplify.

Check

```
5→X: -3→Y: 2X
                10
-3Y+1
                10
X+2Y
                -1
```

The solution of the system is $(5, -3)$.

The solution checks, because $10 = 10$ and $-1 = -1$.

b. (1) $3x + 2y = 1$

(2) $2x - 5y = -2$

In order to eliminate the x-variable, we determine that the least common multiple of the x-coefficients is 6. Therefore, the members of the first equation must be multiplied by 2 and the members of the second equation by -3 to obtain the x-coefficients 6 and -6.

Note: We could have eliminated the y-variable instead by multiplying the first equation by 5 and the second equation by 2.

(1)	$2(3x + 2y) = 2(1)$	Multiply both sides by 2.
(2)	$-3(2x - 5y) = -3(-2)$	Multiply both sides by -3.
(1)	$6x + 4y = 2$	
(2)	$\underline{-6x + 15y = 6}$	
	$19y = 8$	Add corresponding members of the equations.
	$\dfrac{19y}{19} = \dfrac{8}{19}$	Divide both sides by 19.
	$y = \dfrac{8}{19}$	Simplify.

Substitute $\frac{8}{19}$ for y in the first equation to find the x-value of the solution.

(1)	$3x + 2y = 1$	
	$3x + 2\left(\dfrac{8}{19}\right) = 1$	Substitute $\frac{8}{19}$ for y.
	$3x + \dfrac{16}{19} = 1$	Simplify.
	$3x + \dfrac{16}{19} - \dfrac{16}{19} = 1 - \dfrac{16}{19}$	Subtract $\frac{16}{19}$ from both sides.
	$3x = \dfrac{3}{19}$	Simplify.
	$3x\left(\dfrac{1}{3}\right) = \left(\dfrac{3}{19}\right)\left(\dfrac{1}{3}\right)$	Multiply both sides by $\frac{1}{3}$.
	$x = \dfrac{1}{19}$	Simplify.

The solution of the system is $\left(\frac{1}{19}, \frac{8}{19}\right)$.

The check is left to you.

c. (1) $\dfrac{1}{2}x + \dfrac{2}{3}y = 1$

(2) $\dfrac{1}{4}x - \dfrac{1}{5}y = -\dfrac{1}{10}$

Eliminate the fractions in the system of equations by multiplying the members of each equation by the least common multiple of the denominators in the equation.

(1)	$6\left(\dfrac{1}{2}x + \dfrac{2}{3}y\right) = 6(1)$	Multiply by the LCD of 2 and 3.
(2)	$20\left(\dfrac{1}{4}x - \dfrac{1}{5}y\right) = 20\left(-\dfrac{1}{10}\right)$	Multiply by the LCD of 4, 5, and 10.

(1) $3x + 4y = 6$

(2) $\dfrac{5x - 4y = -2}{8x \qquad = \quad 4}$ Add corresponding members of the equations.

$$\frac{8x}{8} = \frac{4}{8}$$ Divide both sides by 8.

$$x = \frac{1}{2}$$ Simplify.

Substitute $\frac{1}{2}$ for x in the first equation to find the y-value of the solution.

(1) $$\frac{1}{2}x + \frac{2}{3}y = 1$$

$$\frac{1}{2}\left(\frac{1}{2}\right) + \frac{2}{3}y = 1$$ Substitute $\frac{1}{2}$ for x.

$$\frac{1}{4} + \frac{2}{3}y = 1$$ Simplify.

$$\frac{1}{4} + \frac{2}{3}y - \frac{1}{4} = 1 - \frac{1}{4}$$ Subtract $\frac{1}{4}$ from both sides.

$$\frac{2}{3}y = \frac{3}{4}$$ Simplify.

$$\frac{3}{2}\left(\frac{2}{3}y\right) = \frac{3}{2}\left(\frac{3}{4}\right)$$ Multiply both sides by $\frac{3}{2}$.

$$y = \frac{9}{8}$$ Simplify.

The solution of the system is $\left(\frac{1}{2}, \frac{9}{8}\right)$.
The check is left to you.

As we have seen in earlier sections of this chapter, a system may have no solution or an infinite number of solutions. Let's see what happens when we apply elimination to systems of these types. Complete the following sets of exercises.

Discovery 6

Inconsistent System of Linear Equations

Solve the given system of linear equations by using elimination.

$$-2x + y = 1$$
$$-4x + 2y = 3$$

Write a rule for determining, by using the elimination method, that a system of linear equations is inconsistent—in other words, or that it has no solution.

A system has no solution if, after elimination, the resulting equation is a contradiction.

Discovery 7

System of Dependent Linear Equations

Solve the given system of linear equations by using elimination.

$$-3x - y = -2$$
$$y = -3x + 2$$

Write a rule for determining, by using the elimination method, that a system of linear equations has dependent equations—in other words, an infinite number of solutions.

A system has an infinite number of solutions if, after elimination, the resulting equation is an identity.

EXAMPLE 2 Solve each system of equations by using elimination. If the system has an infinite number of solutions, describe the solutions.

a. $x + y = -3$ **b.** $y = \dfrac{2}{3}x - \dfrac{1}{3}$

   $x + y = 2$ $3y = 2x - 1$

Algebraic Solution

a. (1) $x + y = -3$

   (2) $x + y = 2$

   (1) $-1(x + y) = -1(-3)$ Multiply both sides of the first equation by -1.

   (2) $x + y = 2$

   (1) $-x - y = 3$

   (2) $\underline{x + y = 2}$

                $0 = 5$ Add corresponding members of the equations.

The result is a contradiction. The system has no solution.

b. (1) $y = \dfrac{2}{3}x - \dfrac{1}{3}$

   (2) $3y = 2x - 1$

Write the first equation in standard form.

$$3(y) = 3\left(\frac{2}{3}x - \frac{1}{3}\right)$$ Multiply by the LCD of 3 to clear the fractions.

$$3y = 2x - 1$$ Simplify.

$$3y - 2x = 2x - 1 - 2x$$ Subtract 2x from both sides.

$$-2x + 3y = -1$$ Simplify and rearrange.

Write the second equation in standard form.

$$3y - 2x = 2x - 1 - 2x$$ Subtract 2x from both sides.

$$-2x + 3y = -1$$ Simplify and rearrange.

In standard form, the system is as follows:

(1) $-2x + 3y = -1$

(2) $-2x + 3y = -1$

(1) $-1(-2x + 3y) = -1(-1)$ Multiply both sides of the first equation by -1.

(2) $-2x + 3y = -1$

(1) $2x - 3y = 1$

(2) $-2x + 3y = -1$

$$0 = 0$$ Add corresponding members of the equations.

The result is an identity. The solutions are all ordered pairs (x, y) that satisfy $y = \frac{2}{3}x - \frac{1}{3}$.

The check is left to you.

The following summary describes the process of algebraically solving a system of linear equations by using elimination:

SOLVING A SYSTEM OF LINEAR EQUATIONS BY USING ELIMINATION

In solving a system of linear equations in two variables, one of three possibilities will occur.

1. **A value for both variables will be determined.**

   One solution exists: the ordered-pair coordinates.

   The system is consistent.

   The equations are independent.

2. **The elimination will result in a contradiction.**

   No solution exists.

   The system is inconsistent.

   The equations are independent.

3. **The elimination will result in an identity.**

   An infinite number of solutions exist. The solution set is the set of all ordered pairs (x, y) that satisfy either equation of the system.

   The system is consistent.

   The equations are dependent.

 6.3.1 Checkup

Solve each system of equations by using the elimination method. Check the solution.

1. $x + y = 10$
   $x - y = 5$

2. $x = 5y + 32$
   $2x + y = -2$

3. $4x + 7y = -3$
   $7x - 2y = 9$

4. $\frac{1}{2}x + \frac{5}{8}y = \frac{9}{8}$
   $\frac{2}{5}x - \frac{1}{2}y = -\frac{7}{10}$

In exercises 5 and 6, solve each system of equations by using the elimination method. If the system has an infinite number of solutions, describe the solution set.

5. $x + 2y = 6$
   $3x = 20 - 6y$

6. $y = 5x + 12$
   $7x + 24 = 2y - 3x$

7. You now know three methods for solving a system of linear equations in two variables: the graphical method, the substitution method, and the elimination method. Explain how you would apply each of these methods to obtain the solution of a system of linear equations.

6.3.2 Modeling the Real World

When two equations are needed to describe the relationship between two variables, we sometimes say that the system consists of **simultaneous equations**, since both equations apply at the same time. The method of elimination is probably the most common way of solving simultaneous equations, unless we can get one equation into a simple enough form so that substitution can be used. In fact, elimination is the method underlying more complicated techniques that we can use to solve systems of three, four, or more simultaneous equations.

EXAMPLE 3 Melinda's class charged admission of $2.50 for adults and $1.25 for students to its play. Melinda counted 10 chairs remaining empty in the auditorium, which holds 500. If the amount collected was $1013.75, how many adults and students attended?

Algebraic Solution

Let a = number of adults attending
c = number of students attending

(1) $a + c = 490$ — The number attending was $500 - 10$ or 490.

(2) $2.50a + 1.25c = 1013.75$ — The cost is the sum of the amount collected for adults, $2.50a$, and the amount collected for students, $1.25c$.

(1) $-1.25(a + c) = -1.25(490)$ — Multiply both sides by -1.25.

(2) $2.50a + 1.25c = 1013.75$

(1) $-1.25a - 1.25c = -612.50$

(2) $\underline{2.50a + 1.25c = 1013.75}$

$1.25a \qquad\quad = 401.25$ — Add corresponding members of the equations.

$\dfrac{1.25a}{1.25} = \dfrac{401.25}{1.25}$ — Divide both sides by 1.25.

$a = 321$ — Simplify.

Substitute 321 for a in the first equation to find the c-value of the solution.

(1) $\qquad\qquad a + c = 490$

$321 + c = 490$ — Substitute 321 for a.

$321 + c - 321 = 490 - 321$ — Subtract 321 from both sides of the equation.

$c = 169$ — Simplify.

The check is left to you.
There were 321 adults and 169 students attending the play.

APPLICATION

According to the 2000 *Forbes* magazine Celebrity 100 list, the combined annual earnings of the highest paid TV writer, Larry David, and TV producer, David E. Kelley, were $360 million. The writer's earnings less twice the producer's earnings were $6 million. Determine the annual earnings of Larry David and David E. Kelley.

(Continued on page 418)

Discussion

Let w = the annual earnings of the TV writer in millions of dollars
$\quad p$ = the annual earnings of the TV producer in millions of dollars
The combined earnings were \$360 million, or $p + w = 360$.
The TV writer earned \$6 million more than twice the earnings of the TV producer, or $w = 2p + 6$.
The system of equations in standard form is

(1) $\quad p + w = 360 \qquad$ or $\qquad p + w = 360$

(2) $\quad w = 2p + 6 \qquad\qquad\qquad \underline{2p - w = -6}$

$$\qquad\qquad\qquad\qquad\qquad 3p \quad\;\; = 354 \qquad \text{Add corresponding members of the equations.}$$

$$\qquad\qquad\qquad\qquad\qquad \frac{3p}{3} = \frac{354}{3} \qquad \text{Divide by 3.}$$

$$\qquad\qquad\qquad\qquad\qquad p = 118 \qquad \text{Simplify.}$$

Substitute 118 for p, and solve for w in the second equation.

$$w = 2p + 6$$
$$w = 2(118) + 6$$
$$w = 242$$

The annual earnings of the highest paid TV writer were \$242 million. The annual earnings of the highest paid TV producer were \$118 million.

 6.3.2 Checkup

1. The Pocahontas circle of the Indian Maidens sold 262 cans of nuts as a fund-raiser. Cashews sold for \$6.50 per can and peanuts for \$4.00 per can. The proceeds from the sale were \$1240.50. Members of the circle lost their tally of how many cans of each were sold. Write and solve a system of equations to help them determine how many of each they sold.

2. According to *Forbes* magazine Celebrity 100 list, the combined earnings of Bruce Willis and Tom Cruise in 2000 were \$113.2 million. The difference in their earnings was \$26.8 million. Given that Bruce Willis earned more than Tom Cruise, write a system of equations and solve them to find the earnings of each.

6.3 Exercises

Solve each system of equations by using the elimination method. Check your solution.

1. $2x - y = -6$
$\quad 5x + y = -8$

2. $5x + y = 8$
$\quad 3x - y = 16$

3. $x + 7y = 19$
$\quad 2y = x - 1$

4. $x + 9y = -12$
$\quad -x + 8y = -5$

5. $5x + y = -24$
$\quad 3x - 2y = 9$

6. $4x - 3y = 26$
$\quad 2x + y = 18$

7. $x + 3y = 2$
$\quad x + 5y = -2$

8. $x + 8y = 37$
$\quad x + 11y = 52$

9. $2x + 7y = 29$
$\quad 4x + 3y = 25$

10. $5x + 11y = -27$
$\quad 10x + 13y = -36$

11. $3x - 5y = 66$
$\quad 4x + 3y = 1$

12. $2x + 7y = 33$
$\quad 3x - 4y = -52$

13. $2x + 9y = 102$
$\quad 5x - 11y = -147$

14. $11x + 5y = 133$
$\quad 15x + 4y = 187$

15. $40x = 23 + 10y$
$\quad 50x + 10y = 94$

16. $15x + 10y = -2$
$\quad 5x = 10y + 18$

17. $5x + 40y = 77$
$\quad 5x + 15y = 17$

18. $4x + 2y = 15$
$\quad -6x + 2y = -30$

19. $10x - 4y = 28$
$\quad 5x + 4y = 35$

20. $2x + y = -7$
$\quad 3x - y = -2$

21. $2x + 8y = 29$

$\quad 13y = 3x + 39$

22. $5x + 15y = -10$

$\quad 19y = 7x + 42$

23. $\frac{1}{4}x + \frac{1}{3}y = \frac{5}{12}$

$\quad \frac{1}{4}x = \frac{1}{3}y - \frac{1}{12}$

24. $\frac{2}{3}x + \frac{3}{5}y = \frac{9}{10}$

$\quad \frac{2}{3}x = \frac{3}{5}y - \frac{1}{10}$

25. $\dfrac{1}{3}x + \dfrac{1}{2}y = -\dfrac{1}{4}$

$\dfrac{1}{6}x - \dfrac{5}{6}y = \dfrac{11}{16}$

26. $\dfrac{3}{7}x + \dfrac{4}{5}y = -\dfrac{1}{6}$

$\dfrac{3}{7}x - \dfrac{8}{15}y = \dfrac{2}{3}$

27. $y = \dfrac{3}{2}x - 9$

$\dfrac{1}{6}x - \dfrac{1}{9}y = 1$

28. $y = \dfrac{6}{5}x + 2$

$\dfrac{3}{5}x - \dfrac{1}{2}y = -1$

29. $\dfrac{4}{5}x - \dfrac{3}{5}y = -1$

$\dfrac{3}{2}x - \dfrac{9}{8}y = 1$

30. $\dfrac{1}{12}x - \dfrac{1}{8}y = -1$

$\dfrac{1}{3}x - \dfrac{1}{2}y = 1$

31. $x - 20y = 70$

$3x + 10y = 70$

32. $x + 35y = 300$

$4x + 15y = 200$

33. $3x + y = 40$

$x + y = 20$

34. $x + 4y = 0$

$x + 7y = -30$

35. $x - y = 300$

$2x - y = -100$

36. $x + y = 300$

$x + 3y = 1400$

37. $10x - 10y = 22$

$y = 2x - 11$

38. $5x + 5y = 10$

$y = 9x - 41$

39. $3.2x + 4.2y = 368$

$4.4x - 2.1y = 128$

40. $1.4x + 2.4y = 225$

$3.4x - 1.6y = 123$

41. $0.05x + 0.10y = 0.75$

$x + y = 11$

42. $0.25x + 0.50y = 5.75$

$x + y = 14$

43. $0.3x + 0.45y = 43.5$

$x + y = 110$

44. $0.55x + 0.25y = 49$

$x + y = 100$

45. $12.50x + 6.50y = 1780$

$x + y = 200$

46. $7.95x + 2.25y = 914.70$

$x + y = 214$

Write a system of equations for each application, and solve the equations by using the elimination method.

47. At Centennial High School's football game, the gate indicated that 683 persons entered. Students were charged $1.50 admission, while nonstudents were charged $5.00 admission. The total receipts at the game were $2645.00. The principal wanted to know how many students attended the game, but unfortunately, the count was not recorded. Can you find the answer for the principal?

48. A church sponsored a bus trip to attend the play *Godspell* at the summer stock theater in a nearby town. Senior citizens were charged $30.00 for the trip, including all expenses, while others were charged $45.00. A total of 56 people took the trip and paid $2190 for the privilege. How many seniors and how many others took the trip?

49. To raise money for a scholarship fund, the faculty and staff at the Community College for Mathematics decided to print a cookbook and a calendar. They sold 50 more cookbooks than calendars and raised $3462.50, selling the cookbooks for $8.50 each and the calendars for $5.00 each. How many of each were sold?

50. For a luncheon, Karlene ordered a tray of ham and cheese from the local deli. The deli advertised the tray as having a weight of 5.5 pounds. The ham sold for $6.00 per pound, and the cheese sold for $4.00 per pound. The total charge for the tray was $29.50. Karlene wanted to know how much ham and how much cheese the deli sold her. Can you help her?

51. The number of adults and children at Midtown Nursery's school play totaled 385. Adults were charged $2 admission, and by mistake, the children were also charged $2 each. The receipts for the play amounted to $770. How many adults and how many children attended the play?

52. Mama Mia's Restaurant offers two lists of items on its menu. The items on the first list are all one price, and the items on the second list are all a second price. If a patron selects one item from each list, the price for the meal is $8.95. If a patron selects two items from each list, the price for the meal is $13.95. Find the price of an item from the first list and the price of an item from the second list.

53. Corinna jogs around a rectangular field that has a perimeter of 620 yards. When she jogs around another field that has the same width, but is 90 yards longer, she jogs a distance of 800 yards. What are the dimensions of the smaller field?

54. Evan can use either of two circular tracks at the gym for his jogging. The radius of one track is 10 yards larger than the radius of the second track. The circumference of the larger track is 20π yards more than the circumference of the smaller track. What is the radius of each track?

55. A triangle has two angles that are equal in size. The third angle equals 90 degrees minus the sum of the measures of the other two angles. What is the size of each angle?

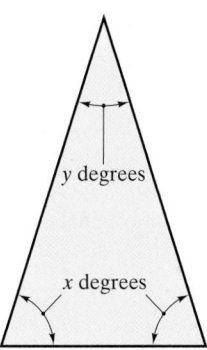

56. Two angles are supplementary. One angle is 30 degrees larger than the other. How large is each angle?

57. The distance from the Sun to the red planet Mars is approximately 4.864×10^7 miles greater than the distance of the Earth from the Sun. The average of the distances from the Sun for the two planets is 1.1728×10^8 miles. Find the approximate distances of the two planets from the Sun.

58. The planet Mercury is approximately 3.6302×10^9 miles closer to the Sun than the planet Pluto. The average distance from the Sun for the two planets is approximately 1.851×10^9 miles. What are the approximate distances of these planets from the Sun?

59. According to data from the U.S. Bureau of Labor Statistics, the median weekly earnings of males exceeded the median weekly earnings of females by $142 in 1998. The two medians totaled $1054. Write a system of equations and solve them to determine the median weekly earnings of males and of females.

60. According to data from the U.S. Bureau of Labor Statistics, the median weekly earnings of black males exceeded the median weekly earnings of black females by $68 in 1998. The combined total of the two medians was $868. Write a system of equations and solve them to determine the median weekly earnings of black males and of black females.

61. According to the *Forbes* magazine Celebrity 100 list, the celebrity earning the most money in 2000 was George Lucas, and the celebrity with the second highest earnings was Oprah Winfrey. Together, they earned $400 million. Lucas earned $50 million less than twice the amount that Winfrey earned. Write a system of equations and solve them to determine the earnings of each.

62. According to the *Forbes* list of the world's richest individuals, the richest person in the world in 2001 was William H. Gates, III. The second-richest person in the world was Warren Edward Buffett. Together, their net worth was $91.0 billion. Gates's worth was $26.4 billion more than Buffet's. Write a system of equations and solve them to determine the worth of each.

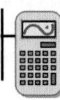

6.3 Calculator Exercises

A variation of the elimination method called the method of alternates can be used to obtain opposite coefficients. Consider the system

(1) $6x - 7y = -21$
(2) $15x - 11y = 6$

Rather than find the least common multiple of 6 and 15, multiply the members of each equation by the coefficient of x in the alternate (other) equation, yielding

(1) $15(6x - 7y) = 15(-21) \quad \rightarrow \quad 90x - 105y = -315$
(2) $6(15x - 11y) = 6(6) \quad \rightarrow \quad 90x - 66y = 36$

Then multiply the members of the second equation by -1 and add.

(1) $90x - 105y = -315$
(2) $\underline{-90x + 66y = -36}$
$-39y = -351$, yielding $y = 9$.

Substituting into one of original equations yields $x = 7$.

Use this method to solve the exercises that follow. Note that the numbers get large and a calculator may help with the arithmetic.

1. $6x + 35y = -52$
$9x - 14y = 55$

2. $12x - 35y = 81$
$15x - 28y = 54$

3. $21a + 10b = -88$
$14a + 15b = 8$

4. $33x + 7y = -3.1$
$6x + 13y = -33.4$

5. $10x + 21y = 19$
$14x - 9y = 1$

6. $8x - 9y = 8$
$12x + 21y = -11$

6.3 Writing Exercises

You now know three methods for solving a system of equations: graphical, substitution, and elimination. Decide which method might be best for each of the exercises that follow. Give reasons for your choices. Then solve the systems.

1. $3x + y = 62$
$5x + 19y = 203$

2. $14x - 3y = -240$
$5x + 3y = -45$

3. $y = 2x - 69$
$y = x - 17$

6.4 More Real-World Models

OBJECTIVES

1 Model real-world situations involving a distance traveled.
2 Model real-world situations involving mixtures and collections.

APPLICATION

The U.S. national debt is about $5.6 trillion. Suppose that the Treasury Department split this debt into two loans. One consists of Treasury bills with an annual interest rate of about 3.6%, and the second loan consists of Treasury constant maturities with an annual interest rate of approximately 4.55%. If about 0.23 trillion dollars is paid in simple interest annually, determine the amount of principal for each loan.

After completing this section, we will discuss this application further. See page 430.

We have now discussed three methods for solving a system of linear equations in two variables: A system can be solved graphically, algebraically by substitution, or algebraically by elimination. We have also discussed real-world situations with each of these methods. In this section, we discuss other applications involving systems of linear equations. Once a system of linear equations is written for a problem, it may be solved with any of the methods we have presented.

Let's review the method we have been using to answer a question about a real-world situation.

MODELING REAL-WORLD SITUATIONS
To model a real-world situation,

* Read and understand the problem.
* Define variables for the unknown quantities, and define other quantities in terms of the defined variables.
* Write equations with the information given.
* Solve the system either graphically or algebraically.
* Check the solution.
* Write an answer for the question asked, in a complete sentence.

6.4.1 Solving Distance-Traveled Models

In previous chapters, we used the formula for distance traveled. Remember, the distance traveled, d, is equal to the rate of motion, r, times the time traveled, t.

$$d = rt$$

We may be given information about more than one situation involving distance traveled. In such a case, we need to solve a system of equations in two variables.

EXAMPLE 1   Romeo and Juliet live 680 miles apart. They plan to meet each other at a designated location somewhere between their homes. Juliet is planning to drive her sports car at an average speed of 65 mph. However, Romeo is driving his pickup truck and towing a boat, so he plans on driving an average of 45 mph. If they both plan to leave at the same time, how many miles from Juliet's home should they plan to meet?

Algebraic Solution

A picture of the trip may help us to determine the equations.

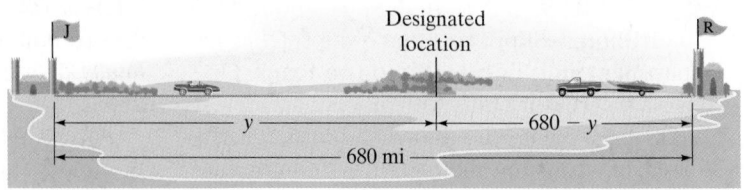

Let y = number of miles from Juliet's home to the meeting place
$680 - y$ = number of miles from Romeo's home to the meeting place
x = time traveled (both drivers travel the same time)

Using the distance-traveled formula, $d = rt$, for each trip, we write the following equations:

$$d = rt$$

(1) Juliet: $\quad\quad\quad y = 65x$

(2) Romeo: $\quad 680 - y = 45x$

We will use substitution to solve this system, since the first equation, $y = 65x$, is already solved for y. Substitute $65x$ for y in the second equation, and solve for x.

(2) $\quad\quad 680 - y = 45x$	
$680 - 65x = 45x$	Substitute 65x for y.
$680 - 65x + 65x = 45x + 65x$	Add 65x to both sides.
$680 = 110x$	Simplify.
$\dfrac{680}{110} = \dfrac{110x}{110}$	Divide both sides by 110.
$6.182 \approx x$	Simplify.

To find the distance from Juliet's home, y, substitute 6.182 for x in the first equation.

(1) $\quad y = 65x$	
$y \approx 65(6.182)$	Substitute 6.182 for x.
$y \approx 401.830$	Simplify.

To check the solution, substitute the solution into both of the original equations.

y	$=$	$65x$	$680 -$	y	$=$	$45x$
401.830		65(6.182)	$680 -$	401.830		45(6.182)
		401.830		278.170		278.190

Remember, we rounded our answer for x. Therefore, the numbers will not check exactly.

The designated location for the meeting place should be about 401.8 miles from Juliet's home.

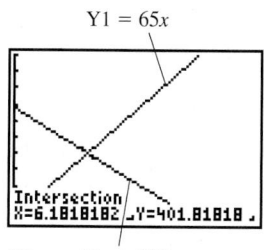

Y1 = 65x

Intersection
X=6.1818182 Y=401.8181

Y2 = −45x + 680

Graphic Solution

Solve the equations for y and graph them.

$$y = 65x$$
$$y = -45x + 680$$

A calculator graph for the solution, (6.182, 401.818), is shown on a window of (0, 20, 10, 0, 1000, 100, 1), using the CALC function to determine the exact intersection. The difference is due to rounding.

The designated location for the meeting place should be about 401.8 miles from Juliet's home.

Sometimes, more than one factor is involved in determining a rate of motion. One such situation involves traveling in the air. High-speed winds that flow between 4 and 6 miles above the earth, mostly from west to east, are called jet streams. Their speeds range between 50 and 250 miles per hour. Over North America there are two (and sometimes three) major jet streams. The force of the wind affects the airspeed by increasing or decreasing the actual speed of an aircraft in relation to the ground. A similar phenomenon occurs in water: Rivers and streams have a current whose speed accelerates or impedes the speed of a boat. In the ocean, streams of water such as the Gulf Stream also exist.

EXAMPLE 2 One evening, Sharon was tracking Delta aircraft between the hubs in Atlanta and Los Angeles, a distance of 1944 miles. The average height above ground of an aircraft was between 5 and 6 miles. The trip from Atlanta to Los Angeles took $5\frac{1}{2}$ hours and the trip from Los Angeles to Atlanta took $3\frac{1}{2}$ hours. Assuming that the planes were traveling at the same average speed when there was no wind, determine the average speed of the planes in still air and the average speed of the jet stream.

Solution

A picture of the trips between Atlanta and Los Angeles may help us determine the equations we need to describe this example.

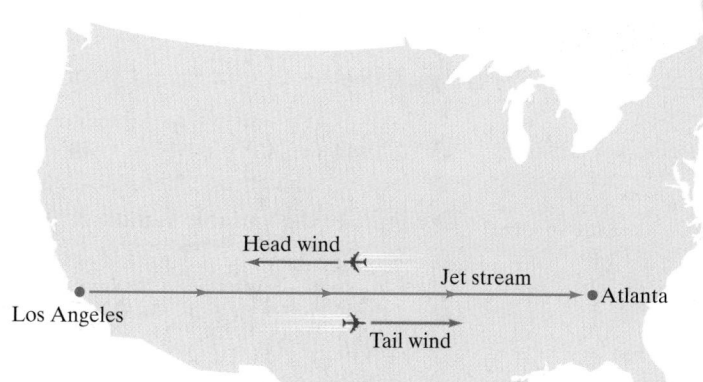

Head wind

Jet stream

Los Angeles

Atlanta

Tail wind

Let x = the speed of the planes in still air
y = the speed of the wind in the jet stream

The speed a plane travels when it is moving against the wind, with a head wind, is the speed of the plane in still air less the speed of the wind.

$$x - y = \text{the speed of the plane in relation to the ground}$$

The speed a plane travels when it is moving in the same direction as the wind, with a tail wind, is the speed of the plane in still air plus the speed of the wind.

$$x + y = \text{the speed of the plane in relation to the ground}$$

 HELPING HAND If an object is traveling in the direction of an additional force, add the rates of motion. If an object is traveling in the opposite direction of an additional force, subtract the rates of motion.

A table may help us organize the information.

	Distance	Rate	Time
Leg of trip to Los Angeles	1944	$x - y$	$5\frac{1}{2}$
Leg of trip to Atlanta	1944	$x + y$	$3\frac{1}{2}$

Using the distance-traveled formula, $d = rt$, for each leg of the trip, we write the following equations:

(1)   Leg of the trip to Los Angeles: $1944 = (x - y)\left(5\frac{1}{2}\right)$

(2)   Leg of the trip to Atlanta: $1944 = (x + y)\left(3\frac{1}{2}\right)$

Simplify using the distributive law.

(1) $1944 = \dfrac{11}{2}x - \dfrac{11}{2}y$

(2) $1944 = \dfrac{7}{2}x + \dfrac{7}{2}y$

Clear the fractions in both equations.

(1) $2(1944) = 2\left(\dfrac{11}{2}x - \dfrac{11}{2}y\right)$ or $3888 = 11x - 11y$

(2) $2(1944) = 2\left(\dfrac{7}{2}x + \dfrac{7}{2}y\right)$ or $3888 = 7x + 7y$

To eliminate the variable y, multiply both sides of each equation.

(1) $7(3888) = 7(11x - 11y)$ Multiply both sides by 7.

(2) $11(3888) = 11(7x + 7y)$ Multiply both sides by 11.

(1) $27216 = 77x - 77y$

(2) $\underline{42768 = 77x + 77y}$

$\quad\quad 69984 = 154x$ Add corresponding members of the equations.

$\dfrac{69984}{154} = \dfrac{154x}{154}$ Divide by 154.

$\quad\quad x \approx 454.44$

Substitute 454.44 for x in the second equation, and solve that equation for y.

$$1944 = \frac{7}{2}x + \frac{7}{2}y$$

$$1944 \approx \frac{7}{2}(454.44) + \frac{7}{2}y$$

$$1944 \approx 1590.54 + \frac{7}{2}y$$

$$1944 - 1590.54 \approx 1590.54 + \frac{7}{2}y - 1590.54$$

$$353.46 \approx \frac{7}{2}y$$

$$\frac{2}{7}(353.46) \approx \frac{2}{7}\left(\frac{7}{2}y\right)$$

$$y \approx 100.99$$

The check is left for you.

The speed of the plane in still air is about 454.44 miles per hour, and the speed of the wind is about 100.99 miles per hour.

6.4.1 Checkup

Write and solve a system of equations for each exercise.

1. Patty left her house to attend her cousin's wedding, traveling up the interstate at 55 miles per hour. After she had been gone for half an hour, her father realized she had forgotten to take her bridal-party dress. He immediately left to overtake her. If he drives at the legal speed of 65 miles per hour, how long will it be until he overtakes her on the interstate?

2. A United Airlines flight traveling from Seattle to Washington, DC, flew at an altitude of about 33,000 feet.

Traveling with the jet stream, the plane arrived approximately 4 hours after departure. A second United Airlines flight traveling from Washington, DC, to Seattle during the same time frame flew against the jet stream at an altitude of 35,000 feet and arrived approximately 5 hours after departure. Assuming that the two planes fly at the same average speed in still air, determine that average speed and the average speed of the wind due to the jet stream. The two cities are approximately 2350 miles apart.

6.4.2 Solving Mixture and Collection Models

Previously, we solved mixture and collection problems by using a linear equation in one variable. A second method is to use a system of linear equations in two variables. The general method is to set up one equation that expresses the amount of each component in the mixture and a second equation that relates the amounts of each component to one another, according to the statement of the problem.

EXAMPLE 3   Susan works in a candy store that stocks two popular Halloween candies. Candy corn sells for $1.25 per pound, and candy witches sell for $2.50 per pound. Susan mixes two pounds of candy corn for every pound of candy witches. A customer wants to buy $25.00 worth of the mixture. How many pounds of each must Susan use in the mixture?

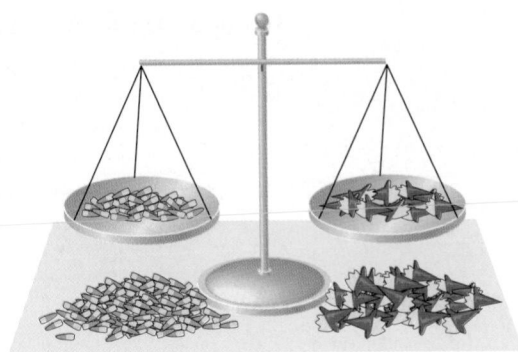

Algebraic Solution

Let c = number of pounds of candy corn
w = number of pounds of candy witches

 HELPING HAND It is convenient to use variable names that will make it easy to remember which variable represents which unknown quantity. In this example, we chose c for corn and w for witches.

Since the mixture contains 2 pounds of candy corn to 1 pound of candy witches, the number of pounds of candy corn, c, equals 2 times the number of pounds of candy witches, w.

$$c = 2w$$

The cost of the mixture will be 1.25 times the number of pounds of candy corn, c, plus 2.50 times the number of pounds of candy witches, w. This will equal the cost of 25.00.

$$1.25c + 2.50w = 25.00$$

The system of equations to be solved is

(1) $c = 2w$

(2) $1.25c + 2.50w = 25.00$

We will use substitution to solve this system because the first equation is already solved for the variable c. Substitute $2w$ for c in the second equation and solve for w.

(2) $1.25c + 2.50w = 25.00$

$1.25(2w) + 2.50w = 25.00$ Substitute $2w$ for c.

$2.50w + 2.50w = 25.00$ Simplify.

$5.00w = 25.00$ Simplify.

$$\frac{5.00w}{5.00} = \frac{25.00}{5.00} \qquad \text{Divide both sides by 5.00.}$$

$w = 5$ Simplify.

Substitute 5 for w in the first equation and solve for c.

(1) $c = 2w$

$c = 2(5)$ Substitute 5 for w.

$c = 10$ Simplify.

To check the solution, substitute it into both of the original equations.

$$\begin{array}{c|c}
c = 2w \\
\hline
10 & 2(5) \\
 & 10
\end{array}
\qquad
\begin{array}{c|c}
1.25c \ + \ 2.50w = 25.00 \\
\hline
1.25(10) + 2.50(5) & 25.00 \\
12.50 \ + \ 12.50 \\
25.00
\end{array}$$

The solution checks.

Susan will need to mix 10 pounds of candy corn with 5 pounds of candy witches in order to sell the mixture for $25.00.

EXAMPLE 4 Su Yung has a collection of 15 antique coins consisting of nickels and dimes. The face value of the coins is $1.10. How many of each denomination does she have?

Algebraic Solution

Let n = number of nickels
d = number of dimes

A table may help us organize the information.

	Nickels	Dimes	Total
Number of coins	n	d	15
Value of collection	$0.05n$	$0.10d$	1.10

Since there is a total of 15 coins, we add the number of nickels, n, and the number of dimes, d.

$$n + d = 15$$

The value of the nickels is 0.05 times the number of nickels, n or $0.05n$. The value of the dimes is 0.10 times the number of dimes, d, or $0.10d$. The value of the collection, 1.10, is equal to the sum of the values of the nickels and dimes.

$$0.05n + 0.10d = 1.10$$

To eliminate the decimals in the equation, multiply both sides of the equation by 100, the LCM of the decimal-place values.

$$100(0.05n + 0.10d) = 100(1.10)$$
$$5n + 10d = 110$$

The system to be solved is

(1) $n + d = 15$
(2) $5n + 10d = 110$

We will use elimination to solve the system, since both equations are written in standard form. To eliminate the variable n, we will multiply both sides of the first equation by -5.

(1) $-5(n + d) = -5(15)$ Multiply both sides by -5.
(2) $5n + 10d = 110$

(1) $\quad -5n - 5d = -75$

(2) $\quad \underline{5n + 10d = 110}$ *Add corresponding members of the equations.*

$\qquad\qquad 5d = 35$

$\qquad\quad \dfrac{5d}{5} = \dfrac{35}{5}$ *Divide both sides by 5.*

$\qquad\qquad d = 7$ *Simplify.*

Substitute 7 for d in the first equation and solve for n.

(1) $\qquad n + d = 15$

$\qquad\quad n + 7 = 15$ *Substitute 7 for d.*

$\qquad n + 7 - 7 = 15 - 7$ *Subtract 7 from both sides.*

$\qquad\qquad n = 8$ *Simplify.*

To check the solution, substitute it into both of the original equations.

$n + d = 15$		$5n + 10d = 110$	
$8 + 7$	15	$5(8) + 10(7)$	110
	15	$40 + 70$	
		110	

The solution checks.

Su Yung has eight nickels and seven dimes.

EXAMPLE 5   Christine needs 500 milliliters of a 22% saline (salt) solution. She has a 10% saline solution and a 30% saline solution. How many milliliters of the 10% saline solution and the 30% saline solution will she need to mix to obtain her desired solution?

Solution

Let x = the number of milliliters of the 10% saline solution

$\quad\;\; y$ = the number of milliliters of the 30% saline solution

A total of 500 milliliters is needed.

(1) $\quad x + y = 500$

The amount of salt in the 10% solution is 10% of x milliliters, or $0.10x$.
The amount of salt in the 30% solution is 30% of y milliliters, or $0.30y$.
The amount of salt in the 22% solution is 22% of 500 milliliters, or $0.22(500)$.

(2) $\quad 0.10x + 0.30y = 0.22(500)$

The system to be solved is

(1) $\quad x + y = 500$ or $x + y = 500$

(2) $\quad 0.10x + 0.30y = 0.22(500)$ or $0.1x + 0.3y = 110$

Eliminate the decimal in the second equation by multiplying by 10.

(2) $\quad 10(0.1x + 0.3y) = 10(110)$

$\qquad\quad x + 3y = 1100$

The system to be solved by elimination is

(1) $\quad x + y = 500$

(2) $\quad x + 3y = 1100$

(1) $\quad -1(x + y) = -1(500)$ *Multiply both sides by −1.*

(2) $\quad x + 3y = 1100$

$$(1) \quad -x - y = -500$$
$$(2) \quad \underline{x + 3y = 1100}$$
$$2y = 600 \qquad \text{Add corresponding members of the equations.}$$
$$\frac{2y}{2} = \frac{600}{2} \qquad \text{Divide both sides by 2.}$$
$$y = 300$$

Substitute 300 for y in the first equation and solve for x.

$$(1) \qquad x + y = 500$$
$$x + 300 = 500$$
$$x + 300 - 300 = 500 - 300$$
$$x = 200$$

The check is left for you to do.

Two hundred milliliters of the 10% saline solution mixed with 300 milliliters of the 30% saline solution will yield 500 milliliters of the 22% saline solution.

EXAMPLE 6   A 40% antifreeze mixture provides freeze-up protection to -10 degrees Fahrenheit. A 70% antifreeze mixture provides freeze-up protection to -62 degrees Fahrenheit. Joe Ben moved to Alaska and needs to lower his freeze-up protection from -10 degrees to -62 degrees. He needs 2 gallons of the 70% antifreeze mixture. How much of the 40% antifreeze mixture must be replaced with pure antifreeze to have a 70% mixture?

Solution

Let x = the number of gallons of pure antifreeze (100% antifreeze)
y = the number of gallons of the 40% antifreeze
The total amount of liquid is 2 gallons.

$$(1) \quad x + y = 2$$

The amount of antifreeze in the pure antifreeze mixture is 100% of the amount, or $1.00x$. The amount of antifreeze in the 40% antifreeze mixture is 40% of the amount, or $0.40y$. The total amount of antifreeze in the 70% antifreeze mixture is 70% of the total amount, or $0.70(2)$.

$$1.00x + 0.40y = 0.70(2)$$
$$(2) \quad 1.00x + 0.40y = 1.40$$

The system of equations is

$$(1) \qquad x + y = 2$$
$$(2) \quad 1.00x + 0.40y = 1.40$$

To solve these equations by substitution, solve the first equation for y and substitute the resulting expression into the second equation.

$$(1) \qquad y = -x + 2$$
$$(2) \qquad 1.00x + 0.40y = 1.40$$
$$1.00x + 0.40(-x + 2) = 1.40$$
$$1.00x - 0.40x + 0.80 = 1.40$$
$$0.60x + 0.80 = 1.40$$

$$0.60x + 0.80 - 0.80 = 1.40 - 0.80$$
$$0.60x = 0.60$$
$$x = 1$$

Using the equation $x + y = 2$, we solve for y.

(1) $x + y = 2$
$$1 + y = 2$$
$$y = 1$$

The check is left for you.

Joe Ben will need to add 1 gallon of pure antifreeze to 1 gallon of a 40% antifreeze mixture to obtain 2 gallons of a 70% antifreeze mixture.

APPLICATION

The U.S. national debt is about $5.6 trillion. Suppose that the Treasury Department split this debt into two loans. One consists of Treasury bills with an annual interest rate of about 3.6%, and the second loan consists of Treasury constant maturities with an annual interest rate of approximately 4.55%. If about 0.23 trillion dollars is paid in simple interest annually, determine the amount of principal for each loan.

Discussion

Let x = the amount, in trillions of dollars, borrowed at an annual rate of 3.6% (Treasury bills)

y = the amount, in trillions of dollars, borrowed at an annual rate of 4.55% (Treasury constant maturities)

Simple interest for 1 year is determined by using the formula $I = Prt$.

$I = x \cdot 0.036 \cdot 1 = 0.036x$ = the amount of interest paid for the Treasury bills

$I = y \cdot 0.0455 \cdot 1 = 0.0455y$ = the amount of interest paid for the Treasury constant maturities

The total interest paid is 0.23 trillion dollars.

(1) $\qquad 0.036x + 0.0455y = 0.23$

The total amount borrowed is 5.6 trillion dollars.

(2) $\qquad x + y = 5.6$

The system to solve is

(1) $0.036x + 0.0455y = 0.23$
(2) $x + y = 5.6$

To eliminate the x, multiply both sides of the second equation by -0.036.

(1) $0.036x + 0.0455y = 0.23$
(2) $-0.036(x + y) = -0.036(5.6)$
(1) $\quad 0.036x + \quad 0.0455y = 0.23$
(2) $\underline{-0.036x + (-0.036y) = -0.2016}$ Add corresponding members of the equations.
$$0.0095y = 0.0284$$
$$\frac{0.0095y}{0.0095} = \frac{0.0284}{0.0095}$$ Divide by 0.0095.
$$y \approx 2.99$$

(Continued on page 431)

Substitute 2.99 for y, and solve for x in the second equation.

$$x + y = 5.6$$
$$x + 2.99 \approx 5.6$$
$$x \approx 2.61$$

Approximately \$2.61 trillion would be in Treasury bills, and about \$2.99 trillion would be in Treasury constant maturities.

 6.4.2 Checkup

Write and solve a system of equations for each mixture and collection exercise.

1. Brian is mixing two types of grass seed to sell as a blend. He must mix x pounds of coarse fescue with y pounds of Kentucky bluegrass to make 100 pounds of mix. The coarse fescue sells for \$0.75 per pound, while the bluegrass sells for \$1.25 per pound. Brian wants to sell the mix for \$1.00 a pound. How much of each should he mix?

2. Rosita had been saving coins to buy Christmas gifts for the family. She had just counted the coins and told her parents there was a total of 498 coins, consisting of 120 nickels and the rest a mix of dimes and quarters. She proudly told them that the coins totaled \$63.75. While the family was out, a burglar entered the house and stole some items, including Rosita's coin collection. The family felt that Rosita was too young to deal with the loss, so they decided to replace the coins in exactly the same mix. How many dimes and quarters did they need to add to 120 nickles to restore Rosita's collection?

3. A chemist needs to mix a 25% glucose solution with a 5% glucose solution to make a new solution that is 10% glucose. If the chemist wants to have 1 liter of the 10% glucose solution, how many liters of the other two should she mix?

4. Recent rates for Series EE savings bonds and for Series HH savings bonds were 4.50% and 4.00%, respectively. What amounts should you invest at each of these interest rates if you wish to earn \$1300 in simple interest after one year on a total investment of \$30,000?

6.4 Exercises

Write and solve a system of equations for each application.

1. Nolan lives in Nashville and Monroe lives in Memphis. The distance between their homes is 220 miles. They plan to meet at a location on l-40 between their homes. Nolan drives his truck at an average speed of 60 miles per hour, and Monroe drives his car at an average speed of 65 miles per hour. If they both leave at the same time, how many miles from Nashville will they meet?

2. Henrieta lives in Houston and Dolores lives in Dallas. The distance between their homes is 240 miles. They plan to meet at a location on l-45 between their homes. Henrietta drives her car at an average speed of 65 miles per hour, and Dolores drives her car at an average speed of 55 miles per hour. If they both leave at the same time, how many miles from Houston will they meet?

3. After Kenny's car broke down, he started walking home. Dolly came by and picked him up. The time he walked was twice as long as the time he rode with Dolly. He was walking at an average rate of 3 miles per hour, and Dolly drove at an average rate of 60 miles per hour. If the total distance to his home was 11 miles when he started walking, how much time did Ken spend walking and how much time riding?

4. On an outing, a troop of Boy Scouts canoed at 12 miles per hour for the first leg of their trip. Then they hiked at 3 miles per hour to reach their destination. It took as many hours for the canoe trip as it did for the hike. If the total distance traveled was 30 miles, find the time they spent canoeing and hiking, and find the distance they covered by canoe.

5. An American Airlines flight travels from Miami to San Francisco in approximately 6 hours, flying at an altitude of 39,000 feet. During the same time, another American Airlines flight travels from San Francisco to Miami in approximately 5 hours, flying at an altitude of 30,200 feet. Assuming that the two planes travel at the same average speed, determine their average speed in still air and the average speed of the wind due to the jet stream. The two cities are approximately 2600 miles apart.

6. A United Airlines flight from Chicago to San Diego arrives 4 hours after departure. During the same time frame, an American Airlines flight from San Diego to Chicago arrives 3.5 hours after departure. The planes fly at altitudes above 30,000 feet. Assuming that they travel at the same average speed, determine their average speed in still air and the average speed of the wind due to the jet

stream. The flight distance between the two cities is approximately 1750 miles.

7. Two Girl Scouts took a canoe trip up a stream. They paddled for 1.5 hours against the stream's current and reached their destination, which was 6 miles from their starting point. After resting for a while, they made the return trip in 45 minutes (0.75 hour). How fast were the girls able to paddle in still water, and what was the average speed of the stream's current?

8. Jack cruised 42 miles downriver in 45 minutes (0.75 hour). His return trip took 1 hour. Determine the average speed of his motorboat in still water and the average speed of the river current.

9. While boating in San Francisco Bay near the Golden Gate Bridge, Joel motored with the tidal current for 15 minutes (0.25 hour) and traveled about 13.4 miles. Traveling against the current for 18 minutes (0.3 hour), he returned to his starting point. Determine the average speed of Joel's motorboat in still water and the average speed of the tidal current.

10. Jenny drove her motorboat with the current near Cape Cod Canal for 20 minutes ($\frac{1}{3}$ hour) and traveled 14.4 miles. Returning to her starting point and traveling against the current took her 24 minutes ($\frac{2}{5}$ hour). Determine her motorboat's average speed in still water and the average speed of the current.

11. Gary's Gourmet Coffee Shop mixes French vanilla coffee, which sells for $9.50 per pound, with hazelnut coffee, which sells for $7.00 per pound. The mix, called Croissant blend, sells for $8.50 per pound. If Gary wishes to make 20 pounds of the blend, how many pounds of each should he use?

12. LaToyia's Tea Shop blended orange spice tea, which sold for $3.50 per pound, with lemon honey tea, which sold for $7.50 per pound. The blend, sold as Spicy Citrus, amounted to 10 pounds of tea that sold for $5.00 per pound. How much of each flavor was used to make the blend?

13. Marian budgeted $250 to buy landscaping plants for her home. She needed 30 plants. If azaleas sell for $5 each and rhododendrons sell for $12 each, how many of each can Marian buy on her budget?

14. Sharon went to the local burger palace to buy burgers and hot dogs for a class picnic. She needed to buy 50 sandwiches for the class with the $95.00 she collected. If hot dogs sell for $1.00 each and burgers sell for $2.50 each, how many of each should she order?

15. The first screening of the latest movie sequel, *Cretaceous Park: The Found World*, had a box office take of $1938 at the local theater. Adults paid $7.50 and children paid $4.50 per ticket. There were four times as many children as adults. How many of these were adults and how many were children?

16. The blockbuster movie, *Car Wars, Part 12: The Princess Drives Back*, opened at the local theater. Adults paid $10.00 and children paid $5.50 per ticket for this show. There were twice as many children as adults at the show, and the box office receipts were $5355.00. How many adults and how many children attended the showing?

17. Philadelphia Treasures offers an assortment of collectable items commemorating the Liberty Bell. This week's assortment features a 1926 postage stamp that sells for $4.00 and a U.S. Franklin half dollar with the Liberty Bell on the back of the coin that sells for $8.00. How many half-dollar coins must be mixed with 100 stamps in order to come out even if the assortment sells for $7.00 per item?

18. Collector's Corner mixes some Cal Ripken, Jr., tribute baseball cards that sell for $2.75 each with some Mike Piazza cards that sell for $1.25 each. Customers can randomly purchase cards for $1.75 each from this mixed bin. How many of the Cal Ripken, Jr., cards should be mixed with 200 of the Mike Piazza cards in order to come out even on the sales? How many cards will be in the mix?

19. Cher works part time as a waitress in one restaurant and part time as a cook in another restaurant. In one week, she worked 15 hours waitressing and 20 hours as a cook and earned $320. In another week, she worked 18 hours waitressing and 24 hours cooking and earned $384. What is her hourly wage on each job?

20. A sale at Denney's Department Store attracted a throng of customers. One customer bought two shirts and four blouses for a total of $132. Another customer bought three shirts and six blouses from the same counters and paid $180. What were the sale prices of the shirts and blouses?

21. Radio station WALG runs a contest to see whether listeners can guess the makeup of a stack of $5 and $10 bills. The station announce that the stack contains 65 bills and is worth $365. The first listener who calls in with the correct number of each wins the money. Can you quickly determine the correct mix?

22. Tillie, the teller at the bank, has a stack of 35 bills. Some are $20 bills and the rest are $50 bills. If the value of the bills is $1300, how many of each does she have?

23. A grapefruit beverage that is 45% concentrated juice is mixed with an orange beverage that is 75% concentrated juice to produce a blend that is 55% concentrated juice. How many gallons of each must be mixed to produce 200 gallons of the blend?

24. A gardener has a solution that is 70% weed killer. He wants to mix this with another solution that is 40% weed killer, to make 5 pints of a solution that will be 50% weed killer. How many pints of each should he mix?

25. A chemist has a solution that is 60% acid and another solution that is 35% acid. How many liters of each should he mix to make 300 liters of a solution that is 50% acid?

26. A gardener has a solution that is 25% fungicide that she wants to mix with another solution that is 40% fungicide. How many pints of each should she mix to make 8 pints of a solution that is 30% fungicide?

27. Al Falfa, a milk farmer, wants to mix milk that is 4.3% butterfat with skim milk that has no butterfat, in order to make milk that is 2% butterfat. How much of each should he mix to make 200 gallons of the 2% butterfat milk? (Round your answer to the nearest gallon.)

28. How many liters of a 45% ammonia solution should be mixed with sterile water containing no ammonia to make 25 liters of a 35% ammonia solution?

29. The coolant in Mabel's car contains 45% antifreeze. How much should be drained and replaced with pure antifreeze in order for Mabel to have 4 gallons of a 60% antifreeze solution?

30. A car mechanic tests the antifreeze in your car and finds that it is 35% antifreeze. How much antifreeze should be drained and replaced with pure antifreeze in order to bring the car up to 5 gallons of 50% antifreeze? (Round your answer to the nearest tenth of a gallon.)

31. A wine maker has one barrel of wine that is 20% alcohol and another barrel that is 12% alcohol. How many gallons of the 12% alcohol wine must he mix with 5 gallons of the 20% alcohol wine to make a batch of wine with a 15% alcohol content? (*Hint*: Let x = the unknown amount of wine and let y = the total amount of the mixed wine.)

32. A pharmacist has one container of medicine with a strength of 45% medication and another container of medicine with a strength of 75% of the same medication. How many ounces of the 45% strength medication should be mixed with 8 ounces of the 75% strength medication to produce a medication with a strength of 60% medication? (See the hint in the previous exercise.)

33. In order to reduce her risk, Catherine split her investment in two mutual funds between one that paid 8.5% simple interest annually and another that paid 7% simple interest annually. She had $10,000 to invest, and at the end of the year she received a total of $752.50 in interest. How much did she invest in each fund?

34. Ali received two student loans for school totaling $10,000. He borrowed the maximum allowed at 7% simple interest and the remainder at 8.25% simple interest. After one year, he owed $725 in interest. How much did Ali borrow at each interest rate?

35. Zelda has inherited $16,500, which she would like to invest. She has an opportunity to invest part of the money in a simple-interest savings account paying 5% per year and the rest in a certificate of deposit paying 7.25% per year. If she wishes to earn $1000 in one year for a vacation next year and also wants to keep some money in the savings account in case she needs it sooner, how much should she invest in each account?

36. In exercise 35, Zelda planned to invest her inheritance. However, she learns that she must pay taxes on the inheritance and will have only $11,000 to invest. She decides to invest the money to earn $600 for a vacation instead. How much should she now invest in each account?

37. Zora plans to reduce her risk by investing her inheritance in two separate accounts. One simple interest certificate account pays 7.25% per year. The second simple interest savings account pays 6% per year. If Zora invests twice as much in the certificate account and needs to earn $1230 in interest, how much should she invest in each account?

38. Roby plans to reduce his risk by investing his inheritance in two separate accounts. One simple interest certificate account pays 7.25% per year. The second simple interest savings account pays 6% per year. If Roby invests $2000 more in the certificate account than in the savings account and needs to earn $1205 in interest, how much should he invest in each account?

39. Princess Fiona needed a short-term loan of $100,000. No one loan company would give her the total amount, so she borrowed $45,000 from one company and $55,000 from another company. The total simple interest on the loans was $6450 for one year. The $45,000 loan charged a simple interest rate of 1% more than the loan rate charged on the $55,000 loan. What were the two interest rates for the loans?

40. D. J. deposited $1500 into one savings account and $1200 into another savings account. The two accounts earned $117.90 in simple interest after a year. The $1200 investment paid a simple-interest rate that was 0.3% greater than the interest rate paid on the $1500 investment. What were the two interest rates for the investments?

6.4 Calculator Exercises

In this chapter, we solved a system of linear equations both graphically and algebraically. However, if we know that we are seeking only integer solutions, we may wish to solve the system numerically. To do so, we must first solve each equation in the system for the dependent variable. Next, we store these equations in the calculator, using the [Y=] key. Then when we view the table, wherever Y1 = Y2, the value of X and of Y1 (or Y2) will represent a coordinate-pair solution of the system.

One application of systems of linear equations in which we may want only integer solutions is the classic "age problem." Consider the following example:

Katie's cousin wants to know Katie's age, so she asks Katie's mom, Brenda. Katie's mom is a math teacher, so she decides to give the cousin the answer by using a math riddle. Brenda tells the cousin, "I am eight times as old as Katie. In 10 years, I will be three times as old as Katie." Use this information to determine how old Katie and Brenda are now.

Solution

Let x = Katie's age now

Let y = Brenda's age now

Brenda is eight times as old as Katie, so

$$y = 8x$$

In 10 years,

$$x + 10 = \text{Katie's age}$$
$$y + 10 = \text{Brenda's age}$$

In 10 years, Brenda will be three times as old as Katie, so $y + 10 = 3(x + 10)$, which simplifies to $y = 3x + 20$

The system of equations is

$$y = 8x$$
$$y = 3x + 20$$

Store these equations in the calculator and view the table.

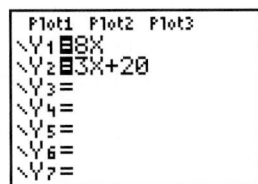 

The table indicates that the solution of the system of equations is (4, 32). Therefore, Katie is now 4 years old and Brenda is 32 years old.

The solution checks, since Brenda is eight times as old as Katie. In 10 years, Katie will be 14 and Brenda will be 42, and then Brenda will be three times as old as Katie.

Use this numerical method to solve the following age problems:

1. Stella is four years younger than three times her son's age. Five years ago, Stella was one year younger than four times her son's age. How old are Stella and her son now?

2. Gulen was daydreaming about her daughter Cecilia. In thinking about their ages, she realized that in 6 years she would be five times as old as Cecilia. In 13 years, she would be three times as old as Cecilia. Can you tell how old Gulen and Cecilia are now?

3. Joe wants to take early retirement in 10 years, when he will be twice as old as his son Paul. Joe is now 25 years older than Paul. How old is Joe now, and how old will he be when he retires?

4. Jenny and Katie bet Susan that she can't solve the riddle about their ages. Jenny is two years younger than Katie. In five years, Katie will be 10% older than Jenny. Help Susan find their ages now.

5. Egan was researching her family history for a school project. In a box of momentos, she found a birthday card from her grandmother to her mom. The card was dated 1965, and the note inside said, "Congratulations, Mary, you are now one-third as old as I am." She found a second card, dated 1977, also from her grandmother to her mother, which read, "Congratulations, Mary, you are now half as old as I am!" When Egan asked her mom how old she was on those two birthdays, her mom (a math teacher) told her to figure it out for herself. How old were her mom and her grandmother in those two years?

6. An antiques dealer found an old dresser and bed with a note attached. The note said that in 1902 the combined age of the bed and dresser was 148 years. It also said that in four more years the dresser would be twice as old as the bed. How old were the two pieces of furniture in 1902?

7. The tiebreaker question in a math contest was an age problem: Fric is three times as old as Frac. In five years, Fric's age will be two years more than twice Frac's age. How old are Fric and Frac now?

6.4 Writing Exercise

Many of the application exercises presented in this section have become classic algebra problems. You will find variations of them in most algebra textbooks. Some students object to these kinds of problems as being contrived and not very practical. For example, students often state that age problems have no practical significance, and one sometimes sees cartoons that joke about distance-traveled problems. Write a short paragraph describing how you feel about these applications. Do you feel they are worthwhile, and if so, why? Do you think they are contrived? Can you suggest better applications?

6.5 Solving Systems of Linear Equations in Three Variables

OBJECTIVES

I Determine whether a given ordered triple is a solution of a system of linear equations in three variables.

2 Solve systems of linear equations in three variables algebraically.

3 Model real-world situations by using systems of linear equations, and solve the equations.

APPLICATION

According to Kirchhoff's laws, the currents I_1, I_2, and I_3 shown in Figure 6.1 are solutions of the system

$$I_1 - I_2 + I_3 = 0$$
$$4I_1 + 2I_2 = 8$$
$$2I_2 + 3I_3 = 7$$

Find the currents, measured in amperes.

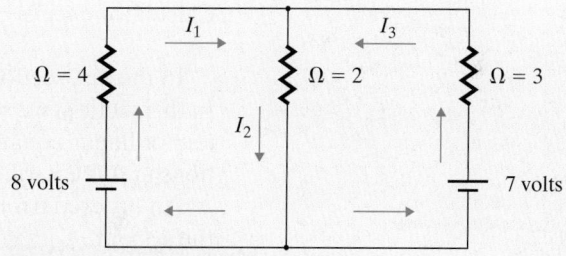

Figure 6.1

After completing this section, we will discuss this application further. See page 443.

6.5.1 Determining an Ordered-Triple Solution of a System of Linear Equations

In Chapter 5, we examined linear equations in two variables. The standard form of a linear equation in two variables is $ax + by = c$, where a and b are not both equal to zero. In this section, we examine a **linear equation in three variables** and systems of linear equations in three variables.

STANDARD FORM FOR A LINEAR EQUATION IN THREE VARIABLES

The standard form for a linear equation in three variables is $ax + by + cz = d$, where $a, b, c,$ and d are real numbers and $a, b,$ and c are not all equal to 0.

A solution of a linear equation in two variables is an ordered pair (x, y) that satisfies the equation. Similarly, a solution of a linear equation in three variables is an **ordered triple** (x, y, z) that satisfies the equation.

SOLUTION OF A LINEAR EQUATION IN THREE VARIABLES

An ordered triple (x, y, z) is a solution of a linear equation in three variables if the values of its coordinates, when substituted for their corresponding variables, result in a true equation.

HELPING HAND If the variables defined are not x, y, and z, the variables are placed in alphabetic order in the ordered triple. For example, an ordered triple could be (k, l, m).

EXAMPLE 1   Determine whether each ordered triple is a solution of the equation $3x - 2y + 8z = -2$.

a. $(2, 0, -1)$ **b.** $(2, 3, 1)$

Solution

a.
$$\begin{array}{c|c} 3x - 2y + 8z = -2 \\ \hline 3(2) - 2(0) + 8(-1) & -2 \\ 6 - 0 + (-8) \\ -2 \end{array}$$

Since $-2 = -2$,
$(2, 0, -1)$ is a solution.

b.
$$\begin{array}{c|c} 3x - 2y + 8z = -2 \\ \hline 3(2) - 2(3) + 8(1) & -2 \\ 6 - 6 + 8 \\ 8 \end{array}$$

Since $8 \neq -2$,
$(2, 3, 1)$ is not a solution.

Check

a. `2→X:0→Y: -1→Z:3X-`
   `2Y+8Z=-2`
                    `1`
b. `2→X:3→Y:1→Z:3X-2`
   `Y+8Z=-2`
                    `0`

Calculator checks confirm the results obtained.

In the last section, we solved systems of linear equations in two variables. In this section, we solve a **system of linear equations in three variables**. A system of linear equations in three variables is a set of at least two linear equations in three variables.

An ordered triple is a solution of the system if it satisfies every equation in the system.

> ### SOLUTION OF A SYSTEM OF LINEAR EQUATIONS IN THREE VARIABLES
> An ordered triple is a solution of a system of linear equations in three variables if it is a solution of every equation in the system.

EXAMPLE 2   Determine whether $(0, 3, -1)$ is a solution of the system

$$3x + 2y - 4z = 10$$
$$2x + 3y + z = 8$$
$$x - 4y + 2z = -14$$

Solution

Substitute the values for the variables.

$$\begin{array}{c|c} 3x + 2y - 4z = 10 \\ \hline 3(0) + 2(3) - 4(-1) & 10 \\ 10 \end{array} \qquad \begin{array}{c|c} 2x + 3y + z = 8 \\ \hline 2(0) + 3(3) + (-1) & 8 \\ 8 \end{array}$$

$$\begin{array}{c|c} x - 4y + 2z = -14 \\ \hline 0 - 4(3) + 2(-1) & -14 \\ -14 \end{array}$$

Check

`0→X:3→Y: -1→Z:3X+`
`2Y-4Z=10`
                 `1`
`2X+3Y+Z=8`
                 `1`
`X-4Y+2Z= -14`
                 `1`

The ordered triple $(0, 3, -1)$ is a solution of the system, because it satisfies all equations in the system.

We graphed ordered pairs on a two-dimensional rectangular coordinate system, or coordinate plane. We graph ordered triples on a three-dimensional rec-

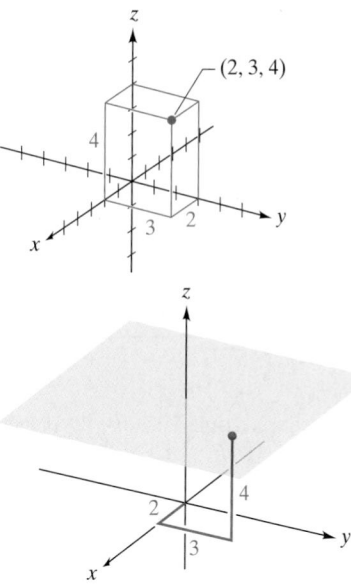

Figure 6.2

tangular coordinate system, or **coordinate space**. A coordinate space is formed by three axes that are perpendicular to each other and that pass through the ordered triple (0, 0, 0). For example, plot the ordered triple (2, 3, 4).

The graph of a linear equation in three variables is a plane that extends without end. To help visualize its location, it is conventional to graph a bounded figure. Even though we do not expect you to graph these bounded figures, one of many planes that contain the ordered triple (2, 3, 4) is shown in Figure 6.2.

Systems of linear equations in three variables have four possible types of solution sets, illustrated as follows:

1. The three planes have one point in common. The ordered-triple coordinates of the common point represents the solution. The system is consistent.

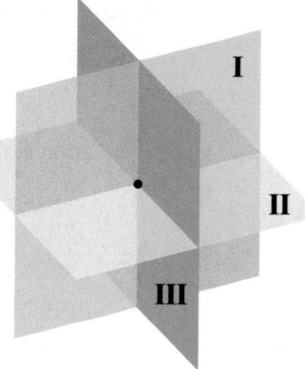

2. The three planes do not have a common intersection. The system has no solution. The system is inconsistent.

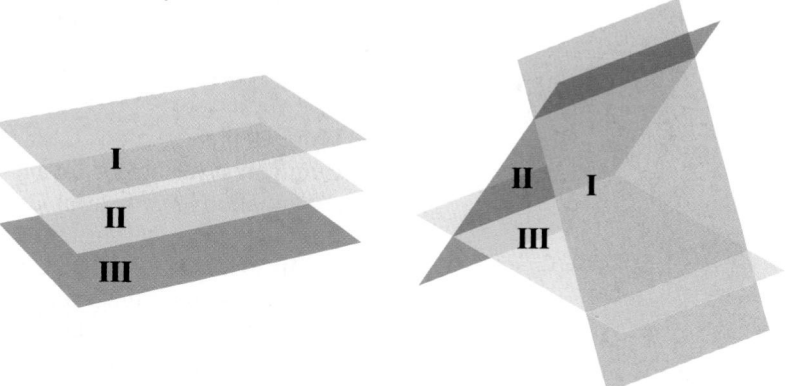

3. The three planes intersect at all points on a line. The system has an infinite number of solutions. The system is consistent.

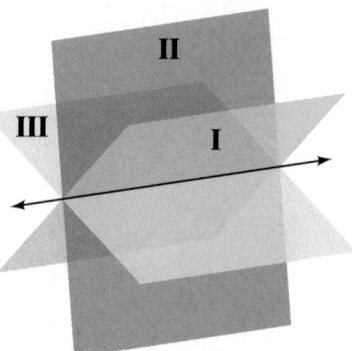

4. The three planes coincide. The system has an infinite number of solutions. The system is consistent, and the equations are dependent.

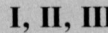

I, II, III

We will not solve these systems graphically. However, this visualization should help you understand the solutions we will obtain algebraically.

6.5.1 Checkup

1. Determine whether the ordered triple $(18, 12, -10)$ is a solution of the equation.

$$\frac{2}{3}x - \frac{3}{4}y + \frac{1}{2}z = -2$$

2. What is the standard form for a linear equation in three variables?

3. Determine whether the ordered triple $(5, -10, 12)$ is a solution of the system.

$$0.5a - 4b + 1.5c = 60.5$$
$$0.3a - 0.2b - 3.4c = -37.3$$
$$1.8a + 2.5b - 4.6c = -71.2$$

4. Can a linear equation in three variables have more than one solution? Explain.

6.5.2 Solving Systems of Linear Equations Algebraically

In this section, we algebraically solve linear equations in three variables, using the elimination method to eliminate a variable and obtain a system of linear equations in two variables.

> **SOLVING A SYSTEM OF LINEAR EQUATIONS IN THREE VARIABLES BY USING ELIMINATION**
> To solve a system of linear equations in three variables by using elimination,
>
> * Write all the equations in standard form.
> * Eliminate a variable from a pair of equations. This results in a linear equation in two variables.
> * Eliminate the *same* variable from a different pair of equations. This results in a second linear equation in two variables.
> * Write a system of linear equations in two variables, using the two equations obtained in the previous steps.
> * Solve the system of linear equations in two variables by any method.
> * Substitute the known values into one of the original equations containing the third variable, and solve for the third variable.
> * Check the solution

EXAMPLE 3 Solve.

$$2x + 3y + z = 7$$
$$x - 2y + z = -4$$
$$-x + 3y - z = 6$$

Solution

$$(1) \quad 2x + 3y + z = 7$$
$$(2) \quad x - 2y + z = -4$$
$$(3) \quad -x + 3y - z = 6$$

We will eliminate the variable z. First use equations (1) and (3), and then use equations (2) and (3).

(1) $\quad 2x + 3y + z = 7$	(2) $\quad x - 2y + z = -4$
(3) $\quad -x + 3y - z = 6$	(3) $\quad -x + 3y - z = 6$
(4) $\quad x + 6y \quad\quad = 13$	(5) $\quad\quad\quad y \quad\quad = 2$

Write a system of equations.

(4) $\quad x + 6y = 13$

(5) $\quad\quad\quad y = 2$

Solve by substituting 2 for y in equation (4).

(4) $\quad x + 6y = 13$

$\quad\quad x + 6(2) = 13$

$\quad\quad x + 12 = 13$

(6) $\quad\quad x = 1$

Substitute 2 for y and 1 for x in equation (2).

Check

(2) $\quad x - 2y + z = -4$

$\quad\quad 1 - 2(2) + z = -4$

$\quad\quad 1 - 4 + z = -4$

$\quad\quad -3 + z = -4$

$\quad\quad z = -1$

```
1→X:2→Y: -1→Z:2X+
3Y+Z=7
                    1
X-2Y+Z= -4
                    1
-X+3Y-Z=6
                    1
```

The ordered-triple solution is $(1, 2, -1)$.

Check the solution as shown.

EXAMPLE 4 Solve.

$$3x + 2y = 0$$
$$4x - z - 4 = 0$$
$$3y + 2z - 3 = 0$$

Solution

Write each equation in standard form.

(1) $\quad 3x + 2y \quad\quad = 0$

(2) $\quad 4x \quad\quad - z = 4$

(3) $\quad\quad\quad 3y + 2z = 3$

The first equation does not have a term containing the variable z. We eliminate z by using equations (2) and (3). Multiply both sides of equation (2) by 2, and add corresponding members of the equations.

(2) $\quad 2(4x - z) = 2(4)$ $\quad$ Yields $\quad\quad 8x - 2z = 8$

(3) $\quad\quad 3y + 2z = 3$ $\quad\quad\quad\quad\quad\quad\quad 3y + 2z = 3$

$\quad\quad\quad\quad\quad\quad\quad\quad\quad\quad\quad\quad$ (4) $\quad 8x + 3y \quad = 11$

Write a system of linear equations in two variables.

(1) $\quad 3x + 2y = 0$

(4) $\quad 8x + 3y = 11$

We eliminate the variable y. Multiply both sides of equation (1) by -3 and both sides of equation (4) by 2. Then add corresponding members of the equations.

(1) $-3(3x + 2y) = -3(0)$ Yields $-9x - 6y = 0$

(4) $2(8x + 3y) = 2(11)$ Yields $\dfrac{16x + 6y = 22}{7x \qquad\;\; = 22}$

$$x = \frac{22}{7}$$

Substitute $\frac{22}{7}$ for x in equation (1), and solve for y.

(1) $\;\;\;3x + \;\;\;\;\; 2y = 0$

$$3\left(\frac{22}{7}\right) + \;\;\;\;\; 2y = 0$$

$$\frac{66}{7} + \;\;\;\;\; 2y = 0$$

$$2y = -\frac{66}{7}$$

$$y = -\frac{33}{7}$$

Substitute $\frac{22}{7}$ for x in equation (2), and solve for z.

(2) $\;\;\;4x - \;\;\;\;\; z = 4$

$$4\left(\frac{22}{7}\right) - \;\;\;\;\; z = 4$$

$$\frac{88}{7} - \;\;\;\;\; z = 4$$

$$-z = -\frac{60}{7}$$

$$z = \frac{60}{7}$$

Check

```
22/7→X: -33/7→Y: 6
0/7→Z:3X+2Y=0
                    1
4X-Z-4=0
                    1
3Y+2Z-3=0
                    1
```

The ordered triple $\left(\frac{22}{7}, -\frac{33}{7}, \frac{60}{7}\right)$ is a solution of the system.

Check the solution on your calculator.

Remember that a system may have no solution or many solutions. The results will appear similar to the corresponding results for equations in two variables.

EXAMPLE 5   Solve.

$$x - 2y + 4z - 1 = 0$$
$$-x - 3y + \;\;z - 8 = 0$$
$$2x - 4y + 8z \qquad = 0$$

Solution

Write each equation in standard form.

(1) $\;\;\;x - 2y + 4z = 1$

(2) $-x - 3y + \;\;z = 8$

(3) $\;\;2x - 4y + 8z = 0$

We eliminate the variable x. Add corresponding members of equations (1) and (2). Multiply both sides of equation (2) by 2, and add corresponding members of the new equation and equation (3).

(1) $\quad x - 2y + 4z = 1$
(2) $\quad -x - 3y + z = 8$
(4) $\quad -5y + 5z = 9$

(2) $\quad 2(-x - 3y + z) = 2(8)$ $\quad$ Yields $\quad\quad -2x - 6y + 2z = 16$
(3) $\quad\quad 2x - 4y + 8z = 0$ $\quad\quad\quad\quad\quad\quad\quad\quad \dfrac{2x - 4y + 8z = 0}{}$
$\quad\quad\quad\quad\quad\quad\quad\quad\quad\quad\quad\quad\quad\quad\quad\quad\quad\quad$ (5) $\quad -10y + 10z = 16$

Write a system of linear equations in two variables.

(4) $\quad -5y + 5z = 9$
(5) $\quad -10y + 10z = 16$

Eliminate the variable y. Multiply both sides of equation (4) by -2, and add corresponding members of the new equation and equation (5).

(4) $\quad -2(-5y + 5z) = -2(9)$ $\quad$ Yields $\quad\quad 10y - 10z = -18$
(5) $\quad\quad -10y + 10z = 16$ $\quad\quad\quad\quad\quad\quad\quad\quad \dfrac{-10y + 10z = 16}{}$
$\quad\quad\quad\quad\quad\quad\quad\quad\quad\quad\quad\quad\quad\quad\quad\quad\quad\quad\quad\quad 0 = -2$

The result is a contradiction. The system has no solution.

EXAMPLE 6 $\quad$ Solve.

$$x + 3y - 2z = 4$$
$$-2x - 6y + 4z = -8$$
$$\frac{1}{4}x + \frac{3}{4}y - \frac{1}{2}z = 1$$

Solution

(1) $\quad x + 3y - 2z = 4$
(2) $\quad -2x - 6y + 4z = -8$
(3) $\quad \dfrac{1}{4}x + \dfrac{3}{4}y - \dfrac{1}{2}z = 1$

Eliminate the variable x. Multiply both sides of equation (1) by 2, and add corresponding members of the resulting equation and equation (2).

(1) $\quad 2(x + 3y - 2z) = 2(4)$ $\quad$ Yields $\quad\quad 2x + 6y - 4z = 8$
(2) $\quad -2x - 6y + 4z = -8$ $\quad\quad\quad\quad\quad\quad\quad\quad \dfrac{-2x - 6y + 4z = -8}{}$
$\quad\quad\quad\quad\quad\quad\quad\quad\quad\quad\quad\quad\quad\quad\quad\quad$ (4) $\quad\quad\quad\quad\quad 0 = 0$

The result is an identity. Equations (1) and (2) are equivalent.
Eliminate the fractions in equation (3) by multiplying both sides of the equation by 4 (the LCD).

(1) $\quad x + 3y - 2z = 4$ $\quad\quad\quad\quad\quad\quad\quad\quad\quad\quad x + 3y - 2z = 4$

(3) $\quad 4\left(\dfrac{1}{4}x + \dfrac{3}{4}y - \dfrac{1}{2}z\right) = 4(1)$ $\quad$ Yields $\quad\quad x + 3y - 2z = 4$

Notice that equations (1) and (3) are equivalent. There is no need to eliminate a variable, since all variables would be eliminated and the result would be an identity.

Since equations (1) and (2) are equivalent and equations (1) and (3) are equivalent, all the equations are equivalent. There are an infinite number of solutions. The solution set is all ordered triples that satisfy $x + 3y - 2z = 4$.

6.5.2 Checkup

Solve exercises 1–3.

1. $2x - 3y + z = -10$
$x - y - z = -5.6$
$3x + y + z = 8.8$

2. $3x + 4y = z + 6$
$x + 3z = y + 3$
$2x - 2y = 10 - 6z$

3. $2x + 4y = 3z + 12$
$0.4x + 0.8y - 0.6z = 2.4$
$x + 2y - 6 = 1.5z$

4. In solving a system of linear equations in three variables algebraically, how can you tell when there is no solution or when there are an infinite number of solutions?

6.5.3 Modeling the Real World

There are many situations in which three unknowns are related in different ways. Setting up a system of linear equations in three variables is helpful in determining a solution for this type of relationship. For example, you may want to combine three different items and set constraints on them.

An ideal algebraic model for this situation would be a system of linear equations in three variables.

EXAMPLE 7 A dietician prepares a dinner for a 27-year-old woman. The meal should have 480 calories, 31 g of fat, and 103 mg of cholesterol. The patient chose roast beef, mashed potatoes, and green beans for her meal. According to the following table, how many servings of each should be prepared?

	Roast Beef (3 oz)	Mashed Potatoes ($\frac{1}{2}$ cup)	Green Beans ($\frac{1}{2}$ cup)
Calories	220	110	20
Fat (in grams)	18	4	0
Cholesterol (in milligrams)	60	13	0

Solution

Let $x =$ the number of 3-oz servings of roast beef
      $y =$ the number of $\frac{1}{2}$-cup servings of mashed potatoes
      $z =$ the number of $\frac{1}{2}$-cup servings of green beans

The total calories are 480.

$$220x + 110y + 20z = 480$$

The total grams of fat are 31.

$$18x + 4y + 0z = 31$$

The total milligrams of cholesterol are 103.

$$60x + 13y + 0z = 103$$

The system is

(1) $220x + 110y + 20z = 480$
(2) $18x + 4y + 0z = 31$
(3) $60x + 13y + 0z = 103$

Since equations (2) and (3) make a system of two variables, solve the system for x and y.

(2) $13(18x + 4y) = 13(31)$ Yields $234x + 52y = 403$
(3) $-4(60x + 13y) = -4(103)$ Yields $\underline{-240x - 52y = -412}$
$$-6x = -9$$
$$x = 1.5$$

Substitute 1.5 for x and solve for y.

(2) $18x + 4y = 31$
$$18(1.5) + 4y = 31$$
$$27 + 4y = 31$$
$$4y = 4$$
$$y = 1$$

Substitute 1.5 for x and 1 for y, and solve for z.

(1) $220x + 110y + 20z = 480$
$$220(1.5) + 110(1) + 20z = 480$$
$$330 + 110 + 20z = 480$$
$$440 + 20z = 480$$
$$20z = 40$$
$$z = 2$$

The dietician should prepare the following:

1.5 servings roast beef

1 serving mashed potatoes

2 servings green beans

APPLICATION

According to Kirchhoff's laws, the currents I_1, I_2, and I_3 are solutions of the system

$$I_1 - I_2 + I_3 = 0$$
$$4I_1 + 2I_2 = 8$$
$$2I_2 + 3I_3 = 7$$

Find the currents, measured in amperes.

Discussion

(1) $1I_1 - 1I_2 + 1I_3 = 0$
(2) $4I_1 + 2I_2 + 0I_3 = 8$
(3) $0I_1 + 2I_2 + 3I_3 = 7$

(Continued on page 444)

Equation (2) has only I_1 and I_2. Therefore, eliminate I_3 from a system containing Equations (1) and (3).

(1) $-3(1I_1 - 1I_2 + 1I_3) = -3(0)$ Yields $\qquad -3I_1 + 3I_2 - 3I_3 = 0$

(3) $0I_1 + 2I_2 + 3I_3 = 7$ $\qquad\qquad\qquad\qquad\quad 0I_1 + 2I_2 + 3I_3 = 7$

$$(4) \quad \overline{-3I_1 + 5I_2 \qquad\quad = 7}$$

Write a new system consisting of Equations (2) and (4), and eliminate I_1.

(2) $3(4I_1 + 2I_2) = 3(8)$ Yields $\qquad 12I_1 + 6I_2 = 24$

(4) $4(-3I_1 + 5I_2) = 4(7)$ Yields $\qquad -12I_1 + 20I_2 = 28$

$$\frac{26I_2}{26} = \frac{52}{26}$$

$$I_2 = 2$$

Substitute into Equation (3).   Substitute into Equation (2).

(3) $\quad 2I_2 + 3I_3 = 7$ $\qquad\qquad$ (2) $\quad 4I_1 + 2I_2 = 8$

$\quad 2(2) + 3I_3 = 7$ $\qquad\qquad\qquad\quad 4I_1 + 2(2) = 8$

$$\qquad I_3 = 1 \qquad\qquad\qquad\qquad\qquad I_1 = 1$$

The solution is $(1, 2, 1)$.

Therefore, $I_1 = 1$ ampere, $I_2 = 2$ amperes, and $I_3 = 1$ ampere.

6.5.3 Checkup

1. In using Kirchhoff's laws to determine the unknown currents in an electrical circuit, the following system of simultaneous equations resulted:

$$I_1 - I_2 - I_3 = 0$$
$$I_1 + 3I_2 = 8$$
$$I_2 - 2I_3 = -1$$

Solve the system for the three currents, I_1, I_2, and I_3, measured in amperes.

2. At the fast-food counter, Edgar ordered a roast beef sandwich, french fries, and a chocolate shake. The place mat on the tray indicated that this combination had 956 calories, 12.88 grams of saturated fat, and 3.92 grams of unsaturated fat, distributed among the constituents as follows:

	Roast beef sandwich	French fries	Thick milk shake
Calories	71 per oz	88 per oz	34 per oz
Saturated fat, grams	0.7 per oz	1.4 per oz	0.49 per oz
Unsaturated fat, grams	0.35 per oz	0.70 per oz	0.035 per oz

How many ounces of each food did Edgar receive?

6.5 Exercises

Determine whether each ordered triple is a solution of the equation $3x + 4y - 5z = 46$.

1. $(2, -3.5, -10.8)$ $\qquad$ **2.** $(-2, 6.5, 5.2)$ $\qquad$ **3.** $(0, 8.5, -2.4)$

4. $(10, -2.5, -5.2)$ $\qquad$ **5.** $(5, -1.5, 7.4)$ $\qquad$ **6.** $(0, 6, -5.2)$

Determine whether each ordered triple is a solution of the equation $\frac{3}{4}a + \frac{1}{3}b = \frac{1}{2}c$.

7. $(8, 12, 10)$ $\qquad\qquad$ **8.** $(-12, 6, -14)$ $\qquad$ **9.** $(-16, 9, -18)$

10. $\left(\dfrac{5}{6}, -\dfrac{9}{16}, \dfrac{7}{8}\right)$ **11.** $\left(\dfrac{2}{3}, -\dfrac{3}{4}, -\dfrac{1}{2}\right)$ **12.** $\left(\dfrac{8}{9}, -2, 0\right)$

Determine whether each ordered triple is a solution of the given system.

13. $(-3, 5, 8)$

$$5x - 3y + 7z = 26$$
$$4x - 5z = -37$$
$$4y + 9z = 92$$

14. $(5, 0, -4)$

$$4x + 8y - 3z = 32$$
$$2x + 5z = 3y - 10$$
$$x - 4y - 7z = 33$$

15. $(14, 78, 21)$

$$x - \dfrac{1}{2}y = z - 46$$
$$x + y + z = 113$$
$$2x - 3y + 4z = -122$$

16. $(8, 10, 5)$

$$0.5x - 4.3y + z = -34$$
$$x + 7.1y - 3.2z = 63$$
$$y - 8.4z = 32$$

Solve.

17. $x - y + z = 14$
$x + y + z = 8$
$x + 2y - 5z = -37$

18. $a + b - c = 6$
$2a + b + c = -10$
$a + 4b - 3c = 9$

19. $\dfrac{1}{3}a + \dfrac{2}{3}b - \dfrac{3}{5}c = -\dfrac{5}{8}$
$2a - \dfrac{1}{3}b - \dfrac{4}{5}c = \dfrac{5}{4}$
$8a + 8b + 8c = 5$

20. $\dfrac{3}{5}x + \dfrac{2}{3}y + z = -\dfrac{1}{15}$
$x - \dfrac{5}{6}y + 2z = -\dfrac{3}{2}$
$3x + 5y - 6z = 8$

21. $4x - 5y - 10z = 63$
$2x + 15z = -108$
$x - 5y - 15z = 99$

22. $4x + 5y - z = 1$
$2x - 10y + 7z = 43$
$6x + 8z = 27$

23. $a - b + c = 385$
$2a - 3b + c = 16$
$4a - 5b + 2c = 503$

24. $p + q + r = 321$
$2p + 3q + r = 33$
$p + 2q + 10r = 582$

25. $5x - 3y + 7z = 8$
$x + 4y - 3z = 11$
$6x + y + 4z = 19$

26. $4x + 2y + 3z = 6$
$3x + y + 2z = 8$
$x + y + z = 5$

27. $8a - 3b + 2c = -29$
$5a + b + 7c = 2$
$3a - 4b - 5c = 30$

28. $4a - 3b + 7c = -15$
$5a + 2b + 13c = 14$
$a + 5b + 6c = 29$

Write a system of linear equations in three variables for each situation, and solve the equations to answer the questions asked.

29. At the Holiday Festival of Lights presented at Greenglade Gardens Resort, adults were charged an admission price of $12.00, children were charged $5.00, and resort guests were charged $3.50. The total number of tickets sold was 1257, and the number of children's tickets sold was 87 more than the total of adults and resort guests. If the total receipts for the festival were $9547, how many tickets were sold to each group of people?

30. Deanna found a box of fine china pieces at an antiques store that matched the set she owned. She paid $53.50 for the box. The box had cups selling for $3.50 each, saucers at $1.00 each, and dinner plates at $4.25 each. The number of cups was equal to the total number of saucers and dinner plates. If there were 18 pieces of china in the box, how many pieces of each kind did Deanna buy?

31. Roderick invested $11,200 in three different accounts in order to distribute his investment risk. Each of the accounts was a simple-interest-bearing account, with one account paying 5% interest annually, another paying 5.5%, and the third paying 6%. The amount invested at 5% was $1400 less than the amount invested at 6%. Roderick's total interest on the investments was $623 after one year. How much did he invest in each account?

32. Siegfried paid 1.5% simple interest each month on his credit card balance, 0.75% simple interest per month on his furniture loan balance, and 0.9% simple interest per month on his car loan balance. His furniture loan balance was five times as large as his credit card balance. His total interest payments for the month amounted to $88.28, and the total amount he owed, not including interest, was $9850. What was the balance on each of Siegfried's loans?

33. Ron collected $184 during a charity drive for a shelter for the homeless. The money consisted entirely of $1, $5, and $10 bills. There were three times as many $1 bills as there were $5 bills. If there were 60 bills, how many of each bill did Ron collect?

34. A packet of stamps sold for $12.25. The package contained $0.32 stamps, $0.03 stamps, and $0.35 stamps. The number of $0.32 stamps was the same as the number of $0.03 stamps. The total number of stamps in the packet was 50. How many of each denomination of stamps were in the packet?

35. Henry's Health Foods Market packages healthful cereal in three sizes. A small box (10.8 oz) sells for $2.39, a medium-sized box (13.5 oz) sells for $2.79, and a large box (20.4 oz) sells for $3.59. Henry's receipts indicate that he sold a total of 145 boxes of the cereal, for a total of $446.55. He sold all but 4 oz of his gross stock of 2350 oz of cereal. How many boxes of each size of cereal did he sell?

36. Bakir's Bakery sells Turkish baklava in three sizes. A package of 6 pieces sells for $3.30, a package of 8 pieces sells for $4.00, and a package of 12 pieces sells for $5.40. At the beginning of the day, Bakir made 60 dozen pieces of baklava. When the day was done, he had sold 77 boxes of the treats and had made $336.00 on the sales. Four pieces of baklava were left. How many of each size of package did Bakir sell?

37. Tenisha invested $7225 to start a portfolio of stocks. She purchased a total of 600 shares. Some shares were gold investments selling for $19.75 each. Others were investments in oil-related industries selling for $9.50 each. The remainder consisted of investments in money market funds selling for $12.00 each. During the first quarter, Tenisha earned a total of $260 in dividends. The gold investment paid $0.50 per share, the oil-related investment $0.10 per share, and the money market funds $0.90 per share. How much did Tenisha invest in each of the three stock funds?

38. Hortense purchased life insurance in units of $1000. Ameritag sold each unit for a premium of $18.00 per year per unit. Bankers Fund sold its units for a premium of $13.20 per year per unit, and Columbia Mutual sold its units for a premium of $15.00 per year per unit. Hortense purchased a mix of 65 units from the three companies at a total annual premium of $942. The companies paid interest each year on the policies. Ameritag paid $1.50 per unit, Bankers Fund paid $0.65 per unit, and Columbia Mutual paid $0.85 per unit. If Hortense received a total of $54.75 in interest payments for one year, how many units of insurance did she purchase from each company?

39. Concetta bakes pastries in her home that she sells to restaurants. She spends $1.25 for each pie, $1.00 for each dozen cookies, and $1.50 for each cake that she bakes. The restaurants pay her $3.50 for each pie, $2.50 for each dozen cookies, and $4.50 for each cake. She bakes three times as many cakes as she does pies. How many of each type of pastry does Concetta bake if she spends $69.50 and receives $200.00 for her efforts?

40. As patient coordinator for a doctor's office, Gretchen schedules patients for visits. Each patient requiring a physical will need 45 minutes, for which the patient will be charged $130. Patients requiring outpatient surgery will need 90 minutes and will be charged $300. Patients requiring diagnostic treatment will need 15 minutes and will be charged $55. Gretchen schedules patients to fill an eight-hour day (480 minutes), for which the total charges are $1595. She schedules one more patient for a physical than she does for surgery. How many of each type of patient does she schedule?

41. In planning Baby Dumpling's meals, the following table is used:

	Cereal (1 oz)	Orange Juice (2 oz)	Milk (2 oz)
Calcium	225 mg	6.8 mg	69.8 mg
Vitamin A	4.5 IU	31.2 IU	116.2 IU
Vitamin C	0.6 mg	35.4 mg	0.6 mg

How many servings of each item should be planned in order for the baby to receive 736 mg of calcium, 505 IU of vitamin A, and 39 mg of vitamin C?

42. Chip will have the items listed in the following table for lunch:

	Hot Dog	Beans (2 oz)	Chips (1 oz)
Protein	6.8 grams	2.7 grams	2 grams
Calcium	11.3 mg	28.4 mg	6.8 mg
Iron	0.8 mg	0.2 mg	0.5 mg

How many servings of each should he have in order to receive 20.3 grams of protein, 64.6 mg of calcium, and 2.8 mg of iron?

6.5 Calculator Exercises

Our calculator can determine the solution of a system of three linear equations in three variables. The calculator uses the mathematical process of matrices to do so. A matrix is a table. Each row of the table contains the coefficients of one of the equations in the system, after the equation is placed into standard form.

The calculator process takes the table and converts it to an equivalent table of coefficients wherein, for each equation, one of the variables has a coefficient of 1 and the other variables in the same equation have coefficients of 0. (This is possible through the properties of equations that you studied

before.) The process results in a table that isolates each of the variables in one of the equations, just as you would do if you had one equation with one variable. By reading each row of the resulting table, you are able to read the solution of the system of equations. It is a lovely process to see once you understand it. To determine the solution of a system of equations by using your calculator,

- Rewrite each equation of the system in standard form, $ax + by + cz = d$.

- Create a table in which each row lists only the coefficients $a, b, c,$ and d, in order.

- Store the table as table A in the calculator.

  [2nd] [MATRX] [▶] [▶] [|] [3] [ENTER] [4]
  [ENTER]

  Type each coefficient and press [ENTER] after each.

- After the table is stored, press [2nd] [QUIT].

- To create the solution table, choose [rref(] under the
  [MATRX] key, [2nd] [MATRX] [▶] [ALPHA] [B].

- Type in the name of table A.

  [2nd] [MATRX] [|] [)] [ENTER]

- The calculator will display a table with 1's and 0's in the
  first three columns and the solution of the system in the
  fourth column. This really is an equivalent system of
  equations (having the same solution as the original system), but with new coefficients a, b, c, and d. If you write
  the new system, you can see its solution.

As an example, solve the system.

$$2x - 3y + z = -9$$
$$x + 2y - 3z = -1$$
$$-2x + y - 4z = -3$$

The table of coefficients is

$$\begin{array}{rrrr} 2 & -3 & 1 & -9 \\ 1 & 2 & -3 & -1 \\ -2 & 1 & -4 & -3 \end{array}$$

*Storing this table in the calculator and obtaining the solution
table are shown in the following diagrams:*

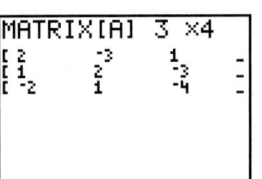

*The solution of the system of equations is given by the new
equivalent system,*

$$1x + 0y + 0z = -1, \quad \text{or} \quad x = -1$$
$$0x + 1y + 0z = 3, \quad \text{or} \quad y = 3$$
$$0x + 0y + 1z = 2, \quad \text{or} \quad z = 2$$

*Thus, the solution is $x = -1$, $y = 3$, and $z = 2$. The solution
also can be read as the ordered triple $(-1, 3, 2)$, given by the
fourth column of the table.*

*When you write the new equivalent system, if one of the equa-
tions is a contradiction, such as $0x + 0y + 0z = 1$, it indicates
that the system has no solution.*

*When you write the new equivalent system, if one of the equa-
tions is an identity, such as $0x + 0y + 0z = 0$, it indicates that
the system has many solutions.*

Use the preceding method to solve the following systems of equations.

1. $x - y - z = 0$
$x + y + z = 4$
$x + y - z = 6$

2. $x + y - z = -4$
$x + 2y + z = -3$
$x - y + z = 6$

3. $2x + 8z = 4$
$4x - y + 8z = 3$
$2x + y = 3$

4. $4x + y - 3z = 2$
$8x - y + 6z = 3$
$y + 9z = 9$

5. $x + 2y + 3z = 6$
$2x - y + z = 2$
$x + 2y - z = 2$

6. $x - 2y + z = 0$
$x - 3y - z = 3$
$2x + y - 4z = 1$

7. $4x + y - z = 30$
$-x + 2y - z = 7$
$9x - 8y + 4z = -3$

8. $3x - y + 3z = 5$
$x + 7y - 2z = 24$
$5x - 6y + 4z = 9$

9. $x - y + z = -2$
$2x + 3y + z = 8$
$3x + 2y + 2z = 10$

10. $3x - 2y + z = -3$
$x - y - 4z = 7$
$4x - 3y - 3z = 4$

6.5 Writing Exercise

In this section, you were shown how to use the elimination
method to solve a system of linear equations in three variables. Explain how you might generalize what you have
learned to solve a system of linear equations in four variables.

CHAPTER 6 SUMMARY

After completing this chapter, you should be able to define in your own words the following key terms:

6.1
consistent
inconsistent
dependent
independent

6.2
substitution method

6.3
elimination method
simultaneous equations

6.5
linear equation in three variables
ordered triple
system of linear equations in three
    variables
coordinate space

Reflections

1. Explain the difference between a solution of a linear equation in two variables and a solution of a system of linear equations in two variables.
2. Describe the method of solving a system of linear equations graphically.
3. What do we mean by an inconsistent system of linear equations?
4. What do we mean by a system of dependent linear equations?
5. What can you say about the solutions of a consistent system of independent linear equations?
6. How do you solve a system of linear equations by using the substitution method?
7. How do you solve a system of linear equations by using the elimination method?
8. In solving a system of linear equations algebraically, how do you know that there is no solution? How do you know that there are many solutions?

CHAPTER 6 SECTION-BY-SECTION REVIEW

6.1

Recall	Examples
Solution of a system of linear equations • An ordered pair that satisfies every equation in the system.	Is (2, 3) a solution of the following system? $x + y = 5$ $2x - y = 1$ Substitute the ordered pair in each equation. $\begin{array}{c c}x + y = 5 & 2x\ -y = 1\\\hline 2+3 \mid 5 & 2(2)\ -3 \mid 1\\ 5 & 4-3\\ & 1\end{array}$ Since $5 = 5$ and $1 = 1$, the ordered pair (2, 3) satisfies both equations and is a solution of the system.
Solve graphically • Graph the equations in the system. • The solution is the coordinates of the ordered-pair intersection of the graphs.	Solve graphically. $y = -x + 5$ $y = 2x - 1$ Graph each equation. The solution is (2, 3).

Determine whether each ordered pair is a solution of the given system of linear equations.

1. $(2, -4); 3x + 2y = -2$
$\qquad 4x - 3y = 20$

2. $\left(\dfrac{17}{7}, \dfrac{8}{7}\right); 2x + y = 6$
$\qquad -x + 3y = 1$

3. $(0.25, -0.45); 4x - 5y = 3$
$\qquad 8x + 5y = 0$

4. $(7, -2); x + 3y = 13$
$\qquad x - y = 5$

5. $\left(\dfrac{2}{3}, \dfrac{2}{3}\right); 3x + 6y = -2$
$\qquad 6x - 3y = 6$

6. $(1.5, -2.4); 6x + 5y = -3$
$\qquad 2x - 10y = 27$

7. $(4, 2); 2(x - 3) = 2$
$\qquad 3(y + 1) = -3$

8. $(-3, 5); x + 5 = 2$
$\qquad 2y - 3 = 7$

Solve each system of equations graphically. If the system has infinitely many solutions, describe the solution set.

9. $2x + y = 17$
$\qquad y = 3x - 18$

10. $3(x + 2) + 1 = -5$
$\qquad 2x - y = -10$

11. $x + 6 = 4$
$\qquad 2(y + 2) = y + 3$

12. $x - 2y = -3$
$\qquad 2x + 4y = 8$

13. $y = 2(x + 2)$
$\qquad 2x - y = 5$

14. $y = \dfrac{3}{2}x - 6$
$\qquad 3x - 2y = 12$

15. $y = 3x - 6$
$\qquad y = -3$

16. $2y - 3 = 1$
$\qquad 5(y - 4) = 10$

17. $2x - 11 = 3$
$\qquad 14 - 2x = x - 7$

18. $y - 1 = 4$
$\qquad 3y - 2 = 13$

19. $2x + 3y = 6$
$\qquad x + y = 1$

20. $y = 2x - 15$
$\qquad y = -3x + 10$

21. $y = 2x + 1$
$\qquad x = 2y + 7$

22. $5y + 6 = 3y + 5$
$\qquad 2(x - 3) + 1 = 4$

Write a system of linear equations to represent each situation, and solve the equations graphically.

23. A survey of 500 voters revealed that 120 people considered themselves independent voters. Of the remaining people surveyed, there were 40 more Republicans than Democrats. However, the survey failed to report how many of each there were. How many more Democrats than independent voters were surveyed?

24. The average hourly earnings of employees in private industry groups has increased since 1990, according to data reported by the U.S. Bureau of Labor Statistics. However, the average hourly earnings of workers in the finance, insurance, and real-estate industry have increased at a faster rate than those for workers in the manufacturing industry. The average hourly earnings of workers in the finance, insurance, and real-estate can be estimated by the function $F(x) = 0.507x + 9.92$, where x is the number of years after 1990. Likewise, the average hourly earnings of workers in manufacturing can be estimated by the function $M(x) = 0.330x + 10.80$. Determine the year in which the average hourly earnings will be the same for both industries.

6.2

Recall	Example
Solve by substitution • Solve one of the equations for a variable • Substitute the expression found in the first step into the other equation, and solve for the second variable. • Substitute the value found for the second variable into one of the original equations, and solve for the remaining variable. • The ordered-pair solution is determined by the values obtained for each variable.	Solve by substitution. $\quad x + y = 5$ $2x - y = 1$ Solve the first equation for y. $y = -x + 5$ Substitute into the second equation and solve. $\qquad\qquad 2x - y = 1$ $\quad 2x - (-x + 5) = 1$ $\qquad 2x + x - 5 = 1$ $\qquad\qquad 3x - 5 = 1$ $\quad 3x - 5 + 5 = 1 + 5$ $\qquad\qquad\qquad 3x = 6$

(Continued on page 450)

Recall	Example
	$$\frac{3x}{3} = \frac{6}{3}$$ $$x = 2$$ Substitute into the first equation and solve. $$x + y = 5$$ $$2 + y = 5$$ $$2 + y - 2 = 5 - 2$$ $$y = 3$$ The solution is $(2, 3)$.

Solve each system of linear equations by using substitution. If a system has an infinite number of solutions, describe the solution set.

25. $8y = 5$
$4x + 8y = 2$

26. $10x - 5y = -14$
$5x - 1 = 0$

27. $4x - y = 5$
$y = 4x + 3$

28. $x - 2y = 71$
$3x - 7y = 275$

29. $2x + 3y = -69$
$2x - 4y = 218$

30. $3(x + 2) - y = -1$
$y = 3x + 7$

31. $x + 8y = 5$
$12x + y = 10$

32. $5x - 3y = 406$
$2x - y = 327$

33. $y = \frac{1}{3}x - 4$
$x - 5y = 0$

34. $y = \frac{1}{2}x + 7$
$y = -\frac{3}{5}x - 4$

35. $3x - y = 10$
$2x - 6y = 26$

36. $x = 160y$
$3x - 440y = 30$

Write a system of linear equations to represent each situation, and solve the equations by using substitution.

37. Two angles are complementary. The second angle is 12 degrees more than twice the first angle. What is the difference in the measures of the two angles?

38. Kidsports Store wants to feature a promotion on Razor scooters. The store plans to sell the scooters for $89.95 each, much lower than the normal price of $100 to $150 each. Kidsports can acquire scooters for sale at a cost of $29.00 each. There will be a fixed cost of $450.00 for advertising and floor space during the promotion.
  a. Write a function for the total cost of acquiring scooters, displaying them, and advertising.
  b. Write a function for the total revenue received from selling scooters.
  c. Write a system of equations, and determine the number of scooters that must be acquired and sold to break even.

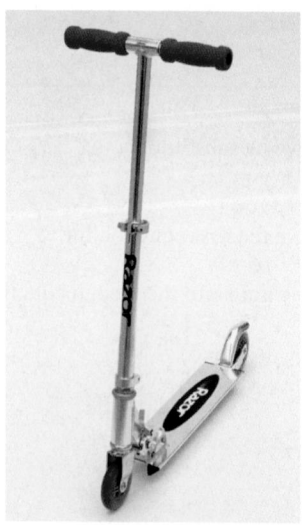

39. In 1998, the Walt Disney Company was ranked as the seventh-largest company in terms of advertising expenditures, and Sears Roebuck and Company was listed as the eight-largest. This was a reversal from the 1991 rankings, when Sears Roebuck and Company ranked higher than the Walt Disney Company in advertising expenditures. Using data published by Competitive Media Reporting, the total spending, in millions of dollars, for advertising for the Walt Disney Company could be estimated by the function $W(x) = 78.1x + 237.8$, where x is the number of years after 1990. Likewise, the total spending for Sears Roebuck and Company could be estimated by the function $S(x) = 40.4x + 406.5$. Determine the year when the two companies spent approximately the same amount for advertising.

6.3

Recall	Example
Solve by elimination • Write both equations in standard form. • Determine the least common multiple of the coefficients of the variable to be eliminated. Multiply the members of the equation by a number to obtain coefficients that have the least common multiple and opposite signs. • Add corresponding members of the equations, eliminating one of the variables. • Solve the resulting equation. • Substitute the value obtained into one of the original equations, and solve that equation. • The ordered-pair solution is determined by the values obtained for each variable.	Solve by elimination. $2x - 3y = 1$ $4x - 5y = 3$ To eliminate the x variable, we need to multiply the first equation by -2 in order to have a coefficient of -4 for x. (1) $\quad -2(2x - 3y) = -2(1)$ (2) $\quad 4x - 5y = 3$ Add corresponding members. (1) $\quad -4x + 6y = -2$ (2) $\quad \underline{4x - 5y = 3}$ $\qquad\qquad y = 1$ Substitute into the first equation and solve. (1) $\quad 2x - 3y = 1$ $\qquad 2x - 3(1) = 1$ $\qquad 2x - 3 = 1$ $\qquad 2x - 3 + 3 = 1 + 3$ $\qquad 2x = 4$ $\qquad \dfrac{2x}{2} = \dfrac{4}{2}$ $\qquad x = 2$ The solution is $(2, 1)$.

Solve each system of equations by using the elimination method. If a system has an infinite number of solutions, describe the solution set.

40. $5x + 3y = -10$
$5x - 3y = 80$

41. $x = 18 - 2y$
$3x + 2y = 30$

42. $\dfrac{1}{2}x + \dfrac{1}{3}y = 3$
$\dfrac{1}{4}x - \dfrac{2}{5}y = -7$

43. $5x + 10y = -55$
$2x - 3y = 6$

44. $2x = 3y + 1$
$15y = 10x + 5$

45. $5x + 7y = 21$
$3x - 2y = 13$

46. $3x + y = 20$
$y = \dfrac{2}{3}x + \dfrac{16}{3}$

47. $7(y - 1) = 5x$
$y = \dfrac{5}{7}x + 1$

48. $3x - 3y = 4$
$9x + 9y = -2$

49. $3x - y = 0$
$2x + 2y = 7$

50. $0.25x + 0.3y = 4$
$0.5x - 0.2y = 4$

51. $0.2x + 0.1y = 5$
$0.02x - 0.01y = 13.5$

52. According to the *Forbes* magazine Celebrity 100 List, in the year 2000 Stephen King's earnings exceeded those of J. K. Rowling, the author of the Harry Potter stories, by 8 million dollars. The combined earnings of the two authors were 80 million dollars. Determine the earnings of each author.

53. A realtor in Chicago is refurbishing a building that is to hold 40 offices. A check of current rates for office rentals indicates that the realtor's competitors are charging about $500 per month for small offices and $1800 per month for large offices. She would like to collect $35,600 per month in rentals to recover her investment and to make a reasonable profit. How many of each size office should she have?

54. The mass of the Earth is approximately 5.328×10^{24} kg greater than the mass of the planet Mars. The total mass of the two planets is approximately 6.612×10^{24} kg. Find the mass of each planet.

6.4

Recall	Example
Model a real-world situation • Read and understand the problem. • Define all variables for the unknown quantities and define other quantities in terms of the defined variables. • Write equations with the information given. • Solve the system. • Check the solution. • In a complete sentence, write an answer to the question asked.	Tony has $20,000 to invest. He wants to diversify his investments by depositing into two accounts, one in mutual bonds at a 7.24% annual interest rate and one in a three-month Treasury bill at a 3.86% annual interest rate. He wants to earn $1279 in interest for one year. Assume that his deposits earn simple interest. How much should Tony deposit in each account? Let x = the amount in mutual bonds at a 7.24% annual rate y = the amount in a three-month Treasury bill at a 3.86% annual rate $$x + y = 20,000$$ $$0.0724x + 0.0386y = 1279$$ Solve by any method. The solution is (15,000, 5000). Tony should invest $15,000 in the mutual bonds and $5,000 in the Treasury bill.

Write and solve a system of linear equations for each situation.

55. A manufacturing plant receives the same component from two suppliers. Supplier A can provide the component at $25 each, while Supplier B can provide the component at $35 each. Under ideal conditions, the plant would buy all of the components from Supplier A, but the plant manager recognizes that it should do business with both suppliers in order to ensure a supply of the components and to promote competition. The budget allows $5500 per month for the component, and the plant needs 200 components each month. How many components should the plant order from each supplier in order to satisfy the budget limitation?

56. A plant-food manufacturer wants to blend a fertilizer that is 10% nitrogen with another that is 5% nitrogen to obtain a blend that is 8% nitrogen. If he wants to make 150 pounds of the blend, how many pounds of each component should he use?

57. A food-processing plant is producing frozen vegetables consisting of a mixture of broccoli and cauliflower. The plant will sell the mixture for 69 cents per pound. The plant sells the broccoli separately for 49 cents a pound and the cauliflower for 99 cents per pound. If it wishes to produce 200 pounds of the mixture, how many pounds of each vegetable should be mixed?

58. A trucking company had to send a truck on a delivery 600 miles away. Part of the route was on an interstate highway where the driver drove at 65 mph. The rest was on rural highways where his speed was 45 mph. If the driver made the trip in 10 hours, how many miles did he drive on the interstate? (*Hint:* If x = hours of interstate driving, then $10 - x$ = hours of highway driving. Likewise, if y = interstate miles, then $600 - y$ = highway miles.)

59. The Buzzards Bay & Vineyard Sound region of Massachusetts has tricky currents that change rapidly and significantly. If a powerboat travels a distance of 20.5 miles with a current for 20 minutes ($\frac{1}{3}$ hour) and returns over the same distance in 21 minutes ($\frac{7}{20}$ hour), determine the boat's average speed in still water and the average speed of the current. Round your answers to the nearest tenth of a mile per hour.

60. One alloy contains 25% copper and another contains 30% copper. How many pounds of the alloy containing the 25% copper must be combined with 40 pounds of the alloy containing 30% copper to form an alloy that contains 27% copper?

61. Dave can lease a car from company A for $150.00 per week plus $0.25 per mile. He can lease the same car from company B for $175.00 per week plus $0.20 per mile. How many miles must he drive during the week in order for the car from company B to be more economical?

62. Julia invested her salary from her last blockbuster movie in two simple-interest funds. Her salary was $10,000,000. One fund paid 4.5% simple interest and the other fund paid 6% simple interest. At the end of the year, Julia received a total of $487,500 in interest payments. How much did she invest in each fund?

6.5

Recall	Examples
Solution of a system of linear equations • An ordered triple that satisfies every equation in the system.	Is $(2, 3, -1)$ a solution of the following system? $$x + y + z = 4$$ $$2x - y + z = 0$$ $$-x + 3y + 2z = 5$$ Substitute the ordered triple into each equation. $\begin{array}{c\|c} x + y + \;z\; = 4 \\ \hline 2 + 3 + (-1) & 4 \\ 4 & \end{array}$ $\begin{array}{c\|c} 2x \;-\; y + \;z\; = 0 \\ \hline 2(2) - 3 + (-1) & 0 \\ 4 \;-\; 3 \;-\; 1 & \\ 0 & \end{array}$ $\begin{array}{c\|c} -x \;+\; 3y \;+\; 2z \;=\; 5 \\ \hline -(2) + 3(3) + 2(-1) & 5 \\ -2 \;+\; 9 \;-\; 2 & \\ 5 & \end{array}$ Since $4 = 4$, $0 = 0$, and $5 = 5$, the ordered pair $(2, 3, -1)$ satisfies all equations and is a solution of the system.

Solve by elimination
• Write all equations in standard form.
• Separate the system into two pairs of equations.
• Eliminate the same variable from each pair of equations.
• Write a system of two variables from the results of the elimination.
• Solve the resulting system of equations.
• Substitute the two values obtained into one of the original equations, and solve that equation for the remaining variable.
• The ordered-triple solution is determined by the values obtained for each variable.
	Solve by elimination.
	(1) $x + y + z = 4$
	(2) $2x - y + z = 0$
	(3) $-x + 3y + 2z = 5$
	Write two pairs of equations.
	(1) $x + y + z = 4$ (2) $2x - y + z = 0$
	(2) $2x - y + z = 0$ (3) $-x + 3y + 2z = 5$
	Eliminate the y from each pair.
	(1) $x + y + \;\;z = \;\;4$ (2) $3(2x - y + z) = 3(0)$
	(2) $\underline{2x - y + \;\;z = 0}$ (3) $-x + 3y + 2z = 5$
	(4) $3x \;\;\;\;\;\; + 2z = 4$
	                     (2) $\;\;6x - 3y + 3z = 0$
	                     (3) $\underline{-x + 3y + 2z = 5}$
	                     (5) $\;\;5x \;\;\;\;\;\; + 5z = 5$
	Write a system of two variables and solve by elimination.
	(4) $-5(3x + 2z) = -5(4)$
	(5) $\;\;\;3(5x + 5z) = 3(5)$
	(4) $-15x - 10z = -20$
	(5) $\underline{\;\;\;15x + 15z = 15}$
	                 $5z = -5$
	                  $z = -1$
	(4) $3x + 2z = 4$
	       $3x + 2(-1) = 4$
	         $3x - 2 = 4$
	             $3x = 6$
	               $x = 2$
	Substitute in an original equation.
	(1) $x + y + z = 4$
	     $2 + y + (-1) = 4$
	         $y + 1 = 4$
	             $y = 3$
	The solution is $(2, 3, -1)$.

Determine whether each ordered triple is a solution of the system.

63. $(3, 4, -2)$

$2x - 3y + 4z = -14$
$5x - 2z = 7$
$3x + 4y = 25$

64. $(0.5, 3.4, -1.2)$

$8x + y + 2z = 5$
$2x - 5y - 5z = -10$
$4x + 2y + 4z = 4$

Solve using the elimination method.

65. $4x + y + z = 7$
$x - y + z = 4$
$2x - y + z = 15$

66. $x + y + z = -1$
$x + y - z = 3$
$3y - z = 4$

67. $x + 2y - z = 3$
$3x - y + z = 6$
$2x - 3y + 2z = 3$

68. $x - 3y + 4z = 3$
$4x + y - z = 8$
$3x + 4y - 5z = 5$

Write a system of linear equations in three variables for each situation, and solve the equations using the elimination method.

69. A charity basketball game charged students $2.50, teachers $5.00, and visitors $7.50 for admission. A total of 140 people attended and $637.50 was raised. The number of students attending was five times the number of teachers. How many of each group attended?

70. Robin had lunch each day at the local deli around the corner from her office. Over a five-day period, she ate sandwiches that had a total fat content of 22 grams, a cholesterol content of 100 mg, and 1446 calories. Her sandwiches were ordered from the following chart:

	Club sandwich	Veggie sandwich	Roast chicken sandwich
Fat, in grams	5	3	6
Cholesterol, in mg	26	0	48
Calories	312	237	348

How many of each sandwich did she eat?

71. Kirchhoff's laws yielded the following system of simultaneous equations for an electrical circuit:

$$I_1 - I_2 + I_3 = 0$$
$$2I_1 + 3I_2 = 11$$
$$3I_2 + 4I_3 = 17$$

Solve the system for the three currents, I_1, I_2, and I_3, measured in amperes.

CHAPTER 6 CHAPTER REVIEW

Solve each system of equations by using the elimination method.

1. $y = \frac{4}{7}x + 2$
$x - 3y = 4$

2. $y = \frac{4}{5}x - 6$
$5y + 2 = 4(x - 7)$

3. $1.25x + 3.5y = -25$
$4.5x + 2.8y = 8$

4. $0.55x - 0.68y = 48$
$-0.51x + 0.68y = 0$

5. $8x + 2y = -6$
$6x - 2y = -78$

6. $2x + 8y = -12$
$y = 2x + 3$

7. $3x + y = 4$
$12x + 4y = 9$

8. $3x - 5y = 11$
$4x + 2y = 13$

9. $\frac{1}{3}x - \frac{2}{5}y = 10$
$\frac{1}{2}x + \frac{4}{5}y = -13$

10. $4x + 8y = 68$
$5x - 3y = -6$

11. $2x + y = 0$
$x + 4y = 3$

12. $5x + 3y = 9$
$x - y = 6$

Solve each system of linear equations graphically.

13. $2y + 2 = y$
$y = -x + 1$

14. $2x + 3 = x$
$2x = 5$

15. $y = 2x + 10$
$3x + y = -10$

16. $2(x - 1) - 4 = 0$
$2x + y = 3$

17. $x + y = 9 - x$
$2x + y = 5$

18. $y = -4x - 3$
$4x + 2y = y - 3$

19. $3x + 4 = 2x$
$x + 7 = 3$

20. $3y + 7 = 2y + 5$
$y + 9 = 7$

21. $3(x - 3) - 1 = 8$
$3y - 10 = 15 - 2y$

22. $y = 3x + 7$
$x - y = -1$

23. $x - 5y = 15$
$x + 5y = -5$

24. $y = -4x - 6$
$y = 3x + 8$

25. $y = x$
$x = 2 - y$

26. $x + 9 = 3$
$3y = 2y - 1$

Solve each system of linear equations by using substitution.

27. $y = \dfrac{5}{11}x - 43$

$x + 2y = -2$

28. $y = \dfrac{3}{4}x + 14$

$y = -\dfrac{2}{3}x - 3$

29. $2x + 3y = 53$

$2x - 4y = -248$

30. $y = 4x - 7$

$2y + 12 = 4x + y + 5$

31. $y = -5x$
$x - y = 1$

32. $5x - 10y = 21$
$5x = -3$

33. $x + 7y = 3$
$4x + y = 9$

34. $5x + 6y = 669$
$x - 2y = 1705$

35. $x - y = 5$
$6x + 2y = 9$

36. $x = 324y$
$3x - 810y = 90$

37. $y = -2x + 7$
$2x + y = 0$

38. $2x - y = 60$
$5x - 3y = 60$

Determine whether each ordered pair is a solution of the given system of linear equations.

39. $\left(\dfrac{5}{6}, -\dfrac{1}{6}\right)$; $4x + 2y = 3$

$x - y = -1$

40. $(-1.6, 3.8)$; $3x + y = -1$

$5x + 5y = 11$

41. $(-4, 2)$; $5x + 7y = -6$

$2x - 7y = -22$

42. $\left(\dfrac{13}{9}, -\dfrac{4}{9}\right)$; $x + y = 1$

$8x - y = 12$

43. $(3, -2)$; $2x - 1 = 5$

$3y + 5 = 2$

44. $(-5, -3)$; $x + 10 = 5$

$3y + 11 = 2$

45. $(-0.44, 1.22)$; $x + 2y = 2$

$2x + 4y = 5$

46. $(-2, -3)$; $2x + 5y = -19$

$3x - 7y = 10$

Write and solve a system of linear equations for each situation.

47. An alloy containing 15% brass is to be combined with an alloy containing 35% brass to form an alloy containing 27% brass. How much of each alloy should be combined to make 200 pounds of the 27% brass alloy? How much more of the 35% alloy will be used than of the 15% alloy?

48. A landlord has 8 one-bedroom apartments and 12 two-bedroom apartments to rent. She wants the two-bedroom apartments to rent for $150 more than the one-bedroom apartments. She wants the total monthly rental income to be $7300. She must determine how much to charge as the monthly rental for each type of apartment. How much will she receive in total rentals each month for all the one-bedroom apartments? How much will she receive in total rentals each month for all the two-bedroom apartments?

49. Two angles are supplementary. The second angle must measure 10 degrees more than the first angle. What does the smaller angle measure?

50. A survey of 700 people was conducted. Forty percent of the men surveyed supported the issue in question, while 70 percent of the women favored it. A total of 400 people favored the issue. How many more women than men were surveyed?

51. Two angles are complementary. One angle measures 12 degrees more than three times the other angle. Find the measures of the angles.

52. A chemist wants to mix a solution containing 15% sulfuric acid with 18 cc of a solution containing 25% sulfuric acid to make a mixture that is 21% sulfuric acid. How much of the 15% concentration should she use? How many cc's of the mixture will she have?

53. A cardiac patient is put on an exercise program of walking and jogging. He reports to his doctor that one day he walked for 30 minutes and jogged for 20 minutes and covered a distance of 3.5 miles. The next day he walked for 40 minutes and jogged for 20 minutes and covered a distance of 4 miles. Assuming that his rates of walking are the same both days and his rates of jogging are the same both days, determine how fast he walked and how fast he jogged, in miles per hour. (*Note:* Convert the times to hours before solving the exercise.)

54. A coffee shop blends gourmet coffee, which sells for $8.50 per pound, with gourmet Dutch chocolate, which sells for $12.50 per pound. The shop makes a mocha blend that sells for $9.50 per pound. How much coffee should be mixed with how much chocolate to make 50 pounds of the blend?

55. Nikki has been offered two part-time jobs while she attends college. One job pays $10.75 per hour and requires her to work on a production line. The other pays $6.50 per hour as a night clerk and allows her to spend her time studying. She can work a total of 20 hours per week without affecting her studies. She wants to earn exactly

$181.00 per week to pay for room and board. How many hours should she work on each job to meet her requirements?

56. A United Airlines flight departs Boston's Logan International Airport and arrives in Los Angeles 5 hours and 45 minutes later. During the same time frame, a United Airlines flight departs Los Angeles International Airport and arrives in Boston 5 hours later. Assuming that the planes travel at the same average speed in still air, determine that average speed, as well as the average speed of the wind caused by the jet stream. The distance between the two cities is 2600 miles.

57. U Rent It will lease a moving van for $39.95 plus $0.15 per mile. Budget Haul It will lease the same type of van for $19.95 plus $0.22 per mile. How many miles must you drive in order for the U Rent It deal to be the less costly choice?

58. A shopkeeper has fixed costs of $75.00 per day and material costs of $2.50 per item produced and sold. The items sell for $6.50 each. How many items must the shopkeeper produce and sell each day in order to break even?

In exercises 59 and 60, solve the system of equations.

59. $x + y + z = 23$
$2x - y + z = -11$
$x + 3y - z = 5$

60. $x + 2y + z = -1$
$x - 2y - 2z = 5$
$x + 6y + z = 2$

61. Using Kirchhoff's laws on an electrical circuit yields the following system of simultaneous equations:

$$I_1 - I_2 - I_3 = 0$$
$$3I_1 + 4I_2 = 18.5$$
$$4I_2 - 6I_3 = -1$$

Solve the system for the three currents, I_1, I_2, and I_3, measured in amperes.

62. Matthew ate fast-food lunches all last week. He selected sandwiches from the following table:

	Turkey Breast	Ham	Hamburger
Fat	4 grams	5 grams	39 grams
Cholesterol	19 mg	28 mg	90 mg
Calories	289 calories	302 calories	640 calories

His total fat intake from the sandwiches was 92 grams, his total cholesterol intake was 255 mg, and his total calories were 2173. How many of each sandwich did he eat?

CHAPTER 6 TEST

A⁺ TEST-TAKING TIPS

Having the right attitude can make the difference between successfully completing a test and failing to do so. Students who say "I've always hated mathematics" or "I've never been good at math" often defeat themselves before they start. Start thinking about ways in which you can overcome "math inability." Pay attention to detail. Don't rush through arithmetic calculations. Check your addition and subtraction. When allowed, repeat the calculation on your calculator. Develop a persistent "don't quit" attitude. Tell yourself that you won't stop working on the test until the time is up. If you are stuck on a problem, start writing on paper what you know. Sometimes the act of writing will unleash thoughts in your mind that will help you solve a problem. Take pleasure in the math exercises you *can* do. This alone may improve your attitude, build up your confidence, and cause you to think more clearly on the test.

1. Determine whether each ordered pair is a solution of the given system of equations.

$(22.8, -64.3); y = -3.5x + 15.5$
$15x + 10y = -301$

Solve each system of equations graphically.

2. $2y - 8 = 0$
$y = 4$

3. $y = \frac{1}{2}x + 3$
$y = \frac{1}{2}x - 5$

4. $y = 5x - 9$
$4x + 8y = 16$

Solve each system of equations by using the substitution method.

5. $x + 2y = 6$
$5x - 11y = -54$

6. $3x + 6y = 12$
$x + 2y = -5$

7. $x + 6y = -2$
$x - 2y = 1$

Solve each system of equations by using the elimination method.

8. $3x + 7y = 11$
   $6x + 14y = 2$

9. $2x + 9y = 16$
   $5y = 8 - x$

10. $5x + 3y = 11$
   $2x + 5y = 7$

In exercises 11–19, write and solve a system of equations for the information given.

11. A factory has a setup cost of $5500 and a cost per item of $65 each to produce. The items sell for $125 each.
   a. Write a function for the total cost of producing the items.
   b. Write a function for the total revenue received for selling the items.
   c. Write a system of equations, and determine the number of items that must be produced and sold to break even.

12. It took Kenny 3 hours to pedal a tandem bicycle built for two to where Dolly was waiting. Dolly's pedaling increased their speed by 4 mph, and the return trip took them only 2 hours. How fast could Kenny pedal alone, and how far did he have to go to find Dolly?

13. A high-speed computer takes 58 nanoseconds to perform six addition and eight multiplication operations. It takes 55 nanoseconds to perform ten additions and five multiplications. How many nanoseconds does the computer take to perform one addition? How many nanoseconds for one multiplication?

14. A candy shop owner is mixing chocolate-covered raisins selling at $1.25 per pound with chocolate-covered peanuts, which sell for $2.00 per pound. He makes a mix to sell at $1.50 per pound. How many pounds of each should he mix in order to make 30 pounds of the mix?

15. Mel is placing a triangular flower garden in one corner of his yard. The corner angle will be 90 degrees. The other two angles will be complementary, and he wants one of them to be twice as large as the other. How large will each angle be?

16. A chemist needs 100 cc of a 30% nitric acid solution. She has a 50% solution and a 10% solution in stock. How many cc of each must she mix in order to make the required solution?

17. Two angles are supplementary. One angle is 15 degrees more than twice the other angle. Determine the measure of each angle.

18. Caitlin's investment broker recommended that she invest money in two mutal funds. Mutual Fund *A* has been earning 9.5% dividends per year, and Mutual Fund *B* has been earning 7% dividends per year. In order to split her risks over the two funds, Caitlin's broker advises her against putting all her money into one of them. Cailtin wants to invest twice as much in the mutual fund *A* account than in the mutual fund *B* account. She would like to realize a total dividend of $780 at the end of the year. How much should she invest in each fund to achieve her goal?

19. Tommy drives his motorboat 36 miles upriver for 36 minutes ($\frac{3}{5}$ hour) and returns to his starting point in 30 minutes. Determine the average speed of his boat in still water and the average speed of the river current.

In exercises 20 and 21, solve the system of equations.

20. $5x - y + 4z = -1$
   $3y + 4z = 0$
   $10x - 7y = 0$

21. $3x - 2y + z = -4$
   $x - y - z = 3$
   $2x - y + 2z = -7$

22. Mike makes bird feeders, birdhouses, and snack tables to sell at craft shows. He spends $2.00 on materials for each feeder, $1.50 for each house, and $3.00 for each table. It takes him 3 hours to make each feeder, 2 hours to make each house, and 4 hours to make each table. In a 40-hour work week, Mike spends a total of $29 on materials, and the number of feeders he makes is one more than the number of tables. How many of each item does Mike make each week?

23. In solving a system of linear equations, one of three outcomes can occur. List the three outcomes, and explain how you would identify them if you were solving the system of equations by means of the graphical method.

Chapter 6

Project

In this chapter, we discussed the effect of the wind resistance of the jet stream on the velocity of an airplane. In this project, we will track two airplanes in flight traveling between the east coast and the west coast of the United States. A possible choice is to and from Atlanta, Georgia, and Los Angeles, California.

Part I

In order to track a flight in progress,

1. Go to the www.trip.com website.

2. Choose "Flight Tracker."

3. Choose "Track a Flight by City and Time."

4. Choose the same airport on the east coast for the departure, the same airport on the west coast for the arrival, and a departure time. When selecting a departure time, remember that there is a three-hour difference between the east coast and west coast time zones. You may have to select several times before you obtain a flight in progress.

5. Record the information from the Web site in the first row of the table in step 7.

6. Choose the same airport on the west coast for the departure, the same airport on the east coast for the arrival, and a departure time. When selecting a departure time, remember that there is a three-hour difference between the east coast and west coast time zones. You may have to select several times before you obtain a flight in progress.

7. Record the information from the Web site in the second row of the following table:

Airline and Flight	Airport	Time	Status	Current Location	Altitude	Instantaneous Speed	Equipment
	Departing						
	Arriving						
	Departing						
	Arriving						

8. Determine the distance (as the crow flies) in miles between the two cities. One Internet site for doing this is www.indo.com/distance/index.html.

9. Use the following guidelines to determine the flight time:

   - The time taken to travel from the east coast to the west coast is three hours more than the difference of Eastern time and Pacific time.
   - The time taken to travel from the west coast to the east coast is three hours less than the difference of Eastern time and Pacific time.

- Write the time as a mixed number. Fractions of an hour can be determined by dividing the number of minutes by 60.

10. Use the distance-traveled formula, $d = rt$, to write a system of equations. Solve the system to determine the average speed of the airplane in flight and the average speed of the wind, assuming that the jet stream is flowing from west to east. Round the speeds to the nearest whole number.

11. Average speeds and instantaneous speeds are not always the same. Compare the average speed of the airplane and the instantaneous speed of the airplane recorded in step 7. If these numbers are different, explain why.

Part II

The point of no return for an aircraft is the point in its flight when the time needed to complete the flight is equal to the time needed to return to the plane's point of origin.

1. For the flight from the east coast to the west coast, write a system of equations needed to determine the time for the point of no return and the distance from the plane's origin. (*Hint:* Let x be the distance from the plane's origin in miles and y be the time in hours of the point of no return.) Sketch a diagram to illustrate this situation.

2. Determine the time of the point of no return for the trip from the west coast to the east coast and the distance the point is from the plane's origin.

3. Compare the locations of the point of no return for each flight.

Chapter 7

Linear Inequalities

In the last few chapters, we studied several ways to solve equations. But sometimes it is useful to work with relations in which one quantity is not necessarily equal to some other quantity. Instead, it can be helpful to know that an algebraic expression is simply less than or greater than some other one.

In this chapter, we will examine linear inequalities and methods of solving them. We will discuss linear inequalities in one variable and in two variables and systems of linear inequalities in two variables. We will see that the same methods used for solving equations—numeric, graphic, and algebraic—work for solving inequalities as well.

Linear inequalities are important in many areas of practical mathematics. In this chapter we will encounter linear inequalities in the areas of science and medicine. We will also use them to determine the range of scores needed to obtain a final grade average.

Keeping within a budget is a familiar and important application of inequalities, both in large businesses and in our personal lives. We will look at several ways of describing costs and budgets in this chapter and show how to solve some kinds of budgetary problems. In the final project for this chapter, you will be ask to set up a small business venture and determine the number of items you must produce and sell to achieve at least a given profit.

7.1 APPLICATION

According to the American Heart Association, high blood pressure (or hypertension) is a major health problem in the world today. In the United States, 50 million people—about one in every four adults—suffer from the condition. Blood pressure classification is based on the average of two or more readings taken at each of two or more screenings. The systolic and diastolic readings, in mm Hg, for the three stages of high blood pressure are listed in the following table:

Category	Systolic	Diastolic
High blood pressure, Stage 3	180 or more	110 or more
High blood pressure, Stage 2	160–179	100–109
High blood pressure, Stage 1	140–159	90–99

For each of these stages, write a linear inequality that represents it. We will discuss this application further. See page 469.

7.1 **Introduction to Linear Inequalities**

OBJECTIVES

1 Identify linear inequalities in one variable.

2 Represent linear inequalities as graphs on a number line.

3 Represent linear inequalities by using interval notation.

4 Model real-world situations by using linear inequalities.

7.1.1 Linear Inequalities in One Variable

In this chapter, we will be solving inequalities of various forms. In this section, we will solve linear inequalities in one variable. A **linear inequality in one variable** (sometimes shortened simply to **linear inequality**) is written by replacing the equals symbol in a linear equation in one variable with an order symbol. The **order symbols** are "greater than" ($>$), "less than" ($<$), "greater than or equal to" ($\geq$), and "less than or equal to" ($\leq$).

> **STANDARD FORM FOR A LINEAR INEQUALITY IN ONE VARIABLE**
> A linear inequality in one variable is an inequality in standard form that can be written in one of the following ways:
>
> $ax + b < 0$
> $ax + b > 0$
> $ax + b \geq 0$
> $ax + b \leq 0$
>
> where a and b are real numbers and $a \neq 0$.

For example, we might have the following inequalities:

1. $2x + 5 < 0$ Standard form: $ax + b < 0$

2. $2x - 5 < 6$ Not standard form

3. $x + 2 + 3(x - 4) < 3x - 8$ Not standard form

The first inequality is in standard form $ax + b < 0$. The other inequalities can be written in this form with algebraic manipulations that we will learn later in the section.

Note that in a linear inequality, the variable is raised to the first power. An example of a nonlinear inequality, which we will solve later in the text, is $2x^2 - 4 < 0$.

Until we learn algebraic manipulations for inequalities, we will identify a linear inequality in one variable, x, as an inequality consisting of two expressions. Each of these expressions can be simplified to the form $ax + b$, where $a \neq 0$ in at least one of the expressions and the coefficient a is not the same in each expression. Using the previous linear examples, we obtain these inequalities:

        $ax + b$ $ax + b$

1. $2x + 5 < 0x + 0$

2. $2x + (-5) < 0x + 6$

3. $4x + (-10) < 3x + (-8)$

EXAMPLE 1   Identify each inequality as linear or nonlinear.

$$\textbf{a.}\ 5x + (x - 2) > 3(x - 2) \qquad \textbf{b.}\ x + 5 < x^3 - 2x$$
$$\textbf{c.}\ \sqrt[4]{x + 2} \geq 3x - 7$$

Solution

a. Linear, because the inequality simplifies to $6x + (-2) > 3x + (-6)$.

b. Nonlinear, because the variable x has an exponent of 3.

c. Nonlinear, because the radical expression contains a variable.

 7.1.1 Checkup

In exercises 1–5, identify each inequality as linear or nonlinear.

1. $4x + 3 \geq 11$

2. $6x - (3x + 2) < 4(x - 1)$

3. $x^2 + 3x < 2x - 5$

4. $\sqrt[3]{x} + 3 \leq 5x + 1$

5. $\dfrac{1}{x} + 3x \geq 2x - 1$

6. What is the difference between a linear equation in one variable and a linear inequality in one variable?

7. How can you decide whether an inequality in one variable is a linear inequality?

7.1.2 Representing Linear Inequalities on a Number Line

In Chapter 1, we discussed order relations involving rational numbers. These numeric inequalities were all true statements. However, just as numeric equations may be true or false, numeric inequalities may also be true or false. For example,

$1 < 5$ is a true inequality because 1 is less than 5.

$-6 > 7$ is a false inequality because -6 is less than 7, not greater than 7.

$-1 \geq -1$ is a true inequality because -1 is equal to -1.

Linear inequalities that include variables may be true or false, depending on the value of the variable being substituted. A solution of a linear inequality is a number that will make the inequality true. For example,

$$x < 5 \text{ is a linear inequality.}$$

There are many possible solutions of this inequality.

4 is a solution because $4 < 5$.

3 is a solution because $3 < 5$.

$4\frac{1}{2}$ is a solution because $4\frac{1}{2} < 5$.

4.75 is a solution because $4.75 < 5$.

In fact, it is impossible to name all the solutions, because there are infinitely many numbers less than 5. Therefore, we say we have a set of solutions. A solution set is a set of all possible solutions. In Chapter 1, we discussed a number-line graph for infinite sets of real numbers. Therefore, we can represent this solution set as a graph on a number line.

Since we want to represent all numbers less than 5, we say that 5 is the *upper bound* of the solution set. We will graph 5 with an open dot because

5 itself is not a solution. All numbers less than 5 are graphed to the left of 5 on the number line.

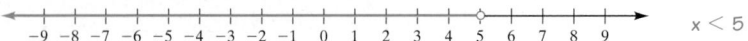

$x < 5$

NUMBER-LINE NOTATION

To graph the solution set of a linear inequality in the form $x < c, x > c, x \le c,$ or $x \ge c$ on a number line,

1. Plot the upper or lower bound, c, of the inequality on the number line.

- Use an open dot if the bound is not included in the solution.
- Use a closed dot if the bound is a solution.

2. Complete the graph, using a solid line.

- Cover all points to the left of the upper bound for a "less than" or "less than or equal to" inequality.
- Cover all points to the right of the lower bound for a "greater than" or "greater than or equal to" inequality.

We can graph a linear inequality on our calculator. The graph of $x < 5$ is shown in Figure 7.1c.

TECHNOLOGY

GRAPHING NUMBER LINES

Graph the solution set of $x < 5$.

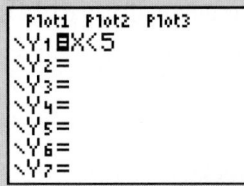

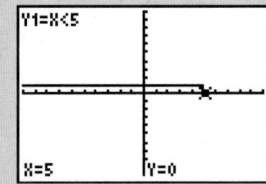

Figure 7.1a   Figure 7.1b   Figure 7.1c

For Figure 7.1a,

Enter the inequality $x < 5$ in Y1. The inequality symbols are found under the TEST menu. The "less than" symbol is option 5.

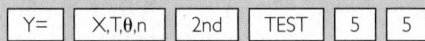

For Figure 7.1b

Graph the number line. The calculator will test the inequality for x-values and graph an ordered pair $(x, 1)$ for a true inequality and $(x, 0)$ for a false inequality.

For Figure 7.1c

Trace and use the arrow keys to display the coordinates graphed. To check the upper bound of a number line, 5, enter the value while tracing.

| TRACE | 5 | ENTER |

If the upper bound is a solution of the inequality, it will have a y-coordinate of 1. If the upper bound is not a solution of the inequality, it will have a y-coordinate of 0. For $x < 5$, the upper bound is not a solution and is graphed as $(5, 0)$.

EXAMPLE 2   Graph the solution set of $x \geq 5$ on a number line.

Solution

Since we want to represent all numbers greater than or equal to 5, we say that 5 is the lower bound of the solution set. We will graph 5 with a closed dot because it is a solution. All numbers greater than 5 are graphed to the right of 5 on the number line.

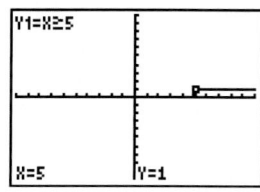

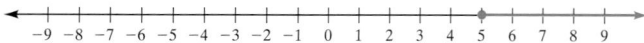

We may combine two linear inequalities into a **compound inequality**.

$$x > -2 \text{ and } x \leq 3$$

The solution set of this compound inequality includes all the real numbers greater than -2 that are also less than or equal to 3. The compound inequality may be written as a **double inequality**.

$$-2 < x \leq 3$$

EXAMPLE 3   Graph the solution set of $-2 < x \leq 3$ on a number line.

Solution

Since we want to represent all numbers between -2 and 3, including 3, we will graph the lower bound, -2, with an open dot because it is not a solution and the upper bound, 3, with a closed dot because it is a solution. All numbers between -2 and 3 are also graphed.

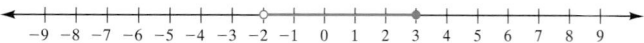

To check the graph of the solution set of a compound inequality on your calculator,

- Partition the inequality into two inequalities.
- Enter the compound inequality using "and," option 1, under "LOGIC" in the TEST menu. See Figure 7.2a

The calculator check is shown in Figure 7.2b.

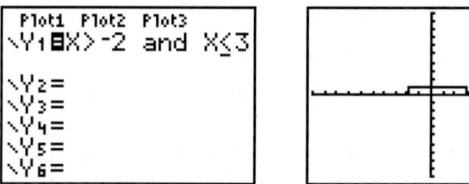

Figure 7.2a   **Figure 7.2b**

7.1.2 Checkup

In exercises 1–3, graph the solution set of each inequality on a number line. Check your graph on your calculator.

1. $x > -2$ **2.** $x \leq 3$ **3.** $1 \leq x < 5$

4. What is the meaning of the open dot used in graphing a linear inequality on a number line? the closed dot?

7.1.3 Representing Linear Inequalities in Interval Notation

A solution set of a linear inequality can also be written in interval notation. **Interval notation** consists of two numbers or symbols representing the lower and upper bounds, if they exist, preceded and followed by parentheses, brackets, or a parenthesis and a bracket, indicating whether the bound is or is not included in the set. If no lower bound exists, we use the symbol $-\infty$ (**negative infinity**) in its place. If no upper bound exists, we use the symbol ∞ (**infinity**) in its place. If the bound is included in the solution, it is preceded by a bracket if it is a lower bound and followed by a bracket if it is an upper bound. (This corresponds to a closed dot on a number line.) If the bound is not included in the set, it is preceded by a parenthesis if it is a lower bound and followed by a parenthesis if it is an upper bound. (This corresponds to an open dot on a number line.) The symbols $-\infty$ and ∞ are not numbers, but indicate that the set continues without bound. Therefore, the symbol $-\infty$ is always preceded by a parenthesis and the symbol ∞ is always followed by a parenthesis.

HELPING HAND The interval notation for representing a solution set of all real numbers is $(-\infty, \infty)$.

For example, write the solution set of $x < 5$ in interval notation.

The solution set includes all numbers less than 5. The upper bound is 5. There is no lower bound, so $-\infty$ will be used in its place. The symbol $-\infty$ must be preceded by a parenthesis. The upper bound, 5, is not included in the solution set, so it is followed by a parenthesis.

$$(-\infty, 5)$$

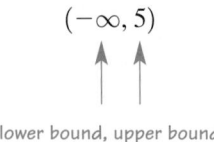

lower bound, upper bound

INTERVAL NOTATION
To write the solution set of a linear inequality in the form $x < c, x > c, x \leq c,$ or $x \geq c,$ using interval notation,

- Determine the bounds for the solutions if they exist. Write the bounds in increasing order, separated by a comma. If no lower bound exists, use the symbol $-\infty$ in its place. If no upper bound exists, use the symbol ∞ in its place.

- Precede or follow each bound of the inequality with a parenthesis or a bracket. A parenthesis is used to denote that the lower or upper bound is not included in the solution set. A bracket is used to denote that the lower or upper bound is included in the solution set.

Remember, the symbols $-\infty$ and ∞ are not numbers. The symbol $-\infty$ is preceded by a parenthesis and the symbol ∞ is followed by a parenthesis.

EXAMPLE 4   Write the solution set of $x \geq 5$ in interval notation.

Solution

The solution set includes all numbers greater than or equal to 5. The lower bound is 5. There is no upper bound, so ∞ will be used in its place. The symbol ∞ must be followed by a parenthesis. The lower bound, 5, is included in the solution set, so it is preceded by a bracket.

The solution set is $[5, \infty)$.

EXAMPLE 5   Write the solution set of $-2 < x \leq 3$ in interval notation.

Solution

The solution set includes all numbers between -2 and 3, including 3. The lower bound is -2. The upper bound is 3. The lower bound, -2, is not included in the solution set, so it is preceded by a parenthesis. The upper bound, 3, is included in the solution set, so it is followed by a bracket.

The solution set is $(-2, 3]$.

The following table summarizes what we have learned in this section by describing some of the most common inequalities you may encounter:

Inequality	Number Line	Interval Notation
$x < a$	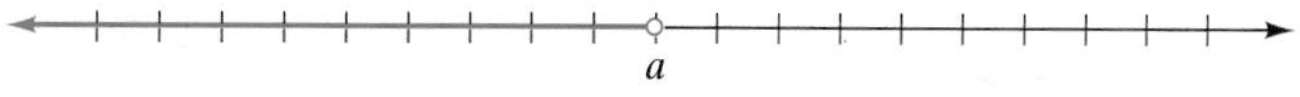	$(-\infty, a)$
$x \leq a$	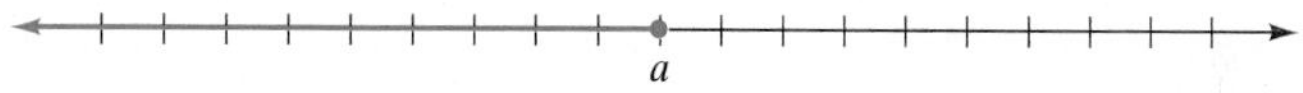	$(-\infty, a]$
$x > a$	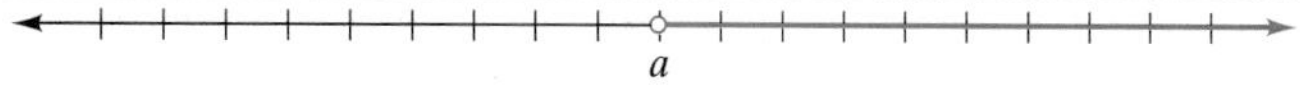	(a, ∞)
$x \geq a$	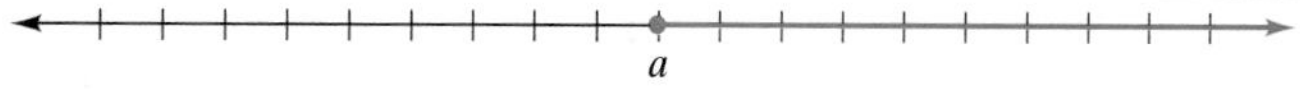	$[a, \infty)$
$a < x < b$	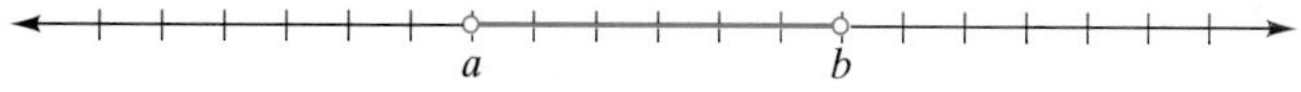	(a, b)
$a \leq x \leq b$	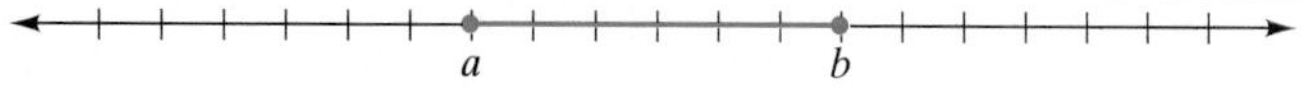	$[a, b]$

EXAMPLE 6   Complete the following table:

Inequality	Number Line	Interval Notation
$x > 5$		
	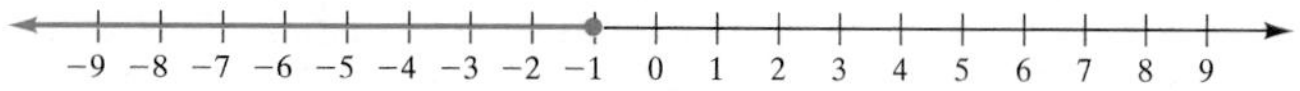	
		$[-3, \infty)$
$-3 \leq x < 0$		

Solution

Inequality	Number Line	Interval Notation
$x > 5$	-9 -8 -7 -6 -5 -4 -3 -2 -1 0 1 2 3 4 5 6 7 8 9	$(5, \infty)$
$x \leq -1$	-9 -8 -7 -6 -5 -4 -3 -2 -1 0 1 2 3 4 5 6 7 8 9	$(-\infty, -1]$
$x \geq -3$	-9 -8 -7 -6 -5 -4 -3 -2 -1 0 1 2 3 4 5 6 7 8 9	$[-3, \infty)$
$-3 \leq x < 0$	-9 -8 -7 -6 -5 -4 -3 -2 -1 0 1 2 3 4 5 6 7 8 9	$[-3, 0)$

7.1.3 Checkup

For each inequality, write the solution set in interval notation.

1. $x > -2$ **2.** $x \leq 3$ **3.** $1 \leq x < 5$

4. Complete the following table.

	Inequality	Number Line	Interval Notation
a.	$x > -3$		
b.		-9 -8 -7 -6 -5 -4 -3 -2 -1 0 1 2 3 4 5 6 7 8 9	
c.			$[-4, 1)$

5. In the solution of a linear inequality written using interval notation, what is the meaning of a bound that is preceded or followed by a parenthesis, and what is the meaning of a bound that is preceded or followed by a bracket?

7.1.4 Modeling the Real World

Inequalities are very useful kinds of relations in real-world situations. They're especially helpful if you have to calculate quantities that must be more or less than a given value, such as what you can spend within a given budget. Sometimes you might want to determine quantities that are *between* two given values, such as how many guests you can invite to a dinner that costs between $100 and $150. In these situations as well, inequalities are the kind of mathematics you need.

EXAMPLE 7 Qu is purchasing begonia tubers over the Internet for her garden. The special offer reads, "Buy 12 tubers of one color for $9.65. Free shipping on orders over $30.00." Write a linear inequality to determine the number of different colors Qu can purchase within her budget of $50.00 and receive free shipping.

Buy **12 tubers of one** color **for $9.65!**

Free shipping on orders over $30.00

Solution

Let x = number of colors Qu can order

The cost is $9.65 per color ordered ($9.65x$). The lower bound is $30.00 and the upper bound is $50.00. The lower bound is not included, because she must spend over $30.00. The upper bound is included, as that is the most she can spend.

$$30.00 < 9.65x \leq 50.00$$

APPLICATION

According to the American Heart Association, high blood pressure (or hypertension) is a major health problem in the world today. In the United States, 50 million people—about one in every four adults—suffer from the condition. Blood pressure classification is based on the average of two or more readings taken at each of two or more screenings. The systolic and diastolic readings, in mm Hg, for the three stages of high blood pressure are listed in the following table:

Category	Systolic	Diastolic
High blood pressure, Stage 3	180 or more	110 or more
High blood pressure, Stage 2	160–179	100–109
High blood pressure, Stage 1	140–159	90–99

For each of these stages, write a linear inequality that represents it.

Discussion

Let s = the systolic reading in mm Hg
    d = the diastolic reading in mm Hg

High blood pressure, stage 3	$s \geq 180$	$d \geq 110$
High blood pressure, stage 2	$160 \leq s \leq 179$	$100 \leq d \leq 109$
High blood pressure, stage 1	$140 \leq s \leq 159$	$90 \leq d \leq 99$

 7.1.4 Checkup

1. Joseph joined a music club that offered a sale in which he could purchase compact discs for $6.99 each, plus a one-time charge of $5.99 for shipping and handling his order. Write a linear inequality in one variable to determine the number of CDs Joseph can order if he can spend no more than $50.00.

2. The American Heart Association lists the following classifications of blood pressure as being within healthy ranges:

Classification	Systolic	Diastolic
High normal range	130–139	85–89
Normal range	Less than 130	Less than 85
Optimal range	Less than 120	Less than 80

Write a linear inequality to represent each of the categories.

7.1 Exercises

Identify each inequality as linear or nonlinear.

1. $3(x + 1) > -(5x - 4)$ **2.** $x + 3 > 2(4x - 1) + 6(x + 2)$ **3.** $4x^2 + 1 < 3x + 2$

4. $\sqrt[3]{x} + 1 \geq x - 1$ **5.** $\frac{5}{7}(a + 2) \geq \frac{3}{7}a + \frac{2}{3}$ **6.** $\frac{2}{3}(b - 6) \leq \frac{1}{3}b - \frac{2}{3}$

7. $0.6x + 2.7 \leq 5.2 - 1.9x$ **8.** $4d - 7 > -4$ **9.** $x + 4(x - 8) > 2(3x + 1)$

10. $4x^3 - 1 > 2x^2 - 5$ **11.** $\sqrt{2x - 7} \leq x + 3$ **12.** $\frac{3}{y} - y < 2$

13. $\frac{5}{x} + 2x > 0$ **14.** $2(x - 1) < -3(x - 2)$ **15.** $2p + 6 < 0$

16. $0.8z - 1.3 \leq 4.7 - 2.6z$

Graph the solution set of each inequality on the number line. Check your results on your calculator. Then write the solution set in interval notation.

17. $x \geq 6$ **18.** $x \leq 9$ **19.** $z < 12$ **20.** $c > -7$

21. $b > -\frac{13}{5}$ **22.** $x < 3\frac{5}{6}$ **23.** $x \leq 4\frac{2}{3}$ **24.** $k \geq \frac{16}{3}$

25. $p \leq 12.59$ **26.** $x \geq 5.1$ **27.** $q > -6.7$ **28.** $R < 45.65$

29. $5 < x < 13$ **30.** $-2 < x < -1$ **31.** $2 < x < 8$ **32.** $3 \leq w \leq 6$

33. $-4 < d \leq 0$ **34.** $0 \leq g < 5$ **35.** $-1 \leq m < 6$ **36.** $2 < x \leq 7$

37. $2 \leq t \leq 7$ **38.** $15 \leq j \leq 30$ **39.** $\frac{2}{5} < s \leq 3\frac{1}{3}$ **40.** $-1\frac{3}{4} \leq a \leq \frac{2}{3}$

41. $-2.5 \leq q < 3.5$ **42.** $-6.5 < y < -1.5$ **43.** $\frac{4}{5} \leq x \leq 4.5$ **44.** $-2.5 < x < \frac{2}{5}$

45. Complete the following table:

	Inequality	Number line	Interval notation
a.	$x < 4.5$		
b.			
c.			$(-\infty, -2)$
d.	$z \geq 5.7$		
e.			
f.			$(2, \infty)$
g.	$2 < y < 8$		
h.			
i.			$[0, 9]$

46. Complete the following table.

	Inequality	Number line	Interval notation
a.			$(-\infty, -9)$
b.		$-9\ -8\ -7\ -6\ -5\ -4\ -3\ -2\ -1\ \ 0\ \ 1\ \ 2\ \ 3\ \ 4\ \ 5\ \ 6\ \ 7\ \ 8\ \ 9$	
c.	$a < \dfrac{1}{2}$		
d.			$[0, \infty)$
e.		$-9\ -8\ -7\ -6\ -5\ -4\ -3\ -2\ -1\ \ 0\ \ 1\ \ 2\ \ 3\ \ 4\ \ 5\ \ 6\ \ 7\ \ 8\ \ 9$	
f.	$b > 2.7$		
g.			$(4, 11)$
h.		$-9\ -8\ -7\ -6\ -5\ -4\ -3\ -2\ -1\ \ 0\ \ 1\ \ 2\ \ 3\ \ 4\ \ 5\ \ 6\ \ 7\ \ 8\ \ 9$	
i.	$-5 \leq c \leq -1$		

47. Seismologists use the Richter scale to classify the magnitude of an earthquake. The following table classifies the magnitudes of less severe earthquakes:

Classification	Richter magnitudes
Generally not felt, but recorded.	Less than 3.5
Often felt, but rarely causes damage.	3.5–5.4
At most slight damage to well-designed buildings and major damage to poorly constructed buildings over small regions.	under 6.0

For each classification, write a linear inequality that represents it.

48. The following table classifies the magnitudes of earthquakes that are destructive:

Classification	Richter magnitudes
Destructive in inhabited areas up to about 100 kilometers across.	6.1–6.9
Major earthquake causing damage over larger areas.	7.0–7.9
Great earthquake causing serious damage in areas several hundred kilometers across.	8.0 or greater

For each classification, write a linear inequality that represents it.

Write a linear inequality in one variable to represent each situation described. Do not attempt to solve the inequality.

49. A car rental firm leases a car for $39.95 plus $0.20 per mile. If your cost of rental is not to exceed $150.00, how many miles can you drive the car?

50. The cost of a production run at a factory is the sum of a setup cost of $300 plus a labor and materials cost of $22 per item produced. What is the number of items that can be produced if the production cost is to be less than $1000?

51. Pete is paid a weekly salary of $450 plus 5% commission on his weekly sales. What must his sales be if he wishes his weekly pay to be more than $800?

52. Josephine pays a weekly rental of $125 for a booth at the local mall. She sells handcrafted baskets there for a price of $15 each. What must be the number of items sold in order for her to realize a profit of at least $250 per week?

53. Sally is planning a shower for her best friend. She has reserved a party room at Le Chien Restaurant, which will charge her $25 for the room and $12.50 per person for lunch. How many people can she invite to the shower if she wishes to spend between $150 and $200 at the restaurant?

54. The local high school is staging *Romeo and Juliet* as a Valentine's Day event. They have spent $350.00 for costumes, props, and so on for the play. They are charging $4.50 per person for admission. How many tickets must they sell in order to realize a profit between $200.00 and $500.00?

7.1 Calculator Exercises

If you have a calculator with split-screen capability, it is often helpful to use the split screen to view an inequality at the same time that you are viewing its graph. As an example, suppose you wish to graph the inequality $x > -3$. The following are the steps used to view this inequality and its graph simultaneously:

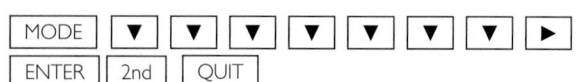

The calculator should exhibit a split screen with the coordinate system above and the home screen below. Now you can enter the instructions to graph the inequality.

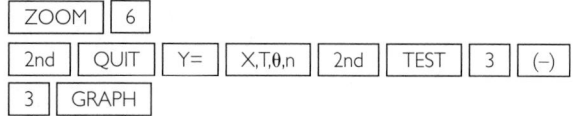

The graph of the inequality is displayed above, and the inequality is displayed below.

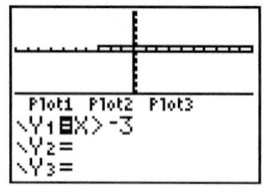

Use the split screen to view the graphs of the given inequalities. Trace the graph to view its bounds.

1. $x < -2$ **2.** $x > -2$

3. $x \leq -2$ **4.** $x \geq -2$

Change the window to decimal scale by entering $\boxed{\text{ZOOM}}$ $\boxed{4}$, and view the following graphs:

5. $-2.5 \leq x \leq 2.5$ **6.** $-2.5 \leq x < 2.5$

7. $-2.5 < x < 2.5$

When you wish to return to a full screen, simply use the $\boxed{\text{MODE}}$ key, move the cursor down to Full, and press $\boxed{\text{ENTER}}$. When you quit the $\boxed{\text{MODE}}$ function, you will be back to the full screen.

7.1 Writing Exercises

A. Key words that imply order relations.

Consider the following list of phrases that imply an order relation, and assign the correct order relation symbol that applies to each phrase:

_____ **1.** At least

_____ **2.** At most

_____ **3.** No more than

_____ **4.** No greater than

_____ **5.** Larger than

_____ **6.** Smaller than

_____ **7.** Not to exceed

_____ **8.** Below

_____ **9.** Above

_____ **10.** Greater than   _____ **11.** Greater than or equal to   _____ **12.** Less than

_____ **13.** Less than or equal to   _____ **14.** No fewer than   _____ **15.** No less than

_____ **16.** No smaller than   _____ **17.** A maximum of   _____ **18.** A minimum of

_____ **19.** Up to   _____ **20.** Down to

After completing the list, suggest at least one additional phrase that matches each of the following order relations:

21. $<$ **22.** $>$ **23.** $\leq$ **24.** $\geq$

Use one of the phrases you suggested to write a real-world application that involves an inequality. Write a complete narrative describing the problem. Then write the linear inequality in one variable that represents the application you described.

B. Alternative notation for graphing inequalities.

In this section, you have used open dots and closed dots to graph an inequality on the number line. You also learned to use parentheses and brackets to indicate the inclusion or exclusion of endpoints in the solution set of an inequality. Consider the following graphs of a linear inequality in one variable:

Number Line	Inequality	Interval notation
1. $\xleftarrow{\quad}$ number line from -5 to 5, bracket at 2 $\rightarrow x$		
2. number line from -5 to 5, mark at 2 $\rightarrow x$		
3. number line from -5 to 5, parenthesis at -1 $\rightarrow x$		
4. number line from -5 to 5, bracket at -4 $\rightarrow x$		
5. number line from -5 to 5, bracket at -3, mark at 2 $\rightarrow x$		

For each graph, write the inequality that corresponds to the graph and write the solution set of the inequality in interval notation. Explain how the graphs differ from those you have seen in this section. Do you think the new notation can be used in place of that shown earlier? Which do you prefer and why?

7.2 Linear Inequalities in One Variable

OBJECTIVES
1 Solve linear inequalities in one variable numerically.
2 Solve linear inequalities in one variable graphically.
3 Solve linear inequalities in one variable algebraically.
4 Model real-world situations by using linear inequalities in one variable, and solve the inequalities.

APPLICATION

The body mass index (BMI) is a measure of weight that takes height into account. The BMI of a person 62 inches tall and weighing x pounds is given by the expression $0.2x - 2$. A doctor may prescribe a certain drug if a person has a BMI of 30 or greater. Determine the weights for which a doctor may prescribe this drug for a person 62 inches in height.

After completing this section, we will discuss this application further. See page 487.

7.2.1 Solving Linear Inequalities Numerically

In previous chapters, we determined whether a number was a solution of an equation by substituting the number for the variable and evaluating the two resulting expressions. If the expressions were equivalent, then the number substituted was called the solution.

We determine a solution of a linear inequality by using the same method of substituting and evaluating. However, we need to determine whether the resulting inequality is true. For example, determine whether 0 is a solution of $2x + 3 \geq x + 5$.

$$
\begin{array}{c|c}
\multicolumn{2}{c}{2x + 3 \geq x + 5} \\
\hline
2(0) + 3 & 0 + 5 \\
3 & 5
\end{array}
$$

The number 0 is not a solution, because the resulting inequality, $3 \geq 5$, is not true.

To find a solution of the inequality, $2x + 3 \geq x + 5$, we continue to substitute values for x until we find a number that results in a true inequality. To do this, it is convenient to use a table, as we did when solving equations numerically.

SOLVING A LINEAR INEQUALITY IN ONE VARIABLE NUMERICALLY

To solve a linear inequality numerically,
Set up an extended table of values.

- The first column is labeled with the name of the independent variable.
- The second column is labeled with the expression on the left side of the inequality.
- The third column is labeled with the expression on the right side of the inequality.

Complete the table.

- Substitute values for the independent variable.
- Evaluate the second and third columns.
- Continue evaluating until the values for the two expressions (the numbers in the second and third columns) result in a true inequality.

The values for the independent variable (the numbers in the first column) used to determine a true inequality are solutions of the inequality.

HELPING HAND Not all solutions may be found numerically. We are limited to those numbers substituted for the independent variable.

EXAMPLE 1 Solve numerically $2x + 3 \geq x + 5$ for integer solutions.

Solution

The following is a sample table for determining the integer solutions:

x	$2x + 3$	$x + 5$	
0	$2(0) + 3$ 3	$0 + 5$ 5	$3 \ne 5$
1	$2(1) + 3$ 5	$1 + 5$ 6	$5 \ne 6$
2	$2(2) + 3$ 7	$2 + 5$ 7	$7 \ge 7$
3	$2(3) + 3$ 9	$3 + 3$ 6	$9 \ge 6$

According to the table, when 2 is substituted for the variable x, the resulting inequality is true $(7 \ge 7)$. Therefore, 2 is a solution of the linear inequality $2x + 3 \ge x + 5$. When 3 is substituted for x, the resulting inequality also is true $(9 \ge 6)$. Therefore, 3 is a solution of the linear inequality $2x + 3 \ge x + 5$. If the table is extended further, more integer solutions will be found. In fact, all integers completing the obvious pattern will result in true inequalities. Therefore, all integers greater than or equal to 2 will be solutions of the inequality.

The table results in only integer solutions, not all solutions. We could have included noninteger solutions, but it would still be impossible to include all solutions.

To solve $2x + 3 \ge x + 5$ numerically on your calculator, set up the table as shown in Figure 7.3a, Figure 7.3b, and Figure 7.3c. The integer solutions are all integers greater than or equal to 2.

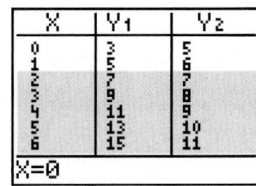

Figure 7.3a   Figure 7.3b   Figure 7.3c

HELPING HAND Although not resulting in all solutions, the numerical method may be used as a visualization to check other methods.

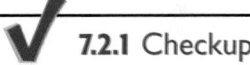 **7.2.1** Checkup

In exercises 1–2, solve numerically for integer solutions.

1. $2x - 5 < 4x - 3$ **2.** $5z + 2 > 3z$

3. What are the limitations of solving a linear inequality numerically?

7.2.2 Solving Linear Inequalities Graphically

A second and more inclusive method of solving a linear inequality is to graph two functions. The functions to be graphed are written by using the expression on the left side of the inequality as a rule for the first function and the expression on the right side as a rule for the second function. For example, for the inequality $2x + 3 \le x + 5$, we write the two functions $y_1 = 2x + 3$ and $y_2 = x + 5$. We then determine the solutions of the inequality $y_1 \le y_2$. Let's explore this method on the graphing calculator. Complete the following set of exercises on your calculator.

Discovery 1

Graphical Solutions of a "Less Than" Linear Inequality in One Variable

To determine the solutions of the inequality $2x + 3 \leq x + 5$, graph the functions $Y1 = 2x + 3$ and $Y2 = x + 5$. Sketch the graphs. Label the point of intersection of the graphs.

1. The intersection of the two graphs is _____.
2. The solution that corresponds to the equality is _____.

Choose the correct answer.

3. To solve a "less than" inequality, $Y1 < Y2$, locate the portion of the Y1 graph *above/below* the Y2 graph. When tracing the graph of Y1, you will find that this is to the *left/right* of the intersection. The *x*-coordinates of the points in this direction are *less than/greater than* the *x*-coordinate of the point of intersection.
4. Combining the solutions found for the equality and the "less than," we determine that the solution set is _____.

The solution set of the inequality $2x + 3 \leq x + 5$ consists of the *x*-coordinate of the point of intersection and the *x*-coordinates of the points to the left of the intersection. Written as an inequality, the solutions are all *x* that satisfy $x \leq 2$.

Correspondingly, we can solve the inequality $2x + 3 \geq x + 5$ graphically. Complete the following set of exercises on your calculator.

Discovery 2

Graphical Solutions of a "Greater Than" Linear Inequality in One Variable

To determine the solutions of the inequality $2x + 3 \geq x + 5$, graph the functions $Y1 = 2x + 3$ and $Y2 = x + 5$. Sketch the graph. Label the point of intersection of the graphs.

1. The intersection of the two graphs is _____.
2. The solution that corresponds to the equality is _____.

Choose the correct answer.

3. To solve a "greater than" inequality, $Y1 > Y2$, locate the portion of the Y1 graph *above/below* the Y2 graph. When tracing the graph of Y1, you will find that this is to the *left/right* of the intersection. The *x*-coordinates of the points in this direction are *less than/greater than* the *x*-coordinate of the point of intersection.
4. Combining the solutions found for the equality and the "greater than," we determine that the solution set is _____.

The solution set of the inequality $2x + 3 \geq x + 5$ consists of the *x*-coordinate of the point of intersection and the *x*-coordinates of the points to the right of the intersection. Written as an inequality, the solution set is all *x* that satisfy $x \geq 2$.

SOLVING A LINEAR INEQUALITY IN ONE VARIABLE GRAPHICALLY

To solve a linear inequality graphically,

- Write two functions, y_1 and y_2, using the expressions on the left and right sides of the inequality, respectively.
- Graph both functions on the same coordinate plane.
- Determine the intersection, if it exists, of the y_1 and y_2 graphs.
- Locate the portion of the y_1 graph below the y_2 graph if the inequality is "less than." Locate the portion of the y_1 graph above the y_2 graph if the inequality is "greater than."
- Determine the x-coordinates for this portion of the graph.
  - If the portion is to the left of the intersection, then x is less than the x-coordinate of the intersection.
  - If the portion is to the right of the intersection, then x is greater than the x-coordinate of the intersection.

The solution set includes either x-values less than or x-values greater than the x-coordinate of the point of intersection. The solution set also includes the x-coordinate of the point of intersection if the inequality includes equality.

EXAMPLE 2   Solve graphically $2x + 3 > x + 5$.

Solution

Graph $y_1 = 2x + 3$ and $y_2 = x + 5$.

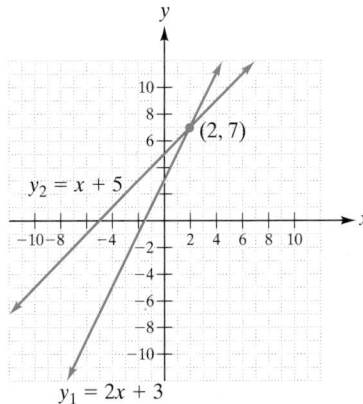
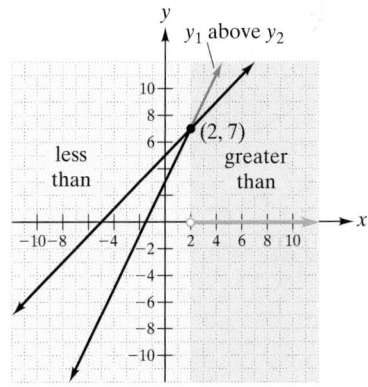

The intersection of the lines is $(2, 7)$. We do not include the x-coordinate of the intersection in the solution, because the inequality does not include equality.

The inequality is "greater than," so the graph of y_1 is above the graph of y_2, to the right of the point of intersection. This is interpreted as an inequality: x is greater than the x-coordinate of the intersection, 2.

The solution set is all x that satisfy $x > 2$, or $(2, \infty)$.

The graphical check is shown in Figure 7.4b.

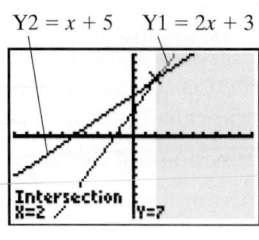

$(-10, 10, -10, 10)$

Figure 7.4a

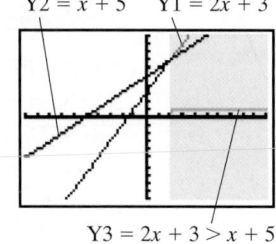

$Y3 = 2x + 3 > x + 5$

$(-10, 10, -10, 10)$

Figure 7.4b

EXAMPLE 3   Solve graphically.

a. $6x - (4x + 8) \leq 7 - 3x$

b. $(5x + 4) - 2(3x + 1) \geq 2(x - 7)$

Calculator/Graphic Solution

a.

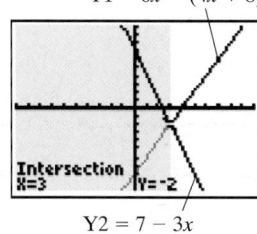

$Y2 = 7 - 3x$

$(-10, 10, -10, 10)$

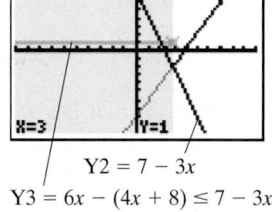

$Y2 = 7 - 3x$

$Y3 = 6x - (4x + 8) \leq 7 - 3x$

$(-10, 10, -10, 10)$

The intersection of the graphs is $(3, -2)$. The inequality is "less than," so we locate the portion of the graph of Y1 that is below the graph of Y2. These points are to the left of the point of intersection and make up the area where x is less than 3. The solution includes the x-coordinate of the point of intersection, because the inequality includes equality. The solution set is all x that satisfy $x \leq 3$, or $(-\infty, 3]$.

b. $(5x + 4) - 2(3x + 1) \geq 2(x - 7)$ does not have an integer-value intersection.

The Trace function will not locate the intersection, so we use Intersect under the CALC function.

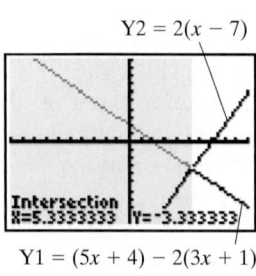

$Y1 = (5x + 4) - 2(3x + 1)$

$(-10, 10, -10, 10)$

The intersection of the two graphs is $(5.\overline{3}, -3.\overline{3})$ or $(\frac{16}{3}, \frac{-10}{3})$. The Y1 graph is above the Y2 graph to the left of the point of intersection. The solution includes the x-coordinate of the intersection point. The solution set is all x that satisfy $x \le \frac{16}{3}$, or $(-\infty, \frac{16}{3}]$.

Remember from Chapter 4 that linear equations may have no solution or may have all-real-number solutions. The same is true for linear inequalities. Graphically solving linear inequalities can result in either of these possibilities.

EXAMPLE 4   Solve graphically.

   a. $4(x - 2) < 4x + 10$ **b.** $3x + 4 \ge 2x + (x + 10)$
   c. $2a + 12 \le 2(a + 3) + 6$

Calculator/Graphic Solution

a.   Y2 = 4x + 10

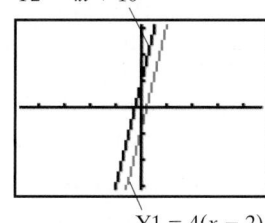

Y1 = 4(x − 2)

$(-47, 47, -31, 31)$

The graphs are parallel, and the graph of Y1 is always below the graph of Y2. Since the inequality is "less than," the entire Y1 graph will determine solutions of the inequality. The solution set is all real numbers or $(-\infty, \infty)$.

b.   Y2 = 2x + (x + 10)

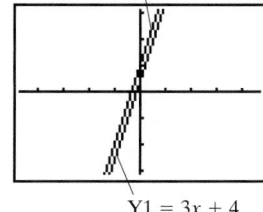

Y1 = 3x + 4

$(-47, 47, -31, 31)$

The graphs are parallel, and the graph of Y1 is always below the graph of Y2. Since the inequality is "greater than" and the Y1 graph will not have any points above the Y2 graph, the inequality has no solution.

c.   Change the variables to x.

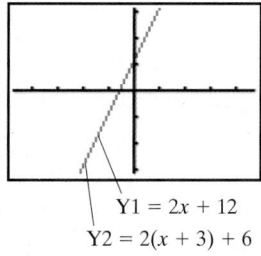

Y1 = 2x + 12
Y2 = 2(x + 3) + 6

$(-47, 47, -31, 31)$

The graphs are the same line. Since the inequality includes an equals sign, all ordered pairs on the line will determine solutions of the inequality. The solution set is the set of all real numbers.

HELPING HAND If the inequality in example 4c did not include an equals sign, no ordered pair would satisfy the inequality, so it would have no solution.

GRAPHICAL SOLUTIONS OF LINEAR INEQUALITIES IN ONE VARIABLE

In solving a linear inequality graphically where Y1 equals the expression on the left side of the inequality and Y2 equals the expression on the right side of the inequality, one of three possibilities will occur.

1. **A solution set exists.**

2. **No solution exists.**
   This occurs for the following conditions:
   Y1 < Y2, and the graph of Y1 is the same as Y2 or always above Y2.
   Y1 > Y2, and the graph of Y1 is the same as Y2 or always below Y2.
   Y1 ≤ Y2, and the graph of Y1 is always above Y2.
   Y1 ≥ Y2, and the graph of Y1 is always below Y2.

3. **A solution set of all real numbers exists.**
   This occurs for the following conditions:
   Y1 < Y2, and the graph of Y1 is always below Y2.
   Y1 > Y2, and the graph of Y1 is always above Y2.
   Y1 ≤ Y2, and the graph of Y1 is the same as Y2 or always below Y2.
   Y1 ≥ Y2, and the graph of Y1 is the same as Y2 or always above Y2.

 7.2.2 Checkup

Solve graphically.

1. $2x - 5 > 4x - 3$ **2.** $5z + 2 \leq 3z$

3. $4(x + 2) - 3(x + 3) < 7 - 3x$

Use the Intersect function on your calculator to find the exact intersection, and determine the solution set.

4. $3(2x - 1) - (2x + 4) < 2(x - 2) + 2$

In exercises 5–7, solve graphically to determine whether there is no solution or the solution set is all real numbers.

5. $3(x + 2) - 4(2x + 1) > 3(2 - x) + 2(1 - x)$

6. $3x - (2x + 5) \geq \frac{1}{2}(2x + 4) - 7$

7. $\frac{1}{2}x + 2 > x - (\frac{1}{2}x + 1)$

8. Explain how you would solve a linear inequality graphically.

9. How would you know from the graphic solution that the solution set is a part of the number line?

10. How would you know from the graphic solution that the solution set is the entire number line or that there is no solution?

7.2.3 Solving Linear Inequalities Algebraically

To solve a linear equation algebraically, we used the addition and multiplication properties of equations (Section 4.2). To solve a linear inequality algebraically, we will need to know similar properties of inequalities. First, let's see if an addition property of inequalities exists. Complete the following set of exercises.

Discovery 3

Addition Property of Inequalities

1. Given the inequality $10 < 12$, add 2 to both expressions and check whether the new expression is also true.

$$
\begin{array}{c}
10 < 12 \\
\hline
\begin{array}{c|c}
10 + 2 & 12 + 2 \\
12 & 14 \quad \text{\small True}
\end{array}
\end{array}
$$

2. Given the inequality $10 < 12$, add -2 to both expressions and check whether the new expression is also true.

3. Given the inequality $10 < 12$, subtract 2 from both expressions and check whether the new expression is also true.

4. Given the inequality $10 < 12$, subtract -2 from both expressions and check whether the new expression is also true.

Write a rule for the addition property of inequalities.

In each of the preceding exercises, we began with a true inequality and added (or subtracted) the same number to (or from) both expressions. The result remained a true inequality. We see that if a number is added to (or subtracted from) both expressions in an inequality, the result is a true inequality.

ADDITION PROPERTY OF INEQUALITIES
Given expressions a, b, and c,

$$\text{if } a < b, \text{ then } a + c < b + c$$
$$\text{if } a > b, \text{ then } a + c > b + c$$
$$\text{if } a \le b, \text{ then } a + c \le b + c$$
$$\text{if } a \ge b, \text{ then } a + c \ge b + c$$

This property holds true for subtraction as well, because subtraction is defined as adding the opposite of a number.

Now let's consider a property of inequalities involving multiplication. Complete the following set of exercises.

Discovery 4

Multiplication Property of Inequalities

1. Given the inequality $10 < 12$, multiply both expressions by 2 and check whether the new expression is also true.

$$
\begin{array}{c}
10 \ < \ 12 \\
\hline
\begin{array}{c|c}
10 \cdot 2 & 12 \cdot 2 \\
20 & 24 \quad \text{\small True}
\end{array}
\end{array}
$$

(Continued on page 482)

2. Given the inequality $10 < 12$, multiply both expressions by -2 and check whether the new expression is also true.

3. Given the inequality $10 < 12$, divide both expressions by 2 and check whether the new expression is also true.

4. Given the inequality $10 < 12$, divide both expressions by -2 and check whether the new expression is also true.

Write a rule for the multiplication property of inequalities.

In each of the preceding exercises, we began with a true inequality and multiplied (or divided) both expressions by the same number. When a positive number was used, the result was a true inequality. When a negative number was used, the result was a false inequality. To make the false inequality true, the inequality symbol must be reversed from "less than" to "greater than" or vice versa.

MULTIPLICATION PROPERTY OF INEQUALITIES

Given expressions a, b, and c with $c > 0$,

$$\text{if } a < b, \text{ then } ac < bc$$
$$\text{if } a > b, \text{ then } ac > bc$$
$$\text{if } a \leq b, \text{ then } ac \leq bc$$
$$\text{if } a \geq b, \text{ then } ac \geq bc$$

Given expressions a, b, and c, with $c < 0$,

$$\text{if } a < b, \text{ then } ac > bc$$
$$\text{if } a > b, \text{ then } ac < bc$$
$$\text{if } a \leq b, \text{ then } ac \geq bc$$
$$\text{if } a \geq b, \text{ then } ac \leq bc$$

This property holds true for division as well, because division is defined as multiplying by the reciprocal of a number.

We are now ready to solve linear inequalities algebraically. To do this, we will need to apply a combination of the properties of inequalities. Since there are several different ways to solve linear inequalities, we will set up a few rules so that, at least in the beginning, we are following the same steps. When we become more sure of ourselves, we may follow these steps in different orders and obtain the same results in the end.

SOLVING A LINEAR INEQUALITY IN ONE VARIABLE ALGEBRAICALLY

To solve a linear inequality by using a combination of properties of inequalities,

- Simplify both expressions in the inequality (preferably, leaving no fractions).

- Isolate the variable in one expression of the inequality (preferably, the expression on the left side) by using the addition property of inequalities.

- Use the addition property of inequalities to isolate the constants in the other expression (preferably, the expression on the right side).
- Use the multiplication property of inequalities to reduce the coefficient of the variable to 1.

Remember to reverse the inequality symbol if a negative number is being used to multiply (or divide) both sides of the inequality.

EXAMPLE 5   Solve. Represent the solution set in the notation given.

a. $2x + 3 > 7$; inequality notation

b. $5x + 4 \geq 6x - 16$; number line

c. $\dfrac{3}{8}\left(x + \dfrac{1}{4}\right) > \dfrac{5}{6}x + 2$; interval notation

Algebraic Solution

a. $2x + 3 > 7$

$2x + 3 - 3 > 7 - 3$ Subtract 3 from both sides.

$2x > 4$ Simplify.

$\dfrac{2x}{2} > \dfrac{4}{2}$ Divide both sides by 2.

$x > 2$

The solution set is the set of all x that satisfy $x > 2$.

b. $5x + 4 \geq 6x - 16$

$5x + 4 - 6x \geq 6x - 16 - 6x$ Subtract 6x from both sides.

$-x + 4 \geq -16$ Simplify.

$-x + 4 - 4 \geq -16 - 4$ Subtract 4 from both sides.

$-x \geq -20$ Simplify.

$\dfrac{-x}{-1} \leq \dfrac{-20}{-1}$ Divide both sides by −1 (the coefficient of x). Remember to reverse the inequality symbol.

$x \leq 20$ Simplify.

A number-line representation is shown.

$$\begin{array}{ccccccccccc} 15 & 16 & 17 & 18 & 19 & 20 & 21 & 22 & 23 & 24 & 25 \end{array}$$

c. $\dfrac{3}{8}\left(x + \dfrac{1}{4}\right) > \dfrac{5}{6}x + 2$

$\dfrac{3}{8}x + \dfrac{3}{32} > \dfrac{5}{6}x + 2$ Distribute.

$96\left(\dfrac{3}{8}x + \dfrac{3}{32}\right) > 96\left(\dfrac{5}{6}x + 2\right)$ Multiply both sides by 96, the LCD of 8, 32, and 6.

$96\left(\dfrac{3}{8}x\right) + 96\left(\dfrac{3}{32}\right) > 96\left(\dfrac{5}{6}x\right) + 96(2)$ Distribute.

$36x + 9 > 80x + 192$ Simplify.

$$36x + 9 - 80x > 80x + 192 - 80x$$ Subtract 80x from both sides.

$$-44x + 9 > 192$$ Simplify.

$$-44x + 9 - 9 > 192 - 9$$ Subtract 9 from both sides.

$$-44x > 183$$ Simplify.

$$\frac{-44x}{-44} < \frac{183}{-44}$$ Divide both sides by -44. Remember to reverse the inequality symbol.

$$x < -\frac{183}{44}$$ Simplify.

The solution set is $\left(-\infty, -\frac{183}{44}\right)$.

EXAMPLE 6   Solve $-2 < 3x - 5 \leq 7$ algebraically.

Algebraic Solution

In order to solve this compound inequality, we will isolate the variable to the middle expression.

$$-2 < 3x - 5 \leq 7$$

$$-2 + 5 < 3x - 5 + 5 \leq 7 + 5$$ Add 5 to all expressions.

$$3 < 3x \leq 12$$ Simplify.

$$\frac{-3}{3} < \frac{3x}{3} \leq \frac{12}{3}$$ Divide all expressions by 3.

$$-1 < x \leq 4$$ Simplify.

As we mentioned earlier, linear equations may have no solution or all-real-number solutions. In Chapter 4, we algebraically determined that a linear equation may have no solution or a solution set of all real numbers.

When we attempted to isolate the variable to one expression, the variable was deleted. The resulting equation was a contradiction. Therefore, the equation had no solution. Similarly, if the resulting inequality is not true, it has no solution.

When we attempted to isolate the variable to one expression, the variable was deleted. The resulting equation was an identity. Therefore, the solution set was the set of all real numbers. A linear inequality has a solution set of all real numbers if the variable is deleted and the resulting inequality is always true.

EXAMPLE 7   Solve algebraically.

a. $4(x + 2) < 4x - 10$ **b.** $3x + 4 \leq 2x + (x + 4)$

Algebraic Solution

a. $$4(x + 2) < 4x - 10$$

$$4x + 8 < 4x - 10$$ Distribute 4.

$$4x + 8 - 4x < 4x - 10 - 4x$$ Subtract 4x from both sides.

$$8 < -10$$ Simplify.

Since this is a false inequality, it has no solution.

b.
$$3x + 4 \le 2x + (x + 4)$$

$$3x + 4 \le 2x + x + 4 \qquad \text{Remove parentheses.}$$

$$3x + 4 \le 3x + 4 \qquad \text{Simplify.}$$

$$3x + 4 - 3x \le 3x + 4 - 3x \qquad \text{Subtract 3x from both sides.}$$

$$4 \le 4 \qquad \text{Simplify.}$$

Since this is a true inequality, the solution set is all real numbers.

ALGEBRAIC SOLUTIONS OF LINEAR INEQUALITIES IN ONE VARIABLE

In solving a linear inequality algebraically, one of three possibilities will occur.

1. **A solution set exists.**

2. **No solution exists.**
   The solution process results in a false inequality.

3. **A solution set exists and it is the set of all real numbers.**
   The solution process results in a true inequality.

7.2.3 Checkup

Solve exercises 1–6 algebraically. Represent the solution in the notation specified.

1. $5x - 8 > 2$; interval notation

2. $3z + 7 \le 7z - 5$; inequality notation

3. $\dfrac{2}{3}\left(x - \dfrac{1}{5}\right) < \dfrac{1}{4}x + 3$; number line

4. $3(x + 2) - 4(2x + 1) > 3(2 - x) + 2(1 - x)$; inequality notation

5. $3x - (2x + 5) \ge \dfrac{1}{2}(2x + 4) - 7$; interval notation

6. $-4 \le \frac{1}{2}x + 3 < 2$; number line

7. In solving a linear inequality algebraically, which property should be used first to isolate the variable to one side of the inequality?

8. How would you know a linear inequality has no solution when you attempt to solve it algebraically?

9. How would you know that the solution set of a linear inequality is the set of all real numbers when you solve it algebraically?

7.2.4 Modeling the Real World

In real-world situations, the terms *less than or equal to* and *greater than or equal to* may be expressed in a variety of ways. For example, two expressions that are used frequently are *at least* (for *greater than or equal to*) and *at most* (for *less than or equal to*). Situations involving these terms or phrases like them can be analyzed as inequalities and then solved graphically or algebraically as we have described. The same rules apply about including a solution point if the relation includes equality and about reversing the direction of an inequality if you need to multiply or divide by a negative number.

EXAMPLE 8 Eata wants to make a *B* in her algebra class. The grading scale states that the range for a *B* is 85–93. Her first four test grades were 93, 100, 88, and 87. Determine the range of grades she must score on her last test in order to make a *B*.

Solution

Let x = fifth test grade

The average of the test grades must be between 85 and 93, inclusive, in order for Eata to make a B.

Algebraic

$$85 \leq \frac{93 + 100 + 88 + 87 + x}{5} \leq 93$$

$$85 \leq \frac{368 + x}{5} \leq 93 \qquad \text{Simplify.}$$

$$5(85) \leq 5\left(\frac{368 + x}{5}\right) \geq 5(93) \qquad \text{Multiply by 5.}$$

$$425 \leq 368 + x \leq 465 \qquad \text{Simplify.}$$

$$425 - 368 \leq 368 + x - 368 \leq 465 - 368 \qquad \text{Subtract 368.}$$

$$57 \leq x \leq 97 \qquad \text{Simplify.}$$

Eata must score between 57 and 97, inclusive, on her fifth test in order to make a B.

Graphic

$$Y2 = \frac{93 + 100 + 88 + 87 + x}{5}$$

$$Y3 = 93$$

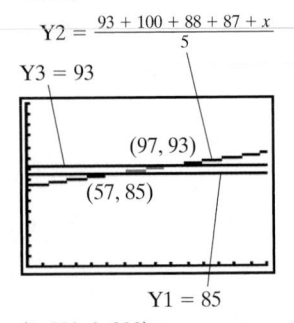

$(97, 93)$

$(57, 85)$

$Y1 = 85$

$(0, 200, 0, 200)$

The intersections are $(57, 85)$ and $(97, 93)$. The portion of the graph of Y2 between Y1 and Y3 is between the intersections. The solution includes the x-coordinates of the intersections. The solution set is the set of all x that satisfy $57 \leq x \leq 97$.

EXAMPLE 9 Jean's Handicrafts sells Christmas wreaths for $25.00 each. Jean estimates that the cost of materials needed to make each wreath is $15.50. The cost of the advertisement is $10.00. If Jean must make a profit of at least $100.00, how many wreaths must she sell?

Solution

Let $\quad x$ = the number of wreaths made and sold
$\quad R(x)$ = the total revenue
$\quad C(x)$ = the total cost
$\quad P(x)$ = the total profit

$$R(x) = 25.00x$$
$$C(x) = 15.50x + 10.00$$

The total profit is the difference of the total revenue and the total cost.

$$P(x) = R(x) - C(x)$$
$$P(x) = 25.00x - (15.50x + 10.00)$$
$$P(x) = 25.00x - 15.50x - 10.00$$
$$P(x) = 9.50x - 10.00$$

Jean must have a profit of at least $100.00, or $P(x) \geq 100.00$.

$$9.50x - 10.00 \geq 100.00$$

Algebraic

$$9.50x - 10.00 \geq 100.00$$

$$9.50x - 10.00 + 10.00 \geq 100.00 + 10.00 \quad \text{Add 10.00.}$$

$$9.50x \geq 110.00 \quad \text{Simplify.}$$

$$\frac{9.50x}{9.50} \geq \frac{110.00}{9.50} \quad \text{Divide by 9.50.}$$

$$x \geq 11.579 \quad \text{(Rounded to the nearest thousandth.)}$$

Jean must make and sell 12 wreaths to make a profit of $100.00.

Graphic

$Y1 = 9.50x - 10.00$

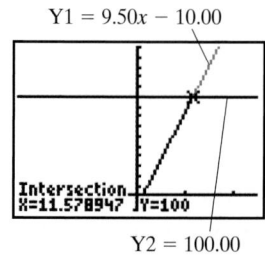

$Y2 = 100.00$

$(-25, 25, -150, 150)$

The solution checks.

APPLICATION

The body mass index (BMI) is a measure of weight that takes height into account. The BMI of a person 62 inches tall and weighing x pounds is given by the expression $0.2x - 2$. A doctor may prescribe a certain drug if a person has a BMI of 30 or greater. Determine the weights for which a doctor may prescribe this drug for a person 62 inches in height.

Write an inequality such that the BMI is greater than or equal to 30.

$$0.2x - 2 \geq 30$$

$$0.2x - 2 + 2 \geq 30 + 2 \quad \text{Add 2 to both sides.}$$

$$0.2x \geq 32 \quad \text{Simplify.}$$

$$\frac{0.2x}{0.2} \geq \frac{32}{0.2} \quad \begin{array}{l}\text{Divide both sides by}\\\text{0.2.}\end{array}$$

$$x \geq 160$$

Discussion

Let x = the weight in pounds of a person 62 inches tall

A doctor may prescribe a certain drug if a person is 62 inches tall and weighs 160 pounds or more.

✓ 7.2.4 Checkup

1. A disk exchange store sells used CD's for $6.99 each. The store buys used CD's for $2.99 each. The average daily cost for having the store open for business is $125.00. Assuming that the store buys and sells the same number of CD's in a day, how many CD's must the store buy and sell in order to make a profit of at least $75.00 per day?

2. After the birth of her baby, Jana counted calories to manage her weight. She wanted to average 1300–1350 calories per day. If her intake for the first six days of the week was 1400, 1100, 1200, 1050, 1450, and 1500 calories, what number of calories can she have on the seventh day to achieve her goal?

3. If a person is 6 feet tall, the body mass index (BMI) is approximated by the expression $0.14x - 1$, where x is the person's weight in pounds. If the person has risk factors for heart disease, medication may be prescribed if his or her BMI is at least 27. Determine the weights for which medication may be prescribed.

7.2 Exercises

Solve numerically for integer solutions.

1. $6(x - 12) < 3x$

2. $3x + 6 > 0$

3. $3(x + 5) + 3 > 3x + 18$

4. $x + 5 > 2(x - 1) - (x - 1)$

5. $3x + 3 > x + 2$

6. $3x + 5 < 2x - 1$

7. $2(4x + 2) + 2x - 7 < 2(5x + 4)$

8. $a + 3(a + 2) < 6(a + 1) - 2a$

Solve graphically.

9. $\dfrac{2}{3}x - \dfrac{2}{3} \geq -\dfrac{3}{4}x - \dfrac{7}{2}$

10. $\dfrac{1}{4}x - \dfrac{5}{2} \leq -\dfrac{1}{2}x - 1$

11. $2(x + 3) - (x - 1) > 8 - (x + 9)$

12. $3(x - 2) - (x - 5) > 7 - (x + 2)$

13. $0.4x - 3.2 \le -0.6x - 0.2$

14. $0.2(2x + 1) \ge -0.2(x - 7)$

15. $2(x + 1) > 3(x - 1) - x$

16. $5 - (2x + 5) > 2(2x - 1) - 2(3x + 1)$

17. $x - (3x + 1) > -x - (x + 1)$

18. $3(x - 1) - 5x > 2(1 - x)$

Solve algebraically. Represent the solution in interval notation.

19. $4x + 12 > 0$

20. $-x - 27 > 11$

21. $-3x + 12 \ge 12$

22. $36 \ge 5x + 8$

23. $-7x - 12 < -26$

24. $-3x + 11 < -10$

25. $15.17 < 5.9x - 4.3$

26. $-4.22c - 0.4 < -21.5$

27. $6.1 > -0.55a + 6.1$

28. $3.05y + 0.09 > 31.2$

29. $2.07z + 4.12 \ge 16.54$

30. $-6.3p + 1.5 \ge -4.8$

31. $-\dfrac{5}{9}b + \dfrac{11}{12} < \dfrac{23}{36}$

32. $-\dfrac{4}{7}z + \dfrac{3}{14} > \dfrac{5}{14}$

33. $156z - 210 > 47z + 662$

34. $728a + 958 < 116a - 878$

35. $-1.05x - 15.41 < 2.55x - 47.09$

36. $21.1x + 0.46 > 10.9x + 0.46$

37. $11x + \dfrac{1}{4} \le 2x + \dfrac{7}{36}$

38. $5p + \dfrac{3}{8} \ge 3p - \dfrac{3}{8}$

39. $\dfrac{2}{5}b - 12 < \dfrac{2}{3}b + 20$

40. $\dfrac{7}{9}x - 15 > \dfrac{4}{9}x - 37$

41. $4x - (3x + 5) < x - 5$

42. $7x + 4 > 3x + 2(2x + 2)$

43. $4x - (3x + 5) \le x - 5$

44. $7x + 4 \ge 3x + 2(2x + 2)$

45. $0.05x + 10.5 < 0.15x - 0.25$

46. $0.01x + 0.11 > 0.47 - 0.09x$

47. $5 < 4 - 3x \le 11$

48. $-2 \le 5x + 3 < 4$

49. $-4 \le 3(x + 1) - 5 \le 7$

50. $2 < 6 - 2(x + 5) < 7$

Write a linear inequality to represent each application and solve.

51. Lee is trying to earn an *A* in his algebra class. To do so, he must have an average of no less than 93 points. He scored 93, 97, 92, 89, and 95 on his first five tests. What range of scores on his last test will earn an *A*?

52. Beckie types student papers for a fee. If her earnings were $38, $62, $56, and $42 for the first four weeks of the term, how much must she earn for the fifth week in order to average more than $50 per week?

53. Luigi is retired, but has a part-time job. He will lose some of his retirement benefits if he earns more than $7500 per year. He makes $9.75 per hour at his job. If he has already earned $5200 this year, how many hours can he work without losing benefits?

54. Bobby rents a stump grinder for $22.00 plus $3.50 per hour. How many hours can he use the grinder and spend no more than $55.00?

55. Elevators must post a weight limit, usually a maximum of 2000 pounds. If the average person weighs 165 pounds, how many can safely ride the elevator?

56. Judy has $120.00 to spend on clothing. She buys a slack suit for $87.50. She would like to buy some sweaters that sell for $12.50 each. How many sweaters can she buy and stay within her budget?

57. Angie has invited 120 people to her wedding reception and plans to serve both vegetarian and meat entrée plates. It will cost her $30.00 per person to serve a vegetarian plate and $35.00 per person to serve a meat entrée plate. If she plans to spend no more than $4000.00 for the dinner, how many meat entrée plates can she order?

58. Hervis leases cars for $25.00 a day and $0.22 per mile. Artz leases cars for $15.00 a day and $0.35 per mile. If you rent a car from Artz, how many miles can you drive per day and still spend less than Hervis would charge?

59. Mike wants to enclose a rectangular garden by using a barn as one of the lengths, with fencing for the other three sides. He wants the length to be 30 feet more than $\frac{3}{4}$ of the width. He has a maximum of 185 feet of fencing and a gate that is 4 feet wide. Find the possible widths for the garden.

60. A rectangular swimming pool is to have a perimeter that does not exceed 240 feet. If the length is to be 15 feet more than twice the width, what widths would satisfy that condition?

61. A photo-processing kiosk charges $0.25 for each print that it produces. The cost of producing the print is $0.08 per print. The overhead cost is $45.00 per day to rent and staff the kiosk. How many prints must be produced each day to realize a profit of at least $50.00 per day?

62. A coffee shop averages a charge of $3.55 per customer for an order. On the average, it costs the shop $1.25 to fill the order. The daily overhead cost of operating the shop is $165.00. How many orders must be sold each day to make a profit of at least $150.00 per day?

63. Daily high temperatures in Nashville, Tennessee, in the month of July were 82, 83, 88, and 92 degrees Fahrenheit. What would the temperature need to register on the fifth day if the average high temperature for the five days was between 82 and 88 degrees Fahrenheit?

64. The closing price of stock for Cisco Systems, Inc., for the first three Fridays in July of 2001 were $16.79, $18.74, and $17.99. What must the closing price be for the next Friday if the average closing price on Friday was between $18.00 and $20.00?

65. A person that is 5 feet 8 inches tall has a body mass index (BMI) that is approximated by the expression $0.146x + 1$, where x is the person's weight in pounds. If the person has no risk factors for heart disease except being over-weight, a doctor will not usually prescribe medication if the BMI is below 30. Determine the weights for which the doctor may not prescribe medication.

66. A person who is 6 feet 2 inches tall has a BMI that is approximated by the expression $0.125x + 0.7$, where x is

the person's weight in pounds. With other risk factors for heart disease, a doctor may choose not to prescribe medication, so long as the person's BMI remains below 27. Determine the weights for which the doctor may choose not to prescribe medication.

67. The total expenditures (federal, state, and local) for social welfare can be estimated by the linear function

$$t(n) = (6.95 \times 10^{10})n + (4.36 \times 10^{11})$$

where n is the number of years since 1980 and $t(n)$ is the total cost in dollars for the year $1980 + n$. Use this function to estimate the number of years in which total expenditures exceed 1.5 trillion dollars (1.5×10^{12}).

68. The total expenditures (state and local) for social welfare can be estimated by the linear function

$$t(n) = (2.95 \times 10^{10})n + (1.63 \times 10^{11})$$

where n is the number of years since 1980 and $t(n)$ is the total cost in dollars for the year $1980 + n$. Use this function to estimate the number of years in which total expenditures exceed 1.0 trillion dollars (1.0×10^{12}).

7.2 Calculator Exercises

In this section, you were shown how to graphically solve an inequality using the calculator. Another way to use your calculator to graphically solve an inequality is to store the expression on the right in Y1 and the expression on the left in Y2 and then graph Y3 = Y1 − Y2 to determine the solution set. If you seek values of x that make Y1 < Y2 (or equivalently, Y1 − Y2 < 0), find the interval where Y3 is below the x-axis (Y3 < 0). If you seek values of x that make Y1 > Y2, find the interval where Y3 is above the x-axis.

For example, use your calculator to solve $2x + 3 > x + 5$ graphically.

• Set the screen to the desired window.

• Enter the expression on the left side of the inequality in Y1.

| Y= | 2 | X,T,θ,n | + | 3 |

• Enter the expression on the right side of the inequality in Y2.

| ▼ | X,T,θ,n | + | 5 |

• Set Y3 equal to Y1 − Y2.

| ▼ | VARS | ▶ | 1 | | 1 | − | VARS | ▶ | 1 |
| 2 |

• Turn off Y1 and Y2.

• Graph Y3.

| GRAPH |

Since we want Y1 > Y2, find where Y3 is above the x-axis. This occurs where $x > 2$, which can be determined by using the | TRACE | key or by using the | 2nd | | CALC | | 2 | keys to find the "zero" of the graph (that is, the point at which the graph crosses the x-axis).

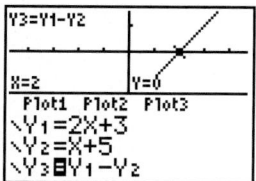

Try this approach on some of the exercises in this section. You must still decide whether to include the bound in your solution set.

7.2 Writing Exercises

Solve the following two inequalities. Discuss the differences in the forms of the inequalities, and explain how these differences affect the solutions of the inequalities. What have you learned from this exercise?

1. $4x - (2x + 5) > 2x - 5$
2. $4x - (2x + 5) \geq 2x - 5$

7.3 Linear Inequalities in Two Variables

OBJECTIVES

1 Identify linear inequalities in two variables.

2 Solve linear inequalities in two variables numerically.

3 Graph linear inequalities in two variables.

4 Graph the two special cases of linear inequalities in two variables.

5 Model real-world situations by using linear inequalities in two variables.

APPLICATION

A portion of the United Parcel Service (UPS) Next Day Air domestic delivery rate chart for customers who receive daily pickup (effective February 5, 2001) is as follows:

Weight, lb	Zone 102	Zone 103	Zone 104	Zone 105
1	14.50	16.50	19.50	21.00
2	15.50	17.50	21.75	23.50
3	16.75	18.50	24.25	26.00
4	17.75	19.75	26.50	28.25
5	18.95	20.75	28.75	30.75
6	19.50	22.00	31.00	33.25
7	20.25	23.25	33.00	35.50
8	21.00	24.25	35.00	38.00
9	21.75	25.25	37.25	40.25
10	22.50	26.25	39.50	42.50

Determine the number of 5-pound packages and the number of 10-pound packages that you can ship to Zone 104 and stay within a shipping budget of $1500.00.

After completing this section, we will discuss this application further. See page 500.

7.3.1 Identifying Linear Inequalities in Two Variables

In this section, we will be solving linear inequalities in two variables. Before we begin, we need to identify a linear inequality in two variables. A **linear inequality in two variables** is written by replacing the equals sign in a linear equation in two variables with an order symbol.

STANDARD FORM FOR A LINEAR INEQUALITY IN TWO VARIABLES

A linear inequality in two variables is an inequality that can be written in one of the forms

$$ax + by < c$$
$$ax + by > c$$
$$ax + by \leq c$$
$$ax + by \geq c$$

where a, b, and c are real numbers and a and b are not both equal to 0.

For example,

Standard form: $ax + by \enspace < c$

1. $2x + 5y < 0$ $a = 2$ $b = 5$ $c = 0$
2. $3x - y < 12$ $a = 3$ $b = -1$ $c = 12$
3. $2x < 8$ $a = 2$ $b = 0$ $c = 8$
4. $3y < -9$ $a = 0$ $b = 3$ $c = -9$

Each of these inequalities is in standard form. Note that each of the variables, x and y, is raised to the first power. This *must* be the case for a linear inequality in two variables.

However, the inequality $y < 5x + 7$ is also a linear inequality in two variables, because we can use the properties of inequalities to rearrange it into standard form. For example,

$$y < 5x + 7$$
$$y - 5x < 5x + 7 - 5x \qquad \text{Subtract 5x from both sides.}$$
$$y - 5x < 7 \qquad \text{Simplify.}$$
$$-5x + y < 7 \qquad \text{Rearrange the left side.}$$

EXAMPLE 1   Identify each inequality as linear or nonlinear. Express each linear inequality in standard form.

a. $2x^2 + 3y > 4$ **b.** $y \geq -3x$ **c.** $x \leq 0$

Solution

a. Nonlinear, because the x-term is squared.

b. Linear

In standard form,

$$y \geq -3x$$
$$y + 3x \geq -3x + 3x$$
$$y + 3x \geq 0$$
$$3x + y \geq 0$$

c. Linear

In standard form, $x \leq 0$.

7.3.1 Checkup

Identify each inequality as linear or nonlinear. Express each linear inequality in standard form.

1. $y > 7x - 10$ **2.** $\frac{1}{3}x < y + 4$

3. $2x + 4 > 5y^2 - 3$ **4.** $5x + 2y < x + y + 10$

5. $3x - 4xy - 6y < 12$ **6.** $y < -\frac{3}{x} + 7$

7.3.2 Solving Linear Inequalities Numerically

To solve a linear inequality in two variables, we will perform the same procedure we used in previous sections. That is, we will substitute the values for the two variables into the inequality and determine whether the inequality is true. A solution of a linear inequality in two variables is an ordered pair that produces a true statement (that is, that satisfies the inequality). For example, in solving $y \leq x + 5$, we can obtain many possible solutions.

$(4, 9)$ is a solution because, when $x = 4$ and $y = 9$, the inequality results in $9 \leq 9$.

$(5, 7)$ is a solution because, when $x = 5$ and $y = 7$, the inequality results in $7 \leq 10$.

$(0.5, 3)$ is a solution because, when $x = 0.5$ and $y = 3$, the inequality results in $3 \leq 5.5$.

To solve the linear inequality numerically, we begin the same way we did with all other numeric solutions: by completing a table of values. For example, in solving $y \leq x + 5$, we write the following table:

x	$y \leq x + 5$	y	
-3	$y \leq -3 + 5$ $y \leq 2$	$y \leq 2$	This means that when x = −3, we have values of y less than or equal to 2.
-2	$y \leq -2 + 5$ $y \leq 3$	$y \leq 3$	This means that when x = −2, we have values of y less than or equal to 3.
-1	$y \leq -1 + 5$ $y \leq 4$	$y \leq 4$	This means that when x = −1, we have values of y less than or equal to 4.
0	$y \leq 0 + 5$ $y \leq 5$	$y \leq 5$	This means that when x = 0, we have values of y less than or equal to 5.
1	$y \leq 1 + 5$ $y \leq 6$	$y \leq 6$	This means that when x = 1, we have values of y less than or equal to 6.
2	$y \leq 2 + 5$ $y \leq 7$	$y \leq 7$	This means that when x = 2, we have values of y less than or equal to 7.

We can see that the number of solutions is infinite. A table of values is not the most convenient method for expressing all the solutions. Therefore, we will not solve any examples with the numeric method.

7.3.2 Checkup

What do we mean by the solution of a linear inequality in two variables?

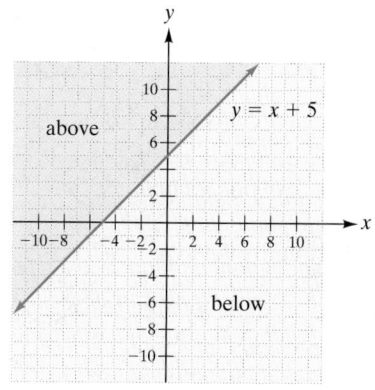

7.3.3 Graphing Linear Inequalities

We used a rectangular coordinate system and graphical methods to illustrate the solutions of linear equations in two variables because the number of solutions was infinite and the solutions could not be listed numerically. We will do the same for linear inequalities in two variables.

To graph an inequality means to create an illustration that represents the solutions of the inequality. Therefore,

1. Every solution of an inequality can be represented by a point on its graph.

2. Every point on a graph represents a solution of its inequality.

The graph of an equation partitions the coordinate plane into three regions: (1) the coordinate pairs that lie on the line that represents the equation; (2) the coordinate pairs above the line; and (3) the coordinate pairs below the line. (For example, see the graph of the equation $y = x + 5$ at the left.)

If we graph the solutions of $y \leq x + 5$ from the table of values found in the previous section, we obtain a series of vertical-line solutions of the inequality. Each line starts at a boundary point on the line representing $y = x + 5$ and extends below the line representing $y = x + 5$.

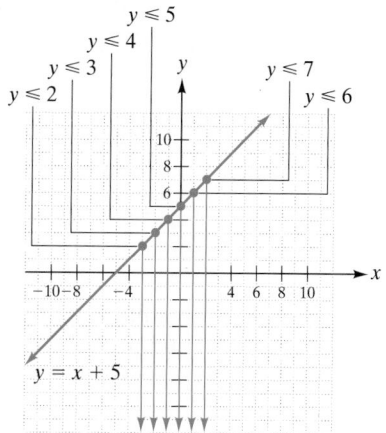

To illustrate the solutions of $y \leq x + 5$, we graph the line for $y = x + 5$ and shade all points below the line.

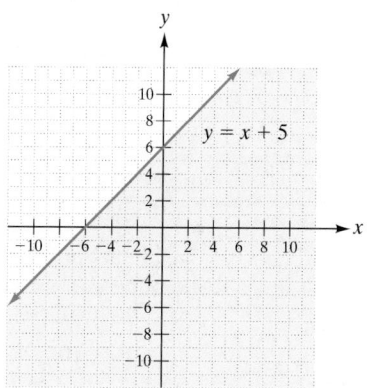

While this method illustrates why we shade a region on one side of the boundary line, it is not the easiest way to graph an inequality. To discover a more straightforward method of graphing a "less than or equal to" inequality, complete the following set of exercises.

Discovery 5

Graphing a "Less Than or Equal To" Linear Inequality in Two Variables

Graph the line determined by the equation $y = 2x + 4$.

1. Use Trace to determine the coordinates of points on the line.
   a. List two of these ordered pairs.
   b. Are the ordered pairs solutions of the inequality $y \le 2x + 4$?

2. Clear the trace, and use the free-moving cursor (arrow keys) to determine ordered pairs above the line.
   a. List two of these ordered pairs.
   b. Are the ordered pairs solutions of the inequality $y \le 2x + 4$?

3. Now use the free-moving cursor (arrow keys) to determine ordered pairs below the line.
   a. List two of these ordered pairs.
   b. Are the ordered pairs solutions of the inequality $y \le 2x + 4$?

Write the rule for graphing a "less than or equal to" linear inequality.

First, we see that the inequality is solved for y. The ordered pairs on the graphed line are solutions of the equation and the "equal to" part of the inequality. We use a solid line to indicate that the ordered pairs on the line are included in the solution. (This is comparable to using a solid dot on a number line for a linear inequality in one variable.) Second, we see that the ordered-pair solutions of the "less than" portion of the inequality are found in the region below the graphed line. We shade this region to indicate that they are included in the solution.

To discover a method for graphing an inequality that does not include equality, complete the following set of exercises.

Discovery 6

Graphing a "Greater Than" Linear Inequality in Two Variables

Graph the line determined by the equation $y = 2x + 4$.

1. Use Trace to determine the coordinates of points on the line.
   a. List two of these ordered pairs.
   b. Are the ordered pairs solutions of the inequality $y > 2x + 4$?

2. Clear the trace, and use the free-moving cursor (arrow keys) to determine ordered pairs above the line.
   a. List two of these ordered pairs.
   b. Are the ordered pairs solutions of the inequality $y > 2x + 4$?

3. Now use the free-moving cursor (arrow keys) to determine ordered pairs below the line.

a. List two of these ordered pairs.
b. Are the ordered pairs solutions of the inequality $y > 2x + 4$?

Write the rule for graphing a "greater than" linear inequality.

First, we see that the inequality is solved for y. The ordered pairs on the graphed line are solutions of the equation, but not solutions of the inequality. We use a dashed line to indicate that these ordered pairs are not included in the solution. (This is comparable to using an open dot on a number line for a linear inequality in one variable.) We also see that the ordered-pair solutions of the "greater than" portion of the inequality are found in the region above the graphed line. We shade this region to indicate that these ordered pairs are included in the solution.

GRAPHING A LINEAR INEQUALITY IN TWO VARIABLES WITH A Y-TERM

To graph a linear inequality in two variables,

- Solve the inequality for y (for example, $y < ax + b$).
- Graph the boundary line determined by the related equation $y = ax + b$.
- Use a solid line when the inequality includes equality.
- Use a dashed line when the inequality does not include equality.
- Shade the correct portion of the coordinate plane determined by the inequality solved for y.
- Shade below the line for a "less than" inequality.
- Shade above the line for a "greater than" inequality.
- Check for the correct shading by choosing a test point in the shaded region to determine whether it is a solution of the inequality.

To graph a linear inequality in two variables on a calculator, we need to do the same as we do by hand.

 **HELPING HAND** The calculator does not graph dashed lines; it always includes the boundary in the graph. It will be up to you not to include the boundary in your graph when that is appropriate. In 7.3 Calculator Exercises, a procedure whereby dashed lines can be displayed is presented.

EXAMPLE 2   Graph.

a. $x - y < -5$ **b.** $6x + 2y \leq -10$

Solution

a. First, solve the inequality for y.

$$x - y < -5$$
$$x - y - x < -5 - x$$
$$-y < -x - 5$$
$$\frac{-y}{-1} > \frac{-x - 5}{-1} \qquad \text{Reverse the inequality when dividing by a negative number.}$$
$$y > x + 5$$

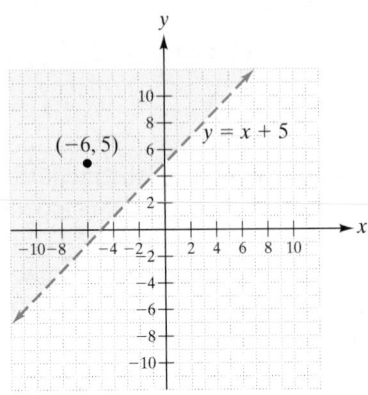

Graph the boundary line determined by $y = x + 5$. This boundary is a dashed line because there is no equality in the inequality. Since the inequality is "greater than," the boundary line is a lower boundary, so we shade above the boundary line. (Remember, after solving for y, we have a "greater than" inequality.)

Check a point in the shaded portion of the graph. A sample ordered pair is $(-6, 5)$. Substituting this solution into the original inequality results in

$$\begin{array}{c|c} x \quad - y < -5 \\ \hline (-6) - 5 & -5 \\ -11 & \end{array}$$

The sample ordered pair is a solution.

The calculator graph is shown in Figure 7.5b.

TECHNOLOGY

GRAPHING A LINEAR INEQUALITY IN TWO VARIABLES

Graph $x - y < -5$.

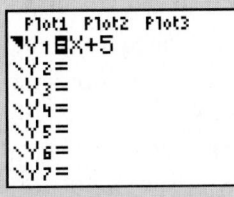

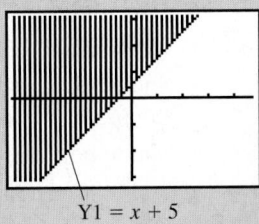

Y1 = x + 5
$(-47, 47, -31, 31)$

Figure 7.5a   Figure 7.5b

First solve for y: $y > x + 5$

For Figure 7.5a,

Enter the boundary equation $y = x + 5$ in Y1.

Since the inequality is "greater than," shade above the boundary line. Use the left arrow key to move left of Y1.

Choose the "shade above" option. The calculator will display a blinking triangle above a diagonal.

| ENTER | ENTER |

For Figure 7.5b,

Graph.

| GRAPH |

The calculator display shows the shaded graph; it does not show a dashed or solid boundary line.

 **HELPING HAND** After entering the boundary equation, to shade above press ENTER twice; to shade below press ENTER three times.

b. Solve the inequality for y.

$$6x + 2y \le -10$$
$$6x + 2y - 6x \le -10 - 6x$$
$$2y \le -6x - 10$$
$$\frac{2y}{2} \le \frac{-6x - 10}{2}$$
$$y \le -3x - 5$$

The boundary line is the graph of the equation $y = -3x - 5$. This boundary is a solid line because the inequality includes equality. The boundary line is an upper boundary for a "less than" inequality, so shade below the boundary.
    Check a point in the shaded region. (This is left for you.)

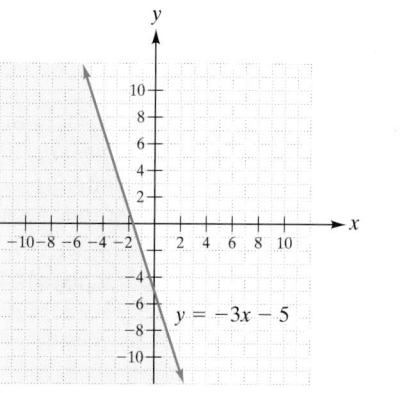

Figure 7.6a

Check

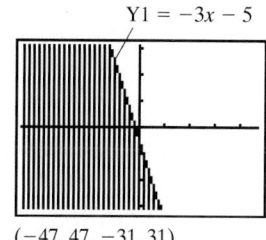

$(-47, 47, -31, 31)$

Figure 7.6b

Since the inequality is a "less than," we want to shade below the boundary line, as shown in Figure 7.6a and 7.6b.

 7.3.3 Checkup

Graph exercises 1 and 2. Check each solution on your calculator.

1. $3x - 2y > 6$ **2.** $y \le -3x + 7$

3. In graphing a linear inequality in two variables, when do you use a dashed boundary line and when do you use a solid one?

4. After graphing a linear inequality in two variables, how should you check that you shaded the proper region of the graph?

7.3.4 Graphing Special Cases of Linear Inequalities

The special cases of linear equations in two variables were the cases when $y = k$ or $x = h$ (Section 5.2). Let's graph inequalities related to each of these cases; that is, we wish to graph the linear inequality $ax + by < c$ when either $a = 0$ or $b = 0$, but not both.

EXAMPLE 3   Graph $y < 5$.

Solution

Graph the boundary line, $y = 5$, with a dashed line. This is a horizontal line. Shade the "less than" portion of the coordinate plane, which is below the upper boundary. Check a point in the shaded region. (This is left for you.)

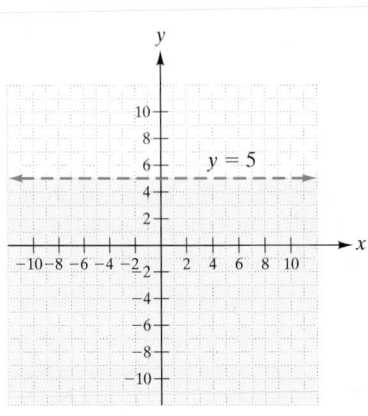

Check

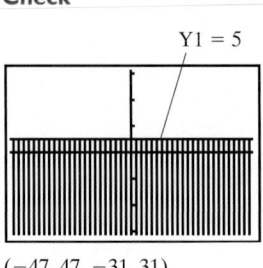

Y1 = 5

$(-47, 47, -31, 31)$

Inequalities with only a y-variable are graphed in the same manner as all other previous inequalities. However, the graph of an inequality not containing a y-variable must be determined in a different manner.

The ordered pairs on the graphed line $x = h$ are solutions of the equality portion of the inequality. The line $x = h$ divides the coordinate plane into two parts. Ordered-pair solutions of "less than" inequalities, $x < h$, are located in the region to the left of the graphed line (the right boundary). Ordered-pair solutions of "greater than" inequalities, $x > h$, are located to the right of the graphed line (the left boundary).

GRAPHING A LINEAR INEQUALITY IN TWO VARIABLES WITH NO Y-TERM
To graph a linear inequality in two variables with no y-term,

- Solve the inequality for x.
- Graph a vertical boundary line for the equation $x = h$.
  - Use a solid line if the inequality includes equality.
  - Use a dashed line if the inequality does not include equality.
- Shade the correct portion of the coordinate plane determined by the solved inequality.
  - Shade to the left of the boundary for a "less than" inequality.
  - Shade to the right of the boundary for a "greater than" inequality.
- Check for the correct shading by choosing a test point in the shaded region to determine whether that point is a solution of the inequality.

Since an inequality with no y-variable (such as $ax < 0$ or $ax > 0$) does not have a related function, we cannot use our previous calculator method to graph the inequality. (See 7.3 Calculator Exercises at the end of this section for a calculator method of solving the inequality).

EXAMPLE 4   Graph $x - 3 < 8$.

Solution

Solve for x.

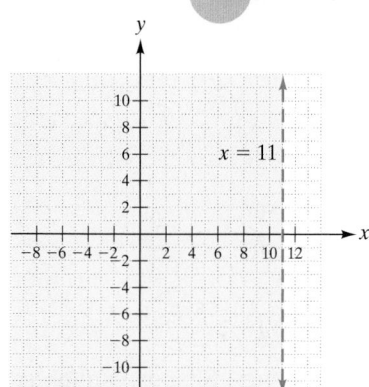

$$x - 3 < 8$$
$$x - 3 + 3 < 8 + 3$$
$$x < 11$$

Graph the boundary line $x = 11$ as a dashed line because $x < 11$ does not include equality. The boundary is a vertical line. Shade to the left of the boundary because the inequality is "less than."

Since $x = 11$ is not a function, we will choose not to use the calculator to graph the inequality.

 7.3.4 Checkup

Graph exercises 1 and 2.

1. $y + 3 > 8$ **2.** $5 - 2x \leq 9$

3. What must be true for a linear inequality in two variables if its graph has a vertical boundary line?

4. What must be true for a linear inequality in two variables if its graph has a horizontal boundary line?

7.3.5 Modeling the Real World

Many situations in the real world involve ranges of values rather than single numbers. For example, in business, you want sales income to be greater than cost, regardless of the actual numbers. You want to spend no more than your budget, whatever the budget is, and you want to produce as many items as you can, whatever that number is. Graphing inequalities can give you a picture of these ranges of values, so you can see whether you're close to your budget or you still have room to spend. As with most real-world situations, remember to check the domain of your variables, so that you don't graph negative values of items or time, for example.

EXAMPLE 5   Shanda is a temporary worker for Make or Break Manufacturing Co. She works at most an eight-hour shift with a 30-minute break. (On slow days, she may work shorter hours.) Shanda has been trained to work on two different assembly lines and may be assigned one or both during her shift. One assembly-line job requires 20 minutes per item. The other job requires 10 minutes per item.

a. Write a linear inequality in two variables needed to determine the numbers of items Shanda can complete during one shift.

b. Graph the inequality. Only positive coordinates found in the first quadrant would make sense in this situation. Why?

c. Determine from the graph two possible combinations of 20-minute and 10-minute items Shanda can produce during one shift.

Solution

a. Let x = number of items assembled requiring 20 minutes per item
y = number of items assembled requiring 10 minutes per item

The time required to assemble the 20-minute items is $20x$.
The time required to assemble the 10-minute items is $10y$.

The total time required to assemble both items is the sum $20x + 10y$. This sum is at most ($\leq$) 480 minutes (8 hours $\times$ 60 minutes per hour) minus the 30-minute break, or a total of 450 minutes.

$$20x + 10y \leq 450$$

b. First, solve the inequality for y.

$$20x + 10y - 20x \leq 450 - 20x$$
$$10y \leq -20x + 450$$
$$\frac{10y}{10} \leq \frac{-20x + 450}{10}$$
$$y \leq -2x + 45$$

Only positive values of x and y make sense, because Shanda cannot assemble a negative number of items.

Graph the boundary line, $y = -2x + 45$. This boundary is included as a solid line in the solution and is an upper boundary. Therefore, the region below the line is shaded.

On your calculator, use a window of $(0, 47, 10, 0, 62, 10, 1)$ to view the first quadrant.

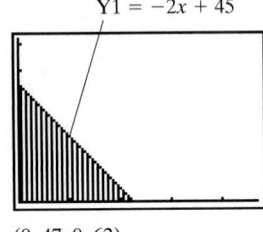

Y1 = −2x + 45

$(0, 47, 0, 62)$

c.

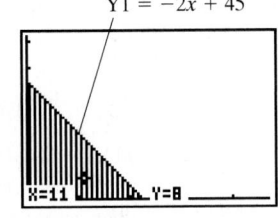

Y1 = −2x + 45   Y1 = −2x + 45

$(0, 47, 0, 62)$ $(0, 47, 0, 62)$

Shanda can produce any combination of 20-minute and 10-minute items determined by the ordered pairs found in the shaded region. For example, $(4, 25)$ and $(11, 8)$ are two ordered pairs in that region.

The ordered pair $(4, 25)$ means that Shanda can produce four 20-minute items and twenty-five 10-minute items during one shift.

The ordered pair $(11, 8)$ means that Shanda can produce eleven 20-minute items and eight 10-minute items during one shift.

APPLICATION

A portion of the United Parcel Service (UPS) Next Day Air domestic delivery rate chart for customers who receive daily pickup (effective February 5, 2001) is as follows:

Weight, lb	Zone 102	Zone 103	Zone 104	Zone 105
1	14.50	16.50	19.50	21.00
2	15.50	17.50	21.75	23.50
3	16.75	18.50	24.25	26.00
4	17.75	19.75	26.50	28.25
5	18.95	20.75	28.75	30.75
6	19.50	22.00	31.00	33.25

Weight, lb	Zone 102	Zone 103	Zone 104	Zone 105
7	20.25	23.25	33.00	35.50
8	21.00	24.25	35.00	38.00
9	21.75	25.25	37.25	40.25
10	22.50	26.25	39.50	42.50

Determine the number of 5-pound packages and the number of 10-pound packages that you can ship to Zone 104 and stay within a shipping budget of $1500.00.

Discussion

Let x = the number of 5-pound packages
$\quad\ y$ = the number of 10-pound packages

According to the table, the cost of shipping a 5-pound package to Zone 104 is $28.75. The cost of shipping a 10-pound package to Zone 104 is $39.50. The total budget for shipping is $1500.00; that is, the total shipping costs should be less than or equal to 1500.00.

$28.75x$ = the cost of shipping the 5-pound packages

$39.50y$ = the cost of shipping the 10-pound packages

$$28.75x + 39.50y \leq 1500.00$$
$$28.75x + 39.50y - 28.75x \leq 1500.00 - 28.75x$$
$$39.50y \leq 1500.00 - 28.75x$$
$$\frac{39.50y}{39.50} \leq \frac{1500.00 - 28.75x}{39.50}$$
$$y \leq \frac{1500.00 - 28.75x}{39.50}$$

Graph the inequality.

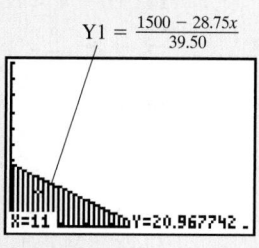

$$Y1 = \frac{1500 - 28.75x}{39.50}$$

(0, 94, 0, 100)

The ordered pairs in the shaded region represent combinations of 5- and 10-pound packages that can be shipped and still allow the shipper to stay within the budget. For example, (11, 20.968) is found in the shaded region. Therefore, eleven 5-pound packages and twenty 10-pound packages can be shipped and still allow the shipper to stay within the budget of $1500.00.

 7.3.5 Checkup

1. Lana is approved for a student loan of no more than $1500 per semester. She figures that she can afford to pay back the loan at the rate of $60 per week. Her parents will contribute $125 occasionally to help her pay back the loan.
   a. Describe Lana's payment plan, using a linear inequality in two variables, where x represents the number of weekly payments Lana makes and y represents the number of occasional payments her parents make on the loan.
   b. Graph the inequality in the first quadrant.
   c. Would a loan such that Lana makes nine weekly payments and her parents make six occasional payments be within the approved loan limit?

   d. Would a loan such that Lana makes 18 weekly payments and her parents make five occasional payments be within the approved loan limit?

2. Use the UPS Next Day Air chart in this section to determine the number of 1-pound packages and 2-pound packages sent to Zone 103 that would overrun a budget of $250.00. Will you overrun the budget by sending thirteen 1-pound packages and nine 2-pound packages to Zone 103? Will you overrun the budget if you send three 1-pound packages and nine 2-pound packages to Zone 103?

7.3 Exercises

Identify each inequality as linear or nonlinear. Express each linear inequality in standard form.

1. $2x + 1.7y > x - 4.6$

2. $y < \sqrt{x - 4} + 9$

3. $y < x^2 + 2x - 3$

4. $-5x + 10y \geq 15$

5. $y > \sqrt{x} + 9$

6. $y > x^3 - 27$

7. $\dfrac{x}{2} - \dfrac{y}{6} > \dfrac{1}{12}$

8. $y \geq -\dfrac{3}{4}x^2 + \dfrac{5}{8}$

9. $4x + 16 \leq y$

10. $3.5y < 4.2x - 2.8$

11. $y \leq -\dfrac{2}{5}x + \dfrac{7}{15}$

12. $\dfrac{x}{6} + \dfrac{y}{3} < 2$

Graph each inequality.

13. $2x + y < 3$

14. $5x - y < 6$

15. $5x - 3y \geq 6$

16. $8x + 7y \geq 14$

17. $y < -\dfrac{3}{4}x + 4$

18. $y < \dfrac{5}{6}x - 3$

19. $y \geq 2.8x - 1.6$

20. $y \geq 4.7 - 1.9x$

21. $2y > 3x - 5$

22. $5y > 3x - 2$

23. $3x + y - 4 \leq x + 2y - 3$

24. $7x - 2y + 8 \leq 5x - 3y + 1$

25. $5y > -20$

26. $3y + 7 > 25$

27. $3x + 6 \leq 9$

28. $2x - 4 \leq 6$

29. $3x + 5y \geq 12$

30. $7x + 9y \geq 63$

31. $-x - y < 7$

32. $x - y > 3$

33. $-3x - 9y > 27$

34. $-12x - 9y > 36$

35. $\dfrac{x}{8} + \dfrac{y}{3} \leq 1$

36. $\dfrac{x}{9} - \dfrac{y}{6} \leq 1$

37. $-\dfrac{4}{7}x + \dfrac{2}{3}y \geq \dfrac{10}{21}$

38. $\dfrac{3}{4}x - \dfrac{5}{6}y \geq \dfrac{13}{24}$

39. $1.8x - 3.2y > 0$

40. $58.2x + 19.4y > 0$

41. $4.6y < 3.5y + 5.94$

42. $8.1y + 16.2 < 72.9y$

43. $x - y > 9$

44. $-x + y > 9$

45. $-x - y > 9$

46. $x + y > 9$

47. $y \leq x$

48. $y \geq -x$

In exercises 49–58, write and graph a linear inequality in two variables that represents the situation presented and answer the questions posed.

49. A Christmas shop sells village pieces for $25.00 each and angel ornaments for $12.00 each. Rita was given a limit of $225.00 to spend to buy decorations for the reception area of her office. What combinations can she buy and stay within the limit imposed? Would she be within the budget if she bought four village pieces and seven angel ornaments? What if she bought seven village pieces and nine angel ornaments?

50. The sale of coupon booklets netted the sixth-grade class $240. The money is to be used to buy books and videos for the class library. If the books sell for $12 each and the videos sell for $18 each, what combinations would be within the amount available? Would the class be able to buy 7 books and six videos? Would they be able to buy 12 books and eight videos?

51. Barbara bought 220 feet of wallpaper border on sale. What are the limits on the dimensions of a rectangular room if she wants to run the border around the top of the walls? Would she have enough for a 70-by-60-foot room? Would she have enough for a 40-by-60-foot room?

52. In exercise 51, the rooms Barbara will decorate have 10-foot ceilings. She also bought enough rolls of wallpaper to cover 2400 square feet of wall space. What are the limits on a rectangular room if Barbara wants to paper all four walls, with no allowances for windows or doors? Does she have enough for a 70-by-60-foot room? Does she have enough for a 40-by-50-foot room?

53. Pablo will be paid $15 per day for demonstrating crafts, plus $12 for each item he sells at a crafts fair. What combinations will allow the artist to earn at least $400? If he works three days and sells 20 items, will he earn at least $400? If he works five days and sells 30 items, will he earn the minimum?

54. Jason is renting a chain saw for $15 per day. He will charge $10 for each tree he cuts down in his neighbors' yards. What combinations of days and trees will allow him to earn at least $75 for the venture? If he cuts 16 trees in three days, will he make his goal of at least $75? What if he cuts 10 trees in four days?

55. At a school fund-raiser, adults were charged $4.50 each and children were charged $2.00 to attend a chili supper. If the school wished to raise at least $250.00, what combinations of ticket sales would assure that goal? Would 40 adult and 25 children's tickets be enough in sales to make the goal? Would 42 adult and 45 children ticket sales be enough?

56. An elevator has a load limit of 2000 pounds. If the average adult weighs 160 pounds and the average child weighs 65 pounds, what combinations would be safe to ride the elevator? Would a group with 5 teachers and 15 children be safe on the elevator? How about a group with 10 teachers and 12 children?

Weight, not over pounds listed	Local, Zones 1 and 2	Zone 3	Zone 4	Zone 5	Zone 6	Zone 7
1	$1.80	$1.83	$1.87	$1.93	$1.99	$2.06
1.5	1.80	1.83	1.87	1.93	1.99	2.06
2	1.84	1.88	1.94	2.02	2.10	2.19
2.5	1.90	1.95	2.00	2.11	2.21	2.33
3	1.94	2.00	2.08	2.20	2.32	2.46
3.5	1.99	2.06	2.15	2.29	2.43	2.60
4	2.03	2.11	2.21	2.37	2.55	2.72

57. A portion of the U.S. Postal Service chart for rates for mailing bound printed matter is shown above.

How many 2.5-pound packages and 4-pound packages can be mailed to Zone 5 without exceeding a mailing budget of $150.00? Can twenty-five 2.5-pound packages and thirty-five 4-pound packages be mailed to Zone 5 without exceeding the budget? Can forty 2.5-pound packages and thirty 4-pound packages be mailed to Zone 5 without exceeding the budget?

58. Using the chart in the preceding exercise, determine the numbers of 4-pound packages that can be shipped to Zones 4 and 7 without exceeding a mailing budget of $150.00. Can twenty 4-pound packages be shipped to Zone 4 along with thirty 4-pound packages to Zone 7 without exceeding the budget? Can thirty-five 4-pound packages be sent to Zone 4 and thirty to Zone 7 without exceeding the budget?

 ## 7.3 Calculator Exercises

If you obtain the applications package for the TI-83 Plus calculator, you can use features under the APPS key to help you graph linear inequalities in two variables. As an example, graph $x - y < 5$.

- First solve the inequality for y: $y > x + 5$
- Enter the inequality into Y1, using the APPS feature:

  | APPS |

  (Note that the "Inequal" option under the APPLICATIONS menu may have a different number than 5

on your calculator. Use the appropriate number to select this application.)

| 5 | ENTER | ALPHA | F4 | ▶ | X,T,θ,n | + |

| 5 |

- Choose a ZOOM setting:

  | ZOOM | 6 |

The calculator will display the following screens as you progress through these steps:

 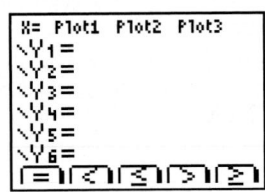

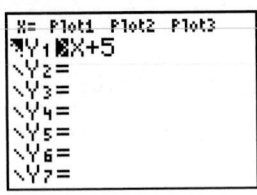 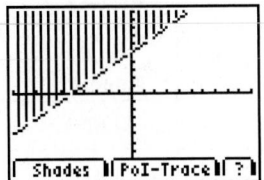

Note that when you use the APPS feature, the calculator will correctly draw a dashed boundary line for an inequality that does not include equality.

The APPS feature will also help you graph inequalities in two variables where the boundary line is a vertical line. As an example, graph $3x - 4 > 5 - (x + 1)$.

• First solve the inequality for x: $x > 2$

• Clear out any equations in ⃞Y=⃞.

• Enter the inequality into X =, using the APPS feature:

⃞APPS⃞

(Choose the appropriate menu number for "Inequal".)

⃞5⃞ ⃞ENTER⃞ ⃞▲⃞ ⃞ENTER⃞ ⃞ALPHA⃞ ⃞F4⃞ ⃞▶⃞ ⃞2⃞

• Select the ZOOM setting:

⃞ZOOM⃞ ⃞6⃞

The following screens are displayed:

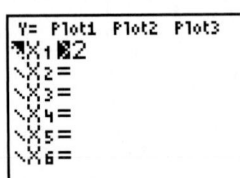 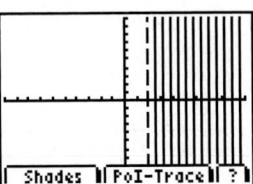

Again, note that the calculator can easily distinguish dashed lines from solid lines when you use the APPS feature.

 ## 7.3 Writing Exercises

In this section, you learned two methods for graphing a linear inequality in two variables: with a pencil and paper and with your calculator. It would be helpful to summarize what you have learned. Complete the following table, and keep it as a reference to help you remember details in graphing linear inequalities in two variables:

Inequality	Boundary	Line	Shading
$y < ax + b$	$y = ax + b$		
$y \leq ax + b$			below
$y > ax + b$		dashed	
$y \geq ax + b$			
$y < c$	$y = c$		
$y \leq c$		solid	
$y > c$			above

Inequality	Boundary	Line	Shading
$y \geq c$			
$x < c$		dashed	
$x \leq c$			left
$x > c$	$x = c$		
$x \geq c$			

Summarize what you have learned by answering the following questions:

1. When should the boundary line be dashed, and when should it be solid?

2. When should you shade above the boundary line, and when should you shade below the boundary line?

3. What does the shaded region represent?

7.4 Systems of Linear Inequalities in Two Variables

OBJECTIVES
1 Graph systems of linear inequalities in two variables.
2 Graph systems of linear inequalities in two variables by using a calculator.
3 Model real-world situations by using systems of linear inequalities.

APPLICATION

Your company packages its product in 5-pound packages and 10-pound packages. At most, 45 packages can be produced daily. You ship your packages by United Parcel Service (UPS) Next Day Air. A portion of the United Parcel Service (UPS) Next Day Air domestic delivery rate chart for customers who receive daily pickup (effective February 5, 2001) is as follows:

Weight, lb	Zone 102	Zone 103	Zone 104	Zone 105
1	14.50	16.50	19.50	21.00
2	15.50	17.50	21.75	23.50
3	16.75	18.50	24.25	26.00
4	17.75	19.75	26.50	28.25
5	18.95	20.75	28.75	30.75
6	19.50	22.00	31.00	33.25
7	20.25	23.25	33.00	35.50
8	21.00	24.25	35.00	38.00
9	21.75	25.25	37.25	40.25
10	22.50	26.25	39.50	42.50

Determine the number of 5-pound packages and the number of 10-pound packages that you can ship to Zone 104 and still stay within a daily shipping budget of $1500.00.

After completing this section, we will discuss this application further. See page 513.

7.4.1 Graphing Systems of Linear Inequalities in Two Variables

We have solved instances of one linear equation in two variables. But sometimes we may need to know the solution of more than one inequality at a time. A **system of linear inequalities in two variables** consists of two or more linear inequalities in two variables. For example, $x + y < 30$ and $2x - 5y > 40$ is a system of linear inequalities. We usually write this system without the word *and*.

$$x + y < 30$$
$$2x - 5y > 40$$

A solution of a system of inequalities is an ordered pair that makes both inequalities true. We have learned to illustrate graphically the solution set of one linear inequality in two variables by using shaded regions. To solve a system of inequalities, we will need to graph each inequality individually and then determine the ordered pairs that make both inequalities true at the same time. These ordered pairs lie in the overlap of the shaded regions.

GRAPHING A SYSTEM OF LINEAR INEQUALITIES IN TWO VARIABLES

To graph the solution of a system of linear inequalities in two variables,

- Graph each inequality individually on the same rectangular coordinate plane.
- Determine all intersections of the boundary lines.
- Determine the region of the coordinate plane that contains solutions of all inequalities (the overlapping shaded regions).
- Check a point in the overlap region to determine whether that point is a solution.

EXAMPLE 1 Graph the following system of linear inequalities in two variables:

$$2x - y < 7$$
$$-x + y \geq 2$$

Solution

a. First, solve each inequality for y.

$$2x - y < 7 \qquad \text{and} \quad -x + y \geq 2$$
$$y > 2x - 7 \qquad\qquad\qquad y \geq x + 2$$

Graph the boundary lines for each inequality. The first inequality has a dashed boundary line (no equality), and the second inequality has a solid boundary line to include the points on the line in the solution set (includes equality).

Determine the intersection of the boundary lines algebraically by solving the related system of equations.

(1) $2x - y = 7$

(2) $-x + y = 2$

By elimination, we obtain

$$x = 9.$$

By substitution, we obtain

$$
\begin{aligned}
(1) \qquad -x + y &= 2 \\
-(9) + y &= 2 \\
-9 + y &= 2 \\
-9 + y + 9 &= 2 + 9 \\
y &= 11
\end{aligned}
$$

The intersection of the boundaries is $(9, 11)$.

b. Shade above both boundaries, because, when solved for y, each inequality was a "greater than" inequality.

The overlap of the shaded regions and the included boundary line shown contains the ordered-pair solutions of the system.

a.

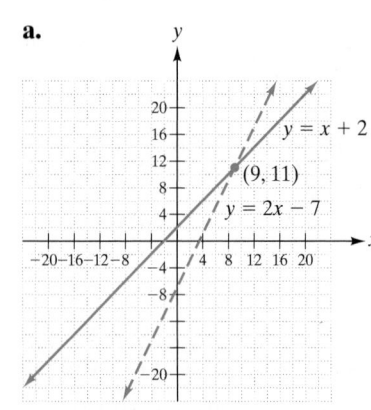

b.

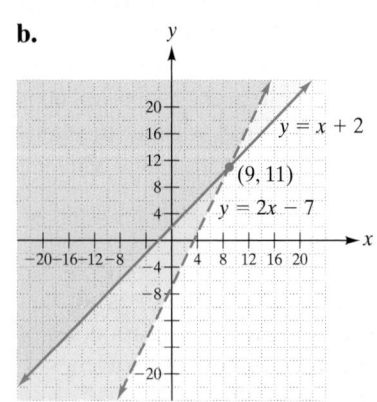

c.

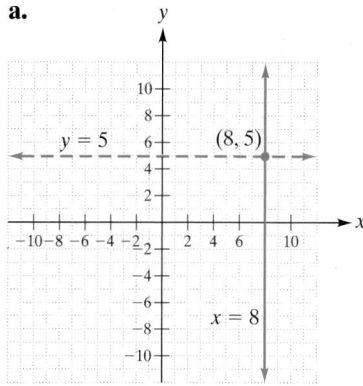

c. Check a point in the overlapping shaded portion. One such point is $(4, 9)$. Check to see whether it is a solution of both inequalities of the system.

$$\begin{array}{ll} 2x - y < 7 & \text{and} & -x + y \geq 2 \\ 2(4) - 9 \mid 7 & & -(4) + 9 \mid 2 \\ 8 - 9 & & -4 + 9 \\ -1 & & 5 \end{array}$$

The point $(4, 9)$ is a solution.

If the system of linear inequalities has inequalities in which one of the variables has a coefficient of 0, the procedure outlined in Example 1 still applies.

EXAMPLE 2 Graph the following system of linear inequalities in two variables:

$$y < 5$$
$$x + 4 \geq 12$$

a.

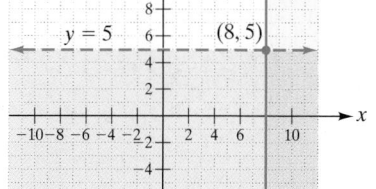

Solution

a. Solve the first inequality for y: $y < 5$
The second inequality does not contain a y and should be solved for x.

$$x + 4 \geq 12$$
$$x + 4 - 4 \geq 12 - 4$$
$$x \geq 8$$

Graph the boundary lines of each inequality. The first line is dashed and the second is solid.
    Determine the intersection of the boundary lines algebraically by solving the related system of equations.

$$y = 5$$
$$x = 8$$

b.

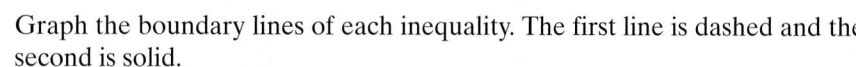

The intersection is the ordered pair $(8, 5)$.

b. Shade below the line $y = 5$ for "less than," and shade to the right of the line $x = 8$ for "greater than."

    The overlap of both shaded regions and the included boundary line shown contains the ordered-pair solutions of the system.

c. Check a point in the shaded portion. One such point is $(10, 3)$. Check to see whether it is a solution of both inequalities of the system.

$$\begin{array}{ll} y < 5 & \text{and} & x + 4 \geq 12 \\ 3 \mid 5 & & 10 + 4 \mid 12 \\ & & 14 \end{array}$$

c.

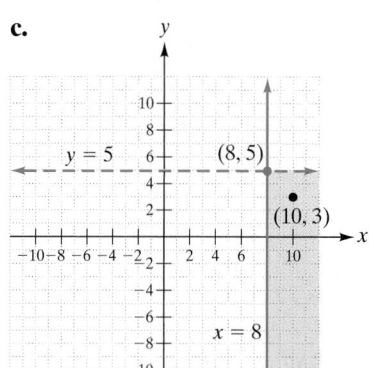

The point $(10, 3)$ is a solution.

Systems of linear inequalities may have more than two inequalities. A solution of such a system is still an ordered pair that makes all the inequalities true. We graphically illustrate the solutions of the system in the same manner as with two inequalities. That is, we graph all the inequalities on the same coordinate system and determine the overlapping shading of the graphs.

This overlapping shading may be hard to determine. It may be easier to graph the boundary lines, then choose a point in each of the regions of the coordinate plane, and check to see whether that point satisfies all the inequalities in the system.

EXAMPLE 3 Graph the following system of linear inequalities in two variables:

$$x + y \leq 10$$
$$y \leq 4$$
$$y \geq 0$$
$$x \geq 0$$

Solution

a. Solve the first inequality for y.

$$y \leq -x + 10$$

Graph the boundary lines of the inequalities. All the boundaries are solid lines because all the inequalities include equality.
    Determine the intersections.

b. Shade the regions for each inequality and determine the overlapping shaded region.

a.

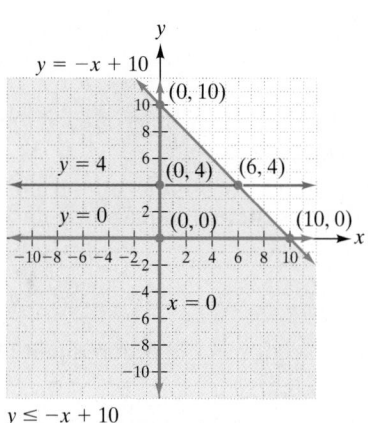

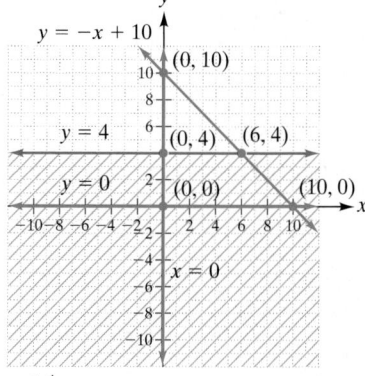

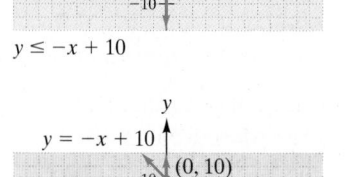

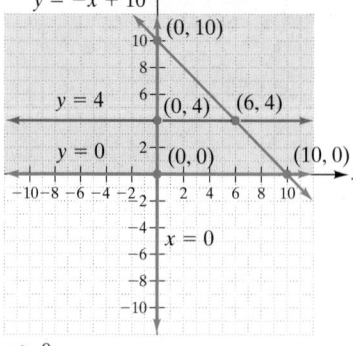

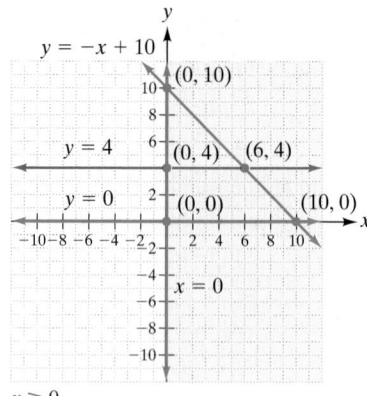

The overlap of all shaded regions and the included boundary lines shown contains the ordered-pair solutions of the system. Because it is so difficult to

b.

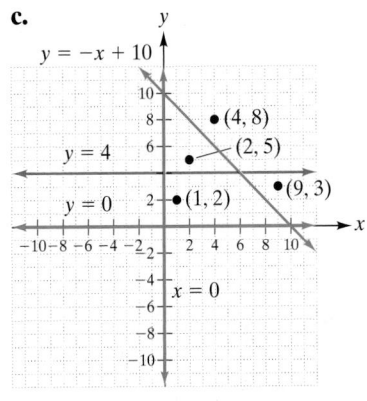

c.

see the overlapping shading when more than two areas are shaded, it is often advisable to graph only the overlap region, as shown in Figure 7.7.

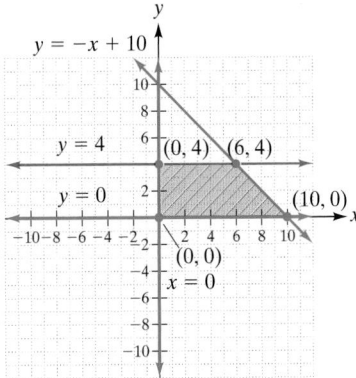

Figure 7.7

c. Alternatively, instead of shading, pick a point in each of the regions defined by the boundaries in the first quadrant. Next, test to determine which of the points you have chosen makes the entire system of inequalities true. Then shade the region which contains that point.

A point in each region is labeled on the graph. Substitute the coordinates of each point into all the inequalities. The ordered pair $(1, 2)$ satisfies all the inequalities. Therefore, shade that region of the first quadrant in which $(1, 2)$ is located. The result is the same as Figure 7.7.

✔ 7.4.1 Checkup

In exercises 1–3, graph each system of linear inequalities.

1. $3x - 5y \leq 5$
$2y + 3 > 7$

2. $y \geq -4x + 3$
$x \geq 1$

3. $x + y < 5$
$y > x + 1$
$x \geq 0$
$y \geq 0$

4. What is the difference between finding the solutions of a linear inequality in two variables and finding the solutions of a system of linear inequalities in two variables?

7.4.2 Graphing Systems of Linear Inequalities by Using a Calculator

To graph a system of linear inequalities in two variables using a calculator, we perform the same steps as we do by hand.

> To graph a system of linear inequalities in two variables with your calculator when both inequalities contain a y-term,
>
> - Solve the inequalities for y.
> - Enter the algebraic expressions in Y1 and Y2.
> - Determine the intersection of the boundaries by tracing or using Intersect under the CALC function.
> - Shade the coordinate plane determined by the first inequality solved for y.
>   - For a "less than" inequality, shade below the boundary line.
>   - For a "greater than" inequality, shade above the boundary line.
> - The second inequality should be entered in the same way as the first.

 **HELPING HAND** The calculator does not graph dashed lines; it includes the boundary in the graph. It will be up to you not to include the boundary in your graph. (See 7.4 Calculator Exercises for further guidance.)

EXAMPLE 4 Use your calculator to graph the following system of linear inequalities in two variables:

$$2x - y < 7$$
$$-x + y \geq 2$$

Solution

Solve the inequalities for y.

$$y > 2x - 7$$
$$y \geq x + 2$$

Enter the boundaries, graph them, and determine their intersection. Let $Y1 = 2x - 7$ and $Y2 = x + 2$, as shown in Figure 7.8a.

Shade above the Y1 graph for a "greater than" inequality. Shade above the Y2 graph for a "greater than or equal to" inequality. The solution set consists of the ordered pairs in the portion of the graph that is shaded twice, as shown in Figure 7.8b.

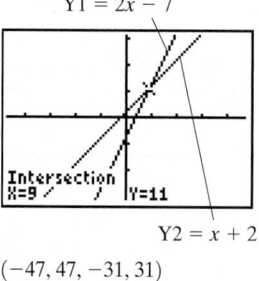

$Y1 = 2x - 7$

$Y2 = x + 2$

$(-47, 47, -31, 31)$

Figure 7.8a

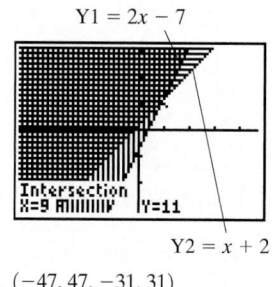

$Y1 = 2x - 7$

$Y2 = x + 2$

$(-47, 47, -31, 31)$

Figure 7.8b

Solving a system of linear inequalities in two variables when one or more of the inequalities contain a y-term with coefficient 0 is more complicated on the calculator. Since the boundary for this type of inequality is not a function, we will not present the calculator procedure here. (See 7.4 Calculator Exercises at the end of this section for instructions.)

When a system of linear inequalities has more than two inequalities, shade the correct portion of the coordinate plane for each inequality that has a boundary line which is a function.

EXAMPLE 5 Use your calculator to graph the following system of linear inequalities in two variables:

$$x + y \leq 10$$
$$y \leq 4$$
$$y \geq 0$$
$$x \geq 0$$

Calculator Solution

We will need to use a close view for this system.

$$(-14.1, 14.1, 1, -12.4, 12.4, 1, 1)$$

Enter the boundaries for the first three inequalities, each of which contains a y-variable.

$$Y1 = -x + 10$$
$$Y2 = 4$$
$$Y3 = 0$$

Because the fourth inequality's boundary, $x = 0$, is not a function, do not graph it. But remember to include it in your description of the solution set of the system of inequalities. (That is, include the y-axis and all points to its right.)

Determine the intersections by tracing. Shade below Y1 and Y2 for a "less than or equal to" inequality. Shade above Y3 for a "greater than or equal to" inequality.

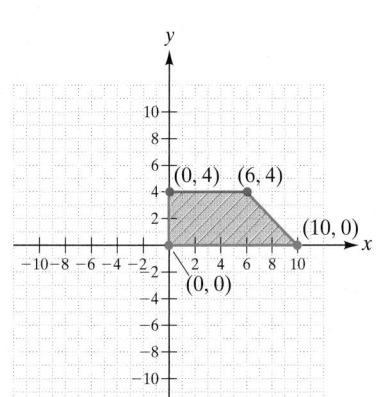

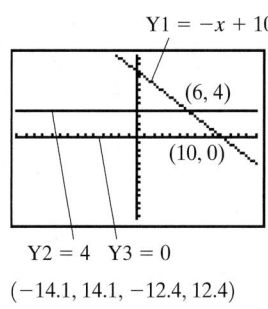

Y1 = −x + 10

(6, 4)

(10, 0)

Y2 = 4   Y3 = 0

(−14.1, 14.1, −12.4, 12.4)

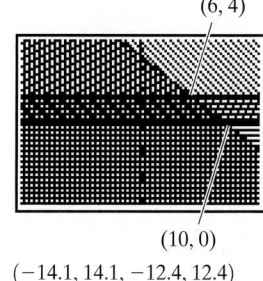

(6, 4)

(10, 0)

(−14.1, 14.1, −12.4, 12.4)

Note: The shaded figure is very difficult to see.

The solution set consists of the ordered pairs that are in the portion of the graph that is shaded three times and that are also on the y-axis or to the right of the y-axis (because $x \geq 0$).

 7.4.2 Checkup

Graph each system of linear inequalities, using your calculator.

1. $y \leq x - 3$

$2x - y \geq -1$

2. $y > -3x + 4$

$x + y > 2$

3. $y \leq \dfrac{2}{3}x + 1$

$y \leq -\dfrac{3}{4}x + 5$

$x \geq 0$

$y \geq 0$

7.4.3 Modeling the Real World

We have seen that systems of inequalities often have many solutions, even an infinite number of solutions. In many real-world situations, we need to restrict the number of solutions to those that make sense, such as restricting solutions to positive values of time. We can do this by adding another inequality to the system, called a **constraint** on the variables. For example, we can add the inequality $t > 0$ to a system involving the variable t for time, so the solutions will include only positive values for t. Or we may want to search a database for unmarried males whose age is over 21 but under 30 and add a constraint that their annual income exceed $100,000. Mathematically, we treat the constraint inequality as just another inequality in the system.

EXAMPLE 6 Donzietta and Kathy make dolls to sell. Donzietta cuts out the patterns and sews and stuffs each doll. This requires 2 hours of work for a rag doll and 6 hours of work for a sculptured doll. Kathy finishes the features and hair of each doll. This requires 3 hours for a rag doll and 2 hours for a sculptured doll. Donzietta plans to work at most 40 hours a week and Kathy at most 32 hours a week.

a. Determine the constraints on producing the dolls for a week.

b. Graph the system of linear inequalities, and determine two possible combinations that satisfy the system.

Solution

a. Let x = number of rag dolls produced per week

y = number of sculptured dolls produced per week

An inequality for Donzietta's contribution is

$$2x + 6y \leq 40$$

An inequality for Kathy's contribution is

$$3x + 2y \leq 32$$

The constraints not stated in the exercise would be that the number of dolls of each kind may not be a negative number.

$$x \geq 0$$
$$y \geq 0$$

Solve each inequality for y.

$$y \leq -\frac{1}{3}x + \frac{20}{3} \qquad \text{Donzietta's contribution}$$

$$y \leq -\frac{3}{2}x + 16 \qquad \text{Kathy's contribution}$$

$$y \geq 0$$

The last constraint is solved for x.

$$x \geq 0$$

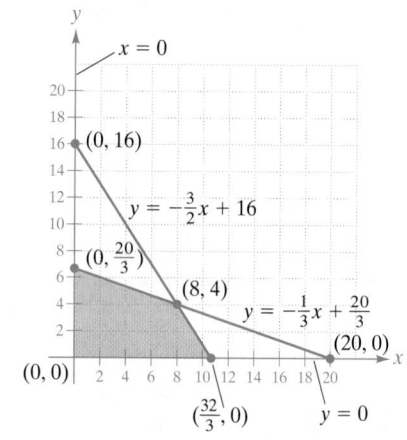

Graph the boundaries with solid lines for equality, and determine the intersections.

Shade each portion of the graph.

For $y \leq -\frac{1}{3}x + \frac{20}{3}$, shade below the boundary line.

For $y \leq -\frac{3}{2}x + 16$, shade below the boundary line.

For $y \geq 0$, shade above the boundary line.

For $x \geq 0$, shade to the right of the boundary line.

b. The solution set consists of all the ordered pairs contained on the boundary lines and in the overlap of the shaded regions. Two possible ordered-pair solutions are (6, 3) and (8, 4).

(6, 3) means that six rag dolls and three sculptured dolls can be produced.

(8, 4) means that eight rag dolls and four sculptured dolls can be produced.

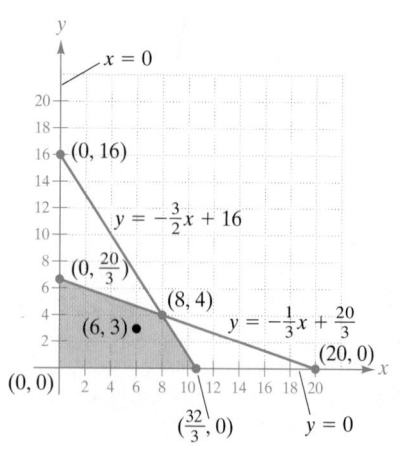

APPLICATION

Your company packages its product in 5-pound packages and 10-pound packages. At most, 45 packages can be produced daily. You ship your packages by United Parcel Service (UPS) Next Day Air. A portion of the United Parcel Service (UPS) Next Day Air domestic delivery rate chart for customers who receive daily pickup (effective February 5, 2001) is as follows:

Weight, lb	Zone 102	Zone 103	Zone 104	Zone 105
1	14.50	16.50	19.50	21.00
2	15.50	17.50	21.75	23.50
3	16.75	18.50	24.25	26.00
4	17.75	19.75	26.50	28.25
5	18.95	20.75	28.75	30.75
6	19.50	22.00	31.00	33.25
7	20.25	23.25	33.00	35.50
8	21.00	24.25	35.00	38.00
9	21.75	25.25	37.25	40.25
10	22.50	26.25	39.50	42.50

Determine the number of 5-pound packages and the number of 10-pound packages that you can ship to Zone 104 and still stay within a daily shipping budget of $1500.00.

Discussion

Let x = the number of 5-pound packages
y = the number of 10-pound packages
We know that we cannot produce or ship a negative number of packages. Therefore, two inequalities of this system are

$$x \geq 0$$
$$y \geq 0$$

If at most 45 packages can be produced daily, a third inequality is

$$x + y \leq 45$$

or

$$y \leq -x + 45$$

According to the table, the cost of shipping a 5-pound package to Zone 104 is $28.75. The cost of shipping a 10-pound package to Zone 104 is $39.50. The total budget for shipping is $1500.00. Thus, the total shipping costs should be less than or equal to 1500.00.

$28.75x$ = the cost of shipping the 5-pound packages
$39.50y$ = the cost of shipping the 10-pound packages

The fourth inequality is

$$28.75x + 39.50y \leq 1500.00$$

which yields

$$28.75x + 39.50y - 28.75x \leq 1500.00 - 28.75x$$
$$39.50y \leq 1500.00 - 28.75x$$
$$\frac{39.50y}{39.50} \leq \frac{1500.00 - 28.75x}{39.50}$$

or

$$y \leq \frac{1500.00 - 28.75x}{39.50}$$

The system of inequalities is thus

$$x \geq 0$$
$$y \geq 0$$
$$y \leq -x + 45$$
$$y \leq \frac{1500.00 - 28.75x}{39.50}$$

(Continued on page 514)

Graph the system of inequalities, and label the point of intersection of the inequalities.

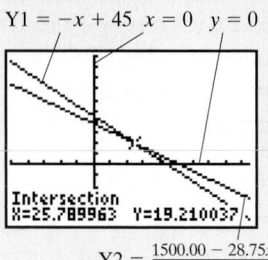

$$Y1 = -x + 45 \quad x = 0 \quad y = 0$$

$$Y2 = \frac{1500.00 - 28.75x}{39.50}$$

$$(-50, 94, -50, 100)$$

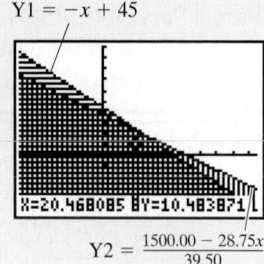

$$Y1 = -x + 45$$

$$Y2 = \frac{1500.00 - 28.75x}{39.50}$$

$$(-50, 94, -50, 100)$$

The ordered pairs in the shaded region represent combinations of 5 and 10 pound packages that can be produced and shipped while still allowing the shipper to stay within the budget. For example, (20.468, 10.484) is found in the shaded region. Therefore, twenty 5-pound packages and ten 10-pound packages can be produced and shipped without exceeding the budget of $1500.00.

Note that the intersection points of the boundary lines are also solutions of the system of inequalities. For example (25.79, 19.21) is an intersection point. Therefore, twenty 5-pound packages and nineteen 10-pound packages can be produced and shipped within the constraints of the budget.

 7.4.3 Checkup

1. Donzietta and Kathy decide to streamline their doll-making operation. They purchase kits that shorten the time required to produce the dolls they make. Donzietta can use precut patterns so that she now spends 1.5 hours of work on each rag doll and 4 hours of work on each sculptured doll. Kathy uses doll heads that have been pre-finished already, so now she can finish a rag doll in 1.5 hours and a sculptured doll in 1 hour. Donzietta changes her plans to work no more than 30 hours per week, and Kathy changes hers to work no more than 24 hours per week. Determine the new constraints on the production of dolls for a week. Graph the system of linear inequalities, and determine two combinations that satisfy the system.

2. Use the UPS Next Day Air chart in this section to determine the number of 1-pound packages and 2-pound packages that can be sent to Zone 103 without exceeding the budget of $250.00 and with the added stipulation that at least 10 packages must be mailed each day. If you ship five 1-pound packages and seven 2-pound packages to Zone 103, will you meet these restrictions? If you ship ten 1-pound packages and ten 2-pound packages to Zone 103, will you meet the restrictions?

7.4 Exercises

Graph each system of linear inequalities. Label the points of intersection.

1. $5x - 3y > 15$
$4x + y > 12$

2. $2x + 3y > -6$
$x - 4y < 8$

3. $2x + 4y \geq 8$
$3x + y \leq -8$

4. $x + y \leq 4$
$x - 2y \leq 7$

5. $y > 3x - 1$
$y > -2x + 4$

6. $y > 3x - 6$
$y > -5x + 2$

7. $y < \frac{3}{4}x + 1$
$y < -\frac{2}{3}x + 1$

8. $y < \frac{1}{2}x + 3$
$y < -\frac{3}{2}x - 1$

9. $3x + 2y < 12$
$y - 2 < 1$

10. $-x + 2y > 4$
$3x - 2 < 4$

11. $3y - 2 > 2y + 2$
$x - 1 < 0$

12. $5x + 10 < -5$
$2y + 7 > 5$

13. $y \leq 3x - 5$
$x > 4 - 2y$

14. $y \geq -2x + 3$
$3x \leq 2y + 8$

15. $y \leq \frac{1}{2}x + 3$
$x + 2y < 6$

16. $y > x + 1$
$x + y < 1$

17. $y \leq 10 - x$
$2y > x + 6$
$y > 0$
$x < 3$

18. $y > \frac{1}{2}x$
$y \leq -2x + 8$
$y \geq 0$
$y \leq 4$
$x \geq 0$

Graph each system of linear inequalities, using your calculator. Determine the points of intersection.

19. $x - 2y > 7$
$5x + 10y < -3$

20. $2x - 3y < 5$
$2y + 6 > 8$

21. $3x - 5y \geq 5$
$10y + 15 > 2$

22. $3x + 5y \geq 15$
$4x - 2y > 7$

23. $y \geq x - 3$
$10x - 5y \geq 14$

24. $y \geq -3x + 4$
$5x - 5y \geq 2$

25. $2x + 9y < 102$
$5x - 11y < -147$

26. $11x + 5y < 133$
$15x + 4y > 187$

27. $3.2x + 4.2y > 368$
$4.4x - 2.1y > 128$

28. $1.4x + 2.4y > 255$
$3.4x - 1.6y > 123$

29. $0.05x + 0.10y < 0.75$
$x + y > 11$

30. $0.25x + 0.50y < 5.75$
$x + y > 14$

For each situation, develop a system of linear inequalities which represents that situation. Graph the system. Determine one possible solution from the solution set.

31. A contractor is staking out an area for a rectangular patio. The perimeter of the patio must be no more than 100 feet. The length must be at least 10 feet more than the width. What are the possible dimensions for the patio?

32. After seeing the design for the patio in exercise 31, the customer decides that while the perimeter still should be no more than 100 feet, the length should be at least twice the width. What are the new possible dimensions for the patio?

33. One investment pays 6% simple interest for a year, while another pays 8% simple interest for a year. If you have no more than $3000 to invest, how much can you invest at each rate in order to earn at least $200 in interest for the year?

34. If you have up to $5000 to invest, part at 6% and part at 8% simple interest for a year, how much can you invest at each rate to earn at least $350 in interest?

35. Hans has two part-time jobs. Because of school, his work is limited to no more than 20 hours per week. He earns $6.50 per hour on the first job and $8.25 on the second job. For what combination of work hours will he earn at

least $150.00 per week? Will he earn at least $150.00 if he works 7 hours on the first job and 10 hours on the second job? How about if he works 5 hours on the first job and 15 hours on the second job?

36. Happy Harry's charges $20 per hour to rent its party room and $5 per guest for snacks and beverages. The minimum number of hours of rental is 1.5. For what combination of the number of guests and the number of hours of rental will the cost be no more than $150? Will 15 guests and a 2-hour rental meet the requirements? How about 25 guests and 2.5 hours of rental?

37. Rosie sells two different entrées at her restaurant. Lasagna costs $1.75 per serving to prepare. Veal parmigiana costs $2.25 per serving to prepare. Past experience indicates that she must prepare at least 50 servings of the lasagna and 25 servings of the veal parmigiana each day. If she wants her cost of preparing the entrées to be no more than $200 per day, what combinations of the number of each entrée will do? If she prepares 60 servings of lasagna and 35 servings of veal parmigiana, will the cost be in the permissible range? What if she prepares 60 servings of lasagna and 50 servings of veal parmigiana?

38. Math Academy is considering contracting out its copying operations. It can pay $0.05 per page to copy the manuscript at the local copy center if it contracts for more than 25,000 pages. Or it can purchase a copy machine for a minimum of $1399, after which the cost would be $0.01 per page copied. For what combination of machine cost and number of copies will it be cheaper to buy the machine and do the copying in-house? Will it be cheaper to do copying in-house if 30,000 copies are needed and the cost of a machine is $1500? What if the academy needs 40,000 copies and the machine costs $1400?

39. A portion of the U.S. Postal Service chart for rates for mailing bound printed matter is as follows:

Weight, not over listed pounds	Local, Zones 1 and 2	Zone 3	Zone 4	Zone 5	Zone 6	Zone 7
1	$1.80	$1.83	$1.87	$1.93	$1.99	$2.06
1.5	1.80	1.83	1.87	1.93	1.99	2.06
2	1.84	1.88	1.94	2.02	2.10	2.19
2.5	1.90	1.95	2.00	2.11	2.21	2.33

(Continued on page 516)

Weight, not over listed pounds	Local, Zones 1 and 2	Zone 3	Zone 4	Zone 5	Zone 6	Zone 7
3	1.94	2.00	2.08	2.20	2.32	2.46
3.5	1.99	2.06	2.15	2.29	2.43	2.60
4	2.03	2.11	2.21	2.37	2.55	2.72

Suppose you must mail some 2.5-pound packages and some 4-pound packages to Zone 5 without exceeding a budget of $150.00. The number of 4-pound packages must be less than twice the number of 2.5-pound packages. Determine the number of packages that can be sent without exceeding the budget or the given restriction. Can fifty 2.5-pound packages and fifty 4-pound packages be mailed to Zone 5 without exceeding the restriction? Can twenty 2.5-pound packages and twenty 4-pound packages be mailed to Zone 5 without exceeding the restrictions?

40. Suppose you must ship 4-pound packages to Postal Service Zones 4 and 7. The number of packages shipped to Zone 7 must always be at least as many as are shipped to Zone 4. Using the chart in the preceding exercise, determine the numbers of 4-pound packages that can be shipped to Zones 4 and 7 without exceeding a mailing budget of $150.00 while still meeting the added restriction. Can thirty 4-pound packages be shipped to Zone 4 along with forty-five 4-pound packages to Zone 7 without exceeding the budget? Can fifteen 4-pound packages be sent to Zone 4 and 35 to Zone 7 without exceeding the budget?

7.4 Calculator Exercises

The APPS feature of the TI-83 Plus calculator enhances the graphs of systems of linear inequalities. Using that feature, you are better able to graph boundary lines that are dashed instead of solid, and you are able to graph boundary lines that are vertical lines. An example follows.

Graph $2x - y < 7$

$\qquad 3x + 6 < 15$

- Solve the first inequality for y: $y > 2x - 7$
- Solve the second inequality for x: $x < 3$
- Store the first inequality in Y1:

$\boxed{\text{APPS}}$

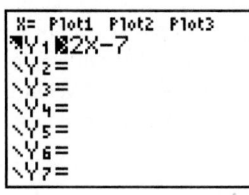 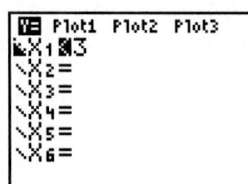

If you wish to see the overlap region more clearly, you may choose a shading option under the Shades menu.

$\boxed{\text{ALPHA}}$ $\boxed{\text{F1}}$ $\boxed{\text{1}}$

This option selects Ineq Intersection and gives you the following screen:

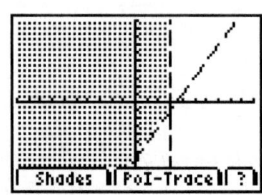

(Choose the option for "Inequal".)

$\boxed{5}$ $\boxed{\text{ENTER}}$ $\boxed{\text{ALPHA}}$ $\boxed{\text{F4}}$ $\boxed{\blacktriangleright}$ $\boxed{2}$ $\boxed{X,T,\theta,n}$
$\boxed{-}$ $\boxed{7}$

- Store the second inequality in X=:

$\boxed{\blacktriangle}$ $\boxed{\blacktriangle}$ $\boxed{\text{ENTER}}$ $\boxed{\text{ALPHA}}$ $\boxed{\text{F2}}$ $\boxed{\blacktriangleright}$ $\boxed{3}$

- Select your ZOOM setting:

$\boxed{\text{ZOOM}}$ $\boxed{6}$

The calculator displays two dashed boundary lines and an overlapping shaded region indicating coordinate pairs that make the system of inequalities true.

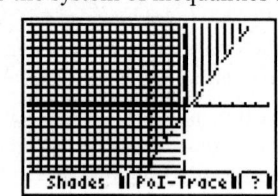

If you wish to determine the intersection point of the boundary lines, enter

$\boxed{\text{ALPHA}}$ $\boxed{\text{F3}}$

and the following screen will show the intersection point, labeled:

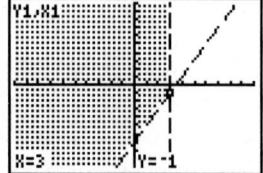

Following are some interesting special cases of systems of linear inequalities that you can graph using the preceding procedure. Describe the solution set of each system.

1. $2x - 3y > -6$
$-2x + 3y > -6$

2. $x + 2y < 6$
$x + 2y > 2$

3. $y > -3x + 2$
$y \le -3x$

4. $y > 2x + 2$
$y \le 2x - 3$

5. $x + 4 > 0$
$3x + 2 < -4$

6. $y - 1 < 0$
$y + 3 > 0$

7. $y < 4 - 2x$
$2x + y \le -1$

8. $y < x - 7$
$x > y + 9$

9. $3y \le x + 6$
$3y - x \ge 6$

10. $y < 2x + 3$
$y - x + 4 > x + 7$

11. $y < 7$
$2y - 3 \le y + 4$

12. $y \ge 3x + 4$
$3x + 5 \le y + 1$

7.4 Writing Exercises

When graphing a system of linear inequalities on your calculator, the choice of an appropriate window is key to being able to see the graph. The window should be large enough to show the intercepts. If any of the inequalities limit the solution space to the first quadrant, the minimum values for x and y should be chosen accordingly. As an example, graph the following system on your calculator using the window indicated.

1. $5.8x + y > 1055.10$
$18.4x + 73.2y < 19{,}361.22$
$x \ge 0$
$y \ge 0$
Window: $(0, 1200, 100, 0, 1200, 100, 1)$

Explain the choice of window suggested for exercise 1. Would you graph the third and fourth inequalities? If not, give reasons. After you understand why these choices were made, describe what you would choose as a suitable window for exercise 2, and which inequalities you would graph. Then graph the system and describe the solution set.

2. $487x + 182y < 51{,}567$
$182x + 487y < 103{,}280$
$x \ge 0$
$y \ge 0$

CHAPTER 7 SUMMARY

After completing this chapter, you should be able to define the following key terms in your own words.

7.1
linear inequality in one variable
order symbols
compound inequality
double inequality
interval notation

negative infinity
infinity

7.3
linear inequality in two variables

7.4
system of linear inequalities in two
variables
constraint

Reflections

1. What is the difference between a linear equation in one variable and a linear inequality in one variable?
2. The solution of a linear inequality can be expressed in interval notation or graphed on the number line. Describe each of these representations.
3. What is the addition property of inequalities? Are there any special considerations you should apply when using this property?
4. What is the multiplication property of inequalities? Are there any special considerations you should apply when using this property?
5. In solving a linear inequality in one variable algebraically, how can you tell that it has no solution? How can you tell that the solution set consists of all real numbers?
6. How does the solution of a linear inequality in two variables differ from the solution of a linear inequality in one variable?
7. How can you show graphically the solution set of a system of linear inequalities in two variables?

CHAPTER 7 SECTION-BY-SECTION REVIEW

7.1

Recall	Examples
Linear inequality in one variable • The standard forms are $ax + b < 0$ $ax + b > 0$ $ax + b \leq 0$ $ax + b \geq 0$ where a and b are real numbers and $a \neq 0$.	The following are linear inequalities in one variable: $2x + 4 < 0$ $x > -2$ $x \leq 3$ $5x + x - 6 \geq -3x$
Number-line notation • Plot the upper or lower bound with an open dot if the bound is not included and a closed dot if the bound is included • Draw a solid line to the left of the upper bound for a "less than" or "less than or equal to" inequality and to the right for a "greater than" or "greater than or equal to" inequality.	Inequality notation · Number-line notation $x < 3$ $x > 3$ $x \leq 3$ $x \geq 3$ $-1 < x \leq 3$
Interval notation • Write the bounds, if they exist, in increasing order, separated by a comma. Use infinity symbols if no bound exists. • Precede or follow each bound with a parenthesis if it is not included in the solution set or a bracket if it is included in the solution set.	Inequality notation · Interval notation $x < 3$ $(-\infty, 3)$ $x > 3$ $(3, \infty)$ $x \leq 3$ $(-\infty, 3]$ $x \geq 3$ $[3, \infty)$ $-1 < x \leq 3$ $(-1, 3]$

Identify each inequality as linear or nonlinear. Write the linear inequalities in standard form.

1. $3x^2 - 2x + 1 < 0$ **2.** $5x - 4 > x + 7$ **3.** $\dfrac{2}{3}x + \dfrac{4}{5} \leq \dfrac{7}{15}$

4. $\sqrt{x} - 3 \geq 6$ **5.** $x - 3 \geq 6$ **6.** $\dfrac{1}{x} - 3x \geq 6$

7. $1.5z - 12.6 < 14.7z$ **8.** $3(a + 2) < 15a - (2a + 1)$

Graph each inequality on a number line. Check your graph on your calculator. Then write the solution set in interval notation.

9. $x < 3$ **10.** $x > -2$ **11.** $x \leq -5$

12. $x \geq -3.5$ **13.** $-2 < a < 4$ **14.** $-1 < b \leq 0$

15. $3 \leq c \leq 5.5$ **16.** $2\dfrac{1}{2} < d < 8$ **17.** $-2.3 \leq f \leq -1\dfrac{1}{3}$

Write a linear inequality in one variable to represent each situation. Do not solve the inequality.

18. One copy machine can make 30 copies per minute, while another machine can make 25 copies per minute. If both machines are used to make copies, how long will it take to make at least 300 copies of a one-page flyer?

19. In designing a roadway, a civil engineer uses the following rules of thumb: The pavement will cost twice the amount of the base material. The sidewalk will cost one-fourth the amount of the pavement. What will the cost of the base material be if the cost of the roadway must be below $200,000?

20. A political campaign manager is deciding whether to use bulk-rate or first-class mailing. The bulk-rate permit costs $125, and the bulk mailing rate is 28 cents per piece. For what number of mailings will the bulk-rate cost be cheaper than the first-class mailing cost of 34 cents per piece?

21. The Fujita Scale (F-Scale) is accepted as the official classification system for tornado damage. The following classifications are used to rate the intensities of less-than-severe tornadoes:

F-Scale number	Intensity	Wind speed
F0	Gale tornado	40–72 mph
F1	Moderate tornado	73–112 mph
F2	Significant tornado	113–157 mph

For each of the categories listed, write a linear inequality that represents that category.

7.2

Recall	Examples
Solving linear inequalities in one variable numerically • Complete a three-column table of values, one column for the independent variable, one for the left side of the inequality, and one for the right side of the inequality. • The solutions are the values of the independent variable that determine a true inequality.	Solve $2x + 1 \leq 3x - 5$ numerically. <table><tr><td>x</td><td>$2x + 1$</td><td>$\leq$</td><td>$3x - 5$</td><td></td></tr><tr><td>5</td><td>11</td><td></td><td>10</td><td>$11 \nleq 10$</td></tr><tr><td>6</td><td>13</td><td></td><td>13</td><td>$13 \leq 13$</td></tr><tr><td>7</td><td>15</td><td></td><td>16</td><td>$15 \leq 16$</td></tr><tr><td>8</td><td>17</td><td></td><td>19</td><td>$17 \leq 19$</td></tr></table> The solutions are all integers greater than or equal to 6.
Solving linear inequalities in one variable graphically • Graph the two functions, y_1 and y_2, respectively defined by the left side of the inequality and the right side of the inequality. • Determine the intersection of the graphs. • Locate the portion of the y_1 graph below the y_2 graph for a "less than" inequality or the portion of the y_1 graph above the y_2 graph for a "greater than" inequality. • If the portion located on the graph is to the left of the intersection point, then x is less than the x-coordinate of the intersection point. If the portion located on the graph is to the right of the intersection point, then x is greater than the x-coordinate of the intersection point. • The solution set includes the x-coordinate of the point of intersection if the inequality includes equality.	Solve $2x + 1 \geq 3x - 5$ graphically. 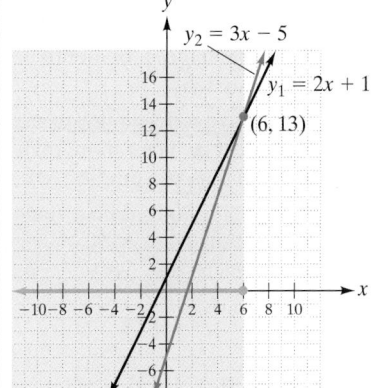 The solution set is the set of all x that satisfy $x \leq 6$.
Solving linear inequalities in one variable algebraically • Simplify both expressions in the inequality (preferably, leaving no fractions). • Isolate the variable on one side of the inequality (preferably, the left side) and the constants on the other side of the inequality by using the addition property of inequalities. • Reduce the coefficient of the variable to 1 by using the multiplication property of inequalities.	Solve $\dfrac{2}{3}x + \dfrac{1}{2} < x - \dfrac{1}{4}$ algebraically. $$\frac{2}{3}x + \frac{1}{2} < x - \frac{1}{4}$$ $$12\left(\frac{2}{3}x + \frac{1}{2}\right) < 12\left(x - \frac{1}{4}\right)$$ $$12\left(\frac{2}{3}x\right) + 12\left(\frac{1}{2}\right) < 12(x) - 12\left(\frac{1}{4}\right)$$ $$8x + 6 < 12x - 3$$ $$8x + 6 - 12x < 12x - 3 - 12x$$ $$-4x + 6 < -3$$ $$-4x + 6 - 6 < -3 - 6$$ $$-4x < -9$$ $$\frac{-4x}{-4} > \frac{-9}{-4}$$ $$x > \frac{9}{4}$$

Solve each inequality numerically for integer solutions.

22. $4x + 7 < 2x - 5$ **23.** $2.4x - 9.6 > 4.8$ **24.** $\frac{3}{5}x - \frac{7}{10} \le \frac{1}{5}x + \frac{1}{2}$ **25.** $\frac{1}{2}x - 2 \ge -\frac{1}{3}x - \frac{11}{3}$

Solve each inequality graphically.

26. $2x - 2 > -x + 4$ **27.** $1.2x + 0.72 \le -2.1x + 8.64$

28. $(x + 6) - 3(x + 1) < (2x + 5) - 2(2x + 3)$ **29.** $(x + 3) + (x + 1) \ge 3(x + 1) - (x - 1)$

Solve each inequality algebraically. Represent the solution in interval notation.

30. $412 + x > 671$

31. $y - \frac{7}{13} < \frac{11}{39}$

32. $3x + 7 < 4x + 21$

33. $14 + 2x < 2x$

34. $8.7x + 4.33 \le -2.4x - 33.41$

35. $6.8z - 9.52 \ge 0$

36. $2(x + 5) - (x + 6) < 2(x + 2)$

37. $3(x - 4) + 2(x + 1) > 5x + 10$

38. $-3 < 4 - 2x \le 0$

39. $8 \le 3(x - 7) + 5 < 20$

For each situation, write and solve an inequality that represents it.

40. Your company has placed a limit on car rentals of no more than $150.00 per trip. If the rental agency charges a flat fee of $49.95 plus $0.18 per mile driven, what range of miles will keep you within budget?

41. Ali must average sales of more than $1500 per month in order to receive his six-month commission. For the first five months, his sales were $2100, $1300, $1650, $1250, and $1725. What should his sales be in the sixth month in order for him to receive his commission?

42. A rectangular flower bed must have a perimeter of no more than 40 feet. If the length must be 4 feet more than the width, what widths would be within the limits?

43. A testing center charges students $35.00 to take a standardized test. The center pays a license fee of $20.00 per student to administer the test. The average daily cost to establish and staff the center is $225.00. How many students must be tested each day to realize a profit of at least $500.00 per day?

44. In order to earn a $B+$ grade for a college mathematics course, a student must have an average of 88 to 92 points. Paul's test scores are 89, 96, 89, 80, and 100. What scores on his last test will earn him a $B+$ in the course?

45. A man who is 5 feet, 6 inches, tall has a body mass index (BMI) given by the expression $0.15x + 2$, where x is his weight in pounds. If the man has other heart risks besides being overweight, a doctor will prescribe medication if his BMI is at least 27. Determine the weights for which a doctor will prescribe medication.

7.3

Recall	Examples

Linear inequality in two variables
• The standard forms are
  $ax + by < c$
  $ax + by > c$
  $ax + by \le c$
  $ax + by \ge c$
where a, b, and c are real numbers and a and b are not both 0.
	Linear inequalities in two variables
	$2x + 4y < 5$
	$x \le 7$
	$y \ge \frac{2}{3}$

Solution of a linear inequality in two variables • An ordered pair is a solution of a linear inequality in two variables if the value of its coordinates satisfy the inequality.	Is $(1, -4)$ a solution of the inequality $x + 2y > -7$? $\begin{array}{c c} x + \ 2y & > -7 \\ \hline 1 + 2(-4) & -7 \\ -7 & \end{array}$ Because $-7 \not> -7$, $(1, -4)$ is not a solution of the inequality.

Recall	Examples
Graph a linear inequality in two variables • If the inequality contains the variable y, solve for y. If not, solve for x. • Graph the boundary line determined by the related equation $y = mx + b$ or $x = c$. • Use a solid line if the boundary is included and a dashed line if the boundary is not included. • Shade below the boundary if the inequality solved for y is a "less than" inequality. Shade above the boundary if the inequality solved for y is a "greater than" inequality. Shade to the left if the inequality solved for x is a "less than" inequality. Shade to the right if the inequality solved for x is a "greater than" inequality.	Graph $-2x + 4y < 1$. Solve for y: $y < \dfrac{1}{2}x + \dfrac{1}{4}$ 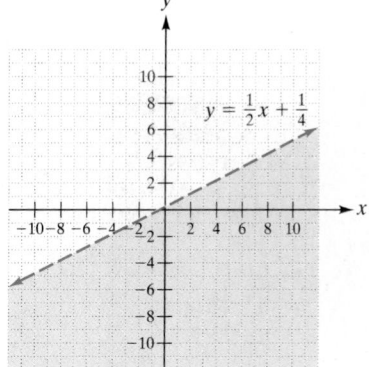 Graph $-2x \geq 6$. Solve for x: $x \leq -3$ 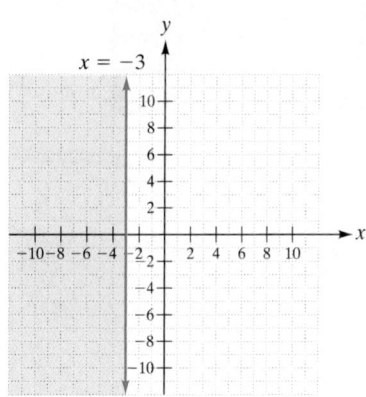

Identify each inequality as linear or nonlinear. Express each linear inequality in standard form.

46. $x + 2y < 12$
47. $y < \dfrac{2}{3}x + \dfrac{5}{9}$
48. $x^2 + y^2 \geq 1$

49. $0.3x + 2.9 > 1.4y$
50. $y \geq \sqrt{x} - 1.44$
51. $y < x^2 + 9$

Graph each inequality.

52. $12x + 6y < 48$
53. $y > \dfrac{3}{5}x - 6$
54. $4y \leq x + 12$
55. $y + 9 \geq 12$

56. $5y - 2 < y - 10$
57. $2x + 16 > x + 18$
58. $8x - 12y > 24$
59. $4.4x + 1.1y \geq 12.1$

60. $7.4x - 14.8y \leq 29.6$
61. $\dfrac{x}{12} + \dfrac{y}{8} > -\dfrac{5}{4}$
62. $-x - y < 2$
63. $y > -9x + 6$

For each situation, write and graph a linear inequality in two variables that represents the situation. From the graph, determine two possible solutions.

64. Oksana has at most $85 to spend on plants for her flower garden. She can buy rhododendrons at $4 a pot and azaleas at $6 a pot. What combinations of plants can she buy and not go over her budget?

65. Rosa gained 5 points for every homework exercise she solved correctly and lost 3 points for every exercise she solved incorrectly or skipped. What combinations would allow her to score at least 80 points on her homework assignment?

66. The U.S. Postal Service rates for priority mail are shown in the following table:

Weight of package	Rate
Up to 1 pound	$3.50
Up to 2 pounds	3.95
Up to 3 pounds	5.20
Up to 4 pounds	6.45
Up to 5 pounds	7.70

Pickup service is available for $10.25 per stop, regardless of the number of pieces. Determine the numbers of 2.5-pound packages and 5-pound packages that can be picked up and shipped without exceeding a shipping budget of $120.00. Can ten 2.5-pound packages and six 5-pound packages be shipped without exceeding the budget? Can twelve 2.5-pound packages and eight 5-pound packages be shipped without exceeding the budget?

7.4

Recall	Examples
Graph a system of linear inequalities • Graph each inequality in the system on the same coordinate plane. • Label all intersections of the boundary lines. • Shade the region of the coordinate plane that contains solutions of all the inequalities. • Check a point in the overlapping region to determine whether that point is a solution.	Graph. $\quad y < 2x + 5$ $\qquad\quad y \geq -3x - 2$ You can verify that the point $(2, 1)$ in the shaded region checks. $\dfrac{y < 2x + 5}{1 \mid 9}$ $\dfrac{y \geq -3x - 2}{1 \mid -8}$

Graph each system of linear inequalities. Label the points of intersection.

67. $2x + y > 10$
$\quad y < 3x - 5$

68. $3(x + 2) + 1 > -5$
$\quad 2x - y < -8$

69. $x + 6 < 4$
$\quad 2(y + 2) > y + 3$

70. $x - 2y \geq 12$
$\quad 2x + 3y < -6$

71. $y < 3x - 6$
$\quad y < -3$

72. $2x + 3y \leq 6$
$\quad x + y \leq 1$

73. $y < 2x - 15$
$\quad y > -3x + 10$

74. $y > 2x + 1$
$\quad x > 2y + 7$

75. $2y + 6 \leq 3y + 5$ **76.** $y < 6 - x$

$2(x - 3) + 1 > 5$ $y < 2x + 1$

77. $y < 25 - \dfrac{1}{4}x$

$y < \dfrac{2}{3}x + 5$

$x \geq 0$ $x \geq 0$

$y \geq 0$ $y \geq 0$

For each situation, write a system of linear inequalities in two variables that represents that situation. Graph the inequalities. Determine one possible solution from the graph.

78. Oksana has at most $85 to purchase potted rhododendrons at $4 each and azaleas at $6 each. She needs at least four more rhododendron plants than azaleas. What possible combinations will meet both criteria?

79. If you have $4000 to invest, part at 5% and part at 6% simple interest for a year, how much can you invest at each rate to earn at least $225 interest?

80. The U.S. Postal Service rates for priority mail are shown in the following table:

Weight of package	Rate
Up to 1 pound	$3.50
Up to 2 pounds	3.95
Up to 3 pounds	5.20
Up to 4 pounds	6.45
Up to 5 pounds	7.70

Pickup service is available for $10.25 per stop, regardless of the number of pieces. Suppose that you must mail 2.5-pound packages and 5-pound packages and the number of 2.5-pound packages must never be more than twice the number of 5-pound packages. Determine the numbers of 2.5- and 5-pound packages that can be picked up and shipped without exceeding a shipping budget of $120.00 or violating the restriction. Can ten 2.5-pound packages and twelve 5-pound packages be shipped? Can seven 2.5-pound packages and six 5-pound packages be shipped?

CHAPTER 7 CHAPTER REVIEW

Graph each inequality on a number line. Write equivalent interval notation for the inequality.

1. $x < -2$ **2.** $x > 7$ **3.** $-1 < x < 3$

4. $x \geq 2.6$ **5.** $x \leq 3\dfrac{2}{3}$ **6.** $-2.4 < x \leq 4\dfrac{1}{3}$

Solve numerically for integer solutions.

7. $5x + 3 < 2x - 9$ **8.** $5.6x - 15.3 > 1.3x + 19.1$ **9.** $\dfrac{1}{6}x + \dfrac{23}{3} \leq \dfrac{13}{6} - \dfrac{5}{3}x$ **10.** $\dfrac{3}{7}x + \dfrac{9}{5} \geq \dfrac{4}{5}x - \dfrac{17}{5}$

Solve each inequality graphically.

11. $5x - 13 > 3(1 - x)$ **12.** $2.1x + 31.71 \leq 8.19 - 3.5x$

13. $3(x - 1) + (x - 1) < 2(x - 1) + 2x - 1$ **14.** $4(x + 2) - 3(x - 5) \leq 5x + 8 - 4(x + 3)$

Solve each inequality algebraically. Represent the solution in interval notation form.

15. $173 - x < 359$ **16.** $z - \dfrac{4}{17} > \dfrac{15}{34}$ **17.** $5x + 4 > 3x + 18$

18. $8x < 8x - 16$ **19.** $3.5x + 19.88 \geq -1.9x + 4.76$ **20.** $2.6y + 9.62 \leq 0$

21. $3(x + 3) < 4(x + 1) - 2(x - 2)$ **22.** $3(x - 3) + 2(x + 2) < 5x + 7$ **23.** $15 < 4(2x - 1) + 5 \leq 25$

Graph each system of linear inequalities. Label points of intersection.

24. $2y + 2 > y$
$\quad y < -x + 1$

25. $2x + 3 > x$
$\quad 2x < 6$

26. $y \geq 2x - 8$
$\quad 3x + y < 8$

27. $2(x - 1) - 4 > 0$
$\quad 2x + y \leq 3$

28. $x + y < 9 - x$
$\quad 2x + y > 5$

29. $3(x - 3) - 1 > 8$
$\quad 3y - 10 < 15 - 2y$

30. $y < 3x + 7$
$\quad x - y > -1$

31. $x - 5y \leq 15$
$\quad x + 5y \leq -5$

32. $y < -4x + 6$
$\quad y > 3x - 8$

33. $y > x$

$\quad x > 2 - y$

34. $x + 9 \leq 3$

$\quad 3y > 2y - 1$

35. $y < -2x + 7$

$\quad y < 2x$
$\quad x \geq 0$
$\quad y \geq 0$

36. $y > \dfrac{3}{4}x - 4$

$\quad y > -\dfrac{2}{3}x + 3$
$\quad x \geq 0$
$\quad y \geq 0$

Graph each linear inequality in two variables.

37. $9x - 5y < 45$

38. $y > \dfrac{4}{3}x - 5$

39. $3y \leq 2x + 9$

40. $y + 13 \geq 8$

41. $-2y - 6 < y - 11$

42. $x - 13 > 3x + 19$

43. $x - 7y > 21$

44. $2.7x + 5.4y \geq 16.2$

45. $4.8x - 1.8y \leq 14.4$

46. $\dfrac{x}{12} + \dfrac{y}{9} > -\dfrac{1}{3}$

47. $x - y < 7$

48. $y > -5x + 7$

49. Based upon data from the U.S. Department of Health and Human Services, body weight classification depends upon a person's height, among other factors. The following table suggests body weight classifications for a person who is 5 feet, 9 inches, tall:

Classification	Weight, lb
Healthy weight	130–165
Moderately overweight	166–190
Severely overweight	Over 190

For each category, write a linear inequality that represents that category.

For each situation, write and solve a linear inequality in one variable that represents that situation. State one possible solution from the solution set.

50. The setup cost for producing gourmet packs of coffee is $255.00. The materials cost $2.50 per pack. The packs will be sold for $12.50 each. How many packs must be sold in order to make a profit of at least $1200.00?

51. Catherine wants to keep her average telephone bill below $42. Her first five months had bills of $45, $36, $52, $48, and $31. What are the limits on her bill for the sixth month?

52. The width of a rectangular street sign has been set at 18 inches. What must the length be if the area of the sign must be more than 600 square inches?

53. Tony has designed a computer software package that he markets through the Internet. He sells the package for $19.95, including shipping and handling. He estimates that he spends $1.45 for each package that he ships. His only overhead charge is a monthly charge of $17.00 for his Web site. How many packages must Tony sell in order to realize a profit of at least $600 per month?

54. Tony tracked his profits for the first five months that he sold his software package. His profits were $344, $434, $254, $705, and $723. What should his profit be for the sixth month if he wishes his average profit to be between $500 and $600?

55. If the U.S. Postal Service charges $2.37 to mail a 4-pound package to Zone 5, how many packages can you send and still remain within a mailing budget of $100.00?

Write and graph a linear inequality in two variables that represents the situation. Determine one possible solution from the graph.

56. A soccer team needs to earn at least 25 points during the season to make the playoffs. If it earns three points for each win, one point for each tie, and no points for a loss, what combinations of wins and ties will land the team in the playoffs?

For each situation, write a system of linear inequalities in two variables that represents the situation. Graph the inequalities. Determine one possible solution from the graph.

57. The U.S. Postal Service charges $2.37 to mail a 4-pound package to Zone 5 and $2.20 to mail a 3-pound package to Zone 5. A company has a mailing budget of $100 per day and has a further restriction that the number of 3-pound packages mailed must never exceed the number of 4-pound packages mailed. Determine how many packages can be mailed and still have the company meet the budget restriction and the added restriction. Can the company send twenty-five 4-pound packages and twelve 3-pound packages and remain within the budget? Can the company send thirty-two 4-pound packages and twenty 3-pound packages and remain within the budget?

58. Farmer McGregor plants oats and wheat on his farm. For conservation purposes, he plants at least twice as many acres of wheat as oats. He can handle up to a total of 540 acres of planting. What combinations of plantings can he realistically consider?

59. A realtor has three efficiency apartments and five regular apartments to rent. The regular apartments will rent for at least $75 more than the efficiency apartments. If the realtor would like to gross at least $6000 per month, what combination of rental rates will meet her wishes?

CHAPTER 7 TEST

 TEST-TAKING TIPS

In attempting to solve a problem on a test, it is helpful to follow some routine procedures. Practice these procedures for homework so that you will be comfortable using them during a test.

- Write down key information about the problem.
- Identify what you need to solve.

- List any formulas or rules you may need for the solution.
- Numerically estimate the answer, using rounded numbers.
- Imagine the same problem with simpler numbers.
- If time permits, solve the problem two different ways (for example, algebraically and graphically) as a check.
- Always check your work.

Identify each inequality as linear or nonlinear. Write the linear inequalities in standard form.

1. $2x - 3y > x + 8$ **2.** $4x^2 + 2x < x - 9$ **3.** $5(x - 3) \geq 4 - (x + 1)$ **4.** $\frac{1}{2}x - 4 \geq y + \frac{3}{8}$

Solve each inequality and represent the solution by using the number line and interval notation if possible.

5. $7(x + 2) - 3(x + 1) > 4(x + 8)$ **6.** $5x + 9 < 2x - 3$

7. $\frac{4}{5}(x - 10) < \frac{1}{5}(4x + 5) + 1$ **8.** $5a - 7 \geq 8a + 1$ **9.** $-2 < 3(x - 4) - 2 \leq 1$

Graph each linear inequality in two variables.

10. $y < -2x + 5$ **11.** $x + 3 > 2x - 1$

Graph each system of linear inequalities in two variables. Label the points of intersection.

12. $x - y > 4$ **13.** $y \geq 4x - 5$ **14.** $3y - 1 > 2$
      $x + 2y > 4$ $y > -3x + 4$ $x - 3 \leq -5$

15. The Fujita Scale described in this chapter classifies tornadoes according to the amount of damage they inflict. The following table lists the classifications of tornadoes that are considered severe or even worse:

F-Scale number	Intensity	Wind speed
F3	Severe tornado	158–206 mph
F4	Devastating tornado	207–260 mph
F5	Incredible tornado	261–318 mph
F6	Inconceivable tornado	319 or higher

For each of the categories listed, write a linear inequality that represents that category.

In exercises 16–18, write and solve a linear inequality in one variable to represent the application. Determine one possible solution from the solution set.

16. The average of three exams must be at least 80 to pass a course. If the first exam score was 83 and the second exam

score was 72, what are the possible scores on the third exam that will enable a student to pass the course?

17. If the last three recorded tornadoes had wind speeds of 168 mph, 172 mph, and 225 mph, what wind speeds for the next tornado will result in an average wind speed for the four tornadoes that is within the range of a severe tornado?

18. Alexandra opened a shop to sell gift baskets. She sells the baskets for an average price of $49.00 each. The materials, food, and gadgets that she uses to create the baskets cost her an average of $23.50 per basket. She rents the storefront, and her overhead costs average $250.00 per week. How many baskets must she sell in order to make a profit of at least $300.00 per week?

In exercise 19, write and graph a linear inequality in two variables to represent the application. Determine one possible solution from the solution set.

19. A Cub Scout earns 5 points for every good deed and 2 points for every activity sheet completed. If a scout must earn at least 20 points in order to receive a medal, what combinations of good deeds and activity sheets will suffice?

In exercise 20, write and graph a system of linear inequalities that represents the application.

20. Alexandra can ship a 10-pound basket to Zone 102 for $22.50 and a basket of the same weight to Zone 103 for $26.25. The number of baskets shipped to Zone 103 must never be less than the number shipped to Zone 102. Alexandra's weekly shipping budget is $500.00? How many baskets can she ship and still meet her budget re-striction and the added restriction? Will she remain with-in budget if she ships 10 baskets to Zone 102 and 15 bas-kets to Zone 103? Will she be within her budget if she ships 5 baskets to Zone 102 and 10 baskets to Zone 103?

21. If you have at most $6000 to invest, part at 4% simple in-terest and part at 8% simple interest, what investment strategies will assure you of earning at least $400 in in-terest for a year?

22. What do we mean by the solution of a system of linear inequalities in two variables?

Project

Part I

You have decided that you need additional income to balance your budget. In order to receive extra income, you plan to produce and sell a product to consumers for a profit.

Write a report on your business venture. Include the following information:

1. The name of your business

2. The product you wish to produce and sell

3. The fixed and variable costs associated with producing and selling your product (include a list of these costs)

4. A cost function

5. The selling price of your product

6. A revenue function

7. The number of items you must produce and sell to break even on your venture

8. The amount of extra income you need and how many items you need to produce to achieve at least that amount

Part II

In this chapter, you have encountered several applications that used real data. For example, one application involved blood pressures, another involved body mass index, and a third involved shipping rates for United Parcel Service deliveries. These data may change over time. To obtain more current data, you may need to research through the library or through internet sites.

Choose an application in this chapter and either verify that the data has not changed or obtain more current data. Reconstruct the application if more current data exists. Write a short report on your findings, listing the sources for your data, and the results of your update of the application, if the data did indeed change.

CHAPTERS 1–7 CUMULATIVE REVIEW

Evaluate.

1. $-(-9)$ **2.** $-|-9|$ **3.** $\sqrt[3]{-\dfrac{27}{64}}$

4. $\sqrt{10}$ **5.** $\left(-\dfrac{9}{14}\right)\left(\dfrac{7}{3}\right)$ **6.** $-\dfrac{3}{8} \div \left(-1\dfrac{2}{3}\right)$

7. $12(-3) \div (-6)(2) \div (-2)$ **8.** $40 + 16 \div 8 - \sqrt{3^2 + 7 \cdot 5} + 5$

9. $\dfrac{2(5^2 - 10) + 4^2 - 1}{8^2 - 2(32)}$

Simplify.

10. $6x - 2(4x - 1)$

11. $4[2(x - 3) + 1] - [5(2x - 4) - 6]$

Consider the relation $y = 2x^2 - 3$.

12. Graph the relation.

13. What is the domain of the relation? The range?

14. Is the relation a function? Justify your answer.

15. Determine the relative minima and the relative maxima.

16. Determine the x-values for which the relation is increasing and decreasing.

17. Determine the x-intercept and the y-intercept.

Solve.

18. $7x - 3 = 5$ **19.** $(x + 3) - 2(3x + 4) = 5$ **20.** $1.2(x + 3) - 4(0.3x + 0.15) = 3$

21. $|2x + 6| = 10$ **22.** $5|x| + 8 = 8$

Solve. Represent the solution as an inequality, using interval notation, and as a graph on a number line.

23. $-12x + 4 \geq -2x + 8$ **24.** $4(x + 3) - 3(x - 2) \leq x - 5$ **25.** $30 < 2x + 10 \leq 70$

Graph. Label enough points to determine the graph.

26. $f(x) = -3x + 4$ **27.** $6x + 3y = 9$ **28.** $2x + 3 = 11$

29. $3y - 2 = 5$ **30.** $y < 4x - 3$

Determine if the graphs of the pair of equations are coinciding lines, parallel lines, only intersecting lines, or both intersecting and perpendicular.

31. $2x + 3y = 21$
$3x - 2y = 2$

Solve the systems of equations.

32. $y = 3x + 4$
$y = 2x - 5$

33. $4x + 2y = 8$
$y = -2x + 4$

Graph the system of inequalities. Determine one possible solution from the graph.

34. $y \geq -x$
$y < 2x + 4$

Determine the slope of a line that satisfies the following conditions.

35. Passes through $(-2, 3)$ and $(-1, -2)$ **36.** Passes through $(6, -5)$ and $(6, 3)$

37. $y = 2x + 3$ **38.** $y = 1.4$

Write an equation for a line that satisfies the following conditions.

39. Passes through $(-2, 3)$ and $(-1, -2)$

40. Passes through $(5, 4)$ and is parallel to the graph of $2x + 3y = 1$.

Write in scientific notation.

41. 5,340,000

Write exercise 42 and 43 in standard notation.

42. 1.2×10^{-4}

43. $-4.783\text{E}-5$

44. Solve for L: $P = 2L + 2W$

45. Given $f(x) = x^2 + 2x - 1$, find $f(-3)$.

46. An investment plan offers 11% simple interest on an investment. How much should Lance invest in order to have $2775 in his account at the end of one year?

47. Two angles are complementary. The larger angle measures 25 degrees more than twice the smaller angle. Find the measures of the two angles.

48. Michael's Coffee Shop sells a special blend called Mike's Favorite. For the blend, Michael mixes Hazelnut Coffee, which sells for $7.50 per pound, with Cinnamon Coffee, which sells for $6.75 per pound. How much of each flavor must he use to create 10 pounds of the Mike's Favorite blend if he sells it for $7.00?

49. April scored 82, 88, 80, and 95 on the first four tests in her Algebra class. There is one more test in the class. In order to earn a B, she must have a test average of at least 85. Determine the score she must earn on the last test in order to earn a B for the semester.

50. Reliable Rentals is running a special. The cost of renting a chain saw is $35 per day plus $1.50 per hour. Write a cost function to represent the cost of renting the chain saw for one day for x hours. What is the cost of renting the chain saw for 12 hours?

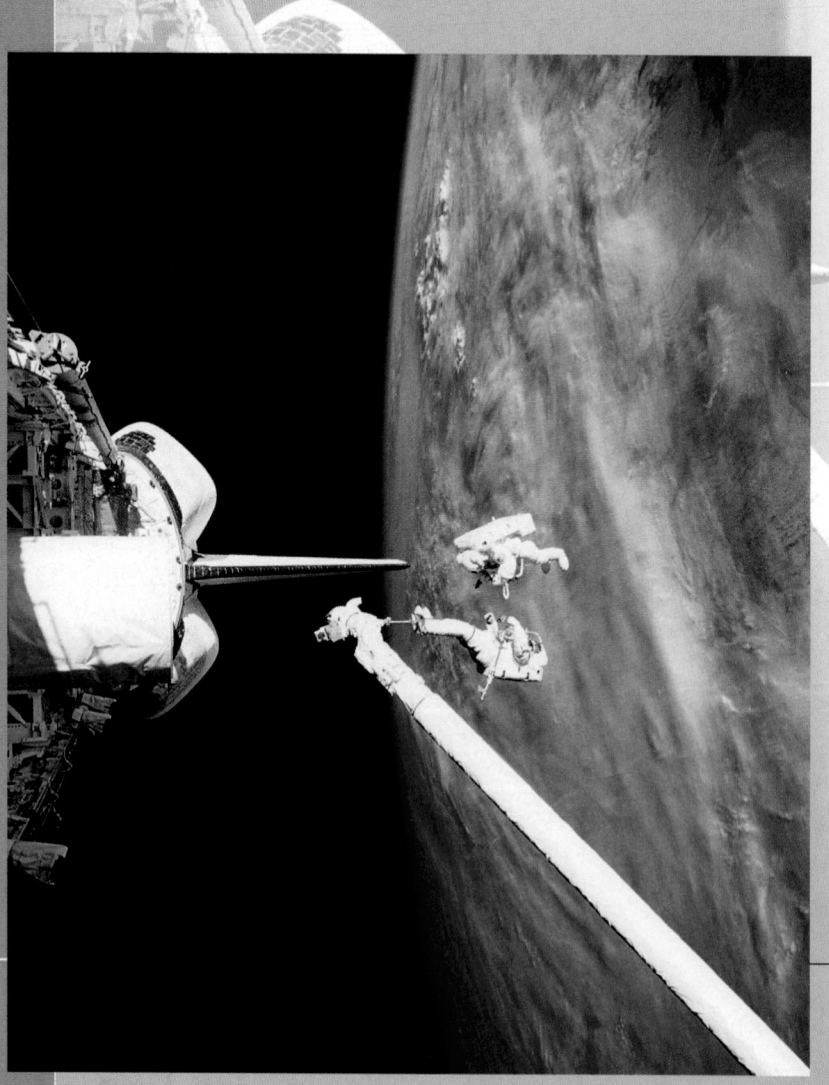

Chapter 8

Polynomial Functions

In this chapter, we return to the idea of functions, which we first discussed in Chapter 3. We examine a particular family of functions, referred to as polynomials, starting in this chapter and continuing for the next few chapters. Polynomials are worth special study because of their importance in so many areas of our daily lives, from calculating costs for repairing or selling homes to determining sales and profits for business items. Even a person's IQ can be determined by evaluating a polynomial.

The motion of an object dropped from a height or thrown into the air is determined by the Earth's gravity. The motion is described by a polynomial function called a quadratic function, which we will discuss in this chapter. We will present several different functions that model the motion of a falling object in typical situations. The fact that the motion of an object due to gravity could be described by a polynomial equation was one of the great scientific and mathematical discoveries of the 17th century and led directly to our present understanding of planetary motion and space travel. We will investigate the motion of a falling object in the project at the end of the chapter.

8.1 APPLICATION

To determine a person's IQ, psychologists ask the person to perform several different tasks, such as completing pictures, arranging pictures, and assembling objects. A sum of the scaled scores reported for each task is then found, and the total scaled score for the performance tasks is related to an IQ score. A polynomial that describes the IQ of a person in the age group 20–24 with a total scaled score of x is $0.006x^2 + 0.824x + 42.706$. Determine the IQ of a person in this age group who received a total scaled score of 60.

We will discuss this application further. See page 538.

8.1 **Introduction to Polynomials**

OBJECTIVES

1 Identify polynomials.

2 Identify the terms of polynomials and classify polynomials by the number of terms they possess.

3 Identify the degree of terms of polynomials and classify polynomials by degree.

4 Write polynomials in one variable in descending and ascending order.

5 Evaluate polynomials.

6 Model real-world situations by using polynomial expressions.

8.1.1 Identifying Polynomials

In Chapter 2, we introduced algebraic expressions. An **algebraic expression** is an expression that contains variables.

The **terms** of an algebraic expression are its addends. There are two types of terms: constant and variable. A **constant term** represents only one number. A **variable term** represents different numbers.

A **numerical coefficient** (or simply, **coefficient**) is the numerical factor of a term.

Two terms are said to be **like terms** if both are constants or if both are variable terms that contain the same variables with the same exponents. We **combine like terms** in order to simplify an algebraic expression. This means that we add the coefficients of the like terms, keeping the variable part intact. For example, $6x^2 - 5x + 3x + 7$ is an algebraic expression in which

$6x^2, -5x,$ and $3x$ are variable terms.
7 is a constant term.
$6, -5, 3,$ and 7 are coefficients of the terms.
$-5x$ and $3x$ are like terms.

By combining like terms, $6x^2 - 5x + 3x + 7 = 6x^2 - 2x + 7$.

In this chapter, we consider algebraic expressions called monomials. A **monomial** is either a constant term or a variable term consisting of one or more variable factors, each having a nonnegative integer exponent. The following are examples of monomials:

2 is a constant term.

$-3x$ is a variable term with a coefficient of -3 and the variable, x, having an exponent of 1.

$4x^2$ is a variable term with a coefficient of 4 and the variable, x, having an exponent of 2.

$-5x^2y^3z$ is a variable term with a coefficient of -5 and three variables x, y, and z, having exponents of 2, 3, and 1, respectively.

Algebraic expressions are not monomials if they have a variable raised to a power other than a nonnegative integer. Thus, expressions with a variable in the denominator of a fraction and expressions with a variable in the radicand of a radical expression are not monomials. We will explain why this is so later in the text. For example, $x^{2/3}$, x^{-5}, $\frac{3}{x^2}$ and $-25\sqrt{4x}$ are not monomials.

A **polynomial** is a monomial or a sum of monomials. Following are some examples:

$x^3 y$ is a monomial.

$x + 2$ is the sum of two monomials: x and 2.

$-7x^2 - 4x$ is the sum of two monomials: $-7x^2$ and $-4x$.

$\frac{1}{3}a + 9b^2 - a^3 b$ is the sum of three monomials: $\frac{1}{3}a$, $9b^2$, and $-a^3 b$.

EXAMPLE 1   Determine whether each expression is a polynomial.

     a. $\dfrac{2}{3}x^2 + \sqrt{5}xy^3$ **b.** $\dfrac{1}{x^2} - 4\sqrt{xy}$

Solution

a. $\frac{2}{3}x^2 + \sqrt{5}xy^3$ is a polynomial, because the two terms, $\frac{2}{3}x^2$ and $\sqrt{5}xy^3$, are monomials, each consisting of a coefficient and variables with nonnegative integer exponents.

b. $\frac{1}{x^2} - 4\sqrt{xy}$ is not a polynomial, because the first term, $\frac{1}{x^2}$, and the second term, $-4\sqrt{xy}$, are not monomials. The first term has a variable in the denominator of a fraction, and the second term has a variable in a radicand.

8.1.1 Checkup

In exercises 1–6, determine whether each expression is a polynomial.

1. $3xy^4$ **2.** $\dfrac{2}{3}x - 3x^2$ **3.** $\dfrac{2}{x} - 5$

4. $3x^{-2} + 4x$ **5.** $\sqrt{3}x + 2$ **6.** $3\sqrt{x} + 2$

7. Explain the difference between the terms of an algebraic expression and the factors of a term.

8. All monomials are polynomials, but not all polynomials are monomials. Explain what this statement means.

9. A polynomial can be simplified by combining its like terms. What are like terms?

8.1.2 Classifying Polynomials by Number of Terms

We can classify a polynomial by the number of its terms. A polynomial with one term is called a monomial. A polynomial with two terms is a **binomial**, and a polynomial with three terms is a **trinomial**. We usually refer to a polynomial with more than three terms simply as a polynomial. Before we classify a polynomial, we must simplify it by combining like terms.

EXAMPLE 2   Classify each polynomial as a monomial, a binomial, a trinomial, or a polynomial.

     a. $6a^2 + 5b^2 + 4ab - a^2 - 2ab + c^2$

     b. $-x^2 + 2xy - 3x + x^2 - y^3$

Solution

First, simplify each polynomial by combining like terms.

a. $6a^2 + 5b^2 + 4ab - a^2 - 2ab + c^2 = 5a^2 + 5b^2 + 2ab + c^2$

There are four terms: $5a^2, 5b^2, 2ab,$ and c^2.

The expression is a polynomial.

b. $-x^2 + 2xy - 3x + x^2 - y^3 = 2xy - 3x - y^3$ Note: $-x^2 + x^2 = 0x^2,$ or 0

There are three terms: $2xy, -3x,$ and $-y^3$.

The expression is a trinomial.

 8.1.2 Checkup

In exercises 1–4, classify each expression as a monomial, a binomial, a trinomial, or a polynomial.

1. $6b^3 + 4b^2c - 3bc^2 - c^3$

2. $3x - 5 + 7x + 12$

3. $5p^2 - 3p + 12 - 5p + 7$ **4.** xyz

5. How can you tell whether a polynomial is a monomial, a binomial, or a trinomial?

8.1.3 Classifying Polynomials by Degree

The **degree of a term** is the sum of the exponents of its variable factors. The **degree of a polynomial** is the largest degree of its variable terms. Remember to simplify the polynomial before determining its degree.

EXAMPLE 3   Determine the degree of each term of each polynomial, and then determine the degree of the polynomial.

a. $6x^2 + 5x^2y^3 - 7x^2$ **b.** $-5x + 6 + 7x - 2x$ **c.** $9x^2y^3z^4$

Solution

First, simplify each polynomial by combining like terms.

a. $6x^2 + 5x^2y^3 - 7x^2 = -x^2 + 5x^2y^3$

The terms are $-x^2$, with a degree of 2, and $5x^2y^3$, with a degree of $2 + 3$, or 5.

The degree of the polynomial is 5, the largest degree of any of its terms.

b. $-5x + 6 + 7x - 2x = 6$

The constant term, 6, can be written as $6x^0$ (because any nonzero number raised to a power of 0 equals 1) and has a degree of 0.

The degree of the polynomial is 0.

c. $9x^2y^3z^4$ already is simplified; it is a monomial.

The polynomial's one term has a degree of $2 + 3 + 4$, or 9.

The degree of the polynomial is 9.

 8.1.3 Checkup

In exercises 1–3, determine the degree of each term of the polynomial, and then determine the degree of the polynomial.

1. $6x^2 - 3xy^3 + 5y^3$ **2.** $5x + 3x^3 - 2x^2 - 3x^3 + 1$

3. $-5x^5y^2z^3 + 2x^2y^3z^4$

4. What is the difference between the degree of a term of a polynomial and the degree of a polynomial?

8.1.4 Writing Polynomials in Descending and Ascending Order

The conventional way to write a polynomial in one variable is to write the terms in order of descending degrees. That is, a polynomial in one variable is written in **descending order** with decreasing values of its variable exponents. Remember that a constant term has a degree of 0 and is always written as the last term. The polynomial $4x^3 - 2x^7 + x^2 - 8$, written in descending order, is $-2x^7 + 4x^3 + x^2 - 8$.

Sometimes it may be desirable to write a polynomial in one variable in ascending order. A polynomial in one variable is written in **ascending order** with increasing values of its variable exponents. The polynomial $4x^3 - 2x^7 + x^2 - 8$, written in ascending order, is $-8 + x^2 + 4x^3 - 2x^7$.

HELPING HAND Take special care when arranging polynomials with terms having negative coefficients. The negative sign must remain with these terms.

EXAMPLE 4 Complete the following table by writing each polynomial in descending and ascending order:

Polynomial	Descending order	Ascending order
$6x^2 + 8x - 7x^4$		
$5 - x^4 + x$		

Solution

Polynomial	Descending order	Ascending order
$6x^2 + 8x - 7x^4$	$-7x^4 + 6x^2 + 8x$	$8x + 6x^2 - 7x^4$
$5 - x^4 + x$	$-x^4 + x + 5$	$5 + x - x^4$

 ### 8.1.4 Checkup

Complete the following table by writing each polynomial in descending and ascending order:

	Polynomial	Descending order	Ascending order
1.	$3 + 2x^3 - 5x - x^2$		
2.	$12 + y^4 + y$		

8.1.5 Evaluating Polynomials

Like an algebraic expression, a polynomial represents different values, depending on the value(s) of its variable(s).

The process to determine the value of a polynomial is called **evaluating** the polynomial.

To evaluate a polynomial,
- Substitute a value for each variable.
- Determine the value of the resulting numeric expression.

To evaluate a polynomial on your calculator,
- Enter the expression with its substituted values.

or
- Store the values of each of the variables and enter the polynomial.

EXAMPLE 5 Evaluate $x^2 + 2xy - 7y^2$

a. for $x = 2$ and $y = -1$. **b.** for $x = \dfrac{3}{4}$ and $y = \dfrac{1}{2}$.

Solution

a. $x^2 + 2xy - 7y^2 = (2)^2 + 2(2)(-1) - 7(-1)^2$ *Substitute 2 for x and −1 for y.*

$\qquad = 4 - 4 - 7$

$\qquad = -7$

b. $x^2 + 2xy - 7y^2 = \left(\dfrac{3}{4}\right)^2 + 2\left(\dfrac{3}{4}\right)\left(\dfrac{1}{2}\right) - 7\left(\dfrac{1}{2}\right)^2$ *Substitute $\tfrac{3}{4}$ for x and $\tfrac{1}{2}$ for y.*

$\qquad = \dfrac{9}{16} + \dfrac{3}{4} - \dfrac{7}{4}$

$\qquad = \dfrac{9}{16} + \dfrac{12}{16} - \dfrac{28}{16}$

$\qquad = -\dfrac{7}{16}$

The two methods of checking with your calculator are shown in Figure 8.1a and Figure 8.1b.

In Figure 8.1a, we substitute the value for the variables and evaluate the polynomial. In Figure 8.1b, we store the values of the variables and enter the polynomial. The calculator is programmed to evaluate the polynomial.

```
(2)²+2(2)(-1)-7(
-1)²
                -7
(3/4)²+2(3/4)(1/
2)-7(1/2)²▶Frac
             -7/16
```

```
2→X: -1→Y:X²+2XY-
7Y²
                -7
3/4→X:1/2→Y:X²+2
XY-7Y²▶Frac
             -7/16
```

Figure 8.1a   **Figure 8.1b**

 **HELPING HAND** When evaluating the same expression more than once, you may recall your last entry by keying [2nd] [ENTRY]. Then edit your entry, using the arrow keys and the delete [DEL] and insert [2nd] [INS] keys.

8.1.5 Checkup

1. Evaluate $2a^2 - 5ab - 9b^2$ for the given values.
  a. $a = -8$ and $b = 2$ **b.** $a = 4$ and $b = -3$

2. Evaluate $5x^3 - 55x^2 + 90x - 246$ for the given values.
  a. $x = 1.4$ **b.** $x = -\dfrac{4}{5}$

8.1.6 Modeling the Real World

Many real-life situations can be described by polynomials, including numerous geometric problems, business situations, and scientific relationships. Many other relationships can be approximated by polynomials, which sometimes makes them easier to work with.

EXAMPLE 6   A rectangular swimming pool is to be enclosed by a fence. The fenced area is a square.

    a. Write a polynomial for the fenced area that is not covered by the pool.

    b. If the pool is 25 feet by 15 feet, and the fence measures 30 feet on a side, determine the fenced area that is not covered by the pool.

Solution

a.

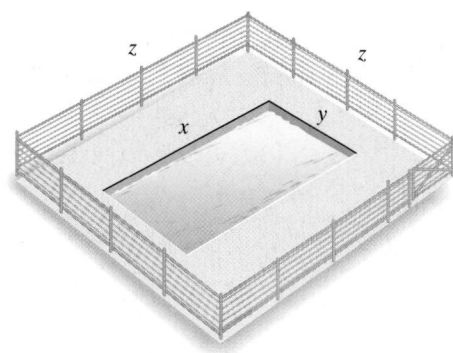

Let x = length of the pool (in feet)

    y = width of the pool (in feet)

    z = length of the side of the fence (in feet)

Fenced area: The area of a square is determined by the product of the two sides, $z \cdot z$, or z^2.

Pool area: The area of a rectangle is determined by the product of the length times the width, xy.

The fenced area not covered by the pool is the difference of the fenced area minus the pool area, or $(z^2 - xy)$ square feet.

b. x = 25 feet, y = 15 feet, and z = 30 feet

$$z^2 - xy = (30)^2 - (25)(15) \qquad \text{Substitute.}$$
$$= 900 - 375$$
$$= 525$$

The fenced area not covered by the pool is 525 square feet.

EXAMPLE 7   Jana decorates sweatshirts for Christy's Crafts in the mall. She is paid $15.00 per sweatshirt and $7.50 per delivery trip. She buys her supplies every delivery trip. She pays $5.41 per sweatshirt, $1.57 for the paint per sweatshirt, and $6.00 in gasoline for each trip.

    a. Write a polynomial that represents Jana's revenue.

    b. Write a polynomial that represents Jana's cost.

    c. If Jana produces 12 sweatshirts and makes one trip to the mall, what is her revenue and cost?

    d. What will be her profit for the 12 sweatshirts in part c?

Solution

Let x = the number of sweatshirts
y = the number of trips to the mall

a. Jana's revenue is \$15.00 per sweatshirt and \$7.50 per trip, or

$$15.00x + 7.50y$$

b. Jana's cost is \$5.41 per sweatshirt, plus \$1.57 per sweatshirt (paint) and \$6.00 per trip, or

$$5.41x + 1.57x + 6.00y = 6.98x + 6.00y$$

c. Evaluate the polynomials representing revenue and cost for $x = 12$ and $y = 1$.

Revenue: $15.00x + 7.50y = 15.00(12) + 7.50(1)$
$$= 187.50$$

Jana's revenue is \$187.50.

Cost: $6.98x + 6.00y = 6.98(12) + 6.00(1)$
$$= 89.76$$

Jana's cost is \$89.76.

d. Profit is the amount of revenue minus the amount of cost.

Profit: $187.50 - 89.76 = 97.74$

Jana's profit is \$97.74.

APPLICATION

To determine a person's IQ, psychologists ask the person to perform several different tasks, such as completing pictures, arranging pictures, and assembling objects. A sum of the scaled scores reported for each task is then found, and the total scaled score for the performance tasks is related to an IQ score. A polynomial that describes the IQ of a person in the age group 20–24 with a total scaled score of x is $0.006x^2 + 0.824x + 42.706$. Determine the IQ of a person in this age group who received a total scaled score of 60.

Discussion

A person's IQ is the value obtained by evaluating the polynomial $0.006x^2 + 0.824x + 42.706$ at a given total scaled score, in this case $x = 60$.

$0.006x^2 + 0.824x + 42.706$
$$= 0.006(60)^2 + 0.824(60) + 42.706$$
$$= 113.746$$
$$\approx 114$$

A person in the age group 20–24 with a total scaled score of 60 has an IQ of about 114.

 8.1.6 Checkup

1. A cabin with a square-shaped base is built upon a rectangular lot.

  a. Write a polynomial for the area of the lot that is not covered by the cabin.

b. If the base of the cabin measures 35 feet on a side, and the lot measures 150 feet by 200 feet, determine the area of the lot that is not covered by the cabin.

2. Mike builds folding tables and birdhouses to sell at a flea market. He can sell the tables for $12 each and the birdhouses for $15 each. The flea market charges him $35.00 each week to rent a booth. He spends $4.25 on materials for each table and $5.85 on materials for each birdhouse. His miscellaneous expenses (travel, packing bags, and so on) cost him $12 each week.

a. Write a polynomial to represent the revenue Mike earns in a week.

b. Write a polynomial to represent his cost for a week.

c. What will be Mike's revenue and cost if he makes and sells 22 tables and 19 birdhouses in a week?

d. What will be his profit in part c?

8.1 Exercises

Determine whether each expression is a polynomial.

1. $2a + 5$

2. $y^{1/3} + y - 1$

3. $x^2 - 3x + 2$

4. $2z^{-2} - 3z^{-1} + 4$

5. $x^{1/2} - 6x - 7$

6. $a^2 + 4a - 17$

7. $5x^2 + 12xy + 2y^2$

8. $\sqrt{b^2 - 4ac}$

9. $\sqrt{5}x - \sqrt{3}y$

10. $2.44x^2 - 1.05x - 15.7$

11. $5\sqrt{x} - 3\sqrt{y}$

12. $\frac{3}{4}x^2 + \frac{5}{8}x - \frac{9}{16}$

13. $\frac{3}{5}x^3 - \frac{2}{3}x^2 + 4x - \frac{7}{10}$

14. $\frac{1}{a} + a + 1$

15. $\frac{4}{x^2} + \frac{1}{x} - \frac{5}{7}$

16. $0.8x - 2.3$

17. $a^{-2} + 17a^{-1} + 13$

18. $\sqrt{6}r + 5$

19. $0.07b^2 - 2.6b + 13.908$

20. $4p^2 - 6pq + q^2$

Classify each expression as a monomial, a binomial, a trinomial, or a polynomial.

21. $3a + 4b - 5c$

22. $3x^3 - 2x^2y + 5xy^2 + 6y^3$

23. $2z^2$

24. $2 - 4b + 7$

25. $x - y$

26. $\frac{1}{7}x + \frac{2}{7}y + \frac{3}{7}z$

27. $4p^4 - 2p^3 + 11p - 57$

28. $b - 2b + 3b - 4b + 5b$

29. $6x^2 - 12 + 8x - 5x^2 + x - 17$

30. $-14x^3$

31. $3b - 4 + 7b$

32. $p + q$

33. $x + 2x + 3x + 4x + 5x$

34. $2a^2 - 6 + a + a^2 - 16a + 6$

35. $\frac{1}{2}x + \frac{2}{3}y - \frac{3}{4}z$

36. $b^2 - 4ac$

Determine the degree of each term of each polynomial and then the degree of the polynomial.

37. $2 - 15c$

38. $-3,298,175$

39. 123

40. $8x^4 + 5x^3y^3 - y^4$

41. $5 + 5x - 4x - x$

42. $b^2 - 4ac$

43. $7x^5 + 2x^2y^2 - 12$

44. $5a + 7$

45. $\pi r^2 + 2\pi rh$

46. $5a^5bc^3 + 6abc^4 + 3ab^2c^2$

47. $4x^2yz^{12} - 8xy^2z^9 + 3x^3y^3z$

48. $6y^2 + 3y - y + y^2 - 2y$

Write each polynomial in descending order.

49. $5 - 2a + 3a^2 + a^3$

50. $4a^9 - 17 + 3a^3 - 2a^6$

51. $\frac{3}{5}x + \frac{4}{5}x^3 - \frac{7}{15}x - \frac{8}{15}x^4$

52. $p^2 - p^5 + p^3 - p^4 - p - 1$

53. $0.1x^3 - 1.72 + 4.6x^2 + 3.06x^4$

54. $7x + 5x^3 + 9x^5 + 15x^7 + 23x^9 + 33$

In exercises 55–60, write each polynomial in ascending order.

55. $7b^2 - 6b^3 + 11 - 2b$

56. $x^2 - 8x^6 + 6x^5 - 4x^3 + x^4$

57. $\frac{2}{7}x + \frac{1}{7}x^5 - \frac{3}{14}x - \frac{5}{14}x^3$

58. $q^2 - q^5 + q^3 - q^4 - q - 1$

59. $0.5x^7 - 2.77 + 3.2x^3 + 9.76x^5$

60. $x^5 - 4x^4 - 2x^3 + 4x^2 - 2x + 4$

61. Evaluate $3x^2 - 4x + 1$ for the given values.
   a. $x = 4$
   b. $x = -2$
   c. $x = 0$

62. Evaluate $3x^3 + 2x^2 + x + 1$ for the given values.
   a. $x = 2$
   b. $x = -1$
   c. $x = 0$

63. Evaluate $-x^2 + 4xy + y^2$ for the given values.
   a. $x = 2, y = 3$
   b. $x = -2, y = -3$
   c. $x = 0, y = 0$

64. Evaluate $-a^3 + a^2b + ab^2 + b^3$ for the given values.
   a. $a = 2, b = 1$
   b. $a = -1, b = -2$
   c. $a = 0, b = 3$

65. Evaluate $1.3m^3 - 2.5m^2 + 3.7m - 4.9$ for the given values.
   a. $m = 1$
   b. $m = 2.5$
   c. $m = -1.5$

66. Evaluate $0.6z^2 + 3.7z - 5.8$ for the given values.
   a. $z = 2$
   b. $z = 4.3$
   c. $z = -2.5$

67. Evaluate $\frac{2}{3}x^2 - x - 3$ for the given values.

   a. $x = -\frac{3}{2}$
   b. $x = \frac{1}{4}$
   c. $x = 3$

68. Evaluate $\frac{3}{4}x^2 - \frac{3}{2}x - \frac{4}{3}$ for the given values.

   a. $x = \frac{2}{3}$
   b. $x = -\frac{2}{3}$
   c. $x = 8$

Geometric formulas were introduced in Chapter 2 and are summarized on the inside book cover. Exercises 69–80 refer to these formulas.

69. Write a polynomial for the perimeter of a triangle whose second side is twice as long as the first side and whose third side is 1 inch longer than the first side. What is the perimeter if the first side measures 8 inches?

70. Write a polynomial for the surface area of a cylinder with a height equal to the radius of its base. What is the surface area of the cylinder if the radius of the base measures 4 inches?

71. Write a polynomial for the surface area of a rectangular solid whose width and length are equal and whose height is 3 meters. What is the surface area of the solid if the width measures 1.5 meters?

72. Write a polynomial for the surface area of a rectangular solid whose width is 5 inches, whose length is 10 inches, and whose height is h inches. What is the surface area if the height is $3\frac{1}{2}$ inches?

73. Write a polynomial for the total area of the three geometric shapes shown in Figure 8.2. Shape A is a square and shapes B and C are rectangles.

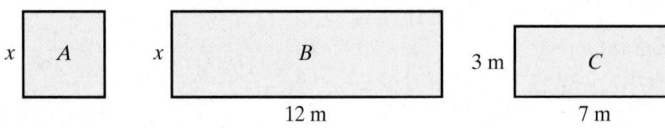

Figure 8.2

74. Write a polynomial for the total area of the three geometric shapes shown in Figure 8.3. Shape A is a right triangle and shapes B and C are squares.

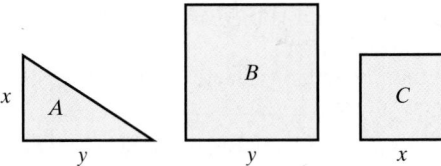

Figure 8.3

75. A rectangular patio measuring 15 feet by 20 feet is placed in a rectangular yard. Write a polynomial for the area of the yard not covered by the patio. What is the area of the yard not covered by the patio if the yard measures 75 feet by 120 feet?

76. A circular swimming pool is placed in a rectangular yard. Write a polynomial for the area of the yard not covered by the pool. What is the area of the yard not covered by the pool if the yard measures 25 meters by 35 meters and the pool has a radius of 4.5 meters?

77. Write a polynomial that represents the cost of replacing a square patio if the replacement cost is $12 per square foot. What will be the cost of replacing a square patio measuring 12 feet on a side?

78. Write a polynomial that represents the cost of replacing a circular patio if the replacement cost is $12 per square foot. What will be the cost of replacing a circular patio with a radius of 12 feet?

79. A contractor charges a flat fee of $200.00 plus $12.00 per square foot to build a deck for a house. The contractor incurs setup costs of $75.00 plus $2.75 per square foot for materials for the project.
   a. Write a polynomial that represents the revenue for building a square deck.
   b. Write a polynomial that represents the cost of building the deck.
   c. What will be the revenue and cost for a deck measuring 15 feet on a side? What will be the profit for building this deck?

80. A contractor charges a flat fee of $350.00 plus $15.00 per square foot to build a deck for a house. The contractor incurs setup costs of $100.00 plus $5.25 per square foot for materials for the project.
   a. Write a polynomial that represents the revenue for building a rectangular deck.
   b. Write a polynomial that represent the cost of building the deck.
   c. What will be the revenue and cost for a deck measuring 15 feet by 18 feet? What will be the profit for building this deck?

81. Use the polynomial that describes the IQ of a person in the age group 20–24 with a total scaled score x, $0.006x^2 + 0.824x + 42.706$, to find the IQ of a person in that age group who received a total scaled score of
   a. 65   **b.** 40   **c.** 50

82. Use the polynomial in the preceding exercise to find the IQ of a person in the given age group who received a total scaled score of
   a. 55   **b.** 70   **c.** 30

83. The tire industry advises consumers that they can optimize the mileage of their tires by using recommended tire pressures. A polynomial that relates the expected mileage of the tires (in thousands of miles) to the inflated tire pressure x (in pounds per square inch) is $-1.143x^2 + 75.214x - 1200$. Use this polynomial to estimate the mileage of the tire if the pressure is maintained at 32 pounds per square inch.

84. Use the polynomial in the preceding exercise to estimate the mileage of a tire whose pressure is maintained at 34 pounds per square inch.

8.1 Calculator Exercises

If you are evaluating a polynomial for several values, it is sometimes helpful to use the List feature of your calculator. As an example, suppose you wish to evaluate the polynomial $0.2x^3 - 1.2x + 6.4$ when x equals the values in the list $\{0, 1, 2, 3, 4, 5\}$. We would store these values in a list and then enter the polynomial by using the list symbol in place of the variable. The keystrokes to do this using the list L1 for x are as follows:

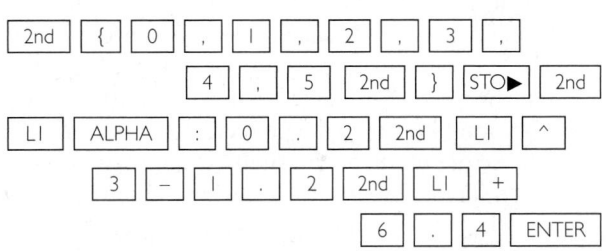

Read the results by using the arrow keys to move the cursor from left to right. The calculator returns the list $\{6.4, 5.4, 5.6, 8.2, 14.4, 25.4\}$, which represents the result of evaluating the polynomial for the list of x-values. You could also have used the TABLE feature of the calculator to do this. However, the list feature is more flexible when you have several variables, as the next example illustrates.

Evaluate $a^2 + 2ab - 3b^2$ when a and b equal the pairs of values in the list $\{(2, 6), (3, 7), (4, 8)\}$, where a is the first value of the pair and b is the second value of the pair. Store the a values in L1 and the b values in L2, maintaining the order of the pairs, and replace a and b by their list notations when you enter the expression. The keystrokes for doing this are as follows:

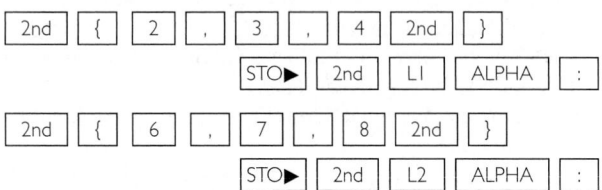

| 2nd | { | 2 | , | 3 | , | 4 | 2nd | } |

| STO▶ | 2nd | LI | ALPHA | : |

| 2nd | { | 6 | , | 7 | , | 8 | 2nd | } |

| STO▶ | 2nd | L2 | ALPHA | : |

| 2nd | LI | x^2 | + | 2 | 2nd | LI | 2nd |

| L2 | − | 3 | 2nd | L2 | x^2 | ENTER |

The calculator returns the list $\{-80, -96, -112\}$ as the results of the evaluation.

Use this method to evaluate the following polynomials for the given values:

1. $3x^3 - 8x^2 + 9x - 12$ for $x = \{-3, -1, 1, 3\}$

2. $4a^3 + 2a^2b + 2ab^2 + b^3$ for $\{(-5, 2), (-3, 0), (-1, -2)\}$

3. $\dfrac{1}{2}x^2 + \dfrac{3}{5}xy - \dfrac{1}{4}y^3$ for the (x, y) pairs
$$\left(\frac{1}{2}, \frac{2}{5}\right), \left(\frac{1}{4}, \frac{3}{5}\right), \left(-\frac{1}{2}, 5\right), \left(0, \frac{2}{5}\right)$$

8.1 Writing Exercises

As you begin the study of polynomials, you will be expanding your knowledge of mathematical models. This would be a good time for you to reflect on your feelings about the study of mathematics. Prepare a short essay describing your experiences to date in studying mathematics. If studying mathematics has been a struggle, try to list reasons why that is so. Do you have personal situations that compete with your studies for your attention? Does your work situation make it difficult to find study time? Have you had a bad past experience in the study of mathematics? Do you have a study group that you can use to help you? Have you developed a negative attitude toward mathematics? If so, can you understand where this attitude originated?

After you have described your current attitudes and situation, reflect on what you can do to make your further study of mathematics a success. What can you do to develop a more positive attitude? What can you do to ensure adequate study time? What can you do to obtain the help you may need with your studies? Try to list a strategy to which you can refer periodically through your studies to keep you on a positive track.

8.2 Polynomial Functions and Their Graphs

OBJECTIVES

1 Create tables of values for polynomial relations.

2 Graph polynomial relations by using sets of ordered pairs.

3 Determine ranges for polynomial relations.

4 Evaluate polynomial functions.

5 Model real-world situations by using polynomial functions.

APPLICATION

Pyrotechnicians must take into account many different relationships in designing a fireworks show. However, if we simplify the projection of a firework shell and consider only the force of gravity on the shell, we can write a polynomial function for the position of the shell. The position function for a 6-inch-diameter shell that is shot directly above the launcher at an initial velocity of 203.5 feet per second and an initial height of 5 feet is $s(t) = -16t^2 + 203.5t + 5$, where $s(t)$ is the height above ground level (in feet) and t is the time in seconds after the shell was shot from the launcher. Determine the range of the function and interpret its meaning.

After completing this section, we will discuss this application further. See page 551.

8.2.1 Creating Tables of Values

In previous chapters of this text, we developed the concept of a relation. A **relation** is a set of ordered pairs. A relation can be written in equation form. The equation then relates a value for an independent variable to a value for a dependent variable. For example, $y = 3x + 2$ relates a value for the independent variable x to a value for the dependent variable y.

A **polynomial relation** equates a polynomial expression in one independent variable to a dependent variable.

> **POLYNOMIAL RELATION**
>
> A polynomial relation is a relation that can be written in the form
> $$y = a_n x^n + a_{n-1} x^{n-1} + a_{n-2} x^{n-2} + \cdots + a_1 x^1 + a_0$$
> where n is the degree of the polynomial; $a_0, a_1, a_2, \ldots, a_n$ are real numbers; and $a_n \neq 0$.

By this definition, all linear equations, $y = ax + b$, are polynomial relations.

Another example of a polynomial relation is $y = x^2 + 2x + 7$. The polynomial $x^2 + 2x + 7$, with the independent variable x, equals the dependent variable y.

A set of all possible values for the independent variable is called the **domain** of the relation. The domain of the relation $y = x^2 + 2x + 7$ is the set of all real numbers.

We can determine an ordered-pair solution of this relation by substituting a value for the independent variable from its domain and obtaining a value for the dependent variable. The corresponding values represent an ordered-pair solution. For example,

$$\text{if } x = 2, \text{ then } y = x^2 + 2x + 7$$
$$y = (2)^2 + 2(2) + 7$$
$$y = 4 + 4 + 7$$
$$y = 15$$

An ordered-pair solution of the relation is $(2, 15)$.

An infinite number of solutions may be found by this method. Previously, we organized the procedure in a table of values. (See Sections 3.1 and 5.1) A **table of values** is a table with a column for the independent variable and a column for the dependent variable. We may want to add a third column between these to show our work.

To complete a table of values,

- Enter a number from the domain in the first column.
- Substitute this value for the independent variable in the second column.
- Evaluate the expression, and enter the result in the third column.

To construct a table of values on your calculator,

- Rename the independent variable x and the dependent variable y, and enter the equation in Y1.

(Continued on page 544)

- Set the calculator to generate the table by entering a minimum value and the amount of the increment to be added to the minimum number. Then set the calculator to perform the evaluations automatically.

An alternative method is to set up the calculator to ask for the x-values.

- Ignore the minimum value and increments. Then set the calculator to ask for the x-values and to perform the evaluations automatically.
- Enter values for x.

EXAMPLE I Create a table of values of possible solutions of the given polynomial relation.

a. $y = x^2 + 2x + 7$ **b.** $y = -2x^3 + 8x - 5$

Solution

a. The domain of $y = x^2 + 2x + 7$ is the set of all real numbers. To determine a table of values, we will use $x = -3, -2, -1, 0,$ and 1.

x	$y = x^2 + 2x + 7$	y
-3	$y = (-3)^2 + 2(-3) + 7$	10
-2	$y = (-2)^2 + 2(-2) + 7$	7
-1	$y = (-1)^2 + 2(-1) + 7$	6
0	$y = (0)^2 + 2(0) + 7$	7
1	$y = (1)^2 + 2(1) + 7$	10

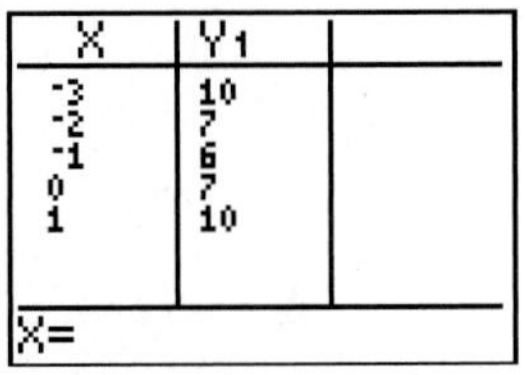

$Y1 = x^2 + 2x + 7$

Figure 8.4a

A calculator version of the table is shown in Figure 8.4a. This table was generated in ASK mode.

The ordered pairs found in the table, $(-3, 10), (-2, 7), (-1, 6), (0, 7),$ and $(1, 10)$, are solutions of the relation $y = x^2 + 2x + 7$.

b. The domain of $y = -2x^3 + 8x - 5$ is all real numbers. To determine a table of values, we will use $x = -3, -2, -1, 0, 1, 2,$ and 3.

x	$y = -2x^3 + 8x - 5$	y
-3	$y = -2(-3)^3 + 8(-3) - 5$	25
-2	$y = -2(-2)^3 + 8(-2) - 5$	-5
-1	$y = -2(-1)^3 + 8(-1) - 5$	-11
0	$y = -2(0)^3 + 8(0) - 5$	-5
1	$y = -2(1)^3 + 8(1) - 5$	1
2	$y = -2(2)^3 + 8(2) - 5$	-5
3	$y = -2(3)^3 + 8(3) - 5$	-35

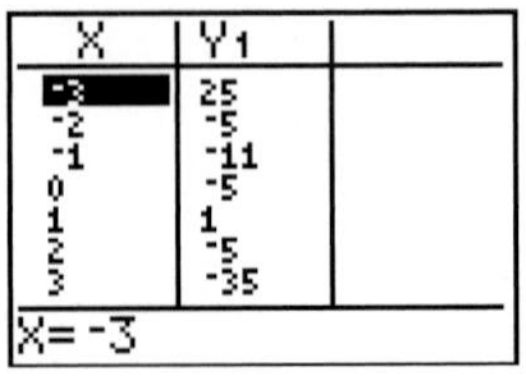

$Y1 = -2x^3 + 8x - 5$

Figure 8.4b

A calculator version of the table is shown in Figure 8.4b. This table was generated in AUTO mode with Tbl Start $= -3$. The ordered pairs found in the table, $(-3, 25), (-2, -5), (-1, -11), (0, -5), (1, 1), (2, -5)$, and $(3, -35)$, are solutions of the relation $y = -2x^3 + 8x - 5$.

8.2.1 Checkup

1. Create a table of values with six possible solutions of the polynomial relation $y = x^3 + 2x^2 - 5x - 6$.

2. Consider the polynomial relation in exercise 1.

   a. Which is the independent variable?

   b. Which is the dependent variable?

   c. What is the domain of the relation?

8.2.2 Graphing Polynomial Relations

Polynomial relations have an infinite number of possible ordered-pair solutions. In order to illustrate these solutions, we use a **graph**.

To graph a polynomial relation means to create an illustration that represents the solutions of the relation. Note that

   1. Every solution of a polynomial relation can be represented by a point on its graph, and

   2. Every point on a graph represents a solution of its associated polynomial relation.

To graph a polynomial relation using ordered pairs,

- Graph ordered-pair solutions found in a table of values.
- Identify a pattern and complete the pattern.
- Label the coordinates of the points graphed.

To graph a polynomial relation on your calculator,

- Select a viewing screen, such as the standard window, integer window, or decimal window.
- Enter the equation in Y1.
- Graph the equation.

EXAMPLE 2   Graph the given polynomial relations.

   a. $y = 2x^2 + 4x + 3$ **b.** $y = x^3 + x^2 + 6$

Solution

a. The domain of $y = 2x^2 + 4x + 3$ is the set of all real numbers. To determine a table of values, we will use $x = -3, -2, -1, 0,$ and 1.

x	$y = 2x^2 + 4x + 3$	y	
-3	$y = 2(-3)^2 + 4(-3) + 3$	9	$(-3, 9)$
-2	$y = 2(-2)^2 + 4(-2) + 3$	3	$(-2, 3)$
-1	$y = 2(-1)^2 + 4(-1) + 3$	1	$(-1, 1)$
0	$y = 2(0)^2 + 4(0) + 3$	3	$(0, 3)$
1	$y = 2(1)^2 + 4(1) + 3$	9	$(1, 9)$

Graph the ordered pairs and connect them with a smooth curve. The calculator check is shown in Figure 8.5a.

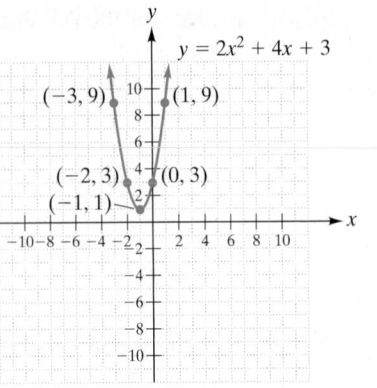
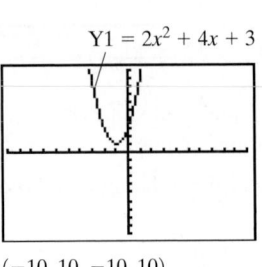

$(-10, 10, -10, 10)$

Figure 8.5a

b. The domain of $y = x^3 + x^2 + 6$ is the set of all real numbers. To determine a table of values, we will use $x = -3, -2, -1, 0, 1,$ and 2.

x	$y = x^3 + x^2 + 6$	y	
-3	$y = (-3)^3 + (-3)^2 + 6$	-12	$(-3, -12)$
-2	$y = (-2)^3 + (-2)^2 + 6$	2	$(-2, 2)$
-1	$y = (-1)^3 + (-1)^2 + 6$	6	$(-1, 6)$
0	$y = (0)^3 + (0)^2 + 6$	6	$(0, 6)$
1	$y = (1)^3 + (1)^2 + 6$	8	$(1, 8)$
2	$y = (2)^3 + (2)^2 + 6$	18	$(2, 18)$

On a coordinate plane, graph the ordered pairs and connect them with a smooth curve. The calculator check is shown in Figure 8.5b.

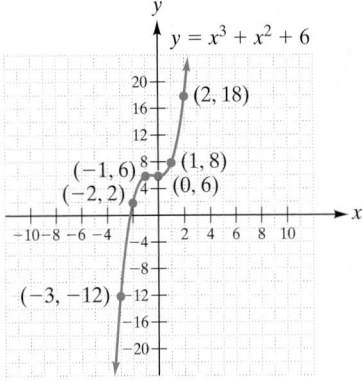
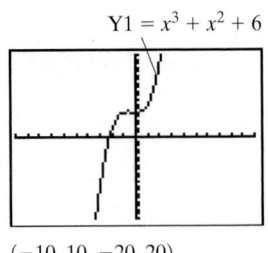

$(-10, 10, -20, 20)$

Figure 8.5b

✓ **8.2.2** Checkup

In exercises 1 and 2, graph the given polynomial relations.

1. $y = x^3 - 5x + 4$ **2.** $y = x^2 - 4x + 2$

3. What are some advantages of representing a polynomial relation with a graph rather than a table?

4. When the domain of a polynomial relation is the set of all real numbers, is it possible to show all solutions of the relation graphically? Explain.

8.2.3 Determining Ranges for Polynomial Relations

The set of all possible values for the dependent variable of a polynomial relation is called the **range** of the relation. To determine the range, we need to consider the relation for all values in its domain. If the domain is infinite, a more effective method of determining the range is to view the graph of the relation.

To determine the range of a polynomial relation from its graph, write a set of the values of the dependent variable (y-coordinates) used in the graph. That is, the range lies between and including the **absolute minimum** of the relation (the smallest value of y) and the **absolute maximum** of the relation (the largest value of y), if they exist.

To determine the range of a polynomial relation from its graph,

- Determine the absolute minimum and the absolute maximum of the relation.
- If the relation has no absolute minimum or absolute maximum, the range is the set of all real numbers.
- If the relation has both an absolute minimum and absolute maximum, the range is the set of all real numbers between and including these values.
- If the relation has no absolute maximum, the range is the set of all real numbers greater than or equal to the absolute minimum.
- If the relation has no absolute minimum, the range is the set of all real numbers less than or equal to the absolute maximum.

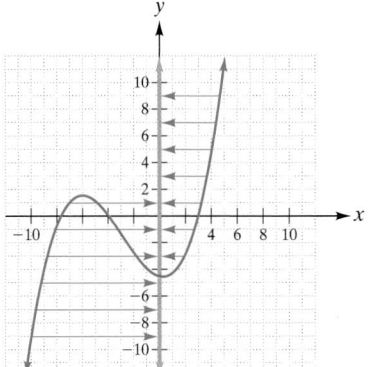

The range is the set of all real numbers, or $(-\infty, \infty)$.

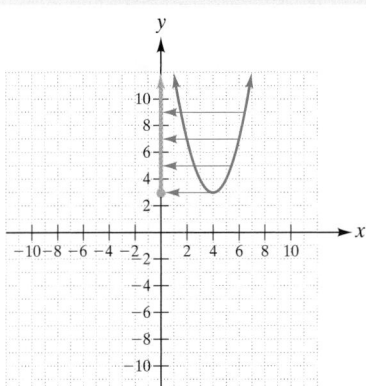

The range is the set of all real numbers greater than or equal to 3, or $y \geq 3$, or $[3, \infty)$.

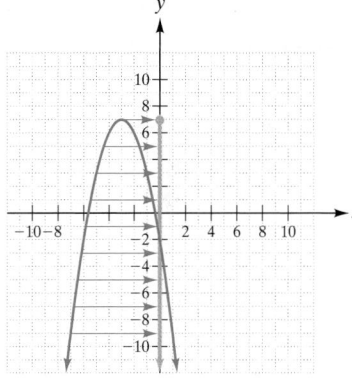

The range is the set of all real numbers less than or equal to 7, or $y \leq 7$, or $(-\infty, 7]$.

To determine the range of a polynomial relation from a calculator graph,

- Trace the graph to determine the y-coordinates of the points. Note the absolute minimum or the absolute maximum.
- Use the CALC function to determine the absolute minimum if it cannot be determined by tracing.
- Use the CALC function to determine the absolute maximum if it cannot be determined by tracing.

HELPING HAND The calculator is an excellent tool to help us determine the range of a relation. However, when we calculate the absolute maximum or minimum, the calculator may not display the exact value. We may need to round the answer.

EXAMPLE 3 Determine the range of each relation graphically.

   a. $y = x^2 + 2x - 4$ **b.** $y = -2x^3 + 5x - 1$ **c.** $y = -x^4 + x - 5$

Calculator Solution

Graph each relation.

a.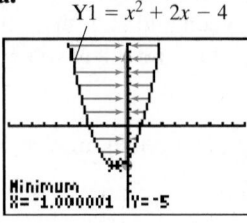

$(-10, 10, -10, 10)$

The y-coordinates have an absolute minimum value of -5. There is no absolute maximum. The range is the set of all real numbers greater than or equal to -5, or $y \geq -5$.

b.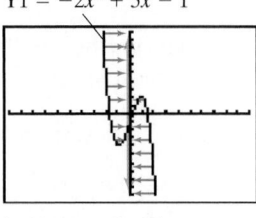

$(-10, 10, -10, 10)$

The y-coordinates have no absolute maximum or absolute minimum value. The range is the set of all real numbers.

c.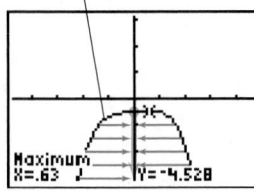

$(-4.7, 4.7, -31, 31)$

The y-coordinates have an absolute maximum value at approximately -4.528. There is no absolute minimum value. The range is the set of all real numbers less than or equal to -4.528. ●

✔ 8.2.3 Checkup

In exercises 1–4, determine the range of each relation graphically.

1. $y = x^3 + 2x^2 - 5x - 6$
2. $y = 8 + 2x - x^2$
3. $y = x^5 - 5x^3 + 4x$
4. $y = 0.2x^4 + 0.1x^3 - 0.5x^2 - 0.8x - 1.2$

5. What is the difference between the range of a polynomial relation and the domain of a polynomial relation?
6. If a polynomial relation has an absolute minimum value, what will be true about the range of the relation?
7. If a polynomial relation has an absolute maximum value, what will be true about the range of the relation?

8.2.4 Evaluating Polynomial Functions

Recall from Chapter 3 that a function is a special type of relation. A **function** relates every element in its domain to only one element in its range. To determine graphically whether a relation is a function, we use the vertical-line test.

> **VERTICAL-LINE TEST**
> If a vertical line can be drawn such that it intersects the graph of a relation more than once, the graph does not represent a function. If the vertical line intersects the graph only once, then the graph represents a function.

Remember that the vertical-line test works because a function can have only one y-value for every x-value. Any graph that crosses a vertical line more than once is a relation but is not a function.

   Are all polynomial relations also functions? To find out, complete the following set of exercises.

Discovery 1

Graphs of Polynomial Functions

Graph each polynomial relation. Draw a vertical line through more than one point on the graph if possible.

1. $y = x$ **2.** $y = x^2$ **3.** $y = x^3$ **4.** $y = x^4$

Are the polynomial relations graphed in exercises 1–4 functions?

It is not possible to draw a vertical line through more than one point on the graphs of these polynomial relations. Therefore, these polynomial relations are *all* functions. In fact, *all* polynomial relations are functions.

To write a function, we use **function notation**. We replace the dependent variable with the name of the function and put the name of the independent variable in parentheses.

POLYNOMIAL FUNCTION

A polynomial function f is a function that can be written in the form

$$f(x) = a_n x^n + a_{n-1} x^{n-1} + a_{n-2} x^{n-2} + \cdots + a_1 x^1 + a_0$$

where n is the degree of the polynomial; $a_0, a_1, a_2, \ldots, a_n$ are real numbers; and $a_n \neq 0$.

HELPING HAND Remember from Chapter 3 that $f(x)$ does *not* mean f times x.

For example, the relation $y = x^2 + 2x + 7$ may be written as the function $f(x) = x^2 + 2x + 7$. We read this as "f of x equals x squared plus 2 times x plus 7."

We may need to evaluate a function for a given value. We use function notation to write the function by replacing the independent variable with the given value. For example, "Evaluate $f(x) = x^2 + 2x + 7$ for $x = 2$" is written "Evaluate $f(2)$, given $f(x) = x^2 + 2x + 7$."

To evaluate a function for a value,

- Substitute the value for the independent variable.
- Evaluate the resulting numeric expression.

EXAMPLE 4 **a.** Evaluate $f(2)$, given $f(x) = x^2 + 2x + 7$.

b. Evaluate $g(-4)$ given $g(x) = x^3 - x^2 + 5x - 6$.

Solution

a. $f(x) = x^2 + 2x + 7$ *Given function*

$f(2) = (2)^2 + 2(2) + 7$ *Substitute 2 for x.*

$f(2) = 4 + 4 + 7$ *Simplify.*

$f(2) = 15$

b. $g(x) = x^3 - x^2 + 5x - 6$ *Given function*

$g(-4) = (-4)^3 - (-4)^2 + 5(-4) - 6$ *Substitute −4 for x.*

$g(-4) = -64 - 16 - 20 - 6$ *Simplify.*

$g(-4) = -106$

EXAMPLE 5   Given the function $g(x) = x^3 + 50x^2 + 100$, evaluate $g(4000)$ on your calculator.

```
(4000)³+50(4000)
²+100
     6.48000001E10
4000→X:X³+50X²+1
00
     6.48000001E10
```

Solution

$g(4000) \approx 6.48 \times 10^{10}$, or approximately 64,800,000,000.

8.2.4 Checkup

1. Evaluate $h(x) = 2x^3 - x^2 - 27x + 36$ at the given values.

   a. $h(0)$ **b.** $h(-1)$ **c.** $h(1)$ **d.** $h(3)$

   e. $h(-4)$ **f.** $h\left(\dfrac{3}{2}\right)$ **g.** $h(0.1)$

2. In evaluating a polynomial function, is it important to know the rules for order of operations? Explain.

8.2.5 Modeling the Real World

Often, certain values of a polynomial function have special importance for real-world situations. For example, the absolute maximum value of a polynomial revenue function is the largest amount of revenue. The absolute maximum value of the polynomial function describing the path of a ball thrown into the air tells how high the ball goes before falling back to earth. Graphs of such polynomial functions are often useful for seeing where or when these special values occur and approximately what the values of the variables are at that point or at that time.

EXAMPLE 6   George makes wooden duck decoys and sells them for $10.00. However, to reduce his inventory before winter, George plans to give a discount of $0.50 for each decoy purchased. If x is the number of decoys a person purchases, then the discount is $0.50 for each decoy purchased, or 0.50x. The purchase price is $10.00 minus the discount, or $10.00 - 0.50x$. The amount of revenue, $R(x)$, is the purchase price, $10.00 - 0.50x$, times the number of decoys purchased, x:

$$R(x) = (10.00 - 0.50x)x, \text{ which becomes } R(x) = 10.00x - 0.50x^2$$

a. Determine the amount of revenue if George sells 5 decoys, 10 decoys, or 20 decoys.

b. What is the range of the revenue function?

c. What does this range tell us about George's discount plan?

Solution

a.
$$R(5) = 10.00(5) - 0.50(5)^2$$
$$R(5) = 37.50$$

If George sells 5 decoys, his revenue is $37.50.

$$R(10) = 10.00(10) - 0.50(10)^2$$
$$R(10) = 50.00$$

If George sells 10 decoys, his revenue is $50.00.

$$R(20) = 10.00(20) - 0.50(20)^2$$
$$R(20) = 0.00$$

If George sells 20 decoys, his revenue is $0.00.

b.

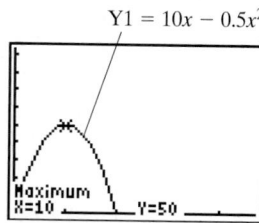

$$Y1 = 10x - 0.5x^2$$

(0, 44, 0, 93)

The range is the set of all real numbers less than or equal to 50.00.

c. The range found shows us that if George gives the $0.50 discount, his largest revenue would be $50.00. (According to part (b), this occurs when he sells 10 decoys.) If George does not limit the discount, he would have a revenue of 0 (with 20 decoys sold) or a negative revenue (with more than 20 decoys sold), which is not a good thing to have. If George wants a more accurate picture of how his discount plan affects his possible profits, he should take into account the cost of manufacturing the decoys and the cost of selling them.

APPLICATION

Pyrotechnicians must take into account many different relationships in designing a fireworks show. However, if we simplify the projection of a firework shell and consider only the force of gravity on the shell, we can write a polynomial function for the position of the shell. The position function for a 6-inch-diameter shell that is shot directly above the launcher at an initial velocity of 203.5 feet per second and an initial height of 5 feet is $s(t) = -16t^2 + 203.5t + 5$, where $s(t)$ is the height above ground level (in feet) and t is the time in seconds after the shell was shot from the launcher. Determine the range of the function and interpret its meaning.

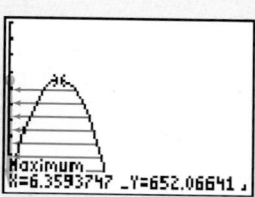

(0, 25, 0, 1000)

Assuming that the shell explodes above the initial height, 5, the function values cannot be less than 5. Therefore, the range of the function is all real numbers between 5 and approximately 652, or $5 \leq y \leq 652$.

Discussion

Let $Y1 = -16x^2 + 203.5x + 5$, and graph the function on your calculator.

The position of the firework shell above the ground will range from 5 to 652 feet, assuming that the shell explodes after 6.36 seconds.

8.2.5 Checkup

1. Consenting Consultants, Inc. (CCI), offers training to a major manufacturer, charging $150 - 4x$ dollars per person, where x is the number of people attending training. (Note that what this amounts to is a discount, depending upon the number of people who attend.) The revenue CCI collects is given by the function

$$R(x) = 150x - 4x^2$$

   a. Determine the amount of revenue CCI will realize if 5 people attend the training, if 10 attend, if 15 attend, if 20 attend, if 25 attend, if 30 attend, if 35 attend, and if 40 attend.

   b. What is the range of the revenue function?
   c. Interpret the range in terms of the discount offered.

2. Using data supplied by the Department of the Army, a statistician developed a function to relate the number of commissioned officers, $n(x)$, to the total strength of the Army, x. The data for the years from 1980 to 2000 yielded the function $n(x) = 0.077x + 31102$. Graph the function, using a window of $(0, 1 \times 10^6, 0, 1 \times 10^5)$. Determine the range of the function and discuss its meaning.

8.2 Exercises

Create a table of values of five possible integer solutions of each polynomial relation.

1. $y = x^3 + 2x^2 - 5x - 6$ **2.** $y = x^3 - 2x^2 - 5x + 6$ **3.** $y = x^2 + 4x + 1$ **4.** $y = x^2 - 4x + 1$

Graph the given polynomial relations. From the graph, determine the range of each relation.

5. $y = 2x - 5$

6. $y = -3x + 2$

7. $y = -x^2 + 6x - 4$

8. $y = -2x^2 - 8x + 1$

9. $y = 2x^2 - 8x + 3$

10. $y = x^2 - 4x - 3$

11. $y = \frac{1}{2}x^2 + 6$

12. $y = \frac{2}{3}x^2 - 4$

13. $y = -\frac{1}{2}x^2 - 3x$

14. $y = -\frac{3}{4}x^2 + 6x$

15. $y = x^3 - 8$

16. $y = x^3 - 1$

17. $y = -\frac{1}{8}x^3 + 1$

18. $y = -\frac{1}{27}x^3 + 1$

19. $y = x^3 - 4x$

20. $y = x^3 - 9x$

21. $y = x^3 + 3x^2 - 10x - 24$

22. $y = x^3 - 7x + 6$

23. $y = \frac{1}{16}x^4 - x^2$

24. $y = \frac{16}{81}x^4 - x^2$

25. $y = -\frac{16}{81}x^4 + x^2$ **26.** $y = -\frac{1}{16}x^4 + x^2$ **27.** $y = -x^4 - x^3 + 11x^2 + 9x - 18$ **28.** $y = x^4 + x^3 - 6x^2 - 4x + 8$

Evaluate the function $f(x) = x^2 + 16x + 64$ at the given values.

29. $f(2)$ **30.** $f(1)$ **31.** $f(-2)$ **32.** $f(-1)$

Evaluate the function $g(x) = 4x^2 - 4x + 1$ at the given values.

33. $g(3)$ **34.** $g(5)$ **35.** $g\left(\frac{3}{2}\right)$ **36.** $g\left(\frac{5}{2}\right)$

Evaluate the function $h(x) = 9x^2 + 12x + 4$ at the given values.

37. $h(-2)$ **38.** $h(-3)$ **39.** $h(-1.5)$ **40.** $h(-1.2)$

Evaluate the function $f(x) = -2x^3 + x^2 - 5x + 8$ at the given values.

41. $f(-2)$ **42.** $f(-1)$ **43.** $f(2)$ **44.** $f(3)$

Evaluate the function $g(x) = 2.7x^3 - 1.5x^2 + 3.5x - 6.7$ at the given values.

45. $g(2)$ **46.** $g(4)$ **47.** $g(-4)$ **48.** $g(-2)$

49. $g(10)$ **50.** $g(20)$ **51.** $g(1)$ **52.** $g(0)$

Evaluate the function $h(x) = \frac{1}{2}x^3 - \frac{3}{4}x^2 + \frac{3}{8}x - \frac{5}{8}$ at the given values. Express your results in fraction notation.

53. $h(2)$ **54.** $h(4)$ **55.** $h(-4)$ **56.** $h(-2)$

57. In order to encourage multiple purchases, Dave's Wholesale Jewelers sells watches at a price that is a function of the number of watches ordered. For x watches ordered, the price will be $150 - 5x$ dollars per watch. Thus, one watch costs $145, two watches cost $140 each, and so forth. The revenue function for an order of x watches is

$$R(x) = 150x - 5x^2$$

a. Determine the revenue if 5 watches are ordered, if 10 watches are ordered, if 15 watches are ordered, if 20 watches are ordered, if 25 watches are ordered, and if 30 watches are ordered.
b. Find the range of the revenue function.
c. Interpret the range in terms of the discount offered.

58. Honest Al rents space at his merchandise mart and offers sliding-scale rates for multimonth rentals. His rate is $300 - 5x$ dollars per month for x months of rental. Thus, the rate for one month's rental is $295, the rate for two months' rental is $290 per month, the rate for three months' rental is $285 per month, and so forth. The total cost for renting the space for x months is

$$C(x) = 300x - 5x^2$$

a. Determine the cost of renting space for 10 months, for 20 months, for 30 months, for 40 months, for 50 months, and for 60 months.
b. Determine the range of the cost function.
c. Interpret the cost function in terms of the discount offered.

59. A statistical study collected data on the length of a Medicare patient's hospital stay and the patient's age. For female patients, a function that relates the length of stay, $S(x)$, to the patient's age, x, is

$$S(x) = -0.011x^2 + 1.91x - 72.54$$

Use this function to estimate a patient's length of stay, given the following ages:

Age (x)	63	65	70	75	80	85
Days ($S(x)$)						

a. The data included information on an 85-year-old woman who had a hospital stay of eight days. How well did the function predict this stay?
b. One 70-year-old woman had a stay of nine days, and another 70-year-old woman had a stay of six days. Did the function predict these stays closely?
c. If you don't think the function predicts well, can you explain why it doesn't?

60. A term life insurance policy lists the following annual premiums for male nonsmokers as a function of their ages:

Age	35	40	45	50
Premium	$385	$505	$770	$1165
Age	55	60	65	70
Premium	$1700	$2645	$4355	$8285

Statistical methods calculate a third-degree polynomial function for this table as

$$P(x) = 0.43x^3 - 58.08x^2 + 2629.48x - 39096$$

where $P(x)$ is the premium and x is the age. Use this function to estimate the premiums for the values of x shown in the following table:

Age (x)	35	40	45	50	55	60	65	70
Premium ($P(x)$)								

a. How well does the function predict the actual premium for a male who is 35 years old?
b. Did the function predict the premium closely for a 70-year-old male?
c. If you don't think the function closely predicts the actual premiums, can you offer a reason that it doesn't?

61. For the Labor Day picnic, an expert pyrotechnician shoots a fireworks rocket from ground level with an initial velocity of 270 feet per second. The position function for the rocket, in feet above ground level, is $s(t) = -16t^2 + 270t$, where t is the time in seconds after the rocket is launched and $0 < t < 8$. State the domain and interpret it. Find the range of the function and interpret it.

62. At the Fourth of July celebration, a professional pyrotechnics expert shoots a rocket straight up from the top of a 20-foot tower. The rocket's initial velocity is 300 feet per second. The position function for the rocket, in feet above ground level, is $s(t) = -16t^2 + 300t + 20$, where t is the time in seconds after the rocket is launched and $0 < t < 10$. State the domain and interpret it. Find the range of the function and interpret it.

63. The real millennium began with a world record parachute jump when 15 parachutists from various countries jumped from the top floor of the Petronas Towers in Malaysia, which stands 1483 feet tall, including a spire. They began their jump in the year 2000 and landed in the year 2001! Their distance from the ground is given by the function $s(t) = -16t^2 + 1242$, where $s(t)$ is the height above ground level (in feet) and t is the time in seconds after they jumped. Determine the range of this function and explain its meaning.

64. While the Petronas Towers are recognized as the tallest buildings in the world, the Sears Tower in Chicago is the highest structure if measured to the highest occupied floor. If a parachutist were permitted to jump from the top of this tower, his or her distance from the ground would be given by the function $s(t) = -16t^2 + 1431$, where $s(t)$ is the height above ground level (in feet) and t is the time in seconds after the parachutist jumped. Determine the range of this function and explain its meaning.

8.2 Calculator Exercises

Part 1. Influence of Degree of Polynomial

Graph the following polynomial functions with a calculator. To view the graph easily, use the decimal viewing window. Then draw a sketch of each graph, labeling the intercepts.

1. $y = x$
2. $y = x^2 + x$
3. $y = x^3 - x$
4. $y = x^4 + 2x^3 - x^2 - 2x$
5. $y = x^5 - 5x^3 + 4x$

Determine the degree of each polynomial.

Compare the graphs. How many times does each graph change direction, either moving up from left to right or moving down from left to right? Compare this number with the degree of the polynomial graphed. Complete the following table:

Polynomial	Degree of polynomial	Number of changes in direction
$y = x$		
$y = x^2 + x$		
$y = x^3 - x$		
$y = x^4 + 2x^3 - x^2 - 2x$		
$y = x^5 - 5x^3 + 4x$		

Discuss the apparent relationship between the degree of the polynomial and the number of times the graph changes direction.

Next, notice which graphs have tails that extend in opposite directions and which graphs have tails that extend in the same direction. Discuss and compare this feature with the degree of the polynomials. Do you see a relationship?

Notice also that when the tails extend in opposite directions, there is at least one value of x that will make the polynomial evaluate to 0. Discuss why this is so.

The degree of the polynomial in a polynomial function can be a useful feature to study to understand what the function's graph will look like.

Part 2. Function Notation on the Calculator

The function notation that you used in this section is also used by the calculator to evaluate functions. You have seen that you can evaluate a function by keying it in with the variable replaced by its value. You have also seen that you can evaluate a function by storing the value of its variable and then keying in the function. A third method, using function notation, is explained next.

To evaluate a function on your calculator by using function notation, first store the function under the Y = menu as Y1. Then type the function notation on the home screen, using Y1 with the value enclosed in parentheses. When you press ENTER, the calculator will display the function value, evaluated at the value specified for the variable.

As an example, evaluate the function $f(x) = x^3 - 2x^2 + 5x - 4$ when $x = 7, -5$, and 2.4.

Key the following instructions into the calculator:

- Store the function in Y = as Y1.

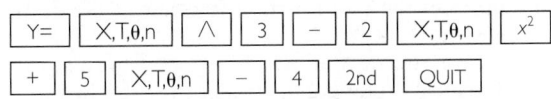

- Key the function notation for the values of x.

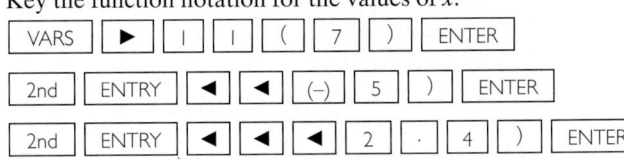

The calculator displays for these instructions are as follows:

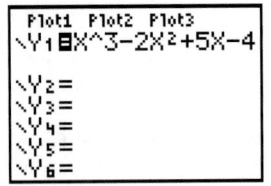

Practice this method on the following exercises:

Evaluate the function $f(x) = 2.5x^3 - 1.6x^2 + 4.2x - 9.3$ for the given values.

1. $f(-2.4)$
2. $f(155.8)$
3. $f(0.944)$

Evaluate the function $h(x) = \frac{2}{3}x^3 + \frac{1}{2}x^2 - 2x + \frac{1}{3}$ for the given values. Use the fraction option of the calculator to display the answer in fractional notation.

4. $h(6)$
5. $h\left(\frac{3}{4}\right)$
6. (-9)

8.2 Writing Exercise

In this section, several examples or exercises mention statistics as an area of mathematics that often uses polynomials in order to explain the relationship between pairs of observations in actual data collected from an application. In the library, browse some statistics texts to find another example where this is done. (*Note:* In the statistics texts, you will find this done under a method called regression analysis.) Don't try to understand the description of how the relation is developed; just take note of the example presented.

Write a short paragraph describing the independent variable and the dependent variable being studied in the example. Also present the relation developed. Experiment with the relation to see if you can substitute values for the independent variable to estimate what the value will be for the dependent variable. Explain what this example means to you. Footnote the text in which you found the example and provide the library call number for the text.

8.3 Quadratic Functions and Their Graphs

OBJECTIVES
1 Identify quadratic functions.
2 Understand the effects of the coefficients of a quadratic function on its graph.
3 Graph quadratic functions.
4 Model real-world situations using graphs of quadratic functions.

APPLICATION

A signal flare is shot upward from a cliff 100 meters high. If the initial velocity of the flare was 91.2 meters per second, a position function for the flare is $s(t) = -4.9t^2 + 91.2t + 100$, where $s(t)$ is measured in meters and time, t, is measured in seconds.

Determine the signal flare's maximum height in meters.

After completing this section, we will discuss this application further. See page 566.

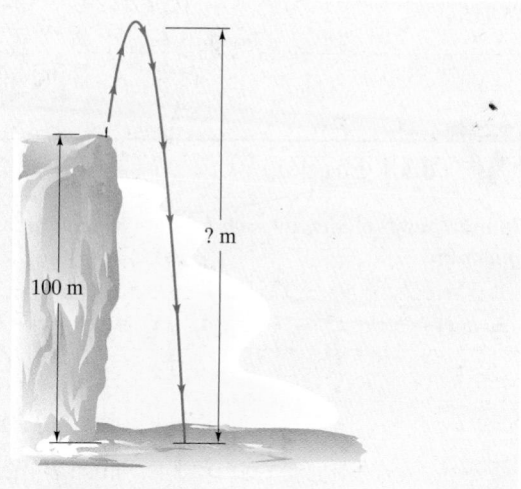

? m

100 m

8.3.1 Identifying Quadratic Functions

In this section, we discuss a special case of a polynomial function, called a quadratic function. A **quadratic function** is a polynomial function with a degree of 2.

STANDARD FORM FOR A QUADRATIC FUNCTION

A quadratic function can be written in the standard form

$$y = ax^2 + bx + c, \text{ where } a \neq 0$$

or

$$f(x) = ax^2 + bx + c, \text{ where } a \neq 0.$$

For example,

Standard form: $y = ax^2 + bx + c$

1. $y = 2x^2 + 3x - 4$ $a = 2, b = 3,$ and $c = -4$
2. $y = x^2 - 2x$ $a = 1, b = -2,$ and $c = 0$
3. $y = -3x^2 + 5$ $a = -3, b = 0,$ and $c = 5$
4. $y = x^2$ $a = 1, b = 0,$ and $c = 0$

The coefficients are identified next to each function. Remember, $a \neq 0$; if $a = 0$, we no longer have a quadratic function. We would then have a linear function of the form $y = bx + c$.

EXAMPLE I Identify each function as quadratic or nonquadratic.

a. $y = x^3 + x^2 - 6x$ **b.** $h(x) = 2x - x^2$ **c.** $y = \dfrac{6}{2x} + 2x - 7$

Solution

a. $y = x^3 + x^2 - 6x$ is not a quadratic function, because $x^3 + x^2 - 6x$ is a third-degree polynomial. (The first term is x cubed.)

b. $h(x) = 2x - x^2$ is a quadratic function, because it can be rearranged into standard form, $h(x) = -x^2 + 2x + 0$.

c. $y = \dfrac{6}{2x} + 2x - 7$ is not a quadratic function, because $\dfrac{6}{2x} + 2x - 7$ is not a polynomial. (A variable is in the denominator of a fraction.)

8.3.1 Checkup

In exercises 1–6, identify each function as quadratic or non-quadratic.

1. $y = -2x^2 + 9x - 125$ **2.** $y = 8x^2 - 12$
3. $h(x) = 5x - 13$ **4.** $f(x) = 3 - 9x - 7x^2$

5. $y = x^3 + x^2 - x$ **6.** $y = -\dfrac{3}{x^2} - 4$

7. What must be true about the coefficient of the squared variable term of a quadratic function? Why is this so?

8.3.2 Understanding the Effects of the Coefficients

Just as the coefficients m and b of a linear function, $y = mx + b$, affect its graph, the values of the coefficients a, b, and c in the standard form of a quadratic function, $y = ax^2 + bx + c$, affect its graph.

All quadratic functions have a U-shaped graph called a **parabola**.

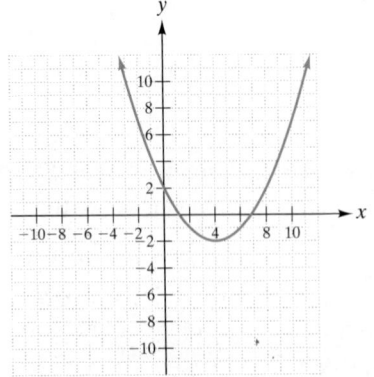

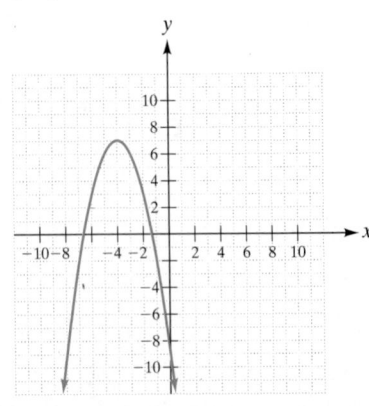

The term ax^2 in the function is called the **quadratic term**. To determine what effect the coefficient a has on the parabola, complete the following set of exercises.

Discovery 2

Effect of the Coefficient *a* on a Quadratic Graph

Sketch the graphs of the given quadratic functions of the form $y = ax^2$, where $b = c = 0$, on the same coordinate plane. Use the decimal window.

1. $y = 0.2x^2$ **2.** $y = x^2$ **3.** $y = 2x^2$

4. $y = -0.2x^2$ **5.** $y = -x^2$ **6.** $y = -2x^2$

Complete the following sentences by choosing the correct words:

7. In exercises 1–3, a is a *positive/negative* number. All of the graphs open *upward/downward*.

8. In exercises 4–6, a is a *positive/negative* number. All of the graphs open *upward/downward*.

9. In exercises 3 and 6, the absolute value of a is greater than 1. The shape of the parabola is *wider/narrower* than the graphs in exercises 2 and 5, in which $a = 1$ or -1.

10. In exercises 1 and 4, the absolute value of a is less than 1. The shape of the parabola is *wider/narrower* than the graphs in exercises 2 and 5, in which $a = 1$ or -1.

The coefficient a of the quadratic term affects the form of the parabola graphed. If the coefficient is positive, the graph opens upward. We say the graph is **concave upward**. If the coefficient is negative, the graph opens downward, or is **concave downward**.

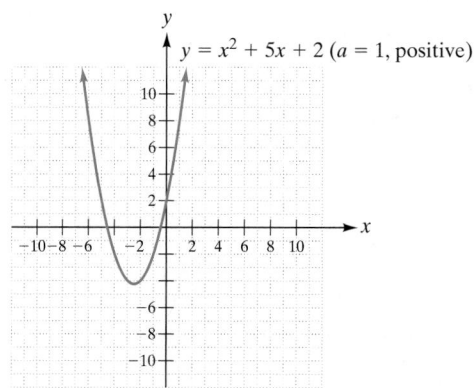

Concave upward

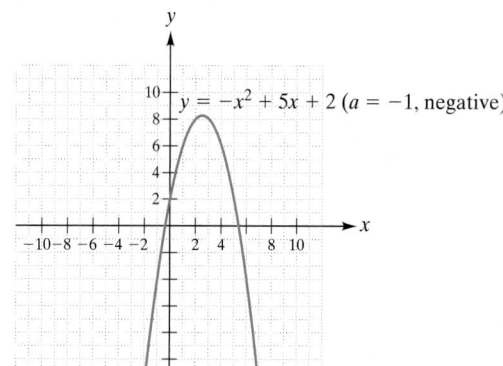

Concave downward

If the absolute value of the coefficient a is greater than 1, the graph is narrower than the graph of a parabola with a coefficient of 1. If the absolute value of the coefficient is less than 1, the graph is wider than the graph of a parabola with a coefficient of 1.

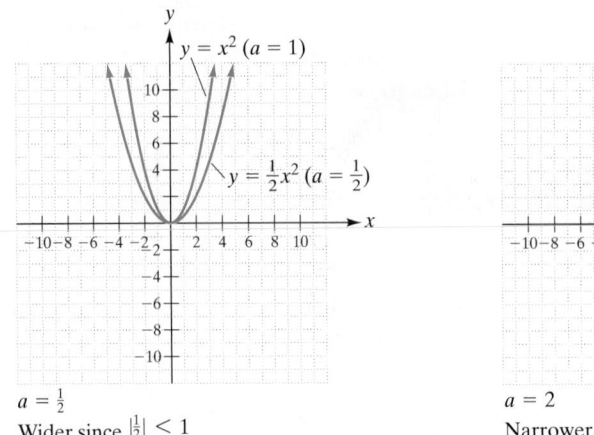

$a = \frac{1}{2}$
Wider since $|\frac{1}{2}| < 1$

$a = 2$
Narrower since $|2| > 1$

Remember from Chapter 3 that a function value is called a **relative maximum** if it is larger than the function values of its neighboring points. The largest relative maximum is called the *absolute maximum*. If a parabola is concave downward, its quadratic function will have an absolute maximum equal to the function value of the highest point of the parabola.

A function value is called a **relative minimum** if it is smaller than the function values of its neighboring points. The smallest relative minimum is called the *absolute minimum*. If a parabola is concave upward, its quadratic function will have an absolute minimum equal to the function value of the lowest point of the parabola.

The highest point on a concave-down parabola and the lowest point on a concave-up parabola are each called the **vertex** of the parabola.

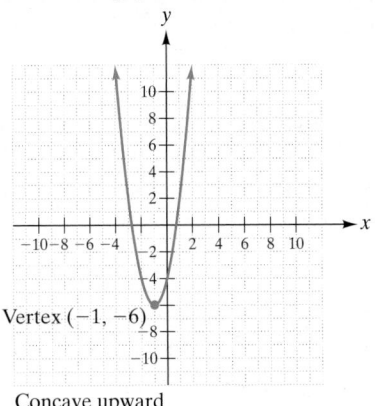

Concave upward
Absolute minimum of -6

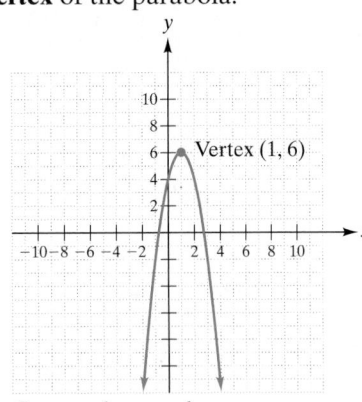

Concave downward
Absolute maximum of 6

The coefficients a and b determine the x-coordinate of the vertex. A formula for this is

$$x = \frac{-b}{2a}$$

We use this x-coordinate to find the y-coordinate of the vertex, which is either the absolute maximum or the absolute minimum of the quadratic function.

To determine the vertex of a quadratic function $y = ax^2 + bx + c$,

- Use the formula for the x-coordinate of the vertex of a quadratic function, which is $x = \frac{-b}{2a}$.
- Find the y-coordinate of the vertex by substituting the value of the x-coordinate into the original function and solving for y.

To determine the vertex of a quadratic function on your calculator,

- Enter the function in Y1.
- Graph the function.
- Trace the graph to determine its highest or lowest point.
- Use the CALC function to determine the minimum or maximum if it cannot be found by tracing.

HELPING HAND The calculator may display the vertex only approximately, so it may not be exact.

EXAMPLE 2 Given the quadratic function $y = 2x^2 + 4x + 3$, determine the coordinates of the vertex of its graph.

Solution

The coefficients are $a = 2$, $b = 4$, and $c = 3$.

$$x = \frac{-b}{2a}$$

$$x = \frac{-(4)}{2(2)} \qquad a = 2, b = 4$$

$$x = -1$$

Check

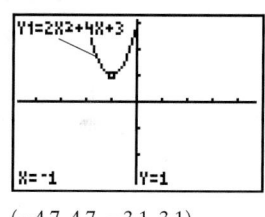

$(-4.7, 4.7, -3.1, 3.1)$

Figure 8.6a

To determine the y-coordinate of the vertex, substitute the value of the x-coordinate into the function and solve for y.

$$y = 2x^2 + 4x + 3$$
$$y = 2(-1)^2 + 4(-1) + 3$$
$$y = 2 - 4 + 3$$
$$y = 1$$

The vertex is $(-1, 1)$.

To determine the vertex on your calculator, see Figures 8.6a and 8.6b. Note that in Figure 8.6a the vertex is located by tracing. However, in Figure 8.6b, the calculated ordered pair is only an approximate location.

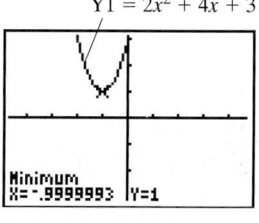

$(-4.7, 4.7, -3.1, 3.1)$

Figure 8.6b

Another feature of a quadratic function's graph is its symmetry. To visualize the meaning of symmetry, complete the following set of exercises on your calculator.

 Discovery 3

Symmetric Graph

1. Consider the graph of $y = x^2$. The vertex of the graph is $(0, 0)$. Complete the table of values for the three integer x-values on either side of $x = 0$, the x-coordinate of the vertex.

(Continued on page 560)

x	y	
-3		
-2		
-1		
0	0	← vertex
1		
2		
3		

2. Graph the function, using the table of values. Label all points graphed. Compare the y-values for the x-values equidistant from $x = 0$.

3. If $x = 1$ or $x = -1$, then $y =$ _____.

4. If $x = 2$ or $x = -2$, then $y =$ _____.

5. If $x = 3$ or $x = -3$, then $y =$ _____.

The graph of $y = x^2$ is a parabola that is symmetric with respect to the y-axis, or to the line graphed by the equation $x = 0$. The y-values that correspond to the x-values equidistant from $x = 0$ are equal. Therefore, if the graph is folded together along the line $x = 0$, the two sides will coincide. We call the vertical line through the vertex the **axis of symmetry** of the parabola.

The constant term c of the quadratic function also affects the graph. To determine the effect of c, complete the following set of exercises on your calculator.

 ## Discovery 4

Effect of the Coefficient c on a Quadratic Graph

Sketch the graph of the given quadratic functions of the form $y = ax^2 + c$, where $b = 0$, on the same coordinate plane. Use the decimal window and label the y-intercept of each graph.

$$y = 2x^2 \qquad y = 2x^2 + 1 \qquad y = 2x^2 - 3$$

1. Write a rule for determining the y-coordinate of the y-intercept of a parabola from its equation.

2. Check your rule for $y = 2x^2 + 3x - 1$.

The y-coordinate of the y-intercept is the constant term. (This is the same as when we were graphing linear equations.) Therefore, the y-coordinate of the y-intercept of the graph of $y = 2x^2 + 3x - 1$ is -1. The y-intercept is $(0, -1)$. We can determine the y-intercept algebraically in the same manner as we did for a linear equation: Substitute 0 for x and solve for y.

For example, given $y = 2x^2 + 3x - 1$, substitute 0 for x.

$$y = 2x^2 + 3x - 1$$
$$y = 2(0)^2 + 3(0) - 1$$
$$y = -1$$

The y-coordinate of the y-intercept is -1, the constant term of the function.

SUMMARY OF THE EFFECTS OF THE COEFFICIENTS OF A QUADRATIC FUNCTION ON ITS GRAPH

The coefficients of a quadratic function written in standard form, $y = ax^2 + bx + c$, affect the graph of the function.

Coefficient a:
If $a > 0$ (positive), then the graph is concave upward.
If $a < 0$ (negative), then the graph is concave downward.

Absolute value of a:
If $|a| > 1$, then the graph is narrower than when $a = 1$.
If $|a| < 1$, then the graph is wider than when $a = 1$.

Coefficient c:
The coefficient c is the y-coordinate of the y-intercept of the graph.
The y-intercept is at $x = 0$, $y = c$ or at $(0, c)$.

Coefficients a and b:
The x-coordinate of the vertex is $\frac{-b}{2a}$.
The axis of symmetry is the line graphed by $x = \frac{-b}{2a}$.

EXAMPLE 3 Given $y = -2x^2 + 6x - 1$, list the properties of its graph.

Solution

The coefficients of $y = -2x^2 + 6x - 1$ are $a = -2$, $b = 6$, and $c = -1$.
  The graph is concave downward, because $a = -2$ (negative).
  The graph is narrow compared to the graph of $y = x^2$, because $a = -2$ and $|-2| = 2 > 1$.
  The y-intercept is $(0, -1)$. (Note that $c = -1$.)
  The x-coordinate of the vertex is

$$x = \frac{-b}{2a} = \frac{-6}{2(-2)} = \frac{-6}{-4} = \frac{3}{2}$$

The y-coordinate of the vertex is

$$y = -2x^2 + 6x - 1$$
$$y = -2\left(\frac{3}{2}\right)^2 + 6\left(\frac{3}{2}\right) - 1$$
$$y = \frac{7}{2}$$

The vertex is $\left(\frac{3}{2}, \frac{7}{2}\right)$.
The axis of symmetry is $x = \frac{3}{2}$.

EXAMPLE 4 Complete the following table:

Function	Coefficients			Properties of graph				
	a	b	c	Wide/narrow	Concave upward/downward	Vertex	Axis of symmetry	y-intercept
$y = 3x^2 + 6$								
$y = -0.5x^2 - 2$								
$y = \frac{1}{4}x^2 - x + 2$								

Solution

Function	Coefficients			Properties of graph				
	a	b	c	Wide/narrow	Concave upward/downward	Vertex	Axis of symmetry	y-intercept
$y = 3x^2 + 6$	3	0	6	narrow	upward	$(0, 6)$	$x = 0$	$(0, 6)$
$y = -0.5x^2 - 2$	-0.5	0	-2	wide	downward	$(0, -2)$	$x = 0$	$(0, -2)$
$y = \frac{1}{4}x^2 - x + 2$	$\frac{1}{4}$	-1	2	wide	upward	$(2, 1)$	$x = 2$	$(0, 2)$

✔ **8.3.2** Checkup

Complete the table. Check your work by viewing the graphs on your calculator.

	Function	Coefficients			Properties of graph				
		a	b	c	Wide/narrow	Concave upward/downward	Vertex	Axis of symmetry	y-intercept
1.	$y = -3x^2 + 2$								
2.	$y = 4x^2 - 8x + 5$								
3.	$y = 0.25x^2 + x - 2$								
4.	$y = \frac{2}{3}x^2 + x$								

5. In graphing the quadratic function $y = ax^2 + bx + c$, How does the value of the coefficient a affect the graph? How does the value of the coefficient b affect the graph? How does the value of the coefficient c affect the graph?

8.3.3 Graphing Quadratic Functions

When we graphed linear functions, we needed two points to determine the linear pattern. Quadratic functions do not have a linear pattern, so we will need more than two solutions to determine a pattern.

To graph a quadratic function,

- Determine the coordinates of the vertex by finding the x-coordinate from the formula $x = \frac{-b}{2a}$, substituting the x-coordinate into the original quadratic function, and solving for y to determine the y-coordinate.
- Construct a table of values by choosing at least two x-values greater than the x-coordinate of the vertex and two corresponding x-values less than the x-coordinate of the vertex.
- Graph the function by plotting the vertex and the set of ordered pairs from the table of values and connecting the points with a smooth curve.

HELPING HAND It helps to graph the axis of symmetry as a dashed vertical line.

EXAMPLE 5 Graph the quadratic function $y = x^2 + 3x + 1$.

Solution

The coefficients are $a = 1$, $b = 3$, and $c = 1$.
First, determine the coordinates of the vertex.
The x-coordinate is found by the formula

$$x = \frac{-b}{2a} = \frac{-3}{2(1)} = -\frac{3}{2}$$

Determine the y-coordinate of the vertex by substituting the value of the x-coordinate into the function and solving for y.

$$y = x^2 + 3x + 1$$
$$y = \left(-\frac{3}{2}\right)^2 + 3\left(-\frac{3}{2}\right) + 1$$
$$y = -\frac{5}{4}$$

The vertex is $\left(-\frac{3}{2}, -\frac{5}{4}\right)$.
The axis of symmetry is the line $x = -\frac{3}{2}$.
Complete a table of values by choosing two x-values less than $-\frac{3}{2}$, such as -2 and -3, and two x-values greater than $-\frac{3}{2}$, such as -1 and 0.

x	$y = x^2 + 3x + 1$	y	
-3	$y = (-3)^2 + 3(-3) + 1$	1	
-2	$y = (-2)^2 + 3(-2) + 1$	-1	$\leftarrow$ vertex $\left(-\frac{3}{2}, -\frac{5}{4}\right)$
-1	$y = (-1)^2 + 3(-1) + 1$	-1	
0	$y = (0)^2 + 3(0) + 1$	1	

Plot the vertex and the ordered pairs found in the table of values. Connect the points with a smooth curve and graph the axis of symmetry.
On your calculator, enter the function in Y1 and graph it.
Note that the decimal screen provides a nicer picture. (See Figure 8.7.)

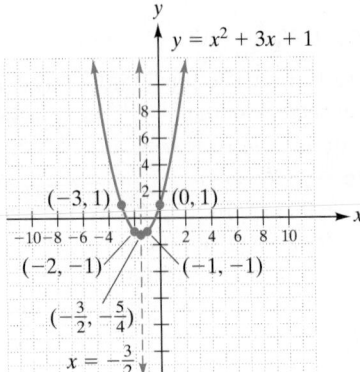

Check

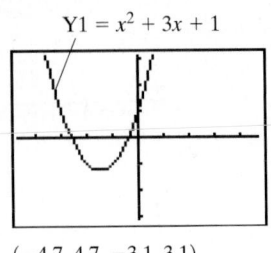

$(-4.7, 4.7, -3.1, 3.1)$

Figure 8.7

EXAMPLE 6 Graph the quadratic function $s(x) = -x^2 + 2x + 1$. Label the vertex. Draw and label the axis of symmetry.

Solution

Given $s(x) = -x^2 + 2x + 1$, we know that $a = -1, b = 2$, and $c = 1$.
Determine the coordinates of the vertex. Use the formula

$$x = \frac{-b}{2a} = \frac{-2}{2(-1)} = 1$$

Substitute 1 for x in the function.

$$s(x) = -x^2 + 2x + 1$$
$$s(1) = -(1)^2 + 2(1) + 1$$
$$s(1) = 2$$

The vertex is $(1, 2)$. The axis of symmetry is the line $x = 1$.
Set up a table of values, using x-values less than and greater than the x-coordinate of the vertex, 1.

x	$s(x) = -x^2 + 2x + 1$	$s(x)$
-1	$s(-1) = -(-1)^2 + 2(-1) + 1$	-2
0	$s(0) = -(0)^2 + 2(0) + 1$	1
2	$s(2) = -(2)^2 + 2(2) + 1$	1
3	$s(3) = -(3)^2 + 2(3) + 1$	-2

← vertex (1, 2)

Graph the vertex and the points found in the table of values. Connect the points with a smooth curve and graph the axis of symmetry.

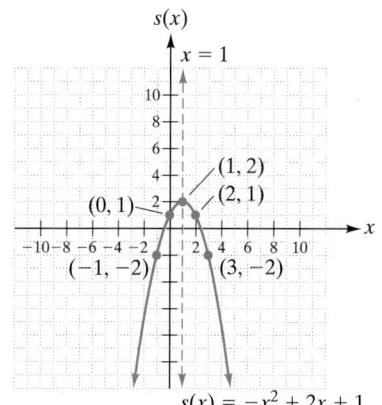

$s(x) = -x^2 + 2x + 1$

Check

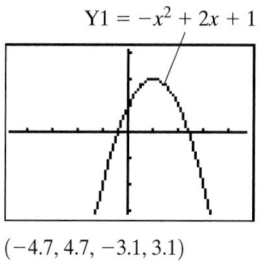

$Y1 = -x^2 + 2x + 1$

$(-4.7, 4.7, -3.1, 3.1)$

The decimal screen provides a nicer picture.
Trace the graph to check the points found in the table of values.

 8.3.3 Checkup

Graph the quadratic functions in exercises 1 and 2. Label the vertex. Draw and label the axis of symmetry.

1. $A(x) = 2x^2 - x - 6$ 2. $y = -x^2 + 12x - 26$

3. In graphing a quadratic function, why should you first find the vertex?

4. After graphing a quadratic function, how can you use the coefficients of the quadratic polynomial to check that your graph is correct?

8.3.4 Modeling the Real World

Quadratic functions model many real-world applications, as we've seen earlier in this chapter. We can use the methods in this section to find the maximum or minimum values of a quadratic function, which is often useful and important information. For example, we can use a quadratic function to determine the profit of a business. The absolute maximum of the function corresponds to the maximum amount of profit.

EXAMPLE 7 A company manufactures and sells x computer printers per month. The monthly profit function is given by $P(x) = 100x - 0.025x^2$, where x is between 0 and 4000 printers. Determine the maximum profit the company can realize and how many printers should be manufactured and sold to achieve this profit.

Solution

$$P(x) = 100x - 0.025x^2$$
or
$$P(x) = -0.025x^2 + 100x$$

Determine the x-value of the vertex by substituting $a = -0.025$, $b = 100$, and $c = 0$.

Check

Let $Y1 = 100x - 0.025x^2$
Graph and determine the absolute maximum value of the function.

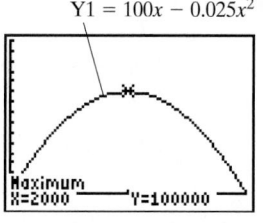

$Y1 = 100x - 0.025x^2$

Maximum
X=2000   Y=100000

$(0, 4000, -10,000, 150,000)$

$$x = -\frac{b}{2a}$$

$$x = \frac{-(100)}{2(-0.025)}$$

$$x = 2000$$

The function $P(x) = 100x - 0.025x^2$ evaluated for $x = 2000$ is

$$P(2000) = 100(2000) - 0.025(2000)^2$$

$$P(2000) = 100,000$$

The vertex is $(2000, 100,000)$. The absolute maximum of the function is 100,000 at $x = 2000$.

The maximum monthly profit for the company is $100,000, when 2000 computer printers are manufactured and sold.

APPLICATION

A signal flare is shot upward from a cliff 100 meters high. If the initial velocity of the flare was 91.2 meters per second, a position function for the flare is $s(t) = -4.9t^2 + 91.2t + 100$, where $s(t)$ is measured in meters and time, t, is measured in seconds. Determine the signal flare's maximum height in meters.

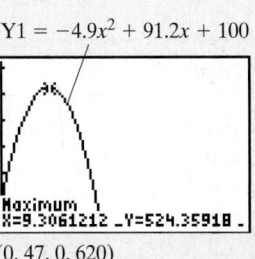

$Y1 = -4.9x^2 + 91.2x + 100$

Maximum
X=9.3061212 _Y=524.35918 _

$(0, 47, 0, 620)$

Discussion

Let $Y1 = -4.9x^2 + 91.2x + 100$

The vertex of the graph is approximately $(9.31, 524.36)$. Therefore, the maximum height of the signal flare is approximately 524 meters, at approximately 9.3 seconds.

 **8.3.4 Checkup**

1. If a water balloon is hurled upward with an initial velocity of 14 meters per second from a window that is 70 meters above the ground, the position formula representing the height of the balloon at time t is given by $s(t) = -4.9t^2 + 14t + 70$. Graph this function. Find the maximum height the balloon will reach.

2. Data provided by the Recording Industry Association of America on the number of compact discs shipped can be used to develop the function $N(x) = -4.9x^2 + 122x + 205$, where $N(x)$ is the number shipped in millions and x is the number of years since 1990. Determine the year in which the number shipped should reach its maximum, assuming that the relation continues to hold. What is the maximum number of compact discs that will be shipped, according to this function? Check your results by graphing the function, using $(0, 50, -100, 1500)$ as the window.

8.3 Exercises

Identify each function as quadratic or nonquadratic.

1. $y = 1 - x - x^2 - x^3$

2. $g(x) = 1.3x^2 - 4.7$

3. $f(x) = 8x + 11$

4. $y = x + 2x^2 - 9$

5. $g(x) = 0.5x^2 + 2.6x - 8.4$

6. $y = 8x^2$

7. $y = -2x^2 - 5x - 7$

8. $r(x) = \frac{1}{x^2} - 3$

9. $a = \pi r^2$

10. $s = 4\pi r^2$

11. $S = 6e^2$

12. $y = 3x^3 + 2x - 5$

13. $y = \frac{7}{x^2} + 3x - 12$

14. $A(x) = \frac{1}{2}x^2$

15. $y = 2x^2 - 2x + 5$

16. $f(x) = 6 - 4x$

In exercises 17 and 18, given the quadratic function, list the properties of its graph.

17. $y = -5x^2 + 10x + 1$ **18.** $y = 6x^2 - 6x - 5$

Complete the table. Check your work on your calculator.

	Function	Coefficients			Properties of graph				
		a	b	c	Wide/narrow	Concave upward/downward	Vertex	Axis of symmetry	y-intercept
19.	$y = 0.6x^2 + 6x - 2$								
20.	$y = -x^2 + 6x - 2$								
21.	$y = 2x^2 + 3x + 5$								
22.	$y = -3x^2 + 6x - 5$								
23.	$y = -\frac{1}{4}x^2 + x - 3$								
24.	$y = \frac{1}{3}x^2 + 2x - 1$								
25.	$f(x) = x^2 + 8x + 1$								
26.	$y = -0.4x^2 + 2.4x - 1.1$								

Graph each quadratic function in exercises 27–42. Label the vertex. Draw and label the axis of symmetry.

27. $f(x) = 2x^2 + 5x - 7$ **28.** $y = 2x^2 + 6x - 5$ **29.** $y = \frac{1}{6}x^2 + 3x + 12$ **30.** $y = \frac{3}{4}x^2 - 6x + 7$

31. $h(x) = 14 + 5x - x^2$ **32.** $f(x) = 11 - 4x + x^2$ **33.** $y = -2x^2 + 8x - 3$ **34.** $y = -3x^2 + 6x + 1$

35. $g(x) = 0.8x^2 - 1.2x$ **36.** $h(x) = 1.2x^2 + 3.6x$ **37.** $y = 0.4x^2$ **38.** $y = -0.7x^2$

39. $y = 3x^2 - 3$ **40.** $y = -6x^2 + 3$ **41.** $f(x) = -0.5x^2 + 3x$ **42.** $g(x) = -0.2x^2 + 4x$

43. Eve threw an apple upward with a speed of 12 feet per second from a height of 24 feet. The position function for the apple is given by $s(t) = -16t^2 + 12t + 24$. Graph the function. Find the maximum height the apple will reach.

44. A football is kicked with a vertical velocity of 60 feet per second from ground level. The position function for the football is given by $s(t) = -16t^2 + 60t$. Graph the function. Determine the coordinates of the vertex of the graph. Interpret these values.

45. Gramps is building Granny a cottage. He wants the foundation to be 280 feet around, but isn't sure what width and length to build. If the width of the foundation is x feet, the area of the foundation is given by $A(x) = 140x - x^2$. Graph the function. Find the vertex. Explain what the coordinates of the vertex indicate.

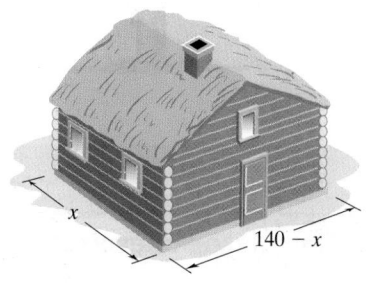

46. Farmer Jones plans to build a small animal pen next to his barn. The pen will be rectangular, with one side formed by the barn. The other three sides will be constructed from 120 feet of fencing. The area of the pen is given by the quadratic function $A(x) = 120x - 2x^2$. Graph the

EXAMPLE 3

Your laboratory partner projects a stone upward from a cliff at a height of 250 feet above the ground. After 1 second, the stone is 324 feet above the base of the cliff, and after 2 seconds the stone is 366 feet above the base of the cliff.

2 second 366 ft

1 second 324 ft

0 second 250 ft

250 ft

a. Write a quadratic function for the height of the stone.

b. Graph the function.

c. Use the graph to predict the length of time required for the stone to hit the ground at the base of the cliff.

d. Use the graph to predict the maximum height of the stone.

Solution

a. Let t = time in seconds

We need to determine the position function, $s(t)$. We have three data points: $(0, 250)$, $(1, 324)$, and $(2, 366)$.

$$s(t) = at^2 + bt + c \qquad \text{Standard form}$$

$(1) \quad 250 = a(0)^2 + b(0) + c \qquad$ Substitute $(0, 250)$.

$(2) \quad 324 = a(1)^2 + b(1) + c \qquad$ Substitute $(1, 324)$.

$(3) \quad 366 = a(2)^2 + b(2) + c \qquad$ Substitute $(2, 366)$.

Simplify each equation.

$(1) \quad 250 = c$

$(2) \quad 324 = a + b + c$

$(3) \quad 366 = 4a + 2b + c$

Since we know $c = 250$, we substitute 250 for c in the simplified equations 2 and 3.

$(2) \quad 324 = a + b + 250$

$(3) \quad 366 = 4a + 2b + 250$

Write the equations in standard form and simplify.

$(2) \quad a + b = 74$

$(3) \quad 2a + b = 58 \qquad$ Divide both expressions by 2.

Solve the remaining system of linear equations. We will use the elimination method in the example.

$(2) \quad -a - b = -74 \qquad$ Multiply both expressions by -1.

$(3) \quad \underline{2a + b = 58}$

$ a = -16$

$ a = -16$

$(2) \quad a + b = 74 \qquad$ Substitute -16 for a in equation (2).

$ -16 + b = 74$

$ b = 90$

Therefore, $a = -16$, $b = 90$, and $c = 250$. Substitute these values into the standard form of the quadratic function.

$$s(t) = at^2 + bt + c$$
$$s(t) = -16t^2 + 90t + 250$$

b. $Y1 = -16x^2 + 90x + 250$

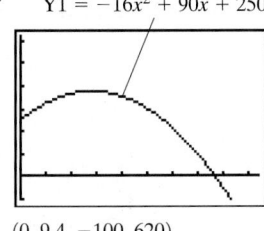

$(0, 9.4, -100, 620)$

c. $Y1 = -16x^2 + 90x + 250$

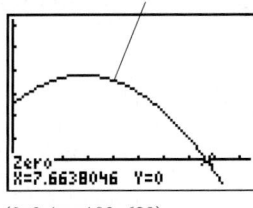

$(0, 9.4, -100, 620)$

c. Trace the graph or calculate a zero to determine when the stone's position is 0 feet above the ground, or $s(t) = 0$.

The stone will hit the ground in approximately 7.7 seconds.

d. $Y1 = -16x^2 + 90x + 250$

$(0, 9.4, -100, 620)$

d. Trace the graph or calculate a maximum to determine the absolute maximum of the function, or the function value of the vertex.

The stone will reach a maximum height of approximately 377 feet.

In the study of statistics, a technique is used to "fit" coordinate pairs to a function. The technique, called **regression analysis**, is intended to statistically analyze data, but also can be used to find the quadratic function that passes through three points in the coordinate plane. Many calculators include this feature. We will utilize it to write a function using real data.

A check of the previous example will result in the same equation, $y = -16x^2 + 90x + 250$, as shown in Figure 8.8c.

TECHNOLOGY

QUADRATIC REGRESSION

Write a quadratic function that contains the three data points $(0, 250)$, $(1, 324)$, and $(2, 366)$.

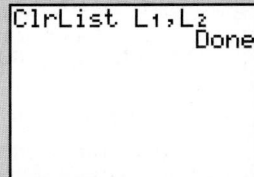

Figure 8.8a

Figure 8.8b

Figure 8.8c

For Figure 8.8a,
Clear the lists where the three data points will be stored.

STAT | 4 | 2nd | LI | , | 2nd | L2 | ENTER

For Figure 8.8b,
Enter the coordinate pairs by choosing option 1, EDIT, under the STAT menu.

(Continued on page 576)

Enter values for x, (0, 1, 2), in L1, and then move to L2 and enter values for y, (250, 324, 366).

| STAT | | I | | 0 | | ENTER | | I | | ENTER | | 2 | | ENTER |

| ► | | 2 | | 5 | | 0 | | ENTER | | 3 | | 2 | | 4 | | ENTER | | 3 | | 6 | | 6 | | ENTER |

For Figure 8.8c,

Exit the EDIT menu.

| 2nd | | QUIT |

Calculate the quadratic function by choosing option 5, QuadReg, under the STAT CALC menu.

| STAT | | ► | | 5 | | 2nd | | LI | | , | | L2 | | ENTER |

Write the function, substituting the given values for a, b, and c into the standard form of the quadratic function. The result is $y = -16x^2 + 90x + 250$.

APPLICATION

The Ohmishima bridge joining the Japanese islands of Ohmishima and Hakatajima is the longest single-span two-hinged solid rib arch bridge in Japan. The lower width of the arch is 297 meters, and the maximum height of the arch is 49 meters. Determine a quadratic function that will model the arch of the bridge.

Discussion

Superimpose a sketch of the bridge on a coordinate plane, placing the origin in the center of the lower width. This will yield 148.5 units ($\frac{1}{2}$ of 297 feet) on each side of the y-axis. Label the three intercepts.

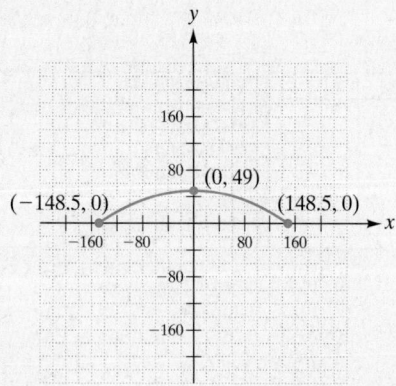

Use the three intercepts, (0, 49), (-148.5, 0), and (148.5, 0), as data points, and determine a quadratic regression function on your calculator.

Write the quadratic function by replacing a, b, and c in the standard form of the quadratic equation.

$$y = -0.002222x^2 + 0x + 49$$
$$y = -0.002222x^2 + 49$$

Check the function by graphing it on your calculator.

Note that the value for a was rounded in our function. Therefore, the x-intercepts are not exact. For accuracy's sake, an engineer would not round values as we did.

✓ 8.4.3 Checkup

1. A message in an attaché case is thrown downward from a plane that is 1500 feet above the water. After 1 second, the case is 1454 feet above the water, and after 2 seconds, it is 1376 feet above the water. Write a quadratic position function for the distance the case is above the water after t seconds. How long will it take the case to reach the water?

2. The Alexander Hamilton Bridge carries approximately 175,000 vehicles per day over the Harlem River in New York City. The length of the main arch of the bridge is 555 feet long, and the clearance at the center of the arch above the mean high water level is 135 feet. Determine a quadratic function that will model the arch of the bridge.

3. The Golden Gate Bridge is a suspension bridge that measures a distance of 4200 feet between its towers, which have a height of 500 feet above the roadway. Assuming that the shape of the main span cables is a parabola, write a quadratic function that will model that shape. (*Hint:* A suspension bridge is different from an arch bridge: The parabola opens upward instead of downward. Superimpose a sketch of the bridge on a coordinate plane, placing the origin at the center of the upper width of the parabola. This will yield x-intercepts at 2100 units on each

side of the y-axis. The y-intercept will be 500 units below the x-axis. Label the three intercepts and then find the function.)

8.4 Exercises

Write the quadratic function for the graph that passes through the given x-intercepts and y-intercept. Graph the function and check the given points.

1. $(2, 0), (-4, 0), (0, -8)$ **2.** $(3, 0), (-1, 0), (0, 3)$ **3.** $(-4, 0), \left(\frac{1}{2}, 0\right), (0, -4)$ **4.** $(-1, 0), \left(\frac{2}{3}, 0\right), (0, -2)$

5. $(0, 3), (3, 0), (-3, 0)$ **6.** $(0, 1), (4, 0), (-4, 0)$ **7.** $(0, -4), (2, 0), (-4, 0)$ **8.** $(0, -2), (3, 0), (-2, 0)$

9. $\left(0, \frac{2}{3}\right), (-2, 0), (2, 0)$ **10.** $\left(0, \frac{1}{4}\right), (1, 0), (-2, 0)$ **11.** $(0, -2), \left(\frac{1}{2}, 0\right), \left(\frac{9}{2}, 0\right)$ **12.** $(0, 1), \left(\frac{1}{3}, 0\right), \left(\frac{-2}{3}, 0\right)$

Write a quadratic function for a graph that passes through the points specified. Graph the function and check the given points.

13. $(2, -15), (-2, -7), (-4, 9)$ **14.** $(2, -8), (4, 10), (-5, -8)$ **15.** $(1, -10), (-2, 5), (2, -7)$

16. $(-1, 10), (1, -12), (3, -10)$ **17.** $(3, 2), (-3, 14), (6, 5)$ **18.** $(2, 1), (-2, -3), (-4, 4)$

19. $(-1, -2), (1, 6), (4, 3)$ **20.** $(2, -3), (-2, 5), (1, 2)$ **21.** $(-1, -3), (-2, -5), (1, -11)$

22. $(2, -3), (-2, -15), (6, -7)$ **23.** $(-2, 4), \left(-3, \frac{5}{2}\right), \left(-1, \frac{13}{2}\right)$ **24.** $\left(5, \frac{13}{3}\right), (6, 4), (3, 7)$

Solve exercises 25–30 by using the position function $s(t) = at^2 + bt + c$. Round your answers to the nearest tenth of a second.

25. A tennis ball is thrown upward from a height of 150 feet. After 1 second, the ball is 146 feet above the ground; after 2 seconds, it is 110 feet above the ground. Write the position function for the ball. How long will it take for the ball to reach the ground?

26. A custard pie is thrown upward from a height of 220 feet. After 1 second, the pie is 222 feet above the ground. After 3 seconds, it is 130 feet above the ground. Write a position function for the pie. How many seconds will it take to reach the ground?

27. A movie stunt dummy is dropped from a height of 400 feet. After 2 seconds, the dummy is 336 feet above the ground. After 4 seconds, it is 144 feet above the ground. Write the position function for the dummy. How many seconds will it take to fall to the ground?

28. A child drops a Beanie Baby from a height of 600 feet. After 2 seconds, the toy is 536 feet above the ground. After 3 seconds, it is 456 feet above the ground. Write the position function for the toy. How many seconds will it take to fall to the ground?

29. An empty fuel cylinder is dropped from a lunar module 2000 feet above the surface of the Moon. After 10 seconds, the cylinder is 1730 feet above the lunar surface. After 20 seconds, it is 920 feet above the lunar surface. Determine the position function for the distance of the cylinder above the surface at any time. After how many seconds will the cylinder touch the surface?

30. A probe is ejected upward from a lunar module 3000 feet above the surface of the Moon. After 10 seconds, the probe is 2980 feet above the lunar surface. After 20 seconds, it is 2420 feet above the lunar surface. Determine the position function for the distance of the probe above the surface at any time. How many seconds will it take for the probe to reach the lunar surface?

31. In 1985, dry natural-gas production in the United States dropped to a low of 20.45 trillion cubic feet. In 1990, production rose to 22.60 trillion cubic feet, and in 1995, it rose to 25.18 cubic feet. If x represents the number of years since 1985 and y represents production, these data points become $(0, 20.45)$, $(5, 22.60)$, and $(10, 25.18)$. Write the quadratic function whose graph contains those points. What would this function predict natural gas production to be for the year 2005? Do you think it is reasonable to assume that production would continue to increase as this function indicates? Explain your answer.

32. In 1985, dry natural-gas production worldwide was 62.17 trillion cubic feet. In 1990, production rose to 73.61 trillion cubic feet, and in 1995, it rose to 77.92 cubic feet. If x represents the number of years since 1985 and y represents production, these data points become $(0, 62.17)$, $(5, 73.61)$, and $(10, 77.92)$. Write the quadratic function whose graph contains those points. What would the function predict production to be for 2005? Do you think this is a reasonable way to predict production? Explain your answer.

33. The *Chicago Tribune* reported that there were 3352 McDonald's franchises in the United States in 1975. In 1985, the number of franchises was 6972. An annual report stated that the number of franchises had grown to 18,299 by 1995. Write a quadratic function whose graph contains these points. Explain how you define the independent variable, x. Use the function to predict the number of franchises in 2005. Do you think the function is a good predictor? Explain your answer.

34. An article in the *Chicago Tribune* noted that the African country of Zimbabwe is a prosperous nation that has experienced a population explosion. The population, in millions, was 5.3 in 1970, 7.1 in 1980, and 11.3 in 2000. Write a quadratic function whose graph contains these points. Use the function to estimate Zimbabwe's population, in millions, for 1990. How does your prediction compare with the actual number of 9.9 reported in the article? Is the function a good estimator? Explain your answer.

35. The Brooklyn Bridge opened to vehicular traffic on May 24, 1883, and remains a spectacular sight to this day. The center span of the suspension bridge between its towers is 1595.5 feet long, and the towers rise 130 feet above the roadway. Approximate the curve of the center span, using a quadratic function.

36. The Mackinac Bridge opened to traffic on November 1, 1957. The bridge is one of the world's longest suspension bridges between cable anchorages. The length of the roadway between the main towers is 3800 feet. The towers rise 350 feet above the roadway. Use a quadratic function to approximate the curve of the center span.

37. The New River Gorge Bridge in West Virginia is an arch bridge with a single span of 1700 feet. The rise of the arch is 360 feet. Approximate the curve of the arch with a quadratic function.

38. The Francis Scott Key Bridge arches over Baltimore Harbor, reaching the height of a 36-story building. The main arch span of the bridge is 1,200 feet long and has a height of 185 feet. Write a quadratic function to approximate the curve of the arch.

8.4 Calculator Exercises

Many models used in mathematics are functions. Some are linear functions, while others are nonlinear. Data gathered from a real-world situation can be used to write mathematical functions to represent the situation. Consider the following data, which come from a production and sales situation:

Number of items produced and sold	5	30	35	40
Cost of production	$155		$335	$365
Revenue from sales	$450	$1200	$1050	
Profit				

1. Make ordered pairs of the data, using the number of items produced and sold as the independent variable and the cost of production as the dependent variable. Plot the data. Does the graph look linear? If so, write a linear function of the form $C(x) = mx + b$ for the relation.

2. Make another set of ordered pairs of data, using the same independent variable, but using revenue from sales as the dependent variable. Plot these data. Does the graph look like a quadratic function? If so, write a quadratic function of the form $R(x) = ax^2 + bx + c$ for the relation.

3. Use the functions you wrote in exercises 1 and 2 to fill in the missing data in the preceding table for cost of production and revenue. Then complete the last line of the table by calculating profit as revenue minus cost of production.

4. Make a third set of ordered pairs of data, using the same independent variable, but using profit as the dependent variable. Graph these pairs of data. Does the relation appear to be a quadratic function? If so, use three of the pairs to write a quadratic function that represents the relation. Check the relation to see whether the fourth pair of data satisfies the function.

5. Find the values of x that make the profit function 0 or less. Discuss what these values represent.

6. Find the coordinates of the vertex of the profit function. Discuss what these values represent.

This is an example of using mathematical models to represent data. Other techniques may be studied to find the "best fit" of a function for a set of data. Once a function has been obtained, it can be used to explain relationships and to answer questions about a relationship. It is a very powerful technique for understanding data. Summarize what the example means to you as a meaningful use of mathematics.

8.4 Writing Exercise

In this section, it was stated that you need two coordinate pairs to determine a linear relationship between two variables and you need three coordinate pairs to determine a quadratic relationship between two variables. Discuss why this is so. Then consider what would happen if you had more than two coordinate pairs and wished to establish a linear relationship. Also, discuss what would happen if you had more than three coordinate pairs and wished to establish a quadratic relationship. What problems could you encounter in each case? How would you deal with these problems?

CHAPTER 8 SUMMARY

After completing this chapter, you should be able to define the following key terms in your own words.

8.1
algebraic expression
terms
constant term
variable term
numerical coefficient
coefficient
like terms
combine like terms
monomial
polynomial
binomial
trinomial
degree of a term
degree of a polynomial
descending order

ascending order
evaluating

8.2
relation
polynomial relation
domain
table of values
graph
range
absolute minimum
absolute maximum
function
function notation

8.3
quadratic function
parabola
quadratic term
concave upward
concave downward
relative maximum
relative minimum
vertex
axis of symmetry

8.4
curve fitting
regression analysis

Reflections

1. In this chapter, you studied polynomial expressions and polynomial functions (or equations). Explain the difference between the two.

2. What is the difference between the degree of a polynomial expression and the degree of a term of a polynomial expression?

3. What is the difference between simplifying a polynomial expression and evaluating a polynomial expression?

4. Explain how the graph of a quadratic function differs from the graph of a linear function.

5. Given a graph, describe the difference between a parabola that is concave upward and one that is concave downward. Given an equation, explain when the graph of the parabola is concave upward and when it is concave downward.

6. The graph of a polynomial function may have relative maxima and an absolute maximum. Alternatively, the graph may have relative minima and an absolute minimum. Explain the difference between the two situations.

7. In the graph of a quadratic function, what is the relationship between the vertex of the parabola and the axis of symmetry?

8. In order to determine the equation of a line from information about points that lie on the line, you must have two coordinate pairs. In order to determine the equation of a parabola from information about points that lie on the parabola, you must have three coordinate pairs. Can you explain why you need an additional coordinate pair for a parabola?

CHAPTER 8 SECTION-BY-SECTION REVIEW

8.1

Recall	Examples

Classification of polynomials
(Always simplify the polynomial first.)
• By number of terms
  One term: monomial
  Two terms: binomial
  Three terms: trinomial
  Four or more terms: polynomial
• By degree
  Determine the degree of each term by adding the exponents of the variable terms. The degree of the polynomial is the largest degree of the terms.
	Classify the polynomial.
	$2x^3y + 8x^2 - x^3y + 4$
	by number of terms and by degree.
	
	First simplify.
	$x^3y + 8x^2 + 4$
	There are three terms:
	$x^3y, 8x^2, 4$
	The degree of each term is as follows:
	$x^3y: 3 + 1 = 4$
	$8x^2: 2$
	4 or $4x^0: 0$
	The polynomial is a fourth-degree trinomial.

Order of terms in a polynomial with one variable • Descending: The terms are written in order from largest to smallest degree. • Ascending: The terms are written in order from smallest to largest degree.	Write in descending and ascending order: $-3x^4 - 2x^3 + 8x^2 - 5x - x^5 + 4$ descending order: $-x^5 - 3x^4 - 2x^3 + 8x^2 - 5x + 4$ ascending order: $4 - 5x + 8x^2 - 2x^3 - 3x^4 - x^5$
Evaluating a polynomial • Substitute values for the variables, and simplify the resulting numeric expression.	Evaluate $-x^2 + 2xy - 4$ for $x = 2$ and $y = -3$. $\begin{aligned} -x^2 + 2xy - 4 &= -(2)^2 + 2(2)(-3) - 4 \\ &= -4 - 12 - 4 \\ &= -20 \end{aligned}$

Determine whether each expression is a polynomial. For those which are polynomials, classify them as monomials, binomials, trinomials, or polynomials.

1. x **2.** $5x - 3$ **3.** $\sqrt{x} + 2$

4. $3x^3 - 4x^2 + x - 1$ **5.** $\dfrac{3}{a} - 2a + 1$ **6.** $3a^4 - 2a^2 + 5$

Determine the degree of each term of each polynomial and then the degree of the polynomial.

7. $5x + 3x^3 - 2$ **8.** $2x^2y + 3xy - 5$ **9.** $x + 9$

10. $0.5a - 3.1a^2 + 9.6a + 3.1a^2$ **11.** $12x^2 + 30x + 3$

Write each polynomial in descending order.

12. $5y^2 + 11y^4 + 12 - 6y + 9y^3$ **13.** $5 - p$ **14.** $\dfrac{1}{4}z^4 + \dfrac{1}{2}z^2 + z + \dfrac{1}{3}z^3 + 1$

15. $0.6b - 2.3b^5 + 1.8 - 9.1b^3$

Evaluate $2x^3 + 11x^2 - 21x - 90$ *for the given values.*

16. $x = 3$ **17.** $x = 0$ **18.** $x = 1$ **19.** $x = -6$ **20.** $x = -\dfrac{5}{2}$

In exercises 21–26, evaluate $a^3 + 2a^2b - 3ab^2 - b^3$ *for the given values.*

21. $a = 0, b = 1$ **22.** $a = -1, b = 0$ **23.** $a = 0, b = 0$ **24.** $a = 1, b = 1$ **25.** $a = -1, b = 1$ **26.** $a = -1, b = -1$

27. Write a polynomial for the perimeter of a rectangle whose length is numerically equal to the square of its width. What is the perimeter if the rectangle has a width of 7 yards?

28. Write a polynomial for the perimeter of a triangle whose second side is three times as long as the first side and whose third side is numerically equal to 1 inch more than the square of the first side. What is the perimeter if the triangle's first side measures 4 inches?

29. Write a polynomial for the total area of the figures shown.

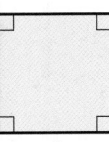 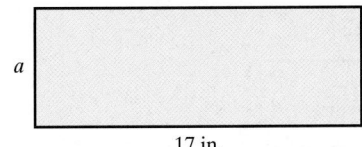 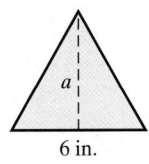

30. Write a polynomial for the shaded area shown.

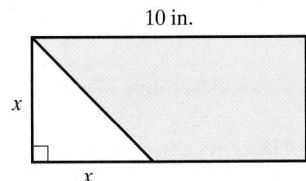

31. A triangular flower garden with a height of x feet and a base of y feet is placed in the center of a square lawn that measures z feet on a side. Write a polynomial for the area of the lawn not covered by the garden. What is this area if the triangle has a height of 6 feet and a base of 10 feet and the lawn measures 80 feet on a side?

32. A contractor charges a flat fee of $500 plus $40 per square foot to build a rectangular room addition onto an existing home. The contractor incurs setup costs of $275 plus a materials cost of $12 per square foot.

  a. Write a polynomial that represents the charge for a room with a length of x feet and a width of y feet.

  b. Write a polynomial that represents the contractor's cost of building the room.

  c. What are the revenue and cost for building a room that is 15 feet wide and 25 feet long? What would be the contractor's profit for taking on the job?

33. Industrial engineers conduct time studies to determine how long it will take a production worker to assemble parts on a production line. One such study attempted to estimate the time (in minutes) it takes a worker with x months of experience to perform a particular task on an automobile assembly line. The polynomial used to estimate the time was $0.01x^2 - 0.67x + 20.09$. Estimate the time it would take a production worker to complete the task if the number of months of experience she has is

  a. 3 months   **b.** 6 months   **c.** 12 months

8.2

Recall	Examples
Graph a polynomial function • Set up a table of values. • Graph the ordered pairs determined by the table of values. • Connect the graphed ordered pairs with a smooth curve.	Graph. $f(x) = x^2 + x - 3$ Set up a table.

x	$f(x) = x^2 + x - 3$	$f(x)$
-4	$f(-4) = (-4)^2 + (-4) - 4$	9
-3	$f(-3) = (-3)^2 + (-3) - 3$	3
-2	$f(-2) = (-2)^2 + (-2) - 3$	-1
-1	$f(-1) = (-1)^2 + (-1) - 3$	-3
0	$f(0) = 0^2 + 0 - 3$	-3
1	$f(1) = 1^2 + 1 - 3$	-1
2	$f(2) = 2^2 + 2 - 3$	3
3	$f(3) = 3^2 + 3 - 3$	9

(Continued on page 582)

Recall	Examples
	$f(x)$ $f(x) = x^2 + x - 3$ $(-4, 9)$ $(3, 9)$ $(-3, 3)$ $(2, 3)$ $(-2, -1)$ $(1, -1)$ $(-1, -3)$ $(0, -3)$
Range of a polynomial function • Determine the absolute minimum and absolute maximum of the function from its graph if they exist. • The range is the set of y-values that lie between the absolute minimum and maximum, inclusive. • If the relation has no absolute maximum, the range is the set of all real numbers greater than or equal to the absolute minimum. • If the relation has no absolute minimum, the range is the set of all real numbers less than or equal to the absolute maximum.	Determine the range of $f(x) = x^2 + x - 3$. $Y1 = x^2 + x - 3$ Minimum X=-.4999995 Y=-3.25 The range is $y \geq -3.25$.
Evaluating a polynomial function • Substitute the x-value and simplify the resulting numeric expression.	Given $f(x) = x^2 + x - 3$, determine $f(-4)$. $f(x) = x^2 + x - 3$ $f(-4) = (-4)^2 + (-4) - 3$ $= 9$

34. Create a table of values for the polynomial relation $y = x^3 - x^2 - 6x$, where x is an integer between -4 and 4.

Graph the given polynomial relations. From the graph, determine the range of each relation.

35. $y = -3x + 5$ **36.** $y = 2x^2 - 2x - 12$

37. $y = x^3 + 2x^2 - 5x - 6$ **38.** $y = x^4 + 2x^3 - 5x^2 - 6x$

In exercises 39–43, evaluate the function $f(x) = 3x^3 - x^2 + 2x - 4$ at the given values.

39. $f(-2)$ **40.** $f(0)$ **41.** $f(2)$ **42.** $f\left(-\dfrac{1}{2}\right)$ **43.** $f(1.7)$

44. Tony's Tees will discount the price of a t-shirt, depending on the number purchased. If a person purchases x shirts, the price per shirt will be $(12 - 0.50x)$ dollars. Tony pays $4 per shirt.
   a. If x shirts are sold, the revenue function for the sale is $R(x) = 12x - 0.50x^2$. Determine the revenue if 5 shirts are sold, if 10 shirts are sold, if 15 shirts are sold, if 20 shirts are sold, and if 25 shirts are sold. What does this tell you about limits that should be placed on the price per shirt?
   b. If x shirts are sold, the total cost of these shirts to Tony is $C(x) = 4x$. What is the cost of 5 shirts, 10 shirts, 15 shirts, and 20 shirts to Tony?

   c. Use the results from parts a and b to determine how much profit will be made from the sale of 5 shirts, 10 shirts, 15 shirts, and 20 shirts.
   d. What would you tell Tony to do in terms of limiting the price per shirt?

45. A statistical study related the cost of milk production, y, to the number of hundreds of gallons produced, x. The relation developed was

$$y = 15{,}800 + 2.2x - 0.001x^2$$

where x is measured in hundreds of gallons and y is measured in dollars. Use the relation to estimate the cost for the following number of gallons produced:

Hundreds of gallons (x)	900	1000	1100	1200	1300	1400
Cost (y)						

What does this relation tell you about the cost of production? If you were the owner of a dairy farm, can you suggest a strategy to contain your cost of production, based upon what this table shows?

46. At its deepest, the Grand Canyon is 6000 vertical feet from rim to river. If a person stands at the edge of the rim and flings a rock upward with a velocity of 40 feet per second, the distance above the river is given by the function $s(t) = -16t^2 + 40t + 6000$, where $s(t)$ is the height above ground level (in feet) and t is the time in seconds after the rock is flung. Determine the range of this function and interpret its meaning.

8.3

Recall	Examples
Standard form for a quadratic function $y = ax^2 + bx + c$ or $f(x) = ax^2 + bx + c$, where $a \neq 0$.	Examples of quadratic functions $y = 2x^2 - 3x + 4$ $g(x) = -\dfrac{1}{2}x^2 + 6x - 2$
Concavity of the graph • Concave upward if $a > 0$. • Concave downward if $a < 0$.	• The graph of $y = 2x^2 - 3x + 4$ is concave upward because the value of a, 2, is positive. • The graph of $g(x) = -\dfrac{1}{2}x^2 + 6x - 2$ is concave downward because the value of a, $-\dfrac{1}{2}$, is negative.
Width of the graph in comparison to the width when $a = 1$. • Narrow graph when $\|a\| > 1$. • Wide graph when $\|a\| < 1$.	• The graph of $y = 2x^2 - 3x + 4$ is narrower than the graph with $a = 1$ because the absolute value of a, 2, is greater than 1. • The graph of $g(x) = -\dfrac{1}{2}x^2 + 6x - 2$ is wider than the graph with $a = 1$ because the absolute value of a, $\dfrac{1}{2}$, is less than 1.
Vertex of the graph • The x-coordinate is $\dfrac{-b}{2a}$. • The y-coordinate is the function evaluated at $x = \dfrac{-b}{2a}$.	The vertex of the graph of $y = 2x^2 - 3x + 4$ is $\left(\dfrac{3}{4}, \dfrac{23}{8}\right)$ because $x = \dfrac{-b}{2a} = \dfrac{-(-3)}{2(2)} = \dfrac{3}{4}$ and $$y = 2x^2 - 3x + 4$$ $$= 2\left(\dfrac{3}{4}\right)^2 - 3\left(\dfrac{3}{4}\right) + 4$$ $$= \dfrac{23}{8}$$
Axis of symmetry • The graph of the equation $x = \dfrac{-b}{2a}$.	The axis of symmetry of the graph of $y = 2x^2 - 3x + 4$ is $x = \dfrac{3}{4}$.
y-intercept • The y-intercept is the ordered pair $(0, c)$.	The y-intercept of the graph of $y = 2x^2 - 3x + 4$ is $(0, 4)$ because $c = 4$.

Identify each equation as quadratic or nonquadratic.

47. $y = x^2 + x + 1$ **48.** $y = x^3 - x - 1$ **49.** $y = \dfrac{5}{x^2} + x + 1$ **50.** $y = x^2 + 4x + 4$

Complete the table. Check by using the calculator.

	Function	Coefficients				Properties of graph				
		a	b	c	Wide/narrow	Concave upward/downward	Vertex	Axis of symmetry	y-intercept
51.	$y = -\dfrac{1}{4}x^2 + \dfrac{1}{2}x + 1$								
52.	$f(x) = -2x^2 + 4x$								
53.	$g(x) = \dfrac{1}{3}x^2 + x$								
54.	$y = 3x^2 - 3x + 1$								

In exercises 55–57, graph each quadratic function. Label the vertex. Draw and label the axis of symmetry.

55. $f(x) = x^2 + 2x - 8$ **56.** $y = -\dfrac{1}{2}x^2 + x - 2$ **57.** $h(x) = 2x^2 - 8$

58. The length of a patio measures 8 feet more than its width. If the width of the patio is denoted by w, its area is given by the function $A(w) = w^2 + 8w$. Graph the function and find its vertex. Does the vertex have any physical meaning?

59. A bridal photographer sets the price of 8-by-10-inch wedding photos by the number of photos ordered. If the bride orders x photos, the total revenue will be $R(x) = 30x - 0.50x^2$ dollars for the sale. Graph the function and find its vertex. Explain what this vertex represents.

60. An egg is thrown upward with a velocity of 60 feet per second from a height of 120 feet. The position function for the egg is given by $s(t) = -16t^2 + 60t + 120$. Determine the vertex of the function and interpret the values.

8.4

Recall	Example
Writing a quadratic function, given three data points	Write a quadratic function for a graph that will pass through $(1, 9)$, $(2, 16)$, and $(3, 25)$.
• Write and simplify three equations, substituting the data points into the standard form of the quadratic equation.	This will result in three equations:
• Solve the resulting system of three equations.	$y = ax^2 + bx + c$
	(1) $\quad 9 = a(1)^2 + b(1) + c$
	(2) $\quad 16 = a(2)^2 + b(2) + c$
	(3) $\quad 25 = a(3)^2 + b(3) + c$

Recall	Example
	Simplify each equation.
	(1) $9 = a + b + c$
	(2) $16 = 4a + 2b + c$
	(3) $25 = 9a + 3b + c$
	First system:
	(1) $9 = a + b + c$
	(2) $16 = 4a + 2b + c$
	(1) $-9 = -a - b - c$
	(2) $\underline{16 = \ \ 4a + 2b + c}$
	(4) $\ \ 7 = \ \ 3a + \ \ b$
	Second system:
	(2) $16 = 4a + 2b + c$
	(3) $25 = 9a + 3b + c$
	(2) $-16 = -4a - 2b - c$
	(3) $\underline{\ \ 25 = \ \ \ \ 9a + 3b + c}$
	(5) $\ \ \ 9 = \ \ \ 5a + \ \ b$
	New system:
	(4) $7 = 3a + b$
	(5) $9 = 5a + b$
	(4) $-7 = -3a - b$
	(5) $\underline{\ \ 9 = \ \ \ 5a + b}$
	$\ \ 2 = \ \ \ 2a$
	$\ \ 1 = \ \ \ \ a$
	Substitute 1 for a in an equation from the new system.
	(4) $7 = 3a + b$
	$7 = 3(1) + b$
	$4 = b$
	Substitute a and b in an original simplified equation.
	$9 = a + b + c$
	$9 = 1 + 4 + c$
	$4 = c$
	Therefore, $a = 1, b = 4$, and $c = 4$. Substitute these values into the standard form of the function.
	$y = ax^2 + bx + c$
	$y = (1)x^2 + (4)x + (4)$
	$y = x^2 + 4x + 4$

In exercises 61–63, write the quadratic function for the graph that passes through the given three points. Check by graphing.

61. $(8, 0), (-3, 0), (0, -24)$ **62.** $(0, 5), (5, 0), \left(-\dfrac{1}{2}, 0\right)$ **63.** $(2, -4), (4, -6), (7, 6)$

64. Use the position function, $s(t) = at^2 + bt + c$, to solve the following application: A hammer is thrown upward from a height of 160 feet. After 1 second, it is 176 feet above the ground. After two seconds, it is 160 feet above the ground. Write the position function for the falling hammer. To the nearest tenth of a second, how long will it take the hammer to reach the ground?

65. In 1910, the number of people age 65 and over in the United States was 4.0 million. In 1950, the number had grown to 12.4 million. In 1990, the number was 31.2 million. Write a quadratic function that relates the number of people age 65 and over to the number of years since

1900. The actual numbers of people age 65 and over were 25.5 million in 1980, 20.1 million in 1970, and 16.7 million in 1960. How well does your quadratic function predict these values? Would you conclude that a quadratic function is a good model to use for these data? What would you predict the number of people age 65 and over to be in the year 2010?

66. The Bixby Creek Bridge is one of the largest single-arch concrete bridges in the world. The main span of the bridge is 320 feet long and 280 feet high. Write a quadratic function to approximate the curve of the arch.

CHAPTER 8 CHAPTER REVIEW

Evaluate $2x^3 - 3x^2 - 29x - 30$ *for the given values.*

1. $x = 5$ **2.** $x = 0$ **3.** $x = 1$ **4.** $x = -2$ **5.** $x = -\dfrac{3}{2}$

Evaluate $2a^3 + 4a^2b - 2ab^2 + b^3$ *for the given values.*

6. $a = -1, b = 0$ **7.** $a = 1, b = 1$ **8.** $a = -1, b = 1$ **9.** $a = -1, b = -1$

10. Classify each expression as a monomial, a binomial, a trinomial, or a polynomial. Determine the degree of each term and the degree of the polynomial. Write the polynomial in descending order.
  a. $5 + 3x^2$ **b.** $15a^2 - 5a^3 + 4 + a$ **c.** $5x^4 + x - 2 + x^5 - 3x^2$

11. Classify each expression as a monomial, a binomial, a trinomial, or a polynomial. Determine the degree of each term and the degree of the polynomial.
  a. $7b + 3b^2 - 4 + 6b$ **b.** $3x^2y - 4xy + 3xy^2 + 5 - 4x^2y^2$ **c.** $4xyz - 2 + 3xyz - 5 - xyz + 7$

In exercises 12–16, evaluate the function $f(x) = 2x^3 - 3x^2 - 23x + 12$ *at the given values.*

12. $f(-3)$ **13.** $f(0)$ **14.** $f(4)$ **15.** $f\left(\dfrac{1}{2}\right)$ **16.** $f(2.2)$

17. Construct a table of values for the polynomial relation $y = 2x^3 + 2x^2 - 12x$, where x is an integer between -4 and 4.

Graph the given polynomial relation. From the graph, determine the range of each relation.

18. $y = 4x - 2$ **19.** $y = 3x^2 + 3x - 6$
20. $y = x^3 - 3x^2 - 13x + 15$ **21.** $y = x^4 - 3x^3 - 13x^2 + 15x$

Graph each quadratic function, labeling the vertex and drawing the axis of symmetry.

22. $f(x) = 2x^2 + 7x - 4$ **23.** $y = \dfrac{1}{4}x^2 - x + 3$ **24.** $h(x) = -x^2 + 9$

Complete the table. Check by using your calculator.

		Coefficients			Properties of graph				
	Function	a	b	c	Wide/narrow	Concave upward/downward	Vertex	Axis of symmetry	y-intercept
25.	$y = \dfrac{1}{3}x^2 + \dfrac{2}{3}x + 1$								
26.	$f(x) = -3x^2 + 6x$								
27.	$y = -\dfrac{1}{4}x^2 + x + 3$								
28.	$g(x) = 2x^2 + 4x - 6$								

In exercises 29–31, write the quadratic function for the graph that passes through the given three points. Check by graphing.

29. $(-2, 12), (-4, 8), (-7, 17)$ **30.** $(9, 0), (-4, 0), (0, -36)$ **31.** $(0, -12), (3, 0), \left(-\dfrac{4}{3}, 0\right)$

32. Write a polynomial for the perimeter of a rectangle whose length has a numerical value that is 5 units more than the cube of the numerical value for its width. What is the perimeter if the width of the rectangle is 3 feet?

33. Write a polynomial for the perimeter of a triangle whose second side has a numerical value that is 3 units less than twice the numerical value of its first side and whose third side has a numerical value that is 7 units less than the square of the numerical value of its first side. What is the perimeter if the first side of the triangle measures 10 centimeters?

34. Write a polynomial for the total area of the figures shown.

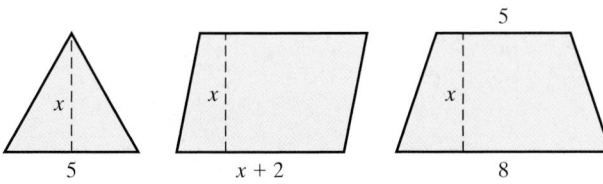

35. Write a polynomial for the shaded area shown.

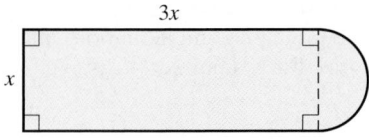

36. A circular pool with a radius of z feet is placed in a rectangular yard with a length of x feet and a width of y feet. Write a polynomial for the area of the yard not covered by the pool. What does this area measure if the pool has a radius of 8 feet and the yard measures 50 feet by 80 feet?

37. Kitchen floor tiles measure 1 foot on a side and sell for $1.50 each. A contractor offers to retile a kitchen floor for a payment of $800. The floor is rectangular.
  a. Write a polynomial that represents the cost of the materials to the contractor.
  b. Write a polynomial that represents the profit (payment minus cost) that the contractor will realize on the job.
  c. What will be the profit for a kitchen whose width is 10 feet and whose length is 15 feet?

38. A reward system for grades has been designed to be progressive. Pop will pay his child for a report card with no failing grades according to the number of A's earned. If the child earns x grades of A, Pop will pay the child $P(x) = 10x + 2x^2$ dollars.
  a. Set up a table of possible values to show how much the child can earn for his report card, given that he is enrolled in seven classes.
  b. Interpret the table. What can you say about the payments in relation to the number of A's?

39. A director of a business school graduate program wanted to relate a student's grade point average (GPA) in the school's master-of-business-administration (MBA) program with the student's score on the standardized test given to business school applicants (GMAT). The director developed the mathematical model

$$y = 1.5 + 0.0082z - 0.0000081z^2$$

where
y is the student's GPA in the MBA program, and
z is the student's score on the GMAT.

a. Using this model, what would you predict a student's GPA to be if her GMAT score was 750?
b. Complete the following table to relate his GPA in the MBA program to his GMAT score:

GMAT score (z)	500	550	600	650	700	750
GPA in MBA (y)						

c. This table should surprise you. What does it say about the relationship between the entrance exam score (GMAT) and the student's performance in graduate school?

40. A supply package is dropped from a lunar module hovering 1500 meters above the surface of the Moon. The position function for the package is $s(t) = -0.8t^2 + 1500$, where t is the number of seconds since the drop. Graph the function. How many seconds will it take for the object to reach the ground?

41. In 1920, the divorce rate per 1000 population was 1.6; in 1940, the divorce rate was 2.0; and in 1970, the divorce rate was 3.5. Write a quadratic function to use as a mathematical model that relates the divorce rate to the number of years since 1900. In 1930, the actual divorce rate was 1.6, and in 1990, it was 4.7. How well does your function predict these values? Would you conclude that the function is a good mathematical model to use to explain the relation between the divorce rate and time? What does the model predict the divorce rate will be for the year 2010?

42. A marketing analyst conducted a study to determine how sales (number of units sold) might be affected by the price x (in cents) of a product. She determined that the polynomial $0.046x^2 - 10.98x + 729.87$ estimates the number of greeting cards sold when the price is x cents per card. How many cards would be sold when the price is 99 cents per card? What if the price is $1.19 per card?

43. The deepest gorge on Earth is the Yarlung Tsangpo River Canyon in Tibet, which is more than 3 miles deep at its lowest point. If a rock were to be tossed downward with an initial velocity of 60 feet per second and were to fall directly into the river, its distance above the river would be given by the function $s(t) = -16t^2 - 60t + 19,500$, where $s(t)$ is the height above ground level (in feet) and t is the time in seconds after the rock was tossed. Determine the range of this function and interpret its meaning.

44. As a fund-raiser, a school is selling coupon books that give discounts to the purchaser at various stores and restaurants in the community. The school offers reduced prices for multiple purchases in order to increase sales. If a customer purchases x coupon books, the revenue to the school is given by $R(x) = 15x - x^2$ dollars. Determine the maximum revenue the school can receive from one person's multiple purchase and how many books must be sold to receive this maximum revenue.

45. The Hell Gate Bridge over the East River in New York City provided a much-needed railroad connection when it was completed in 1916. The bridge has an arch that spans 977.5 feet and rises 170 feet from the tracks. Write a quadratic function to approximate the curve of the arch.

CHAPTER 8 TEST

TEST-TAKING TIPS

In reviewing material for a test, it is important to read carefully and deliberately. Do not read your mathematics text the same way that you read a history book, a novel, or a newspaper. You must read a mathematics text slowly, absorbing each and every word. You may have to reread the material several times until it begins to make sense. Each word or symbol is important because mathematics texts have many thoughts condensed into a few statements.

Do not try to memorize illustrative examples. You will be overwhelmed with memorization, and the further you go, the more difficult memorizing will be. Much of mathematics is based upon a few fundamental principles (for example, the vertical line test) and definitions (for example, range). Concentrate on these principles and definitions, and commit them to memory. Try to see how each example is just a reapplication of the principles and definitions, and you won't have to memorize the examples.

Classify each expression as a monomial, a binomial, a trinomial, or a polynomial.

1. $123x^2y^3z$ **2.** $3a^3 + 5a^2b + 7ab^2 + 9b^3$ **3.** $2 - c$

Determine the degree of each polynomial.

4. $13 - 3x^3 + 6x - 0.5x^5 + 1.7x^2 - x^4$ **5.** $5x^2y^3 + 3xy^2 - y + 17$

Write the polynomial in descending order.

6. $15 + 3x^4 - 7x + x^5 + 9x^2 + 21x$

Write the polynomial in ascending order.

7. $a - \dfrac{2}{3}a^3 - \dfrac{5}{6}a^2 - \dfrac{4}{9}$

Evaluate $a^2 + 3ab^3 - 7b^2 - b - 6$ for the given values.

8. $a = 0, b = 3$ **9.** $a = -2, b = 0$ **10.** $a = 2, b = -3$

11. Carpeting for a living room sells for $16.50 per square yard, with an installation charge of $75.00. Write a polynomial for the total cost of carpeting a rectangular room. What will the cost be for a room that measures 15 feet by 22 feet, which is 5 yards by $7\frac{1}{3}$ yards?

Given $g(x) = 3x^2 + 7x - 6$, find

12. $g(-3)$ **13.** $g(0)$ **14.** $g(1)$

Consider the relation $y = \frac{1}{2}x^2 - 2x - 6$.

15. Find the vertex.   **16.** Graph the relation.   **17.** Determine the range of the relation graphically.

18. Is the relation a function? Justify your answer.

Complete the following table:

	Coefficients			Properties of graph				
Function	*a*	*b*	*c*	**Wide/narrow**	**Concave upward/downward**	**Vertex**	**Axis of symmetry**	***y*-intercept**
19. $y = \dfrac{1}{2}x^2 + 2x + 3$								
20. $y = 3x^2 - 3x + \dfrac{1}{4}$								

In exercises 21 and 22, write the quadratic function for the graph that passes through the given points.

21. $(0, 8), (2, 0), (4, 0)$ **22.** $(-1, 6), (1, 4), (2, 9)$

23. A researcher used data from the U.S. Bureau of Labor Statistics to develop a polynomial that estimates the average annual earnings of males who had completed x years of schooling, where x varied from 9th grade $(x = 9)$ through college $(x = 16)$. The earnings are estimated by the polynomial $1020x^2 - 19{,}195x + 114{,}446$ (dollars). Determine the average earnings of a male who completes an associate's degree program $(x = 14)$. How many more dollars per year can he be expected to earn over a male who has only a high school education $(x = 12)$?

24. Construction on the Tower of London began in 1078 by William of Normandy. It was completed 20 years later, rising nearly 100 feet high. The tower held many famous prisoners, including Sir Thomas More and Anne Boleyn. If they had attempted to drop a note from the top of the tower, its distance above the ground could be approximated by the function $s(t) = -16t^2 + 100$, where $s(t)$ was the height above ground level (in feet) and t was the time in seconds. Determine the range of this function and interpret its meaning.

25. A supplier of ergonomically designed computer keyboards offers a multiple-purchase discount to a firm that plans to purchase them for its employees. If the firm buys x keyboards, the supplier will realize a profit of $P(x) = 75x - 2x^2$ dollars, after subtracting cost from revenue. State the vertex of the graph of this function. Explain what the coordinates of the vertex indicate. What is the maximum profit the supplier can realize, and how many keyboards should be sold to achieve this profit?

26. The Bayonne Bridge is one of the longest steel arch bridges in the world and links Bayonne, New Jersey, with Staten Island, New York. The length of the arch span is 1,675 feet and the height of the arch is 325 feet at the center. Write a quadratic function to approximate the curve of the arch.

27. Define what is meant by the term *quadratic function*. Describe the special features of such a function. Explain how you would sketch the graph of a quadratic function.

Chapter 8

Project

Part I

In this chapter, we have given several examples of objects being dropped or propelled upward or downward. In this project, we will use the Tl-83 Plus and a Texas Instrument Calculator-Based Ranger (CBR) to collect our own data involving dropping and tossing a pillow.

You will need the CBR, the Ranger program, and an object (a small pillow is easy to use).

1. Run the Ranger program.

   a. Connect the CBR to the TI-83 Plus.

   b. Under programs, select RANGER.

   c. Under MAIN MENU, select 1:SETUP/SAMPLE.

   d. With the up or down arrow, select a line, and press ENTER to change the settings. The screen settings should read as follows:

REAL TIME:	YES
TIME(S):	15
DISPLAY:	DIST
BEGIN ON:	[ENTER]
SMOOTHING:	NONE
UNITS:	FEET

   When the settings are correct, use the arrow to move the cursor to the top of the page, and press ENTER to start.

   e. Press ENTER to start the program.

   f. Place the CBR on the floor.

   g. Press ENTER again to start data collection. The CBR will begin to make clicking noises. The calculator will collect data points every 0.2 second.

   h. From a height of about six feet directly above the CBR, drop a pillow on top of it.

   i. When the data have been collected, press ENTER to view the PLOT MENU. Choose 2:SELECT DOMAIN. You will be prompted to choose the left and right bounds of your data set. You will need to eliminate any points before the pillow dropped and after it hit the floor. You may need to do this more than once. You may need to repeat the sample if your data do not appear to be correct. Press 5 to exit. The calculator now has the time in seconds stored in L1 and the distance above the CBR stored in L2.

   j. Calculate the quadratic regression equation for your data. Use the data points to write a quadratic function $s(t)$ representing the height above the motion detector in feet, with t representing the time after the release of the pillow.

2. Repeat the steps in exercise 1, but this time, from a height of about 5 feet, toss the pillow *upward* so that it will fall on top of the motion detector.

We have discussed the formula $s(t) = -16t^2 + v_0 t + s_0$, where $s(t)$ is the vertical distance in feet, t is time in seconds, v_0 is the initial velocity, and s_0 is the initial height. How do your two quadratic equations compare with this theoretical formula? Can you explain why your equation is not in this exact form?

Part II

In this chapter, you encountered several exercises that used real data. For example, in Section 8.4, exercises 31 through 34 each ask you to write a quadratic function and predict future data.

Search the Internet or library for a similar set of data. List the source of your data. Use the data to write an exercise. Give enough information to write three data points, ask for a quadratic function to model the data, and ask for a prediction of future data.

Chapter 9

Exponents and Polynomials

We've seen how to write and evaluate polynomials and how to graph them; our next step is to learn how to work with them. We will look at how to add, subtract, multiply, and divide polynomials. However, before we work with polynomial operations, we will need to look at the rules of exponents.

Scientific notation is an application of exponential expressions and the rules of exponents. Many measurements are either so large or so small that we use scientific notation to write them easily and the rules of exponents to multiply and divide them easily. For example, the distance between the Earth and a star is so large that we resort to scientific notation to write it.

We will also continue to look at geometric applications of polynomials. We will see that many such applications may be described by polynomials. It is important to be able to simplify these polynomials by adding, subtracting, multiplying, and dividing. This will make it easier to work with the polynomials when we are ready to solve polynomial equations.

Raising a polynomial to a power (expanding the polynomial) is sometimes a long and time-consuming process. In the project at the end of the chapter, we will discuss a method that will help us accomplish this expansion more quickly. The method was first published in 1665 and was used even before that time.

9.1 APPLICATION

In a study of the white-tailed deer population in the state of Nebraska, 23 tagged deer were recovered.

a. Determine the area over which the white-tailed deer ranged if an average movement of x miles was recorded.

b. In one extreme case, a deer was recovered at about three times the average distance. Determine the area over which this white-tailed deer ranged.

c. Determine the ratio of the area covered by the deer in part b to the area covered by the deer in part a.

d. If the average movement of a deer was 38 miles, determine the area ranged over by the deer for this average and the area ranged over by the deer in the extreme case.

We will discuss this application further. See page 603.

9.1 Operations Involving Nonnegative Exponents

1 Rewrite exponential expressions by using their definitions.

2 Simplify exponential expressions by using the product rule.

3 Simplify exponential expressions by using the quotient rule.

4 Simplify exponential expressions by using the power-to-a-power rule, product-to-a-power rule, and quotient-to-a-power rule.

5 Simplify exponential expressions by using more than one rule for exponents.

6 Check simplified exponential expressions by using a calculator.

7 Model real-world situations by using exponential expressions.

9.1.1 Defining Exponential Expressions

We discussed numeric exponential expressions in previous chapters. An exponential expression is an expression that has a base and an exponent. We evaluate an exponential expression by writing it in expanded form and using repeated multiplication. By convention, we do not write an exponent of 1, although we might do so in an intermediate step in solving a problem. In Chapter 1, we defined an exponential expression with a nonzero base and a zero exponent as equal to 1.

$$4^3 = 4 \cdot 4 \cdot 4 = 64$$
$$4^1 = 4$$
$$4^0 = 1$$

Remember, 0^0 is indeterminate. We will explain why this is true later in the chapter. See page 597.

We define variable exponential expressions in the same manner as numeric exponential expressions.

INTEGER EXPONENTS

For any base a and nonnegative integer exponent n,

$$a^n = a \cdot a \cdot a \cdot \ldots \cdot a \quad n \text{ factors}$$
$$a^1 = a$$
$$a^0 = 1, \text{ where } a \neq 0$$

The base a may be a constant, a variable, or an expression.

For example,

$$x^4 = x \cdot x \cdot x \cdot x$$
$$x^1 = x$$

For all $x \neq 0$, $x^0 = 1$.

EXAMPLE I   Write the given expressions in expanded form.

a. $-2x^4$ **b.** $(-2x)^4$ **c.** a^3b^2 **d.** mn^0 **e.** $(x + 2)^2$

Solution

a. $-2x^4 = -2 \cdot x \cdot x \cdot x \cdot x$

b. $(-2x)^4 = (-2x)(-2x)(-2x)(-2x)$

c. $a^3b^2 = a \cdot a \cdot a \cdot b \cdot b$

d. $mn^0 = m \cdot 1 = m$ Remember, $n^0 = 1$.

e. $(x + 2)^2 = (x + 2)(x + 2)$

HELPING HAND It is very important to determine the base of an exponential expression. In Example 1a, the base is x. In Example 1b, the base is $-2x$.

 9.1.1 Checkup

In exercises 1–5, write the expressions in expanded form.

1. 3^2x^3 **2.** $(-5y)^3$ **3.** $-5y^3$ **4.** $(a-b)^4$ **5.** p^0q^4

6. Explain the difference between the terms *expanded form* and *exponential expression*.

9.1.2 The Product Rule

In order to perform multiplication involving exponential expressions, we need to discover a few rules. Use your calculator to complete the following set of exercises to determine a rule for multiplication of exponential expressions with the same base.

 Discovery 1

Multiplication of Exponential Expressions with the Same Base

Evaluate each expression, and compare the results obtained in the first column with the corresponding results in the second column.

1. a. $4^3 \cdot 4^2 =$ _____   **b.** $4^5 =$ _____

2. a. $(-2)^4 \cdot (-2)^2 =$ _____   **b.** $(-2)^6 =$ _____

3. a. $\left(\dfrac{3}{4}\right)^2 \cdot \left(\dfrac{3}{4}\right) =$ _____   **b.** $\left(\dfrac{3}{4}\right)^3 =$ _____

Write a rule for multiplication of two exponential expressions with the same base.

4. Use the rule to simplify $x^4 \cdot x^3 =$ _____.

In the first column, the base of each factor is the same. In the second column, the base is the same as the bases of the factors in the first column, and the exponent is the sum of the factors' exponents. The results are the same. Therefore, to multiply exponential expressions with like bases, add the exponents.

Using this rule, we see that the product of $x^4 \cdot x^3$ is x^{4+3}, or x^7. We can illustrate the rule by rewriting each expression as a product of its factors and simplifying.

$$x^4 \cdot x^3 = (x \cdot x \cdot x \cdot x) \cdot (x \cdot x \cdot x)$$
$$= x \cdot x \cdot x \cdot x \cdot x \cdot x \cdot x$$
$$= x^7$$

PRODUCT RULE FOR EXPONENTS

For any base a and integer exponents m and n,

$$a^m \cdot a^n = a^{m+n}$$

The base a may be a constant, a variable, or an expression.

EXAMPLE 2 Simplify.

a. $y^3 \cdot y$ **b.** $(a + b)^4(a + b)$ **c.** $-2x^3 \cdot x^2 \cdot x$

Solution

a. $y^3 \cdot y = y^{3+1} = y^4$

b. $(a + b)^4(a + b) = (a + b)^{4+1} = (a + b)^5$

c. $-2x^3 \cdot x^2 \cdot x = -2x^{3+2+1} = -2x^6$

9.1.2 Checkup

Simplify exercises 1–4.

1. $a^3 \cdot a^6$ **2.** $z \cdot z^6$ **3.** $(x + y)(x + y)^7$ **4.** $\frac{1}{2}a^2 \cdot a \cdot a^5$

5. What is the difference between $x^2 \cdot x^3$ and $x^2 + x^3$? Which of these expressions can be simplified by using the product rule?

9.1.3 The Quotient Rule

Before we begin to discuss division of polynomials, we must discover rules of exponents involving division. Use your calculator to complete the following set of exercises to determine a rule for division of exponential expressions with the same base.

Discovery 2

Division of Exponential Expressions with the Same Base

Evaluate each expression, and compare the results obtained in the first column with the corresponding results in the second column.

1. a. $\dfrac{4^5}{4^2} = $ _____   **b.** $4^3 = $ _____

2. a. $\dfrac{(-2)^4}{(-2)^2} = $ _____   **b.** $(-2)^2 = $ _____

3. a. $\dfrac{\left(\frac{3}{4}\right)^3}{\left(\frac{3}{4}\right)} = $ _____   **b.** $\left(\frac{3}{4}\right)^2 = $ _____

4. a. $\dfrac{0^5}{0^5} = $ _____   **b.** $0^0 = $ _____

Write a rule for division of two exponential expressions with the same base.

5. Use the rule to simplify $\dfrac{x^7}{x^3} = $ _____.

In the first column, the bases of the divisor and dividend are the same. In the second column, the base is the same as the bases of the divisor and dividend, and the exponent is the difference of the exponents in the quotient. The results are the same. Therefore, to divide exponential expressions with like bases, subtract the exponent in the denominator from the exponent in the numerator.

Using this rule, we find that the quotient of $\dfrac{x^7}{x^3}$ is x^{7-3}, or x^4. We can illustrate the rule by rewriting each expression as a product of its factors and simplifying.

$$\frac{x^7}{x^3} = \frac{x \cdot x \cdot x \cdot x \cdot x \cdot x \cdot x}{x \cdot x \cdot x}$$
$$= x \cdot x \cdot x \cdot x$$
$$= x^4$$

Note: In the discovery box, we evaluated $\frac{0^5}{0^5} = \frac{0}{0}$, which we determined to be indeterminate in Chapter 1. Now, using the rule we just discovered, we see that $\frac{0^5}{0^5} = 0^{5-5} = 0^0$. It follows that 0^0 is indeterminate.

QUOTIENT RULE FOR EXPONENTS
For any nonzero base a and integer exponents m and n,

$$\frac{a^m}{a^n} = a^{m-n}$$

The base a may be a constant, a variable, or an expression.

EXAMPLE 3 Simplify.

a. $\dfrac{y^3}{y}$ **b.** $\dfrac{(a+b)^4}{(a+b)}$ **c.** $\dfrac{-6b^2}{18b}$

Solution

a. $\dfrac{y^3}{y} = y^{3-1} = y^2$ **b.** $\dfrac{(a+b)^4}{(a+b)} = (a+b)^{4-1} = (a+b)^3$

c. $\dfrac{-6b^2}{18b} = \left(\dfrac{-6}{18}\right)\left(\dfrac{b^2}{b}\right) = \dfrac{-1}{3}b^{2-1} = -\dfrac{1}{3}b$

Note: The last expression can be written as $-\frac{b}{3}$.

9.1.3 Checkup

Simplify exercises 1–3.

1. $\dfrac{a^{12}}{a^8}$ **2.** $\dfrac{22z^4}{121z}$ **3.** $\dfrac{(x+5)^6}{(x+5)^4}$

4. What is the difference between $\dfrac{x^5}{x^2}$ and $x^5 - x^2$? Which can be simplified by using the quotient rule for exponents?

9.1.4 Power Rules

Sometimes, we may need to simplify an exponential expression raised to a power. Use your calculator to complete the following set of exercises to determine a rule for simplifying an exponential expression raised to a power.

Discovery 3

Exponential Expressions Raised to a Power

Evaluate each expression, and compare the results obtained in the first column with the corresponding results in the second column.

1. a. $(4^3)^2 = $ _____   **b.** $4^6 = $ _____

2. a. $[(-2)^2]^4 = $ _____   **b.** $(-2)^8 = $ _____

3. a. $\left[\left(\dfrac{3}{4}\right)^2\right]^3 = $ _____   **b.** $\left(\dfrac{3}{4}\right)^6 = $ _____

Write a rule for raising an exponential expression to a power.

4. Use the rule to simplify $(x^4)^3 = $ _____.

In the second column, the base is the same as the base in the first column, and the exponent is the product of the exponents in the first column. The results are the same. Therefore, to raise an exponential expression to a power, multiply the exponents.

With this rule, $(x^4)^3$ is $x^{4\cdot3}$, or x^{12}. We can illustrate the rule by rewriting the expression as a product of its factors and simplifying.

$$
\begin{aligned}
(x^4)^3 &= (x \cdot x \cdot x \cdot x)^3 \\
&= (x \cdot x \cdot x \cdot x) \cdot (x \cdot x \cdot x \cdot x) \cdot (x \cdot x \cdot x \cdot x) \\
&= x \cdot x \cdot x \cdot x \cdot x \cdot x \cdot x \cdot x \cdot x \cdot x \cdot x \cdot x \\
&= x^{12}
\end{aligned}
$$

POWER-TO-A-POWER RULE FOR EXPONENTS

For any base a and integer exponents m and n,

$$(a^m)^n = a^{mn}$$

The base a may be a constant, a variable, or an expression.

We also need rules to simplify a product or a quotient raised to a power. Use your calculator to complete the following set of exercises to determine a rule for simplifying a product raised to a power.

Discovery 4

Products Raised to a Power

Evaluate each expression, and compare the results obtained in the first column with the corresponding results in the second column.

1. a. $(4 \cdot 2)^3 = $ _____   **b.** $4^3 \cdot 2^3 = $ _____

2. a. $(-2 \cdot 3)^2 = $ _____   **b.** $(-2)^2 \cdot 3^2 = $ _____

3. a. $\left(\dfrac{3}{4} \cdot \dfrac{2}{5}\right)^3 = $ _____   **b.** $\left(\dfrac{3}{4}\right)^3 \cdot \left(\dfrac{2}{5}\right)^3 = $ _____

Write a rule for a product raised to a power.

4. Use the rule to simplify $(xy)^4 = $ _____.

In the first column, we determine a product and then raise the product to a power. In the second column, we raise each factor to a power and then multiply the result. The results are the same. Therefore, raising a product to a power is equivalent to multiplying the factors raised to a power.

Applying this rule, we obtain $(xy)^4 = x^4y^4$. We can illustrate the rule by rewriting the exponential expression as its factors, rearranging the factors by means of the commutative and associative properties, and simplifying.

$$(xy)^4 = (xy)(xy)(xy)(xy)$$
$$= x \cdot y \cdot x \cdot y \cdot x \cdot y \cdot x \cdot y$$
$$= x \cdot x \cdot x \cdot x \cdot y \cdot y \cdot y \cdot y$$
$$= x^4 y^4$$

PRODUCT-TO-A-POWER RULE

For any factors a and b and integer exponent m,

$$(ab)^m = a^m b^m$$

The factors a and b may be constants, variables, or expressions.

The result for a quotient raised to a power is determined in a similar way. For example,

$$\left(\frac{x}{y}\right)^4 = \left(\frac{x}{y}\right)\left(\frac{x}{y}\right)\left(\frac{x}{y}\right)\left(\frac{x}{y}\right) = \frac{x^4}{y^4}$$

QUOTIENT-TO-A-POWER RULE

For any dividend a, nonzero divisor b, and integer exponent m,

$$\left(\frac{a}{b}\right)^m = \frac{a^m}{b^m}$$

The dividend a and divisor b may be constants, variables, or expressions.

EXAMPLE 4   Simplify.

a. $(-2x)^4$ **b.** $(-2x^3)^4$ **c.** $\left(\dfrac{5y}{z}\right)^3$

Solution

a. $(-2x)^4 = (-2)^4 x^4 = 16x^4$ Product-to-a-power rule

b. $(-2x^3)^4 = (-2)^4 (x^3)^4$ Product-to-a-power rule

$$= 16x^{3 \cdot 4}$$ Power-to-a-power rule

$$= 16x^{12}$$

c. $\left(\dfrac{5y}{z}\right)^3 = \dfrac{(5y)^3}{z^3}$ Quotient-to-a-power rule

$$= \frac{5^3 y^3}{z^3}$$ Product-to-a-power rule

$$= \frac{125 y^3}{z^3}$$

9.1.4 Checkup

Simplify exercises 1–4.

1. $(-3a)^5$ **2.** $(ab)^{18}$ **3.** $(-4x^8)^3$ **4.** $\left(\dfrac{-2p}{q}\right)^4$

5. What is the difference between $2x^3$ and $(2x)^3$? Which of the two expressions can be simplified with the product-to-a-power rule?

9.1.5 Combining Rules

Now that we have learned the rules for exponents, we will apply more than one rule to simplify an exponential expression. The order in which we apply the rules may vary to obtain the same simplification.

EXAMPLE 5 Simplify.

a. $-2x^2y \cdot 3x^3y^3$ **b.** $\dfrac{12x^4y}{3x^2y}$ **c.** $\dfrac{(a^3b)^2}{a^2b}$ **d.** $\left(\dfrac{2m^3}{3mn^2}\right)^2$

Solution

a. $-2x^2y \cdot 3x^3y^3 = (-2 \cdot 3)(x^2 \cdot x^3)(y \cdot y^3)$

$\qquad\qquad\qquad = -6x^{2+3}y^{1+3} \qquad$ Product rule

$\qquad\qquad\qquad = -6x^5y^4$

b. $\dfrac{12x^4y}{3x^2y} = \left(\dfrac{12}{3}\right)\left(\dfrac{x^4}{x^2}\right)\left(\dfrac{y}{y}\right)$

$\qquad\qquad = 4x^{4-2}y^{1-1} \qquad$ Quotient rule

$\qquad\qquad = 4x^2y^0 \qquad$ Definition: $y^0 = 1$

$\qquad\qquad = 4x^2$

c. $\dfrac{(a^3b)^2}{a^2b} = \dfrac{(a^3)^2b^2}{a^2b} \qquad$ Product-to-a-power rule

$\qquad\quad = \dfrac{a^{3\cdot2}b^2}{a^2b} \qquad$ Power-to-a-power rule

$\qquad\quad = \dfrac{a^6b^2}{a^2b}$

$\qquad\quad = a^{6-2}b^{2-1} \qquad$ Quotient rule

$\qquad\quad = a^4b$

d. $\left(\dfrac{2m^3}{3mn^2}\right)^2 = \dfrac{(2m^3)^2}{(3mn^2)^2} \qquad$ Quotient-to-a-power rule

$\qquad\qquad\quad = \dfrac{2^2(m^3)^2}{3^2m^2(n^2)^2} \qquad$ Product-to-a-power rule

$\qquad\qquad\quad = \dfrac{4m^{3\cdot2}}{9m^2n^{2\cdot2}} \qquad$ Power-to-a-power rule

$\qquad\qquad\quad = \dfrac{4m^6}{9m^2n^4} \qquad$ $\frac{m^6}{m^2} = m^{6-2} = m^4$

$\qquad\qquad\quad = \dfrac{4m^4}{9n^4}$

A second method is to simplify the original quotient before applying the quotient-to-a-power rule. Remember, $\frac{m^3}{m} = m^{3-1} = m^2$.

$$\left(\frac{2m^3}{3mn^2}\right)^2 = \left(\frac{2m^2}{3n^2}\right)^2 = \frac{(2m^2)^2}{(3n^2)^2} = \frac{2^2(m^2)^2}{3^2(n^2)^2} = \frac{4m^4}{9n^4}$$

9.1.5 Checkup

Simplify exercises 1–4.

1. $\frac{1}{2}a^2b^3\left(-\frac{4}{5}ab\right)$

2. $\frac{21p^4q^2}{7p^4q}$

3. $\frac{(-x^2)^3}{2x}$

4. $\left(\frac{3a^4b}{4a^2b}\right)^3$

5. What is the difference between $(a^m)^n$ and $a^m(a^n)$?

9.1.6 Checking Simplified Expressions

The calculator can be used to check for equivalence of simplified expressions if the expressions involve a single variable.

TECHNOLOGY

CALCULATOR CHECK FOR EQUIVALENCE OF SIMPLIFIED EXPRESSIONS

Is $\left(\frac{3}{5}x^3\right)^2 = \frac{9}{25}x^6$?

Numeric Check

Graphic Check

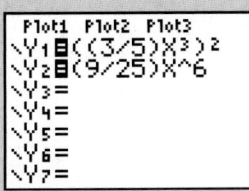

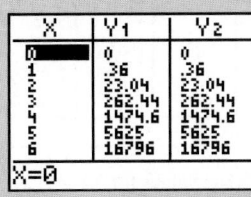

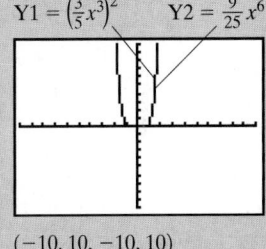

$(-10, 10, -10, 10)$

Figure 9.1a

Figure 9.1b

Figure 9.1c

For Figure 9.1a,

Enter the original expression, $\left(\frac{3}{5}x^3\right)^2$, in Y1.

((3 ÷ 5) X,T,θ,n MATH 3) x^2

Enter the simplified expression, $\frac{9}{25}x^6$, in Y2.

(9 ÷ 2 5) X,T,θ,n ^ 6

For Figure 9.1b,

To check the equivalence numerically, view the table of values of Y1, the original expression, and Y2, the simplified expression with whole number values for x.

(Continued on page 602)

| 2nd | TBLSET | 0 | ENTER | 1 | ENTER | ENTER | ▼ | ENTER | 2nd | TABLE |

Check the values for Y1, the original expression, and Y2, the simplified expression. If the simplification was correct, corresponding values should be equal. (Note that we could have looked at a table of values for other x-values.)

For Figure 9.1c,

To check the equivalence graphically, view the graphs of Y1, the original expression, and Y2, the simplified expression.

| ZOOM | 6 |

The graphs should coincide. Check by using the trace feature of the calculator or by setting the graph of Y2 to be drawn with a bubble.

EXAMPLE 6   Simplify. Check numerically and graphically.

$$[(-3x)^2]^3$$

Solution

$$[(-3x)^2]^3 = [(-3)^2x^2]^3 \quad \text{Product-to-a-power rule}$$
$$= [9x^2]^3 \quad \text{Product-to-a-power rule}$$
$$= 9^3(x^2)^3 \quad \text{Power-to-a-power rule}$$
$$= 729x^6 \quad \text{Power-to-a-power rule}$$

Numeric Check

X	Y1	Y2
0	0	0
1	729	729
2	46656	46656
3	531441	531441
4	2.99E6	2.99E6
5	1.14E7	1.14E7
6	3.4E7	3.4E7

X=0

The values for Y1 = Y2.
The simplification is correct.

Graphic Check

Y1 = $[(-3x)^2]^3$ Y2 = $729x^6$

$(-4.7, 4.7, -3.1, 3.1)$
The graphs of Y1 and Y2 are the same.
The simplification is correct.

9.1.6 Checkup

In exercises 1–5, check the following simplifications numerically and graphically to determine whether they are correct.

1. $x^2 \cdot x^5 = x^7$ **2.** $x^2 \cdot x^5 = x^{10}$ **3.** $(2x^2)^3 = 6x^5$

4. $(3x^3)^2 = 9x^6$ **5.** $\left(\dfrac{6x^2}{2x^3}\right)^3 = \dfrac{27}{x^3}$

6. When checking a simplification numerically and graphically, are you able to see all the values the variable can assume? Explain.

9.1.7 Modeling the Real World

Exponential expressions occur frequently in all kinds of real-world situations. As always, you must remember to include the units of real quantities in your calculations; that is, raising the quantity 3 feet to the third power gives you, not 27 feet, but 27 *cubic* feet (ft^3).

$$(3 \text{ ft})^3 = 3^3 \text{ ft}^3 = 27 \text{ ft}^3$$

EXAMPLE 7   Determine the area of a square mail room with sides of length s feet. A publishing company is expanding into new offices, and the new mail room will still be a square, but with sides of length $2s$. Determine the area of the new mail room, and compare it with the area of the original mail room.

Solution

Let s = length of the side of the original mail room

The area is found by squaring the length of the side of a square. The original mail room is s^2 square feet.

Let $2s$ = length of the side of the new mailroom

The area is found by squaring the length of the side of a square. The new mailroom is $(2s)^2 = 2^2 s^2 = 4s^2$ square feet.

Since the new mailroom is $4s^2$ and the original mailroom is s^2, the area of the new room is $\frac{4s^2}{s^2} = 4$ times the area of the original room.

APPLICATION

In a study of the white-tailed deer population in the state of Nebraska, 23 tagged deer were recovered.

a. Determine the area over which the white-tailed deer ranged if an average movement of x miles was recorded.

b. In one extreme case, a deer was recovered at about three times the average distance. Determine the area over which this white-tailed deer ranged.

c. Determine the ratio of the area covered by the deer in part b to the area covered by the deer in part a.

d. If the average movement of a deer was 38 miles, determine the area ranged over by the deer for this average and the area ranged over by the deer in the extreme case.

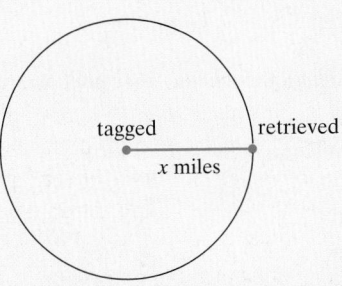

Discussion

a. Let x = the number of miles of the average movement of deer. It is reasonable to assume that the area of movement is a circle with a radius of x miles, yielding an area of πx^2 square miles.

b. Let $3x$ = the number of miles of the movement of the deer in the extreme case. The area is a circle with a radius of $3x$ miles.

$$\pi(3x)^2 = \pi 9x^2 = 9\pi x^2 \text{ square miles}$$

c. The ratio of the area covered by the deer in part b to the area covered by the deer in part a is

$$\frac{9\pi x^2}{\pi x^2} = \frac{9}{1} = 9$$

In other words, in the extreme case, the deer ranged over an area nine times that ranged over by the average deer.

d. If x = 38 miles, then the average area was

$$\pi x^2 = \pi(38)^2 = 1444\pi$$

or approximately 4536 square miles.

In the extreme case, the deer's area was

$$9\pi x^2 = 9\pi(38)^2 = 9 \cdot \pi \cdot 1444 = 12{,}996\pi$$

or approximately 40,828 square miles (nine times the range of the average deer).

9.1.7 Checkup

A box has the shape of a cube, with each side measuring x inches. Determine the volume of the box. If another box is to be made with each side measuring half the length of the original box, determine the volume of the new box. Compare the volume of the original box with that of the new box. What are the two volumes if the first box has a side measuring 1.5 meters on a side?

9.1 Exercises

Write in expanded form.

1. $-3x^4$

2. $-4c^3$

3. $(-3x)^4$

4. $(-4c)^3$

5. $a^3b^0c^5$

6. x^2yz^0

7. $\left(\dfrac{3}{4}\right)^3 \cdot x^2$

8. $\left(\dfrac{2}{3}\right)^2 \cdot x^3$

9. $5(x + y)^2$

10. $-4(p - q)^2$

Simplify.

11. $x^5 \cdot x^8$

12. $a^9 \cdot a^{14}$

13. $y \cdot y^{13}$

14. $b^{23} \cdot b$

15. $\dfrac{1}{2}x^2 \cdot x \cdot x^5$

16. $\dfrac{1}{3}y^2 \cdot y^3 \cdot y$

17. $(x + y)^4(x + y)^2$

18. $(x - y)^5(x - y)^3$

19. $(x + 3)^2(x + 3)$

20. $(x + 9)^{12}(x + 9)$

21. $\dfrac{p^{11}}{p^6}$

22. $\dfrac{a^5}{a^3}$

23. $\dfrac{54q^7}{18q}$

24. $\dfrac{39t^8}{13t}$

25. $\dfrac{(2x - 3)^8}{(2x - 3)^3}$

26. $\dfrac{(4x + 7)^9}{(4x + 7)^2}$

27. $\dfrac{-3(p + q)^2}{9(p + q)}$

28. $\dfrac{-5(xy + 2)^4}{10(xy + 2)}$

29. $(a^5)^6$

30. $(m^7)^4$

31. $(-3x)^4$

32. $(-5y)^3$

33. $(abc)^{21}$

34. $(xyz)^9$

35. $(5m^3)^3$

36. $(3k^4)^4$

37. $[(x + y)^3]^2$

38. $[(a + 2b)^4]^3$

39. $[(a - b)^4]^1$

40. $[(c - 3d)^5]^1$

41. $(x^2)^0$

42. $(x^0)^3$

43. $\left(\dfrac{b}{d}\right)^4$

44. $\left(\dfrac{m}{n}\right)^5$

45. $\left(\dfrac{3b}{c}\right)^4$

46. $\left(\dfrac{5y}{z}\right)^4$

47. $\left(\dfrac{-d}{2c}\right)^6$

48. $\left(\dfrac{-m}{3n}\right)^4$

49. $-7a^3b \cdot 5a^2b^4$

50. $-8c^4d^2 \cdot 3cd^5$

51. $\dfrac{15x^3y^2}{3xy^2}$

52. $\dfrac{18ab^2}{3ab}$

53. $\dfrac{-27ab^2c^3}{15bc}$

54. $\dfrac{-27p^3q^2s}{6ps}$

55. $\dfrac{-4x(4 - x)^4}{2(4 - x)}$

56. $\dfrac{5t^2(15 - 4t)^7}{15t(15 - 4t)^3}$

57. $\dfrac{(p^5q^7)^3}{p^6q}$

58. $\dfrac{(k^4m^6)^2}{k^3m}$

59. $\left(\dfrac{4x^3}{y^2}\right)^2$

60. $\left(\dfrac{5a^2}{7b}\right)^2$

61. $\left(\dfrac{-3p^4q^2}{p^2q}\right)^3$

62. $\left(\dfrac{-5c^2d^3}{cd^2}\right)^2$

63. $[(2a)^2]^5$

64. $[(-3t)^2]^2$

65. $\left[\left(\dfrac{x}{2y}\right)^2\right]^3$

66. $\left[\left(\dfrac{3z}{w}\right)^2\right]^3$

In exercises 67–78, check the simplifications numerically and graphically to determine whether they are correct.

67. $x^3 \cdot x^4 = x^{12}$

68. $x^2 \cdot x^5 = x^{10}$

69. $x^3 \cdot x^4 = x^7$

70. $x^2 \cdot x^5 = x^7$

71. $(3x^4)^2 = 9x^8$

72. $(5x^3)^2 = 25x^6$

73. $(3x^4)^2 = 9x^6$

74. $(5x^3)^2 = 25x^5$

75. $\left(\dfrac{8x^3}{2x^6}\right)^2 = \dfrac{16}{x^6}$

76. $\left(\dfrac{9x^2}{3x^5}\right)^2 = \dfrac{9}{x^6}$

77. $\left(\dfrac{8x^3}{2x^6}\right)^2 = \dfrac{16}{x^5}$

78. $\left(\dfrac{9x^2}{3x^5}\right)^2 = \dfrac{9}{x^9}$

79. A child's square play area is enlarged by increasing each side by a factor of five. (That is, each side is multiplied by 5.) Write exponential expressions for the areas of the original play area and the enlarged play area. Compare the areas of the two squares. What are the two areas if the original area had a side of 6 feet?

80. A department store's square display area is enlarged by increasing each side by a factor of 2.5. Write exponential expressions for the areas of the original and the enlarged display areas. Compare the two areas. What are the two areas if the original area had a side of 12 feet?

81. A storage bin in the shape of a cube is enlarged by increasing each side by a factor of four. Write exponential expressions for the volume of the original bin and the enlarged bin. Compare the volumes of the two bins. What are the two volumes if the original bin had a side of 1.5 feet?

82. A block of ice with the shape of a cube begins to melt. Its size is decreased so that each side is 0.8 of its original length. Write exponential expressions for the volumes of the original cube of ice and the smaller cube of ice. Compare the volumes of the two cubes. What are the two volumes if the larger cube measured 22 inches on a side?

83. A hot-air balloon has the shape of a sphere with a radius of x feet. A second hot-air balloon also has the shape of a sphere, but with a radius that is twice the length of the first balloon. Write exponential expressions for the volumes of the two balloons. Compare the volumes of the balloons. What are the two volumes if the smaller balloon has a radius of 4 feet?

84. A company sells various sizes of exercise balls used in physical fitness conditioning. The company recommends a ball with a radius of x centimeters for a person who is less than 5 feet, 6 inches tall and one with a radius that is 1.5 times as large for a person that is over 6 feet, 5 inches tall. Write exponential expressions for the amount of leather it would take to make the exercise balls (the surface area of each ball). Compare the two surface areas. If the smaller exercise ball has a radius of 27 centimeters, what are the two surface areas?

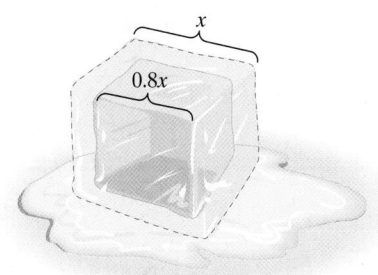

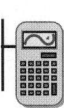

9.1 Calculator Exercises

In this section, you learned how to check to see whether you had properly simplified an expression by using the TABLE or GRAPH features of your calculator. You were instructed to store the original expression in Y1 and the simplified expression in Y2 and then to view the table of values to see whether Y1 was equal to Y2 for each value of the variable, or to view the graphs to see whether they coincided. However, there is another way to use the TABLE or GRAPH features to check expressions.

Instead of storing the two expressions in Y1 and Y2, store both of them in Y1, using an equals sign, since they are presumed to be equivalent. Then, when you view the table, you

should see a column of values all equal to 1, indicating that the two expressions are indeed equal to each other. If you see any values of 0 in the table for Y1, this indicates that the two expressions are not equal. Likewise, if you trace along the graph, you should see a horizontal line at Y1 = 1 for a correct simplification and a horizontal line at Y1 = 0 for an incorrect simplification. The calculator screens shown next are for the same example used in the text.

Is $\left(\dfrac{3}{5}x^3\right)^2 = \dfrac{9}{25}x^6$?

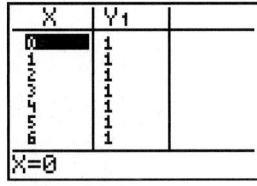

$(-10, 10, 1, -10, 10, 1)$

Use this method to check the following expressions for equivalence:

1. $x^4 \cdot x^5 = x^{20}$?   **2.** $x^4 \cdot x^5 = x^9$?   **3.** $(x^4)^5 = x^{20}$?

4. $(x^4)^5 = x^9$?   **5.** $(3x^4)^2 = 9x^8$?   **6.** $(3x^4)^2 = 9x^6$?

9.1 Writing Exercises

The rules of exponents sometimes yield results that may run contrary to your intuition. To help you deal with this, explain briefly why the following statements are true, and illustrate each with an example.

1. A negative number raised to an even power results in a positive number.

2. A negative number raised to an odd power results in a negative number.

3. When two exponential expressions have the same base, but different exponents, you may properly write their product as the base raised to the power of the sum of the exponents.

4. When two exponential expressions have different bases, but the same exponent, you may *not* properly write their product as the product of the bases raised to the power of the sum of the exponents.

9.2 Operations Involving Negative Exponents

OBJECTIVES

1 Write exponential expressions with negative exponents as equivalent expressions with positive exponents.

2 Simplify exponential expressions with negative exponents by using the rules of exponents.

3 Model real-world situations by using scientific notation.

APPLICATION

An FM radio station broadcasts at a frequency of 102.1 MHz, or 102.1×10^6 cycles per second. Write this frequency in scientific notation. What is the wavelength of these radio waves in meters?

After completing this section, we will discuss this application further. See page 612.

 9.2.1 Writing with Positive Exponents

In the last section, we limited our discussion of exponents to nonnegative integers. However, in Chapter 1 we defined exponential expressions with a numeric base raised to a negative exponent. To evaluate an exponential expression with a nonzero base raised to a negative exponent, rewrite the expression as an exponential expression with the reciprocal of the base and the opposite value of the exponent (a positive number). That is,

$$10^{-1} = \left(\frac{1}{10}\right)^1 = \frac{1}{10^1} = \frac{1}{10} \qquad \left(\frac{1}{10}\right)^{-2} = \left(\frac{10}{1}\right)^2 = 10^2 = 100$$

Remember that 0 raised to a negative exponent is undefined.

$$0^{-2} = \left(\frac{1}{0}\right)^2 = \frac{1^2}{0^2} = \frac{1}{0} \qquad \text{Division by zero is undefined.}$$

While we recognize that expressions with negative exponents are not polynomials, our work with polynomials can result in such expressions. Therefore, in this section, we will complete our discussion of exponential expressions involving integer exponents by discussing negative integer exponents with variable bases. Using the rules of exponents in the last section, we might obtain an expression involving negative exponents. For example, if we use the quotient rule, to simplify $\frac{x^3}{x^5}$, we obtain

$$\frac{x^3}{x^5} = x^{3-5} = x^{-2}$$

However, if we simplify $\frac{x^3}{x^5}$ by writing it first in expanded form, we obtain

$$\frac{x^3}{x^5} = \frac{x \cdot x \cdot x}{x \cdot x \cdot x \cdot x \cdot x} = \frac{1}{x \cdot x} = \frac{1}{x^2}$$

Therefore, $x^{-2} = \frac{1}{x^2}$. The same result is obtained by using the procedure that we used with numeric bases.

For all $x \neq 0$,

$$x^{-4} = \left(\frac{x}{1}\right)^{-4} = \left(\frac{1}{x}\right)^4 = \frac{1}{x^4}$$

$$\frac{1}{x^{-4}} = \frac{1}{\left(\frac{1}{x}\right)^4} = \frac{1}{\frac{1}{x^4}} = 1 \div \frac{1}{x^4} = 1 \cdot \frac{x^4}{1} = x^4$$

HELPING HAND As long as only factors are involved, a base with a negative exponent in the numerator is placed in the denominator and given a positive exponent equal to the absolute value of the original exponent. Also, a base with a negative exponent in the denominator is placed in the numerator and given a positive exponent equal to the absolute value of the original exponent.

$$\frac{a^{-n}}{b^{-m}} = \frac{b^m}{a^n}$$

In this text, we will assume that variables represent nonzero values, so that the given expressions are defined.

INTEGER EXPONENTS
For any base a and nonnegative integer exponent n,

$$a^{-n} = \frac{1}{a^n}, \text{ where } a \neq 0$$

$$\frac{1}{a^{-n}} = a^n, \text{ where } a \neq 0$$

The base a may be a constant, a variable, or an expression.

EXAMPLE 1 Write equivalent expressions with positive exponents.

a. $\dfrac{x^{-2}}{y^{-3}}$ **b.** $\dfrac{x^3 y^{-1}}{3^{-2} z^2}$ **c.** $a^3 b^{-2}$

Solution

a. $\dfrac{x^{-2}}{y^{-3}} = \dfrac{y^3}{x^2}$

Move the factor x^{-2} from the numerator into the denominator as the factor x^2.
Move the factor y^{-3} from the denominator into the numerator as the factor y^3.

b. $\dfrac{x^3 y^{-1}}{3^{-2}z^2} = \dfrac{3^2 x^3}{yz^2}$ Move the factor y^{-1} from the numerator into the denominator as the factor y. Move the factor 3^{-2} from the denominator into the numerator as the factor 3^2.

$$= \dfrac{9x^3}{yz^2}$$ Evaluate 3^2.

c. $a^3 b^{-2} = \dfrac{a^3}{b^2}$ Move the factor b^{-2} into the denominator as the factor b^2.

9.2.1 Checkup

In exercises 1–3, write each expression with positive exponents.

1. $\dfrac{x^{-4}}{y^{-2}}$ **2.** $\dfrac{5^{-2}a^4}{2^{-3}b^{-3}}$ **3.** $p^{-5}q^5$

4. If an exponential expression has factors with negative exponents, is it always possible to rewrite the expression with nonnegative exponents? Explain.

9.2.2 Simplifying Expressions

Assuming that the variables represent nonzero values so that the given expressions are defined, all of the rules of exponents that we introduced in the previous section are valid for expressions having negative exponents. A summary of these rules follows.

SUMMARY OF RULES FOR EXPONENTS

For any real numbers a and b and integers m and n,

$a^1 = a$ Exponent of one

$a^0 = 1, a \neq 0$ Exponent of zero

$a^m \cdot a^n = a^{m+n}$ Product rule for exponents

$\dfrac{a^m}{a^n} = a^{m-n}, a \neq 0$ Quotient rule for exponents

$(a^m)^n = a^{mn}$ Power-to-a-power rule for exponents

$(ab)^m = a^m b^m$ Product-to-a-power rule for exponents

$\left(\dfrac{a}{b}\right)^m = \dfrac{a^m}{b^m}, b \neq 0$ Quotient-to-a-power rule for exponents

---

EXAMPLE 2   Simplify and write with positive exponents.

a. $(2x^4)(-3x^{-2})$ **b.** $\dfrac{4x^{-2}}{8x^{-3}}$

Solution

a. $(2x^4)(-3x^{-2}) = [2 \cdot (-3)](x^4 x^{-2})$

$$= -6x^{4+(-2)}$$ Product rule for exponents

$$= -6x^2$$

b. $\dfrac{4x^{-2}}{8x^{-3}} = \left(\dfrac{4}{8}\right)\left(\dfrac{x^{-2}}{x^{-3}}\right)$

$$= \dfrac{1}{2}x^{-2-(-3)}$$ Quotient rule for exponents

$$= \frac{1}{2}x^1$$

$$= \frac{1}{2}x, \quad \text{or} \quad \frac{x}{2}$$

The checks for 2a and 2b are as follows:

2a. Numeric Check

X	Y₁	Y₂
0	ERROR	0
1	-6	-6
2	-24	-24
3	-54	-54
4	-96	-96
5	-150	-150
6	-216	-216

X=0

$Y1 = (2x^4)(-3x^{-2})$
$Y2 = -6x^2$

2b. Graphic Check

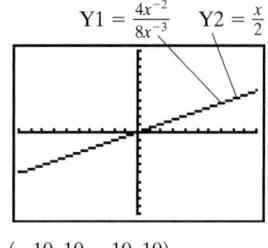

$Y1 = \frac{4x^{-2}}{8x^{-3}}$ $Y2 = \frac{x}{2}$

$(-10, 10, -10, 10)$

EXAMPLE 3   Simplify and write with positive exponents.

a. $(3x^{-3}y^4)^{-2}$ **b.** $(3x^2yz)(-4xy^{-2}z^{-1})$

c. $\dfrac{3^{-2}x^{-3}y}{3^{-1}xy^{-4}}$ **d.** $\left(\dfrac{-2a^3b^2}{ab^{-2}}\right)^3$

Solution

a. $(3x^{-3}y^4)^{-2} = 3^{-2}(x^{-3})^{-2}(y^4)^{-2}$ Product-to-a-power rule

$$= \frac{1}{3^2}x^6y^{-8}$$ Power-to-a-power rule

$$= \frac{1}{9} \cdot x^6 \cdot \frac{1}{y^8}$$ Write with positive exponents.

$$= \frac{x^6}{9y^8}$$

b. $(3x^2yz)(-4xy^{-2}z^{-1}) = [3 \cdot (-4)](x^2x)(yy^{-2})(zz^{-1})$

$$= -12x^{2+1}y^{1+(-2)}z^{1+(-1)}$$ Product rule for exponents

$$= -12x^3y^{-1}z^0$$

$$= \frac{-12x^3}{y}$$ Write with positive exponents. Remember, $z^0 = 1$.

c. $\dfrac{3^{-2}x^{-3}y}{3^{-1}x\,y^{-4}} = \left(\dfrac{3^{-2}}{3^{-1}}\right)\left(\dfrac{x^{-3}}{x}\right)\left(\dfrac{y}{y^{-4}}\right)$

$$= 3^{-2-(-1)}x^{-3-1}y^{1-(-4)}$$ Quotient rule for exponents

$$= 3^{-1}x^{-4}y^5$$

$$= \frac{1}{3} \cdot \frac{1}{x^4} \cdot y^5$$

$$= \frac{y^5}{3x^4}$$

d. $\left(\dfrac{-2a^3b^2}{ab^{-2}}\right)^3 = \dfrac{(-2)^3(a^3)^3(b^2)^3}{a^3(b^{-2})^3}$

Product-to-a-power rule
Quotient-to-a-power rule

$$= \dfrac{-8a^9b^6}{a^3b^{-6}}$$

Power-to-a-power rule

$$= -8 \cdot \dfrac{a^9}{a^3} \cdot \dfrac{b^6}{b^{-6}}$$

$$= -8a^6b^{12}$$

Quotient rule for exponents

The quotient-to-a power rule may be generalized to the case where the exponent is negative. If the exponent is negative and we apply the rule, we obtain

$$\left(\frac{x}{y}\right)^{-4} = \frac{x^{-4}}{y^{-4}} = \frac{y^4}{x^4}$$

Since $\dfrac{y^4}{x^4}$ also equals $\left(\dfrac{y}{x}\right)^4$, it follows that $\left(\dfrac{x}{y}\right)^{-4} = \left(\dfrac{y}{x}\right)^4$.

The result is the reciprocal of the quotient, raised to the positive power. We will write a special rule for this case.

QUOTIENT-TO-A-NEGATIVE-POWER RULE
For any nonzero dividend and divisor a and b and integer exponent $m > 0$,

$$\left(\frac{a}{b}\right)^{-m} = \left(\frac{b}{a}\right)^m = \frac{b^m}{a^m}$$

The elements a and b may be constants, variables, or expressions.

For example,

$$\left(\frac{3}{4}\right)^{-2} = \left(\frac{4}{3}\right)^2 = \frac{4^2}{3^2} = \frac{16}{9} \qquad \left(\frac{m}{n}\right)^{-3} = \left(\frac{n}{m}\right)^3 = \frac{n^3}{m^3}$$

EXAMPLE 4 Simplify and write with positive exponents.

a. $\left(\dfrac{m^{-3}}{n^{-2}}\right)^{-4}$ **b.** $\left(\dfrac{4xy^{-2}}{16x^{-3}y}\right)^{-2}$

Solution

a. $\left(\dfrac{m^{-3}}{n^{-2}}\right)^{-4} = \dfrac{(m^{-3})^{-4}}{(n^{-2})^{-4}}$

Quotient-to-a-power rule

$$= \dfrac{m^{12}}{n^8}$$

Power-to-a-power rule

b. Before using the rules of exponents, simplify the quotient by writing it with positive exponents.

$$\left(\frac{4xy^{-2}}{16x^{-3}y}\right)^{-2} = \left(\frac{x \cdot x^3}{4y \cdot y^2}\right)^{-2}$$

$$= \left(\frac{x^4}{4y^3}\right)^{-2}$$

$$= \left(\frac{4y^3}{x^4} \right)^2 \qquad \text{Quotient-to-a-negative-power rule}$$

$$= \frac{4^2(y^3)^2}{(x^4)^2} \qquad \text{Quotient-to-a-power rule}$$
$$\qquad\qquad\qquad \text{Product-to-a-power rule}$$

$$= \frac{16y^6}{x^8} \qquad \text{Power-to-a-power rule}$$

 HELPING HAND As in the case of simplifying positive exponents, the order the rules of exponents are applied may vary. The results will be the same.

✓ 9.2.2 Checkup

In exercises 1–8, simplify and write with only positive exponents.

1. $\left(-\frac{2}{3}z^6 \right)\left(-\frac{3}{4}z^{-8} \right)$ **2.** $\frac{15b^{-5}}{3b^{-4}}$ **3.** $(2a^{-1}b^3)^{-1}$

7. $\left(\frac{h^{-4}}{k^{-2}} \right)^{-3}$ **8.** $\left(\frac{8c^{-2}d}{12cd^{-3}} \right)^{-3}$

4. $\left(\frac{2}{3}a^3bc^{-2} \right)\left(\frac{3}{4}ab^{-4}c^2 \right)$ **5.** $\frac{5^{-3}p^2q^{-4}}{5^{-6}pq^{-2}}$ **6.** $\left(\frac{-x^2y^3}{5xy^{-1}} \right)^3$

9. All monomials are exponential expressions, but not all exponential expressions are monomials. Explain.

9.2.3 Modeling the Real World

In Chapter 1, we discussed scientific notation. Scientific notation is used to write both very large numbers and very small numbers, which we encounter often in the field of science. Recall the following equivalences:

Standard Notation	Scientific Notation	Calculator Notation
2,340,000	2.34×10^6	2.34E6
0.000058	5.8×10^{-5}	5.8E−5

At times, it is necessary to perform operations on large or small numbers. The rules of exponents allow us to do so.

EXAMPLE 5 The Milky Way galaxy contains roughly 100 billion stars, one of which is the Sun. The galaxy is arranged in a huge disklike structure with a diameter of about 1×10^5 light-years. A light-year, the distance light travels in a year at the speed of 1.86×10^5 miles per second, is about 5.9×10^{12} miles. Approximate the length of the diameter of the Milky Way in miles.

Solution

To determine the length in miles, we will multiply the length in light-years by the number of miles in a light-year.

$$(1 \times 10^5)(5.9 \times 10^{12}) = (1 \times 5.9) \times (10^5 \times 10^{12})$$
$$= 5.9 \times (10^{5+12})$$
$$= 5.9 \times 10^{17}$$

Check

We can check this result on the calculator by entering the problem two different ways.

The diameter of the Milky Way is approximately 5.9×10^{17} miles.

APPLICATION

An FM radio station broadcasts at a frequency of 102.1 MHz, or 102.1×10^6 cycles per second. Write this frequency in scientific notation. What is the wavelength of these radio waves in meters?

Discussion

$$102.1 \times 10^6 = (1.021 \times 10^2) \times 10^6$$
$$= 1.021 \times (10^2 \times 10^6)$$
$$= 1.021 \times 10^8$$

The speed of light has a constant value of 2.998×10^8 meters per second. To determine the wavelength, divide the speed of light by the number of cycles per second.

$$\frac{2.998 \times 10^8}{102.1 \times 10^6} = \frac{2.998 \times 10^8}{1.021 \times 10^8}$$
$$= \left(\frac{2.998}{1.021}\right)\left(\frac{10^8}{10^8}\right)$$
$$\approx 2.936 \times 10^0$$
$$\approx 2.936$$

The wavelength of the waves is about 2.936 meters.

```
(2.998E8)/(102.1
E6)
       2.936336925
```

✓ 9.2.3 Checkup

It takes moonlight about $1\frac{1}{4}$ seconds to reach the Earth. If light travels at the rate of 1.86×10^5 miles per second, use the distance formula to estimate the distance to the moon in miles. Write your answer in scientific notation. Given that there are 5.28×10^3 feet in each mile, estimate the distance in feet, using scientific notation.

9.2 Exercises

In exercises 1-54, simplify. Write with positive exponents.

1. p^{-3}

2. q^{-2}

3. $\dfrac{1}{q^{-5}}$

4. $\dfrac{1}{p^{-3}}$

5. $\dfrac{p^{-3}}{q^{-5}}$

6. $\dfrac{q^{-2}}{p^{-3}}$

7. $\dfrac{d^4}{c^{-3}}$

8. $\dfrac{c^2}{d^{-5}}$

9. $\dfrac{5^{-2}h^3}{2^{-4}k^{-4}}$

10. $\dfrac{3^{-2}x^2}{4^{-3}y^{-3}}$

11. $p^{-3}q^5$

12. p^3q^{-2}

13. $(5a^3)(-4a^{-1})$

14. $(-6h^4)(-5h^{-2})$

15. $\dfrac{4m^{-2}}{16m^{-3}}$

16. $\dfrac{2y^{-3}}{12y^{-4}}$

17. $\dfrac{4^{-3}n^{-2}}{3^{-4}n^{-3}}$

18. $\dfrac{2^{-5}x^{-3}}{5^{-2}x^{-4}}$

19. $(c^{-4})^{-2}$

20. $(p^{-3})^{-2}$

21. $(5a^{-2}b^2)^{-3}$

22. $(5m^{-4}n^2)^{-2}$

23. $(-6p^3q)(7p^{-3}q^{-4})$

24. $(13x^3y)(3x^{-2}y^2)$

25. $(1.4x^2)(4.3x^3y^{-2})$

26. $(5.7yz)(2.7y^2z^{-1})$

27. $\left(\dfrac{3}{7}x^2y^{-1}\right)\left(\dfrac{14}{15}x^{-1}y^4\right)$

28. $\left(\dfrac{5}{9}a^3b^{-4}\right)\left(\dfrac{12}{25}a^{-2}b^2\right)$

29. $(3x^5yz)(-7xy^{-4}z^{-1})$

30. $(-2m^3pq^2)(4m^{-1}p^{-1}q)$

31. $\dfrac{m^{-2}}{m^5}$

32. $\dfrac{k^{-1}}{k}$

33. $\dfrac{-7c^{-5}}{21c^{-7}}$

34. $\dfrac{14d^{-2}}{20d^{-4}}$

35. $\dfrac{8x^2y}{2x^{-1}y^{-3}}$

36. $\dfrac{12cd^3}{4c^{-1}d^{-2}}$

37. $\dfrac{5^{-3}h^{-1}k}{5^{-2}hk^{-6}}$

38. $\dfrac{9^{-4}u^{-2}v^2}{9^{-3}u^3v^{-1}}$

39. $\dfrac{121a^{-3}b^3}{11a^2b^{-2}}$

40. $\dfrac{17a^{-7}b^{-4}}{51a^{-2}b^{-7}}$

41. $\left(\dfrac{a^{-3}b}{2ab^{-3}}\right)^3$

42. $\left(\dfrac{-cd^{-3}}{4c^{-2}d}\right)^3$

43. $\left(\dfrac{-4p^3q^2}{2pq^{-3}}\right)^5$

44. $\left(\dfrac{-8km^3}{4k^{-4}m^2}\right)^5$

45. $\left(\dfrac{a}{b}\right)^{-3}$

46. $\left(\dfrac{s}{t}\right)^{-1}$

47. $\left(\dfrac{4x}{y}\right)^{-3}$ **48.** $\left(\dfrac{3a}{b}\right)^{-5}$ **49.** $\left(\dfrac{-3p^2}{q^{-2}}\right)^{-4}$ **50.** $\left(\dfrac{-5c^3}{d^{-1}}\right)^{-2}$

51. $\left(\dfrac{y^{-1}}{z^{-2}}\right)^{-3}$ **52.** $\left(\dfrac{d^{-3}}{e^{-4}}\right)^{-2}$ **53.** $\left(\dfrac{5a^{-2}b}{25ab^{-3}}\right)^{-3}$ **54.** $\left(\dfrac{15s^2t^{-3}}{3s^{-1}t^4}\right)^{-4}$

55. The Sun is approximately 2.8×10^4 light years away from the center of the Milky Way galaxy. Given that a light year is about 5.9×10^{12} miles, approximate the distance in miles from the Sun to the center of the Milky Way galaxy.

56. If all the stars in the galaxy were placed the same distance from Earth, Deneb would be the brightest star in the sky. Deneb is about 3.23×10^3 light years away from Earth. Approximate this distance in miles.

57. A country music radio station broadcasts at a frequency of 107.7 MHz, or 107.7×10^6 cycles per second. Determine the wavelength of these radio waves in meters, by dividing the speed of light, 2.998×10^8 meters per second, by the number of cycles per second.

58. A classic rock radio station broadcasts at a frequency of 103.5 MHz. Using the preceding exercise as a guide, determine the wavelength of these radio waves in meters.

9.2 Calculator Exercises

If you make a mistake when you are simplifying an exponential expression that has several steps, you may have difficulty finding the mistake. In 9.1 Calculator Exercises, you were shown how to check that your simplified expression was equivalent to the original expression. If you made a mistake, you can check for equivalence step-by-step to find where the mistake was made. For example, consider the following simplification, which has a mistake:

$$\left[\left(\dfrac{-2x^{-1}}{3x^{-2}}\right)^2\right]^{-3} = \left(\dfrac{-2x^{-1}}{3x^{-2}}\right)^{-6} =$$

$$\left(\dfrac{3x^{-2}}{-2x^{-1}}\right)^6 = \left(\dfrac{-2x}{3x^2}\right)^6 = \dfrac{64x^6}{729x^8} = \dfrac{64}{729x^2}$$

If you check the original expression against the rightmost simplification, you will see that the two aren't equivalent.

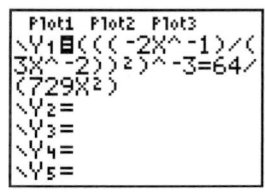

However, if you check step-by-step, you can see that the mistake occurred in the third step.

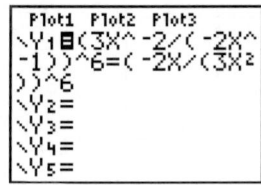

When you recheck your work, you see that

$$\left(\dfrac{3x^{-2}}{-2x^{-1}}\right)^6 = \left(\dfrac{-3x}{2x^2}\right)^6 = \dfrac{729x^6}{64x^{12}} = \dfrac{729}{64x^6}$$

and this simplification is equivalent to the original expression.

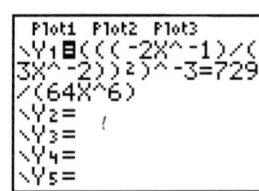 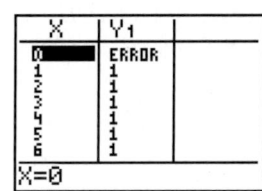

Use the preceding method to determine whether there is an error in each of the following exercises and in which step the error occurs:

1. $\left(\dfrac{3}{5}x^3\right)\left(\dfrac{5}{7}x^{-2}\right) = \left(\dfrac{3}{5}\cdot\dfrac{5}{7}\right)(x^3 \cdot x^{-2}) = \left(\dfrac{3}{7}\right)(x^{-6}) = \dfrac{3}{7x^6}$

2. $(2x^{-3})^{-4} = 2^{-4}(x^{-3})^{-4} = \left(\dfrac{1}{2^4}\right)(x^{-7}) = \dfrac{1}{16x^7}$

3. $\left(\dfrac{x^{-4}}{3x}\right)^{-2} = \dfrac{(x^{-4})^{-2}}{3^{-2}x^{-2}} = \dfrac{x^8}{-6x^{-2}} = \dfrac{x^8x^2}{-6} = -\dfrac{x^{10}}{6}$

9.2 Writing Exercise

In the preceding calculator exercises, the TABLE check displays an error message when x is equal to zero. Explain why this is so.

9.3 Polynomial Addition and Subtraction

OBJECTIVES

1 Add polynomials.
2 Subtract polynomials.
3 Model real-world situations by using polynomials.

APPLICATION

Pyrotechnicians for the famous Boom's Day festival plan the grand finale with a large display of aerial shell fireworks. Assuming that only gravity has an effect on the shells, the height above ground level for an 8-inch shell shot at a velocity of 235 feet per second from a height of 5 feet is given by the function $s(t) = -16t^2 + 235t + 5$, where t is time in seconds. The height above ground level for a 10-inch shell shot at a velocity of 263 feet per second from a height of 5 feet is given by the function $s(t) = -16t^2 + 263t + 5$, where t is again time in seconds. Determine the difference in the height of the two shells after t seconds if the shells are shot at the same time.

After completing this section, we will discuss this application further. See page 617.

9.3.1 Adding Polynomials

To add two algebraic expressions, we used the distributive law to remove the parentheses and combined like terms. We will do the same for polynomials.

ADDITION OF POLYNOMIALS

To add polynomials enclosed in a set of parentheses,

- Use the distributive law to remove the parentheses.
- Combine like terms.

EXAMPLE I   Add the polynomials $(2x^3 - 3x^2 + x - 5)$ and $(5x^3 - 3x^2 + 7)$.

Solution

$(2x^3 - 3x^2 + x - 5) + (5x^3 - 3x^2 + 7)$

$= 2x^3 - 3x^2 + x - 5 + 5x^3 - 3x^2 + 7$ Use the distributive law to remove the parentheses.

$= 7x^3 - 6x^2 + x + 2$ Combine like terms.

Sometimes it is easier to see the like terms if we align the polynomials in columns with like terms aligned. We can write missing addends with a coefficient of 0 to keep the columns complete.

$$\begin{array}{r} 2x^3 - 3x^2 + x - 5 \\ + 5x^3 - 3x^2 + 0x + 7 \\ \hline 7x^3 - 6x^2 + x + 2 \end{array}$$

Check

X	Y₁	Y₂
0	2	2
1	4	4
2	36	36
3	140	140
4	358	358
5	732	732
6	1304	1304

X=0

Y1 = $(2x^3 - 3x^2 + x - 5) + (5x^3 - 3x^2 + 7)$
Y2 = $7x^3 - 6x^2 + x + 2$

EXAMPLE 2   Add.

$$(6x^2y + 4x^2 - xy^2 + xy) + (5xy + 6x^2 - x^2y^3)$$

Solution

$(6x^2y + 4x^2 - xy^2 + xy) + (5xy + 6x^2 - x^2y^3)$

$= 6x^2y + 4x^2 - xy^2 + xy + 5xy + 6x^2 - x^2y^3$ Use the distributive law to remove the parentheses.

$= 6x^2y - x^2y^3 - xy^2 + 6xy + 10x^2$ Combine like terms.

Aligned in columns, the result is as follows:

$$
\begin{aligned}
&6x^2y + 4x^2 - xy^2 + xy \\
&\underline{+ 6x^2 + 5xy - x^2y^3} \\
&6x^2y + 10x^2 - xy^2 + 6xy - x^2y^3
\end{aligned}
$$

Align like terms.

 9.3.1 Checkup

In exercises 1 and 2, add as indicated.

1. Add $(3x^3 + 2x^2 - 5x + 7)$ and $(5x^3 - 5x^2 + 2 + 4x)$.

2. $(3x^3 + 2x^2y - 4xy^2 + y^3) + (4y^3 - 3xy^2 - 3x^2y + 4x^3)$

3. Explain what is meant by "combining like terms."

9.3.2 Subtracting Polynomials

To subtract algebraic expressions, we used the distributive law to remove the parentheses by taking the opposite of the subtrahend, added the opposite of the subtrahend to the minuend, and combined like terms. We will do the same for polynomials.

> **SUBTRACTION OF POLYNOMIALS**
>
> To subtract polynomials enclosed in a set of parentheses,
>
> • Remove the parentheses, remembering to take the opposite of the terms within the subtrahend.
> • Combine like terms.

EXAMPLE 3   Subtract the polynomial $(5x^3 - 3x^2 + 7)$ from the polynomial $(2x^3 - 3x^2 + x - 5)$.

Solution

$(2x^3 - 3x^2 + x - 5) - (5x^3 - 3x^2 + 7)$

$= 2x^3 - 3x^2 + x - 5 - 5x^3 + 3x^2 - 7$ Remove the parentheses. (Take the opposite of the subtrahend.)

$= -3x^3 + x - 12$ Combine like terms.

To use columnar subtraction, align like terms. However, remember to take the opposite of the subtrahend before adding.

$$
\begin{aligned}
&2x^3 - 3x^2 + x - 5 \\
&\underline{- (5x^3 - 3x^2 + 0x + 7)}
\end{aligned}
\quad \text{or} \quad
\begin{aligned}
&2x^3 - 3x^2 + x - 5 \\
&\underline{-5x^3 + 3x^2 - 0x - 7} \\
&-3x^3 + x - 12
\end{aligned}
$$

Change signs and add.

Check

X	Y₁	Y₂
0	-12	-12
1	-14	-14
2	-34	-34
3	-90	-90
4	-200	-200
5	-382	-382
6	-654	-654

X=0

Y1 = $(2x^3 - 3x^2 + x - 5) -$ $(5x^3 - 3x^2 + 7)$
Y2 = $-3x^3 + x - 12$

EXAMPLE 4 Subtract.

$$(6x^2y + 4x^2 - xy^2 + xy) - (5xy + 6x^2 - x^2y^3)$$

Solution

$$(6x^2y + 4x^2 - xy^2 + xy) - (5xy + 6x^2 - x^2y^3)$$
$$= 6x^2y + 4x^2 - xy^2 + xy - 5xy - 6x^2 + x^2y^3 \qquad \text{Take the opposite of the subtrahend.}$$

$$= 6x^2y - 2x^2 + x^2y^3 - xy^2 - 4xy \qquad \text{Combine like terms.}$$

Aligned in columns, the result is as follows:

$$
\begin{array}{l}
6x^2y + 4x^2 - xy^2 + xy \\
-(6x^2 + 5xy - x^2y^3)
\end{array}
\quad \text{or} \quad
\begin{array}{l}
6x^2y + 4x^2 - xy^2 + xy \\
- 6x^2 - 5xy + x^2y^3 \\
\hline
6x^2y - 2x^2 - xy^2 - 4xy + x^2y^3
\end{array}
$$

9.3.2 Checkup

In exercises 1 and 2, subtract as indicated.

1. Subtract $(5a^2 - 2 + 6a)$ from $(6a + 13 - 2a^2 + a^3)$.

2. $(4a^4 + 3a^3b - 7a^2b^2 + 9ab^3 - 3b^4) - (2a^2b^2 + 4ab^3 - 5b^4)$

3. In subtracting polynomials, which polynomial is the subtrahend, and why must you take the opposite of each term of the subtrahend when removing its parentheses? Why do you not have to perform this operation in adding two polynomials?

9.3.3 Modeling the Real World

Many business relations may be defined as a polynomial. Three such relations are the revenue function, the total cost function, and the profit function. We discussed the relationship of these functions to one another in previous chapters. We know that the total cost is found by adding all of the costs of manufacturing and advertising the product. We also know that the profit may be found by subtracting the total cost of an item from the revenue received from the same number of items. Therefore, it is very important to know how to perform the operations of addition and subtraction of polynomials to be able to work with these functions.

EXAMPLE 5 The cost of manufacturing x items is given by the function $m(x) = 4.95x + 231$. The cost of advertising and distributing for the same x items is given by the function $a(x) = 0.50x + 125$. Write a total cost function for x items. Each item sells for $12.00. Write a revenue function, $R(x)$. Write a profit function for x items.

Solution

The total cost of x items, $C(x)$, is the sum of the manufacturing costs and the advertising and distributing costs.

$$C(x) = m(x) + a(x)$$
$$C(x) = (4.95x + 231) + (0.50x + 125)$$
$$C(x) = 5.45x + 356$$

The revenue received for x items, $R(x)$, is the product of the selling price and the number of items, x.

$$R(x) = 12.00x$$

The profit on x items, $P(x)$, is the difference of the revenue and the total cost.

$$P(x) = R(x) - C(x)$$
$$P(x) = 12.00x - (5.45x + 356)$$
$$P(x) = 12.00x - 5.45x - 356 \qquad \text{Change the signs to subtract.}$$
$$P(x) = 6.55x - 356$$

APPLICATION

Pyrotechnicians for the famous Boom's Day festival plan the grand finale with a large display of aerial shell fireworks. Assuming that only gravity has an effect on the shells, the height above ground level for an 8-inch shell shot at a velocity of 235 feet per second from a height of 5 feet is given by the function $s(t) = -16t^2 + 235t + 5$, where t is time in seconds. The height above ground level for a 10-inch shell shot at a velocity of 263 feet per second from a height of 5 feet is given by the function $s(t) = -16t^2 + 263t + 5$, where t is again time in seconds. Determine the difference in the height of the two shells after t seconds if the shells are shot at the same time.

Discussion

The difference in height of the two shells is

$$(-16t^2 + 263t + 5) - (-16t^2 + 235t + 5)$$
$$= -16t^2 + 263t + 5 + 16t^2 - 235t - 5$$
$$\qquad \text{Change the signs to subtract.}$$
$$= 28t$$

The difference in the height is $28t$ feet, within the domains of the two functions.

✓ 9.3.3 Checkup

1. Sybil sells box lunches from a booth in the town square. She pays the town council $15.00 per day to use the booth and $0.25 per box lunch for permission to sell on the square. Her box lunches cost $1.25 each to prepare. She sells the box lunches for $4.50 each.

   a. Write a daily total cost function, $C(x)$.
   b. Write a daily revenue function, $R(x)$.
   c. Write a daily profit function, $P(x)$.

2. A projectile is shot upwards with an initial velocity of 220 feet per second from a tower that is 60 feet above ground.

The distance of the projectile from the ground is given by the polynomial function $s(t) = -16t^2 + 220t + 60$, where t is the number of seconds after launch. Another projectile is shot upwards at the same time from ground level with an initial velocity of 200 feet per second. The distance of this projectile from the ground is given by the polynomial function $s(t) = -16t^2 + 200t$. Determine the difference in heights of the two projectiles after t seconds.

9.3 Exercises

Add the polynomials as indicated.

1. Add $(9x^2 - 17x + 31)$ and $(2x^4 + 3x^2 + 12)$.

2. Add $(b^3 + 2b^2 + 5b)$ and $(4b^2 - b + 2)$.

3. $(5x^4 + 6x + 3x^3 - 2x^2 - 12) + (4x^4 + 21 - 8x^2 - 9x)$

4. $(-z^3 + z + z^2 + 1) + (2z^3 + 5 + 4z)$

5. Add $(5x^2y - 3xy^2 + 6y^3)$ and $(15x^3 - 8x^2y + 3xy^2)$.

6. Add $(3a^3 + 5a^2b - 2ab^2 + 6b^3)$ and $(b^3 + ab^2 + a^2b + a^3)$.

7. Add $(6 - 7a^3 + 3a^2 - 5a)$, $(6a + 8a^3 + 2)$, and $(5a^2 - 8a - 9)$.

8. Add $(x + 5 + x^2)$, $(2x^2 - 4x)$, and $(3x + 7)$.

9. $\left(\frac{2}{3}y^4 + \frac{1}{6}y^3 + 3y^2 - \frac{1}{3}y + \frac{5}{9}\right) + \left(\frac{7}{3}y^3 - \frac{8}{9}y^2 + \frac{5}{6}y - 3\right)$

10. $\left(\frac{5}{6}x^3 + \frac{17}{24} - \frac{1}{2}x^2 - \frac{3}{4}x\right) + \left(\frac{7}{8} + \frac{1}{4}x^3 - \frac{1}{6}x^2 + \frac{1}{2}x\right)$

11. $(12.07x^3 + 8.6x^2 - 3.19x + 14) + (6.7x^3 - 9.83x^2 + 7x - 4.265)$

12. $(5.1y^2 - 3.6y + 0.8y^3 - 3.7) + (4 - 0.8y - 0.1y^3 + 1.1y^2)$

13. Add $(4756a^3 - 3219a^2 - 1816a + 2083)$ and $(361a^3 + 54217a^2 + 12)$.

14. Add $(509b - 471b^3 + 211 + 54b^2)$ and $(471b^3 - 509b - 4b^2 - 11)$.

15. $(3a + 4b) + (5b + 6c)$

16. $(5x + 7y) + (-6y + 11z)$

Subtract the polynomials as indicated.

17. $(5z^3 + 27z^2 - 35z + 42) - (3z^3 + 16z - 72)$ **18.** $(a^3 - 5a^2 - 4a + 10) - (a^3 - 2a^2 + a - 5)$

19. $(a^5 - 9) - (a^5 - a^4 + a^3 - a^2 + a - 9)$ **20.** $(b^9 + b^7 + b^5 + b^3 + b + 15) - (b^9 + 15)$

21. Subtract $(12x + 7 + 9x^2)$ from $(16x^2 - 32 + 9x)$.

22. Subtract $(11y^3 + 3 + y^2 - 6y)$ from $(6y^3 - 8 + 2y + y^2)$.

23. $(13a^3 - 6a^2 + 11)$ minus $(12a - 3 + 18a^3)$ **24.** $(15 + 3x + 6x^3)$ minus $(5x^3 + 2x^2 + 6)$

25. $(42x^3 + 17x^2y + 3xy^2 + 23y^3)$ decreased by $(47x^2y + 12y^3)$

26. $(51p^3 + 18p^2q - 7pq^2 + 41q^3)$ decreased by $(9pq^2 - 8p^2q - 10)$

27. $(4a + 7c)$ decreased by $(2b + 6d)$ **28.** $(12x + 17y)$ decreased by $(12y - 8z)$

29. $\left(\frac{5}{7}x^2 + \frac{8}{21}x - \frac{11}{14}\right) - \left(\frac{1}{2}x^2 + \frac{5}{6}x + \frac{19}{42}\right)$ **30.** $\left(\frac{2}{3}x^3 - \frac{17}{24}x + \frac{5}{8}x^2\right) - \left(\frac{3}{8}x^3 + \frac{1}{6}x^2 - \frac{11}{24}x + \frac{1}{3}\right)$

31. Subtract $(12.2x^3 - 0.1x^2y + 0.78xy^2 + 13.07y^3)$ from $(21.2x^3 + 0.9x^2y - 13.22xy^2 + 81.07y^3)$.

32. Subtract $(4.6x^2 + 9.3xy + 17.02y^2)$ from $(7.08x^2 - 3.21xy + 8.27y^2)$.

33. $(5062z^2 - 106z + 8295)$ minus $(379z^2 + 4297z + 1108)$

34. $(476b^3 + 178b^2 - 471b + 972)$ minus $(562b^3 + 873b - 619)$

35. Philbert's Pots has a production process with a fixed setup cost of $200.00 each time a production run is made. It costs $2.50 for the labor to produce a single pot. The cost of materials is $2.00 per pot.

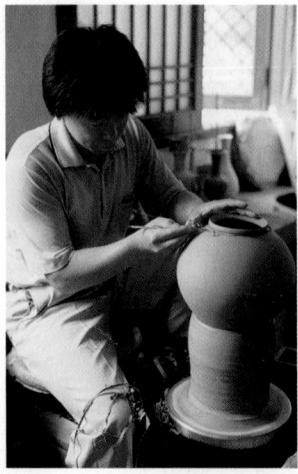

a. Determine the cost function $C(x)$ for producing a run of x pots.

b. If the pots can be sold for $13.50 each, determine the revenue function $R(x)$ for selling x pots.

c. Determine the profit function, $P(x)$.

d. What is the profit if 20 pots are produced and sold? What if 30 pots are produced and sold?

36. America's Best Cellular, Incorporated, has a production process to manufacture ABC cellular phones with a setup cost of $500. The cost of labor and materials for each phone is $20.

a. Determine the cost function for producing a run of x cellular phones.

b. The cellular phones sell for a price of $50. Determine the revenue function for selling x phones.

c. Determine the profit function.

d. What is the profit if 75 phones are sold? What is it if 100 phones are sold?

37. Shonda bakes gourmet cakes and sells them as a business venture. Her records show that the cost of baking x cakes is given by the function $b(x) = 3.50x + 25$. The cost of delivering x cakes to the customer is given by $d(x) = 1.50x + 5$. Write a total cost function for x cakes. The cakes sell for \$28.50 each. Write a revenue function, $r(x)$. Write a profit function for x cakes baked and sold.

38. Faouzi prints and assembles trivia desk calendars and sells them. The cost of printing and assembling x calendars is given by the function $a(x) = 125 + 1.25x$. The cost of advertising and distributing the calendars is given by $d(x) = 0.55x + 55$. Write a total cost function for printing and distributing x calendars. The calendars sell for \$11.95 each. Write the revenue function, $r(x)$. Write a profit function for x calendars distributed and sold.

39. At a sporting event, a fireworks shell is shot upward with an initial velocity of 240 feet per second from the top of the stadium, a height of 100 feet. At the same time, an-other shell is shot upward from ground level with an initial velocity of 250 feet per second. After t seconds, the height above ground level for the first shell is given by the function $s(t) = -16t^2 + 240t + 100$, and the height above ground level for the second shell is given by $s(t) = -16t^2 + 250t$. Determine the difference in the height of the shells after t seconds.

40. A signal flare is shot upward at a velocity of 180 feet per second from a tower that is 40 feet above ground. A second flare is shot upward from ground level at a velocity of 150 feet per second. After t seconds, the height above ground level of the first flare is given by $s(t) = -16t^2 + 180t + 40$, and the height above ground level of the second flare is given by $s(t) = -16t^2 + 150t$. Determine the difference in the height of the flares after t seconds if they are shot at the same time.

 9.3 Calculator Exercises

When you are checking your addition and subtraction of polynomials on the calculator, you may want to store the original polynomial in Y1 and the resulting polynomial in Y2. Rather than checking that the TABLE values are equal for Y1 and Y2 or checking that the graphs coincide, you can store Y1 = Y2 in Y3 instead and check to see whether the TABLE values of Y3 are always equal to 1. Likewise, you can check to see that tracing the graph of Y3 is a horizontal line, where Y3 = 1. An example follows.

Is $(3x^3 - x^2 + 4x - 5) + (3x^2 - 5x + 7) = $
$$3x^3 + 2x^2 - x + 2?$$

The screens for this method of checking follow.

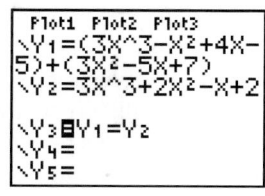 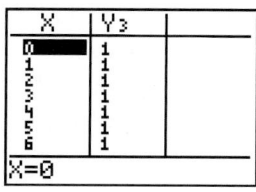 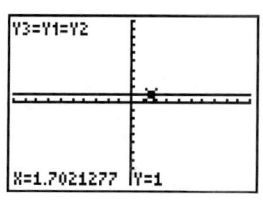

$(-10, 10, 1, -10, 10, 1)$

Note that the expressions in Y1 and Y2 have been "turned off" so that they will not be displayed on the table or graph. To turn them off, move the cursor over the equals sign for Y1 and for Y2, and press ENTER. This will remove the black box around the equals sign, which indicates that it is turned off.

Note also that, to type the expression into Y3, you will need to press the VARS key, move the cursor rightwards to Y-VARS, select 1 for Function, and then select 1 for Y1 or 2 for Y2.

Perform the following additions and subtractions of polynomials, and then check your results on the calculator, using the method just set out.

1. Add $(2x^4 + 4x^2 - 3x + 12)$ and $(3x^3 - 5x^2 + 7x - 9)$.

2. Add $(5.6x^3 + 1.17x^2 - 0.45x + 2.6)$ and $(4.1x^4 - 0.3x^2 + 1.9x + 2.33)$.

3. Subtract $(5x^5 - 3x^2 + 7x - 4)$ from $(3x^4 + 2x^3 - 3x^2 - 2x + 8)$.

4. Subtract $\left(\frac{2}{3}x^2 + \frac{5}{6}x - \frac{4}{9}\right)$ from $\left(\frac{5}{6}x^3 + \frac{1}{2}x^2 - \frac{7}{9}x + 4\right)$.

 **9.3 Writing Exercises**

In this section, you were asked to subtract polynomials. The directions were stated several different ways. You must be careful that you correctly determine which polynomial is being subtracted from which. For example, if you are asked to subtract $(5x^2 - 6)$ from $(11x^2 - 2x + 3)$, which of the following expressions is the correct one?

1. $5x^2 - 6 - 11x^2 - 2x + 3$
2. $11x^2 - 2x + 3 - 5x^2 - 6$
3. $(5x^2 - 6) - (11x^2 - 2x + 3)$
4. $(11x^2 - 2x + 3) - (5x^2 - 6)$

Explain why you think a particular expression is the correct one. Then describe the error(s) in the other three expressions.

9.4 Polynomial Multiplication

OBJECTIVES

1 Multiply by a monomial.

2 Multiply two binomials.

3 Multiply polynomials of two or more terms.

4 Multiply polynomials, resulting in special products.

5 Model real-world situations by using polynomials.

APPLICATION

The design of a swimming pool calls for its length to be twice its width. A 5-foot concrete walkway surrounds the pool. Determine a function, $A(w)$, for the area covered by the walkway.

After completing this section, we will discuss this application further. See page 626.

9.4.1 Multiplying with Monomials

We are ready to begin multiplication of polynomials. We start by multiplying two monomials. Remember that we multiplied algebraic expressions with one term by using the commutative and associative properties to rearrange the factors. We also used the rules for exponents discussed in Section 9.1 to help simplify the products. Recall, too, that we multiplied expressions when one factor had more than one term, using the distributive law. We do the same when multiplying a polynomial by a monomial.

PRODUCTS WITH A MONOMIAL FACTOR

To multiply a monomial by a monomial, use the commutative and associative properties to rearrange factors. Simplify, using the rules of exponents.

To multiply a polynomial with more than one term by a monomial, use the distributive law. Simplify, using the rules of exponents.

EXAMPLE 1 Multiply.

a. $(5x^2 y)(-3xy)$ **b.** $2x^2(x - y + 2z)$

Solution

Note that we did monomial multiplication in Section 9.1.

a.
$$\begin{aligned}
(5x^2 y)(-3xy) &= 5 \cdot x^2 \cdot y \cdot -3 \cdot x \cdot y \\
&= 5 \cdot -3 \cdot x^2 \cdot x \cdot y \cdot y \\
&= -15 \cdot x^{2+1} \cdot y^{1+1} \qquad \text{Product rule} \\
&= -15x^3 y^2
\end{aligned}$$

b. $2x^2(x - y + 2z) = 2x^2 \cdot x - 2x^2 \cdot y + 2x^2 \cdot 2z$ Distribute $2x^2$.

$$= 2x^3 - 2x^2y + 4x^2z$$ Product rule

If we multiply in columnar form, we write the following:

$$x - y + 2z$$
$$2x^2$$
$$\overline{2x^3 - 2x^2y + 4x^2z}$$

 9.4.1 Checkup

In exercises 1–2, multiply.

1. $5a^2b^4(-5a^3bc)$

2. $-4m^3(m^2n - 5mn^2 + n^3)$

3. When multiplying two polynomials, when would you use only the commutative and associative laws? For what type of polynomial multiplication would you need to use the distributive law?

9.4.2 Multiplying Binomials

To multiply binomials, we need to use the distributive law. For example,

$$(x + 2)(x + 3) = (x + 2)x + (x + 2)3$$ Distribute $(x + 2)$.
$$= x^2 + 2x + 3x + 6$$ Distribute x and 3.
$$= x^2 + 5x + 6$$ Combine like terms.

or $(x + 2)(x + 3) = x(x + 3) + 2(x + 3)$ Distribute $(x + 3)$.
$$= x^2 + 3x + 2x + 6$$ Distribute x and 2.
$$= x^2 + 5x + 6$$ Combine like terms.

Check

$Y1 = (x + 2)(x + 3)$

or $x + 2$ Vertical method
$$x + 3$$
$$\overline{3x + 6}$$ Distribute 3.
$$x^2 + 2x$$ Distribute x.
$$\overline{x^2 + 5x + 6}$$ Combine like terms.

All three methods result in the same polynomial, $x^2 + 5x + 6$.

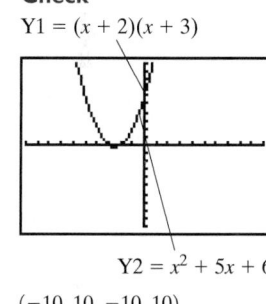

$Y2 = x^2 + 5x + 6$

$(-10, 10, -10, 10)$

PRODUCTS WITH TWO BINOMIAL FACTORS

$$(a + b)(c + d) = a(c + d) + b(c + d)$$
$$= ac + ad + bc + bd$$

For example,

$$(a + b)(c + d) = a(c + d) + b(c + d)$$

$$(x + 4)(x - 5) = x(x - 5) + 4(x - 5)$$
$$= x^2 - 5x + 4x - 20$$
$$= x^2 - x - 20$$

This process is sometimes called the **FOIL** method. The name comes from the following labels:

| first
term | last
term | first
term | last
term | First
terms | Outside
terms | Inside
terms | Last
terms |

$$(a \;+\; b) \qquad (c \;+\; d) \;=\; ac \;+\; ad \;+\; bc \;+\; bd$$

| outside
term | inside
term | inside
term | outside
term |

For example,

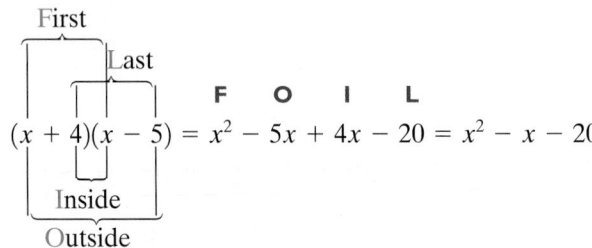

$$\text{F} \quad \text{O} \quad \text{I} \quad \text{L}$$
$$(x + 4)(x - 5) = x^2 - 5x + 4x - 20 = x^2 - x - 20$$

EXAMPLE 2 Multiply.

a. $(2x + 3)(4x - 1)$ **b.** $(x + 3)(x - 3)$ **c.** $(x + 5)^2$

Solution

$$\qquad\qquad\qquad\quad \text{F} \quad\; \text{O} \quad\; \text{I} \quad\; \text{L}$$
a. $(2x + 3)(4x - 1) = 8x^2 - 2x + 12x - 3$
$$= 8x^2 + 10x - 3 \qquad \text{Combine like terms.}$$

$$\qquad\qquad\qquad\quad \text{F} \quad\; \text{O} \quad\; \text{I} \quad\; \text{L}$$
b. $(x + 3)(x - 3) = x^2 - 3x + 3x - 9$
$$= x^2 + 0x - 9 \qquad \text{Combine like terms.}$$
$$= x^2 - 9$$

c. $(x + 5)^2 = (x + 5)(x + 5) \qquad \text{Expand.}$
$$= x^2 + 5x + 5x + 25 \qquad \text{FOIL}$$
$$= x^2 + 10x + 25 \qquad \text{Combine like terms.}$$

✔ 9.4.2 Checkup

In exercises 1–3, multiply.

1. $(a + 7)(7a + 1)$ **2.** $(2y - 1)(2y + 1)$

3. $(x - 5)^2$

4. In multiplying one binomial expression by another, you can use the FOIL method or the distributive law. Which do you prefer to use, and why?

9.4.3 Multiplying Polynomials

To multiply two polynomial factors, pick one factor and distribute its terms over all the terms of the second polynomial. We can do this several ways—for example,

$$(x - 4)(x^2 + 2x - 1)$$
$$= x(x^2 + 2x - 1) - 4(x^2 + 2x - 1) \qquad \text{Distribute } (x^2 + 2x - 1).$$
$$= x^3 + 2x^2 - x - 4x^2 - 8x + 4 \qquad \text{Distribute } x \text{ and } -4.$$
$$= x^3 - 2x^2 - 9x + 4 \qquad \text{Combine like terms.}$$

or

$$
\begin{array}{r}
x^2 + 2x - 1 \qquad \text{Vertical method} \\
x - 4 \\
\hline
-4x^2 - 8x + 4 \qquad \text{Distribute } -4. \\
x^3 + 2x^2 - x \qquad \text{Distribute } x. \\
\hline
x^3 - 2x^2 - 9x + 4
\end{array}
$$

PRODUCTS WITH POLYNOMIAL FACTORS
To multiply two polynomial factors, use the distributive law to distribute the terms of one factor over the terms of the second factor.

EXAMPLE 3 Multiply.

a. $(a^2 + 2a + 5)(a^2 - a + 4)$ **b.** $(x - 2)^3$ **c.** $(x - 2)(x^2 + 2x + 4)$

Solution

a. $(a^2 + 2a + 5)(a^2 - a + 4)$
$$= a^2(a^2 - a + 4) + 2a(a^2 - a + 4) + 5(a^2 - a + 4) \qquad \text{Distribute.}$$
$$= a^4 - a^3 + 4a^2 + 2a^3 - 2a^2 + 8a + 5a^2 - 5a + 20 \qquad \text{Distribute.}$$
$$= a^4 + a^3 + 7a^2 + 3a + 20 \qquad \text{Simplify.}$$

b. $(x - 2)^3 = (x - 2)(x - 2)(x - 2)$ Expand.
$$= [x(x - 2) - 2(x - 2)](x - 2) \qquad \text{Distribute } (x - 2).$$
$$= (x^2 - 2x - 2x + 4)(x - 2) \qquad \text{Distribute } x \text{ and } -2.$$
$$= (x^2 - 4x + 4)(x - 2) \qquad \text{Combine like terms.}$$
$$= x(x^2 - 4x + 4) - 2(x^2 - 4x + 4) \qquad \text{Distribute } (x^2 - 4x + 4).$$
$$= x^3 - 4x^2 + 4x - 2x^2 + 8x - 8 \qquad \text{Distribute } x \text{ and } -2.$$
$$= x^3 - 6x^2 + 12x - 8 \qquad \text{Combine like terms.}$$

c. $(x - 2)(x^2 + 2x + 4)$
$$= x(x^2 + 2x + 4) - 2(x^2 + 2x + 4) \qquad \text{Distribute } (x - 2).$$
$$= x^3 + 2x^2 + 4x - 2x^2 - 4x - 8 \qquad \text{Distribute } x \text{ and } -2.$$
$$= x^3 - 8 \qquad \text{Simplify.}$$

 9.4.3 Checkup

In exercises 1–3, multiply.

1. $(x + 2)(3x^2 - 2x - 8)$

2. $(x + 1)^3$

3. $(x + 1)(x^2 - x + 1)$

4. You have seen two methods for multiplying polynomials. One is to use the distributive property repeatedly and the other is to perform vertical multiplication. Which would you choose to use, and why?

9.4.4 Special Products

In previous examples, we determined certain products that we now identify as **special products**.

The **product of a sum and difference of the same two terms** is considered a special product. To determine a rule for this product, complete the following set of exercises.

Discovery 5

Product of the Sum and Difference of the Same Two Terms

Multiply.

   1. $(x + 5)(x - 5)$ **2.** $(3x + 1)(3x - 1)$ **3.** $(x + y)(x - y)$

Write a rule for determining by inspection the product of the sum and difference of the same two terms.

The product of the sum and difference of the same two terms is the difference of the square of the first term and the square of the second term.

PRODUCT OF THE SUM AND DIFFERENCE OF THE SAME TWO TERMS

$$(a + b)(a - b) = a^2 - b^2$$

We know this is true by determining the product algebraically.

$$(a + b)(a - b) = a^2 - ab + ab - b^2$$
$$= a^2 - b^2$$

For example, multiply $(x + 2)(x - 2)$.

$$(a + b)(a - b) = a^2 - b^2$$

$$(x + 2)(x - 2) = x^2 - 2^2 = x^2 - 4 \qquad a = x, b = 2$$

The **square of a binomial** is also considered a special product. To determine a rule for this product, complete the following set of exercises.

Discovery 6

Square of a Binomial

Rewrite in expanded form and multiply.

   1. $(x + 5)^2$ **2.** $(3x - 1)^2$ **3.** $(x + y)^2$ **4.** $(x - y)^2$

Write a rule for determining by inspection the square of a binomial.

The square of a binomial that is a sum of two terms is the square of the first term, plus two times the product of the first and last terms, plus the square of the last term. The square of a binomial that is a difference of two terms is the square of the first term, minus two times the product of the first and last terms, plus the square of the last term.

SQUARE OF A BINOMIAL

$$(a + b)^2 = a^2 + 2ab + b^2$$
$$(a - b)^2 = a^2 - 2ab + b^2$$

We can also determine these products algebraically.

$$(a + b)^2 = (a + b)(a + b) \qquad (a - b)^2 = (a - b)(a - b)$$
$$= a^2 + ab + ab + b^2 \qquad\qquad = a^2 - ab - ab + b^2$$
$$= a^2 + 2ab + b^2 \qquad\qquad = a^2 - 2ab + b^2$$

For example, multiply $(x + 2)^2$ and $(x - 2)^2$.

$$(a + b)^2 = a^2 + 2 \quad a \quad b + b^2$$

$$(x + 2)^2 = x^2 + 2 \cdot x \cdot 2 + 2^2 = x^2 + 4x + 4 \qquad a = x, b = 2$$

$$(a - b)^2 = a^2 - 2 \quad a \quad b + b^2$$

$$(x - 2)^2 = x^2 - 2 \cdot x \cdot 2 + 2^2 = x^2 - 4x + 4 \qquad a = x, b = 2$$

HELPING HAND These special products are helpful to know, but can be found without memorizing the formulas simply by using the FOIL method of multiplying two binomials.

EXAMPLE 4 Multiply.

a. $(2x - 5)(2x + 5)$ **b.** $(2x + 5)^2$

Solution

a. This is the product of the sum and difference of the same two terms. The first term is $2x$ and the second term is 5.

$$(a + b)(a - b) = a^2 - b^2$$

$$(2x - 5)(2x + 5) = (2x)^2 - 5^2 \qquad a = 2x, b = 5$$
$$= 4x^2 - 25$$

b. This is the square of a binomial. The binomial is the sum of $2x$ and 5.

$$(a + b)^2 = a^2 + 2 \quad a \quad b + b^2$$

$$(2x + 5)^2 = (2x)^2 + 2(2x)(5) + 5^2 \qquad a = 2x, b = 5$$
$$= 4x^2 + 20x + 25$$

9.4.4 Checkup

In exercises 1–3, multiply. When using the special-products forms, always state what a and b are.

1. $(3y - 7)(3y + 7)$
2. $(3y - 7)^2$
3. $(3y + 7)^2$

4. You can choose to learn the forms for special products, or you can choose to obtain the results by just applying the general rules for polynomial multiplication as described in previous sections of this chapter. Which do you prefer? Why?

9.4.5 Modeling the Real World

We have seen that many important quantities are expressed as products. Geometric areas are products of the sides of a rectangle, the square of the side of a square, or the product of the base and the height of a parallelogram. Volumes are products of three dimensions, such as the width, length, and height of a rectangular solid or the third power of the side of a cube. Calculations of these kinds of quantities often involve multiplying polynomials.

EXAMPLE 5 A rectangular garden plot has a length that is twice its width. Jay plans to increase the length by 5 feet and decrease the width by 2 feet. Write a polynomial representation of the area of the new garden plot in terms of the original width, w.

Solution

$$\text{Let } w = \text{original width (in feet)}$$
$$2w = \text{original length (in feet)}$$

$$w - 2 = \text{new width (in feet)}$$
$$2w + 5 = \text{new length (in feet)}$$

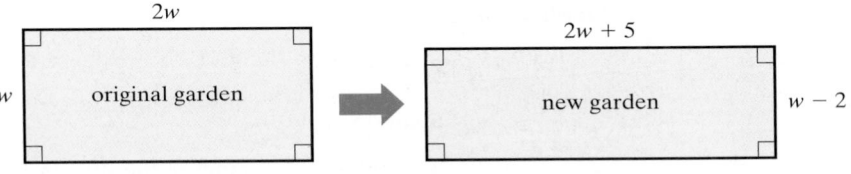

The area is the product of the length and width.

$$(2w + 5)(w - 2) = 2w^2 - 4w + 5w - 10$$
$$= 2w^2 + w - 10$$

The new area is $(2w^2 + w - 10)$ square feet.

APPLICATION

The design of a swimming pool calls for its length to be twice its width. A 5-foot concrete walkway surrounds the pool. Determine a function, $A(w)$, for the area covered by the walkway.

Discussion

Let w = the width of the pool in feet
$2w$ = length of the pool in feet

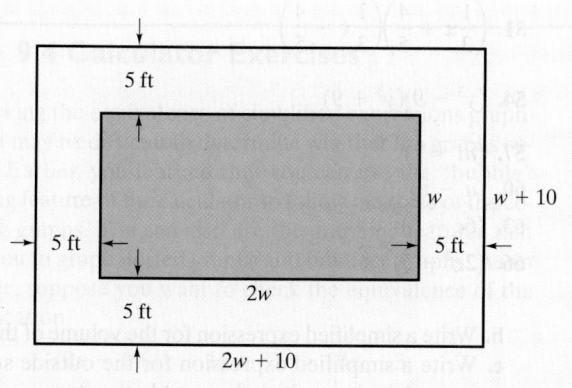

The walkway extends 5 feet on each side of the pool. Therefore, the width of the pool and walkway is $(w + 10)$ feet. The length of the pool and walkway is $(2w + 10)$ feet.

The area of the pool and walkway is the length times the width.

$$(w + 10)(2w + 10) =$$
$$(2w^2 + 30w + 100) \text{ square feet}$$

The area of the pool is the width times the length, $w(2w)$, or $2w^2$ square feet.

A function for the area of the walkway, in square feet, is the combined area of the pool and walkway minus the area of the pool.

$$A(w) = (2w^2 + 30w + 100) - (2w^2)$$
$$A(w) = 30w + 100$$

 9.4.5 Checkup

1. Jay has another garden plot for flowers. It is a square garden. He wishes to make it rectangular by increasing one side by 4 feet to become the width and by enlarging the other side to 3 feet more than 2 times its original measure to become the length. Write a polynomial expression to represent the area of the new flower garden in terms of its original side, s. What is the area of the new garden if it measured 5 feet on a side before the increase?

2. A rectangular garden has a length that is three times its width. A brick walkway around the garden measures 2 feet across. Determine a function for the area covered by the walkway.

9.4 Exercises

Multiply.

1. $(8ab^2)(-2a^3b)$

2. $(-5p^3q)(-3p^2q^2)$

3. $-2x(3x - y + 2z)$

4. $-a(4a - 3b + 2c)$

5. $2a^3(3a + 2b - c)$

6. $3x^2(2x - 3y + z)$

7. $(x + 4)(x + 2)$

8. $(a + 6)(a + 7)$

9. $(5 + x)(3 + 2x)$

10. $(4 + z)(1 + 7z)$

11. $(2x + 5)(3y - 2)$

12. $(5a - 2)(3b + 4)$

13. $(3x + 4y)(x - 2y)$

14. $(2a + 3b)(a - 4b)$

15. $(a - 2.4)(5a + 3.8)$

16. $(p - 1.7)(4p + 2.9)$

17. $(2x + 1.1)(3y + 3.2)$

18. $(5m - 4.9)(4n + 2.7)$

19. $\left(a + \dfrac{2}{3}\right)\left(a + \dfrac{1}{3}\right)$

20. $\left(x - \dfrac{1}{5}\right)\left(x + \dfrac{3}{5}\right)$

21. $(2x^2 - 3)(x^2 + 4)$

22. $(5x^3 + 1)(x^3 - 3)$

23. $(4x^2 + 3)(2x + 1)$

24. $(6x^3 - 5)(4x^2 - 3)$

25. $(2x + 3y^2)(3x^2 - 5y)$

26. $(4a^2 + 3b)(6a - 5b^2)$

27. $(3a^2 + 5b^3)(a^2 + b^3)$

28. $(7x^4 - 2y^3)(4x^4 - y^3)$

29. $(x + 4)(x^2 - 4x + 16)$

30. $(x - 1)(x^2 + x + 1)$

31. $(3x - 2)(2x^2 - 5x - 3)$

32. $(4x + 3)(x^2 - 5x - 2)$

33. $(x^2 + x + 1)(x^2 + 2x + 3)$

34. $(a^2 + a - 2)(a^2 - 2a + 1)$

35. $(a + b + c)^2$

36. $(a - b - c)^2$

37. $(z + 3)^3$

38. $(r + 2)^3$

39. $(3a - 2b)^3$

40. $(2x - 3y)^3$

41. $(x - 5)(x + 5)$

42. $(y + 12)(y - 12)$

43. $(3m + 7)(3m - 7)$

44. $(5p - 4)(5p + 4)$

45. $(2a + 3b)(2a - 3b)$

46. $(9p - 2q)(9p + 2q)$

47. $(4x - 1.5)(4x + 1.5)$

48. $(3z + 2.5)(3z - 2.5)$

9.5 Polynomial Division

OBJECTIVES

1 Divide by a monomial.

2 Divide by a polynomial.

3 Model real-world situations by using polynomials.

APPLICATION

Sio signed a contract for her new job. According to the contract, she is to make $20,000 the first year and receive an increase of 5% per year for the next two years. Sio's total salary for three years is given by the function $s(x) = \frac{20,000(1 - x^3)}{1 - x}$, where x is the sum of 1 and the percentage increase, converted to a decimal, or 1.05. Write an equivalent function by simplifying $s(x)$. Determine Sio's total salary for three years, using the original function and the simplified function.

After completing this section, we will discuss this application further. See page 636.

9.5.1 Dividing with Monomials

We have already divided algebraic expressions by a single-term expression. To divide two monomials, we use the commutative and associative properties to rearrange the factors. We also use the rules for exponents to help simplify the quotient, as in Section 9.1. *We will assume that variables represent non-zero values, so that the expressions are defined.*

To divide a polynomial of more than one term by a monomial, we use the distributive law, as we did with algebraic expressions.

QUOTIENTS WITH A MONOMIAL DIVISOR

To divide a monomial by a monomial, use the commutative and associative properties to rearrange factors. Simplify by using the rules of exponents.

To divide a polynomial with more than one term by a monomial, use the distributive law. Again, simplify by using the rules of exponents.

EXAMPLE 1 Divide. Express your answer with positive exponents.

a. $\dfrac{6x^5y^4z}{-2xy^2z}$ **b.** $\dfrac{4x^2y^3z - 2xy^2z}{2xyz}$ **c.** $\dfrac{2a^2 + 6a^2b - 4b^2}{2a}$

Solution

a. Note that we simplified this type of example in Section 9.1.

$$\frac{6x^5y^4z}{-2xy^2z} = \frac{6}{-2} \cdot \frac{x^5}{x} \cdot \frac{y^4}{y^2} \cdot \frac{z}{z}$$

$$= -3 \cdot x^{5-1} \cdot y^{4-2} \cdot z^{1-1} \qquad \text{Quotient rule}$$

$$= -3x^4y^2z^0 \qquad \text{Remember, } z^0 = 1.$$

$$= -3x^4y^2$$

b. $\dfrac{4x^2y^3z - 2xy^2z}{2xyz} = \dfrac{4x^2y^3z}{2xyz} - \dfrac{2xy^2z}{2xyz}$ Distribute 2xyz.

$$= \frac{4}{2} \cdot \frac{x^2}{x} \cdot \frac{y^3}{y} \cdot \frac{z}{z} - \frac{2}{2} \cdot \frac{x}{x} \cdot \frac{y^2}{y} \cdot \frac{z}{z}$$

$$= 2x^{2-1}y^{3-1}z^{1-1} - 1x^{1-1}y^{2-1}z^{1-1} \qquad \text{Quotient rule}$$

$$= 2x^1y^2z^0 - 1x^0y^1z^0$$

$$= 2xy^2 - y$$

c. $\dfrac{2a^2 + 6a^2b - 4b^2}{2a} = \dfrac{2a^2}{2a} + \dfrac{6a^2b}{2a} - \dfrac{4b^2}{2a}$ Distribute 2a.

$$= 1a^{2-1} + 3a^{2-1}b - 2a^{-1}b^2 \qquad \text{Rules of exponents}$$

$$= a + 3ab - \frac{2b^2}{a} \qquad \text{Simplify.}$$

9.5.1 Checkup

In exercises 1–3, divide. Express your answers with positive exponents.

1. $\dfrac{27a^4b^5c^3}{9a^2b^2c^2}$

2. $\dfrac{18x^2y^2 + 3xy^3}{-3xy}$

3. $\dfrac{12x^3 + 6x^2y - 9xy^2 - 2y^3}{4x^2}$

4. In dividing a polynomial by a monomial, when would you need to use the distributive law?

9.5.2 Dividing Polynomials

To divide a polynomial by a polynomial other than a monomial, we use long division. Polynomial long division follows the same steps as numeric long division. Review the steps of long division, as shown in the following example:

$$\begin{array}{r} 12 \\ 56\overline{)675} \\ -56 \\ \hline 115 \\ -112 \\ \hline 3 \end{array}$$

Divide: $67 \div 56 = 1$. Place 1 in the quotient.
Multiply: $1(56) = 56$.
Subtract: $67 - 56 = 11$. Bring down the 5.
Divide: $115 \div 56 = 2$. Place 2 in the quotient.
Multiply: $2(56) = 112$.
Subtract: $115 - 112 = 3$. The remainder is 3.

The quotient is $12\frac{3}{56}\left(\frac{\text{remainder}}{\text{divisor}}\right)$. We can check division.

$$56 \cdot 12 + 3 = 675$$

divisor · quotient + remainder = dividend

EXAMPLE 2   Divide $8x^2 + 10x - 3$ by $2x + 3$.

Solution

The dividend is $8x^2 + 10x - 3$. The divisor is $2x + 3$.

$$
\begin{array}{r}
4x \\
2x + 3 \overline{) 8x^2 + 10x - 3}
\end{array}
$$

Divide the first term of the dividend by the first term of the divisor. $\frac{8x^2}{2x} = 4x$

$$
\begin{array}{r}
4x \\
2x + 3 \overline{) 8x^2 + 10x - 3} \\
8x^2 + 12x
\end{array}
$$

Multiply the partial quotient

by the divisor. $4x(2x + 3) = 8x^2 + 12x$

$$
\begin{array}{r}
4x \\
2x + 3 \overline{) 8x^2 + 10x - 3} \\
-(8x^2 + 12x) \\
\hline
-2x - 3
\end{array}
$$

Subtract (change signs and add), and bring down the next term. $(8x^2 + 10x) - (8x^2 + 12x) =$

$8x^2 + 10x - 8x^2 - 12x = -2x$

$$
\begin{array}{r}
4x - 1 \\
2x + 3 \overline{) 8x^2 + 10x - 3} \\
-(8x^2 + 12x) \\
\hline
-2x - 3
\end{array}
$$

Divide the first term of the new dividend by the first term of the divisor, $\frac{-2x}{2x} = -1$

$$
\begin{array}{r}
4x - 1 \\
2x + 3 \overline{) 8x^2 + 10x - 3} \\
-(8x^2 + 12x) \\
\hline
-2x - 3 \\
-2x - 3
\end{array}
$$

Multiply the partial quotient

by the divisor. $-1(2x + 3) = -2x - 3$

$$
\begin{array}{r}
4x - 1 \\
2x + 3 \overline{) 8x^2 + 10x - 3} \\
-(8x^2 + 12x) \\
\hline
-2x - 3 \\
-(-2x - 3) \\
\hline
0
\end{array}
$$

Subtract (change signs and add).

$(-2x - 3) - (-2x - 3) = -2x - 3 + 2x + 3 = 0$

Check

X	Y₁	Y₂
0	-1	-1
1	3	3
2	7	7
3	11	11
4	15	15
5	19	19
6	23	23

X=0

The quotient is $4x - 1$.
The remainder is 0.

We can check by multiplying.

$$(2x + 3)(4x - 1) = 8x^2 - 2x + 12x - 3 \quad \text{divisor} \cdot \text{quotient}$$

$$= 8x^2 + 10x - 3 \quad = \text{dividend}$$

$Y1 = \frac{8x^2 + 10x - 3}{2x + 3}$

$Y2 = 4x - 1$

EXAMPLE 3 Divide $\dfrac{4x^3 + 3x + 5}{2x - 3}$.

Solution

You will need to write any missing addends with 0 coefficients in order to keep the columns straight. There is no x^2 term, so we need to add $0x^2$ in order to write in decreasing order.

$$
2x - 3 \overline{) 4x^3 + 0x^2 + 3x + 5}
$$

Write as long division. Write missing addend with 0 coefficient.

$$
\begin{array}{r}
2x^2 \\
2x - 3 \overline{) 4x^3 + 0x^2 + 3x + 5}
\end{array}
$$

Divide the first term of the dividend by the first term of the divisor. $\frac{4x^3}{2x} = 2x^2$

$$
\begin{array}{r}
2x^2 \\
2x - 3 \overline{) 4x^3 + 0x^2 + 3x + 5} \\
4x^3 - 6x^2
\end{array}
$$

Multiply the partial quotient by the divisor.

$2x^2(2x - 3) = 4x^3 - 6x^2$

$$\begin{array}{r} 2x^2 \\ 2x - 3 \overline{\smash{)}\, 4x^3 + 0x^2 + 3x + 5} \\ \underline{-(4x^3 - 6x^2)} \\ 6x^2 + 3x \end{array}$$

Subtract (change signs and add), and bring down the next term. $(4x^3 + 0x^2) - (4x^3 - 6x^2) =$ $4x^3 + 0x^2 - 4x^3 + 6x^2 = 6x^2$

$$\begin{array}{r} 2x^2 + 3x \\ 2x - 3 \overline{\smash{)}\, 4x^3 + 0x^2 + 3x + 5} \\ \underline{-(4x^3 - 6x^2)} \\ 6x^2 + 3x \end{array}$$

Divide the first term of the new dividend by the first term of the divisor. $\frac{6x^2}{2x} = 3x$

$$\begin{array}{r} 2x^2 + 3x \\ 2x - 3 \overline{\smash{)}\, 4x^3 + 0x^2 + 3x + 5} \\ \underline{-(4x^3 - 6x^2)} \\ 6x^2 + 3x \\ 6x^2 - 9x \end{array}$$

Multiply the partial quotient by the divisor. $3x(2x - 3) = 6x^2 - 9x$

$$\begin{array}{r} 2x^2 + 3x \\ 2x - 3 \overline{\smash{)}\, 4x^3 + 0x^2 + 3x + 5} \\ \underline{-(4x^3 - 6x^2)} \\ 6x^2 + 3x \\ \underline{-(6x^2 - 9x)} \\ 12x + 5 \end{array}$$

Subtract (change signs and add), and bring down the next term. $(6x^2 + 3x) - (6x^2 - 9x) = 6x^2 + 3x - 6x^2 + 9x = 12x$

$$\begin{array}{r} 2x^2 + 3x + 6 \\ 2x - 3 \overline{\smash{)}\, 4x^3 + 0x^2 + 3x + 5} \\ \underline{-(4x^3 - 6x^2)} \\ 6x^2 + 3x \\ \underline{-(6x^2 - 9x)} \\ 12x + 5 \end{array}$$

Divide the first term of the new dividend by the first term of the divisor. $\frac{12x}{2x} = 6$

$$\begin{array}{r} 2x^2 + 3x + 6 \\ 2x - 3 \overline{\smash{)}\, 4x^3 + 0x^2 + 3x + 5} \\ \underline{-(4x^3 - 6x^2)} \\ 6x^2 + 3x \\ \underline{-(6x^2 - 9x)} \\ 12x + 5 \\ 12x - 18 \end{array}$$

Multiply the partial quotient by the divisor. $6(2x - 3) = 12x - 18$

$$\begin{array}{r} 2x^2 + 3x + 6 + \dfrac{23}{2x - 3} \\ 2x - 3 \overline{\smash{)}\, 4x^3 + 0x^2 + 3x + 5} \\ \underline{-(4x^3 - 6x^2)} \\ 6x^2 + 3x \\ \underline{-(6x^2 - 9x)} \\ 12x + 5 \\ \underline{-(12x - 18)} \\ 23 \end{array}$$

Subtract (change signs and add). Write the remainder as a fraction in the quotient.

$(12x + 5) - (12x - 18) =$
$12x + 5 - 12x + 18 = 23$

The quotient is $2x^2 + 3x + 6 + \frac{23}{2x - 3}$.

Check: $(2x - 3)(2x^2 + 3x + 6) + 23$

divisor   quotient   remainder

$$= 2x(2x^2 + 3x + 6) - 3(2x^2 + 3x + 6) + 23$$
$$= 4x^3 + 6x^2 + 12x - 6x^2 - 9x - 18 + 23$$
$$= 4x^3 + 3x + 5$$

dividend

EXAMPLE 4   Divide $\dfrac{27x^3 - 1}{3x - 1}$.

Solution

The coefficients of the x^2 and x terms of the dividend are 0.

$$\begin{array}{r} 9x^2 \\ 3x - 1 \overline{)27x^3 + 0x^2 + 0x - 1} \\ -(27x^3 - 9x^2) \\ \hline 9x^2 + 0x \end{array}$$

Divide the first term of the dividend by the first term of the divisor. Multiply the partial quotient by the divisor. Subtract, and bring down the next term.

$$\begin{array}{r} 9x^2 + 3x \\ 3x - 1 \overline{)27x^3 + 0x^2 + 0x - 1} \\ -(27x^3 - 9x^2) \\ \hline 9x^2 + 0x \\ -(9x^2 - 3x) \\ \hline 3x - 1 \end{array}$$

Divide the first term of the dividend by the first term of the divisor. Multiply the partial quotient by the divisor. Subtract, and bring down the next term.

$$\begin{array}{r} 9x^2 + 3x + 1 \\ 3x - 1 \overline{)27x^3 + 0x^2 + 0x - 1} \\ -(27x^3 - 9x^2) \\ \hline 9x^2 + 0x \\ -(9x^2 - 3x) \\ \hline 3x - 1 \\ -(3x - 1) \\ \hline 0 \end{array}$$

Divide the first term of the dividend by the first term of the divisor. Multiply the partial quotient by the divisor. Subtract, and bring down the next term.

The quotient is $9x^2 + 3x + 1$. Check: $(3x - 1)(9x^2 + 3x + 1) = 27x^3 - 1$

9.5.2 Checkup

In exercises 1–3, divide. Check your answer.

1. Divide $(10x^2 + 7x - 12)$ by $(2x + 3)$.
2. $\dfrac{8x^3 + 16x^2 + 27}{2x + 5}$
3. $\dfrac{8x^3 + 27}{2x + 3}$

4. In dividing one polynomial by another, when will there be a remainder? What does the remainder represent?

9.5.3 Modeling the Real World

We've already mentioned that polynomial functions and exponential expressions are both very common in modeling real-world situations. Therefore, performing mathematical operations with these functions and expressions becomes an important part of working with applications of mathematics. Any operation you can do with numbers you can also do with polynomials. The only difference is that polynomials don't reduce to simple terms quite as easily as real numbers do.

EXAMPLE 5   Alfredo Travel Company offers a spring tour of Lutz's Gardens. The regular price is $99.95 per package. A reduction of $1.00 per package purchased is given for groups. The bus expenses are estimated to be $105.00, the driver is paid $300.00, and the tickets to the gardens are $19.95 per person. The bus can carry up to 30 people.

   a. Determine a total cost function, $C(x)$.
   b. Determine a revenue function, $R(x)$.
   c. Determine a profit function, $P(x)$.
   d. Determine an average profit (profit per package) function, $A(x)$.
   e. If 23 members of Springfield Garden Club purchase a tour package, what is the profit per package for Alfredo Travel Company?

Solution

Let x = the number of tour packages purchased

   a. The total cost is $105.00, plus $300.00 (that is, $405.00), plus $19.95 per ticket (number of tour packages purchased).

$$C(x) = 19.95x + 405$$

   b. The reduction in the package is $1.00 times the number of packages purchased, or $1.00x$. The price of the package is $99.95 minus the reduction, or $(99.95 - 1.00x)$. The revenue is the price of the package times the number of packages purchased, or $(99.95 - 1.00x)x$.

$$R(x) = (99.95 - 1.00x)x$$
$$R(x) = 99.95x - x^2$$

   c. The profit is the revenue minus the total cost.

$$P(x) = (99.95x - x^2) - (19.95x + 405)$$
$$P(x) = 99.95x - x^2 - 19.95x - 405$$
$$P(x) = -x^2 + 80x - 405$$

   d. The average profit (profit per package) is determined by dividing the profit by the number of packages purchased.

$$A(x) = \frac{-x^2 + 80x - 405}{x}$$
$$A(x) = -x + 80 - \frac{405}{x}$$

   e. If $x = 23$, find $A(23)$.

$$A(23) = -(23) + 80 - \frac{405}{(23)}$$
$$A(23) \approx 39.39$$

The average profit for Alfredo Travel Company is $39.39 per package if 23 packages are purchased.

APPLICATION

Sio signed a contract for her new job. According to the contract, she is to make $20,000 the first year and receive an increase of 5% per year for the next two years. Sio's total salary for three years is given by the function $s(x) = \frac{20,000(1 - x^3)}{1 - x}$, where x is the sum of 1 and the percentage increase, converted to a decimal, or 1.05. Write an equivalent function by simplifying $s(x)$. Determine Sio's total salary for three years, using the original function and the simplified function.

Discussion

$$s(x) = \frac{20,000(1 - x^3)}{1 - x}$$

$$s(x) = \frac{20,000 - 20,000x^3}{1 - x}$$

Divide.

Write in decreasing order:

$$
\begin{array}{r}
20,000x^2 + 20,000x + 20,000 \\
-x + 1 \overline{)\,-20,000x^3 + 0x^2 + 0x + 20,000} \\
\underline{-(-20,000x^3 + 20,000x^2)} \\
-20,000x^2 + 0x \\
\underline{-(-20,000x^2 + 20,000x)} \\
-20,000x + 20,000 \\
\underline{-(-20,000x + 20,000)} \\
0
\end{array}
$$

Therefore,

$$s(x) = 20,000x^2 + 20,000x + 20,000$$

To determine Sio's total salary for three years, we need to evaluate each function with $x = 1.05$.

original function

$$s(x) = \frac{20,000(1 - x^3)}{1 - x}$$

$$s(1.05) = \frac{20,000(1 - 1.05^3)}{1 - 1.05}$$

$$s(1.05) = 63,050$$

simplified function

$$s(x) = 20,000x^2 + 20,000x + 20,000$$

$$s(1.05) = 20,000(1.05)^2 + 20,000(1.05) + 20,000$$

$$s(1.05) = 63,050$$

Sio will receive $63,050 over the three-year period.

9.5.3 Checkup

1. A hotel offered a package deal for a weekend stay that included tickets to dinner and a touring production of a Broadway play. The price of the package was $259.00 per person. A reduction of $5.00 per person was offered for group purchases. The hotel incurred a fixed cost of $200 paid to a travel agent to arrange a group excursion and spent $25.00 per person for the dinner and $50.00 per person for the tickets to the play. Assume that x people in a group purchase the package.

a. Determine a total cost function, $C(x)$, for the group's excursion.

b. Determine a revenue function, $R(x)$, for the group's excursion.

c. Determine a profit function, $P(x)$, for the group's excursion.

d. Determine an average profit function, $A(x)$, for the group's excursion.

e. If 10 people in a group purchased the package, what was the average profit per package for the excursion?

2. At his birth, Benjamin received a gift of $5000. For the next three years, he received an additional gift on his birthday that was $5000 increased by 10% per year. The total amount of money Benjamin received from his benefactor is given by the function $g(y) = \frac{5000(1 - y^4)}{1 - y}$, where y is the sum of the percentage increase (converted to decimal form) and 1. Write a function equivalent to $g(y)$ by simplifying it. Using the original function and the simplified function, determine the total amount Benjamin received.

9.5 Exercises

In exercises 1–40, divide as indicated, and express your results with positive exponents only.

1. $\dfrac{20a^4b^2c}{-5a^2bc}$

2. $\dfrac{18x^7y^2z^2}{-9x^3yz^2}$

3. $\dfrac{6x^4y^6z^2 + 18x^2y^3z}{3x^2y^3z}$

4. $\dfrac{15x^5y^5z^3 + 10x^3y^5z^2}{5x^3y^4z^2}$

5. $\dfrac{3p^3 - 9p^2q + 6pq^2 - 12q^3}{3pq}$

6. $\dfrac{7c^3 - 21c^2d + 14cd^2 - 35d^3}{7cd}$

7. $\dfrac{6x^3 + 12x^2 - 18x}{3x}$

8. $\dfrac{21a^4 - 14a^2 + 42}{7a^2}$

9. $\dfrac{4x^2 - 2x + 12}{8x}$

10. $\dfrac{25z^4 - 10z^3 + 45z^2 + 15z}{5z}$

11. $\dfrac{3.72x^4 - 6.96x^2 + 1.08}{1.2x^2}$

12. $\dfrac{7.578m - 10.525m^2 + 12.63m^3}{-4.21m^3}$

13. $\dfrac{9x^4 + 6x^3y - 18x^2y^2 - 24xy^3 + 72y^4}{-3xy}$

14. $\dfrac{12a^3 - 30a^2b - 24ab^2 + 6b^3}{6a^2b}$

15. Divide $15x^2 - 8x - 12$ by $3x + 2$.

16. Divide $12x^2 - 19x - 21$ by $4x + 3$.

17. $(2x^2 + 9x - 35) \div (x + 7)$

18. $(3x^2 + x - 44) \div (x + 4)$

19. $\dfrac{5x^2 - 24x - 36}{x - 6}$

20. $\dfrac{6x^2 - 25x - 25}{x - 5}$

21. $(3y^2 + 19y - 20) \div (y + 8)$

22. $(5z^2 + 3z - 10) \div (z + 2)$

23. $\dfrac{6a^2 - 5a - 30}{2a - 5}$

24. $\dfrac{12b^2 - 13b - 15}{3b - 7}$

25. $(5x^3 - 14x^2 - 3x + 2) \div (5x + 2)$ ✓

26. $(12x^3 - 11x^2 + 17x - 10) \div (3x - 2)$

27. $\dfrac{8x^3 + 8x - 5}{2x - 1}$

28. $\dfrac{9x^3 + 17x - 14}{3x - 2}$

29. $(9a^2 - 49) \div (3a + 7)$

30. $(49b^2 - 121) \div (7b + 11)$

31. $\dfrac{5x^2 + 30x + 9}{5x + 3}$

32. $\dfrac{36y^2 + 60y + 25}{6y + 5}$

33. $(16z^2 - 88z + 121) \div (4z - 11)$

34. $(64p^2 - 112p + 49) \div (8p - 7)$

35. $\dfrac{x^3 + 27}{x + 3}$

36. $\dfrac{y^3 + 343}{y + 7}$

37. $\dfrac{a^3 - 125}{a - 5}$

38. $\dfrac{b^3 - 27}{b - 3}$

39. Divide $64x^3 + 27$ by $4x + 3$.

40. Divide $125z^3 - 8$ by $5z - 2$.

41. Jennifer established a business that buys and resells used graphing calculators. She advertised in the local paper that she was in the market to buy and sell the calculators. She spent $45 for her advertisement. She bought used calculators for $35 each and resold them for $50 each.

a. Determine a total cost function for Jennifer's operation if she buys x calculators.

b. Determine a revenue function if she sells all the calculators she buys.

c. Determine a profit function.

d. Determine an average profit (per calculator) function.

e. What will be Jennifer's average profit if she sells 30 calculators?

42. Karla formed a company to design Web pages for clients. She spent $1000 to advertise in business journals that she provided such a service. She spent about $75 in labor and materials for each Web page she designed. She charged her clients $350 for each Web page she designed for them.

a. Determine a total cost function for Karla's operation if she designs x pages.

b. Determine a revenue function.

c. Determine a profit function.

d. Determine an average profit (per Web page) function.

e. What will be Karla's average profit if she designs 10 Web pages?

43. Scott enrolled in a college program that would give him a bachelor's degree and a master's degree after five years of combined study. To help him through the program, his parents gave him $6000 the first year and increased that amount by 10% each subsequent year. A function for the total amount of money his parents gave Scott is given by $m(t) = \frac{6000(1 - t^5)}{1 - t}$, where t is the sum of 1 and the percentage increase (converted to decimal form) each year. Write a simplified expression for this function by performing the division. Determine the total amount of money Scott received by first evaluating the original function and then evaluating the simplified function. What would be the total amount if Scott's parents had increased each year's payment by 15%?

44. To encourage her son to do well in high school, Sharon promised to give him $10 for his first A, and double the amount each time he earned another A. She was surprised to see that her son earned six A's during the grading period. She paid him $d(t) = \frac{10(1 - x^6)}{1 - x}$ dollars, where x is the factor for increasing the amount. Write a simplified ex-

pression for this function by performing the division. Evaluate the original function and the simplified function, given that $x = 2$, since Sharon doubled the amount each time. What would the total amount have been if she had tripled the amount each time?

45. To encourage a client to continue to rent equipment from his company, Randy offers a percentage discount on the charge for each additional week of rental. A client pays $250 to rent a piece of equipment for the first week, and the amount is reduced 10% each subsequent week of rental. If the client rents the equipment for five weeks, a function for the total rental cost is $c(n) = \frac{250(1 - n^5)}{1 - n}$, where n is the difference of 1 and the percent reduction (converted to a decimal). Write a simplified expression for this function by performing the division. Evaluate the original function and the simplified function for the 10% reduction.

46. Nelsy operates a collector basket business out of her home. Recently, she discovered that over the previous six weeks her sales declined steadily at a rate of 8% per week. If Nelsy's first week's sales were $550, a function for the total amount of her sales over the six-week period is $s(w) = \frac{550(1 - w^6)}{1 - w}$, where w is the difference of 1 and the percentage decline (converted to decimal form). Write a simplified expression for this function by performing the division. Evaluate the original function and the simplified function for the 8% decline.

9.5 Calculator Exercises

Using the table feature of your calculator, you easily can examine functions that involve polynomial division to see what effect varying the values of the variable will have on the end result. Following are two exercises that compare the pressure exerted on a floor by a man or a woman, compared with that exerted by an elephant. You will be surprised at the results!

1. Pressure is measured in units of force per unit of area. For example, if an elephant weighs 7 tons and each of its four feet is considered to be a circle with a radius of 8 inches, then the pressure the elephant exerts on the floor would be 14,000 pounds divided by four times the area of one of the elephant's foot-pads, $4(\pi 8^2)$, or approximately 17.5 pounds per square inch. By contrast, a woman weighing 110 pounds and wearing high heels would exert a pressure of 110 divided by two times the area of her heel. (We make the simplifying assumption that her weight is distributed on her heels and not on the soles of her shoes.) Write a polynomial division expression for the pressure per square inch exerted by the woman's weight. Set up a table of values to evaluate this expression for a heel radius that varies from 0.25 to 1 inch, in increments of 0.25 inch. For what radius of the heel is the woman's pressure on the floor approximately the same as the elephant's?

2. Assuming that a man's weight is fully distributed on his circular heels, the pressure exerted on the floor by a man weighing 175 pounds would be equal to his weight divided by the area of both of his heels. Write a polynomial division expression for the pressure per square inch exerted by the man's weight. Construct a table of values for a heel radius that varies from 1 to 1.5 inches, in increments of 0.05 inch. For what radius of the heel is the man's pressure on the floor approximately the same as the pressure of an elephant, as described in exercise 1?

9.5 Writing Exercises

Refer to the writing exercise in Section 9.4. There you were asked to match various forms of polynomial multiplication operations to the resulting polynomials. For example, if you determined that $(A - B)(A + B) = A^2 - B^2$, then you should see that $\frac{A^2 - B^2}{A + B} = A - B$. In this way, division can be related to multiplication.

Write a short summary of how you can use the relationships that you determined from the writing exercise in Section 9.4 to help you perform polynomial divisions. When will these

relationships not be useful? See if you can use the relationships to perform the following divisions more easily.

1. $(a^2 + 2ab + b^2) \div (a + b)$ 2. $(x^3 - y^3) \div (x - y)$
3. $(x^3 + 1) \div (x^2 - x + 1)$ 4. $(a^2 - b^2) \div (a + b)$
5. $(x^2 - 16x + 64) \div (x - 8)$ 6. $(z^3 + 125) \div (z + 5)$
7. $(b^3 - 1) \div (b - 1)$ 8. $(x^6 - y^6) \div (x^3 - y^3)$
9. $(4p^2 - 9q^2) \div (2p + 3q)$ 10. $(8x^3 - 729) \div (2x - 9)$

CHAPTER 9 SUMMARY

After completing this chapter, you should be able to define in your own words the following key terms.

9.4
special products
product of the sum and difference of
    the same two terms
square of a binomial

Reflections

1. Explain the difference between the expanded form and the exponential form of an algebraic expression.
2. If an algebraic expression has an exponential expression with a positive exponent greater than 1, how can you evaluate the exponential expression?
3. If an algebraic expression has an exponential expression with an exponent equal to 1, how can you evaluate the exponential expression?
4. If an algebraic expression has an exponential expression with an exponent equal to 0, how can you evaluate the exponential expression?
5. If an algebraic expression has an exponential expression with a negative exponent, how can you evaluate the exponential expression?
6. State the product rule for exponents and the power-to-a-power rule for exponents. How do they differ from one another?
7. In adding two polynomials, what do you do to simplify the result?
8. How does subtracting two polynomials differ from adding two polynomials?
9. What properties and laws of the real-number system are important to the process of multiplying two polynomials?
10. How do you divide a polynomial expression by a monomial expression?
11. How do you divide a polynomial expression by another polynomial expression when neither is a monomial?

CHAPTER 9 SECTION-BY-SECTION REVIEW

9.1

Recall	Exercises
Integer exponents $a^n = a \cdot a \cdot a \cdot \ldots \cdot a$ a factors $a^1 = a$ $a^0 = 1$, where $a \neq 0$	Simplify. Assume that all expressions are defined. $t^4 = t \cdot t \cdot t \cdot t$ $t^1 = t$ $t^0 = 1$

(Continued on page 640)

Recall	Exercises
Rules for exponents $a^m \cdot a^n = a^{m+n}$ Product rule $\dfrac{a^m}{a^n} = a^{m-n}, a \neq 0$ Quotient rule $(a^m)^n = a^{mn}$ Power-to-a-power rule $(ab)^m = a^m b^m$ Product-to-a-power rule $\left(\dfrac{a}{b}\right)^m = \dfrac{a^m}{b^m}, b \neq 0$ Quotient-to-a-power rule	Simplify. Assume that all expressions are defined. $t^3 t^6 = t^{3+6} = t^9$ $\dfrac{t^6}{t^3} = t^{6-3} = t^3$ $(t^3)^6 = t^{3\cdot6} = t^{18}$ $(st)^3 = s^3 t^3$ $\left(\dfrac{s}{t}\right)^2 = \dfrac{s^2}{t^2}$

Write in expanded form.

1. $-5c^2$ **2.** $(-5c)^2$ **3.** $4(x+y)^2 z^0$ **4.** $\left(\dfrac{2}{3}\right)^4 x^2$

Simplify exercises 5–23.

5. $a^2 \cdot a^7$ **6.** $(p+q)^3(p+q)$ **7.** $\dfrac{2}{3} b^3 \cdot b \cdot b^4$ **8.** $\dfrac{t^{12}}{t^9}$

9. $\dfrac{24a^4}{3a^2}$ **10.** $\dfrac{(x+3y)^5}{(x+3y)^4}$ **11.** $(cd)^{22}$ **12.** $(-2a)^4$

13. $(-2a)^5$ **14.** $(3x^3)^2$ **15.** $[(a+b)^2]^5$ **16.** $(a^3)^0$

17. $\left(\dfrac{-2x}{3z}\right)^3$ **18.** $\left(\dfrac{-4d}{e}\right)^3$ **19.** $-2xy^2 \cdot 4x^3y^5$ **20.** $\dfrac{72x^4y^5z^3}{-24x^2y^4z^3}$

21. $\dfrac{(x^2y^3)^4}{x^3y^5}$ **22.** $\left(\dfrac{2p^2q}{pq^2}\right)^3$ **23.** $\left[\left(\dfrac{m}{4n}\right)^2\right]^2$

24. A square is decreased in size by reducing each side to one-fourth of its original length. Write expressions for the areas of the original square and the reduced square. Compare the two areas.

25. A circular garden is increased in size by doubling its radius. Write expressions for the area of the original garden and the area of the enlarged garden. Compare the two areas. What is the area of the enlarged garden if the original garden had a radius of 6 feet?

9.2

Recall	Examples
Integer exponents $a^{-n} = \dfrac{1}{a^n}$, where $a \neq 0$ $\dfrac{1}{a^{-n}} = a^n$, where $a \neq 0$	Simplify. Assume that all expressions are defined. $t^{-3} = \dfrac{1}{t^3}$ $\dfrac{1}{t^{-2}} = t^2$ $\dfrac{t^{-3}}{t^{-4}} = \dfrac{t^4}{t^3} = t$
Rules for negative exponents are the same as rules for positive exponents.	Simplify. Write with positive exponents. $t^{-3}t^5 = t^{-3+5} = t^2$ $\dfrac{t^6}{t^{-3}} = t^{6-(-3)} = t^9$ $(t^{-3})^6 = t^{-3\cdot6} = t^{-18} = \dfrac{1}{t^{18}}$ $(st)^{-3} = s^{-3}t^{-3} = \dfrac{1}{s^3 t^3}$
Quotient-to-a-negative-power rule $\left(\dfrac{a}{b}\right)^{-m} = \left(\dfrac{b}{a}\right)^m = \dfrac{b^m}{a^m}$	$\left(\dfrac{s}{t}\right)^{-2} = \dfrac{s^{-2}}{t^{-2}} = \dfrac{t^2}{s^2}$

Simplify exercises 26–35. Write with positive exponents.

26. $\dfrac{3^{-2}h^{-4}}{k^{-2}}$ **27.** c^8d^{-5} **28.** $(-b^2)(-3b^{-5})$ **29.** $\dfrac{5z^{-3}}{15z^{-7}}$

30. $(5a^{-4}b^2)^{-3}$ **31.** $\left(\dfrac{2}{3}p^3q^{-2}\right)\left(\dfrac{3}{5}p^{-5}q^{-3}\right)$ **32.** $\dfrac{144x^{-4}y^{-3}}{12x^2y^{-4}}$ **33.** $\left(\dfrac{27a^5b^{-3}}{9a^2b^7}\right)^2$

34. $\left(\dfrac{a^{-3}}{b^{-4}}\right)^{-2}$ **35.** $\left(\dfrac{2a^{-2}b}{24a^5b^{-2}}\right)^{-2}$

36. The star Pollux is 3.37×10^1 light years away from the Earth. Given that a light year is about 5.9×10^{12} miles, approximate the distance of Pollux from Earth.

9.3

Recall	Examples
Addition of polynomials • Use the distributive law to remove parentheses. • Combine like terms.	Add. $(x^3 + 2x^2 - x + 4) + (3x^2 - 6x + 2)$ $= x^3 + 2x^2 - x + 4 + 3x^2 - 6x + 2$ $= x^3 + 5x^2 - 7x + 6$
Subtraction of polynomials • Use the distributive law to remove parentheses by taking the opposite of the subtrahend. • Combine like terms.	Subtract. $(x^3 + 2x^2 - x + 4) - (3x^2 - 6x + 2)$ $= x^3 + 2x^2 - x + 4 - 3x^2 + 6x - 2$ $= x^3 - x^2 + 5x + 2$

Add the polynomials.

37. Add $(5x^4 + 3x^3 + 6x - 3)$ and $(4x^3 + 5x^2 + 7)$.   **38.** $(3.57z^3 - 2.08z^2 + 8.77z - 1.99) + (4.73 - 2.98z + 5.64z^2)$

In exercises 39–41, subtract the polynomials as indicated.

39. Subtract $(a^4 + 2a^3 + 3a^2 + 4a + 5)$ from $(5a^4 + a^3 + a^2 + a + 1)$.

40. $(65z^4 + 27z^2 + 36)$ decreased by $(16z^3 + 8z + 12)$

41. $\left(\dfrac{5}{8}b^4 + \dfrac{7}{8}b^3 - \dfrac{3}{4}b^2 + \dfrac{1}{2}b - \dfrac{1}{4}\right)$ minus $\left(\dfrac{1}{2}b^4 + \dfrac{3}{8}b^2 + \dfrac{1}{8}b\right)$

42. The cost of producing a certain item consists of a fixed cost of $10.00 to set up the run and $3.50 per item for labor and materials. The items sell for $10.00 each.
   a. Write the total cost function for producing x items, $C(x)$.
   b. Write a revenue function for selling x items, $R(x)$.
   c. Write the profit function, $P(x)$.
   d. What is the profit if 10 items are produced and sold? What if 25 items are produced and sold?

43. Danielle produces prints of her watercolor artwork and frames the prints for sale. The cost of printing and framing x copies of a piece of art is given by the function $f(x) = 150 + 25x$. Danielle's cost of packaging and mailing x copies is given by the function $m(x) = 25 + 4.50x$. Write a total cost function for framing and mailing x copies. Danielle charges $125 for each framed print mailed to a customer. Write a revenue function, $r(x)$. Write a profit function for x prints framed and mailed to customers.

9.4

Recall	Examples
Multiply polynomials • To multiply a monomial by a monomial, use the rules of exponents. • To multiply a polynomial with more than one term by a monomial, use the distributive law and the rules for exponents. • To multiply two binomials, use the FOIL method. • To multiply two polynomials with more than two terms, use the distributive law and the rules for exponents.	Multiply. $2x^3 \cdot 5xy^2 = 10x^4y^2$ $2x^3(x^2 + 3x - 5) = 2x^5 + 6x^4 - 10x^3$ $(x + 3)(2x - 1) = 2x^2 - x + 6x - 3$ $\qquad = 2x^2 + 5x - 3$ $(x - 2)(x^2 + 3x - 1) = x(x^2 + 3x - 1) - 2(x^2 + 3x - 1)$ $\qquad = x^3 + 3x^2 - x - 2x^2 - 6x + 2$ $\qquad = x^3 + x^2 - 7x + 2$

(Continued on page 642)

Recall	Examples
Product of the sum and difference of the same two terms $(a + b)(a - b) = a^2 - b^2$	Multiply. $\begin{aligned}(x + 3)(x - 3) &= x^2 - 3^2 \\ &= x^2 - 9\end{aligned}$
Square of a binomial $(a + b) = a^2 + 2ab + b^2$ $(a - b) = a^2 - 2ab + b^2$	Multiply. $\begin{aligned}(x + 2) &= x^2 + 2(x)(2) + 2^2 \\ &= x^2 + 4x + 4\end{aligned}$ $\begin{aligned}(x - 3) &= x^2 - 2(x)(3) + 3^2 \\ &= x^2 - 6x + 9\end{aligned}$

In exercises 44–62, multiply as indicated.

44. $-3a^3b(5ab^3)$ **45.** $(-6.9x^3z^4)(3.4xz^3)$ **46.** $6x^3(3x^2 + 2x - 7)$

47. $-7a(3a^6 + a^4 - 2a^2)$ **48.** $4a^2(2a - 3b + c)$ **49.** $(p + 6)(p - 9)$

50. $(5x - 2)(x + 11)$ **51.** $(y - 1.8)(y + 3.4)$ **52.** $(2x + 1)(3x^2 + 5x - 4)$

53. $(a - 3)(a^2 + 3a + 9)$ **54.** $(x^2 + 2x + 3)(x^2 - x + 5)$ **55.** $(z^2 + 2z - 3)^2$

56. $(b - 4)^3$ **57.** $(2x - 5)(2x + 5)$ **58.** $\left(\frac{4}{5}x - \frac{1}{2}\right)\left(\frac{4}{5}x + \frac{1}{2}\right)$

59. $(z^2 - 10)(z^2 + 10)$ **60.** $(y + 9)^2$

62. $(x^3 + 3)^2$ **61.** $(3x - 5)^2$

63. The length of a box is 3 inches more than the height. The width of the box is 3 inches less than the height.
   a. Write expressions for the length, width, and height of the box.
   b. Write an expression for the volume of the box.
   c. Write an expression for the surface area of the box, including the top.

64. Andre plans to enlarge the deck on his house. The current deck is three times as long as it is wide. Andre plans to double the width and add 9 feet to the length.
   a. Write a polynomial for the current area of the deck.
   b. Write a polynomial for the area of the planned enlarged deck.
   c. What is the difference of the planned area and the current area?

9.5

Recall	Examples
Divide polynomials • To divide a monomial by a monomial, use the rules for exponents. • To divide a polynomial with more than one term by a monomial, use the distributive law and the rules for exponents. • To divide two polynomials with more than one term, use the steps of long division.	Divide. $\frac{6x^2}{3x} = \frac{6}{3} \cdot \frac{x^2}{x} = 2x$ $\frac{6x^2 + 3x - 2}{3x} = \frac{6x^2}{3x} + \frac{3x}{3x} - \frac{2}{3x}$ $= 2x + 1 - \frac{2}{3x}$ $2x + 1 + \frac{x - 4}{2x + 1}$ $2x + 1\overline{)4x^2 + 5x - 3}$ $\underline{-(4x^2 + 2x)}$ $3x - 3$ $\underline{-(2x + 1)}$ $x - 4$

In exercises 65–73, divide as indicated and express the results with positive exponents only.

65. $\dfrac{144x^6y^3z^4}{2x^5y^2z^6}$ **66.** $\dfrac{6ab^2 + 18a^2b}{3a^2b^2}$ **67.** $\dfrac{15b^4 - 10b^3 - 25b^2 + 5b}{-5b}$ **68.** Divide $15x^2 + 19x - 56$ by $5x - 7$.

69. $(8x^2 + 2x - 19) \div (2x + 3)$ **70.** $\dfrac{x^3 - 44x + 35}{x + 7}$ **71.** $\dfrac{z^3 - 8}{z - 2}$

72. $(16a^2 - 25) \div (4a + 5)$ **73.** $(4x^2 + 36x + 81) \div (2x + 9)$

74. An owner of a produce stand has fixed costs of $35.00 per day to operate the stand. During October, he sells only pumpkins. He purchases the pumpkins from local farmers at a cost of $1.25 each. In turn, he sells the pumpkins for $3.75 each.
- **a.** Determine a total cost function for the daily operation of the stand if the owner buys x pumpkins to sell.
- **b.** Determine a revenue function for selling the x pumpkins.
- **c.** Determine a daily profit function.
- **d.** Determine an average daily profit (per pumpkin) function.
- **e.** What will be the average profit if the owner sells 25 pumpkins in a day?

75. The charge to rent a piece of machinery was $2000 for the first year. Every year thereafter, the charge was reduced by 15%. The total charge for renting the machinery for four years is given by the function $c(t) = \dfrac{2000(1 - t^4)}{1 - t}$, where t is equal to 1 minus the percent reduction. Determine a simplified function by performing the division. Evaluate the original function and the simplified function to determine the total charge for four years.

CHAPTER 9 CHAPTER REVIEW

Simplify. Write with only positive exponents.

1. $m^{-7}n^5$

2. $(3a^{-5}b^4)^{-2}$

3. $\left(\dfrac{21x^4y^{-3}}{7x^3y^2}\right)^2$

4. $\dfrac{5^{-3}t^3}{4^{-3}s^4}$

5. $\dfrac{128a^{-3}b^{-2}}{32a^2b^{-5}}$

6. $\left(\dfrac{2}{5}x^2y^{-3}\right)\left(\dfrac{5}{7}x^{-5}y^{-1}\right)$

7. $\left(\dfrac{15h^2k^5}{3h^{-1}k^7}\right)^{-2}$

8. $s^2 \cdot s^8$

9. $(m + n)^5(m + n)$

10. $(3y^2)^3$

11. $\dfrac{c^9}{c^6}$

12. $(x^0)^8$

13. $-5a^2b^3 \cdot 3ab^4$

14. $\dfrac{1}{2}z^4 \cdot z \cdot z^2$

15. $(b^5)^7$

16. $(-3d)^6$

17. $(-3d)^3$

18. $(x^4y^5z)^4$

19. $\dfrac{25b^5}{5b^3}$

20. $\dfrac{(a + 2b)^9}{(a + 2b)^2}$

21. $\dfrac{64p^5q^7}{-4p^2q}$

22. $[(4x)^2]^3$

23. $\left[\left(\dfrac{c}{2d}\right)^3\right]^2$

24. $\left(\dfrac{-3a^2}{4b}\right)^3$

25. $\dfrac{(3a^3)^3}{54a^7}$

26. $\left(\dfrac{21a^6b^2}{7a^3b^6}\right)^2$

Add the polynomials as indicated.

27. Add $(2y^2 + 4y - 3)$ and $(8y^2 - 12y + 15)$.

28. $\left(\dfrac{3}{8}a^3 + \dfrac{3}{4}a^2 - \dfrac{5}{8}a + \dfrac{1}{4}\right) + \left(\dfrac{3}{4}a^3 + \dfrac{7}{16}a - \dfrac{5}{8}a^2 + \dfrac{11}{16}\right)$

29. Add $(4.9z^3 - 6.82z^2 + 12z - 11.07)$ and $(4.6 - 1.83z + 4.9z^2)$.

Subtract the polynomials as indicated.

30. $(2x^3 + 6x^2 - 9x + 13) - (4x^3 + 17x^2 - x + 6)$

31. Subtract $(5a^4 + 4a^3 + 3a^2 + 2a + 1)$ from $(6a^4 + 3a^2 + 4a + 5)$.

32. $(117z^4 + 43z^2 + 88)$ decreased by $(18z^3 + 50z + 32)$

Multiply.

33. $11x^5y^6(13x^2y^6)$

34. $-6x(2x^4 - 4x^3 + 6x^2 + 8x - 10)$

35. $9a(4a^5 + 2a^3 - 3a)$

36. $(m - 11)(m + 11)$

37. $(z - 8)^2$

38. $(4a - 7)(a + 13)$

39. $(13 - x)(13 + x)$

40. $(b^3 + 4)^2$

41. $(t + 2)^3$

42. $(7x - 2)(x^2 + 4x - 3)$ **43.** $(p + q + r)^2$ **44.** $(b - 4)(b^2 + 4b + 16)$

45. $(y + 5)(y - 9)$ **46.** $(x^2 + 3x + 1)(x^2 - x + 4)$

Divide. Express your results with positive exponents.

47. $\dfrac{81x^5y^7z}{27x^2y^6z^2}$ **48.** $\dfrac{14b^4 - 21b^3 - 35b^2 + 28b}{-7b}$ **49.** $\dfrac{16cd^2 + 8c^2d}{4c^2d^2}$

50. $\dfrac{24x^2 - 37x - 5}{3x - 5}$ **51.** $(2x^2 - 13x - 50) \div (2x + 5)$ **52.** $\dfrac{27a^3 - 8}{3a - 2}$

53. $(25x^3 + 39x - 8) \div (5x - 1)$

Write exercises 54 and 55 in expanded form.

54. $-3d^4$ **55.** $(-3d)^4$

56. Shondra opened a day-care center. Her daily cost to operate the center was $35.00. She spent $2.75 per child each day for supplies and snacks. She charged her clients $10.00 per day for each child left with the center.
  a. Write a total cost function for caring for x children on a given day.
  b. Write a revenue function if x children are cared for on a given day.
  c. Write a profit function for the day.
  d. Write the average profit function per child.
  e. What will be the average profit per child if the center has 10 children being cared for on a given day?

57. Reggie tutored math students. To encourage them to hire him in groups, he offered a discount. The price for tutoring one student was $25.00 per hour. If a second student attended the tutoring session, the price was reduced 5%. Each subsequent student also received the 5% discount. Reggie left it to the students to figure out how they would divide the cost among themselves. The total amount of money Reggie received for each hour of tutoring five students is given by the function $m(s) = \frac{25(1 - s^5)}{1 - s}$, where s is 1 minus the percentage discount (in decimal form). Write a simplified expression for $m(s)$ by performing the division. Evaluate the original function and the simplified function, given that the discount is 5%.

58. Stars that are a distance of more than 300 light-years away from Earth have uncertainty in the measure of their distances. Earlier in the chapter, it was said that the star Deneb would be the brightest star in the sky if all stars were the same distance from the Earth. The accepted distance of Deneb from Earth is 3,227.7 light-years. With the uncertainty taken into account, this distance may really be only about 1,600 light-years from Earth. Given that a light-year is about 5.9×10^{12} miles, approximate the distance of 1,600 light-years in miles.

59. The length of a rectangular box is 3 inches more than twice the height of the box. The width of the box is equal to the height.
  a. Write expressions for the length, width, and height of the box.
  b. Write an expression for the volume of the box.
  c. Write an expression for the surface area of the box.
  d. What is the volume of the box if its height is 8 inches?
  e. What is the surface area of the box if its height is 8 inches?

60. A triangle has a base that is 5 cm more than twice its height.
  a. Write a polynomial that represents the area of the triangle if its height is x cm.
  b. What is the area of the triangle if its height is 12 cm?

61. A square patio measures x feet on a side. It will be enlarged to a width that is double the length of a current side and a length that is 5 feet more than the new width.
  a. Write a polynomial for the current area of the patio.
  b. Write a polynomial for the enlarged area of the patio.
  c. What is the difference of the enlarged area and the original area?

CHAPTER 9 TEST

A^+ ### TEST-TAKING TIPS

When studying for a test, you should try to improve your notes. Read your notes and textbook with pencil in hand. Identify and label what you read according to categories. This chapter had many different methods for manipulating polynomials. If you organize your notes by the various methods and work examples for each method, you will more likely recall these methods during a test. Try to think of various kinds of exercises to which you would apply each method. Compare the methods to see what the differences are. List key points that dictate the method to use. If something confuses you, place a big question mark next to it, and try to get help with the question before taking the test. Then go back and read the confusing material again to see if the help you got reduced your confusion.

Simplify and write with positive exponents.

1. $(2x - 1)(2x - 1)^8$

2. $3x^2y^0z^{-3}$

3. $m^{-9}n^4$

4. $(-2x^2y^{-1})^{-3}$

5. $\dfrac{a^{12}}{a^5}$

6. $\left[\dfrac{(2p)^3}{3q}\right]^{-2}$

7. $\dfrac{24a^2b^{-3}c}{3ab^2c^{-2}}$

8. $\dfrac{(a^2b^4)^3}{a^2b^5c}$

9. Add $(3y^2 + 16 - 7y + 5y^3)$ and $(7y + 6y^2 + 4y^4 - 11)$.

10. Subtract $(2x^4 - 4x^2 - 2x - 9)$ from $(7x^5 + 23x^3 + 17x^2 - 39)$.

Multiply.

11. $(-2p^3q^{-5}r^2)(5.7p^6q^7r)$

12. $-4t(2t^3 - 3t^2 - 8t + 6)$

13. $(9 - 5d)(9 + 5d)$

14. $(3x + 4)(5x - 7)$

15. $(4z - 3)(2z^2 - z + 5)$

16. $(x + 3)^2$

In exercises 17–18, divide. Express your results with positive exponents.

17. $\dfrac{15x^6 + 25x^5 - 5x^3}{5x^2}$

18. $\dfrac{15x^2 + x - 28}{3x - 4}$

19. The width of a box is 4 inches more than the height. The length is twice the height.
  a. Write a polynomial expression for the volume of the box.
  b. Write a polynomial expression for the surface area of the box, including its top.
  c. What are the volume and surface area if the box has a height of 5 inches?

20. Dallie set up a small business providing fruit baskets to customers. The weekly cost for her shop was $235.00. She spent $5.75 per basket for the basket, fruit, and decorations. She charged her customers $25.00 per basket.

  a. Write a total cost function if Dallie produced x baskets per week.
  b. Write a revenue function if she sold x baskets per week.
  c. Write a profit function for the week.
  d. Write the average profit function per basket.
  e. What will be the average profit per basket if Dallie sold 45 baskets in one week?

21. Explain the difference between $x^5 \cdot x^8$ and $(x^5)^8$. Simplify each expression.

Project

Part I

How would you simplify $(x + y)^3$? One way would be to multiply the binomial $(x + y)$ by itself three times. That would be very time consuming and prone to error. Surely, there must be an easier way! You will be happy to know that there is. The famous mathematician Blaise Pascal employed such a method in 1653, called his arithmetical triangle. The first account of his method was printed in 1665. To employ the method, you must construct a triangle as follows:

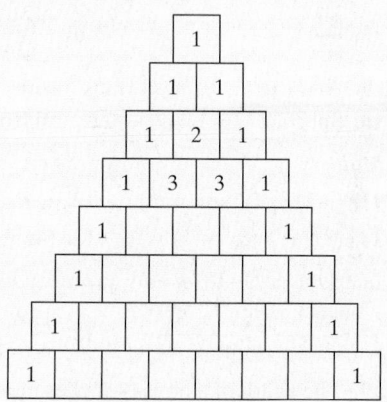

The entry in a particular cell of the triangle is formed by adding the two numbers directly above the cell. Then the numbers along a row are used to determine the coefficients of each term in the simplification of $(x + y)^n$ (that is, any power of a binomial). The top row is the coefficient of $(x + y)^0$, the row below that yields the coefficients for $(x + y)^1$, and so forth. Thus,

$(x + y)^0 = 1$ since the first row contains $\{1\}$.
$(x + y)^1 = x + y$, since the second row contains $\{1, 1\}$
$(x + y)^2 = x^2 + 2xy + y^2$, since the third row contains $\{1, 2, 1\}$
$(x + y)^3 = x^3 + 3x^2y + 3xy^2 + y^3$, since the fourth row contains $\{1, 3, 3, 1\}$

Note that the exponents of x and y in each term sum to the power that you are expanding. So always start with x to that power as the first term, and then reduce the exponent of x by one and increase the exponent of y by one for each subsequent term of the expansion.

One thing you should know for sure: $(x + y)^n \neq x^n + y^n$!

1. Write the following power expansions, using the remaining rows of the triangle.

   a. $(x + y)^4$ **b.** $(x + y)^5$ **c.** $(x + y)^6$ **d.** $(x + y)^7$

2. Now add two more rows to the triangle. Then use these to write the power expansion of $(x + y)^9$.

Pascal's triangle will also work for other binomial power expansions. For example, if we replace y by -1,

$$(x - 1)^4 = x^4 + 4x^3(-1)^1 + 6x^2(-1)^2 + 4x^1(-1)^3 + (-1)^4$$
$$= x^4 - 4x^3 + 6x^2 - 4x + 1$$

3. See if you can use Pascal's triangle to determine the following binomial powers:

   a. $(x + 1)^4$ **b.** $(x - 3)^5$ **c.** $(2x + 1)^3$ **d.** $(x - 3y)^4$ **e.** $(2x - 5y)^6$

Part II

Many Internet sites present interesting facts about Pascal's triangle. For example, if you were to take all the cells containing odd numbers and color them one color, and all the cells containing even numbers and color them another color, interesting patterns emerge. Search the Internet for a site that studies Pascal's triangle. Try to find a site that contains a large triangle grid, and print it out. If the cells of the triangle are not completely filled out, complete the calculations for the cells. Then color all the even-numbered cells one color.

1. One of the patterns you should observe is symmetry. Write a short explanation of what this means to you.

2. Another pattern that is present is something called a *fractal*. Research this term, and explain what it means in relation to your triangle. Visit the companion website to obtain a program called Sierpins, that will create a fractal on your calculator.

Part III

Pascal's triangle was studied by others beside Pascal. Omar Khayyám was said to have studied it. Many centuries earlier, it was described by a Chinese mathematician, Yang Hui. Similar triangular arrangements were known to the Arabs about the same time that the Chinese were using it. Find a reference on the history of mathematics, and report on a civilization that was known to use Pascal's triangle. Document your reference. Identify any individuals, and describe what is known about the civilization's use of the triangle.

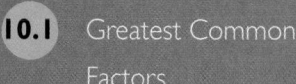

Chapter 10

Factoring

In this chapter, we continue our discussion of polynomials. We will discuss methods of factoring a polynomial—that is, writing the polynomial as a product of factors. This will turn out to be very important for simplifying complicated polynomial expressions and making them easier to work with when we solve equations and inequalities. Factoring is also the basis for working with rational expressions and equations.

Geometric figures often involve areas that can be described by polynomial expressions. Designers and architects must work with these expressions as they draw plans for the efficient use of space in a home, an office, or a manufacturing plant. Engineers and construction workers also use polynomials as a check on the dimensions written on plans and lists of materials. The opening photo shows the Rose Center for Earth and Space, which contains the Hayden Sphere having a diameter of 87 feet enclosed within a cubic-shaped building, 95 feet on each edge. We will describe some of these applications in each section of the chapter. We will conclude the chapter with a project illustrating the use of polynomials in determining the area of rectangular figures.

10.1 APPLICATION

The official dimensions of an Olympic-size pool are 50 meters by 21 meters. If the width of a walkway around the pool is x meters on all sides, write a polynomial expression that a designer could use to determine the area of the walkway. Factor the polynomial and discuss the meaning of the factors. If a customer wanted to install a 10-meter walkway around his Olympic-size pool, how many square meters of material would he need?

We will discuss this application further. See page 658.

10.1 Greatest Common Factors

OBJECTIVES

1 Determine greatest common factors for sets of monomials.

2 Factor out greatest common monomial factors from polynomials.

3 Factor out greatest common polynomial factors from polynomials.

4 Factor polynomials by grouping.

5 Model real-world situations by factoring out greatest common factors from polynomials.

10.1.1 Determining the Greatest Common Factor

In Chapter 1, we stated that two numbers which are multiplied together are called factors of the resulting product. For example, since $2 \cdot 6 = 12$, then 2 and 6 are factors of 12. Also, we may describe a factor of a given number as a number that divides into the given number evenly, or with a remainder of 0. For example, 2 is a factor of 12 because 2 divides into 12 with a remainder of 0.

To determine the integer factors of a number, divide the number by all the integers whose absolute value is less than or equal to the number itself. The integers that divide evenly are factors.

For example, to determine the positive integer factors of 24, we divide 24 by all the positive integers from 1 to 24. The positive integer factors of 24 are 1, 2, 3, 4, 6, 8, 12, and 24. Since we are repeatedly dividing by consecutive integers, the calculator will do this quickly for us by means of a table.

HELPING HAND Remember, there is no need to look beyond the numbers whose absolute value is the number itself.

T E C H N O L O G Y

INTEGER FACTORS

Determine the integer factors of 24.

X	Y₁
1	24
2	12
3	8
4	6
5	4.8
6	4
7	3.4286
X=1	

X	Y₁
-7	-3.429
-6	-4
-5	-4.8
-4	-6
-3	-8
-2	-12
-1	-24
X=-7	

Figure 10.1a   Figure 10.1b

For Figure 10.1a,

Set up the table.

Set the first column for x equal to integer values starting at 1.

| 2nd | TBLSET | 1 | ENTER | 1 | ENTER | ▼ | ENTER |

Set the second column for the number, 24, divided by the integer in the first column, x.

| Y= | 2 | 4 | ÷ | X,T,θ,n |

View the table to find the integer factors in the first column that correspond to an integer in the second column.

2nd		TABLE

Move down the table to determine additional positive integer factors of 24.

For Figure 10.1b,

Move up the table to determine the negative integer factors of 24.

The integer factors of 24 are $-24, -12, -8, -6, -4, -3, -2, -1, 1, 2, 3, 4, 6, 8, 12,$ and 24.

The **greatest common factor (GCF)** of a set of positive integers is the largest factor common to all the numbers.

To find the GCF of a set of numbers, determine the factors for each number in the set, and then choose the largest factor common to all the sets of factors.

For example, determine the GCF for 24, 36, and 72.

The factors of 24 are 1, 2, 3, 4, 6, 8, 12, and 24.

The factors of 36 are 1, 2, 3, 4, 6, 9, 12, 18, and 36.

The factors of 72 are 1, 2, 3, 4, 6, 8, 9, 12, 18, 24, 36, and 72.

greatest common factor = 12

The GCF of 24, 36, and 72 is 12.

To find the GCF of a set of monomials with variable factors, determine the product of the common variable factors, each having the smallest exponent common to that variable.

For example, determine the GCF of a^3bc^2 and ac^4.

The variables a and c are common to both monomials. The smallest exponent of a common to both monomials is 1, and the smallest exponent of c common to both monomials is 2.

The GCF of a^3bc^2 and ac^4 is the product $a^1 \cdot c^2$, or ac^2.

This process leads to an alternative method of finding the GCF of a set of numbers. Before we discuss this method, we need the following definition: A **prime number** is a counting number greater than 1 that has exactly two different counting-number factors, namely, 1 and itself. The prime numbers less than 30 are 2, 3, 5, 7, 11, 13, 17, 19, 23, 29.

To factor a number *completely*, we write the number as a product of prime numbers.

For example, factor 24 completely.

$$24 = 2 \cdot 2 \cdot 2 \cdot 3 = 2^3 \cdot 3$$

To find the GCF of a set of numbers, factor each number in the set completely, and then determine the product of the common prime factors, each having the smallest exponent for that prime number.

For example, determine the GCF of 24, 36, and 72.

Write the prime factorization of each number.

$$24 = 2 \cdot 2 \cdot 2 \cdot 3 = 2^3 \cdot 3$$
$$36 = 2 \cdot 2 \cdot 3 \cdot 3 = 2^2 \cdot 3^2$$
$$72 = 2 \cdot 2 \cdot 2 \cdot 3 \cdot 3 = 2^3 \cdot 3^2$$

The prime factors common to all three numbers are 2 and 3. The smallest exponent of 2 is 2. The smallest exponent of 3 is 1.

The GCF of 24, 36, and 72 is the product $2^2 \cdot 3^1$, or 12.

DETERMINING THE GREATEST COMMON FACTOR

To find the GCF of a set of monomials, determine the product of

- the GCF of the coefficients;
- the GCF of each variable.

EXAMPLE 1 Determine, for each set of monomials, its greatest common monomial factor.

a. $12x^2y^4z^3$ and $30xy^3z^2$ **b.** $132a^2b^2c^2$ and $72a^2b^3$

Solution

a. $12x^2y^4z^3$ and $30xy^3z^2$

The factors of 12 and 30 are small, and we should see that the GCF of the coefficients is 6.

The GCF of the variable factors contains the smallest exponent common to each, or xy^3z^2.

The GCF of $12x^2y^4z^3$ and $30xy^3z^2$ is $6xy^3z^2$.

b. $132a^2b^2c^2$ and $72a^2b^3$

The coefficients 132 and 72 are large, so we will use the alternative method to find their GCF.

$$132 = 2^2 \cdot 3 \cdot 11 \qquad \text{Factor each number completely.}$$
$$72 = 2^3 \cdot 3^2$$

The GCF of the coefficients is $2^2 \cdot 3 = 12$.
The GCF of the variable factors is a^2b^2.
The GCF of $132a^2b^2c^2$ and $72a^2b^3$ is $12a^2b^2$.

10.1.1 Checkup

For each set of monomials, determine its greatest common monomial factor.

1. $12a^2b^3c$; $18a^3bc^2$ **2.** $84x^3y^5z^2$; $210x^7y^3z^2$ **3.** $132pq$; $143p^3q^2r$; $77p^2r^3$

10.1.2 Factoring Out Greatest Common Monomial Factors

In Chapter 1, we reversed the distributive law in order to factor numeric expressions. The distributive law states that

$$a(b + c) = ab + ac, \quad \text{so that} \quad 2(3 + 4) = 2 \cdot 3 + 2 \cdot 4$$

To reverse the law and factor, we write

$$ab + ac = a(b + c), \quad \text{so that} \quad 2 \cdot 3 + 2 \cdot 4 = 2(3 + 4)$$

We are now ready to do the same for polynomials. (Later on, we will use this technique to simplify complicated rational expressions and equations.) To factor a polynomial completely, we first determine the greatest common factor (GCF) of its terms and then use the distributive law.

EXAMPLE 2 Factor $2x^2 + 4x + 6$.

Solution

$$2x^2 + 4x + 6 = 2 \cdot x^2 + 2 \cdot 2x + 2 \cdot 3$$

> The GCF is 2. Write each term as a product of the GCF and its remaining factor.

$$= 2(x^2 + 2x + 3)$$

> Use the distributive law.

We can check all factoring by multiplying the factors and obtaining the original polynomial or by a numeric or graphic check on the calculator if the expression contains only one variable. This will verify that the product of the factors is equivalent to the original polynomial, but it will not verify that the polynomial is factored completely.

Numeric Check

X	Y1	Y2
-3	12	12
-2	6	6
-1	4	4
0	6	6
1	12	12
2	22	22
3	36	36

X = -3

Y1 = $2x^2 + 4x + 6$
Y2 = $2(x^2 + 2x + 3)$
The table confirms that
Y1 = Y2.

Graphic Check

Y1 = $2x^2 + 4x + 6$

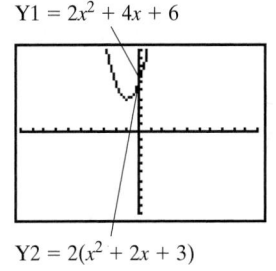

Y2 = $2(x^2 + 2x + 3)$

$(-10, 10, -10, 10)$
The graph confirms that the lines are coinciding.

A polynomial that is factored completely has no common factors in its terms. Thus, the polynomial $12x - 6y = 2(6x - 3y)$ is not factored completely, because the terms in the polynomial factor $6x - 3y$ have a GCF, 3, and may be factored again. Therefore,

$$12x - 6y = 2(6x - 3y)$$
$$= 2 \cdot 3(2x - y)$$
$$= 6(2x - y)$$

Now $12x - 6y = 6(2x - y)$ is factored completely, because the terms in the binomial $2x - y$ do not have a common factor.

FACTORING OUT THE GCF FROM A POLYNOMIAL
To factor out the GCF from a polynomial,

- Determine the GCF of all the terms. If the GCF is 1, there is no need to continue.
- Write each term as a product of the GCF and another factor.
- Reverse the distributive law, and write a product of the GCF and a polynomial.

HELPING HAND If the first term of the polynomial is negative, write a product of the opposite of the GCF and a polynomial.

EXAMPLE 3   Factor completely. Check the factors.

a. $-10x^4y^2 - 15xy^3 - 10x^2y^3$ **b.** $28a^3b^4 + 42a^4 - 14a^2b^5$

Solution

a. $-10x^4y^2 - 15xy^3 - 10x^2y^3$

$= (-5xy^2) \cdot 2x^3 + (-5xy^2) \cdot 3y + (-5xy^2) \cdot 2xy$ The GCF is 5xy², but the first term is negative, so we factor out −5xy².

$= -5xy^2(2x^3 + 3y + 2xy)$ Use the distributive law.

The check is left for you.

b.   Try to factor out the GCF mentally. Determine the polynomial by asking yourself what factor is needed to determine the original product.

$28a^3b^4 + 42a^4 - 14a^2b^5 = 14a^2($ $+$ $-$ $)$ The GCF is 14a². Write each term as a product of the GCF and a factor.

$= 14a^2(2ab^4 + 3a^2 - b^5)$ $14a^2 \cdot 2ab^4 = 28a^3b^4$
$14a^2 \cdot 3a^2 = 42a^4$
$14a^2 \cdot -b^5 = -14a^2 b^5$

The check is left for you.

10.1.2 Checkup

Factor exercises 1–2 completely. Check.

1. $6a^3b^2 - 12a^2b^3 + 24a^2b^2$

2. $-8x^2y^3z - 24x^3y^2 + 36xyz$

3. How can you check to see that you have correctly factored out a greatest common factor from a polynomial?

4. After you have factored out a common factor from a polynomial, what should you do to be sure that you have factored out the greatest common factor?

10.1.3 Factoring Out Greatest Common Polynomial Factors

A polynomial may have a polynomial other than a monomial as its greatest common factor.

EXAMPLE 4   Factor $2x(x + 1) + 3(x + 1)$.

Solution

$2x(x + 1) + 3(x + 1)$ has two terms, $2x(x + 1)$ and $3(x + 1)$, and a GCF of $(x + 1)$.

$2x(x + 1) + 3(x + 1) = (x + 1)(2x + 3)$ Use the distributive law.

Check to see that the product of the factors is equivalent to the simplification of the original polynomial. That is,

$$(x + 1)(2x + 3) = 2x^2 + 3x + 2x + 3 = 2x^2 + 5x + 3$$

is equivalent to the original polynomial,

$$2x(x + 1) + 3(x + 1) = 2x^2 + 2x + 3x + 3 = 2x^2 + 5x + 3$$

Since the polynomial is in one variable, we can check numerically or graphically.

Numeric Check

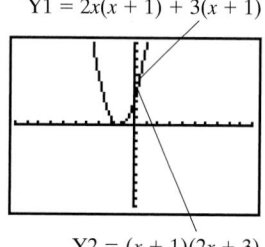

Y1 = 2x(x + 1) + 3(x + 1)
Y2 = (x + 1)(2x + 3)
The table confirms Y1 = Y2.

Graphic Check

Y1 = 2x(x + 1) + 3(x + 1)

Y2 = (x + 1)(2x + 3)
(−10, 10, −10, 10)
The graphs of Y1 and Y2 are equivalent.

EXAMPLE 5 Factor completely and check.

a. $3x(x^2 + 1) + 4(x^2 + 1)$ **b.** $5x^2(2x + 3y) - (2x + 3y)$

Solution

a. $3x(x^2 + 1) + 4(x^2 + 1) = (x^2 + 1)(3x + 4)$ The GCF is $(x^2 + 1)$.

The check is left for you.

b. $5x^2(2x + 3y) - (2x + 3y) = (2x + 3y)(5x^2 - 1)$ The GCF is $(2x + 3y)$.
    Remember that $-(2x + 3y)$
    is the same as $-1(2x + 3y)$.

The multiplication check is left for you. A numeric or graphic check is not possible, because the polynomial has two variables.

✓ **10.1.3** Checkup

Factor exercises 1–2 completely and check.

1. $7a(2a + b) + 4b(2a + b)$
2. $4x^2(3x - 2) - 3(3x - 2)$

10.1.4 Factoring by Grouping

A polynomial, such as the polynomial $2ax^2 + 4bx + axy + 2by$, may not appear to have a greatest common factor. However, notice that if we rewrite this polynomial as a sum of two binomials, each binomial has a GCF.

$$2ax^2 + 4bx + axy + 2by$$
$$(2ax^2 + 4bx) + (axy + 2by)$$

The first binomial, $2ax^2 + 4bx$, has a GCF of $2x$. The second binomial, $axy + 2by$, has a GCF of y. Factor out the corresponding GCF from each binomial.

$$2x(ax + 2b) + y(ax + 2b)$$

Now the two terms $2x(ax + 2b)$ and $y(ax + 2b)$ have a GCF of $(ax + 2b)$. Factoring this GCF out of each term, we obtain

$$(ax + 2b)(2x + y)$$

This process is called **factoring by grouping**.
Remember to check your factoring.

FACTORING A POLYNOMIAL WITH FOUR TERMS BY GROUPING

To factor a polynomial with four terms by grouping,

- Factor out the GCF from all of the terms.
- Rewrite the polynomial as a sum of two binomials.
- Factor out the corresponding GCF from each of the two binomials.
- Determine that the binomial in each group is the GCF of the resulting terms.
- Factor out the common binomial.
- If no common factor is found, a different grouping may be needed or the polynomial may not factor.

EXAMPLE 6   Factor completely and check.

a. $2x^2 - 10x + 7x - 35$ **b.** $3x^4 + 6x^2 - x^2 - 2$
c. $6x^2 - 3x + 12x - 6$

Solution

a. $2x^2 - 10x + 7x - 35$

$= (2x^2 - 10x) + (7x - 35)$ Group the terms.

$= 2x(x - 5) + 7(x - 5)$ Factor out the GCF ($2x$ and 7) from each group.

$= (x - 5)(2x + 7)$ Factor out $(x - 5)$ from each term.

The check is left for you.

b. $3x^4 + 6x^2 - x^2 - 2$

$= (3x^4 + 6x^2) + (-x^2 - 2)$ Group the terms.

$= 3x^2(x^2 + 2) + (-1)(x^2 + 2)$ Factor out the GCF from each group ($3x^2$ and -1).

$= (x^2 + 2)(3x^2 - 1)$ Factor out $(x^2 + 2)$ from the remaining terms.

The check is left for you.

   c. $6x^2 - 3x + 12x - 6$

$$= 3(2x^2 - x + 4x - 2) \qquad \text{Factor out the GCF (3).}$$
$$= 3[(2x^2 - x) + (4x - 2)] \qquad \text{Group the terms.}$$
$$= 3[x(2x - 1) + 2(2x - 1)] \qquad \text{Factor out the GCF of each group.}$$
$$= 3(2x - 1)(x + 2) \qquad \text{Factor out } (2x - 1) \text{ from the remaining terms.}$$

   The check is left for you.

10.1.4 Checkup

Factor completely and check.

  1. $6x^2 - 10x + 9x - 15$ **2.** $2y^4 + 2y^2 - 5y^2 - 5$ **3.** $15a^3 - 20a^2 + 105a^2 - 140a$ **4.** $ax + ay + bx + by$

10.1.5 Modeling the Real World

It sometimes happens that the simplest way to model a complicated real-world situation gives rise to a complicated polynomial expression. But once you have the polynomial written down, you should always think about ways to factor it. This will make any operations you need to perform with the polynomial simpler and will help you avoid errors.

EXAMPLE 7   Casey's father decided he would give his son a surprise birthday present. He wrote the following message:

> Every day for a week you will receive a gift of $5 more than you received the day before.

   a. Write a polynomial for the total amount Casey will receive if he receives x dollars the first day. Simplify the polynomial.

   b. Factor out the GCF from your simplified expression in part a.

   c. Explain in your own words the meaning of the binomial factor.

> Suppose Casey receives $10 the first day from his father.

   d. Evaluate the polynomial in part a to determine the total amount Casey will receive in a week.

   e. Evaluate the binomial factor found in part b.

   f. Show that the value obtained in part e has the meaning given in part c.

Solution

   a. Let $x =$ amount received the first day

      Casey will receive

$$x + (x + 5) + (x + 10) + (x + 15) + (x + 20) + (x + 25)$$
$$+ (x + 30) \text{ or } (7x + 105) \text{ dollars}$$

   b. $7x + 105 = 7(x + 15)$

   c. The binomial $(x + 15)$ is the average daily amount in dollars that Casey will receive for each of the seven days.

   d. $7x + 105 = 7(10) + 105 = 70 + 105 = 175$

      If Casey receives $10 on the first day, he will receive $175 in a week from his father.

e. $x + 15 = 10 + 15 = 25$

f. Casey will receive an average of $25 per day for a week. This will amount to

$$\frac{7 \text{ days}}{1 \text{ week}} \cdot \frac{\$25}{1 \text{ day}} = \frac{\$175}{1 \text{ week}}$$

or $175 for the week, the same amount found in part d.

APPLICATION

The official dimensions of an Olympic-size pool are 50 meters by 21 meters. If the width of a walkway around the pool is x meters on all sides, write a polynomial expression that a designer could use to determine the area of the walkway. Factor the polynomial and discuss the meaning of the factors. If a customer wanted to install a 10-meter walkway around his Olympic-size pool, how many square meters of material would he need?

Discussion

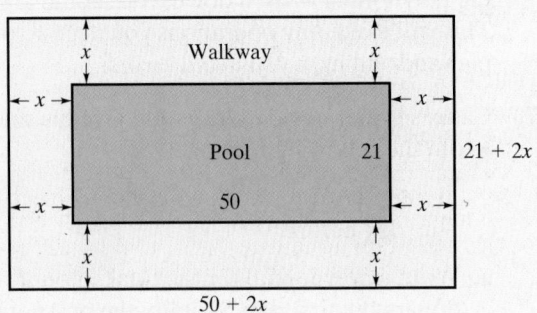

50 + 2x

Pool and walkway:

$21 + 2x = $ width
$50 + 2x = $ length
$(21 + 2x)(50 + 2x) = $ area

Pool:

$21 = $ width
$50 = $ length
$(21)(50) = $ area

The area of the walkway is the difference of the area of the pool and walkway and the area of the pool.

$$(21 + 2x)(50 + 2x) - (21)(50) = 1050 + 42x + 100x + 4x^2 - 1050$$
$$= 142x + 4x^2$$

The area of the walkway is $142x + 4x^2$ square meters.

$$142x + 4x^2 = 2x(71 + 2x) \qquad \text{Factor.}$$

The area of the walkway is equivalent to a rectangle $2x$ meters by $(71 + 2x)$ meters.

If the walkway is 10 meters wide, the area not covered by the walkway is

$$142x + 4x^2 = 142(10) + 4(10)^2 \quad \text{or} \quad 2x(71 + 2x) = (2 \cdot 10)(71 + 2 \cdot 10)$$
$$= 1820 \qquad\qquad\qquad\qquad = 1820$$

The area of the walkway is 1820 square meters.

10.1.5 Checkup

1. Katie gave up smoking as her New Year's resolution. Her dad told her that as an incentive, every month for six months he would give her $12 more than he gave her the month before if she continued refraining from smoking.

   a. Write a polynomial for the total amount of money Katie will receive if she receives x dollars after not smoking for six months.
   b. Factor the polynomial and explain what each factor represents.
   c. If her dad gives her $50 the first month, how much will Katie receive if she doesn't smoke for the six months?
   d. What is her average monthly amount received for not smoking? Does this check with your explanation in part b?

2. Many new office buildings have an atrium in the center that opens up to the full height of the building. One such building has an atrium that measures 150 feet long and 100 feet wide. The atrium is surrounded by offices that add an additional x feet to all sides of the atrium. Write a polynomial for the area of the floor space of the offices. Factor the polynomial and discuss the meaning of the factors. If the offices add an additional 40 feet to all sides of the atrium, how many square feet of area do the offices occupy?

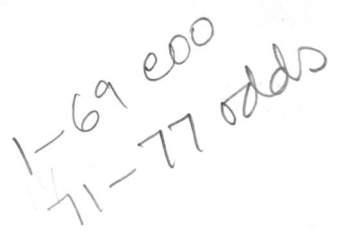

10.1 Exercises

For each set of monomials, determine the greatest common monomial factor.

1. $60a^2b^4c^3$; $50a^3bc^2$
2. $45x^2yz$; $90x^3yz^2$
3. $252x^3y^4$; $180x^2z$
4. $126u^3v^4$; $315u^5v^2$
5. $45pq^4$; $135r^4s$
6. $80c^5d^2$; $24a^2b^2$
7. $63xyz$; $98xyz$
8. $40a^2bc$; $250a^2bc$
9. $60a^2b^3c^3$; $90ab^2c$; $150a^2b^4c^2$
10. $120x^2y^4z^3$; $270x^2y^2z$; $750xy^2z$

Factor exercises 11–70 completely and check.

11. $4x + 12y$
12. $-7x + 21y$
13. $8x^3 - 4x^2 + 12x - 24$
14. $9d^5 - 12d^3 + 21d + 24$
15. $3a^4 - 5a^3 + 7a^2$
16. $-4p^5 - 9p^4 - 11p^3$
17. $-3x^5 - 9x^4 - 12x^3$
18. $5m^3 - 15m^2 + 30m$
19. $7x^4y^2 - 3x^2y^2 + 9x^2y^4$
20. $-3u^3v + 4u^2v^2 - 8uv^3$
21. $8a^5b^3c + 4a^4b^2 + 16a^3c$
22. $7x^3y^5 - 21x^2y^4 + 63xy^3$
23. $66u^3v^4 - 88u^4v^3$
24. $-39c^2d^3 + 52c^3d^2$
25. $3x^3 + 5y^4$
26. $p^4 + 7q^2$
27. $5x(x + 3) - 4(x + 3)$
28. $3y(y + 7) - 5(y + 7)$
29. $x(2x + y) + 2y(2x + y)$
30. $3a(a + 3b) + b(a + 3b)$
31. $6x^2 + 10x + 21x + 35$
32. $15x^2 + 6x + 10x + 4$
33. $x^2 + 8x + x + 8$
34. $y^2 + 4y + y + 4$
35. $2a^2 + 3a - 2a - 3$
36. $2z^2 + 7z - 2z - 7$
37. $2x^2 + xy + 4xy + 2y^2$
38. $3p^2 + 9pq + pq + 3q^2$
39. $x^2 + xy - xy - y^2$
40. $m^2 + mn - mn - n^2$
41. $10xy - 55y + 24x - 132$
42. $28xy + 21y - 36x - 27$
43. $12ac + 3bc + 4ad + bd$
44. $10mp + 5np + 4mq + 2nq$
45. $2x^2y^2 + 3xy - 8xy - 12$
46. $7a^2b^2 - 2ab + 21ab - 6$
47. $-x^2 - 3x - xy - 3y$
48. $-a^2 - ab - 7a - 7b$
49. $x^4 + x^2y^2 + 2x^2y^2 + 2y^4$
50. $3p^4 - 3p^2q^2 + p^2q^2 - q^4$
51. $ac + bc + ad + bd$
52. $xy + y^2 + zx + zy$
53. $8x^2 + 4x + 24x + 12$
54. $21y^2 + 14y + 84y + 56$
55. $u^4 + u^3v - 2u^3v - 2u^2v^2$
56. $a^4 + a^3b + 2a^3b + 2a^2b^2$
57. $6a^4 + 6a^3b^2 + 6a^3b + 6a^2b^3$
58. $5c^5 + 5c^4d^2 + 5c^3d + 5c^2d^3$
59. $-4x^3 - 12x^2 - 2x^2 - 6x$
60. $-10y^3 + 50y^2 - 15y^2 + 75y$
61. $4x^2 + 6xz + 6xz + 9z^2$
62. $9u^2 + 12uv + 12uv + 16v^2$
63. $36a^2 - 30ab - 30ab + 25b^2$
64. $16x^2 + 12xy + 12xy + 9y^2$
65. $5m^3 + 5m^2n + 5m^2n + 5mn^2$
66. $7p^3 - 7p^2q - 7p^2q + 7pq^2$
67. $2x^3 + 3x + 8x^2 + 12$
68. $5c^4 + 7c - 15c^3 - 21$
69. $5a^2x + 2b^2x + 15a^2y + 6b^2y$
70. $2a^2c + 3b^2c + 14a^2d + 21b^2d$

71. Amy's dad promised her that for each soccer goal she made, he would give her $1 more than he gave her for her last soccer goal. She made nine goals during the season.

a. Write a polynomial for the total amount of money Amy will receive if she receives x dollars for her first goal.
b. Factor out the GCF from the simplified expression in part a.
c. Explain what the factors in part b represent.
d. If Amy's dad gave her $10 for her first goal, determine the total amount of money Amy received. Check your explanation in part c, using $10 for x.

72. Wynona started a program to quit smoking. She decided to limit herself to two fewer cigarettes than her limit from the day before. She was able to hold to the program.

a. Write a polynomial for the total number of cigarettes Wynona smoked during the first week if she smoked c cigarettes the first day.
 b. Factor out the GCF from the simplified expression in part a.
c. Explain what the factors in part b represent.
d. If Wynona smoked 20 cigarettes the first day, determine the total number of cigarettes she smoked the first week. Does your explanation in part c check, using that number? How many cigarettes did she smoke on the seventh day of the week?

73. In a sports program with n teams, it can be shown that the number of times each team must play every other team exactly two times in the program is given by $n^2 - n$.
a. Factor the expression completely.
b. Evaluate the original expression and the factored expression for a program that has 21 teams in it.
c. Which expression was easier to evaluate and why?

74. It can be shown that in a sports program with n teams, the number of ways in which you can have three teams finish in the top three positions is given by $n^3 - 2n^2 - n^2 + 2n$.
a. Factor this expression completely.
b. Evaluate the original expression and the factored expression for a program that has 11 teams in it.
c. Which expression was easier to evaluate and why?

75. The area of a rectangle is given by the expression $x^2 + 7x - 3x - 21$. Factor this expression, using grouping to determine algebraic expressions for the rectangle's width and length.

76. The area of a square is given by the expression $4x^2 + 6x + 6x + 9$. Factor this expression, using grouping to determine an algebraic expression for the length of the square's sides.

77. An employee parking lot measures 50 feet by 200 feet. It is surrounded by a green strip of grass and shrubs that adds x feet to all of its sides. Write a polynomial for the area of the green strip. Factor the polynomial and discuss the meaning of the factors. If the green strip adds 10 feet to all sides of the parking lot, how many square feet of greenery is there?

78. A painting measures 30 inches in length and 24 inches in width. The painting is surrounded by a frame that adds x inches to each side. Write a polynomial for the area of the frame facing out from the painting. Factor the polynomial and discuss the meaning of the factors. If the frame adds 5 inches to each side of the painting, how many square inches of frame are added to the face of the painting.

10.1 Calculator Exercises

Part 1. Finding the Greatest Common Factor by Using the Calculator

The TI-83 Plus has an instruction under the CATALOG key that will find the greatest common divisor (gcd) of two numbers. This will be the greatest common factor of the two numbers. To use the instruction, press [2nd] [CATALOG], which lists the catalog of all instructions. Then scroll down the catalog listing until the cursor is pointed at gcd, and press [ENTER]. This will place the instruction on the home screen, after which you can type in the two numbers for which you want the greatest common factor. When you press [ENTER],

the calculator will display the GCF. For example, if you wish the GCF of the numbers 56 and 14, the following screens find it:

 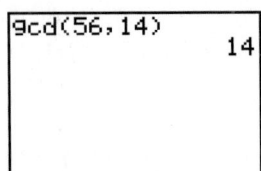

If you need to find the GCF of three or more numbers, enter them in pairs, and find the GCF of each pair. Then select the smallest GCF of all the pairs, which will be the GCF of all the numbers. Can you explain why?

For example, find the GCF of 56, 14, and 24.

```
gcd(56,14)
                14
gcd(56,24)
                 8
gcd(14,24)
                 2
```

The GCF of 56, 14, and 24 is 2, since that is the largest number that divides all three numbers evenly.

Use this calculator feature to find the greatest common factors in the following exercises.

1. $105x^2yz^3$, $147xy^2z^2$ **2.** $104ab^2c^2$, $65b^3c$, $143a^2c^2$

3. $108x^3$, $96x^2$, $72x^5$

4. $64abc$, $128abc^2$, $192ab^2c$, $224a^2bc$

Part 2. Factoring a Number into Prime Factors by Using the Calculator

Graphing calculators can be programmed to perform certain operations. The program[1] that follows will factor a number into a product of prime factors. To place your calculator into a program edit mode, enter PRGM ▶ ▶ to New, and then press the ENTER key. This will place the calculator into a mode so that you can enter a program. At each step, key in the instructions, and press ENTER to move to the next step.

If you make a mistake, use the arrow keys to move back up and correct it. Many of the instructions use keys with which you are already familiar. Any other programming instruction can be found by pressing the PRGM key, using the arrow keys to find the instruction, and then pressing the ENTER key to select the instruction. Enter the following program:

PROGRAM:PRIME

Input N

N→M

1→C

2→D

Lbl 1

M/D→B

[1]By permission of Nan Burwell, Pellissippi State Technical Community College.

If fpart(B)=0

Goto 2

C+2→D

D→C

If D≥($\sqrt{}$(M)+1)

Goto 3

Goto 1

Lbl 2

ipart(B)→M

Disp D

Pause

If M=1

Goto 3

Goto 1

Lbl 3

If M≠1

Disp M

If M=N

Disp "PRIME"

Disp "DONE"

Note: fPart (is found under the MATH key.

iPart (is found under the MATH key.

The equation sign and the inequality signs are found under the 2nd TEST keys.

To run the program, enter PRGM ▼ to go to the appropriate program, and then press ENTER once to select the program and again to execute the program. At the question mark prompt, key in the number to be factored, and press ENTER repeatedly until the calculator indicates that the factoring is done. If you press ENTER again, the program will be restarted so that you can enter another number.

Use the program to factor the following numbers completely:

1. 30   **2.** 108   **3.** 525   **4.** 1287   **5.** 1547   **6.** 4500

10.1 Writing Exercises

Gennifer was asked to completely factor the polynomials listed. In each case, the answer she gave was incorrect. Explain what is wrong with each result.

1. $-8a - 4b = -4(2a - b)$

2. $5x^2 + 10x = 5(x^2 + 2x)$

3. $4x^2 + 3x + 4x + 3 = (4x + 3)(x + 0)$

4. $4y^3 - 6y^2 = 2^2 \cdot y^3 - 2 \cdot 3 \cdot y^2$

10.2 Factoring Perfect-Square Trinomials, Differences of Squares, and Sums and Differences of Cubes

OBJECTIVES

1 Factor a perfect-square trinomial.

2 Factor a difference of two squares.

3 Factor a sum or difference of two cubes.

4 Model real-world situations by factoring special products.

APPLICATION

A certified report for Quality Jewelry reports that an emerald-cut square diamond of 4.22 carats is worth $59,360. To make the diamond more lustrous, two sides of the face are decreased by x millimeters. The surface area of the face, in square millimeters, of this more lustrous diamond is given by the polynomial $88.36 - 19.8x + x^2$. Factor the polynomial, and determine the length of the sides of the original diamond.

After completing this section, we will discuss this application further. See page 668.

In the previous chapter, we determined the product of two polynomials. We identified certain of these products as special products. If we can identify these products, we can determine their factors quickly.

10.2.1 Factoring a Perfect-Square Trinomial

One special product is the square of a binomial.

$$(a + b)^2 = a^2 + 2ab + b^2$$
$$(a - b)^2 = a^2 - 2ab + b^2$$

We can reverse this operation by factoring the trinomial into a square of a binomial.

FACTORING A PERFECT-SQUARE TRINOMIAL
To factor a perfect-square trinomial, we write

$$a^2 + 2ab + b^2 = (a + b)^2$$
$$a^2 - 2ab + b^2 = (a - b)^2$$

In order to do this factoring, we need to recognize when a polynomial is a perfect-square trinomial. In a perfect-square trinomial, the first and last terms are perfect squares, and the middle term is either positive or negative twice the product of the terms that are squared—that is, $a^2 + 2ab + b^2$ or $a^2 - 2ab + b^2$. The binomial factor of a perfect-square trinomial is $(a + b)^2$ or $(a - b)^2$. For example,

$$a^2 + 2ab + b^2 = (a + b)^2$$

$$x^2 + 10x + 25 = (x)^2 + 2(x)(5) + (5)^2 = (x + 5)^2$$

$$a^2 - 2ab + b^2 = (a - b)^2$$

$$x^2 - 10x + 25 = (x)^2 - 2(x)(5) + (5)^2 = (x - 5)^2$$

We can check factoring by multiplying the factors, the result should be the polynomial that was factored. We can also check numerically or graphically on the calculator.

Numeric Check

X	Y₁	Y₂
-3	64	64
-2	49	49
-1	36	36
0	25	25
1	16	16
2	9	9
3	4	4

X= -3

$Y1 = x^2 - 10x + 25$
$Y2 = (x - 5)^2$
The table shows Y1 = Y2.

Graphic Check

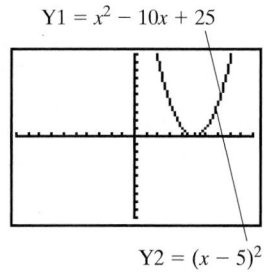

$Y1 = x^2 - 10x + 25$

$Y2 = (x - 5)^2$
$(-10, 10, -10, 10)$
The graphs are coinciding.

EXAMPLE I   Factor completely and check.

a. $4x^2 + 12xy + 9y^2$ **b.** $3x^2 - 24xy + 48y^2$ **c.** $m^2 + 12m + 144$
d. $x^4 + 18x^2 + 81$

Solution

a. $4x^2 + 12xy + 9y^2 = (2x)^2 + 2(2x)(3y) + (3y)^2$ Determine that $a = 2x$ and $b = 3y$.

$$= (2x + 3y)^2$$

The check is left for you.

b. $3x^2 - 24xy + 48y^2 = 3(x^2 - 8xy + 16y^2)$ Factor out a common factor.

$$= 3[(x)^2 - 2(x)(4y) + (4y)^2]$$ Determine that $a = x$ and $b = 4y$.

$$= 3(x - 4y)^2$$

The check is left for you.

c. $m^2 + 12m + 144$ does not factor. The first and last terms are perfect squares: m^2 and 12^2. However, the middle term is not $2(m)(12)$.

d. $x^4 + 18x^2 + 81 = (x^2)^2 + 2(x^2)(9) + (9)^2$ Determine that $a = x^2$ and $b = 9$.

$\qquad\qquad\qquad\qquad = (x^2 + 9)^2$

The check is left for you.

 10.2.1 Checkup

Factor exercises 1–5 completely and check.

1. $z^2 + 12z + 36$ **2.** $49a^2 + 42ab + 9b^2$ **3.** $5x^2 - 60x + 180$ **4.** $p^2 - 8p + 64$ **5.** $y^4 + 8y^2 + 16$

6. What should you look for in a polynomial in order to factor it as a perfect-square trinomial?

10.2.2 Factoring a Difference of Two Squares

One special product consists of two factors, one the sum and the other the difference of the same two terms.

$$(a + b)(a - b) = a^2 - b^2$$

If we reverse this operation, we obtain two binomial factors.

> **FACTORING THE DIFFERENCE OF TWO SQUARES**
> To factor the difference of two squares, we write
> $$a^2 - b^2 = (a + b)(a - b)$$

In order to do this factoring, we need to recognize a polynomial as the difference of two squares. The difference of two squares is a binomial consisting of the subtraction of two perfect-square terms—that is, $a^2 - b^2$. The factors of a difference of two squares are $(a + b)$ and $(a - b)$.

For example, factor $x^2 - 25$ completely.

$$a^2 \quad - \quad b^2 \;=\; (a + b)(a - b)$$

$$x^2 - 25 = (x)^2 - (5)^2 = (x + 5)(x - 5)$$

The factors may be checked, as with the previous objective.

EXAMPLE 2   Factor completely and check.

a. $4x^2 - 9y^2$ **b.** $3x^2 - 48$ **c.** $144 + x^2y^2$
d. $m^4 - 81$

Solution

a. $4x^2 - 9y^2 = (2x)^2 - (3y)^2$ Determine that $a = 2x$ and $b = 3y$.

$\qquad\qquad\quad = (2x + 3y)(2x - 3y)$ Write a product of the sum and difference of a and b.

The check is left for you.

b. $3x^2 - 48 = 3(x^2 - 16)$ *Factor out the common factor.*

$\qquad\qquad = 3[(x)^2 - (4)^2]$ *Determine that $a = x$ and $b = 4$.*

$\qquad\qquad = 3(x + 4)(x - 4)$ *Write the special product as a product of the sum and difference of a and b.*

The check is left for you.

c. $144 + x^2y^2$ will not factor. This is not a difference of squares, but a *sum* of squares.

d. Remember that a variable term is a perfect square when its exponent is an even number. That is, x^4 is a perfect square of x^2 because $x^4 = (x^2)^2$, and x^6 is a perfect square of x^3 because $x^6 = (x^3)^2$.

Therefore, we can factor the difference of two squares if the variable terms have even exponents.

$m^4 - 81 = (m^2)^2 - (9)^2$ *Determine that $a = m^2$ and $b = 9$.*

$\qquad\qquad = (m^2 + 9)(m^2 - 9)$ *Write a product of the sum and difference of a and b.*

$\qquad\qquad = (m^2 + 9)[(m)^2 - (3)^2]$ *Determine the perfect-square terms of the second factor.*

$\qquad\qquad = (m^2 + 9)(m + 3)(m - 3)$ *Write the second factor as a product of a sum and a difference.*

The check is left for you.

10.2.2 Checkup

Factor exercises 1–5 completely and check.

1. $y^2 - 4$ **2.** $25x^2 - 16y^2$ **3.** $2z^2 - 18$

4. $a^2 + b^2$ **5.** $c^4 - 16$

6. What are the things you should look for in a polynomial in order to factor it as the difference of two squares?

10.2.3 Factoring a Sum or Difference of Cubes

A third type of polynomial that we need to be able to factor is the product of a binomial and a trinomial. Complete the following set of exercises to determine this product and to discover a pattern you can use to factor it.

Discovery 1

Factoring a Sum or Difference of Two Cubes

Multiply.

1. $(x + 5)(x^2 - 5x + 25)$ **2.** $(x - 5)(x^2 + 5x + 25)$

Write a rule to factor the sum or difference of two cubes.

The first product is a sum of two cubes. The second product is a difference of two cubes. In order to factor these polynomials, we turn these results around and write a binomial and trinomial factor.

FACTORING THE SUM OR DIFFERENCE OF TWO CUBES

To factor the sum or difference of two cubes, we write

$$a^3 + b^3 = (a + b)(a^2 - ab + b^2)$$
$$a^3 - b^3 = (a - b)(a^2 + ab + b^2)$$

HELPING HAND The binomial factor has the same sign as the polynomial being factored. The middle term of the trinomial has the opposite sign of the polynomial being factored. The last term in the trinomial is always positive.

$$a^3 + b^3 = (a + b)(a^2 - ab + b^2)$$
$$S \qquad\ O \quad\ AP$$
$$a^3 - b^3 = (a - b)(a^2 + ab + b^2)$$
$$S \qquad\ O \quad\ AP$$

Remember SOAP.
S = same
O = opposite
AP = always positive

If you cannot remember this, divide the sum of two cubes, $a^3 + b^3$, by $a + b$ and the difference of two cubes, $a^3 - b^3$, by $a - b$ to obtain the trinomial factor.

For example, factor $x^3 - 27$ completely.

$$a^3 - b^3 = (a - b)(a^2 + a\,b + b^2)$$
$$x^3 - 27 = (x)^3 - (3)^3 = (x - 3)[x^2 + x(3) + (3)^2]$$
$$= (x - 3)(x^2 + 3x + 9)$$

The factors may be checked by multiplying, or they may be checked numerically or graphically on your calculator.

EXAMPLE 3 Factor completely and check.

a. $m^3 + 64$ **b.** $27x^3 - 125y^3$

Solution

a. $m^3 + 64 = (m)^3 + (4)^3$ Determine $a = m$ and $b = 4$.

$ = (m + 4)[m^2 - 4m + (4)^2]$ Write a product of a binomial factor
$S \quad\quad O \quad\ AP$ and a trinomial factor.

$ = (m + 4)(m^2 - 4m + 16)$ Simplify.

The check is left for you.

b. $27x^3 - 125y^3 = (3x)^3 - (5y)^3$ Determine $a = 3x$ and $b = 5y$.

$ = (3x - 5y)[(3x)^2 + (3x)(5y) + (5y)^2]$ Write a product of
$S \qquad\quad O \qquad\quad AP$ a binomial and
a trinomial.

$ = (3x - 5y)(9x^2 + 15xy + 25y^2)$ Simplify.

The check is left for you.

10.2.3 Checkup

Factor exercises 1–3 completely and check.

1. $a^3 + 8$ **2.** $a^3 - 8$ **3.** $8x^3 + 27y^3$

4. In factoring a polynomial, how would you recognize it to be the sum or difference of two cubes?

10.2.4 Modeling the Real World

Many applications involving the area of a figure are modeled by polynomials. If the shape of the original figure is a rectangle, then factoring these polynomials will result in the length and the width of the figure. This will enable us to work with figures when the dimensions vary and are written in terms of a variable.

EXAMPLE 4   Lynn wants to redesign the square deck on her house, using as much of the outside railing and all of the usable carpet as possible. However, she cannot use 9 square feet of the outdoor carpet because it has grease stains from the grill.

a. Determine the possible dimensions of Lynn's new deck in terms of the original side of length x feet.

b. If Lynn's original deck was 12 feet by 12 feet, determine the new deck's dimensions.

c. Using the dimensions found in part b, draw two possible decks.

d. Determine the amount of railing for each deck in part c.

e. If only the original railing is used, determine whether either drawing in part c is possible.

Solution

a. Since the length of the original deck (carpet) is x feet, the area of the original square deck is x^2.

  The new deck (carpet) area is 9 square feet less than the original area, or $x^2 - 9$.

  To determine the dimensions of the new deck, we must factor the area to determine the dimensions that were multiplied together.

$$x^2 - 9 = (x)^2 - (3)^2 = (x + 3)(x - 3)$$

The resulting dimensions are $(x + 3)$ feet and $(x - 3)$ feet.

b. If $x = 12$, then the new deck's dimensions are 15 feet $(x + 3)$ by 9 feet $(x - 3)$.

c.

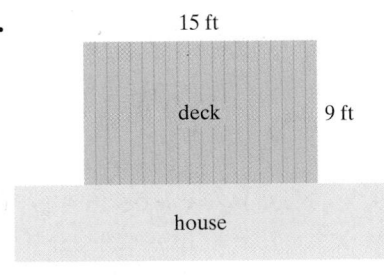

d. The first drawing will need 9 + 15 + 9, or 33, feet of railing. The second drawing will need 15 + 9 + 15, or 39, feet of railing.

e. The original deck had 12 + 12 + 12, or 36, feet of railing. The first drawing in part c shows the only possible deck using these conditions.

APPLICATION

A certified report for Quality Jewelry reports that an emerald-cut square diamond of 4.22 carats is worth $59,360. To make the diamond more lustrous, two sides of the face are decreased by x millimeters. The surface area of the face, in square millimeters, of this more lustrous diamond is given by the polynomial $88.36 - 19.8x + x^2$. Factor the polynomial, and determine the length of the sides of the original diamond.

Discussion

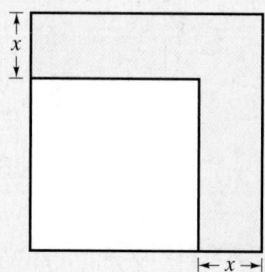

The polynomial is a perfect-square trinomial, because $9.4^2 = 88.36$ and $2(9.4) = 19.8$.

$$88.36 - 19.8x + x^2 = (9.4 - x)^2 \quad \text{Factor.}$$

The diamond face is $9.4 - x$ millimeters on a side, or 9.4 millimeters minus x millimeters. Therefore, the original diamond was 9.4 millimeters on a side.

 10.2.4 Checkup

1. Lynn has a square plot in her yard, measuring 16 feet on a side, that she wants to make into an outdoor sitting area. She plans to put a square patio in the plot for furniture, with the remainder of the plot landscaped as a garden.
   a. Determine a polynomial for the area of the plot available for landscaping if the patio measures p feet on a side.
   b. Factor the polynomial for the area. Use the factors to determine the dimensions of an equivalent rectangular garden area if the patio will be 7 feet on a side.
   c. Lynn sees an advertisement for garden plants which states that the package contains enough plants for a 10-by-20-foot garden. Will the package be large enough for her garden?

2. Marge decided to salvage an antique quilt that had damage along one side. To maintain the integrity of the design, she had to trim each side of the quilt. The original shape of the quilt was square, and an equal width was trimmed from each side. Given that Marge trimmed x feet from each side, a polynomial representing the new area of the quilt is $81 - 18x + x^2$ square feet. Factor the polynomial, and determine the length of each side of the original quilt. If 1.25 feet were trimmed from each side, use the factored expression to determine the area of the trimmed quilt.

10.2 Exercises

In exercises 1–58, factor completely. Check by multiplication, by a table of values, or graphically.

1. $x^2 + 4x + 4$

2. $p^2 + 14p + 49$

3. $16z^2 + 40z + 25$

4. $36x^2 + 84x + 49$

5. $x^2 + 13x + 169$

6. $y^2 + 12y + 144$

7. $x^2 - 10x + 25$

8. $y^2 - 16y + 64$

9. $36z^2 - 60z + 25$

10. $64m^2 - 48m + 9$

11. $c^2 - 16d + 16d^2$

12. $z^2 - 20z + 25$

13. $a^4 - 32a^2 + 256$

14. $b^4 - 98b^2 + 2401$

15. $16x^4 - 72x^2 + 81$

16. $625y^4 - 200y^2 + 16$

17. $3x^2 + 24x + 48$

18. $16x^2 + 32x + 16$

19. $2a^2 - 12a + 18$

20. $5x^2 - 70x + 245$

21. $p^3 + 2p^2q + pq^2$ **22.** $y^2z^2 + 4yz^3 + 4z^4$ **23.** $2p^5 - 4p^3q^2 + 2pq^4$ **24.** $5u^5 - 40u^3v^2 + 80uv^4$

25. $m^5 + 2m^3n^2 + mn^4$ **26.** $p^7 + 2p^5q^2 + p^3q^4$ **27.** $x^2 - 100$ **28.** $y^2 - 64$

29. $121 - c^2$ **30.** $196 - b^2$ **31.** $49a^2 - 4$ **32.** $25z^2 - 9$

33. $25 - 4y^2$ **34.** $64 - 9x^2$ **35.** $16u^2 - 9v^2$ **36.** $36a^2 - 25b^2$

37. $7z^2 - 28$ **38.** $8x^2 - 8$ **39.** $25 + 4p^2$ **40.** $9y^2 + 4$

41. $x^4 - 625$ **42.** $a^4 - 1296$ **43.** $256 - z^4$ **44.** $625 - b^4$

45. $x^8 - 1$ **46.** $y^8 - 256$ **47.** $x^3 - 27$ **48.** $z^3 - 343$

49. $a^3 + 64$ **50.** $m^3 + 1$ **51.** $27x^3 + 64y^3$ **52.** $125p^3 + 27q^3$

53. $8p^3 - 125q^3$ **54.** $27u^3 - 8v^3$ **55.** $p^4 + 64pq^3$ **56.** $125x^3y + y^4$

57. $81u^4 - 3uv^3$ **58.** $2a^3b - 54b^4$

59. A farmer leases you a plot of land for a garden. The plot is square, but it contains a smaller square, 15 feet on a side, that cannot be used.

   a. Write a polynomial for the area of the land you will be able to garden, assuming that the plot is x feet on a side.

   b. Factor the polynomial to find the dimensions of a rectangular plot with an equivalent area.

   c. If the square plot is 100 feet on a side, would you have more garden than another rectangular plot that measures 85 by 100 feet and that is available for the same rental fee? Explain.

60. A machinist must fabricate a square piece of metal that is y inches on a side and that has a triangle cut out of it with a base of x inches and a height of one-half the base.

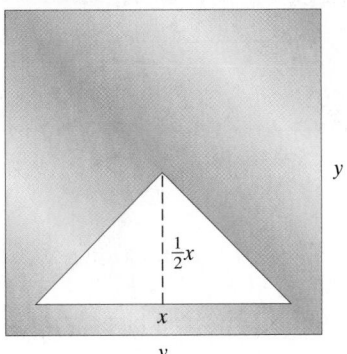

a. Write a polynomial for the area of the metal square with the triangle removed.

b. Factor the polynomial to find the dimensions of a rectangular piece of metal that would have an equivalent area.

c. If the square is 12 inches on a side and the triangle has a base of 4 inches, what are the dimensions of a rectangle with an equivalent area?

61. For a rock concert, promoters built a square stage and surrounded it with a mosh pit that was x feet on each side. A polynomial that represents the area of the stage and mosh pit is $1225 + 140x + 4x^2$ square feet. Factor the polynomial and determine the dimensions of the stage. If the mosh pit was 25 feet wide on each side of the stage, use the factored expression to find the area of the stage and mosh pit. Find the area of the stage, and use the two areas to determine the area of the mosh pit.

62. A landscaper is developing a design for a customer's yard that has the shape of a square. She plans to border the yard with a shrub trim that will be x feet wide. A polynomial that represents the central area of the yard not covered with the shrub trim is $4225 - 260x + 4x^2$ square feet. Factor the polynomial and determine the dimensions of the yard. If the shrub trim is 5 feet wide, use the factored expression to find the central area. Find the area of the yard, and use the two areas to determine the area of the shrub trim.

10.2 Calculator Exercises

Students often forget that a calculator can help them when they are factoring the difference of two squares or the sum or difference of two cubes. Many times, students are stumped and cannot see that a polynomial has a perfect-square value or a perfect-cube value. Remember, you can always use the square-root function or the cube-root function on your calculator to help recognize these forms. Determine whether the following are perfect squares or cubes by using your calculator (be careful, some may be both):

1. 841   **2.** 42,875   **3.** 361   **4.** 729

Now that you know you can check your calculator for perfect squares and cubes, use the calculator to help you factor the following polynomials:

5. $x^2 - 1225$ **6.** $a^3 - 6859$ **7.** $256y^2 - 441$

8. $343b^3 - 1728$ **9.** $z^2 - 729$ **10.** $z^3 - 729$

10.2 Writing Exercise

Many students make the mistake of believing that $(a + b)^2 = a^2 + b^2$ and that $(a - b)^2 = a^2 - b^2$, both of which are not true for all values of a and b. Based upon what you have learned in this section about special products, explain why these two mistaken beliefs are not true. The mis-

takes are made so often that they have a special name: Some call them the "Freshman's Dream," because they are usually made by first-year college students just beginning their math studies. Resolve now that you will not be one of those who make these two common mistakes.

10.3 Factoring Trinomials by Using the Trial-and-Error Method

OBJECTIVES

1 Factor trinomials of the form $ax^2 + bx + c, a = 1$.

2 Factor trinomials of the form $ax^2 + bx + c, a \neq 1$.

3 Model real-world situations by factoring trinomials.

APPLICATION

According to the *USA Volleyball Rule Book*, the standard playing court must be surrounded by an area (free zone) with a recommended minimum width of 2 meters. The area, in square meters, of the playing court with a free zone of x meters in width is given by the polynomial $162 + 54x + 4x^2$. Factor the polynomial and determine the dimensions of the playing court.

After completing this section, we will discuss this application further. See page 680.

In this section, we continue our discussion of factoring. We will now factor a quadratic trinomial: a trinomial of the form $ax^2 + bx + c$, with $a \neq 0, b \neq 0$, and $c \neq 0$. The **leading coefficient** of the quadratic trinomial is a, the coefficient of the quadratic term, x^2.

10.3.1 Factoring Quadratic Trinomials with $a = 1$

To determine a process for factoring a quadratic trinomial, we need to find a pattern.

We can see this pattern by using variables instead of numbers and multiplying.

$$
\begin{array}{cccc}
\text{F} & \text{O} & \text{I} & \text{L}
\end{array}
$$

$$(x + m)(x + n) = x^2 + nx + mx + mn \quad \text{Use the FOIL method.}$$

$$= x^2 + (n + m)x + mn \quad \text{Factor a common factor out of the middle two terms.}$$

$$= x^2 + \quad bx \quad + \quad c$$

This technique shows that $n + m = b$ and $mn = c$. In other words, look for two numbers whose product is c and whose sum is b.

FACTORING A QUADRATIC TRINOMIAL WITH A LEADING COEFFICIENT OF 1

To factor a quadratic trinomial with a leading coefficient $a = 1$, write

$$x^2 + bx + c = (x + m)(x + n), \text{ where } mn = c \text{ and } n + m = b$$

EXAMPLE 1   Factor $x^2 + x - 6$ completely.

Solution

$x^2 + x - 6$ $a = 1, b = 1, \text{ and } c = -6$

First, determine the factors of c, -6. Next, add the pairs of factors to determine the sum of b, 1. It is easier to do this in a chart so that we do not omit any possibilities.

Factor	Factor	Sum of Factors
1	-6	-5
-1	6	5
2	-3	-1
-2	3	1 $\leftarrow b$

Choose the last pair of factors, -2 and 3, because their sum is 1 (b).

$$x^2 + x - 6 = (x - 2)(x + 3)$$

We should always check the factors.

Check

Multiplication

$(x - 2)(x + 3) = x^2 + 3x - 2x - 6$
$ = x^2 + x - 6$

Numeric

X	Y1	Y2
-3	0	0
-2	-4	-4
-1	-6	-6
0	-6	-6
1	-4	-4
2	0	0
3	6	6

X = -3

Y1 = $x^2 + x - 6$
Y2 = $(x - 2)(x + 3)$
The table shows that
Y1 = Y2.

Graphic

Y1 = $x^2 + x - 6$

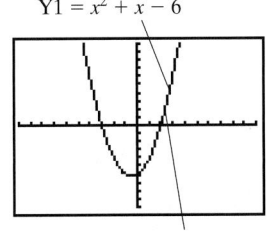

Y2 = $(x - 2)(x + 3)$
$(-10, 10, -10, 10)$
The graph shows that Y1 and Y2 are coinciding lines.

We could set up the previous chart of factors on the calculator as shown in Figure 10.2b.

T E C H N O L O G Y

CHART OF FACTORS

Determine the factors of -6 (ac) that add to 1(b).

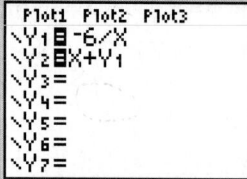

Figure 10.2a   Figure 10.2b

For Figure 10.2a,

Set up the table.

Set the first column for x equal to integer values starting at 1.

| 2nd | | TBLSET | | 1 | | ENTER | | 1 | | ENTER | | ▼ | | ENTER |

Set the second column, Y1, equal to ac, -6, divided by the integer in the first column, x.

| Y= | | (−) | | 6 | | ÷ | | X,T,θ,n |

Set the third column, Y2, equal to the sum of the first two columns, x and Y1.

| ▼ | | X,T,θ,n | | + | | VARS | | ▶ | | 1 | | 1 |

For Figure 10.2b,

View the table.

| 2nd | | TABLE |

Find a value in the Y2 column that is equal to b, 1. The factors of ac, -6, that have a sum of b, 1, are the integer values of x and Y1 corresponding to that value of b in Y2.

The factors of -6 that have a sum of 1 are -2 and 3.

To save time in choosing possible factors, we need to see a pattern. Complete the following set of exercises.

Discovery 2

Determining Signs for the Factors of c

Factor the following quadratic trinomials of the form $x^2 + bx + c$.

1. $x^2 + 5x + 6$

If c is positive and b is positive, then the factors of c are both

_____.

2. $x^2 - 5x + 6$

If c is positive and b is negative, then the factors of c are both

_____.

3. $x^2 + x - 6$

If c is negative and b is positive, then the factors of c have different signs. The factor with the larger absolute value is _____.

4. $x^2 - x - 6$

If c is negative and b is negative, then the factors of c have different signs. The factor with the larger absolute value is _____.

Write a rule to determine when the factors are positive or negative.

SIGNS OF FACTORS

Given $x^2 + bx + c$, the sign of the factors are determined in the following way:

If c is positive, then the factors both have the same sign as b.
If c is negative, then the factors have different signs. (The factor with the larger absolute value has the same sign as b.)

EXAMPLE 2 Factor completely and check.

a. $z^2 + 11z + 24$ **b.** $x^2 + 4x - 7$ **c.** $2x^2 + 10x + 8$

d. $x^4 + 13x^2 + 36$

Solution

a. $z^2 + 11z + 24$ $a = 1, b = 11,$ and $c = 24$

Note: c is positive, so both factors have the sign of b, which is positive.
Write these factors of 24 (c):

Factor	Factor	Sum of Factors
1	24	25
2	12	14
3	8	11 $\leftarrow b$
4	6	10

X	Y1	Y2
1	24	25
2	12	14
3	8	11
4	6	10
5	4.8	9.8
6	4	10
7	3.4286	10.429

X=3

Y1 = 24/x
Y2 = x + Y1

$z^2 + 11z + 24 = (z + 3)(z + 8)$

The check is left for you. *Note:* The variable z must be changed to x in order to enter the functions on your calculator.

b. $x^2 + 4x - 7$ $a = 1, b = 4,$ and $c = -7$

Note: c is negative, so the factors have different signs. The factor with the larger absolute value must be positive. Write these factors of -7 (c):

Factor	Factor	Sum of Factors
-1	7	6

The possible factors of -7 do not add to the desired sum of 4.
Therefore, $x^2 + 4x - 7$ does not factor.

c. $2x^2 + 10x + 8$

First, factor out the common factor.

$$2(x^2 + 5x + 4) \qquad a = 1, b = 5, \text{ and } c = 4$$

Note: c is positive, so both factors must have the sign of b, positive.
Write these factors of 4 (c):

Factor	Factor	Sum of Factors
1	4	5 ← b
2	2	4

$$2x^2 + 10x + 8 = 2(x^2 + 5x + 4)$$
$$= 2(x + 1)(x + 4)$$

The check is left for you. *Note:* If you multiply, multiply all three factors to check.

d. The process of factoring used in this section can be applied to trinomials that are not quadratic. Any polynomial that can be written in the form $y^2 + by + c$, where y is x^2, x^3, and so forth, can be factored in a similar manner.

$x^4 + 13x^2 + 36$ is not a quadratic trinomial, because the degree of the polynomial is 4. However, it is a quadratic-like trinomial, with $y = x^2$ and $y^2 = (x^2)^2 = x^4$. Note that $a = 1$, $b = 13$, and $c = 36$.

Note: c is positive, so both factors must have the sign of b, which is positive.

Factor	Factor	Sum of Factors
1	36	37
2	18	20
3	12	15
4	9	13 ← b

The first term in the binomials is x^2.

$$x^4 + 13x^2 + 36 = (x^2 + 4)(x^2 + 9)$$

Check by multiplying:

$$(x^2 + 4)(x^2 + 9) = x^4 + 9x^2 + 4x^2 + 36$$
$$= x^4 + 13x^2 + 36$$

10.3.1 Checkup

Factor exercises 1–4 completely and check.

1. $b^2 - 12b + 35$ **2.** $c^2 - 5c - 24$

3. $6d^2 + 60d + 126$ **4.** $y^4 - 2y^2 - 3$

5. If a trinomial is quadratic-like, what must be true about the exponent of each of the three terms?

6. What do we mean by the quadratic term of a trinomial?

10.3.2 Factoring Quadratic Trinomials with $a \neq 1$

In this section, we factor a quadratic trinomial, $ax^2 + bx + c$, where $a \neq 1$. First, let's review the multiplication problem that results in such a polynomial.

$$\overset{\text{F}\quad\text{O}\quad\text{I}\quad\text{L}}{(x + 3)(2x - 1)} = 2x^2 - x + 6x - 3 = 2x^2 + 5x - 3$$

We want to reverse this multiplication by factoring the trinomial into two binomials. We call this the **trial-and-error method** because we make a guess and check to see whether it is correct.

EXAMPLE 3 Factor $2x^2 + 5x - 3$ completely.

Solution

The product of the first terms in each binomial factor must equal $2x^2$. The product of the last terms in each binomial factor must equal -3.

$$2x^2 + 5x - 3 = (\_\_\_)(\_\_\_)$$

First, determine the possible factors that result in each binomial factor.

$$2x^2 = x \cdot 2x \qquad -3 = 1 \cdot -3$$
$$= -1 \cdot 3$$

Next, fill in the binomial factors with all the possible combinations, and multiply to find the middle term of the trinomial. Remember that the middle term is the sum of the product of the outer terms and the product of the inner terms.

$$2x^2 + 5x - 3 = (\_\_\_\_)(\_\_\_\_)$$

	Middle term
$(\ x + 1\ )(2x - 3\ )$	$-3x + 2x = -x$
$(\ x - 1\ )(2x + 3\ )$	$3x - 2x = x$
$(2x + 1\ )(\ x - 3\ )$	$-6x + x = -5x$
$(2x - 1\ )(\ x + 3\ )$	$6x - x = 5x$

Check

Y1 = $2x^2 + 5x - 3$

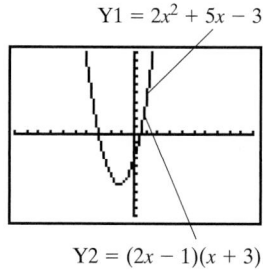

Y2 = $(2x - 1)(x + 3)$

$(10, 10, -10, 10)$

The factors $(2x - 1)(x + 3)$ result in the middle term $5x$. Multiply the two binomials to check your factors completely.

$$(2x - 1)(x + 3) = 2x^2 + 6x - x - 3 = 2x^2 + 5x - 3$$

A graphic check will result in equivalent graphs for the two functions Y1 = $2x^2 + 5x - 3$ and Y2 = $(2x - 1)(x + 3)$.

Therefore, $2x^2 + 5x - 3 = (2x - 1)(x + 3)$.

FACTORING A QUADRATIC TRINOMIAL WITH A LEADING COEFFICIENT NOT EQUAL TO 1 BY USING THE TRIAL-AND-ERROR METHOD

To factor a quadratic trinomial by using the trial-and-error method,

- Factor out the GCF from all terms.
- Determine the possible factors of the first term and the last term.
- Write a set of binomial factors for each combination of the factors.
- Multiply the factors to determine which set results in the middle term of the trinomial.

Check the factors by multiplying to obtain the original trinomial or by entering into your calculator the original polynomial and the factors as separate functions and determining that the table of values have equivalent Y1 and Y2 columns or the graphs are equivalent.

EXAMPLE 4   Factor $6x^2 + 19x + 8$ completely.

Solution

First, determine the factors for the first and last terms.

$$6x^2 = x \cdot 6x \qquad\qquad 8 = 1 \cdot 8$$
$$= 2x \cdot 3x \qquad\qquad = -1 \cdot -8$$
$$= -x \cdot -6x \qquad\qquad = 2 \cdot 4$$
$$= -2x \cdot -3x \qquad\qquad = -2 \cdot -4$$

Before we fill in the binomial factors, observe that we need to use only the positive factors, because the trinomial has all positive terms. This eliminates half of the combinations.

$$6x^2 = x \cdot 6x \qquad 8 = 1 \cdot 8$$
$$= 2x \cdot 3x \qquad\quad = 2 \cdot 4$$

Fill in the binomial factors with all the combinations, and multiply to find the middle term of the original trinomial.

$$6x^2 + 19x + 8 = (\underline{\quad\quad})(\underline{\quad\quad})$$

	Middle term
$(\ x + \underline{1}\ )(6x + \underline{8}\ )$	$8x + 6x = 14x$
$(\ x + \underline{2}\ )(6x + \underline{4}\ )$	$4x + 12x = 16x$
$(6x + \underline{1}\ )(\ x + \underline{8}\ )$	$48x + x = 49x$
$(6x + \underline{2}\ )(\ x + \underline{4}\ )$	$24x + 2x = 26x$
$(2x + \underline{1}\ )(3x + \underline{8}\ )$	$16x + 3x = 19x$
$(2x + \underline{2}\ )(3x + \underline{4}\ )$	$8x + 6x = 14x$
$(3x + \underline{1}\ )(2x + \underline{8}\ )$	$24x + 2x = 26x$
$(3x + \underline{2}\ )(2x + \underline{4}\ )$	$12x + 4x = 16x$

The factors $(2x + 1)$ and $(3x + 8)$, when multiplied together, result in the correct middle term. Multiply the factors to check.

$$(2x + 1)(3x + 8) = 6x^2 + 16x + 3x + 8 = 6x^2 + 19x + 8$$

A numeric check with $Y1 = 6x^2 + 19x + 8$ and $Y2 = (2x + 1)(3x + 8)$ results in $Y1 = Y2$.

Therefore, $6x^2 + 19x + 8 = (2x + 1)(3x + 8)$.

Check

X	Y1	Y2
-3	5	5
-2	-6	-6
-1	-5	-5
0	8	8
1	33	33
2	70	70
3	119	119

X= -3

$Y1 = 6x^2 + 19x + 8$
$Y2 = (2x + 1)(3x + 8)$

HELPING HAND We could have eliminated more of the combinations if we had been more observant. For example, six of the combinations had a common factor in one of the binomials. We know that that is not possible, because the original trinomial did not have a common factor. If we had observed this fact, we would have had only two possible combinations: $(6x + 1)(x + 8)$ and $(2x + 1)(3x + 8)$.

EXAMPLE 5   Factor completely.

a. $8x^2 - 26x + 15$ **b.** $6x^2 - 9x - 10$ **c.** $12x^2 - 14xy - 6y^2$
d. $3x^4 + 2x^2 - 8$

Solution

a. $8x^2 - 26x + 15$

First, determine the possible factors of the first and last terms.

$$8x^2 = x \cdot 8x \qquad 15 = 1 \cdot 15$$
$$= 2x \cdot 4x \qquad\quad = -1 \cdot -15$$
$$= 3 \cdot 5$$
$$= -3 \cdot -5$$

Next, fill in the binomial factors with all the possible combinations, and multiply to find the correct middle term in the original trinomial.

Before we begin, observe that the middle term is negative. Therefore, the positive factors of 15 are not possible combinations, as they will result in only positive middle terms. Knowing this reduces the possibilities.

$$8x^2 = x \cdot 8x \qquad 15 = -1 \cdot -15$$
$$= 2x \cdot 4x \qquad\quad = -3 \cdot -5$$

$8x^2 - 26x + 15 = (\underline{\quad}\ \underline{\quad})(\underline{\quad}\ \underline{\quad})$ Middle term

$(\ x - 1\ )(8x - 15)$	$-15x - 8x = -23x$
$(\ x - 3\ )(8x - 5\ )$	$-5x - 24x = -29x$
$(8x - 1\ )(\ x - 15)$	$-120x - x = -121x$
$(8x - 3\ )(\ x - 5\ )$	$-40x - 3x = -43x$
$(2x - 1\ )(4x - 15)$	$-30x - 4x = -34x$
$(2x - 3\ )(4x - 5\ )$	$-10x - 12x = -22x$
$(4x - 1\ )(2x - 15)$	$-60x - 2x = -62x$
$(4x - 3\ )(2x - 5\ )$	$-20x - 6x = -26x$

The factors $(4x - 3)$ and $(2x - 5)$, when multiplied together, result in the middle term, $-26x$.

$$8x^2 - 26x + 15 = (4x - 3)(2x - 5)$$

The check is left for you.

b. $6x^2 - 9x - 10$

First, determine the possible factors of the first and last terms.

$$6x^2 = x \cdot 6x \qquad -10 = 1 \cdot -10$$
$$= 2x \cdot 3x \qquad\qquad = -1 \cdot 10$$
$$= 2 \cdot -5$$
$$= -2 \cdot 5$$

Next, fill in the binomial factors with all the possible combinations, and multiply to find the correct middle term in the original trinomial.

$6x^2 - 9x - 10 = (\underline{\quad}\ \underline{\quad})(\underline{\quad}\ \underline{\quad})$ Middle term

$(\ x + 1\ )(6x - 10)$	
$(\ x - 1\ )(6x + 10)$	
$(\ x + 2\ )(6x - 5\ )$	$-5x + 12x = 7x$
$(\ x - 2\ )(6x + 5\ )$	$5x - 12x = -7x$
$(6x + 1\ )(\ x - 10)$	$-60x + x = -59x$
$(6x - 1\ )(\ x + 10)$	$60x - x = 59x$

$$\underline{(6x + \underline{2}\ )(\ x - \underline{5}\ )}$$
$$\underline{(6x - \underline{2}\ )(\ x + \underline{5}\ )}$$
$$\underline{(2x + \underline{1}\ )(3x - \underline{10})} \quad {\small -20x + 3x = -17x}$$
$$\underline{(2x - \underline{1}\ )(3x + \underline{10})} \quad {\small 20x - 3x = 17x}$$
$$\underline{(2x + \underline{2}\ )(3x - \underline{5}\ )}$$
$$\underline{(2x - \underline{2}\ )(3x + \underline{5}\ )}$$
$$\underline{(3x + \underline{1}\ )(2x - \underline{10})}$$
$$\underline{(3x - \underline{1}\ )(2x + \underline{10})}$$
$$\underline{(3x + \underline{2}\ )(2x - \underline{5}\ )} \quad {\small -15x + 4x = -11x}$$
$$\underline{(3x - \underline{2}\ )(2x + \underline{5}\ )} \quad {\small 15x - 4x = 11x}$$

Note: We did not find the middle term for the binomial factors with common factors.

No factors result in the middle term.

The trinomial $6x^2 - 9x - 10$ does not factor.

c. $12x^2 - 14xy - 6y^2$

First, factor out the greatest common factor, 2.
$$12x^2 - 14xy - 6y^2 = 2(6x^2 - 7xy - 3y^2)$$

Now determine the possible factors of the first and last terms.

$$6x^2 = x \cdot 6x \qquad -3y^2 = y \cdot -3y$$
$$= 2x \cdot 3x \qquad = -y \cdot 3y$$

Next, fill in the binomial factors with all the possible combinations, and multiply to find the correct middle term in the original trinomial.

$$6x^2 - 7xy - 3y^2 = (\underline{\quad}\ \underline{\quad})(\underline{\quad}\ \underline{\quad}) \qquad {\small \text{Middle term}}$$
$$\underline{(\ x + \underline{y})(6x - \underline{3y})}$$
$$\underline{(\ x - \underline{y})(6x + \underline{3y})}$$
$$\underline{(6x + \underline{y})(\ x - \underline{3y})} \quad {\small -18xy + xy = -17xy}$$
$$\underline{(6x - \underline{y})(\ x + \underline{3y})} \quad {\small 18xy - xy = 17xy}$$
$$\underline{(2x + \underline{y})(3x - \underline{3y})}$$
$$\underline{(2x - \underline{y})(3x + \underline{3y})}$$
$$\underline{(3x + \underline{y})(2x - \underline{3y})} \quad {\small -9xy + 2xy = -7xy}$$
$$\underline{(3x - \underline{y})(2x + \underline{3y})} \quad {\small 9xy - 2xy = 7xy}$$

Note: The binomial factors with common factors are not possible solutions. The factors $(3x + y)$ and $(2x - 3y)$, when multiplied together, result in the correct middle term.

$$12x^2 - 14xy - 6y^2 = 2(6x^2 - 7xy - 3y^2)$$
$$= 2(3x + y)(2x - 3y)$$

The check is left for you.

d. $3x^4 + 2x^2 - 8$

Quadratic-like trinomials are also found with the leading coefficient not equal to 1. Just as before, we factor them with the same procedure we used for quadratic trinomials.

First, determine the possible factors of the first and last terms. Since the middle term is a factor of x^2, we will need to factor the first term $3x^4$ into factors of x^2.

$$3x^4 = x^2 \cdot 3x^2 \qquad -8 = 1 \cdot -8$$
$$= -1 \cdot 8$$
$$= 2 \cdot -4$$
$$= -2 \cdot 4$$

Next, fill in the binomial factors with all the possible combinations, and multiply to find the correct middle term in the original trinomial.

$$3x^4 + 2x^2 - 8 = (\underline{\quad}\ \underline{\quad})(\underline{\quad}\ \underline{\quad}) \qquad \text{Middle term}$$

$$(x^2 + \underline{1})(3x^2 - \underline{8}) \qquad -8x^2 + 3x^2 = -5x^2$$
$$(x^2 - \underline{1})(3x^2 + \underline{8}) \qquad 8x^2 - 3x^2 = 5x^2$$
$$(x^2 + \underline{2})(3x^2 - \underline{4}) \qquad -4x^2 + 6x^2 = 2x^2$$

We do not need to continue, since we have found the needed middle term. The factors $(x^2 + 2)$ and $(3x^2 - 4)$, when multiplied together result in the correct middle term.

$$3x^4 + 2x^2 - 8 = (x^2 + 2)(3x^2 - 4)$$

The check is left for you.

10.3.2 Checkup

Factor exercises 1–4 completely and check.

1. $2x^2 - 9x - 5$ **2.** $15x^2 + 7x + 2$

3. $30x^2 + 65x + 30$ **4.** $6x^4 - 23x^2 + 20$

5. In this section, it was stated that if the original trinomial being factored does not have a common factor, then neither of the binomial factors can have a common factor. Explain why this is so.

10.3.3 Modeling the Real World

Trinomials are used to model many real-world situations, such as the equation of motion for a projectile or the equation describing the impact that occurs in some kinds of collisions. In some cases, the trinomial can be factored by using the methods described in this section, enabling us to solve the equations directly. In geometry problems, if the area of a figure is described by a trinomial, factoring it can help us determine the dimensions of the figure.

EXAMPLE 6

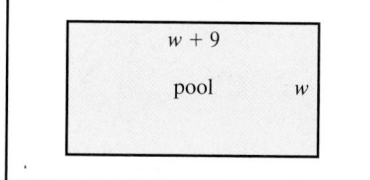

Mandy has designed a swimming pool area for a customer. The length of the pool is to be 9 feet more than the width. The pool is to have a concrete walk around it, with equal width on opposite sides. However, the walk is wider along one set of sides.

Determine the two walk widths if Mandy's figures show that the total area covered by the walk and the pool is $(w^2 + 18w + 56)$ square feet, where w is the pool width.

Solution

To determine the outside dimensions of the walks, find the factors that are multiplied to obtain the total area, $w^2 + 18w + 56$.

$$w^2 + 18w + 56 = (w + 4)(w + 14)$$

The outside dimensions are $(w + 4)$ feet and $(w + 14)$ feet.

The outside width is $(w + 4)$. To determine the width of the walk, we subtract the width of the pool, w, and divide by 2. (The walks along each side are equal.)

$$\frac{(w + 4) - w}{2} = \frac{4}{2} = 2$$

The walk is 2 feet wide along the pool's length.

e length is $(w + 14)$. To determine the width of the walk on this side, we subtract the length of the pool, $w + 9$, and divide by 2.

$$\frac{(w + 14) - (w + 9)}{2} = \frac{5}{2} = 2\frac{1}{2}$$

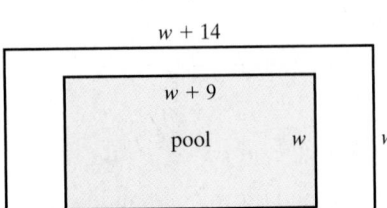

The walk is $2\frac{1}{2}$ feet along the pool's width.

APPLICATION

According to the *USA Volleyball Rule Book*, the standard playing court must be surrounded by an area (free zone) with a recommended minimum width of 2 meters. The area, in square meters, of the playing court with a free zone of x meters in width is given by the polynomial $162 + 54x + 4x^2$. Factor the polynomial and determine the dimensions of the playing court.

Discussion

The dimensions of the area that includes the playing court and the free zone can be found by factoring.

$$162 + 54x + 4x^2 = (18 + 2x)(9 + 2x)$$

The total area is $(18 + 2x)$ meters by $(9 + 2x)$ meters.

To determine the dimensions of the playing court, we need to subtract $2x$ from the total length and

$2x$ from the total width. We must subtract $2x$ because the free zone adds x meters on each side of the playing court, or $2x$.

$$(18 + 2x) - 2x = 18$$
$$(9 + 2x) - 2x = 9$$

The area of the playing court is 18 meters by 9 meters.

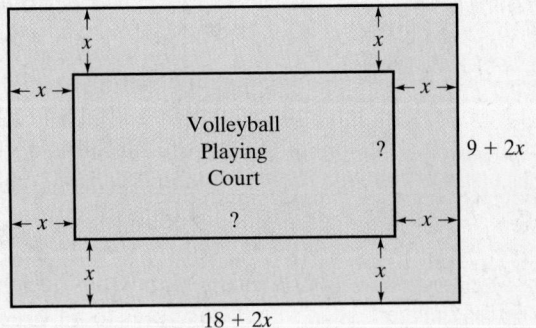

10.3.3 Checkup

1. Mandy discovered that she had reversed the numbers on the constant term of the trinomial in Example 6. The trinomial should have been $w^2 + 18w + 65$. Rework the problem to find what the two widths of the walk should then be.

2. The recommended layout for a tennis court on which singles play states that the playing court must be surrounded by a free area such that the baseline at each end is a specified distance from a fixed obstruction and the sideline is a specified distance from a fixed obstruction.

The area, in square feet, of the playing area with a free zone of x feet from the baseline and x feet from a sideline is given by the polynomial $4x^2 + 210x + 2106$. Factor the polynomial, and determine the dimensions of the playing area of the tennis court.

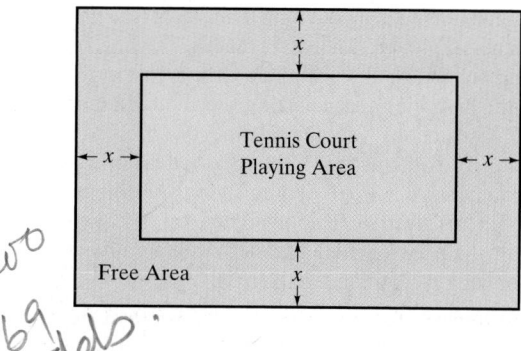

10.3 Exercises

In exercises 1–64, factor completely and check.

1. $x^2 + 14x + 45$ **2.** $p^2 + 14p + 48$ **3.** $y^2 - 15y + 56$ **4.** $u^2 + 11u - 26$

5. $p^2 - 9p - 36$ **6.** $v^2 - 12v - 13$ **7.** $z^2 + 6z + 12$ **8.** $m^2 - 9m + 15$

9. $x^4 + 25x^2 + 144$ **10.** $b^4 + 29b^2 + 100$ **11.** $x^4 - 2x^2 - 3$ **12.** $d^4 + 18d^2 - 175$

13. $3a^2 + 48a + 165$ **14.** $7b^2 - 70b + 168$ **15.** $4c^2 + 44c - 104$ **16.** $7p^2 - 7p - 140$

17. $x^2 - 11xy + 24y^2$ **18.** $a^2 - 17ab + 72b^2$ **19.** $x^2 + 11xy - 12y^2$ **20.** $a^2 - 9ab - 90b^2$

21. $-3a^2 - 15ab - 18b^2$ **22.** $-5x^2 + 35xy - 60y^2$ **23.** $-2x^2 + 14xy + 36y^2$ **24.** $-6x^2 + 24xy + 126y^2$

25. $3x^2 + 10x + 3$ **26.** $4x^2 + 12x + 7$ **27.** $2x^2 - 15x + 7$ **28.** $3x^2 - 34x + 11$

29. $3x^2 - x - 2$ **30.** $5x^2 - 2x - 3$ **31.** $5m^2 + 9m - 2$ **32.** $7p^2 + 34p - 5$

33. $2m^2 + 7m - 3$ **34.** $3t^2 - 12t + 4$ **35.** $4a^2 + 25a + 6$ **36.** $10c^2 + 41c + 4$

37. $9d^2 - 13d + 4$ **38.** $8b^2 - 17b + 9$ **39.** $6x^2 - 23x - 4$ **40.** $4x^2 - 31x - 8$

41. $8y^2 + 7y - 18$ **42.** $6z^2 + 19z - 20$ **43.** $6b^2 + 17b + 12$ **44.** $15a^2 + 28a + 12$

45. $20x^2 - 31x + 12$ **46.** $14z^2 + 13z - 12$ **47.** $18x^2 - 9x - 20$ **48.** $8x^2 - 22x - 63$

49. $18p^2 - 57p - 21$ **50.** $-63x^2 - 30x + 48$ **51.** $2x^4 + 11x^2 + 9$ **52.** $24x^4 - 22x^2 + 3$

53. $4m^4 + 13m^2 - 12$ **54.** $5z^4 + 39z^2 - 54$ **55.** $5x - 6x^2 + 56$ **56.** $21 - 5b - 4b^2$

57. $6x^2 + 5xy - 6y^2$ **58.** $14a^2 + 25ab + 6b^2$ **59.** $4u^2 - 39uv + 56v^2$ **60.** $3x^2 - 20xy + 32y^2$

61. $9x^4 + 13x^2y^2 + 4y^4$ **62.** $16x^4 + 41x^2y^2 + 25y^4$ **63.** $10x^2y^2 + xy - 21$ **64.** $21p^2q^2 - 2pq - 8$

65. The length of a rectangle is 8 inches more than its width. After both the length and width have been increased, the area of the larger rectangle is given by $(w^2 + 14w + 24)$ square inches, where w is the original width.
 a. Factor the polynomial to find the increased width and length.
 b. By how much was the width increased?
 c. By how much was the length increased?

66. A rectangular poster design has a width that is 5 inches less than its length. The width is increased while the length is decreased. The area of the new poster is given by $(x^2 - 4x + 3)$ square inches, where x is the original length.
 a. Factor the polynomial to find the new width and length.
 b. By how much was the width increased?
 c. By how much was the length decreased?

67. A right triangle has its two legs increased by the same amount. After the increase, the area of the triangle is equal to the polynomial $(x^2 + \frac{21}{2}x + 20)$ square inches, where x is the length of the original triangle's small side.
 a. Determine the lengths of the legs of the enlarged triangle as a function of x.
 b. By how much were the legs increased?
 c. What is the expression for the length of the long leg of the original triangle?

68. A right triangle has its two legs decreased by the same amount. After the reduction, the area of the triangle is equal to the polynomial $(\frac{3}{2}x^2 - 12x + 24)$ square inches, where x is the length of the original triangle's small leg.
 a. Determine the lengths of the legs of the reduced triangle as a function of x.
 b. By how much were the legs decreased?
 c. What is the expression for the length of the long leg of the original triangle?

69. The recommended layout for a tennis court on which tournament doubles are held states that the playing court must be surrounded by a free area. The free area is defined such that the baseline at each end is a specified distance from a fixed obstruction. Also, the sideline is a specified distance from a fixed obstruction. If the specified distance is x feet, the area, in square feet, of the playing area, including the free zone, is given by the polynomial $4x^2 + 228x + 2808$. Factor the polynomial, and determine the dimensions of the playing area of the tennis court.

70. Requirements for the markings of a football field state that the end zones should be 10 yards long, but can be shortened if the field itself is too short. One football stadium, Sanford Stadium at the University of Georgia, is famous for being surrounded by English privet hedges, and games there are referred to as being played "between the hedges." Suppose this field has end zones that are x yards long and sidelines that are $2x$ yards wide. A polynomial that represents the area of the football field is $8x^2 + 506x + 5300$ square yards. Factor the polynomial, and determine the dimensions of the playing area of the field.

10.3 Calculator Exercises

It is sometimes difficult to factor a trinomial by using the trial-and-error method if the coefficients are not simple. In these situations, you can use a different calculator technique to find the factors. The technique is based upon the concepts you learned in an earlier chapter concerning the x-intercepts of a graph. You learned that the x-intercepts of a graph represent the points at which the y-coordinate equals zero. You can use this information to help you factor a trinomial.

Suppose you wish to factor the trinomial $40x^2 + 91x - 153$. After trying several combinations of factors that do not work, you decide to use the graphing feature of your calculator to help you. The steps involved are as follows:

1. Define a function with the polynomial.
$$y = 40x^2 + 91x - 153$$

2. Store the function using the $\boxed{\text{Y=}}$ key, and graph the function on the calculator. Note that you will need to adjust the window appropriately to see the graph well.

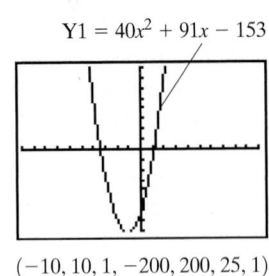

$(-10, 10, 1, -200, 200, 25, 1)$

3. Find the x-intercepts of the graph by using the zero option under the $\boxed{\text{2nd}}$ $\boxed{\text{CALC}}$ key.

After finding each x-intercept, quit the graph and use the Fraction option under the $\boxed{\text{MATH}}$ key to convert the x-values to fractions. (When you key $\boxed{\text{X,T,θ,n}}$, the intercept value will be stored there.)

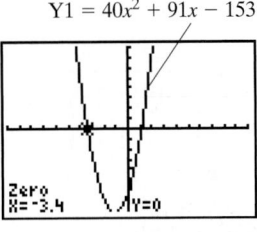
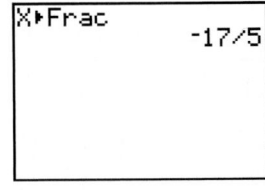
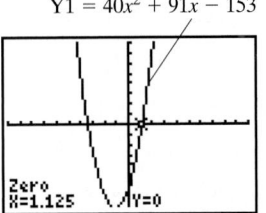
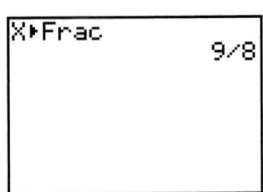

Note that if the trinomial factors into two binomial factors, the x-values of the x-intercepts must be rational numbers and should always be able to be written as fractions.

$$x = 1.125 \text{ is equivalent to } x = \frac{9}{8}$$

$$x = -3.4 \text{ is equivalent to } x = \frac{-17}{5}$$

4. Take the two equations solved for x, and rewrite them in standard form, $ax + b = 0$.

$$x = \frac{9}{8} \qquad\qquad x = \frac{-17}{5}$$

$$8x = 9 \qquad\qquad 5x = -17$$

$$8x - 9 = 0 \quad \text{and} \quad 5x + 17 = 0$$

5. The final two expressions on the left sides of the equations in step 4 are the factors of your original polynomial. Thus, $40x^2 + 91x - 153 = (8x - 9)(5x + 17)$. You can check this answer by multiplication.

Use the preceding method to factor the following trinomials:

1. $40x^2 + 21x - 180$ 2. $80x^2 - 232x + 117$
3. $108x^2 - 177x + 55$ 4. $180x^2 + 327x + 143$

10.3 Writing Exercise

Ned factored $-3x^2 - 15x + 312$ into $-3(x - 8)(x + 13)$. Lamar factored the same trinomial into $3(13 + x)(8 - x)$. Check to see whether the two answers multiply to the same trinomial. If they do, discuss how the two students may have obtained different answers, and try to identify the properties of real numbers that tell us that the two are equivalent expressions. State which answer you prefer and explain why.

10.4 Factoring Trinomials by Using the Grouping Method

OBJECTIVES

1 Factor trinomials of the form $ax^2 + bx + c$.

2 Model real-world situations by factoring polynomials.

APPLICATION

Delaware's Fred Rust Arena has an ice surface that is classified as an Olympic-sized surface. The length of the surface is twice its width. If a surrounding area of equal width borders the ice surface, then a polynomial for the area of the enclosed surface is $2x^2 + 15x + 25$ square feet. Factor the polynomial and determine the width of the border. If the width of the ice surface is 100 feet, evaluate the factors to determine the area of the enclosed surface and its dimensions.

After completing this section, we will discuss this application further. See page 689.

10.4.1 Factoring Quadratic Trinomials by Using the *ac* Method

Sometimes, another method may be easier to use to factor quadratic trinomials when the leading coefficient does not equal 1. This is an algorithm called the *ac* method. The **ac method** reverses the FOIL method by rewriting the middle term as a sum, using factors of the product of the coefficients *a* and *c* in the trinomial $ax^2 + bx + c$. For example,

$$\begin{array}{cccc} \text{F} & \text{O} & \text{I} & \text{L} \\ \downarrow & \downarrow & \downarrow & \downarrow \end{array}$$
$$(x + 3)(2x - 1) = 2x^2 - x + 6x - 3 = 2x^2 + 5x - 3$$

We want to reverse this process entirely and obtain

$$2x^2 + 5x - 3 = 2x^2 - x + 6x - 3 = (x + 3)(2x - 1)$$

To do this, first determine the product of the coefficients *a* and *c*. Since $a = 2$ and $c = -3$, the product $ac = 2(-3) = -6$.

Next, determine the factors of $ac(-6)$ whose sum is b (5).

This is the same procedure we used in factoring a quadratic trinomial with a leading coefficient of 1.

Factor	Factor	Sum of Factors	
1	−6	−5	
−1	6	5	← b
2	−3	−1	
−2	3	1	

X	Y1	Y2
-1	6	5
0	ERROR	ERROR
1	-6	-5
2	-3	-1
3	-2	1
4	-1.5	2.5
5	-1.2	3.8

X=-1

Y1 = (2 · −3)/x
Y2 = x + Y1

Remember, we can set up the table in the calculator as we did in the previous section. The only difference is that we set the second column to be the product (ac) divided by x.

We will use the factors $−1$ and 6, because their sum is the same as b, 5. Rewrite the middle term as a sum, using $−1$ and 6 as the new coefficients; that is,

$$\overset{5x}{\downarrow}$$

$$2x^2 + 5x − 3 = 2x^2 − x + 6x − 3$$

Factor the resulting polynomial, using grouping.

$$(2x^2 − x) + (6x − 3) = x(2x − 1) + 3(2x − 1) \qquad \text{Factor a common factor from each group.}$$

$$= (2x − 1)(x + 3) \qquad \text{Factor a common binomial from each term.}$$

Check the results.

Check

Multiplication

$$(2x − 1)(x + 3) = 2x^2 + 6x − x − 3$$
$$= 2x^2 + 5x − 3$$

Numeric

X	Y1	Y2
-3	0	0
-2	-5	-5
-1	-6	-6
0	-3	-3
1	4	4
2	15	15
3	30	30

X=-3

Y1 = $2x^2 + 5x − 3$
Y2 = $(2x − 1)(x + 3)$
The table shows that
Y1 = Y2.

Graphic

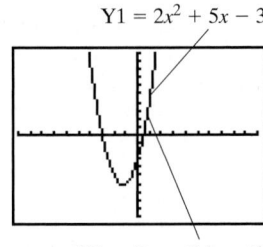

Y1 = $2x^2 + 5x − 3$

Y2 = $(2x − 1)(x + 3)$
$(−10, 10, −10, 10)$
The graph shows that Y1 and Y2 are coinciding lines.

FACTORING A QUADRATIC TRINOMIAL WITH A LEADING COEFFICIENT NOT EQUAL TO 1 BY USING THE AC METHOD

To factor a quadratic trinomial by using the ac method,

- Factor out the GCF from all terms.
- Determine the product ac.
- Determine those factors m and n of the product ac whose sum is b.
- Rewrite the trinomial as a four-term polynomial,
  $ax^2 + mx + nx + c$.
- Factor the resulting polynomial by using grouping.

Check the factors by multiplying to obtain the original trinomial or by entering into your calculator the original polynomial and the product of the factors as separate functions and determining that the table of values have equivalent Y1 and Y2 columns or the graphs are equivalent.

EXAMPLE 1 Factor $6x^2 + 19x + 8$ completely.

Solution

$6x^2 + 19x + 8$
Since $a = 6$, $b = 19$, and $c = 8$, it follows that $ac = 6 \cdot 8 = 48$.
    Using a table, we determine the factors of the product that result in a sum of 19 (b). Remember, since ac and b are both positive, we will not need to use the negative factors of 48.

Factor	Factor	Sum of Factors
1	48	49
2	24	26
3	16	19 ←b
4	12	16
6	8	14

X	Y₁	Y₂
1	48	49
2	24	26
3	16	19
4	12	16
5	9.6	14.6
6	8	14
7	6.8571	13.857

X=3

$Y1 = (6 \cdot 8) \div x$
$Y2 = x + Y1$

The calculator can help us find the factors.
    The factors 3 and 16 result in a sum of 19.
    Rewrite the trinomial, using these factors as coefficients of the two new middle terms.

$$6x^2 + 19x + 8 = 6x^2 + 3x + 16x + 8$$
$$= (6x^2 + 3x) + (16x + 8) \qquad \text{Group.}$$
$$= 3x(2x + 1) + 8(2x + 1) \qquad \text{Factor out a common factor from each group.}$$
$$= (2x + 1)(3x + 8) \qquad \text{Factor out a common binomial factor from each term.}$$

The check is left for you.
$6x^2 + 19x + 8 = (2x + 1)(3x + 8)$

EXAMPLE 2 Factor completely.

a. $8x^2 - 26xy + 15y^2$ **b.** $6x^2 - 9x - 10$ **c.** $12x^2 - 14x - 6$
d. $3x^4 + 2x^2 - 8$

Solution

a. $8x^2 - 26xy + 15y^2$

Since $a = 8$, $b = -26$, and $c = 15$, it follows that $ac = 8 \cdot 15 = 120$.

Using a table, determine the factors of the product that result in a sum of -26 (b). We will need to use only the negative factors of ac, because ac is positive and b is negative. *Note:* You may want to use the calculator table.

Factor	Factor	Sum of Factors
-1	-120	-121
-2	-60	-62
-3	-40	-43
-4	-30	-34
-5	-24	-29
-6	-20	-26 ←b
-8	-15	-23
-10	-12	-22

The factors -6 and -20 result in a sum of -26. Rewrite the trinomial, using these factors as coefficients of the two new middle terms.

$$8x^2 - 26xy + 15y^2 = 8x^2 - 6xy - 20xy + 15y^2$$

$$= (8x^2 - 6xy) + (-20xy + 15y^2) \qquad \text{Group.}$$

$$= 2x(4x - 3y) + (-5y)(4x - 3y) \qquad \text{Factor out common factors.}$$

$$= (4x - 3y)(2x - 5y) \qquad \text{Factor out common binomial factor.}$$

The check is left for you.

$$8x^2 - 26xy + 15y^2 = (4x - 3y)(2x - 5y)$$

b. $6x^2 - 9x - 10$

Since $a = 6, b = -9$, and $c = -10$, it follows that $ac = 6 \cdot -10 = -60$.

Using a table, determine the factors of the product that result in a sum of -9 (b). *Note:* You may want to use the calculator table.

Factor	Factor	Sum of Factors
1	−60	−59
−1	60	59
2	−30	−28
−2	30	28
3	−20	−17
−3	20	17
4	−15	−11
−4	15	11
5	−12	−7
−5	12	7
6	−10	−4
−6	10	4

No set of factors results in a sum of -9. The trinomial does not factor.

$$6x^2 - 9x - 10 \text{ does not factor.}$$

c. $12x^2 - 14x - 6$

First, factor out the common factor of 2.

$$12x^2 - 14x - 6 = 2(6x^2 - 7x - 3)$$

Since $a = 6, b = -7$, and $c = -3$, it follows that $ac = 6 \cdot -3 = -18$. Using a table, determine the factors of the product that result in a sum of -7 (b). *Note:* You may want to use the calculator table.

Factor	Factor	Sum of Factors	
1	−18	−17	
−1	18	17	
2	−9	−7	←b
−2	9	7	
3	−6	−3	
−3	6	3	

The factors 2 and -9 result in a sum of -7. Rewrite the trinomial, using these factors as coefficients of the two new middle terms.

$$12x^2 - 14x - 6 = 2(6x^2 - 7x - 3)$$

$$= 2(6x^2 + 2x - 9x - 3)$$

$$= 2[(6x^2 + 2x) + (-9x - 3)] \qquad \text{Group.}$$

$$= 2[2x(3x + 1) - 3(3x + 1)] \quad \text{Factor out common factors.}$$

$$= 2(3x + 1)(2x - 3) \quad \text{Factor out a common binomial factor.}$$

The check is left for you.

$$12x^2 - 14x - 6 = 2(3x + 1)(2x - 3)$$

d. $3x^4 + 2x^2 - 8$ is not a quadratic trinomial. However, the *ac* method may still be used to factor this trinomial.

Since $a = 3$, $b = 2$, and $c = -8$, it follows that $ac = 3 \cdot -8 = -24$. Using a table, determine the factors of the product that result in a sum of 2 (*b*). *Note:* You may want to use the calculator table.

Factor	Factor	Sum of Factors
1	−24	−23
−1	24	23
2	−12	−10
−2	12	10
3	−8	−5
−3	8	5
4	−6	−2
−4	6	2 ← *b*

The factors −4 and 6 result in a sum of 2. Rewrite the trinomial, using these factors as coefficients of the two new middle terms.

$$3x^4 + 2x^2 - 8 = 3x^4 - 4x^2 + 6x^2 - 8$$

$$= (3x^4 - 4x^2) + (6x^2 - 8) \quad \text{Group.}$$

$$= x^2(3x^2 - 4) + 2(3x^2 - 4) \quad \text{Factor out common factors.}$$

$$= (3x^2 - 4)(x^2 + 2) \quad \text{Factor out a common binomial factor.}$$

The check is left for you.

$$3x^4 + 2x^2 - 8 = (3x^2 - 4)(x^2 + 2)$$

 **HELPING HAND** Both the trial-and-error method and the *ac* method result in the same factors. As you work exercises, you will develop a sense regarding which method is easier for you to use. Sometimes, one method is shorter than the other. Other times, either method will be about as difficult.

✓ 10.4.1 Checkup

Factor exercises 1–4 completely, using the ac method of factoring a trinomial with a leading coefficient not equal to 1.

1. $2x^2 - 9x - 5$

2. $15x^2 + 7x + 2$

3. $30x^2 + 65x + 30$

4. $6x^4 - 23x^2 + 20$

5. If you had to factor a trinomial whose leading coefficient was not equal to 1, would you first try the trial-and-error method or the *ac* method? Explain your choice.

10.4.2 Modeling the Real World

In previous sections, we factored polynomials representing the area of a rectangle to determine its dimensions. This same procedure may be used to determine the dimensions of other geometric figures as well. However, it may be necessary to factor the polynomial into more than two factors.

EXAMPLE 3 Carri has a triangular flower bed in the corner of her yard where the driveway enters the street at a 90-degree angle. The length of the flower bed along the driveway is twice the length of the flower bed along the street. She plans to increase the length of the sides of her flower bed along the driveway and street by the same number of feet. She has determined that the area of the new flower bed will be $\left(x^2 + \frac{15}{2}x + \frac{25}{2}\right)$ square feet, given that x is the original number of feet on the street side of the flower bed.

a. Determine the lengths of the sides of the new flower bed in terms of x.

b. By how many feet does Carri plan to increase each side?

Solution

a. The area of a triangle is found by the formula $A = \frac{1}{2}bh$. Since the triangle is a right triangle, the base and height are the lengths of the sides along the driveway and the street, respectively. To determine the lengths of these sides, we must factor the area given into the form $\frac{1}{2}bh$.

$$
\begin{array}{ccc}
A & \frac{1}{2} & bh \\
\downarrow & \downarrow & \downarrow
\end{array}
$$

$$x^2 + \frac{15}{2}x + \frac{25}{2} = \frac{1}{2}(2x^2 + 15x + 25) \qquad \text{Factor out the } \frac{1}{2}.$$

$$= \frac{1}{2}(x + 5)(2x + 5) \qquad \text{Factor the } bh \text{ into two binomials.}$$

The sides of the new triangular bed are $(x + 5)$ feet and $(2x + 5)$ feet.

b. Since the original dimensions of the flower bed were x feet and $2x$ feet, Carri plans to add 5 feet to each side.

APPLICATION

Delaware's Fred Rust Arena has an ice surface that is classified as an Olympic-sized surface. The length of the surface is twice its width. If a surrounding area of equal width borders the ice surface, then a polynomial for the area of the enclosed surface is $2x^2 + 15x + 25$ square feet. Factor the polynomial and determine the width of the border. If the width of the ice surface is 100 feet, evaluate the factors to determine the area of the enclosed surface and its dimensions.

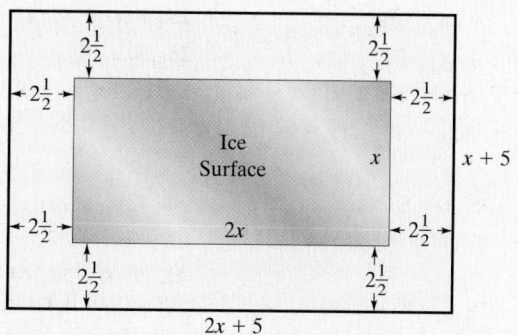

Discussion

$ac = 2(25) = 50$

10 and 5 are factors of 50 that add to 15.

$2x^2 + 15x + 25 = 2x^2 + 10x + 5x + 25$

$\qquad\qquad\qquad = (2x^2 + 10x) + (5x + 25)$

$\qquad\qquad\qquad = 2x(x + 5) + 5(x + 5)$

$\qquad\qquad\qquad = (2x + 5)(x + 5)$

Given the factors, five feet is added to the length and width or the walkway is $2\frac{1}{2}$ feet wide.

Let $x = 100$ feet and evaluate the factors.

$(2x + 5)(x + 5) = (2 \cdot 100 + 5)(100 + 5)$

$\qquad\qquad\qquad = 21{,}525$

The total area enclosed is 21,525 square feet.

Note that the factors 205 feet and 105 feet are the dimensions of the enclosed surface.

We can check to see that the original polynomial gave us the same value for the area.

$2x^2 + 15x + 25 = 2(100)^2 + 15(100) + 25$

$\qquad\qquad\qquad = 21{,}525$

 10.4.2 Checkup

1. Carri has another triangular garden, on the other side of the driveway, that has to be reduced. The length of the flower bed along the drive is three times the length along the street. She cut both sides of the triangle by the same amount. The reduced garden has an area equal to the polynomial $\frac{3}{2}x^2 - 8x + 8$, where x is the length of the small side of the original garden.

   a. Determine the sides of the reduced garden in terms of x.

   b. By how many feet did Carri reduce each side of the garden?

2. Most hockey games are played on a National Hockey League–sized rink. The length of the ice surface for these rinks is 30 feet more than twice the width of the surface. If a surrounding area of equal width borders the ice surface, a polynomial for the total area (border and ice surface) is $2x^2 + 75x + 675$. Factor the polynomial to determine expressions for the width and length of the total area. Determine the width of the border. If the width of the ice surface is 85 feet, evaluate the factors to determine the length and the width of the total area, and determine the total area.

10.4 Exercises

In exercises 1–54, use the ac method to factor completely. Check the factors by multiplying, either numerically or graphically.

1. $3x^2 + 19x + 20$

2. $5x^2 + 19x + 12$

3. $6x^2 + 31x + 35$

4. $28x^2 + 29x + 6$

5. $10q^2 + 27q + 11$

6. $8p^2 + 30p + 13$

7. $4x^2 + 14x + 45$

8. $25x^2 + 25x + 16$

9. $20z^2 - 51z + 12$

10. $40y^2 - 31y + 6$

11. $6m^2 - 13m + 6$

12. $10k^2 - 29k + 10$

13. $56p^2 + 13p - 3$

14. $72x^2 + 23x - 4$

15. $15p^2 - 26p + 8$

16. $20q^2 - 49q + 35$

17. $32x^2 - 12x - 5$

18. $24x^2 - 14x - 3$

19. $28k^2 - 19k - 99$

20. $56k^2 - 23k - 45$

21. $24m^2 + 77m - 117$

22. $30x^2 + 47x - 60$

23. $40x^2 - 148x + 28$

24. $54x^2 - 84x + 16$

25. $30x^2 + 87x + 30$

26. $30x^2 + 65x + 30$

27. $16x^2 + 42x - 18$

28. $18x^2 + 64x - 32$

29. $-18x^2 + 75x - 72$

30. $-24x^2 + 76x - 40$

31. $16x^2 - 52x - 80$

32. $24y^2 - 6y - 66$

33. $6x^3 + 25x^2 + 25x$

34. $10y^3 + 29y^2 + 35y$

35. $6x^4 + 7x^2 + 2$

36. $10x^4 + 17x^2 + 3$

37. $8x^4 + 46x^2 + 63$

38. $12y^4 + 28y^2 + 15$

39. $6x^4 - 19x^2 + 15$

40. $12x^4 - 41x^2 + 35$

41. $12x^4 - 17x^2 + 6$

42. $6x^4 - 29x^2 + 35$

43. $8m^4 - 2m^2 - 15$

44. $12p^4 - 8p^2 - 15$

45. $15x^4 + 14x^2 - 16$

46. $18x^4 + 19x^2 - 12$

47. $60y^4 + 57y^2 - 84$

48. $56x^4 + 46x^2 - 30$

49. $-24x^4 + 32x^2 - 10$

50. $-45y^4 - 66y^2 - 24$

51. $6x^2 - 7xy - 20y^2$

52. $21x^2 - 22xy - 8y^2$

53. $40p^2 - 67pq + 28q^2$

54. $54p^2 - 57pq + 10q^2$

55. Shannon had a triangular-shaped patio behind her house. The height of the triangle was three-fourths the length of the base. She increased both the height and the base of the triangle by the same amount. As a result, the area of the enlarged patio became $6x^2 + \frac{49}{2}x + \frac{49}{2}$ square feet, where the base of the original patio was $3x$ feet. Determine the base and the height of the enlarged patio in terms of x. By how many feet did Shannon increase the height and the base?

56. Tyler constructed a walkway along one side of a triangular corner of his yard. Before constructing the walkway, the triangular yard had a base that was five times as large as its height. After Tyler constructed the walkway, the area of the triangular yard was $\frac{5}{2}x^2 - 18x + 18$ square feet. Factor this polynomial to determine the dimensions of the yard. By how many feet was each side of the triangular yard reduced?

57. For doubles play in badminton, the recommended free space surrounding the court is at least 2 meters on all four sides. If the length of a badminton court is 1 meter more than twice its width, and the free space on each side is of equal width, the area of the court and the free space can be approximated by the polynomial $2x^2 + 19x + 42$ square meters. Factor the polynomial and determine the width of the free space. If the width of the court is approximately 6 meters, evaluate the factors to determine the length and width of the total area, including the free space, and the amount of that area.

58. The recommended playing surface for an English standard billiard table has a length that is twice the width. If the table has a free space of equal width around each side, a polynomial for the total area is $2x^2 + 21x + 49$ square meters. Factor the polynomial and determine the width of the free space. If the width of the playing area is 1.8 meters, evaluate the factors to determine the dimensions of the total area, as well as the area itself.

10.4 Calculator Exercises

In using the ac method, if you store the values of a, b, and c from the trinomial $ax^2 + bx + c$ in your calculator, you can proceed from one trinomial to another without having to reenter the expressions for Y1 and Y2. The following example illustrates the procedure:

Factor $85x^2 - 91x - 72$ using the ac method.

a. Store values for a, b, and c.

| 8 | 5 | STO▶ | ALPHA | A | ALPHA | : |

| (−) | 9 | 1 | STO▶ | ALPHA | B | ALPHA |

| : | (−) | 7 | 2 | STO▶ | ALPHA | C |

| ENTER |

b. Set Y1 = $\frac{ac}{x}$.

| Y= | ALPHA | A | × | ALPHA | C | ÷ |

| X,T,θ,n | ▼ |

c. Set Y2 to test for the values of m and n that make $m + n = b$ a true statement or Y2 = x + Y1 = B.

d. Set your table as before and view the table, scanning for a 1 in the Y2 column.

In this case, Y2 is 1 when the values of m and n are 45 and −136, respectively. Now you can use the grouping method to show that

$$85x^2 - 91x - 72 = 85x^2 - 136x + 45x - 72$$
$$= (85x^2 - 136x) + (45x - 72)$$
$$= 17x(5x - 8) + 9(5x - 8)$$
$$= (17x + 9)(5x - 8)$$

For subsequent problems, you need only store the new values for a, b, and c and go directly to the table to find the new values of m and n.

Note that when you are scanning the table for the pair of numbers, there is some logic to follow in deciding whether to scan in the negative-integer direction or the positive-integer direction. If the coefficient b is negative and c is positive, you will be searching for two negative integers and will want to scan in that direction. If the coefficient c is negative, you may scan in either direction, since one of the integers will be positive and the other negative. If the coefficients b and c are both positive, you need only scan in the positive-integer direction, since the two integers will be positive.

Use this procedure to factor the following polynomials by the ac method. Be sure to factor completely.

1. $96x^2 - 16x - 2$ 2. $24x^2 + 7x - 55$
3. $32x^2 + 102x + 81$ 4. $72x^2 - 99x + 34$
5. $4p^4 + 109p^2 + 225$

10.4 Writing Exercise

Chuck used the *ac* method to factor the trinomial $6x^2 + 3x - 30$ as shown.

Study Chuck's work, and explain why he has not correctly factored the trinomial completely.

$$6x^2 + 3x - 30 = 6x^2 + 15x - 12x - 30$$
$$= 3x(2x + 5) - 6(2x + 5)$$
$$= (3x - 6)(2x + 5)$$

10.5 General Strategies for Factoring

OBJECTIVES   1 Factor any given polynomial completely.
2 Model real-world situations by factoring polynomials.

APPLICATION

The first telephone receivers had a cone-shaped mouthpiece. The volume of the mouthpiece was represented by the polynomial $(\frac{1}{3}\pi x^3 + \pi x^2)$ cubic inches, where x is the length of the radius in inches. Determine a polynomial that expresses the height in terms of the radius.

After completing this section, we will discuss this application further. See page 696.

10.5.1 Factoring Polynomials

In the previous sections, we discussed methods of factoring polynomials. We now discuss a general strategy to use when any given polynomial is to be factored. The following steps should be employed to factor a polynomial completely.

✔ **STEPS TO FACTOR A POLYNOMIAL COMPLETELY**

1. Factor out the greatest common factor from all terms. Factor out -1 if the leading coefficient is negative.

Determine the type of polynomial remaining, and factor as indicated next for each. Repeat steps 2–4 until all polynomial factors (with the exception of a common monomial factor) do not factor.

2. Binomial—Difference of two squares:

$$a^2 - b^2 = (a + b)(a - b)$$

Sum or difference of two cubes:

$$a^3 + b^3 = (a + b)(a^2 - ab + b^2)$$
$$a^3 - b^3 = (a - b)(a^2 + ab + b^2)$$

Otherwise, the polynomial will not factor.

3. Trinomial—Perfect-square trinomial:

$$a^2 + 2ab + b^2 = (a + b)^2$$
$$a^2 - 2ab + b^2 = (a - b)^2$$

Quadratic trinomial with leading coefficient of 1:

$$x^2 + bx + c = (x + m)(x + n)$$

where $mn = c$ and $n + m = b$
Quadratic trinomial with leading coefficient greater than 1: trial-and-error method or ac method
Otherwise, the polynomial will not factor.

4. Four-term polynomial—Factor by grouping
Otherwise, the polynomial will not factor.

Check the factors by multiplying to obtain the original polynomial or by entering into your calculator the original polynomial and the product of the factors as separate functions and determining that the tables of values have equivalent Y1 and Y2 columns or the graphs are equivalent.

EXAMPLE 1 Factor completely and check.

a. $3y^2 - 243$ **b.** $16y^4 - 200y^2 + 625$

c. $6x^2 + 16x - 70$ **d.** $5x^3 + 12x^2 + 7x$

e. $2a^3b - 8a^2b^2 - 3a^2b^2 + 12ab^3$ **f.** $4x^2 + 14x + 49$

g. $15 + 8x + x^2$

Solution

a. $3y^2 - 243$

The GCF is 3. Therefore, factor out 3 from all terms.

$$3y^2 - 243 = 3(y^2 - 81)$$

The remaining factor is a binomial; in fact, it is a difference of squares:
$y^2 - 81 = (y)^2 - (9)^2 = (y + 9)(y - 9)$.

$$3y^2 - 243 = 3(y^2 - 81)$$
$$= 3(y + 9)(y - 9)$$

The remaining factors are binomials. However, neither is a difference of squares or a difference of cubes. The binomials will not factor.

The check is left for you.

$$3y^2 - 243 = 3(y + 9)(y - 9)$$

b. $16y^4 - 200y^2 + 625$

There are no common factors. The polynomial is a trinomial; in fact, it is a perfect-square trinomial: $(4y^2)^2 - 2(4y^2)(25) + (25)^2 = (4y^2 - 25)^2$.

$$16y^4 - 200y^2 + 625 = (4y^2 - 25)^2$$

The remaining polynomial factors are binomials. Both are differences of squares, $4y^2 - 25 = (2y)^2 - (5)^2 = (2y + 5)(2y - 5)$, so we write the factors twice.

$$16y^4 - 200y^2 + 625 = (4y^2 - 25)^2$$
$$= (2y + 5)(2y - 5)(2y + 5)(2y - 5)$$
$$= (2y + 5)^2(2y - 5)^2$$

The remaining factors are binomials that will not factor.

The check is left for you.

$$16y^4 - 200y^2 + 625 = (2y + 5)^2(2y - 5)^2$$

c. $6x^2 + 16x - 70$

The GCF is 2. Therefore, factor 2 out of all terms.

$$6x^2 + 16x - 70 = 2(3x^2 + 8x - 35)$$

The remaining factor is a trinomial with $a > 1$. Since $a = 3$ and $c = -35$ are not both prime, it will be easier to use the ac method: $ac = -105$.

Note: You may want to use the calculator table.

Factor	Factor	Sum of Factors
1	-105	-104
-1	105	104
3	-35	-32
-3	35	32
5	-21	-16
-5	21	16
7	-15	-8
-7	15	8 ← b

Rewrite the middle term of the trinomial, $8x$, as $-7x + 15x$, and factor by grouping.

$$6x^2 + 16x - 70 = 2(3x^2 + 8x - 35)$$
$$= 2(3x^2 - 7x + 15x - 35)$$
$$= 2[(3x^2 - 7x) + (15x - 35)] \quad \text{Group terms.}$$
$$= 2[x(3x - 7) + 5(3x - 7)] \quad \text{Factor out a common factor from each group.}$$
$$= 2(3x - 7)(x + 5) \quad \text{Factor out a common binomial factor from each term.}$$

The remaining binomial factors will not factor.

The check is left for you.

$$6x^2 + 16x - 70 = 2(3x - 7)(x + 5)$$

d. $5x^3 + 12x^2 + 7x$

The GCF is x. Factor x out of each term.

$$5x^3 + 12x^2 + 7x = x(5x^2 + 12x + 7)$$

The remaining factor is a trinomial with $a > 1$. Since $a = 5$ and $c = 7$ are prime, we will use the trial-and-error method. Possible factors for the first and last terms are $5x^2 = x \cdot 5x$ and $7 = 1 \cdot 7$. Since $b = 12$ is a positive number, we will use only positive factors of 7. Filling in the blanks with possibilities and checking for the correct middle term will result in the following:

	Middle term
$x(\underline{} \ \underline{})(\underline{} \ \underline{})$	
$x(\underline{x} + \underline{1})(\underline{5x} + \underline{7})$	$7x + 5x = 12x$
$x(\underline{x} + \underline{7})(\underline{5x} + \underline{1})$	$x + 35x = 36x$

We choose $x(x + 1)(5x + 7)$, because it results in the correct middle term.

$$5x^3 + 12x^2 + 7x = x(5x^2 + 12x + 7)$$
$$= x(x + 1)(5x + 7)$$

The remaining binomial factors will not factor.

The check is left for you.

$$5x^3 + 12x^2 + 7x = x(x + 1)(5x + 7)$$

e. $2a^3b - 8a^2b^2 - 3a^2b^2 + 12ab^3$

The GCF is ab. Factor ab out of each term.

$$2a^3b - 8a^2b^2 - 3a^2b^2 + 12ab^3 = ab(2a^2 - 8ab - 3ab + 12b^2)$$

The remaining polynomial has four terms. We will factor it by grouping.

$$2a^3b - 8a^2b^2 - 3a^2b^2 + 12ab^3 = ab(2a^2 - 8ab - 3ab + 12b^2)$$
$$= ab[(2a^2 - 8ab) + (-3ab + 12b^2)]$$

Group terms.

$$= ab[2a(a - 4b) + (-3b)(a - 4b)]$$

Factor out a common factor from each group.

$$= ab(a - 4b)(2a - 3b)$$

Factor out a common binomial factor from each term.

The remaining binomial factors will not factor.

The check is left for you.

$$2a^3b - 8^2b^2 - 3a^2b^2 + 12ab^3 = ab(a - 4b)(2a - 3b)$$

f. $4x^2 + 14x + 49$

There are no common factors. The polynomial is a trinomial. It looks similar to a perfect-square trinomial, but upon closer inspection, we see that it is not.

$$4x^2 + 14x + 49 \neq (2x)^2 + 2(2x)(7) + (7)^2$$
$$\uparrow$$
$$28x$$

The polynomial is a trinomial with $a > 1$. Since $a = 4$ and $c = 49$ are not prime, we will use the ac method: $ac = 196$. We will use a table to determine the factors, m and n, of the product 196 that will result in a sum of 14 (b). Since b is positive, we will only use the positive factors of 196. *Note:* You may want to use the calculator table.

Factor	Factor	Sum of Factors
1	196	197
2	98	100
4	49	53
7	28	35
14	14	28

Since no set of factors results in a sum of 14, the trinomial will not factor. $4x^2 + 14x + 49$ will not factor.

g. $15 + 8x + x^2$

There are no common factors. The polynomial is a trinomial that is not written in standard form. We will factor this trinomial by trial and error. Possible factors for the first and last terms are as follows:

$$15 = 1 \cdot 15 \qquad x^2 = x \cdot x$$
$$= 3 \cdot 5$$

Since $b = 8$ is a positive number, we will use only positive factors of 15. Filling in the blanks with possibilities and checking for the correct middle term results in the following:

$$(\_\_\_)(\_\_\_) \qquad \text{Middle term}$$
$$(\underline{1} + x)(\underline{15} + x) \qquad x + 15x = 16x$$
$$(\underline{3} + x)(\underline{5} + x) \qquad 3x + 5x = 8x$$

We choose $(3 + x)(5 + x)$, because this results in $8x$ for the middle term.

$$15 + 8x + x^2 = (3 + x)(5 + x)$$

The check is left for you.

$$15 + 8x + x^2 = (3 + x)(5 + x)$$

10.5.1 Checkup

Factor exercises 1–7 completely and check.

1. $10z^2 + 35z - 150$ **2.** $6a^3 + 26a^2 + 8a$ **3.** $4x^2 + 15x + 25$

4. $12x^3 - 60x^2y - 6x^2y + 30xy^2$ **5.** $-6a^2 + 24$ **6.** $81x^4 - 72x^2 + 16$

7. $12 - 17h - 5h^2$

8. What is the first thing you should always look for in factoring a polynomial?

9. If you are factoring a binomial, for what forms should you be on the lookout?

10. In factoring a trinomial, for what forms should you be looking?

11. In factoring a polynomial with four terms, what method should you try to use?

10.5.2 Modeling the Real World

Many mathematical relationships cannot be described exactly by polynomials. However, they often can be approximated by polynomials, especially if we are interested in only a small domain of the variable. Because of this capability to approximate, factoring and simplifying polynomials is a very useful technique to learn.

Many business applications involving revenue may be represented by a polynomial having a GCF of the variable representing the number of items sold. When this occurs, the common factor represents the number of items sold, and the second factor represents the cost of each item sold. Therefore, factoring the polynomial allows us to determine the selling price in terms of the number of items sold.

EXAMPLE 2 A tour agency offers special group package rates. One such offer is to give one free ticket if certain conditions are met. Another possibility is to reduce the advertised cost of the ticket. A ticket agent uses a formula to determine that the revenue from a tour is represented by the polynomial $(149.95x - 0.99x^2)$, where x is the number of people participating. Factor the polynomial, and determine whether a free ticket was offered or the advertised cost of the tickets was reduced.

Solution

x = number of people participating

$$149.95x - 0.99x^2 = x(149.95 - x) \qquad \text{\textit{Factor out the common factor.}}$$

The revenue is the product of x participants and $(149.95 - x)$ dollars, or $(149.95 - 1x)$ dollars.

Therefore, the advertised cost of the ticket ($149.95) is reduced and is $149.95 minus $1 per participant.

APPLICATION

The first telephone receivers had a cone-shaped mouthpiece. The volume of the mouthpiece was represented by the polynomial $(\frac{1}{3}\pi x^3 + \pi x^2)$ cubic inches, where x is the length of the radius in inches. Determine a polynomial that expresses the height in terms of the radius.

Discussion

x is the length of the radius in inches.

$\frac{1}{3}\pi x^3 + \pi x^2$ is the volume of the mouthpiece.

The formula for the volume of a cone is $\frac{1}{3}\pi r^2 h$. To determine the height, factor out $\frac{1}{3}\pi x^2$ from the polynomial representing the volume.

$$\frac{1}{3}\pi x^3 + \pi x^2 = \frac{1}{3}\pi x^2(x + 3)$$

Comparing the formula for the volume with the factored polynomial, we have

$$\frac{1}{3}\pi r^2 \quad h$$
$$\downarrow \qquad \downarrow$$
$$\frac{1}{3}\pi x^2(x + 3)$$

The radius is x inches and the height is $(x + 3)$ inches.

✔ 10.5.2 Checkup

1. A supplier of computer software offers a special deal to customers who purchase multiple copies of their software. The total price of purchasing x copies of the software is given as $(349.95x - 8.95x^2)$ dollars. Factor this polynomial. From the factors, explain how the supplier figured the special deal.

2. A tray is made from a rectangular sheet of metal by cutting out squares from each corner and folding up the edges. The volume of the tray is determined to be $(35x - 24x^2 + 4x^3)$ cubic inches, where x is the measure of the side of the squares removed from each corner. Factor the expression for the volume to obtain expressions for the height, width, and length of the tray. What are the dimensions of the tray if the squares cut from the corners measure 1 inch on a side?

1–31 e00
41 odds
33

10.5 Exercises

In exercises 1–32, use the general strategy to factor completely. Check.

1. $3y^3 + 3y^2 + 3y$
2. $10x^3 + 35x^2 - 5x$
3. $10abc^2 + 15abc - 20ab$
4. $6xyz^2 + 10xyz - 6xy$
5. $-40a^2 - 24ab + 48ac$
6. $-28x^2 - 35xy + 21xz$
7. $108x^5 - 75x^3$
8. $32u^3 - 98u$
9. $-200x^3y + 32xy^3$
10. $-18a^3b + 128ab^3$
11. $x^3 - 16x^2 + 64x$
12. $z^3 - 10z^2 + 25z$
13. $12u^2v + 36uv^2 + 27v^3$
14. $50x^3 + 40x^2y + 8xy^2$
15. $3x^4 - 48x^2 + 7x^2 - 112$
16. $2x^4 - 18x^2 + 5x^2 - 45$
17. $4p^4 - 37p^2 + 9$
18. $9z^4 - 40z^2 + 16$
19. $2 + 8y - 42y^2$
20. $3 + 6y - 72y^2$
21. $4u^2v^2 + 36uv + 56$
22. $5a^2b^2 + 35ab + 60$
23. $32x^3 - 64x^2 - 28x^2 + 56x$
24. $15x^3 - 18x^2 - 60x^2 + 72x$
25. $12x^4 + 26x^3 - 30x^2$
26. $15z^4 + 10z^3 - 40z^2$
27. $x^6 - 5x^3y^3 + 3x^3y^3 - 15y^6$
28. $p^6 + 2p^3q^3 - 5p^3q^3 - 10q^6$
29. $x^4 - 5x^3 + 8x^2 - 40x$
30. $2y^4 + 6y^3 + 8y^2 + 24y$
31. $1 - k^8$
32. $256s^8 - 1$

33. To encourage sales, Big Ed's Pizza Palace offers discounts on multiple orders. If a person orders x pizzas, the total price is $(14.75x - 0.95x^2)$ dollars. Factor this polynomial. From the factors, explain how Big Ed figured the discount. Should Big Ed put a limit on the number of pizzas a person may order? Explain.

34. A music club offers its members a special deal on music CD's. If a person orders x CD's from the club, the total price, including shipping and handling, is $(15.70x - 2.24x^2)$ dollars. Factor this polynomial. Explain how the special deal works. Should the music club put a limit on the number of CD's a member may order? Explain.

35. Because of a shortage of fashion dolls for the holiday season, Tallmart stores charge a surcharge on purchases of more than one doll. If a person purchases x dolls, the total price is $(19.95x + 2.99x^2)$ dollars. Factor this polynomial. Explain how Tallmart figured the surcharge.

36. Food Tiger stores have a sale on Thanksgiving turkeys. To discourage excessive purchases by any customer, they charge a surcharge on purchases of more than one turkey. If a customer purchases x turkeys, the total price is $(4.95x + 2x^2)$ dollars. Factor this polynomial and explain how the surcharge is administered.

37. A sheet of posterboard has its corners cut out and sides folded up to form an open box. The volume of the box is $135x - 48x^2 + 4x^3$ cubic inches, where x is the length of the sides of the squares cut from the corners. Factor the polynomial to determine the length, width, and height of the box. What are the dimensions and volume of the box if x is 2 inches?

38. An open box is formed by cutting out the corners of a piece of tin and folding up the sides. The volume of the box is $x^3 + 30x^2 + 200x$ cubic inches, where x is the length of the sides of the squares cut from the corners. Factor the polynomial to determine expressions for the length, width, and height of the box. What are the dimensions and volume of the box if x is 5 inches?

39. The volume of a cylindrical hatbox is $\pi(4x^3 - 20x^2 + 25x)$ cubic inches, where x is the height of the box. Factor the polynomial to determine an expression for the radius of the box. What are the dimensions and volume of the box if x is 8 inches?

40. The volume of a cylindrical waste container is $\pi(x^3 - 16x^2 + 64x)$ cubic inches, where x is the height of the container. Factor the polynomial to determine the radius and diameter of the container. What are the dimensions and volume of the container if x is 15 inches?

41. In his drafting class, Dennis had to draw different perspective views of a cube with a smaller cube carved out of one corner. The original cube measured x inches on a side. Dennis figured that the volume of the remaining object was $x^3 - 216$ cubic inches. Factor this polynomial. Do the factors have any interpretation? What are the dimensions of the cube that was carved out of the larger cube?

42. For another project, Dennis had to draw different perspective views of an object that consisted of two different-sized cubes attached on one face. The smaller cube measured x inches on a side. Dennis determined that the total volume of the object was $x^3 + 13,824$ cubic inches. Factor this polynomial. Do the factors have any interpretation? What are the dimensions of the larger cube?

10.5 Calculator Exercises

In 10.3 Calculator Exercises, you were shown how to use the x-intercepts of a graph to help you factor a difficult trinomial. Actually, this method will work for *any* polynomial that can be factored, even when some of the methods we have studied will not. As an example, suppose you wish to factor the polynomial $40x^3 + 122x^2 + 79x - 21$.

Follow these steps:

1. Define a function $y = 40x^3 + 122x^2 + 79x - 21$.

2. Graph the function, using an appropriate window, such as $(-2, 2, 1, -10, 10, 1, 1)$. Note that since we are interested only in finding x-intercepts, the window need not be large enough to show all minima and maxima.

3. Determine the x-values of the x-intercepts.

$x = 0.2$ $x = -1.5$ $x = -1.75$

Y1 = $40x^3 + 122x^2 + 79x - 21$ Y1 = $40x^3 + 122x^2 + 79x - 21$ Y1 = $40x^3 + 122x^2 + 79x - 21$

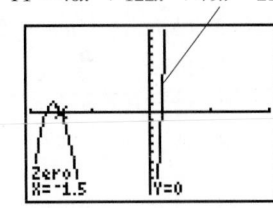

4. Convert the x-values to fractions, and write the resulting equations in standard form.

$$x = \frac{1}{5} \qquad x = -\frac{3}{2} \qquad x = -\frac{7}{4}$$

$$5x - 1 = 0 \qquad 2x + 3 = 0 \qquad 4x + 7 = 0$$

Try the method on the following polynomials:

1. $16x^3 + 4x^2 - 264x + 189$

2. $50x^3 + 25x^2 - 18x - 9$

3. $90x^3 + 213x^2 - 53x - 140$

4. $18x^3 - 45x^2 - 128x + 320$

5. The polynomial factors into the product of the expressions on the left of the equations in step 4.

$$40x^3 + 122x^2 + 79x - 21 = (5x - 1)(2x + 3)(4x + 7)$$

Check your results by multiplying, check them numerically with a table of values, or check them graphically.

10.5 Writing Exercise

Perform the following two multiplications, which involve squares of binomials:

1. $(a^n + b^n)^2$ **2.** $(a^n - b^n)^2$

You should see that the results are two more recognizable forms that could be helpful in factoring. Write a description of how these two results, when turned around, could help you factor some trinomials.

Use your description to factor the trinomials that follow as illustrations of what you have described. Once you do the initial factoring, be sure to check to see whether the factors can be factored further.

3. $x^6 + 2x^3y^3 + y^6$ **4.** $p^8 - 2p^4q^4 + q^8$

CHAPTER 10 SUMMARY

After completing this chapter, you should be able to define the following key terms in your own words.

10.1
greatest common factor (GCF)
prime number
factoring by grouping

10.3
leading coefficient
trial-and-error method

10.4
ac method

Reflections

1. In factoring out a common factor from a polynomial, why do you always seek the greatest common factor?

2. How does a greatest common factor differ from a least common denominator?

3. Explain the difference between a monomial factor and a polynomial factor.

4. Other than a common factor, which can be factored—the difference of two squares or the sum of two squares? Describe the process.

5. Other than a common factor, which can be factored—the difference of two cubes or the sum of two cubes? Describe the process.

6. What is a perfect-square trinomial?

7. Explain what the *ac* method of factoring a trinomial enables you to do.

8. After factoring a polynomial, how can you tell that you factored it correctly?

9. Compare the operation of multiplying polynomials with that of factoring a polynomial.

CHAPTER 10 SECTION-BY-SECTION REVIEW

10.1

Recall	Examples
Factoring out a GCF • Determine the GCF of all terms. If the leading coefficient is negative, use a negative GCF. • Write each term as a product of the GCF and another factor. • Reverse the distributive law to write a product of the GCF and a polynomial.	Factor. $12x^2 + 4x - 6 = 2(6x^2 + 2x - 3)$ $-24a^2b^3 - 6ab^2 + 12b = -6b(4a^2b^2 + ab - 2)$
Factoring by grouping (four terms) • Factor the GCF out of all terms. • Write the polynomial as a sum of two binomials. • Factor each binomial. • Determine the binomial GCF of the resulting terms. • Write a product of the binomial GCF and another binomial.	Factor. $6x^2 - 12x + 10x - 20 = 2(3x^2 - 6x + 5x - 10)$ $= 2[(3x^2 - 6x) + (5x - 10)]$ $= 2[3x(x - 2) + 5(x - 2)]$ $= 2[(x - 2)(3x + 5)]$ $= 2(x - 2)(3x + 5)$

Factor exercises 1–5 completely and check.

1. $20a^6 - 28a^4 + 44a^2$ 　　　**2.** $22u^3v^2 + 22u^2v^3$ 　　　**3.** $3x^3 + 3x + x^2 + 1$

4. $7a^4 + 7a^2b^2 + 7a^2b^2 + 7b^4$ 　　　**5.** $15ac + 18ad + 20bc + 24bd$

6. It can be shown that the sum of the natural numbers from 1 to n equals $\frac{1}{2}n^2 + \frac{1}{2}n$.
  a. Factor the expression completely.
  b. Evaluate the original expression and the factored expression for the sum of the natural numbers from 1 to 12.
  c. Which expression was easier to evaluate and why?

7. The area of a rectangle is numerically equal to the polynomial

$$2x^2 - 6x + 5x - 15$$

Factor the expression, using the grouping method to determine algebraic expressions for the rectangle's width and length.

8. Because of a shortage of labor in a resort town, the owners of a fast-food restaurant offered to increase their employees' monthly salary by $10 for each additional month they remain employed. If an employee earns x dollars the first month, write a polynomial for her total pay for seven months. Factor out the GCF from the simplified expression. Interpret what the factors represent. If the employee's beginning salary is $960, determine her total pay for the seven months. Check your interpretation for this beginning salary.

9. A backyard patio design has a rectangular-shaped center made from tiles. The center measures 24 feet by 15 feet. This central section is surrounded by a brick border that adds x feet to each side of the patio. Write a polynomial that represents the area of the brick border. Factor the polynomial and discuss the meaning of the factors. If the brick border adds 3 feet to each side of the patio, how many square feet of brick border are there?

10.2

Recall	Examples
Factoring a perfect-square trinomial • $a^2 + 2ab + b^2 = (a + b)^2$ • $a^2 - 2ab + b^2 = (a - b)^2$	Factor. $x^2 + 10x + 25 = (x + 5)^2$ $4x^2 - 12x + 9 = (2x - 3)^2$
Factoring the difference of two squares • $a^2 - b^2 = (a + b)(a - b)$	Factor. $9x^2 - 4y^2 = (3x + 2y)(3x - 2y)$
Factoring the difference of two cubes • $a^3 - b^3 = (a - b)(a^2 + ab + b^2)$	Factor. $x^3 - 8 = (x - 2)(x^2 + 2x + 4)$
Factoring the sum of two cubes • $a^3 + b^3 = (a + b)(a^2 - ab + b^2)$	Factor. $27x^3 + y^3 = (3x + y)(9x^2 - 3xy + y^2)$

Factor exercises 10–27 completely and check.

10. $p^2 + 12p + 36$ **11.** $q^2 - 16q + 64$ **12.** $9x^2 + 30x + 25$ **13.** $49y^2 - 112y + 64$

14. $a^2 - 9a + 81$ **15.** $x^7 + 6x^5y + 9x^3y^2$ **16.** $x^2 - 169$ **17.** $625 - a^2$

18. $12x^2 - 75$ **19.** $p^2 - q^2$ **20.** $p^2 + q^2$ **21.** $9x^2 - 25y^2$

22. $16x^4 - 81$ **23.** $x^8 - 1$ **24.** $c^3 + 27$ **25.** $c^3 - 27$

26. $8z^3 - 125$ **27.** $5h^3 + 40k^3$

28. A design for a school flag is a square with a smaller square 2 feet on a side in its corner. The small square has an icon of the school's mascot in it. The small square is light blue, and the remainder of the large square is navy blue.
  a. Write a polynomial for the area of the flag that is navy blue.
  b. Factor the polynomial to determine the dimensions of a rectangle with the same area as the area in navy blue.
  c. If the large square of the flag is 5 feet on a side, would it require more navy-blue material than a rectangular piece of navy-blue material that is 3 feet by 8 feet? Explain.

29. A circular pond is surrounded by grass and an exercise path that adds x feet to the radius of the circle. A polynomial that represents the area of the pond and exercise space is $\pi(x^2 + 200x + 10000)$ square feet. Factor the polynomial and determine the radius of the pond. If the grass and exercise path add 20 feet to the radius of the pond and exercise space, use the factored expression to find the area of the pond and exercise space. Find the area of the pond. Use these two areas to find the area of the exercise space.

10.3

Recall	Examples
Factoring a trinomial with a leading coefficient of 1, using the trial-and-error method • $x^2 + bx + c = (x + m)(x + n)$, where $mn = c$ and $n + m = b$.	Factor. $x^2 + 7x + 12 = (x + 4)(x + 3)$
Factoring a quadratic trinomial with a leading coefficient not equal to 1, using the trial-and-error method • Factor out the GCF from all terms. • Determine the possible factors of the first term and the third term. • Write a set of binomial factors for each combination. • Multiply the factors to determine which set results in the original trinomial.	Factor. $4x^2 + 16x + 15$ $\begin{aligned}4x^2 &= x \cdot 4x & 15 &= 1 \cdot 15 \\ &= 2x \cdot 2x & &= 3 \cdot 5\end{aligned}$ Both factors are positive. Check each combination. $4x^2 + 16x + 15 = (2x + 3)(2x + 5)$

Use the trial-and-error method in exercises 30–40 to factor completely. Check.

30. $z^2 + 2z - 99$ **31.** $p^2 + 5pq - 66q^2$ **32.** $6a^2 + 96a + 234$ **33.** $x^4 + 8x^2 + 15$

34. $4q^3 - 28q^2 - 240q$ **35.** $x^2y^2 - 4xy - 117$ **36.** $-7x^2 + 98x - 168$ **37.** $2x^2 - 11x + 5$

38. $6x^2 + 17x + 5$ **39.** $28a^2b^2 + 91ab + 21$ **40.** $-45x^3 - 102x^2 + 48x$

41. The base of a triangle measures twice its height. After the base and the height are increased, the enlarged triangle has an area given by the expression

$$2x^2 + 10x + \frac{21}{2}$$

where x is the height of the original triangle.
  a. Factor the expression to produce expressions for the base and height of the enlarged triangle.
  b. By how much was the base of the original triangle increased?
  c. By how much was the height of the original triangle increased?

42. A rectangle has a width that is 6 inches less than its length. After the width and length are increased, the larger rectangle has an area given by $(x^2 + 17x + 30)$ square inches, where x is the length of the original rectangle.
  a. Factor the polynomial to find the expressions for the increased length and width.
  b. By how much was the length increased?
  c. By how much was the width increased?

43. For Olympic tournaments and world championships, the size of the basketball court is prescribed and must have a minimum clear space on each side. If x meters are allowed on each side, a polynomial for the total area needed is given by $420 + 86x + 4x^2$ square meters. Factor the polynomial and determine the dimensions of the basketball court itself.

10.4

Recall	Example
Factoring a quadratic trinomial by using the *ac* method • Factor out the GCF from all terms. • Determine the product *ac*. • Determine the factors whose sum is the coefficient of the middle term *b*. • Write the polynomial as a sum of four terms, using the factors as coefficients of the two terms that replace the middle term. • Factor by grouping.	Factor. $12x^2 - 2x - 4 = 2(6x^2 - x - 2)$ $ac = (6)(-2) = -12$ The factors of -12 that result in a sum of $-1, b$, are -4 and 3. $12x^2 - 2x - 4 = 2(6x^2 - x - 2)$ $\qquad = 2[(6x^2 - 4x) + (3x - 2)]$ $\qquad = 2[2x(3x - 2) + 1(3x - 2)]$ $\qquad = 2[(3x - 2)(2x + 1)]$ $\qquad = 2(3x - 2)(2x + 1)$

In exercises 44–57, use the ac method to factor completely. Check the factors.

44. $3x^2 + 23x + 4$ **45.** $8x^2 + 22x + 9$ **46.** $15y^2 - 29y + 12$

47. $10x^2 + 29x + 24$ **48.** $12y^2 - 13y - 14$ **49.** $8x^4 + 22x^2 + 15$

50. $10x^4 + 9x^2 - 9$ **51.** $12x^2 - 35x + 9$ **52.** $6p^4 - 29p^2 - 28$

53. $24x^2 + 76x + 32$ **54.** $-40x^4 - 30x^2 + 175$ **55.** $6x^2 - 7xy - 20y^2$

56. $12x^2 + 40xy + 25y^2$ **57.** $20x^2 - 27xy + 9y^2$

58. A bowling lane is about 19 times as long as it is wide. Usually, 4.5 meters are added to each end of the lane for an approach and for collecting the pins, and 1.5 meters are added to each side of the lane for ball returns and gutters. A polynomial representing the total area covered by the lane and the added area is $(19x^2 + 66x + 27)$ square meters. Factor the polynomial. If the width of the lane is approximately 1 meter, evaluate the factors to determine the dimensions of the total area and the area itself.

10.5

Recall	Examples
Factoring a polynomial • Factor out the GCF from all terms. • Binomials must be the difference of two squares, the difference of two cubes, or the sum of two cubes. • Trinomials may be perfect-square trinomials or may be factored by trial and error or by the *ac* method. • Four-term polynomials must be factored by grouping.	Factor the following polynomials: $3x^2 - 6x + 6 = 3(x^2 - 2x + 2)$ $x^2 - 16 = (x + 4)(x - 4)$ $x^3 - 64 = (x - 4)(x^2 + 4x + 16)$ $x^3 + 64 = (x + 4)(x^2 - 4x + 16)$ $x^2 + 8x + 16 = (x + 4)^2$ $x^2 + 8x + 15 = (x + 5)(x + 3)$ $2x^2 + x - 15 = 2x^2 - 5x + 6x - 15$ $\qquad = (2x^2 - 5x) + (6x - 15)$ $\qquad = x(2x - 5) + 3(2x - 5)$ $\qquad = (2x - 5)(x + 3)$ $6x^2 - 15x + 2x - 5 = (6x^2 - 15x) + (2x - 5)$ $\qquad = 3x(2x - 5) + 1(2x - 5)$ $\qquad = (2x - 5)(3x + 1)$

In exercises 59–66, use the general strategy to factor completely. Check.

59. $-12x^3 + 60x^2y - 75xy^2$ **60.** $7x^4 + 7x^3 + 7x^2$ **61.** $12x^3 - 243x$ **62.** $32x^3 + 32x^2 + 8x$

63. $24x^3 - 14x^2 - 90x$ **64.** $256x^4 - 288x^2 + 81$ **65.** $36x^4 - 25x^2 + 4$ **66.** $2x^4 + 14x^3 - 8x^2 - 56x$

67. The site for a memorial statue is a rectangular piece of land whose length is four times its width. The base of the statue is a square measuring 5 feet on a side.
  a. If x is the width of the site, write a polynomial for the area of land that will not be covered by the statue.
  b. Factor the polynomial to obtain the dimensions of a rectangular plot with an area equivalent to the uncovered area.

  c. If the site has a width of 80 feet, use the factors to find the dimensions of a rectangle with an area equivalent to the uncovered area.

68. A caterer offers a discount for catering a dinner to the planners of a conference for mathematics instructors. If x instructors attend the dinner, the total cost to the planners is $(19.95x - 0.10x^2)$ dollars. Factor this polynomial to determine the price charged for each instructor, and explain how the discount is figured.

CHAPTER 10 CHAPTER REVIEW

Factor exercises 1–37 completely and check.

1. $z^2 + 9z - 90$

2. $a^2 - 18a + 72$

3. $x^2 + 14xy + 45y^2$

4. $5a^2 + 70a + 245$

5. $2a^2 + 8ab + 12b^2$

6. $x^4 + 10x^2 + 21$

7. $3q^3 - 33q^2 - 126q$

8. $-6x^2 + 42x + 360$

9. $10 + 7x + x^2$

10. $x^2 - 289$

11. $x^3 - 1$

12. $4x^2 - 64$

13. $u^2 + v^2$

14. $36x^2 - 49y^2$

15. $81x^4 - 1$

16. $p^2 + 22p + 121$

17. $q^2 - 30q + 225$

18. $27a^3 + 64b^3$

19. $27a^2b^2 - 72ab + 48$

20. $-50x^3 + 120x^2y - 72xy^2$

21. $8x^4 - 2x^3 + 6x^2 - 12x$

22. $35u^3v^2 + 25u^2v^3$

23. $2x^3 + 10x + x^2 + 5$

24. $m^2 - 2mn - 8mn + 16n^2$

25. $4a^4 + 8a^2b^2 + 8a^2b^2 + 16b^4$

26. $2x^2 - 13x + 11$

27. $24ac + 20ad + 18bc + 15bd$

28. $7x^2 - 19x - 6$

29. $10x^2 - 11x - 6$

30. $36a^2 + 66a + 24$

31. $-30x^2 - 28x + 32$

32. $12x^4 + 13x^2 + 3$

33. $54x^3 + 36x^2 + 6x$

34. $81x^4 - 72x^2 + 16$

35. $4x^4 - 61x^2 + 225$

36. $12x^3 + 18x^2 - 30x^2 - 45x$

37. $3x^4 + 15x^3 - 27x^2 - 135x$

38. The area of a rectangle in square feet is given by the polynomial $8x^2 - 2x - 3$. Factor this polynomial to determine algebraic expressions for the rectangle's width and length.

39. The width of a rectangle is 2 inches less than its length. After the width and length are increased, the larger rectangle has an area given by $(6x^2 - 10x - 4)$ square inches, where x is the length of the original rectangle.
 a. Factor the polynomial to find expressions for the increased length and width.
 b. By how much was the length increased?
 c. By how much was the width increased?

40. The base of a triangle is twice its height. After the base and the height are increased, the enlarged triangle has an area in square feet given by the polynomial
$$x^2 + 12x + 32$$
where x is the height of the original triangle.
 a. Factor the polynomial to produce expressions for the base and height of the enlarged triangle.
 b. By how much was the base of the original triangle increased?
 c. By how much was the height of the original triangle increased?

41. A judge levies a progressively higher penalty each time a person is cited for littering. A person that has been cited x times for littering pays a total penalty of $(50x + 25x^2)$ dollars. Factor this polynomial to determine the penalty paid for a given citation, and explain how the judge determined the penalty.

42. The sheets of a desk calendar are square with a smaller square 3 inches on a side inside. The small square has a daily cartoon in it. The remainder of the large square is white with the day's date on it.
 a. Write a polynomial for the area of the sheet not covered by the cartoon.
 b. Factor the polynomial to determine the dimensions of a rectangle with the same area as that not covered by the cartoon.
 c. If the large square measures 5 inches on a side, is the area not covered by the cartoon larger than the area covered by the cartoon?

43. In the sport of archery, the face of the target is circular. If the face includes a border that is x mm wide around the scoring region, a polynomial for the total area facing the archer is $\pi(372100 + 1220x + x^2)$ square mm. Factor the polynomial to determine the radius of the face of the target. What is the area of the face?

44. A quilt has a central design that is 68 inches wide and 88 inches long. The border around the center adds x inches to each side of the quilt. Write a polynomial that represents the area of the quilt's border. Factor the polynomial and discuss the meaning of the factors. If the border adds 16 inches to each side of the quilt, how many square inches of border are there?

CHAPTER 10 TEST

TEST-TAKING TIPS

Doing well on tests requires perseverance in the classroom. Attend all classes and take full notes. Research shows that successful students never cut class and usually record at least 64% of what is discussed in class. Failing students record half as much and often miss class. Missing one class puts you behind in the course by at least two lessons. Can you see why? It is important to stay current. Do not allow yourself to fall behind in homework or study, or the entire course will become an effort and a struggle to catch up. Participate in the class by asking questions. Instructors welcome students' questions. If necessary, see the instructor outside of class if you do not understand a lesson and can't get your questions answered in class. Do not delay in getting help with difficult concepts.

Factor exercises 1–12 completely.

1. $81a^3 + 54a^2 + 9a$
2. $p^3 + 125$
3. $-8a^4b^2 - 36a^3b^3 - 16a^2b^4$
4. $a(a^2 + b^2) - 5b(a^2 + b^2)$
5. $15x^2 - 21xy + 10xy - 14y^2$
6. $64a^2 - 49b^2$
7. $25x^2 - 70x + 49$
8. $3x^3 - 27x^2 + 24x$
9. $x^2 - 4xy - 21y^2$
10. $14x^2 + 25x + 9$
11. $4x^4 + 27x^2 - 7$
12. $x^2 + 8x + 14$

13. A box with a height of x inches has a volume of $(2x^3 + 5x^2 - 3x)$ cubic inches. Factor the polynomial to determine expressions for the width and length of the box.

14. The base of a triangle is 7 inches more than its height. After the base and height are increased by the same amount, the area of the new triangle is $(\frac{1}{2}x^2 + \frac{17}{2}x + 30)$ square inches.
    a. Factor the polynomial to produce expressions for the base and height of the enlarged triangle.
    b. By how much were the base and height increased?
    c. What was the expression for the base of the original triangle?

15. Edgar received progressively more reward money from his parents each time he passed an algebra test. When Edgar passed x tests, the total amount of money he had received was $(20x + 5x^2)$ dollars. Factor this polynomial, and explain how Edgar's parents determined his reward for passing a given test.

16. For a school science project, Reggie designed a packing crate to protect a raw egg that was to be dropped from a height of 20 feet. The objective was to protect the egg so that it would not shatter. Reggie placed the egg in a small box in the shape of a cube 2 inches on a side. This box in turn was placed in a bigger crate that was also shaped like a cube. The larger crate was filled with gelatin to absorb the impact of the drop.
    a. Write a polynomial for the volume of the crate, not including the small cube.
    b. Factor the polynomial. Do the factors have any physical interpretation? Explain.
    c. If the crate measures 12 inches on a side, how many cubic inches of gelatin will it take to fill the crate along with the boxed egg?

17. A potter designs a serving tray that has a circular center surrounded by a border that is x inches wide. The surface area of the tray, including the border, is $\pi(x^2 + 18x + 81)$ square inches. Factor this polynomial to determine the radius of the center of the tray. If the border is 2 inches wide, determine the surface area of the tray, including the border. Then determine the area of the center, and use the two areas to find the area of the border.

18. David factored the following polynomial as shown:
$$12x^3 + 28x^2 - 27x - 63 = (12x^3 + 28x^2) + (-27x - 63)$$
$$= 4x^2(3x + 7) - 9(3x + 7)$$
$$= (3x + 7)(4x^2 - 9)$$
    a. Identify the method David used to factor the polynomial.
    b. Explain what is wrong with David's solution, and describe how you would correct it.
    c. What is the correct solution?

Project

In this chapter, we have used geometric figures as applications of factoring. In this project, we will illustrate some of these applications. In particular, we will explore the special products. To illustrate why the special products factor as they do, we want to visualize the rectangular objects that are formed.*

Part I

We know that $a^2 + 2ab + b^2 = (a + b)^2$. We can demonstrate this fact geometrically using the square in the illustration.

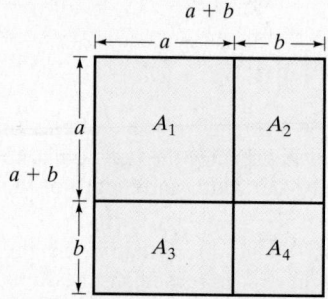

1. Determine the area of the shaded square with sides $(a + b)$.

2. Determine the sum of the areas of the four numbered sections A_1, A_2, A_3, and A_4.

$$A_1 + A_2 + A_3 + A_4$$

3. The areas found in exercises 1 and 2 are equivalent. Write an equation to represent this finding.

4. Choose a small positive integer value for a and a different value for b. Show that these values are solutions of the equation in exercise 3.

5. On graph paper, draw the square in the illustration, using the values chosen for a and b, from exercise 4.

Part II

We know that $a^2 - 2ab + b^2 = (a - b)^2$. We can demonstrate this fact geometrically using the square in the illustration.

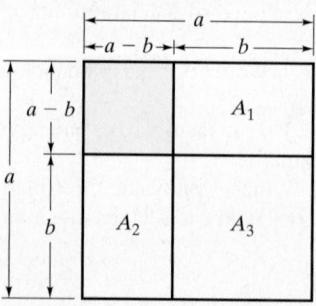

*Instructors may use Cuisinaire® Algebra Tiles for this project.

1. Determine the area of the shaded square with sides $(a - b)$.

2. Determine the difference of the entire area of the figure, A, and the sum of the areas of the three numbered sections.

$$A - (A_1 + A_2 + A_3)$$

3. The areas found in exercises 1 and 2 are equivalent. Write an equation to represent this finding.

4. Choose a small positive integer value for a and a different value for b. Show that these values are solutions of the equation in exercise 3.

5. On graph paper, draw the square in the illustration, using the values chosen for a and b, from exercise 4.

Part III

We know that $a^2 - b^2 = (a + b)(a - b)$. We can demonstrate this fact geometrically using the rectangle in the illustration.

1. Determine the area of the shaded rectangle with sides $(a + b)$ and $(a - b)$.

2. Determine the difference of the entire area of the figure, A, and the sum of the areas of the two numbered sections.

$$A - (A_1 + A_2)$$

3. The areas found in exercises 1 and 2 are equivalent. Write an equation to represent this finding.

4. Choose a small positive integer value for a and a different value for b. Show that these values are solutions of the equation in exercise 3.

5. On graph paper, draw the rectangle in the illustration, using the values chosen for a and b, from exercise 4.

Chapter 11

Now that we have learned how to work with polynomial expressions, including adding, subtracting, and factoring them, we can proceed to solve polynomial equations and inequalities. In this chapter, we will examine numeric, graphic, and algebraic methods for solving polynomial equations, particularly quadratic equations and inequalities. Because the most precise of these methods is the algebraic approach, we will look at several ways of solving equations algebraically, including factoring, using the principle of square roots, completing the square, and using the quadratic formula.

Economists utilize many types of polynomial equations to model trends in sales and to forecast future sales. The information that is learned can be used to set interest rates for business loans or for deciding whether a company needs to borrow money in a short-term transaction. In this chapter, we will look at some economic models that involve quadratic equations and show how they illustrate the methods of solution we describe.

Another important application of quadratic equations is the Pythagorean theorem, named after the Greek philosopher and mathematician Pythagoras of Samos. We will explore many different uses for this theorem in the applications and the project for the chapter.

Quadratic and Other Polynomial Equations and Inequalities

11.1 APPLICATION

Sales for Ameritrade Holding increased rapidly after the company went public in 1997. Using data from the company's annual financial reports dating from 1993, we find that the amount of sales in millions of dollars, y, is given by $y = 6.75x^2 - 7.5x + 27.5$, where x is the number of years after 1993. In 1998, sales were \$164.1 million. Assuming that the equation continues to apply, determine when the company will double this amount.

We will discuss this application further. See page 715.

11.1 Solving Equations Numerically and Graphically

11.1.1 Identifying Polynomial Equations in One Variable

In this chapter, we will solve polynomial equations in one variable. A **polynomial equation in one variable**, or, simply, a **polynomial equation**, is an equation that relates two polynomials. In an earlier chapter, we discussed polynomial equations in one variable with a degree of 1. These equations were of the form $ax + b = 0$, where $a \neq 0$. We called such equations linear equations in one variable, or, again simply, linear equations.

In this chapter, we discuss other polynomial equations. A polynomial equation with a degree of 2 is called a **quadratic equation in one variable**, or a **quadratic equation**. A polynomial equation with a degree of 3 is called a **cubic equation in one variable**, or a **cubic equation**.

STANDARD FORMS FOR POLYNOMIAL EQUATIONS

Given real numbers a, b, and c, where $a \neq 0$,

A linear equation is written in the form

$$ax + b = 0$$

A quadratic equation is written in the form

$$ax^2 + bx + c = 0$$

A cubic equation is written in the form

$$ax^3 + bx^2 + cx + d = 0$$

In general, a polynomial equation in one variable is written in the form

$$P(x) = 0$$

where $P(x)$ is a polynomial.

EXAMPLE 1 Determine whether each equation is a polynomial equation. Identify each polynomial equation as quadratic or cubic when applicable.

a. $3x^2 + 5x^6 - 7x^{-3} = 4$ **b.** $6x^2 + 8 - 7x^3 = 12$

c. $3x = 2x^2$ **d.** $\dfrac{1}{4}x^4 = \dfrac{2}{3}x^4 + \dfrac{7}{8}x^6$

Solution

a. $3x^2 + 5x^6 - 7x^{-3} = 4$ is not a polynomial equation, because x has an exponent of -3, which is not a positive integer.

b. $6x^2 + 8 - 7x^3 = 12$, or, equivalently, $7x^3 - 6x^2 + 4 = 0$, is a cubic polynomial equation.

   c. $3x = 2x^2$, or, equivalently, $2x^2 - 3x = 0$, is a quadratic polynomial equation.

   d. $\frac{1}{4}x^4 = \frac{2}{3}x^4 + \frac{7}{8}x^6$, or, equivalently, $\frac{7}{8}x^6 + \frac{5}{12}x^4 = 0$, is a polynomial equation.

11.1.1 Checkup

In exercises 1–5, determine whether each equation is a polynomial equation. Identify each polynomial equation as quadratic or cubic when applicable.

  1. $3x^2 - 5 = 4x + 2$ **2.** $x^{-2} + 4x - 7 = 0$

  3. $2\sqrt{x} + 2x - 3 = 0$ **4.** $1.2x^3 - 3.7 = 5.6$

  5. $5x^5 - 2x^2 = 3x + 7$

  6. What is the difference between a polynomial expression and a polynomial equation?

  7. What is meant by the degree of a polynomial equation?

11.1.2 Solving Polynomial Equations Numerically

In an earlier chapter, we determined whether a number was a solution of a linear equation by substituting the number into the variable and evaluating the two resulting expressions. If the resulting equation was true, the number substituted was called the solution of the original equation. We use the same procedure for a polynomial equation. The set of all possible solutions of an equation is called the solution set of the equation. For example, given $x^2 - 3 = -x + 3$, determine whether 2 is a solution.

$$\begin{array}{c|c} x^2 - 3 = & -x + 3 \\ \hline (2)^2 - 3 & -(2) + 3 \\ 4 - 3 & 1 \\ 1 & \end{array}$$

The number 2 is thus a solution, because the resulting equation, $1 = 1$, is true.

A linear equation that is not an identity has at most one solution. However, other polynomial equations may have more than one solution and not be identities.

In order to find other solutions of the equation $x^2 - 3 = -x + 3$, continue to substitute values for x and determine whether the resulting equations are true. A table of values is helpful in organizing this method.

SOLVING A POLYNOMIAL EQUATION NUMERICALLY

To solve a polynomial equation numerically,
Set up an extended table of values as follows:

* The first column is labeled with the independent variable.
* The second column is labeled with the expression on the left side of the equation.
* The third column is labeled with the expression on the right side of the equation.

Complete the table:

* Substitute values for the independent variable.
* Evaluate the second and third columns.

(Continued on page 710)

- Continue until values for the two expressions (the numbers in the second and third columns) are equal.

The values for the independent variable (the number in the first column) that result in equivalent expressions are the solutions.

For example, a sample table to determine another solution of $x^2 - 3 = -x + 3$ could appear as follows:

x	$x^2 - 3$	$-x + 3$	
-4	$(-4)^2 - 3$ $16 \ - 3$ 13	$-(-4), + 3$ $4 \ + 3$ 7	$13 > 7$, so -4 is not a solution.
-3	$(-3)^2 - 3$ $9 \ - 3$ 6	$-(-3) + 3$ $3 \ + 3$ 6	$6 = 6$, so -3 is a solution.
-2	$(-2)^2 - 3$ $4 \ - 3$ 1	$-(-2) + 3$ $2 \ + 3$ 5	$1 < 5$, so -2 is not a solution.

According to the table, when -3 is substituted for the variable x, the two expressions are equivalent ($6 = 6$). Therefore, -3 is a second solution of the polynomial equation.

To solve $x^2 - 3 = -x + 3$ numerically for integer solutions on your calculator,

$$\text{Let } Y1 = x^2 - 3 \quad \text{and} \quad Y2 = -x + 3$$

The solutions are the x-values that determine equal Y1 and Y2 values. The two solutions from the table are -3 and 2, as shown in Figure 11.1c.

Figure 11.1a   Figure 11.1b   Figure 11.1c

Both linear equations and polynomial equations may have noninteger solutions, no solution, or an infinite number of solutions. However, when solved numerically, certain polynomial equations may have a different result than that found previously with linear equations. To see these different results, complete the following example.

EXAMPLE 2   Solve numerically if possible.

a. $a^3 + a^2 - 4a + 2 = 2a^2 + 2a + 2$

b. $2(x^2 + 3x - 5) = 2x^2 + 2(3x - 5)$

c. $x^3 + 5x^2 + 2x - 7 = x^3 + 5x^2 + 2$

d. $2x^2 + 7x - 4 = 2(x^2 + 3x + 1) + x$

e. $x^2 + 2x + 4 = 3x^2 + 6x + 12$

a.

X	Y1	Y2	
-2	6	6	6 = 6
-1	6	2	6 > 2
0	2	2	2 = 2
1	0	6	0 < 6
2	6	14	6 < 14
3	26	26	26 = 26
4	66	42	66 > 42

X=4

$Y1 = x^3 + x^2 - 4x + 2$
$Y2 = 2x^2 + 2x + 2$

b.

X	Y1	Y2	
0	-10	-10	-10 = -10
1	-2	-2	-2 = -2
2	10	10	10 = 10
3	26	26	26 = 26
4	46	46	46 = 46
5	70	70	70 = 70
6	98	98	98 = 98

X=0

$Y1 = 2(x^2 + 3x - 5)$
$Y2 = 2x^2 + 2(3x - 5)$

c.

X	Y1	Y2	
0	-7	2	-7 < 2
1	1	8	1 < 8
2	25	30	25 < 30
3	71	74	71 < 74
4	145	146	145 < 146
5	253	252	253 > 252
6	401	398	401 > 398

X=0

$Y1 = x^3 + 5x^2 + 2x - 7$
$Y2 = x^3 + 5x^2 + 2$

d.

X	Y1	Y2	
0	-4	2	-4 < 2
1	5	11	5 < 11
2	18	24	18 < 24
3	35	41	35 < 41
4	56	62	56 < 62
5	81	87	81 < 87
6	110	116	110 < 116

X=0

$Y1 = 2x^2 + 7x - 4$
$Y2 = 2(x^2 + 3x + 1) + x$

e.

X	Y1	Y2	
0	4	12	4 < 12
1	7	21	7 < 21
2	12	36	12 < 36
3	19	57	19 < 57
4	28	84	28 < 84
5	39	117	39 < 117
6	52	156	52 < 156

X=0

$Y1 = x^2 + 2x + 4$
$Y2 = 3x^2 + 6x + 12$

Calculator Numeric Solution

a. $a^3 + a^2 - 4a + 2 = 2a^2 + 2a + 2$

Rewrite the equation as $x^3 + x^2 - 4x + 2 = 2x^2 + 2x + 2$.

The solutions are $-2, 0$, and 3, because when $-2, 0$, and 3 are substituted for the variable in both expressions, the results are equivalent.

b. $2(x^2 + 3x - 5) = 2x^2 + 2(3x - 5)$

The expressions are equal for all x-values in the table. If we rewrite this equation in standard form, we obtain an identity, $0 = 0$. The permissible replacements for each expression are all real numbers. The solution set is the set of all real numbers.

c. $x^3 + 5x^2 + 2x - 7 = x^3 + 5x^2 + 2$

The expression on the left is less than the expression on the right for $x = 4$ and greater than the expression on the right for $x = 5$. A noninteger solution lies somewhere between 4 and 5.

Note: We will find the exact solution later in this section.

d. $2x^2 + 7x - 4 = 2(x^2 + 3x + 1) + x$

The expression on the left is always less than the expression on the right. Note that the expression on the left is always 6 less than the expression on the right $Y2 - Y1 = 6$. It appears that the expressions will never be equal. If we rewrite the equation in standard form, we obtain $-6 = 0$, a contradiction. Therefore, the original equation has no solution.

e. $x^2 + 2x + 4 = 3x^2 + 6x + 12$

The expression on the left is always less than the expression on the right. The expressions do not appear to be equal. Note that $Y2 - Y1$ changes values. If we rewrite the equation in standard form, we obtain $2x^2 + 4x + 8 = 0$. This equation has no real-number solution. We will determine the solution of the equation in Chapter 13. ●

HELPING HAND Note that we cannot determine the solutions of the equations in parts d and e, because in part d there is no solution and in part e there is no real-number solution. When we write the equations in standard form, we see that an equation with no solution results in a contradiction, whereas an equation with no real-number solution does not result in a contradiction.

NUMERICAL SOLUTIONS OF A POLYNOMIAL EQUATION

To solve a polynomial equation for integer solutions numerically, set up an extended table of values. In analyzing the table, you will find that one of five possibilities will occur.

Integer solutions exist. The solutions are the integers in the first column that correspond to equal values in the second and third columns.

(Continued on page 712)

Noninteger solutions exist. In comparing corresponding values in the second and third columns, it is found that the order changes from "less than" to "greater than," or from "greater than" to "less than." A noninteger solution is between the two integers in the first column that indicate this change.

No solution exists. In comparing corresponding values in the second and third columns, it is found that one column is always less than the value in the other column. There is a constant difference between the values in the second column and the third column. The equation is a contradiction.

No real-number solution exists. In comparing corresponding values in the second and third columns, it is found that the value in one column is always less than the value in the other column. There is no constant difference between the values in the second and third column. The equation is not a contradiction.

An infinite number of solutions exist. In comparing corresponding values in the second and third columns, it is found that the values are equal. The equation is an identity. The solution set consists of all numbers for which the equation is defined.

In conclusion, a polynomial equation may be solved numerically by using a table of values. However, if the solution is noninteger, it will be difficult to find by that method. Also, it is difficult to know whether all possible solutions have been found.

11.1.2 Checkup

Solve exercises 1–5 numerically.

1. $b^3 + 3b^2 + 2b = 2b^2 + 6b + 4$

2. $6x^2 + 3x = 2x + 1$

3. $(3x - 4)(x + 1) = 3x^2 - x - 1$

4. $x^2 + 4x + 5 = 2x^2 + 8(x + 1) + 2$

5. $2(x^2 + 2x - 1) - x(x - 1) = 3x(x + 1) - 2(x^2 - x + 1)$

6. Explain the difference between integer solutions of a polynomial equation and noninteger solutions.

7. What are some of the limitations of the numerical method in solving a polynomial equation?

11.1.3 Solving Polynomial Equations Graphically

A second way to determine a real-number solution of an equation is to graph two functions. The functions to be graphed are written by using each expression in the equation as a rule for one of the functions.

SOLVING A POLYNOMIAL EQUATION GRAPHICALLY
To solve a polynomial equation graphically,

• Write two functions, using each expression in the equation as a rule.
• Graph both functions on the same coordinate plane by plotting points found in a table of values and connecting the points with a smooth curve to include all values in the domains of the functions.

The solutions of the equation are the *x*-coordinates of the points of intersection of the two graphs.

The *y*-coordinates of the points of intersection of the two graphs are the values obtained for both expressions when the equation is evaluated with the solutions.

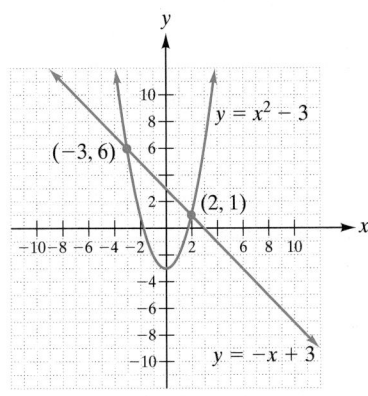

$$Y1 = x^2 - 3$$

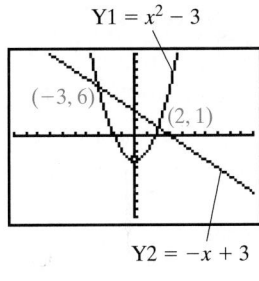

$$Y2 = -x + 3$$

$$(-10, 10, -10, 10)$$

Figure 11.2

For example, solve $x^2 - 3 = -x + 3$ graphically.

Set up a table of values for each function.

x	$y = x^2 - 3$	y	x	$y = -x + 3$	y
-1	$y = (-1)^2 - 3$	-2	-1	$y = -(-1) + 3$	4
0	$y = (0)^2 - 3$	-3	0	$y = -(0) + 3$	3
1	$y = (1)^2 - 3$	-2	1	$y = -(1) + 3$	2
2	$y = (2)^2 - 3$	1	2	$y = -(2) + 3$	1

When plotted, the points result in the graph shown at the left.

The two solutions are the x-coordinates of the intersections, -3 and 2.

Solve $x^2 - 3 = -x + 3$ graphically on your calculator.

The intersection of the graphs may be found by using the Intersect option under the CALC menu. The solutions of the equation are the x-coordinates of the points of intersection of the two graphs.

The y-coordinates of the points of intersection of the two graphs are the values obtained for both expressions when the equation is evaluated with the solutions.

The two solutions are the x-coordinates of the intersections, -3 and 2, as shown in Figure 11.2.

Remember that polynomial equations may have a finite number of solutions, an infinite number of solutions, no solutions, or no real-number solutions. Thus, solving polynomial equations graphically may yield any of these possibilities.

EXAMPLE 3   Solve graphically if possible.

a. $x^3 + x^2 - 4x + 2 = 2x^2 + 2x + 2$

b. $4x^2 - 3x - 3 = x$

c. $2(x^2 + 3x - 5) = 2x^2 + 2(3x - 5)$

d. $x^2 + 2x + 4 = 3x^2 + 6x + 12$

e. $2x^2 + 7x - 4 = 2(x^2 + 3x + 1) + x$

Calculator Graphic Solution

Define the two expressions as rules for functions, and enter them into your calculator.

a. $x^3 + x^2 - 4x + 2 = 2x^2 + 2x + 2$

The three solutions are the x-coordinates of the intersections: -2, 0, and 3.

b. $4x^2 - 3x - 3 = x$

The two solutions are the x-coordinates of the intersections, -0.5 and 1.5, as shown in Figure 11.3a and Figure 11.3b. Intersect, under the CALC menu, was used to find the intersection of the graphs.

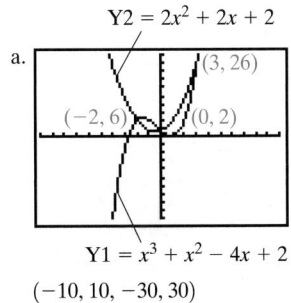

$$Y2 = 2x^2 + 2x + 2$$

a.

$$Y1 = x^3 + x^2 - 4x + 2$$

$$(-10, 10, -30, 30)$$

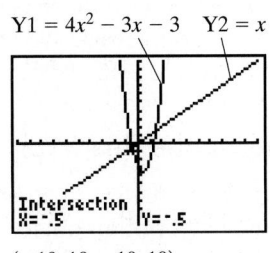

$$Y1 = 4x^2 - 3x - 3 \quad Y2 = x$$ $\qquad$ $$Y1 = 4x^2 - 3x - 3 \quad Y2 = x$$

Intersection   X=-.5   Y=-.5 $\qquad$ Intersection   X=1.5   Y=1.5

$$(-10, 10, -10, 10)$$ $\qquad\qquad$ $$(-10, 10, -10, 10)$$

Figure 11.3a $\qquad\qquad\qquad$ **Figure 11.3b**

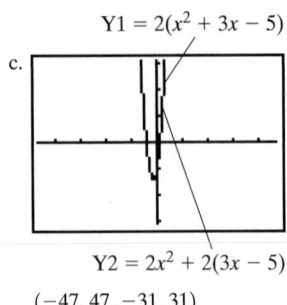

Y1 = 2(x² + 3x − 5)

c.

Y2 = 2x² + 2(3x − 5)

(−47, 47, −31, 31)

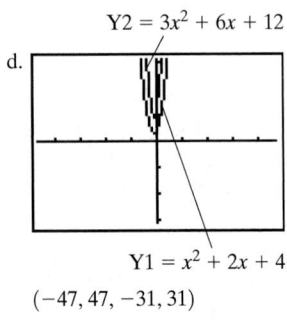

Y2 = 3x² + 6x + 12

d.

Y1 = x² + 2x + 4

(−47, 47, −31, 31)

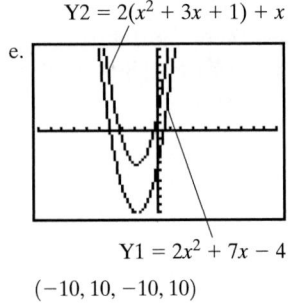

Y2 = 2(x² + 3x + 1) + x

e.

Y1 = 2x² + 7x − 4

(−10, 10, −10, 10)

c. $2(x^2 + 3x - 5) = 2x^2 + 2(3x - 5)$

The graphs appear to be the same. Rewriting the equation results in $0 = 0$, an identity. The solution set is the set of all real numbers (the common domain of the graphed functions).

d. $x^2 + 2x + 4 = 3x^2 + 6x + 12$

The graphs do not appear to intersect. The standard form of the equation is $2x^2 + 4x + 8 = 0$. The equation is not a contradiction. There is no real-number solution.

e. $2x^2 + 7x - 4 = 2(x^2 + 3x + 1) + x$

The graphs do not appear to intersect. Rewriting the equation results in a contradiction, $-6 = 0$. The equation has no solution.

GRAPHICAL SOLUTIONS OF POLYNOMIAL EQUATIONS

To solve a polynomial equation graphically, graph the two functions defined by the expressions on the left and right sides of the equation. In analyzing the graphs, you will find that one of four possibilities will occur.

One or more solutions exist. The graphs intersect. The solutions are the x-coordinates of the points of intersection.

No solution exists. The graphs do not appear to intersect. The original equation is a contradiction.

No real-number solution exists. The graphs do not appear to intersect. The original equation may be written in standard form and is not a contradiction.

An infinite number of solutions exist.

The graphs appear to coincide. The original equation is an identity. The solution set consists of all the numbers in the common domain of the graphed functions.

In conclusion, we solve a polynomial equation graphically by graphing two functions. On a calculator, this method may be used to find noninteger solutions. Also, with this method we are better able to find all the solutions because we can see the number of points of intersection.

11.1.3 Checkup

Solve exercises 1–5 graphically.

1. $x^3 + 2x^2 = x + 2$

2. $20x^2 + 4x = 15x + 3$

3. $(3x - 2)(x - 1) = 3x^2 - 5(x + 1)$

4. $x^2 - 2(3x - 5) = 3(x^2 - 6x) + 30$

5. $3(2x^3 + 2x^2) - (x + 2) + x^2 = 6x^3 + 7x^2 - x - 2$

6. In solving polynomial equations, we sometimes conclude that there are no solutions of the equation and other times determine that there are no real-number solutions of the equation. What do you think is the difference between these two statements?

7. Once you graphically obtain the solutions of a polynomial equation, how can you check to be sure they are the solutions?

11.1.4 Modeling the Real World

An application of a quadratic equation is the **vertical-position equation**, used to find the height of an object that was dropped or projected into the air. The height s of the object (in feet) is found by using the equation $s = -16t^2 + v_0t + s_0$, where t is the time (in seconds), v_0 is the initial velocity, and s_0 is the initial height. The initial velocity is 0 feet per second if the object is dropped, is positive if the object is thrown upward and is negative if the object is thrown downward. Galileo discovered this formula in the late 1500s.

EXAMPLE 4

According to legend, Galileo simultaneously dropped two balls of different weights from the Leaning Tower of Pisa in Italy, to see if they fell at the same rate or different rates. The Greek philosopher Aristotle had said that the heavier object would fall faster, and this was accepted as the truth for 2000 years. Galileo showed that if you neglect air resistance, the balls fall at the same rate. If the balls fell 179 feet, how long did it take them to hit the ground?

Graphic Solution

Substitute values into the vertical-position equation. The balls hit the ground when they are 0 feet above ground level, or when $s = 0$ feet, $v_0 = 0$ feet per second (the objects were dropped from rest), and $s_0 = 179$ feet.

$$s = -16t^2 + v_0t + s_0$$
$$0 = -16t^2 + 0t + 179$$
$$0 = -16t^2 + 179$$

The intersections of $Y1 = 0$ and $Y2 = -16x^2 + 179$ occur on the x-axis at approximately $(3.34, 0)$ and $(-3.34, 0)$. The negative value of the x-intercept is not valid for this situation, because we cannot have a negative value for time.

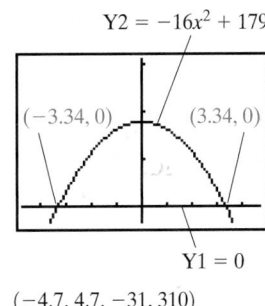

$$(-4.7, 4.7, -31, 310)$$

The objects were in the air for about 3.34 seconds.

APPLICATION

Sales for Ameritrade Holding increased rapidly after the company went public in 1997. Using data from the company's annual financial reports dating from 1993, we find that the amount of sales in millions of dollars, y, is given by $y = 6.75x^2 - 7.5x + 27.5$, where x is the number of years after 1993. In 1998, sales were $164.1 million. Assuming that the equation continues to apply, determine when the company will double this amount.

Discussion

Let x = the number of years after 1993
    y = the amount of annual sales in millions of dollars
To double the amount of sales in 1998, the amount of sales will be $164.1 \cdot 2 = 328.2$ million dollars.

$$y = 6.75x^2 - 7.5x + 27.5$$
$$328.2 = 6.75x^2 - 7.5x + 27.5 \quad \text{Substitute } 328.2 \text{ for y.}$$

(Continued on page 716)

Solve by graphing.

$Y2 = 6.75x^2 - 7.5x + 27.5$

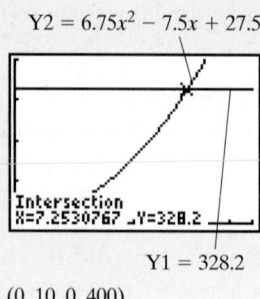

$Y1 = 328.2$

$(0, 10, 0, 400)$

The point of intersection is about (7.25, 328.2). The solution is 7.25.

Sales will reach double the 1998 amount of 164.1 million dollars 7.25 years after 1993 (1993 + 7.25 = 2000.25), or in the year 2001.

 11.1.4 Checkup

1. The Empire State Building is 1250 feet tall. If King Kong dropped a banana from the top of the building, how long would it take for the banana to hit the ground?

2. Per capita personal income has increased steadily, according to data published by the Bureau of Economic Development of the U.S. Department of Commerce. Using the four most recently available years of data, one can estimate the per capita personal income for residents of the state of California by the equation $y = 0.175x^2 + 1.305x + 26.655$, where x is the number of years since 1997 and y is the income in thousands of dollars. In 1997, the per capita personal income for residents of California was reported to be \$26,555. Assuming that the equation continues to apply, determine the year in which the per capita personal income will have increased by \$15,000 over the 1997 value.

11.1 Exercises

Determine whether each equation is a polynomial equation. Identify each polynomial equation as quadratic or cubic, where applicable.

1. $3x^3 - 2x^2 + x = 5$

2. $3y - 2y^{-1} + 4 = 0$

3. $3\sqrt{y} + y - 4 = 0$

4. $5.7x^3 - 1.9x^2 = 8.6x - 3.5$

5. $\frac{1}{4}x^4 + 3x^2 - \frac{3}{4} = 0$

6. $5(z - 1)^3 = 4(z - 1)^2$

7. $4(x - 2)(x + 7) = 16$

8. $3x + 5\sqrt{x} = 17$

9. $3x^{-2} - 5x = 4x^2$

10. $4a^3 + a^2 - 3a + 2 = 0$

11. $1.7x^2 + 3.2x = 5.7$

12. $\frac{2}{3}x^2 - \frac{5}{9}x = \frac{1}{6}$

Solve numerically.

13. $x^2 + 8 = 6x$

14. $x^2 - 6 = -x$

15. $4x^3 = x + 1$

16. $x^2 + x = 15$

17. $x^2 - 7 = x^2 + 3$

18. $x^2 - 3x + 2 = x(x - 3) + 7$

19. $x^2 + 5x + 1 = 1 + x(5 + x)$

20. $x(7 - x^2) + 6 = 6 + 7x - x^3$

21. $x^2 - 2x + 6 = 12 - 4x + 2x^2$

22. $x^2 + 2x + 4 = 1 - (x + 1)^2$

23. $\frac{1}{2}x^2 - x = 6 - 3x$

24. $\frac{1}{2}x^2 + 2x = \frac{3}{2}x + 6$

25. $2 - 0.2x^2 = 0.6x$

26. $6 + 2x - 0.4x^2 = 1.2x$

Solve graphically and check.

27. $x^2 - 3 = 6$

28. $x^2 - 8 = 8$

29. $x^2 - 3 = 2x$

30. $x^2 + 2x + 9 = 1 - 4x$

31. $x^3 = 4x$

32. $\frac{1}{2}x^3 = 4$

33. $x^2 - 3x - 10 = 0$

34. $x^3 + 3x^2 - x - 3 = 0$

35. $x^2 - 2x + 1 = x^2 - 2x - 3$

36. $3 - x^2 = 8 - x$

37. $x^2 + 1 = 3x^2 + 3$

38. $3 - x^2 = 6 - x^2$

39. $x(x + 3) = x^2 + 3x$

40. $(x + 1)^2 = x^2 + 2x + 1$

41. $4x^2 - x^3 = x^2 - 4x$

42. $x^3 - 6x + 2 = 2 - x^2$

43. $x^2 - 10x + 30 = \frac{1}{2}x^2 - 5x + 15$

44. $\frac{1}{2}x^2 + 5 = -\frac{1}{3}x^2 - 2$

45. $x^3 - 2x^2 + 1 = x^3 - 2x^2 + 9$

46. $x^2 - 3 = x^2 - 1$

47. $x(x^2 - 3) - 5(x + 1) = x^3 - 8x - 5$

48. $(x - 2)(x^2 + 2x) = x^3 - 4x$

49. $4x^2 = 9$

50. $9x^2 = 16$

51. $10x^3 - 7x^2 - 4x = 3x - 4$

52. $4x^3 - 8x^2 = 7x - 5$

53. $x^2 - 0.9x - 10.36 = 0$

54. $x^2 - 5x + 3.36 = 0$

55. $x^3 + 3.7x^2 = 1.74x + 7.56$

56. $x^3 + 0.1x^2 + 5 = 6.02x + 6.2$

In exercises 57–62, use the position equation, $s = -16t^2 + v_0 t + s_0$, to solve the problem presented.

57. A tightrope walker drops her hat from 40 feet above the ground. How many seconds will it take for the hat to hit the ground?

58. A gardener drops his pruning shears while trimming a tree. The gardener is 16 feet above the ground. How long will it take for the shears to hit the ground?

59. A tightrope walker throws a silver dagger vertically downward from 40 feet above the ground. If the initial velocity of the throw is 5 feet per second, how many seconds will it take for the dagger to hit the ground? (*Note:* $v_0 = -5$, since the object is thrown downward. If it had been thrown upward, the initial velocity would have been positive.)

60. A gardener tosses a pruned branch vertically downward to the ground from a height of 16 feet. If the initial velocity of the toss is 2 feet per second, how long will it take for the branch to hit the ground?

61. A tightrope walker throws a baton vertically upward at a velocity of 5 feet per second. If he is 40 feet above the ground, how many seconds will it take for the baton to hit the ground?

62. A gardener tosses a hammer vertically upward to shoo away a squirrel in a tree. The hammer is released at a height of 6 feet above the ground with an initial velocity of 10 feet per second. The gardener misses both the squirrel and the tree. How long will it take for the hammer to hit the ground?

63. Using data published by the Internal Revenue Service for the years from 1995 to 1999, a statistician developed an equation to estimate the total collections from corporate income taxes. The statistician used the equation $y = -2.143x^2 + 19.371x + 173.314$, where x is the number of years since 1995 and y is the total taxes in billions of dollars. The 1995 tax collection totaled \$175 billion. Using the preceding equation, determine the year in which the total collections will decrease by \$25 billion, assuming that the relation given in the equation continues to apply. From the graph, does it appear likely that the relation will continue to apply? Explain.

64. Using data published by the Internal Revenue Service for the years from 1995 to 1999, a statistician developed an equation to estimate the total collections from estate and gift taxes. The statistician used the equation $y = 0.286x^2 + 2.217x + 15.071$, where x is the number of years since 1995 and y is the total taxes in billions of dollars. If the 1995 collections totaled \$15 billion, determine the year in which the collections would be triple this amount, assuming that the relation given in the equation continues to apply. Do you think it is reasonable to assume that the relation will continue to apply? Explain.

65. Per capita personal-income data by states are published by the Bureau of Economic Development of the U.S. Department of Commerce. Using the four most recently available years of data, one can estimate the per capita personal income of residents of the state of Hawaii by the equation $y = 0.175x^2 + 0.165x + 25.715$, where x is the number of years since 1997 and y is the income in thousands of dollars. In 1997, the per capita personal income for residents of Hawaii was reported to be \$25,700. Assuming that the equation continues to apply, determine the year in which the per capita personal income will become twice the 1997 value. Does this result seem reasonable to you? Explain.

66. Using the data published by the Bureau of Economic Development, the per capita personal income of residents of the state of New York can be estimated by the equation

$y = 0.25x^3 - 1.1x^2 + 2.65x + 29.7$, where x is the number of years since 1997 and y is the income in thousands of dollars. In 1997, the per capita personal income of residents of New York was reported to be $29,700. Assuming that the equation continues to apply, determine the year in which the per capita personal income will become double the 1997 value. Does this result seem reasonable to you? Explain.

67. A statistical study attempted to relate the monthly sales (y) in a sales territory to the number of sales representatives (x) assigned to the territory. The polynomial equation derived from the study was $y = 11.55 + 20.67x - 1.35x^2$, where sales are recorded in thousands of dollars. How many sales representatives should be assigned to a territory if the company wants monthly sales to reach $100,000?

68. Using the statistical model described in exercise 67, how many sales representatives should be assigned to a territory if the company wants monthly sales to reach $75,000?

11.1 Calculator Exercises

Part 1. Using the ZBox Command to Improve a Graph's View

In solving a polynomial equation graphically, it is often necessary to experiment with different settings of the window to be able to see the graph clearly. As an example, solve the equation

$$x^3 + 5.2x^2 = 36.9 + 4.49x$$

using the following instructions:

1. Graph the expressions on the left and right sides of the equation, using the decimal window. The graph appears to have the shape of a parabola, which is the shape of a quadratic function. But the expressions are cubic and linear, indicating that the decimal window is not a good view.

2. Next, change to a standard window. More of the graph appears, but it still does not seem complete.

3. Change to an integer window. Much more of the graph appears, but it seems squeezed horizontally.

4. Change Xmin and Xmax by dividing each setting by 10.

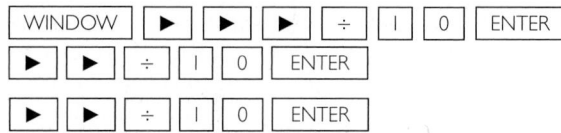

This will spread out the graph. Then view the graph. GRAPH This improves the graph, but it looks as if the window is still too small.

5. Change Ymin and Ymax by multiplying each setting by 2.

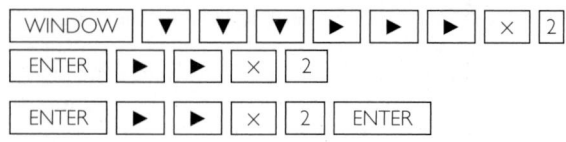

This will expand the graph in the y direction. Then view the graph. GRAPH Now you see that the graph consists of a curve that is intersected by a straight line at two or more points.

6. Find the rightmost intersection. One solution of the equation is $x = 2.5$.

7. The leftmost intersection is still not clearly shown. Use ZOOM 1 to box the region where the two graphs

meet. Move the cursor to the left upper corner of your desired box and press ENTER . Then use the arrow keys right and down to create the box. Be sure that the box includes the entire portion of the graphs where the intersection occurs. Press ENTER again, and the calculator will enlarge the graph to fit the box.

8. Now you can clearly see that the line intersects the curve at two points. This would be difficult to see otherwise. Find the two intersection points. The other two solutions are $x = -3.6$ and $x = -4.1$. After completing this exercise, restore the calculator window to the standard setting, ZOOM 6 .

Use the preceding approach to solve the following equations graphically (you can use multipliers or divisors other than those used in the example you just completed).

1. $x^3 + 5x^2 + 4x + 20 = 5x^2 + 25x$; verify the solutions $x = 1$, $x = 4$, and $x = -5$.

2. $x^3 + 3x^2 - 3x = x^2 - 2x + 2$; verify the solutions $x = -2$, $x = -1$, and $x = 1$.

3. $x^3 - 2x^2 = 30x - x^2$; verify the solutions $x = -5$, $x = 0$, and $x = 6$.

Part 2. Solving Graphically by First Writing Equations in Standard Form

Another helpful technique for solving a polynomial equation graphically is to simplify the equation by placing it in standard form. To do this, move all terms to the left side of the equation, using the properties of equations. The equations will now be of the form

$$ax^2 + bx + c = 0$$
$$ax^3 + bx^2 + cx + d = 0$$

and so on. Then store the left side of the equation in Y1 and set Y2 = 0 (since the right side of the equation is 0). When you graph the equations, their solutions will be the x-coordinates of the points of intersection of the graphs of Y1 and Y2. However, since the graph of Y2 = 0 is the x-axis, the solutions will also be the x-coordinates of the x-intercepts of the graph of Y1. Try this method on the four equations in Part 1 to verify the solutions listed there.

11.1 Writing Exercise

When graphing the expressions on the left and right sides of a polynomial equation as two functions, you can identify both intercepts and intersection points. Define "intercepts" and "intersection points," pointing out the differences and similarities between them. Draw an example of a graph of the expressions on the left and right sides of a polynomial equation, labeling the intercepts and the intersection points. Can the intercepts and the intersection points occur at the same location? Explain.

11.2 Solving Equations Algebraically by Factoring

OBJECTIVES

1 Solve polynomial equations by using the zero factor property.

2 Solve polynomial equations by factoring.

3 Model real-world situations by using polynomial equations, and solve the equations by factoring.

APPLICATION

According to annual reports for Texas Instruments, Inc., the company's gross profit for the years 1998–2000 are given in the following table:

	2000	1999	1998
Gross profit in billions of dollars	5.8	4.5	3.1

a. Let x = the number of years after 1998 and y = the gross profit in billions of dollars.

Determine a quadratic function for the gross profit.

b. Use the quadratic function found in part a, and determine the year in which the gross profit will reach $10 billion.

After completing this section, we will discuss this application further. See page 725.

In the preceding section, we used numeric and graphic methods to solve polynomial equations. While such methods work, they are sometimes limited in their usefulness. Therefore, we need an algebraic method to solve these equations. One such method is to solve by factoring.

11.2.1 Solving Equations Using the Zero Factor Property

We begin by reviewing a property of real numbers that we discussed in Chapter 1. The multiplication property of zero stated that the product of a real number and 0 is 0. We can relate this property to another property called the zero factor property. The *zero factor property* states that if a product is 0, then one or both of the factors must be 0.

For example, $6 \cdot 0 = 0, 0 \cdot (-5) = 0$, or $0 \cdot 0 = 0$.

ZERO FACTOR PROPERTY
If $ab = 0$, then either $a = 0, b = 0$, or both.

We can use this property to solve an equation of the form $P(x) = 0$ when $P(x)$ is in factored form, such as when $P(x) = (x + 5)(x - 2)$.

For example, solve $(x + 5)(x - 2) = 0$.
If $(x + 5)(x - 2) = 0$, then $x + 5 = 0$, $x - 2 = 0$, or both.

$$\underset{a}{\uparrow} \quad \underset{b}{\uparrow} \qquad \underset{a}{\uparrow} \quad \underset{b}{\uparrow}$$

We can now determine the solutions by solving the linear equations.

$$x + 5 = 0 \qquad \text{or} \qquad x - 2 = 0$$
$$x + 5 - 5 = 0 - 5 \qquad\qquad x - 2 + 2 = 0 + 2$$
$$x = -5 \qquad\qquad x = 2$$

The solutions of the equation $(x + 5)(x - 2) = 0$ are -5 and 2.
We can check the solutions by substituting them into the original equation.

Let x = −5

$(x + 5)(x - 2) = 0$
$(-5 + 5)(-5 - 2)$ | 0
$(0) \qquad (-7)$
0

Let x = 2

$(x + 5)(x - 2) = 0$
$(2 + 5)(2 - 2)$ | 0
$(7) \qquad (0)$
0

Both values result in true equations. Therefore, the solutions are -5 and 2.

EXAMPLE 1 Solve and check.

a. $(2x + 3)(5x - 4) = 0$ **b.** $x(x + 6)(3x - 2) = 0$

Algebraic Solution

a. $(2x + 3)(5x - 4) = 0$

Use the zero factor property.

$$2x + 3 = 0 \qquad \text{or} \qquad 5x - 4 = 0$$
$$2x + 3 - 3 = 0 - 3 \qquad\qquad 5x - 4 + 4 = 0 + 4$$
$$2x = -3 \qquad\qquad\qquad 5x = 4$$
$$x = -\frac{3}{2} \qquad\qquad\qquad x = \frac{4}{5}$$

The solutions of the equation are $-\frac{3}{2}$ and $\frac{4}{5}$.

b. $x(x + 6)(3x - 2) = 0$

We now have three factors equal to 0. The zero factor property can be expanded to more than two factors. Therefore, we set all three factors equal to 0.

$$x = 0 \quad \text{or} \quad x + 6 = 0 \quad \text{or} \quad 3x - 2 = 0$$
$$x = -6 \qquad\qquad x = \frac{2}{3}$$

The solutions of the equation are 0, -6, and $\frac{2}{3}$.

Substitution Check

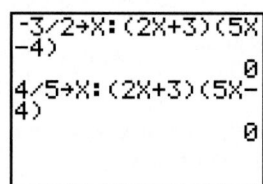

Check the solutions on your calculator by substituting the values for the variable in the expression on the left side of the equation. The result should equal 0, the expression on the right side of the equation.

After substituting the solutions into the expression on the left, we find that the results are 0. The solutions check.

Graphic Check

$$Y1 = x(x + 6)(3x - 2)$$

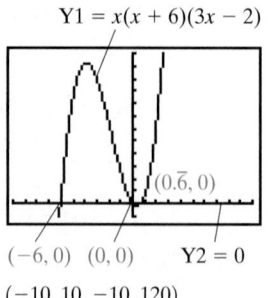

$(-6, 0) \quad (0, 0) \qquad Y2 = 0$

$(-10, 10, -10, 120)$

A graphic check on the calculator results in three solutions: 0, -6, and $0.\overline{6}$ (or $\frac{2}{3}$).

11.2.1 Checkup

Solve and check.

1. $(x + 7)(4x - 3) = 0$ **2.** $x(4x - 5)(x - 1) = 0$

11.2.2 Solving Equations by Factoring

As we know, most polynomial equations are not written in the factored form given in Example 1. We may have to use the algebra skills from previous chapters to manipulate the equation into that form.

SOLVING A POLYNOMIAL EQUATION ALGEBRAICALLY BY FACTORING

To solve a polynomial equation algebraically by factoring,

- Write the equation in the standard form $P(x) = 0$.
- Factor $P(x)$.
- Set each factor equal to 0 and solve for the variable.
- Check the solutions by substitution or by solving numerically or graphically.

For example, solve $x^2 + x - 6 = 0$.

Graphic Check

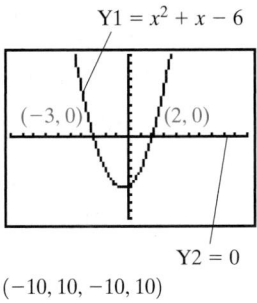

$$x^2 + x - 6 = 0 \qquad \text{Standard form}$$
$$(x + 3)(x - 2) = 0 \qquad \text{Factor.}$$
$$x + 3 = 0 \quad \text{or} \quad x - 2 = 0 \quad \text{Zero factor property}$$
$$x = -3 \qquad\qquad x = 2$$

Y1 = $x^2 + x - 6$

$(-3, 0)$ $(2, 0)$

Y2 = 0
$(-10, 10, -10, 10)$

Check the solution by graphing the two functions determined by each side of the original equation. Let $Y1 = x^2 + x - 6$ and $Y2 = 0$.

 **HELPING HAND** The second function, $Y2 = 0$, is the x-axis and cannot be seen as a line. Therefore, the function does not need to be graphed to determine the solution. We can graph Y1 alone and determine the x-intercepts.

The x-coordinates of the x-intercepts—that is, the intersections of the graph of Y1 and the x-axis—are -3 and 2. The solutions are -3 and 2.

If we cannot trace and find the intersection of the graph with the x-axis, the calculator will calculate it for us. Under the CALC function, choose Zero. Set an interval by choosing a left bound (a point on the graph to the left of the x-intercept) and a right bound (a point on the graph to the right of the x-intercept). Then select a guess between the two bounds. The calculator will display the x-intercept. The solutions (or **roots**) are the x-coordinates of each x-intercept.

EXAMPLE 2   Solve and check.

a. $x^2 - 9 = 0$ **b.** $4x^2 + 20x + 25 = 0$
c. $-10x^2 - 41x + 77 = 0$

Algebraic Solution

a.
$$x^2 - 9 = 0 \qquad \text{Standard form}$$
$$(x + 3)(x - 3) = 0 \qquad \text{Factor.}$$
$$x + 3 = 0 \quad \text{or} \quad x - 3 = 0 \qquad \text{Zero factor property}$$
$$x = -3 \qquad\qquad x = 3$$

The solutions are -3 and 3.

Graphic Check

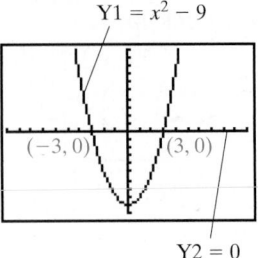

$(-10, 10, -10, 10)$

The graphic check results in the graph of $Y1 = x^2 - 9$.
The x-coordinates of the x-intercepts are -3 and 3, the solutions.

b.
$$4x^2 + 20x + 25 = 0 \qquad \text{Standard form}$$
$$(2x + 5)^2 = 0 \qquad \text{Factor.}$$

Set each factor equal to 0 and solve. Since the factor is squared, both factors are $2x + 5$. There is no need to solve two equations.

$$2x + 5 = 0$$
$$x = -\frac{5}{2}$$

The solution is $-\frac{5}{2}$. It is called a **double root**.

Graphic Check

$Y1 = 4x^2 + 20x + 25$

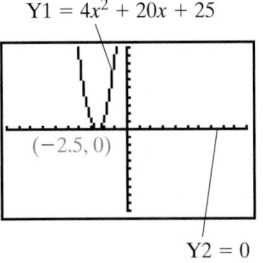

$(-10, 10, -10, 10)$

The graphic check results in the graph of $Y1 = 4x^2 + 20x + 25$. There is only one x-intercept. The x-coordinate of the x-intercept, $-\frac{5}{2}$, is the solution. You need to use Zero under the CALC function to find the intercept.

 **HELPING HAND** A double root will always occur when the graphic check of a quadratic equation in standard form results in one x-intercept.

c.
$$-10x^2 - 41x + 77 = 0 \qquad \text{Standard form}$$
$$-1(2x + 11)(5x - 7) = 0 \qquad \text{Factor.}$$

Set each factor equal to 0 and solve. The common factor -1 cannot equal 0, because it is a constant.

$$2x + 11 = 0 \quad \text{or} \quad 5x - 7 = 0$$
$$x = -\frac{11}{2} \qquad\qquad x = \frac{7}{5}$$

The solutions are $-\frac{11}{2}$ and $\frac{7}{5}$.

Substitution Check

```
-11/2→X: -10X²-41
X+77
               0
7/5→X: -10X²-41X+
77
               0
```

Both solutions, $-\frac{11}{2}$ and $\frac{7}{5}$, when substituted for the variable, result in expressions equal to 0, the second expression value.

EXAMPLE 3 Solve and check.

a. $6x^2 + 14x = 3x + 7$ b. $(x + 3)(x - 7) = -9$

Algebraic Solution

a. $6x^2 + 14x = 3x + 7$

Neither side of the equation is 0. Use the properties of equations to write the equation in standard form.

$6x^2 + 14x - 3x - 7 = 3x + 7 - 3x - 7$

$6x^2 + 11x - 7 = 0$ *Standard form.*

$(3x + 7)(2x - 1) = 0$ *Factor.*

$3x + 7 = 0$ or $2x - 1 = 0$ *Zero factor property*

$$x = -\frac{7}{3} \qquad\qquad x = \frac{1}{2}$$

The solutions are $-\frac{7}{3}$ and $\frac{1}{2}$.

Graphic Check

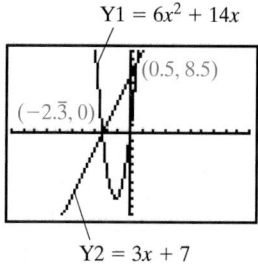

Y1 = $6x^2 + 14x$

(0.5, 8.5)

$(-2.\overline{3}, 0)$

Y2 = $3x + 7$

$(-10, 10, -10, 10)$

The graphic check shows the intersections of Y1 = $6x^2 + 14x$ and Y2 = $3x + 7$ are $(-2.\overline{3}, 0)$ and $(0.5, 8.5)$. The solutions are $-2.\overline{3}$ and 0.5 or $-\frac{7}{3}$ and $\frac{1}{2}$.

b. $(x + 3)(x - 7) = -9$

Neither side of the equation is 0. Multiply the expression on the left and then use the properties of equations to write the equation in standard form.

$x^2 - 4x - 21 = -9$

$x^2 - 4x - 12 = 0$ *Standard form*

$(x + 2)(x - 6) = 0$ *Factor.*

$x + 2 = 0$ or $x - 6 = 0$ *Zero factor property*

$x = -2 \qquad\qquad x = 6$

The solutions are -2 and 6.

Graphic Check

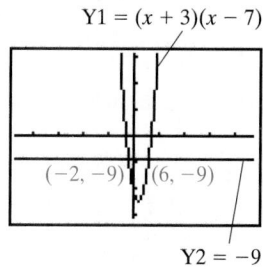

Y1 = $(x + 3)(x - 7)$

$(-2, -9)$ $(6, -9)$

Y2 = -9

$(-47, 47, -31, 31)$

The graphic check results in the intersection of the graphs Y1 = $(x + 3)(x - 7)$ and Y2 = -9. The solutions are -2 and 6.

 11.2.2 Checkup

Solve exercises 1–4 and check.

1. $a^2 - 81 = 0$

2. $16x^2 + 24x + 9 = 0$

3. $4x^2 - 5x = 12x - 15$

4. $(x + 4)(x - 2) = 16$

5. Before you can use the zero factor property to solve a polynomial equation, what must be true about the form of the equation?

11.2.3 Modeling the Real World

Polynomial equations are commonly applied to geometry problems. Use the formulas given to you in earlier chapters, and solve them with the methods of this section. Be careful to evaluate the solutions to see whether they make sense within the constraints of the problem. Solutions of the equation sometimes may not make physical sense.

EXAMPLE 4 LaChung plans to make an open rectangular box from a flat piece of tin. He needs the length to be 5 feet more than the width. He plans to cut out square corners 3 feet on a side for the height of the box. He needs the box to hold 198 cubic feet of mulch.

a. Find the dimensions of the box.

b. Find the dimensions of the piece of tin needed.

Algebraic Solution

a. Let x = width of the box

$x + 5$ = length of the box

The volume formula for a rectangular solid is $V = LWH$. Substitute values or expressions for the variables L, W, H, and V.

$$V = LWH$$
$$198 = (x + 5)(x)(3)$$
$$198 = 3x^2 + 15x \qquad \text{Simplify.}$$
$$3x^2 + 15x - 198 = 0 \qquad \text{Standard form}$$
$$3(x^2 + 5x - 66) = 0 \qquad \text{Factor out the GCF.}$$
$$3(x - 6)(x + 11) = 0 \qquad \text{Factor the trinomial.}$$

$x - 6 = 0$ or $x + 11 = 0$	Zero factor property
$x = 6$ $\qquad$ $x = -11$	Solve.

Since x is the width of the box, $x = -11$ is not an appropriate choice. Therefore, the width is 6 feet.

The length is $x + 5 = 6 + 5 = 11$ feet.

The box's dimensions are 11 feet by 6 feet by 3 feet.

b. The length of the tin must be 11 feet plus 6 feet (two 3-foot corner sections), or 17 feet.

The width of the tin must be 6 feet plus 6 feet (again, two 3-foot corner sections), or 12 feet.

EXAMPLE 5 Phyllis plans to build a storage shed for her garden tools and lawn mower. She wants the length of the shed to be twice the width and the height to be 3 feet more than the width. Phyllis has enough materials to cover 340 square feet. She would like to use all the materials. Determine the possible dimensions of the shed.

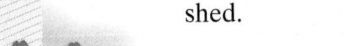

Algebraic Solution

Let x = width

$2x$ = length

$x + 3$ = height

The surface area of a rectangular solid is given by the formula

$$SA = 2LW + 2WH + 2LH$$

$$340 = 2(2x)(x) + 2(x)(x + 3) + 2(2x)(x + 3) \qquad \text{Substitute expressions.}$$
$$340 = 4x^2 + 2x^2 + 6x + 4x^2 + 12x \qquad \text{Simplify.}$$
$$340 = 10x^2 + 18x$$

$$10x^2 + 18x - 340 = 0 \qquad \text{Standard form}$$
$$2(5x^2 + 9x - 170) = 0 \qquad \text{Factor out the GCF.}$$
$$2(5x^2 - 25x + 34x - 170) = 0 \qquad \text{Factor, using the ac method.}$$
$$2[5x(x - 5) + 34(x - 5)] = 0$$
$$2(x - 5)(5x + 34) = 0$$
$$x - 5 = 0 \qquad \text{or} \qquad 5x + 34 = 0 \qquad \text{Zero factor property}$$
$$x = 5 \qquad\qquad x = -\frac{34}{5}$$

The width of the shed is 5 feet. (A negative width is not possible, so we discard the solution $-\frac{34}{5}$ feet.) The length is twice the width, or 10 feet. The height is 3 feet more than the width, or 8 feet.

APPLICATION

According to annual reports for Texas Instruments, Inc., the company's gross profit for the years 1998–2000 are given in the following table:

	2000	1999	1998
Gross profit in billions of dollars	5.8	4.5	3.1

a. Let x = the number of years after 1998 and y = the gross profit in billions of dollars. Determine a quadratic function for the gross profit.

b. Use the quadratic function found in part a, and determine the year in which the gross profit will reach $10 billion.

Discussion

a. Let x = the number of years after 1998

y = the gross profit in billions of dollars

The ordered pairs from the table are $(0, 3.1)$, $(1, 4.5)$, and $(2, 5.8)$.

Use the quadratic regression function on your calculator to determine an equation.

(*Note:* Directions for writing a quadratic function were given in Section 8.4.)

```
L1      L2      L3    2
0       3.1     ------
1       4.5
2       5.8
------  ------

L2(4) =
```

```
QuadReg
y=ax²+bx+c
a=⁻.05
b=1.45
c=3.1
```

The quadratic equation is $y = -0.05x^2 + 1.45x + 3.1$.

b. To determine when the gross profit reaches $10 billion, let $y = 10$.

$$10 = -0.05x^2 + 1.45x + 3.1$$
$$0.05x^2 - 1.45x + 6.9 = 0 \qquad \text{Write in standard form.}$$
$$5x^2 - 145x + 690 = 0 \qquad \text{Multiply by the LCD of the decimals, or 100.}$$
$$5(x^2 - 29x + 138) = 0 \qquad \text{Factor out the common factor of 5.}$$
$$5(x - 6)(x - 23) = 0 \qquad \text{Factor.}$$
$$x - 6 = 0 \quad \text{or} \quad x - 23 = 0 \qquad \text{Zero factor property.}$$
$$x = 6 \qquad\qquad x = 23$$

(Continued on page 726)

The gross profit will reach $10 billion 6 years after 1998 (that is, 2004) and 23 years after 1998 (that is, 2021). Note that 23 years beyond the starting year is very far into the future and is likely not a reliable prediction.

It is more reasonable to assume that if the relation continues to hold, gross profit will reach $10 million in the year 2004.

11.2.3 Checkup

1. For her art class, Trekisa needs to make a tray from a piece of poster board. From a rectangular piece of board, she must cut squares x inches on a side out of each corner, fold up the sides, and join them to make the tray. Her assignment is to make a tray that is 7 inches long, with a width that is 2 inches more than three times its height. The tray must have a volume of 35 cubic inches.
   a. Determine the dimensions of the tray.
   b. Determine the dimensions of the rectangular piece of poster board.

2. Justin must make a box from balsa wood for a project in his architecture class. He is allowed to use 40 square inches of balsa wood for the project. The height of the box must be the same as its width. The box must have a length that is 2 inches more than the width. What will be the dimensions of the box?

3. Annual reports for a computer-manufacturing company listed the gross profit for the years 1998–2000 as follows:

Year	1998	1999	2000
Gross profit in billions of dollars	6.44	6.60	6.68

a. An analyst believed that profits would reach a maximum and then begin to fall. A quadratic function could be used to represent this belief. Let $x =$ the number of years after 1998 and $y =$ the gross profit in billions of dollars. Determine a quadratic function for the gross profit.
b. Use the quadratic function found in part a and the zero factor property to determine the year when the gross profit will drop to $5 billion.
c. Do you think that your quadratic function is a reliable predictor? Explain.

11.2 Exercises

Solve exercises 1–44 and check.

1. $(x + 6)(x + 11) = 0$

2. $(p + 9)(p + 13) = 0$

3. $\left(\frac{3}{5}x - \frac{9}{20}\right)\left(x + \frac{2}{3}\right) = 0$

4. $\left(\frac{2}{3}x + \frac{4}{9}\right)\left(x - \frac{7}{9}\right) = 0$

5. $3x(x + 9)(2x - 5) = 0$

6. $9x(x - 4)(6x + 1) = 0$

7. $(7x - 49)(49x - 7) = 0$

8. $(5x - 45)(45x - 5) = 0$

9. $(4x + 3)(2x - 9)(x + 6) = 0$

10. $(2z + 7)(3z - 8)(z - 7) = 0$

11. $(0.2x + 6.8)(1.3x - 1.69) = 0$

12. $(1.4x + 4.2)(0.7x - 2.8) = 0$

13. $0 = x^2 + 10x + 24$

14. $z^2 + 13z + 40 = 0$

15. $x^2 + 33 = 14x$

16. $52 - 17x + x^2 = 0$

17. $4x^2 + 5x + 24x + 30 = 0$

18. $7x^2 + 8x + 14x + 16 = 0$

19. $5x^2 + 3x = 8$

20. $7x = 3x^2 - 20$

21. $15x^2 = 35x$

22. $63x = 18x^2$

23. $18x^2 - 3x = 5 - 30x$

24. $8x^2 + 12x = 14x + 21$

25. $16x^2 + 72x + 81 = 0$

26. $49x^2 - 28x + 4 = 0$

27. $4x^2 + 25x + 18 = 5x - 7$

28. $9x^2 + 29x + 8 = 5x - 8$

29. $b^2 + 7 = 71$

30. $c^2 - 5 = 139$

31. $9z^2 = 25$

32. $64p^2 = 81$

33. $(x + 1)^2 = 49$

34. $(x - 2)^2 = 36$

35. $(x - 3)(x - 2) = 42$

36. $(x - 5)(x + 4) = 22$

37. $x^2 + (x + 3)^2 = 225$

38. $(x + 5)^2 + x^2 = 625$

39. $x^3 + 7x^2 - 9x - 63 = 0$

40. $z^3 + 5z^2 - 16z - 80 = 0$

41. $18x^3 + 45x^2 - 50x - 125 = 0$

42. $12p^3 + 36p^2 - 147p - 441 = 0$

43. $3x^3 - 3x^2 + 12x - 12 = 0$

44. $2x^3 - 16x^2 + 3x - 24 = 0$

45. Phil is designing a water trough for his ranch. The trough will be in the shape of a rectangular box and must hold 72 cubic feet of water. Phil wants it to be 3 feet high, with a width that is 1 foot more than half its length.
  a. If the length is denoted by x, write an expression for the volume of the trough.
  b. Write an equation in terms of the volume, and solve the equation for the dimensions of the trough.

46. A rectangular jewelry box is designed to have a height of 4 centimeters. The length of the box is 1 centimeter more than eight times its width.
  a. If the width is denoted by x, write an expression for the volume of the box.
  b. Write an equation for the volume of the box when it will hold 820 cubic centimeters, and solve the equation for the dimensions of the box.

47. Michelle is designing drawers for a cabinet. She wants each drawer to have a width that is three times its height and a length that is 2 inches more than four times the height. She will need wood for the four sides of the drawer and the bottom.
  a. Write an expression for the surface area of the five sides for which Michelle must obtain wood.
  b. If Michelle figures that she needs 700 square inches of wood for the drawer, what are its dimensions?

48. A cardboard chute in the shape of a rectangular box is constructed with a width that is five times its height and a length that is 2 units more than seven times its height. The chute is open on its two ends.

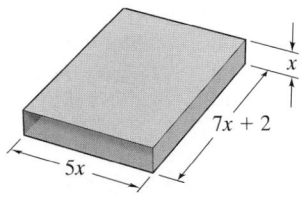

  a. Write an expression for the surface area of the chute.
  b. If the outer surface of the chute has an area of 828 square inches, what are the dimensions of the chute?

49. Steve designs a sail for his sailboat. The sail will be triangular, but not a right triangle. He wants the height to be 4 feet more than the base.
  a. Write an expression for the area of the sail.
  b. If the area of the sail is to be 30 square feet, what are the base and height of the sail?

50. The base of a triangular garden is 2 meters more than five times its height. The garden covers 260 square meters of ground.
  a. Write an expression for the area of the garden.
  b. What do the base and height of the garden measure?

51. Claims adjusters are concerned with establishing who was at fault in an accident. To help them decide, they use skid marks to determine the speeds of the vehicles. The formula is

$$V^2 = 30FS$$

where V is the velocity of the vehicle (in miles per hour), F is the coefficient of friction of the road in decimal form, and S is the skid length. Write an equation involving velocity if a car skidded 243 feet and the coefficient of friction of the road was 40%. Solve the equation by factoring to determine the velocity of the vehicle.

52. Using the formula from exercise 51, write an equation for the velocity of a car that skidded 250 feet when the coefficient of friction of the road was 27%. Solve the equation by factoring to determine the velocity of the car.

53. Annual reports for a manufacturer of electronic equipment listed the profit for the years 1998–2000 as follows:

Year	1998	1999	2000
Profit in millions of dollars	2.90	2.72	2.30

  a. A statistician decided that a quadratic function could be used to represent these data. Let $x = $ the number of years after 1998 and $y = $ the profit in millions of dollars. Determine a quadratic function for the profit.
  b. Use the quadratic function found in part a and the zero factor property to determine the year when the profit will drop to $2 million.
  c. Do you think the function is a reliable predictor for this exercise? Explain.

54. Annual reports for a manufacturer of athletic shoes listed the profit for the years 1998–2000 as follows:

Year	1998	1999	2000
Profit in millions of dollars	3.66	4.14	4.44

  a. The chief executive officer (CEO) of the company believed that profits would increase for a period of time and then begin to fall as the line of shoes became less popular. A quadratic function could be used to represent this belief. Let $x = $ the number of years after 1998 and $y = $ the profit in millions of dollars. Determine a quadratic function for the profit.
  b. Use the quadratic function found in part a and the zero factor property to determine the year when the profit will drop to $3 million.
  c. Do you think the function is a reliable predictor for this exercise? Explain.

11.2 Calculator Exercises

You have learned to use the regression feature of your calculator to write a quadratic equation, given three points through which the function passes. You can also use this feature to write a polynomial equation of higher degree when you are given more points through which the function passes. As an example, suppose you wish to find the cubic equation that passes through the points $(0, 0)$, $(1, 2)$, $(-1, -6)$, and $(2, 18)$. To do so, complete the following steps:

• Clear List 1 and List 2 of the calculator.

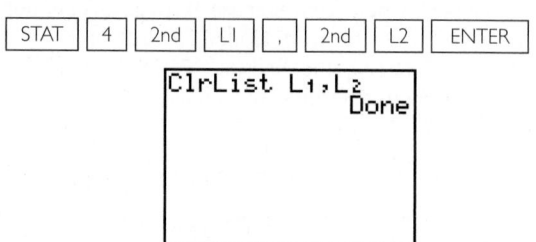

• Store the x-values of the ordered pairs in List 1 and the y-values of the ordered pairs in List 2.

Use the cursor to move to the appropriate list, and then key in the values, pressing ENTER after each, until you have entered the pairs of values into the lists. Be sure the pairs match in the lists.

L1	L2	L3	2
0	0		
1	2		
-1	-6		
2	18		
------	------		

L2(5) =

• Calculate the cubic equation for the points and store the equation in Y1.

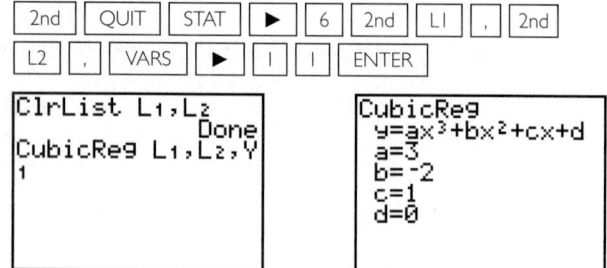

CubicReg
y=ax³+bx²+cx+d
a=3
b=-2
c=1
d=0

• To view the plotted coordinate pairs and the equation, use the [STAT PLOT] feature of the calculator.

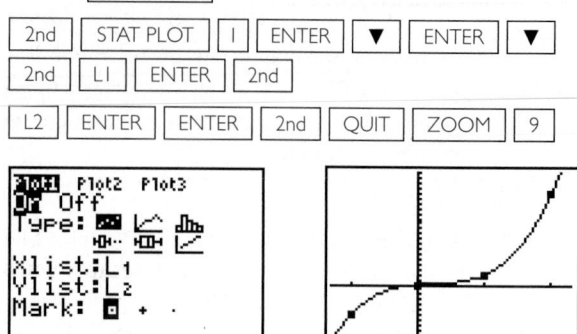

Note that when you use this feature to find a polynomial equation, the number of coordinate pairs through which the function passes must be at least one more than the degree of the polynomial that you are writing. That is, to find a quadratic equation, you must have three coordinate pairs; for a cubic equation, you must have four coordinate pairs; and for a fourth-degree equation, you must have five coordinate pairs. The calculator will accept only up to a fourth-degree equation. To calculate such an equation, you would follow the same steps as you just did, except that when you choose the calculate option, you would choose "QuartReg" instead of "CubicReg."

Use the preceding method to answer the following questions (remember that when you are keying in the values of the coordinate pairs, you can clear the previous set of values stored in Lists L1 and L2 by moving the cursor to the top of the list, pressing [CLEAR], and then moving the cursor back down to the location of the first value in the list):

1. Find the cubic equation that passes through $(0, -4)$, $(1, -2)$, $(2, 2)$, and $(-1, -10)$.

2. Find the quartic equation (that is, fourth-degree equation) that passes through $(0, 1)$, $(1, 1)$, $(2, 11)$, $(-1, 5)$, and $(-2, 31)$.

3. Find the cubic equation that passes through $(0, -6)$, $(1, -7)$, $(5, 9)$, and $(-1, 15)$.

4. Find the quartic equation that passes through $(0, 6)$, $(1, -0.7)$, $(2, -1.6)$, $(3, 20.1)$, and $(-1, 13.7)$.

11.2 Writing Exercise

In solving the equation

$$a^2 - 9a = 0$$

Chantel factored the left side, used the zero factor property, and stated that the solutions of the equation were $a = 0$ and $a = 9$. In contrast, Holly just divided both sides of the equation by a and solved the resulting equation to get only one solution, $a = 9$. Which do you think is the correct way to solve the equation? Explain.

11.3 Solving Quadratic Equations by Using Square Roots

OBJECTIVES

1 Simplify square-root expressions by using the product rule for square roots.

2 Simplify square-root expressions by using the quotient rule for square roots.

3 Solve quadratic equations in one variable algebraically by using the principle of square roots.

4 Model real-world situations by using quadratic equations, and solve the equations by using the principle of square roots.

APPLICATION

The compound-interest formula $A = P(1 + r)^t$ is used to determine the compounded amount A, given a principal P at an interest rate per period r for t periods. This formula can lead to various polynomial equations, including quadratic equations.

Due to the bank policy on education savings accounts, at age 18 Tommy must withdraw his ed-

ucation fund of $7237. He plans to reinvest all of his fund at an annual interest rate such that this money will increase to a total of $8000 in two years. Determine the annual interest rate at which he must reinvest his funds.

After completing this section, we will discuss this application further. See page 738.

In Section 11.1, methods for solving polynomial equations were presented. The methods will work for polynomial equations of any degree. However, quadratic equations are a special type of polynomial equation that have been studied more extensively. Additional methods for solving these special polynomial equations are presented in the next section.

11.3.1 Simplifying Square Roots by Using the Product Rule

In the previous section, we learned that some quadratic equations had solutions which we could not find by using the process of factoring. Therefore, we need to identify another algebraic method of finding these solutions.

Before we begin, we will review some terminology associated with a square-root expression. A square-root expression is of the form $\sqrt{a}$, where the symbol $\sqrt{}$ is called a radical and a is the radicand. In Chapter 1, we defined $\sqrt{a}$ as the principal square root and $-\sqrt{a}$ as the negative square root. Always remember that a square root of a negative number is not defined in the real-number system.

$$\sqrt{36} = 6 \text{ because } (6)^2 = 36$$
$$-\sqrt{36} = -6 \text{ because } (-6)^2 = 36$$
$$\sqrt{-36} \text{ is not a real number.}$$

EVALUATING SQUARE-ROOT EXPRESSIONS
To evaluate a principal square root, observe that $\sqrt{c^2} = |c|$.
To evaluate a negative square root, observe that $-\sqrt{c^2} = -|c|$.

If the radicand cannot be written as the square of a number, we approximate the answer. We use the calculator to do this.

One property of square roots involves multiplication. To discover this product rule, use your calculator to complete the following exercises.

Discovery 1

Multiplication of Square Roots

Evaluate each expression on your calculator, and compare the results obtained in the first column with the corresponding results in the second column.

1. a. $\sqrt{3} \cdot \sqrt{7} \approx$ _____   **b.** $\sqrt{21} \approx$ _____

2. a. $\sqrt{2} \cdot \sqrt{3} \approx$ ____   **b.** $\sqrt{6} \approx$ ____

Write a rule for multiplying square roots.

The value of the expression in the first column is equivalent to the expression in the second column. The first value is the product of square roots. The second value is the square root of the product of the radicands in the first expression. Therefore, to multiply square roots, we multiply the radicands and then take the square root of the product.

PRODUCT RULE FOR SQUARE ROOTS
For any real numbers $\sqrt{a}$ and $\sqrt{b}$,

$$\sqrt{a} \cdot \sqrt{b} = \sqrt{ab}$$

To simplify square roots, we read the product rule from right to left: $\sqrt{ab} = \sqrt{a} \cdot \sqrt{b}$. For example, to simplify a square root, we rewrite the radicand as a product of a perfect square (preferably the largest possible perfect-square factor) and another factor. We then reverse the product rule by writing a product of square roots. Finally, we simplify the perfect square root. For example,

$$\sqrt{24} = \sqrt{4 \cdot 6} = \sqrt{4}\sqrt{6} = \sqrt{2^2}\sqrt{6} = 2\sqrt{6}$$

Check

```
2√(6)
          4.898979486
√(24)
          4.898979486
```

To check this result on your calculator, first evaluate the square-root expression found algebraically and then evaluate the given square-root expression. The results should be the same.

EXAMPLE I   Simplify.

a. $\sqrt{48}$ **b.** $\sqrt{128}$ **c.** $-\sqrt{72}$

Solution

a. $\sqrt{48} = \sqrt{16 \cdot 3} = \sqrt{16}\sqrt{3} = 4\sqrt{3}$ $16 = 4^2$

HELPING HAND If we had not used the largest perfect-square factor, 16, we would need to simplify the expressions twice.

$$\sqrt{48} = \sqrt{4 \cdot 12} = \sqrt{4}\sqrt{12} = 2\sqrt{12} = 2\sqrt{4 \cdot 3} = 2\sqrt{4}\sqrt{3}$$
$$= 2 \cdot 2\sqrt{3} = 4\sqrt{3}$$

b. $\sqrt{128} = \sqrt{64 \cdot 2} = 8\sqrt{2}$ $64 = 8^2$

c. $-\sqrt{72} = -\sqrt{36} \cdot \sqrt{2} = -6\sqrt{2}$ $36 = 6^2$

EXAMPLE 2   Simplify.

a. $-2 + \sqrt{24}$ b. $3 - \sqrt{27}$ c. $\dfrac{4 + \sqrt{32}}{4}$ d. $\dfrac{-6 - \sqrt{28}}{-4}$

Solution

a. $-2 + \sqrt{24} = -2 + \sqrt{4 \cdot 6} = -2 + 2\sqrt{6}$

b. $3 - \sqrt{27} = 3 - \sqrt{9 \cdot 3} = 3 - 3\sqrt{3}$

c. $\dfrac{4 + \sqrt{32}}{4} = \dfrac{4 + \sqrt{16 \cdot 2}}{4} = \dfrac{4 + 4\sqrt{2}}{4}$

$\qquad = \dfrac{\overset{1}{\cancel{4}}\,(1 + \sqrt{2})}{\underset{1}{\cancel{4}}}$ Factor the numerator and divide out the 4.

$\qquad = 1 + \sqrt{2}$

d. $\dfrac{-6 - \sqrt{28}}{-4} = \dfrac{-6 - \sqrt{4 \cdot 7}}{-4} = \dfrac{-6 - 2\sqrt{7}}{-4}$

$\qquad = \dfrac{\overset{1}{-2}\,(3 + \sqrt{7})}{\underset{2}{-\cancel{4}}}$ Factor the numerator and divide out the -2.

$\qquad = \dfrac{3 + \sqrt{7}}{2}$

 11.3.1 Checkup

Simplify exercises 1–4 without using a calculator. Then check the results on your calculator.

1. $\sqrt{98}$ **2.** $-3 + \sqrt{32}$ **3.** $\dfrac{5 + \sqrt{75}}{5}$ **4.** $\dfrac{-6 - \sqrt{45}}{-3}$

5. Explain the difference between finding the square root of a number and squaring a number.

11.3.2 Simplifying Square Roots by Using the Quotient Rule

Square roots also have a property involving quotients. To see what it is, complete the following set of exercises.

 Discovery 2

Division of Square Roots

Evaluate each expression on your calculator, and compare the results obtained in the first column with the corresponding results in the second column.

1. a. $\dfrac{\sqrt{9}}{\sqrt{3}} \approx$ _____   **b.** $\sqrt{3} \approx$ _____

2. a. $\dfrac{\sqrt{6}}{\sqrt{3}} \approx$ _____   **b.** $\sqrt{2} \approx$ _____

Write a rule for dividing square roots.

The value of the expression in the first column is equivalent to the expression in the second column. The first value is the quotient of square roots. The second value is the square root of the quotient of the radicands in the first expression. Therefore, to divide square roots, we divide the radicands and then take the square root of the quotient.

QUOTIENT RULE FOR SQUARE ROOTS
For any real numbers $\sqrt{a}$ and $\sqrt{b}$, where $b \neq 0$,

$$\frac{\sqrt{a}}{\sqrt{b}} = \sqrt{\frac{a}{b}}$$

We will show that this rule is true in a later chapter.

To simplify square roots, we read the quotient rule from right to left: $\sqrt{\frac{a}{b}} = \frac{\sqrt{a}}{\sqrt{b}}$. For example, to evaluate a square root having a fractional radicand, we reverse the quotient rule by writing a quotient of square roots. Then we simplify each square root, using the product rule. For example,

$$\sqrt{\frac{5}{36}} = \frac{\sqrt{5}}{\sqrt{36}} = \frac{\sqrt{5}}{6}$$

We must be careful when writing the final result. If the result is a fraction, it is conventional to write the denominator without radicals. The process of changing the fraction to an equivalent form without a radical denominator is called **rationalizing the denominator**. That is, we make the denominator a rational number.

For example, simplify

$$\sqrt{\frac{36}{5}} = \frac{\sqrt{36}}{\sqrt{5}} = \frac{6}{\sqrt{5}}$$

To rationalize the denominator, we use the multiplication property of 1. That is, we multiply the numerator and denominator by the same value. We choose a value that results in the denominator having a perfect-square radicand, so that we can simplify it. Since the denominator is $\sqrt{5}$ and we want a perfect-square radicand, we can multiply the denominator by itself: $\sqrt{5}\sqrt{5} = \sqrt{25} = 5$. Therefore, rationalizing the denominator results in

$$\frac{6}{\sqrt{5}} = \frac{6}{\sqrt{5}} \cdot \frac{\sqrt{5}}{\sqrt{5}} = \frac{6\sqrt{5}}{5}$$

EXAMPLE 3 Simplify without a calculator. Rationalize all denominators. Check your results on your calculator.

a. $\sqrt{\frac{3}{4}}$ **b.** $\sqrt{\frac{48}{50}}$

Solution

a. $\sqrt{\frac{3}{4}} = \frac{\sqrt{3}}{\sqrt{4}} = \frac{\sqrt{3}}{2}$ *Quotient rule*

Check

```
√(3)/2
          .8660254038
√(3/4)
          .8660254038
```

The results check.

b. $\sqrt{\dfrac{48}{50}} = \dfrac{\sqrt{48}}{\sqrt{50}} = \dfrac{\sqrt{16}\sqrt{3}}{\sqrt{25}\sqrt{2}} = \dfrac{4\sqrt{3}}{5\sqrt{2}}$ Quotient rule

$= \dfrac{4\sqrt{3}}{5\sqrt{2}} \cdot \dfrac{\sqrt{2}}{\sqrt{2}} = \dfrac{4\sqrt{6}}{5 \cdot 2}$ Rationalize the denominator.

$= \dfrac{4\sqrt{6}}{10} = \dfrac{2\sqrt{6}}{5}$ Simplify.

```
(2√(6))/5
        .9797958971
√(48/50)
        .9797958971
```

The results check.

A second method would be to reduce the fractional radicand before using the quotient rule.

$$\sqrt{\dfrac{48}{50}} = \sqrt{\dfrac{24}{25}} = \dfrac{\sqrt{24}}{\sqrt{25}} = \dfrac{\sqrt{4}\sqrt{6}}{\sqrt{25}} = \dfrac{2\sqrt{6}}{5}$$

 11.3.2 Checkup

Simplify exercises 1–3 without a calculator. Rationalize all denominators. Check your results on your calculator.

1. $\sqrt{\dfrac{7}{64}}$ **2.** $\sqrt{\dfrac{64}{7}}$ **3.** $\sqrt{\dfrac{250}{45}}$

4. What does it mean to rationalize the denominator of a quotient?

11.3.3 Solving Quadratic Equations by Using the Principle of Square Roots

We are now ready to examine a second algebraic method for solving a quadratic equation. In the previous section, we found a solution of an equation such as $x^2 = 9$ by factoring. For example,

$$x^2 = 9$$
$$x^2 - 9 = 0$$
$$(x - 3)(x + 3) = 0$$
$$x - 3 = 0 \quad \text{or} \quad x + 3 = 0$$
$$x = 3 \qquad\qquad x = -3$$

The solutions are 3 and -3, which we can combine by writing ± 3.

Another way to solve $x^2 = 9$ is to determine values for x that can be squared to equal 9. There are two such numbers: 3 and -3.

We are actually taking the square root of both sides of the equation in order to solve for the variable.

$$x^2 = 9$$
$$\sqrt{x^2} = \sqrt{9} \qquad \text{Take the square root of both sides.}$$

$$|x| = 3 \qquad \text{Evaluate.}$$
$$x = 3 \quad \text{or} \quad x = -3 \qquad \text{Solve the absolute-value equation.}$$

As mentioned previously, this pair of solutions may also be written as $x = \pm 3$.

PRINCIPLE OF SQUARE ROOTS
For any positive number b, if $a^2 = b$, then $a = \sqrt{b}$ or $a = -\sqrt{b}$.

We can also use the principle of square roots to determine solutions of equations such as $x^2 = 5$, in which the squared variable does not equal a perfect square.

$$x^2 = 5$$
$$x = \sqrt{5} \quad \text{or} \quad x = -\sqrt{5}$$

We could not solve, for example, the equivalent equation $x^2 - 5 = 0$ by factoring over the rational numbers.

SOLVING A QUADRATIC EQUATION BY USING THE PRINCIPLE OF SQUARE ROOTS

To solve a quadratic equation of the form $ax^2 + bx + c = 0$ when $b = 0$ (that is, $ax^2 + c = 0$),

- Solve for x^2.
- Apply the principle of square roots.
- Solve the resulting equations.

EXAMPLE 4 Solve and check.

a. $4x^2 - 9 = 0$ **b.** $2x^2 + 4 = 4$ **c.** $x^2 + 6 = 1$
d. $(x + 2)^2 = 9$ **e.** $(x + 3)^2 - 3 = 2$ **f.** $x^2 - 4x + 4 = 5$
g. $3x^2 - 16 = 0$

Algebraic Solution

a. $4x^2 - 9 = 0$

$$x^2 = \frac{9}{4} \qquad \text{Solve for } x^2.$$

$$x = \sqrt{\frac{9}{4}} \quad \text{or} \quad x = -\sqrt{\frac{9}{4}} \qquad \text{Principle of square roots}$$

$$x = \frac{3}{2} \qquad\qquad x = -\frac{3}{2}$$

Check by substitution.

$4x^2 - 9 = 0$		$4x^2 - 9 = 0$	
$4\left(\dfrac{3}{2}\right)^2 - 9$	0	$4\left(-\dfrac{3}{2}\right)^2 - 9$	0
$4\left(\dfrac{9}{4}\right) - 9$		$4\left(\dfrac{9}{4}\right) - 9$	
$9 \quad - 9$		$9 \quad - 9$	
0		0	

The solutions are $\pm\frac{3}{2}$.

b. $2x^2 + 4 = 4$ Solve for x^2.

$$x^2 = 0 \qquad \text{Principle of}$$
$$x = 0 \qquad \text{square roots}$$

The solution is 0, because 0 is neither positive nor negative.

Check

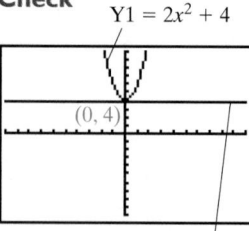

Y1 = $2x^2 + 4$

(0, 4)

Y2 = 4

$(-10, 10, -10, 10)$

The solution is 0, the x-coordinate of the intersection.

c. $x^2 + 6 = 1$

$x^2 = -5$ Solve for x^2.

The solution is not a real number, because there is no real number whose square is negative.

Check

Check by graphing. The graphs do not intersect. In standard form, the equation is $x^2 + 5 = 0$. There is no real-number solution.

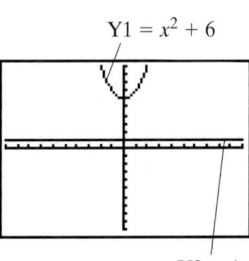

$Y1 = x^2 + 6$

$Y2 = 1$

$(-10, 10, -10, 10)$

d. This is a variable expression squared.

$$(x + 2)^2 = 9$$

$x + 2 = \sqrt{9}$ or $x + 2 = -\sqrt{9}$ Principle of square roots

$x + 2 = 3$ $\qquad$ $x + 2 = -3$

$x = 1$ $\qquad\qquad$ $x = -5$

The solutions are 1 and -5.

The check is left for you.

e. $(x + 3)^2 - 3 = 2$

$(x + 3)^2 = 5$ Solve for the squared expression $(x + 3)$.

$x + 3 = \sqrt{5}$ or $x + 3 = -\sqrt{5}$ Principle of square roots

$x = -3 + \sqrt{5}$ $\qquad$ $x = -3 - \sqrt{5}$

$x \approx -0.764$ $\qquad$ $x \approx -5.236$

The solutions are $-3 \pm \sqrt{5}$, or approximately -0.764 and -5.236.

The check is left for you.

f. The expression on the left is a perfect-square trinomial.

$x^2 - 4x + 4 = 5$

$(x - 2)^2 = 5$ Factor into a binomial square.

$x - 2 = \sqrt{5}$ or $x - 2 = -\sqrt{5}$ Principle of square roots

$x = 2 + \sqrt{5}$ $\qquad$ $x = 2 - \sqrt{5}$

$x \approx 4.236$ $\qquad$ $x \approx -0.236$

The solutions are $2 \pm \sqrt{5}$, or approximately 4.236 and -0.236.

The check is left for you.

g. $3x^2 - 16 = 0$

$3x^2 = 16$

$x^2 = \dfrac{16}{3}$ Solve for x^2.

$x = \pm\sqrt{\dfrac{16}{3}}$ Principle of square roots

$x = \pm\dfrac{4}{\sqrt{3}} \cdot \dfrac{\sqrt{3}}{\sqrt{3}}$ Simplify and rationalize the denominator.

$x = \pm\dfrac{4\sqrt{3}}{3}$

The solutions are $\pm\frac{4\sqrt{3}}{3}$, or approximately ± 2.309. The check is left for you.

11.3.3 Checkup

Solve exercises 1–7 and check.

1. $x^2 - 7 = 9$ **2.** $9x^2 - 25 = 0$

3. $x^2 + 12 = 3$ **4.** $(x - 4)^2 = 16$

5. $(x - 1)^2 + 5 = 8$ **6.** $x^2 + 2x + 1 = 6$

7. $2x^2 - 49 = 0$

8. How can you use the principle of square roots to solve a quadratic equation?

11.3.4 Modeling the Real World

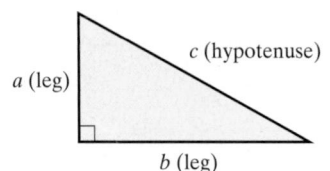

c (hypotenuse)

a (leg)

b (leg)

Another geometric application uses a formula that relates the lengths of three sides of a right triangle. A right triangle is a triangle that contains a 90° angle (**right angle**), denoted by the ⌐ symbol at the angle. The **legs** of a right triangle are the two sides that form the right angle. The **hypotenuse** is the side opposite the right angle.

The **Pythagorean theorem** states that the sum of the squares of the lengths of the legs of a right triangle is equal to the square of the length of the hypotenuse.

In other words, $a^2 + b^2 = c^2$, where a and b are the lengths of the legs of a right triangle and c is the length of the hypotenuse.

EXAMPLE 5 The Leaning Tower of Pisa is 190 feet tall, but was built on unstable ground. It began to tip while under construction in 1173. By 1990, the tower was 15 feet out of perpendicular. That year, construction began to correct the tilt. In 2001, the construction was complete. The tower is now 13.5 feet out of perpendicular. Determine the vertical distance of the top of the tower above the ground in 1990 and 2001.

Algebraic Solution

Let x = height

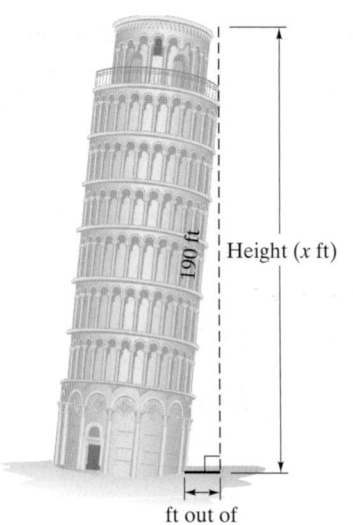

190 ft

Height (x ft)

ft out of perpendicular

1990	2001	
$a^2 + b^2 = c^2$	$a^2 + b^2 = c^2$	
$(15)^2 + x^2 = (190)^2$	$(13.5)^2 + x^2 = (190)^2$	Substitute.
$x^2 = (190)^2 - (15)^2$	$x^2 = (190)^2 - (13.5)^2$	Solve for x^2.
$x^2 = 35{,}875$	$x^2 = 35{,}917.75$	
$x = \pm\sqrt{35{,}875}$	$x = \pm\sqrt{35{,}917.75}$	Principle of square roots
$x \approx 189.41$	$x \approx 189.52$	A negative answer is not possible.

In 1990, the top of the Leaning Tower of Pisa was about 189.41 feet above the ground. In 2001, the top was about 189.52 feet above the ground.

EXAMPLE 6 A large boat is towing a smaller boat with a rope that spans 50 feet between hookups. If the rope is attached to the smaller boat at a point 10 feet below the level of attachment to the larger boat (see Figure 11.4), what is the distance between the boats?

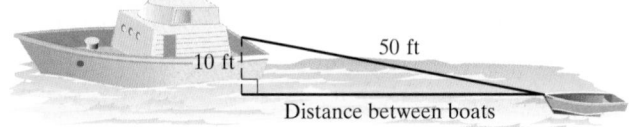

Figure 11.4

Algebraic Solution
Let x = the distance between the boats

$a^2 + b^2 = c^2$	Pythagorean theorem
$x^2 + 10^2 = 50^2$	Substitute x for a, 10 for b, and 50 for c.
$x^2 + 100 = 2500$	Simplify.
$x^2 = 2500 - 100$	Solve for x^2.
$x^2 = 2400$	Simplify.
$x = \pm\sqrt{2400}$	Principle of square roots
$x = \sqrt{400 \cdot 6}$	Only the positive root is possible, because we are solving for a distance.
$x = 20\sqrt{6}$	Simplify.
$x \approx 48.99$	

The boats are $20\sqrt{6}$ feet apart, or approximately 49 feet apart.

EXAMPLE 7 One of the most memorable games in World Series history occurred on October 21, 1975. Carlton Fisk of the Boston Red Sox ended Game 6 at Fenway Park with a 12th-inning home run over the wall called the Green Monster. The Green Monster is 315 feet from home plate.

If Carlton hit the baseball at a height of 3 feet and the ball crossed the Green Monster at a point 318 feet away in a straight line from the bat, approximate the height of the ball above the ground as it crossed the wall.

Algebraic Solution
Let x = the height of the ball as it crossed the Green Monster

We see that we have a right triangle.

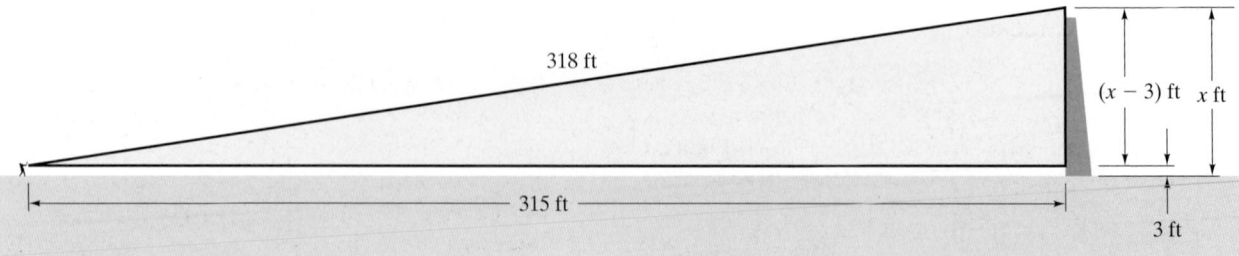

$$a^2 + b^2 = c^2 \qquad \text{Pythagorean theorem}$$
$$(x - 3)^2 + 315^2 = 318^2 \qquad \text{Substitute.}$$
$$(x - 3)^2 = 318^2 - 315^2 \qquad \text{Isolate the squared term.}$$
$$(x - 3)^2 = 1899 \qquad \text{Simplify.}$$
$$x - 3 = \pm\sqrt{1899} \qquad \text{Principle of square roots}$$
$$x = \pm\sqrt{1899} + 3$$
$$x \approx 46.6 \qquad \text{A negative solution does not represent a height.}$$

The ball crossed the Green Monster at a height of approximately 46.6 feet.

APPLICATION

The compound-interest formula $A = P(1 + r)^t$ is used to determine the compounded amount A, given a principal P at an interest rate per period r for t periods. This formula can lead to various polynomial equations, including quadratic equations.

Due to the bank policy on education savings accounts, at age 18 Tommy must withdraw his education fund of $7237. He plans to reinvest all of his fund at an annual interest rate such that this money will increase to a total of $8000 in two years. Determine the annual interest rate at which he must reinvest his funds.

Discussion

Let x = interest rate

$$A = P(1 + x)^t \qquad \text{Compound-interest formula}$$
$$8000 = 7237(1 + x)^2 \qquad A = 8000, P = 7237, t = 2$$
$$\frac{8000}{7237} = (1 + x)^2 \qquad \text{Solve for the squared term.}$$
$$(1 + x)^2 = \frac{8000}{7237}$$
$$1 + x = \sqrt{\frac{8000}{7237}} \qquad \text{or} \qquad 1 + x = -\sqrt{\frac{8000}{7237}} \qquad \text{Principle of square roots}$$
$$x = -1 + \sqrt{\frac{8000}{7237}} \qquad x = -1 - \sqrt{\frac{8000}{7237}}$$
$$x \approx 0.0514 \qquad \qquad \text{This is a negative solution.}$$

Tommy must reinvest his funds at approximately 5.14%. (Only the positive solution is used; the negative solution is not applicable.)

11.3.4 Checkup

1. A flagpole that is 20 feet tall tilts so that it is 1.5 feet out of perpendicular. Determine the pole's height.

2. A 3.5-foot metal rod is used to hook a disabled vehicle to a tow truck. The two hookups are separated by a horizontal distance of 2.8 feet. What is the vertical distance between the hookups?

3. A Little League player hit a line drive that just barely cleared a fence that was 100 feet away from home plate.

If the ball crossed the fence at a point 101 feet away in a straight line from the bat, estimate the height of the fence. Assume that the player hit the ball at a height of 2 feet.

4. Tommy is considering investing his money in an aggressive mutual fund instead of in the bank. If he invests the $7237 that he has on hand for two years, what equivalent annual interest rate must he realize in order to have the funds grow to $10,000?

11.3 Exercises

Simplify without using a calculator. Check your results on your calculator.

1. $\sqrt{63}$
2. $\sqrt{80}$
3. $\sqrt{243}$
4. $\sqrt{75}$
5. $\sqrt{147}$
6. $-\sqrt{128}$
7. $-\sqrt{125}$
8. $\sqrt{27}$
9. $-3 + \sqrt{20}$
10. $-4 + \sqrt{28}$
11. $2 - \sqrt{50}$
12. $7 - \sqrt{32}$
13. $\dfrac{4 - \sqrt{48}}{4}$
14. $\dfrac{3 - \sqrt{18}}{3}$
15. $\dfrac{-6 + \sqrt{45}}{3}$
16. $\dfrac{-4 + \sqrt{24}}{2}$
17. $\dfrac{-5 - \sqrt{18}}{-3}$
18. $\dfrac{-3 - \sqrt{40}}{-2}$
19. $\dfrac{-6 - \sqrt{54}}{-3}$
20. $\dfrac{-4 - \sqrt{44}}{-2}$
21. $\sqrt{\dfrac{16}{5}}$
22. $\sqrt{\dfrac{9}{7}}$
23. $\sqrt{\dfrac{50}{48}}$
24. $\sqrt{\dfrac{45}{24}}$

Solve and check.

25. $x^2 = 144$
26. $x^2 = 121$
27. $a^2 = 13$
28. $b^2 = 15$
29. $q^2 = 98$
30. $p^2 = 200$
31. $2x^2 - 32 = 0$
32. $3x^2 - 27 = 0$
33. $4x^2 - 25 = 0$
34. $16y^2 - 49 = 0$
35. $9x^2 = 2$
36. $64x^2 - 1 = 0$
37. $3x^2 + 4 = 6$
38. $5x^2 + 4 = 11$
39. $m^2 + 7 = 5$
40. $12 + v^2 = 5$
41. $(x - 5)^2 = 0$
42. $(2x - 7)^2 = 0$
43. $(z - 7)^2 = 4$
44. $(x + 6)^2 = 9$
45. $(4a - 3)^2 = 4$
46. $(3b - 5)^2 = 1$
47. $x^2 = 1.69$
48. $y^2 = 2.89$
49. $(x + 3)^2 - 1 = 3$
50. $(x - 6)^2 - 5 = 20$
51. $2(m - 4)^2 - 6 = 12$
52. $4(n + 9)^2 + 3 = 19$
53. $x^2 + 10x + 25 = 9$
54. $x^2 - 14x + 49 = 100$
55. $9x^2 - 6x + 1 = 144$
56. $25x^2 + 10x + 1 = 64$
57. $(x - 7)^2 - 5 = 1$
58. $(x - 12)^2 + 4 = 9$
59. $(2x + 1)^2 - 3 = 7$
60. $(4x - 5)^2 + 7 = 9$
61. $(x + 3)^2 - 6 = 6$
62. $(x - 15)^2 + 6 = 18$
63. $5x^2 - 4 = 0$
64. $7x^2 - 9 = 0$
65. $2a^2 - 13 = 12$
66. $11b^2 - 8 = 8$

Use the Pythagorean theorem to write a quadratic equation. Then solve the equation by using the principle of square roots.

67. A wheelchair ramp has a length of 61 inches. The horizontal distance of the ramp measures 60 inches. What is the vertical distance of the ramp?

68. Trent's kite is flying on 82 feet of string. His dad is standing directly below the kite and is 18 feet away from Trent. How high is the kite?

69. A traffic-control helicopter uses radar to determine that a car on a straight highway is 5000 feet away. The helicopter is 3000 feet above the highway. How far along the ground is the car from the helicopter?

70. A wire supporting a radio tower is attached to the top of the tower and to the ground. The wire is 130 meters long and it is attached to the ground 50 meters from the base of the tower. How tall is the tower?

71. The gable end of a roof is a right triangle with a span of 50 feet. The distance from the peak of the roof to either eave is the same. Find this distance.

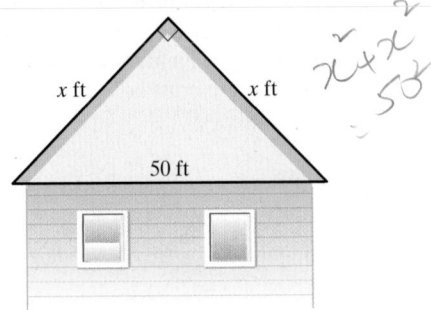

x ft   *x* ft

50 ft

72. The diagonal of a square is 20 cm long. What is the perimeter of the square?

73. A bridge that is 1 mile long expands by 1 foot during hot weather. If the bridge did not have adequate expansion joints, how high would the center of the bridge rise?

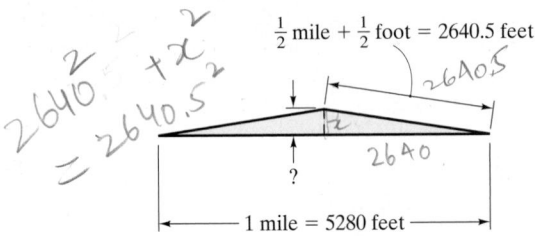

$\frac{1}{2}$ mile + $\frac{1}{2}$ foot = 2640.5 feet

?

1 mile = 5280 feet

74. A driveway is 50 feet long. During hot weather, it would expand 2 inches. If the driveway did not have expansion joints, how high would the center of the drive rise?

75. A soccer player kicked a goal from a distance of 25 feet along the ground to the goal. The ball entered the goal just below the horizontal bar, and the distance from there to the player's foot was 26 feet. How high was the ball as it crossed the goal line? If the goal was 8 feet high, by how many feet did the ball miss the goal's horizontal bar?

76. A soccer player kicked a ball from a distance of 30 feet along the ground to the goal. The ball missed the goal, crossing just above the horizontal bar, and the distance from there to the player's foot was 31.2 feet. How high was the ball as it crossed over the goal? If the goal was 8 feet high, by how many feet did the ball clear the bar?

Use the compound-interest formula $A = P(1 + r)^t$ to write a quadratic equation, and then solve the equation by using the principle of square roots.

77. If a $4000 investment compounded to $4708.90 after two years, determine the annual interest rate of the investment.

78. An investment of $6000 compounded to $6678.15 after two years. What was the annual interest rate of the investment?

79. At what annual interest rate must $5000 be invested in order for it to compound to $6000 after two years?

80. In order for $1200 to compound to $1600 in two years, what must be the annual interest rate of the investment?

81. Jean invests $7500 for two years in order to have money to add a sunroom onto her home. If she wants to have $9500 for the sunroom, determine the annual compound interest rate she must seek in order to meet her goal.

82. Brenda inherited $20,000 and invested it in a mutual fund for two years. At the end of that time, her investment had grown to $28,085. Determine the annual compound interest rate that would yield that amount of money.

Use the vertical-position formula to write a quadratic equation, and solve the equation by using the principle of square roots.

83. A water balloon is dropped from a 400-foot tower. How long will it take for the balloon to reach the ground?

84. A brick falls from the top ledge of a building that is 576 feet tall. How long will it take for the brick to reach the ground?

85. How long will it take a sky diver to descend from 12,000 feet to 5000 feet?

86. How long will it take a sky diver to descend from 11,000 feet to 3500 feet?

Write a quadratic equation and use the principle of square roots to solve it.

87. A company compiled data on its salespeople. The data were used to relate a salesperson's travel cost to the number of accounts handled. The mathematical model developed from the data estimated a salesperson's annual travel cost as $C(x) = 0.044x^2 + 4700$, where x is the number of accounts the salesperson manages. Use the model to estimate the number of accounts a salesperson should have if her annual travel costs are $9000.

88. Use the mathematical model in exercise 87 to estimate the number of accounts a salesperson should have if his annual travel costs are $6000.

11.3 Calculator Exercises

A popular algebra problem deals with sizes of pizza. In order to double the size (area) of a pizza, you must increase its diameter by a factor of $\sqrt{2}$; that is, $d_2 = \sqrt{2}\, d_1$, where d_2 is the diameter of the pizza with doubled area and d_1 is the diameter of the original pizza. Thus, a pizza that is about twice as large as a 5-inch pizza will be 7 inches in diameter. With this in mind, complete the following table, rounding answers to the nearest inch.

Original Pizza Diameter	Twice-as-Large Pizza Diameter
5 inches	7 inches
6 inches	
7 inches	
8 inches	
9 inches	
10 inches	
11 inches	
12 inches	

Take the table with you the next time you go out for pizza, and use it for price comparisons!

11.3 Writing Exercises

1. Rationalize the expression $\frac{7}{\sqrt{2}}$. Without using your calculator, do you know what a decimal approximation of $\sqrt{2}$ is? It is approximately equal to 1.414. If you had to evaluate the given expression without using your calculator, but instead using the preceding decimal approximation, would you prefer to evaluate the given expression or the expression after you rationalized it? Which would be easier to do with pencil and paper? Why?

2. Suppose you wanted to add two expressions, $\frac{7}{\sqrt{2}}$ and $\frac{1}{\sqrt{3}}$. Rationalize each of these expressions. Which would be easier to add, the original expressions or the rationalized expressions? State the reasons for your choice. Can you write an exact expression for the sum?

11.4 Solving Quadratic Equations by Completing the Square

OBJECTIVES

1 Complete perfect-square trinomials.

2 Solve quadratic equations by completing the square.

3 Model real-world situations by using quadratic equations, and solve the equations by completing the square.

APPLICATION

The American Automobile Manufacturers Association (AAMA) tabulates information annually. According to its records, the number of U.S. passenger cars produced (in millions of automobiles) is as listed in the following table:

Year	Production in millions
1995	6.4
1996	6.1
1997	5.9

a. Statisticians predicted that the decreasing trend would be followed by an increase in production. Therefore, write a quadratic equation for y million automobiles produced x years after 1995.

b. Using the equation in part a, determine the number of years after 1995 when 6 million automobiles will be produced. Interpret the solutions.

After completing this section, we will discuss this application further. See page 746.

11.4.1 Completing Perfect-Square Trinomials

As long as a variable expression is a perfect-square trinomial, which can be written as a binomial squared, we can solve an equation by using the principle of square roots. However, we may have an equation, such as $x^2 + 6x = 2$, with a variable expression, $x^2 + 6x$, that is not a perfect-square trinomial. We need to develop a method that will enable us to write an equivalent equation with a perfect-square trinomial.

First, we need to remember that a perfect-square trinomial is in the form $a^2 + 2ab + b^2$. Therefore, using this form to write a trinomial square in terms of x, we substitute x for a.

$$a^2 + 2ab + b^2$$
$$x^2 + 2xb + b^2 \qquad \text{Substitute x for a.}$$
or
$$x^2 + 2bx + b^2 \qquad \text{Rewrite the term 2xb.}$$

The coefficient of x^2 is 1 and the coefficient of x is $2b$. Therefore, to determine a value for b, we divide the coefficient of x by 2. Once we have the value for b, we square it and add it to the trinomial to make the trinomial a perfect square.

EXAMPLE 1 Determine what value must be added to each of the following expressions to obtain a perfect-square trinomial.

a. $x^2 + 6x$ **b.** $x^2 - 3x$ **c.** $x^2 + \dfrac{2}{3}x$

Solution

a. $x^2 + 6x$

The coefficient of x^2 is 1 and the coefficient of x is 6. Dividing 6 by 2, we obtain a value for b, $\frac{6}{2} = 3$. Therefore, $b^2 = (3)^2 = 9$. We need to add 9 to the expression to obtain a perfect-square trinomial: $x^2 + 6x + 9$, or $(x + 3)^2$.

b. $x^2 - 3x$

The coefficient of x^2 is 1 and the coefficient of x is -3. Dividing -3 by 2, we obtain a value for b, $-\frac{3}{2}$. Therefore, $b^2 = (-\frac{3}{2})^2 = \frac{9}{4}$. We need to add $\frac{9}{4}$ to the expression to obtain a perfect-square trinomial: $x^2 - 3x + \frac{9}{4}$, or $(x - \frac{3}{2})^2$.

c. $x^2 + \dfrac{2}{3}x$

The coefficient of x^2 is 1 and the coefficient of x is $\frac{2}{3}$. Dividing by 2 (or multiplying by $\frac{1}{2}$), we obtain a value for b, $\frac{2}{3} \cdot \frac{1}{2} = \frac{1}{3}$. Therefore, $b^2 = (\frac{1}{3})^2 = \frac{1}{9}$. We need to add $\frac{1}{9}$ to the expression to obtain a perfect-square trinomial: $x^2 + \frac{2}{3}x + \frac{1}{9}$, or $(x + \frac{1}{3})^2$. ●

✓ 11.4.1 Checkup

In exercises 1–3, determine what value must be added to each expression to obtain a perfect-square trinomial.

1. $x^2 + 8x$ **2.** $x^2 - 7x$ **3.** $x^2 - \dfrac{4}{5}x$

4. Explain what is meant by a perfect-square trinomial.

11.4.2 Solving Quadratic Equations by Completing the Square

Now we are ready to solve the equation we introduced at the beginning of this section, $x^2 + 6x = 2$. To do this, we will use a process called completing the square. **Completing the square** is a procedure used to determine a solution of an equation by rewriting the equation as a perfect-square trinomial equal to a rational number. For example, solve $x^2 + 6x = 2$ by completing the square.

$$x^2 + 6x = 2$$
$$x^2 + 6x + 9 = 2 + 9$$

Add 9 to both sides because that is the value of b^2 needed to complete the trinomial square on the left side of the equation.

$$(x + 3)^2 = 11$$

Rewrite the trinomial square as a binomial square.

$$x + 3 = \sqrt{11} \quad \text{or} \quad x + 3 = -\sqrt{11}$$

Principle of square roots

$$x = -3 + \sqrt{11} \qquad x = -3 - \sqrt{11}$$
$$x \approx 0.317 \qquad x \approx -6.317$$

The solutions are $-3 \pm \sqrt{11}$, or approximately 0.317 and -6.317.

SOLVING A QUADRATIC EQUATION BY COMPLETING THE SQUARE

To solve a quadratic equation by completing the square,

- Isolate the variable terms on one side of the equation.
- Divide both sides of the equation by the coefficient of x^2. (This step is not needed if the coefficient is 1.)
- Determine the value needed to complete the square by dividing the coefficient of x by 2 and squaring the result.
- Add the value obtained to both sides of the equation.
- Rewrite the trinomial as a binomial squared.
- Use the principle of square roots to determine the possible solutions, and solve.

EXAMPLE 2 Solve and check.

a. $x^2 - 5x + 2 = 5$ **b.** $x^2 + 4x + 6 = 2$
c. $x^2 + x + 2 = 0$

Algebraic Solution

a. $x^2 - 5x + 2 = 5$
$$x^2 - 5x = 3$$

Isolate the variable terms.

$$x^2 - 5x + \frac{25}{4} = 3 + \frac{25}{4}$$

Add the value needed to complete the square: $(-\frac{5}{2})^2 = \frac{25}{4}$.

(Continued on page 744)

$$\left(x - \frac{5}{2}\right)^2 = \frac{37}{4}$$ *Binomial squared*

$$x - \frac{5}{2} = \sqrt{\frac{37}{4}} \quad \text{or} \quad x - \frac{5}{2} = -\sqrt{\frac{37}{4}}$$ *Principle of square roots*

$$x - \frac{5}{2} = \frac{\sqrt{37}}{2} \qquad\qquad x - \frac{5}{2} = -\frac{\sqrt{37}}{2}$$

$$x = \frac{5}{2} + \frac{\sqrt{37}}{2} \qquad\qquad x = \frac{5}{2} - \frac{\sqrt{37}}{2}$$

$$x = \frac{5 + \sqrt{37}}{2} \qquad\qquad x = \frac{5 - \sqrt{37}}{2}$$

$$x \approx 5.541 \qquad\qquad x \approx -0.541$$

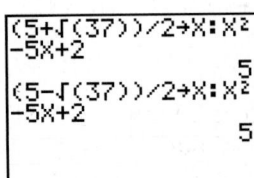

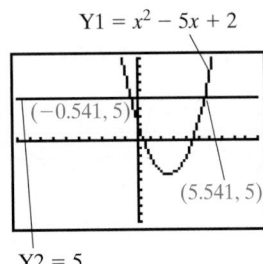

Check by substitution on your calculator. When the solutions are substituted for the variable, the expression equals 5, the expression on the right.

The solutions are $\frac{5 \pm \sqrt{37}}{2}$, or approximately 5.541 and −0.541.

Solve by graphing. Let $Y1 = x^2 - 5x + 2$ and $Y2 = 5$.

The solutions are approximately 5.541 and −0.541.

b. $x^2 + 4x + 6 = 2$

$$x^2 + 4x = -4$$ *Isolate the variable terms.*

$$x^2 + 4x + 4 = -4 + 4$$ *Add the value needed to complete the square: $\left(\frac{4}{2}\right)^2 = 2^2 = 4$.*

$$(x + 2)^2 = 0$$ *Binomial squared*

$$x + 2 = 0$$ *Principle of square roots*

$$x = -2$$ *Only one root equals 0.*

The solution is −2 (a double root).

The check is left for you.

c. $x^2 + x + 2 = 0$

$$x^2 + x = -2$$ *Isolate the variable terms.*

$$x^2 + x + \frac{1}{4} = -2 + \frac{1}{4}$$ *Add the value needed to complete the square: $\left(\frac{1}{2}\right)^2 = \frac{1}{4}$.*

$$\left(x + \frac{1}{2}\right)^2 = -\frac{7}{4}$$

The solution is not a real number, because there is no real number whose square is negative.

✓ 11.4.2 Checkup

Solve exercises 1–4 by completing the square.

1. $x^2 + 8x + 6 = 8$ **2.** $x^2 - 16x + 65 = 1$

3. $x^2 - 2x + 5 = 0$ **4.** $5x^2 - 4x - 5 = 0$

5. In order to use the method of completing the square to solve a quadratic equation, what must be true of the coefficient of the squared term? Is it always possible to write an equivalent equation that meets this condition? Explain.

11.4.3 Modeling the Real World

Completing the square is a very good general method for solving quadratic equations. It works with equations that can't be factored and with those that can be factored. Remember that, when using the method of completing the square, you should always start by isolating the terms containing the variable.

EXAMPLE 3 Go Away travel agency is preparing a special sale. The usual cost of a ticket is $150. The agency needs to collect $1000 to break even on the event. The agency wants to offer a special plan by reducing the ticket cost. If Go Away advertises a reduction of $1 per person purchasing a ticket, determine the number of people that need to purchase this special offer so that the travel agency will break even.

Algebraic Solution

Let x = number of people purchasing tickets
The ticket cost is $150 - 1x$, or $150 - x$.

The amount the agency will collect is the product of the number of people purchasing a ticket and the cost per ticket.

$$x(150 - x) = 1000$$
$$150x - x^2 = 1000$$
$$x^2 - 150x = -1000 \qquad \text{Rearrange and divide both sides by } -1.$$

$$x^2 - 150x + (-75)^2 = -1000 + (-75)^2 \qquad \text{Add to both sides.}$$
$$(x - 75)^2 = 4625 \qquad (-\tfrac{150}{2})^2 = (-75)^2$$
$$x - 75 = \pm\sqrt{4625} \qquad \text{Principle of square roots.}$$
$$x = 75 \pm \sqrt{4625}$$
$$x \approx 7 \quad \text{or} \quad x \approx 143$$

The agency will break even when 7 or 143 people purchase tickets. It is unlikely that the agency will sell tickets for $150 - $143, or $7 each. The logical answer is that they need to sell 7 tickets. (They most likely will limit the discount so that the cost does not fall below a certain number of dollars.)

APPLICATION

The American Automobile Manufacturers Association (AAMA) tabulates information annually. According to its records, the number of U.S. passenger cars produced (in millions of automobiles) is as listed in the following table:

Year	Production in millions
1995	6.4
1996	6.1
1997	5.9

a. Statisticians predicted that the decreasing trend would be followed by an increase in production. Therefore, write a quadratic equation for y million automobiles produced x years after 1995.

b. Using the equation in part a, determine the number of years after 1995 when 6 million automobiles will be produced. Interpret the solutions.

Discussion

a. Let x = the number of years after 1995

   y = the number of millions of automobiles produced

Use quadratic regression on your calculator to write a quadratic equation for the data points $(0, 6.4)$, $(1, 6.1)$, and $(2, 5.9)$.

```
L3      L4      L5      4
0       6.4     ------
1       6.1
2       5.9
------  ------
L4(4) =
```

```
QuadReg
y=ax²+bx+c
a=.05
b=-.35
c=6.4
```

The quadratic equation for the number of millions of automobiles produced x years after 1995 is $y = 0.05x^2 - 0.35x + 6.4$.

b. $y = 0.05x^2 - 0.35x + 6.4$

$6 = 0.05x^2 - 0.35x + 6.4$ Let $y = 6$ (million automobiles)

$0.05x^2 - 0.35x + 0.4 = 0$ Standard form

$5x^2 - 35x + 40 = 0$ Multiply both sides by 100.

$x^2 - 7x + 8 = 0$ Divide both sides by 5, the coefficient of x^2.

$x^2 - 7x = -8$ Isolate the variable terms.

$x^2 - 7x + \dfrac{49}{4} = -8 + \dfrac{49}{4}$ Add the value needed to complete the square: $(\frac{-7}{2})^2 = \frac{49}{4}$.

$\left(x - \dfrac{7}{2}\right)^2 = \dfrac{17}{4}$ Binomial squared

$x - \dfrac{7}{2} = \sqrt{\dfrac{17}{4}}$ or $x - \dfrac{7}{2} = -\sqrt{\dfrac{17}{4}}$ Principle of square roots

$x = \dfrac{7}{2} + \dfrac{\sqrt{17}}{2}$ $x = \dfrac{7}{2} - \dfrac{\sqrt{17}}{2}$

$x \approx 5.6$ $x \approx 1.4$

According to the equation, the production level will be 6 million automobiles 1.4 years after 1995, or $1995 + 1.4 = 1996.4 \approx 1997$, and 5.6 years after 1995, or $1995 + 5.6 = 2000.6 \approx 2001$.

11.4.3 Checkup

1. A retailer usually sells an item for $60. To promote sales, the price of the item is reduced by $2 times the number of items purchased. The retailer determines that each sale must be $200 to break even on the promotion. Determine the number of items that must be sold to each purchaser in order for the retailer to break even.

2. According to the AAMA, U.S. retail sales of all passenger cars (in millions of automobiles) is as listed in the following table:

Year	Retail sales in millions
1995	8.6
1996	8.5
1997	8.3

a. Statisticians predicted that the decreasing trend would continue. Write a quadratic equation for y million automobiles produced x years after 1995.

b. Using the equation in part a, determine the number of years after 1995 when the retail sales will be 5 million automobiles. Interpret the solutions.

11.4 Exercises

Determine what value must be added to each expression to obtain a trinomial square.

1. $x^2 + 18x$ **2.** $x^2 - 8x$ **3.** $x^2 - 9x$ **4.** $x^2 + 5x$

5. $x^2 + \dfrac{3}{4}x$ **6.** $x^2 + \dfrac{3}{5}x$ **7.** $x^2 + x$ **8.** $x^2 - x$

9. $x^2 - 6x$ **10.** $x^2 + 8x$ **11.** $x^2 + 9x$ **12.** $x^2 - 5x$

13. $x^2 + \dfrac{8}{9}x$ **14.** $x^2 + \dfrac{6}{7}x$ **15.** $x^2 - 14x$ **16.** $x^2 + 10x$

Solve by completing the square.

17. $x^2 + 6x - 20 = 35$ **18.** $x^2 - 8x - 25 = 40$ **19.** $x^2 - 3x = 28$ **20.** $x^2 + 3x = 40$

21. $x^2 + \dfrac{4}{7}x + \dfrac{3}{49} = 0$ **22.** $x^2 + \dfrac{6}{5}x + \dfrac{8}{25} = 0$ **23.** $x^2 + x - 30 = 60$ **24.** $x^2 - x - 16 = 40$

25. $x^2 - 6x = 2$ **26.** $x^2 + 8x = 5$ **27.** $x^2 + 9x = 1$ **28.** $x^2 - 5x = 2$

29. $x^2 + \dfrac{8}{9}x = 2$ **30.** $x^2 + \dfrac{6}{7}x = 1$ **31.** $x^2 - x - 5 = 0$ **32.** $x^2 + x - 10 = 0$

33. $x^2 + x + 10 = 0$ **34.** $x^2 - x + 6 = 0$ **35.** $2x^2 + 6x - 1 = 0$ **36.** $4x^2 + 12x - 3 = 0$

37. $3x^2 + x - 7 = 0$ **38.** $5x^2 + 2x - 3 = 0$ **39.** $x^2 - 14x + 55 = 6$ **40.** $x^2 + 10x + 35 = 10$

41. $4x^2 - 20x + 30 = 5$ **42.** $9x^2 + 12x + 10 = 6$ **43.** $\dfrac{1}{2}x^2 + 5x - 2 = 0$ **44.** $\dfrac{1}{3}x^2 - 2x + 1 = 0$

45. $\dfrac{1}{3}x^2 + \dfrac{1}{9}x - \dfrac{1}{6} = 0$ **46.** $\dfrac{1}{2}x^2 - \dfrac{1}{6}x - \dfrac{3}{4} = 0$ **47.** $\dfrac{2}{3}x^2 - 2x - \dfrac{5}{6} = 0$ **48.** $\dfrac{3}{5}x^2 - 6x - \dfrac{9}{10} = 0$

Determine a quadratic equation for each situation, and solve the equation by completing the square.

49. The regular cost of a theater ticket is $36.00. A discount of $1.00 per person is offered for group purchases. If the theater wants a breakeven of $250 on a group sale, how many tickets should be sold to a group?

50. A stadium ordinarily charges $18 per person to attend a sporting event. A discount of $1.00 per person is offered for group purchases. If the stadium wants a breakeven of

$50 on a group sale, how many tickets should be sold to a group?

51. A restaurant ordinarily charges $19.50 for a steak dinner. For holiday group parties, the restaurant offers a discount of $0.25 per person from this price. Determine the number of persons needed in a group to realize a breakeven point of $200.

52. An accounting firm ordinarily is charged $75.00 licensing fee for each employee that uses a copyrighted accounting software package. The supplier offers a discount of $1.50 per copy for multiple licenses. If the supplier wants a breakeven point of $900, how many licenses must the accounting firm purchase?

53. Model cars sell for $20 each. To encourage sales, a discount is given so that the price is reduced by $1 times the number of cars purchased. How many cars can be purchased for $75? The dealer does not intend to sell model cars for less than $12 each.

54. To discourage customers from hoarding Bean Buddies the local store ordinarily sells them for $5 each, but adds a surcharge of 50 cents times the number of dolls purchased to the price of each doll. (A purchase of one doll costs $5.50, a purchase of two dolls costs $6.00 each, a purchase of three dolls costs $6.50 each, and so on.) Write a quadratic function for the cost of purchasing x dolls. How many Bean Buddies can be purchased for $100?

55. The demand for a product is related to its price, x, by the polynomial function $D(x) = \frac{1}{3}x^2 - 64x + 3100$, where x is any price up to $50. At what value should the price be set in order to have a demand of 2000 units for the product? Round your answer to the nearest dollar.

56. Using the same demand function as in exercise 55, at what value should the price be set in order to have a demand of 1000 units for the product? Round your answer to the nearest dollar.

57. A large construction company studied the relationship between the size of a contract, x (in millions of dollars), and the cost to the company for preparing the bid for the contract, C(x) (in thousands of dollars). A statistical analysis yielded the mathematical model $C(x) = 0.11x^2 + 3.08x + 11$. If the company spends $55,000 to prepare a bid, what should be the approximate size of the contract?

58. Using the same relationship as in exercise 57, what should be the approximate size of a contract if the cost of preparing the bid is $40,000?

59. The length of a rectangle is 4 inches more than its width. The area of the rectangle is 117 square inches. Find its dimensions.

60. The base of a triangle measures 8 cm more than its height. The area of the triangle is 90 square centimeters. Find the height and base of the triangle.

61. A foundation for a shed is to be rectangular, with a width that is 5 feet less than the length. The area of the foundation is to be approximately 85 square feet. Find the length and width of the foundation to the nearest tenth of a foot.

62. One leg of a right triangle is 8 feet longer than the other leg. The hypotenuse of the triangle is approximately 25 feet long. Find the lengths of the legs of the triangle to the nearest hundredth of a foot.

In exercises 63 and 64, use the vertical-position formula $s = -16t^2 + v_0t + s_0$ to write a quadratic equation.

63. Starting 16 feet above the ground, a ball is thrown upward with an initial velocity of 32 feet per second.
    a. Write a quadratic equation for the time it will take the ball to reach the ground. (*Note:* Since the ball is thrown upward, the velocity will be positive; if the ball had been thrown downward, the velocity would be negative.)
    b. Complete the square to find an exact expression for the time it will take the ball to reach the ground.
    c. Estimate the time to the nearest tenth of a second.

64. From a height of 8 feet, a projectile is fired upward with an initial velocity of 88 feet per second.

    a. Write a quadratic equation for the time it will take the projectile to reach the ground.
    b. Complete the square to find an exact expression for the time it will take the projectile to reach the ground.
    c. Estimate the time to the nearest tenth of a second.

65. According to the annual reports of American Eagle Outfitters, the number of its employees, in thousands of people, is as listed in the following table:

Year	Thousands of employees
1997	5.4
1998	6.7
1999	7.6

    a. Statisticians predicted that the trend would first increase and then decrease. Write a quadratic equation for the number of employees, y, in thousands, x years after 1997.
    b. Using the equation in part a, determine the number of years after 1997 when the number of employees will return to the 1997 level of 5.4 thousand. Interpret the solutions.

66. According to the annual reports of Abercrombie & Finch Company, the number of its employees, in thousands of people, is as listed in the following table:

Year	Thousands of employees
1997	4.9
1998	6.7
1999	9.5

a. Statisticians predicted that the increasing trend would continue. Write a quadratic equation for the number of employees, y, in thousands, x years after 1997.

b. Using the equation in part a, determine the number of years after 1997 when the number of employees will increase to 20 thousand. Interpret the solutions.

11.4 Calculator Exercises

Try to solve the quadratic equations that follow by completing the square. Then graph the equations by storing the left side in Y1 and the right side in Y2. In some cases you can find a real-number solution of the equation, and in other cases you cannot. Do you see a connection between those for which you cannot find a real-number solution and their graphs? Is there a connection between those for which you can find a real-number solution and their graphs? Now study the solutions you obtained by completing the square. Do you see any connection between the graphs when you find a rational real-number solution? between the graphs when you find an irrational real-number solution?

1. $x^2 + 8x + 15 = 0$ **2.** $x^2 + 4x + 7 = 0$

3. $x^2 + x - 1 = 0$ **4.** $2x^2 + 11x + 12 = 0$

5. $2x^2 + 6x + 7 = 0$ **6.** $8x^2 + 8x - 5 = 0$

11.4 Writing Exercise

You have seen that some quadratic equations can be solved by factoring and applying the zero factor property, while others can be solved by completing the square to make a square trinomial. If you are faced with solving a quadratic equation algebraically, discuss which method you would try to apply first, explaining the reasons for your choice. Then discuss a situation in which the factoring method might not be best and you would need to try completing the square. Finally, discuss when completing the square might not be best. You may wish to illustrate each situation with a sample exercise from this section.

11.5 Solving Quadratic Equations by Using the Quadratic Formula

OBJECTIVES

1 Solve quadratic equations by using the quadratic formula.

2 Determine the number and type of solutions of quadratic equations by using the discriminant.

3 Model real-world situations by using quadratic equations, and solve the equations by using the quadratic formula.

APPLICATION

Water in a stream moves fastest at the surface and slowest at the bottom. A hydrologist measured the velocity of a 2-meter-deep stream at different depths and determined that the velocity of the stream, in meters per second, could be approximated by the equation $y = -0.058x^2 + 0.206x + 0.063$, where x is the depth of the stream in meters.

a. If the average velocity of the stream was 0.191 meter per second, determine the depth at which this average velocity occurred.

b. Hydrologists use the "four-tenths rule" to determine the average velocity of a stream at a point. That is, the average velocity of a stream is approximately the velocity of the stream at a depth of four-tenths of the stream's depth. Compare the approximation obtained by the four-tenths rule with the approximation obtained in part a.

After completing this section, we will discuss this application further. See page 757.

11.5.1 Solving Quadratic Equations by Using the Quadratic Formula

When we have to repeat a process many times, it often saves time to develop a formula for the process. Therefore, instead of completing the square to solve a quadratic equation, we may want to use a formula. To develop such a formula we begin with the standard form of a quadratic equation in one variable, $ax^2 + bx + c = 0$, and repeat the process of completing the square, using letters instead of numbers and remembering that a, b, and c represent real numbers.

$$ax^2 + bx + c = 0$$
$$ax^2 + bx = -c \qquad \text{Isolate the variable terms.}$$
$$x^2 + \frac{b}{a}x = \frac{-c}{a} \qquad \text{Divide by the coefficient of } x^2.$$

Determine the value needed to complete the square. Multiply the coefficient of x by $\frac{1}{2}$ and square the result.

$$\frac{b}{a} \cdot \frac{1}{2} = \frac{b}{2a}$$
$$\left(\frac{b}{2a}\right)^2 = \frac{b^2}{4a^2}$$

Add $\frac{b^2}{4a^2}$ to both sides.

$$x^2 + \frac{b}{a}x + \frac{b^2}{4a^2} = \frac{-c}{a} + \frac{b^2}{4a^2}$$

Simplify the right side.

$$\frac{-c}{a} + \frac{b^2}{4a^2} = \frac{-4ac}{4a^2} + \frac{b^2}{4a^2} = \frac{-4ac + b^2}{4a^2} = \frac{b^2 - 4ac}{4a^2}$$

Write the left side as a binomial squared: $(x + \frac{b}{2a})^2$.

$$\left(x + \frac{b}{2a}\right)^2 = \frac{b^2 - 4ac}{4a^2} \qquad \text{Binomial squared.}$$

Use the principle of square roots to determine the solutions.

$$x + \frac{b}{2a} = \pm\sqrt{\frac{b^2 - 4ac}{4a^2}}$$
$$x + \frac{b}{2a} = \pm\frac{\sqrt{b^2 - 4ac}}{2a}$$
$$x = \frac{-b}{2a} \pm \frac{\sqrt{b^2 - 4ac}}{2a}$$
$$x = \frac{-b \pm \sqrt{b^2 - 4ac}}{2a}$$

In writing this formula, be careful to write just one fraction for the right side of the equation.

We can use the preceding formula to solve any quadratic equation. To do this, first write the quadratic equation in standard form: $ax^2 + bx + c = 0$, with real numbers a, b, and c and $a \neq 0$. Then determine the values to substitute for a, b, and c.

QUADRATIC FORMULA

The quadratic formula used for solving a quadratic equation in standard form, $ax^2 + bx + c = 0$, with real numbers a, b, and c and $a \neq 0$, is

$$x = \frac{-b \pm \sqrt{b^2 - 4ac}}{2a}$$

EXAMPLE 1 Solve and check.

a. $2x^2 + 3x - 5 = 0$ **b.** $2x^2 + 3x - 4 = 0$ **c.** $2x^2 + 4x = -2$

d. $2y^2 - 5y + 4 = 0$ **e.** $x^2 + 6x + 1 = 0$

Algebraic Solution

a. $2x^2 + 3x - 5 = 0$

$x = \dfrac{-b \pm \sqrt{b^2 - 4ac}}{2a}$ $a = 2, b = 3,$ and $c = -5$

$x = \dfrac{-(3) \pm \sqrt{(3)^2 - 4(2)(-5)}}{2(2)}$ Substitute values for a, b, and c.

$x = \dfrac{-3 \pm \sqrt{9 + 40}}{4}$

$x = \dfrac{-3 \pm \sqrt{49}}{4}$

$x = \dfrac{-3 \pm 7}{4}$

$x = \dfrac{-3 + 7}{4}$ or $x = \dfrac{-3 - 7}{4}$

$x = \dfrac{4}{4}$ $x = -\dfrac{10}{4}$

$x = 1$ $x = -\dfrac{5}{2}$

The solutions are 1 and $-\frac{5}{2}$.

Note: These solutions could have been determined by factoring.

Graphic Check

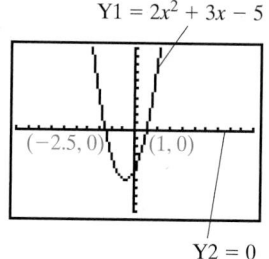

Y1 = $2x^2 + 3x - 5$

$(-2.5, 0)$ $(1, 0)$

Y2 = 0

The solutions are -2.5 and 1.

b. $2x^2 + 3x - 4 = 0$

$x = \dfrac{-b \pm \sqrt{b^2 - 4ac}}{2a}$ $a = 2, b = 3,$ and $c = -4$

$x = \dfrac{-(3) \pm \sqrt{(3)^2 - 4(2)(-4)}}{2(2)}$

$x = \dfrac{-3 \pm \sqrt{9 + 32}}{4}$

$x = \dfrac{-3 \pm \sqrt{41}}{4}$

$x = \dfrac{-3 + \sqrt{41}}{4}$ or $x = \dfrac{-3 - \sqrt{41}}{4}$

$x \approx 0.851$ $x \approx -2.351$

Substitution Check

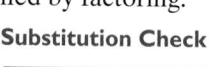

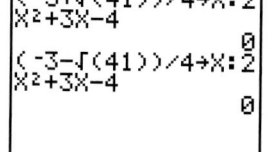

The solutions are $\frac{-3 \pm \sqrt{41}}{4}$, or approximately 0.851 and -2.351.

Note: These solutions could not have been found by factoring.

If we substitute the approximate solutions, we will not obtain an exact check, because $0.001402 \neq 0$.

c. $2x^2 + 4x = -2$

$$2x^2 + 4x + 2 = 0 \qquad \text{Standard form}$$

$$x = \frac{-b \pm \sqrt{b^2 - 4ac}}{2a} \qquad a = 2, b = 4, \text{ and } c = 2$$

$$x = \frac{-(4) \pm \sqrt{(4)^2 - 4(2)(2)}}{2(2)}$$

$$x = \frac{-4 \pm \sqrt{16 - 16}}{4}$$

$$x = \frac{-4 \pm \sqrt{0}}{4}$$

$$x = \frac{-4}{4}$$

$$x = -1$$

The solution is -1.
The check is left for you.

d. $2y^2 - 5y + 4 = 0 \qquad$ This is a quadratic equation in terms of y.

$$y = \frac{-b \pm \sqrt{b^2 - 4ac}}{2a} \qquad a = 2, b = -5, \text{ and } c = 4$$

$$y = \frac{-(-5) \pm \sqrt{(-5)^2 - 4(2)(4)}}{2(2)}$$

$$y = \frac{5 \pm \sqrt{25 - 32}}{4}$$

$$y = \frac{5 \pm \sqrt{-7}}{4}$$

$\sqrt{-7}$ is not a real number. Therefore, the expressions $\frac{5 \pm \sqrt{-7}}{4}$ do not represent real numbers. The equation has no real-number solution.

e. $x^2 + 6x + 1 = 0 \qquad \text{Standard form}$

$$x = \frac{-b \pm \sqrt{b^2 - 4ac}}{2a} \qquad a = 1, b = 6, \text{ and } c = 1$$

$$x = \frac{-(6) \pm \sqrt{(6)^2 - 4(1)(1)}}{2(1)}$$

$$x = \frac{-6 \pm \sqrt{36 - 4}}{2}$$

$$x = \frac{-6 \pm \sqrt{32}}{2}$$

$$x = \frac{-6 \pm 4\sqrt{2}}{2}$$

$$x = \frac{\overset{1}{2}(-3 \pm 2\sqrt{2})}{\underset{1}{2}} \qquad \begin{array}{l}\text{Factor the numerator and divide} \\ \text{out the 2.}\end{array}$$

$$x = -3 \pm 2\sqrt{2}$$

The solutions are $-3 \pm 2\sqrt{2}$, or approximately -0.172 and -5.828.
The check is left for you.

11.5.1 Checkup

Solve exercises 1–5 by using the quadratic formula, and check your solutions.

1. $2x^2 + x - 6 = 0$
2. $z^2 + z = 7$
3. $3(3p^2 - 4p) = -4$
4. $3k^2 + 2k + 5 = 3k$
5. $x^2 + 4x - 23 = 0$

6. The quadratic formula appears to be a single rule, yet sometimes there are two separate solutions, sometimes only one double root, and sometimes no real number. Explain how this is possible.

7. What does the symbol $\pm$ mean?

11.5.2 Determining Characteristics of Solutions by Using the Discriminant

In previous examples, the solutions of a quadratic equation varied in both number (two, one, or none) and type (rational, irrational, or no real number). At times, it is necessary only to determine the type and number of solutions. For example, if it is known that no real-number solution exists, there may not be any reason to attempt to solve the equation. Is there any way to determine the number and kind of solutions of an equation without solving the equation? See if you can find an answer by completing the following set of exercises, using the quadratic formula.

Discovery 3

Characteristics of the Quadratic-Formula Solution

In Example 1, we obtained the solutions of the given equations. Determine a value for the radicand, $b^2 - 4ac$, of the quadratic formula.

1. $2x^2 + 3x - 5 = 0$

$x = 1$ or $x = -\dfrac{5}{2}$

(two rational solutions)

$b^2 - 4ac = $ _____

2. $2x^2 + 3x - 4 = 0$

$x = \dfrac{-3 + \sqrt{41}}{4}$ or $x = \dfrac{-3 - \sqrt{41}}{4}$

(two irrational solutions)

$b^2 - 4ac = $ _____

3. $2x^2 + 4x + 2 = 0$

$x = -1$

(one rational solution)

$b^2 - 4ac = $ _____

4. $2y^2 - 5y + 4 = 0$

(no real-number solutions)

$b^2 - 4ac = $ _____

Write a rule for determining the number of and type of solutions of a quadratic equation by using $b^2 - 4ac$.

The radicand $b^2 - 4ac$ determines the characteristics of the solutions of a quadratic equation. If the radicand is a perfect square ($\neq 0$), the equation has two rational-number solutions. If the radicand is 0, the equation has one rational-number solution. If the radicand is a positive number that is not a perfect square, the equation has two irrational-number solutions. If the radicand is a negative number, the equation has no real-number solutions.

We can understand why this is true if we look at the entire quadratic formula, $x = \dfrac{-b \pm \sqrt{b^2 - 4ac}}{2a}$. The square root simplifies to a positive rational number if the radicand is a perfect square. In the formula, this results in a rational number being added to or subtracted from another rational number and then divided by a second rational number. The result is always two rational numbers.

The square root simplifies to 0 if the radicand is 0. In the formula, this results in 0 being added to or subtracted from another rational number and then divided by a second rational number. The result is always one rational number.

The square root remains if the radicand is a positive number that is not a perfect square. This results in an irrational number being added to or subtracted from a rational number and then divided by a second rational number. The result is always two irrational numbers.

The square root is not defined in the real-number system if the radicand is a negative number. In that case, we cannot obtain a real-number solution of the equation.

We call the radicand $b^2 - 4ac$ the **discriminant** of the quadratic equation $ax^2 + bx + c = 0$.

DETERMINING THE CHARACTERISTICS OF THE SOLUTION OF A QUADRATIC EQUATION BY USING THE DISCRIMINANT

Determine the value of the discriminant $b^2 - 4ac$, of a quadratic equation in standard form, $ax^2 + bx + c = 0$, with rational numbers a, b, and c and $a \neq 0$. One of four possibilities occur.

Discriminant	Number and Type of Real-Number Solutions
0	One rational-number solution
Perfect square, not equal to 0	Two rational-number solutions
Positive number, not a perfect square	Two irrational-number solutions
Negative number	No real-number solution

EXAMPLE 2 Determine the characteristics of the solution of each quadratic equation. Do not solve the equation.

a. $15x^2 + 26x = 12$ **b.** $\dfrac{3}{4}x^2 - \dfrac{2}{3}x + \dfrac{1}{9} = 0$

c. $0.3x^2 - 0.6x + 0.3 = 0$ **d.** $3x^2 - 7x + 12 = 0$

Solution

a. $\quad 15x^2 + 26x = 12$

$\quad\quad 15x^2 + 26x - 12 = 0$ *Standard form*

$\quad\quad\quad b^2 - 4ac = (26)^2 - 4(15)(-12)$ *a = 15, b = 26, and c = −12*

$\quad\quad\quad\quad\quad\quad = 1396$

The equation has two irrational solutions, because 1396 is positive, but not a perfect square.

b. $\dfrac{3}{4}x^2 - \dfrac{2}{3}x + \dfrac{1}{9} = 0$

$\quad\quad b^2 - 4ac = \left(-\dfrac{2}{3}\right)^2 - 4\left(\dfrac{3}{4}\right)\left(\dfrac{1}{9}\right)$ *a = ¾, b = −⅔, c = ⅑*

$\quad\quad\quad\quad\quad = \dfrac{1}{9}$

The equation has two rational solutions, because $\frac{1}{9}$ is a perfect square.

HELPING HAND If you prefer not to work with fractions, write an equivalent equation by multiplying by the LCD of 36. The new equivalent equation, $27x^2 - 24x + 4 = 0$ will have a discriminant value of 144, a perfect square as well.

c. $0.3x^2 - 0.6x + 0.3 = 0$

$$b^2 - 4ac = (-0.6)^2 - 4(0.3)(0.3) \qquad a = 0.3, b = -0.6, c = 0.3$$
$$= 0$$

The equation has one rational solution, because the discriminant is 0.

d. $3x^2 - 7x + 12 = 0$

$$b^2 - 4ac = (-7)^2 - 4(3)(12) \qquad a = 3, b = -7, \text{ and } c = 12$$
$$= 49 - 144$$
$$= -95$$

The equation has no real-number solution, because the discriminant is negative.

11.5.2 Checkup

In exercises 1–4, determine the characteristics of the solution(s) of each quadratic equation. Do not solve the equation.

1. $0.1x^2 + x + 2.3 = 0$ **2.** $25x^2 - 60x + 36 = 0$

3. $3x^2 - 4x = 32$ **4.** $\frac{1}{2}x^2 + 3x + 17 = 0$

5. Under what conditions will a quadratic equation have rational numbers as solutions; irrational numbers as solutions; no real-number solution?

11.5.3 Modeling the Real World

We now have four algebraic methods (factoring, square root principle, completing the square, and quadratic formula), as well as a numeric method and a graphic method, for solving quadratic equations in one variable. First, let's review the strengths and weaknesses of these methods.

Method	Strengths	Weaknesses
Numeric	Easy to set up for integer values on a calculator	Time consuming if completed by hand Will determine solutions only for the input values (usually integers)
Graphic	Easy to set up on a calculator Real-number solutions will always be found if they exist	Appropriate range may be difficult to determine; graph may be hard to read Solution will be rounded for irrational and repeating fraction values
Algebraic	Exact solutions will be found if they exist	Time consuming at times

After reviewing this list, we likely will want to rely more heavily on algebraic methods than on the numeric and graphic methods. Now let us review the steps involved in solving a quadratic equation algebraically.

SOLVING A QUADRATIC EQUATION IN ONE VARIABLE ALGEBRAICALLY

To solve a quadratic equation in one variable, $ax^2 + bx + c = 0$, algebraically, select one of the following methods:

- If $b = 0$—that is, if the quadratic equation has no linear (x) term— solve for x^2 and use the principle of square roots to solve the equation.
- Factor $ax^2 + bx + c$ and, if possible, use the zero factor property to solve the equation.
- If the quadratic expression will not factor, or if it is difficult to factor, use the quadratic formula or complete the square to solve the equation.

EXAMPLE 3   Chloe is planning a Christmas sale. She usually sells a doll cradle for $40. To encourage customers to purchase more than one cradle, she plans to reduce the price of the cradle by $1 dollar per cradle purchased. In order to make the cradles, she purchased a saw for $150. She estimated that the cost of materials for each cradle is $5.

a. Write a revenue function.

b. Write a total cost function.

c. Write a profit function.

d. Determine the number of doll cradles that Chloe must make and sell to break even, and find the selling price for that number of cradles.

Solution

a. Let x = the number of doll cradles sold

$R(x)$ = the total revenue

The selling price of the cradles is $40 minus $1 per cradle sold, or $40 - 1x$. The revenue is the product of the number of cradles sold and the selling price. Thus,

$$R(x) = x(40 - x)$$
$$R(x) = 40x - x^2$$

b. Let x = the number of cradles made

$C(x)$ = the total cost of making and selling the cradles

The total cost is $5 per cradle, plus $150 for the saw.

$$C(x) = 5x + 150$$

c. Let $P(x)$ = the profit for making and selling the cradles

The profit is the difference of the revenue and the total cost.

$$P(x) = R(x) - C(x)$$
$$P(x) = (40x - x^2) - (5x + 150)$$
$$P(x) = 40x - x^2 - 5x - 150$$
$$P(x) = -x^2 + 35x - 150$$

d. Chloe will break even when the profit is 0 dollars, or $P(x) = 0$.

$$-x^2 + 35x - 150 = 0 \qquad a = -1, b = 35, c = -150$$

$$x = \frac{-b \pm \sqrt{b^2 - 4ac}}{2a} \qquad \text{Quadratic formula}$$

$$x = \frac{-(35) \pm \sqrt{(35)^2 - 4(-1)(-150)}}{2(-1)}$$

$$x = \frac{-35 \pm \sqrt{1225 - 600}}{-2}$$

$$x = \frac{-35 \pm \sqrt{625}}{-2}$$

$$x = \frac{-35 \pm 25}{-2}$$

$$x = \frac{-35 + 25}{-2} \qquad \text{or} \qquad x = \frac{-35 - 25}{-2}$$

$$x = 5 \qquad\qquad\qquad x = 30$$

$$40 - x = 40 - 5 = 35 \qquad 40 - x = 40 - 30 = 10 \qquad \text{Selling price}$$

Chloe will break even when she sells 5 cradles for 35 dollars or when she sells 30 cradles for 10 dollars. She would not sell cradles for 10 dollars, so the correct answer is that the break-even point occurs when Chloe sells 5 cradles.

APPLICATION

Water in a stream moves fastest at the surface and slowest at the bottom. A hydrologist measured the velocity of a 2-meter-deep stream at different depths and determined that the velocity of the stream, in meters per second, could be approximated by the equation $y = -0.058x^2 + 0.206x + 0.063$, where x is the depth of the stream in meters.

a. If the average velocity of the stream was 0.191 meter per second, determine the depth at which this average velocity occurred.

b. Hydrologists use the "four-tenths rule" to determine the average velocity of a stream at a point. That is, the average velocity of a stream is approximately the velocity of the stream at a depth of four-tenths of the stream's depth. Compare the approximation obtained by the four-tenths rule with the approximation obtained in part a.

Discussion

a. Let y = the velocity of the stream in meters per second
x = the stream's depth in meters

$$y = -0.058x^2 + 0.206x + 0.063$$
$$0.191 = -0.058x^2 + 0.206x + 0.063 \qquad \text{Let y = 0.191 meter per second}$$

Write the equation in standard form and solve using the quadratic formula.

$$-0.058x^2 + 0.206x - 0.128 = 0 \qquad a = -0.058, b = 0.206, c = -0.128$$

(Continued on page 758)

$$x = \frac{-b \pm \sqrt{b^2 - 4ac}}{2a} \qquad \text{Quadratic formula}$$

$$x = \frac{-(0.206) \pm \sqrt{(0.206)^2 - 4(-0.058)(-0.128)}}{2(-0.058)}$$

$$x = \frac{-0.206 \pm \sqrt{0.01274}}{-0.116}$$

$$x = \frac{-0.206 + \sqrt{0.01274}}{-0.116} \qquad \text{or} \qquad x = \frac{-0.206 - \sqrt{0.01274}}{-0.116}$$

$$x \approx 0.8 \qquad\qquad\qquad\qquad x \approx 2.7$$

Since the stream is 2 meters deep, the average velocity of the stream occurred at a depth of about 0.8 meter.

b. The four-tenths rule states that the average velocity occurs at 0.4 of the depth (2m).

$$(0.4)(2) = 0.8$$

The equation in part a is a good approximation, because the same value, 0.8 meter, was obtained.

11.5.3 Checkup

Set up a quadratic equation to represent each situation, and then use the quadratic formula to solve the equation.

1. A manufacturer supplies cabinets at wholesale price to a dealer. The price per cabinet depends on the number of cabinets the dealer orders. The listed price of $150 per cabinet will be reduced by $5 per cabinet ordered. It costs the manufacturer $65 to make a cabinet, and there is a fixed delivery cost of $150 for each order.

  a. Write a revenue function for an order of cabinets.

  b. Write a cost function for an order of cabinets.

  c. Write a profit function for an order of cabinets.

  d. Determine the number of cabinets that must be ordered to break even on an order. Interpret the break-even point(s).

2. Using data from the New York Stock Exchange (NYSE) fact book for the years 1995 through 1999, one can relate the number of shares of stocks listed on the NYSE to the market value of the stocks by the quadratic equation $y = 1.5x^2 - 8x + 151$. The variable y represents the number of shares listed (in billions), and the variable x represents the market value of the shares listed (in trillions of dollars). Use this relation and the quadratic formula to determine for what market value the number of stocks listed would equal 200 billion shares.

11.5 Exercises

Use the quadratic formula to find the exact solution; then check your solution.

1. $x^2 - 12x + 27 = 0$

2. $x^2 - 12x + 35 = 0$

3. $2x^2 + 3x - 15 = 2x + 6$

4. $3x^2 + 3x - 10 = x + 6$

5. $2z^2 + 11z + 5 = 0$

6. $3y^2 + 31y + 36 = 0$

7. $8(2p^2 - p + 1) = 7$

8. $5(5q^2 + 4q + 1) = 1$

9. $x^2 + 3x + 4 = 0$

10. $x^2 + x + 5 = 0$

11. $-5a^2 + 4a - 7 = 0$

12. $-3b^2 + b - 5 = 0$

13. $v^2 - 5v + 2 = 0$

14. $t^2 - 8t + 10 = 0$

15. $x(x - 4) + 1 = 0$

16. $x(x + 6) + 7 = 0$

17. $4x(x + 1) + 6 = 6 - x$

18. $3x(x - 3) - 2 = x - 2$

19. $3d^2 + 10 = 17$

20. $2r^2 + 4 = 19$

21. $16m = m^2 + 55$

22. $72 = -k^2 - 18k$

23. $(x - 4)(x + 4) = 2(x - 4)$

24. $6(x + 3) = (x - 3)(x + 3)$

25. $x^2 - 6.3x + 7.2 = 0$

26. $x^2 + 3.7x + 1.6 = 0$

27. $1.8 - 5.6x - x^2 = 0$

28. $6.2 + 1.8x - x^2 = 0$

29. $a^2 + 24.01 = 9.8a$

30. $b^2 + 14.2b = -50.41$

31. $1.7z^2 + 1.3z + 5.6 = 0$

32. $2.9t^2 - 5.2t + 8.1 = 0$

Evaluate the discriminant in exercises 33–48 to determine the characteristics of the solution(s) of each quadratic equation. Do not solve the equation.

33. $x^2 - 11x + 24 = 0$ **34.** $x^2 + 8x + 21 = 0$ **35.** $a^2 + 12a + 36 = 0$ **36.** $b^2 - 3b - 8 = 0$

37. $z^2 = 4z - 5$ **38.** $y^2 + 81 = 18y$ **39.** $6x^2 - 11x - 7 = 0$ **40.** $5x^2 - 125 = 0$

41. $7p^2 - 15 = 0$ **42.** $10q^2 - 7q - 12 = 0$ **43.** $1 - 5x - 4x^2 = 0$ **44.** $6x - 5 = 3x^2$

45. $z - 1.25 = 0.2z^2$ **46.** $2.1 - 1.3t - 4.3t^2 = 0$ **47.** $0.3x - 2.8 = 1.7x^2$ **48.** $x^2 = 8.6x - 18.49$

49. A gourmet meat shop that ships frozen steaks to customers charges $9 per pound. To encourage larger orders, the shop promises to reduce the price by $0.10 for every pound of steak ordered. The shop spends $6 per pound for the steak and spends $7.50 for packaging and shipping each order.
   a. Write a revenue function for an order of steaks.
   b. Write a cost function for an order of steaks.
   c. Write a profit function for an order of steaks.
   d. Determine the number of pounds of steak that must be ordered to break even on an order. Interpret the break-even point(s).

50. Jennah sells hair ribbons through a mail-order business in her home. She ordinarily sells the ribbons for $3.25 each, but reduces this price by $0.05 per ribbon ordered to encourage multiple orders. Her hair ribbons cost $2.00 each to make, and packaging and shipping costs $3.00 per order.
   a. Write a revenue function for an order of hair ribbons.
   b. Write a cost function for an order of hair ribbons.
   c. Write a profit function for an order of hair ribbons.
   d. Determine the number of hair ribbons that must be ordered to break even on an order. Interpret the break-even point(s).

51. A promoter rents and decorates a party hall for $600 a night. He charges guests $50 per person to attend a New Year's Eve party. He would ordinarily pay a caterer $20 per person to provide food and beverages for the party. However, the caterer offers to reduce the price per person by $0.25 for each person attending, to gain the promoter's business.
   a. Write a revenue function for the amount of money the promoter will collect for the party.
   b. Write a cost function for the amount of money the promoter will spend on the party.
   c. Write a profit function for this promotion.
   d. Determine the number of guests that must attend to break even on the party. Interpret the break-even point(s).

52. A consulting firm charges customers $250 per day for services. The firm hires freelance consultants and pays them a flat amount of $500 to work a job. In addition, the firm would ordinarily pay the freelancers $75 per day, but because the consultants prefer multiple-day assignments, the firm agrees to reduce this daily payment by $5 for each day the job lasts.
   a. Write a revenue function for the amount of money the firm earns for a consulting job.
   b. Write a cost function for the amount of money the firm pays the freelance consultant to work the job.
   c. Write a profit function for one job that the firm arranges.

   d. Determine the number of days that a job must last for the firm to break even. Interpret the break-even point(s).

53. The U.S. Bureau of the Census reported that the world population first reached 1 billion people in the year 1804. Since then, it has grown to 6 billion in the year 1999. A quadratic equation that relates the world population size y (in billions of people) to the number of years since 1800, x, is $y = 0.00026x^2 - 0.028x + 1.13$. Use this equation and the quadratic formula to determine the number of years since 1800 that it will take to reach a world population of 7 billion people. Translate your answer into the year in which the world population may reach that number.

54. According to the Energy Information Administration, total energy consumption for the world in 1986 was 313.48 quadrillion British thermal units (Btu's). By 1998, consumption was estimated to be 377.72 quadrillion Btu's. The data for the intervening years appears to follow a parabolic curve that can be estimated by the quadratic equation $y = -0.34x^2 + 13.3x + 246$, where the variable y is the worldwide consumption in quadrillions of Btu's and the variable x is the number of years since 1980. This equation suggests that worldwide energy consumption will start to decrease soon, which would be environmentally desirable if true. Use the equation and the quadratic formula to determine when the consumption might drop to 350 Btu's. Translate your answer into the year in which this could happen if the equation is valid.

55. A statistical study of the consumer price index (CPI) for the years 1950 to 1999 yielded the quadratic function

$$y = 0.066x^2 - 0.145x + 22.81$$

where y is the CPI (a percentage) and x is the number of years since 1950. Use the quadratic formula to determine the year after 1950 in which the CPI was approximately 100%.

56. Using the function in exercise 55, determine the year after 1950 in which the CPI was approximately 125%.

57. A loading ramp is to be built from the ground to the loading dock. The cross section of the ramp is a right triangle with a base that is 6 feet more than the height and a hypotenuse that is 12 feet more than the height. Use the Pythagorean theorem to write a polynomial equation. Find the height, base, and hypotenuse of the ramp.

58. Marge and Gretchen are cutting out right triangles for a quilt they are making. The instructions call for the height of the triangles to be $\frac{3}{4}$ of the base and the hypotenuse to be $\frac{1}{2}$ inch longer than the base. Write a polynomial equation. Find the lengths of the three sides of the triangles.

11.5 Calculator Exercises

The calculator can be helpful in applying the quadratic formula to solve a quadratic equation. We will store the formula for the discriminant in Y8 and the two formulas for the solutions in Y9 and Y0. Then we can just store values for a, b, and c in the calculator and view Y8, Y9, and Y0 to see the results for the quadratic formula. The steps to do so follow:

• Store the formula for the discriminant in Y8.

Y=	▼	▼	▼	▼	▼	▼	▼
ALPHA	B	x^2	−	4	ALPHA	A
ALPHA	C	ENTER

• Store the first solution formula in Y9.

(	(−)	ALPHA	B	+	2nd	√
ALPHA	B	x^2	−	4	ALPHA	A
ALPHA	C	)	)	÷	(	2
ALPHA	A	)	ENTER

• Store the second solution formula in Y0.

(	(−)	ALPHA	B	−	2nd	√
ALPHA	B	x^2	−	4	ALPHA	A
ALPHA	C	)	)	÷	(	2	ALPHA	A
)	2nd	QUIT

• Store the values for a, b, and c of the quadratic equation in the A, B, and C locations of the calculator. For example, to solve the quadratic equation, $2x^2 + 9x - 35 = 0$, store 2 in A, 9 in B and −35 in C.

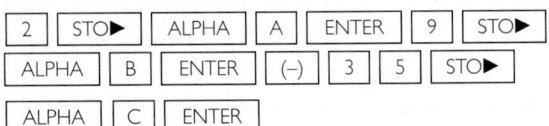

• Recall Y8 to see the value of the discriminant.

| VARS | ▶ | 1 | 8 | ENTER |

The calculator will display 361, which is a perfect square of 19, indicating that the equation has two rational solutions.

• Recall Y9 and Y0 to see the two solutions of the equation.

| VARS | ▶ | 1 | 9 | ENTER | VARS | ▶ |
| 1 | 0 | ENTER |

The two solutions are $x = 2.5$ and $x = -7$.

To solve another equation, it is only necessary to store new values for A, B, and C and then recall Y8, Y9, and Y0. The formulas will remain stored in the calculator until you delete them from these three Y locations.

You should make Y8, Y9, and Y0 inactive when you are storing formulas there. Simply move the cursor over the equals sign under Y= and press ENTER . The box around the equals sign should disappear, and now these locations will not appear in any tables or graphs that you subsequently create.

Use this method to complete the following table. Remember to place the equation in standard form to determine the values of a, b, and c. When the solution is an irrational number, approximate the solution as a decimal to the nearest thousandth.

	Equation	Value of Discriminant	Types of Roots (Rational, Irrational, Not Real)	Number of Unlike Roots	Roots
1.	$x^2 + 6 = 5x$	1	rational	2	2, 3
2.	$9x^2 + 6x = -1$				
3.	$2x^2 + 1 = 7x$				
4.	$x^2 + 6x = -10$				
5.	$x^2 = 6 - x$				
6.	$5x^2 - 6x = 0$				
7.	$x^2 + 0.36 = 1.2x$				
8.	$1.7x^2 + x + 1.9 = 0$				
9.	$1.5x^2 + 1.2x = 3.6$				

	Equation	Value of Discriminant	Types of Roots (Rational, Irrational, Not Real)	Number of Unlike Roots	Roots
10.	$\frac{1}{4}x^2 + x = \frac{1}{8}$				
11.	$x^2 - \frac{1}{6}x = \frac{1}{6}$				
12.	$\frac{1}{5}x^2 + \frac{2}{3}x = -\frac{7}{8}$				

11.5 Writing Exercises

1. The term *discriminant* is an important term in this section. It is a derivative of the word *discriminate*. Look up this word in a dictionary, and select the definition that most fits its use in this section. Then use the definition to explain why the discriminant is so important to the study of quadratic equations presented here.

2. Although the arithmetic may become more tedious when you solve a quadratic equation whose coefficients are fractions, the quadratic formula still applies. You can choose to solve the equation with the fractions, or you can choose to use the least common denominator to first clear the equation of fractions and then apply the quadratic formula.

   First solve the equations with the fractions. Then clear the equations of fractions before solving. After solving the equations both ways, review your results. Is it better to clear the fractions before solving the equations? Explain.

   a. $\frac{5}{6}x^2 - \frac{4}{9}x - \frac{2}{3} = 0$ **b.** $\frac{1}{4} - \frac{1}{2}x - \frac{1}{9}x^2 = 0$

11.6 More Real-World Models

OBJECTIVES   **1** Solve formulas for a variable.

2 Model real-world situations by recognizing special triangles.

APPLICATION

We can extend the Pythagorean theorem to three dimensions. The square of the length of the diagonal of a rectangular solid, d, is the sum of the squares of the length of the three sides of the solid, a, b, and c.

a. Determine a formula for the length of the diagonal of a rectangular solid, d.

b. According to the *Guinness Book of World Records*, the smallest video camera was manufactured in the United States by Super Circuits in the year 2000. The camera measures 1.62 centimeters on each side. Use the formula generated in part a to determine the length of the diagonal of the camera's interior.

We will discuss this application further in this section. See page 763.

11.6.1 Solving Formulas for Variables

We can use the process for solving polynomial equations to solve formulas for variables.

EXAMPLE 1 Solve the formula for the area of a circle, $A = \pi r^2$, for r.

Solution

$$A = \pi r^2 \qquad \text{Quadratic equation in terms of } r$$

$$\frac{A}{\pi} = r^2 \qquad \text{Divide both sides by } \pi.$$

$$r = \sqrt{\frac{A}{\pi}} \qquad \begin{array}{l}\text{Principle of square roots}\\ \text{Use the positive root for length of } r.\end{array}$$

EXAMPLE 2 Solve for the indicated variable.

a. Vertical-position formula, $s = -16t^2 + v_0t$, for t.

b. Cylinder surface area formula, $S = 2\pi r^2 + 2\pi rh$, for r.

Solution

a. Solve $s = -16t^2 + v_0t$ for t.

$$16t^2 - v_0t + s = 0 \qquad \begin{array}{l}\text{Quadratic equation in terms of } t\\ a = 16, b = -v_0, c = s\end{array}$$

$$t = \frac{-b \pm \sqrt{b^2 - 4ac}}{2a} \qquad \text{Quadratic formula}$$

$$t = \frac{-(-v_0) \pm \sqrt{(-v_0)^2 - 4(16)(s)}}{2(16)}$$

$$t = \frac{v_0 \pm \sqrt{v_0^2 - 64s}}{32}$$

b. Solve $S = 2\pi r^2 + 2\pi rh$ for r.

$$2\pi r^2 + 2\pi rh - S = 0 \qquad \begin{array}{l}\text{Quadratic equation in terms of } r\\ a = 2\pi, b = 2\pi h, c = -S\end{array}$$

$$r = \frac{-b \pm \sqrt{b^2 - 4ac}}{2a} \qquad \text{Quadratic formula}$$

$$r = \frac{-(2\pi h) \pm \sqrt{(2\pi h)^2 - 4(2\pi)(-S)}}{2(2\pi)}$$

$$r = \frac{-2\pi h \pm \sqrt{4\pi^2 h^2 + 8\pi S}}{4\pi}$$

$$r = \frac{-2\pi h \pm \sqrt{4(\pi^2 h^2 + 2\pi S)}}{4\pi}$$

$$r = \frac{-2\pi h \pm 2\sqrt{\pi^2 h^2 + 2\pi S}}{4\pi}$$

$$r = \frac{\overset{1}{2}(-\pi h \pm \sqrt{\pi^2 h^2 + 2\pi S})}{2\pi}$$

$$r = \frac{-\pi h \pm \sqrt{\pi^2 h^2 + 2\pi S}}{2\pi}$$

APPLICATION

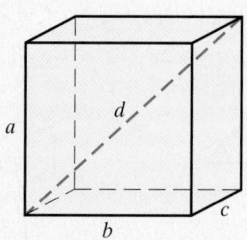

We can extend the Pythagorean theorem to three dimensions. The square of the length of the diagonal of a rectangular solid, d, is the sum of the squares of the lengths of the three sides of the solid, a, b, and c.

a. Determine a formula for the length of the diagonal of a rectangular solid, d. That is, solve the given formula for d.

b. According to the *Guinness Book of World Records*, the smallest video camera was manufactured in the United States by Super Circuits in the year 2000. The camera measures 1.62 centimeters on each side. Use the formula generated in part a to determine the length of the diagonal of the camera's interior.

Discussion

a.
$$a^2 + b^2 + c^2 = d^2$$
$$\sqrt{a^2 + b^2 + c^2} = \sqrt{d^2} \quad \text{Take the square root of both sides.}$$
$$\sqrt{a^2 + b^2 + c^2} = d$$

or

$$d = \sqrt{a^2 + b^2 + c^2}$$

b. Substitute 1.62 for each of sides a, b, and c.
$$d = \sqrt{(1.62)^2 + (1.62)^2 + (1.62)^2}$$
$$d \approx 2.806$$

The diagonal of the tiny camera is about 2.806 centimeters.

 ### 11.6.1 Checkup

Solve each formula for the variable specified.

1. Area of a square, $A = s^2$, for the length of the side s.

2. Pythagorean theorem, $a^2 + b^2 = c^2$, for the length of side a.

3. Surface area of an open canister, $A = \pi r^2 + 2\pi rh$, for the radius r.

4. In August 1998, the world's smallest nightclub was opened by four British men. It was called the "Miniscule of Sound" and had a maximum capacity of 14 people, including the DJ! The nightclub was 8 feet by 4 feet by 8 feet in dimensions. Determine the length of a diagonal from one corner of the ceiling to the opposite corner of the floor.

11.6.2 Special Triangles

Now we are ready to apply the Pythagorean theorem to some special triangles. First let's consider an isosceles right triangle. An **isosceles right triangle** has two **congruent** (equal) legs and two congruent (45°) angles opposite the congruent legs.

To determine the length of the hypotenuse in terms of the length of the congruent legs, we solve the Pythagorean theorem for c, substituting a for the length of each leg.

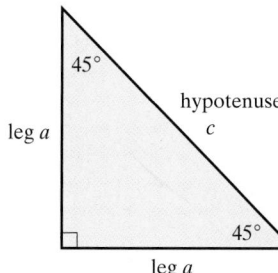

$$a^2 + b^2 = c^2$$
$$a^2 + a^2 = c^2 \quad \text{Substitute.}$$
$$2a^2 = c^2 \quad \text{Simplify.}$$
$$\pm\sqrt{2a^2} = c \quad \text{Principle of square roots}$$
$$\pm a\sqrt{2} = c \quad \text{Simplify.}$$

Therefore, the hypotenuse of an isosceles right triangle is $a\sqrt{2}$.

ISOSCELES RIGHT TRIANGLE (45°–45°–90°)
The length c of the hypotenuse of an isosceles right triangle is the length a of the congruent leg times $\sqrt{2}$.

$$c = a\sqrt{2}$$

For example, determine the hypotenuse of an isosceles right triangle if the length of each congruent leg is 5 cm.

$$c = a\sqrt{2}$$
$$c = 5\sqrt{2} \qquad \text{Substitute.}$$

The hypotenuse is $5\sqrt{2}$ cm, or approximately 7.071 cm.

EXAMPLE 3   Determine the length of each congruent leg of an isosceles right triangle if the hypotenuse is 5 cm.

Algebraic Solution

Substitute 5 for c in the formula and solve for a.

$$c = a\sqrt{2}$$
$$5 = a\sqrt{2} \qquad \text{Substitute.}$$
$$\frac{5}{\sqrt{2}} = \frac{a\sqrt{2}}{\sqrt{2}} \qquad \text{Divide both sides by } \sqrt{2}.$$
$$\frac{5}{\sqrt{2}} = a \qquad \text{Simplify.}$$
$$\frac{5}{\sqrt{2}} \cdot \frac{\sqrt{2}}{\sqrt{2}} = a \qquad \text{Rationalize the denominator.}$$
$$\frac{5\sqrt{2}}{2} = a \qquad \text{Simplify.}$$

The congruent legs are $\frac{5\sqrt{2}}{2}$ cm, or approximately 3.536 cm.

EXAMPLE 4   A single-engine airplane is flying at a low altitude. The pilot takes a reading on the power lines he is approaching. The angle between the horizontal and the power lines is 45°. The distance reading is 100 feet. If the pilot needs to clear the power lines at a height of 75 feet due to local regulations, will he need to adjust his present course?

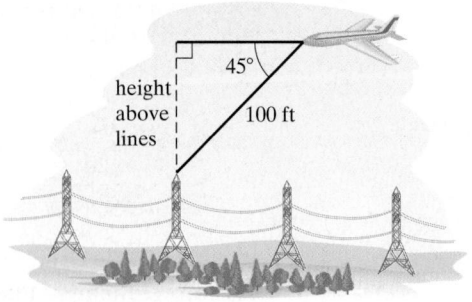

Algebraic Solution

Let x = height above power lines
The hypotenuse is 100 feet.

$$c = a\sqrt{2} \qquad c = 100, a = x$$
$$100 = x\sqrt{2}$$
$$\frac{100}{\sqrt{2}} = x$$
$$\frac{100}{\sqrt{2}} \cdot \frac{\sqrt{2}}{\sqrt{2}} = x$$
$$\frac{100\sqrt{2}}{2} = x$$
$$50\sqrt{2} = x$$
$$x \approx 70.71 \text{ feet}$$

The pilot must adjust his altitude in order to clear the power lines at 75 feet.

Now let's consider an equilateral triangle. An **equilateral triangle** has three congruent (equal) sides and three congruent (60°) angles. If we draw an altitude for this triangle, it will bisect an angle and the side opposite the angle. Then we will have two congruent right triangles. The angles will measure 30°–60°–90°, and the side opposite the 90° angle will be twice the length of the side opposite the 30° angle.

To determine the length of the second leg (the one opposite the 60° angle), we solve the Pythagorean theorem for b, substituting a for the length of the shorter leg and $2a$ for the length of the hypotenuse.

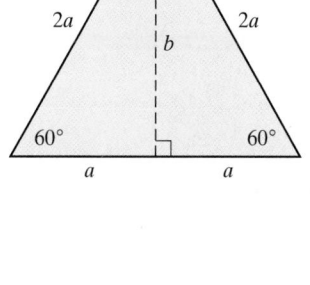

$$a^2 + b^2 = c^2$$
$$a^2 + b^2 = (2a)^2 \qquad \text{Substitute.}$$
$$a^2 + b^2 = 4a^2 \qquad \text{Simplify.}$$
$$b^2 = 4a^2 - a^2 \qquad \text{Subtract } a^2 \text{ from both sides.}$$
$$b^2 = 3a^2 \qquad \text{Simplify.}$$
$$b = \pm\sqrt{3a^2} \qquad \text{Principle of square roots}$$
$$b = \pm a\sqrt{3} \qquad \text{Simplify.}$$

Therefore, the side opposite the 60° angle of a 30°–60°–90° right triangle is $a\sqrt{3}$.

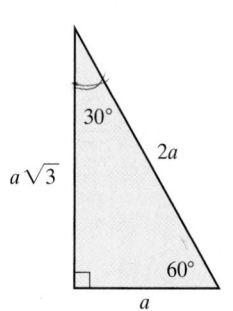

RIGHT TRIANGLE (30°–60°–90°)
The length c of the hypotenuse of a 30°–60°–90° triangle is twice the length a of the shorter side.

$$c = 2a$$

The length b of the longer leg of a 30°–60°–90° triangle is the length a of the shorter leg times $\sqrt{3}$.

$$b = a\sqrt{3}$$

For example, determine the length of the longer leg of a 30°–60°–90° triangle if the shorter leg is 5 inches.

Substitute 5 for a in the formula and solve for b.

$$b = a\sqrt{3}$$
$$b = 5\sqrt{3} \qquad \text{Substitute.}$$

The longer leg is $5\sqrt{3}$ inches, or approximately 8.660 inches.

EXAMPLE 5 Determine the length of the hypotenuse of a 30°–60°–90° triangle if the length of the side opposite the 60° angle is 5 inches.

Algebraic Solution

Use the formula for the longer side, $b = a\sqrt{3}$, substitute 5 for b, and solve for the shorter side, a.

$$b = a\sqrt{3}$$
$$5 = a\sqrt{3} \qquad \text{Substitute.}$$
$$\frac{5}{\sqrt{3}} = \frac{a\sqrt{3}}{\sqrt{3}} \qquad \text{Divide both sides by } \sqrt{3}.$$
$$\frac{5}{\sqrt{3}} = a \qquad \text{Simplify.}$$
$$\frac{5}{\sqrt{3}} \cdot \frac{\sqrt{3}}{\sqrt{3}} = a \qquad \text{Rationalize the denominator.}$$
$$\frac{5\sqrt{3}}{3} = a \qquad \text{Simplify.}$$

Now substitute for a in the formula for the hypotenuse, $c = 2a$.

$$c = 2a$$
$$c = 2\left(\frac{5\sqrt{3}}{3}\right) \qquad \text{Substitute.}$$
$$c = \frac{10\sqrt{3}}{3} \qquad \text{Simplify.}$$

Therefore, the hypotenuse is $\frac{10\sqrt{3}}{3}$ inches, or approximately 5.774 inches.

EXAMPLE 6 It is recommended that an amateur diver stay within 30 feet of the surface of the water. An amateur diver carrying a homing device is alone in the water. At an angle from the water surface of 60°, a boat receiving the diver's signal measures the distance to the diver to be 40 feet. Is the diver out of the safety range?

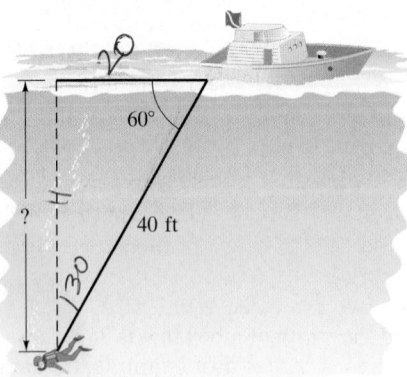

Algebraic Solution

The hypotenuse of the triangle (c) is 40 feet. We are looking for the side opposite the 60° angle (b).

$$c = 2a$$
$$40 = 2a \qquad \text{Substitute.}$$
$$20 = a$$

The shorter side is 20 feet.

$$b = a\sqrt{3}$$
$$b = 20\sqrt{3} \qquad \text{Substitute.}$$
$$b \approx 34.64$$

The diver is approximately 34.64 feet below the water surface. He is below the safety level for amateur divers.

HELPING HAND Remember that the special-triangle formulas are derived from the Pythagorean theorem. Therefore, if you do not remember them, you can always use the Pythagorean theorem to solve the problem. It will just take more steps.

✔ 11.6.2 Checkup

1. Determine the length of each congruent leg of an isosceles right triangle whose hypotenuse measures 6 mm.

2. A ramp makes a 30° angle with the floor. If the vertical distance of the ramp is 3 feet, find the horizontal distance of the ramp and the ramp's length.

3. Describe the features of an isosceles right triangle.

4. Describe the features of an equilateral triangle.

5. Describe the features of a 30°–60°–90° triangle.

11.6 Exercises

In exercises 1–12, solve each formula for the specified variable.

1. $E = mc^2$ for c

2. $A = 4\pi r^2$ for r

3. $V = \pi r^2 h$ for r

4. $F = \dfrac{mv^2}{R}$ for v

5. $E = \dfrac{1}{2}mv^2$ for v

6. $A = \dfrac{\pi r^2 S}{360}$ for r

7. $C = \dfrac{n(n-1)}{2}$ for n

8. $S = \dfrac{n(n+1)}{2}$ for n

9. $x^2 - y^2 = c^2$ for x

10. $x^2 + y^2 = r^2$ for x

11. $A = P(1 + r)^2$ for r

12. $S = 16t^2 + gt$ for t

13. The length of a diagonal, D, inside a rectangular box is related to the dimensions of the box by the Pythagorean theorem, $D^2 = L^2 + W^2 + H^2$, where L, W, and H are, respectively, the length, width, and height of the box. Solve this equation for L. Use the equation to find the length of a box that is 8 inches high, is 16 inches wide, and has a diagonal that is 24 inches long.

14. Solve the formula in the preceding exercise for W. Use the equation to find the width of a box that is 24 inches long, is 24 inches high, and has a diagonal that is 36 inches long.

Use the properties of an isosceles right triangle to solve. Round your answers to the nearest hundredth.

15. Determine the length of the hypotenuse of an isosceles right triangle whose legs measure 3.75 meters.

16. What is the length of the hypotenuse of an isosceles right triangle whose legs measure 6.55 feet?

17. What will be the lengths of the legs of an isosceles right triangle whose hypotenuse measures 7.6 feet?

18. If the hypotenuse of an isosceles right triangle measures 14.9 centimeters, what are the lengths of its legs?

19. The top of a telephone pole is anchored with a guy wire that makes a 45-degree angle with the ground. If the pole is 20 feet high, how long is the guy wire?

20. A guy wire makes a 45-degree angle from the ground to the top of a pole. The wire is anchored 15 feet away from the base of the pole. How long is the wire?

21. A wire runs from the top of a building to the ground at a 45-degree angle. If the wire is 42.5 feet long, how high is the building?

22. A wire at a boot camp runs from the top of a tower to the ground at a 45-degree angle. If the wire is 145 feet long, how high is the tower?

23. An amateur diver is expected to stay within 30 feet of the surface of the water. His homing device sends a signal back to the boat at an angle of 45° from the water level and from a distance of 40 feet. Is the diver within the expected depth?

24. The amateur diver in exercise 23 moves further in the water and now sends a signal back to the boat at an angle of 45° from the water level, but from a distance of 50 feet. Is the diver within the expected depth?

In the exercises that follow, assume that the right triangles are 30°–60°–90° triangles, and use the properties of such triangles to solve. Round your answers to the nearest tenth.

25. Determine the lengths of the unknown sides of a triangle if the side opposite the 60° angle measures 18.6 inches.

26. In a triangle, if the side opposite the 60° angle measures 2.19 centimeters, find the lengths of the other sides of the triangle.

27. Given that the hypotenuse of a triangle measures 22 inches, what are the lengths of the legs?

28. What are the measures of the legs of a triangle if the hypotenuse measures 34 centimeters?

29. If the short leg of a triangle measures 21 centimeters, what are the lengths of the other sides?

30. What will be the lengths of the other sides when the length of the short leg of a triangle measures 8.2 inches?

31. An airplane is flying at an altitude of 4000 feet and is approaching the runway at a 30-degree angle. Find the horizontal distance to the set-down point on the runway and the slanted distance from the plane to the runway.

32. A kite is 120 feet above the ground. The string to the kite makes a 60-degree angle with the ground. Find the horizontal distance between the kite and its flyer and the length of the string between the kite and its flyer.

33. A guy wire reaches from the top of a pole to the ground at a 60-degree angle. The distance between the base of the pole and the point at which the wire is anchored to the ground measures 8 feet. Find the height of the pole and the length of the wire.

34. A telephone pole is 13 meters high. The pole is anchored with a guy wire that makes a 60-degree angle with the ground. How much wire is needed, and how far from the base of the pole should it be anchored?

11.6 Calculator Exercises

Remember that you can check your calculations for special triangles by using the Pythagorean theorem and your calculator. In Example 3 of this section, you determined that if the hypotenuse of an isosceles right triangle was $c = 5$ cm, then the length of each congruent leg was $a = \frac{5\sqrt{2}}{2}$ cm. To check this on the calculator, store the values for a and c, and type the Pythagorean theorem when $a = b$. If the calculator displays 1, the theorem holds; if the calculator displays 0, the theorem does not hold and an error has been made. The screen for this example follows:

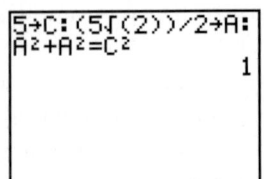

To check Example 5, store 5 for b, $\frac{5\sqrt{3}}{3}$ for a, and $\frac{10\sqrt{3}}{3}$ for c. The screen for this example follows:

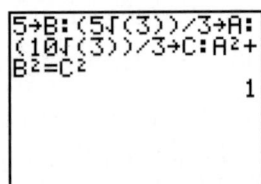

Use the method to check the following special-triangle calculations:

1. Given an isosceles right triangle with legs of length a and hypotenuse of length c, check the following pairs of values:

   a. $a = 8, c = 8\sqrt{2}$ **b.** $a = 12, c = 12\sqrt{3}$ **c.** $a = 6\sqrt{2}, c = 12$ **d.** $a = \dfrac{8\sqrt{3}}{3}, c = 8$

2. Given a $30°–60°–90°$ triangle with legs of length a and b and hypotenuse of length c, check the following triples of values:

   a. $a = 5, b = 5\sqrt{3}, c = 10$ **b.** $a = 10, b = 10\sqrt{3}, c = 5$ **c.** $a = 5, b = 5\sqrt{2}, c = 10$ **d.** $a = 5, b = 10, c = 5\sqrt{3}$
   e. $a = 10, b = 10\sqrt{3}, c = 20$

11.6 Writing Exercises

In this section, you were given formulas to be used to determine the lengths of the sides of special right triangles. Unless you are very careful, it is easy to make a mistake in trying to recall which formula to use. You may confuse which formula to assign to which side of the triangle, or you may not remember whether to use a factor of $\sqrt{2}$ or of $\sqrt{3}$. However, one of the helping hands in this section suggested that you can always fall back on the Pythagorean theorem to calculate the lengths of the sides of these special triangles. Explain how you would determine all sides of the following triangles:

1. An isosceles right triangle when you know the length of one of the congruent legs.

2. An isosceles right triangle when you know the length of the hypotenuse.

3. A $30°–60°–90°$ triangle when you know the length of the shortest leg.

4. A $30°–60°–90°$ triangle when you know the length of the longest leg.

5. A $30°–60°–90°$ triangle when you know the length of the hypotenuse.

6. An equilateral triangle when you know the height of the triangle.

11.7 Solving Quadratic Inequalities

OBJECTIVES

1 Identify quadratic inequalities in one variable.

2 Solve quadratic inequalities numerically.

3 Solve quadratic inequalities graphically.

4 Solve quadratic inequalities algebraically.

5 Model real-world situations by using quadratic inequalities, and solving the equations.

APPLICATION

Kaiser Foundation Health Plan reported its net income (sales minus expenses) for the years from 1987 to 1995. The company CEO believes that the firm has done well if its net income is at least $600 million. A function used to estimate the annual net income in millions of dollars is $I(x) = -28x^2 + 338x - 317$, where x is the number of years after 1987. Determine the years for which the CEO believes the company has done well.

After completing this section, we will discuss this application further. See page 781.

Earlier in this text, we solved linear inequalities in one variable. Recall that the solution set for a linear inequality contains an infinite number of solutions. Therefore, we describe the set as an inequality or in interval notation. For example, we might have the following descriptions:

Inequality	*Interval*	*Inequality*	*Interval*
$x < 1$	$(-\infty, 1)$	$x \le 1$	$(-\infty, 1]$
$x > 1$	$(1, \infty)$	$x \ge 1$	$[1, \infty)$
$4 < x < 5$	$(4, 5)$	$4 \le x < 5$	$[4, 5)$
$4 < x \le 5$	$(4, 5]$	$4 \le x \le 5$	$[4, 5]$

11.7.1 Identifying Quadratic Inequalities in One Variable

In this section, we will be solving quadratic inequalities in one variable. A **quadratic inequality in one variable**, or, simply, a **quadratic inequality**, is written by replacing the equals sign in a quadratic equation in one variable with an order symbol.

> ### STANDARD FORM FOR A QUADRATIC INEQUALITY IN ONE VARIABLE
>
> A quadratic inequality in standard form is written in one of the forms
>
> $$ax^2 + bx + c < 0$$
> $$ax^2 + bx + c > 0$$
> $$ax^2 + bx + c \le 0$$
> $$ax^2 + bx + c \ge 0$$
>
> where a, b, and c are real numbers and $a \ne 0$.

EXAMPLE 1   Determine whether each inequality is a quadratic inequality.

a. $2x^2 + 3x - 5 < 0$ **b.** $x^2 + 5x - 7 \le 2x^2 + 4x + 15$

c. $0.3x^3 + 5.1x^2 > x - 1.2$ **d.** $6x^{-2} + 5x^{-1} < 0$

Solution

a. $2x^2 + 3x - 5 < 0$ is a quadratic inequality.

b. $x^2 + 5x - 7 \le 2x^2 + 4x + 15$ is a quadratic inequality, because it simplifies to $-x^2 + x - 22 \le 0$.

c. $0.3x^3 + 5.1x^2 > x - 1.2$ is not a quadratic inequality, because it has a third-degree term (exponent of 3).

d. $6x^{-2} + 5x^{-1} < 0$ is not a quadratic inequality, because it has variables with negative exponents.

11.7.1 Checkup

In exercises 1–4, determine whether each inequality is a quadratic inequality.

1. $2x + 3 \le x^2 - 1$

2. $\frac{2}{3}x^2 - 3x \ge \frac{1}{4}$

3. $3x + 2x^{-2} > 4x^{-1} + 8$ **4.** $1.2\sqrt{x} - 3.3x^2 \le 2.5x$

5. Explain the difference between a quadratic equation and a quadratic inequality.

11.7.2 Solving Quadratic Inequalities Numerically

We solved linear inequalities numerically for integer solutions by using a table of values. We will use the same method to solve a quadratic inequality in one variable for integer solutions.

SOLVING A QUADRATIC INEQUALITY NUMERICALLY
To solve a quadratic inequality numerically for integer solutions, set up an extended table of values.

- The first column is labeled with the name of the independent variable.
- The second column is labeled with the expression on the left side of the inequality.
- The third column is labeled with the expression on the right side of the inequality.

Complete the table.

- Substitute values for the independent variable.
- Evaluate the second and third columns.
- Continue until values for the two expressions (the numbers in the second and third column) result in a true inequality.

The values for the independent variable (the numbers in the first column) that result in a true inequality are solutions of the inequality.

HELPING HAND Not all solutions may be found by this method. We are limited to the numbers substituted for the independent variable.

EXAMPLE 2   Solve $x^2 + 2x - 1 \geq 2$ numerically.

Numeric Solution

Set up a table of values.

x	$x^2 + 2x - 1$	2	
-5	14	2	$14 \geq 2$
-4	7	2	$7 \geq 2$
-3	2	2	$2 \geq 2$
-2	-1	2	$-1 \not\geq 2$
-1	-2	2	$-2 \not\geq 2$
0	-1	2	$-1 \not\geq 2$
1	2	2	$2 \geq 2$
2	7	2	$7 \geq 2$
3	14	2	$14 \geq 2$

Calculator Numeric Solution

X	Y1	Y2	
-4	7	2	$7 \geq 2$
-3	2	2	$2 \geq 2$
-2	-1	2	$-1 \not\geq 2$
-1	-2	2	$-2 \not\geq 2$
0	-1	2	$-1 \not\geq 2$
1	2	2	$2 \geq 2$
2	7	2	$7 \geq 2$

X= -4

Y1 = $x^2 + 2x - 1$
Y2 = 2

A calculator table will find the integer solutions numerically.

 We obtain the same solution set: all integers less than or equal to -3 or all integers greater than or equal to 1, as shown in the figure.

According to the table, when -5, -4, and -3 are substituted into the inequality, the result is a true statement. Also, when 1, 2, and 3 are substituted into the inequality, a true statement results. Therefore, there appear to be two sets of solutions: all integers less than or equal to -3 is the first set, and all integers greater than or equal to 1 is the second set.

HELPING HAND This method is not recommended, as it does not result in a complete solution set, only an integer solution set. However, as a visualization, it may be used to check other methods.

11.7.2 Checkup

Solve numerically for integer solutions.

1. $x^2 + 2x - 7 \geq 3x + 5$ **2.** $2x^2 + 5x - 3 < 0$ **3.** $-x^2 + 4 < x$

11.7.3 Solving Quadratic Inequalities Graphically

A second and more inclusive method for determining the solutions of an inequality is to graph two functions, y_1 and y_2. We will use the two expressions in the inequality as the rules of the functions. If we are solving a "less than" $(y_1 < y_2)$ inequality, we choose the x-values that correspond to the portion of the graph for which y_1 is below y_2. If we are solving a "greater than" $(y_1 > y_2)$ inequality, we choose the x-values that correspond to the portion of the graph for which y_1 is above y_2.

SOLVING A QUADRATIC INEQUALITY GRAPHICALLY
To solve a quadratic inequality graphically,

- Write two functions, y_1 and y_2, using the expressions on the left and right sides of the inequality as rules.
- Graph both functions on the same coordinate plane.
- Determine the intersections, if they exist, of the y_1 and y_2 graphs.
- Locate the portion of the y_1 graph below the y_2 graph if the inequality is "less than." Locate the portion of the y_1 graph above the y_2 graph if the inequality is "greater than."
- Determine the x-coordinates for this portion of the graph.
- If the portion is to the left of the intersection, then x is less than the x-coordinate of the intersection.
- If the portion is to the right of the intersection, then x is greater than the x-coordinate of the intersection.
- If the portion is between the left and right intersections, then x is greater than the x-coordinate of the left intersection and less than the x-coordinate of the right intersection.

The solution set of the inequality includes the x-coordinates of the points of intersection of the two graphs if the inequality contains equality.

EXAMPLE 3   Solve graphically $x^2 + 2x - 1 \geq 2$.

Graphic Solution

Let $Y1 = x^2 + 2x - 1$ and $Y2 = 2$

The solution contains the x-coordinates of the points of intersection, -3 and 1, because the inequality contains "equal to." The graph of Y1 is above the graph of Y2 to the left of the intersection point $(-3, 2)$ and to the right of the intersection point $(1, 2)$, as shown in Figures 11.5a, 11.5b, and 11.5c.

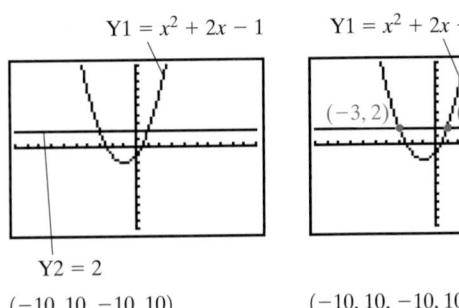

Figure 11.5a

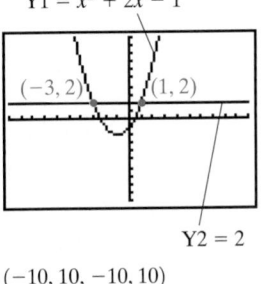

Figure 11.5b

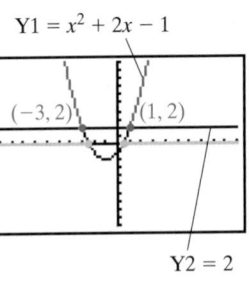

Figure 11.5c

Therefore, the solution is $x \le -3$ or $x \ge 1$. The solution set may be written in interval notation as $(-\infty, -3]$ or $[1, \infty)$. In mathematics, we write this as $(-\infty, -3] \cup [1, \infty)$, where the symbol $\cup$ is called **union** and means "or."

EXAMPLE 4 Solve graphically.

a. $x^2 + 3x - 4 \le 2x + 2$ b. $3x^2 - x - 2 > 0$

c. $-x^2 + 6x > 6$

Graphic Solution

a. $x^2 + 3x - 4 \le 2x + 2$

Let $Y1 = x^2 + 3x - 4$ and $Y2 = 2x + 2$

The solution contains both x-coordinates of the intersections, because the inequality contains "equal to." Since the inequality also contains "less than," we determine that Y1 is below Y2 between the points of intersection.

Therefore, the solution set is $-3 \le x \le 2$, or $[-3, 2]$.

b. $3x^2 - x - 2 > 0$

Let $Y1 = 3x^2 - x - 2$ and $Y2 = 0$ (The latter will graph a horizontal line on the x-axis.)

The solution does not contain the x-coordinates of the points of intersection. Since the inequality is "greater than," we determine that Y1 is above Y2 to the left of the intersection point $\left(-\frac{2}{3}, 0\right)$ and to the right of the intersection point $(1, 0)$.

Therefore, the solution set is $x < -\frac{2}{3}$, or $x > 1$. In interval notation, the solution is $\left(-\infty, -\frac{2}{3}\right) \cup (1, \infty)$.

Numeric Check

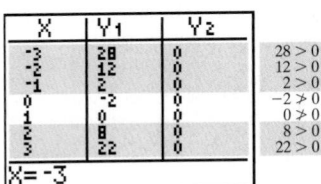

$Y1 = x^2 + 3x - 4$
$Y2 = 2x + 2$
The solution checks.

$Y1 = 3x^2 - x - 2$
$Y2 = 0$
The solution checks.

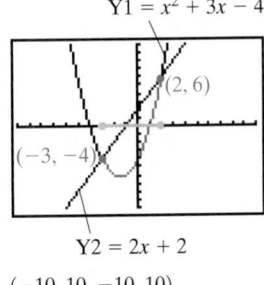

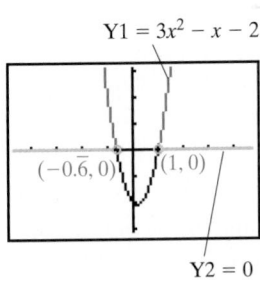

Y2 = 6

(4.73, 6)

(1.27, 6)

Y1 = $-x^2 + 6x$

$(-10, 10, -10, 10)$

c. $-x^2 + 6x > 6$

Let Y1 = $-x^2 + 6x$ and Y2 = 6

The solution does not contain the x-coordinates of the points of intersection. Since the inequality is "greater than," we determine that Y1 is above Y2 between the approximate points of intersection, $(1.27, 6)$ and $(4.73, 6)$.

Therefore, the approximate solution set is $1.27 < x < 4.73$. In interval notation, the solution is approximately $(1.27, 4.73)$.

X	Y1	Y2
-1	-7	6
0	0	6
1	5	6
2	8	6
3	9	6
4	8	6
5	5	6

X = -1

Y1 = $-x^2 + 6x$
Y2 = 6
The solution checks.

-7 ≯ 6
0 ≯ 6
5 ≯ 6
8 > 6
9 > 6
8 > 6
5 ≯ 6

There are special cases of linear inequalities that have no solution or for which the solution is all real numbers. There are also special cases of quadratic inequalities. In fact, several types of solutions are possible. To see what they are, complete the following set of exercises on your calculator.

Discovery 4

Special Cases of Quadratic Inequalities

Graph the functions that follow. Label points of intersection if possible. Determine the solutions of the given inequalities.

1. $f(x) = -x^2 - 5$
$g(x) = 0$
 a. $-x^2 - 5 > 0$
 b. $-x^2 - 5 \geq 0$
 c. $-x^2 - 5 < 0$
 d. $-x^2 - 5 \leq 0$

2. $h(x) = x^2 + 5x + 8$
$j(x) = 0$
 a. $x^2 + 5x + 8 > 0$
 b. $x^2 + 5x + 8 \geq 0$
 c. $x^2 + 5x + 8 < 0$
 d. $x^2 + 5x + 8 \leq 0$

3. $y_1 = -x^2 - 10x - 25$
$y_2 = 0$
 a. $-x^2 - 10x - 25 > 0$
 b. $-x^2 - 10x - 25 \geq 0$
 c. $-x^2 - 10x - 25 < 0$
 d. $-x^2 - 10x - 25 \leq 0$

4. $y_1 = 2x^2 - 12x + 18$
$y_2 = 0$
 a. $2x^2 - 12x + 18 > 0$
 b. $2x^2 - 12x + 18 \geq 0$
 c. $2x^2 - 12x + 18 < 0$
 d. $2x^2 - 12x + 18 \leq 0$

Discovery 4 illustrates that some quadratic inequalities have no solution, while others have a solution set consisting of all real numbers. Still other solution sets contain only one value or all real numbers except one value.

EXAMPLE 5 Solve graphically.

a. $2x^2 \geq x^2$ **b.** $x^2 - 4x + 7 > -x^2 + 4x - 1$
c. $x^2 + 2 \leq 2$

Graphic Solution

Numeric Check

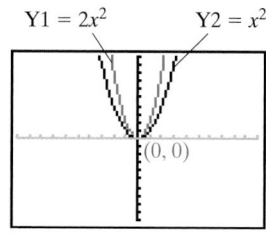

$Y1 = 2x^2$ $Y2 = x^2$

$(0, 0)$

$(-10, 10, -10, 10)$

a. $2x^2 \geq x^2$

Let $Y1 = 2x^2$ and $Y2 = x^2$

The solution contains the x-coordinate of the intersection point, $(0, 0)$, because the inequality contains equality. Since the inequality is "greater than," we determine that Y1 is above Y2 at all points to the left and to the right of the intersection. Therefore, the solution set is the set of all real numbers.

X	Y1	Y2	
-3	18	9	$18 \geq 9$
-2	8	4	$8 \geq 4$
-1	2	1	$2 \geq 1$
0	0	0	$0 \geq 0$
1	2	1	$2 \geq 1$
2	8	4	$8 \geq 4$
3	18	9	$18 \geq 9$

X= -3

$Y1 = 2x^2$
$Y2 = x^2$
The values of Y1 are always greater than or equal to their Y2 values.

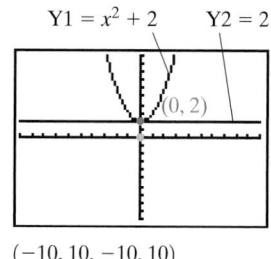

$Y1 = x^2 - 4x + 7$

$(2, 3)$

$Y2 = -x^2 + 4x - 1$

$(-10, 10, -10, 10)$

b. $x^2 - 4x + 7 > -x^2 + 4x - 1$

Let $Y1 = x^2 - 4x + 7$ and $Y2 = -x^2 + 4x - 1$

The solution does not contain the x-coordinate of the intersection point, $(2, 3)$, because the inequality does not contain equality. Since the inequality is "greater than," we determine that Y1 is above Y2 at all points to the left of the intersection point and all points to the right of the intersection point.

Therefore, the solution set is the set of all real numbers except 2, or $x \neq 2$. In interval notation, the solution is $(-\infty, 2) \cup (2, \infty)$.

X	Y1	Y2	
-2	19	-13	$19 > -13$
-1	12	-6	$12 > -6$
0	7	-1	$7 > -1$
1	4	2	$4 > 2$
2	3	3	$3 \not> 3$
3	4	2	$4 > 2$
4	7	-1	$7 > -1$

X= -2

$Y1 = x^2 - 4x + 7$
$Y2 = -x^2 + 4x - 1$
The values of Y1 are always greater than the corresponding values for Y2 except when $x = 2$. The solution set is all real numbers except 2.

$Y1 = x^2 + 2$ $Y2 = 2$

$(0, 2)$

$(-10, 10, -10, 10)$

c. $x^2 + 2 \leq 2$

Let $Y1 = x^2 + 2$ and $Y2 = 2$

The solution contains the x-coordinate of the intersection point, $(0, 2)$, because the inequality contains "equal to." Since the inequality is "less than," we determine that Y1 is not below Y2.

Therefore, the only solution is 0.

X	Y1	Y2	
-2	6	2	$6 \not\leq 2$
-1	3	2	$3 \not\leq 2$
0	2	2	$2 \leq 2$
1	3	2	$3 \not\leq 2$
2	6	2	$6 \not\leq 2$
3	11	2	$11 \not\leq 2$
4	18	2	$18 \not\leq 2$

X= -2

$Y1 = x^2 + 2$
$Y2 = 2$
The solution is 0, the only value with corresponding equal Y1 and Y2 values. ●

 11.7.3 Checkup

Solve exercises 1–6 graphically.

1. $x^2 + 2x - 7 \geq 3x + 5$ **2.** $2x^2 + 5x - 3 < 0$

3. $-x^2 + 4 < x$ **4.** $\dfrac{1}{2}x^2 + 2 \geq x^2 + 2$

5. $x^2 - 4x + 1 > -3$

6. $x^2 - 4x + 7 \leq -5 + 8x - 2x^2$

7. Describe what you might see if you graphically solved a quadratic inequality with the following results:
 a. The solution set consists of all real numbers.
 b. The solution set consists of a single value.
 c. The solution set consists of an interval on the number line.
 d. The inequality has no real-number solution.

11.7.4 Solving Quadratic Inequalities Algebraically

Several methods are available for solving a quadratic inequality algebraically. We will illustrate one of them.

When we solved quadratic inequalities graphically, we first determined the points of intersection of the graphs. Algebraically, we need to determine these points by solving a quadratic equation. Remember, we have several methods available to do this. We call these points of equality **critical points**—endpoints of possible interval solutions. After determining the possible intervals for the solution, we need test only one point in each interval. If one point in an interval is a solution, all points in the interval are solutions.

SOLVING A QUADRATIC INEQUALITY ALGEBRAICALLY

To solve a quadratic inequality in one variable algebraically,

- Determine the critical points (points of equality) by solving a quadratic equation and replacing the inequality symbol with an equals sign. Use the original inequality to determine whether the points of equality are in the solution.
- Determine the possible intervals for the solution by using the critical points as endpoints.
- Test one value in each interval for a true statement. If the inequality is true for the value tested, then the inequality is true for all values in the interval.

Write the solution set as a union of all possible intervals.

EXAMPLE 6 Solve algebraically.

a. $x^2 + 3x - 4 \le 2x + 2$ **b.** $3x^2 - x - 2 > 0$ **c.** $-x^2 + 6x > 3$

Algebraic Solution

a. $x^2 + 3x - 4 \le 2x + 2$

Determine the critical points.

$$x^2 + 3x - 4 = 2x + 2$$
$$x^2 + x - 6 = 0$$
$$(x + 3)(x - 2) = 0$$
$$x = -3 \quad \text{or} \quad x = 2$$

The inequality contains equality, so the critical points are solutions. We include them in the intervals.

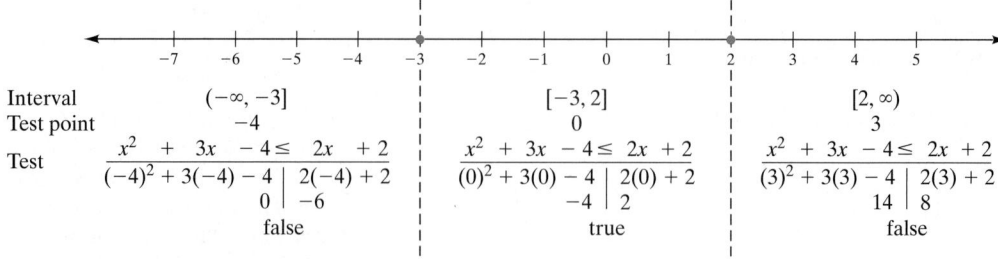

Interval	$(-\infty, -3]$	$[-3, 2]$	$[2, \infty)$
Test point	-4	0	3
Test	$x^2 + 3x - 4 \le 2x + 2$	$x^2 + 3x - 4 \le 2x + 2$	$x^2 + 3x - 4 \le 2x + 2$
	$(-4)^2 + 3(-4) - 4 \mid 2(-4) + 2$	$(0)^2 + 3(0) - 4 \mid 2(0) + 2$	$(3)^2 + 3(3) - 4 \mid 2(3) + 2$
	$0 \mid -6$	$-4 \mid 2$	$14 \mid 8$
	false	true	false

Therefore, the solution is $[-3, 2]$.

b. $3x^2 - x - 2 > 0$

Determine the critical points.

$$3x^2 - x - 2 = 0$$
$$(3x + 2)(x - 1) = 0$$
$$x = -\frac{2}{3} \quad \text{or} \quad x = 1$$

The inequality does not contain equality. We do not include the critical points as solutions.

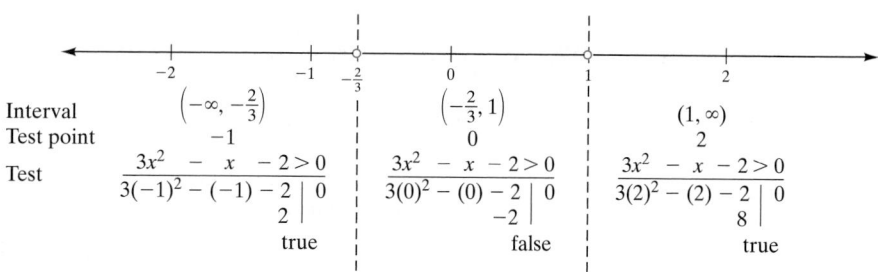

Interval	$\left(-\infty, -\frac{2}{3}\right)$	$\left(-\frac{2}{3}, 1\right)$	$(1, \infty)$
Test point	-1	0	2
Test	$3x^2 - x - 2 > 0$	$3x^2 - x - 2 > 0$	$3x^2 - x - 2 > 0$
	$3(-1)^2 - (-1) - 2 \mid 0$	$3(0)^2 - (0) - 2 \mid 0$	$3(2)^2 - (2) - 2 \mid 0$
	2	-2	8
	true	false	true

Therefore, the solution is $\left(-\infty, -\frac{2}{3}\right) \cup (1, \infty)$.

c. $-x^2 + 6x > 3$

Determine the critical points.

$$-x^2 + 6x = 3$$
$$-x^2 + 6x - 3 = 0$$

This equation does not factor. We need to use the quadratic formula. Use $a = -1, b = 6,$ and $c = -3$.

$$x = \frac{-b \pm \sqrt{b^2 - 4ac}}{2a}$$

$$x = \frac{-(6) \pm \sqrt{(6)^2 - 4(-1)(-3)}}{2(-1)}$$

$$x = \frac{-6 \pm \sqrt{24}}{-2}$$

$$x = \frac{-6 \pm 2\sqrt{6}}{-2}$$

$$x = \frac{\overset{1}{-2}(3 \pm \sqrt{6})}{\underset{1}{-2}}$$

$$x = 3 \pm \sqrt{6}$$

$$x \approx 0.551 \quad \text{or} \quad x \approx 5.450$$

The inequality does not contain equality. We do not include the critical points as solutions.

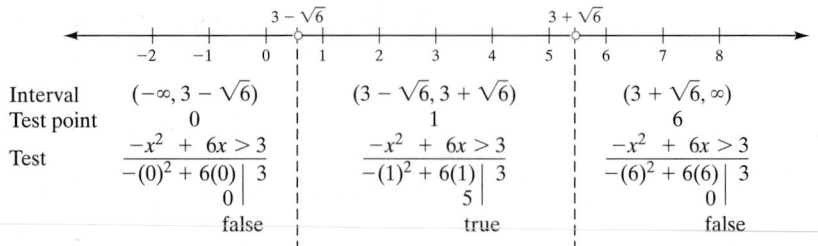

Therefore, the solution is $(3 - \sqrt{6}, 3 + \sqrt{6})$.

The special cases of quadratic inequalities solved graphically in Example 5 may also be solved algebraically.

EXAMPLE 7 Solve algebraically.

a. $2x^2 \geq x^2$ **b.** $x^2 - 4x + 7 > -x^2 + 4x - 1$

c. $x^2 + 2 \leq 2$

Algebraic Solution

a. $2x^2 \geq x^2$

Determine the critical points.

$$2x^2 = x^2$$
$$x^2 = 0$$
$$x = 0$$

The inequality contains equality. We include the critical point as a solution.

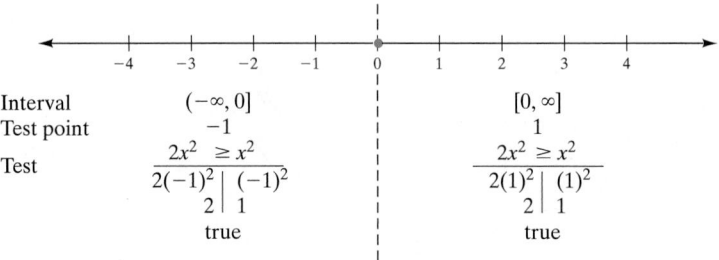

Therefore, the solution is $(-\infty, 0] \cup [0, \infty)$, which is equivalent to $(-\infty, \infty)$, or the set of all real numbers.

b. $x^2 - 4x + 7 > -x^2 + 4x - 1$

Determine the critical points.

$$x^2 - 4x + 7 = -x^2 + 4x - 1$$
$$2x^2 - 8x + 8 = 0$$
$$2(x^2 - 4x + 4) = 0$$
$$2(x - 2)^2 = 0$$
$$x - 2 = 0$$
$$x = 2$$

The inequality does not contain equality. We do not include the critical point as a solution.

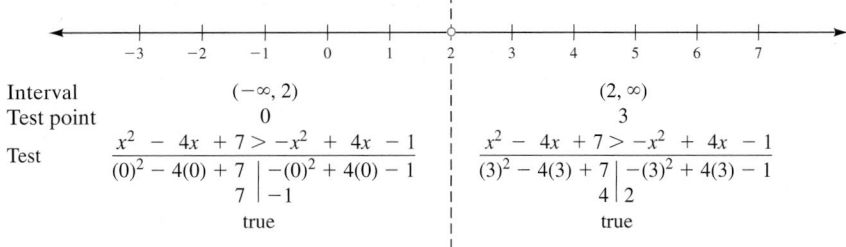

Interval	$(-\infty, 2)$	$(2, \infty)$
Test point	0	3
Test	$\dfrac{x^2 - 4x + 7 > -x^2 + 4x - 1}{(0)^2 - 4(0) + 7 \mid -(0)^2 + 4(0) - 1}$	$\dfrac{x^2 - 4x + 7 > -x^2 + 4x - 1}{(3)^2 - 4(3) + 7 \mid -(3)^2 + 4(3) - 1}$
	$7 \mid -1$	$4 \mid 2$
	true	true

Therefore, the solution is $(-\infty, 2) \cup (2, \infty)$, or $x \neq 2$.

c. $x^2 + 2 \leq 2$

Determine the critical points.

$$x^2 + 2 = 2$$
$$x^2 = 0$$
$$x = 0$$

The inequality contains equality. We include the critical point as a solution.

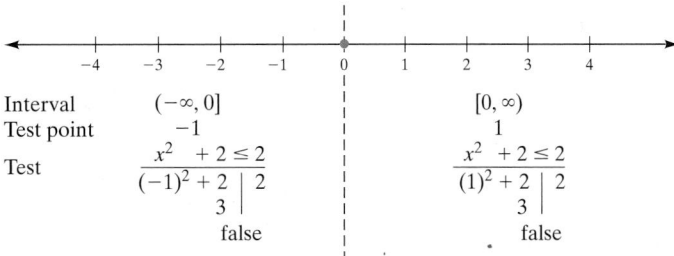

Interval	$(-\infty, 0]$	$[0, \infty)$
Test point	-1	1
Test	$\dfrac{x^2 + 2 \leq 2}{(-1)^2 + 2 \mid 2}$	$\dfrac{x^2 + 2 \leq 2}{(1)^2 + 2 \mid 2}$
	3	3
	false	false

Therefore, the solution is neither possible interval. However, the critical point is a solution. The only solution is 0.

 11.7.4 Checkup

Solve exercises 1–6 algebraically.

1. $x^2 + 2x - 7 \geq 3x + 5$ **2.** $2x^2 + 5x - 3 < 0$

3. $-x^2 + 4 < x$ **4.** $\dfrac{1}{2}x^2 + 2 \geq x^2 + 2$

5. $x^2 - 4x + 1 > -3$

6. $x^2 - 4x + 7 \leq -5 + 8x - 2x^2$

7. How do you determine the critical points of a quadratic inequality?

8. Why is it important to know the critical points of a quadratic inequality when solving the inequality algebraically?

11.7.5 Modeling the Real World

Inequalities often give us important information about the quantities related by a quadratic equation. If the equation describes the height of a thrown or falling object, an inequality might tell us when the object is higher or lower than a given height. If the equation describes a cost or sales function, an inequality might tell us the range of numbers of items we need to sell to make a profit. For these reasons, inequalities are very useful in business and science.

EXAMPLE 8 In a movie, a crash dummy is projected upward at a speed of 100 feet per second from the top of a 200-foot cliff. Determine the length of time the dummy is above the point of origin.

Algebraic Solution

Use the vertical-position function $s(t) = -16t^2 + v_0 t + s_0$ for time t, v_0 as the initial velocity, and s_0 as the initial height to write a function. Let $v_0 = 100$ and $s_0 = 200$. The position function then becomes

$$s(t) = -16t^2 + 100t + 200$$

Use this position function to write an inequality. We want to know when the position is above (that is, greater than) the point of origin, 200 feet. The dummy will be above the point of origin when $s(t) > 200$.

$$-16t^2 + 100t + 200 > 200$$

Determine the critical points.

$$-16t^2 + 100t + 200 = 200$$
$$-16t^2 + 100t = 0$$
$$-4t(4t - 25) = 0$$

$$-4t = 0 \qquad \text{or} \qquad 4t - 25 = 0$$

$$t = 0 \qquad\qquad\qquad t = \frac{25}{4}$$

The inequality does not contain equality, so the critical points are not solutions.

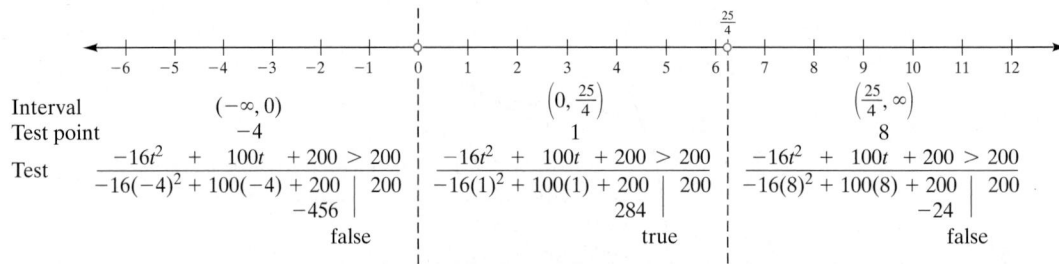

Interval	$(-\infty, 0)$	$\left(0, \frac{25}{4}\right)$	$\left(\frac{25}{4}, \infty\right)$
Test point	-4	1	8
Test	$\dfrac{-16t^2 + 100t + 200 > 200}{-16(-4)^2 + 100(-4) + 200 \mid 200}$	$\dfrac{-16t^2 + 100t + 200 > 200}{-16(1)^2 + 100(1) + 200 \mid 200}$	$\dfrac{-16t^2 + 100t + 200 > 200}{-16(8)^2 + 100(8) + 200 \mid 200}$
	-456	284	-24
	false	true	false

The solution is $\left(0, \frac{25}{4}\right)$ or $0 < t < \frac{25}{4}$.

The dummy is above the point of origin between 0 and $\frac{25}{4}$ seconds, or between 0 and 6.25 seconds, after projection.

EXAMPLE 9 Elisa sells fruit drinks at the local fair. She has determined from past summers that the cost for x days of operation can be estimated by the cost function $C(x) = 2x^2 - 20x + 240$. Determine the number of days she can operate the stand at a cost of no more than $200.

Numeric Solution

We need to determine when the cost function is less than or equal to 200: $C(x) \le 200$.

$$2x^2 - 20x + 240 \le 200$$

Use the inequality $2x^2 - 20x + 240 \le 200$.

X	Y1	Y2	
2	208	200	$208 \not\le 200$
3	198	200	$198 \le 200$
4	192	200	$192 \le 200$
5	190	200	$190 \le 200$
6	192	200	$192 \le 200$
7	198	200	$198 \le 200$
8	208	200	$208 \not\le 200$
X=2			

Since we need to determine only integer solutions, set up a table of values for x days. Let $Y1 = 2x^2 - 20x + 240$ and $Y2 = 200$.

According to Figure 11.6, Elisa should operate her stand for at least three days but not more than seven days, to keep her cost at $200 or less. ●

Figure 11.6

APPLICATION

Kaiser Foundation Health Plan reported its net income (sales minus expenses) for the years from 1987 to 1995. The company CEO believes that the firm has done well if its net income is at least $600 million. A function used to estimate the annual net income in millions of dollars is $I(x) = -28x^2 + 338x - 317$, where x is the number of years after 1987. Determine the years for which the CEO believes the company has done well.

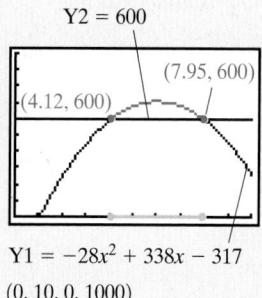

Y2 = 600

(4.12, 600)   (7.95, 600)

Y1 = $-28x^2 + 338x - 317$

(0, 10, 0, 1000)

Discussion

Let x = number of years after 1987

$I(x)$ = annual net income in millions of dollars

$I(x) = -28x^2 + 338x - 317$

$-28x^2 + 338x - 317 \ge 600$ Net income is at least 600.

Let Y1 = $-28x^2 + 338x - 317$ and Y2 = 600

The graph of the function representing the net income is above the graph of the constant function for x-values of 5, 6, and 7 years after 1987. This represents the years 1992, 1993, and 1994.

Check

A numeric table of values confirms this solution.

X	Y1	Y2
2	247	600
3	445	600
4	587	600
5	673	600
6	703	600
7	677	600
8	595	600
X=2		

Y1 = $-28x^2 + 338x - 317$
Y2 = 600

✔ 11.7.5 Checkup

1. A baseball is thrown upward at a speed of 50 feet per second from the top of a tower that is 120 feet above the ground. Determine the length of time the baseball is no lower than the top of the tower.

2. The wholesale price of a computer is $600. The minimum order a dealer can place is 10 computers. For every additional computer above 10, the price per computer to the dealer is reduced by $20. If x represents the number of computers above 10 that are ordered, the price per computer will be $600 - 20x$, and the number ordered will be $10 + x$. The total revenue from the sale will be $R(x) = (600 - 20x)(10 + x)$. How many computers must be ordered for the revenue to be at least $7500?

3. Using data from the *Universal Almanac*, the growth in population of the Mexico City urban area can be approximated by the function $y = 0.005x^2 + 0.234x + 2.6$, where y is measured in millions of people and x is the number of years after 1950. Using this function, in what years would you expect the population of the Mexico City area to exceed 30 million people?

11.7 Exercises

Determine whether each inequality is a quadratic inequality.

1. $2x^2 - 5 < 4x + 1$

2. $5a - 3 \geq 2a^{-2} + 7$

3. $1.3z - 2.8 > 4.3z^2$

4. $\dfrac{2}{3}x^3 + \dfrac{1}{7}x < \dfrac{5}{7}$

5. $4x^{-2} + 5x^{-1} \geq 2x + 1$

6. $x^2 > 0$

7. $\dfrac{1}{3}x^2 - \dfrac{5}{6}x > \dfrac{3}{4}$

8. $3x - 5 > x^2 + 8$

9. $2x^3 + 4 \leq x^2 + x$

10. $\dfrac{5}{9}x - \dfrac{2}{9}x^2 > \dfrac{2}{3}\sqrt{x} - \dfrac{5}{6}$

11. $3\sqrt{x} + 5x < x^2 - 5$

12. $5.4x^2 + 4.1x - 0.9 \geq 0$

Solve for integer solutions numerically.

13. $x^2 + 3x - 7 \geq x + 8$

14. $x^2 + 4x - 3 \leq 2x + 5$

15. $3x^2 + 9x + 7 < 11 - 2x$

16. $2x^2 + 4x - 8 < 2 + 3x$

17. $4x^2 - 13x - 7 > 18x + 1$

18. $3x^2 + 9x - 5 > 2x + 1$

Solve graphically.

19. $x^2 + 2x + 1 > 4$

20. $x^2 - 8x + 16 \geq 9$

21. $2x^2 + x - 3 < 7$

22. $x^2 - 2x - 8 \leq 2x - 3$

23. $24 + 5x - x^2 < -x + 8$

24. $14 + 5x - x^2 < -2x + 6$

25. $2x^2 - 5 \leq x^2 + 4$

26. $x^2 + 3x - 10 > 10 - 3x - x^2$

27. $\dfrac{1}{4}x^2 - 3 \geq x^2 + 2$

28. $5 - \dfrac{1}{5}x^2 \leq 1 - x^2$

29. $x^2 + 6x + 10 \geq -x^2 - 6x - 8$

30. $-x^2 - 4x \geq 4$

31. $x^2 - 4x + 8 \geq \dfrac{1}{4}x^2 - x + 5$

32. $x^2 - 8x + 15 \geq -x^2 + 8x - 17$

Solve exercises 33–54 algebraically.

33. $x^2 - 2x - 15 \geq 9$

34. $x^2 - 2x - 24 \leq 11$

35. $8 + 2x - x^2 > -7$

36. $10 + 3x - x^2 > -8$

37. $4x^2 - 12x + 9 \leq 4$

38. $4x^2 + 20x + 25 \geq 9$

39. $6 + 5x - x^2 < -8$

40. $20 + x - x^2 > -10$

41. $x^2 + 3 \geq -x + 15$

42. $\dfrac{1}{2}x^2 + 2 \leq 2x + 8$

43. $10 - 2x^2 < 4x + 4$

44. $7 - x^2 < -2x - 1$

45. $x^2 - x - 6 \leq 12 + 2x - 2x^2$

46. $2x^2 - 11x + 5 \geq -10 + 22x - 4x^2$

47. $x^2 - 3x + \dfrac{21}{4} \geq 2x^2 - 6x + \dfrac{15}{2}$

48. $-x^2 - 5x - \dfrac{17}{4} \leq -\dfrac{1}{5}x^2 - x + \dfrac{3}{4}$

49. $x^2 - x - \dfrac{7}{4} > -2x^2 + 2x - \dfrac{5}{2}$

50. $x^2 + 6x + 5 < -x^2 - 6x - 13$

51. $x^2 + 4x + 8 \geq 4$

52. $-x^2 + 8x - 14 \geq 2$

53. $2 + 4x - x^2 > 10 - 4x + x^2$

54. $-7 - 4x - \dfrac{1}{2}x^2 < x^2 + 8x + 17$

55. An egg is thrown upward from a stand that is 25 feet above the ground. The initial speed of the egg is 30 feet per second. For what length of time is the egg higher than the stand?

56. An arrow is projected upward with an initial velocity of 40 feet per second from a height of 220 feet above the ground. Determine the length of time the arrow is higher than the height at which it was released.

57. A rock is hurled upward with an initial velocity of 75 feet per second from a cliff that is 1200 feet high. For what length of time is the height of the rock no more than 800 feet?

58. A golf ball is hit upward from a ledge that is 600 feet high. The initial speed of the ball is 60 feet per second. Determine the length of time during which the height of the ball is no more than 200 feet.

59. The cost to operate a booth at a flea market for x days is given by the cost function $C(x) = 3x^2 + x + 35$. How many days can the booth be operated at a cost of no more than $100?

60. The cost to run a stand at a farmers' market for x weeks is given by the cost function $C(x) = 5x^2 + 25x + 50$. For how many weeks can the stand be run with a cost of less than $250?

61. A statistical study related the hourly compensation costs for production workers in Japan as $y = 0.01x^2 + 0.57x + 2.27$, where y is the cost in U.S. dollars and x is the number of years after 1975. Use this function to estimate the years in which the hourly costs exceeded $10.

62. Use the function in exercise 61 to determine the years in which the hourly cost was below $6.75.

63. The statistical study mentioned in exercise 61 also related the hourly compensation costs for U.S. production workers as $y = -0.007x^2 + 0.68x + 6.51$, where x is the number of years after 1975. Use this function to estimate the years in which the hourly costs exceeded $15.

64. Use the function in exercise 63 to determine the years in which the hourly costs were below $10.

11.7 Calculator Exercises

A. Using the TEST Function to Graphically Solve a Quadratic Inequality

You may use the TEST function of the calculator to determine whether a quadratic inequality is true or false. The calculator denotes a true statement as "1" and a false statement as "0." Using the TEST function, you can graph the number-line solution of an inequality. As an example, suppose you wish to solve the inequality $x^2 + 5x - 6 > 0$.

• Set the screen to integer coordinates for x and tenths coordinates for y.

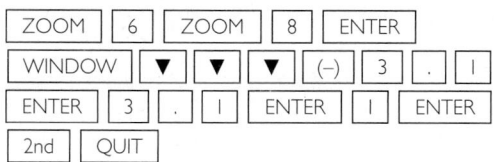

• Enter the inequality in Y1. The inequality symbols are found under the TEST menu.

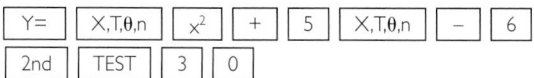

• Graph.

GRAPH

The calculator will graph a y-coordinate of 1 for all true statements and a 0 for all false statements. This will appear as a line segment on the screen. You can imagine the graph as a number-line representation of the solution intervals for the inequality. You will have to use your understanding of the endpoints to correctly interpret the graph. Use the procedure to check the solutions of the exercises in 11.7 Exercises.

B. Using the TEST Function to Solve a Quadratic Inequality Numerically

You have been shown how to use the TABLE function to solve a quadratic inequality numerically. Sometimes it is confusing to compare Y1 and Y2 to determine the correct order relation. There is a way to reduce this confusion. As an example, solve $2 - 3x - x^2 > x - 5$ numerically for integer solutions.

• Store the inequality Y2 > Y3 in Y1, the left side of the inequality in Y2, and the right side of the inequality in Y3. (Doing this will make it easier to scan for values of x that make the inequality true.)

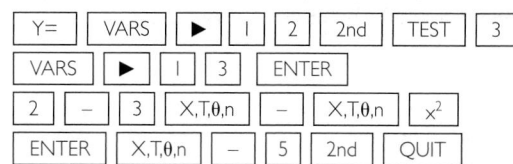

• Set the TBLSET to start at 0 and increment by 1 automatically.
• View the table, scanning up and down for various x-values. Wherever Y1 is equal to 1, the corresponding x-value is an integer solution of the inequality; wherever Y1 is equal to 0, the corresponding x-value is not a solution.

With this method, you no longer have to compare the Y2 and Y3 values to determine whether the x-value makes the inequality true or not. You see this immediately by reading the Y1 value. Use this procedure to check the solutions of exercises 13–18 in 11.7 Exercises.

11.7 Writing Exercise

When the graphical method is used to solve a quadratic inequality, the graph of the expression on the left, stored in Y1, may not intersect the graph of the expression on the right, stored in Y2. However, this does not necessarily mean that the inequality has no solutions. Sketch situations in which this is so, and discuss conditions under which the inequality may still have a solution.

CHAPTER 11 SUMMARY

After completing this chapter, you should be able to define the following key terms in your own words.

11.1
polynomial equation in one variable
polynomial equation
quadratic equation in one variable
quadratic equation
cubic equation in one variable
cubic equation
vertical-position equation

11.2
root
double root

11.3
rationalizing the denominator
right angle
legs
hypotenuse
Pythagorean theorem

11.4
completing the square

11.5
quadratic formula
discriminant

11.6
isosceles right triangle
congruent
equilateral triangle

11.7
quadratic inequality in one variable
quadratic inequality
union
critical points

Reflections

1. Define what is meant by a polynomial equation. What is the difference between a polynomial equation of degree two and one of degree three?
2. Explain how the zero factor property of real numbers can help you solve a polynomial equation algebraically.
3. Explain how the principle of square roots can help you solve a quadratic equation algebraically.
4. What is the quadratic formula and how is it used?
5. When can the Pythagorean theorem be used in modeling real-world situations?
6. What is the difference between a quadratic equation and a quadratic inequality? How do the solutions to each differ?

CHAPTER 11 SECTION-BY-SECTION REVIEW

11.1

Recall	Examples
A polynomial equation in one variable is written in the form $P(x) = 0$, where $P(x)$ is a polynomial.	$2x^3 + x^2 - 4x = 0$ Cubic equation $x^2 + x + 5 = 0$ Quadratic equation $-2x^4 + x^2 + 1 = 0$ Polynomial equation
Solving numerically • Complete a three-column table of values: one column for the independent variable, one for the left side of the equation, and one for the right side of the equation. • The solution is the value for the independent variable that results in equal values for the left and right sides of the equation.	Solve $x^2 - 2 = 2x + 1$ numerically. | x	$x^2 - 2$	$= 2x + 1$
---	---	---
-1	-1	-1
0	-2	1
1	-1	3
2	2	5
3	7	7	 The solutions are -1 and 3.

Recall	Example
Solving graphically • Graph the two functions defined by the left side of the equation and the right side of the equation. • The solution is the x-coordinate of the point of intersection of the two graphs.	Solve $x^2 - 2 = 2x + 1$ graphically. 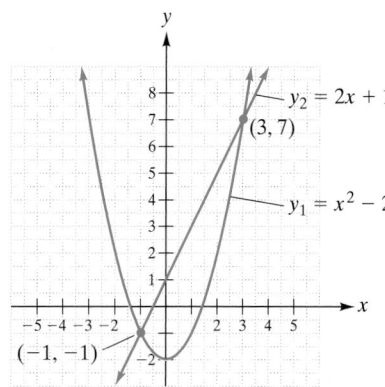 The solutions are -1 and 3.

Solve numerically for integer solutions.

1. $2x^2 + 5x - 16 = 7x + 8$ **2.** $\dfrac{1}{3}x^2 + 5x + 6 = x - 3$ **3.** $0.7x^2 - 2.5x + 4.6 = 1.7x + 1.1$

Solve exercises 4–9 graphically.

4. $x^2 - 4x + 4 = 9$ **5.** $x^2 - 6 = \dfrac{1}{4}x^2 + 6$ **6.** $x^2 + 3x - 5 = 7 - 2x - x^2$

7. $-0.3x^2 + 4x + 5 = 2.2x - 11.5$ **8.** $x^2 + 8x + 8 = 0$ **9.** $2x^3 - 18x = 3x^2 - 27$

10. An amateur scientist drops a lead weight from a platform 96 feet above the ground. How many seconds will it take for the weight to reach the ground?

11. Using data from the Motion Picture Association of America, the box office receipts (in billions of dollars) for the years 1995–1999 can be approximated by $y = 0.014x^2 + 0.42x + 5.49$, where x is the number of years since 1995. Box office receipts for 1995 were approximately $5.5 billion. Assuming that the equation still holds, determine the year in which the box office receipts will be double the 1995 receipts.

11.2

Recall	Example
Solving algebraically by factoring • Write the equation in standard form, $P(x) = 0$. • Factor $P(x)$. • Set each factor equal to 0 and solve. • Check the solution(s).	Solve. $\quad x^2 - 2 = 2x + 1$ $\quad x^2 - 2x - 3 = 0$ $\quad (x - 3)(x + 1) = 0$ $x - 3 = 0 \quad$ or $\quad x + 1 = 0$ $\quad x = 3 \qquad\qquad x = -1$ Check the solution.

Solve exercises 12–19 algebraically by factoring.

12. $x^2 - 5 = x + 1$ **13.** $6x^2 - x - 77 = 0$ **14.** $2x + 10 = x^2 - 5x + 2$

15. $x^2 - 7x - 60 = 0$ **16.** $3x^2 + 5x - 3 = 1 + 5x + 3x^2$ **17.** $6x^2 - 8x = 9x - 12$

18. $\dfrac{1}{4}x^2 + \dfrac{3}{2}x - \dfrac{19}{8} = \dfrac{3}{8}x + 2$ **19.** $9x^2 - 49 = 0$

20. Emma purchased 300 square feet of remnant carpeting to carpet a playroom in the bonus room of her home. She wants to lay the carpet in an area that has a length that is 5 feet more than the width, and she wants to use all of the carpeting. Write a polynomial equation and solve it by factoring to determine the dimensions of the carpeted area.

21. The chief financial officer of a fund-raising organization listed the donations received for the last three years:

Year	1998	1999	2000
Donations in millions of dollars	6.44	6.50	6.44

a. The officer used a quadratic function to represent the donations received per year. Let x = the number of years after 1998 and y = the donations in millions of dollars. Determine a quadratic function for the profit.
b. Use the quadratic function and the zero factor property to determine the year when the donations will drop to $5 million.

11.3

Recall	Examples
Product Rule for Square Roots For any real numbers $\sqrt{a}$ and $\sqrt{b}$, $\sqrt{a} \cdot \sqrt{b} = \sqrt{ab}$.	Simplify, using the product rule. $\sqrt{28} = \sqrt{4 \cdot 7} = \sqrt{4} \cdot \sqrt{7} = 2\sqrt{7}$ $\dfrac{-2 + \sqrt{24}}{4} = \dfrac{-2 + \sqrt{4 \cdot 6}}{4}$ $= \dfrac{-2 + 2\sqrt{6}}{4}$ $= \dfrac{\overset{1}{2}(-1 + \sqrt{6})}{\underset{2}{4}}$ $= \dfrac{-1 + \sqrt{6}}{2}$
Quotient Rule for Square Roots For any real numbers $\sqrt{a}$ and $\sqrt{b}$ where $b \neq 0$, $\frac{\sqrt{a}}{\sqrt{b}} = \sqrt{\frac{a}{b}}$.	Simplify, using the quotient rule. $\sqrt{\dfrac{49}{2}} = \dfrac{\sqrt{49}}{\sqrt{2}} = \dfrac{7}{\sqrt{2}}$ Rationalize the denominator. $= \dfrac{7 \cdot \sqrt{2}}{\sqrt{2} \cdot \sqrt{2}} = \dfrac{7\sqrt{2}}{\sqrt{4}} = \dfrac{7\sqrt{2}}{2}$
Solving algebraically, using the principle of square roots, when $ax^2 + c = 0$ • Solve for x^2. • Apply the principle of square roots. • Solve the resulting equations. • Check the solution(s).	Solve. $2x^2 - 9 = 7$ $2x^2 - 9 = 7$ $2x^2 = 16$ $x^2 = 8$ $x = \sqrt{8}$ or $x = -\sqrt{8}$ $x = 2\sqrt{2}$ $\quad x = -2\sqrt{2}$ Check the solution.

Simplify without using a calculator.

22. $-\sqrt{49}$ **23.** $\sqrt{200}$ **24.** $\sqrt{32}$ **25.** $\sqrt{8}$

26. $-\sqrt{72}$ **27.** $\dfrac{-6 - \sqrt{50}}{5}$ **28.** $\dfrac{14 + \sqrt{98}}{14}$ **29.** $\dfrac{3 - \sqrt{27}}{6}$

30. $\dfrac{-4 + \sqrt{20}}{-2}$ **31.** $\sqrt{\dfrac{36}{5}}$ **32.** $\sqrt{\dfrac{40}{32}}$

Solve exercises 33–39 by using the principle of square roots.

33. $4x^2 - 100 = 0$ **34.** $a^2 + 5 = 9$ **35.** $x^2 + 3 = 15$ **36.** $15 + b^2 = 7$
37. $(x - 4)^2 = 9$ **38.** $x^2 + 18x + 81 = 16$ **39.** $(z - 2)^2 + 5 = 7$

40. A boat ramp has a length of 35 feet. If the vertical distance of the ramp is 15 feet, what is the horizontal distance?

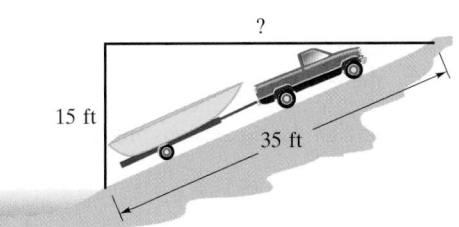

?

15 ft

35 ft

41. Stephanie received a length-of-service bonus of $2,000 from her employer and invested it in a money market fund for two years. At the end of that period, her investment had grown to $2,332.80. Use the compound-interest formula to write a quadratic equation, and then solve the equation, using the principle of square roots, to determine the annual compound-interest rate that resulted in the growth of Stephanie's investment.

42. Modern sky divers typically free-fall from 12,000 feet above the ground until 2500 feet, when they open their parachutes. They can maneuver in free fall by controlling the position of their bodies. Use the vertical-position formula to write an equation, and then solve the equation, using the principle of square roots, to determine the maximum length of time the sky divers would be free-falling.

11.4

Recall	Example
Solving algebraically by completing the square • Isolate the variable terms. • Divide both sides by the leading coefficient. • Add the square of one-half of the coefficient of the x term to both sides. • Rewrite the trinomial as a binomial squared. • Use the principle of square roots to solve the resulting equations. • Check the solution(s).	Solve. $2x^2 + 5x - 1 = 0$ $2x^2 + 5x - 1 = 0$ $2x^2 + 5x = 1$ $x^2 + \dfrac{5}{2}x = \dfrac{1}{2}$ $x^2 + \dfrac{5}{2}x + \dfrac{25}{16} = \dfrac{1}{2} + \dfrac{25}{16}$ $\left(x + \dfrac{5}{4}\right)^2 = \dfrac{33}{16}$ $x + \dfrac{5}{4} = \sqrt{\dfrac{33}{16}}$ or $x + \dfrac{5}{4} = -\sqrt{\dfrac{33}{16}}$ $x + \dfrac{5}{4} = \dfrac{\sqrt{33}}{4}$ $x + \dfrac{5}{4} = -\dfrac{\sqrt{33}}{4}$ $x = -\dfrac{5}{4} + \dfrac{\sqrt{33}}{4}$ $x = -\dfrac{5}{4} - \dfrac{\sqrt{33}}{4}$ $x = \dfrac{-5 + \sqrt{33}}{4}$ $x = \dfrac{-5 - \sqrt{33}}{4}$ Check the solution.

Solve exercises 43–45 by completing the square.

43. $p^2 - 4p - 96 = 0$ **44.** $2x^2 - 5x - 12 = 0$ **45.** $x^2 + \dfrac{1}{2}x = 1$

46. According to the annual reports of Casio Computer Company, the firm's cost of goods sold, in billions of dollars, for the last three years is as listed in the following table:

Year	Cost of goods in billions of dollars
1999	2.6
2000	2.7
2001	2.5

a. Statisticians predicted that the trend would continue to decrease. Write a quadratic equation for the cost of goods in y billion dollars in terms of x years after 1999.

b. Using the equation in part a, determine the number of years after 1999 when the cost of goods will be $2 billion. Interpret the solutions.

47. A retailer sells figurines for $20 each. In order to promote sales, the price of each figurine is reduced by $1 times the number of figurines purchased. The retailer de-

termines that each sale must be for $100 in order for her to break even on the promotion. Determine the number of figurines that must be sold to each purchaser in order for the retailer to break even.

48. The length of a rectangle is 8 inches more than twice its width. The area of the rectangle is 90 square inches. Write a quadratic equation and solve it by completing the square, to find the dimensions of the rectangle.

11.5

Recall	Example
Solving algebraically by using the quadratic formula • Write the equation in standard form, $ax^2 + bx + c = 0$. • Use the quadratic formula, $x = \dfrac{-b \pm \sqrt{b^2 - 4ac}}{2a}$ • Check the solution(s).	Solve. $2x^2 + 5x - 1 = 0$ $$a = 2, b = 5, c = -1$$ $$x = \frac{-b \pm \sqrt{b^2 - 4ac}}{2a}$$ $$x = \frac{-(5) \pm \sqrt{(5)^2 - 4(2)(-1)}}{2(2)}$$ $$x = \frac{-5 \pm \sqrt{25 + 8}}{4}$$ $$x = \frac{-5 \pm \sqrt{33}}{4}$$ Check the solution.
Determining the characteristics of the solution of a quadratic equation by using the discriminant First, determine the value of the discriminant, $b^2 - 4ac$, for the quadratic equation in standard form. • If the discriminant is 0, the equation has one rational solution.	Determine the characteristics of the solution of the given quadratic equation by using the discriminant. $x^2 + 2x + 1 = 0$ $b^2 - 4ac = (2)^2 - 4(1)(1) = 0$ The equation has one rational solution.
• If the discriminant is a perfect square not equal to 0, the equation has two rational solutions.	$x^2 - x - 2 = 0$ $b^2 - 4ac = (-1)^2 - 4(1)(-2) = 9$ The equation has two rational solutions.
• If the discriminant is a positive number that is not a perfect square, the equation has two irrational solutions.	$x^2 + 3x + 1 = 0$ $b^2 - 4ac = (3)^2 - 4(1)(1) = 5$ The equation has two irrational solutions.
• If the discriminant is a negative number, the equation has no real-number solution.	$x^2 - 2x + 3 = 0$ $b^2 - 4ac = (-2)^2 - 4(1)(3) = -8$ The equation has no real-number solution.

Use the quadratic formula to find exact solutions.

49. $x^2 - 2x - 63 = 0$ **50.** $x^2 = 3x + 3$ **51.** $25x^2 + 1 = 10x$ **52.** $z^2 + \dfrac{7}{20}z - \dfrac{3}{10} = 0$

53. $x^2 + 2.1x - 10.8 = 0$ **54.** $y^2 - 5y + 12 = 0$ **55.** $x^2 - 10x + 6 = 0$ **56.** $3a^2 - 4a - 12 = 0$

In exercises 57–60, evaluate the discriminant to determine the characteristics of the solution(s) of each quadratic equation. Do not solve the equation.

57. $x^2 + 2x + 10 = 0$ **58.** $x^2 + 20x + 55 = 2x - 26$

59. $x^2 = 10x + 75$ **60.** $x = x^2 - 11$

61. Mary Claus sells Christmas Rum Cakes out of her home. Ordinarily, she charges a customer $6.50 for each cake, but to encourage multiple purchases, Mary reduces the price per cake by $0.50 for each cake ordered. She spends $2.50 to make a cake and $3.95 to package and ship an order to the customer.
 a. Write a revenue function for an order of x cakes.
 b. Write a cost function for an order of x cakes.
 c. Write a profit function for an order of x cakes.
 d. Determine the number of cakes that must be ordered for Mary to break even on an order. Interpret the break-even point(s).

62. Using data from the Energy Information Administration, one can estimate world nuclear power generation by the equation $y = -4.1x^2 + 163x + 693$, where y is the world's total generation of nuclear energy in billions of kilowatt-hours and x is the number of years after 1980. Use the equation and the quadratic formula to determine when the world's total generation first reached 2,000 billion kilowatt-hours. Translate this solution to the year in which that happened.

11.6

Recall	Examples
To solve formulas for a specified variable, use any of the algebraic methods.	Solve. $\quad s = 4.9t^2 + 10 \quad$ for t $s = 4.9t^2 + 10$ $s - 10 = 4.9t^2$ $\dfrac{s - 10}{4.9} = t^2$ $t = \pm\sqrt{\dfrac{s-10}{4.9}}$
The length c of the hypotenuse of an isosceles right triangle is the length a of a congruent leg times $\sqrt{2}$. $c = a\sqrt{2}$	Determine the length of the hypotenuse of an isosceles triangle, given that each of the congruent legs measures 4 centimeters. $c = a\sqrt{2}$ $c = 4\sqrt{2}$ The hypotenuse is $4\sqrt{2}$ centimeters.
The length c of the hypotenuse of a 30°–60°–90° triangle is twice the length a of the smaller leg. $c = 2a$	Determine the length of the hypotenuse of a 30°–60°–90° triangle, given that the smaller leg measures 3 centimeters. $c = 2a$ $c = 2(3)$ $c = 6$ The hypotenuse is 6 centimeters.
The length b of the longer leg of a 30°–60°–90° triangle is the length a of the shorter leg times $\sqrt{3}$. $b = a\sqrt{3}$	Determine the length of the longer leg of a 30°–60°–90° triangle, given that the shorter leg measures 3 centimeters. $b = a\sqrt{3}$ $b = 3\sqrt{3}$ The longer leg is $3\sqrt{3}$ centimeters.

In exercises 63–66, solve for the variable specified.

63. $s = -16t^2 + 50 \quad$ for t **64.** $A = \dfrac{1}{4}\pi d^2 \quad$ for d

65. $a = bx^2 + c \quad$ for x **66.** $s = -16t^2 + v_0t + s_0 \quad$ for t

67. The length D of a diagonal inside a rectangular box is related to the dimensions of the box by the Pythagorean theorem, $D^2 = L^2 + W^2 + H^2$, where L, W, and H are, respectively, the length, width, and height of the box. Solve this equation for H. Use the equation to find the height of a box that is 8 inches long, is 4 inches wide, and has a diagonal that is 12 inches long.

68. A rope attached to the top of a pole makes a 45-degree angle with the ground. If the pole is 7 feet high, how long is the rope?

69. Assume that a right triangle is a 30°–60°–90° triangle. Determine the lengths of the sides of the triangle if the hypotenuse measures 20 inches.

11.7

Recall	Examples
Solving quadratic inequalities numerically • Complete a three-column table of values: one column for the independent variable, one for the left side of the inequality, and one for the right side of the inequality. • The solution set consists of all values of the independent variable that result in true inequalities.	Solve $x^2 - 2 \leq 2x + 1$ numerically. <table><tr><td>x</td><td>$x^2 - 2$</td><td>$\leq$</td><td>$2x + 1$</td></tr><tr><td>-1</td><td>-1</td><td></td><td>-1</td></tr><tr><td>0</td><td>-2</td><td></td><td>1</td></tr><tr><td>1</td><td>-1</td><td></td><td>3</td></tr><tr><td>2</td><td>2</td><td></td><td>5</td></tr><tr><td>3</td><td>7</td><td></td><td>7</td></tr></table> The solutions are all integers between -1 and 3, inclusive.
Solving quadratic inequalities graphically • Graph the two functions defined by the left side of the inequality and the right side of the inequality. • Determine the portion of the y_1 graph that lies above the y_2 graph for "greater than" inequalities or the portion of the y_1 graph that lies below the y_2 graph for "less than" inequalities. • Determine the x-coordinates for the appropriate portion of the graph. • The solution set consists of the x-coordinate of the intersection if the order relation includes equality.	Solve $x^2 - 2 \leq 2x + 1$ graphically. 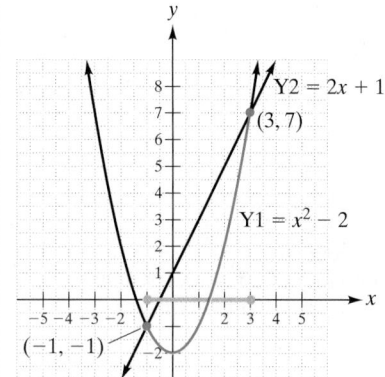 The solution set is the set of all real numbers between -1 and 3, inclusive, or $[-1, 3]$.
Solving algebraically • Determine the critical numbers by solving the quadratic equation formed by replacing the inequality sign by an equals sign. • Determine the possible intervals for the solution by using the critical numbers as endpoints. • Test one value in each interval for a true statement. If the inequality is true for the value tested, then the inequality is true for all values in the interval.	Solve $x^2 - 2 < 2x + 1$ algebraically. Solve the associated equation: $x^2 - 2 = 2x + 1$ $x^2 - 2x - 3 = 0$ $(x - 3)(x + 1) = 0$ $x - 3 = 0 \quad$ or $\quad x + 1 = 0$ $x = 3 \qquad\qquad x = -1$ The critical points are 3 and -1. The inequality does not include equality so we do not include 3 and -1 as solutions. 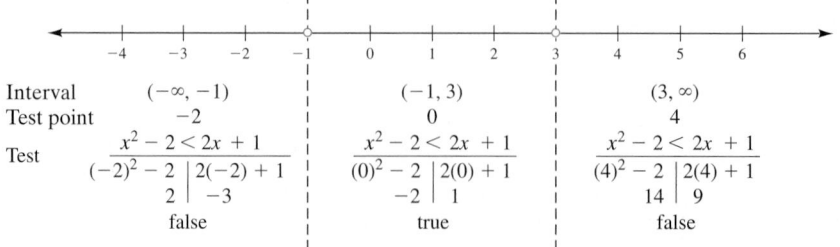 The solution set is $(-1, 3)$. Note that this example differs from the two above in its order relation. Therefore, the solution set is different.

Solve numerically for integer solutions.

70. $x^2 + 6x - 9 \leq 4x + 15$

71. $2x^2 + 3x - 8 > 6 + 4x - x^2$

Solve graphically.

72. $x^2 + 2x - 3 \geq 5$

73. $5 - x^2 \geq -4$

74. $x^2 - 2x - 1 < x + 3$

75. $x^2 + 3 < 1 - x^2$

76. $x^2 - 6x + 11 > -\dfrac{1}{2}x^2 + 3x - \dfrac{5}{2}$

77. $x^2 - 8x + 18 \leq \dfrac{1}{4}x^2 - 2x + 6$

78. $10 - 6x + x^2 < -4 + 6x - x^2$

Solve exercises 79–85 algebraically.

79. $x^2 - 4x + 4 < \dfrac{1}{4}x^2 - x + 3$

80. $x^2 + x - 6 \leq 12 - 2x - 2x^2$

81. $3 - z^2 > -2z$

82. $x^2 + 6x + 11 > -x^2 - 6x - 7$

83. $1 - 2x^2 \leq 1 - x^2$

84. $x^2 + 4x - 2 < -\dfrac{1}{2}x^2 - 2x - 8$

85. $p^2 - 6p - 1 < 0$

86. The cost to rent a booth at a carnival for x days is given by the cost function $C(x) = 3x^2 + 25x + 50$. How many days can the booth be rented for a cost of no more than $200?

87. A dart is shot upward from a height 30 feet above the ground. The initial speed of the dart is 25 feet per second. For what length of time is the dart above the ground?

CHAPTER 11   CHAPTER REVIEW

Simplify without using a calculator.

1. $\sqrt{50}$

2. $-\sqrt{48}$

3. $\dfrac{-3 + \sqrt{18}}{3}$

4. $\dfrac{4 - \sqrt{80}}{-2}$

5. $\sqrt{\dfrac{14}{25}}$

6. $\sqrt{\dfrac{75}{18}}$

Solve graphically.

7. $x^2 - 4 = 2x - 1$

8. $3 - x^2 = -3 - \dfrac{1}{3}x^2$

9. $x^2 + 6x + 9 = 4$

10. $15x^2 + 13x - 72 = 0$

11. $2x + 10 = x^2 + 4x + 7$

12. $x^2 + 9x + 7 = 0$

13. $\dfrac{1}{6}x^2 - 2x + 7 = 9 - \dfrac{1}{4}x^2$

14. $x^2 - 6x - 40 = 0$

Solve numerically for integer solutions.

15. $2x^2 - x - 28 = x^2 + 3x + 17$

16. $3x^2 - 5x - 10 = 7x + 5$

Solve algebraically.

17. $12x^2 + 9x = 8x + 6$

18. $x^2 + 3x - 88 = 0$

19. $5x^2 + 14x - 6 = 8 - 19x$

20. $49x^2 - 16 = 0$

21. $5x^2 - 180 = 0$

22. $11 + b^2 = 2$

23. $(x - 9)^2 = 16$

24. $(p - 7)^2 + 6 = 9$

25. $z^2 + 12z = -33$

26. $m^2 + 7m + 12 = 9$

27. $b^2 + 2.4b - 4.32 = 0$

28. $7q^2 - 7q - 78 = (8q^2 - 3q - 50) - (q^2 + 4q + 28)$

29. $x^2 + 20x + 84 = 0$

30. $2y^2 + 3y - 5 = 0$

31. $4x^2 + 16x - 2 = 0$

Use the quadratic formula to find the exact solution.

32. $4 + 8x + x^2 = 0$

33. $x^2 = 7x + 2$

34. $36q^2 + 25 = 60q$

35. $b^2 - 7b + 16 = 0$

36. $z^2 + 6z - 55 = 0$

37. $4m^2 + 7m = 36$

Solve for the variable specified.

38. $a = \dfrac{x(x + 1)}{2}$ for x

39. $x^2 + y^2 + z^2 = r^2$ for y

40. $A = 4\pi r^2$ for r

Solve numerically for integer solutions.

41. $x^2 + 2x - 7 \leq 4x + 8$

42. $x^2 + 9x - 11 \leq 4x + 3$

Solve algebraically.

43. $z^2 + 5z - 6 < 12 - 10z - 2z^2$

44. $m^2 - 15 > 2m$

45. $x^2 + 2x - 1 > -x^2 - 2x + 1$

46. $p^2 - 8p - 2 > 0$

47. $2x^2 - 5x - 11 \le 19 + 8x - x^2$

48. $3q^2 + 2q - 17 \ge 5q + 19$

Solve exercises 49–56 graphically.

49. $x^2 - 4x - 1 \ge x + 5$

50. $2 - x^2 > 5 + x^2$

51. $x^2 - 2x - 3 < 12$

52. $7 - x^2 > -9$

53. $x^2 + 4x - 2 > -5 - 4x - x^2$

54. $-x^2 - 10x - 27 < x^2 + 10x + 23$

55. $-x^2 + 6x - 7 > -\dfrac{1}{3}x^2 + 2x - 4$

56. $-x^2 + 6x - 6 \ge -\dfrac{1}{3}x^2 + 2x$

57. A student drops a notebook from 160 feet above ground level. How many seconds will it take for the notebook to reach the ground?

58. The cost of producing x items is given by the mathematical model $C(x) = 5x^2 + 75x + 875$. How many items can be produced for approximately \$3500?

59. Find the time it will take for a rock to reach the ground if it is shot upward from a slingshot with an initial velocity of $v_0 = 64$ feet per second from a height of $s_0 = 32$ feet.

60. How long will it take a sky diver to free-fall from 11,000 feet to 4000 feet?

61. The hypotenuse of a right triangle measures 42.5 meters, and one of its legs measures 34 meters. What is the measure of the other leg?

62. The area of a rectangle is 144 square cm. Its length measures 6 cm less than three times its width. Find the dimensions of the rectangle.

In exercises 63 and 64, use the compound-interest formula $A = P(1 + r)^t$ to set up a quadratic equation for each situation, and then solve the equation.

63. What must the annual interest rate be for \$1200 to compound to a total amount of \$1323 after two years?

64. At what annual interest rate must \$1200 be invested in order for it to compound to a total amount of \$1500 after two years?

65. Angela sells Easter baskets for \$10.50 per arrangement. To encourage multiple purchases, she reduces the price per arrangement by \$0.50 for each arrangement ordered. She spends \$7.50 on each arrangement and \$3.00 to deliver an order of the baskets.
  a. Write a revenue function for an order of x baskets.
  b. Write a cost function for an order of x baskets.
  c. Write a profit function for an order of x baskets.
  d. Determine the number of baskets that Angela must sell to break even on an order. Interpret the break-even point(s).

66. The two legs of a right triangle measure 20 inches and 48 inches. What is the length of the hypotenuse?

67. The U.S. Department of Health and Human Services reported that cigarette consumption among persons 18 and older dropped from 2.817 thousand per capita in 1990 to 2.146 thousand per capita in 1999. With data inserted for intervening years, the drop follows the quadratic function $y = -0.01x^2 + 0.01x + 2.8$, where x is the number of years after 1990 and y is the per capita consumption in thousands of cigarettes. Use this function to determine the year in which per capita consumption would drop to 1,000 cigarettes ($y = 1$).

68. A statistical study of the number of purchases of an office software product listed the following data:

Year	1998	1999	2000
Number sold in millions of units	3.55	4.36	4.81

  a. The statistician believes that sales will follow a quadratic function and will begin to decrease after a period of time. Determine a quadratic function for the number of units sold.
  b. Use the quadratic function and the zero property of equations to determine the year when sales will drop to 4 million units.

69. What are the lengths of the legs of an isosceles right triangle whose hypotenuse measures 14.21 centimeters?

70. A rope attached to the top of a pole makes a 45-degree angle with the ground. If the pole is 11 feet high, how long is the rope?

71. Assume that a right triangle is a 30°–60°–90° triangle. Determine the lengths of the sides of the triangle if the hypotenuse measures 32 inches.

72. The cost to rent a booth at a carnival for x days is given by the cost function $C(x) = 5x^2 + 20x + 100$. How many days can the booth be rented for a cost of no more than \$500?

CHAPTER 11 TEST

TEST-TAKING TIPS

Many students have failed a test at some point in their academic life. It is important that you respond to such a failure properly. Don't blame your teacher, your background, your past performance, your personal deficiencies, or other difficulties in your personal life. These are just excuses for the lack of success. Instead, make up your mind that you will do whatever it takes to master the skills being taught. Learn from your mistakes. If you can, meet with your instructor to review the test, emphasizing the problems you missed. Try to see how you should have approached the problem. If your mistakes are careless ones, think of steps you can take to avoid them in the future. For example, if you make mistakes in transcribing a problem to your paper, make a vow always to check what you transcribe before you begin working the problem. If you make arithmetic mistakes, stop doing the arithmetic in your head. Instead, do the work on paper, and use a calculator whenever you are permitted to do so. Remember that doing your work neatly and in an orderly manner will help you understand the problem. The secret to being successful at mathematics is to be patient and to persevere.

Solve.

1. $x^2 - 4x - 9 = 2x + 7$

2. $2x^3 + 9x^2 - 23x - 66 = 0$

3. $a^2 + 2a - 5 = 2(a^2 + a - 2)$

4. $x^2 - 6x + 13 = 0$

5. $2x^2 - 7x + 3 = 2x - 1$

6. $x^2 - 5x + 4 = 7 - 3x - x^2$

7. $x^2 - x - 6 \le 6$

8. $(x + 3)(x - 2) < 3(x + 3)$

9. $2x^2 + 7x - 9 \ge 4x + 11$

In exercises 10–11, evaluate the discriminant, and then find the exact roots if possible.

10. $x^2 - 9x + 9 = 7x - 5$

11. $x^2 - 3x + 12 = 0$

12. Solve $s = at^2 + c$ for t.

13. Solve. $x^2 - 2 = x + 4$
   a. numerically
   b. graphically
   c. algebraically

14. The hypotenuse of a $30°–60°–90°$ triangle measures 28 inches. Find the exact length of the two legs of the triangle. Then approximate the lengths to the nearest tenth of an inch if they are irrational numbers.

15. Keanu free-falls from 11,500 feet above the ground until he reaches 2600 feet, when he opens his parachute. Use the vertical-position equation, $s = -16t^2 + v_0 t + s_0$, to determine the length of time Keanu was free-falling.

16. Use the compound-interest formula, $A = P(1 + r)^t$, to solve the following interest problem: What must the annual interest rate be for $1400 to compound to a total amount of $1573 in two years?

17. Nicole sells Valentine flower arrangements for $12.50 per arrangement. To encourage multiple purchases, she reduces the price per arrangement by $0.50 for each arrangement ordered. She spends $9.50 on each arrangement and $3.95 to package and deliver an order of the arrangements.
   a. Write a revenue function for an order of x arrangements.
   b. Write a cost function for an order of x arrangements.
   c. Write a profit function for an order of x arrangements.
   d. Determine the number of arrangements that must be ordered to break even on an order. Interpret the break-even point(s).

18. The total net income for Texas Instruments, Inc., is listed in the following table:

Year	Net income in billions of dollars
1998	0.4
1999	1.4
2000	3.1

   a. Statisticians predicted that the increasing trend would continue. Write a quadratic equation for the net income in y billion dollars in terms of x years after 1998.
   b. Using the equation in part a, determine the number of years after 1998 when the net income will be $5 billion.

19. A gangplank to board a ship is 40 feet long. It touches the ship at a height of 10 feet above the dock. What is the horizontal distance between the ship and the end of the gangplank?

20. Explain how to use the value of the discriminant of a quadratic equation to describe the characteristics of the solution(s) of the equation.

Chapter 11

Project

In this chapter, we have applied the Pythagorean theorem to many different applications that we may encounter in our daily lives. In this project, we will explore other extensions of this very important theorem.

Part I

This activity is designed to illustrate why the Pythagorean theorem is true. It is based on an actual proof attributed to Pythagoras.
The following figure illustrates the fact that $a^2 + b^2 = c^2$.

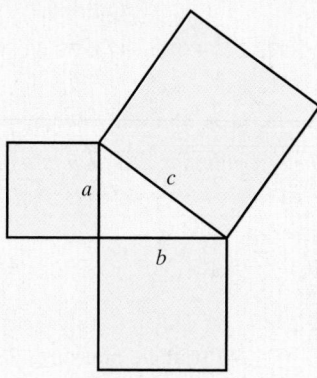

Figure 11.6

1. Determine the area of each of the squares.

2. According to the Pythagorean theorem, the sum of the areas of the smaller two squares is equal to the area of the larger square. Write an equation for this statement, using the areas you found in exercise 1. Your equation should be the Pythagorean theorem.

3. Let $a = 6$, $b = 8$, and $c = 10$. Calculate the area of the squares, and verify that the Pythagorean theorem is correct for these values.

4. Draw Figure 11.6 on graph paper, using the values $a = 6$, $b = 8$, and $c = 10$. Cut up and reassemble the two small squares to form the larger square.

Part II

An iterative example of the use of the Pythagorean theorem is demonstrated in the following figure:

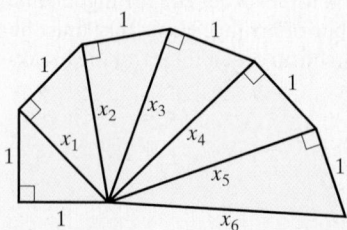

Determine the exact length of x_1, then x_2, then x_3, and so on, until you get a value for x_6. Do not approximate the lengths.

Part III

The equation $x^2 + y^2 = z^2$, where x, y, and z are integers, is called a Diophantine equation. A Greek mathematician, Diophantus of Alexandria, proved that any set of three integers that satisfy this equation has the form

$$x = a^2 - b^2 \quad y = 2ab \quad \text{and} \quad z = a^2 + b^2$$

where a and b are integers.

This Diophantine equation is a special case of the Pythagorean theorem that was used in the chapter to solve application problems. We will examine such equations further.

1. Complete the following table of values for the previous Diophantine equation:

Case	a	b	$x = a^2 - b^2$	$y = 2ab$	$z = a^2 + b^2$	Diophantine triple
1.	2	1	3	4	5	3, 4, 5
2.	3	1				
3.	3	2				
4.	4	1				
5.	4	2				
6.	4	3				
7.	5	1				
8.	5	2				
9.	5	3				
10.	5	4				

2. Check each row of the table to be sure that the Diophantine triple does in fact satisfy the Pythagorean theorem. Summarize what your check revealed.

3. Which of the cases have Diophantine triples that are just multiples of a smaller Diophantine triple. What does this observation indicate to you about finding Diophantine triples that satisfy the Pythagorean theorem?

4. How many different Diophantine triples do you think can be found that satisfy the Pythagorean theorem?

5. Substitute the Diophantine expressions for x, y, and z into the Pythagorean theorem to see if they satisfy the theorem.

6. Program your calculator to create Diophantine triples.

```
PROGRAM:DIOPHANT
Disp "A?, A>1"
Input A
seq(A,X,1,A-1,1) →L1
seq(X,X,1,A-1,1) →L2
L1²-L2²→L3
2*L1*L2→L4
L1²+L2²→L5
Disp "PRESS STAT 1"
Disp "TO VIEW L3,L4,L5"
```

7. Use the program from step 6 to create Diophantine triples when $a = 10$ and b assumes integer values from 1 to 9.

a	b	x, L3	y, L4	z, L5	Diophantine triple
10	1				
10	2				
10	3				
10	4				
10	5				
10	6				
10	7				
10	8				
10	9				

8. Check whether all the triples in step 7 satisfy the Pythagorean theorem. Report the results of your check.

Part IV

To finish our examination of the Pythagorean theorem and its related Diophantine triples, research the mathematician Pythagoras of Samos or Diophantus of Alexandria in the library or by using the Internet. Write a one-page summary of interesting facts that you discover about either of these famous mathematicians.

CHAPTERS 1–11 CUMULATIVE REVIEW

Simplify exercises 1–12. Write all answers with positive exponents.

1. $(2a^2 + 3ab - 4b^2) + (a^2 + ab - b^2)$
2. $(1.5x^2 + 2.3xy - y^2) - (0.5x^2 + 1.3xy - y^2)$

3. $(3x + 1)(2x - 6)$
4. $(5x + 1)^2$
5. $(2x + 3)(2x - 3)$
6. $(2.1xy^2z)(3xy^3z^2)$
7. $6x^2y^{-1}z^0$

8. $\dfrac{2x^{-1}y}{4xy^2}$
9. $\left[\dfrac{(-2a)^3}{5b}\right]^{-2}$
10. $(-2p^2q^{-3}r)(4p^3q^5r)$
11. $\dfrac{2x^2 + 4x - 6}{2x}$
12. $\dfrac{x^2 + 4x - 5}{x + 5}$

13. Given $f(x) = -x^2 + 2x - 1$, evaluate $f(-3)$.
14. Write 5.6×10^{-3} in standard notation.

15. Write -3.4 E6 in standard notation.
16. Solve $A = \frac{1}{2}bh$ for b.
17. Solve $s = -16t^2 + 5$ for t.

Factor completely.

18. $x^2 + 2x - 15$
19. $10s^2 + 7s - 12$
20. $9x^2 - 16$
21. $4x^2 + 12x + 9$
22. $a^3 + 27$
23. $2a^4 - 32$

24. *Solve.*

$2x - 4y = 1$
$x + \ y = 6$

25. *Graph and determine one solution.*

$y < 2x + 5$
$y \geq -2x - 1$

Solve.

26. $x^2 - 4x + 7 = 2x - 1$
27. $x^2 - 5x - 12 = 0$
28. $3x^2 + 6x = 9$
29. $2x^2 + 4x - 5 = 2x^2 + 4x + 3$

30. $2x^2 - 5x + 10 = x^2 - 2$
31. $(x - 2)(2x + 1) = 4$
32. $2x + 4 = 5x - 1$
33. $2(x + 3) + 4 = (x + 4) + (x + 6)$

34. $5x - 2(2.5x - 1) = 4$
35. $|2x - 1| = 4$
36. $3|x| - 2 = 7$

Solve exercises 37–39. Write the solution set in interval notation.

37. $2x - 5 > 3x + 4$
38. $x^2 - 4x - 5 \geq x + 1$
39. $3t^2 - 4t < 2t^2 + 1$

40. Consider the relation $y = -3x + 4$.
  a. Graph the relation.
  b. Is the relation a function? Justify your answer.
  c. Determine the x-values for which the relation is increasing and the x-values for which it is decreasing.
  d. What is the domain of the relation? The range?
  e. Determine the x-intercept and the y-intercept algebraically.

41. Consider the relation $y = x^2 + 4x - 1$.
  a. Determine the x-intercept and the y-intercept algebraically.
  b. Determine the vertex algebraically.
  c. Graph enough points to determine the curve. Label the vertex and the axis of symmetry.
  d. What is the domain of the relation? The range?

42. Write a linear equation that passes through $(3, -2)$ and $(4, 1)$.

43. Write a linear equation that passes through $(-1, 2)$ and is perpendicular to $2x + 3y = 1$.

44. Raynoc plans to sell a mixture of peanuts and cashews at his shop. He usually sells peanuts for $5.00 per pound and cashews for $12.00 per pound. How many pounds of each should he mix to sell 7 pounds of the mixture for $10.00?

45. The crystal ball at Times Square dropped 77 feet in 60 seconds at midnight of the year 2002, through a controlled drop. If the ball were to free-fall 77 feet, how long would it take to drop that distance?

46. An investment opportunity pays 12% simple interest per year. How much should Georgette invest in order to have a total amount of $16,800 at the end of the year to purchase a classic Corvette?

47. Annual reports for a manufacturer of mountain bikes listed the profit for the years 1999–2001 as follows:

Year	1999	2000	2001
Profit in million dollars	9.5	9.8	8.9

  a. An accountant decided that a quadratic function could be used to represent the dwindling profits. Let $x = $ the number of years after 1999 and $y = $ the profit in millions of dollars. Determine a quadratic function for the profit.
  b. Use the quadratic function found in part a and the zero factor property to determine the year when the profit will drop to $3.5 million.

48. Sam sells sparklers for the Fourth of July celebration. Ordinarily, the packages sell for $3.50 each. To encourage multiple purchases, Sam reduces the price of each package by $0.25 per package ordered. He spends $2.00 per package on supplies and $1.50 for shipping and handling each order.
  a. Write a revenue function for an order of x packages.
  b. Write a cost function for an order of x packages.
  c. Write a profit function for an order of x packages.
  d. Determine the number of packages that must be ordered to break even on an order. Interpret the break-even point(s).

49. Determine the length of the legs of an isosceles triangle if the hypotenuse measures 8 meters.

50. Janet scored 96, 82, and 94 on the first three tests in her history class. In order to make a B in the course, she must have a test average between 85 and 90, inclusive. Determine the scores she must make on her last test in order to earn a B.

Chapter 12

Rational Expressions, Functions, and Equations

In this chapter, we define rational expressions and functions and perform operations on them. Since rational functions are simply ratios of two polynomials, the discussion follows naturally from the preceding chapters on polynomial expressions and functions. Once we've established the basic techniques for working with rational functions, we can show how to use them to help solve equations.

Equations involving rational functions are extremely important in virtually all areas of applied mathematics. They are central to the basic description of electrical circuits, arising in all areas of electronic and computer design. Economic theory and business practice involve rational functions, as do medical practice, biology, and chemistry. We will discuss only the fundamental ideas of rational functions in this chapter, but you should be aware that they underlie the mathematical descriptions of most areas of modern technology. The project concluding this chapter explores two different mathematical models involving rational expressions in the areas of business and psychology.

12.1 APPLICATION

According to the U.S. Bureau of the Census, the number of civilians in the labor force who are over 16 years of age and the number in the labor force who are unemployed are as listed in the following table:

Year	Unemployed labor force (millions of people)	Labor force (millions of people)	Unemployment rate
1997	6.7	136.3	
1998	6.2	137.7	
1999	5.9	139.4	
2000	5.7	140.9	

a. Determine the unemployment rate for each year.

b. The number of people unemployed in the labor force (in millions) may be approximated by $U(x) = -0.33x + 6.62$, where x is the number of years after 1997. The number of people in the labor force (in millions) may be approximated by $L(x) = 1.6x + 136.1$, where x is the number of years after 1997. Determine a function $R(x)$ that approximates the unemployment rate x years after 1997.

(Continued on page 800)

c. Use the formula from part b to approximate the unemployment rate for the years 1997, 1998, 1999, and 2000.

d. Compare the approximations with the actual values in the table. Would you consider this function a good predictor of the unemployment rate?

After completing this section, we will discuss this application further. See page 810.

12.1 Rational Expressions and Functions

OBJECTIVES

1 Identify rational expressions.

2 Determine the domain of rational functions.

3 Graph rational functions by using a set of ordered pairs.

4 Model real-world situations by using rational functions.

12.1.1 Rational Expressions

In Chapter 1, we defined a **rational number** to be a number that can be written as a ratio $\frac{a}{b}$, where a and b are integers with $b \neq 0$.

$$\frac{5}{8} \qquad \text{rational number (fraction)}$$

$$0.625 \qquad \text{rational number (decimal)}$$

$$-17 \qquad \text{rational number (integer)}$$

 **HELPING HAND** Remember that all integers are rational numbers, because they can be written with a denominator of 1. For example, $-17 = \frac{-17}{1}$.

In this chapter, we discuss rational expressions. A **rational expression** can be written as a ratio $\frac{A}{B}$, where A and B are polynomials with $B \neq 0$. As we continue our discussion of rational expressions with variables in the denominator, we assume that the denominator does not equal 0.

RATIONAL EXPRESSION
A rational expression can be written in the form

$$\frac{A}{B}$$

where A and B are polynomials with $B \neq 0$.

$$\frac{x + y}{2x} \qquad \text{rational expression (algebraic fraction)}$$

$$6x + 4 \qquad \text{rational expression (polynomial)}$$

Other kinds of expressions, such as $x^2 + \frac{5}{2z}$, are also rational expressions. To see that this is true, we need to manipulate the expression. Since we do not yet have these skills, we will identify a rational expression as a sum of terms, each of which is a rational expression. The term x^2 can be written as a rational expression in the form $\frac{x^2}{1}$, and the term $\frac{5}{2z}$ is already in ratio form. Therefore, $x^2 + \frac{5}{2z}$ is a rational expression.

EXAMPLE 1 Identify each expression as rational or nonrational.

a. $\dfrac{12 + \sqrt{x}}{x}$ **b.** $x^5 y + 9z$

c. $\dfrac{x^3 + x^2 - 3}{x - 5}$ **d.** a^{-3}

Solution

a. $\frac{12 + \sqrt{x}}{x}$ is not a rational expression. The expression $12 + \sqrt{x}$ is not a polynomial, because it contains a square-root expression with a variable radicand, $\sqrt{x}$. Therefore, $\frac{12 + \sqrt{x}}{x}$ is not a ratio of two polynomials.

b. $x^5 y + 9z$ is a polynomial and a rational expression, because it may be written as $\frac{x^5 y + 9z}{1}$, or $\frac{x^5 y}{1} + \frac{9z}{1}$.

c. $\frac{x^3 + x^2 - 3}{x - 5}$ is a rational expression, because $(x^3 + x^2 - 3)$ and $(x - 5)$ are both polynomials.

d. a^{-3} is a rational expression, because it can be written as $\frac{1}{a^3}$.

Note: Example 1b is a polynomial and also a rational expression. According to the definition of rational expressions, all polynomial expressions are rational expressions. However, since we have already studied polynomial expressions, we will limit our main discussion in this chapter to nonpolynomial rational expressions.

12.1.1 Checkup

In exercises 1–6, identify each expression as rational or nonrational.

1. $\dfrac{x^2 + 2x + 1}{x + 1}$ **2.** $\dfrac{3\sqrt[5]{x} + 2x}{x + 3}$ **3.** $x^2 + \dfrac{9}{x - 3}$

4. $4x^3 - 2x^2 + x - 5$ **5.** $4x^{-2}$ **6.** $\dfrac{2xy}{z}$

7. All polynomials are rational expressions, but not all rational expressions are polynomials. Explain why.

12.1.2 Determining the Domain of Rational Functions

A special type of relation called a rational relation equates a rational expression in one independent variable to a dependent variable. Some rational relations are also functions and, therefore, may be written using function notation.

> **RATIONAL FUNCTION**
> A rational function can be written in the form
>
> $$f(x) = \dfrac{A(x)}{B(x)}$$
>
> where $A(x)$ and $B(x)$ are polynomial functions with $B(x) \neq 0$.

Examples of **rational functions** are as follows:

$$f(x) = \frac{6}{x} \qquad\qquad g(x) = \frac{x + 2}{x + 1}$$

$$h(x) = \frac{x^2 + 2x - 15}{x + 5} \qquad j(x) = \frac{5}{x} + \frac{4}{x^2}$$

The domain of a polynomial function is all real numbers. However, that may not be true for a rational function. Sometimes, substituting certain values for the independent variable in a rational expression results in undefined values for the fractional term. Such numbers must be excluded from the domain. We do this by identifying them as restrictions on the domain, or **restricted values**. We need to be able to determine these restricted values. Complete the following set of exercises to determine a method for finding the restrictions on the domain of a rational function.

 Discovery 1

Restricted Values

1. Complete the tables of values for the given rational functions.

    a. $f(x) = \dfrac{6}{x}$ **b.** $g(x) = \dfrac{x + 2}{x - 1}$ **c.** $h(x) = \dfrac{x^2 - x - 2}{x - 2}$

x	f(x)		x	g(x)		x	h(x)
-3			-3			-3	
-2			-2			-2	
-1			-1			-1	
0			0			0	
1			1			1	
2			2			2	
3			3			3	

Determine the restricted values of each function.

2. Graph the given rational functions on a calculator screen $(-9.4, 9.4, 1, -6.2, 6.2, 1, 1)$. Trace each function's graph. What happens when you reach the point on a graph for the restricted x-values found in exercise 1?

Write a rule for determining the restricted values of a rational function by viewing its table of values.

    Write a rule for determining the restricted values of a rational function by viewing its graph.

    Write a rule for determining the restricted values of a rational function algebraically.

    The restrictions on the domain of a rational function are the numbers that result in undefined values for the rational expression. On the function's graph, the restricted values (x-values) of the function do not have corresponding function values. This occurs because the function cannot be evaluated when the denominator of the rational expression is 0. Therefore, to determine the restricted values algebraically, we set the expression in the denominator of the rational expression equal to 0 and solve for x.

We have discovered three methods—numeric, graphic, and algebraic—of determining the restricted values from the domain of a rational function. The numeric and graphic methods have limitations on their effectiveness, so they should be used only as a check of the algebraic method.

> **To determine the restricted values of a rational function, algebraically**
>
> - Set the expression in the denominator equal to 0.
> - Solve for the independent variable.
>
> This method will find all the restricted values.

To determine the restricted integer values of a rational function numerically on your calculator, set up a table of integer values for the function. The restricted integer values are the x-values that result in errors in the table. Note that this table does not show all the restricted values.

To determine the restricted values of a rational function graphically on your calculator, graph the function and trace the graph. The x-values that have no corresponding y-values are restricted values for the rational function. Note that this method will determine only specific x-values, depending upon the calculator window setting.

EXAMPLE 2 Determine the restricted values, and then state the domain, of the rational function $f(x) = \frac{x^2 - 4}{x - 2}$.

Solution

Algebraic

Set the expression in the denominator equal to 0 and solve.

$$x - 2 = 0$$
$$x = 2$$

Numeric

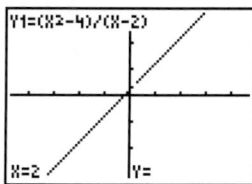

$$Y1 = \frac{x^2 - 4}{x - 2}$$

The function is not defined at $x = 2$.

Graphic

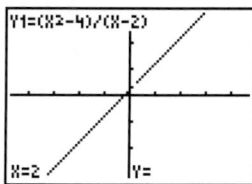

$(-47, 47, -31, 31)$

Trace the graph and ask for the value of $x = 2$.

The graph has a "hole" at $x = 2$.

The restricted value is 2.

Therefore, the domain is the set of all real numbers not equal to 2.

$$x \neq 2 \qquad \text{Inequality notation}$$
$$(-\infty, 2) \cup (2, \infty) \qquad \text{Interval notation}$$

 **HELPING HAND** The union symbol ($\cup$) joins the two intervals and is read "or," meaning that the elements in the domain are in the first interval or in the second interval (or both).

EXAMPLE 3 Determine algebraically the restricted values, and then state the domain, of the given rational functions. Check your answer numerically or graphically.

a. $h(x) = \dfrac{x+4}{x^2-x-20}$ **b.** $j(x) = \dfrac{6}{x^2+12}$ **c.** $m(x) = \dfrac{3x+1}{5x-8}$

Solution

Algebraic

a. $h(x) = \dfrac{x+4}{x^2-x-20}$

To determine the restricted values, set the denominator equal to 0 and solve for x by factoring.

$$x^2 - x - 20 = 0$$
$$(x+4)(x-5) = 0$$
$$x+4 = 0 \quad \text{or} \quad x-5 = 0$$
$$x = -4 \qquad\qquad x = 5$$

The restricted values are -4 and 5.

Domain: All real numbers not equal to -4 or 5

$$x \neq -4 \quad \text{or} \quad x \neq 5$$
$$(-\infty, -4) \cup (-4, 5) \cup (5, \infty)$$

b. $j(x) = \dfrac{6}{x^2+12}$

Algebraic

To determine the restricted values, set the denominator equal to 0 and solve for x.

$$x^2 + 12 = 0$$

The expression on the left does not factor. Also, note that if we subtract 12 from both sides, we obtain $x^2 = -12$. There is no real number whose square is a negative number. Therefore, the function has no restricted values in the real-number system.

Domain: All real numbers

$$(-\infty, \infty)$$

Numeric Check

Since the restricted values are integers, check the answer numerically. Let $Y1 = \dfrac{x+4}{x^2-x-20}$ and view the table.

X	Y1
-4	ERROR
-3	-.125
-2	-.1429
-1	-.1667
0	-.2
1	-.25
2	-.3333

X= -1

X	Y1
-1	-.1667
0	-.2
1	-.25
2	-.3333
3	-.5
4	-1
5	ERROR

X= -1

The function values for $x = -4$ and $x = 5$ are undefined.
     The restricted values are -4 and 5. The answer checks.

Graphic Check

The graphed function does not appear to have any "holes" in it and hence likely has no restrictions.

$$Y1 = \dfrac{6}{x^2+12}$$

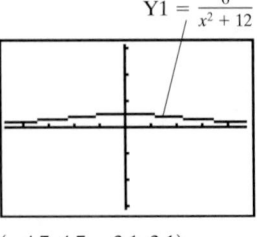

$(-4.7, 4.7, -3.1, 3.1)$

c. $m(x) = \dfrac{3x + 1}{5x - 8}$

Algebraic

To determine the restricted values, set the denominator equal to 0 and solve for x.

$$5x - 8 = 0$$
$$5x = 8$$
$$x = \frac{8}{5}$$

The restricted value is $\frac{8}{5}$.

Domain: All real numbers not equal to $\frac{8}{5}$

$$x \neq \frac{8}{5}$$
$$(-\infty, \tfrac{8}{5}) \cup (\tfrac{8}{5}, \infty)$$

Graphic Check

Since the restricted value is not an integer, we cannot use a table of integer values. Check graphically. Ask for the value of $x = 1.6$.

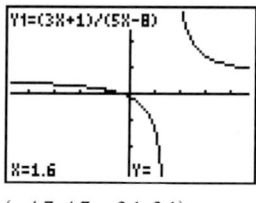

$(-4.7, 4.7, -3.1, 3.1)$

The function is not defined for $x = 1.6$, or $\frac{8}{5}$. The answer checks.

 12.1.2 Checkup

In exercises 1–4, determine the restricted values algebraically and then state the domain of the given rational functions. Check your answer numerically or graphically.

1. $s(x) = \dfrac{x^2 + x - 12}{x + 4}$ **2.** $g(x) = \dfrac{x + 8}{x^2 + x - 12}$

3. $C(x) = \dfrac{8x}{x^2 + 1}$ **4.** $a(b) = \dfrac{4}{2b - 3}$

5. What does it mean to say that a rational function has restricted values?

12.1.3 Graphing Rational Functions

Rational functions have an infinite number of possible solutions. In order to illustrate these solutions, we use a graph. When we graphed linear functions, we needed two points to determine the linear pattern and one additional point to be used as a check. To graph quadratic functions, we determined the vertex and several points to the left and right of it in order to see the pattern. To graph a rational function, we determine the restricted values. These values will not have corresponding function values. Then we graph several points to the left and right of each restricted value to see a pattern. This may be a very lengthy process by hand.

To graph a rational function,

- Determine the restrictions on the domain.
- Determine a table of values by choosing x-values less than and greater than the restricted values. Pay close attention to the x-values that are close to the restricted values.
- Graph the function by plotting the sample set of ordered pairs from the table of values and connecting the points with a smooth curve determined by the pattern seen. Be careful not to cross the vertical line where x is equal to a restricted value, because the restricted values do not have corresponding function values.

Note: The calculator TABLE function in Ask mode will set up a table for you.

HELPING HAND The calculator window setting is very important in determining the "picture" you receive. You should carefully consider your choices.

EXAMPLE 4 Graph the rational functions.

a. $y = \dfrac{x^2 - x - 15}{x + 3}$ **b.** $y = \dfrac{5}{x} + \dfrac{4}{x^2}$ **c.** $p(x) = \dfrac{x^2 - 4}{x - 2}$

Solution

a. $y = \dfrac{x^2 - x - 15}{x + 3}$

The restricted value is -3. Set up a table of values, using x-values less than and greater than -3.

x	y
-6	-9
-5	-7.5
-4	-5
-3.5	-1.5
-3.1	22.9
-3.01	292.99

x	y
-2.99	-307
-2.9	-36.9
-2.5	-12.5
-2	-9
-1	-6.5
0	-5
1	-3.75
2	-2.6
3	-1.5
4	-0.4286

Graph the ordered pairs found in the table of values, and connect the points with a smooth curve. Do not connect across the vertical line $x = -3$.

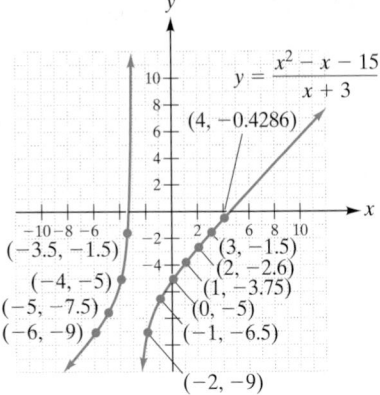

Calculator Solution

a. Enter the equation in Y1, graph it, and choose the best viewing screen. Note that the standard screen adds an extra line in the graph. Note also that the graph appears to be defined at $x = -3$, a restricted value.

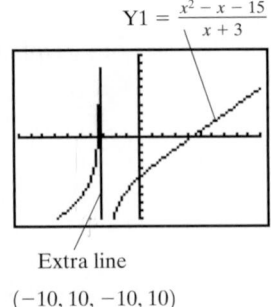

$$Y1 = \frac{x^2 - x - 15}{x + 3}$$

Extra line

$(-10, 10, -10, 10)$

The extra line is due to the fact that the calculator is in connected mode. It connected the two parts of the graph across the vertical line $x = -3$. To graph only points, change the calculator to dot mode, as shown in Figure 12.1c.

T E C H N O L O G Y

GRAPHING RATIONAL FUNCTIONS

Graph $y = \dfrac{x^2 - x - 15}{x + 3}$.

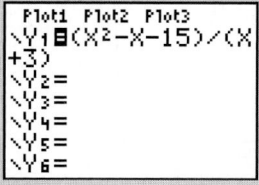

Figure 12.1a

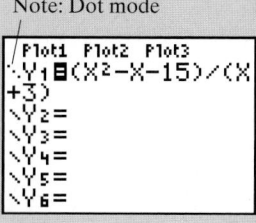

Note: Dot mode

Figure 12.1b

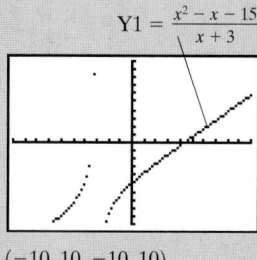

$Y1 = \dfrac{x^2 - x - 15}{x + 3}$

$(-10, 10, -10, 10)$

Figure 12.1c

For Figure 12.1a,

Enter the function, $y = \frac{x^2 - x - 15}{x + 3}$, in the $Y =$ menu. Remember to group both the numerator and denominator with a set of parentheses.

| Y= | (| X,T,θ,n | x² | − | X,T,θ,n | − | 1 | 5 |) | ÷ | (| X,T,θ,n | + |

| 3 |) |

For Figure 12.1b,

Set the calculator to dot mode.

Move the arrow to the left of Y1 = and press | ENTER | six times.

For Figure 12.1c,

Graph the function on a standard window.

| ZOOM | 6 |

Note that the calculator only graphed points on the graph and did not connect them. You will need to connect these points when you draw the graph on paper.

b. $y = \dfrac{5}{x} + \dfrac{4}{x^2}$

The restricted value is 0. Therefore, complete a table of values by choosing x-values less than 0 and x-values greater than 0.

x	y	x	y
−6	−0.722	0.01	40500
−5	−0.84	0.1	450
−4	−1	0.5	26
−3	−1.222	1	9
−2	−1.5	2	3.5
−1	−1	3	2.111
−0.5	6	4	1.5
−0.1	350	5	1.16
−0.01	39500	6	0.944

(Continued on page 808)

Graph the ordered pairs found in the table of values, and connect the points with a smooth curve. Do *not* connect across the vertical line when x is equal to the restricted value, or $x = 0$.

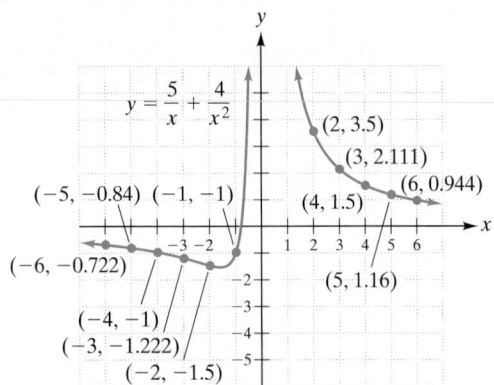

Calculator Solution

On your calculator, graph in dot mode the rational function $y = \frac{5}{x} + \frac{4}{x^2}$.

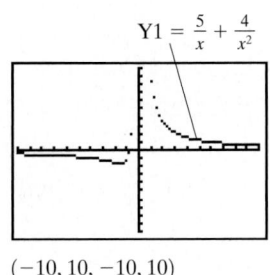

$(-10, 10, -10, 10)$

c. $p(x) = \dfrac{x^2 - 4}{x - 2}$

The restricted value is 2. Set up a table of values, using x-values less than and greater than 2.

x	$p(x)$	x	$p(x)$
-1	1	2.01	4.01
0	2	2.1	4.1
1	3	2.5	4.5
1.5	3.5	3	5
1.9	3.9	4	6
1.99	3.99	5	7

Graph the ordered pairs found in the table of values, and connect the points with a smooth curve. Again, do not cross the vertical line $x = 2$.

There is a "hole" in the graph at $x = 2$. Thus, this point should be enclosed with an open dot to show that it is not included in the linear graph.

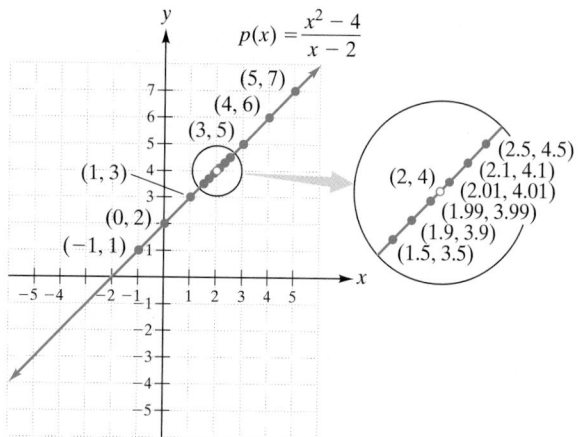

$$p(x) = \frac{x^2 - 4}{x - 2}$$

Calculator Solution

HELPING HAND It is always a good idea to look at your table of values as well as your calculator graph. Your calculator graphs may be misleading without proper settings.

Enter the equation in Y1, graph it, and choose the best viewing screen.
   Note: The standard screen does not show the "hole," because it is an integer value for x.

$$Y1 = \frac{x^2 - 4}{x - 2}$$

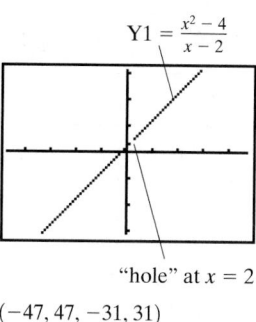

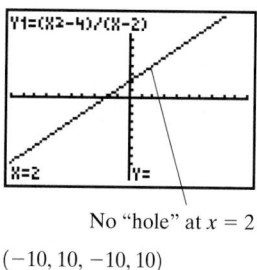

"hole" at $x = 2$ No "hole" at $x = 2$

$(-47, 47, -31, 31)$ $(-10, 10, -10, 10)$

 12.1.3 Checkup

Graph the rational functions in exercises 1–4.

1. $y = \dfrac{8}{x^2} + \dfrac{24}{x^3}$

2. $h(x) = \dfrac{x^2 + 8x + 15}{x + 3}$

3. $g(x) = \dfrac{x^2 - 5x + 8}{x - 2}$

4. $c(x) = \dfrac{x}{x^2 + 3}$

5. Do you think a graph presents a more complete description of the solutions of a rational function than a table of values does? Explain.

12.1.4 Modeling the Real World

Many business and economic applications are modeled by rational functions. Tables of values and graphs of these functions help us analyze the situations modeled and make appropriate business decisions.

EXAMPLE 5   A printer is purchased for $5000 with an additional service contract that costs $400 the first year and increases $100 per year thereafter. The total cost of the copier for x years is given by $C(x) = 5000 + 350x + 50x^2$.

a. Determine the average cost per year for x years.

b. Graph the average-cost function for the domain $1 \le x \le 20$. Determine the minimum average cost per year.

Solution

a. The average cost per year for x years is calculated by dividing the total cost by the number of years.

Let x = the number of years

$$C(x) = \text{the total cost}$$

$$C_{ave}(x) = \text{the average cost per year}$$

$$C_{ave}(x) = \frac{C(x)}{x}$$

$$C_{ave}(x) = \frac{5000 + 350x + 50x^2}{x}$$

b.

$$Y1 = \frac{5000 + 350x + 50x^2}{x}$$

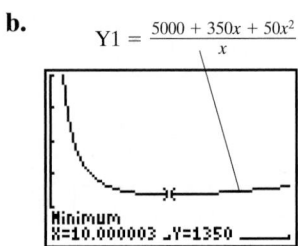

Minimum
X=10.000003 Y=1350

(0, 20, 0, 5000)

Note that the function is decreasing and then increasing. The relative minimum of the graph is located at (10, 1350). The minimum average cost per year is $1350. This minimum value occurs when the printer has been in use 10 years.

In business analysis, the year in which the average cost per year is at its minimum is referred to as the replacement time. Therefore, the printer should be replaced when it has been in use 10 years.

APPLICATION

According to the U.S. Bureau of the Census, the number of civilians in the labor force who are over 16 years of age and the number in the labor force who are unemployed are as listed in the following table:

Year	Unemployed labor force (millions of people)	Labor force (millions of people)	Unemployment rate
1997	6.7	136.3	
1998	6.2	137.7	
1999	5.9	139.4	
2000	5.7	140.9	

a. Determine the unemployment rate for each year.

Determine a function $R(x)$ that approximates the unemployment rate x years after 1997.

b. The number of people unemployed in the labor force (in millions) may be approximated by $U(x) = -0.33x + 6.62$, where x is the number of years after 1997. The number of people in the labor force (in millions) may be approximated by $L(x) = 1.6x + 136.1$, where x is the number of years after 1997.

c. Use the formula from part b to approximate the unemployment rate for the years 1997, 1998, 1999, and 2000.

d. Compare the approximations with the actual values in the table. Would you consider this function a good predictor of the unemployment rate?

Discussion

a. In order to determine the unemployment rate, we divide the number of people unemployed by the total number in the labor force.

$$1997: \frac{6.7}{136.3} \approx 0.049 = 4.9\% \qquad 1998: \frac{6.2}{137.7} \approx 0.045 = 4.5\%$$

$$1999: \frac{5.9}{139.4} \approx 0.042 = 4.2\% \qquad 2000: \frac{5.7}{140.9} \approx 0.04 = 4\%$$

b. $R(x) = \dfrac{U(x)}{L(x)}$

$R(x) = \dfrac{-0.33x + 6.62}{1.6x + 136.1}$

c.

X	Y1
0	.04864
1	.04568
2	.04279
3	.03996
4	.03719
5	.03449
6	.03185

X=0

$x = 0$: 1997: $0.049 = 4.9\%$ $x = 1$: 1998: $0.046 = 4.6\%$
$x = 2$: 1999: $0.043 = 4.3\%$ $x = 3$: 2000: $0.04 = 4\%$

Y1 = $(-0.33x + 6.62)/(1.6x + 136.1)$

d. According to the table, the approximate unemployment rates are very close to the rates determined in part a. The function is a good predictor.

12.1.4 Checkup

1. Paul purchased a big-screen television set for $2500. He also purchased a maintenance agreement for the set that costs $75 for the first year and increases $50 per year thereafter. The total cost of the television and maintenance agreement is $C(x) = 2500 + 50x + 25x^2$ for x years.

a. Determine the average cost per year if Paul keeps the set and the agreement for x years.

b. Graph the average-cost function for the domain $1 \le x \le 20$. Determine the minimum average cost per year.

c. What advice would you give to Paul?

2. Using data from the U.S. Bureau of the Census, the following table shows the number of civilians over 16 years of age in the labor force and the number employed in nonfarm employment:

Year	Nonfarm employment (millions of people)	Labor force (millions of people)	Proportion of the labor force in nonfarm employment	$P(x)$
1997	122.7	136.3		
1998	125.9	137.7		
1999	128.8	139.4		
2000	131.4	140.9		

a. Calculate the proportion of the labor force in nonfarm employment for each year.

b. The number of people employed in nonfarm employment (in millions) may be approximated by $N(x) = 2.9x + 122.9$, where x is the number of years after 1997. The number of people in the labor force (in millions) may be approximated by $L(x) = 1.6x + 136.1$, where x is the number of years after 1997. Write a rational function $P(x)$ that approx-

imates the proportion of the labor force in nonfarm employment x years after 1997.

c. Use the function from part b to approximate the proportion of the labor force in nonfarm employment for the years 1997 through 2000.

d. Compare the approximations from part c with the proportions in part a. Is the function from part b a good predictor of the proportion of nonfarm employment?

12.1 Exercises

Determine whether each expression is rational or nonrational.

1. $\dfrac{x^2 + 10x + 25}{x + 5}$

2. $\dfrac{x + 4\sqrt{x} + 7}{x - 2}$

3. $z - 4 - \dfrac{25}{z - 4}$

4. $0.8a^2 - 0.2a + 1.2$

5. $\dfrac{x + 3}{\sqrt{x} - 2}$

6. $6x - 3x^{-1}$

7. $6x^2 + 17x - 3$

8. $\dfrac{\sqrt{x} + 3}{5x}$

9. $\dfrac{11}{3xy}$

10. $\dfrac{\sqrt{x^2 + 4}}{x - 6}$

11. $3x^{-2} + 4x^{-1} + 3 - x$

12. $\dfrac{1}{2}x^2 - \dfrac{3}{4}x + \dfrac{7}{8}$

13. $\dfrac{0.5x + 7}{1.2x - 4}$

14. $a + 7 - \dfrac{6}{a + 1}$

15. $\dfrac{2 - 3\sqrt{x}}{2x}$

16. $\dfrac{4}{7 - xy}$

Determine the restricted values algebraically, and then state the domain of each rational function. Check your answer numerically or graphically.

17. $y = \dfrac{3 - x}{x}$

18. $y = \dfrac{1 - x}{x^2 + 8x + 15}$

19. $h(x) = \dfrac{x - 5}{x^2 - 11x + 30}$

20. $g(x) = \dfrac{4x^2 + 1}{2x^2 + 6x}$

21. $C(x) = \dfrac{x + 7}{x^2 - 4}$

22. $y = \dfrac{5x^2 - 2}{x^3 - 4x^2 - x + 4}$

23. $y = \dfrac{2x^2 + 7}{x^2 - 5x}$

24. $b(x) = \dfrac{x + 7}{5x}$

25. $f(x) = \dfrac{x + 2}{x^3 - 2x^2 + 4x - 8}$

26. $y = \dfrac{55}{x^2 + 4x + 4}$

27. $f(x) = \dfrac{7}{2x^2 + 17x + 35}$

28. $y = \dfrac{12}{25x^2 + 30x + 9}$

29. $y = \dfrac{x^2 - 3x}{8x^2 + 6x - 9}$

30. $F(x) = \dfrac{x^3 - 8}{4x^3 + 20x^2 - 64x - 320}$

31. $y = \dfrac{4x}{4x^2 + 36x + 81}$

32. $g(x) = \dfrac{7x - 3}{14x^2 + 25x + 9}$

33. $h(x) = \dfrac{6 - x}{x^2 - 22x + 121}$

34. $y = \dfrac{5x + 2}{x^2 + 3x + 4}$

35. $y = \dfrac{3x^2}{9x^2 - 25}$

36. $y = \dfrac{9x^3}{6x^2 + 13x - 15}$

37. $C(x) = \dfrac{5x^3 + 1}{2x^3 + 4x^2 - 18x - 36}$

38. $y = \dfrac{2x^2 + 1}{16x^2 - 49}$

39. $y = \dfrac{3x + 8}{2x^2 + 5x + 7}$

40. $f(x) = \dfrac{9 - 2x}{9x^2 - 24x + 16}$

41. $y = \dfrac{x + 4}{2x + 3}$

42. $g(x) = \dfrac{8x^2 + 19}{10x^2 + 3x - 18}$

In exercises 43–62, graph each rational function.

43. $y = \dfrac{6}{x} + \dfrac{9}{x^2}$

44. $y = \dfrac{25x^2 - 4}{5x + 2}$

45. $h(x) = \dfrac{x^2 - 49}{x - 7}$

46. $g(x) = \dfrac{4x}{x^3 + 1}$

47. $y = \dfrac{2x}{x^2 + 3}$

48. $f(x) = \dfrac{9}{x} - \dfrac{36}{x^2}$

49. $y = \dfrac{6x}{x^3 + 8}$

50. $g(x) = \dfrac{2x^2 + 5}{3x + 1}$

51. $f(x) = \dfrac{x^2 + 1}{x - 3}$

52. $z(x) = \dfrac{6x^2 - x - 35}{2x - 5}$

53. $R(x) = \dfrac{2x^2 + 7x + 6}{x + 2}$

54. $y = \dfrac{x}{x^2 + 1}$

55. $y = \dfrac{8}{x}$

56. $y = \dfrac{1 - x}{x + 1}$

57. $y = \dfrac{x - 5}{x + 3}$

58. $g(x) = \dfrac{x^2 + 3}{x^2 + x - 2}$

59. $f(x) = \dfrac{2x + 3}{x^2 - 2x - 3}$

60. $c(x) = \dfrac{x^2 + x - 2}{x + 2}$

61. $g(x) = \dfrac{2x^2 - 5x + 2}{x - 2}$

62. $y = -\dfrac{4}{x}$

63. The purchase price of a new heating-and-cooling system is \$6500. The manufacturer offers a maintenance agreement for the system that can be purchased for \$125 the first year and that increases by \$50 per year thereafter. The total cost of the system and maintenance agreement is $C(x) = 6500 + 100x + 25x^2$ for x years.
   a. Determine the average cost per year if the system is kept for x years.
   b. Graph the average-cost function for the domain $0 \le x \le 40$. Determine the minimum average cost per year.
   c. What advice would you give the customer?

64. Acme Machines sells a high-speed copy machine for \$10,000. The maintenance agreement for the machine costs \$1200 for the first year and increases by \$600 per year thereafter. The total cost of the copier and maintenance agreement is $C(x) = 10{,}000 + 900x + 300x^2$ for x years of use.
   a. Determine the average cost per year if the copier is kept for x years.
   b. Graph the average-cost function for the domain $0 \le x \le 20$. Determine the minimum average cost per year.
   c. What recommendation would you make to the customer?

65. Using data from the U.S. Bureau of the Census, the following table shows crude-oil production figures (in millions of barrels) for the latest available years:

Year	Domestic	Total	Proportion from domestic	P(x)
1995	2,394	5,037		
1998	2,282	5,460		
1999	2,147	5,334		

   a. Calculate the proportion of the total crude-oil production that results from domestic production for each year.
   b. The number of barrels of domestic crude-oil production (in millions) may be approximated by $D(x) = -56.12x + 2405$, where x is the number of years after 1995. The total number of barrels of crude oil production, domestic and imported (in millions), may be approximated by $T(x) = 89.65x + 5068$, where x is the number of years after 1995. Write a rational function $P(x)$ that approximates the proportion of crude-oil production from domestic sources x years after 1995.
   c. Use the function from part b to approximate the proportion of crude oil production from domestic sources for the years 1995, 1998, and 1999.
   d. Compare the approximations from part c with the proportions in part a. Is the function from part b a good predictor of the proportion of domestic crude-oil production?

66. Using data from the U.S. Bureau of the Census, the following table shows the number of families (in millions) with children under 18 years of age and the total number of families (in millions) for the latest available years:

Year	Number with children under 18 years	Total number	Proportion with children under 18 years	$P(x)$
1995	34.3	69.3		
1998	34.8	70.9		
1999	34.6	71.5		

a. Calculate the proportion of families with children under 18 years of age for each year.

b. The number of families (in millions) with children under 18 years of age may be approximated by $N(x) = 0.10x + 34.3$, where x is the number of years after 1995. The total number of families (in millions) may be approximated by $T(x) = 0.55x + 69.3$, where x is the number of years after 1995. Write a rational function $P(x)$ that approximates the proportion of families with children under 18 years of age x years after 1995.

c. Use the function from part b to approximate the proportion of families with children under 18 years of age for the years 1995, 1998, and 1999.

d. Compare the approximations from part c with the proportions in part a. Is the function from part b a good predictor of the proportion of families with children under 18 years of age?

12.1 Calculator Exercises

When graphing rational functions with your calculator, you must carefully choose a window that provides a good picture of the graph. To do so, you may need to use a window setting other than the options available under the ZOOM key. Remember that you can choose any window you wish by using the WINDOW key.

As an example, graph $y = \frac{6x + 13}{x + 1}$. First, store the function in Y1 and set Y1 to graph in dot mode. The first screen shows the graph in the decimal window, ZOOM 4. Not much of the graph is visible. The second screen is the integer screen, ZOOM 8 ENTER. Here, the graph is too compressed, with much of that part of it near the origin unclear.

The third screen is the standard screen, ZOOM 6. This screen is more revealing around the origin, but the y-axis needs to be extended. Do this by pressing WINDOW and changing Ymin and Ymax to double their size. At the same time, set Yscl to 0 so that tick marks on the y-axis won't interfere with your view. This is the fourth screen. With it, you can more clearly discern the behavior of the function, particularly around the y-axis. Finally, remember that you can always use the ZOOM BOX option to enlarge a particular area of the graph that you need to see in more detail. The four screens are as follows:

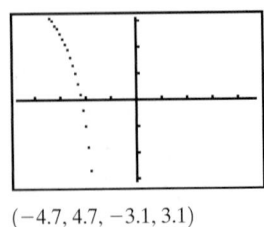

$(-4.7, 4.7, -3.1, 3.1)$

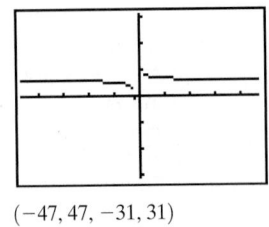

$(-47, 47, -31, 31)$

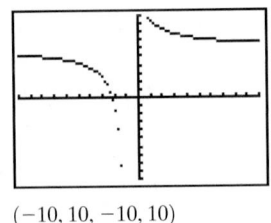

$(-10, 10, -10, 10)$

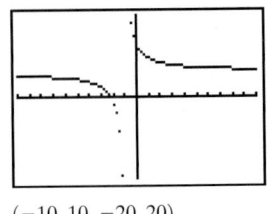

$(-10, 10, -20, 20)$

Find an appropriate window for the following functions:

1. $f(x) = \dfrac{x^3 + 5}{x^2 - 3}$ **2.** $y = \dfrac{1}{x^3 - 1} + \dfrac{1}{x^2}$ **3.** $g(x) = \dfrac{3}{x - 2} + \dfrac{5}{x^2 + 1}$

12.1 Writing Exercise

In the study of statistics, the *average* of a set of observations (that is, numbers) is calculated as the total of the observations divided by the number of observations in the set. Symbolically,

$$\text{ave} = \frac{\text{total}}{n}$$

where n is the number of observations in the set. Consider the restrictions on the divisor in this rational expression. Obviously, n cannot equal 0. What does this mean in terms of collecting observations and calculating averages? Does it make sense that zero should be a restricted value? What minimum number of observations is required to calculate an average?

Also in the study of statistics, the *variability* in a set of observations is calculated as the total of squared deviations from the average, divided by the number of observations less 1. Don't be concerned if you don't understand all the terminology in this definition. We are really interested only in what is happening with the divisor. Symbolically,

$$\text{var} = \frac{\text{total of squares}}{n - 1}$$

What values are restricted on the divisor in this rational expression? Can you measure variability in numbers if you have zero observations? What if you have only one observation? Again, do the restrictions make sense in the context of the application? Write a short summary of your understanding of the questions raised in this application.

12.2 Multiplication and Division of Rational Expressions

OBJECTIVES

1 Simplify rational expressions.

2 Multiply rational expressions.

3 Divide rational expressions.

4 Model real-world situations by using multiplication and division of rational expressions.

APPLICATION

Sio signed a contract for her new job. According to the contract, she is to make $20,000 the first year and receive an increase of 5% per year for the next two years. Sio's total salary for three years is given by the function $s(x) = \frac{20,000 - 20,000x^3}{1 - x}$, where x is the sum of 1 and the percentage increase converted to a decimal, or 1.05. Write an equivalent function by simplifying $s(x)$. Determine Sio's total salary for three years, using the original function and the simplified function.

After completing this section, we will discuss this application further. See page 824.

12.2.1 Simplifying Rational Expressions

We are now ready to simplify rational expressions that are algebraic fractions. We simplify these expressions by means of the same procedures we use with numeric fractions.

To write equivalent rational expressions, we need to review a property of real numbers discussed in Chapter 1. The multiplication identity property states that the product of a real number and 1 is the real number itself. We use this property to write equivalent rational numbers, which are numbers that have the same value.

To write an equivalent numeric fraction in simplest form, we factor out the greatest common factor (GCF) of the numerator and denominator, rewrite the GCF as 1, and eliminate the factor of 1 by using the identity property. For example, simplify $\frac{6}{14}$.

$$\frac{6}{14} = \frac{2 \cdot 3}{2 \cdot 7} \quad \text{Factor out the GCF, 2.}$$

$$= \frac{2}{2} \cdot \frac{3}{7}$$

$$= 1 \cdot \frac{3}{7} \quad \text{Rewrite the GCF as 1.}$$

$$= \frac{3}{7}$$

The multiplication identity property also holds for rational expressions. It allows us to write equivalent rational expressions, using the same procedure as with numeric fractions. **Equivalent rational expressions** have the same value for any permissible replacement of the independent variable. Complete the following set of exercises to see the meaning of this definition.

 Discovery 2

Simplified Expressions and Restricted Values

To simplify $\frac{6x^2}{14x}$, complete the following steps:

$$\frac{6x^2}{14x} = \frac{2x \cdot 3x}{2x \cdot 7} \quad \text{Factor out the GCF, 2x.}$$

$$= \frac{2x}{2x} \cdot \frac{3x}{7}$$

$$= 1 \cdot \frac{3x}{7} \quad \text{Rewrite the GCF as 1.}$$

$$= \frac{3x}{7}$$

Check the equivalence of the expressions $\frac{6x^2}{14x}$ and $\frac{3x}{7}$ by graphing. Let $Y1 = \frac{6x^2}{14x}$ and $Y2 = \frac{3x}{7}$.

1. Both graphs appear to be coinciding. However, look at a table of values for $x = -3, -2, -1, 0, 1, 2,$ and 3. What do you see?

2. Determine the restricted values for each expression.

3. Explain why the two expressions are equivalent for all values except when $x = 0$.

Note that the original expression to be simplified had a restricted value of $x = 0$, but the simplified expression did not have a restricted value. The permissible values for x in the original expression are all real numbers except 0. The permissible values for x in the simplified expression are all real numbers. Therefore, the two expressions are said to be equivalent expressions for their permissible values.

Another property of multiplication that is often useful in simplifying rational expressions is the multiplication property of -1. The multiplication property of -1 states that the product of a real number and -1 is the opposite of the number itself. This is useful in simplifying rational expressions when the numerator and denominator are opposites.

HELPING HAND Two polynomials are opposites if every term of one polynomial corresponds to an opposite term in the second polynomial.

For example, simplify $\frac{x-1}{1-x}$.

The numerator and denominator are opposites because x and $-x$ are corresponding terms, as are 1 and -1. Therefore, if we rewrite the denominator as a product of -1 and its opposite, we can factor out a 1 from the expression.

$$\frac{x-1}{1-x} = \frac{x-1}{-1(-1+x)}$$ *Rewrite the denominator as a product of -1 and its opposite.*

$$= \frac{x-1}{-1(x-1)}$$ *Rearrange the binomial factor in the denominator.*

$$= \frac{x-1}{x-1} \cdot \frac{1}{-1}$$

$$= 1 \cdot -1$$ *Rewrite the GCF as 1.*

$$= -1$$

If the numerator or denominator polynomials contain more than one term, it is necessary to factor them in order to find the GCF.

SIMPLIFYING RATIONAL EXPRESSIONS
To simplify rational expressions,

- Factor the numerator and denominator.
- Write a product of two rational expressions, with one factor containing the GCF of the numerator and denominator and the other containing the remaining factors.
- Rewrite the factor containing the GCF as 1.
- Multiply, obtaining

$$\frac{A}{B} = \frac{k \cdot C}{k \cdot D} = \frac{k}{k} \cdot \frac{C}{D} = 1 \cdot \frac{C}{D} = \frac{C}{D}$$

where A, B, C, and D are polynomials, $B \neq 0$, $D \neq 0$, and k is the GCF of A and B.

HELPING HAND As in previous chapters, we can check the simplification of one variable expressions numerically or graphically.

EXAMPLE 1 Simplify.

a. $\dfrac{5x + 15}{10x - 25}$ b. $\dfrac{x^2 - 5x - 14}{x^2 - 8x + 7}$ c. $\dfrac{16 - x^4}{x^3 - 2x^2 + 4x - 8}$

Solution

a. $\dfrac{5x + 15}{10x - 25} = \dfrac{5(x + 3)}{5(2x - 5)}$ *Factor the numerator and denominator.*

$\qquad\qquad = \dfrac{5}{5} \cdot \dfrac{x + 3}{2x - 5}$ *Write a product, with one factor containing the GCF of the numerator and denominator and the other containing the remaining factors.*

$\qquad\qquad = 1 \cdot \dfrac{x + 3}{2x - 5}$ *Rewrite the GCF as 1.*

$\qquad\qquad = \dfrac{x + 3}{2x - 5}$

Numeric Check

As in previous chapters, we can check our simplification on the calculator by entering the simplified expression and the original expression. All Y1 and Y2 values should be equal.

X	Y1	Y2
0	-.6	-.6
1	-1.333	-1.333
2	-5	-5
3	6	6
4	2.3333	2.3333
5	1.6	1.6
6	1.2857	1.2857

X=0

$Y1 = \dfrac{x + 3}{2x - 5}$

$Y2 = \dfrac{5x + 15}{10x - 25}$

b. $\dfrac{x^2 - 5x - 14}{x^2 - 8x + 7} = \dfrac{(x + 2)(x - 7)}{(x - 1)(x - 7)}$ *Factor the numerator and denominator.*

$\qquad\qquad = \dfrac{x - 7}{x - 7} \cdot \dfrac{x + 2}{x - 1}$ *Write a product, with one factor containing the GCF of the numerator and denominator and the other containing the remaining factors.*

$\qquad\qquad = 1 \cdot \dfrac{x + 2}{x - 1}$ *Rewrite the GCF as 1.*

$\qquad\qquad = \dfrac{x + 2}{x - 1}$ *Multiply.*

Graphic Check

We can check our simplification graphically by entering Y1 = the simplified expression and Y2 = the original expression. The graphs should coincide.

$Y1 = \frac{x + 2}{x - 1}$

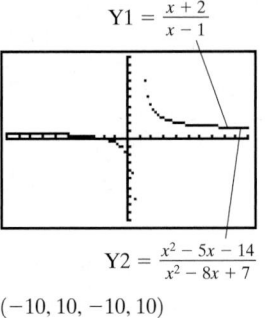

$Y2 = \frac{x^2 - 5x - 14}{x^2 - 8x + 7}$

$(-10, 10, -10, 10)$

c. $\dfrac{16 - x^4}{x^3 - 2x^2 + 4x - 8} = \dfrac{(2 - x)(2 + x)(4 + x^2)}{(x - 2)(x^2 + 4)}$ *Factor the numerator and denominator.*

$\qquad\qquad = \dfrac{-1(x - 2)(2 + x)(4 + x^2)}{(x - 2)(x^2 + 4)}$ *Rewrite the factor $2 - x$ as $-1(x - 2)$.*

$\qquad\qquad = \dfrac{(x - 2)(4 + x^2)}{(x - 2)(x^2 + 4)} \cdot \dfrac{-1(2 + x)}{1}$ *Write a product, with one factor containing the GCF of the numerator and denominator and the other containing the remaining factors. Note: $4 + x^2 = x^2 + 4$.*

$\qquad\qquad = 1 \cdot \dfrac{-1(2 + x)}{1}$ *Rewrite the GCF as 1.*

$\qquad\qquad = -1(2 + x) \quad \text{or} \quad -2 - x$ *Multiply.*

The check is left for you.

Note: You may be aware of a process called **cancellation** that may be used to reduce the number of steps in simplifying rational expressions. However, cancellation must be used properly and with great care. The process involves dividing out common *factors* from the numerator and denominator. For example, Example 1b could have been worked as follows:

$$\frac{x^2 - 5x - 14}{x^2 - 8x + 7} = \frac{(x + 2)(x - 7)}{(x - 1)(x - 7)}$$ Factor the numerator and denominator.

$$= \frac{(x + 2)\overset{1}{(x - 7)}}{(x - 1)\underset{1}{(x - 7)}}$$ Cancel or divide out a common factor of $(x - 7)$. Note: $\frac{x-7}{x-7} = 1$.

$$= \frac{x + 2}{x - 1}$$

 **HELPING HAND** The remaining *x-terms* cannot be canceled. We can cancel or divide out *factors* only. We cannot cancel terms.

CAUTION
To simplify a rational expression by cancellation, the numerator and denominator must be factored. Only factors can be canceled or divided out.

 12.2.1 Checkup

Simplify exercises 1–4.

1. $\dfrac{12x^2y^2z}{28x^3yz}$ **2.** $\dfrac{3a - 6}{9a + 12}$ **3.** $\dfrac{x^2 - 6x + 8}{x^2 - 10x + 24}$

4. $\dfrac{25 - x^2}{x^3 - 5x^2 + 7x - 35}$

5. What is the difference between a factor and a term of a polynomial?

6. When simplifying a rational expression, you can cancel only factors, not terms. Why?

12.2.2 Multiplying Rational Expressions

To multiply numeric fractions, we multiply the numerators together, multiply the denominators together, and, if necessary, simplify the resulting fraction. For example,

$$\frac{3}{4} \cdot \frac{5}{6} = \frac{3 \cdot 5}{4 \cdot 6} = \frac{15}{24} = \frac{3}{3} \cdot \frac{5}{8} = 1 \cdot \frac{5}{8} = \frac{5}{8}$$

We use the same procedure for multiplication of rational expressions.

$$\frac{6x}{5} \cdot \frac{3y}{4x} = \frac{6x \cdot 3y}{5 \cdot 4x} = \frac{18xy}{20x} = \frac{2x}{2x} \cdot \frac{9y}{10} = 1 \cdot \frac{9y}{10} = \frac{9y}{10}$$

MULTIPLYING RATIONAL EXPRESSIONS
To multiply rational expressions,

- Multiply the numerators.
- Multiply the denominators.
- Factor the numerator and denominator.

(Continued on page 820)

• Simplify the resulting rational expression, as in

$$\frac{A}{B} \cdot \frac{C}{D} = \frac{AC}{BD}$$

where A, B, C, and D are polynomials with $B \neq 0$ and $D \neq 0$.

EXAMPLE 2 Multiply and simplify.

a. $\dfrac{x+4}{x-5} \cdot \dfrac{x}{x+1}$ **b.** $\dfrac{2x-3}{x^2-9} \cdot \dfrac{x-3}{6x^2-7x-3}$

Solution

a. $\dfrac{x+4}{x-5} \cdot \dfrac{x}{x+1} = \dfrac{(x+4)x}{(x-5)(x+1)}$ Multiply numerators and denominators.

$$= \frac{x^2+4x}{x^2-4x-5}$$

After multiplying and simplifying, the result may have a product of factors in the numerator and denominator. It is not necessary to complete the multiplication of these factors.

Numeric Check

X	Y₁	Y₂
-1	ERROR	ERROR
0	0	0
1	-.625	-.625
2	-1.333	-1.333
3	-2.625	-2.625
4	-6.4	-6.4
5	ERROR	ERROR

X= -1

$Y1 = \dfrac{x^2+4x}{x^2-4x-5}$

$Y2 = \dfrac{x+4}{x-5} \cdot \dfrac{x}{x+1}$

Note that for all x-values, the corresponding Y1 and Y2 are equal.

HELPING HAND For ease in reading and to ensure that the calculator performs the operations correctly, place fractions in parentheses.

b. $\dfrac{2x-3}{x^2-9} \cdot \dfrac{x-3}{6x^2-7x-3}$

$$= \frac{(2x-3)(x-3)}{(x^2-9)(6x^2-7x-3)}$$ Multiply numerators and denominators.

$$= \frac{(2x-3)(x-3)}{(x-3)(x+3)(2x-3)(3x+1)}$$ Factor.

$$= \frac{(2x-3)(x-3)}{(2x-3)(x-3)} \cdot \frac{1}{(x+3)(3x+1)}$$ Factor out the GCF.

$$= \frac{1}{(x+3)(3x+1)}$$ Multiply.

The check is left for you.

Just as we may cancel when simplifying algebraic fractions, we may cancel with care when multiplying algebraic fractions. To do this, we simplify by cancellation before we multiply.

To multiply rational expressions by canceling,

- Factor the numerators and denominators.
- Cancel, or divide out, common factors in pairs—one from the numerator and one from the denominator.
- Multiply the remaining factors.

The product will not need to be simplified after all common factors are canceled. For example, the product in Example 2b may be found by this method.

$$\frac{2x - 3}{x^2 - 9} \cdot \frac{x - 3}{6x^2 - 7x - 3}$$

$$= \frac{2x - 3}{(x + 3)(x - 3)} \cdot \frac{x - 3}{(2x - 3)(3x + 1)} \qquad \text{Factor.}$$

$$= \frac{2x - 3}{(x + 3)(x - 3)} \cdot \frac{x - 3}{(2x - 3)(3x + 1)} \qquad \text{Cancel out factors.}$$

$$= \frac{1}{(x + 3)(3x + 1)} \qquad \text{Multiply.}$$

The check is left for you.

 12.2.2 Checkup

Multiply and simplify.

1. $\dfrac{8x}{15y} \cdot \dfrac{3y^3}{4x^3}$ **2.** $\dfrac{x - 1}{x + 1} \cdot \dfrac{x + 4}{x^2}$ **3.** $\dfrac{10x + 4}{x^2 - 16} \cdot \dfrac{x + 4}{5x^2 - 18x - 8}$

12.2.3 Dividing Rational Expressions

To divide numeric fractions, we multiply by the reciprocal of the divisor and, if necessary, simplify the resulting fraction. For example,

$$\frac{3}{4} \div \frac{5}{6} = \frac{3}{4} \cdot \frac{6}{5} = \frac{18}{20} = \frac{2}{2} \cdot \frac{9}{10} = 1 \cdot \frac{9}{10} = \frac{9}{10}$$

We use the same procedure to divide rational expressions.

$$\frac{15a^3b}{2cd} \div \frac{20ac^2}{3d} = \frac{15a^3b}{2cd} \cdot \frac{3d}{20ac^2} = \frac{45a^3bd}{40ac^3d} = \frac{5ad}{5ad} \cdot \frac{9a^2b}{8c^3} = 1 \cdot \frac{9a^2b}{8c^3} = \frac{9a^2b}{8c^3}$$

DIVIDING RATIONAL EXPRESSIONS
To divide rational expressions,

- Multiply by the reciprocal of the divisor.
- Factor the numerator and denominator.
- Simplify the resulting rational expression, as in

$$\frac{A}{B} \div \frac{C}{D} = \frac{A}{B} \cdot \frac{D}{C} = \frac{AD}{BC}$$

where A, B, C, and D are polynomials with $B \neq 0$, $D \neq 0$, and $C \neq 0$.

EXAMPLE 3 Divide and simplify.

$$\textbf{a.} \ \frac{2x + 1}{2x^2 - 8} \div \frac{2x + 1}{x^4 - 16} \qquad \textbf{b.} \ \frac{x^2 + 8x + 12}{x^2 + 12} \div (x + 6)$$

Solution

Numeric Check

a. $\dfrac{2x + 1}{2x^2 - 8} \div \dfrac{2x + 1}{x^4 - 16}$

$$= \frac{2x + 1}{2x^2 - 8} \cdot \frac{x^4 - 16}{2x + 1}$$

Rewrite as multiplication by reciprocal of divisor.

$$= \frac{(2x + 1)(x^4 - 16)}{(2x^2 - 8)(2x + 1)}$$

Multiply numerators and denominators.

$$= \frac{(2x + 1)(x^2 + 4)(x + 2)(x - 2)}{2(x + 2)(x - 2)(2x + 1)}$$

Factor.

$$= \frac{(2x + 1)(x + 2)(x - 2)}{(2x + 1)(x + 2)(x - 2)} \cdot \frac{x^2 + 4}{2}$$

Factor out the GCF.

$$= \frac{x^2 + 4}{2}$$

Multiply.

X	Y1	Y2
-1	2.5	2.5
0	2	2
1	2.5	2.5
2	4	ERROR
3	6.5	6.5
4	10	10
5	14.5	14.5

X = -1

$$Y1 = \frac{x^2 + 4}{2}$$

$$Y2 = \frac{2x + 1}{2x^2 - 8} \div \frac{2x + 1}{x^4 - 16}$$

Note that Y1 and Y2 values are equal for all values in the domain of the functions.

b. $\dfrac{x^2 + 8x + 12}{x^2 + 12} \div (x + 6)$

$$= \frac{x^2 + 8x + 12}{x^2 + 12} \cdot \frac{1}{x + 6}$$

Rewrite as multiplication by the reciprocal of the divisor.

$$= \frac{x^2 + 8x + 12}{(x^2 + 12)(x + 6)}$$

Multiply numerators and denominators.

$$= \frac{(x + 6)(x + 2)}{(x^2 + 12)(x + 6)}$$

Factor.

$$= \frac{x + 6}{x + 6} \cdot \frac{x + 2}{x^2 + 12}$$

Factor out the GCF.

$$= \frac{x + 2}{x^2 + 12}$$

Multiply.

The check is left for you.

We may also cancel, with care, when dividing rational expressions.

To divide rational expressions by canceling,

- Multiply by the reciprocal of the divisor.
- Factor the numerators and denominators.
- Cancel, or divide out, common factors in pairs—one from the numerator and one from the denominator.
- Multiply the remaining factors.

The product will not need to be simplified after all common factors are canceled.

For Example 3a,

$$\frac{2x + 1}{2x^2 - 8} \div \frac{2x + 1}{x^4 - 16}$$

$$= \frac{2x + 1}{2x^2 - 8} \cdot \frac{x^4 - 16}{2x + 1} \qquad \text{Rewrite as multiplication.}$$

$$= \frac{2x + 1}{2(x + 2)(x - 2)} \cdot \frac{(x^2 + 4)(x + 2)(x - 2)}{2x + 1} \qquad \text{Factor.}$$

$$= \frac{\cancel{2x + 1}}{2\cancel{(x + 2)}\cancel{(x - 2)}} \cdot \frac{(x^2 + 4)\cancel{(x + 2)}\cancel{(x - 2)}}{\cancel{2x + 1}} \qquad \text{Cancel out common factors.}$$

$$= \frac{x^2 + 4}{2} \qquad \text{Multiply.}$$

 **HELPING HAND** Remember, we cancel common factors. This means that we must be canceling in a *multiplication* problem. Therefore, *always* change the division problem to multiplication *before* canceling factors.

 12.2.3 Checkup

Divide and simplify exercises 1–3.

1. $\dfrac{25x^3 y^2}{z^2} \div \dfrac{20xy^3}{z^3}$ **2.** $\dfrac{5a + 7}{3x^2 - 3} \div \dfrac{10a + 14}{9x^4 - 9}$

3. $\dfrac{5x^2 + 19x - 4}{x^2 - 4} \div (5x - 1)$

4. Why must you first change a division problem to a multiplication problem before performing any cancellation of factors?

12.2.4 Modeling the Real World

Rational expressions occur in many models of real-world problems. One common example comes from the distance-traveled formula, $d = rt$, where d is the distance traveled by a moving object, r is the average rate, or speed, of the object (which we assume to be constant), and t is the time of travel. If we solve this equation for the time t, we get the rational expression $t = \frac{d}{r}$.

EXAMPLE 4 Balto the rescue dog traveled a distance of 125 miles.

a. Write an expression for the time traveled if Balto traveled at a speed of x mph.

b. Write an expression for the new distance Balto traveled if he kept the time constant, but increased his average speed by 10 mph.

Solution

a. Let x = rate in miles per hour

$$t = \frac{d}{r} \qquad \text{Distance-traveled formula, solved for } t$$

$$t = \frac{125}{x} \qquad \text{Substitute 125 for } d \text{ and } x \text{ for } r.$$

An expression for the time Balto traveled is $\frac{125}{x}$ hours.

b. Since Balto increased his average speed by 10 mph, his new speed (rate) is $x + 10$. The time is still $\frac{125}{x}$.

$$d = rt \qquad \text{\small Distance-traveled formula}$$

$$d = (x + 10)\left(\frac{125}{x}\right)$$

$$d = \left(\frac{x + 10}{1}\right)\left(\frac{125}{x}\right)$$

$$d = \frac{125(x + 10)}{x}$$

An expression for the new distance Balto traveled is $\frac{125(x + 10)}{x}$ miles.

APPLICATION

Sio signed a contract for her new job. According to the contract, she is to make $20,000 the first year and receive an increase of 5% per year for the next two years. Sio's total salary for three years is given by the function $s(x) = \frac{20,000 - 20,000x^3}{1 - x}$, where x is the sum of 1 and the percentage increase converted to a decimal, or 1.05. Write an equivalent function by simplifying $s(x)$. Determine Sio's total salary for three years, using the original function and the simplified function.

Discussion

$$s(x) = \frac{20,000 - 20,000x^3}{1 - x}$$

$$s(x) = \frac{20,000(1 - x^3)}{1 - x}$$

$$s(x) = \frac{20,000(\overset{1}{\cancel{1 - x}})(1 + x + x^2)}{\underset{1}{\cancel{1 - x}}}$$

$$s(x) = 20,000(1 + x + x^2)$$
$$s(x) = 20,000 + 20,000x + 20,000x^2$$

Evaluate each function for $x = 1.05$.

Original function

$$s(x) = \frac{20,000 - 20,000x^3}{1 - x}$$

$$s(1.05) = \frac{20,000 - 20,000(1.05)^3}{1 - 1.05}$$

$$s(1.05) = 63,050$$

Simplified function

$$s(x) = 20,000 + 20,000x + 20,000x^2$$
$$s(1.05) = 20,000 + 20,000(1.05) + 20,000(1.05)^2$$
$$s(1.05) = 63,050$$

Sio will receive $63,050 over the three-year period. (This simplified function was determined in Section 9.5 by division.)

12.2.4 Checkup

1. Reza traveled 150 miles.

   a. Write an expression for his average rate of speed if it took x hours for the trip.

   b. If Reza were to double his rate of speed and increase his time by 1 hour, write an expression for the new distance he would travel.

2. When Benjamin enrolled at a community college, he received a scholarship of $1500. The following year, the scholarship was increased by 20%. The total amount of money Benjamin received for his two years of study is given by $s(y) = \frac{1500 - 1500y^2}{1 - y}$, where y is the sum of the percentage increase (converted to decimal form) and 1. Write an equivalent function by simplifying. Using the original function and the simplified function, determine the total amount Benjamin received.

12.2 Exercises

Simplify.

1. $\dfrac{36x^3y^2z}{54x^2y^5}$

2. $\dfrac{21x^3y^4}{35x^5yz^2}$

3. $\dfrac{-56a^3b^2}{84a^2b^2c}$

4. $\dfrac{99ab^5c}{-72a^3b^2}$

5. $\dfrac{9x-27}{18x-6}$

6. $\dfrac{8x+16}{24x+12}$

7. $\dfrac{8-z}{z-8}$

8. $\dfrac{x-y}{y-x}$

9. $\dfrac{5x^2y^3+10xy^4}{15xy^2-20x^2y^2}$

10. $\dfrac{15x^2y-9x^3y}{6x^2y^2+3xy^3}$

11. $\dfrac{-6x+9y-3z}{3x-21y+3z}$

12. $\dfrac{-7x+28y-14z}{14x-7y+7z}$

13. $\dfrac{x^2-8x+12}{x^2-10x+24}$

14. $\dfrac{x^2-6x+9}{x^2-10x+21}$

15. $\dfrac{x^2-x-6}{x^2+9x+14}$

16. $\dfrac{x^2+12x+32}{x^2-x-20}$

17. $\dfrac{3x^2+7x+2}{3x^2-11x-4}$

18. $\dfrac{2x^2+11x+5}{2x^2-17x-9}$

19. $\dfrac{6x^2-x-2}{4x^2-4x-3}$

20. $\dfrac{10x^2+7x-12}{8x^2+2x-15}$

21. $\dfrac{8x^2+12x+4}{24x^2-40x-16}$

22. $\dfrac{12x^2+6x-6}{27x^2+45x-18}$

23. $\dfrac{2x^2+xy-y^2}{3x^2+4xy+y^2}$

24. $\dfrac{5x^2-6xy+y^2}{4x^2-3xy-y^2}$

25. $\dfrac{64-x^2}{x^3+2x^2-64x-128}$

26. $\dfrac{25-x^2}{x^3+3x^2-25x-75}$

27. $\dfrac{2x^2-7x-15}{x^3-5x^2+11x-55}$

28. $\dfrac{3x^2-17x-28}{x^3-7x^2+9x-63}$

Multiply and simplify.

29. $\dfrac{12a^2b^2}{35cd^2}\cdot\dfrac{49c^2d^2}{27ab^3}$

30. $\dfrac{15a^3b^2}{16c^3d^5}\cdot\dfrac{24c^3d^3}{25ab^4}$

31. $\dfrac{-2x^3}{5y^2}\cdot\dfrac{15xy^2}{8x^2}$

32. $\dfrac{12x}{15y^3}\cdot\dfrac{-5xy^5}{26x^4}$

33. $(2.6x^2)\cdot\dfrac{3.1x}{y^2}$

34. $\dfrac{7x^2y}{6z}\cdot 9$

35. $\dfrac{1}{3}\cdot\dfrac{2a^2}{3b}$

36. $\dfrac{7a^3}{3b^2}\cdot\dfrac{2}{5}$

37. $\dfrac{x+5}{x}\cdot\dfrac{x^3}{x+4}$

38. $\dfrac{x^3}{x-11}\cdot\dfrac{x+15}{x^2}$

39. $\dfrac{3x+2}{x+4}\cdot\dfrac{x-4}{3x+2}$

40. $\dfrac{x+7}{4x+3}\cdot\dfrac{4x+3}{x-7}$

41. $\dfrac{x+7}{2x+1}\cdot\dfrac{5x+2}{x}$

42. $\dfrac{3x+5}{x}\cdot\dfrac{x+4}{2x+9}$

43. $\dfrac{x^2-x-6}{2x^2+3x+1}\cdot\dfrac{2x^2-x-1}{x^2+5x+6}$

44. $\dfrac{x^2+8x+15}{3x^2-5x-2}\cdot\dfrac{3x^2+7x+2}{x^2-2x-15}$

45. $\dfrac{x^2-xy-2y^2}{x^2-9y^2}\cdot\dfrac{x^2-3xy}{x^2-y^2}$

46. $\dfrac{xy-5y^2}{x^2-4y^2}\cdot\dfrac{x^2-xy-2y^2}{x^2-25y^2}$

Divide and simplify exercises 47–70.

47. $\dfrac{54a^3b^2}{c}\div\dfrac{36ab^2}{c}$

48. $\dfrac{49a^4b^2}{c^2}\div\dfrac{56a^3b^3}{c^2}$

49. $\dfrac{-21x^2y^3}{z}\div\dfrac{7xy^2}{z^2}$

50. $\dfrac{33x^3y}{z^3}\div\dfrac{-11x^5y^3}{z^2}$

51. $(32ab^2)\div\dfrac{8ab^3}{c}$

52. $(21a^2b^4)\div\dfrac{7ab^2}{c^2}$

53. $\dfrac{21x^2y^2}{z^3}\div(-3xy^3)$

54. $\dfrac{-34x^3y}{z^2}\div(17x^2y^4)$

55. $\dfrac{a+4}{9a^2-1}\div\dfrac{5a+20}{3a+1}$

56. $\dfrac{7a+21}{36a^2-1}\div\dfrac{a+3}{6a-1}$

57. $\dfrac{x^2-16}{5y-2}\div\dfrac{x+4}{25y^2-4}$

58. $\dfrac{x^2-49}{4y+3}\div\dfrac{x+7}{16y^2-9}$

59. $\dfrac{2x^2-5x-3}{x^2-5x-14}\div\dfrac{3x^2+10x+3}{x^2-4x-21}$

60. $\dfrac{x^2-2x-8}{2x^2+x-3}\div\dfrac{3x^2+8x+4}{2x^2-5x-12}$

61. $\dfrac{8x^2+2x-3}{4x^2-1}\div(4x+3)$

62. $\dfrac{15x^2+16x+4}{9x^2-4}\div(5x+2)$

63. $(x+9)\div\dfrac{2x^2+19x+9}{3x+4}$

64. $(x+11)\div\dfrac{2x^2+17x-55}{5x-2}$

65. $\dfrac{1}{3x-5}\div\dfrac{1}{x+7}$

66. $\dfrac{5}{4x+3}\div\dfrac{7}{2x+1}$

67. $\dfrac{x^2+3xy}{7xy+2y^2}\div\dfrac{x^2+4xy+3y^2}{7x^2-5xy-2y^2}$

68. $\dfrac{x^2+7xy+10y^2}{3x^2-4xy-4y^2}\div\dfrac{x^2+5xy}{3xy+2y^2}$

69. $\dfrac{4x+8y}{3x-12y}\div\dfrac{4x+4y}{6x+6y}$

70. $\dfrac{5x+35y}{6x+18y}\div\dfrac{5x-10y}{8x+24y}$

71. Katarina drove a distance of 170 miles. If her average rate of travel (speed) was x miles per hour, write a rational expression for her time of travel. If she had traveled for the same amount of time, but had decreased her average speed by 10 mph, write an expression for the distance she would have traveled. How far would she have traveled at the reduced speed if her original speed was 50 miles per hour?

72. Brian traveled 155 miles. Write an expression for his time of travel if his average speed was x mph. If he travels the same length of time, but increases his speed by 8 mph, write an expression for the distance he would travel. How far would he travel at the increased speed if his original speed was 45 miles per hour?

73. To encourage her grandson to do well in college, Sharon promised him $500 for completing his first semester of a full-time course load. She promised to increase the amount by 20% for each subsequent semester of a full-time course load that he completed. The total amount that the grandson received was $m(x) = \frac{500 - 500x^3}{1 - x}$, where x is the sum of 1 and the percentage increase (converted to decimal form). Write an equivalent function by simplifying this rational function. Using the original function and the simplified function, determine the total amount that the grandson received.

74. Alexandra is a sales representative for a pharmaceutical company. Her sales during the first week of the month totaled $50,000 and increased by 25% each week for the next two weeks. Her total sales for the three-week period were $s(x) = \frac{50,000 - 50,000x^3}{1 - x}$, where x is the sum of 1 and the percentage increase (converted to decimal form). Write an equivalent function by simplifying this rational function. Using the original function and the simplified function, determine the total amount of Alexandra's sales.

75. A box has a volume given by the formula $V = LWH$. Solve this formula for L. If the volume of the box is $15x^3 + 7x^2 - 2x$ cubic inches, its height is x inches, and its width is $3x + 2$ inches, use the new formula to determine an expression for the length in terms of the height x. If the height of the box measures 4 inches, what are the box's other dimensions?

76. A cylindrical can has a volume given by the formula $V = \pi r^2 h$. Solve this formula for h. If a can has a volume of $6\pi x^3 - 5\pi x^2$ cubic inches and its radius is x inches, use the new formula to determine an expression for the height of the can in terms of the radius x. If the radius of the can measures 1.5 inches, what is its height?

77. The fixed cost of producing an item is $300 and the variable cost is $5 per item produced. The revenue received for selling the item is $25 per item.
   a. Write a cost function and a revenue function for producing and selling x items.
   b. Write an expression for the ratio of revenue to cost for x items, and simplify the expression.
   c. If a production run produces 40 items, what is the value of the revenue-to-cost ratio?

78. The fixed cost of producing an item is $175 and the variable cost is $25 per item produced. The revenue received for selling the item is $75 per item.
   a. Write a cost function and a revenue function for producing and selling x items.
   b. Write an expression for the ratio of cost to revenue for x items, and simplify the expression.
   c. If a production run produces 50 items, what is the value of the cost-to-revenue ratio?

12.2 Calculator Exercises

In this section, you have used functions of the form $f(x) = \frac{k(1 - x^n)}{(1 - x)}$, where k is a constant and n is the integer 2 or 3. This function was also used in earlier chapters. It is an interesting function, as you have seen. When n equals 2 or 3, you can factor the numerator, and one of the factors is the denominator, so you can simplify the function after factoring.

Let's explore the function on the calculator. To make the exploration easier, we will set k equal to 1 and $f(x)$ equal to y. Then the function becomes $y = \frac{(1 - x^n)}{(1 - x)}$.

One property of this function is that when you specify a value for n, the denominator always divides evenly into the numerator. The resulting function has degree that is one less than n.

Complete the table that follows. Use your calculator to sketch a graph of the original function and the simplified function for each value of n. To expedite the graphing, store the function as Y1 = $(1 - X \wedge N)/(1 - X)$, and store the various values for N before viewing each graph. The graphs can best be viewed with a window of $(-4.7, 4.7, 1, -31, 31, 10, 1)$. When sketching the graphs, be sure to show the restricted values as holes or asymptotes. Use long division to find the simplified functions. Can you discern a pattern?

Value for n	Original function	Simplified function
2	$y = \dfrac{(1 - x^2)}{(1 - x)}$	$y = 1 + x$
3	$y = \dfrac{(1 - x^3)}{(1 - x)}$	$y = 1 + x + x^2$
4	$y = \dfrac{(1 - x^4)}{(1 - x)}$	
5		
6		

12.2 Writing Exercise

The rational expression $\frac{x^3 + 3x^2 - 4x - 12}{x^3 + 2x^2 - 9x - 18}$ simplifies to $\frac{x - 2}{x - 3}$. Prove to yourself that this is so. The original expression has more restricted values than the simplified expression. Determine what the restricted values are for both expressions. Finally, discuss the implications of this difference in restricted values for the two expressions. If you simplify a rational expression, which should you use as the restricted values in the domain of the function, those from the simplified expression or those from the original expression? Explain the reasons for your answer.

12.3 Addition and Subtraction of Rational Expressions

OBJECTIVES

1 Add and subtract rational expressions with like denominators.

2 Determine the least common denominator for a set of rational expressions.

3 Write equivalent rational expressions, using the least common denominator as the new denominator.

4 Add and subtract rational expressions with unlike denominators.

5 Model real-world situations by using expressions involving addition and subtraction of rational expressions.

APPLICATION

Julio travels from City A to City B at an average rate of 40 miles per hour. He makes the return trip at an average rate of 60 miles per hour. Determine his average rate for the round-trip.

After completing this section, we will discuss this application further. See page 839.

12.3.1 Adding and Subtracting Rational Expressions with Like Denominators

Now that we can multiply rational expressions, we are ready to add and subtract them. To add or subtract numeric fractions with like denominators, we add or subtract the numerators in order to obtain the numerator of the answer. The denominator of the answer is the like denominator. We simplify the result if necessary. For example,

$$\frac{3}{10} + \frac{1}{10} = \frac{3+1}{10} = \frac{4}{10} = \frac{2}{2} \cdot \frac{2}{5} = 1 \cdot \frac{2}{5} = \frac{2}{5}$$

$$\frac{3}{10} - \frac{1}{10} = \frac{3-1}{10} = \frac{2}{10} = \frac{2}{2} \cdot \frac{1}{5} = 1 \cdot \frac{1}{5} = \frac{1}{5}$$

We use the same procedure for adding and subtracting rational expressions.

$$\frac{2x}{5y} + \frac{x^2}{5y} = \frac{2x + x^2}{5y} = \frac{x(2+x)}{5y}$$

$$\frac{7ab}{2a} - \frac{-3a^2}{2a} = \frac{7ab - (-3a^2)}{2a} = \frac{7ab + 3a^2}{2a}$$

$$= \frac{a(7b + 3a)}{2a} = \frac{a}{a} \cdot \frac{7b + 3a}{2} = \frac{7b + 3a}{2}$$

ADDING AND SUBTRACTING RATIONAL EXPRESSIONS WITH LIKE DENOMINATORS

To add rational expressions,

- Add the numerators.
- Use the like denominator for the denominator of the sum, as in

$$\frac{A}{B} + \frac{C}{B} = \frac{A + C}{B}$$

where A, B, and C are polynomials with $B \neq 0$.

To subtract rational expressions,

- Subtract the numerators.
- Use the like denominator for the denominator of the difference, as in

$$\frac{A}{B} - \frac{C}{B} = \frac{A - C}{B}$$

where A, B, and C are polynomials with $B \neq 0$.

Remember to simplify the answer if possible.

EXAMPLE 1 Add or subtract and simplify.

a. $\dfrac{2x}{x+3} + \dfrac{6}{x+3}$ **b.** $\dfrac{3x}{3x^2 - 13x + 14} - \dfrac{7}{3x^2 - 13x + 14}$

c. $\dfrac{x+4}{x^2 + 3x - 4} - \dfrac{x-1}{x^2 + 3x - 4}$

Solution

Numeric Check

a. $\dfrac{2x}{x+3} + \dfrac{6}{x+3} = \dfrac{2x+6}{x+3}$ Add numerators.

$= \dfrac{2(x+3)}{x+3}$ Factor.

$= 2$ Simplify.

X	Y₁	Y₂
-3	2	ERROR
-2	2	2
-1	2	2
0	2	2
1	2	2
2	2	2
3	2	2

X= -3

Y1 = 2

Y2 = $\dfrac{2x}{x+3} + \dfrac{6}{x+3}$

b. $\dfrac{3x}{3x^2 - 13x + 14} - \dfrac{7}{3x^2 - 13x + 14}$

$= \dfrac{3x-7}{3x^2 - 13x + 14}$ Subtract numerators.

$= \dfrac{3x-7}{(3x-7)(x-2)}$ Factor.

$= \dfrac{1}{x-2}$ Simplify.

Note that all Y1 and Y2 values are equal for the x-values in the domain of the functions.

The check is left for you.

c. $\dfrac{x+4}{x^2 + 3x - 4} - \dfrac{x-1}{x^2 + 3x - 4}$

$= \dfrac{(x+4) - (x-1)}{x^2 + 3x - 4}$ Subtract numerators.

$= \dfrac{5}{x^2 + 3x - 4}$

The check is left for you.

 HELPING HAND Be careful when subtracting numerators with more than one term. It is helpful to use parentheses to group the numerators, as in Example 1c.

✓ **12.3.1** Checkup

Add or subtract and simplify exercises 1–4.

1. $\dfrac{5xy}{3z} - \dfrac{7x}{3z}$ **2.** $\dfrac{3x}{x+7} + \dfrac{21}{x+7}$

3. $\dfrac{x+5}{2x^2 + 5x - 12} + \dfrac{x-8}{2x^2 + 5x - 12}$

4. $\dfrac{2x+7}{x^2 + 3x - 18} - \dfrac{x+7}{x^2 + 3x - 18}$

5. What must be true before you can add or subtract rational expressions?

6. After adding or subtracting rational expressions, what should you do to the result (in addition to checking your work)?

12.3.2 Determining the Least Common Denominator

In order to add or subtract numeric fractions, we must verify that the fractions have like denominators. If they do not, we must change them to equivalent fractions with a common denominator. The **least common denominator (LCD)** is the smallest number of which each denominator is a factor. That is, the LCD is the smallest number divisible by each of the denominators.

To determine the least common denominator of a set of fractions, first factor each of the denominators into its prime factors. Next, write a product consisting of each different denominator factor, using the largest exponent that occurred for each factor.

For example, determine the LCD of $\frac{7}{12}$ and $\frac{1}{30}$.

Factor the denominators.

$$12 = 2^2 \cdot 3$$
$$30 = 2 \cdot 3 \cdot 5$$

The different denominator factors are 2, 3, and 5; the greatest exponent of 2 is 2, of 3 is 1, and of 5 is 1.

Multiply.

$$\text{LCD} = 2^2 \cdot 3 \cdot 5 = 60$$

We use the same procedure for finding the LCD of rational expressions.

For example, determine the least common denominator of $\frac{7x}{8xy^3}$ and $\frac{3z}{20x^2y}$.

Factor the denominators.

$$8xy^3 = 2^3 \cdot x \cdot y^3$$
$$20x^2y = 2^2 \cdot 5 \cdot x^2 \cdot y$$

The different denominator factors are 2, 5, x, and y. The greatest exponent of 2 is 3, of 5 is 1, of x is 2, and of y is 3.

Multiply.

$$\text{LCD} = 2^3 \cdot 5 \cdot x^2 \cdot y^3 = 40x^2y^3$$

DETERMINING THE LEAST COMMON DENOMINATOR
To determine the least common denominator of a set of rational expressions,

- Factor each denominator into its prime factors.
- Write a product consisting of each different factor, using the greatest exponent that occurs for each.

EXAMPLE 2 Determine the least common denominator in each of the following sets of rational expressions:

a. $\frac{3x}{x+3}$ and $\frac{2x}{x+2}$ **b.** $\frac{x+5}{4x^2-9}$ and $\frac{3x-1}{2x^2-x-3}$

c. $\frac{2}{x-2}$ and $\frac{-2}{-x+2}$

Solution

a. $\frac{3x}{x+3}$ and $\frac{2x}{x+2}$

The denominators do not factor. Therefore, we use both denominators as factors for the LCD.

$$\text{LCD} = (x+3)(x+2)$$

There is no need to multiply these factors.

b. $\dfrac{x+5}{4x^2-9}$ and $\dfrac{3x-1}{2x^2-x-3}$

Factor the denominators.

$$4x^2 - 9 = (2x+3)(2x-3)$$
$$2x^2 - x - 3 = (2x-3)(x+1)$$
$$\text{LCD} = (2x+3)(2x-3)(x+1)$$

c. $\dfrac{2}{x-2}$ and $\dfrac{-2}{-x+2}$

Notice that the denominators are opposites. Therefore, either denominator may be considered the LCD.

$$\text{LCD} = x-2 \quad \text{or} \quad \text{LCD} = -x+2$$

12.3.2 Checkup

In exercises 1–4, determine the least common denominator in each set of rational expressions.

1. $\dfrac{5x}{12x^2y}$ and $\dfrac{4z}{15xy^3}$

2. $\dfrac{5}{x+4}$ and $\dfrac{7x}{x-5}$

3. $\dfrac{x-1}{25x^2-1}$ and $\dfrac{2x+5}{15x^2-2x-1}$

4. $\dfrac{7}{x-5}$ and $\dfrac{8}{5-x}$

5. What is the meaning of the word *least* in relation to finding the least common denominator of a set of rational expressions?

6. To find the least common denominator of a set of rational expressions, you should first factor the expressions into prime factors. What is the meaning of the word *prime*?

12.3.3 Writing Equivalent Rational Expressions with the LCD

Now that we can determine the LCD of a set of rational expressions, we are ready to change the expressions to equivalent expressions, with the LCD as the new denominator of each. We use the multiplication identity property to do this.

Recall the procedure for writing equivalent numeric fractions. First, we must determine the factor needed to multiply the original denominator in order to obtain the new denominator. Next, we multiply the original numeric fraction by a factor of 1 (using the needed factor as the numerator and denominator).

For example, write a numeric fraction equivalent to $\frac{3}{4}$, but with a denominator of 20.

$$\frac{3}{4} = \frac{?}{20} \qquad \text{Since } 20 = 4 \cdot 5, \text{ we need a factor of 5.}$$

$$\frac{3}{4} = \frac{3}{4} \cdot \frac{5}{5} = \frac{15}{20} \qquad \text{Multiply by a factor of 1, } \tfrac{5}{5}.$$

We use the same procedure to write equivalent rational expressions. For example, write a rational number equivalent to $\frac{7x}{8xy^3}$, but with a denominator of $40x^2y^3$.

$$\frac{7x}{8xy^3} = \frac{?}{40x^2y^3}$$

Factor the denominators.

$$40x^2y^3 = 2^3 \cdot 5 \cdot x^2 \cdot y^3$$
$$8xy^3 = 2^3 \cdot \quad x \cdot y^3$$

HELPING HAND It is easier to see the "needed" factors if we arrange the factors in the same order, leaving a space for the missing factors, such as 5.

The given denominator, $8xy^3$, must be multiplied by $5x$ to obtain the desired denominator, $40x^2y^3$.

$$\frac{7x}{8xy^3} \cdot \frac{5x}{5x} = \frac{35x^2}{40x^2y^3}$$

DETERMINING EQUIVALENT RATIONAL EXPRESSIONS

To determine an equivalent rational expression,

- Factor the denominator of the original rational expression and the new desired denominator (LCD).
- Determine the factors needed to multiply the original denominator to obtain the desired denominator.
- Multiply the original rational expression by a factor of 1, using the factors found in the preceding step as the numerator and denominator.

EXAMPLE 3 Write an equivalent rational expression with the given LCD.

a. $\dfrac{3x}{x + 3}$; LCD $= x^2 + 5x + 6$

b. $\dfrac{x + 5}{4x^2 - 9}$; LCD $= 4x^4 + 4x^3 - 9x^2 - 9x$

c. $\dfrac{2}{x - 2}$; LCD $= 2 - x$

Solution

a. $\dfrac{3x}{x + 3}$; LCD $= x^2 + 5x + 6$

Factor the denominators.

$$x^2 + 5x + 6 = (x + 3)(x + 2)$$
$$x + 3 \qquad = x + 3 \qquad\qquad \text{\small x + 3 does not factor.}$$

The given denominator must be multiplied by $(x + 2)$ to obtain the desired denominator.

$$\frac{3x}{x + 3} \cdot \frac{x + 2}{x + 2} = \frac{3x^2 + 6x}{x^2 + 5x + 6}$$

b. $\dfrac{x + 5}{4x^2 - 9}$; LCD $= 4x^4 + 4x^3 - 9x^2 - 9x$

Factor the denominators.

$$4x^4 + 4x^3 - 9x^2 - 9x = x(2x + 3)(2x - 3)(x + 1)$$
$$4x^2 - 9 \qquad\qquad = (2x + 3)(2x - 3)$$

The given denominator must be multiplied by $x(x + 1)$ to obtain the desired denominator.

$$\frac{x + 5}{4x^2 - 9} \cdot \frac{x(x + 1)}{x(x + 1)} = \frac{x^3 + 6x^2 + 5x}{4x^4 + 4x^3 - 9x^2 - 9x}$$

c. $\frac{2}{x - 2}$; LCD $= 2 - x$

The denominators do not factor. However, they are opposites. Therefore, the given denominator must be multiplied by -1 to obtain the desired denominator.

$$\frac{2}{x - 2} \cdot \frac{-1}{-1} = \frac{-2}{-x + 2} \quad \text{or} \quad \frac{-2}{2 - x}$$

12.3.3 Checkup

Write an equivalent rational expression with the given least common denominator.

1. $\frac{-4x}{3xy^2}$; LCD $= 12x^2y^2$ **2.** $\frac{2x}{x + 5}$; LCD $= x^2 + 9x + 20$

3. $\frac{2x + 1}{x^2 - 16}$; LCD $= x^3 + 7x^2 - 16x - 112$

4. $\frac{7}{x - 5}$; LCD $= 5x - x^2$

12.3.4 Adding and Subtracting Rational Expressions with Unlike Denominators

To add or subtract numeric fractions with unlike denominators, we change the fractions to equivalent fractions with like denominators and add or subtract. For example, to add $\frac{7}{12} + \frac{1}{30}$, we use the LCD, 60, that we found earlier.

$$\frac{7}{12} + \frac{1}{30} = \frac{7}{12} \cdot \frac{5}{5} + \frac{1}{30} \cdot \frac{2}{2} \qquad \text{Change to like denominators.}$$

$$= \frac{35}{60} + \frac{2}{60}$$

$$= \frac{37}{60} \qquad \text{Add numerators.}$$

We use the same procedure to add and subtract rational expressions. For example,

$$\frac{7x}{8xy^3} + \frac{3x^2}{20x^2y} = \frac{7x}{8xy^3} \cdot \frac{5x}{5x} + \frac{3x^2}{20x^2y} \cdot \frac{2y^2}{2y^2} \qquad \begin{array}{l}\text{Change to like denominators.}\\ \text{We found the LCD, } 40x^2y^3\text{, earlier.}\end{array}$$

$$= \frac{35x^2}{40x^2y^3} + \frac{6x^2y^2}{40x^2y^3}$$

$$= \frac{35x^2 + 6x^2y^2}{40x^2y^3} \qquad \text{Add numerators.}$$

$$= \frac{35 + 6y^2}{40y^3} \cdot \frac{x^2}{x^2} \qquad \text{Simplify.}$$

$$= \frac{35 + 6y^2}{40y^3}$$

ADDING AND SUBTRACTING RATIONAL EXPRESSIONS WITH UNLIKE DENOMINATORS

To add rational expressions,

- Change the rational expressions to equivalent expressions with like denominators.
- Add the numerators.
- Use the like denominator for the denominator of the sum, as in

$$\frac{A}{B} + \frac{C}{B} = \frac{A + C}{B}$$

where A, B, and C are polynomials with $B \neq 0$.

To subtract rational expressions,

- Change the rational expressions to equivalent expressions with like denominators.
- Subtract the numerators.
- Use the like denominator for the denominator of the difference, as in

$$\frac{A}{B} - \frac{C}{B} = \frac{A - C}{B}$$

where A, B, and C are polynomials with $B \neq 0$.

Remember to simplify the answer if possible.

EXAMPLE 4 Add or subtract and simplify.

a. $\dfrac{3x}{x + 3} + \dfrac{2x}{x + 2}$ **b.** $\dfrac{5}{x + 4} + \dfrac{x - 4}{x^2 + 8x + 16}$

c. $\dfrac{2}{x - 2} + \dfrac{2}{2 - x}$ **d.** $\dfrac{x + 5}{4x^2 - 9} - \dfrac{x - 1}{2x^2 - x - 3}$

e. $\dfrac{x}{x + 3} - \dfrac{2x}{x - 3} + \dfrac{x^2 + 10x + 3}{x^2 - 9}$

Solution

a. $\dfrac{3x}{x + 3} + \dfrac{2x}{x + 2}$

Neither denominator factors. LCD $= (x + 3)(x + 2)$. Multiply the first rational expression by $\frac{x + 2}{x + 2}$ and the second rational expression by $\frac{x + 3}{x + 3}$.

$= \dfrac{3x(x + 2)}{(x + 3)(x + 2)} + \dfrac{2x(x + 3)}{(x + 2)(x + 3)}$

$= \dfrac{3x^2 + 6x}{(x + 3)(x + 2)} + \dfrac{2x^2 + 6x}{(x + 2)(x + 3)}$ 　Multiply each numerator. Do not multiply the denominator until after you add.

$= \dfrac{5x^2 + 12x}{(x + 3)(x + 2)}$ 　Add numerators.

$= \dfrac{x(5x + 12)}{(x + 3)(x + 2)}$ 　Factor.

Numeric Check

X	Y1	Y2
-3	ERROR	ERROR
-2	ERROR	ERROR
-1	-3.5	-3.5
0	0	0
1	1.4167	1.4167
2	2.2	2.2
3	2.7	2.7

X= -3

$Y1 = \dfrac{x(5x + 12)}{(x + 3)(x + 2)}$

$Y2 = \dfrac{3x}{x + 3} + \dfrac{2x}{x + 2}$

For all x-values in the domain of the functions, $Y1 = Y2$.

b. $\dfrac{5}{x+4} + \dfrac{x-4}{x^2+8x+16} = \dfrac{5}{x+4} + \dfrac{x-4}{(x+4)^2}$ *Factor the denominators.*

The largest power of the common factor, $x + 4$, is 2. Therefore, LCD $= (x+4)^2$. Multiply the first expression by $\frac{x+4}{x+4}$.

$$= \dfrac{5(x+4)}{(x+4)(x+4)} + \dfrac{x-4}{(x+4)^2}$$

$$= \dfrac{5x+20}{(x+4)^2} + \dfrac{x-4}{(x+4)^2} \qquad \text{\textit{Multiply.}}$$

$$= \dfrac{6x+16}{(x+4)^2} \qquad \text{\textit{Add numerators.}}$$

$$= \dfrac{2(3x+8)}{(x+4)^2} \qquad \text{\textit{Factor to simplify.}}$$

The check is left for you.

c. $\dfrac{2}{x-2} + \dfrac{2}{2-x}$

The denominators do not factor. However, they are opposites: $x - 2$ and $2 - x$. Therefore, either denominator may be used as the LCD. We will use $x - 2$. Multiply the second expression by $\frac{-1}{-1}$.

$$= \dfrac{2}{x-2} + \dfrac{2(-1)}{(2-x)(-1)}$$

$$= \dfrac{2}{x-2} + \dfrac{-2}{x-2} \qquad \text{\textit{Multiply.}}$$

$$= \dfrac{2-2}{x-2} \qquad \text{\textit{Subtract numerators.}}$$

$$= \dfrac{0}{x-2} = 0 \qquad \text{\textit{Simplify.}}$$

The check is left for you.

d. $\dfrac{x+5}{4x^2-9} - \dfrac{x-1}{2x^2-x-3} = \dfrac{(x+5)}{(2x+3)(2x-3)} - \dfrac{(x-1)}{(2x-3)(x+1)}$

Factor the denominators.

$$4x^2 - 9 \quad = (2x+3)(2x-3)$$
$$2x^2 - x - 3 = \quad\quad (2x-3)(x+1)$$

LCD $= (2x+3)(2x-3)(x+1)$. Multiply the first expression by $\frac{x+1}{x+1}$ and the second expression by $\frac{2x+3}{2x+3}$.

$$= \frac{(x+5)(x+1)}{(2x+3)(2x-3)(x+1)} - \frac{(x-1)(2x+3)}{(2x-3)(x+1)(2x+3)}$$

$$= \frac{x^2+6x+5}{(2x+3)(2x-3)(x+1)} - \frac{2x^2+x-3}{(2x-3)(x+1)(2x+3)} \qquad \text{Multiply.}$$

$$= \frac{(x^2+6x+5) - (2x^2+x-3)}{(2x+3)(2x-3)(x+1)} \qquad \text{Subtract}$$
numerators.

$$= \frac{x^2+6x+5-2x^2-x+3}{(2x+3)(2x-3)(x+1)}$$

$$= \frac{-x^2+5x+8}{(2x+3)(2x-3)(x+1)} \qquad \text{Simplify.}$$

The check is left for you.

e. $\dfrac{x}{x+3} - \dfrac{2x}{x-3} + \dfrac{x^2+10x+3}{x^2-9} = \dfrac{x}{x+3} - \dfrac{2x}{x-3} + \dfrac{x^2+10x+3}{(x+3)(x-3)}$

Factor the denominators. LCD $= (x+3)(x-3)$. Multiply the first expression by $\frac{x-3}{x-3}$ and the second expression by $\frac{x+3}{x+3}$.

$$= \frac{x(x-3)}{(x+3)(x-3)} - \frac{2x(x+3)}{(x-3)(x+3)} + \frac{x^2+10x+3}{(x+3)(x-3)}$$

$$= \frac{x^2-3x}{(x+3)(x-3)} - \frac{2x^2+6x}{(x-3)(x+3)} + \frac{x^2+10x+3}{(x+3)(x-3)} \qquad \text{Multiply.}$$

$$= \frac{(x^2-3x) - (2x^2+6x) + (x^2+10x+3)}{(x+3)(x-3)} \qquad \text{Subtract and}$$
add numerators.

$$= \frac{x^2-3x-2x^2-6x+x^2+10x+3}{(x+3)(x-3)}$$

$$= \frac{x+3}{(x+3)(x-3)} \qquad \text{Simplify.}$$

$$= \frac{1}{x-3} \qquad \text{Simplify.}$$

The check is left for you.

12.3.4 Checkup

Add or subtract and simplify.

1. $\dfrac{5}{6x^2y} + \dfrac{9}{2xy^2}$

2. $\dfrac{2x}{x-3} + \dfrac{5x}{x-8}$

3. $\dfrac{5}{x-4} - \dfrac{3}{4-x}$

4. $\dfrac{x}{x^2+10x+21} + \dfrac{3}{x^2+8x+15}$

5. $\dfrac{x}{x-4} - \dfrac{3}{x+4} + \dfrac{6x}{x^2-16}$

12.3.5 Modeling the Real World

Geometric formulas are often written as rational expressions. For example, the area of a rectangle is $A = LW$, where L is the rectangle's length and W is its width. We can solve for L to obtain $L = \frac{A}{W}$, which is a rational expression.

This formula is useful for modeling practical situations in which you want to determine the dimensions of some rectangular object with a given area.

EXAMPLE 5 Two rectangular bins are arranged side by side, as shown in the figures that follow. The area of the smaller bin is 25 square feet and the area of the larger bin is 36 square feet. Determine the total length of the two bins in terms of the width W using the formula $L = \frac{A}{W}$.

a.

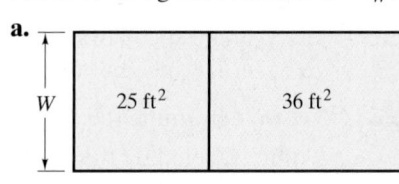

b.

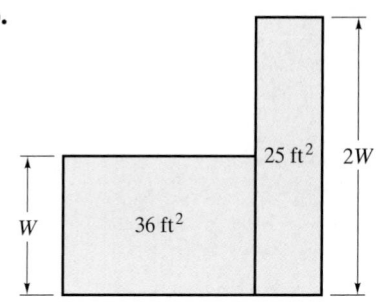

c.

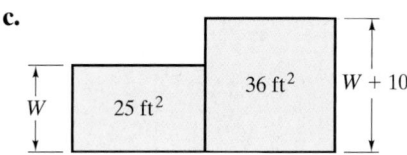

Solution

a. The length of the smaller bin is $\frac{25}{W}$ feet. The length of the larger bin is $\frac{36}{W}$ feet. The total length is $\frac{25}{W} + \frac{36}{W} = \frac{61}{W}$ feet.

b. The length of the smaller bin is $\frac{25}{2W}$ feet. The length of the larger bin is $\frac{36}{W}$ feet. The total length is $\frac{25}{2W} + \frac{36}{W} = \frac{25}{2W} + \frac{72}{2W} = \frac{97}{2W}$ feet.

c. The length of the smaller bin is $\frac{25}{W}$ feet. The length of the larger bin is $\frac{36}{W + 10}$ feet. The total length is $\frac{25}{W} + \frac{36}{W + 10} = \frac{25W + 250}{W(W + 10)} + \frac{36W}{W(W + 10)} = \frac{61W + 250}{W(W + 10)}$ feet.

EXAMPLE 6 Margaret breaks her leg during her vacation and requires surgery. She wants to have the surgery at home. The hospital she is in is 3 miles from the airport. The flight home is 625 miles. The air attendant states that the average speed of a plane is three times the average speed of ground transportation in an ambulance.

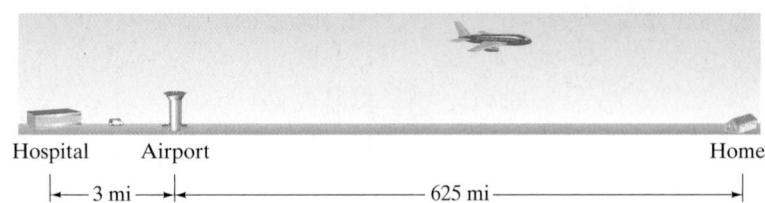

a. Write a function $T(x)$ for the total time traveled to arrive at Margaret's home airport.

b. Graph the function.

c. If the ambulance's speed from the hospital to the airport was 65 mph, estimate, to the nearest hundredth of an hour, the total time traveled to Margaret's home airport.

d. If Margaret travels 4 hours to reach her home airport, estimate, to the nearest mile per hour, the average speed of the ambulance from the hospital to the airport.

Solution

a. Let $x =$ the average speed of the ambulance
$3x =$ the average speed of the plane

Since $t = \frac{d}{r}$, we determine the following from the problem:

$\frac{3}{x} =$ time traveled by the ambulance

$\frac{625}{3x} =$ time traveled by the plane

The total time traveled is the sum of the two times, or $T(x) = \frac{3}{x} + \frac{625}{3x}$.

b. $Y1 = \frac{3}{x} + \frac{625}{3x}$

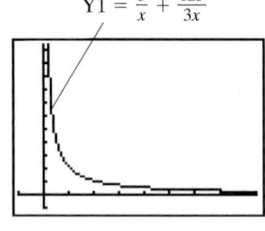

$(-10, 84, -10, 114)$

c. Evaluate $T(65)$. According to the graph, when x is 65, $T(x)$ is approximately 3.25.

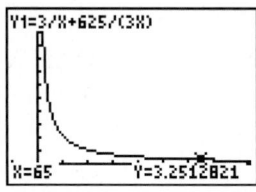

$(-10, 84, -10, 114)$

The total time traveled is approximately 3.25 hours when the ambulance speed is 65 mph.

d. Trace the calculator graph. When $T(x) \approx 4$, x is approximately 53.

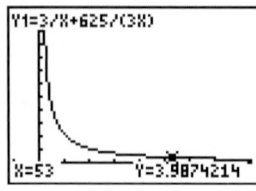

$(-10, 84, -10, 114)$

The average ambulance speed is approximately 53 mph when the total time traveled is 4 hours.

APPLICATION

Julio travels from City A to City B at an average rate of 40 miles per hour. He makes the return trip at an average rate of 60 miles per hour. Determine his average rate for the round-trip.

Discussion

Let x = the distance between City A and City B
$x + x = 2x$ = the total distance traveled for the round-trip
Since $t = \frac{d}{r}$, we can determine the time traveled.
$\frac{x}{40}$ = the time traveled from City A to City B
$\frac{x}{60}$ = the time traveled from City B to City A
$\frac{x}{40} + \frac{x}{60}$ = the total time traveled

$$\frac{x}{40} + \frac{x}{60} = \frac{3x}{120} + \frac{2x}{120} \qquad \text{LCD is 120.}$$

$$= \frac{5x}{120}$$

$$= \frac{x}{24}$$

The average rate for the entire trip is given by the total distance traveled divided by the total time traveled.

$$2x \div \frac{x}{24} = \frac{2\overset{1}{x}}{1} \cdot \frac{24}{\underset{1}{x}}$$

$$= 48$$

The average rate of travel for the round-trip was 48 miles per hour. Note that this is in contrast to the average of the two speeds, or 50 miles per hour.

✓ 12.3.5 Checkup

1. **a.** The width of a rectangle is given by $\frac{x-1}{5x}$ yards, and its length is given by $\frac{x+1}{3x}$ yards. Find an expression for the perimeter of the rectangle.
   b. A second rectangle has a width given by $\frac{x+1}{2x}$ yards and a length given by $\frac{3x+2}{x}$ yards. Find an expression for its perimeter.
   c. Find the difference between the two perimeters.
2. A trucking company makes a delivery 270 miles away. The truck travels 60 miles on rural roads and 210 miles on interstate highways. The average speed of the truck on the interstate is 10 miles per hour less than twice its average speed on the rural roads.
   a. Letting x be the average speed on the rural roads, write a function $T(x)$ for the total time of the delivery trip.

   b. Graph the function on your calculator.
   c. If the average speed on the rural road was 35 mph, use the graph to estimate to the nearest tenth of an hour the total time of the delivery trip.
   d. If the total time of the trip was 6 hours, use the graph to find the average speed on the rural roads, and calculate the average speed on the interstate.
3. A plane travels against the wind at an average speed of 200 mph. For the return trip traveling with the wind, the plane averages 300 mph. Determine the average speed of the plane for the round-trip.

12.3 Exercises

Add or subtract and simplify.

1. $\dfrac{5}{x} + \dfrac{8}{x}$

2. $\dfrac{3}{x} + \dfrac{9}{x}$

3. $\dfrac{2x}{5xy} + \dfrac{3x}{5xy}$

4. $\dfrac{b}{4ab} + \dfrac{3b}{4ab}$

5. $\dfrac{x-1}{x+6} + \dfrac{x+1}{x+6}$

6. $\dfrac{x-5}{x+12} + \dfrac{2x+5}{x+12}$

7. $\dfrac{2b}{b^2 - 25} + \dfrac{10}{b^2 - 25}$

8. $\dfrac{21}{z^2 - 49} + \dfrac{3z}{z^2 - 49}$

9. $\dfrac{x - 5}{x + 3} + \dfrac{x + 5}{x + 3} + \dfrac{x + 9}{x + 3}$

10. $\dfrac{x - 2}{2x + 1} + \dfrac{3x + 1}{2x + 1} + \dfrac{4x + 5}{2x + 1}$

11. $\dfrac{4}{d} - \dfrac{9}{d}$

12. $\dfrac{3}{b} - \dfrac{11}{b}$

13. $\dfrac{9xy}{7x^2y} - \dfrac{2xy}{7x^2y}$

14. $\dfrac{2xy}{5xy^2} - \dfrac{7xy}{5xy^2}$

15. $\dfrac{2x - 3}{x + 9} - \dfrac{x - 3}{x + 9}$

16. $\dfrac{5x - 4}{x + 8} - \dfrac{2x - 1}{x + 8}$

17. $\dfrac{4x - 13}{x - 5} - \dfrac{2x - 3}{x - 5}$

18. $\dfrac{3x + 5}{x - 6} - \dfrac{x + 17}{x - 6}$

19. $\dfrac{x + 3}{x + 11} - \dfrac{2x - 8}{x + 11}$

20. $\dfrac{x + 2}{x + 13} - \dfrac{3x - 7}{x + 13}$

21. $\dfrac{z^2}{z + 4} - \dfrac{16}{z + 4}$

22. $\dfrac{c^2}{c + 7} - \dfrac{49}{c + 7}$

23. $\dfrac{3c}{c^2 - 9} - \dfrac{9}{c^2 - 9}$

24. $\dfrac{5b}{b^2 - 4} - \dfrac{10}{b^2 - 4}$

25. $\dfrac{5x + 1}{2x + 3} - \dfrac{x + 2}{2x + 3} - \dfrac{2x - 5}{2x + 3}$

26. $\dfrac{2x + 5}{3x - 4} - \dfrac{x + 4}{3x - 4} - \dfrac{3x - 2}{3x - 4}$

27. $\dfrac{5}{x} + \dfrac{8}{-x}$

28. $\dfrac{3}{-x} + \dfrac{9}{x}$

29. $\dfrac{3x - 2}{x - 5} + \dfrac{x + 8}{5 - x}$

30. $\dfrac{x + 4}{7 - x} + \dfrac{4x - 17}{x - 7}$

31. $\dfrac{a^2}{a - 3} + \dfrac{9}{3 - a}$

32. $\dfrac{25}{5 - c} + \dfrac{c^2}{c - 5}$

33. $\dfrac{5x}{x^2 - 9} + \dfrac{15}{9 - x^2}$

34. $\dfrac{3x - 4}{x^2 - 1} + \dfrac{3 - 4x}{1 - x^2}$

35. $\dfrac{3}{2x} + \dfrac{7}{6x^2}$

36. $\dfrac{9}{5x^2} + \dfrac{3}{10x^3}$

37. $\dfrac{3x + 1}{x - 6} + \dfrac{x - 4}{2x - 12}$

38. $\dfrac{2x - 3}{x + 9} + \dfrac{x + 2}{2x + 18}$

39. $\dfrac{2x}{2x - 3} + \dfrac{6x + 9}{4x^2 - 9}$

40. $\dfrac{2x - 1}{5x + 2} + \dfrac{4 - 10x}{25x^2 - 4}$

41. $\dfrac{7}{3x - 5} + \dfrac{4}{3x + 5}$

42. $\dfrac{5}{2x - 3} + \dfrac{7}{2x + 3}$

43. $\dfrac{7}{x - 5} + \dfrac{x + 1}{x^2 - 10x + 25}$

44. $\dfrac{2}{x - 8} + \dfrac{3x + 4}{x^2 - 16x + 64}$

45. $\dfrac{b + 8}{b} + \dfrac{b}{b + 8}$

46. $\dfrac{a - 3}{2a} + \dfrac{2a}{a - 3}$

47. $\dfrac{x - 2}{4x + 12} + \dfrac{2x + 1}{x^2 - 9}$

48. $\dfrac{x - 1}{x^2 - 4} + \dfrac{2x - 1}{3x + 6}$

49. $\dfrac{x + 7}{x^2 + 2x - 15} + \dfrac{8 - x}{x^2 + 9x + 20}$

50. $\dfrac{3}{x^2 + 8x + 7} + \dfrac{4 - x}{x^2 + 5x - 14}$

51. $\dfrac{3a - b}{a^2 b} + \dfrac{a + 2b}{ab^2}$

52. $\dfrac{p - q}{pq^2} + \dfrac{p + q}{p^2 q}$

53. $\dfrac{3x}{x - 2y} + \dfrac{4y}{2y - x}$

54. $\dfrac{x}{5x - y} + \dfrac{3y}{y - 5x}$

55. $\dfrac{5x + 2y}{x^2 - 9y^2} + \dfrac{6}{x + 3y}$

56. $\dfrac{2x + y}{4x^2 - 9y^2} + \dfrac{7}{2x - 3y}$

57. $\dfrac{8}{x^2 + 6xz + 9z^2} + \dfrac{2}{x^2 - 9z^2}$

58. $\dfrac{5}{x^2 - 4y^2} + \dfrac{3}{x^2 + 4xy + 4y^2}$

59. $\dfrac{7x + 5}{x - 2} - \dfrac{x - 1}{2 - x}$

60. $\dfrac{2x - 3}{x - 7} - \dfrac{x + 4}{7 - x}$

61. $\dfrac{3x}{x^2 - 4} - \dfrac{6}{4 - x^2}$

62. $\dfrac{5a}{a^2 - 36} - \dfrac{30}{36 - a^2}$

63. $\dfrac{6x - 3}{3x + 15} - \dfrac{x - 1}{x + 5}$

64. $\dfrac{3x + 7}{2x + 3} - \dfrac{x + 4}{6x + 9}$

65. $\dfrac{7}{2x - 5} - \dfrac{9}{2x + 5}$

66. $\dfrac{9}{3x - 1} - \dfrac{4}{3x + 1}$

67. $\dfrac{6}{x^2 + 12x + 36} - \dfrac{2x + 1}{x + 6}$

68. $\dfrac{3}{x^2 - 14x + 49} - \dfrac{x - 4}{x - 7}$

69. $\dfrac{x}{8 - x} - \dfrac{8 - x}{x}$

70. $\dfrac{x - 10}{x} - \dfrac{x}{x - 10}$

71. $\dfrac{x + 2}{3x + 12} - \dfrac{x - 4}{x^2 - 16}$

72. $\dfrac{7}{x^2 - 9} - \dfrac{9}{x^2 + 3x}$

73. $\dfrac{11}{b^2 - 9} - \dfrac{7}{2b^2 - b - 15}$

74. $\dfrac{14}{a^2 - 25} - \dfrac{9}{2a^2 - 9a - 5}$

75. $\dfrac{7}{5ab^2} - \dfrac{15}{10a^2 b} - \dfrac{3}{2b}$

76. $\dfrac{5}{16x^2 y} - \dfrac{2}{24xy} - \dfrac{7}{6xy^2}$

77. $\dfrac{4p - q}{p + 3q} - \dfrac{p + 2q}{p + 3q}$

78. $\dfrac{2x - z}{x + 5z} - \dfrac{x - 4z}{x + 5z}$

79. $\dfrac{7}{x^2 + 6xy + 9y^2} - \dfrac{2}{x^2 - 9y^2}$ **80.** $\dfrac{8}{x^2 + 10xy + 25y^2} - \dfrac{3}{x^2 - 25y^2}$

Add or subtract and simplify exercises 81–92.

81. $\dfrac{x}{x + 5} - \dfrac{4}{x - 5} + \dfrac{10x}{x^2 - 25}$

82. $\dfrac{6z}{z^2 - 81} - \dfrac{3}{z - 9} + \dfrac{2z}{z + 9}$

83. $1 + \dfrac{x - 9}{x + 10}$

84. $3 + \dfrac{x}{3x + 1}$

85. $5 - \dfrac{4}{2x + 7}$

86. $1 - \dfrac{x}{x + 4}$

87. $\dfrac{2x - 3}{x + 7} - 1$

88. $\dfrac{5x + 8}{9x - 2} - 4$

89. $\dfrac{3}{x + 3} + x + 5$

90. $x + 3 + \dfrac{2}{x + 1}$

91. $2x + 1 - \dfrac{3}{x - 4}$

92. $\dfrac{5x}{4x - 1} - 5x$

93. In Figure 12.2 (a), the smaller rectangle has an area of 20 square feet, and the larger rectangle has an area of 45 square feet. Given the lengths shown, find an expression for the sum of the two widths of the rectangles in terms of L. Repeat the exercise for Figure 12.2(b).

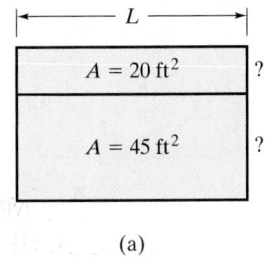

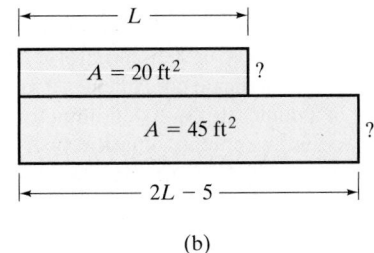

(a)   (b)

Figure 12.2

94. In Figure 12.3 (a), the smaller rectangular box has a volume of 60 cubic inches, while the larger rectangular box has a volume of 200 cubic inches. Given the lengths shown, find an expression for the difference in the two widths in terms of L. Repeat the exercise for Figure 12.3(b).

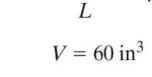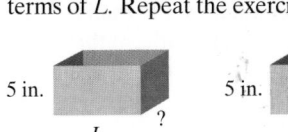

5 in.   5 in.

L $2L$

$V = 60\ \text{in}^3$ $V = 200\ \text{in}^3$

(a)

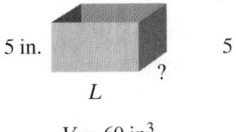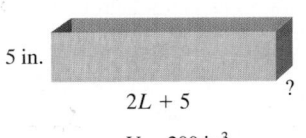

5 in.   5 in.

L $2L + 5$

$V = 60\ \text{in}^3$ $V = 200\ \text{in}^3$

(b)

Figure 12.3

95. The width of a rectangle is given by $\frac{24}{x - 1}$ feet, and its length is given by $\frac{72}{x + 1}$ feet. Find an expression for the perimeter of the rectangle.

96. The base of a trapezoid measures $\frac{60}{x + 3}$ inches, its top measures $\frac{42}{x + 3}$ inches, and its sides measure $\frac{9}{x}$ inches and $\frac{21 + x}{2x}$ inches. Find an expression for the perimeter of the trapezoid.

97. Joey has entered an amateur triathlon competition. One leg of the race is a 26-mile stretch in which he will run. Another leg of the race is a 110-mile stretch in which he will bicycle. Joey figures that he can bike 20 miles per hour faster than he can run.
a. Letting x be his average running speed, write a function for the total time it will take Joey to complete the two legs of the race.
b. Graph the function.
c. If Joey's average running speed is 9 miles per hour, use the graph to estimate, to the nearest tenth of an hour, the total time to complete the two legs of the race.
d. If the total time to complete the two legs of the race is 8 hours, use the graph to estimate Joey's average running speed, and then determine his biking speed.

98. In Joey's amateur triathlon competition, another leg of the competition is a swim race. For this competition, he must swim 1 mile upriver and then return. The speed of the river current is 2 miles per hour.
a. Letting x be Joey's swimming speed in still water, write a function for the total time it will take him to swim up the river and back.
b. Graph the function in the window $(0, 9.4, 1, 0, 6.2, 1, 1)$.
c. If Joey's swimming speed in still water is 2.5 mph, how long will it take him to complete this leg of the competition? Round your answer to the nearest tenth of an hour.
d. If Joey wants to complete the swimming leg of the competition in 1.5 hours, how fast should he be able to swim in still water?

99. A boat travels upriver at an average speed of 25 mph and returns downriver at an average speed of 35 mph. Determine the average speed of the boat for the entire round-trip.

100. One member of a speed ski team averaged 125 mph down the slopes. The other member of the team averaged 100 mph down the slopes. Determine the average speed for the team for both runs down the slopes.

12.3 Calculator Exercise

In this section, you were shown how to check your addition and subtraction of rational functions by using the table (numeric-check) feature of your calculator. One problem that often occurs in doing this is that a student will incorrectly place parentheses when storing the original expressions and the simplified expressions into the calculator.

As an example, suppose you were checking the simplification $\frac{3x + 2}{x + 1} + \frac{2x + 3}{x + 1} = 5$.

When you enter the left expression, be sure to place parentheses around both numerators and both denominators. If you fail to do so, the table will not indicate equivalence. The only time that you can key in a numerator or a denominator without parentheses is when there is no mathematical operation contained within the expression. Even then, it is a safe-

ty measure to always enclose numerators and denominators in parentheses. Also, when the numerator or denominator is a product of expressions in parentheses, you must use an additional pair of parentheses to enclose the entire product.

Practice checking the following equivalent expressions:

1. $\dfrac{x + 4}{2x} + \dfrac{3x - 4}{2x} = 2$

2. $\dfrac{5}{3 - x} - \dfrac{x - 8}{x - 3} = -1$

3. $\dfrac{3x - 5}{2x + 1} + \dfrac{4x + 7}{2x + 1} = \dfrac{7x + 2}{2x + 1}$

4. $\dfrac{4}{x + 3} - \dfrac{5}{x + 2} = \dfrac{-x - 7}{(x + 3)(x + 2)}$

5. $\dfrac{x + 1}{x^2 - x - 6} + \dfrac{2x - 1}{x^2 + 5x + 6} = \dfrac{3x^2 - 3x + 6}{(x - 3)(x + 2)(x + 3)}$

12.3 Writing Exercise

Part 1.

Many mathematical models use rational expressions to explain the behavior of physical objects. Listed here are several simple mathematical models that use rational expressions:

1. If two resistors R_1 and R_2 (measured in ohms) are connected in parallel, then the combined resistance R_T (in ohms) is given by $\frac{1}{R_T} = \frac{1}{R_1} + \frac{1}{R_2}$.

2. The combined capacitance C of two capacitors C_1 and C_2 connected in series is given by $\frac{1}{C} = \frac{1}{C_1} + \frac{1}{C_2}$.

3. If two blood vessels share common endpoints (connected in parallel), and if r_1 and r_2 are the resistances of the two vessels to the flow of blood (measured in dynes), then the combined resistance of the vessels is given by $\frac{1}{r} = \frac{1}{r_1} + \frac{1}{r_2}$.

4. The force F required to stretch a spring is given by $F = kx$, where x is the distance the spring stretches and k is the spring constant of the spring. If two springs with spring constants k_1 and k_2 are connected in sequence, the spring constant K of the combination is given by $\frac{1}{K} = \frac{1}{k_1} + \frac{1}{k_2}$.

5. The focal length f of a lens is related to the distances q and p from the lens of the object and of the image by the formula $\frac{1}{f} = \frac{1}{p} + \frac{1}{q}$.

6. If two lenses with focal lengths f_1 and f_2 are placed in contact with one another, the focal length of the combination is given by $\frac{1}{f} = \frac{1}{f_1} + \frac{1}{f_2}$.

Search in the library for a reference that discusses one of these applications. Or, if you prefer, find a reference on another mathematical model that involves rational expressions. Describe the model you find. If the reference contains an example that uses the model, summarize the example in a brief written report. Be sure to include the title, author, publisher, and call number of the reference you found.

Part 2.

Now that you know how to add, subtract, multiply, and divide rational expressions, you should be able to figure out the steps needed to simplify a complex rational expression.

Suppose you were asked to simplify the expression, $\dfrac{\frac{2}{x} + \frac{x}{2}}{\frac{y}{3} - \frac{3}{y}}$. List the steps you would follow to do this.

Hints: Can you simplify the numerator? What about the denominator? Can you rewrite the expression as a division exercise?

Now see if you can use your steps to simplify the following expressions:

1. $\dfrac{\dfrac{3xy}{2z^2}}{\dfrac{9x^3y}{8z}}$

2. $\dfrac{2x + \dfrac{1}{x}}{5x}$

3. $\dfrac{\dfrac{a + b}{a - b}}{\dfrac{a^2 + 2ab + b^2}{a^2 - b^2}}$

4. $\dfrac{\dfrac{2x}{5} - \dfrac{5}{x}}{\dfrac{2x}{5} + \dfrac{1}{15x}}$

5. $\dfrac{\dfrac{1}{a^2} + \dfrac{2}{ab}}{\dfrac{3}{ab^2} + \dfrac{1}{b}}$

12.4 Solving Rational Equations in One Variable

OBJECTIVES

1 Identify rational equations in one variable.

2 Solve rational equations algebraically.

3 Differentiate between a rational equation and a rational expression.

4 Model real-world situations by using rational equations.

APPLICATION

The maximum variable height (in feet) of an object viewed at a distance of 100 feet from a $\frac{1}{2}$-inch (4.8-millimeter) camera lens combined with an x-millimeter focal length is given by $h(x) = \frac{480}{x}$. Determine the focal length of a camera that Mariam will need to take a picture of a Christmas tree that is 20 feet tall.

After completing this section, we will discuss this application further. See page 852.

12.4.1 Rational Equations in One Variable

A **rational equation in one variable**, or, simply, **rational equation**, is an equation that can be written in the form of a rational expression, containing one variable, that is equal to 0. We are now ready to solve such equations.

STANDARD FORM FOR A RATIONAL EQUATION IN ONE VARIABLE

A rational equation in one variable is written in the form

$$R(x) = 0$$

where $R(x)$ is a rational expression.

For example, the following are rational equations in one variable:

$$\frac{2}{x} = \frac{x}{8} \qquad \text{Standard form, } \frac{2}{x} - \frac{x}{8} = 0$$

$$\frac{x+2}{x+1} = \frac{x+1}{x} \qquad \text{Standard form, } \frac{x+2}{x+1} - \frac{x+1}{x} = 0$$

$$\frac{5}{x} + \frac{x-5}{3x} - \frac{x^2-9}{x+3} = 0 \qquad \text{Standard form}$$

EXAMPLE 1 Determine whether each equation is rational.

a. $3x^2 + 5x^6 - 7x^3 = 0$ **b.** $6x^{-2} + 8 - 7x^{-3} = 12$

c. $x^{3/4} = 2x^2$

Solution

a. $3x^2 + 5x^6 - 7x^3 = 0$ is a polynomial equation, but it is also a rational equation, because each term can be written as a rational expression with a denominator of 1.

$$\frac{3x^2}{1} + \frac{5x^6}{1} - \frac{7x^3}{1} = 0$$

b. $6x^{-2} + 8 - 7x^{-3} = 12$ is a rational equation, because an expression that contains a variable with a negative integer exponent can be written as a rational expression.

$$6x^{-2} = \frac{6}{x^2} \quad \text{and} \quad 7x^{-3} = \frac{7}{x^3}$$

Therefore, $\frac{6}{x^2} + 8 - \frac{7}{x^3} - 12 = 0$.

c. $x^{3/4} = 2x^2$ is not a rational equation, because an expression containing a variable base with a fractional exponent cannot be written as a rational expression.

 12.4.1 Checkup

In exercises 1–6, determine whether each equation is rational.

1. $\dfrac{1}{\sqrt{x}} = \dfrac{x}{2}$ **2.** $4z^2 - 3z - 1 = 0$ **3.** $a + 2 = a^{1/2}$

4. $\dfrac{5}{x} + 3 = \dfrac{x}{7}$ **5.** $\dfrac{3}{2x} + \dfrac{1}{x-2} = 5$ **6.** $3m^{-2} + 4m^{-1} = -6$

7. Explain the difference between a rational equation and a rational expression.

12.4.2 Solving Rational Equations Algebraically

In previous chapters, we solved linear and quadratic equations numerically and graphically. The same procedures are used to solve rational equations numerically and graphically. We will use these procedures to check our solutions when we solve rational equations algebraically.

We are now ready to solve rational equations in one variable algebraically. We have already solved linear equations with fractional coefficients. To simplify these equations, we applied the multiplication property of equations. The multiplication property of equations states that if both sides of an equation are multiplied by the same nonzero expression, the result is an equivalent equation. We used this property to clear the fractions by multiplying by the least common denominator of all the fractions. For example, solve $\frac{1}{2}x + \frac{3}{4}x = \frac{2}{3}x + 7$.

$$\frac{1}{2}x + \frac{3}{4}x = \frac{2}{3}x + 7 \qquad \text{The LCD is 12.}$$

$$12\left(\frac{1}{2}x + \frac{3}{4}x\right) = 12\left(\frac{2}{3}x + 7\right) \qquad \text{Multiply by 12.}$$

$$\overset{6}{\cancel{12}}\left(\frac{1}{2}x\right) + \overset{3}{\cancel{12}}\left(\frac{3}{4}x\right) = \overset{4}{\cancel{12}}\left(\frac{2}{3}x\right) + 12(7) \qquad \text{Distribute 12.}$$

$$6x + 9x = 8x + 84$$

$$15x = 8x + 84$$

$$7x = 84 \qquad \text{Subtract 8x.}$$

$$x = 12 \qquad \text{Divide by 7.}$$

We use the same procedure for rational equations. However, we need to be aware of the restricted values for the equation.

SOLVING A RATIONAL EQUATION ALGEBRAICALLY
To solve a rational equation in one variable algebraically,

- Determine the restricted values for the equation.
- Multiply both sides of the equation by the least common denominator of all the terms.
- Solve the resulting equation.
- Discard any solutions that are restricted values.

Check the solution by substituting or by solving numerically or graphically.

EXAMPLE 2   Solve $\frac{2}{x} + \frac{4}{2x} = \frac{3}{x} + 1$.

Algebraic Solution

$$\frac{2}{x} + \frac{4}{2x} = \frac{3}{x} + 1 \qquad \text{The restricted value is 0.}$$
$$\text{The LCD is } 2x.$$

$$2x\left(\frac{2}{x} + \frac{4}{2x}\right) = 2x\left(\frac{3}{x} + 1\right) \qquad \text{Multiply by } 2x.$$

$$\overset{1}{2x}\left(\frac{2}{x}\right) + \overset{1}{2x}\left(\frac{4}{2x}\right) = \overset{1}{2x}\left(\frac{3}{x}\right) + 2x(1) \qquad \text{Distribute } 2x.$$

$$4 + 4 = 6 + 2x$$
$$8 = 6 + 2x$$
$$2 = 2x \qquad \text{Subtract 6.}$$
$$1 = x \qquad \text{Divide by 2.}$$

Substitution Check

$$\frac{2}{x} + \frac{4}{2x} = \frac{3}{x} + 1$$

$$\frac{2}{1} + \frac{4}{2(1)} \quad \bigg| \quad \frac{3}{1} + 1$$

$$2 + 2 \quad \bigg| \quad 3 + 1$$

$$4 \quad \bigg| \quad 4$$

The solution is 1, because when 1 is substituted for x, the resulting expressions are equal.

Numeric Check

X	Y1	Y2
-3	-1.333	0
-2	-2	-.5
-1	-4	-2
0	ERROR	ERROR
1	4	4
2	2	2.5
3	1.3333	2

X= -3

$$Y1 = \frac{2}{x} + \frac{4}{2x}$$

$$Y2 = \frac{3}{x} + 1$$

The solution is 1, because this value results in Y1 = Y2.

Graphic Check

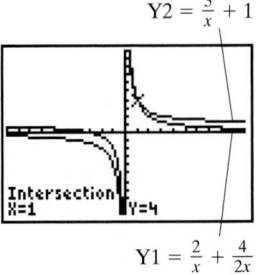

$$Y2 = \frac{3}{x} + 1$$

Intersection
X=1   Y=4

$$Y1 = \frac{2}{x} + \frac{4}{2x}$$

$(-10, 10, -10, 10)$

The intersection is $(1,4)$. The solution is 1.

EXAMPLE 3   Solve.

a. $\dfrac{a}{a-2} = \dfrac{a+1}{a-3}$ b. $x^{-2} - 2 = 2$

c. $1 - \dfrac{4}{x+2} = \dfrac{2x}{x+2}$ d. $\dfrac{2}{x+3} - \dfrac{x+1}{x^2+5x+6} = \dfrac{4}{x^2-4}$

Algebraic Solution

a.

$$\frac{a}{a-2} = \frac{a+1}{a-3}$$

Restricted values are 2 and 3. The LCD is $(a-2)(a-3)$.

$$(a-2)(a-3)\left(\frac{a}{a-2}\right) = (a-2)(a-3)\left(\frac{a+1}{a-3}\right)$$ Multiply by the LCD.

$$(a-3)a = (a-2)(a+1)$$ Simplify.

$$a^2 - 3a = a^2 - a - 2$$ Simplify.

$$-2a = -2$$ Subtract a^2 and add a.

$$a = 1$$ Divide by -2.

b.

$$x^{-2} - 2 = 2$$

$$\frac{1}{x^2} - 2 = 2$$

Rewrite without the negative exponent. The restricted value is 0. The LCD is x^2.

$$x^2\left(\frac{1}{x^2} - 2\right) = 2x^2$$ Multiply by x^2.

$$x^2\left(\frac{1}{x^2}\right) - 2x^2 = 2x^2$$ Distribute x^2.

$$1 - 4x^2 = 0$$ Subtract $2x^2$.

$$(1 + 2x)(1 - 2x) = 0$$ Factor.

$$1 + 2x = 0 \quad \text{or} \quad 1 - 2x = 0$$ Use the zero factor property.

$$x = -\frac{1}{2} \qquad x = \frac{1}{2}$$ Solve.

c.

$$1 - \frac{4}{x+2} = \frac{2x}{x+2}$$

The restricted value is -2. The LCD is $(x+2)$.

$$(x+2)\left(1 - \frac{4}{x+2}\right) = (x+2)\left(\frac{2x}{x+2}\right)$$ Multiply by the LCD.

$$(x+2)(1) - (x+2)\left(\frac{4}{x+2}\right) = (x+2)\left(\frac{2x}{x+2}\right)$$ Distribute the LCD.

$$x + 2 - 4 = 2x$$

$$x - 2 = 2x$$

$$-2 = x$$ Subtract x.

We determined that -2 is a restricted value. Therefore, it cannot be a solution.

Numeric Check

Since the solution is an integer, check by solving numerically. Substitute x for a.

X	Y1	Y2
-3	.6	.33333
-2	.5	.2
-1	.33333	0
0	0	-.3333
1	-1	-1
2	ERROR	-3
3	3	ERROR

X=-3

$$Y1 = \frac{x}{x-2}$$

$$Y2 = \frac{x+1}{x-3}$$

The solution is 1, because this value results in $Y1 = Y2$.

Graphic Check

Check by graphing on a calculator decimal screen.

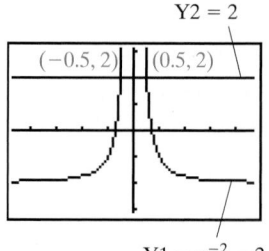

$(-4.7, 4.7, -3.1, 3.1)$

The intersections are $\left(-\frac{1}{2}, 2\right)$ and $\left(\frac{1}{2}, 2\right)$. The solutions are $-\frac{1}{2}$ and $\frac{1}{2}$.

Numeric Check

X	Y1	Y2
-3	5	6
-2	ERROR	ERROR
-1	-3	-2
0	-1	0
1	-.3333	.66667
2	0	1
3	.2	1.2

X=-3

$$Y1 = 1 - \frac{4}{x+2}$$

$$Y2 = \frac{2x}{x+2}$$

Note that when $x = -2$, Y1 and Y2 are undefined.

We call -2 an **extraneous solution**. This is the reason it is very important to determine the restricted values for the equation first. The given equation has no solution.

d. $\dfrac{2}{x+3} - \dfrac{x+1}{x^2+5x+6} = \dfrac{4}{x^2-4}$

Factor each denominator. Restricted values are -3, -2, and 2.

$\dfrac{2}{x+3} - \dfrac{x+1}{(x+3)(x+2)} = \dfrac{4}{(x+2)(x-2)}$

$(x+2)(x-2)(x+3)\left(\dfrac{2}{x+3} - \dfrac{x+1}{(x+3)(x+2)}\right)$

$= (x+2)(x-2)(x+3)\left(\dfrac{4}{(x+2)(x-2)}\right)$

Multiply by the LCD, $(x+2)(x-2)(x+3)$.

$(x+2)(x-2)\overset{1}{\cancel{(x+3)}}\left(\dfrac{2}{\underset{1}{\cancel{x+3}}}\right) - \overset{1}{\cancel{(x+2)}}(x-2)\overset{1}{\cancel{(x+3)}}\left(\dfrac{x+1}{\underset{1}{\cancel{(x+3)}}\underset{1}{\cancel{(x+2)}}}\right)$

$= \overset{1}{\cancel{(x+2)}}\overset{1}{\cancel{(x-2)}}(x+3)\left(\dfrac{4}{\underset{1}{\cancel{(x+2)}}\underset{1}{\cancel{(x-2)}}}\right)$

Distribute the LCD.

$(x+2)(x-2)2 - (x-2)(x+1) = (x+3)4$

$(2x^2-8) - (x^2-x-2) = 4x+12$

$x^2 + x - 6 = 4x + 12$

$x^2 - 3x - 18 = 0$ Subtract 4x and 12.

$(x-6)(x+3) = 0$ Factor.

$x - 6 = 0 \quad \text{or} \quad x + 3 = 0$ Use the zero factor property.

$x = 6 \qquad\qquad x = -3$ Solve.

The solution is 6. The value -3 is a restricted value and not a solution.

Numeric Check

X	Y1	Y2
-3	ERROR	.8
-2	ERROR	ERROR
-1	1	-1.333
0	.5	-1
1	.33333	-1.333
2	.25	ERROR
3	.2	.8

X=-3

X	Y1	Y2
1	.33333	-1.333
2	.25	ERROR
3	.2	.8
4	.16667	.33333
5	.14286	.19048
6	.125	.125
7	.11111	.08889

X=1

$Y1 = \dfrac{2}{x+3} - \dfrac{x+1}{x^2+5x+6}$

$Y2 = \dfrac{4}{x^2-4}$

Note that when $x = -3$, Y1 is not defined $x = -3$ is an extraneous solution.
When $x = 6$, Y1 = Y2. The solution is 6.

An alternative method may be used to solve Example 3a. The rational equation to be solved, $\frac{a}{a-2} = \frac{a+1}{a-3}$, consists of two expressions, each of which is an algebraic fraction: $\frac{a}{a-2}$ and $\frac{a+1}{a-3}$. This special type of rational equation is called a **proportion**. To solve a proportion, we can equate the cross products; that is, we multiply the numerator of one fraction and the denominator of the other fraction and solve the remaining equation.

$$\dfrac{a}{a-2} \bowtie \dfrac{a+1}{a-3}$$

$a(a-3) = (a+1)(a-2)$ Equate the cross products.

$a^2 - 3a = a^2 - a - 2$ Note: This is the same as the equation in the previous solution.

$-3a = -a - 2$ Subtract a^2 from both sides.

$-2a = -2$ Add a to both sides.

$a = 1$ Divide both sides by -2.

Some rational equations are contradictions and have no solution. Other rational equations are identities and have an infinite number of solutions. We determine these situations algebraically the same way as we did for linear and quadratic equations.

EXAMPLE 4   Solve.

a. $\dfrac{3x - 15}{6 + 3x} = \dfrac{x - 10}{x + 2}$ b. $\dfrac{3}{x} = \dfrac{6}{2x}$

Algebraic Solution

a.
$$\dfrac{3x - 15}{6 + 3x} = \dfrac{x - 10}{x + 2}$$ The restricted value is -2.

$$\dfrac{3x - 15}{3(2 + x)} = \dfrac{x - 10}{x + 2}$$ Factor the denominators.

$$3(x + 2)\left(\dfrac{3x - 15}{3(2 + x)}\right) = 3(x + 2)\left(\dfrac{x - 10}{x + 2}\right)$$ Multiply by the LCD, $3(x + 2)$.

$$3x - 15 = 3x - 30$$

$$-15 = -30$$ Subtract 3x.

Since $-15 \neq -30$, the equation is a contradiction. It has no solution.

b.
$$\dfrac{3}{x} \bowtie \dfrac{6}{2x}$$ The restricted value is 0.

$$3(2x) = 6x$$ Equate the cross products.

$$6x = 6x$$ Subtract 6x.

$$0 = 0$$

Since $0 = 0$, the equation is an identity. The solution set is the set of all real numbers not equal to 0. (0 is the restricted value for the equation.)

ALGEBRAIC SOLUTIONS OF RATIONAL EQUATIONS
To solve a rational equation algebraically, determine the restricted values, multiply both sides of the equation by the LCD of all fractional terms, and solve the resulting equation. One of three possibilities will occur.

One or more solutions exist.
No solution exists. The solution process results in a contradiction.
An infinite number of solutions exist. The solution set is the set of all real numbers not equal to the restricted values. The solution process results in an identity.

12.4.2 Checkup

Solve exercises 1–5 and check.

1. $\dfrac{7}{2x} + \dfrac{3}{x} = 2 - \dfrac{3}{2x}$ 2. $\dfrac{x - 10}{x - 1} = \dfrac{10 - 2x}{2 - 2x}$

3. $\dfrac{2x}{x - 3} = \dfrac{6}{x - 3} + 1$ 4. $\dfrac{6}{3x + 3} = \dfrac{2}{x + 1}$

5. $\dfrac{x + 2}{x - 3} - \dfrac{15}{x^2 - x - 6} = \dfrac{2}{x - 3}$

6. How can you tell whether a rational equation has extraneous solutions?

7. In solving a rational equation algebraically, how can you tell whether the equation has no solutions?

8. In solving a rational equation algebraically, how can you tell whether the equation has all real numbers as solutions, excluding restricted values?

12.4.3 Differentiating between Expressions and Equations

One of the most common mistakes made in dealing with the topics of this chapter is a student becoming confused between simplifying a rational expression and solving a rational equation.

$$\text{Equation} \quad \frac{2}{x} + \frac{3}{x} = 1$$

$$\text{Expression} \quad \frac{2}{x} + \frac{3}{x} - 1$$

Solve equation.

$$x\left(\frac{2}{x} + \frac{3}{x}\right) = x(1)$$

$$\overset{1}{x}\left(\frac{2}{x}\right) + \overset{1}{x}\left(\frac{3}{x}\right) = x$$

$$2 + 3 = x$$

$$5 = x$$

The solution of the equation is 5.

Simplify expression.

$$\frac{2}{x} + \frac{3}{x} - 1 = \frac{2}{x} + \frac{3}{x} - \frac{x}{x}$$

$$= \frac{5 - x}{x}$$

The simplification of the expression is $\frac{5-x}{x}$.

Caution: Expressions multiplied by the LCD will not result in equivalent expressions.

$$\frac{2}{x} + \frac{3}{x} - 1 \neq x\left(\frac{2}{x} + \frac{3}{x} - 1\right)$$

$$\neq \overset{1}{x}\left(\frac{2}{x}\right) + \overset{1}{x}\left(\frac{3}{x}\right) - x(1)$$

$$\neq 2 + 3 - x$$

$$\neq 5 - x$$

This result is not equivalent to the simplification $\frac{5-x}{x}$ previously found. Therefore, we cannot simplify a rational expression by multiplying it by the LCD.

HELPING HAND You can *solve* a rational *equation* by multiplying both sides by the LCD. However, you cannot *simplify* a rational *expression* by multiplying it by the LCD.

EXAMPLE 5 Determine which of the following is an equation and which an expression. Solve the given equation and simplify the given expression.

a. $\dfrac{x+1}{x-5} - \dfrac{x-5}{x-1}$ **b.** $\dfrac{x+1}{x-5} = \dfrac{x-5}{x-1}$

Solution

a. $\dfrac{x+1}{x-5} - \dfrac{x-5}{x-1}$ is an expression.

Simplify.

$$\frac{x+1}{x-5} - \frac{x-5}{x-1} = \frac{(x+1)(x-1)}{(x-5)(x-1)} - \frac{(x-5)(x-5)}{(x-1)(x-5)} \qquad \text{LCD is } (x-5)(x-1).$$

$$= \frac{x^2-1}{(x-5)(x-1)} - \frac{x^2-10x+25}{(x-1)(x-5)}$$

$$= \frac{(x^2-1)-(x^2-10x+25)}{(x-5)(x-1)}$$

$$= \frac{x^2-1-x^2+10x-25}{(x-5)(x-1)}$$

$$= \frac{10x-26}{(x-5)(x-1)}$$

$$= \frac{2(5x-13)}{(x-5)(x-1)}$$

b. $\dfrac{x+1}{x-5} = \dfrac{x-5}{x-1}$ is an equation.

Solve.

$$\frac{x+1}{x-5} \bowtie \frac{x-5}{x-1} \qquad \text{The restricted values are 5 and 1.}$$

$$(x+1)(x-1) = (x-5)(x-5) \qquad \text{Equate the cross products.}$$

$$x^2-1 = x^2-10x+25$$

$$-1 = -10x+25$$

$$-26 = -10x$$

$$x = \frac{26}{10}$$

$$x = \frac{13}{5}$$

12.4.3 Checkup

1. Determine which of the following is an equation and which an expression. Solve the given equation and simplify the given expression.

   a. $\dfrac{2x}{x-4} - 1 + \dfrac{9}{x-4}$ **b.** $\dfrac{2x}{x-4} - 1 = \dfrac{9}{x-4}$

2. Explain how you can tell the difference between a rational equation and a rational expression.

3. What mathematical operations can you perform on a rational expression?

4. What mathematical operations can you perform on a rational equation?

12.4.4 Modeling the Real World

Recall the simple-interest formula $I = PRT$, where I is the amount of simple interest, P is the amount of principal, R is the rate of interest, and T is the amount of time. When this formula is solved for the principal, it gives rise to the rational equation $P = \frac{I}{RT}$. This relationship is another example of writing rational expressions to model real-world situations.

EXAMPLE 6 At the end of a two-year investment of $2000, Chantell had earned simple interest of $125 on one investment and $130 on a second investment. She remembered that the second investment earned 0.25% higher interest than the first investment. What were the two simple-interest rates?

Algebraic Solution

Let $\qquad x =$ interest rate of the first investment (in decimal form)

$x + 0.0025 =$ interest rate of the second investment

The time T is two years.

The first investment interest was $125.

The second investment interest was $130.

Since $P = \frac{I}{RT}$,

$$P_1 = \frac{125}{x(2)} \qquad \text{First investment principal}$$

$$P_2 = \frac{130}{(x + 0.0025)2} \qquad \begin{array}{l}\text{Second investment} \\ \text{principal}\end{array}$$

The total principal was $2000. Therefore, $P_1 + P_2 = 2000$.

$$\frac{125}{x(2)} + \frac{130}{(x + 0.0025)2} = 2000$$

$$\frac{125}{2x} + \frac{130}{2x + 0.005} = 2000$$

$$125(2x + 0.005) + 130(2x) = 2000(2x)(2x + 0.005) \qquad \text{LCD} = 2x(2x + 0.005)$$

$$250x + 0.625 + 260x = 8000x^2 + 20x \qquad \text{Write in standard form.}$$

$$8000x^2 - 490x - 0.625 = 0$$

$$x = \frac{-b \pm \sqrt{b^2 - 4ac}}{2a} \qquad \text{Quadratic formula}$$

$$x = \frac{-(-490) \pm \sqrt{(-490)^2 - 4(8000)(-0.625)}}{2(8000)} \qquad \begin{array}{l}a = 8000, \\ b = -490, \\ c = -0.625\end{array}$$

$$x = \frac{490 \pm \sqrt{240,100 + 20,000}}{16,000}$$

$$x = \frac{490 \pm \sqrt{260,100}}{16,000}$$

$$x = \frac{490 \pm 510}{16,000}$$

$$x = \frac{1000}{16,000} \quad \text{or} \quad x = \frac{-20}{16,000}$$

$$x = 0.0625 \quad \text{or} \quad x = -0.00125$$

Graphic Solution

Check graphically.

$Y2 = 2000$

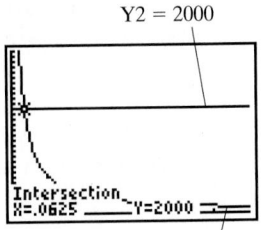

Intersection
X=.0625 Y=2000

$$Y1 = \frac{125}{2x} + \frac{130}{2(x + 0.0025)}$$

$(0, 1.175, 0, 3100)$

The x-coordinate of the point of intersection of the graphs is the solution: 0.0625.

Check these possible solutions graphically by using a window of $(0, 1.175, 0.1, 0, 3100, 100, 1)$.

The x-coordinate of the point of intersection of the graphs is the solution: 0.0625. The value -0.00125 is not a realistic solution.

Chantell earned an interest rate of 0.0625, or 6.25%, on the first investment and an interest rate of $0.0625 + 0.0025 = 0.65$, or 6.5%, on the second investment.

APPLICATION

The maximum variable height (in feet) of an object viewed at a distance of 100 feet from a $\frac{1}{2}$-inch (4.8-millimeter) camera lens combined with an x-millimeter focal length is given by $h(x) = \frac{480}{x}$. Determine the focal length of a camera that Mariam will need to take a picture of a Christmas tree that is 20 feet tall.

Discussion

Let x = the focal length in millimeters
$h(x)$ = the maximum variable height of an object

$$h(x) = \frac{480}{x}$$

$$20 = \frac{480}{x} \qquad \text{Substitute 20 for } h(x).$$

$$20 \cdot x = \left(\frac{480}{x}\right)x \qquad \text{Multiply both sides by } x.$$

$$20x = 480$$

$$\frac{20x}{20} = \frac{480}{20} \qquad \text{Divide both sides by 20.}$$

$$x = 24$$

Mariam will need a camera with a 24-millimeter focal lens.

✓ 12.4.4 Checkup

1. At the end of a three-year investment period, Akim earned $225 in simple interest. He also earned $225 on a second investment for a two-year period. The second investment offered a reduced interest rate, since the investment period was shorter. The simple-interest rate was 0.5% less than that for the longer period. If Akim invested a total of $4000 in the two investments, what were the two simple-interest rates?

2. The maximum variable height (in feet) of an object viewed at a distance of 200 feet from a $\frac{2}{3}$-inch (6.6-millimeter) camera lens combined with an x = millimeter focal length is given by $h(x) = \frac{1320}{x}$. Determine the focal length needed to take a picture of a building that is 66 feet high.

12.4 Exercises

Determine whether each equation is rational.

1. $x^{-3} + 2x^{-2} - x^{-1} = 5$

2. $c^{-1/3} + 5 = 2c$

3. $\dfrac{1}{x+7} + \dfrac{1}{x} = 5$

4. $\dfrac{1}{5+x} = \dfrac{\sqrt{2x-1}}{x^2}$

5. $v^{1/2} + 3v = 10$

6. $\dfrac{4.7}{x} - \dfrac{1.2}{x^2} = 2.6$

7. $\dfrac{\sqrt{2x+7}}{x} = \dfrac{1}{3-x}$

8. $\dfrac{\sqrt{5}}{2h^2} = 4 - \dfrac{\sqrt{3}}{h}$

9. $\dfrac{1}{3x^2} + \dfrac{5}{6x} = \dfrac{8}{9}$

10. $3x^{-2} = 5 + x^{-1}$

11. $\dfrac{\sqrt{3}}{x^2} + \dfrac{\sqrt{2}}{x} = 7$

12. $\dfrac{1}{4x+1} + \dfrac{4}{3x} = 7$

Solve exercises 13–84 algebraically and check numerically, graphically, or by substitution.

13. $\dfrac{1}{a} = \dfrac{5}{9} + \dfrac{2}{3}$

14. $\dfrac{5}{7} - \dfrac{3}{b} = \dfrac{1}{3}$

15. $\dfrac{11}{x} - \dfrac{3}{4} = \dfrac{5}{8}$

16. $\dfrac{3}{x} = \dfrac{5}{8} - \dfrac{3}{4}$

17. $1 - \dfrac{8}{x} = \dfrac{10}{x}$

18. $1 - \dfrac{12}{x} = \dfrac{2}{x}$

19. $\dfrac{1}{2} = \dfrac{5}{2x} - \dfrac{6}{x}$

20. $\dfrac{1}{3} = \dfrac{3}{2x} - \dfrac{1}{x}$

21. $2 = \dfrac{6}{x-2}$

22. $3 = \dfrac{4}{x-5}$

23. $\dfrac{x-4}{2x+3} = \dfrac{5}{21}$

24. $\dfrac{x-1}{3x+2} = \dfrac{6}{13}$

25. $\dfrac{1}{x-6} = \dfrac{17}{2x+3}$

26. $\dfrac{17}{2x-5} = \dfrac{47}{4x+3}$

27. $\dfrac{8}{2x-3} + \dfrac{4}{x} = 0$

28. $\dfrac{17}{5x+2} - \dfrac{2}{x} = 0$

29. $\dfrac{2x + 3}{x + 4} = \dfrac{5}{x + 4}$

30. $\dfrac{3x + 1}{x + 7} = \dfrac{4}{x - 2}$

31. $\dfrac{16}{x} = x$

32. $\dfrac{81}{x} = x$

33. $\dfrac{z}{7} - \dfrac{7}{z} = 0$

34. $\dfrac{2}{y} = \dfrac{y}{32}$

35. $\dfrac{10}{k} + 3 = k$

36. $m = \dfrac{28}{m} - 3$

37. $\dfrac{x}{4} + \dfrac{4}{x} = 3 - \dfrac{4}{x}$

38. $\dfrac{x}{4} + \dfrac{1}{2x} = \dfrac{7}{4} - \dfrac{1}{x}$

39. $\dfrac{y}{5} - \dfrac{21}{5y} = 1 + \dfrac{3}{y}$

40. $\dfrac{z}{5} - \dfrac{2}{z} = \dfrac{2}{5} + \dfrac{1}{z}$

41. $\dfrac{x + 7}{x + 9} = \dfrac{x}{x + 1}$

42. $\dfrac{x + 8}{x + 4} = \dfrac{x + 1}{x - 1}$

43. $\dfrac{x - 6}{x + 5} = \dfrac{x - 4}{3x + 2}$

44. $\dfrac{x + 3}{4x - 1} = \dfrac{x + 7}{1 - 4x}$

45. $x^{-1} + \dfrac{2}{3} = \dfrac{3}{5}$

46. $x^{-1} + \dfrac{3}{7} = \dfrac{5}{8}$

47. $x^{-2} - 6 = 3$

48. $x^{-2} + 3 = 19$

49. $1 - 4x^{-1} = 60x^{-2}$

50. $14x^{-2} = 1 + 5x^{-1}$

51. $\dfrac{9}{x - 7} + 3 = \dfrac{2x - 5}{x - 7}$

52. $\dfrac{5}{x - 5} + 4 = \dfrac{3x - 10}{x - 5}$

53. $\dfrac{x + 2}{3} = \dfrac{x^2 - x + 7}{3x - 6}$

54. $\dfrac{5x - 2}{3} = \dfrac{5x^2 + 14x + 2}{3x + 9}$

55. $\dfrac{x - 3}{x + 4} = \dfrac{14}{x^2 + 6x + 8}$

56. $\dfrac{z + 4}{z + 1} = \dfrac{30}{z^2 - 2z - 3}$

57. $\dfrac{25}{p^2 + p - 12} = \dfrac{p + 4}{p - 3}$

58. $\dfrac{q + 3}{q - 4} = \dfrac{42}{q^2 - 2q - 8}$

59. $\dfrac{m - 2}{m + 1} = \dfrac{7m + 22}{m^2 + 6m + 5}$

60. $\dfrac{x + 6}{x - 4} = \dfrac{5x - 3}{x^2 - 7x + 12}$

61. $\dfrac{1 - 4x}{1 - x} + 2 = \dfrac{6x - 3}{x - 1}$

62. $\dfrac{2x - 1}{x - 1} - 1 = \dfrac{5x - 7}{x - 1} - 4$

63. $2 + \dfrac{z + 2}{2z + 1} = \dfrac{3z + 6}{2z + 1} + 1$

64. $\dfrac{x + 5}{x - 6} + \dfrac{1}{2} = \dfrac{3x + 4}{2x - 12}$

65. $\dfrac{1}{x + 5} + \dfrac{21}{x^2 - 25} = 2$

66. $\dfrac{1}{x - 9} + \dfrac{1}{8} = \dfrac{3}{x^2 - 81}$

67. $\dfrac{x + 5}{x - 3} = \dfrac{1 + x}{2 - x}$

68. $\dfrac{x + 9}{x - 2} = \dfrac{3 + x}{7 - x}$

69. $\dfrac{1}{x + 3} + \dfrac{1}{x} = 2$

70. $\dfrac{1}{x + 2} - \dfrac{1}{x} = 3$

71. $2w^{-3} + 2w^{-2} - w^{-1} = 1$

72. $15v^{-3} - 5v^{-2} + 1 = 3v^{-1}$

73. $\dfrac{1}{3u^2} + \dfrac{5}{6u} = \dfrac{8}{9}$

74. $\dfrac{3}{5x^2} + \dfrac{2}{x} = \dfrac{7}{11}$

75. $\dfrac{3}{x - 5} + \dfrac{3}{x^2 - 25} = \dfrac{3x + 1}{x^2 - 25}$

76. $\dfrac{3}{x + 4} + \dfrac{2}{x + 2} = \dfrac{5x + 9}{x^2 + 6x + 8}$

77. $\dfrac{3x + 7}{x - 1} = \dfrac{3x - 3}{x - 1}$

78. $\dfrac{1}{t + 5} - \dfrac{1}{t + 3} = \dfrac{-6}{t^2 + 8t + 15}$

79. $\dfrac{b + 4}{b - 2} = \dfrac{b^2 + 3b - 4}{b^2 - 3b + 2}$

80. $\dfrac{x - 5}{2x + 3} = \dfrac{x^2 - 7x + 10}{2x^2 - x - 6}$

81. $\dfrac{a - 3}{a - 1} + \dfrac{3}{a + 1} = \dfrac{(a + 3)(a - 2)}{a^2 - 1}$

82. $\dfrac{5}{b - 3} - \dfrac{2}{b^2 - 9} = \dfrac{5b + 13}{b^2 - 9}$

83. $\dfrac{x}{2} - 1 + \dfrac{4}{x} = \dfrac{3}{2x} - \dfrac{1}{2}$

84. $x - 1 + \dfrac{2}{x} = \dfrac{1}{2} - \dfrac{1}{2x}$

85. To diversify her savings, Marianne placed part of her money into one account that paid her $286 simple interest for two years and the remainder in another account that paid her $117 simple interest for two years. The first account was able to offer her 2% more simple interest than the second account could. If she invested a total of $3500, what were the annual interest rates she received?

86. Roger invested some of his savings in an account that paid him 6% simple interest annually, amounting to $420. He invested other savings in an account that paid him 5% simple interest annually, amounting to $367.50. The money in the second account was invested for 18 months longer than that for the first account. What were the two time periods for the investments if the total invested in both accounts was $5600?

87. Two blood vessels in a human body are connected in parallel. The combined resistance R of the blood vessels to the flow of blood is related to the resistances r_1 and r_2 of the individual vessels by the mathematical model $\dfrac{1}{R} = \dfrac{1}{r_1} + \dfrac{1}{r_2}$. If the combined resistance is 11.5 dynes and the resistance of the first vessel is 20.5 dynes, what is the resistance of the second vessel, to the nearest tenth of a dyne?

flow of blood →

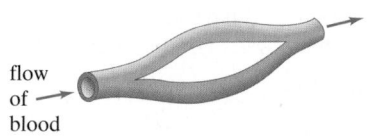

88. When two springs are connected in sequence, the spring constant of the combination K is related to the spring constants k_1 and k_2 of the individual springs by the mathematical model $\frac{1}{K} = \frac{1}{k_1} + \frac{1}{k_2}$. If the constant for the combination is 1.5 pounds per inch and the constant for the first spring is 6 pounds per inch, what is the constant for the second spring?

89. The maximum variable height (in feet) of an object viewed at a distance of 100 feet from a $\frac{1}{3}$-inch (3.6-millimeter) camera lens combined with an $x = $ millimeter focal length is given by $h(x) = \frac{360}{x}$. Determine the focal length needed to take a picture of a 6-foot-tall man.

90. The maximum variable height (in feet) of an object viewed at a distance of 200 feet from a $\frac{1}{4}$-inch (2.7-millimeter) camera lens combined with an $x = $ millimeter focal length is given by $h(x) = \frac{540}{x}$. Determine the focal length needed to take a picture of a billboard that is 18 feet high.

 ## 12.4 Calculator Exercises

The calculator contains a "solve" feature that can be used to find solutions of rational equations. To use this feature, first store the left side of the equation in Y1 and the right side of the equation in Y2. Then key in the expression

$$\text{solve}\ (Y1 - Y2, x, n)$$

where n is an initial guess close to where you might expect the solution to be, and "solve" is a function selected from the calculator's catalog of functions. The keystrokes used to solve the equation $\frac{7}{2x} + \frac{3}{x} = 2 - \frac{3}{2x}$ are as follows:

1. Store the two sides of the equation in Y1 and Y2.

2. Enter $\boxed{\text{2nd}}$ $\boxed{\text{CATALOG}}$ and move the arrow down to the "solve(" line. Press $\boxed{\text{ENTER}}$.

3. Complete the parentheses by entering Y1 − Y2, x, n, and).

$\boxed{\text{VARS}}$ $\boxed{\blacktriangleright}$ $\boxed{1}$ $\boxed{1}$ $\boxed{-}$ $\boxed{\text{VARS}}$ $\boxed{\blacktriangleright}$ $\boxed{1}$ $\boxed{2}$
$\boxed{,}$ $\boxed{\text{X,T,}\theta\text{,n}}$ $\boxed{,}$ $\boxed{3}$ $\boxed{)}$ $\boxed{\text{ENTER}}$

Here, 3 is our initial guess at the solution. The calculator returns a 4, which is the actual solution.

Once you have entered these keystrokes, you can reuse them for a new equation without having to repeat them.

4. Change what you have stored in Y1 and Y2, using the new left-hand and right-hand expressions of the new equation. Press $\boxed{\text{2nd}}$ $\boxed{\text{QUIT}}$.

5. Press $\boxed{\text{2nd}}$ $\boxed{\text{ENTRY}}$, which will repeat the "solve" instruction, and change the initial guess appropriately. Press $\boxed{\text{ENTER}}$, and the calculator will yield a solution of the new equation.

Try this approach on the exercises that follow. The shortcomings of the method are that you must know how many solutions to seek and you must know approximately where the solutions are. If your initial guess is far from the actual solution, the calculator will time out and give you an error message, since it uses an iterative approach to find the solution. These shortcomings underscore the need to know how to solve a rational equation algebraically.

1. $\dfrac{4x + 3}{3x - 1} - \dfrac{3}{4} = \dfrac{4}{5}$

2. $\dfrac{5}{c} + c = \dfrac{21}{2}$

3. $\dfrac{17}{x} = 6 + \dfrac{5}{x}$

4. $\dfrac{3}{k} = \dfrac{17}{4} - \dfrac{2}{5k}$

5. $\dfrac{h}{7} = \dfrac{3h}{5} - \dfrac{16}{21}$

6. $\dfrac{18}{5 - z} = \dfrac{8 - z}{2 - z}$

12.4 Writing Exercises

1. In solving a rational equation algebraically, sometimes the result is an identity. You might assume that the set of all real numbers would then be the solution set of the equation. However, that is not always true; there may be numbers that do not satisfy the equation. Write a short paragraph describing what you must consider when a rational equation results in an identity. Describe what qualifications you must make on your statement of the solutions in this situation.

2. In this section, you were cautioned to not confuse a rational equation with a rational expression. Define what we mean by these two terms. Explain what mathematical operations can be performed on each. Describe how you distinguish one from the other.

(12.5) More Real-World Models

OBJECTIVES
1 Solve real-world models involving rates.
2 Solve real-world models involving proportions.
3 Solve real-world models involving geometric proportions.
4 Solve formulas for a specified variable.

APPLICATION

A formula for the total resistance R in a parallel circuit is $R^{-1} = r_1^{-1} + r_2^{-1}$, where R is the total resistance, r_1 is the resistance of the first resistor, and r_2 is the resistance of the second resistor.

a. Solve the formula for R.

b. A typical lightbulb has a resistance of about 100 ohms. An iron has a resistance of 15 ohms. If a lightbulb and an iron are connected in a parallel circuit, determine the total resistance.

After completing this section, we will discuss this application further. See page 862.

We now know three methods for solving a rational equation: numeric, graphic, and algebraic. We have also solved several types of real-world models with each of these methods. Many other types of models also can be solved with the methods.

In this section, we discuss some common types of real-world models involving rational equations. Remember, there is no set way to work all problems. Also, once a rational equation is written for the problem, it may be solved with any of the methods we have learned.

Let's review the method we have been using to answer a question about a real-world model.

SOLVING REAL-WORLD MODELS

To solve an application problem,

- Read and understand the problem.
- Define a variable for the unknown quantity, and define other quantities in terms of the variable if possible.
- Write an equation with the given information.
- Solve the equation either numerically, graphically, or algebraically.
- Check the solution.
- Write an answer to the question in a complete sentence.

12.5.1 Solving Rate Problems

A ratio is the quotient of two quantities. A **rate** is a ratio (such as miles per hour) used to compare different kinds of measurements. Rate of work is one type of problem that leads to solving a rational equation. This type of problem is usually referred to as a work problem.

Before we begin, we need to be able to determine a rate of work. We know that if a painter needs four hours to paint a room, then $\frac{1}{4}$ of the job is completed in one hour. In two hours, $\frac{2}{4} = \frac{1}{2}$ of the job is completed. If we use this information, we determine the rate of work by solving any one of the following equations:

Let x = the rate of work

time · rate = amount of job completed

$$1 \cdot x = \frac{1}{4}$$ $\frac{1}{4}$ of job completed

$$2 \cdot x = \frac{1}{2}$$ $\frac{1}{2}$ of job completed

$$3 \cdot x = \frac{3}{4}$$ $\frac{3}{4}$ of job completed

$$4 \cdot x = 1$$ entire job completed

All of the equations solved for x result in $x = \frac{1}{4}$. Therefore, the rate of work is $\frac{1}{4}$ job per hour.

RATE OF WORK

If a task can be completed in x hours, then the rate of work is $\frac{1}{x}$ of the task per hour.

Next, we need to understand that if more than one person or machine is working on a task, the total rate of work is the sum of the individual rates. For example, suppose two painters are painting a room. The first painter paints at a rate of $\frac{1}{4}$ room per hour, and his helper paints at a rate of $\frac{1}{6}$ room per hour. Determine the total rate of work.

The sum of the two rates is

$$\frac{1}{4} + \frac{1}{6} = \frac{3}{12} + \frac{2}{12} = \frac{5}{12}$$

Therefore, the total rate of work is $\frac{5}{12}$ room per hour.

EXAMPLE 1 Michelangelo working alone can paint a room in four hours. His helper, Leonardo, would need six hours to do the job alone. Determine the number of hours required for the two painters to paint the room together.

Algebraic Solution

Let x = number of hours needed to paint the room together

$\frac{1}{4}$ is the rate of work for Michelangelo.

$\frac{1}{6}$ is the rate of work for Leonardo.

$\frac{1}{x}$ is the rate of work for both painters working together.

Graphic Check

Check your solution graphically.

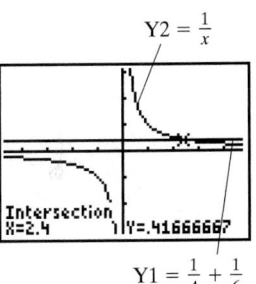

$Y1 = \frac{1}{4} + \frac{1}{6}$

$(-4.7, 4.7, -3.1, 3.1)$

The solution is 2.4.

The sum of the individual rates equals the total rate.

$$\frac{1}{4} + \frac{1}{6} = \frac{1}{x} \qquad \text{\small The restricted value is 0.}$$

$$12x\left(\frac{1}{4} + \frac{1}{6}\right) = 12x\left(\frac{1}{x}\right) \qquad \text{\small Multiply by the LCD.}$$

$$12x\left(\frac{1}{4}\right) + 12x\left(\frac{1}{6}\right) = 12x\left(\frac{1}{x}\right) \qquad \text{\small Distribute the LCD.}$$

$$3x + 2x = 12$$

$$5x = 12$$

$$x = \frac{12}{5} \quad \text{or} \quad 2\frac{2}{5}$$

It would take $2\frac{2}{5}$ hours for the painters to paint the room if they work together. Check your solution graphically on your calculator. The solution is 2.4 or $2\frac{2}{5}$.

EXAMPLE 2 Normally, it takes three minutes to fill a sink. However, today the drain does not close properly and allows water to drain as it is being filled. Therefore, the sink is filled at the end of five minutes. If the water is turned off, determine the number of minutes that water will remain in the sink.

Algebraic Solution

Let x = number of minutes it takes to drain the sink

$\frac{1}{3}$ is the rate of work for filling the sink.

$\frac{1}{x}$ is the rate of work for draining the sink.

$\frac{1}{5}$ is the rate of work for filling and draining at the same time.

Graphic Check

Check your solution graphically.

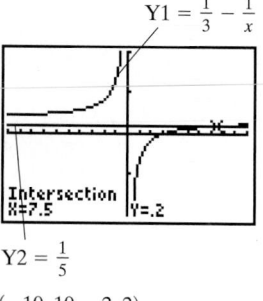

$Y1 = \frac{1}{3} - \frac{1}{x}$

$Y2 = \frac{1}{5}$

$(-10, 10, -2, 2)$

The difference of the individual rates equals the combined rate.

The solution is 7.5.

$$\frac{1}{3} - \frac{1}{x} = \frac{1}{5}$$

The restricted value is 0.

$$15x\left(\frac{1}{3} - \frac{1}{x}\right) = 15x\left(\frac{1}{5}\right)$$

Multiply by the LCD.

$$15x\left(\frac{1}{3}\right) - 15x\left(\frac{1}{x}\right) = 15x\left(\frac{1}{5}\right)$$

Distribute the LCD.

$$5x - 15 = 3x$$

$$2x = 15$$

$$x = \frac{15}{2} \quad \text{or} \quad 7\frac{1}{2}$$

It takes $7\frac{1}{2}$ minutes to drain the sink. Therefore, water will remain in the sink for $7\frac{1}{2}$ minutes.

 12.5.1 Checkup

1. Merle can mow the lawn around his farmhouse in 2 hours. His wife, Glenna, can do the same job in 1.6 hours. Assuming that they have two lawn mowers and can work together without getting in each other's way, how long will it take them to mow the lawn together?

2. Water is being pumped out of a storage tank at a rate that would empty the tank in two hours. At the same time, water is being pumped into the tank at an unknown rate. If an empty tank can be filled in five hours while water is simultaneously being pumped out, how long would it take to fill an empty tank if no water is pumped out?

12.5.2 Solving Proportion Problems

Two ratios (rates) that are equal in value may be written as a rational equation called a proportion. In order for the two ratios to be equal, the numerator and the denominator of each must be written in the same order. For example, $\frac{\text{cost}}{\text{weight}} = \frac{\text{cost}}{\text{weight}}$, but $\frac{\text{cost}}{\text{weight}} \neq \frac{\text{weight}}{\text{cost}}$.

EXAMPLE 3 Justin is practicing batting a baseball with his mother. He has hit 10 out of 20 pitches. His goal is to hit 60% of the pitches. How many consecutive pitches must he hit to raise his average to 60%?

Algebraic Solution	Numeric Check

Algebraic Solution

Let x = number of consecutive pitches he must hit

The current ratio of hits to pitches $\left(\frac{\text{hits}}{\text{pitches}}\right)$ is $\frac{10}{20}$.

The ratio after x consecutive hits $\left(\frac{\text{hits}}{\text{pitches}}\right)$ is $\frac{10 + x}{20 + x}$.

The desired ratio of hits to pitches is 60%, or $\frac{60}{100}$.

The proportion is as follows:

$$\frac{10 + x}{20 + x} = \frac{60}{100}$$

$$1000 + 100x = 1200 + 60x$$ Equate the cross products.

$$40x = 200$$

$$x = 5$$

Numeric Check

Check your solution numerically.

X	Y₁	Y₂
0	.5	.6
1	.52381	.6
2	.54545	.6
3	.56522	.6
4	.58333	.6
5	.6	.6
6	.61538	.6

X=6

$$Y1 = \frac{10 + x}{20 + x}$$

$$Y2 = \frac{60}{100}$$

The restricted value is −20.

The solution is 5.

Justin must hit five consecutive pitches to raise his average to 60%.

Check your solution numerically on your calculator. The solution is 5.

12.5.2 Checkup

Michelle has scored 6 out of 10 baskets from the free-throw line. How many more consecutive baskets must she make in order to bring her average up to 75%?

12.5.3 Solving Geometric Proportions

Geometric figures can be proportional. One example of proportional figures is similar triangles. **Similar triangles** are triangles with congruent corresponding angles and proportional corresponding sides. The symbol for similar is ~. For example,

$\triangle ABC \sim \triangle DEF$ is read "triangle ABC is similar to triangle DEF."

In Figure 12.4, sides AB and DE, BC and EF, and AC and DF are corresponding sides. **Corresponding sides** are in the same position in the figure. **Corresponding angles** are A and D, B and E, and C and F.

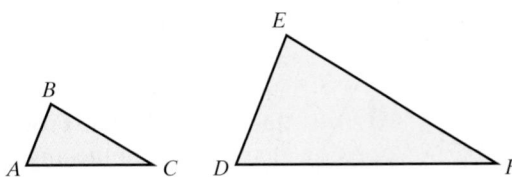

Figure 12.4

The ratio of the measure of any side of one triangle and the measure of the corresponding side on the second triangle are always constant. For example,

$$\frac{AB}{DE} = \frac{BC}{EF} = \frac{AC}{DF}$$

 **HELPING HAND** Compare sides from the same triangle in the same order in the ratio $\frac{\triangle ABC}{\triangle DEF}$.

Therefore, if two sides of one triangle and one corresponding side of a second triangle are known, the second corresponding side may be found by using a proportion.

EXAMPLE 4 Given that $\triangle ABC \sim \triangle DEF$, determine the length of the unknown labeled side.

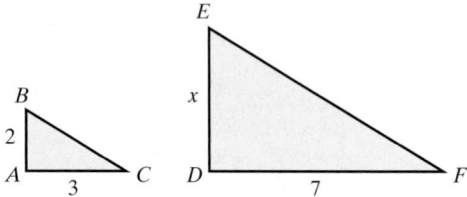

Algebraic Solution

Since $\triangle ABC \sim \triangle DEF$, we have three pairs of corresponding sides that are proportional. However, only two pairs of corresponding sides have their dimensions labeled.

$$\frac{AB}{DE} = \frac{AC}{DF} \qquad \text{Compare corresponding sides } (\tfrac{\triangle ABC}{\triangle DEF}).$$

$$\frac{2}{x} = \frac{3}{7} \qquad \text{Substitute values.}$$

$$3x = 14$$

$$x = \frac{14}{3}$$

The length of the unknown side is $\frac{14}{3}$.

Graphic Check

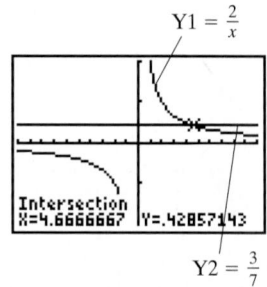

$Y1 = \frac{2}{x}$

Intersection
X=4.6666667 Y=.42857143

$Y2 = \frac{3}{7}$

$(-10, 10, -2, 2)$

Graphing the problem on a calculator results in the solution of approximately 4.667.

EXAMPLE 5 To determine the height of a tree, Sam stood next to it and had Kyle measure the length of his shadow and the tree's shadow. Use Figure 12.5 to determine the height of the tree.

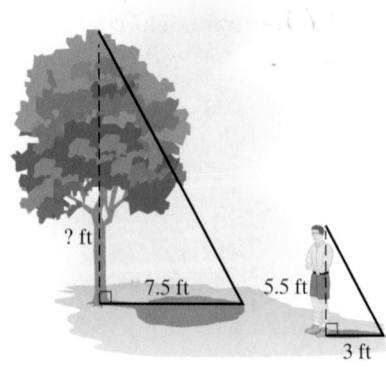

Figure 12.5

Algebraic Solution

We first must determine whether the triangles formed are similar—that is, whether the measures of corresponding angles are equal.

We know that the tree and Sam both form right angles with the ground. From geometry, we know that the angles formed by the sun striking the tree and Sam are equal. From geometry, we also know that the sum of the angles of a triangle is 180 degrees and the sums of the two known angles are equal. Hence, the third angles must be equal.

Graphic Check

Graphing the problem on a calculator results in the solution 13.75. Note that the equation is actually a linear equation.

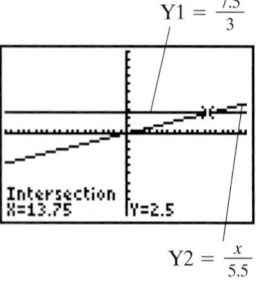

$Y1 = \frac{7.5}{3}$

Intersection
X=13.75   Y=2.5

$Y2 = \frac{x}{5.5}$

$(-20, 20, -10, 10)$

The corresponding sides are therefore proportional, because the corresponding angles are congruent. Let x equal the height of the tree.

$$\frac{7.5}{3} = \frac{x}{5.5}$$ *Compare corresponding sides*

$$3x = 41.25$$

$$x = 13.75$$

The height of the tree is 13.75 feet.

 12.5.3 Checkup

1. $\triangle ABC \sim \triangle DEF$. Use proportions to find the lengths of the unknown sides.

2. A tower stands 20 feet high and casts a shadow of 18 feet. The combined shadow cast by the tower and an antenna mounted on its top is 27 feet. Find the height of the antenna.

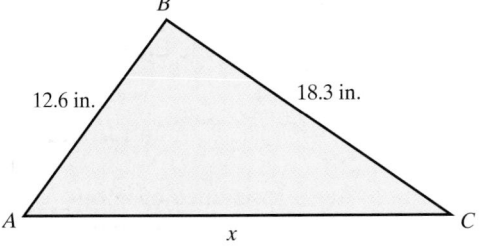
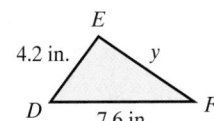

12.5.4 Solving Formulas

Many applications involve the use of formulas, which frequently are rational equations with several variables. Often, we must solve the formula for the variable of interest. We use the same procedure to solve formulas that we use to solve a rational equation.

EXAMPLE 6 **a.** The ideal-gas law states that, given the initial pressure P_1, initial volume V_1, initial temperature T_1, final pressure P_2, final volume V_2, and final temperature T_2, $\dfrac{P_1 V_1}{T_1} = \dfrac{P_2 V_2}{T_2}$. Solve for the initial temperature T_1.

b. Coulomb's law is used to determine the force F between two particles, given the net charge q_1 and q_2 on the particles, the distance d between the particles, and a proportional constant k. Solve $F = \dfrac{kq_1q_2}{d^2}$ for d.

Solution

a. $\dfrac{P_1 V_1}{T_1} = \dfrac{P_2 V_2}{T_2}$

$\dfrac{P_1 V_1 T_2}{P_2 V_2} = \dfrac{P_2 V_2 T_1}{P_2 V_2}$ *Equate cross products.*
Divide by $P_2 V_2$.

$\dfrac{P_1 V_1 T_2}{P_2 V_2} = T_1$ *Simplify.*

or

$$T_1 = \frac{P_1 V_1 T_2}{P_2 V_2}$$

b.
$$F = \frac{kq_1q_2}{d^2}$$

$$d^2(F) = d^2\left(\frac{kq_1q_2}{d^2}\right) \qquad \text{Multiply by the LCD, } d^2.$$

$$\frac{d^2F}{F} = \frac{kq_1q_2}{F} \qquad \text{Divide by F.}$$

$$d^2 = \frac{kq_1q_2}{F}$$

$$\sqrt{d^2} = \sqrt{\frac{kq_1q_2}{F}} \qquad \text{Take the square root.}$$

$$d = \pm\sqrt{\frac{kq_1q_2}{F}}$$

APPLICATION

A formula for the total resistance R in a parallel circuit is $R^{-1} = r_1^{-1} + r_2^{-1}$, where R is the total resistance, r_1 is the resistance of the first resistor, and r_2 is the resistance of the second resistor.

a. Solve the formula for R.
b. A typical lightbulb has a resistance of about 100 ohms. An iron has a resistance of 15 ohms. If a lightbulb and an iron are connected in a parallel circuit, determine the total resistance.

Discussion

a.
$$R^{-1} = r_1^{-1} + r_2^{-1}$$

$$\frac{1}{R} = \frac{1}{r_1} + \frac{1}{r_2} \qquad \text{Write as a rational equation.}$$

$$\frac{1}{R}(Rr_1r_2) = \left(\frac{1}{r_1} + \frac{1}{r_2}\right)(Rr_1r_2) \qquad \text{Multiply by the LCD, } Rr_1r_2.$$

$$\frac{1}{R}(Rr_1r_2) = \left(\frac{1}{r_1}\right)(Rr_1r_2) + \left(\frac{1}{r_2}\right)(Rr_1r_2) \qquad \text{Distributive property}$$

$$r_1r_2 = Rr_2 + Rr_1$$

$$r_1r_2 = R(r_2 + r_1) \qquad \text{Factor.}$$

$$R = \frac{r_1r_2}{r_1 + r_2} \qquad \text{Divide both sides by } (r_1 + r_2).$$

b. $R = \dfrac{r_1r_2}{r_1 + r_2}$

$$R = \frac{(100)(15)}{100 + 15} \qquad \text{Substitute 100 for } r_1 \text{ and 15 for } r_2.$$

$$R = \frac{1500}{115}$$

$$R = \frac{300}{23}$$

The total resistance is $\frac{300}{23}$ ohms.

12.5.4 Checkup

1. In physics, the gravitational attraction between two spheres is given by $F = \dfrac{Gm_1 m_2}{D^2}$, where G is the universal gravitational constant, D is the distance between the centers of the spheres, and m_1 and m_2 are the masses of the two spheres. Solve this formula for D.

2. Two resistors are wired in parallel, and the total resistance of the circuit is 12 ohms. Find the resistance of the two resistors when the resistance of the second resistor is 10 ohms more than the resistance of the first resistor.

12.5 Exercises

1. Miriam can type a report in 3 hours. Saul can type the same report in 4.2 hours. Assuming that they can work together to get the report done, how long will it take them?

2. Paul can wax his car in 45 minutes. His big brother John can do the job in 30 minutes. If they work together, how long will it take them to wax Paul's car?

3. It takes Jacques twice as long to clean the house as it does his wife, Simone. When they work together, they can clean the house in $3\frac{1}{3}$ hours. How long does it take each of them to do the job alone?

4. When Mary Lynn cleans the pool, it takes her an hour longer than it does when Joe cleans the pool. If they can clean the pool in $1\frac{1}{2}$ hours when they work together, how long does it take each of them working alone?

5. Work records show that Felix averages nine hours to do a certain plumbing job, while Oscar averages six hours to do the same job. How long would you expect them to take to do the job together?

6. JoAnne can grade a set of test papers in 15 minutes. Bob can grade the same number of test papers in 25 minutes. How long will it take them to grade the set of papers working together?

7. A tank is filled by two separate pipelines. When the red pipeline is used alone to fill the tank, it takes four hours. When the blue pipeline is used alone to fill the tank, it takes seven hours. How long would it take if both pipelines were used together to fill the tank?

8. A tank can be drained by two separate drains. When the large drain is used alone, it takes 5 hours to drain the tank. When the smaller drain is used alone, it takes 7.5 hours to drain the tank. If both drains are used together, how long will it take to drain the tank?

9. A gas supply line is used to fill storage tanks. Gas line A can fill the tank in $\frac{2}{3}$ of the time it takes gas line B to fill the tank. Together, the lines can fill the tank in $4\frac{4}{5}$ hours. How long does it take each line alone to fill the tank?

10. Two water supply lines can be used to fill the neighborhood pool. Used alone, the second line takes an hour and a half longer than the first line to fill the pool. Together, they can fill the pool in $4\frac{1}{2}$ hours. How long does it take each line used alone to fill the pool?

11. Ricki has earned 45 points out of 60 extra-credit points in her math class. If she completes all of the additional assignments for full credit, how many additional extra-credit points must there be if she wants her average for extra credit to be 80%?

12. A 2.5-liter mixture of diluted vinegar was made by combining 0.8 liter of vinegar with distilled water. How much more pure vinegar must be added to the mixture to bring the percent of vinegar in the mixture up to 40%?

13. The student nature club has spent $120 of its $300 in funds preparing for an Earth Day celebration. The club's budget allocated 45% of all funds to be spent on the celebration. If the club does not want to touch its remaining funds, but instead has a fund-raiser to raise additional funds, how much money should be raised and spent on the celebration to bring the percentage up to 45% of all funds?

14. Felix has raised $75 for his club's fund-raiser. So far, all the members have raised a total of $200, including Felix's contribution. Felix wants to make a last-ditch effort to raise enough additional money so that his contribution will be 50% of the total raised. If no other members raise additional funds, how much more must Felix raise to bring his percentage up to 50% of the total?

In Figure 12.6, $\triangle PQR \sim \triangle XYZ$. Use these triangles for exercises 15, 17, and 19, and find the measures of the unknown sides of the similar triangles from the information provided. In Figure 12.7, $\triangle JKL \sim \triangle MNO$. Use these triangles for exercises 16, 18, and 20, and find the measures of the unknown sides of the similar triangles from the information given.

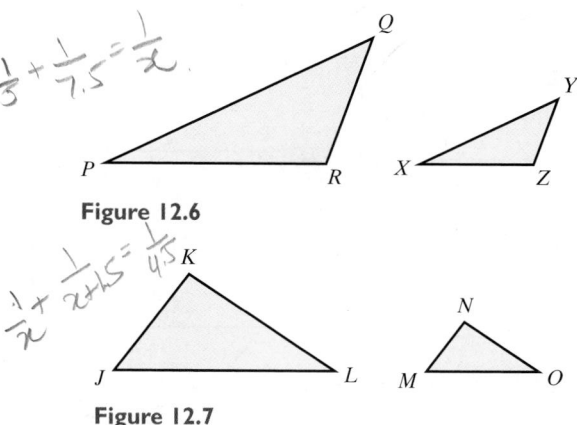

Figure 12.6

Figure 12.7

15. $\overline{PQ}$ measures 24 inches, $\overline{QR}$ measures 12 inches, $\overline{XZ}$ measures 3 inches, and $\overline{YZ}$ measures 2 inches.

16. $\overline{JK}$ measures 7 mm, $\overline{JL}$ measures 35 mm, $\overline{NO}$ measures 8 mm, and $\overline{MO}$ measures 10 mm.

17. $\overline{QR}$ measures 4.4 cm, $\overline{PR}$ measures 5.875 cm, $\overline{XY}$ measures 4.03 cm, and $\overline{XZ}$ measures 2.35 cm.

18. $\overline{KL}$ measures 32.4 yards, $\overline{JL}$ measures 43.92 yards, $\overline{MN}$ measures 6.9 yards, and $\overline{NO}$ measures 13.5 yards.

19. $\overline{PQ}$ measures $10\frac{5}{16}$ feet, $\overline{PR}$ measures $5\frac{5}{8}$ feet, $\overline{XY}$ measures $4\frac{1}{8}$ feet, and $\overline{YZ}$ measures $1\frac{3}{4}$ feet.

20. $\overline{JK}$ measures $9\frac{7}{18}$ feet, $\overline{KL}$ measures $20\frac{1}{24}$ feet, $\overline{MN}$ measures $4\frac{1}{3}$ feet, and $\overline{MO}$ measures $13\frac{1}{2}$ feet.

21. Bridgette was walking along a beach edged by a cliff. She wondered how high the cliff was, so she paced the distance from its base to the tip of its shadow. The shadow measured 35 paces, so she estimated it to be 35 feet long. Then she estimated her shadow to be 4 feet long. If Bridgette is 5.5 feet tall, what is the estimated height of the cliff, based on this information?

22. In late afternoon, a 6-foot lamppost casts a 14-foot shadow. How high is a flagpole if it casts a 35-foot shadow?

23. In physics, the electrical attraction between two charged spheres (with opposite signs) is given by $F = \dfrac{kq_1q_2}{D^2}$ where k is the proportionality constant, D is the distance between the centers of the spheres, and q_1 and q_2 are the charges on the two spheres. Solve this formula for D.

24. Kepler's third law for planetary orbits states that $T = \dfrac{4\pi^2a^3}{MG}$, where T is the orbital period of a planet, a is the length of the longer axis of the elliptical orbit, M is the mass of the sun, and G is the gravitational constant from Newton's law of gravitation. Solve this formula for G.

25. A circuit has two resistors wired in parallel. The resistance of one resistor is 5 ohms less than twice the resistance of the other resistor. The total resistance of the circuit is 6 ohms. Find the resistance of each resistor.

26. Two resistors in a circuit are wired in parallel. The resistance of the second resistor is 5 ohms more than three times the resistance of the first resistor. The total resistance of the circuit is 4 ohms. Find the resistance of each resistor.

12.5 Calculator Exercises

While the emphasis of this chapter was on solving rational equations algebraically, you were shown how to check your solutions graphically. To do so, you stored the left expression of the equation in Y1 and the right expression of the equation in Y2, graphed the two equations, and found where the two graphs intersected. With rational equations, the graphs sometimes are difficult to see and the intersection points are not obvious. One remedy for this situation is to define Y3 to be equal to the difference of Y1 and Y2. Then, whenever Y3 equals zero, you have found a value such that Y1 = Y2, which is the solution of the equation. These values will occur wherever the graph of Y3 crosses the x-axis—that is, at x-intercepts.

As an example, solve $\dfrac{6x + 13}{x + 1} = \dfrac{7}{x^2(x + 1)} - \dfrac{4}{x^2}$. Letting Y1 $= \dfrac{6x + 13}{x + 1}$ and Y2 $= \dfrac{7}{x^2(x + 1)} - \dfrac{4}{x^2}$ define Y3 = Y1 − Y2. The graph of Y1 and Y2 is shown on the first screen, and the graph of Y3 is shown on the second screen:

(Remember, you can turn a graph off or on by moving the cursor to the equals sign and pressing [ENTER]. When the box around the equals sign is missing, the graph is turned off. In the first screen, Y3 is turned off; in the second, Y1 and Y2 are turned off.) From the second screen, it is easier to see that there are two x-intercepts, indicating two solutions of the equation. These are found by using the zero option under the [CALC] function. Verify that the solutions are $x = 0.333\ldots$ and $x = -1.5$.

Try this method on the following rational equations:

1. $\dfrac{2x + 1}{x} = 1 + \dfrac{12}{x^2}$ 2. $\dfrac{x + 1}{x + 3} = \dfrac{8}{2x}$

3. $\dfrac{1}{x} + \dfrac{1}{x + 2} = \dfrac{5}{12}$

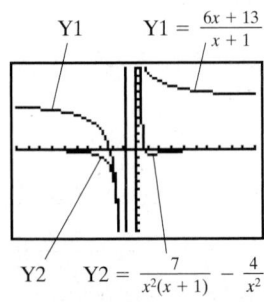

Y1   Y1 $= \dfrac{6x + 13}{x + 1}$

Y2   Y2 $= \dfrac{7}{x^2(x + 1)} - \dfrac{4}{x^2}$

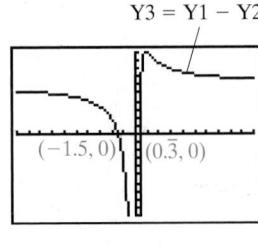

Y3 = Y1 − Y2

$(-1.5, 0)$ $(0.\overline{3}, 0)$

12.5 Writing Exercises

In this chapter, you learned that in some instances you can use cross products to solve a rational equation. Consider each of the equations that follow. For which equations can you use cross products? Explain your reasons.

1. $\dfrac{2}{x} = \dfrac{6}{x+1}$ **2.** $\dfrac{2}{x} = \dfrac{6}{x} + 1$

3. $x^2 - \dfrac{1}{x} = \dfrac{x+1}{3}$ **4.** $\dfrac{x^2 - 1}{x} = \dfrac{x+1}{3}$

CHAPTER 12 SUMMARY

After completing this chapter, you should be able to define the following key terms in your own words.

12.1
rational number
rational expression
rational function
restricted values

12.2
equivalent rational expressions
cancellation

12.3
least common denominator

12.4
rational equation in one variable
extraneous solution
proportion

12.5
rate
similar triangles
corresponding sides
corresponding angles

Reflections

1. What is the difference between a polynomial expression and a rational expression?
2. What does it mean to cancel when multiplying rational expressions? Can you cancel when dividing rational expressions? Explain.
3. What must be true before you can add or subtract two rational expressions?
4. What is the difference between a rational equation and a rational expression?
5. How do restricted values affect the solutions of a rational equation?
6. In solving a rational equation algebraically, how can you tell when it has no solutions? How can you tell when it has many solutions? How can you tell when it has extraneous solutions?

CHAPTER 12 SECTION-BY-SECTION REVIEW

12.1

Recall	Examples
A rational function can be written as a ratio of two polynomials.	$f(x) = \dfrac{2x+1}{x-5}$ is a rational function.
The domain of a rational function is the set of all real numbers except the function's restricted values.	Determine the domain of $f(x) = \dfrac{2x+1}{x-5}$. The restricted values are determined when $x - 5 = 0$ $x = 5$ The domain is the set of all real numbers not equal to 5, or $x \neq 5$, or $(-\infty, 5) \cup (5, \infty)$.

(Continued on page 866)

Recall	Examples
Graph a rational function • Determine the restricted value(s). • Determine a table of values by choosing the *x*-values less than, and the *x*-values greater than, the restricted values. • Plot the points and connect them with a smooth curve. Do not cross the vertical line where *x* equals the restricted value.	Graph $f(x) = \dfrac{2x + 1}{x - 5}$. The restricted value is 5.

x	*f(x)*	*x*	*f(x)*
−5	0.9	5.1	112
−2	0.4	5.5	24
0	−0.2	6	13
2	−1.7	7	7.5
4	−9	10	4.2
4.5	−20	15	3.1
4.9	−108		

Recall	Examples
Graph a rational function on a calculator • Determine the restricted value(s). • Set the calculator to Dot MODE. • Enter the function in the Y = menu. • Graph on an appropriate window.	Graph $f(x) = \dfrac{2x + 1}{x - 5}$.

$Y1 = \dfrac{2x + 1}{x - 5}$

$(-20, 20, -20, 20)$

Determine whether each expression is rational or nonrational.

1. $3 + \dfrac{x - 5}{x^2 + 6x + 9}$ **2.** $x^2 - \dfrac{\sqrt{x} + 3x}{x + 1}$ **3.** $5x^2 - 3x + 2 - 4x^{-1}$ **4.** $2x + 5$

Determine the restricted values algebraically, and then state the domain of each rational equation.

5. $y = \dfrac{-3}{x^2}$ **6.** $f(x) = \dfrac{2x + 7}{x^2 + 5x - 24}$ **7.** $y = \dfrac{5}{4x^2 + 9}$ **8.** $y = \dfrac{5}{4x^2 - 9}$

In exercises 9–12, graph each rational function.

9. $y = \dfrac{180}{x^2} + \dfrac{30}{x}$ **10.** $g(x) = \dfrac{x^2 - x - 2}{x - 2}$ **11.** $p(x) = 1 - \dfrac{4}{x} - \dfrac{21}{x^2}$ **12.** $y = \dfrac{x^2 - 3x - 10}{x^2 - x - 12}$

13. The purchase price of a computerized cash register is $1800. The manufacturer provides a maintenance agree-ment on the register that is priced at $125 for the first year and that increases by $50 per year each year thereafter.

The total cost of the cash register and maintenance agreement is given by $C(x) = 1800 + 100x + 25x^2$ if the cash register is used and maintained for x years.

a. Determine the average cost per year if the cash register is kept for x years.

b. Graph the average-cost function for the domain $0 \le x \le 20$.

Determine the minimum average cost per year.

c. What advice on maintaining the cash register would you give this business?

12.2

Recall	Examples
Simplifying rational expressions • Factor the numerator and denominator. • Simplify the results, as in $\dfrac{A}{B} = \dfrac{\overset{1}{k} \cdot C}{\underset{1}{k} \cdot D} = \dfrac{C}{D}$, where A, B, C, and D are polynomials and $B \ne 0$, $D \ne 0$, and k is the GCF of A and B.	Simplify. $\dfrac{x^2 + 2x}{x^2 + x - 2}$ $\dfrac{x^2 + 2x}{x^2 + x - 2} = \dfrac{x(x + 2)}{(x + 2)(x - 1)}$ $= \dfrac{x}{x - 1}$
Multiplying rational expressions • Multiply the numerators and multiply the denominators, as in $\dfrac{A}{B} \cdot \dfrac{C}{D} = \dfrac{AC}{BD}$, where A, B, C, and D are polynomials and B and $D \ne 0$. • Simplify the results.	Multiply. $\dfrac{x}{x - 1} \cdot \dfrac{x^2 - 1}{2x^2 + 3x + 1}$ $\dfrac{x}{x - 1} \cdot \dfrac{x^2 - 1}{2x^2 + 3x + 1} = \dfrac{x}{x - 1} \cdot \dfrac{(x - 1)(x + 1)}{(2x + 1)(x + 1)}$ $= \dfrac{x}{2x + 1}$
Dividing rational expressions • Change the division problem to an equivalent multiplication problem, such as $\dfrac{A}{B} \div \dfrac{C}{D} = \dfrac{A}{B} \cdot \dfrac{D}{C} = \dfrac{AD}{BC}$, where A, B, C, and D are polynomials and $B \ne 0$, $C \ne 0$, and $D \ne 0$. • Simplify the results.	Divide. $\dfrac{x + 2}{x^2 - 2x + 1} \div \dfrac{x}{x^2 + 3x - 4}$ $\dfrac{x + 2}{x^2 - 2x + 1} \div \dfrac{x}{x^2 + 3x - 4}$ $= \dfrac{x + 2}{x^2 - 2x + 1} \cdot \dfrac{x^2 + 3x - 4}{x}$ $= \dfrac{x + 2}{(x - 1)(x - 1)} \cdot \dfrac{(x - 1)(x + 4)}{x}$ $= \dfrac{(x + 2)(x + 4)}{x(x - 1)}$

Simplify.

14. $\dfrac{-56x^3yz}{126x^2y^3z}$ **15.** $\dfrac{5p - 15}{15p + 25}$ **16.** $\dfrac{15x - 3x^2}{6x^3 - 30x^2}$ **17.** $\dfrac{2x^2 + x - 15}{4x^2 + 13x + 3}$ **18.** $\dfrac{b^2 - 49}{b^3 + 7b^2 + 2b + 14}$

Multiply and simplify.

19. $\dfrac{9x^3}{35y} \cdot \dfrac{-5y^2}{6x^5}$ **20.** $\dfrac{a + 5}{a - 4} \cdot (a + 1)$ **21.** $\dfrac{4m - 4}{m^2 - 25} \cdot \dfrac{m - 5}{2m^2 + 5m - 7}$

22. $\dfrac{a - 5}{a + 4} \cdot \dfrac{a + 4}{5 - a}$ **23.** $\dfrac{7x^2y}{5y - 3x} \cdot \dfrac{3x - 5y}{42xy^3}$ **24.** $\dfrac{2m + 6}{9m + 45} \cdot \dfrac{-3m - 15}{10m + 30}$

In exercises 25–33, divide and simplify.

25. $\dfrac{14z^5}{25x^2y^3} \div \dfrac{-7z^3}{15xy^4}$ **26.** $\dfrac{27a^2b}{8cd} \div \dfrac{9c^2d}{16ab^2}$ **27.** $\dfrac{x^2 + 4x + 3}{5x^2 + 5} \div \dfrac{x^2 - 3x + 2}{x^4 - 1}$

28. $\dfrac{3x^2y}{z^3} \div (2xyz)$ **29.** $(21a^2bc) \div \dfrac{3ab}{2c}$ **30.** $\dfrac{x^2 - 4}{x^2 + 4} \div (x - 2)$

31. $(2x^2 - x - 3) \div \dfrac{2x^2 - 7x + 6}{2x^2 - x - 6}$ **32.** $\dfrac{a + 3}{a + 6} \div \dfrac{a + 2}{a + 4}$ **33.** $\dfrac{x - 11}{x - 3} \div \dfrac{11 - x}{3 - x}$

34. Mario traveled 220 miles. If the trip took t hours, write an expression for his average rate of speed. If Mario cuts his average speed in half and adds two hours to his time, write an expression for the new distance he would travel.

35. Given the formula for the volume of a box, $V = LWH$, solve for the height H. A box has a volume given by $4x^3 + 12x^2$ cubic inches. The base of the box is a square, with each side measuring $2x$ inches. Use the new formula to find the height of the box.

36. The fixed cost (setup cost) to produce an item is $500, and the variable cost per item produced is $40. The revenue received for selling each item produced is $80 per item.
   a. Write a cost function and a revenue function for producing and selling x items.
   b. Write an expression for the ratio of cost to revenue for x items, and simplify the expression.

12.3

Recall	Examples
Adding and subtracting rational expressions with like denominators • Add or subtract the numerators, as in $$\dfrac{A}{B} + \dfrac{C}{B} = \dfrac{A + C}{B}$$ $$\dfrac{A}{B} - \dfrac{C}{B} = \dfrac{A - C}{B}$$ where A, B, and C are polynomials and $B \neq 0$. • Simplify the results.	Add. $\dfrac{5}{x} + \dfrac{x + 2}{x}$ $$\dfrac{5}{x} + \dfrac{x + 2}{x} = \dfrac{x + 7}{x}$$ Subtract. $\dfrac{x}{x - 3} - \dfrac{x + 1}{x - 3}$ $$\dfrac{x}{x - 3} - \dfrac{x + 1}{x - 3} = \dfrac{x - (x + 1)}{x - 3}$$ $$= \dfrac{x - x - 1}{x - 3}$$ $$= \dfrac{-1}{x - 3}$$
Adding and subtracting rational expressions with unlike denominators • Determine the LCD of the denominators of all the fractions. • Change the fractions to equivalent fractions with the LCD as the denominator. • Add or subtract by using the procedure for like denominators.	Add. $\dfrac{2x}{x - 3} + \dfrac{1}{x + 3}$ $$\dfrac{2x}{x - 3} + \dfrac{1}{x + 3} = \dfrac{2x(x + 3)}{(x - 3)(x + 3)} + \dfrac{1(x - 3)}{(x + 3)(x - 3)}$$ $$= \dfrac{2x^2 + 6x + x - 3}{(x - 3)(x + 3)}$$ $$= \dfrac{2x^2 + 7x - 3}{x - 3}$$ Subtract. $\dfrac{2}{x} - \dfrac{x - 1}{x + 4}$ $$\dfrac{2}{x} - \dfrac{x - 1}{x + 4} = \dfrac{2(x + 4)}{x(x + 4)} - \dfrac{x(x - 1)}{x(x + 4)}$$ $$= \dfrac{2x + 8 - x^2 + x}{x(x + 4)}$$ $$= \dfrac{-x^2 + 3x + 8}{x(x + 4)}$$

In exercises 37–50, add or subtract, and simplify.

37. $\dfrac{3y}{5x} + \dfrac{7y^2}{5x}$ **38.** $\dfrac{5b^3}{7a} - \dfrac{2b}{7a}$ **39.** $\dfrac{x - 8}{2x - 5} + \dfrac{3x - 2}{2x - 5}$

40. $\dfrac{10x^2 + 3x + 5}{4x^2 - 9} - \dfrac{2 - 7x}{4x^2 - 9}$ **41.** $\dfrac{4}{5x^3y} + \dfrac{7}{15x^2y^2}$ **42.** $\dfrac{5x}{x - 8} + \dfrac{2x}{x + 4}$

43. $\dfrac{8}{2x - 1} - \dfrac{4}{x + 5}$ **44.** $\dfrac{7}{x - y} + \dfrac{4}{y - x}$ **45.** $\dfrac{4x}{9x^2 - 16} + \dfrac{2}{15x - 20}$

46. $\dfrac{5}{2x^2 - 5x - 3} + \dfrac{7}{3x^2 - 11x + 6}$ **47.** $\dfrac{7}{2x - 18} - \dfrac{3x + 4}{x^2 - 81}$ **48.** $\dfrac{8}{2x^2 + x - 6} - \dfrac{2}{3x^2 + 4x - 4}$

49. $\dfrac{7}{x + 5} + \dfrac{6}{x - 5} + \dfrac{8}{5 - x}$ **50.** $\dfrac{7x}{x + 2} - \dfrac{11}{3x} - \dfrac{5}{6}$

51. The lengths of the sides of a triangle are given by the rational expressions $\dfrac{x - 1}{x + 1}$, $\dfrac{x - 3}{x}$, and $\dfrac{x + 1}{2x}$. Add the lengths to find an expression for the perimeter of the triangle.

52. The perimeter of a rectangle is given by the rational expression $\dfrac{14x^2 + 14x + 6}{x(x + 1)}$. If the width of the rectangle is given by the expression $\dfrac{3x}{x + 1}$, use subtraction to find an expression for the length of the rectangle.

53. Bob and Gretchen took turns driving on a rural highway. Bob's average speed was 10 miles per hour greater than Gretchen's average speed. Bob drove 125 miles of the trip and Gretchen drove 80 miles of the trip.

a. Letting x be Gretchen's average speed, write a function $t(x)$ for the total time of the trip.

b. Graph the function.

c. If Gretchen's average speed was 30 miles per hour, estimate the total time of the trip, to the nearest tenth of an hour.

d. If the time of the trip was 4.5 hours, what were Gretchen's average speed and Bob's average speed?

12.4

Recall	Examples
Rational equation in one variable • A rational equation in one variable is written in the form $R(x) = 0$, where $R(x)$ is a rational expression.	Rational equations in one variable are $\dfrac{2x + 1}{x} = 0$ $6 + \dfrac{5}{2x} = \dfrac{x - 1}{x + 3}$ which can be written as $R(x) = 0$.
Solving a rational equation • Determine the restricted values. • Multiply both sides by the LCD of all of the terms. • Solve the resulting equation. • Discard any solutions that are restricted values. • Check the solution by substituting, solving numerically, or solving graphically.	Solve. $\dfrac{2x^2}{2x + 1} + 3 = \dfrac{x}{2x + 1}$ Restricted values: $2x + 1 = 0$ $$x = -\dfrac{1}{2}$$ $\dfrac{2x^2}{2x + 1} + 3 = \dfrac{x}{2x + 1}$ $(2x + 1)\left(\dfrac{2x^2}{2x + 1} + 3\right) = (2x + 1)\left(\dfrac{x}{2x + 1}\right)$ $(2x + 1)\left(\dfrac{2x^2}{2x + 1}\right) + (2x + 1)(3) = (2x + 1)\left(\dfrac{x}{2x + 1}\right)$ $$2x^2 + 6x + 3 = x$$ $$2x^2 + 5x + 3 = 0$$ $$(2x + 3)(x + 1) = 0$$ $$2x + 3 = 0 \quad \text{or} \quad x + 1 = 0$$ $$x = -\dfrac{3}{2} \qquad x = -1$$ Check the solutions.

Determine whether each equation is rational.

54. $\dfrac{2x + 3}{6} = \dfrac{5}{\sqrt{x} - 1}$ **55.** $3x - 7 = 4x^{1/3}$ **56.** $5x^{-2} + 3x^{-1} + 4 = 18$ **57.** $\dfrac{2}{x} - \dfrac{5}{x - 3} = 17$

Solve exercises 58–73 algebraically.

58. $x + 3 = \dfrac{3x - 26}{2 - x}$ **59.** $\dfrac{3}{x + 3} = \dfrac{6}{2x + 6}$ **60.** $\dfrac{7 - x}{2 - x} = \dfrac{x + 9}{x - 2}$

61. $\dfrac{4}{3x} = 1 - \dfrac{17}{3x}$ **62.** $\dfrac{5}{2x + 4} = \dfrac{2x + 1}{2}$ **63.** $\dfrac{3x + 7}{2x - 4} = \dfrac{11}{3}$

64. $\dfrac{z}{50} - \dfrac{2}{z} = 0$ **65.** $b = 1 + \dfrac{30}{b}$ **66.** $a + \dfrac{13}{a} = -\dfrac{3}{a} - 8$

67. $1 = 12p^{-2} + p^{-1}$ **68.** $\dfrac{x - 2}{x - 6} + 3 = \dfrac{3x - 14}{x - 6}$ **69.** $\dfrac{3}{m + 4} + \dfrac{7}{m^2 - 16} = 2$

70. $\dfrac{2}{(3x + 5) - (2x + 1)} = \dfrac{8}{3(x + 5) + (x + 1)}$ **71.** $\dfrac{7}{x + 3} + \dfrac{2x - 1}{x^2 + x - 6} = \dfrac{5x - 2}{x^2 - 4}$

72. $1 + \dfrac{1}{x^2} = \dfrac{5}{x}$ **73.** $\dfrac{5x}{3} + \dfrac{2}{3} = \dfrac{x - 1}{x}$

74. Hope invested part of her annual bonus into one account that paid her $300 simple interest for one year and the remainder in a second account that paid her $560 simple interest for one year. The second account paid her 1% more interest than the first account. If she invested a total of $13,000, what were the annual interest rates that she received?

12.5

Recall	Example
Modeling real-world situations	Jamie has 15 out of 20 questions correct on her computer quiz. How many consecutive questions must she answer correctly to have an average of 80%.
• Define a variable.	Let x = the number of correct consecutive questions The current ratio is $\frac{15}{20}$. The ratio after x correct consecutive questions is $\frac{15 + x}{20 + x}$. The desired ratio is 80%, or $\frac{80}{100} = \frac{4}{5}$.
• Write an equation.	$\dfrac{15 + x}{20 + x} = \dfrac{4}{5}$
• Solve the equation.	$75 + 5x = 80 + 4x$ $5 = x$
• Check the solution.	Check: $\dfrac{15 + x}{20 + x} = \dfrac{4}{5}$ $\begin{array}{c\|c} \dfrac{15 + (5)}{20 + (5)} & \dfrac{4}{5} \\ \dfrac{20}{25} & \\ \dfrac{4}{5} & \end{array}$
• Write a one-sentence answer to the question asked.	Jamie must answer the next five questions correctly in order to have an average of 80%.

75. Norman can install drywall in an average-sized room in six hours. His assistant can do the room in eight hours. How long will it take them to do the room working together?

76. Lucy and Ethel are packaging candy on an assembly line. It takes Lucy 10 minutes longer to pack a crate of candy than it takes Ethel to do the same. If they work together to pack the crate, they can do so in 30 minutes. How long does it take Lucy and Ethel each to do the job alone?

77. One production line can fill a barrel full of chocolate drops in 10 minutes. Another production line can fill the barrel in 15 minutes. How long will it take both lines to fill a barrel working together?

78. One production line takes twice as long to produce widgets as does a high-speed production line. When orders backlog, both production lines must be run. If it takes both lines working together $3\frac{1}{2}$ hours to produce an order, how long would it take each line working alone?

79. Maya and her roommate placed equal amounts of money into their joint account to pay bills. Maya placed $85 into the account, and, upon counting the money, she finds that the total is $210. Her roommate has evidently added money to the account. How much more money must Maya add to the account to make her share 50%?

80. A diluted alcohol solution is made by mixing 1.5 cups of pure grain alcohol with 12 cups of water. How many more cups of pure alcohol should be added to the solution to make it a 25% alcohol solution?

81. Given $\triangle GHI \sim \triangle JKL$.

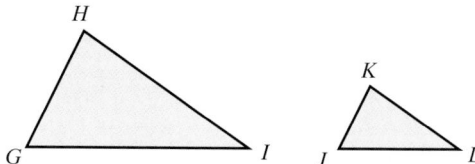

Use the triangles to find the measures of the missing sides of the similar triangles, given the following information:
a. $\overline{GH}$ measures 4 inches, $\overline{HI}$ measures 8 inches, $\overline{JK}$ measures 7 inches, and $\overline{JL}$ measures 21 inches.
b. $\overline{GH}$ measures $3\frac{1}{2}$ feet, $\overline{GI}$ measures 13 feet, $\overline{JK}$ measures $2\frac{5}{8}$ feet, and $\overline{KL}$ measures $5\frac{13}{16}$ feet.
c. $\overline{HI}$ measures 9.22 cm, $\overline{GI}$ measures 14.08 cm, $\overline{KL}$ measures 2.305 cm, and $\overline{JK}$ measures 1.88 cm.

82. A surveyor sighted along similar triangles to determine the height of a bluff. He was standing 120 feet away from the bluff and 15 feet away from his surveying partner, who was holding the surveyor's rod. The top of the 6-foot rod was in line with the top of the bluff from the partner's point of view. How high was the bluff?

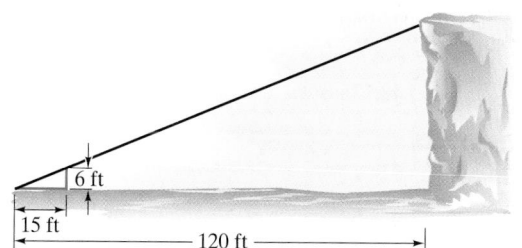

6 ft
15 ft
120 ft

83. In the "Case of the Musgrave Ritual," Sherlock Holmes needs to know where the shadow of an elm tree falls at a certain time of day "when the sun is over the oak tree." However, the tree was struck by lightning and is gone. Fortunately, the owner of the estate knows that the elm was 64 feet tall, because he had used trigonometry to determine that before the tree fell. Holmes quickly fashioned a 6-foot rod and measured its shadow to be 9 feet. He then determined what the length of the shadow of the oak tree would have been. It was elementary, my dear student. What was the length of the tree's shadow?

84. Two resistors are in a parallel circuit. One resistor has a resistance that is 5 ohms more than twice the resistance of the other resistor. The total resistance of the circuit is 30 ohms. Find the resistances of the two resistors, to the nearest ohm.

85. If two lenses with focal lengths f_1 and f_2 are placed in contact with one another, the focal length of the combination is given by $\frac{1}{f} = \frac{1}{f_1} + \frac{1}{f_2}$. Solve for f.

CHAPTER 12 CHAPTER REVIEW

Graph each rational function.

1. $y = \dfrac{45}{x^2} + \dfrac{15}{x}$

2. $g(x) = \dfrac{x+5}{x^2+5x+4}$

3. $p(x) = \dfrac{2x^2-x-3}{x+1}$

4. $y = \dfrac{x^2-5x-6}{x^2+4x-5}$

Determine the restricted values algebraically, and then state the domain of each rational equation.

5. $y = \dfrac{5}{9x^2+25}$

6. $y = \dfrac{5}{9x^2-25}$

7. $g(x) = \dfrac{x+5}{3x}$

8. $h(x) = \dfrac{5x-7}{3x^2+2x-8}$

Simplify.

9. $\dfrac{3x^2+19x+20}{2x^2+13x+15}$

10. $\dfrac{k^2-25}{k^3-5k^2+6k-30}$

11. $\dfrac{-72x^4y^2z}{96xy^6z}$

12. $\dfrac{7q-21}{14q+28}$

13. $\dfrac{12x^2-6x^3}{9x^3-36x}$

Add or subtract, and simplify.

14. $\dfrac{7}{3x+2} - \dfrac{2}{x-3}$

15. $\dfrac{14}{a-b} + \dfrac{9}{b-a}$

16. $\dfrac{7b}{3a} + \dfrac{5b^2}{3a}$

17. $\dfrac{3p^2}{4m} - \dfrac{2p}{4m}$

18. $\dfrac{5}{4x^2y^3} + \dfrac{3}{6x^2y}$

19. $\dfrac{x}{x+3} + \dfrac{4x}{x-5}$

20. $\dfrac{2x}{4x^2-25} + \dfrac{1}{6x-15}$

21. $\dfrac{7}{2x^2+5x-3} + \dfrac{5}{3x^2+11x+6}$

22. $\dfrac{3z-5}{2z-7} + \dfrac{z-9}{2z-7}$

23. $\dfrac{10x^2 - 8x}{25x^2 - 16} - \dfrac{5x - 4}{25x^2 - 16}$

24. $\dfrac{3}{x + 6} + \dfrac{2}{x - 6} + \dfrac{4}{6 - x}$

25. $\dfrac{x}{x + 1} - \dfrac{3}{2x} - \dfrac{3}{4}$

26. $\dfrac{5}{2x + 10} - \dfrac{2x + 1}{x^2 - 25}$

27. $\dfrac{3}{4x^2 - 17x + 15} - \dfrac{2}{5x^2 - 19x + 12}$

Multiply and simplify.

28. $\dfrac{m^2 - 36}{2m^2 + 9m + 10} \cdot \dfrac{3m + 6}{m - 6}$

29. $\dfrac{z - 3}{z - 7} \cdot \dfrac{7 - z}{z - 3}$

30. $\dfrac{8a^4}{21b} \cdot \dfrac{-35b^3}{12a^7}$

31. $(2c + 3) \cdot \dfrac{c + 7}{c + 3}$

32. $\dfrac{3b - 4a}{24ab^3} \cdot \dfrac{8a^2b}{4a - 3b}$

33. $\dfrac{5p + 20}{12p + 96} \cdot \dfrac{-9p - 72}{15p + 60}$

Divide and simplify.

34. $\dfrac{22g^8}{15h^3k^2} \div \dfrac{-11g^4}{10hk^3}$

35. $\dfrac{18p^2q^2}{25m^2n} \div \dfrac{3mn^3}{5pq^5}$

36. $(15m^3np^2) \div \dfrac{2mn}{5p}$

37. $\dfrac{4x^2 - 9}{4x^2 + 9} \div (2x + 3)$

38. $\dfrac{x^2 - x - 6}{2x^2 + 8} \div \dfrac{x^2 - 5x + 6}{x^4 - 16}$

39. $\dfrac{5a^2b^3}{c^5} \div (3a^2bc^2)$

40. $\dfrac{z - 3}{z - 6} \div \dfrac{z + 4}{z + 8}$

41. $\dfrac{k - 7}{k - 12} \div \dfrac{7 - k}{12 - k}$

42. $(2x^2 + 9x - 5) \div \dfrac{2x^2 + 3x - 2}{2x^2 + 5x + 2}$

Solve exercises 43–64 algebraically.

43. $\dfrac{4}{7} - \dfrac{1}{y} = \dfrac{9}{28}$

44. $\dfrac{4}{x - 1} = \dfrac{5x - 16}{7}$

45. $2 - \dfrac{4}{x + 3} = \dfrac{x + 2}{x + 3}$

46. $\dfrac{5}{m + 3} + \dfrac{8}{m^2 - 9} = \dfrac{3}{5}$

47. $\dfrac{16}{2x + 6} = \dfrac{5}{x}$

48. $35 = p^{-2} + 2p^{-1}$

49. $\dfrac{x - 6}{x - 4} = \dfrac{x + 7}{2(x + 3) - (x + 10)}$

50. $\dfrac{a}{a - 7} = \dfrac{a + 1}{a - 2}$

51. $1 - \dfrac{6}{x} = \dfrac{6}{x} - \dfrac{1}{2} + \dfrac{21}{2x}$

52. $1 + 7x^{-1} = 3$

53. $\dfrac{10}{x + 5} - \dfrac{24}{x^2 + x - 20} = \dfrac{1}{x^2 - 16}$

54. $\dfrac{2}{(x + 3) + (x - 9)} = \dfrac{3}{2(x + 4) + (x - 17)}$

55. $x + 5 = \dfrac{x - 11}{2 - x}$

56. $5 = \dfrac{3(x + 5)}{x + 3} + \dfrac{4x}{2x + 6}$

57. $\dfrac{1}{2} - \dfrac{3}{2x} = \dfrac{1}{2} + \dfrac{1}{2x} - \dfrac{2}{x}$

58. $5x = \dfrac{9x + 4}{x + 2}$

59. $\dfrac{x}{3} = \dfrac{27}{x}$

60. $\dfrac{4 - x}{3 - x} = \dfrac{x + 11}{x - 3}$

61. $\dfrac{x + 8}{2} = \dfrac{2x + 11}{x + 1}$

62. $x + 1 = \dfrac{x^2 + 5x - 12}{3x - 1}$

63. $\dfrac{2x}{x + 4} = 1 - \dfrac{8}{x + 4}$

64. $\dfrac{x - 4}{x - 8} = \dfrac{6x}{x^2 - 4x - 32}$

65. Dave can hang wallpaper in a room and finish in 10 hours. Joan can do the same room and finish in 7 hours. How long will it take them to paper the room working together?

66. A diluted glue mixture is made by mixing 2 pints of glue with 7 pints of distilled water. How many more pints of glue are needed to bring the mixture up to a 30% glue solution?

67. Antonio drove 120 miles in t hours. Write an expression for his average rate of speed. If he reduces his average speed to $\frac{2}{3}$ of his original speed and adds 1 hour to his time, write an expression for the new distance he could travel.

68. A tree casts a shadow of 18 feet. At the same time of day, a mailbox that is 4 feet high casts a shadow of 3 feet. How high is the tree?

69. Given $V = LWH$ as the volume of a box, solve for H. A box has a width of x inches and a length that is five times the width. If the volume of the box is $10x^3 + 5x^2$ cubic inches, use the new formula to write an expression for the height of the box.

70. One production line takes $1\frac{1}{2}$ hours longer to fill an order than a second production line takes. When both lines are used, the order can be filled in $3\frac{1}{3}$ hours. How long does it take each line working alone to fill the order?

71. The lengths of the sides of a triangle are given by the expressions $\frac{6}{x-3}$, 3, and $\frac{x-1}{x}$. Find an expression for the perimeter of the triangle.

72. Latoya has two roommates. Each pays one-third of the bills. Latoya has added $70 to the amount her roommates collected, and the total was $250. How much more does Latoya have to add to the money to make her contribution one-third of the total?

73. Two resistors are connected in parallel. The total resistance of the circuit is 25 ohms. If one resistor has a resistance that is 12 ohms more than the resistance of the other resistor, what are the two resistances, to the nearest ohm?

74. The combined capacitance C of two capacitors C_1 and C_2 connected in series is given by $\frac{1}{C} = \frac{1}{C_1} + \frac{1}{C_2}$. Solve for C.

CHAPTER 12 TEST

TEST-TAKING TIPS

Many students have been conditioned to think that they should use only their textbook when they need a reference to help them with their math difficulties. This is not true. Even though your instructor may emphasize the text, remember that you can always seek other explanations of the same material in other textbooks. When you have difficulty understanding a concept, go to the library and check other math textbooks to see how they present the material. You may find that you more readily understand the examples in another text. Or you may find that another author has a manner of explanation that fits better with your style of learning. When students decide to check another text, it is often an eye-opener for them, and they find, to their delight, that they have a much better grasp of the material they are studying. Also, other texts will provide you with many more exercises that you can use to rehearse and prepare for the test.

Determine the restricted values, and then state the domain of each rational equation.

1. $y = \dfrac{x - 7}{2x^2 + 6x}$

2. $g(x) = \dfrac{x^2 - 4x - 21}{x^2 - 5x - 36}$

3. Graph the rational function $f(x) = \dfrac{x + 4}{x - 2}$.

Simplify.

4. $\dfrac{24x + 56}{9x^2 - 49}$

5. $\dfrac{2z^3 - 18z}{z^3 + 3z^2 + 9z + 27}$

Perform the operations and simplify.

6. $\dfrac{5}{x^2 + 4x} + \dfrac{2}{x^2 + x - 12}$

7. $\dfrac{5x^2 - 40x}{20x^2 + 30x} \div \dfrac{x^2 - 64}{2x^2 + 19x + 24}$

8. $\dfrac{2x + 1}{x^2 + 5x} - \dfrac{x - 4}{x^2 + 10x + 25}$

9. $\dfrac{24x^2y}{x^2 - xy - 2y^2} \cdot \dfrac{3x^2 - 12xy + 12y^2}{16x^3y}$

Solve exercises 10–18.

10. $x - 5 = \dfrac{5 - x}{x}$

11. $\dfrac{16}{x - 3} = \dfrac{x - 3}{x}$

12. $\dfrac{b}{5} + \dfrac{1}{2} = \dfrac{1}{5} - \dfrac{11}{10b}$

13. $\dfrac{3}{m} - \dfrac{m}{75} = 0$

14. $6 + 14x^{-1} + 4x^{-2} = 0$

15. $\dfrac{x^2 + 6x + 9}{4x + 20} = \dfrac{x + 6}{x + 5}$

16. $\dfrac{2}{x + 2} = \dfrac{3}{x + 3}$

17. $\dfrac{x - 5}{x + 1} = 1 - \dfrac{6}{x + 1}$

18. $\dfrac{5 - x}{3 - x} = \dfrac{x + 9}{x - 3}$

19. Given $z = \frac{x - m}{s}$, solve for s.

20. Chris wants to estimate the height of a tree that casts a shadow measuring 32.5 feet. She knows that a 7-foot pole in the yard casts a shadow of 10.5 feet. What is the height of the tree?

21. Jack can rake the leaves in his yard in 6 hours. When Jill helps him, they can rake the leaves in 2 hours and 24 minutes (that is, $2\frac{2}{5}$ hours). How long would it take Jill to rake the leaves alone?

22. Explain why a rational equation can be solved algebraically by multiplying both sides of the equation by the least common denominator (LCD), but a rational expression cannot be simplified by multiplying by the LCD.

Project

Part I

In this project, we will first examine rational functions that occur when considering averages. We have seen that we can often express the cost of producing x items as a polynomial function. For example, if there is a setup cost of $500 to prepare for production and a variable cost of $12 per item produced, the cost of producing x items is given by the function $c(x) = 12x + 500$. The average cost of producing x items is determined by dividing this cost function by the number of items produced: $c_{ave}(x) = \frac{12x + 500}{x}$.

1. Graph this function, using a window of $(0, 940, 100, 0, 62, 10, 1)$. Sketch the graph.

2. Answer the following questions:
   a. Does the average cost increase or decrease as the number of items produced increases? Explain what the graph indicates.
   b. As x gets arbitrarily large, what value does the average cost approach? Trace along the curve to explore this.
   c. Do you think there is a limit to how low the average cost will drop as x gets larger? Explain.

3. Complete the second column of the following table of values (the $p_{ave}(x)$ column will be completed later):

x	$c_{ave}(x)$	$p_{ave}(x)$
100		
200		
300		
400		
500		
1000		
1500		
2000		
2500		
3000		

4. Will the average cost of producing x items ever become less than $12? Explain what you think is happening. This is an example of a limit, a concept that you will encounter in many mathematics applications.

5. Next, suppose that all the items produced can be sold for $50 each. Write a revenue function $r(x)$.

6. Use the cost and revenue functions to define a profit function $p(x)$.

7. Finally, write an average-profit function $p_{ave}(x) = \frac{p(x)}{x}$.

8. Graph the average-profit function, using the same window as in step 1, and sketch its graph.

9. Complete the table of values in step 3, using the average-profit function from step 7.

10. Describe the behavior of the average-profit function as the number of items produced and sold increases.

11. Discuss what this project has shown you with regard to using mathematics to model a real-world process and understand how the process behaves.

Part II

Another rational function was proposed by L. L. Thurstone to model the number of successful acts per unit of time that one could accomplish after a given number of practice sessions. His model was $f(x) = \frac{a(x + b)}{(x + b) + c}$, where $f(x)$ represents the number of successful attempts after x rehearsals. Using $a = 40, b = 1$, and $c = 2$, we can hypothesize an example of Thurstone's model as

$$n(x) = \frac{40(x + 1)}{x + 3}$$

where $n(x)$ is the number of words a particular person can read per minute and x is the number of weeks that the person has been practicing. (Note: This hypothetical example is not based on experimental data.)

Use the latter model to answer the following questions:

1. How does $n(x)$ behave as x increases? Sketch a graph of the function.

2. Is there a practical limit to the number of words per minute that the person can read as the number of weeks of practice grows progressively larger?

3. Interpret what Thurstone's model illustrates in this application.

Part III

Search the literature or the Internet for information on another real-world model that uses rational functions. You may find such an application in areas such as learning theory (L. L. Thurstone's work), physiology (the research of W. O. Fems and J. Marsh on muscle contraction), general business (amortization formulas for repaying a loan), physics (acceleration is a function of force divided by mass), electronics (Ahmdahl's law for determining speedup in computer processing), etc.

Once you have found an application, write a short description of the model used, explaining what its behavior illustrates, and document your reference source.

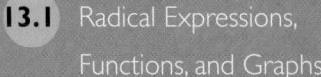

Chapter 13

Radical Expressions, Functions, and Equations

We have discussed radicals and numbers with exponents in several previous chapters. In this chapter, we examine algebraic expressions that involve radicals and rational exponents. We use the properties of radicals and rational exponents to see how to perform mathematical operations on these expressions. Then we study radical equations in one variable and use standard numerical, graphical, and algebraic methods to solve these equations. We complete our discussion with the topic of imaginary numbers and the complex number system.

Radical equations occur in many common real-life situations. For example, the geometric formulas for determining the dimensions of a square or circular object with a given area involve square roots. The geometric formulas for determining the dimensions of a cubic or spherical object from a given volume involve cube roots. Aside from geometry, many kinds of data from statistical analysis can be described by radical expressions. Also, many relations involving acceleration due to gravity use square roots. We will use these kinds of expressions to illustrate concepts throughout this chapter.

In the project at the end of the chapter, we will discuss a more complicated relation involving rational exponents: "windchill" formulas that meterologists use in their weather forecasts.

13.1 APPLICATION

According to annual reports, the number of employees, in thousands, of the Sony Corporation can be estimated by the function $e(x) = 137\sqrt[6]{x}$, where x is the number of years after 1994.

a. Graph the function.

b. Use the graph to determine whether the function is increasing or decreasing.

c. Assuming that the trend shown on the graph does not change, approximate the number of people that will be employed by Sony in 2014.

We will discuss this application further. See page 888.

13.1 Radical Expressions, Functions, and Graphs

OBJECTIVES

1 Evaluate radical expressions with a real-number radicand.
2 Identify radical functions.
3 Determine the domain of radical functions.
4 Graph radical functions.
5 Model real-world situations with expressions and functions involving radicals.

13.1.1 Evaluating Radical Expressions

In Chapter 1, we defined a radical expression as an expression that can be written in the form $\sqrt[n]{a}$, where n (an integer greater than 1) is the index (the plural is *indices*) and a is the radicand. The symbol $\sqrt{}$ is called the radical sign. If the index is 2, it is conventional not to write it.

We revisited square-root expressions in Chapter 11 and simplified square roots with numerical radicands. Now we will discuss radical expressions having indices of 3 and greater. The process of evaluating radicals with even indices differs from that of evaluating radicals with odd indices. We call radicals with even indices even roots. Radicals with odd indices are called odd roots.

Square roots are examples of even roots. In general, to evaluate even roots, we follow the same rules as for evaluating square roots.

To evaluate even roots,
- The radicand of an even root cannot be a negative number, because all real numbers are positive when raised to an even power. (We discuss this case in Section 13.7.)
- The principal root, $\sqrt[n]{a}$, is always a positive value.
- The negative root, $-\sqrt[n]{a}$, is always a negative value.

For example,
$\sqrt[4]{-2}$ does not represent a real number.
$\sqrt{-2}$ does not represent a real number.
The principal root $\sqrt[4]{81} = 3$, because $3^4 = 81$.
The principal square root $\sqrt{81} = 9$, because $9^2 = 81$.
The negative root $-\sqrt[4]{81} = -3$, because $-\sqrt[4]{81}$ is the opposite of the principal root $\sqrt[4]{81}$.
The negative square root $-\sqrt{81} = -9$, because $-\sqrt{81}$ is the opposite of the principal root $\sqrt{81}$.
To evaluate odd roots, we do not have to worry about cases such as principal roots, negative roots, or nonreal numbers. For example,

$\sqrt[3]{64} = 4$, because $4^3 = 64$. $\qquad$ $\sqrt[5]{32} = 2$, because $2^5 = 32$.
$\sqrt[3]{-64} = -4$, because $(-4)^3 = -64$. $\quad$ $\sqrt[5]{-32} = -2$, because $(-2)^5 = -32$.

EVALUATING RADICAL EXPRESSIONS WITH NUMERICAL RADICANDS
To evaluate a radical expression $\sqrt[n]{a}$,
- If n is an integer >1 and $a = 0$, then the result is 0.
- If n is an even integer >1 and $a < 0$, then the result is not a real number.
- If n is an even integer >1 and $a > 0$, then $\sqrt[n]{a} = |b|$ if $b^n = a$.
- If n is an odd integer >1, then $\sqrt[n]{a} = b$ if $b^n = a$.

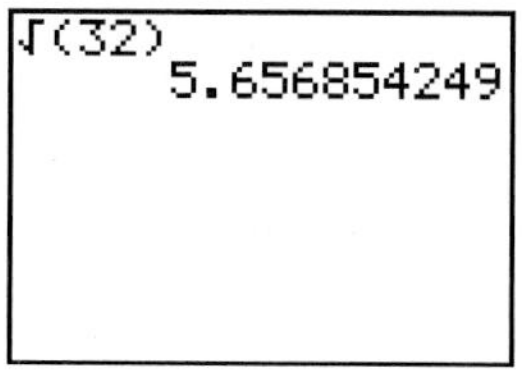

```
√(32)
        5.656854249
```

If the radicand cannot be written as a power of the index, we approximate the answer. We can use a calculator to do this. For example, $\sqrt{32} \approx 5.657$, because $(5.657)^2 \approx 32$.

EXAMPLE 1 Evaluate. Round your results to the nearest thousandth if necessary.

a. $\sqrt[3]{-12}$ **b.** $\sqrt[4]{16}$
c. $-\sqrt[4]{16}$ **d.** $\sqrt[4]{-16}$

Solution

a. $\sqrt[3]{-12} \approx -2.289$, because $(-2.289)^3 \approx -12$.
b. $\sqrt[4]{16} = 2$, because $2^4 = 16$.
c. $-\sqrt[4]{16} = -2$.

 Since $2^4 = 16$, it follows that $-\sqrt[4]{16} = -(2) = -2$. This is the negative root.
d. $\sqrt[4]{-16}$ is not a real number, because the radicand is negative and the root is even.

See Figures 13.1a and 13.1b for calculator solutions.

T E C H N O L O G Y

ROOTS WITH INDICES GREATER THAN 2

Evaluate.　　**a.** $\sqrt[3]{-12}$　**b.** $\sqrt[4]{16}$　**c.** $-\sqrt[4]{16}$　**d.** $\sqrt[4]{-16}$

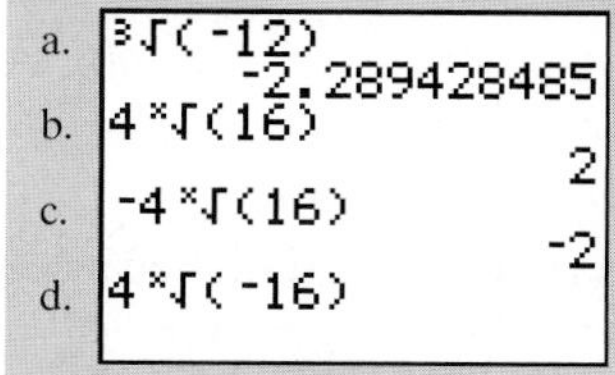

```
a. ³√(-12)
          -2.289428485
b. 4 ˣ√(16)
              2
c. -4 ˣ√(16)
             -2
d. 4 ˣ√(-16)
```

```
ERR:NONREAL ANS
1▊Quit
2:Goto
```

Figure 13.1a 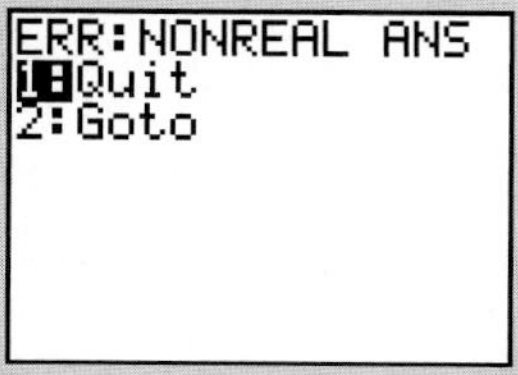   Figure 13.1b

For Figure 13.1a,

The cube root is option 4 under the MATH menu.

a. | MATH | | 4 | | (-) | | 1 | | 2 | |) | | ENTER |

 The xth root is option 5 under the MATH menu. Enter the index before the xth root. Enclose the radicand in a set of parentheses.

b. | 4 | | MATH | | 5 | | (| | 1 | | 6 | |) | | ENTER |
c. | − | | 4 | | MATH | | 5 | | (| | 1 | | 6 | |) | | ENTER |
d. | 4 | | MATH | | 5 | | (| | − | | 1 | | 6 | |) |

For Figure 13.1b,

| ENTER |

The error message "NONREAL ANS" means that we are attempting to take an even root of a negative number.

HELPING HAND Do not attempt to enter xth roots as a product of the index and the square root $\sqrt[4]{16} \neq 4\sqrt{16}$.

13.1.1 Checkup

Evaluate exercises 1–6. Round results to the nearest thousandth if necessary.

1. $\sqrt[4]{625}$ 2. $-\sqrt[4]{625}$ 3. $\sqrt[4]{-625}$
4. $\sqrt[3]{-343}$ 5. $\sqrt[3]{343}$ 6. $\sqrt[5]{117}$

7. Explain the difference between finding the square root of a number and squaring a number.

8. When a negative number is multiplied as a factor an even number of times, what will be the sign of the result? Use this fact to explain why the radicand of an even root must be a positive number if the solution is to be a real number.

9. When a negative number is multiplied as a factor an odd number of times, what will be the sign of the result? Use this fact to explain why the radicand of an odd root may be a positive number or a negative number.

13.1.2 Identifying Radical Functions

A **radical relation** equates a radical expression in one independent variable to a dependent variable—for example, $y = \sqrt{x}$ or $y = \sqrt[3]{2x + 5}$.

Some radical relations are also **radical functions** and can be written in function notation, such as $f(x) = \sqrt{x}$ and $g(x) = \sqrt[3]{2x + 5}$.

> **RADICAL FUNCTION**
> A radical function can be written in the form
> $$f(x) = A(x)$$
> where $A(x)$ is a radical expression with the variable x in the radicand.

EXAMPLE 2   State whether each function is a radical function.

a. $f(x) = \sqrt[3]{5}x$ **b.** $m(x) = 5\sqrt[3]{x}$ **c.** $p(t) = (\sqrt{t + 3})^5$

Solution

a. $f(x) = \sqrt[3]{5}x$ is not a radical function, because the radicand does not contain a variable. This is a linear equation.

b. $m(x) = 5\sqrt[3]{x}$ is a radical function, because the radicand contains the independent variable x.

c. $p(t) = (\sqrt{t + 3})^5$ is a radical function, because the radicand, $t + 3$, contains the independent variable t.

13.1.2 Checkup

State whether each function in exercises 1–3 is a radical function.

1. $f(x) = 2\sqrt{x} + 5$ 2. $y = \sqrt{2x + 5}$

3. $y = (\sqrt[3]{x} + 1)^2$

4. How does a radical function differ from a polynomial function?

13.1.3 Determining the Domain of a Radical Function

In Chapter 12, we saw that rational functions had values excluded from their domains. We called such restrictions on the domain *restricted values*. Radical functions may have values restricted from their domains as well. Sometimes, substituting values for the independent variable in a radical expression results in values that are not real numbers. (In this section, we are concerned with only the real-number system. Another number system is discussed in Section 13.7.) Complete the following set of exercises to determine a method for finding these restricted values.

 Discovery 1

Restricted Values

1. Complete the following tables of values for the given radical functions:

a. $f(x) = \sqrt{x}$

x	f(x)
-3	
-2	
-1	
0	
1	

b. $g(x) = \sqrt[4]{x + 1}$

x	g(x)
-3	
-2	
-1	
0	
1	

c. $h(x) = \sqrt[3]{x}$

x	h(x)
-3	
-2	
-1	
0	
1	

d. $j(x) = \sqrt[5]{x + 1}$

x	j(x)
-3	
-2	
-1	
0	
1	

2. Determine the restricted values of each function if possible.

3. The functions in exercises 1a and 1b are defined by even roots. The functions in exercises 1c and 1d are defined by odd roots. Which functions have restricted values?

4. Graph the given radical functions with restricted values on a decimal calculator screen. Trace each function's graph. What happens when you reach the restricted x-values you found in exercise 2?

5. Write a rule for determining the restricted values of a radical function by viewing its table of values.

6. Write a rule for determining the restricted values of a radical function by viewing its graph.

7. Write a rule for determining the restricted values of a radical function algebraically.

Only the radical functions defined by even roots have restricted values. In viewing a table of values, the restricted values of a radical function are the numbers that display "ERROR" on a calculator. When we view the function's graph, we see that the restricted values of the function do not have corresponding function values. Since these restricted values occur when the radicand of an even root is less than 0, we can determine the restricted values of even roots algebraically by setting the radicand less than 0 and solving for the independent variable.

We have discovered three methods—numeric, graphic, and algebraic—for determining the restrictions on the domain of a radical function. The numeric and graphic methods have limitations on their effectiveness. They should be used primarily to check the algebraic method.

> To determine algebraically the restricted values of a radical function with an even index,
>
> - Set the expression in the radicand less than 0.
> - Solve for the independent variable.

This method will find all the restricted values.

To check the restricted integer values of a radical function numerically on your calculator, set up a table of values.

Restricted integer values are the x-values that result in nonreal values (ERROR in the table). This method will only determine integer values.

To check the restricted values of a radical function graphically on your calculator, graph the function. The x-values that have no corresponding y-values are restricted values for the radical function. The x-values found by this method will vary according to the window setting of your calculator.

EXAMPLE 3   Determine the restricted values, and then state the domain of the radical function $y = \sqrt[4]{3x} + 1$.

Algebraic Solution

$3x < 0$

$x < 0$

The restricted values are all real numbers less than 0.

   The domain is all real numbers greater than or equal to 0.

$x \geq 0$ Inequality notation

$[0, \infty)$ Interval notation

Numeric Solution

$Y1 = \sqrt[4]{3x} + 1$

X	Y1	
-3	ERROR	
-2	ERROR	
-1	ERROR	
0	1	
1	2.3161	
2	2.5651	
3	2.7321	

X=0

The function is not defined for x-values less than 0.

The restricted values are all real numbers less than 0.

   The domain is all real numbers greater than or equal to 0.

$x \geq 0$ Inequality notation

$[0, \infty)$ Interval notation

Graphic Solution

$Y1 = \sqrt[4]{3x} + 1$

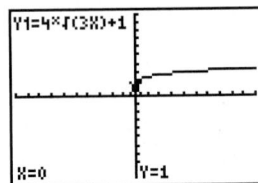

Y1=4×√(3X)+1

X=0   Y=1

$(-10, 10, -10, 10)$
The graph of the function does not exist for x-values less than 0.

The restricted values are all real numbers less than 0.

   The domain is all real numbers greater than or equal to 0.

$x \geq 0$ Inequality notation

$[0, \infty)$ Interval notation

HELPING HAND   If we want to find the domain directly, without going through the separate step of finding the restricted values, we set the radicand greater than or equal to 0 and solve for the independent variable. For example, $3x \geq 0$, or $x \geq 0$.

EXAMPLE 4 Determine the restricted values algebraically if they exist, and then state the domains of the given radical functions. Check your answers numerically or graphically.

a. $y = \sqrt[3]{x + 5} + \sqrt[3]{x - 8}$ **b.** $f(x) = \sqrt{x + 5} + \sqrt{x - 8}$

Solution

a. $y = \sqrt[3]{x + 5} + \sqrt[3]{x - 8}$

Algebraic Solution

There are no restricted values, because the index of the radical, 3, is odd.

Domain: All real numbers

$(-\infty, \infty)$

Graphic Check

$Y1 = \sqrt[3]{x + 5} + \sqrt[3]{x - 8}$

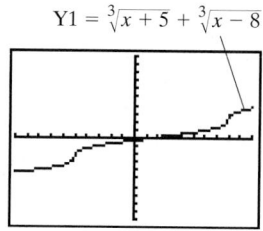

$(-10, 10, -10, 10)$

Since there are no restricted values, check graphically.

The graph appears to be continuous and has no "holes" for restricted values.

b. $f(x) = \sqrt{x + 5} + \sqrt{x - 8}$

Algebraic Solution

There are restricted values, because the index of the radical expressions, 2, is even. To determine the restricted values, set both of the radicands less than 0 and solve for x.

$$x + 5 < 0 \quad \text{or} \quad x - 8 < 0$$
$$x < -5 \quad\quad\quad x < 8$$

According to these inequalities, the restricted values are all real numbers less than -5 or all real numbers less than 8. Hence, the restricted values are all real numbers less than 8, since this restriction covers both inequalities.

Domain: All real numbers greater than or equal to 8

$x \geq 8$

$[8, \infty)$

Numeric Check

X	Y1
4	ERROR
5	ERROR
6	ERROR
7	ERROR
8	3.6056
9	4.7417
10	5.2872

X=8

$Y1 = \sqrt{x + 5} + \sqrt{x - 8}$

The table shows that restricted integer values are less than 8.

13.1.3 Checkup

In exercises 1–3, determine the restricted values and then state the domains of the given functions.

1. $y = \sqrt[3]{x + 2}$ 2. $y = \sqrt[4]{3x + 1} - 2$
3. $g(x) = \sqrt{x + 1} + \sqrt{x - 3}$
4. What values must be restricted from the domain of a radical function?

5. When will a radical function have restrictions on its domain?
6. When you graph a radical function on your calculator, how can you tell whether there are restrictions on its domain?

13.1.4 Graphing Radical Functions

Radical functions have an infinite number of possible solutions. We use a graph to illustrate these solutions. As with any function, to graph a radical function, we determine the domain and then use those values to plot ordered-pair solutions. We need to determine enough ordered pairs so that we can see the pattern. This may be a very lengthy process by hand.

To graph a radical function,

- Determine the restricted values of the domain.
- Determine a table of values by choosing enough x-values in the domain to see a pattern. Pay close attention to the x-values near the restricted values.
- Graph the function by plotting the sample set of ordered pairs from the table of values and connecting the points with a smooth curve determined by the pattern seen. Be careful not to cross the vertical line where x is equal to a restricted value, because the restricted values do not have corresponding function values.

Note: The calculator TABLE function in Ask mode will set up this table.

To graph a radical function on your calculator,

- Enter the equation in Y1.
- Choose an appropriate viewing screen (integer, decimal, standard, or set your own values), and graph the equation.
- Determine individual solutions. Trace and move the cursor, using the left and right arrow keys.

 **HELPING HAND**   The calculator window setting is very important to the picture you receive. You should consider your choices carefully.

EXAMPLE 5   Graph each radical function.

a. $y = 2\sqrt{3x + 10}$ **b.** $m(x) = \sqrt[3]{2x - 9}$

Solution

a. $y = 2\sqrt{3x + 10}$

Restricted value are

$$3x + 10 < 0$$
$$3x + 10 - 10 < 0 - 10$$
$$3x < -10$$
$$x < -\frac{10}{3}$$

Complete a table of values by choosing x-values in the domain, which is the set of all real numbers greater than or equal to $-\frac{10}{3}$.

x	y
$-\dfrac{10}{3}$	0
-3.2	1.2649
-3.1	1.6733
-3	2
-2	4
-1	5.2915
0	6.3246
1	7.2111
2	8
3	8.7178

Graph the ordered pairs found in the table of values, connecting the points with a smooth curve.

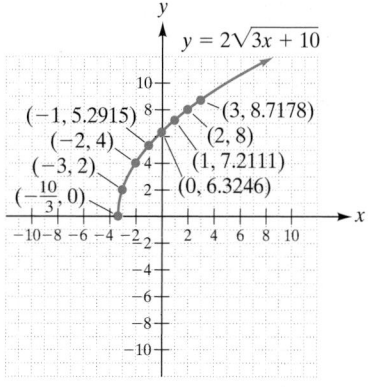

Use your calculator to graph the radical function $y = 2\sqrt{3x + 10}$.

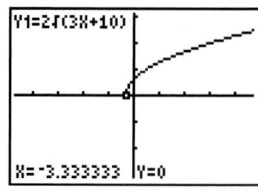

$(-47, 47, -31, 31)$

b. $m(x) = \sqrt[3]{2x - 9}$

The domain of $m(x)$ is the set of all real numbers. Using x-values in the domain, set up a table of values.

x	y
-2	-2.351
-1	-2.224
0	-2.08
1	-1.913
2	-1.71
3	-1.442
4	-1
5	1
6	1.4422
7	1.71
8	1.9129

Graph the ordered pairs found in the table of values, and connect the points with a smooth curve.

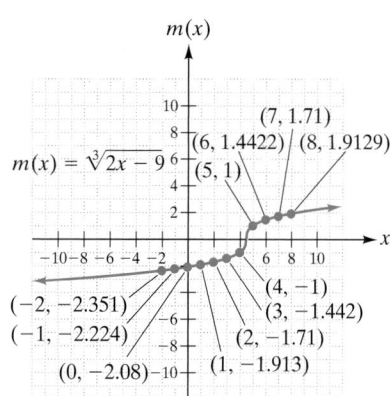

Use your calculator to graph $m(x) = \sqrt[3]{2x - 9}$.

$Y1 = \sqrt[3]{2x - 9}$

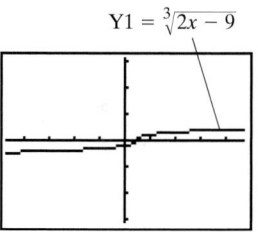

$(-47, 47, -31, 31)$

13.1.4 Checkup

In exercises 1–3, graph each radical function.

1. $y = 2\sqrt{x} + 2$ **2.** $f(x) = \sqrt[3]{2x - 3}$
3. $y = \sqrt[4]{2x + 3}$

4. List the steps you would follow to graph a radical function with pencil and paper. What would you do differently if the function has restrictions on its domain?

13.1.5 Modeling the Real World

In geometry, many area formulas involve squaring a variable. For example, we square the length of a side, s, to obtain the area of a square: $A = s^2$. We can find the exact length of the side by solving the formula for s. When we do, we get $s = \sqrt{A}$. Many volume formulas involve cubing a variable, such as the length of an edge, s. For example, the volume of a cube is $V = s^3$. We can find the exact length of the edge by solving for s. When we do, we obtain $s = \sqrt[3]{V}$.

EXAMPLE 6

a. The volume V of a sphere with radius of length r is $V = \frac{4}{3}\pi r^3$. Write a formula for r in terms of V.

b. The geodesic dome of the Spaceship Earth at the Epcot Center has a volume of 2.2×10^6 cubic feet. Assuming that the dome is a sphere, what is its diameter?

Solution

a. $V = \dfrac{4}{3}\pi r^3$

$3V = 4\pi r^3$ Multiply both sides by 3.

$\dfrac{3V}{4\pi} = r^3$ Divide both sides by 4π.

$\sqrt[3]{\dfrac{3V}{4\pi}} = r$ Take the cube root of both sides.

or $r = \sqrt[3]{\dfrac{3V}{4\pi}}$

b. We will use the formula derived in part a to find the radius of the dome.

$$r = \sqrt[3]{\frac{3V}{4\pi}}$$

$$r = \sqrt[3]{\frac{3(2.2 \times 10^6)}{4\pi}}$$

$$r \approx 80.68$$

The diameter is twice the radius, or $d = 2r$.

$$d = 2r$$
$$d \approx 2(80.68)$$
$$d \approx 161.36$$

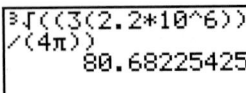

The diameter of the geodesic dome is approximately 161.36 feet.

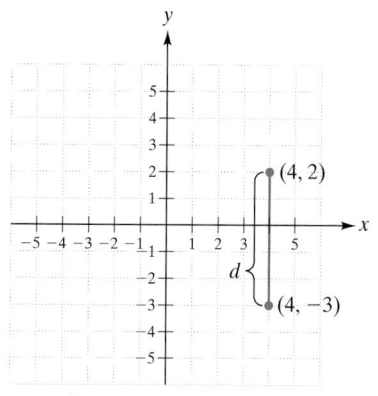

An application of a radical expression is used to determine the distance between two points in a rectangular coordinate system.

First, we must determine the distance between two points that lie on the same vertical or horizontal line. To determine the distance between two points that lie on a vertical line, we take the absolute value of the difference between the two y-coordinates of the points. For example, the vertical distance between $(4, 2)$ and $(4, -3)$ is

$$|2 - (-3)| = |2 + 3| = |5| = 5$$

The distance between the two points is 5.

To find the distance between two points that lie on a horizontal line, we take the absolute value of the difference between the two x-coordinates of the points. For example, the horizontal distance between $(1, 3)$ and $(4, 3)$ is

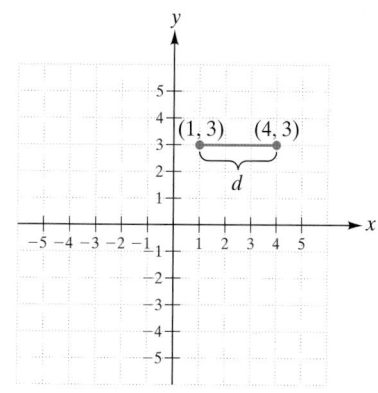

$$|1 - 4| = |-3| = 3$$

The distance between the two points is 3.

Next, to determine the distance between two points that do not lie on a horizontal or vertical line, we apply the Pythagorean theorem. (See Figure 13.2.)

$$\text{Let } a = |x_2 - x_1| \text{ and } b = |y_2 - y_1|$$
$$a^2 + b^2 = c^2$$
$$(x_2 - x_1)^2 + (y_2 - y_1)^2 = d^2 \qquad \textit{Substitute.}$$

Note: We do not need the absolute-value symbols, because we are squaring the number. It does not matter whether the number is positive or negative; squaring will always result in a positive number.

$$\sqrt{(x_2 - x_1)^2 + (y_2 - y_1)^2} = d \qquad \textit{Principle of powers}$$

We do not write the negative square root, because we cannot have a negative distance.

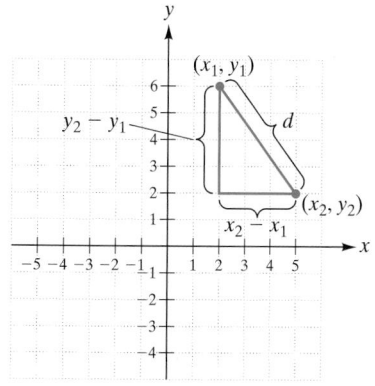

DISTANCE FORMULA

The distance d between any two points on a coordinate plane (x_1, y_1) and (x_2, y_2) is

$$d = \sqrt{(x_2 - x_1)^2 + (y_2 - y_1)^2}$$

Figure 13.2

EXAMPLE 7   Determine the distance between the points $(2, 3)$ and $(5, 1)$.

Solution

Let $x_1 = 2$, $y_1 = 3$, $x_2 = 5$, and $y_2 = 1$

$$d = \sqrt{(x_2 - x_1)^2 + (y_2 - y_1)^2}$$
$$d = \sqrt{(5 - 2)^2 + (1 - 3)^2} \qquad \textit{Substitute.}$$
$$d = \sqrt{3^2 + (-2)^2}$$
$$d = \sqrt{9 + 4}$$
$$d = \sqrt{13}$$

The distance between the two points is $\sqrt{13}$, or approximately 3.606.

APPLICATION

According to annual reports, the number of employees, in thousands, of the Sony Corporation can be estimated by the function $e(x) = 137\sqrt[6]{x}$, where x is the number of years after 1994.

a. Graph the function.

b. Use the graph to determine whether the function is increasing or decreasing.

c. Assuming that the trend shown on the graph does not change, approximate the number of people that will be employed by Sony in 2014.

Discussion

a. The restricted values are $x < 0$.

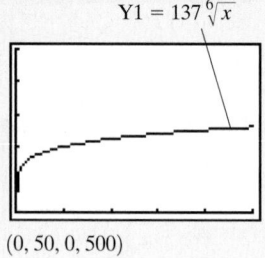

$$Y1 = 137\sqrt[6]{x}$$

(0, 50, 0, 500)

b. The function is increasing, because the graph rises from left to right.

c. The year 2014 is 20 years after 1994. Therefore, let $x = 20$.

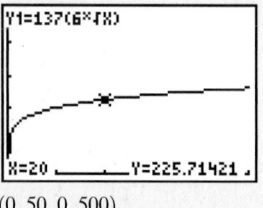

(0, 50, 0, 500)

According to the graph, Sony will employ about 225.714 thousand people, or 225,714, in 2014.

✓ 13.1.5 Checkup

1. The area covered by the dome of Saint Peter's Basilica in Rome is approximately 19,360 square feet. Write a formula for the radius of a circle in terms of its area A. Use the formula to estimate the radius of the dome. What is the dome's diameter?

2. Find the distance between the two points $(5, -2)$ and $(5, 7)$.

3. Find the distance between the two points $(-4, -5)$ and $(3, -5)$.

4. Find the distance between the two points $(-2, 1)$ and $(3, 4)$.

5. According to annual reports, the number of employees, in thousands, of the Amway Corporation can be estimated by the function $e(x) = 6.1\sqrt[3]{x}$, where x is the number of years after 1990.
   a. Graph the function.
   b. Use the graph to determine whether the function is increasing or decreasing.
   c. Assuming that the trend shown on the graph does not change, approximate the number of people that will be employed by Amway in 2010.

13.1 Exercises

Round, to the nearest thousandth if necessary.

1. $-\sqrt[4]{1296}$ **2.** $-\sqrt[4]{2401}$ **3.** $\sqrt[4]{-1296}$ **4.** $\sqrt[4]{-2401}$

5. $\sqrt[3]{216}$ **6.** $\sqrt[3]{1331}$ **7.** $\sqrt[3]{-216}$ **8.** $\sqrt[3]{-1331}$

9. $-\sqrt[3]{216}$ **10.** $-\sqrt[3]{1331}$ **11.** $\sqrt{110}$ **12.** $\sqrt{200}$

13. $-\sqrt{110}$ **14.** $-\sqrt{200}$ **15.** $\sqrt{-110}$ **16.** $\sqrt{-200}$

17. $\sqrt[3]{110}$ **18.** $\sqrt[3]{200}$ **19.** $-\sqrt[3]{110}$ **20.** $-\sqrt[3]{200}$

21. $\sqrt[5]{20}$ **22.** $\sqrt[6]{25}$ **23.** $\sqrt[3]{12.5}$ **24.** $\sqrt[4]{72.8}$

Identify each function as radical or nonradical.

25. $y = \sqrt{5x} + 7$ **26.** $f(x) = 5\sqrt{4x-7} + 8$ **27.** $f(x) = 2\sqrt{3x+1} - 5$ **28.** $y = 13 - \sqrt{7x}$

Determine the restricted values, and then state the domain of each function.

29. $y = \sqrt{x-3}$ **30.** $y = 5 + 4\sqrt{x}$ **31.** $y = \sqrt[3]{x-3}$

32. $y = \sqrt{3x-1} - \sqrt{x+2}$ **33.** $y = 2\sqrt{x} + 3$ **34.** $y = 2 + \sqrt[3]{3x+5}$

35. $y = \sqrt{2x+1} - \sqrt{x-3}$ **36.** $f(x) = 2 + \sqrt{3x+5}$ **37.** $y = (\sqrt[3]{5x+9})^4$

38. $y = (\sqrt[5]{3-4x})^2$ **39.** $y = (\sqrt[4]{3x+2})^5$ **40.** $y = (\sqrt{7x-5})^3$

In exercises 41–50, create a table of values and graph the functions on a Cartesian coordinate plane. Check, using your calculator.

41. $y = \sqrt{x}$ **42.** $y = 3\sqrt{x}$ **43.** $y = \sqrt[3]{x} - 5$ **44.** $y = 2\sqrt[3]{x} + 5$

45. $y = 5\sqrt{x} - 10$ **46.** $y = 6 - 3\sqrt{x}$ **47.** $f(x) = \sqrt{x-5}$ **48.** $h(x) = \sqrt{2x+5}$

49. $g(x) = \sqrt[3]{8x}$ **50.** $p(x) = -\sqrt[3]{4x}$

51. A painting is in the shape of a square. Given the area of the painting, write the equation for the length of a side. Find the length of a side in each situation.
 a. The painting is a mural whose area is 289 square feet.
 b. The painting is a canvas whose area is 90.25 square inches.
 c. The painting is a locket whose area is 115 square millimeters.

52. A canvas cover has a square shape. Find the length of a side in each situation.
 a. The canvas covers an area of 529 square yards.
 b. The canvas covers an area of 110.25 square centimeters.
 c. The canvas covers an area of 45 square feet.

53. A puddle of standing water has the shape of a circle. Given the area of the puddle, write the equation for the length of its radius. Find the exact length of the radius of the puddle in each situation. Then approximate the length to the nearest thousandth.
 a. The puddle has an area of 196π square inches.
 b. The puddle has an area of 49 square yards.
 c. The puddle has an area of 108 square meters.

54. Find the exact length of the radius of a circular metal ring that encloses the area cited. Then approximate the length to the nearest thousandth.
 a. The metal ring encloses an area of 225π square decimeters.
 b. The metal ring encloses an area of 361 square feet.

 c. The metal ring encloses an area of 200 square centimeters.

55. A wooden box is shaped like a cube. Given the volume of the box, write the equation for the length of a side. Find the length of a side of the box in each situation. Round your answers to the nearest hundredth.
 a. The volume of the box is 1728 cubic inches.
 b. The volume of the box is 120 cubic meters.

56. A bundle of paper has the shape of a cube. Find the length of a side of the bundle in each situation. Round your answers to the nearest hundredth.
 a. The volume of the bundle is 9261 cubic centimeters.
 b. The volume of the bundle is 500 cubic feet.

57. A balloon has a spherical shape. Given the volume of the balloon, write equations for the length of its radius and of its diameter. Find the radius and diameter in each situation.
 a. The volume of the balloon is 288π cubic decimeters.
 b. The volume of the balloon is 250 cubic yards.

58. A mesh wire cage has the shape of a sphere. Find the radius and diameter of the cage for each situation.
 a. The cage encloses a volume of 972π cubic inches.
 b. The cage encloses a volume of 1000 cubic meters.

59. The area covered by the dome of St. Paul's Cathedral in London is approximately 8200 square feet. Using the formula for the circular area covered by a dome, write the equation for the length of the radius of the circle. What is the radius of the dome? What is its diameter?

60. The area of the square floor of the central building of the Taj Mahal is approximately 377,000 square feet. Using the formula for the area of a square, write the equation for the length of a side of the floor. What is the length of one of its sides?

In exercises 61–66, find the distance on the coordinate plane between the given points.

61. $(8, -3)$ and $(8, 2)$ **62.** $(-2, 3)$ and $(-2, 11)$

63. $(-3, 6)$ and $(5, 6)$ **64.** $(-9, -1)$ and $(-2, -1)$

65. $(-3, -4)$ and $(2, 5)$ **66.** $(-5, 7)$ and $(3, -4)$

67. The market value of data communications equipment, in millions of dollars, is approximated by the function $d(x) = 9.4\sqrt[4]{x}$, where x is the number of years after 1993.
  a. Graph the function.
  b. Use the graph to determine whether the function is increasing or decreasing.
  c. Assuming that the trend shown on the graph continues, approximate the 2010 market value of data communications equipment.

68. The market value of car rentals, in millions of dollars, is approximated by the function $r(x) = 26\sqrt[4]{x}$, where x is the number of years after 1993.
  a. Graph the function.
  b. Use the graph to determine whether the function is increasing or decreasing.
  c. Assuming that the function holds for all time, approximate the 2010 market value of car rentals.

69. In statistical analyses, data are often summarized in tables called frequency tables. In designing such a table, rules of thumb are used to decide on the number of rows in the table. One such rule is that r, the number of rows in the table, should be equal to the square root of n, the number of pieces of data collected. Graph this relation for values of n ranging from 0 to 400. How many rows should you have in a frequency table for 250 pieces of data?

70. In statistics, if a set of data consists of the integers from 1 to n, the variability of the numbers is given by

$$s = \sqrt{\frac{n(n + 1)}{12}}$$

Graph this relation for values of n ranging from 1 to 250. What is the variability of the integers from 1 to 150?

13.1 Calculator Exercises

In graphing a radical function in which the radicand is a perfect square, an interesting result occurs. To see this, first store $f(x) = \sqrt{x^2 + 6x + 9}$ in Y1 on your calculator, and graph $f(x)$. Notice that the graph never drops below the x-axis. Can you explain why?

Now store $g(x) = x + 3$ in Y2 on your calculator, and graph that function. What do you notice about this graph? You should see that the graphs coincide for certain values of x. Determine the interval for which the graphs coincide. Are the two functions equivalent?

Now store $h(x) = |x + 3|$ in Y3, and set Y3 to use the bubble test when it graphs. (Remember, to use the bubble, move

the cursor to the left of Y3, and press ENTER repeatedly until you see the "−0" icon.) Does this graph coincide with either $f(x)$ or $g(x)$? Does it suggest how you can simplify $f(x)$?

Next, consider the following simplification of $f(x)$:

$$f(x) = \sqrt{x^2 + 6x + 9} = \sqrt{(x + 3)^2} = x + 3$$

What is wrong with this simplified function? Your calculator graphs should suggest that $f(x)$ does not always simplify as shown. Can you properly state how $f(x)$ simplifies?

Use the same approach to explore the following functions:

1. $f(x) = \sqrt{x^2 - 4x + 4}$ **2.** $f(x) = \sqrt{4x^2 - 12x + 9}$

13.1 Writing Exercise

For certain values of x, $\sqrt{x^2} = x$, and for certain other values of x, $\sqrt{x^2} = -x$. Identify the values of x that make each statement true. Then explain how you determined your answer.

13.2 Rational Exponents

OBJECTIVES

1 Evaluate expressions of the form $a^{1/n}$, where a is a real number and n is a positive or negative integer.

2 Evaluate expressions of the form $a^{m/n}$, where a is a real number, m is a positive or negative integer, and n is a positive integer.

3 Graph a function defined as an expression having a rational exponent.

4 Model real-world situations by using expressions involving rational exponents.

APPLICATION

A major concern for a person or animal in cold weather is frostbite—damage to body tissue caused by being frozen. Since wind and cold affect the rate of heat lost through exposed skin, knowing the time it takes to become frostbitten is very important.

According to Environment Canada, the time to frostbite in t minutes is related to the wind speed v in km/h at a height of 10 m and with the actual air temperature $T°C$ by the formula

$$t = \{[-24.5(0.667v + 4.8)] + 2111\} \cdot (-4.8 - T)^{-1.668}$$

If the wind speed is a constant 40 km/h and the temperature is forecast to drop from $-25°C$ to $-50°C$, determine the time to frostbite at $5°C$ intervals.

After completing this section, we will discuss this application further. See page 899.

13.2.1 Evaluating Expressions Involving $a^{1/n}$

In Chapter 1, we evaluated expressions having exponents. In this chapter, we have evaluated expressions involving radicals. There is a relationship between these two types of expressions. Let's see if we can discover the relationship with our calculators.

Discovery 2

Radical and Rational Exponents

Evaluate each expression, and compare the results obtained in the left column with the corresponding results in the right column.

1. **a.** $\sqrt{3} =$ _____   **b.** $3^{1/2} =$ _____
2. **a.** $\sqrt[3]{3} =$ _____   **b.** $3^{1/3} =$ _____
3. **a.** $\sqrt[4]{3} =$ _____   **b.** $3^{1/4} =$ _____

Write a rule for writing a radical expression as an exponential expression.

In the column on the left, we evaluate a radical expression. In the column on the right, we evaluate an exponential expression consisting of a base that is the same as the radicand of the radical expression on the left and an exponent that is the reciprocal of the index of the radical expression on the left. The results are the same in both columns.

We define an expression with a positive rational exponent having a numerator of 1 to be a radical expression.

EXPRESSIONS WITH RATIONAL EXPONENTS HAVING A NUMERATOR OF 1

$$a^{1/n} = \sqrt[n]{a}$$

In this equation, a is a real number and n is a positive integer >1.

EXAMPLE 1 Write the following expressions as radical expressions, and evaluate them if possible.

a. $625^{1/4}$ **b.** $-625^{1/4}$ **c.** $(-625)^{1/4}$
d. $1024^{1/5}$ **e.** $-1024^{1/5}$ **f.** $(-1024)^{1/5}$

Solution

a. $625^{1/4} = \sqrt[4]{625} = 5$ $5^4 = 625$

b. $-625^{1/4} = -\sqrt[4]{625} = -5$

The negative sign is not part of the base and therefore indicates a negative root.

Check

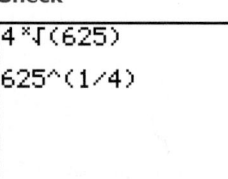

Figure 13.3

c. $(-625)^{1/4} = \sqrt[4]{-625}$, which is not a real number.

The radicand is negative and the index is even.

d. $1024^{1/5} = \sqrt[5]{1024} = 4$ $4^5 = 1024$

e. $-1024^{1/5} = -\sqrt[5]{1024} = -4$

The negative sign is not part of base and therefore indicates a negative root.

f. $(-1024)^{1/5} = \sqrt[5]{-1024} = -4$ $(-4)^5 = -1024$

To check your solution, enter the radical expression and the expression having a rational exponent. The results should be the same. The check for Example 1a is shown in Figure 13.3.

We can also define a negative rational exponent as a radical expression.

EXPRESSIONS WITH NEGATIVE RATIONAL EXPONENTS HAVING A NUMERATOR OF 1

$$a^{-1/n} = \frac{1}{a^{1/n}} = \frac{1}{\sqrt[n]{a}}, a \neq 0$$

In these equations, a is a real number and n is a positive integer > 1.

EXAMPLE 2   Write the following expressions as radical expressions, and evaluate them if possible. Express your answers as fractions.

a. $625^{-1/4}$ **b.** $-625^{-1/4}$ **c.** $(-625)^{-1/4}$

d. $1024^{-1/5}$ **e.** $-1024^{-1/5}$ **f.** $(-1024)^{-1/5}$

Solution

Check

a. $625^{-1/4} = \dfrac{1}{625^{1/4}} = \dfrac{1}{\sqrt[4]{625}} = \dfrac{1}{5}$ $5^4 = 625$

b. $-625^{-1/4} = -\dfrac{1}{625^{1/4}} = -\dfrac{1}{\sqrt[4]{625}} = -\dfrac{1}{5}$ $5^4 = 625$

c. $(-625)^{-1/4} = \dfrac{1}{(-625)^{1/4}} = \dfrac{1}{\sqrt[4]{-625}}$, which is not a real number.

The radicand is negative and the index is even.

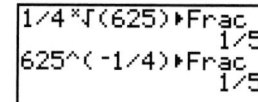

Figure 13.4

d. $1024^{-1/5} = \dfrac{1}{1024^{1/5}} = \dfrac{1}{\sqrt[5]{1024}} = \dfrac{1}{4}$ $4^5 = 1024$

e. $-1024^{-1/5} = -\dfrac{1}{1024^{1/5}} = -\dfrac{1}{\sqrt[5]{1024}} = -\dfrac{1}{4}$ $4^5 = 1024$

f. $(-1024)^{-1/5} = \dfrac{1}{(-1024)^{1/5}} = \dfrac{1}{\sqrt[5]{-1024}} = \dfrac{1}{-4} = -\dfrac{1}{4}$ $(-4)^5 = -1024$

The check for Example 2a is shown in Figure 13.4.

If the radicand cannot be written as a power of the index, approximate the answer. Use your calculator to do this.

EXAMPLE 3   Evaluate. Round to the nearest thousandth.

```
625^(1/5)
        3.623898318
-1024^(-1/4)
       -.1767766953
```

a. $625^{1/5}$ **b.** $-1024^{-1/4}$

Calculator Solution

a. $625^{1/5} \approx 3.624$ **b.** $-1024^{-1/4} \approx -0.177$

✓ 13.2.1 Checkup

Write as a radical expression, and evaluate if possible.

1. $16^{1/4}$ **2.** $-16^{1/4}$ **3.** $(-16)^{1/4}$

Write as a radical expression, and evaluate if possible. Express your answer as a fraction.

4. $125^{-1/3}$ **5.** $-125^{-1/3}$ **6.** $(-125)^{-1/3}$

Evaluate exercises 7 and 8 to the nearest thousandth.

7. $500^{1/4}$ **8.** $650^{-1/3}$

9. Describe the relationship between a radical expression and an expression with a rational exponent whose numerator is equal to 1.

13.2.2 Evaluating Expressions Involving $a^{m/n}$

Some expressions, such as $4^{3/2}$, have a rational exponent with a numerator not equal to 1. We need to be able to evaluate these expressions. Complete the following set of exercises with your calculator to discover a rule for writing an equivalent radical expression from an expression containing a rational exponent.

Discovery 3

More Radicals and Rational Exponents

Evaluate each expression, and compare the results obtained in each row.

1. **a.** $4^{3/2} =$ _____   **b.** $(\sqrt{4})^3 =$ _____   **c.** $\sqrt{4^3} =$ _____
2. **a.** $64^{2/3} =$ _____   **b.** $(\sqrt[3]{64})^2 =$ _____   **c.** $\sqrt[3]{64^2} =$ _____
3. **a.** $8^{4/3} =$ _____   **b.** $(\sqrt[3]{8})^4 =$ _____   **c.** $\sqrt[3]{8^4} =$ _____
4. **a.** $(-8)^{4/3} =$ _____   **b.** $(\sqrt[3]{-8})^4 =$ _____   **c.** $\sqrt[3]{(-8)^4} =$ _____

Write a rule for converting an expression with a rational exponent to a radical expression.

In each row, the two radical expressions have the same value as the expression with a rational exponent. (Note: On some calculators, this is not true. If your calculator is one of these, see 13.2 Calculator Exercises.) As we already know, the base of the expression with a rational exponent is the radicand of the radical expression, and the denominator of the rational exponent is the index of the radical expression. The numerator of the rational exponent is the power to which the radical expression is raised, or the power to which the radicand is raised.

We can confirm these statements by rewriting the radical expression as an expression with a rational exponent and simplifying by means of the power-to-a-power rule for exponents. For example, in the first Discovery 3 exercise,

$$(\sqrt{4})^3 = (4^{1/2})^3 = 4^{3/2} \quad \text{or} \quad \sqrt{4^3} = (4^3)^{1/2} = 4^{3/2}$$

Therefore, $4^{3/2} = (\sqrt{4})^3 = \sqrt{4^3}$.

In general,

$$(\sqrt[n]{a})^m = (a^{1/n})^m = a^{m/n} \quad \text{or} \quad \sqrt[n]{a^m} = (a^m)^{1/n} = a^{m/n}$$

Therefore, $a^{m/n} = (\sqrt[n]{a})^m = \sqrt[n]{a^m}$.

RATIONAL EXPONENTS

$$a^{m/n} = \sqrt[n]{a^m} \quad \text{or} \quad (\sqrt[n]{a})^m$$

In this equation, a is a real number, $a^{1/n}$ is defined, m and n have no common factors other than 1, m is an integer $\neq 0$, and n is a positive integer > 1.

HELPING HAND Whether one takes the power or the root first does not matter.

EXAMPLE 4 Write as an equivalent radical expression, and evaluate.

$$\textbf{a. } 4^{3/2} \qquad \textbf{b. } (-32)^{3/5} \qquad \textbf{c. } 12^{2/5} \qquad \textbf{d. } 16^{-3/4} \qquad \textbf{e. } -32^{-4/5}$$

Solution

a. $4^{3/2} = (\sqrt{4})^3 = 2^3 = 8 \quad$ or $\quad 4^{3/2} = \sqrt{4^3} = \sqrt{64} = 8$

b. $(-32)^{3/5} = (\sqrt[5]{-32})^3 = (-2)^3 = -8 \quad$ or $\quad \sqrt[5]{(-32)^3} = \sqrt[5]{-32{,}768} = -8$

c. $12^{2/5} = (\sqrt[5]{12})^2 \approx 2.702 \qquad$ Calculate on your calculator, because 12 cannot
be written as a power of the index 5.

d. $16^{-3/4} = \dfrac{1}{16^{3/4}} = \dfrac{1}{(\sqrt[4]{16})^3} = \dfrac{1}{(2)^3} = \dfrac{1}{8}$

or

$$16^{-3/4} = \dfrac{1}{16^{3/4}} = \dfrac{1}{\sqrt[4]{16^3}} = \dfrac{1}{\sqrt[4]{4096}} = \dfrac{1}{8}$$

e. $-32^{-4/5} = -\dfrac{1}{32^{4/5}} = -\dfrac{1}{(\sqrt[5]{32})^4} = -\dfrac{1}{2^4} = -\dfrac{1}{16} \qquad$ Write as a positive
exponent first.

Check

```
((√(4))³
                    8
4^(3/2)
                    8
```

Figure 13.5

To check on your calculator, enter the radical expression and the exponential expression. The check for Example 4a is shown in Figure 13.5.

✓ 13.2.2 Checkup

Rewrite exercises 1–4 as radical expressions, and evaluate.

1. $16^{3/2}$ **2.** $(-125)^{5/3}$ **3.** $15^{3/2}$ **4.** $-125^{-5/3}$

5. Explain how the numerator and the denominator of a rational exponent are used to transform an expression into a radical expression.

13.2.3 Graphing a Radical Function

Since a radical expression can be expressed as an expression with a rational exponent, a radical function such as $f(x) = \sqrt{x}$ can also be written as $f(x) = x^{1/2}$. We determine the domain of a function defined by an expression involving a rational exponent as if it were defined by a radical expression. Remember that radical functions with odd indices do not have restricted values. Therefore, if the rational exponent has a denominator that is an odd integer, the function has no restriction on its domain.

> To determine algebraically the restricted values of a function defined by an expression involving rational exponents with an even denominator (or an even index),
>
> • Set the base of the expression involving rational exponents less than 0.
> • Solve for the independent variable.

EXAMPLE 5   Determine the restricted values algebraically if they exist, and then state the domain of each function. Check your answer numerically or graphically.

a. $y = (x - 7)^{2/5}$ **b.** $g(x) = (3x + 4)^{3/4}$

Solution

a. $y = (x - 7)^{2/5}$

The function has no restricted values, because the denominator of the rational exponent, 5, is odd.

Domain: All real numbers

$(-\infty, \infty)$

Graphic Check

Since there are no restricted values, check graphically.

$Y1 = (x - 7)^{2/5}$

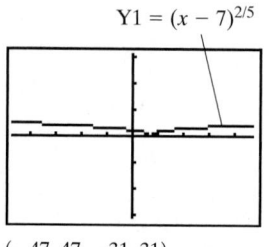

$(-47, 47, -31, 31)$

The graph appears to be continuous and has no "holes" for restricted values.

b. $g(x) = (3x + 4)^{3/4}$

The function has restricted values, because the denominator of the rational exponent is even. To determine the restricted values, set the base of the expression less than 0 and solve for x.

$$3x + 4 < 0$$

$$x < -\frac{4}{3}$$

The restricted values are all real numbers less than $-\frac{4}{3}$.

Domain: All real numbers greater than or equal to $-\frac{4}{3}$

$x \geq -\frac{4}{3}$

$[-\frac{4}{3}, \infty)$

Numeric Check

Since the restricted values are less than a fraction, check the solution numerically by selecting x-values close to $-\frac{4}{3}$. This will not be an exact check.

X	Y1
-1.6	ERROR
-1.5	ERROR
-1.4	ERROR
-1.3	.17783
-1.2	.50297
-1.1	.76529
-1	1

X=-1.6

$Y1 = (3x + 4)^{3/4}$

Viewing the table, we see that x-values less than -1.3 are restricted. It appears that the restricted values are $x < -1.3^{3/4}$.

We graph a radical function defined by an expression involving rational exponents in the same manner as we do other radical functions.

To graph a radical function defined by an expression involving rational exponents,

- Determine the restricted values.
- Determine a table of values by choosing enough points to determine the curve. Pay attention to the x-values that are close to the restricted values.
- Graph the points. Do not graph a point having an x-value equal to a restricted value.

EXAMPLE 6 Graph each radical function.

$$\text{a. } y = -3x^{3/4} \qquad \text{b. } g(x) = 5(x + 1)^{2/3}$$

Solution

a. $y = -3x^{3/4}$

Set up a table of values by using x-values in the domain, which is the set of all real numbers greater than or equal to 0.

x	y
0	0
0.1	-0.5335
0.5	-1.784
1	-3
2	-5.045
3	-6.839
4	-8.485
5	-10.03
6	-11.5
7	-12.91

Graph the ordered pairs found in the table of values, and connect the points with a smooth curve.

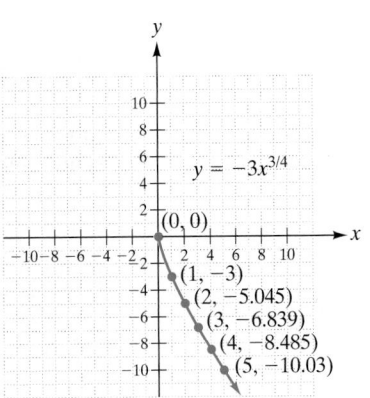

Calculator Check

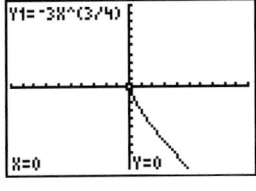

$(-10, 10, -10, 10)$

b. $g(x) = 5(x + 1)^{2/3}$

Set up a table of values by using x-values in the domain, the set of all real numbers.

x	y
-3	7.937
-2	5
-1	0
0	5
1	7.937
2	10.4

Graph the ordered pairs found in the table of values, and connect the points with a smooth curve.

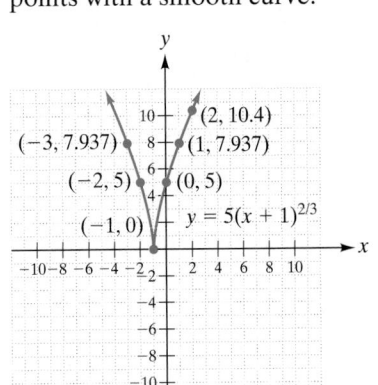

Calculator Check

$Y1 = 5(x + 1)^{2/3}$

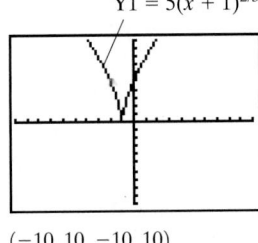

$(-10, 10, -10, 10)$

13.2.3 Checkup

Determine the restricted values algebraically if they exist, and then state the domain of each function. Check your answer numerically or graphically.

1. $y = (x + 2)^{1/2}$ **2.** $f(x) = (2x - 8)^{4/3}$

Graph the functions given in exercises 3 and 4 on a Cartesian coordinate system, and check your graphs with your calculator.

3. $y = -2x^{1/5}$ **4.** $f(x) = 2(x - 1)^{3/2}$

5. List the steps that you should follow if you are graphing a radical function defined by an expression involving rational exponents.

13.2.4 Modeling the Real World

We have seen that formulas can give rise to radical functions, often written as expressions involving rational exponents. Frequently, a table of values or a graph of the radical function makes it easier to see a bigger "picture" of what is happening.

EXAMPLE 7   The absolute luminosity (brightness) of a main-sequence star (a star similar to our Sun) is proportional to the 3.5 power of its mass. The function $L = kM^{3.5}$, where M is the mass of a star and k is the constant of proportionality, can be used to compare the absolute luminosity L of main-sequence stars.

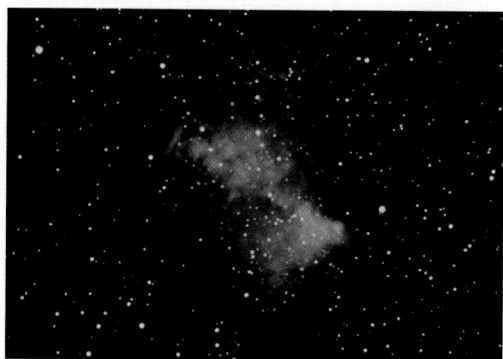

Define the mass of the Sun as $M_\odot$. If another main-sequence star has a mass that is x times the mass of the Sun, its mass is given by $xM_\odot$, and its luminosity is given by $L = k(xM_\odot)^{3.5} = kx^{3.5}M_\odot^{3.5}$. To compare this star's brightness with the Sun's, form the ratio of luminosity measures, $y = \dfrac{L_{\text{star}}}{L_{\text{Sun}}} = \dfrac{kx^{3.5}M_\odot^{3.5}}{kM_\odot^{3.5}} = x^{3.5}$. This gives us a function, $y = x^{3.5}$, that can be used to compare the luminosity of a main-sequence star whose mass is x times that of the Sun with the luminosity of the Sun.

a. Use the comparison function to set up a table of values to compare main-sequence stars whose masses are 1, 10, and 100 times the mass of the Sun $(M_\odot)$.

b. Graph the comparison function, and specify whether it is increasing or decreasing.

c. If the mass of a star is 10 times the mass of another star, approximately how much brighter (to the nearest thousand units) will the larger star be?

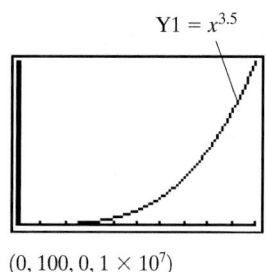

Solution

a. Let Y1 $= x^{3.5}$, and use the Ask mode of TblSet to generate the table of values.

b.

Y1 $= x^{3.5}$

(0, 100, 0, 1 × 10⁷)

The function is increasing, because its graph rises from left to right.

c. In comparing the absolute luminosities of two stars of mass $M = 1M_\odot$ and $M = 10M_\odot$, we see that the second star has 10 times more mass than the first.

The brightness increases from 1 to about 3162.

$$\frac{3162}{1} = 3162 \approx 3000$$

The star with mass $10M_\odot$ is about 3000 times as bright as the star with $1M_\odot$.

When we compare $M = 10M_\odot$ and $M = 100M_\odot$ (the mass increased 10 times), we find that the luminosity increases from 3162 to 10,000,000.

$$\frac{10,000,000}{3162} \approx 3163 \approx 3000$$

Similarly, the brightness increases about 3000 times.

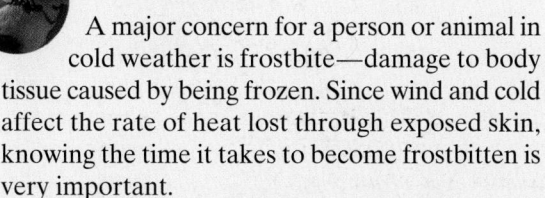

APPLICATION

A major concern for a person or animal in cold weather is frostbite—damage to body tissue caused by being frozen. Since wind and cold affect the rate of heat lost through exposed skin, knowing the time it takes to become frostbitten is very important.

According to Environment Canada, the time to frostbite in t minutes is related to the wind speed, v in km/h at a height of 10 m and the actual air temperature $T°C$ by the formula

$$t = \{[-24.5(0.667v + 4.8)] + 2111\} \cdot (-4.8 - T)^{-1.668}$$

If the wind speed is a constant 40 km/h and the temperature is forecast to drop from $-25°C$ to $-50°C$, determine the time to frostbite at 5°C intervals.

Discussion

Environment Canada states that the time in minutes to frostbite is related to the wind speed and air temperature. If the wind speed is constant, we can simplify the relation by substituting the wind speed of 40 km/h into the equation.

$$t = \{[-24.5(0.667v + 4.8)] + 2111\} \cdot (-4.8 - T)^{-1.668}$$
$$t = \{[-24.5(0.667 \cdot 40 + 4.8)] + 2111\} \cdot (-4.8 - T)^{-1.668} \quad \text{Use order of operations to evaluate.}$$

$$t = 1339.74(-4.8 - T)^{-1.668}$$

(Continued on page 900)

Set up a table of values on your calculator. Use the ASK mode, or set the table difference to -5. (The air temperature is dropping.)

X	Y1	
-25	8.9062	
-30	6.1586	
-35	4.5537	
-40	3.5269	
-45	2.828	
-50	2.3241	
X=		

$$Y1 = 1339.74(-4.8 - x)^{-1.668}$$

According to the table, if the wind speed is a constant 40 km/h, then when the air temperature is $-25°C$, exposed skin will become frostbitten in nine minutes, at $-30°C$ in six minutes, at $-35°C$ in five minutes, at $-40°C$ in four minutes, at $-45°C$ in three minutes, and at $-50°C$ in two minutes.

 13.2.4 Checkup

1. Use the function $y = x^{3.5}$ from Example 7 to compare the absolute luminosities of main-sequence stars with that of the Sun. Assume that the stars have masses of 1, 4, 16, and 64 times the mass of the Sun. Use the results to determine how the absolute brightness of a main-sequence star changes as its mass is increased by a factor of four.

2. Given the formula for the time to frostbite, in minutes, $t = \{[-24.5(0.667v + 4.8)] + 2111\} \cdot (-4.8 - T)^{-1.668}$,

assume that the wind speed v is 30 km/h. With this value for v, simplify the equation by using the order of operations. With the simplified equation, set up a table of values on your calculator to determine the time to frostbite for air temperatures varying from $-15°C$ to $-50°C$, in increments of $-5°C$. Interpret the results of your table.

13.2 Exercises

Write as a radical expression, and evaluate if possible.

1. $36^{1/2}$ **2.** $100^{1/2}$ **3.** $216^{1/3}$ **4.** $729^{1/3}$

5. $-81^{1/4}$ **6.** $-625^{1/4}$ **7.** $(-81)^{1/4}$ **8.** $(-625)^{1/4}$

9. $-8^{1/3}$ **10.** $-27^{1/3}$ **11.** $(-8)^{1/3}$ **12.** $(-27)^{1/3}$

Write as a radical expression, and evaluate if possible. Express your answer as a fraction.

13. $36^{-1/2}$ **14.** $100^{-1/2}$ **15.** $216^{-1/3}$ **16.** $729^{-1/3}$

17. $81^{-1/4}$ **18.** $625^{-1/4}$ **19.** $-81^{-1/4}$ **20.** $(-625)^{-1/4}$

21. $8^{-1/3}$ **22.** $27^{-1/3}$ **23.** $(-8)^{-1/3}$ **24.** $(-27)^{-1/3}$

Evaluate to the nearest thousandth.

25. $522^{1/2}$ **26.** $478^{1/3}$ **27.** $522^{1/4}$ **28.** $478^{1/5}$

29. $522^{-1/3}$ **30.** $478^{-1/2}$ **31.** $522^{-1/5}$ **32.** $478^{-1/4}$

Write as a radical expression, and evaluate if possible. Express your results as integers or fractions.

33. $27^{4/3}$ **34.** $25^{3/2}$ **35.** $(-27)^{4/3}$ **36.** $(-25)^{3/2}$

37. $27^{-4/3}$ **38.** $25^{-3/2}$ **39.** $-27^{-4/3}$ **40.** $-25^{-3/2}$

41. $4^{5/2}$ **42.** $9^{3/2}$ **43.** $8^{7/3}$ **44.** $125^{5/3}$

45. $4^{-7/2}$ **46.** $16^{-5/2}$ **47.** $-16^{3/2}$ **48.** $-64^{3/2}$

49. $(-9)^{3/2}$ **50.** $(-4)^{3/2}$ **51.** $-81^{-3/2}$ **52.** $-36^{-3/2}$

Approximate to the nearest thousandth if possible.

53. $28^{5/4}$ **54.** $36^{5/4}$ **55.** $(-21)^{4/3}$ **56.** $(-15)^{2/3}$

57. $5^{-2/3}$ **58.** $7^{-2/3}$ **59.** $-42^{-2/3}$ **60.** $-36^{-2/3}$

61. $(-88)^{3/8}$ **62.** $(-66)^{5/6}$

Determine the restricted values, and then state the domain of each function.

63. $f(x) = (5 - 4x)^{3/4}$ **64.** $f(x) = (7 - 2x)^{2/3}$

65. $g(x) = (5 - 4x)^{2/3}$ **66.** $h(x) = (7 - 2x)^{3/4}$

Graph the functions in exercises 67–74 on a Cartesian coordinate system, and check the graphs with your calculator.

67. $y = x^{3/4} + 2$ **68.** $y = 2 - x^{3/4}$ **69.** $y = 3 - x^{2/3}$ **70.** $y = x^{2/3} - 3$

71. $y = x^{4/5} + 1$ **72.** $y = x^{2/5} + 1$ **73.** $y = 5(x - 4)^{2/3}$ **74.** $y = 3(4 - x)^{2/3}$

75. Claims adjusters and highway patrol officers use the lengths of skid marks to determine the speed of a vehicle involved in an accident. The velocity in miles per hour is estimated as $V = (12S)^{1/2}$, where S is the length of the skid mark in feet, and the coefficient of friction of the road is assumed to be 40%. Graph this function for values of the domain between 0 and 300 feet. If a car leaves a skid mark that is 250 feet long, what would you estimate the speed of the car to be?

76. If the coefficient of friction on a road is 35%, the velocity in miles per hour of a car that leaves a skid mark of S feet is given by the mathematical model $V = (10.5S)^{1/2}$. Graph this function for values of the domain between 0 and 300 feet. If a car leaves a skid mark that is 200 feet long, what would you estimate the speed of the car to be?

13.2 Calculator Exercises

Consider the exponential expression $(-32)^{3/5}$. Expressions of this type have the following features:

- The base is negative.
- The rational exponent has an odd-numbered denominator.
- The rational exponent has a numerator that is not equal to 1.

Some calculators are not programmed to correctly evaluate exponential expressions that have these features. You should enter the expression, as given, into your calculator and see whether the calculator returns a result of -8. You can verify that this is the correct result by rewriting the expression as a radical expression and evaluating. If your calculator does not

give you this result, you will need to rewrite the expression in one of the following ways and key the new expression into your calculator:

- $[(-32)^{1/5}]^3$
- $[(-32)^3]^{1/5}$
- $(\sqrt[5]{-32})^3$
- $\sqrt[5]{(-32)^3}$

You should experiment with your calculator and use the form that returns the correct result.

1. Evaluate $(-27)^{2/3}$.

2. Rewrite the expression in exercise 1 four different ways and check for equivalence with the original expression.

13.2 Writing Exercise

In the definition of rational exponents,

$$a^{m/n} = \sqrt[n]{a^m} = (\sqrt[n]{a})^m$$

it was stated that n is a positive integer greater than 1.

1. Consider what would happen if $n = 1$. Describe what this would do to the expression $a^{m/n}$, and state why the value $n = 1$ is excluded from the definition.

2. Next, consider what would happen if $n = 0$. What would this do to the expression $a^{m/n}$, and why is $n = 0$ excluded from the definition?

3. Explain why we don't need to consider negative integer values for n.

4. Finally, when considering an expression of the form $a^{m/n}$, you should express the exponent $\frac{m}{n}$ in lowest terms. Why?

13.3 Properties of Rational Exponents

OBJECTIVES

1 Evaluate exponential expressions with rational exponents by using properties of exponents.

2 Simplify exponential expressions with rational exponents by using properties of exponents.

3 Model real-world situations by using exponential expressions containing rational exponents.

APPLICATION

The Stefan–Boltzmann law states that the temperature of a star is equal to the expression $\left(\frac{F}{\sigma}\right)^{1/4}$, where F is the energy flux (amount of energy that leaves the surface) of the star and the Greek letter σ (pronounced *sigma*) is a constant. Write an expression for the total temperature of two stars if one star's energy flux is twice the other.

After completing this section, we will discuss this application further. See page 906.

13.3.1 Evaluating Rational Exponents

The properties that we used earlier with integer exponents also apply to rational exponents. Let's review these properties.

PROPERTIES OF RATIONAL EXPONENTS

$a^{-n} = \dfrac{1}{a^n}$	Definition of a negative exponent
$\dfrac{1}{a^{-n}} = a^n$	Definition of a negative exponent
$a^m \cdot a^n = a^{m+n}$	Product rule
$\dfrac{a^m}{a^n} = a^{m-n}$	Quotient rule
$(a^m)^n = a^{mn}$	Power-to-a-power rule
$(ab)^m = a^m b^m$	Product-to-a-power rule
$\left(\dfrac{a}{b}\right)^m = \dfrac{a^m}{b^m}$	Quotient-to-a-power rule
$\left(\dfrac{a}{b}\right)^{-m} = \left(\dfrac{b}{a}\right)^m$	Quotient-to-a-negative-power rule

In these expressions, a and b are nonzero real expressions such that both sides of the equation represent real numbers, and m and n are rational expressions.

EXAMPLE 1 Use the properties of exponents to evaluate the given expressions. Check by means of your calculator.

a. $25^{3/4} \cdot 25^{-1/4}$ **b.** $\dfrac{16^{3/8}}{16^{1/8}}$ **c.** $2^{1/2} \cdot 8^{1/2}$ **d.** $\left(\dfrac{3^5}{2^{10}}\right)^{-2/5}$

Solution

a. $25^{3/4} \cdot 25^{-1/4} = 25^{3/4+(-1/4)} = 25^{1/2} = 5$ *Product rule*

b. $\dfrac{16^{3/8}}{16^{1/8}} = 16^{3/8-1/8} = 16^{1/4} = 2$ *Quotient rule*

c. $2^{1/2} \cdot 8^{1/2} = (2 \cdot 8)^{1/2} = 16^{1/2} = 4$ *Product-to-a-power rule*

d. $\left(\dfrac{3^5}{2^{10}}\right)^{-2/5} = \dfrac{(3^5)^{-2/5}}{(2^{10})^{-2/5}}$ *Quotient-to-a-power rule*

$ = \dfrac{3^{(5)(-2/5)}}{2^{(10)(-2/5)}} = \dfrac{3^{-2}}{2^{-4}}$ *Power-to-a-power rule*

$ = \dfrac{2^4}{3^2} = \dfrac{16}{9}$ *Definition of a negative exponent*

A second way to evaluate this expression is to use the quotient-to-a-negative-power rule first and then use the quotient-to-a-power rule.

$$\left(\frac{3^5}{2^{10}}\right)^{-2/5} = \left(\frac{2^{10}}{3^5}\right)^{2/5} = \frac{2^{10(2/5)}}{3^{5(2/5)}} = \frac{2^4}{3^2} = \frac{16}{9}$$

To check on your calculator, carefully enter the expressions into it. Remember to enclose the rational exponents in parentheses.

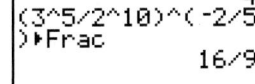

a. `25^(3/4)*25^(-1/4)`
   `                5`
b. `16^(3/8)/16^(1/8)`
   `                2`

c. `2^(1/2)*8^(1/2)`
   `              4`
d. `(3^5/2^10)^(-2/5)`
   `)▶Frac`
   `            16/9`

 13.3.1 Checkup

Use the properties of exponents to evaluate exercises 1–4. Check, using your calculator.

1. $36^{5/6} \cdot 36^{-1/3}$ **2.** $\dfrac{8^{7/12}}{8^{1/4}}$ **3.** $3^{1/3} \cdot 9^{1/3}$ **4.** $\left(\dfrac{5^2}{3^4}\right)^{-3/2}$

5. Explain the difference between an integer exponent and a rational exponent. Give examples of each.

13.3.2 Simplifying Expressions with Rational Exponents

To simplify an exponential expression with a rational exponent and a variable base, we use the same rules we used for evaluating an exponential expression with a numerical base.

However, suppose we are evaluating an expression whose exponent has an even-numbered denominator, such as $x^{1/2}$. Then we must be sure that the variable represents only nonnegative numbers, because an even root is not defined for a negative number.

Similarly, suppose we are evaluating an expression whose exponent is negative, such as $x^{-1/3}$. Then we must be sure that the variable represents only nonzero numbers, because we rewrite negative exponents as the reciprocal of the base and a positive power [for example, $\left(\frac{1}{x}\right)^{1/3}$]. If the variable x is zero, the reciprocal $\frac{1}{x}$ is undefined.

In this text, we will assume that the variable base does not represent values for which the expression is undefined.

EXAMPLE 2   Simplify, using the properties of exponents. Write each result with positive exponents.

a. $x^{1/2}x^{-1/4}$ **b.** $\dfrac{y^{5/8}}{y^{1/4}}$ **c.** $(xy)^{3/4}$ **d.** $\left(\dfrac{x}{y}\right)^{-2/3}$

Solution

a. $x^{1/2}x^{-1/4} = x^{1/2+(-1/4)} = x^{1/4}$ Product rule

b. $\dfrac{y^{5/8}}{y^{1/4}} = y^{5/8-1/4} = y^{3/8}$ Quotient rule

c. $(xy)^{3/4} = x^{3/4}y^{3/4}$ Product-to-a-power rule

d. $\left(\dfrac{x}{y}\right)^{-2/3} = \dfrac{x^{-2/3}}{y^{-2/3}}$ Quotient-to-a-power rule

$\qquad = \dfrac{y^{2/3}}{x^{2/3}}$ Definition of a negative exponent

A second way to evaluate the last expression is to use the quotient-to-a-negative-power rule first and then use the quotient-to-a-power rule.

$$\left(\frac{x}{y}\right)^{-2/3} = \left(\frac{y}{x}\right)^{2/3} = \frac{y^{2/3}}{x^{2/3}}$$

Using combinations of the properties of exponents allows us to simplify more complicated expressions. When we combine the properties of exponents, we may simplify the expression in different orders of steps and still obtain the correct results. Therefore, do not be concerned if you apply the properties of exponents using a different order from what you see in the examples, as long as you obtain the same answers.

EXAMPLE 3   Simplify. Write the result, using positive exponents.

a. $(25x^4y^3)^{1/2}$ **b.** $(2x^{3/4}y)(3x^{2/3}y^{5/8})$ **c.** $\left(\dfrac{5x^{1/10}}{x^{-2/5}}\right)^5$ **d.** $n^{1/3}(n^{2/3} + n^{3/4})$

Solution

a. $(25x^4y^3)^{1/2} = 25^{1/2}(x^4)^{1/2}(y^3)^{1/2}$ Product-to-a-power rule

$\qquad = 25^{1/2}x^{4\cdot1/2}y^{3\cdot1/2}$ Power-to-a-power rule

$\qquad = 5x^2y^{3/2}$

b. $(2x^{3/4}y)(3x^{2/3}y^{5/8}) = (2 \cdot 3)x^{3/4+2/3}y^{1+5/8}$ Product rule

$\qquad = 6x^{17/12}y^{13/8}$

c. $\left(\dfrac{5x^{1/10}}{x^{-2/5}}\right)^5 = \dfrac{(5x^{1/10})^5}{(x^{-2/5})^5}$ Quotient-to-a-power rule

$\qquad = \dfrac{5^5x^{(1/10)5}}{x^{(-2/5)5}}$ Product-to-a-power rule, power-to-a-power rule

$\qquad = \dfrac{5^5x^{1/2}}{x^{-2}}$

$$= 5^5 x^{1/2-(-2)} \qquad \text{Quotient rule}$$
$$= 5^5 x^{5/2} \quad \text{or} \quad 3125x^{5/2}$$

d. $n^{1/3}(n^{2/3} + n^{3/4}) = n^{1/3}n^{2/3} + n^{1/3}n^{3/4} \qquad \text{Distribute } n^{1/3}.$
$$= n^{1/3+2/3} + n^{1/3+3/4} \qquad \text{Product rule}$$
$$= n + n^{13/12}$$

13.3.2 Checkup

Simplify exercises 1–7. Write the results, using positive exponents.

1. $(x^9)^{1/3}$ **2.** $x^{1/3} \cdot x^{-1/6}$ **3.** $\dfrac{z^{2/3}}{z^{1/6}}$ **4.** $\left(\dfrac{p}{q}\right)^{-3/5}$

5. $(4a^2b^3)^{1/2}$ **6.** $\left(\dfrac{8x^{2/3}}{4x^{-2/3}}\right)^3$ **7.** $a^{1/2}(a^{1/4} - a^{2/3})$

8. In the expression $x^{3/4}$, what restrictions are placed on values of x because of the even denominator in the exponent?

9. In the expression $x^{-2/3}$, what restrictions are placed on the values of x because of the negative exponent?

10. Explain why it is important to be aware of restrictions on the base of an expression with rational exponents whenever conditions such as those illustrated in exercises 8 and 9 are present.

13.3.3 Modeling the Real World

In our daily lives, we often need to increase or decrease the size of objects that are in the shape of geometric figures. For instance, we might want to increase the size of a room or decrease the size of a fenced area for a pet. We may know the desired area and must determine the dimensions. Usually, we need to evaluate exponential expressions in order to determine these lengths.

EXAMPLE 4 The formula $s = A^{1/2}$ is used to determine the length of the side of a square, given the area A.

a. If the area of a square doubles, how is the length of a side affected? (The ratio of the areas is 2 to 1. What is the ratio of the new side to the original side?)
b. If the area of a square triples, how is the length of a side affected? (The ratio of the areas is 3 to 1. What is the ratio of the new side to the original side?)
c. If the area of a square is multiplied by a factor of x, determine the ratio of the length of a new side to the length of the original side.

Solution

a. Let $A =$ the original area of the square
     $s_1 =$ the length of the original side

Then $2A =$ the area of the new square (the area doubled)
     $s_2 =$ the length of the new side

Using the formula $s = A^{1/2}$, we have

$$s_1 = A^{1/2} \quad \text{for the original side}$$
$$s_2 = (2A)^{1/2} = 2^{1/2}A^{1/2} \quad \text{for the new side}$$

The ratio of the new side to the original side is

$$\frac{s_2}{s_1} = \frac{2^{1/2}A^{1/2}}{A^{1/2}} = \frac{2^{1/2}}{1}$$

or approximately $\frac{1.414}{1}$, when the ratio of the new area to the original area is $\frac{2}{1}$.

b. Let $A =$ the original area of the square

$s_1 =$ the length of the original side

Then $3A =$ the area of the new square (the area tripled)

$s_3 =$ the length of the new side

Using the formula $s = A^{1/2}$, we have

$$s_1 = A^{1/2} \quad \text{for the original side}$$
$$s_3 = (3A)^{1/2} = 3^{1/2}A^{1/2} \quad \text{for the new side}$$

The ratio of the new side to the original side is

$$\frac{s_3}{s_1} = \frac{3^{1/2}A^{1/2}}{A^{1/2}} = \frac{3^{1/2}}{1}$$

or approximately $\frac{1.732}{1}$, when the ratio of the new area to the original area is $\frac{3}{1}$.

c. Let $A =$ the original area of the square

$s_1 =$ the length of the original side

Then $xA =$ the area of the new square (the area is multiplied by a factor of x)

$s_4 =$ the length of the new side

Using the formula $s = A^{1/2}$, we have

$$s_1 = A^{1/2} \quad \text{for the original side}$$
$$s_4 = (xA)^{1/2} = x^{1/2}A^{1/2} \quad \text{for the new side}$$

The ratio of the new side to the original side is

$$\frac{s_4}{s_1} = \frac{x^{1/2}A^{1/2}}{A^{1/2}} = \frac{x^{1/2}}{1}$$

or $\frac{\sqrt{x}}{1}$, when the ratio of the new area to the original area is $\frac{x}{1}$.

APPLICATION

The Stefan–Boltzmann law states that the temperature of a star is equal to the expression $\left(\frac{F}{\sigma}\right)^{1/4}$, where F is the energy flux (amount of energy that leaves the surface) of the star and the Greek letter σ (pronounced *sigma*) is a constant. Write an expression for the total temperature of two stars if one star's energy flux is twice the other.

Discussion

Let $x =$ the amount of energy flux of the first star

$2x =$ the amount of energy flux of the second star

$$\left(\frac{x}{\sigma}\right)^{1/4} = \text{the temperature of the first star}$$

$$\left(\frac{2x}{\sigma}\right)^{1/4} = \text{the temperature of the second star}$$

The total temperature is the sum of the individual temperatures.

$$\left(\frac{x}{\sigma}\right)^{1/4} + \left(\frac{2x}{\sigma}\right)^{1/4} = \frac{x^{1/4}}{\sigma^{1/4}} + \frac{2^{1/4}x^{1/4}}{\sigma^{1/4}}$$

Quotient-to-a-power rule

$$= \frac{x^{1/4} + 2^{1/4}x^{1/4}}{\sigma^{1/4}}$$

Add rational expressions.

$$= \frac{x^{1/4}(1 + 2^{1/4})}{\sigma^{1/4}}$$

Factor the numerator.

$$= (1 + 2^{1/4})\frac{x^{1/4}}{\sigma^{1/4}}$$

Rewrite product as factors.

$$= (1 + 2^{1/4})\left(\frac{x}{\sigma}\right)^{1/4}$$

Quotient-to-a-power rule

The total temperature of the two stars is approximately $2.189\left(\frac{x}{\sigma}\right)^{1/4}$, or about 2.189 times the temperature of the first star.

$$\approx 2.189\left(\frac{x}{\sigma}\right)^{1/4}$$

Evaluate.

 ### 13.3.3 Checkup

1. The formula $r = \left(\frac{A}{\pi}\right)^{1/2}$ is used to determine the length of the radius of a circle whose area is A.
   a. If the area of a circle is doubled, how is the radius of the circle affected? (The ratio of the area is 2 to 1; what is the ratio of the new radius to the original radius?)
   b. If the area of the circle is tripled, how is the radius of the circle affected? (The ratio of the area is 3 to 1; what is the ratio of the new radius to the original radius?)
   c. If the area of the circle is multiplied by a constant C, determine the ratio of the lengths of the radii.

2. A tank of water has an opening near its base. The flow of water from the opening has a velocity (in feet per second) given by the expression $v = (64.4h)^{1/2}$, where h is the height of the water above the opening.
   a. Write an expression for the velocity of the water flow when the height of water in the tank is doubled.
   b. Determine an expression for the difference between the velocities—that is, the velocity of the water flow when its height is doubled, less the velocity of the water flow from the original height.

13.3 Exercises

Evaluate by using the properties of exponents. Check, using a calculator.

1. $9^{-1/2}$
2. $81^{-1/4}$
3. $\dfrac{1}{8^{-1/3}}$
4. $\dfrac{1}{16^{-1/4}}$

5. $27^{1/3} \cdot 27^{4/3}$
6. $64^{1/4} \cdot 64^{5/4}$
7. $\dfrac{32^{4/5}}{32^{1/5}}$
8. $\dfrac{64^{5/6}}{64^{1/3}}$

9. $(125^{1/2})^{2/3}$
10. $(256^{1/3})^{3/4}$
11. $3^{1/2} \cdot 12^{1/2}$
12. $5^{1/3} \cdot 25^{1/3}$

13. $\left(\dfrac{8}{27}\right)^{1/3}$
14. $\left(\dfrac{27}{125}\right)^{2/3}$
15. $\left(\dfrac{9}{16}\right)^{-1/2}$
16. $\left(\dfrac{81}{16}\right)^{-3/4}$

17. $[(2^{1/2})(8^{1/2})]^{-3/2}$
18. $[(2^{1/2})(32^{1/2})]^{-4/3}$

Simplify exercises 19–64. Write the results, using positive exponents.

19. $x^{-2/3}$
20. $x^{-5/6}$
21. $\dfrac{12}{y^{-3/4}}$

22. $\dfrac{2}{y^{-4/7}}$
23. $z^{2/3} \cdot z^{3/4}$
24. $z^{3/4} \cdot z^{7/8}$

25. $\dfrac{p^{5/6}}{p^{2/3}}$
26. $\dfrac{x^{5/9}}{x^{2/3}}$
27. $(b^{3/4})^{8/9}$

28. $(z^{2/5})^{15/16}$
29. $(x^2 y)^{3/4}$
30. $(pq^3)^{5/6}$

31. $\left(\dfrac{x^2}{y^3}\right)^{5/6}$
32. $\left(\dfrac{a^4}{b^3}\right)^{7/12}$
33. $\left(\dfrac{a}{b}\right)^{-1/3}$

34. $\left(\dfrac{m}{n}\right)^{-3/4}$
35. $c^{3/5} \cdot c^{-4/5}$
36. $k^{5/7} \cdot k^{-3/7}$

37. $(8a^5 b^6)^{2/3}$
38. $(16c^{8/9} d^2)^{3/4}$
39. $(5a^{3/4} b^{2/5})(2a^{1/3} b^{2/5})$

40. $(3x^{1/5}y^{1/3})(-2x^{3/5}y^{1/4})$

41. $\left(\dfrac{m^{3/7}}{2m^{-2/7}}\right)^2$

42. $\left(\dfrac{4z^{5/9}}{z^{-2/9}}\right)^3$

43. $(3a^{1/3}b^2c^{1/6})^{-2}(2a^{4/3}b^{1/4}c^{5/6})^3$

44. $(2x^{4/9}y^{3/4}z)^{-3}(2x^{2/3}y^{1/4}z^3)^2$

45. $x^{1/4}(x^{2/3} - x^{4/5})$

46. $z^{2/3}(z^{1/4} - z^{1/3})$

47. $2x^{2/5}(x^{1/5} + 3y^{2/5})$

48. $5a^{2/7}(3a^{1/7} - b^{4/7})$

49. $a^{1/4}b^{1/3}(3 - a^{1/2}b^{5/6})$

50. $x^{1/4}y^{3/4}(x^{1/2}y^{3/8} - 4)$

51. $(x^{1/2} - y^{1/2})(x^{1/3} + y^{1/3})$

52. $(c^{2/3} - d^{2/3})(c^{1/2} + d^{1/2})$

53. $(x^{1/4} + y^{1/4})(x^{1/4} - y^{1/4})$

54. $(p^{3/4} - q^{3/4})(p^{3/4} + q^{3/4})$

55. $(x^2 + y^2)(x^{1/2} - y^{1/2})$

56. $(a^{2/3} + b^{2/3})(a - b)$

57. $(x^{1/3} + 2)(x^{1/3} - 2)$

58. $(4 - z^{1/4})(4 + z^{1/4})$

59. $(x^{1/4} + 2)^2$

60. $(x^{1/3} + 3)^2$

61. $(x - y^{1/2})^2$

62. $(x^{1/2} - y)^2$

63. $(a^{1/2} + b^{1/2})^2$

64. $(c^{1/2} - d^{1/2})^2$

65. The length s of the side of a cube is related to the volume V of the cube by the formula $s = V^{1/3}$. How should you increase the length of a side in order to increase the volume of the cube to eight times its original volume?

66. The length s of the side of a cube is related to the surface area A of the cube by the formula $s = \left(\frac{A}{6}\right)^{1/2}$. How should you increase the length of a side in order to increase the surface area of the cube to nine times its original area?

67. The time t (in seconds) it would take an object to fall to the ground from a distance d (in feet) above the ground is given by the formula $t = \left(\frac{d}{16.1}\right)^{1/2}$. If you triple the height from which you drop the object, how does this change the time it will take the object to fall to the ground? (*Hint:* Determine the ratio of the times at given distances of d and $3d$.)

68. Using the formula from exercise 67, what will be the change in the time it will take the object to fall to the

ground if you quadruple the height from which you drop the object?

69. The time t (in seconds) it would take an object to fall to the ground from a distance d (in meters) above the ground is given by the formula $t = \left(\frac{d}{4.9}\right)^{1/2}$. If you triple the height from which you drop the object, how long it will take the object to fall to the ground? How does this answer compare with that of exercise 67? Does the change of the unit of measure from feet to meters affect the change in time?

70. Using the formula from exercise 69, what will be the change in the time it will take the object to fall to the ground if you quadruple the height from which you drop the object? How does this answer compare with that of exercise 68? Does the change of the unit of measure from feet to meters affect the change in time?

13.3 Calculator Exercises

Remember that the calculator can help you check whether you have simplified expressions with rational exponents correctly when the expressions involve only a single variable. Store the original expression in Y1 and the simplified expression in Y2. You can create a table of values and check for equivalence, as you have seen before. But you can also use the graphing feature along with the bubble test. To do so, after storing the expressions in Y1 and Y2, move the cursor to the extreme left of Y2 and press ENTER four times to select the symbol "−0". Doing this will cause a bubble to trace the graph of Y2. Then, when you press GRAPH, the calculator will first

graph Y1, and if Y2 is an equivalent graph, the bubble will trace along the graph for Y1. This indicates that the graphs coincide and the two expressions are equivalent; otherwise, the two expressions are not equivalent.

Simplify the following expressions, using only positive exponents, and check by graphing with the bubble feature just explained.

1. $(x^3)^{2/3}$

2. $(x^3)^{-2/3}$

3. $\dfrac{x^4}{x^{9/4}}$

4. $\dfrac{x^2}{x^{7/2}}$

5. $x^{3/4}(x^{3/2} - x^{2/3})$

6. $(x^{1/3} + 1)(x^{1/3} - 1)$

13.3 Writing Exercise

In the exercises in this section, you were asked to simplify the expression $(a^{1/2} + b^{1/2})^2$. If you simplified this expression correctly, you saw that the result was *not* $(a + b)$. This is an im-

portant point to remember. Explain why the result is not $(a + b)$. Would you expect $(c^{1/2} - d^{1/2})^2$ to be equal to $(c - d)$? Explain.

13.4 Properties of Radicals

OBJECTIVES

1 Simplify radical expressions by using the product rule for radicals.

2 Simplify radical expressions by using the quotient rule for radicals.

3 Model real-world situations by using expressions containing radical expressions.

APPLICATION

The most complete fossil (95%) of a *Tyrannosaurus rex*, Sue, is on display at the Field Museum in Chicago, Illinois. Its height at the hips is 13 feet.

a. Model the leg motion as that of a pendulum, and determine the length of time in seconds during which the leg completed one back-and-forth motion.

b. Dinosaur tracks have been uncovered in the roofs of coal mines where peat in the tracks was reduced to coal. The distance between the prints of the same foot (which constitutes a back-and-forth movement) measures about 18.6 feet. Use this distance to estimate Sue's walking speed.

After completing this section, we will discuss this application further. See page 917.

Since radical expressions are equivalent to exponential expressions with rational exponents, certain properties of radicals correspond to properties of exponents. In Chapter 11, we defined two such properties for square roots: the product rule and the quotient rule. We now extend these definitions to all radical expressions.

13.4.1 Simplifying Radical Expressions by Using the Product Rule

The product rule for square roots states that for any real numbers $\sqrt{a}$ and $\sqrt{b}$, $\sqrt{a} \cdot \sqrt{b} = \sqrt{a \cdot b}$. For example, $\sqrt{3} \cdot \sqrt{2} = \sqrt{3 \cdot 2} = \sqrt{6}$. This rule holds for all radical expressions.

PRODUCT RULE FOR RADICALS

For any real numbers $\sqrt[n]{a}$ and $\sqrt[n]{b}$ and any integer index n greater than 1,

$$\sqrt[n]{a} \cdot \sqrt[n]{b} = \sqrt[n]{ab}$$

We can show that this statement is true if we rewrite the radical expressions as exponential expressions with rational exponents and apply the product-to-a-power rule for exponents. That is, we would have the following:

$$\sqrt[3]{-3} \cdot \sqrt[3]{8} = (-3)^{1/3} \cdot 8^{1/3} = (-3 \cdot 8)^{1/3} = (-24)^{1/3} = \sqrt[3]{-24}$$
$$\sqrt[4]{6} \cdot \sqrt[4]{8} = 6^{1/4} \cdot 8^{1/4} = (6 \cdot 8)^{1/4} = 48^{1/4} = \sqrt[4]{48}$$

For the general rule, $\sqrt[n]{a} \cdot \sqrt[n]{b} = a^{1/n}b^{1/n} = (ab)^{1/n} = \sqrt[n]{ab}$.

SIMPLIFYING RADICAL EXPRESSIONS — CONDITION I

If a radical expression has a radicand with a perfect nth factor and an index of n, it can be simplified by using the product rule.

$$\sqrt[n]{ab} = \sqrt[n]{a} \cdot \sqrt[n]{b}$$

To simplify an nth root,

- Factor the radicand. One factor must be a perfect nth power.
- Reverse the product rule to simplify.

EXAMPLE I   Simplify without a calculator.

  a. $\sqrt{48}$ **b.** $\sqrt[3]{54}$ **c.** $\sqrt[4]{80}$ **d.** $\sqrt[5]{-96}$

Solution

a. $\sqrt{48} = \sqrt{16 \cdot 3}$ 16 is the largest perfect-square factor of 48.

   $= \sqrt{16} \cdot \sqrt{3}$ Product rule

   $= 4\sqrt{3}$

Check

```
4√(3)
            6.92820323
√(48)
            6.92820323
```

Figure 13.6

HELPING HAND If we had not used the largest perfect square factor, 16, but instead used a factor of 4, we would need to simplify the expression twice.

$$\sqrt{48} = \sqrt{4 \cdot 12} = \sqrt{4}\sqrt{12} = 2\sqrt{12} = 2\sqrt{4 \cdot 3} = 2\sqrt{4}\sqrt{3}$$
$$= 2 \cdot 2\sqrt{3} = 4\sqrt{3}$$

b. $\sqrt[3]{54} = \sqrt[3]{27 \cdot 2}$ 27 is the largest perfect-cube factor of 54.

   $= \sqrt[3]{27} \cdot \sqrt[3]{2}$ Product rule

   $= 3\sqrt[3]{2}$

c. $\sqrt[4]{80} = \sqrt[4]{16 \cdot 5}$ 16 is the largest fourth-power factor of 80.

   $= \sqrt[4]{16} \cdot \sqrt[4]{5}$ Product rule

   $= 2\sqrt[4]{5}$

d. $\sqrt[5]{-96} = \sqrt[5]{-32 \cdot 3}$ Use a negative factor.
                                              -32 is a fifth-power factor of -96.

   $= \sqrt[5]{-32} \cdot \sqrt[5]{3}$ Product rule

   $= -2\sqrt[5]{3}$

To check these results on your calculator, first evaluate the result found algebraically and then evaluate the given radical expression. The two values should be the same. The check for Example 1a is shown in Figure 13.6.

Before we expand the preceding method of simplification to variable radicands, we must remember that evaluating even roots always results in a positive number. Therefore, $\sqrt{x^2} = |x|$ and $\sqrt[4]{x^4} = |x|$. If we assume that no radicand is formed by raising negative quantities to even powers, there will be no need to write these results as absolute values.

In this text, we will assume that no radicand is formed by raising negative quantities to even roots. Therefore, the results will not be written as absolute values.

We simplify radical expressions with variable radicands by using the same procedure. In addition to rewriting the numerical coefficient as a product with a perfect power factor, we also rewrite the variables as products with a perfect power factor. (A perfect power factor is a factor whose exponent is a multiple of the index of the radical.)

Perfect squares of variables must have an exponent that is a multiple of 2.

Perfect Square

$$x^2, \text{ because } x^2 = (x)^2$$
$$x^4, \text{ because } x^4 = (x^2)^2$$
$$x^6, \text{ because } x^6 = (x^3)^2$$

Likewise, perfect cubes of variables must have an exponent that is a multiple of 3.

Perfect cube

$$x^3, \text{ because } x^3 = (x)^3$$
$$x^6, \text{ because } x^6 = (x^2)^3$$
$$x^9, \text{ because } x^9 = (x^3)^3$$

Therefore, a perfect nth powers must have an exponent that is a multiple of n.

EXAMPLE 2 Simplify.

a. $\sqrt{x^5}$ **b.** $\sqrt{x^4 y^7}$ **c.** $\sqrt[3]{a^4 b^8}$

Solution

a. To determine the exponent of the largest perfect-root factor of a variable, determine the largest multiple of the index that is less than or equal to the variable exponent. For example, the largest perfect-square factor for x^5 is x^4, because 4 is the largest multiple of 2 less than the exponent 5. Therefore, $x^5 = x^4 \cdot x$.

$$\sqrt{x^5} = \sqrt{x^4 \cdot x} \qquad \text{\small x^4 is the largest perfect-square factor of x^5, $x^4 = (x^2)^2$.}$$
$$= \sqrt{x^4} \cdot \sqrt{x} \qquad \text{\small Product rule}$$
$$= x^2 \sqrt{x}$$

b. $\sqrt{x^4 \cdot y^7} = \sqrt{x^4 \cdot y^6 \cdot y} \qquad \text{\small $x^4 = (x^2)^2$ and $y^6 = (y^3)^2$}$
$$= \sqrt{x^4 \cdot y^6} \sqrt{y} \qquad \text{\small Product rule}$$
$$= x^2 y^3 \sqrt{y}$$

c. $\sqrt[3]{a^4 b^8} = \sqrt[3]{a^3 \cdot a \cdot b^6 \cdot b^2} \qquad \text{\small $a^3 = (a)^3$ and $b^6 = (b^2)^3$}$
$$= \sqrt[3]{a^3 b^6} \sqrt[3]{ab^2} \qquad \text{\small Product rule}$$
$$= ab^2 \sqrt[3]{ab^2}$$

EXAMPLE 3 Simplify.

a. $\sqrt{50 x y^{15} z^4}$ **b.** $\sqrt[3]{64 a^3 b^6 c^{14}}$ **c.** $\sqrt{3x^2 + 12x + 12}$

Solution

a. $\sqrt{50xy^{15}z^4} = \sqrt{25 \cdot 2 \cdot x \cdot y^{14} \cdot y \cdot z^4}$ Find the perfect-square factors of the radicand.

$$= \sqrt{25y^{14}z^4}\,\sqrt{2xy}$$ Product rule

$$= 5y^7z^2\sqrt{2xy}$$

b. $\sqrt[3]{64a^3b^6c^{14}} = \sqrt[3]{64 \cdot a^3 \cdot b^6 \cdot c^{12} \cdot c^2}$ Find the perfect-cube factors of the radicand.

$$= \sqrt[3]{64a^3b^6c^{12}}\,\sqrt[3]{c^2}$$ Product rule

$$= 4ab^2c^4\sqrt[3]{c^2}$$

c. $\sqrt{3x^2 + 12x + 12} = \sqrt{3(x^2 + 4x + 4)}$ This radicand has terms, not factors. To simplify, factor the radicand.

$$= \sqrt{3(x + 2)^2}$$

$$= \sqrt{(x + 2)^2}\,\sqrt{3}$$

$$= (x + 2)\sqrt{3}$$

Note: Our assumption that no radicand is formed by raising negative quantities to even roots allows us to write the result, $|x + 2|\sqrt{3}$, without absolute values.

EXAMPLE 4   Multiply and simplify.

a. $(2x\sqrt{3y})(4\sqrt{6xy})$ **b.** $\sqrt{2x - 6} \cdot \sqrt{x - 3}$

c. $2\sqrt[3]{4a^2b^5} \cdot 3ab\sqrt[3]{12ab^2}$

Solution

a. $(2x\sqrt{3y})(4\sqrt{6xy}) = 2x \cdot 4\sqrt{3y \cdot 6xy}$ Multiply nonradical factors and radical factors.

$$= 8x\sqrt{18xy^2}$$

$$= 8x\sqrt{9 \cdot 2 \cdot x \cdot y^2}$$ Factor.

$$= 8x \cdot 3y\sqrt{2x}$$ Simplify.

$$= 24xy\sqrt{2x}$$

b. $\sqrt{2x - 6} \cdot \sqrt{x - 3} = \sqrt{(2x - 6)(x - 3)}$

$$= \sqrt{2(x - 3)(x - 3)}$$ Factor.

$$= \sqrt{2(x - 3)^2}$$

$$= (x - 3)\sqrt{2}$$ Simplify.

c. $2\sqrt[3]{4a^2b^5} \cdot 3ab\sqrt[3]{12ab^2}$

$$= 2 \cdot 3ab\sqrt[3]{4a^2b^5 \cdot 12ab^2}$$ Multiply nonradical factors and radical factors.

$$= 6ab\sqrt[3]{48a^3b^7}$$

$$= 6ab\sqrt[3]{8 \cdot 6 \cdot a^3 \cdot b^6 \cdot b}$$ Factor.

$$= 6ab \cdot 2ab^2\sqrt[3]{6b}$$ Simplify.

$$= 12a^2b^3\sqrt[3]{6b}$$

✓ 13.4.1 Checkup

Simplify without a calculator. Check the results on your calculator.

1. $\sqrt{192}$ 2. $\sqrt[3]{-40}$ 3. $\sqrt[4]{162}$ 4. $\sqrt[3]{x^5 y^{12}}$

Simplify exercises 5–9.

5. $\sqrt{98x^4 y^3 z^9}$ 6. $\sqrt{20x^2 + 20x + 5}$

7. $(3a\sqrt{5b})(2\sqrt{5ab})$ 8. $\sqrt{5x + 10} \cdot \sqrt{2x + 4}$
9. $4x\sqrt[3]{12xy^7} \cdot 3y\sqrt[3]{36x^2 y^4}$
10. What does the term "perfect nth-root factor of a radicand" mean?

13.4.2 Simplifying Radical Expressions by Using the Quotient Rule

The quotient rule for square roots states that for any real numbers $\sqrt{a}$ and $\sqrt{b}$, $\frac{\sqrt{a}}{\sqrt{b}} = \sqrt{\frac{a}{b}}$. For example, $\frac{\sqrt{6}}{\sqrt{2}} = \sqrt{\frac{6}{2}} = \sqrt{3}$. We can extend this rule for all radical expressions.

QUOTIENT RULE FOR RADICALS
For any real numbers $\sqrt[n]{a}$ and $\sqrt[n]{b}$ and any integer index n greater than 1,

$$\frac{\sqrt[n]{a}}{\sqrt[n]{b}} = \sqrt[n]{\frac{a}{b}}$$

We can show that this statement is true if we rewrite the radical expressions as exponential expressions with rational exponents and apply the quotient rule for exponents. That is, we would have the following:

$$\frac{\sqrt[3]{8}}{\sqrt[3]{2}} = \frac{8^{1/3}}{2^{1/3}} = \left(\frac{8}{2}\right)^{1/3} = \sqrt[3]{\frac{8}{2}} = \sqrt[3]{4}$$

$$\frac{\sqrt[4]{16}}{\sqrt[4]{2}} = \frac{16^{1/4}}{2^{1/4}} = \left(\frac{16}{2}\right)^{1/4} = \sqrt[4]{\frac{16}{2}} = \sqrt[4]{8}$$

For the general rule, $\dfrac{\sqrt[n]{a}}{\sqrt[n]{b}} = \dfrac{a^{1/n}}{b^{1/n}} = \left(\dfrac{a}{b}\right)^{1/n} = \sqrt[n]{\dfrac{a}{b}}$

SIMPLIFYING RADICAL EXPRESSIONS — CONDITION 2
If a radical expression has a fractional radicand, it can be simplified.

$$\sqrt[n]{\frac{a}{b}} = \frac{\sqrt[n]{a}}{\sqrt[n]{b}}$$

For example,

$$\sqrt{\frac{3}{4}} = \frac{\sqrt{3}}{\sqrt{4}} = \frac{\sqrt{3}}{2}$$

We must be careful when writing the final result. If the result is a fraction, it is conventional to write the *denominator* without radicals. The process of changing the fraction to an equivalent form without a radical denominator is called **rationalizing the denominator**. That is, we make the denominator a rational number. Rationalizing the denominator makes it easier to add or subtract radical expressions because the denominators are rational numbers and not radicals.

> ### SIMPLIFYING RADICAL EXPRESSIONS — CONDITION 3
> If a radical expression has a denominator containing a radical, it can be rationalized.

For example,

$$\sqrt{\frac{4}{3}} = \frac{\sqrt{4}}{\sqrt{3}} = \frac{2}{\sqrt{3}}$$

is not fully simplified. To rationalize the denominator, we use the multiplication property of 1; that is, we multiply the numerator and denominator by the same value. We choose a value that will result in the denominator's being a perfect root, so that we can simplify it. Since the denominator is $\sqrt{3}$ and we want a perfect-square root, we can multiply the numerator and denominator of the fraction by $\sqrt{3}$.

$$\frac{2}{\sqrt{3}} \cdot \frac{\sqrt{3}}{\sqrt{3}} = \frac{2\sqrt{3}}{\sqrt{9}} = \frac{2\sqrt{3}}{3}$$

We evaluate other roots by using the same process.

EXAMPLE 5 Simplify without a calculator. Rationalize all denominators.

a. $\sqrt{\dfrac{63}{50}}$ **b.** $\sqrt[3]{\dfrac{5}{8}}$ **c.** $\sqrt[3]{\dfrac{8}{5}}$

Solution

a. $\sqrt{\dfrac{63}{50}} = \dfrac{\sqrt{63}}{\sqrt{50}} = \dfrac{\sqrt{9}\sqrt{7}}{\sqrt{25}\sqrt{2}} = \dfrac{3\sqrt{7}}{5\sqrt{2}}$ Quotient rule

$= \dfrac{3\sqrt{7}}{5\sqrt{2}} \cdot \dfrac{\sqrt{2}}{\sqrt{2}} = \dfrac{3\sqrt{14}}{5\sqrt{4}} = \dfrac{3\sqrt{14}}{5 \cdot 2} = \dfrac{3\sqrt{14}}{10}$ Rationalize the denominator.

b. $\sqrt[3]{\dfrac{5}{8}} = \dfrac{\sqrt[3]{5}}{\sqrt[3]{8}} = \dfrac{\sqrt[3]{5}}{2}$ Quotient rule

c. $\sqrt[3]{\dfrac{8}{5}} = \dfrac{\sqrt[3]{8}}{\sqrt[3]{5}} = \dfrac{2}{\sqrt[3]{5}}$ Quotient rule

$= \dfrac{2}{\sqrt[3]{5}} \cdot \dfrac{\sqrt[3]{5^2}}{\sqrt[3]{5^2}} = \dfrac{2\sqrt[3]{5^2}}{\sqrt[3]{5^3}} = \dfrac{2\sqrt[3]{25}}{5}$ To obtain a perfect-cube root in the denominator, multiply by $(\sqrt[3]{5})^2$ or $\sqrt[3]{5^2}$.

Check

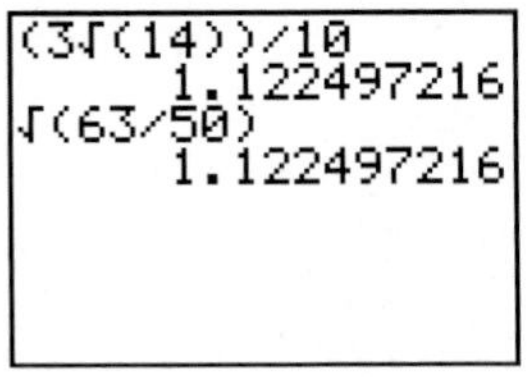

```
(3√(14))/10
          1.122497216
√(63/50)
          1.122497216
```

Figure 13.7

To check these results on your calculator, first evaluate the result found algebraically, and then evaluate the given radical expression. The two values should be the same. The check for Example 5a is shown in Figure 13.7.

EXAMPLE 6 Simplify. Rationalize all denominators.

$$\textbf{a.}\ \sqrt{\dfrac{16ab}{3c}} \qquad \textbf{b.}\ \sqrt{\dfrac{72x^3y^2}{15xy}} \qquad \textbf{c.}\ \sqrt[3]{\dfrac{35m^2}{12np^2}}$$

Solution

a.
$$\sqrt{\dfrac{16ab}{3c}} = \dfrac{\sqrt{16ab}}{\sqrt{3c}} = \dfrac{4\sqrt{ab}}{\sqrt{3c}} \qquad \textit{Quotient rule}$$

$$= \dfrac{4\sqrt{ab}}{\sqrt{3c}} \cdot \dfrac{\sqrt{3c}}{\sqrt{3c}} = \dfrac{4\sqrt{3abc}}{\sqrt{(3c)^2}} = \dfrac{4\sqrt{3abc}}{3c} \qquad \begin{array}{l}\textit{Rationalize the}\\ \textit{denominator.}\end{array}$$

b. First, simplify the radicand before applying the quotient rule.

$$\sqrt{\dfrac{72x^3y^2}{15xy}} = \sqrt{\dfrac{24x^2y}{5}} = \dfrac{\sqrt{24x^2y}}{\sqrt{5}} \qquad \textit{Quotient rule}$$

$$= \dfrac{\sqrt{24x^2y}}{\sqrt{5}} \cdot \dfrac{\sqrt{5}}{\sqrt{5}} = \dfrac{\sqrt{120x^2y}}{\sqrt{5^2}} = \dfrac{\sqrt{4x^2 \cdot 30y}}{\sqrt{5^2}} = \dfrac{2x\sqrt{30y}}{5}$$

$$\begin{array}{r}\textit{Rationalize the}\\ \textit{denominator.}\end{array}$$

c.
$$\sqrt[3]{\dfrac{35m^2}{12np^2}} = \dfrac{\sqrt[3]{35m^2}}{\sqrt[3]{12np^2}} \qquad \textit{Quotient rule}$$

To determine the factor needed to rationalize the denominator, we first determine the smallest perfect cube that is divisible by the numerical factor of the radicand, 12. The smallest perfect cube divisible by 12 is $216 = 6^3$. 216 divided by 12 is 18. Therefore, 18 is the needed numerical factor. Each variable factor must also be a perfect cube. We determine the variable factors needed to be n^2 and p. The cube root of $18n^2p$ is the factor needed to rationalize the denominator.

$$= \dfrac{\sqrt[3]{35m^2}}{\sqrt[3]{12np^2}} \cdot \dfrac{\sqrt[3]{18n^2p}}{\sqrt[3]{18n^2p}} = \dfrac{\sqrt[3]{630m^2n^2p}}{\sqrt[3]{216n^3p^3}} = \dfrac{\sqrt[3]{630m^2n^2p}}{6np}$$

We also use the quotient rule for radicals to divide radical expressions with the same index.

EXAMPLE 7 Divide and simplify. Rationalize all denominators.

$$\textbf{a.}\ \dfrac{\sqrt{3y}}{\sqrt{6xy}} \qquad \textbf{b.}\ \dfrac{2a\sqrt{5a}}{4\sqrt{6ab^2}} \qquad \textbf{c.}\ \dfrac{\sqrt[3]{4a^2b^5}}{\sqrt[3]{12ab^2}}$$

Solution

a.
$$\dfrac{\sqrt{3y}}{\sqrt{6xy}} = \sqrt{\dfrac{3y}{6xy}} = \sqrt{\dfrac{1}{2x}} \qquad \textit{Quotient rule}$$

$$= \sqrt{\dfrac{1}{2x}} \cdot \sqrt{\dfrac{2x}{2x}} = \sqrt{\dfrac{2x}{(2x)^2}} = \dfrac{\sqrt{2x}}{2x} \qquad \textit{Rationalize the denominator.}$$

b. $\dfrac{2a\sqrt{5a}}{4\sqrt{6ab^2}} = \dfrac{a\sqrt{5a}}{2\sqrt{6ab^2}} = \dfrac{a\sqrt{5a}}{2b\sqrt{6a}}$ 　　　　　Simplify.

$\qquad = \dfrac{a}{2b}\sqrt{\dfrac{5a}{6a}} = \dfrac{a}{2b}\sqrt{\dfrac{5}{6}}$ 　　　　　Quotient rule

$\qquad = \dfrac{a}{2b}\sqrt{\dfrac{5}{6}} \cdot \sqrt{\dfrac{6}{6}} = \dfrac{a}{2b}\sqrt{\dfrac{30}{36}} = \dfrac{a}{2b} \cdot \dfrac{\sqrt{30}}{6} = \dfrac{a\sqrt{30}}{12b}$ 　Rationalize the denominator.

c. $\dfrac{\sqrt[3]{4a^2b^5}}{\sqrt[3]{12ab^2}} = \sqrt[3]{\dfrac{4a^2b^5}{12ab^2}} = \sqrt[3]{\dfrac{ab^3}{3}} = b\sqrt[3]{\dfrac{a}{3}}$ 　　　Quotient rule

$\qquad = b\sqrt[3]{\dfrac{a}{3}} \cdot \sqrt[3]{\dfrac{3^2}{3^2}} = b\sqrt[3]{\dfrac{9a}{3^3}} = \dfrac{b\sqrt[3]{9a}}{3}$ 　　Rationalize the denominator. ●

13.4.2 Checkup

Simplify without a calculator. Rationalize all denominators.
Check your results on your calculator.

1. $\sqrt{\dfrac{27}{20}}$ 　　**2.** $\sqrt[3]{\dfrac{27}{2}}$

Simplify exercises 3–7. Rationalize all denominators.

3. $\sqrt{\dfrac{25xy}{2z}}$ 　**4.** $\sqrt{\dfrac{18a^2b^3}{14ab^2}}$ 　**5.** $\sqrt[3]{\dfrac{6x}{10y^2z}}$

6. $\dfrac{4b\sqrt{3ab}}{12\sqrt{15a^2b}}$ 　**7.** $\dfrac{\sqrt[3]{2c^3d}}{\sqrt[3]{8cd^4}}$

8. What does it mean to rationalize the denominator of a radical expression?

13.4.3 Modeling the Real World

Another example of a radical expression we may encounter is found in a formula for the length of time (in seconds) that it takes for a pendulum to swing from one side of its vertical rest position to the other and back (called the *period* of the pendulum). This time, T, is determined by the acceleration due to gravity, g, and the length of its suspension, L, according to the formula $T = 2\pi\sqrt{\dfrac{L}{g}}$. The motion of a pendulum in a grandfather clock can be modeled by this formula. However, with some simplifying assumptions, many other kinds of motion can also be modeled with the formula, such as the swaying of a skyscraper in a strong wind or the swinging of a person's leg while walking.

EXAMPLE 8 Yueling practices daily for a 10-km walk. She swings each arm in a forward-and-back motion as she walks. If the acceleration due to gravity is about 32 ft/sec² and the length of her arm is 2 feet, determine the time it takes Yueling's arm to complete the forward-and-back motion and return to its original starting position, assuming that her arms swing in a pendulum motion and not by her own efforts.

Solution

We will model the motion of Yueling's as a pendulum and use the formula $T = 2\pi\sqrt{\dfrac{L}{g}}$.

$$T = 2\pi\sqrt{\dfrac{L}{g}}$$

$$T = 2\pi\sqrt{\dfrac{2}{32}} \qquad \text{Substitute 32 for } g \text{ and 2 for } L.$$

$$T = 2\pi\sqrt{\dfrac{1}{16}} \qquad \text{Simplify the radicand.}$$

$$T = 2\pi \cdot \dfrac{1}{4} \qquad \text{Simplify the radical factor.}$$

$$T = \dfrac{\pi}{2} \qquad \text{Multiply and simplify.}$$

The forward-and-back motion of Yueling's arm takes $\frac{\pi}{2}$ seconds, or approximately 1.57 seconds.

Note that Yueling's arm is not a true pendulum, because it may swing faster or slower due to her own effort.

APPLICATION

The most complete fossil (95%) of a *Tyrannosaurus rex*, Sue, is on display at the Field Museum in Chicago, Illinois. Its height at the hips is 13 feet.

a. Model the leg motion as that of a pendulum, and determine the length of time in seconds during which the leg completed one back-and-forth motion.

b. Dinosaur tracks have been uncovered in the roofs of coal mines where peat in the tracks were reduced to coal. The distance between consecutive prints of the same foot measures 9.3 feet. Twice this distance constitutes a back-and-forth movement measuring about 18.6 feet. Use this distance to estimate Sue's walking speed.

Solution

a.
$$T = 2\pi\sqrt{\dfrac{L}{g}} \qquad \text{Pendulum formula}$$

$$T = 2\pi\sqrt{\dfrac{13}{32}} \qquad \text{Substitute 32 for } g \text{ and 13 for } L.$$

$$T = 2\pi\dfrac{\sqrt{13}}{\sqrt{32}} \qquad \text{Quotient rule}$$

$$T = 2\pi\dfrac{\sqrt{13}}{\sqrt{16 \cdot 2}}$$

$$T = 2\pi\dfrac{\sqrt{13}}{4\sqrt{2}} \qquad \text{Product rule}$$

$$T = 2\pi\dfrac{\sqrt{13}}{4\sqrt{2}} \cdot \dfrac{\sqrt{2}}{\sqrt{2}} \qquad \text{Rationalize the denominator.}$$

$$T = 2\pi\dfrac{\sqrt{26}}{4\sqrt{4}}$$

$$T = 2\pi\dfrac{\sqrt{26}}{4 \cdot 2}$$

$$T = \overset{1}{2}\pi\dfrac{\sqrt{26}}{8}$$

$$T = \dfrac{\pi\sqrt{26}}{4} \approx 4$$

One back-and-forth motion required $\frac{\pi\sqrt{26}}{4}$ seconds, or approximately 4 seconds. Sue's leg is not a true pendulum, because the dinosaur could have swing it faster or slower due to its own effort. However, this model gives us an approximate time that we can use.

b.
$$r = \dfrac{d}{t} \qquad \text{Distance-traveled formula}$$

$$r \approx \dfrac{18.6 \text{ ft}}{4 \text{ sec}} \qquad \text{Substitute.}$$

$$r \approx 4.65 \dfrac{\text{ft}}{\text{sec}}$$

Sue's approximate running time is about 4.65 ft/sec.

Converting this figure to miles per hour yields

$$4.65 \dfrac{\text{feet}}{\text{second}} \cdot 60 \dfrac{\text{second}}{\text{minute}} \cdot 60 \dfrac{\text{minute}}{\text{hour}} \cdot \dfrac{1 \text{ mile}}{5280 \text{ feet}} \approx 3.17 \dfrac{\text{miles}}{\text{hour}}$$

13.4.3 Checkup

1. The pendulum in a grandfather clock is 16 inches long ($1\frac{1}{3}$ feet long). What is the period of the pendulum?

2. The distance from Steve's heel to his hip is about 3 feet. Determine the length of time in seconds during which his leg does one back-and-forth swing, or one stride. The dis-

tance between footprints of Steve's same leg is about 6 feet. Determine his walking speed. How does Steve's speed compare with the *Tyrannosaurus rex*'s speed in the application in this section?

13.4 Exercises

Simplify without a calculator. Check your results on your calculator.

1. $\sqrt{28}$

2. $\sqrt{75}$

3. $\sqrt[3]{-686}$

4. $\sqrt[3]{-192}$

5. $\sqrt[4]{112}$

6. $\sqrt[4]{7203}$

7. $\sqrt[5]{-6250}$

8. $\sqrt[5]{-1944}$

Simplify.

9. $\sqrt{20x^4y^3z^2}$

10. $\sqrt{12x^3y^9z^8}$

11. $\sqrt[3]{72m^5n^9}$

12. $\sqrt[3]{108p^7q^8}$

13. $\sqrt[4]{162x^4y^5}$

14. $\sqrt[4]{80a^6b^8}$

15. $\sqrt[5]{486a^6b^{10}c^2}$

16. $\sqrt[5]{729x^2y^8z^{15}}$

17. $\sqrt{5x^2 + 30x + 45}$

18. $\sqrt{3x^3 + 6x^2 + 3x}$

Multiply and simplify.

19. $\sqrt{7x} \cdot \sqrt{14y}$

20. $\sqrt{6x} \cdot \sqrt{12y}$

21. $2\sqrt{7xy} \cdot x\sqrt{14y}$

22. $-3\sqrt{7pq} \cdot 2\sqrt{3p^4q}$

23. $\sqrt{7x + 14y} \cdot \sqrt{3x + 6y}$

24. $\sqrt{p^2 + 2pq} \cdot \sqrt{3p + 6q}$

25. $\sqrt{x + 2} \cdot \sqrt{x^2 + 3x + 2}$

26. $\sqrt{x - 7} \cdot \sqrt{x^2 - 6x - 7}$

27. $\sqrt{x^2 + x - 2} \cdot \sqrt{x^2 + 3x - 4}$

28. $\sqrt{x^2 + 6x + 8} \cdot \sqrt{x^2 - 16}$

29. $\sqrt[3]{-4x^4y^2} \cdot \sqrt[3]{6xy}$

30. $\sqrt[3]{-25x^4y} \cdot \sqrt[3]{10x^2y^7}$

Simplify without a calculator. Rationalize all denominators. Check your results on your calculator.

31. $-\sqrt{\dfrac{36}{49}}$

32. $-\sqrt{\dfrac{25}{81}}$

33. $\sqrt{\dfrac{1210}{1440}}$

34. $\sqrt{\dfrac{1690}{1960}}$

35. $-\sqrt{\dfrac{8}{27}}$

36. $-\sqrt{\dfrac{27}{125}}$

37. $\sqrt[3]{\dfrac{4}{125}}$

38. $\sqrt[3]{\dfrac{5}{512}}$

39. $\sqrt[3]{-\dfrac{9}{25}}$

40. $\sqrt[3]{-\dfrac{10}{49}}$

41. $\sqrt[5]{\dfrac{1}{16}}$

42. $\sqrt[5]{\dfrac{1}{81}}$

Simplify. Rationalize all denominators.

43. $\sqrt{\dfrac{4x^2}{9y^2}}$

44. $\sqrt{\dfrac{25x}{49y^2}}$

45. $\sqrt{\dfrac{3x}{25y^2}}$

46. $\sqrt{\dfrac{5p}{49q^2}}$

47. $\sqrt{\dfrac{4xy}{5z}}$

48. $\sqrt{\dfrac{36ab}{7c}}$

49. $\sqrt{\dfrac{2x^2}{6}}$

50. $\sqrt{\dfrac{7z^4}{35}}$

51. $\sqrt[3]{\dfrac{27x^3}{y^6}}$

52. $\sqrt[3]{\dfrac{-64x^6y^3}{z^9}}$

53. $\sqrt[3]{\dfrac{3}{25x^2}}$

54. $\sqrt[3]{\dfrac{7}{36a^2}}$

55. $\sqrt[3]{\dfrac{3x^2y}{5xy^2z}}$

56. $\sqrt[3]{\dfrac{4a^2b^2c}{7ab^2c^2}}$

Divide and simplify. Rationalize all denominators.

57. $\dfrac{\sqrt{5}}{\sqrt{x}}$

58. $\dfrac{\sqrt{10}}{\sqrt{y}}$

59. $\dfrac{\sqrt{4x}}{\sqrt{8xy}}$

60. $\dfrac{\sqrt{5a}}{\sqrt{15ab}}$

61. $\dfrac{\sqrt{z^3}}{\sqrt{8}}$

62. $\dfrac{\sqrt{p^3}}{\sqrt{27}}$

63. $\dfrac{\sqrt{ab^4}}{\sqrt{a^2b}}$

64. $\dfrac{\sqrt{c^2d}}{\sqrt{c^3d^4}}$

65. $\dfrac{3x\sqrt{5x^2y}}{6\sqrt{15x^3y}}$

66. $\dfrac{5p\sqrt{3pq^2}}{15\sqrt{6pq^3}}$

67. $\dfrac{\sqrt[3]{3}}{\sqrt[3]{x^2}}$

68. $\dfrac{\sqrt[3]{7}}{\sqrt[3]{a^2}}$

69. $\dfrac{\sqrt[3]{3x}}{\sqrt[3]{12xy}}$

70. $\dfrac{\sqrt[3]{6b}}{\sqrt[3]{24ab}}$

71. $\dfrac{\sqrt[3]{-8xy^2z^2}}{\sqrt[3]{3x^2y^2z}}$

72. $\dfrac{\sqrt[3]{-27a^2bc^4}}{\sqrt[3]{5a^2b^2c^3}}$

73. Find the period of a pendulum whose length is 7 feet.

74. Find the period of a pendulum whose length is 12 feet.

75. Use the pendulum formula to model the motion of a man's leg as he walks. If his leg measures 3 feet from his heel to his hip, determine the time it takes his leg to complete the forward-and-back motion to return to its starting position.

76. A boy's leg measures 1.6 feet from his heel to his hip. Use the pendulum formula to determine the time it takes his leg to complete the forward-and-back motion to return to its starting position as he walks.

77. The distance between Eydie's heel and her hip is about 2.5 feet. Determine the number of seconds of her stride—that is, the time it takes for a leg to do one back-and-forth swing. The distance between footprints of her same leg is about 4 feet. What is Eydie's walking speed?

78. Gargantua, the latest movie monster, has a leg length of about 25 feet. What is the time required for him to take a stride? The distance between his running footprints for the same leg is about 60 feet. How fast is Gargantua running?

13.4 Calculator Exercises

When given a radical expression to simplify, you must be able to determine the largest perfect power factor of the radicand that you can extract from the radical expression. For example, when you simplified $\sqrt[3]{24}$, you had to know that 8 was the largest perfect-cube factor of 24. That was not too difficult to figure out, but suppose you had to simplify $\sqrt[3]{24,565}$. It would be very difficult to determine the largest perfect cube factor of 24,565. The table feature of the calculator can be used to help you find this factor. The steps to do so are as follows:

* In Y1, store x^p, where p will be the exponent of the power that you are seeking.

| Y= | X,T,Θ,n | ^ | ALPHA | P | ENTER |

* In Y2, store $N/Y1$, where N will be the radicand of the radical expression.

| ALPHA | N | ÷ | VARS | ▶ | 1 | 1 | 2nd | QUIT |

* Store the index of the radicand in P and the radicand in N.

| 3 | STO▶ | ALPHA | P | ENTER | 2 | 4 | 5 | 6 |
| 5 | STO▶ | ALPHA | N | ENTER |

* Set the table feature to start at 2 and increment by 1 in the Auto mode.

| 2nd | TBLSET | 2 | ENTER | 1 | ENTER | ENTER |
| ▼ | ENTER | 2nd | QUIT |

* View the table.

| 2nd | TABLE |

* Scroll down the table until you find the first integer value in the Y2 column. The value in the X column will be the base of the perfect power factor, the value in the Y1 column will be the perfect power factor, and the value in the Y2 column will be the other factor.

X	Y1	Y2
12	1728	14.216
13	2197	11.181
14	2744	8.9523
15	3375	7.2785
16	4096	5.9973
17	4913	5
18	5832	4.2121

X=17

* From the screen, we can see that

$$24,565 = 4913 \cdot 5 = 17^3 \cdot 5$$

so

$$\sqrt[3]{24,565} = \sqrt[3]{17^3 \cdot 5} = 17\sqrt[3]{5}$$

Use this method to simplify the following radicals:

1. $\sqrt[3]{13,718}$ **2.** $\sqrt[4]{199,927}$

3. $\sqrt[7]{-1664}$ **4.** $\sqrt{9251}$

13.4 Writing Exercises

The following examples illustrate the use of rational exponents to simplify radical expressions involving different indices. In each of the examples, an incorrect step has been made, which yields a result that is not equivalent to the original expression. You may use the TABLE feature to see that they are not equivalent to the original expression. Check each step to see where the mistake is made during simplification. Describe the incorrect step.

1. $\sqrt[3]{x^2} \sqrt[5]{x^3} = x^{2/3} \cdot x^{3/5} = x^{(2+3)/(3+5)} = x^{5/8} = \sqrt[8]{x^5}$

2. $\sqrt[3]{x^2} \sqrt[5]{x^3} = x^{3/2} \cdot x^{5/3} = x^{9/6 + 10/6} = x^{19/6} = \sqrt[6]{x^{19}} = \sqrt[6]{x^{18} \cdot x} = x^3 \sqrt[6]{x}$

3. $\dfrac{\sqrt[5]{x^3} \cdot}{\sqrt[3]{x^2}} = \dfrac{x^{3/5}}{x^{2/3}} = x^{3/5 - 2/3} = x^{(3-2)/(5-3)} = x^{1/2} = \sqrt{x}$

13.5 Operations on Radicals

OBJECTIVES

1 Add and subtract radical expressions.

2 Multiply radical expressions.

3 Divide radical expressions, and rationalize a denominator having two terms.

4 Model real-world situations by using expressions involving the addition or subtraction of radicals.

APPLICATION

The Hephaesteion was a temple built by the Greeks in the fifth century B.C. The ratio of its width to its height is approximated by the expression $\frac{2}{\sqrt{5}-1}$. This number is called the *golden ratio*. Early Greeks believed that rectangles with this ratio were aesthetically pleasing to the eye. Rationalize the denominator of the expression.

After completing this section, we will discuss this application further. See page 926.

13.5.1 Adding and Subtracting Radical Expressions

We are now ready to add and subtract radical expressions. We need to develop a property of radicals that we can use in addition and subtraction. Let's discuss addition and subtraction of like radicals. **Like radicals** have the same index and radicand. The expression $2\sqrt{3} + 5\sqrt{3}$ contains two terms with like radicals, $2\sqrt{3}$ and $5\sqrt{3}$. The coefficients of the radicals in the expression are 2 and 5, respectively.

To add (or subtract) like radicals, we add (or subtract) the coefficients of each like radical.

$$2\sqrt{3} + 5\sqrt{3} = (2 + 5)\sqrt{3} = 7\sqrt{3}$$

ADDITION AND SUBTRACTION OF LIKE RADICALS

To add or subtract like radicals, use the respective formulas

$$a\sqrt[n]{x} + b\sqrt[n]{x} = (a + b)\sqrt[n]{x}$$

$$a\sqrt[n]{x} - b\sqrt[n]{x} = (a - b)\sqrt[n]{x}$$

We can show that this statement is true by using the distributive law, as we did when we combined like terms. For example,

$$2x + x = 2x + 1x = (2 + 1)x = 3x \qquad \text{Like terms}$$
$$2\sqrt{x} + \sqrt{x} = 2\sqrt{x} + 1\sqrt{x} = (2 + 1)\sqrt{x} = 3\sqrt{x} \qquad \text{Like radicals}$$

HELPING HAND Just as we cannot add or subtract unlike terms, such as $2x + y$, we cannot add or subtract unlike radicals, such as $2\sqrt{x} + \sqrt{y}$.

EXAMPLE 1 Add or subtract as indicated.

    a. $2\sqrt{xy} + 3\sqrt[3]{xy} - 4\sqrt[3]{xy} - \sqrt{xy}$

    b. $3x\sqrt{y} - 5x\sqrt{y} + 4x\sqrt{y}$

    c. $5a\sqrt[3]{abc} + 4\sqrt[3]{abc}$

Solution

    a. $2\sqrt{xy} + 3\sqrt[3]{xy} - 4\sqrt[3]{xy} - \sqrt{xy}$

$$= 2\sqrt{xy} - \sqrt{xy} + 3\sqrt[3]{xy} - 4\sqrt[3]{xy} \qquad \text{Rearrange terms.}$$
$$= (2 - 1)\sqrt{xy} + (3 - 4)\sqrt[3]{xy} \qquad \text{Subtract coefficients.}$$
$$= \sqrt{xy} - \sqrt[3]{xy}$$

    b. $3x\sqrt{y} - 5x\sqrt{y} + 4x\sqrt{y}$

$$= (3x - 5x + 4x)\sqrt{y} \qquad \text{Add and subtract coefficients.}$$
$$= 2x\sqrt{y}$$

    c. $5a\sqrt[3]{abc} + 4\sqrt[3]{abc} = (5a + 4)\sqrt[3]{abc} \qquad \text{Add coefficients.}$

Sometimes we need to simplify the radical expressions to have like radicals.

EXAMPLE 2 Add or subtract as indicated.

    a. $5\sqrt{32} - \sqrt{72}$ **b.** $3ab\sqrt{ab} + 2b\sqrt{a^3b} - a\sqrt{4ab^3}$

    c. $\sqrt{2} + \sqrt{\dfrac{1}{2}}$

Solution

    a. $5\sqrt{32} - \sqrt{72} = 5 \cdot 4\sqrt{2} - 6\sqrt{2} \qquad \text{Simplify radical expressions.}$

$$= 20\sqrt{2} - 6\sqrt{2} \qquad \text{Subtract coefficients.}$$
$$= 14\sqrt{2}$$

    b. $3ab\sqrt{ab} + 2b\sqrt{a^3b} - a\sqrt{4ab^3}$

$$= 3ab\sqrt{ab} + 2ab\sqrt{ab} - 2ab\sqrt{ab} \qquad \text{Simplify radical expressions.}$$
$$= (3ab + 2ab - 2ab)\sqrt{ab} \qquad \text{Add and subtract coefficients.}$$
$$= 3ab\sqrt{ab}$$

    c. $\sqrt{2} + \sqrt{\dfrac{1}{2}} = \sqrt{2} + \dfrac{\sqrt{1}}{\sqrt{2}} \qquad \text{Quotient rule}$

$$= \sqrt{2} + \dfrac{\sqrt{1}}{\sqrt{2}} \cdot \dfrac{\sqrt{2}}{\sqrt{2}} \qquad \text{Rationalize the denominator.}$$
$$= \sqrt{2} + \dfrac{\sqrt{2}}{2}$$
$$= \sqrt{2} + \dfrac{1}{2}\sqrt{2}$$
$$= \left(1 + \dfrac{1}{2}\right)\sqrt{2} \qquad \text{Add coefficients.}$$
$$= \dfrac{3}{2}\sqrt{2}$$

13.5.1 Checkup

In exercises 1–6, add or subtract as indicated.

1. $3\sqrt{ab} + 5\sqrt[4]{ab} - 4\sqrt{ab} + 2\sqrt[4]{ab}$

2. $3p\sqrt{q} - 7p\sqrt{q} + 8p\sqrt{q}$ 3. $5m\sqrt[3]{mn} + 2\sqrt[3]{mn}$

4. $2\sqrt{48} + \sqrt{75}$ 5. $6xy\sqrt{xy} + 5y\sqrt{x^3y} - 4x\sqrt{9xy^3}$

6. $\sqrt{3} - \sqrt{\dfrac{1}{3}}$

7. Define what is meant by the term *like radicals*. Give examples of like radicals and unlike radicals.

8. Explain what a coefficient of a radical term is.

9. What must you do to the radical terms of an addition or subtraction problem before collecting like terms? Why?

13.5.2 Multiplying Radicals

We have already multiplied radical expressions with one term. We use the properties of radicals to simplify these expressions. We are now ready to expand our multiplication of radicals to radical expressions that have more than one term.

To multiply, we use the distributive law, as we did to multiply polynomial expressions with more than one term. For example,

$$2(x + 3) = 2x + 6 \qquad \text{Polynomial expression}$$
$$\sqrt{2}(\sqrt{x} + \sqrt{3}) = \sqrt{2x} + \sqrt{6} \qquad \text{Radical expression}$$
$$(x + 3)(x - 4) = x^2 - 4x + 3x - 12 \qquad \text{Polynomial expression}$$
$$= x^2 - x - 12$$
$$(\sqrt{x} + 3)(\sqrt{x} - 4) = \sqrt{x^2} - 4\sqrt{x} + 3\sqrt{x} - 12 \qquad \text{Radical expression}$$
$$= x - \sqrt{x} - 12$$

EXAMPLE 3   Multiply and simplify.

a. $3\sqrt[3]{x}(4\sqrt[3]{x^2} - 2)$ **b.** $(\sqrt{2} - \sqrt{xy})(\sqrt{5} + \sqrt{xy})$

c. $(\sqrt{2} + \sqrt{3})^2$ **d.** $(\sqrt{x} - 2)^2$

Solution

a. $3\sqrt[3]{x}(4\sqrt[3]{x^2} - 2) = 12\sqrt[3]{x^3} - 6\sqrt[3]{x}$ Distributive law

$$= 12x - 6\sqrt[3]{x}$$

b. $(\sqrt{2} - \sqrt{xy})(\sqrt{5} + \sqrt{xy})$

$$= \sqrt{10} + \sqrt{2xy} - \sqrt{5xy} - \sqrt{(xy)^2} \qquad \text{FOIL method}$$
$$= \sqrt{10} + \sqrt{2xy} - \sqrt{5xy} - xy$$

c. $(\sqrt{2} + \sqrt{3})^2$

$$= (\sqrt{2} + \sqrt{3})(\sqrt{2} + \sqrt{3}) \qquad \text{Expand factors.}$$
$$= \sqrt{4} + \sqrt{6} + \sqrt{6} + \sqrt{9} \qquad \text{FOIL method}$$
$$= 2 + 2\sqrt{6} + 3$$
$$= 5 + 2\sqrt{6}$$

We should have noticed that the given expression was a special product, $(a + b)^2 = a^2 + 2ab + b^2$. Then,

$$(\sqrt{2} + \sqrt{3})^2 = (\sqrt{2})^2 + 2\sqrt{2}\sqrt{3} + (\sqrt{3})^2 = 2 + 2\sqrt{6} + 3$$
$$= 5 + 2\sqrt{6}$$

d. $(\sqrt{x} - 2)^2 = (\sqrt{x})^2 - 2 \cdot \sqrt{x} \cdot 2 + 2^2$ $(a - b)^2 = a^2 - 2ab + b^2$

$$= x - 4\sqrt{x} + 4$$

The product of the sum and difference of two terms is the difference of their squares.

$$(a + b)(a - b) = a^2 - b^2$$

The expressions $a + b$ and $a - b$ are called **conjugates** of each other. When we multiply conjugates that contain square roots, we obtain an interesting result. Complete the following set of exercise to discover this result.

Discovery 4

Conjugates

Multiply and simplify.

1. $(\sqrt{2} + \sqrt{3})(\sqrt{2} - \sqrt{3})$ **2.** $(2 + \sqrt{x})(2 - \sqrt{x})$

What did you notice about the products?

In these exercises, we noticed that the factors contained square roots, but the product did not contain a square root.

We can see why this is true if we multiply $(\sqrt{x} + \sqrt{y})(\sqrt{x} - \sqrt{y})$, using the special product $(a + b)(a - b) = a^2 - b^2$. We obtain

$$(\sqrt{x} + \sqrt{y})(\sqrt{x} - \sqrt{y}) = (\sqrt{x})^2 - (\sqrt{y})^2$$
$$= x - y$$

This kind of simplification works only for radicals involving square roots. We will use conjugates in the next objective to rationalize denominators.

EXAMPLE 4   Multiply. $(\sqrt{x} - 2\sqrt{2})(\sqrt{x} + 2\sqrt{2})$

Solution

$$(\sqrt{x} - 2\sqrt{2})(\sqrt{x} + 2\sqrt{2}) = (\sqrt{x})^2 - (2\sqrt{2})^2 \qquad {\scriptstyle (a - b)(a + b) = a^2 - b^2}$$
$$= x - 8$$

13.5.2 Checkup

Multiply exercises 1–6 and simplify. Where possible, check for equivalence of your results, using your calculator.

1. $7\sqrt[3]{x}(5\sqrt[3]{x^2} + 4\sqrt[3]{x})$ **2.** $(\sqrt{3} - \sqrt{ab})(\sqrt{7} + \sqrt{ab})$

3. $(\sqrt{10} + \sqrt{5})(\sqrt{10} - \sqrt{5})$

4. $(11 + 3\sqrt{z})(11 - 3\sqrt{z})$ **5.** $(\sqrt{10} + \sqrt{5})^2$

6. $(\sqrt{z} - 5)^2$

7. Describe what we mean by a conjugate of a radical expression. Give an example.

8. What special result is obtained when you multiply conjugate expressions?

13.5.3 Dividing Radicals

We have already divided radical expressions with one term. We used the properties of radicals to simplify these expressions. We are now ready to expand our division of radicals to radical expressions that have more than one term.

To divide a radical expression by a radical expression with a single term, we use the distributive law, as we did to divide polynomial expressions. For example,

$$\frac{x+6}{2} = \frac{x}{2} + \frac{6}{2}$$

Polynomial expression

$$= \frac{1}{2}x + 3$$

Simplify.

$$\frac{\sqrt{x} + \sqrt{6}}{\sqrt{2}} = \frac{\sqrt{x}}{\sqrt{2}} + \frac{\sqrt{6}}{\sqrt{2}}$$

Radical expression

$$= \frac{\sqrt{x}}{\sqrt{2}} \cdot \frac{\sqrt{2}}{\sqrt{2}} + \sqrt{3}$$

Simplify and rationalize the denominator.

$$= \frac{\sqrt{2x}}{2} + \sqrt{3} \quad \text{or} \quad \frac{1}{2}\sqrt{2x} + \sqrt{3}$$

To divide a radical expression with square roots by a radical expression containing two terms with square roots, we use the multiplication property of 1 and conjugates. Remember that the product of conjugates with at least one square-root term results in an expression without square roots. Therefore, this division process will rationalize the denominator.

EXAMPLE 5 Divide and simplify.

a. $\dfrac{\sqrt{5}}{\sqrt{3} + \sqrt{2}}$ **b.** $\dfrac{2\sqrt{x} + \sqrt{5}}{\sqrt{x} - 2\sqrt{2}}$ **c.** $\dfrac{\sqrt{3} + \sqrt{7}}{\sqrt{3} - \sqrt{7}}$

Solution

a. $\dfrac{\sqrt{5}}{\sqrt{3} + \sqrt{2}} = \dfrac{\sqrt{5}}{\sqrt{3} + \sqrt{2}} \cdot \dfrac{\sqrt{3} - \sqrt{2}}{\sqrt{3} - \sqrt{2}}$

Multiplication by 1
Use the conjugate of the denominator as numerator and denominator of the factor of 1.

$$= \frac{\sqrt{5}(\sqrt{3} - \sqrt{2})}{(\sqrt{3} + \sqrt{2})(\sqrt{3} - \sqrt{2})}$$

Multiply.

$$= \frac{\sqrt{15} - \sqrt{10}}{(\sqrt{3})^2 - (\sqrt{2})^2}$$

Distribute and simplify.

$$= \frac{\sqrt{15} - \sqrt{10}}{1}$$

$$= \sqrt{15} - \sqrt{10}$$

b. $\dfrac{2\sqrt{x} + \sqrt{5}}{\sqrt{x} - 2\sqrt{2}}$

$$= \frac{2\sqrt{x} + \sqrt{5}}{\sqrt{x} - 2\sqrt{2}} \cdot \frac{\sqrt{x} + 2\sqrt{2}}{\sqrt{x} + 2\sqrt{2}}$$

Multiplication by 1
Use the conjugate of the denominator as numerator and denominator of the factor of 1.

$$= \frac{(2\sqrt{x} + \sqrt{5})(\sqrt{x} + 2\sqrt{2})}{(\sqrt{x} - 2\sqrt{2})(\sqrt{x} + 2\sqrt{2})}$$

Multiply.

$$= \frac{2\sqrt{x^2} + 4\sqrt{2x} + \sqrt{5x} + 2\sqrt{10}}{(\sqrt{x})^2 - (2\sqrt{2})^2}$$

Use the FOIL method and simplify. See Example 4.

$$= \frac{2x + 4\sqrt{2x} + \sqrt{5x} + 2\sqrt{10}}{x - 8}$$

c. $\dfrac{\sqrt{3} + \sqrt{7}}{\sqrt{3} - \sqrt{7}} = \dfrac{\sqrt{3} + \sqrt{7}}{\sqrt{3} - \sqrt{7}} \cdot \dfrac{\sqrt{3} + \sqrt{7}}{\sqrt{3} + \sqrt{7}}$

Multiplication by 1
Use the conjugate of the denominator as numerator and denominator of the factor of 1.

$$= \dfrac{(\sqrt{3} + \sqrt{7})(\sqrt{3} + \sqrt{7})}{(\sqrt{3} - \sqrt{7})(\sqrt{3} + \sqrt{7})}$$

Multiply.

$$= \dfrac{(\sqrt{3})^2 + 2\sqrt{21} + (\sqrt{7})^2}{(\sqrt{3})^2 - (\sqrt{7})^2}$$

FOIL method

$$= \dfrac{3 + 2\sqrt{21} + 7}{3 - 7}$$

$$= \dfrac{10 + 2\sqrt{21}}{-4}$$

$$= \dfrac{2(5 + \sqrt{21})}{2(-2)}$$

Factor the numerator and denominator to simplify the fraction.

$$= \dfrac{5 + \sqrt{21}}{-2}$$

$$= \dfrac{5 + \sqrt{21}}{-2} \cdot \dfrac{-1}{-1}$$

Multiplication by 1
Use −1 as the numerator and denominator in order to eliminate a negative denominator.

$$= \dfrac{-5 - \sqrt{21}}{2}$$

13.5.3 Checkup

Divide exercises 1–4 and simplify. Check for equivalence of your result by using your calculator.

1. $\dfrac{\sqrt{6x} + \sqrt{15}}{\sqrt{3}}$

2. $\dfrac{\sqrt{13}}{\sqrt{5} + \sqrt{2}}$

3. $\dfrac{\sqrt{6} + \sqrt{2}}{\sqrt{6} - \sqrt{2}}$

4. $\dfrac{8\sqrt{x} - \sqrt{3}}{\sqrt{x} + 2\sqrt{3}}$

5. Suppose one radical expression involving square roots is being divided by another radical expression with two terms involving square roots. What is accomplished by multiplying both the numerator and the denominator by the conjugate of the denominator? Why would this be desirable?

13.5.4 Modeling the Real World

When we use geometric formulas involving radicals, we encounter values that are not perfect-square radicands, but are irrational numbers instead. Evaluating complicated radical expressions can be tricky, even with a calculator; it is usually a good idea to simplify these radical expressions by using the rules for radicals we've discussed in this section.

EXAMPLE 6   Fernando plans to increase the size of a square fenced area in his yard. The present area is 48 square feet. He plans to increase the fenced area to 108 square feet and keep the shape as a square. Determine the amount of additional fencing that Fernando needs to purchase. (Round your answer to the nearest foot.)

Solution

Determine the length of the sides of the present area and the new area.

Present area: $s = \sqrt{A}$ New area: $s = \sqrt{A}$

$s = \sqrt{48}$ $s = \sqrt{108}$

$s = \sqrt{16 \cdot 3}$ $s = \sqrt{36 \cdot 3}$

$s = 4\sqrt{3}$ $s = 6\sqrt{3}$

Determine the amount of fencing for both areas. Use the perimeter formula.

Present area: $P = 4s$ New area: $P = 4s$

$P = 4 \cdot 4\sqrt{3}$ $P = 4 \cdot 6\sqrt{3}$

$P = 16\sqrt{3}$ $P = 24\sqrt{3}$

The difference in the new perimeter and the present perimeter is the amount of fencing needed.

$$24\sqrt{3} - 16\sqrt{3} = 8\sqrt{3} \approx 14$$

Fernando needs $8\sqrt{3}$ feet of additional fencing, or approximately 14 feet of fencing.

APPLICATION

The Hephaesteion was a temple built by the Greeks in the fifth century B.C. The ratio of its width to its height is approximated by the expression $\frac{2}{\sqrt{5} - 1}$. This number is called the golden ratio. Early Greeks believed that rectangles with this ratio were aesthetically pleasing to the eye. Rationalize the denominator of the expression.

Discussion

$$\frac{2}{\sqrt{5} - 1} = \frac{2}{\sqrt{5} - 1} \cdot \frac{\sqrt{5} + 1}{\sqrt{5} + 1}$$

Use the conjugate of the denominator as the numerator and denominator of the factor of 1.

$$= \frac{2(\sqrt{5} + 1)}{(\sqrt{5} - 1)(\sqrt{5} + 1)}$$

Multiply.

$$= \frac{2(\sqrt{5} + 1)}{4}$$

$$= \frac{\sqrt{5} + 1}{2}$$

The golden ratio is $\frac{\sqrt{5} + 1}{2}$.

Note: The project in Chapter 4 discusses the golden ratio in more detail.

✔ 13.5.4 Checkup

Sandra purchased enough baseboard to surround her sunporch floor, which was a square measuring 350 square feet. Julie purchased baseboard for her sunporch, which was also a square, but measuring 224 square feet. What is the difference in the amounts of baseboard the women purchased? Ignore cuts in the baseboard for doors.

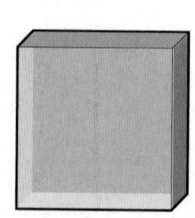

Sandra

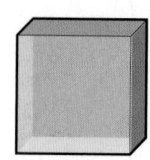

Julie

13.5 Exercises

Add or subtract as indicated.

1. $2\sqrt{28} - \sqrt{63}$

2. $5\sqrt{24} - \sqrt{54}$

3. $\sqrt{10} - \sqrt{\dfrac{1}{10}}$

4. $\sqrt{13} + \sqrt{\dfrac{1}{13}}$

5. $\sqrt[3]{24} - 4\sqrt[3]{3}$

6. $2\sqrt[3]{250} + \sqrt[3]{54}$

7. $5\sqrt{75} - 2\sqrt{27} + \sqrt{48}$

8. $4\sqrt{80} - 2\sqrt{45} + \sqrt{20}$

9. $5\sqrt{x} + 9\sqrt{x}$

10. $8\sqrt{bc} - 10\sqrt{bc}$

11. $\sqrt{25x} + \sqrt{36x}$

12. $\sqrt{64p} + \sqrt{144p}$

13. $\sqrt{8x^3} - \sqrt{50x^3}$

14. $\sqrt{48y^5} - \sqrt{75y^5}$

15. $\sqrt{9a} + \sqrt{16a^3}$

16. $\sqrt{121b^3} - \sqrt{16b}$

17. $7a\sqrt{b} + 9a\sqrt{b} - 2a\sqrt{b}$

18. $2x\sqrt{y} - 5x\sqrt{y} + 4x\sqrt{y}$

19. $5\sqrt{pq} - 4\sqrt[3]{pq} + 2\sqrt{pq} + 11\sqrt[3]{pq}$

20. $4\sqrt[3]{ab} - 3\sqrt{ab} - 5\sqrt{ab} + 2\sqrt[3]{ab}$

21. $7y\sqrt{x^3y} + 3x\sqrt{xy^3} - 4xy\sqrt{xy}$

22. $3d\sqrt{c^3d} - 8c\sqrt{cd^3} + 2cd\sqrt{cd}$

23. $2x\sqrt[3]{x^2y^4z} - 3y\sqrt[3]{x^5yz}$

24. $6x\sqrt[4]{xy^5z^2} + y\sqrt[4]{x^5yz^2}$

Multiply and simplify.

25. $\sqrt{7}(\sqrt{5} - \sqrt{7})$

26. $\sqrt{5}(\sqrt{6} + \sqrt{5})$

27. $\sqrt{3}(\sqrt{x} - \sqrt{5})$

28. $2\sqrt{5}(\sqrt{7} + \sqrt{a})$

29. $3\sqrt{a}(2\sqrt{a} - 5)$

30. $8\sqrt{c}(3\sqrt{c} + 4)$

31. $2\sqrt[3]{x}(4\sqrt[3]{x^2} - 6\sqrt[3]{x})$

32. $7\sqrt[3]{a}(5\sqrt[3]{a} - 2\sqrt[3]{a^2})$

33. $(\sqrt{3} - 5\sqrt{6})(2\sqrt{3} + \sqrt{8})$

34. $(3\sqrt{6} + 2)(\sqrt{5} - 4\sqrt{3})$

35. $(\sqrt{3} - \sqrt{x})(\sqrt{2} + \sqrt{x})$

36. $(\sqrt{2} - \sqrt{z})(\sqrt{5} + \sqrt{z})$

37. $(5 - \sqrt{6})(5 + \sqrt{6})$

38. $(\sqrt{11} + \sqrt{12})(\sqrt{11} - \sqrt{12})$

39. $(12 + \sqrt{p})(12 - \sqrt{p})$

40. $(13 - \sqrt{q})(13 + \sqrt{q})$

41. $(\sqrt{2x} + \sqrt{3y})(\sqrt{2x} - \sqrt{3y})$

42. $(\sqrt{6x} + \sqrt{3y})(\sqrt{6x} - \sqrt{3y})$

43. $(\sqrt{a} + 4)^2$

44. $(\sqrt{bc} + 5)^2$

45. $(3\sqrt{b} - 2)^2$

46. $(9\sqrt{c} - 3)^2$

47. $(\sqrt{x} - \sqrt{y})^2$

48. $(\sqrt{2x} - \sqrt{3y})^2$

Divide and simplify.

49. $\dfrac{\sqrt{21x} - \sqrt{14}}{\sqrt{7}}$

50. $\dfrac{\sqrt{10x} + \sqrt{30}}{\sqrt{5}}$

51. $\dfrac{\sqrt{a} - 12}{\sqrt{a}}$

52. $\dfrac{9 + \sqrt{b}}{\sqrt{b}}$

53. $\dfrac{\sqrt{x} + \sqrt{y} + \sqrt{z}}{\sqrt{x}}$

54. $\dfrac{\sqrt{ab} - \sqrt{ac} + \sqrt{bc}}{\sqrt{b}}$

55. $\dfrac{18}{\sqrt{6} + \sqrt{3}}$

56. $\dfrac{24}{\sqrt{10} + \sqrt{2}}$

57. $\dfrac{\sqrt{3} + \sqrt{2}}{\sqrt{3} - \sqrt{2}}$

58. $\dfrac{\sqrt{3} - \sqrt{2}}{\sqrt{3} + \sqrt{2}}$

59. $\dfrac{3x}{\sqrt{x} - 2}$

60. $\dfrac{4w}{3 - \sqrt{w}}$

61. $\dfrac{3b - 4}{\sqrt{3b} - 2}$

62. $\dfrac{ab - 1}{\sqrt{ab} + 1}$

63. $\dfrac{\sqrt{x} + 3}{\sqrt{x} - 3}$

64. $\dfrac{6 - \sqrt{p}}{6 + \sqrt{p}}$

65. $\dfrac{3\sqrt{x} - 4}{4 - 3\sqrt{x}}$

66. $\dfrac{3 - 5\sqrt{x}}{5\sqrt{x} - 3}$

67. One painting on a square canvas measures 490 square inches. A second painting on a square canvas measures 810 square inches. How many more inches of frame will the second painting require than that required for the first painting?

68. The footprint of one computer is a square measuring 150 square inches. The footprint of a second computer is also a square, but it measures 216 square inches. By how many inches does the perimeter of the footprint for the second computer exceed that of the first? (Note: The footprint of a computer is the base upon which it rests.)

69. A box in the shape of a cube has a volume of 1750 cubic inches. It contains a smaller box shaped like a cube, with a volume of 896 cubic inches. Determine the difference in the perimeters of the bases of the two boxes.

70. A pair of stacking tables in the shapes of cubes consists of one whose volume is 2000 cubic inches and another whose volume is 6750 cubic inches. Determine the difference in the perimeters of the bases of the two tables.

13.5 Calculator Exercises

Operations with radicals represent another situation in which the calculator can be helpful in checking your results if the radical expressions contain only a single variable. By storing the original expression in Y1 and your simplified result in Y2 and comparing the graphs for equivalence, you can see whether your result is correct. Likewise, you can use the TABLE function to explore the equivalence of the two expressions. These methods have already been demonstrated for you in the text.

However, if you store the original expression in Y2 and your simplified result in Y3, then you can define Y1 to be Y2 = Y3, using the TEST key to do so. Then, when you do a TABLE check of your simplification, if the Y1 column contains values of 1, your simplification is equivalent to the original expression. If the column for Y1 contains values of zero,

your simplifications is not equivalent to the original expression, and you need to check it again.

The screens to check that $\frac{x-4}{\sqrt{x}-2} = \sqrt{x}+2$ is the correct simplification are as follows:

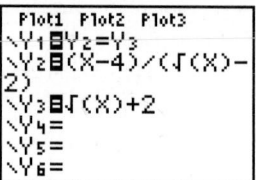

As you can see, the two expressions are equivalent, except for restricted values.

Simplify the following expressions, and use the method described to check your results for equivalence.

1. $(3\sqrt[3]{x} + 5\sqrt{x}) + (\sqrt[3]{x} - 3\sqrt{x}) + (4\sqrt{x} + 7)$
2. $(2\sqrt{x} + 7) - (\sqrt{x} + 9)$
3. $2\sqrt[3]{x}(3\sqrt[3]{2x} + 5)$
4. $(2\sqrt{x} + 3\sqrt{2x})(\sqrt{x} + \sqrt{2x})$
5. $\dfrac{x-16}{\sqrt{x}+4}$
6. $\dfrac{21}{\sqrt{x}-7}$
7. $(3\sqrt{x} + 5)(\sqrt{x} - 7) + 3\sqrt{x} - 6$

13.5 Writing Exercise

Sometimes, in upper-level mathematics courses, you may be asked to rationalize the numerator of a radical expression. Suppose you had the expression $\frac{\sqrt{x}+5}{\sqrt{x}-2}$. Explain what it would

mean to rationalize the numerator. Then list the steps you would follow to do so. Use your list to rationalize the numerator of the expression presented here.

13.6 Solving Radical Equations in One Variable

OBJECTIVES

1 Identify radical equations in one variable.

2 Solve a square-root equation algebraically.

3 Solve a radical equation algebraically.

4 Solve an equation involving rational exponents algebraically.

5 Solve real-world situations that are modeled by radical equations, including those using the distance formula.

APPLICATION

An explosion of energy E (in ft-lb) creates a crater whose diameter d (in feet) is found using the formula $d = kE^{1/3}$, where k is a constant.

a. The explosion of 1 ton of TNT has an energy of 3.1×10^9 ft-lb and creates a crater with a diameter of 30 feet. Determine a value for the constant k.

b. The Barringer meteor crater near Winslow, Arizona, is 4000 feet in diameter. Determine the energy of its explosion. (Use the value for k found in part a.)

After completing this section, we will discuss this application further. See page 938.

13.6.1 Radical Equations in One Variable

We are now ready to solve radical equations in one variable. A **radical equation in one variable**, or, simply, **radical equation**, can be written in the form of an expression that contains radicals with variables in the radicands and that is equal to 0. Since an exponential expression with a rational exponent is equivalent to a radical expression, a radical equation may also be written in the form of an exponential expression that has a variable base with a rational exponent and that is equal to 0.

> **STANDARD FORM FOR A RADICAL EQUATION IN ONE VARIABLE**
>
> A radical equation in one variable is written in the form
>
> $$D(x) = 0$$
>
> where $D(x)$ is an expression containing radicals with a variable in the radicand or an exponential expression with a variable base and a rational exponent.

EXAMPLE 1 State whether each of the following equations is a radical equation:

a. $\sqrt[5]{x + 3} = 2x$ **b.** $\sqrt{3}x + x = 0$ **c.** $x^{3/4} = 2x^2$

Solution

a. $\sqrt[5]{x + 3} = 2x$ is a radical equation, because $\sqrt[5]{x + 3} - 2x = 0$ and the radicand, $x + 3$, contains a variable.

b. $\sqrt{3}x + x = 0$ is not a radical equation, because the radicand does not contain a variable.

c. $x^{3/4} = 2x^2$ is a radical equation, because $x^{3/4} - 2x^2 = 0$ is equivalent to $\sqrt[4]{x^3} - 2x^2 = 0$ and the radicand, x^3, contains a variable.

13.6.1 Checkup

In exercises 1–3, state whether each of the equations is a radical equation.

1. $\sqrt{5} + x = 7$ **2.** $\sqrt[3]{2x} = x + 3$ **3.** $5x - 3 = x^{2/3}$

4. What is the difference between a radical equation and a radical expression?

13.6.2 Solving Square-Root Equations

In previous chapters, we solved equations numerically and graphically. The same procedures are used to solve radical equations numerically and graphically. We will use these procedures to check our solutions when we solve radical equations algebraically.

To solve algebraically an equation containing a radical expression, we must determine a process that reverses the effect of taking a root of a radicand.

If we begin with an equation that is true and we raise both expressions contained in the equation to the same power, the result is a true equation.

> **PRINCIPLE OF POWERS**
> If $a = b$, then $a^n = b^n$.

We use this principle to solve algebraically equations in one variable that contain radicals and rational exponents.

SOLVING A SQUARE-ROOT EQUATION ALGEBRAICALLY
To solve an equation algebraically by using the principle of powers,

- Determine the restricted values of the independent variable for each radical expression.
- Isolate a square-root term on one side of the equation.
- Square both sides of the equation.
- Solve the resulting equation. (If the resulting equation contains a square root, this process must be repeated.)
- Discard any solutions that are restricted values.

Check the solutions numerically, graphically, or by substitution.

EXAMPLE 2   Solve algebraically and check by substitution.

a. $\sqrt{3x + 1} = 4$ **b.** $\sqrt{2x} = -6$ **c.** $\sqrt{x + 6} = x$

Algebraic Solution

a. $\sqrt{3x + 1} = 4$

Restricted values are all real numbers less than $-\frac{1}{3}$.

$(\sqrt{3x + 1})^2 = 4^2$ Square both sides.

$3x + 1 = 16$

$3x + 1 - 1 = 16 - 1$

$3x = 15$

$x = 5$

Substitution Check

$$\sqrt{3x + 1} = 4$$

$$\begin{array}{c|c} \sqrt{3(5) + 1} & 4 \\ \sqrt{16} & \\ 4 & \end{array}$$

The solution, $x = 5$, checks.

b. $\sqrt{2x} = -6$

$\sqrt{2x}$ is a principal root and represents a nonnegative number. Therefore, the expression $\sqrt{2x}$ cannot equal the negative number -6. We say that the equation has no real-number solution.

c. $\sqrt{x + 6} = x$

Restricted values are all real numbers less than -6.

$(\sqrt{x + 6})^2 = x^2$ Square both sides.

$x + 6 = x^2$

$x^2 - x - 6 = 0$ Standard form for a polynomial equation.

$(x - 3)(x + 2) = 0$ Factor.

$x = 3$ or $x = -2$

Substitution Check

$$\sqrt{x + 6} = x$$

$$\begin{array}{c|c} \sqrt{(3) + 6} & 3 \\ 3 & \end{array}$$

The solution is 3.

$$\sqrt{x + 6} = x$$

$$\begin{array}{c|c} \sqrt{(-2) + 6} & -2 \\ 2 & \end{array}$$

Since $2 \neq -2$, -2 is not a solution.

Check the solutions by substitution.

In Example 1c, applying the principle of powers resulted in a true equation, but one that is not equivalent to the original equation. A simplified example to illustrate this result follows.

$$x = 1 \qquad \text{(This equation has one solution, 1.)}$$

Applying the principle of powers, we can square both sides of the equation.

$$x^2 = 1^2 \text{ or } x^2 = 1 \text{ (This equation has two solutions, 1 and } -1.)$$

Not all solutions of the squared equation are solutions of the original equation. However, all solutions of the original equation *are* solutions of the squared equation. This means that we have an *extraneous solution* of -1 after squaring the original equation.

 **HELPING HAND** We must always check the final solutions to determine whether they are solutions of the original equation.

EXAMPLE 3   Solve algebraically and check numerically, graphically, or by substitution.

   a. $\sqrt{3x + 10} - x = 4$ **b.** $2\sqrt{5x} + 1 = 11$
   c. $\sqrt{5a + 4} = \sqrt{4a - 3}$ **d.** $\sqrt{x} + 2 = \sqrt{x + 5}$

Algebraic Solution

a. $\sqrt{3x + 10} - x = 4$

               Restricted values of the radical expression are all real numbers less than $\frac{-10}{3}$.

$$\sqrt{3x + 10} = x + 4 \qquad \text{\textit{Add x to both sides.}}$$
$$(\sqrt{3x + 10})^2 = (x + 4)^2 \qquad \text{\textit{Square both sides.}}$$
$$3x + 10 = x^2 + 8x + 16$$
$$x^2 + 5x + 6 = 0 \qquad \text{\textit{Set the quadratic expression equal to 0.}}$$

$$(x + 3)(x + 2) = 0 \qquad \text{\textit{Factor.}}$$
$$x = -3 \quad \text{or} \quad x = -2 \qquad \text{\textit{Set each factor equal to 0 and solve.}}$$

Numeric Check

Since both solutions are integers, check numerically on your calculator.

X	Y1	Y2
-4	ERROR	4
-3	4	4
-2	4	4
-1	3.6458	4
0	3.1623	4
1	2.6056	4
2	2	4

X= -3

Y1 = $\sqrt{3x + 10} - x$
Y2 = 4

The solutions are -3 and -2.

b. $2\sqrt{5x} + 1 = 11$ *Restricted values of the radical expression are all real numbers less than 0.*

$$2\sqrt{5x} = 10 \qquad \text{\textit{Subtract 1 from both sides.}}$$
$$\frac{2\sqrt{5x}}{2} = \frac{10}{2} \qquad \text{\textit{Divide both sides by 2.}}$$
$$\sqrt{5x} = 5$$
$$(\sqrt{5x})^2 = 5^2 \qquad \text{\textit{Square both sides.}}$$
$$5x = 25$$
$$x = 5$$

Substitution Check

$$2\sqrt{5x} + 1 = 11$$

$$\begin{array}{c|c} 2\sqrt{5(5)} + 1 & 11 \\ 2\sqrt{25} + 1 & \\ 2 \cdot 5 + 1 & \\ & 11 \end{array}$$

The solution is 5.

c. $\sqrt{5a + 4} = \sqrt{4a - 3}$

Restricted values of the radical expressions are all real numbers less than $\frac{3}{4}$.

$(\sqrt{5a + 4})^2 = (\sqrt{4a - 3})^2$ *Square both sides.*

$5a + 4 = 4a - 3$

$a = -7$

The possible solution is a restricted value. The original equation has no real-number solution.

Graphic Check

$Y1 = \sqrt{5x + 4}$

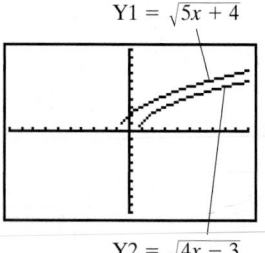

$Y2 = \sqrt{4x - 3}$

$(-10, 10, -10, 10)$

The graphs do not intersect. There are no real number solutions.

d. $\sqrt{x} + 2 = \sqrt{x + 5}$

Restricted values of the radical expressions are all real numbers less than 0.

$(\sqrt{x} + 2)^2 = (\sqrt{x + 5})^2$ *Square both sides.*

$x + 4\sqrt{x} + 4 = x + 5$

$4\sqrt{x} = 1$ *Isolate the radical term.*

$(4\sqrt{x})^2 = (1)^2$ *Square both sides.*

$16x = 1$

$x = \dfrac{1}{16}$

Substitution Check

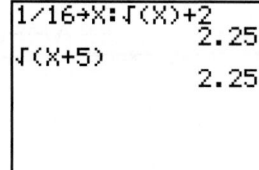

The solution is $\frac{1}{16}$.

Radical equations may have no solution or have infinitely many solutions.

EXAMPLE 4   Solve algebraically and check graphically.

  a. $\sqrt{2x} + 3 = \sqrt{2x} - 5$ **b.** $2\sqrt{2a + 1} = \sqrt{8a + 4}$

Algebraic Solution

a. $\sqrt{2x} + 3 = \sqrt{2x} - 5$

Restricted values of the radical expressions are all real numbers less than 0.

$3 = -5$ *Isolate the radical expression by subtracting $\sqrt{2x}$.*

This is a contradiction. There is no solution.

Graphic Check

$Y1 = \sqrt{2x} + 3$

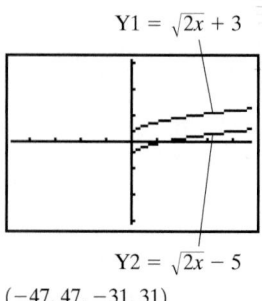

$Y2 = \sqrt{2x} - 5$

$(-47, 47, -31, 31)$

The graphs do not appear to intersect. There is no solution.

b. $2\sqrt{2a + 1} = \sqrt{8a + 4}$

Restricted values of the radical expressions are all real numbers less than $-\frac{1}{2}$.

$$(2\sqrt{2a + 1})^2 = (\sqrt{8a + 4})^2$$ *Square both sides.*
$$4(2a + 1) = 8a + 4$$
$$8a + 4 = 8a + 4$$
$$4 = 4$$ *Subtract 8a from both sides.*

This is an identity. The solution set is all real numbers greater than or equal to $-\frac{1}{2}$. (Real numbers less than $-\frac{1}{2}$ are restricted values.)

Graphic Check

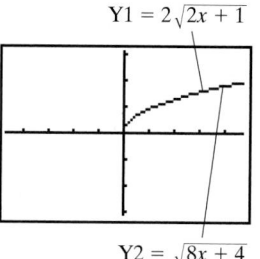

$Y1 = 2\sqrt{2x + 1}$

$Y2 = \sqrt{8x + 4}$

$(-47, 47, -31, 31)$

The graphs are coinciding. The equation has infinitely many solutions. The solution set is all real numbers greater than or equal to $-\frac{1}{2}$.

   13.6.2 Checkup

Solve exercises 1–6 algebraically and check.

1. $\sqrt{x + 12} = x$

2. $x - \sqrt{5x - 16} = 2$

3. $3\sqrt{2x} + 2 = 8$

4. $\sqrt{2z + 1} = \sqrt{z - 4}$

5. $\sqrt{x + 9} = \sqrt{x} + 1$

6. $\sqrt{x + 1} + 2 = \sqrt{x + 1} - 3$

7. The principle of powers states that if an equation is true and if you raise the expressions on the left and right sides of the equation to the same power, you obtain another equation that is true. Does this mean that the new equation has the same solutions as the original equation? Explain your answer.

13.6.3 Solving Other Radical Equations

To solve other radical equations, we use the same procedure as with square-root equations, except we raise each expression to the power of the index.

> **SOLVING A RADICAL EQUATION ALGEBRAICALLY**
> To solve an equation algebraically by using the principle of powers,
>
> - Determine the restricted values for the independent variable for each radical expression.
> - Isolate a radical term on one side of the equation.
> - Raise both sides of the equation to a power that is the same as the index of the isolated radical term—that is, to the second power for square roots, to the third power for cube roots, and so on.
> - Solve the resulting equation. (If the resulting equation contains a radical, this process must be repeated.)
>
> Check the solutions numerically, graphically, or by substitution.

EXAMPLE 5   Solve algebraically and check numerically, graphically, or by substitution.

a. $\sqrt[4]{x - 5} = 2$ **b.** $\sqrt[3]{x^2 + 12x} + 5 = 9$

4

Algebraic Solution

a. $\sqrt[4]{x-5} = 2$ *Restricted values for the radical expression are all real numbers less than 5.*

$(\sqrt[4]{x-5})^4 = (2)^4$ *Raise to the fourth power on both sides.*

$x - 5 = 16$

$x = 21$

Substitution Check

$$\sqrt[4]{x-5} = 2$$

$$\begin{array}{c|c} \sqrt[4]{(21)-5} & 2 \\ \sqrt[4]{16} & \\ 2 & \end{array}$$

The solution is 21.

b. $\sqrt[3]{x^2 + 12x} + 5 = 9$ *There are no restricted values. The index, 3, is odd.*

$\sqrt[3]{x^2 + 12x} = 4$ *Subtract 5 from both sides.*

$(\sqrt[3]{x^2 + 12x})^3 = (4)^3$ *Cube both sides.*

$x^2 + 12x = 64$ *Simplify.*

$x^2 + 12x - 64 = 0$ *Standard form of a quadratic equation*

$(x + 16)(x - 4) = 0$ *Factor.*

$x = -16$ or $x = 4$ *Solve.*

Graphic Check

$Y1 = \sqrt[3]{x^2 + 12x} + 5$

$(-16, 9)$ $(4, 9)$

$Y2 = 9$

$(-47, 47, -31, 31)$

The x-coordinates of the intersections are -16 and 4.
The solutions are -16 and 4.

13.6.3 Checkup

Solve exercises 1–3 algebraically and check.

1. $\sqrt[3]{4x} = 2$ **2.** $\sqrt[4]{x-3} = 2$ **3.** $\sqrt[3]{x^2 + 6x} + 2 = 5$

4. When using the principle of powers to solve a radical equation, how do you decide what power to apply to the expressions on the left and right sides of the original radical equation?

13.6.4 Solving an Equation with Rational Exponents

To solve an equation involving an expression with a rational exponent, we use the principle of powers. To obtain an integer exponent, we raise both expressions in the equation to the denominator of the fractional exponent. That is, if we have $x^{2/3}$, we raise it to a power of 3, $(x^{2/3})^3$. By the power-to-a-power property of exponents, we can simplify this expression to $(x^{2/3})^3 = x^{(2/3)(3)} = x^2$.

SOLVING AN EQUATION WITH A RATIONAL EXPONENT ALGEBRAICALLY

To solve an equation algebraically by using the principle of powers,

- Determine the restricted values of the independent variable.
- Isolate a term with an exponential expression on one side of the equation.
- Raise both expressions in the equation to the power of the denominator of the exponent.
- Solve the resulting equation. (If the resulting equation contains a rational exponential expression, this process must be repeated.)

Check the solutions numerically, graphically, or by substitution.

For example, solve $x^{2/3} = 4$.

$$x^{2/3} = 4 \qquad \text{There are no restricted values.}$$
$$(x^{2/3})^3 = 4^3 \qquad \text{Raise both sides to a power of 3.}$$
$$x^2 = 64$$
$$x = \pm 8 \qquad \text{Principle of square roots}$$

Check by substitution.

$$\begin{array}{c|c} x^{2/3} = 4 \\ \hline (8)^{2/3} & 4 \\ 4 & \end{array}$$

$$\begin{array}{c|c} x^{2/3} = 4 \\ \hline (-8)^{2/3} & 4 \\ 4 & \end{array}$$

The solution is 8.   The solution is -8.

In the example, we used the principle of square roots to solve the equation $x^2 = 64$ and obtain $x = \pm 8$. Before we continue, we need to expand the principle so that it is a rule for all roots. We need to consider two different cases: when the exponent is even and when the exponent is odd.

> **PRINCIPLE OF ROOTS**
> *Case 1.* If n is a positive even integer, b is a positive number, and $a^n = b$, then $a = \sqrt[n]{b}$ or $a = -\sqrt[n]{b}$.
> *Case 2.* If n is a positive even integer, b is a negative number, and $a^n = b$, then a is not a real number.
> *Case 3.* If n is a positive even integer, b is 0, and $a^n = b$, then $a = 0$.
> *Case 4.* If n is a positive odd integer, b is a real number, and $a^n = b$, then $a = \sqrt[n]{b}$.

The principle of roots will allow us to solve equations such as the following:

Case 1. $x^4 = 16$ Case 2. $x^4 = -16$ Case 3. $x^4 = 0$
$\qquad x = \pm\sqrt[4]{16}$ x is not a real number. $x = 0$
$\qquad x = \pm 2$

Case 4. $x^3 = 8$ or $x^3 = -8$
$\qquad x = \sqrt[3]{8}$ $x = \sqrt[3]{-8}$
$\qquad x = 2$ $x = -2$

EXAMPLE 6   Solve algebraically and check numerically, graphically, or by substitution.

     a. $(x + 2)^{3/4} = 8$ **b.** $(x + 2)^{4/3} = 3$ **c.** $(x + 2)^{2/3} = -3$

Algebraic Solution

a. $(x + 2)^{3/4} = 8$ Restricted values are all real numbers less than -2.

$[(x + 2)^{3/4}]^4 = (8)^4$ Raise both sides to a power of 4.

$\qquad (x + 2)^3 = 4096$

$\qquad\quad x + 2 = \sqrt[3]{4096}$ Principle of roots

$\qquad\quad x + 2 = 16$

$\qquad\qquad\quad x = 14$

Graphic Check

$Y1 = (x + 2)^{3/4}$

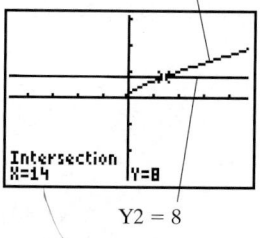

$Y2 = 8$

$(-47, 47, -31, 31)$

The x-coordinate of the intersection is 14.

       The solution is 14.

b. $(x + 2)^{4/3} = 3$ There are no restricted values.   **Substitution Check**

$[(x + 2)^{4/3}]^3 = 3^3$ Raise both sides to a power of 3.

$(x + 2)^4 = 27$

$x + 2 = \pm\sqrt[4]{27}$ Principle of roots

$x = -2 \pm \sqrt[4]{27}$ Subtract 2 from both sides.

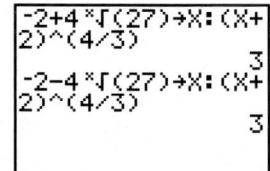

The substitutions of the possible solutions result in 3, the right side of the equation.
The solutions are $-2 \pm \sqrt[4]{27}$.

c. $(x + 2)^{2/3} = -3$ There are no restricted values.   **Graphic Check**

$[(x + 2)^{2/3}]^3 = (-3)^3$ Raise both sides to a power of 3.

$(x + 2)^2 = -27$

There is no real-number solution of the equation, because, according to the principle of roots, the square of $(x + 2)$ cannot be a negative number.

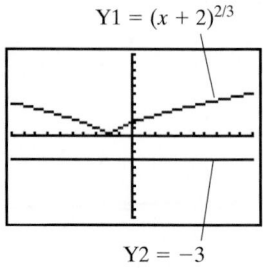

Y1 = $(x + 2)^{2/3}$

Y2 = -3

$(-10, 10, -10, 10)$

The graphs do not intersect. There is no real-number solution.

 13.6.4 Checkup

Solve exercises 1–3 algebraically and check.

1. $x^{3/5} = 8$ **2.** $(x + 5)^{3/2} = 27$ **3.** $(3 - x)^{2/3} + 3 = 2$

4. To solve an equation with rational exponents, you may need to use the principle of powers and then the principle of roots. Explain the difference between these two principles.

13.6.5 Modeling the Real World

Radical equations are used to describe many situations involving the acceleration due to gravity. In this section, we will discuss one such situation. Dancers and basketball players have great jumping abilities and seem to "hang in the air." The relationship between their hang time t and the vertical height of the jump, d, is given by the equation $t = 2\sqrt{\frac{2d}{g}}$, where g is the acceleration due to gravity.

EXAMPLE 7   **a.** Given that the acceleration due to gravity is 32 feet per second squared, write an equation for the hang time t.

b. According to the *Guinness Book of Records*, Rune Almen of Sweden made the highest high jump on record from a standing position. His hang time was recorded as 1.248 seconds. Determine the vertical height of his jump.

Solution

a. Use the given equation for hang time, substitute 32 for g, and then simplify.

$$t = 2\sqrt{\frac{2d}{g}}$$

$$t = 2\sqrt{\frac{2d}{32}}$$

$$t = 2\sqrt{\frac{d}{16}} \qquad \text{Simplify.}$$

$$t = 2\left(\frac{\sqrt{d}}{\sqrt{16}}\right) \qquad \text{Quotient rule}$$

$$t = 2\left(\frac{\sqrt{d}}{4}\right) \qquad \text{Simplify.}$$

$$t = \frac{\sqrt{d}}{2}$$

An equation for the hang time t (in seconds) is $t = \frac{\sqrt{d}}{2}$, where d is the vertical height in feet.

b. Let x = the vertical height in feet

$$1.248 = \frac{\sqrt{x}}{2} \qquad t = 1.248$$

$$2(1.248) = 2\left(\frac{\sqrt{x}}{2}\right) \qquad \text{Multiply both sides by 2.}$$

$$2.496 = \sqrt{x}$$

$$(2.496)^2 = (\sqrt{x})^2 \qquad \text{Square both sides.}$$

$$6.230016 = x$$

Rune's record vertical jump was about 6.23 feet.

EXAMPLE 8 Determine the possible y-coordinates of a point 5 units from $(-1, 4)$ with an x-coordinate of 2.

Solution

Recall that the distance between two points (x_1, y_1) and (x_2, y_2) can be calculated as $d = \sqrt{(x_2 - x_1)^2 + (y_2 - y_1)^2}$.
Let $x_1 = 2$ and y = the missing y-coordinate
Let $x_2 = -1$ and $y_2 = 4$

$$d = \sqrt{(x_2 - x_1)^2 + (y_2 - y_1)^2}$$

$$5 = \sqrt{(-1 - 2)^2 + (4 - y)^2} \qquad \text{Substitute 5 for } d, (2, y) \text{ for } (x_1, y_1), \text{ and}$$
$$\qquad\qquad (-1, 4) \text{ for } (x_2, y_2).$$

$$5 = \sqrt{(-3)^2 + (4 - y)^2}$$

$$5 = \sqrt{9 + 16 - 8y + y^2}$$

$$5 = \sqrt{25 - 8y + y^2}$$

$$5^2 = (\sqrt{25 - 8y + y^2})^2 \qquad \text{Square both sides.}$$

$$25 = 25 - 8y + y^2$$

$$y^2 - 8y = 0 \qquad \text{Standard form for a quadratic equation}$$
$$y(y - 8) = 0 \qquad \text{Factor.}$$
$$y = 0 \quad \text{or} \quad y = 8 \qquad \text{Solve.}$$

The two possible points are $(2, 0)$ and $(2, 8)$.

APPLICATION

An explosion of energy E (in ft-lb) creates a crater whose diameter d (in feet) is found using the formula $d = kE^{1/3}$, where k is a constant.

a. The explosion of 1 ton of TNT has an energy of 3.1×10^9 ft-lb and creates a crater with a diameter of 30 feet. Determine a value for the constant k.

b. The Barringer meteor crater near Winslow, Arizona, is 4000 feet in diameter. Determine the energy of its explosion. (Use the value for k found in part a.)

Discussion

a. Use the given formula.

$$d = kE^{1/3}$$
$$30 = k(3.1 \times 10^9)^{1/3} \qquad \text{Substitute 30 for } d \text{ and } 3.1 \times 10^9 \text{ for } E.$$
$$k = \frac{30}{(3.1 \times 10^9)^{1/3}} \qquad \text{Solve for } k.$$
$$k \approx 0.02$$

b.
$$d = kE^{1/3}$$
$$4000 = 0.02E^{1/3} \qquad \text{Substitute 4000 for } d \text{ and } 0.02 \text{ for } k.$$
$$\frac{4000}{0.02} = E^{1/3} \qquad \text{Divide both sides by 0.02.}$$
$$200,000,000 = E^{1/3}$$
$$(200,000)^3 = (E^{1/3})^3 \qquad \text{Cube both sides.}$$
$$8,000,000,000,000,000 = E$$
$$\text{or } E = 8 \times 10^{15}$$

The energy of the meteor explosion that caused the crater near Winslow was approximately 8×10^{15} ft-lb.

✓ 13.6.5 Checkup

1. According to the *Guinness Book of Records*, Gete Bjordalshakka (Norway) made the highest high jump on record for women from a standing position. Her hang time was recorded as 1.116 seconds. Determine the vertical height of her jump.

2. Find the possible y-coordinates of a point that is 10 units away from the point $(6, 5)$ and has an x-coordinate of -2.

3. Wolf Creek Crater in Australia is about 2950 feet in diameter. Determine the energy of its explosion. (Use the formula $d = 0.02E^{1/3}$, where d is the diameter in feet and E is the energy in ft-lb.)

13.6 Exercises

Solve exercises 1–90 algebraically and check.

1. $\sqrt{x} - 3 = 5$
2. $\sqrt{x} - 5 = 2$
3. $\sqrt{x} + 1.7 = 4.5$
4. $\sqrt{x} - 2.4 = 3.8$
5. $\sqrt{x} + 8 = 5$
6. $\sqrt{x} + 11 = 10$

7. $\sqrt{3x} - 4 = 2$

8. $\sqrt{7x} - 2 = 12$

9. $9 - 3\sqrt{2z} = 0$

10. $15 - 5\sqrt{6z} = 0$

11. $3\sqrt{6x} + 1 = 10$

12. $4\sqrt{3x} + 3 = 15$

13. $\sqrt{x + 5} = 3$

14. $\sqrt{3 + x} = 5$

15. $\sqrt{2x + 5} + 4 = 9$

16. $\sqrt{3x + 1} + 2 = 5$

17. $\sqrt{3x + 4} - 2 = 6$

18. $\sqrt{5x + 4} - 3 = 5$

19. $\sqrt{5 - 4x} = x$

20. $\sqrt{x + 30} = x$

21. $\sqrt{6x - 5} = \sqrt{4x + 5}$

22. $\sqrt{3x + 7} = \sqrt{5x - 2}$

23. $3\sqrt{x + 2} = \sqrt{x + 10}$

24. $4\sqrt{x + 3} = \sqrt{x + 18}$

25. $x = 3 + 2\sqrt{x - 4}$

26. $x = 3 + 2\sqrt{2x - 10}$

27. $x = \sqrt{21 - x} - x^2 + 3$

28. $\sqrt{(4x + 5)(x + 1)} = 5 + x$

29. $\sqrt{x^2 - 14x + 49} = 7 - x$

30. $\sqrt{4x^2 - 12x + 9} = 3 - 2x$

31. $x = \sqrt{2x + 7}$

32. $\sqrt{8 - 3x} + x = 0$

33. $5 + \sqrt{25 - 2x} = x$

34. $7 + \sqrt{49 - 2x} = x$

35. $\sqrt{x} - 3 = \sqrt{x - 27}$

36. $\sqrt{x} - 8 = \sqrt{x - 64}$

37. $\sqrt{x - 7} = 7 - \sqrt{x}$

38. $\sqrt{x + 15} = 3 + \sqrt{x}$

39. $\sqrt{2x + 5} - 7 = \sqrt{5 + 2x} + 3$

40. $\sqrt{4x + 3} - 5 = \sqrt{3 + 4x} + 9$

41. $\sqrt{2x + 3} = 1 - \sqrt{x + 5}$

42. $\sqrt{3x - 2} = 1 + \sqrt{x + 3}$

43. $\sqrt[3]{3x} = -6$

44. $\sqrt[3]{6x} = -3$

45. $\sqrt[4]{2x} = 10$

46. $\sqrt[4]{2x - 10} = 4$

47. $\sqrt[5]{2x} = -2$

48. $\sqrt[5]{9x} = -3$

49. $\sqrt[3]{x + 3} = 1$

50. $\sqrt[3]{x - 7} = -5$

51. $\sqrt[4]{x + 1} = 5$

52. $\sqrt[4]{x + 3} = 6$

53. $\sqrt[3]{2x + 1} = 3$

54. $\sqrt[3]{3x + 2} = -4$

55. $\sqrt[4]{x - 5} = 2$

56. $\sqrt[4]{x - 7} = 3$

57. $\sqrt[5]{2x - 5} = 3$

58. $\sqrt[5]{3x - 1} = 2$

59. $\sqrt[3]{x^2 + 7x + 7} = 9$

60. $\sqrt[4]{x^2 + 24x + 2} = 5$

61. $\sqrt[4]{3x - 5} = \sqrt[4]{2x + 4}$

62. $\sqrt[4]{5x - 7} = \sqrt[4]{x + 5}$

63. $\sqrt[4]{5x - 2} = 2\sqrt[4]{3}$

64. $\sqrt[3]{3x + 6} = 3\sqrt[3]{2}$

65. $x = \sqrt[4]{18x^2 - 81}$

66. $x = \sqrt[4]{8x^2 - 16}$

67. $\sqrt[4]{2x} + \sqrt[4]{3x} = 0$

68. $\sqrt[3]{5x} + \sqrt[3]{x} = 0$

69. $\sqrt[4]{3x - 5} + \sqrt[4]{x - 3} = 0$

70. $\sqrt[4]{x + 3} + \sqrt[4]{2x + 1} = 0$

71. $\sqrt[3]{3x - 5} + \sqrt[3]{x - 3} = 0$

72. $\sqrt[5]{5x - 2} + \sqrt[3]{3x - 2} = 0$

73. $\sqrt[4]{7x - 1} - \sqrt[4]{x + 11} = 0$

74. $\sqrt[4]{8x - 1} - \sqrt[4]{x + 6} = 0$

75. $x^{4/3} = 16$

76. $x^{2/3} = 25$

77. $(x + 6)^{2/5} = 4$

78. $(x + 8)^{2/3} = 16$

79. $(x - 4)^{3/4} = 27$

80. $(x - 9)^{3/2} = 64$

81. $x^{-2/3} = 4$

82. $x^{-3/4} = 27$

83. $(x - 5)^{-3/4} = 27$

84. $(4x + 5)^{3/4} = 125$

85. $(5x - 3)^{-2/3} = \dfrac{1}{25}$

86. $(6x - 1)^{-3/4} = \dfrac{1}{64}$

87. $x^{2/3} + 5 = 3$

88. $x^{4/5} + 6 = 4$

89. $x^{3/4} - 7 = 1$

90. $x^{3/5} - 6 = 21$

91. Michael "Wild Thing" Wilson of the Harlem Globetrotters slam-dunked a regulation basketball 12 feet at the Conseco Fieldhouse in Indianapolis, Indiana, in April of 2000. His hang time for the shot was about 1.16 seconds. Determine the vertical distance of his jump. Given that he is 6 feet, 7 inches, tall, does this hang time added to his height equal a 12-foot high basket?

92. Given the hang-time formula $t = 2\sqrt{\frac{2d}{g}}$ and the acceleration due to gravity of 9.8 meters per second squared, write an equation for t. Use the equation to determine the distance in meters of Michael "Wild Thing" Wilson's vertical jump if his hang time was about 1.16 seconds.

93. Find the possible y-coordinates of a point that has an x-coordinate of 8 and is 13 units away from $(3, 4)$.

94. Find the possible y-coordinates of a point that has an x-coordinate of -13 and is 17 units away from $(2, -3)$.

95. Find the possible x-coordinates of a point that has a y-coordinate of -1 and is 10 units away from the point $(-2, 5)$.

96. Find the possible x-coordinates of a point that has a y-coordinate of 1 and is 5 units away from $(1, -3)$.

97. Find the possible y-coordinates of a point that has an x-coordinate of 6 and is 5 units away from $(-2, 1)$.

98. Find the possible x-coordinates of a point that has a y-coordinate of 1 and is 2 units away from $(1, 9)$.

99. One of the largest known craters believed to have been produced by a meteorite was discovered in northwestern Quebec, Canada. The circular pit is approximately 2.5 miles in diameter, or 13,200 feet. Use the formula presented in this section to determine the energy of the explosion when the meteor struck.

100. Use the formula in this section to determine the energy of the explosion of a meteor that creates a crater having a diameter of 5000 feet.

101. Civil engineers relate the crushing load L (in tons) for a square wooden pillar that has a thickness T (in inches) and a height H (in feet) by using the mathematical model $T = \left(\dfrac{LH^2}{25}\right)^{1/4}$. If a square wooden pillar has a thickness of 4 inches and a height of 8 feet, what is its crushing load?

102. Use the formula in exercise 101 to find the crushing load of a wooden pillar that is 6 inches thick and 10 feet tall.

13.6 Calculator Exercises

Some radical equations may be difficult or impossible to solve algebraically. When this is the case, you can use your calculator to solve the equations graphically. Even so, you will need to be creative in how you do that. First of all, you may have to explore different window settings to get a good picture of what is happening with the graphs. You may have to increase the range of values on the x-axis while reducing the range of values on the y-axis in order to see the graphs well. You can easily do this by adjusting the WINDOW settings. Also, you may have to try different strategies for graphing. The following are two strategies to try:

Strategy 1. Store the left side of the equation in Y1 and the right side of the equation in Y2, and graph the equations

to see where the graphs intersect. The x-values of the intersection points will be the solutions of the equations.

Strategy 2. Store the left side of the equation in Y2 and the right side of the equation in Y3, and turn off these two functions by moving the cursor to the equals sign under Y= and pressing ENTER. Then, define Y1 to equal Y2 − Y3, using the VARS to key this definition into Y1. Next, graph Y1 and use the zero function under the CALC key to find the x-intercepts of Y1. The x-values of these intercepts represent solutions of the radical equation.

Use one of the preceding strategies on your calculator to solve the following equations:

1. $(\sqrt[4]{2x+1})(\sqrt[3]{x-13}) = 9$

2. $\sqrt[5]{3x+2} - \sqrt[4]{8x+1} = -1$

3. $\dfrac{\sqrt{3x+4}}{\sqrt[4]{x-4}} = \sqrt[3]{4x-16}$

4. $\sqrt[3]{2x+1} = \sqrt[4]{3x^2}$

5. $x^{3/2} + 2x^{1/2} - 7 = 0$

6. $3x^{2/3} - 5x^{1/3} - 9 = 0$

13.6 Writing Exercise

You have learned methods for solving many different types of equations, including the following:

1. linear equations, by isolating the variable;

2. polynomial equations, by factoring and using the zero factor rule;

3. rational equations, by multiplying through by the least common denominator; and

4. radical and exponential equations, by applying the power principle and the root principle.

In some cases, you may have a combination of these types of equations, which you can solve by creatively applying the techniques you have learned. Discuss how you would approach the following exercises, and then try to solve the equations using the foregoing methods.

1. $\sqrt{m + \dfrac{1}{m}} = \dfrac{5\sqrt{2}}{7}$

2. $\dfrac{\sqrt{5k+1}}{\sqrt{5k-1}} = \dfrac{\sqrt{51}}{7}$

3. $\dfrac{\sqrt{3}}{\sqrt{2z-3}} = \sqrt{\dfrac{3}{z}}$

4. $\sqrt{2 + \sqrt{p}} = 4$

5. $\dfrac{\sqrt{a-7}}{2} = \dfrac{4}{\sqrt{a+5}}$

6. $\sqrt{\dfrac{1}{t} + \dfrac{1}{t+2}} = \dfrac{\sqrt{15}}{6}$

13.7 Equations with Imaginary-Number Solutions

OBJECTIVES

1 Write imaginary numbers.

2 Perform operations on imaginary numbers.

3 Perform operations on complex numbers.

4 Solve equations having an imaginary solution.

5 Model real-world situations by using complex numbers.

APPLICATION

In studying the motion of biological cells and viruses, we must solve the equation $3.0 = 2.5p - 12.6p^2$. Determine the solution set of the equation.

After completing this section, we will discuss this application further. See page 953.

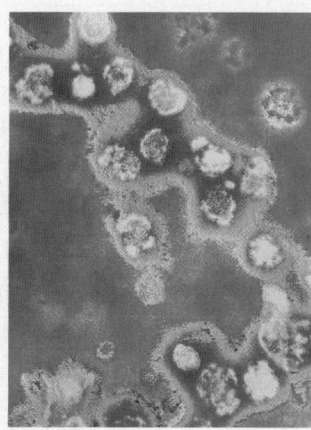

In this section, we complete our discussion of solving equations that have non-real-number solutions.

The key to working with square roots of negative numbers is to isolate the factor -1 in the radicand. Then we can proceed to work with the rest of the number, carrying that factor along as if it were a variable. We illustrate this idea by first introducing and operating with imaginary numbers. Then we combine imaginary numbers with real numbers to form complex numbers and work with them.

13.7.1 Writing Imaginary Numbers

First, we must define a number to be the principal square root of -1. This number is the **imaginary unit i**.

DEFINITION OF i
The imaginary unit $i = \sqrt{-1}$.
Therefore, $i^2 = -1$.

Using this definition and the product rule for square roots, we can define the square root of any negative real number in terms of i. For example, write $\sqrt{-5}$ in terms of i.

$$\sqrt{-5} = \sqrt{-1 \cdot 5} \qquad \text{Isolate a factor of } -1 \text{ in the radicand.}$$
$$= \sqrt{-1} \cdot \sqrt{5} \qquad \text{Product rule for square roots}$$
$$= i\sqrt{5} \qquad \text{Substitute } i \text{ for } \sqrt{-1}.$$

This process leads us to the definition of the square root of a negative real number in terms of i.

SQUARE ROOT OF A NEGATIVE NUMBER
$\sqrt{-b} = i\sqrt{b}$, where b is any positive real number.

HELPING HAND Be careful when writing the expression $\sqrt{b}i$ with the square root first. It can be hard to see that i is not under the square-root symbol. Therefore, we recommend that you write the expression with the i first, as $i\sqrt{b}$.

The calculator can be set to write the square root of a negative number in terms of i. For example, $\sqrt{-5}$ is written in terms of i as shown in Figures 13.8a and 13.8b.

TECHNOLOGY

IMAGINARY NUMBERS

Write $\sqrt{-5}$ in terms of i.

For Figure 13.8a,

Change the calculator to $a + bi$ mode by highlighting "$a + bi$" under the MODE menu.

| MODE | ▼ | ▼ | ▼ | ▼ | ▼ | ▼ | ▶ | ENTER |

Return to the home screen.

| 2nd | QUIT |

Enter the square-root expression to be written in terms of i.

| 2nd | √ | (−) | 5 |) | ENTER |

This is an approximate decimal value for $i\sqrt{5}$. Enter this value to check.

| 2nd | i | 2nd | √ | 5 |) | ENTER |

```
√(-5)
          2.236067977i
i√(5)
          2.236067977i
```
Figure 13.8a

For Figure 13.8b,

For ease in reading approximate values, set the calculator to the number of decimal places you want to round to. For example, round to two decimal places the values found in Figure 13.8a.

Set the calculator to round to two decimal places.

| MODE | ▼ | ▶ | ▶ | ▶ | ENTER | 2nd | QUIT |

Reenter the square-root expression.

| 2nd | √ | (−) | 5 |) | ENTER |

Reenter $i\sqrt{5}$ to check.

| 2nd | i | 2nd | √ | 5 |) | ENTER |

```
√(-5)
                 2.24i
i√(5)
                 2.24i
```
Figure 13.8b

Note: You could recall the previous entries by using | 2nd | ENTRY |

EXAMPLE I   Write in terms of i.

a. $\sqrt{-36}$ **b.** $\sqrt{-\dfrac{9}{16}}$ **c.** $\sqrt{-8}$

Solution

a. $\sqrt{-36} = i\sqrt{36}$ *Write the square root of a negative number in terms of i.*

$\quad\quad = 6i$ $\sqrt{36} = 6$

b. $\sqrt{-\dfrac{9}{16}} = i\sqrt{\dfrac{9}{16}}$ *Write the square root of a negative number in terms of i.*

$= \dfrac{3}{4}i$ $\sqrt{\frac{9}{16}} = \frac{3}{4}$

c. $\sqrt{-8} = i\sqrt{8}$ *Write the square root of a negative number in terms of i.*

$= 2i\sqrt{2}$ $\sqrt{8} = 2\sqrt{2}$

Check these solutions with your calculator. Round approximate values to two decimal places, as shown in Figures 13.9a and 13.9b.

a. √(-36)
                    6.00i
b. √(-9/16)▶Frac
                    3/4i

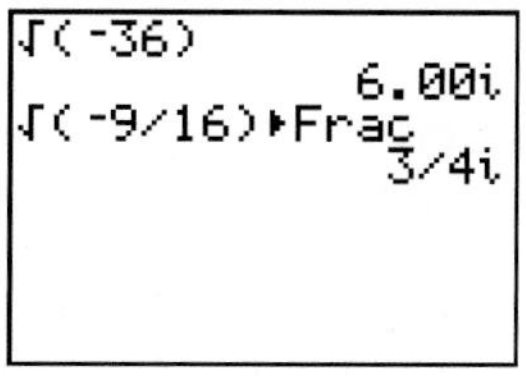

Figure 13.9a

c. √(-8)
                    2.83i
2i√(2)
                    2.83i

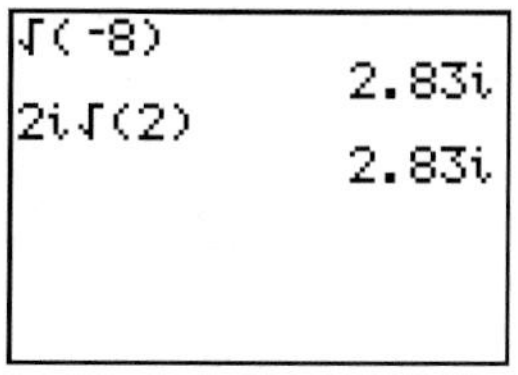

Figure 13.9b

 13.7.1 Checkup

Write the expressions in exercises 1–3 in terms of i.

1. $\sqrt{-49}$ **2.** $\sqrt{-\dfrac{4}{81}}$ **3.** $\sqrt{-27}$

4. Explain the difference between the following two radicals:
   a. $-\sqrt{16}$ **b.** $\sqrt{-16}$
5. What is an imaginary number?

13.7.2 Operations on Imaginary Numbers

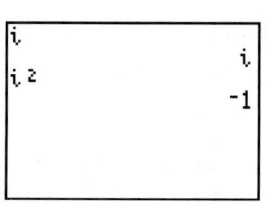

Figure 13.10

When we perform operations with imaginary numbers, we must first write all square roots of negative numbers in terms of i. Even though the symbol i is *not* a variable, we can then treat it as a variable when performing operations.

Before we begin, we need to evaluate i raised to powers greater than 2. For example, evaluate i^1 and i^2 on your calculator, as shown in Figure 13.10.

Complete the following set of exercises with your calculator to discover the results of raising i to powers greater than 2.

 Discovery 5

Powers of i

1. Evaluate the following expressions:
   $i^1 = \underline{\quad i \quad}$ $i^2 = \underline{\quad -1 \quad}$
   a. $i^3 = \underline{\qquad}$ **b.** $i^4 = \underline{\qquad}$
   c. $i^5 = \underline{\qquad}$ **d.** $i^6 = \underline{\qquad}$
2. Look for a pattern and predict the following:
   $i^7 = \underline{\qquad}$ *(Continued on page 944)*

Check your prediction on your calculator. Were you correct? The calculator value is written in scientific notation. If you evaluate the calculator value, it should be approximately $-i$.

3. Predict a value for i^8 and check your results again.

$$i^8 = \underline{\hspace{2cm}}$$

Look for a pattern and then write a rule for evaluating i to any power.

The values for the powers of i repeat the pattern $i, -1, -i, 1, i, \ldots$.

POWERS OF i

To determine the value of i raised to a power, divide the power by 4.

- If the remainder is 1, the value of the expression is i.
- If the remainder is 2, the value of the expression is -1.
- If the remainder is 3, the value of the expression is $-i$.
- If the remainder is 0, the value of the expression is 1.

We can see algebraically that this pattern holds:

$$i^1 = i$$
$$i^2 = -1$$
$$i^3 = (i^2)(i) = (-1)(i) = -i$$
$$i^4 = (i^3)(i) = (-i)(i) = -i^2 = -(-1) = 1$$
$$i^5 = (i^4)(i) = (1)(i) = i$$
$$\vdots$$

We can add, subtract, multiply, and divide imaginary numbers.

Addition	$ai + bi = (a + b)i$	Add the coefficients of i.
Subtraction	$ai - bi = (a - b)i$	Subtract the coefficients of i.
Multiplication	$(ai)(bi) = (ab)i^2$	Multiply the coefficients of i. Remember that $i \cdot i = i^2$.
	$= (ab)(-1)$	Substitute -1 for i^2.
	$= -ab$	Simplify.
Division	$\dfrac{ai}{bi} = \dfrac{a}{b}$	Simplify.

HELPING HAND We do not need to memorize these patterns. Simply perform the operations, treating i as a variable.

For example,

$$\sqrt{-2}\sqrt{-18} = i\sqrt{2} \cdot i\sqrt{18}$$ Write each square root of a negative number in terms of i.

$$= i^2\sqrt{36}$$ Product rule for radicals

$$= -1 \cdot 6$$ Substitute -1 for i^2.

$$= -6$$

 **HELPING HAND** Remember to write the square roots of a negative number in terms of i before you perform operations with imaginary numbers. For example, $\sqrt{-2} \cdot \sqrt{-18} \neq \sqrt{36}$.

EXAMPLE 2 Perform the indicated operation.

a. $\sqrt{-27} + \sqrt{-12}$ b. $2\sqrt{-63} - 3\sqrt{-28}$ c. $\dfrac{\sqrt{-25}}{\sqrt{-16}}$

d. $\sqrt{-2}\sqrt{-3}\sqrt{-5}$ e. $(\sqrt{-3} + \sqrt{-4})(\sqrt{-2} - \sqrt{-6})$

Solution

Check

a. $\sqrt{-27} + \sqrt{-12} = i\sqrt{27} + i\sqrt{12}$ Write the square root of a negative number in terms of i.

$= 3i\sqrt{3} + 2i\sqrt{3}$ Simplify the radicals.

$= 5i\sqrt{3}$

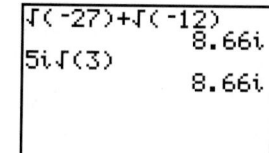

Figure 13.11

b. $2\sqrt{-63} - 3\sqrt{-28} = 2i\sqrt{63} - 3i\sqrt{28}$ Write the square root of a negative number in terms of i.

$= 2i \cdot 3\sqrt{7} - 3i \cdot 2\sqrt{7}$ Simplify the radicals.

$= 6i\sqrt{7} - 6i\sqrt{7}$ Multiply.

$= 0$

c. $\dfrac{\sqrt{-25}}{\sqrt{-16}} = \dfrac{i\sqrt{25}}{i\sqrt{16}}$ Write the square root of a negative number in terms of i.

$= \dfrac{5i}{4i}$ Simplify the radicals.

$= \dfrac{5}{4}$ Simplify.

d. $\sqrt{-2}\sqrt{-3}\sqrt{-5} = i\sqrt{2} \cdot i\sqrt{3} \cdot i\sqrt{5}$ Write the square root of a negative number in terms of i.

$= i^3\sqrt{30}$ Product rule for square roots

$= (-i)(\sqrt{30})$ Substitute $-i$ for i^3.

$= -i\sqrt{30}$

e. $(\sqrt{-3} + \sqrt{-4})(\sqrt{-2} - \sqrt{-6})$

$= (i\sqrt{3} + i\sqrt{4})(i\sqrt{2} - i\sqrt{6})$ Write the square root of a negative number in terms of i.

$= i^2\sqrt{6} - i^2\sqrt{18} + i^2\sqrt{8} - i^2\sqrt{24}$ FOIL method

$= i^2\sqrt{6} - 3i^2\sqrt{2} + 2i^2\sqrt{2} - 2i^2\sqrt{6}$ Simplify the radicals.

$= (-1)\sqrt{6} - 3(-1)\sqrt{2} + 2(-1)\sqrt{2} - 2(-1)\sqrt{6}$ Substitute -1 for i^2.

$= -\sqrt{6} + 3\sqrt{2} - 2\sqrt{2} + 2\sqrt{6}$ Multiply.

$= \sqrt{6} + \sqrt{2}$

We can check these operations on the calculator. Set your calculator to round to two decimal places and to $a + bi$ mode, and then enter the expression. Answers with radicals are approximated and can be checked by entering the exact answer and comparing the two results. The check for Example 2a is shown in Figure 13.11.

13.7.2 Checkup

In exercises 1–5, perform the indicated operation.

1. $\sqrt{-4} + \sqrt{-64}$

2. $5\sqrt{-20} - 3\sqrt{-125}$

3. $\dfrac{\sqrt{-100}}{\sqrt{-121}}$

4. $\sqrt{-6}\,\sqrt{-10}\,\sqrt{-15}$

5. $(\sqrt{-5} - \sqrt{-6})(\sqrt{-2} - \sqrt{-15})$

6. Is the imaginary number i a variable? Explain.

7. How is the imaginary number i treated the same as a variable?

13.7.3 Operations on Complex Numbers

As we performed operations on imaginary numbers in the last section, we obtained values expressed in different forms. For example, some of the results were real numbers and others were expressions that contained the imaginary number i. Together they make up the **complex-number system**.

> **STANDARD FORM FOR A COMPLEX NUMBER**
> A complex number is written in the standard form
>
> $$a + bi$$
>
> where a and b are real numbers.
>
> If $b = 0$, then $a + 0i = a$ is a real number.
> If $b \neq 0$, then $a + bi$ is an imaginary number.
> If $a = 0$ and $b \neq 0$, then $0 + bi = bi$ is a pure imaginary number.
>
> We call a the real part, and b the imaginary part, of the complex number.

In sum, the set of complex numbers consists of two mutually exclusive subsets: real numbers and imaginary numbers. Remember that two sets are mutually exclusive when any number that belongs to one set does not belong to the other set. To visualize the set of complex numbers, we add the set of imaginary numbers to the set of real numbers that we viewed in Figure 1.12, obtaining the following diagram:

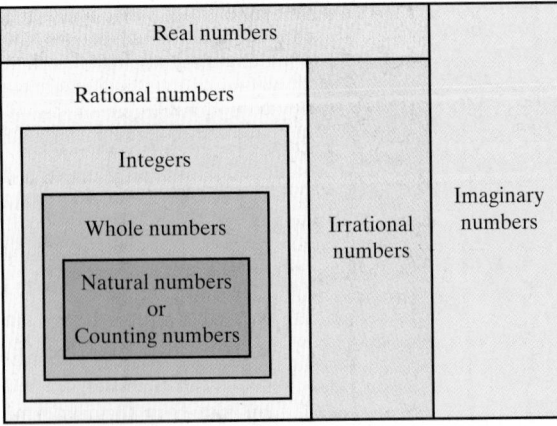

Two complex numbers $a + bi$ and $c + di$ are equal if and only if $a = c$ and $b = d$. For example,

$\sqrt{4} + \sqrt{-3}$ and $2 + i\sqrt{3}$ are equal, because $\sqrt{4} = 2$ and $\sqrt{-3} = i\sqrt{3}$.

We can add and subtract complex numbers by adding the real parts and the imaginary parts separately.

$$(a + bi) + (c + di) = (a + c) + (b + d)i$$
$$(a + bi) - (c + di) = (a - c) + (b - d)i$$

Again, we treat i as a variable and perform the operations in the same manner as adding and subtracting binomials.

EXAMPLE 3   Add or subtract. Write your answer in standard form for a complex number.

    a. $(2 + 3i) + (-4 + 5i)$ **b.** $7 - (2 + i)$ **c.** $-2i - (3 - 2i)$

Solution   **Check**

a. $(2 + 3i) + (-4 + 5i) = [2 + (-4)] + (3 + 5)i$ a.

$$= -2 + 8i$$

b. $7 - (2 + i) = 7 - 2 - i = 5 - i$

c. $-2i - (3 - 2i) = -2i - 3 + 2i = -3$

```
a. (2+3i)+(-4+5i)
                -2+8i
b. 7-(2+i)
                 5-i
c. -2i-(3-2i)
                  -3
```

Calculator checks for Examples 3a, 3b, and 3c are shown in the figure.   ●

We multiply complex numbers by using the distributive property. Remember, this leads to the FOIL method for binomials.

HELPING HAND Remember to simplify powers of i. Write the products in standard form for complex numbers.

EXAMPLE 4   Multiply. Write your answer in standard form for a complex number.

    a. $2(3 - 4i)$ **b.** $3i(6 + i)$ **c.** $(-5 + 2i)(7 - 6i)$

Solution

a. $2(3 - 4i) = 6 - 8i$ Distributive property

b. $3i(6 + i) = 18i + 3i^2$ Distributive property

$$= 18i + 3(-1)$$
$$= 18i - 3$$
$$= -3 + 18i$$

c. $(-5 + 2i)(7 - 6i) = -35 + 30i + 14i - 12i^2$ FOIL method

$$= -35 + 30i + 14i - 12(-1)$$
$$= -35 + 30i + 14i + 12$$
$$= -23 + 44i$$

Check the solutions with your calculator, as in Example 3.   ●

When we multiplied the sum and difference of the same two terms using binomials, we obtained a special product. Similarly, the product of $a + bi$ and $a - bi$ results in a special product.

The product is a real number, the sum of the squares of a and b. Complex numbers of the form $a + bi$ and $a - bi$ are called **complex conjugates**.

PRODUCT OF COMPLEX CONJUGATES

$$(a + bi)(a - bi) = a^2 + b^2$$

We can verify this result by multiplying $a + bi$ and $a - bi$ algebraically.

$$\begin{aligned}(a + bi)(a - bi) &= a^2 - abi + abi - b^2i^2 \\ &= a^2 - abi + abi - b^2(-1) \\ &= a^2 - abi + abi + b^2 \\ &= a^2 + b^2\end{aligned}$$

HELPING HAND Be careful not to confuse this product with the special product for real numbers $(a + b)(a - b) = a^2 - b^2$.

The following pairs of complex numbers are examples of complex conjugates.

$2 + 3i$	and	$2 - 3i$
$-4 - i$	and	$-4 + i$
5 (or $5 + 0i$)	and	5 (or $5 - 0i$)
$2i$ (or $0 + 2i$)	and	$-2i$ (or $0 - 2i$)

To divide a complex number by a nonzero real number, we use the distributive property, as we did with binomials.

$$\frac{a + bi}{c} = \frac{a}{c} + \frac{bi}{c} = \frac{a}{c} + \frac{b}{c}i$$

However, to divide two complex numbers $\frac{a + bi}{c + di}$, we multiply the numerator and denominator by the conjugate of the denominator, $c - di$.

$$\frac{a + bi}{c + di} = \frac{(a + bi)(c - di)}{(c + di)(c - di)}$$

EXAMPLE 5 Divide. Write your answer in standard form for a complex number.

a. $\dfrac{3 + 9i}{3}$ **b.** $\dfrac{2 - 3i}{4 + i}$ **c.** $\dfrac{-2 + 4i}{5i}$

Solution

a. $\dfrac{3 + 9i}{3} = \dfrac{3}{3} + \dfrac{9i}{3}$ Distributive property

$= 1 + 3i$

Check this solution with your calculator.

b. $\dfrac{2 - 3i}{4 + i} = \dfrac{(2 - 3i)(4 - i)}{(4 + i)(4 - i)}$ Multiply the numerator and denominator by the conjugate of the denominator.

$$= \frac{8 - 2i - 12i + 3i^2}{16 + 1}$$

$(a + bi)(a - bi) = a^2 + b^2$, with $a = 4$ and $b = 1$

$$= \frac{8 - 2i - 12i + 3(-1)}{16 + 1}$$

$$= \frac{8 - 2i - 12i - 3}{17}$$

$$= \frac{5 - 14i}{17}$$

$$= \frac{5}{17} - \frac{14}{17}i$$ Standard form

Check this solution your calculator.

c. $\dfrac{-2 + 4i}{5i} = \dfrac{(-2 + 4i)(-5i)}{(5i)(-5i)}$ Conjugate of the denominator, $5i$, or $0 + 5i$, is $0 - 5i$, or $-5i$.

$$= \frac{10i - 20i^2}{25}$$ $(5i)(-5i) = -25i^2 = -25(-1) = 25$

$$= \frac{10i - 20(-1)}{25}$$

$$= \frac{10i + 20}{25}$$

$$= \frac{10}{25}i + \frac{20}{25}$$

$$= \frac{2}{5}i + \frac{4}{5}$$

$$= \frac{4}{5} + \frac{2}{5}i$$ Standard form

Check this solution with your calculator.

 13.7.3 Checkup

Add or subtract.

1. $(15 - 12i) + (11 + 7i)$ **2.** $11 - (5 - 3i)$

3. $21i - (8 + i)$

Multiply.

4. $i(12 + 3i)$ **5.** $(8 - 4i)(9 + 2i)$ **6.** $(7 + 3i)(7 - 3i)$

Divide exercises 7–9.

7. $\dfrac{24 - 18i}{6}$ **8.** $\dfrac{7 - 8i}{2i}$ **9.** $\dfrac{5}{8 - i}$

10. Explain the difference between the conjugate of a radical expression and the conjugate of a complex number. How are these two kinds of conjugates used?

13.7.4 Solving Equations Having Imaginary-Number Solutions

Some quadratic equations can be solved by using the principle of square roots. That is,

for any positive number b, if $a^2 = b$, then $a = \pm\sqrt{b}$.

This principle limits us to a positive number b in order to obtain real-number solutions. For example, solve $x^2 = 1$.

$$x^2 = 1$$
$$x = \pm 1$$

To solve the equation $x^2 = -1$, we need a new principle.

PRINCIPLE OF SQUARE ROOTS FOR IMAGINARY-NUMBER SOLUTIONS
For any negative number b, if $a^2 = b$, then $a = \pm i\sqrt{b}$.

Now we can solve the equation $x^2 = -1$.

$$x^2 = -1$$
$$x = \pm i\sqrt{1}$$
$$x = \pm i$$

We can check the solutions by substitution.

Let $x = i$ $\qquad$ Let $x = -i$

$$
\begin{array}{c|c}
x^2 = -1 & \\
\hline
(i)^2 & -1 \\
-1 &
\end{array}
\qquad
\begin{array}{c|c}
x^2 = -1 & \\
\hline
(-i)^2 & -1 \\
i^2 & \\
-1 &
\end{array}
$$

Both solutions check.

EXAMPLE 6 Solve.

a. $2x^2 + 16 = 0$ $\quad$ **b.** $(x + 3)^2 + 16 = 0$ $\quad$ **c.** $3(2x - 3)^2 = -15$

Solution

Substitution Check

a. $2x^2 + 16 = 0$

$$2x^2 = -16 \qquad \text{Subtract 16 from both sides.}$$

$$\frac{2x^2}{2} = -\frac{16}{2} \qquad \text{Divide both sides by 2.}$$

$$x^2 = -8 \qquad \text{Simplify.}$$

$$x = \pm i\sqrt{8} \qquad \begin{array}{l}\text{Principle of square roots} \\ \text{for imaginary-number} \\ \text{solutions}\end{array}$$

$$x = \pm 2i\sqrt{2} \qquad \text{Simplify.}$$

```
2i√(2)→X:2X²+16
                  0
-2i√(2)→X:2X²+16
                  0
```

The substitution of the possible solutions result in 0, the right side of the equation. The solutions are $\pm 2i\sqrt{2}$.

b. $(x + 3)^2 + 16 = 0$

$$(x + 3)^2 = -16 \qquad \text{Subtract 16 from both sides.}$$

$$x + 3 = \pm i\sqrt{16} \qquad \begin{array}{l}\text{Principle of square roots for imaginary-number} \\ \text{solutions}\end{array}$$

$$x + 3 = \pm 4i \qquad \text{Simplify.}$$

$$x = -3 \pm 4i \qquad \text{Subtract 3 from both sides.}$$

The check is left for you.

c. $3(2x - 3)^2 = -15$

$(2x - 3)^2 = -5$ Divide both sides by 3.

$2x - 3 = \pm i\sqrt{5}$ Principle of square roots for imaginary-number solutions

$2x = 3 \pm i\sqrt{5}$ Add 3 to both sides.

$x = \dfrac{3 \pm i\sqrt{5}}{2}$ Divide both sides by 2.

$x = \dfrac{3}{2} \pm \dfrac{\sqrt{5}}{2} i$ Standard form

The check is left for you.

A second method of solving quadratic equations in the standard form, $ax^2 + bx + c = 0$, is to use the quadratic formula,

$$x = \frac{-b \pm \sqrt{b^2 - 4ac}}{2a}$$

When we solved quadratic equations with the use of the quadratic formula, we found that some equations had no real-number solutions. This was the result when the discriminant was a negative number. Now we can see that these solutions are imaginary numbers.

EXAMPLE 7   Solve.

     a. $3x^2 + 8x + 15 = 0$ **b.** $\dfrac{y - 2}{y - 5} = \dfrac{4}{y + 4}$

Solution   **Check**

a. $3x^2 + 8x + 15 = 0$

$x = \dfrac{-b \pm \sqrt{b^2 - 4ac}}{2a}$ Quadratic formula

$x = \dfrac{-(8) \pm \sqrt{(8)^2 - 4(3)(15)}}{2(3)}$ $a = 3, b = 8,$ and $c = 15$

$x = \dfrac{-8 \pm \sqrt{64 - 180}}{6}$

$x = \dfrac{-8 \pm \sqrt{-116}}{6}$

$x = \dfrac{-8 \pm i\sqrt{116}}{6}$ Write the square root of a negative number in terms of i.

$x = \dfrac{-8 \pm i\sqrt{4 \cdot 29}}{6}$ Simplify the radical.

$x = \dfrac{-8 \pm 2i\sqrt{29}}{6}$

$x = -\dfrac{8}{6} \pm \dfrac{2i\sqrt{29}}{6}$

$x = -\dfrac{4}{3} \pm \dfrac{\sqrt{29}}{3} i$

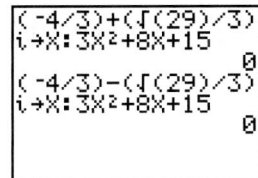

Check the two imaginary-number solutions by substitution on your calculator.

13.7 Exercises

Write in terms of i.

1. $\sqrt{-100}$ **2.** $\sqrt{-144}$ **3.** $\sqrt{-\dfrac{16}{49}}$ **4.** $\sqrt{-\dfrac{36}{121}}$

5. $\sqrt{-32}$ **6.** $\sqrt{-75}$ **7.** $2\sqrt{-50}$ **8.** $-3\sqrt{-98}$

Perform the indicated operations.

9. $7\sqrt{-36} + 9\sqrt{-4}$ **10.** $5\sqrt{-16} + 2\sqrt{-9}$ **11.** $2\sqrt{-121} - 3\sqrt{-9}$ **12.** $3\sqrt{-144} - 2\sqrt{-25}$

13. $\dfrac{\sqrt{-144}}{\sqrt{-225}}$ **14.** $\dfrac{\sqrt{-64}}{\sqrt{-100}}$ **15.** $\sqrt{-5}\sqrt{-8}\sqrt{-10}$ **16.** $\sqrt{-3}\sqrt{-5}\sqrt{-15}$

17. $(\sqrt{-6} + \sqrt{-4})(\sqrt{-3} - \sqrt{-8})$ **18.** $(\sqrt{-3} + \sqrt{-10})(\sqrt{-5} - \sqrt{-6})$ **19.** $(\sqrt{-5} - \sqrt{-7})(\sqrt{-5} + \sqrt{-7})$

20. $(\sqrt{-3} + \sqrt{-11})(\sqrt{-3} - \sqrt{-11})$

Add or subtract. Write your answer in standard form for a complex number.

21. $(3 + 5i) + (-8 - i)$ **22.** $(9 - 17i) + (-11 + 8i)$ **23.** $(6 + 2i) - (3 + 3i)$ **24.** $(14 + 3i) - (5 - 6i)$

25. $(4 - 5i) - (3 - 5i)$ **26.** $(6 + 8i) - (6 + 2i)$ **27.** $\left(\dfrac{1}{2} + 5i\right) + \left(3 - \dfrac{2}{3}i\right)$ **28.** $\left(\dfrac{3}{8} + 6i\right) + \left(\dfrac{1}{8} - 7i\right)$

29. $(4.5 + 6.7i) - (2.88 - 4.68i)$ **30.** $(12.7 - 3.63i) + (4.6 + 12.7i)$ **31.** $(2\sqrt{3} - i\sqrt{2}) + (5\sqrt{3} + 2i\sqrt{2})$

32. $(5\sqrt{5} + i\sqrt{3}) - (7\sqrt{5} + 8i\sqrt{3})$

Multiply. Write your answer in standard form for a complex number.

33. $(4 - 3i)(6 + 5i)$ **34.** $(9 + 6i)(5 - 4i)$ **35.** $(5 + 7i)(5 - 7i)$

36. $(11 - 3i)(11 + 3i)$ **37.** $(-6 - i)(-6 + i)$ **38.** $(-5 + 2i)(-5 - 2i)$

39. $(0.4 - 3.1i)(5.7 + 0.8i)$ **40.** $(2.5 + 3.4i)(1.2 - 0.6i)$ **41.** $\left(\dfrac{3}{5} + \dfrac{1}{2}i\right)\left(\dfrac{2}{5} - \dfrac{2}{3}i\right)$

42. $\left(\dfrac{3}{4} - \dfrac{2}{3}i\right)\left(\dfrac{3}{5} + \dfrac{1}{3}i\right)$ **43.** $i\sqrt{2}(\sqrt{2} + i\sqrt{3})$ **44.** $-3i\sqrt{3}(\sqrt{6} - i\sqrt{15})$

45. $(\sqrt{3} - i\sqrt{5})(\sqrt{3} + i\sqrt{5})$ **46.** $(\sqrt{7} + 2i\sqrt{6})(\sqrt{7} - 2i\sqrt{6})$

Divide. Write your answer in standard form for a complex number.

47. $\dfrac{7 + 21i}{7}$ **48.** $\dfrac{16 + 48i}{8}$ **49.** $\dfrac{-4 + 7i}{3 - i}$ **50.** $\dfrac{47 + 13i}{5 + 2i}$

51. $\dfrac{6 - 5i}{2i}$ **52.** $\dfrac{12 + 9i}{-6i}$ **53.** $\dfrac{16.2 - 13.5i}{2.7i}$ **54.** $\dfrac{9.03 + 6.45i}{4.3i}$

55. $\dfrac{21.2 - 10.4i}{4.5 + i}$ **56.** $\dfrac{-19.7 + 21.5i}{1 + 7.4i}$ **57.** $\dfrac{\sqrt{6} + i\sqrt{14}}{i\sqrt{2}}$ **58.** $\dfrac{\sqrt{15} - i\sqrt{21}}{i\sqrt{3}}$

59. $\dfrac{2 + 5i\sqrt{6}}{2\sqrt{3} + i\sqrt{2}}$ **60.** $\dfrac{23 + 4i\sqrt{10}}{\sqrt{5} + 2i\sqrt{2}}$

Solve exercises 61–100.

61. $a^2 + 7 = 0$ **62.** $b^2 + 11 = 0$ **63.** $z^2 + 5 = 1$ **64.** $m^2 + 18 = 2$

65. $3p^2 + 75 = 0$ **66.** $4q^2 + 64 = 0$ **67.** $4d^2 + 12 = 0$ **68.** $5c^2 + 10 = 0$

69. $(t + 1)^2 + 9 = 0$ **70.** $(s - 3)^2 + 36 = 0$ **71.** $4(x - 5)^2 + 22 = 2$ **72.** $2(x + 3)^2 + 21 = 5$

73. $4(2x + 5)^2 = -16$ **74.** $3(2y - 1)^2 = -75$ **75.** $2(z + 2.5)^2 + 10.58 = 0$ **76.** $3(y - 1.4)^2 + 7.68 = 0$

c. $3(2x - 3)^2 = -15$

$\quad\quad (2x - 3)^2 = -5$ Divide both sides by 3.

$\quad\quad 2x - 3 = \pm i\sqrt{5}$ Principle of square roots for imaginary-number solutions

$\quad\quad\quad\quad 2x = 3 \pm i\sqrt{5}$ Add 3 to both sides.

$\quad\quad\quad\quad x = \dfrac{3 \pm i\sqrt{5}}{2}$ Divide both sides by 2.

$\quad\quad\quad\quad x = \dfrac{3}{2} \pm \dfrac{\sqrt{5}}{2}i$ Standard form

The check is left for you.

A second method of solving quadratic equations in the standard form, $ax^2 + bx + c = 0$, is to use the quadratic formula,

$$x = \frac{-b \pm \sqrt{b^2 - 4ac}}{2a}$$

When we solved quadratic equations with the use of the quadratic formula, we found that some equations had no real-number solutions. This was the result when the discriminant was a negative number. Now we can see that these solutions are imaginary numbers.

EXAMPLE 7   Solve.

a. $3x^2 + 8x + 15 = 0$ **b.** $\dfrac{y - 2}{y - 5} = \dfrac{4}{y + 4}$

Solution

a. $3x^2 + 8x + 15 = 0$

$\quad x = \dfrac{-b \pm \sqrt{b^2 - 4ac}}{2a}$ Quadratic formula

$\quad x = \dfrac{-(8) \pm \sqrt{(8)^2 - 4(3)(15)}}{2(3)}$ $a = 3$, $b = 8$, and $c = 15$

$\quad x = \dfrac{-8 \pm \sqrt{64 - 180}}{6}$

$\quad x = \dfrac{-8 \pm \sqrt{-116}}{6}$

$\quad x = \dfrac{-8 \pm i\sqrt{116}}{6}$ Write the square root of a negative number in terms of i.

$\quad x = \dfrac{-8 \pm i\sqrt{4 \cdot 29}}{6}$ Simplify the radical.

$\quad x = \dfrac{-8 \pm 2i\sqrt{29}}{6}$

$\quad x = -\dfrac{8}{6} \pm \dfrac{2i\sqrt{29}}{6}$

$\quad x = -\dfrac{4}{3} \pm \dfrac{\sqrt{29}}{3}i$

Check

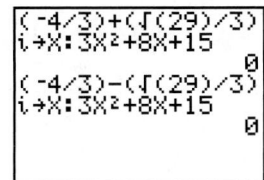

Check the two imaginary-number solutions by substitution on your calculator.

b.
$$\frac{y-2}{y-5} = \frac{4}{y+4}$$

Check

$$y^2 + 4y - 2y - 8 = 4y - 20 \qquad \text{Cross multiply.}$$

$$y^2 - 2y + 12 = 0 \qquad \text{Standard form of a quadratic equation (in terms of y)}$$

```
1+i√(11)→Y:(Y-2)
/(Y-5)
           .56-.37i
4/(Y+4)
           .56-.37i
```

$$y = \frac{-b \pm \sqrt{b^2 - 4ac}}{2a} \qquad \text{Quadratic formula}$$

$$y = \frac{-(-2) \pm \sqrt{(-2)^2 - 4(1)(12)}}{2(1)} \qquad a = 1, b = -2, \text{ and } c = 12$$

Check the two imaginary-number solutions by substitution on your calculator. The check of $1 + i\sqrt{11}$ is shown.

$$y = \frac{2 \pm \sqrt{4 - 48}}{2}$$

$$y = \frac{2 \pm \sqrt{-44}}{2}$$

$$y = \frac{2 \pm i\sqrt{44}}{2} \qquad \text{Write the square root of a negative number in terms of i.}$$

$$y = \frac{2 \pm i\sqrt{4 \cdot 11}}{2} \qquad \text{Simplify the radical.}$$

$$y = \frac{2 \pm 2i\sqrt{11}}{2}$$

$$y = \frac{2}{2} \pm \frac{2i\sqrt{11}}{2}$$

$$y = 1 \pm i\sqrt{11}$$

✓ 13.7.4 Checkup

Solve exercises 1–5.

1. $5x^2 + 20 = 0$

2. $(x - 4)^2 + 27 = 0$

3. $-6(3z + 1)^2 = 12$

4. $5x^2 + 3x + 9 = 0$

5. $\dfrac{x + 4}{x - 6} = \dfrac{2}{x + 8}$

6. In the quadratic formula, what is the discriminant?

7. If the solutions of a quadratic equation are imaginary numbers, what must be the sign of the discriminant?

13.7.5 Modeling the Real World

In previous chapters, we examined an important formula called Ohm's law, $V = IR$, that applies to direct-current (DC) circuits, such as the circuit form when you connect a battery. However, there is another kind of electric circuit, called an alternating-current (AC) circuit, that is used to provide the electric power in houses, offices, and industry. The electrical quantities in an AC circuit—for example, voltage and current—can be described by complex numbers.

In an AC circuit, Ohm's law takes on a slightly different form, $V = IZ$. The quantity V is still called the voltage and the quantity I is still called the current, although both may be complex numbers. The quantity Z is called the impedance and includes other electrical properties that we won't discuss, as well as the resistance R, which is the same as in DC circuits. When engineers work with the AC form of Ohm's law, they often have to determine the magnitude of the voltage, which is denoted as $|V|$. If the voltage is given in the complex form $a + bi$, the magnitude of the voltage is the principal square root of the sum of the squares of the real part and the imaginary part, or $|V| = \sqrt{a^2 + b^2}$. We will solve a couple of problems by means of these equations in order to give you practice in working with complex numbers.

EXAMPLE 8   In a particular circuit, the current, I, is $2 - 3i$ amperes and the impedance, Z, is $6 + 2i$ ohms. Find the magnitude of the total voltage, $|V|$, across this part of the circuit.

Solution

The formula for total voltage is $V = IZ$.

$$V = IZ$$
$$V = (2 - 3i)(6 + 2i)$$
$$V = 12 + 4i - 18i - 6i^2$$
$$V = 12 + 4i - 18i - 6(-1)$$
$$V = 12 + 4i - 18i + 6$$
$$V = 18 - 14i$$

The total voltage is $18 - 14i$ volts.
   The magnitude of the total voltage is $|V| = \sqrt{a^2 + b^2}$, where $a = 18$ and $b = -14$.

$$|V| = \sqrt{(18)^2 + (-14)^2} \approx 22.8$$
Use a calculator to approximate to two decimal places.

The magnitude of the total voltage is approximately 22.8 volts.

```
√((18)²+(-14)²)
             22.80
```

APPLICATION

In studying the motion of biological cells and viruses, we must solve the equation $3.0 = 2.5p - 12.6p^2$. Determine the solution set of the equation.

Discussion

$$3.0 = 2.5p - 12.6p^2$$
$$12.6p^2 - 2.5p + 3.0 = 0$$
Standard form of a quadratic equation

$$p = \frac{-b \pm \sqrt{b^2 - 4ac}}{2a}$$
Quadratic formula

$$p = \frac{-(-2.5) \pm \sqrt{(-2.5)^2 - 4(12.6)(3.0)}}{2(12.6)}$$
$a = 12.6$, $b = -2.5$, and $c = 3.0$

$$p = \frac{2.5 \pm \sqrt{-144.95}}{25.2}$$

$$p = \frac{2.5 \pm i\sqrt{144.95}}{25.2}$$
Write the square root of a negative number in terms of i.

 13.7.5 Checkup

Find the magnitude of the total voltage across a circuit when the current is $4 - 2i$ amperes and the impedance is $5 + 3i$ ohms.

13.7 Exercises

Write in terms of i.

1. $\sqrt{-100}$ 2. $\sqrt{-144}$ 3. $\sqrt{-\dfrac{16}{49}}$ 4. $\sqrt{-\dfrac{36}{121}}$

5. $\sqrt{-32}$ 6. $\sqrt{-75}$ 7. $2\sqrt{-50}$ 8. $-3\sqrt{-98}$

Perform the indicated operations.

9. $7\sqrt{-36} + 9\sqrt{-4}$ 10. $5\sqrt{-16} + 2\sqrt{-9}$ 11. $2\sqrt{-121} - 3\sqrt{-9}$ 12. $3\sqrt{-144} - 2\sqrt{-25}$

13. $\dfrac{\sqrt{-144}}{\sqrt{-225}}$ 14. $\dfrac{\sqrt{-64}}{\sqrt{-100}}$ 15. $\sqrt{-5}\sqrt{-8}\sqrt{-10}$ 16. $\sqrt{-3}\sqrt{-5}\sqrt{-15}$

17. $(\sqrt{-6} + \sqrt{-4})(\sqrt{-3} - \sqrt{-8})$ 18. $(\sqrt{-3} + \sqrt{-10})(\sqrt{-5} - \sqrt{-6})$ 19. $(\sqrt{-5} - \sqrt{-7})(\sqrt{-5} + \sqrt{-7})$

20. $(\sqrt{-3} + \sqrt{-11})(\sqrt{-3} - \sqrt{-11})$

Add or subtract. Write your answer in standard form for a complex number.

21. $(3 + 5i) + (-8 - i)$ 22. $(9 - 17i) + (-11 + 8i)$ 23. $(6 + 2i) - (3 + 3i)$ 24. $(14 + 3i) - (5 - 6i)$

25. $(4 - 5i) - (3 - 5i)$ 26. $(6 + 8i) - (6 + 2i)$ 27. $\left(\dfrac{1}{2} + 5i\right) + \left(3 - \dfrac{2}{3}i\right)$ 28. $\left(\dfrac{3}{8} + 6i\right) + \left(\dfrac{1}{8} - 7i\right)$

29. $(4.5 + 6.7i) - (2.88 - 4.68i)$ 30. $(12.7 - 3.63i) + (4.6 + 12.7i)$ 31. $(2\sqrt{3} - i\sqrt{2}) + (5\sqrt{3} + 2i\sqrt{2})$

32. $(5\sqrt{5} + i\sqrt{3}) - (7\sqrt{5} + 8i\sqrt{3})$

Multiply. Write your answer in standard form for a complex number.

33. $(4 - 3i)(6 + 5i)$ 34. $(9 + 6i)(5 - 4i)$ 35. $(5 + 7i)(5 - 7i)$

36. $(11 - 3i)(11 + 3i)$ 37. $(-6 - i)(-6 + i)$ 38. $(-5 + 2i)(-5 - 2i)$

39. $(0.4 - 3.1i)(5.7 + 0.8i)$ 40. $(2.5 + 3.4i)(1.2 - 0.6i)$ 41. $\left(\dfrac{3}{5} + \dfrac{1}{2}i\right)\left(\dfrac{2}{5} - \dfrac{2}{3}i\right)$

42. $\left(\dfrac{3}{4} - \dfrac{2}{3}i\right)\left(\dfrac{3}{5} + \dfrac{1}{3}i\right)$ 43. $i\sqrt{2}(\sqrt{2} + i\sqrt{3})$ 44. $-3i\sqrt{3}(\sqrt{6} - i\sqrt{15})$

45. $(\sqrt{3} - i\sqrt{5})(\sqrt{3} + i\sqrt{5})$ 46. $(\sqrt{7} + 2i\sqrt{6})(\sqrt{7} - 2i\sqrt{6})$

Divide. Write your answer in standard form for a complex number.

47. $\dfrac{7 + 21i}{7}$ 48. $\dfrac{16 + 48i}{8}$ 49. $\dfrac{-4 + 7i}{3 - i}$ 50. $\dfrac{47 + 13i}{5 + 2i}$

51. $\dfrac{6 - 5i}{2i}$ 52. $\dfrac{12 + 9i}{-6i}$ 53. $\dfrac{16.2 - 13.5i}{2.7i}$ 54. $\dfrac{9.03 + 6.45i}{4.3i}$

55. $\dfrac{21.2 - 10.4i}{4.5 + i}$ 56. $\dfrac{-19.7 + 21.5i}{1 + 7.4i}$ 57. $\dfrac{\sqrt{6} + i\sqrt{14}}{i\sqrt{2}}$ 58. $\dfrac{\sqrt{15} - i\sqrt{21}}{i\sqrt{3}}$

59. $\dfrac{2 + 5i\sqrt{6}}{2\sqrt{3} + i\sqrt{2}}$ 60. $\dfrac{23 + 4i\sqrt{10}}{\sqrt{5} + 2i\sqrt{2}}$

Solve exercises 61–100.

61. $a^2 + 7 = 0$ 62. $b^2 + 11 = 0$ 63. $z^2 + 5 = 1$ 64. $m^2 + 18 = 2$

65. $3p^2 + 75 = 0$ 66. $4q^2 + 64 = 0$ 67. $4d^2 + 12 = 0$ 68. $5c^2 + 10 = 0$

69. $(t + 1)^2 + 9 = 0$ 70. $(s - 3)^2 + 36 = 0$ 71. $4(x - 5)^2 + 22 = 2$ 72. $2(x + 3)^2 + 21 = 5$

73. $4(2x + 5)^2 = -16$ 74. $3(2y - 1)^2 = -75$ 75. $2(z + 2.5)^2 + 10.58 = 0$ 76. $3(y - 1.4)^2 + 7.68 = 0$

77. $\left(b - \dfrac{1}{2}\right)^2 + \dfrac{1}{4} = 0$

78. $\left(c + \dfrac{2}{3}\right)^2 + \dfrac{9}{16} = 0$

79. $x^2 + 2x + 4 = 0$

80. $x^2 - 6x + 14 = 0$

81. $b^2 - 10b + 27 = 0$

82. $x^2 + 14x + 51 = 0$

83. $4y^2 + 4y + 5 = 0$

84. $9x^2 + 24x + 17 = 0$

85. $9p^2 - 12p + 8 = 0$

86. $16z^2 - 8z + 7 = 0$

87. $x^2 - 2.4x + 3.44 = 0$

88. $m^2 + 4.2m + 7.41 = 0$

89. $2y^2 - 2y + 1.22 = 0$

90. $3z^2 + 3.6z + 1.35 = 0$

91. $x^2 - \dfrac{2}{3}x + \dfrac{13}{36} = 0$

92. $x^2 + \dfrac{1}{2}x + \dfrac{25}{144} = 0$

93. $4z^2 + \dfrac{4}{3}z + \dfrac{2}{9} = 0$

94. $9x^2 + 3x + \dfrac{13}{16} = 0$

95. $\dfrac{x}{x - 2} = \dfrac{5}{x + 3}$

96. $\dfrac{a}{a - 9} = \dfrac{4}{a + 7}$

97. $\dfrac{y + 7}{y - 3} = \dfrac{6}{y + 4}$

98. $\dfrac{b + 5}{b - 6} = \dfrac{8}{b + 1}$

99. $\dfrac{z - 5}{z(z + 3)} = \dfrac{4}{5}$

100. $\dfrac{c - 9}{c(c + 3)} = \dfrac{2}{3}$

101. Find the magnitude of the total voltage V across a circuit if the current I is $3 + 5i$ amperes and the impedance Z is $5 + 4i$ ohms. Use the formula $V = IZ$.

102. What is the total voltage V across a circuit when the current I is $5 - i$ amperes and the impedance Z is $8 + 2i$ ohms?

103. Use the formula $I = \dfrac{V}{Z}$ to find the magnitude of the current I across a circuit when the voltage V is $18 + i$ volts and the impedance Z is $2 + 3i$ ohms.

104. Find the magnitude of the current I across a circuit when the voltage V is $40 + 44i$ volts and the impedance Z is $6 + 4i$ ohms.

105. In analyzing the laminar flow of fluids, energy loss is calculated with the use of the quantity $K = 1.00 - 2.67r + r^2$, where r is the ratio of cross-sectional areas of the fluid. Let $K = -1.75$ and solve the resulting equation for r.

106. Using the relation in exercise 105, let $K = -2.50$, and solve the resulting equation for r.

13.7 Calculator Exercises

Now that you have learned to set your calculator to complex-number mode, you can investigate how the calculator handles roots of negative numbers when the index of the radical is something other than 2. Use the complex-number setting for the following exercises:

Simplify.

1. $\sqrt{-25}$ **2.** $\sqrt[3]{-27}$ **3.** $\sqrt[4]{-64}$

4. $\sqrt[5]{-32}$ **5.** $\sqrt[4]{-324}$ **6.** $\sqrt[4]{-40{,}000}$

When you have your calculator in complex-number mode, you can use the fraction key to change decimal values to fractions. For example, convert $0.25 - 0.8i$ to fractional notation by entering the expression, followed by

| MATH | ▶ | ENTER |

which results in the expression $\frac{1}{4} - \frac{4}{5}i$.

Convert the following complex numbers to fractional notation:

7. $1.5 - 3.5i$ **8.** $0.4 + 3.2i$ **9.** $-1.8i$

The calculator also can determine the magnitude of a complex number, as required in Example 8 in this section. Remember that the magnitude of $a + bi$ is calculated as $\sqrt{a^2 + b^2}$. For example, to find the magnitude of $3 + 4i$, enter

| MATH | ▶ | ▶ | 5 | 3 | + | 4 | 2nd | i |) |

| ENTER |

The calculator returns a value of 5 for the magnitude of $3 + 4i$. Use this feature to find the magnitude of the following complex numbers:

10. $6 + 8i$ **11.** $6 - 8i$ **12.** $2 - 5i$

13. 22 **14.** $-7i$ **15.** $i\sqrt{3}$

13.7 Writing Exercise

Fractal geometry is an area of mathematics that uses complex numbers to draw extraordinary pictures called fractals. While you may not have the math skills to fully understand the principles that underlie the process, you can still enjoy the amazing pictures obtained. In the library, find a reference on fractals. You may want to search for some of the most popular drawings, such as the Mandelbrot set or the Sierpinski triangle.

Write a short description of what you have found. Include the title, author, and library call number of the text you reference. Some texts provide programs that can be entered into a graphing calculator to reproduce the designs. You may want to try to do so.

CHAPTER 13 SUMMARY

After completing this chapter, you should be able to define the following key terms in your own words.

13.1
radical relation
radical function

13.4
rationalizing the denominator

13.5
like radicals
conjugates

13.6
radical equation in one variable
radical equation

13.7
imaginary unit i
complex-number system
complex conjugates

Reflections

1. What is the relationship between a radical expression with a variable in its radicand and a variable expression with a rational exponent?

2. How can you determine the restricted values of a radical function?

3. What does it mean to "rationalize a radical expression"?

4. In operating with radical expressions involving square roots, it is sometimes important to be able to determine the conjugate of an expression. What is the conjugate of a radical expression? When is it necessary to work with the conjugate of a radical expression?

5. When solving a radical equation algebraically, how can you tell whether a solution is extraneous?

6. What is the difference between the real-number system and the complex-number system?

7. Explain the process of adding and subtracting two complex numbers.

8. Can you use the FOIL method to multiply two complex numbers? If so, explain how.

9. How can you form the conjugate of a complex number?

10. How is the conjugate of a complex number used to perform division of complex numbers?

11. When you use the quadratic formula to solve a quadratic equation, how can you tell from the nature of the discriminant that the solutions will be imaginary numbers?

CHAPTER 13 SECTION-BY-SECTION REVIEW

In these exercises, we will assume that the variable base of an exponential expression or a variable radicand does not represent values for which the expression is not defined.

13.1

Recall	Examples
A radical function can be written in the form $f(x) = A(x)$, where $A(x)$ is a radical expression with the variable x in the radicand.	Some radical functions are $f(x) = \sqrt{x + 4}$ $g(x) = \sqrt[3]{2x - 1}$ $y = \sqrt[4]{x} - 2$ $y = \left(\sqrt[3]{x}\right)^2$
The domain of a radical function having an odd index is the set of all real numbers. The domain of a radical function having an even index is the set of all real numbers except the values restricted from the domain. The radicand cannot be negative.	Determine the domain of $f(x) = \sqrt{2x + 1}$. The restricted values are determined by $2x + 1 < 0$ $2x < -1$ $x < -\dfrac{1}{2}$ The domain is the set of all real numbers greater than or equal to $-\frac{1}{2}$, $x \geq -\frac{1}{2}$, or $\left[-\frac{1}{2}, \infty\right)$.
Graph of a radical function • Determine the restricted values. • Determine a table of values by choosing the x-values in the domain of the function. • Plot the points and connect them with a smooth curve.	Graph $f(x) = \sqrt{2x + 1}$. The restricted values are $x < -\frac{1}{2}$. Table of values

x	$f(x) = \sqrt{2x + 1}$	$f(x)$
$-\dfrac{1}{2}$	$f\left(-\dfrac{1}{2}\right) = \sqrt{2\left(-\dfrac{1}{2}\right) + 1}$	0
0	$f(0) = \sqrt{2(0) + 1}$	1
4	$f(4) = \sqrt{2(4) + 1}$	3
5	$f(5) = \sqrt{2(5) + 1}$	$\sqrt{11}$

| Distance formula
The distance between two points (x_1, y_1) and (x_2, y_2), on a coordinate plane is $d = \sqrt{(x_2 - x_1)^2 + (y_2 - y_1)^2}$. | Determine the distance between the points $(2, 5)$ and $(-1, 3)$.
Let $x_1 = 2$, $y_1 = 5$, $x_2 = -1$, and $y_2 = 3$
$d = \sqrt{(x_2 - x_1)^2 + (y_2 - y_1)^2}$
$d = \sqrt{(-1 - 2)^2 + (3 - 5)^2}$
$d = \sqrt{(-3)^2 + (-2)^2}$
$d = \sqrt{13}$ |

Evaluate. Round to the nearest thousandth if necessary.

1. $\sqrt{225}$

2. $\sqrt{2.89}$

3. $\sqrt{\dfrac{49}{64}}$

4. $\sqrt[3]{-\dfrac{27}{125}}$

5. $-\sqrt[4]{1296}$

6. $\sqrt[4]{-1296}$

7. $\sqrt{150}$

8. $\sqrt[3]{-35}$

9. $\sqrt[5]{125}$

Determine the restricted values and then state the domain of each radical function.

10. $y = \sqrt{2x - 7}$

11. $y = \sqrt[3]{2x - 7}$

Use a table of values to graph each radical function in exercises 12 and 13.

12. $f(x) = 2\sqrt{x}$

13. $y = \sqrt[3]{3x - 4}$

14. In statistics, the variability of the integers 1 to n is calculated as $s = \sqrt{\dfrac{n(n + 1)}{12}}$. Calculate the variability of the integers from 1 to 23.

In exercises 15–18, find the distance on the coordinate plane between the points for the coordinate pairs listed.

15. $(4, -3)$ and $(4, 7)$

16. $(-2, -4)$ and $(7, -4)$

17. $(6, 2)$ and $(9, 6)$

18. $(-3, -1)$ and $(4, 8)$

19. Find the length of a side of a square picture frame that encloses an area of 121 square inches.

20. Find the length of the side of a cube-shaped box whose volume is $\frac{125}{216}$ cubic yards.

21. Find the radius and diameter of a rubber ball whose volume is 65.45 cubic centimeters.

22. According to *U.S. Market Trends and Forecasts*, the market value of athletic footwear, in millions of dollars, is approximated by the function $v(x) = 13\sqrt[10]{x}$, where x is the number of years after 1993.
a. Graph the function.
b. Use the graph to determine whether the function is increasing or decreasing.
c. Assuming that the trend shown in the graph continues, approximate the 2010 market value of athletic footwear.

13.2

Recall	Examples
Expressions with rational exponents can be written as radical expressions. $$a^{1/n} = \sqrt[n]{a}$$ $$a^{-1/n} = \frac{1}{a^{1/n}} = \frac{1}{\sqrt[n]{a}}$$ $$a^{m/n} = \sqrt[n]{a^m} = (\sqrt[n]{a})^m$$	Write as an equivalent radical expression. $$x^{1/3} = \sqrt[3]{x}$$ $$x^{-1/2} = \frac{1}{x^{1/2}} = \frac{1}{\sqrt{x}}$$ $$x^{3/4} = \sqrt[4]{x^3} = (\sqrt[4]{x})^3$$
Graph of a function having a rational exponent • Determine the restricted values. • Determine a table of values by choosing the x-values in the domain of the function. • Plot the points and connect them with a smooth curve.	Graph. $g(x) = (x - 1)^{3/4}$ The restricted values are determined by $$x - 1 < 0$$ $$x < 1$$ The domain is the set of all real numbers greater than or equal to 1, $x \geq 1$, or $[1, \infty)$.

x	$g(x) = (x - 1)^{3/4}$	$g(x)$
1	$g(1) = (1 - 1)^{3/4}$	0
2	$g(2) = (2 - 1)^{3/4}$	1
6	$g(6) = (6 - 1)^{3/4}$	3.3
17	$g(17) = (17 - 1)^{3/4}$	8

Recall	Examples
	$g(x)$ graph with points $(2,1)$, $(6,3.3)$, $(17,8)$, $(1,0)$, $g(x)=(x-1)^{3/4}$

In exercises 23–30, write each expression as a radical expression, and evaluate if possible.

23. $(-64)^{1/3}$ **24.** $-16^{1/4}$ **25.** $(-16)^{1/4}$ **26.** $-64^{-1/3}$

27. $-64^{4/3}$ **28.** $(-64)^{3/2}$ **29.** $16^{-3/4}$ **30.** $(-16)^{-3/4}$

Determine the restricted values and then state the domain of each radical function.

31. $y=(4x+9)^{3/4}$ **32.** $f(x)=2x^{1/3}+2$

Use a table of values to graph each radical function in exercises 33 and 34.

33. $g(x)=x^{3/2}+1$ **34.** $y=(2x+1)^{1/3}$

35. When the coefficient of friction on a road is 30%, the velocity V, in miles per hour, of a car that leaves a skid mark of S feet is given by the mathematical model $V=(9s)^{1/2}$. Graph this function for values of S between 0 and 200 feet. What would you estimate the speed of a car to be if it leaves a skid mark that is 150 feet long?

13.3

Recall	Examples
All the properties of integer exponents apply to rational exponents. • Product rule • Quotient rule • Power-to-a-power rule • Product-to-a-power rule • Quotient-to-a-power rule	Simplify, using the properties of exponents. $x^{-1/3}x^{2/3}=x^{-1/3+2/3}=x^{1/3}$ $\dfrac{y^{3/4}}{y^{1/2}}=y^{3/4-1/2}=y^{3/4-2/4}=y^{1/4}$ $(9ab)^{1/2}=9^{1/2}a^{1/2}b^{1/2}=3a^{1/2}b^{1/2}$ $\left(\dfrac{x}{y}\right)^{-3/5}=\dfrac{x^{-3/5}}{y^{-3/5}}=\dfrac{y^{3/5}}{x^{3/5}}$

Evaluate, using the properties of exponents.

36. $\dfrac{1}{4^{-1/2}}$ **37.** $8^{2/3}\cdot8^{5/3}$ **38.** $\dfrac{9^{7/3}}{9^{2/3}}$

39. $\left(\dfrac{4}{9}\right)^{-3/2}$ **40.** $[(3^{1/2})(27^{1/2})]^{-3/2}$ **41.** $(64^{2/3})^{3/4}$

Simplify exercises 42–48. Write your results using positive exponents.

42. $\dfrac{x^{2/3}}{x^{5/6}}$ **43.** $y^{2/5}\cdot y^{-3/10}$ **44.** $(z^{3/5})^{5/9}$

45. $\left(\dfrac{a^3}{b^6}\right)^{-5/12}$ **46.** $(8x^6y^9)^{4/3}$ **47.** $(2a^{3/4}b^{1/3})(3a^{1/3}b^{2/3})$ **48.** $x^{1/3}(x^{2/3}-x^{1/3})$

49. How should the length of the side of a square be changed to increase the area of the square to five times its original size?

13.4

Recall	Examples
Product rule for radicals $$\sqrt[n]{a} \cdot \sqrt[n]{b} = \sqrt[n]{ab}$$	Simplify, using the product rule. $$\sqrt{18xy^3z^5} = \sqrt{9 \cdot 2 \cdot x \cdot y^2 \cdot y \cdot z^4 \cdot z}$$ $$= \sqrt{9y^2z^4}\sqrt{2xyz}$$ $$= 3yz^2\sqrt{2xyz}$$
Quotient rule for radicals $$\frac{\sqrt[n]{a}}{\sqrt[n]{b}} = \sqrt[n]{\frac{a}{b}}$$	Simplify, using the quotient rule. $$\sqrt[3]{\frac{27x^3y}{z^3}} = \frac{\sqrt[3]{27x^3y}}{\sqrt[3]{z^3}} = \frac{3x\sqrt[3]{y}}{z}$$
Rationalize the denominators • If the denominator is one term, multiply the numerator and denominator by the value needed to make the denominator a perfect root.	Rationalize the denominator. $$\frac{2\sqrt{xy}}{\sqrt{5}} \cdot \frac{\sqrt{5}}{\sqrt{5}} = \frac{2\sqrt{5xy}}{5}$$

Simplify without a calculator. Check the results with your calculator.

50. $\sqrt{6300}$ **51.** $\sqrt[3]{-320}$ **52.** $\sqrt[4]{162}$

Simplify.

53. $\sqrt{45x^4y^7z^2}$ **54.** $\sqrt[3]{-64x^2y^7}$ **55.** $\sqrt{12x^2 - 36x + 27}$

Multiply and simplify.

56. $\sqrt{2x} \cdot \sqrt{8x^3}$ **57.** $\sqrt[3]{-2a^2b} \cdot \sqrt[3]{20a^2b^5}$ **58.** $\sqrt{x^2 - 2xy} \cdot \sqrt{3x - 6y}$

Simplify without a calculator, rationalizing all denominators.

59. $\sqrt{\dfrac{25}{64}}$ **60.** $\sqrt{\dfrac{13}{289}}$ **61.** $\sqrt{\dfrac{27}{343}}$

62. $\sqrt[3]{\dfrac{6}{25}}$ **63.** $\dfrac{\sqrt{50}}{\sqrt{60}}$

Simplify and rationalize in exercises 64–67.

64. $\sqrt{\dfrac{25a^2}{64b^4}}$ **65.** $\sqrt{\dfrac{16m}{5}}$ **66.** $\sqrt[3]{\dfrac{3z^3}{4x^2y}}$ **67.** $\dfrac{4a\sqrt{5ab^2}}{12\sqrt{10ab^3}}$

68. Find the period of a pendulum whose length is 3 feet.

13.5

Recall	Examples
Addition and subtraction $$a\sqrt[n]{x} + b\sqrt[n]{x} = (a + b)\sqrt[n]{x}$$ $$a\sqrt[n]{x} - b\sqrt[n]{x} = (a - b)\sqrt[n]{x}$$	Add or subtract. $$2\sqrt[3]{x} + 3\sqrt[3]{x} = (2 + 3)\sqrt[3]{x} = 5\sqrt[3]{x}$$ $$\sqrt{x} - 3\sqrt{x} = (1 - 3)\sqrt{x} = -2\sqrt{x}$$
Multiplication • Use the distributive property. • If the factors are like roots, use the product rule to simplify.	Multiply. $$2\sqrt[3]{x}(3\sqrt[3]{x} - 1) = 6\sqrt[3]{x^2} - 2\sqrt[3]{x}$$ $$(\sqrt{2} + \sqrt{x})(3\sqrt{5} - \sqrt{x}) = 3\sqrt{10} - \sqrt{2x} + 3\sqrt{5x} - x$$

Recall	Examples
Division • If the divisor (denominator) is one term, distribute it over the terms of the numerator. • If the divisor (denominator) is two terms containing square roots, multiply the numerator and denominator by the conjugate of the denominator.	Divide. $$\dfrac{2 + \sqrt{xy}}{\sqrt{x}} = \dfrac{2}{\sqrt{x}} + \dfrac{\sqrt{xy}}{\sqrt{x}}$$ $$= \dfrac{2}{\sqrt{x}} \cdot \dfrac{\sqrt{x}}{\sqrt{x}} + \dfrac{\sqrt{y}}{1}$$ $$= \dfrac{2\sqrt{x}}{x} + \sqrt{y}$$ $$\dfrac{2 + \sqrt{x}}{3 - \sqrt{x}} = \dfrac{2 + \sqrt{x}}{3 - \sqrt{x}} \cdot \dfrac{3 + \sqrt{x}}{3 + \sqrt{x}}$$ $$= \dfrac{6 + 5\sqrt{x} + x}{9 - x}$$

Add or subtract as indicated.

69. $3\sqrt{5} - 2\sqrt{5} + 7\sqrt{5} - \sqrt{5}$

70. $7\sqrt{11} + 2\sqrt{44}$

71. $\sqrt{15} + \sqrt{\dfrac{1}{15}}$

72. $5\sqrt[3]{16} - \sqrt[3]{54}$

73. $\sqrt{49x} - \sqrt{25x}$

74. $4b\sqrt{a^3b} - 7a\sqrt{ab^3} + 8ab\sqrt{ab}$

Multiply and simplify.

75. $\sqrt{7}(\sqrt{14} - \sqrt{7})$

76. $\sqrt{2a}(6 + \sqrt{2a})$

77. $(2 - \sqrt{5})(4 + \sqrt{5})$

78. $(\sqrt{5x} + \sqrt{7y})(\sqrt{5x} - \sqrt{7y})$

79. $(\sqrt{x} + 8)^2$

80. $3\sqrt[3]{x}(2\sqrt[3]{x^2} + \sqrt[3]{x})$

Divide exercises 81–86 and simplify.

81. $\dfrac{\sqrt{15x} - \sqrt{30}}{\sqrt{5}}$

82. $\dfrac{\sqrt{z} - 5}{\sqrt{z}}$

83. $\dfrac{24}{\sqrt{5} + 2}$

84. $\dfrac{\sqrt{x} + 2}{\sqrt{x} - 2}$

85. $\dfrac{2\sqrt{x} - 5}{5 - 2\sqrt{x}}$

86. $\dfrac{2x - 9}{\sqrt{2x} - 3}$

87. One square mirror measures 720 square inches. A larger square mirror measures 1620 square inches. How much more framing material will the second mirror require than that required for the first mirror?

13.6

Recall	Examples
Solving a radical equation algebraically • Determine the restricted values. • Isolate the radical term to one side of the equation. • Raise both sides of the equation to the power of the index. • Solve the resulting equation. • Check the solution(s) numerically, graphically, or by substitution.	Solve. $4 + \sqrt{x + 2} = x$ The restricted values are determined by $x + 2 < 0$ $\qquad\qquad x < -2$ Solve. $$4 + \sqrt{x + 2} = x$$ $$\sqrt{x + 2} = x - 4$$ $$(\sqrt{x + 2})^2 = (x - 4)^2$$ $$x + 2 = x^2 - 8x + 16$$ $$x^2 - 9x + 14 = 0$$ $$x = \dfrac{-b \pm \sqrt{b^2 - 4ac}}{2a}$$ $$x = \dfrac{-(-9) \pm \sqrt{(-9)^2 - 4(1)(14)}}{2(1)}$$ $$x = \dfrac{9 \pm \sqrt{25}}{2}$$ $$x = \dfrac{9 \pm 5}{2}$$ $$x = 7, x = 2$$ Every solution must be checked. See the following check.

(Continued on page 962)

Recall	Examples
	Graphic check
	Y1 = 4 + $\sqrt{x + 2}$ Y2 = x
	$(-10, 10, -10, 10)$
	The intersection is $(7, 7)$.
	The solution is 7. Note that $x = 2$ is an extraneous solution.
Solving algebraically a radical equation having a rational exponent • Determine the restricted values. • Isolate the term having the rational exponent to one side of the equation. • Raise both sides of the equation to the power of the denominator. • Solve the resulting equation. • Check the solution(s) numerically, graphically, or by substitution.	Solve. $(x + 1)^{2/3} = 2$ There are no restricted values, because the power has an odd denominator, 3. Solve. $(x + 1)^{2/3} = 2$ $((x + 1)^{2/3})^3 = 2^3$ $(x + 1)^2 = 8$ $\sqrt{(x + 1)^2} = \sqrt{8}$ $x + 1 = \pm 2\sqrt{2}$ $x = -1 \pm 2\sqrt{2}$ Check by substitution. The solutions are $-1 \pm 2\sqrt{2}$.

In exercises 88–97, solve algebraically and check.

88. $2\sqrt{3x} + 3 = 15$ **89.** $\sqrt{4x - 3} = x - 2$ **90.** $2\sqrt{x + 5} = \sqrt{5x + 9}$

91. $\sqrt{x + 6} + 3 = \sqrt{5x - 1}$ **92.** $\sqrt{x + 8} = \sqrt{x - 2}$ **93.** $\sqrt[3]{2x} = -4$

94. $\sqrt[4]{x + 3} = 3$ **95.** $(x + 7)^{2/5} = 4$ **96.** $x^{-3/2} = 8$

97. $2x^{2/3} - 5 = 1$

98. Find the possible y-coordinates of a point that is 5 units away from the point $(3, -2)$ and that has an x-coordinate of 6.

99. If the coefficient of friction on a road is 35%, the velocity V, in miles per hour, of a car that leaves a skid mark of S feet is given by the mathematical model $V = \sqrt{10.5S}$. What would be the length of the skid mark of a car traveling 40 miles per hour if the car had to stop suddenly?

100. The time t in seconds that it would take an object to fall a distance d in feet is given by the expression $t = \sqrt{\frac{2d}{32.2}}$. From what height did a peach drop if it took 6 seconds to fall to the ground?

13.7

Recall	Examples
Square root of a negative number, $\sqrt{-b} = i\sqrt{b}$.	Write in terms of i. $\sqrt{-72} = i\sqrt{72} = i\sqrt{36 \cdot 2} = 6i\sqrt{2}$
Addition and subtraction of complex numbers $(a + bi) + (c + di) = (a + c) + (b + d)i$ $(a + bi) - (c + di) = (a - c) + (b - d)i$	Add or subtract. $(-1 + 3i) + (2 + 4i) = (-1 + 2) + (3 + 4)i$ $\qquad\qquad\qquad\quad = 1 + 7i$ $(-2 - 3i) - (4 + i) = (-2 - 4) + (-3 - 1)i$ $\qquad\qquad\qquad\quad = -6 - 4i$
Multiplication of complex numbers • Use the distributive property to simplify.	Multiply. $\begin{aligned} 3i(-2 + 4i) &= -6i + 12i^2 \\ &= -6i + 12(-1) \\ &= -6i - 12 \\ &= -12 - 6i \end{aligned}$ $\begin{aligned} (2 + i)(-2 - i) &= -4 - 2i - 2i - i^2 \\ &= -4 - 4i - (-1) \\ &= -4 - 4i + 1 \\ &= -3 - 4i \end{aligned}$
Division of complex numbers • If the divisor (denominator) is one term, distribute it over the terms of the numerator. • If the divisor (denominator) is two terms, multiply the numerator and denominator by the conjugate of the denominator.	Divide. $\begin{aligned} \frac{2 - 5i}{2} &= \frac{2}{2} - \frac{5i}{2} \\ &= 1 - \frac{5}{2}i \end{aligned}$ $\begin{aligned} \frac{2 + i}{3 - i} &= \frac{2 + i}{3 - i} \cdot \frac{3 + i}{3 + i} \\ &= \frac{6 + 5i + i^2}{9 - i^2} \\ &= \frac{6 + 5i - 1}{9 - (-1)} \\ &= \frac{5 + 5i}{10} \\ &= \frac{5}{10} + \frac{5i}{10} \\ &= \frac{1}{2} + \frac{1}{2}i \end{aligned}$
| Solve an equation involving imaginary-number roots<br>• Use the quadratic formula to solve.<br>• Check the solution (s) numerically or by substitution. | Solve $x^2 + 2x + 3 = 0$.<br>$a = 1, b = 2$, and $c = 3$<br>$x = \dfrac{-b \pm \sqrt{b^2 - 4ac}}{2a}$<br>$x = \dfrac{-(2) \pm \sqrt{(2)^2 - 4(1)(3)}}{2(1)}$<br>$x = \dfrac{-2 \pm \sqrt{-8}}{2}$<br>$x = \dfrac{-2 \pm 2i\sqrt{2}}{2}$<br>$x = -\dfrac{2}{2} \pm \dfrac{2i\sqrt{2}}{2}$<br>$x = -1 \pm i\sqrt{2}$<br>Check by substitution.<br><br>```
-1+i√(2)→X:X²+2X
+3
                  0
-1-i√(2)→X:X²+2X
+3
                  0
```<br>The solutions are $-1 \pm i\sqrt{2}$. |
```

Write in terms of i.

101. $\sqrt{-64}$

102. $\sqrt{-\dfrac{25}{49}}$

103. $\sqrt{-6.25}$

104. $\sqrt{-50}$

105. $\sqrt{-\dfrac{81}{2}}$

106. $-5\sqrt{-32}$

Perform the indicated operations.

107. $\sqrt{-25} + \sqrt{-49}$

108. $\sqrt{-144} - \sqrt{-64}$

109. $2\sqrt{-36} + 6\sqrt{-81}$

110. $11\sqrt{-4} - 3\sqrt{-36}$

111. $\dfrac{\sqrt{-169}}{\sqrt{-225}}$

112. $\sqrt{-2}\,\sqrt{-6}\,\sqrt{-75}$

113. $\sqrt{-2}(\sqrt{-6} + \sqrt{75})$

114. $(\sqrt{-2} + \sqrt{-18})(\sqrt{-4} - \sqrt{-9})$

115. $(\sqrt{-7} + \sqrt{-11})(\sqrt{-7} - \sqrt{-11})$

Perform the indicated operations. Write your answer in standard form.

116. $(17 + 4i) + (12 - 6i)$

117. $(12 - 3i) - (1 + i)$

118. $(2 + 5i)(-3 + 4i)$

119. $(3 + 11i)(3 - 11i)$

120. $\dfrac{12 + 15i}{3}$

121. $\dfrac{14 - 21i}{-7i}$

122. $\dfrac{41 + i}{5 - 2i}$

123. $(5\sqrt{7} + 8i\sqrt{13}) + (3\sqrt{7} - 2i\sqrt{13})$

124. $i\sqrt{6}(\sqrt{3} - i\sqrt{6})$

125. $\left(\dfrac{2}{3} - \dfrac{3}{5}i\right)\left(\dfrac{1}{2} + \dfrac{1}{3}i\right)$

126. $\dfrac{18 - 11i\sqrt{6}}{3\sqrt{2} + i\sqrt{3}}$

Solve exercises 127–142.

127. $z^2 + 10 = 1$

128. $5t^2 + 125 = 0$

129. $3a^2 + 33 = 0$

130. $(r - 5)^2 + 36 = 0$

131. $7(x + 2)^2 + 23 = 2$

132. $3(4x + 1)^2 = -27$

133. $\left(m + \dfrac{2}{5}\right)^2 + \dfrac{16}{25} = 0$

134. $x^2 - 10x + 29 = 0$

135. $4y^2 + 4y + 10 = 0$

136. $z^2 + 16z + 17 = 0$

137. $9x^2 - 12x + 40 = 0$

138. $4x^2 + 4x + 5.84 = 0$

139. $x^2 - \dfrac{3}{2}x + \dfrac{5}{8} = 0$

140. $\dfrac{x}{x - 5} = \dfrac{10}{x + 1}$

141. $\dfrac{y + 4}{y - 8} = \dfrac{9}{y + 6}$

142. $\dfrac{b - 8}{b(b + 7)} = \dfrac{2}{3}$

143. What is the magnitude of the total voltage V across a circuit if the current I is $5 + 4i$ amperes and the impedance Z is $9 - 2i$ ohms? Use the formula $V = IZ$. Round your answer to the nearest tenth of a volt.

144. Use the formula $Z = \dfrac{V}{I}$ to find the magnitude of the impedance Z when the voltage V is $33 + 19i$ volts and the current I is $5 + 2i$ amperes. Express your answer as a simplified radical.

CHAPTER 13 CHAPTER REVIEW

Evaluate. Round to the nearest thousandth if necessary.

1. $\sqrt{196}$

2. $\sqrt{4.41}$

3. $\sqrt[3]{-\dfrac{27}{64}}$

4. $\sqrt[4]{256}$

5. $\sqrt[4]{-256}$

6. $\sqrt[3]{-29}$

Write as a radical expression, and evaluate if possible.

7. $121^{1/2}$

8. $(-125)^{1/3}$

9. $-81^{1/4}$

10. $(-81)^{1/4}$

11. $729^{5/6}$

12. $(-729)^{5/6}$

13. $(-8)^{-2/3}$

14. $(-81)^{-3/4}$

Use the properties of exponents to evaluate.

15. $\dfrac{1}{9^{-1/2}}$

16. $27^{2/3} \cdot 27^{5/3}$

17. $\dfrac{6^{7/3}}{6^{2/3}}$

18. $(36^{2/3})^{3/4}$

19. $\left(\dfrac{9}{16}\right)^{-3/2}$

20. $[(3^{1/2})(12^{1/2})]^{-3/2}$

Simplify without a calculator, rationalizing all denominators.

21. $-\sqrt{\dfrac{72}{98}}$

22. $\sqrt{\dfrac{15}{144}}$

23. $\sqrt{\dfrac{50}{338}}$

24. $\sqrt[3]{-\dfrac{64}{125}}$

25. $\sqrt{150}$

26. $\sqrt[3]{-448}$

27. $\sqrt[4]{48}$

28. $\sqrt[5]{-486}$

Perform the indicated operations and simplify.

29. $\sqrt{15} \cdot \sqrt{35}$

30. $\sqrt{\dfrac{3}{7}} - \sqrt{\dfrac{7}{3}}$

31. $6\sqrt{13} + 3\sqrt{52}$

32. $4\sqrt{7} - 3\sqrt{7} + 11\sqrt{7} - \sqrt{7}$

33. $\sqrt{3}(\sqrt{6} - \sqrt{3})$

34. $(7 - \sqrt{3})(9 + \sqrt{3})$

35. $(\sqrt{10} - \sqrt{5})(\sqrt{10} + \sqrt{5})$

36. $6\sqrt[3]{375} + 2\sqrt[3]{24}$

Simplify. Write your results using positive exponents.

37. $\dfrac{x^{2/3}}{x^{3/4}}$

38. $z^{3/5} \cdot z^{-3/2}$

39. $\left(\dfrac{p^4}{q^8}\right)^{-5/16}$

40. $(27x^9y^{12})^{4/3}$

41. $\sqrt{72x^6y^3z^4}$

42. $\sqrt[3]{-64x^2y^7}$

43. $2\sqrt{3xy} \cdot y\sqrt{6x}$

44. $\sqrt[3]{-3a^5b^5} \cdot \sqrt[3]{18ab^2}$

45. $\sqrt{10x + 35y} \cdot \sqrt{2x + 7y}$

46. $\sqrt{\dfrac{36x^2}{49y^4}}$

47. $\sqrt{\dfrac{25z}{3}}$

48. $\sqrt[3]{\dfrac{-27a}{b^3}}$

49. $\dfrac{\sqrt{6a}}{\sqrt{12ab}}$

50. $\dfrac{\sqrt{x^3y^2}}{\sqrt{x^4y^5}}$

51. $\sqrt{121x} - \sqrt{81x}$

52. $16d\sqrt{c^3d} - 11c\sqrt{cd^3} + 9cd\sqrt{cd}$

53. $(\sqrt{a} + 9)^2$

54. $\sqrt{6a}(3 + \sqrt{6a})$

55. $(\sqrt{3x} + \sqrt{2y})(\sqrt{3x} - \sqrt{2y})$

56. $2\sqrt[3]{x^2}(3\sqrt[3]{x} + 5\sqrt[3]{x^2})$

57. $\dfrac{\sqrt{18x} - \sqrt{42}}{\sqrt{6}}$

58. $\dfrac{\sqrt{m} - 9}{\sqrt{m}}$

59. $\dfrac{18}{\sqrt{7} + 2}$

60. $\dfrac{3 + \sqrt{x}}{3 - \sqrt{x}}$

61. $\dfrac{7 - 3\sqrt{a}}{3\sqrt{a} - 7}$

62. $\dfrac{25 - 7x}{5 - \sqrt{7x}}$

Graph each radical function. State the restricted values on the domain.

63. $f(x) = 5 - \sqrt{x}$

64. $y = x^{3/2} - 3$

Solve.

65. $(x - 1)^{3/2} = \sqrt[3]{2x + 5}$

66. $(x + 4)^{3/5} = (5x - 1)^{2/3}$

67. $\sqrt{x} + 2 = 11$

68. $3\sqrt{5x} + 7 = 52$

69. $\sqrt{3x + 1} = x - 3$

70. $2\sqrt{x + 9} = \sqrt{9x + 1}$

71. $\sqrt{2x + 1} + 2 = \sqrt{6x + 1}$

72. $\sqrt{10 - x} = \sqrt{3 - x}$

73. $\sqrt[3]{3x} = -6$

74. $\sqrt[4]{x + 2} = 3$

75. $(x + 5)^{2/5} = 9$

76. $x^{-3/2} = 27$

77. $3x^{2/3} - 7 = 2$

Write in terms of i.

78. $\sqrt{-169}$

79. $\sqrt{-\dfrac{64}{81}}$

80. $\sqrt{-\dfrac{25}{6}}$

81. $10\sqrt{-80}$

82. $\sqrt{-6.25}$

83. $-\sqrt{-108}$

Perform the indicated operations. Write your answer in standard form.

84. $(22.3 - 1.33i) + (8.55 - 2.9i)$

85. $\sqrt{-64} + \sqrt{-4}$

86. $\sqrt{-196} - \sqrt{-100}$

87. $\left(\frac{3}{4} + \frac{1}{2}i\right)\left(\frac{4}{9} - 2i\right)$

88. $-1.5i(2.9 - 4i)$

89. $7\sqrt{-25} + 3\sqrt{-64}$

90. $\sqrt{-2}\sqrt{-6}\sqrt{-48}$

91. $(\sqrt{-3} - \sqrt{-10})(\sqrt{-5} + \sqrt{-6})$

92. $(\sqrt{-13} + \sqrt{-17})(\sqrt{-13} - \sqrt{-17})$

93. $(21 - i) - (7 + i)$

94. $(9 - 3i)(-9 + 3i)$

95. $\frac{18 - 27i}{9i}$

96. $\frac{32 - 9i}{2 - 3i}$

97. $\left(\frac{5}{6} - 7i\right) + \left(\frac{5}{6} + i\right)$

98. $\frac{12i + \sqrt{10}}{2\sqrt{5} + i\sqrt{2}}$

99. $\frac{\sqrt{-289}}{\sqrt{-841}}$

Solve exercises 100–107.

100. $\frac{d - 11}{d(d + 9)} = \frac{5}{8}$

101. $(m + 7)^2 + 25 = 0$

102. $4(y - 3)^2 + 29 = 5$

103. $x^2 - 16x + 67 = 0$

104. $9y^2 + 30y + 32 = 0$

105. $a^2 + 100 = 19$

106. $6b^2 + 78 = 0$

107. $\frac{y + 9}{y + 1} = \frac{5}{y - 7}$

108. Use the formula $Z = \frac{V}{I}$ to find the magnitude of the current I when the voltage V is $30 + 52i$ volts and the impedance Z is $9 + 5i$ ohms. Round your answer to the nearest tenth of an ampere.

109. Find the length of a side of a square napkin whose area is 529 square inches.

110. In statistics, the variability of the integers 1 to n is calculated as $s = \sqrt{\frac{n(n + 1)}{12}}$. Calculate the variability of the integers from 1 to 27.

111. The time t in seconds it would take a pear to fall a distance d in feet is given by the expression $t = \sqrt{\frac{2d}{32.2}}$. From what height did the pear fall if it took 4 seconds to reach the ground?

112. Using the formula for the period T of a pendulum of length L, $T = 2\pi\sqrt{\frac{L}{32}}$, find the period of a pendulum whose length is 3.5 feet.

Find the distance on the coordinate plane between the points for the coordinate pairs listed in exercises 113–116.

113. $(6, -3)$ and $(6, 7)$

114. $(-4, 8)$ and $(2, 8)$

115. $(5, -1)$ and $(9, 2)$

116. $(5, 7)$ and $(-2, 10)$

117. Find the possible x-coordinates of a point that has a y-coordinate of -2 and is 4 units away from $(2, 8)$.

118. Find the possible y-coordinates of a point that has an x-cordinate of 7 and is 5 units away from $(3, 2)$.

CHAPTER 13 TEST

A^+ **TEST-TAKING TIPS**

Doing well on tests requires perseverance in the classroom. Attend all classes and take detailed class notes. Research shows that successful students never cut class and usually record at least 64% of what is discussed in class. Failing students record half as much and often miss class. Missing one class puts you behind in the course by at least two lessons. Can you see why? It is important to stay current. Do not allow yourself to fall behind in homework or study, or the entire course will become an effort and a struggle to catch up. Participate in the class by asking questions. Instructors welcome students' questions. See the instructor outside of class if you do not understand a lesson and can't get your questions answered in class. Do not delay in getting help with difficult concepts.

Evaluate if possible. Round decimal answers to the nearest thousandth.

1. $\sqrt{196}$

2. $\sqrt[4]{21}$

3. $\sqrt{-25}$

4. $\sqrt[3]{\frac{-8}{125}}$

Use the properties of exponents to evaluate the given expressions.

5. $16^{5/2} \cdot 16^{-1/4}$

6. $(81^{4/3})^{3/8}$

7. $\left(\frac{8}{27}\right)^{-2/3}$

In exercises 8–15, simplify. Express your results as a single radical expression whenever possible. Rationalize all denominators. (Assume that the variable base of an exponential expression or a variable radicand does not represent values for which the expression is not defined.)

8. $\dfrac{x^{4/5}}{x^{3/10}}$

9. $(\sqrt{2x} - \sqrt{5y})(\sqrt{2x} + \sqrt{5y})$

10. $\sqrt{6x^3y} \cdot \sqrt{15xy^2}$

11. $\sqrt{\dfrac{3z^3}{2xy^2}}$

12. $\dfrac{\sqrt{50xy^3}}{\sqrt{98x^3y^2}}$

13. $\sqrt{81x} + 2\sqrt{25x} - 3\sqrt{16x}$

14. $5\sqrt[3]{x}(2\sqrt[3]{x^2} + 3\sqrt[3]{x})$

15. $\dfrac{\sqrt{x} + 4}{\sqrt{x} + 1}$

Solve exercises 16–19.

16. $(x + 4)^{1/3} + 3 = 2$

17. $3\sqrt{x} - 5 = 13$

18. $\sqrt{3x + 1} = x - 1$

19. $\sqrt{5x + 1} - 2 = \sqrt{x + 1}$

20. Graph the radical function $f(x) = \sqrt{x + 7} - 3$, and state the restricted values on the domain.

Perform the indicated operations. Write your answer in standard form.

21. $3\sqrt{-100} - 2\sqrt{-49}$

22. $\sqrt{-2}\sqrt{-7}\sqrt{-56}$

23. $\sqrt{-2}(\sqrt{-7} - \sqrt{56})$

24. $\dfrac{\sqrt{-225}}{\sqrt{-289}}$

25. $(\sqrt{-6} - \sqrt{-13})(\sqrt{-6} + \sqrt{-13})$

26. $(8 + 2i)(-5 + i)$

27. $\dfrac{29 + 29i}{5 - 2i}$

28. $(22.47 - 13.6i) - (12.2 - 7.32i)$

29. $i\sqrt{15}(\sqrt{3} + i\sqrt{5})$

30. $\dfrac{44 + 16i}{2i}$

Solve exercises 31–36.

31. $x^2 + 196 = 0$

32. $(t + 2)^2 + 15 = 0$

33. $2z^2 + 5 = 2.12$

34. $x^2 - 6x + 17 = 0$

35. $9x^2 - 6x + 7 = 0$

36. $\dfrac{y - 2}{y(y + 1)} = \dfrac{4}{7}$

In exercises 37 and 38, find the distance on the coordinate plane between the points for the coordinate pairs listed.

37. $(3, -2)$ and $(3, 5)$

38. $(2, -3)$ and $(5, 4)$

39. A square tablecloth measures 2000 square inches. A larger square tablecloth measures 3125 square inches. How much more border will the larger tablecloth require over that required by the smaller tablecloth?

40. Explain how you would determine the restricted values in the domain of a radical function in which the index of the radical is an even number. Use an example to illustrate your explanation. Then explain how you would describe the domain of a radical function.

41. What is the magnitude of the total voltage V across a circuit if the current I is $8 + 3i$ amperes and the impedance Z is $11 - 3i$ ohms? Use the formula $V = IZ$, and round your answer to the nearest tenth of a volt.

42. Explain the difference between the real-number system and the complex-number system.

Project

Many mathematical equations—or models, as they are sometimes referred to—involve radical expressions. In this project, we will investigate one of these models. The model helps us understand the calculation of windchill factors. We will investigate the model using English and metric systems of measurement.

Part I

The term "windchill" was first used by an Antarctic explorer, Paul Siple, in 1939. He described the chilling effect of the wind when combined with a low air temperature on the human skin. Siple and Charles Passel were responsible for the first windchill formulas, which were used until the winter of 2001. At that time, the United States and Canada revised the formulas after extensive experimentation revealed some of their shortcomings.

One form of the Siple–Passel windchill formula is

$$T = 0.0817(3.71v^{1/2} + 5.81 - 0.25v)(t - 91.4) + 91.4$$

where t is the actual temperature in Fahrenheit degrees and v is the wind velocity in miles per hour.

One form of the revised windchill formula is

$$T = 35.74 + 0.6215t - 35.75v^{0.16} + 0.4275v^{0.16}$$

where t is the actual temperature in Fahrenheit degrees and v is the wind velocity in miles per hour.

1. Write both formulas for the windchill factor, given an air temperature of 32°F.

2. Complete the following table, which calculates the windchill factor for both formulas if the temperature measures 32°F and the wind velocity, in miles per hour, is as shown in the table:

Wind speed v, mph	10	15	20	25	30	35	40
Siple–Passel windchill factor T, °F							
Revised windchill factor T, °F							

3. Describe what happens to the measured temperature of 32°F as the wind speed increases. That is, what do you "feel" is happening to the temperature?

4. Graph both equations in step 1 on the same coordinate plane.

5. Describe the difference in the windchill factor obtained by the two equations.

6. Repeat the first five steps for a measured air temperature of 0°F.

7. Compare the tables and graphs you have produced, and describe what happens when the wind speed is held constant and the temperature is dropping.

Part II

Canada uses the metric system of measurement. Therefore, a metric version of the Siple–Passel formulas is also available.

One form of the Siple–Passel windchill formula is given by the equation

$$T = 0.045(5.27v^{1/2} + 10.45 - 0.28v)(t - 33) + 33$$

where t is the actual temperature in Celsius degrees and v is the wind velocity in kilometers per hour.

One form of the revised windchill formula is given by the equation

$$T = 13.12 + 0.6215t - 11.37v^{0.16} + 0.3965v^{0.16}$$

where t is the actual temperature in Celsius degrees and v is the wind velocity in kilometers per hour.

Repeat steps 1 through 7, using 0°C and −10°C as the air temperature. Compare your findings from Part I and Part II.

Part III

Additional adjustments are being proposed to the revised Siple–Passel formulas in order to take account of factors that reflect one's comfort level with the temperature and humidity. Find a reference that discusses one of these concepts, and write a short summary of your findings. Be sure to document your source.

CHAPTERS 1–13 CUMULATIVE REVIEW

Simplify and write with positive exponents.

1. $2^0 x^{-1} y^2$

2. $(3x^{1/2} y^{3/4})^2$

3. $\left(\dfrac{(2s)^2}{3t} \right)^{-3}$

Simplify.

4. $(2x^3 + 6x^2 y - 2xy^2 + y^3) + (3x^3 + 4xy^2 - y^3)$

5. $(1.2a^2 - 3.6ab + b^2) - (4a^2 + 2.71ab - 3.4b^2)$

6. $(6.8m^2 n)(-2mn^2 p)$

7. $(3a - b)(2a + 4b)$

8. $(2x + 3)(2x - 3)$

9. $(2x + 3)^2$

10. $\dfrac{-15x^2 y^3 z}{3xyz}$

11. $\dfrac{2m^2 n + 4mn - 8n^2}{2m^2 n}$

12. $\dfrac{x^2 + 2x - 3}{x^2 - 9}$

13. $\dfrac{x}{x + 2} + \dfrac{3}{x - 2}$

14. $\dfrac{1}{x + 2} - \dfrac{3}{x - 3} - \dfrac{x + 3}{x^2 + x - 12}$

15. $\dfrac{2a - 5}{a + 2} \cdot \dfrac{2a^2 + 3a - 2}{4a^2 - 25}$

16. $\dfrac{2x}{2x^2 + 9x + 4} \div \dfrac{4x^2 y}{x^2 + 9x + 20}$

17. $\sqrt{8y} + \sqrt{2y} - 2\sqrt{18y}$

18. $\sqrt[3]{25a^4 b} \cdot \sqrt[3]{5ab}$

19. $(\sqrt{3} - \sqrt{5})(\sqrt{3} + \sqrt{5})$

20. $\dfrac{\sqrt{21abc^3}}{\sqrt{7a^3 b}}$

21. $\dfrac{\sqrt{x} + 2}{\sqrt{x} - 1}$

22. $x^{1/2}(x^{2/3} - x^{3/4})$

Factor exercises 23–25 completely if possible.

23. $16a^2 - 25b^2$

24. $x^2 - 2x - 8$

25. $3x^2 - 9x - 30$

26. Given $f(x) = -x^2 + 3x - 1$, find $f(-2)$.

Graph and label as indicated. Determine the domain and range of each function.

27. $f(x) = -3.2x + 1$; three points on the graph

28. $y = x^2 + 4x + 6$; vertex, y-intercept, enough points to determine the curve, and the axis of symmetry

29. $g(x) = -2x^2 + x + 1$; vertex, x-intercept, y-intercept, enough points to determine the curve, and the axis of symmetry

30. $y = \dfrac{x^2 - 4}{x + 2}$; enough points to determine the curve

31. $h(x) = \dfrac{2}{x}$; enough points to determine the curve

32. $y = 2\sqrt{x}$; enough points to determine the curve

33. $a(x) = x^{2/3} + 2$; enough points to determine the curve

Solve.

34. $2(x + 3) - 4(x - 1) = x - 2$

35. $x^2 - 3x + 9 = 0$

36. $x^2 + 2x - 15 = 0$

37. $2t^2 + 3t = 4$

38. $\dfrac{x + 1}{x - 1} = \dfrac{x + 3}{x + 2}$

39. $\dfrac{2}{x^2} + \dfrac{3}{2x^2} = \dfrac{1}{8}$

40. $\sqrt{2x - 3} + 5 = 9$

Solve. Write the solution set in interval notation.

41. $2x^2 - 9x - 18 < 0$

Solve.

42. $3x - 4y = 3$
$\quad\; x + y = 8$

In exercises 43 and 44, write an equation of the line that satisfies the given conditions.

43. Passes through the points $(3, -5)$ and $(-4, 2)$

44. Perpendicular to $2x + 3y = 5$ and passes through $(-2, 1)$

45. Write a quadratic equation for a curve that passes through the points $(0, -4)$, $(-4, 0)$, and $(1, 0)$.

46. The diagonal of a rectangle measures 20 feet. The length of the rectangle is 4 feet more than the width. Find the dimensions of the rectangle.

47. Determine the possible y-coordinates of a point 5 units from $(-3, 4)$ with an x-coordinate of 0.

48. Kevin can rake the leaves in his yard in 5 hours. His sister Elizabeth can rake the leaves in her yard in 4 hours. It is 12:00 noon. Can Kevin and Elizabeth, working together, be through raking leaves before their mother arrives home at 2:30 P.M.?

49. A small business spends $1000 on production equipment. Each item produced costs $0.55 to produce and is sold for $1.25. How many items must be sold before the business breaks even?

50. A sky diver free-falls from a height of 10,000 feet above the ground. He opens his parachute at a height of 2500 feet. Use the vertical-position equation to determine the length of time the sky diver is free-falling.

Chapter 14

Exponential and Logarithmic Functions and Equations

In this chapter, we introduce two new types of functions: the exponential function and the logarithmic function. We show how to use the properties of exponents to perform operations with these functions and how to solve equations involving them. Exponential and logarithmic functions have a special relationship: Each is the inverse function of the other. We begin the chapter by discussing what inverse functions are, when they exist, and how we can find them.

Exponential and logarithmic functions are used to model many important real-world situations in business and science. Exponential models of growth and decay describe situations as varied as compound interest, population growth, and radioactive decay. Logarithms are used to describe the magnitudes of earthquakes, sound levels, and the acidity of a chemical solution. We discuss some of these applications throughout the chapter. We conclude with projects illustrating exponential and logarithmic models.

14.1 APPLICATION

Currency exchange rates vary daily. According to a recent report, a function for determining the number of European euros given x United States dollars was $f(x) = 1.1173x$. Determine the inverse function $f^{-1}(x)$ and explain its meaning.

We will discuss this application further. See page 983.

14.1 Inverse Functions

OBJECTIVES

1 Determine the inverse of a relation.
2 Determine whether a function is a one-to-one function.
3 Graph inverse functions.
4 Determine the inverse of a function.
5 Model real-world situations by using inverse functions.

14.1.1 Determining the Inverse of a Relation

In Chapter 3, we defined a relation as a set of ordered pairs. For example, the set of ordered pairs a biologist obtains when Celsius temperatures are converted to Fahrenheit temperatures,

$$T = \{(20, 68), (21, 69.8), (22, 71.6), (23, 73.4)\}$$

is a relation.

If we reverse the order in each ordered pair, another useful relation results. For example, the set of ordered pairs a biologist obtains when Fahrenheit temperatures are converted to Celsius temperatures,

$$\{(68, 20), (69.8, 21), (71.6, 22), (73.4, 23)\}$$

is also a relation.

This second relation is called the **inverse** of the relation T and is denoted T^{-1}.

$$T = \{(20, 68), (21, 69.8), (22, 71.6), (23, 73.4)\}$$
$$T^{-1} = \{(68, 20), (69.8, 21), (71.6, 22), (73.4, 23)\}$$

 **HELPING HAND** Do *not* confuse the inverse notation T^{-1} with exponential notation. The -1 is *not* an exponent.

> **INVERSE OF A RELATION WRITTEN AS A SET OF ORDERED PAIRS**
> To determine the inverse of a relation written as a set of ordered pairs, interchange the coordinates of the independent and dependent variables in the original relation.

EXAMPLE 1 Determine the inverses of the following relations:

a. $f = \{(3, 6), (5, 10), (4, 8), (0, 0)\}$
b. $A = \{(1, 2), (-2, 3), (4, -5), (0, 2)\}$

Solution

a. $f = \{(3, 6), (5, 10), (4, 8), (0, 0)\}$
$f^{-1} = \{(6, 3), (10, 5), (8, 4), (0, 0)\}$

b. $A = \{(1, 2), (-2, 3), (4, -5), (0, 2)\}$
$A^{-1} = \{(2, 1), (3, -2), (-5, 4), (2, 0)\}$

Recall that a relation can also be written as an equation in two variables. To determine the inverse of a relation expressed in this form, we follow the same procedure as with ordered pairs: Interchange the independent and dependent variables. For example, given x = Celsius temperature, y = Fahrenheit temperature, and the relation $y = \frac{9}{5}x + 32$, determine the inverse relation.

Interchange the variables. Let x = Fahrenheit temperature and y = Celsius temperature.

$$x = \frac{9}{5}y + 32$$

Solve for y.

$$x - 32 = \frac{9}{5}y$$

$$\frac{5}{9}(x - 32) = y \quad \text{or} \quad y = \frac{5}{9}(x - 32)$$

INVERSE OF A RELATION WRITTEN AS AN EQUATION

To determine the inverse of a relation written as an equation, interchange the independent and dependent variables in the original equation and solve for the new dependent variable.

EXAMPLE 2   Determine the inverse relations.

a. $y = 3x + 5$ **b.** $y = x^2 - 2$

Solution

a. $y = 3x + 5$

$x = 3y + 5$ Interchange the variables.

$x - 5 = 3y$ Solve for y.

$\dfrac{x - 5}{3} = y \quad \text{or} \quad y = \dfrac{1}{3}x - \dfrac{5}{3}$

b. $y = x^2 - 2$

$x = y^2 - 2$ Interchange the variables.

$x + 2 = y^2$ Solve for y.

$\pm\sqrt{x + 2} = y \quad \text{or} \quad y = \pm\sqrt{x + 2}$

14.1.1 Checkup

In exercises 1–4, determine the inverse of each relation.

1. $g = \{(1, 0.5), (2, 1), (3, 1.5), (4, 2), (5, 2.5), (6, 3)\}$ **2.** $K = \left\{\left(\frac{1}{2}, 3\right), \left(\frac{2}{3}, 4\right), \left(\frac{3}{4}, 5\right), \left(\frac{4}{5}, 4\right), \left(\frac{5}{6}, 3\right)\right\}$

3. $y = -4x + 1$ **4.** $y = 3x^2 + 6$

5. The inverse of the function $f(x)$ is denoted by $f^{-1}(x)$. What does the notation $[f(x)]^{-1}$ indicate? Describe the differences in the two notations.

14.1.2 Determining One-to-One Functions

In Chapter 3, we also defined a function. A function is a relation such that every element in the domain corresponds to only one element in the range. In Example 1, we determined the inverse relation of the two functions f and A.

$$f = \{(3, 6), (5, 10), (4, 8), (0, 0)\}$$
$$A = \{(1, 2), (-2, 3), (4, -5), (0, 2)\}$$

The inverse of the function f is a function.

$$f^{-1} = \{(6, 3), (10, 5), (8, 4), (0, 0)\}$$

However, the inverse of the function A is not a function, because an element in the domain of the inverse relation, 2, is paired with two different elements in the range, 1 and 0.

$$A^{-1} = \{(2, 1), (3, -2), (-5, 4), (2, 0)\}$$

In order for each element in the domain of the inverse to be paired with only one element in the range, each element in the range of the function must be paired with only one element in the domain. The function is then called a **one-to-one function**. The inverse of a one-to-one function is a function.

> **ONE-TO-ONE FUNCTION**
> A function is a one-to-one function if no two different ordered pairs have the same second coordinate.

HELPING HAND A one-to-one function is a set of ordered pairs in which no two different ordered pairs have the same first coordinate or the same second coordinate.

EXAMPLE 3 Determine whether the following relations are functions. If so, is the function one-to-one?

a. $h = \{(2.75, 3.25), (0.5, 3.25), (-2.25, 2.75)\}$
b. $R = \{(0, 1.5), (1, 2.5), (3, 4.5)\}$
c. $j = \{(3, 5), (3, -5), (2, 6), (2, -6)\}$

Solution

a. $h = \{(2.75, 3.25), (0.5, 3.25), (-2.25, 2.75)\}$
The relation h is a function, because each ordered pair has a different first coordinate.
The function h is not one-to-one, because the second coordinate, 3.25, is repeated in two ordered pairs. The inverse of h is not a function.

b. $R = \{(0, 1.5), (1, 2.5), (3, 4.5)\}$
The relation R is a function because each ordered pair has a different first coordinate.
The function R is one-to-one, because no two ordered pairs have the same second coordinate. The inverse of R is a function.

c. $j = \{(3, 5), (3, -5), (2, 6), (2, -6)\}$
The relation j is not a function, because the first coordinates, 3 and 2, are repeated in two different ordered pairs.

In Example 2, we determined the inverse of two functions written in equation form:

a. $y = 3x + 5$ **b.** $y = x^2 - 2$

The inverse of the first function, $y = \frac{1}{3}x - \frac{5}{3}$, is a function. However, the inverse of the second function, $y = \pm\sqrt{x + 2}$, is not a function. If we graph each of the inverses, we can see that one is a function and that one is not a function by using the vertical-line test.

a.

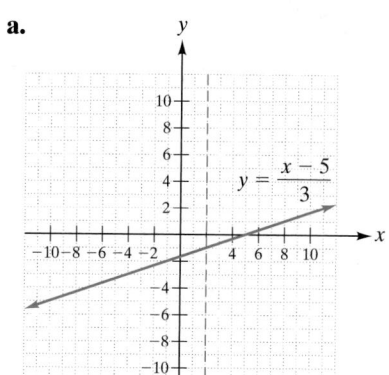

b.

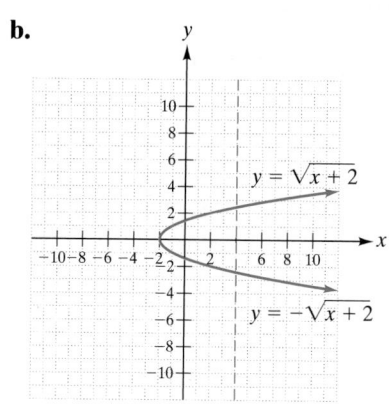

Note: All sample vertical lines drawn for part a intersect the graph once. A sample vertical line is drawn for part b that intersects the graph more than once.

We can also determine whether the inverse of a function is a function from the graph of the function itself. Complete the following set of exercises to see when a function is one-to-one, or has an inverse that is a function.

Discovery 1

Graphs of One-to-One Functions

In the following graphs of one-to-one functions, draw a horizontal line through more than one point on the graph if possible:

1.

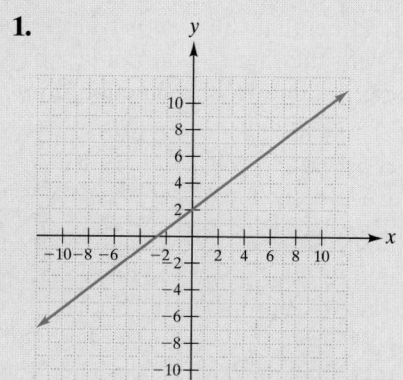

2.

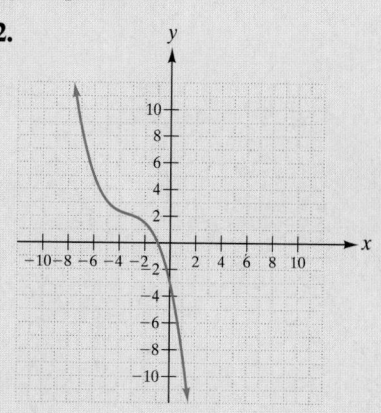

(Continued on page 978)

Write a rule for determining a one-to-one function from the graph of a function by drawing a horizontal line through points on the graph.

Check your rule on the following graph of a function, which is not one-to-one:

3.

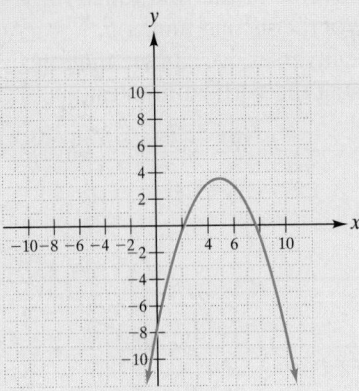

All possible horizontal lines cross the graphs of the one-to-one functions only once. It is possible to draw a horizontal line that crosses the function more than once if that function is not one-to-one.

HORIZONTAL-LINE TEST

If a horizontal line can be drawn such that it intersects the graph of a function more than once, the graph does not represent a one-to-one function. If the horizontal line intersects the graph only once, then the graph represents a one-to-one function.

We know that this rule is true because if we graph two distinct ordered pairs with the same y-coordinate, they will lie on the same horizontal line. However, a one-to-one function does not have two ordered pairs with the same y-coordinate.

HELPING HAND According to the vertical-line test for functions, it is not possible to draw a vertical line that intersects the graph of a function more than once. According to the horizontal-line test for one-to-one functions, it is not possible to draw a horizontal line that intersects the graph of a one-to-one function more than once. Therefore, it is not possible to draw a vertical or a horizontal line that intersects the graph of a one-to-one function more than once.

EXAMPLE 4   Determine whether the following graphs represent one-to-one functions:

a.

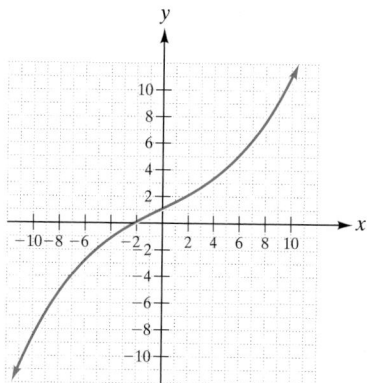

b.

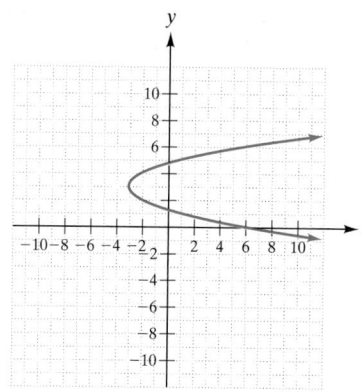

c.

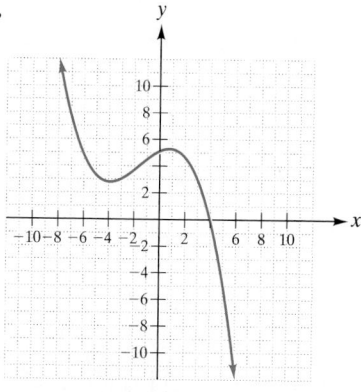

Solution

a. The graph represents a one-to-one function, because all possible vertical and horizontal lines cross the graph only once.

b.

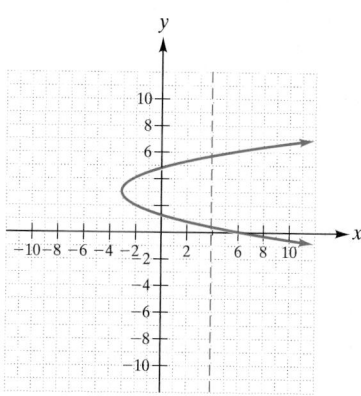

The graph does not represent a function, because a vertical line can be drawn that intersects the graph more than once.

c.

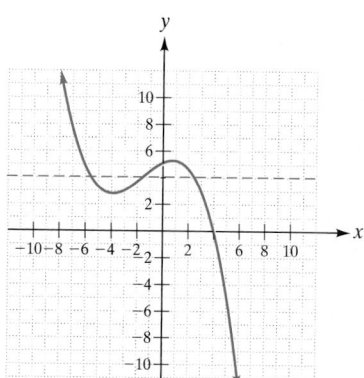

The graph is a function, because it passes the vertical-line test, but a horizontal line can be drawn that intersects the graph more than once. Therefore, it is not a one-to-one function.

14.1.2 Checkup

Determine whether the relations are functions. If so, is the function one-to-one?

1. $B = \{(0, -3), (2, 1), (4, 5), (6, 9), (8, 13)\}$ **2.** $G = \{(-2, 5), (-1, 2), (0, 1), (1, 2), (2, 5)\}$

3. $K = \{(2.1, 4.2), (-2.1, 4.2), (2.1, 6.3)\}$

Determine whether the graphs in exercises 4–6 represent one-to-one functions.

4.
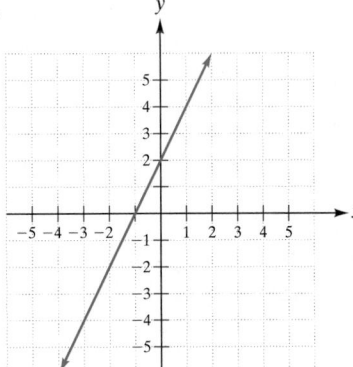

5.
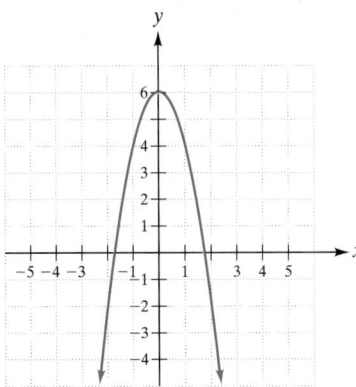

6.
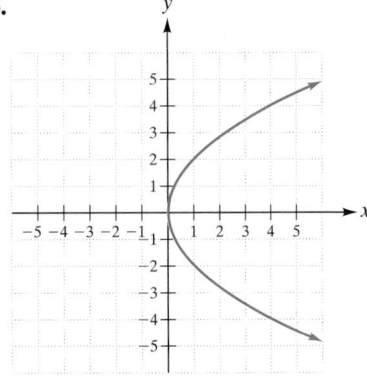

7. For what purpose is the vertical-line test used?

8. For what purpose is the horizontal-line test used?

9. Every one-to-one function is a relation, but not every relation is a one-to-one function. Explain why this is so.

14.1.3 Graphing Inverse Functions

We interchange the coordinates of the ordered pairs of a one-to-one function to obtain its inverse function. Therefore, the domain and range of a one-to-one function interchange, as do the domain and range of its inverse function. If we graph a one-to-one function and its inverse function on the same coordinate plane, as well as graph the line $y = x$, we obtain an interesting picture. For example,

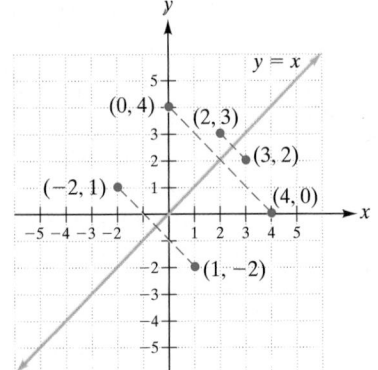

Graph the function $f = \{(2, 3), (0, 4), (-2, 1)\}$.

Graph the inverse function $f^{-1} = \{(3, 2), (4, 0), (1, -2)\}$.

Graph the equation $y = x$.

When we graph the function f and its inverse function f^{-1} on the same coordinate plane, we see that the points graphed are the same distance, on either side, from the line representing the equation $y = x$. This line is called the **line of symmetry**. Visualizing the line of symmetry can help us determine whether we have graphed a function and its inverse function correctly.

HELPING HAND If you fold your graph paper along the line of symmetry, the graph of the function and the graph of its inverse should match.

EXAMPLE 5 On your calculator, graph the following function and its inverse, as well as the line of symmetry $y = x$:

$$f(x) = x + 4$$
$$f^{-1}(x) = x - 4$$

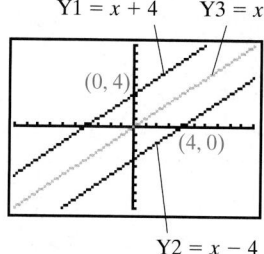

$(-10, 10, -10, 10)$

Solution

Note that the graphs for $y = x + 4$ and $y = x - 4$ appear to be symmetric about the line $y = x$.

HELPING HAND You need to be careful, because looks are sometimes deceiving. However, you can trace the two lines to determine that the coordinates of the ordered pairs interchange. For example, the ordered pair $(0, 4)$ lies on the graph of $y = x + 4$ and the ordered pair $(4, 0)$ lies on the graph of the inverse $y = x - 4$.

14.1.3 Checkup

1. On your calculator, graph the following function, its inverse, and the line of symmetry $y = x$:

$$f(x) = 2x - 4 \quad \text{and} \quad f^{-1}(x) = \frac{1}{2}x + 2$$

2. What does the line of symmetry $y = x$ mean to you?

14.1.4 Determining an Inverse Function

If a one-to-one function $f(x)$ is defined by an equation, we can determine the inverse function $f^{-1}(x)$ in the same way we obtained an inverse for a relation.

> To determine the inverse of a function $f(x)$,
>
> - Replace $f(x)$ with y. ✓
> - Interchange x and y.
> - Solve the resulting equation for y.
> - Replace y with $f^{-1}(x)$.

EXAMPLE 6 Determine the inverse function $f^{-1}(x)$. Check the inverse with your calculator.

a. $f(x) = \sqrt[3]{2x}$ **b.** $f(x) = \dfrac{2}{3x} + 4$

Solution

a. $f(x) = \sqrt[3]{2x}$

$y = \sqrt[3]{2x}$ Replace f(x) with y.

$x = \sqrt[3]{2y}$ Interchange x and y.

$x^3 = 2y$ Solve for y.

$y = \dfrac{1}{2}x^3$

$f^{-1}(x) = \dfrac{1}{2}x^3$ Replace y with $f^{-1}(x)$.

Calculator Check

Graph the function, its inverse, and the line of symmetry to check visually.

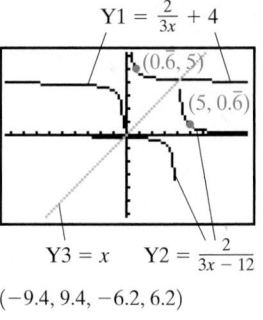

$Y2 = \frac{1}{2}x^3$ $Y1 = \sqrt[3]{2x}$

$Y3 = x$

$(-4.7, 4.7, -3.1, 3.1)$

b. $f(x) = \dfrac{2}{3x} + 4$

$y = \dfrac{2}{3x} + 4$ Replace f(x) with y.

$x = \dfrac{2}{3y} + 4$ Interchange x and y.

$x - 4 = \dfrac{2}{3y}$ Subtract 4 from both sides.

$3y(x - 4) = 3y\left(\dfrac{2}{3y}\right)$ Multiply both sides by 3y.

$3xy - 12y = 2$ Simplify.

$y(3x - 12) = 2$ Factor out a common factor of y.

$\dfrac{y(3x - 12)}{3x - 12} = \dfrac{2}{3x - 12}$ Divide both sides by $(3x - 12)$.

$y = \dfrac{2}{3x - 12}$ Simplify.

Calculator Check

The calculator check is very difficult to visualize. Check to see whether the coordinates are reversed in the ordered pairs of the two functions.

$Y1 = \dfrac{2}{3x} + 4$

$(0.6, 5)$

$(5, 0.\overline{6})$

$Y3 = x$ $Y2 = \dfrac{2}{3x - 12}$

$(-9.4, 9.4, -6.2, 6.2)$

14.1.4 Checkup

In exercises 1 and 2, determine the inverse of each function. Then graph the function, its inverse, and the line of symmetry.

1. $f(x) = -2x + 6$ **2.** $g(x) = x^3$

3. When is the inverse of a function not a function?

4. How can you test to find whether a given function has an inverse that is a function?

14.1.5 Modeling the Real World

We have seen many times in this text that important relations in business, science, and social studies are functions. We can write inverse functions for many of them, just as we've learned to do in this section. It is often helpful to interpret the meaning of these inverse functions in terms of the original situation.

This change in your point of view can sometimes provide new insights into your mathematical model.

EXAMPLE 7 Marvin is paid a salary of $300 per week, plus a commission of 6% of his total sales.

a. Write a function $I(x)$ that represents Marvin's weekly income.

b. Determine the inverse function $I^{-1}(x)$.

c. Interpret the meaning of the inverse function $I^{-1}(x)$.

Solution

a. Marvin's income is $300, plus 6% of his weekly sales.

Let x = the amount of Marvin's weekly sales

$I(x)$ = weekly income

$I(x) = 300 + 0.06x$

b.
$$I(x) = 300 + 0.06x$$
$$y = 300 + 0.06x \qquad \text{Let } y = I(x).$$
$$x = 300 + 0.06y \qquad \text{Interchange } x \text{ and } y.$$
$$x - 300 = 0.06y \qquad \text{Solve for } y.$$
$$\frac{x - 300}{0.06} = y \qquad \text{Divide both sides by 0.06.}$$
$$\frac{50}{3}x - 5000 = y$$
$$I^{-1}(x) = \frac{50}{3}x - 5000 \qquad \text{Replace } y \text{ with } I^{-1}(x).$$

c. The inverse function $I^{-1}(x)$ represents the amount of Marvin's weekly sales in terms of his total pay for the week.

APPLICATION

Currency exchange rates vary daily. According to a recent report, a function for determining the number of European euros, given x U.S. dollars, was $f(x) = 1.1173x$. Determine the inverse function $f^{-1}(x)$ and explain its meaning.

Discussion

$$f(x) = 1.1173x$$
$$y = 1.1173x \qquad \text{Substitute } y \text{ for } f(x).$$
$$x = 1.1173y \qquad \text{Interchange } x \text{ and } y.$$

$$\frac{x}{1.1173} = \frac{1.1173y}{1.1173} \qquad \text{Solve for } y.$$
$$\frac{1}{1.1173}x = y$$
$$0.8950x \approx y$$
$$y \approx 0.8950x$$
$$f^{-1}(x) \approx 0.8950x \qquad \text{Substitute } f^{-1}(x) \text{ for } y.$$

The inverse function is used to determine the number of U.S. dollars, given x European euros.

14.1.5 Checkup

1. Dimitri invests $1200 in a fund that pays 5% simple interest per year.
   a. Write a function $I(x)$ that represents the interest the fund will earn after x years.
   b. Determine the inverse function $I^{-1}(x)$.
   c. Interpret the meaning of the inverse function.

2. A function for determining the number of British pounds, given x U.S. dollars is $f(x) = 0.6919x$. Determine the inverse function and explain its meaning.

14.1 Exercises

Determine the inverse of each relation.

1. $h = \{(-3, 5), (-2, 4), (-1, 3), (0, 2), (1, 1), (2, 0)\}$
2. $d = \{(5, 25), (4, 16), (3, 9), (2, 4), (1, 1), (0, 0)\}$
3. $A = \{(3, 6), (2, 7), (1, 8), (0, 8), (-1, 7), (-2, 6)\}$
4. $J = \{(-4, 1), (3, 2), (-2, 1), (1, 2), (0, 1), (-1, 2)\}$

5. $y = 2x - 8$
6. $y = 3x + 6$
7. $y = -3x + 2$
8. $y = 5 - 2x$
9. $y = \frac{3}{4}x + 9$

10. $y = -\frac{2}{3}x + \frac{4}{3}$
11. $y = 0.125x - 2.5$
12. $y = 1.8 - 0.4x$
13. $y = x^2 - 2$
14. $y = x^3 + 1$

Determine whether the given relations are functions. If so, is the function one-to-one?

15. $k = \{(5, 2), (10, 4), (15, 8), (20, 16), (25, 32)\}$
16. $j = \{(-8, 3), (-9, 4), (-10, 5), (-11, 4), (-12, 3)\}$
17. $Q = \{(-2, 0), (-4, 2), (-6, 4), (-8, 2), (-10, 0)\}$
18. $P = \{(1, 16), (2, 15), (3, 14), (4, 13), (5, 12)\}$
19. $A = \{(-1, 0), (0, 1), (-1, 2), (0, 3)\}$
20. $B = \{(6, 5), (-6, 3), (6, 7), (-6, 4)\}$

Determine whether each graph represents a one-to-one function.

21.

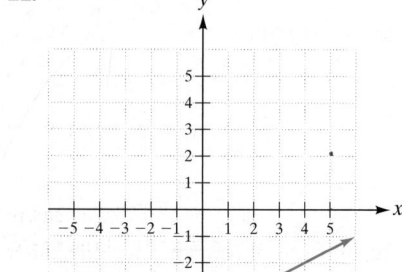

22.

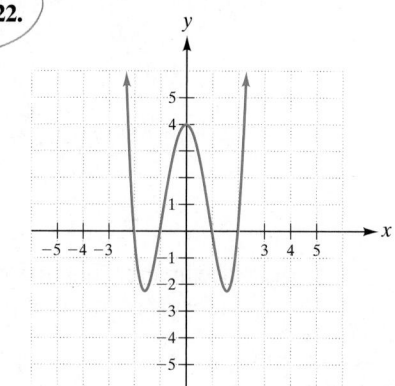

23.

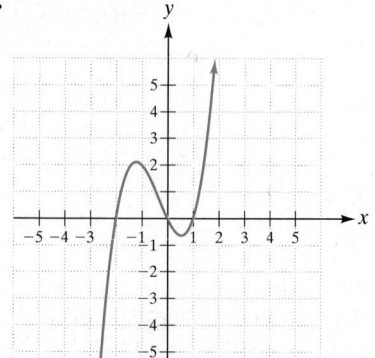

24.

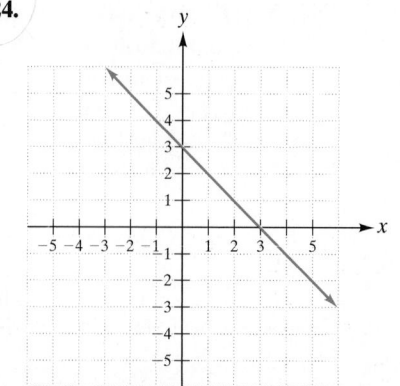

25.

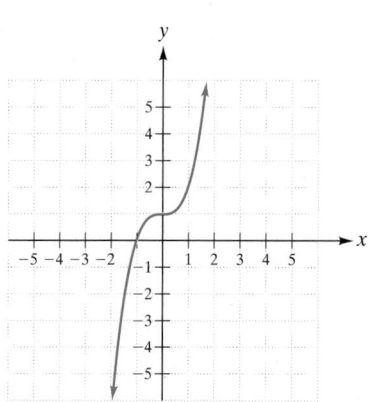

26.

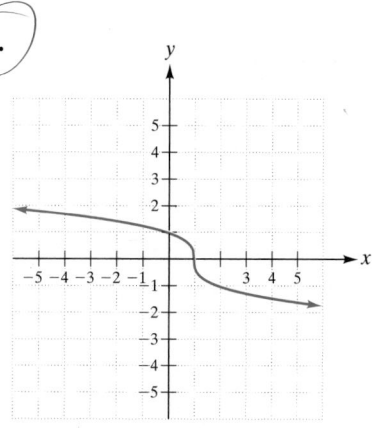

Determine the inverse of each function. Then use your calculator to graph the function, its inverse, and the line of symmetry, y = x.

27. $g(x) = 3x - 6$ **28.** $f(x) = -4x + 2$ **29.** $y = \frac{2}{3}x - 4$ **30.** $y = -\frac{2}{5}x + 4$

31. $h(x) = x^2 - 1$ **32.** $r(x) = 2 - x^2$ **33.** $y = \frac{1}{3}x^3 - 4$ **34.** $y = \frac{1}{4}x^3 + 2$

Write a function that represents each situation. Then determine the inverse of the function. Intepret what the inverse of the function represents.

35. Julia is paid $1500 per month, plus a commission of 2% of the value of all stock transactions she handles. What is Julia's monthly income?

36. Harvey earns $125 per week, plus an average of 15% of his sales in tips waiting tables. What are his weekly earnings?

37. Antonio drives at an average speed of 55 miles per hour. What distance does he travel in a given amount of time?

38. Grace operates a machine that cuts 30 pairs of jeans per hour. How many pairs of jeans can she cut in a given amount of time?

39. Priscilla borrowed $2000 from her parents. They agreed that until she repays the loan, they will charge her 0.5% simple interest per month. What is the total amount Priscilla will repay?

40. Kimberly purchased a bond for $1500. The bond pays 0.4% simple interest per month. How much will Kimberly earn for the bond?

14.1 Calculator Exercises

The calculator can draw the inverse of a function without your having to derive the inverse algebraically. As an example, graph the function $y = \frac{1}{2}x^3 + 3$. Do so by storing the function in Y1, and graph it with the standard window setting. Now draw the inverse by entering the following:

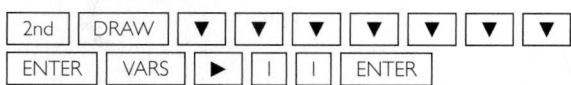

The calculator will draw the inverse of the function on the same graph as the function. Once you have algebraically determined the inverse, you can check it by first graphing the function (stored in Y1), then graphing your algebraically de-

rived inverse (stored in Y2), and finally instructing the calculator to draw the inverse of the original function. If the graph of your inverse and the calculator's inverse coincide, you have correctly determined the inverse. (Note: You will not be able to trace the two graphs to verify that they coincide, but you can check visually.) In the following exercises, find the inverse algebraically, graph the function and its inverse, and then check, using this feature of your calculator just discussed:

1. $y = 0.3x^2 - 4$ **2.** $f(x) = 0.1x^3 + 2x$

3. $g(x) = 5\sqrt{x} - 1.5$ **4.** $y = \frac{1}{x} + 2$

14.1 Writing Exercise

In this section, you were cautioned not to confuse the notation $f^{-1}(x)$ with the notation $[f(x)]^{-1}$. Discuss what each notation means, using an example for each. Then consult a

dictionary to find definitions for the words *inverse* and *reciprocal*, and relate these definitions to what you have studied.

14.2 Exponential Functions

OBJECTIVES
1 Identify exponential functions.
2 Evaluate exponential functions.
3 Graph exponential functions.
4 Graph the natural exponential function.
5 Model real-world situations by using exponential functions.

APPLICATION

The population of the United States (in millions) may be approximated by the exponential function $f(x) = 250(1.01)^x$, where x is the number of years after 1990.

a. Graph this function.

b. According to the graph, what is the estimated population of the United States in the year 2010?

c. According to U.S. Bureau of the Census Projections, the U.S. population will be between 291 million and 311 million in 2010. Compare your results in part b with these projections.

After completing this section, we will discuss this application further. See page 994.

14.2.1 Identifying Exponential Functions

In previous chapters, we studied polynomial functions, rational functions, and radical functions. Each of these functions can be written with a term with a variable base and a rational-number exponent.

$$f(x) = x^2 \qquad g(x) = x^{-2} \qquad h(x) = x^{1/2}$$

Polynomial function   Rational function   Radical function

In this section, we examine a function that has a term with a rational-number base and a variable exponent, the reverse of what we previously studied.

$$f(x) = 2^x \qquad g(x) = 2^{-x} \qquad h(x) = \left(\frac{1}{2}\right)^x$$

We call each of these functions an **exponential function**.

EXPONENTIAL FUNCTION

An exponential function f can be written in the form

$$f(x) = a^x$$

where $a > 0$, $a \neq 1$, and x is any real number.

Note the restrictions on the base a. The base must be positive and not equal to 1. A base of 1 results in the function $f(x) = 1^x$ or $f(x) = 1$, a constant function.

EXAMPLE 1 For each function, state whether it is an exponential function or not.

a. $y = 5^x$ **b.** $f(x) = x^{-5/3}$ **c.** $g(x) = 0^x$

Solution

a. $y = 5^x$ is an exponential function, because it has a rational-number base and a variable exponent.

b. $f(x) = x^{-5/3}$ is not an exponential function, because the base is a variable. In fact, this is a radical function.

c. $g(x) = 0^x$ is not an exponential function, because it has a base of 0.

 14.2.1 Checkup

For each function, state whether it is an exponential function or not.

1. $h(x) = x^{2/3}$ **2.** $k(x) = \left(\dfrac{2}{3}\right)^x$ **3.** $y = 1^x$

14.2.2 Evaluating Exponential Functions

To evaluate an exponential function, we substitute the value of the independent variable and evaluate the resulting numeric expression. This expression will be an exponential expression containing a real-number exponent. We already know how to evaluate an exponential expression with a rational exponent. For example,

$$a^3 = a \cdot a \cdot a \qquad a^{-2} = \frac{1}{a^2} \qquad a^{3/4} = (\sqrt[4]{a})^3$$

Therefore,

$$2^3 = 2 \cdot 2 \cdot 2 = 8 \qquad 5^{-2} = \frac{1}{5^2} = \frac{1}{25} \qquad 16^{3/4} = (\sqrt[4]{16})^3 = 2^3 = 8$$

We also may need to evaluate an exponential expression with an irrational exponent, such as $a^{\sqrt{3}}$. For example, evaluate $2^{\sqrt{3}}$.

Since $\sqrt{3} \approx 1.732$, we can approximate $2^{\sqrt{3}}$ by evaluating $2^{1.7}$, $2^{1.73}$, or $2^{1.732}$, and so on, depending on the accuracy we require.

However, since we are using a calculator to perform these operations, we can also enter the expression with an irrational exponent and obtain an approximation with the accuracy of our calculator display.

```
2^1.7
       3.249009585
2^1.73
       3.317278183
2^1.732
       3.321880096
```

```
2^√(3)
       3.321997085
```

EXAMPLE 2 Evaluate the function $g(x) = 8^x$ at the given values.

a. $g(2)$ **b.** $g(-2)$ **c.** $g\left(\dfrac{2}{3}\right)$ **d.** $g(\sqrt{2})$

Solution

```
8^√(2)
       18.93050099
```

a. $g(2) = 8^2 = 64$ **b.** $g(-2) = 8^{-2} = \dfrac{1}{8^2} = \dfrac{1}{64}$

c. $g\left(\dfrac{2}{3}\right) = 8^{2/3} = (\sqrt[3]{8})^2 = 2^2 = 4$ **d.** $g(\sqrt{2}) = 8^{\sqrt{2}} \approx 18.93$

14.2.2 Checkup

In exercises 1–4, evaluate the function $h(x) = 4^x$ *at the given values.*

1. $h(3)$ **2.** $h(-3)$ **3.** $h\left(\dfrac{3}{2}\right)$ **4.** $h(\sqrt{3})$

5. Are there any real-number values of x for which the function $h(x) = 4^x$ cannot be evaluated? (*Hint:* Consider each of the subsets of numbers that make up the real-number system.)

14.2.3 Graphing Exponential Functions

We graph a function by first determining the domain of the function and then evaluating the function for values in its domain. We do not need to determine the domain of an exponential function, because it is given to us in the definition; that is, x is a real number. Therefore, the domain is the set of all real numbers.

To graph an exponential function in the form $f(x) = a^x$, set up a table of values with real-number values for x, plot the points, and connect them with a smooth curve.

EXAMPLE 3   Graph $f(x) = 8^x$. Determine the function's domain and range, the graph's y-intercept and x-intercept, and whether the function is increasing or decreasing.

Solution

Set up a table of values.

x	$f(x)$
-3	$\dfrac{1}{512}$
-2	$\dfrac{1}{64}$
-1	$\dfrac{1}{8}$
0	1
1	8
2	64
3	512

Plot the points and connect them with a smooth curve.

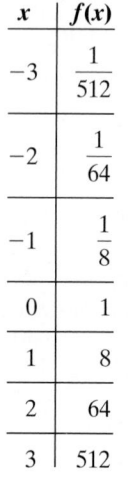

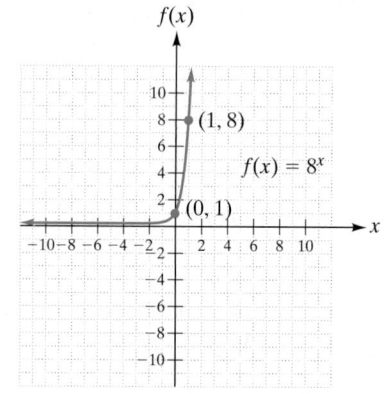

Calculator Check

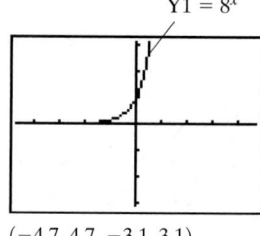

$Y1 = 8^x$

$(-4.7, 4.7, -3.1, 3.1)$

The domain of $f(x)$ is the set of all real numbers. The range of $f(x)$ is the set of all real numbers greater than 0. The graph's y-intercept is $(0, 1)$. The graph does not have an x-intercept. The function is increasing.

Exponential functions can also be written in the form $f(x) = a^{-x}$, where $a > 0$ and $a \neq 1$. These functions are equivalent to other functions written in the form $f(x) = a^x$, where $a > 0$ and $a \neq 1$. Complete the following set of exercises on your calculator to determine the relationship between the two forms.

Discovery 2

Equivalent Exponential Functions

1. Graph each exponential function on a decimal window.

   a. $f(x) = 2^x$ **b.** $f(x) = 2^{-x}$

   c. $f(x) = \left(\frac{1}{2}\right)^x$ **d.** $f(x) = \left(\frac{1}{2}\right)^{-x}$

2. Match the graphs of the functions.

Write a rule for determining equivalent functions.

The functions $f(x) = 2^{-x}$ and $f(x) = \left(\frac{1}{2}\right)^x$ are equivalent. The functions $f(x) = 2^x$ and $f(x) = \left(\frac{1}{2}\right)^{-x}$ are also equivalent.

Using the properties of exponents, we can show algebraically that $a^{-x} = \left(\frac{1}{a}\right)^x$ and $a^x = \left(\frac{1}{a}\right)^{-x}$, where $a > 0$, $a \neq 1$, and x is a rational number.

$$a^{-x} = (a^{-1})^x \qquad \text{Power-to-a-power rule}$$
$$= \left(\frac{1}{a}\right)^x \qquad \text{Definition of a negative exponent}$$

$$a^x = (a^{-1})^{-x} \qquad \text{Power-to-a-power rule}$$
$$= \left(\frac{1}{a}\right)^{-x} \qquad \text{Definition of a negative exponent}$$

EXAMPLE 4   Graph the function $f(x) = \left(\frac{1}{8}\right)^x = 8^{-x}$. Determine the function's domain and range, the graph's y-intercept and x-intercept, and whether the function is increasing or decreasing.

Solution

Set up a table of values.

Plot the points and connect them with a smooth curve.

Calculator Check

$$Y1 = \left(\frac{1}{8}\right)^x = 8^{-x}$$

x	$f(x)$
-3	512
-2	64
-1	8
0	1
1	$\frac{1}{8}$
2	$\frac{1}{64}$
3	$\frac{1}{512}$

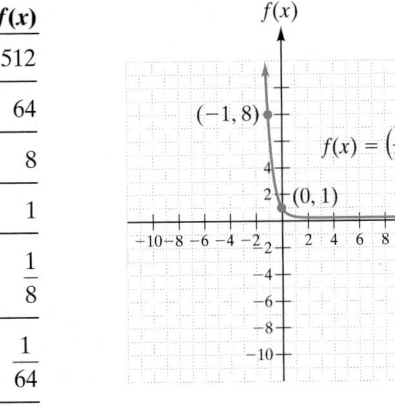

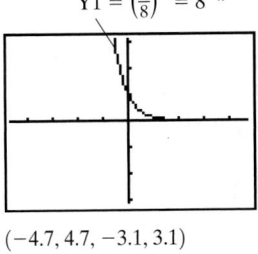

$(-4.7, 4.7, -3.1, 3.1)$

The domain of $f(x)$ is the set of all real numbers. The range of $f(x)$ is the set of all real numbers greater than 0. The graph's y-intercept is $(0, 1)$. The graph does not have an x-intercept. The function is decreasing.

Exponential functions and their graphs have several common characteristics. Some of these characteristics are common to all exponential functions, and others are common to certain sets of functions. Two such characteristics are determined by the value of the base a. Complete the following set of exercises to discover the common characteristics of exponential functions and their graphs.

Discovery 3

Effect of the Base a on an Exponential Graph

1. Sketch the graphs of the given exponential functions of the form $f(x) = a^x$, where $a > 0$, on the same coordinate plane. Use the decimal window.

   a. $f(x) = 2^x$ **b.** $f(x) = 5^x$ **c.** $f(x) = 10^x$
   d. $f(x) = \left(\dfrac{1}{2}\right)^x$ **e.** $f(x) = \left(\dfrac{1}{5}\right)^x$ **f.** $f(x) = \left(\dfrac{1}{10}\right)^x$

   Use your graphs to complete the following exercises:
2. Determine the domain of each function.
3. Determine the range of each function.
4. Are all the functions one-to-one?
5. Determine the y-intercept of each function.
6. Determine the x-intercept of each function.

Choose the correct answer.

7. In exercise 1a–1c, $a > 1$. The function is *increasing/decreasing*. The larger the value of a, the *steeper/shallower* the graph.
8. In exercises 1d–1f, $0 < a < 1$. The function is *increasing/decreasing*. The smaller the value of a (as a approaches 0), the *steeper/shallower* the graph.

The exponential function $y = a^x$, where $a > 1$, is always increasing. The larger the value of a, the steeper is the function's graph. The exponential function $y = a^x$, where $0 < a < 1$, is always decreasing. The smaller the value of a, the steeper is the function's graph. All exponential functions are one-to-one and have a domain of all real numbers and a range of all positive real numbers, and their graphs have a y-intercept of $(0, 1)$. Exponential functions have no x-intercept.

Let's summarize what we have learned about the exponential function $f(x) = a^x$ and its corresponding graph.

CHARACTERISTICS OF AN EXPONENTIAL FUNCTION AND ITS GRAPH

For the exponential function $f(x) = a^x$ and its graph, we have the following characteristics:

Domain	All real numbers
Range	All positive real numbers

The function is one-to-one.

y-intercept	$(0, 1)$
x-intercept	There is no x-intercept. The graph of the function approaches, but does not reach, the x-axis.

If $a > 1$, then the function is always increasing.
If $0 < a < 1$, then the function is always decreasing.

14.2.3 Checkup

In exercises 1 and 2, graph each exponential function.
  1. $y = 8^{2x}$ **2.** $y = 8^{-2x}$

3. Why are negative numbers excluded from the range of the exponential function $y = a^x, a > 0$?

14.2.4 The Natural Exponential Function

In all of the examples of exponential functions we have discussed, we chose to use a rational-number base. However, we could have chosen an irrational-number base. In fact, many applications of exponential functions involve a special irrational number as the base. This number occurs often in nature and in human activity, showing up in the description of continuously compounded interest, the decay rates of radioactive elements, and even the growth rates of bacterial populations.

The irrational number is denoted by the symbol e and is called the **natural base**. Rounded to five decimal places, the value of e is 2.71828. The calculator gives us a more accurate value to nine decimal places, as shown in Figure 14.1.

T E C H N O L O G Y

NATURAL BASE, e

Calculate the natural base e to nine decimal places.

```
e
        2.718281828
e^(1)
        2.718281828
```

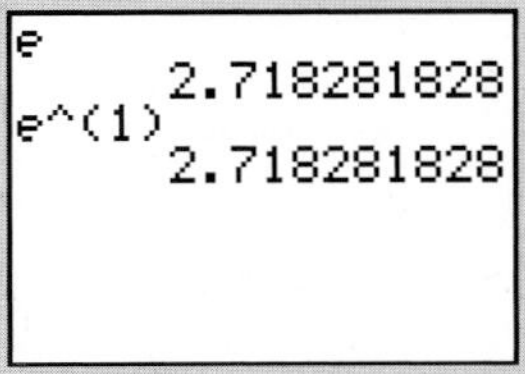

Figure 14.1

For Figure 14.1,
Enter e.

| 2nd | | e | | ENTER |

If e is a base of an expression, the calculator allows a more convenient method. Enter e^x followed by the exponent. Remember to close the parentheses. To evaluate e, enter e^1.

| 2nd | | e^x | | 1 | |) | | ENTER |

The irrational number e is used to describe the **natural exponential function** $f(x) = e^x$. Remember, the base e is a positive number greater than 1. Therefore, the domain of the function is the set of all real numbers. The range is the set of all positive real numbers. The graph of the function has a *y*-intercept at $(0, 1)$ and is always increasing. Calculate a table of values on your calculator, and use the coordinates to graph the function. Check your graph with your calculator.

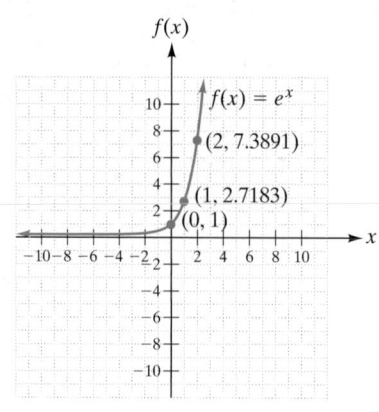

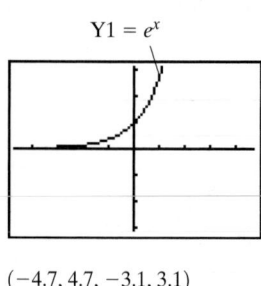

$(-4.7, 4.7, -3.1, 3.1)$

NATURAL EXPONENTIAL FUNCTION

The natural exponential function is defined as

$$f(x) = e^x$$

where *x* is a real number.

EXAMPLE 5 Graph the function $f(x) = 2e^x$.

Solution

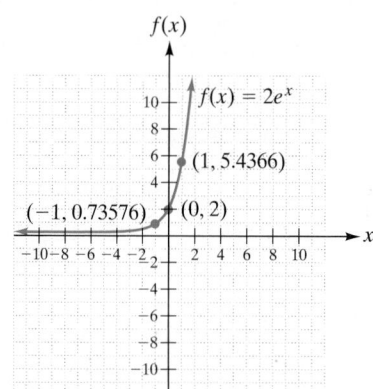

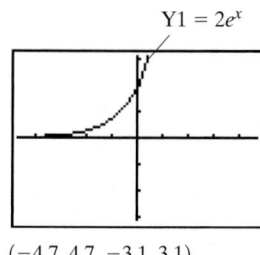

$(-4.7, 4.7, -3.1, 3.1)$

14.2.4 Checkup

In exercises 1 and 2, graph each function and label the y-intercept.

1. $y = 0.5e^x$ **2.** $y = e^{-x}$

3. The number *e* is an irrational number. What does this mean?

14.2.5 Modeling the Real World

In section 14.2.1, we explained that an exponential function $y = a^x$ is different from a polynomial function $y = x^a$.

As an example, the exponential function $y = 2^x$ increases at a faster rate than the polynomial function $y = x^2$. Therefore, we use an exponential function to describe situations that change quickly over a large range of values of the domain.

Exponential functions model many types of phenomena in the real world. These phenomena range from the calculation of one's balance in a savings account, to the determination of the amount owed on a loan, to the prediction of numbers of individuals in a population.

In Chapter 2, we first discussed the calculation of interest. Remember the formula for simple interest, $I = Prt$, where P is the amount of principal, r is the rate of interest per period, and t is the number of periods. Compound interest is based on taking the interest accumulated in one period and adding it to the principal before determining the interest received the next period.

Let $t = 1$ year. At the end of one year, the amount is the principal plus the interest.

$$P + Prt = P + Pr(1)$$
$$= P + Pr$$
$$= P(1 + r)$$

If this amount is left at the same interest rate for one more year, the total amount is the last amount plus the interest.

$$P(1 + r) + [P(1 + r)]rt = P(1 + r) + [P(1 + r)]r(1)$$
$$= P(1 + r) + Pr(1 + r)$$
$$= P(1 + r)(1 + r)$$
$$= P(1 + r)^2$$

If we continue the pattern, we see that at the end of three years the amount is $P(1 + r)^3$. Therefore, we can arrive at the formula given for the compounded amount

$$A = P(1 + r)^t$$

where P = principal, r = interest per period, and t = number of periods.

EXAMPLE 6   Alex plans to invest $10,000 at an annual compounded rate of 8%. Determine the amount of his investment at the end of two years.

Solution

Use the compound-interest formula with $P = 10,000$, $r = 0.08$, and $t = 2$.

$$A = P(1 + r)^t$$
$$A = 10,000(1 + 0.08)^2$$
$$A = 10,000(1.08)^2$$
$$A = 11,664$$

Alex will have $11,664 in his account at the end of two years.

```
10000(1.08)²
           11664
```

Continuously compounded interest is compound interest with very short periods, so that the addition of interest occurs on a continuous basis. A formula for the continuously compounded amount is

$$A = Pe^{rt}$$

where P = principal, t = number of years, and r = annual interest rate compounded continuously.

EXAMPLE 7   Alex plans to invest $10,000 at a rate of 8% compounded continuously. Determine the amount of his investment at the end of two years.

Solution

Use the continuously compounded interest formula with $P = 10,000$, $r = 0.08$, and $t = 2$.

```
10000→P:0.08→R:2
→T:Pe^(RT)
        11735.10871
```

$$A = Pe^{rt}$$
$$A = 10,000e^{(0.08)(2)}$$
$$A = 10,000e^{0.16}$$
$$A \approx 11735.11$$

Alex will have approximately $11,735.11 in his account at the end of two years.

APPLICATION

The population of the United States (in millions) may be approximated by the exponential function $f(x) = 250(1.01)^x$, where x is the number of years after 1990.

a. Graph this function.

b. According to the graph, what is the estimated population of the United States in the year 2010?

c. According to U.S. Bureau of the Census Projections, the U.S. population will be between 291 million and 311 million in 2010. Compare your results in part b with these projections.

Discussion

a.

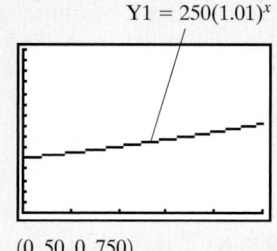 $Y1 = 250(1.01)^x$

$(0, 50, 0, 750)$

b. Let $x = 2010 - 1990 = 20$

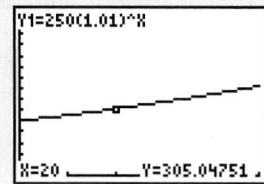

$(0, 50, 0, 750)$

According to the model, the population of the United States will be approximately 305 million in the year 2010.

c. A population of 305 million falls within the projected range.

14.2.5 Checkup

1. Cindy invests $5000 at an annual compounded interest rate of 6%. How much will she have after four years?

2. If Cindy had invested the $5000 at a continuously compounded rate of 6%, how much would she have after four years?

3. The population of the world (in billions) may be approximated by the exponential function $y = 5.2975(1.014)^x$, where x is the number of years after 1990. Graph this function. Use the function to estimate the population in the year 2010. According to projections, the world population will be approximately 6,840,423,256 in 2010. Compare your results with this figure.

1-47 evo
49-55 odds

14.2 Exercises

For each function, state whether it is an exponential function or not.

1. $y = 7^x$
2. $d(x) = x^{1.5}$
3. $f(x) = 0.3^x$
4. $y = 22.3^x$
5. $g(x) = x^{0.3}$
6. $y = x^4$
7. $y = 1.57^x$
8. $y = 12^x$
9. $y = x^{-3}$
10. $T(x) = \left(\dfrac{5}{8}\right)^x$
11. $R(x) = \left(\dfrac{2}{3}\right)^x$
12. $b(x) = 1.5^x$

Evaluate the function $g(x) = 16^x$ at the given values.

13. $g(3)$
14. $g(4)$
15. $g(-2)$
16. $g(-1)$
17. $g\left(\dfrac{1}{2}\right)$
18. $g\left(\dfrac{3}{4}\right)$
19. $g(\sqrt{2})$
20. $g(0)$

Evaluate the function $h(x) = 0.64^x$ at the given values.

21. $h(0.3)$
22. $h(1.5)$
23. $h\left(-\dfrac{1}{3}\right)$
24. $h\left(\dfrac{5}{6}\right)$
25. $h(-\sqrt{2})$
26. $h(\sqrt{5})$

Graph each exponential function.

27. $f(x) = 4^x$
28. $C(x) = 3^x$
29. $g(x) = 4^{-x}$
30. $D(x) = 3^{-2x}$
31. $h(x) = 4^{2x}$
32. $F(x) = 3^{2x-1}$
33. $j(x) = 4^{(1/2)x}$
34. $K(x) = 3^{x/4}$
35. $k(x) = 4^{x-1}$
36. $L(x) = 3^x + 1$
37. $m(x) = 4^x - 1$
38. $M(x) = 3^{x+1} - 1$
39. $y = e^{(1/2)x}$
40. $y = e^{0.2x}$
41. $y = \dfrac{1}{2}e^x$
42. $y = 0.2e^x$
43. $y = e^{(-1/2)x}$
44. $y = e^{-0.2x}$
45. $y = e^x + \dfrac{1}{2}$
46. $y = e^{0.2x} + 0.2$
47. $y = \dfrac{1}{2}e^{(1/2)x}$
48. $y = 0.2e^{0.2x}$

t → # of periods
$A = P(1+r)^t$
r → rate/period
$A = Pe^{rt}$

In exercises 49–52, write a function for the total value of the investment. Graph the function and answer the question.

49. Jerome received his annual bonus of $6000, which he invested at an annually compounded interest rate of 5.5%. How much money was invested after seven years?

50. Marissa received a $3000 commission for landing a book adoption with a major school. She invested the money at an interest rate of 6.2%, compounded annually. What was her investment worth after four years?

51. Jolene inherited $8000, which she invested at a continuously compounded interest rate of 4.8%. How much did she earn after five years?

52. Marty's regional sales netted her a commission of $4500. If she invests the money at a continuously compounded interest rate of 7.5%, how much will the investment be worth after 10 years?

53. The number of cable television subscribers can be estimated by the mathematical model $y = 49.75(1.03)^x$,

where x is the number of years after 1990 and y is the number of subscribers, in millions. Graph this function. What does the model predict the number of subscribers will be in the year 2010?

54. According to data from the U.S. Health Care Financing Administration, the national health expenditures can be estimated by the mathematical model

$$y = 779.03(1.05)^x$$

where x is the number of years after 1990 and y is the total expenditures in billions of dollars. Graph this function. What does the model predict expenditures will be in the year 2010?

55. The operating revenue for the cellular telephone industry can be estimated by the model $y = 6.002e^{1.95x}$, where x is the number of years after 1990 and y is the operating revenue in millions of dollars. Graph this function. What does the model predict the operating revenue will be in the year 2015?

56. The number of wireless service providers can be estimated by the model $y = 907e^{0.08x}$, where x is the number of years after 1994 and y is the number of wireless providers. Graph this function. What does the model predict the number of wireless service providers will be in the year 2015?

14.2 Calculator Exercises

The calculator allows you to easily explore more complicated mathematical models that involve exponential expressions. Graph the following functions and study the effects of the operations indicated:

Could you have predicted any of the results you obtained? Which rules of exponents apply to these exercises? See if you can use the rules of exponents to write a simplified form of some of the functions listed. Then check your results by graphing.

1. **a.** $f(x) = e^x$ **b.** $g(x) = e^{-x}$

2. **a.** $h(x) = e^x + e^{-x}$ **b.** $j(x) = e^x - e^{-x}$

3. **a.** $k(x) = (e^x)(e^{-x})$ **b.** $m(x) = e^x \div e^{-x}$

4. **a.** $F(x) = e^{(1/2)x}$ **b.** $G(x) = e^{(-1/2)x}$

5. **a.** $H(x) = e^{(1/2)x} + e^{(-1/2)x}$ **b.** $J(x) = e^{(1/2)x} - e^{(-1/2)x}$

6. **a.** $K(x) = (e^{(1/2)x})(e^{(-1/2)x})$ **b.** $M(x) = (e^{(1/2)x}) \div (e^{(-1/2)x})$

14.2 Writing Exercise

The definition of an exponential function $f(x) = a^x$ specifies that a must be positive and must not equal 1. Consider an exponential equation with a negative base, such as $y = (-2)^x$. Set up a table of values, using integer values of x. What do you notice about the values of y? Does this equation behave like an exponential function? Can you understand why the base of an exponential expression cannot be negative? Explain.

Next, consider an exponential function with a base of 1, for example, $y = 1^x$. What does the table of values look like

for this equation? Can you see why it is excluded from the discussion of exponential functions? Explain.

Finally, consider the functions $f(x) = -2^x$, $g(x) = (-2)^x$, and $h(x) = 2^x$. Are any of these functions the same? Which of the functions is an exponential function? Graph each function separately, using a decimal window. Trace the graphs. What do the graphs show? Does this example further confirm the rule that the base of an exponential function must be positive? Explain.

14.3 Logarithmic Functions

OBJECTIVES

1 Identify logarithmic functions.

2 Evaluate logarithms.

3 Graph logarithmic functions.

4 Model real-world situations by using logarithmic functions.

APPLICATION

The magnitude R, measured on the Richter scale, of an earthquake of intensity I is defined by the function $R = \log \frac{I}{I_0}$, where I_0 is the intensity of a very small vibration in the Earth used as a standard. The San Francisco earthquake of 1906 had a magnitude of 8.3 on the Richter scale. Eighty-three years later, in 1989, San Francisco had another earthquake, this time measuring 7.1.

a. Determine the intensity of each earthquake.

b. The difference on the Richter scale of the 1906 and 1989 earthquakes is 1.2. Determine the ratio of the intensity of the 1906 earthquake to the 1989 earthquake.

After completing this section, we will discuss this application further. See page 1008.

14.3.1 Identifying Logarithmic Functions

In section 14.1, we introduced the concept of the inverse of a function. The horizontal-line test states that if it is not possible to draw a horizontal line which intersects the graph of a function more than once, then the graph represents a one-to-one function. A one-to-one function has an inverse function.

According to the horizontal-line test, all exponential functions are one-to-one and, therefore, have inverse functions.

Let's determine the inverse function of the exponential function.

$f(x) = a^x$	Exponential function
$y = a^x$	Replace $f(x)$ with y.
$x = a^y$	Interchange x and y.
$y = $ the power to which we raise a to obtain x	Solve for y.

In order to write this relationship mathematically, we define the **logarithm of x with base a** to be the power to which we raise a to get x.

LOGARITHM OF X WITH BASE A

The logarithm of x with base a is written in the form $\log_a x$ and is defined as

$$y = \log_a x \quad \text{if and only if} \quad x = a^y$$

where a and x are positive real numbers and $a \neq 1$.

Now we can write the inverse function of the exponential function $f(x) = a^x$ as

$$y = \log_a x$$

or

$$f^{-1}(x) = \log_a x$$

One example of an exponential function and its inverse function is

$$g(x) = 2^x \quad \text{and} \quad g^{-1}(x) = \log_2 x$$

We have now identified a new function called the **logarithmic function**, which is the inverse of the exponential function.

> ### LOGARITHMIC FUNCTION WITH BASE A
> The logarithmic function with base a can be written in the form
>
> $$f(x) = \log_a x$$
>
> where a and x are positive real numbers and $a \neq 0$.

EXAMPLE 1   Determine the logarithmic function that is the inverse of the given exponential function.

a. $f(x) = 5^x$ **b.** $g(x) = 10^x$

Solution

a. Given $f(x) = 5^x$, $f^{-1}(x) = \log_5 x$. $a = 5$

b. Given $g(x) = 10^x$, $g^{-1}(x) = \log_{10} x$. $a = 10$

14.3.1 Checkup

In exercises 1 and 2, determine the logarithmic function that is the inverse of the given exponential function.

1. $f(x) = 3^x$

2. $h(x) = 13^x$

3. What is wrong with the statement, "The inverse function of the exponential function $F(x) = 7^x$ is the logarithmic function $F^{-1}(x) = \log_x 7$"? How can you guard against making this mistake?

14.3.2 Evaluating Logarithms

According to the definition of the logarithm of x with base a, the two equations $y = \log_a x$ and $x = a^y$ are equivalent. For example,

$$y = \log_a x \qquad x = a^y$$
$$3 = \log_2 8 \qquad 8 = 2^3$$

$y = 3$

Therefore, by definition, *a logarithm is an exponent.*

To evaluate a logarithm $\log_a x$, ask the question, "To what power must a be raised to obtain x?" For example, evaluate $\log_3 9$.

The power to which 3 must be raised to obtain 9 is 2. Therefore, since $3^2 = 9$, then $\log_3 9 = 2$.

EXAMPLE 2 Evaluate.

a. $\log_2 32$ **b.** $\log_4 2$ **c.** $\log_3 \dfrac{1}{3}$

Solution

a. $\log_2 32$

The power to which 2 must be raised to obtain 32 is 5. Therefore, since $2^5 = 32$, it follows that $\log_2 32 = 5$.

b. $\log_4 2$

The power to which 4 must be raised to obtain 2 is $\frac{1}{2}$. Therefore, since $4^{1/2} = 2$, it follows that $\log_4 2 = \frac{1}{2}$.

c. $\log_3 \dfrac{1}{3}$

The power to which 3 must be raised to obtain $\frac{1}{3}$ is -1. Therefore, since $3^{-1} = \frac{1}{3}$, it follows that $\log_3 \frac{1}{3} = -1$.

When evaluating logarithms, we sometimes encounter certain special cases that are properties of all logarithms. Complete the following set of exercises to identify these properties.

Discovery 4

Properties of Logarithms

Determine the following logarithms in the form $\log_a x$:

1. a. $\log_3 1$ **b.** $\log_5 1$

2. a. $\log_3 3$ **b.** $\log_5 5$

3. a. $\log_3(3^2)$ **b.** $\log_5(5^2)$

4. a. $\log_3(-1)$ **b.** $\log_5(-1)$

5. a. $\log_3 0$ **b.** $\log_5 0$

6. In exercise 1, $x = 1$. The logarithms are _____.

7. In exercise 2, $x = a$, the base. The logarithms are _____.

8. In exercise 3, $x = a^2$, the base squared. The logarithms are _____.

9. In exercise 4, $x = -1$. The logarithms are _____.

10. In exercise 5, $x = 0$. The logarithms are _____.

The properties discovered in these exercises come directly from the definition of a logarithm.

PROPERTIES OF LOGARITHMS

For any positive real numbers a and x, where $a \neq 1$, logarithms have the following properties.

$\log_a 1 = 0$	because $a^0 = 1$.
$\log_a a = 1$	because $a^1 = a$.
$\log_a a^m = m$	because $a^m = a^m$.
$\log_a(-1)$ is undefined	because a is a positive number and there is no power to which a can be raised to obtain a negative number.
$\log_a 0$ is undefined	because a is a positive number and there is no power to which a can be raised to obtain 0.

The base of a logarithm can be any positive real number except 1. However, one base that appears more often than other rational numbers is the base 10. A base-10 logarithm is called the **common logarithm**.

COMMON LOGARITHM

The common logarithm of a number x is $\log_{10} x$ or $\log x$.

HELPING HAND When we write $\log x$, the base is understood to be 10.

The calculator has a special key for evaluating the common logarithm. For example, evaluate $\log_{10} 100$, and check your results on your calculator.

$$\log_{10} 100 = 2, \text{ because } 10^2 = 100, \text{ as shown in Figure 14.2.}$$

TECHNOLOGY

THE COMMON LOGARITHM

Evaluate $\log_{10} 100$.

```
log(100)
              2
```

Figure 14.2

For Figure 14.2,
Enter the common-logarithm function, log, followed by 100. Remember to close the set of parentheses.

| LOG | 1 | 0 | 0 |) | ENTER |

EXAMPLE 3 Evaluate and check on your calculator.

a. log 1   **b.** log 0.01   **c.** log 20   **d.** log 0

Solution

a. log 1 = 0   *Because $10^0 = 1$.*

b. log 0.01 = −2   *Because $10^{-2} = \frac{1}{10^2} = \frac{1}{100} = 0.01$.*

c. log 20 ≈ 1.301   *You must evaluate this on your calculator, because 10 cannot easily be raised to a power to obtain 20.*

d. log 0 is undefined.

Calculator Check

```
a. log(1)
                    0
b. log(0.01)
                   -2
c. log(20)
           1.301029996
d. log(0)
```

```
d. ERR:DOMAIN
   1:Quit
   2:Goto
```

 **HELPING HAND** The calculator will return an error message when the logarithm is undefined.

The logarithm with base e is called the **natural logarithm**. It is written using the special symbol ln x, and is read "el en of x."

NATURAL LOGARITHM
The natural logarithm of a number x is $\log_e x$ or $\ln x$.

All the properties of logarithms listed in this section also hold for natural logarithms.

PROPERTIES OF NATURAL LOGARITHMS
For any positive real number x, natural logarithms have the following properties:

ln 1 = 0	because $e^0 = 1$.
ln e = 1	because $e^1 = e$.
ln e^m = m	because $e^m = e^m$.
ln (−1) is undefined	because e is a positive number and there is no power to which e can be raised to obtain a negative number.
ln 0 is undefined	because e is a positive number and there is no power to which e can be raised to obtain 0.

The calculator has a special key for evaluating the natural logarithm. For example, evaluate ln e^2, and check your results on your calculator.

$\ln e^2 = 2$, because a property of logarithms states that $\ln e^m = m$ and in this case $m = 2$, as shown in Figure 14.3.

TECHNOLOGY

THE NATURAL LOGARITHM

Evaluate $\ln e^2$.

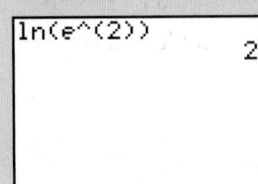

Figure 14.3

For Figure 14.3,

Enter the natural logarithm, ln, followed by e^x and the value of the exponent, 2. Remember to close the both sets of parentheses.

| LN | 2nd | e^x | 2 |) |) | ENTER |

EXAMPLE 4   Evaluate $\ln \frac{1}{e^2}$ and check on your calculator.

Solution

$$\ln \frac{1}{e^2} = \ln e^{-2} = -2$$

Calculator Check

```
ln(1/e^(2))
              -2
ln(e^(-2))
              -2
```

To evaluate logarithms with bases other than 10 and e on the calculator, we need to introduce the **change-of-base formula**. It can be shown that $\log_a x$ can be written as a quotient of logarithms with a base other than a.

CHANGE-OF-BASE FORMULA

For any positive real numbers a, b, and x, where $a \neq 1$ and $b \neq 1$,

$$\log_a x = \frac{\log_b x}{\log_b a} \quad \text{or} \quad \log_a x = \frac{\ln x}{\ln a}$$

To evaluate logarithms with bases other than 10 and e,

• Use the change-of-base formula with common logarithms.

$$\log_a x = \frac{\log_{10} x}{\log_{10} a}$$

• Use the change-of-base formula with natural logarithms.

$$\log_a x = \frac{\ln x}{\ln a}$$

HELPING HAND Note that the original logarithm has base a and the quotient has log a or ln a in the denominator (base). This will help you to remember the location of a in the quotient.

EXAMPLE 5 Evaluate $\log_5 12$.

    a. Use the change-of-base formula and natural logarithms.
    b. Use the change-of-base formula and common logarithms.

Solution

a. $\log_5 12 = \dfrac{\ln 12}{\ln 5} \approx 1.544$

b. $\log_5 12 = \dfrac{\log 12}{\log 5} \approx 1.544$

Calculator Check

```
ln(12)/ln(5)
            1.543959311
log(12)/log(5)
            1.543959311
```

Note: We obtain the same value in 5a and 5b.

 14.3.2 Checkup

Evaluate.

1. $\log_2 16$ **2.** $\log_4 \dfrac{1}{64}$

Evaluate exercises 3–7 and check with your calculator.

3. $\log 1000$ **4.** $\log 0.001$ **5.** $\log (-1)$

6. $\ln e^4$ **7.** $\ln \dfrac{1}{e^4}$

8. Evaluate $\log_8 24$, using
    a. natural logarithms.
    b. common logarithms.

9. Explain the difference between the common logarithm of a number and the natural logarithm of a number. Are these the only two forms of logarithms we can use?

14.3.3 Graphing Logarithmic Functions

We are now ready to graph a logarithmic function. We illustrate two methods of setting up a table of values for the function.

    The first method involves setting up a table of values for the inverse of the logarithmic function (the exponential function), which is usually easier to evaluate than the logarithmic function. Reversing the coordinates will then determine the table of values for the logarithmic function.

To set up a table of values for a logarithmic function,

- Identify the inverse function (the exponential function).
- Set up a table of values for the exponential function.
- Reverse the coordinates in the ordered pairs to determine the ordered pairs of the logarithmic function.

EXAMPLE 6 Graph the logarithmic function $f(x) = \log_8 x$.

The inverse function is the exponential function $f^{-1}(x) = 8^x$. The domain of this inverse function is the set of all real numbers. (We set up a table of values for this function in Section 14.2.)

$f^{-1}(x) = 8^x$

x	$f^{-1}(x)$
-3	$\dfrac{1}{512}$
-2	$\dfrac{1}{64}$
-1	$\dfrac{1}{8}$
0	1
1	8
2	64
3	512

Reverse coordinates →

$f(x) = \log_8 x$

x	$f(x)$
$\dfrac{1}{512}$	-3
$\dfrac{1}{64}$	-2
$\dfrac{1}{8}$	-1
1	0
8	1
64	2
512	3

Note: We graph both the exponential function and the logarithmic function. We also graph $y = x$, the line of symmetry, to illustrate that the graphs represent inverse functions.

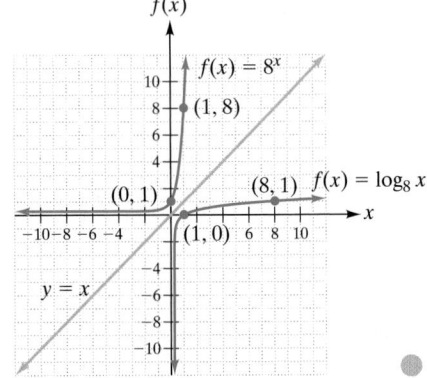

The second method of completing a table of values for the logarithmic function involves substituting values for the independent variable and evaluating the function. First, we determine the domain of the logarithmic function. Since it is the inverse of an exponential function, we reverse the domain and range of the two functions.

Exponential Function			Logarithmic Function
Domain	←	all real numbers →	Range
Range	←	all positive real numbers →	Domain

Therefore, the domain of a logarithmic function is the set of all positive real numbers.

To complete the table of values, we may need to use a calculator. Also, remember to use the change-of-base formula when needed.

To set up a table of values for a logarithmic function,

• Substitute positive real-number values from the domain of the function for the independent variable.
• Determine values of the dependent variable by evaluating the resulting expression.

EXAMPLE 7 Graph the logarithmic function $f(x) = \log_8 x$.

Use your calculator and the change-of-base formula to determine a table of values for the function. Let $f(x) = \log_8 x = \frac{\ln x}{\ln 8}$.

x	$f(x)$
0.001	-3.322
0.01	-2.215
0.1	-1.107
0.5	-0.333
1	0
2	0.333
3	0.528
4	0.667
5	0.774
6	0.862

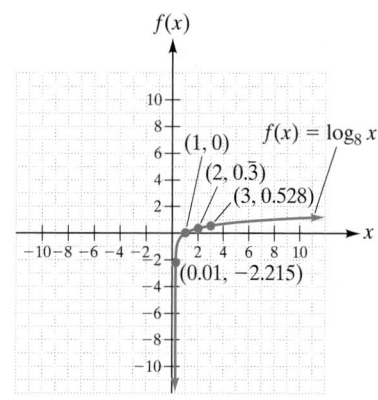

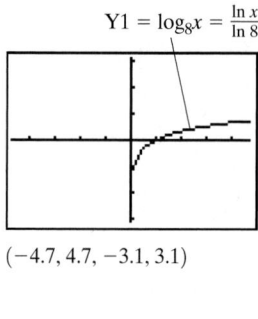

$(-4.7, 4.7, -3.1, 3.1)$

We also have logarithmic functions identified by a common logarithm and a natural logarithm.

COMMON LOGARITHMIC FUNCTION

The common logarithmic function is written in the form

$$f(x) = \log_{10} x \quad \text{or} \quad f(x) = \log x$$

where x is a positive real number.

NATURAL LOGARITHMIC FUNCTION

The natural logarithmic function is defined as $f(x) = \log_e x$ and is written in the form

$$f(x) = \ln x$$

where x is a positive real number.

EXAMPLE 8 Graph.

a. $y = \log x$ **b.** $g(x) = \ln x$

Solution

a. For $y = \log x$ we illustrate the first graphical method.

$$f^{-1}(x) = 10^x \qquad f(x) = \log x$$

x	$f^{-1}(x)$
-3	0.001
-2	0.01
-1	0.1
0	1
1	10
2	100
3	1000

Reverse coordinates →

x	$f(x)$
0.001	-3
0.01	-2
0.1	-1
1	0
10	1
100	2
1000	3

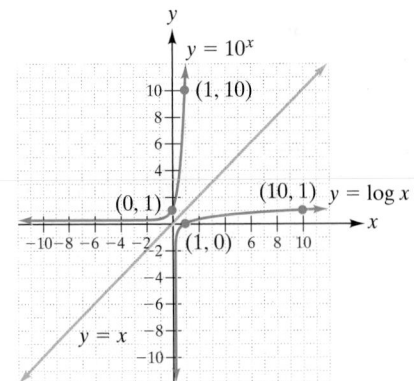

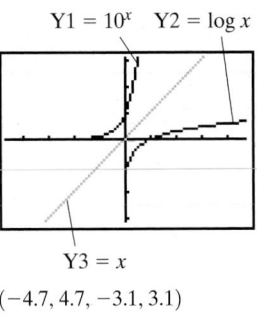

Y1 = 10^x Y2 = log x

Y3 = x

$(-4.7, 4.7, -3.1, 3.1)$

b. For $g(x) = \ln x$, we illustrate the second graphical method.

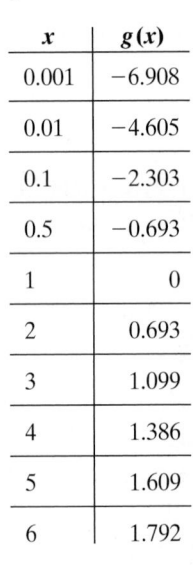

x	$g(x)$
0.001	-6.908
0.01	-4.605
0.1	-2.303
0.5	-0.693
1	0
2	0.693
3	1.099
4	1.386
5	1.609
6	1.792

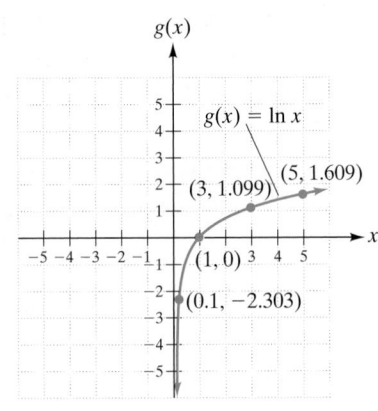

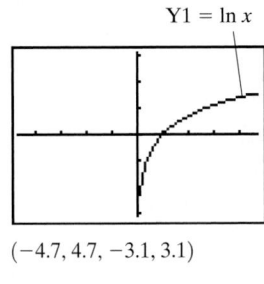

Y1 = ln x

$(-4.7, 4.7, -3.1, 3.1)$

We can summarize the characteristics of a logarithmic function and its graph by remembering that a logarithmic function is an inverse of an exponential function. Therefore, the domain and range are interchanged.

CHARACTERISTICS OF A LOGARITHMIC FUNCTION AND ITS GRAPH

Given the logarithmic function $f(x) = \log_a x$ and its graph, we have the following characteristics:

Domain	All positive real numbers
Range	All real numbers

The function is one-to-one.

y-intercept	There is no y-intercept. The graph of the function approaches, but does not reach, the y-axis.
x-intercept	$(1, 0)$

If $a > 1$, then the function is always increasing.
If $0 < a < 1$, then the function is always decreasing.

✓ 14.3.3 Checkup

In exercises 1–3, use a table of values to graph each logarithmic function.

1. $f(x) = \log_3 x$ **2.** $g(x) = \log(3x)$ **3.** $h(x) = \ln(3x)$

4. Explain the relationship between the domain and the range of an exponential function and the domain and the range of its inverse function—that is, the corresponding logarithmic function.

14.3.4 Modeling the Real World

Logarithmic functions are used to model many different real-world situations, especially when the independent variable can take a wide range of values. The reason for this is that the values of the logarithmic function change slowly over a large range of values of the domain. Examples of logarithmic functions are found in many of the sciences. For instance, the magnitude of an earthquake is measured on a Richter scale, which uses a logarithmic function to describe motion from millionths of an inch to tens of feet. The loudness of a sound, from the merest whisper to the roar of thunder, is measured in decibels, which are defined by a logarithmic function.

Another important application of common logarithms is found in chemistry. The pH value of a substance is its measure of acidity or alkalinity. Pure water has a pH of 7.0. A pH greater than 7.0 represents an alkaline substance, and a pH value less than 7.0 represents an acidic substance.

The pH of a substance is defined by the function

$$pH = -\log[H^+]$$

where $[H^+]$ is the hydrogen ion concentration in moles per liter.

EXAMPLE 9   Orange juice has a hydrogen ion concentration of 6.3×10^{-4} moles per liter. Determine the pH of orange juice.

Calculator Solution

Substitute 6.3×10^{-4} for $[H^+]$ in the pH formula.

$$pH = -\log[H^+]$$
$$pH = -\log(6.3 \times 10^{-4})$$
$$pH \approx 3.2$$

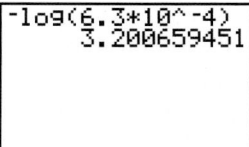

EXAMPLE 10   Crackers have a pH of 8.4. Determine their hydrogen ion concentration.

Solution

Substitute 8.4 into the formula for pH.

$$pH = -\log[H^+]$$
$$8.4 = -\log[H^+]$$
$$-8.4 = \log[H^+] \qquad \text{Divide both sides by } -1.$$
$$10^{-8.4} = [H^+] \qquad \text{Write as an exponential equation.}$$
$$4.0 \times 10^{-9} \approx [H^+]$$

```
10^(-8.4)
   3.981071706E-9
```

Note: This calculator value is rounded to one decimal place.

APPLICATION

The magnitude R, measured on the Richter scale, of an earthquake of intensity I is defined by the function $R = \log \frac{I}{I_0}$, where I_0 is the intensity of a very small vibration in the Earth used as a standard. The San Francisco earthquake of 1906 had a magnitude of 8.3 on the Richter scale. Eighty-three years later, in 1989, San Francisco had another earthquake, this time measuring 7.1.

a. Determine the intensity of each earthquake.

b. The difference on the Richter scale of the 1906 and 1989 earthquakes is 1.2. Determine the ratio of the intensity of the 1906 earthquake to the 1989 earthquake.

Discussion

a. Use the formula $R = \log \dfrac{I}{I_0}$ with $R = 8.3$ and $R = 7.1$.

$$R = \log \frac{I}{I_0} \qquad\qquad R = \log \frac{I}{I_0}$$
$$8.3 = \log \frac{I}{I_0} \qquad\qquad 7.1 = \log \frac{I}{I_0} \qquad \text{Substitute values.}$$
$$10^{8.3} = \frac{I}{I_0} \qquad\qquad 10^{7.1} = \frac{I}{I_0} \qquad \text{Write exponential equations.}$$
$$I = 10^{8.3} I_0 \qquad\qquad I = 10^{7.1} I_0$$

The intensity of the 1906 earthquake was $10^{8.3} I_0$, and the intensity of the 1989 earthquake was $10^{7.1} I_0$.

b. The ratio of the intensity of the 1906 earthquake to the intensity of the 1989 earthquake is $\frac{10^{8.3} I_0}{10^{7.1} I_0} = \frac{10^{8.3}}{10^{7.1}} \approx 15.8$.

The intensity of the 1906 earthquake was approximately 15.8 times as great as that of the 1989 earthquake.

Note: These magnitudes have been revised. See the 14.3.4 Checkup for additional information.

✓ 14.3.4 Checkup

1. Pure water has a pH of 7.0. What is its hydrogen ion concentration?

2. Until 1979, an earthquake of magnitude 8.5 on the Richter scale was thought to be the most powerful possible. Since then, improvements in seismic measuring techniques have enabled scientists to refine the scale. Now 9.5 is considered to be the practical limit to measuring an earthquake's magnitude. As a result, the magnitude of the Alaskan earthquake of 1964 has been revised upward to 9.2 on the Richter scale. How does its intensity compare with that of the 1989 San Francisco earthquake, which had a magnitude of 7.1 on the Richter scale?

3. The refinement of the Richter scale also downgraded the magnitude of the 1906 San Francisco earthquake from 8.3 to 7.9. How has the intensity of this earthquake been affected by the downgrade?

14.3 Exercises

Determine the inverse logarithmic function for each exponential function. (Note: k and b are constants.)

1. $f(x) = 11^x$ **2.** $h(x) = 15^x$ **3.** $g(x) = 6^x$ **4.** $p(x) = 29^x$

5. $H(x) = k^x$ **6.** $J(x) = b^x$

Evaluate.

7. $\log_2 64$ **8.** $\log_2 128$ **9.** $\log_3 243$ **10.** $\log_3 729$

11. $\log_4 256$ **12.** $\log_5 125$ **13.** $\log_2 \dfrac{1}{8}$ **14.** $\log_4 \dfrac{1}{16}$

15. $\log_4 0.25$ **16.** $\log_2 0.125$ **17.** $\log_3 \dfrac{1}{81}$ **18.** $\log_5 \dfrac{1}{125}$

Evaluate and check with your calculator.

19. $\log 10$ **20.** $\log 10,000$ **21.** $\log 0.0001$ **22.** $\log 0.000001$

23. $\ln e^3$ **24.** $\ln e^7$ **25.** $\ln e^{-5}$ **26.** $\ln e^{-8}$

27. $\ln \dfrac{1}{e^5}$ **28.** $\ln \dfrac{1}{e^7}$

Evaluate, using your calculator.

29. $\log 15$ **30.** $\log 23$ **31.** $\log \dfrac{1}{12}$ **32.** $\log \dfrac{1}{19}$

33. $\log 1.35$ **34.** $\log 12.5$ **35.** $\ln 14$ **36.** $\ln 27$

37. $\ln 2.85$ **38.** $\ln 11.6$ **39.** $\ln \dfrac{1}{5}$ **40.** $\ln \dfrac{1}{15}$

Evaluate each logarithm, using common logarithms to change the base.

41. $\log_4 12$ **42.** $\log_5 28$ **43.** $\log_2 10$ **44.** $\log_4 100$

45. $\log_5 2.88$ **46.** $\log_3 14.7$ **47.** $\log_5 \dfrac{2}{3}$ **48.** $\log_3 \dfrac{3}{5}$

Evaluate each logarithm, using natural logarithms to change the base.

49. $\log_8 15$ **50.** $\log_6 35$ **51.** $\log_3 5.9$ **52.** $\log_7 3.75$

53. $\log_5 \dfrac{1}{7}$ **54.** $\log_{12} \dfrac{1}{3}$

In exercises 55–60, create a table of values and graph the function.

55. $f(x) = \log_5 x$ **56.** $F(x) = \log_7 x$ **57.** $g(x) = \log(x + 2)$ **58.** $G(x) = \log(x - 2)$

59. $h(x) = \ln(x + 2)$ **60.** $H(x) = \ln(x - 2)$

61. Rainwater typically has a pH of 6.2. Determine its hydrogen ion concentration.

62. Acid rain in many parts of the world has a pH value as low as 3. What is the hydrogen ion concentration of this level of acid rain?

63. Determine the pH value of seawater if its hydrogen ion concentration is 3.2×10^{-9} moles per liter.

64. What is the pH value of milk, whose hydrogen ion concentration is 4×10^{-7} moles per liter?

65. An estimated 50,000 earthquakes of magnitude 3.5 occur worldwide each year. How do these compare in intensity with the 1906 San Francisco earthquake, which had a magnitude of 8.3 (before the revision mentioned in the previous Checkup) on the Richter scale?

66. Every year, there are approximately 800 earthquakes worldwide with a magnitude of 5.5 on the Richter scale. How do these compare in intensity with the 1989 San Francisco earthquake, which had a magnitude of 7.1 on the Richter scale?

14.3 Calculator Exercises

Use your calculator to graph the following logarithmic functions:

1. $f(x) = \log x$　　　　**2.** $g(x) = \log(x + 1)$　　　　**3.** $h(x) = \log x + \log(x + 1)$

4. $k(x) = \log[x(x + 1)]$　　**5.** $j(x) = \log x - \log(x + 1)$　　**6.** $m(x) = \log\left(\dfrac{x}{x + 1}\right)$

Compare the graphs of exercises 3 and 4. Then compare the graphs of exercises 5 and 6. In each pair of exercises, do the graphs appear to be the same? How can you check this using your calculator? If the graphs are the same, does that suggest a relationship to the logarithm of a product or the logarithm of a quotient? How would you state this relationship? Are there any restrictions on the domains of the functions?

14.3 Writing Exercises

1. Explain the difference between $\log_3 5$ and $\log_5 3$. Evaluate these logarithms, and use your calculator to help you see the difference.

2. Explain the difference between the functions $y = \log x + 2$ and $y = \log(x + 2)$. Graph the functions on your calculator to help you see the difference between them.

3. How does the domain of the function $y = \log x + 2$ differ from the domain of the function $y = \log(x + 2)$?

14.4 Properties of Logarithms

OBJECTIVES

1 Rewrite logarithmic expressions by using the properties of logarithms.

2 Model real-world situations by using logarithms.

APPLICATION

If you change the speakers in your stereo system, going from large speakers to smaller ones, there will be a peak response.

To determine the peak response, you must evaluate the expression

$$20[\log 2.6 + \log Q_s + 0.35(\log V_a - \log V_b)]$$

where V_a = volume of the larger speaker, V_b = volume of the smaller speaker, and Q_s is the Q of the system (the ratio of the reactance to the resistance of the circuit). Condense this logarithmic expression.

After completing this section, we will discuss this application further. See page 1014.

14.4.1 Using Properties of Logarithms

In section 14.3, we discussed properties of logarithms that were derived from the definition of logarithms. In order to rewrite logarithmic expressions, we need to expand our list of properties. Since we know that logarithms are exponents, by definition, we should expect to have properties of logarithms that correspond to the properties of exponents that we discussed in Chapter 9. That is,

$$a^m a^n = a^{m+n} \qquad \text{Product rule}$$

$$\frac{a^m}{a^n} = a^{m-n} \qquad \text{Quotient rule}$$

$$(a^m)^n = a^{mn} \qquad \text{Power-to-a-power rule}$$

Complete the following set of exercises on your calculator to discover these properties.

Discovery 5

Product, Quotient, and Power Rules of Logarithms

Approximate each expression, and compare the results obtained in the left column with the corresponding results in the right column.

1. Product rule

    a. $\log(3 \cdot 4) \approx$ _____   **b.** $\log 3 + \log 4 \approx$ _____

    c. $\ln(2 \cdot 5) \approx$ _____   **d.** $\ln 2 + \ln 5 \approx$ _____

2. Quotient rule

    a. $\log \dfrac{3}{4} \approx$ _____   **b.** $\log 3 - \log 4 \approx$ _____

    c. $\ln \dfrac{2}{5} \approx$ _____   **d.** $\ln 2 - \ln 5 \approx$ _____

3. Power rule

    a. $\log 3^4 \approx$ _____   **b.** $4 \log 3 \approx$ _____

    c. $\ln 2^5 \approx$ _____   **d.** $5 \ln 2 \approx$ _____

Write the following rules of logarithms:

4. Product rule

5. Quotient rule

6. Power rule

For the product rule, the logarithm of a product is equal to the sum of the logarithms of the factors. For the quotient rule, the logarithm of a quotient is equal to the difference of the logarithm of the numerator and the logarithm of the denominator. For the power rule, the logarithm of a base raised to a power is equal to the product of the power and the logarithm of the base.

PRODUCT, QUOTIENT, AND POWER RULES OF LOGARITHMS

Assume that m and n are any real number, variable, or expression, where $m > 0$ and $n > 0$. Also, assume that $a > 0$, $a \neq 1$, and c is a real number.

	Logarithm with Base a	Natural Logarithm
Product rule	$\log_a mn = \log_a m + \log_a n$	$\ln mn = \ln m + \ln n$
Quotient rule	$\log_a \dfrac{m}{n} = \log_a m - \log_a n$	$\ln \dfrac{m}{n} = \ln m - \ln n$
Power rule	$\log_a m^c = c \log_a m$	$\ln m^c = c \ln m$

HELPING HAND There is no property of logarithms that can be used to simplify the log of a sum, $\log_a(m + n)$, or the log of a difference, $\log_a(m - n)$.

We can use the properties of logarithms stated in this section and Section 14.3 to rewrite logarithms. We will be working with variable expressions as well as numeric ones.

We will assume that the variables do not represent values for which the expression is not defined.

First, we **expand** a logarithm by using the properties stated previously and writing a sum, a difference, or a product.

EXAMPLE 1 Expand each logarithmic expression.

a. $\log 3x$ **b.** $\log_2 \dfrac{x}{3}$ **c.** $\log_5 x^3$

Solution

a. $\log 3x = \log 3 + \log x$ Product rule

b. $\log_2 \dfrac{x}{3} = \log_2 x - \log_2 3$ Quotient rule

c. $\log_5 x^3 = 3 \log_5 x$ Power rule

We **condense** a logarithm by using the properties stated previously, reading right to left. In other words, we write a product, quotient, or power.

EXAMPLE 2 Condense each logarithmic expression.

a. $\ln 2 - \ln x$ **b.** $\ln 2 + \ln x + \ln y$ **c.** $2 \ln y$

Solution

a. $\ln 2 - \ln x = \ln \dfrac{2}{x}$ Quotient rule

b. $\ln 2 + \ln x + \ln y = \ln 2xy$ Product rule

c. $2 \ln y = \ln y^2$ Power rule

We can combine the rules of exponents to rewrite more complicated expressions. Remember, the logarithms must have the same base in order to apply the rules of logarithms.

EXAMPLE 3 Expand the logarithmic expression $\log_3 \sqrt{3x}$.

Solution

$$\log_3 \sqrt{3x} = \log_3 (3x)^{1/2} \qquad \text{Definition of square root}$$

$$= \frac{1}{2} \log_3 3x \qquad \text{Power rule}$$

$$= \frac{1}{2}(\log_3 3 + \log_3 x) \qquad \text{Product rule}$$

$$= \frac{1}{2} \log_3 3 + \frac{1}{2} \log_3 x \qquad \text{Distribute } \tfrac{1}{2}.$$

$$= \frac{1}{2}(1) + \frac{1}{2} \log_3 x \qquad \log_3 3 = 1$$

$$= \frac{1}{2} + \frac{1}{2} \log_3 x$$

EXAMPLE 4   Condense the logarithmic expression $2 \ln(2x + 3) + \ln(x + 3) - \ln x$.

Solution

$$2 \ln(2x + 3) + \ln(x + 3) - \ln x$$
$$= \ln(2x + 3)^2 + \ln(x + 3) - \ln x \qquad \text{Power rule}$$
$$= \ln(2x + 3)^2(x + 3) - \ln x \qquad \text{Product rule}$$
$$= \ln \frac{(2x + 3)^2(x + 3)}{x} \qquad \text{Quotient rule}$$

14.4.1 Checkup

Expand each logarithmic expression.

1. $\log_5 11x$ **2.** $\ln \dfrac{x}{y}$

In exercises 3 and 4, condense each logarithmic expression.

3. $\log x + \log y - \log z$ **4.** $-3 \ln x$

5. Expand the logarithmic expression $\log_5 5x^2$.
6. What does it mean to expand a logarithmic expression?
7. What does it mean to condense a logarithmic expression?

14.4.2 Modeling the Real World

Condensing logarithmic expressions is an important skill needed to solve logarithmic equations. Condensing logarithms also may simplify the evaluation of expressions by enabling us to calculate a single logarithm instead of calculating more than one logarithm in an expression. The accuracy of our results in evaluating one expression (rounding once) may be greater than when we evaluate an expression in which we round more than one value.

EXAMPLE 5   The power gain, in decibels, of an electronic device is determined by the logarithmic expression $10(\log P_0 - \log P_i)$, where P_0 is the output power, in watts, and P_i is the input power, in watts.

a. Condense the expression, using the rules of logarithms.

b. Determine the power gain of an amplifier with an output of 25.0 W and an input of 0.625 W.

Solution

a. $10(\log P_0 - \log P_i) = 10 \left[\log \left(\dfrac{P_0}{P_i} \right) \right] \qquad$ Quotient rule

$$= \log \left(\dfrac{P_0}{P_i} \right)^{10} \qquad \text{Power rule}$$

b. $\log \left(\dfrac{P_0}{P_i} \right)^{10} = \log \left(\dfrac{25.0}{0.625} \right)^{10} \approx 16.021 \text{ dB}$

APPLICATION

If you change the speakers in your stereo system, going from large speakers to smaller ones, there will be a peak response.

To determine the peak response, you must evaluate the expression

$$20[\log 2.6 + \log Q_s + 0.35(\log V_a - \log V_b)]$$

where V_a = volume of the larger speaker, V_b = volume of the smaller speaker, and Q_s is the Q of the system (the ratio of the reactance to the resistance of the circuit). Condense this logarithmic expression.

Discussion

$$20[\log 2.6 + \log Q_s + 0.35(\log V_a - \log V_b)]$$

$$= 20\left[\log 2.6 + \log Q_s + 0.35 \log\left(\frac{V_a}{V_b}\right)\right] \qquad \text{Quotient rule}$$

$$= 20\left[\log 2.6 + \log Q_s + \log\left(\frac{V_a}{V_b}\right)^{0.35}\right] \qquad \text{Power rule}$$

$$= 20 \log\left[(2.6)(Q_s)\left(\frac{V_a}{V_b}\right)^{0.35}\right] \qquad \text{Product rule}$$

$$= \log\left[2.6Q_s\left(\frac{V_a}{V_b}\right)^{0.35}\right]^{20} \qquad \text{Power rule}$$

 14.4.2 Checkup

In exercises 1 and 2, use the rules of logarithms to expand the right side of each application from geometry.

1. $\ln A = \ln(\pi r^2)$ **2.** $\log A = \log\left(\frac{1}{2}bh\right)$

3. Light passing through a transparent solution follows the model

$$y = c(\ln I_0 - \ln I)$$

where c is a constant related to the solution, I_0 is the intensity of the light striking the solution, and I is the intensity at a depth of y units in the solution.

a. Use the rules of logarithms to condense the right side of the model.

b. If $c = 8.5$ for a solution, find the depth y when the intensity at the surface is three times the intensity at a depth of y units.

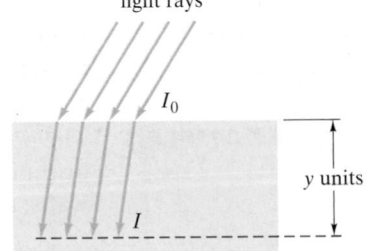

light rays

I_0

y units

I

1–37 odds

1–37 odds

14.4 Exercises

Expand each logarithmic expression.

1. $\log 12a$

2. $\ln 25c$

3. $\ln x^3$

4. $\log z^3$

5. $\log_5 \frac{x}{5}$

6. $\log_7 \frac{7}{y}$

7. $\log \frac{2x^2}{y}$

8. $\log \frac{x^2}{5y^3}$

9. $\log_3 x^3 y^2$

10. $\log_2 a^2 b^4$

11. $\ln \sqrt[3]{xy^2}$

12. $\log \sqrt[4]{x^3 y}$

13. $\log \frac{\sqrt{2x}}{\sqrt[3]{y}}$

14. $\ln \frac{\sqrt[3]{3x^2}}{\sqrt{2y}}$

15. $\log_3 3a$

16. $\log_5 5z$

17. $\log_5 10xy$

18. $\log_3 \frac{xy}{6}$

19. $\log_a(ab^2)$

20. $\log_6(3ab^2)$

Condense each logarithmic expression.

21. $\log x + \log(x + 5)$

22. $\log(x - 1) + \log(x + 1)$

23. $2 \ln x + 3 \ln y$

24. $3 \ln b - 2 \ln c$

25. $2 \log_3(x + 3) - \log_3(x - 1)$

26. $5 \log_5(a - 5) + 2 \log_5 a$

27. $\frac{1}{2} \ln x - \frac{1}{5} \ln(x + 1)$

28. $\frac{1}{4} \log z - \frac{1}{3} \log(z + 5)$

29. $\log xy - \log xz$

30. $2 \log pq - \log p$

In exercises 31 and 32, use the mathematical model for power gain, $G = \log\left(\frac{P_0}{P_i}\right)^{10}$, where P_0 is the output power in watts and P_i is the input power in watts.

31. Determine the power gain G, in decibels, for an amplifier with an output P_0 of 20 watts and an input P_i of 1.5 watts.

32. What is the power gain G, in decibels, of an amplifier with an output P_0 of 30 watts and an input P_i of 2 watts?

In exercises 33 and 34, use the mathematical model for light passing through a transparent solution, $y = c \ln \frac{I_0}{I}$, where I_0 is the intensity of light striking the solution, I is the intensity at a depth of y units in the solution, and c is a related constant.

33. If $c = 12$ for a liquid, find the depth y when the surface intensity I_0 is 3.5 times the intensity at a depth of y units.

34. If $c = 10$ for a liquid, at what depth y will the surface intensity be six times the intensity at a depth of y units?

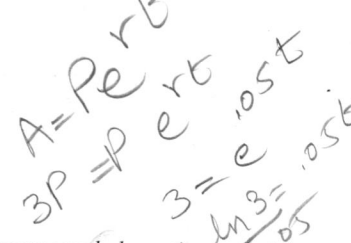

If money in an account is compounded continuously, the amount in the account will grow to k times the original investment in $t = \frac{\ln k}{r}$ years when the annual rate of interest is r. Use this model for the following exercises:

35. How long will it take the amount invested to triple if the annual interest rate is 5%?

36. At an annual interest rate of 4.5%, how long will it take an investment to double in value?

37. How long will it take an investment to increase to four times its original value if the annual interest rate is 9%?

38. If the annual interest rate is 10%, how long will it take an investment to increase by a factor of 3.5 times its original value?

14.4 Calculator Exercises

You were told that the log of a sum or a difference does not usually simplify. To investigate this claim, graph the functions in exercises 1–4 on your calculator, using a decimal window.

1. $\log 3x$

2. $\log x + \log 2x$

3. $\log(x^2 - x)$

4. $\log x^2 - \log x$

5. Compare the graphs of exercises 1 and 2. Are they equivalent?

6. Compare the graphs of exercises 3 and 4. Are they equivalent?

7. What is wrong with the statement $\log(A + B) = \log A + \log B$?

8. What is wrong with the statement $\log(A - B) = \log A - \log B$?

14.4 Writing Exercise

One of the most useful applications of exponential and logarithmic expressions is the determination of the half-life of chemical elements. Find a library or Internet reference that discusses half-life, and write a short summary of what you find. Be sure to cite your reference.

14.5 Solving Exponential and Logarithmic Equations

OBJECTIVES

1 Identify exponential and logarithmic equations in one variable.

2 Solve exponential and logarithmic equations in one variable algebraically.

3 Model real-world situations by using exponential and logarithmic equations in one variable.

APPLICATION

According to Newton's law of cooling, the temperature T of a body t minutes after it is placed in surroundings having a constant temperature T_0 is determined by the function $T(t) = T_0 + Ce^{-kt}$, where C is the difference between the initial temperature and the surrounding temperature and k is the cooling constant.

a. Boiling water ($100°C$) is placed in a freezer at $0°C$. After 25 minutes, the water temperature is $50°C$. Determine the constant of cooling.

b. Determine the temperature of the water after 2 hours.

After completing this section, we will discuss this application further. See page 1024.

14.5.1 Exponential and Logarithmic Equations in One Variable

If we take the logarithm of an expression, we call the expression the **argument** of the logarithm. For example, the expression $(2x + 1)$ is the argument of $\log(2x + 1)$. A **logarithmic equation in one variable**, or, simply, **logarithmic equation**, is an equation with a logarithmic expression having a variable in the argument of the logarithm. For example, the following are logarithmic equations:

$$\log_3 x = 5 \qquad \ln(x + 1) = \ln 5$$

An **exponential equation in one variable**, or, simply, **exponential equation**, is an equation with an exponential expression with a variable exponent. For example, the following are exponential equations:

$$4^x = 26 \qquad e^x = e^{2x-3}$$

EXAMPLE 1 Identify each equation as exponential, logarithmic, or neither.

a. $\ln 3 = x + 4$ b. $10e^x = 100$ c. $x^5 + 6 = \log 10$

Solution

a. $\ln 3 = x + 4$ is neither a logarithmic nor an exponential equation. The logarithmic argument does not contain a variable; therefore, $\ln 3$ represents a number. The equation is a linear equation.

b. $10e^x = 100$ is an exponential equation, because the exponent of the exponential expression contains a variable.

c. $x^5 + 6 = \log 10$ is neither a logarithmic nor an exponential equation. The exponential expression does not have a variable exponent, and the logarithmic expression is a number. The equation is a polynomial equation.

14.5.1 Checkup

In exercises 1–4, identify each equation as exponential, logarithmic, or neither.

1. $8 - x = e^2$ **2.** $2 + \log x = 5$

3. $2^x = 4^{x+3}$ **4.** $\log 5 = x - 1$

5. What is the difference between an exponential equation in one variable and a logarithmic equation in one variable?

14.5.2 Solving Equations

We are now ready to solve exponential and logarithmic equations algebraically. To solve these equations, we need to discuss certain properties. Complete the following set of exercises to discover two such properties.

Discovery 6

Properties of Exponential and Logarithmic Equations

Complete the following statements with "true" or "false":

1. a. If $2 = 2$ is true, then $5^2 = 5^2$ is _____.

  b. If $2 = 2$ is true, then $12^2 = 12^2$ is _____.

  c. If $2 = 2$ is true, then $e^2 = e^2$ is _____.

Write a rule for determining a true equation using exponentials.

2. a. If $2 = 2$ is true, then $\log_5 2 = \log_5 2$ is _____.

  b. If $2 = 2$ is true, then $\log_{12} 2 = \log_{12} 2$ is _____.

  c. If $2 = 2$ is true, then $\ln 2 = \ln 2$ is _____.

Write a rule for determining a true equation using logarithms.

If we begin with an equation that is true and write an exponential expression for each side of the equation, using the same base and the original expressions as exponents, the result is a true equation. If we begin with an equation that is true and take the same logarithm of both expressions contained in the equation, the result is a true equation.

PROPERTIES OF EXPONENTIAL AND LOGARITHMIC EQUATIONS

For a positive real number $a \neq 1$ and real numbers x and y, the following properties hold:

1. a. If $a^x = a^y$, then $x = y$.

  b. If $x = y$ then $a^x = a^y$.

For positive real numbers x, y, and $a \neq 1$, the following properties hold:

2. a. If $\log_a x = \log_a y$, then $x = y$.

  b. If $x = y$, then $\log_a x = \log_a y$.

First, we will discuss solving exponential equations. One example is an exponential equation that is written with both sides having exponential expressions with the same base. We use property 1a to solve this exponential equation by equating the exponents. For example, solve $4^x = 4^3$.

$$4^x = 4^3$$

$$x = 3 \qquad \text{Equate the exponents of both sides.}$$

Another example is an exponential equation that cannot be written with both sides having exponential expressions with the same base. To solve this type of equation, we choose a logarithm, usually log or ln, and equate the logarithms of both sides, using property 2b. For example, solve $4^x = 5$.

$$4^x = 5$$

$$\ln 4^x = \ln 5 \qquad \text{Equate the natural logarithms of both sides.}$$

$$x \ln 4 = \ln 5 \qquad \text{Power rule}$$

$$x = \frac{\ln 5}{\ln 4} \qquad \text{Divide both sides by ln 4.}$$

$$x \approx 1.16$$

SOLVING AN EXPONENTIAL EQUATION ALGEBRAICALLY

To solve an exponential equation algebraically,

- Write both sides of the equation as exponential expressions with the same base, equate the exponents, and solve for the variable.

or

- Isolate the exponential expression, take the logarithms of both sides, and solve for the variable.

EXAMPLE 2 Solve algebraically. Check your solution by substitution, numerically, or graphically.

a. $2^x = 32$ **b.** $e^x = 72$ **c.** $5e^{2x} = 8$

Solution

a. $2^x = 32$

$2^x = 2^5$ — Rewrite 32 with the same base as the expression on the left, or 2: $32 = 2^5$.

$x = 5$ Equate the exponents of both sides.

Substitution Check

$$2^x = 32$$

$$2^5 \mid 32$$

$$32 \mid$$

Since $32 = 32$ the solution checks.

Graphic Check

b. $e^x = 72$

$\ln e^x = \ln 72$ Equate the natural logarithms of both sides.

$x = \ln 72$ Property of natural logarithms

$x \approx 4.277$

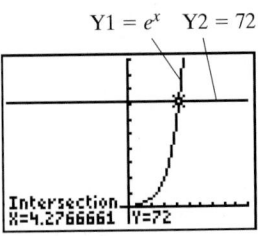

$$Y1 = e^x \quad Y2 = 72$$

Intersection
X=4.2766661 Y=72

$(-10, 10, -10, 100)$

The solution is about 4.277 because the x-coordinate of the intersection is about 4.277.

c. $5e^{2x} = 8$

$e^{2x} = \dfrac{8}{5}$ Divide both sides by 5.

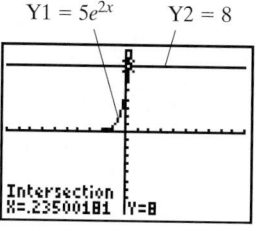

$Y1 = 5e^{2x}$ $Y2 = 8$

$\ln e^{2x} = \ln \dfrac{8}{5}$ Equate the natural logarithms of both sides.

$2x = \ln \dfrac{8}{5}$ Property of natural logarithms, $\ln e^{2x} = 2x.$

Intersection
X=.23500181 Y=8

$(-10, 10, -10, 10)$

$x = \dfrac{\ln \dfrac{8}{5}}{2}$ Divide both sides by 2.

The solution is about 0.235 because the x-coordinate of the intersection is about 0.235.

$x \approx 0.235$

We use property 2a to solve logarithmic equations by equating the arguments of logarithms with the same base. For example, solve $\ln(x + 3) = \ln 2$.

$\ln(x + 3) = \ln 2$

$x + 3 = 2$ Equate the arguments of the logarithms of both sides.

$x = -1$ Substract 3 from both sides.

To solve a logarithmic equation that cannot be written with the same base, we isolate the logarithmic expression, use the definition of logarithms to write an exponential equation, and solve for the variable. For example, solve $\log_7(2x) = 6$.

$\log_7(2x) = 6$

$7^6 = 2x$ Definition of logarithms

$\dfrac{7^6}{2} = x$ Divide both sides by 2.

$x = 58824.5$

SOLVING A LOGARITHMIC EQUATION ALGEBRAICALLY

To solve a logarithmic equation algebraically,

- Determine the restricted values.
- Write both sides of the equation with logarithmic expressions with the same base, equate the arguments, and solve for the variable.

or

- Isolate the logarithmic expression, use the definition of logarithms to write an exponential equation, and solve for the variable.

EXAMPLE 3 Solve algebraically. Check your solution by substitution, numerically, or graphically.

a. $\log x + \log(x - 2) = \log 8$ **b.** $3 \ln 2x = 9$

Algebraic Solution

a. Logarithms are defined for positive values only. Therefore, the restricted values are $x \le 0$ and $x - 2 \le 0$ or $x \le -2$.

$$\log x + \log(x - 2) = \log 8$$

$$\log[x(x - 2)] = \log 8 \qquad \text{Product rule.}$$

$$x(x - 2) = 8 \qquad \begin{array}{l}\text{Equate the} \\ \text{arguments of} \\ \text{the logarithms} \\ \text{of both sides.}\end{array}$$

$$x^2 - 2x - 8 = 0$$

$$(x - 4)(x + 2) = 0$$

$$x - 4 = 0 \quad \text{or} \quad x + 2 = 0$$

$$x = 4 \qquad\qquad x = -2$$

Since $x = -2$ is a restricted value, the only solution is 4.

Graphic Check

$Y2 = \log 8$ $Y1 = \log x + \log (x - 2)$

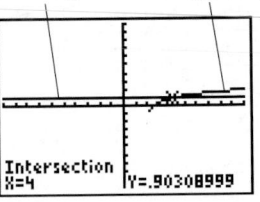

$(-10, 10, -10, 10)$

According to the graph, there is only one solution, 4. Note that -2 is a restricted value and cannot be a solution.

b. The restricted values are $2x \le 0$, or $x \le 0$.

$$3 \ln 2x = 9$$

$$\ln 2x = 3 \qquad \text{Divide both sides by 3.}$$

$$e^3 = 2x \qquad \begin{array}{l}\text{Definition of natural} \\ \text{logarithms.}\end{array}$$

$$\frac{e^3}{2} = x \qquad \text{Divide both sides by 2.}$$

$$x \approx 10.043$$

Graphic Check

$Y2 = 9$ $Y1 = 3 \ln 2x$

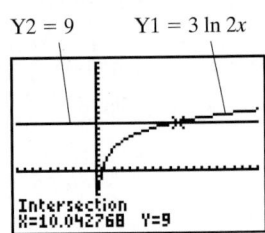

$(-10, 20, -10, 20)$

The solution is about 10.043 because the x-coordinate of the intersection is about 10.043.

An exponential equation or a logarithmic equation may not have a solution, or it may have more than one solution. The first of these alternatives occurs if the equation is a contradiction, the second if it is an identity.

EXAMPLE 4 Solve $\log x^4 - \log x = \log x^3$.

Solution

$$\log x^4 - \log x = \log x^3$$

$$\log\left(\frac{x^4}{x}\right) = \log x^3 \qquad \text{Quotient rule.}$$

$$\log x^3 = \log x^3 \qquad \text{Simplify.}$$

$$x^3 = x^3$$

Equate the arguments of the logarithms of both sides.

$$x^3 - x^3 = x^3 - x^3$$

Subtract x^3 from both sides.

$$0 = 0$$

This is an identity. The equation thus has an infinite number of solutions. The solution set is the set of all positive real numbers, the permissible values of the variables in the equation.

ALGEBRAIC SOLUTIONS OF AN EXPONENTIAL OR LOGARITHMIC EQUATION

To solve an exponential or a logarithmic equation algebraically, determine the restricted values and solve the equation. One of three possibilities will occur.

One or more solutions exist.

No solution exists. The solution process results in a contradiction.

An infinite number of solutions exist. The solution set is the set of all real numbers not equal to the restricted values. The solution process results in an identity.

  14.5.2 Checkup

Solve exercises 1–7 algebraically.

1. $3^x = 81$

2. $e^x = 3$

3. $2e^x = 7$

4. $\ln 2 + 2 \ln x = \ln(6 - x)$

5. $5 + \log 3x = 7$

6. $\log 4x^2 = \log x + \log 4x$

7. $2^{x(x+2)} = 2^{(x-1)^2}$

8. What does it mean to solve an exponential or a logarithmic equation in one variable by the algebraic method?

9. What advantage is there in using the algebraic method over the numerical or graphical method of solving an exponential or a logarithmic equation?

14.5.3 Modeling the Real World

In section 14.2, we determined the balance in an account earning continuously compounded interest. This is one example of a quantity that increases over time according to the **exponential growth** model. A similar **exponential decay** model represents a quantity that decreases steadily over time. Examples of these models are found in biology, chemistry, and business, as well as other physical and social sciences.

The exponential growth model is represented by the function $A(t) = A_0 e^{kt}$, where $A(t)$ is the amount of a substance present at time t, A_0 is the initial amount of the substance present, and k represents the growth constant $(k > 0)$. The exponential decay model is represented by the same function, except that the growth factor is negative $(k < 0)$.

EXAMPLE 5 According to the U.S. Bureau of the Census, the population of the United States in 1999 was 273 million people. The population is projected to increase to 404 million people in 2050.

a. Determine the growth constant.

b. Write an exponential growth function to model the U.S. population.

c. Use the function in part b to estimate the U.S. population in 2025.

Solution

a. $A(t) = A_0 e^{kt}$

Exponential growth function; t = number of years after 1999, $A(t) = 404$, $A_0 = 273$, $t = 2050 - 1999 = 51$

$404 = 273 e^{k(51)}$

$\dfrac{404}{273} = e^{51k}$

Divide both sides by 273.

$\ln \dfrac{404}{273} = \ln e^{51k}$

Equate the natural logarithms of both sides.

$\ln \dfrac{404}{273} = 51k$

Inverse property of natural logarithms

$k = \dfrac{1}{51} \ln \dfrac{404}{273}$

Multiply both sides by $\frac{1}{51}$.

$k \approx 0.008$

The growth constant is about 0.008.

b. $A(t) = A_0 e^{kt}$

Exponential growth function; t = number of years after 1999

$A(t) \approx 273 e^{0.008t}$

$A(t)$ = population in millions, $A_0 = 273$ million, $k \approx 0.008$, t = number of years after 1999

c. $A(t) \approx 273 e^{0.008t}$

$A(26) \approx 273 e^{0.008(26)}$

Substitute 26 for $t(2025 - 1999)$.

$A(26) \approx 336$

The population of the United States will be approximately 336 million in 2025.

EXAMPLE 6 The gross national product has risen exponentially since 1990. The gross national product can be approximated by the exponential growth model $A(t) = A_0 e^{0.0412t}$, where t is the number of years after 1990 and A_0 is the gross national product for 1990. Determine when the 1990 gross national product will double.

Solution

Let A_0 = the gross national product in 1990

$2A_0$ = double the 1990 gross national product

Therefore, let $A(t) = 2A_0$ and solve for t.

$A(t) = A_0 e^{0.0412t}$

$2A_0 = A_0 e^{0.0412t}$

$2 = e^{0.0412t}$

Divide both sides by A_0.

$\ln 2 = \ln e^{0.0412t}$

Equate the natural logarithms of both sides.

$\ln 2 = 0.0412t$

Inverse property of natural logarithms

$t = \dfrac{\ln 2}{0.0412}$

Divide both sides by 0.0412.

$t \approx 16.8$

The 1990 gross national product will double in about 17 years.

If a quantity grows exponentially, the amount of time it takes to increase to twice its original amount is called the **doubling time**. Note that the amount originally present does not affect this time.

If a quantity decays exponentially, the amount of time it takes to diminish to half of its original amount is called the **half-life**. As with the doubling time, the amount originally present does not affect this time.

EXAMPLE 7   Carbon-14 (C-14) is a radioactive form of carbon found in all living organisms. After an organism dies, the C-14 in it begins to decay. The rate of decay appears to be steady. The half-life of C-14 is about 5730 years. The exponential decay function can be used to approximate the amount of C-14 present t years after an organism dies.

a. Determine the decay factor.

b. If a fossil is found with one-third the C-14 of its living form, determine the time since the organism died.

Solution

a. A_0 = the original amount of C-14 present

$\dfrac{1}{2} A_0$ = half the original amount of C-14 present

Therefore, let $A(t) = \frac{1}{2} A_0$ and $t = 5730$, and then solve for k.

$$A(t) = A_0 e^{kt}$$

$$\frac{1}{2} A_0 = A_0 e^{k \cdot 5730}$$

$$\frac{1}{2} = e^{5730k} \qquad \text{Divide both sides by } A_0.$$

$$\ln \frac{1}{2} = \ln e^{5730k} \qquad \text{Equate the natural logarithms of both sides.}$$

$$\ln \frac{1}{2} = 5730k \qquad \text{Property of natural logarithms}$$

$$k = \left(\ln \frac{1}{2} \right)\left(\frac{1}{5730} \right) \qquad \text{Multiply both sides by } \tfrac{1}{5730}.$$

$$k \approx -0.000121$$

The decay constant is about -0.000121.

b. Let $\frac{1}{3} A_0$ = one-third of the amount of the original C-14

$$A(t) = A_0 e^{kt}$$

$$\frac{1}{3} A_0 \approx A_0 e^{-0.000121t}$$

$$\frac{1}{3} \approx e^{-0.000121t} \qquad \text{Divide both sides by } A_0.$$

$$\ln \frac{1}{3} \approx \ln e^{-0.000121t} \qquad \text{Equate the natural logarithms of both sides.}$$

$$\ln \frac{1}{3} \approx -0.000121t \qquad \text{Property of natural logarithms.}$$

$$t \approx \left(\ln \frac{1}{3} \right)\left(-\frac{1}{0.000121} \right) \qquad \text{Multiply both sides by } -\frac{1}{0.000121}.$$

$$t \approx 9079.44$$

The organism has been dead about 9079 years.

APPLICATION

According to Newton's law of cooling, the temperature T of a body t minutes after it is placed in surroundings having a constant temperature T_0 is determined by the function $T(t) = T_0 + Ce^{-kt}$, where C is the difference between the initial temperature and the surrounding temperature and k is the cooling constant.

a. Boiling water (100°C) is placed in a freezer at 0°C. After 25 minutes, the water temperature is 50°C. Determine the constant of cooling.

b. Determine the temperature of the water after 2 hours.

Discussion

a. $T(t) = T_0 + Ce^{-kt}$

$$50 = 0 + 100e^{-k \cdot 25} \qquad \text{Substitute 50 for } T(t), \text{ 0 for } T_0, \text{ 100 for } C, \text{ and 25 for } t.$$

$$50 = 100e^{-25k}$$

$$\frac{1}{2} = e^{-25k} \qquad \text{Divide both sides by 100.}$$

$$\ln \frac{1}{2} = \ln e^{-25k} \qquad \text{Equate the natural logarithms of both sides.}$$

$$\ln \frac{1}{2} = -25k \qquad \text{Inverse property of natural logarithms}$$

$$k = -\frac{1}{25} \ln \frac{1}{2} \qquad \text{Multiply both sides by } -\frac{1}{25}.$$

$$k \approx 0.0277$$

The constant of cooling is about 0.0277.

b. $T(120) \approx 0 + 100e^{-0.0277(120)}$ Substitute 120 minutes for t, 0 for T_0, 100 for C, and -0.0277 for k.

$$T(120) \approx 3.6$$

The temperature is about 3.6°C after 2 hours.

14.5.3 Checkup

1. The U.S. Census Bureau estimated the total female population of the United States to be about 128 million in 1990. This population had grown to about 139 million in 1999. Use the exponential growth model to develop a mathematical model that estimates the total female population t years after 1990. What does the model predict the total female population will be in 2010?

2. The gross national product growth model is

$$A(t) = A_0 e^{0.0412t}$$

where t is the number of years after 1990 and A_0 is the gross national product for 1990. When will the gross national product triple in value?

3. Radium-226 has a half-life of 1620 years. Determine the decay factor of radium-226. How long will it take for a store of radium-226 to diminish to 90 percent of its original amount?

4. Use Newton's law of cooling to determine the time it would take boiling water to cool to 20°C when placed in a freezer at 0°C.

14.5 Exercises

Identify each equation as exponential, logarithmic, or neither.

1. $-3 + \log x^2 = 1$
2. $2e^x - 5 = 3$
3. $e^3 + x = 2x - 1$
4. $\log x^3 + 3 = 5$
5. $e^{2x} - 1 = 5$
6. $\log 8 + x = \log 10$
7. $x = \log 4 + \log 7$
8. $x + e^2 = 1$

Solve exercises 9–70.

9. $5^{2x} = 625$
10. $3^{x+2} = 81$
11. $e^t = 2$
12. $e^{3t} + 5 = 7$
13. $5^x = 1$
14. $3^m = 0.5$
15. $3e^{-x} = 2$
16. $e^{-2x} + 1 = 5$
17. $(2^{3x})(2^5) = 2^8$
18. $(5^x)(5^{x+3}) = 5^4$
19. $2^x = 2^{3x-6}$
20. $e^{3x-1} = e^{2x+3}$
21. $5^{2x(x-2)} = 5^{3(x-1)}$
22. $7^{a^2+4} = 7^{5a}$
23. $e^{x(x-2)} = e^{x+10}$
24. $3^{2x(x+6)} = 3^{5(x+3)}$
25. $3^{x+x(x-3)} = 3^{(x-1)^2}$
26. $5^{(x+3)^2} = 5^{x(x+6)}$
27. $8^{5(x-3)} = 8^{4(x-2)}8^{(x-7)}$
28. $e^{a+7} = e^{3a+4}e^{3-2a}$
29. $3^{3x} = 9^{x+1}$
30. $5^{x+7} = 25^{x-2}$
31. $5^{2b+5} = 5^4$
32. $7^{3c-1} = 7^{c+9}$
33. $3^{2x^2} \cdot 3^{5x} = 27$
34. $5^{3x^2} \cdot 5^{5x} = 25$
35. $e^{0.5x} = 4$
36. $10^{1.5x} = 8$
37. $(3.5)^x = 12$
38. $(4.3)^z - 5 = 1$
39. $\log_5 c = 5$
40. $3 \log_2 b = 12$
41. $\log x = 0$
42. $\log x = 1$
43. $\ln z = 0$
44. $\ln y = 2$
45. $\log_3 k = 2$
46. $2 \log_5 x = 8$
47. $\log_5 x = -3$
48. $\log_3 k + 5 = 1$
49. $\log_3(a + 2) = 2$
50. $\log_2(3a - 7) = 3$
51. $2 \ln x = 1$
52. $-5 \ln x = 10$
53. $\log x + 4 = 2$
54. $3 \log x - 5 = 1$
55. $5 \ln z + 4 = 4$
56. $\ln x + 2 = 3$
57. $\ln x + \ln(x + 1) = \ln 12$
58. $\ln x + \ln(2x - 5) = \ln 25$
59. $2 \ln(p - 5) = \ln 9$
60. $\ln 16 = 2 \ln(x - 7)$
61. $\log 2x + \log 3x = \log 6 + 2 \log x$
62. $3 \log x - \log x = 2 \log x$
63. $\ln x + \ln(x + 2) = 2 \ln(x + 2)$
64. $2 \ln(x + 2) = \ln(x + 2) + \ln 2$
65. $\log(x + 4) - \log x = \log 2$
66. $\log(x + 3) - \log x = \log 5$
67. $\log x - \log(x + 6) = -1$
68. $\log_2 x - \log_2(x + 6) = -2$
69. $\log_3(x + 16) - \log_3 x = 2$
70. $\ln(x^2 - 3x + 2) - \ln(x - 1) = \ln 5$

71. Expenditures for public elementary and secondary schools totaled $218 billion in 1990 and grew exponentially to $328 billion in 1998. Use the exponential growth function to model the growth in expenditures since 1990. If this growth continues, what will be the estimated expenditures in the year 2010?

72. The U.S. Bureau of the Census reported the total male population of the United States to be about 121 million in 1990. It reported that that population had grown to 133 million in 1999. Use the exponential growth function to model the growth in the male population since 1990. What does the model predict the male population will be in the year 2010?

73. Per capita income in the United States has grown exponentially over the last decade. Per capita income can be estimated by the growth model $A(t) = A_0 e^{0.0416t}$, where t is the number of years after 1990 and A_0 is per capita income for 1990. If the trend continues, in what year will per capita income double its 1990 value?

74. Using the growth model in exercise 73, in what year will per capita income increase to 1.5 times its value in 1990?

75. Uranium-238 has a half-life of 4.5 billion years. Determine the decay factor of uranium-238. How long will it take for a sample of uranium-238 to diminish to 98% of its original amount?

76. Using the information in exercise 75, determine how long it will take for a sample of uranium-238 to diminish to 90% of its original amount.

14.5 Calculator Exercises

Using your calculator, solve the equation $e^{x(x+2)} = e^{(x+1)^2}$ graphically. If you enter the expressions as shown in the following diagram, you may notice that the calculator indicates that the graphs cross at $x \approx -1.414214$.

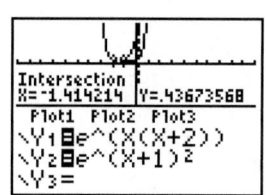

$(-10, 10, -10, 10)$

However, if you attempt to solve the same equation algebraically, you see that it has no solution. The reason may be an inappropriate placement of parentheses in entering the expressions into the calculator. Let's explore the ways you can enter the right-hand expression into your calculator. When you press the e^x key, the calculator starts a set of parentheses for you. If you follow this initial parenthesis with the expression $x + 1$, close the parentheses, and square the expression, the calculator assumes that the resulting expression is equivalent to the expression $(e^{(x+1)})^2$. To instruct the calculator that the expression $(x + 1)$ is to be squared, you must enclose the squared expression in another pair of parentheses: $e^{((x+1)^2)}$. To verify this,

enter all three expressions into your calculator, graph each separately, and then compare the graphs. The keystrokes needed to enter all three are as follows:

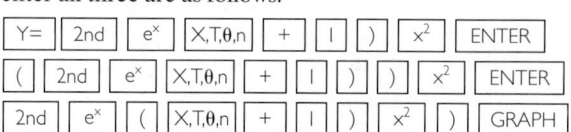

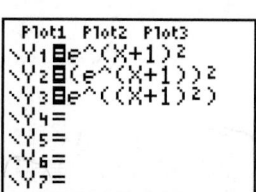

Notice that the graphs of Y1 and Y2 are identical. This indicates that when you square an exponent of e, the calculator assumes that the entire exponential expression is to be squared. The graph of Y3 is different from the other two. This indicates that if you want the exponent of e to be squared, you must use an additional set of parentheses to enclose the entire squared exponent. Be careful that you do this. Remember that it is safer to use an additional set of parentheses to be sure that the calculator behaves as you wish. Just make certain that the parentheses do not change the meaning of your expression.

14.5 Writing Exercise

In this section, you saw examples of exponential growth models and exponential decay models. Describe the two models and state when it would be appropriate to use either. Then

describe a situation that would be an appropriate application of each model.

CHAPTER 14 SUMMARY

After completing this chapter, you should be able to define the following key terms in your own words.

14.1
inverse
one-to-one function
line of symmetry

14.2
exponential function
natural base
natural exponential function

14.3
logarithm of *x* with base *a*
logarithmic function
common logarithm
natural logarithm
change-of-base formula

14.4
expand
condense

14.5
argument
logarithmic equation in one variable
exponential equation in one variable
exponential growth
exponential decay
doubling time
half-life

Reflections

1. What do we mean by the inverse of a relation?
2. How can you construct the inverse of a relation?
3. What is a one-to-one function?
4. What is an exponential function?
5. What is a logarithmic function?
6. What are the two most frequently used bases for a logarithmic function?
7. State the following rules of logarithms:
   a. The product rule
   b. The quotient rule
   c. The power rule

CHAPTER 14 SECTION-BY-SECTION REVIEW

14.1

Recall	Examples
Inverse of a relation • If the relation is a set of ordered pairs, interchange the coordinates of the original ordered pairs. • If the relation is written as an equation, interchange the independent and dependent variables and solve for the new dependent variable.	Determine the inverse of the given relations. $A = \{(1, 2), (3, 5), (-2, 1)\}$ $A^{-1} = \{(2, 1), (5, 3), (1, -2)\}$ $6y = 2x - 4$ $6x = 2y - 4$ $6x + 4 = 2y$ $3x + 2 = y$ or $y = 3x + 2$
One-to-one function • If the relation is a set of ordered pairs, no two different ordered pairs should have the same first coordinate or the same second coordinate.	Determine whether the given relations are one-to-one functions. $A = \{(2, -2), (3, 4), (1, 2)\}$ is a one-to-one function, because no two values are repeated for the first or second coordinates.

(Continued on page 1028)

Recall	Examples
One-to-one function (*continued*) • If the relation is described by a graph, the graph must pass both the vertical-line test and the horizontal-line test.	 According to the graph, the relation is not a one-to-one function, because it fails the horizontal line test.
Graph of a function and its inverse • Determine the inverse of a function. • Graph the function and its inverse. • Graph the line of symmetry, $y = x$.	Graph $f(x) = x^3 - 4$ and its inverse. Determine the inverse. $$f(x) = x^3 - 4$$ $$x = y^3 - 4$$ $$y = \sqrt[3]{x + 4}$$ $$f^{-1}(x) = \sqrt[3]{x + 4}$$

Determine the inverse of each relation.

1. $h = \{(2, 4.5), (4, 3.5), (6, 2.5), (8, 1.5), (10, 0.5)\}$

2. $y = 3x - 9$

3. $y = x^2 + 1$

4. $y = x^3 - 1$

Determine whether the inverse of each function is a function.

5. $P = \{(9, 1), (8, 2), (7, 3), (6, 1), (5, 2), (4, 3)\}$

6. $Q = \{(1, 9), (2, 8), (3, 7), (4, 6), (5, 5), (6, 4), (7, 3)\}$

Determine whether each graph represents a one-to-one function.

7.

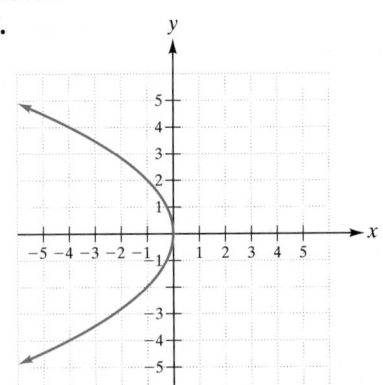

8.

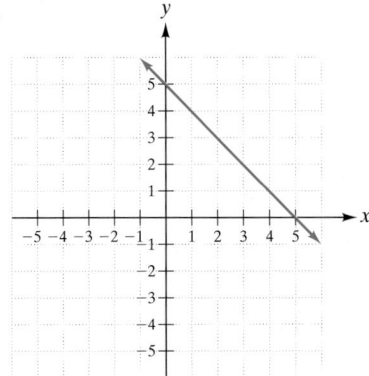

9.

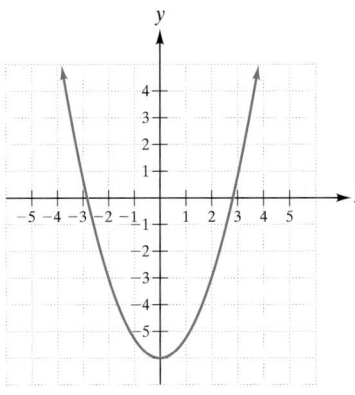

Determine the inverse of each function. Sketch the graph of the function, its inverse, and the line of symmetry.

10. $f(x) = \dfrac{3}{4}x - 3$ **11.** $y = x^3 + 8$

For each situation, write a function that represents it. Determine the inverse of the function and interpret what the inverse function represents.

12. Motomo is paid \$570 per week, plus 3% commission on the value of all sales she handles. What is Motomo's weekly income?

13. Valecia can complete an average of 15 algebra problems in one hour. What is the average number of problems she can complete in a given number of hours?

14.2

Recall	Examples
Exponential functions can be written as $f(x) = a^x$, where $a > 0$ and $a \neq 1$.	Examples of exponential functions are $f(x) = 5^x$ $g(x) = 5^{-x}$
Graph an exponential function • Set up a table of values. • Plot the ordered pairs and connect them with a smooth curve. The y-intercept is always $(0, 1)$.	Graph $g(x) = 5^{-x}$. <table><tr><td>X</td><td>Y1</td></tr><tr><td>-2</td><td>25</td></tr><tr><td>-1</td><td>5</td></tr><tr><td>0</td><td>1</td></tr><tr><td>1</td><td>.2</td></tr><tr><td>2</td><td>.04</td></tr><tr><td>3</td><td>.008</td></tr><tr><td>4</td><td>.0016</td></tr></table> X=-2 $g(x)$ $(-1, 5)$ $g(x) = 5^{-x}$ $(0, 1)$ $(1, 0.2)$ $(3, 0.008)$

(Continued on page 1030)

Recall	Examples
The natural exponential function is written as $f(x) = e^x$, where e is the natural base.	Examples of natural exponential functions are $f(x) = e^x$ $g(x) = e^{-x}$
Compound interest The compounded amount is $A = P(1 + r)^t$, where P = principal, r = interest rate per period, and t = number of periods.	Determine the amount of an investment of $5000 at an annual compounded rate of 6% at the end of three years. $A = P(1 + r)^t$ $A = 5000(1 + 0.06)^3$ $A = 5955.08$ There will be $5955.08 in the account at the end of three years.

Determine whether each function is an exponential function.

14. $f(x) = 2^x$ **15.** $y = x^2$

Evaluate the function $g(x) = 1.21^x$ at the given values.

16. $g(2)$ **17.** $g(-1)$ **18.** $g(0)$

19. $g\left(\dfrac{1}{2}\right)$ **20.** $g(0.3)$ **21.** $g(\sqrt{2})$

In exercises 22–24, graph each function.

22. $f(x) = 5^x - 1$ **23.** $g(x) = 9^{(1/2)x}$ **24.** $y = 2e^{0.5x}$

25. Paul invested his holiday bonus of $1000 into an account that paid 4% interest compounded annually. Write a function for the total value of his investment. Graph the function. How much was the investment worth after four years?

26. The mathematical model $y = 330.53(1.06)^x$, where x is the number of years after 1990 and y is total expenditures in billions of dollars, predicts national health expenditures, by the U.S. government. What does the model predict that expenditures will be for the year 2010?

14.3

Recall	Examples
Logarithm of x with base a $y = \log_a x$ if and only if $x = a^y$	Determine the inverse function of $f(x) = 4^x$ $f^{-1}(x) = \log_4 x$
Change-of-base formula $\log_a x = \dfrac{\log_b x}{\log_b a} = \dfrac{\ln x}{\ln a}$	Evaluate $\log_3 6$. $\log_3 6 = \dfrac{\ln 6}{\ln 3} \approx 1.631$

Recall	Examples
Graph a logarithmic function Method 1 • Determine the inverse exponential function. • Set up a table of values for the exponential function. • Interchange the coordinates of the exponential function. • Plot the points and connect them with a smooth curve. The x-intercept is always $(1, 0)$.	Graph. $g(x) = \log_2 x$ Method 1 $g^{-1}(x) = 2^x$

x	$g^{-1}(x)$		x	$g(x)$
-3	0.125		0.125	-3
-2	0.25		0.25	-2
-1	0.5		0.5	-1
0	1		1	0
1	2		2	1
2	4		4	2
3	8		8	3

Recall	Examples
Graph a logarithmic function Method 2 • Set up a table of values. • Plot the ordered pairs and connect them with a smooth curve. The x-intercept is always $(1, 0)$.	Graph. $g(x) = \log_2 x$ Method 2

x	$g(x)$
0.125	-3
0.25	-2
0.5	-1
1	0
2	1
4	2
8	3

For each exponential function, determine the logarithmic function that is its inverse.

27. $h(x) = 8^x$ **28.** $A(x) = Ae^{kx}$

Evaluate.

29. $\log 100$ **30.** $\log_2 16$ **31.** $\log_3 \dfrac{1}{27}$

32. $\log 1.5$ **33.** $\ln e^4$ **34.** $\ln 10$

35. $\ln e^{-2}$ **36.** $\log \dfrac{3}{5}$ **37.** $\log_5 15$

Create a table of values for each function and graph.

38. $y = \log_3 x$ **39.** $f(x) = \ln(x - 2)$

The pH value of a substance is defined by the function

$$\text{pH} = -\log[\text{H}^+]$$

where $[\text{H}^+]$ *is the hydrogen ion concentration in moles per liter. Use this relation to answer the following questions:*

40. The hydrogen ion concentration of lemon juice is 6.2×10^{-3} moles per liter. What is the pH of lemon juice?

41. The pH value of blood is typically 7.4. What is the hydrogen ion concentration of blood?

14.4

Recall	Examples
Product rule • $\log_a mn = \log_a m + \log_a n$ • $\ln mn = \ln m + \ln n$	Expand. $\log_3 5x$ $\log_3 5x = \log_3 5 + \log_3 x$ Condense. $\log_3 6 + \log_3 2$ $\log_3 6 + \log_3 2 = \log_3 12$
Quotient rule • $\log_a \dfrac{m}{n} = \log_a m - \log_a n$ • $\ln \dfrac{m}{n} = \ln m - \ln n$	Expand. $\ln \dfrac{2x}{5}$ $\ln \dfrac{2x}{5} = \ln 2x - \ln 5 = \ln 2 + \ln x - \ln 5$ Condense. $\ln 6 - \ln 2$ $\ln 6 - \ln 2 = \ln \frac{6}{2} = \ln 3$
Power rule • $\log_a m^c = c \log_a m$ • $\ln m^c = c \ln m$	Expand. $\log_3 x^5$ $\log_3 x^5 = 5 \log_3 x$ Condense. $4 \ln 3$ $4 \ln 3 = \ln 3^4 = \ln 81$

Write each logarithmic expression in expanded form.

42. $\log 2x^4$ **43.** $\log_7 \dfrac{7x}{y}$ **44.** $\ln 25x^2 y^3 z$ **45.** $\ln \dfrac{\sqrt[3]{2x^2 y}}{\sqrt{yz}}$

In exercises 46–49, write each logarithmic expression in condensed form.

46. $\log(x-3) + \log(x+3)$ **47.** $5 \ln x - 3 \ln y$

48. $\dfrac{1}{3} \log x - \dfrac{1}{2} \log y$ **49.** $\log ab - \log bc + \log cd$

50. Money invested at a continuously compounded interest rate r will grow to k times the original amount invested in t years, where $t = \frac{\ln k}{r}$. How long will it take an investment to triple in value if the annual interest rate is 6%?

14.5

Recall	Examples
Solving an exponential equation • Write both sides of the equation as an exponential with the same base, equate the exponents, and solve for the variable. or • Isolate the exponential expression, take the logarithm of both sides, and solve for the variable.	Solve. $2^x = 32$ $2^x = 32$ $2^x = 2^5$ $x = 5$ Solve. $e^{2x} = 4$ $e^{2x} = 4$ $\ln e^{2x} = \ln 4$ $2x = \ln 4$ $x = \dfrac{\ln 4}{2}$ $x \approx 0.693$

Recall	Examples
Solving a logarithmic equation • Determine the restricted values. • Write both sides of the equation as a logarithm with the same base, equate the arguments, and solve for the variable.	Solve. $\ln x^2 = \ln x + \ln 4$ Restricted values are $x \leq 0$. $\ln x^2 = \ln x + \ln 4$ $\ln x^2 = \ln 4x$ $x^2 = 4x$ $x^2 - 4x = 0$ $x(x - 4) = 0$ $x = 0, x = 4$ The solution is $x = 4$. (0 is a restricted value.)
or • Isolate the logarithmic expression, raise both sides to the same power, and solve for the variable.	Solve. $\log(x + 3) = 2$ Restricted values are $x + 3 \leq 0$, or $x \leq -3$. $\log(x + 3) = 2$ $10^{\log(x+3)} = 10^2$ $x + 3 = 10^2$ $x = 10^2 - 3$ $x = 97$

Identify each equation as exponential, logarithmic, or neither.

51. $e^x + 5 = 7$ **52.** $\ln 7 - x = \ln 3$ **53.** $\log x^2 + 2 = \log x$

54. $x = e^3 - 1$ **55.** $2^x + 7 = 2^{3x}$ **56.** $y = \log_2 5 - e^2$

Solve exercises 57–75.

57. $3^{x+1} = 243$ **58.** $e^{-5t} - 2 = 3$ **59.** $12^a = 1$ **60.** $(3^{2x})(3^5) = 3^8$

61. $e^{2x(x-3)} = e^{2(x-2)(x-1)}$ **62.** $4^{x^2+2x} = 64$ **63.** $5^{2x-3} = 5^{x+9} \cdot 5^{x-12}$ **64.** $10^{2.3x} = 4$

65. $\log_7 a = 49$ **66.** $5 \log_3 x = 45$ **67.** $\ln z = 1$ **68.** $\log b = 0$

69. $\log k = -3$ **70.** $5 + \log_2 x = 3$ **71.** $3 \ln x = 1$ **72.** $\ln(2x^2 + 7x) = \ln 15$

73. $2 \log(x - 2) = \log 9$ **74.** $\log x + \log(x + 4) = 2 \log(x + 2)$ **75.** $7 \ln x - \ln x^3 = 4 \ln x$

76. Census data reported that the resident population 85 years and older was 3.022 million in 1990 and had grown exponentially to 4.177 million in 1999. Use the growth function to model the growth in the resident population 85 years and older since 1990. If this growth continues, what will be the estimated resident population 85 years and older in 2010. The projected resident population 85 years and older that year is 5.786 million. Compare your model's projections with this number.

77. The decay function for carbon-14 (C-14) is approximated as $A(t) = A_0 e^{kt}$, where $k \approx -0.000121$. If skeletal remains are found with two-thirds of the C-14 of the living form, determine the time since the body's demise.

CHAPTER 14 REVIEW

Evaluate.

1. $\log 0.001$ **2.** $\log_7 343$ **3.** $\log_2 \dfrac{1}{64}$

4. $\log 5.1$ **5.** $\ln e^{-3}$ **6.** $\ln 100$

7. $\ln e$ **8.** $\ln \dfrac{4}{7}$ **9.** $\log_3 36$

Write each logarithmic expression in expanded form.

10. $\log 3x^2 y$ **11.** $\log_3 \dfrac{9a}{b}$ **12.** $\ln 100p^3 q^2 r$ **13.** $\log \dfrac{\sqrt[5]{6x^3}}{\sqrt{xy}}$

Write each logarithmic expression in condensed form.

14. $\log(x^2 - 9) - \log(x + 3)$

15. $2 \log c - 5 \log d$

16. $\dfrac{1}{2} \ln x - 3 \ln y$

17. $\log 2xy - \log xz + \log 3yz$

Determine the inverse of each relation.

18. $y = 5 - x$

19. $y = 1 - x^2$

20. $y = 4 - x^3$

21. $y = 2 - \dfrac{4}{5}x$

22. $m(x) = 5^x$

23. $G(x) = ab^x$, a and b are constants.

Evaluate the function $g(t) = 1 + \left(\dfrac{4}{9}\right)^t$ at the given values.

24. $g(3)$

25. $g(-1)$

26. $g(0)$

27. $g\left(\dfrac{1}{2}\right)$

28. $g(1.5)$

29. $g(-\sqrt{3})$

Graph each function.

30. $m(x) = 3^x - 2$

31. $k(x) = 8^{(1/3)x}$

32. $y = -2e^{0.4x}$

33. $y = \log_2 x - 2$

34. $g(x) = \log(x + 1)$

Solve.

35. $\log_6 b = 216$

36. $2 \log_3 z = 6$

37. $\log z = 1$

38. $\ln b = 0$

39. $\log 2m = -4$

40. $-1 + \log_3 x = 4$

41. $2 \log x = 1$

42. $\log(2x^2 + 3x) = \log 9$

43. $2 \ln (x + 3) = \ln 4$

44. $\log 4x + \log(x + 1) = 2 \log(2x + 1)$

45. $2 \log a + \log a^4 = 6 \log a$

46. $5^{x-3} = 125$

47. $e^{-t} + 4 = 5$

48. $7^a = e$

49. $(6^7)(6^{3x}) = 6^2$

50. $e^{x(x+6)} = e^{(x+5)(x+1)}$

51. $2^{x^2} \cdot 2^{-3x} = 16$

52. $7^{2x-6} = (7^{3x-11})(7^{5-x})$

53. $10^{1.8x} = 9$

In exercises 54 and 55, write a function that represents each situation. Determine the inverse of the function and interpret what the inverse function represents.

54. Diana earns \$1200 per month, plus 2.5% commission on all of her sales at a furniture gallery. What is Diana's monthly income?

55. Carlyle is averaging 95 kilometers per hour on his auto trip through Germany. How far can he travel in a given number of hours?

56. Kristin invested her inheritance of \$8500 into an account that paid 6.5% interest compounded annually. Write a function for the total value of her investment. Graph the function. How much was the investment worth after 10 years?

57. The mathematical model $y = 52.28(1.03)^x$, where x is the number of years after 1990 and y is the number of households in millions, predicts the number of households with cable TV in the United States. What does the model predict the number of households with cable TV will be for the year 2010, assuming the same rate of growth? Does it seem reasonable to assume that the number of cable TV systems will continue to grow this way?

The pH value of a substance is defined by the function

$$\text{pH} = -\log[\text{H}^+]$$

where $[\text{H}^+]$ is the hydrogen ion concentration in moles per liter.

   Use this relation to answer exercises 58 and 59.

58. The hydrogen ion concentration of apple juice is 1.6×10^{-4} moles per liter. What is the pH of apple juice?

59. The pH value of gastric juice is typically 1.6. What is the hydrogen ion concentration of gastric juice?

60. Money invested at a continuously compounded interest rate r will grow to k times the original amount invested in t years, where $t = \frac{\ln k}{r}$. How long will it take an investment to double in value if the annual interest rate is 5.6%?

61. Revenues for public elementary and secondary schools totaled \$218 billion in 1990 and grew exponentially to \$343 billion in 1998. Use the growth function $A(t) = A_0 e^{kt}$ to model the growth in revenues since 1980. If this growth continues, what will be the estimated revenues in the years 2005 and 2010?

62. The decay function for carbon-14 (C-14) is approximated as $A(t) = A_0 e^{kt}$, where $k \approx -0.000121$. If the remains of a living organism are found with 40% of the C-14 of its living form, determine the time since the organism died.

CHAPTER 14 TEST

TEST-TAKING TIPS

If math is your most challenging subject, make sure to study it before you study all other subjects. Do not leave it for last. You must study math when you are most alert and fresh. Then you will do better and you will recall more of what you have studied. It has been shown that a few minutes' break every half-hour while you are studying will refresh you and help you retain more information. Begin each study session with a review of previous material studied. This will keep your math skills polished. Read your text before attempting to do the assigned exercises. Always check your answers with those provided in the text. If you cannot reconcile an answer with that in the text, be sure to mark the exercise so that you can ask about it in class. Try to look ahead at some of the material that will be presented in the next class, so that that material will not be difficult to understand when your instructor presents it.

1. Graph the function $f(x) = x^3$, its inverse, and the line of symmetry.
2. Graph $f(x) = 2^x + 1$.
3. Graph $g(x) = \ln(x + 2)$.
4. Find the inverse of the function $F(x) = 2x - 8$.
5. Evaluate $g(x) = 16^{(1/2)x}$ at the given values.
   a. $g(-1)$ **b.** $g(0)$ **c.** $g(4)$
   d. $g(0.3)$ **e.** $g(\sqrt{3})$

Write the expanded form of each expression.

6. $\log 3x^2 y^3 z$ 7. $\log_3 \dfrac{9a^2}{b}$

Write the condensed form of each expression.

8. $2 \log x + \log(x - 4)$ 9. $\dfrac{1}{2} \ln x - 2 \ln y$

10. Solve. $9^x = 3^{x+1}$
    a. numerically **b.** graphically **c.** algebraically

Solve exercises 11–14.

11. $(2^{3x})(2^{-2}) = 8$ 12. $2 \ln a = 3.1$
13. $2 \log x + \log x^4 = 6 \log x$ 14. $9^x = 5$

15. Pedro earns $400 per week, plus a commission of 3% of all sales he makes at the appliance store.
    a. Write a function for his total weekly income.
    b. Determine the inverse of the function.
    c. Interpret what the inverse of the function represents.

16. Kate discovered a bank deposit book that her deceased grandfather had owned. The bankbook indicated that he had made a single deposit of $2000 into an account that paid 5% interest compounded annually. Write a function for the amount of money that is in the account after x years. What was the investment worth after 20 years?

17. Money invested at a continuously compounded interest rate r will grow to k times the original amount invested in t years, where $t = \frac{\ln k}{r}$. How long will it take an investment to double in value if the annual interest rate is 7.5%?

18. According to U.S. Bureau of Economic Analysis, the personal income per capita was $19,614 in 1990 and had grown exponentially to $28,525 in 1998. Use the growth function to model the growth in personal income. If this growth continues, what will be the estimated personal income per capita in 2010?

19. What do we mean by the common logarithm of a number?

Project

Part I

In this chapter, we discussed Newton's law of cooling. According to Sir Isaac Newton, objects warmer than their surroundings will eventually cool to a common temperature with the surroundings. The cooling rate depends on how much hotter the object is than its surroundings.

You will need a Calculator Base Laboratory (CBL) with temperature probe, your TI-83 Plus calculator, a cooler with ice, and boiling water in a cup.*

1. Enter the COOLTEMP program from the Prentice Hall companion Web site.

2. Connect the TI-83 Plus calculator to the CBL that has a temperature probe in Chan 1.

3. Place the temperature probe in the ice chest and close the lid. Press "mode" to determine the temperature of the surroundings inside the ice chest, and record this temperature as T_0.

4. Place the temperature probe in a cup of boiling water. Allow the temperature probe to warm to the temperature of the water.

5. Place the cup of water with the probe in the ice chest. Run the program, following the directions on the screen. Allow the program to run for 15 minutes.

The COOLTEMP program is designed to record the temperature the instant the program begins and every 60 seconds for 30 minutes. It will store the number of minutes, x, in L_3 and the temperature T in degrees Celsius in L_4. The program will also graph the results on your calculator.

1. Write an equation for the recorded temperature T. Use the STAT CALC command "ExpReg L_3, L_5" to determine an exponential regression model in the form $T = a*b^x$ for the data recorded.

2. Enter the equation from Step 1 in Y1, and graph the equation. Does this equation appear in order to "fit" the recorded points?

3. According to Newton's law of cooling, the temperature T of a body in t minutes after it is placed in surroundings having a constant temperature T_0, is determined by the function $T = T_0 + Ce^{-kt}$, where C is the difference between the initial temperature and the surrounding temperature and k is the cooling constant.

    a. Let T_0 equal the temperature inside the ice chest, T equal the temperature of the water after cooling 30 minutes, and t equal 30 minutes. Determine the constant of cooling, k.

    b. Write an equation for T.

*If boiling water is not available, you may use ice water and measure warming.

c. Store the equation in Y2, and graph it. Does this equation appear to "fit" the recorded points?

4. Compare the equations found in step 1 and step 3. Which equation appears to be a better model? Can you explain why the equations differ?

Part II

In this chapter, we used an exponential function to approximate the population of the United States. The world population may also be determined with the use of an exponential function.

1. Search the Internet or your library to find the population of the world every year from 1991 to the present.

2. Enter these data in your calculator. Let L_1 be the number of years after 1990 and L_2 be the world population.

3. Calculate an exponential equation, using the STAT CALC command "ExpReg L_1, L_2."

4. Enter the equation found in step 3 in Y1, and graph it. Does this equation appear to "fit" the recorded points?

5. Use the equation in step 3 to approximate the world population in 2010.

6. Search the Internet or your library to find the projected population of the world in 2010. Compare your results in step 5 with the projections. If the results differ, explain why?

Part III

Many situations fit into the exponential growth model. Some of these include the spread of the AIDS virus, the growth of bacteria, sales of compact discs, sales of computers, and values of collectibles such as art or sport cards. Collect data on a variable that may be experiencing exponential growth. You may search the Internet or your library.

1. Develop a mathematical model that represents the growth.

2. Check your model against the data you collected.

3. Use the model to predict a future value of the variable.

Write a summary of your findings.

Part IV

Many situations may fit into a logarithmic model. As you saw in this chapter, the magnitude R of an earthquake of intensity I is defined by the function $R = \log \frac{I}{I_0}$, where I_0 is the intensity of a very small vibration in the Earth used as a standard. Collect data on two major earthquakes. You will need to know the magnitude R measured on the Richter scale for each. Place the magnitude data into the model and solve for the intensity of each earthquake, measured in terms of I_0. Then form the ratio of the two intensities and interpret the result. How do the two earthquakes compare in their intensities?

CHAPTERS 1–14 CUMULATIVE REVIEW

Simplify and write with positive exponents.

1. $5^{-2}x^0y^2$ **2.** $(-2x^{-3}y^2)^2$ **3.** $\left(\dfrac{4x^{3/4}}{x^{-1/4}}\right)^2$

Simplify.

4. $(3.2a^2 - 2.6ab + 1.7b^2) - (4a^2 + 2.6ab - b^2)$ **5.** $(4.5x^2y)(-2y^2z)$ **6.** $(3x + 2y)(3x - 2y)$

7. $(2x + 6)(2x - 3)$ **8.** $(x - 4)^2$ **9.** $\dfrac{25x^2y^4z^2}{-5xyz}$

10. $\dfrac{2r^2s + 4rs - 8s^2}{2rs}$ **11.** $\dfrac{x^2 - x - 6}{x^2 - 9}$ **12.** $\dfrac{x}{x + 2} + \dfrac{3}{x - 2} + \dfrac{x - 1}{x^2 - 4}$

13. $\dfrac{1}{x + 4} - \dfrac{x}{x - 2}$ **14.** $\dfrac{2m - 3}{m + 1} \cdot \dfrac{2m^2 + 5m + 3}{4m^2 - 9}$ **15.** $\dfrac{16x}{y^3 + y^2 - 12y} \div \dfrac{4x^2y}{y^2 - y - 20}$

16. $\sqrt{18x} - \sqrt{50x} + 4\sqrt{8x}$ **17.** $\sqrt[3]{16a^2b^2} \cdot \sqrt[3]{8ab^2}$ **18.** $(\sqrt{2} - \sqrt{3})(\sqrt{2} + \sqrt{3})$

19. $\dfrac{\sqrt{32xy^2z}}{\sqrt{8x^3y}}$ **20.** $\dfrac{\sqrt{x} + 4}{\sqrt{x} - 3}$ **21.** $x^{2/3}(x^{3/4} - x^{1/3})$

Factor completely if possible.

22. $25m^2 - 36n^2$ **23.** $2x^2 - 4x - 16$ **24.** $x^2 + 6x - 6$

Graph and label as indicated. Determine the domain and range of each function.

25. $f(x) = \dfrac{2}{3}x + 1$; three points on the graph

26. $y = x^2 + 5x + 4$; vertex, y-intercept, x-intercept, enough points to determine the curve, and the axis of symmetry

27. $y = \dfrac{x^2 + 4}{x + 2}$; enough points to determine the curve   **28.** $h(x) = \sqrt{x - 5}$; enough points to determine the curve

29. $g(x) = x^{2/3} + 3$; enough points to determine the curve   **30.** $y = 3^x$; enough points to determine the curve

31. $f(x) = \ln(x - 3)$; enough points to determine the curve

Solve.

32. $2(x - 5) - (x + 4) = 2x - 8$ **33.** $2(x + 3.1) = (x - 4.2) + (x - 6)$ **34.** $x^2 - 2x - 12 = 12$ **35.** $x^2 - 5x + 9 = 0$

36. $\dfrac{x + 1}{x + 2} = \dfrac{x - 1}{x + 3}$ **37.** $4\sqrt{x + 1} - 14 = -2$ **38.** $e^{2x-4} = e^{x(x-2)}$ **39.** $2\log_3 x = \log_3 5$

Solve. Write the solution set in interval notation.

40. $2x - 3 < 3(x - 4)$ **41.** $x^2 + 2x \geq 2$

Solve.

42. $3x - 2y = 4$
    $x + y = 8$

In exercises 43 and 44, write an equation of a line that satisfies the given conditions.

43. Passes through the points $(3, -1)$ and $(-2, 2)$

44. Is perpendicular to $x + 4y = 2.3$ and passes through $(-2, 1)$

45. Write a quadratic equation for a curve that passes through the points $(0, -6)$, $(-4, 0)$, and $(1, 0)$.

46. Given $f(x) = \frac{2}{3}x + 6$, find $f^{-1}(x)$.

47. Happy Recipe Company bought a rebuilt copier for $525 to reproduce its latest recipe book. If it costs $5 per book for materials to print the books and each book sells for $15, determine the break-even point—that is, the point at which the revenue equals the cost of the copies.

48. Nathan has 15 feet of landscaping timbers to place diagonally across his rectangular flower garden. If he wants to use all of the timbers and have the length of the garden be 3 feet more than the width, determine the dimensions of the garden.

49. A tank is drained by two separate drains. When the drain on the left is used alone, it takes 4 hours to drain the tank. When the drain on the right is used alone, it takes 6.5 hours to drain the tank. How long will it take to drain the tank if both drains are used? (Round your answer to the nearest tenth of an hour.)

50. If money is compounded continuously, the amount in the account will grow to k times the original amount invested in $t = \frac{\ln k}{r}$ years when the annual rate of interest is r. How long will it take the amount invested to triple if the annual interest rate is 4.5%?

Photo Credits

CHAPTER 1 **CO** Maria Stenzel/National Geographic Society Image Collection; **p. 15** Alison Wright/Stock Boston; **p. 26** NASA Headquarters; **p. 26** James M. McCann/Photo Researchers, Inc.; **p. 42** Andy Lyons/Getty Images, Inc.; **p. 44** © David Lees/CORBIS; **p. 55** Jessica Wecker/Photo Researchers, Inc. **p. 58** Topham/The Image Works; **p. 68** Calvin Larsen/Photo Researchers, Inc.; **p. 90** Paramount/Picture Desk, Inc./Kobal Collection

CHAPTER 2 **CO** Charles Graham/eStock Photography LLC; **p. 103** Paolo Koch/Photo Researchers Inc.; **p. 124** SuperStock, Inc.; **p. 131** David Taylor/Getty Images, Inc.; **p. 134** 2001 The Longaberger® Company

CHAPTER 3 **CO** Bob Daemmrich Photography, Inc.; **p. 154** Roy Morsch/Corbis/Stock Market; **p. 155** AP/Wide World Photos

CHAPTER 4 **CO** Mary Jelliffe/The Ancient Art & Architecture Collection Ltd.; **p. 232** John Elk, III/Stock Boston; **p. 234** Susanne Buckler/Getty Images, Inc.; **p. 256** Stacy Pick/Stock Boston; **p. 264** Kent Knudson/Stock Boston; **p. 266** Alan Schein/Corbis/Stock Market; **p. 285** Boyd Norton/The Image Works; **p. 293** W. Cody/ CORBIS; **p. 297** Chris Bensley/Stock Boston

CHAPTER 5 **CO** Chad Ehlers/International Stock Photography Ltd.; **p. 317** A. Ramey/PhotoEdit; **p. 341** Paul J. Sutton/Duomo Photography Incorporated; **p. 350** © Bettman/CORBIS; **p. 363** Monique Salaber/NASA, The Image Works

CHAPTER 6 **CO** Eliot Cohen; **p. 398** Michael Ventura/International Stock Photography Ltd.; **p. 450** David Young-Wolff/PhotoEdit

CHAPTER 7 **CO** Matthew Borkoski/Stock Boston; **p. 490** AP/Wide World Photos; **p. 502** Elise Amendola/AP/Wide World Photos; **p. 515** Chris Takagi; **p. 522** David Young-Wolff/PhotoEdit

CHAPTER 8 **CO** NASA/Johnson Space Center/The Image Works; **p. 550** A. Gurmankin/M. Morina/Unicorn Stock Photos; **p. 569** Kyodo News International, Inc.; **p. 577** John Wang/Getty Images, Inc./PhotoDisc, Inc. **p. 578** E.R. Degginger/Color-Pic, Inc.; **p. 584** Richard Renaldi

CHAPTER 9 **CO** Randy Faris/CORBIS; **p. 614** Kent Wood/Peter Arnold, Inc.; **p. 618** Paul Griffin/Stock Boston; **p. 638** Jeff Vanuga/CORBIS; **p. 646** Salaber/The Image Works

CHAPTER 10 **CO** Rafael Macia/Photo Researchers, Inc.; **p. 660** Roberto Soncin Gerometta/Lonely Planet Images/Photo 20-20; **p. 670** Curtis Martin/Lonely Planet Images/Photo 20-20; **p. 691** Getty Images, Inc.

CHAPTER 11 **CO** SuperStock, Inc.; **p. 715** Pictures Colour Library, Ltd./eStock Photography, LLC; **p. 717** (top) Robert Frerck/Odyssey Productions/Chicago; **p. 717** (bottom) Robert Frerck/Odyssey Productions/Chicago; **p. 737** Bill Horsman/Stock Boston; **p. 739** Ron Sherman/Stock Boston; **p. 748** David Bartruff/Getty Images, Inc.; **p. 761** U.S. Supercircuits; **p. 787** Guy Sauvage/Agence Vandystadt/Getty Images, Inc.

CHAPTER 12 **CO** SuperStock, Inc.; **p. 827** Tony Freeman/PhotoEdit

CHAPTER 13 **CO** AP/Wide World Photos; **p. 886** Eunice Harris/Photo Researchers, Inc.; **p. 888** Romer/Explorer/Photo Researchers, Inc.; **p. 890** Robert Frerck/Odyssey Productions/Chicago; **p. 890** Will and Deni McIntyre/Photo Researchers, Inc.; **p. 891** AP/Wide World Photos; **p. 898** Joe Sohm/Chromosohm, Stock Boston; **p. 909** SuperStock, Inc.; **p. 920** Ronald Sheridan/The Ancient Art & Architecture Collection Ltd.; **p. 928** Francois Gohier/Photo Researchers, Inc.; **p. 941** National Institute for Biological Standards and Control, England/Science Photo Library/Photo Researcher, Inc.; **p. 955** Adam Hart-Davis/Science Photo Library/Photo Researchers, Inc.; **p. 956** (left) Gregory Sams/Science Photo Library; **p. 956** (right) Gregory Sams/Science Photo Library

CHAPTER 14 **CO** David Bartruff/Stock Boston; **p. 986** Al Stephenson/Woodfin Camp & Associates; **p. 997** Tom Chargin/Tom Chargin Photography; **p. 1007** Linc Cornell/Stock Boston; **p. 1009** Simon Fraser/Science Photo Library/Photo Researchers, Inc.; **p. 1023** Sinclair Stammers/Science Photo Library, Photo Researchers, Inc.; **p. 1026** Ann States/CORBIS/SABA Press Photos, Inc.

Answers to Selected Exercises

Chapter 1

SECTION 1.1 EXERCISES

1. (a) integer, rational **(b)** natural, whole, integer, rational **(c)** natural, whole, integer, rational **(d)** rational **(e)** rational
2. (a) natural, whole, integer, rational **(b)** rational **(c)** natural, whole, integer, rational **(d)** rational **(e)** rational
3. (a) whole, integer, rational **(b)** rational **(c)** rational **(d)** rational **(e)** rational **4. (a)** natural, whole, integer, rational
(b) integer, rational **(c)** rational **(d)** rational **(e)** rational **5.** 5 and 6 **6.** -2 and -1
7.

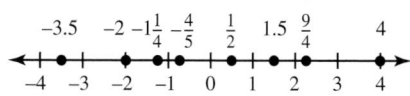

8.

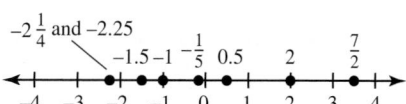

9. $<$ **10.** $<$ **11.** $>$

12. $<$ **13.** $>$ **14.** $>$ **15.** $<$ **16.** $<$ **17.** $<$ **18.** $>$ **19.** $>$ **20.** $<$ **21.** $=$ **22.** $=$ **23.** $>$ **24.** $<$
25. $<$ **26.** $>$ **27.** $=$ **28.** $=$ **29.** $>$ **30.** $<$ **31.** $=$ **32.** $=$ **33.** $=$ **34.** $=$ **35.** $0.295 < 0.7$

36. $0.06 > -2.56$ **37.** $-5 < -1 \le 0$ **38.** $20 \le 24 < 25$ **39.** $0.4 = \frac{2}{5}$ **40.** $-\frac{43}{50} = -0.86$ **41.** 15.34 **42.** 25

43. 15.34 **44.** 25 **45.** $3\frac{1}{3}$ **46.** 25 **47.** -23 **48.** -25 **49.** -23 **50.** -25 **51.** -15 **52.** $-\frac{1}{2}$ **53.** 25

54. 35 **55.** $-\frac{3}{2}$ **56.** $-\frac{5}{4}$ **57. (a)** -26.97 **(b)** $|-26.97|$ **58. (a)** -5.82 **(b)** $|-5.82|$ **59.** $+45; -115; +200$

60. $+1.25; -0.75; -2; +1.75; +1.50$ **61.** -14 **62.** -58 **63. (a)** $+5$ **(b)** -5 **(c)** -2 **64. (a)** $+100$ **(b)** -200

65. $|-70|$ **66.** $|-15|$

1.1 CALCULATOR EXERCISES

1.

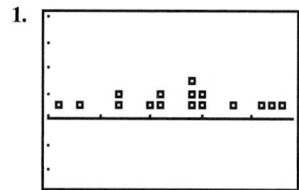

2.

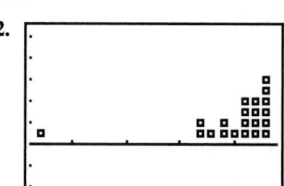

SECTION 1.2 EXERCISES

1. 2 **2.** 1 **3.** -11 **4.** -5 **5.** -2 **6.** -2 **7.** -547 **8.** 682 **9.** 6.66 **10.** 3.16 **11.** -1.3 **12.** -2.8

13. 1.37 **14.** 2.09 **15.** -3.37 **16.** -11.4 **17.** $-\frac{11}{10}$ **18.** $-\frac{19}{28}$ **19.** $-\frac{11}{18}$ **20.** $-\frac{7}{24}$ **21.** $\frac{4}{9}$ **22.** $\frac{1}{2}$ **23.** $\frac{11}{12}$

24. $2\frac{13}{15}$ **25.** -16 **26.** -12 **27.** -7 **28.** -6 **29.** 18 **30.** 16 **31.** -547 **32.** 682 **33.** 3.7 **34.** 5.71

35. 1.17 **36.** 4.4 **37.** -1.26 **38.** -3.04 **39.** $-\frac{1}{10}$ **40.** $\frac{5}{28}$ **41.** $-\frac{17}{18}$ **42.** $-\frac{11}{24}$ **43.** $\frac{13}{9}$ **44.** $\frac{7}{6}$ **45.** $-\frac{18}{7}$

46. $-\frac{7}{5}$ **47.** $-3\frac{1}{5}$ **48.** $-3\frac{1}{3}$ **49.** $-1\frac{1}{3}$ **50.** $-1\frac{3}{4}$ **51.** 7 **52.** 90 **53.** $-\frac{5}{12}$ **54.** $-\frac{17}{20}$ **55.** 43 **56.** 29 **57.** 9.21

58. 16.32 **59.** $-\frac{7}{12}$ **60.** $\frac{25}{12}$ **61.** 1008 **62.** 7479 **63.** -4 **64.** 234 **65.** $-\frac{1}{5}$ **66.** $-\frac{1}{15}$ **67.** 15.35 **68.** -2.81

69. $29,786,000 + 2,881,000 = 32,667,000$; the projected population is 32,667,000. **70.** $1,003,000 - 15,000 = 988,000$; the projected population is 988,000. **71.** $98 - (-90) = 188$; the range is 188°F. **72.** $44 - (-56) = 100$; the range is 100°F.
73. $130 - (-180) = 310$; the change in the mean surface temperature is 310°C. **74.** $350 - (-170) = 520$; the change in the mean surface temperature is 520°C. **75.** $20,320 - (-282) = 20,602$; the range is 20,602 feet. **76.** $163 - (-2) = 165$; the range is 165 meters. **77.** $\frac{1}{2} + \frac{1}{4} + 1 + 1 + \frac{1}{2} = 3\frac{1}{4}$; the recipe requires $3\frac{1}{4}$ cups of dry ingredients. **78.** $1\frac{1}{4} + \frac{3}{4} + \frac{1}{2} + 2\frac{1}{4} = 4\frac{3}{4}$; the pattern uses $4\frac{3}{4}$ yards of material. **79.** $-255 + 375 + (-575) + 1525 = 1070$; Karin is $1070 above her quota because 1070 is a positive number.

80. $385 + 285 + (-555) + (-405) + 265 + 575 = 550$; Karin is $550 above her quota.
81. $1500 + 150 + 150 + 150 + (-75) + (-500) + (-200) + 12 = 1187$; the net balance in Lindsay's account is $1187.
82. $500 + 75 + 50 + 100 + (-125) + 35 = 635$; the balance in the account is $635.
83. $19.95 + 19.95 + 19.95 + (-25) + (-59.27) + (-19.95) = -44.37$; Rosie lost $44.37 that morning.
84. $39.95 + 27.95 + 19.99 + (-175) + (-29.95) + (-14.95) = -132.01$; Richard lost $132.01 that hour.
85. $8.4 - 0.8 - 0.7 - 1.1 = 5.8$; the length of A is 5.8 inches.   **86.** $3.4 - 0.3 - 0.3 = 2.8$; the length of B is 2.8 inches; $3.4 - 1.4 = 2.0$; twice the length of C is 2 inches. C is 1 inch.

1.2 CALCULATOR EXERCISES

1. $7\dfrac{6}{7}$ **2.** $-2\dfrac{69}{113}$ **3.** $2\dfrac{253}{487}$ **4.** $-5\dfrac{13}{19}$

SECTION 1.3 EXERCISES

1. 24   **2.** 28   **3.** -15 **4.** -36 **5.** -20 **6.** -48 **7.** -135 **8.** -224 **9.** 128   **10.** 605   **11.** 0   **12.** 0

13. 0.34   **14.** 68.4   **15.** 7.29   **16.** 5   **17.** -12.5 **18.** -1.083 **19.** $\dfrac{5}{24}$ **20.** $\dfrac{22}{51}$ **21.** $-1\dfrac{1}{4}$ **22.** 1   **23.** $-\dfrac{3}{28}$

24. $-\dfrac{25}{51}$ **25.** -5 **26.** -8 **27.** 8   **28.** 5   **29.** 9   **30.** 8   **31.** 0   **32.** 0   **33.** -200 **34.** -90 **35.** 8.5

36. 0.05   **37.** 1   **38.** -1 **39.** -0.2 **40.** -0.5 **41.** -0.8 **42.** -0.005 **43.** $\dfrac{96}{125}$ **44.** $\dfrac{33}{34}$ **45.** $-2\dfrac{2}{9}$ **46.** $-6\dfrac{3}{5}$

47. $\dfrac{7}{9}$ **48.** $\dfrac{18}{25}$ **49.** $-\dfrac{64}{21}$ **50.** $-\dfrac{27}{17}$ **51.** undefined   **52.** undefined   **53.** 1   **54.** $-\dfrac{4}{9}$ **55.** $\dfrac{2}{9}$ **56.** $-\dfrac{7}{16}$ **57.** -42

58. -14 **59.** 4800   **60.** 60,000   **61.** $-\dfrac{4}{75}$ **62.** $\dfrac{4}{75}$ **63.** 1.144   **64.** 0.638   **65.** 0   **66.** 0   **67.** -800 **68.** -540

69. 0.486   **70.** 9.3   **71.** $\dfrac{4}{3}$ **72.** 1   **73.** -3 **74.** 20   **75.** -1.7 **76.** 12.7   **77.** 19.2   **78.** 6.6

79. $(-0.22)(645) = -141.90$; Sara has $141.90 deducted every two weeks; $(6)(-141.9) = -851.40$; this amounts to $851.40 every 12 weeks.

80. $-5 \times 9 \times 6 = -270$; Ron will bet $270 in a week.   **81.** $(0.05)(40)\left(1\dfrac{1}{2}\right)(3) = 9$; Sammy earns $9 for his work.

82. $(0.25)(20)(6) = 30$; Melanie earns $30 for her troop.   **83.** $(50)\left(8\dfrac{1}{2}\right)(4) = 1700$; George drove approximately 1700 miles.

84. $\left(9\dfrac{1}{2}\right)\left(1\dfrac{3}{4}\right)(5) = 83\dfrac{1}{8}$; Michele will hike $83\dfrac{1}{8}$ miles.   **85.** $(24)(0.59)(14) = 198.24$; the total retail value for the vegetables is $198.24.
86. $(12)(18)(12)(-0.35) = -907.2$; the cost the distributor incurs for this order is $907.20.   **87.** $659 \div 19.4 \approx 33.97$; Al will use approximately 34 gallons of gas.   **88.** $659 \div 38 \approx 17.34$; Al's gas mileage on the trip was approximately 17.34 mpg.
89. $47,500 \div 40 = 1187.5$; 1187 bottles can be filled.   **90.** $2800 \times 16 \div 98 \approx 457.14$; 457 boxes can be filled.

91. $16,000,000 \text{ seconds} \times \dfrac{1 \text{ minute}}{60 \text{ seconds}} \times \dfrac{1 \text{ hour}}{60 \text{ minutes}} \times \dfrac{1 \text{ day}}{24 \text{ hours}} \approx 185.2$; it will take a little over 185 days.
92. $20,000,000 \div 60 \div 60 \div 24 \div 7 \approx 33.1$; it will take a little over 33 weeks.
93. $335 \div (3 \times 20) \approx 5.6$; Billie will need six CD cabinets to store her collection.
94. $650 \div (3 \times 125) = 1.7$; Bruce will need two CD cabinets to store his collection.
95. $4,040,374,000 \div 4 \approx 1,010,093,500$; the estimated expenditure per quarter is $1,010,093,500.
96. $(50 \div 1350) \times 1000 \approx 37.04$; it would take Phillipe Petit approximately 37 minutes to complete a 1000-foot walk.
97. $0.65 \times 15,000 = 9750$; invest $9750 in U.S. stocks.
98. $0.2 \times 45,000 = 9000$; invest $9000 in global stocks.

99. $20 \div 3 = 6.7$; it takes 6.7 minutes per drawing. $6.7\dfrac{\text{minutes}}{\text{drawing}} \times 17 \text{ drawings} = 113.9$; it will take approximately 114 minutes to complete 17 drawings.   **100.** $45 \text{ rivets} \div 8 \text{ feet} \approx 5.6$ rivets per foot. $5.6\dfrac{\text{rivets}}{\text{foot}} \times 5 \text{ feet} = 28$; it will require 28 rivets.

1.3 CALCULATOR EXERCISES

1. 20   **2.** 20   **3.** -27 **4.** -27 **5.** 16   **6.** 16   **7.** 5   **8.** 5

SECTION 1.4 EXERCISES

1. $3 \cdot 3 \cdot 3 \cdot 3 = 81$ **2.** $2 \cdot 2 \cdot 2 \cdot 2 \cdot 2 \cdot 2 = 64$ **3.** $(-3)(-3)(-3)(-3) = 81$ **4.** $(-2)(-2)(-2)(-2)(-2)(-2) = 64$
5. $-(3 \cdot 3 \cdot 3 \cdot 3) = -81$ **6.** $-(2 \cdot 2 \cdot 2 \cdot 2 \cdot 2 \cdot 2) = -64$ **7.** $(-4)(-4)(-4) = -64$ **8.** $(-3)(-3)(-3)(-3)(-3) = -243$
9. $-(4 \cdot 4 \cdot 4) = -64$ **10.** $-(3 \cdot 3 \cdot 3 \cdot 3 \cdot 3) = -243$ **11.** 6.25   **12.** 0.16   **13.** $-\dfrac{9}{49}$ **14.** $-\dfrac{16}{81}$ **15.** $2\dfrac{10}{27}$ **16.** $3\dfrac{3}{8}$

17. 102.4947066   **18.** 1,640,401.445   **19.** 0   **20.** 0   **21.** 1   **22.** 1   **23.** −1   **24.** −1   **25.** 1256   **26.** $-\dfrac{4}{17}$ **27.** 1

28. 1   **29.** −1   **30.** −1   **31.** 1   **32.** 1   **33.** 6   **34.** 9   **35.** 16   **36.** 18   **37.** −5   **38.** −7   **39.** 0.8   **40.** 0.9

41. $-\dfrac{4}{3}$ **42.** $-\dfrac{6}{7}$ **43.** 1   **44.** −1   **45.** 0   **46.** 0   **47.** not a real number   **48.** not a real number

49. between 3 and 4, or approximately 3.162   **50.** between 4 and 5, or approximately 4.690

51. between −2 and −1, or approximately −1.732   **52.** between −4 and −3, or approximately −3.742

53. 4   **54.** 2   **55.** 12   **56.** 21   **57.** 10.726   **58.** 16.288   **59.** −5   **60.** −4   **61.** $\dfrac{1}{2}$ **62.** $-\dfrac{9}{11}$

63.

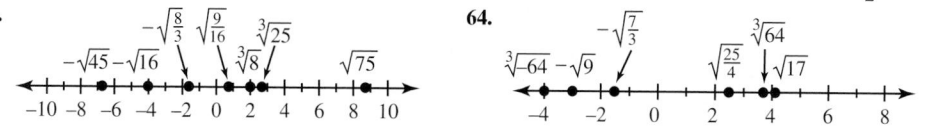

64.

65. Complete the following table:

Description	Number line
All real numbers between −1.5 and 3, including −1.5	−1.5; closed circle at −1.5, open circle at 3, shaded between, on scale −3 −2 −1 0 1 2 3
All real numbers between −1.5 and 3, including 3	−1.5; open circle at −1.5, closed circle at 3, shaded between, on scale −3 −2 −1 0 1 2 3
All real numbers between −1.5 and 3, inclusive	−1.5; closed circle at −1.5, closed circle at 3, shaded between, on scale −3 −2 −1 0 1 2 3
All real numbers between −1.5 and 3	−1.5; open circle at −1.5, open circle at 3, shaded between, on scale −3 −2 −1 0 1 2 3
All real numbers greater than −1.5	−1.5; open circle at −1.5, shaded to the right, on scale −3 −2 −1 0 1 2 3
All real numbers less than or equal to 3	closed circle at 3, shaded to the left, on scale −3 −2 −1 0 1 2 3

66. Complete the following table:

Description	Number line
All real numbers between −1 and 2.5, including −1	2.5; closed circle at −1, open circle at 2.5, shaded between, on scale −3 −2 −1 0 1 2 3
All real numbers between −1 and 2.5, including 2.5	2.5; open circle at −1, closed circle at 2.5, shaded between, on scale −3 −2 −1 0 1 2 3
All real numbers between −1 and 2.5, inclusive	2.5; closed circle at −1, closed circle at 2.5, shaded between, on scale −3 −2 −1 0 1 2 3
All real numbers between −1 and 2.5	2.5; open circle at −1, open circle at 2.5, shaded between, on scale −3 −2 −1 0 1 2 3
All real numbers greater than or equal to 2.5	2.5; closed circle at 2.5, shaded to the right, on scale −3 −2 −1 0 1 2 3
All real numbers less than or equal to −1	closed circle at −1, shaded to the left, on scale −3 −2 −1 0 1 2 3

67. $(755)^2 = 570{,}025$; the Great Pyramid covers 570,025 ft^2 of ground.   **68.** Light blue = 36 ft^2; white = 64 ft^2; dark blue = 125 ft^2

69. Jennie will need about 9.434 feet of logs.   **70.** Jennie will need 17 feet of logs.   **71.** Each side is approximately 1.442 feet.
72. Each side is approximately 2.466 feet.   **73.** Each side is approximately 3.464 feet.   **74.** The pond is approximately 7.071 feet by 7.071 feet by 2 feet.   **75.** Each side is approximately 614.003 feet.   **76.** The diagonal is approximately 868.332 feet.   **77.** The radius is approximately 78.583 feet.   **78.** The radius is approximately 339.921 feet.

1.4 CALCULATOR EXERCISES

Part I:

Number, n	Square, n^2	Cube, n^3
1	1	1
2	4	8
3	9	27
4	16	64
5	25	125
6	36	216
7	49	343
8	64	512
9	81	729
10	100	1000
11	121	1331
12	144	1728
13	169	2197
14	196	2744
15	225	3375
16	256	4096
17	289	4913
18	324	5832
19	361	6859
20	400	8000

1. 14   **2.** 23   **3.** 17   **4.** 21

Part II.

1. 8.949   **2.** -13.930 **3.** 1.331   **4.** $\frac{2}{3}$, or .667   **5.** 1.1399, or 1.140   **6.** -3.3

SECTION 1.5 EXERCISES

1. $\left(\frac{1}{2}\right)^1 = \frac{1}{2}$ **2.** $\left(\frac{1}{5}\right)^1 = \frac{1}{5}$ **3.** $(2)^2 = 4$ **4.** $(5)^2 = 25$ **5.** $\left(-\frac{1}{2}\right)^1 = -\frac{1}{2}$

6. $\left(-\frac{1}{5}\right)^1 = -\frac{1}{5}$ **7.** $(-2)^2 = 4$ **8.** $(-5)^2 = 25$

	Standard Notation	Scientific Notation	Calculator Notation
9.	23,450,000,000	2.345×10^{10}	2.345E10
10.	18,300,000,000,000,000,000	1.83×10^{19}	1.83E19
11.	-0.000006591	-6.591×10^{-6}	-6.591E-6
12.	-0.00000000072193	-7.2193×10^{-10}	-7.2193E-10
13.	3.6943	3.6943×10^{0}	3.6943E0
14.	9.98031	9.98031×10^{0}	9.98031E0
15.	$-711,030$	-7.1103×10^{5}	-7.1103E5
16.	$-100,500,000,000$	-1.005×10^{11}	-1.005E11
17.	0.01966	1.966×10^{-2}	1.966E-2
18.	0.00000555	5.555×10^{-6}	5.555E-6
19.	-9.95	-9.95×10^{0}	-9.95E0
20.	-8.103	-8.103×10^{0}	-8.103E0
21.	27,000,000	2.7×10^{7}	2.7E7
22.	5,470,000,000	5.47×10^{9}	5.47E9
23.	-0.00030303	-3.0303×10^{-4}	-3.0303E-4
24.	-0.00000000511	-5.11×10^{-9}	-5.11E-9
25.	1.26	1.26×10^{0}	1.26E0
26.	-8.81	-8.81×10^{0}	-8.81E0

27. 1,231,000,000,000   **28.** 351,700,000,000   **29.** 8.793×10^{11}; 879,300,000,000   **30.** 2.857×10^{-1}; 0.2857
31. 1.3175×10^{7}; 2.68008×10^{8}; 0.04915898 crime per person, or 4.9 crimes per 100 people.   **32.** 1.635×10^{6}; 0.006100564 crime per person, or 6 crimes per 1000 people.   **33.** 3.025×10^{7}; 2.70561×10^{8}; 11.2% of the population was Hispanic.
34. 2.315×10^{6}; 2.68008×10^{8}; deaths were 0.864% of the US population.   **35.** 2,650,000,000 Christmas cards were sold.
36. 925,000,000 Valentine's Day cards were sold.   **37.** Each pyramid at Giza would weigh over 1.2×10^{10} pounds.   **38.** 2×10^{6}; the Egyptians had to move 2×10^{6} blocks.   **39.** 2.36016×10^{7}; the Great Wall of China is 2.36016×10^{7} feet long.   **40.** 1.0979641×10^{7}; the Grand Coulee Dam contains 1.0979641×10^{7} cubic yards of concrete.   **41.** 1.628×10^{-9}; the perimeter is 1.628×10^{-9} meters.
42. 1.212×10^{-9}; the perimeter is 1.212×10^{-9} meters.   **43.** 3.029×10^{-10}; the length of an edge is 3.029×10^{-10} meters.
44. 4.070×10^{-10}; the length of an edge is 4.070×10^{-10} meters.

1.5 CALCULATOR EXERCISES

1. 6×10^{9} **2.** 6×10^{-4} **3.** 1.963×10^{12} **4.** 1.628×10^{13}

SECTION 1.6 EXERCISES

1. $\left(\dfrac{3}{8}\right)\left(\dfrac{5}{7}\right) - \left(\dfrac{3}{8}\right)\left(\dfrac{1}{9}\right)$ **2.** $-23(31) + (-23)(25)$ **3.** $2.7(-1.5) + 2.7(3.2)$ **4.** $\left(1\dfrac{1}{4}\right)\left(5\dfrac{1}{2}\right) - \left(1\dfrac{1}{4}\right)\left(3\dfrac{1}{8}\right)$ **5.** $\dfrac{217}{7} - \dfrac{175}{7}$

6. $\dfrac{-78}{6} + \dfrac{108}{6}$ **7.** $-15 - 19.3$ **8.** $-5.7 - 0.06$ **9.** $\dfrac{6}{7} + \dfrac{5}{9}$ **10.** $51 + 19$ **11.** $-1\dfrac{1}{7} + 2\dfrac{1}{5}$ **12.** $-89 + 17$

13. $-19.37 - 15.043$ **14.** $-2\dfrac{1}{4} - 3\dfrac{5}{8}$ **15.** $15(17 + 23)$ **16.** $81(22 + 8)$ **17.** $-3(14 + 21)$ **18.** $-13(21 + 16)$

19. $6(1.2 - 1.3)$ **20.** $5(4.5 - 7)$ **21.** $\dfrac{2}{3}(3 - 9)$ **22.** $\dfrac{3}{4}(2 - 10)$ **23.** -79 **24.** -129 **25.** 1   **26.** 30   **27.** -5
28. -17 **29.** 3   **30.** 2   **31.** -139 **32.** -106 **33.** 121   **34.** -11 **35.** -22.15 **36.** -3 **37.** 78   **38.** 66

39. $\dfrac{64}{135}$ **40.** $\dfrac{27}{80}$ **41.** $\dfrac{73}{72}$ **42.** $\dfrac{1}{3}$ **43.** -53 **44.** -221 **45.** 24 **46.** 115 **47.** 2 **48.** 4 **49.** 10 **50.** 8

51. 0 **52.** 0 **53.** undefined **54.** undefined **55.** The recommended weight is 115.28 pounds. **56.** The recommended weight is 145.64 pounds. **57.** The recommended weight is 185.9 pounds. **58.** The recommended weight is 160.6 pounds. **59.** The current will be approximately 0.173 ampere. **60.** The current will be approximately 0.289 ampere. **61.** The speed was approximately 31.6 mph. **62.** The speed was approximately 49 mph. **63.** The speed was approximately 56 mph. **64.** Yes, the car was going faster than 65 mph. **65.** The flow rate is approximately 42.264 gallons per minute. **66.** The flow rate is approximately 84 gallons per minute. **67.** The average of the index return is -18.8. **68.** The average daily low temperature for the week is 0.1°C.

1.6 CALCULATOR EXERCISES

1. 27 **2.** $-\dfrac{13}{15}$

Reflections

1–9. Answers will vary.

CHAPTER 1 SECTION-BY-SECTION REVIEW

1. -15: integer, rational; 1000: natural, whole, integer, rational; 0: whole, integer, rational; -2000: integer, rational; 13: natural, whole, integer, rational; $-\dfrac{15}{17}$: rational; 12.97: rational; $3\dfrac{5}{8}$: rational; $\dfrac{12}{4} = 3$: natural, whole, integer, rational

2. a.

3. $>$ **4.** $<$ **5.** $=$ **6.** $>$ **7.** $<$ **8.** $<$ **9.** $>$ **10.** $<$ **11.** $<$

b.

12. $=$ **13.** $<$ **14.** $=$ **15.** $-100 \le -49 < -23$ **16.** $\dfrac{13}{17} > \dfrac{252}{529}$ **17.** $-\dfrac{1729}{50} = -34.58$ **18.** $-\dfrac{17}{33}$ **19.** 67 **20.** 32.698

21. -257 **22.** $0, -5, -2, -3$ **23.** $-125{,}000{,}000$ **24.** -14 **25.** -4529 **26.** $-\dfrac{2}{3}$ **27.** $\dfrac{6403}{3225}$ **28.** $-335{,}597$

29. 14.017 **30.** $\dfrac{58}{63}$ **31.** -44.668 **32.** -123 **33.** 0 **34.** 26 **35.** $-\dfrac{53}{56}$ **36.** -2.4 **37.** -3.97 **38.** 0 **39.** $-1\dfrac{2}{3}$

40. $-2{,}977{,}581$ **41.** -2564 **42.** $\dfrac{47}{165}$ **43.** 124.92 **44.** 945 **45.** 28.9 **46.** 3 **47.** $\dfrac{43}{12}$ **48.** -47 **49.** 7.75

50. $1\dfrac{43}{45}$ **51.** $735.66 + (-276.12) + (-187.05) + (-68.57) + 75.00 + 185.00 + 50 + (-4.65) + (-12.00) = 497.27$; Cleta has $497.27 left in her account at the end of the month. **52.** $14{,}494 - (-282) = 14{,}776$; the range of the elevation between these two points is 14,776 feet.

53. 78 **54.** -105 **55.** $\dfrac{37}{7}$ **56.** -0.922 **57.** 0 **58.** -25 **59.** $\dfrac{11}{570}$ **60.** 8.4048 **61.** 3,477,285 **62.** $-\dfrac{2}{13}$

63. $-638{,}775$ **64.** 0 **65.** $-64{,}000$ **66.** 7.04 **67.** $-\dfrac{5}{6}$ **68.** 0 **69.** -55 **70.** 26 **71.** $-\dfrac{5}{6}$ **72.** -4.59 **73.** 0

74. undefined **75.** 13.7 **76.** -537 **77.** 219 **78.** $1\dfrac{5}{22}$ **79.** $\dfrac{65}{119}$ **80.** 6.94 **81.** -100 **82.** -9.24 **83.** $-\dfrac{13}{17}$

84. 83 **85.** 10.3 **86.** $0.83(125) = 103.75$; you donated $103.75 to program services. **87.** $\dfrac{1}{7}(192{,}631{,}000) \approx 27{,}519{,}000$; the number of people who could not find the U.S. on a world map is approximately 27,519,000. **88.** $345 \div 18.7 \approx 18.45$; Clarence's average mileage is approximately 18.45 mpg. **89.** 256 **90.** -243 **91.** 81 **92.** 1.44 **93.** $\dfrac{64}{27}$ **94.** 0 **95.** 1 **96.** 1 **97.** -1

98. $\dfrac{729}{4096}$ **99.** $\dfrac{16807}{243}$ **100.** 1 **101.** -1 **102.** -15 **103.** 3.079 **104.** 1 **105.** 1 **106.** indeterminate **107.** 1

108. 0 **109.** 9 **110.** 0.8 **111.** $\dfrac{3}{5}$ or 0.6 **112.** -7 **113.** not a real number **114.** 1.13 **115.** -21.7 **116.** $-\dfrac{24}{37}$

117. 3.873 **118.** 3 **119.** 0.5 **120.** $\dfrac{4}{9}$ **121.** -30 **122.** -2.154 **123.**

124. Number Lines

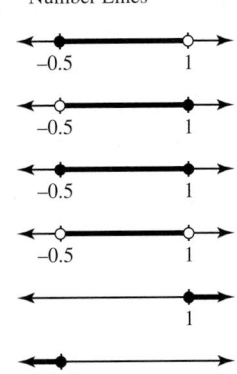

125. The diamond is 27 meters on a side. **126.** The box is $4\frac{1}{2}$ inches on a side.

127. $\left(\frac{1}{4}\right)^1 = \frac{1}{4}$ **128.** $4^1 = 4$ **129.** $4^2 = 16$ **130.** $\left(-\frac{1}{4}\right)^2 = \frac{1}{16}$ **131.** 1.89×10^{-7}; 1.89E-7

132. -2.7085×10^{10}; -2.7085E10 **133.** 589,000,000,000; 5.89E11

134. -0.00007093; -7.093×10^{-5} **135.** 6.821×10^{11} **136.** 911,900,000,000; 2.298×10^{11}; U.S. imports exceed exports by 229,800,000,000. **137.** 0.000008; 4×10^{-6}; the radius of the blood cell is 4×10^{-6} meters. **138.** $-2.6(-1.9) + (-2.6)(3.2)$ **139.** $\frac{5}{6}\left(-\frac{3}{5}\right) + \frac{5}{6}\left(-\frac{4}{15}\right)$

140. $\frac{1687}{7} - \frac{1372}{7}$ **141.** $-2.7 - 3.09$ **142.** $-32 - (-51)$ **143.** $21(18 + 42)$

144. $5(97 - 17)$ **145.** $5.6(18 - 8)$ **146.** $\frac{3}{4}(121 - 161)$ **147.** -334 **148.** 117 **149.** 29

150. -118 **151.** 21 **152.** 0 **153.** undefined **154.** The lean body weight is 140 pounds.
155. The average of the median weekly earnings is $545.25. **156.** $2(7 + 5) + 2\sqrt{7^2 + 5^2}$; Mary will need 41.2 feet of lace.

CHAPTER 1 CHAPTER REVIEW

1. 3.554×10^{17}; 3.554 E17 **2.** -9.2×10^{-5}; -9.2 E-5 **3.** 0.00000000000794; 7.94 E-12 **4.** $-68,760$; -6.876×10^4

5. $42(73 - 23)$ **6.** $-31(18 - 42)$ **7.** $>$ **8.** $<$ **9.** $>$ **10.** $<$ **11.** $>$ **12.** $=$ **13.** $<$ **14.** $>$ **15.** $>$ **16.** $-\frac{9}{25}$

17. 102 **18.** -0.085 **19.** -1457 **20.** 0.17 **21.** $-1\frac{3}{8}$ **22.** 12,147 **23.** 0.055 **24.** $\frac{159}{455}$ **25.** -182 **26.** -2.48

27. $-\frac{3}{11}$ **28.** 1.03 **29.** $\frac{3}{25}$ **30.** 4545 **31.** $\frac{2}{11}$ **32.** -2 **33.** 0 **34.** 88 **35.** $\frac{1}{16}$ **36.** 31.6316 **37.** 4,800,000

38. $-\frac{1}{3}$ **39.** indeterminate **40.** undefined **41.** -56.8 **42.** 8.2 **43.** $\frac{34}{55}$ **44.** 2.5 **45.** 204 **46.** $\frac{15}{28}$ **47.** 248

48. -20 **49.** 267 **50.** 50 **51.** 10 **52.** 0 **53.** undefined **54.** 169 **55.** 5252.335 **56.** -125 **57.** 625

58. $-2\frac{93}{125}$ **59.** 1 **60.** -1 **61.** 81 **62.** $-\frac{1}{243}$ **63.** 19.4481 **64.** 1 **65.** 1 **66.** -1 **67.** 22 **68.** -22

69. -22 **70.** indeterminate **71.** 0 **72.** 1 **73.** 65.02 **74.** 1 **75.** 1 **76.** 1 **77.** $\frac{1}{100}$ **78.** $\frac{1}{100}$ **79.** 10

80. -10 **81.** 50 **82.** not a real number **83.** $-\frac{21}{38}$ **84.** 1.2 **85.** -0.112 **86.** -47.282 **87.** -4 **88.** $\frac{9}{4}$ **89.** -6.5

90. -2.924 **91.** 1.6 **92.** -1.395 **93.** $\frac{3}{4}$ **94.** -2.56 **95.** -33.38 **96.** $7\frac{17}{42}$ **97.** $-35,640$ **98.** $-\frac{2}{5}$ **99.** 0

100. -16.5 **101.** -6 **102.** $74 \leq 75 < 76$ **103.** $\frac{1}{7} > -\frac{2}{5}$ **104.** $\frac{13}{25} = 0.52$

105. a.

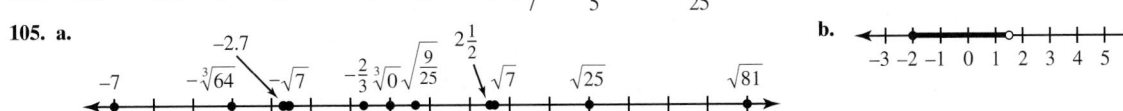

106. Approximately 117 credits run per minute. **107.** It takes $7\frac{1}{2}$ gallons of water to fill ten 10-gallon hats.

108. Captain Picard commanded 582 more people than Captain Kirk. Captain Janeway commanded 303 fewer people than Captain Kirk.
109. 5952 footballs will be needed. **110.** The average of the four indexes is -92.185. **111.** The length of each side is 5.070×10^{-10} meters.

CHAPTER 1 TEST

1. $<$ **2.** $>$ **3.** $=$ **4.** $\frac{1}{5} \leq \frac{1}{4} < \frac{5}{16}$ **5.** $-2.3 > -3.2$ **6.** $0.06 = \frac{3}{50}$ **7.** 13.37 **8.** -20 **9.** -36 **10.** $\frac{3}{95}$

11. -3.602 **12.** $\frac{2}{5}$ **13.** -4409 **14.** -45.78 **15.** $-17,128,272$ **16.** -1011.55 **17.** 91 **18.** $-\frac{4}{3}$ **19.** undefined

20. 0 **21.** 2 **22.** -14.45 **23.** 0 **24.** $-\frac{1}{42}$ **25.** 5.1 **26.** -3

27. a.

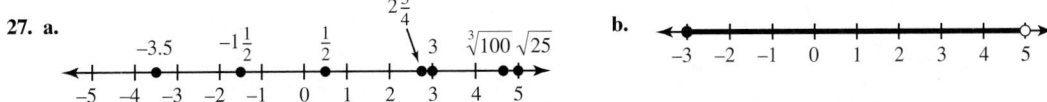

28. 5.239×10^{15} **29.** -2.03×10^{-6} **30.** 2.25 **31.** $\dfrac{256}{81}$ **32.** 1 **33.** 0 **34.** -64 **35.** -4.008 **36.** indeterminate

37. $-\dfrac{1}{10}$ **38.** 1 **39.** $\dfrac{6}{11}$ **40.** -1.897 **41.** -72.2 **42.** 11 **43.** 20 **44.** 0 **45.** $9.5^2 - 68$; 22.25; the garden will have 22.25 ft^2 in excess. **46.** 0.0378; the milk contains 0.0378 liters of fat; there are 0.004725 liters of fat per serving. **47.** 7.9176×10^{12}; the GDP of the U.S. exceeds that of Russsia by 7,917,600,000,000 U.S. dollars. **48.** If the signs are the same, the product is positive. If the signs are different, the product is negative. **49.** If the sum of the negative signs is even, the product is positive. If the sum of the negative signs is odd, the product is negative.

Chapter 2

SECTION 2.1 EXERCISES

1. Let $p =$ the total price; $\dfrac{3}{4}p$ **2.** Let $p =$ the amount invested; $0.06p$ **3.** Let $x =$ the total amount; $\dfrac{x}{15}$ **4.** Let $c =$ the count; $\dfrac{55}{c}$

5. Let $x =$ a number; $2.5x - \dfrac{19.59}{x}$ **6.** Let $x =$ a number; $x^2 + 3x$ **7.** Let $x =$ a number; $12x - 25$ **8.** Let $x =$ a number; $8x - 3$

9. Let $c =$ the cost of a chair; $3c + 80$ **10.** Let $a =$ the first grade, $b =$ the second grade, $c =$ the third grade; $\dfrac{a + b + c}{3}$

11. Let $L =$ the length of a rectangle; $\dfrac{1}{3}(L - 5)$ **12.** Let $L =$ the length of a side of a square; $2L - 5$ **13.** Let $x =$ a number; $5x + 4$

14. Let $N =$ the number of adults; $\dfrac{1}{3}N + 6$ **15.** Let $x =$ a number; $y =$ another number; $x + y + 2xy$ **16.** Let $x =$ a number;

$3x - \dfrac{10}{x}$ **17.** Let $x =$ a number; $\dfrac{2x}{x + 5}$ **18.** Let $x =$ a number; $y =$ another number; $xy - 8$ **19.** Let $x =$ the dollars invested in the first fund; $10{,}000 - x$ **20.** Let $a =$ the number of adults; $456 - a$ **21. (a)** 3 **(b)** -3 **(c)** 9 **(d)** -9 **22. (a)** 1.5 **(b)** -1.5

(c) 2.25 **(d)** -2.25 **23.** -10 **24.** -34 **25.** 7 **26.** 13 **27.** -3.1 **28.** 1.82 **29.** 54.21 **30.** 20.7 **31.** -51

32. -81 **33.** $\dfrac{41}{2}$ **34.** $34\dfrac{1}{3}$ **35.** 392 **36.** 67.2 **37.** $\dfrac{64}{15}$ **38.** 66 **39.** 14.28 **40.** $\dfrac{5}{3}$ **41.** 44 **42.** 57 **43.** 20.25

44. 14.76 **45.** 9 **46.** 13 **47.** 21 **48.** 15 **49.** 1.7 **50.** 4.8 **51.** 10.7 **52.** 13.8 **53.** 5 **54.** 2 **55.** $\dfrac{3}{2}$ **56.** $\dfrac{1}{3}$

57. Let $a =$ the number of adult tickets, $c =$ the number of children's tickets; $6a + 2c$; they raised $2170. **58.** Let $x =$ the number of hours on the job, $y =$ the number of hours on the second job; $9.50x + 12.25y$; Katie earned $289.50. **59.** Let $c =$ the number of computers sold, $a =$ the number of pieces of ancillary equipment sold; $200c + 50a - 750$; Pablo's profit is $3650. **60.** Let $T =$ the number of hours tutoring, $A =$ the number of hours administering tests; $1200 + 8T + 6A$; Shondra earned $2380. **61.** Let $D =$ the diameter of a circle; πD; $\pi(12) \approx 37.7$; the circumference is about 37.7 feet. **62.** Let $r =$ the radius of a circle; $2\pi r$; $2\pi(3) \approx 18.8$; the circumference is about 18.8 feet. **63.** Let $d =$ the number of days, $m =$ the number of miles; $64.99d + 0.29m$; 882.42; the cost is $882.42. **64.** Let $d =$ the number of days, $m =$ the number of miles; $76.99d + 0.29m$; 281.72; the cost is $281.72. **65.** Let $s =$ the length of a side of the fan; the area is s^2 and the perimeter is $4s$; the area is 441 square inches and the perimeter is 84 inches.

66. Let $b =$ the base of the triangle, $h =$ the height of the triangle, a and $c =$ the lengths of the other two sides of the triangle; the area is $\dfrac{1}{2}bh$ and the perimeter is $a + b + c$; the area is 65 square inches and the perimeter is 38 inches. **67.** Let $p =$ the retail price of the coat; the tax is $0.085p$ and the total price is $p + 0.085p$ or $1.085p$; the tax is $12.67 and the total price is $161.67. **68.** Let $b =$ the bill; the tip is $0.15b$ and the total bill with tip is $b + 0.15b$ or $1.15b$; the tip is $13.14 and the total bill is $100.72.

2.1 CALCULATOR EXERCISES

1. 130 **2.** 1 **3.** 9 **4.** 61 **5.** 0.25 **6.** $\dfrac{289}{49}$

1. 256 **2.** 1.96 **3.** $\dfrac{441}{25}$ **4.** 121 **5.** 0.49 **6.** $\dfrac{529}{49}$

SECTION 2.2 EXERCISES

1. (a) 4 **(b)** 12 **(c)** $2x^2, -6x, x$ **(d)** $2, -6, 1, 12$ **(e)** $-6x, x$ **2. (a)** 5 **(b)** 1 **(c)** $3y^5, y^3, -y, 4y^3$ **(d)** $3, 1, -1, 1, 4$ **(e)** $y^3, 4y^3$
3. (a) 3 **(b)** none **(c)** $3.4a, -11.2b, -0.3a$ **(d)** $3.4, -11.2, -0.3$ **(e)** $3.4a, -0.3a$
4. (a) 4 **(b)** none **(c)** $\dfrac{2}{5}m, \dfrac{1}{10}n, -\dfrac{3}{10}m, \dfrac{1}{5}n$ **(d)** $\dfrac{2}{5}, \dfrac{1}{10}, -\dfrac{3}{10}, \dfrac{1}{5}$ **(e)** $\dfrac{2}{5}m$ and $-\dfrac{3}{10}m$; $\dfrac{1}{10}n$ and $\dfrac{1}{5}n$
5. (a) 2 **(b)** none **(c)** $3m(n - 5)$; $6(n - 5)$ **(d)** 3, 6 **(e)** none
6. (a) 2 **(b)** none **(c)** $3p(p + q)$; $-q(p + q)$ **(d)** $3, -1$ **(e)** none
7. (a) 4 **(b)** 7 **(c)** $x^2, 3xy, -y^2$ **(d)** $1, 3, -1, 7$ **(e)** none **8. (a)** 4 **(b)** -3 **(c)** $a^2, b^2, 6ab$ **(d)** $1, -1, 6, -3$ **(e)** none

9. $x + 14$ **10.** $52a + 109$ **11.** $x^3 + x + 6$ **12.** $7h^3 - 2h^2 - 9$ **13.** $1.87a + 6.78b$ **14.** $17.348z - 44.004x$ **15.** $-\frac{1}{2}x + \frac{1}{2}$

16. $\frac{13}{8}z + \frac{7}{32}$ **17.** $7x^3 - 2x^2y - 4xy^2 + 9y^3$ **18.** $10d^5 - 10d^3e^2 + 14d^2e^3 - 16e^5$ **19.** $\frac{65}{42}x + \frac{19}{42}y$ **20.** $-\frac{118}{99}a + \frac{134}{99}b$

21. $3.6x + 4.8$ **22.** $6.7y - 4.2$ **23.** $66x - 79$ **24.** $25 - 10t$ **25.** $-72 - 43y$ **26.** $718z - 91$ **27.** $-x + 6y + 3z$
28. $40a + 33b + 14c$ **29.** $2.3y + 0.8z$ **30.** $-1.92t - 12.8r$ **31.** 0 **32.** $2x - 2y$ **33.** 0 **34.** 0 **35.** $2a$ **36.** $-10m$
37. $-30x - 45$ **38.** $-60z - 12$ **39.** $20z + 56$ **40.** $2a + 410$ **41.** $7.7x - 16.06$ **42.** $3.96a - 2.88$ **43.** $-84m + 49$

44. $-12z + 330$ **45.** $20b - 9$ **46.** $-15y + 40$ **47.** $3x + 5$ **48.** $-6m - 3n$ **49.** $-3b + \frac{1}{2}c$ **50.** $0.3m - n$

51. $12a + 3b - \frac{115}{8}c$ **52.** $-15x - \frac{17}{5}y + 25z$ **53.** $-73a + 78b - 60c$ **54.** $-51x + 252y - 249z$ **55.** $40.3x - 6.7y$

56. $65.8p - 18.2q$ **57.** $\frac{5}{12}p - \frac{55}{84}q$ **58.** $-\frac{11}{24}a - \frac{1}{18}b$ **59.** $11 + 4x + 4y$ **60.** $-51x + 31y - 19$

61. $-24a - 30b - 17c + 18$ **62.** $-117x - 77y - 31z + 33$ **63.** $6x + 60y - 132z$ **64.** $4a + 26b - 64c$ **65.** $7a + 8c$
66. $6x - 13y$ **67.** $0.7m - 8.4n$ **68.** $-7.48a + 13.9b$ **69.** $261.99x + 0.29y + 100$; the trip expenses were \$1329.21.
70. $227.99d + 0.29m + 150$; the total expense was \$1051.47.   **71.** $124h + 44$; the cost is \$912.   **72.** $150.45h$; the contract is \$3610.80.
73. The cost is $13.50 + 0.90c$ dollars; the cost is \$238.50.   **74.** The profit is $23 + 30h$ dollars; the profit is \$128.   **75.** The profit is
$2.75 + 1.50w$ dollars; the profit is \$8.75.   **76.** The profit is $7 + 1.50p$ dollars; the profit is \$175.   **77.** The total cost is $10.7892x$ dollars;
Laurie's change is $50 - 10.7892x$ dollars; Laurie's cost is \$32.37 and her change is \$17.63.   **78.** Mary Lynn's profit is $10.50x - 18.50$ dollars;
her profit is \$212.50.

2.2 CALCULATOR EXERCISES

1. $-5.342x - 11.8105$ **2.** $1492x - 3996$ **3.** $-\frac{21}{260}x - \frac{33}{52}y$ **4.** $-1\frac{32}{75}x + 2\frac{13}{50}$ **5.** $0.91x + 1.0035$

6. $-1018.3646x - 119.3616$ **7.** $x + 3$

SECTION 2.3 EXERCISES

1. equation   **2.** expression   **3.** expression   **4.** equation   **5.** equation   **6.** expression   **7.** equation   **8.** expression
9. expression   **10.** equation   **11.** yes   **12.** no   **13.** no   **14.** no   **15.** no   **16.** yes   **17.** yes   **18.** yes   **19.** yes
20. no   **21.** no   **22.** yes   **23.** yes   **24.** no   **25.** Let $x =$ a number; $x + 6 = 15$ **26.** Let $x =$ a number; $2x = 12$

27. Let $c =$ the number of children, $a =$ the number of adults; $2c - 5 = a + 2$ **28.** Let $c =$ the annual cost; $\frac{c}{12} = 2200$

29. Let $n =$ a number; $n^2 - 21 = 100$ **30.** Let $n =$ a number; $2n - 100 = n + 50$ **31.** Let $n =$ a number; $2(n + 5^2) = n + 100$

32. Let $n =$ a number; $\frac{1}{2}n + 60 = 200$ **33.** Let $x =$ a number; $2(x + 2) = 4 + 2x$ **34.** Let $x =$ a number; $3(x - 6) = 3x - 18$

35. Let $x =$ a number; $17 + \frac{x}{2} = 4 + 3x$ **36.** Let $x =$ a number; $x^2 + x = 2x$ **37.** Let $P =$ the perimeter, $d =$ the diameter, and $r =$

the radius; $P = d + \pi r$ **38.** Let $r =$ average rate of speed, $d =$ the distance, $t =$ the time; $r = \frac{d}{t}$ **39.** $I = A - P$ **40.** $A = P + I$

41. Let $n =$ the number of nickels, $d =$ the number of dimes; $n + d = 2n$ **42.** Let $q =$ the number of quarters; $d =$ the number of dimes;
$q = 2d + 5$ **43.** Let $x =$ the amount invested at 5%, $y =$ the amount invested at 7%; $0.05x + 0.07y = 176$ **44.** Let $x =$ the amount
invested at 8%, $y =$ the amount invested at 10%; $0.08x + 0.10y = 256$ **45.** Let $x =$ the measure of one angle, $y =$ the measure of a
second angle; $x = 3y + 10$ **46.** Let $x =$ the length of one side of a triangle, $y =$ the length of a second side of a triangle; $x = 2y + 6$

47. $H = \frac{NI}{L}$ **48.** $n = \frac{m}{M}$

2.3 CALCULATOR EXERCISES

1. 0, which means false.   **2.** 1, which means true.   **3.** $10 + \frac{42}{7}x = 4x$ **4.** $10 + \frac{42}{7x} = 4x$

SECTION 2.4 EXERCISES

1. 468 in²; 104 in.   **2.** 600 mm²; 160 mm   **3.** 24 cm²; 20 cm   **4.** 40 ft²; 26 ft   **5.** $\frac{25}{4}$ ft²; 10 ft   **6.** 31.36 m²; 22.4 m

7. 3060 m²; 242 m   **8.** 3705 in²; 368 in.   **9.** 86.6 in²; 33.0 in.   **10.** 23.8 ft²; 17.3 ft
11. The area of carpet needed is 20 yd², and the cost will be \$139.80.   **12.** The triangle will contain an area of 54 ft².
13. The area of coverage is 2704 in², and the amount of fringe needed for the perimeter is 208 in.
14. The deck contained an area of 225 ft², and Marcos will need 56 ft of latticework.
15. They roam over an area of about 855.3 mi², and the circumference of this area is approximately 103.7 miles.
16. They roam over an area of about 20,106.2 mi², and the circumference of this area is approximately 502.7 miles.

17. The area of the standard lot is 6000 ft^2, and the perimeter of the lot is 320 ft.
18. The square footage of this area is about 78.5 ft^2, and the border of the area will be 31.4 ft.
19. 15 ft^3; 46 ft^2 **20.** 72 ft^3; 108 ft^2 **21.** 421.875 in^3; 337.5 in^2 **22.** 551.368 cm^3; 403.44 cm^2 **23.** 35.3 in^3; 61.3 in^2
24. 117.8 in^3; 133.5 in^2 **25.** 4189 cm^3; 1257 cm^2 **26.** 524 in^3; 314 in^2 **27.** 0.049 ft^3; 0.817 ft^2 **28.** 33.51 cm^3; 64.38 cm^2
29. The box will hold 16 ft^3 of toys. Jim painted 40 ft^2 of surface area, using 2 pints of paint.
30. The cube contains 15.625 ft^3 of ice. The surface area is 37.5 ft^2.   **31.** The tank will hold 4188.8 in^3 of gas. The surface area is 1256.6 in^2.
32. The volume of the crate is 16 ft^3. The outer surface contains 40 ft^2 of wood.
33. The case contains 5832 in^3, with a surface area of 1944 in^2.   **34.** It will take 1005.3 in^3 of pellets.; the surface area is 603.2 in^2.
35. The volume of the paperweight is 89.8 in^3.   **36.** The volume of the bag is 84.8 in^3.   **37.** The third angle is 85°.   **38.** The third angle is 60°.   **39.** The radius of the ball is 47.625 inches, and the volume equals 452,474 cubic inches. The surface area equals 28,502 square inches, and the area covered by the mirrors is 27,600 square inches.   **40.** The volume of the Destiny laboratory is 4310 ft^3, and the volume of the room is 1040 ft^3. The Destiny laboratory contains the volume of approximately 4 rooms.   **41.** 25°   **42.** 19°
43. 115°   **44.** 56°   **45.** 79°   **46.** 45°   **47.** The pitch of the roof is 40°.   **48.** The angle formed with the driveway is 20°.
49. The other angle is 135°.   **50.** The other angle is 120°.   **51.** The temperature is 77°F.   **52.** The temperature is 87.8°F.
53. The temperature is $33\frac{1}{3}$°C.   **54.** The temperature is 18.9°C.   **55.** The temperature is 212°F.   **56.** The temperature is 1945.4°F.
57. The temperature is 203°F.   **58.** The temperature is 5370.8°F.   **59.** The temperature is −38.87°C.   **60.** The temperature is 356.58°C.   **61.** The temperature is 28.9°C.   **62.** The temperature is 22.8°C.   **63.** JoAnne paid $162.50 in interest.   **64.** Jack paid $385 in interest.   **65.** $2.92 of interest is earned.   **66.** The interest will be $2 and the total will be $102.   **67.** $48 in interest is paid.
68. The simple interest is $144.   **69.** The distance covered was 522.5 miles.   **70.** The trip was 511.5 miles.   **71.** He drove 223.806 miles.   **72.** He drove a distance of 335.214 miles.

2.4 CALCULATOR EXERCISES

Students should develop other programs for various formulas.

CHAPTER 2 SECTION-BY-SECTION REVIEW

Reflections

1–7. Answers will vary.

Exercises

In exercises 1–3, let x = the number.

1. (a) $55x + 4$ **(b)** $55(x + 4)$ **2. (a)** $\frac{3}{4}(x + 35)$ **(b)** $\frac{3}{4}x + 35$ **3. (a)** $2x - 20$ **(b)** $20 - 2x$ **4.** $2500 + 275n$ **5.** $\frac{650}{h}$

6. $200 + 5.5k$ **7.** $0.4x$; $x - 0.4x$ or $0.6x$ **8.** $758 - x$ **9.** $0.25x$; $x - 0.25x$ or $0.75x$ **10.** $\frac{2}{3}$ **11.** $-\frac{2}{3}$ **12.** $\frac{4}{9}$ **13.** $-\frac{4}{9}$

14. $\frac{2}{3}$ **15.** 15   **16.** −48   **17.** 15   **18.** 7.5   **19.** $\frac{4}{15}$ **20.** undefined   **21.** 77.48   **22.** 323

23. (a) Let L = length, W = width, H = height, $V = LWH$ **(b)** $V = 2,491,776$ ft^3
24. terms = 4; constants = none; variable terms = $3x, -2y, 4x, 9y$; coefficients = 3, −2, 4, 9; like terms = $3x$ and $4x$, $-2y$ and $9y$
25. terms = 4; constants = none; variable terms = $2a^2, -a, 3a^2, -5a^3$; coefficients = 2, −1, 3, −5; like terms = $2a^2$ and $3a^2$
26. terms = 3; constants = 5.1; variable terms = $2.4x, 6.2x$; coefficients = 2.4, 5.1, 6.2; like terms = $2.4x, 6.2x$
27. terms = 2; constants = none; variable terms = $4a(a + b), -b(a + b)$; coefficients = 4, −1; like terms = none

28. $0.2z$ **29.** $24x - 42$ **30.** $\frac{1}{12}x + \frac{7}{8}y$ **31.** $9x^2 + 28xy + 5y^2 - 2$ **32.** $7a + 8b$ **33.** $-7.6a + 9.4b$ **34.** $-90x + 28y$

35. $6x^2 - 34xy$ **36.** $3a - 4b + \frac{5}{3}c$ **37.** $129x + 3y$ **38.** $-12x - 21y - 31$ **39.** $10a - 26b - 3c + 15$

40. $4x + 2$; total number of push-ups is 182.   **41.** $335d + 0.2m$; $1409   **42.** $2.75(x + y) - (1.25x + 1.5y) = 1.5x + 1.25y$; the net profit is $930.   **43.** $20 - 5.3x$; Katie received $4.10 in change.   **44.** $8.75x - 52.65$; Margaret's net profit is $227.35.

45. expression   **46.** equation   **47.** no   **48.** yes   **49.** yes   **50.** yes   **51.** no   **52.** Let x = the number, $5 + 4x = 65 + \frac{x}{4}$
53. Let x = the number of pamphlets, $1500 = 1.35x$ **54.** Let x = the dollars invested in one account, y = the dollars invested in another account, $0.06x + 0.08y = 1500$ **55.** Let x = the dollars invested in one account, y = the dollars invested in another account,
$x + y = 30,000$ **56.** $f = \frac{\sqrt{\frac{T}{m}}}{2L}$ **57.** 208 m^2; 68 m   **58.** 880 in^2; 128 in.   **59.** 225 cm^2; 60 cm   **60.** 65 m^2; 34 m
61. 467.59 ft^2; 76.65 ft   **62.** 6615 in^3; 2478 in^2 **63.** 3112.136 mm^3; 1278.96 mm^2 **64.** 54,965.3 in^3; 8143.01 in^2
65. 145,124.7 cm^3; 13,355.04 cm^2 **66.** 117.3 cm^3; 151.6 cm^2 **67.** The other angle is 122°.   **68.** The other angle is 32°.
69. The third angle is 25°.   **70.** Dan should order 5100 ft^2 of sod and 252 ft of fencing.   **71.** The garden will have about 197.9 ft^2.
72. The two 10-in. pizzas are the better deal.   **73.** The truck will hold 80 ft^3.   **74.** The surface area is approximately 113.1 ft^2.

75. The total amount of the loan is $956.25   **76.** The temperature is 122°F.   **77.** The temperature is $26\frac{2}{3}$°C.

78. LuAnn traveled 356.5 miles.

CHAPTER 2 CHAPTER REVIEW

1. terms = 4; constants = 23; variable terms = $12x$, y, $-z$; coefficients = 12, 1, -1, 23; like terms = none
2. terms = 3; constants = 75; variable terms = $3(a - 2)$, $5(b - 4)$; coefficients = 3, 5; like terms = none
3. terms = 5; constants = 12, -18; variable terms = $-7x$, $14x$, x; coefficients = 12, -7, 14, -18, 1; like terms = 12 and -18; $-7x$, $14x$, x
4. terms = 5; constants = none; variable terms = b^2, $2b$, $-3b^2$, $6b$, b^3; coefficients = 1, 2, -3, 6, 1; like terms = b^2 and $-3b^2$; $2b$ and $6b$
5. 324   **6.** -324 **7.** 324   **8.** -324 **9.** -5.23 **10.** -12 **11.** not a real number   **12.** 0   **13.** 35   **14.** undefined

15. 2   **16.** $-\frac{5}{4}$ **17.** 3   **18.** yes   **19.** no   **20.** yes   **21.** yes   **22.** no   **23.** no   **24.** $17h$ **25.** $8m + 10$

26. $27x - 8 + 15y$ **27.** $15x^4 + x^3 - 7x^2 - 46x - 1$ **28.** $10.9a + 3.4b$ **29.** $-27y + 15$ **30.** $4g + 4$

31. $-x - 2y - 15z$ **32.** $22a + 33b$ **33.** $8x - 12y + \frac{30}{13}z$ **34.** $3x - 4y + 6z$ **35.** $35.9x - 49.3y$

36. $-144a + 132b - 252$ **37.** $-30x + 45$ **38.** $100 - 19.95x$; the change back is $40.15.
39. $0.89x + 0.49y$; the total cost is $4.54.   **40.** Let x = a number, $x^2 + 2x = x + 306$
41. Let x = the number of hours, $225 + 45x$; Lakeetha will make $5625 in earnings.
42. $500 + 130n$; Carmen has $2450 deposited after 15 weeks; she has a $1825 balance after the withdrawal.

43. $112.5x + 275$; the total earned was $2525.   **44.** Chum's pool will hold 1413.7 ft^3 of water.   **45.** The temperature is $35\frac{5}{9}$°C.

46. You will have $31,600.67 over 10 years.   **47.** Randy bicycled 16.25 miles.   **48.** The can's volume is about 12.44 in^3.; the can's

surface area is about 31.91 in^2.   **49.** $1.50x - 175$; the profit is $200.   **50.** $v = \frac{ax}{x + k}$ **51.** The volume is about 0.005 cubic inches.

52. equation   **53.** expression   **54.** expression   **55.** equation   **56.** The complementary angle is 5°.   **57.** The supplementary
angle is 137°.   **58.** The third angle is 91°.

CHAPTER 2 TEST

1. $3500 - x$ **2.** $0.08x$; $x + 0.08x$ or $1.08x$; the total cost is $16.20.   **3.** Let l = length, w = width, $l = 3w - 5$ **4.** $z = \frac{x - m}{s}$
5. 9   **6.** -8 **7.** 36   **8.** -36 **9.** 36   **10.** eight terms   **11.** y^3, $-5y^2$, $15y$, $7y^2$, $4y$, $6y^3$ **12.** -3, -12

13. 1, -5, 15, -3, 7, -12, 4, 6   **14.** y^3 and $6y^3$, $-5y^2$ and $7y^2$, $15y$ and $4y$, -3 and -12 **15.** $\frac{5}{6}x + \frac{29}{18}y - \frac{5}{9}$ **16.** $6p - q$

17. $-5x + 9$ **18.** $-30x + 10$ **19.** no   **20.** yes   **21.** The toolbox's volume is 12 ft^3; the outside surface area is 34 ft^2
22. Tracy's account will have $6400 in 40 years.   **23.** The supplementary angle is 102°.   **24.** The area is 15 square meters. The
perimeter is 17 meters.   **25.** The temperature is 77°F.   **26.** Answer will vary. Possible answer: The area is the space contained within
the boundaries, whereas the perimeter is the distance around the boundaries.

Chapter 3

SECTION 3.1 EXERCISES

1.

x	y
-2	-6
-1	-1
0	4
1	9
2	14
3	19

2.

x	y
-3	18
-2	10
-1	2
0	-6
1	-14

3.

x	y
-15	-11
-10	-8
-5	-5
0	-2
5	1
10	4
15	7

4.

x	y
-18	-11
-9	-4
0	3
9	10
18	17
27	24
36	31

5.

x	y
-2	-3
-1	-0.7
0	1.6
1	3.9
2	6.2

6.

x	y
-3	5.2
-1	-4.4
0	-9.2
1	-14
3	-23.6

7.

x	y
-1	2
-4	1
-7	0
-10	-1
-13	-2

8.

x	y
-10	-2
-4	-1
2	0
8	1
14	2

For problems 9–16, answers will vary. One possibility is given.

9.

x	y
-2	-20
-1	-14
0	-8
1	-2
2	4

10.

x	y
-2	37
-1	26
0	15
1	4
2	-7

11.

x	y
-14	-6
-7	-4
0	-2
7	0
14	2

12.

x	y
-16	11
-8	8
0	5
8	2
16	-1

13.

x	y
-2	11.3
-1	6.7
0	2.1
1	-2.5
2	-7.1

14.

x	y
-2	-22
-1	-11.4
0	-0.8
1	9.8
2	20.4

15.

x	y
-6	-5
-2	-2
0	$-\frac{1}{2}$
2	1
6	4

16.

x	y
-11	-6
-3	-3
5	0
13	3
21	6

17.

x	y
-4	-61
-2	-37
0	-13
2	11
4	35

18.

x	y
-3	39
-1	21
1	3
3	-15

19.

y	z
-6	3
-3	4
0	5
3	6
6	7

20.

q	p
-16	-18
-8	-11
0	-4
8	3
16	10

21.

b	a
-2	-22.7
-1	-8.5
0	5.7
1	19.9
2	34.1

22.

n	m
-3	-9.4
-2	-7.5
-1	-5.6
0	-3.7
1	-1.8

23.

x	y
-3	10
-2	3
-1	0
0	1
1	6
2	15
3	28

24.

x	y
-2	-44
-1	-24
0	-10
1	-2
2	0
3	-4
4	-14

25.

x	y
-2	14
-1	-5
0	-12
1	-7
2	10

26.

x	y
5	-63
6	-100
7	-145
8	-198
9	-259

27.

x	y
-3	$\frac{1}{2}$
-1	-2
1	undefined
3	8

28.

x	y
-1	$\frac{7}{4}$
1	$-\frac{3}{2}$
3	undefined
5	$\frac{23}{2}$

29. (a) Let d = distance, t = time, $d = 55t$

(b)

t	d
4	220
8	440
12	660
16	880
20	1100

30. (a) Let C = total cost, p = number of place settings, $C = 15.50p$

(b)

p	C
2	31
4	62
6	93
8	124
10	155
12	186

31.

H	V
1	8
3	24
5	40
7	56
9	72

32.

s	V
1	1
2	8
3	27
4	64
5	125
6	216

33.

t	I
1	225
2	450
3	675
4	900
5	1125
6	1350
7	1575
8	1800
9	2025
10	2250
11	2475
12	2700

34.

t	I
1	210
3	630
5	1050
7	1470
9	1890
11	2310
13	2730
15	3150

35.

R	I
1	9
2	4.5
3	3
4	2.25
5	1.8
6	1.5
7	1.286
8	1.125
9	1

36.

R	V
1	5
2	10
3	15
4	20
5	25
6	30
7	35
8	40
9	45
10	50

37. $(-2, 1.6), (-1, 2.8), (0, 4), (1, 5.2), (2, 6.4)$ **38.** $(-4, 7.2), (-2, 6.6), (0, 6), (2, 5.4), (4, 4.8)$ **39.** $\left(\frac{1}{6}, \frac{5}{6}\right), \left(\frac{1}{5}, \frac{4}{5}\right), \left(\frac{1}{4}, \frac{3}{4}\right), \left(\frac{1}{3}, \frac{2}{3}\right), \left(\frac{1}{2}, \frac{1}{2}\right)$

40. $(1, 1), (2, 3), (3, 6), (4, 10), (5, 15)$ **41.** $(2, 8), (4, 16), (6, 24), (8, 32), (10, 40)$ **42.** $(4, 6), (6, 4), (8, 2), (9.5, 0.5)$

43. $(1990, 110), (1991, 66), (1992, 22), (1993, 83), (1994, 175), (1995, 198), (1996, 225), (1997, 244), (1998, 238)$

44. $(1990, 80.90), (1992, 68.68), (1993, 61.48), (1994, 56.21), (1995, 51.00), (1996, 47.70), (1997, 42.78), (1998, 39.43)$

45. domain: $\{3, 5, 7, 9\}$; range: $\{15.8, 17.8, 19.8, 21.8\}$ **46.** domain: $\{11, 15, 19, 23\}$; range: $\{8, 12, 16, 20\}$

47. domain: $\{4\}$; range: $\{-3, -1, 1, 3, \ldots\}$ **48.** domain: $\{3, 2, 1, 0, -1, \ldots\}$; range: $\{-1\}$

49. domain: $\{2\}$; range: $\{\ldots, -2, -1, 0, 1, 2, \ldots\}$ **50.** domain: $\{\ldots, -2, -1, 0, 1, 2, \ldots\}$; range: $\{0\}$

51. domain: $\{2, 4, 6\}$; range: $\{3, 11, 19\}$ **52.** domain: $\{10, 20, 30\}$; range: $\{35, 65, 95\}$

53. domain: $\{0, 1, 2, 3, 4, 5, 6\}$; range: $\{0, 1, 2, 3, 4, 5, 6\}$

54. domain: $\{-4, -3, -2, -1, 0, 1, 2, 3, 4\}$; range: $\{-12, -11, -10, -9, -8, -7, -6, -5, -4\}$

55. domain: $\{0, 0.5, 1, 1.5, 2, 2.5, 3, 3.5, 4\}$; range: $\{1, 1.25, 2, 3.25, 5, 7.25, 10, 13.25, 17\}$

56. domain: $\{0, 0.5, 1, 1.5, 2, 2.5, 3, 3.5, 4\}$; range: $\{-2, -1.75, -1, -0.25, 2, 4.25, 7, 10.25, 14\}$

57. domain: $\{2, 3, 6, 11, 18, 27\}$; range: $\{0, 1, 2, 3, 4, 5\}$ **58.** domain: $\{-2, -1, 2, 7, 14, 23\}$; range: $\{0, 1, 2, 3, 4, 5\}$

59. domain: $\{-3, -1, 1, 3, 5, 7\}$; range: $\left\{-6, -2, -1\frac{1}{5}, 1\frac{1}{5}, 2, 6\right\}$ **60.** domain: $\{-3, -1, 1, 3, 5, 7\}$; range: $\left\{-10, -3\frac{1}{3}, 1\frac{3}{7}, 2, 3\frac{1}{3}, 10\right\}$

61.

t	s
0	50
0.5	46
1	34
1.5	14

62.

t	s
0	0
1	209
2	386
3	531
4	644
5	725
6	774
7	791
8	776
9	729
10	650
11	539
12	396
13	221
14	14

63. $(1, 65), (2, 130), (3, 195), (4, 260)$ **64.** $(2.5, 78), (2.75, 70.91), (3, 65), (3.25, 60), (3.5, 55.71)$

65. (a) Let n = number of weeks, R = rental cost, $R = 175n + 225$

(b)

n	R ($)
1	400
2	575
3	750
4	925

66. (a) Let h = number of hours, R = rental cost, $R = 75h + 300$

(b)

h	R ($)
2	450
2.5	487.50
3	525
3.5	562.50
4	600

SECTION 3.2 EXERCISES

1. 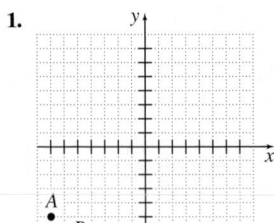 **2.** 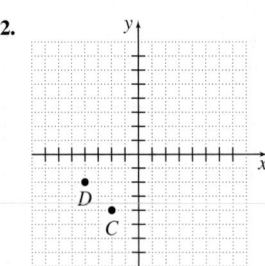 **3.** 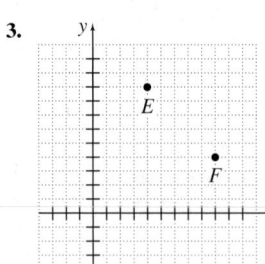 **4.**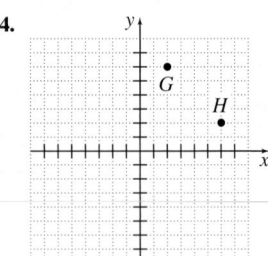

5. 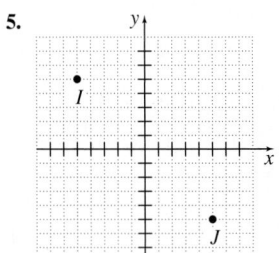 **6.** 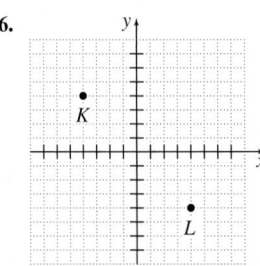 **7.** 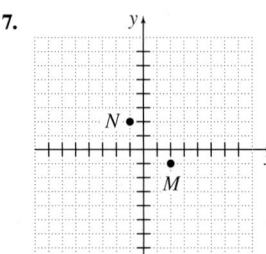 **8.**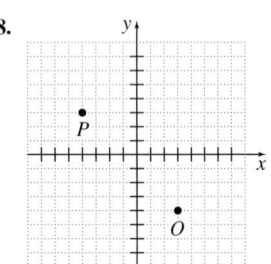

9. 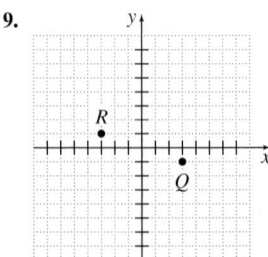 **10.** 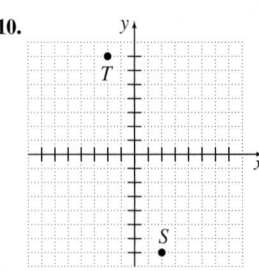 **11.** 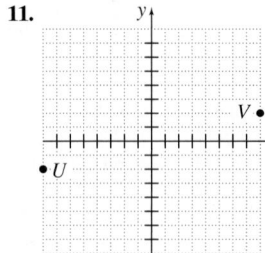 **12.**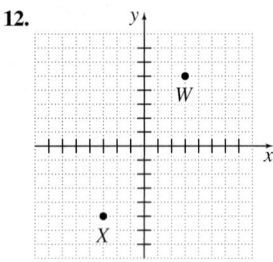

13. 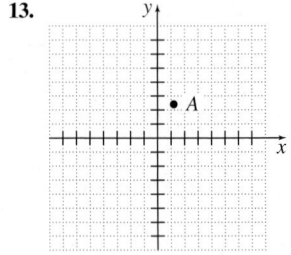 **14.** 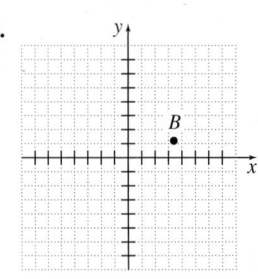 **15.** 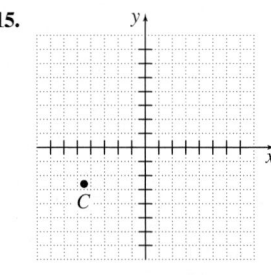 **16.**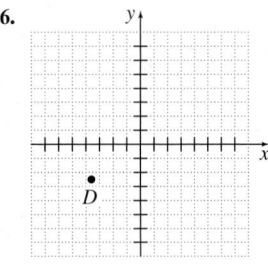

17. 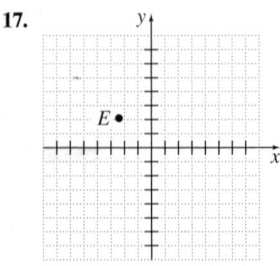 **18.** 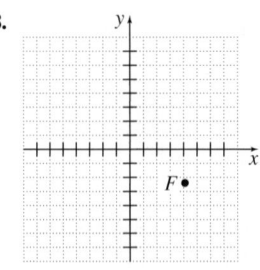 **19.** 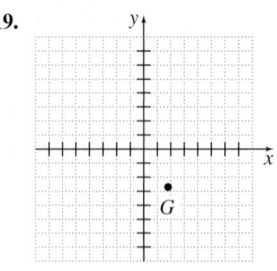 **20.**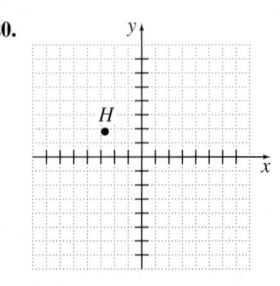

21. quadrant II   **22.** quadrant III   **23.** quadrant I   **24.** quadrant IV   **25.** quadrant III   **26.** quadrant II   **27.** y-axis
28. x-axis   **29.** quadrant IV   **30.** quadrant I   **31.** x-axis   **32.** y-axis   **33.** x-axis   **34.** origin   **35.** quadrant I
36. quadrant II   **37.** y-axis   **38.** x-axis   **39.** quadrant IV   **40.** quadrant III   **41.** quadrant III   **42.** x-axis   **43.** y-axis
44. quadrant IV   **45.** $A(8, 2)$, $B(-9, 7)$, $C(0, 0)$, $D(-2, -3)$, $E(0, 4)$, $F(5, -6)$, $G(-5, 0)$
46. $A(-7, -6)$, $B(2, 4)$, $C(-5, 2)$, $D(0, -9)$, $E(5, -5)$, $F(9, -7)$, $G(6, 0)$

47.
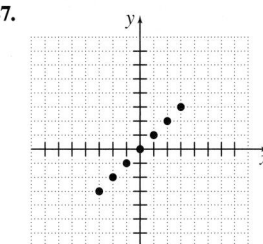

48.
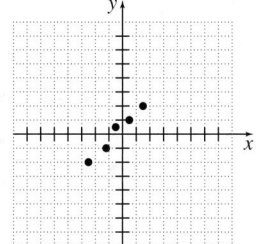

49.
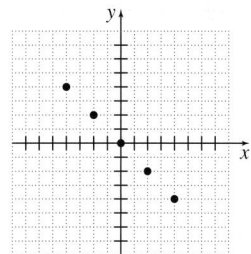

50.
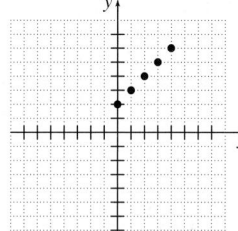

51.
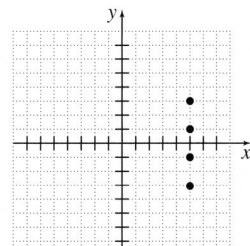

52.
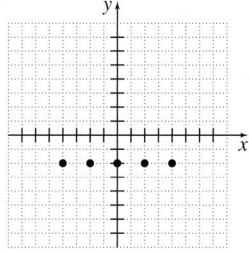

53.
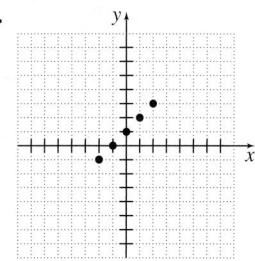

54.
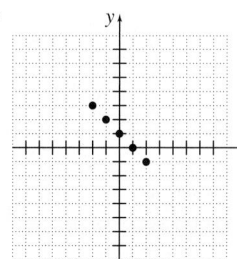

55.
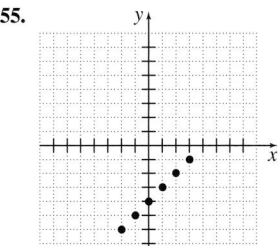

56.
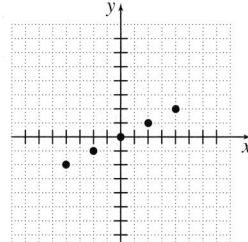

57.
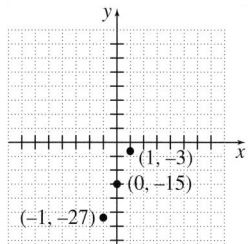

58.
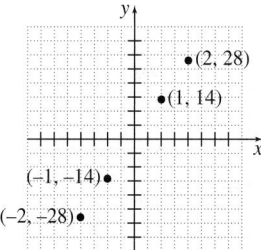

59.
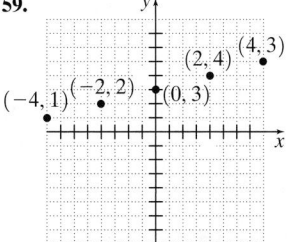

60.
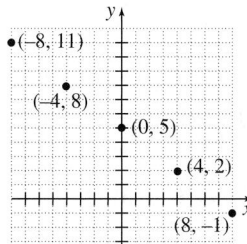

61.
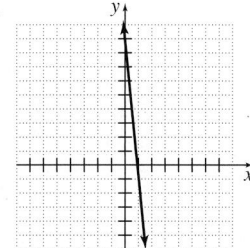

62.
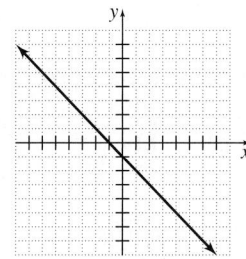

63.
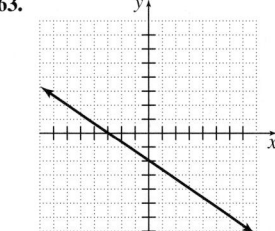

64.
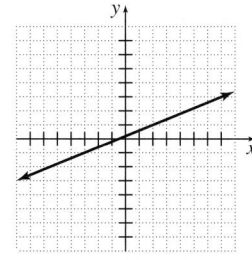

65.
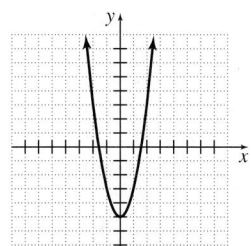

66.
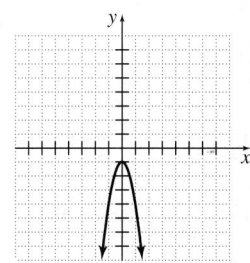

67.
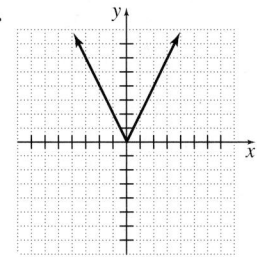

68.
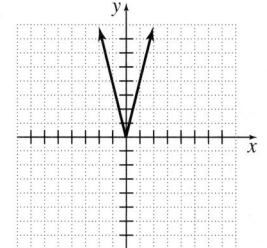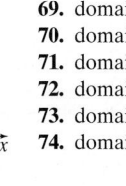

69. domain: all real numbers; range: all real numbers ≤ -1

70. domain: all real numbers ≥ 3; range: all real numbers

71. domain: all real numbers; range: all real numbers ≥ 3

72. domain: $\{3\}$; range: all real numbers

73. domain: $0 \leq x \leq 40$; range: $37 \leq y \leq 61$, approximately

74. domain: $0 \leq x \leq 10$; range: $350 \leq y \leq 440$, approximately

75. (a) 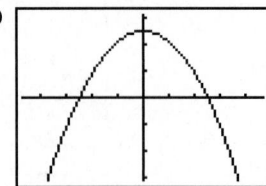 **(b)** domain: $0 \le x \le 2.5$; range: $0 \le y \le 100$ **76. (a)** 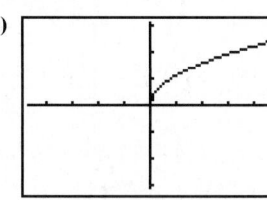 **(b)** domain: $x \ge 0$; range: $y \ge 0$

3.2 CALCULATOR EXERCISES

A 1. Integer setting   **2.** Decimal setting   **3.** Standard setting

B 1. $Y1 = 0.6x - 1.2$ **2.** $Y1 = -0.5x + 2.2$ **3.** $Y1 = |x| - 2$

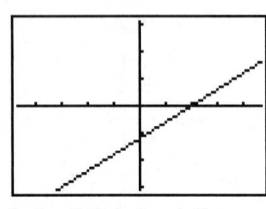 $(-2, -2.4), (-1, -1.8),$ $(0, -1.2), (1, -0.6), (2, 0)$

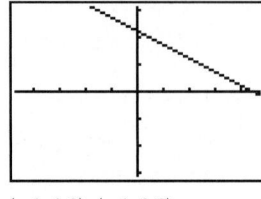 $(-2, 3.2), (-1, 2.7),$ $(0, 2.2), (1, 1.7), (2, 1.2)$

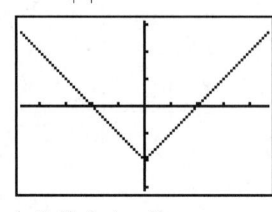 $(-2, 0), (-1, -1),$ $(0, -2), (1, -1), (2, 0)$

4. $Y1 = |x - 2|$ **5.** $Y1 = x - 2$

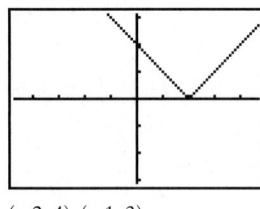 $(-2, 4), (-1, 3),$ $(0, 2), (1, 1), (2, 0)$

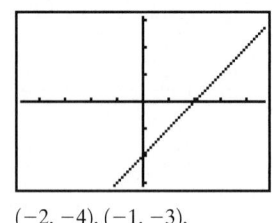 $(-2, -4), (-1, -3),$ $(0, -2), (1, -1), (2, 0)$

6. exercise 3: [MATH] [▶] [|] [X,T,θ,n] [)] [−] [2];

exercise 4: [MATH] [▶] [|] [X,T,θ,n] [−] [2] [)];

exercise 5: [X,T,θ,n] [−] [2]

SECTION 3.3 EXERCISES

1. not a function   **2.** function   **3.** function   **4.** function   **5.** not a function   **6.** not a function   **7.** function   **8.** function
9. function   **10.** not a function   **11.** not a function   **12.** function   **13.** function   **14.** not a function   **15.** 112   **16.** 1012
17. -128 **18.** -328 **19.** 60   **20.** -36 **21.** 7   **22.** 14   **23.** $20a + 12$ **24.** $-20a + 12$ **25.** $20h + 52$ **26.** $20h - 28$
27. $20a - 68$ **28.** $-20a + 92$ **29.** $20a + 20h + 12$ **30.** $20x + 20h + 12$ **31.** 75   **32.** 3   **33.** 53   **34.** 11   **35.** 11.82
36. 4.62   **37.** $\dfrac{173}{25}$ **38.** $\dfrac{93}{25}$ **39.** $\dfrac{203}{25}$ **40.** $147\dfrac{1}{2}$ **41.** $2b^2 - 4b + 5$ **42.** $2b^2 + 4b + 5$ **43.** 6   **44.** 9   **45.** 24
46. 12   **47.** 4.5   **48.** 8.7   **49.** 22.5   **50.** 9.03   **51.** 7   **52.** 11   **53.** 23   **54.** 5   **55.** 31   **56.** 28   **57.** 24
58. 23   **59.** not a real number   **60.** not a real number   **61.** 25.5   **62.** 25.6   **63.** 24.5   **64.** 25.2
65. $f(x) = 1500 + 35x$; the production run costs \$15,500.   **66.** $f(x) = 25,000 + 550x$; the cost of production is \$1,675,000.
67. $f(x) = 125x - 470$; the profit is \$49,530.   **68.** $f(x) = 1200x - 185$; the profit will be \$59,815.
69. $f(x) = 39 + 25x$; the charge for renting a truck for three days will be \$114.   **70.** $f(x) = 475 + 165x$; Susie's week's pay will be \$1135.
71. $f(x) = 2.5 + x$; the charge is \$9.50.   **72.** $f(x) = 140 + 18.5x$; there will be a \$1527.50 charge for a party of 75 guests.
73. $f(x) = 175x - 3x^2$; if 22 customers make reservations for the tour, the company will make \$2398.
74. $f(x) = (375 + 25x)(40 - x)$; the total monthly rental receipts will be \$16,650.
75. Let $x =$ the number of CD's sold over 5, $f(x) = 25 + 4x$, the cost of 12 used CD's will be \$53.   **76.** Let $x =$ the number of comics
sold over 10, $f(x) = 35 + 1.5x$; the cost of 18 comics will be \$47.   **77.** Approximately 67.34% of households owned their own homes
in 2001.   **78.** The anticrime budget for the year 2001 was approximately \$24.07 billion. Answers will vary.

3.3 CALCULATOR EXERCISES

exercises 1–6   exercises 7–12   exercises 13–18

X	Y1
65	278916
-83	-5.6E5
3.1416	45.017
1.4142	7.2426
5634	1.8E11
-3.142	-23.28

$Y1 = X^3 + X^2 + X + 1$

X	Y1
-8	4
0	4
-4	0
.8	4.8
-6.3	2.3
.75	4.75

$Y1 = \sqrt{(X^2 + 8X + 16)}$

X	Y1
10	1
-5	-.5
20	.33333
5.5	10
5	ERROR
.2	-1.042

$Y1 = 5/(X-5)$

SECTION 3.4 EXERCISES

1. (a) $(-2, 0), (6, 0)$ **(b)** $(0, 3)$ **(c)** 4 **(d)** none **(e)** $x < 2$ **(f)** $x > 2$
2. (a) $(-1, 0), (3, 0)$ **(b)** $(0, -3)$ **(c)** none **(d)** -4 **(e)** $x > 1$ **(f)** $x < 1$
3. (a) $(3, 0), (-5, 0)$ **(b)** $(0, -3)$ **(c)** none **(d)** -4 **(e)** $x > -1$ **(f)** $x < -1$
4. (a) $(4, 0), (-8, 0)$ **(b)** $(0, 4)$ **(c)** 6 **(d)** none **(e)** $x < -2$ **(f)** $x > -2$

5. $y = 3x - 6$

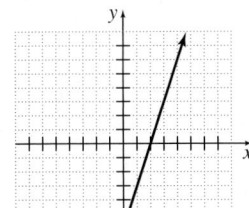

x-intercept: $(2, 0)$
y-intercept: $(0, -6)$

6. $y = 4x + 8$

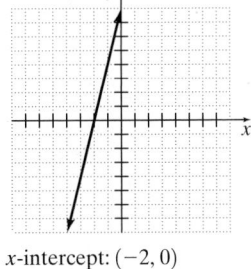

x-intercept: $(-2, 0)$
y-intercept: $(0, 8)$

7. $y = \frac{1}{2}x + 1$

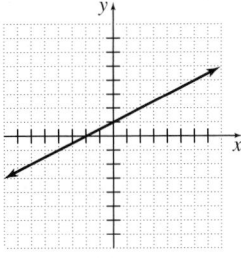

x-intercept: $(-2, 0)$
y-intercept: $(0, 1)$

8. $y = \frac{2}{3}x + 4$

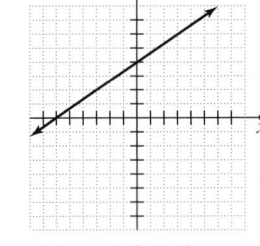

x-intercept: $(-6, 0)$
y-intercept: $(0, 4)$

9. $y = 1.2x - 6$

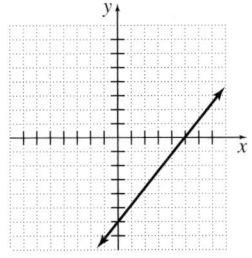

x-intercept: $(5, 0)$
y-intercept: $(0, -6)$

10. $y = 0.2x + 1$

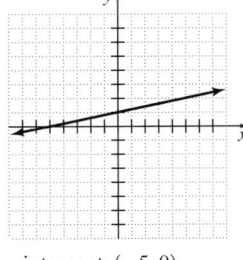

x-intercept: $(-5, 0)$
y-intercept: $(0, 1)$

11. $f(x) = -12x + 24$

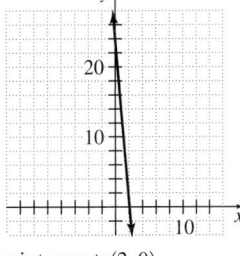

x-intercept: $(2, 0)$
y-intercept: $(0, 24)$

12. $F(x) = 15x - 45$

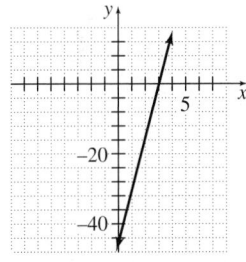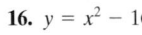

x-intercept: $(3, 0)$
y-intercept: $(0, -45)$

13. $f(x) = 9x + 15$

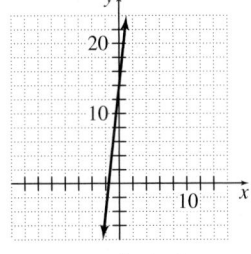

x-intercept: $\left(-\frac{5}{3}, 0\right)$
y-intercept: $(0, 15)$

14. $G(x) = -8x + 36$

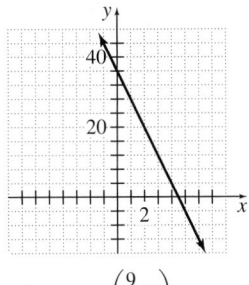

x-intercept: $\left(\frac{9}{2}, 0\right)$
y-intercept: $(0, 36)$

15. $y = x^2 - 9$

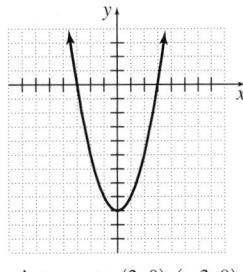

x-intercepts: $(3, 0), (-3, 0)$
y-intercept: $(0, -9)$

16. $y = x^2 - 16$

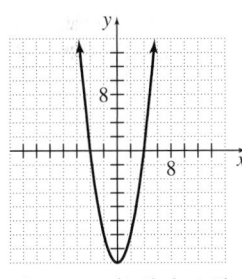

x-intercept: $(4, 0), (-4, 0)$
y-intercept: $(0, -16)$

17. $y = x^2 + 6x + 9$

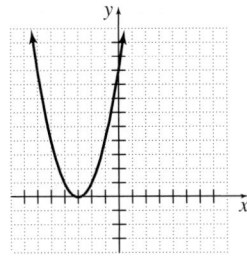

x-intercept: $(-3, 0)$
y-intercept: $(0, 9)$

18. $y = x^2 - 4x + 4$

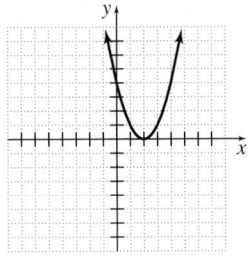

x-intercept: $(2, 0)$
y-intercept: $(0, 4)$

19. $y = 4x^2 + 4x + 1$

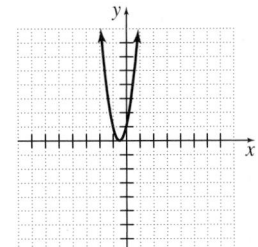

x-intercept: $\left(-\frac{1}{2}, 0\right)$
y-intercept: $(0, 1)$

20. $y = 4x^2 - 12x + 9$

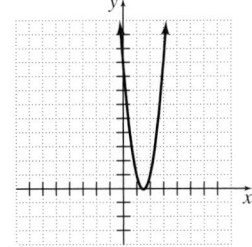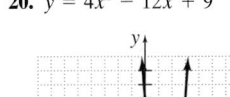

x-intercept: $\left(\frac{3}{2}, 0\right)$
y-intercept: $(0, 9)$

21. $g(x) = x^2 + 10x - 3$

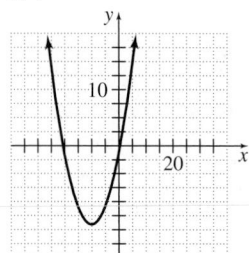

x-intercepts: $(-10.29, 0)$, $(0.29, 0)$
y-intercept: $(0, -3)$

22. $f(x) = 0.4x^2 - 0.4x - 6.5$

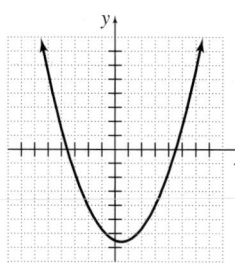

x-intercepts: $(-3.56, 0)$, $(4.56, 0)$
y-intercept: $(0, -6.5)$

23. $H(x) = x^2 - 5x - 24$

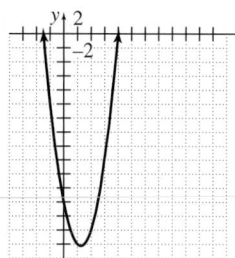

x-intercepts: $(-3, 0), (8, 0)$
y-intercept: $(0, -24)$

24. $g(x) = 2x^2 + 13x - 70$

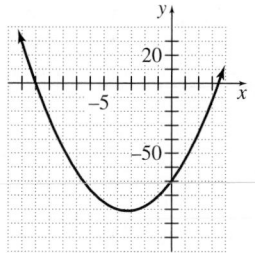

x-intercepts: $(-10, 0), (3.5, 0)$
y-intercept: $(0, -70)$

25. $y = x^3 + x^2 - 2x$

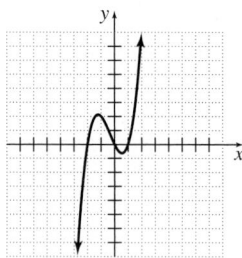

x-intercepts: $(-2, 0)$, $(0, 0), (1, 0)$
y-intercept: $(0, 0)$

26. $y = x^3 + 4x^2 + 3x$

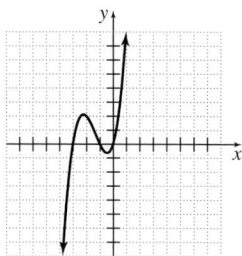

x-intercepts: $(-3, 0), (-1, 0)$, $(0, 0)$
y-intercept: $(0, 0)$

27. $f(x) = x^3 + 2x^2 - x - 2$

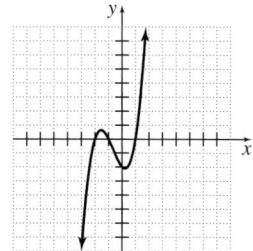

x-intercepts: $(-2, 0)$, $(-1, 0), (1, 0)$
y-intercept: $(0, -2)$

28. $f(x) = x^3 + x^2 - 4x - 4$

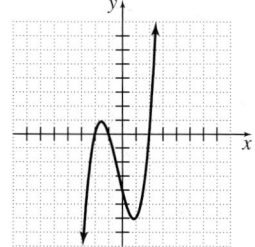

x-intercepts: $(-2, 0)$, $(-1, 0), (2, 0)$
y-intercept: $(0, -4)$

29. $h(x) = |x| - 6$

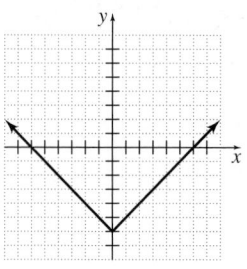

x-intercepts: $(-6, 0)$, $(6, 0)$
y-intercept: $(0, -6)$

30. $f(x) = |2x| - 6$

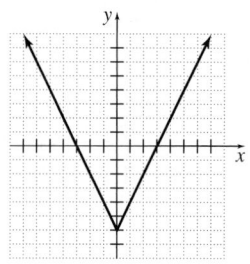

x-intercepts: $(-3, 0), (3, 0)$
y-intercept: $(0, -6)$

31. $y = |2x - 3| - 1$

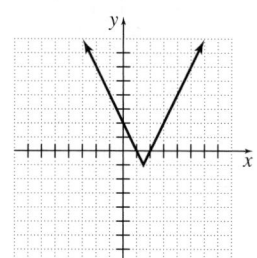

x-intercepts: $(1, 0), (2, 0)$
y-intercept: $(0, 2)$

32. $y = 5 - |3x + 1|$

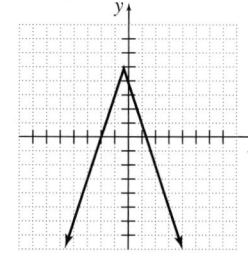

x-intercepts: $(-2, 0)$, $\left(1\frac{1}{3}, 0\right)$
y-intercept: $(0, 4)$

33. $y = |x^2 - 2| - 1$

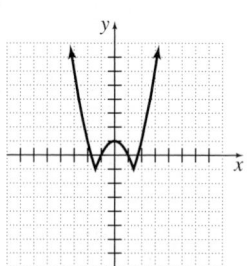

x-intercept: $(-1.73, 0)$, $(-1, 0), (1, 0), (1.73, 0)$
y-intercept: $(0, 1)$

34. $y = |x^2 - 3| - 1$

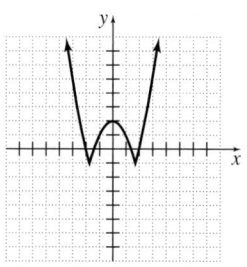

x-intercept: $(-2, 0), (2, 0)$, $(-1.41, 0), (1.41, 0)$
y-intercept: $(0, 2)$

35. $y = 2x + 8$

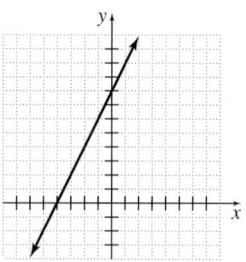

increasing for all
x-values

36. $y = 4 - x$

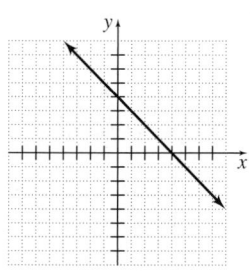

decreasing for all
x-values

37. $f(x) = 3 - 2x$

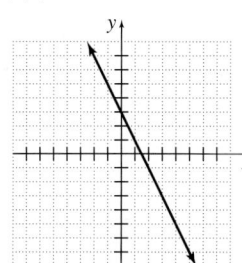

decreasing for all x-values

38. $g(x) = x - 2$

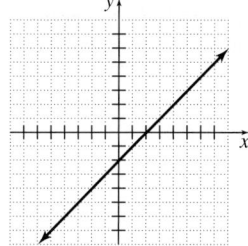

increasing for all x-values

39. $y = 1 - x^2$

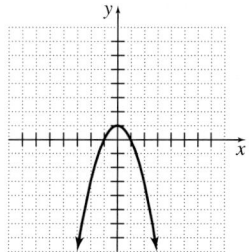

increasing for $x < 0$
decreasing for $x > 0$
relative maximum is 1

40. $y = x^2 + 1$

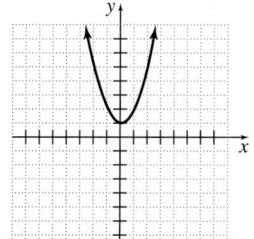

increasing for $x > 0$
decreasing for $x < 0$
relative minimum is 1

41. $g(x) = x^2 + 4x + 3$

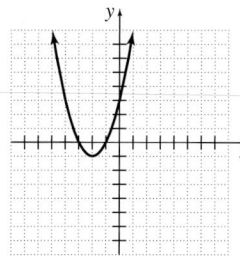

increasing for $x > -2$
decreasing for $x < -2$
relative minimum is -1

42. $h(x) = 4x - 5 - x^2$

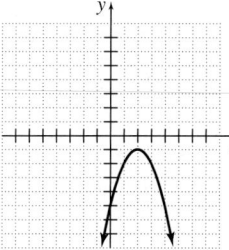

increasing for $x < 2$
decreasing for $x > 2$
relative maximum is -1

43. $y = |x + 3|$

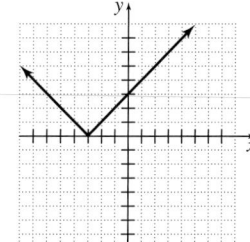

increasing for $x > -3$
decreasing for $x < -3$
relative minimum is 0

44. $y = |x| + 3$

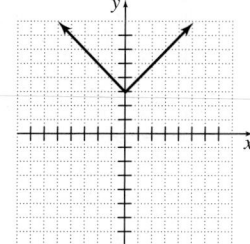

increasing for $x > 0$
decreasing for $x < 0$
relative minimum is 3

45. $f(x) = -|x + 3|$

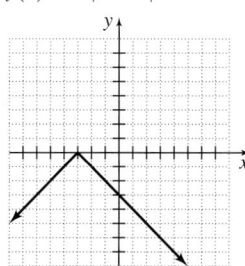

increasing for $x < -3$,
decreasing for $x > -3$
relative maximum is 0

46. $p(x) = -|x| + 3$

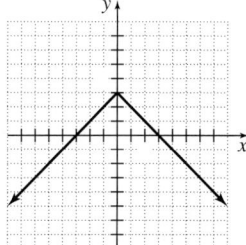

increasing for $x < 0$,
decreasing for $x > 0$
relative maximum is 3

47. $(4, 7)$ **48.** $(8, 1)$ **49.** $(-2, 3)$ **50.** $(-1, -2)$
51. $(-0.75, 5.75)$ **52.** $(-2.\overline{6}, 10.\overline{6})$ **53.** $(3.8, 12)$ **54.** $(3, 5)$
55. $(-1, 3), (1, 3)$ **56.** $(-5, 9), (5, 9)$ **57.** $(1, 3), (3, 11)$
58. $(-4, -2), (-1, 4)$ **59.** $(-2, -1), (4, 2)$
60. $(-3, -3), (6, 6)$ **61.** $(-7, 2), (7, 2)$ **62.** $(4, 4), (-12, 4)$

63. increasing: $0 < x < 20$
decreasing: $x > 20$
relative maximum is 4000

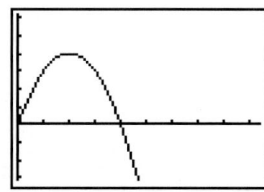

64. increasing: $0 < x < 50$
decreasing: $x > 50$
relative maximum is 2500

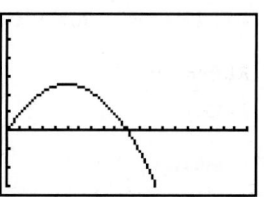

65. Let $p(x) =$ profit
$p(x) = 5x - 50$
no relative maximum or
minimum
no decreasing values
increasing for $x > 0$

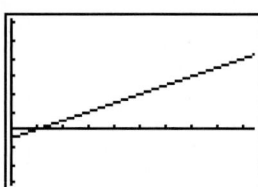

66. Let $f(x) =$ remaining money
$f(x) = 50,000 - 5000x$
decreasing values: $x > 0$
no relative maximum or
minimum
no increasing values

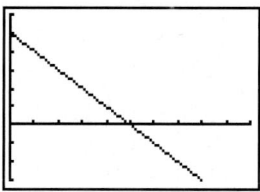

67. (a) $f(x) = 4x + 50$
  (c) Y1 = $4x + 50$
    Y2 = $10x$

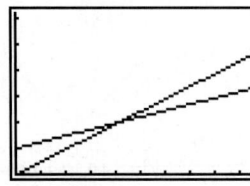

  intersection at $(8.\overline{3}, 83.\overline{3})$

(b) $g(x) = 10x$
(d) At $x = 8.3$ containers, the cost for producing and the cost for selling are the same at 83.33. This means that nine containers must be sold to cover the costs of production.

68. (a) $f(x) = 12x + 100$
  (c) Y1 = $12x + 100$
    Y2 = $24x$

  intersection at $(8.\overline{3}, 200)$

(b) $g(x) = 24x$
(d) At $x = 8.\overline{3}$ carvings, the cost and revenue are equal at 200. This means that nine carvings must be sold to cover the costs of production.

69. (a) $f(x) = 200 + 50x; g(x) = 75x$
  (b) Y1 = $200 + 50x$
    Y2 = $75x$

  intersection at $(8, 600)$

  (c) At eight credit hours, the pay is the same: $600.

70. (a) $f(x) = 25 + 10x; g(x) = 15x$
  (b) Y1 = $25 + 10x$
    Y2 = $15x$

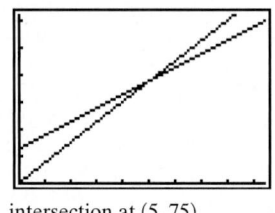

  intersection at $(5, 75)$

  (c) At five lawns, the pay is the same: $75.

71. x-intercept is approximately $(2.5, 0)$ This means that Yahoo! broke even in 1997.
72. x-intercepts are approximately $(0.3, 0), (4.7, 0),$ and $(5.4, 0)$. This means that Northwest Airlines broke even in 1993, 1997, and 1998.

3.4 CALCULATOR EXERCISES

For problems 1–3, graphs will vary.   **1.** $(-4.\overline{6}, 1.\overline{4})$ and $(6, 5)$ **2.** $(-5, -1)$ and $(5, -1)$ **3.** $(-6, 18)$ and $(6, 18)$

CHAPTER 3 SECTION-BY-SECTION REVIEW

Reflections

1–8. Answers will vary.

Exercises

1.

a	b
−3	13
−2	11
−1	9
0	7
1	5
2	3
3	1

2.

x	y
9	10
6	8
3	6
0	4
−3	2
−6	0
−9	−2

3.

x	y
−3	−2.4
−2	−2
−1	−1.6
0	−1.2
1	−0.8
2	−0.4
3	0

4.

x	y
−18	−30,190
−7	−1898
0	−22
6	962
21	45,002
22.5	55,457.375

5.

x	y
−2	21
−1	12
0	5
1	0
2	3
3	4

6.

x	y
−2.7	7.994
−1.9	−4.054
−0.6	−13.804
0	−14.2
0.8	−10.696
1.5	−3.85
2.4	10.136

7. Answers will vary. Possible answer:

x	y
−2	7.7
−1	6.1
0	4.5
1	2.9
2	1.3

8. Answers will vary. Possible answer:

x	y
−4	−12
−2	−9
0	−6
2	−3
4	0

9. Answers will vary. Possible answer:

x	y
−4	17
−2	13
0	9
2	5
4	1

10.

x	y
−5	207
−3	91
−1	15
1	−21
3	−17
5	27

11.

x	y
−15	−1
−10	2
−5	5
0	8
5	11
10	14
15	17

12.

x	y
−3	−64.2
−2	−47.1
−1	−30
0	−12.9
1	4.2
2	21.3
3	38.4

13. (a) Let x = gallons of gas
C = total cost
$C = 1.649x$

(b)

x	C
5	$8.25
10	$16.49
15	$24.74
20	$32.98

14.

r	A
4	50.265
6	113.097
8	201.062
10	314.159

15.

a	b
10	80
20	70
30	60
40	50
45	45

16.

t	I
2	$240
3	$360
4	$480

17.

F	C
−23	−30.6
−14	−25.6
0	−17.8
41	5
50	10
59	15
100	37.8

18. $(−10, 37), (−5, 22), (0, 7), (5, −8), (10, −23)$
19. $(−8, 0), (−7, 1), (−4, 2), (1, 3), (8, 4)$
20. $(−6, −5), (−3, −3), (0, −1), (3, 1), (6, 3)$
21. $(2, 1), (4, 2), (6, 3), (8, 4), (10, 5)$
22. $(1993, 86{,}900), (1994, 70{,}100), (1995, 92{,}500), (1996, 98{,}600), (1997, 119{,}900)$
23. domain: $\{1, 3, 5, 7, 9\}$; range: $\{2, 6, 10, 14, 18\}$
24. domain: $\{\ldots, −6, −4, −2, 0, 2, 4, 6, \ldots\}$; range: $\{\ldots, 6, 4, 2, 0, −2, −4, −6, \ldots\}$
25. domain: $\{−5, −4, −3, −2, −1\}$; range: $\{−1, 3, 7, 11, 15\}$
26. domain: $\{0, 0.5, 1, 1.5, 2\}$; range: $\{2.5, 2.75, 3.5, 4.75, 6.5\}$

27. domain: $\{−5, −4, −3, −2, −1, 0\}$; range: $\{0, 1, \sqrt{2}, \sqrt{3}, 2, \sqrt{5}\}$ **28.** domain: $\{−4, −2, 0, 2, 4\}$; range: $\{2.4, 4, 12, −12, −4\}$

29.

L	T
1	1.11
8	3.14
16	4.44
24	5.44
32	6.28

30.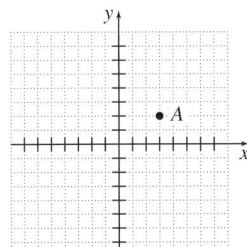

31.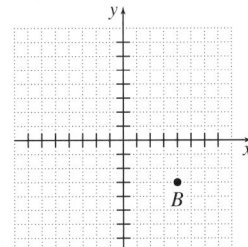

32.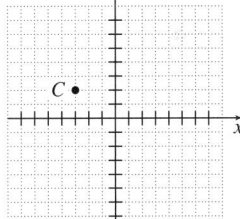

33.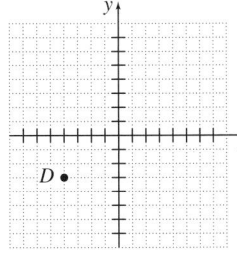

34.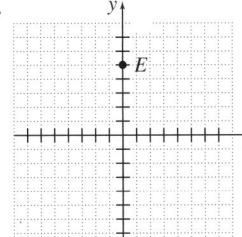

35.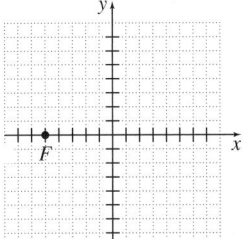

36. $A = (5, 3); B = (−2, −5); C = (2, −2);$
$D = (−3, 5); E = (5, 0); F = (0, −4);$
$G = (0, 0)$
37. quadrant I   **38.** quadrant III   **39.** quadrant IV   **40.** quadrant II   **41.** x-axis   **42.** y-axis   **43.** origin

44.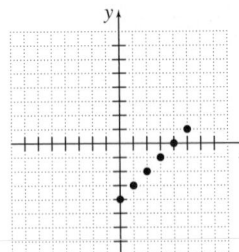

45.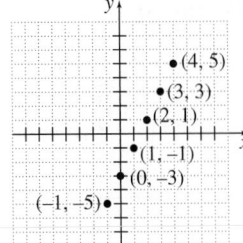

46.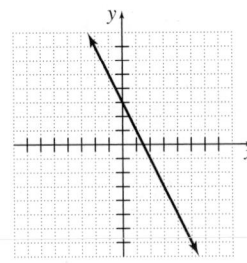

47.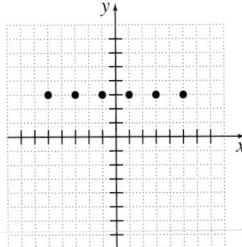

48. domain: all real numbers; range: all real numbers ≤ 2

49. domain: all real numbers; range: $\{-3\}$

50. domain: $1980 \leq x \leq 1998$; range: $71 \leq y \leq 90$, approximately

51. (a) $y = 10x + 40$
 (c) domain: $\{1,2,3,4,\dots\}$
  range: $\{50,60,70,80,\dots\}$

(b)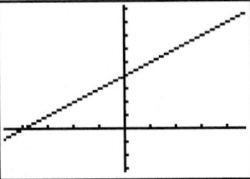

52. (a) $y = x^2$
 (c) domain: $x > 0$
  range: $y > 0$

(b)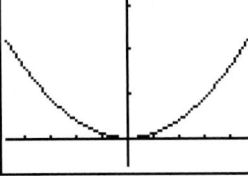

53. not a function **54.** function **55.** not a function **56.** function **57.** -39 **58.** 97 **59.** 3 **60.** 27.8 **61.** $1 - 4h$

62. $4b + 13$ **63.** 44 **64.** 14 **65.** -2.25 **66.** $5a^2 + a - 4$ **67.** $5a^2 - a - 4$ **68.** $-\dfrac{63}{16}$ **69.** -2 **70.** 0 **71.** 6

72. Let x = number of widgets, $C(x)$ = cost, $C(x) = 4500 + 17x$ **73.** Let x = number attending, $T(x)$ = total charge, $T(x) = 1500 + 125x$
$C(1200) = \$24,900$ $T(20) = \$4000$

74. Let x = number of painted faces, $P(x)$ = profit, $P(x) = 1.50x - 15$
$P(135) = \$187.50$

75. Net sales are $\$164.9$ billion.

76. (a) none **(b)** $(0, 7)$ **(c)** none **(d)** 1 **(e)** $x > 3$ **(f)** $x < 3$

77. $y = 3x + 9$

78. $y = \dfrac{3}{4}x - 9$

79. $y = x^2 - 0.36$

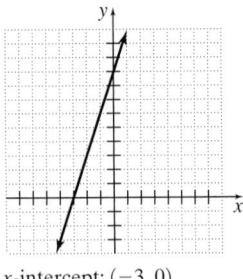

x-intercept: $(-3, 0)$
y-intercept: $(0, 9)$

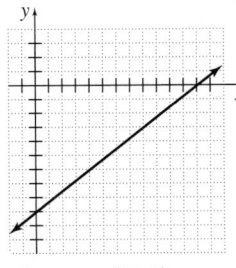

x-intercept: $(12, 0)$
y-intercept: $(0, -9)$

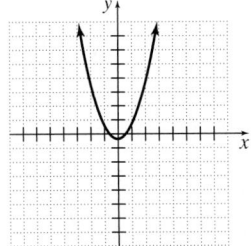

x-intercept: $(-0.6, 0), (0.6, 0)$
y-intercept: $(0, -0.36)$

80. $y = |x| - 4$

81. $h(x) = 6 - 2x$

82. $y = 3 - x^2$

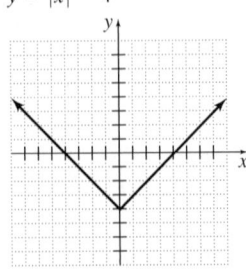

x-intercept: $(-4, 0), (4, 0)$
y-intercept: $(0, -4)$

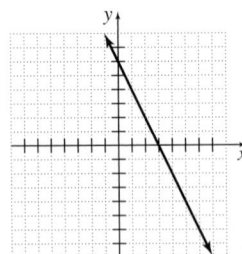

decreasing for all values of x

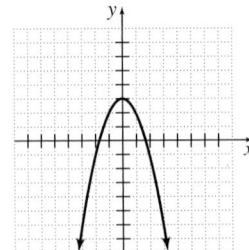

increasing for $x < 0$
decreasing for $x > 0$
relative maximum is 3

83. $y = |x| + 2$

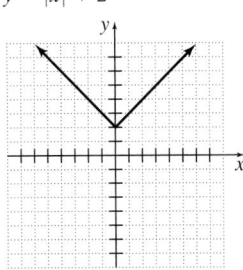

increasing for $x > 0$
decreasing for $x < 0$
relative minimum is 2

84. $y = |x^2 - 1|$

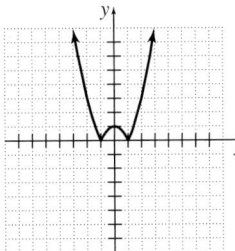

increasing for
$-1 < x < 0, x > 1$
decreasing for $x < -1$,
$0 < x < 1$
relative maximum is 1
relative minima are $0, 0$

85. $(3, 4)$ **86.** $(-2, -2), (3, 3)$ **87.** $(-7, 2), (-3, 2)$

88.

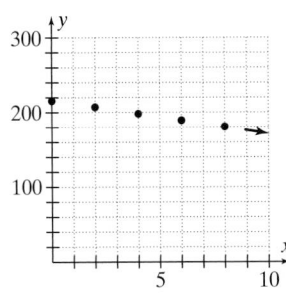

increasing for $x > 0$
no relative maximum or minimum
no decreasing values

89. Let $R(x)$ = revenue, $R(x) = 10.45x$

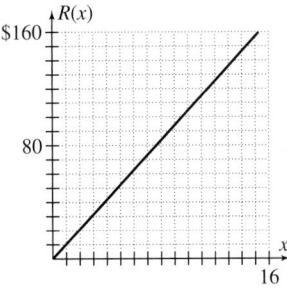

no relative minimum or maximum
increasing for all values of $x > 0$

90. Let $f(x)$ = balance
$f(x) = 216 - 4.5x$

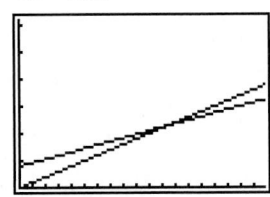

no relative minimum or maximum
decreasing for all values of $x > 0$

91. (a) $f(x) = 400 + 65x$
$g(x) = 100x$
(b) $Y1 = 400 + 65x$
$Y2 = 100x$

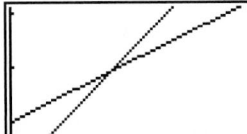

intersect at approximately $(11.4, 1143)$

(c) At about 11.4 credit hours, the stipend is the
same at about \$1143 for both options.

92. (a) $f(x) = 500 + 12x$
$g(x) = 25x$
(b) $Y1 = 500 + 12x$
$Y2 = 25x$

intersect at approximately $(38.5, 961.5)$

(c) At about 38.5 items, the cost to produce and the
revenue are equal at about \$961.50 each. This
means that 39 items must be sold to break even.

CHAPTER 3 CHAPTER REVIEW

1. 0   **2.** 18   **3.** 7.2   **4.** 11.7   **5.** $b + 9$ **6.** $-h + 8$ **7.** 0   **8.** 0   **9.** -6.25 **10.** $v^2 - 3v - 4$ **11.** $v^2 + 3v - 4$

12. $-\dfrac{14}{9}$ **13.** 4   **14.** 8   **15.** 16   **16.** function   **17.** not a function

18. $y = 4.8x - 1.2$

19. $y = \dfrac{2}{5}x + 4$

20. $y = x^2 - 1.21$

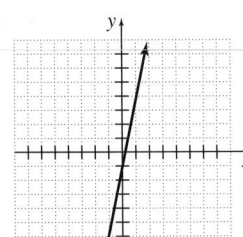

x-intercept at $(0.25, 0)$
y-intercept at $(0, -1.2)$
increasing for all x-values

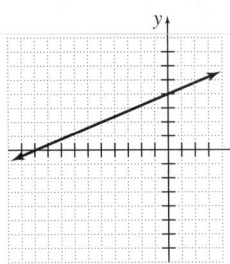

x-intercept at $(-10, 0)$
y-intercept at $(0, 4)$
increasing for all x-values

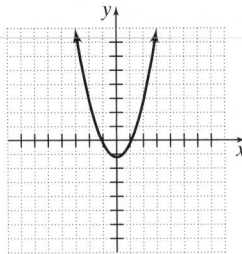

x-intercepts at $(-1.1, 0)$
and $(1.1, 0)$
y-intercepts at $(0, -1.21)$
increasing for $x > 0$
decreasing for $x < 0$
relative minimum is -1.21

21. $y = 2 - |x|$

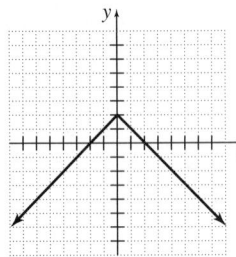

x-intercepts at $(-2, 0), (2, 0)$
y-intercept at $(0, 2)$
increasing for $x < 0$
decreasing for $x > 0$
relative maximum is 2

22. $(-3, -4)$
23. $(0, 0), (3, 9)$
24. $(-1, 2), (3, 6)$
25. domain: $\{2, 4, 6, 8, 10\}$; range: $\{1, 2, 3, 4, 5\}$
26. domain: $\{\ldots, -6, -4, -2, 0, 2, 4, 6, \ldots\}$; range: $\{3\}$
27. domain: $\{-5, -4, -3, -2, -1\}$; range: $\{25, 16, 9, 4, 1\}$
28. domain: $\{0, 0.5, 1, 1.5, 2\}$; range: $\{-1.5, -1.25, -0.5, 0.75, 2.5\}$
29. $(-6, 60), (-3, 36), (0, 12), (3, -12), (6, -36)$
30. $(3, 1), (2, 2), (1, \sqrt{7}), (0, \sqrt{10}), (-1, \sqrt{13}), (-2, 4)$
31. $(-7, 1), (0, 5), (7, 9), (14, 13), (21, 17)$

32.

x	y
-4	119
-2	33
0	-5
2	5
4	63

33.

x	y
-12	-14
-8	-11
-4	-8
0	-5
4	-2
8	1
12	4

34.

x	y
-2	25.2
-1	20.5
0	15.8
1	11.1
2	6.4

35.

x	y
-2	35
$-\frac{3}{4}$	0
0	-15
$\frac{3}{4}$	-25.5
5	0

36.

x	y
-6	95
-3	20
0	1
3	32
6	119
9	260

37.

x	y
-3.7	63.014
-2.2	26.504
-0.7	10.694
0	10.4
0.8	15.584
2.3	41.174
3.8	87.464

38. Answers will vary.
Possible answer:

x	y
-1	-6.1
0	-1.6
1	2.9

39. Answers will vary.
Possible answer:

x	y
-4	2
0	3
4	4

40. Answers will vary.
Possible answer:

x	y
-3	19
0	10
3	1

41. $\left(\dfrac{1}{4}, 1.5708\right), \left(\dfrac{1}{2}, 3.1416\right), (1, 6.2832), \left(\dfrac{3}{2}, 9.4248\right), (2, 12.566)$

42.

s	A
3	9
5	25

43.

a	b
30	150
60	120
90	90
120	60
150	30

44.

t	I
2	240
3	360
4	480

45.

C	F
−10	14
−5	23
0	32
5	41
10	50
15	59
20	68
25	77

46. (a) Let x = number of disks, C = cost

$C = 7.95x$

(b)

x	C
1	7.95
2	15.90
3	23.85
4	31.80
5	39.75

47. Let x = number of items, $C(x)$ = total cost, $C(x) = 2500 + 12x$; the cost is $22,300.
48. Let x = number of hours, $C(x)$ = cost, $C(x) = 15 + 2x$; the cost is $35.
49. Let x = number of people, $C(x)$ = total charges, $C(x) = 275 + 9.50x$; the total charges are $1557.50.
50. Let x = number of admissions, $P(x)$ = profit, $P(x) = 4x - 185$; the profit is $1055.
51. Let $F(x)$ = amount of fluid remaining
$F(x) = 250 - 3.5x$

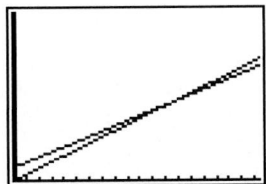

decreasing for $x > 0$

52. Let $F(x)$ = total amount in savings
$F(x) = 1000 + 50x$

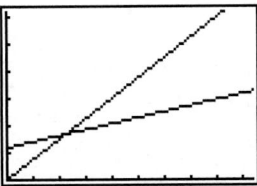

increasing for $x > 0$

53. (a) Let x = number of years
$f(x)$ = first option amount
$g(x)$ = second option amount
$f(x) = 25{,}000 + 5000x$
$g(x) = 6000x$
(b) Y1 $= 25{,}000 + 5000x$
Y2 $= 6000x$

intersect at (25, 150,000)

(c) At 25 years, the money received is the same: $150,000.

54. (a) Let $f(x)$ = total acquisition cost
$g(x)$ = revenue
$f(x) = 22x + 600$
$g(x) = 75x$
(b) Y1 $= 22x + 600$
Y2 $= 75x$

intersect at approximately (11.32, 849.06)

(c) At about 11.32 appliances, the total acquisition cost and total revenue are equal at about $849.06. This means that 12 appliances must be sold to break even.

CHAPTER 3 TEST

1.

x	y
−9	0
0	−9
3	60

2. (a)

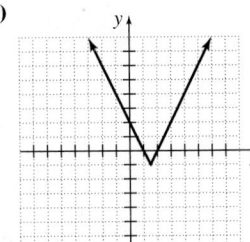

(b) none
(c) −1
(d) $x > 1.5$
(e) $x < 1.5$
(f) $(2, 0)$ and $(1, 0)$
(g) $(0, 2)$

3. (a) no, because it does not pass the vertical-line test. **(b)** $x \geq -4$ **(c)** all real numbers
4. $A = (1, 2); B = (-2, -4); C = (-5, 3); D = (2, -5); E = (0, -2)$ **5.** quadrant IV

6.

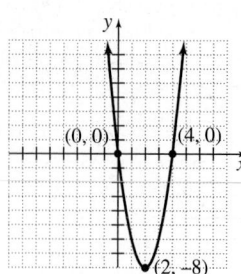

7.

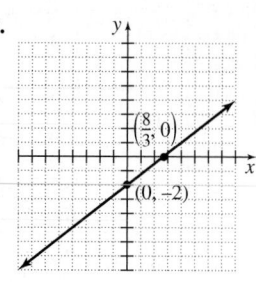

8.

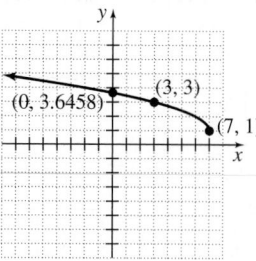

9. 8   **10.** 3   **11.** $\frac{1}{2}a + 6$ **12.** $\frac{1}{2}a + 7$ **13.** Let $C(x) = $ cost of production, $C(x) = 450 + 21.50x$ **14.** The cost is $5825.

15. $(2, -4)$ **16.** Answers will vary.

CHAPTERS 1–3 CUMULATIVE REVIEW

1. 0, 12   **2.** 0, 12   **3.** $-\frac{2}{3}; 0; 12; 1\frac{4}{5}; -0.33$ **4.** $\sqrt{7}$ **5.** $>$ **6.** $>$ **7.** $<$

8.

$$-\sqrt{25} \quad -3.1 \quad -\frac{1}{2} \quad \sqrt{5} \quad 2\frac{3}{4}$$

number line from -9 to 9

9. -15 **10.** -2.56 **11.** 3   **12.** $-\frac{31}{24}$ **13.** $-1\frac{37}{40}$ **14.** $\frac{63}{640}$

15. -62.208 **16.** 0   **17.** 96   **18.** 31   **19.** -6.2 **20.** 0   **21.** -8 **22.** $\frac{4}{5}$ **23.** -1.095 **24.** not a real number

25. 1.051   **26.** 1   **27.** 1   **28.** indeterminate   **29.** -4096 **30.** 4096   **31.** 100   **32.** 3.05×10^{-6} **33.** -4.2356×10^{6}
34. 0.0356   **35.** 678,000,000   **36.** 7   **37. (a)** 6 **(b)** $a^3, -2a^2, a, -2a^3, 7a$ **(c)** -5 **(d)** $-a^3 - 2a^2 + 8a - 5$ **38.** $4y + z$

39. $\frac{3}{4}y - \frac{43}{48}$ **40.** $-6x - 12$ **41.** No   **42.**

graph of a parabola

43. all real numbers; all real numbers ≥ 3
44. Yes, since all possible vertical lines cross the graph a maximum of one time.
45. The relative minimum is 3.
46. $x > 0$
47. (a) -2 **(b)** $\frac{1}{3}h - 4$
48. The volume is 13.78125 ft³.
49. Kelsie's interest is $110; the total investment is $610.

50. Let $C(x) = $ cost, $C(x) = 35 + 2.80x$; the cost of producing 150 ornaments in one production run is $455.

Chapter 4

SECTION 4.1 EXERCISES

1. linear   **2.** linear   **3.** nonlinear   **4.** linear   **5.** linear   **6.** nonlinear   **7.** nonlinear   **8.** nonlinear   **9.** linear

10. linear   **11.** $2x - 7 = x + 2; x = 9$ **12.** $\frac{1}{2}x - 3 = x - 1; x = -4$ **13.** $0.5x + 1.25 = 0.5(x + 2.5); x = $ all real numbers

14. $\frac{2}{3}x + 1 = x + 2 - \frac{1}{3}(x + 3); x = $ all real numbers   **15.** $x - (4.5 - 0.5x) = 1.5(x + 2)$; no solution

16. $x + 3(x - 5) = 2(2x - 3)$; no solution   **17.** $\frac{1}{3}x + 1 = \frac{3}{2}x - 1$; noninteger between 1 and 2

18. $x - 1 = 3x + 4$; noninteger between -2 and -3 **19.** 14   **20.** 5   **21.** -6 **22.** -8 **23.** 7   **24.** 0
25. all real numbers   **26.** 0.75   **27.** no solution   **28.** no solution   **29.** no solution
30. all real numbers   **31.** 0.5   **32.** all real numbers

33. $3x + 2 = 4 - x; x = \frac{1}{2}$ **34.** $0.4x - 1.5 = 1 - 0.6x; x = 2.5$ **35.** $\frac{1}{2}x + 5 = 4 - 0.5(6 - x)$; no solution

36. $x + \frac{1}{3}(x + 6) = \frac{4}{3}(x + 3)$; no solution   **37.** -3 **38.** all real numbers   **39.** no solution   **40.** 2   **41.** all real numbers

42. −1   **43.** 15   **44.** all real numbers   **45.** no solution   **46.** −8   **47.** 3   **48.** approximately 0.727   **49.** all real numbers
50. no solution   **51.** The number of miles is 80.   **52.** The number of pages is 20.   **53.** The factory should produce 20 pairs of shoes.
54. The number of hours is 8.   **55.** She can spend $29.   **56.** She should lose 1 pound.   **57.** The dimensions are 5 feet by 8 feet.
58. The lengths are 2.8 inches and 11.2 inches.   **59.** Any number of rolls will have the same charge for both.   **60.** There is no number
of hours for which the fee will be equal.   **61.** The seventh day temperature is 78°F.   **62.** The fifth day temperature is 51°F.

4.1 CALCULATOR EXERCISES

A 1. 22   **2.** −12   **3.** 250   **4.** −500   **B** Students should experiment with same exercises.

SECTION 4.2 EXERCISES

1. 18   **2.** −104   **3.** 116   **4.** 76   **5.** −7.98   **6.** −4.606   **7.** 0   **8.** $\frac{14}{9}$ **9.** −3   **10.** 50   **11.** −16.85   **12.** 22.6

13. $-\frac{14}{5}$ **14.** $\frac{32}{7}$ **15.** 36   **16.** 123   **17.** $-\frac{17}{16}$ **18.** $\frac{17}{30}$ **19.** −90   **20.** 0   **21.** 81   **22.** −62   **23.** −0.02

24. −3.7   **25.** $\frac{2}{5}$ **26.** −16   **27.** 4.88   **28.** −69.2   **29.** −57   **30.** $16\frac{1}{5}$ **31.** 3   **32.** 17   **33.** −2.398

34. −10   **35.** $\frac{1}{3}$ **36.** $-\frac{5}{7}$ **37.** 14   **38.** −41   **39.** −1   **40.** 6.3   **41.** 0   **42.** −1   **43.** Six servings remain in the box.

44. He can have 700 calories.   **45.** His gross pay was $2351.58.   **46.** It was marked down $41.46.   **47.** She should borrow $1\frac{3}{4}$ cups.

48. He can work $4\frac{5}{12}$ hours.   **49.** There was $4.37 sales tax.   **50.** It was reduced by 14,574 people.   **51.** She must buy 45 feet; no,
she will not have enough.   **52.** The loan value is $93,000.   **53.** There were 480 paid admissions.   **54.** The price was about $250.06.
55. They must sell 580 packets.   **56.** It will take 18 months.   **57.** The estate was worth $75,400.   **58.** The price was about $5.99.
59. The height is 17.5 feet.   **60.** The height must be 5 feet.   **61.** The height must be about 6 feet.   **62.** The radius is about 16 inches.
63. You must place $6400 into savings.   **64.** The average speed was about 65.8 mph.   **65.** The quarterly profits were $63,650.
66. The jackpot was $45,990,000.   **67.** Her sales were $1350.   **68.** Her sales were $120,000.   **69.** Sixteen dogs is the maximum
number allowed.   **70.** There are 109 mountains worldwide that are over 24,000 feet high.   **71.** The 1999 estimated population is
272,000,000 people.   **72.** The height of the volcano is 19,400 ft.   **73.** A single hair is 100,000 nanometers wide.   **74.** The temperature
of the Sun is approximately 3,167,000°F.   **75.** The mechanic needs about 5.7 gallons of solution.   **76.** The gardener needs about
2.5 pints of solution.

4.2 CALCULATOR EXERCISES

1. about 461.2 feet   **2.** about 55.3 inches   **3.** $2\frac{1}{8}$ per share   **4.** 6 pieces   **5.** about 12 gallons   **6.** about 123 square feet

7. 62.5 grams   **8.** 60 grams; does not agree; difference may be round-off error   **9.** about 16.7 feet per second

SECTION 4.3 EXERCISES

1. −2   **2.** 3   **3.** −44   **4.** 12   **5.** 0   **6.** 4.5   **7.** 3.3   **8.** 5   **9.** 7.2   **10.** 1   **11.** $\frac{1}{2}$ **12.** $\frac{2}{3}$ **13.** 2   **14.** −1

15. 30   **16.** −2   **17.** 1   **18.** 0   **19.** 8   **20.** −3   **21.** all real numbers   **22.** all real numbers   **23.** no solution
24. no solution   **25.** all real numbers   **26.** no solution   **27.** −6.6   **28.** all real numbers   **29.** no solution   **30.** 6

31. no solution   **32.** 8   **33.** 6   **34.** all real numbers   **35.** no solution   **36.** 10   **37.** $\frac{10}{9}$ **38.** $\frac{28}{3}$ **39.** $-\frac{1}{18}$ **40.** $\frac{9}{5}$

41. −120   **42.** −66   **43.** $-\frac{8}{13}$ **44.** $-\frac{3}{8}$ **45.** 107.5   **46.** 3.6   **47.** 0   **48.** 8.8   **49.** −0.05   **50.** 70

51. You could drive about 1250 miles.   **52.** You could drive 1727.5 miles.   **53.** The monthly payments would be $85.50.

54. The monthly charge would be $140.   **55.** There were 12 liters of 30% solution.   **56.** There were $2\frac{1}{2}$ gallons of 40% solution.

57. Her brother's average weekly earnings are $352.55.   **58.** His profit last year was $8988.   **59.** Compensation is the same by the two
companies for all levels of sales.   **60.** The costs are the same for any number of days.   **61.** The plans will never pay the same.
62. There is no number of miles for which the two plans cost the same.   **63.** $6250 is the additional amount necessary to meet Nathan's
budget.   **64.** $10,500 is the additional amount necessary to meet Max's budget.   **65.** 301 chirps per minute is the number for that
temperature.   **66.** 228 chirps per minute is the number for that temperature.

4.3 CALCULATOR EXERCISES

1. The sales were $652.80.   **2.** She should administer 3.5 tablets.   **3.** He needs three sheets.

SECTION 4.4 EXERCISES

1. $s = \dfrac{P}{4}$ **2.** $b = \dfrac{A}{h}$ **3.** $d = \dfrac{C}{\pi}$ **4.** $r = \dfrac{C}{2\pi}$ **5.** $L = \dfrac{V}{WH}$ **6.** $W = \dfrac{V}{LH}$ **7.** $L = \dfrac{S - 2WH}{2W + 2H}$ **8.** $H = \dfrac{S - 2LW}{2L + 2W}$

9. $h = \dfrac{V}{\pi r^2}$ **10.** $h = \dfrac{S - 2\pi r^2}{2\pi r}$ **11.** $P = \dfrac{I}{rt}$ **12.** $r = \dfrac{I}{Pt}$ **13.** $g = \dfrac{v}{t}$ **14.** $t = \dfrac{v}{g}$ **15.** $R = \dfrac{V}{I}$ **16.** $m = \dfrac{E}{c^2}$

17. $m = x - zs$ **18.** $s = \dfrac{x - m}{z}$ **19.** $y = -\dfrac{4}{3}x$ **20.** $y = -\dfrac{1}{3}x$ **21.** $y = \dfrac{1}{2}x$ **22.** $y = \dfrac{1}{3}x$ **23.** $y = -x$ **24.** $y = x$

25. $y = -\dfrac{5}{4}x + 5$ **26.** $y = -x - 3$ **27.** $y = -x - 7$ **28.** $y = 13x - 13$ **29.** $y = \dfrac{1}{2}x + 2$ **30.** $y = -\dfrac{8}{7}x - 8$

31. $y = x - 1$ **32.** $y = 3x - 2$ **33.** $y = 4x - 19$ **34.** $y = -4x + 13$ **35.** $y = -2x + 8$ **36.** $y = 3x + 35$

37. $y = -x - 6$ **38.** $y = -5x - 62$ **39.** $y = \dfrac{2}{3}x + 10$ **40.** $y = -\dfrac{3}{4}x$ **41.** $y = -\dfrac{2}{3}x - \dfrac{1}{3}$ **42.** $y = \dfrac{4}{9}x + \dfrac{1}{6}$

43. $P = 200 + 85m$; $m = \dfrac{1}{85}P - \dfrac{40}{17}$; It will take 24 months to pay off \$2240. It will take 12 months to pay off \$1200.

44. $c = 75 + 35h$; $h = \dfrac{1}{35}c - \dfrac{15}{17}$; Richard worked 12 hours for \$495; Richard worked 16.5 hours for \$652.50.

45. $T = 22c + 12.50$; $c = \dfrac{1}{22}T - \dfrac{25}{44}$; she can spend about \$1.02 on each student for a \$35 party; she can spend about \$1.70 on each student for a \$50 party.

46. $T = 175 + 14c$; $c = \dfrac{1}{14}T - \dfrac{25}{2}$; they can spend about \$35.71 for each member if they raise \$675; they can spend about \$58.92 on each member if they raise \$1000.

47. $B = 75 + 3T$; $T = \dfrac{1}{3}B - 25$; Ted's weekly earnings are about \$216.67 when his boss averages \$725 per week; Ted's weekly earnings are \$400 when his boss averages \$1275 per week.

48. $M = 1.5F - 0.75$; $F = \dfrac{2}{3}M + \dfrac{1}{2}$; females would average \$8.80 per hour if males earned \$12.45 per hour; females would average \$14.50 per hour if males earned \$21 per hour.

49. $C = 85 + 185d$; $d = \dfrac{C - 85}{185}$. With costs limited to \$270, the equipment can be rented for one day. With costs limited to \$825, the equipment can be rented for four days.

50. $d = 40h + 50$; $h = \dfrac{d - 50}{40}$. If the distance to Fairbanks is 510 miles, then it will take about 11.5 hours. If the distance to Anchorage is 130 miles, then it will take about 2 hours.

51. $V = 15h$; $h = \dfrac{V}{15}$; the height should be 4 feet for a volume of 60 cubic feet. The height should be $6\dfrac{2}{3}$ feet for a volume of 100 cubic feet.

52. $S = 30 + 16h$; $h = \dfrac{1}{16}S - \dfrac{15}{8}$; the height would be 3.5 feet if the total surface area were 86 square feet; the height would be 5.625 feet if the total surface area were 120 square feet.

4.4 CALCULATOR EXERCISES

The amount P to invest when $A = \$10,000$ at $t = 5, 7, 10,$ and 12 is listed in the following table:

t	$P(\$)$
5	7985.20
7	7297.90
10	6376.30
12	5827.50

The amount P to invest when $A = \$25,000$ at $t = 5, 7, 10,$ and 12 is listed in the following table:

t	$P(\$)$
5	17,617
7	15,316
10	12,415
12	10,793

As more time passes or increases, the amount of money that must be invested decreases.

SECTION 4.5 EXERCISES

1. The doses are 6 grains, 8 grains, and 10 grains.   **2.** The doses are 16 cc, 15 cc, 14 cc, 13 cc, and 12 cc.   **3.** The number awarded at each stage is three prizes, five prizes, seven prizes, and nine prizes.   **4.** The number awarded each day are 2 prizes, 4 prizes, 6 prizes, 8 prizes, and 10 prizes.   **5.** The lowest grade was 81 points and the highest grade was 88 points.   **6.** The ratings are $-4, -2, 0, 2$, and 4.

7. Each side measures $9\frac{3}{4}$ inches.   **8.** The sides measure 2.5 cm, 5 cm, and 4 cm.   **9.** The sides measure 6 feet, 6 feet, and 4 feet.

10. The sides measure $15\frac{1}{5}$ meters, $15\frac{1}{5}$ meters, and $7\frac{3}{5}$ meters.   **11.** The dimensions are 85 yards by 115 yards.

12. Each side measures $\frac{1}{4}$ inch.   **13.** The dimensions are 95 cm by 52.25 cm.   **14.** The dimensions are 8 inches by 18 inches.

15. The dimensions should be 8 feet by 40 feet; it will cover 320 square feet of yard.   **16.** Each side would be 24 feet; yes, 576 ft^2 of area is greater than 320 ft^2 of area.   **17.** The measure of the angle behind the fence is 42° and the other two angles are 90° and 48°. **18.** The supplement of the angle is 15° and the other angles measure 90° and 75°.   **19.** The amount borrowed was $4000; the interest was $500.   **20.** The amount borrowed was $1200; the interest was $108.   **21.** You must invest about $4587.16.   **22.** About $11,160.71 should be invested.   **23.** He should invest about $455,000.   **24.** About $7421.15 should be invested.   **25.** $9500 was invested at 8%; $5500 was invested at 6.5%.   **26.** About $7494.74 was invested at 9.2%; about $7505.26 was invested at 3.5%.

27. $A = 1.07P$; $P = \dfrac{A}{1.07}$; about $1261.68 should be invested to have $1350; about $2336.45 should be invested to have $2500.

28. $P = 1.11L$; $L = \dfrac{P}{1.11}$; you will receive $1200 as a loan if the payback is $1332; you will receive about $720.72 as a loan if the payback is $800.   **29.** The original price was $85.   **30.** The original price was about $129.99; the amount of the reduction was about $19.50. **31.** The original cost was about $12.47.   **32.** The dealer paid $200.   **33.** The markup percentage was 80%.   **34.** The markup percentage was 40%.   **35.** The regular price is $260.   **36.** The regular price is $73.75.   **37.** The SRP should be about $19,230.77. **38.** The SRP should be about $10,582.35.   **39.** The original price is about $59.96 if the sale price is $53.96; the original price is about $109.94 if the sale price is $98.95.   **40.** The subtotal is about $25 when the total cost is $27.19; the subtotal is $132 when the total cost is $143.55.   **41.** His hourly wage before the increase was about $13.45.   **42.** Her salary before the increase was about $30,660.38. **43.** The bill before the gratuity was added was about $124.56.   **44.** The bill before the gratuity was added was about $228.65. The gratuity was $34.30.   **45.** Each angle labeled x is 72 degrees.   **46.** Each angle labeled x is 72 degrees and each angle labeled y is 54 degrees.

4.5 CALCULATOR EXERCISES

1. $\{122, 131, 140, 149\}$ **2.** $-17.77777778; -3.888888889; 10; 23.88888889; 37.77777778$ **3.** $\{75, 60, 45, 30\}$ **4.** $\{60, 90, 120, 150\}$

SECTION 4.6 EXERCISES

1. -9 and 9 **2.** -9 and 9 **3.** -4 and 2 **4.** 2 and 4 **5.** -3 and 1 **6.** 1 and 5 **7.** -2 and 0 **8.** -3 and -1 **9.** -47 and 5 **10.** -10 and 48 **11.** -41 and 3 **12.** -19 and 17 **13.** -3.92 and 8.26 **14.** 21.931 and 23.729 **15.** -1.133 and 0.333 **16.** -7.4375 and 5.6875 **17.** no solution   **18.** -6 **19.** 32 **20.** no solution   **21.** -8.677 and -1.333 **22.** -0.5 and 5.5 **23.** -138 and 138 **24.** -2400 and 2400 **25.** no solution   **26.** -0.009 and 0.009 **27.** -41.67 and 41.67

28. no solution   **29.** $-14\frac{5}{9}$ and $14\frac{5}{9}$ **30.** $-\frac{11}{12}$ and $\frac{11}{12}$ **31.** -864 and -292 **32.** -774 and 168 **33.** -1221 and 2663

34. 527 and 775 **35.** 5 and 53 **36.** $-\frac{7}{2}$ and $\frac{43}{2}$ **37.** no solution   **38.** no solution   **39.** -42 and 70 **40.** $-\frac{73}{2}$ and $\frac{67}{2}$

41. -6 and 2 **42.** $-\frac{33}{2}$ and $\frac{1}{2}$ **43.** -16 and -8 **44.** $\frac{3}{2}$ and $\frac{13}{2}$ **45.** The maximum depth of Lake Superior is either 513 feet or 1333 feet; the maximum depth is 1333 feet.   **46.** The maximum depth of the Caspian Sea is either 3363 feet or 7267 feet; the Caspian Sea is 3363 feet deep.   **47.** The minimum and maximum heights the clothes will fit are 5 feet, 3 inches, and 6 feet, 3 inches.   **48.** The limits are 5 feet, 9 inches, and 6 feet, 5 inches.   **49.** The minimum and maximum percentages are 39% and 45%.   **50.** The minimum and maximum percentages are 43.5% and 52.5%.   **51.** The range is from 130 pounds to 134 pounds.   **52.** The range is from 0.95 pound to 1.05 pounds.   **53.** The lengths range from $5\frac{1}{4}$ inches to $5\frac{3}{4}$ inches.   **54.** The lengths range from 25.375 inches 25.625 inches.

55. The possible values for the extreme score are 49 and 93.   **56.** The range of scores is from 38 points to 94 points.   **57.** The pairs of ages are 35 and 47 or 47 and 59.   **58.** The monthly sales for the other brand are either 9,600,000 bottles or 14,600,000 bottles.

4.6 CALCULATOR EXERCISES

1. 0.5 **2.** no solution   **3.** 4 and -13 **4.** 1.5 **5.** -7.5 and 7.5

CHAPTER 4 SECTION-BY-SECTION REVIEW

Reflections

1–8. Answers will vary.

Exercises

1. nonlinear **2.** linear **3.** linear **4.** nonlinear **5.** linear **6.** nonlinear **7.** $\frac{3}{4}(x+7)-5=\frac{1}{3}(x+12)$; $x=9$

8. -6 **9.** 6 **10.** 3 **11.** noninteger between -3 and -2 **12.** all real numbers **13.** no solution

14. $2.3(x-5.6)+4=3x-11.3$; $x=3.5$ **15.** 2 **16.** -2 **17.** all real numbers **18.** no solution **19.** 2.4

20. The two offers are equivalent at 10 hours. **21.** The fourth-week donation should be $2165.

22. The sides measure 7.75 feet, 7.75 feet, and 10.75 feet. **23.** 26 **24.** $\frac{32}{39}$ **25.** -2 **26.** 2.933 **27.** -68 **28.** -38.5

29. 105 **30.** $\frac{16}{25}$ **31.** 12 **32.** -2.98 **33.** 3.5 **34.** There were 58 passes given. **35.** There were 280 graduates.

36. They must sell 5334 books. **37.** The total proceeds were $42,630. **38.** $1000=0.0005c$; $c=2,000,000$; the current achieved was 2 million amperes per square centimeter. **39.** He needs approximately 11.43 quarts of solution.

40. -14 **41.** 2 **42.** -3.4 **43.** $-\frac{11}{6}$ **44.** all real numbers **45.** no solution **46.** You can drive 909 miles.

47. The annual depreciation is about $17,857. **48.** The offers are the same at 15 hours.

49. Erin needs to invest $11,000 in the second fund. **50.** $h=\frac{2A}{b+B}$ **51.** $W=\frac{S-2LH}{2L+2H}$ **52.** $y=\frac{6}{5}x-\frac{22}{15}$

53. $A=6000+8000n$; $n=\frac{A-6000}{8000}$; It will last 9 years if the amount is $78,000; it will last 15 years if the amount is $126,000.

54. The lengths should be 10 inches, 12 inches, and 14 inches. **55.** The complement is $42°$ and the two other angles measure $69°$ each.
56. The dimensions are 30 feet by 90 feet. **57.** It should be invested for six years. **58.** You should invest $10,000.
59. The suit's original price should have been $262.50. **60.** The artist was paid $25. **61.** 7 **62.** 8 and 16 **63.** no solution

64. $-\frac{1}{2}$ and $\frac{15}{2}$ **65.** 1 and 13 **66.** no solution **67.** -2 and 4 **68.** The permissible limits on the part are 62.75 mm

and 62.83 mm. **69.** The limits on the percentage of voters are 45% and 53%.

CHAPTER 4 CHAPTER REVIEW

1. all real numbers **2.** no solution **3.** -4 **4.** 3 **5.** 3.1 **6.** 2 **7.** noninteger between -4 and -3 **8.** -3
9. all real numbers **10.** -2 **11.** no solution **12.** linear **13.** nonlinear **14.** nonlinear **15.** linear **16.** linear

17. nonlinear **18.** -348 **19.** $\frac{3}{10}$ **20.** 32 **21.** -72.3 **22.** -14 **23.** -14.59 **24.** 7 **25.** -444 **26.** $-\frac{2}{3}$

27. 7.49 **28.** 8.7 **29.** 6.4 **30.** $-\frac{1}{8}$ **31.** $\frac{11}{7}$ **32.** 6 **33.** 1.5 **34.** all real numbers **35.** no solution **36.** -3

37. 0 and 8 **38.** no solution **39.** $-\frac{14}{5}$ and 2 **40.** 4 and 10 **41.** -5 and 3 **42.** $y=\frac{2}{3}x+12$ **43.** $y=-\frac{1}{6}x+\frac{1}{4}$

44. $h=\frac{S-2\pi r^2}{2\pi r}$ **45.** $P=\frac{I}{RT}$ **46.** Two possible heights are 517 ft and 1023 ft; 1023 ft is the correct height. **47.** The limits are
67 and 77 points. **48.** The subtotal was about $65.58; the total bill was about $70.99. **49.** He expects to receive $14,125.
50. She worked 1250 hours. **51.** An employee needs 120 houses. **52.** The annual depreciation is $11,750. **53.** She needs an
additional $414. **54.** The pieces should be 5 inches, 6 inches, 7 inches, and 8 inches. **55.** The other angles measure $59°$ each.
The measure of the supplement is $121°$. **56.** The original price was about $139.94. **57.** The simple-interest rate should be 6.5%.
58. You should invest $12,500. **59.** The dimensions are 16.8 feet by 25.2 feet. **60.** The price before tax was about $299; sales tax was

about $26.16. **61.** $A=40+55.5h$; $h=\frac{A-40}{55.5}$; a job that costs $178.75 lasts 2.5 hours; a job that costs $95.50 lasts 1 hour.

62. The angle made with the width of the rug is $55°$. **63.** The mechanic needs $3\frac{1}{3}$ gallons of the solution.

CHAPTER 4 TEST

1. linear **2.** nonlinear **3.** nonlinear **4.** linear **5.** no solution **6.** -3 **7.** -2.079 **8.** all real numbers

9. $1\frac{3}{4}$ **10.** -10 and 4 **11.** no solution **12.** The pieces should be cut into sections measuring 14 inches, 15 inches, and 16 inches.

13. His monthly payments will be $135. **14.** The price before it went on sale was about $239.93. **15.** $W=\frac{P-2L}{2}$; the

width is 7.6 inches. **16.** The other angles measure $69°$ each. The measure of the supplement is $111°$. **17.** 1500 liters of 60% apple
juice must be added. **18.** The two payment plans will be equal for $7500 of sales. **19.** The two plans will cost the same if the job lasts
10 hours. **20.** The limits on the true percent are 49% and 55%. **21.** Answers will vary.

Chapter 5

SECTION 5.1 EXERCISES

1. linear; $\sqrt{5}x + 7y = 35$ **2.** linear; $\sqrt{3}x + 2y = 6$ **3.** nonlinear   **4.** nonlinear   **5.** nonlinear   **6.** nonlinear
7. linear; $2x = 5$ **8.** linear; $5y = 14$

9. Answers will vary.
Possible answer:

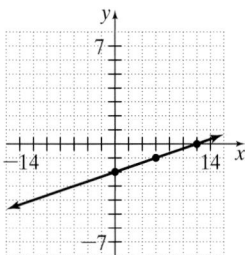

$(0, -2), (12, 0)$, and $(6, -1)$
are three possible solutions.

10. Answers will vary.
Possible answer:

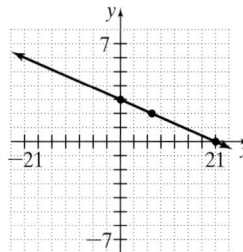

$(0, 3), (21, 0)$, and $(7, 2)$
are three possible solutions.

11. Answers will vary.
Possible answer:

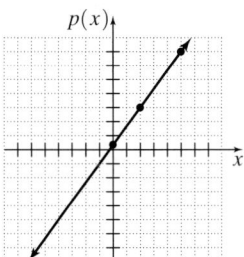

$\left(0, \dfrac{1}{3}\right), (2, 3)$, and $(5, 7)$
are three possible solutions.

12. Answers will vary.
Possible answer:

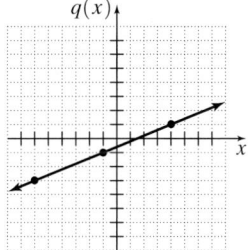

$(-6, -3), (-1, -1)$, and $(4, 1)$
are three possible solutions.

13. Answers will vary.
Possible answer:

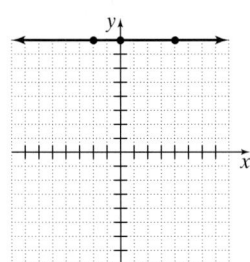

$(0, 8), (4, 8)$, and $(-2, 8)$
are three possible solutions.

14. Answers will vary.
Possible answer:

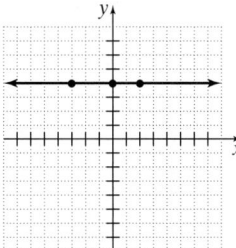

$(0, 4), (2, 4)$, and $(-3, 4)$
are three possible solutions.

15. Answers will vary.
Possible answer:

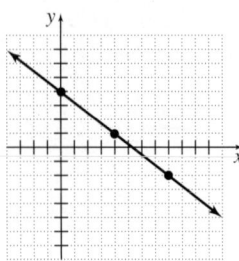

$(0, 4), (4, 1)$, and $(8, -2)$
are three possible solutions.

16. Answers will vary.
Possible answer:

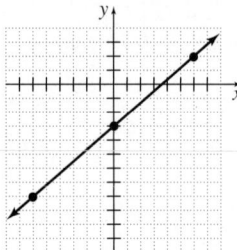

$(-6, -8), (0, -3)$, and $(6, 2)$
are three possible solutions.

17. Answers will vary.
Possible answer:

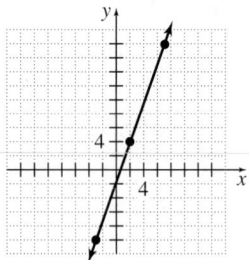

$(-3, -10), (2, 4)$, and $(7, 18)$
are three possible solutions.

18. Answers will vary.
Possible answer:

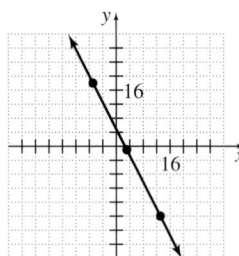

$(-7, 18), (3, -1)$, and $(13, -20)$
are three possible solutions.

19. Answers will vary.
Possible answer:

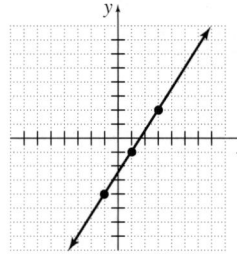

$(-1, -4), (1, -1)$, and $(3, 2)$
are three possible solutions.

20. Answers will vary.
Possible answer:

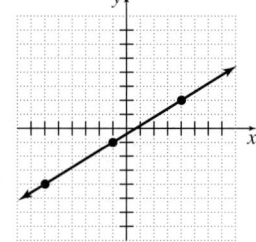

$(-6, -4), (-1, -1)$, and $(4, 2)$
are three possible solutions.

21. Answers will vary.
Possible answer:

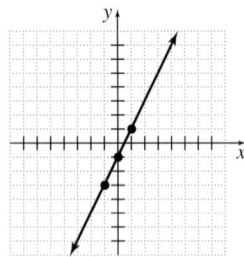

$(-1, -3), (0, -1)$, and $(1, 1)$
are three possible solutions.

22. Answers will vary.
Possible answer:

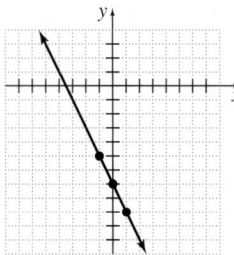

$(-1, -5), (0, -7)$, and $(1, -9)$
are three possible solutions.

23. Answers will vary.
Possible answer:

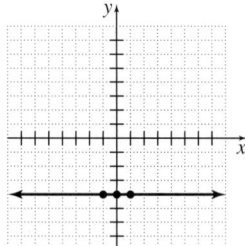

$(-1, -4), (0, -4)$, and $(1, -4)$
are three possible solutions.

24. Answers will vary.
Possible answer:

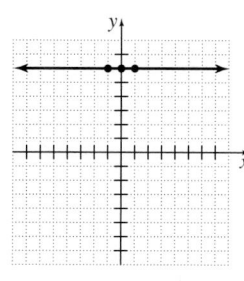

$(-1, 6)$, $(0, 6)$, and $(1, 6)$
are three possible solutions.

25. x-intercept: $(-2, 0)$
y-intercept: $(0, 4)$

26. x-intercept: $(-5, 0)$
y-intercept: $(0, -5)$

27. x-intercept: $(0, 0)$
y-intercept: $(0, 0)$

28. x-intercept: $(0, 0)$
y-intercept: $(0, 0)$

29. x-intercept: $(3, 0)$
y-intercept: none

30. x-intercept: none
y-intercept: $(0, -4)$

31. x-intercept: $(4, 0)$

y-intercept: $\left(0, \dfrac{12}{5}\right)$

32. x-intercept: $(9, 0)$

y-intercept: $(0, 7)$

33. x-intercept: $\left(\dfrac{7}{2}, 0\right)$

y-intercept: $(0, -2)$

34. x-intercept: $(31, 0)$

y-intercept: $(0, -31)$

35. x-intercept: $\left(-\dfrac{27}{2}, 0\right)$

y-intercept: $(0, -3)$

36. x-intercept: $(-3, 0)$

y-intercept: $\left(0, -\dfrac{9}{2}\right)$

37. x-intercept: $(6, 0)$
y-intercept: $(0, 4)$

38. x-intercept: $(-9, 0)$
y-intercept: $(0, -15)$

39. x-intercept: $(0, 0)$
y-intercept: $(0, 0)$

40. x-intercept: $(0, 0)$
y-intercept: $(0, 0)$

41. x-intercept: $(10, 0)$
y-intercept: none

42. x-intercept: $(5, 0)$
y-intercept: none

43. x-intercept: none
y-intercept: $(0, 11)$

44. x-intercept: none
y-intercept: $(0, -5)$

45. $(0, -24)$ **46.** $(0, 2)$ **47.** $(0, -15)$ **48.** $(0, -16)$ **49.** $(0, 0)$ **50.** $(0, 0)$ **51.** $(0, -2)$ **52.** $(0, 3)$ **53.** $(0, 0)$
54. $(0, 0)$ **55.** $(0, 9)$ **56.** $(0, -8)$ **57.** $(0, 0)$ **58.** $(0, 0)$

59.

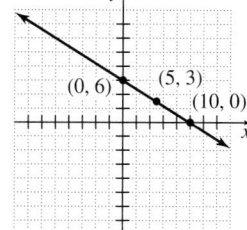

60.

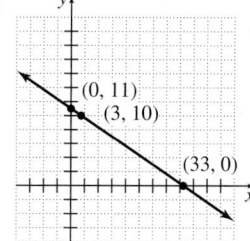

61.

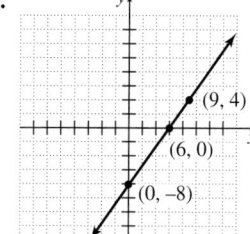

62.

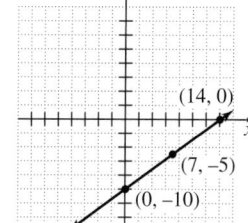

63.

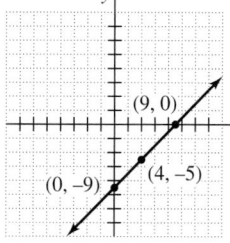

64.

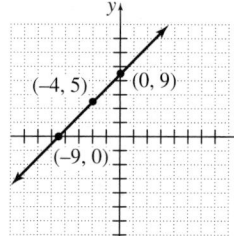

65.

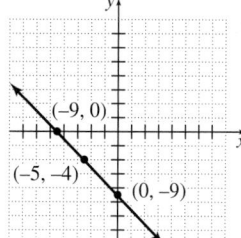

66.

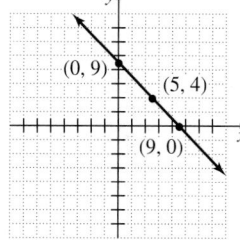

67.

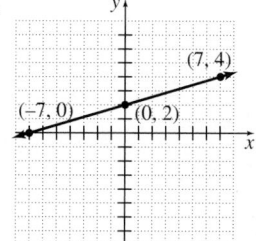

68.

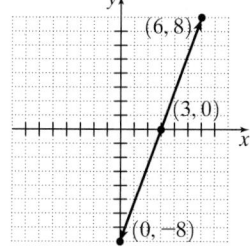

69. (a) $(1, 0), (2, 5), (4, 15)$

(b)

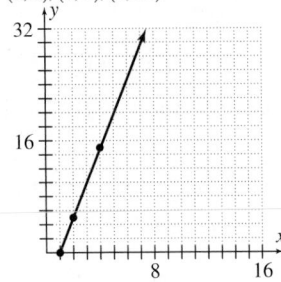

(c) A crew of five people would pack 20 boxes per minute.
(d) Answers will vary. Possible answer:
There is probably a maximum number of boxes that can be packed by a large number of people, given the size of the packing plant.

71. (a) $(2, 6), (3.5, 10.5), (10.5, 31.5)$

(b)

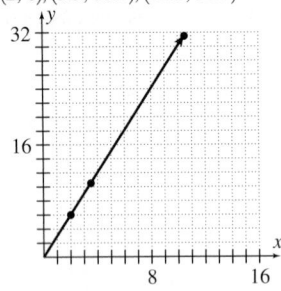

(c) The border would be 12 inches.
(d) Yes, all equilateral triangles have three sides of equal measure.

73. (a) The cost would be $325.
(b) The cost would be $362.50.
(c) yes   **(d)**

x	$D(x)$
0	250
5	268.75
10	287.50
15	306.25
20	325
25	343.75
30	362.50
35	381.25
40	400
45	418.75
50	437.50

75. Let x = number of customers, $P(x)$ = profit,
$P(x) = 14.25x - 1000$ Intercepts: $(0, -1000), (70.175, 0)$
If no one goes to the water park, then the park loses $1000. If 71 people go to the water park, then the park breaks even or makes a tiny profit of $11.75.

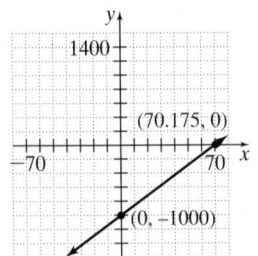

70. (a) $(0, 100), (6, 85), (10, 75)$

(b)

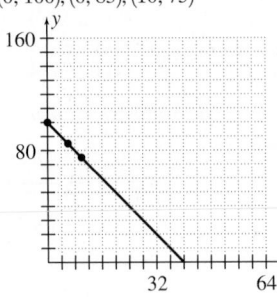

(c) The score for missing 12 questions would be 70.
(d) domain: 0 to 40

72. (a) $(15, 50), (25, 70), (10, 40)$

(b)

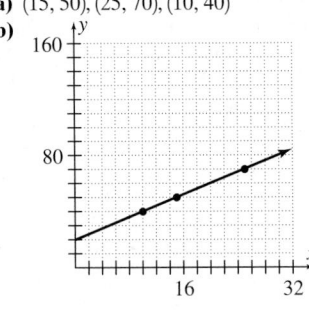

(c) The perimeter would be 60 cm.
(d) Yes, it is a linear equation.

74. (a) She will earn $600.
(b) She will earn $425.
(c) yes   **(d)**

x	$P(x)$
0	300
2	350
4	400
6	450
8	500
10	550
12	600
14	650
16	700
18	750
20	800

76. Let x = number of games, $P(x)$ = profit,
$P(x) = 1.35x - 200$ Intercepts: $(0, -200), (148.15, 0)$
If there are no games bowled, then there is a $200 loss. If 148 games are bowled, then the bowling alley just about breaks even.

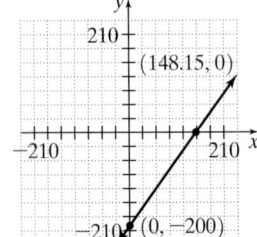

77. $d = 1944 - 425.7t$ Intercepts: $(0, 1944), (4.57, 0)$
At the start of the trip, $t = 0$, the distance between Atlanta and Los Angeles is 1944 miles. At the end of the trip, $d = 0$, the time it took to reach the destination was 4 hours and 0.57 of an hour (equal to 34 minutes). From the graph, it takes just a little over 2 hours to fly 1000 miles from Atlanta.

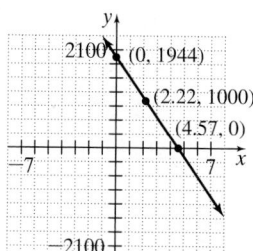

78. $d = 3471 - 389.3t$ Intercepts: $(0, 3471) (8.9, 0)$
At the start of the trip, $t = 0$, the distance between New York City and London is 3471 miles. At the end of the trip, $d = 0$, the time it took to reach the destination was 8 hours and 0.9 of an hour (equal to 54 minutes). From the graph, it takes almost 4 hours to fly 2000 miles from New York City.

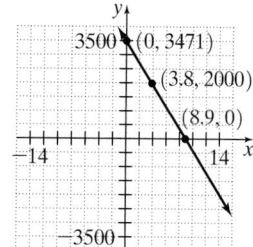

5.1 CALCULATOR EXERCISES

1. $x = \dfrac{1}{3}$

2. $x = 8$

3. $x = -3$

SECTION 5.2 EXERCISES

1. $m = -\dfrac{3}{2}$; the graph is a function that is decreasing.   **2.** $m =$ undefined; the graph is not a function.   **3.** $m = 0$; the graph is a function that is constant.   **4.** $m = \dfrac{7}{2}$; the graph is a function that is increasing.   **5.** $m = \dfrac{1}{6}$; the graph is a function that is increasing.

6. $m = 0$; the graph is a function that is constant.   **7.** $m =$ undefined; the graph is not a function.   **8.** $m = -\dfrac{5}{2}$; the graph is a function that is decreasing.   **9.** $\dfrac{2}{3}$ **10.** $-\dfrac{9}{5}$ **11.** 0   **12.** undefined   **13.** undefined   **14.** $\dfrac{15}{7}$ **15.** -2 **16.** 0   **17.** $-\dfrac{4}{5}$

18. $\dfrac{5}{3}$ **19.** -6 **20.** 0.4   **21.** $\dfrac{19}{12}$ **22.** $-\dfrac{63}{44}$ **23.** The slope is 21 and the y-intercept is $(0, 15)$.   **24.** The slope is -19 and the y-intercept is $(0, 28)$.   **25.** The slope is 5.95 and the y-intercept is $(0, -2.01)$.   **26.** The slope is -3.6 and the y-intercept is $(0, 14.8)$.

27. The slope is -1255 and the y-intercept is $(0, 85,600)$.   **28.** The slope is 45 and the y-intercept is $(0, 1250)$.   **29.** The slope is 4 and the y-intercept is $(0, -16)$.   **30.** The slope is -8 and the y-intercept is $(0, 13)$.   **31.** The slope is 0 and the y-intercept is $(0, -2)$.

32. The slope is 0 and the y-intercept is $(0, -22)$.   **33.** The slope is $\dfrac{5}{2}$ and the y-intercept is $\left(0, -\dfrac{7}{2}\right)$.   **34.** The slope is $\dfrac{4}{3}$ and the y-intercept is $\left(0, \dfrac{5}{3}\right)$.   **35.** The slope is undefined and there is no y-intercept.   **36.** The slope is undefined and there is no y-intercept.

37.

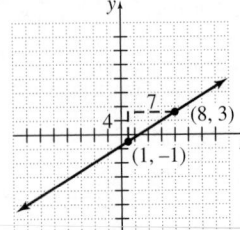

38.

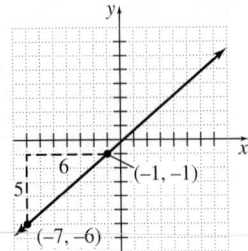

39.

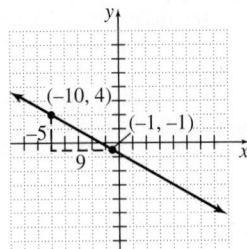

40.

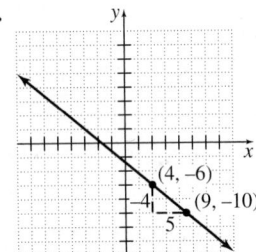

41.

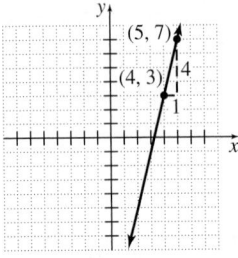

42.

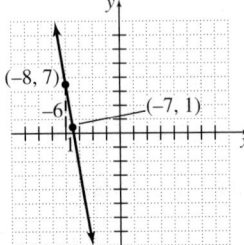

43.

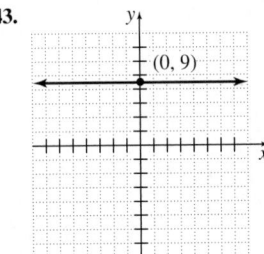

44.

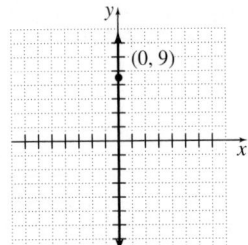

45.

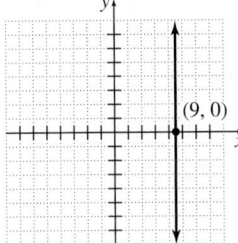

46.

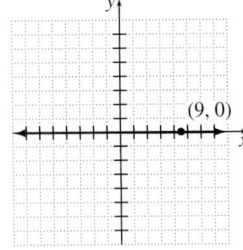

47.

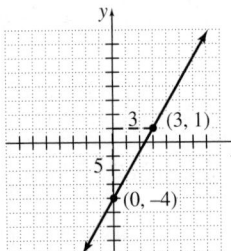

48.

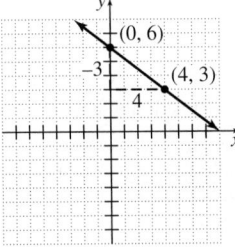

49.

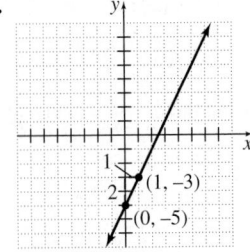

50.

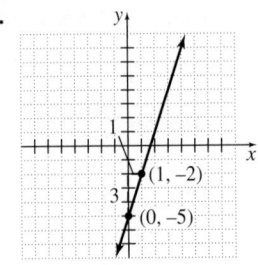

51.

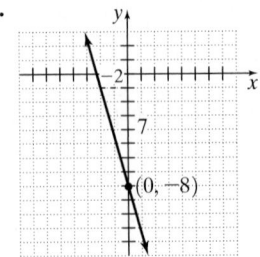

52.

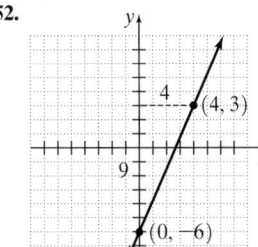

53.

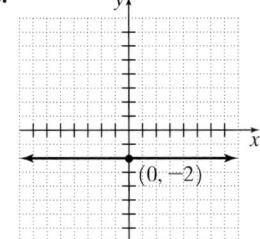

54.

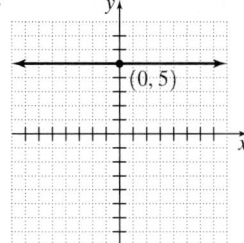

55.

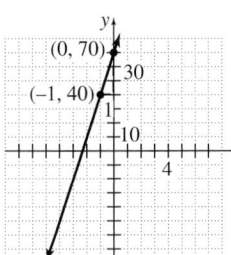

56.

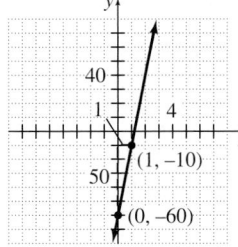

57.

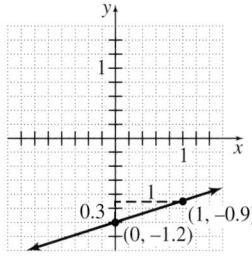

58.

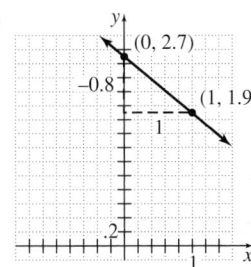

59. The grade of the advertised terrain is 35%.
60. The grade of the terrain is 25%.
61. The pitch of the roof is 27.5%.
62. The pitch of the roof is approximately 42%.
63. Graph b
64. Graph c
65. Graph a
66. Graph b
67. $3960 per year
68. $3350 per year
69. $7.237 billion per year
70. An increase of 0.27 year per year

71. (a)

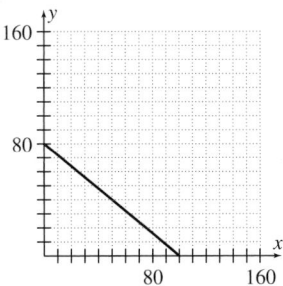

(b) At $x = 10$, $y = 72$:
   $x = 20$, $y = 64$
   $x = 40$, $y = 48$
   $x = 64$, $y = 28.8$
(c) Demand decreases as price increases.
(d) At a price of $100 or more, demand is 0.

72. (a)

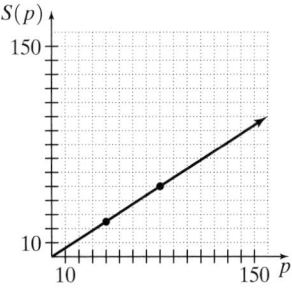

(b) At $x = 10$, $y = 6.25$:
   $x = 20$, $y = 12.5$
   $x = 40$, $y = 25$
   $x = 64$, $y = 40$
(c) As price increases, supply increases.
(d) No; answers will vary.

5.2 CALCULATOR EXERCISES

Equation	$y = mx + b$		Conclusions	
	m	b	Graph's inclination, ↗ or ↘	Graph's y-intercept
$y = 3x + 6$	3	6	↗	6
$y = -2x + 7$	-2	7	↘	7
$y = -x - 3$	-1	-3	↘	-3
$y = 4x - 1$	4	-1	↗	-1
$5x - 3y = 9$	$\dfrac{5}{3}$	-3	↗	-3
$4x + 5y = 10$	$-\dfrac{4}{5}$	2	↘	2
$y = \dfrac{7}{8}x - \dfrac{3}{4}$	$\dfrac{7}{8}$	$-\dfrac{3}{4}$	↗	$-\dfrac{3}{4}$
$y = -1.7x + 3.2$	-1.7	3.2	↘	3.2

SECTION 5.3 EXERCISES

1. intersecting and perpendicular   **2.** intersecting and perpendicular   **3.** parallel   **4.** parallel   **5.** only intersecting
6. only intersecting   **7.** coinciding   **8.** coinciding   **9.** only intersecting   **10.** only intersecting   **11.** parallel   **12.** coinciding
13. intersecting and perpendicular   **14.** intersecting and perpendicular   **15.** parallel   **16.** only intersecting   **17.** parallel
18. parallel   **19.** coinciding   **20.** parallel   **21.** coinciding   **22.** only intersecting   **23.** only intersecting   **24.** parallel

25. only intersecting   **26.** parallel   **27. (a)** $y = 0.25x + 3.5$ **(b)** $y = 0.25x$ **(c)** no break-even point   **(d)** $y = \dfrac{1}{3}x$

(e) At 42 candy bars, Brook will break even.   **(f)** At 10 candy bars, Brook will break even.
(g) At about 5.45 candy bars, Brook will break even. At 6 candy bars, Brook will start making a profit.   **(h)** Answers will vary.
28. (a) $y = 285 + 200x$ **(b)** $y = 200x$ **(c)** no break-even point   **(d)** $y = 300x$ **(e)** At 2.85 radios, Joe will break even. At 3 radios, Joe will make a profit.   **(f)** At 3.5 radios, Joe will break even. At 4 radios, Joe will make a profit.   **(g)** Answers will vary.
29. $y_1 = 35 + 0.25x, m = 0.25, b = 35;$
$y_2 = 60; m = 0, b = 60;$
their graphs will intersect because the slopes are not equal.;

30. $y_1 = 65, m = 0, b = 65;$
$y_2 = 15 + 10x, m = 10, b = 15;$
their graphs will intersect because the slopes are not equal.;

The intersection is (100, 60). At 100 miles, the prices are equal at \$60 per day.

The intersection is (5, 65). At 5 hours, the costs will be equal at \$65.

31. $y_1 = 10x; m = 10, b = 0; y_2 = 15(x - 4), m = 15, b = -60; y_3 = 150; m - 0, b = 150; y_1$ and y_2 intersect at (12, 120). At 12 seconds, Speedie will catch Archie at a distance of 120 feet, or before the end zone 150 feet away.
32. $y_1 = 18x; y_2 = 21.5(x - 0.75);$ at a distance of about 82.9 miles, Tom will overtake Victor.
33. Yes. The intersection point is (0.69, 1031.7). In mid-1990, the number of returns filed by each type of corporation was the same.
34. Yes. The intersection point is $(-10, 17.1)$. However, since x is the number of employees, it cannot be negative, so the graphs do not intersect within the domains of the functions.
35. $y_1 = 50 + 2x, m = 2, b = 50; y_2 = 75, m = 0, b = 75; x = 8, m =$ undefined, $b =$ none; yes, y_1 intersects with $x = 8$ and y_2 intersects with $x = 8$ because each has a different slope. $x = 8$ is perpendicular to y_2 because one is vertical and the other is horizontal. The intersection of y_1 and $x = 8$ is (8, 66). At 8 years, she will receive \$66. The intersection of y_2 and $x = 8$ is (8, 75). At 8 years, she will receive \$75. The intersection of y_1 and y_2 is (12.5, 75). At $12\frac{1}{2}$ years she will receive the same either way, \$75.

36. $y_1 = 0.26x$, $m = 0.26$, $b = 0$; $y_2 = 8 + 0.02x$, $m = 0.02$, $b = 8$; $x = 30$, $m =$ undefined, $b =$ none; yes, $x = 30$ will intersect each y_1 and y_2 because they have different slopes. No, $x = 30$ is not perpendicular to either y_1 or y_2. y_1 and y_2 intersect at $(33\frac{1}{3}, 8\frac{2}{3})$. At $33\frac{1}{3}$ words, the cost is the same either way, \$8.67. y_1 and $x = 30$ intersect at $(30, 7.8)$. At 30 words, the cost is \$7.80. y_2 and $x = 30$ intersect at $(30, 8.6)$. At 30 words, the cost is \$8.60.

5.3 CALCULATOR EXERCISES

1. $(-10, 10, 1, -10, 10, 1)$ **2.** $(-47, 47, 10, -31, 31, 10)$ **3.** Yes   **4.** Yes   **5.** Intersecting; $(6, 2250)$ **6.** Parallel
7. Intersecting; $(14.4, 1800)$

SECTION 5.4 EXERCISES

1. $y = \frac{3}{2}x - 1$ **2.** $y = \frac{1}{2}x + 1$ **3.** $x = -5$ **4.** $x = 4\frac{1}{2}$ **5.** $y = -3x$ **6.** $y = 3\frac{1}{3}$ **7.** $y = -\frac{3}{2}$ **8.** $y = -\frac{1}{5}x - 3$

9. $y = -\frac{2}{5}x + 4$ **10.** $y = -\frac{1}{7}x - 9$ **11.** $y = \frac{5}{9}x$ **12.** $y = \frac{1}{7}x - 1$ **13.** $y = 4x - \frac{3}{4}$ **14.** $y = 11x + \frac{1}{2}$

15. $y = -4.1x + 0.5$ **16.** $y = -6.2x - 2.2$ **17.** $y = -33$ **18.** $y = -4x$ **19.** $y = \frac{2}{3}x - 5$ **20.** $y = -2x + 8$

21. $y = -3x + 4$ **22.** $y = \frac{4}{3}x + 5$ **23.** $y = -1.7x + 3.6$ **24.** $y = 1.4x - 3.3$ **25.** $y = -\frac{3}{2}x - \frac{1}{2}$ **26.** $y = 6$

27. $x = -1$ **28.** $y = -5x + 11$ **29.** $y = 2x + 3$ **30.** $y = \frac{2}{7}x + \frac{20}{7}$ **31.** $y = 2$ **32.** $x = 2$ **33.** $y = \frac{5}{14}x + \frac{51}{14}$

34. $y = \frac{14}{27}x - \frac{8}{27}$ **35.** $y = -\frac{3}{2}x$ **36.** $y = -\frac{9}{2}x + \frac{67}{4}$ **37.** $y = -\frac{42}{13}x + \frac{21}{13}$ **38.** $y = \frac{35}{11}x + \frac{129}{11}$ **39.** $y = x + 0.4$

40. $y = -19x + 25$ **41.** $y = \frac{3}{8}x + 4$ **42.** $y = -\frac{5}{3}x + 12$ **43.** $y = -\frac{1}{2}x + 2$ **44.** $y = 2x - 3$ **45.** $y = 3x - 5$

46. $y = -3x + 2$ **47.** $y = 3x - \frac{13}{6}$ **48.** $y = 2x - \frac{3}{2}$ **49.** $y = -1.2x + 2.8$ **50.** $y = -0.8x + 4.2$ **51.** $y = -\frac{1}{2}x$

52. $y = \frac{1}{3}x$ **53.** $y = -\frac{1}{3}x + 7$ **54.** $y = 0.17x + 2.4$ **55.** $y = -\frac{1}{5}x - 27$ **56.** $y = -\frac{1}{2}x + 3$ **57.** $y = \frac{3}{2}x$ **58.** $y = -\frac{4}{5}x$

59. $y = \frac{2}{3}x - 2$ **60.** $y = -\frac{4}{5}x - \frac{8}{5}$

61. (a) $m = -6239$ **(b)** $y = -6239x + 77,103$ **(c)** 58,386; yes   **(d)** In the year 2005 the number of cases will be about 8,474.   **(e)** The cases will be close to zero in 2007. It is hard to predict. The trend might not continue.   **62. (a)** $m = -389$ **(b)** $y = -389x + 13,802$
(c) 12,635. Fairly close.   **(d)** In the year 2005, the number of cases will be about 9,523.   **(e)** Within 8 years, the number of cases dropped by about 3000, so in 24 years, in 2029, you would think you would be at 0 cases. It will probably happen sooner, though, since the decline gets larger every year.   **63.** $y = x + 273$; a Kelvin temperature of 373 corresponds to 100°C.
64. $y = x + 460$; the Rankine temperature that corresponds to 75°F is 535.   **65.** $y = 2.6x + 12.8$ **66.** $y = -3.8x + 21.6$
67. $y = -0.005x + 52.5$ **68.** $y = -0.008x + 55$; answers will vary.   **69.** $y = 1000x - 1000$; the prediction is 9000 sales.

70. $y = -\frac{81}{140}x + \frac{1515}{28}$; the prediction is about 13.6 years.   **71.** $y = 120x + 5470$; the predicted enrollment in 1996 is 6190 students; yes, this is a good estimate; the predicted enrollment in 2010 would be 7870 students.   **72.** $y = -250,000x + 5,125,000$; the predicted business is \$5,125,000 in 1990; this is a good estimate; the predicted business would be \$125,000 in 2010.

73. (a) $y = 3.78x + 57.1$ **(b)** The economic loss for 1997 was about \$121.36 billion; this estimate is \$2.34 billion lower than the actual loss, but relatively close.   **(c)** In 2005, the economic loss will be about \$151.6 billion.
74. (a) $y = 6x + 271$ **(b)** The estimated expenditure for 1998 was about \$319 trillion; this estimate is \$5 trillion lower than the actual expenditure, but relatively close. In 2005, the estimated expenditure will be about \$361 trillion.

5.4 CALCULATOR EXERCISES

1. $y = 55x + 265$ **2.** $y = 32x - 863$ **3.** $y = 2.670054945x + 6.4$ **4.** $y = 5x - 0.75$ **5.** $y = 3.5x + 38.5$ **6.** $y = -\frac{2}{3}x + \frac{26}{3}$

Reflections

1–8. Answers will vary.

CHAPTER 5 SECTION-BY-SECTION REVIEW

1. linear; $0.6x - y = -2.3$ **2.** nonlinear   **3.** linear; $4x - 2y = 2$ **4.** linear; $6y = 19$

5. $(-1, 10), (0, 8), (4, 0)$

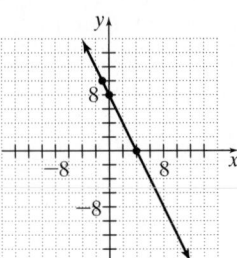

6. $(0, -7), (13, 1), (-13, -15)$

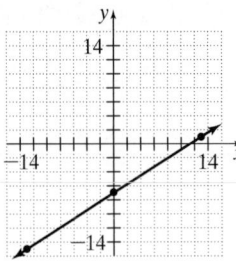

7. $(0, -9), (5, -9), (-5, -9)$

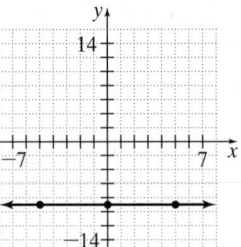

8. Answers will vary.
Possible answer:

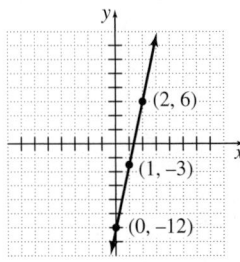

$(0, -12), (1, -3),$ and $(2, 6)$
are three possible solutions.

9. Answers will vary.
Possible answer:

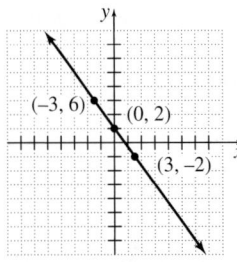

$(-3, 6), (0, 2),$ and $(3, -2)$
are three possible solutions.

10. Answers will vary.
Possible answer:

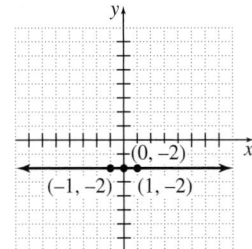

$(-1, -2), (0, -2),$ and $(1, -2)$
are three possible solutions.

11. Answers will vary.
Possible answer:

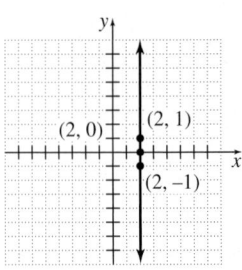

$(2, -1), (2, 0),$ and $(2, 1)$
are three possible solutions.

12. x-intercept: $(-3, 0)$; y-intercept: $(0, -2)$
13. x-intercept: $(-10, 0)$; y-intercept: $(0, 4)$
14. $(0, -3)$
15. $(3, 0)$
16. $(0, 4)$
17. $(0, 0)$
18. $(0, -10)$

19.

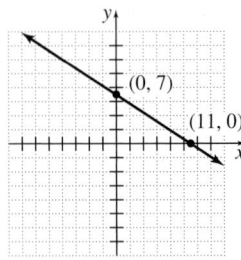

20.

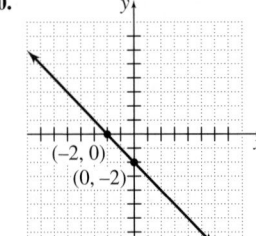

21.

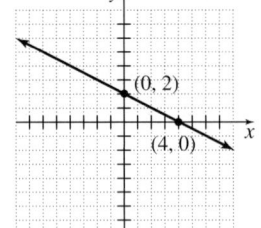

22. (a)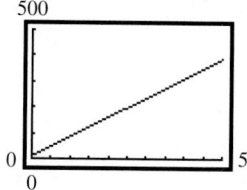

(b) She will receive $235 for a job that is 30 pages long.

23. Let x = the number of copies, $P(x)$ = profit, $P(x) = 0.02x - 25$

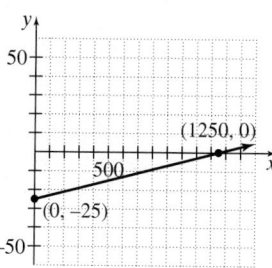

When no copies are made, there is a loss of $25. When 1250 copies are made, the center breaks even.

24. It takes about 1.9 hours to fly 1000 miles from Chicago.

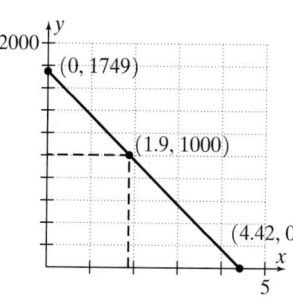

25. (a) $y = 5200 - 100x$

(b) The x-intercept is $(52, 0)$. At 52 weeks, the balance is $0; the y-intercept is $(0, 5200)$. At 0 weeks, the balance is $5200.

(c)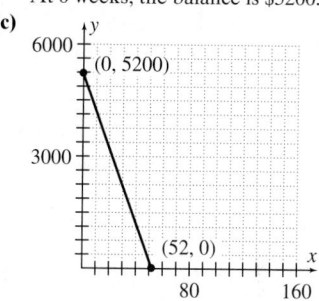

(d) There is $2000 in the account.

26. 0; yes; constant   **27.** $\frac{9}{2}$; yes; increasing   **28.** $-\frac{8}{5}$; yes; decreasing   **29.** undefined; no   **30.** $\frac{1}{2}$ **31.** $-\frac{5}{3}$ **32.** 0

33. undefined   **34.** The slope is 23 and the y-intercept is $(0, -51)$.   **35.** The slope is $-\frac{6}{5}$ and the y-intercept is $\left(0, \frac{12}{5}\right)$.

36. The slope is 0 and the y-intercept is $\left(0, -\frac{9}{2}\right)$.

37. 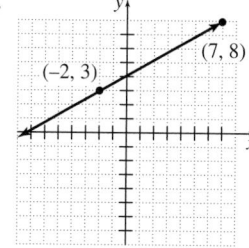   **38.** 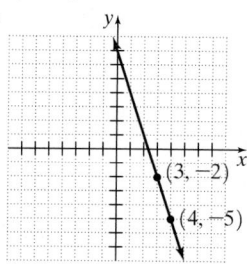   **39.** 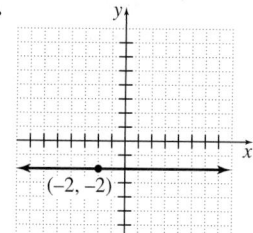   **40.**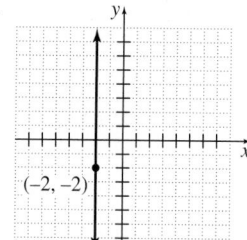

41. 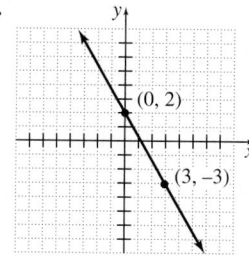   **42.** 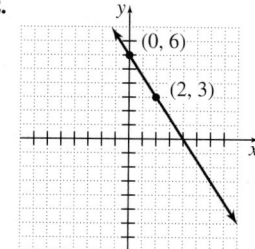   **43.** 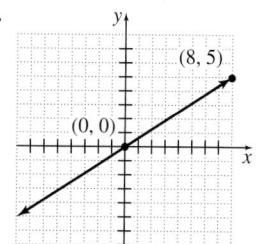   **44.**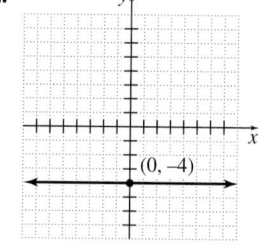

45. The grade is 2.5%.   **46.** The depreciation is $121.25 per year.   **47.** graph c   **48.** graph a   **49.** only intersecting
50. coinciding   **51.** parallel   **52.** parallel   **53.** intersecting and perpendicular   **54.** only intersecting
55. intersecting and perpendicular   **56.** intersecting and perpendicular

57. (a) $y = 35x + 85$ **(b)** $y = 35x$ **(c)** 250 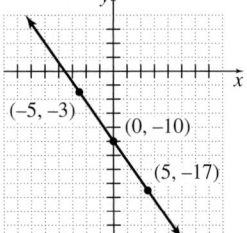 ; there is no break-even point. **(d)** $y = 60x$

(e) 250 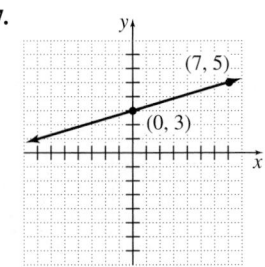 ; the break-even point is at (3.4, 204). He will start making a profit when he sells the fourth calculator. **(f)** J. R. should sell the calculators for $60 each.

58. Parallel. The slope is 15.9 for both equations. **59.** $y = -\dfrac{1}{4}x + 1$ **60.** $y = -2x + 3$ **61.** $y = \dfrac{3}{5}x - 2$ **62.** $y = -3.5$

63. $x = 2.6$ **64.** $y = -5x + 7$ **65.** $y = \dfrac{3}{11}x + \dfrac{28}{11}$ **66.** $y = \dfrac{1}{3}x + \dfrac{14}{3}$ **67.** $y = 4x - 3$ **68.** $y = -\dfrac{1}{2}x + 5$

69. (a) $y = \dfrac{5}{2}x + 10$ **(b)** The prediction is 22.5 thousands of records sold. This is not very close to the actual sales.

70. (a) slope is 1.9 **(b)** $y = 1.9x + 90.4$ **(c)** In 1998, there was about 124.6 million non-farm employees. The estimate was 1.2 million less. **(d)** In 2005, there will be about 137.9 million non-farm employees.

CHAPTER 5 CHAPTER REVIEW

1. Answers will vary. Possible answer:

2. Answers will vary. Possible answer:

3. Answers will vary. Possible answer:

4.

5.

6.

7.

8.

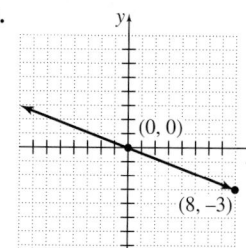

9.

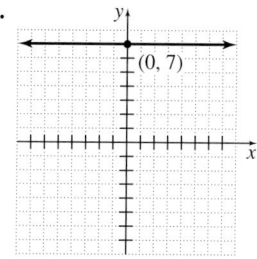

10.

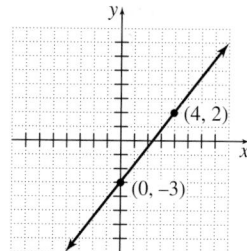

11.

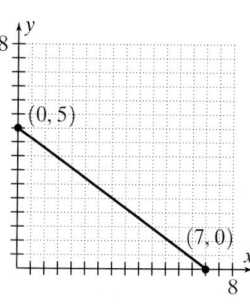

12.

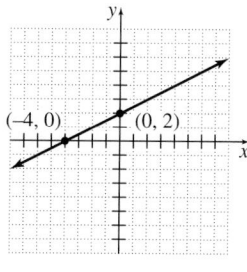

13.

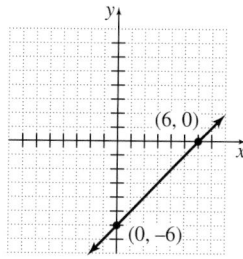

14. only intersecting   **15.** only intersecting   **16.** parallel   **17.** intersecting and perpendicular   **18.** coinciding
19. intersecting and perpendicular   **20.** parallel
21. $(-1, 8), (0, 8), (1, 8)$ **22.** $(-9, 0), (0, 8), (9, 16)$ **23.** $(-11, -17), (0, -8), (11, 1)$ **24.** linear; $x - y = 0$

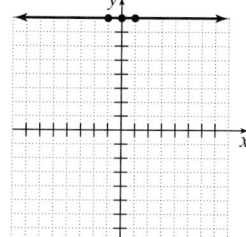

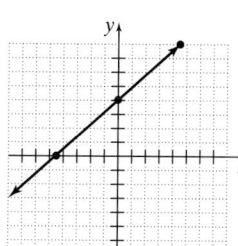

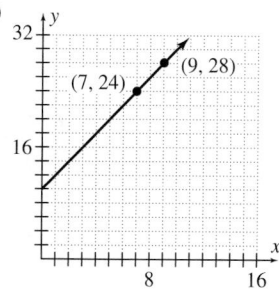

25. nonlinear   **26.** linear; $1.3x - y = 0.5$ **27.** linear; $15x - y = 0$ **28.** linear; $5x - 8y = -21$ **29.** nonlinear
30. The slope is undefined; there is no y-intercept.   **31.** The slope is $-\frac{3}{2}$; the y-intercept is $(0, -5)$.   **32.** The slope is 3; the y-intercept
is $(0, -2)$.   **33.** The slope is 0; the y-intercept is $(0, 0)$.   **34.** The slope is 13; the y-intercept is $(0, -15)$.   **35.** The slope is -5.03; the
y-intercept is $(0, 7.92)$.   **36.** 0   **37.** undefined   **38.** -3 **39.** $\dfrac{13}{6}$ **40.** $y = 4x - 18$ **41.** $y = -\dfrac{1}{4}x - \dfrac{5}{4}$
42. $y = -\dfrac{2}{3}x + 5$ **43.** $x = 4.1$ **44.** $y = -3x - 8$ **45.** $y = 8$ **46.** $y = 4x + 3.2$ **47.** $y = 3x - 2$ **48.** $y = 2x + 2$

49. (a)

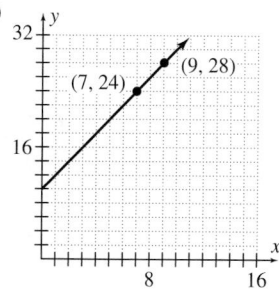

(b) See the graph in part a; Frank
would receive $26 for 8 innings.
Frank would receive $30 for
10 innings.

50. (a) $y = -5x + 180$
(b) The x-intercept is $(36, 0)$. After 36 hours,
the tank is empty; the y-intercept is $(0, 180)$.
After 0 hours, the tank has 180 gallons.
(c)

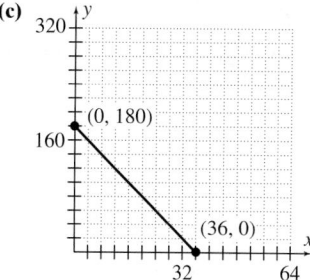

(d) After 16 hours, 100 gallons remain in the tank.
51. (a) $y = 20 + 50x$ **(b)** The trainer would earn $120.   **(c)** He would earn $95.

52. (a) $c(x) = 0.75x + 225$ **(b)** $r(x) = 2x$
(c)

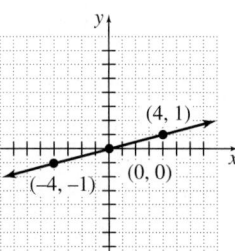

The break-even point is $(180,360)$.
Revenue equals cost at 180 bows.

(d) No, 150 is less than 180; yes, 200 is greater than 180; she must sell at least 180 bows.

53. The grade of the hill is 48%. **54.** The pitch is 16.7%. **55.** The average rate of change is $13 per year. **56.** graph a
57. (a) slope = 33 **58.** Let x = the number of pictures, $P(x)$ = profit, $P(x) = 33.50x - 35$ **59.** graph c
(b) $y = 1361 + 33x$
(c)

Year	Predicted	Actual
1994	1427	1414
1995	1460	1428
1996	1493	1479
1997	1526	1501

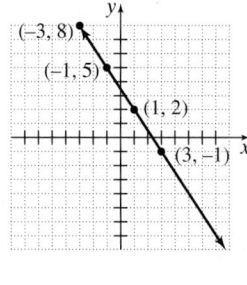

(d) In 2005, the emission level will be about 1790 million metric tons.

When no pictures are framed, Beckie lost $35.
When 1 picture is framed, Beckie will almost break even.

CHAPTER 5 TEST

1. linear **2.** linear **3.** nonlinear **4.** nonlinear **5.** Answers will vary. Possible answer: **6.**

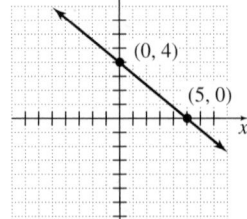

7.

8.

9.

10.

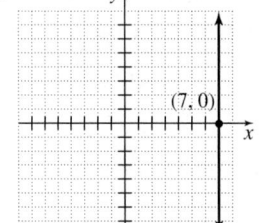

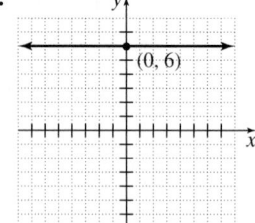

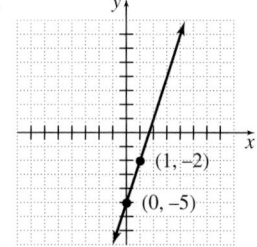

11.

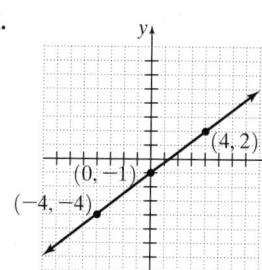

12. 0　**13.** $-\dfrac{2}{7}$　**14.** 1　**15.** undefined　**16.** only intersecting　**17.** parallel

18. coinciding　**19.** intersecting and perpendicular　**20.** The slope is $\dfrac{3}{4}$; the y-intercept is $(0, 3)$.

21. The slope is -1; the y-intercept is $(0, 3)$.　**22.** $y = -\dfrac{5}{3}x + \dfrac{13}{3}$　**23.** $y = 9x + 7$

24. $y = 6x + 8$　**25.** $y = \dfrac{1}{3}x + \dfrac{7}{3}$　**26.** $(0, -5), (2, -1), (4, 3)$; slope is 2; increasing function

27. The depreciation is about $666.67 per year.
28. The average rate of change was about 6.3 acres per year.

29. $y = 1450 - 150x$　**30.** Answers will vary.

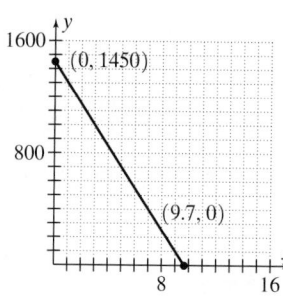

Before any payments, the balance is $1450. After
10 payments, the balance is zero. At $x = 4$, $y = 850$;
after 4 months, his remaining loan is $850.

CHAPTER 1–5 CUMULATIVE REVIEW EXERCISES

1. (a) $<$　**(b)** $=$　**(c)** $>$
2.

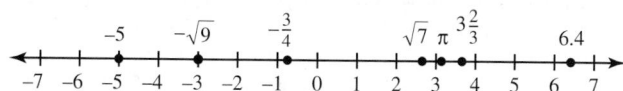

3. 1.02　**4.** -12　**5.** $\dfrac{7}{9}$　**6.** -1.581　**7.** not a real number　**8.** 3　**9.** $-\dfrac{43}{35}$　**10.** 392　**11.** 5　**12.** 15　**13.** -16.98

14. $\dfrac{5}{36}$　**15.** 1　**16.** -81　**17.** 81　**18.** $\dfrac{9}{4}$　**19.** $8a - b$　**20.** $11x + 16y$

21. (a) 6　**(b)** $x^3, -2x^2, 7x, -5x^3, 2x$　**(c)** -4　**(d)** $1, -2, 7, -5, 2, -4$　**(e)** $7x, 2x; x^3, -5x^3$
22. yes　**23.**

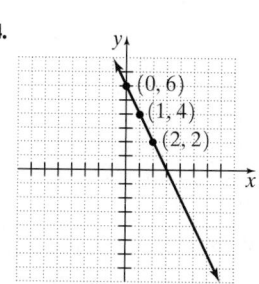

24. Domain is all real numbers; range is all $y \geq -1$.
25. Yes. It passes the vertical-line test.
26. The relation is increasing for $x > 0$.
27. The minimum is -1. There is no maximum.
28. (a) -5　**(b)** $-5 + 3h$
29. $x = -1$
30. $x = -\dfrac{7}{3}$
31. $x =$ all real numbers
32. no solution
33. $x = \dfrac{7}{5}$ or $x = -1$

34.

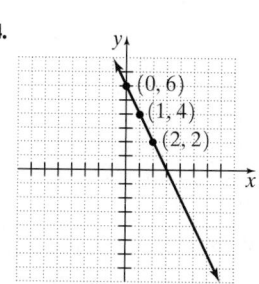

35.

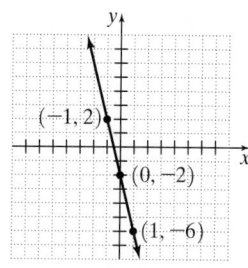

36.

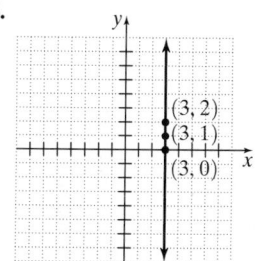

37.

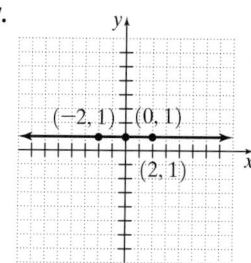

38. parallel   **39.** intersecting and perpendicular   **40.** 4   **41.** 3   **42.** undefined   **43.** $y = -\frac{3}{2}x - \frac{1}{2}$ **44.** $y = -\frac{1}{4}x + \frac{7}{2}$

45. $z = 3A - x - y$ **46.** Let x = original price; $0.7x = 87.49$; $x = 124.99$; The original price was $124.99.

47. Let x = amount borrowed; $1612.50 = x + 0.075x$; $x = 1500$; The amount borrowed was $1500.

48. Let x = the number of CD's; $300 + 2.50x = 15x$
Break-even point is 24 CD's.
If 20 CD's are made, then there will be a loss.
If 30 CD's are made, then there will be a profit.

49. (a) 400
(b) $y = 400x + 7500$
(c) The number of students in 2000 will be about 9900 students.
(d) The number of students in 2010 will be about 13,500; this seems reasonable for a rapidly growing area.

50. The amount of depreciation is $6500 per year.

Chapter 6

SECTION 6.1 EXERCISES

1. solution   **2.** solution   **3.** solution   **4.** not a solution   **5.** not a solution   **6.** not a solution   **7.** not a solution
8. solution   **9.** not a solution   **10.** not a solution   **11.** $(0, -2)$ **12.** all ordered pairs (x, y) that satisfy $x + 3y = 3$ **13.** $(5, -1)$
14. $(3, 0)$ **15.** no solution   **16.** $(-4, 4)$ **17.** all ordered pairs (x, y) that satisfy $x - y = -1$ **18.** no solution   **19.** $\left(3, \frac{9}{2}\right)$
20. $\left(\frac{1}{2}, \frac{2}{3}\right)$ **21.** $(1, 2)$ **22.** all ordered pairs (x, y) that satisfy $y = -\frac{2}{3}x - 2$ **23.** no solution   **24.** $(1, -3)$
25. all ordered pairs (x, y) that satisfy $y = \frac{1}{2}x + 2$ **26.** no solution   **27.** $(2, 3)$ **28.** no solution   **29.** $(1, 4)$ **30.** $(2, 3)$
31. no solution   **32.** $(-3, -1)$ **33.** $\left(\frac{1}{6}, -\frac{1}{2}\right)$ **34.** $\left(\frac{1}{4}, \frac{5}{2}\right)$
35. all ordered pairs (x, y) that satisfy $y = x - 7$ **36.** $(0, 3)$ **37.** no solution   **38.** no solution
39. (a) First system for 200 miles or less: Let y = cost; Turtle Rental: $y = 39.95$; Snail Rental: $y = 79.95$; Second system for more than 200 miles: Let x = number of miles over 200, y = cost; Turtle Rental: $y = 39.95 + 0.25x$, Snail Rental: $y = 79.95$.   (b) There is no solution for the first system. The second system solution is $(160, 79.95)$.   (c) Snail Rental and Turtle Rental will cost the same for $200 + 160$ or 360 miles.   (d) Snail Rental will cost less for 300 miles.   (e) Turtle Rental will cost less for 600 miles.
40. (a) Let x = number of miles, y = cost; Turtle Rental: $y = 49.95$, Snail Rental: $y = 19.95 + 0.40x$ (b) The solution is $(120, 49.95)$.
(c) Snail Rental and Turtle Rental will cost the same for 120 miles.   (d) Turtle Rental will cost less for 300 miles.   (e) Turtle Rental will cost less for 600 miles.
41. (a) $y = 125 + 0.23x$, $y = 0.34x$;   (b) The solution is $(1136.36, 386.36)$.   (c) It will be more cost effective to use pre-sorted mailing for more than 1136 pieces.   (d) It will be more cost effective to use first class mailing for 1136 or less pieces.
42. (a) With coupon book: $y = 50 + 13.5x$; without coupon book: $y = 17.5x$ (b) $(12.5, 218.75)$
(c) It will be beneficial to purchase the coupon book if you will purchase 13 or more dinners.
(d) It will be more cost effective not to purchase the coupon book if you will purchase 12 or fewer dinners; the two options are equal at the intersection of 12.5 dinners, with a cost of $218.75.
43. $x + y = 500$; $y = 3x$; the car payment is $125 and the rent payment is $375.
44. $330x + 450y = 7800$; $x + y = 20$; there should be 10 one-bedroom apartments and 10 two-bedroom apartments.
45. $x + y = 40$; $7x + 8.2y = 298$; she should work 15 hours at the job that pays $8.20 per hour and 25 hours at the job paying $7.00 per hour.
46. $18x + 22y = 316$; $y = 3 + x$; his wages are $6.25 per hour and $9.25 per hour.   **47.** $L = 2W - 25$; $2L + 2W = 550$; the width is 100 feet and the length is 175 feet.   **48.** $W = \frac{1}{2}L + 5$; $2W + 2L = 310$; the width is 55 feet and the length is 100 feet.
49. $y = 2x + 1$; $x + y = 22$; Shania had 7 records and Reba had 15 records.   **50.** $y = x + 5$; $x + y = 21$; there were eight cups and 13 saucers.   **51.** The participation will be equal in 2022.   **52.** The number of travelers will be equal in 1990.

6.1 CALCULATOR EXERCISES

1. $x = 3$, $y = -5$ **2.** $x = -\frac{7}{5}$, $y = 12$ **3.** $x = 8$, $y = 9$ **4.** $x = \frac{9}{4}$, $y = -\frac{7}{2}$

SECTION 6.2 EXERCISES

1. $(12, -7)$ **2.** $(4, 1)$ **3.** $(5, 0)$ **4.** $\left(\frac{5}{2}, \frac{3}{2}\right)$ **5.** $(-5, -5)$ **6.** $(1, -1)$ **7.** no solution   **8.** all ordered pairs (x, y) that satisfy $y = 2x + 3$ **9.** $(16, 13)$ **10.** $(2, -2)$ **11.** $\left(-\frac{13}{5}, -\frac{8}{5}\right)$ **12.** $(4, 9)$ **13.** $(-8, -1)$ **14.** no solution   **15.** $\left(\frac{3}{2}, \frac{5}{2}\right)$
16. $(-3, -7)$ **17.** all ordered pairs (x, y) that satisfy $y = -3x + 2$ **18.** $(4, -7)$ **19.** $(-4, 5)$ **20.** $(0, -2)$ **21.** $(4, 11)$

22. $(5, -1)$ **23.** $(2, -15)$ **24.** no solution   **25.** all ordered pairs (x, y) that satisfy $x - y = -1$ **26.** all ordered pairs (x, y) that satisfy $x + 3y = 3$ **27.** $(1, 2)$ **28.** $(1, -3)$ **29.** $(-2, 2)$ **30.** $(8, 7)$ **31.** all ordered pairs (x, y) that satisfy $y = \frac{1}{2}x + 2$

32. all ordered pairs (x, y) that satisfy $x = -\frac{3}{2}y - 3$ **33.** no solution   **34.** no solution   **35.** $(2, 3)$ **36.** $(2, 3)$ **37.** $(2, 1)$

38. $(2, -1)$ **39.** all ordered pairs (x, y) that satisfy $y = x - 7$ **40.** $\left(\frac{3}{2}, \frac{9}{4}\right)$ **41.** no solution   **42.** no solution   **43.** $(14, -5)$

44. $(9, 2)$ **45.** $x + y = 90; y = 4x - 10$; the angles measure $20°$ and $70°$.
46. $x + y = 180; y - x = 40$; the angles measure $70°$ and $110°$.
47. $x + x + y = 180; y = x + x + 20$; the angles measure $40°, 40°,$ and $100°$.
48. $x + x + y = 180; y = x + 15$; the angles measure $55°, 55°,$ and $70°$.
49. $R = 2r + 5; 2\pi R = 283$; the radius of each measures 20 inches and 45 inches.
50. $2\pi R = 163; R = 3r - 10$; the radius of each measures 12 cm and 26 cm.
51. $x + y = 15; y = 25 - x$; there is no solution.
52. $x + y = 12; y = 12 - x$; the solutions are all ordered pairs (x, y) that satisfy $x + y = 12$, where $x =$ acres of alfalfa and $y =$ acres of wheat.
53. (a) $C(x) = 2500 + 22x$ **(b)** $R(x) = 49x$
    (c) $C(x) = 2500 + 22x$
        $R(x) = 49x$
        They must sell 93 ovens to break even.
54. (a) $R(x) = 89x$ **(b)** $C(x) = 3600 + 35x$
    (c) $R(x) = 89x$
        $C(x) = 3600 + 35x$
        They must sell 67 fans to break even.
55. The net stock values will be equal in 1999.   **56.** The net stock values will be equal in 1995.   **57.** $x + y = 1.9563 \times 10^{12}$; $y = x + 5.31 \times 10^{10}$; The amounts are $\$9.516 \times 10^{11}$ from individual income taxes and $\$1.0047 \times 10^{12}$ from other sources.
58. $\frac{x + y}{2} = 2.75 \times 10^{-10}; y - x = 4.5 \times 10^{-10}$; the diameters are 5×10^{-11} m for helium and 5×10^{-10} m for cesium.

6.2 CALCULATOR EXERCISES

1. $(153.75, 225.85)$ **2.** $(0.2, 5)$ **3.** $\left(\frac{13}{21}, \frac{17}{15}\right)$ **4.** $\left(\frac{13}{7}, \frac{8}{17}\right)$

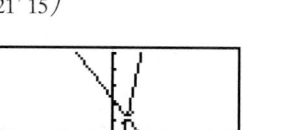

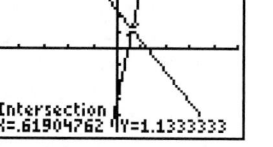

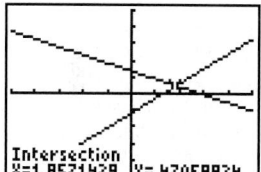

SECTION 6.3 EXERCISES

1. $(-2, 2)$ **2.** $(3, -7)$ **3.** $(5, 2)$ **4.** $(-3, -1)$ **5.** $(-3, -9)$ **6.** $(8, 2)$ **7.** $(8, -2)$ **8.** $(-3, 5)$ **9.** $(4, 3)$ **10.** $(-1, -2)$
11. $(7, -9)$ **12.** $(-8, 7)$ **13.** $(-3, 12)$ **14.** $(13, -2)$ **15.** $\left(\frac{13}{10}, \frac{29}{10}\right)$ **16.** $\left(\frac{4}{5}, -\frac{7}{5}\right)$ **17.** $\left(-\frac{19}{5}, \frac{12}{5}\right)$ **18.** $\left(\frac{9}{2}, -\frac{3}{2}\right)$
19. $\left(\frac{21}{5}, \frac{7}{2}\right)$ **20.** $\left(-\frac{9}{5}, -\frac{17}{5}\right)$ **21.** $\left(\frac{13}{10}, \frac{33}{10}\right)$ **22.** $\left(-\frac{41}{10}, \frac{7}{10}\right)$ **23.** $\left(\frac{2}{3}, \frac{3}{4}\right)$ **24.** $\left(\frac{3}{5}, \frac{5}{6}\right)$ **25.** $\left(\frac{3}{8}, -\frac{3}{4}\right)$ **26.** $\left(\frac{7}{9}, -\frac{5}{8}\right)$
27. all ordered pairs (x, y) that satisfy $y = \frac{3}{2}x - 9$ **28.** all ordered pairs (x, y) that satisfy $y = \frac{6}{5}x + 2$ **29.** no solution
30. no solution   **31.** $(30, -2)$ **32.** $(20, 8)$ **33.** $(10, 10)$ **34.** $(40, -10)$ **35.** $(-400, -700)$ **36.** $(-250, 550)$
37. $\left(\frac{44}{5}, \frac{33}{5}\right)$ **38.** $\left(\frac{43}{10}, -\frac{23}{10}\right)$ **39.** $(52, 48)$ **40.** $(63, 57)$ **41.** $(7, 4)$ **42.** $(5, 9)$ **43.** $(40, 70)$ **44.** $(80, 20)$ **45.** $(80, 120)$
46. $(76, 138)$ **47.** $x + y = 683; 1.5x + 5y = 2645$; there were 220 students who attended the game.
48. $x + y = 56; 30x + 45y = 2190$; there were 22 senior citizens and 34 others.
49. $x - y = 50; 8.5x + 5y = 3462.5$; they sold 275 cookbooks and 225 calendars.
50. $x + y = 5.5; 6x + 4y = 29.5$; they sold her 3.75 pounds of ham and 1.75 pounds of cheese.
51. $x + y = 385; 2x + 2y = 770$; the solution is any number of adults x and any number of children y, where $x + y = 385$.
52. $x + y = 8.95; 2x + 2y = 13.95$; there is no solution.
53. $2x + 2y = 620; 2x + 2(y + 90) = 800$; the solutions are any number of yards x and any number of yards y, where $x + y = 310$.
54. $R - r = 10; 2\pi R = 2\pi r + 20\pi$; the solutions are any radius r and any radius R, where $R - r = 10$.

55. $x + x + y = 180$; $y = 90 - 2x$; there is no solution.

56. $x + y = 180$; $y - x = 30$; the angles are 75° and 105°.

57. $y = x + 4.864 \times 10^7$; $\dfrac{x + y}{2} = 1.1728 \times 10^8$; the distances from the sun are 9.296×10^7 miles for Earth and 1.416×10^8 miles for Mars.

58. $y = 3.6302 \times 10^9 + x$; $\dfrac{x + y}{2} = 1.851 \times 10^9$; the distances from the sun are 3.59×10^7 miles for Mercury and 3.661×10^9 miles for

Pluto.

59. $M - F = 142$
$M + F = 1054$
$M = 598$; $F = 456$
The median weekly earnings
are $598 for males and
$456 for females.

60. $M - F = 68$
$M + F = 868$
$M = 468$; $F = 400$
The median weekly earnings
are $468 for black males, and
$400 for black females.

61. $2W - 50 = L$
$W + L = 400$
$W = 150$ million; $L = 250$ million
The earnings were $150 million for
Oprah Winfrey and $250 million
for George Lucas.

62. $B + 26.4 = G$
$B + G = 91$
$B = 32.3$ billion; $G = 58.7$ billion
William Gates' worth was $58.7 billion
and Warren Buffet's worth was
$32.3 billion.

6.3 CALCULATOR EXERCISES

1. $(3, -2)$ **2.** $(-2, -3)$ **3.** $(-8, 8)$ **4.** $(0.5, -2.8)$ **5.** $\left(\dfrac{1}{2}, \dfrac{2}{3}\right)$ **6.** $\left(\dfrac{1}{4}, -\dfrac{2}{3}\right)$

SECTION 6.4 EXERCISES

1. Let x = time driving, y = distance from Nashville; $y = 60x$, $220 - y = 65x$; They will meet at a distance of 105.6 miles from Nashville.

2. Let x = time driving, y = distance from Houston; $y = 65x$, $240 - y = 55x$; They will meet at a distance of 130 miles from Houston.

3. Let x = time riding, y = distance walking; $y = 6x$, $11 - y = 60x$; He walked for $\dfrac{1}{3}$ hour and rode for $\dfrac{1}{6}$ hour.

4. Let x = time canoeing, y = distance canoeing; $y = 12x$, $30 - y = 3x$; They canoed for 2 hours and hiked for 2 hours. They canoed 24 miles.

5. Let x = average speed of planes in still air, y = average wind speed; $6(x - y) = 2600$, $5(x + y) = 2600$; The average plane speed in still air is about 476.7 miles per hour and the average wind speed is about 43.3 miles per hour.

6. Let x = average speed of planes in still air, y = average wind speed; $4(x - y) = 1750$, $3.5(x + y) = 1750$; The average plane speed in still air is 468.75 mph and the average wind speed is 31.25 mph.

7. Let x = average speed of paddling in still water, y = average current speed; $1.5(x - y) = 6$, $0.75(x + y) = 6$; The average speed of paddling was 6 mph and the average current speed was 2 mph.

8. Let x = average speed of the boat in still water, y = average speed of river; $0.75(x + y) = 42$, $1(x - y) = 42$; The average speed of the boat in still water is 49 mph and the average speed of the river is 7 mph.

9. Let x = average speed of the boat in still water, y = average speed of the current; $0.25(x + y) = 13.4$, $0.3(x - y) = 13.4$; The average speed of the boat in still water is about 49.13 mph and the average speed of the current is about 4.47 mph.

10. Let x = average speed of the boat in still water and y = average speed of the current; $\dfrac{1}{3}(x + y) = 14.4$, $\dfrac{2}{3}(x - y) = 14.4$; The average speed of the boat in still water is about 32.4 mph and the average speed of the current was 0.8 mph.

11. Let x = pounds of French vanilla coffee, y = pounds of hazelnut coffee; $x + y = 20$, $9.50x + 7.00y = 8.50(20)$; He should use 12 pounds of French vanilla coffee and 8 pounds of hazelnut coffee.

12. Let x = pounds of orange spice tea and y = pounds of lemon honey tea; $x + y = 10$, $3.50x + 7.50y = 5.00(10)$; The blend contained 6.25 pounds of orange spice tea and 3.75 pounds of lemon honey tea.

13. Let x = the number of azalea plants and y = the number of rhododendron plants; $x + y = 30$, $5x + 12y = 250$; She can buy 16 azalea plants and 14 rhododendron plants.

14. Let x = the number of burgers and y = the number of hot dogs; $x + y = 50$, $2.50x + 1.00y = 95.00$; She could order 30 burgers and 20 hot dogs.

15. Let x = the number of adults and y = the number of children; $y = 4x$, $7.50x + 4.50y = 1938$; There were 76 adults and 304 children.

16. Let x = the number of adults and y = the number of children; $y = 2x$, $10.00x + 5.50y = 5355$; There were 255 adults and 510 children.

17. Let x = the number of coins and y = the total number of items in the assortment; $y = x + 100$, $4(100) + 8x = 7y$; The assortment will need 300 half-dollar coins.

18. Let x = the number of Cal Ripken cards and y = the total number of cards in the bin; $y = x + 200$, $2.75x + 1.25(200) = 1.75y$; You will need 100 Cal Ripken, Jr. cards, and this will make a total of 300 cards in the mix.

19. Let x = hourly wage on job 1 and y = hourly wage on job 2; $15x + 20y = 320$, $18x + 24y = 384$; Many solutions; any pair of hours (x, y) where $15x + 20y = 320$ will satisfy the system.

20. Let x = sale price of shirts and y = sale price of blouses; $2x + 4y = 132$, $3x + 6y = 180$; No solution; there are no sale prices that will satisfy this system of equations.

21. Let x = number of $5 bills and y = number of $10 bills: $x + y = 65, 5x + 10y = 365$; There were 57 $5 bills and 8 $10 bills.

22. Let x = number of $20 bills and y = number of $50 bills; $x + y = 35, 20x + 50y = 1300$; There are 15 $20 bills and 20 $50 bills.

23. Let x = gallons of grapefruit beverage and y = gallons of orange beverage; $x + y = 200, 0.45x + 0.75y = 0.55(200)$; They must mix $133\frac{1}{3}$ gallons of grapefruit beverage with $66\frac{2}{3}$ gallons of orange beverage.

24. Let x = pints of 70% solution and y = pints of 40% solution; $x + y = 5, 0.70x + 0.40y = 0.50(5)$; He should mix $1\frac{2}{3}$ pints of 70% solution and $3\frac{1}{3}$ pints of 40% solution.

25. Let x = liters of 60% acid and y = liters of 35% acid; $x + y = 300, 0.60x + 0.35y = 0.50(300)$; He should mix 180 liters of 60% acid with 120 liters of 35% acid.

26. Let x = pints of 25% solution and y = pints of 40% solution; $x + y = 8, 0.25x + 0.40y = 0.30(8)$; She should mix $5\frac{1}{3}$ pints of 25% solution with $2\frac{2}{3}$ pints of 40% solution.

27. Let x = gallons of 4.3% milk and y = gallons of skim milk; $x + y = 200, 0.043x + 0y = 0.02(200)$; He should mix about 93 gallons of 4.3% milk with about 107 gallons of skim milk.

28. Let x = liters of 45% solution and y = liters of sterile water; $x + y = 25, 0.45x + 0y = 0.35(25)$; Mix $19\frac{4}{9}$ liters of 45% solution with $5\frac{5}{9}$ liters of sterile water.

29. Let x = gallons of 45% antifreeze and y = gallons of pure antifreeze; $x + y = 4, 0.45x + 1.00y = 0.60(4)$; Mabel should drain off about 1.1 gallons and replace with about 1.1 gallons of pure antifreeze.

30. Let x = gallons of 35% antifreeze and y = gallons of pure antifreeze; $x + y = 5, 0.35x + 1.00y = 0.50(5)$; The mechanic should drain off about 1.2 gallons and replace with about 1.2 gallons of pure antifreeze.

31. Let x = gallons of 12% wine and y = total gallons of 15% wine; $5 + x = y, 0.20(5) + 0.12x = 0.15y$; He should mix $8\frac{1}{3}$ gallons of 12% wine with the 5 gallons of 20% wine to make the 15% wine.

32. Let x = ounces of 45% medication and y = ounces of the blended 60% medication; $x + 8 = y, 0.45x + 0.75(8) = 0.60y$; He should mix 8 ounces of the 45% medication with 8 ounces of the 75% to produce 16 ounces of the 60% medication.

33. Let x = amount invested at 8.5% and y = amount invested at 7%; $x + y = 10,000, 0.085x + 0.07y = 752.50$; She invested $3500 at 8.5% and $6500 at 7%.

34. Let x = amount borrowed at 7% and y = amount borrowed at 8.25%; $x + y = 10,000, 0.07x + 0.0825y = 725$; He borrowed $8000 at 7% and $2000 at 8.25%.

35. Let x = amount invested at 5% and y = amount invested at 7.25%; $x + y = 16,500, 0.05x + 0.0725y = 1000$; She should invest $8722.22 at 5% and $7777.78 at 7.25%.

36. Let x = amount invested at 5% and y = amount invested at 7.25%; $x + y = 11,000, 0.05x + 0.0725y = 600$; She should invest $8777.78 at 5% and $2222.22 at 7.25%.

37. Let x = amount in certificates and y = amount in savings; $x = 2y, 0.0725x + 0.06y = 1230$; She should invest $12,000 in certificates and $6000 in savings.

38. Let x = amount in certificates and y = amount in savings; $x = y + 2000, 0.0725x + 0.06y = 1205$; He should invest $10,000 in certificates and $8000 in savings.

39. Let x = interest rate for $45,000 loan and y = interest rate for $55,000 loan; $x = y + 0.01, 45,000x + 55,000y = 6450$; The interest rate was 7% for the $45,000 loan and 6% for the $55,000 loan.

40. Let x = interest rate for $1500 account and y = interest rate for $1200 account; $y = x + 0.003, 1500x + 1200y = 117.90$; The interest rate was about 4.23% for the $1500 investment and about 4.53% for the $1200 investment.

6.4 CALCULATOR EXERCISES

1. Stella is 32 years old, and her son is 12 years old.   **2.** Gulen is 29 years old, and Cecilia is 1 year old.
3. Joe is 40 years old, and he will retire at 50 years old.   **4.** Jenny is 15 years old, and Katie is 17 years old.
5. On those two birthdays, Mom was 12 and 24 years old. Grandmother was 36 and 48 years old.
6. The dresser was 100 years old, and the bed was 48 years old.   **7.** Fric is 21 years old, and Frac is 7 years old.

SECTION 6.5 EXERCISES

1. solution   **2.** not a solution   **3.** solution   **4.** solution   **5.** not a solution   **6.** not a solution   **7.** not a solution
8. solution   **9.** solution   **10.** solution   **11.** not a solution   **12.** solution   **13.** not a solution   **14.** solution   **15.** solution

16. not a solution   **17.** $(4, -3, 7)$ **18.** $(2, -5, -9)$ **19.** $\left(\frac{3}{4}, -\frac{3}{4}, \frac{5}{8}\right)$ **20.** $\left(\frac{1}{3}, \frac{3}{5}, -\frac{2}{3}\right)$ **21.** $\left(0, \frac{9}{5}, -\frac{36}{5}\right)$ **22.** $\left(\frac{9}{2}, -\frac{17}{5}, 0\right)$

23. $(573, 471, 283)$ **24.** $(756, -522, 87)$ **25.** infinite number of solutions   **26.** no solution   **27.** no solution
28. infinite number of solutions

29. $12x + 5y + 3.5z = 9547$
$x + y + z = 1257$
$x - y + z = -87$
There were 487 adult tickets, 672 child tickets, and 98 resort guest tickets.

30. $3.5c + s + 4.25p = 53.5$
$c - s - p = 0$
$c + s + p = 18$
Deanna bought nine cups, five saucers, and four dinner plates.

31. $x + y + z = 11,200$
$x - z = -1400$
$0.05x + 0.055y + 0.06z = 623$
There are an infinite number of solutions.

32. $0.015c + 0.0075f + 0.009a = 88.28$
$5c - f = 0$
$c + f + a = 9850$
Siegfried's balances are $246.67 on credit card loan, $1233.33 on furniture loan, and $8370 on car loan.

33. $x + 5y + 10z = 184$
$x - 3y = 0$
$x + y + z = 60$
Ron collected 39 $1 bills, 13 $5 bills, and eight $10 bills.

34. $0.32x + 0.03y + 0.35z = 12.25$
$x - y = 0$
$x + y + z = 50$
The packet contains 15 $0.32 stamps, 15 $0.03 stamps, and 20 $0.35 stamps.

35. $s + m + l = 145$
$2.39s + 2.79m + 3.59l = 446.55$
$10.8s + 13.5m + 20.4l = 2350 - 4$
There were 35 small boxes, 40 medium-sized boxes, and 70 large boxes.

36. $s + e + t = 77$
$6s + 8e + 12t = 60(12) - 4$
$3.3s + 4e + 5.4t = 336$
There is no solution.

37. $g + l + m = 600$
$19.75g + 9.50l + 12m = 7225$
$0.50g + 0.10l + 0.90m = 260$
Tenisha invested $1975 in gold, $2850 in oil, and $2400 in the money market.

38. $A + B + C = 65$
$18A + 13.2B + 15C = 942$
$1.5A + 0.65B + 0.85C = 54.75$
Hortense purchased 10 units of Ameritag, 35 units of Bankers Fund, and 20 units of Columbia Mutual.

39. $1.25x + y + 1.5z = 69.5$
$3.5x + 2.5y + 4.5z = 200$
$-3x + z = 0$
Concetta bakes 10 pies, 12 dozen cookies, and 30 cakes.

40. $45p + 90s + 15d = 480$
$130p + 300s + 55d = 1595$
$p - s = 1$
Gretchen schedules three physicals, two surgeries, and 11 diagnostic treatments.

41. $225C + 6.8J + 69.8M = 736$
$4.5C + 31.2J + 116.2M = 505$
$0.6C + 35.4J + 0.6M = 39$
The plan should include two servings (2 oz) of cereal, one serving (2 oz) of orange juice, and four servings (8 oz) of milk.

42. $6.8H + 2.7B + 2C = 20.3$
$11.3H + 28.4B + 6.8C = 64.6$
$0.8H + 0.2B + 0.5C = 2.8$
Chip should have two servings of hot dogs, one serving of beans, and two servings of chips.

6.5 CALCULATOR EXERCISES

1. $(2, 3, -1)$ **2.** $(1, -3, 2)$ **3.** $(0.5, 2, 0.375)$ **4.** $\left(\frac{1}{4}, 3, \frac{2}{3}\right)$ **5.** $(1, 1, 1)$ **6.** $(-1, -1, -1)$ **7.** $(5, 2, -8)$ **8.** $(7, 1, -5)$
9. no solution   **10.** many solutions; all ordered triples that satisfy $x - y - 4z = 7$.

CHAPTER 6 SECTION-BY-SECTION REVIEW

1. solution   **2.** solution   **3.** not a solution   **4.** not a solution   **5.** not a solution   **6.** solution   **7.** not a solution

8. solution   **9.** $(7, 3)$ **10.** $(-4, 2)$ **11.** $(-2, -1)$ **12.** $\left(\frac{1}{2}, \frac{7}{4}\right)$ **13.** no solution

14. all ordered pairs (x, y) that satisfy $y = \frac{3}{2}x - 6$ **15.** $(1, -3)$ **16.** no solution

17. all ordered pairs (x, y) that satisfy $2x - 11 = 3$ or $x = 7$ **18.** all ordered pairs (x, y) that satisfy $y - 1 = 4$ or $y = 5$ **19.** $(-3, 4)$

20. $(5, -5)$ **21.** $(-3, -5)$ **22.** $\left(\frac{9}{2}, -\frac{1}{2}\right)$ **23.** $x + y = 500 - 120$; $y = x + 40$; there were 50 more Democrats than Independents.

24. $x = 4.97$; the earnings will be the same in 1995.   **25.** $\left(-\frac{3}{4}, \frac{5}{8}\right)$ **26.** $\left(\frac{1}{5}, \frac{16}{5}\right)$ **27.** no solution   **28.** $(-53, -62)$

29. $(27, -41)$ **30.** all ordered pairs (x, y) that satisfy $y = 3x + 7$ **31.** $\left(\frac{15}{19}, \frac{10}{19}\right)$ **32.** $(575, 823)$ **33.** $(30, 6)$

34. $(-10, 2)$ **35.** $\left(\frac{17}{8}, -\frac{29}{8}\right)$ **36.** $\left(120, \frac{3}{4}\right)$ **37.** $x + y = 90$; $y = 2x + 12$; the difference in the angles is 38°.

38. (a) $C(x) = 29x + 450$ **(b)** $R(x) = 89.95x$
(c) $C(x) = 29x + 450$
$R(x) = 89.95x$
$x = 7.38$
Therefore, they must acquire and sell eight scooters to break even.

39. $x = 4.47$. Therefore, the spending would be about equal in 1995.   **40.** $(7, -15)$ **41.** $(6, 6)$ **42.** $(-4, 15)$ **43.** $(-3, -4)$

44. no solution   **45.** $\left(\frac{133}{31}, -\frac{2}{31}\right)$ **46.** $(4, 8)$ **47.** all ordered pairs (x, y) that satisfy $y = \frac{5}{7}x + 1$ **48.** $\left(\frac{5}{9}, -\frac{7}{9}\right)$ **49.** $\left(\frac{7}{8}, \frac{21}{8}\right)$

50. $(10, 5)$ **51.** $(350, -650)$ **52.** King earned $44 million; Rowling earned $36 million.

53. She should have 28 small offices and 12 large offices.
54. The mass of the Earth is 5.97×10^{24} kg; the mass of Mars is 6.42×10^{23} kg.
55. $25x + 35y = 5500$; $x + y = 200$; the plant should order 150 components from Supplier A and 50 components from Supplier B.
56. $x + y = 150$; $0.1x + 0.05y = 0.08(150)$; he should use 90 pounds of 10% nitrogen with 60 pounds of 5% nitrogen.
57. $0.49x + 0.99y = 0.69(200)$; $x + y = 200$; they should mix 120 pounds of broccoli with 80 pounds of cauliflower.
58. $x + y = 10$; $65x + 45y = 600$; he drove 487.5 miles on the interstate.   **59.** $20.5 = (x - y)\frac{7}{20}$, $20.5 = (x + y)\frac{1}{3}$; the boat's speed is about 60.0 mph. The water's speed is about 1.5 mph.   **60.** $y = x + 40$; $0.25x + 0.30(40) = 0.27y$; the 27% copper alloy contains 60 pounds of the alloy containing 25% copper.   **61.** $y = 150 + 0.25x$; $y = 175 + 0.2x$; he must drive more than 500 miles.   **62.** $x + y = 10,000,000$; $0.045x + 0.06y = 487,500$; $7,500,000$ was invested in the 4.5% interest fund, and $2,500,000$ in the 6% interest fund.   **63.** not a solution
64. solution   **65.** $(11, -15, -22)$ **66.** $\left(\frac{1}{3}, \frac{2}{3}, -2\right)$ **67.** infinite number of solutions   **68.** infinite number of solutions
69.
$$s + t + v = 140$$
$$2.5s + 5t + 7.5v = 637.5$$
$$s - 5t = 0$$
The attendees were 75 students, 15 teachers, and 50 visitors.

70.
$$5C + 3V + 6R = 22$$
$$26C + 0V + 48R = 100$$
$$312C + 237V + 348R = 1446$$
She ate two club sandwiches, two veggie sandwiches, and one roast chicken sandwich.

71. The currents are $I_1 = 1$ ampere, $I_2 = 3$ amperes, and $I_3 = 2$ amperes.

CHAPTER 6 CHAPTER REVIEW

1. $(-14, -6)$ **2.** all ordered pairs (x, y) that satisfy $y = \frac{4}{5}x - 6$ **3.** $(8, -10)$ **4.** $(1200, 900)$ **5.** $(-6, 21)$ **6.** $(-2, -1)$
7. no solution   **8.** $\left(\frac{87}{26}, -\frac{5}{26}\right)$ **9.** $(6, -20)$ **10.** $(3, 7)$ **11.** $\left(-\frac{3}{7}, \frac{6}{7}\right)$ **12.** $\left(\frac{27}{8}, -\frac{21}{8}\right)$ **13.** $(3, -2)$ **14.** no solution
15. $(-4, 2)$ **16.** $(3, -3)$ **17.** no solution   **18.** all ordered pairs (x, y) that satisfy $y = -4x - 3$
19. all ordered pairs (x, y) that satisfy $x + 7 = 3$ or $x = -4$ **20.** all ordered pairs (x, y) that satisfy $y + 9 = 7$ or $y = -2$ **21.** $(6, 5)$
22. $(-3, -2)$ **23.** $(5, -2)$ **24.** $(-2, 2)$ **25.** $(1, 1)$ **26.** $(-6, -1)$ **27.** $(44, -23)$ **28.** $(-12, 5)$ **29.** $(-38, 43)$
30. all ordered pairs (x, y) that satisfy $y = 4x - 7$ **31.** $\left(\frac{1}{6}, -\frac{5}{6}\right)$ **32.** $\left(-\frac{3}{5}, -\frac{12}{5}\right)$ **33.** $\left(\frac{20}{9}, \frac{1}{9}\right)$ **34.** $(723, -491)$
35. $\left(\frac{19}{8}, -\frac{21}{8}\right)$ **36.** $\left(180, \frac{5}{9}\right)$ **37.** no solution   **38.** $(120, 180)$ **39.** not a solution   **40.** solution   **41.** solution
42. solution   **43.** not a solution   **44.** solution   **45.** not a solution   **46.** not a solution
47. $x + y = 200$; $0.15x + 0.35y = 0.27(200)$; 80 pounds of 15% brass; 120 pounds of 35% brass; 40 pounds more of the 35% alloy will be used.   **48.** $8x + 12y = 7300$; $y = x + 150$; she will receive $2200 for all of the one-bedroom apartments and $5100 for all of the two-bedroom apartments.   **49.** $x + y = 180$; $y = x + 10$; the smaller angle measures 85°.
50. $x + y = 700$; $0.7x + 0.4y = 400$; there were 100 more women surveyed.
51. $x + y = 90$; $y = 3x + 12$; the angles measure 19.5° and 70.5°.
52. $x + 18 = y$; $0.15x + 0.25(18) = 0.21y$; she should use 12 cc of the 15% solution to make 30 cc of the mixture.
53. $3.5 = \frac{1}{2}x + \frac{1}{3}y$; $4 = \frac{2}{3}x + \frac{1}{3}y$; he walked at 3 mph and jogged at 6 mph.
54. $x + y = 50$; $8.5x + 12.5y = 9.5(50)$; the shop should mix 37.5 pounds of gourmet coffee with 12.5 pounds of dutch chocolate.
55. $10.75x + 6.5y = 181$; $x + y = 20$; she should work 12 hours at $10.75 per hour and 8 hours at $6.50 per hour.
56. $5.75(x - y) = 2600$; $5(x + y) = 2600$; the air speed x equals 486 mph, and the wind speed y is 34 mph.
57. $U(x) = 39.95 + 0.15x$; $B(x) = 19.95 + 0.22x$; you must drive at least 286 miles in order for U Rent It to be less costly.
58. $C(x) = 7.5 + 2.5x$; $R(x) = 6.5x$; the shopkeeper must produce and sell at least 19 items in order not to lose money.
59. $(-10, 12, 21)$ **60.** $\left(\frac{1}{2}, \frac{3}{4}, -3\right)$
61. The three currents are $I_1 = 3.5$ amperes, $I_2 = 2$ amperes, and $I_3 = 1.5$ amperes.
62.
$$4t + 5h + 39b = 92$$
$$19t + 28h + 90b = 255$$
$$289t + 302h + 640b = 2173$$
Matthew ate one turkey sandwich, two ham sandwiches, and two hamburgers.

CHAPTER 6 TEST

1. solution   **2.** all ordered pairs (x, y) that satisfy $y = 4$ **3.** no solution   **4.** $(2, 1)$ **5.** $(-2, 4)$ **6.** no solution
7. $\left(\frac{1}{4}, -\frac{3}{8}\right)$ **8.** no solution   **9.** $(8, 0)$ **10.** $\left(\frac{34}{19}, \frac{13}{19}\right)$

11. (a) $C(x) = 5500 + 65x$ **(b)** $R(x) = 125x$
(c) $C(x) = 5500 + 65x$
$R(x) = 125x$
$x = 91.67$

Therefore, 92 items must be produced and sold in order to break even.

12. $y = 3x; y = 2(x + 4)$; Kenny's rate was 8 mph and the distance to Dolly was 24 miles.

13. $6x + 8y = 58; 10x + 5y = 55$; the computer takes 3 nanoseconds per addition and 5 nanoseconds per multiplication.

14. $x + y = 30; 1.25x + 2y = 1.5(30)$; he should mix 20 pounds of raisins with 10 pounds of peanuts.

15. $x + y = 90; y = 2x$; the angles measure 30° and 60°.

16. $x + y = 100; 0.5x + 0.1y = 0.3(100)$; she should mix 50 cc of 50% solution and 50 cc of 10% solution.

17. $x + y = 180; y = 2x + 15$; the two angles measure 125° and 55°. **18.** $0.095x + 0.07y = 780; x = 2y$; Caitlin invested $6000 in Fund A, and $3000 in Fund B.

19. $36 = \frac{3}{5}(x - y); 36 = \frac{1}{2}(x + y)$; the boat's speed is 66 mph; the speed of the current is 6 mph.

20. $\left(\frac{7}{5}, 2, -\frac{3}{2}\right)$ **21.** Infinite number of solutions; all triple that satisfies $x - y - z = 3$. **22.** Mike makes four bird feeders, eight birdhouses, and three snack tables. **23.** One solution exists; no solution exists; an infinite number of solutions exist. Answers will vary.

Chapter 7

SECTION 7.1 EXERCISES

1. linear **2.** linear **3.** nonlinear **4.** nonlinear **5.** linear **6.** linear **7.** linear **8.** linear **9.** linear
10. nonlinear **11.** nonlinear **12.** nonlinear **13.** nonlinear **14.** linear **15.** linear **16.** linear

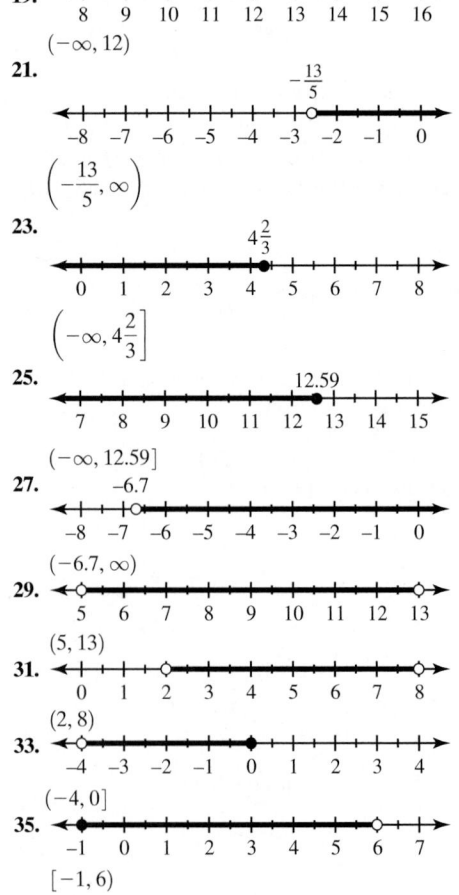

17.
$[6, \infty)$

18.
$(-\infty, 9]$

19.
$(-\infty, 12)$

20.
$(-7, \infty)$

21.
$\left(-\frac{13}{5}, \infty\right)$

22.
$\left(-\infty, 3\frac{5}{6}\right)$

23.
$\left(-\infty, 4\frac{2}{3}\right]$

24.
$\left[\frac{16}{3}, \infty\right)$

25.
$(-\infty, 12.59]$

26.
$[5.1, \infty)$

27.
$(-6.7, \infty)$

28.
$(-\infty, 45.65)$

29.
$(5, 13)$

30.
$(-2, -1)$

31.
$(2, 8)$

32.
$[3, 6]$

33.
$(-4, 0]$

34.
$[0, 5)$

35.
$[-1, 6)$

36.
$(2, 7]$

37.
[2, 7]

38.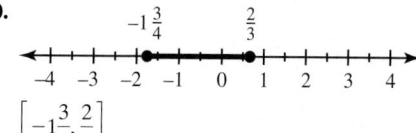
[15, 30]

39.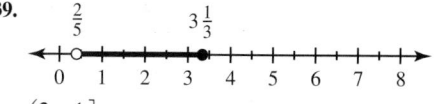
$\left(\frac{2}{5}, 3\frac{1}{3}\right]$

40.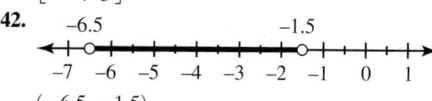
$\left[-1\frac{3}{4}, \frac{2}{3}\right]$

41.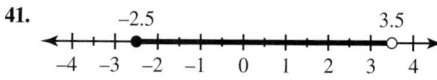
[-2.5, 3.5)

42.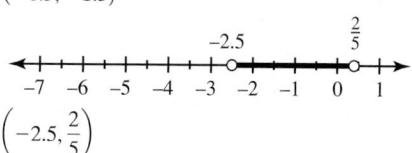
(-6.5, -1.5)

43.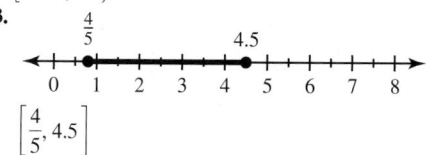
$\left[\frac{4}{5}, 4.5\right]$

44.
$\left(-2.5, \frac{2}{5}\right)$

45. (a) 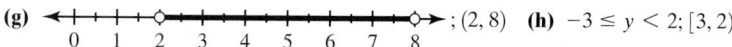 ; $(-\infty, 4.5)$ **(b)** $x \le 3; (-\infty, 3]$

(c) $x < -2;$ 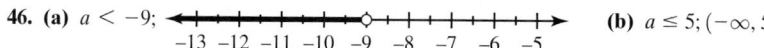 **(d)** 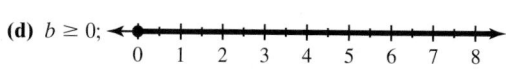 ; $[5.7, \infty)$

(e) $z > -7; (-7, \infty)$ **(f)** $z > 2;$

(g) 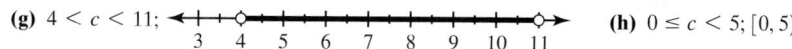 ; (2, 8) **(h)** $-3 \le y < 2; [3, 2)$

(i) $0 \le y \le 9;$

46. (a) $a < -9;$ **(b)** $a \le 5; (-\infty, 5]$

(c) ; $\left(-\infty, \frac{1}{2}\right)$ **(d)** $b \ge 0;$

(e) $b > -4; (-4, \infty)$ **(f)** ; $(2.7, \infty)$

(g) $4 < c < 11;$ **(h)** $0 \le c < 5; [0, 5)$

(i) ; $[-5, -1]$

47. (a) Let r = magnitude, $r < 3.5$ **(b)** $3.5 \le r \le 5.4$ **(c)** $r < 6.0$
48. (a) Let r = magnitude, $6.1 \le r \le 6.9$ **(b)** $7.0 \le r \le 7.9$ **(c)** $r \ge 8.0$
49. $39.95 + 0.2x \le 150$ **50.** $300 + 22x < 1000$ **51.** $800 < 450 + 0.05x$
52. $15x - 125 \ge 250$ **53.** $150 \le 25 + 12.5x \le 200$ **54.** $200 \le 4.5x - 350 \le 500$

7.1 CALCULATOR EXERCISES

1.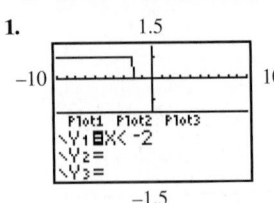

2.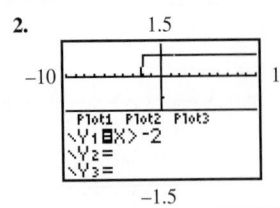

3.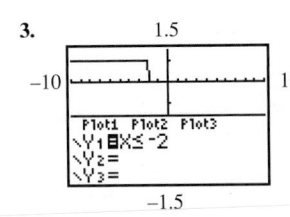

4.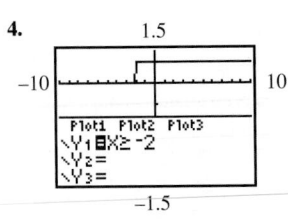

5.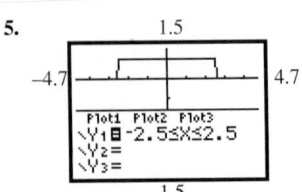

6.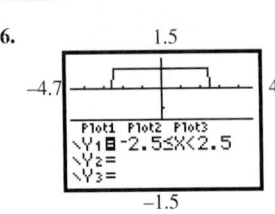

7.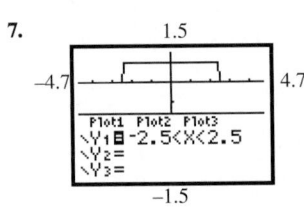

SECTION 7.2 EXERCISES

1. integers less than 24   **2.** integers greater than -2 **3.** no solution   **4.** all integers   **5.** integers greater than $\frac{1}{2}$

6. integers less than -6 **7.** all integers   **8.** no solution   **9.** $x \geq -2$ **10.** $x \leq 2$ **11.** $x > -4$ **12.** $x > 2$ **13.** $x \leq 3$
14. $x \geq 2$ **15.** all real numbers   **16.** all real numbers   **17.** no solution   **18.** no solution   **19.** $(-3, \infty)$ **20.** $(-\infty, -38)$

21. $(-\infty, 0]$ **22.** $\left(-\infty, \frac{28}{5}\right]$ **23.** $(2, \infty)$ **24.** $(7, \infty)$ **25.** $(3.3, \infty)$ **26.** $(5, \infty)$ **27.** $(0, \infty)$ **28.** $(10.2, \infty)$ **29.** $[6, \infty)$

30. $(-\infty, 1]$ **31.** $\left(\frac{1}{2}, \infty\right)$ **32.** $\left(-\infty, -\frac{1}{4}\right)$ **33.** $(8, \infty)$ **34.** $(-\infty, -3)$ **35.** $(8.8, \infty)$ **36.** $(0, \infty)$ **37.** $\left(-\infty, -\frac{1}{162}\right]$

38. $\left[-\frac{3}{8}, \infty\right)$ **39.** $(-120, \infty)$ **40.** $(-66, \infty)$ **41.** no solution   **42.** no solution   **43.** all real numbers

44. all real numbers   **45.** $(107.5, \infty)$ **46.** $(3.6, \infty)$ **47.** $\left[\frac{-7}{3}, -\frac{1}{3}\right)$ **48.** $\left[-1, \frac{1}{5}\right)$ **49.** $[-2/3, 3]$

50. $(-5.5, -3)$ **51.** $\frac{93 + 97 + 92 + 89 + 95 + x}{6} \geq 93$; Lee must score 92 or better.   **52.** $\frac{38 + 62 + 56 + 42 + x}{5} > 50$; she must

earn more than \$52.   **53.** $9.75x + 5200 \leq 7500$; he can work 235 hours or less.   **54.** $22 + 3.5x \leq 55$; he can use the grinder 9 hours or

less.   **55.** $165x \leq 2000$; the number of people can be 12 or less.   **56.** $12.5x + 87.5 \leq 120$; she can buy one or two sweaters.

57. $35x + 30(120 - x) \leq 4000$; She can order at most 80 meat entrées.   **58.** $25 + 0.22x > 15 + 0.35x$; you can drive 76 miles or less.

59. $2x + \frac{3}{4}x + 30 - 4 \leq 185$; the width must be 57 feet or less.   **60.** $2x + 2(2x + 15) \leq 240$; the width must be 35 feet or less.

61. $0.17x - 45 \geq 50$; at least 559 prints must be produced each day.
62. $3.55x - (1.25x + 165) \geq 150$; at least 137 orders must be sold each day.

63. $82 < \frac{82 + 83 + 88 + 92 + x}{5} < 88$; The temperature for the fifth day must be between $65°$ and $95°$.

64. $18 < \frac{16.79 + 18.74 + 17.99 + x}{4} < 20$; The closing price must be between \$18.48 and \$26.48.

65. $0.146x + 1 < 30$; The weight must be less than 198.6 pounds.
66. $0.125x + 0.7 < 27$; The weight must be less than 210.4 pounds.
67. $(6.95 \cdot 10^{10})n + 4.36 \cdot 10^{11} > 1.5 \cdot 10^{12}$; The expenditure exceeds this amount 16 years after 1980, or 1996.
68. $(2.95 \cdot 10^{10})n + 1.63 \cdot 10^{11} > 1.0 \cdot 10^{12}$; The expenditure exceeds this amount 29 years after 1980, or 2009.

7.2 CALCULATOR EXERCISE

Answers will vary.

SECTION 7.3 EXERCISES

1. linear; $x + 1.7y > -4.6$ **2.** nonlinear   **3.** nonlinear   **4.** linear; $-5x + 10y \geq 15$ **5.** nonlinear   **6.** nonlinear
7. linear; $6x - 2y > 1$ **8.** nonlinear   **9.** linear; $4x - y \leq -16$ **10.** linear; $-4.2x + 3.5y < -2.8$
11. linear; $6x + 15y \leq 7$ **12.** linear; $x + 2y < 12$

13. $y < -2x + 3$

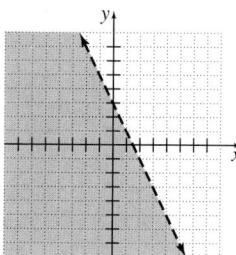

14. $y > 5x - 6$

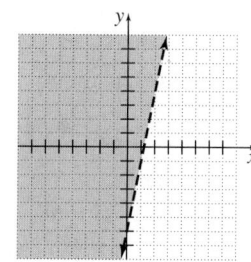

15. $y \leq \frac{5}{3}x - 2$

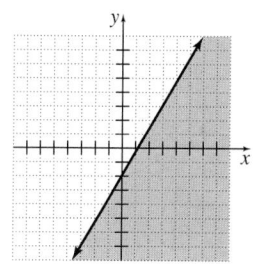

16. $y \geq -\frac{8}{7}x + 2$

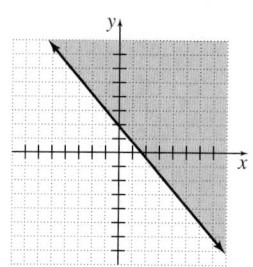

17. $y < -\frac{3}{4}x + 4$

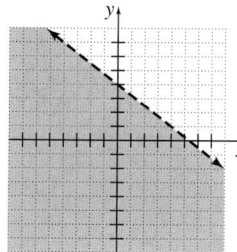

18. $y < \frac{5}{6}x - 3$

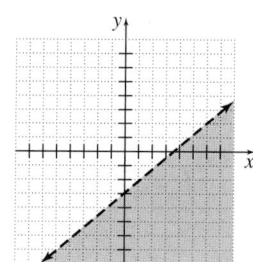

19. $y \geq 2.8x - 1.6$

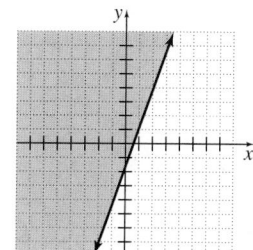

20. $y \geq -1.9x + 4.7$

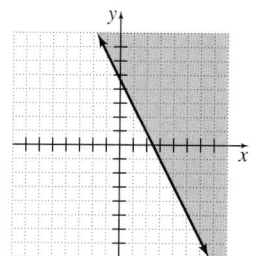

21. $y > \frac{3}{2}x - \frac{5}{2}$

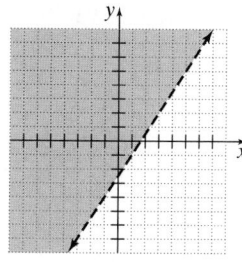

22. $y > \frac{3}{5}x - \frac{2}{5}$

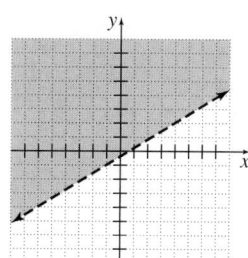

23. $y \geq 2x - 1$

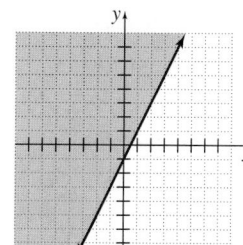

24. $y \leq -2x - 7$

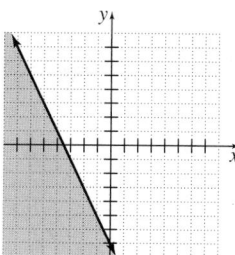

25. $y > -4$

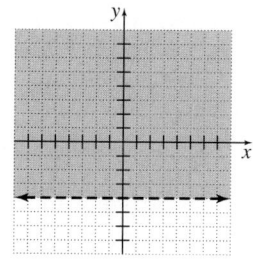

26. $y > 6$

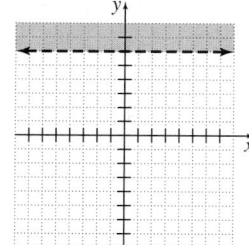

27. $x \leq 1$

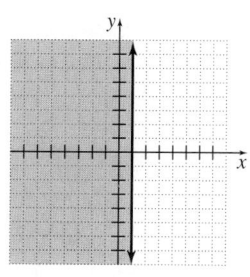

28. $x \leq 5$

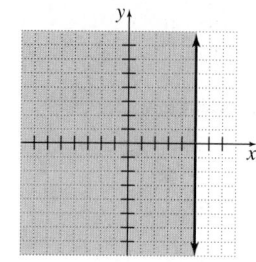

29. $y \geq -\frac{3}{5}x + \frac{12}{5}$

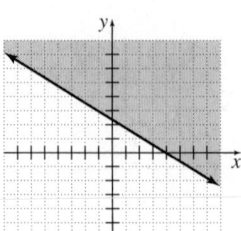

30. $y \geq -\frac{7}{9}x + 7$

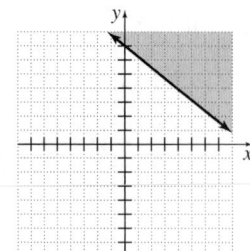

31. $y > -x - 7$

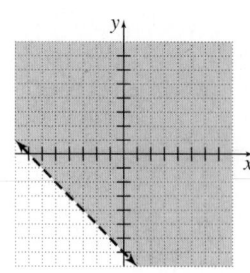

32. $y < x - 3$

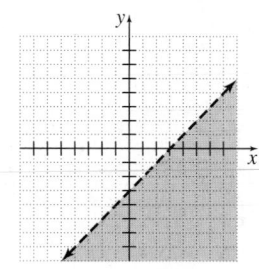

33. $y < -\frac{1}{3}x - 3$

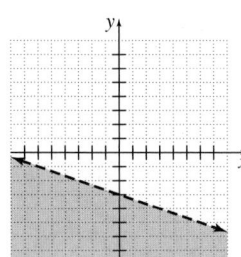

34. $y < -\frac{4}{3}x - 4$

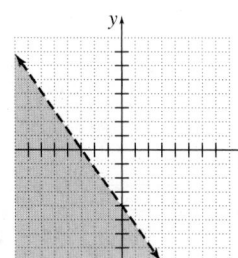

35. $y \leq -\frac{3}{8}x + 3$

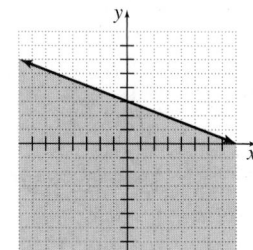

36. $y \geq \frac{2}{3}x - 6$

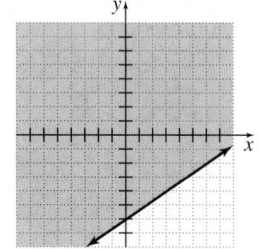

37. $y \geq \frac{6}{7}x + \frac{5}{7}$

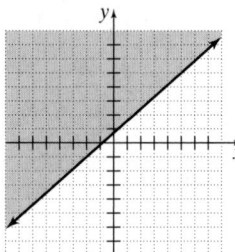

38. $y \leq \frac{9}{10}x - \frac{13}{20}$

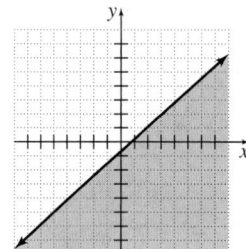

39. $y < 0.5625x$

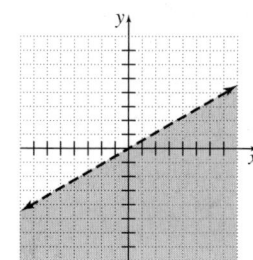

40. $y > -3x$

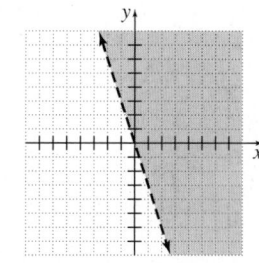

41. $y < 5.4$

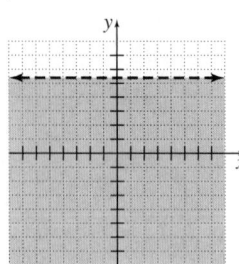

42. $y > 0.25$

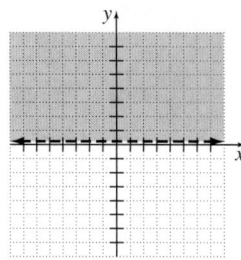

43. $y < x - 9$

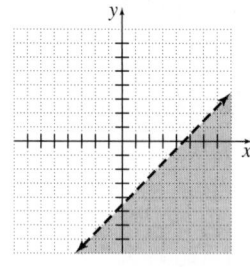

44. $y > x + 9$

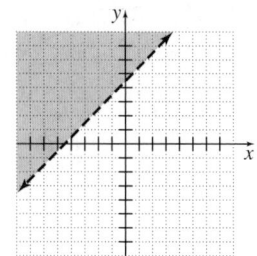

45. $y < -x - 9$

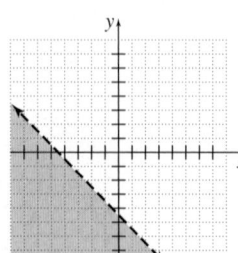

46. $y > -x + 9$

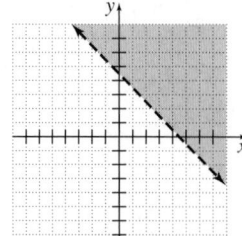

47. $y \leq x$

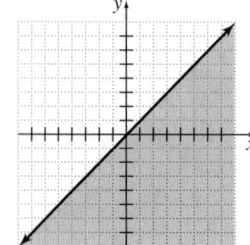

48. $y \geq -x$

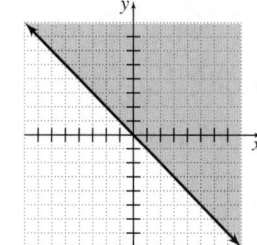

49. $25x + 12y \le 225$

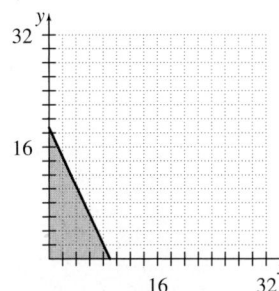

ordered pairs in the shaded region; yes; no

50. $12x + 18y \le 240$

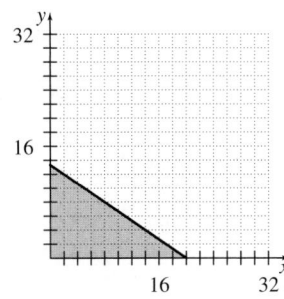

ordered pairs in the shaded region; yes; no

51. $2x + 2y \le 220$

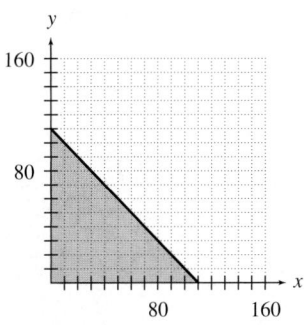

ordered pairs in the shaded region; no; yes

52. $2(10x + 10y) \le 2400$

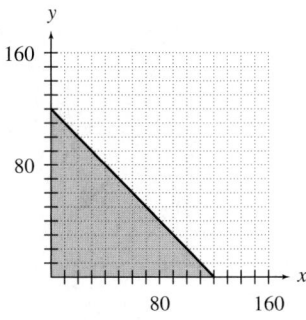

ordered pairs in the shaded region; no; yes

53. $15x + 12y \ge 400$

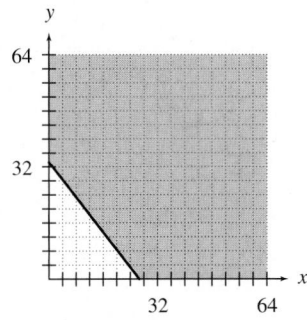

ordered pairs in the shaded region; no; yes

54. $15x + 10y \ge 75$

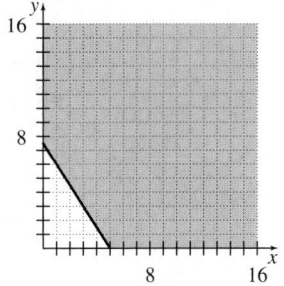

ordered pairs in the shaded region; yes; yes

55. $4.5x + 2y \ge 250$

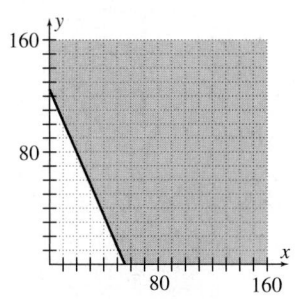

ordered pairs in the shaded region; no; yes

56. $160x + 65y \le 2000$

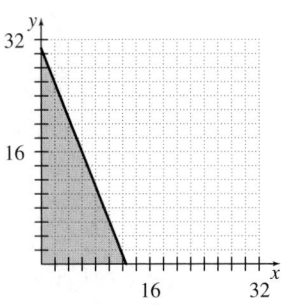

ordered pairs in the shaded region; yes; no

57. $2.11x + 2.37y \le 150$

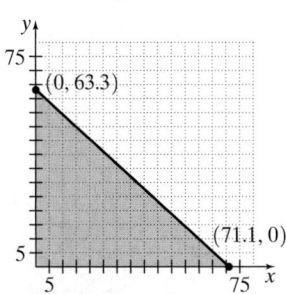

(0, 63.3)

(71.1, 0)

ordered pairs in the shaded region; yes; no

58. $2.21x + 2.72y \le 150.00$

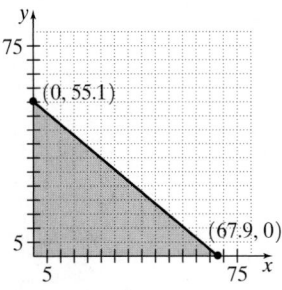

(0, 55.1)

(67.9, 0)

ordered pairs in the shaded region; yes; no

7.3 CALCULATOR EXERCISES

Answers will vary.

SECTION 7.4 EXERCISES

1.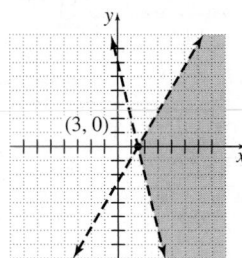

2.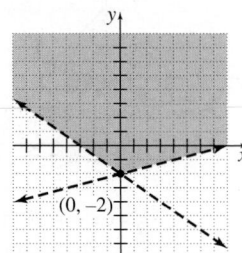

3.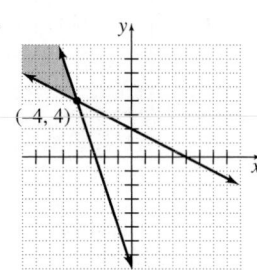

4.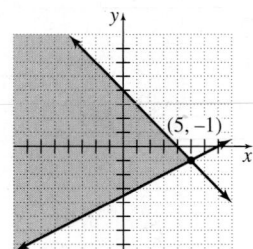

5.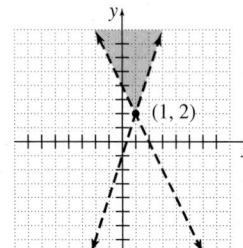

6.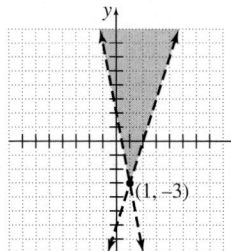

7.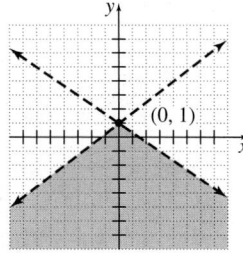

8.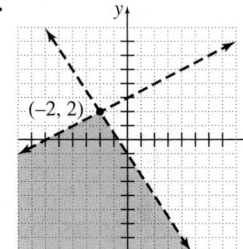

9.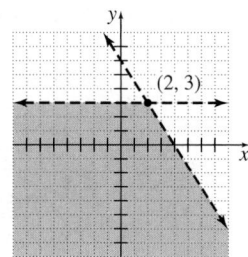

10.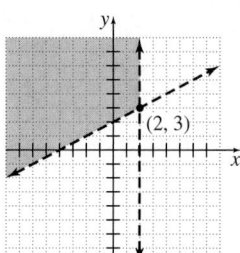

11.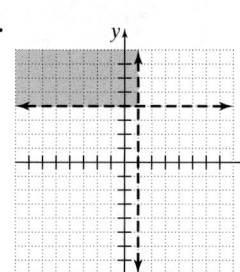

12.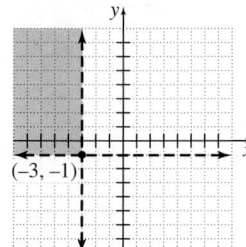

13.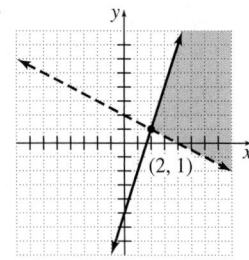

14.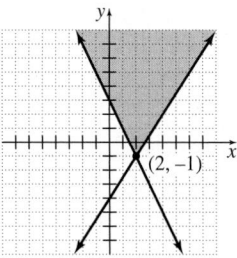

15.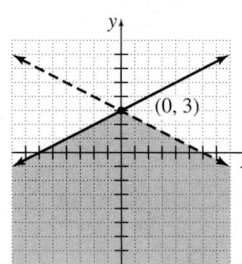

16.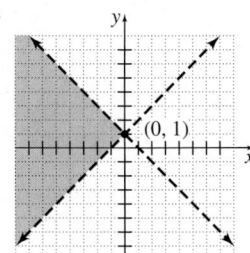

17.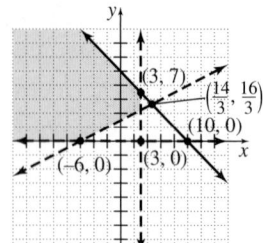

18.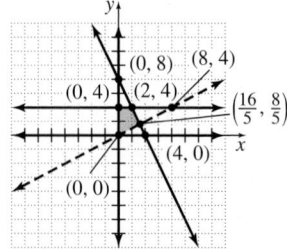

19.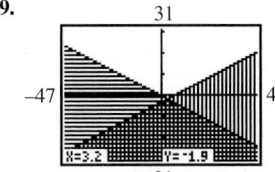

20.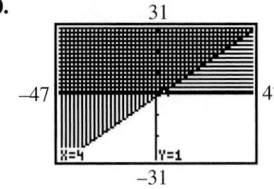

21.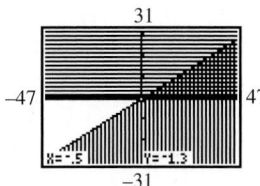

22.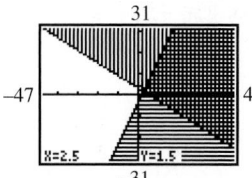

23.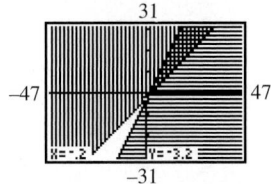

24.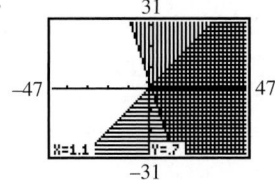

25.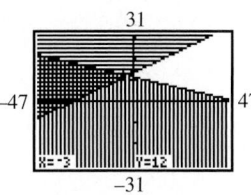

26.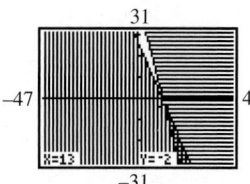

27.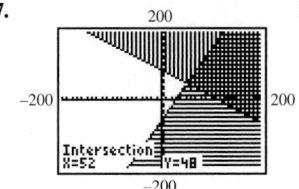

28.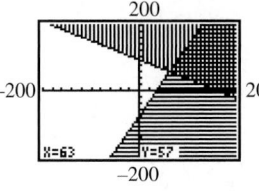

29.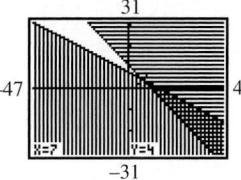

30.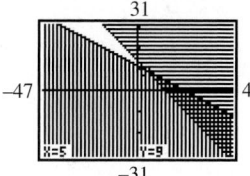

31. $2x + 2y \leq 100$
$y \geq x + 10$
$x \geq 0$
$y \geq 0$

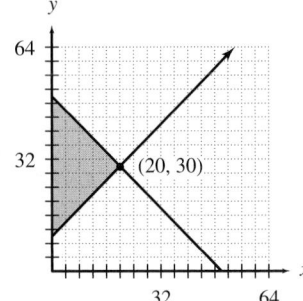

possible answer:
10 feet by 30 feet

32. $2x + 2y \leq 100$
$y \geq 2x$
$x \geq 0$
$y \geq 0$

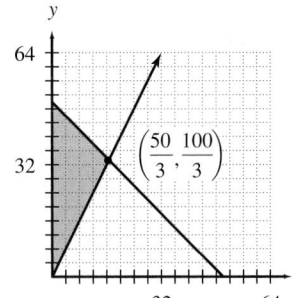

possible answer:
5 feet by 25 feet

33. $x + y \leq 3000$
$0.06x + 0.08y \geq 200$
$x \geq 0$
$x \geq 0$

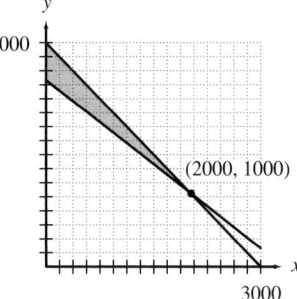

possible answer:
$1000 at 6%; $2000 at 8%

34. $x + y \leq 5000$
$0.06x + 0.08y \geq 350$
$x \geq 0$
$y \geq 0$

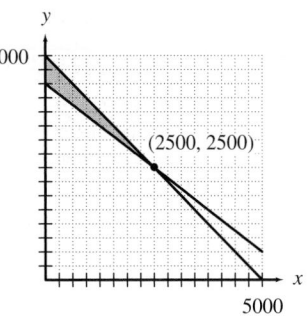

possible answer:
$1000 at 6%; $4000 at 8%

35. $x + y \leq 20$
$6.5x + 8.25y \geq 150$
$x \geq 0$
$y \geq 0$

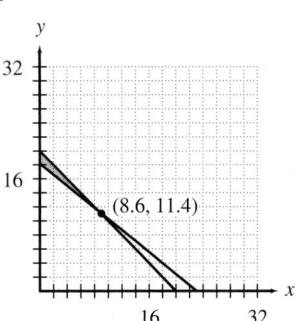

possible answer: He could work 6 hours on the
first job and 14 hours on the second job; no; yes

36. $20x + 5y \le 150$
$x \ge 1.5$
$y \ge 0$

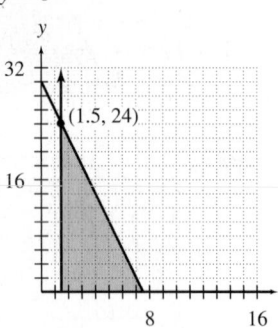

possible answer:
3 hours and 10 guests; yes; no

37. $1.75x + 2.25y \le 200$
$x \ge 50$
$y \ge 25$

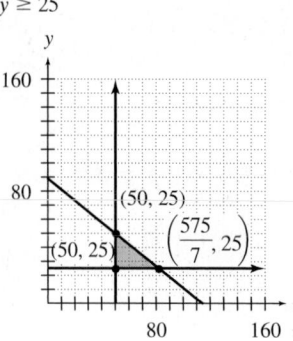

possible answer:
75 servings of lasagna and
30 servings of veal; yes; no

38. $0.05x > 0.01x + y$
$x \ge 25{,}000$
$y \ge 1399$

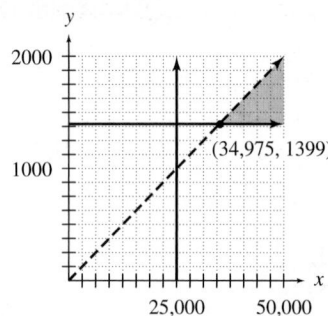

possible answer:
45,000 copies and $1600 machine cost;
no; yes

39. $2.11x + 2.37y \le 150$
$y < 2x$

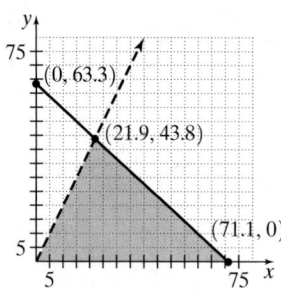

possible answer:
40 packages weighing
2.5 pounds and 48 packages
weighing 4 pounds; no; yes.

40. $2.21x + 2.72y \le 150$
$y \ge x$

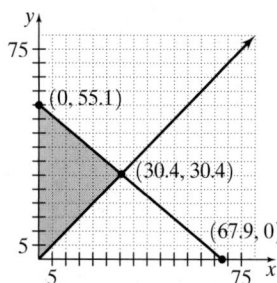

possible answer:
18 packages from zone 4 and
30 packages from zone 7;
no; yes.

7.4 CALCULATOR EXERCISES

1.

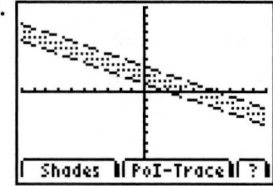

The solution is all ordered
pairs contained in the
shaded region.

2.

The solution is all ordered
pairs contained in the
shaded region.

3.

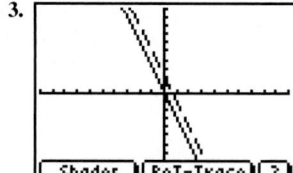

There are no ordered pairs
that satisfy this system of
linear inequalities.

4.

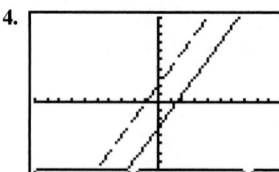

There are no ordered pairs
that satisfy this system of
linear inequalities.

5.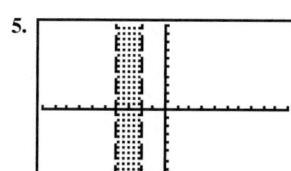

The solution is all ordered pairs contained in the shaded region.

6.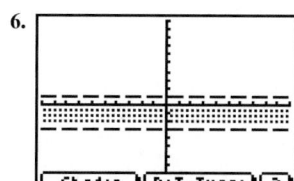

The solution is all ordered pairs contained in the shaded region.

7.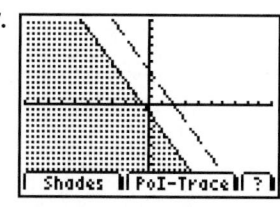

The solution is all ordered pairs below and including the boundary line.

8.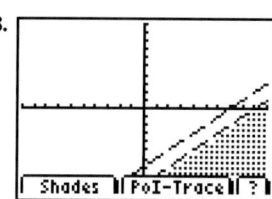

The solution is all ordered pairs below the boundary line $y = x - 9$.

9.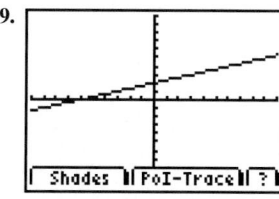

The solution is all ordered pairs on the line $3y = x + 6$ or $y = \frac{1}{3}x + 2$.

10.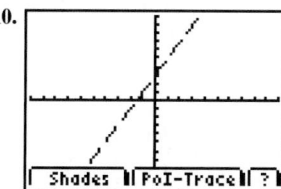

There are no ordered pairs that satisfy this system of linear inequalities.

11.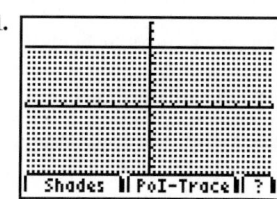

The solution is all ordered pairs below the line $y = 7$. Note, the line is solid but is not part of the solution.

12.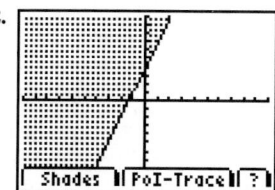

The solution is all ordered pairs above and including the boundary line $y = 3x + 4$.

Reflections

1–7. Answers will vary.

CHAPTER 7 SECTION-BY-SECTION REVIEW

1. nonlinear   **2.** linear; $4x - 11 > 0$ **3.** linear; $\frac{2}{3}x + \frac{1}{3} \le 0$ **4.** nonlinear   **5.** linear; $x - 9 \ge 0$ **6.** nonlinear

7. linear; $-13.2z - 12.6 < 0$ **8.** linear; $-10a + 7 < 0$

9. [number line from 0 to 8, open circle at 3]
$(-\infty, 3)$

10. [number line from −8 to 0, open circle at −2]
$(-2, \infty)$

11. [number line from −8 to 0, closed circle at −5]
$(-\infty, -5]$

12. [number line with −3.5 marked, closed circle at −3.5]
$[-3.5, \infty)$

13. [number line from −4 to 4, open circle at −2 and 4]
$(-2, 4)$

14. [number line from −4 to 4, open circle at −1, closed circle at 0]
$(-1, 0]$

15. [number line from 0 to 8, 5.5 marked, closed circle at 3 and 5.5]
$[3, 5.5]$

16. [number line from 0 to 8, $2\frac{1}{2}$ marked, open circle at $2\frac{1}{2}$ and 8]
$\left(2\frac{1}{2}, 8\right)$

17.

[number line from −4 to 4, −2.3 and $-1\frac{1}{3}$ marked, closed circles]

$\left[-2.3, -1\frac{1}{3}\right]$

18. $(30 + 25)x \ge 300$

19. $x + 2x + \frac{1}{4}(2x) < 200{,}000$ **20.** $125 + 0.28x < 0.34x$ **21. (a)** $F0; 40 \le f \le 72$ **(b)** $F1; 73 \le f \le 112$ **(c)** $F2; 113 \le f \le 157$

22. integers less than -6 **23.** integers greater than 6   **24.** integers less than or equal to 3   **25.** integers greater than or equal to -2

26. $x > 2$
$(2, \infty)$

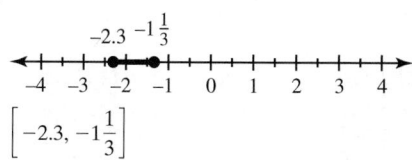

27. $x \le 2.4$
$(-\infty, 2.4]$

[number line from 0 to 8, 2.4 marked, closed circle at 2.4]

28. no solution   **29.** all real numbers
$(-\infty, \infty)$

30. $(259, \infty)$ **31.** $\left(-\infty, \dfrac{32}{39}\right)$ **32.** $(-14, \infty)$ **33.** no solution   **34.** $(-\infty, -3.4]$ **35.** $[1.4, \infty)$ **36.** $(0, \infty)$

37. no solution   **38.** $[2, 3.5)$ **39.** $[8, 12)$

40. $49.95 + 0.18x \le 150$; the number of miles driven should be less than or equal to 555 miles.

41. $\dfrac{2100 + 1300 + 1650 + 1250 + 1725 + x}{6} > 1500$; his sales should be greater than \$975.

42. $2x + 2(x + 4) \le 40$; the width can be no more than 8 feet.   **43.** $35x - (20x + 225) \ge 500$; they must have at least 49 students per day.

44. $88 \le \dfrac{89 + 96 + 89 + 80 + 100 + x}{6} \le 92$; the student must score 74 to 98 on the last test.

45. $0.15x + 2 \ge 27$; the weight must be at least $166\dfrac{2}{3}$ pounds.

46. linear; $x + 2y < 12$ **47.** linear; $6x - 9y > -5$ **48.** nonlinear   **49.** linear; $3x - 14y > -29$

50. nonlinear   **51.** nonlinear

52. $y < -2x + 8$ **53.** $y > \dfrac{3}{5}x - 6$ **54.** $y \le \dfrac{1}{4}x + 3$ **55.** $y \ge 3$

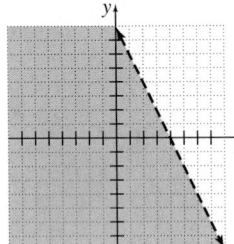   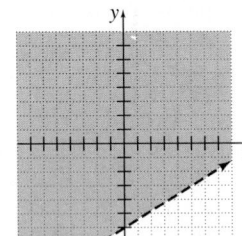   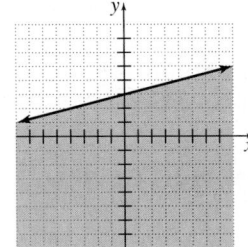   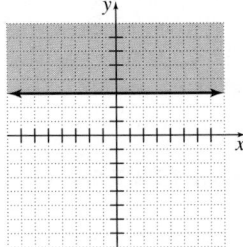

56. $y < -2$ **57.** $x > 2$ **58.** $y < \dfrac{2}{3}x - 2$ **59.** $y \ge -4x + 11$

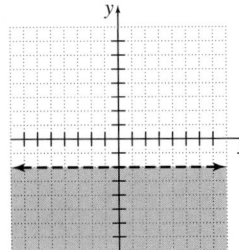   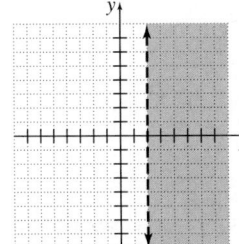   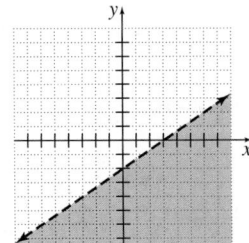   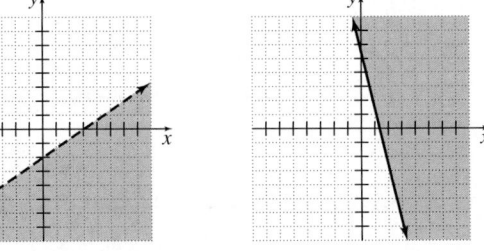

60. $y \ge 0.5x - 2$ **61.** $y > -\dfrac{2}{3}x - 10$ **62.** $y > -x - 2$ **63.** $y > -9x + 6$

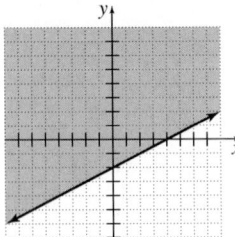   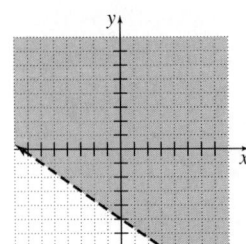   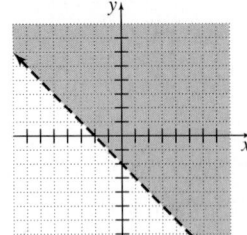   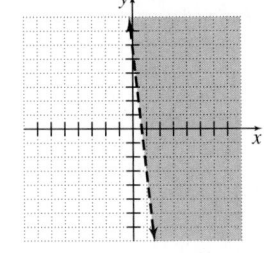

64. $4x + 6y \leq 85$

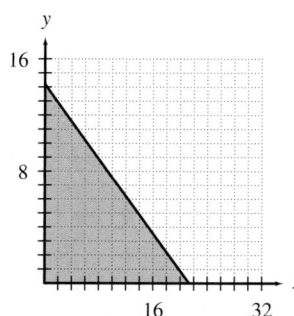

possible answer: 15 rhododendrons and 2 azaleas, or 10 rhododendrons and 5 azaleas

65. $5x - 3y \geq 80$

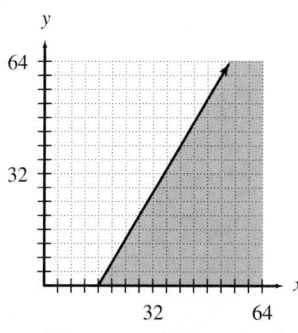

possible answer: 40 correct and 3 incorrect, or 30 correct and 0 incorrect

66. $5.20x + 7.70y + 10.25 \leq 120$

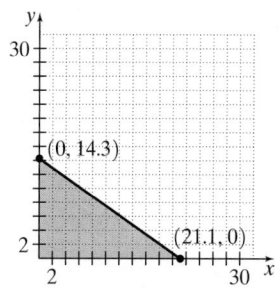

possible answer: 5 packages weighing 2.5 pounds and 8 packages weighing 5 pounds; eight 2.5 pound packages and six 5 pound packages; yes; no.

67.

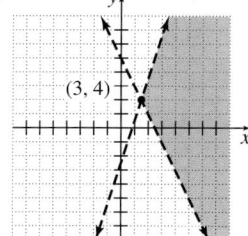

68.

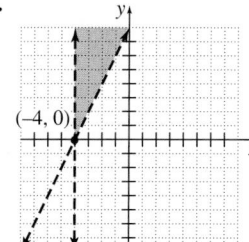

69.

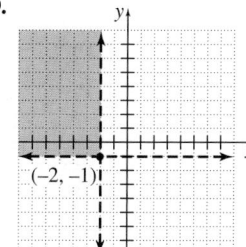

70.

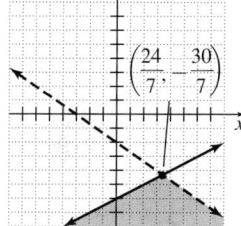

71.

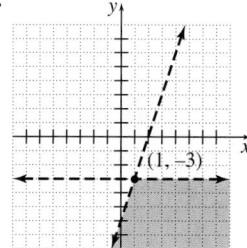

72.

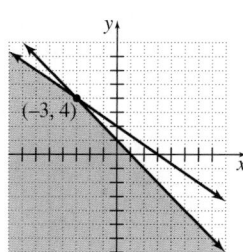

73.

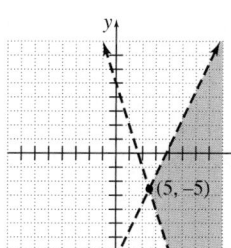

74.

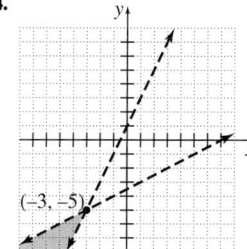

75.

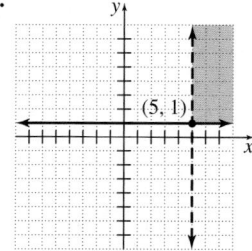

76.

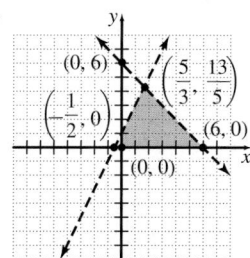

77.

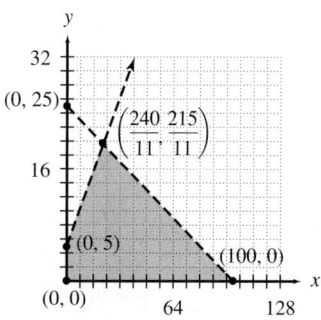

78. $4x + 6y \leq 85$
$y + 4 \leq x$
$x \geq 0$
$y \geq 0$

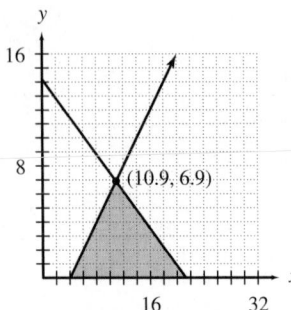

possible answer: 15 rhododendrons and 2 azaleas

79. $0.05x + 0.06y \geq 225$
$x + y \leq 4000$
$x \geq 0$
$y \geq 0$

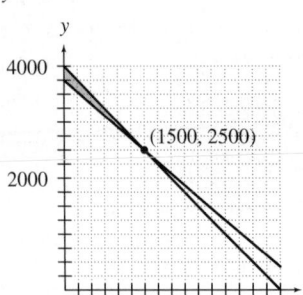

possible answer: $200 at 5% and $3800 at 6%

80. $5.20x + 7.70y + 10.25 \leq 120$
$x \leq 2y$

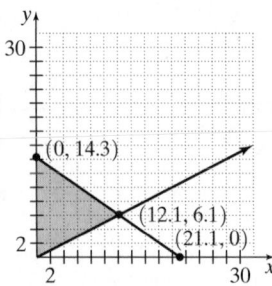

possible answer: 9 packages weighing 2.5 pounds each and 7 packages weighing 5 pounds each; no; yes.

CHAPTER 7 CHAPTER REVIEW

1.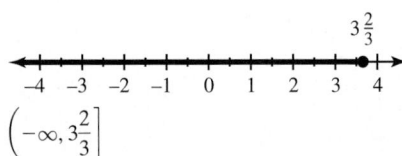
$(-\infty, -2)$

2.
$(7, \infty)$

3.
$(-1, 3)$

4.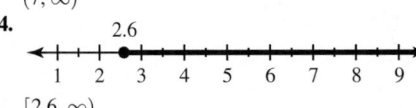
$[2.6, \infty)$

5.
$\left(-\infty, 3\frac{2}{3}\right]$

6.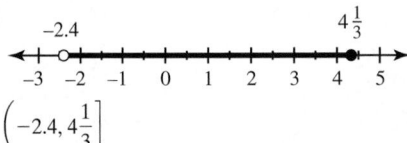
$\left(-2.4, 4\frac{1}{3}\right]$

7. integers less than -4 **8.** integers greater than 8   **9.** integers less than or equal to -3 **10.** integers less than or equal to 14

11. $x > 2$
$(2, \infty)$

12. $x \leq -4.2$
$(-\infty, -4.2]$

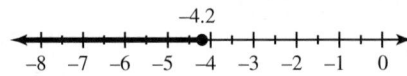

13. all real numbers
$(-\infty, \infty)$

14. no solution
15. $(-186, \infty)$

16. $\left(\frac{23}{34}, \infty\right)$ **17.** $(7, \infty)$ **18.** no solution   **19.** $[-2.8, \infty)$ **20.** $(-\infty, -3.7]$ **21.** $(-\infty, -1)$ **22.** $(-\infty, \infty)$ **23.** $\left(\frac{7}{4}, 3\right]$

24.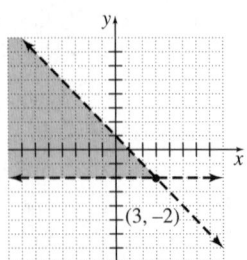

25.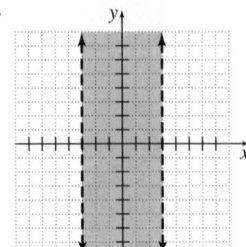

26.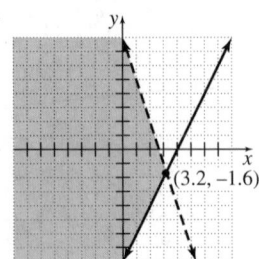

27.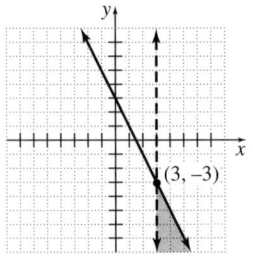

28.
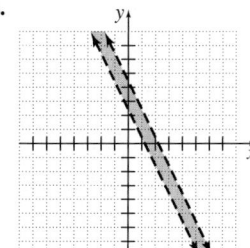

29.
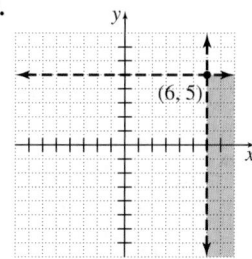
(6, 5)

30.
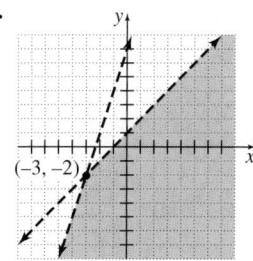
(−3, −2)

31.
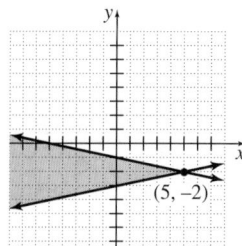
(5, −2)

32.
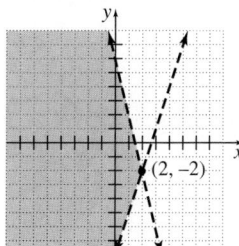
(2, −2)

33.
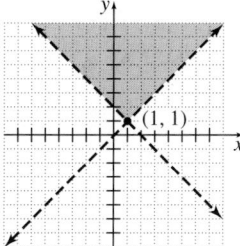
(1, 1)

34.
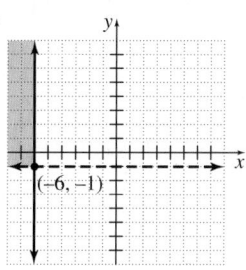
(−6, −1)

35.
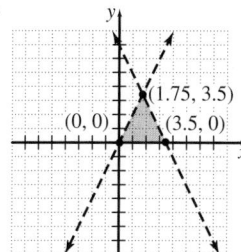
(1.75, 3.5)
(0, 0)   (3.5, 0)

36.
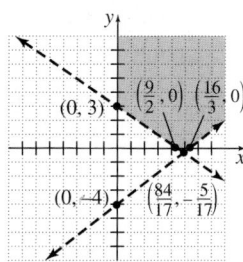
(0, 3) $\left(\frac{9}{2}, 0\right)$ $\left(\frac{16}{3}, 0\right)$
(0, −4) $\left(\frac{84}{17}, -\frac{5}{17}\right)$

37. $y > \frac{9}{5}x - 9$
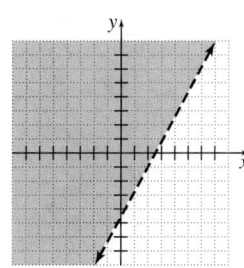

38. $y > \frac{4}{3}x - 5$
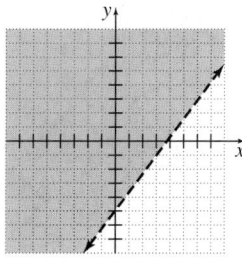

39. $y \le \frac{2}{3}x + 3$
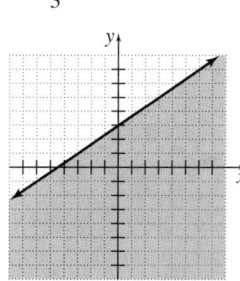

40. $y \ge -5$
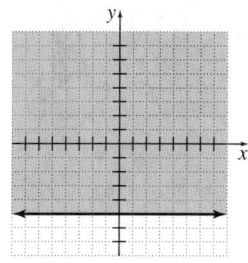

41. $y > \frac{5}{3}$
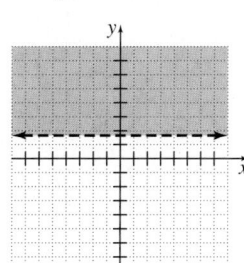

42. $x < -16$
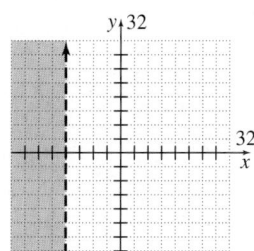
32
32

43. $y < \frac{1}{7}x - 3$
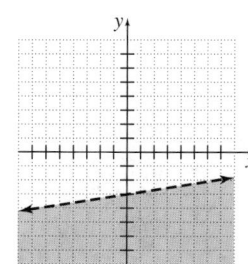

44. $y \ge -0.5x + 3$
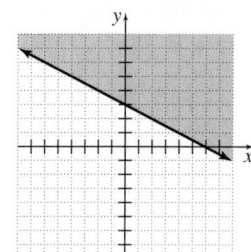

45. $y \geq \dfrac{8}{3}x - 8$

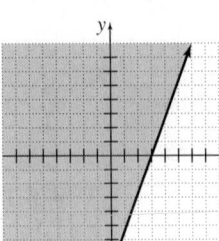

46. $y > -\dfrac{3}{4}x - 3$

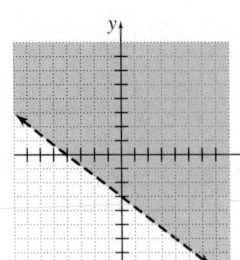

47. $y > x - 7$

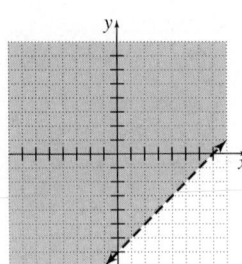

48. $y > -5x + 7$

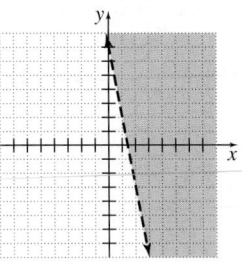

49. $130 \leq w \leq 165$; $166 \leq w \leq 190$; $w > 190$

50. $12.5x - (255 + 2.5x) \geq 1200$; to make a profit, at least 146 packs must be sold; possible answer: There were 200 packs sold.

51. $\dfrac{45 + 36 + 52 + 48 + 31 + x}{6} < 42$; her sixth phone bill must be less than \$40; possible answer: Her phone bill should be \$35.

52. $18x > 600$; the length must be greater than $33\dfrac{1}{3}$ inches; possible answer: The length is 40 inches.

53. $19.95x - (1.45x + 17) \geq 600$; they must sell at least 34 packages; possible answer: sell 50 packages.

54. $500 < \dfrac{344 + 434 + 254 + 705 + 723 + x}{6} < 600$; the profit should be between \$540 and \$1140; possible answer: profit of \$1000.

55. $2.37x \leq 100$; you can send no more than 42 packages; possible answer: send 40 packages.

56. $3x + y \geq 25$

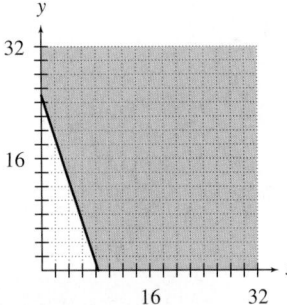

possible answer:
10 wins and five ties

57. $2.37x + 2.20y \leq 100$
$y < x$

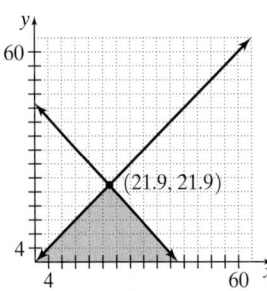

possible answer:
send twenty 4 pound packages and
nineteen 3 pound packages; yes; no

58. $y \geq 2x$
$x + y \leq 540$
$x \geq 0$
$y \geq 0$

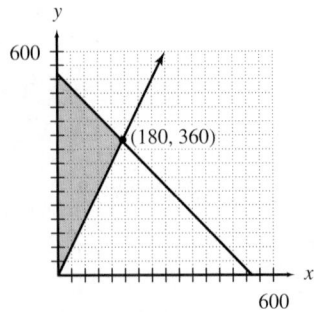

possible answer:
100 acres of oats and
400 acres of wheat

59. $3x + 5y \geq 6000$
$y \geq x + 75$
$x \geq 0$
$y \geq 0$

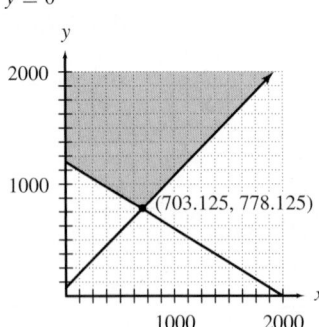

possible answer:
\$1000 for an efficiency;
\$1500 for a regular apartment

CHAPTER 7 TEST

1. linear; $x - 3y > 8$ **2.** nonlinear   **3.** linear; $6x - 18 \geq 0$ **4.** linear; $\frac{1}{2}x - y \geq 4\frac{3}{8}$ **5.** no solution

6.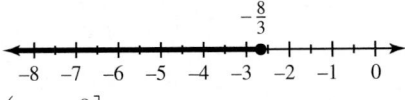

$(-\infty, -4)$

7. ◄——————————————►

$(-\infty, \infty)$

8.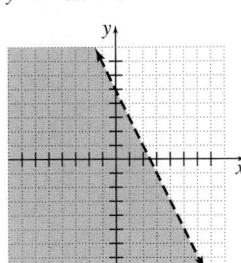

$\left(-\infty, -\frac{8}{3}\right]$

9. ◄—+—+—+—+—+—●—○—+—+—►
-1 0 1 2 3 4 5 6 7

$(4, 5]$

10. $y < -2x + 5$ **11.** $x < 4$

12.

13.

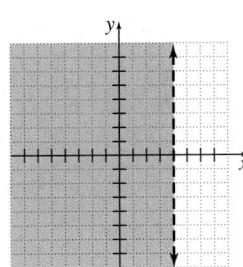

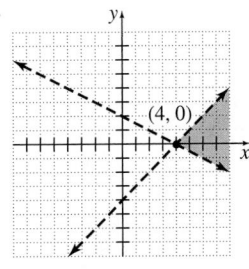

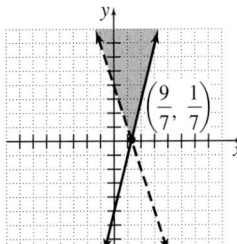

14.   **15.** $F3; 158 \leq f \leq 206$
$F4; 207 \leq f \leq 260$
$F5; 261 \leq f \leq 318$
$F6; f \geq 319$

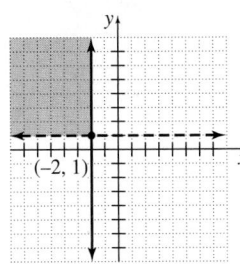

16. $\dfrac{83 + 72 + x}{3} \geq 80$; the score must be at least 85; possible answer: The score is 90 points.

17. $158 \leq \dfrac{565 + f}{4} \leq 206$; the fourth tornado's wind speed must be between 67 mph and 259 mph inclusive; possible answer: wind speed is 200 mph.

18. $49x - (23.5x + 250) \geq 300$; she must sell at least 22 baskets; possible answer: she sells 25 baskets.

19. $5x + 2y \geq 20$ **20.** $22.50x + 26.25y \leq 500$ **21.** $0.04x + 0.08y \geq 400$ **22.** Answers will vary.
$y \geq x$ $x + y \leq 6000$
$x \geq 0$
$y \geq 0$

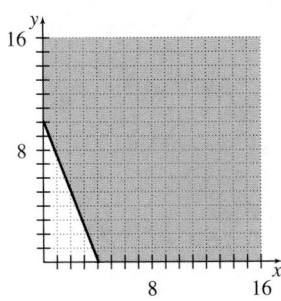

possible answer:
five good deeds and
five activity sheets

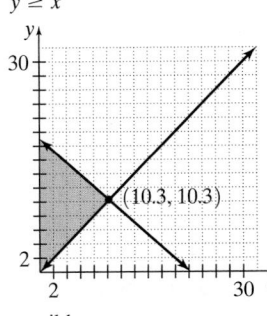

possible answer:
ship 4 baskets to zone 102
and 11 baskets to zone
103; no; yes

possible answer:
$500 at 4% and $5500 at 8%

CUMULATIVE REVIEW CHAPTERS 1–7

1. 9 **2.** −9 **3.** $-\dfrac{3}{4}$ **4.** 3.16 **5.** $-\dfrac{3}{2}$ **6.** $\dfrac{9}{40}$ **7.** −6 **8.** 35 **9.** undefined **10.** $-2x + 2$ **11.** $-2x + 6$

12.

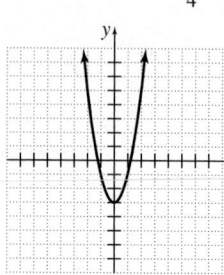

13. The domain of the relation is all real numbers. The range of the relation is all real numbers ≥ -3.

14. The relation is a function, since all possible vertical lines cross the graph a maximum of one time.

15. The relative minimum value is −3 at $x = 0$. There is no relative maximum.

16. The relation is increasing for all $x > 0$ and decreasing for all $x < 0$.

17. The x-intercepts are $(-1.225, 0)$ and $(1.225, 0)$. The y-intercept is $(0, -3)$.

18. $\dfrac{8}{7}$ **19.** −2 **20.** All real numbers **21.** 2 and −8 **22.** $x = 0$

23. $x \leq -\dfrac{2}{5}; \left(-\infty, -\dfrac{2}{5}\right]$ **24.** no solution **25.** $10 < x \leq 30$ $(10, 30]$

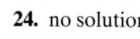

26. 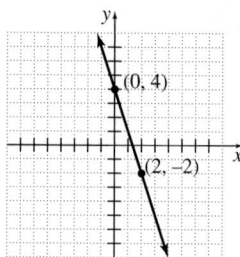 **27.** 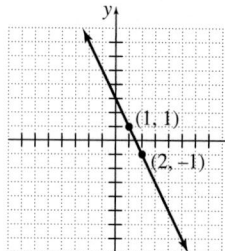 **28.** 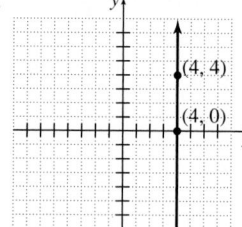 **29.**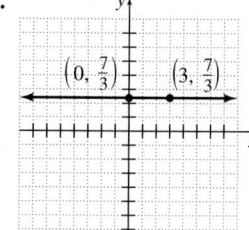

30.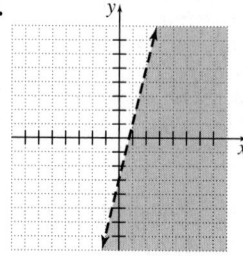

31. intersecting and perpendicular lines.
32. $(-9, -23)$
33. all ordered pairs that satisfy $y = -2x + 4$.

34.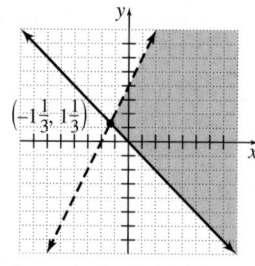

One solution is $(1, 0)$.

35. −5 **36.** undefined **37.** 2 **38.** 0 **39.** $y = -5x - 7$ **40.** $y = -\dfrac{2}{3}x + \dfrac{22}{3}$ **41.** 5.34×10^{6} **42.** 0.00012

43. −0.00004783 **44.** $L = \dfrac{P}{2} - W$ **45.** 2 **46.** Lance should invest \$2,500.

47. The smaller angle measures $21\dfrac{2}{3}°$ and the larger angle measures $68\dfrac{1}{3}°$.

48. Mike should mix $3\dfrac{1}{3}$ pounds of hazelnut coffee and $6\dfrac{2}{3}$ pounds of cinnamon coffee.

49. April must score at least 80 on the last test to get a B in her algebra class.
50. $c(x) = 35 + 1.50x$; the cost of renting a chain saw for 12 hours is \$53.

Chapter 8

SECTION 8.1 EXERCISES

1. yes   **2.** no   **3.** yes   **4.** no   **5.** no   **6.** yes   **7.** yes   **8.** no   **9.** yes   **10.** yes   **11.** no   **12.** yes   **13.** yes
14. no   **15.** no   **16.** yes   **17.** no   **18.** yes   **19.** yes   **20.** yes   **21.** trinomial   **22.** polynomial   **23.** monomial
24. binomial   **25.** binomial   **26.** trinomial   **27.** polynomial   **28.** monomial   **29.** trinomial   **30.** monomial
31. binomial   **32.** binomial   **33.** monomial   **34.** binomial   **35.** trinomial   **36.** binomial   **37.** degrees of terms: 0, 1; degree
of polynomial: 1   **38.** degree of term: 0; degree of polynomial: 0   **39.** degree of term: 0; degree of polynomial: 0
40. degrees of terms: 4, 6; degree of polynomial: 6   **41.** degree of term: 0, degree of polynomial: 0
42. degrees of terms: 2, 2; degree of polynomial: 2   **43.** degrees of terms: 5, 4, 0; degree of polynomial: 5
44. degrees of terms: 1, 0; degree of polynomial: 1   **45.** degrees of terms: 2, 2; degree of polynomial: 2
46. degrees of terms: 9, 6, 5; degree of polynomial: 9   **47.** degrees of terms: 15, 12, 7; degree of polynomial: 15
48. degree of term: 2; degree of polynomial: 2   **49.** $a^3 + 3a^2 - 2a + 5$ **50.** $4a^9 - 2a^6 + 3a^3 - 17$ **51.** $-\frac{8}{15}x^4 + \frac{4}{5}x^3 + \frac{2}{15}x$
52. $-p^5 - p^4 + p^3 + p^2 - p - 1$ **53.** $3.06x^4 + 0.1x^3 + 4.6x^2 - 1.72$ **54.** $23x^9 + 15x^7 + 9x^5 + 5x^3 + 7x + 33$
55. $11 - 2b + 7b^2 - 6b^3$ **56.** $x^2 - 4x^3 + x^4 + 6x^5 - 8x^6$ **57.** $\frac{1}{14}x - \frac{5}{14}x^3 + \frac{1}{7}x^5$ **58.** $-1 - q + q^2 + q^3 - q^4 - q^5$
59. $-2.77 + 3.2x^3 + 9.76x^5 + 0.5x^7$ **60.** $4 - 2x + 4x^2 - 2x^3 - 4x^4 + x^5$ **61. (a)** 33   **(b)** .21   **(c)** 1
62. (a) 35   **(b)** -1 **(c)** 1   **63. (a)** 29   **(b)** 29   **(c)** 0   **64. (a)** -1 **(b)** -13 **(c)** 27
65. (a) -2.4 **(b)** 9.0375   **(c)** -20.4625 **66. (a)** 4   **(b)** 21.204   **(c)** -11.3 **67. (a)** 0   **(b)** $-\frac{77}{24}$ **(c)** 0
68. (a) -2 **(b)** 0   **(c)** $34\frac{2}{3}$ **69.** The perimeter measures $4x + 1$ units; the perimeter measures 33 inches.
70. The surface area measures $4\pi x^2$ square units; the surface area measures 64π in^2.
71. The surface area measures $2x^2 + 12x$ square units; the surface area measures 22.5 m^2.
72. The surface area measures $30h + 100$ square units; the surface area measures 205 in^2.
73. The total area is $x^2 + 12x + 21$ m^2.   **74.** The total area is $x^2 + \frac{1}{2}xy + y^2$ square units.
75. The area not covered by the patio is $lw - 300$ ft^2; the area not covered by the patio is 8700 ft^2.
76. The area not covered by the pool is $lw - \pi r^2$ square units; the area not covered by the pool is approximately 811.4 m^2.
77. The replacement cost is $12x^2$ dollars; the replacement cost is $1728.
78. The replacement cost is $12\pi r^2$ dollars; the replacement cost is about $5428.67.
79. (a) Revenue $= 200 + 12x^2$ **(b)** Cost $= 75 + 2.75x^2$ **(c)** The revenue is $2900; the cost is $693.75; the profit is $2206.25.
80. (a) Revenue $= 350 + 15lw$ **(b)** Cost $= 100 + 5.25lw$ **(c)** The revenue is $4400; the cost is $1517.50. The profit is $2882.50.
81. (a) The IQ is 122.   **(b)** The IQ is 85.   **(c)** The IQ is 99.   **82. (a)** The IQ is 106.   **(b)** The IQ is 130.   **(c)** The IQ is 73.
83. The expected mileage is 36,416 miles.   **84.** The expected mileage is 35,968 miles.

8.1 CALCULATOR EXERCISES

1. $\{-192, -32, -8, 24\}$ **2.** $\{-432, -108, -24\}$
3. $\{0.229, 0.06725, -32.625, -0.016\}$ or $\left\{\dfrac{229}{1000}, \dfrac{269}{4000}, -\dfrac{261}{8}, -\dfrac{2}{125}\right\}$

SECTION 8.2 EXERCISES

1.

x	y
-2	4
-1	0
0	-6
1	-8
2	0

2.

x	y
-2	0
-1	8
0	6
1	0
2	-4

3.

x	y
-2	-3
-1	-2
0	1
1	6
2	13

4.

x	y
-2	13
-1	6
0	1
1	-2
2	-3

5. all real numbers   **6.** all real numbers   **7.** $y \le 5$ **8.** $y \le 9$ **9.** $y \ge -5$ **10.** $y \ge -7$ **11.** $y \ge 6$ **12.** $y \ge -4$
13. $y \le 4.5$ **14.** $y \le 12$ **15.** all real numbers   **16.** all real numbers   **17.** all real numbers   **18.** all real numbers
19. all real numbers   **20.** all real numbers   **21.** all real numbers   **22.** all real numbers   **23.** $y \ge -4$ **24.** $y \ge -1.27$ (approx.)
25. $y \le 1.27$ (approx.)   **26.** $y \le 4$ **27.** $y \le 20.97$ (approx.)   **28.** $y \ge -3.12$ (approx.)   **29.** 100   **30.** 81   **31.** 36   **32.** 49
33. 25   **34.** 81   **35.** 4   **36.** 16   **37.** 16   **38.** 49   **39.** 6.25   **40.** 2.56   **41.** 38   **42.** 16   **43.** -14 **44.** -52

45. 15.9   **46.** 156.1   **47.** -217.5 **48.** -41.3 **49.** 2578.3   **50.** 21,063.3   **51.** -2 **52.** -6.7 **53.** $\dfrac{9}{8}$ **54.** $\dfrac{167}{8}$
55. $-\dfrac{364}{8}$ **56.** $-\dfrac{67}{8}$

57. (a) When 5, 10, 15, 20, 25, and 30 watches are ordered, the revenue is $625, $1000, $1125, $1000, $625, and $0 respectively.
(b) The range is all non-negative real numbers less than or equal to 1125.   **(c)** This range shows us that the maximum revenue would be $1125.
58. (a) When the space is rented for 10, 20, 30, 40, 50, and 60 months, the cost is $2500, $4000, $4500, $4000, $2500, and $0, respectively.
(b) The range is all real numbers less than or equal to 4500.   **(c)** This range shows us that the maximum cost would be $4500, and there is a value of x for which the cost is $0.

59.

Age (x)	63	65	70	75	80	85
Days ($S(x)$)	4.131	5.135	7.26	8.835	9.86	10.335

(a) The function predicted a stay of 10 days for an 85-year-old woman, which is not a very good prediction of the actual data for an 8-day stay for a woman that age.   **(b)** The function predicted a stay of 7 days for a 70-year-old woman, which is a fairly good prediction of the actual data for a 6-day stay, but not as good for a 9-day stay.   **(c)** Answers will vary.

60.

Age (x)	35	40	45	50	55	60	65	70
Premium ($P(x)$)	224.05	675.20	802.35	928	1374.65	2464.80	4520.95	7865.60

(a) The prediction is not very good.   **(b)** The prediction is relatively close.   **(c)** Answers will vary.
61. The domain is all real numbers between 0 and 8; the rocket is in the air for 8 seconds before it explodes. It explodes before it reaches the maximum value of the function; the range is all real numbers less than or equal to 1136; the maximum height will be 1136 feet.
62. The domain is all real numbers between 0 and 10; the rocket is in the air for 10 seconds before it explodes; the range is all real numbers less than or equal to 1426.25; the maximum height will be at 1426.25 feet.
63. The range is all non-negative real numbers less than or equal to 1242. The parachutists jumped from a height of 1242 feet.
64. The range is all non-negative real numbers less than or equal to 1431. The parachutists would jump from a height of 1431 feet.

8.2 CALCULATOR EXERCISES

Part I.

1. $y = x$
 x int $= (0, 0)$
 y int $= (0, 0)$

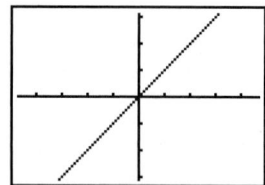

2. $y = x^2 + x$
 x int $= (-1, 0)$ and $(0, 0)$
 y int $= (0, 0)$

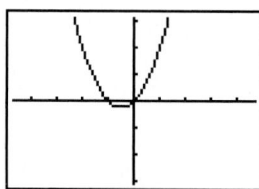

3. $y = x^3 - x$
 x int $= (1, 0)(-1, 0)(0, 0)$
 y int $= (0, 0)$

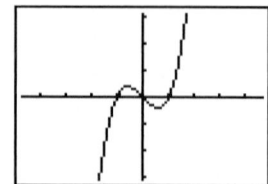

4. $y = x^4 + 2x^3 - x^2 - 2x$
 x int $= (1, 0)\ (0, 0)(-1, 0)(-2, 0)$
 y int $= (0, 0)$

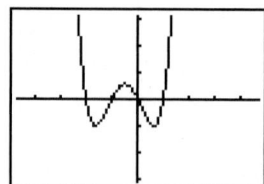

5. $y = x^5 - 5x^3 + 4x$

x int $= (-2, 0)(-1, 0)(0, 0)(1, 0)(2, 0)$

y int $= (0, 0)$

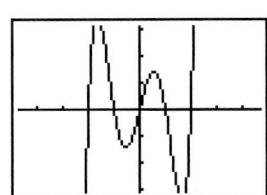

	Degree	# of changes
$y = x$	1	0
$y = x^2 + x$	2	1
$y = x^3 - x$	3	2
$y = x^4 + 2x^3 - x^2 - 2x$	4	3
$y = x^5 - 5x^3 + 4x$	5	4

Part 2.

1. -63.156 **2.** 9416390.016 **3.** -4.65793664 **4.** $\dfrac{451}{3}$ **5.** $-\dfrac{29}{48}$ **6.** $-\dfrac{2563}{6}$

SECTION 8.3 EXERCISES

1. nonquadratic   **2.** quadratic   **3.** nonquadratic   **4.** quadratic   **5.** quadratic   **6.** quadratic   **7.** quadratic
8. nonquadratic   **9.** quadratic   **10.** quadratic   **11.** quadratic   **12.** nonquadratic   **13.** nonquadratic
14. quadratic   **15.** quadratic   **16.** nonquadratic
17. $a = -5$ $b = 10$ $c = 1$
The graph is narrow compared with that of $y = x^2$ and concave downward. The vertex is $(1, 6)$. The axis of symmetry is $x = 1$.
The y-intercept is $(0, 1)$.
18. $a = 6$ $b = -6$ $c = -5$

The graph is narrow compared with that of $y = x^2$ and concave upward. The vertex is $\left(\dfrac{1}{2}, -6\dfrac{1}{2}\right)$.

The axis of symmetry is $x = \dfrac{1}{2}$. The y-intercept is $(0, -5)$.

19–26.

	a	b	c	graph wide/narrow	graph concave upward/downward	graph vertex	axis of symmetry	y-intercept
19.	0.6	6	-2	wide	upward	$(-5, -17)$	$x = -5$	$(0, -2)$
20.	-1	6	-2	neither	downward	$(3, 7)$	$x = 3$	$(0, -2)$
21.	2	3	5	narrow	upward	$(-0.75, 3.875)$	$x = -0.75$	$(0, 5)$
22.	-3	6	-5	narrow	downward	$(1, -2)$	$x = 1$	$(0, -5)$
23.	$-\dfrac{1}{4}$	1	-3	wide	downward	$(2, -2)$	$x = 2$	$(0, -3)$
24.	$\dfrac{1}{3}$	2	-1	wide	upward	$(-3, -4)$	$x = -3$	$(0, -1)$
25.	1	8	1	neither	upward	$(-4, -15)$	$x = -4$	$(0, 1)$
26.	-0.4	2.4	-1.1	wide	downward	$(3, 2.5)$	$x = 3$	$(0, -1.1)$

27.

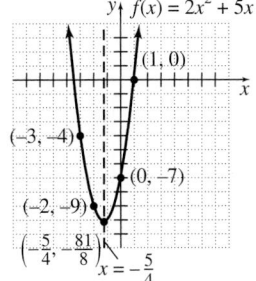

$f(x) = 2x^2 + 5x - 7$
$(1, 0)$
$(-3, -4)$
$(0, -7)$
$(-2, -9)$
$\left(-\dfrac{5}{4}, -\dfrac{81}{8}\right)$
$x = -\dfrac{5}{4}$

28.

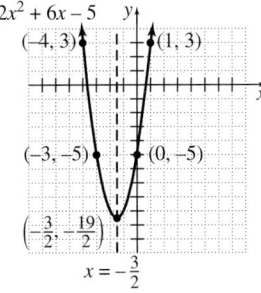

$y = 2x^2 + 6x - 5$
$(-4, 3)$
$(1, 3)$
$(-3, -5)$
$(0, -5)$
$\left(-\dfrac{3}{2}, -\dfrac{19}{2}\right)$
$x = -\dfrac{3}{2}$

29.

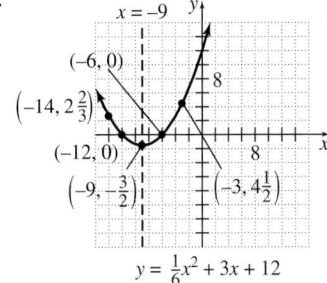

$x = -9$
$(-6, 0)$
8
$\left(-14, 2\dfrac{2}{3}\right)$
$(-12, 0)$
$\left(-9, -\dfrac{3}{2}\right)$
$\left(-3, 4\dfrac{1}{2}\right)$
$y = \dfrac{1}{6}x^2 + 3x + 12$

30.

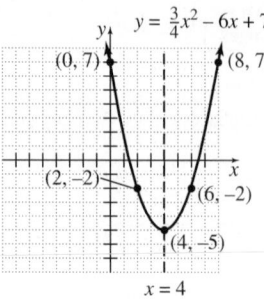

$y = \frac{3}{4}x^2 - 6x + 7$
(0, 7), (8, 7)
(2, -2), (6, -2)
(4, -5)
$x = 4$

31.

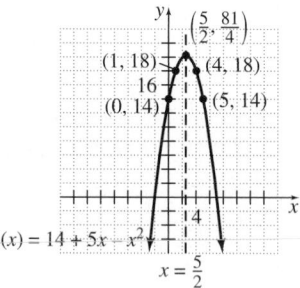

$\left(\frac{5}{2}, \frac{81}{4}\right)$
(1, 18), (4, 18)
16
(0, 14), (5, 14)
4
$h(x) = 14 + 5x - x^2$
$x = \frac{5}{2}$

32.

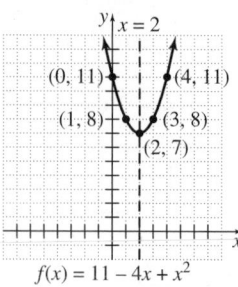

$x = 2$
(0, 11), (4, 11)
(1, 8), (3, 8)
(2, 7)
$f(x) = 11 - 4x + x^2$

33.

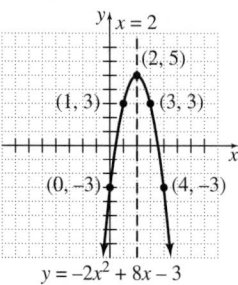

$x = 2$
(2, 5)
(1, 3), (3, 3)
(0, -3), (4, -3)
$y = -2x^2 + 8x - 3$

34.

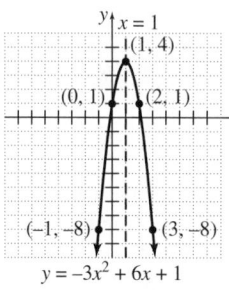

$x = 1$
(1, 4)
(0, 1), (2, 1)
(-1, -8), (3, -8)
$y = -3x^2 + 6x + 1$

35.

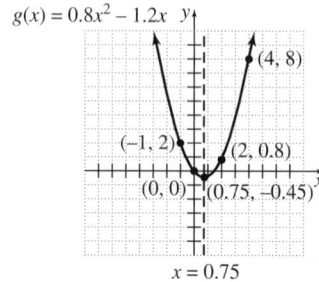

$g(x) = 0.8x^2 - 1.2x$
(4, 8)
(-1, 2), (2, 0.8)
(0, 0), (0.75, -0.45)
$x = 0.75$

36.

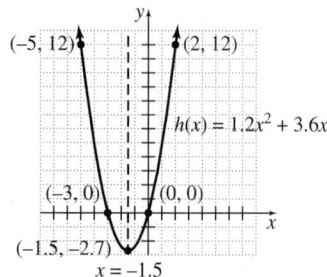

(-5, 12), (2, 12)
$h(x) = 1.2x^2 + 3.6x$
(-3, 0), (0, 0)
(-1.5, -2.7)
$x = -1.5$

37.

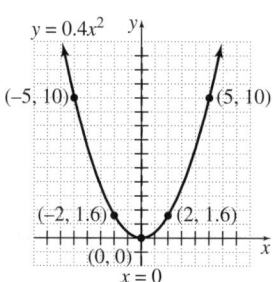

$y = 0.4x^2$
(-5, 10), (5, 10)
(-2, 1.6), (2, 1.6)
(0, 0)
$x = 0$

38.

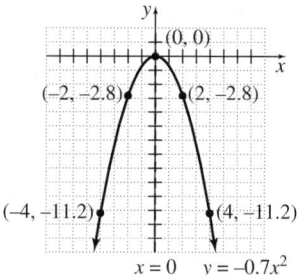

(0, 0)
(-2, -2.8), (2, -2.8)
(-4, -11.2), (4, -11.2)
$x = 0$ $y = -0.7x^2$

39.

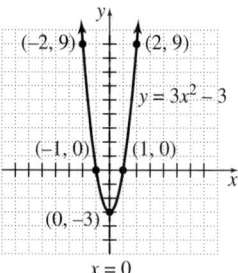

(-2, 9), (2, 9)
$y = 3x^2 - 3$
(-1, 0), (1, 0)
(0, -3)
$x = 0$

40.

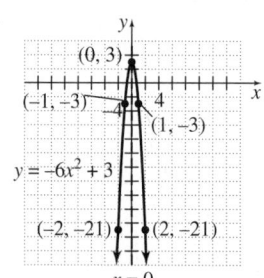

(0, 3)
(-1, -3), (1, -3)
4
$y = -6x^2 + 3$
(-2, -21), (2, -21)
$x = 0$

41.

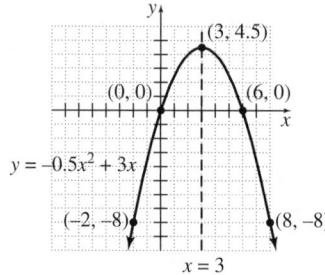

(3, 4.5)
(0, 0), (6, 0)
$y = -0.5x^2 + 3x$
(-2, -8), (8, -8)
$x = 3$

42.

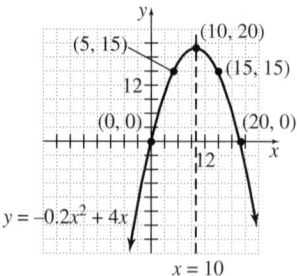

(5, 15), (10, 20)
(15, 15)
12
(0, 0), (20, 0)
12
$y = -0.2x^2 + 4x$
$x = 10$

43. $Y1 = -16x^2 + 12x + 24$

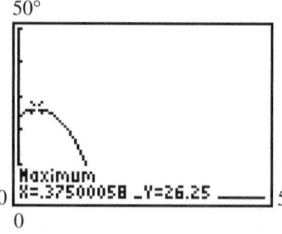

50°
Maximum
X=.3750005B Y=26.25
0
0   5

The maximun height the
apple will reach is 26.25 feet.

44. $Y1 = -16x^2 + 60x$

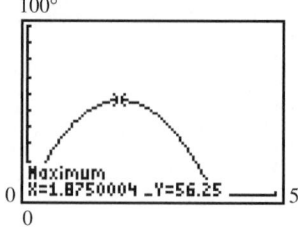

100°
Maximum
X=1.B750004 Y=56.25
0
0   5

The vertex is (1.875, 56.25);
the football will reach a maximum
height of 56.25 feet after
1.875 seconds.

45. $Y1 = 140x - x^2$

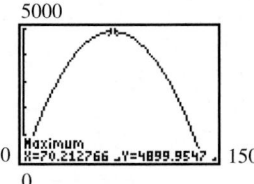

The vertex is $(70, 4900)$; the room has a maximum area of 4900 ft^2 when the width is 70 feet.

46. $Y1 = 120x - 2x^2$

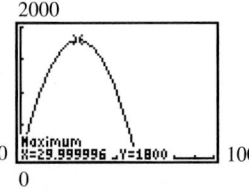

The vertex is $(30, 1800)$; the pen has a maximum area of 1800 ft^2 when the length of two sides are each 30 feet.

47. $Y1 = 10x - x^2$

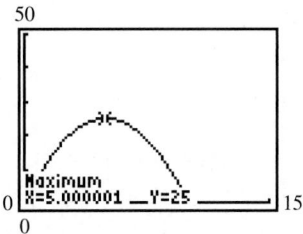

The vertex is $(5, 25)$; the triangle has a maximum area of 25 square inches when the height is 5 inches.

48. $Y1 = x^2 + 10x$

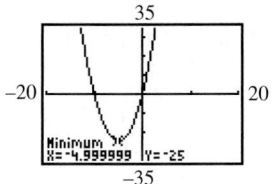

The vertex is $(-5, -25)$; no, the vertex does not have any physical meaning.

49. $Y1 = 46x - x^2$

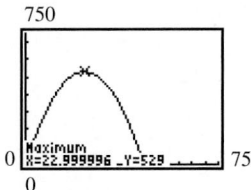

The vertex is $(23, 529)$; at $x = \$23$, the revenue will be at a maximum of \$529. Yes, the seller should limit the number of dolls sold.

50. $Y1 = 250x - 25x^2$

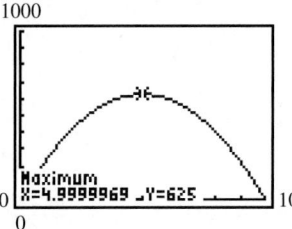

When the number of travelers equals 5, they will maximize their revenue at \$625. Yes, they should limit their number.

51. $Y1 = 6x - 0.05x^2 - 5$

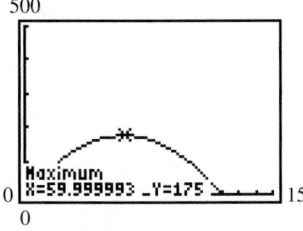

The maximum profit would be \$175, when they sell 60 discs.

52. $Y1 = 16000x - 200x^2 - 1000$

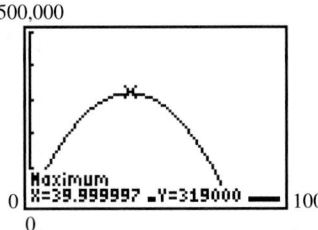

The maximum profit would be \$319,000, when they sell 40 pieces of major equipment.

8.3 CALCULATOR EXERCISES

1.

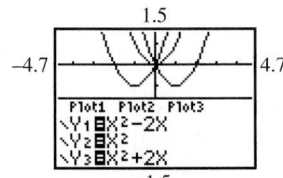

2.

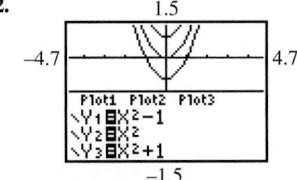

3.

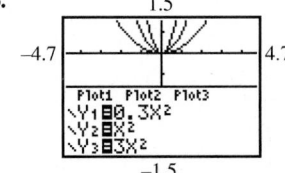

4.

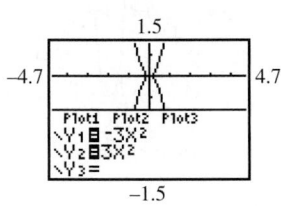

5.

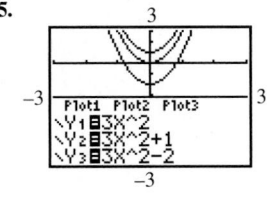

6.

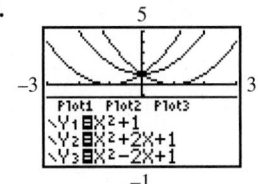

Answers may vary.

SECTION 8.4 EXERCISES

1. $y = x^2 + 2x - 8$

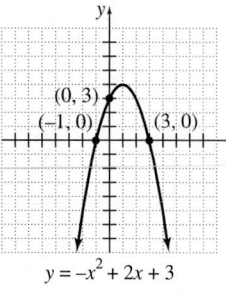

$y = x^2 + 2x - 8$

2. $y = -x^2 + 2x + 3$

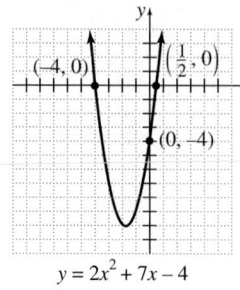

$y = -x^2 + 2x + 3$

3. $y = 2x^2 + 7x - 4$

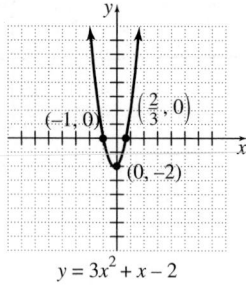

$y = 2x^2 + 7x - 4$

4. $y = 3x^2 + x - 2$

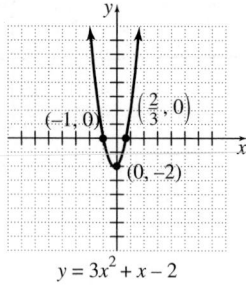

$y = 3x^2 + x - 2$

5. $y = -\frac{1}{3}x^2 + 3$

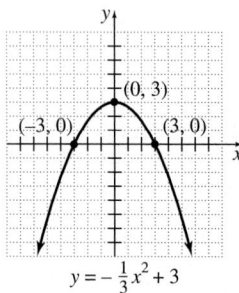

$y = -\frac{1}{3}x^2 + 3$

6. $y = -\frac{1}{16}x^2 + 1$

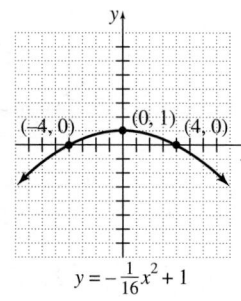

$y = -\frac{1}{16}x^2 + 1$

7. $y = \frac{1}{2}x^2 + x - 4$

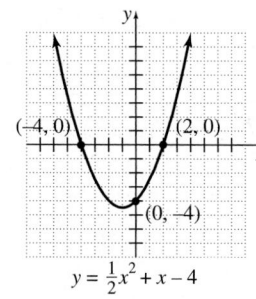

$y = \frac{1}{2}x^2 + x - 4$

8. $y = \frac{1}{3}x^2 - \frac{1}{3}x - 2$

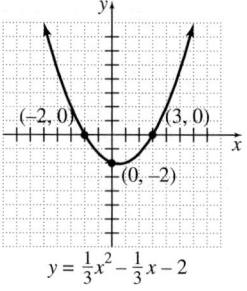

$y = \frac{1}{3}x^2 - \frac{1}{3}x - 2$

9. $y = -\frac{1}{6}x^2 + \frac{2}{3}$

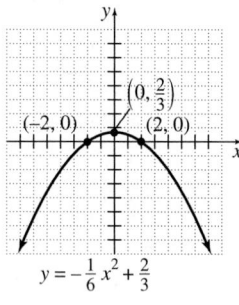

$y = -\frac{1}{6}x^2 + \frac{2}{3}$

10. $y = -\frac{1}{8}x^2 - \frac{1}{8}x + \frac{1}{4}$

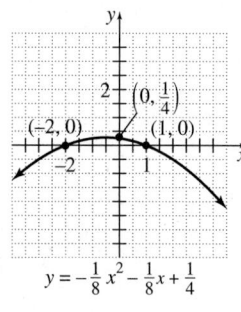

$y = -\frac{1}{8}x^2 - \frac{1}{8}x + \frac{1}{4}$

11. $y = -\frac{8}{9}x^2 + \frac{40}{9}x - 2$

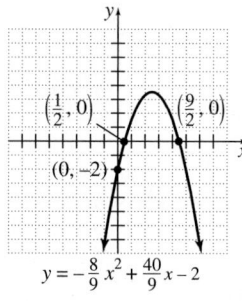

$y = -\frac{8}{9}x^2 + \frac{40}{9}x - 2$

12. $y = -\frac{9}{2}x^2 - \frac{3}{2}x + 1$

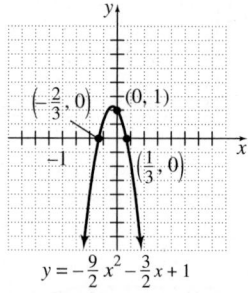

$y = -\frac{9}{2}x^2 - \frac{3}{2}x + 1$

13. $y = x^2 - 2x - 15$

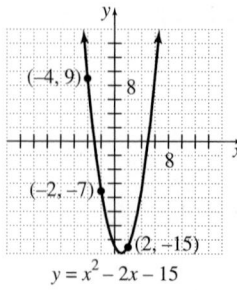

$y = x^2 - 2x - 15$

14. $y = x^2 + 3x - 18$

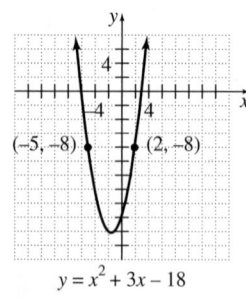

$y = x^2 + 3x - 18$

15. $y = 2x^2 - 3x - 9$

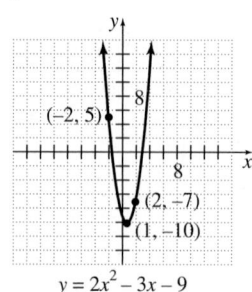

$y = 2x^2 - 3x - 9$

16. $y = 3x^2 - 11x - 4$

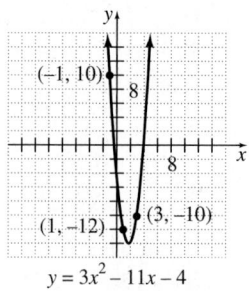

$y = 3x^2 - 11x - 4$

17. $y = \frac{1}{3}x^2 - 2x + 5$

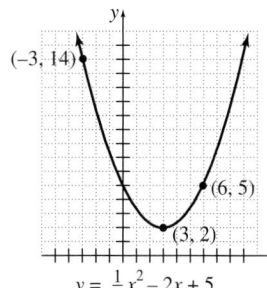

$y = \frac{1}{3}x^2 - 2x + 5$

18. $y = \frac{3}{4}x^2 + x - 4$

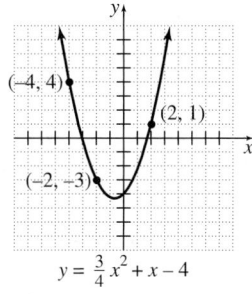

$y = \frac{3}{4}x^2 + x - 4$

19. $y = -x^2 + 4x + 3$

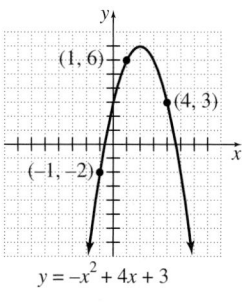

$y = -x^2 + 4x + 3$

20. $y = -x^2 - 2x + 5$

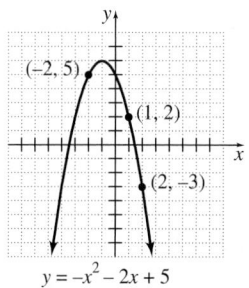

$y = -x^2 - 2x + 5$

21. $y = -2x^2 - 4x - 5$

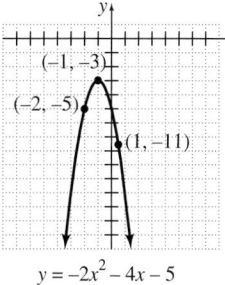

$y = -2x^2 - 4x - 5$

22. $y = -\frac{1}{2}x^2 + 3x - 7$

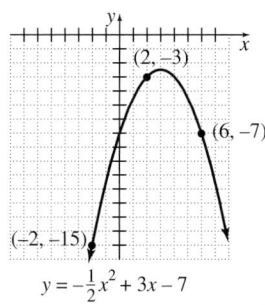

$y = -\frac{1}{2}x^2 + 3x - 7$

23. $y = \frac{1}{2}x^2 + 4x + 10$

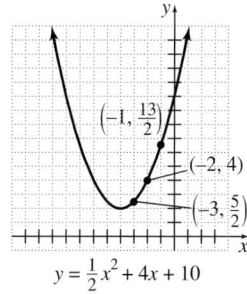

$y = \frac{1}{2}x^2 + 4x + 10$

24. $y = \frac{1}{3}x^2 - 4x + 16$

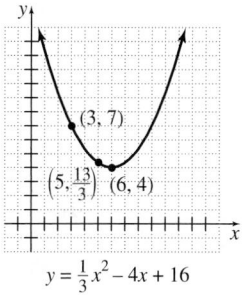

$y = \frac{1}{3}x^2 - 4x + 16$

25. $s(t) = -16t^2 + 12t + 150$
The ball will hit the ground in approximately 3.5 seconds.

26. $s(t) = -16t^2 + 18t + 220$
The pie will hit the ground in approximately 4.3 seconds.

27. $s(t) = -16t^2 + 400$
The dummy will hit the ground in 5 seconds.

28. $s(t) = -16t^2 + 600$
The toy will hit the ground in approximately 6.1 seconds.

29. $s(t) = -2.7t^2 + 2000$
The package will touch the surface at approximately 27.2 seconds.

30. $s(t) = -2.7t^2 + 25t + 3000$
The probe will reach the surface in approximately 38.3 seconds.

31. $f(x) = .0086x^2 + .387x + 20.45$
In the year 2005, it will produce 31.63 ft³. Answers will vary.

32. $f(x) = -0.1426x^2 + 3.001x + 62.17$
In the year 2005, it will produce 65.15 trillion ft³. Answers will vary.

33. $f(x) = 38.535x^2 - 23.35x + 33.52$, where x represents the number of years since 1975. In 2005, there will be 37,333 franchises. Answers will vary.

34. $f(x) = 0.001x^2 + 0.17x + 5.3$
For 1990, it is estimated to be 9.1 million, close to the actual number of 9.9 million.

35. $y = 0.0002x^2$

36. $y = 0.000097x^2$

37. $y = -0.0005x^2 + 360$

38. $y = -0.0005x^2 + 185$

8.4 CALCULATOR EXERCISES

1. $(5, 155) (35, 335) (40, 365)$; yes; $c(x) = 6x + 125$
2. $(5, 450) (30, 1200) (35, 1050)$; yes; $R(x) = -2x^2 + 100x + 0$
3.

Number of items produced and sold	5	30	35	40
Cost of production	$155	$305	$335	$365
Revenue from sales	$450	$1200	$1050	$800
Profit	$295	$895	$715	$435

*Answers in bold

4. $(5, 295) (30, 895) (35, 715)$; yes; $P(x) = -2x^2 + 94x - 125$
5. When the value of x is greater than 46, the profit becomes less than zero.
6. $(23.5, 979.5)$
When approximately 23 items are produced, $979.50 profit is achieved.

CHAPTER 8 SECTION-BY-SECTION REVIEW

1. yes; monomial   **2.** yes; binomial   **3.** no   **4.** yes; polynomial   **5.** no   **6.** yes; trinomial

7. The degrees of the terms are 1, 3, 0; the degree of the polynomial is 3.

8. The degrees of the terms are 3, 2, 0; the degree of the polynomial is 3.

9. The degrees of the terms are 1, 0; the degree of the polynomial is 1.   **10.** The degree of the term is 1; the degree of the polynomial is 1.

11. The degrees of the terms are 2, 1, 0; the degree of the polynomial is 2.   **12.** $11y^4 + 9y^3 + 5y^2 - 6y + 12$ **13.** $-p + 5$

14. $\frac{1}{4}z^4 + \frac{1}{3}z^3 + \frac{1}{2}z^2 + z + 1$ **15.** $-2.3b^5 - 9.1b^3 + 0.6b + 1.8$ **16.** 0   **17.** -90 **18.** -98 **19.** 0   **20.** 0   **21.** -1

22. -1 **23.** 0   **24.** -1 **25.** 3   **26.** 1   **27.** The perimeter is $2w + 2w^2$ units; the perimeter is 112 yards.

28. The perimeter is $x^2 + 4x + 1$ units; the perimeter is 33 inches.   **29.** The total area is $a^2 + 20a$ in².

30. The shaded area is $-\frac{1}{2}x^2 + 10x$ in².   **31.** The area of the lawn not covered by the garden is $z^2 - \frac{1}{2}xy$ square feet; the area is 6370 ft².

32. (a) The charge for a room is $500 + 40xy$ dollars.   **(b)** The cost to build a room is $275 + 12xy$ dollars.   **(c)** The revenue is $15,500; the cost is $4775; the net return is $10,725.

33. (a) It would take 18.17 minutes.   **(b)** It would take 16.43 minutes.   **(c)** It would take 13.49 minutes.

34.

x	y
-3	-18
-2	0
-1	4
0	0
1	-6
2	-8
3	0

35. The range is all real numbers.

36. The range is all real numbers greater than or equal to -12.5.

37. The range is all real numbers.

38. The range is all real numbers greater than or equal to -9.

39. $f(-2) = -36$ **40.** $f(0) = -4$ **41.** $f(2) = 20$

42. $f\left(-\frac{1}{2}\right) = -\frac{45}{8}$ **43.** $f(1.7) = 11.249$

44. (a) The revenue for 5, 10, 15, 20, and 25 shirts is $47.50, $70, $67.50, $40, and $-$12.50, respectively. Answers will vary.   **(b)** The cost for 5, 10, 15, and 20 shirts is $20, $40, $60, and $80, respectively.   **(c)** The profit from 5, 10, 15, and 20 shirts is $27.50, $30, $7.50, and $-$40, respectively.   **(d)** Answers will vary.

45.

Hundreds of Gallons (x)	900	1000	1100	1200	1300	1400
Cost (y)	16,970	17,000	17,010	17,000	16,970	16,920

; answers will vary.

46. The range is all real numbers less than or equal to 6025. The maximum height of the rock was 6025 feet.

47. quadratic   **48.** nonquadratic   **49.** nonquadratic   **50.** quadratic

51–54.

	a	b	c	graph wide/narrow	graph concave upward/downward	graph vertex	axis of symmetry	y-intercept
51.	$-\frac{1}{4}$	$\frac{1}{2}$	1	wide	downward	$(1, \frac{5}{4})$	$x = 1$	$(0, 1)$
52.	-2	4	0	narrow	downward	$(1, 2)$	$x = 1$	$(0, 0)$
53.	$\frac{1}{3}$	1	0	wide	upward	$(-\frac{3}{2}, -\frac{3}{4})$	$x = -\frac{3}{2}$	$(0, 0)$
54.	3	-3	1	narrow	upward	$(\frac{1}{2}, \frac{1}{4})$	$x = \frac{1}{2}$	$(0, 1)$

55.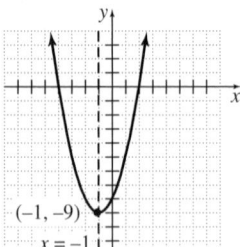

56.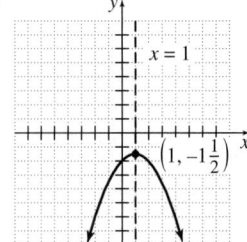

57.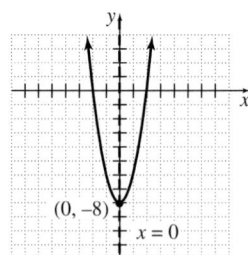

58.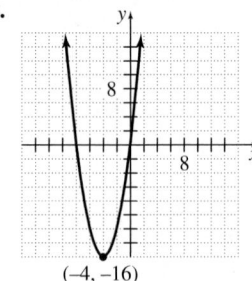

$(-4, -16)$; no, the vertex has no physical meaning.

59.

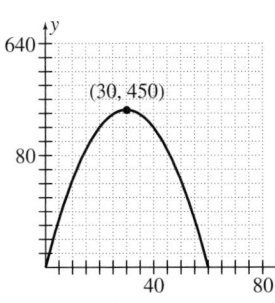

The vertex is (30, 450); the revenue is at a maximum of $450 when 30 photos are ordered.

60. The vertex is (1.875, 176.25); the egg reaches a maximum height of 176.25 feet at 1.875 seconds.
61. $y = x^2 - 5x - 24$ **62.** $y = -2x^2 + 9x + 5$ **63.** $y = x^2 - 7x + 6$
64. $s(t) = -16t^2 + 32t + 160$; the hammer will reach the ground in approximately 4.3 seconds.
65. $f(x) = 0.00325x^2 + 0.015x + 3.525$; in 1980, 1970, and 1960, the predictions are 25.525 million, 20.5 million, and 16.125 million, respectively; answers will vary; in the year 2010, the prediction is 44.5 million.
66. $f(x) = 0.0109375x^2 + 280$

CHAPTER 8 CHAPTER REVIEW

1. 0   **2.** -30 **3.** -60 **4.** 0   **5.** 0   **6.** -2 **7.** 5   **8.** 5   **9.** -5
10. (a) binomial; degree of each term is 0, 2; degree of polynomial is 2; $3x^2 + 5$ **(b)** polynomial; degree of each term is 2, 3, 0, 1; degree of polynomial is 3; $-5a^3 + 15a^2 + a + 4$ **(c)** polynomial; degree of each term is 4, 1, 0, 5, 2; degree of polynomial is 5; $x^5 + 5x^4 - 3x^2 + x - 2$
11. (a) trinomial; degree of each term is 2, 1, 0; degree of polynomial is 2.   **(b)** polynomial; degree of each term is 3, 2, 3, 0, 4; degree of polynomial is 4.   **(c)** monomial; degree of term is 3; degree of polynomial is 3.
12. 0   **13.** 12   **14.** 0   **15.** 0   **16.** -31.824
17.

x	y
-3	0
-2	16
-1	12
0	0
1	-8
2	0
3	36

18. The range is all real numbers.
19. The range is all real numbers greater than or equal to -6.75.
20. The range is all real numbers.
21. The range is all real numbers greater than or equal to -87.04.

22. $x = -\dfrac{7}{4}$

23. $x = 2$

24. $x = 0$

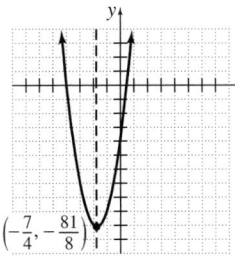

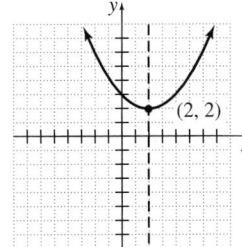

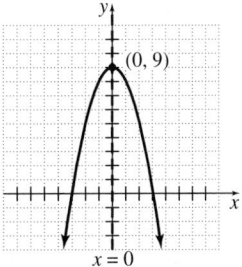

25–28.

	a	b	c	graph wide/narrow	graph concave upward/downward	graph vertex	axis of symmetry	y-intercept
25.	$\frac{1}{3}$	$\frac{2}{3}$	1	wide	upward	$(-1, \frac{2}{3})$	$x = -1$	$(0, 1)$
26.	-3	6	0	narrow	downward	$(1, 3)$	$x = 1$	$(0, 0)$
27.	$-\frac{1}{4}$	1	3	wide	downward	$(2, 4)$	$x = 2$	$(0, 3)$
28.	2	4	-6	narrow	upward	$(-1, -8)$	$x = -1$	$(0, -6)$

29. $y = x^2 + 8x + 24$ **30.** $y = x^2 - 5x - 36$ **31.** $y = 3x^2 - 5x - 12$ **32.** The polynomial for the perimeter is $2w^3 + 2w + 10$ units; the perimeter is 70 feet.   **33.** The perimeter is $x^2 + 3x - 10$ units; the perimeter is 120 cm.

34. The polynomial for the total area is $x^2 + 11x$ square units.   **35.** The polynomial for the shaded area is $\dfrac{24 + \pi}{8}x^2$ square units.

36. The polynomial is $xy - \pi z^2$ square feet; the area measures approximately 3798.9 ft².

37. (a) The cost is $1.5lw$ dollars.   **(b)** The profit is $800 - 1.5lw$ dollars.   **(c)** The profit will be $575.

38. (a)

x	$P(x)$
0	0
1	12
2	28
3	48
4	72
5	100
6	132
7	168

(b) Answers will vary.

39. (a) The predicted GPA is 3.09375.

(b)

GMAT score, z	500	550	600	650	700	750
GPA in MBA, y	3.575	3.55975	3.504	3.40775	3.271	3.09375

(c) As the entrance exam score goes up, the student's predicted performance goes down.

40. $Y1 = -0.8x^2 + 1500$
$Y2 = 0$

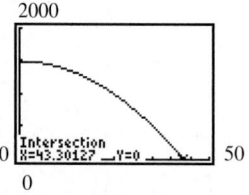

The object will reach the ground in approximately 43.3 seconds.

41. $f(x) = 0.0006x^2 - 0.016x + 1.68$; in 1930 and 1990, the predicted divorce rate per 1000 population is 1.74 and 5.1, respectively; answers will vary; in the year 2010, the prediction is 7.18 per 1000 population.

42. (a) They can sell about 94 cards at 99 cents each.
(b) They can sell about 75 cards at $1.19 each.

43. The range is all real numbers less than or equal to 19,500. The rock had a maximum height of 19,500 feet.

44. The vertex is at $(7.5, 56.25)$. The maximum revenue is $56 when 7 or 8 books are sold.

45. $y = -0.000712x^2 + 170$.

CHAPTER 8 TEST

1. monomial   **2.** polynomial   **3.** binomial   **4.** 5   **5.** 5   **6.** $x^5 + 3x^4 + 9x^2 + 14x + 15$ **7.** $-\dfrac{4}{9} + a - \dfrac{5}{6}a^2 - \dfrac{2}{3}a^3$

8. -72 **9.** -2 **10.** -224 **11.** The total cost will be $16.5lw + 75$ dollars; the total cost will be $680.   **12.** 0   **13.** -6 **14.** 4

15. The vertex is $(2, -8)$.   **16.**

17. The range is all real numbers greater than or equal to -8.
18. Yes; the graph of the relation passes the vertical line test.

19–20.

	a	b	c	graph wide/narrow	graph concave upward/downward	graph vertex	axis of symmetry	y-intercept
19.	$\frac{1}{2}$	2	3	wide	upward	$(-2, 1)$	$x = -2$	$(0, 3)$
20.	3	-3	$\frac{1}{4}$	narrow	upward	$(\frac{1}{2}, -\frac{1}{2})$	$x = \frac{1}{2}$	$(0, \frac{1}{4})$

21. $y = x^2 - 6x + 8$ **22.** $y = 2x^2 - x + 3$ **23. (a)** His estimated earnings are $45,636 with an associate's degree.   **(b)** He will earn $14,650 more than a male high school graduate.   **24.** The range is all real numbers less than or equal to 100. The paper had a maximum height of 100 feet.   **25.** The vertex is $(18.75, 703.125)$. $703 is the maximum profit earned by selling 19 keyboards.
26. $y = -0.000463x^2 + 325$ **27.** Answers will vary.

Chapter 9

SECTION 9.1 EXERCISES

1. $-3 \cdot x \cdot x \cdot x \cdot x$ **2.** $-4 \cdot c \cdot c \cdot c$ **3.** $(-3x)(-3x)(-3x)(-3x)$ **4.** $(-4c)(-4c)(-4c)$ **5.** $a \cdot a \cdot a \cdot c \cdot c \cdot c \cdot c \cdot c$

6. $x \cdot x \cdot y$ **7.** $\frac{3}{4} \cdot \frac{3}{4} \cdot \frac{3}{4} \cdot x \cdot x$ **8.** $\left(\frac{2}{3}\right)\left(\frac{2}{3}\right) \cdot x \cdot x \cdot x$ **9.** $5(x + y)(x + y)$ **10.** $-4(p - q)(p - q)$ **11.** x^{13} **12.** a^{23}

13. y^{14} **14.** b^{24} **15.** $\frac{1}{2}x^8$ **16.** $\frac{1}{3}y^6$ **17.** $(x + y)^6$ **18.** $(x - y)^8$ **19.** $(x + 3)^3$ **20.** $(x + 9)^{13}$ **21.** p^5 **22.** a^2

23. $3q^6$ **24.** $3t^7$ **25.** $(2x - 3)^5$ **26.** $(4x + 7)^7$ **27.** $-\frac{1}{3}(p + q)$ **28.** $-\frac{1}{2}(xy + 2)^3$ **29.** a^{30} **30.** m^{28} **31.** $81x^4$

32. $-125y^3$ **33.** $a^{21}b^{21}c^{21}$ **34.** $x^9y^9z^9$ **35.** $125m^9$ **36.** $81k^{16}$ **37.** $(x + y)^6$ **38.** $(a + 2b)^{12}$ **39.** $(a - b)^4$

40. $(c - 3d)^5$ **41.** 1 **42.** 1 **43.** $\frac{b^4}{d^4}$ **44.** $\frac{m^5}{n^5}$ **45.** $\frac{81h^4}{c^4}$ **46.** $\frac{625y^4}{z^4}$ **47.** $\frac{d^6}{64c^6}$ **48.** $\frac{m^4}{81n^4}$ **49.** $-35a^5b^5$

50. $-24c^5d^7$ **51.** $5x^2$ **52.** $6b$ **53.** $\frac{-9abc^2}{5}$ **54.** $\frac{-9p^2q^2}{2}$ **55.** $-2x(4 - x)^3$ **56.** $\frac{t(15 - 4t)^4}{3}$ **57.** p^9q^{20} **58.** k^5m^{11}

59. $\frac{16x^6}{y^4}$ **60.** $\frac{25a^4}{49b^2}$ **61.** $-27p^6q^3$ **62.** $25c^2d^2$ **63.** $1024a^{10}$ **64.** $81t^4$ **65.** $\frac{x^6}{64y^6}$ **66.** $\frac{729z^6}{w^6}$ **67.** Incorrect

68. Incorrect **69.** Correct **70.** Correct **71.** Correct **72.** Correct **73.** Incorrect **74.** Incorrect **75.** Correct
76. Correct **77.** Incorrect **78.** Incorrect
79. The original area is x^2 square units; the enlarged area is $25x^2$ square units; the enlarged area is 25 times bigger; if the original side is 6 feet, the original area and enlarged area are 36 ft^2 and 900 ft^2, respectively.
80. The original area is x^2 square units; the enlarged area is $6.25x^2$ square units; the enlarged area is 6.25 times bigger; if the original side is 12 feet, then the original area and enlarged area are 144 ft^2 and 900 ft^2, respectively.
81. The volume of the original bin is x^3 cubic units; the volume of the enlarged bin is $64x^3$ cubic units; the volume of the enlarged bin is 64 times greater; if the original side measures 1.5 feet, then the volumes of the original bin and enlarged bin are 3.375 ft^3 and 216 ft^3, respectively.
82. The volume of the original block is x^3 cubic units; the volume of the smaller block is $0.512x^3$ cubic units; the volume of the smaller block is 0.512 times the volume of the larger block; if the original side is 22 inches, then the volume of the original block and smaller block are 10,648 in^3 and 5451.776 in^3, respectively.
83. The volume of the first balloon is $\frac{4}{3}\pi x^3$; the volume of the second balloon is $\frac{32}{3}\pi x^3$; the volume of the second balloon is 8 times the volume of the first balloon. If the smaller balloon has a radius of 4 feet, the volume is about 268 cubic feet and the volume of the second balloon is about 2145 cubic feet.
84. The amount of leather for the smaller ball is $4\pi x^2$ and for the large ball is $9\pi x^2$; the amount of leather for the large ball is 2.25 times that for the small ball. If the radius of the small ball is 27 cm, the amount of the leather is about 9161 square centimeters for the small ball and 20,612 square centimeters for the large ball.

9.1 CALCULATOR EXERCISES

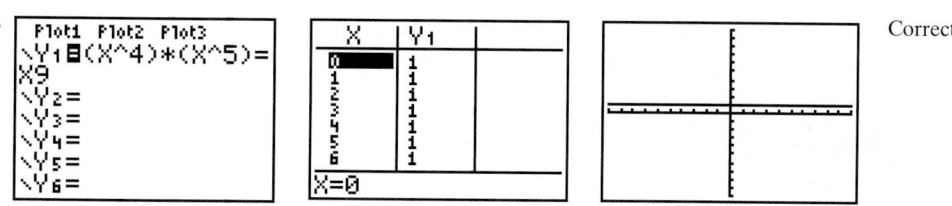

1. Incorrect

2. Correct

3. 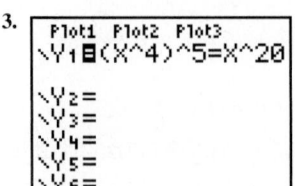 Correct

4. 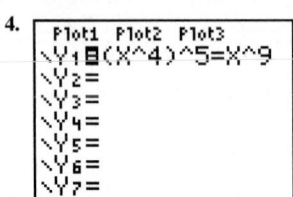 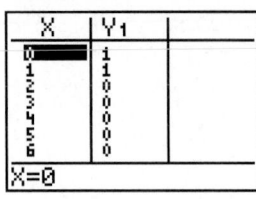 Incorrect

5. 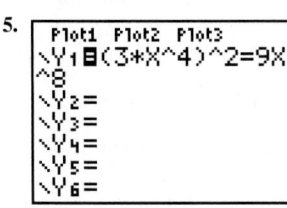 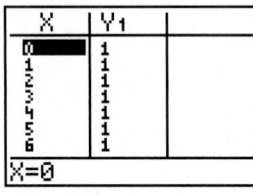 Correct

6. 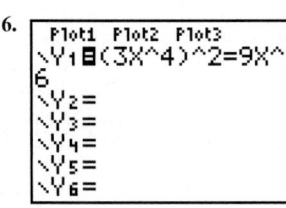 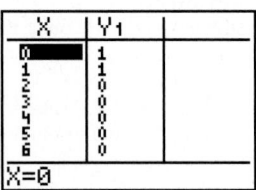 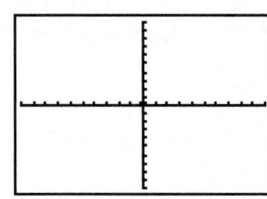 Incorrect

SECTION 9.2 EXERCISES

1. $\dfrac{1}{p^3}$ **2.** $\dfrac{1}{q^2}$ **3.** q^5 **4.** p^3 **5.** $\dfrac{q^5}{p^3}$ **6.** $\dfrac{p^3}{q^2}$ **7.** d^4c^3 **8.** c^2d^5 **9.** $\dfrac{16h^3k^4}{25}$ **10.** $\dfrac{64x^2y^3}{9}$ **11.** $\dfrac{q^5}{p^3}$ **12.** $\dfrac{p^3}{q^2}$

13. $-20a^2$ **14.** $30h^2$ **15.** $\dfrac{m}{4}$ **16.** $\dfrac{y}{6}$ **17.** $\dfrac{81n}{64}$ **18.** $\dfrac{25x}{32}$ **19.** c^8 **20.** p^6 **21.** $\dfrac{a^6}{125b^6}$ **22.** $\dfrac{m^8}{25n^4}$ **23.** $\dfrac{-42}{q^3}$

24. $39xy^3$ **25.** $\dfrac{6.02x^5}{y^2}$ **26.** $15.39y^3$ **27.** $\dfrac{2xy^3}{5}$ **28.** $\dfrac{4a}{15b^2}$ **29.** $\dfrac{-21x^6}{y^3}$ **30.** $-8m^2q^3$ **31.** $\dfrac{1}{m^7}$ **32.** $\dfrac{1}{k^2}$ **33.** $\dfrac{-c^2}{3}$

34. $\dfrac{7d^2}{10}$ **35.** $4x^3y^4$ **36.** $3c^2d^5$ **37.** $\dfrac{k^7}{5h^2}$ **38.** $\dfrac{v^3}{9u^5}$ **39.** $\dfrac{11b^5}{a^5}$ **40.** $\dfrac{b^3}{3a^5}$ **41.** $\dfrac{b^{12}}{8a^{12}}$ **42.** $\dfrac{-c^9}{64d^{12}}$ **43.** $-32p^{10}q^{25}$

44. $-32k^{25}m^5$ **45.** $\dfrac{b^3}{a^3}$ **46.** $\dfrac{t}{s}$ **47.** $\dfrac{y^3}{64x^3}$ **48.** $\dfrac{b^5}{243a^5}$ **49.** $\dfrac{1}{81p^8q^8}$ **50.** $\dfrac{1}{25c^6d^2}$ **51.** $\dfrac{y^3}{z^6}$ **52.** $\dfrac{d^6}{e^8}$ **53.** $\dfrac{125a^9}{b^{12}}$

54. $\dfrac{t^{28}}{625s^{12}}$ **55.** The distance is about 1.652×10^{17} miles. **56.** The distance is about 1.9057×10^{16} miles.

57. The wavelength is 2.784 meters. **58.** The wavelength is 2.8966 meters.

9.2 CALCULATOR EXERCISES

1. 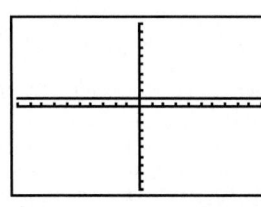 Error in second step

2. 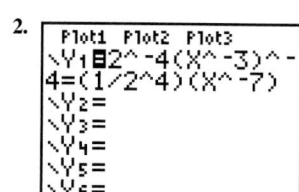 Error in second step

3. 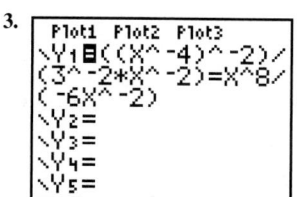 Error in second step

SECTION 9.3 EXERCISES

1. $2x^4 + 12x^2 - 17x + 43$ **2.** $b^3 + 6b^2 + 4b + 2$ **3.** $9x^4 + 3x^3 - 10x^2 - 3x + 9$ **4.** $z^3 + z^2 + 5z + 6$

5. $-3x^2y + 6y^3 + 15x^3$ **6.** $4a^3 + 6a^2b - ab^2 + 7b^3$ **7.** $a^3 + 8a^2 - 7a - 1$ **8.** $3x^2 + 12$ **9.** $\frac{2}{3}y^4 + \frac{5}{2}y^3 + \frac{19}{9}y^2 + \frac{1}{2}y - \frac{22}{9}$

10. $\frac{13}{12}x^3 - \frac{2}{3}x^2 - \frac{1}{4}x + \frac{19}{12}$ **11.** $18.77x^3 - 1.23x^2 + 3.81x + 9.735$ **12.** $0.7y^3 + 6.2y^2 - 4.4y + 0.3$

13. $5117a^3 + 50998a^2 - 1816a + 2095$ **14.** $50b^2 + 200$ **15.** $3a + 9b + 6c$ **16.** $5x + y + 11z$ **17.** $2z^3 + 27z^2 - 51z + 114$
18. $-3a^2 - 5a + 15$ **19.** $a^4 - a^3 + a^2 - a$ **20.** $b^7 + b^5 + b^3 + b$ **21.** $7x^2 - 3x - 39$ **22.** $-5y^3 + 8y - 11$
23. $-5a^3 - 6a^2 - 12a + 14$ **24.** $x^3 - 2x^2 + 3x + 9$ **25.** $42x^3 - 30x^2y + 3xy^2 + 11y^3$ **26.** $51p^3 + 26p^2q - 16pq^2 + 41q^3 + 10$

27. $4a - 2b + 7c - 6d$ **28.** $12x + 5y + 8z$ **29.** $\frac{3}{14}x^2 - \frac{19}{42}x - \frac{26}{21}$ **30.** $\frac{7}{24}x^3 + \frac{11}{24}x^2 - \frac{1}{4}x - \frac{1}{3}$

31. $9x^3 + x^2y - 14xy^2 + 68y^3$ **32.** $2.48x^2 - 12.51xy - 8.75y^2$ **33.** $4683z^2 - 4403z + 7187$
34. $-86b^3 + 178b^2 - 1344b + 1591$ **35. (a)** $C(x) = 200 + 4.5x$ **(b)** $R(x) = 13.5x$ **(c)** $P(x) = 9x - 200$
(d) The profit is $-\$20$ if 20 pots are sold (that is a loss of \$20) and \$70 if 30 pots are sold.
36. (a) $C(x) = 500 + 20x$ **(b)** $R(x) = 50x$ **(c)** $P(x) = 30x - 500$
(d) The profit is \$1750 if 75 phones are sold and \$2500 if 100 phones are sold.
37. $C(x) = 5.00x + 30$;
    $R(x) = 28.50x$;
    $P(x) = 23.50x - 30$
38. $C(x) = 180 + 1.80x$;
    $R(x) = 11.95x$;
    $P(x) = 10.15x - 180$
39. The difference is $(-10t + 100)$ feet.   **40.** The difference is $(30t + 40)$ feet.

9.3 CALCULATOR EXERCISES

1.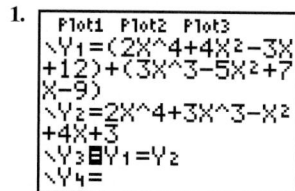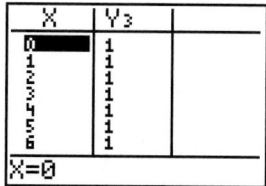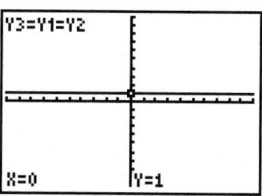

2.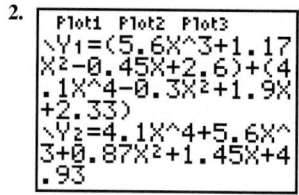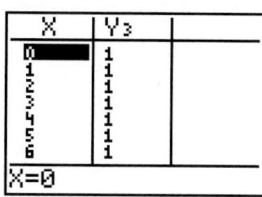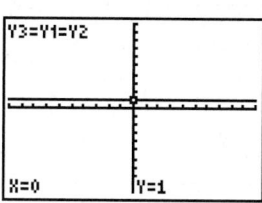

3.

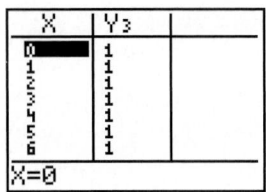 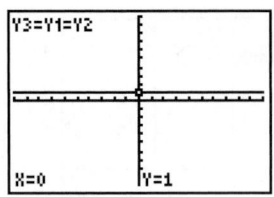

4.

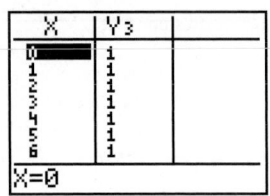 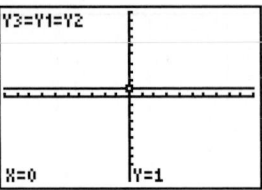

SECTION 9.4 EXERCISES

1. $-16a^4b^3$ **2.** $15p^5q^3$ **3.** $-6x^2 + 2xy - 4xz$ **4.** $-4a^2 + 3ab - 2ac$ **5.** $6a^4 + 4a^3b - 2a^3c$ **6.** $6x^3 - 9x^2y + 3x^2z$
7. $x^2 + 6x + 8$ **8.** $a^2 + 13a + 42$ **9.** $2x^2 + 13x + 15$ **10.** $7z^2 + 29z + 4$ **11.** $6xy - 4x + 15y - 10$
12. $15ab + 20a - 6b - 8$ **13.** $3x^2 - 2xy - 8y^2$ **14.** $2a^2 - 5ab - 12b^2$ **15.** $5a^2 - 8.2a - 9.12$ **16.** $4p^2 - 3.9p - 4.93$

17. $6xy + 6.4x + 3.3y + 3.52$ **18.** $20mn + 13.5m - 19.6n - 13.23$ **19.** $a^2 + a + \dfrac{2}{9}$ **20.** $x^2 + \dfrac{2}{5}x - \dfrac{3}{25}$ **21.** $2x^4 + 5x^2 - 12$

22. $5x^6 - 14x^3 - 3$ **23.** $8x^3 + 4x^2 + 6x + 3$ **24.** $24x^5 - 18x^3 - 20x^2 + 15$ **25.** $6x^3 - 10xy + 9x^2y^2 - 15y^3$
26. $24a^3 - 20a^2b^2 + 18ab - 15b^3$ **27.** $3a^4 + 8a^2b^3 + 5b^6$ **28.** $28x^8 - 15x^4y^3 + 2y^6$ **29.** $x^3 + 64$ **30.** $x^3 - 1$
31. $6x^3 - 19x^2 + x + 6$ **32.** $4x^3 - 17x^2 - 23x - 6$ **33.** $x^4 + 3x^3 + 6x^2 + 5x + 3$ **34.** $a^4 - a^3 - 3a^2 + 5a - 2$
35. $a^2 + b^2 + c^2 + 2ab + 2bc + 2ac$ **36.** $a^2 + b^2 + c^2 - 2ab - 2ac + 2bc$ **37.** $z^3 + 9z^2 + 27z + 27$ **38.** $r^3 + 6r^2 + 12r + 8$
39. $27a^3 - 54a^2b + 36ab^2 - 8b^3$ **40.** $8x^3 - 36x^2y + 54xy^2 - 27y^3$ **41.** $x^2 - 25$ **42.** $y^2 - 144$ **43.** $9m^2 - 49$ **44.** $25p^2 - 16$

45. $4a^2 - 9b^2$ **46.** $81p^2 - 4q^2$ **47.** $16x^2 - 2.25$ **48.** $9z^2 - 6.25$ **49.** $\dfrac{4}{25}x^2 - 1$ **50.** $\dfrac{16}{49}y^2 - 4$ **51.** $\dfrac{1}{9}x^2 - \dfrac{16}{25}$

52. $\dfrac{25}{49}y^2 - \dfrac{9}{16}$ **53.** $x^4 - 49$ **54.** $y^6 - 81$ **55.** $4x^6 - 25y^2$ **56.** $25a^{10} - 4b^4$ **57.** $m^2 + 14m + 49$ **58.** $p^2 + 12p + 36$
59. $x^2 - 2xy + y^2$ **60.** $u^2 - 2uv + v^2$ **61.** $4p^2 + 36pq + 81q^2$ **62.** $25m^2 - 110mn + 121n^2$ **63.** $36c^2 - 60c + 25$
64. $64d^2 + 48d + 9$ **65.** $9x^6 + 12x^3 + 4$ **66.** $4z^4 - 20z^2 + 25$ **67.** $4x^4 - 12x^2y^3 + 9y^6$ **68.** $25x^6 + 30x^3y^2 + 9y^4$
69. (a) length: $(18 - 2x)$ in.; width: $(12 - 2x)$ in.; height: (x) in. **(b)** The volume is $(4x^3 - 60x^2 + 216x)$ in³.
(c) The surface area is $(-4x^2 + 216)$ in².
70. (a) length: $(10 - 2y)$ in.; width: $(10 - 2y)$ in.; height: y in. **(b)** The volume is $(4y^3 - 40y^2 + 100y)$ in³.
(c) The surface area is $(-4y^2 + 100)$ in².
71. (a) (πx^2) ft² **(b)** $[\pi(x - 5)^2]$ ft² **(c)** The area of the deck is $(10\pi x - 25\pi)$ ft².
72. (a) (36π) ft² **(b)** $[\pi(6 - x)^2]$ ft² **(c)** The area of the deck is $(12\pi x - \pi x^2)$ ft².

9.4 CALCULATOR EXERCISES

1. not equivalent; $4x^2 - 1$ **2.** not equivalent; $x^2 - 1$ **3.** equivalent **4.** equivalent **5.** not equivalent; $x^2 - \dfrac{35}{6}x - 1$

6. equivalent **7.** equivalent **8.** not equivalent; $0.3x^3 - 1.4x^2 - 0.3x + 1.4$

SECTION 9.5 EXERCISES

1. $-4a^2b$ **2.** $-2x^4y$ **3.** $2x^2y^3z + 6$ **4.** $3x^2yz + 2y$ **5.** $\dfrac{p^2}{q} - 3p + 2q - \dfrac{4q^2}{p}$ **6.** $\dfrac{c^2}{d} - 3c + 2d - \dfrac{5d^2}{c}$ **7.** $2x^2 + 4x - 6$

8. $3a^2 - 2 + \dfrac{6}{a^2}$ **9.** $\dfrac{x}{2} - \dfrac{1}{4} + \dfrac{3}{2x}$ **10.** $5z^3 - 2z^2 + 9z + 3$ **11.** $3.1x^2 - 5.8 + \dfrac{0.9}{x^2}$ **12.** $-\dfrac{1.8}{m^2} + \dfrac{2.5}{m} - 3$

13. $\dfrac{-3x^3}{y} - 2x^2 + 6xy + 8y^2 - \dfrac{24y^3}{x}$ **14.** $\dfrac{2a}{b} - 5 - \dfrac{4b}{a} + \dfrac{b^2}{a^2}$ **15.** $5x - 6$ **16.** $3x - 7$ **17.** $2x - 5$ **18.** $3x - 11$

19. $5x + 6$ **20.** $6x + 5$ **21.** $3y - 5 + \dfrac{20}{y + 8}$ **22.** $5z - 7 + \dfrac{4}{z + 2}$ **23.** $3a + 5 - \dfrac{5}{2a - 5}$ **24.** $4b + 5 + \dfrac{20}{3b - 7}$

25. $x^2 - 3.2x + 0.68 + \dfrac{0.64}{5x + 2}$ **26.** $4x^2 - x + 5$ **27.** $4x^2 + 2x + 5$ **28.** $3x^2 + 2x + 7$ **29.** $3a - 7$ **30.** $7b - 11$

31. $x + 5.4 - \dfrac{7.2}{5x + 3}$ **32.** $6y + 5$ **33.** $4z - 11$ **34.** $8p - 7$ **35.** $x^2 - 3x + 9$ **36.** $y^2 - 7y + 49$ **37.** $a^2 + 5a + 25$

38. $b^2 + 3b + 9$ **39.** $16x^2 - 12x + 9$ **40.** $25z^2 + 10z + 4$

41. (a) $C(x) = 35x + 45$
 (b) $R(x) = 50x$
 (c) $P(x) = 15x - 45$
 (d) $A(x) = 15 - \dfrac{45}{x}$
 (e) The average profit is \$13.50 per calculator.

42. (a) $C(x) = 75x + 1000$
 (b) $R(x) = 350x$
 (c) $P(x) = 275x - 1000$
 (d) $A(x) = 275 - \dfrac{1000}{x}$
 (e) The average profit is \$175 per web page.

43. $6000t^4 + 6000t^3 + 6000t^2 + 6000t + 6000$; he will receive about \$36,631 for a 10% increase and about \$40,454 for a 15% increase.
44. $10x^5 + 10x^4 + 10x^3 + 10x^2 + 10x + 10$; Sharon paid \$630 when the amount was doubled and \$3640 when the amount was tripled.
45. $250n^4 + 250n^3 + 250n^2 + 250n + 250$; the total rental cost was about \$1024.
46. $550w^5 + 550w^4 + 550w^3 + 550w^2 + 550w + 550$; the total sales were about \$2706.

9.5 CALCULATOR EXERCISES

1. $\dfrac{55}{\pi r^2}$ pounds per square inch

r (inch)	Pounds per square inch
0.25	280.1
0.50	70.0
0.75	31.1
1.00	17.5

For a radius of 1 inch, the woman's pressure on the floor approximates the elephant's.

2. $\dfrac{175}{2\pi r^2}$ pounds per square inch

r (inches)	Pounds per square inch
1.0	27.9
1.05	25.3
1.10	23.0
1.15	21.1
1.20	19.3
1.25	17.8
1.30	16.5
1.35	15.3
1.40	14.2
1.45	13.2
1.50	12.4

For a radius of 1.25 inches, the pressure is approximately the same as the elephant's.

CHAPTER 9 SECTION-BY-SECTION REVIEW

Reflections 1–11

Answers will vary.

1. $-5 \cdot c \cdot c$ **2.** $(-5c)(-5c)$ **3.** $4 \cdot (x + y) \cdot (x + y)$ **4.** $\dfrac{2}{3} \cdot \dfrac{2}{3} \cdot \dfrac{2}{3} \cdot \dfrac{2}{3} \cdot x \cdot x$ **5.** a^9 **6.** $(p + q)^4$ **7.** $\dfrac{2}{3}b^8$ **8.** t^3

9. $8a^2$ **10.** $x + 3y$ **11.** $c^{22}d^{22}$ **12.** $16a^4$ **13.** $-32a^5$ **14.** $9x^6$ **15.** $(a + b)^{10}$ **16.** 1 **17.** $\dfrac{-8x^3}{27z^3}$ **18.** $\dfrac{-64d^3}{e^3}$

19. $-8x^4y^7$ **20.** $-3x^2y$ **21.** x^5y^7 **22.** $\dfrac{8p^3}{q^3}$ **23.** $\dfrac{m^4}{256n^4}$

24. Area of original square: x^2 square units; area of reduced square: $\dfrac{x^2}{16}$ square units; the smaller area is $\dfrac{1}{16}$ of the original area.

25. Area of original garden: πr^2 square units; area of enlarged garden: $4\pi r^2$ square units; the enlarged garden has an area that is 4 times that of the original garden; the garden has an area of 144π ft^2.

26. $\dfrac{k^2}{9h^4}$ **27.** $\dfrac{c^8}{d^5}$ **28.** $\dfrac{3}{b^3}$ **29.** $\dfrac{z^4}{3}$ **30.** $\dfrac{a^{12}}{125b^6}$ **31.** $\dfrac{2}{5p^2q^5}$ **32.** $\dfrac{12y}{x^6}$ **33.** $\dfrac{9a^6}{b^{20}}$ **34.** $\dfrac{a^6}{b^8}$ **35.** $\dfrac{144a^{14}}{b^6}$

36. The distance is about 1.9883×10^{14} miles.   **37.** $5x^4 + 7x^3 + 5x^2 + 6x + 4$ **38.** $3.57z^3 + 3.56z^2 + 5.79z + 2.74$

39. $4a^4 - a^3 - 2a^2 - 3a - 4$ **40.** $65z^4 - 16z^3 + 27z^2 - 8z + 24$ **41.** $\dfrac{1}{8}b^4 + \dfrac{7}{8}b^3 - \dfrac{9}{8}b^2 + \dfrac{3}{8}b - \dfrac{1}{4}$

42. (a) $C(x) = 10 + 3.5x$ **(b)** $R(x) = 10x$ **(c)** $P(x) = 6.5x - 10$ **(d)** The profit is \$55 for 10 items and \$152.50 for 25 items.

43. $C(x) = 175 + 29.5x$
$R(x) = 125x$
$P(x) = 95.5x - 175$
44. $-15a^4b^4$ **45.** $-23.46x^4z^7$ **46.** $18x^5 + 12x^4 - 42x^3$ **47.** $-21a^7 - 7a^5 + 14a^3$ **48.** $8a^3 - 12a^2b + 4a^2c$
49. $p^2 - 3p - 54$ **50.** $5x^2 + 53x - 22$ **51.** $y^2 + 1.6y - 6.12$ **52.** $6x^3 + 13x^2 - 3x - 4$ **53.** $a^3 - 27$
54. $x^4 + x^3 + 6x^2 + 7x + 15$ **55.** $z^4 + 4z^3 - 2z^2 - 12z + 9$ **56.** $b^3 - 12b^2 + 48b - 64$ **57.** $4x^2 - 25$ **58.** $\frac{16}{25}x^2 - \frac{1}{4}$
59. $z^4 - 100$ **60.** $y^2 + 18y + 81$ **61.** $9x^2 - 30x + 25$ **62.** $x^6 + 6x^3 + 9$
63. (a) length: $(x + 3)$ in.; width: $(x - 3)$ in.; height: x in.   **(b)** The volume is $(x^3 - 9x)$ in^3.   **(c)** The surface area is $(6x^2 - 18)$ in^2.
64. (a) The current area is $3x^2$ ft^2.   **(b)** The new area is $(6x^2 + 18x)$ ft^2.   **(c)** The difference is $(3x^2 + 18x)$ square feet.

65. $\frac{72xy}{z^2}$ **66.** $\frac{2}{a} + \frac{6}{b}$ **67.** $-3b^3 + 2b^2 + 5b - 1$ **68.** $3x + 8$ **69.** $4x - 5 - \frac{4}{2x + 3}$ **70.** $x^2 - 7x + 5$ **71.** $z^2 + 2z + 4$
72. $4a - 5$ **73.** $2x + 9$
74. (a) $C(x) = 35.00 + 1.25x$
    (b) $R(x) = 3.75x$
    (c) $P(x) = 2.50x - 35.00$
    (d) $A(x) = 2.50 - \dfrac{35.00}{x}$
    (e) The average profit is \$1.10 per pumpkin.
75. $2000t^3 + 2000t^2 + 2000t + 2000$; the total charge is about \$6373.

CHAPTER 9 CHAPTER REVIEW

1. $\frac{n^5}{m^7}$ **2.** $\frac{a^{10}}{9b^8}$ **3.** $\frac{9x^2}{y^{10}}$ **4.** $\frac{64t^3}{125s^4}$ **5.** $\frac{4b^3}{a^5}$ **6.** $\frac{2}{7x^3y^4}$ **7.** $\frac{k^4}{25h^6}$ **8.** s^{10} **9.** $(m + n)^6$ **10.** $27y^6$ **11.** c^3 **12.** 1
13. $-15a^3b^7$ **14.** $\frac{z^7}{2}$ **15.** b^{35} **16.** $729d^6$ **17.** $-27d^3$ **18.** $x^{16}y^{20}z^4$ **19.** $5b^2$ **20.** $(a + 2b)^7$ **21.** $-16p^3q^6$
22. $4096x^6$ **23.** $\frac{c^6}{64d^6}$ **24.** $\frac{-27a^6}{64b^3}$ **25.** $\frac{a^2}{2}$ **26.** $\frac{9a^6}{b^8}$ **27.** $10y^2 - 8y + 12$ **28.** $\frac{9}{8}a^3 + \frac{1}{8}a^2 - \frac{3}{16}a + \frac{15}{16}$
29. $4.9z^3 - 1.92z^2 + 10.17z - 6.47$ **30.** $-2x^3 - 11x^2 - 8x + 7$ **31.** $a^4 - 4a^3 + 2a + 4$ **32.** $117z^4 - 18z^3 + 43z^2 - 50z + 56$
33. $143x^7y^{12}$ **34.** $-12x^5 + 24x^4 - 36x^3 - 48x^2 + 60x$ **35.** $36a^6 + 18a^4 - 27a^2$ **36.** $m^2 - 121$ **37.** $z^2 - 16z + 64$
38. $4a^2 + 45a - 91$ **39.** $169 - x^2$ **40.** $b^6 + 8b^3 + 16$ **41.** $t^3 + 6t^2 + 12t + 8$ **42.** $7x^3 + 26x^2 - 29x + 6$
43. $p^2 + q^2 + r^2 + 2pq + 2qr + 2pr$ **44.** $b^3 - 64$ **45.** $y^2 - 4y - 45$ **46.** $x^4 + 2x^3 + 2x^2 + 11x + 4$ **47.** $\frac{3x^3y}{z}$
48. $-2b^3 + 3b^2 + 5b - 4$ **49.** $\frac{4}{c} + \frac{2}{d}$ **50.** $8x + 1$ **51.** $x - 9 - \frac{5}{2x + 5}$ **52.** $9a^2 + 6a + 4$ **53.** $5x^2 + x + 8$
54. $-3 \cdot d \cdot d \cdot d \cdot d$ **55.** $-3 \cdot -3 \cdot -3 \cdot -3 \cdot d \cdot d \cdot d \cdot d$
56. (a) $C(x) = 35.00 + 2.75x$
    (b) $R(x) = 10x$
    (c) $P(x) = 7.25x - 35.00$
    (d) $A(x) = 7.25 - \dfrac{35.00}{x}$
    (e) The average profit is \$3.75 per child.
57. $25s^4 + 25s^3 + 25s^2 + 25s + 25$; the total amount received is about \$113.11.   **58.** The distance is 9.44×10^{15} miles.
59. (a) length: $(2x + 3)$ in.; width: x in.; height: x in.   **(b)** The volume is $(2x^3 + 3x^2)$ in^3.   **(c)** The surface area is $(10x^2 + 12x)$ in^2.
    (d) The volume is 1216 in^3.   **(e)** The surface area is 736 in^2.
60. (a) The area is $\left(x^2 + \dfrac{5}{2}x\right)$ cm^2.   **(b)** The area is 174 cm^2.
61. (a) The current area is x^2 ft^2.   **(b)** The new area will be $(4x^2 + 10x)$ ft^2.   **(c)** The difference is $(3x^2 + 10x)$ square feet.

CHAPTER 9 TEST

1. $(2x - 1)^9$ **2.** $\frac{3x^2}{z^3}$ **3.** $\frac{n^4}{m^9}$ **4.** $-\frac{y^3}{8x^6}$ **5.** a^7 **6.** $\frac{9q^2}{64p^6}$ **7.** $\frac{8ac^3}{b^5}$ **8.** $\frac{a^4b^7}{c}$ **9.** $4y^4 + 5y^3 + 9y^2 + 5$
10. $7x^5 - 2x^4 + 23x^3 + 21x^2 + 2x - 30$ **11.** $-11.4p^9q^2r^3$ **12.** $-8t^4 + 12t^3 + 32t^2 - 24t$ **13.** $81 - 25d^2$ **14.** $15x^2 - x - 28$
15. $8z^3 - 10z^2 + 23z - 15$ **16.** $x^2 + 6x + 9$ **17.** $3x^4 + 5x^3 - x$ **18.** $5x + 7$
19. (a) The volume is $(2x^3 + 8x^2)$ in^3.   **(b)** The surface area is $(10x^2 + 24x)$ in^2.   **(c)** The volume is 450 in^3; the surface area is 370 in^2.
20. (a) $C(x) = 5.75x + 235$ **(b)** $R(x) = 25x$ **(c)** $P(x) = 19.25x - 235$ **(d)** $A(x) = 19.25 - \dfrac{235}{x}$ **(e)** The average profit is \$14.03
per basket.   **21.** Answers will vary.

Chapter 10

SECTION 10.1 EXERCISES

1. $10a^2bc^2$ **2.** $45x^2yz$ **3.** $36x^2$ **4.** $63u^3v^2$ **5.** 45 **6.** 8 **7.** $7xyz$ **8.** $10a^2bc$ **9.** $30ab^2c$
10. $30xy^2z$ **11.** $4(x+3y)$ **12.** $-7(x-3y)$ **13.** $4(2x^3-x^2+3x-6)$ **14.** $3(3d^5-4d^3+7d+8)$ **15.** $a^2(3a^2-5a+7)$
16. $-p^3(4p^2+9p+11)$ **17.** $-3x^3(x^2+3x+4)$ **18.** $5m(m^2-3m+6)$ **19.** $x^2y^2(7x^2-3+9y^2)$ **20.** $-uv(3u^2-4uv+8v^2)$
21. $4a^3(2a^2b^3c+ab^2+4c)$ **22.** $7xy^3(x^2y^2-3xy+9)$ **23.** $22u^3v^3(3v-4u)$ **24.** $-13c^2d^2(3d-4c)$ **25.** does not factor
26. does not factor   **27.** $(x+3)(5x-4)$ **28.** $(y+7)(3y-5)$ **29.** $(2x+y)(x+2y)$ **30.** $(a+3b)(3a+b)$
31. $(3x+5)(2x+7)$ **32.** $(5x+2)(3x+2)$ **33.** $(x+8)(x+1)$ **34.** $(y+4)(y+1)$ **35.** $(2a+3)(a-1)$
36. $(2z+7)(z-1)$ **37.** $(2x+y)(x+2y)$ **38.** $(p+3q)(3p+q)$ **39.** $(x+y)(x-y)$ **40.** $(m+n)(m-n)$
41. $(2x-11)(5y+12)$ **42.** $(4x+3)(7y-9)$ **43.** $(4a+b)(3c+d)$ **44.** $(2m+n)(5p+2q)$ **45.** $(2xy+3)(xy-4)$
46. $(7ab-2)(ab+3)$ **47.** $-1(x+3)(x+y)$ **48.** $-1(a+b)(a+7)$ **49.** $(x^2+y^2)(x^2+2y^2)$ **50.** $(p+q)(p-q)(3p^2+q^2)$
51. $(a+b)(c+d)$ **52.** $(x+y)(y+z)$ **53.** $4(2x+1)(x+3)$ **54.** $7(3y+2)(y+4)$ **55.** $u^2(u+v)(u-2v)$
56. $a^2(a+b)(a+2b)$ **57.** $6a^2(a+b^2)(a+b)$ **58.** $5c^2(c+d^2)(c^2+d)$ **59.** $-2x(x+3)(2x+1)$ **60.** $-5y(y-5)(2y+3)$
61. $(2x+3z)^2$ **62.** $(3u+4v)^2$ **63.** $(6a-5b)^2$ **64.** $(4x+3y)^2$ **65.** $5m(m+n)^2$ **66.** $7p(p-q)^2$ **67.** $(2x^2+3)(x+4)$
68. $(5c^3+7)(c-3)$ **69.** $(5a^2+2b^2)(x+3y)$ **70.** $(2a^2+3b^2)(c+7d)$
71. **(a)** She will receive $(9x+36)$ dollars.   **(b)** $9(x+4)$.   **(c)** The binomial $(x+4)$ is the average amount in dollars that she will receive for each of her nine goals.   **(d)** She will receive $126.
72. **(a)** She smoked $(7c-42)$ cigarettes the first week.   **(b)** $7(c-6)$ **(c)** The binomial $(c-6)$ represents the average number of cigarettes she smoked for each of seven days.   **(d)** She smoked 98 cigarettes; yes; she smoked eight cigarettes on the seventh.
73. **(a)** $n(n-1)$ **(b)** Each equals 420 times.   **(c)** The factored expression was easier to evaluate; answers will vary.
74. **(a)** $n(n-2)(n-1)$ **(b)** Each equals 990 times.   **(c)** The factored expression was easier to evaluate; answers will vary.
75. The rectangle's length is $(x+7)$ units and width is $(x-3)$ units.
76. The length of the square's side is $(2x+3)$ units.
77. **(a)** The area of the green strip is $(4x^2+500x)$ square feet.   **(b)** $4x(x+125)$; the area is equivalent to the area of a rectangle $4x$ feet by $(x+125)$ feet.   **(c)** The area is 5400 ft^2.
78. **(a)** The area of the frame is $(4x^2+108x)$ square inches.   **(b)** $4x(x+27)$; the area is equivalent to the area of a rectangle $4x$ inches by $(x+27)$ inches.   **(c)** The area is 640 in^2.

10.1 CALCULATOR EXERCISES

Part 1

1. $21xyz^2$ **2.** $13c$ **3.** $12x^2$ **4.** $32abc$

Part 2

1. $30 = 2 \cdot 3 \cdot 5$ **2.** $108 = 2^2 \cdot 3^3$ **3.** $525 = 3 \cdot 5^2 \cdot 7$ **4.** $1287 = 3^2 \cdot 11 \cdot 13$ **5.** $1547 = 7 \cdot 13 \cdot 17$ **6.** $4500 = 2^2 \cdot 3^2 \cdot 5^3$

SECTION 10.2 EXERCISES

1. $(x+2)^2$ **2.** $(p+7)^2$ **3.** $(4z+5)^2$ **4.** $(6x+7)^2$ **5.** does not factor   **6.** does not factor   **7.** $(x-5)^2$ **8.** $(y-8)^2$
9. $(6z-5)^2$ **10.** $(8m-3)^2$ **11.** does not factor   **12.** does not factor   **13.** $(a+4)^2(a-4)^2$ **14.** $(b+7)^2(b-7)^2$
15. $(2x+3)^2(2x-3)^2$ **16.** $(5y+2)^2(5y-2)^2$ **17.** $3(x+4)^2$ **18.** $16(x+1)^2$ **19.** $2(a-3)^2$ **20.** $5(x-7)^2$
21. $p(p+q)^2$ **22.** $z^2(y+2z)^2$ **23.** $2p(p+q)^2(p-q)^2$ **24.** $5u(u+2v)^2(u-2v)^2$ **25.** $m(m^2+n^2)^2$ **26.** $p^3(p^2+q^2)^2$
27. $(x+10)(x-10)$ **28.** $(y+8)(y-8)$ **29.** $(11+c)(11-c)$ **30.** $(14+b)(14-b)$ **31.** $(7a+2)(7a-2)$
32. $(5z+3)(5z-3)$ **33.** $(5+2y)(5-2y)$ **34.** $(8+3x)(8-3x)$ **35.** $(4u+3v)(4u-3v)$ **36.** $(6a+5b)(6a-5b)$
37. $7(z+2)(z-2)$ **38.** $8(x+1)(x-1)$ **39.** does not factor   **40.** does not factor   **41.** $(x^2+25)(x+5)(x-5)$
42. $(a^2+36)(a+6)(a-6)$ **43.** $(16+z^2)(4+z)(4-z)$ **44.** $(25+b^2)(5+b)(5-b)$ **45.** $(x^4+1)(x^2+1)(x+1)(x-1)$
46. $(y^4+16)(y^2+4)(y+2)(y-2)$ **47.** $(x-3)(x^2+3x+9)$ **48.** $(z-7)(z^2+7z+49)$ **49.** $(a+4)(a^2-4a+16)$
50. $(m+1)(m^2-m+1)$ **51.** $(3x+4y)(9x^2-12xy+16y^2)$ **52.** $(5p+3q)(25p^2-15pq+9q^2)$
53. $(2p-5q)(4p^2+10pq+25q^2)$ **54.** $(3u-2v)(9u^2+6uv+4v^2)$ **55.** $p(p+4q)(p^2-4pq+16q^2)$
56. $y(5x+y)(25x^2-5xy+y^2)$ **57.** $3u(3u-v)(9u^2+3uv+v^2)$ **58.** $2b(a-3b)(a^2+3ab+9b^2)$
59. **(a)** The garden area is (x^2-225) ft^2.   **(b)** The rectangular plot has dimensions of $(x+15)$ ft by $(x-15)$ ft.
    (c) Yes, the plot that is 100 ft on each side has a larger garden area than the 85-by-100-foot plot.
60. **(a)** The area is $\left(y^2 - \dfrac{1}{4}x^2\right)$ in^2.   **(b)** An equivalent area is that of a rectangle measuring $\left(y+\dfrac{1}{2}x\right)$ in. by $\left(y-\dfrac{1}{2}x\right)$ in.
    (c) The dimensions are 14 in. by 10 in.
61. $(35+2x)(35+2x)$; the dimensions of the stage are 35 feet by 35 feet.
    The area of the stage and mosh pit is 7225 ft^2.
    The area of the stage is 1225 ft^2.
    The area of the mosh pit is 6000 ft^2.

62. $(65 - 2x)(65 - 2x)$; the dimensions of the yard are 65 feet by 65 feet.
The central area of the yard is 3025 ft^2.
The area of the yard is 4225 ft^2.
The area of the shrub trim is 1200 ft^2.

10.2 CALCULATOR EXERCISES

1. perfect square   **2.** perfect cube   **3.** perfect square   **4.** perfect square and perfect cube   **5.** $(x - 35)(x + 35)$
6. $(a - 19)(a^2 + 19a + 361)$ **7.** $(16y - 21)(16y + 21)$ **8.** $(7b - 12)(49b^2 + 84b + 144)$ **9.** $(z - 27)(z + 27)$
10. $(z - 9)(z + 9z + 81)$

SECTION 10.3 EXERCISES

1. $(x + 5)(x + 9)$ **2.** $(p + 6)(p + 8)$ **3.** $(y - 7)(y - 8)$ **4.** $(u - 2)(u + 13)$ **5.** $(p + 3)(p - 12)$ **6.** $(v + 1)(v - 13)$
7. does not factor   **8.** does not factor   **9.** $(x^2 + 9)(x^2 + 16)$ **10.** $(b^2 + 4)(b^2 + 25)$ **11.** $(x^2 + 1)(x^2 - 3)$
12. $(d^2 - 7)(d^2 + 25)$ **13.** $3(a + 5)(a + 11)$ **14.** $7(b - 4)(b - 6)$ **15.** $4(c - 2)(c + 13)$ **16.** $7(p + 4)(p - 5)$
17. $(x - 3y)(x - 8y)$ **18.** $(a - 8b)(a - 9b)$ **19.** $(x - y)(x + 12y)$ **20.** $(a + 6b)(a - 15b)$ **21.** $-3(a + 2b)(a + 3b)$
22. $-5(x - 3y)(x - 4y)$ **23.** $-2(x + 2y)(x - 9y)$ **24.** $-6(x + 3y)(x - 7y)$ **25.** $(3x + 1)(x + 3)$ **26.** does not factor
27. $(x - 7)(2x - 1)$ **28.** $(x - 11)(3x - 1)$ **29.** $(x - 1)(3x + 2)$ **30.** $(x - 1)(5x + 3)$ **31.** $(m + 2)(5m - 1)$
32. $(p + 5)(7p - 1)$ **33.** does not factor   **34.** does not factor   **35.** $(a + 6)(4a + 1)$ **36.** $(c + 4)(10c + 1)$
37. $(d - 1)(9d - 4)$ **38.** $(b - 1)(8b - 9)$ **39.** $(x - 4)(6x + 1)$ **40.** $(x - 8)(4x + 1)$ **41.** $(y + 2)(8y - 9)$
42. $(z + 4)(6z - 5)$ **43.** $(2b + 3)(3b + 4)$ **44.** $(5a + 6)(3a + 2)$ **45.** $(5x - 4)(4x - 3)$ **46.** $(2z + 3)(7z - 4)$
47. $(3x - 4)(6x + 5)$ **48.** $(2x - 9)(4x + 7)$ **49.** $3(2p - 7)(3p + 1)$ **50.** $-3(7x + 8)(3x - 2)$ **51.** $(2x^2 + 9)(x^2 + 1)$
52. $(4x^2 - 3)(6x^2 - 1)$ **53.** $(m^2 + 4)(4m^2 - 3)$ **54.** $(z^2 + 9)(5z^2 - 6)$ **55.** $-1(2x - 7)(3x + 8)$ **56.** $-1(b + 3)(4b - 7)$
57. $(2x + 3y)(3x - 2y)$ **58.** $(2a + 3b)(7a + 2b)$ **59.** $(u - 8v)(4u - 7v)$ **60.** $(x - 4y)(3x - 8y)$ **61.** $(x^2 + y^2)(9x^2 + 4y^2)$
62. $(x^2 + y^2)(16x^2 + 25y^2)$ **63.** $(2xy + 3)(5xy - 7)$ **64.** $(3pq - 2)(7pq + 4)$
65. **(a)** The increased width is $(w + 2)$ in. and the increased length is $(w + 12)$ in.   **(b)** The width was increased by 2 in.
(c) The length was increased by 4 in.
66. **(a)** The new width is $(x - 3)$ in. and the new length is $(x - 1)$ in.   **(b)** The width was increased by 2 in.
(c) The length was decreased by 1 in.
67. **(a)** The lengths of the legs are $(x + 8)$ in. and $(2x + 5)$ in.   **(b)** They were each increased by 8 in.   **(c)** The expression is $(2x - 3)$ in.
68. **(a)** The lengths of the legs are $(x - 4)$ in. and $(3x - 12)$ in.   **(b)** They were each decreased by 4 in.   **(c)** The expression is $(3x - 8)$ in.
69. $4(x + 18)(x + 139)$ or $(2x + 36)(2x + 78)$; the dimensions of the playing area are 36 feet by 78 feet.
70. $2(4x + 53)(x + 50)$ or $(4x + 53)(2x + 100)$; the dimensions of the playing area are 53 yards by 100 yards.

10.3 CALCULATOR EXERCISES

1. $(5x + 12)(8x - 15)$ **2.** $(4x - 9)(20x - 13)$ **3.** $(9x - 11)(12x - 5)$ **4.** $(12x + 13)(15x + 11)$

SECTION 10.4 EXERCISES

1. $(3x + 4)(x + 5)$ **2.** $(5x + 4)(x + 3)$ **3.** $(3x + 5)(2x + 7)$ **4.** $(7x + 2)(4x + 3)$ **5.** $(5q + 11)(2q + 1)$
6. $(4p + 13)(2p + 1)$ **7.** cannot be factored   **8.** cannot be factored   **9.** cannot be factored   **10.** $(5y - 2)(8y - 3)$
11. $(3m - 2)(2m - 3)$ **12.** $(5k - 2)(2k - 5)$ **13.** $(7p - 1)(8p + 3)$ **14.** $(8x - 1)(9x + 4)$ **15.** $(5p - 2)(3p - 4)$
16. cannot be factored   **17.** $(4x + 1)(8x - 5)$ **18.** $(6x + 1)(4x - 3)$ **19.** $(4k - 9)(7k + 11)$ **20.** $(8k - 9)(7k + 5)$
21. $(8m - 9)(3m + 13)$ **22.** $(5x + 12)(6x - 5)$ **23.** $(5x - 1)(8x - 28)$ **24.** $(9x - 2)(6x - 8)$ **25.** $(6x + 15)(5x + 2)$
26. $(15x + 10)(2x + 3)$ **27.** $(8x - 3)(2x + 6)$ **28.** $(9x - 4)(2x + 8)$ **29.** $-3(2x - 3)(3x - 8)$ **30.** $-4(2x - 5)(3x - 2)$
31. $4(4x^2 - 13x - 20)$ **32.** $6(4y^2 - y - 11)$ **33.** $x(2x + 5)(3x + 5)$ **34.** $y(10y^2 + 29y + 35)$ **35.** $(3x^2 + 2)(2x^2 + 1)$
36. $(5x^2 + 1)(2x^2 + 3)$ **37.** $(4x^2 + 9)(2x^2 + 7)$ **38.** $(6y^2 + 5)(2y^2 + 3)$ **39.** $(3x^2 - 5)(2x^2 - 3)$ **40.** $(3x^2 - 5)(4x^2 - 7)$
41. $(3x^2 - 2)(4x^2 - 3)$ **42.** $(3x^2 - 7)(2x^2 - 5)$ **43.** $(4m^2 + 5)(2m^2 - 3)$ **44.** $(6p^2 + 5)(2p^2 - 3)$ **45.** $(3x^2 - 2)(5x^2 + 8)$
46. $(9x^2 - 4)(2x^2 + 3)$ **47.** $3(4y^2 + 7)(5y^2 - 4)$ **48.** $2(7x^2 - 3)(4x^2 + 5)$ **49.** $-2(6x^2 - 5)(2x^2 - 1)$ **50.** $-3(5y^2 + 4)(3y^2 + 2)$
51. $(3x + 4y)(2x - 5y)$ **52.** $(3x - 4y)(7x + 2y)$ **53.** $(8p - 7q)(5p - 4q)$ **54.** $(6p - 5q)(9p - 2q)$

55. $\frac{1}{2}(3x + 7)(4x + 7)$; the enlarged triangle has a height of $(3x + 7)$ feet and a base of $(4x + 7)$ feet; the base and height were increased by

7 feet.

56. $\frac{1}{2}(5x - 6)(x - 6)$; the reduced triangle has a height of $(x - 6)$ feet and a base of $(5x - 6)$ feet; the base and height were reduced

by 6 feet.

57. $(2x + 7)(x + 6)$; the length is $(2x + 7)$ meters and the width is $(x + 6)$ meters which indicates that the free space added to each side of
the width is 3 meters; since $x = 6$, the dimensions of the total area are 12 meters by 19 meters with an area of 228 square meters.

58. $(2x + 7)(x + 7)$; the length is $(2x + 7)$ meters and the width is $(x + 7)$ meters which indicates that the free space added to each side is
3.5 meters; since $x = 1.8$, the dimensions of the total area are 8.8 meters by 10.6 meters with an area of 93.28 square meters.

10.4 CALCULATOR EXERCISES

1. $2(4x-1)(12x+1)$ **2.** $(3x+5)(8x-11)$ **3.** $(2x+3)(16x+27)$ **4.** $(3x-2)(24x-17)$ **5.** $(p^2+25)(4p^2+9)$

SECTION 10.5 EXERCISES

1. $3y(y^2+y+1)$ **2.** $5x(2x^2+7x-1)$ **3.** $5ab(2c^2+3c-4)$ **4.** $2xy(3z^2+5z-3)$ **5.** $-8a(5a+3b-6c)$
6. $-7x(4x+5y-3z)$ **7.** $3x^3(6x+5)(6x-5)$ **8.** $2u(4u+7)(4u-7)$ **9.** $-8xy(5x+2y)(5x-2y)$
10. $-2ab(3a+8b)(3a-8b)$ **11.** $x(x-8)^2$ **12.** $z(z-5)^2$ **13.** $3v(2u+3v)^2$ **14.** $2x(5x+2y)^2$ **15.** $(x+4)(x-4)(3x^2+7)$
16. $(x+3)(x-3)(2x^2+5)$ **17.** $(p+3)(p-3)(2p+1)(2p-1)$ **18.** $(z+2)(z-2)(3z+2)(3z-2)$ **19.** $-2(3y-1)(7y+1)$
20. $-3(4y-1)(6y+1)$ **21.** $4(uv+2)(uv+7)$ **22.** $5(ab+3)(ab+4)$ **23.** $4x(x-2)(8x-7)$ **24.** $3x(5x-6)(x-4)$
25. $2x^2(x+3)(6x-5)$ **26.** $5z^2(z+2)(3z-4)$ **27.** $(x^3-5y^3)(x^3+3y^3)$ **28.** $(p^3+2q^3)(p^3-5q^3)$ **29.** $x(x-5)(x^2+8)$
30. $2y(y+3)(y^2+4)$ **31.** $(1+k^4)(1+k^2)(1+k)(1-k)$ **32.** $(16s^4+1)(4s^2+1)(2s+1)(2s-1)$
33. $x(14.75-0.95x)$; the regular price is \$14.75; for every additional pizza, a discount of \$0.95 will be given; yes.
34. $x(15.70-2.24x)$; CD's regularly cost \$15.70; for every additional CD, a discount of \$2.24 will be given; yes.
35. $x(19.95+2.99x)$; a doll costs \$19.95; for every doll purchased, a surcharge of \$2.99 will be given.
36. $x(4.95+2x)$; a turkey costs \$4.95; for every turkey purchased, x, a charge of \$2 will be given.
37. $x(9-2x)(15-2x)$; the dimensions are x inches by $(9-2x)$ inches by $(15-2x)$ inches; if $x=2$, the dimensions are 2 inches by 5 inches by 11 inches with a volume of 110 cubic inches.
38. $x(x+20)(x+10)$; the dimensions are x inches by $(x+20)$ inches by $(x+10)$ inches; if $x=5$, the dimensions are 5 inches by 25 inches by 15 inches with a volume of 1875 cubic inches.
39. $\pi x(2x-5)^2$; the radius is $(2x-5)$ inches; if $x=8$, the radius is 11 inches and the volume is 968π cubic inches or about 3041 cubic inches.
40. $\pi x(x-8)^2$; the radius is $(x-8)$ inches and the diameter is $(2x-16)$ inches; if $x=15$, the radius is 7 inches, the diameter of 14 inches, the height is 15 inches and the volume is 735π cubic inches or about 2309 cubic inches.
41. $(x-6)(x^2+6x+36)$; no, since $V=e^3$ for a cube; each side of the smaller cube was 6 inches.
42. $(x+24)(x^2-24x+576)$; no; the larger cube is 24 inches on each side.

10.5 CALCULATOR EXERCISES

1. $(2x+9)(4x-3)(2x-7)$ **2.** $(2x+1)(5x-3)(5x+3)$ **3.** $(3x+7)(5x-4)(6x+5)$ **4.** $(3x+8)(3x-8)(2x-5)$

CHAPTER 10 SECTION-BY-SECTION REVIEW

Reflections

1–9. Answers will vary.
1. $4a^2(5a^4-7a^2+11)$ **2.** $22u^2v^2(u+v)$ **3.** $(x^2+1)(3x+1)$ **4.** $7(a^2+b^2)^2$ **5.** $(5c+6d)(3a+4b)$
6. (a) $\frac{1}{2}n(n+1)$ **(b)** The sum is 78. **(c)** The factored expression is easier to evaluate; answers will vary.
7. The width is $(x-3)$ units and the length is $(2x+5)$ units.
8. $x+(x+10)+(x+20)+(x+30)+(x+40)+(x+50)+(x+60)$; $7(x+30)$; the employee earns an average of $(x+30)$ dollars for each of 7 months; the salary is \$6930.
9. The area is $(4x^2+78x)$ square feet; $2x(2x+39)$; the area of the border is equivalent to the area of a rectangle $2x$ feet by $(2x+39)$ feet; if $x=3$, the area is 270 square feet.
10. $(p+6)^2$ **11.** $(q-8)^2$ **12.** $(3x+5)^2$ **13.** $(7y-8)^2$ **14.** does not factor **15.** $x^3(x^2+3y)^2$ **16.** $(x-13)(x+13)$
17. $(25-a)(25+a)$ **18.** $3(2x+5)(2x-5)$ **19.** $(p-q)(p+q)$ **20.** does not factor **21.** $(3x-5y)(3x+5y)$
22. $(4x^2+9)(2x+3)(2x-3)$ **23.** $(x^4+1)(x^2+1)(x+1)(x-1)$ **24.** $(c+3)(c^2-3c+9)$ **25.** $(c-3)(c^2+3c+9)$
26. $(2z-5)(4z^2+10z+25)$ **27.** $5(h+2k)(h^2-2hk+4k^2)$
28. (a) x^2-4 **(b)** $(x+2)(x-2)$ **(c)** No, because some of the rectangle is filled in with light blue material.
29. $\pi(x+100)^2$; the radius is $(x+100)$ feet; the area of the pond and exercise space is about 45,239 square feet; the area of the pond is about 31,416 square feet; the area of the exercise space is about 13,823 square feet.
30. $(z-9)(z+11)$ **31.** $(p-6q)(p+11q)$ **32.** $6(a+3)(a+13)$ **33.** $(x^2+3)(x^2+5)$
34. $4q(q+5)(q-12)$ **35.** $(xy+9)(xy-13)$ **36.** $-7(x-2)(x-12)$ **37.** $(x-5)(2x-1)$
38. $(2x+5)(3x+1)$ **39.** $7(ab+3)(4ab+1)$ **40.** $-3x(3x+8)(5x-2)$
41. (a) The base is $(2x+7)$ units and the height is $(2x+3)$ units. **(b)** The base was increased by 7 units.
(c) The height was increased by $(x+3)$ units.
42. (a) The new length is $(x+15)$ units and the new width is $(x+2)$ units. **(b)** It was increased by 15 units.
(c) It was increased by 8 units.
43. $2(2x+15)(x+14)$ or $(2x+15)(2x+28)$; the dimensions of the basketball court are 15 meters by 28 meters.
44. cannot be factored **45.** $(4x+9)(2x+1)$ **46.** $(5y-3)(3y-4)$ **47.** cannot be factored **48.** $(3y+2)(4y-7)$
49. $(4x^2+5)(2x^2+3)$ **50.** $(5x^2-3)(2x^2+3)$ **51.** cannot be factored **52.** cannot be factored **53.** $4(2x+1)(3x+8)$
54. $-5(4x^2-7)(2x^2+5)$ **55.** $(3x+4y)(2x-5y)$ **56.** $(6x+5y)(2x+5y)$ **57.** $(4x-3y)(5x-3y)$

58. $(19x + 9)(x + 3)$; the dimensions of the total area are 4 meters by 28 meters; the total area is 112 square meters; the dimensions of the lane are 1 meter by 19 meters with an area of 19 square meters.
59. $-3x(2x - 5y)^2$ **60.** $7x^2(x^2 + x + 1)$ **61.** $3x(2x + 9)(2x - 9)$ **62.** $8x(2x + 1)^2$ **63.** $2x(4x - 9)(3x + 5)$
64. $(4x + 3)^2(4x - 3)^2$ **65.** $(3x + 2)(3x - 2)(2x + 1)(2x - 1)$ **66.** $2x(x + 7)(x - 2)(x + 2)$
67. (a) The land area not covered is $(4x^2 - 25)$ ft^2. **(b)** The dimensions would be $(2x + 5)$ ft by $(2x - 5)$ ft.
  (c) The dimensions would be 165 ft by 155 ft.
68. $x(19.95 - 0.10x)$; for every instructor, x, there is a 10¢ discount.

CHAPTER 10 CHAPTER REVIEW

1. $(z + 15)(z - 6)$ **2.** $(a - 6)(a - 12)$ **3.** $(x + 9y)(x + 5y)$ **4.** $5(a + 7)^2$ **5.** $2(a^2 + 4ab + 6b^2)$
6. $(x^2 + 3)(x^2 + 7)$ **7.** $3q(q + 3)(q - 14)$ **8.** $-6(x + 5)(x - 12)$ **9.** $(x + 2)(x + 5)$ **10.** $(x + 17)(x - 17)$
11. $(x - 1)(x^2 + x + 1)$ **12.** $4(x + 4)(x - 4)$ **13.** does not factor **14.** $(6x + 7y)(6x - 7y)$ **15.** $(9x^2 + 1)(3x + 1)(3x - 1)$
16. $(p + 11)^2$ **17.** $(q - 15)^2$ **18.** $(3a + 4b)(9a^2 - 12ab + 16b^2)$ **19.** $3(3ab - 4)^2$ **20.** $-2x(5x - 6y)^2$
21. $2x(4x^3 - x^2 + 3x - 6)$ **22.** $5u^2v^2(7u + 5v)$ **23.** $(x^2 + 5)(2x + 1)$ **24.** $(m - 2n)(m - 8n)$ **25.** $4(a^2 + 2b^2)^2$
26. $(x - 1)(2x - 11)$ **27.** $(6c + 5d)(4a + 3b)$ **28.** $(x - 3)(7x + 2)$ **29.** $(2x - 3)(5x + 2)$ **30.** $6(2a + 1)(3a + 4)$
31. $-2(3x - 2)(5x + 8)$ **32.** $(3x^2 + 1)(4x^2 + 3)$ **33.** $6x(3x + 1)^2$ **34.** $(3x + 2)^2(3x - 2)^2$ **35.** $(2x + 5)(2x - 5)(x + 3)(x - 3)$
36. $3x(2x + 3)(2x - 5)$ **37.** $3x(x + 5)(x + 3)(x - 3)$ **38.** The width is $(2x + 1)$ units and the length is $(4x - 3)$ units.
39. (a) The new length is $(3x + 1)$ in. and the new width is $(2x - 4)$ in. **(b)** The length was increased by $(2x + 1)$ in.
  (c) The width was increased by $(x - 2)$ in.
40. (a) The base is $(2x + 16)$ units and the height is $(x + 4)$ units. **(b)** The base was increased by 16 units.
  (c) The height was increased by 4 units.
41. $x(50 + 25x)$; a judge sets a fine of \$75 for the first penalty and then adds on \$25 for x number of times cited in the future.
42. (a) $x^2 - 9$ **(b)** $(x + 3)(x - 3)$ **(c)** Yes
43. $\pi(610 + x)^2$; the radius is $(610 + x)$mm; the area is $372{,}100\pi$ mm^2 or about 1,168,987 mm^2.
44. $312x + 4x^2$; $4x(78 + x)$ or $2x(156 + 2x)$; if $x = 16$, the area is 6016 square inches.

CHAPTER 10 TEST

1. $9a(3a + 1)^2$ **2.** $(p + 5)(p^2 - 5p + 25)$ **3.** $-4a^2b^2(a + 4b)(2a + b)$ **4.** $(a^2 + b^2)(a - 5b)$
5. $(5x - 7y)(3x + 2y)$ **6.** $(8a + 7b)(8a - 7b)$ **7.** $(5x - 7)^2$ **8.** $3x(x - 1)(x - 8)$
9. $(x + 3y)(x - 7y)$ **10.** $(2x + 1)(7x + 9)$ **11.** $(x^2 + 7)(2x + 1)(2x - 1)$ **12.** does not factor
13. $x(2x - 1)(x + 3)$; the width is $(2x - 1)$ inches; the length is $(x + 3)$ inches.

14. (a) $\frac{1}{2}(x + 12)(x + 5)$; the base is $(x + 12)$ inches and the height is $(x + 5)$ inches.
  (b) The base and height were increased by 5 inches each. **(c)** The base of the original triangle was $(x + 7)$ inches.

15. $5x(4 + x)$ or $x(20 + 5x)$; his parents gave him \$25 for his first test and then added \$5 for x tests he passed.
16. (a) $s^3 - 8$ **(b)** $(s - 2)(s^2 + 2s + 4)$; no since $V = e^3$.
  (c) You will need 1720 cubic inches of gelatin to fill the larger crate and boxed egg.
17. $\pi(x + 9)^2$; the radius of the center tray is 9 inches; the total area is 121π square inches or about 380 square inches; the area of the center is 81π square inches or about 254 square inches; the area of the border is 40π square inches or about 126 square inches.
18. (a) factor by grouping **(b)** He did not factor completely. **(c)** $(3x + 7)(2x + 3)(2x - 3)$

Chapter 11

SECTION 11.1 EXERCISES

1. polynomial; cubic **2.** not a polynomial **3.** not a polynomial **4.** polynomial; cubic **5.** polynomial **6.** polynomial; cubic
7. polynomial; quadratic **8.** not a polynomial **9.** not a polynomial **10.** polynomial; cubic **11.** polynomial; quadratic
12. polynomial; quadratic **13.** 2 and 4 **14.** -3 and 2 **15.** non-integer between 0 and 1. **16.** non-integers between -4 and -5
and between 3 and 4. **17.** no solution **18.** no solution **19.** all real numbers **20.** all real numbers
21. no real-number solution **22.** no real-number solution **23.** -6 and 2 **24.** -4 and 3 **25.** -5 and 2 **26.** -3 and 5
27. -3 and 3 **28.** -4 and 4 **29.** -1 and 3 **30.** -4 and -2 **31.** $-2, 0,$ and 2 **32.** 2 **33.** -2 and 5
34. $-3, -1,$ and 1 **35.** no solution **36.** no real-number solution **37.** no real-number solution **38.** no solution
39. all real numbers **40.** all real numbers **41.** $-1, 0,$ and 4 **42.** $-3, 0,$ and 2 **43.** no real-number solution
44. no solution **45.** no solution **46.** no solution **47.** all real numbers **48.** all real numbers
49. $-\frac{3}{2}$ and $\frac{3}{2}$ **50.** $-\frac{4}{3}$ and $\frac{4}{3}$ **51.** $-0.8, 0.5,$ and 1 **52.** $-1, 0.5,$ and 2.5 **53.** -2.8 and 3.7 **54.** 0.8 and 4.2
55. $-3.6, -1.5,$ and 1.4 **56.** $-2.4, -0.2,$ and 2.5 **57.** It will hit the ground in approximately 1.58 seconds.
58. They will hit the ground in 1 second. **59.** The dagger will hit the ground in approximately 1.43 seconds.
60. The branch will hit the ground in approximately 0.94 second. **61.** It will hit the ground in 1.75 seconds.

62. The hammer will hit the ground in 1 second. **63.** The total collection will decrease by \$25 billion in the year 2006; no
64. The collections will triple the amount in the year 2003; no **65.** The per capita personal income will become twice the 1997 value in the year 2009; yes **66.** The per capita personal income will double the 1997 value in the year 2003; yes
67. There is no real-number solution. **68.** There should be between 5 and 11 sales reps.

11.1 CALCULATOR EXERCISES

Students should verify solutions with the calculator.

SECTION 11.2 EXERCISES

1. -11 and -6 **2.** -13 and -9 **3.** $-\frac{2}{3}$ and $\frac{3}{4}$ **4.** $-\frac{2}{3}$ and $\frac{7}{9}$ **5.** $-9, 0,$ and $\frac{5}{2}$ **6.** $-\frac{1}{6}, 0,$ and 4 **7.** $\frac{1}{7}$ and 7 **8.** $\frac{1}{9}$ and 9

9. $-6, -\frac{3}{4},$ and $\frac{9}{2}$ **10.** $-\frac{7}{2}, \frac{8}{3},$ and 7 **11.** -34 and 1.3 **12.** -3 and 4 **13.** -6 and -4 **14.** -8 and -5 **15.** 3 and 11

16. 4 and 13 **17.** -6 and $-\frac{5}{4}$ **18.** -2 and $-\frac{8}{7}$ **19.** $-\frac{8}{5}$ and 1 **20.** $-\frac{5}{3}$ and 4 **21.** 0 and $\frac{7}{3}$ **22.** 0 and $\frac{7}{2}$

23. $-\frac{5}{3}$ and $\frac{1}{6}$ **24.** $-\frac{3}{2}$ and $\frac{7}{4}$ **25.** $-\frac{9}{4}$ **26.** $\frac{2}{7}$ **27.** $-\frac{5}{2}$ **28.** $-\frac{4}{3}$ **29.** -8 and 8 **30.** -12 and 12 **31.** $-\frac{5}{3}$ and $\frac{5}{3}$

32. $-\frac{9}{8}$ and $\frac{9}{8}$ **33.** -8 and 6 **34.** -4 and 8 **35.** -4 and 9 **36.** -6 and 7 **37.** -12 and 9 **38.** -20 and 15

39. $-7, -3,$ and 3 **40.** $-5, -4,$ and 4 **41.** $-\frac{5}{2}, -\frac{5}{3},$ and $\frac{5}{3}$ **42.** $-3.5, -3,$ and 3.5 **43.** 1 **44.** 8

45. (a) The volume is $\left(\frac{3}{2}x^2 + 3x\right)$ ft^3. (b) The dimensions are 6 ft by 3 ft by 4 ft.
46. (a) The volume is $(32x^2 + 4x)$ cm^3. (b) The dimensions are 5 cm by 4 cm by 41 cm.
47. (a) The surface area is $(26x^2 + 10x)$ in^2. (b) The dimensions are 15 in. by 5 in. by 22 in.
48. (a) The surface area is $(84x^2 + 24x)$ units^2. (b) The dimensions are 3 in. by 15 in. by 23 in.

49. (a) The area is $\left(\frac{1}{2}x^2 + 2x\right)$ ft^2. (b) The dimensions are 6 ft base, 10 ft height.

50. (a) The area is $\left(\frac{5}{2}x^2 + x\right)$ m^2. (b) The dimensions are 10 m height, 52 m base.

51. $V^2 = 30(0.40)(243)$; the vehicle was traveling at 54 mph.
52. $V^2 = 30(0.27)(250)$; the vehicle was traveling at 45 mph.
53. (a) $y = -0.12x^2 - 0.06x + 2.9$ (b) In 2001, the profit will drop to \$2 million. (c) Answers will vary.
54. (a) $y = -0.09x^2 + 0.57x + 3.66$ (b) In 2004, the profit will drop to \$3 million. (c) Answers will vary.

11.2 CALCULATOR EXERCISES

1. $y = x^3 - 2x^2 + 3x - 4$ **2.** $y = x^4 - x^3 + x^2 - x + 1$ **3.** $y = -1.5x^3 + 10x^2 - 9.5x - 6$ **4.** $y = 0.5x^4 - 0.2x^3 - 7x + 6$

SECTION 11.3 EXERCISES

1. $3\sqrt{7}$ **2.** $4\sqrt{5}$ **3.** $9\sqrt{3}$ **4.** $5\sqrt{3}$ **5.** $7\sqrt{3}$ **6.** $-8\sqrt{2}$ **7.** $-5\sqrt{5}$ **8.** $3\sqrt{3}$ **9.** $-3 + 2\sqrt{5}$ **10.** $-4 + 2\sqrt{7}$

11. $2 - 5\sqrt{2}$ **12.** $7 - 4\sqrt{2}$ **13.** $1 - \sqrt{3}$ **14.** $1 - \sqrt{2}$ **15.** $-2 + \sqrt{5}$ **16.** $-2 + \sqrt{6}$ **17.** $\frac{5}{3} + \sqrt{2}$ **18.** $\frac{3}{2} + \sqrt{10}$

19. $2 + \sqrt{6}$ **20.** $2 + \sqrt{11}$ **21.** $\frac{4\sqrt{5}}{5}$ **22.** $\frac{3\sqrt{7}}{7}$ **23.** $\frac{5\sqrt{6}}{12}$ **24.** $\frac{\sqrt{30}}{4}$ **25.** ± 12 **26.** ± 11 **27.** $\pm\sqrt{13}$

28. $\pm\sqrt{15}$ **29.** $\pm 7\sqrt{2}$ **30.** $\pm 10\sqrt{2}$ **31.** ± 4 **32.** ± 3 **33.** $\pm\frac{5}{2}$ **34.** $\pm\frac{7}{4}$ **35.** $\pm\frac{\sqrt{2}}{3}$ **36.** $\pm\frac{1}{8}$ **37.** $\pm\frac{\sqrt{6}}{3}$

38. $\pm\frac{\sqrt{35}}{5}$ **39.** no real-number solution **40.** no real-number solution **41.** 5 **42.** $\frac{7}{2}$ **43.** 5 and 9 **44.** -9 and -3

45. $\frac{1}{4}$ and $\frac{5}{4}$ **46.** $\frac{4}{3}$ and 2 **47.** ± 1.3 **48.** ± 1.7 **49.** -5 and -1 **50.** 1 and 11 **51.** 1 and 7 **52.** -11 and -7

53. -8 and -2 **54.** -3 and 17 **55.** $-\frac{11}{3}$ and $\frac{13}{3}$ **56.** $-\frac{9}{5}$ and $\frac{7}{5}$ **57.** $7 \pm \sqrt{6}$ **58.** $12 \pm \sqrt{5}$ **59.** $\frac{-1 \pm \sqrt{10}}{2}$

60. $\frac{5 \pm \sqrt{2}}{4}$ **61.** $-3 \pm 2\sqrt{3}$ **62.** $15 \pm 2\sqrt{3}$ **63.** $\pm\frac{2\sqrt{5}}{5}$ **64.** $\pm\frac{3\sqrt{7}}{7}$ **65.** $\pm\frac{5\sqrt{2}}{2}$ **66.** $\pm\frac{4\sqrt{11}}{11}$

67. It is 11 inches high. **68.** The kite is 80 feet high. **69.** The distance is 4000 feet. **70.** The tower is 120 meters tall.
71. The distance is $25\sqrt{2}$, or approximately 35.36 feet. **72.** The perimeter is $40\sqrt{2}$, or approximately 56.57 cm.
73. It would rise approximately 51.4 feet. **74.** It would rise about 24.5 inches. **75.** The ball was about 7.14 feet high. The ball missed the bar by 0.86 feet. **76.** The ball was about 8.57 feet high. The ball cleared the bar by 0.57 feet.

77. The annual interest rate was 8.5%. **78.** The annual interest rate was 5.5%. **79.** The annual interest rate was 9.54%. **80.** The annual interest rate was 15.47%. **81.** The annual interest rate was 12.55%. **82.** The annual interest rate was 18.5%. **83.** It will take 5 seconds for the balloon to hit the ground. **84.** It will reach the ground in 6 seconds. **85.** It will take $\dfrac{5\sqrt{70}}{2}$, or approximately 20.9 seconds. **86.** It will take $\dfrac{25\sqrt{3}}{2}$, or approximately 21.65 seconds. **87.** She manages about 313 accounts. **88.** He manages about 172 accounts.

11.3 CALCULATOR EXERCISES

7 in., 8 in., 10 in., 11 in., 13 in., 14 in., 16 in., 17 in.

SECTION 11.4 EXERCISES

1. 81 **2.** 16 **3.** $\dfrac{81}{4}$ **4.** $\dfrac{25}{4}$ **5.** $\dfrac{9}{64}$ **6.** $\dfrac{9}{100}$ **7.** $\dfrac{1}{4}$ **8.** $\dfrac{1}{4}$ **9.** 9 **10.** 16 **11.** $\dfrac{81}{4}$ **12.** $\dfrac{25}{4}$ **13.** $\dfrac{16}{81}$ **14.** $\dfrac{9}{49}$

15. 49 **16.** 25 **17.** -11 and 5 **18.** -5 and 13 **19.** -4 and 7 **20.** -8 and 5 **21.** $-\dfrac{3}{7}$ and $-\dfrac{1}{7}$ **22.** $-\dfrac{4}{5}$ and $-\dfrac{2}{5}$

23. -10 and 9 **24.** -7 and 8 **25.** $3 \pm \sqrt{11}$ **26.** $-4 \pm \sqrt{21}$ **27.** $\dfrac{-9 \pm \sqrt{85}}{2}$ **28.** $\dfrac{5 \pm \sqrt{33}}{2}$ **29.** $\dfrac{-4 \pm \sqrt{178}}{9}$

30. $\dfrac{-3 \pm \sqrt{58}}{7}$ **31.** $\dfrac{1 \pm \sqrt{21}}{2}$ **32.** $\dfrac{-1 \pm \sqrt{41}}{2}$ **33.** no real-number solution **34.** no real-number solution

35. $\dfrac{-3 \pm \sqrt{11}}{2}$ **36.** $\dfrac{-3 \pm 2\sqrt{3}}{2}$ **37.** $\dfrac{-1 \pm \sqrt{85}}{6}$ **38.** -1 and $\dfrac{3}{5}$ **39.** 7 **40.** -5 **41.** $\dfrac{5}{2}$ **42.** $-\dfrac{2}{3}$ **43.** $-5 \pm \sqrt{29}$

44. $3 \pm \sqrt{6}$ **45.** $\dfrac{-1 \pm \sqrt{19}}{6}$ **46.** $\dfrac{1 \pm \sqrt{55}}{6}$ **47.** $\dfrac{3 \pm \sqrt{14}}{2}$ **48.** $\dfrac{10 \pm \sqrt{106}}{2}$

49. They break even when 10 tickets or 27 tickets are sold. **50.** They break even when 4 tickets or 15 tickets are sold. **51.** They break even when 13 people or 66 people are in the group. **52.** They break even when 20 licenses or 30 licenses are purchased. **53.** The number of cars is 5. **54.** $f(x) = 0.5x^2 + 5x$; a total of 10 dolls can be purchased. **55.** The price should be set at $19. **56.** The price should be set at $42. **57.** The contract should be about $10,400,000. **58.** The contract should be for about $7,400,000. **59.** The width is 9 in. and the length is 13 in. **60.** The height is 10 cm and the base is 18 cm. **61.** The length is 12.1 ft and the width is 7.1 ft. **62.** The lengths of the legs are 13.22 ft and 21.22 ft. **63. (a)** $0 = -16t^2 + 32t + 16$ **(b)** It will take $1 + \sqrt{2}$ seconds. **(c)** It will take approximately 2.4 seconds.

64. (a) $0 = -16t^2 + 88t + 8$ **(b)** It will take $\dfrac{11 \pm \sqrt{129}}{4}$ seconds. **(c)** It will take approximately 5.6 seconds.

65. (a) $y = -0.2x^2 + 1.5x + 5.4$ **(b)** The number of employees will return to the 1997 level in the year 2005.
66. (a) $y = 0.5x^2 + 1.3x + 4.9$ **(b)** The number of employees will increase to 20 thousand in the year 2002.

11.4 CALCULATOR EXERCISES

1. -5 and -3 **2.** no real-number solution **3.** approximately -1.618 and 0.618 **4.** -4 and -1.5 **5.** no real-number solution **6.** approximately -1.435 and 0.435

SECTION 11.5 EXERCISES

1. 3 and 9 **2.** 5 and 7 **3.** $-\dfrac{7}{2}$ and 3 **4.** $-\dfrac{8}{3}$ and 2 **5.** -5 and $-\dfrac{1}{2}$ **6.** -9 and $-\dfrac{4}{3}$ **7.** $\dfrac{1}{4}$ **8.** $-\dfrac{2}{5}$

9. no real-number solution **10.** no real-number solution **11.** no real-number solution **12.** no real-number solution

13. $\dfrac{5 \pm \sqrt{17}}{2}$ **14.** $4 \pm \sqrt{6}$ **15.** $2 \pm \sqrt{3}$ **16.** $-3 \pm \sqrt{2}$ **17.** $-\dfrac{5}{4}$ and 0 **18.** 0 and $\dfrac{10}{3}$ **19.** $\pm\dfrac{\sqrt{21}}{3}$ **20.** $\pm\dfrac{\sqrt{30}}{2}$

21. 5 and 11 **22.** -12 and -6 **23.** -2 and 4 **24.** -3 and 9 **25.** 1.5 and 4.8 **26.** -3.2 and -0.5 **27.** $-2.8 \pm \sqrt{9.64}$
28. $0.9 \pm \sqrt{7.01}$ **29.** 4.9 **30.** -7.1 **31.** no real-number solution **32.** no real-number solution **33.** two rational solutions
34. no real-number solution **35.** one rational solution **36.** two irrational solutions **37.** no real-number solution
38. one rational solution **39.** two rational solutions **40.** two rational solutions **41.** two irrational solutions
42. two rational solutions **43.** two irrational solutions **44.** no real-number solution **45.** one rational solution
46. two irrational solutions **47.** no real-number solution **48.** one rational solution **49. (a)** $R(x) = x(9 - 0.1x)$
(b) $C(x) = 7.50 + 6x$ **(c)** $P(x) = -0.1x^2 + 3x - 7.50$ **(d)** He breaks even when he sells 2.75 pounds at $8.73 per pound or 27.2 pounds at $6.28 per pound.
50. (a) $R(x) = x(3.25 - 0.05x)$ **(b)** $C(x) = 3.00 + 2x$ **(c)** $P(x) = -0.05x^2 + 1.25x - 3.00$
(d) She breaks even when she sells 3 ribbons at $3.10 each or 23 ribbons at $2.10 each.
51. (a) $R(x) = 50x$ **(b)** $C(x) = x(20 - 0.25x) + 600$ **(c)** $P(x) = 0.25x^2 + 30x - 600$
(d) The promoter breaks even when there are 18 guests at $15.50 each.

52. **(a)** $R(x) = 250x$ **(b)** $C(x) = x(75 - 5x) + 500$ **(c)** $P(x) = 5x^2 + 175x - 500$ **(d)** The firm breaks even when the job lasts about 3 days with a variable cost of $60 per day.
53. The population reaches 7 billion at 214 years after 1800. In the year 2014, the population will reach 7 billion people.
54. The consumption is expected to be 350 BTU's at 10.8 years after 1980 and 283 years after 1980.
55. The year was 1986.   **56.** The year was 1991.
57. $(x + 12)^2 = x^2 + (x + 6)^2$; The measurements are 18 ft height, 24 ft base, and 30 ft hypotenuse.
58. $\left(x + \frac{1}{2}\right)^2 = x^2 + \left(\frac{3}{4}x\right)^2$; The measurements are 2 in. base, $1\frac{1}{2}$ in. height, and $2\frac{1}{2}$ in. hypotenuse.

11.5 CALCULATOR EXERCISES

Equation	Value of Discriminant	Types of Roots	Number of Unlike Roots	Roots
1. $x^2 + 6 = 5x$	1	rational	2	2, 3
2. $9x^2 + 6x = -1$	0	rational	1	$-\frac{1}{3}$
3. $2x^2 + 1 = 7x$	41	irrational	2	0.149, 3.351
4. $x^2 + 6x = -10$	-4	not real	2	not real
5. $x^2 = 6 - x$	25	rational	2	$-3, 2$
6. $5x^2 - 6x = 0$	36	rational	2	0, 1.2
7. $x^2 + 0.36 = 1.2x$	0	rational	1	0.6
8. $1.7x^2 + x + 1.9 = 0$	-11.92	not real	2	not real
9. $1.5x^2 + 1.2x = 3.6$	23.04	rational	2	$-2, 1.2$
10. $\frac{1}{4}x^2 + x = \frac{1}{8}$	1.125	irrational	2	$-4.121, 0.121$
11. $x^2 - \frac{1}{6}x = \frac{1}{6}$	$0.69\overline{4}$	rational	2	$-0.\overline{3}, 0.5$
12. $\frac{1}{5}x^2 + \frac{2}{3}x = -\frac{7}{8}$	$-0.2\overline{5}$	not real	2	not real

SECTION 11.6 EXERCISES

1. $c = \pm\dfrac{\sqrt{Em}}{m}$ **2.** $r = \pm\dfrac{\sqrt{A\pi}}{2\pi}$ **3.** $r = \pm\dfrac{\sqrt{V\pi h}}{\pi h}$ **4.** $v = \pm\dfrac{\sqrt{FRm}}{m}$ **5.** $v = \pm\dfrac{\sqrt{2Em}}{m}$ **6.** $r = \pm\dfrac{6\sqrt{10A\pi S}}{\pi S}$

7. $n = \dfrac{1 \pm \sqrt{1 + 8C}}{2}$ **8.** $n = \dfrac{-1 \pm \sqrt{1 + 8S}}{2}$ **9.** $x = \pm\sqrt{y^2 + c^2}$ **10.** $x = \pm\sqrt{r^2 - y^2}$ **11.** $r = -1 \pm \dfrac{\sqrt{AP}}{P}$

12. $t = \dfrac{-g \pm \sqrt{g^2 + 64S}}{32}$ **13.** $L = \sqrt{D^2 - W^2 - H^2}$; the length is 16 inches.   **14.** $W = \sqrt{D^2 - L^2 - H^2}$; the width is 12 inches.
15. The hypotenuse measures approximately 5.30 m.   **16.** The hypotenuse measures approximately 9.26 feet.
17. The lengths are approximately 5.37 feet.   **18.** The lengths are approximately 10.54 cm.   **19.** The wire is approximately 28.28 feet.
20. The wire is approximately 21.21 feet.   **21.** The building is approximately 30.05 feet tall.
22. The tower is approximately 102.53 feet high.   **23.** Yes, 28.28 feet is less than 30 feet.   **24.** No, 35.36 feet is greater than 30 feet.
25. The lengths are approximately 10.7 in. and 21.5 in.   **26.** The lengths are approximately 1.3 cm and 2.5 cm.
27. The lengths are 11 in. and approximately 19.1 in.   **28.** The lengths are 17 cm and approximately 29.4 cm.
29. The lengths are 42 cm and approximately 36.4 cm.   **30.** The lengths are 16.4 in. and approximately 14.2 in.
31. The horizontal distance is about 6928.2 ft and the slanted distance is 8000 ft.
32. The horizontal distance is approximately 69.3 ft and the length of string is approximately 138.6 ft.
33. The height of the pole is approximately 13.9 ft and the length of the wire is 16 ft.
34. The amount of wire is approximately 15.0 m, and it should be anchored approximately 7.5 m from the base of the pole.

11.6 CALCULATOR EXERCISES

1. (a) yes **(b)** no **(c)** yes **(d)** no   **2. (a)** yes **(b)** no **(c)** no **(d)** no **(e)** yes

SECTION 11.7 EXERCISES

1. quadratic inequality   **2.** not a quadratic inequality   **3.** quadratic inequality   **4.** not a quadratic inequality
5. not a quadratic inequality   **6.** quadratic inequality   **7.** quadratic inequality   **8.** quadratic inequality
9. not a quadratic inequality   **10.** not a quadratic inequality   **11.** not a quadratic inequality   **12.** quadratic inequality
13. all integers less than or equal to -5 and greater than or equal to 3
14. all integers greater than or equal to -4 and less than or equal to 2   **15.** all integers greater than -4 and less than 1
16. all integers greater than -3 and less than 2   **17.** all integers less than 0 or greater than 8
18. all integers less than -3 or greater than 0   **19.** $(-\infty, -3) \cup (1, \infty)$ **20.** $(-\infty, 1] \cup [7, \infty)$
21. $(-2.5, 2)$ **22.** $[-1, 5]$ **23.** $(-\infty, -2) \cup (8, \infty)$ **24.** $(-\infty, -1) \cup (8, \infty)$ **25.** $[-3, 3]$ **26.** $(-\infty, -5) \cup (2, \infty)$
27. no solution   **28.** no solution   **29.** $(-\infty, \infty)$ **30.** -2 **31.** $(-\infty, \infty)$ **32.** $(-\infty, \infty)$ **33.** $(-\infty, -4] \cup [6, \infty)$
34. $[-5, 7]$ **35.** $(-3, 5)$ **36.** $(-3, 6)$ **37.** $\left[\dfrac{1}{2}, \dfrac{5}{2}\right]$ **38.** $(-\infty, -4] \cup [1, \infty)$ **39.** $(-\infty, -2) \cup (7, \infty)$ **40.** $(-5, 6)$
41. $(-\infty, -4] \cup [3, \infty)$ **42.** $[-2, 6]$ **43.** $(-\infty, -3) \cup (1, \infty)$ **44.** $(-\infty, -2) \cup (4, \infty)$ **45.** $[-2, 3]$ **46.** $\left(-\infty, \dfrac{1}{2}\right] \cup [5, \infty)$
47. $\dfrac{3}{2}$ **48.** $(-\infty, \infty)$ **49.** $\left(-\infty, \dfrac{1}{2}\right) \cup \left(\dfrac{1}{2}, \infty\right)$ **50.** no solution   **51.** $(-\infty, \infty)$ **52.** 4   **53.** no solution
54. $(-\infty, -4) \cup (-4, \infty)$ **55.** The egg is higher than the stand between 0 and 1.875 seconds.
56. The arrow is above the point of origin between 0 and 2.5 seconds.
57. The rock is no more than 800 feet from about 7.866 seconds to 11.316 seconds.
58. The ball is no more than 200 feet from about 7.215 seconds to 8.279 seconds.
59. The booth can be operated from one day to four days.   **60.** The stand can be operated from one week to four weeks.
61. The years are from 1987 on.   **62.** The years were from 1970 to 1981.   **63.** The years are from 1990 on.
64. The years were from 1975 to 1980.

11.7 CALCULATOR EXERCISES

Answers will vary.

CHAPTER 11 SECTION-BY-SECTION REVIEW

Reflections

1–6. Answers will vary.

Exercises

1. -3 and 4   **2.** -9 and -3 **3.** 1 and 5   **4.** -1 and 5   **5.** -4 and 4   **6.** -4 and 1.5   **7.** -5 and 11
8. approximately -6.83 and -1.17 **9.** $-3, 3,$ and 1.5   **10.** It will reach the ground in approximately 2.45 seconds.
11. They will double in the year 2005.   **12.** -2 and 3   **13.** $-\dfrac{7}{2}$ and $\dfrac{11}{3}$ **14.** -1 and 8   **15.** -5 and 12   **16.** no solution
17. $\dfrac{4}{3}$ and $\dfrac{3}{2}$ **18.** $\dfrac{5}{2}$ and -7 **19.** $\pm\dfrac{7}{3}$ **20.** $5w + w^2 = 300$; The dimensions are 15 feet by 20 feet.
21. (a) $y = -0.06x^2 + 0.12x + 6.44$ **(b)** The donations will drop to $5 million in the year 2004.   **22.** -7 **23.** $10\sqrt{2}$ **24.** $4\sqrt{2}$
25. $2\sqrt{2}$ **26.** $-6\sqrt{2}$ **27.** $-\dfrac{6}{5} - \sqrt{2}$ **28.** $1 + \dfrac{\sqrt{2}}{2}$ **29.** $\dfrac{1 - \sqrt{3}}{2}$ **30.** $2 - \sqrt{5}$ **31.** $\dfrac{6\sqrt{5}}{5}$ **32.** $\dfrac{\sqrt{5}}{2}$ **33.** ± 5
34. ± 2 **35.** $\pm 2\sqrt{3}$ **36.** no real-number solution   **37.** 1 and 7   **38.** -13 and -5 **39.** $2 \pm \sqrt{2}$
40. The horizontal distance is $10\sqrt{10}$ ft, or approximately 31.6 ft.   **41.** The interest rate is 8%.
42. They would free-fall about 24 seconds, with wind resistance and maneuvering ignored.
43. -8 and 12   **44.** $-\dfrac{3}{2}$ and 4   **45.** $\dfrac{-1 \pm \sqrt{17}}{4}$
46. (a) $y = -0.15x^2 + 0.25x + 2.6$ **(b)** The cost of the goods will be $2 billion in the year 2002.
47. They will break even selling 10 figurines.   **48.** $x^2 + 4x - 45 = 0$; the dimensions are 5 in. by 18 in.
49. -7 and 9   **50.** $\dfrac{3 \pm \sqrt{21}}{2}$ **51.** $\dfrac{1}{5}$ **52.** $-\dfrac{3}{4}$ and $\dfrac{2}{5}$ **53.** -4.5 and 2.4
54. no real-number solution   **55.** $5 \pm \sqrt{19}$ **56.** $\dfrac{2 \pm 2\sqrt{10}}{3}$ **57.** no real-number solution

58. one rational solution   **59.** two rational solutions   **60.** two irrational solutions   **61. (a)** $R(x) = x(6.50 - 0.50x)$

(b) $C(x) = 2.50x + 3.95$ **(c)** $P(x) = -0.5x^2 + 4x - 3.95$ **(d)** The break-even points are sales of two cakes at $5.50 each.

62. The world's total generation first reaches 2000 billion kwh in the year 1992.

63. $s = -16t^2 + 50; t = \dfrac{\pm\sqrt{50 - s}}{4}$ **64.** $d = \dfrac{2\sqrt{\pi A}}{\pi}$ **65.** $x = \pm\dfrac{\sqrt{b(a - c)}}{b}$ **66.** $t = \dfrac{-v_0 \pm \sqrt{(v_0)^2 + 64(s_0 - s)}}{-32}$

67. $H = \sqrt{D^2 - L^2 - W^2}$; the height is 8 inches.   **68.** The rope is $7\sqrt{2}$ ft, or approximately 9.9 feet.

69. The lengths of the sides are 10 in. and $10\sqrt{3}$ in., or approximately 17.3 in.

70. all integers greater than or equal to -6 and less than or equal to 4   **71.** all integers less than -2 or greater than 2

72. $(-\infty, -4] \cup [2, \infty)$ **73.** $[-3, 3]$ **74.** $(-1, 4)$ **75.** no solution   **76.** $(-\infty, 3) \cup (3, \infty)$ **77.** 4   **78.** $(1.59, 4.41)$

79. $\left(2 - \dfrac{2}{3}\sqrt{6}, 2 + \dfrac{2}{3}\sqrt{6}\right)$ **80.** $[-3, 2]$ **81.** $(-1, 3)$ **82.** $(-\infty, -3) \cup (-3, \infty)$ **83.** $(-\infty, \infty)$ **84.** no solution

85. $(3 - \sqrt{10}, 3 + \sqrt{10})$ **86.** The booth can be rented up to four days.   **87.** The dart is above the ground from 0 seconds to approximately 2.358 seconds.

CHAPTER 11 CHAPTER REVIEW

1. $5\sqrt{2}$ **2.** $-4\sqrt{3}$ **3.** $-1 + \sqrt{2}$ **4.** $-2 + 2\sqrt{5}$ **5.** $\dfrac{\sqrt{14}}{5}$ **6.** $\dfrac{5\sqrt{6}}{6}$ **7.** -1 and 3   **8.** -3 and 3   **9.** -5 and -1

10. $-2\dfrac{2}{3}$ and $1\dfrac{4}{5}$ **11.** -3 and 1   **12.** approximately -8.14 and -0.86 **13.** approximately -0.85 and 5.65   **14.** -4 and 10

15. -5 and 9   **16.** -1 and 5   **17.** $-\dfrac{3}{4}$ and $\dfrac{2}{3}$ **18.** -11 and 8   **19.** -7 and $\dfrac{2}{5}$ **20.** $\pm\dfrac{4}{7}$ **21.** ± 6

22. no real-number solution   **23.** 5 and 13   **24.** $7 \pm \sqrt{3}$ **25.** $-6 \pm \sqrt{3}$ **26.** $\dfrac{-7 \pm \sqrt{37}}{2}$ **27.** -3.6 and 1.2

28. all real numbers   **29.** -14 and -6 **30.** $-\dfrac{5}{2}$ and 1   **31.** $\dfrac{-4 \pm 3\sqrt{2}}{2}$ **32.** $-4 \pm 2\sqrt{3}$ **33.** $\dfrac{7 \pm \sqrt{57}}{2}$ **34.** $\dfrac{5}{6}$

35. no real-number solution   **36.** -11 and 5   **37.** -4 and $\dfrac{9}{4}$ **38.** $x = \dfrac{-1 \pm \sqrt{1 + 8a}}{2}$ **39.** $y = \sqrt{r^2 - x^2 - z^2}$

40. $r = \pm\dfrac{\sqrt{A\pi}}{2\pi}$ **41.** all integers greater than or equal to -3 and less than or equal to 5

42. all integers greater than or equal to -7 and less than or equal to 2   **43.** $(-6, 1)$ **44.** $(-\infty, -3) \cup (5, \infty)$

45. $(-\infty, -1 - \sqrt{2}) \cup (-1 + \sqrt{2}, \infty)$ **46.** $(-\infty, 4 - 3\sqrt{2}) \cup (4 + 3\sqrt{2}, \infty)$ **47.** $\left[-\dfrac{5}{3}, 6\right]$ **48.** $(-\infty, -3] \cup [4, \infty)$

49. $(-\infty, -1] \cup [6, \infty)$ **50.** no solution   **51.** $(-3, 5)$ **52.** $(-4, 4)$ **53.** $(-\infty, -3.58) \cup (-0.42, \infty)$ **54.** $(-\infty, -5) \cup (-5, \infty)$

55. $(0.88, 5.12)$ **56.** 3   **57.** The notebook will reach the ground in $\sqrt{10}$ seconds or approximately 3.16 seconds.

58. The number is about 16 items.   **59.** It will reach the ground in $2 + \sqrt{6}$ seconds or approximately 4.45 seconds.

60. It will take $\sqrt{437.5}$ seconds or approximately 20.92 seconds.   **61.** The other leg is 25.5 meters.

62. The dimensions are 8 cm by 18 cm.   **63.** The interest rate is 5%.   **64.** The interest rate is approximately 11.8%.

65. (a) $R(x) = x(10.50 - 0.50x)$ **(b)** $C(x) = 7.50x + 3.00$ **(c)** $P(x) = -0.5x^2 + 3x - 3$ **(d)** The break-even point occurs by selling

two baskets at $9.50.   **66.** The hypotenuse is 52 inches.   **67.** The per capita consumption will drop to 1000 cigarettes in the year 2004.

68. (a) $y = -0.18x^2 + 0.99x + 3.55$ **(b)** The sales will drop to 4 million units in the year 2003.

69. They each measure $7.105\sqrt{2}$ cm, or approximately 10 cm.   **70.** The rope is $11\sqrt{2}$ feet long, or approximately 15.6 feet.

71. The sides measure 16 in. and $16\sqrt{3}$ in., or approximately 27.7 in.   **72.** The booth can be rented up to seven days.

CHAPTER 11 TEST

1. -2 and 8   **2.** $-5.5, -2,$ and 3   **3.** no real number soution   **4.** no real-number solution   **5.** 0.5 and 4   **6.** -0.82 and 1.82

7. $[-3, 4]$ **8.** all integers greater than -3 and less than 5   **9.** $(-\infty, -4] \cup \left[\dfrac{5}{2}, \infty\right)$ **10.** 200; two irrational solutions; $8 \pm 5\sqrt{2}$

11. -39; no real-number solutions   **12.** $t = \pm\dfrac{\sqrt{a(s - c)}}{a}$ **13. (a)** -2 and 3   **(b)** -2 and 3   **(c)** -2 and 3

14. The legs are 14 in. and $14\sqrt{3}$ in., or approximately 24.2 in.   **15.** Keanu was free-falling for about 24 seconds.

16. The interest rate must be approximately 6%.   **17. (a)** $R(x) = x(12.50 - 0.50x)$ **(b)** $C(x) = 9.50x + 3.95$

(c) $P(x) = -0.50x^2 + 3x - 3.95$ **(d)** Break-even point occurs when two arrangements are sold at $11.50 each.

18. (a) $y = 0.35x^2 + 0.65x + 0.4$ **(b)** The net income will be $5 billion in the year 2001.

19. The horizontal distance is $10\sqrt{15}$ feet or about 38.7 feet.   **20.** Answers will vary.

CHAPTER 1-11 CUMULATIVE REVIEW

1. $3a^2 + 4ab - 5b^2$ **2.** $x^2 + xy$ **3.** $6x^2 - 16x + 6$ **4.** $25x^2 + 10x + 1$ **5.** $4x^2 - 9$ **6.** $6.3x^2y^5z^3$ **7.** $\dfrac{6x^2}{y}$ **8.** $\dfrac{1}{2x^2y}$

9. $\dfrac{25b^2}{64a^6}$ **10.** $-8p^5q^2r^2$ **11.** $x + 2 - \dfrac{3}{x}$ **12.** $(x - 1)$ **13.** -16 **14.** $.0056$ **15.** $-3,400,000$ **16.** $\dfrac{2A}{h} = b$

17. $t = \dfrac{\pm\sqrt{5 - s}}{4}$ **18.** $(x + 5)(x - 3)$ **19.** $(5s - 4)(2s + 3)$ **20.** $(3x - 4)(3x + 4)$ **21.** $(2x + 3)^2$ **22.** $(a + 3)(a^2 - 3a + 9)$

23. $2(a + 2)(a - 2)(a^2 + 4)$ **24.** $x = \dfrac{25}{6}$ $y = \dfrac{11}{6}$ **25.** possible solution: $(1, 2)$

26. 4 and 2   **27.** $2.5 \pm \sqrt{18.25}$ **28.** -3 and 1
29. no solution   **30.** no real-number solution
31. 2.64 and -1.14 **32.** $\dfrac{5}{3}$ **33.** all real numbers
34. no solution   **35.** $-\dfrac{3}{2}$ and $\dfrac{5}{2}$ **36.** -3 and 3
37. $(-\infty, 9)$ **38.** $(-\infty, -1] \cup [6, \infty)$

39. $(2 - \sqrt{5}, 2 + \sqrt{5})$ **40. (a)**   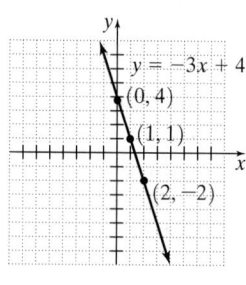   **(b)** Yes; it passes the vertical-line test   **(c)** Decreasing for all x
(d) Domain is all real numbers; range is all real numbers.
(e) x int: $\left(\dfrac{4}{3}, 0\right)$
y int: $(0, 4)$

41. (a) x int: $(-2 \pm \sqrt{5}, 0)$ y int: $(0, -1)$ **(b)** $(-2, 5)$ **(c)**
(d) Domain is all real numbers; range is $y \geq -5$.
42. $y = 3x - 11$ **43.** $y = \dfrac{3}{2}x + \dfrac{7}{2}$
44. He should mix 2 pounds of peanuts with 5 pounds of cashews.
45. It will take about 2.2 seconds to drop 77 feet.
46. She should invest $15,000.
47. (a) $y = -0.6x^2 + 0.9x + 9.5$ **(b)** The profit will drop to $3.5 million in the year 2003.
48. (a) $R(x) = x(3.50 - 0.25x)$ **(b)** $C(x) = 2x + 1.50$
(c) $P(x) = -0.25x^2 + 1.50x - 1.50$
(d) The break-even points are about 2 and 5. He must sell two sparklers at $3.00 each.
49. The legs are both $4\sqrt{2}$ meters or about 5.7 meters long.   **50.** Janet must score between 68 and 88 inclusive.

Chapter 12

SECTION 12.1 EXERCISES

1. rational expression   **2.** not a rational expression   **3.** rational expression   **4.** rational expression   **5.** not a rational expression
6. rational expression   **7.** rational expression   **8.** not a rational expression   **9.** rational expression
10. not a rational expression   **11.** rational expression   **12.** rational expression   **13.** rational expression   **14.** rational expression
15. not a rational expression   **16.** rational expression   **17.** $0; x \neq 0$ **18.** -5 and $-3; x \neq -5, x \neq -3$ **19.** 5 and 6; $x \neq 5, x \neq 6$
20. -3 and $0; x \neq -3, x \neq 0$ **21.** -2 and $2; x \neq -2, x \neq 2$ **22.** $-1, 1,$ and $4; x \neq -1, x \neq 1, x \neq 4$ **23.** 0 and $5x \neq 0; x \neq 5$

24. $0; x \neq 0$ **25.** $2; x \neq 2$ **26.** $-2; x \neq -2$ **27.** -5 and $-\dfrac{7}{2}; x \neq -5; x \neq -\dfrac{7}{2}$ **28.** $-\dfrac{3}{5}; x \neq -\dfrac{3}{5}$

29. $-\dfrac{3}{2}$ and $\dfrac{3}{4}; x \neq -\dfrac{3}{2}, x \neq \dfrac{3}{4}$ **30.** $-5, -4,$ and $4; x \neq -5, x \neq -4, x \neq 4$ **31.** $-\dfrac{9}{2}; x \neq -\dfrac{9}{2}$ **32.** $-\dfrac{9}{7}$ and $-\dfrac{1}{2}; x \neq -\dfrac{9}{7}, x \neq -\dfrac{1}{2}$

33. $11; x \neq 11$ **34.** no restricted values; all real numbers   **35.** $-\dfrac{5}{3}$ and $\dfrac{5}{3}; x \neq -\dfrac{5}{3}, x \neq \dfrac{5}{3}$ **36.** -3 and $\dfrac{5}{6}; x \neq -3, x \neq \dfrac{5}{6}$

37. $-3, -2,$ and $3; x \neq -3, x \neq -2, x \neq 3$ **38.** $-\dfrac{7}{4}$ and $\dfrac{7}{4}; x \neq -\dfrac{7}{4}, x \neq \dfrac{7}{4}$ **39.** no restricted values; all real numbers

40. $\dfrac{4}{3}; x \neq \dfrac{4}{3}$ **41.** $-\dfrac{3}{2}; x \neq -\dfrac{3}{2}$ **42.** $-\dfrac{3}{2}$ and $\dfrac{6}{5}; x \neq -\dfrac{3}{2}, x \neq \dfrac{6}{5}$

43. $y = \dfrac{6}{x} + \dfrac{9}{x^2}$

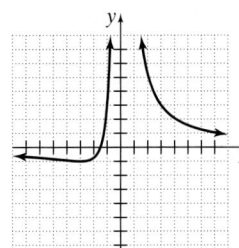

44. $y = \dfrac{25x^2 - 4}{5x + 2}$

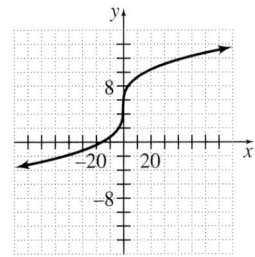

45. $h(x) = \dfrac{x^2 - 49}{x - 7}$

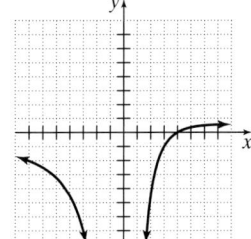

46. $g(x) = \dfrac{4x}{x^3 + 1}$

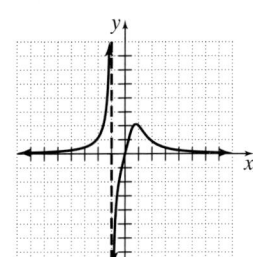

47. $y = \dfrac{2x}{x^2 + 3}$

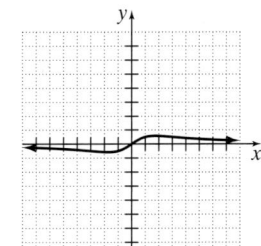

48. $f(x) = \dfrac{9}{x} - \dfrac{36}{x^2}$

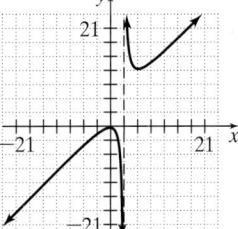

49. $y = \dfrac{6x}{x^3 + 8}$

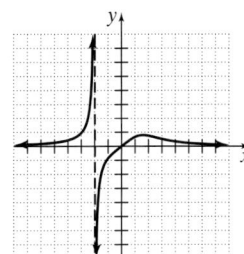

50. $g(x) = \dfrac{2x^2 + 5}{3x + 1}$

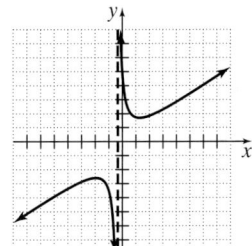

51. $f(x) = \dfrac{x^2 + 1}{x - 3}$

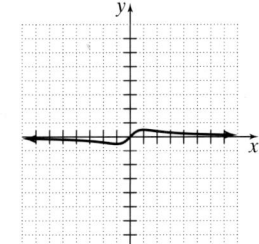

52. $z(x) = \dfrac{6x^2 - x - 35}{2x - 5}$

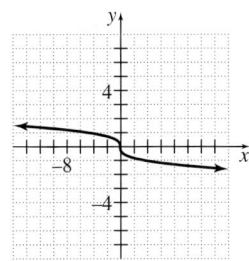

53. $R(x) = \dfrac{2x^2 + 7x + 6}{x + 2}$

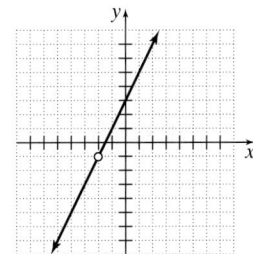

54. $y = \dfrac{x}{x^2 + 1}$

55.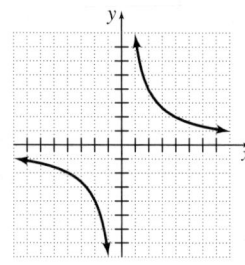

56.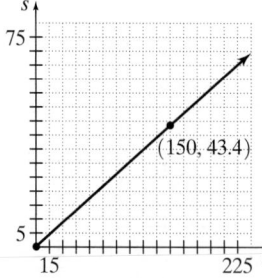

57.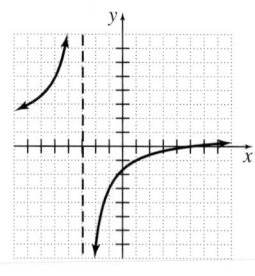

58.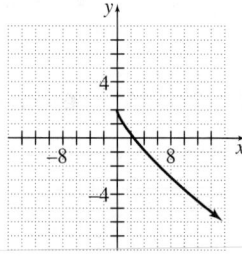

59.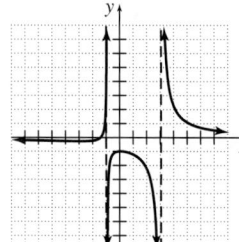

60.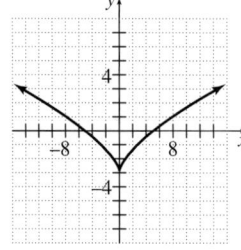

61.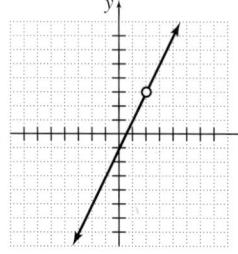

62.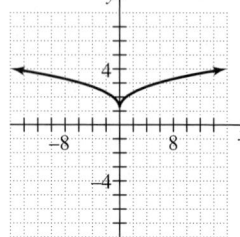

63. (a) $C_{ave}(x) = \dfrac{6500 + 100x + 25x^2}{x}$

(b) The minimum average cost per year would be about \$906.25, when $x = 16$.

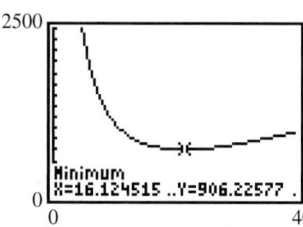

(c) The customer would want to buy a new system after 16 years.

64. (a) $C_{ave}(x) = \dfrac{10,000 + 900x + 300x^2}{x}$

(b) The minimum average cost per year would be about \$4364.15, when $x = 5.7$.

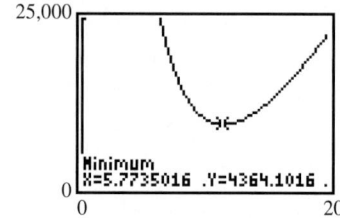

(c) After about five years and eight months, the customer would want to buy a new copier.

65. (a) 1995: 0.475
1998: 0.418
1999: 0.403

(b) $P(x) = \dfrac{D(x)}{T(x)}$

$P(x) = \dfrac{-56.12x + 2405}{89.65x + 5068}$

(c)

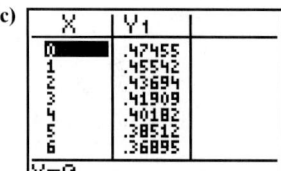

$x = 0$, 1995: 0.475
$x = 3$, 1998: 0.419
$x = 4$, 1999: 0.402

(d) The approximations from part c are close to the proportions in part a. The function is therefore a good predictor.

66. (a) 1995: 0.495
1998: 0.491
1999: 0.484

(b) $P(x) = \dfrac{N(x)}{T(x)} = \dfrac{0.10x + 34.3}{0.55x + 69.3}$

(c)

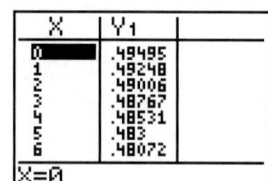

$x = 0$, 1995: 0.495
$x = 3$, 1998: 0.488
$x = 4$, 1999: 0.485

(d) This function is a fair predictor, since approximations for $x = 0$ and $x = 4$ are fairly close. The approximation for $x = 3$ is a bit off.

12.1 CALCULATOR EXERCISES

1. (window: $-5, 5, 1, -10, 10, 5, 1$)　　**2.** (window: $-2, 2, 1, -10, 10, 5, 1$)　　**3.** (window: $-2, 2, 1, -5, 5, 1, 1$)

SECTION 12.2 EXERCISES

1. $\dfrac{2xz}{3y^3}$　**2.** $\dfrac{3y^3}{5x^2z^2}$　**3.** $-\dfrac{2a}{3c}$　**4.** $-\dfrac{11b^3c}{8a^2}$　**5.** $\dfrac{3(x-3)}{2(3x-1)}$　**6.** $\dfrac{2(x+2)}{3(2x+1)}$　**7.** -1　**8.** -1　**9.** $\dfrac{y(x+2y)}{3-4x}$　**10.** $\dfrac{x(5-3x)}{y(2x+y)}$

11. $\dfrac{-2x+3y-z}{x-7y+z}$　**12.** $\dfrac{-x+4y-2z}{2x-y+z}$　**13.** $\dfrac{x-2}{x-4}$　**14.** $\dfrac{x-3}{x-7}$　**15.** $\dfrac{x-3}{x+7}$　**16.** $\dfrac{x+8}{x-5}$　**17.** $\dfrac{x+2}{x-4}$　**18.** $\dfrac{x+5}{x-9}$

19. $\dfrac{3x-2}{2x-3}$　**20.** $\dfrac{5x-4}{4x-5}$　**21.** $\dfrac{(x+1)(2x+1)}{2(x-2)(3x+1)}$　**22.** $\dfrac{2(x+1)(2x-1)}{3(x+2)(3x-1)}$　**23.** $\dfrac{2x-y}{3x+y}$　**24.** $\dfrac{5x-y}{4x+y}$　**25.** $-\dfrac{1}{x+2}$

26. $-\dfrac{1}{x+3}$　**27.** $\dfrac{2x+3}{x^2+11}$　**28.** $\dfrac{3x+4}{x^2+9}$　**29.** $\dfrac{28ac}{45b}$　**30.** $\dfrac{9a^2}{10b^2d^2}$　**31.** $-\dfrac{3x^2}{4}$　**32.** $-\dfrac{2y^2}{13x^2}$　**33.** $\dfrac{8.06x^3}{y^2}$　**34.** $\dfrac{21x^2y}{2z}$

35. $\dfrac{2a^2}{9b}$　**36.** $\dfrac{14a^3}{15b^2}$　**37.** $\dfrac{x^2(x+5)}{x+4}$　**38.** $\dfrac{x(x+15)}{x-11}$　**39.** $\dfrac{x-4}{x+4}$　**40.** $\dfrac{x+7}{x-7}$　**41.** $\dfrac{(x+7)(5x+2)}{x(2x+1)}$　**42.** $\dfrac{(3x+5)(x+4)}{x(2x+9)}$

43. $\dfrac{(x-3)(x-1)}{(x+1)(x+3)}$　**44.** $\dfrac{(x+5)(x+2)}{(x-2)(x-5)}$　**45.** $\dfrac{x(x-2y)}{(x+3y)(x-y)}$　**46.** $\dfrac{y(x+y)}{(x+2y)(x+5y)}$　**47.** $\dfrac{3a^2}{2}$　**48.** $\dfrac{7a}{8b}$　**49.** $-3xyz$

50. $-\dfrac{3}{x^2y^2z}$　**51.** $\dfrac{4c}{b}$　**52.** $3ab^2c^2$　**53.** $-\dfrac{7x}{yz^3}$　**54.** $-\dfrac{2x}{y^3z^2}$　**55.** $\dfrac{1}{5(3a-1)}$　**56.** $\dfrac{7}{6a+1}$　**57.** $(x-4)(5y+2)$

58. $(x-7)(4y-3)$　**59.** $\dfrac{(2x+1)(x-3)}{(x+2)(3x+1)}$　**60.** $\dfrac{(x-4)^2}{(x-1)(3x+2)}$　**61.** $\dfrac{1}{2x+1}$　**62.** $\dfrac{1}{3x-2}$　**63.** $\dfrac{3x+4}{2x+1}$　**64.** $\dfrac{5x-2}{2x-5}$

65. $\dfrac{x+7}{3x-5}$　**66.** $\dfrac{5(2x+1)}{7(4x+3)}$　**67.** $\dfrac{x(x-y)}{y(x+y)}$　**68.** $\dfrac{y(x+2y)}{x(x-2y)}$　**69.** $\dfrac{2(x+2y)}{x-4y}$　**70.** $\dfrac{4(x+7y)}{3(x-2y)}$

71. $\dfrac{170}{x}$ hours; $(x-10)\left(\dfrac{170}{x}\right)$ miles; the distance is 136 miles.　**72.** $\dfrac{155}{x}$ hours; $(x+8)\left(\dfrac{155}{x}\right)$ miles; the distance is $182\dfrac{5}{9}$ miles.

73. $m(x) = 500(1 + x + x^2)$; the total amount that Sharon's grandson received was $1820.
74. $s(x) = 50{,}000(1 + x + x^2)$; the total amount of Alexandra's sales was $190,625.

75. $L = \dfrac{V}{WH}$; the length is $(5x-1)$ inches; height: 4 inches; width: 14 inches; length: 19 inches

76. $h = \dfrac{V}{\pi r^2}$; the height is $(6x-5)$ inches; the height is 4 inches.

77. **(a)** $C(x) = 300 + 5x$; $R(x) = 25x$　**(b)** $\dfrac{5x}{60+x}$　**(c)** The ratio is 2.

78. **(a)** $C(x) = 175 + 25x$; $R(x) = 75x$　**(b)** $\dfrac{7+x}{3x}$　**(c)** The ratio is $\dfrac{57}{150}$.

12.2 CALCULATOR EXERCISES

Value of n	Original function	Simplified function
2	$y = \dfrac{(1-x^2)}{(1-x)}$	$y = 1 + x$
3	$y = \dfrac{(1-x^3)}{(1-x)}$	$y = 1 + x + x^2$
4	$y = \dfrac{(1-x^4)}{(1-x)}$	$y = 1 + x + x^2 + x^3$
5	$y = \dfrac{(1-x^5)}{(1-x)}$	$y = 1 + x + x^2 + x^3 + x^4$
6	$y = \dfrac{(1-x^6)}{(1-x)}$	$y = 1 + x + x^2 + x^3 + x^4 + x^5$

SECTION 12.3 EXERCISES

1. $\dfrac{13}{x}$　**2.** $\dfrac{12}{x}$　**3.** $\dfrac{1}{y}$　**4.** $\dfrac{1}{a}$　**5.** $\dfrac{2x}{x+6}$　**6.** $\dfrac{3x}{x+12}$　**7.** $\dfrac{2}{b-5}$　**8.** $\dfrac{3}{z-7}$　**9.** 3　**10.** 4　**11.** $-\dfrac{5}{d}$　**12.** $-\dfrac{8}{b}$

13. $\dfrac{1}{x}$ **14.** $-\dfrac{1}{y}$ **15.** $\dfrac{x}{x+9}$ **16.** $\dfrac{3(x-1)}{x+8}$ **17.** 2 **18.** 2 **19.** $\dfrac{-x+11}{x+11}$ **20.** $\dfrac{-2x+9}{x+13}$ **21.** $z-4$ **22.** $c-7$

23. $\dfrac{3}{c+3}$ **24.** $\dfrac{5}{b+2}$ **25.** $\dfrac{2(x+2)}{2x+3}$ **26.** $\dfrac{-2x+3}{3x-4}$ **27.** $-\dfrac{3}{x}$ **28.** $\dfrac{6}{x}$ **29.** 2 **30.** 3 **31.** $a+3$ **32.** $5+c$

33. $\dfrac{5}{x+3}$ **34.** $\dfrac{7}{x+1}$ **35.** $\dfrac{9x+7}{6x^2}$ **36.** $\dfrac{3(6x+1)}{10x^3}$ **37.** $\dfrac{7x-2}{2(x-6)}$ **38.** $\dfrac{5x-4}{2(x+9)}$ **39.** $\dfrac{2x+3}{2x-3}$ **40.** $\dfrac{2x-3}{5x+2}$

41. $\dfrac{3(11x+5)}{(3x-5)(3x+5)}$ **42.** $\dfrac{6(4x-1)}{(2x-3)(2x+3)}$ **43.** $\dfrac{2(4x-17)}{(x-5)^2}$ **44.** $\dfrac{5x-12}{(x-8)^2}$ **45.** $\dfrac{2(b^2+8b+32)}{b(b+8)}$ **46.** $\dfrac{5a^2-6a+9}{2a(a-3)}$

47. $\dfrac{x^2+3x+10}{4(x+3)(x-3)}$ **48.** $\dfrac{2x^2-2x-1}{3(x+2)(x-2)}$ **49.** $\dfrac{2(11x+2)}{(x+5)(x-3)(x+4)}$ **50.** $\dfrac{-x^2+6x-2}{(x+7)(x+1)(x-2)}$ **51.** $\dfrac{a^2+5ab-b^2}{a^2b^2}$

52. $\dfrac{p^2+q^2}{p^2q^2}$ **53.** $\dfrac{3x-4y}{x-2y}$ **54.** $\dfrac{x-3y}{5x-y}$ **55.** $\dfrac{11x-16y}{(x-3y)(x+3y)}$ **56.** $\dfrac{2(8x+11y)}{(2x-3y)(2x+3y)}$ **57.** $\dfrac{2(5x-9z)}{(x+3z)^2(x-3z)}$

58. $\dfrac{4(2x+y)}{(x+2y)^2(x-2y)}$ **59.** $\dfrac{4(2x+1)}{x-2}$ **60.** $\dfrac{3x+1}{x-7}$ **61.** $\dfrac{3}{x-2}$ **62.** $\dfrac{5}{a-6}$ **63.** $\dfrac{x}{x+5}$ **64.** $\dfrac{8x+17}{3(2x+3)}$

65. $\dfrac{-4(x-20)}{(2x-5)(2x+5)}$ **66.** $\dfrac{15x+13}{(3x-1)(3x+1)}$ **67.** $\dfrac{-x(2x+13)}{(x+6)^2}$ **68.** $\dfrac{-x^2+11x-25}{(x-7)^2}$ **69.** $\dfrac{16(x-4)}{x(8-x)}$ **70.** $\dfrac{-20(x-5)}{x(x-10)}$

71. $\dfrac{x-1}{3(x+4)}$ **72.** $\dfrac{-2x+27}{x(x+3)(x-3)}$ **73.** $\dfrac{15b+34}{(b+3)(b-3)(2b+5)}$ **74.** $\dfrac{19a-31}{(a-5)(a+5)(2a+1)}$ **75.** $\dfrac{14a-15b-15a^2b}{10a^2b^2}$

76. $\dfrac{15y-4xy-56x}{48x^2y^2}$ **77.** $\dfrac{3(p-q)}{p+3q}$ **78.** $\dfrac{x+3z}{x+5z}$ **79.** $\dfrac{5x-27y}{(x+3y)^2(x-3y)}$ **80.** $\dfrac{5(x-11y)}{(x+5y)^2(x-5y)}$ **81.** $\dfrac{x-4}{x-5}$ **82.** $\dfrac{2z+3}{z+9}$

83. $\dfrac{2x+1}{x+10}$ **84.** $\dfrac{10x+3}{3x+1}$ **85.** $\dfrac{10x+31}{2x+7}$ **86.** $\dfrac{4}{x+4}$ **87.** $\dfrac{x-10}{x+7}$ **88.** $\dfrac{-31x+16}{9x-2}$ **89.** $\dfrac{x^2+8x+18}{x+3}$ **90.** $\dfrac{x^2+4x+5}{x+1}$

91. $\dfrac{2x^2-7x-7}{x-4}$ **92.** $\dfrac{-10x(2x-1)}{4x-1}$ **93. (a)** The total width is $\dfrac{65}{L}$ feet. **(b)** The total width is $\dfrac{5(17L-20)}{L(2L-5)}$ feet.

94. (a) The difference is $\dfrac{8}{L}$ inches. **(b)** The difference is $\dfrac{4(4L-15)}{L(2L+5)}$ inches. **95.** The perimeter is $\dfrac{96(2x-1)}{(x-1)(x+1)}$ feet.

96. The perimeter is $\dfrac{x^2+246x+117}{2x(x+3)}$ inches.

97. (a) $T(x) = \dfrac{26}{x} + \dfrac{110}{x+20}$
(b)
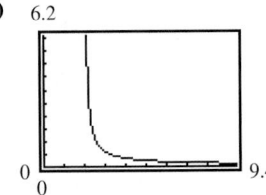

98. (a) $T(x) = \dfrac{1}{x-2} + \dfrac{1}{x+2}$
(b) 6.2

(c) The total time is approximately 6.7 hours.
(d) The average speeds for running and bicycling are approximately 6.6 mph and 26.6 mph, respectively.

(c) It will take approximately 2.2 hours.
(d) He should swim at approximately 2.8 mph.

99. The average speed is 29.2 mph.   **100.** The average speed is 111.1 mph.

12.3 CALCULATOR EXERCISES

Students should use the method described to check the equivalent expressions.

SECTION 12.4 EXERCISES

1. rational   **2.** not rational   **3.** rational   **4.** not rational   **5.** not rational   **6.** rational   **7.** not rational   **8.** rational

9. rational   **10.** rational   **11.** rational   **12.** rational   **13.** $\dfrac{9}{11}$ **14.** $\dfrac{63}{8}$ **15.** 8   **16.** -24 **17.** 18   **18.** 14

19. -7 **20.** $\dfrac{3}{2}$ **21.** 5   **22.** $\dfrac{19}{3}$ **23.** 9   **24.** -5 **25.** 7   **26.** 11   **27.** $\dfrac{3}{4}$ **28.** $\dfrac{4}{7}$ **29.** 1   **30.** -2 and 5

31. ± 4 **32.** ± 9 **33.** ± 7 **34.** ± 8 **35.** -2 and 5   **36.** -7 and 4   **37.** 4 and 8   **38.** 1 and 6   **39.** -4 and 9

40. -3 and 5   **41.** 7   **42.** 6   **43.** $\dfrac{1}{2}$ and 8   **44.** -5 **45.** -15 **46.** $\dfrac{56}{11}$ **47.** $\pm\dfrac{1}{3}$ **48.** $\pm\dfrac{1}{4}$ **49.** -6 and 10

50. -7 and 2   **51.** no solution   **52.** no solution   **53.** 11   **54.** -8 **55.** 5   **56.** -7 and 6   **57.** -9 and 1   **58.** -9

59. -4 and 8   **60.** -3 and 5   **61.** all real numbers not equal to 1   **62.** no solution   **63.** no solution

64. all real numbers not equal to 6   **65.** $-\dfrac{11}{2}$ and 6   **66.** -11 and 3   **67.** approximately -2.8117 and 2.3117

68. approximately -6.6714 and 5.1714 **69.** approximately -2.5811 and 0.5811 **70.** approximately -1.5774 and -0.4226

71. $w = -1$ and $w = \pm\sqrt{2}$ **72.** $v = 3$ and $v = \pm\sqrt{5}$ **73.** approximately -0.3024 and 1.2399

74. approximately -0.2758 and 3.4187 **75.** no solution   **76.** no solution   **77.** no solution   **78.** no solution

79. all real numbers except $b = 2$ and $b = 1$.   **80.** all real numbers except $x = -\dfrac{3}{2}$ and $x = 2$.   **81.** all integers not equal to -1 or 1.

82. all integers not equal to -3 or 3.   **83.** non-real solutions   **84.** non-real solutions   **85.** The interest on the first account is 6.5% and on the second account is 4.5%.   **86.** The first investment was for 2 years and the second for 3.5 years.   **87.** The resistance of the second vessel is 26.2 dynes.   **88.** The constant for the second spring is 2 pounds per square inch.   **89.** The focal length needed is 60 millimeters.   **90.** The focal length needed is 30 millimeters.

12.4 CALCULATOR EXERCISES

1. $x = 7$ **2.** $c = 10$ and $c = \dfrac{1}{2}$ **3.** $x = 2$ **4.** $k = \dfrac{4}{5}$ **5.** $h = \dfrac{5}{3}$ **6.** $z = -4$ and $z = -1$

SECTION 12.5 EXERCISES

1. It will take them 1.75 hours working together.   **2.** It will take them 18 minutes working together.   **3.** It takes Jacques 10 hours and Simone 5 hours.   **4.** It will take Joe about 2.58 hours and Mary Lynne about 3.58 hours.   **5.** It will take 3.6 hours working together.
6. It will take 9.375 minutes working together.   **7.** It will take approximately 2.55 hours for both pipes together.
8. It will take 3 hours to drain when both are used.   **9.** It takes Line A 8 hours and Line B 12 hours.   **10.** It will take the first line about 8.3 hours and the second line about 9.8 hours.   **11.** There must be an additional 15 points.

12. Another $\dfrac{1}{3}$ liter of pure vinegar should be added.   **13.** They should raise and spend $27.27.   **14.** He must raise another $50.

15. $\overline{XY}$ measures 4 inches; $\overline{PR}$ measures 18 inches   **16.** $\overline{MN}$ measures 2 mm; $\overline{KL}$ measures 28 mm   **17.** $\overline{PQ}$ measures 10.075 cm;

$\overline{YZ}$ measures 1.76 cm   **18.** $\overline{JK}$ measures 16.56 yards; $\overline{MO}$ measures 18.3 yards   **19.** $\overline{QR}$ measures $4\dfrac{3}{8}$ feet; $\overline{XZ}$ measures $2\dfrac{1}{4}$ feet

20. $\overline{JL}$ measures $29\dfrac{1}{4}$ feet; $\overline{NO}$ measures $9\dfrac{1}{4}$ feet   **21.** The estimated height is 48.125 feet.   **22.** The flag pole is 15 feet high.

23. $D = \sqrt{\dfrac{Kq_1q_2}{F}}$ **24.** $G = \dfrac{4\pi^2 a^3}{MT^2}$ **25.** The resistance of the resistors is 10 ohms and 15 ohms, respectively.
26. The resistance of the resistors is 5 ohms and 20 ohms, respectively.

12.5 CALCULATOR EXERCISES

1. $x = -4$ and $x = 3$ **2.** $x = \dfrac{3 \pm \sqrt{57}}{2}$ or $x \approx 5.27$ and $x \approx -2.27$ **3.** $x = -\dfrac{6}{5}$ and $x = 4$

CHAPTER 12 SECTION-BY-SECTION REVIEW

Reflections

1–6. Answers will vary.

Exercises

1. rational expression   **2.** nonrational expression   **3.** rational expression   **4.** rational expression   **5.** $0; x \neq 0$

6. -8 and 3; $x \neq -8, x \neq 3$ **7.** no restricted values; all real numbers   **8.** $\pm\dfrac{3}{2}; x \neq -\dfrac{3}{2}, x \neq \dfrac{3}{2}$

9. $y = \dfrac{180}{x^2} + \dfrac{30}{x}$ **10.** $g(x) = \dfrac{x^2 - x - 2}{x - 2}$ **11.**   **12.**

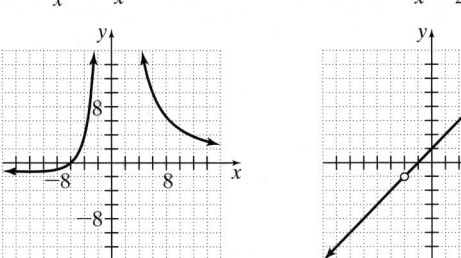

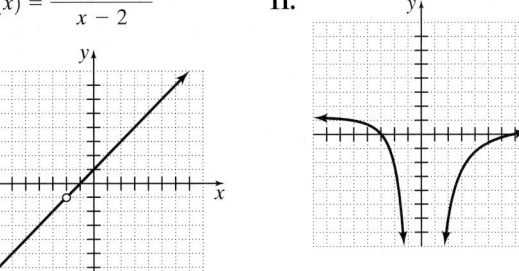

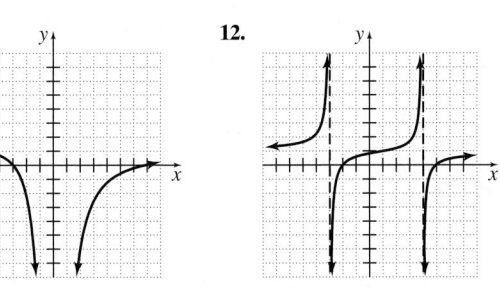

13. (a) $C_{ave}(x) = \dfrac{1800 + 100x + 25x^2}{x}$

(b) 5000

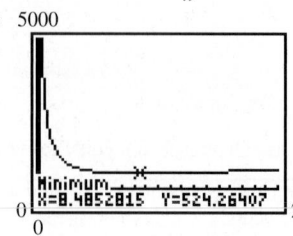

The minimum is 524.27 at $x = 8.5$. This means that the minimum average cost per year is \$524.27.

(c) After $8\frac{1}{2}$ years, the cash register should be replaced.

14. $-\dfrac{4x}{9y^2}$ **15.** $\dfrac{p - 3}{3p + 5}$ **16.** $-\dfrac{1}{2x}$ **17.** $\dfrac{2x - 5}{4x + 1}$ **18.** $\dfrac{b - 7}{b^2 + 2}$ **19.** $-\dfrac{3y}{14x^2}$

20. $\dfrac{(a + 5)(a + 1)}{a - 4}$ **21.** $\dfrac{4}{(m + 5)(2m + 7)}$ **22.** -1 **23.** $-\dfrac{x}{6y^2}$ **24.** $-\dfrac{1}{15}$

25. $-\dfrac{6yz^2}{5x}$ **26.** $\dfrac{6a^3b^3}{c^3d^2}$ **27.** $\dfrac{(x + 3)(x + 1)^2}{5(x - 2)}$ **28.** $\dfrac{3x}{2z^4}$ **29.** $14ac^2$ **30.** $\dfrac{x + 2}{x^2 + 4}$

31. $(x + 1)(2x + 3)$ **32.** $\dfrac{(a + 3)(a + 4)}{(a + 6)(a + 2)}$ **33.** 1

34. His average speed is $\dfrac{220}{t}$ mph; his new distance is $\dfrac{110(t + 2)}{t}$ miles.

35. $H = \dfrac{V}{LW}$; the height is $(x + 3)$ inches.

36. (a) $C(x) = 500 + 40x$; $R(x) = 80x$ **(b)** The ratio is $\dfrac{25 + 2x}{4x}$.

37. $\dfrac{y(3 + 7y)}{5x}$ **38.** $\dfrac{b(5b^2 - 2)}{7a}$ **39.** 2   **40.** $\dfrac{10x^2 + 10x + 3}{(2x - 3)(2x + 3)}$ **41.** $\dfrac{12y + 7x}{15x^3y^2}$ **42.** $\dfrac{x(7x + 4)}{(x - 8)(x + 4)}$ **43.** $\dfrac{44}{(2x - 1)(x + 5)}$

44. $\dfrac{3}{x - y}$ **45.** $\dfrac{2(13x + 4)}{5(3x - 4)(3x + 4)}$ **46.** $\dfrac{29x - 3}{(2x + 1)(x - 3)(3x - 2)}$ **47.** $\dfrac{x + 55}{2(x - 9)(x + 9)}$ **48.** $\dfrac{10(2x - 1)}{(2x - 3)(x + 2)(3x - 2)}$

49. $\dfrac{5(x - 9)}{(x + 5)(x - 5)}$ **50.** $\dfrac{37x^2 - 32x - 44}{6x(x + 2)}$ **51.** The perimeter is $\dfrac{5x^2 - 4x - 5}{2x(x + 1)}$ units.   **52.** The length is $\dfrac{4x + 3}{x}$ units.

53. (a) $t(x) = \dfrac{125}{10 + x} + \dfrac{80}{x}$ **(b)**

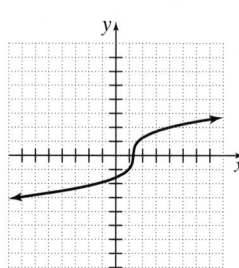

(c) The trip took 5.8 hours.
(d) Gretchen's speed was 40 mph and Bob's speed was 50 mph.

54. not rational   **55.** not rational   **56.** rational   **57.** rational   **58.** -8 and 4   **59.** all integers not equal to -3 **60.** no solution

61. 7   **62.** -3 and $\dfrac{1}{2}$ **63.** 5   **64.** ± 10 **65.** -5 and 6   **66.** -4 **67.** -3 and 4   **68.** no solution   **69.** -3 and $\dfrac{9}{2}$

70. all real numbers except -4 **71.** $-\dfrac{3}{2}$ and 4   **72.** $\dfrac{5 \pm \sqrt{21}}{2}$ **73.** non-real solutions   **74.** The first interest rate was 6%.

The second interest rate was 7%.   **75.** It will take them $\dfrac{24}{7}$ hours, or approximately 3.4 hours.   **76.** It will take Viv 55.4 minutes and

Lucy 65.4 minutes.   **77.** It will take 6 minutes working together.   **78.** It will take the high-speed line $5\dfrac{1}{4}$ hours and the other line

$10\dfrac{1}{2}$ hours, working alone.   **79.** She should add another \$40.   **80.** Another two cups should be added.   **81. (a)** $\overline{GI}$ measures 12 in;

$\overline{KL}$ measures 14 in.   **(b)** $\overline{HI}$ measures $7\dfrac{3}{4}$ feet; $\overline{JL}$ measures $9\dfrac{3}{4}$ feet   **(c)** $\overline{GH}$ measures 7.52 feet; $\overline{JL}$ measures 3.52 feet

82. The bluff is 48 feet high.   **83.** The tree's shadow was 96 feet long.   **84.** The resistance of the two resistors are 44 ohms and 93 ohms.

85. $f = \dfrac{f_1 f_2}{f_2 + f_1}$

CHAPTER 12 CHAPTER REVIEW

1. $y = \dfrac{45}{x^2} + \dfrac{15}{x}$

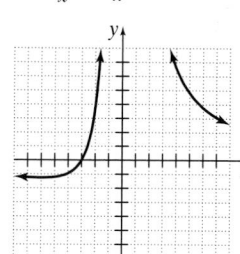

2. $g(x) = \dfrac{x + 5}{x^2 + 5x + 4}$

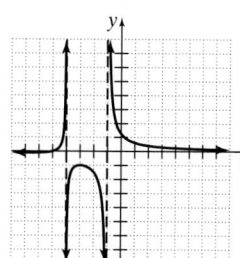

3. $p(x) = \dfrac{2x^2 - x - 3}{(x + 1)}$

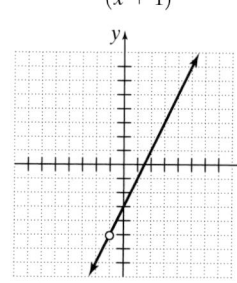

4.

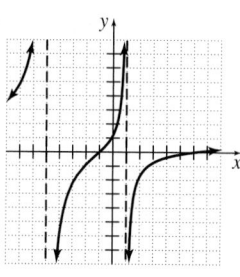

5. no restricted values; all real numbers **6.** $\pm\dfrac{5}{3}; x \neq -\dfrac{5}{3}, x \neq \dfrac{5}{3}$ **7.** $0; x \neq 0$ **8.** -2 and $\dfrac{4}{3}; x \neq -2, x \neq \dfrac{4}{3}$

9. $\dfrac{3x + 4}{2x + 3}$ **10.** $\dfrac{k + 5}{k^2 + 6}$ **11.** $-\dfrac{3x^3}{4y^4}$ **12.** $\dfrac{q - 3}{2(q + 2)}$ **13.** $\dfrac{-2x}{3(x + 2)}$ **14.** $\dfrac{x - 25}{(3x + 2)(x - 3)}$ **15.** $\dfrac{5}{a - b}$ **16.** $\dfrac{b(7 + 5b)}{3a}$

17. $\dfrac{p(3p - 2)}{4m}$ **18.** $\dfrac{5 + 2y^2}{4x^2y^3}$ **19.** $\dfrac{x(5x + 7)}{(x + 3)(x - 5)}$ **20.** $\dfrac{8x + 5}{3(2x - 5)(2x + 5)}$ **21.** $\dfrac{31x + 9}{(2x - 1)(x + 3)(3x + 2)}$ **22.** 2 **23.** $\dfrac{2x - 1}{5x + 4}$

24. $\dfrac{x - 30}{(x + 6)(x - 6)}$ **25.** $\dfrac{x^2 - 9x - 6}{4x(x + 1)}$ **26.** $\dfrac{x - 27}{2(x + 5)(x - 5)}$ **27.** $\dfrac{7x - 2}{(4x - 5)(x - 3)(5x - 4)}$ **28.** $\dfrac{3(m + 6)}{2m + 5}$ **29.** -1

30. $-\dfrac{10b^2}{9a^3}$ **31.** $\dfrac{(2c + 3)(c + 7)}{c + 3}$ **32.** $-\dfrac{a}{3b^2}$ **33.** $-\dfrac{1}{4}$ **34.** $-\dfrac{4g^4k}{3h^2}$ **35.** $\dfrac{6p^3q^7}{5m^3n^4}$ **36.** $\dfrac{75m^2p^3}{2}$ **37.** $\dfrac{2x - 3}{4x^2 + 9}$

38. $\dfrac{(x + 2)^2}{2}$ **39.** $\dfrac{5b^2}{3c^7}$ **40.** $\dfrac{(z - 3)(z + 8)}{(z - 6)(z + 4)}$ **41.** 1 **42.** $(x + 5)(2x + 1)$ **43.** 4 **44.** approximately -0.51 and 4.71

45. 0 **46.** $\dfrac{1}{3}$ and 8 **47.** 5 **48.** $\dfrac{1}{7}$ and $\dfrac{1}{5}$ **49.** no solution **50.** $-\dfrac{7}{4}$ **51.** 15 **52.** $\dfrac{7}{2}$

53. approximately -4.01 and 6.51 **54.** all real numbers except 3 **55.** -7 and 3 **56.** all real numbers except -3

57. all real numbers except 0. **58.** -1 and $\dfrac{4}{5}$ **59.** -9 and 9 **60.** no solution **61.** -7 and 2 **62.** non-real solutions

63. -4 **64.** 8 and -2 **65.** It will take them $\dfrac{70}{17}$ hours, or approximately 4.1 hours. **66.** Another 1 pint of glue is needed.

67. His average rate of speed is $\dfrac{120}{t}$ mph; the new distance is $\dfrac{80(t + 1)}{t}$ miles. **68.** The tree is 24 feet high.

69. $H = \dfrac{V}{LW}$; the height is $(2x + 1)$ inches. **70.** It would take the first line 6 hours and the second line 7.5 hours, working alone.

71. The perimeter is $\dfrac{(4x - 3)(x - 1)}{x(x - 3)}$ units. **72.** She needs an additional $20.

73. Each resistor is approximately 45 ohms and 57 ohms. **74.** $c = \dfrac{c_1c_2}{c_2 + c_1}$

CHAPTER 12 TEST

1. -3 and $0; x \neq -3, x \neq 0$ **2.** -4 and $9; x \neq -4, x \neq 9$ **3.**

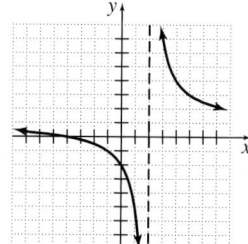

6. $\dfrac{7x - 15}{x(x + 4)(x - 3)}$ **7.** $\dfrac{1}{2}$ **8.** $\dfrac{x^2 + 15x + 5}{x(x + 5)^2}$ **4.** $\dfrac{8}{3x - 7}$ **5.** $\dfrac{2z(z - 3)}{z^2 + 9}$

9. $\dfrac{9(x - 2y)}{2x(x + y)}$ **10.** -1 and 5 **11.** $11 \pm 4\sqrt{7}$

12. no real-number solutions **13.** ± 15 **14.** -2 and $-\dfrac{1}{3}$

15. 3 **16.** 0 **17.** all real numbers except -1.

18. no solution **19.** $s = \dfrac{x - m}{z}$ **20.** The height is $21\dfrac{2}{3}$ feet. **21.** Jill can rake the leaves in four hours. **22.** Answers will vary.

Chapter 13

SECTION 13.1 EXERCISES

1. -6 **2.** -7 **3.** not a real number   **4.** not a real number   **5.** 6   **6.** 11   **7.** -6 **8.** -11 **9.** -6 **10.** -11
11. 10.488   **12.** 14.142   **13.** -10.488 **14.** -14.142 **15.** not a real number   **16.** not a real number   **17.** 4.791   **18.** 5.848
19. -4.791 **20.** -5.848 **21.** 1.821   **22.** 1.710   **23.** 2.321   **24.** 2.921   **25.** nonradical   **26.** radical   **27.** radical
28. nonradical   **29.** all real numbers less than 3; $[3, \infty)$ **30.** all real numbers less than 0; $[0, \infty)$ **31.** no restricted values; $(-\infty, \infty)$

32. all real numbers less than $\dfrac{1}{3}$; $\left[\dfrac{1}{3}, \infty\right)$ **33.** all real numbers less than 0; $[0, \infty)$ **34.** no restricted values; $(-\infty, \infty)$

35. all real numbers less than 3; $[3, \infty)$ **36.** all real numbers less than $-\dfrac{5}{3}$; $\left[-\dfrac{5}{3}, \infty\right)$ **37.** no restricted values; $(-\infty, \infty)$

38. no restricted values; $(-\infty, \infty)$ **39.** all real numbers less than $-\dfrac{2}{3}$; $\left[-\dfrac{2}{3}, \infty\right)$ **40.** all real numbers less than $\dfrac{5}{7}$; $\left[\dfrac{5}{7}, \infty\right)$

41. 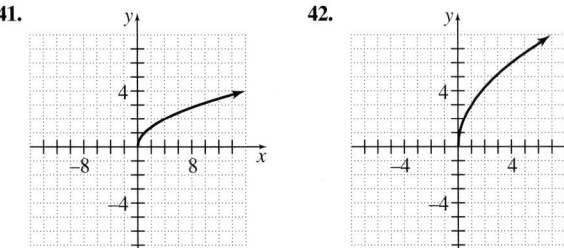   **42.** 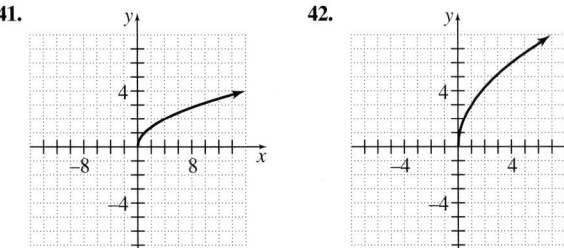   **43.** 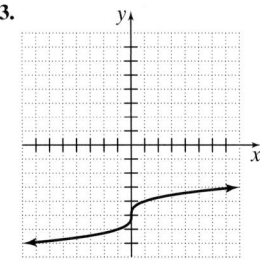   **44.**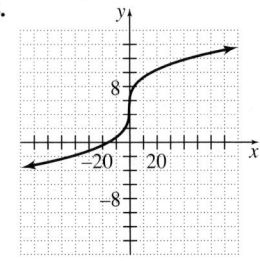

45. 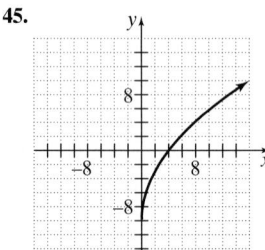   **46.** 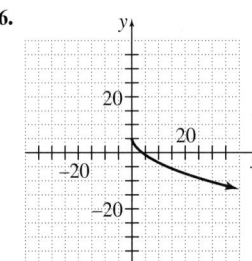   **47.** 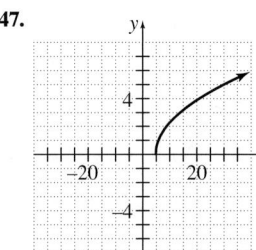   **48.**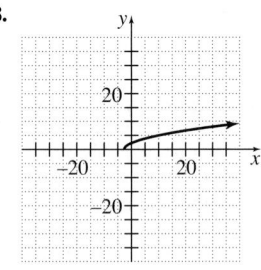

49. 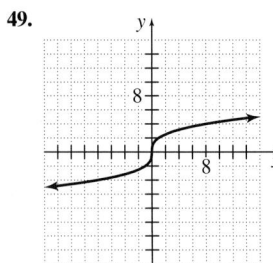   **50.**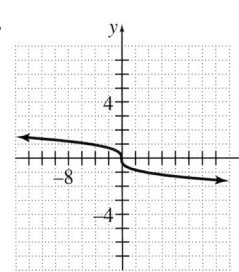

51. $s = \sqrt{A}$ **(a)** The length of the side is 17 ft.   **(b)** The length of the side is 9.5 inches.   **(c)** The length of the side is approximately 10.724 mm

52. $r = \sqrt{\dfrac{A}{\pi}}$ **(a)** The length of the side is 23 yards.   **(b)** The length of the side is 10.5 cm.   **(c)** The length of the side is approximately 6.708 feet.

53. (a) The length of the radius is 14 inches.   **(b)** The length of the radius is $\sqrt{\dfrac{49}{\pi}} \approx 3.949$ yards.
(c) The length of the radius is $\sqrt{\dfrac{108}{\pi}} \approx 5.863$ meters.

54. (a) The length of the radius is 15 dm.   **(b)** The length of the radius is $\dfrac{19}{\sqrt{\pi}} \approx 10.720$ ft
(c) The length of the radius is $\sqrt{\dfrac{200}{\pi}} \approx 7.979$ cm.

55. $e = \sqrt[3]{V}$ **(a)** The length of the side is 12 inches.   **(b)** The length of the side is 4.93 meters.
56. (a) The length of the side is 21 cm.   **(b)** The length of the side is 7.94 feet.

57. $r = \sqrt[3]{\dfrac{3V}{4\pi}}$; $d = 2\sqrt[3]{\dfrac{3V}{4\pi}}$ **(a)** The radius is 6 dm and the diameter is 12 dm.
(b) The radius is approximately 3.908 yards and the diameter is approximately 7.816 yards.

58. (a) The radius is 9 inches and the diameter is 18 inches.
(b) The radius is approximately 6.204 meters and the diameter is approximately 12.407 meters.

59. $r = \sqrt{\dfrac{8200}{\pi}}$ The radius and diameter are approximately 51.090 ft. and 102.179 ft., respectively.

60. $s = \sqrt{377,000}$ The side is approximately 614 feet.

61. 5 units   **62.** 8 units   **63.** 8 units   **64.** 7 units   **65.** $\sqrt{106}$ units   **66.** $\sqrt{185}$ units

67. (a) $d(x) = 9.4\sqrt[4]{x}$ **68. (a)** $r(x) = 26\sqrt[4]{x}$

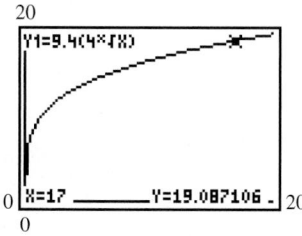

(b) The function is increasing.

(c) Assuming that the trend continues, the 2010 market value of data communications equipment would be about $19 million.

(b) The function is increasing.

(c) If the trend holds, the 2010 market value of car rentals will be about $53 million.

69.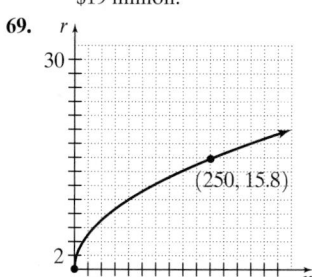

There should be about 16 rows in the table.

70.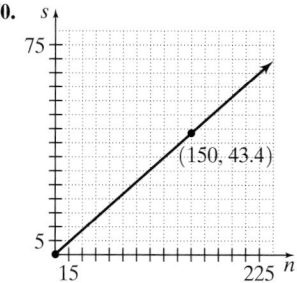

The variability is about 43.

13.1 CALCULATOR EXERCISES

1. $f(x)$ simplifies to $f(x) = |x - 2|$.   **2.** $f(x)$ simplifies to $f(x) = |2x - 3|$.

SECTION 13.2 EXERCISES

1. $\sqrt{36} = 6$ **2.** $\sqrt{100} = 10$ **3.** $\sqrt[3]{216} = 6$ **4.** $\sqrt[3]{729} = 9$ **5.** $-\sqrt[4]{81} = -3$ **6.** $-\sqrt[4]{625} = -5$

7. $\sqrt[4]{-81}$, not a real number   **8.** $\sqrt[4]{-625}$, not a real number   **9.** $-\sqrt[3]{8} = -2$ **10.** $-\sqrt[3]{27} = -3$ **11.** $\sqrt[3]{-8} = -2$

12. $\sqrt[3]{-27} = -3$ **13.** $\dfrac{1}{\sqrt{36}} = \dfrac{1}{6}$ **14.** $\dfrac{1}{\sqrt{100}} = \dfrac{1}{10}$ **15.** $\dfrac{1}{\sqrt[3]{216}} = \dfrac{1}{6}$ **16.** $\dfrac{1}{\sqrt[3]{729}} = \dfrac{1}{9}$ **17.** $\dfrac{1}{\sqrt[4]{81}} = \dfrac{1}{3}$ **18** $\dfrac{1}{\sqrt[4]{625}} = \dfrac{1}{5}$

19. $-\dfrac{1}{\sqrt[4]{81}} = -\dfrac{1}{3}$ **20.** $\dfrac{1}{\sqrt[4]{-625}}$, not a real number   **21.** $\dfrac{1}{\sqrt[3]{8}} = \dfrac{1}{2}$ **22.** $\dfrac{1}{\sqrt[3]{27}} = \dfrac{1}{3}$ **23.** $\dfrac{1}{\sqrt[3]{-8}} = -\dfrac{1}{2}$ **24.** $\dfrac{1}{\sqrt[3]{-27}} = -\dfrac{1}{3}$

25. $\sqrt{522} \approx 22.847$ **26.** $\sqrt[3]{478} \approx 7.819$ **27.** $\sqrt[4]{522} \approx 4.780$ **28.** $\sqrt[5]{478} \approx 3.435$ **29.** $\dfrac{1}{\sqrt[3]{522}} \approx 0.124$ **30.** $\dfrac{1}{\sqrt{478}} \approx 0.046$

31. $\dfrac{1}{\sqrt[5]{522}} \approx 0.286$ **32.** $\dfrac{1}{\sqrt[4]{478}} \approx 0.214$ **33.** $(\sqrt[3]{27})^4 = 81$ **34.** $(\sqrt{25})^3 = 125$ **35.** $(\sqrt[3]{-27})^4 = 81$

36. $(\sqrt{-25})^3$, not a real number   **37.** $\dfrac{1}{(\sqrt[3]{27})^4} = \dfrac{1}{81}$ **38.** $\dfrac{1}{(\sqrt{25})^3} = \dfrac{1}{125}$ **39.** $-\dfrac{1}{(\sqrt[3]{27})^4} = -\dfrac{1}{81}$ **40.** $-\dfrac{1}{(\sqrt{25})^3} = -\dfrac{1}{125}$

41. $(\sqrt{4})^5 = 32$ **42.** $(\sqrt{9})^3 = 27$ **43.** $(\sqrt[3]{8})^7 = 128$ **44.** $(\sqrt[3]{125})^5 = 3125$ **45.** $\dfrac{1}{(\sqrt{4})^7} = \dfrac{1}{128}$ **46.** $\dfrac{1}{(\sqrt{16})^5} = \dfrac{1}{1024}$

47. $-(\sqrt{16})^3 = -64$ **48.** $-(\sqrt{64})^3 = -512$ **49.** $(\sqrt{-9})^3$, not a real number   **50.** $(\sqrt{-4})^3$, not a real number

51. $-\dfrac{1}{(\sqrt{81})^3} = -\dfrac{1}{729}$ **52.** $-\dfrac{1}{(\sqrt{36})^3} = -\dfrac{1}{216}$ **53.** $(\sqrt[4]{28})^5 \approx 64.409$ **54.** $(\sqrt[4]{36})^5 \approx 88.182$ **55.** $(\sqrt[3]{-21})^4 \approx 57.937$

56. $(\sqrt[3]{-15})^2 \approx 6.082$ **57.** $\dfrac{1}{(\sqrt[3]{5})^2} \approx 0.342$ **58.** $\dfrac{1}{(\sqrt[3]{7})^2} \approx 0.273$ **59.** $-\dfrac{1}{(\sqrt[3]{42})^2} \approx -0.083$ **60.** $-\dfrac{1}{(\sqrt[3]{36})^2} \approx -0.092$

61. $(\sqrt[8]{-88})^3$, not a real number   **62.** $(\sqrt[6]{-66})^5$, not a real number

63. The restricted values are all real numbers greater than $\frac{5}{4}$. The domain of the function is $\left(-\infty, \frac{5}{4}\right]$.

64. There are no restricted values. The domain is all real numbers, or $(-\infty, \infty)$.

65. There are no restricted values. The domain is all real numbers, or $(-\infty, \infty)$.

66. The restricted values are all real numbers greater than $\frac{7}{2}$. The domain of the function is $\left(-\infty, \frac{7}{2}\right]$.

67. 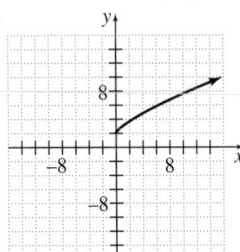 **68.** 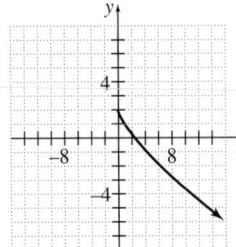 **69.** 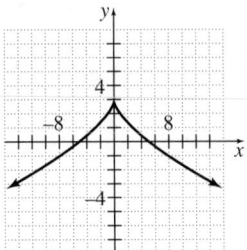 **70.**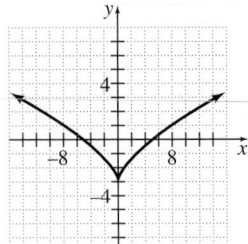

71. 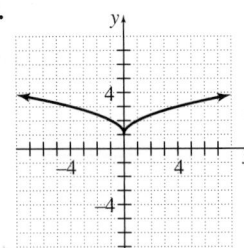 **72.** 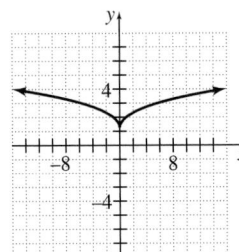 **73.** 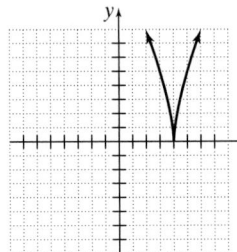 **74.**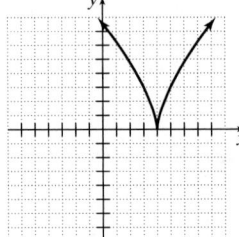

75.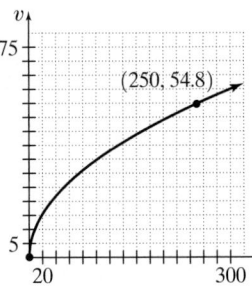

The velocity should be about 55 mph.

76.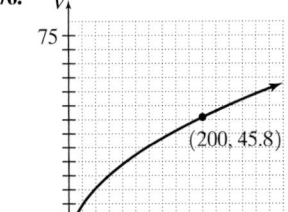

The velocity should be about 46 mph.

13.2 CALCULATOR EXERCISES

1. 9   **2.** $(-27)^{2/3}, (\sqrt[3]{(-27)})^2, ((-27)^{1/3})^2, \sqrt[3]{(-27)^2}$

SECTION 13.3 EXERCISES

1. $\frac{1}{3}$ **2.** $\frac{1}{3}$ **3.** 2   **4.** 2   **5.** 243   **6.** 512   **7.** 8   **8.** 8   **9.** 5   **10.** 4   **11.** 6   **12.** 5   **13.** $\frac{2}{3}$ **14.** $\frac{9}{25}$

15. $\frac{4}{3}$ **16.** $\frac{8}{27}$ **17.** $\frac{1}{8}$ **18.** $\frac{1}{16}$ **19.** $\frac{1}{x^{2/3}}$ **20.** $\frac{1}{x^{5/6}}$ **21.** $12y^{3/4}$ **22.** $2y^{4/7}$ **23.** $z^{17/12}$ **24.** $z^{13/8}$ **25.** $p^{1/6}$

26. $\frac{1}{x^{1/9}}$ **27.** $b^{2/3}$ **28.** $z^{3/8}$ **29.** $x^{3/2}y^{3/4}$ **30.** $p^{5/6}q^{5/2}$ **31.** $\frac{x^{5/3}}{y^{5/2}}$ **32.** $\frac{a^{7/3}}{b^{7/4}}$ **33.** $\frac{b^{1/3}}{a^{1/3}}$ **34.** $\frac{n^{3/4}}{m^{3/4}}$ **35.** $\frac{1}{c^{1/5}}$

36. $k^{2/7}$ **37.** $4a^{10/3}b^4$ **38.** $8c^{2/3}d^{3/2}$ **39.** $10a^{13/12}b^{4/5}$ **40.** $-6x^{4/5}y^{7/12}$ **41.** $\frac{m^{10/7}}{4}$ **42.** $64z^{7/3}$ **43.** $\frac{8a^{10/3}c^{13/6}}{9b^{13/4}}$

44. $\frac{z^3}{2y^{7/4}}$ **45.** $x^{11/12} - x^{21/20}$ **46.** $z^{11/12} - z$ **47.** $2x^{3/5} + 6x^{2/5}y^{2/5}$ **48.** $15a^{3/7} - 5a^{2/7}b^{4/7}$ **49.** $3a^{1/4}b^{1/3} - a^{3/4}b^{7/6}$

50. $x^{3/4}y^{9/8} - 4x^{1/4}y^{3/4}$ **51.** $x^{5/6} + x^{1/2}y^{1/3} - x^{1/3}y^{1/2} - y^{5/6}$ **52.** $c^{7/6} + c^{2/3}d^{1/2} - c^{1/2}d^{2/3} - d^{7/6}$ **53.** $x^{1/2} - y^{1/2}$ **54.** $p^{3/2} - q^{3/2}$

55. $x^{5/2} - x^2y^{1/2} + x^{1/2}y^2 - y^{5/2}$ **56.** $a^{5/3} - a^{2/3}b + ab^{2/3} - b^{5/3}$ **57.** $x^{2/3} - 4$ **58.** $16 - z^{1/2}$ **59.** $x^{1/2} + 4x^{1/4} + 4$

60. $x^{2/3} + 6x^{1/3} + 9$ **61.** $x^2 - 2xy^{1/2} + y$ **62.** $x - 2x^{1/2}y + y^2$ **63.** $a + 2a^{1/2}b^{1/2} + b$ **64.** $c - 2c^{1/2}d^{1/2} + d$

65. The side length should be doubled.   **66.** The side length should be tripled.

67. The time is increased by a factor of $\sqrt{3}$, or the time is about $1.732t$ seconds.
68. The time is increased by a factor of 2, or the time is $2t$ seconds. **69.** The answer is the same as that for Section Exercise 67. The change from feet to meters does not affect the change in time.
70. The answer is the same as that for Section Exercise 68. The change from feet to meters does not affect the change in time.

13.3 CALCULATOR EXERCISES

1. x^2 **2.** $\dfrac{1}{x^2}$ **3.** $x^{7/4}$ **4.** $\dfrac{1}{x^{3/2}}$ **5.** $x^{9/4} - x^{17/12}$ **6.** $x^{2/3} - 1$

SECTION 13.4 EXERCISES

1. $2\sqrt{7}$ **2.** $5\sqrt{3}$ **3.** $-7\sqrt[3]{2}$ **4.** $-4\sqrt[3]{3}$ **5.** $2\sqrt[4]{7}$ **6.** $7\sqrt[4]{3}$ **7.** $-5\sqrt[5]{2}$ **8.** $-3\sqrt[5]{8}$ **9.** $2x^2yz\sqrt{5y}$
10. $2xy^4z^4\sqrt{3xy}$ **11.** $2mn^3\sqrt[3]{9m^2}$ **12.** $3p^2q^2\sqrt[3]{4pq^2}$ **13.** $3xy\sqrt[4]{2y}$ **14.** $2ab^2\sqrt[4]{5a^2}$ **15.** $3ab^2\sqrt[5]{2ac^2}$ **16.** $3yz^3\sqrt[5]{3x^2y^3}$
17. $(x+3)\sqrt{5}$ **18.** $(x+1)\sqrt{3x}$ **19.** $7\sqrt{2xy}$ **20.** $6\sqrt{2xy}$ **21.** $14xy\sqrt{2x}$ **22.** $-6p^2q\sqrt{21p}$ **23.** $(x+2y)\sqrt{21}$
24. $(p+2q)\sqrt{3p}$ **25.** $(x+2)\sqrt{x+1}$ **26.** $(x-7)\sqrt{x+1}$ **27.** $(x-1)\sqrt{(x+2)(x+4)}$ **28.** $(x+4)\sqrt{(x+2)(x-4)}$
29. $-2xy\sqrt[3]{3x^2}$ **30.** $-5x^2y^2\sqrt[3]{2y^2}$ **31.** $-\dfrac{6}{7}$ **32.** $-\dfrac{5}{9}$ **33.** $\dfrac{11}{12}$ **34.** $\dfrac{13}{14}$ **35.** $-\dfrac{2\sqrt{6}}{9}$ **36.** $\dfrac{3\sqrt{15}}{25}$ **37.** $\dfrac{\sqrt[3]{4}}{5}$
38. $\dfrac{\sqrt[3]{5}}{8}$ **39.** $-\dfrac{\sqrt[3]{45}}{5}$ **40.** $-\dfrac{\sqrt[3]{70}}{7}$ **41.** $\dfrac{\sqrt[5]{2}}{2}$ **42.** $\dfrac{\sqrt[5]{3}}{3}$ **43.** $\dfrac{2x}{3y}$ **44.** $\dfrac{5\sqrt{x}}{7y}$ **45.** $\dfrac{\sqrt{3x}}{5y}$ **46.** $\dfrac{\sqrt{5p}}{7q}$ **47.** $\dfrac{2\sqrt{5xyz}}{5z}$
48. $\dfrac{6\sqrt{7abc}}{7c}$ **49.** $\dfrac{x\sqrt{3}}{3}$ **50.** $\dfrac{z^2\sqrt{5}}{5}$ **51.** $\dfrac{3x}{y^2}$ **52.** $\dfrac{-4x^2y}{z^3}$ **53.** $\dfrac{\sqrt[3]{15x}}{5x}$ **54.** $\dfrac{\sqrt[3]{42a}}{6a}$ **55.** $\dfrac{\sqrt[3]{75xy^2z^2}}{5yz}$ **56.** $\dfrac{\sqrt[3]{196ac^2}}{7c}$
57. $\dfrac{\sqrt{5x}}{x}$ **58.** $\dfrac{\sqrt{10y}}{y}$ **59.** $\dfrac{\sqrt{2y}}{2y}$ **60.** $\dfrac{\sqrt{3b}}{3b}$ **61.** $\dfrac{z\sqrt{2z}}{4}$ **62.** $\dfrac{p\sqrt{3p}}{9}$ **63.** $\dfrac{b\sqrt{ab}}{a}$ **64.** $\dfrac{\sqrt{cd}}{cd^2}$ **65.** $\dfrac{\sqrt{3x}}{6}$ **66.** $\dfrac{p\sqrt{2q}}{6q}$
67. $\dfrac{\sqrt[3]{3x}}{x}$ **68.** $\dfrac{\sqrt[3]{7a}}{a}$ **69.** $\dfrac{\sqrt[3]{2y^2}}{2y}$ **70.** $\dfrac{\sqrt[3]{2a^2}}{2a}$ **71.** $\dfrac{-2\sqrt[3]{9x^2z}}{3x}$ **72.** $\dfrac{-3\sqrt[3]{25b^2c}}{5b}$
73. The period is approximately 2.9 seconds. **74.** The period is approximately 3.8 seconds. **75.** It takes approximately 1.92 seconds.
76. It takes approximately 1.4 seconds. **77.** The leg takes 1.756 seconds; Eydie's speed is about 4.56 feet per second.
78. The stride takes 5.554 seconds; Gargantua is running about 21.6 feet per second.

13.4 CALCULATOR EXERCISES

1. $19\sqrt[3]{2}$ **2.** $13\sqrt[4]{7}$ **3.** $-2\sqrt[6]{13}$ **4.** $29\sqrt{11}$

SECTION 13.5 EXERCISES

1. $\sqrt{7}$ **2.** $7\sqrt{6}$ **3.** $\dfrac{9}{10}\sqrt{10}$ **4.** $\dfrac{14}{13}\sqrt{13}$ **5.** $-2\sqrt[3]{3}$ **6.** $13\sqrt[3]{2}$ **7.** $23\sqrt{3}$ **8.** $12\sqrt{5}$ **9.** $14\sqrt{x}$ **10.** $-2\sqrt{bc}$
11. $11\sqrt{x}$ **12.** $20\sqrt{p}$ **13.** $-3x\sqrt{2x}$ **14.** $-y^2\sqrt{3y}$ **15.** $(3+4a)\sqrt{a}$ **16.** $(11b-4)\sqrt{b}$ **17.** $14a\sqrt{b}$ **18.** $x\sqrt{y}$
19. $7\sqrt{pq} + 7\sqrt[3]{pq}$ **20.** $6\sqrt[3]{ab} - 8\sqrt{ab}$ **21.** $6xy\sqrt{xy}$ **22.** $-cd\sqrt{cd}$ **23.** $-xy\sqrt[3]{x^2yz}$ **24.** $7xy\sqrt[4]{xyz^2}$ **25.** $\sqrt{35} - 7$
26. $\sqrt{30} + 5$ **27.** $\sqrt{3x} - \sqrt{15}$ **28.** $2\sqrt{35} + 2\sqrt{5a}$ **29.** $6a - 15\sqrt{a}$ **30.** $24c + 32\sqrt{c}$ **31.** $8x - 12\sqrt[3]{x^2}$
32. $35\sqrt[3]{a^2} - 14a$ **33.** $6 + 2\sqrt{6} - 30\sqrt{2} - 20\sqrt{3}$ **34.** $3\sqrt{30} - 36\sqrt{2} + 2\sqrt{5} - 8\sqrt{3}$ **35.** $\sqrt{6} + \sqrt{3x} - \sqrt{2x} - x$
36. $\sqrt{10} + \sqrt{2z} - \sqrt{5z} - z$ **37.** 19 **38.** -1 **39.** $144 - p$ **40.** $169 - q$ **41.** $2x - 3y$ **42.** $6x - 3y$
43. $a + 8\sqrt{a} + 16$ **44.** $bc + 10\sqrt{bc} + 25$ **45.** $9b - 12\sqrt{b} + 4$ **46.** $81c - 54\sqrt{c} + 9$ **47.** $x - 2\sqrt{xy} + y$
48. $2x - 2\sqrt{6xy} + 3y$ **49.** $\sqrt{3x} - \sqrt{2}$ **50.** $\sqrt{2x} + \sqrt{6}$ **51.** $1 - \dfrac{12\sqrt{a}}{a}$ **52.** $\dfrac{9\sqrt{b}}{b} + 1$ **53.** $1 + \dfrac{\sqrt{xy}}{x} + \dfrac{\sqrt{xz}}{x}$
54. $\sqrt{a} - \dfrac{\sqrt{abc}}{b} + \sqrt{c}$ **55.** $6(\sqrt{6} - \sqrt{3})$ **56.** $3(\sqrt{10} - \sqrt{2})$ **57.** $5 + 2\sqrt{6}$ **58.** $5 - 2\sqrt{6}$ **59.** $\dfrac{3x\sqrt{x} + 6x}{x - 4}$
60. $\dfrac{12w + 4w\sqrt{w}}{9 - w}$ **61.** $\sqrt{3b} + 2$ **62.** $\sqrt{ab} - 1$ **63.** $\dfrac{x + 6\sqrt{x} + 9}{x - 9}$ **64.** $\dfrac{36 - 12\sqrt{p} + p}{36 - p}$ **65.** -1 **66.** -1
67. It will require $8\sqrt{10} \approx 25.3$ inches more of material. **68.** It is $4\sqrt{6} \approx 9.8$ inches larger.
69. The difference is $4\sqrt[3]{14} \approx 9.6$ inches. **70.** The difference is $20\sqrt[3]{2} \approx 25.2$ inches.

13.5 CALCULATOR EXERCISES

Students should check the results by one of the methods presented.

1. $4\sqrt[3]{x} + 6\sqrt{x} + 7$ **2.** $\sqrt{x} - 2$ **3.** $6\sqrt[3]{2x^2} + 10\sqrt[3]{x}$ **4.** $8x + 5x\sqrt{2}$ **5.** $\sqrt{x} - 4$ **6.** $\dfrac{21\sqrt{x} + 147}{x - 49}$ **7.** $3x - 13\sqrt{x} - 41$

SECTION 13.6 EXERCISES

1. 64 **2.** 49 **3.** 7.84 **4.** 38.44 **5.** no solution **6.** no solution **7.** 12 **8.** 28 **9.** $\dfrac{9}{2}$ **10.** $\dfrac{3}{2}$ **11.** $\dfrac{3}{2}$ **12.** 3

13. 4   **14.** 22   **15.** 10   **16.** $\dfrac{8}{3}$ **17.** 20   **18.** 12   **19.** 1   **20.** 6   **21.** 5   **22.** $\dfrac{9}{2}$ **23.** -1 **24.** -2 **25.** 5

26. 7   **27.** 4   **28.** -2.42 and 2.75 **29.** all real numbers less than or equal to 7   **30.** all real numbers less than or equal to $\dfrac{3}{2}$

31. approximately 3.83   **32.** -4.70 **33.** 8   **34.** 12   **35.** 36   **36.** 64   **37.** 16   **38.** 1   **39.** no solution

40. no solution   **41.** no solution   **42.** 6   **43.** -72 **44.** $-\dfrac{9}{2}$ **45.** 5000   **46.** 133   **47.** -16 **48.** -27 **49.** -8

50. 8   **51.** 256   **52.** 81   **53.** 13   **54.** -22 **55.** 21   **56.** 88   **57.** 124   **58.** 11   **59.** -8 and 1   **60.** -27 and 3

61. 9   **62.** 3   **63.** 10   **64.** 52   **65.** 3   **66.** 2   **67.** 0   **68.** 0   **69.** no solution   **70.** no solution   **71.** 2   **72.** $\dfrac{1}{2}$

73. 2   **74.** 1   **75.** ± 8 **76.** ± 125 **77.** -38 and 26   **78.** -72 and 56   **79.** 85   **80.** 25   **81.** $\pm\dfrac{1}{8}$ **82.** $\dfrac{1}{81}$ **83.** $\dfrac{406}{81}$

84. 155   **85.** $-\dfrac{122}{5}$ and $\dfrac{128}{5}$ **86.** $\dfrac{257}{6}$ **87.** no real-number solution   **88.** no real-number solution   **89.** 16   **90.** 243

91. The vertical distance is 5.3824 feet. His height of $6.58\overline{3}$ feet added to this jump just about equals a 12-foot basket at 11.97 feet.

92. $t = 2\sqrt{\dfrac{d}{4.9}}$; the distance would be 1.64 meters.   **93.** -8 and 16   **94.** -10 and 5   **95.** -10 and 6   **96.** 4 and -2

97. no solution   **98.** no solution   **99.** The energy was approximately 2.875×10^{17} ft-lb.   **100.** The energy was 1.5625×10^{16} ft-lb.
101. The crushing load was 100 tons.   **102.** The crushing load was 324 tons.

13.6 CALCULATOR EXERCISES

1. 40   **2.** 10   **3.** 20   **4.** -0.307 and 1.412   **5.** 2.462   **6.** -1.291 and 20.920

SECTION 13.7 EXERCISES

1. $10i$ **2.** $12i$ **3.** $\dfrac{4}{7}i$ **4.** $\dfrac{6}{11}i$ **5.** $4i\sqrt{2}$ **6.** $5i\sqrt{3}$ **7.** $10i\sqrt{2}$ **8.** $-21i\sqrt{2}$ **9.** $60i$ **10.** $26i$ **11.** $13i$ **12.** $26i$

13. $\dfrac{4}{5}$ **14.** $\dfrac{4}{5}$ **15.** $-20i$ **16.** $-15i$ **17.** $\sqrt{2} + 2\sqrt{3}$ **18.** $\sqrt{15} - 2\sqrt{2}$ **19.** 2   **20.** 8   **21.** $-5 + 4i$ **22.** $-2 - 9i$

23. $3 - i$ **24.** $9 + 9i$ **25.** 1   **26.** $6i$ **27.** $\dfrac{7}{2} + \dfrac{13}{3}i$ **28.** $\dfrac{1}{2} - i$ **29.** $1.62 + 11.38i$ **30.** $17.3 + 9.07i$ **31.** $7\sqrt{3} + i\sqrt{2}$

32. $-2\sqrt{5} - 7i\sqrt{3}$ **33.** $39 + 2i$ **34.** $69 - 6i$ **35.** 74   **36.** 130   **37.** 37   **38.** 29   **39.** $4.76 - 17.35i$ **40.** $5.04 + 2.58i$

41. $\dfrac{43}{75} - \dfrac{1}{5}i$ **42.** $\dfrac{121}{180} - \dfrac{3}{20}i$ **43.** $-\sqrt{6} + 2i$ **44.** $-9\sqrt{5} - 9i\sqrt{2}$ **45.** 8   **46.** 31   **47.** $1 + 3i$ **48.** $2 + 6i$

49. $-\dfrac{19}{10} + \dfrac{17}{10}i$ **50.** $9 - i$ **51.** $-\dfrac{5}{2} - 3i$ **52.** $-\dfrac{3}{2} + 2i$ **53.** $-5 - 6i$ **54.** $1.5 - 2.1i$ **55.** $4 - 3.2i$ **56.** $2.5 + 3i$

57. $\sqrt{7} - i\sqrt{3}$ **58.** $-\sqrt{7} - i\sqrt{5}$ **59.** $\sqrt{3} + 2i\sqrt{2}$ **60.** $3\sqrt{5} - 2i\sqrt{2}$ **61.** $a = \pm i\sqrt{7}$ **62.** $b = \pm i\sqrt{11}$ **63.** $z = \pm 2i$

64. $m = \pm 4i$ **65.** $p = \pm 5i$ **66.** $q = \pm 4i$ **67.** $d = \pm i\sqrt{3}$ **68.** $c = \pm i\sqrt{2}$ **69.** $t = -1 \pm 3i$ **70.** $s = 3 \pm 6i$

71. $x = 5 \pm i\sqrt{5}$ **72.** $x = -3 \pm 2i\sqrt{2}$ **73.** $x = -\dfrac{5}{2} \pm i$ **74.** $y = \dfrac{1}{2} \pm \dfrac{5}{2}i$ **75.** $z = -2.5 \pm 2.3i$ **76.** $y = 1.4 \pm 1.6i$

77. $b = \dfrac{1}{2} \pm \dfrac{1}{2}i$ **78.** $c = -\dfrac{2}{3} \pm \dfrac{3}{4}i$ **79.** $x = -1 \pm i\sqrt{3}$ **80.** $x = 3 \pm i\sqrt{5}$ **81.** $b = 5 \pm i\sqrt{2}$ **82.** $x = -7 \pm i\sqrt{2}$

83. $y = -\dfrac{1}{2} \pm i$ **84.** $x = -\dfrac{4}{3} \pm \dfrac{1}{3}i$ **85.** $p = \dfrac{2}{3} \pm \dfrac{2}{3}i$ **86.** $z = \dfrac{1}{4} \pm \dfrac{\sqrt{6}}{4}i$ **87.** $x = 1.2 \pm i\sqrt{2}$ **88.** $m = -2.1 \pm i\sqrt{3}$

89. $y = \dfrac{1}{2} \pm \dfrac{3}{5}i$ **90.** $z = -0.6 \pm 0.3i$ **91.** $x = \dfrac{1}{3} \pm \dfrac{1}{2}i$ **92.** $x = -\dfrac{1}{4} \pm \dfrac{1}{3}i$ **93.** $z = -\dfrac{1}{6} \pm \dfrac{1}{6}i$ **94.** $x = -\dfrac{1}{6} \pm \dfrac{1}{4}i$

95. $x = 1 \pm 3i$ **96.** $a = -\dfrac{3}{2} \pm \dfrac{3\sqrt{15}}{2}i$ **97.** $y = -\dfrac{5}{2} \pm \dfrac{\sqrt{159}}{2}i$ **98.** $b = 1 \pm 2i\sqrt{13}$ **99.** $z = -\dfrac{7}{8} \pm \dfrac{3i\sqrt{39}}{8}$

100. $c = -\dfrac{3}{4} \pm \dfrac{3i\sqrt{23}}{4}$ **101.** The magnitude of the total voltage is approximately 37.336 volts.

102. The total voltage is $42 + 2i$ volts.   **103.** The magnitude of the current is 5 amperes.

104. The magnitude of the current is approximately 8.2 amperes.   **105.** $r = 1.335 \pm \dfrac{i\sqrt{3.8711}}{2}$ **106.** $r = 1.335 \pm \dfrac{i\sqrt{6.8711}}{2}$

13.7 CALCULATOR EXERCISES

1. $5i$ **2.** -3 **3.** $2 + 2i$ **4.** -2 **5.** $3 + 3i$ **6.** $10 + 10i$ **7.** $\dfrac{3}{2} - \dfrac{7}{2}i$ **8.** $\dfrac{2}{5} + \dfrac{16}{5}i$

9. $-\dfrac{9}{5}i$ **10.** 10   **11.** 10   **12.** 5.39   **13.** 22   **14.** 7   **15.** 1.73

CHAPTER 13 SECTION-BY-SECTION REVIEW

Reflections

1–11: Answers will vary.

1. 15   **2.** 1.7   **3.** $\dfrac{7}{8}$ **4.** $-\dfrac{3}{5}$ **5.** -6 **6.** not a real number   **7.** 12.247   **8.** -3.271 **9.** 2.627

10. all real numbers less than $\dfrac{7}{2}$; $\left[\dfrac{7}{2}, \infty\right)$ **11.** no restricted values; $(-\infty, \infty)$

12. $f(x) = 2\sqrt{x}$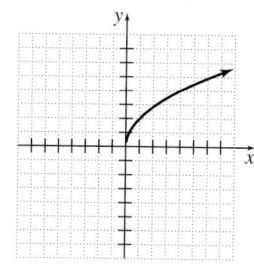

13. $y = \sqrt[3]{3x - 4}$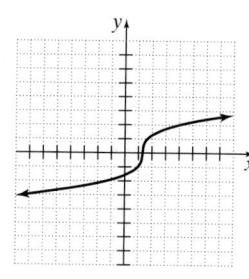

14. The variability is 6.78.   **15.** 10 units   **16.** 9 units

17. 5 units   **18.** about 11.4 units   **19.** 11 inches   **20.** $\dfrac{5}{6}$ yard

21. The approximate radius is 2.5 cm and the diameter is 5 cm.

22. (a) $v(x) = 13\sqrt[10]{x}$

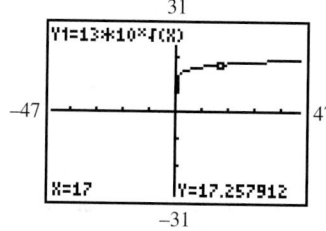

(b) The function is increasing.
(c) The 2010 market value of athletic footwear would be $17.26 million.

23. $\sqrt[3]{-64} = -4$ **24.** $-\sqrt[4]{16} = -2$ **25.** $\sqrt[4]{-16}$, not a real number   **26.** $-\dfrac{1}{\sqrt[3]{64}} = -\dfrac{1}{4}$ **27.** $-(\sqrt[3]{64})^4 = -256$

28. $(\sqrt{-64})^3$, not a real number   **29.** $\dfrac{1}{(\sqrt[4]{16})^3} = \dfrac{1}{8}$ **30.** $\dfrac{1}{(\sqrt[4]{-16})^3}$, not a real number

31. The restricted values are all real numbers less than $-\dfrac{9}{4}$. The domain is all real numbers greater than or equal to $-\dfrac{9}{4}$, or $\left[-\dfrac{9}{4}, \infty\right)$.

32. There are no restricted values. The domain is all real numbers, or $(-\infty, \infty)$.

33. $g(x) = x^{3/2} + 1$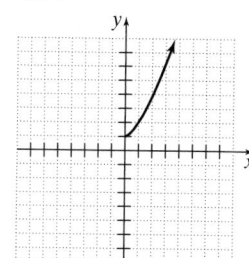

34. $y = (2x + 1)^{1/3}$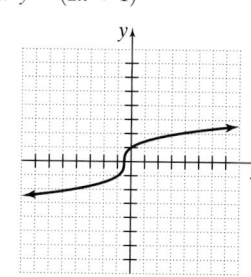

35.
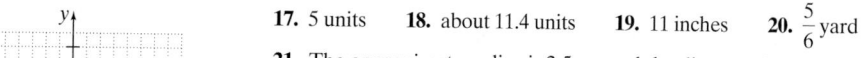
The approximate speed is 36.7 mph.

36. 2   **37.** 128

38. $(\sqrt[3]{9})^5 \approx 38.9$ **39.** $\dfrac{27}{8}$ **40.** $\dfrac{1}{27}$ **41.** 8   **42.** $\dfrac{1}{x^{1/6}}$ **43.** $y^{1/10}$ **44.** $z^{1/3}$ **45.** $\dfrac{b^{5/2}}{a^{5/4}}$ **46.** $16x^8 y^{12}$ **47.** $6a^{13/12}b$

48. $x - x^{2/3}$ **49.** The new side should be $\sqrt{5}$ times larger than the old side.   **50.** $30\sqrt{7}$ **51.** $-4\sqrt[3]{5}$ **52.** $3\sqrt[4]{2}$

53. $3x^2 y^3 z\sqrt{5y}$ **54.** $-4y^2\sqrt[3]{x^2 y}$ **55.** $(2x - 3)\sqrt{3}$ **56.** $4x^2$ **57.** $-2ab^2\sqrt[3]{5a}$ **58.** $(x - 2y)\sqrt{3x}$ **59.** $\dfrac{5}{8}$ **60.** $\dfrac{\sqrt{13}}{17}$

61. $\dfrac{3\sqrt{21}}{49}$ **62.** $\dfrac{\sqrt[3]{30}}{5}$ **63.** $\dfrac{\sqrt{30}}{6}$ **64.** $\dfrac{5a}{8b^2}$ **65.** $\dfrac{4\sqrt{5m}}{5}$ **66.** $\dfrac{z\sqrt[3]{6xy^2}}{2xy}$ **67.** $\dfrac{a\sqrt{2b}}{6b}$ **68.** The period is approximately

1.9 seconds.   **69.** $7\sqrt{5}$ **70.** $11\sqrt{11}$ **71.** $\dfrac{16}{15}\sqrt{15}$ **72.** $7\sqrt[3]{2}$ **73.** $2\sqrt{x}$ **74.** $5ab\sqrt{ab}$ **75.** $7\sqrt{2} - 7$ **76.** $6\sqrt{2a} + 2a$

77. $3 - 2\sqrt{5}$ **78.** $5x - 7y$ **79.** $x + 16\sqrt{x} + 64$ **80.** $6x + 3\sqrt[3]{x^2}$ **81.** $\sqrt{3x} - \sqrt{6}$ **82.** $1 - \dfrac{5\sqrt{z}}{z}$ **83.** $24\sqrt{5} - 48$

84. $\dfrac{x + 4\sqrt{x} + 4}{x - 4}$ **85.** -1 **86.** $\sqrt{2x} + 3$ **87.** It will require $24\sqrt{5}$ inches more material.   **88.** 12   **89.** 7   **90.** 11

91. 10   **92.** no solution   **93.** -32 **94.** 78   **95.** -39 and 25   **96.** $\dfrac{1}{4}$ **97.** $\pm 3\sqrt{3}$ **98.** -6 and 2

99. The skid mark would be approximately 152.4 feet long.   **100.** It fell from a height of 579.6 feet.   **101.** $8i$ **102.** $\dfrac{5}{7}i$ **103.** $2.5i$

104. $5i\sqrt{2}$ **105.** $\dfrac{9i\sqrt{2}}{2}$ **106.** $-20i\sqrt{2}$ **107.** $12i$ **108.** $4i$ **109.** $66i$ **110.** $4i$ **111.** $\dfrac{13}{15}$ **112.** $-30i$

113. $-2\sqrt{3} + 5i\sqrt{6}$ **114.** $4\sqrt{2}$ **115.** 4   **116.** $29 - 2i$ **117.** $11 - 4i$ **118.** $-26 - 7i$ **119.** 130   **120.** $4 + 5i$

121. $3 + 2i$ **122.** $7 + 3i$ **123.** $8\sqrt{7} + 6i\sqrt{3}$ **124.** $6 + 3i\sqrt{2}$ **125.** $\dfrac{8}{15} - \dfrac{7}{90}i$ **126.** $\sqrt{2} - 4i\sqrt{3}$

127. $z = \pm 3i$ **128.** $t = \pm 5i$ **129.** $a = \pm i\sqrt{11}$ **130.** $r = 5 \pm 6i$ **131.** $x = -2 \pm i\sqrt{3}$ **132.** $x = -\dfrac{1}{4} \pm \dfrac{3}{4}i$

133. $m = -\dfrac{2}{5} \pm \dfrac{4}{5}i$ **134.** $x = 5 \pm 2i$ **135.** $y = -\dfrac{1}{2} \pm \dfrac{3}{2}i$ **136.** $z = -8 \pm \sqrt{47}$ **137.** $x = \dfrac{2}{3} \pm 2i$ **138.** $x - 0.5 \pm 1.1i$

139. $x = \dfrac{3}{4} \pm \dfrac{1}{4}i$ **140.** $x = \dfrac{9}{2} \pm \dfrac{i\sqrt{119}}{2}$ **141.** $y = -\dfrac{1}{2} \pm \dfrac{i\sqrt{383}}{2}$ **142.** $b = -\dfrac{11}{4} \pm \dfrac{i\sqrt{71}}{4}$

143. The magnitude of the total voltage is 59.0 volts.   **144.** The magnitude of the impedance is $5\sqrt{2}$ ohms.

CHAPTER 13 CHAPTER REVIEW

1. 14   **2.** 2.1   **3.** $-\dfrac{3}{4}$ **4.** 4   **5.** not a real number   **6.** -3.072 **7.** $\sqrt{121} = 11$ **8.** $\sqrt[3]{-125} = -5$ **9.** $-\sqrt[4]{81} = -3$

10. $\sqrt[4]{-81}$, not a real number   **11.** $(\sqrt[6]{729})^5 = 243$ **12.** $(\sqrt[6]{-729})^5$, not a real number   **13.** $\dfrac{1}{(\sqrt[3]{-8})^2} = \dfrac{1}{4}$

14. $\dfrac{1}{(\sqrt[4]{-81})^3}$, not a real number   **15.** 3   **16.** 2187   **17.** $(\sqrt[3]{6})^5 \approx 19.812$ **18.** 6   **19.** $\dfrac{64}{27}$ **20.** $\dfrac{1}{(\sqrt{6})^3} \approx 0.068$ **21.** $-\dfrac{6}{7}$

22. $\dfrac{\sqrt{15}}{12}$ **23.** $\dfrac{5}{13}$ **24.** $-\dfrac{4}{5}$ **25.** $5\sqrt{6}$ **26.** $-4\sqrt[3]{7}$ **27.** $2\sqrt[4]{3}$ **28.** $-3\sqrt[5]{2}$ **29.** $5\sqrt{21}$ **30.** $-\dfrac{4}{21}\sqrt{21}$ **31.** $12\sqrt{13}$

32. $11\sqrt{7}$ **33.** $3\sqrt{2} - 3$ **34.** $60 - 2\sqrt{3}$ **35.** 5   **36.** $34\sqrt[3]{3}$ **37.** $\dfrac{1}{x^{1/12}}$ **38.** $\dfrac{1}{z^{9/10}}$ **39.** $\dfrac{q^{5/2}}{p^{5/4}}$ **40.** $81x^{12}y^{16}$

41. $6x^3yz^2\sqrt{2y}$ **42.** $-4y^2\sqrt[3]{x^2y}$ **43.** $6xy\sqrt{2y}$ **44.** $-3a^2b^2\sqrt[3]{2b}$ **45.** $(2x + 7y)\sqrt{5}$ **46.** $\dfrac{6x}{7y^2}$ **47.** $\dfrac{5\sqrt{3z}}{3}$ **48.** $\dfrac{-3\sqrt[3]{a}}{b}$

49. $\dfrac{\sqrt{2b}}{2b}$ **50.** $\dfrac{\sqrt{xy}}{xy^2}$ **51.** $2\sqrt{x}$ **52.** $14cd\sqrt{cd}$ **53.** $a + 18\sqrt{a} + 81$ **54.** $3\sqrt{6a} + 6a$ **55.** $3x - 2y$ **56.** $6x + 10x\sqrt[3]{x}$

57. $\sqrt{3x} - \sqrt{7}$ **58.** $1 - \dfrac{9\sqrt{m}}{m}$ **59.** $6(\sqrt{7} - 2)$ **60.** $\dfrac{9 + 6\sqrt{x} + x}{9 - x}$ **61.** -1 **62.** $5 + \sqrt{7x}$

63. Restricted values are all real numbers less than 0.

64. Restricted values are all real numbers less than 0.

65. 2.681   **66.** -0.429 and 1.061   **67.** 81   **68.** 45   **69.** 8   **70.** 7   **71.** 4   **72.** no solution   **73.** -72

74. 79   **75.** -248 and 238   **76.** $\dfrac{1}{9}$ **77.** $\pm 3\sqrt{3}$

78. $13i$ **79.** $\dfrac{8}{9}i$ **80.** $\dfrac{5i\sqrt{6}}{6}$ **81.** $40i\sqrt{5}$ **82.** $2.5i$

83. $-6i\sqrt{3}$ **84.** $30.85 - 4.23i$ **85.** $10i$ **86.** $4i$

87. $\dfrac{4}{3} - \dfrac{23}{18}i$ **88.** $-6 - 4.35i$ **89.** $59i$ **90.** $-24i$

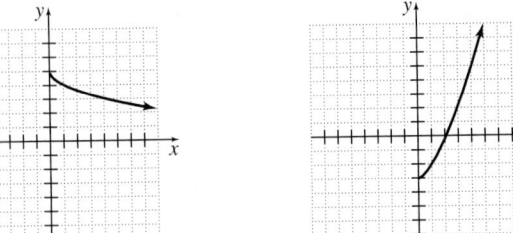

91. $\sqrt{15} + 8\sqrt{2}$ **92.** 4   **93.** $14 - 2i$ **94.** $-72 + 54i$ **95.** $-3 - 2i$ **96.** $7 + 6i$ **97.** $\dfrac{5}{3} - 6i$ **98.** $\sqrt{2} + i\sqrt{5}$

99. $\dfrac{17}{29}$ **100.** $\dfrac{53}{10} \pm \dfrac{\sqrt{1049}}{10}$ **101.** $m = -7 \pm 5i$ **102.** $y = 3 \pm i\sqrt{6}$ **103.** $x = 8 \pm i\sqrt{3}$ **104.** $y = -\dfrac{5}{3} \pm \dfrac{i\sqrt{7}}{3}$

105. $a = \pm 9i$ **106.** $b = \pm i\sqrt{13}$ **107.** $y = \dfrac{3}{2} \pm \dfrac{\sqrt{281}}{2}$ **108.** The magnitude of the current is approximately 5.8 amperes.

109. The side is 23 inches.　**110.** The variability is approximately 7.937.　**111.** It fell from a height of 257.6 feet.
112. The period is approximately 2.08 seconds.　**113.** 10 units　**114.** 6 units　**115.** 5 units　**116.** $\sqrt{58}$ units　**117.** no solution
118. -1 and 5

CHAPTER 13 TEST

1. 14　**2.** 2.141　**3.** not a real number　**4.** $-\dfrac{2}{5}$　**5.** 512　**6.** 9　**7.** $\dfrac{9}{4}$　**8.** $\sqrt{x}$　**9.** $2x - 5y$　**10.** $3x^2y\sqrt{10y}$

11. $\dfrac{z\sqrt{6xz}}{2xy}$　**12.** $\dfrac{5\sqrt{y}}{7x}$　**13.** $7\sqrt{x}$　**14.** $10x + 15\sqrt[3]{x^2}$　**15.** $\dfrac{x + 3\sqrt{x} - 4}{x - 1}$　**16.** -5　**17.** 36　**18.** 5　**19.** 3

20. Restricted values are all real numbers less than -7.

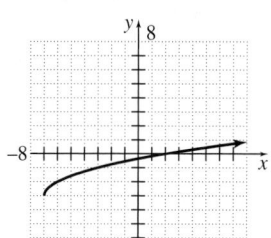

21. $16i$　**22.** $-28i$　**23.** $-\sqrt{14} - 4i\sqrt{7}$　**24.** $\dfrac{15}{17}$　**25.** 7　**26.** $-42 - 2i$　**27.** $3 + 7i$

28. $10.27 - 6.28i$　**29.** $-5\sqrt{3} + 3i\sqrt{5}$　**30.** $8 - 22i$　**31.** $x = \pm14i$　**32.** $t = -2 \pm i\sqrt{15}$

33. $z = \pm1.2i$　**34.** $x = 3 \pm 2i\sqrt{2}$　**35.** $x = \dfrac{1}{3} \pm \dfrac{i\sqrt{6}}{3}$　**36.** $y = \dfrac{3}{8} \pm \dfrac{i\sqrt{215}}{8}$　**37.** 7 units

38. $\sqrt{58}$ units　**39.** It will require $20\sqrt{5}$ inches more border.　**40.** Answers will vary.
41. The magnitude of the total voltage is approximately 97.4 volts.　**42.** Answers will vary.

CHAPTERS 1–13 CUMULATIVE REVIEW EXERCISES

1. $\dfrac{y^2}{x}$　**2.** $9xy^{3/2}$　**3.** $\dfrac{27t^3}{64s^6}$　**4.** $5x^3 + 6x^2y + 2xy^2$　**5.** $-2.8a^2 - 6.31ab + 4.4b^2$　**6.** $-13.6m^3n^3p$　**7.** $6a^2 + 10ab - 4b^2$

8. $4x^2 - 9$　**9.** $4x^2 + 12x + 9$　**10.** $-5xy^2$　**11.** $1 + \dfrac{2}{m} - \dfrac{4n}{m^2}$　**12.** $\dfrac{x - 1}{x - 3}$　**13.** $\dfrac{x^2 + x + 6}{(x + 2)(x - 2)}$　**14.** $-\dfrac{3x^2 + 22x + 42}{(x + 2)(x - 3)(x + 4)}$

15. $\dfrac{2a - 1}{2a + 5}$　**16.** $\dfrac{x + 5}{2xy(2x + 1)}$　**17.** $-3\sqrt{2y}$　**18.** $5a\sqrt[3]{a^2b^2}$　**19.** -2　**20.** $\dfrac{c}{a}\sqrt{3c}$　**21.** $\dfrac{x + 3\sqrt{x} + 2}{x - 1}$　**22.** $x^{7/6} - x^{5/4}$

23. $(4a + 5b)(4a - 5b)$　**24.** $(x - 4)(x + 2)$　**25.** $3(x - 5)(x + 2)$　**26.** -11

27.

(−2, 7.4)
(0, 1)
(2, −5.4)

Domain: all real numbers
Range: all real numbers

28.

(0, 6)
(−1, 3)
(−2, 2)
$x = -2$

Domain: all real numbers
Range: all real numbers ≥ 2

29.

$\left(\dfrac{1}{4}, \dfrac{9}{8}\right)$
(0, 1)
$\left(-\dfrac{1}{2}, 0\right)$　(1, 0)
(2, −5)
$x = \dfrac{1}{4}$

Domain: all real numbers
Range: all real numbers $\leq \dfrac{9}{8}$

30.

(0, −2)　(2, 0)

Domain: all real numbers $\neq -2$
Range: all real numbers $\neq -4$

31.

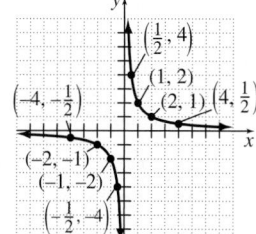

$\left(\dfrac{1}{2}, 4\right)$
(1, 2)
$\left(-4, -\dfrac{1}{2}\right)$　(2, 1)　$\left(4, \dfrac{1}{2}\right)$
(−2, −1)
(−1, −2)
$\left(-\dfrac{1}{2}, -4\right)$

Domain: all real numbers $\neq 0$
Range: all real numbers $\neq 0$

32.

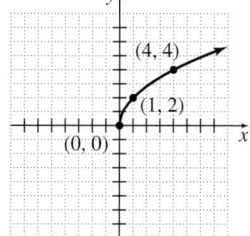

(4, 4)
(1, 2)
(0, 0)

Domain: all real numbers ≥ 0
Range: all real numbers ≥ 0

33.

(−8, 6)　(8, 6)
(−1, 3)　(1, 3)
(0, 2)

Domain: all real numbers
Range: all real numbers ≥ 2

34. 4　**35.** $\dfrac{3 \pm 3i\sqrt{3}}{2}$

36. -5 and 3　**37.** $\dfrac{-3 \pm \sqrt{41}}{4}$

38. -5　**39.** $\pm2\sqrt{7}$

40. $\dfrac{19}{2}$　**41.** $\left(-\dfrac{3}{2}, 6\right)$

42. (5, 3)　**43.** $y = -x - 2$

44. $y = \dfrac{3}{2}x + 4$ **45.** $y = x^2 + 3x - 4$ **46.** The rectangle is 12 feet by 16 feet.   **47.** The possible y-coordinates are 0 or 8.

48. Yes, they can finish raking the leaves.   **49.** The company must sell at least 1429 items in order to break even.
50. The skydiver was free-falling for about 21.65 seconds.

Chapter 14

SECTION 14.1 EXERCISES

1. $h^{-1} = \{(5, -3), (4, -2), (3, -1), (2, 0), (1, 1), (0, 2)\}$ **2.** $d^{-1} = \{(25, 5), (16, 4), (9, 3), (4, 2), (1, 1), (0, 0)\}$
3. $A^{-1} = \{(6, 3), (7, 2), (8, 1), (8, 0), (7, -1), (6, -2)\}$ **4.** $J^{-1} = \{(1, -4), (2, 3), (1, -2), (2, 1), (1, 0), (2, -1)\}$
5. $y = \dfrac{1}{2}x + 4$ **6.** $y = \dfrac{1}{3}x - 2$ **7.** $y = -\dfrac{1}{3}x + \dfrac{2}{3}$ **8.** $y = -\dfrac{1}{2}x + \dfrac{5}{2}$ **9.** $y = \dfrac{4}{3}x - 12$ **10.** $y = -\dfrac{3}{2}x + 2$ **11.** $y = 8x + 20$
12. $y = -2.5x + 4.5$ **13.** $y = \pm\sqrt{x + 2}$ **14.** $y = \sqrt[3]{x} - 1$ **15.** function; yes   **16.** not a function   **17.** not a function
18. function; yes   **19.** not a function   **20.** not a function   **21.** yes   **22.** no   **23.** no   **24.** yes   **25.** yes   **26.** yes
27. $g^{-1}(x) = \dfrac{1}{3}x + 2$ **28.** $f^{-1}(x) = -\dfrac{1}{4}x + \dfrac{1}{2}$ **29.** $y^{-1} = \dfrac{3}{2}x + 6$

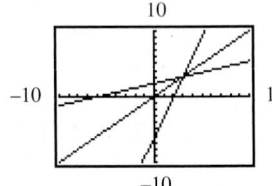   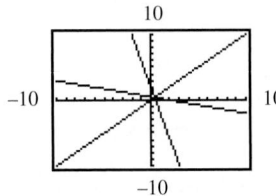   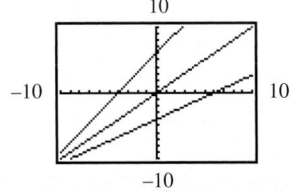

30. $y^{-1} = -\dfrac{5}{2}x + 10$ **31.** $h^{-1}(x) = \pm\sqrt{x + 1}$ **32.** $r^{-1}(x) = \pm\sqrt{2 - x}$

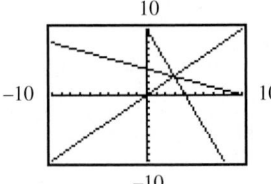   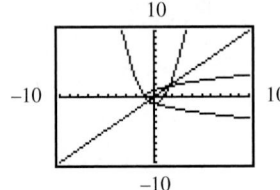   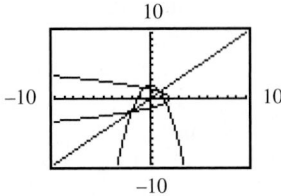

33. $y^{-1} = \sqrt[3]{3x + 12}$ **34.** $y^{-1} = \sqrt[3]{4x - 8}$ **35.** $f(x) = 1500 + 0.02x$
$\qquad\qquad\qquad\qquad\qquad\qquad\qquad\qquad\qquad\qquad\qquad f^{-1}(x) = 50x - 75{,}000$
The inverse function represents the amount of Julia's transactions in terms of her monthly income.

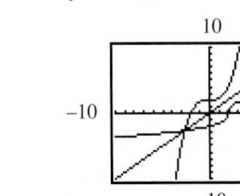   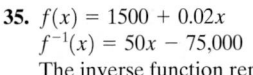

36. $f(x) = 125 + 0.15x$ **37.** $f(x) = 55x$ **38.** $f(x) = 30x$ **39.** $f(x) = 2000 + 10x$
$f^{-1}(x) = 6\dfrac{2}{3}x - 833\dfrac{1}{3}$ $f^{-1}(x) = \dfrac{x}{55}$ $f^{-1}(x) = \dfrac{x}{30}$ $f^{-1}(x) = \dfrac{1}{10}x - 200$

The inverse function represents the amount of Harvey's sales in terms of his weekly income.   The inverse function represents the time in terms of the distance traveled.   The inverse function represents the number of hours in terms of the number of pairs in jeans.   The inverse function represents the number of months in terms of the total amount to repay.

40. $f(x) = 1500 + 6x$; $f^{-1}(x) = \dfrac{1}{6}x - 250$; the inverse function represents the number of months in terms of the amount earned.

14.1 CALCULATOR EXERCISES

1. $y^{-1} = \pm\sqrt{\dfrac{x+4}{0.3}}$ **2.** $f^{-1}(x) = \sqrt[3]{\dfrac{x-2}{0.1}}$ **3.** $g^{-1}(x) = \left(\dfrac{x+1.5}{5}\right)^2$ **4.** $y^{-1} = \dfrac{1}{x-2}$

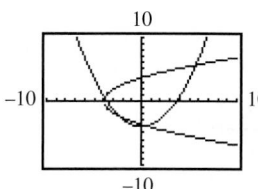 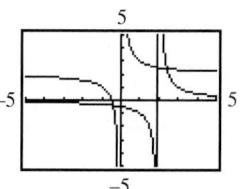

SECTION 14.2 EXERCISES

1. exponential   **2.** not exponential   **3.** exponential   **4.** exponential   **5.** not exponential   **6.** not exponential   **7.** exponential

8. exponential   **9.** not exponential   **10.** exponential   **11.** exponential   **12.** exponential   **13.** 4096   **14.** 65,536   **15.** $\dfrac{1}{256}$

16. $\dfrac{1}{16}$ **17.** 4   **18.** 8   **19.** ≈ 50.453 **20.** 1   **21.** ≈ 0.875 **22.** 0.512   **23.** ≈ 1.160 **24.** ≈ 0.689 **25.** ≈ 1.880 **26.** ≈ 0.369

27. $Y1 = 4^x$ **28.** $Y1 = 3^x$ **29.** $Y1 = 4^{-x}$

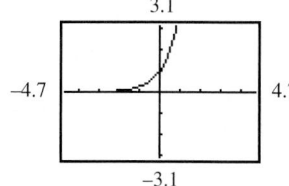 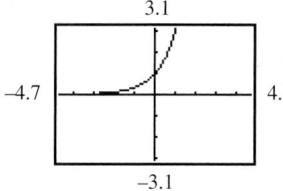 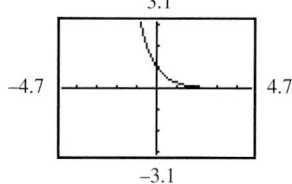

30. $Y1 = 3^{-2x}$ **31.** $Y1 = 4^{2x}$ **32.** $Y1 = 3^{2x-1}$

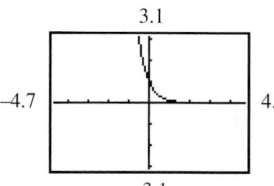 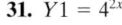 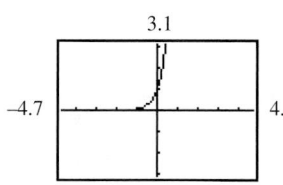 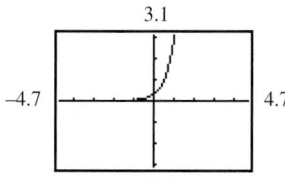

33. $Y1 = 4^{1/2x}$ **34.** $Y1 = 3^{x/4}$ **35.** $Y1 = 4^{x-1}$

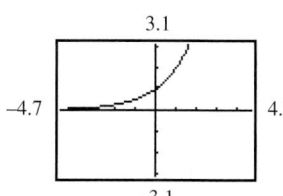 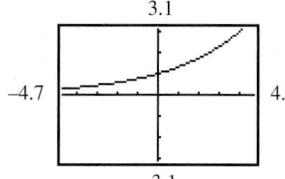 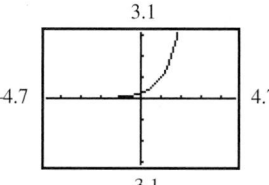

36. $Y1 = 3^x + 1$ **37.** $Y1 = 4^x - 1$ **38.** $Y1 = 3^{x+1}$

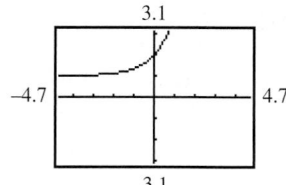 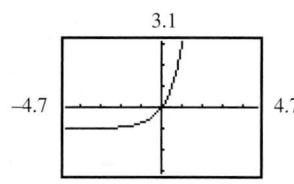 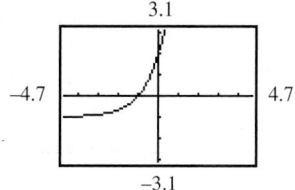

39. $Y1 = e^{1/2x}$

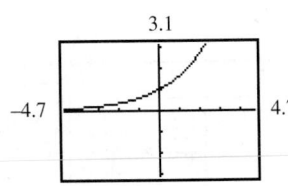

40. $Y1 = e^{0.2x}$

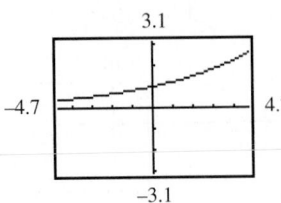

41. $Y1 = \frac{1}{2}e^x$

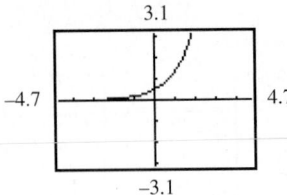

42. $Y1 = 0.2e^x$

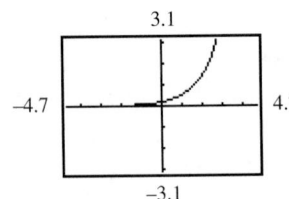

43. $Y1 = e^{-1/2x}$

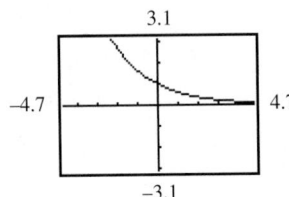

44. $Y1 = e^{-0.2x}$

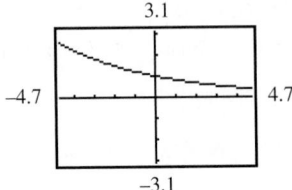

45. $Y1 = e^x + \frac{1}{2}$

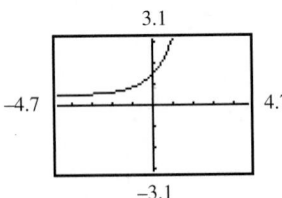

46. $Y1 = e^{0.2x} + 0.2$

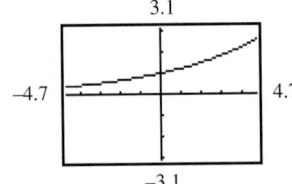

47. $Y1 = \frac{1}{2}e^{1/2x}$

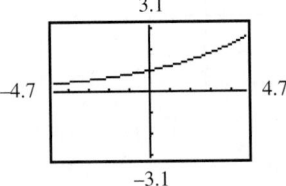

48. $Y1 = 0.2e^{0.2x}$

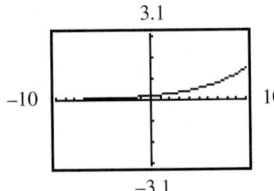

49. $A = 6000(1.055)^t$

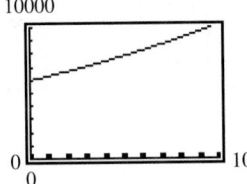

The investment was worth
$8728.08.

50. $A = 3000(1.062)^t$

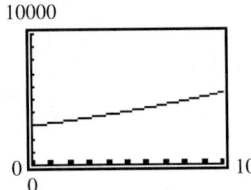

The investment was worth
$3816.10.

51. $A = 8000e^{0.048t}$

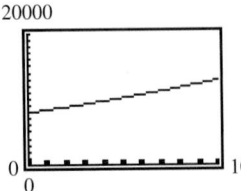

She will have earned $2169.99.

52. $A = 4500e^{0.075t}$

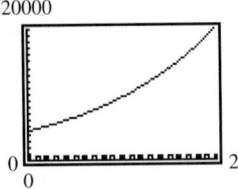

The investment will be worth
$9526.50

53.

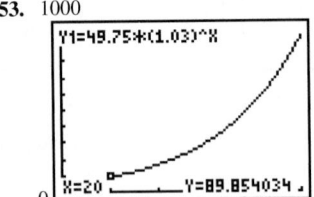

The number of subscribers will be
about 90 million in the year 2010.

54.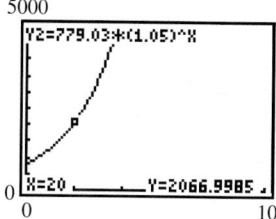

Expenditures will be approximately $2067 billion in the year 2010.

55.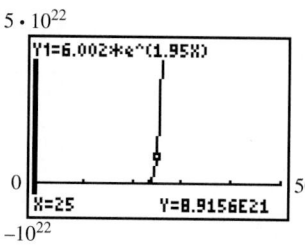

The model predicts that in the year 2015 the operating revenue will be 8.92×10^{21} million.

56.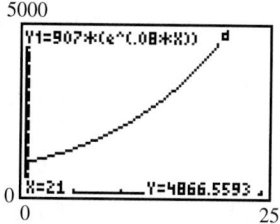

In the year 2015, the number of wireless providers will be approximately 4867.

14.2 CALCULATOR EXERCISES

1. (a)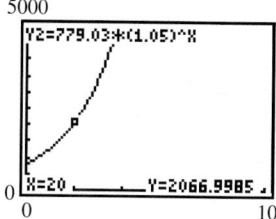

(b)

2. (a)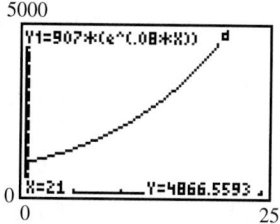

(b)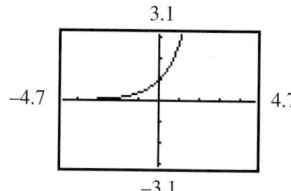

3. (a)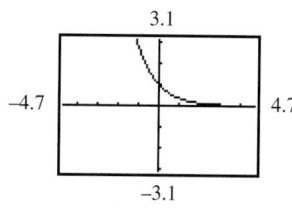

(b)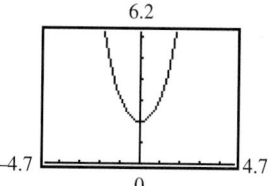

4. (a)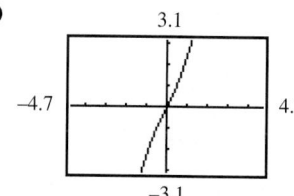

(b)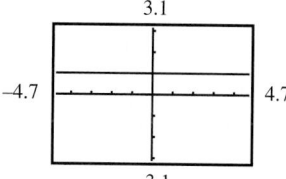

5. (a)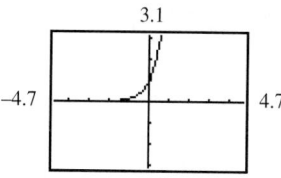

(b)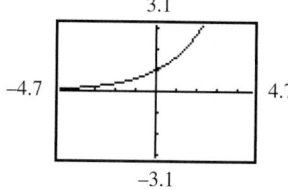

6. (a)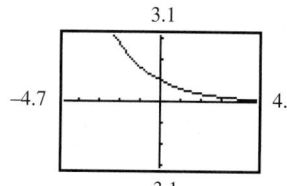

(b)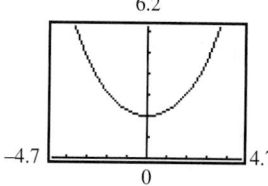

SECTION 14.3 EXERCISES

1. $f^{-1}(x) = \log_{11}x$ **2.** $h^{-1}(x) = \log_{15}x$ **3.** $g^{-1}(x) = \log_6 x$ **4.** $p^{-1}(x) = \log_{29}x$ **5.** $H^{-1}(x) = \log_k x$ **6.** $J^{-1}(x) = \log_b x$ **7.** 6
8. 7 **9.** 5 **10.** 6 **11.** 4 **12.** 3 **13.** -3 **14.** -2 **15.** -1 **16.** -3 **17.** -4 **18.** -3 **19.** 1 **20.** 4
21. -4 **22.** -6 **23.** 3 **24.** 7 **25.** -5 **26.** -8 **27.** -5 **28.** -7 **29.** ≈ 1.176 **30.** ≈ 1.362 **31.** ≈ -1.079
32. ≈ -1.279 **33.** ≈ 0.130 **34.** ≈ 1.097 **35.** ≈ 2.639 **36.** ≈ 3.296 **37.** ≈ 1.047 **38.** ≈ 2.451 **39.** ≈ -1.609 **40.** ≈ -2.708
41. ≈ 1.792 **42.** ≈ 2.070 **43.** ≈ 3.322 **44.** ≈ 3.322 **45.** ≈ 0.657 **46.** ≈ 2.447 **47.** ≈ -0.252 **48.** ≈ -0.465 **49.** ≈ 1.302
50. ≈ 1.984 **51.** ≈ 1.616 **52.** ≈ 0.679 **53.** ≈ -1.209 **54.** ≈ -0.442

55. $Y1 = \dfrac{\log x}{\log 5}$

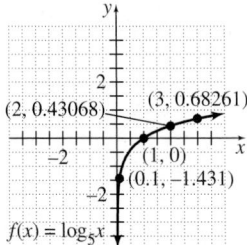

56. $Y1 = \dfrac{\log x}{\log 7}$

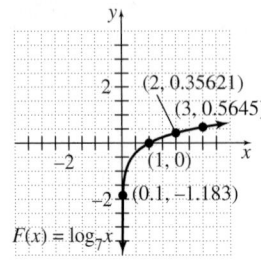

57. $Y1 = \log(x + 2)$

X	Y1
-1.99	-2
-1.5	-.301
-1	0
0	.30103
1	.47712
2	.60206
3	.69897

Y1◻log(X+2)

(graph, continued) $g(x) = \log(x + 2)$ with points (0, 0.30103), (2, 0.60206), (−1, 0), (−1.99, −2)

58. $Y1 = \log(x - 2)$

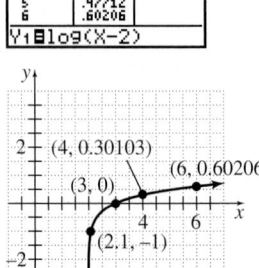

59. $Y1 = \ln(x + 2)$

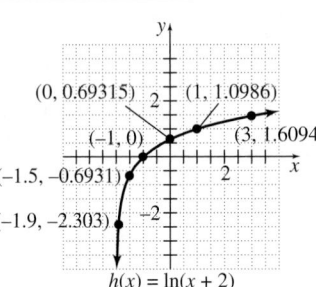

60. $Y1 = \ln(x - 2)$

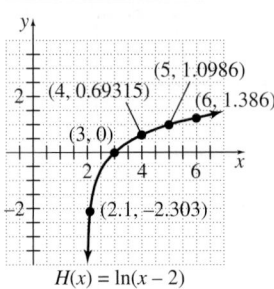

61. The hydrogen ion concentration is 0.0000006 mole per liter.   **62.** The hydrogen ion concentration is 0.001 mole per liter.
63. The pH is approximately 8.5.   **64.** The pH is approximately 6.4.   **65.** The larger earthquake's intensity was 63095.7 times as great as the smaller earthquake's intensity.   **66.** The larger earthquake's intensity was 39.8 times as great as the smaller earthquake's intensity.

14.3 CALCULATOR EXERCISES

1.
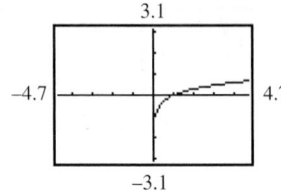

2.
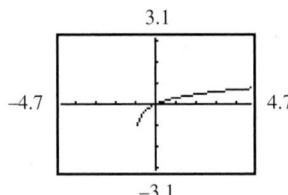

3.
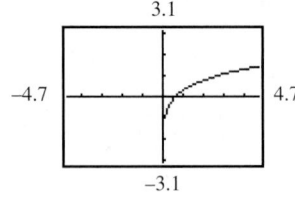

4.
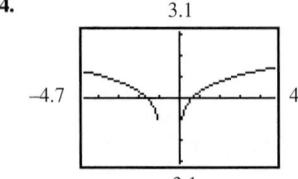

5.
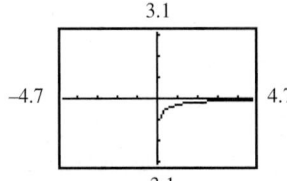

6.
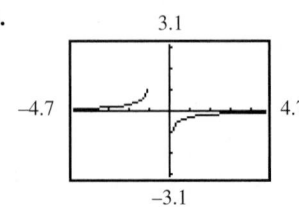

SECTION 14.4 EXERCISES

1. $\log 12 + \log a$ **2.** $\ln 25 + \ln c$ **3.** $3 \ln x$ **4.** $3 \log z$ **5.** $\log_5 x - 1$ **6.** $1 - \log_7 y$

7. $\log 2 + 2 \log x - \log y$ **8.** $2 \log x - \log 5 - 3 \log y$ **9.** $3 \log_3 x + 2 \log_3 y$ **10.** $2 \log_2 a + 4 \log_2 b$

11. $\dfrac{1}{3}\ln x + \dfrac{2}{3}\ln y$ **12.** $\dfrac{3}{4}\log x + \dfrac{1}{4}\log y$ **13.** $\dfrac{1}{2}\log 2 + \dfrac{1}{2}\log x - \dfrac{1}{3}\log y$

14. $\dfrac{1}{3}\ln 3 + \dfrac{2}{3}\ln x - \dfrac{1}{2}\ln 2 - \dfrac{1}{2}\ln y$ **15.** $1 + \log_3 a$ **16.** $1 + \log_5 z$ **17.** $\log_5 10 + \log_5 x + \log_5 y$

18. $\log_3 x + \log_3 y - \log_3 6$ **19.** $1 + 2 \log_a b$ **20.** $\log_6 3 + \log_6 a + 2 \log_6 b$ **21.** $\log x(x + 5)$

22. $\log (x^2 - 1)$ **23.** $\ln x^2 y^3$ **24.** $\ln \dfrac{b^3}{c^2}$ **25.** $\log_3 \dfrac{(x + 3)^2}{x - 1}$ **26.** $\log_5(a - 5)^5 a^2$ **27.** $\ln \dfrac{\sqrt{x}}{\sqrt[5]{x + 1}}$

28. $\log \dfrac{\sqrt[4]{z}}{\sqrt[3]{z + 5}}$ **29.** $\log \dfrac{y}{z}$ **30.** $\log pq^2$ **31.** The power gain is approximately 11.249 decibels.

32. The power gain is approximately 11.761 decibels.   **33.** The depth is approximately 15.033 units.

34. The depth is approximately 17.918 units.   **35.** It will take approximately 22 years.

36. It will take approximately 15.4 years.   **37.** It will take approximately 15.4 years.   **38.** It will take approximately $12\dfrac{1}{2}$ years.

14.4 CALCULATOR EXERCISES

1.
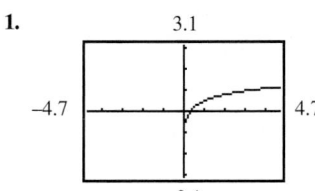

2.
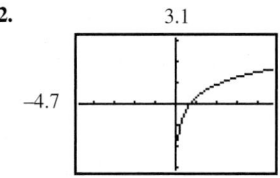

3.
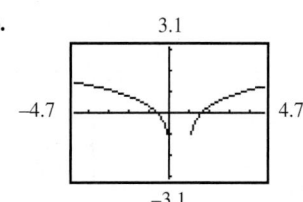

4.
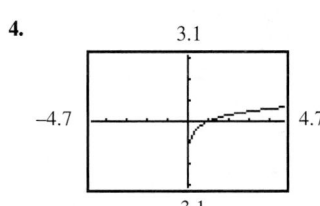

5. no   **6.** no

7. The logarithm of a sum is not equivalent to the sum of the logarithms.

8. The logarithm of a difference is not equivalent to the difference of the logarithms.

SECTION 14.5 EXERCISES

1. logarithmic   **2.** exponential   **3.** neither   **4.** logarithmic   **5.** exponential   **6.** neither   **7.** neither   **8.** neither

9. 2   **10.** 2   **11.** ≈ 0.693 **12.** ≈ 0.231 **13.** 0   **14.** ≈ -0.631 **15.** ≈ 0.405 **16.** ≈ -0.693 **17.** 1   **18.** $\dfrac{1}{2}$

19. 3   **20.** 4   **21.** $\dfrac{1}{2}$ and 3   **22.** 1 and 4   **23.** -2 and 5   **24.** -5 and $\dfrac{3}{2}$ **25.** no solution   **26.** no solution

27. all real numbers   **28.** all real numbers   **29.** 2   **30.** 11   **31.** $-\dfrac{1}{2}$ **32.** 5   **33.** -3 and $\dfrac{1}{2}$ **34.** -2 and $\dfrac{1}{3}$ **35.** ≈ 2.773

36. ≈ 0.602 **37.** ≈ 1.984 **38.** ≈ 1.228 **39.** 3125   **40.** 16   **41.** 1   **42.** 10   **43.** 1   **44.** ≈ 7.389 **45.** 9   **46.** 625

47. $\dfrac{1}{125}$ **48.** $\dfrac{1}{81}$ **49.** 7   **50.** 5   **51.** ≈ 1.649 **52.** ≈ 0.135 **53.** $\dfrac{1}{100}$ **54.** 100   **55.** 1   **56.** ≈ 2.718 **57.** 3

58. 5   **59.** 8   **60.** 11   **61.** all real numbers greater than 0   **62.** all real numbers greater than 0   **63.** no solution   **64.** 0

65. 4   **66.** $\dfrac{3}{4}$ **67.** $\dfrac{2}{3}$ **68.** 2   **69.** 2   **70.** 7

71. $A(t) = 218e^{0.051t}$

In the year 2010, the estimated expenditures will be about $604.6 billion.

72. $A(t) = 121e^{0.011t}$

In the year 2010, the male population will be about 150.8 million.

73. It will double its value in the year 2007.   **74.** It will be 1.5 times its value in the year 2000.   **75.** The decay factor ≈ -0.154;

it will take approximately 0.131 billion years.   **76.** It will take approximately 0.684 billion years.

14.5 CALCULATOR EXERCISES

Students should read about graphing techniques.

CHAPTER 14 SECTION-BY-SECTION REVIEW

Reflections

1–7. Answers will vary.

Exercises

1. $h^{-1} = \{(4.5, 2), (3.5, 4), (2.5, 6), (1.5, 8), (0.5, 10)\}$ **2.** $y^{-1} = \frac{1}{3}x + 3$ **3.** $y^{-1} = \pm\sqrt{x-1}$ **4.** $y^{-1} = \sqrt[3]{x+1}$ **5.** not a function

6. function   **7.** no   **8.** yes   **9.** no   **10.** $f^{-1}(x) = \frac{4}{3}x + 4$ **11.** $y^{-1} = \sqrt[3]{x-8}$

12. $f(x) = 570 + 0.03x$;
$f^{-1}(x) = 33.\overline{3}x - 19{,}000$, which represents the value of all sales in terms of Motomo's weekly salary.

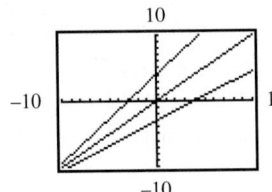
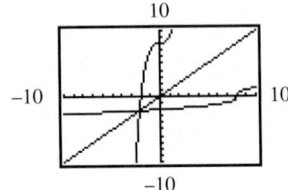

13. $f(x) = 15x$; $f^{-1}(x) = \dfrac{x}{15}$, which represents the number of hours in terms of the number of problems solved.

14. exponential   **15.** not exponential   **16.** 1.4641   **17.** ≈ 0.826 **18.** 1   **19.** 1.1   **20.** ≈ 1.059 **21.** ≈ 1.309

22. 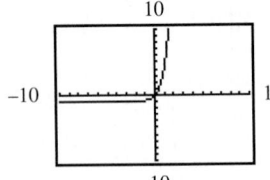   **23.** 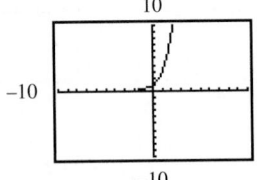   **24.**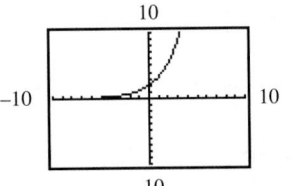

25. $A = 1000(1.04)^t$

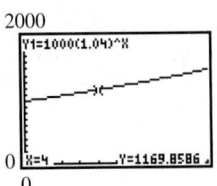

26. The estimated expenditures predicted for the year 2010 will be about $1060.1 billion.

27. $h^{-1}(x) = \log_8 x$ **28.** $A^{-1}(x) = \frac{1}{k}\ln\frac{x}{A}$ **29.** 2   **30.** 4   **31.** -3 **32.** ≈ 0.176

33. 4   **34.** ≈ 2.303 **35.** -2 **36.** ≈ -0.222 **37.** ≈ 1.683

38. $Y1 = \dfrac{\log x}{\log 3}$

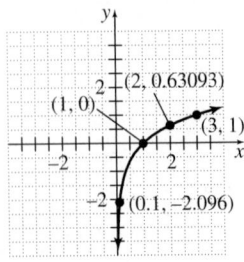

39. $Y1 = \ln(x-2)$

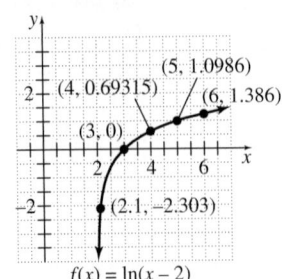

$f(x) = \ln(x-2)$

40. The pH is approximately 2.208.
41. The hydrogen ion concentration is 3.98×10^{-8} mole per liter.
42. $4\log 2 + 4\log x$ **43.** $1 + \log_7 x - \log_7 y$
44. $\ln 25 + 2\ln x + 3\ln y + \ln z$
45. $\frac{1}{3}\ln 2 + \frac{2}{3}\ln x + \frac{1}{3}\ln y - \frac{1}{2}\ln y - \frac{1}{2}\ln z$

46. $\log (x^2 - 9)$ **47.** $\ln\dfrac{x^5}{y^3}$ **48.** $\log\dfrac{\sqrt[3]{x}}{\sqrt{y}}$ **49.** $\log ad$

50. It will take approximately 18.3 years.   **51.** exponential
52. neither   **53.** logarithmic   **54.** neither
55. exponential   **56.** neither   **57.** 4   **58.** ≈ -0.322

59. 0   **60.** $\dfrac{3}{2}$ **61.** no solution   **62.** -3 and 1

63. all real numbers   **64.** ≈ 0.262 **65.** 2.569×10^{41}

66. 19,683   **67.** ≈ 2.718 **68.** 1   **69.** $\dfrac{1}{1000}$ **70.** $\dfrac{1}{4}$

71. ≈ 1.396 **72.** -5 and $\dfrac{3}{2}$ **73.** 5   **74.** no solution

75. all real numbers greater than 0   **76.** $A(t) = 3.022e^{0.036t}$ The model projects that in 2010 the resident population for those 85 years and older will be about 6.208 million. This is higher than the given number.   **77.** The time since the skeleton's demise is 3351 years.

CHAPTER 14 CHAPTER REVIEW

1. -3 **2.** 3   **3.** -6 **4.** ≈ 0.708 **5.** -3 **6.** ≈ 4.605 **7.** 1   **8.** ≈ -0.560 **9.** ≈ 3.262 **10.** $\log 3 + 2\log x + \log y$

11. $2 + \log_3 a - \log_3 b$ **12.** $\ln 100 + 3\ln p + 2\ln q + \ln r$ **13.** $\frac{1}{5}\log 6 + \frac{3}{5}\log x - \frac{1}{2}\log x - \frac{1}{2}\log y$ **14.** $\log(x - 3)$ **15.** $\log \dfrac{c^2}{d^5}$

16. $\ln \dfrac{\sqrt{x}}{y^3}$ **17.** $\log 6y^2$ **18.** $y^{-1} = 5 - x$ **19.** $y^{-1} = \pm\sqrt{1 - x}$ **20.** $y^{-1} = \sqrt[3]{4 - x}$ **21.** $y^{-1} = -\dfrac{5}{4}x + \dfrac{5}{2}$

22. $m^{-1}(x) = \log_5 x$ **23.** $G^{-1}(x) = \log_b\left(\dfrac{x}{a}\right)$ **24.** ≈ 1.088 **25.** $\dfrac{13}{4}$ **26.** 2   **27.** $1\dfrac{2}{3}$ **28.** ≈ 1.296 **29.** ≈ 5.074

30.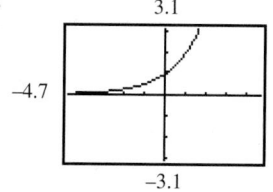

31.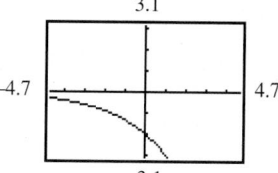

32.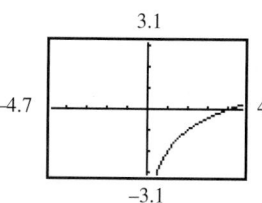

33.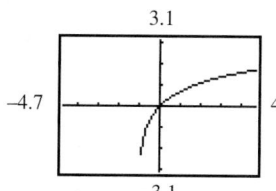

34.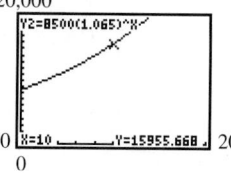

35. 6^{216} **36.** 27   **37.** 10   **38.** 1

39. $\dfrac{1}{20{,}000}$ **40.** 243   **41.** ≈ 3.162

42. -3 and $\dfrac{3}{2}$ **43.** -1 **44.** no solution

45. all real numbers greater than 0   **46.** 6

47. 0   **48.** ≈ 0.514 **49.** $-\dfrac{5}{3}$ **50.** no solution   **51.** -1 and 4   **52.** all real numbers   **53.** ≈ 0.530

54. $f(x) = 1200 + 0.025x$; $f^{-1}(x) = 40x - 48{,}000$; the inverse function represents her total sales in terms of her monthly income.

55. $f(x) = 95x$
$f^{-1}(x) = \dfrac{x}{95}$;
The inverse function represents the number of hours in terms of the distance traveled.

56. $A = 8500(1.065)^t$

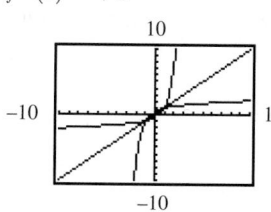

The investment was worth $15,955.67.

57. The model predicts that about 94.4 million households will have cable TV in the United States in 2010. It does not seem reasonable that the same rate of growth would occur between 1990 and 2010.
58. The pH is approximately 3.796.
59. The hydrogen ion concentration is approximately 2.5×10^{-2} mole per liter.
60. It will take approximately 12.4 years.
61. $A(t) = 218e^{0.057t}$ In the year 2005, the estimated revenue would be about $513 billion. In 2010, it would be about $682 billion.
62. It has been about 7573 years since the organism died.

CHAPTER 14 TEST

1. $f(x) = x^3$
$f^{-1}(x) = \sqrt[3]{x}$

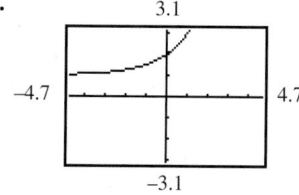

2.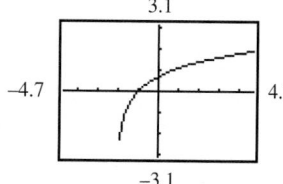

3.

4. $F^{-1}(x) = \dfrac{1}{2}x + 4$ **5.** (a) $\dfrac{1}{4}$ (b) 1   (c) 256   (d) ≈ 1.516 (e) ≈ 11.036 **6.** $\log 3 + 2\log x + 3\log y + \log z$

7. $2 + 2 \log_3 a - \log_3 b$ **8.** $\log x^2(x - 4)$ **9.** $\ln \dfrac{\sqrt{x}}{y^2}$ **10. (a)** 1   **(b)** 1   **(c)** 1   **11.** $\dfrac{5}{3}$ **12.** ≈ 4.771

13. all real numbers greater than 0   **14.** ≈ 0.732 **15. (a)** $f(x) = 400 + 0.03x$ **(b)** $f^{-1}(x) = 33.\overline{3}x - 13{,}333.\overline{3}$ **(c)** The inverse function represents the value of weekly sales in terms of his weekly income.

16. $f(x) = 2000(1.05)^x$; the investment was worth approximately \$5306.60.   **17.** It will take approximately 9.24 years.

18. $A(t) = 19614e^{0.047t}$ The estimated personal income per capita will be about \$50,211 in 2010.   **19.** Answers will vary.

CHAPTERS 1–14 CUMULATIVE REVIEW EXERCISES

1. $\dfrac{y^2}{25}$ **2.** $\dfrac{4y^4}{x^6}$ **3.** $16x^2$ **4.** $-0.8a^2 - 5.2ab + 2.7b^2$ **5.** $-9x^2y^3z$ **6.** $9x^2 - 4y^2$ **7.** $4x^2 + 6x - 18$

8. $x^2 - 8x + 16$ **9.** $-5xy^3z$ **10.** $r + 2 - \dfrac{4s}{r}$ **11.** $\dfrac{x + 2}{x + 3}$ **12.** $\dfrac{x^2 + 2x + 5}{(x + 2)(x - 2)}$ **13.** $-\dfrac{(x + 2)(x + 1)}{(x + 4)(x - 2)}$

14. 1   **15.** $\dfrac{4(y - 5)}{xy^2(y - 3)}$ **16.** $6\sqrt{2x}$ **17.** $4ab\sqrt[3]{2b}$ **18.** -1 **19.** $\dfrac{2}{x}\sqrt{yz}$ **20.** $\dfrac{x + 7\sqrt{x} + 12}{x - 9}$

21. $x^{17/12} - x$ **22.** $(5m + 6n)(5m - 6n)$ **23.** $2(x - 4)(x + 2)$ **24.** does not factor

25.

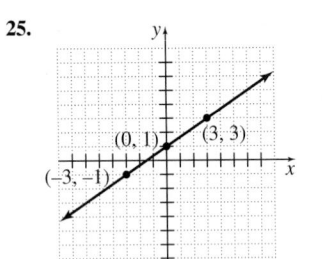

Domain: all real numbers
Range: all real numbers

26.

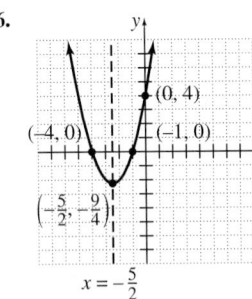

$x = -\dfrac{5}{2}$

Domain: all real numbers

Range: all real numbers $\geq -\dfrac{9}{4}$

27.

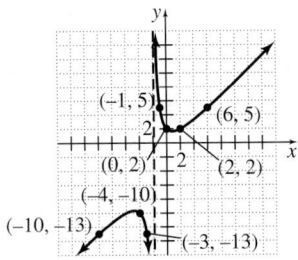

Domain: all real numbers $\neq -2$
Range: all real numbers ≤ -9.66 or ≥ 1.66

28.

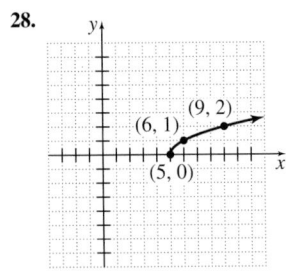

Domain: all real numbers ≥ 5
Range: all real numbers ≥ 0

29.

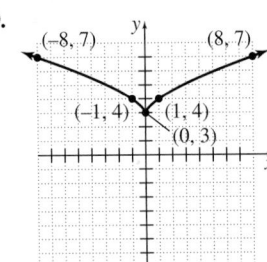

Domain: all real numbers
Range: all real numbers ≥ 3

30.

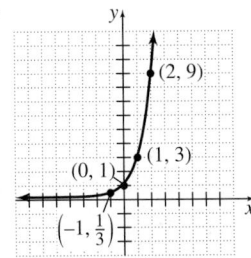

Domain: all real numbers
Range: all real numbers > 0

31.

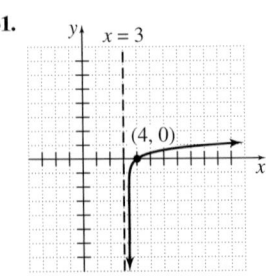

Domain: all real numbers > 3
Range: all real numbers

32. -6 **33.** no solution   **34.** 6 and -4 **35.** $\dfrac{5}{2} \pm \dfrac{\sqrt{11}}{2}i$ **36.** $-\dfrac{5}{3}$ **37.** 8   **38.** 2

39. $\sqrt{5}$ **40.** $(9, \infty)$ **41.** $(-\infty, -1 - \sqrt{3}] \cup [-1 + \sqrt{3}, \infty)$ **42.** $(4, 4)$ **43.** $y = -\dfrac{3}{5}x + \dfrac{4}{5}$

44. $y = 4x + 9$ **45.** $y = \dfrac{3}{2}x^2 + \dfrac{9}{2}x - 6$ **46.** $f^{-1}(x) = \dfrac{3}{2}x - 9$

47. The company must sell 53 books in order to break even.

48. The dimensions are 9 feet by 12 feet.

49. It will take the drains about 2.5 hours to drain the tank.

50. It will take about 24.4 years for the money to triple.

Answers to Checkup Exercises

Chapter 1

SECTION 1.1

1.1.1 Checkup

1. $15, \frac{6}{3}$ (or 2), 1 billion   **2.** all of the numbers   **3.** $15, -3, 0, \frac{6}{3}$ (or 2), 1 billion, -180 **4.** $15, 0, \frac{6}{3}$ (or 2), 1 billion

1.1.2 Checkup

1.

1.1.3 Checkup

1. $>$ **2.** $=$ **3.** $>$ **4.** $15 > 5$ **5.** $-2 < 6$ **6.** $\frac{11}{4} = 2.75$ **7.** $-3 \le 0 < 4$

1.1.4 Checkup

1. 15   **2.** 3.3   **3.** $\frac{2}{7}$

1.1.5 Checkup

1. $-3\frac{1}{3}$ **2.** $\frac{1}{2}$ **3.** 15   **4.** -35 **5.** $-\frac{4}{7}$

1.1.6 Checkup

1. $|-80| = 80$ **2.** -0.6% **3.** $-20; -26$

SECTION 1.2

1.2.1 Checkup

1. 34   **2.** -5.75 **3.** 4.5   **4.** -4 **5.** $\frac{1}{28}$

1.2.2 Checkup

1. -24 **2.** 34   **3.** -11.75 **4.** -4.5 **5.** $-\frac{41}{28}$ **6.** 4

1.2.3 Checkup

1. 7   **2.** $-\frac{3}{10}$ **3.** 22.2

1.2.4 Checkup

1. $897.63 + 355 + 572 - 120 - 300 - 185.23 - 104.50 - 231.97 - (-231.97) - 10$; the current balance in Beverly's account is $1104.90.

2. The December precipitation was $\frac{21}{100}$ inches lower than the average.

SECTION 1.3

1.3.1 Checkup

1. 72   **2.** $\frac{1}{4}$ **3.** -0.144 **4.** -3

A-119

1.3.2 Checkup

1. (a) 3 **(b)** −3 **(c)** $\frac{3}{5}$ **2. (a)** −4.1 **(b)** 36 **(c)** 0 **3. (a)** $\frac{4}{9}$ **(b)** $-\frac{27}{16}$ **(c)** undefined

1.3.3 Checkup

1. −48 **2.** $-\frac{64}{21}$ **3.** 0 **4.** 125 **5.** −11.76

1.3.4 Checkup

1. 6 **2.** 31

1.3.5 Checkup

1. The rate of transmission is 175 miles/hour. You can send a message 437.5 miles. **2.** The size of the black population is about 35,427,786 people. **3.** The density of Australia is about 6.46 people per square mile. **4.** The rate is 7.5 cases/nurse. The ward can handle 90 cases.

SECTION 1.4

1.4.1 Checkup

1. 1.69 **2.** 0 **3.** $-\frac{32}{3125}$ **4. (a)** 36 **(b)** −36 **(c)** −216 **(d)** −216 **5. (a)** 7 **(b)** −7 **(c)** −7 **(d)** 1 **(e)** 1 **(f)** −1

1.4.2 Checkup

1. (a) 7 **(b)** 0.9 **(c)** $\frac{5}{6}$ **(d)** −4 **(e)** not a real number

2. (a) 4 and 5; 4.123105626 **(b)** −3 and −4; −3.872983346

3. (a) 5 **(b)** −4 **(c)** $\frac{2}{5}$ **(d)** 2.080

1.4.3 Checkup

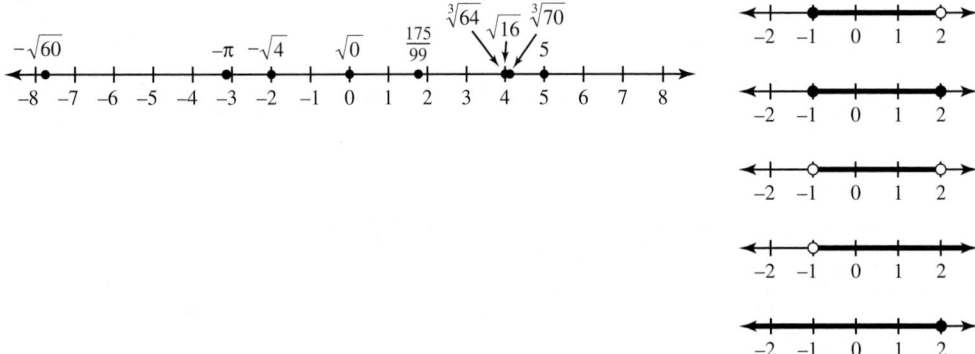

1.4.4 Checkup

1. The length of an edge of Rubik's Revenge is about 7.34 centimeters; the length of an edge of the center square is about 3.67 centimeters; the volume of the center cube is 49.375 cubic centimeters. **2.** The measure of each edge is about 37.56 feet. **3.** The area of the Go board is 292.41 square inches.

SECTION 1.5

1.5.1 Checkup

1. $\left(\frac{1}{3}\right)^1 = \frac{1}{3}$ **2.** $3^2 = 9$ **3.** $\left(-\frac{1}{3}\right)^1 = -\frac{1}{3}$ **4.** $(-3)^2 = 9$

1.5.2 Checkup

	Standard Notation	Scientific Notation	Calculator Notation
1.	65,000,000	6.5×10^7	6.5E7
2.	-0.000312	-3.12×10^{-4}	$-3.12\text{E-}4$
3.	$-83,300,000$	-8.33×10^7	$-8.33\text{E}7$
4.	0.0093	9.3×10^{-3}	$9.3\text{E-}3$
5.	0.00365	3.65×10^{-3}	$3.65\text{E-}3$
6.	$-2,340$	-2.34×10^3	$-2.34\text{E}3$

1.5.3 Checkup

1. 299,792,458; 300 million meters per second would be a useful approximation.
2. $(6.75 \times 10^4)(2) = 1.35 \times 10^5$; it would take about 1.35×10^5 calories to change 250 g of water at 100°C to steam at 100°C.
3. The approximate mass of the Moon is 7.3×10^{22} tons.

SECTION 1.6

1.6.1 Checkup

1. $5(3) + 5(7)$ **2.** $17(5) - 25(5)$ **3.** $\dfrac{-46}{2} + \dfrac{62}{2}$ **4.** $5 - 9$ **5.** $5(1.2 + 1.8) = 15$ **6.** $(17 - 23)\dfrac{1}{3} = -2$

1.6.2 Checkup

1. 3   **2.** -97 **3.** 2   **4.** -4 **5.** -65

1.6.3 Checkup

1. His weight is 170.72 pounds.   **2.** The average is about 1,066,966 short tons.

Chapter 2

SECTION 2.1

2.1.1 Checkup

1. Let b = length of base; h = height; $\dfrac{1}{2}bh$ **2.** Let p = previous price; $p + 100$ **3.** Let n = a number; $8n - 6$

2.1.2 Checkup

1. (a) 5   **(b)** 4   **(c)** $\dfrac{10}{9}$
2. (a) -24 **(b)** 24   **(c)** -24 **(d)** 288   **(e)** -288

2.1.3 Checkup

1. $110 \cdot d = 110 \cdot 3.5 = 385$; The cost of the generator is \$385.

2. Let l = length; w = width; $\sqrt{l^2 + w^2}$; $\sqrt{100^2 + \left(\dfrac{160}{3}\right)^2} = \dfrac{340}{3} = 113\dfrac{1}{3}$; the diagonal is $113\dfrac{1}{3}$ yards.

SECTION 2.2

2.2.1 Checkup

1. terms—3; variable terms—$3y^2$, $9y$; constant terms—8; coefficients—3, 9, 8; like terms—none
2. terms—2; variable terms—$3p(p + 8)$, $-5(p + 8)$; constant terms—none; coefficients—3, -5; like terms—none
3. terms—6; variable terms—$3x^2$, $3xy$, $6y$, $-5xy$, $7y$; constant terms—-2; coefficients—3, 3, 6, -5, 7, -2; like terms—$3xy$, $-5xy$; $6y$, $7y$

2.2.2 Checkup

1. $2x^3 + 2x^2 - 4x$ **2.** $6xy - 2yz$ **3.** $\dfrac{25x}{12} + \dfrac{13y}{12}$

2.2.3 Checkup

1. $50p + q$ **2.** $-18a - 14b - 4c$ **3.** $20p + 204$ **4.** $7x + 53$

2.2.4 Checkup

1. Her profit for the month is $415x - 225$ dollars; her profit for six paintings is \$2265.
2. Let l = length; w = width; $2(l + w) = 2l + 2w$; the perimeter is 284 inches.

SECTION 2.3

2.3.1 Checkup

1. equation   **2.** expression   **3.** expression   **4.** equation

2.3.2 Checkup

1. yes   **2.** no

2.3.3 Checkup

Let n = number of cartridges; $15n + 1500 = 25n$

2.3.4 Checkup

1. Let x = number of students moved, $49x = 141$ **2.** $R = \dfrac{V}{I}$

SECTION 2.4

2.4.1 Checkup

1. The exact area is 64π square yards, or approximately 201.1 square yards.
2. The exact circumference is 16π yards, or approximately 50.3 yards.

2.4.2 Checkup

1. The surface area is 5329π square feet, or approximately 16,741.5 square feet.

2.4.3 Checkup

1. The third angle measures $55°$.   **2.** The second angle measures $31°$.

2.4.4 Checkup

1. The temperature at the center of the Earth's core could be $3982°C$.   **2.** The temperatures on the Moon range from $273°F$ to $-274°F$.
3. The simple interest is \$1700.   **4.** The distance is equal to 8760 ft, or 1.66 miles.

2.4.5 Checkup

1. The equatorial circumference of Mercury is approximately 15,331 km.
2. The volume is equal to 1537.7 ft^3, and the surface area is equal to 885.4 ft^2.

Chapter 3

SECTION 3.1

3.1.1 Checkup

1.

F	86	77	68	59	50	41	32
C	30	25	20	15	10	5	0

2.

b	-3	$-\frac{1}{6}$	0	$\frac{2}{3}$	2	3	5
a	25	8	7	3	-5	-11	-23

3.1.2 Checkup

1. Temperature ordered pairs: (86, 30), (77, 25), (68, 20), (59, 15), (50, 10), (41, 5), (32, 0); ordered pairs for (b, a): $(-3, 25)$, $\left(-\dfrac{1}{6}, 8\right)$, $(0, 7)$, $\left(\dfrac{2}{3}, 3\right)$, $(2, -5)$, $(3, -11)$, $(5, -23)$

3.1.3 Checkup

1. Domain is {5, 10, 15}; range is {15, 30, 45}.
2. Domain is {0, 1, 2, 3, …}, or the set of whole numbers; range is {4, 6, 8, 10, …}, or the set of all even integers 4 or larger.
3. Domain is {0.5, 1.5, 2.5}; range is {15.5, 16.5, 17.5}.

3.1.4 Checkup

1. (a) $r = 410 + 131n$

(b)

n	2	3	4	5
r	672	803	934	1065

(c) $(2, 672), (3, 803), (4, 934), (5, 1065)$

2. If the height is $h = \{29.5, 38.5, 47.5\}$ inches, the corresponding volume is $V = \{8142, 10{,}626, 13{,}110\}$ cubic inches.

3. (1990, 99,200), (1991, 106,800), (1992, 107,400), (1993, 109,700), (1994, 113,600), (1995, 113,800), (1996, 121,700), (1997, 124,100)

SECTION 3.2

3.2.1 Checkup

1.

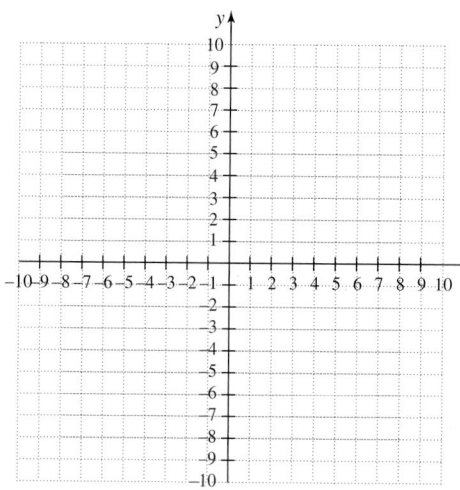

The graph corresponds to the setting of ZStandard.

ZOOM 6

2.

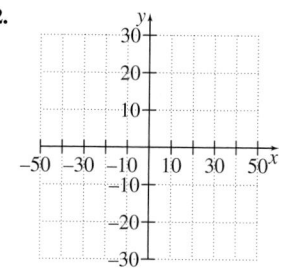

The graph corresponds to the setting of ZInteger.

ZOOM 6 ZOOM 8 ENTER

3.2.2 Checkup

1–2.

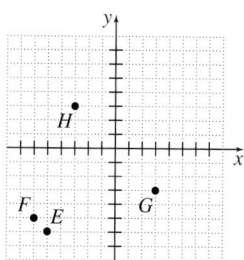

3. Answers may vary.
4. $A(-5, -2), B(5, 5), C(-1, 1), D(0, 4), E(4, -3), F(-3, 0)$
5. (a) II **(b)** I **(c)** IV **(d)** III **(e)** origin **(f)** x-axis **(g)** y-axis

6.

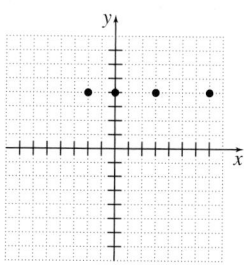

7.

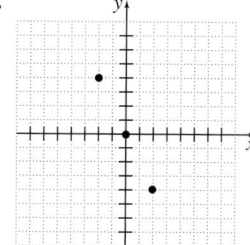

8.

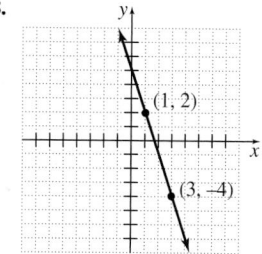

9.

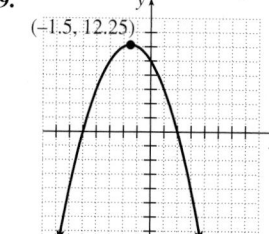

3.2.3 Checkup

1. Domain is $\{-3\}$; range is all real numbers.
2. Domain is all real numbers ≥ -4; range is all real numbers.
3. Domain is all real numbers between -5 and 5 inclusive; range is all real numbers between -5 and 5 inclusive.

3.2.4 Checkup

1. (a)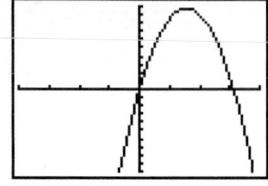

(b) Domain is from 0 to 15.625; range is all real numbers between 0 and 977 approximately.
2. Domain is $\{1, 2, 3, 4, 5, 6, 7\}$ or day of the week; range is $\{60, 65, 68, 70, 64, 55, 50\}$ or number of nondefective parts produced each day.

SECTION 3.3

3.3.1 Checkup

1. function **2.** function **3.** not a function **4.** not a function

3.3.2 Checkup

1. function **2.** function **3.** not a function **4.** not a function

3.3.3 Checkup

1. $f(4)$ is not a real number. $f(-4) = -8$
2. $h(1) = -3; h(0) = -8; h(-1) = -17$
3. $g(b) = 4b - 8; g(1) = -4; g(b + 1) = 4b - 4$

3.3.4 Checkup

1. (a) $c(d) = 5 + 10d$ **(b)** $c(4) = 45$
2. The amount of income tax for the year 2000 is $5,870,000,000.

SECTION 3.4

3.4.1 Checkup

1. x-intercept is $(5, 0)$; y-intercept is $(0, 3)$. **2.** x-intercepts are $(-2, 0)$, $(-1, 0)$, and $(1, 0)$; y-intercept is $(0, -1)$. **3.** x-intercepts are $(2, 0)$ and $(-2, 0)$; y-intercept is $(0, -4)$. **4.** x-intercepts: $(-2, 0)$, $(1, 0)$, $(2.25, 0)$; y-intercept: $(0, 3)$.

3.4.2 Checkup

1. (a) relative maximum of 5 at $x = -3$ **(b)** no relative minimum **(c)** $x < -3$ **(d)** $x > -3$
2. (a) relative maximum of 0 at $x = -1$ **(b)** no relative minimum **(c)** $x < -1$ **(d)** $x > -1$
3. (a) no relative maximum **(b)** relative minimum of -1 at $x = 1$ **(c)** $x > 1$ **(d)** $x < 1$
4. (a) relative maximum of 45.9375 at $x = 1.5$ **(b)** relative minima of -40 at $x = -1$ and 24 at $x = 3$ **(c)** $-1 < x < 1.5$ and $x > 3$
(d) $x < -1$ and $1.5 < x < 3$

3.4.3 Checkup

1. $(1, 0)$ and $(4, 3)$ **2.** $(5, -5)$ **3.** $(-1, -3)$ and $(2.5, 2.25)$

3.4.4 Checkup

1. (a) $C(x) = 0.60x + 35$
(b) $R(x) = 1.00x$
(c) The break-even point is $x = 87.5$, which means that with 88 quarters, they will begin to make a profit.

2. The x-intercept is approximately $(5.2, 0)$. This means that sometime early in 1997, the surplus was 0.

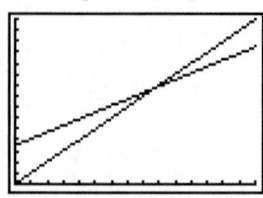

Chapter 4

SECTION 4.1

4.1.1 Checkup

1. linear **2.** nonlinear **3.** linear **4.** nonlinear **5.** nonlinear

4.1.2 Checkup

1. -2 **2.** noninteger between 2 and 3 **3.** 2 **4.** noninteger between 12 and 13

4.1.3 Checkup

1. $x = -2$

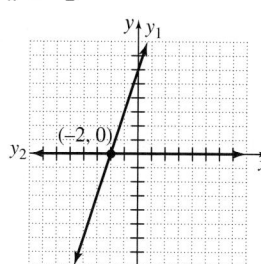

$y_1 = 3x + 6$
$y_2 = 0$

2. $b = 2.6$

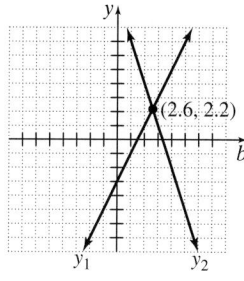

$y_1 = 2b - 3$
$y_2 = 10 - 3b$

3. $x = 2$

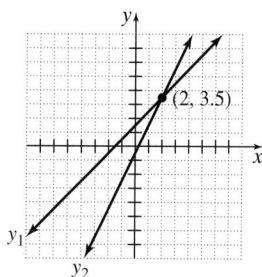

$y_1 = \frac{1}{2}(3 + 2x)$

$y_2 = 2x - \frac{1}{2}$

4. $x = 12.5$

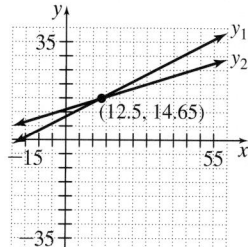

$y_1 = 8.40 + 0.50x$
$y_2 = 10.90 + 0.30x$

4.1.4 Checkup

1. all real numbers

x	y_1	y_2
-3	-10	-10
-2	-8	-8
-1	-6	-6
0	-4	-4
1	-2	-2
2	0	0
3	2	2

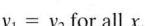

$y_1 = (x - 1) + (x - 3)$
$y_2 = 2(x - 2)$

$y_1 = y_2$ for all x.

2. no solution

x	y_1	y_2
-3	-7	-15
-2	-3	-11
-1	1	-7
0	5	-3
1	9	1
2	13	5
3	17	9

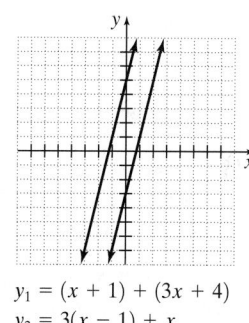

$y_1 = (x + 1) + (3x + 4)$
$y_2 = 3(x - 1) + x$

$y_1 - y_2 = 8$ for all x.

4.1.5 Checkup

1. The cost of production will equal the revenue received when 10 baskets are produced and sold.
2. Phillipe must score a 96 in order to achieve an average of 90.

SECTION 4.2

4.2.1 Checkup

1. $\frac{6}{35}$ **2.** 15 **3.** 22

4.2.2 Checkup

1. -325 **2.** -33 **3.** $\frac{2}{15}$ **4.** 5

4.2.3 Checkup

1. On Earth, the astronaut and his suit weigh about 360 pounds.   **2.** Big Bird is 98 inches tall.   **3.** Each side of Cheryl's garden will be 9.5 feet long.   **4.** The height of the triangle is $5\frac{1}{2}$ feet.

SECTION 4.3

4.3.1 Checkup

1. $-\dfrac{13}{2}$ **2.** any real number   **3.** no solution   **4.** -3 **5.** 13

4.3.2 Checkup

1. The monthly payments will be $75.
2. The break-even point is any real number, since the cost will always equal the revenue.
3. The mixture used 10 cc's of the 40% alcohol solution.

SECTION 4.4

4.4.1 Checkup

1. $h = \dfrac{2A}{b}$ **2.** $a = 4A - b - c - d$ **3.** $y = -2x + \dfrac{5}{2}$ **4.** $y = \dfrac{7}{3}x + 9$

4.4.2 Checkup

1. (a) $c = 0.12x + 49.95$ **(b)** $x = \dfrac{25}{3}c - \dfrac{1665}{4}$

    (c) You can drive approximately 1250 miles if your vacation budget is $200.
    (d) For $150, you can drive approximately 834 miles; for $250, approximately 1667 miles; for $500, approximately 3750 miles.

2. $C = \dfrac{5}{9}(F - 32)$

SECTION 4.5

4.5.1 Checkup

1. The integers are 12, 14, and 16.
2. The lengths Jim needs to cut are 3 inches, 4 inches, 5 inches, 6 inches, and 7 inches.

4.5.2 Checkup

The sandbox will have two sides of 3.5 feet and a third side of 5 feet.

4.5.3 Checkup

1. You will actually receive $2752.29. The interest paid on the loan will be $247.71.
2. Zeke will borrow $4000 at 7% and $1000 at 8.5%.

4.5.4 Checkup

1. The room charge before adding the surcharge is $62.50.
2. One of the roads must be rotated 34°.

SECTION 4.6

4.6.1 Checkup

1. linear   **2.** linear   **3.** nonlinear   **4.** linear

4.6.2 Checkup

1. 1 and $-1.66\overline{6}$ **2.** 1 and -1 **3.** 5 and -7 **4.** no solutions   **5.** $-1.66\overline{6}$ and 3

4.6.3 Checkup

1. 9   **2.** 1 and $-1.66\overline{6}$ **3.** 1 and -1 **4.** 5 and -7 **5.** no solutions   **6.** $-1.66\overline{6}$ and 3

4.6.4 Checkup

$|72 - h| = 3$; $h = 69$ and $h = 75$; the smallest height for drivers will be 69 inches (or 5 feet, 9 inches), and the largest height will be 75 inches (or 6 feet, 3 inches).

Chapter 5

SECTION 5.1

5.1.1 Checkup

1. nonlinear   **2.** linear; $y = -\dfrac{3}{8}$ **3.** linear; $8.2x - y = 3.6$

4. linear; $\sqrt{3}x - \sqrt{2}y = 0$ **5.** nonlinear

5.1.2 Checkup

1.

x	y
6	0
6	1
6	2

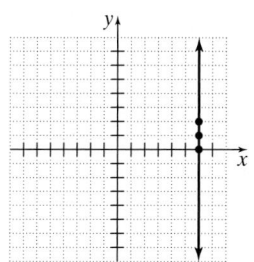

2.

x	y
0	−1
1	3
2	7

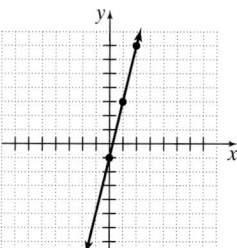

3.

x	y
0	2
5	5
−5	−1

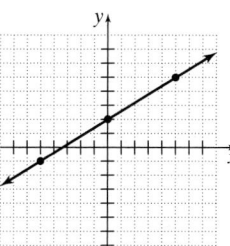

4.

x	y
2	−7
4	−10
6	−13

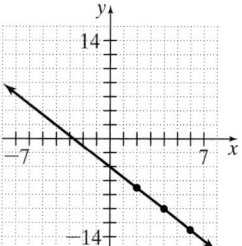

5.

x	y
0	4
2	2
4	0

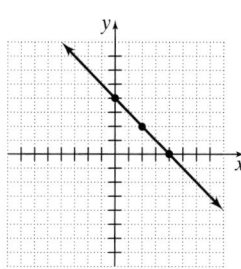

6.

x	y
0	5
3	3
6	1

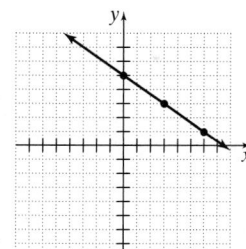

5.1.3 Checkup

1. The x-intercept is $(2, 0)$ and the y-intercept is $(0, 12)$.   **2.** The x-intercept is $\left(-\dfrac{11}{5}, 0\right)$ and the y-intercept is $(0, 11)$.

3. $y = 7x - 15$; the y-intercept is $(0, -15)$.   **4.** $y = -\dfrac{4}{3}x + 8$; the y-intercept is $(0, 8)$.

5.1.4 Checkup

1. The x-intercept is $(16, 0)$.   **2.** The x- and y-intercepts occur at the origin, $(0, 0)$.   **3.** The y-intercept is $\left(0, \dfrac{3}{5}\right)$.

5.1.5 Checkup

1.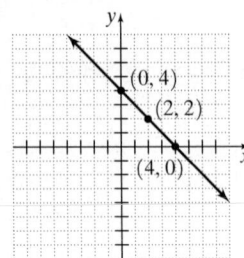

2.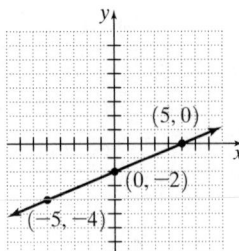

5.1.6 Checkup

1. (a) $(1, 3), (3, 7), (5, 11)$

(b)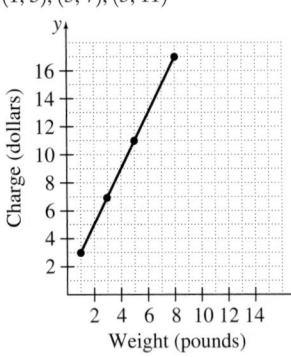

(c) The cost for an 8-pound package is $17.

(d) Answers may vary.

2. The intercepts of the graph are $(0, 5645)$ and $(3.8, 0)$. This means that at $t = 0$ the plane hasn't moved yet, so the distance d is 5645 miles. When the distance is 0, the time is approximately 3 hours and 48 minutes, or 3.8 hours. The time t it takes to fly 3500 miles is approximately 1.44 hours.

3. The linear function for the profit that Don will realize is $y = 140x - 700$, with y as profit and x as the number of students in the class. The intercepts are $(0, -700)$ and $(5, 0)$. When there are no students in his class $(x = 0)$, Don loses $700. With five students in his class, Don breaks even, or has a profit of 0.

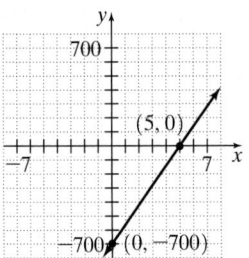

SECTION 5.2

5.2.1 Checkup

1. The slope is zero, and the graph is a function that is constant.
2. The slope is $\frac{2}{5}$, and the graph is a function that is increasing.
3. The slope is -3, and the graph is a function that is decreasing.
4. The slope is undefined, and the graph is not a function.

5.2.2 Checkup

1. $m = \frac{3}{2}$ **2.** $m = 0$ **3.** $m = $ undefined

5.2.3 Checkup

1. $m = -\frac{7}{11}$; y-intercept is $(0, 13)$.   **2.** $m = -\frac{8}{3}$; y-intercept is $(0, 4)$.   **3.** $m = 0$; y-intercept is $(0, -6)$.
4. $m = $ undefined; there is no y-intercept.

5.2.4 Checkup

1. 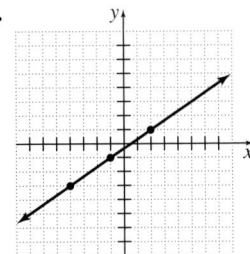 **2.** 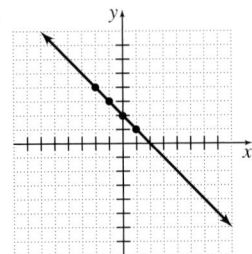 **3.** 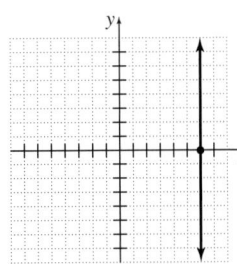 **4.**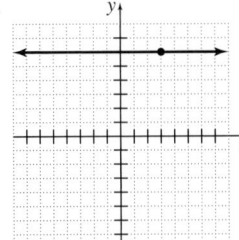

5. 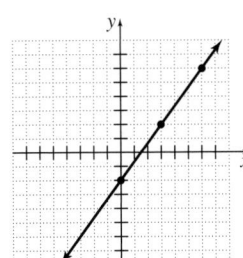 **6.** 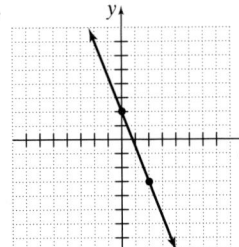 **7.**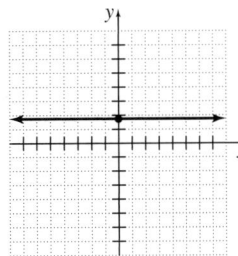

5.2.5 Checkup

1. The grade is 40%.   **2.** The average rate of change is 122 stations per year.
3. Graph c. The slope of the third portion is twice that of the first portion. The rest period was 4 minutes.
4. The depreciation is $9000 per year.   **5.** graph c

SECTION 5.3

5.3.1 Checkup

1. intersecting and perpendicular   **2.** parallel   **3.** intersecting   **4.** parallel   **5.** perpendicular   **6.** coinciding

7. If $m_1 = -\dfrac{1}{m_2}$, then the lines are perpendicular, but if $m_1 \neq m_2$, then the lines intersect.

8.

Situation	Relationship between slopes m_1 and m_2	Relationship between y-intercept values b_1 and b_2
Graphs coincide	$m_1 = m_2$	$b_1 = b_2$
Graphs are parallel	$m_1 = m_2$	$b_1 \neq b_2$
Graphs intersect only	$m_1 \neq m_2$ and $m_1 \cdot m_2 \neq -1$	None
Graphs intersect and are perpendicular	$m_1 \cdot m_2 = -1$	None

5.3.2 Checkup

1. (a) Let x = the number of lawns for which Tim cares. $y = 8.50x + 6.50x + 85$ **(b)** $y = 20x$
   (c) Since the slopes are different, the lines intersect and there is a break-even point.
   (d) If the revenue function is $y = 15x$, the slopes are the same, but the y-intercepts are different and the lines are parallel.
   Tim will never break even.
2. No, their intersection would be to the left of $x = 0$.

SECTION 5.4

5.4.1 Checkup

1. $y = -\dfrac{1}{4}x + 2$ **2.** $x = 2$ **3.** $y = -3$ **4.** $y = 5x - 4$ **5.** $y = 4x - 1$ **6.** $y = -0.4x + 3.2$

5.4.2 Checkup

1. $y = 2x + 3$ **2.** $y = \dfrac{3}{4}x - \dfrac{9}{2}$

5.4.3 Checkup

1. $y = \dfrac{2}{3}x + \dfrac{5}{3}$ **2.** $y = \dfrac{3}{2}$ **3.** $x = -\dfrac{5}{2}$

5.4.4 Checkup

1. $y = -\dfrac{3}{2}x + \dfrac{13}{2}$ **2.** $y = -\dfrac{5}{2}x + \dfrac{3}{2}$

5.4.5 Checkup

1. (a) $m = -0.371$ **(b)** $y = -0.371x + 37$ **(c)** 35.1; yes   **(d)** 33.29
2. (a) $m = \dfrac{5}{9}$ **(b)** $C = \dfrac{5}{9}(F - 32)$ **(c)** 24°

Chapter 6

SECTION 6.1

6.1.1 Checkup

1. The ordered pair is not a solution.   **2.** The ordered pair is a solution.

6.1.2 Checkup

1. $(1, 1)$ **2.** $(4, -5)$ **3.** Infinitely many solutions; any ordered pair satisfying $y = 2x + 3$ **4.** no solution   **5.** Answers will vary.
6. No; the graphs of two consistent and independent equations will be intersecting lines. The graphs of two independent and inconsistent equations will be parallel lines.

6.1.3 Checkup

1. At 291.5 miles, both offers are equal.
2. Let x = the width of the current garden and y = its length; $2x + 2y = 64$; $4x + 2y + 30 = 118$; the solution is $(12, 20)$. The current garden measures 12 feet by 20 feet; the future garden will be 24 feet by 35 feet.   **3.** In the year 2019, the percent of males and females will be equal.

SECTION 6.2

6.2.1 Checkup

1. $\left(\dfrac{1}{4}, -\dfrac{9}{2}\right)$ **2.** $\left(\dfrac{2}{11}, -\dfrac{26}{11}\right)$ **3.** infinitely many solutions; any coordinate pair that satisfies $y = 5x - 3$ **4.** no solution
5. If the system results in an identity, then there will be many solutions, a consistent system, and dependent equations. If the system results in a contradiction, then there will be no solutions, an inconsistent system, and independent equations.

6.2.2 Checkup

1. The two angles measure 125° and 55°.
2. (a) $C(x) = 245 + 2.50x$ **(b)** $R(x) = 20x$
(c) $y = 245 + 2.5x$
    $y = 20x$
    Deanna will break even when 14 cakes are produced and sold.
3. The number of workers will be equal in the year 2016.

SECTION 6.3

6.3.1 Checkup

1. $(7.5, 2.5)$ **2.** $(2, -6)$ **3.** $(1, -1)$ **4.** $\left(\dfrac{1}{4}, \dfrac{8}{5}\right)$ **5.** no solution
6. infinitely many solutions; any coordinate pair that satisfies $y = 5x + 12$

6.3.2 Checkup

1. Let x = the number of cans of cashews sold and y = the number of cans of peanuts sold; $x + y = 262$; $6.50x + 4.00y = 1240.50$; the solution of the system of equations is $(77, 185)$. The Indian Maidens sold 77 cans of cashews and 185 cans of peanuts.

2. Let B = Bruce Willis's earnings; Let T = Tom Cruise's earnings
   $B + T = 113.2$
   $B - T = 26.8$
   Bruce Willis made \$70 million, and Tom Cruise made \$43.2 million.

SECTION 6.4

6.4.1 Checkup

1. Let x = Patty's time on the road and y = the distance she traveled when her dad caught up with her. Her dad's distance would also be y, and his time on the road is 0.5 hour less than Patty's time; $y = 55x$; $y = 65(x - 0.5)$; the solution is $(3.25, 178.75)$. Patty's dad must travel for $3.25 - 0.5 = 2.75$ hours before catching up with Patty. They will have traveled 178.75 miles before he catches her.
2. Let x = average speed of planes in still air, y = average speed of wind; $4(x + y) = 2350$, $5(x - y) = 2350$; the average speed of the planes in still air is 528.75 mph; the average speed of the wind is 58.75 mph.

6.4.2 Checkup

1. Let x = the number of pounds of coarse fescue in the mix and y = the number of pounds of Kentucky blue grass in the mix; $x + y = 100$; $0.75x + 1.25y = 1.00(100)$; the solution is $(50, 50)$. The mix must contain 50 pounds of each type of grass seed.
2. Let x = the number of dimes Rosita saved and y = the number of quarters; $x + y + 120 = 498$; $0.05(120) + 0.10x + 0.25y = 63.75$; the solution is $(245, 133)$. Rosita had saved 245 dimes and 133 quarters.
3. Let x = the amount of 25% solution and y = the amount of 5% solution in the mix; $x + y = 1$; $0.25x + 0.05y = 0.10(1)$; the solution is $\left(\dfrac{1}{4}, \dfrac{3}{4}\right)$. The mix should contain 0.25 liter of the 25% solution and 0.75 liter of the 75% solution.
4. Let x = the amount invested at 4.50%, y = the amount invested at 4.00%; $x + y = 30{,}000$; $0.045x + 0.04y = 1300$; invest \$20,000 at 4.50% and \$10,000 at 4.00%.

SECTION 6.5

6.5.1 Checkup

1. yes   **2.** Answers will vary.   **3.** yes   **4.** Answers will vary.

6.5.2 Checkup

1. $(0.8, 4.5, 1.9)$ **2.** no solution   **3.** infinite number of solutions

6.5.3 Checkup

1. The three currents are $I_1 = 3$ amperes, $I_2 = \dfrac{5}{3}$ amperes, and $I_3 = \dfrac{4}{3}$ amperes.
2. Edgar received 4 ounces of a roast beef sandwich, 3 ounces of french fries, and 12 ounces of milk shake.

Chapter 7

SECTION 7.1

7.1.1 Checkup

1. linear   **2.** linear   **3.** nonlinear   **4.** nonlinear   **5.** nonlinear

7.1.2 Checkup

1. 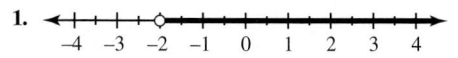   **2.**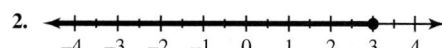

3.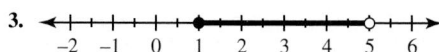

7.1.3 Checkup

1. $(-2, \infty)$ **2.** $(-\infty, 3]$ **3.** $[1, 5)$
4. (a) ; $(-3, \infty)$ **(b)** $x \leq 0$; $(-\infty, 0]$

(c) $-4 \leq x < 1$;

7.1.4 Checkup

1. Let x = the number of CD's Joseph can order; $6.99x + 5.99 \leq 50.00$
2. Let s = systolic pressure, d = diastolic pressure; $130 \leq s \leq 139, 85 \leq d \leq 89, s < 130, d < 85, s < 120, d < 80.$

SECTION 7.2

7.2.1 Checkup

1. $x > -1$ **2.** $z > -1$

7.2.2 Checkup

1. $x < -1$ $(-\infty, -1)$

2. $x \leq -1$ $(-\infty, -1]$

3. $x < 2$ $(-\infty, 2)$

4. $(-\infty, 2.5)$ **5.** no solution   **6.** all real numbers   **7.** all real numbers

7.2.3 Checkup

1. $(2, \infty)$ **2.** $z \geq 3$ **3.**
7.52
4. no solution   **5.** $(-\infty, \infty)$

6.

7.2.4 Checkup

1. The store must buy and sell at least 50 CD's.   **2.** She can have between 1400 and 1750 calories, inclusive.
3. The weight must be at least 200 pounds.

SECTION 7.3

7.3.1 Checkup

1. linear   **2.** linear   **3.** nonlinear   **4.** linear   **5.** nonlinear   **6.** nonlinear

7.3.3 Checkup

1. **2.**

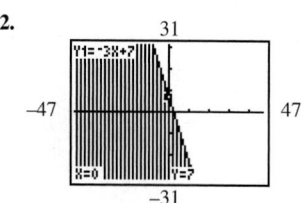

7.3.4 Checkup

1. **2.**

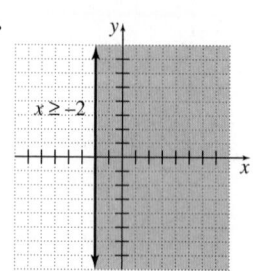

$x \geq -2$

7.3.5 Checkup

1. (a) Let x = the number of payments Lana makes and y = the number of payments her parents make; $60x + 125y \le 1500$

(b)

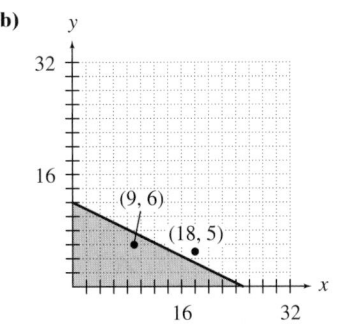

(c) Yes, see the graph.
(d) No, see the graph.

2. Yes. No.

SECTION 7.4

7.4.1 Checkup

1.

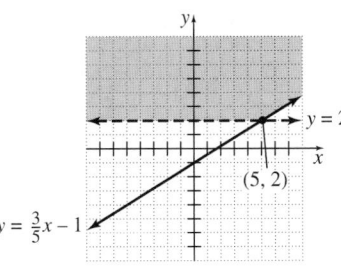

2.

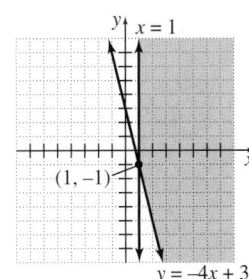

3.

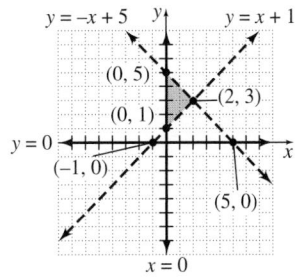

7.4.2 Checkup

1. $Y1 = x - 3$
$Y2 = 2x + 1$

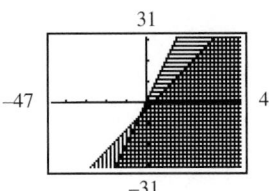

2. $Y1 = -3x + 4$
$Y2 = -x + 2$

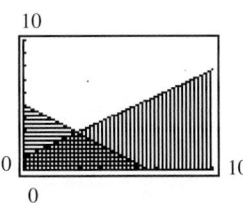

3. Note: Set window to first quadrant to achieve last two inequalities.

$Y1 = \dfrac{2}{3}x + 1$

$Y2 = -\dfrac{3}{4}x + 5$

7.4.3 Checkup

1. Let x = the number of rag dolls and y = the number of sculptured dolls produced; $1.5x + 4y \le 30$; $1.5x + y \le 24$; ; possible combinations: $(5, 5)$ indicates that they produce 5 of each doll; $(10, 2)$ indicates that they produce 10 rag dolls and 2 sculptured dolls; see points on figure.

2. Yes, No.

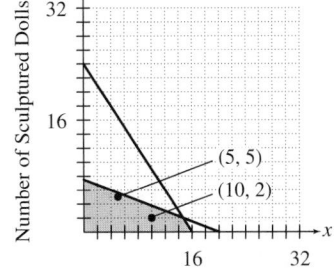

Chapter 8

SECTION 8.1

8.1.1 Checkup

1. polynomial   **2.** polynomial   **3.** not a polynomial   **4.** not a polynomial   **5.** polynomial   **6.** not a polynomial

8.1.2 Checkup

1. polynomial   **2.** binomial   **3.** trinomial   **4.** monomial

8.1.3 Checkup

1. The degrees of the terms are 2, 4, and 3; the degree of the polynomial is 4.
2. The polynomial simplifies to $-2x^2 + 5x + 1$; the degrees of the terms are 2, 1, and 0; the degree of the polynomial is 2.
3. The degrees of the terms are 10 and 9; the degree of the polynomial is 10.

8.1.4 Checkup

descending order	ascending order
1. $2x^3 - x^2 - 5x + 3$	$3 - 5x - x^2 + 2x^3$
2. $y^4 + y + 12$	$12 + y + y^4$

8.1.5 Checkup

1. (a) 172 **(b)** 11   **2. (a)** -214.08 **(b)** -355.76 or $-\dfrac{8894}{25}$

8.1.6 Checkup

1. Let x = length of a side of the cabin, a = length of the lot, b = width of the lot, A = area not covered by the cabin.   **(a)** $A = ab - x^2$
(b) If $x = 35$, $a = 200$, and $b = 150$, then $A = 28{,}775$ square feet.
2. Let x = the number of tables made and sold, y = the number of birdhouses made and sold, R = the revenue received from sales, and
C = the cost of making the items.   **(a)** $R = 12x + 15y$ **(b)** $C = 47 + 4.25x + 5.85y$
(c) If $x = 22$, $y = 19$, $R = \$549$, $C = \$251.65$ **(d)** The profit is $297.35.

SECTION 8.2

8.2.1 Checkup

1.

x	y
-3	0
-2	4
-1	0
0	-6
1	-8
2	0

2. (a) x **(b)** y **(c)** all real numbers

8.2.2 Checkup

1.

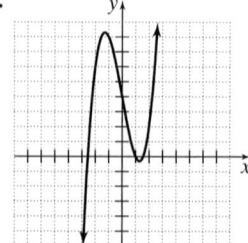

2.

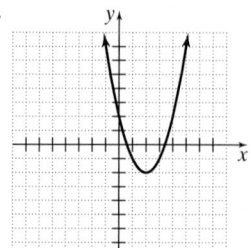

8.2.3 Checkup

1. all real numbers   **2.** all real numbers less than or equal to 9   **3.** all real numbers
4. all real numbers greater than or equal to -2.3 (approx.)

8.2.4 Checkup

1. (a) 36   **(b)** 60   **(c)** 10   **(d)** 0   **(e)** 0   **(f)** 0   **(g)** 33.292

8.2.5 Checkup

1. (a)

No. attending	Revenue
5	$650
10	$1100
15	$1350
20	$1400
25	$1250
30	$900
35	$350
40	$-$400

(b) The revenue will be at most $1406.25.   **(c)** Answers will vary.

2. All real numbers greater than 31,102. As the strength of the army increases, so does the number of commissioned officers.

SECTION 8.3

8.3.1 Checkup

1. quadratic   **2.** quadratic   **3.** nonquadratic   **4.** quadratic   **5.** nonquadratic   **6.** nonquadratic

8.3.2 Checkup

	a	b	c	w/n	u/d	vertex	axis	y int
1.	-3	0	2	narrow	down	$(0, 2)$	$x = 0$	$(0, 2)$
2.	4	-8	5	narrow	up	$(1, 1)$	$x = 1$	$(0, 5)$
3.	0.25	1	-2	wide	up	$(-2, -3)$	$x = -2$	$(0, -2)$
4.	$\dfrac{2}{3}$	1	0	wide	up	$\left(-\dfrac{3}{4}, -\dfrac{3}{8}\right)$	$x = -\dfrac{3}{4}$	$(0, 0)$

8.3.3 Checkup

1.

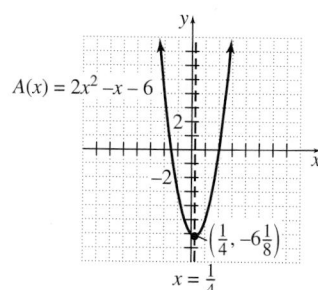

$A(x) = 2x^2 - x - 6$

$\left(\dfrac{1}{4}, -6\dfrac{1}{8}\right)$

$x = \dfrac{1}{4}$

2.

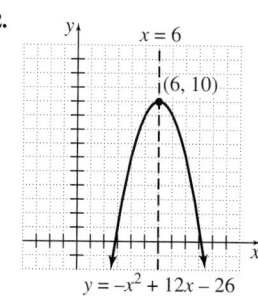

$x = 6$

$(6, 10)$

$y = -x^2 + 12x - 26$

8.3.4 Checkup

1.

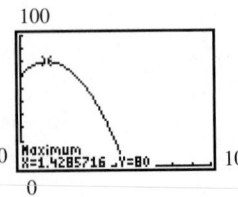

2.

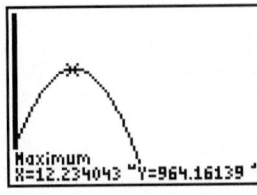

The maximum height the balloon reaches is approximately 80 meters. Tracing on the graph, we readily see that the balloon will return to the ground in approximately 5.5 seconds.

In the 13th year, 2003, the maximum number of compact discs shipped will be approximately 964 discs.

SECTION 8.4

8.4.1 Checkup

1. $y = 0.75x^2 + 2.75x - 1$

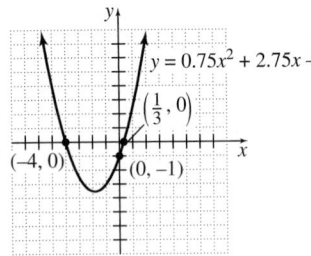

8.4.2 Checkup

1. $y = x^2 + x - 6$

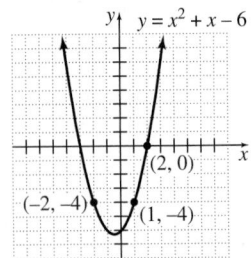

8.4.3 Checkup

1. Let t = time in seconds after release and $s(t)$ = position of the case in feet above the water. $s(t) = -16t^2 - 30t + 1500$; the case will reach the water in approximately 8.8 seconds.

2. $y = -0.00175x^2 + 135$ **3.** $y = 0.000113x^2 - 500$

Chapter 9

SECTION 9.1

9.1.1 Checkup

1. $3 \cdot 3 \cdot x \cdot x \cdot x$ **2.** $(-5y)(-5y)(-5y)$ **3.** $-5 \cdot y \cdot y \cdot y$ **4.** $(a - b)(a - b)(a - b)(a - b)$ **5.** $q \cdot q \cdot q \cdot q$

9.1.2 Checkup

1. a^9 **2.** z^7 **3.** $(x + y)^8$ **4.** $\dfrac{a^8}{2}$

9.1.3 Checkup

1. a^4 **2.** $\dfrac{2}{11}z^3$ **3.** $(x + 5)^2$

9.1.4 Checkup

1. $-243a^5$ **2.** $a^{18}b^{18}$ **3.** $-64x^{24}$ **4.** $\dfrac{16p^4}{q^4}$

9.1.5 Checkup

1. $\dfrac{-2a^3b^4}{5}$ **2.** $3q$ **3.** $\dfrac{-x^5}{2}$ **4.** $\dfrac{27a^6}{64}$ **5.** $a^{mn}; a^{m+n}$

9.1.6 Checkup

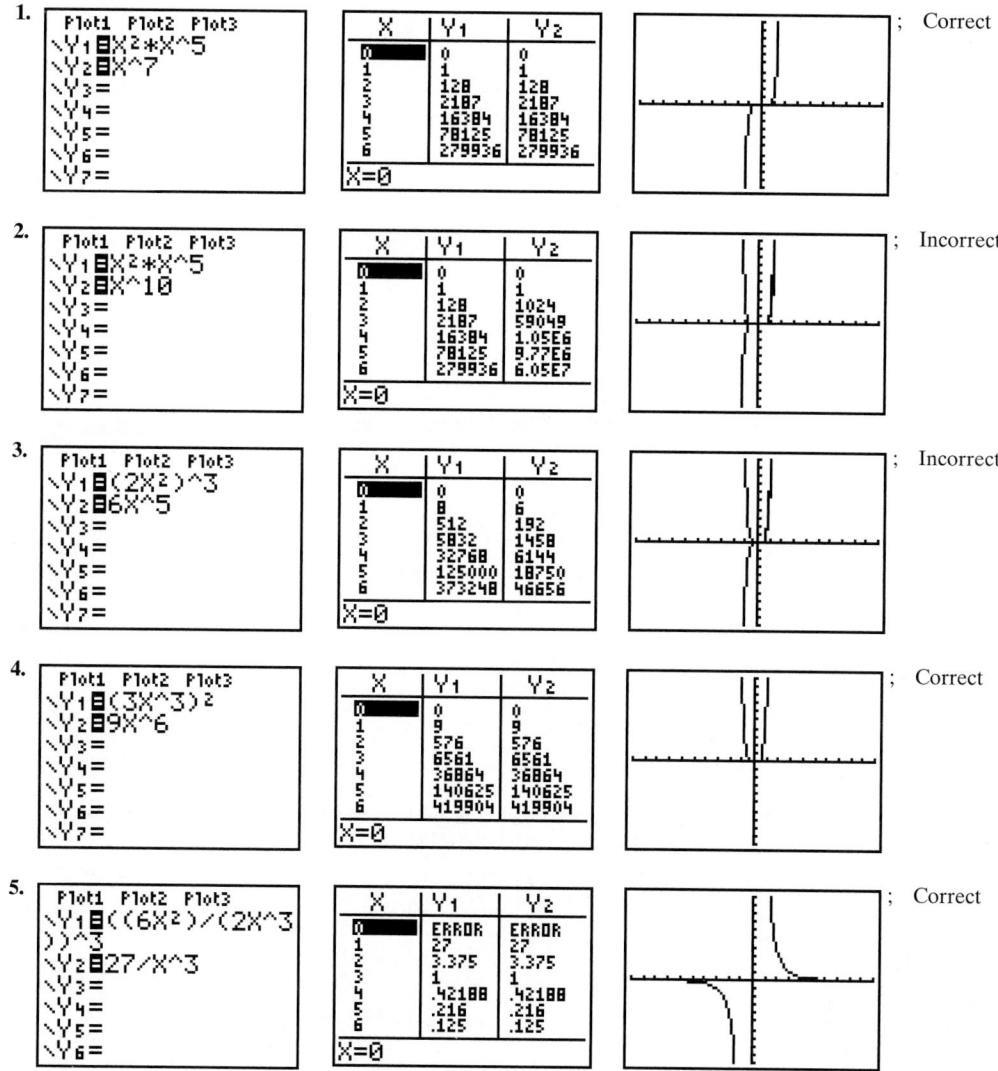

1. ; Correct

2. ; Incorrect

3. ; Incorrect

4. ; Correct

5. ; Correct

9.1.7 Checkup

The volume of the original box is x^3. The volume of the new box is $\dfrac{1}{8}x^3$. The volume of the new box is one-eighth as large as the volume of the original box. If the original box measures 1.5 meters on a side, its volume is 3.375 cubic meters, and the volume of the new box is 0.421875 cubic meters.

SECTION 9.2

9.2.1 Checkup

1. $\dfrac{y^2}{x^4}$ **2.** $\dfrac{8a^4b^3}{25}$ **3.** $\dfrac{q^5}{p^5}$

9.2.2 Checkup

1. $\dfrac{1}{2z^2}$ **2.** $\dfrac{5}{b}$ **3.** $\dfrac{a}{2b^3}$ **4.** $\dfrac{a^4}{2b^3}$ **5.** $\dfrac{125p}{q^2}$ **6.** $\dfrac{-x^3\,y^{12}}{125}$ **7.** $\dfrac{h^{12}}{k^6}$ **8.** $\dfrac{27c^9}{8d^{12}}$

9.2.3 Checkup

1. The distance to the moon is 2.325×10^5 miles, or 1.2276×10^9 feet.

SECTION 9.3

9.3.1 Checkup

1. $8x^3 - 3x^2 - x + 9$ **2.** $7x^3 - x^2y - 7xy^2 + 5y^3$

9.3.2 Checkup

1. $a^3 - 7a^2 + 15$ **2.** $4a^4 + 3a^3b - 9a^2b^2 + 5ab^3 + 2b^4$

9.3.3 Checkup

1. Let $x = $ the number of box lunches made and sold in a day. **(a)** $C(x) = 15 + 1.5x$ **(b)** $R(x) = 4.50x$ **(c)** $P(x) = 3x - 15$
2. The difference in the heights is when $(20t + 60)$ feet.

SECTION 9.4

9.4.1 Checkup

1. $-25a^5b^5c$ **2.** $-4m^5n + 20m^4n^2 - 4m^3n^3$

9.4.2 Checkup

1. $7a^2 + 50a + 7$ **2.** $4y^2 - 1$ **3.** $x^2 - 10x + 25$

9.4.3 Checkup

1. $3x^3 + 4x^2 - 12x - 16$ **2.** $x^3 + 3x^2 + 3x + 1$ **3.** $x^3 + 1$

9.4.4 Checkup

1. $9y^2 - 49$ **2.** $9y^2 - 42y + 49$ **3.** $9y^2 + 42y + 49$

9.4.5 Checkup

1. Let $A(s)$ be the area of the new garden. $A(s) = 2s^2 + 11s + 12$. If the garden measured 5 feet on a side, the area of the new garden will be 117 square feet.
2. Let x be the width of the garden; $3x$ is the length of the garden; $x + 4$ is the width of the garden and walkway, and $3x + 4$ is the length of the garden and walkway. If $A(x)$ is the area of the walkway, $A(x) = (3x + 4)(x + 4) - 3x(x)$. Thus, $A(x) = 16x + 16$.

SECTION 9.5

9.5.1 Checkup

1. $3a^2b^3c$ **2.** $-6xy - y^2$ **3.** $3x + \dfrac{3y}{2} - \dfrac{9y^2}{4x} - \dfrac{y^3}{2x^2}$

9.5.2 Checkup

1. $5x - 4$ **2.** $4x^2 - 2x + 5 + \dfrac{2}{2x + 5}$ **3.** $4x^2 - 6x + 9$

9.5.3 Checkup

1. (a) $C(x) = 200 + 75x$ **(b)** $R(x) = 259.00x - 5.00x^2$ **(c)** $P(x) = -5x^2 + 184x - 200$

(d) $A(x) = -5x + 184 - \dfrac{200}{x}$ **(e)** The average profit is $114.00 per package.

2. $g(y) = 5000y^3 + 5000y^2 + 5000y + 5000$
Benjamin received a total of $23,205.00.

Chapter 10

SECTION 10.1

10.1.1 Checkup

1. $6a^2bc$ **2.** $42x^3y^3z^2$ **3.** $11p$

10.1.2 Checkup

1. $6a^2b^2(a - 2b + 4)$ **2.** $-4xy(2xy^2z + 6x^2y - 9z)$

10.1.3 Checkup

1. $(2a + b)(7a + 4b)$ **2.** $(3x - 2)(4x^2 - 3)$

10.1.4 Checkup

1. $(3x - 5)(2x + 3)$ **2.** $(y^2 + 1)(2y^2 - 5)$ **3.** $5a(3a - 4)(a + 7)$ **4.** $(x + y)(a + b)$

10.1.5 Checkup

1. **(a)** Let $x =$ the amount Katie received the first month. Let $T(x) =$ the total amount of money Katie will receive. $T(x) = 6x + 180$.
(b) $T(x) = 6(x + 30)$; $(x + 30)$ is the average monthly amount Katie will receive if she stops smoking for six months.
(c) $T(50) = 480$; Katie will receive $480.
(d) Katie's average monthly receipt is $80, and this does check with the interpretation from part b.
2. **(a)** $(100 + 2x)(150 + 2x) - (150)(100)$; $4x(125 + x)$; the offices' area is 26,400 ft^2.

SECTION 10.2

10.2.1 Checkup

1. $(z + 6)^2$ **2.** $(7a + 3b)^2$ **3.** $5(x - 6)^2$ **4.** Does not factor.   **5.** $(y^2 + 4)^2$

10.2.2 Checkup

1. $(y - 2)(y + 2)$ **2.** $(5x - 4y)(5x + 4y)$ **3.** $2(z - 3)(z + 3)$ **4.** Does not factor.
5. $(c - 2)(c + 2)(c^2 + 4)$

10.2.3 Checkup

1. $(a + 2)(a^2 - 2a + 4)$ **2.** $(a - 2)(a^2 + 2a + 4)$ **3.** $(2x + 3y)(4x^2 - 6y + 9y^2)$

10.2.4 Checkup

1. **(a)** Let $P =$ the length of a side of the square patio; $A(p) =$ the area to be landscaped. $A(p) = 16^2 - p^2$.   **(b)** $A(p) = (16 - p)(16 + p)$.
An equivalent rectangular area will be one that is 9 feet by 23 feet.   **(c)** The package covers an area of 200 square feet. The landscaped
area will be 207 square feet. The package is not large enough.
2. $(9 - x)^2$; the original quilt was 9 feet by 9 feet; the area of the trimmed quilt is 60.0625 ft^2.

SECTION 10.3

10.3.1 Checkup

1. $(b - 7)(b - 5)$ **2.** $(c - 8)(c + 3)$ **3.** $6(d + 7)(d + 3)$ **4.** $(y^2 - 3)(y^2 + 1)$

10.3.2 Checkup

1. $(2x + 1)(x - 5)$ **2.** Does not factor.   **3.** $5(2x + 3)(3x + 2)$ **4.** $(2x^2 - 5)(3x^2 - 4)$

10.3.3 Checkup

1. Since the polynomial factors as $(w + 13)(w + 5)$, the outside dimensions are $(w + 5)$ feet wide and $(w + 13)$ feet long. Comparing these dimensions with the inside width of w feet and length of $w + 9$ ft, we find that the walk is 2.5 feet wide along the pool's length and 2 feet wide along its width.

2. $2(x + 39)(2x + 27)$ or $(2x + 78)(2x + 27)$; the dimensions of the playing area are 27 feet by 78 feet.

SECTION 10.4

10.4.1 Checkup

1–4. These have the same answers as in 10.3.2 checkup.

10.4.2 Checkup

1. (a) The polynomial factors as $\frac{1}{2}(3x - 4)(x - 4)$. The sides of the reduced garden measure $(3x - 4)$ and $(x - 4)$ feet.

(b) Carri reduced each of the two sides of the garden by 4 feet.

2. $(2x + 45)(x + 15)$; the width is x feet and the length is $2x + 30$ feet; the border is one-half of 15 or 7.5 feet wide; the width is 85 feet, the length is 200 feet, and the total area is 21,500 square feet.

SECTION 10.5

10.5.1 Checkup

1. $5(2z - 5)(z + 6)$ **2.** $2a(3a + 1)(a + 4)$ **3.** Does not factor.   **4.** $6x(x - 5y)(2x - y)$ **5.** $-6(a - 2)(a + 2)$
6. $(3x - 2)^2(3x + 2)^2$ **7.** $(4 + h)(3 - 5h)$ or $-1(5h - 3)(h + 4)$

10.5.2 Checkup

1. $x(349.95 - 8.95x)$; For every copy of software purchased, a discount of \$8.95 will be given.
2. The polynomial factors as $x(5 + 2x)(7 + 2x)$. Therefore, the height is x inches, the width is $5 + 2x$ inches, and the length is $7 + 2x$ inches. If the height is 1 inch, the width is 7 inches and the length is 9 inches.

Chapter 11

SECTION 11.1

11.1.1 Checkup

1. a quadratic polynomial   **2.** not a polynomial   **3.** not a polynomial   **4.** a cubic polynomial   **5.** a polynomial

11.1.2 Checkup

1. $-2, -1$, and 2   **2.** noninteger solutions between -1 and 0 and between 0 and 1   **3.** contradiction, no solution
4. not a contradiction, no real-number solution   **5.** The solution set is all real numbers.

11.1.3 Checkup

1. $-2, -1$, and 1   **2.** -0.2 and 0.75   **3.** no solution, contradiction   **4.** no real-number solution, not a contradiction
5. The solution set is all real numbers, identity.

11.1.4 Checkup

1. The banana would hit the ground after approximately 8.8 seconds.
2. In 2004, per capita income will have increased by \$15,000 over the 1997 value.

SECTION 11.2

11.2.1 Checkup

1. -7 and $\frac{3}{4}$ **2.** 0, 1, and $\frac{5}{4}$

11.2.2 Checkup

1. -9 and 9   **2.** $-\frac{3}{4}$ **3.** $\frac{5}{4}$ and 3   **4.** -6 and 4

11.2.3 Checkup

1. (a) The dimensions of the tray are 1 inch by 5 inches by 7 inches.   **(b)** The dimensions of the posterboard are 7 inches by 9 inches.
2. The dimensions of the box are 2 inches by 2 inches by 4 inches.   **3. (a)** $y = -0.04x^2 + 0.2x + 6.44$ **(b)** The gross profit will drop to $5 billion in the year 2007.   **(c)** Answers will vary.

SECTION 11.3

11.3.1 Checkup

1. $7\sqrt{2}$ **2.** $-3 + 4\sqrt{2}$ **3.** $1 + \sqrt{3}$ **4.** $2 + \sqrt{5}$

11.3.2 Checkup

1. $\dfrac{\sqrt{7}}{8}$ **2.** $\dfrac{8\sqrt{7}}{7}$ **3.** $\dfrac{5\sqrt{2}}{3}$

11.3.3 Checkup

1. ± 4 **2.** $\pm \dfrac{5}{3}$ **3.** No real number.   **4.** 0 and 8   **5.** $1 \pm \sqrt{3}$

6. $-1 \pm \sqrt{6}$ **7.** $\pm \dfrac{7\sqrt{2}}{2}$

11.3.4 Checkup

1. The pole's height is about 19.94 feet.   **2.** The vertical distance between the two hookups is 2.1 feet.   **3.** The fence is approximately 16.2 feet high.   **4.** Tommy must realize an equivalent annual interest rate of 17.5%.

SECTION 11.4

11.4.1 Checkup

1. 16   **2.** $\dfrac{49}{4}$ **3.** $\dfrac{4}{25}$

11.4.2 Checkup

1. $x = -4 \pm 3\sqrt{2}$ **2.** $x = 8$ **3.** no real solution

4. $x = \dfrac{2 \pm \sqrt{29}}{5}$

11.4.3 Checkup

1. To break-even, the retailer must sell 4 or 26 items.
2. (a) $y = -0.05x^2 - 0.05x + 8.6$ **(b)** In the year 2003, the sales will be 5 million automobiles.

SECTION 11.5

11.5.1 Checkup

1. $x = -2, x = 1.5$ **2.** $z = \dfrac{-1 \pm \sqrt{29}}{2}$ **3.** $p = \dfrac{2}{3}$ **4.** no real-number solutions   **5.** $x = -2 \pm 3\sqrt{3}$

11.5.2 Checkup

1. Discriminant is 0.08; there are two irrational solutions.   **2.** Discriminant is zero; there is one rational solution.
3. Discriminant is 400; there are two rational solutions.   **4.** Discriminant is -25; there is no real-number solution.

11.5.3 Checkup

1. (a) $R(x) = x(150 - 5x)$ **(b)** $C(x) = 150 + 65x$ **(c)** $P(x) = -5x^2 + 85x - 150$ **(d)** They will break even when 2 cabinets are sold for $140 each or 15 cabinets are sold for $65 each.   **2.** The number listed will equal 200 billion shares when the market value is about $9 trillion.

SECTION 11.6

11.6.1 Checkup

1. $s = \sqrt{A}$ **2.** $a = \sqrt{c^2 - b^2}$ **3.** $r = \dfrac{-\pi h + \sqrt{\pi^2 h^2 + \pi A}}{\pi}$ or $r = -h + \sqrt{h^2 + \dfrac{A}{\pi}}$ **4.** The diagonal measures 12 ft.

11.6.2 Checkup

1. The congruent legs are $3\sqrt{2}$, or approximately 4.2 mm long.   **2.** The horizontal distance of the ramp is $3\sqrt{3}$ feet (or approximately 5.2 feet), and the ramp's length is 6 feet.

SECTION 11.7

11.7.1 Checkup

1. yes   **2.** yes   **3.** no   **4.** no

11.7.2 Checkup

1. All integers less than or equal to -3 or all integers greater than or equal to 4.   **2.** The integers between -3 and 1 exclusive; that is, $x = -2, -1,$ and 0.   **3.** All integers less than -2 or all integers greater than 1.

11.7.3 Checkup

1. $x \le -3$ or $x \ge 4$ or $(-\infty, -3] \cup [4, \infty)$ **2.** $-3 < x < \dfrac{1}{2}$ **3.** $x < -2.562$ or $x > 1.562$ **4.** $x = 0$

5. $x \ne 2$ or $(-\infty, 2) \cup (2, \infty)$ **6.** $x = 2$

11.7.4 Checkup

same answers as in 11.7.3.

11.7.5 Checkup

1. Let t = the time the baseball is no lower than the top of the tower, $0 \le t \le 3.125$; the baseball is no lower than the top of the tower for the first 3.125 seconds of its flight.
2. If the revenue must be at least $7500, then the value of x is between 5 and 15 inclusive. Therefore, the number of computers ordered must be between 15 and 25 inclusive.   **3.** In the year 2005, Mexico City's population is predicted to exceed 30 million people.

Chapter 12

SECTION 12.1

12.1.1 Checkup

1. a rational expression   **2.** not a rational expression   **3.** a rational expression   **4.** a rational expression
5. a rational expression   **6.** a rational expression

12.1.2 Checkup

1. The restricted value is -4. The domain is all real numbers not equal to -4.
2. The restricted values are -4 and 3. The domain is all real numbers except -4 and 3.
3. There are no restricted values. The domain is all real numbers.
4. The restricted value is 1.5. The domain is all real numbers except 1.5.

12.1.3 Checkup

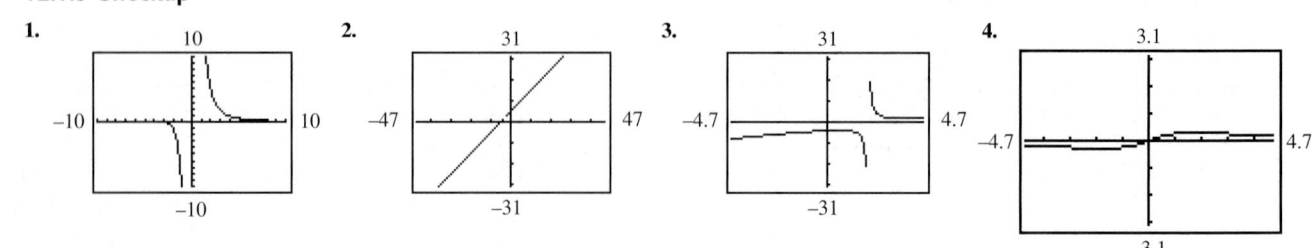

12.1.4 Checkup

1. (a) $C_{ave}(x) = \dfrac{2500 + 50x + 25x^2}{x}$ **(b)** The minimum average cost per year is $550.00.   **(c)** Paul should replace the set after 10 years.

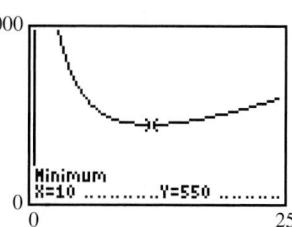

2. (a) 1997: 0.900   **(b)** $P(x) = \dfrac{N(x)}{L(x)} = \dfrac{2.9x + 122.9}{1.6x + 136.1}$ **(c)** $x = 0$: 1997: 0.903
1998: 0.914
1999: 0.924
2000: 0.933

$x = 1$: 1998: 0.914
$x = 2$: 1999: 0.924
$x = 3$: 2000: 0.934

X	Y1
0	.90301
1	.91358
2	.92391
3	.934
4	.94386
5	.9535
6	.96294

X=0

(d) According to the table in part c, the approximate proportions are very close to those in part a. Therefore, the function is a good predictor.

SECTION 12.2

12.2.1 Checkup

1. $\dfrac{3y}{7x}$ **2.** $\dfrac{a - 2}{3a + 4}$ **3.** $\dfrac{x - 2}{x - 6}$ **4.** $-\dfrac{5 + x}{x^2 + 7}$

12.2.2 Checkup

1. $\dfrac{2y^2}{5x^2}$ **2.** $\dfrac{(x - 1)(x + 4)}{x^2(x + 1)} = \dfrac{x^2 + 3x - 4}{x^3 + x^2}$ **3.** $\dfrac{2}{(x - 4)^2} = \dfrac{2}{x^2 - 8x + 16}$

12.2.3 Checkup

1. $\dfrac{5x^2z}{4y}$ **2.** $\dfrac{3(x^2 + 1)}{2} = \dfrac{3x^2 + 3}{2}$ **3.** $\dfrac{x + 4}{(x - 2)(x + 2)} = \dfrac{x + 4}{x^2 - 4}$

12.2.4 Checkup

1. (a) Let $R(x) =$ Reza's average speed. $R(x) = \dfrac{150}{x}$.   **(b)** New rate $= \dfrac{300}{x}$. New time $= x + 1$. New distance $= \dfrac{300(x + 1)}{x}$.

2. $s(y) = 1500(1 + y)$; the total amount that Benjamin received was $3300.

SECTION 12.3

12.3.1 Checkup

1. $\dfrac{5xy - 7x}{3z}$ or $\dfrac{x(5y - 7)}{3z}$ **2.** 3   **3.** $\dfrac{1}{x + 4}$ **4.** $\dfrac{x}{(x + 6)(x - 3)}$

12.3.2 Checkup

1. $60x^2y^3$ **2.** $(x + 4)(x - 5)$ **3.** $(5x - 1)(5x + 1)(3x - 1)$ **4.** $-1(x - 5)$ or $5 - x$

12.3.3 Checkup

1. $\dfrac{-16x^2}{12x^2y^2}$ **2.** $\dfrac{2x(x + 4)}{(x + 5)(x + 4)}$ or $\dfrac{2x^2 + 8x}{x^2 + 9x + 20}$ **3.** $\dfrac{(2x + 1)(x + 7)}{(x^2 - 16)(x + 7)}$ or $\dfrac{2x^2 + 15x + 7}{x^3 + 7x^2 - 16x - 112}$ **4.** $\dfrac{-7x}{-x(x - 5)}$ or $\dfrac{-7x}{5x - x^2}$

12.3.4 Checkup

1. $\dfrac{27x + 5y}{6x^2y^2}$ **2.** $\dfrac{7x^2 - 31x}{x^2 - 11x + 24}$ or $\dfrac{x(7x - 31)}{(x - 3)(x - 8)}$ **3.** $\dfrac{8}{x - 4}$ **4.** $\dfrac{x^2 + 8x + 21}{(x + 3)(x + 5)(x + 7)}$ **5.** $\dfrac{x + 3}{x - 4}$

12.3.5 Checkup

1. (a) $\dfrac{16x + 4}{15x}$ **(b)** $\dfrac{7x + 5}{x}$ **(c)** The second perimeter is larger than the first by $\dfrac{89x + 71}{15x}$.

2. (a) $T(x) = \dfrac{60}{x} + \dfrac{210}{2x - 10}$ **(b)**

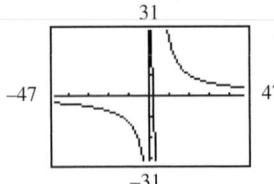

(c)

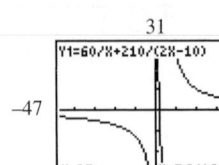

The total time of the delivery trip is approximately 5.2 hours.

(d)

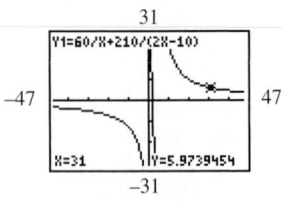

The average speed on the rural roads was 31 mph and on the interstate was 52 mph.

3. The average speed of the plane for the round-trip was 240 mph.

SECTION 12.4

12.4.1 Checkup

1. not a rational equation.   **2.** a rational equation.   **3.** not a rational equation.   **4.** a rational equation.   **5.** a rational equation.
6. a rational equation.

12.4.2 Checkup

1. 4   **2.** contradiction, no solution   **3.** The solution is extraneous, since it is the same as the restricted value.
4. Identity; solution is all real numbers except -1.   **5.** -5 (3 is an extraneous solution.)

12.4.3 Checkup

1. (a) expression; $\dfrac{x + 13}{x - 4}$ **(b)** equation; $x = 5$

12.4.4 Checkup

1. Let x = the interest rate for the three-year period. Then $\dfrac{225}{3x} + \dfrac{225}{2(x - 0.005)} = 4000$ and $x = 0.05$. The interest rate for the three-year investment was 5% and for the two year investment was 4.5%.

2. The focal length needed is 20 millimeters.

SECTION 12.5

12.5.1 Checkup

1. Let x = the time it will take to do the lawn together. Then $\dfrac{1}{2} + \dfrac{1}{1.6} = \dfrac{1}{x}$. $x = \dfrac{8}{9}$. It will take them $\dfrac{8}{9}$ of an hour to do the lawn together.

2. Let x = the time to fill an empty tank. Then $\dfrac{1}{5} = \dfrac{1}{x} - \dfrac{1}{2}$. $x = \dfrac{10}{7}$. It will take $1\dfrac{3}{7}$ hours to fill an empty tank if no water is pumped out.

12.5.2 Checkup

Let x = the number of consecutive baskets she must make. Then $\dfrac{6 + x}{10 + x} = \dfrac{75}{100}$. $x = 6$. Michelle must make six consecutive baskets to average 75%.

12.5.3 Checkup

1. Side AC measures 22.8 inches. Side EF measures 6.1 inches. **2.** The antenna is 10 feet tall.

12.5.4 Checkup

1. $D = \sqrt{\dfrac{Gm_1m_2}{F}}$

2. Let x = the first resistor's resistance. Then $\dfrac{1}{12} = \dfrac{1}{x} + \dfrac{1}{x + 10}$. $x = 20$. The two resistances are 20 ohms and 30 ohms.

Chapter 13

SECTION 13.1

13.1.1 Checkup

1. 5 **2.** -5 **3.** not a real number **4.** -7 **5.** 7 **6.** 2.592

13.1.2 Checkup

1. radical **2.** nonradical **3.** radical

13.1.3 Checkup

1. There are no restricted values. The domain is all real numbers. $(-\infty, \infty)$
2. The restricted values are all real numbers less than $-\frac{1}{3}$. The domain is all real numbers greater than or equal to $-\frac{1}{3}$, or $[-\frac{1}{3}, \infty)$.
3. The restricted values are all real numbers less than 3. The domain is all real numbers greater than or equal to 3, or $[3, \infty)$.

13.1.4 Checkup

1. 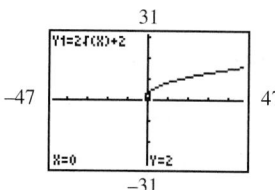 **2.** 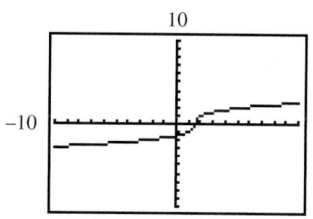 **3.**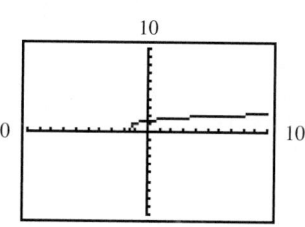

13.1.5 Checkup

1. $r = \sqrt{\dfrac{A}{\pi}} = \sqrt{\dfrac{19{,}360}{\pi}} \approx 78.5$; the radius of the dome is approximately 78.5 feet. The diameter is approximately 157 feet.
2. 9 **3.** 7 **4.** $\sqrt{34} \approx 5.831$
5. **(a)** $e(x) = 6. \sqrt[3]{x}$

(b) The function is increasing for all $x > 0$.
(c) In the year 2010, there will be approximately 17,000 people working for Amway if the trend on the graph remains the same.

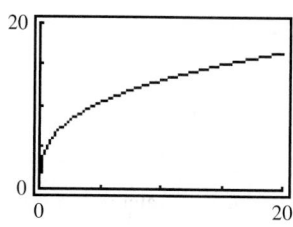

SECTION 13.2

13.2.1 Checkup

1. $\sqrt[4]{16} = 2$ **2.** $-\sqrt[4]{16} = -2$ **3.** $\sqrt[4]{-16}$, not a real number **4.** $\dfrac{1}{\sqrt[3]{125}} = \dfrac{1}{5}$ **5.** $-\dfrac{1}{\sqrt[3]{125}} = \dfrac{1}{5}$ **6.** $\dfrac{1}{\sqrt[3]{-125}} = -\dfrac{1}{5}$
7. 4.729 **8.** 0.115

13.2.2 Checkup

1. $(\sqrt[3]{16})^3 = \sqrt[3]{16^3} = 64$ **2.** $(\sqrt[3]{-125})^5 = \sqrt[3]{(-125)^5} = -3125$ **3.** $(\sqrt{15})^3 = \sqrt{15^3} \approx 58.095$

4. $-\dfrac{1}{(\sqrt[3]{125})^5} = -\dfrac{1}{\sqrt[3]{125^5}} = -\dfrac{1}{3125}$; note: $-3.2E - 4$ on the calculator

13.2.3 Checkup

1. The restricted values are all real numbers less than -2. The domain is all real numbers greater than or equal to -2, or $[-2, \infty)$.

2. There are no restricted values.   **3.** $y = -2x^{1/5}$ **4.** $f(x) = 2(x-1)^{3/2}$
The domain is all real numbers, or $(-\infty, \infty)$.

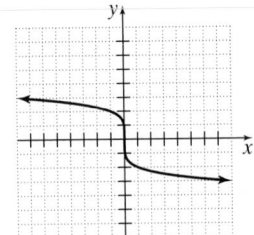
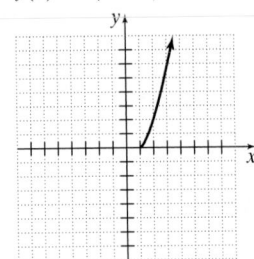

13.2.4 Checkup

1.

x	y
1	1
4	128
16	16384
64	2,097,152

128
128
128

The absolute brightness increases 128 times as the mass is increased by a factor of 4.

2.

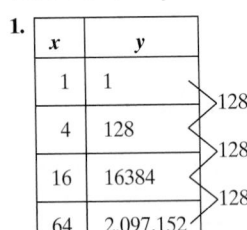

As the temperature decreases by 5°C, the time it takes for frostbite to occur decreases exponentially.

SECTION 13.3

13.3.1 Checkup

1. 6   **2.** 2   **3.** 3   **4.** $\dfrac{729}{125}$

13.3.2 Checkup

1. x^3 **2.** $x^{1/6}$ **3.** $z^{1/2}$ **4.** $\dfrac{q^{3/5}}{p^{3/5}}$ **5.** $2ab^{3/2}$ **6.** $8x^4$ **7.** $a^{3/4} - a^{7/6}$

13.3.3 Checkup

1. **(a)** The radius increases by a factor of $\sqrt{2}$.   **(b)** The radius increases by a factor of $\sqrt{3}$.
(c) The ratio of the length of the larger radius to the smaller is 2.
2. **(a)** $v = (128.8h)^{y_2}$ **(b)** $v_2 - v_1 = (\sqrt{128.8} - \sqrt{64.4})h^{y_2} \approx 3.324\sqrt{h}$

SECTION 13.4

13.4.1 Checkup

1. $8\sqrt{3}$ **2.** $-2\sqrt[3]{5}$ **3.** $3\sqrt[4]{2}$ **4.** $xy^4\sqrt[3]{x^2}$ **5.** $7x^2yz^4\sqrt{2yz}$ **6.** $(2x+1)\sqrt{5}$ **7.** $30ab\sqrt{a}$ **8.** $(x+2)\sqrt{10}$
9. $72x^2y^4\sqrt[3]{2y^2}$

13.4.2 Checkup

1. $\dfrac{3\sqrt{15}}{10}$ **2.** $\dfrac{3\sqrt[3]{4}}{2}$ **3.** $\dfrac{5\sqrt{2xyz}}{2z}$ **4.** $\dfrac{3\sqrt{7ab}}{7}$ **5.** $\dfrac{\sqrt[3]{75xyz^2}}{5yz}$ **6.** $\dfrac{b\sqrt{5a}}{15a}$ **7.** $\dfrac{\sqrt[3]{2c^2}}{2d}$

13.4.3 Checkup

1. The period of the pendulum is 1.28 seconds.
2. The length of time for one stride is approximately 1.9238 seconds. Steve's walking speed is approximately 6.24 feet per second.

SECTION 13.5

13.5.1 Checkup

1. $-\sqrt{ab} + 7\sqrt[4]{ab}$ **2.** $4p\sqrt{q}$ **3.** $(5m + 2)\sqrt[3]{mn}$ **4.** $13\sqrt{3}$ **5.** $-xy\sqrt{xy}$ **6.** $\frac{2}{3}\sqrt{3}$

13.5.2 Checkup

1. $35x + 28\sqrt[3]{x^2}$ **2.** $\sqrt{21} + \sqrt{3ab} - \sqrt{7ab} - ab$ **3.** 5 **4.** $121 - z$ **5.** $15 + 10\sqrt{2}$ **6.** $z - 10\sqrt{z} + 25$

13.5.3 Checkup

1. $\sqrt{2x} + \sqrt{5}$ **2.** $\dfrac{\sqrt{65} - \sqrt{26}}{3}$ **3.** $2 + \sqrt{3}$ **4.** $\dfrac{8x - 17\sqrt{3x} + 6}{x - 12}$

13.5.4 Checkup

Let s = the length of a side of the square floor and A = the area of the square floor, $s = \sqrt{A}$. The perimeter of the floor is $p = 4s = 4\sqrt{A}$. The perimeter of Sandra's floor is $p_1 = 4\sqrt{350}$ and of Julie's floor is $p_2 = 4\sqrt{224}$. The difference is $d = p_1 - p_2 = 4\sqrt{350} - 4\sqrt{224} = 4\sqrt{14}$. This is approximately 15 feet.

SECTION 13.6

13.6.1 Checkup

1. nonradical **2.** radical **3.** radical

13.6.2 Checkup

1. 4. Note: $x = -3$ is an extraneous solution. **2.** 5 and 4 **3.** 2 **4.** no solution; $z = -5$ is an extraneous solution.

5. 16 **6.** no solution; $x = \dfrac{21}{4}$ is an extraneous solution.

13.6.3 Checkup

1. 2 **2.** 19 **3.** -9 and 3

13.6.4 Checkup

1. 32 **2.** 4 **3.** There is no real-number solution.

13.6.5 Checkup

1. Gete's vertical jump was about 4.98 feet. **2.** $y = -1, y = 11$ **3.** The energy of the Wolf Creek explosion is 3.21×10^{15} ft-lb.

SECTION 13.7

13.7.1 Checkup

1. $7i$ **2.** $\dfrac{2}{9}i$ **3.** $3i\sqrt{3}$ **4.** **(a)** -4 **(b)** $4i$; answers may vary.

13.7.2 Checkup

1. $10i$ **2.** $-5i\sqrt{5}$ **3.** $\dfrac{10}{11}$ **4.** $-30i$ **5.** $7\sqrt{3} - 4\sqrt{10}$

13.7.3 Checkup

1. $26 - 5i$ **2.** $6 + 3i$ **3.** $-8 + 20i$ **4.** $-3 + 12i$ **5.** $80 - 20i$ **6.** 58 **7.** $4 - 3i$ **8.** $-4 - \dfrac{7}{2}i$ **9.** $\dfrac{8}{13} + \dfrac{1}{13}i$

13.7.4 Checkup

1. $x = \pm 2i$ **2.** $x = 4 \pm 3i\sqrt{3}$ **3.** $z = -\dfrac{1}{3} \pm \dfrac{i\sqrt{2}}{3}$ **4.** $x = \dfrac{-3 \pm 3i\sqrt{19}}{10}$ **5.** $x = -5 \pm i\sqrt{19}$

13.7.5 Checkup

The magnitude of the total voltage across the circuit is approximately 26.1 volts.

Chapter 14

SECTION 14.1

14.1.1 Checkup

1. $g^{-1} = \{(0.5, 1), (1, 2), (1.5, 3), (2, 4), (2.5, 5), (3, 6)\}$ **2.** $K^{-1} = \left\{\left(3, \frac{1}{2}\right), \left(4, \frac{2}{3}\right), \left(5, \frac{3}{4}\right), \left(4, \frac{4}{5}\right), \left(3, \frac{5}{6}\right)\right\}$

3. $y = -\dfrac{1}{4}x + \dfrac{1}{4}$ **4.** $y = \pm\sqrt{\dfrac{1}{3}x - 2}$

14.1.2 Checkup

1. This is a one-to-one function.   **2.** This is a function, but it is not one-to-one.   **3.** This is not a function.   **4.** The graph represents a one-to-one function.   **5.** The graph does not represent a one-to-one function, since it fails the horizontal-line test.
6. The graph does not represent a one-to-one function, since it fails the vertical-line test.

14.1.3 Checkup

1.

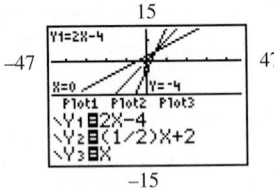

14.1.4 Checkup

1. $f^{-1}(x) = -\dfrac{1}{2}x + 3$ **2.** $g^{-1}(x) = \sqrt[3]{x}$

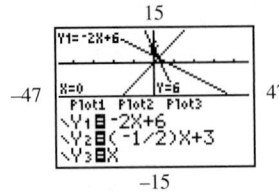 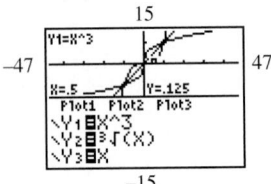

14.1.5 Checkup

1. (a) $I(x) = 60x$ **(b)** $I^{-1}(x) = \dfrac{x}{60}$ **(c)** The inverse function represents the number of years over which Dimitri must invest $1200 at 5%
simple interest per year in order to earn x dollars in interest.
2. $f^{-1}(x) = 1.4453x$; the inverse function represents the number of U.S. dollars, given x number of British pounds.

SECTION 14.2

14.2.1 Checkup

1. not an exponential function   **2.** exponential function   **3.** not an exponential function

14.2.2 Checkup

1. 64   **2.** $\dfrac{1}{64}$ or 0.015625   **3.** 8   **4.** approximately 11.036

14.2.3 Checkup

1. $y = 8^{2x}$

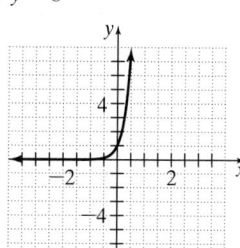

2. $y = 8^{-2x}$

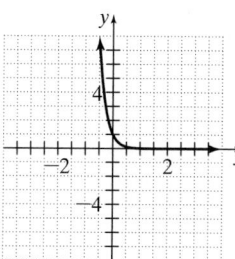

14.2.4 Checkup

1.

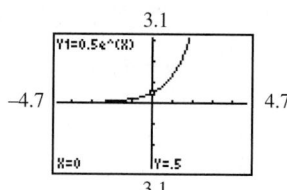

2.

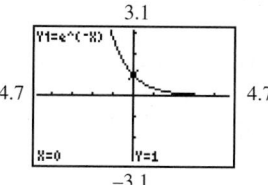

14.2.5 Checkup

1. After four years, Cindy will have $6312.38.

2. After four years, Cindy would have had $6356.25.

3. $y = 5.2975(1.014)^x$

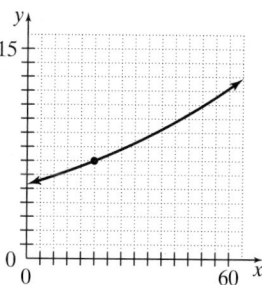

At the year 2010, $x = 20$; therefore, $y = 7$ billion. The estimate is just slightly higher than the given projection for 2010.

SECTION 14.3

14.3.1 Checkup

1. $f^{-1}(x) = \log_3 x$ **2.** $h^{-1}(x) = \log_{13} x$

14.3.2 Checkup

1. 4   **2.** -3 **3.** 3   **4.** -3 **5.** undefined   **6.** 4   **7.** -4 **8. (a)** 1.52832 **(b)** 1.52832

14.3.3 Checkup

1.

x	$f(x)$
0.001	-6.288
0.01	-4.192
0.1	-2.096
0.5	-0.6309
1	0
2	0.6309
3	1
4	1.2619
5	1.465

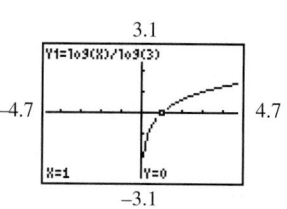

2.

x	$g(x)$
0.001	-2.523
0.01	-1.523
0.1	-0.5229
0.5	0.1761
1	0.4771
2	0.7782
3	0.9542
4	1.0792
5	1.1761

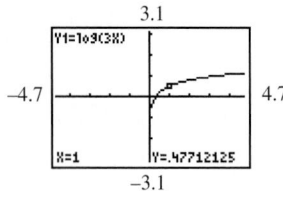

3.

x	$h(x)$
0.001	-5.809
0.01	3.507
0.1	-1.204
0.5	0.4055
1	1.0986
2	1.7918
3	2.1972
4	2.4849
5	2.7081

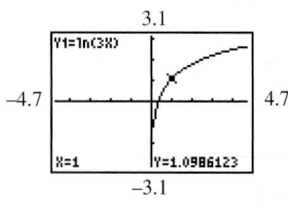

14.3.4 Checkup

1. The concentration of hydrogen in pure water is 10^{-7} mole per liter.   **2.** The ratio of the intensity of the 1964 Alaskan earthquake to the intensity of the 1989 San Francisco earthquake is $\dfrac{10^{9.2}}{10^{7.1}}$, or approximately 126. The Alaskan quake was 126 times as great as the 1989 San Francisco quake.   **3.** The ratio of the intensity of the downgraded earthquake to the original intensity is $\dfrac{10^{7.9}}{10^{8.3}}$, or approximately 0.40. The intensity is downgraded to 0.40 of its original intensity.

SECTION 14.4

14.4.1 Checkup

1. $\log_5 11 + \log_5 x$ **2.** $\ln x - \ln y$ **3.** $\log\left(\dfrac{xy}{z}\right)$ **4.** $\ln(x^{-3}) = \ln\left(\dfrac{1}{x^3}\right)$ **5.** $1 + 2\log_5 x$

14.4.2 Checkup

1. $\ln A = \ln \pi + 2\ln r$ **2.** $\log A = \log b + \log h - \log 2$ **3. (a)** $y = c\ln\left(\dfrac{I_0}{I}\right)$ **(b)** The depth is 9.3 units below the surface.

SECTION 14.5

14.5.1 Checkup

1. neither   **2.** logarithmic   **3.** exponential   **4.** neither

14.5.2 Checkup

1. 4   **2.** $\ln(3)$ **3.** $\ln\left(\dfrac{7}{2}\right)$ **4.** $\dfrac{3}{2}$ **5.** $\dfrac{100}{3}$ **6.** all real numbers   **7.** no solution

14.5.3 Checkup

1. Let $A(t)$ represent the total female population t years after 1990. Then $A(t) = 128e^{0.009t}$. In the year 2010, $t = 20$, and the total female population of the United States is predicted to be 153.244 million.

2. The gross national product will triple in value in 28 years, according to this prediction.

3. $A(t) = A_0 e^{-0.000428t}$. When $A(t) = 0.9A_0$, $t = \dfrac{\ln(0.9)}{-0.000428}$, or $t \approx 246$; the radium-226 will diminish to 90 percent of its original amount in 246 years.

4. Let $T(t)$ represent the temperature of the liquid after t minutes in the freezer. The solution is $t = \dfrac{\ln\left(\dfrac{1}{5}\right)}{-0.0277}$, or $t \approx 58.1$. The water will cool to 20°C after 58 minutes.

Index

Slope of a Line (m)

$$m = \frac{y_2 - y_1}{x_2 - x_1}$$

(x_1, y_1) and (x_2, y_2) are coordinates of two points on a line

Linear Equation in Two Variables

Standard Form

$ax + by = c$

a, b, and c are real numbers and a and b are not both equal to zero

Slope–Intercept Form

$y = mx + b$

m is the slope of the graphed line, and b is the y-coordinate of the y-intercept of the graph

Point–Slope Form

$y - y_1 = m(x - x_1)$

m is the slope of the graphed line, and (x_1, y_1) are coordinates of a point on the line

Special Cases

$y = k$

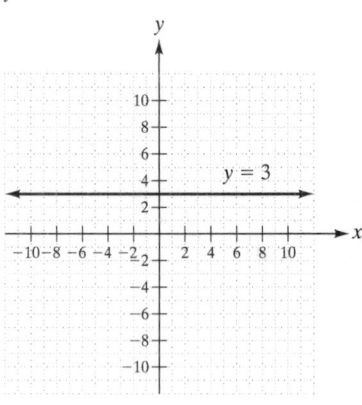

$x = h$

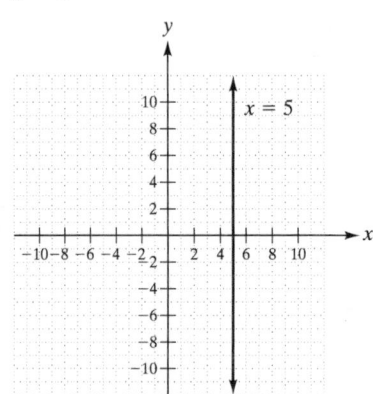

Standard Form for a Quadratic Function

$f(x) = ax^2 + bx + c$ a, b, and c are real numbers and $a \neq 0$

Calculator Windows

This text uses the following notation to identify the calculator window setting:

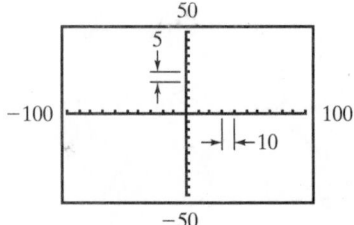

$(-100, 100, 10, -50, 50, 5, 1)$
(x minimum value, x maximum value, x scale, y minimum value, y maximum value, y scale, x resolution) .

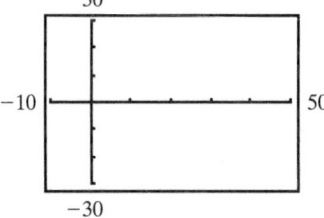

$(-10, 50, -30, 30)$
(x minimum value, x maximum value, y minimum value, y maximum value)

Sample Functions

Linear Function

$f(x) = ax, a > 0$

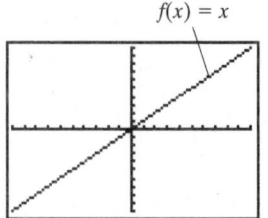

$(-10, 10, -10, 10)$

$f(x) = ax, a < 0$

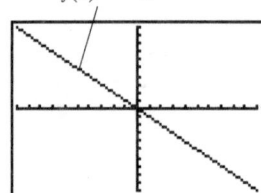

$(-10, 10, -10, 10)$